2009 46th ACM/IEEE Design Automation Conference

(DAC 2009)

San Francisco, CA, USA
26 – 31 July 2009

Pages 1-471

IEEE Catalog Number: CFP09DAC-PRT
ISBN: 978-1-60558-497-3

Copyright © 2009, ACM
All Rights Reserved

****This publication is a representation of what appears in the IEEE Digital Libraries. Some format issues inherent in the e-media version may also appear in this print version.*

IEEE Catalog Number: CFP09DAC-PRT
ISBN 13:   978-1-60558-497-3
Library of Congress No.:   85-644924
ISSN: 0738-100X

Additional Copies of This Publication Are Available From:

Curran Associates, Inc
57 Morehouse Lane
Red Hook, NY 12571 USA
Phone: (845) 758-0400
Fax: (845) 758-2633
E-mail: curran@proceedings.com

TABLE OF CONTENTS

SESSION 12 – DESIGN INTEGRITY CHALLENGES

SESSION 13 – PANEL

SESSION 14 – SPECIAL SESSION:
VERIFYING AN SOC MONSTER: WHOSE JOB IS IT ANYWAY?

SESSION 15 – TIMING SIMULATION: OPTIMIZED EMBEDDED SOFTWARE AND MPSOCS

SESSION 16 – ADVANCES IN EMBEDDED SYSTEM MODELING AND OPTIMIZATION

SESSION 23 – ANALOG/RF SIMULATION AND STATISTICAL MODELING

SESSION 24 – RECENT ADVANCES IN TIMING, ECO AND LOGIC OPTIMIZATION

SESSION 25 – PANEL

SESSION 26 – SPECIAL SESSION: COMPUTATION IN THE POST-TURING ERA

SESSION 27 – ADVANCES IN PHYSICAL SYNTHESIS

SESSION 39 – EMBEDDED SYSTEM DESIGN FOR LOW-POWER

SESSION 40 – HARDWARE AUTHENTICATION, CHARACTERIZATION AND TRUSTED DESIGN

SESSION 41 – TARGETED TEST AND DIAGNOSIS

SESSION 42 – CHALLENGES OF MEMORY-AWARE DESIGN FOR EMBEDDED SYSTEMS

<u>SESSION 54 – MODEL ORDER REDUCTION TECHNIQUES AND APPLICATIONS</u>

GENERAL CHAIR'S MESSAGE

Welcome to San Francisco and the 46th Design Automation Conference!

For thousands of us across the electronic design and design automation fields, DAC is a week of intense networking and learning. DAC is the one event each year where the entire life cycle of IC design technology is on display: new challenges in panels and special sessions; new technologies in research papers; new solutions in the exhibit suites of EDA startups; new insights into tools and methodologies at the Exhibitor Forum and the new User Track; new integrated reference flows from leading foundries and IP providers; continuing education in the form of full-day tutorials; ... The opportunities for professional growth and new connections are truly endless. This is why business leaders, market leaders, technical leaders and thought leaders all converge at DAC.

The technical conference introduces DAC's new User Track – a three-day track of over 80 papers and posters that are written by, selected by, and targeted at users of design tools. In the User Track, attendees learn from their colleagues in leading-edge design organizations about the latest solutions to critical design and methodology challenges.

DAC's exhibition showcases approximately 200 companies, including all of the largest EDA vendors. The Exhibitor Forum theater returns with a full three-day schedule of focused technical presentations from exhibitors. The Best of DAC Contest also returns with additional Attendees' Choice categories: now you can vote on the web for your favorite vendors and new products. New this year in the exhibition is IC Design Central (ICDC) – a cluster of exhibitors and a presentation theater that brings together the entire ecosystem for SOC enablement. IP providers, design service providers, and foundry service providers can all be found in the ICDC.

DAC is also the meeting place for many EDA- and design-related organizations, workshops and events. Who meets at DAC? Standards bodies, such as Accellera and Si2. Roadmapping efforts, such as the ITRS Design Technology working group. Established symposia, such as SASP, NANOARCH, MSE, SLIP, DFM&Y, and HOST. Workshops on topics ranging from UML to low-power design to bio-design automation. A total of 19 workshops and colocated events will take place at DAC.

DAC 2009 is proud to present a stellar lineup of keynote and plenary sessions. In response to the recent upheavals in the macroeconomy as well as in the IC design and design automation industries, DAC opens its program with a special Monday afternoon panel that brings together the CEOs of the leading EDA companies. Attendees will hear our industry leaders' views on market, business, and technology futures, as well as the overall outlook for the EDA industry. DAC is also proud to present keynote presentations from outstanding industry leaders: Fu-Chieh Hsu, Vice President, Design and Technology Platform, TSMC; and William Dally, Chief Scientist, NVIDIA and Willard R. and Inez Kerr Bell Professor of Computer Science at Stanford University. Finally, a plenary panel on Thursday, moderated by Walden C. Rhines of Mentor Graphics Corp. will address the questions of "How Green is My Silicon Valley" and what 'green' means for our industry.

On behalf of DAC's sponsoring organizations and the many hundreds of volunteers who make DAC happen each year, I warmly welcome you to DAC 2009!

Best regards,

Andrew B. Kahng
General Chair, 46th Design Automation Conference

EXECUTIVE COMMITTEE

GENERAL CHAIR
Andrew B. Kahng
Univ. of California, San Diego
CSE & ECE Depts.
Mailcode #0404
9500 Gilman Dr.
La Jolla, CA 92093-0404
+1-858-822-4884
abk@cs.ucsd.edu

VICE/FINANCE CHAIR
Sachin Sapatnekar
Univ. of Minnesota
4-174 EE/CSci Building
200 Union St. SE
Minneapolis, MN 55455
+1-612-625-0025
sachin@ece.umn.edu

**TECHNICAL PROGRAM
CO-CHAIR**
Patrick Groeneveld
Magma Design Automation, Inc.
1650 Technology Dr.
San Jose, CA 95110
+1-408-565-7654
patrick@magma-da.com

**TECHNICAL PROGRAM
CO-CHAIR**
Narendra Shenoy
Synopsys, Inc.
700 E. Middlefield Rd.
Mountain View, CA 94043
+1-650-584-1734
narendra.shenoy@synopsys.com

EDA INDUSTRY CHAIR
Yervant Zorian
Virage Logic Corp.
47100 Bayside Pkwy.
Fremont, CA 94538
+1-510-360-8035
yervant.zorian@viragelogic.com

PAST CHAIR
Limor Fix
Intel Research Pittsburgh
4720 Forbes Ave., Ste. 410
Mailstop CM2
Pittsburgh, PA 15213
+1-412-297-4021
limor.fix@intel.com

TUTORIAL CHAIR
Dennis Sylvester
Univ. of Michigan
Electrical Engineering &
    Computer Science
1301 Beal Ave.
Ann Arbor, MI 48109
+1-734-615-8783
dennis@eecs.umich.edu

PANEL CHAIR
Greg Spirakis
Consultant
San Jose, CA
+1-408-218-5577
gspiraki@pacbell.net

DESIGN COMMUNITY CHAIR
Soha Hassoun
Tufts Univ.
161 College Ave.
Medford, MA 02155
+1-617-627-5177
soha@cs.tufts.edu

NEW INITIATIVES CHAIR
Leon Stok
IBM Corp.
2070 Route 52
Hopewell Jct., NY 12533
+1-845-892-5262
leonstok@us.ibm.com

PUBLICITY CHAIR
Nanette V. Collins
Nanette V. Collins Marketing &
    Public Relations
37 Symphony Rd., Unit A
Boston, MA 02115
+1-617-437-1822
nanette@nvc.com

**EUROPE/MIDDLE EAST
REPRESENTATIVE**
Georges Gielen
ESAT-MICAS
Katholieke Univ. Leuven
Kasteelpark Arenberg 10
Leuven, B-3001 Belgium
+32-16-321047
georges.gielen@esat.kuleuven.be

**ASIA/SOUTH PACIFIC
REPRESENTATIVE**
Kazutoshi Wakabayashi
NEC Corp.
1753 Shimonumabe, Nakahara-Ku
Kawasaki, 211-8666 Japan
+81-435-9486
wakaba@bl.jp.nec.com

ACM/SIGDA REPRESENTATIVE
Diana Marculescu
Carnegie Mellon Univ.
ECE Dept. – HH #2124
5000 Forbes Ave.
Pittsburgh, PA 15213
+1-412-268-1167
dianam@ece.cmu.edu

**IEEE/CASS/CANDE/CEDA
REPRESENTATIVE**
Ramesh Chandra
Qualcomm, Inc.
6455 Lusk Blvd.
San Diego, CA 92121
+1-858-658-4315
ramesh.chandra@qualcomm.com

**EDA CONSORTIUM
REPRESENTATIVE**
Anne Cirkel
Mentor Graphics Corp.
8005 SW Boeckman Rd.
Wilsonville, OR 97070
+1-503-685-7934
anne_cirkel@mentor.com

CONFERENCE MANAGER
Kevin Lepine
MP Associates, Inc.
1721 Boxelder St., Ste. 107
Louisville, CO 80027
+1-303-530-4562
kevin@mpassociates.com

EXHIBITS MANAGER
Lee Wood
MP Associates, Inc.
1721 Boxelder St., Ste. 107
Louisville, CO 80027
+1-303-530-4562
lee@mpassociates.com

TECHNICAL PROGRAM COMMITTEE

Patrick Groeneveld
Technical Program Co-Chair
Magma Design Automation, Inc.
San Jose, CA

Rob Aitken
ARM Ltd.
Sunnyvale, CA

Syed Alam
Everspin Technologies, Inc.
Austin, TX

Kunihiro Asada
Univ. of Tokyo
Tokyo, Japan

Kaustav Banerjee
Univ. of California
Santa Barbara, CA

Kia Bazargan
Univ. of Minnesota
Minneapolis, MN

Valeria Bertacco
Univ. of Michigan
Ann Arbor, MI

Davide Bertozzi
Univ. di Ferrara
Ferrara, Italy

Clive Bittlestone
Texas Instruments, Inc.
Dallas, TX

Per Bjesse
Synopsys, Inc.
Hillsboro, OR

Eric Bracken
Ansoft Corp.
Pittsburgh, PA

Yu (Kevin) Cao
Arizona State Univ.
Tempe, AZ

Luca Carloni
Columbia Univ.
New York, NY

Shih-Chieh Chang
National Tsing-Hua Univ.
Hsinchu, Taiwan

Karam S. Chatha
Arizona State Univ.
Tempe, AZ

Narendra Shenoy
Technical Program Co-Chair
Synopsys, Inc.
Mountain View, CA

Deming Chen
Univ. of Illinois, Urbana-Champaign
Urbana, IL

Wu-Tung Cheng
Mentor Graphics Corp.
Wilsonville, OR

Eli Chiprout
Intel Corp.
Hillsboro, OR

Olivier Coudert
Magma Design Automation, Inc.
San Jose, CA

Luca Daniel
Massachusetts Institute of Technology
Cambridge, MA

Abhijit Davare
Intel Corp.
Hillsboro, OR

Azadeh Davoodi
Univ. of Wisconsin
Madison, WI

Bjorn De Sutter
Ghent Univ.
Ghent, Belgium

Andre DeHon
Univ. of Pennsylvania
Philadelphia, PA

Alper Demir
Koç Univ.
Sariyer-Istanbul, Turkey

Rainer Doemer
Univ. of California
Irvine, CA

Rolf Drechsler
Univ. Bremen
Bremen, Germany

Petru Eles
Linköping Univ.
Linkoping, Sweden

Peter Feldmann
IBM Corp.
Yorktown Hts., NY

Alberto Ferrari
Parades S.c.a.r.l.
Rome, Italy

Masahiro Fujita
Univ. of Tokyo
Tokyo, Japan

Nagib Z. Hakim
Intel Corp.
Santa Clara, CA

Masanori Hashimoto
Osaka Univ.
Osaka, Japan

Graham R. Hellestrand
Embedded Systems Technology, Inc.
San Carlos, CA

Jaeha Kim
Stanford Univ.
Stanford, CA

Farinaz Koushanfar
Rice Univ.
Houston, TX

David S. Kung
IBM Corp.
Yorktown Hts., NY

Marcello Lajolo
NEC Corp.
Princeton, NJ

Guy Lemieux
Univ. of British Columbia
Vancouver, BC, Canada

Tao Lin
Magma Design Automation, Inc.
San Jose, CA

Jinfeng Liu
Synopsys, Inc.
Mountain View, CA

Igor L. Markov
Univ. of Michigan
Ann Arbor, MI

Nobu Matsumoto
Toshiba Corp.
Kawasaki, Japan

Raj S. Mitra
Texas Instruments, Inc.
Bangalore, India

Ravi Nanavati
Bluespec, Inc.
Waltham, MA

Vijaykrishnan Narayanan
Pennsylvania State Univ.
University Park, PA

Nicola Nicolici
McMaster Univ.
Hamilton, ON, Canada

Raymond Nijssen
Achronix Semiconductor Corp.
San Jose, CA

Ralph H. Otten
Technische Univ. Eindhoven
Eindhoven, The Netherlands

Mustafa Ozdal
Intel Corp.
Hillsboro, OR

Sule Ozev
Arizona State Univ.
Tempe, AZ

David Z. Pan
Univ. of Texas
Austin, TX

Davide Pandini
STMicroelectronics
Agrate Brianza, Italy

Barry M. Pangrle
Consultant
San Jose, CA

Sridevan Parameswaran
Univ. of New South Wales
Sydney, Australia

Alessandro Pinto
United Technologies
Berkeley, CA

Miodrag Potkonjak
Univ. of California
Los Angeles, CA

Ruchir Puri
IBM Corp.
Yorktown Hts., NY

Anand Raghunathan
Purdue Univ.
West Lafayette, IN

John Regehr
Univ. of Utah
Salt Lake City, UT

Prashant Saxena
Synopsys, Inc.
Hillsboro, OR

Patrick R. Schaumont
Virginia Polytechnic Institute and State Univ.
Blacksburg, VA

Youngsoo Shin
KAIST
Daejeon, Republic of Korea

Eli Singerman
Intel Corp.
Haifa, Israel

Ankur Srivastava
Univ. of Maryland
College Park, MD

Mircea Stan
Univ. of Virginia
Charlottesville, VA

Jürgen Teich
Friedrich-Alexander Univ.
Erlangen-Nürnberg
Erlangen, Germany

Frank Vahid
Univ. of California
Riverside, CA

Gerd Vandersteen
Vrije Univ. Brussel
Brussels, Belgium

Paul Villarrubia
IBM Corp.
Austin, TX

Jens Vygen
Univ. Bonn
Bonn, Germany

Janet Meiling Wang
Univ. of Arizona
Tucson, AZ

Yosinori Watanabe
Cadence Design Systems, Inc.
Berkeley, CA

Hai Zhou
Northwestern Univ.
Evanston, IL

Vladimir Zolotov
IBM Corp.
Yorktown Hts., NY

PANEL COMMITTEES

Greg Spirakis
Panel Chair
Consultant
San Jose, CA

Technical Panel Committee

Jason Cong
Univ. of California
Los Angeles, CA

Eshel Haritan
CoWare, Inc.
San Jose, CA

Andreas Kuehlmann
Cadence Design Systems, Inc.
Berkeley, CA

Nagaraj NS
Texas Instruments, Inc.
Dallas, TX

Ruchir Puri
IBM Corp.
Yorktown Hts., NY

Juan C. Rey
Mentor Graphics Corp.
San Jose, CA

Hiroyuki Yagi
STARC
Yokohama, Japan

Pavilion Panel Committee

Sabina Burns
Virage Logic Corp.
Fremont, CA

Rich Goldman
Synopsys, Inc.
Mountain View, CA

Dave Kelf
Sigmatix, Inc.
Stoneham, MA

David Lin
Denali Software, Inc.
Sunnyvale, CA

Jim Lipman
Sidense Corp.
Livermore, CA

Tiffany Sparks
Chartered Semiconductor Mfg., Inc.
Milpitas, CA

Yatin Trivedi
Synopsys, Inc.
Mountain View, CA

TUTORIAL COMMITTEE

Dennis Sylvester
Tutorial Chair
Univ. of Michigan
Ann Arbor, MI

Rob Aitken
ARM Ltd.
Sunnyvale, CA

Harry Foster
Mentor Graphics Corp.
San Jose, CA

David Kung
IBM Corp.
Yorktown Hts., NY

EXHIBITOR LIAISON COMMITTEE

Yervant Zorian
Exhibitor Liaison Chair
Virage Logic Corp.
Fremont, CA

Peggy Aycinena
EDA Confidential
San Mateo, CA

Dagmar Berendes
ThinkBold Corporate Communications, Inc.
Campbell, CA

Jana Burke
Mentor Graphics Corp.
Tuscon, AZ

Chuck Byers
Consultant
San Jose, CA

Anne Cirkel
Mentor Graphics Corp.
Wilsonville, OR

Craig Cochran
ChipVision Design Systems
San Jose, CA

Nanette V. Collins
Nanette V. Collins Marketing & Public Relations
Boston, MA

George Harper
Bluespec, Inc.
Waltham, MA

Jill Jacobs
MOD Marketing and Events LLC
San Jose, CA

Andrew B. Kahng
Univ. of California, San Diego
La Jolla, CA

Drew McGrady
The MathWorks, Inc.
Natick, MA

Gabe Moretti
Gabe on EDA
Venice, FL

Herta Schreiner
Synopsys, Inc.
Mountain View, CA

Rob van Blommestein
SpringSoft, Inc.
San Jose, CA

Lee Wood
MP Associates, Inc.
Louisville, CO

STRATEGY COMMITTEE

Sachin Sapatnekar
Strategy Chair
Univ. of Minnesota
Minneapolis, MN

Clive Bittlestone
Texas Instruments, Inc.
Dallas, TX

Dennis Brophy
Mentor Graphics Corp.
Wilsonville, OR

Anne Cirkel
Mentor Graphics Corp.
Wilsonville, OR

Joe Civello
Agilent
Santa Rosa, CA

Michelle Clancy
Cayenne Communications
Washington, NC

Nannette Collins
Nannete V. Collins Marketing & Public Relations
Boston, MA

Limor Fix
Intel Research Pittsburgh
Pittsburgh, PA

Bob Gardner
EDA Consortium
San Jose, CA

Mike Gianfagna
Atrenta, Inc.
San Jose, CA

Patrick Groeneveld
Magma Design Automation, Inc.
San Jose, CA

Charlie Kahle
QThink, Inc.
San Diego, CA

Andrew B. Kahng
Univ. of California, San Diego
La Jolla, CA

Ken Karnofsky
The MathWorks, Inc.
Natick, MA

David Lin
Denali Software, Inc.
Sunnyvale, CA

Diana Marculescu
Carnegie Mellon Univ.
Pittsburgh, PA

Gabe Moretti
Gabe on EDA
Venice, FL

Shishpal Rawat
Intel Corp.
Folsom, CA

Rob A. Rutenbar
Carnegie Mellon Univ.
Pittsburgh, PA

Scott Sandler
SpringSoft, Inc.
San Jose, CA

Gary Smith
Gary Smith EDA
Santa Clara, CA

Greg Spirakis
Consultant
San Jose, CA

Leon Stok
IBM Corp.
Hopewell Jct., NY

Steve Trimberger
Xilinx, Inc.
San Jose, CA

Yatin Trivedi
Synopsys, Inc.
Mountain View, CA

Nancy Weiss
Synopsys, Inc.
Mountain View, CA

Lee Wood
MP Associates, Inc.
Louisville, CO

Andrew Yang
Apache Design Solutions, Inc.
Mountain View, CA

Yervant Zorian
Virage Logic Corp.
Fremont, CA

USER TRACK COMMITTEE

Leon Stok
User Track Co-Chair
IBM Corp.
Hopewell Jct., NY

Jarrod Brooks
Cypress Semiconductor
Lexington, KY

Nitin Chawla
STMicroelectronics
Greater Noida, India

Kyu-Myung Choi
Samsung
Yongin City, Republic of Korea

Sorin Dobre
Qualcomm, Inc.
San Diego, CA

Toshio Fujisawa
Toshiba Corp.
Kawasaki, Japan

Wolfgang Grimm
Qimonda AG
Munich, Germany

Bryan Heard
QThink, Inc.
San Diego, CA

Matt Moore
IBM Corp.
Research Triangle Park, NC

Uwe Müller
Qimonda AG
Munich, Germany

Nobuyuki Nishiguchi
STARC
Yokohama, Japan

Soha Hassoun
User Track Co-Chair
Tufts Univ.
Medford, MA

Peter Nord
ST-Ericsson
Lund, Sweden

Srinivas Nori
Microsoft Corp.
Mountain View, CA

Robert Patti
Tezzaron Semiconductor Corp.
Naperville, IL

David Peterman
Texas Instruments, Inc.
Dallas, TX

Thierry Pfirsch
Alcatel-Lucent
Illkirch, France

Doug Quist
NVIDIA Corp.
Santa Clara, CA

Milton Ribeiro
Bematech S/A
Curitiba, Brazil

Doug Stiles
Microsoft Corp.
Mountain View, CA

Alicia Strang
Marvell Semiconductor, Inc.
Aliso Viejo, CA

Raj R. Varada
Intel Corp.
Santa Clara, CA

STUDENT DESIGN CONTEST JUDGES

Bill Bowhill
DAC/ISSCC SDC Co-Chair
Intel Corp.
Hudson, MA

Michael Attig
Xilinx, Inc.
San Jose, CA

Fabio Campi
STMicroelectronics
Bologna, Italy

Shine Chung
Taiwan Semiconductor Manufacturing Co., Ltd.
Hsinchu, Taiwan

Claude Gauthier
Prism Circuits
Los Gatos, CA

Brian Ginsburg
Texas Instruments, Inc.
Dallas, TX

David Harris
Harvey Mudd College
Claremont, CA

Ron Ho
Sun Microsystems Labs
Menlo Park, CA

Chris H. Kim
Univ. of Minnesota
Minneapolis, MN

Jaejoon Kim
IBM Corp.
Yorktown Hts., NY

Sandip Kundu
Univ. of Massachusetts
Amherst, MA

Byunghoo Jung
DAC/ISSCC SDC Co-Chair
Purdue Univ.
West Lafayette, IN

Saraju Mohanty
Univ. of N. Texas
Denton, TX

Jun Ohta
Nara Institute of Science and Technology
Nara, Japan

Vojin Oklobdzija
Univ. of Texas, Dallas
Richardson, TX

Mondira (Mandy) Pant
Intel Corp.
Westborough, MA

Vijay Raghunathan
Purdue Univ.
West Lafayette, IN

Bud Taddiken
Microtune, Inc.
Plano, TX

Marc Tiebout
Infineon Technologies AG
Villach, Austria

Peter Wilson
Univ. of Southampton
Southampton, United Kingdom

Changsik Yoo
Hanyang Univ.
Seoul, Republic of Korea

KEYNOTE PANEL

Aart de Geus
CEO and
Chairman of the Board
Synopsys, Inc.

Walden C. Rhines
CEO and
Chairman of the Board
Mentor Graphics Corp.

Lip-Bu Tan
President and CEO
Cadence Design
Systems, Inc.

Futures for EDA: The CEO View

The IC design and EDA ecosystem has witnessed a number of structural and financial upheavals in recent years. Perennial challenges—new technologies and markets, R&D investment levels, business and revenue models, and interoperability——are magnified by the current economic downturn. This year, DAC will open its program with a special Monday afternoon panel that brings together the CEOs of the leading EDA companies. The panelists will give their views on market, business and technology futures, as well as the overall outlook for the EDA industry.

SPECIAL PLENARY PANEL

Walden C. Rhines
EDA Consortium, Chair
Mentor Graphics Corp.
Wilsonville, OR

How Green Is My Silicon Valley

There has been a lot of talk about 'green' technology, and how it can save our economy and the electronics industry. But what is really being done about it in Silicon Valley—the spawning ground for all things high-tech?

This panel of experts—drawn from technology heavyweights, governmental agencies, venture capitalists and start-ups—will give their views on what 'green' means for our industry. For some, it means low-power chip and system design. Others focus on completely new solutions that go well beyond traditional electronics to help lower our collective carbon footprint. Green also means leverage of existing technologies in new eco-friendly applications ranging from water management to smokestack emission monitoring.

Innovation is the only answer to this, the preeminent challenge of the 21st century. And innovation is what Silicon Valley and the global high-tech industry are all about. Come hear the best and brightest talk about their favorite new color and how it will take our industry and the larger economy into a brighter, more sustainable future.

TUESDAY KEYNOTE ADDRESS

Fu-Chieh Hsu
Vice President, Design and Technology Platform,
Taiwan Semiconductor Manufacturing Company, Ltd.,
Hsinchu, Taiwan

Overcoming the New Design Complexity Barrier: Alignment of Technology and Business Models

Abstract: The semiconductor industry today faces a new design complexity barrier. Convergence of product features brings unprecedented complexities on process technology, architectural and gate-count. Integration and cost drivers bring 3-D stacking with through-silicon vias. Heterogeneity brings high-voltage devices, image sensors, and MEMS into a single product design optimization. From IP migration, to variability modeling and mitigation, to NRE cost, the new paradigm is that these are no longer discrete issues addressable by point-tool solutions. Overcoming this new design complexity barrier requires breakthrough technologies and integrated EDA solutions.

Funding the development of new EDA solutions must make business sense. But while traditional ROI assessments are based on "one company-centric" business models, the next stage of the industry's evolution requires ROI and business models that acknowledge coexistence, and even collaboration, of multiple companies. This is especially true during this unprecedented worldwide financial earthquake. A win-win-win future will depend on the emergence of an innovative business model that is realized collaboratively by key players across the IC design and manufacturing ecosystem. Fortunately, recent initiatives have the promise to meet this need.

Biography: Dr. Fu-Chieh Hsu has served as Vice President of Design and Technology Platform for Taiwan Semiconductor Manufacturing Company, Ltd. since April 2006. He is responsible for all design service operations at TSMC and works with the Marketing and R&D departments to provide customers with technology platform solutions.

Dr. Hsu founded Monolithic System Technology, Inc. (MoSys) in 1991 and served as its Chairman and Chief Executive Officer until retiring at the end of 2004. He was Chairman and President of Myson Technology Inc. (now Myson Century Inc.) from 1990 to 1991. Prior to that, Dr. Hsu worked at Integrated Device Technology, Inc. as Chief Technology Officer and Vice President as well as other senior positions. Dr. Hsu also served at Hewlett-Packard Labs.

Dr. Hsu has published or contributed to more than 40 papers and also holds 55 U.S. patents. Dr. Hsu received his Bachelor of Science degree in electrical engineering from National Taiwan Univ. in 1978, and Master of Science and Ph.D. degrees in electrical engineering and computer sciences from Univ. of California, Berkeley, in 1981 and 1983, respectively.

WEDNESDAY KEYNOTE ADDRESS

William J. Dally
Chief Scientist and Senior Vice President of Research,
NVIDIA Corp., Santa Clara, CA
Willard R. and Inez Kerr Bell Professor of Engineering,
Stanford Univ., Palo Alto, CA

The End of Denial Architecture and the Rise of Throughput Computing

Abstract: Throughput-optimized processors, such as graphics processing units (GPUs) have scaled at historic rates in recent years, and continue to do so along a design trajectory that is largely unhindered by conventional dogmas and legacies. These processors recognize that two critical aspects of machine organization are key to performance: parallel execution and hierarchical memory organization. Conventional processors, which present an illusion of sequential execution and uniform, flat memory, find their performance increasing only slowly over time, and their evolution is at an end. In contrast, throughput processors embrace, rather than deny, parallelism and memory hierarchy to realize large performance and efficiency advantages. Throughput processors have hundreds of cores today and will have thousands of cores by 2015. They will deliver most of the performance, and most of the user value, in future computer systems.

This talk will discuss some of the challenges and opportunities in the architecture and programming of future throughput processors, as it relates to the EDA world. First, CAD tools, flows, and methodologies are clearly crucial to the design of these processors, and these must adapt to support such designs. Second, in this changing landscape, CAD tools must need to evolve to run on throughput processors. In throughput processors, performance derives from the parallelism available from the plentiful arithmetic units, and efficiency derives from locality, overcoming restrictions stemming from communication bandwidth bottlenecks that dominate cost, performance, and power.   This talk will discuss exploitation of parallelism and locality with examples drawn from the Imagine and Merrimac projects, from NVIDIA GPUs, and from three generations of stream programming systems.

Biography: William J. Dally is Chief Scientist and Senior Vice President of Research at NVIDIA Corp. and the Willard R. and Inez Kerr Bell Professor of Engineering at Stanford University. Dally and his group have developed system architecture, network architecture, signaling, routing, and synchronization technology that can be found in most large parallel computers today. While at Bell Labs Bill contributed to the BELLMAC32 microprocessor and designed the MARS hardware accelerator. At Caltech he designed the MOSSIM Simulation Engine and the Torus Routing Chip which pioneered wormhole routing and virtual-channel flow control. While a Professor of Electrical Engineering and Computer Science at the Massachusetts Institute of Technology his group built the J-Machine and the M-Machine, experimental parallel computer systems that pioneered the separation of mechanisms from programming models and demonstrated very low overhead synchronization and communication mechanisms. At Stanford Univ.his group has developed the Imagine processor, which introduced the concepts of stream processing and partitioned register organizations. Bill has worked with Cray Research and Intel Corp. to incorporate many of these innovations in commercial parallel computers, with Avici Systems to incorporate this technology into Internet routers, co-founded Velio Communications to commercialize high-speed signaling technology, and co-founded Stream Processors, Inc. to commercialize stream processor technology. He is a Member of the National Academy of Engineering, a Fellow of the IEEE, a Fellow of the ACM, and a Fellow of the American Academy of Arts and Sciences. He has received numerous honors including the IEEE Seymour Cray Award and the ACM Maurice Wilkes award. He currently leads projects on computer architecture, network architecture, and programming systems. He has published over 200 papers in these areas, holds over 50 issued patents, and is an author of the textbooks, *Digital Systems Engineering and Principles and Practices of Interconnection Networks.*

AWARDS

Marie R. Pistilli Women in EDA Achievement Award

Telle Whitney – *President and CEO, Anita Borg Institute for Women and Technology, Palo Alto, CA*
For her significant contributions in helping women advance in the field of EDA technology.

P.O. Pistilli Undergraduate Scholarships for Advancement in Computer Science and Electrical Engineering

The objective of the P.O. Pistilli Scholarship program is to increase the pool of professionals in Electrical Engineering, Computer Engineering, and Computer Science from under-represented groups (women, African American, Hispanic, Native American, and physically challenged). In 1989, ACM Special Interest Group on Design Automation (SIGDA) began providing the program. Beginning in 1993, the Design Automation Conference provided the funds for the scholarship and SIGDA continues to administer the program for DAC. DAC normally funds two or more $4000 scholarships, renewable up to five years, to graduating high school seniors.

The 2009 recipient is:
Cesar A. Torres, Jr.
Attending Stanford University, majoring in Computer Science.

For more information about the P.O. Pistilli scholarship, contact Andrew B. Kahng, *Univ. of California, San Diego, CSE & ECE Depts., Mailcode #0404, 9500 Gilman Dr., La Jolla, CA 92093-0404. Email: abk@cs.ucsd.edu*

A. Richard Newton Graduate Scholarships

The DAC Executive Committee has chosen to name our existing DAC Graduate Scholarships after the late Professor A. Richard Newton. We feel that supporting young faculty and graduate research is an appropriate way to honor his vision and carry out some of his goals. Each year the Design Automation Conference sponsors several $24,000 scholarships to support graduate research and study in Design Automation (DA), with emphasis in "design and test automation of electronic and computer systems". Each scholarship is awarded directly to a University for the Faculty Investigator to expend in direct support of one or more DA graduate students. The criteria for granting such a scholarship expanded in 1996 to include financial need. The criteria are: the academic credentials of the student(s); the quality and applicability of the proposed research; the impact of the award on the DA program at the institution; and financial need. Preference is given to institutions that are trying to establish new DA research programs.

Advisor: **Shiyan Hu** – *Michigan Technological Univ., Houghton, MA*
Student: **Chen Liao** – *Michigan Technological Univ., Houghton, MA*
Project: High Performance Placement and Routing Techniques for Digital Microfluidic Biochip Design

2008 Phil Kaufman Award for Distinguished Contributions to EDA Sponsored by the EDA Consortium and IEEE Council on EDA

Dr. Aart J. de Geus – *Chairman, CEO and Co-Founder of Synopsys, Inc., Mountain View, CA*
Aart de Geus is the recipient of the prestigious 2008 Phil Kaufman Award for his significant impact on EDA and IC design through productizing advanced techniques beginning with logic optimization and synthesis.

2009 IEEE Emanuel R. Piore Award

Dr. David J. DeWitt - *Microsoft Corp., Madison, WI*
For fundamental contributions to the architecture, algorithms, and implementation of innovative database systems.

IEEE Transactions on Computer-Aided Design 2009 Donald O. Pederson Best Paper Award

Hristo Nikolov, Todor Stefanov, Ed F. Deprettere – *Leiden Univ., Leiden, The Netherlands*
For the paper entitled, *Systematic and Automated Multiprocessor System Design, Programming, and Implementation.* IEEE Transactions on Computer-Aided Design of Integrated Circuits and Systems, vol. 27, no. 3, pp. 542-555, March 2008.

IEEE Circuits and Systems Society 2009 Industrial Pioneer Award

Paul E. Jacobs – *Qualcomm CDMA Technologies, San Diego, CA*
Paul Jacobs' vision of wireless data services has been pioneering and transformational
in the worldwide cellular industry. At Qualcomm, starting from the development of the first Palm OS® based Smartphone, and now in his capacity as CEO, he has grown company revenues to over $10B through this vision.

IEEE Circuits and Systems Society 2009 Vitold Belevitch Award

Ronald A. Rohrer – *SystemIC, Bend, OR*
For advancing circuit theory to solve practical circuit design problems with everlasting impact on the global electronic industry.

IEEE Circuits and Systems Society 2009 VLSI Transactions Best Paper Award

Nagarajan Ranganathan, Justin E. Harlow III, Mahalingam Venkataraman – *Univ. of S. Florida, Tampa, FL*
For the paper entitled, *A Fuzzy Optimization Approach for Variation Aware Power Minimization During Gate Sizing.* IEEE Transactions on Very Large Scale Integration (VLSI) Systems, vol. 16, no. 8, pp. 975-984, August 2008.

2009 IEEE Fellow

Steven Nowick – *Columbia Univ., New York, NY*
For contributions to asynchronous and mixed-timing integrated circuits and systems.

Joel Phillips – *Cadence Research Labs, Berkeley, CA*
For contributions to numerical techniques in the design of high-frequency and radio frequency systems.

ACM Fellow

Jason (Jingsheng) Cong – *Univ. of California, Los Angeles, CA*
For contributions to electronic design automation.

Jonathan S. Rose – *Univ. of Toronto, Toronto, ON, Canada*
For contributions to the architecture and computer-aided design of field-programmable gate arrays (FPGAs).

Rob A. Rutenbar – *Carnegie Mellon Univ., Pittsburgh, PA*
For contributions to computer-aided design tools for mixed-signal integrated circuits.

SIGDA Distinguished Service Award

Nikil Dutt – *Univ. of California, Irvine, CA*
For contributions to ACM's Special Interest Group on Design Automation during the past fifteen years as a SIGDA officer, coordinator of the University Booth in its early years, and most recently, as Editor-in-Chief of the ACM Transactions on Design Automation of Electronic Systems.

ACM Outstanding PhD Dissertation Award in EDA

Student: **Kai-Hui Chang** – *Univ. of Michigan, Ann Arbor, MI*
Advisors: **Igor Markov, Valeria Bertacco** – *Univ. of Michigan, Ann Arbor, MI*
For the dissertation, *Functional Design Error Diagnosis, Correction and Layout Repair of Digital Circuits.*

SIGDA Outstanding New Faculty Award

Yu (Kevin) Cao – *Arizona State Univ., Tempe, AZ*
For a junior faculty member early in his/her academic career who demonstrates outstanding potential as an educator and/or researcher in the field of electronic design automation.

ACM Transactions on Design Automation of Electronic Systems (TODAES) 2009 Best Paper Award

Sivaram Gopalakrishnan, Priyank Kalla – *Univ. of Utah, Salt Lake City, UT*
Optimization of Polynomial Datapaths Using Finite Ring Algebra. ACM Transactions on Design Automation of Electronic Systems (TODAES), Volume 12, Issue 4 (September 2007).

46th DAC Best Paper Candidates

Seven papers were nominated by the Technical Program Committee as DAC Best Paper Candidates. Final decisions will be made after the papers are presented at the conference. The awards for the best paper will be presented at 12:00pm on Thursday, July 30, in Gateway Ballroom, prior to the Special Plenary Panel.

10.3 Generating Test Programs to Cover Pipeline Interactions

Thanh Nga Dang – *National Univ. of Singapore, Singapore*
Abhik Roychoudhury – *National Univ. of Singapore, Singapore*
Tulika Mitra – *National Univ. of Singapore, Singapore*
Prabhat Mishra – *Univ. of Florida, Gainesville, FL*

12.1 GPU Friendly Fast Poisson Solver for Structured Power Grid Network Analysis

Jin Shi – *Tsinghua Univ., Beijing, China*
Yici Cai – *Tsinghua Univ., Beijing, China*
Wenting Hou – *Synopsys, Inc., Beijing, China*
Liwei Ma – *Synopsys, Inc., Beijing, China*
Sheldon X.-D. Tan – *Univ. of California, Riverside, CA*
Pei-Hsin Ho – *Synopsys, Inc., Hillsboro, OR*
Xiaoyi Wang – *Tsinghua Univ., Beijing, China*

15.1 An Efficient Approach for System-Level Timing Simulation of Compiler-Optimized Embedded Software

Zhonglei Wang – *Technische Univ. München, Munich, Germany*
Andreas Herkersdorf – *Technische Univ. München, Munich, Germany*

18.2 A Computing Origami: Folding Streams in FPGAs

Andrei Hagiescu – *National Univ. of Singapore, Singapore*
Weng-Fai Wong – *National Univ. of Singapore, Singapore*
David Bacon – *IBM Corp., Hawthorne, NY*
Rodric Rabbah – *IBM Corp., Hawthorne, NY*

22.1 Statistical Multi-Layer Process Space Coverage for At-Speed Test

Jinjun Xiong – *IBM Corp., Yorktown Hts., NY*
Yiyu Shi – *Univ. of California, Los Angeles, CA*
Vladimir Zolotov – *IBM Corp., Yorktown Hts., NY*
Chandu Visweswariah – *IBM Corp., Yorktown Hts., NY*

23.2 A Robust and Efficient Harmonic Balance (HB) Using Direct Solution of HB Jacobian

Amit Mehrotra – *Berkeley Design Automation, Santa Clara, CA*
Abhishek Somani – *Berkeley Design Automation, Bangalore, India*

24.3 New Spare Cell Design for IR Drop Minimization in Engineering Change Order

Hsien-Te Chen – *National Tsing-Hua Univ., Hsinchu, Taiwan*
Chieh-Chun Chang – *National Tsing-Hua Univ., Hsinchu, Taiwan*
TingTing Hwang – *National Tsing-Hua Univ., Hsinchu, Taiwan*

2009 DAC/ISSCC Student Design Contest Winners

The Student Design Contest promotes excellence in the design of electronic systems by providing a competition for graduate and undergraduate students at universities and colleges. It is co-organized by DAC and ISSCC. This year we received over 60 submissions in three categories: operational systems, operational chips and conceptual designs based on simulation. Nine award winners were selected. The Student Design Contest is co-sponsored by DAC, industry sponsors and ISSCC.

Awards will be presented at the DAC Pavilion on the exhibit floor, Booth 1928, Monday, July 27 from 5:00 – 6:00pm. The ceremony will include brief overview of presentations from each winning project team.

Smart Memories Polymorphic Chip Multiprocessor

Ofer Shacham, Zain Asgar, Han Chen, Amin Firoozshahian, Rehan Hameed, Christos Kozyrakis, Wajahat Qadeer, Stephen Richardson, Alexandre Solomatnikov, Don Stark, Megan Wachs, Mark Horowitz – *Stanford Univ., Palo Alto, CA*

Large-Scale SRAM Variability Characterization Chip in 45nm CMOS

Zheng Guo, Andrew Carlson, Liang-Teck Pang, Kenneth Duong, Tsu-Jae King Liu, Borivoje Nikolic - *University of California, Berkeley, CA*

A 600MS/s 30mW 0.13μm CMOS ADC Array Achieving over 60dB SFDR with Adaptive Digital Equalization

Wenbo Liu, Yun Chiu – *Univ. of Illinois, Urbana-Champaign, Urbana, IL*

A 212MPixels/s 4096x2160p Multiview Video Encoder Chip for 3D/Quad HDTV Applications

Li-Fu Ding, Wei-Yin Chen, Pei-Kuei Tsung, Tzu-Der Chuang, Pai-Heng Hsiao, Yu-Han Chen, Shao-Yi Chien, Liang-Gee Chen – *National Taiwan Univ., Taipei, Taiwan*

A Fully-Automated Process Characterization Macro for Gate Dielectric Breakdown

John Keane, Shrinivas Venkatraman, Chris H. Kim – *Univ. of Minnesota, Minneapolis, MN*
Paulo F. Butzen – *Univ. Estadual do Rio Grande do Sul, Porto Alegre, Brazil*

A Heterogeneous MPSOC with Hardware Supported Dynamic Task Scheduling for Software Defined Radio

Torsten Limberg, Markus Winter, Marcel Bimberg, Reimund Klemm, Marcos Tavares, Holger Eisenreich, Georg Ellguth, Jens-Uwe Schlüßler, Emil Matus, Gerhard Fettweis – *Technische Univ. Dresden, Dresden, Germany*
Hendrik Ahlendorf – *ZMD AG, Dresden, Germany*

An Ultrasensitive CMOS Magnetic Biosensor Array for Point-Of-Care (POC) Microarray Application

Hua Wang, Ali Hajimiri – *California Institute of Technology, Pasadena, CA*

Phoenix: An Ultra-Low-Power Processor for Cubic Millimeter Sensor Systems

Mingoo Seok, Scott Hanson, Yu-Shiang Lin, Zhiyoong Foo, Daeyeon Kim, Yoonmyung Lee, Nurrachman Liu, Dennis Sylvester, David Blaauw – *Univ. of Michigan, Ann Arbor, MI*

A 1.2 V 26 mW Configurable Multiuser Mobile MIMO-OFDM/-OFDMA Baseband Processor

Jung-Mao Lin, Hsin-Yi Yu, Yu-Jen Wu, Hsi-Pin Ma – *National Tsing-Hua Univ., Hsinchu, Taiwan*

REVIEWERS

A total of 682 manuscripts were submitted to the 46th DAC. The Conference Executive and Technical Program Committees wish to acknowledge the time and effort spent by the following people who reviewed these manuscripts. Our thanks to all of those who participated and contributed to the success of the conference.

Mark Aagaard
Magdy Abadir
Yael Abarbanel
Rawan Abdel-Khalek
Samar Abdi
Saurabh Adya
Vishwani Agrawal
Ali Ahmadinia
Suhail Ahmed
Usman Ahmed
Rami Ahola
Shoaib Akram
Mohammad Al Faruque
Yousra Alkabani
Bartomeu Alorda
Charles Alpert
Said Al-Sarawi
Behnam Amelifard
Chirayu S. Amin
Michael Anderson
Nikos Andrikos
Josef Angermeier
Will Archer
Laurent Arditi
Farhadur Arifin
Krit Athikulwongse
William Au
Pietro Babighian
Mustafa Badaroglu
Xiaoliang Bai
Luz Balado
Felice Balarin
James (Yong-Chan) Ban
David Baneres
Ansuman Banerjee
Francisco Barat
Sandro Bartolini
Christoph Bartoschek
Faik Baskaya
Prasenjit Basu
Shabbir Batterywala
Lars Bauer
Erwin Behnen
Dan Benua
Janick Bergeron
Jay Bhadra
Sandeep Bhatia
Bishnupriya Bhattacharya

Koustav Bhattacharya
Sambuddha Bhattacharya
Alizadeh Marafeh Bijan
Jesse Bingham
Heather Bittlestone
Stephen Blythe
Srinivas Bodapati
Thomas Boesch
Cristiana Bolchini
Bradley Neil Bond
Talal Bonny
Samir Boubezari
Ulrich Brenner
Philip Brisk
Annette Bunker
Doron Bustan
Paulo Butzen
Lukai Cai
Paul Campbell
Fabio Campi
Joan Carletta
Dan Carmi
Luigi Carro
Francky Catthoor
Eduard Cerny
Wander Cesario
Krishnendu Chakrabarty
Ashutosh Chakraborty
Herwin Chan
Vikas Chandra
Henry Chang
Ning Chang
Yao-Wen Chang
Nisha Checka
Gang Chen
Genlong Chen
Howard Chen
Hung-Ming Chen
Mingsong Chen
Tai-Chen Chen
Tsung-Hao Chen
Weiwei Chen
Xi Chen
Yang Chen
Ying-Yu Chen
Yu Chen
Zhimin Chen
Chen-Yong Cher

Eric Cheung
Patrick Chiang
Scott Erick Chilstedt
Darius Chiu
Minsik Cho
Kiyoung Choi
Pai H. Chou
Chris Chu
Jason Clemons
Mike Clinton
Rebecca Collins
Nicola Concer
Erika Cota
Scott A. Cromar
Hamed Dadgour
Rahul Dani
Animesh Datta
Scott Davidson
Aurelia De Colle
Fernando DeBernardinis
Andrew DeOrio
Onur Derin
Veerle Desmet
Marc DeWilde
Nagu Dhanwada
Michail Dimopoulos
Duo Ding
Eugene Ding
Quang Sy Dinh
Chen Dong
Xiangyu Dong
Frederic Doucet
Steven Drager
Xiaogang Du
Jianjun Duan
Hritam Dutta
Wolfgang Ecker
Stephen A. Edwards
Stephan Eggersgluess
Cindy Eisner
Tarek Ali El Moselhy
Reouven Elbaz
Abe Elfadel
Peeter Ellervee
Ahmed El-Rayis
Heiko Falk
Joachim Falk
Jia-Wei Fang

Morteza Fayyazi
Chunyang Feng
Alberto Ferrante
Goerschwin Fey
Leandro Fiorin
Paulo Flores
William Fornaciari
Cristiano Forzan
Luca Fossati
Ranan Fraer
Mark Fredrickson
Stefan Frehse
Martin Frerichs
Qiang Fu
Franco Fummi
Altaf Abdul Gaffar
Shanghua Gao
Youxin Gao
Alok Garg
Rajesh Garg
Annajirao Garimella
Lovic Gauthier
Aman Gayasen
Christian Genz
Andreas Gerstlauer
Amir Masoud Gharehbaghi
AmirHossein GholamiPour
Roberto Giorgi
Jens Gladigau
Michael Glaß
Amit Goel
Neeraj Goel
Shahin Golshan
Min Gong
Maziar Goudarzi
Elena Gramatova
David Grant
Matthias Gries
Joel Grodstein
Daniel Grosse
John Grout
Wolfgang Guenther
Zhihua Gui
Eric Xu Guo
Ruifeng Guo
Xu Guo
Pallav Gupta
Matthew Guthaus
Peter Hallschmid
Andrzej Handkiewicz
Frank Hannig
David Hathaway
Christian Haubelt
Chun He
Michael Healy
Stephan Held

Mona Hella
Sabine Helwig
Dwight Hill
Mokhtar Hirech
C. Richard Ho
Tsung-Yi Ho
Chih-Jen Hsiao
Michael Hsiao
Yu-Chung Hsiao
Jin Hu
Jiun-Lang Huang
Michael Huang
Shih-Hsu Huang
Wei Huang
Yu Huang
William Hung
Walid Ibrahim
Michiko Inoue
Tsuyoshi Iwagaki
Juergen Jaeger
Ashok Jagannathan
Shah Jahinuzzaman
Alok Jain
Himanshu Jain
Andhi Janapsatya
Wooyoung Jang
Geert Janssen
Alvin Jee
Christophe Jego
Jie-Hong Roland Jiang
Le Jin
Tongdan Jin
Minge Jing
Tong Jing
Bijoy Jose
Knut Just
Lalitha Mohana Kalyani
    Garimella
Gila Kamhi
Teja Karkhanis
Chandramouli V. Kashyap
Srinivas Katkoori
Masamichi Kawarabayashi
Mahesh Ketkar
Zurab Khasidashvili
Yasin Khatami
Chris Kim
Dae Hyun Kim
Kyoung Park Kim
Tae-Hyoung Kim
Youngsoo Kim
Sebastian Kinder
Dimitrij Kissler
Seok-Bum Ko
Dirk Koch
Karl Koehler

Alfred Koelbl
Yoshihisa Kojima
Alex Kondratyev
Hessam Kooti
Dmitry Korchemny
Joeseph Kozhaya
Andrzej Krasniewski
Victor Kravets
Rainer Kress
Franz Kreupl
Ratna Krishnamoorthy
Harsha Krishnamurthy
Smita Krishnaswamy
Ulrich Kuehne
Akash Kumar
Anurag Kumar
Shashi Kumar
Masanori Kurimoto
Eren Kursun
Young-Su Kwon
Bo-Cheng Lai
Liyang Lai
Vahid Lari
Alf Larsson
Erik Larsson
Viet Le
Jae Ho Lee
Junghoon Lee
Kangmin Lee
Ren-Jie Lee
Yeonbok Lee
Young-Joon Lee
Miriam Leeser
Alex Levin
Cheng-Hong Li
Dexin Li
Hong Li
Huawei Li
Peng Li
Xianfeng Li
Xin Li
Zhuo Li
Jiye Liang
Michael Lifshitz
Sung Lim
Chieh Lin
Xijiang Lin
Nicola L'Insalata
Chien-Nan Liu
Frank Liu
Xiaodong Liu
Xun Liu
Zhu Liu
Marisa Lopez-Vallejo
Yinghai Lu
Zhonghai Lu

Gregory M. Lucas
Martin Lukasiewycz
Slobodan Lukovic
Xu Luo
Yuchun Ma
Alberto Macii
Juan Antonio Maestro
Ali Mahdoum
Zohaib Mahmood
Pongstorn Maidee
Abhranil Maiti
Mehrdad Majzoobi
Sohaib Majzoub
Pui-In Mak
Wai-Kei Mak
Rakesh Malik
Yossi Malka
Hans Manhaeve
Mohammad Mansour
Giovanni Mariani
Ewout Martens
Thomas Martin
Maurizio Martina
Guido Masera
Jens Maßberg
Takshi Matsumoto
Peter Maurer
Dennis McCain
Judy McCullen
Kevin McCullen
Farhad Mehdipour
Richard Membarth
Seda Ogrenci Memik
Noel Menezes
Vinod Menezes
William Migatz
Paolo Miliozzi
Salvador Mir
Alan Mishchenko
Prabhat Mishra
Tania Mishra
Michale Moffitt
Nagm Mohamed
Maria Molina
Andrea Molino
Matteo Monchiero
Jose Monteiro
Robert Montoye
In-Ho Moon
Francisco Assis Moreira do
    Nascimento
Vasily Moshnyaga
Anna Moss
Saibal Mukhopadhyay
Dirk Müller
Brian Mulvaney

Satyakiran Munaga
Marcello Mura
Indira Nair
Tsuneo Nakata
Gi-Joon Nam
Alberto Nannarelli
Jagan Narasimhan
Sridhar Narayanan
Ramesh Narayanaswamy
Ognen Nastov
Ekanathan Palamadai
    Natarajan
Suriyaprakash Natarajan
Venki Natarajan
Leyla Nazhandali
Kyle Nelson
Eduardo Wanderley Netto
Dimitris Nikolos
Tasuku Nishihara
Franc Novak
Ian O'Connor
Kimihiro Ogawa
Umit Y. Ogras
Dongkeun Oh
Satoshi Ohtake
Mwaffaq Otoom
Jin Ouyang
Rohit Pagariya
Gus Palau
Gianluca Palermo
Murthy Palla
Luigi Palopoli
Chengzhi Pan
Xiaoda Pan
Rajendran Panda
Sujan Pandey
Chris Papachristou
Chris Papademetrious
Alexandros
    Papakonstantinou
Yegna Parasuram
Sudeep Pasricha
Krutartha Patel
Mohit Pathak
Marek Patyra
Jorgen Peddersen
Chi-Chen Peng
Gregory Peterson
Fedor Pikus
Christian Pilato
Sebastien Pillement
Stephen Plaza
Massimo Poncino
Mikhail Popovich
Katalin Popovici
Mario Porrmann

Mihalis Psarakis
Madhura Purnaprajna
Masood Qazi
Haifeng Qian
Katherine Qiang
Xiaoke Qin
Stephen Quay
Stefano Quer
Martin Radetzki
Jon Rafkind
Vijay Raghunathan
Karthik Rajagopal
Balaji Raman
Nagarajan Ranganathan
Jagdish Rao
Seid Hadi Rasouli
Kaushik Ravindran
Sherief Reda
Felix Reimann
Andre Reis
Haoxing Ren
Javier Resano
Michael A. Riepe
Danilo Rimondi
Tero Rissa
Fredy Rivera
Alberto Rosti
Frederic Rousseau
Abinash Roy
Jarrod Roy
Subir Roy
Hans Sahm
Biranchinath Sahu
Kewal Saluja
Pablo Sanchez
Marcos Sanchez-Elez
Ranganathan
    Sankaralingam
Tsutomu Sasao
Toshinori Sato
Sneh Saurabh
John Savage
Ioannis Savidis
Gunar Schirner
Thomas Schlichter
Ulf Schlichtmann
Klaus Schneider
Robert Schwencker
Andrew Seawright
Jaume Segura
Alper Sen
Simha Sethumadhavan
Kenshu Seto
Muhammad Shafique
Lesley Shannon
Manish Sharma

Namrata Shekhar
Rupesh Shelar
Eric Sherk
Kaijian Shi
Pengyu Shi
Sean Xiaokang Shi
Yiyu Shi
Youhua Shi
Dongwan Shin
Jeonghee Shin
Hamid Shojaei
Emily Shriver
Carlos Silva
L. Miguel Silveira
Eric Simpson
Amit Singh
Amith Singhee
Saurabh Sinha
Satish Sivaswamy
Kevin Skadron
Iouliia Skliarova
Valery Sklyarov
Gerard Smit
Mathias Soeken
Mani Soma
Sampada Sonalkar
Kin Cheong Sou
Ashish Srivastava
Zoran Stamenković
Ted Stanion
Balsha Stanisic
Steve Start
Todor Stefanov
Vladimir Stojanovic
Martin Streubühr
Markus Struzyna
Sander Stuijk
Prasad Subramaniam
Sri Raga Sudha Garimella
Andre Suelflow
Mohd-Shahiman Sulaiman
Jian Sun
Jin Sun
Krishnan Sundaresan
Savithri Sundareswaran
Stephen Sunter
Peter Sutton
Cliff Sze
Frank Szorc
Taraneh Taghavi
Murali Talupur
Yutaka Tamiya
Genichi Tanaka
Huaxing Tang

Xiaoyong Tang
Hideo Tanida
Alexander Taubin
John Taylor
Himanshu Thapliyal
Michael Theobald
Brandon Thompson
Daniel Tille
Praveen Tiwari
Saurabh Tiwary
Hiroyuki Tomiyama
Jelena Trajkovic
Louise Trevillyan
Anne-Marie Trullemans-
   Anckaert
Antonino Tumeo
Emre Tuncer
Mark Tuttle
Raimund Ubar
Carlos Valderrama
Werner van Almsick
Arjan van Genderen
Tom Vander Aa
Hans Vandierendonck
Harsh Vardhan
Dmitry G. Vasilyev
Andreas Veneris
Mahalingam Venkataraman
G. Venkatesh
Oreste Villa
Tiziano Villa
Massimo Violante
Ines Viskic
Kugan Vivekanandarajah
Ilya Wagner
Muhammad Aqeel Wahlah
Robert Walker
Steven Walstra
Lu Wan
Evelyn Wang
Wei-Shen Wang
Weixun Wang
Xiaofang Wang
Yu Wang
Yuankai Wang
Hiroshi Watanabe
Liqiong Wei
John-David Wellman
Yen-Cheng Wen
Xiaoqing Wen Wen
Stefan Wildermann
Gustavo Wilke
Robert Wille
Michael Wirthlin

Jens Peter Wittenburg
Francis Wolff
Peng Wu
Tai-Hsuan Wu
Hans-Joachim Wunderlich
Dong Xiang
Hua Xiang
Lin Xie
Yuan Xie
Lluís Ribas Xirgo
Chuan Xu
Qiang Xu
Qinwei Xu
Masaaki Yamada
Guang Yang
Jae-Seok Yang
Xiaojian Yang
Xuejun Yang
Young Jin Yoon
Hiroaki Yoshida
Guo Yu
Hao Yu
Peng Yu
Xiaochun Yu
Kun Yuan
Lin Yuan
Xin Yuan
Ping-Hung Yuh
Joseph Zambreno
Arek Zawada
Christian Zebelein
Jindrich Zejda
Jun Zeng
Xuan Zeng
Yaping Zhan
Bin Zhang
Chun Zhang
Chunhui Zhang
Lei Zhang
Liang Zhang
Sushu Zhang
Yilin Zhang
Lu Zhao
Xin Zhao
Yanling Zhi
Qi Zhu
Daniel Ziener
Tobias Ziermann
Zeljko Zilic
Lyudmila Zinchenko
Vladimir Zolotov
Ke Zong
Wei Zou

System Prototypes: Virtual, Hardware or Hybrid?

Organizers

Tom Borgstrom
Synopsys
Mountain View, CA

Eshel Haritan
CoWare
San Jose, CA

Chair

Ron Wilson
EDN
San Jose, CA

David Abada
Amicus Wireless
Sunnyvale, CA

Ramesh Chandra
Qualcomm
San Diego, CA

Chuck Cruse
LSI Corporation
Colorado Springs, CO

Andrew Dauman
Synopsys
Sunnyvale, CA

Olivier Mielo
ST-Ericsson
Sophia Antipolis, France

Achim Nohl
CoWare
Aachen, Germany

PANEL SUMMARY

Almost all SoC designs today use hardware prototyping at some point of the development cycle to perform hardware/software validation or test interfaces to real-world stimulus. Recently, virtual prototypes have emerged as a way to run system level tests and perform hardware/software co-simulation before silicon or hardware prototypes are available. Does the emergence of virtual prototyping mean the end for FPGA-based prototypes? Will the hardware prototype continue to live on as an integral part of the SoC design cycle? Will hybrids of virtual and hardware prototypes be the answer? Join our expert panelists in a spirited discussion on the best approach for speeding the verification and validation of complex SoC using system prototypes. Virtual, hardware or hybrid – which will reign supreme?

Categories and Subject Descriptors

B [Hardware]; B.7 [Integrated Circuits]; B.7.2 [Design Aids]; D [Software]; D.2 [Software Engineering]; D.2.4 [Software/Program Verification]; D.2.5 [Testing and Debugging]

General Terms

Measurement, Performance, Design, Economics, Verification.

Keywords

Virtual prototype, virtual platform, rapid prototype, FPGA prototype, system prototype, SystemC TLM, embedded software, hardware/software co-verification, system validation

Permission to make digital or hard copies of part or all of this work for personal or classroom use is granted without fee provided that copies are not made or distributed for profit or commercial advantage and that copies bear this notice and the full citation on the first page. To copy otherwise, to republish, to post on servers or to redistribute to lists, requires prior specific permission and/or a fee.
DAC'09, July 26-31, 2009, San Francisco, California, USA

PANELIST VIEWPOINTS

David Abada: When it comes to system level validation and embedded SW development, FPGA-based rapid prototyping platform proves to be a unique SOC/ASIC verification tool. I believe such systems are superior to compute-based prototyping tools, especially at final development phases where exhaustive system level validation tests are performed.

FPGA-based verification platforms provide SOC development and verification engineers with the following:

- Modeling accuracy. Implementation is a direct synthesis of final RTL with time tried accuracy, hence having minimal modeling risk. In the cases where FPGA implementation deviates from final ASIC implementation due to technology differences, the deviations are typically minor, adding little accuracy risk. This is also important to the SW developers where access to a close-to-final SOC platform further minimizes characterization and performance modeling risks.
- System level integration. FPGA-based prototyping platforms allow system level integration and tests with *actual* hardware typically external to the SOC. Components such as memories, communications controllers etc., can be added to the FPGA prototype, increasing test coverage and further reducing design risks.
- Near/real-time behavior. FPGA-based prototyping platforms enable functional tests with *actual* third party equipment/applications, exposing the design to real-world stimuli. In addition to extensive validation opportunities, such a system also provides an accurate environment for various SOC performance metrics, such as real-time throughput etc.
- Debugging. FPGA-based systems can be made to allow the observing of internal nodes using both hardware and SW tools. This allows a fairly quick and effective bug resolution.
- Portability. FPGA-based prototyping systems can be enclosed in box and transported to locations where physical

presence is mandatory such as customer's lab, applications in remote areas (i.e. base-stations), certification labs etc. Early SOC tests at a customer's site demonstrating near-final behavior are invaluable for both design validation and boosting the business success of final product.

As the saying goes "there is nothing like the real thing", FPGA-based prototype can be a very close equivalent to final implementation, offering comprehensive system validation environment and a valuable pre-silicon SW development environment.

Ramesh Chandra: The complexity of current SoC designs both in terms of architecture and functionality requires significant modeling and validation early in the design cycle. The overall product design cycles and cost of bugs in complex designs also demands more upfront validation. Virtual platforms with fast and efficient C models have provided an excellent framework for HW-SW co-design development and validation. Hardware prototype that have been widely used for pre-silicon validation of HW with applications will continue to provide for and augment the verification quality of complex IPs and SoCs. Complexity of some of the IPs and lack of full co-relation between C models and designs will invariably leave issues uncovered with virtual platforms and will be adequately addressed with HW prototyping.

Chuck Cruse: System Prototypes: Virtual, Hardware or Hybrid? The answer is yes - all of the above. The hybrid is not necessarily a final solution, but an intermediate step from a pure virtual platform to a final hardware prototype.

One of the promises of virtual prototyping is to minimize the effort and risk "step-function" classically seen in hardware/software integration. The hope is that the virtual platform could also facilitate the birthing of the hardware prototype. Historically, developing an FPGA based prototype has presented a step-function task, similar to (and including) hardware/software integration. The virtual prototyping environment may serve as an excellent vehicle for verifying the functional correctness of a partitioned FPGA prototype, before it is realized on a PCB.

A cascaded process is being investigated which starts with a virtual platform targeting the ASIC, followed closely by a virtual representation of the partitioned FPGA hardware prototype. A similar approach has succeeded in the past for a simulation environment with the RTL based DUT, where a partitioned FPGA solution was executed. Starting with the pure virtual model of the FPGA target, a progressive set of hybrid configurations could be implemented, ultimately resulting in a standalone ASIC prototype. The potential time savings using this approach would maximize the effectiveness of the hardware prototype in a normally narrow window of opportunity.

Andrew Dauman: Traditional verification methodologies have been sufficient and appropriate for several decades. With System on Chip designs finally being a reality, with increasing software content, increasing IP usage, and the integration of a wide range of peripherals (interfaces) all in one design, there is a requirement for an evolution in verification methodologies – System Prototyping. Verification of the components is no longer sufficient to ensure correct functionality and fulfillment of system requirements. For one technology to be able to supplant another in verification, it must at least fully replace all the capabilities of its predecessor. Ideally, it does so and brings additional benefits, in completeness of the solution or perhaps in performance metrics.

In the comparison of virtual versus physical (hardware) prototyping, we will find that each offers unique benefits, neither is a superset. Virtual prototypes enable software development earlier in the schedule (before RTL is fixed) with minimal distribution costs while FPGA-based prototypes enable software development and system validation prior to silicon availability with minimal extra modeling requirements. It is the characteristics of the desired verification methodology that will drive the use of virtual platforms, physical prototyping or hybrids.

With the focus then on methodology and not verification technology, there are some universal truths in verification: users want faster verification, comprehensive verification, and are increasingly verifying systems including embedded software, not just silicon, in the prototype. As each approach offers unique value, there may be many applications where selective use of a each prototype approach offers the most comprehensive verification with pragmatic schedules and costs. A hybrid approach offers fulfillment of the broadest possible set of requirements.

Olivier Mielo: Virtual Platforms are emerging as an alternative solution for SOC validation but can they completely replace HW prototyping? Do we have enough accuracy in a Virtual Platform?

Virtual platforms are modeled using object oriented programming languages (SystemC/C++). The designer inherits the full power of object oriented languages including flexibility, visibility, object specialization and incremental refinement from initial "draft phase" to a fully refined platform. This makes Virtual Platform very attractive in contrast to hardware prototype that requires synthesis, timing optimizations and debugging without much visibility for every revision.

Since RTL is too slow to be a reasonable solution for system level tasks there is a need for a faster model. The right solution however depends of the validation use case:

- **Virtual Platforms** are useful for functional verification where timing accuracy is less important and many software layers (like mobile phone protocol stack) are simulated. Timing can be added to Virtual Platforms but more the timing is accurate more the speed is slow down
- **Hardware prototypes** are very useful where timing is involved and when interfaces between cores or cores and IPs are being verified.

978-1-60558-497-3/09 $25.00 © 2009 ACM

What to do with only HW for step by step debugging on processors without breaking the real time flow?

- Post silicon validation may benefit from a **hybrid approach**. It will combine the accuracy and speed of the hardware with the flexibility and visibility of the virtual platform. The post silicon bugs found in field testing are often very tricky. For such cases ones the user can benefits from all available methods.

Achim Nohl: System prototyping has a long history within the design of electronic products. The evolution and innovation of system prototyping methods has been driven by the ever increasing design complexity. Due to the growing heterogeneity of hardware and software concepts used in a single electronic product, not a single prototyping solution can serve all needs at once. The selection of the right prototyping method should be made by the quality and productivity gain it can deliver during the design and maintenance of an electronic product. Therefore, different objectives of the various hardware and software design tasks have to be evaluated with the characteristics of the prototyping methods such as the effort of prototyping, performance, accuracy, visibility, controllability and connectivity. However, it would unrealistic to assume that different prototypes can be created for all the different hardware and software design tasks. Here, again a pragmatic trade-off need to be made between the value and the cost required for the prototyping. Going forward, a major requirement of a prototyping solution will be that it serves as an integration environment allowing different prototyping means to cooperate with each other. Depending on the design task and the focus of this tasks, sub-systems should be replaceable by simulation models at different levels of abstraction (untimed, approximately timed, cycle accurate, register transfer level), or hardware emulators (FPGA etc.), or even existing hardware. Virtual platforms and the TLM 2.0 interoperability standard serve as such an integrative prototyping environment.

978-1-60558-497-3/09 $25.00 © 2009 ACM

Circuit Techniques for Dynamic Variation Tolerance

Keith Bowman, James Tschanz, Chris Wilkerson, Shih-Lien Lu,
Tanay Karnik, Vivek De, and Shekhar Borkar

Intel Corporation, Hillsboro, OR

keith.a.bowman@intel.com

Abstract

Three circuit techniques for dynamic variation tolerance are presented: (i) Sensors with adaptive voltage and frequency circuits, (ii) Tunable replica circuits for timing-error prediction with error recovery, and (iii) Embedded error-detection sequential circuits with error recovery. These circuits mitigate the clock frequency guardbands for dynamic variations, thus improving microprocessor performance and energy-efficiency. These circuits are described with a focus on the different trade-offs in guardband reduction and design overhead. Opportunities for CAD to further enhance microprocessor performance and energy efficiency are offered.

Categories & Subject Description:

B.7.1   [Integrated Circuits]: Types and design styles

General Terms: Design, performance, reliability

Keywords: Dynamic variations, parameter variations, variation sensors, replica paths, timing errors, error detection, error-detection sequential, error recovery, error correction, variation-tolerant circuits, resilient circuits

1. Introduction

Dynamic variations in supply voltage (V_{CC}), temperature, transistor aging, cross-coupling capacitance, and multiple-input switching adversely impact the clock frequency (F_{CLK}) and energy efficiency of microprocessors. Conventional designs build guardbands into the operating F_{CLK} to ensure correct functionality within the presence of worst-case dynamic variations over the microprocessor lifetime. Consequently, these inflexible designs cannot exploit opportunities for higher performance by increasing F_{CLK} or lower energy by lowering V_{CC} during favorable operating conditions and lack of aging degradation. Since most systems usually operate at nominal conditions where worst-case scenarios rarely occur, these infrequent dynamic variations severely limit the performance and energy efficiency of conventional microprocessor designs.

In this paper, dynamic variation sources are reviewed and three circuit techniques are described to reduce the F_{CLK} guardbands for dynamic variations to improve microprocessor performance and energy efficiency. The variation-tolerant circuit techniques provide different trade-offs in guardband reduction and design overhead. CAD research opportunities to further enhance microprocessor performance and energy efficiency are offered.

Permission to make digital or hard copies of part or all of this work for personal or classroom use is granted without fee provided that copies are not made or distributed for profit or commercial advantage and that copies bear this notice and the full citation on the first page. To copy otherwise, to republish, to post on servers or to redistribute to lists, requires prior specific permission and/or a fee.

DAC'09, July 26-31, 2009, San Francisco, California, USA

2. Dynamic Variation Sources

The primary sources of dynamic variations include: (i) V_{CC} droops, (ii) temperature changes, (iii) transistor drain current aging and recovery, (iv) cross-coupling capacitance changes, and (v) multiple inputs switching (MIS) in logic gates. V_{CC} droops are induced from abrupt changes in switching activity, resulting in large current transients in the power delivery system. The magnitude and duration of V_{CC} droops depend on the interaction of capacitive and inductive parasitics at the board, package, and die levels with changes in current demand [1]. V_{CC} droops contain high frequency (i.e., fast changing) and low frequency (i.e., slow changing) components and occur locally and globally across the die. Temperature variations depend on workload, environmental conditions, and the heat-removal capability of the package. Temperature changes occur at a relatively slow time scale with local hot spots on the die.

Transistor drive current degrades slowly over time as a function of gate bias and temperature conditions. As an example in Fig. 1, path delay degradation is measured across different circuit types from a 45nm test-chip at various intervals of accelerated DC stress [2]. Since the path delay reduction is similar for different circuit types, a circuit path containing simple inverter components sufficiently tracks the worst-case DC stress on critical paths.

Cross-coupling capacitance between adjacent wires can induce a data-dependent RC delay on the wire, which varies as adjacent wires switch in the same or opposite direction. MIS affects logic gate delay depending on the input transitions and arrival times. Path delay variations resulting from cross-coupling capacitance and MIS are fast-changing and local.

3. Sensors with Adaptive Voltage & F_{CLK}

As illustrated in Fig. 2 for a generic microprocessor with N pipeline stages, on-die sensors coupled with dynamic voltage and frequency (DVF) control circuits [2]-[5] can adjust F_{CLK}, V_{CC}, or

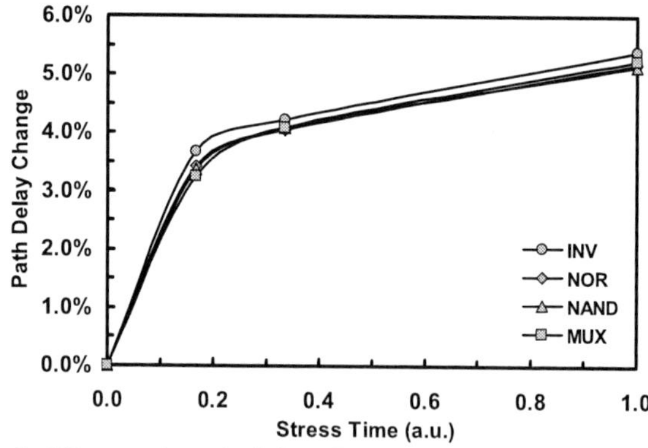

Fig. 1: Measured path delay degradation versus accelerated DC stress time for various circuit types [2].

978-1-60558-497-3/09 $25.00 © 2009 ACM

body bias in response to dynamic variations. Thermal sensors [4] are inserted at local hot spots on the die to measure worst-case temperature. V_{CC} droop sensors [1] detect global V_{CC} fluctuations observed across the die which affect all circuit paths. Although the actual drain current change from aging and recovery depends on the local transistor activity factor and state probability, aging sensors [2], [5]-[9] can track critical path degradation from worst-case DC stress to capture the effects and aging and recovery during power up/down sequences and sleep modes [2]. Since this design requires time to detect and respond to dynamic variations to avoid actual timing violations, this circuit technique reduces the F_{CLK} guardbands for slow-changing global variations, resulting in higher average F_{CLK}. Alternatively, the average F_{CLK} benefits may be converted to lower average energy by decreasing V_{CC}.

As an example of this circuit technique, measurements from a 90nm test-chip [5] demonstrate the dynamic response to temperature changes in Fig. 3. The test-chip [5] contains thermal and V_{CC} droop sensors as well as adaptive F_{CLK} and body biasing circuits. Starting at time zero in Fig. 3, the test-chip operates at an F_{CLK} and body bias targeted for 60°C. Body bias is optimized for minimum total power at a given F_{CLK}. As temperature increases, the thermal sensor detects the temperature change and communicates this change to the adaptive control circuits. Since the test-chip performance degrades at higher temperature, the adaptive control circuits respond by lowering F_{CLK} to ensure correct operation at the higher temperature. Furthermore, the body bias is also lowered to counteract the increasing leakage current at higher temperatures. As temperature decreases, F_{CLK} and body bias are adjusted to the original values to maintain higher performance and energy efficiency.

The disadvantages of on-die sensors and adaptive circuits include the inability to respond to fast-changing variations such as high-frequency V_{CC} droops [1] or local path-level variations.

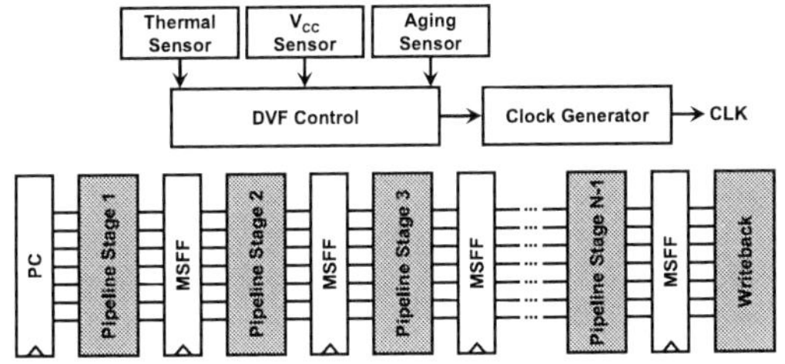

Fig. 2: **Sensors with dynamic voltage and frequency (DVF) control circuits.**

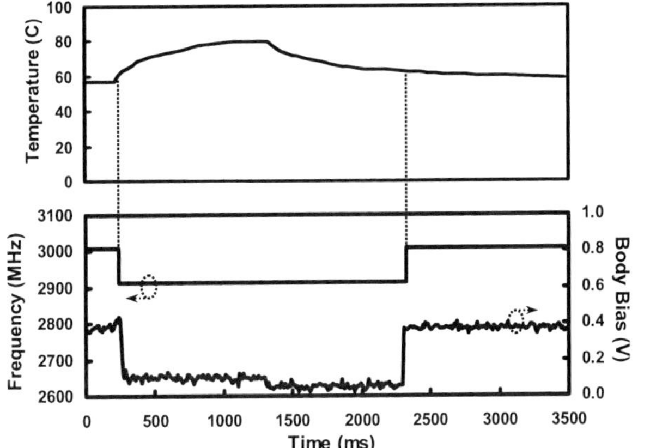

Fig. 3: **Measured dynamic response in clock frequency and forward body bias to on-die temperature variations [5].**

Furthermore, analog and digital sensors as well as adaptive circuits require substantial calibration time per die, leading to increased testing costs. Although sensors may be tuned during test to reduce the impact of within-die variations on sensor accuracy, an F_{CLK} guardband is still necessary to ensure coverage across a wide range of V_{CC} and temperature conditions as well as for transistor aging.

4. Tunable Replica Circuits with Recovery

In comparison to the previous circuit technique in Section 3, F_{CLK} guardbands are further reduced by combining tunable replica circuits (TRC) for timing-error prediction with error recovery [2] as illustrated in Fig. 4. This scheme removes the response time constraint imposed on the previous technique, thus eliminating the F_{CLK} guardbands for both slow and fast global dynamic variations. TRCs, which toggle each clock cycle, are inserted adjacent to each pipeline stage and calibrated to track local critical path delays. In contrast to previous TRC designs with time-to-digital converters [10], the TRCs in Fig. 4 drive an error-detection sequential (EDS) [11]-[17] to detect late timing transitions from dynamic variations. The TRC and the local pipeline stage use the same V_{CC} and clock, enabling the TRC to detect V_{CC} droops at finer granularity and to capture the clock-to-data relationship per pipeline stage.

TRCs are designed with different logic stages (e.g., inverters, NANDs, NORs, pass gates, and repeated interconnects). In Fig. 5, path delay sensitivities to V_{CC} droop and temperature are measured across various circuit types. At high V_{CC}, a combination of inverter and repeated interconnect delay components is adequate for the TRC to track critical path delay variations [10]. From Fig. 5, additional delay components (e.g., stacked gates and pass gates) or re-tuning may be required to track variations across wide ranges of V_{CC} and temperature, where delay sensitivities increase and the inverse delay dependence on temperature is observed.

When dynamic variations induce a late timing transition in the TRC, the EDS generates an error signal, which is pipelined to the writeback stage and the error control unit (ECU). Valid control logic at the writeback stage invalidates erroneous data. The ECU replays the appropriate instruction and signals the DVF control to halve F_{CLK} to ensure correct operation during the replay even if dynamic variations persist [17]. After the replayed instruction finishes, the ECU sends a reset signal to validate data at the writeback stage and signals the DVF control to resume at target F_{CLK}. By detecting and correcting timing errors, proper logic functionality is maintained. Since additional clock cycles are

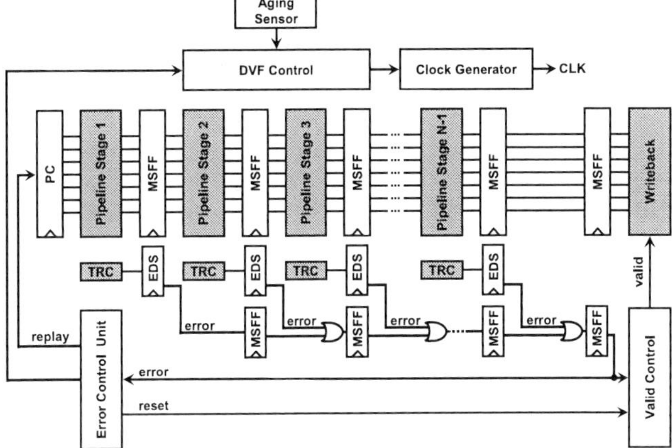

Fig. 4: **Tunable replica circuits (TRCs) for error prediction with error recovery.**

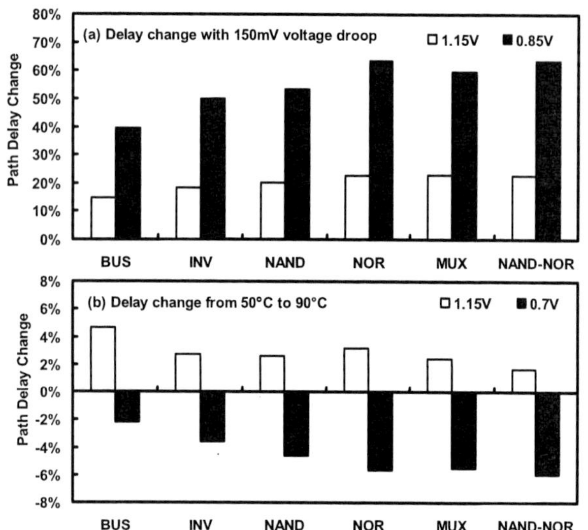

Fig. 5: Measured path delay change for various circuit types due to (a) 150mV voltage droop and (b) temperature change from 50°C to 90°C [2].

required for error recovery, instructions per cycle (IPC) reduce as errors occur. Assuming infrequent timing errors from dynamic variations, the IPC reduction is relatively small as compared to the F_{CLK} gains from removing the guardbands for slow and fast global dynamic variations, resulting in higher performance. As an alternative to performance benefits, the F_{CLK} gains may be traded-off for energy efficiency by reducing V_{CC}.

The disadvantages of the TRC with recovery include signaling errors and activating recovery when an actual error did not occur. In addition, TRCs cannot detect path-level dynamic variations and TRCs require a delay guardband to ensure that the TRC per pipeline is slower than the critical paths across wide ranges of V_{CC} and temperature. Although the digital TRCs are less complex than the analog sensors in Section 3, post-silicon tuning of TRCs impact testing costs. The recovery design requires an additional stabilization pipeline stage to accommodate the 1-cycle latency for propagating error signals in the N-1 pipeline stage [14]-[17]. This stabilization stage ensures that instructions are valid before committing the state at writeback.

5. Embedded Error Detection with Recovery

As illustrated in Fig. 6, embedding EDS circuits [11]-[17] in all critical paths eliminates the F_{CLK} guardbands for fast and slow as well as global and local dynamic variations. In contrast to the circuit techniques in Sections 3 and 4, this design does not require post-silicon tuning. Embedded EDS circuits detect late transitions in the actual critical paths. Error signals from each EDS per pipeline stage are combined via an OR tree to generate a single error signal. The recovery circuits in Fig. 4 and Fig. 6 are similar. When dynamic variations induce an actual timing error, the error is detected and corrected to maintain proper system functionality. In addition to removing the F_{CLK} guardbands for dynamic variations, further F_{CLK} benefits are possible by exploiting path-activation probabilities. If the slowest paths on the die are infrequently activated, the F_{CLK} may increase higher than the critical path operating frequency. When these critical paths are activated, the timing error is detected and corrected.

In Fig. 7, measurements from a 65nm test-chip [17] with embedded EDS and recovery circuits demonstrate the capability to

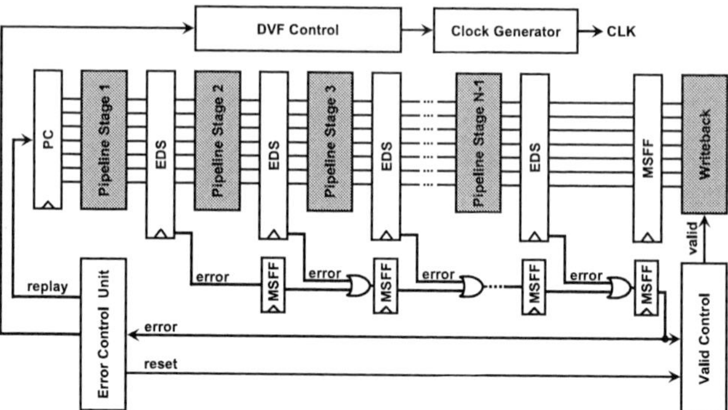

Fig. 6: Embedded error-detection sequential (EDS) circuits with error recovery.

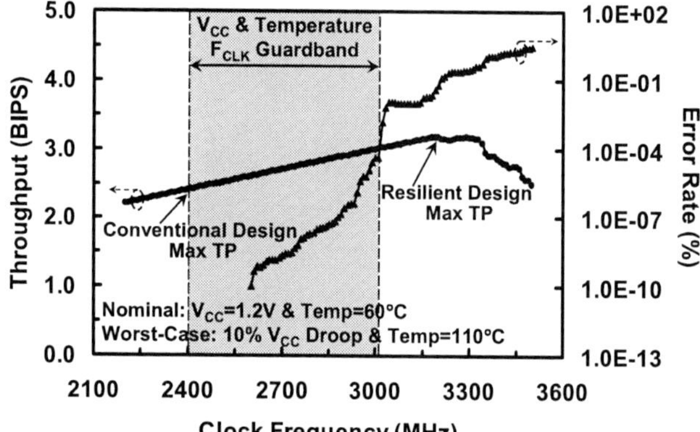

Fig. 7: Measured throughput (TP) and error rate versus clock frequency (F_{CLK}) for a 65nm resilient circuit test-chip with embedded EDS and recovery circuits [17].

eliminate the F_{CLK} guardbands for dynamic V_{CC} and temperature variations as well as to exploit path-activation probabilities for maximizing throughput. For a given F_{CLK}, error rate depends on the path histogram, as dictated by design optimization, as well as path-activation probabilities and environmental variations, as determined by workloads. In Fig. 7, throughput increases linearly as F_{CLK} increases with no errors. Once errors occur, IPC reduces as a function of error rate and recovery time. Since V_{CC} droop events are infrequent, throughput gains continue as F_{CLK} increases into the V_{CC} and temperature guardband region. When F_{CLK} reaches 3020MHz, the first path failure occurs under nominal conditions, resulting in a sharp error rate increase. Since the path-activation probability is low for slow paths in this workload, further throughput gains are achieved at higher F_{CLK}. The maximum throughput of 3.17 billion instructions per second (BIPS) corresponds to a 3200MHz F_{CLK}. Increasing F_{CLK} further leads to a larger error rate, where IPC reduction outweighs F_{CLK} gains. The maximum throughput to guarantee correct functionality within the presence of worst-case dynamic V_{CC} and temperature variations for a conventional design is 2.4BIPS, corresponding to an F_{CLK} of 2400MHz. From Fig. 7, a resilient design with embedded EDS and recovery circuits enables 25% throughput gain over a conventional design by eliminating the F_{CLK} guardband for dynamic V_{CC} and temperature variations and an additional 7% throughput increase from exploiting path-activation probabilities.

The disadvantages of embedded EDS circuits with recovery include the additional clock energy overhead for EDS circuits. Although datapath metastability has previously been the primary concern for EDS circuits, this issue has been addressed in recent EDS implementations [16]-[17]. The fundamental trade-off in

978-1-60558-497-3/09 $25.00 © 2009 ACM

6

embedded EDS circuits is the maximum path delay (max-delay) constraint versus the minimum path delay (min-delay) constraint. The fraction of the cycle time in which late timing transitions can be detected in embedded EDS circuits directly penalizes the min-delay constraint. Although this trade-off may not be advantageous for microprocessors with deep pipelines (i.e., small number of logic stages between sequentials) and corresponding stringent min-delay requirements, the microarchitecture for recent microprocessors has moved towards shallow pipelines (i.e., large number of logic stages between sequentials) to improve energy efficiency [18]-[19]. Microprocessors with shallow pipelines greatly relax the min-delay requirements as compared to a deep pipeline design, enabling a more effective trade-off of max-delay improvement for min-delay penalty. Furthermore, embedded EDS circuits require duty-cycle control circuits to maintain a constant high clock phase delay (i.e., error-detection window) at low and high F_{CLK} for min-delay protection. As described for the recovery design in Section 4, an additional stabilization pipeline stage is required for recovery.

6. Opportunities for CAD

Recommendations are offered for CAD research to further enhance the performance and energy efficiency of dynamic variation-tolerant circuit techniques. First, a rigorous analysis of microprocessor path-activation probabilities is desired across a range of typical workloads. This exploration would provide insight to the potential gains of exploiting activation probabilities in critical paths through embedded EDS and recovery circuits. Second, the path-activation probabilities introduce a new variable for timing optimization. As illustrated through a conceptual example in Fig. 8(a), all paths in a conventional timing optimization must satisfy a target cycle time. For embedded EDS and recovery circuits, timing optimization is based on combining the path-delay histogram with path-activation probabilities. As described in Fig. 8(b), infrequently activated paths could have a relaxed timing constraint as compared to frequently activated paths to reduce energy. Third, less stringent assumptions for cross-coupling capacitance and MIS could be applied in timing analysis for energy savings. Pre-silicon design methodologies often apply pessimistic assumptions for cross-coupling capacitance and MIS even though these events may rarely occur, leading to an over-design for many paths, and consequently, high energy costs. Allowing embedded EDS and recovery circuits to detect and correct late transitions from infrequent cross-coupling and MIS events, the timing constraints for cross-coupling capacitance and MIS could be relaxed, resulting in lower energy. Finally, the additional circuit features for dynamic variation tolerance introduce new placement and routing complexities, warranting an investigation in floorplanning optimization.

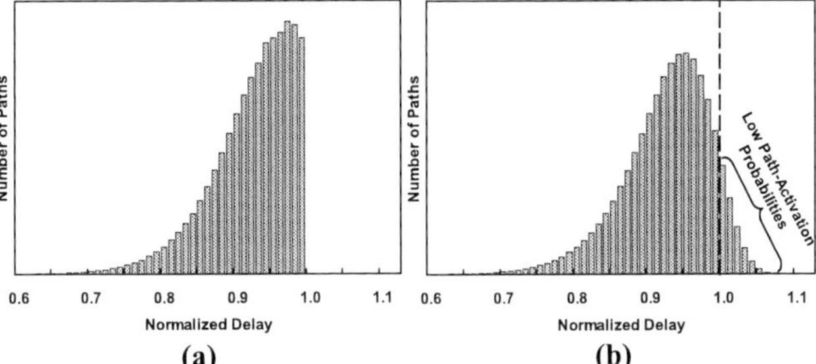

Fig. 8: Conceptual path-delay histograms optimized for (a) conventional and (b) embedded EDS with recovery circuits.

7. Conclusion

Dynamic variation sensors with adaptive voltage and frequency circuits, tunable replica circuits for timing-error prediction with error recovery, and embedded error-detection sequential (EDS) circuits with error recovery are three circuit techniques that reduce the clock frequency guardbands for dynamic variations to improve microprocessor performance and energy efficiency. These circuits provide unique trade-offs in guardband removal and overhead. CAD research opportunities to further enhance performance and energy efficiency include: (i) Explore path-activation probabilities across various workloads, (ii) Optimize timing by coupling the path-delay histogram with path-activation probabilities, (iii) Relax cross-coupling capacitance and multiple-input switching constraints in timing analysis, and (iv) Optimize floorplanning to accommodate additional circuit features.

8. References

[1]   A. Muhtaroglu, et al., "On-Die Droop Detector for Analog Sensing of Power Supply Noise," *IEEE J. Solid-State Circuits*, pp. 651-660, Apr. 2004.

[2]   J. Tschanz, et al., "Tunable Replica Circuits and Adaptive Voltage-Frequency Techniques for Dynamic Voltage, Temperature, and Aging Variation Tolerance," in *IEEE Symp. VLSI Circuits*, June 2009.

[3]   T. Fischer, et al., "A 90-nm Variable Frequency Clock System for a Power-Managed Itanium Architecture Processor," *IEEE J. Solid-State Circuits*, pp. 218-228, Jan. 2006.

[4]   R. McGowen, et al., "Power and Temperature Control on a 90-nm Itanium Family Processor," *IEEE J. Solid-State Circuits*, pp. 229-237, Jan. 2006.

[5]   J. Tschanz, et al., "Adaptive Frequency and Biasing Techniques for Tolerance to Dynamic Temperature-Voltage Variations and Aging," in *IEEE ISSCC Dig. Tech. Papers*, Feb. 2007, pp. 292-293.

[6]   J. Keane, et al., "An On-Chip NBTI Sensor for Measuring PMOS Threshold Voltage Degradation," in *Proc. ISLPED*, Aug. 2007, pp. 189-194.

[7]   E. Karl, et al., "Compact In-Situ Sensors for Monitoring Negative-Bias-Temperature-Instability Effect and Oxide Degradation," in *IEEE ISSCC Dig. Tech. Papers*, Feb. 2008, pp. 410-411.

[8]   T. Kim, et al., "Silicon Odometer: An On-chip Reliability Monitor for Measuring Frequency Degradation of Digital Circuits," *IEEE J. Solid-State Circuits*, pp. 874-880, Apr. 2008.

[9]   A. Cabe, et al., "Small Embeddable NBTI Sensors (SENS) for Tracking On-Chip Performance Decay," in *IEEE ISQED*, Mar. 2009, pp. 1-6.

[10] A. Drake, et al., "A Distributed Critical-Path Timing Monitor for a 65nm High-Performance Microprocessor," in *IEEE ISSCC Dig. Tech. Papers*, Feb. 2007, pp. 398-399.

[11] P. Franco and E. J. McCluskey, "Delay Testing of Digital Circuits by Output Waveform Analysis," in *Proc. IEEE Intl. Test Conf.*, Oct. 1991, pp. 798-807.

[12] P. Franco and E. J. McCluskey, "On-Line Testing of Digital Circuits," in *Proc. IEEE VLSI Test Symp.*, Apr. 1994, pp. 167-173.

[13] M. Nicolaidis, "Time Redundancy Based Soft-Error Tolerance to Rescue Nanometer Technologies," in *Proc. IEEE VLSI Test Symp.*, Apr. 1999, pp. 86-94.

[14] D. Ernst, et al., "Razor: A Low-Power Pipeline Based on Circuit-Level Timing Speculation," in *Proc. IEEE/ACM Intl. Symp. Microarchitecture (MICRO-36)*, Dec. 2003, pp. 7-18.

[15] S. Das, et al., "A Self-Tuning DVS Processor Using Delay-Error Detection and Correction," *IEEE J. Solid-State Circuits*, pp. 792-804, Apr. 2006.

[16] S. Das, et al., "Razor II: In Situ Error Detection and Correction for PVT and SER Tolerance," *IEEE J. Solid-State Circuits*, pp. 32-48, Jan. 2009.

[17] K. A. Bowman, et al., "Energy-Efficient and Metastability-Immune Resilient Circuits for Dynamic Variation Tolerance," *IEEE J. Solid-State Circuits*, pp. 49-63, Jan. 2009.

[18] V. Srinivasan, et al., "Optimizing Pipelines for Power and Performance," in *Proc. Intl. Symp. Microarchitecture (MICRO-35)*, Nov. 2002, pp. 333-344.

[19] A. Hartstein and T. R. Puzak, "The Optimum Pipeline Depth Considering Both Power and Performance," *ACM Trans. Arch. and Code Opt. (TACO)*, pp. 369-388, Dec. 2004.

Enabling Adaptability Through Elastic Clocks

Emre Tuncer

Elastix Corporation
Los Gatos, CA, USA

emre@elastix-corp.com

Jordi Cortadella

Universitat Politècnica de Catalunya
Barcelona, Spain

Luciano Lavagno

Politecnico di Torino
Torino, Italy

ABSTRACT

Power and performance benefits of scaling are lost to worst case margins as uncertainty of device characteristics is increasing. Adaptive techniques can dynamically adjust the margins required to tolerate variability and recover a significant part of the benefits lost due to worst-case conditions. Additionally, the stringent timing requirements for the synthesis of low-skew clock trees involve higher power consumption, and limit the adaptability to varying operating conditions. This paper introduces an elastic clocking scheme as an adaptive technique to confront variability and provide substantial power savings by dynamically adjusting to operating conditions. The synthesis and sign-off analysis of the elastic clocks is fully automated. Changes to the design flow and sign-off analysis of elastic clocks are addressed by automation of design flow support.

Categories and Subject Descriptors

B.8.2 Performance Analysis and Design Aids.

General Terms

Design, Reliability, Economics.

Keywords

Adaptive voltage scaling, desynchronization, GALS, low power design.

1. INTRODUCTION

Increasing process variability and decreasing operating voltage, as feature sizes scale down, reduce potential power-performance gains. Statistical design methods can reduce overdesign due to unrealistic worst-case assumptions [1], [6]. Large volume parts can be binned and sold at different price points to recover some portion of performance versus yield trade-off. However, binning is not applicable to ASICs due to commercial reasons, and statistical timing analysis does not address the margins needed for environmental variations such as temperature and voltage changes.

Permission to make digital or hard copies of part or all of this work for personal or classroom use is granted without fee provided that copies are not made or distributed for profit or commercial advantage and that copies bear this notice and the full citation on the first page. To copy otherwise, to republish, to post on servers or to redistribute to lists, requires prior specific permission and/or a fee.
DAC'09, July 26-31, 2009, San Francisco, California, USA

Ad-hoc recovery of design margins is common place in today's world. Off-the-shelf processor parts can be over-clocked well beyond their rated speeds by employing sophisticated cooling. By the same token, reducing the supply voltage to run a fast part at the specified frequency can save energy and power, which are becoming a primary concern for all electronic systems. Adaptive Voltage Scaling (AVS) provides this capability by sensing on-chip conditions dynamically and reducing or increasing the supply voltage to run the part at the required speed [2]. The power gains in AVS, however, are limited by the ability of predicting data path delays across the variation space [3].

AVS addresses static (process) or slowly varying (temperature, and to some degree aging) variations. The response time of the voltage regulation loop is usually hundreds or thousands of clock cycles. Cycle-to-cycle variations, such as IR-drop due to dynamic loads, must be handled by increasing the margins, thus reducing the achievable power gains.

Fine-grained application of AVS to individual cores or blocks in an SOC further improves power gains, based on load and performance requirements. However, the clock skew due to voltage domain crossing quickly becomes the limiting factor for performance, and increases the hold time fixing overhead. A solution to overcome this limitation is the adoption of asynchronous communication techniques between blocks [4]. The GALS (Globally Asynchronous Locally Synchronous) approach provides the flexibility to have each block driven by its own separate clock, and possibly supply voltage, while still enabling safe communication with other blocks. The main drawback of a GALS approach is the synchronization latency required to cross different clock domains, which may have a significant impact on the performance of the system.

Elastic clocks, where the period is dynamically adjusted to data path delays at the current operating conditions, provide the ability to minimize AVS margins due to IR-drop and clock skew. They also reduce latency in inter-block communication due to the asynchronous nature of the local clock controller protocol. Elastic clocks are implemented in a synchronous design flow through the desynchronization process.

2. DESYNCHRONIZATION

The separation between functionality and performance has always been a cornerstone of digital circuit design, enabling the development of tools that support functional specification, using snthesizable Verilog or VHDL; logic synthesis; physical design; equivalence checking and static timing analysis. Even testing schemes based on coupling of full stuck-at functional testing with limited at-speed performance testing, benefit from this separation.

978-1-60558-497-3/09 $25.00 © 2009 ACM

Asynchronous circuits are interesting for the performance, power, modularity and Electro-Magnetic Interference properties. However, they so far suffered from a fundamental violation of the separation between functionality and performance, which required new tools and languages to be developed and learned, thus making their widespread adoption virtually impossible. In recent years, extensive work has been done on a local clock signal generation technique known as "desynchronization" [5], [9].

The desynchronization process is automated by a tool that essentially augments Clock Tree Synthesis in a standard design flow, by creating localized clocks, and synthesizing handshake logic between the asynchronous controllers that generate them, thus creating elastic clocks. This control layer contains matched delays to guarantee the appropriate synchronization timing (setup and hold) between all pairs of communicating registers.

Desynchronization retains the separation between functionality and timing throughout the design flow, because it actually does not modify the logic and registers (apart from some transformations of flip-flops into latch pairs). In fact, the elastic circuit as a result of desynchronization still works using the notion of "cycles", except that now the cycles are determined by local combinational logic delays (including the effects of PVT variations) and handshaking, rather than by an external PLL or synchronous clock reference. As such, it permits the re-use of IP blocks, design tools and design flows that were originally created for synchronous design, but provides the advantages of asynchronous circuit to a "traditional" designer.

3. ELASTIC CLOCKS

The control layer, and the corresponding delay elements that generate the clocking for different partitions of the circuit, configure the elastic clocks that dynamically adjust their frequency to the data path delays. The delay elements are located close to the associated data path blocks, increasing the spatial correlation between the data path and the delay elements. Hence, the margins required to cover the variability can be reduced. The control layer and the delay elements can immediately react to the cycle-to-cycle variations that affect the data path, such as IR-drop and voltage ripple, and can delay the clock pulses as needed. This guarantees that the circuit operates correctly in a wider range of static or dynamic variations than traditional synchronous AVS.

The delay lines are synthesized such that each pair of communicating registers is always clocked to ensure setup and hold times. The false paths are ignored, as in usual Static Timing Analysis, and do not contribute to the synthesis of the delay lines. Multi-cycle paths require special attention. During calculation of the longest delay, the delays of the multi-cycle paths are normalized with the multi-cycle factor, and they contribute based on their slack. As a simplified example, if the minimum slack path between a pair of registers is due to a multi-cycle path of factor 3, then the delay line is synthesized such that it has one third the delay of the multi-cycle path.

Elastic clocks are generated on a per clock domain basis, multiple clock domains are handled by desynchronizing each domain separately. In this scheme, asynchronous clock domains are handled without any extra steps. The communication between asynchronous clock domains is left untouched through insertion of elastic clocks. Synchronous clock domains require explicit handshake signals to control data transfer between them.

The area overhead of elastic clocks is minimized by transforming only the interface flip-flops into master/slave latches. Interface flip-flops are defined as the registers that receive input from other partitions. Internal flip-flops have logic only from their partition in their transitive-fanin. Enable trees are automatically created for the master latches. The slave latches are connected to the slave enable trees, which also clock the original flip-flops that are internal to the partitions. The overhead of the master enable trees is minimized by appropriately choosing the number and location of the required master latches. The control logic has a constant overhead of a few hundred gates per partition; hence the relative area increase depends on the complexity of the partition connections which is mirrored in the control layer. The overall area overhead for a partition size of two hundred thousand gates is about 2%.

Scan testing is supported with the elastic clocks. At the tester non-overlapping clocks applied to transformed master/slave latches, hence operating them synchronously. The non-overlapping clocks can be generated internally from the test clock or provided externally. Overall, the elastic clocks use the same scan-based methodology and ATPG flows as before, and no change is necessary for testing flows.

Clock gating is a widely used technique to reduce dynamic power. During desynchronization, existing clock gating circuits are copied to slave enable trees. If a gating element controls registers across multiple slave enables trees, the gating element is cloned. The data path delays include clock gating checks and matched delay lines are adjusted to represent clock gating signal arrival times, if they become critical.

4. DESIGN FLOW

The insertion of elastic clocks is done after placement of a gate-level netlist, just before clock tree synthesis (Figure 1). The circuit is first partitioned into a set of logic blocks and elastic clocks are created for each block. Handshake signals are inserted to synchronize elastic clocks that drive the registers of each communicating block. The period of each elastic clock is determined by the post-placement timing characteristics of the corresponding block.

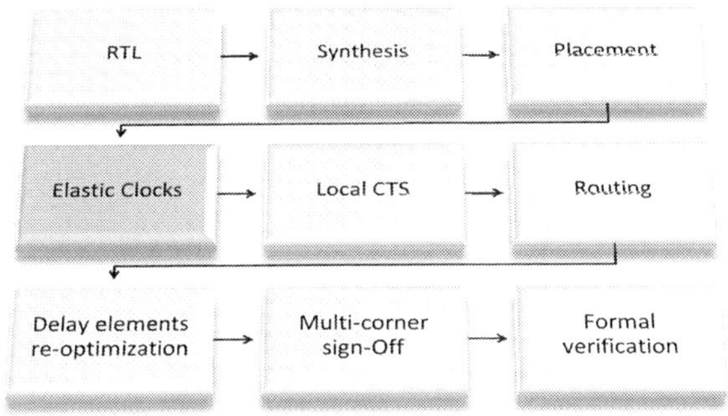

Figure 1. Elastic Clocks design flow

New timing constraints are generated automatically during the insertion of elastic clocks, to define new local clocks and constrain delay elements ("Delay" in Figure 2) for proper optimization. The delay elements have to match data path delays

978-1-60558-497-3/09 $25.00 © 2009 ACM

at all sign-off corners. The margins between the delay elements and the corresponding data path delays can be optimized using statistical methods [7].

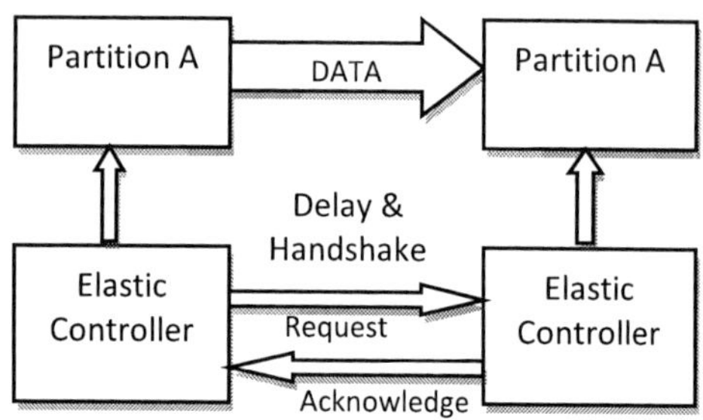

Figure 2. Elastic clock connections

During sign-off, explicit timing checks are done between the data path and the delay elements at all corners, to guarantee the correct clocking of the logic to meet the setup and hold conditions. In Figure 2, this corresponds to verifying that the data path delay between partitions A and B is always smaller than the delay between the associated controllers ("Delay") at all design corners, including the clock-tree insertion delays in each partition and the setup times of the receiving registers. These checks are performed using standard industrial sign-off tools by means of automatically generated scripts.

At the top level, each elastic block is connected to the desynchronized on-chip network asynchronously, by using the handshake signals. The handshake signals are used to "sense" the operating conditions and drive a voltage controller circuitry. This provides a unique opportunity to apply AVS where the sensing circuit is also an elastic clock generator, thus reducing the stringent clock skew requirements across multiple cores or blocks and reducing margins (e.g. due to cycle-to-cycle IR-drop). These signals are also used to control the voltage for a fine-grain AVS [8].

Figure 3 illustrates a connection scheme, where the clock controllers generate the elastic clocks for individual blocks and provide handshake signals ("HS") to the voltage controller and the asynchronous on-chip interconnect. The voltage controller drives a voltage regulator module to dynamically adjust the voltage, according to the relative timing of the handshake signals between the interconnect network and the elastic block. A purely asynchronous communication between the elastic blocks and the on-chip interconnect significantly reduces the communication latency.

5. CONCLUSION

An AVS scheme using elastic clocks eliminates stringent clock skew requirements across multiple cores and blocks, and reduces margins due to cycle-to-cycle variations. The asynchronous nature of elastic clocks avoids the latency penalty introduced by GALS schemes, and makes fine grain voltage scaling a possibility without performance overhead.

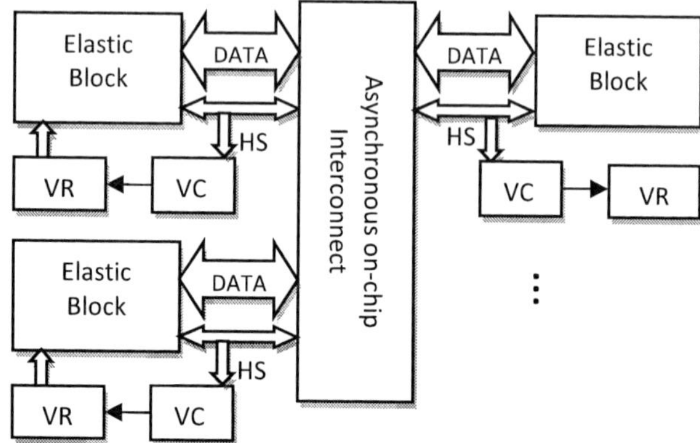

Figure 3. Scheme for voltage regulation

6. REFERENCES

[1] Visweswariah, C., "Statistical analysis and optimization in the presence of gate and interconnect delay variations," in *Proceeding of the 2006 international Workshop on System-Level interconnect Prediction*. Munich, Germany, March 04-05, 2006.

[2] Dhar, S., Maksimović, D., and Kranzen, B. Closed-loop adaptive voltage scaling controller for standard-cell ASICs. In *Proceedings of the 2002 international Symposium on Low Power Electronics and Design* (Monterey, California, USA, August 12 - 14, 2002). ISLPED'02. ACM, New York, NY, 103-107.

[3] Elgebaly, M. and Sachdev, M. Variation-aware adaptive voltage scaling system. *IEEE Trans. Very Large Scale Integr. Syst.* 15, 5 (May. 2007), 560-571.

[4] Iyer, A. and Marculescu, D. Power and performance evaluation of globally asynchronous locally synchronous processors. In *Proceedings of the 29th Annual international Symposium on Computer Architecture* (Anchorage, Alaska, May 25 - 29, 2002). International Symposium on Computer Architecture. IEEE Computer Society, Washington, DC, 158-168.

[5] J. Cortadella, A. Kondratyev, L. Lavagno, and C. Sotiriou, "De-synchronization: synthesis of asynchronous circuits from synchronous specifications," *IEEE Transactions on Computer-Aided Design*, vol. 25, no. 10, pp. 1904--1921, Oct. 2006.

[6] Cao, Y. and Clark, L. T. Mapping statistical process variations toward circuit performance variability: an analytical modeling approach. In *Proceedings of the 42nd Annual Conference on Design Automation* (Anaheim, California, USA, June 13 - 17, 2005). DAC'05. ACM, New York, NY, 658-663.

[7] Liu, Q. and Sapatnekar, S. S. Synthesizing a representative critical path for post-silicon delay prediction. In *Proceedings of the 2009 international Symposium on Physical Design* (San Diego, California, USA, March 29 - April 01, 2009). ISPD '09. ACM, New York, NY, 183-190.

[8] J. Cortadella, V. Singhal, E. Tuncer, and L. Lavagno. A variability-aware scheme for high-performance asynchronous circuit voltage regulation. US patent application. November 2008

[9] Andrikos, N., Lavagno, L., Pandini, D., and Sotiriou, C. P. A fully-automated desynchronization flow for synchronous circuits. In Proceedings of the 44th Annual Conference on Design Automation (San Diego, California, June 04 - 08, 2007). DAC '07. ACM, New York, NY, 982-985.

978-1-60558-497-3/09 $25.00 © 2009 ACM

Addressing Design Margins through Error-tolerant Circuits

Shidhartha Das[1], David Blaauw[2], David Bull[1], Krisztián Flautner[1] and Rob Aitken[1]
[1]ARM Ltd., Cambridge, U.K
[2]Department of EECS, University of Michigan, Ann Arbor, U.S.A

ABSTRACT

We review adaptive design techniques with particular emphasis on error-tolerant techniques. We compare and contrast traditional adaptive approaches with error-tolerant techniques and analyze the margins eliminated by each of them. We discuss the applications of the latter to on-chip communication and signal-processing. Finally, we focus on a specific example of an error-tolerant technique for general-purpose computing called Razor.

I. INTRODUCTION

Increasingly, design margins are required to address rising PVT variations leading to substantial power and performance losses. Adaptive techniques mitigate the impact of margins by dynamically tuning circuit parameters to compensate for variations. However, such techniques cannot track localized transients, the margins for which can be significant, especially at advanced process nodes. This has motivated recent research efforts into, so-called, error-tolerant approaches for dynamic compensation. In such techniques, intermittent timing errors during circuit operation are detected and recovered from. Allowing circuits to fail enables elimination of worst-case margins leading to significant improvements in energy-efficiency and performance.

II. CATEGORIZING VARIATIONS

At smaller geometries, inter- and intra-die process variations worsen due to inherent limitations in accurately controlling the manufacturing process. Environmental uncertainties such as power supply droops, temperature hot-spots, coupling noise and clock jitter as well as transistor ageing contribute to performance variations of transistors. Variations can be classified as local (e.g. temperature hot-spots) or global (e.g. ambient temperature fluctuations) based on their spatial reach. Depending on their temporal rate of change, variations can be categorized as slow-changing (e.g. process variations and ageing effects) or fast-changing (coupling noise). Compensating for such variations requires operating at higher voltages or lower frequencies to account for unforeseen slow-down caused due to worst-case PVT variations. This process of margining ensures correctness at the expense of higher power and performance impact.

III. TRADITIONAL ADAPTIVE TECHNIQUES

Instead of operating at a single operating point, adaptive techniques tune voltage and frequency as determined by silicon and operating conditions. The traditional adaptive design approaches [1-4], or the "always-correct" techniques, use a pre-characterized look-up table or "canary" circuits to predict the failure limit of a chip and operate the system close to the predicted limit.

A. Look-up table based approach

In the look-up table based approach, the processor is pre-frequency for a given supply voltage.

Permission to make digital or hard copies of part or all of this work for personal or classroom use is granted without fee provided that copies are not made or distributed for profit or commercial advantage and that copies bear this notice and the full citation on the first page. To copy otherwise, to republish, to post on servers or to redistribute to lists, requires prior specific permission and/or a fee.
DAC'09, July 26-31, 2009, San Francisco, California, USA

The safe voltage-frequency pairs are obtained by performing conventional timing analysis on the processor. Typically, the operating frequency is decided based on the deadline under which a given computational task needs to be completed. Accordingly, the supply voltage corresponding to the frequency requirement is "dialed in". The table look-up approach exploits periods of low CPU utilization by dynamically scaling voltage and frequency, thereby leading to energy savings. However, its reliance on conventional timing analysis performed at the combination of worst-case process, voltage and temperature corners implies that none of the safety margins are eliminated at a particular operating point.

B. Canary-circuits based approach

An alternative approach relies on the use of the so-called "canary" circuits to predict the failure point [1-4]. Canary circuits are typically implemented as delay chains which approximate the critical path of the processor. Voltage and frequency are scaled to the extent that this replica-delay path fails to meet timing. The replica-path tracks the critical-path delay across inter-die process variations and global fluctuations in supply voltage and temperature, thereby eliminating margins due to global PVT variations. However, the on-die location of the critical-path and its replica differs. Consequently, margins are added to the replica-path in order to budget for delay mismatches due to intra-die process and local variations in temperature and supply voltage. Margins are also required to address fast-changing transient effects, such as coupling noise, which are difficult to respond to in time using this approach. Furthermore, mismatches in the scaling characteristics of both paths require additional safety margins. These margins ensure that the processor still operates correctly even at the point of failure of the replica-path.

IV. ERROR-TOLERANT TECHNIQUES

As process technology scales, the local variations worsen thereby undermining the efficacy of traditional adaptive techniques. "Error-tolerant" techniques address local variations by scaling voltage and frequency till the point where the processor incurs timing errors. Error-detection circuits flag such an occurrence and engage a recovery mechanism to restore correct state. This eliminates all worst-case safety margins and enables significantly improved performance and energy efficiency over the traditional techniques. Their relative complexity makes the general applicability of such systems difficult. However, they are naturally amenable for communications and signal-processing where existing mechanisms can be overloaded to detect and correct timing errors.

Worm et al. [5] apply this concept to self-calibrating on-chip interconnects wherein bit-transfers occur at voltages below the safe

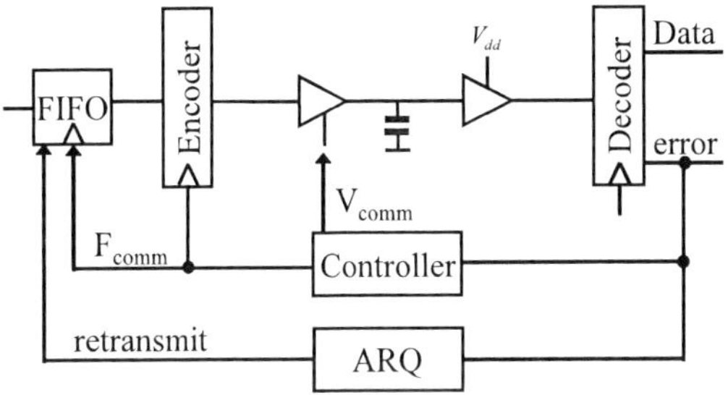

Figure 1 Self-calibrating interconnects

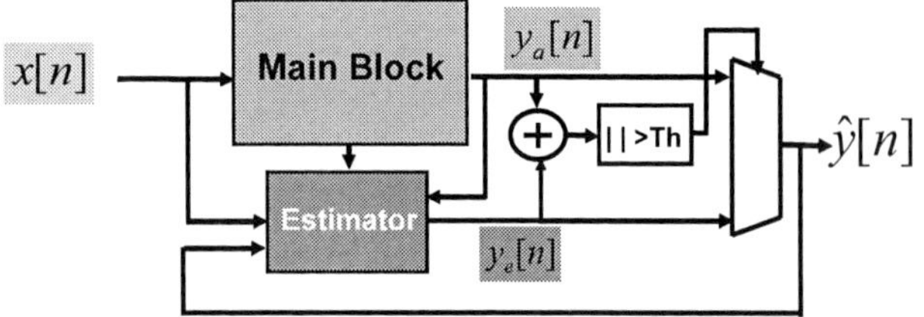

Figure 2 Algorithmic Noise Tolerance

limit. This enables bit-transfers at the lowest possible operating voltage while still guaranteeing the required performance and the targeted Bit Error Rate (BER). Error-detection occurs by encoding data words with so-called self synchronizing codes before transmission. The receiver is augmented with a checker unit that decodes the received code word and flags timing errors. Correction occurs by requesting re-transmission, as shown in figure 2. Furthermore, an additional controller obtains feedback from the checker and accordingly adjusts the voltage and the frequency of the transmission. By reacting to the error-rates, the controller is able to adapt to the operating conditions and thus eliminate worst-case safety margins. This improves the energy efficiency of the on-chip busses with negligible BER degradation.

Algorithmic Noise Tolerance [6] by Shanbhag et al. uses voltage-overscaling to significantly reduce energy consumption of processing blocks such as FIR filters while incurring intermittent timing errors. The main block is voltage scaled beyond the point of failure. Error-detection occurs by comparing the output of the main processor block against an estimator block which computes correct result based on previous history. Error-correction occurs by overwriting the result of the main block with that of the estimator block. The estimator block is significantly cheaper in terms of area and power as compared to the main block which is being voltage-scaled. At low error-rates, the benefits of aggressive scaling on the main block compensates for the overhead of correction, leading to significant energy savings.

V. ERROR-TOLERANT TECHNIQUES FOR GENERAL-PURPOSE COMPUTING

In general-purpose computing, timing errors should not corrupt the

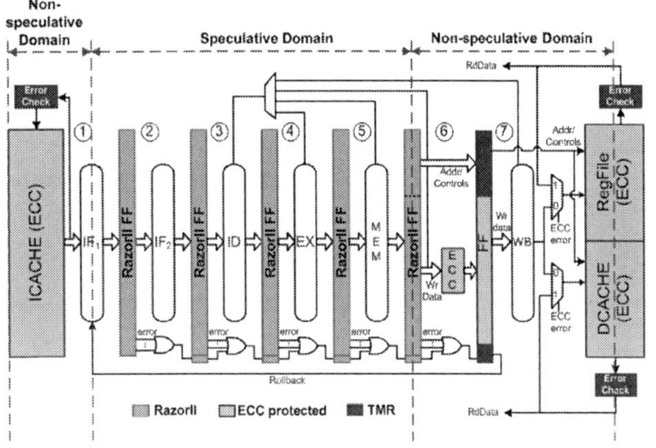

Figure 3 RazorII processor pipeline diagram

committed architectural state. We proposed Razor [7] as the first "error-tolerant" adaptive technique applied to general-purpose computing. Razor uses a delay-error tolerant flip-flop which detects timing errors by flagging spurious transitions on critical-path endpoints [8]. Recovery is achieved through a conventional architectural replay mechanism. This enables the supply voltage to be scaled to the point of first failure (PoFF) of a die for a given frequency. Thus, all margins due to global and local PVT variations

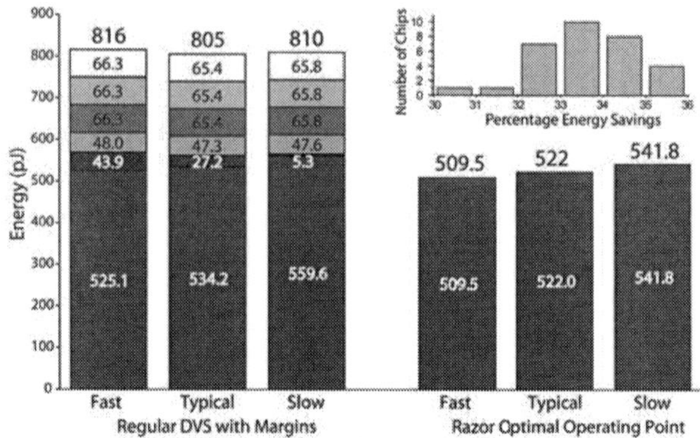

Figure 4 RazorII energy savings

are eliminated, resulting in significant energy savings. In addition, the supply voltage can be scaled even lower than the PoFF into the sub-critical region, deliberately tolerating a targeted error rate. Razor error-detection also enables tolerance to logic and register SER.

We designed and implemented a Razor-enabled 64-bit processor implemented in 0.13μm technology. The architecture (Figure 3) is divided into a pipeline with speculative state protected using RazorII flip-flops (FF), and a non-speculative memory and register file protected by ECC or triple-module redundancy (TMR). The 7th stage was designed to be non-timing critical to stabilize the pipeline state. In the event of an error, the pipeline is flushed and the failing instruction is re-executed. In case of repeatedly failing instructions, the error controller switches the clock frequency by half for 8 cycles. Figure 4 shows the measured energy dissipation for 3 die when operating at 0.04% error rate. Gains were 33.1 to 37.5% compared to the energy when the supply voltage is elevated to ensure correct operation for all 31 fabricated die at 85C with 10% margin for wearout, supply fluctuation and safety.

Bowman et al. [9] developed a similar approach for high-performance chips. Instead of keeping frequency fixed and reducing the supply voltage, they keep the supply voltage constant and use error-detection to improve throughput by 25-32% at the same operating voltage.

VI. CONCLUSION

In this paper, we compared traditional adaptive techniques with error-tolerant techniques and discussed the margins eliminated by them. We discussed a specific "error-tolerant" approach for general-purpose computing called Razor.

REFERENCES

[1] A. Drake, et al., ISSCC, 2007
[2] T. Burd, et al. ,JSSC, Vol. 35, No. 11, 2000
[3] M. Nakai, et al., JSSC, Vol. 40, No. 1, 2005
[4] K. Nowka, et al., JSSC, Vol. 37, No. 11, 2002
[5] Worm et al., *TVLSI*, Vol. 13, No. 1, January 2005.
[6] R. Hegde and N. R. Shanbhag, *JSSC*, Vol.39, No. 2, February 2004.
[7] S. Das, et al., JSSC, Vol. 41, No. 4, 2006
[8] Blaauw et al., ISSCC 2008
[9] Bowman et al., ISSCC 2008

Worst-Case Aggressor-Victim Alignment with Current-Source Driver Models

Ravikishore Gandikota
University of Michigan

Li Ding
Synopsys, CA

Peivand Tehrani
Synopsys, CA

David Blaauw
University of Michigan

ABSTRACT

Crosstalk delay-noise which occurs due to the simultaneous transitions of victim and aggressor drivers is very sensitive to their mutual alignment. Hence, during static noise analysis, it is crucial to identify the worst-case victim-aggressor alignment which results in the maximum delay-noise. Although several approaches have been proposed to obtain the worst-case aggressor alignment, most of them compute only the worst-case stage delay of the victim. However, in reality it is essential to compute the worst-case combined delay of victim stage and victim receiver gate [5, 9]. We propose a heuristic approach to compute the worst-case aggressor alignment which maximizes the victim receiver output arrival time. In this work, we use a novel *cumulative gate overdrive voltage* ($CGOV$) metric to model the victim receiver output transition. HSPICE simulations, performed on industrial nets to validate the proposed methodology, show an average error of 1.7% in delay-noise when compared to the worst-case alignment obtained by an exhaustive sweeping.

Categories and Subject Descriptors

B.7.2 [**Integrated Circuits**]: Design Aids; B.8.2 [**Performance and Reliability**]: Performance Analysis and Design Aids

General Terms

Algorithms, Design

Keywords

Crosstalk, delay noise, CSM

1. INTRODUCTION

Continuous scaling of device dimensions in the nano-meter regime has led to several key challenges in the timing verification of circuits. As the spacing between adjacent wires continues to shrink, the coupling capacitance now dominates the total wire capacitance. Furthermore, as technology advances, we are seeing increasing chip frequencies and decreasing voltage margins. All of the above trends exacerbate crosstalk noise which occurs due to the charge transfer between simultaneously-switching capacitively-coupled inter-

Permission to make digital or hard copies of part or all of this work for personal or classroom use is granted without fee provided that copies are not made or distributed for profit or commercial advantage and that copies bear this notice and the full citation on the first page. To copy otherwise, to republish, to post on servers or to redistribute to lists, requires prior specific permission and/or a fee.

DAC'09, July 26-31, 2009, San Francisco, California, USA

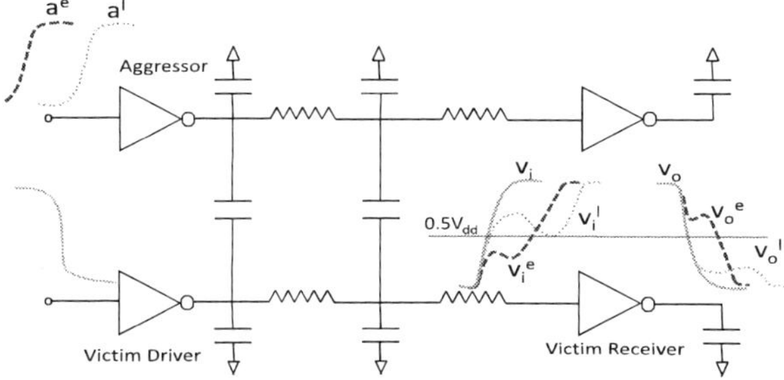

Figure 1: Aggressor alignment maximizes the victim receiver input delay

connects. Therefore, it has become imperative to accurately model the impact of crosstalk noise on circuit delay while performing signoff timing analysis of nano-meter designs.

Besides crosstalk noise, we have several non-linear effects that must be modeled in order to accurately estimate the gate and interconnect delays. These effects include multiple-input switching, resistive and inductive interconnects, power-grid noise and non-linear receiver gate capacitances, etc. Traditionally, cell delays have been computed by using slew/capacitance based look-up tables or k-factor equations, where the output load is usually a lumped capacitance and the input transition is approximated by a ramp or other characterization waveform shapes [1, 2, 3]. In this framework, an iterative effective capacitance ($Ceff$) based technique was developed in [4] to model the resistive shielding effect observed in distributed interconnects. In [5, 11], the authors obtain a more accurate estimate of the crosstalk noise bump by accounting for the non-linearity of victim driver resistance. However, the above modifications are mostly ad-hoc in nature and may not be very accurate when they are all combined together to perform the timing verification of nano-meter designs. In contrast, current source-based models (CSM) have emerged as a more fundamental approach for performing timing analysis since they are independent of pre-characterized ramp input waveforms and lumped capacitive output loading. In [7], a CSM was proposed where the output current depends on the DC voltage levels of the input and the output pins and an extra capacitance was added to the output pin to account for the transient effects. In [9], it was shown that the CSM can be effectively used for performing crosstalk noise analysis.

Traditionally, the objective of delay-noise analysis has been to maximize (or minimize for MIN analysis) the victim stage delay. Under the linear superposition assumption, it was shown in [10] that the victim stage delay is maximized, for a rising victim transition, if the peak of the coupling noise bump (Vp) is aligned at the point where the noiseless victim waveform crosses the $0.5Vdd + Vp$ voltage level. However, the true objective of delay-noise analysis is not to maximize the victim stage delay, but to maximize the combined sum of the victim stage delay and the victim receiver gate delay. In other words, the worst-case alignment of aggressors should maximize the output arrival time of the victim receiver gate. We show with an example that maximizing the former quantity does not guarantee maximization of the latter. In Figure 1, we have a capacitively-coupled victim-aggressor network with a lightly loaded victim receiver gate. The victim input transition is fixed and the alignment of the aggressor input transition is varied. The noisy victim waveforms v_i^e and v_i^l correspond to the two aggressor transitions a_i^e and a_i^l respectively. It is can be noted that v_i^l has a later 50% crossing time (t_{50} or *arrival time*) and consequently a greater stage delay than v_i^e. However, the victim transition v_i^l (with the maximum stage delay) results in the output transition v_o^l for which the coupling noise arrives too late after it has finished switching. In comparison, the earlier victim transition v_i^e, which has a lesser interconnect delay, results in the victim output transition v_o^e having the latest arrival time. Hence, maximizing only the victim stage delay can sometimes be inaccurate and the worst-case aggressor alignment should be computed such that it maximizes the victim receiver output arrival time.

It was shown in [5] that the worst-case aggressor alignment is a non-linear function of victim slew rate, noise bump and victim receiver output loading. They proposed the use of pre-characterized look-up tables to identify the worst-case aggressor alignment which however requires additional overhead in cell library characterization. In [9], techniques of constrained optimization were used to obtain the worst-case aggressor-victim alignment with the objective of maximizing the victim output arrival time. However, this may require multiple non-linear simulations and can be expensive in terms of runtime overhead. In this work, we present a heuristic method to find the worst-case aggressor-victim alignment considering non-linear CMOS drivers. We propose the *cumulative gate overdrive voltage* ($CGOV$) metric as a proxy for the total victim receiver output current. We know that the rate of the victim receiver output transition is proportional to the amount of current sourced by the victim receiver gate. Hence, the alignment with the lowest $CGOV$ will result in the slowest output transition having the latest output arrival time. Using the $CGOV$ metric, we propose a heuristic approach to compute the worst-case aggressor alignment which maximizes the victim receiver output arrival time. Since the victim receiver output arrival time is estimated without actually simulating the output waveform, the proposed approach proves to be very runtime efficient. HSPICE simulations performed on industrial nets to validate the proposed methodology show an average error of 1.7% in delay-noise when compared to the worst-case alignment obtained by an exhaustive sweep.

2. PROBLEM DESCRIPTION

In this section, we analyze the problem of computing the worst-case crosstalk delay in a static timing analysis (STA) framework, where every net has a timing window representing the period within the clock cycle within which the net can switch. In [14], it was shown that victim alignment at the latest point in its timing window was optimal and always resulted in the latest victim output arrival time. In this work, we focus on the problem of computing the alignment of aggressors relative to the victim transition such that they satisfy their respective timing window constraints and produce the maximum delay-noise at victim output transition.

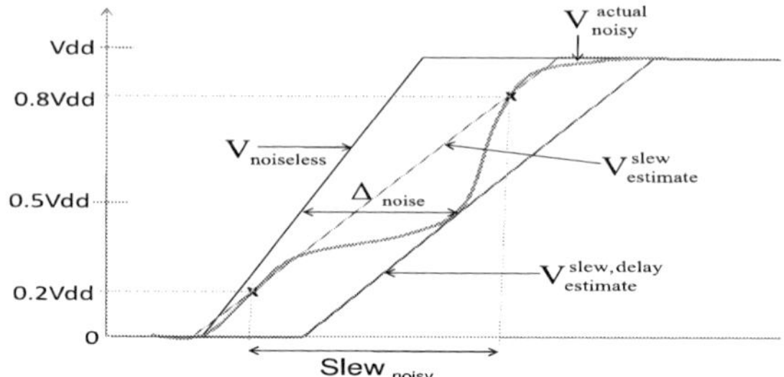

Figure 2: Waveforms at victim receiver input

It must be noted that the problem of computing delay-noise and timing windows are mutually dependent, since delay-noise depends on aggressor timing windows and timing windows are a function of the delay-noise coupled onto them. However, it was shown in [15, 16, 17] that this *chicken-and-egg* problem could be solved by updating the delay-noise and timing windows iteratively. In this paper, we consider the problem of computing the worst-case delay-noise, given the aggressor timing windows at some iteration of the outer loop.

As seen earlier, the worst-case delay-noise should maximize the victim output arrival time and not just the victim stage delay. The computation of victim output transition requires two steps, first the noisy waveform is computed at victim receiver input, second the victim receiver gate is simulated with the noisy input waveform. In cases where the victim net has a significant amount of coupling, the shape of the noisy victim waveform (e.g. V^{actual}_{noisy} in Figure 2) can be quite different from a ramp. [1] Such noisy victim waveforms waveforms cannot be used directly as inputs while simulating the victim receiver gate with traditional look-up table based cell delay models which are characterized with ramp input waveforms. Instead, an *equivalent ramp* is often fitted to noisy victim waveform and then used as the input transition for the victim receiver gate. Several heuristic approaches can be used to fit an equivalent ramp signal. One approach matches the slew rate of the noisy victim transition by fitting a ramp (i.e. $V^{slew}_{estimate}$ in Figure 2) through the 20-80% VDD voltage trip points of V^{actual}_{noisy}. For more conservative delay-noise analysis, the above ramp can be delayed (i.e. $V^{slew,delay}_{estimate}$) such that its arrival time matches that of V^{actual}_{noisy}. It can be seen in the figure that $V^{slew,delay}_{estimate}$ under-estimates the actual waveform V^{actual}_{noisy} and results in a pessimistic victim output arrival time. Alternatively, the authors in [12] propose the use of an equivalent ramp which is closest to the noisy victim waveform in a weighted least-square sense. In [13], the authors obtain a *transition quantity* by integrating the area beneath the noisy victim waveform and use it as a metric to obtain an equivalent ramp waveform.

With traditional look-up table based delay models, it is not possible to compute the *exact* victim output waveform with arbitrary input waveforms. Although, aggressor alignment which maximizes only the victim stage delay can be optimistic, the use of equivalent ramp such as $V^{slew,delay}_{estimate}$ often guarantees an extra pessimism in delay-noise. Overall, the use of traditional look-up table based delay models can lead to erroneous results due to the inherent modeling approximations associated with it. This is especially true when there is a significant amount of coupling noise and the shape of the victim waveform significantly differs from a

---

[1] For simplicity of discussion, we will use a ramp as the representative waveform for all characterization waveform shapes.

978-1-60558-497-3/09 $25.00 © 2009 ACM

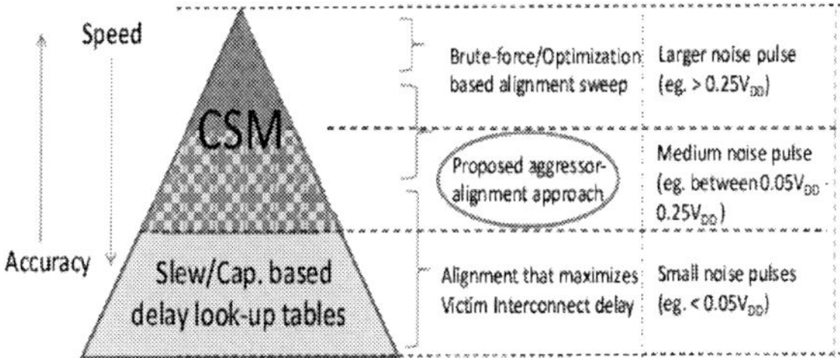

Figure 3: Proposed Methodology

ramp signal. In contrast, using current source driver models (CSMs), accurate victim output waveform can be computed even with arbitrary noisy input waveforms (e.g. V_{noisy}^{actual}). We already know that the aggressor alignment which maximizes the victim stage delay may not necessarily maximize the output arrival time. Therefore, to prevent *optimistic* delay-noise analysis in the CSM framework, we must instead find the alignment that maximizes the output arrival time.

Although, CSMs provide better accuracy over look-up table based delay models, they are computationally more expensive. Also, for victim nets with relatively small amount of coupling noise, delay-noise analysis with traditional models provide sufficient accuracy. Hence, in order to maximize the accuracy of delay-noise engine without significantly adding a runtime penalty, we propose the following noise analysis methodology (as shown in Figure 3). For the commonly occurring case of very small (e.g. $\leq 5\%$ VDD) coupling noise bumps, we propose to use look-up table based delay models which are very fast and provide reasonably accuracy. Note that if the coupling noise bump is very small, then the amount of delay noise would correspondingly be small. Therefore, we do not require a very accurate delay noise engine in this region and the aggressor-alignment can be computed very quickly by maximizing only the victim stage delay [16]. However, for victim nets with relatively larger coupling noise, we propose the use of CSMs to accurately model the noisy victim waveforms and the worst-case alignment of aggressors is then computed such that it maximizes the victim output arrival time. In this work, we present a heuristic approach which accurately computes the worst-case aggressor alignment for relatively larger noise bumps (e.g. [5% , 25%] VDD). For even larger noise bumps (e.g. $\geq$ 25% VDD), we suggest the use of more accurate optimization techniques to compute the the the worst-case alignment. However, with modern place-and-route tools, it is uncommon to have a large number of victim nets with such high amounts of coupling noise. Hence, we need to use the computationally expensive optimization engine only on a very small fraction of the entire design. Overall, we believe that the proposed methodology produces accurate results with very fast runtime.

The rest of the paper is organized as follows: In Section 3, we first explain the metric which can be used as a proxy for the total victim receiver output current. In Section 4, we then see how this metric is used to compute the worst-case aggressor alignment for both MIN and MAX analysis. In Section 5, we present results and compare the delay-noise obtained by using our proposed methodology with that obtained by doing worst-case input alignment. We finally conclude this paper in Section 6.

3. CUMULATIVE GATE OVERDRIVE VOLTAGE

In this section, we propose a metric to model the total victim receiver output current. The key observation is that the rate of victim output transition directly depends on the amount of current sourced by the receiver gate. Hence, we can use the total victim receiver output current as a proxy for the victim output transition. Therefore, we can solve for the *latest* occurring victim output transition without even simulating the victim output response. We first summarize the relationship between input voltage and the output current sourced by a CMOS inverter for a rising output transition

- The output current sourced by the driver is negligible if input gate voltage (V_{inp}) is less than the threshold voltage (V_{th})

- The output current sourced by the pull-up network of the driver is proportional to the gate *overdrive* voltage $(V_{inp} - V_{th})^\alpha$, for some $\alpha \in (1, 2)$ [18]

In this work, the total output current sourced by the CMOS driver is modeled by the *cumulative gate overdrive voltage* ($CGOV$) which is defined as the area between the input waveform and the gate threshold voltage [13] as shown in Figure 4.

$$CGOV = \int_{t_{inp}}^{t_{out}} (V_{inp} - V_{th})^\alpha \, dt \qquad (1)$$

Note that t_{inp} is the time when V_{inp} crosses the gate threshold voltage V_{th} which is a function of the threshold voltages of the transistors in the pull-up (pull-down) stack for a rising (falling) output transition. Similarly t_{out} is the time when the output waveform V_{out} crosses the target voltage level (V_{th}^{out}) of the output loading gate. Therefore, $CGOV$ actually models the total output current that is required to switch the output waveform to the level of the target voltage level V_{th}^{out}.

One can note that $CGOV$ for a certain gate is only a function of the gate output loading since it tracks the amount of output current that must be sourced for the output transition to switch upto a certain voltage level. In order to accurately compute $CGOV$, we need to model the dependence of output current on the gate output voltage and must also account for the effects of parasitic Miller coupling between gate input and output nodes. However, we see that even when the above mentioned second order effects are not modeled in equation (1), the $CGOV$ tracks the arrival time of noisy victim output transition very closely.

Consider the aggressor-victim coupled network shown in Figure 1. First, we fix the victim receiver and sweep all the other coupled-circuit parameters, such as victim/aggressor slew rates, driver sizes, victim interconnect coupling/ground capacitance etc. by randomly assigning their values. Then, we perform HSPICE simulations on each circuit and obtain

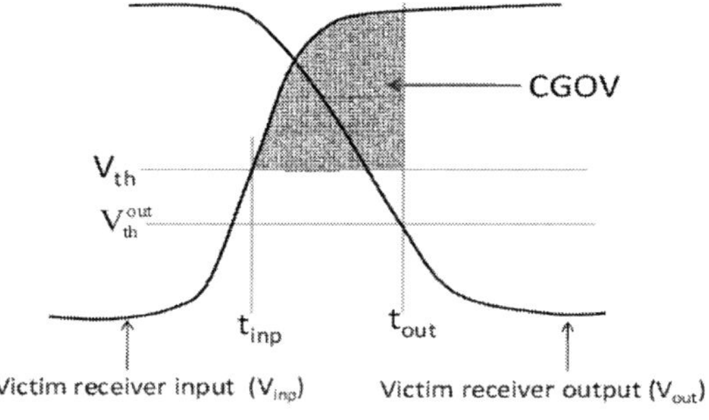

Figure 4: Cumulative Gate Overdrive Voltage

978-1-60558-497-3/09 $25.00 © 2009 ACM

the corresponding noiseless v_i and the noisy v_i^l victim transitions. The histogram of ($\approx$ 1500 data points) noise peaks is shown in Figure 5. Finally, for every circuit, the respective $CGOV$ values were calculated by using equation (1) and integrating up to the arrival times of the corresponding victim output transitions, v_i and v_o, respectively. The gate threshold voltages were assumed to be 0.5VDD. Shown in Figure 5, is a scatter plot of $CGOV$ values obtained for the circuits. On the X(Y) axis, we plot the percentage error of $CGOV$ values with respect to the mean for the corresponding noiseless(noisy) victim transitions.

Since, the slew rate of v_i determines how fast the output v_o transitions, across all the circuits, we obtain different slew rates of the output transition v_o. We know that the output current has a dependence on the output voltage which is not modeled by $CGOV$ and can lead to errors in the computed $CGOV$ values. As seen in Figure 5, the magnitude of error denoted by the spread of 10% ([-3%,7%]) around the mean $CGOV$ is quite small. Also, recall that the objective of computing $CGOV$ is to use it for estimating the noisy victim output arrival time. Therefore, it is necessary for the $CGOV$ values of the noiseless v_i and noisy victim transitions v_i^l to track each other very closely. It can be seen in the figure that the maximum error in the corresponding $CGOV$ values across all circuits (denoted by the vertical distance from the 45° inclined line) is only 1.23%. Hence, we conclude that the $CGOV$ metric is not very sensitive to coupling noise and remains fairly constant irrespective of the shape of the input transition. This observation allows us to compute $CGOV$ for the noiseless victim transition and use the same $CGOV$ value in tracking the noisy victim output arrival time.

4. WORST-CASE AGGRESSOR ALIGNMENT

A brute-force approach of computing the worst-case aggressor alignment would be sweeping the aggressor transition within its timing window and choosing that transition which results in the latest victim output arrival time. However, this requires multiple non-linear simulations of the coupled victim-aggressor network and is prohibitively expensive. In this section, we show how the worst-case aggressor alignment can be computed more efficiently using the $CGOV$ metric by using only a single non-linear simulation to obtain the noise bump. It was also seen earlier that $CGOV$ is very robust and is fairly insensitive to the aggressor alignment. Therefore, instead of performing multiple nonlinear simulations by sweeping the aggressor transition and simulating the victim output response, we propose to compute

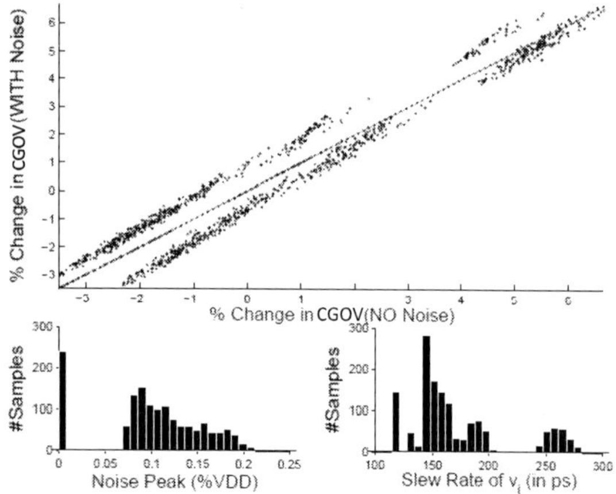

Figure 5: Scatter plot of $CGOV$.

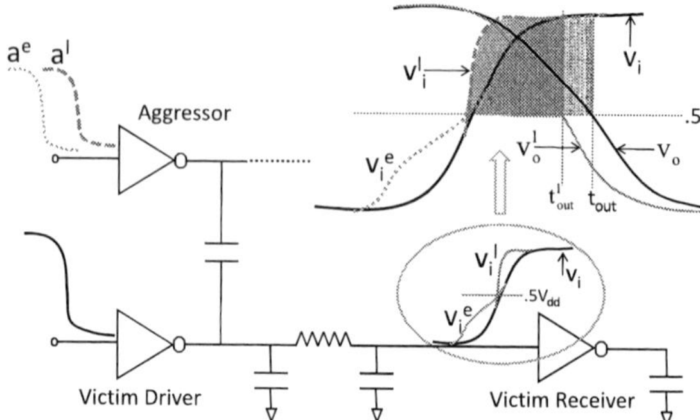

Figure 6: Worst-case alignment for MIN analysis.

$CGOV$ and use it as a proxy for the victim output arrival time.

Consider the case when both the victim and the aggressor drivers have a falling input transition as shown in Figure 6. We perform MIN analysis where we seek to find the aggressor transition which results in the earliest possible victim output arrival time. In order to do that, we first simulate the noiseless victim input and output transitions, v_i and v_o, assuming no switching at the aggressor input. We then compute $CGOV$ for the noiseless victim transition using Equation (1) and integrating up to the noiseless output arrival time t_{out}. In the previous section we showed that $CGOV$ is fairly insensitive to the input waveform shape. Hence, the $CGOV$ for noisy victim waveforms (e.g. v_i^l) can be assumed to be the same as that computed earlier for the noiseless transition v_i. Next, the noisy victim receiver input waveform (e.g. v_i^l) is obtained with a falling transition at the aggressor input. The victim receiver output arrival time can be estimated without actually simulating the victim output response. The unknown victim output arrival time (t_{out}^l) can instead be obtained by using v_i^l and integrating using Equation (1) until the $CGOV$ matches that of the noiseless waveform.

In Figure 6, we see the two noisy victim transitions, v_i^e and v_o^l, corresponding to the early (a^e) and late (a^l) aggressor transitions. It can be seen that the noise aligns early for the victim transition v_i^e and does not really affect the victim waveform above the threshold voltage level (assumed to be 0.5VDD in this example). On the other hand, the noise due to a^l aligns later and affects the victim waveform above the threshold voltage level and forces the output to transition faster. In this example, it can be seen that the later aggressor transition (a^l) results in a faster victim output transition v_o^l having an earlier output arrival time t_{out}^l. Finally, the worst-case aggressor alignment in MIN analysis can be chosen among all feasible aggressor transitions such that it results in the earliest arrival time at the output of the victim receiver. It is easy to see that the same technique can be used in the MAX analysis with mutually opposite aggressor-victim transitions. In this case, we sweep the aggressor transition within its timing window and use the $CGOV$ metric to choose the alignment which results in the latest arrival time (t_{out}) at the victim receiver output.

It is key to note that the proposed approach does not employ a non-linear CSM engine to simulate the victim receiver output response within every iteration. Instead, it computes $CGOV$ and uses it as a proxy for the victim receiver output arrival time. Therefore, this leads to a substantial speedup over approaches which employ expensive non-linear CSM based simulations. We will now see how further speed-ups can be obtained by using a linear superposition based frame-

978-1-60558-497-3/09 $25.00 © 2009 ACM

work to compute the noisy victim receiver input waveforms corresponding to each aggressor alignment.

The proposed algorithm requires the enumeration of aggressor alignment within its timing window. With the use of optimization techniques, the number of such enumerations are typically small. Nevertheless, for every aggressor alignment, we would still need to perform expensive non-linear simulations using the CSMs to obtain the accurate noisy victim waveforms. Hence, to significantly reduce the computation overhead, we use the principle of linear superposition and combine the noiseless victim waveform with the coupling noise to obtain an estimate of the noisy victim response. It is well-known that the use of linear superposition can lead to an underestimation of the coupling noise peak, if the change in noise peak due to the non-linearity of victim driver is not modeled. However, it accurately estimates the pulse width of the noise waveform. Hence, although linear superposition under-estimates the noise peak, it doesn't necessarily under-estimate the time-to-peak of the noise which is needed to estimate the alignment. Therefore, the $CGOV$ metric is not very sensitive to the non-linearity of victim driver resistance. In the results section, we show that the average error introduced by the superposition assumption is typically very less ($\approx 1.7\%$). Therefore, we make an engineering decision and use the principle of superposition to compute the noisy victim waveforms. Since, non-linear simulation using CSMs are performed only once in order to obtain the coupling noise and noiseless victim waveform, the proposed alignment approach is overall very fast.

In this paper, we have so far analyzed the case when the victim is coupled to a single aggressor. However, in a typical circuit, the victim net is often coupled to more than a single aggressor. In such cases, it is necessary to find the worst-case alignment of all the aggressors that are coupled to the victim such that it results in the maximum victim output arrival time. A heuristic often used to compute the relatively alignment among aggressors is that they are aligned such that all the noise peaks coincide and produces the largest cumulative noise peak. The cumulative noise bump is then optimally aligned with the victim waveform. In contrast, any other alignment among aggressors would result in a combined noise bump with a smaller noise peak and a wider pulse width. However, it was shown in [5] that using the noise bump with the largest noise peak resulted in an error which was less than 5% across exhaustive SPICE simulations. Therefore, in this work we have focused on the alignment of the cumulative noise bump with the victim transition such that the victim receiver output arrival time is maximized.

5. RESULTS

In this section, we will show experimental results that verify the accuracy and effectiveness of our proposed approach for computing the worst-case aggressor alignment. All experiments were performed on the fully coupled victim-aggressor circuit shown in the Figure 7 in $65nm$ technology node. The non-linear CMOS gates were simulated with detailed internal RC parasitics extracted from an industrial library. For the interconnect wire load model in the $65nm$ technology node [20], we assume a wire resistance $R = 0.5\Omega/\mu m$ and a wire ground capacitance $C_g = 0.2fF/\mu m$. We know that the coupling capacitance C_c is a function of the spacing and the relative amount of overlap between the aggressor-victim nets. If the victim-aggressor nets are routed very closely in the same metal layer, then coupling capacitance could account for a significant portion of the total wiring capacitance. Conversely, the magnitude of the coupling capacitance would be small if the victim-aggressor nets

Table 1: **Parameters of the experimental victim-aggressor circuit**

Parameters	Set of Parameter Values
Aggressor Driver Strength	(2X , 12X)
Aggressor Driver Input Slew (ps)	(10, 200)
Victim Driver Strength	(2X , 8X , 12X)
Victim Driver Input Slew (ps)	(10, 50, 100, 200)
Victim Interconnect Length (μm)	(5, 50, 100, 200)
Coupling scaling factor k	(0.5 , 1, 1.5, 2)
Victim Receiver Load (fF)	(1 , 10 , 50 , 200)

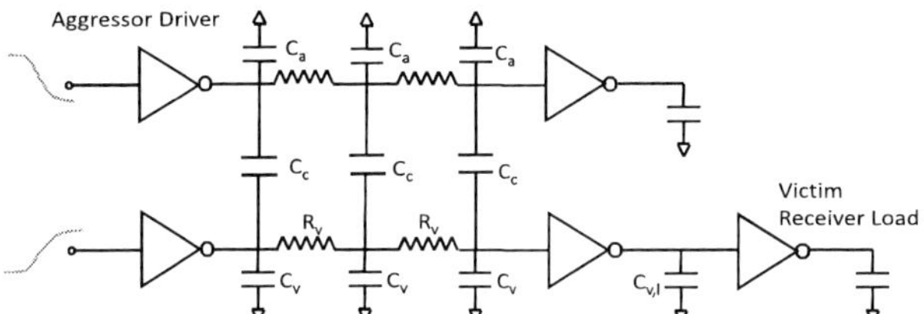

Figure 7: Experimental Victim-Aggressor Circuit

are routed at a relatively farther distance from each other. Therefore, in our experiments we treat the coupling capacitance C_c as a variable which is obtained by appropriately scaling the ground capacitance. In other words, $C_c = k * C_g$, where k is a scaling factor which ranges from $[0.5, 2]$ in our experiments.

The proposed approach of finding the worst-case aggressor alignment is validated on the victim-aggressor coupled circuit shown in Figure 7. However, in reality the representative circuit can have several variable parameters such as the type/strength of the victim-aggressor driver and receiver gates, the input slew rates, the interconnect length, the coupling capacitance scaling factor (k), the amount of receiver loadings, etc. In order to adequately sample the above mentioned parameter space, we perform a total of about 1500 simulations by sweeping the parameter values (shown in Table 1).

In each simulation, we compare the accuracy of our proposed approach with the *golden* aggressor alignment obtained by using HSPICE based brute-force enumeration. The aggressor transition was enumerated with a discretization step size of $2ps$ and HSPICE simulations were performed to simulate the victim receiver output response for every aggressor alignment. Finally the aggressor transition which resulted in the latest victim output arrival time was reported. In comparison, the proposed approach computed the worst-case aggressor alignment by using the $CGOV$ metric to predict the victim output arrival time. Once, the aggressor alignment is computed, HSPICE was used to simulate the victim receiver output response. In order to evaluate the accuracy, we express the difference in the victim receiver output arrival time (w.r.t. golden) as a percentage of the worst-case delay noise at the victim receiver output. A histogram of the percentage error in delay noise obtained across all simulations is shown in Figure 8 for MAX analysis, when the victim receiver gate is an inverter. It can be seen that the proposed approach accurately estimates the worst-case aggressor alignment since we observe an average error of 0.75% across all simulations.

Earlier we claimed that aggressor alignment which maximizes the victim receiver *input* arrival time does not necessarily maximize the victim receiver *output* arrival time. To validate the above claim, we performed a similar brute-force enumeration and computed the aggressor alignment which

978-1-60558-497-3/09 $25.00 © 2009 ACM

maximizes the victim receiver input arrival time (referred to *INP Align* in Table2). Using the above computed aggressor alignment, we simulate the victim receiver output transition and finally compare the percentage error in the output arrival time w.r.t golden obtained earlier. The histogram of the errors for inverter receiver gate (Input alignment in Figure 8) shows that about 20% of the cases report an error of 100% in delay noise. These cases occur when the input alignment results in a coupling noise which aligns too late with the victim transition (as shown in Figure 1). Overall, it can be seen that the proposed approach performs much better than the brute-force input alignment enumeration. This establishes the significance of considering the victim receiver gate in the computation of the worst-case aggressor alignment.

We repeat the above experiment for different victim receiver gates and in Table 2 we show the average % error in delay noise w.r.t. golden. It can be seen that the proposed approach accurately computes the worst-case aggressor alignment across different receiver gates. It can be noted that even with the stack-effect, the proposed approach is more accurate than the input alignment enumeration. Across all simulations, the average error of the proposed approach is 1.7% compared to an error 8.49% obtained with the input enumeration approach.

6. CONCLUSIONS

In this paper, it was seen that worst-case aggressor alignment must be computed such that it always maximizes the victim receiver output arrival time. In order to model the victim receiver output waveform, we define and use the cumulative gate overdrive voltage ($CGOV$) metric to model the total victim receiver output current. Since, the victim receiver output transition directly depends on the amount of current sourced by the receiver gate, the alignment with the lowest $CGOV$ would correspondingly lead to the slowest transition having the maximum delay. HSPICE simulations, performed on industrial nets to validate the proposed methodology, show an average relative error of 1.7% in delay-noise when compared to the worst-case alignment obtained by an exhaustive sweep.

7. REFERENCES

[1] J. Qian, S. Pullela, and L. T. Pillage. Modeling the effective capacitance of RC interconnect, *In IEEE Trans. on CAD*, pages 1526-1535, 1994.

[2] F. Dartu, N. Menezes, J. Qian, and L. T. Pillage. A gate-delay model for high speed cmos circuits, *In Proc. DAC*, pages 576-580, 1994.

[3] R. Arunachalam, F. Dartu, and L. T. Pileggi. CMOS gate delay models for general RLC loading, *In Proc. ICCD*, pages 576-580, 1994.

[4] F. Dartu, N. Menezes, and L. T. Pileggi. Performance computation for precharacterized CMOS gates with RC loads, *In IEEE Trans. on CAD*, pages 544-553, 1996.

[5] D. Blaauw, S. Sirichotiyakul, and C. Oh. Driver modeling and alignment for worst-case delay noise, *In IEEE Trans. on VLSI*, 11(2), April 2003.

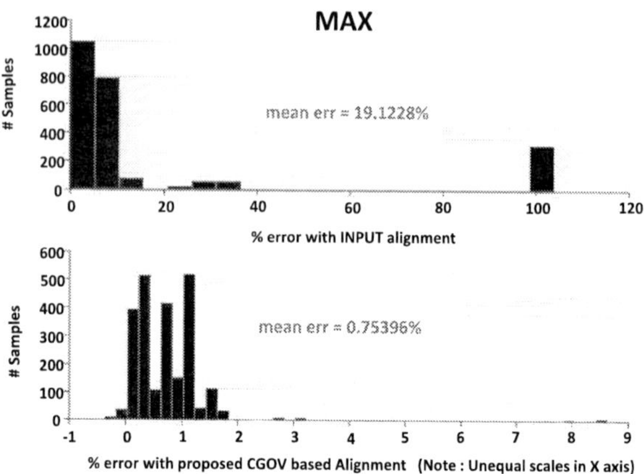

Figure 8: Histogram of % error in delay noise with INVX receiver gate for MAX analysis

[6] S. K. Gupta L. C. Chen and M. A. Breur. A new gate delay model for simultaneous switching and its applications, In *Proc. DAC*, pages 289-294, 2001.

[7] J. F. Croix and D. F. Wong. Blade and Razor: Cell and Interconnect Delay Analysis Using Current-Based Model, In *Proc. DAC*, pages 386-389, 2003.

[8] C. Amin, C. Kashyap, N. Menezes, K. Kilpak, and E. Chiprout. A multi-port current source model for multiple-input switching effects in CMOS library cells, In *Proc. DAC*, pages 247-252, 2006.

[9] I. Keller, K. Tseng, and N. Verghese. A robust cell-level crosstalk delay change analysis, In *Proc. ICCAD*, 2004.

[10] P. D. Gross, R. Arunachalam, K. Rajagopalan, and L. T. Pileggi. Determination of wrost-case aggressor alignment for delay calculation, In *In Proc. ICCAD*, pages 212-219, 1998.

[11] V. Rajappan and S. S. Sapatnekar. An Efficient Algorithm for Calculating the Worst-case Delay due to Crosstalk, In *In Proc. ICCD*, pages 76, 2003.

[12] M. Hashimoto, Y. Yamada and H. Onodera. Equivalent waveform propagation for static timing analysis, *In IEEE Trans. on CAD*, 23(4), April 2004.

[13] L. Ding, P. Tehrani and A. Kasnavi. Determining Equivalent Waveforms for Distorted Waveforms, *U.S. Patent 7272807*, 2007.

[14] R. Gandikota, K. Chopra, D. Blaauw, D. Sylvester, M. Becer and J. Geada. Victim alignment in crosstalk aware timing analysis. *In Proc. ICCAD*, pages 698-704, 2007.

[15] S. S. Sapatnekar. A Timing Model Incorporating the Effect of Crosstalk on Delay and its Application to Optimal Channel Routing, *In IEEE Trans. on CAD*, pages 550-559, 2000.

[16] R. Arunachalam, K. Rajagopal, and L. T. Pilleggi. TACO: Timing analysis with coupling, *In Proc. DAC*, pages 266-269, 2000.

[17] H. Zhou, N. Shenoy and W. Nicholls. Timing Analysis with Crosstalk is a Fixpoint on a Complete Lattice, *In Proc. DAC*, pages 714-719, 2001.

[18] H. Im, M. Song, T. Hiramoto and T. Sakurai. Physical insight into fractional power dependence of saturation current on gate voltage in advanced short-channel MOSFETs (alpha-power law model), *In Proc. ISLPED*, pages 13-18, 2002.

[19] K. Agarwal, Y. Cao, T. Sato, D. Sylvester and C. Hu. Efficient Generation of Delay Change Curves for Noise-Aware Static Timing Analysis, *In Proc. ASPDAC*, pages 77-84, 2002.

[20] N. Lu. Statistical and Corner Modeling of Interconnect Resistance and Capacitance, *In Proc. CICC*, pages 853-856, 2006.

Table 2: Percentage errors in delay-noise wrt golden

	RISE MAX		FALL MAX		RISE MIN		FALL MIN	
Receiver	CGOV Align	INP Align	CGOV Align	INP Align	CGOV Align	INP Align	CGOV Align	INP Align
INV	0.77	7.24	0.75	19.12	-2.32	-6.66	-2.10	-6.36
NAND4X	0.35	7.34	4.41	5.11	-2.99	-12.87	-3.07	-4.50
NOR4X	3.85	5.36	1.38	8.36	-3.35	-7.39	-2.23	-12.31
BUF	0.26	8.34	0.31	8.42	-1.67	-7.20	-1.34	-9.87
AND4X	0.29	8.36	1.86	8.19	-1.51	-6.14	-1.07	-8.75
OR4X	1.46	7.63	0.37	9.92	-0.99	-6.06	-2.23	-12.40

978-1-60558-497-3/09 $25.00 © 2009 ACM

A Moment-Based Effective Characterization Waveform
For Static Timing Analysis

David D. Ling, Chandu Visweswariah and Peter Feldmann
IBM T.J. Watson Research Center, Yorktown Heights, NY 10598

Soroush Abbaspour
IBM Systems and Technology Group, Hopewell Junction, NY 12533

Abstract---Static timing analysis of VLSI circuits relies on tabular models of logic gates obtained during library characterization. At characterization time logic gates are simulated at the circuit level with a range of input waveforms (e.g., saturated ramps with different slews) and various output loads. At timing analysis time the same gates are driven by input waveforms that differ from the class of characterization waveforms. This paper proposes a method for mapping the waveforms that arise in static timing analysis to members of the class of waveforms used to characterize gate timing performance during library characterization. The method is based on the moments of the input waveform, which describe concisely the salient features of the waveform. The mapping between the input waveform and the effective characterization waveform is accomplished by positing functional relationships between the input waveform's moments and the parameters of the characterization waveform. The unknown coefficients of this functional relationship are determined by minimizing the worst case error of the output waveform over a representative set of input waveforms and gate loads. The technique requires no change to the library characterization procedure, and minimal change to the static timing tool with little to no additional computational burden on the timer.

Categories and Subject Descriptors
B.7.2 [INTEGRATED CIRCUITS]: Design Aids
General Terms
Algorithms
Keywords
Current Source Model, Timing Analysis

I. INTRODUCTION

Modern VLSI circuits exhibit digital signal waveforms that increasingly depart from the ideal ramp-like shapes that are at the foundation of static timing analysis. The fundamental reason for waveform complexity is the aggressively scaled devices used in logic gates and the increased predominance of

Permission to make digital or hard copies of part or all of this work for personal or classroom use is granted without fee provided that copies are not made or distributed for profit or commercial advantage and that copies bear this notice and the full citation on the first page. To copy otherwise, to republish, to post on servers or to redistribute to lists, requires prior specific permission and/or a fee.
DAC'09, July 26-31, 2009, San Francisco, California, USA

interconnect parasitic effects. Static timing tools need to fully account for these increasingly complex waveforms in order to retain the required accuracy,

Timing analysis tools have undergone significant improvement in their ability to compute realistic waveforms. At the logic gate output nodes, current source models, e.g. CCS and ECSM, combined with reduced-order models of interconnect produce almost SPICE-accurate waveforms [1,2]. Even the traditional effective capacitance based techniques no longer rely just on the ramp with the delay and output slew predicted by the table model. Instead the delay and slew are post-processed into a linearized Thevenin model [3] which is analyzed in conjunction with the interconnect. The realistic waveforms produced at the gate output are then propagated through the interconnect model resulting in realistic waveforms at the input nodes of all fan-out gates.

Unfortunately, most gate timing models used in static timing analysis, both the traditional and the emerging current-source models, are based on table look-up and the input waveform is characterized by only a single number: slew. In order to propagate the signal to the next stage through fan-out gates described by such models, we need to revert the painstakingly computed almost SPICE-accurate waveforms to simple ramp-like waveforms described by only one slew number that can be used to look-up the table model [4,5]. This step represents today the weakest link, with the potentially largest loss of accuracy in static timing analysis. In Section II, we provide an example of three different input waveforms with the same arrival time and slew that result in gate delays that differ by 16%. This problem needs to be addressed, preferably without changing either the gate characterization process or the static timing tool.

There are three categories of solutions to address this problem. The first utilizes more advanced models that can evaluate the propagation of arbitrary waveforms through the gates, e.g., transistor level timing, CSM VIVO models [6,7,8]. A second category attempts to bring the characterization waveforms closer to the realistic signals computed during timing analysis. These characterization waveforms replace simple ramps by more realistic functions, e.g., Sine-squared, or pre-driver models [9]. The third category attempts to match the arbitrary input waveform computed during the timing analysis with the best available characterization waveform (the effective waveform) which then can be employed in conjunction with traditional gate models to predict the waveform that propagates through the gate [4,10,11].

The first category of solutions is not widely used in the context of gate level timing analysis since it requires more advanced modeling and longer analysis times. For the second category of solutions, the input characterization waveforms are assumed to be non-ramp, e.g., using Sine-squared waveforms or a pre-driver method [9]. These characterization techniques might help to reduce the inaccuracy of using ramp waveforms as the inputs during characterization. However, as we previously argued, the real waveforms that arise in timing analysis may not look like characterization waveforms, whatever choice of waveform one makes for characterization. Thus, there remains a need to map the actual waveform to a characterization waveform in order to improve accuracy. Such a mapping is the subject of this paper, and hence the approach we propose belongs in the third category.

There have been different techniques proposed to approximate the arbitrary waveform at the input of a cell with an effective waveform. The point-based technique, the most widely used, generally passes an equivalent line through the $0.5V_{dd}$ crossing point of the voltage waveform and chooses the 10-90% transition time of this line equal to that of the arbitrary waveform [4]. Clearly, it is possible to revise these techniques to use a different reference point than the $0.5V_{dd}$ crossing point or calculate the slew differently (e.g., 20-80% instead of 10-90%), or even use different rules for different types of gates. Although these modifications may improve the accuracy in certain cases, we show that the point-based technique still incurs an unnecessarily high loss of accuracy.

The least squared error-based technique [4] finds the effective slew such that the sum of the squares of the sampled differences (for p sampling points in the range of interest) between a ramp with an effective slew and the real waveform is minimized. A weighted least squared error-based technique has also been suggested in [4]. This technique multiplies each squared term by a weight factor where the weight is equal to the derivative of the output waveform divided by the derivate of the input waveform in the time domain. However, this is a circular problem for timing analysis. In other words, the effective input slew can not be calculated until the output waveform is calculated and the output waveform can not be determined until the input waveform is calculated.

Note that the problem of approximating a waveform affected by coupling and propagation noise with an effective ramp waveform is also studied in [10,12]. These techniques do not carry over to the timing problem, which is the focus of this paper.

II. GATE CHARACTERIZATION WAVEFORMS

For static timing analysis to be efficient, the various constituent gates of a design must have their timing properties pre-characterized. A typical gate library consists of a thousand or more gates, each representing a different logic function, drive strength, transistor topology, etc. For each of these gates,

the timing properties of the gate are characterized by running repeated SPICE simulations for different input waveforms and different capacitive loads. The resulting gate delays and output slews are then tabulated (or fit) as functions of input slew and output load capacitance.

In this paper, we will refer to the waveforms used to drive the gates in the library characterization process as *characterization waveforms*. The choice of characterization waveform can vary according to library, technology and design style. The simplest choice is a saturated ramp waveform. Another choice is the sine-squared waveform. Many ASIC libraries use the sine-squared waveform for library characterization. Both the saturated ramp and the sine-squared waveforms have the desirable property that there is a parameterized analytic description of the waveform, and characteristics such as the 50% crossing time and the slew can be expressed analytically in terms of the waveform parameters.

Another choice for the characterization waveform is to use the output waveform obtained from simulation by driving a specially chosen pre-driver (inverter or other simple gate) with a saturated ramp input of small fixed slew. The slew of the output waveform is controlled by varying the capacitive load between the pre-driver and the gate being characterized. Such simulated waveforms do not have a closed-form analytic representation, but there is a one-to-one correspondence between the load capacitance and the waveform slew. These simulated waveforms are attributed with either the slew or the load capacitance.

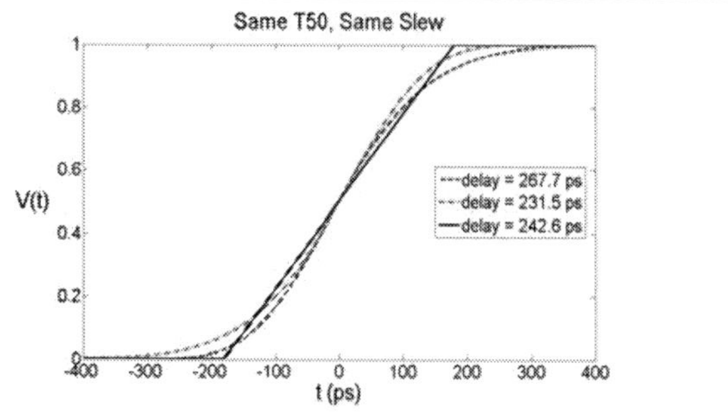

Fig. 1. Three different waveforms with identical values of T_{50} and *Slew* (10-90%). In the standard method, all three input waveforms would be assigned the same delay value (in this example, for an inverter characterized with a saturated ramp, all three would be assigned delay = 242.6 ps). However, the actual delays vary from 231.5 ps for the red (dot-dashed) curve to 267.7 ps for the blue (dashed) curve -- a variation of 16%.

Once the library is characterized, static timing analysis tools rely on the pre-computed and tabulated timing properties to time each gate in the design. In particular, the input waveform appearing at the switching input of a gate under static timing analysis (which we denote as a *timing waveform*) must be mapped to a characterization waveform. The standard approach to this mapping is to measure the 50% crossing time and the slew of the timing waveform and choose the characterization waveform that matches these values. The delay and output slew of the gate are then obtained from either function evaluation or table interpolation.

978-1-60558-497-3/09 $25.00 © 2009 ACM

One shortcoming of the standard approach is that since the timing properties of the gate are treated as dependent on input slew and load capacitance only, all timing waveforms with the same slew will result in the same gate delay and output slew values. Figure 1 shows three different timing waveforms with the same 10-90% slew value. These three waveforms result in actual delays (through an inverter) that vary up to 16%, yet the standard approach would assign the same delay value to all three timing waveforms. This paper focuses on trying to reduce the inaccuracy associated with the standard approach.

III. THE OPTIMAL RAMP

In this section we consider an idealized problem: Suppose our choice of characterization waveform is a saturated ramp. Given an arbitrary timing waveform, how do we best map it to a characterization waveform? The solution presented in this section is impractical, but helps point us to a practical approach that is presented in later sections.

For both the saturated ramp and the sine-squared waveform, we can express the characterization waveform in terms of the 50% crossing time and slew as $v_C(t - T_{50}, Slew)$. Let us denote the timing waveform as $v_T(t)$. The mapping problem becomes one of finding functionals F_T and F_S that map the timing waveform to values of:

$$\begin{aligned} T_{50} &= F_T[v_T(t)] \\ Slew &= F_S[v_T(t)] \end{aligned} \tag{1}$$

We formulate the solution as an optimization problem. Suppose we drive a gate G (under some load condition) by the timing waveform $v_T(t)$ and also by the characterization waveform $v_C(t - T_{50}, Slew)$. Let $u_T(t)$ be the output waveform when G is driven by $v_T(t)$. Similarly, let $u_C(t - T_{50}, Slew)$ be the output waveform when G is driven by $v_C(t - T_{50}, Slew)$. We shall treat T_{50} and $Slew$ as tunable parameters and select the values that minimize the L1 norm error between $u_T(t)$ and $u_C(t - T_{50}, Slew)$:

$$E_{L1}(T_{50}, Slew) = \int_{-\infty}^{\infty} dt \left| u_T(t) - u_C(t - T_{50}, Slew) \right|. \tag{2}$$

The solution to this two-parameter minimization problem is obtained by invoking a nonlinear optimizer called DEFT (DErivative-Free Tuner). DEFT is a tuner that is tailored to small optimization problems which require the accuracy of SPICE and for which gradients are not available. DEFT makes use of a nonlinear optimization package called DFO (Derivative-Free Optimizer) [13]. DFO incrementally builds successively more accurate quadratic models of the objective function and each of the constraints as it carries out the optimization in a trust-region framework [14,15]. The quadratic problem is repeatedly solved subject to simple and trust-region bounds. Model maintenance, consisting of deleting sample points, adding new sample points and re-formulating the quadratic model from scratch is periodically applied to evolve the accuracy of the model even in the absence of gradients.

DEFT repeatedly invokes SPICE as a black box function evaluator with the latest solution vector from DFO at each iteration. Fast re-run capabilities of SPICE are exploited via a programming interface to improve efficiency. DEFT has a failure recovery mechanism whereby "electrical failure" is detected if, for example, a voltage crossing does not occur in the transient simulation. In such cases, DFO cuts back its step and tries to keep away from the regions of the search space that caused electrical failure.

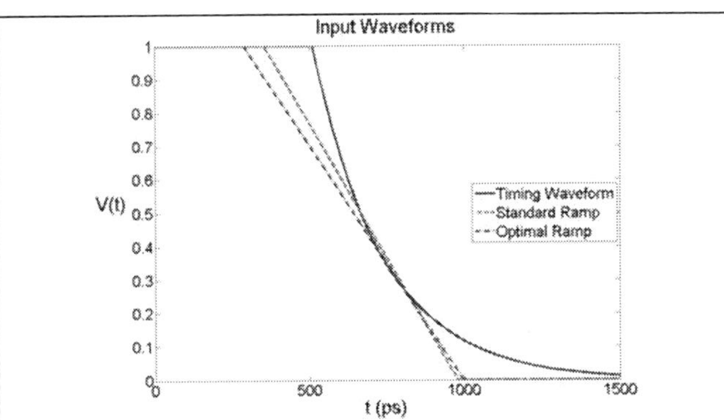

Fig. 2. A typical timing waveform is shown in black (solid), the standard effective ramp shown in red (dashed), and the Optimal Ramp shown in blue (dot-dashed).

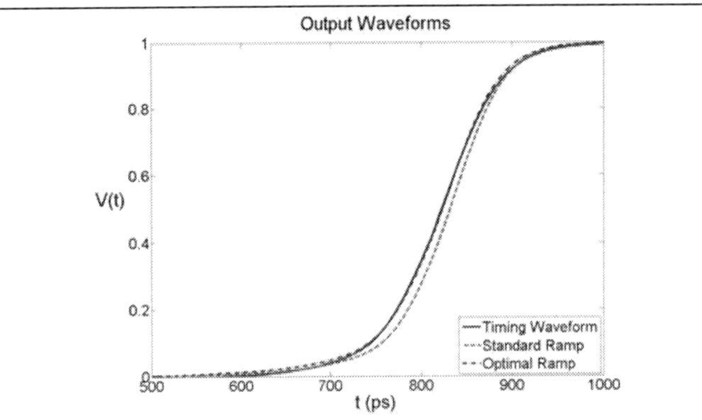

Fig. 3. The output waveforms of an inverter driven by the input waveforms in Fig. 2. The output waveform for the original timing waveform is shown in black (solid), that for the standard ramp in red (dashed), and that for the Optimal Ramp in blue (dot-dashed).

We have applied this approach to an inverter gate for a variety of input waveforms and gate loads. Figures 2 and 3 show the results for one case of an inverter being driven by a waveform obtained as the output of another inverter. In Fig. 2, the timing waveform is shown in black (solid), while the standard ramp is shown in red (dashed), and the optimal ramp is shown in blue (dot-dashed). Note that the optimal ramp does

978-1-60558-497-3/09 $25.00 © 2009 ACM

21

not match either the 50% crossing time or the slew of the timing waveform. This observation is critical to the success of the method proposed in later sections. Figure 3 shows the output waveforms for each of the three input waveforms from Fig. 2. As expected, the optimal ramp results in closer agreement with the exact output waveform than does the standard ramp.

By definition, the optimal ramp results in the *most accurate* representation of the *output* waveform over the entire space of input ramp waveforms. The timing accuracy of the optimal ramp is significantly improved over the standard ramp. However, the procedure we have just described is inherently impractical for timing analysis as it requires a full circuit-level optimization within the delay calculator. On the other hand, it reveals the potential for significant accuracy improvement through the use of more sophisticated methods to map timing waveforms into the characterization set. In the next section we describe a practical method to achieve much of this accuracy improvement.

III. AN EFFECTIVE CHARACTERIZATION WAVEFORM

In this section we propose a practical method for mapping an arbitrary timing waveform to a characterization waveform. We focus here on static timing analysis, i.e., we do not address the effect of crosstalk noise on timing. We assume that the timing waveform is either purely monotonic or nearly monotonic.

For a monotonic waveform $v(t)$ representing a rising (or falling) transition between ground and V_{dd}, the quantity

$$P(t) = \frac{\dot{v}(t)}{v(\infty) - v(-\infty)} \qquad (3)$$

behaves like a probability density function (pdf), i.e., $P(t) \geq 0$ for all t, and the area under $P(t)$ is one. We can then define the moments of our timing waveform in terms of the moments of the pdf $P(t)$:

$$m_n[v(t)] \equiv \int_{-\infty}^{\infty} dt\, P(t)\, t^n \; . \qquad (4)$$

The first moment, m_1, locates the waveform in time – it is the mean time of the pdf $P(t)$ (in this terminology, the 50% crossing time is the median). Thus, if a waveform is shifted forward in time by an amount T, its first moment will be transformed as $m_1 \rightarrow m_1 + T$.

For higher-order moments ($n \geq 2$), we prefer to use the central moments, which are defined as:

$$\mu_n[v(t)] \equiv \int_{-\infty}^{\infty} dt\, P(t)\, (t - m_1)^n \; . \qquad (5)$$

The second central moment, μ_2, represents the variance of the pdf (effectively, the width of the waveform transition and therefore related to the slew). The third and fourth central moments represent the skewness (asymmetry) and the kurtosis (likelihood of extreme values) of the pdf. Given all of the moments of a waveform, one can synthesize the waveform itself. For the purposes of timing analysis, we expect a waveform will be well-characterized by relatively few moments, and hence moments rather than crossing times are a compact way of capturing the salient features of a waveform.

The central moments have the property of being invariant under time translation – if one shifts the waveform in time, the central moments do not change. They therefore describe the shape of the waveform only and not its position in time.

It is worth observing that the two curved waveforms in Fig. 1 have identical moments, except for having μ_3 values of opposite sign. Evidently, the skewness of an input waveform can have a significant impact on the output waveform it produces.

The numerical computation of the waveform moments is efficient compared to the calculation of crossing times required to determine T_{50} and *Slew*. The moment computation involves integration over the waveform, while the crossing time calculation involves a root-finding procedure.

Our approach to mapping a timing waveform to a characterization waveform is as follows: Given the timing waveform, $v_T(t)$, calculate its moments m_1, μ_2, μ_3, μ_4, etc. Then, use the moments to compute a 50% crossing time and a slew from:

$$\begin{aligned}
T_{50} &= F_T[v_T(t)] = m_1 + \alpha_2 \sqrt{\mu_2} + \alpha_3 \sqrt[3]{\mu_3} + \cdots \\
Slew &= F_S[v_T(t)] = \quad\;\; \beta_2 \sqrt{\mu_2} + \beta_3 \sqrt[3]{\mu_3} + \cdots
\end{aligned} \qquad (6)$$

Once we have the values of T_{50} and *Slew*, we can use $v_C(t - T_{50}, Slew)$ as our effective characterization waveform (ECW).

The roots of the moments are taken in Eq. (6) so that the coefficients $\{\alpha_2, \beta_2, \alpha_3, \beta_3, \ldots\}$ are dimensionless. Moreover, time translation symmetry demands that the coefficients of m_1 in the expressions for T_{50} and *Slew* be 1 and 0, respectively.

It remains for us to determine the coefficients $\{\alpha_2, \beta_2, \alpha_3, \beta_3, \ldots\}$. We shall include moments up to third order and omit all higher order terms in Eq. (6). This leaves four unknown dimensionless constants to be determined. To this end, we select a sample set of N actual timing waveforms and a set of M output loads. This results in a total of NM total combinations of input waveform and output load. For each combination, we simulate in SPICE the gate G driven by the actual timing waveform and by the ECW determined by Eq. (6). We then measure the L1 norm error between the two output waveforms and solve the following minimax optimization problem:

978-1-60558-497-3/09 $25.00 © 2009 ACM

$$\min_{\{\alpha_2,\alpha_3,\beta_2,\beta_3\}} \max_{\{i,j\}} E_{L1}^{ij}$$

$$E_{L1}^{ij} \equiv \int_{-\infty}^{\infty} dt \left| u_T^{ij}(t) - u_C^j\left(t - T_{50}^i, Slew^i\right) \right| \tag{7}$$

$$T_{50}^i = m_1(i) + \alpha_2 \sqrt{\mu_2(i)} + \alpha_3 \sqrt[3]{\mu_3(i)}$$

$$Slew^i = \beta_2 \sqrt{\mu_2(i)} + \beta_3 \sqrt[3]{\mu_3(i)}$$

where the index i refers to the timing waveform and the index j refers to the output load. Thus, u_T^{ij} is the output waveform produced by the ith timing waveform and the jth output load. Similarly, $u_C^j(t - T_{50}, Slew)$ is the output waveform produced by driving the gate with the jth output load by the characterization waveform having parameters T_{50} and $Slew$. For the ith timing waveform, the moments are $m_1(i)$, $\mu_2(i)$, etc. The minimax optimization problem is solved by invoking the DEFT optimizer as described in the previous section.

A minimax approach is chosen (rather than minimization of a mean-square L1 norm error) so as to focus the effort of the optimizer on improving the worst outlier (the largest L1 norm error).

After the optimization solution is obtained, the resulting coefficients $\{\alpha_2, \beta_2, \alpha_3, \beta_3, ...\}$ define a means of obtaining an effective T_{50} and $Slew$ for any timing waveform driving gate G. In principal, this procedure should be repeated for each gate in the library. However, there is the intriguing possibility that one set of coefficients may be adequate to time many different gates (or perhaps all). In such an event, the application of this method would require very little additional labor.

IV. RESULTS

In this section we present the results of applying the moment-based ECW method to several gates.

First, we apply the method as outlined in Eq. (7) to an inverter. The sample set of timing waveforms are an $N=20$ subset of waveforms extracted from a static timing run on a 65 nm ASIC design, with slews ranging from 10 ps to 640 ps. The sample loads consist of an $M=8$ set of loads ranging from 20 fF to 100 fF. The solution of the minimax optimization yields the coefficients shown in Table I.

TABLE I
COEFFICIENTS FOR AN INVERTER

α_2	α_3	β_2	β_3
-0.127	-0.032	3.040	0.061

Figure 4 shows a comparison in delay errors between the standard ramp representation of the timing waveforms and the moment-based representation resulting from the minimax optimization. The standard ramp representation results in delay errors ranging between -45 ps to +20 ps. The moment-

based representation yields significantly improved delay errors, with errors ranging from -12 ps to +16 ps. The distribution of delay errors is much tighter and clustered around zero delay error, rather than widely distributed as in the standard ramp case. (Similar results are obtained for sine-squared characterization waveforms.)

It is worth stressing that the minimax optimization did not minimize the worst case delay error, but rather the worst case L1 norm error -- a measure of the error in the output waveform over all time. There was no guarantee that individual delay errors would improve relative to the standard ramp. And yet we did see a significant reduction in delay error.

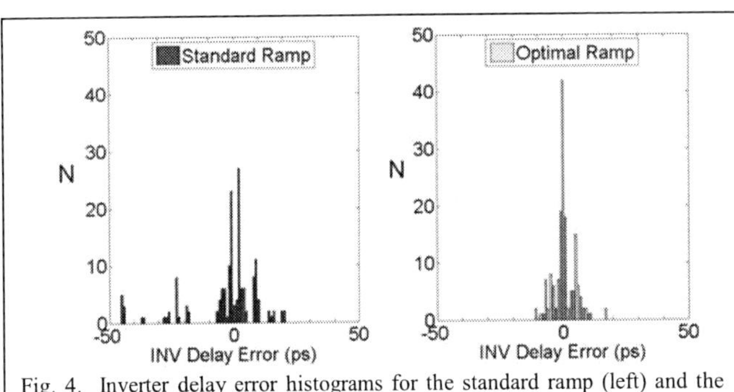

Fig. 4. Inverter delay error histograms for the standard ramp (left) and the moment-based effective ramp (right).

The results in Fig. 4 raise the question of whether the ECW we have obtained will produce accurate results on a broader set of input waveforms and output loads. To answer this question, we applied the same coefficients to the inverter driven by 98 different timing waveforms and loaded with 27 different output loads (a total of 2646 cases). We did not change the coefficients used to compute the ECW. The resulting delay errors are shown in Fig. 5. The spread in error for the moment-based ECW is a little wider than in the smaller sample set, but it is still significantly smaller than the spread for the standard ramp.

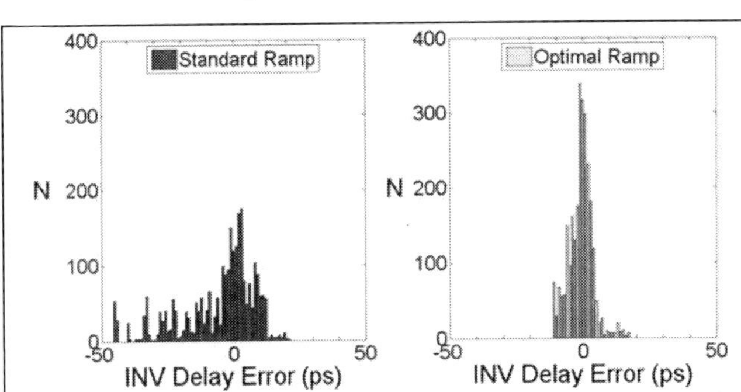

Fig. 5. Inverter delay error histograms for the standard ramp (left) and the moment-based effective ramp (right) for 2646 cases built from a larger set of timing waveforms (98) and gate loads (27).

The results shown to this point were based on applying the moment-based ECW method to an inverter gate to obtain coefficients expected to give good results for the original

978-1-60558-497-3/09 $25.00 © 2009 ACM

inverter. Is it necessary to repeat this procedure for each gate in the library? That is, does each gate require its own set of coefficients $\{\alpha_2, \beta_2, \alpha_3, \beta_3, \ldots\}$? In principle, one could make the calculation of the coefficients part of the gate characterization process for each gate. But, perhaps it is possible that one set of coefficients will work for all (or at least many) of the gates in the library. To explore this point, we carried out simulations for both NOR and NAND gates, using the coefficients shown in Table I. The delay error results for the NOR gate are shown in Fig. 6 (and the results for the NAND gate are essentially identical). As was the case for the inverter, the delay errors for the NOR gate show a significant improvement in the width of the delay error distribution for the moment-based ECW. These results suggest that one set of coefficients may work for many different gates, thereby reducing the amount of additional characterization effort required to leverage the moment-based ECW method.

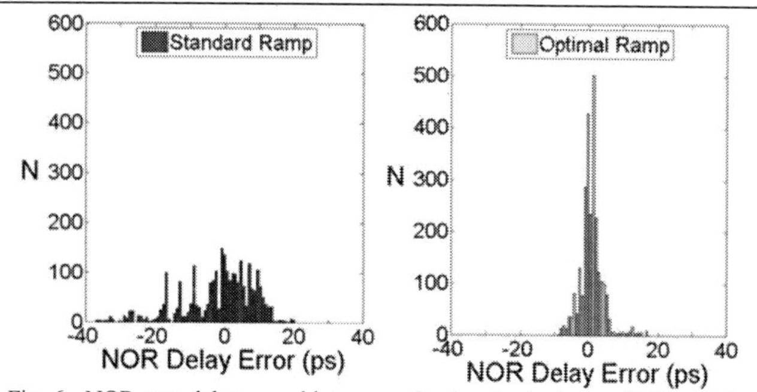

Fig. 6. NOR gate delay error histograms for the standard ramp (left) and the moment-based effective ramp (right) using the coefficients obtained for an inverter.

We conclude by revisiting the waveforms shown in Fig. 1. Table II compares the delays obtained with the moment-based ECW method to the exact values and the values obtained with the standard method. The method proposed here accounts for most of the delay variation among the waveforms.

SUMMARY AND CONCLUSIONS

This paper presents a method for mapping a timing waveform to an effective characterization waveform based on the moments of the timing waveform. The method does not require any change to the library characterization process and only minimal changes to the static timing tool with little or no additional computational burden on the timer. The method can be applied to any choice of characterization waveform (not just a ramp or a sine-squared), including simulation-based waveforms. The errors in gate timing are demonstrated to be significantly reduced in the case of a ramp input.

The functional relationship between T_{50} and *Slew* and the waveform moments need not be the choice made in Eq. (6). Any function of the moments may be used, with the coefficients in the function determined through the same minimax process described in Section III. One could also keep additional moments in order to increase the accuracy.

As gate loads evolved from pure capacitance to RC interconnect, the notion of an effective capacitance was introduced in order to prolong the utility of timing analysis based on gate models constructed for purely capacitive loads. Similarly, in the deep sub-micron regime, timing waveforms often bear little resemblance to the waveforms used to characterize gates. An effective characterization waveform method may likewise enable us to continue to use the same characterization strategy as in the past with improved accuracy.

TABLE II
DELAY RESULTS

Waveform	Exact	Standard	Optimal
Blue (dashed)	267.7	242.6	259.0
Red (dot-dashed)	231.5	242.6	232.5

REFERENCES

[1] Composite Current Source Model, Synopsys web resource, www.synopsys.com/products/libertyccs/libertyccs.html.

[2] Effective Current Source Model, Cadence web resource, www.cadence.com/partners/languages/ecsm.aspx.

[3] J. Qian, S. Pullela and L. T. Pillage, *"Modeling the effective capacitance for the RC interconnect of CMOS gates,"* IEEE Transactions on Computer Aided Design of Integrated Circuits and Systems, vol. 13, pp. 1526--1535, 1994.

[4] M. Hashimoto, Y. Yamada and H. Onodera, *"Equivalent waveform propagation for static timing analysis,"* Proc. of IEEE International Conference on Computer Aided Design, pp. 169--175, 2003.

[5] P. Feldmann, S. Abbaspour, D. Sinha, G. Schaeffer, R. Banerji and H. Gupta, *"Driver waveform computation for timing analysis with multiple voltage threshold driver models,"* Proc. of ACM/IEEE Design Automation Conference, pp. 425--428, 2008.

[6] I. Keller, K. H. Tam and V. Kariat, *"Challenges in gate level modeling for delay and SI at 65nm and below,"* Proc. of ACM/IEEE Design Automation Conference, pp. 468--473, 2008.

[7] S. Raja, F. Varadi, M. R. Becer and J. Geada, *"Transistor level gate modeling for accurate and fast timing, noise, and power analysis,"* Proc. of ACM/IEEE Design Automation Conference, pp. 456--461, 2008.

[8] N. Menezes, C. V. Kashyap and C. S. Amin, *"A true electrical cell model for timing, noise, and power grid verification,"* Proc. of ACM/IEEE Design Automation Conference, pp. 462--467, 2008.

[9] *"Motivations and methodology for nanometer library characterization"*, Cadence technical paper, www.cadence.com.

[10] S. Nazarian, M. Pedram, E. Tuncer and T. Lin, *"Sensitivity-based gate delay propagation in static timing analysis,"* Proc. International Symposium on Quality of Electronic Design, pp. 536--541, 2005.

[11] D. Blaauw, V. Zolotov and S. Sundareswaran, "Slope propagation in static timing analysis," *IEEE Trans. On Computer Aided-Design of Integ. Cir. & Sys.*, pp. 1180-1195, 2002.

[12] A. Jain, D. Blaauw and V. Zolotov, *"Accurate delay computation for noisy waveform shapes,"* Proc. of IEEE International Conference on Computer-Aided Design, pp. 946--952, 2005.

[13] A. R. Conn, K. Scheinberg and Ph. L. Toint, *"A derivative-free optimization algorithm in practice,"* Proceedings of the 7th AIAA/USAF/NASA/ISSMO Symposium on Multidisciplinary Analysis and Optimization, 1988, St. Louis, MO.

[14] A. R. Conn, K. Scheinberg and Ph. L. Toint, *"On the convergence of derivative-free methods for unconstrained optimization,"* Approximation Theory and Optimization, tribute to M. J. D. Powell, ed. M. D. Buhmann and A. Iserles, Cambridge University Press, pp. 83--108, 1997.

[15] A. R. Conn, K. Scheinberg and Ph. L. Toint, *"Recent progress in unconstrained nonlinear optimization without derivatives,"* Journal of Mathematical Programming, vol. 79, pp. 397--414, 1997.

A False-Path Aware Formal Static Timing Analyzer Considering Simultaneous Input Transitions

Shihheng Tsai and Chung-Yang (Ric) Huang

Graduate Institute of Electronics Engineering

National Taiwan University, Taipei, Taiwan

ABSTRACT

Timing closure has always been the biggest bottleneck in the modern VLSI design flow. Traditional timing verification techniques such as Static Timing Analysis (STA) are usually too conservative or sometimes too optimistic. This inaccuracy may lead to an unnecessary procrastination of time to market or even silicon failure. It is mainly due to the inability to detect false paths and handle multiple-input-transitioning effects in the timing analysis process. In this paper, we proposed a novel Formal Static Timing Analysis (FSTA) technique which can model the multiple-input transitioning effects, detect the false paths, and generate an input transition pattern for the true critical path at the same time. This is achieved by tightly integrating a state-of-the-art Boolean Satisfiability (SAT) solver with a STA engine, under a specialized multiple-input-transition timing library. Our experiments compare the FSTA engine with the traditional STA and random simulation techniques. The results show that our approach greatly outperforms random simulation while obtaining more accurate timing analysis results than STA.

Categories and Subject Descriptors: B.7.2 [**Integrated Circuits**] Design Aids -- *Verification, Simulation.* B.8.2 [**Performance and Reliability**] Performance Analysis and Design Aids

General Terms: *Algorithm, Performance, Reliability*

Keywords: *Critical path selection, multiple input transitioning, false path, static timing analysis, formal method*

1. INTRODUCTION

For the continuously increasing VLSI design complexity, timing closure has become the biggest bottleneck in the entire design flow. Among the various research attempts to resolve the timing closure problem, *Static Timing Analysis* (STA) has been the sole winner due to its extremely fast run time and acceptable accuracy. Nowadays, STA is an essential step in either early-stage timing analysis or post-layout timing verification.

However, there are two major inheriting deficiencies in STA. The first is the inability to automatically detect false paths. A path is called false in a circuit if the transition on its input cannot be propagated along the path and observed on the output [1]. If a path is false, its delay can be ignored during the critical path analysis. According to [2], there could be more than 30% of paths in a circuit that are false. Therefore, it is very likely that the critical path found by STA is false and the corresponding circuit delay is

Permission to make digital or hard copies of part or all of this work for personal or classroom use is granted without fee provided that copies are not made or distributed for profit or commercial advantage and that copies bear this notice and the full citation on the first page. To copy otherwise, to republish, to post on servers or to redistribute to lists, requires prior specific permission and/or a fee.
DAC'09, July 26-31, 2009, San Francisco, California, USA

unrealistic and over-pessimistic.

The other deficiency in STA is the negligence of the simultaneous input transitions. *Single input transition* (SIT) is a basic hypothesis of STA, where only one input node of a circuit element can switch (rise or fall) at a time. The gate delay model is then built under SIT assumption. However, in real circuits, simultaneously *multiple input transition* (MIT) is more likely to happen, especially for the patterns excited by a bank of flip-flops. It is stated that neglecting MIT may lead to a 100% error on the gate delay and output slew in a single stage [3] [4]. Moreover, timing analysis with SIT model should encounter a higher rate of false paths because the side inputs of its propagation path are assumed constants and thus are more difficult to be justified. Therefore, we should consider MIT when the accurate timing analysis is required.

Previous researches on false path problems mainly focus on two aspects: false path detection [1] and false-path-aware timing analysis [5][6][7]. The false path detection algorithms usually model the false path condition as a test pattern generation or Boolean satisfiability problem, and utilize ATPG or formal verification engines to verify it. On the other hand, the false-path-aware timing analyzers take the explicitly or implicitly specified false paths as the inputs, and apply techniques such as node splitting or additional delay element restoring to identify the true critical path. Recently, the work [8] and several commercial tools are able to integrate the false path detection and false-path-aware timing analysis into an automatic flow. However, none of them take the MIT model into account. Therefore, as we will describe in Section2, their timing analysis results can be either too pessimistic or optimistic.

In [9] and [10], the authors defined and discussed the MIT effects and thus pushed the STA min/max delay to a safer and more conservative bound. The accuracy was further improved by applying current source models in [3] and [4]. However, these works mainly focused on the impacts of the MIT timing modeling on a single or a few gates. They did not address how the circuit delay can be calculated under such model. In addition, without considering the potential false paths in the MIT-based timing analysis, the critical path delay may still be too pessimistic.

In this paper, we propose a novel Formal Static Timing Analysis (FSTA) algorithm in which false path and multiple-input-transitioning effects are considered at the same time. This is achieved by incorporating state-of-the-art Boolean Satisfiability (SAT) solver and MIT timing library under a timing analysis engine. Our algorithm can automatically search for a *true* critical path under the MIT timing model and at the end, generate an input transition pattern that can be simulated to validate and debug the critical path delay. Our experimental results on various benchmark circuits show that the traditional STA is indeed too pessimistic. Many of the timing critical paths it reports are either false or with delays that are over-estimated in reality. In average, the true path delays generated by our FSTA are about 92.2% and 101.6% when

978-1-60558-497-3/09 $25.00 © 2009 ACM

compared to those false and true path delays reported by STA, respectively. In addition, our results are 7.61% more critical with a much shorter run time than those by the simulation-based approach.

The rest of the paper is organized as follows: we will first introduce the MIT timing model and definitions in Section 2. Section 3 illustrates our program flow and algorithm implementation. The results are shown and discussed in Section 4, followed by the conclusion in Section 5.

2. Multiple Input Transition Timing Analysis

In this section, we will define the multiple-input-transition timing model and describe how it is constructed and utilized in our timing analysis framework.

2.1 MIT Timing Model

We will use a 2-input gate to illustrate the multiple-input-transition timing model as shown in Figure 1. We define the arrival time arr_i of an input signal i as the time when its voltage crosses the 50% point of the transition, and its transition time TR_i as the time difference between the 10% and 90% points. Assume a 2-input logic gate Y with inputs a and b as shown in Figure 1(a), the arrival time difference $Darr_y$ (or simply $Darr$) is defined as "arr_a-arr_b". It is clear that positive $Darr$ means signal A arrives after B, and negative when A switches before B.

We call the input signals transiting "simultaneously" if $|Darr_y| < TR_a + TR_b$. As shown in Figure 1(b), let the waveform of pin a be fixed in the middle and that of pin b be sliding from left to right. It can be shown that when the waveform of b lies between the $\pm(TR_a + TR_b)$ boundaries, both transitions will overlap and the multiple-input-transitioning effects should be considered. On the contrary, when $|Darr_y| \geq (TR_a + TR_b)$, the output waveform will be mostly determined by one of the inputs. The above discussion also works for the extreme cases when $TR_a \gg TR_b$ or $TR_b \gg TR_a$.

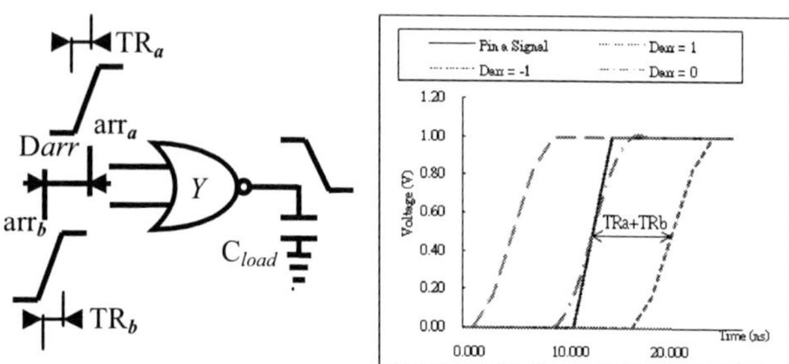

(a) Parameters of MIT model   **(b) Simultaneous transition waveform**

Figure 1. Simultaneous Multiple Input Transition

The *normalized Darr*, or just called *Darr* for short, is the original *Darr* divided by $(TR_a + TR_b)$. Therefore, the valid MIT effects mentioned above will have Darr between -1 and 1. In the remaining article, we will use the *normalized Darr* for convenience.

Please note that simultaneous input transitions may either increase or decrease the cell delay, when compared to the single transition case. For example, when both inputs of an AND gate are switching from low to high (i.e. to its input non-controlling value), its delay is actually exceeding that of a single rising transition. On the contrary, when there are simultaneous falling transitions at the inputs of an AND gate, its output delay will be smaller. This phenomenon applies to almost all the gate types. In our FSTA algorithm, we will use it to guide the critical path search process for transitions with more critical timing.

However, according to our HSPICE simulations, the cell delay characteristics function is much more complicated. Figure 2 are the simulation results for a 2-input NAND gates with simultaneous input falling transitions. Figure 2(a) is when both inputs have transition time = 0.15ns, and in Figure 2(b) one input has transition time = 0.15ns and the other = 0.60ns. We can observe that: (1) The waveform is not symmetric and thus the lowest delay does not necessarily occur when $Darr = 0$. (2) The waveform is not monotonically increasing on both sides of the valley point(i.e. the point with smallest delay). For example, in Figure 2(a) there are some "bumps" near the valley, and in Figure 2(b) the delay is actually decreasing towards $Darr = 1$.

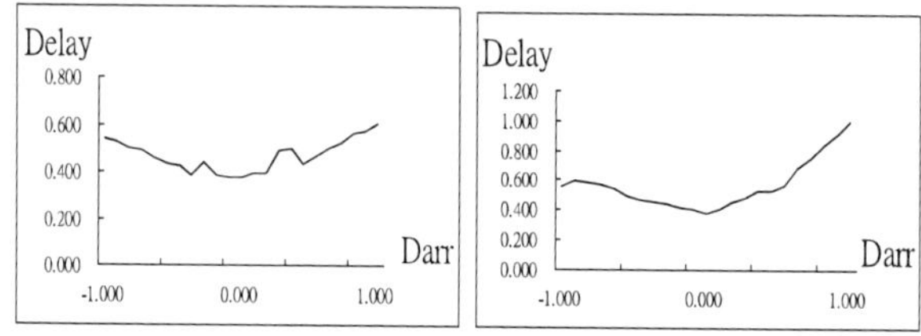

(a) $TR_a = TR_b = 0.15$ns   **(b) $TR_a = 0.15$ns, $TR_b = 0.60$ns**
Figure 2. Cell Delay versus Arrival Time Difference

2.2 MIT Timing Library

Due to the complicated characteristics function of MIT timing model, we adopt the table-lookup method for computing the cell delay as in the traditional STA library. However, unlike STA where the cell delay is determined by two factors: input transition time and output capacitance, our MIT timing model defines the cell delay by the following four parameters: TR_a, TR_b, $Darr$, and C_{load}. Obviously, this will require a different timing library from the traditional STA.

To construct the MIT library, we conducted several SPICE experiments, each of which is for a different combination of the above four parameters. For example, given a 2-input gate, we need to run totally (N_a * N_b * Ndarr * Ncout * 4) SPICE simulations, where N_a, N_b, Ndarr, and Ncout are the number of sampling points for TR_a, TR_b, $Darr$, and C_{load} in the look-up table, respectively. The typical values for them are between 5 and 7. The number 4 at the end refers to the four different tables for output cell rise_delay, fall_delay, rise_transition, and fall_transition. The above combinations lead to around $4*5^4 \sim 4*7^4 = 2,500 \sim 9,604$ entries in the library for a single cell. When more-input gates are taken into account, this number will grow exponentially due to the complex pin combinations. Therefore, we need to simplify the model by considering the 2 or 3 most critical input pins and treating the others as constant values. This simplification, however, still produces much more accurate results than the SIT model and is valid for most of the real cases. Therefore, in order to simplify our experiment, we map all gates (except for inverters) of our circuits into 2-input gates.

In our implementation, we actually store the 4-dimensional (4-D) lookup tables as various 2-D tables. Each 2-D table is associated with a combination of TR_a and TR_b values, and within each table the delay is looked up by $Darr$ and C_{load}. Please note that when the TR_a and TR_b values do not exactly match any of the 2-D tables, four of the adjacent tables will be examined and the delay will be the interpolation of the corresponding cells in these tables.

In addition, when $|Darr|$ gets larger, the MIT effect is asymptotic to the SIT model. Therefore, we set the delay in the MIT library to SIT value when $|Darr| \geq 2$.

978-1-60558-497-3/09 $25.00 © 2009 ACM

3. False Path Analysis

We will illustrate the false path detection rules and our formal static timing analysis (FSTA) algorithms on searching timing critical true paths in this section.

3.1 False Path Formulation

Given a signal s in a circuit, we denote the symbols $*s$ and $s*$ to be the logic values of s in the previous and current cycles, respectively. Let $\vec{v}_p$ and $\vec{v}_c$ be the input patterns in the previous and currently clock cycles, respectively. If we simulate these two patterns, we will have the logic values on s as $*s(\vec{v}_p)$ and $s*(\vec{v}_c)$.

We can then define the *transition function* T for signal s as:

$$T_s(\vec{v}_p, \vec{v}_c) = {}*s(\vec{v}_p) \oplus s*(\vec{v}_c) \tag{1}$$

Clearly, $T_s = 1$ if and only if the input transition $(\vec{v}_p, \vec{v}_c)$ causes a switch on signal s.

A path π in a circuit is a connected sequence of signals { s_1, s_2, ... s_n } from the circuit input s_1 to the circuit output s_n. We say the path π is a *true path* if and only if the following "*true path constraint* TP" can be satisfied:

$$\text{TP}(\pi): \exists (\vec{v}_p, \vec{v}_c) \cdot \prod_{i=1}^{i=n} T_{si} \tag{2}$$

In other words, there exists an input pattern pair $(\vec{v}_p, \vec{v}_c)$ such that a transition on the primary input (i.e. $T_{s1} = 1$) can be propagated along the path and be observed on the primary output (i.e. $T_{sn} = 1$).

On the other hand, we say that a path π is false if and only if Equation (2) can never be satisfied.

Figure 3 demonstrates a simple example for a false path. The path { G1, G2, G3 } is annotated with a sequence of possible transitions. It is clear that the rising transition on primary input can go from G1 to G2, but will be blocked by G3. Similarly, the falling transition will also be blocked. Therefore, the path is a false path. It can be verified that the above Equation (2) can not be satisfied in this example.

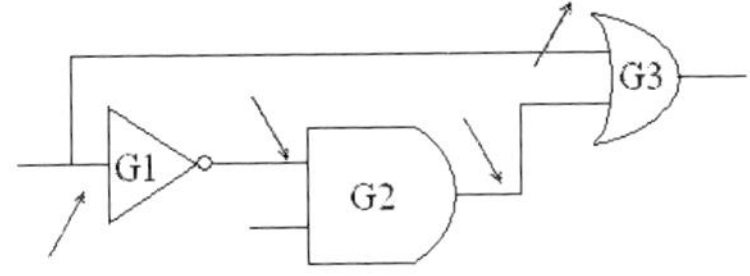

Figure 3. A False Path Example

Please note that the above true and false path constraints are defined under the multiple-input-transition (MIT) timing model. For the single-input-transition (SIT) model, we need to add the "*side-pin-constant constraint*", which assures the side pin signals remain constant values.

Definition 1: Given a signal s on a path π, the *side pins* of s, denoted as $sp(s)$, are the pins that fanout to the same gate as s, but do not belong to π. Moreover, for the single transition to propagate through s, all the side pins must be kept as non-controlling values[1]

---

[1] The non-controlling value for a gate is the input value that does not solely determine the output value of this gate. For example, the non-controlling values for the AND and OR gates are 1 and 0, respectively.

at both previous and current clock cycles. We call this side-pin constraint of s, or say $spc(s)$.

The true path constraint for the SIT model can then be modified from Equation (2) as:

$$\text{TPS}(\pi): \exists (\vec{v}_p, \vec{v}_c) \cdot \prod_{i=1}^{i=n} T_{s_i} \cdot spc(s_i) \tag{3}$$

From Equations (2) and (3), we can learn that the true path constraint for SIT model is more stringent that that of MIT model. Therefore, if a path is proven to be false path in the MIT model, it will not be a true path for the SIT model.

Note that a transition can be propagated through a binate gate (i.e. XOR gate) only if there is a single input transition. For example, given a 2-input XOR gate, a transition can go through this gate as long as its side pin remains constant (0 or 1) in both clock cycles. However, if both pins of the XOR gate have transitions, the output will be evaluated as a stable value as shown in Table 1, even though it may produce a glitch, which is not our concern in this work. Therefore, in our true critical path searching procedure, we will exclude the binate gates when considering simultaneous input transitions.

Table 1. Output Logic of XOR Gate with MIT model

Input signal	Rise	Fall
Rise	Zero	One
Fall	One	Zero

3.2 False Path Checking

We implemented the false path checker by constructing the true path constraint and solving it with a Boolean Satisfiability (SAT) solver. The false path checker has two modes. The first one is the full path SIT mode whose purpose is to examine whether the critical path selected by STA is sensitizable (i.e. a true path). The Boolean constraint in this mode is generated according to Equation (3) and solved by SAT in a single run.

The second mode is an incremental mode that adds the signal transition function (i.e. Equation (1)) one at a time along the path under validation. We modify the Equation (2) to the "*partial true path constraint*" as:

$$\text{TP}\big|_x^y(\pi): \exists (\vec{v}_p, \vec{v}_c) \cdot \prod_{i=x}^{i=y} T_{si} \text{, where } x \geq 1 \text{ and } y \leq n \tag{4}$$

More specifically, we incrementally add the transition functions from primary output to primary input and call the SAT solver to solve the partial true path constraints. If the partial true path constraint becomes false when we reach a signal s_k, we can early terminate the checking process and conclude that the path will never be a true path. We call the signal s_k the "*false element*".

Figure 4 demonstrates this process. Suppose we stop at the false element on the left, we can deduce that the partial path before this false element must be true. Therefore, there must exist a path from one of the side pins of this partial path that is true. As we will present in Section 4.2, we will pick one of these candidate pins and continue the search for the true critical path.

Note that the above incremental mode works only from the backward direction. If the Boolean constraints are added one by one from PI, no partial true path can be concluded since the observability is not guaranteed by the partial constraints.

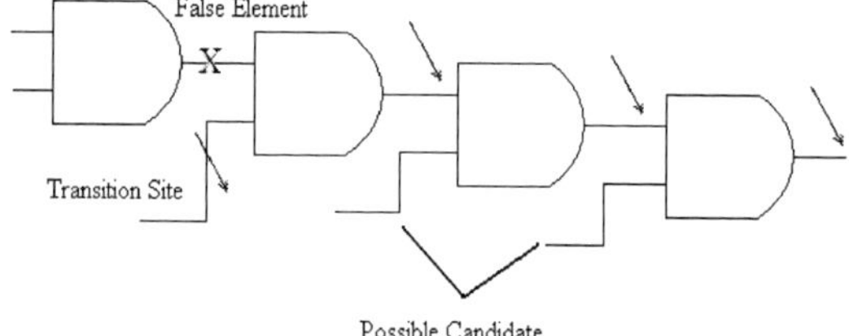

Figure 4. Incremental Mode of False Path Checker

4. FSTA Framework and Implementation

In this section, we will demonstrate our framework and implementation. Figure 5 shows the overview of our timing analysis framework. It consists of several parts. The top level includes a circuit parser, STA engine and STA libraries. Our STA engine is a traditional block-based delay calculator and the STA libraries are generated under single transition model. The output of the first stage is the STA candidate critical path with STA delay information of the circuit. The false path checker examines whether the selected path meets the true path constraint by a state-of-the-art SAT solver. If the result turns out to be a false path, the path searching handler will manage to seek for another critical delay candidate path. Once a true candidate critical path is found, the simulator will compute the corresponding delay with given input pattern pairs using MIT libraries which are derived from HSPICE simulation as mentioned in section 2. The MIT Guider tries to push the delay bound by making the side pin signals to transit simultaneously. The details of MIT Guider will be illustrated in the next section. In addition, we may continue the path searching progress to iteratively refine the candidate solutions until the user specified terminating condition is fulfilled.

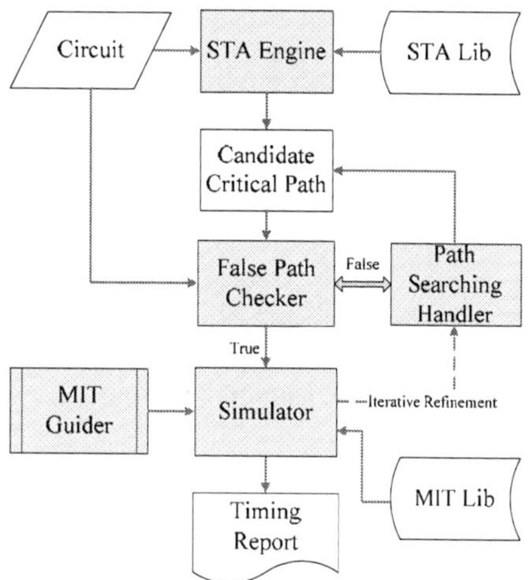

Figure 5. The Overview of FSTA

4.1 Program Flow

First we run static timing analysis to generate the most critical delay path and the estimated delays of each circuit cells with transition times. Note that the timing information here is considered under single transition and the path sensibility is not taken into account yet.

After the STA critical path is examined by the false path checker, we define our program to do one of the following:

1. If the selected candidate path turns out to be true, we use the MIT guider to push the delay bound.

2. If the candidate path turns out to be false, we search the neighborhood of this critical path or other paths with close delay value.

In the former situation, its STA delay is sensitizable and the remaining problem becomes: whether there are any MIT patterns that can make the path delay worse. Hence, we apply the general rules that concluded in section 2.1 to guide the pattern searching. This will be shown in section 4.3.

In the latter one, the computed STA delay is an unrealistic number which is neither lower bound (i.e. the corresponding signal transitions are non-sensitizable) nor upper bound (there might be some path with worse delay when considering multiple input transitions.) of the circuit critical delay. However, the STA delay values could be good reference so that we utilize it in our path searching techniques as shown in section 4.2.

Until a timing critical path is found to be true, the result of the false path checker would be transformed into input pattern pairs and fed into a simulator together with our MIT Libraries. Finally, the output timing report will include the critical path we found, the corresponding delay, and also the generated pattern that can be used for timing validation. Moreover, the path searching handler can iteratively refine the true circuit delay by deriving the next critical candidates until all the paths are enumerated.

4.2 Path Searching Handler

We propose a novel heuristics to handle the path searching process. First, we attach the *delay difference* (DD) to each gate in the STA run. The delay difference here refers to the difference of delays between the winner and loser signals of the MAX competition in STA process. To be specific, assume there is a k-input gate with k signal arrival time ar_i and k transition time tr_i respectively, where $i = 1 \sim k$. The k possible cell delays are then computed by their corresponding timing arc delay function DF_i of input transition time tr_i and output capacitance c. While the STA chooses the $worstcasedelay = \mathrm{MAX}\left(\{(ar_i + DF_i(tr_i,c))|_{i=1 \sim k}\}\right)$ to be delay of the output signal, we attach the delay differences $DD = worstcasedelay - (ar_i + DF_i(tr_i,c))$ to the corresponding circuit elements. Although in our implementation, only 2 or less input gates are considered, this technique could certainly be generalized to multiple input logic cells. Since these numbers are all computed in STA, this step costs almost no extra computation effort.

The circuit elements in the true part of the candidate path are then sorted by their delay differences. Intuitively, picking up the smaller delay difference element would lose less delay value when the path is "bended" at the cell (i.e. reconnect to the selected cell's side pin), unless this new path is false. Indeed, if we assume that MIT affects the cell delays by a portion, which is of the same probability density function in all circuit elements, it turns out that the smaller DD side pins will have higher probability to lead to critical paths. Hence, we choose circuit elements with ascending DD order one at a time, and reconnect the path to their side pins. Afterward, the false path checker examines the true path constraints on the resulting partial paths. When the Boolean SAT solver returns satisfiable, we keep back tracing from this side pin and push circuit element into the new candidate path. The backtrack priority of input pins is according to STA timing information (higher delay value) while true path property is guaranteed.

We enumerate all the possible bended paths until the delay difference of the bending site is too large for MIT to recover the

978-1-60558-497-3/09 $25.00 © 2009 ACM

gap. In our program the stopping criteria is set to 50% of STA critical path delay. If all the delays of these bended paths are not critical enough, the same process will be executed on the next candidate critical path acquired from another PO with the second worst delay. This progress continues until we find the path delay greater than the STA bound or all possible candidates are examined. The overall path searching handler algorithm is shown in Algorithm 1.

This greedy algorithm guides us to find the more probable critical paths in the neighborhood. Although this is a heuristic, the probability that a path with low priority occurs to have worst path delay is quite small. The efficiency is shown in our experimental results.

Algorithm 1: Path Searching Handler

```
PathSearchHandler(){
    sortedPoList = sort POs by STA arrival time;
    for j = (0 … |PO|-1) // for each cell in sortedPoList
        candPath = STACriPathTrace(sortedPoList[j]);
        bound = candPath->getSTADelay();
        length= FalsePathCheckInc(candPath); // incremental check
        if ( switchPath(length) == true)   return; // found solution
}
bool switchPath (length){ // true part length
    cellArray = sort true part of critical path by delay difference;
    for (i=0 … length-1 ){
        if (switch at cellArray[i] is available) // switch to side pin
            MITGuider(); // illustrated in 4.2
            MaxArrTime = Simulator; // simulate patterns
    if (MaxArrTime > bound ) return true;
    else return false;
}
```

4.3 Multiple Input Transition Guider

The principle in the multiple input transition (MIT) guider is very similar to the path searching handler in the previous section. After the candidate path is determined by the path searching handler or STA true path, we calculate and store the *MIT potential delay difference* for each candidate path element.

The MIT potential delay difference is determined by 4 parameters as in MIT timing model. The discrepancy is that the arrival time difference Darr and input transition times here are estimated values by STA instead of the real ones. The reason why these values provide good reference is that larger STA arrival time difference implies smaller probability of simultaneous signal switching. Thus we filter out those side pin transitions with low probability in resulting worse path delay. Also because the candidate path is examined to be true by false path checker, these transition and arrival time values can be reliable when MIT occurs.

The MIT general rules are described in section 2 and we rewrite here as:

$$(Delay_{new} < Delay_{ori} \&\& Slew_{new} < Slew_{ori}) \mid input = controlling$$

$$(Delay_{new} > Delay_{ori} \&\& Slew_{new} > Slew_{ori}) \mid input \mathrel{!}= controlling$$

Since we are searching for the worst delay bound, the MIT guider tends to make side pins transition when input signals are non-controlling, and make side pins remain constant when controlling signals are applied at the cell inputs. Algorithm 2 demonstrates the implementation of the algorithm.

Algorithm 2: MIT Guider

```
MITGuider(){
    for (i=0 … |criPath|-1)
        criPath[i]->setMITPossibleDD(Tr_a,Tr_b,Darr,C_load);
        // values by STA
    sortArray = criElement sorted by PossibleDD;
    for (i=0…|criPath|-1)
        cell = sortArray[i];
        if (cell == 2-input-cell)
            addMITConstraint(cell);
            if (cell->getInputSignalType() = controlling)
                cell->makeSidePinConstant();
            else   cell->makeSidePinTransition();
            falsePathCheck();
            NewArrTime = Simulator();
}
```

5. Experimental Result

We conduct our experiments on ISCAS85 benchmark circuits on a Linux workstation with 3.2GHz CPU and 2GB RAM.

We compare our FSTA algorithm with both STA and random simulation approaches, and the results are shown in Table 2. The first two columns are the circuit names and the critical path delays reported by STA. The third column shows if the STA critical path is true. Note that 8 out of 11 testcases are proven false by our FSTA engine. The worst case delays (divided by the STA delays) of random simulation and the corresponding run time (in seconds) are shown in columns 4 and 5. Note that the run time reported here is the time when the worst delay is encountered instead of total run time. We set the terminating condition of random simulation as 3×10^6 consecutive iterations with the found worst delay unchanged. We present the true path delays and total program run time by our FSTA algorithm in the 6th and 7th columns, respectively. And the last column summarizes the FSTA improvement ratio over the random simulation in term of the critical path delays.

For the three true path cases, we can generate the input transition patterns with extremely fast run time and demonstrate that, in average, the critical path delay reported by STA is 1.6% less than that by our FSTA. This STA pessimistic result is due to the negligence of the multiple input transitioning effects. On the other hand, for the eight false path cases, the STA delay bound is not realistic since STA did not exclude the false paths during the delay calculation. Our algorithm can find the true critical paths very efficiently and show that the delays are about 94.8% of the STA false path delays in average. Please note that our result always comes with an input transition pattern which can be simulation and validated with the MIT timing library.

We also perform random simulation on these test cases. The results show that our algorithm can identify true critical paths that are 7.61% more in delays with 32 times faster in runtime. Moreover, the advantages over random simulation are more evident when the design scale increases. For small circuits, random simulation approaches to the real bound very fast since the total number of combinations is very limited. But for larger circuits such as C6288 and C7552, our work outperform it by 25% ~ 50% with much smaller run time. Note that there are 3.73×10^{15} paths in our synthesized C6288 circuit, which is an enormous number and hence it's impossible to enumerate the path space exhaustively.

Figure 6 depicts the growth of delays computed by our FSTA and random simulation for the circuit C7552. We see that random simulation climbs slowly while our FSTA framework grows much faster and saturates much earlier.

978-1-60558-497-3/09 $25.00 © 2009 ACM

Table 2. ISCAS85 benchmark result

Circuit	STA data		Random		FSTA		FSTA/ Random
	STA Delay	T/F	Delay Ratio	Run Time	Delay Ratio	Run Time	Improve Ratio (%)
C17	5.424	T	1.047	0	1.047	0	+0
C432	60.78	F	0.844	1507.7	0.889	2.5	+5.25
C499	53.31	F	1.013	4739.7	1.001	18.3	-1.18
C880	62.51	T	0.943	2972.1	1.000	0.88	+6.02
C1355	54.47	F	0.994	3757.9	1.014	22.5	+2.02
C1908	74.24	F	0.779	286.14	0.834	83.4	+7.23
C2670	70.73	T	0.944	6349.2	1.000	1.01	+5.98
C3540	84.95	F	0.981	22654.5	0.937	494.7	-4.35
C5315	71.67	F	0.861	4965.6	0.931	134.72	+8.12
C6288	205.4	F	0.666	5739.0	0.839	682.29	+25.86
C7552	111.0	F	0.621	5458.0	0.937	1741.1	+50.93
Average			0.881	5311.8	0.948	163.8	+7.61

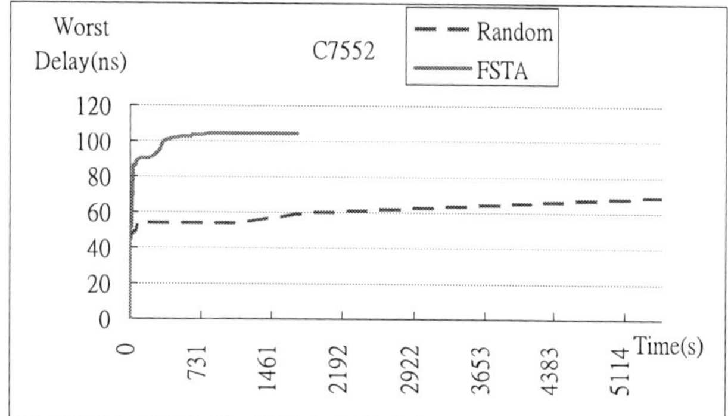

Figure 6. FSTA and Random Delay Curves of C7552

6. Conclusion and Future Works

In this paper we proposed a novel formal static timing analysis (FSTA) technique that can identify false paths with the multiple-input-transition (MIT) timing models. Our experimental results on various benchmark circuits show that the traditional STA can be optimistic or pessimistic. Our FSTA can approach to the real delay bound fast under the MIT timing model, and generate input transition patterns for timing verification and debugging.

In the future, we will extend our framework considering process variations by adapting statistical or multi-corner multi-mode delay models. In addition, we will enhance the 2-input MIT timing library for multiple-input cells by further simplifications. Also, we will attempt to fully integrate our timing engine with the SAT solver in order to reduce the overhead caused by the frequent interactions with the external SAT solver.

7. REFERENCES

[1] D.H. Du, S.H. Yen, and S.Ghantz, "On the General False Path Problem in Timing Analysis," in Proceedings of the 26th ACM/IEEE conference on Design Automation, pp. 556-560, 1989

[2] J.Zeng, M. Abadir, J. Abraham, "False Timing Path Identification Using ATPG Techniques and Delay-Based Information," in Proceedings of Design Automation Conference, 2002

[3] C. Amin, C. Kashyap, N. Menezes, K. Killpack and Eli. Chipout, "A Multi-port Current Source Model for Multiple-Input Switching Effects in CMOS Library Cells," in Proceedings of Design Automation Conference, 2006.

[4] B. Amelifard, S. Hatami, H. Fatemi, M. Pedram, "A Current Source Model for CMOS Logic Cells Considering Multiple Input Switching and Stack Effect," in Proceedings of Design, Automation and Test in Europe, 2008.

[5] Y. Kukimoto, R. K. Brayton, "Timing-Safe False Path Removal for Combinational Modules," in Proceedings of IEEE/ACM International Conference on Computer-Aided Design, pp. 544-549 Nov. 1999.

[6] E. Goldberg, A. Saldanha, "Timing Analysis with Implicitly Specified False Paths," VLSI Design, 2000. 13th International Conference on 2000, pp. 518-522

[7] S. Zhou, B. Yao, H. Chen,y. Zhu, M. Hutton, T.Collins and S. Srinivasan, "Improving the Efficiency of Static Timing Analysis with False Paths," in Proceedings of International Conference on Computer Aided Design, pp.527-531, 2005

[8] J. J. Liou, A. Krstic, L.C. Wang and K. T. Cheng, "False-Path-Aware Statistical Timing Analysis and Efficient Path Selection for Delay Testing and Timing Validation," in Proceedings of International Conference of Computer Aided Design, pp.566-569, 2002

[9] V. Chandramouli and K.A. Sakallah, "Modeling the Effects of Temporal Proximity of Input Transitions on Gate Propagation Delay and Transition Time," in Proceedings of Design Automation Conference, 1996.

[10] L.C. Chen, S.K. Gupta and M.A. Breuer, "A New Gate Delay Model for Simultaneous Switching and Its Applications," in Proceedings of Design Automation Conference, pp.617-622, 1996.

[11] D. Blaauw, V. Zolotov, and S. Sundareswaran, "Slope Propagation in Static Timing Analysis," IEEE Transactions on Computer-Aided Design of Integrated Circuits and Systems, Vol. 21, No. 10, Oct, 2002.

[12] S. T. Huang, T. M. Parng, J. M. Shyu, "A New Approach to Solving False Path Problem in Timing Analysis," in Proceedings of International Conference on Computer-Aided Design, pp. 216-219, Nov. 1991.

[13] E. Bolender, H.M. Lipp, "Timing Verification: A New Understanding of False Paths," in Proceedings of Design Automation European Conference, pp. 383 – 387, Mar, 1992.

[14] J. Bhadra, M. S. Abadir, J. A. Abraham, "A Quick and Inexpensive Method to Identify False Critical Paths Using ATPG Techniques: an Experiment with a PowerPC Microprocessor" in Proceedings of IEEE 2000 Custom Integrated Circuits Conference, pp. 71-74, May. 2000.

[15] J. Zeng, M. S. Abadir, J. Bhadra, J. A. Abraham, "Full chip false timing path identification: applications to the PowerPC microprocessors" in Proceedings of Design, Automation and Test in Europe, pp. 514-518, 2001.

978-1-60558-497-3/09 $25.00 © 2009 ACM

Way Stealing: Cache-assisted Automatic Instruction Set Extensions

Theo Kluter[†]
ties.kluter@epfl.ch

Philip Brisk[†]
philip.brisk@epfl.ch

Paolo Ienne[†]
paolo.ienne@epfl.ch

Edoardo Charbon[‡§]
e.charbon@tudelft.nl

Ecole Polytechnique Fédérale de Lausanne (EPFL)
[†]School of Computer and Communication Sciences
[‡]School of Engineering
CH–1015 Lausanne, Switzerland

Delft University of Technology
[§]Circuits and Systems Group
NL–2600 AA Delft, The Netherlands

ABSTRACT

This paper introduces *Way Stealing*, a simple architectural modification to a cache-based processor to increase data bandwidth to and from application-specific *Instruction Set Extensions (ISEs)*. Way Stealing provides more bandwidth to the ISE-logic than the register file alone and does not require expensive coherence protocols, as it does not add memory elements to the processor. When enhanced with Way Stealing, ISE identification flows detect more opportunities for acceleration than prior methods; consequently, Way Stealing can accelerate applications to up to $3.7\times$, whilst reducing the memory sub-system energy consumption by up to 67%, despite data-cache related restrictions.

Categories and Subject Descriptors

B.3.2 [**Memory Structures**]: Design Styles; D.3.4 [**Programming Languages**]: Processors—*Optimization*

General Terms

Design, Performance

Keywords

Application-Specific Processors, Instruction Set Extensions, Way Stealing, Memory Coherence, Automatic Identification

1. INTRODUCTION

The performance requirements of embedded systems must continuously grow within a stringent cost and energy envelope. As a consequence, algorithms to automatically identify application-specific custom *Instruction Set Extensions (ISEs)* have been proposed in recent years [5]. These ISEs effectively extract maximal performance from relatively simple cores.

Permission to make digital or hard copies of part or all of this work for personal or classroom use is granted without fee provided that copies are not made or distributed for profit or commercial advantage and that copies bear this notice and the full citation on the first page. To copy otherwise, to republish, to post on servers or to redistribute to lists, requires prior specific permission and/or a fee.
DAC'09, July 26-31, 2009, San Francisco, California, USA

Unfortunately, the data bandwidth between the main processor and the ISE logic is limited—indeed a classical formulation of the ISE identification problem uses register-port availability as a constraint [9]. To mitigate this problem, *Architecturally Visible Storage (AVS)* uses local memory elements to intrinsically increase the data bandwidth, effectively bypassing the processor itself [1]. AVS systems may use either flip-flops to store scalar variables or local memories to store arrays.

AVS suffers from classic coherence problems in cache based systems; correctness can be ensured by embedding the AVS memories in coherence protocols, such as those available in high-end embedded multiprocessor systems [8]. Using a coherence protocol to facilitate AVS in a single-processor system unfortunately imposes a high cost in terms of energy and silicon real estate: namely the tag array and a more complex cache state machine (see Figure 1).

Way Stealing, the contribution of this work, addresses the aforementioned concerns. Way stealing is a simple architectural modification to a cache-based processor (see Figure 2). Way stealing extends several existing paths in the processor data-cache interface (see Figure 3) to increase bandwidth to and from ISEs, similar in principle to coherent AVS, but without the overhead of a coherence protocol. Some small modifications to ISE identification algorithms to account for Way Stealing are presented as well.

The rest of the paper is organized as follows: Section 2 details the related work in the domain. Section 3 introduces our concept of Way Stealing, the specific problems that one could encounter by its introduction, and brings effective and efficient solutions to all of them. We show in Section 4 how Way Stealing can be integrated in a framework that automatically identifies custom instruction set extensions. In Section 6 we address several benchmarks, and show the effectiveness of ISE with Way Stealing, by using our FPGA-based emulation platform and ISE identification framework as described in Section 5. Section 7 concludes the paper.

2. RELATED WORK

Biswas *et al.* [1] introduced techniques to automatically identify ISEs that can read and write directly to an AVS local memory; this allowed for potentially larger ISEs with greater speedups than prior methods that forbade the inclusion of memory accesses into ISEs. The compiler inserted DMA transfers into the program to move data between main memory and the AVS. Kluter *et al.* [8] observed that this approach could lead to incorrect results because coherence between the AVS and L1-cache was not maintained; to correct the situation, the AVS and L1-cache were integrated into a coherence

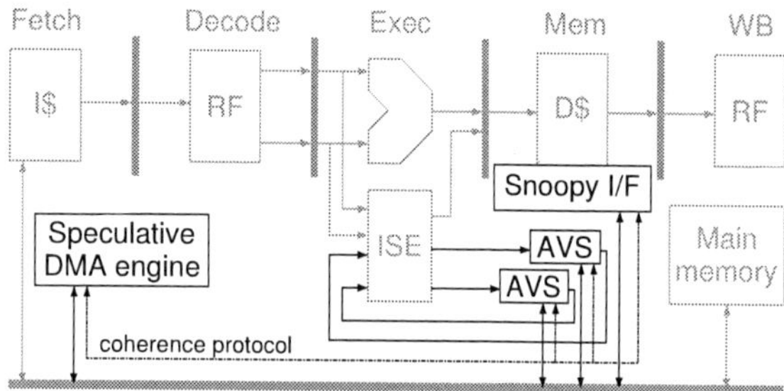

Figure 1: Current state-of-the-art Automatic Instruction Set Extension algorithms provide high bandwidth to the ISE-logic by adding Architecturally Visible Storage, however, they require extensive hardware added to a standard processor pipeline [1, 8].

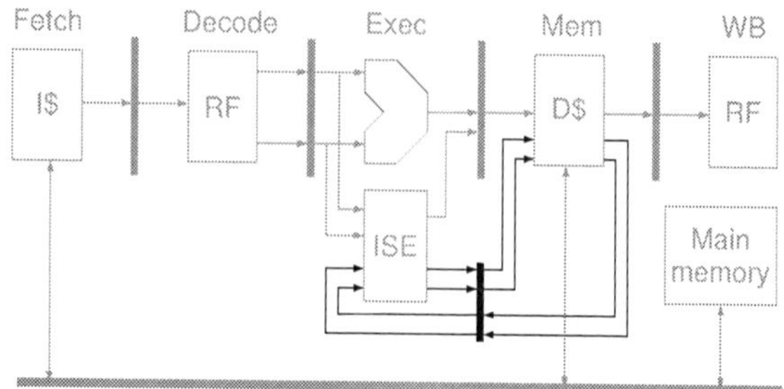

Figure 2: Automatic Instruction Set Extension utilizing Way Stealing, as proposed in this work, provide high bandwidth to the ISE-logic with only simple architectural modification to the data cache and insertion of pipeline registers.

protocol. Coherence protocols are typically meant for multiprocessor systems: their area overhead and impact on performance and power consumption due to increased bus traffic are significant. *Way Stealing*, achieves coherence but without the overhead of a protocol. Its drawback is that the number of AVS memories is limited by the number of ways in the cache; the coherence-based approach, in contrast, may allocate as many AVS memories as desired.

Other methods, orthogonal to AVS, have been proposed to increase the bandwidth between the processor core and ISEs. Cong *et al.* [2], for example, introduced shadow registers, which are written by the processor's ALU in parallel with the main register file; however, this only increases ISE input bandwidth, with no affect on the output bandwidth. Their ISEs were k-input cones, with at most k inputs and one output; our custom instructions may have any number of inputs and outputs, such as a 3-input, 3-output ISE used for color space conversion in the CJPEG benchmark; their method could not identify this ISE as it is not a k-input cone.

Jayaseelan *et al.* [6] used the forwarding mechanism to resolve data hazards in a pipelined processor to increase the I/O bandwidth to ISEs from 2-inputs and 1-output (the register file) to 4-inputs and 2-outputs; Karuri *et al.* [7] proposed the use of VLIW-style clustered register files, where each bank has 2 inputs and 1 output. These approaches are compatible with Way Stealing, and could be used if the ISE reads and/or writes many scalar variables; Way Stealing should be used when ISE data originates from arrays.

Many reconfigurable cache architectures have been proposed in the past. Ranganathan *et al.* [11] argued that cache associativity should be configurable at runtime to meet varying needs across disparate workloads. One of their proposed configurations places the data/tag SRAM under compiler control, similar in principle to Way Stealing. Their evaluation, however, focused on instruction reuse, while ours focuses on increase performance through ISEs.

3. MORE BANDWIDTH: WAY STEALING

In an n-way set-associative data cache, each way contains separate data and tag memories. All memories are activated in parallel during a lookup. After *hit* detection, the valid data is selected by an n-input multiplexor, and is sent to the processor for storage in the register file (i.e. Figure 3). Prior to the multiplexor, there are n values read in parallel, which, in principle, could be rerouted to the ISE (see Figure 2).

Writing is similar; n values can be written at once. The write behavior of a data cache is normally implemented by a write decoder, which, in turn, is activated by the *hit* signals. By extending the write decoder, providing n data line from the ISE logic to the

n data memories, and placing a multiplexor in front of each data memory, the write bandwidth can be extended similarly (see Figure 2).

3.1 Preload and Locking

Way Stealing needs a preload-lock and preload-unlock custom instruction to prevent extensive multiplexing and tag comparison in the data cache. At compile time, these instructions load and lock (unlock) a given array in a pre-determined way of the cache. These instructions require simple extensions to the cache state machine.

During execution of an application, the cache page replacement policy determines which way of the cache will store a given array; in many cases the array will be scattered across different ways. Thus, a compiler cannot feasibly predict the state of the cache prior to executing an ISE. The compiler, therefore, must statically select the ways that hold each array that will be read or written by an ISE at runtime. Additionally, the compiler must ensure that these arrays are not evicted until after the program section containing the ISEs finishes its execution.

The preload-lock and preload-unlock instructions are similar to cache prefetching and software cache-line locking [3]. Our preloading scheme, however, differs from prefetching, as the compiler, rather than the dynamic replacement policy, selects the ways that hold each array. Three distinct situations may occur:

1. Similar to prefetching, the data does not reside in the cache at all, and is loaded using the normal cache load/evict procedure (shown as step 1 in Figure 4). Contrary to prefetching, where the replacement policy selects the way dynamically, the preload instruction statically selects the way, and then locks it, possibly impeding the replacement policy.

2. Similar to prefetching, in case the data already resides in the cache, and in the correct way, nothing has to be done (shown as step 2 in Figure 4).

3. Contrary to prefetching, the data may already reside in the cache, but in a different way than specified by the preload instruction (shown as step 3 in Figure 4). For normal cache operation there may only be one copy of a given data structure in the cache. By loading the data in the specified way, the preload instruction would violate this rule, impeding correct cache behavior; moreover, if the data already in the cache was in the modified state, the preload operation would create an inter-way coherence violation. One approach to solve the problem is to evict the data in the target cache, copy it from the source way to the target way, and then invalidate the source way. A second approach, which we have adopted,

978-1-60558-497-3/09 $25.00 © 2009 ACM

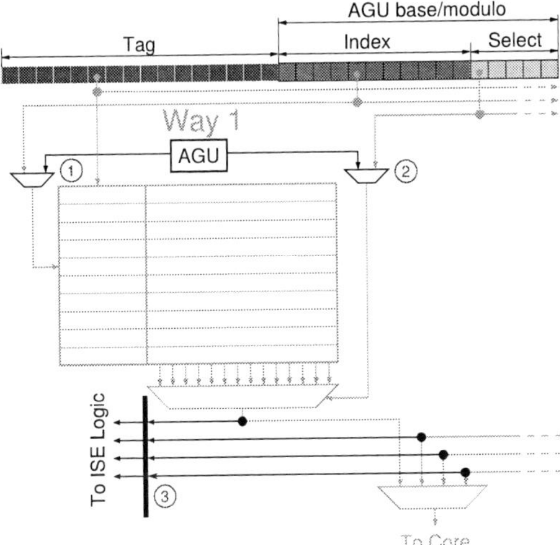

Figure 3: Block diagram of the read path of a standard n-way set-associative data cache. The Way Stealing modifications are shown in black. Multiplexor (1) may be on the critical path. Multiplexor (2) is out of the critical path. The pipeline registers (3) restrict the additional load due to long lines to the ISE logic—they only add an extra pipeline register input load to the read path. The write path is similar and omitted in this figure.

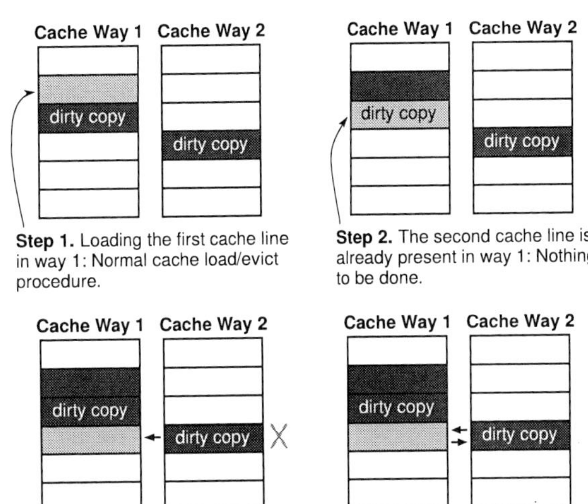

Step 1. Loading the first cache line in way 1: Normal cache load/evict procedure.

Step 2. The second cache line is already present in way 1: Nothing to be done.

Step 3. The third cache line resides in the wrong way, we can either copy and invalidate (left) or swap the contents of the cache lines of both ways (right). Simply loading the contents in way 1 would result in a inter cache coherence problem and/or violate the cache rule that only 1 copy of a data structure may reside in the cache for proper operation.

Figure 4: The different steps of preloading a data structure in way 1 having a potential risk for inter cache coherence problems.

is to swap the contents of the source and destination ways, including the tags. Although the second approach seems more complex, it is faster than the first because it does not initiate a cache line writeback.

3.2 Address Generation Units

In normal cache behavior, all ways from the cache are addressed identically. The address provided by the process is divided into a *Tag*, an *Index* and a *Select*, as shown on the top of Figure 3. The *Tag* determines a hit or miss in a given way through a comparison against the tag memories; the *Index* selects a cache line; and the *Select* picks the correct data in the selected cache line. Using this addressing scheme for Way Stealing would reduce usability in two ways. Firstly, all arrays used in an ISE must be aligned such that the *Index* and *Select* parts of their start addresses are identical; secondly, the access pattern of the data structures must be identical as well. The first restriction can be enforced by padding; however the second restriction limits the algorithmic behavior which an ISE may exhibit.

To alleviate these restrictions, Way Stealing introduces an *Address Generation Unit (AGU)*, much like those found in DSPs, for each way of the cache. Each AGU contains a count register (A_{count}), a stride register (A_{stride}), a mask register (A_{mask}), and a base register (A_{base}). The preload instruction ensures that the array is already present in a given way. This reduces the total addressing of the AGU to the *Index* and *Select* parts, as shown in Figure 3. The preload instruction stores the start address (*Index, Select*) of the data structure into A_{base}. Figure 5 depicts the addressing scheme that the AGU can perform. Although simple, this addressing scheme is easy to automatically detect and is representative of most embedded applications.

3.3 Zero Overhead Loops

Way Stealing requires pipeline registers, shown in Figure 2 and Figure 3, which minimize the increase in the processor's critical path; however, they impose a multi-cycle execution phase of an

ISE, as depicted on the left of Figure 6. The program execution, shown on the left hand side, can be accelerated using pipelining, as shown on the right hand side.

Pipelining the path from the data cache to the ISEs is similar to zero overhead loops instructions. The branch and increment phase of a program are removed to amortize their cost on each iteration. Using zero overhead loops in conjunction with Way Stealing amortizes the read delay on each iteration, and only adds one initial read delay at the beginning of each loop. Figure 6 illustrate the benefits of this approach.

3.4 Coherence

Way Stealing is intrinsically coherent, unlike AVS [1, 8], as it introduces no new memory elements into the processor. Coherence protocols in multiprocessor systems can be extended to accommodate Way Stealing as well, but doing so is beyond the scope of this paper.

3.5 Way Stealing Restrictions

To ensure proper cache operation, only one copy of an array may reside in the cache; therefore, only one data lane, as depicted in Figure 2 can be reserved for each array. Parallel reads and writes are only possible for distinct data structures. Secondly, as the capacity of a way is fixed, special care must be taken for data structures that exceed this size.

4. AUTO ISE WITH WAY STEALING

This section describes a Way Stealing-aware ISE identification flow. ISE selection algorithms, in general, need to model the cost of operations on a normal RISC pipeline in contrast to the cost of the operations executing on an ISE-enhanced pipeline. The concept of Way Stealing introduces extra costs normally not considered in ISE identification flows. The primary challenge is to model the costs of preload and lock/unlock instructions when evaluating an ISE, while accounting for the restrictions listed in the preceding section. Furthermore, Way Stealing requires some specific compiler analyses in the identification process.

978-1-60558-497-3/09 $25.00 © 2009 ACM

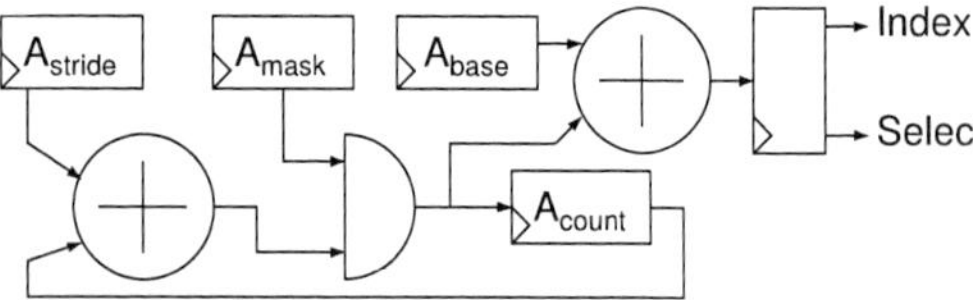

Figure 5: The functional block diagram of the Address Generation Unit (AGU) shown in Figure 3.

4.1 Modeling the Cost of Way Stealing

Way Stealing has tangible benefits, but also incurs some costs, which must be modeled accurately in order to determine when and where it is best applied.

The first cost occurs when a preload instruction swaps cache lines, as depicted in step 3 of Figure 4, which we assume takes N_{swap} cycles. If the arrays occupy α cache lines, then the maximum preload cost is defined by $\lambda_{preload} = \alpha \cdot N_{swap} + 2\alpha$. This cost is only incurred when a swap occurs; the cost of a cache-line load is identical for both normal operation and Way Stealing, and can thus be omitted. At compile time, we cannot determine how many cache lines must be swapped; the estimate, conservatively, is based on worst-case assumptions. The extra factor 2α accounts for cycling through all cache lines during the preload instruction, and unlocking of α cache lines after executing an ISE. The preload cost $\lambda_{preload}$ replaces the λ_{DMA} in the algorithm described in [1]; however, $\lambda_{preload}$ is orders of magnitude smaller than λ_{DMA}, as no memory accesses are required.

The second cost, λ_{store}, is the cost of store instructions. RISC based processors execute arithmetic operations on the contents of registers. To move the result to the memory hierarchy, the contents of a register needs to be moved from the register file back to the data cache by using a store instruction. The cost related to this store instruction is $\lambda_{store} = 1$ processor cycle. However, if the store instruction(s) is/are selected to be part of an ISE with Way Stealing and takes place on one of the data lanes depicted in Figure 2, than λ_{store} reduces to 0. In this case the arithmetic operation is directly performed on the data cache contents, bypassing the register file.

The final cost, λ_{load}, is the cost of load instructions. Similar to the store instruction the contents of the data cache has to be moved to the register file by (a) load instruction(s) with an associated cost of $\lambda_{load} = 1$ processor cycle. However, when the load instruction(s) is/are selected to be part of an ISE with Way stealing and are allocated to the data lanes depicted in Figure 2, than the λ_{load} should reduce to 0. However, λ_{load} does not become zero, because the pipeline registers inserted in the path between the data-cache and ISE logic increases λ_{load} from 1 cycle to 2.

4.2 ISE Identification Flow

The flow to find ISEs in conjunction with Way Stealing consists of ten phases:

1. **Data structure disambiguation**: The compiler attempts to disambiguate all data structures in the program [1]. All load instructions to arrays that are not disambiguated are marked as forbidden; no ISE may include a forbidden operation.

2. **Loop detection**: The loop hierarchy is computed [10]. All loops with a data-dependent loop count and those that cannot be perfectly nested are marked as forbidden.

3. **Zero overhead loop detection**: The innermost (leaf) loops in the hierarchy identified in the preceding step are examined for compatibility with zero overhead Way Stealing, described

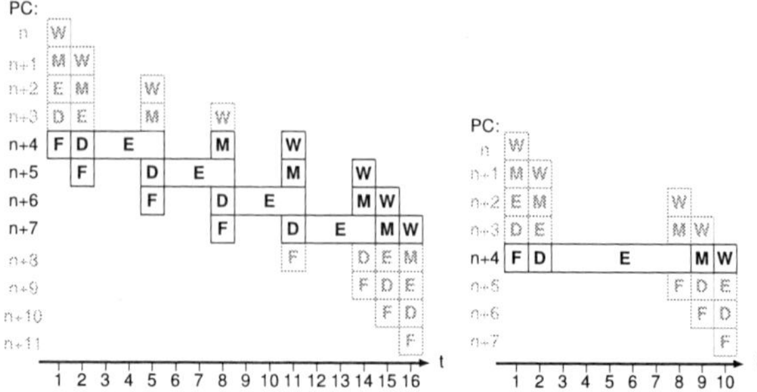

Figure 6: (Left) The execution of a program segment with a unrolled loop of four custom instructions applying Way Stealing. Due to inserted pipeline registers, the read of the data-cache takes 2 cycles, extending the execution phase to 3 cycles. (Right) By introducing the concept of zero overhead loops in the custom instruction, the data-cache read can be pipelined and the same program segment executes faster.

in Section 3.3. If the loop has a speedup, this loop is selected as ISE candidate and marked with a *zero overhead* annotation. To prevent the ISE identification step 5 from selecting parts of the loop as an ISE candidate, all operations in the loop are marked as forbidden.

4. **Access pattern detection**: For each loop, the access pattern to each disambiguated data structure is analyzed and matched with the AGU's capability. Load instructions that follow an incompatible access pattern are marked as forbidden.

5. **ISE identification**: ISEs are identified using a known algorithm, e.g., [9]. Loads and stores may be included in the search, and their costs are annotated with λ_{load} or λ_{store} as described in Section 4.1. All ISE candidates including load and/or store instructions are marked with a *Way Stealing* annotation.

6. **Way Stealing Filtering**: All ISE candidates whose data lane or array requirements exceed the number of cache ways, or whose arrays exceed the capacity of a way, are removed from consideration. Transformations such as loop unrolling must be made aware of these restrictions to be used successfully in conjunction with Way Stealing. The loops selected in phase 3 are added as potential ISE candidates.

7. **Preload Annotation**: $\lambda_{preload}$ is added to the cost of each ISE candidate with a *Way Stealing* annotation. The cost for configuring the AGUs is calculated and added to each candidate as well. After the update, all ISEs no longer having any speedup are removed.

8. **Custom instruction selection**: The best m ISE candidates are selected; m is a designer-specific parameter.

9. **Preload instruction insertion**: Preload and unlock instructions are inserted, always outside the body of leaf loops. In all other situations, they are directly placed before and after the ISEs. More sophisticated placement algorithms are beyond the scope of this paper. AGU setup instructions are inserted similarly.

10. **Custom instruction insertion**: ISEs are inserted at the appropriate locations.

978-1-60558-497-3/09 $25.00 © 2009 ACM

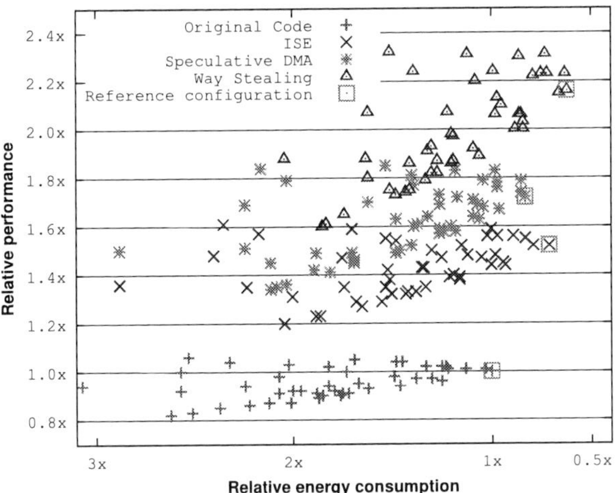

Figure 7: Design space exploration of the CJPEGV2-DATA1 testbench for the different processor specializations. The cache configuration delivering the best performance per energy for the original system is chosen as reference point.

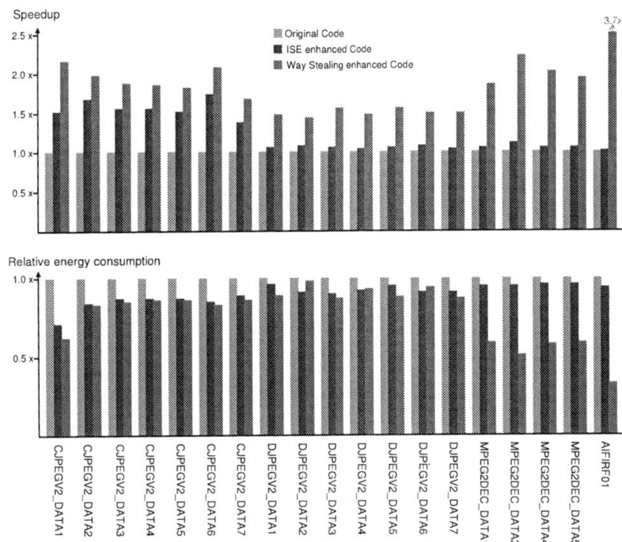

Figure 8: (Top) The speed up of several EEMBC benchmarks and data sets applying traditional ISEs and Way Stealing relative to the run-time of the original source code. (Bottom) The corresponding relative energy consumption graph of the same benchmarks as shown on the top.

4.3 Example of Generated Code

To demonstrate the overhead introduced by Way Stealing, following code snippet is used:

```
for (i=0; i<10; i++)
    b[i] = ise1(a[i],c[i]);
```

After the complete identification phase, the code-snippet is transformed into the following Way Stealing assembly code:

```
preload b,w1,10*4    // load b[] into way 1
preload a,w2,10*4    // load a[] into way 2
preload c,w3,10*4    // load c[] into way 3
setall  Amask,-1     // all Amask regs = 0xFF
setall  Acount,0     // all Acount regs = 0
setall  Astride,1    // all Astride regs = 1
ise1    w1,w2,w3,10  // perform ISE
unlock  b,w1,10*4    // unlock way 1
unlock  a,w2,10*4    // unlock way 2
unlock  c,w3,10*4    // unlock way 3
```

5. EXPERIMENTAL SETUP

We used modified an internally developed research compiler to perform ISE identification using Way Stealing. We also integrated prior relevant algorithms [1, 8] into the same environment. Our target was an OpenRISC-compatible FPGA-based emulation platform with all of the architectural changes required to support Way Stealing, as described in Section 3.

We selected several applications from the EEMBC benchmark suite [4] to evaluate Way Stealing. For each benchmark, the compiler generated VHDL models of the ISE, with or without Way Stealing, which were then added to the FPGA-based emulation platform. The compiler also generates modified C-code that includes calls to the appropriate ISEs, including prefetching and lock (unlock) instructions. The modified C-code was cross-compiled using a gcc 3.4.4 toolchain based on "newlib" for the OpenRISC.

The FPGA-based emulation platform has software controllable 16 kB instruction and data caches. Our experiments used an 2, 4, 8, and 16 kB 1-way, 2-way, and 4-way set associative instruction cache with a *Least Recently Used (LRU)* replacement policy. We used a 2, 4, 8, and 16 kB 4-way set associative data cache with an LRU replacement policy that was enhanced to support Way Stealing.

For all benchmarks, we imposed a 4:2 input-output constraint on the ISE identification algorithm when Way Stealing was not used, which is consistent with the register file restrictions; when Way Stealing was used, we increased the input-output constraint to 6:5 to account for the greater bandwidth.

We used CACTI [12] to determine the read/write energy consumption for different cache configurations in a 90 nm technology. The energy consumed by the Way Stealing read/write accesses is conservatively overestimated by assuming that each read/write access consumes the same amount of energy as a normal cache read access. The external memory and bus-access read/write energy consumption is estimated to be 792pJ per access. The energy values reported here only include the dynamic energy consumed in the memory sub-system; this model does not include processor and leakage energy.

For all experiments we performed a cache design space exploration on the original system (as shown in Figure 7). We determined the cache configuration with the best performance-energy product as the reference configuration for our experiments.

6. EXPERIMENTAL RESULTS

Figure 8 shows the relative execution time and energy consumption of several EEMBC benchmarks augmented with Way Stealing custom instructions with respect to the original code. Figure 9 shows a detailed graph of the CJPEG benchmark, shown as the first set of bars in Figure 8, and includes comparison with traditional ISEs [9] and speculative DMA [8]. Both Figure 8 and Figure 9 only represent the results relative to the reference configuration.

Figure 8 shows clearly that Way Stealing consumes significantly less energy and achieves tangible speedups compared to the baseline. The reduction of energy consumption due to four effects, the first of which is common to all ISE methods, and the latter three of which are specific to Way Stealing: (1) replacing several instruc-

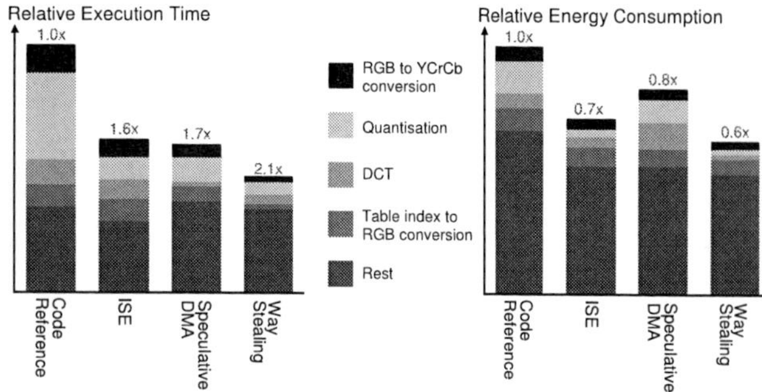

Figure 9: Detailed graph of the CJPEG benchmark shown in Figure 8. (Left) Way Stealing provides more opportunities for ISE detection due to its reduced overhead cost; however, on some kernels (like the DCT) the state-of-the-art outperforms Way Stealing. (Right) Way Stealing does not require expensive data structure moves to and from an AVS, consuming significantly less energy.

tions by one ISEs reduces the number of instruction cache fetches; (2) allocating one data structure to each way reduces data cache thrashing and capacity misses for data structures that exceed the capacity of a single way; (3) parallelizing cache read/write accesses reduces the total number of cache access; and (4) the use of zero overhead loop instructions removes a significant number of instruction cache accesses.

Figure 9 compares Way Stealing to both traditional ISEs and ISEs augmented with AVS [1] and speculative DMA to ensure coherence [8]. The *DCT* kernel benefits significantly from storing an 8×8-byte matrix in an AVS with eight read and eight write ports to the pipelined DCT logic (an ISE). Unfortunately, since each data structure may reside in just one way of the cache, Way Stealing can only provide a single read and write port, regardless of associativity; this sequentializes the loads and stores, inhibiting the acceleration of this particular kernel compared to coherent AVS.

Second, among all kernels shown in Figure 9, Way Stealing finds more opportunities for acceleration compared to existing methods, with or without coherent AVS. This results from the reduced overhead of $\lambda_{preload}$ compared to the λ_{DMA} required for DMA transfers when AVS is used. The use of zero overhead loops in conjunction with Way Stealing provides additional advantages that traditional ISEs, whose execution bodies are restricted to convex data flow subgraphs, cannot.

Lastly, Figure 9 shows that *Speculative DMA* consumes significantly more energy in the quantization kernel than the other three scenarios. *Speculative DMA* increases the miss rate in the data cache due to the DMA transfers required to facilitate coherent AVS; this is typical of other benchmarks as well. Way Stealing avoids these extra misses as it access the data directly from the cache.

7. CONCLUSIONS

Way Stealing is a simple architectural modification to a cache-based embedded processor that significantly increases data bandwidth to and from ISEs. Our results, derived by a cycle-accurate FPGA-based emulation platform, shows that ISEs enhanced with Way Stealing improve performance and reduces power consumption compared to the state-of-the-art. Due to restrictions imposed by the cache, Way Stealing does not achieve the best performance on all kernels compared to AVS, but ensures coherence at a much cheaper cost, and does not require the instantiation of new AVS memories. For these reasons, we believe that Way Stealing should be used instead of coherent AVS for modern embedded systems where energy efficiency and reduced cost trump raw performance.

8. ACKNOWLEDGMENTS

The authors thank Bhargava Shastry for his extensive work in generating the experimental results, and Xilinx for their generous donation of FPGA devices through their *XUP* program.

9. REFERENCES

[1] P. Biswas, N. Dutt, L. Pozzi, and P. Ienne. Introduction of architecturally visible storage in instruction set extensions. *IEEE Transactions on Computer-Aided Design of Integrated Circuits and Systems*, CAD-26(3):435–46, Mar. 2007.

[2] J. Cong, G. Han, and Z. Zhang. Architecture and compiler optimizations for data bandwidth improvement in configurable embedded processors. *IEEE Transactions on Very Large Scale Integration (VLSI) Systems*, 14(9):986–97, Sept. 2006.

[3] J. A. Fisher, P. Faraboschi, and C. Young. *Embedded Computing: A VLIW Approach to Architecture, Compilers and Tools*. Morgan Kaufmann, San Francisco, Calif., 2005.

[4] T. R. Halfhill. EEMBC releases first benchmarks. *Microprocessor Report*, 1 May 2000.

[5] P. Ienne and R. Leupers, editors. *Customizable Embedded Processors—Design Technologies and Applications*. Systems on Silicon Series. Morgan Kaufmann, San Mateo, Calif., 2006.

[6] R. Jayaseelan, H. Liu, and T. Mitra. Exploiting forwarding to improve data bandwidth of instruction-set extensions. In *Proceedings of the 43rd Design Automation Conference*, pages 43–48, San Francisco, Calif., July 2006.

[7] K. Karuri, A. Chattopadhyay, M. Hohenauer, R. Leupers, G. Ascheid, and H. Meyr. Increasing data-bandwidth to instruction-set extensions through register clustering. In *Proceedings of the International Conference on Computer Aided Design*, pages 166–71, San Jose, Calif., Nov. 2007.

[8] T. Kluter, P. Brisk, P. Ienne, and E. Charbon. Speculative DMA for Architecturally Visible Storage in Instruction Set Extensions. In *Proceedings of the International Conference on Hardware/Software Codesign and System Synthesis*, pages 243–48, Atlanta, Ga., Oct. 2008.

[9] L. Pozzi, K. Atasu, and P. Ienne. Exact and approximate algorithms for the extension of embedded processor instruction sets. *IEEE Transactions on Computer-Aided Design of Integrated Circuits and Systems*, CAD-25(7):1209–29, July 2006.

[10] G. Ramalingam. On loops, dominators, and dominance frontiers. *ACM Transactions on Programming Languages and Systems (TOPLAS)*, 24(5):455–90, Sept. 2002.

[11] P. Ranganathan, S. V. Adve, and N. P. Jouppi. Reconfigurable caches and their application to media processing. In *Proceedings of the 27th Annual International Symposium on Computer Architecture*, pages 214–24, Vancouver, June 2000.

[12] D. Tarjan, S. Thoziyoor, and N. P. Jouppi. CACTI 4.0. Technical Report HPL-2006-86, Hewlett-Packard Development Company, Palo Alto, Calif., June 2006.

978-1-60558-497-3/09 $25.00 © 2009 ACM

SysCOLA: A Framework for Co-Development of Automotive Software and System Platform

Zhonglei Wang, Andreas Herkersdorf
Lehrstuhl für Integrierte Systeme
Technische Universität München
Arcisstr. 21, 80290 München, Germany
{zhonglei.wang, herkersdorf}@tum.de

Wolfgang Haberl
Institut für Informatik
Technische Universität München
Boltzmannstr. 3, 85748 Garching, Germany
haberl@in.tum.de

Martin Wechs
BMW Forschung und Technik GmbH
Hanauerstr. 46, 80992 München, Germany
martin.wechs@bmw.de

ABSTRACT

A modeling language with formal semantics is able to capture a system's functionality unambiguously, without concerning implementation details. Such a formal language is well-suited for a design process that employs formal techniques and supports hardware/software synthesis. On the other hand, SystemC is a widely used system level design language with hardware-oriented modeling features. It provides a desirable simulation framework for system architecture design and exploration. This paper presents a design framework, called *SysCOLA*, that makes use of the unique advantages of both a new formal modeling language, *COLA*, and SystemC, and allows for parallel development of application software and system platform. In SysCOLA, function design and architecture exploration are done in the COLA based modeling environment and the SystemC based virtual prototyping environment, respectively. Our concepts of *abstract platform* and *virtual platform abstraction layer* facilitate the orthogonalization of functionality and architecture by means of mapping and integration in the respective environments. As SysCOLA is targeted at the automotive domain, the whole design approach is showcased using a case study of designing an automotive system.

Categories and Subject Descriptors

I.6.5 [**Simulation and Modeling**]: Model Development—*Modeling methodologies*; C.4 [**Performance of Systems**]: Modeling Techniques

General Terms

Design, Languages, Performance

Keywords

System modeling, COLA, Virtual prototyping, SystemC

Permission to make digital or hard copies of part or all of this work for personal or classroom use is granted without fee provided that copies are not made or distributed for profit or commercial advantage and that copies bear this notice and the full citation on the first page. To copy otherwise, to republish, to post on servers or to redistribute to lists, requires prior specific permission and/or a fee.
DAC'09, July 26-31, 2009, San Francisco, California, USA

1. INTRODUCTION

Today, the complexity of embedded systems is ever increasing. Many embedded systems are multiprocessor or distributed systems, where applications are partitioned into several tasks and allocated to a set of processing elements, which are interconnected by a bus system and cooperate to realize the system's functionality. Due to the growing time-to-market pressure, traditional system development methods, that software design starts after the hardware platform is ready for use, are not suitable anymore. Instead, a design method that allows for parallel development of both software and target platform is more desirable. This design method should also be able to tackle the ever-increasing design complexity of embedded systems.

In this paper, we introduce a design framework, SysCOLA, that covers a high level design process and allows parallel development of application software and target system platform. Here, the term *system platform* represents the whole platform lying under the application software layer, including a *software platform* and a *hardware platform*. The software platform contains software layers that provide services required for realizing the designated functionality.

1.1 Design Process Overview

The proposed design process is divided into three stages, namely system modeling, virtual prototyping, and system realization, as depicted in Figure 1. There exist different understandings of the term *virtual prototyping*. In this paper, it is defined as a process of generating an executable model that emulates the whole system under design. System modeling and virtual prototyping are carried out in two separate environments of SysCOLA. A newly developed formal language, *the component language* (COLA) serves as the formal foundation of the modeling environment, while the virtual prototyping environment is based on SystemC. Corresponding to the three design stages, three abstraction levels of a system platform are defined: the *abstract platform* at the model level, the *virtual platform* in SystemC and finally the implemented *real platform*.

In the modeling environment, the system's functionality is captured with an application model. The model is refined and then partitioned into *clusters*, which are tasks from an operational point of view. The system platform is modeled

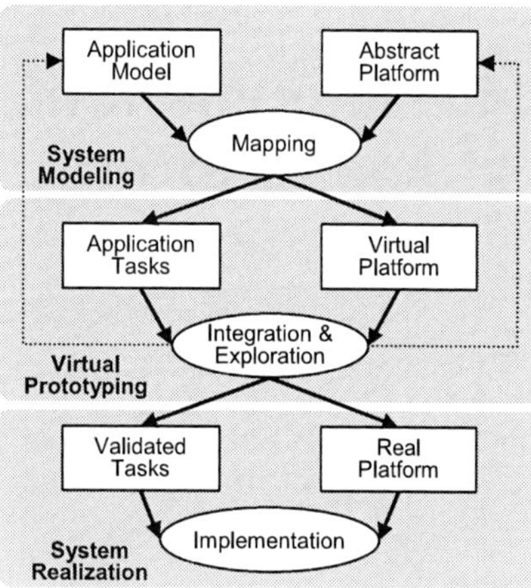

Figure 1: Design Process

as an abstract platform, allowing for automatic mapping of the software clusters and automatic code generation. However, the abstract platform is specified solely with a set of resource figures and thus cannot provide an insight into the system architecture for design space exploration. Therefore, it is only used to find suitable topologies of the system platform and fix some parameters. Rather, the detailed exploration is done on a virtual platform, using a simulative approach. Different from the abstract platform, the virtual platform is an executable SystemC model that captures the dynamic behavior of the system platform under design and can execute the generated software tasks. It gives an environment for functional validation and allows early system integration as well as performance evaluation. For system realization, the real platform is implemented, according to design decisions made on the virtual platform, and the validated tasks are then executed on the finished real platform.

Within SysCOLA, the application software and the system platform are developed in parallel, in a tightly coupled loop. Software errors found by testing on the virtual platform or a change of the virtual platform architecture may lead to a change at the model level. Then, newly generated code is tested again on the virtual platform. Such iterations reduce the possibility of errors in the final system and avoid costly redesigns.

1.2 Related Work

There exist many high level design frameworks that cover both application modeling and system platform design, such as SystemClick [12], CoFluent Studio [2], and Metropolis [6]. These frameworks as well as ours are based on the similar ideas but are different from each other in terms of target application domains and conceptual approaches. SystemClick is specially targeted at WLAN design. Unlike COLA, the modeling language Click, used in SystemClick, defines the causal order of tasks and their resource requirements, but does not deal with specifying their implementation. Tasks are implemented by hand. CoFluent Studio and other purely SystemC-based design solutions model the system's functionality using C/C++ semantics. They rule out formal techniques and hinder automated synthesis. Simulink-based solutions like [9] have the same limitation. Among these design frameworks, the most similar one to ours is Metropolis, which supports both function and architecture modeling in an integrated environment. In contrast, we prefer separation of function design and architecture exploration

in two different environments to make use of the unique advantages of a formal language like COLA and a system level design language like SystemC. The orthogonalization of functionality and architecture is achieved by means of mapping and integration in the respective environments. Also, some key concepts like abstract platform, logical communication, and virtual platform abstraction layer (VPAL) make our methodology different from Metropolis's.

The AUTOSAR project [1] also focuses on automotive systems design and has similar objectives as ours. However, it is aimed more at defining a standard for automotive systems design, while we focus on establishing a concrete design framework. When needed, SysCOLA can be adapted to generate AUTOSAR compatible designs.

There are also many research activities or commercial tools that focus on one or more aspects of a design flow. It is out of the scope of this paper to give an exhaustive list of all the related works. Instead, some survey papers like [7] can be referred to.

The rest of the paper is organized as follows: Section 2 presents the SysCOLA framework in detail. A case study is shown in Section 3. Finally, Section 4 concludes this paper.

2. THE SYSCOLA FRAMEWORK

SysCOLA employs techniques of graphical modeling, model-level simulation, formal verification, code generation, etc., supported by a well-integrated tool-chain in the modeling environment. Within this environment a system's functionality is modeled, verified and then realized in hardware or software. We focus more on the software-based implementation because of its flexibility and low cost. The main work in the virtual prototyping environment is to generate a virtual prototype that emulates the system under design. With such a virtual prototype at hand, the functionality of the generated tasks is validated and the design space of the system platform is explored. The concept of VPAL is introduced to facilitate integration of the application software and the virtual platform and reduce design effort.

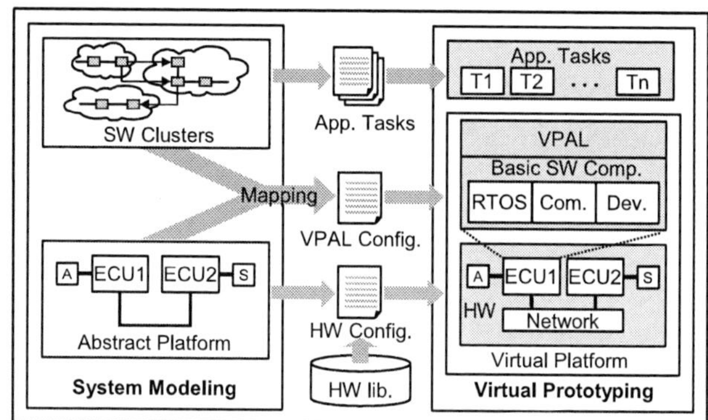

Figure 2: The SysCOLA Framework

Figure 2 shows the architecture of the SysCOLA framework. A prototype of the framework was implemented based on the Eclipse platform. As shown in Figure 2, the application behavior and important information captured in the COLA model are forwarded to the virtual prototyping environment in the form of application tasks and configuration files. A detailed introduction to both design environments is given in the following sub-sections.

978-1-60558-497-3/09 $25.00 © 2009 ACM

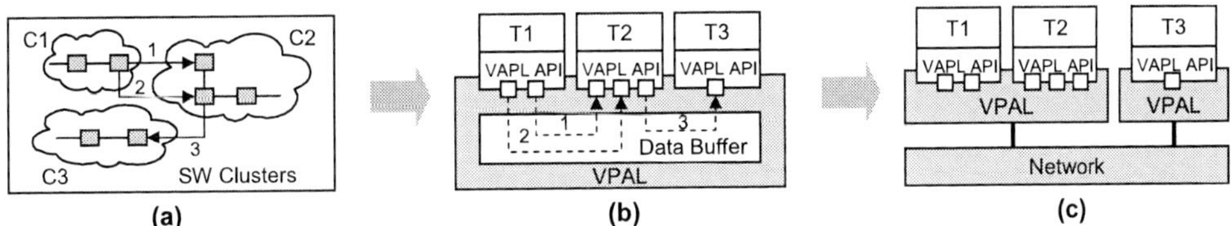

Figure 3: Logical Communication

2.1 System Modeling

2.1.1 Modeling Language and Semantics

Model-driven development, based on a formally defined language, allows for a clear and unambiguous description of a system's functionality, without concerning implementation details. To set up a modeling environment, one critical issue is to find a suitable model of computation, which can capture the target system's characteristics efficiently. In this paper, we aim at automotive control systems as our target systems. *Data flow*, as used e.g. in MATLAB/Simulink, is the state-of-the-art concept for models in this domain. Unfortunately, the tools like MATLAB/Simulink lack either formally defined semantics or integrated tool support throughout the entire development process. Therefore, a new modeling language, COLA, that follows the paradigm of synchronous data flow was developed as the formal foundation of our modeling environment. COLA offers an easily understandable graphical syntax and a formally defined semantics. We describe details about COLA in [11].

2.1.2 Application Modeling

In COLA, design starts from modeling functional requirements in a *feature model*. The feature model is converted to a *functional model* throughout several steps of model transformation and rearrangement. In a functional model, the data flows are expressed by networks of COLA *units*. A unit can be further composed hierarchically, or occurs in terms of *blocks* that define the basic (arithmetic and boolean) operations. The communication between units is realized by *channels*. In addition to the hierarchy of networks, COLA provides *automata* to express control flows in a functional model. With automata complex networks of units can be partitioned into disjoint operating modes, the activation of which depends on the input events.

COLA allows for modeling in a component-based flavor, which facilitates model reusability. The correctness of modeling and model refinement is checked by model-level simulation [13] and formal verification.

2.1.3 Clustering and Logical Communication

Before mapping model entities to computing nodes of the platform, units of the functional model are grouped into *clusters*—the model representation of software tasks, each annotated with resource requirements on each possible processing element the task might be mapped to. A clustering is derived from an optimized software architecture w.r.t. reusability, maintainability, design guidelines, documentation and others.

To enable software development independent of the system platform, we introduce the notion of *logical communication*. According to the concept of logical communication, two clusters exchange data by means of a logical address (i.e., data id) assigned to the channel between them, as shown in Figure 3(a). After mapped to a networked sys-tem, this logical communication is retained. The generated software tasks communicate with each other still by logical addresses, without caring whether they are located at the same node or not. The realization of the logical communication on a concrete platform is introduced later.

2.1.4 Automatic Mapping and Scheduling

An abstract platform is constructed using the hardware modeling capabilities of COLA. The purpose of the abstract platform is to specify the topology and capabilities of the target platform and thus give a basis for mapping of clusters. We have developed an automatic mapping algorithm, which takes the application clusters and the abstract platform as its input and tries to find a valid mapping with the resource requirements of the tasks and the platform capabilities as the foundation of its decision. We describe this mapping algorithm in [10].

The execution order of the tasks can be decided either at runtime by a dynamic scheduling algorithm or pre-runtime by a static scheduling algorithm. To guarantee that the resulting system behaves the same as modeled, we propose to use static scheduling. An algorithm for static scheduling of COLA tasks is also introduce in [10].

2.1.5 Code Generation

After finding a proper mapping that fulfills the requirements, C code is generated [8]. COLA also allows hardware implementation of the modeled functionality for higher performance [15]. The code generation relies on a set of template-like translation rules. As COLA has a slender syntax, the number of translation rules is small. The translation rules are well documented in [8] and [15]. As an example, the task generated from the cluster C2 (Figure 3) is shown in Figure 4. It reads the input data from the logical addresses 1 and 2 and sends the output data to the logical address 3.

```
1. void T2(){
2.   Int in_0, in_1; Int out_0;
3.   receive_data(&in_0, sizeof(in_0), 1);
4.   receive_data(&in_1, sizeof(in_1), 2);
5.   ... // behavioral code
6.   send_data(&out_0, sizeof(out_0), 3);
7. }
```

Figure 4: An Example of Generated Code

2.2 Virtual Prototyping

2.2.1 Virtual Platform Abstraction Layer

The VPAL is aimed at wrapping the whole virtual platform. Using this layer, the virtual platform can be regarded as a functional entity that provides a set of services, from the application's perspective. The software tasks use these services by calling a fixed VPAL API. In this way, the tasks can be simulated on different virtual platforms without the need of changing any code. Only adaptation of the VPAL

978-1-60558-497-3/09 $25.00 © 2009 ACM

to the new platform is needed. To achieve a small and simple VPAL API, the VPAL provides four basic functions, as shown in Figure 5. The API is easily extendable. The functions *send_data()* and *receive_data()* are defined for sending and receiving data, respectively. We define a task as a *stateful* task, if the execution of this task depends on its history (*state*). This state data is read at the beginning of the task's execution and saved in the end, by calling *read_state()* and *save_state()*, respectively.

```
send_data(void *buf, ssize_t len, unsigned short int dataid);
receive_data(void *buf, ssize_t len, unsigned short int dataid);
read_state(void *buf, ssize_t len, unsigned short int taskid);
save_state(void *buf, ssize_t len, unsigned short int taskid);
```

Figure 5: VPAL API

It is one of the duties of the VPAL to retain the logical communication between software tasks in the virtual platform. It can be realized by a shared data buffer, where the tasks exchange data using logical addresses, as shown in Figure 3(b). To go one step further, the logical communication between tasks on different nodes is mapped to the physical communication that is emulated by a network simulation model, as shown in Figure 3(c). To set up exactly the same logical communication as the model level, configuration information generated from the model is used. Using this configuration information, the VPAL is able to allocate buffer space during the startup phase, initialize sensor and actuator hardware for operation, and distinguish between local and remote communications. A segment of configuration information is shown in Figure 6. It is used to configure the VPAL for the task shown in Figure 4.

```
<node id="1">
 <task id="2" name="T2">
  <data id="1" channel="c_in_0" name="in_0" type="INT" len="4"/>
  <data id="2" channel="c_in_1" name="in_1" type="INT" len="4"/>
  <data id="3" channel="c_out" name="out_0" type="INT" len="4"/>
 </task>
 ...
</node>
```

Figure 6: VPAL Configuration Information

2.2.2 System Platform Design

The system platform design can be regarded as the process of building and refining a virtual platform. It starts with defining the critical platform services in the VPAL, including communication mechanisms, device interaction, and scheduling. These services are then refined and realized in the corresponding basic software components. For example, a RTOS is chosen and modeled to take charge of scheduling and device drivers are developed to interact with devices. The VPAL lies between the application tasks and the basic software components and controls the interaction between these software components. Meanwhile, the hardware platform is designed to realize these services in hardware components including processing nodes, sensors and actuators, interconnected by a network. During the design, both functional and timing simulations are needed. Functional simulation is used to verify each change or refinement of the virtual platform, while timing simulation provides performance statistics to explore the system architecture.

To which level a platform component should be refined depends on how many details we need to explore. If a new component is to be developed, its model should be refined down to implementation. In contrast, if we decide to use an existing one, it is unnecessary to simulate all the details as accurate as the real one. To achieve low simulation complexity and high simulation speed, we just need an abstract model that captures the most important features of the existing component and is accurate enough to make important design decisions.

2.2.3 Functional Simulation

In SysCOLA, functional simulation can be performed early during modeling, by means of model-level simulation [13]. This model-level simulation tests a functional model before mapping and code generation and thus cannot take the underlying platform into account. Different from the model-level simulation, the simulation on a virtual platform allows for testing of both generated code and the services provided by the target platform.

Thanks to the VPAL, the details of the virtual platform are transparent from the application software's point of view, and thus the software tasks can run on the virtual platform at an arbitrary abstraction level. Validation of application software is made possible early in the design cycle, even when there is only the VPAL available. The refinement of the virtual platform can be verified by comparing functional simulation results from the refined virtual platform and an earlier one. For example, the network design can be verified by comparing simulation results from the setup shown in Figure 3(b) and the one shown in Figure 3(c).

2.2.4 Performance Simulation

During embedded systems design, performance is one of the most important factors to be considered. If performance requirements cannot be met, this could lead to system failure. It is necessary to handle performance issues from early design phases until system implementation.

We estimate the WCET of each task, which serves as a basis for automatic mapping and scheduling. However, these static analyses have limited ability to capture real workload scenarios and often result in an over-designed system. Therefore, we use performance simulation to validate whether the performance requirements are really satisfied and show how pessimistic the static analyses are.

In [14, 16], we propose source code level and *intermediate source code* (ISC) level approaches for software timing simulation. Using our tools, each software task is converted to a timed simulation model in SystemC. In comparison to instruction set simulation and binary level simulation, our approaches allow for simulation at much higher speeds without compromising accuracy [14].

The temporal behavior of the system platform can also be abstracted with delay values, to be annotated into the functional simulation model. For example, a network model is annotated with delays, determined with respect to the bus protocol and the data length. Similarly, each RTOS overhead due to, e.g. scheduling or context switches, is also associated with a delay. Such delay values can be obtained by means of measurement. The simulation models of these system components are linked with the simulation models of software tasks to generate a simulator of the whole system, which we call a virtual prototype. Depending on the simulation results, system architecture will then be adjusted and refined iteratively. With the stepwise refinement of the virtual platform, the granularity of the timing values can also be refined, down to cycle-accurate if necessary.

978-1-60558-497-3/09 $25.00 © 2009 ACM

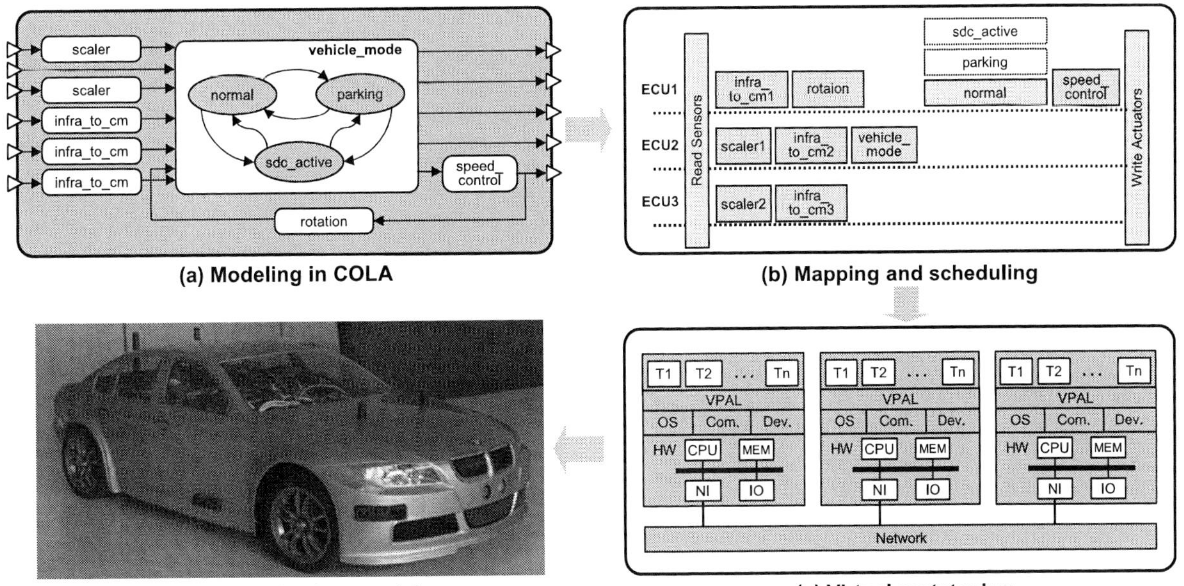

(d) Implementation in a model car

Figure 7: Case Study

3. CASE STUDY

The presented approach has been evaluated using a case study. As the framework is targeted at the automotive domain, we chose an upcoming feature of premium cars, namely an automatic parking system, to imitate the concerns and requirements of automotive design. We chose a model car with a scale of 1:10 as basis for system realization.

3.1 Functionality and Design Space

The intended functionality included (i) manual steering via remote control, (ii) automatic discovery and parking into a parking space of sufficient length, and (iii) ability to run alongside a wall while going around all convexities.

In principle, the design space for the designated functionality could be very large. However, in our case study, the design space was constrained by the availability of resources and our research interests. We chose Gumstix micro computers [3] as processing nodes. An Ethernet network with RTnet protocol [4] was chosen as the communication fabric to study the real-time capability of the Ethernet. RTnet is based on Xenomai [5], which implements a RTOS emulation framework on Linux. These constraints reduce the design space but don't hinder the demonstration of the key concepts. The same methodology is surely applicable for exploring a larger design space.

3.2 System Modeling

The discussed functionality was transferred into a COLA model. Eleven distributable clusters were defined, as shown in Figure 7(a). The system was separated into three operating modes, namely **normal**, **parking**, and **sdc_active**, each corresponding to one defined function. In the **normal** mode, the user can modify speed and direction using the remote control. The **parking** mode makes the model car run in parallel to a wall and search for a gap of sufficient size to park. In the **sdc_active mode**, the car just runs parallel to a given wall and adjusts its distance to all convexities found. The mode cluster **vehicle_mode** controls the transitions between the three operating modes, which are triggered using the remote control.

After specification of the software model and an abstract platform, the clusters were distributed onto available computing nodes and a scheduling plan was generated automatically. Figure 7(b) shows an exemplified result of mapping and scheduling on a network of three computing nodes. As shown, the scheduling cycle starts from reading input data from sensors and ends with writing output data to actuators. The tasks are scheduled in between, with the starting time explicitly specified. The tasks that can be executed in parallel are mapped to different nodes. Whereas, **normal**, **parking**, and **sdc_active** must be delayed until the mode decision is made by **vehicle_mode**. As they execute exclusively, they are mapped to the same node. In the same way, mapping and scheduling were repeated for two-node and four-node systems. At last, software tasks and configuration files were generated from the model.

3.3 Virtual Prototyping

As the design space was narrowed down by the constraints discussed before, the work in this phase was significantly eased, including: (i) refining VPAL services down to implementation, (ii) writing device drivers, (iii) deciding the number of computing nodes, (iv) system integration, (v) functional validation and performance evaluation.

By integrating the generated code into a virtual platform, an executable virtual prototype is generated for functional validation. By the time of functional validation, the virtual platform was at a high level of abstraction, containing the VPAL and the functional models of RTOS, devices and network. Figure 8(a) and (b) show the exemplified simulation results. The two simulations, each run with 1000 test cases, take only 1.92 and 1.98 seconds, respectively, with useful logging information generated for error detection. The first simulation tested the system in the **parking** mode, while the second one tested the system in the **sdc_active** mode. The distances from the wall to the right side and the front-right side of the car were calculated using randomly gener-

978-1-60558-497-3/09 $25.00 © 2009 ACM

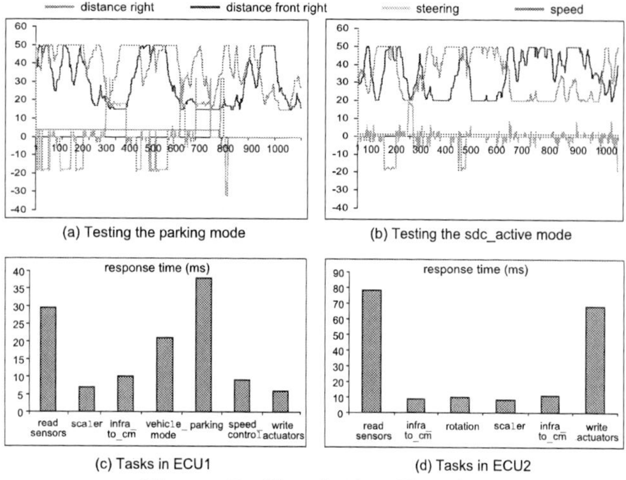

Figure 8: Simulation Results

ated data representing the sensor data. Both parking and steering algorithms depend on the two distances. The distances from the car to the front, back and left obstacles were set constants in the value range of safe distances. In both figures, we can see the changes of the steering value with respect to the distance values. In Figure 8(a), we can also see the parking behavior after around 700 simulation steps.

The parking system must react to the environment fast enough, so we defined that the scheduling cycle could not extend 200 ms. To decide the number of ECUs needed to satisfy this performance requirement, the virtual platform was refined for fine-grained timing annotation and performance simulation. To connect all the sensors, at least two ECUs were needed. Figure 8(c) and (d) show the respective maximal response times of all the tasks mapped to two ECUs, obtained by simulation with 10000 test cases. The simulation took less than one minute. According to the simulation results, the scheduling cycle was adjusted, which resulted in a cycle length of 238 ms. In the same way, the systems with 3 ECUs and 4 ECUs were evaluated and resulted in cycle lengths of 178.5 ms and 145 ms, respectively. The increase of the number of ECUs cannot improve the performance anymore, because there were five tasks that must run sequentially due to data dependencies. For higher performance and lower cost, these five tasks should be allocated to an ECU with higher computing power, while the other tasks could run on cheaper ECUs. Another solution is to reduce the overhead of basic software components. According to the exploration result, we decided to implement the three-node system. After all the high-level design decisions were made, the VPAL was refined and implemented as a middleware. All the device drivers were also implemented.

3.4 System Realization

For system realization, three Gumstix micro computers were embedded into the model car shown in Figure 7(d). The infrared and supersonic sensors were used for distance sensing and connected via serial and I2C bus, respectively. A cellular phone was connected via bluetooth and served as the user interface for remote control. Control of steering, motor, and light system was realized by an interface board, equipped with a serial interface. Executing the application code, the middleware and the basic software components on the real platform confirmed the simulation results obtained in the virtual prototyping phase.

4. CONCLUSIONS

In the paper, we presented the SysCOLA framework, which consists of two design environments. The concepts of logical communication, abstract platform and VPAL were introduced to facilitate parallel development of application software and system platform. In the modeling environment of SysCOLA, a system's functionality is captured in a COLA model, verified and validated using formal techniques, and finally realized in software by means of code generation. An *abstract platform* provides an abstract view of the system architecture and serves as the basis for automatic mapping of the software model onto the system platform. The system platform is designed in the form of a *virtual platform*, in the SystemC based virtual prototyping environment. The VPAL wraps the virtual platform and enables early system integration. The whole design process was showcased using a case study of designing an automotive parking system.

5. REFERENCES

[1] http://www.autosar.org.
[2] http://www.cofluentdesign.com.
[3] http://www.gumstix.com.
[4] http://www.rtnet.org.
[5] http://www.xenomai.org.
[6] F. Balarin, Y. Watanabe, H. Hsieh, L. Lavagno, C. Passerone, and A. Sangiovanni-Vincentelli. Metropolis: An integrated electronic system design environment. *Computer*, pages 45–52, 2003.
[7] D. Densmore, R. Passerone, and A. Sangiovanni-Vincentelli. A platform-based taxonomy for ESL design. *IEEE Des. Test*, pages 359–374, 2006.
[8] W. Haberl, M. Tautschnig, and U. Baumgarten. From COLA Models to Distributed Embedded Systems Code. *IAENG International Journal of Computer Science*, 2008.
[9] K. Huang and et al. Simulink-based MPSoC design flow: case study of motion-jpeg and h.264. In *Proceedings of the Design Automation Conference*, pages 39–42, 2007.
[10] S. Kugele, W. Haberl, M. Tautschnig, and M. Wechs. Optimizing automatic deployment using non-functional requirement annotations. In *Leveraging Applications of Formal Methods, Verification and Validation*, 2008.
[11] S. Kugele, M. Tautschnig, A. Bauer, C. Schallhart, S. Merenda, W. Haberl, C. Kühnel, F. Müller, Z. Wang, D. Wild, S. Rittmann, and M. Wechs. COLA – The component language. Technical Report TUM-I0714, Technische Universität München, Sept. 2007.
[12] C. Sauer, M. Gries, and H.-P. Löb. SystemClick: a domain-specific framework for early exploration using functional performance models. In *Proceedings of the Design Automation Conference*, pages 480–485, 2008.
[13] Z. Wang, W. Haberl, S. Kugele, and M. Tautschnig. Automatic generation of SystemC models from component-based designs for early design validation and performance analysis. In *Proceedings of the 7th international workshop on software and performance*, 2008.
[14] Z. Wang and A. Herkersdorf. An efficient approach for system-level timing simulation of compiler-optimized embedded software. In *Proceedings of the 46th Design Automation Conference (DAC'09)*, July 2009.
[15] Z. Wang, S. Merenda, M. Tautschnig, and A. Herkersdorf. A model driven development approach for implementing reactive systems in hardware. In *Proceedings of International Forum on Specification and Design Languages (FDL'08)*, pages 197–202, September 2008.
[16] Z. Wang, A. Sanchez, and A. Herkersdorf. SciSim: A Software Performance Estimation Framework using Source Code Instrumentation. In *Proceedings of the 7th international workshop on software and performance*, 2008.

978-1-60558-497-3/09 $25.00 © 2009 ACM

Designing Heterogeneous ECU Networks via Compact Architecture Encoding and Hybrid Timing Analysis[*]

Michael Glaß[†], Martin Lukasiewycz[†], Jürgen Teich[†], Unmesh D. Bordoloi[‡], Samarjit Chakraborty[*]

[†]University of Erlangen-Nuremberg,
Germany
{glass,martin.lukasiewycz,teich}@cs.fau.de

[‡]Verimag, France
unmesh.bordoloi@imag.fr

[*]Technical University of Munich,
Germany
samarjit@tum.de

ABSTRACT

In this paper, a design method for automotive architectures is proposed. The two main technical contributions are (i) a novel hardware/software architecture *encoding* that unifies a number of design steps, i.e., resource allocation, process binding, message routing, scheduling, and parameter estimation for the processor and bus schedulers, and (ii) a hybrid scheme that allows different timing analysis techniques to be applied to different bus protocols (viz., CAN and FlexRay) within the same architecture in order to derive global performance estimates such as end-to-end delays of messages. The use of the compact encoding technique substantially reduces the underlying search space, and the hybrid timing analysis scheme allows the combination of known timing analysis techniques from the real-time systems domain. The proposed techniques were combined into a tool-chain and a real-life case study to illustrate their advantages.

Categories and Subject Descriptors

C.3 [**Special-purpose and application-based systems**]: Real-time and embedded systems

General Terms

Design, Performance

Keywords

Automotive, Design Space Exploration, Timing Analysis

1. INTRODUCTION AND MOTIVATION

Today, it is fairly common for high-end cars to have more than 80 different electronic control units (ECUs) running a variety of distributed control applications and communicating via multiple buses and gateways implementing different communication protocols, e.g., CAN, LIN, and FlexRay. Optimally designing such a complex and heterogeneous ECU network poses several challenges, which has led to a lot of recent interest in developing design and analysis techniques specifically directed towards the automotive domain. Design methodologies for such complex embedded systems typically follow a series of design steps such as *resource allocation*, *process binding*, *message routing*, *scheduling* and *parameter estimation* for different processor and bus schedulers. Since a number of design constraints, e,g., real-time properties of safety-critical applications cannot be verified until all the design steps are complete and a concrete implementation is derived, such that multi-phase approaches hardly come up with global optimal solutions.
Contributions: This paper adresses the general design space exploration problem, cf. Fig. 1, by proposing an architecture encoding scheme that allows a number of hardware/software co-design tasks – such as resource allocation, process binding, message routing, scheduling, etc. – to be represented as linear constraints with binary variables. As a result, the design space exploration problem

---

[*]Supported in part by the German Science Foundation (DFG), SFB 694

Permission to make digital or hard copies of part or all of this work for personal or classroom use is granted without fee provided that copies are not made or distributed for profit or commercial advantage and that copies bear this notice and the full citation on the first page. To copy otherwise, to republish, to post on servers or to redistribute to lists, requires prior specific permission and/or a fee.
DAC'09, July 26-31, 2009, San Francisco, California, USA

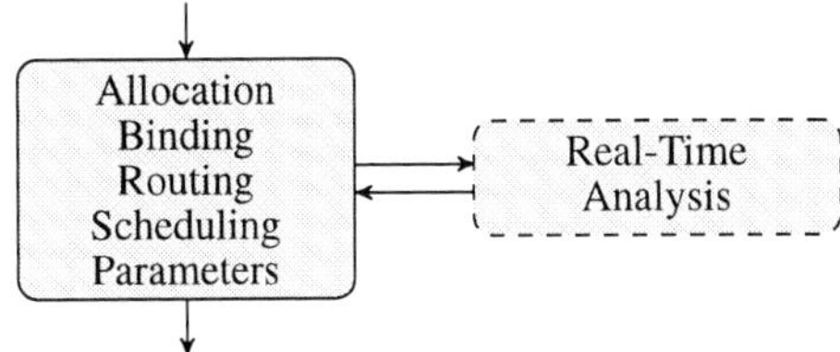

Figure 1: Proposed design space exploration approach.

can now be formulated as a *binary search problem*, and the drawbacks associated with multi-phase design approaches can largely be avoided. The encoding scheme also substantially reduces the underlying search space and solutions to the resulting search problem are complete feasible implementations which respect all system constraints. As a result, it is now possible to use combinations of binary search and efficient heuristics for quickly solving multi-objective optimization problems arising in real-life automotive designs.

The presented techniques are geared towards automotive architectures, consisting of a large and heterogeneous collection of ECUs, buses, and gateways running distributed control applications and processing several message streams connecting sensors to actuators. The optimization schemes clearly need to rely on timing *analysis* and performance *evaluation* techniques. Given the heterogeneous nature of the architectures in the automotive domain, having a single general-purpose timing analysis technique for all resources is difficult and often impossible. For example, different bus protocols such as CAN and FlexRay require very different modeling and analysis techniques. To get around this problem, a hybrid analysis technique is proposed which allows results from different timing analysis techniques to be composed together. In particular, it is shown how response time analysis techniques – which use a system of recurrence relations – for CAN protocol can be used with timing analysis techniques for FlexRay to derive overall timing properties of architectures containing buses of both these types.

Both, the architecture encoding scheme, as well as the hybrid timing analysis technique have been implemented into a tool-chain which is capable of analyzing real-life case studies from the automotive electronics domain. The utility of the proposed approach is illustrated through a detailed case study later in this paper.

Related Work: Resource allocation and process binding strategies are studied in [1, 11], mostly based on either exact approaches like *Integer Linear Programs* (ILPs) or using meta-heuristics like *Evolutionary Algorithms* (EAs). Process and message scheduling as well as the parameter estimation has been thoroughly studied in approaches like [4]. There exist design space exploration approaches that unify some or all steps in a single exploration tool, cf., e.g., [16]. However, these approaches are not able to restrict their search space to the set of feasible implementations only. This leads to serious drawbacks in case few feasible implementations exist due to stringent constraints. Real-time analysis of automotive networks has been studied in [6, 14]. For the sake of flexibility and extensibility, the work at hand proposes a jitter propagation model that utilizes existing analysis techniques, depending on the individual resource type and used scheduler.

2. DESIGN SPACE EXPLORATION

This section presents the proposed one-step design space exploration approach for automotive networks. After an introduction of the used exploration model, the encoding of the tasks of resource allocation, process binding, message routing, scheduling, and parameter estimation as a binary search problem based on linear constraints is presented. The combination of the binary search problem and a meta-heuristic to ensure a fast convergence to the optimal

---

implementations with respect to multiple objectives completes this section.

The exploration model is defined by a *specification* that consists of an *application* and an *architecture*. From this specification, various *implementations* can be derived by defining the *allocation* of the architecture and the *mapping* of the application, the *routing* of messages and the used *parameters*. The specification consists of an *architecture graph* G_R and an *application graph* G_T : The architecture is given by a directed graph $G_R(R, E_R)$. The vertices R represent resources such as ECUs ($R_{ECU} \subset R$), buses like FlexRay ($R_{FR} \subset R$) or CAN ($R_{CAN} \subset R$), communication controllers ($R_{CC} \subset R$), as well as sensors and actuators. The directed edges E_R indicate available communication connections between two resources. The application is given by a bipartite directed graph $G_T(T, E_T)$ with $T = P \cup M$. The vertices T are either processes $p \in P$ or messages $m \in M$. Each edge $e \in E_T$ connects a vertex in P to one in M, or vice versa. Each process can have multiple incoming edges that indicate the data dependencies to communication information of the predecessor message. On the other hand, each message has exactly one predecessor process as the sender, but a process can of course have multiple successor messages. To allow multicasts, each message can have multiple successor processes. Each process $p \in P$ can be implemented on a unique set of resources $R_p \subseteq R$. Each message $m \in M$ can be routed on a subset of resources from R_m with $R_m \subseteq R$. An implementation consists of the *allocation graph* G_A that is deduced from the architecture graph and a function i that maps the application onto the allocation graph. The allocation is a directed graph $G_A(A, E_A)$ that is an induced subgraph of the architecture graph G_R. The allocation contains all resources that are available in the current implementation and the edges are induced from the graph G_R. Each process $p \in P$ is bound to exactly one allocated resource $i(p)$ such that $i(p) \in (A \cap R_p)$. Each message in $m \in M$ is routed on a tree that is a subgraph of the allocation such that $i(m) \subseteq G_A$ with all vertices in R_m. These bindings and routings have to be performed such that all data dependencies given by the following two conditions are satisfied:

1. For each message $m \in M$, the root of the routing has to equal the binding of the predecessor sender process $p \in P$. It holds:

$$\forall (p, m) \in E_T : root(i(m)) = i(p)$$

2. For each process $p \in P$ the routings of the predecessor message $m \in M$ have to be routed on the same resource as the binding of process p. The following holds:

$$\forall (c, m) \in E_T : i(p) \in i(m)$$

An implementation is *feasible* if all requirements regarding the process and message mapping, the data dependencies, and the system constraints are fulfilled.

With the definition of a feasible implementation, the task of the *design space exploration* can be formulated as the following multi-objective optimization problem:

DEFINITION 1 (DESIGN SPACE EXPLORATION).
optimize $f(\mathbf{x})$
subject to:
    $\mathbf{x}$ is a feasible *implementation*

In real-world problems, the objective function f consists of multiple functions including also non-linear equations. In single-objective optimization, the feasible set of networks is totally ordered, whereas in multi–objective optimization problems, the feasible set is only partially ordered and, thus, there is generally not only one global optimum, but a set of *Pareto solutions*. A Pareto-optimal solution is better in at least one objective when compared to any other feasible solution.

2.1 Model Encoding

In the following, a binary search problem is defined such that a solution $\mathbf{x}$ corresponds to a *feasible* implementation x, regarding the allocation of resources, the binding of processes, the routing of messages, and the parameter set. The symbolic encoding consists of the following binary variables:

- $\mathbf{r}$ - one variable for each resource $r \in R$ indicating whether this resource is allocated (1) or not (0).

- $\mathbf{p_r}$ - one variable for each pair of process $p \in P$ and available resources $r \in R_p$, indicating whether the process is bound onto the resource r (1) or not (0).

- $\mathbf{m_r}$ - one variable for each message $m \in M$ and available resources $r \in R_m$, indicating whether m is routed over the resource r (1) or not (0).

- $\mathbf{m_{r,n}}$ - one variable for each message and resource pair indicating on which communication step $n \in \mathbb{N}$ (messages are propagated in steps) a message is routed over the resource.

The linear constraints are formulated as follows:

$$\forall p \in P : \qquad \sum_{r \in R_p} \mathbf{p_r} = 1 \qquad (1a)$$

$$\forall m \in M : \qquad \sum_{r \in R_m} \mathbf{m_{r,0}} = 1 \qquad (1b)$$

$$\forall m \in M, p \in \{\tilde{p}|(\tilde{p}, m) \in E_T\}, r \in R_p \cap R_m :$$
$$\mathbf{p_r} - \mathbf{m_{r,0}} = 0 \qquad (1c)$$

$$\forall p \in P, m \in \{\tilde{m}|(\tilde{m}, p) \in E_T\}, r \in R_p \cap R_m :$$
$$\mathbf{m_r} - \mathbf{p_r} \geq 0 \qquad (1d)$$

$$\forall m \in M, r \in R_m : \qquad \sum_{i=1}^{n} \mathbf{m_{r,i}} \leq 1 \qquad (1e)$$

$$\sum_{i=1}^{n} \mathbf{m_{r,i}} - \mathbf{m_r} \geq 0 \qquad (1f)$$

$$\forall m \in M, r \in R_m, i = \{1, ..., n\} :$$
$$\mathbf{m_r} - \mathbf{m_{r,i}} \geq 0 \qquad (1g)$$

$$\forall m \in M, r \in R_m, i = \{1, ..., n-1\} :$$
$$-\mathbf{m_{r,i+1}} + \sum_{\tilde{r} \in R_m \wedge e=(\tilde{r},r) \in E_R} \mathbf{m_{\tilde{r},i}} \geq 0 \qquad (1h)$$

$$\forall p \in P, r \in R_p : \qquad \mathbf{r} - \mathbf{p_r} \geq 0 \qquad (1i)$$

$$\forall m \in M, r \in R_m : \qquad \mathbf{r} - \mathbf{m_r} \geq 0 \qquad (1j)$$

$$\forall r \in R : \quad -\mathbf{r} + \sum_{m \in M \wedge r \in R_m} \mathbf{m_r} + \sum_{p \in P \wedge r \in R_p} \mathbf{p_r} \geq 0 \qquad (1k)$$

Equation (1a) ensures that each process is bound exactly once. Equations (1b) and (1c) imply that each message has exactly one root that equals the used resource of the predecessor process. Analogously, for each process the predecessor messages have to be routed on the corresponding resources as stated in Equation (1d). Equation (1e) ensures that a message can pass a resource at most once such that no cycles occur in the route of a message. A message has to exist in one communication step on a resource in order to be correctly routed on this resource as implied by the Equations (1f) and (1g). Equation (1h) states that a communication is only possible between adjacent resources. The Equations (1i) and (1j) imply that a process or message, respectively, is bound or routed on an allocated resource only. On the other hand, Equation (1k) states that a resource is only allocated if at least one process is bound or a message is routed on this resource.

Besides the constraints arising from resource allocation, process binding, and message routing, additional constraints regarding the system and the parameters of its components have to be respected. In this work, an encoding for system constraints that depend on the chosen parameters of a system component is proposed. In particular, this encoding carries out the task of parameter estimation implicitly.

In the automotive area, stringent bus load constraints are applied to the used CAN buses. The maximal rational load of a CAN bus is manufacturer-dependent and commonly between 0.4 and 0.6 [12]. The capacity of a high-speed CAN is $64,000$ byte/s and the low-speed CAN has a capacity of $16,000$ byte/s. The constraints are formulated as follows:

$$\forall r \in R_{CAN} :$$

$$\sum_{m \in M} \lceil \frac{\sigma_{CAN}(m)}{\rho(m)} \rceil \cdot \mathbf{m_r} \leq \lfloor \lambda(r) \cdot 64000 \rfloor \cdot \mathbf{r_{hi}} + \lfloor \lambda(r) \cdot 16000 \rfloor \cdot \mathbf{r_{lo}}$$
$$(2a)$$

$$\mathbf{r_{hi}} + \mathbf{r_{lo}} = 1 \qquad (2b)$$

978-1-60558-497-3/09 $25.00 © 2009 ACM

44

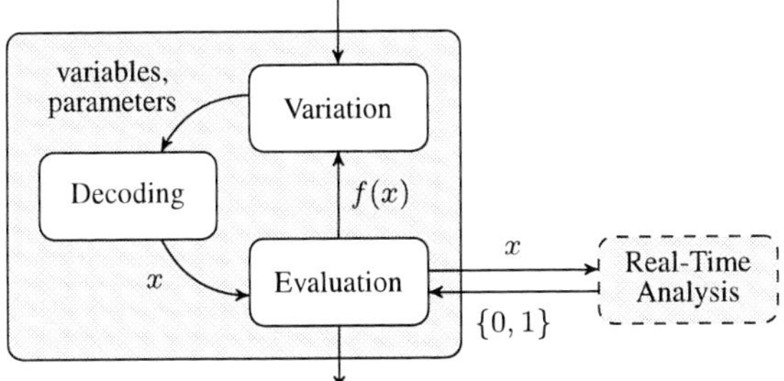

Figure 2: Proposed design space optimization flow.

- $\sigma_{CAN} : M \to \mathbb{N}$ - size of bytes of each message (including additional CAN overhead per message)
- $\rho : M \to \mathbb{N}$ - period of a message
- $\lambda : R_{CAN} \to [0; 1]$ - maximal rational load of the bus

Equation (2a) states that the load of a CAN bus does not exceed the allowed bus load. Equation (2b) ensures that exactly one CAN variant, high-speed or low-speed, is chosen.

With respect to energy consumption, ECU processors can be set to different operation modes leading to different characteristics regarding computational capacity and energy consumption. The constraints are formulated as follows:
$\forall r \in R_{ECU}$:

$$\sum_{p \in P} \left\lceil \frac{\sigma_{ECU}(p)}{\rho(p)} \right\rceil \cdot \mathbf{p_r} \le \sum_{o \in \mu(r)} \lfloor \lambda(r) \cdot \gamma(r, o) \rfloor \cdot \mathbf{r_o} \qquad (3a)$$

$$\sum_{o \in \mu(r)} \mathbf{r_o} = 1 \qquad (3b)$$

- $\sigma_{ECU} : P \times R_{ECU} \to \mathbb{N}$ - instructions of a process on an ECU
- $\rho : P \to \mathbb{N}$ - period of a process
- $\mu : R_{ECU} \to 2^O$ - set of operation modes of an ECU
- $\gamma : R_{ECU} \times O \to \mathbb{N}$ - the computational capacity of an ECU under a given mode in instructions per second
- $\lambda_{ECU} : R_{ECU} \to [0; 1]$ - the maximal utilization of an ECU

Equation (3a) states that the computational load of an ECU bus does not exceed the allowed utilization, Equation (3b) ensures that exactly one operation mode is chosen.

Given a single solution $\mathbf{x}$ of this search problem, the corresponding implementation x is deduced by constructing the allocation from the $\mathbf{r}$ variables, the binding for each process from the $\mathbf{p_r}$ variables, the routing of the message from the $\mathbf{m_r}$ and $\mathbf{m_{r,n}}$ variables, and the system parameters from the corresponding variables.

2.2 Optimization

Common optimization approaches that are based on *Integer Linear Programs*, cf. [11], or *Evolutionary Algorithms*, cf. [8], only are either restricted to a single linear objective function or do not perform well on optimization problems with many constraints and few feasible solutions. With the linear constraints using binary variables introduced previously, the design space exploration problem as stated in Def. 1 can be carried out efficiently by using a heuristic *SAT decoding* optimization approach based on [9]. This hybrid optimization approach based on an *Evolutionary Algorithm* and a *PB (pseudo Boolean) solver* allows the optimization of multiple conflicting and non-linear objectives under linear constraints in a binary search space, cf. Fig. 2. In particular, the approach *varies* the variables of the binary search problem as well as additional parameters used, e.g., for process and message priorities or scheduling policies. These additional parameters are optimized in parallel to the binary search problem if they do not affect the feasibility of the implementation. The *decoding* solves the binary search problem with respect to the varied variables and, thus, gathers feasible implementations. Decoded implementations are *evaluated* to determine their objectives and their real-time properties are verified. The results from the evaluation guide the variable and parameter variation in the next iteration. The optimization approach iteratively improves found implementations such that with a higher runtime and more evaluated implementations, respectively, the quality of the results increases.

3. HYBRID TIMING ANALYSIS

Next, a timing analysis is presented that determines the real-time properties for an application by handling each process and message separately and only considering the jitter propagation. The distinction between processes and messages allows a specific analysis which has an effect on the response time delay of multicast and multihop routed messages. The jitter propagation model allows a component-wise analysis either with a common method based on recurrence functions, e.g., [7, 14], or with more sophisticated methods like the *Real-Time Calculus* (RTC) [5, 13].

3.1 Function Timing Analysis

Each process $p \in P$ is implemented on a single resource $i(p) \in R$ such that the worst case response time a_p is calculated dependent on the scheduler of this resource. On the other hand, each message $m \in M$ is implemented on the tree $i(m) \in G_A$ such that the response $a_{m,r}$ has to be determined for each resource in the graph $i(m)$. The end-to-end latency of one path of the application π with $\pi \subseteq T$ is determined as follows:

$$a_\pi = \sum_{p \in \pi \cap P} a_p + \sum_{m \in \pi \cap M} \sum_{r \in route(m)} a_{m,r} \qquad (4a)$$

with

$$route(m) = (i(p), ..., i(\tilde{p})) \subseteq i(m) \text{ and } (p, m), (m, \tilde{p}) \in G_T \qquad (4b)$$

This means that the end-to-end latency is given by the sum of all process execution times and message response times that arise on the path π.

Given a function of the application $G_F \subseteq G_T$ which is described by a directed acyclic graph, the worst case execution time of the function is:

$$a_{G_F} = \max_{\pi \in \{\text{all paths in } G_F\}} a_\pi \qquad (5)$$

Instead of enumerating all paths in G_F and searching for the maximal a_π, the evaluation is performed efficiently by applying the *Bellman-Ford algorithm* [2] to the graph G_F with the cost for the nodes being a_p or $a_{m,r}$, respectively. Thus, the complexity of Equation (5) is $O(|F| \cdot |E_F|)$ with F being the nodes of G_F and E_F the edges, respectively.

The real-time analysis of an application requires the process execution times a_p and messages response times $a_{m,r}$. For this purpose, a jitter propagation model also becomes necessary. Here, the jitter is the unwanted variation between two consecutive periodic processes executions or messages. The output jitter of the process execution is j_p for $p \in P$ and a message is $j_{m,r}$ for $m \in M$ and the resource $r \in R$. The input jitter j_p^* of a process $p \in p$ is system dependent if this process is time-triggered or the maximum of the jitter values of the predecessor messages $\max(j_{pred(p),i(p)})$ if the process is event-triggered, respectively. The input jitter $j_{m,r}^*$ of a message $m \in M$ with the predecessor task p with $(p, m) \in E_T$ on the resource $r \in R$ is calculated as follows:

$$j_{m,r}^* = \begin{cases} j_p, & \text{if } r = root(i(m)) \\ j_{m,pred(r)}, & \text{else} \end{cases} \qquad (6)$$

Where $pred(r)$ denotes the predecessor resource of the resource r on the routing tree $i(m)$. Taking these jitter values into account allows an exact and flexible real-time analysis independent of the underlying methodology. With the given jitter values, each process execution time a_p and messages response time $a_{m,r}$ can be calculated independently as shown exemplarily in the following subsections.

3.1.1 Process Execution Time (Priority Scheduling)

Given a priority-based scheduler with preemption, the worst case response time and jitter is calculated as follows:

$$a_p = c_p + \sum_{\tilde{p} \in hp(p)} \left\lceil \frac{a_p + j_p^*}{\tau_{\tilde{p}}} \right\rceil \cdot c_{\tilde{p}} \qquad (7a)$$

$$j_p = a_p - c_p \qquad (7b)$$

Where c_p ($c_{\tilde{p}}$) is the required time for the execution of the process p ($\tilde{p}$), $\tau_{\tilde{p}}$ is the period of the process $\tilde{p}$, and the function $hp(p)$

978-1-60558-497-3/09 $25.00 © 2009 ACM

determines all processes that are implemented on the same resource as p and have a higher priority.

3.1.2 Message Response Time on CAN (ET)

For an event-triggered (ET) CAN bus $r \in R_{CAN}$, one message $m \in M$ induces the following response time and jitter:

$$a_{m,r} = c_{m,r} + b_{m,r} + \sum_{\tilde{m} \in hp(m)} \left\lceil \frac{a_{m,r} + j^*_{m,r} - c_{m,r}}{\tau_{\tilde{m},r}} \right\rceil \cdot c_{\tilde{m},r} \tag{8a}$$

$$j_{m,r} = a_{m,r} + c_{m,r} \tag{8b}$$

Here, $c_{m,r}$ $(c_{\tilde{m},r})$ is the transmission time of the message m $(\tilde{m})$ on the bus r, $\tau_{\tilde{m}}$ is the period of the message $\tilde{m}$, and the function $hp(m)$ determines all message that are routed over the CAN bus r and have a higher priority than m. In contrast to the priority-based scheduler, preemption is not possible on the CAN bus and $b_{m,r}$ is added which denotes the maximal transmission time of one lower priority message.

3.1.3 Message Response Time on FlexRay (TT)

Each message $m \in M$ that is routed on the time-triggered (TT) *static segment* of the FlexRay bus $r \in R_{FR}$ induces the response time and jitter as follows:

$$a_{m,r} = c_{m,r} + s_{m,r} \tag{9a}$$

$$j_{m,r} = c_{m,r} + \left\lceil \frac{j^*_{m,r}}{s_{m,r}} \right\rceil \cdot s_{m,r} \tag{9b}$$

Where $c_{m,r}$ denotes the transmission time for a static slot and $s_{m,r}$ the maximal time difference (usually a multiple of the cycle time) between two static slots that transmit the message m.

The analysis of the *dynamic segment* of the FlexRay bus is based on the demand-bound criteria approach which is well-known in the real-time scheduling literature [3]. This is approach is adapted to specifically model the dynamic segment of the FlexRay protocol. The model and the associated analysis mechanisms have been integrated into a publicly-available Matlab-based tool called the Real-Time Calculus Toolbox [15].

4. CASE STUDY

The methodology presented in this paper, cf. Fig. 2, is realized as a tool-chain: The proposed exploration is implemented using the OPT4J framework [10] while the real-time analysis is based on recurrence relations and the RTC toolbox [15]. The methodology was applied to a case study, modeling a typical automotive subnetwork: The network architecture consists of 15 ECUs, connected via two CAN buses, one FlexRay bus, and a central gateway. The 9 sensors, and 5 actuators are connected via LIN buses to the ECUs. The application consisting of four functions, an *adaptive cruise control* (ACC), a *brake-by-wire* (BW), an *air conditioning function* (C1), and a *multimedia control* (C2), with 46 processes and 42 messages in total has to be mapped onto the given architecture.

The subnetwork is optimized in terms of the monetary cost in Euro(€) and energy consumption in Watts. Given constraints are the maximal load 40% for the CAN bus, the maximal utilization 95% for ECUs as well as real-time constraints regarding the end-to-end delay from the sensors to the corresponding actuators, given in Table 1.

The optimization of this subnetwork was performed by an exploration with the presented approach including a parallel optimization of priorities of the processes and messages as well as the scheduling of the messages on the static or dynamic segment of the FlexRay bus. The exploration was performed with an Evolutionary Algorithm meta-heuristic, improving the implementations iteratively by 5075 objective evaluations within 4 hours and 43 minutes on an Intel Core 2 Quad 2.66 GHz with 3GB RAM. Only 21 seconds are spent on the variation and decoding while the remaining time accounts for evaluation, in particular the real-time analysis due to a delaying file interface of the RTC.

The results of the optimization are given in Table 1. Given is a hand-made reference implementation with the monetary cost 214.40€ and an energy consumption of 432.5 Watts that respects

	energy(W)	cost(€)	ACC(ms)	BW(ms)	C1(ms)	C2(ms)
deadline	-	-	100	50	30	50
ref.	432.5	214.4	99.7	30.2	11.8	32.8
impl. 1	394.9	189.1	43.1	37.5	17.3	39.6
impl. 2	396.2	184.7	41.2	34.0	15.1	49.0
impl. 3	399.5	182.8	41.2	34.0	12.5	49.0
impl. 4	416.7	182.3	38.1	28.9	13.9	39.7

Table 1: Detailed results of the best found implementations.

all real-time constraints. Four non-dominated high quality implementations are found improving the reference implementation in both objectives, monetary cost and energy consumption. The found implementations allow to decrease the monetary cost by 11.8% to 15% while decreasing the energy consumption by about 3.6% to 8.7% at the same time. These results are obtained by binding the processes such that a higher utilization is achieved and some ECUs become redundant and can be removed from the implementation. At the same time, the priorities of the messages and processes are varied such that the real-time constraints are still respected.

5. CONCLUSION

The paper presents a unified design space exploration approach for real-time automotive networks that performs the design steps of resource allocation, process binding, message routing, scheduling, and parameter optimization. A combination of a binary search problem and a meta-heuristic allows to reduce the search space to the feasible implementations only while achieving a fast convergence to the optimal implementations. For the verification of the real-time constraints, a hybrid timing analysis is introduced that combines the benefits of known analysis techniques that each fit best for particular resource types like buses, ECUs, etc. A case study modeling an automotive subsystem shows the capability of the proposed methodology.

6. REFERENCES

[1] N. Banerjee and R. Kumar. Multiobjective network design for realistic traffic models. In *Proceedings of GECCO '07*, pages 1904–1911, 2007.

[2] R. Bellman. On a routing problem. *Quarterly of Applied Mathematics*, 16:87–90, 1958.

[3] G. Buttazzo. *Hard Real-Time Computing Systems: Predictable Scheduling Algorithms and Applications*. Springer, 2005.

[4] A. Davare, Q. Zhu, M. D. Natale, C. Pinello, S. Kanajan, and A. Sangiovanni-Vincentelli. Period optimization for hard real-time distributed automotive systems. In *Proceedings of DAC '07*, pages 278–283, 2007.

[5] A. Hagiescu, U. D. Bordoloi, S. Chakraborty, P. Sampath, P. V. V. Ganesan, and S. Ramesh. Performance analysis of flexray-based ECU networks. In *Proceedings of DAC '07*, pages 284–289, 2007.

[6] A. Hamann, R. Racu, and R. Ernst. Formal methods for automotive platform analysis and optimization. In *Proceedings Future Trends in Automotive Electronics and Tool Integration Workshop (DATE Conference)*, 2006.

[7] M. G. Harbour, M. H. Klein, and J. P. Lehoczky. Timing analysis for fixed-priority scheduling of hard real-time systems. *IEEE Trans. Softw. Eng.*, 20(1):13–28, 1994.

[8] R. Kumar, P. K. Singh, and P. P. Chakrabarti. Multiobjective EA approach for improved quality of solutions for spanning tree problem. In *Proceedings of EMO '05*, pages 811–825, 2005.

[9] M. Lukasiewycz, M. Glaß, C. Haubelt, and J. Teich. Sat-decoding in evolutionary algorithms for discrete constrained optimization problems. In *Proceedings of CEC '07*, pages 935–942, 2007.

[10] Opt4J. The optimization framework for java. http://www.opt4j.org/, Version 1.5.

[11] D. Rajan and A. Atamtürk. A directed cycle-based column-and-cut generation method for capacitated survivable network design. *Networks*, 43(4):201–211, 2004.

[12] K. Richter and R. Ernst. How OEMs and suppliers can face the network integration challenges. In *Proceedings of DATE '06*, pages 183–188, 2006.

[13] L. Thiele, S. Chakraborty, and M. Naedele. Real-time calculus for scheduling hard real-time systems. In *Proceedings of ISCAS '00*, pages 101–104, 2000.

[14] K. Tindell, A. Burns, and A. Wellings. Calculating controller area network (CAN) message response times. *Control Engineering Practice*, 3:1163–1169, 1995.

[15] E. Wandeler and L. Thiele. Real-Time Calculus (RTC) Toolbox. http://www.mpa.ethz.ch/Rtctoolbox, 2006.

[16] H. Zeng, A. Davare, A. Sangiovanni-Vincentelli, S. Sonalkar, S. Kanajan, and C. Pinello. Design space exploration of automotive platforms in metropolis. In *SAE Congress*, 2006.

978-1-60558-497-3/09 $25.00 © 2009 ACM

Optimizing Throughput of Power- and Thermal-Constrained Multicore Processors Using DVFS and Per-Core Power-Gating

Jungseob Lee and Nam Sung Kim

Department of Electrical and Computer Engineering, University of Wisconsin - Madison, WI, U.S.A.

{jslee9, nskim3} at wisc dot edu

ABSTRACT

Process variability from a range of sources is growing as technology scales below 65nm, resulting in increasingly nonuniform transistor delay and leakage power both within a die and across dies. As a result, the negative impact of process variations on the maximum operating frequency and the total power consumption of a processor is expected to worsen. Meanwhile, manufacturers have integrated more cores in a single die, substantially improving the throughput of a processor running highly-parallel applications. However, many existing applications do not have high enough parallelism to exploit multiple cores in a die. In this paper, first, we analyze the throughput impact of applying per-core power gating and dynamic voltage and frequency scaling to power- and thermal-constrained multicore processors. To optimize the throughput of the multicore processors running applications with limited parallelism, we exploit power- and thermal-headroom resulted from power-gated idle cores, allowing active cores to increase operating frequency through supply voltage scaling. Our analysis using a 32nm predictive technology model shows that optimizing the number of active cores and operating frequency within power, thermal, and supply voltage scaling limits improves the throughput of a 16-core processor by ~16%. Furthermore, we extend our throughput analysis and optimization to consider the impact of within-die process variations leading to core-to-core frequency (and leakage power) variations in a multicore processor. Our analysis shows that exploiting core-to-core frequency variations improves the throughput of a 16-core processor by ~75%.

Categories and Subject Descriptors

C.4 [Performance of Systems]: Design studies

General Terms

Design, Performance

Keywords

DVFS, power gating, multicore processor

1. INTRODUCTION

Today we are at an inflection point in the computing landscape as we enter the multicore era. Recently, manufacturers have announced multicore processors because technology scaling has allowed more devices in a single die. Moreover, manufacturer roadmap promises to repeatedly double the number of cores per die to continue Moore's Law. With technology scaling, however, manufactured dies exhibit a large spread of *maximum operating frequency* (F_{max}) and *leakage power* due to process variations that can be classified into two categories: *die-to-die* (D2D) and *within-die* (WID). D2D variations affect all transistors on a die equally, while WID variations induce different electrical characteristics across a die. This leads to increasingly nonuniform transistor delay and leakage both within a die and across dies. Furthermore, it is expected that individual cores are becoming small enough that spatially correlated WID process variations manifest themselves as *core-to-core* (C2C) F_{max} and leakage power variations in future multicore processors.

K. Bowman et al. and J. Tschanz et al. evaluated the impact of D2D and WID variations on the F_{max} distribution of single core processors [1][2]. K. Bowman, et al. also presented an analytical throughput model for *globally-clocked* multicore processors and the sensitivity of several processor designs' throughputs to process variations [3]. E. Humenay et al. modeled the impact of WID variations on C2C F_{max} variations in multicore processors and analyzed the benefit of *adaptive body-biasing* (ABB) and *adaptive voltage scaling* (AVS) [4]. S. Herbert et al. presented an analytical model for the throughput of *frequency-island* (FI) multicore processors, and quantified the performance benefit of the FI design style across a range of multicore processor designs [5]. R. Rao et al. provided analytical thermal and power models, and analyzed the impacts of a thermal constraint on multicore processor performance [6]. J. Donald et al. presented power-performance trade-off analysis under the presence of process variations, and provided a method to predict an optimal cut-off point for turning-off extra cores in multicore processors [7]. M. Hill et al. showed that an optimal core size for a given die size providing the highest throughput depends on the level of parallelism in applications [8]. Finally, Woo et al. analyzed energy-efficient core size for many-core processors considering parallelism in applications [9].

As an alternative to optimize the throughput of multicore processors for a wide range of parallelism, we can use *per-core power-gating* (PCPG) and *dynamic voltage and frequency scaling* (DVFS); both of them are very commonly used in many commercial multicore processors to improve the energy and power efficiency of multicore processors. A multicore processor with many, smaller cores provides much higher throughput than one with a few, larger cores for applications with high parallelism. However, for applications with limited parallelism, a multicore processor with many, smaller cores suffers from significantly lower throughput than one with a fewer, larger cores due to lower single-thread performance of the smaller cores. To improve the single-thread performance of the smaller cores (thus the overall throughput of the multicore processor), F_{max} can be increased through *supply voltage* (V_{DD}) scaling. In recent multicore processors, however, increasing V_{DD} (and F_{max}) is often limited by power and thermal constraints rather than a maximum V_{DD} ($V_{DD,max}$) constraint. Note that manufactured dies with the same multicore processor design can have different power and thermal constraints depending on a target market segment: server, desktop, and mobile; the power and thermal constraints are often determined by the capacity of supply voltage regulators and cooling solutions in computing platforms customized for each market segment.

When a multicore processor with many, smaller cores runs applications with limited parallelism, extra power- and thermal-headroom can be provided for the multicore processor after idle cores are disabled by PCPG. This allows one or more active cores to increase V_{DD} and F_{max} (thus single-thread performance) within power and thermal constraints. Furthermore, WID C2C F_{max} variations can be also exploited to improve the single-thread performance of a multicore processor with many, smaller cores through F_{max} increase; smaller cores exhibit a more relative F_{max} spread among the cores in a die [5]. When all cores are active in a globally-clocked multicore processor, F_{max} is limited by the slowest core in a die. However, we need only a few number of active cores due to limited parallelism in applications, allowing a multicore processor to pick faster cores for running applications. Then the slowest core among the chosen faster cores determines the processor's F_{max} that is much faster than the original F_{max} determined by the slowest core in a die. However, note that F_{max} increase of faster cores can be more severely limited by power and

Permission to make digital or hard copies of part or all of this work for personal or classroom use is granted without fee provided that copies are not made or distributed for profit or commercial advantage and that copies bear this notice and the full citation on the first page. To copy otherwise, to republish, to post on servers or to redistribute to lists, requires prior specific permission and/or a fee.

DAC'09, July 26-31, 2009, San Francisco, California, USA

978-1-60558-497-3/09 $25.00 © 2009 ACM

thermal constraints, since faster ones often consume much more leakage power than slower ones.

In this paper, we analyze and optimize the throughput of power- and thermal-constrained multicore processors using the first-order approximated models that allow early architectural analysis. First, we analyze the throughput impact of applying PCPG and DVFS simultaneously to power- and thermal- constrained multicore processors. To maximize the throughput using PCPG and DVFS for a given multicore configuration (core size) and parallelism in applications, we optimize 1) the number of active (or idle) cores that determine available power- and thermal-headroom and 2) V_{DD} and F_{max} that affect the single-thread performance of the active cores within power and thermal constraints. Second, we extend our analysis and optimization to consider the WID process variations in a multicore processor. To optimize the throughput using PCPG and DVFS, we exploit WID C2C F_{max} variations by picking faster cores for active ones that can maximize the F_{max} of the multicore processor within power and thermal constraints.

The remainder of this paper is organized as follows: Section 2 presents an overview of Amdahl's Law to model the throughput of multicore processors. Section 3 analyzes and optimizes the throughput of power- and thermal-constrained multicore processors. Section 4 extends analysis and optimization to consider WID process variations. Section 5 details experimental methodology. Section 6 concludes the study.

2. THROUGHPUT ANALYSIS USING AMDAHL'S LAW

According to *Amdahl's Law*, when a fraction F of a program's execution time was infinitely parallelizable with no overhead and the remaining fraction, $1 - F$, was totally sequential, the throughput speedup on N processors is governed by:

$$speedup = \left((1 - F) + \frac{F}{N}\right)^{-1} \tag{1}$$

To apply Amdahl's Law to a multicore processor, a cost model for the number and performance of cores is needed and a simple hardware model from [8] is adopted in this paper: 1) a multicore processor of given die size and technology generation can contain at most *N base core equivalents* (BCEs), where a single BCE implements the baseline core; 2) (micro-) architects have techniques for using the resources of multiple BCEs to create a richer core with greater sequential performance. Let the performance of an 1-BCE core be 1 and architects can expend the resources of R BCEs to create a rich core with sequential performance *Perf(R)*. The multicore performance model shown in [8] allows *Perf(R)* to be an arbitrary function, but basically assumes $Perf(R) = \sqrt{R}$; the efforts that devote R BCEs resources will result in performance $\sqrt{R}$. Thus, architects can double performance at a cost of 4 BCEs, triple it for 9 BCEs, etc. Similar functions, e.g., $^{1.5}\sqrt{R}$ was explored in [8], but no important change to the results was found.

A multicore processor on a die requires that all its cores have the same cost. A multicore processor die with a resource budget of N=16 BCEs, for example, can support 16 cores of 1 BCE each, or in general, N/R cores of R BCEs each (rounded down to an integer number of cores). Figure 1-(a) shows a cartoon of a possible multicore processor with four 4-BCE cores with a resource budget of N=16 BCEs. Under Amdahl's Law, the throughput speedup of a multicore processor (relative to using one 1-BCE core) depends on F, total die resources in BCEs (N), and the BCE resources (R) devoted to increase the performance of each core. The processor uses one core to execute sequentially at performance *Perf(R)*. It uses all N/R cores to execute in parallel at performance *Perf(R)×N/R*. Overall, the throughput speedup is represented as follows:

$$speedup = \left(\frac{1 - F}{Perf(R)} + \frac{F \cdot R}{Perf(R) \cdot N}\right)^{-1} \tag{2}$$

Figure 1-(b) shows the throughput speedup of multicore processors versus F and R for N=16 multicore processors. Depending on parallelism in applications, a multicore processor

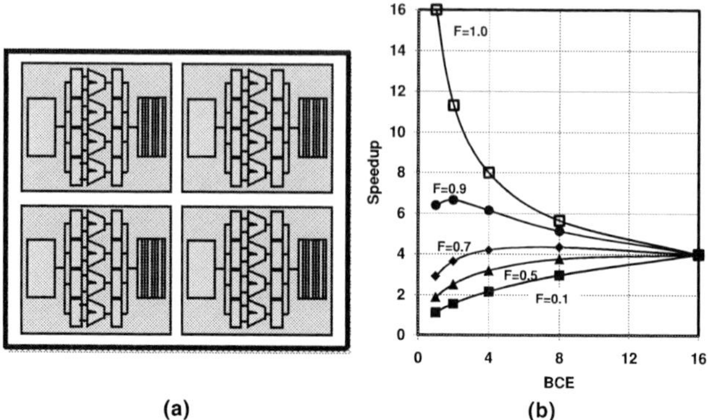

Figure 1. A multicore processors with four 4-BCE cores in (a); the cartoon is from [8]. Throughput speedup of multicore processors versus BCE for N=16 in (b).

(a)   (b)

typically shows either immediate performance improvement or performance loss as it uses more powerful cores.

3. THROUGHPUT OPTIMIZATION USING DVFS AND PER-CORE POWER-GATING

When parallelism in an application is limited, one or more idle cores disabled by PCPG can also provide extra power- and thermal-headroom in power- and thermal-constrained multicore processors. This, in turn, allows active cores to increase F_{max} through V_{DD} scaling, improving the single-thread performance of the active cores (thus the throughput of the multicore processor). In this section, we analyze the throughput impact of applying PCPG and DVFS to power- and thermal-constrained multicore processors. Then we maximize the throughput at given parallelism by optimizing 1) the number of active (and idle) cores and 2) V_{DD} and F_{max}, while satisfying the power and thermal constraints.

We can approximate the throughput speedup of a multicore processor by extending (2) in Section 2 as follows:

$$speedup = \left(\frac{1 - F}{f(V_{DD}) \cdot Perf(R)} + \frac{F}{f(V_{DD}) \cdot Perf(R) \cdot n}\right)^{-1} \tag{3}$$

where n is the number of active cores; and $f(V_{DD})$ is the frequency scaling factor at V_{DD}, normalized to the F_{max} at $V_{DD,TDP}$ — the maximum V_{DD} limited by *thermal design power* (TDP) with all the cores running simultaneously at sustainable maximum power and performance (i.e., a multicore processor uses up all the power- and thermal-budget at the TDP point). Since we can approximate the total power of a multicore processor at the TDP point — P_{TDP} as follows:

$$P_{TDP} = P_{dyn,\,TDP} + P_{leak,\,TDP} \tag{4}$$

$$P_{dyn,\,TDP} = \frac{N}{R} \cdot (C_{eff} \cdot R) \cdot V_{DD,\,TDP}^2 \cdot F_{max}(V_{DD,\,TDP}) \tag{5}$$

$$P_{leak,\,TDP} = \frac{N}{R} \cdot (I_{leak}(V_{DD,\,TDP}) \cdot R) \cdot V_{DD,\,TDP} \tag{6}$$

where $P_{dyn,\,TDP}$ and $P_{leak,\,TDP}$ are *dynamic power* (P_{dyn}) and *active leakage power* (P_{leak}) at $V_{DD,\,TDP}$; N/R is the number of cores on a processor; C_{eff} is the effective switching capacitance of a baseline 1-BCE core; and $F_{max}(V_{DD,\,TDP})$ and $I_{leak}(V_{DD,\,TDP})$ are F_{max} and leakage current of a baseline core at $V_{DD,\,TDP}$. Then we can approximate P_{dyn} and P_{leak} of a multicore processor as follows:

$$P_{dyn} = \frac{n}{N/R} \cdot f(V_{DD}) \cdot \left(\frac{V_{DD}}{V_{DD,\,TDP}}\right)^2 \cdot P_{dyn,\,TDP} \tag{7}$$

$$P_{leak} = \frac{n}{N/R} \cdot l(V_{DD}) \cdot \left(\frac{V_{DD}}{V_{DD,\,TDP}}\right) \cdot P_{leak,\,TDP} \tag{8}$$

where $l(V_{DD})$ is the I_{leak} scaling factor at V_{DD}, normalized to the leakage current at $V_{DD,\,TDP}$. Then we can approximate the total power of a multicore processor as follows:

978-1-60558-497-3/09 $25.00 © 2009 ACM

$$P_{tot} = P_{dyn} + P_{leak} = \frac{n}{N/R} \qquad (9)$$

$$\cdot \left\{ f(V_{DD}) \cdot \left(\frac{V_{DD}}{V_{DD,TDP}} \right)^2 + L \cdot l(V_{DD}) \cdot \left(\frac{V}{V_{DD,TDP}} \right) \right\} \cdot P_{dyn,TDP}$$

where L is the P_{leak} ratio to P_{dyn} at $V_{DD,TDP}$ ($P_{leak,TDP} / P_{dyn,TDP}$); see Section 5 for experimental methodology and parameters for generating f and l.

With the presented throughput speedup and power models, our objective is to maximize the speedup at given R and F by optimizing 1) the number of active cores and 2) V_{DD} (and corresponding F_{max}), while satisfying the $V_{DD,\ max}$, P_{TDP}, and $T_{j,max}$ constraints as follows:

Objective:

$$maximize(speedup) \qquad (10)$$

Constraints:

$$V \le V_{DD,max}, P_{tot} \le P_{TDP}, T_j \le T_{j,\,max} \qquad (11)$$

To maximize the throughput speedup, we can increase both f and n while satisfying the constraints. As we increase f by raising V_{DD}, P_{dyn} and P_{leak} grow more than quadratically and exponentially, respectively. This super-linearly increasing power consumption in exchange for faster F_{max} rapidly uses up available power- and thermal-headroom provided by the power-gated idle cores, limiting the speedup. Meanwhile, as we increase n, available power- and thermal-headroom decreases, because the number of power-gated idle cores decreases, limiting f increase. Therefore, for given F and R, there must be the optimal number of active cores and corresponding V_{DD} and F_{max}.

Table 1 shows the throughput speedup improvement in percentage and the number of active cores of multicore processors after the optimization using PCPG and DVFS for a wide range of parallelism. The 1) white-, 2) light-gray-, and 3) dark-gray-color cells in Table 1 imply that the improvement is limited by 1) $V_{DD,max}$, 2) P_{TDP} or $T_{j,\,max}$, and 3) both P_{TDP} and $T_{j,\,max}$. We assumed that L is 0.3/0.7 at $V_{DD,TDP}$; we also analyzed the throughput speedup improvement with $L = 0.4/0.6$, but the difference was less than 1%. When $V_{DD,TDP}$ is set to 0.8V and F is 0, the throughput speedup improvement for the 16-core configuration was 16%. Note that the speedup in this case was limited by a $V_{DD,max}$ constraint. However, as F increases, the throughput speedup improvement diminishes, because more cores are active (thus less power- and thermal-headroom to increase V_{DD} and F_{max} of the active cores). Hence, the improvement is often limited by either P_{TDP} or $T_{j,\,max}$ constraint. For example, if F is equal to or greater than 0.7, we could not improve the throughput speedup due to lack of power- and thermal-headroom to increase V_{DD} and F_{max}. As the core size increases (thus the number of cores per die decreases), the improvement also diminishes, because the power consumption of larger cores increases more steeply than smaller one (i.e., larger ones have less power- and thermal-headroom than smaller ones to

Table 1. Throughput speedup improvement (TSI) in % and the number of active cores (*n*) of multicore processors.

F	The number of total cores per die							
	16 (1R)		8 (2R)		4 (4R)		2 (8R)	
	TSI	*n*	TSI	*n*	TSI	*n*	TSI	*n*
0.0	15.6	1	15.6	1	12.5	1	8.4	1
0.1	12.8	4	9.5	2	6.0	2	3.0	1
0.2	10.0	4	5.7	4	3.1	3	0.0	2
0.3	7.3	6	3.9	5	1.8	3	0.0	2
0.4	5.1	9	2.4	6	0.3	3	0.0	2
0.5	3.9	10	1.4	6	0.0	4	0.0	2
0.6	2.2	10	0.0	8	0.0	4	0.0	2
0.7	1.0	13	0.0	8	0.0	4	0.0	2
0.8	0.0	16	0.0	8	0.0	4	0.0	2

(Row-label column: $V_{DD,TDP} = 0.8V$)

increase V_{DD} and F_{max}). Furthermore, as F increases, the optimal number of active cores increases (i.e., more active cores at lower V_{DD} and F_{max} is more beneficial in improving the throughput speedup). On average, the optimization applying PCPG and DVFS simultaneously improves the throughput speedup by 9.1%, 6.4%, and 3.9% for the 16-, 8-, and 4-core processors when $V_{DD,TDP}$ is 0.8V and F is between 0~0.5.

4. THROUGHPUT OPTIMIZATION WITH WID C2C F_{MAX} AND P_{LEAK} VARIATIONS

As mentioned in Section 1, individual cores are becoming small enough that spatially correlated WID process variations manifest themselves as C2C F_{max} and P_{leak} variations. In a globally-clocked multicore processor, F_{max} is often limited by the slowest core, although other cores can run at faster F_{max}. In this section, we also exploit C2C F_{max} variations to optimize the throughput speedup in conjunction with the throughput speedup improvement method presented in Section 3; we pick an optimal set of active cores that can maximize F_{max} (and the resulting throughput speedup) of a multicore processor within power and thermal constraints.

Figure 2-(a) shows a systematic V_{th} variation map of a 16-core processor; see Section 5 for experimental methodology and parameters for generating the variation map. Each rectangle represents a core, and each pair of numbers in a rectangle in Figure 2-(b) corresponds to F_{max} (normalized to the slowest core F_{max}) and P_{leak} (normalized to the least leaky core P_{leak}) of a core, respectively. Some cores are ~1.65× faster than the F_{max} of the slowest core while ~3.3× leakier than the P_{leak} of the least leaky core in a multicore processor die; shorter L_{eff} and lower V_{th} due to WID process variations make cores faster and leakier. Assuming that a multicore processor is globally-clocked and all the cores are running, the normalized F_{max} of the multicore processor is 1, which is limited by the F_{max} of the slowest core. However, when parallelism in an application is limited, only a few number of active cores are needed. Hence, to improve the single-thread performance of the active core(s) (thus the throughput of the multicore processor), faster cores can be picked for running the application and V_{DD} and F_{max} can be optimized accordingly. For example, we can pick the fastest 4 cores; the normalized F_{max} of the cores are 1.65, 1.56, and 1.53 and there are two cores with 1.56 F_{max} in Figure 2-(b). Then the normalized F_{max} of the multicore processor running these 4 cores will be 1.53. As a result, picking the fastest 4 cores gives 50%, 46%, 42%, 37%, and 30% more throughput speedup than running all the cores when R is 1 and F is 0.1, 0.2, 0.3, 0.4, and 0.5, respectively. Note that faster cores often consume much more P_{leak}. Hence, depending on parallelism in applications, C2C P_{leak} variations also significantly impact on picking an optimal set of fast active cores due to power and thermal constraints.

When $f_i(V_{DD})$ is the frequency scaling factor of core i at V_{DD} (normalized to the F_{max} at $V_{DD,TDP}$), f in (3) and (7) is decided by f_i of the slowest core among the picked n active cores in N/R cores. To approximate P_{dyn}, we can use (7) where f is replaced with f_i of the slowest core among the picked n cores. To estimate P_{leak}, we can modify (8), since each core consumes different P_{leak} due to WID process variations. Then the total P_{leak} of a multicore processor can be approximated as follows:

Figure 2. A WID V_{th} variation map for a 16-core processor in (a). The corresponding F_{max} and P_{leak} map in (b).

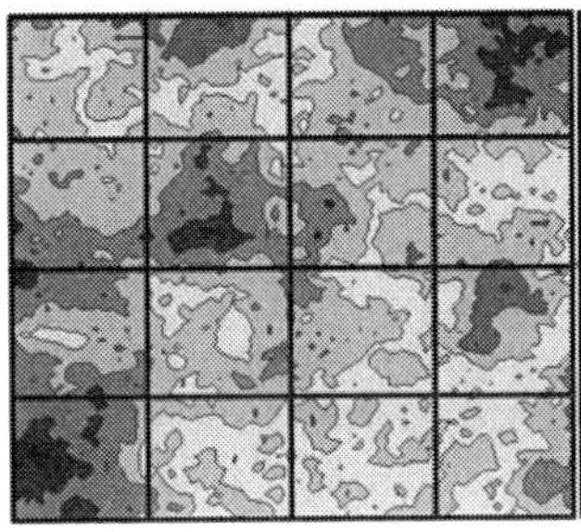

1.12, 1.41	1.00, 1.08	1.30, 2.04	1.65, 2.97
1.56, 2.28	1.56, 2.78	1.18, 1.77	1.19, 1.20
1.52, 2.46	1.37, 1.77	1.26, 1.62	1.04, 1.00
1.53, 3.30	1.26, 1.37	1.20, 1.18	1.31, 1.66

(a) (b)

978-1-60558-497-3/09 $25.00 © 2009 ACM

Table 2. Avg. throughput speedup improvement (TSI) in % and the avg. number of active cores (n) of multicore processors.

F	The number of total cores per die							
	16 (1R)		8 (2R)		4 (4R)		2 (8R)	
	TSI	n	TSI	n	TSI	n	TSI	n
0.0	74.5	1	53.2	1	34.0	1	14.8	1
0.1	59.7	1	40.3	1	24.0	1	9.1	1
0.2	49.5	2	30.8	2	15.0	1	3.9	1
0.3	41.8	3	24.3	3	9.4	2	1.6	1
0.4	35.4	4	18.7	3	5.3	2	0.0	2
0.5	29.9	6	14.0	4	2.5	2	0.0	2
0.6	24.2	6	8.9	4	0.6	2	0.0	2
0.7	17.9	8	4.5	5	0.0	4	0.0	2
0.8	11.0	10	1.3	5	0.0	4	0.0	2
0.9	3.3	11	0.1	5	0.0	4	0.0	2

(The leftmost column of the table is labeled $V_{DD,TDP} = 0.8V$.)

$$P_{leak} = \sum_{i=1}^{n} l_i(V_{DD}) \cdot \left(\frac{V_{DD}}{V_{DD,TDP}} \right) \cdot P_{leak,\,TDP} \qquad (12)$$

where $l_i(V_{DD})$ is the I_{leak} scaling factor of core i at V_{DD} (normalized to the I_{leak} of the least leaky R-BCE core at $V_{DD,TDP}$) among the picked n cores; see Section 5 for experimental methodology and parameters for generating f_i and l_i.

Table 2 shows the average improvement of the throughput speedup of multicore processors in percentage and the average number of active cores of multicore processors; the average improvement was obtained from 100 die samples (variation maps) that were generated through Monte-Carlo simulations; the average number of active cores — n was rounded to a nearest integer value. Each cell color representing a limiting constraint (e.g., $T_{j,\,max}$) is determined by majority samples; for example, 75% of samples for give F are limited by a $T_{j,\,max}$ constraint, then the corresponding cell is marked with light-gray color. Exploiting WID C2C F_{max} variations along with applying PCPG and DFVS simultaneously provides much higher throughput speedup for a wider range of parallelism than the results shown in Table 1. We also have noticeable throughput speedup improvement even for higher F, because disabling one or two very slow cores lead to significantly higher processor's F_{max}; higher F_{max} with less number of active cores often results in higher throughput speedup than lower F_{max} with more number of active cores. As the number of cores in a die increases, we have a more relative F_{max} spread among the cores as noted in [5]. This provides more opportunities in exploiting WID C2C F_{max} variations, leading to higher relative throughput speedup for a multicore processor with smaller cores. When F is 0, the throughput speedup of a 16-core processor was limited by a $V_{DD,max}$ constraint in Table 1. However, when we consider WID C2C variations, the fastest core (that is usually picked when F is 0) usually consume much more P_{leak} than slower ones. As a result, the throughput speedup is limited by $T_{j,max}$ rather than $V_{DD,max}$ although exploiting WID C2C F_{max} variations still provides much higher improvement than the results shown in Table 1. On average, the optimization exploiting WID C2C F_{max} variations improves the throughput speedup by 34.7%, 19.6%, and 9.1% for the 16-, 8-, and 4-core processors when $V_{DD,TDP}$ is 0.8V and F is between 0~0.9.

5. EXPERIMENTAL METHODOLOGY

We generated f, frequency scaling factor in Section 3 by measuring frequency of a 24-stage FO4 inverter chain for a range of V_{DD} values (0.6~1.05V); we used the 32nm PTM model for metal gate/high-k CMOS (V2.0) and HSPICE for the measurement. We also generated l, leakage scaling factor from a dummy circuit; we populated a large number of NOT (50%), NAND (30%), and NOR (20%) gates, and measured the leakage current of the dummy circuit; each gate excluding INVs had the various numbers of inputs (2~4) and randomly selected input states were applied to measure the leakage current for the same range of V_{DD} that is used to obtain

f. In addition, we generated spatially correlated V_{th} and L_{eff} maps through the models shown in [5] for Section 4. To model multicore processors, we used the following parameters for 100 35mm² dies: WID correlation distance coefficient, ϕ (0.5); WID V_{th} variation $\sigma_{V_{th}}^{sys}$ (6.4%); and D2D variation $\sigma_{V_{th}}^{D2D}$ (5.0%) as presented in [5]. We broke each variation map into 80×80 grid points, and obtained a pair of V_{th} and L_{eff} values from each grid point that is modeled with a 24-stage FO4 inverter chain for frequency and a dummy circuit for leakage. Then we applied the corresponding pair of V_{th} and L_{eff} values to the 32nm predictive technology model. Note that f_i is decided by the slowest grid point in the core [5], while l_i is obtained from the leakage current of all the grid points in each core. We set P_{TDP} at $V_{DD,TDP}$ to 120W for a thermal simulation purpose, which is typical for server class multicore processors. We also set L to 0.3/0.7; L can be parameterized to explore a different design or technology. Then we can estimate P_{dyn}, P_{leak}, and P_{tot} for the range of V_{DD} (e.g., using (7), (8), and (9)). We compute T_j of each core in a die sample using 1) HotSpot [10] with 0.3K/W for the convection resistance [6], 2) the calculated P_{tot} of each core in the die, and 3) the given die size (35mm²). We assume that T_{jmax} is 100°C since 120W power consumption across a die leads to $T_j = \sim 100$°C in steady state with the aforementioned parameters.

6. CONCLUSION

Our work presented throughput analysis and optimization of power- and thermal-constrained multicore processors by formulating the first-order approximated throughput and power models in conjunction with the widely-used thermal model. The first-order analytical model presented in this study enables early architectural trade-off analysis, although detailed physical design parameters are not available. When parallelism in an application is limited, PCPG was applied to one or more idle cores, providing extra power- and thermal-headroom in a multicore processor. Then V_{DD} (thus F_{max}) was increased using DVFS, improving the throughput speedup within V_{DD} scaling, power, and thermal limits. Our results showed that optimizing the number of active cores and F_{max} (through V_{DD} scaling) improved the throughput of a power- and thermal-constrained 16-core processor by ~16% for a range of parallelism; combining PCPG with DVFS benefits multicore processors with many, small cores that suffer from lower single-thread performance than ones with a few, larger cores when parallelism in an application is limited. We also extended our throughput speedup analysis and optimization considering WID process variations. Our analysis shows that exploiting C2C F_{max} variations along with applying PCPG and DVFS simultaneously improved the throughput speedup of globally-clocked multicore processors where F_{max} is limited by the slowest core in a die. On average, 16-core processors exploiting WID C2C F_{max} variations showed the throughput speedup improvement by ~75% for a range of parallelism.

REFERENCES

[1] K. Bowman et al., "Impact of die-to-die and within-die parameter fluctuations on the maximum clock frequency distribution for gigascale integrations," IEEE JSSC, Feb. 2002.

[2] J. Tschanz et al., "Adaptive body bias for reducing impacts of die-to-die and within-die parameter variations on microprocessor frequency and leakage," IEEE JSSC, Nov. 2002.

[3] K. Bowman et al., "Impact of die-to-die and within-die parameter variations on the throughput distribution of multicore processors," in Proc. IEEE ISLPED, Aug. 2007.

[4] E. Humenay et al., "Impact of process variations on multicore performance symmetry," in Proc. IEEE DATE, Apr. 2007.

[5] S. Herbert et al., "Characterizing chip-multiprocessor variability-tolerance," in Proc. IEEE DAC, Jun. 2008.

[6] R. Rao et al., "Throughput of multi-core processors under thermal constraints," in Proc. IEEE ISLPED, Aug. 2007.

[7] J. Donald et al., "Power efficiency for variation-tolerant multicore processors," in Proc. IEEE ISLPED, Aug. 2006.

[8] M. Hill et al., "Amdahl's law in the multicore era," IEEE Computer, Jul. 2008.

[9] D. Woo et al., "Extending Amdahl's Law for energy-efficient computing in the many-core era," IEEE Computer, Dec. 2008.

[10] HotSpot, http://lava.cs.virginia.edu/HotSpot/index.htm.

978-1-60558-497-3/09 $25.00 © 2009 ACM

Design Automation for a 3DIC FFT Processor for Synthetic Aperture Radar: A Case Study

Thorlindur Thorolfsson
trthorol@ncsu.edu

Kiran Gonsalves
kgonsal@ncsu.edu

Paul D. Franzon
paulf@ncsu.edu

Department of Electrical and Computer Engineering
North Carolina State University Box 7911
Raleigh, NC 27695

ABSTRACT

This work discusses a 1024-point, memory-on-logic 3DIC FFT processor for synthetic aperture radar (SAR), sent to fabrication in the 180 nm MIT Lincoln Labs 3D FDSOI 1.5 V process[12] along with the design flow required to realize it with off-the-shelf commercial 2D tools. The work shows how the vertical dimension can be exploited for novel memory architecture tradeoffs that are not feasible in 2D, reducing the energy consumed per memory operation in the FFT by 60.3%. In comparison to its 2D counterpart, the SAR FFT processor exhibits a 53.0% decrease in average wire length, a 24.6% increase in maximum operating frequency and a 25.3% decrease in total silicon area.

Categories and Subject Descriptors

B.7.1 [**Integrated Circuits**]: Types and Design Styles; C.4 [**Performance Of Systems**]: Design studies

General Terms

Design

Keywords

SAR, 3DIC, TSV, FFT

1. INTRODUCTION

New developments in fabrication technology allow vertical integration using 3D thru-silicon vias. Vertical integration has the potential to cut wire length drastically for standard cell designs as reported by Davis et al.[5]. This is important because as designs move to smaller feature sizes the wires will increasingly dominate the delay and the power budgets of digital logic circuits[7]. 3D integration has several major obstacles, including increased thermal densities[10], increased test costs, and the lack of commercial EDA tool support for 3DICs[2]. It has been shown that using custom 3D placement and routing tools, a 28-51% reduction in total

wire length can be achieved[4]. In this paper we demonstrate one way in which an application-specific-processor can be re-architectured to take advantage of 3DIC technology. In this paper an FFT engine designed for use in a Synthetic Aperture Radar (SAR) processor is used as a case study. We re-architectured the baseline design to reduce power and total area simultaneously by taking advantage of 3DIC with through-silicon via (TSV) technology. Furthermore, by separating the logic and memory layers we simplify the tool flow so that the design can be executed easily using extensions of current 2D CAD tools. 3DIC stacking is used to interconnect the power-optimized memory to the logic tier while reducing overall area. The design has been sent to fabrication at MIT Lincoln Laboratory but fabrication of the chip has not yet been completed.

The paper is organized in the following manner. Section 2 describes the algorithm on which the SAR FFT processor is based. Section 3 describes the architecture of the SAR FFT processor. Section 4 describes the tool flow and manufacturing process. Finally, Section 5 compares the metrics of the 3D implementation to a 2D equivalent.

2. SAR ALGORITHM

Synthetic aperture radar, unlike most radar, is used for imaging. While conventional images are formed using the visible spectrum, SAR images are formed using the radio region of the spectrum. A tremendous amount of digital signal processing and memory bandwidth is required to form a SAR image. The required digital signal processing and memory bandwidth increase exponentially with the desired image resolution. This makes a SAR FFT processor an excellent candidate to demonstrate the memory bandwidth benefits that 3D integrated circuits can provide. The image-forming algorithm used for the SAR FFT processor is derived from the one used in the RASSP[6] project and based on the Range Doppler Algorithm[8]. The steps required to form the SAR image along, with the portion of the floating point operations performed by each of the steps (for 30 cm imaging resolution), are shown in Table 1. It is important to note the majority of all the floating point operations are FFT/IFFT operations, which occur in steps 2, 3 and 5.

3. ARCHITECTURE

As we have shown in Table 1 and discussed in Section 2, the majority of SAR processing involves computing FFTs to some degree. As a result, the main objective of the design is to efficiently calculate the FFTs used in the SAR algo-

Permission to make digital or hard copies of part or all of this work for personal or classroom use is granted without fee provided that copies are not made or distributed for profit or commercial advantage and that copies bear this notice and the full citation on the first page. To copy otherwise, to republish, to post on servers or to redistribute to lists, requires prior specific permission and/or a fee.
DAC'09, July 26-31, 2009, San Francisco, California, USA

978-1-60558-497-3/09 $25.00 © 2009 ACM

Table 1: The Steps in the SAR Algorithm.

Step	%
1)Range Low Pass FIR Filtering	35.6%
2)Range Fast Fourier Transform	12.3%
3)Azimuth Fast Fourier Transform	22.3%
4)Azimuth Complex Multiply	3.8%
5)Azimuth Inverse Fast Fourier Transform	26.0%
FFT Steps Combined (Steps 2, 3 & 5)	60.6%

rithm. We use a radix-2 Cooley-Tukey FFT [3] for all the FFT calculations in the processor. A radix-2 FFT has a data dependency that resembles a hypercube. This hypercube data dependency can be exploited in two ways. First, a radix-2 FFT will process two memory locations every cycle, one of which will have odd parity while the other will have even parity. As a result we can split the processing memory into two independent memory groups that never need to be accessed at the same time. Second, we can sub-divide the even and odd groups into smaller subgroups where each processing element is only connected to the absolute minimum number of memory locations required to successfully compute the FFT. Furthermore, in this subdivision each memory subgroup is not accessed by more than one processing element at the same time. The benefit of splitting the memories into smaller subgroups is that smaller memories are faster, and since each memory subgroup can be accessed simultaneously, the system can perform a greater number of reads and writes per cycle. Conversely, a single memory will require less area as only one set of peripheral logic (write driver and sense amp) is required. We use Cacti 4.1[13] to assess the architectural tradeoff, by comparing the properties of a single 8 kByte memory to sixteen 512 Byte memories. The memory-core area savings of using a single memory would have been 67.6%. By using multiple smaller memories, the energy per read is reduced by 60.8% (from 68.205 to 26.718 pJ), the energy per write is reduced by 57.6% (from 14.48 to 6.142 pJ) and the memory bandwidth is increased by 854.9% (13.4 to 128.4 GBps). The number of wires interconnecting the memory to logic is increased from 150 to 2272 wires. 3DIC stacking is used to minimize the area impact of the added wires. Furthermore, the single memory will have a shorter and simpler interconnect structure between the logic and the memory. This tradeoff is illustrated in Figure 1.

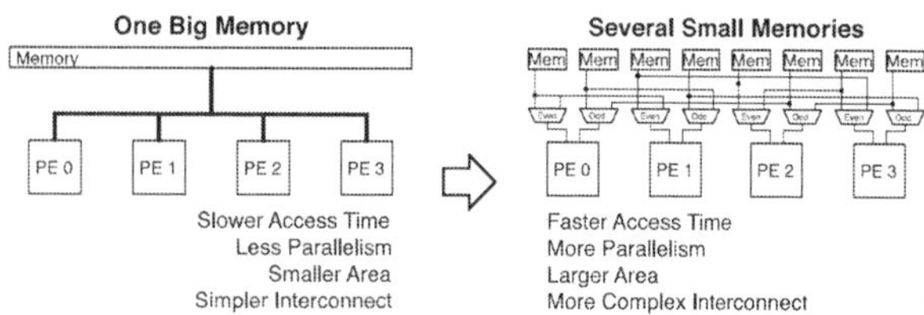

Figure 1: The memory design tradeoffs.

Overall, the architecture we fabricated can process a 1024-pixel wide image using an FFT of the same width. Each pixel/data point in the FFT has a precision of 32 imaginary and 32 real bits. Computing an N-point FFT requires using $N/2$ FFT twiddle factors of the same precision. To minimize the actual storage of the FFT twiddle factors we utilize two optimizations. First, we use trigonometric properties to reduce the number of twiddle factors stored[11] from $N/2$ to $N/8 + 1$. For the 1024 point FFT this effectively reduces the number of twiddle factors stored from 512 to 129. Second, if any bit is the same for all the words, that bit is hard coded into the processing element rather being stored in the ROM. This optimization effectively reduces the number of bits required to store the twiddle factors from 64 to 52. We use the memory-dividing scheme described above to divide the processing memory in 32 smaller memories (16 even and 16 odd). Furthermore, every single memory is dual-ported (one read and one write port). Overall, this allows the system to perform 32 memory accesses per cycle (16 reads and 16 writes), completing a 1024-point FFT in 653 cycles assuming five pipeline stages. Each of the different components of the architecture are described below and illustrated in Figure 3.

The system consists of four different components, eight processing elements, one controller, thirty two SRAMs, and eight ROMs. The processing elements are the core of the system, implementing the FFT butterfly with four floating point multipliers and six addition/subtraction units. The internal structure of the processing element is shown in Figure 2. The controller orchestrates the overall operation of the system, by setting the addresses and read enables of the memories. The controller requires very little communication with the processing elements, only three signals per processing element. The SRAMs implement the main processing memory using 8-transistor dual ported SRAMs. The ROMs store the FFT twiddle factors and are implemented as single ported NOR type ROMs.

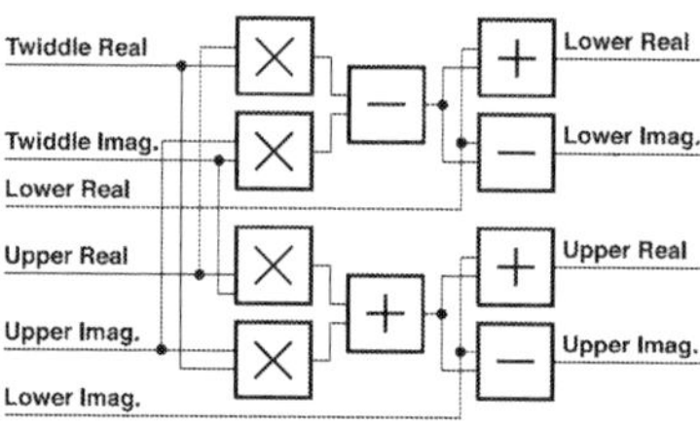

Figure 2: The internal structure of a processing element.

The SAR FFT processor architecture is a good example of a design that has a significant number of heavily shared and interconnected resources. For this reason it can be expected to benefit significantly from 3D integration due to long wires in the interconnect between these resources (memories and processing elements).

4. IMPLEMENTATION AND TOOL FLOW

In this section we discuss the design, implementation, tool flow and manufacturing process used to create the system. Before the design flow is explained, it is important to understand the manufacturing process. The MIT Lincoln Labs' manufacturing process is a three tier, 180 nm wafer scale 3D integration process[9, 1]. It features a 1.5 V low power fully depleted silicon on insulator CMOS technology with one layer of polysilicon, three metal layers per tier and a back-metal layer between the top two tiers, with an additional metal layer on top of the entire stack. The bottom tier is named A, the middle tier B and the top tier C. Tier

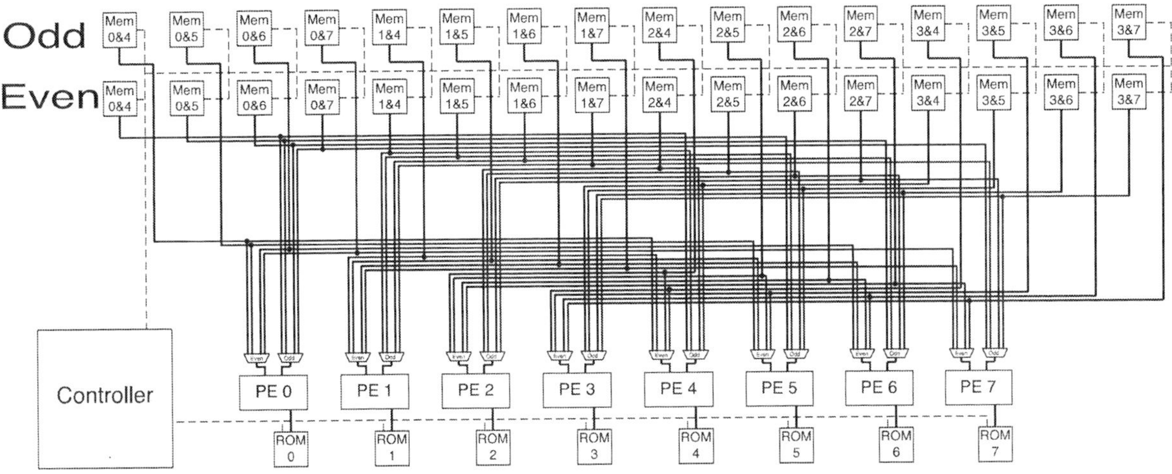

Figure 3: The SAR FFT processor architecture.

A is closest to the heat sink. Tier C is the only tier which has off-chip inputs and outputs. Tiers B and C face down, while tier A faces up. Figure 4 shows a side view of the process with the silicon-thru vias and the orientation of the tiers shown. In this process the dimensions of a single thru-silicon via are 2.5×2.5 μm and the smallest pitch the vias can be placed on is 3.9 μm.

Overall, the design is a mix between standard cell and full custom design. The processing elements and controller are coded in Verilog, while the memories (SRAMs and ROMs) are implemented using full custom design. One of the benefits of doing the memories in full custom, rather than using an off-the-shelf memory generator, is that it allows the thru-silicon vias to be implemented on the outside edges of the memories. This simplifies the flow as the thru-silicon vias get placed along with the memory. This, however, is not the case for the 24 logic-to-logic vias which must have their position predetermined and are then placed in the final assembly stage.

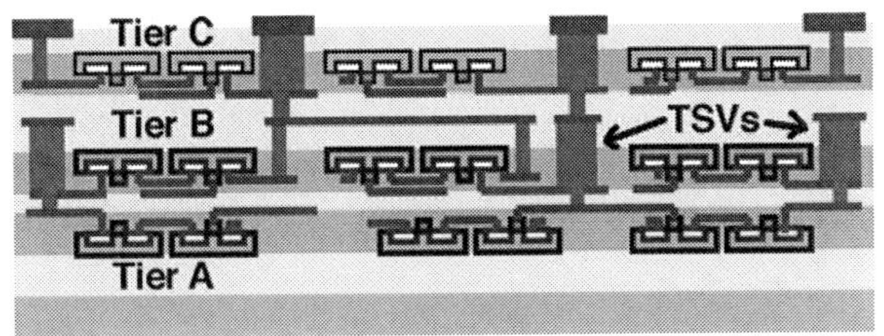

Figure 4: A side view of the MIT Lincoln Labs' process with the silicon-thru vias and tier orientation shown.

Figure 5 shows the complete design flow. The first step in the design flow is 3D floorplanning, partitioning and selecting the locations for the memories. In the 3D floorplanning phase, the main objective is to get the memories as close as possible to the processing elements that use them. We define PE0, PE1, PE2 and PE3 to be the lower numbered processing elements and PE4, PE5, PE6 and PE7 to be the upper numbered processing elements. Figure 3 shows that every

memory is connected to one lower numbered PE and one upper numbered PE. To exploit this connectivity we partition the system so that the controller and the memories are on the middle tier (tier B), with the upper numbered PEs and their respective twiddle factor ROMs placed on tier A and the lower numbered PEs along with their ROMs placed on tier C. This partitioning scheme guarantees that a memory is never more than one tier away from the processing elements that are connected to it. This means the memory is also on the same tier as the controller that sets its address lines. On the middle tier we have thirty-two memories and one controller to place. To accomplish this, we use an 11×3 grid. We place the controller in the center location of the grid in the middle tier. For the remaining memories we use a Python constraints package to generate an optimal memory placement based on the distance a given memory is to the two processing elements that use it. The resulting floorplan is shown in Figure 8. In the system there are a total of 8280 thru-silicon vias 4128 of those vias connect the logic on tier A to the memories on tier B, another 4128 connect the logic on tier C to the memories on tier B, the remaining 24 thru vias connect the controller to the processing elements.

The next step in the design flow is synthesis, which was accomplished using a standard cell library based on the IIT-SoC library from the Illinois Institute of Technology. Each tier is synthesized separately in Synopsys Design Compiler. After synthesis, we perform static timing analysis and add an additional pipeline stage to the processing elements until adding another pipeline stage to the processing elements does not result in any overall speed increase. The optimal pre-place and route pipeline depth for the system was discovered to be five stages for this manufacturing process and standard cells, yielding a maximum operation frequency of 196 MHz (without parasitics).

After synthesis, we perform place and route. This stage deviates the most from a conventional 2D flow. In order to successfully complete place and route, the global information about the placement of the memories and the thru-silicon vias is required. Using standard string and file manipulation functions built into the TCL interpreter in Encounter, the thru-via and pin locations can easily be extracted by parsing the DEF files of the custom memories designs. Using

978-1-60558-497-3/09 $25.00 © 2009 ACM

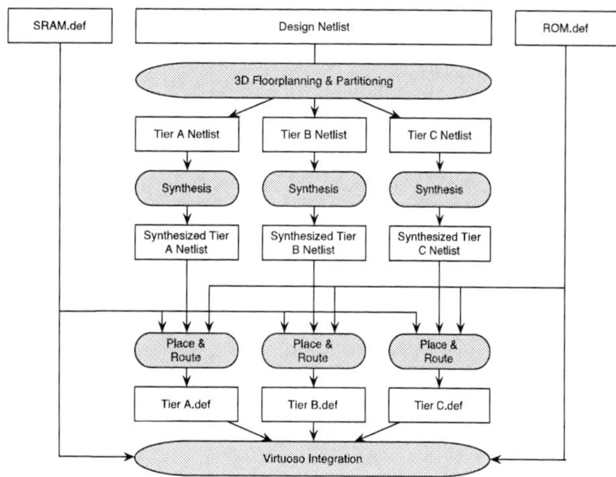

Figure 5: The design flow.

the information from the DEF file, routing and placement is blocked over the areas of the memories and inter-tier silicon-thru via location. Normal placement is then performed, followed by clock tree synthesis. Due to the fact that the process only allows three metal layers per tier, the clock tree is not routed before regular routing, which is common for processes with a greater number of metals. Instead the clock tree is routed along with all other routing. This causes more clock skew than would have occurred if a greater number of metal layers had been available. After clock tree synthesis, the "preassignPin" command is then used to place virtual input/output pins directly on top of the thru-vias on the edge of the memories. Encounter then performs routing as normal, connecting the standard cells, clock tree and virtual pins (effectively performing 3D routing). After place and route, the design along with its parasitics is imported into PrimeTime and post-place and route timing analysis is performed. In this step it is important to make sure that each tier has no setup or hold violations. It is also important to make sure that signals that travel between tiers have no setup or hold violations either. This step is greatly simplified due to the fact there are very few logic-to-logic vias (24) and the remaining signals are either data pins to the SRAMs or address pins to the twiddle factor ROMs.

Finally, all the tiers are imported separately into Virtuoso. In Virtuoso the three tiers and the full-custom memories are combined. For the 24 signals that connect the controller to processing elements, the through-silicon vias are placed by hand, the rest are placed automatically as part of the memory. The reason the 24 TSVs were placed by hand is that since there were so few of them it was quicker to place them by hand then to write a script to do so. However, this process can easily by scripted using Skill code in Virtuoso as the location of all through-silicon vias are known. Furthermore, scripting this process would be necessary for other 3D designs that contain a greater number of logic to logic through-silicon vias. The power and ground rings of the three tiers are then combined into 3D meshes, by placing thru-silicon vias all along the perimeter. Due to the fact that Encounter routed over the power and ground rings in a some areas, a few thru-silicon vias along the perimeter had to be removed to avoid shorts. All in all, we managed to

fit 4554 power and ground vias between tiers A and B and 4800 vias between tier B and C. The next step was to place the input and output pads and perform a final DRC and LVS. Furthermore, since the process being used is an SOI process, we added extra power and ground decoupling capacitors where ever there was room left over to compensate for the limited native decoupling in SOI. Figure 6 shows the three tiers stacked, along with the thru-silicon vias.

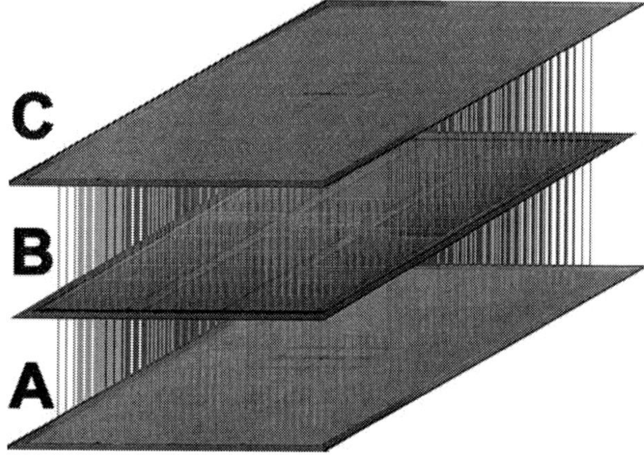

Figure 6: The 3D SAR FFT processor with thru silicon vias drawn in.

5. RESULTS

To quantify the improvements of the 3D circuit over its 2D counterpart, we place and route the design in 2D. In order to ensure a fair comparison between the two circuits, the circuit is not resynthesized, instead the same synthesis output is used. For the comparison, a literally identical floorplan is used. This floorplan is essentially the floorplan of tier B expanded with the ROMs placed in similar locations to the 3D version, shown in Figure 9. Due to increased congestion, the 2D design does not route successfully with the same area as its 3D counterpart (4.8×4.8 mm). To remedy this, the area used for place and route is grown until the design routes without any design rule violations. Compared to the 3D version, the total area used must be expanded significantly from $3 \times 2.6 \times 3$ mm for the 3D circuit to 5.6×5.6 mm for its 2D counterpart, which is 25.3% increase in total area. To get just core placement area, we exclude the power and ground rings from the total area (0.1 mm on every side) and the comparison becomes $3 \times 2.8 \times 2.4$ mm versus 5.4×5.4 mm. The area discrepancy between the total area and the core area illustrates an interesting point: given the same total area and same power and ground ring width, a 3D design will devote more area to the power and ground rings. The next metric examined was net length. We extract all net information directly from Encounter, combining the information from all the different tiers. As expected, the average wire length decreased drastically from 836.0 μm down to 392.9 μm. This is a 53.0% decrease. Similarly, the total wire length decreased from 19.107 m to 8.238 m, a total of 53%. A histogram of the wire lengths is shown in Figure 7.

In order to gather the speed and power metrics of the design, we have to extract the parasitics and characterize the switching activity of the design. The parasitics are ex-

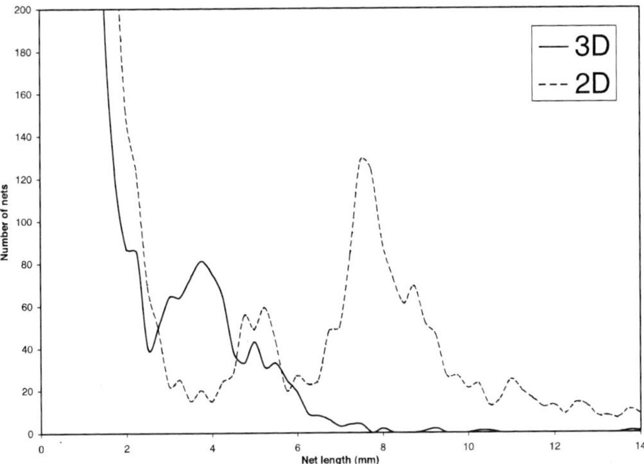

Figure 7: Histogram of wire lengths of the SAR FFT processor for both the 2D and 3D versions (bin size = 250μm).

Table 2: Comparison between the 2D and 3D metrics of the SAR FFT along with read and write energy from Cacti.

Metric	2D	3D	%
Total Area (mm^2)	31.36	23.40	25.3%
Core Area (mm^2)	29.16	20.16	30.9%
Mean Net Length (μm)	836.0	392.9	53.0%
Total Wire Length (m)	19.107	8.238	56.9%
Max Speed (MHz)	63.7	79.4	24.6%
Critical Path (ns)	15.7	12.6	19.7%
Logic Power @ 63.7MHz (mW)	340.0	324.9	4.4%
Logic Power @ 79.4 MHz (mW)	——	409.2	——
FFT Logic Energy (μJ)	3.552	3.366	5.2%

Table 3: Read and write energy from Cacti comparing the un-optimized to the optimized design.

Metric	Divided	Undivided	%
Bandwidth ($GBps$)	13.4	128.4	854.9%
Energy Per Write (pJ)	14.48	6.142	57.6%
Energy Per Read (pJ)	68.205	26.718	60.8%
Memory Wires ($\#$)	150	2272	-1414.7%

tracted into a SPEF file using Encounter. The switching activity is generated by simulating an FFT test bench in Mentor Graphics Modelsim and exporting the resulting activity of the test bench to a SAIF file. Both files were then read into Synopsys PrimeTime. In PrimeTime the clock period was increased to the fastest clock that did not cause any setup violations, to determine the maximum operating frequency. The 3D design simulated correctly at 79.4 MHz (12.6ns), whereas the 2D design simulated correctly at 63.7 MHz (15.7ns). This is a 24.6% increase in maximum operating frequency and a 19.7% improvement in clock speed. As these numbers may seem a bit slow for the given technology node, it is important to keep two points in mind. First, the process only has three metal layers which limits clock tree routing causing more skew than would occur if more metals were available. Second, the standard cell library does not have the multi-adder cells that many commercial libraries have, which would have helped increase the maximum operating frequency. Finally, using both the SPEF and the SAIF file, power dissipation numbers (excluding power dissipated in the memories) were generated using PrimeTime. For the 3D design the power dissipation is determined for both the maximum operating frequency of the 2D and the 3D design, while for the 2D design the power dissipation is only determined for its own maximum operating frequency. At an operating frequency of 79.4 MHz the 3D design dissipates 409.2 mW. Operating at 63.7 MHz the 3D design dissipates 324.9 mW and the 2D design dissipates 340.0 mW. This is a 4.4% improvement. Using the power numbers of both circuits operating at maximum frequency, we compute the energy (excluding memory accesses) required per 1024-point FFT. The energy required for the completing the FFT in 3D is 3.366 μJ as opposed to 3.552 μJ for the 2D version which is a 5.2% improvement. The results are summarized in Table 2, followed by a summary of the memory tradeoffs in Section 3 in Table 3.

6. CONCLUSIONS

The main point of this paper is not necessarily to compare the 2D and 3D implementations of the same design but also to show how a system can be re-optimized in 3D in ways that are not available in 2D. Furthermore, the system can be realized with use of commercial 2D tools. Thus it is not necessary to use 3D tools. The 3D optimized design permits a single large memory to be broken down into multiple smaller memories to reduce the energy consumption in memory operations per FFT by 60.3%. This memory re-optimization would not be suited to a 2D design due to the high interconnect cost. In the 2D design, the increase in interconnect area is greater than the increase in memory area. Case in point, the 2D implementation of the archicture on the left side of Figure 1 is significantly worse than the 3D implementation of the archictecture to the right of Figure 1 in all metrics - power, performance and area. Finally, comparing the 2D and 3D implementations of the SAR FFT processor, we show an average wire length reduction of 53.0%, an overall wire length reduction of 56.9%, a 24.6% increase in maximum operating frequency, a 5.2% reduction in energy per FFT and a 30.9% reduction in area.

7. ACKNOWLEDGMENTS

This project was funded by DARPA under contract FA8650-04-C-7127, and contract FA8650-04-C-7120 both managed by AFRL. Additional funding was provided by Semiconductor Research Corporation. The authors would like to thank MIT Lincoln Labs for providing access to their FD-SOI technology and Magnus Halldorsson at Reykjavik University for help with the memory partitioning approach.

8. REFERENCES

[1] J. Burns, B. Aull, C. Chen, C.-L. Chen, C. Keast,

978-1-60558-497-3/09 $25.00 © 2009 ACM

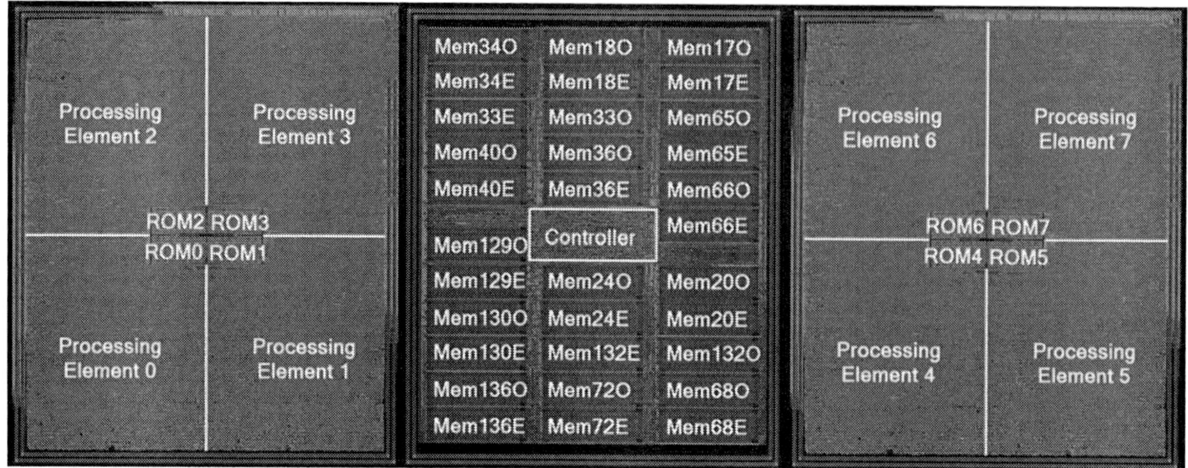

Figure 8: The 3D floorplan.

J. Knecht, V. Suntharalingam, K. Warner, P. Wyatt, and D. Yost. A wafer-scale 3-D circuit integration technology. *IEEE Transactions on Electron Devices*, 53(10):2507–2516, October 2006.

[2] P. Clarke. Eda's big three unready for 3d chip packaging. *EE Times Asia Online*, October 2007.

[3] J. W. Cooley and J. W. Tukey. An algorithm for the machine calculation of complex fourier series. *Mathematics of Computation*, 19(90):297–301, 1965.

[4] S. Das, A. Chandrakasan, and R. Reif. Design tools for 3-d integrated circuits. In *ASPDAC: Proceedings of the 2003 conference on Asia South Pacific design automation*, pages 53–56, New York, NY, USA, 2003. ACM.

[5] W. R. Davis, J. Wilson, S. Mick, J. Xu, H. Hua, C. Mineo, A. M. Sule, M. Steer, and P. D. Franzon. Demystifying 3D ICs: The Pros and Cons of Going Vertical. *IEEE Design And Test of Computers*, 22(6):498–510, Nov.-Dec. 2005.

[6] C. Hein, J. Pridgen, and W. Kline. RASSP Virtual Prototyping of DSP Systems. *Design Automation Conference DAC 97*, pages 492–497, 1997.

[7] R. Ho, K. Mai, and M. Horowitz. The Future of Wires. *Proceedings of the IEEE*, 89(4):490–504, 2001.

[8] M. Jin and C. Wu. SAR correlation algorithm which accommodates large-range migration. *IEEE Transactions on Geoscience and Remote Sensing*, 22(6):592–597, 1984.

[9] Massachusetts Institute of Technology Lincoln Labs. *MITLL Low-Power FDSOI CMOS Process Design Guide*, revision 2008:6 edition, September 2008.

[10] A. Rahman and R. Reif. Thermal analysis of three-dimensional (3-d) integrated circuits (ics). *Interconnect Technology Conference, 2001. Proceedings of the IEEE 2001 International*, pages 157–159, 2001.

[11] T. Sansaloni, A. Pérez-Pascual, V. Torres, and J. Valls. Scheme for Reducing the Storage Requirements of FFT Twiddle Factors on FPGAs. *The Journal of VLSI Signal Processing*, 47(2):183–187, 2007.

[12] V. Suntharalingam, R. Berger, J. Burns, C. Chen, C. Keast, J. Knecht, R. Lambert, K. Newcomb, D. O'Mara, D. Rathman, D. Shaver, A. Soares, C. Stevenson, B. Tyrrell, K. Warner, B. Wheeler, D.-R. Yost, and D. Young. Megapixel cmos image sensor fabricated in three-dimensional integrated circuit technology. *Solid-State Circuits Conference, 2005. Digest of Technical Papers. ISSCC. 2005 IEEE International*, pages 356–357 Vol. 1, Feb. 2005.

[13] S. Wilton and N. Jouppi. CACTI: an enhanced cache access and cycle time model. *Solid-State Circuits, IEEE Journal of*, 31(5):677–688, 1996.

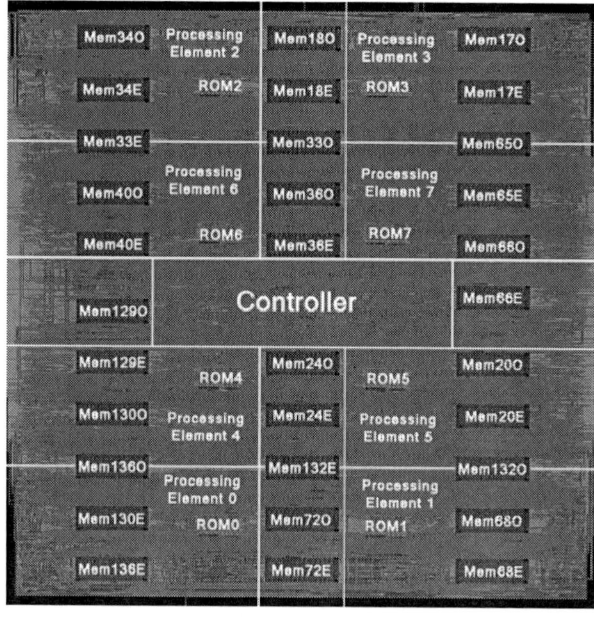

Figure 9: The 2D floorplan for comparison.

978-1-60558-497-3/09 $25.00 © 2009 ACM

Selective Wordline Voltage Boosting for Caches to Manage Yield under Process Variations

Yan Pan[†], Joonho Kong[‡], Serkan Ozdemir[†], Gokhan Memik[†], Sung Woo Chung[‡]

[†]Northwestern University
2145 Sheridan Road, Evanston, IL
{panyan,s-ozdemir,g-memik}@northwestern.edu

[‡]Korea University
Anam-dong Seongbuk-Gu, Seoul, 136-713 Korea
{luisfigo77,swchung}@korea.ac.kr

ABSTRACT

One of the most important hurdles of technology scaling is process variations, i.e., variations in device characteristics. Process variations cause large fluctuations in performance and power consumption in the manufactured chips. In addition, these fluctuations cause reductions in the chip yields. In this work, we present an analysis of a representative high-performance processor architecture and show that the caches have the highest probability of causing yield losses under process variations. We then propose a novel selective wordline voltage boosting mechanism that aims at reducing the latency of the cache lines that are affected by process variations. We show that our approach can eliminate over 80% of the yield losses under medium level of variations, while incurring less than 1% per-access energy overhead on average and less than 4.5% area overhead.

Categories and Subject Descriptors

B.3.2 [**Memory Structures**]: Design Styles - *cache memories*

General Terms

Performance, Design, Reliability

Keywords

Cache, Process Variations, Yield, Access Time Failure, Selective Wordline Voltage Boosting

1. INTRODUCTION

One of the major challenges faced by deep submicron technologies is process variations, which adversely affect performance and power consumption, and hurt yield [30]. Among the various factors that affect yield, parametric yield losses are the most dominant factor; with smaller technologies, parametric yield losses constitute more than 50% of the losses [21].

Although various design components are affected by process variations, caches are the most vulnerable. The 6T SRAM cell represents the finest feature sizes for a fabrication technology; in fact, foundries usually use 6T SRAM arrays for technology qualification. In addition, according to FMAX theory [4], components with high number of parallel and shallow critical paths, such as caches, are most susceptible to process variations. Caches also occupy a large share of chip area; hence they face larger variations. Any single failing SRAM cell can directly affect yield if no preventive techniques are adopted.

There are three kinds of SRAM cell failures: unstable read, unstable write, and access time failures. As we will further discuss

Permission to make digital or hard copies of part or all of this work for personal or classroom use is granted without fee provided that copies are not made or distributed for profit or commercial advantage and that copies bear this notice and the full citation on the first page. To copy otherwise, to republish, to post on servers or to redistribute to lists, requires prior specific permission and/or a fee.
DAC'09, July 26-31, 2009, San Francisco, California, USA

in Section 2, among these failures types, the access time failures are the most crucial [1]. Access time failure means the actual access time to the cache exceeds the pre-defined access time. As variation in threshold voltages amplifies, the probability of the access time failures increases [1]. In case of higher level caches (level 2 and 3), NUCA cache architecture [13] can be adopted to endure different cache access times due to process variations and hence eliminate the effect of increased access times on yields. Another conventional way to mitigate process variations in caches is to employ redundancy schemes [5][8][11][19][25][29][32]. However, such schemes have considerable overhead. First, the redundant lines cause significant area overhead. In addition, the number of redundant cache lines may have to be increased significantly to cover all faulty lines under severe process variations, particularly considering that the redundant lines themselves are prone to process variations.

In this work, we propose a simple, yet efficient, technique to mitigate process variations by selective voltage boosting. By selectively boosting the voltage of components on the failing critical paths, the cache access times can be significantly reduced, while still maintaining small power and area overheads. The proposed technique can be applied to any SRAM array structure, however, in this work, we focus on level 1 (L1) caches since their access times have a direct impact on the chip yield.

The rest of this paper is organized as follows. In Section 2, we present our analysis showing that caches are indeed the most susceptible components to process variations. We demonstrate our proposed technique in Section 3, while the evaluation results in terms of yield, power, and area costs are provided in Section 4. A review of related works is presented in Section 5 and we conclude our paper in Section 6.

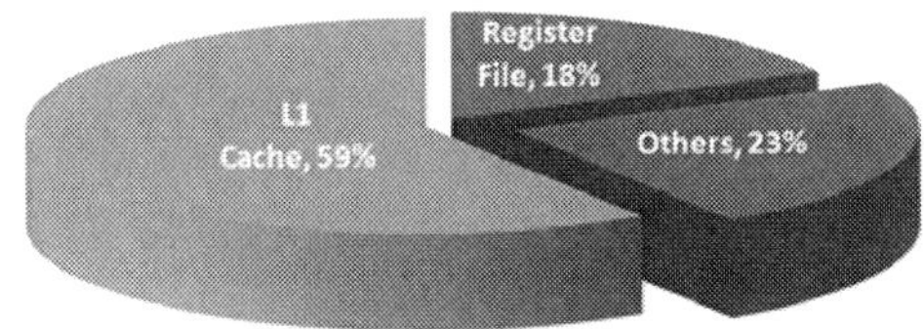

Figure 1. Distribution of critical paths in a representative processor architecture.

Table 1. Probability of cache failures w.r.t. σV_{th} [2]. $P(x)$ indicates the probability of failure 'x' occurring.

σV_{th}	P(read failure)	P(write failure)	P(access time failure)	P(cache failure)
20	1E-4	1E-4	*1E-4*	1E-4
30	2E-4	2E-4	*5E-4*	8E-4
40	1E-3	6E-4	*3E-3*	4E-3
50	5.8E-3	3.8E-3	*1.4E-2*	2.2E-2

2. ANALYSIS OF YIELD LOSSES

To analyze the impact of process variations on a processor architecture, we carried out a Monte Carlo study on 1000 chips modeled after an Alpha 21364 (EV7) processor using the framework described in Section 4.1. We considered the critical paths in each pipeline stage and run simulations for the branch predictor, register rename unit, issue queue, register file, integer execution unit, and L1 data cache [15]. Figure 1 shows the distribution of the critical paths among various pipeline components. Our results reveal that 59% of the critical path lies in the L1 caches. This is not a surprising result, as previous work [4] has shown that structures with many independent paths with low logic depth are most likely affected by parametric variations. Thus, we focus on implementing our voltage boosting scheme only on caches. Although it is not depicted here, L2/L3 caches are also highly affected by process variations. However, since they are usually not on the processor critical path, other techniques such as NUCA can be adopted to alleviate the increase in access latencies in L2/L3 caches.

SRAM cells have been used as primary storage in microprocessors. Currently, they are widely used in caches, register files, etc. due to many advantages of SRAM cells (such as fast access speed, no need to refresh, and stability). However, as process variations become severe, SRAM cells become vulnerable in terms of their stability and access speed. Table 1 depicts the relation between the probabilities of cache failures and the standard deviation of V_{th} (σV_{th}) [2]. When σV_{th} is 20mV, the probability of three kinds of cache failure is 1E-4. However, as the σV_{th} is increased, the access time failure becomes dominant. Furthermore, the situation gets worse in the cache line granularity. Liang et al. [17] simulated line-level failure rate for a cache with 32 Bytes linesize and observed a 64% cache line failure in 32nm technology. Although redundant cache lines can replace these faulty cache lines, 64% cache line failure will be hard to recover from. This calls for a multitude of techniques to mitigate cache access time failures under process variations. These results also suggest that we can expect a significant yield improvement when we alleviate the cache access time failure.

3. SELECTIVE VOLTAGE BOOSTING

3.1 Design philosophy

3.1.1 Delay Reduction by Boosted Voltage

In a synchronized design, the longest pipeline stage determines the frequency of the processor. If this frequency is below the pre-determined minimum processor frequency, the processor becomes parametric yield loss, despite the fact that it might be functional. The main goal in this work is to reduce the cache access latency if it causes a parametric yield loss. Specifically, we propose to save the failing paths by boosting their power supply voltage.

The relation between power supply voltage and delay is described by Alpha Law [26]:

$$Delay \quad \propto \quad \frac{V_{dd}}{(V_{dd} - V_{th})^{\alpha}} \qquad (1)$$

Here 'α' is a technology dependent constant that is greater than 1. Hence, without tuning the V_{th} in the circuit, the delay can be reduced by boosting the V_{dd}.

3.1.2 Selective Wordline Voltage Boosting

The reduced delay by boosting supply voltage does not come for free. It incurs increased dynamic and static power. Hence, careful design compromises must be made to achieve an efficient

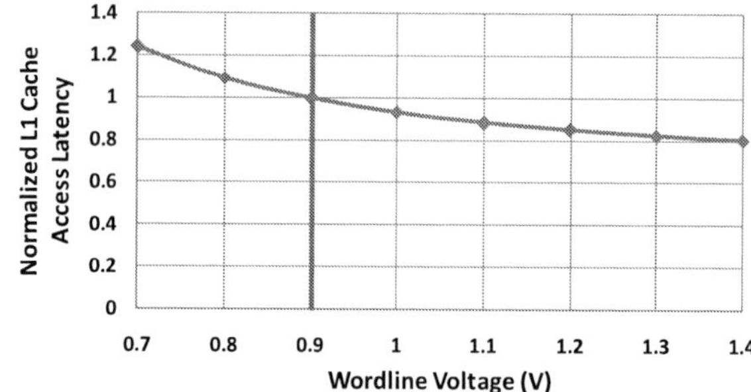

Figure 2. The relation between wordline voltage and cache delay. The results are obtained using SPICE simulations.

implementation. In our design, we propose to *only boost the voltage on selected wordlines*. There are several reasons for this design decision. First, the delay on the long wordlines is a major component of the overall cache access time. Boosted supply voltage in its drivers can directly reduce this delay. Second, boosted wordline voltage helps to enhance read current that goes through the pass gates (PG) of the cells, hence helps to discharge the bitlines faster. Third, the leakage power is low on wordlines as they only affect gate leakage to the PG's. Thus, boosted voltage will not significantly increase the overall leakage power of the cache. Finally, failing cells are naturally grouped into rows that share the same wordline. Therefore, selective wordline boosting effectively implements boosting of only the failing rows.

To estimate the efficiency of boosting only the wordline voltage, we simulated the access latency versus various wordline voltages using the framework described in Section 4.1. The results are shown in Figure 2. As the wordline voltage is increased from 0.9V to 1.3V, the L1 cache access latency is reduced by 18%. This result shows that the wordline voltage has the potential of being effectively used to reduce the access latency of caches.

3.2 Overall architecture

With the aim of boosting the voltage on selected wordlines, we propose the architecture shown in Figure 3. Only the circuits associated with a single wordline is illustrated as the remaining circuit design is not affected by our optimization. During chip testing, the rows in the L1 cache that fail the timing test are marked in a special purpose register file, where a single fault bit is dedicated for each wordline. In addition to the dedicated register, by using off-chip EEPROM, this failure information can be kept across system power down. The fault bits are used to drive the power selection circuit for the wordline buffers. Failing wordlines

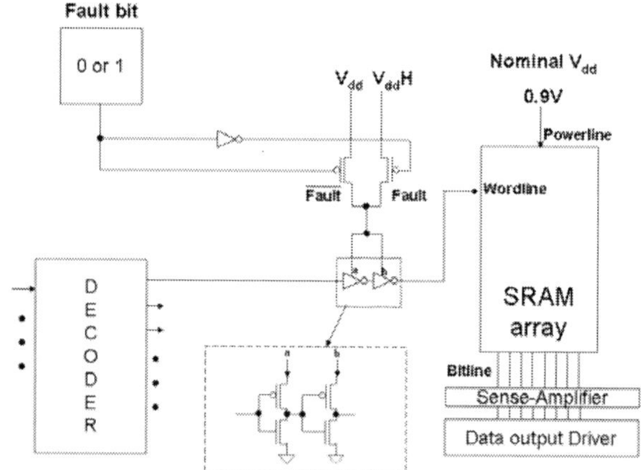

Figure 3. Proposed wordline voltage boosting architecture.

will adopt higher power supply voltage ($V_{dd}H$) in its drivers (the inverters), while passing wordlines will be fed with the nominal voltage (V_{dd}). The remaining components in the L1 cache all work under nominal power supply. In our evaluation, we target a 32KB L1 cache with wordline/bitline slicing, where 1024 separate wordlines are present. Thus 128 Bytes of fault bit storage (1024 bits), together with 2048 power gating PMOS's are added to the cache. Utilizing fault bits in our cache architecture makes sure that only wordlines of the faulty rows are raised high, hence the dynamic and static power overhead of our approach remains low. The distribution of $V_{dd}H$ could increase power routing complexity. However, this can be alleviated if techniques in [14] is used.

As the decoders are still working at the nominal voltage, while the wordline buffers can be raised to $V_{dd}H$, there is a voltage mismatch. However, there is no need to employ a level shifter. To achieve full rail ($0\sim V_{dd}H$) signal on the wordline, the supply voltage of both inverters are boosted to $V_{dd}H$. There is still increased leakage going through the PMOS of the first inverter due to incomplete shut-off of the PMOS (i.e., when input is V_{dd}, supply is $V_{dd}H$). We fully modeled this leakage and evaluated it to be minimal as compared to the total leakage in the L1 cache which is dominated by the SRAM array leakage (a detailed leakage overhead is presented in Section 4.1). Hence we chose not to adopt level shifters to avoid extra area overhead and additional latency on the critical path.

4. EVALUATION

4.1 Evaluation Methodology

To evaluate the efficiency of our proposed technique, we built a SPICE simulation framework that models process variations. In addition, a detailed architectural simulation framework (described in Section 4.1.3) is adopted to model the effect of our approach on the processor.

4.1.1 Circuit Model for L1 Cache

We adopt the floorplan and circuit structure of a 32KB L1 cache as described in CACTI 5.3 [35]. One critical path for the data array is shown in Figure 4.

A path consists of address H-Tree, address pre-decoder, address post-decoder, wordline driver, wordline, SRAM cell, bitline, precharge circuit, sense amplifier, multiplexer, and output data H-Tree. H-Trees, wordlines, and bitlines are modeled as RC lines, with repeaters inserted in the H-Trees. The nominal parameter values for the MOSFETs and interconnects are extracted from 45nm PTM [23]. The dimensions of the SRAM cells are based on the design by Hamzaoglu et al. [11]. The SPICE model without process variations is calibrated using CACTI 5.3.

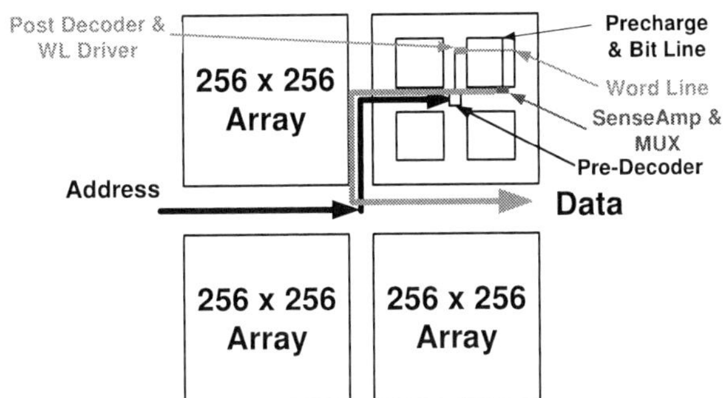

Figure 4. L1 Cache data array components: components on one path are shown.

Table 2. Nominal and 3σ variation values for each source of process variations modeled.

	Gate Length	Threshold Voltage	Metal Width	Metal Thickness	ILD Thickness
Nominal Value	45 nm	220 mV	0.25 μm	0.55 μm	2.5 nm
Low Var. [%]	±6.6	±12	±22	±22	±23
Medium Var. [%]	±10	±18	±33	±33	±35
High Var. [%]	±13.3	±24	±44	±44	±46

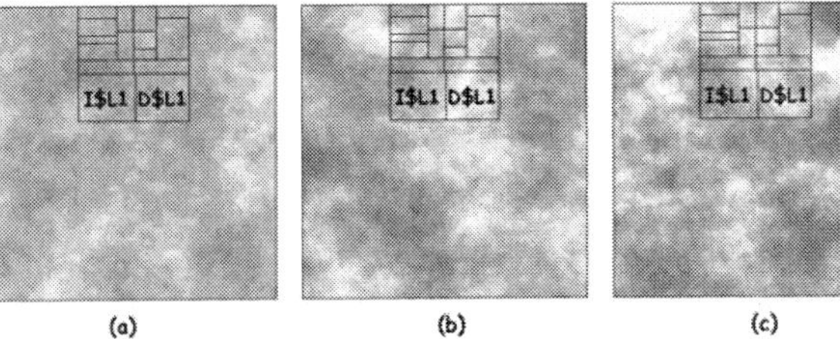

Figure 5. V_{th} variation maps for (a) low, (b) medium, and (c) high process variations.

4.1.2 Process Variation Modeling

Processes like sub-wavelength lithography and aggressive technology scaling result in statistical variations in circuit parameters such as gate-oxide thickness, channel length, and Random Doping Effects (RDE) [3]. These parametric variations can be classified into die-to-die (D2D) variations and within-die (WID) variations. D2D variation refers to the variation in process parameters across dies and wafers, whereas WID variation takes place in device features within a single die. Parametric variations can be of two categories: spatially-correlated (systematic) variations where devices close to each other have a higher probability of observing a similar variation level, and random (uncorrelated) variations causing random differences between various devices within a die. In this work, we model both systematic and random parametric variations.

To effectively model parametric variations, we account for five different variation parameters: metal thickness (T), inter-layer dielectric thickness (ILD or H), line-width (W) on interconnects, gate length (L_{gate}) and threshold voltage (V_{th}) for the MOS devices. We use the variation limits given by Nassif [20] as shown in Table 2 as our base case (Medium Variation). We also study two other cases where process variations are less and more severe (Low and High Variation, respectively); the parameter variations for those cases are also given in Table 2. Three sample maps corresponding to the V_{th} in the three process variation levels are presented in Figure 5. Note that high process variation corresponds to a larger span of colors on the map, and vice versa. To take into account the spatial correlation, we use a range factor (φ) in the two dimensional layout of the chip. Thus, each process parameter can be expressed as a function of its mean (μ), standard deviation (σ), and the range (φ) values. In this work, we used a

Table 3. Normalized dynamic energy and leakage power consumption of our proposed logic.

Scheme	Dynamic energy	Leakage power
Baseline 0.9V	1.0000	1.0000
Selective 1.0V	1.0216	1.0064
Selective 1.1V	1.0419	1.0076
Selective 1.2V	1.0685	1.0092
Selective 1.3V	1.0977	1.0125
Whole 1.0V	1.3649	1.6172

range factor of 0.2. With this background, we have generated a spatial map of parameter values using the VARIUS process variation model [27] implemented in the R statistical tool [34]. We divide the circuit floorplan into a grid of 5000 x 5000 points. Our framework then picks the process variation values on this grid and maps them to the RC line sections and MOSFETS in the SPICE model for simulation.

We use the Monte Carlo method to model a batch of chips and study the impact of process variations on cache yield. We simulated a total of 1000 chips for each process variation severity level. The cache access latency under process variations is simulated across all the paths, together with dynamic and leakage power statistics.

4.1.3 Architectural Simulation

To evaluate the energy consumption impact, we extracted cache access traces of SPEC2000 benchmark applications using SimpleScalar 3.0 Alpha simulator [31]. Using per access energy consumption and leakage power from the SPICE simulations, we calculated overall cache energy consumption including both leakage and dynamic energy for L1 data and instruction caches. In the following section, we present results for eight representative applications (applu, crafty, fma3d, gcc, gzip, mcf, mesa, and twolf). SimPoint [28] is used to improve the accuracy of the simulations. Data and instruction caches are 32KB in size, 2-way set associative, and have 32-byte linesizes. The clock frequency of the simulated processor is set to be 3.5 GHz.

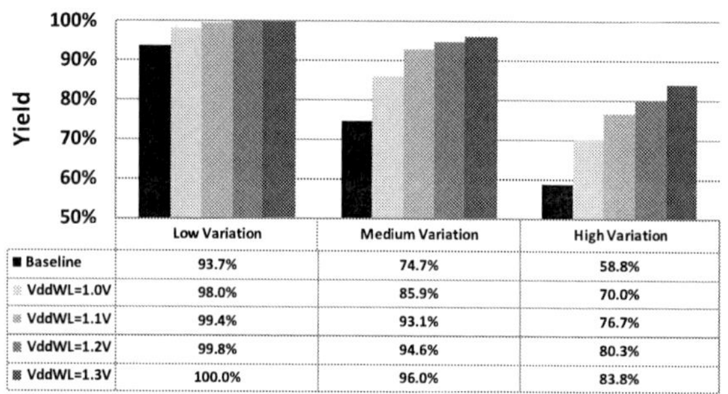

	Low Variation	Medium Variation	High Variation
■ Baseline	93.7%	74.7%	58.8%
VddWL=1.0V	98.0%	85.9%	70.0%
VddWL=1.1V	99.4%	93.1%	76.7%
▨ VddWL=1.2V	99.8%	94.6%	80.3%
■ VddWL=1.3V	100.0%	96.0%	83.8%

Figure 6. Yield Improvement.

4.2 Yield Enhancement

The parametric yield result for L1 data and instruction caches for the 1000 simulated chips are shown in Figure 6.

We simulated the latency of the top 128 critical paths of the L1 data and instruction caches under low, medium, and high process variations as described in the previous section. We assume the timing specification (T_{cutoff}) of the caches to be the mean latency ($\mu_{Tmedium}$) for the medium variation case plus half its standard deviation ($\sigma_{Tmedium}$). That is:

$$T_{cutoff} = \mu_{Tmedium} + 0.5\ \sigma_{Tmedium}.$$

Any cache with

$$Max(T_{pathi} \mid i=1\ to\ 128) > T_{cutoff}$$

is considered to be a yield loss. The same T_{cutoff} is used for all the three process variation scenarios. Hence, for a baseline cache without boosted wordline voltage, higher yield is seen when the process variations are low, while significant yield losses are experienced when the process variations are high. Across the variation scenarios, raising the wordline voltage can significantly improve the parametric yield: for the medium variation case, when the wordline voltage is raised to 1.0V, cache yield is improved by 10.8%, which corresponds to a 45.0% loss reduction.

For the same variation levels, when the wordline voltage is raised to 1.3V, cache yield is improved by 28.5%, corresponding to an 85.2% reduction in the yield losses. For severe process variations, our scheme improves the yield level by up to 42.5%, corresponding to over 60% reduction in yield losses. For weaker process variations, our scheme can eliminate all the yield losses.

Based on our simulations across all the 1000 chips, 11.7% of the critical paths (rows) failed the timing requirement for the medium process variation scenario, on average. However, due to the spatial correlation in the process variations, the failing rows are not evenly distributed. Some of the failing chips have a larger number of failing rows, whereas others have few. Although redundancy schemes may save some of these chips, it alone will not be enough to improve the yield levels considerably particularly under severe process variations. However, our technique allows any number of wordlines to be raised to higher voltage, thus has the potential to improve the yield even with cases of massive failures. In fact, redundancy schemes can be used on top of our technique to recover particle related failures to further improve the overall yield.

4.3 Cost

The proposed design improves the cache access time at the price of increased energy and area consumption.

4.3.1 Energy consumption

In our proposed technique, we only raise the supply voltage of the wordline drivers for failing lines to minimize the energy overhead. Table 3 shows the normalized dynamic and static energy consumption for each scheme. Baseline scheme uses the nominal voltage (0.9V) for wordline voltage. Selective 1.0~1.3V schemes represent our proposed technique where only the wordline driver supply voltage is raised to the corresponding level. As a reference, we also show a case where the power supply voltage of the whole cache is raised to 1.0V (denoted by *Whole 1.0V*). The reported dynamic energy numbers correspond to reading a 64-bit word in a boosted line. For boosted lines, there is a small dynamic power overhead (2.1%~9.8%) compared with the baseline scheme as shown in Table 3. Note that only the originally failing lines face this overhead. Compared to raising the supply voltage of the whole cache to 1.0V, even the Selective 1.3V scheme consumes 80.4% less dynamic power. As described in Section 4.2, about 11.7% of the cache wordlines need to be boosted. Hence, the dynamic power consumption overhead in practice is even smaller (0.98% on average for the Selective 1.3V scheme). In addition, assuming 100 wordlines (10% of the complete set of wordlines) need to be boosted, we calculated the leakage overhead of our approach. The leakage overhead is also negligible (1.25% for

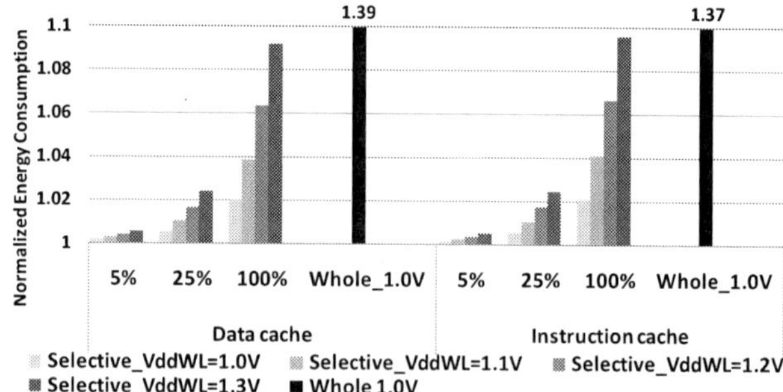

Figure 7. Geometric mean of normalized energy consumption in L1 caches across SPEC2K benchmarks. All of the schemes are normalized to baseline scheme using the nominal voltage.

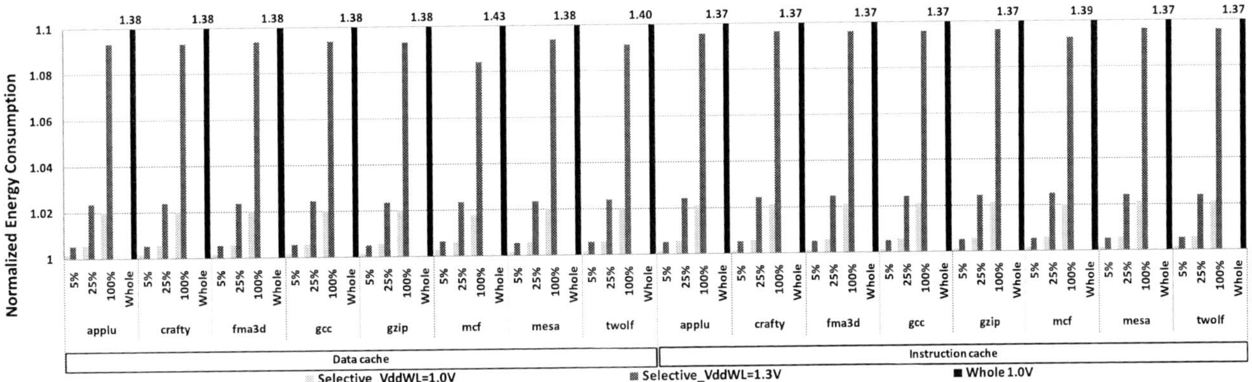

Figure 8. Normalized energy consumption in L1 caches when executing SPEC2K benchmark suite. All of the schemes are normalized to baseline scheme which uses the nominal voltage (0.9V).

Selective 1.3V) since the wordline leakage constitutes only a small fraction of the total leakage, while the sub-threshold leakage in the SRAM array dominates the cache leakage.

The actual power consumption impact varies depending on the chip sample as well as the application executed since the fault bitmap is different across chip samples and cache access patterns are different among applications. For example, there will not be any energy overhead if we do not access the boosted cache lines. Figure 7 shows the geometric mean of normalized energy consumption for the five schemes across the eight selected SPEC2000 applications. To gather these results, we generate 100 chips with randomly distributed failing cache rows. For example, with the 5% failure rate, we randomly select 5% of the rows for each of the 100 chips. Then, we simulate the eight applications monitoring the fraction of accesses that are made to failing rows. Using this information, we calculate the energy overhead of our proposed schemes. The x-axis shows the failure rate. A reference scheme with the whole cache power supply voltage raised to 1.0V is also shown (*Whole 1.0V*). In general, the dynamic power overhead is proportional to the number of boosted wordlines. For a case of 25% wordlines of data cache boosted to 1.0V and 1.3V, the geometric mean of the energy overhead is 0.55% and 2.3%, respectively. In the extreme case of having 100% boosted wordlines, the overhead can be as high as 9.2%. This is still drastically lower than that (39%) of raising the whole cache power supply to 1.0V. Instruction cache shows similar trends.

Figure 8 shows energy consumption of the L1 cache for specific applications. We present results for three schemes: "selective 1.0V", "selective 1.3V", and "whole 1.0V". As shown in Figure 8, "selective 1.0V" and "selective 1.3V" with 25% of the wordlines boosted has at most 0.58% and 2.3% energy overhead (for mcf application) compared to the baseline scheme, respectively. The studied applications show very similar behaviors. Overall, our selective schemes are much more energy-efficient, when compared to raising the supply voltage of the whole cache, regardless of which application is running on the microprocessor.

4.3.2 Area

Our proposed technique requires extra storage for the fault bits, inverters for logic generation, and power gating PMOS transistors. The power gating PMOS transistors take 3.54% of total L1 cache area. This overhead is comparable to similar techniques such as Gated-V_{dd} [22]. The storage needed for the fault bits is limited as only 128 Bytes (1024 bits) for a 32KB cache, which is less than

0.4% of the cache capacity. The total area cost can be conservatively estimated to be below 4.5% of total cache area.

Further area reduction can be achieved by increasing the granularity of control. By sharing the power gating PMOS across n wordlines, the area cost can be significantly reduced to one *n-th* of the above estimated value. This comes at the cost of extra power consumption for cases with sporadic failing lines.

5. RELATED WORK

There have been many studies on robust microprocessor design under process variations. Agarwal et al. [1][2] analyzed the cache operation failures under process variation and proposed process variation tolerant cache architecture, where the cache access failure is shown to be a dominant factor as σV_{th} is increased. Their proposed cache architecture utilizes remapping of column mux when there are faulty cache lines. Though their architecture improves yield significantly, it is not suitable for severe process variation environments because the reduced number of available cache lines leads to large performance overhead. The IBM Cell [24] and Power6 [9] processor both use increased array supply voltage to improve stability and read performance. However, their technique exercises a much coarser control granularity and they raise SRAM cell supply while we proposed to have fine-grain wordline voltage boosting. Chen et al. [6] compared three types of SRAM design under yield constraints. They compared these three types of SRAM designs (differential 6T, single-ended 6T, and 8T SRAM) in terms of energy and area considering transistor sizing for their robustness. Mutyam et al. [18] proposed process variation tolerant block rearrangement technique. This technique rearranges cache blocks which have similar access latency. It tries to minimize performance overhead incurred by process variation. However, their technique has performance overheads since their technique does not reduce the latency of faulty cache line but only allocates the cache lines that have similar latency in a same set through the block rearranging scheme.

In the circuit level, many techniques have been proposed to maintain the yield under process variation. Chen et al. [7] compared Adaptive Body Bias (ABB) to Adaptive Supply Voltage (ASV). Though both ABB and ASV can be used to reach sufficient yield, adopting either ABB or ASV is enough for yield improvement according to their analysis. They also concluded that ASV is simpler to implement than ABB. Tschanz et al. [36] introduced variation tolerant circuit techniques: ABB and adaptive V_{dd}. They suggested that using both ABB and adaptive V_{dd} is efficient with respect to variation tolerance. Li et al. [16] proposed

978-1-60558-497-3/09 $25.00 © 2009 ACM

ASV technique which can supply V_{dd} adaptively in the processor pipeline. Though their technique reduces power consumption under the yield constraints, it is not applicable to caches where timing is critical as they confessed. Gregg et al. [10] proposed Individual Well Adaptive Body Biasing (IWABB). By adopting their circuitry, many biasing modes are available after fabrication considering the characteristics of each chip. Using their design flow adopting IWABB, yield can be significantly improved. However, their technique is based on ABB, which is more complicated to implement than adjusting V_{dd}. Teodorescu et al. [33] proposed Dynamic Fine-Grain Body Biasing (D-FGBB). Compared to conventional body biasing techniques that are statically applied to the fabricated chips, they proposed a dynamic body biasing technique which provides more flexibility. However, they do not exploit the special features of SRAM arrays.

6. CONCLUSION

As process technology scales down continuously and more transistors are packed in a limited die area, parametric yield losses have become an important problem for manufacturers. In this paper, we proposed a novel yield-aware selective wordline voltage boosting technique for caches. By boosting the voltage on selected wordlines, the overall yield of the cache can be significantly improved. SPICE simulations show over 80% reductions in yield losses, with average per-access energy overhead of less than 1%. The area overhead is well below 4.5% and can be further reduced by increasing the control granularity.

7. ACKNOWLEDGEMENT

This work was in part supported by the Korea Science and Engineering Foundation (KOSEF) grant funded by the Korea government (MEST) (No. R01-2007-000-20750-0); US National Science Foundation (NSF) grants CNS-0551639, IIS-0536994, CCF-0747201, and CCF-0541337; DoE CAREER Award DEFG02-05ER25691; and by Wissner-Slivka Chair funds.

8. REFERENCES

[1] A. Agarwal, et al., "A process-tolerant cache architecture for improved yield in nanoscale technologies", *IEEE Transaction on VLSI Systems, vol. 13*, pp. 27-38, 2005.

[2] A. Agarwal, et al., "Process variation in embedded memories: failure analysis and variation aware architecture", *IEEE Jnl. of Solid-State Circuits, vol. 40*, 2005.

[3] S. Borkar, et al., "Parameter variations and impact on circuits and microarchitecture". *In Proc. of DAC*, 2003.

[4] K. A. Bowman, et al., "Impact of Die-to-Die and Within-Die Parameter Fluctuations on the Maximum Clock Frequency Distribution for Gigascale Integration". *IEEE Jnl. of Solid-State Circuits, vol. 37*, 2002.

[5] W. Bryg and J. Alabado, "The UltraSPARC T1 Processor - Reliability, Availability, and Serviceability."

[6] G. K. Chen, et al., "Yield-driven near-threshold SRAM design". *In Proc. of Int'l Conference on Computer Aided Design*, 2007.

[7] T. Chen and S. Naffziger. "Comparison of Adaptive Body Bias (ABB) and Adaptive Supply Voltage (ASV) for Improving Delay and Leakage under the Presence of Process Variation". *IEEE Transaction on VLSI Systems, vol. 11*, pp. 888-899, 2003.

[8] A. Das, S. Ozdemir, G. Memik, J. Zambreno, and A. N. Choudhary, "Microarchitectures for Managing Chip Revenues under Process Variations". *Computer Architecture Letters, vol. 6*, pp. 29-32, 2007.

[9] J. Friedrich, et al., "Design of the Power6 Microprocessor", In ISSCC, 2007.

[10] J. Gregg and T. W. Chen. "Post Silicon Power/Performance Optimization in the Presence of Process Variations Using Individual Well Adaptive Body Biasing (IWABB)". *In proceedings of IEEE Int'l Symposium on Quality Electronic Design*, 2007.

[11] F. Hamzaoglu, et al., "A 153Mb-SRAM Design with Dynamic Stability Enhancement and Leakage Reduction in 45nm High-k Metal-Gate CMOS Technology", *In Proc. of Int'l Solid State Circuits Conference*, 2008.

[12] N. P. Jouppi, "Improving direct-mapped cache performance by the addition of a small fully-associative cache and prefetch buffers", *ACM SIGARCH Computer Architecture News, vol. 18*, 1990.

[13] C. Kim, D. Burger, and S. W. Keckler. "An adaptive, non-uniform cache structure for wire-delay dominated on-chip caches". *In Proc. of Int'l Conference on Architectural Support for Programming Languages and Operating Systems*, 2002.

[14] N. Kim, et al., "Single-VDD and single-VT super-drowsy techniques for low-leakage high-performance instruction caches", *In Proc. of Int'l Symposium on Low Power Electronics and Design*, 2004.

[15] K. Krewell, Alpha EV7 Processor: A High-Performance Tradition Continues, Apr. 2002.

[16] H. Li, et al., "SAVS: a self-adaptive variable supply-voltage technique for process- tolerant and power-efficient multi-issue superscalar processor design". *In Proc. of ASPDAC*, 2006.

[17] X. Liang, et al., "Process Variation Tolerant 3T1D-Based Cache Architectures". *In Proc. of IEEE/ACM Int'l Symposium on Microarchitecture*, 2007.

[18] M. Mutyam and N. Vijaykrishnan. "Working with process variation aware caches". *In proc. of DATE*, 2007.

[19] S. Naffziger, et al., "The Implementation of the Itanium 2 Microprocessor," *IEEE Jnl. of Solid-State Circuits, vol. 37*, 2002.

[20] S. R. Nassif. "Modeling and Analysis of Manufacturing Variations". *In Proc. of IEEE Conf. on Custom Integrated Circuits*, 2001.

[21] S. Ozdemir, D. Sinha, G. Memik, J. Adams, and H. Zhou, "Yield-Aware Cache Architectures". *In Proc. of IEEE/ACM Int'l Symposium on Microarchitecture*, pp. 15-25, 2006.

[22] M. Powell, et al., "Gated-Vdd: A circuit technique to reduce leakage in deep-submicron cache memories", ISLPED, pp. 90-95, 2000.

[23] Predictive Technology Model (PTM), Arizona State University. http://www.eas.asu.edu/~ptm/

[24] M. Riley, et al., "Implementation of the 65nm Cell Broadband Engine", *In Proc. of IEEE Custom Integrated Circuits Conf.*, 2007.

[25] B. F. Romanescu, et al., "Reducing the Impact of Intra-Core Process Variability with Criticality-Based Resource Allocation and Prefetching". *In Proc. of ACM Int'l Conference on Computing Frontiers*, 2008.

[26] T. Sakurai and A. R. Newton. "Alpha-Power Law MOSFET Model and its Applications to CMOS Inverter Delay and Other Formulas", *IEEE Jnl. of Solid-State Circuits, vol. 25*, pp. 584-594, 1990.

[27] S. R. Sarangi, et al., "VARIUS: A Model of Process Variation and Resulting Timing Errors for Microarchitects", *IEEE Transactions on Semiconductor Manufacturing, vol. 21*, pp. 3-13, 2008.

[28] T. Sherwood, et al., "Automatically characterizing large scale program behavior". *In Proc. of Int'l Conference on Architectural Support for Programming Languages and Operating Systems*, 2002.

[29] P. Shivakumar, et al., "Exploiting microarchitectural redundancy for defect tolerance". *In Proc. of IEEE Int'l Conference on Computer Design*, 2003.

[30] SIA. International Technology Roadmap for Semiconductors, 2005. Available at http://public.itrs.net.

[31] SimpleScalar toolset. http://www.simplescalar.com.

[32] G. S. Sohi, "Cache Memory Organization to Enhance the Yield of High Performance VLSI Processors". *IEEE Transaction on Computers, vol. 38*, pp. 484-492, 1989.

[33] R. Teodorescu, et al., "Mitigating Parameter Variation with Dynamic Fine-Grain Body Biasing". *In Proc. of IEEE/ACM Int'l Symposium on Microarchitecture*, pp. 27-42, 2007.

[34] The R Project for Statistical Computing. Available from: http://www.r-project.org.

[35] S. Thoziyoor, et al., "CACTI 5 Technical Report. HP Labs".

[36] J. Tschanz, K. A. Bowman, and V. De. "Variation-tolerant circuits: circuit solutions and techniques". *In Proc. of Design Automation Conference*, pp. 762-763, 2005.

[37] B. Zhai, et al., "The limit of dynamic voltage scaling and insomniac dynamic voltage scaling", *IEEE Transaction on VLSI Systems, vol. 13*, pp. 1239-1252, 2005

978-1-60558-497-3/09 $25.00 © 2009 ACM

Double Patterning Lithography Friendly Detailed Routing with Redundant Via Consideration*

Kun Yuan, Katrina Lu† David Z. Pan
ECE Dept. Univ. of Texas at Austin, Austin, TX 78712
{kyuan, yiotse}@cerc.utexas.edu, dpan@ece.utexas.edu

ABSTRACT

In double patterning lithography (DPL), coloring conflict and stitch minimization are the two main challenges. Post layout decomposition algorithm [1] [2]may not be enough to achieve high quality solution for DPL-unfriendly designs, due to complex 2D patterns in lower metal layers. Therefore, DPL-friendliness is needed at routing stage [3]. Another key yield improvement technique is redundant via insertion [4] [5]. However, this would increase the complexity in DPL-compliance. To make designs manufacturable in DPL, we should not insert a redundant via if it results in coloring conflict. This paper is the first work to consider DPL and redundant via together. We have developed two algorithms, *post-routing DPL-aware insertion* and *DPL-friendly routing with redundant via consideration* to take into account redundant via DPL-compliance. Experimental results show that, compared to a DPL-aware optimization flow without redundant via consideration, we can improve insertion rate by 43% while still achieving zero coloring conflicts. Moreover, we can reduce the number of vias and stitches by 9% and 17% respectively.

Categories and Subject Descriptors

B.7.2 [**Hardware, Integrated Circuit**]: Design Aids

General Terms

Algorithms, Design

Keywords

Detailed Routing, Double Patterning, Redundant Via

1. INTRODUCTION

Double patterning lithography [6] is considered as a most likely solution for 32nm/22nm technology. In DPL, the original layout is decomposed into two masks (colors), e.g., BLACK and GRAY. However, this introduces the challenges of minimizing conflict and stitch. Most researches [1] [2] focus on post layout decomposition but it may be not enough for poor designs, especially due to complex 2D patterns in lower metal layers. DPL-friendliness is needed from design side. Recently, Cho et al [3] proposed the first DPL-friendly detailed routing for highly decomposable routing.

Another key yield improvement technique is the redundant via insertion [4, 5]. This could introduce complexity in DPL compliance. The challenge comes from the metal which is used to cover the via and the redundant via in both layers. It is commonly referred as extra metal. Besides not violating the minimum spacing min_{sp} rule, minimum coloring spacing min_{dp} needs to be satisfied as well to avoid conflicts. Fig. 1 (a) shows a motivational example, where the top figures are metal2 and the bottoms are for metal1. As Fig. 1 (a) indicates, E1 and E2 are the extra metals. To minimize the number of stitches, we prefer assigning them the same color as the metal the via touches in corresponding layer, which is referred as *stitch-free extension*. However, this may cause conflicts due to the coloring assignment of existing layout.

*This work is supported in part by NSF, SRC, Sun, Qualcomm and equipment donations from Intel.

†Katrina Lu is now affiliated with Intel Corporation, Hillsboro OR.

Permission to make digital or hard copies of part or all of this work for personal or classroom use is granted without fee provided that copies are not made or distributed for profit or commercial advantage and that copies bear this notice and the full citation on the first page. To copy otherwise, to republish, to post on servers or to redistribute to lists, requires prior specific permission and/or a fee.
DAC'09, July 26-31, 2009, San Francisco, California, USA

In Fig. 1 (b), the *stitch-free extension* will introduce a conflict in both metal1 and metal2. To resolve this issue, as Fig. 1 (c) illustrates, we can flip the coloring of the extra metal E2 by introducing an additional stitch. However, it is not always possible to remove the conflict, such as the one in metal1. In such case, we need to modify the layout and move one of the three features.

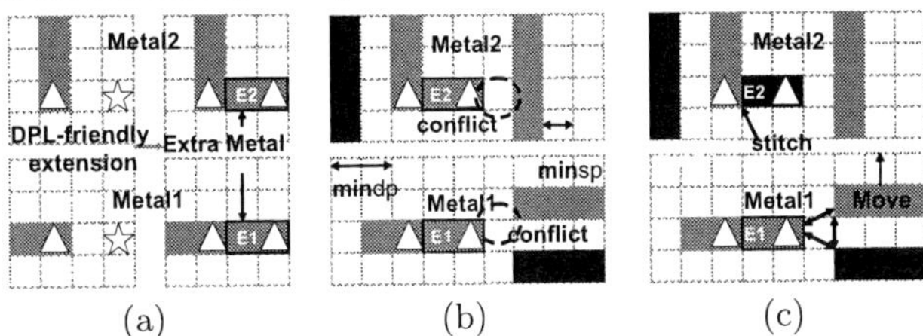

Figure 1: This figure illustrates redundant via DPL-compliance problem. A via or redundant via goes through the corresponding triangles in metal1 and metal2. The star denotes the feasible redundant via candidate without violating min_{sp}. E1 and E2 are the extra metals, enclosed by solid rectangles.

In this paper, we consider DPL and redundant via insertion together. We develop two algorithms, *post-routing DPL-aware insertion* and *DPL-friendly routing with redundant via consideration*, to take into account redundant via DPL-compliance. The experimental results are very promising and show the effectiveness of our proposed algorithms.

2. PRELIMINARIES AND PREVIOUS WORK

2.1 Preliminaries

Our algorithm works on routing grids. The min_{sp} and min_{dp} for metals and via cuts are taken as one-grid and two-grid size respectively, but they can be easily extended to other values. Only metal1 and metal2 are in our discussion, because the detailed routing and DPL are mainly for low metal layers.

We explain some key concepts in Fig 2. If a redundant via candidate for a via does not violate min_{sp} in the existing layout, we refer it as *feasible*. If a via does not have any *feasible* candidate, it is *dead*, otherwise *alive*. If two *feasible* candidates of different vias can not be inserted at the same time due to min_{sp} constraints, there is an *external conflict* between them. As Fig. 2 (a) shows, there are three vias A, B, and C. Via C is a *dead* via. A and B are *alive*, and their *feasible* candidates are A1 and B1-B3 respectively. There is an *external conflict* between A1 and B1. Instead, we can select A1 and B2 as redundant via solutions without violating min_{sp}. However, under DPL, when min_{dp} is additionally considered, the *feasible* candidates may result in conflicts or extra stitches. We refer to a *feasible* candidate as *DPL-feasible*, if its *stitch-free extension* configuration will not cause conflicts under the pre determined layout and coloring. Basically, a *DPL-feasible* candidate has a DPL-friendly, conflict and stitch free, redundant via solution. If a via does not have any *DPL-feasible* candidate, it is *DPL-dead*, otherwise *DPL-alive*. With the coloring assignment in Fig. 2 (b), if the same solutions A1 and B2 are selected, it will produce two *coloring conflicts* as Fig. 2 (c) illustrates when *stitch-free extension* is applied. A is a *DPL-dead* via because its only *feasible* candidate A1 is not *DPL-feasible*.

2.2 Previous work on DPL-friendly routing

To improve layout decomposability in metal layers, Cho [3] et al. proposed a DPL-friendly simultaneous routing and decomposition using *coloring path* and *color shadow* techniques. A two bit-variable is introduced for each grid to keep track of the colorability information, which is one of the four states {BG, $B\overline{G}$,

978-1-60558-497-3/09 $25.00 © 2009 ACM

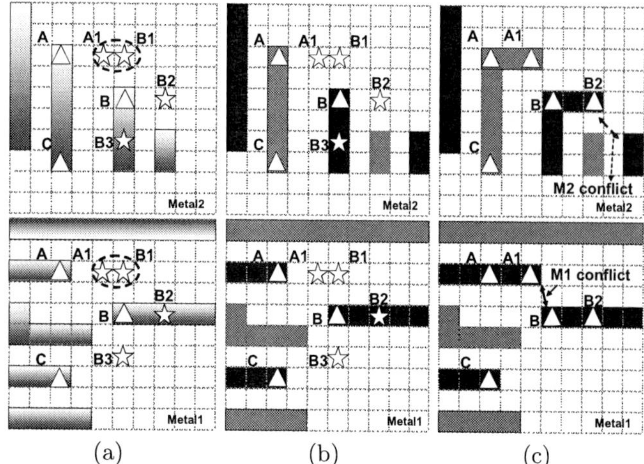

(a)   (b)   (c)

Figure 2: These figures illustrate the basic concepts of *feasible* redundant via candidate, *dead* and *alive* via. *DPL-feasible* candidate, *DPL-dead* and *DPL-alive* via are also explained.

$\overline{BG}$, $\overline{\overline{BG}}$}. As a preprocessing, all the routing blockages will be colored by a layout decomposition algorithm and *color shadow* will be performed to generate the starting grid state map for all the routing grids. In *color shadow*, the state of surrounding grids within min_{dp} distance from the colored layout will be assigned. For example, grids near a GRAY grid will have either $B\overline{G}$ or $\overline{BG}$ states.

When each net is to be routed, other than the traditional costs, they also apply DPL awareness penalty, based on current routing solution and grid states. As an example, Fig. 3 (a) shows the existing configuration, and DPL-friendly routing will be performed for net S-T. Assuming the path in Fig. 3 (b) is picked due to possible *conflict and stitch free* wiring, the grids along the path will be colored through *coloring path* according to its grid states. In *coloring path*, a grid with $B\overline{G}$ and $\overline{BG}$ will be deterministically colored as BLACK and GRAY, and there will be a grid of conflict if the state is $\overline{\overline{BG}}$. For a grid in BG state, which has the freedom to be in either BLACK or GRAY, it will be assigned to the nearest color along the path to minimize the number of stitches. After coloring a path, we will apply *coloring shadow* for it as well. Fig. 3 (c) shows the colored new route with updated state map. These three steps, DPL-friendly routing, *coloring path*, and *coloring shadow*, repeat for all the nets.

However, the work [3] did not handle redundant via DPL compliance. As an example, Fig. 4(a) shows the routing result for new net S-T in Fig. 3(c) plus detailed metal2 information. Suppose via C has a single *feasible* redundant candidate C1, and A1-A2 and B1-B2 are the *feasible* ones for A and B respectively. Under the existing coloring assignment, as Fig. 4(b) illustrates, A2/B2 is not *DPL-feasible* because of the $\overline{BG}$ grid locations in metal1. A1 and C1 will also be in trouble as Fig. 4(b) shows. C1 is not *DPL-feasible* due to the coloring of M2 in the new S-T path. On the other side, A1 will have conflict with the existing layout in metal2 if its configuration is *stitch-free extension*. In this case, both A and C are *DPL-dead*. If we simply flip the coloring of the entire M2 in the S-T routing path to save *DPL-feasible candidates* A1 and C1, it will make B *DPL-dead* in turn. As another solution, M2 might be split to make all vias *DPL-alive* but at a cost of an extra stitch. This example illustrates the fact that, their algorithm has difficulty producing high quality solution for both redundant via insertion and DPL.

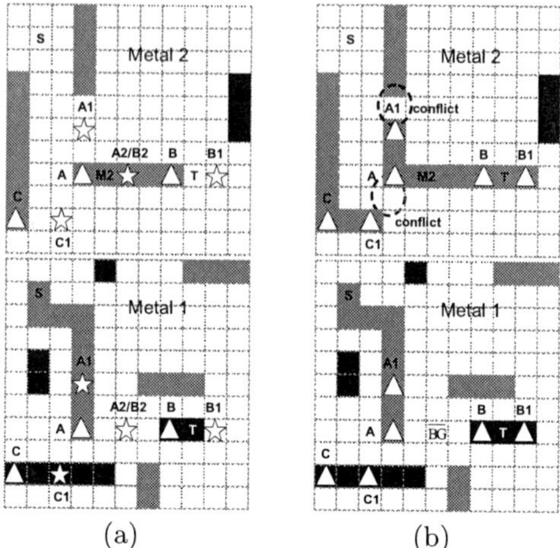

(a)   (b)

Figure 4: This example shows the limitation of the previous work [3] for handling redundant via DPL-compliance.

3. POST ROUTING DPL-AWARE REDUNDANT VIA INSERTION

In our post routing redundant via insertion, the existing layout and coloring will not be modified to honor the optimization result from routing. We do not allow any coloring conflict resulting from insertion because it is not manufacturable.

Problem Formulation 1 *post-routing DPL-aware redundant via insertion:* Without modifying the existing layout and coloring assignment, maximize the redundant via insertion rate while introducing as few as possible stitches and zero coloring conflicts.

Inspired by the work [5], we formulate an integer linear programming algorithm to perform insertion and coloring at the same time. To enable simultaneous optimization, for each *feasible* redundant via candidate of the via shown in Fig. 5 (a), we first define four types of *potential configurations* depending on the coloring of the extra metal. Fig. 5 (b)-(e) show the examples for the *feasible* candidate on the right side, where (b) is the *stitch-free extension* case while the other three all result in stitches by flipping extra metal colors in one or both layers. We disable the configuration that flips the coloring of existing routing wires to honor pre determined solution. That is to say, the configuration similar to Fig. 5 (f) will be forbidden because the coloring of existing metal2 is flipped. In our redundant via solutions, only the above four *potential configurations* are included for maintaining the uniformity of extra metal coloring. Other configurations with some stitches inserted in the middle of the extra metal can be easily extended.

Our ILP formulation is as follows and the notations can be found in Table 1.

$$maximize:$$

$$\sum_{\forall r_{ij} \in R} r_{ij} - \lambda \sum_{\forall p_{ij}^k \in P} s_{ij}^k \cdot o_{ij}^k \tag{1}$$

$$where \quad \sum_{f_{ij} \, of \, v_i} r_{ij} \le 1 \quad \forall v_i \in V \tag{2}$$

$$r_{ij} + r_{mn} \le 1 \quad \forall (f_{ij}, f_{mn}) \in EF \tag{3}$$

$$\sum_{p_{ij}^k \, of \, f_{ij}} o_{ij}^k = r_{ij} \quad \forall f_{ij} \in F \tag{4}$$

$$o_{ij}^k = 0 \quad \forall p_{ij}^k \in CP \tag{5}$$

$$o_{ij}^k + o_{mn}^l \le 1 \quad \forall (p_{ij}^k, p_{mn}^l) \in ECP \tag{6}$$

$$o_{ij}^k = 0 \quad \forall p_{ij}^k \notin B_{ij} \cup CP, \quad \forall f_{ij} \in NFP \tag{7}$$

The objective function (1) is to maximize the insertion rate while minimizing the number of stitches the *potential configuration* introduces. λ is used to tune the relative importance between stitches with respect to insertion rate.

Constraint (2) implies that at most one *feasible* candidate will be picked for each via. The *external conflict* is avoided by applying Constraint (3). Constraint (4) guarantees only one *potential*

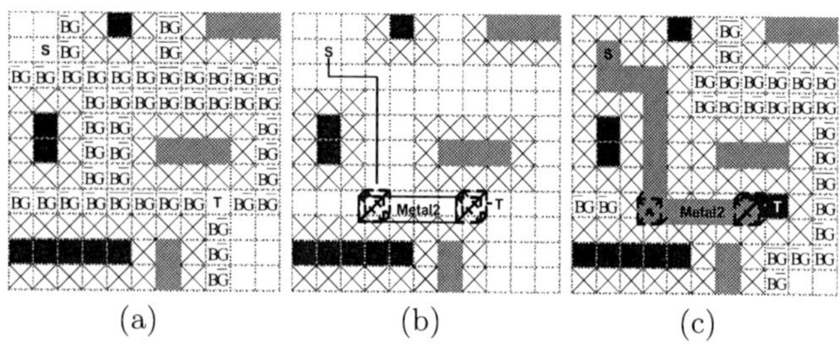

(a)   (b)   (c)

Figure 3: This example shows the main idea of [3]. The objects are layering in metal1 by default if not specially notated. The checked boxes are the blockages due to min_{sp}. Except BG, the state is shown in the grid.

978-1-60558-497-3/09 $25.00 © 2009 ACM

64

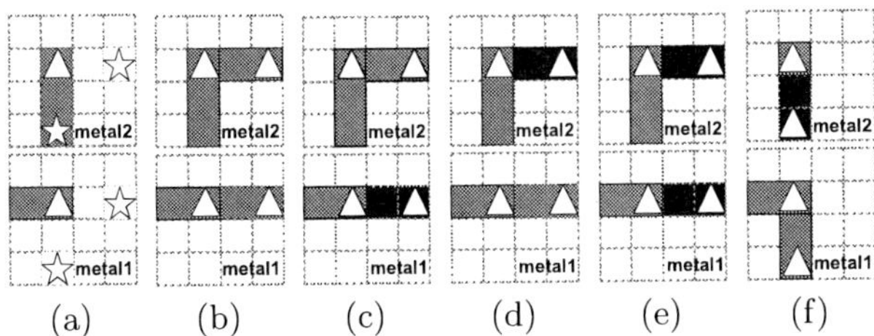

(a)   (b)   (c)   (d)   (e)   (f)

Figure 5: This example illustrates the *potential configurations* for redundant via insertion in DPL.

Table 1: Notation

v_i	the ith via with at least one feasible redundant via locations
V	the set of all v_i
f_{ij}	the jth feasible redundant via location of v_i
F	the set of all f_{ij}
r_{ij}	binary variable. $r_{ij} = 1$ if f_{ij} is chosen
R	the set of all r_{ij}
(f_{ij}, f_{mn})	f_{ij}, f_{mn} are a pair of feasible redundant candidates for different vias v_i and v_m ($i \neq m$), which can not be inserted simultaneously due to *external conflict*
EF	the set of all (f_{ij}, f_{mn})
p_{ij}^k	the kth *potential configuration* of f_{ij}
P	the set of all p_{ij}^k
s_{ij}^k	number of stitches p_{ij}^k will introduce
o_{ij}^k	binary variable. $o_{ij}^k = 1$ if p_{ij}^k is chosen
(p_{ij}^k, p_{mn}^l)	a pair of p_{ij}^k, p_{mn}^l for different vias v_i and v_m ($i \neq m$) will cause coloring conflicts to each other
ECP	the set of all (p_{ij}^k, p_{mn}^l)
CP	the set of all p_{ij}^k, which will cause coloring conflict with existing layout and coloring assignment
B_{ij}	the set of best *potential configurations* for f_{ij} which have minimum number of stitches and do not cause coloring conflict with existing layout and coloring assignment
NFP	the set of all f_{ij} whose p_{ij}^k is NOT involved in any ECP

configuration will be selected for each inserted redundant via location. Constraint (5) and (6) are used to prevent the coloring conflict due to insertion. Constraint (5) disables the *potential configuration* which will cause coloring conflict with the existing layout, and Constraints (6) is used to avoid the situation when a pair of redundant vias cause coloring conflict between each other. Constraint (7) prunes the sub optimal *potential configurations* of certain *feasible* candidates. If a candidate can not have conflict with existing layout or other redundant vias, we will not pick its *potential configurations* which do not have minimum number of stitches. We also adopt speed-up techniques similar to the work [5].

4. DPL-FRIENDLY DETAILED ROUTING WITH REDUNDANT VIA CONSIDERATION

The post-routing DPL-aware redundant via insertion may not be enough for DPL-unfriendly designs. If we only do DPL-friendly routing as [3], we might end up with a lot of *DPL-dead* vias, which do not have *conflict and stitch free* redundant via solutions. Therefore, it is in high demand to consider redundant via and DPL together during routing. Our formal problem statement is as follows.

Problem Formulation 2 *DPL-friendly Detailed Routing with Redundant Via Consideration:* Given an input netlist, perform simultaneous routing and coloring to minimize the number of *DPL-dead* vias while maintaining highly decomposable wiring path and other design objectives.

In our algorithm, we will first present *via color shadow* in Section 4.1 to penalize *DPL-dead* vias during routing. Moreover, for larger optimization space and better solution quality, we propose *equivalent transformation* in Section 4.2.

4.1 Via Color Shadow

Eliminating *DPL-dead* vias during routing is more complicated than dealing with *dead* vias as in [7]. Besides min_{sp}, the coloring

of the layout has to be taken into account as well. In consequence, we need to have different routing costs to predict and penalize the *DPL-dead* vias, according to various coloring configurations. These costs will be updated each time after we find a path and assign its coloring.

Our algorithm is presented in Algorithm 1. First of all, we should avoid hurting the *DPL-alive* vias in the existing layout when routing and coloring the new nets. We propose a penalty pair $(g.Vcost(B), g.Vcost(G))$ for each grid g in line 1-10. The value of the pair reflects the cost of coloring the grid as BLACK and GRAY respectively. The cost will be higher if more *DPL-feasible* candidates of the existing vias will be killed due to the grid state. Meanwhile in line 11-16 , we would like to avoid generating a *DPL-dead* via when routing a new net. To determine whether a potential via v is *DPL-dead*, we need to know the coloring of the grids it links in both layers. Suppose the higher and the lower grids which v links are $v.h$ and $v.l$ respectively, and their colors are $v.ch$ and $v.cl$, we introduce four costs $v.Vcost(v.ch, v.cl)$ for each possible via v in the routing graph. These are to account for the potential *DPL-dead* cases, when ch and cl can be either BLACK(B) or GRAY(G).

The proposed *DPL-dead* via avoidance costs can be penalized during rouging based on the associated grid states. It is straightforward for these grids with a state except BG. However, it is tricky to deal with BG grid because of a chicken and egg problem. BG can be assigned either BLACK or GRAY. If we do not consider its exact coloring during routing as in [3], we have no idea which one out of these proposed costs should be penalized. We can certainly apply some estimated cost for BG grid but the solution quality will be deteriorated. This motivates us to perform simultaneous coloring and routing for BG grids.

4.2 Equivalent Transformation

In the previous work [3], when a grid is BG, its exact coloring will not be considered during routing. It will be picked as either BLACK or GRAY greedily after the path is determined. This narrows the optimization space. Moreover, as discussed above, we also need detailed coloring information for BG grid on the fly to better eliminate *DPL-dead* vias. Therefore, we would like to have certain *look ahead* capability to distinguish different situations, when the BG grid is in BLACK or GRAY.

Our idea is to replace each BG grid with two equivalent grids, as Fig. 6 illustrates, which are *brothers* to each other. These two grids have the state of $B\overline{G}$ and $\overline{B}G$, and are used to track the different cases when this original BG is colored as BLACK and GRAY respectively. During routing, we will apply penalty on the equivalent grids rather than the original BG grid.

The underlying routing graph is changed accordingly as well, as shown in Fig. 6 (b). Both the new grids link to the grids the BG one connects. Because the two equivalent grids are physically same, in any found path, only one of them should be there. To ensure this, first of all, in the updated routing graph, there should be no routing edge between the equivalent grids.

Algorithm 1 via color shadow

Require: a colored path p
Ensure: *DPL-dead* via avoidance cost assignment
1: **for** each *DPL-alive via* v in p **do**
2:      **for** each *DPL-feasible* candidate rvc of v **do**
3:          **for** each $g \in Gt$, which Gt denotes all the available routing grids within min_{dp} from rvc **do**
4:              **if** g is $BLACK$, it will have conflict with the *stitch-free extension* of rvc **then**
5:                  $g.Vcost(B) \mathrel{+}= f_1$
6:              **end if**
7:              Similar when g is $GRAY$
8:          **end for**
9:      **end for**
10: **end for**
11: **for** each $v \in V$, which V denotes all the available via locations within min_{dp} from p **do**
12:      **if** v will be *DPL-dead* when the coloring $(v.ch, v.cl)$ of its up and down grids is $(BLACK, BLACK)$ **then**
13:          $v.Vcost(B, B) \mathrel{+}= C_2$
14:      **end if**
15:      Similar when $(v.ch, v.cl)$ has other coloring configurations
16: **end for**

978-1-60558-497-3/09 $25.00 © 2009 ACM

65

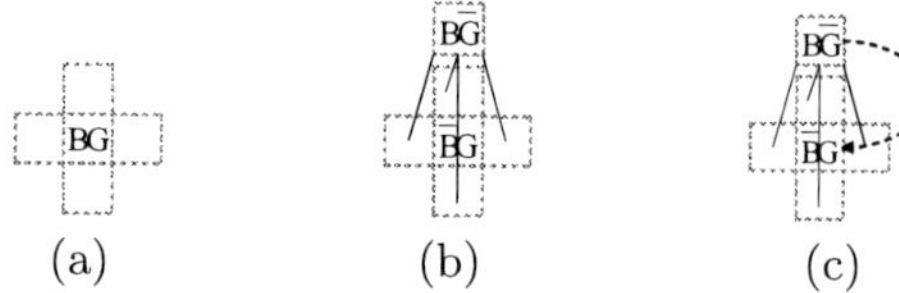

(a)   (b)   (c)

Figure 6: This example illustrates equivalent transformation. The solid line means there is a routing edge connecting the two grids.

However, there will still be an *illegal loop problem* coming from above *equivalent transformation*. During the path propagation we may still generate a route linking the equivalent grids through a set of other grids as Fig. 6 (c). This configuration is illegal because the selection of $B\overline{G}$ and $\overline{B}G$ is exclusive as we just discussed. Our solution for this problem is simple but effective. Whenever we would like to update a cost of an equivalent grid during routing, we will trace back to see whether its ancestors contain the *brother* grid. If so, it returns true and this updating request will not be executed.

5. EXPERIMENTAL RESULT

We implement our algorithm in C++ and test four scaled benchmarks on Intel Core 3.0GHz Linux machine with 32G RAM. Glpk [8] is applied as the solver for integer linear programming. Our *post-routing DPL-aware insertion* is denoted as **POSTDPL**, and λ is set as 0.4. On the other side, our *DPL-friendly routing with redundant via consideration* is referred as **DPRRV**. The parameters α, β and γ are set as the same as the previous work [3]: α and γ are 9 and 6 respectively while $\beta >> 10$. We set both C_1 and C_2 100.

For comparative reason, we also implement a DPL-aware optimization flow without considering redundant via compliance **DPR+POST**. In **DPR**, *DPL-friendly routing* [3] is first performed. For fair comparison, we also apply the techniques in [7] to reduce the number of *dead* vias. The reason is that in our routing algorithm **DPRRV**, *via cost shadow* can eliminate *dead* vias as a side effect. The *DPL-dead* via it primarily removes is a super set of *dead* via. In the following post-routing via insertion **POST**, we first apply [5] to maximize the redundant via insertion rate. For each inserted one, we will then try to select the best *potential configuration* to eliminate the resulting coloring conflict with minimum stitches introduced. After this, if there are still coloring conflicts due to insertion, we will remove the corresponding redundant vias.

Table 2-4 presents our experiment results. For all the tables, "WL" is the total wirelength of metal1 and metal2 with the unit *um*. "CPU" is the total runtime by second. "via" and "DPDV" are the number of vias and *DPL-dead* vias respectively, and "DPDV%" is the ratio of "DPDV' over "via". "ST(v)" is the number of additional stitches caused by post-routing via insertion. "ST(t)" is the total number of stitches in the final layout, including both routing wires and redundant vias. "rv" shows the number of inserted redundant vias without causing any coloring conflict. "rv%" is the insertion rate, which is the ratio of "rv" over "via". The "total" row shows the sum or average for all four test cases. "ratio" row is calculated with respect to corresponding "total" item in Table 2. All of our experiments achieve 100% routability and zero coloring conflicts.

Table 2 shows the result of DPL-aware optimization flow without considering redundant via compliance, **DPR+POST**. **DPR** only focuses on obtaining zero coloring conflicts and minimizing the number of stitches for routing wires. However, as we find out, it produces many *DPL-dead* vias, i.e, there are a large portion of vias, averagely 6.9%, which do not have *conflict and stitch free* redundant via solutions. Therefore, if we simply apply **POST** to insert redundant vias, our insertion rate is only 69.4% averagely when no resulting coloring conflict is allowed. These experimental results demonstrate the strong demand for considering redundant via insertion and DPL together.

To find better DPL-friendly redundant via solution, we first replace **POST** by our approach **POSTDPL** in post-routing phase after applying **DPR**. This is denoted as **DPR+ POSTDPL** and the results are shown in Table 3. By performing simultaneous insertion and coloring, **POSTDPL** can increase the insertion rate to 94.1% averagely without causing any coloring conflict. Compared to **DPR+POST**, although we introduce a few more stitches due to insertion, the number is relative small compared

Table 2: DPR+POST: The DPL flow without considering redundant via

ckt	WL	CPU	via	DPDV	DPDV%	ST(v)	ST(t)	rv	rv%
C1	0.67	12	313	26	8.3	4	41	209	66.8
C2	1.67	64	757	74	9.8	3	86	519	68.6
C3	4.06	184	1004	64	6.3	13	293	696	69.3
C4	8.95	680	2062	123	6.0	21	565	1447	70.2
total	15.35	940	4136	287	6.9	41	985	2871	69.4
ratio	1	1	1	1	1	1	1	1	1

to the total number of stitches in the final layout. Moreover, the runtime overhead is very little. On the other side, because we also apply **DPR** without redundant via DPL-compliance in this experiment, a significant number of *DPL-dead* vias are still there. This implies a noticeable improvement space. It is highly desirable to eliminate these unfriendly vias from design side.

Table 3: DPR+POSTDPL: Result for post-routing DPL-awareness insertion

ckt	WL	CPU	via	DPDV	DPDV%	ST(v)	ST(t)	rv	rv%
C1	0.67	12	313	26	8.3	3	40	288	92.0
C2	1.67	66	757	74	9.8	9	92	691	91.3
C3	4.06	188	1004	64	6.3	19	299	950	94.6
C4	8.95	683	2062	123	6.0	29	573	1962	95.2
total	15.35	949	4136	287	6.9	60	1004	3891	94.1
ratio	1	1.01	1	1	1	1.46	1.02	1.36	1.36

To eliminate those vias which do not have *DPL-feasible* redundant via solution during design, we will further replace **DPR** by **DPRRV** before applying **POSTDPL**. This is referred as **DPRRV+POSTDPL**. As expected, our approach achieves averagely 89% less *DPL-dead* vias. The insertion rate is improved to 99.3% averagely, with zero conflicts and 73% less stitches introduced. Moreover, we are able to reduce the number of total stitches in final layout by 17%, including both routing wires and redundant vias. The reason is that by doing *equivalent transformation*, the coloring of BG grid is planned during our detailed routing. It has higher possibility to conduct global optimization. Because of the same reason, we are also able to reduce the number of vias by 9%, while **DPR** has to use more vias to resolve potential coloring conflicts. As the last mention, all these improvements in our approach are obtained with little overhead on wirelength and runtime.

Table 4: DPRRV+POSTDPL: Result for DPL-friendly Post-Routing and detailed routing with redundant via consideration

ckt	WL	CPU	via	DPDV	DPDV%	ST(v)	ST(t)	rv	rv%
C1	0.67	17	283	6	2.1	2	34	279	98.6
C2	1.68	90	654	13	2.0	4	61	640	97.9
C3	4.10	277	930	3	0.3	2	240	928	99.8
C4	9.03	750	1915	9	0.5	3	482	1908	99.6
total	15.48	1134	3782	31	0.8	11	817	3755	99.3
ratio	1.008	1.21	0.91	0.11	0.12	0.27	0.83	NA	1.43

6. CONCLUSION

In this paper, we have developed two algorithms to take into account redundant via DPL-compliance in post-routing and during-routing stages respectively. Experimental results are very promising.

7. REFERENCES

[1] Andrew B. Kahng et al. Layout decomposition for double patterning lithography. In *Proc. of ICCAD*, November 2008.

[2] Kun Yuan et al. Double patterning layout decomposition for simultaneous conflict and stitch minimization. In *Proc. of ISPD*, March 2009.

[3] Minsik Cho et al. Double patterning technology friendly detailed routing. In *Proc. of ICCAD*, November 2008.

[4] Huang-Yu Chen et al. Novel Full-Chip Gridless Routing Considering Double-Via Insertion. In *Proc. of DAC*, Jul 2006.

[5] Kuang-Yao Lee et al. Optimal post-routing redundant via insertion. In *Proc. of ISPD*, April 2008.

[6] Mircea Dusa et al. Pitch doubling through dual-patterning lithography challenges in integration and litho budgets. In *Proc. of SPIE*, March 2007.

[7] G. Xu et al. Redundant-Via Enhanced Maze Routing for Yield Improvement. In *Proc. Asia and South Pacific Design Automation Conf.*, Jan 2005.

[8] http://www.gnu.org/software/glpk/glpk.html/.

978-1-60558-497-3/09 $25.00 © 2009 ACM

Use of Lithography Simulation for the Calibration of Equation-Based Design Rule Checks

David Abercrombie	Fedor Pikus	Cosmin Cazan
Mentor Graphics Corp.	Mentor Graphics Corp.	Mentor Graphics Corp.
8005 SW Boeckman Rd.	8005 SW Boeckman Rd.	8005 SW Boeckman Rd.
Wilsonville, OR 97070	Wilsonville, OR 97070	Wilsonville, OR 97070
1-503-685-1446	1-503-685-4857	1-503-685-0307
david_abercrombie@mentor.com	fedor_pikus@mentor.com	cosmin_cazan@mentor.com

ABSTRACT

Designers using one-dimension measurements in nanometer designs can't readily identify features prone to excessive variation during processing. Process simulation provides high resolution checking, but requires significant computing resources. Equation-based design rule checking (eqDRC) offers the DRC performance with capture of complex process issues using multi-dimensional equations. One challenge to adopting eqDRC is the definition and calibration of the equations. We will show how a lithographic simulator can be used to define and calibrate an eqDRC equation.

Categories and Subject Descriptors

B.7.2 **[Hardware : Integrated Circuits : Design Aids]**

General Terms

Measurement, Design, Verification.

Keywords

DFM, DRC, verification, equation based DRC, Leff, LFD, lithography, manufacturability

1. INTRODUCTION

In modern semiconductor processing, the patterns imaged on the wafer do not match the drawn shapes in the layout. Rectangular shapes in the design database are significantly distorted during lithography and etch processing. One such distortion is often called "corner rounding," because sharp corners in the rectangular shapes of the design database become rounded on the wafer. For designs in nanometer process nodes, this phenomenon becomes even more problematic as the layout feature size becomes smaller than the wavelength of the light used for lithography [1].

In this paper, the effect of polysilicon corner rounding on the gate length (L) of a transistor gate (Figure 1) is examined.

Permission to make digital or hard copies of part or all of this work for personal or classroom use is granted without fee provided that copies are not made or distributed for profit or commercial advantage and that copies bear this notice and the full citation on the first page. To copy otherwise, to republish, to post on servers or to redistribute to lists, requires prior specific permission and/or a fee.
DAC'09, July 26-31, 2009, San Francisco, California, USA

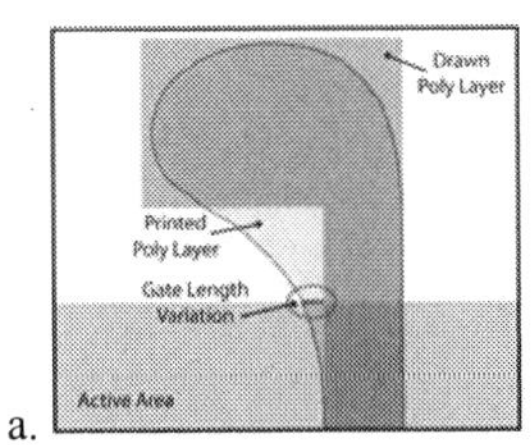
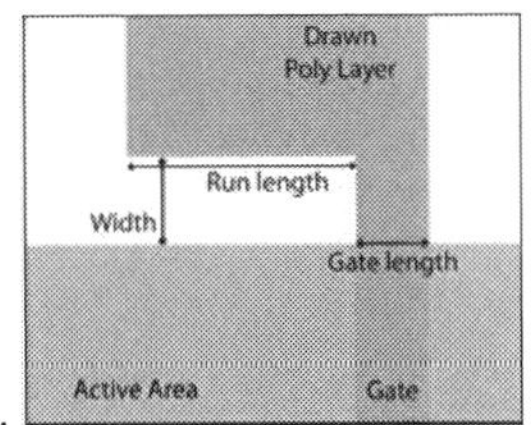

Figure 1: a) Gate length variation due to corner rounding, b) Characteristics of cornered gates

Since altering the gate length of a transistor introduces undesirable parametric effects, eliminating gate length variation will improve design reliability and performance. Many methods have been proposed using LFD techniques to analyze parametric effects caused by lithography effects [1]. In this paper, we will use lithographic corner rounding of polysilicon as an example of a complex multi-dimensional process issue, and show how a lithography simulator, such as Calibre® LFD, can be used to derive and calibrate equation-based design rules. Using Calibre LFD to simulate lithography effects on a systematically varied set of simple layout test structures will allow us to 1) extract variation measurements, 2) derive empirical fitting equations to approximate the effect, and 3) use both the equation-based DRC and Calibre LFD on a real layout to compare the equation-based approach to the simulation results. Equation-based DRC will enhance and complement the use of LFD software by allowing the designer to run equation-based rule checks similar to traditional DRC. The designer would be able to run these checks quickly throughout the design process, as necessary. When further stages of the design process are reached, equation-based DRC can be used to identify the areas where lithography simulations may be appropriate or necessary.

2. PROBLEM FORMULATION

The characteristics of the polysilicon rounding depend on several factors that are traditionally modeled using highly accurate, but compute-intensive, lithography simulators. Although every design shape within an optical radius of a given shape has an influence on the final printed shape, for this particular issue, we hypothesized that the two most important factors that are directly proportional to the induced gate length variation by corner rounding are the width between the polysilicon corner and active area, and the run length of the polysilicon layer corner (Figure 1). We suggest that, given these two parameters, the gate length variation due to the curvature of the printed polysilicon layer can be closely approximated with an equation-based design rule that estimates the printed results.

In this experiment, we will use equation-based DRC (eqDRC) to implement an equation-based approximation of this effect as a high-speed alternative to full lithography simulation. We will use Calibre LFD for both the calibration and validation processes of the eqDRC formulas in this experiment. Once the equation-based rule is calibrated, it can be used when a first-order approximation is useful at full chip level, or to narrow down the most sensitive locations for detailed simulation.

3. EQUATION DERIVATION

By understanding the behavior of the L-shaped polysilicon bend in terms of the **width** and **run-length** (as previously defined), we should be able to predict the gate variation incurred, and, potentially, the electrical effects suffered by the transistor. Moreover, we should be able to predict the minimum width and run-length at which the curvature ceases to affect the effective length of the gate. The process of creating and calibrating this equation-based rule is outlined below.

1. Create varied test structures and simulate printed results with lithography simulator

2. Analyze the results of simulation to understand correlation between parameters (width and run-length) and induced gate length variation

3. Derive an equation model to fit the observed behavior

3.1 Test Structure Simulation

By examining a 45 nm design, it was found that polysilicon shapes incurring corner rounding could be generally divided into L-shapes and U-shapes. To create a correlation between the two parameters and the gate length variation, numerous test structures were created with a run-length between 140nm and 320 nm for U-shapes and between 10nm and 100nm for L-shapes, and width values between 30nm and 100nm for all shapes.

We created a series of test structures to measure the gate length variation for different width values at different run-length values using Calibre LFD simulation with a 45nm LFD kit provided by the foundry. Figure 2 shows how the test structure matrix for L-shapes was drawn in the test layout Figure 2 shows how the test structure matrix for L-shapes was drawn in the test layout.

Figure 2: Polysilicon test structure
(gray is active area; orange is the polysilicon layer)

To obtain the gate length variation incurred by each test structure, the test layout was simulated with a 45 nm Calibre LFD Kit provided by the foundry. The simulated structures were analyzed to determine how much the channel gate length changed, and the numerical results were extracted.

3.2 Correlation Analysis

To fit an equation-based rule, we first needed to understand the qualitative effect of the width and run-length on the gate length variation. Having extracted the results from the LFD simulation of the test structures, we analyzed the correlation between width, run-length, and gate length variation to achieve our goal: mathematically approximating the gate length variation using only two-dimensional parameters. The correlation analysis was done individually on each shape type, since the observed behavior for each shape was significantly different.

For L-shape structures, as seen in Figure 3, a 3D correlation graph revealed that the relationship between gate length variation and width was non-linear, and as the space between the polysilicon corner and the active area increased, the gate length variation decreased. The correlation between the run-length and gate length variation was more complex. Qualitative analysis of the correlation graph revealed that, as the run-length of the polysilicon corner increased, the gate length variation also increased in a seemingly linear fashion, but only up to a point, after which the run-length value dipped slightly, then seemed to stop affecting the gate length variation. The peak variation, found at 95 nm, was due to OPC fragmentation points at that same dimension.

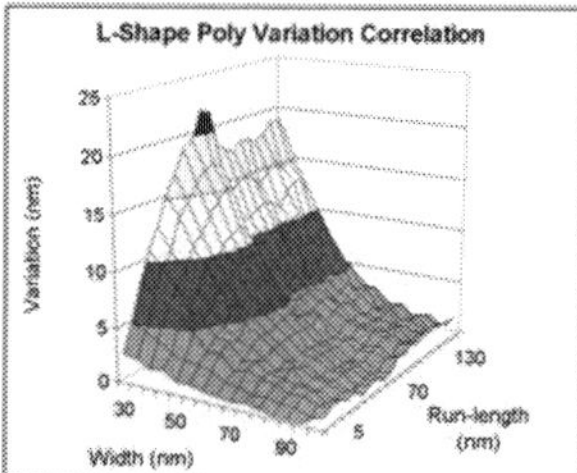

Figure 3: L-shape correlation graph

Since only two values of run-length were considered for U-shape structures, the correlation analysis was only concerned with the relationship between the width parameter and gate length variation. Qualitative analysis for both values of run-length showed that, as the width gets larger, the gate length variation gets exponentially smaller. The difference between the two run-lengths is fairly small, but the larger run-length (320 nm) appears to produce a slightly larger gate length variation, especially at small widths.

3.3 Derive Equation

Once the general correlation between the parameters and gate length variation was established, the next step was establishing the correct function types for replicating the observed qualitative behavior. Since our equation was to be based purely on empirical data, a least squares approach was taken. This approach involves determining the "best fit" line (Figure 4) where the error is the sum of the squares of the differences between the values on the approximating line and the empirical values [2].

Letting $F(x_i)$ denote the ith predicted value and v_i the ith empirical value requires that $F(x_i)$ equation constants be found that minimize the quantity:

$$\sum_{i=0}^{n} \left[v_i - F(x_i) \right]^2$$

For the U-shape collected data, we obtained good results using exponential fit of the form:

$$GLV = Ae^{(B*x)} \qquad \text{(where x = width)}$$
$$GLV = \text{gate length variation}$$

The equation below was minimized and equation constants found.

$$\sum_{i=0}^{n}\left[v_i - Ae^{(B*x_i)}\right]^2$$

$$GLV_{140\,nm} = 73.44\,e^{(-0.046*x)} \qquad \text{(where x = width)}$$
$$GLV_{320\,nm} = 81.588\,e^{(-0.0482*x)} \qquad \text{(where x = width)}$$

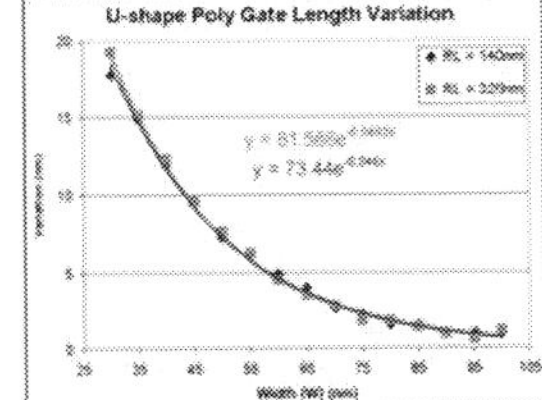

Figure 4: U-shape polysilicon equation model

For the L-shape data, in terms of the width, a 2nd order polynomial equation was found to be the best fit (Figure 5). In terms of the run-length, as generalized in Figure 5, the data set seems to be best described by two separate equations (Figure 6). Fitting the data revealed that the best-fit was two linear equations.

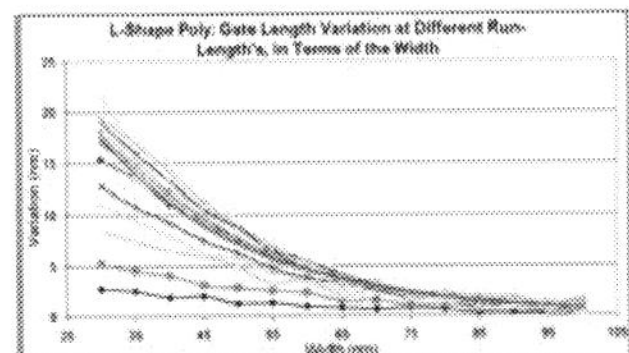

Figure 5: L-shape polysilicon width correlation

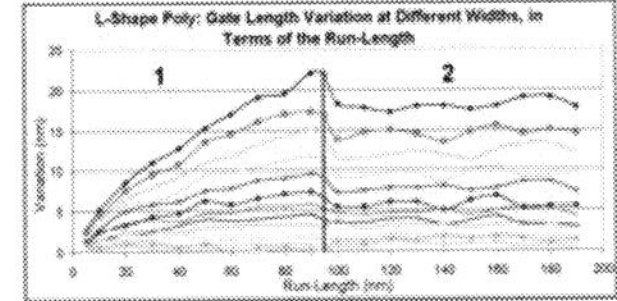

Figure 6: L-shape polysilicon run-length correlation

Thus, the combined model was an equation in terms of both the width and run-length values. To approximate the least-squares fit, the modeling equation was adjusted to:

$$\sum_{i=0}^{n}\left[v_i - F(x_i,y_i)\right]^2$$

(where x = run length and y = width)
$$F(x_i,y_i) = (Ax_i + B)\times(Cy_i^2 + Dy_i + E)$$

The equation was solved for the smallest possible quantity and the finalized equations were defined as follows:

L-shape for range width = 0 – 95 nm:
$$F(x_i,y_i) = (0.12x_i + 3.07)\times(0.00037\,y_i^2 - 0.068\,y_i + 3.22)$$

L-shape for range run-length > 95 nm:
$$F(x_i,y_i) = (0.004x_i + 5.91)\times(0.00075\,y_i^2 - 0.13\,y_i + 5.97)$$

4. EXPERIMENTAL RESULTS

To evaluate the accuracy of our empirically-based equation rule, an equation-based DRC deck was configured to find each cornered gate in a 45 nm production design, measure the run-lengths and widths of each corner, and apply the correct formulas. Each gate length variation predicted was compared to the nominal value from the LFD simulation. In Figure 7, the "Equality Line" represents the points on the plot at which the **LFD simulated** and **equation-predicted** values are equal to each other. In a perfect fit, all data points would be positioned on the "Equality Line".

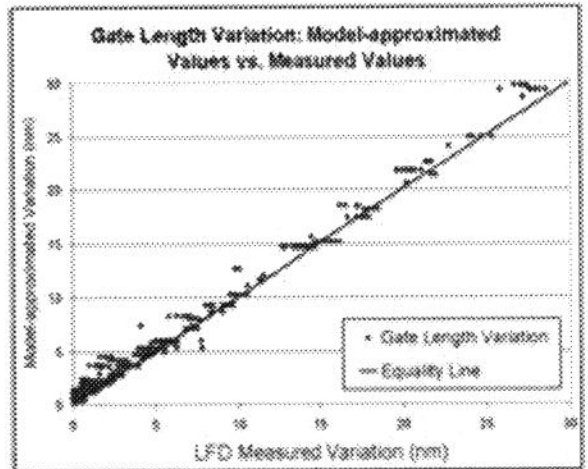

Figure 7: Predicted equation values vs. LFD-measured values

As observed, the calculated values are highly accurate in predicting the first order effects of corner rounding. In the plot, horizontal data sets represent gate length variations that had the same width and run-length, but where other higher order effects produced further gate length variation. Figure 8 shows that the majority of the residual error is smaller than 1nm, with the entire range fitting between –2.5 nm and 4 nm.

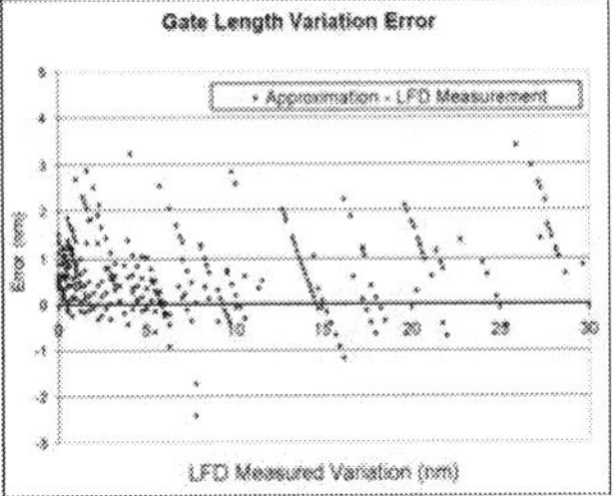

Figure 8: Gate length variation model error

Equation-based DRC can be used in DRC decks to quickly identify and display information about possible issues in the design, as shown in Figure 9.

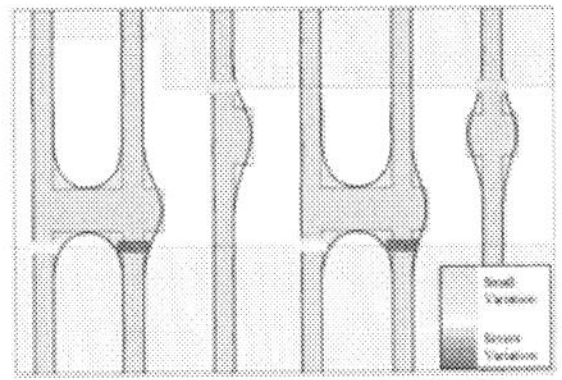

Figure 9: Using the equation-based model to identify design issues

This experiment has shown the process by which an equation-based rule can be derived to approximate the effective gate length increase of a transistor gate due to L-poly corner rounding. Furthermore, it has also shown how this approximation can be used to identify areas that are most sensitive to this phenomenon, or to approximate the first order effect at the full-chip level when extremely accurate simulation is not necessary. It is important to note, however, that this concept can be extended and applied for many other purposes.

978-1-60558-497-3/09 $25.00 © 2009 ACM

To show the usefulness of a numerically-based approximation for the designer, a rule was devised to determine the dimensions at which it would be safe to draw U-shape gates. Since the design was limited to U-shape gates with run-lengths of 140 nm and 320 nm, the only change a designer could make to minimize gate length variation was in the width parameter of the gate structure. Given the equations developed for approximating the gate length variation of U-shape structures, it is possible to inversely solve for the desired width at which the gate length variation is negligible.

$$GLV = Ae^{(B*x)} \Rightarrow x = \frac{\left(\ln\dfrac{GLV}{A}\right)}{B}$$

(where x = width)

Figure 10 shows an example of the types of information and fix suggestions that can be generated using an equation-based rule. The rule check displays the gate length variation incurred by this particular gate, along with how much variation occurred on each side of the gate. It also displays a target maximum gate length variation considered to be acceptable (2 nm), and displays the value at which the width between the polysilicon corner and active area needs to be set, such that we achieve this minimal gate length variation.

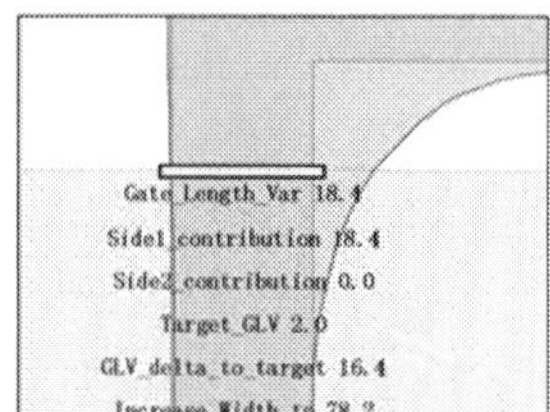

Figure 10: Corrections suggested by equation-based rule check (all units in nm)

Defining a negligible gate length variation to be 2 nm, and solving for the desired width, we get:

- Desired width for 140 nm U-shape gate = 78.2 nm
- Desired width for 320 nm U-shape gate = 76.9 nm

This simple example demonstrates how an equation-based rule can be used to offer quick solutions to designers during the design process. Rather than relying on a compute-intensive LFD simulation, the designer can run the equation-based checks in a matter of minutes, and the equation-based rule check can provide accurate and straightforward fix suggestions.

5. POTENTIAL DESIGN FLOW

In the previous sections, we have shown how calibrating a simple equation-based rule can yield fairly accurate results. In this section, we expand on how this can be directly useful to the designer for drawing designs that can be manufactured, while limiting the design time necessary for this process. Figure 11 shows how design optimization can largely be achieved by using equation-based rules.

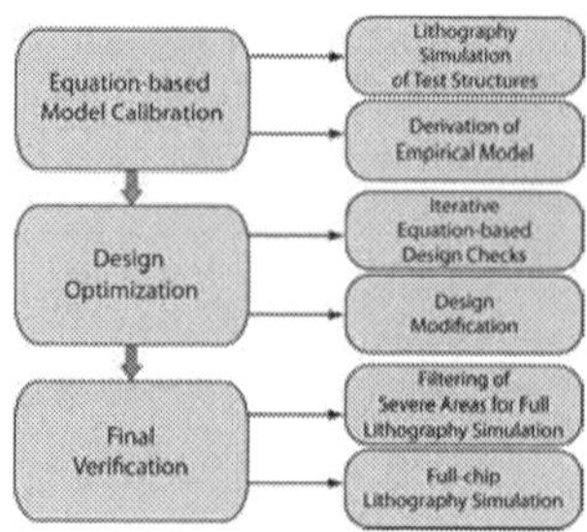

Figure 11: Potential design flow diagram

For iterative checks during all stages of design, the equation-based check approach offers an effective and time-saving alternative. The design used to validate the equation rule in this experiment was found to run 300 times faster using equation-based DRC compared to a lithography simulation run. Given the relative accuracy of the equation-based rule approach, it can be safely used to filter out more severe design areas where full lithography simulation is required. Figure 12 illustrates how design areas can be targeted by variation severity.

Figure 12: Design area by severity (red denotes most severe rule deviations)

In the final verification stage, full-lithography simulation is still necessary for accurate validation of the design.

6. CONCLUSIONS

In this paper, we have shown how lithography simulation can be used for the calibration and validation of equation-based design rules. We have shown how such rules can help designers improve design practices to eliminate manufacturability issues during the design process with a high level of accuracy, compared to the golden simulation sign-off. This experiment shows how these simple and time-efficient equation-based rules can even be constructed using the data provided by simulation tools and models provided by the foundries, providing valuable information and actionable feedback to the designer.

7. REFERENCES

[1] R. S. Fathy, M. Al-Imam, et al. "Litho Aware Method for Circuit Timing/Power Analysis through Process". Proc. SPIE 6521, 65210O (2007)

[2] Burden, R. L., Faires D. and Reynolds A. C. (1981): Numerical Analysis, (2nd ed.), PWS Publishers. Boston, Massachusetts.

Carbon Nanotube Circuits in the Presence of Carbon Nanotube Density Variations

Jie Zhang   Nishant Patil   Arash Hazeghi   Subhasish Mitra

Departments of Electrical Engineering and Computer Science, Stanford University, Stanford, CA

Abstract

Carbon Nanotubes (CNTs) are grown using chemical synthesis. As a result, it is extremely difficult to ensure exact positioning and uniform density of CNTs. Density variations in CNT growth can compromise reliability of Carbon Nanotube Field Effect Transistor (CNFET) circuits, and result in increased delay variations. A parameterized model for CNT density variations is presented based on experimental data extracted from aligned CNT growth. This model is used to quantify the impact of such variations on design metrics such as noise margin and delay variations of CNFET circuits. Finally, we analyze correlation that exists in aligned CNT growth, and demonstrate how the reliability of CNFET circuits can be significantly improved by taking advantage of such correlation.

Categories and Subject Descriptors

B.7 [Hardware]: Integrated Circuits

General Terms

Design, Performance, Reliability

Keywords

Carbon Nanotube, CNT, CNT Density Variation, CNT Correlation.

1. Introduction

Carbon Nanotube Field Effect Transistors (CNFETs) show promise as extensions to Silicon CMOS. Ideal CNFET circuits show 20X Energy-Delay Product (EDP) advantage over 16 nm Silicon CMOS [Patil 09]. This analysis assumes that CNFETs consist of multiple aligned semiconducting Carbon Nanotubes (CNTs) with a uniform density of 250 CNTs/μm [Deng 07a]. However, current CNT synthesis processes are far from being perfect:

1) A third of grown CNTs are metallic [Kang 07], creating source-drain shorts in CNFETs causing excessive leakage and reduced noise margins in CNFET circuits. Hence, metallic CNTs (m-CNTs) must be removed [Zhang 06, Collins 01]. However, current m-CNT removal techniques are not perfect as they do not remove all m-CNTs and also inadvertently remove some s-CNTs.

2) Although CNT growth on quartz yields a large fraction (> 99%) of aligned CNTs [Kang 07, Patil 08a], there exists a non-negligible fraction of mis-positioned CNTs which may cause incorrect logic functionality [Patil 08b]. Layout design principles described in [Patil 08b, Bobba 09] can enable CNFET circuits immune to such mis-positioned CNTs.

3) The average density of CNTs obtained today is between 10-50 CNTs/μm [Kocabas 07]. Advances in CNT synthesis are essential to improve the average density from this value to the required density of 250 CNTs/μm. However, mere increase in average CNT density is not enough. CNT density variations, resulting from the lack of precise control of CNT location during synthesis, also pose serious challenges to circuit design using CNFETs. CNFETs fabricated using CNTs of density variations not only have large variations in their performance, but also have a significant probability of complete failure in cases when there is no CNT present in the CNFET. The presence of m-

CNTs can introduce additional variations in performance (delay) and increased probability of failure.

In this paper, we characterize CNT density variations using Scanning Electron Microscopy (SEM) and Atomic Force Microscopy (AFM) images (e.g., Fig. 1.1) of aligned CNTs grown on quartz. The impact of CNT density variations on CNFET circuits is analyzed. Furthermore, since grown CNTs are aligned over long distances (>100 μm), correlation exists between CNT counts of CNFETs at different locations. We show that such correlation in CNT count can be effectively utilized to design CNFET circuits with significantly improved noise margin characteristics and reduced delay variations.

The key contributions of this paper are:

1) Quantitative analysis of the impact of non-uniform CNT density distribution on CNFET circuit performance.

2) Characterization of CNT density distributions from SEM and AFM images of aligned CNT growth samples.

3) A parameterized model based on renewal theory [Cox 62] for CNT density variations fitted to experimental data.

4) Special layout design guidelines for CNFET-based circuits in the presence of CNT density variations by utilizing correlation.

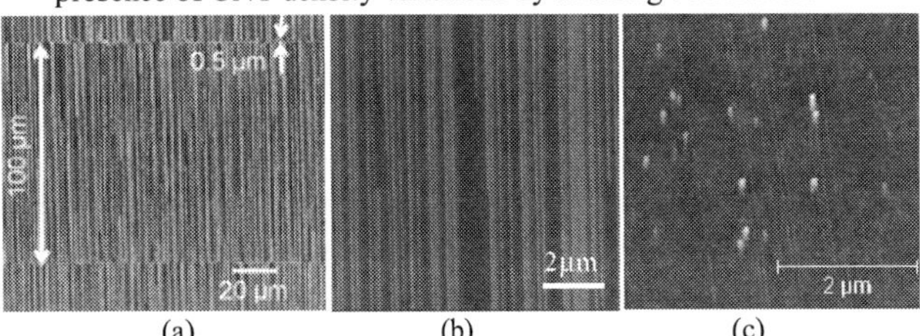

Figure 1.1. (a) SEM image of aligned CNTs on quartz with catalyst stripes of size 0.5 μm. SEM (b) and AFM (c) images of aligned CNTs between catalyst stripes.

2. Characterization and Modeling of CNT Density Variations

In order to characterize CNT density variations, we analyzed SEM and AFM images (Fig. 1.1) of aligned CNTs grown on single-crystal quartz wafers. Although different techniques exist for CNT synthesis [Dai 02], aligned CNT growth on quartz has been considered preferable for electronic device applications [Kang 07]. In our experiments, aligned CNTs were grown on single-crystal quartz wafers using Fe nanoparticles as the catalyst patterned at predefined stripes using lithography as shown in Fig. 1.1a [Kang 07, Patil 08a]. We performed image processing on the SEM and AFM images (e.g. Fig. 1.1b and 1.1c) to extract the positions of ~1,500 CNTs. Furthermore, a parameterized analytical model for CNT count distribution is derived in Sec. 2.2 and 2.3, and fitted to the experimental data. We assume that the CNT density distribution does not vary across the CNT sample.

2.1. CNT Count and Spacing Distributions

For the purpose of CNT count modeling, we represent a CNFET (Fig. 2.1b) as a rectangular box with width W and length L^1, with CNTs aligned along the L direction (Fig. 2.1a). *CNT count* is defined as the number of CNTs that completely bridge the upper and the lower sides of the box (and is, hence, equal to the number of CNTs that connect source and drain contacts in the corresponding CNFET). We denote CNT count by $N(W, L)$ since it is a function of width (W) and length (L). $N(W, L)$ is a random variable that varies depending on the location of the box (or equivalently, the CNFET).

Permission to make digital or hard copies of part or all of this work for personal or classroom use is granted without fee provided that copies are not made or distributed for profit or commercial advantage and that copies bear this notice and the full citation on the first page. To copy otherwise, to republish, to post on servers or to redistribute to lists, requires prior specific permission and/or a fee.
DAC'09, July 26-31, 2009, San Francisco, California, USA

---

¹ The length (L) of the box is equal to sum of CNT channel length and the lengths of the source and drain doped CNTs regions.

To simplify the analysis, we consider the 1-D limit of the function $N(W, L)$ as L approaches 0:

$$N(W) = \lim_{L \to 0} N(W, L) \qquad (2.1)$$

Such simplification is acceptable because, for relatively aligned CNT growth and small channel-length[2] (Length(CNT) >> L) CNFETs, the difference between the number of CNTs passing through the upper side of the box and the number of CNTs passing through the lower side of the box is negligible.

CNT spacing, denoted by S, is defined as the distance between neighboring CNTs measured perpendicular to the direction of L as L approaches 0. Figure 2.2a shows the CNT spacing distribution extracted from SEM images of CNTs (e.g., Fig. 1.1) by considering spacing between all possible pairs of CNTs in the images.

To study the possible correlation between CNT spacings inside a box or CNFET, we number the CNTs in the box from 1 to N. The CNT immediately outside the box to the left is numbered 0. Let S_j ($j = 1, 2, 3...$) be the spacing random variables from the $(j-1)^{th}$ CNT to the jth CNT (Fig. 2.2b). Because the exact location of the box is arbitrary, all random variables S_j ($j = 1, 2, 3...$) follow the same distribution as shown in Fig. 2.2a but can potentially be correlated with each other. However, both correlation coefficient calculation and chi-squared independence test [Ross 05] performed on experimentally extracted data show that these random variables are in fact highly independent.

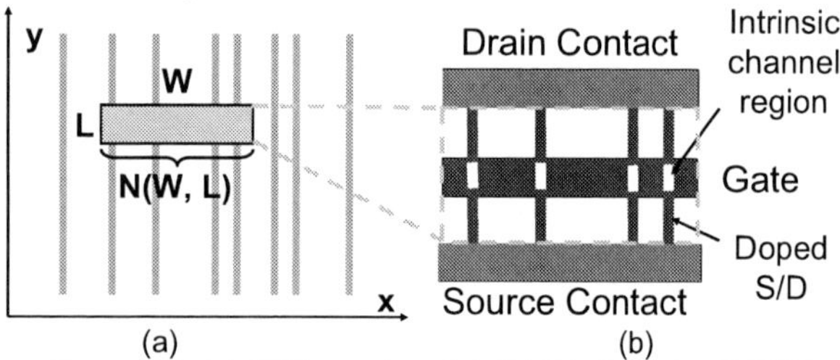

(a)   (b)

Figure 2.1. (a) CNT count model. (b) 2D view of a CNFET.

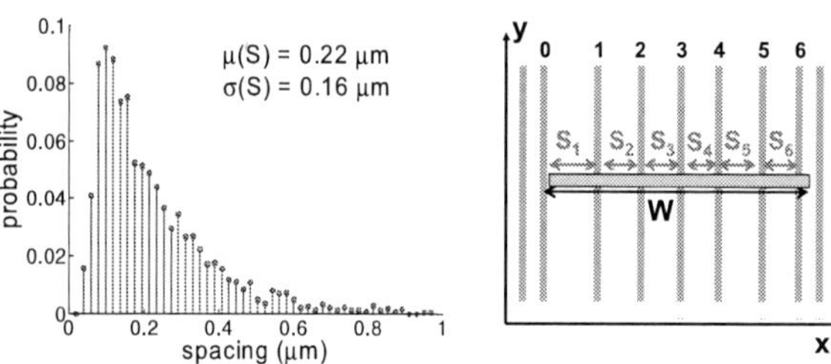

Figure 2.2. (a) CNT spacing distribution. (b) Model for correlation between CNT spacing

2.2. CNT Count Distribution from Spacing Distribution

The distribution of CNT count, $N(W)$, is important for characterizing the statistical performance of CNFETs because it decides the drive current of a CNFET. However, extracting the distribution of $N(W)$ experimentally can be time-consuming since this distribution is dependent on W. Spacing distribution, on the other hand, is width independent and easy to extract experimentally. Renewal theory [Cox 62, Cameron 98] can be applied to derive the distribution of $N(W)$ from CNT spacing distribution.

Based on the results from Sec. 2.1, we assume that CNT spacing, S_j ($j = 1, 2, 3 ...$) is identical and independently distributed (i.i.d), with

probability density function (pdf) $f_S(s)$, cumulative distribution function (cdf) $F_S(s)$, mean μ_S and standard deviation σ_S.

Let SS_n denote the sum of n successive CNT spacings:

$$SS_n = \sum_{j=1}^{n} S_j \qquad (2.2)$$

The pdf of SS_n can be extracted experimentally or derived by convolving (denoted by the "$*$" sign) the individual distributions:

$$f_{SS_n}(w) = f_{S1}(w) * f_{S2}(w) * ... * f_{Sn}(w) \qquad (2.3)$$

where $f_{S1}(s) = f_{S2}(s) = ... = f_{Sn}(s) = f_S(s)$.

First, we will consider the case when the left side of the box is to the right of CNT 0 by an infinitesimal amount (Fig. 2.2). In that case,

$$\text{Prob}\{N(W) <= n\} = \text{Prob}\{SS_{n+1} > W\} = 1 - F_{SS_{n+1}}(W) \qquad (2.4)$$

where Prob(I) represents the probability of event I, and $F_{SSn}(W)$ the cumulative distribution of SS_n. From (2.4), the probability distribution of $N(W)$ is

$$\text{Prob}\{N(W) = n\} = F_{SS_n}(W) - F_{SS_{n+1}}(W) \qquad (2.5)$$

$F_{SS_0}(W)$ is a constant 1, and (2.5) holds for all non-negative integers n.

Equation (2.5) calculates the distribution of $N(W)$ for the cases when the left side of the box is to the right of CNT 0 by an infinitesimal amount as shown in Fig. 2.2a. In reality, the left side of the box corresponding to a CNFET can be at any random position relative to CNT 0. In this case, the distribution of the spacing between the left side of the box and the first CNT inside the box, denoted by S_1^*, is not necessarily equal to S_1. [Cox 62] found that the pdf of S_1^* is given by

$$f_{S1*}(t) = \frac{1}{\mu_S} \int_t^{\infty} f_{S1}(x) dx \qquad (2.6)$$

If (2.6) is used in place of $f_{S1}(w)$ in (2.3), a more strict result can be derived in place of (2.5) [Cox 62].

2.3. Asymptotic CNT Count Distribution

The methodology discussed in Sec. 2.2 can be used to derive the distribution $N(W)$ for arbitrary W. When $W \to \infty$, the asymptotic distribution of $N(W)$ can be shown [Cox 62] to follow a Gaussian distribution as a result of the central limit theorem. The asymptotic mean and variance of $N(W)$ can be derived [Cox 62] as

$$\lim_{W \to \infty} \mu[N(W)] = \frac{W}{\mu_S} \qquad (2.7)$$

$$\lim_{W \to \infty} \sigma^2[N(W)] = \frac{W\sigma_S^2}{\mu_S^3} \qquad (2.8)$$

Therefore, we have

$$\lim_{W \to \infty} N(W) \sim Gauss(\frac{W}{\mu_S}, \frac{W\sigma_S^2}{\mu_S^3}) \qquad (2.9)$$

In practice, we find that W does not need to be very large to reach this asymptotic limit. Figure 2.3 shows the experimentally extracted count distribution N for a box of $W = 2$ μm along with the predicted distributions derived by both the exact calculation (2.5) and Gaussian approximation (2.9). Both predictions, i.e., (2.5) and (2.9), produce similar results compared to the experimentally extracted distribution. In the example of Fig. 2.3, the average CNT count under a gate of 2-μm width is only 8.9. As will be shown in Sec.3, the minimum width of CNFETs (or equivalently, the minimum CNT count in a CNFET) is constrained by the presence of m-CNTs, and CNFETs with average CNT count < 8.9 are hardly practical in reality. This indicates that the Gaussian approximation can be appropriate for most practical sizes of CNFETs.

---

[2] Aligned CNT growth processes produce CNTs with lengths > 100 μm [Kang 07]. Note that, the scalability of CNFET technology is determined by the channel length of CNFETs rather than the length of CNTs themselves.

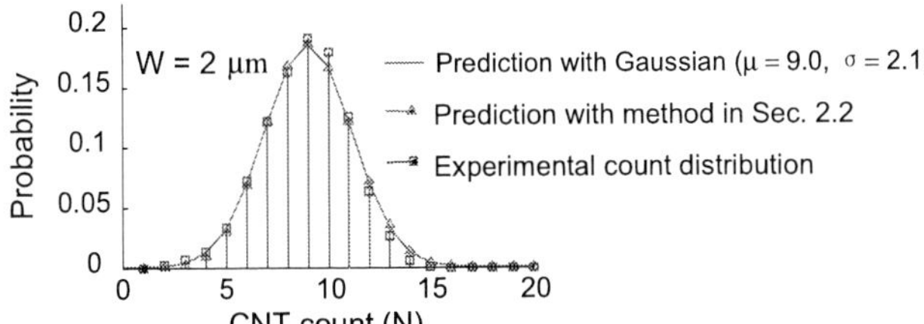

Figure 2.3. Experimentally extracted CNT count distribution compared to predicted distributions.

The asymptotic approximation can greatly simplify the characterization for CNT density variation, because the only parameters needed from the CNT synthesis technology are the mean (μ_S) and variance (σ_S) of CNT spacing, as shown in (2.9).

2.4. Count and Spacing Distributions for Semiconducting CNTs

The discussion in Sec. 2.1 to 2.3 applies to all CNTs, regardless of their types (s- or m-CNTs). For practical VLSI circuit applications, more than 99.99% of grown m-CNTs must be removed [Zhang 08], and the distributions of interest after m-CNT removal are the spacing and count distributions of the s-CNTs. We derive such distributions by assuming that the probability of any CNT being an m-CNT (s-CNT) is p_m (p_s), with $p_m + p_s = 1$, independent of the types of any of its neighboring CNTs.

Consider the spacing between two s-CNTs separated by a random number of m-CNTs. We label the first s-CNT as CNT 0 and the subsequent s-CNT as CNT K (with K-1 m-CNTs lying between the two s-CNTs). According to the above assumptions, K is a geometrically distributed random variable [Ross 01]. The spacing between the two s-CNTs can be modeled as the following stochastic sum of the original CNT spacing distribution:

$$S_{s-CNT} = \sum_{i=1}^{K} S_i \quad with \; K \sim Geometric(p_s) \qquad (2.10)$$

In general, this s-CNT spacing distribution can be derived from its moment generating function, which is the composite function of the moment generating functions of K and S. But for the asymptotic case described in Sec. 2.3, the derivation can be greatly simplified because only mean and variance are needed to characterize the desired s-CNT spacing distribution (because it is Gaussian):

$$\mu(S_{s-CNT}) = \mu(K)\mu(S) \qquad (2.11)$$

$$\sigma^2(S_{s-CNT}) = \sigma^2(K)\mu^2(S) + \mu(K)\sigma^2(S) \qquad (2.12)$$

where $\mu(K) = \dfrac{1}{p_s}$ and $\sigma^2(K) = \dfrac{1-p_s}{p_s}$.

2.5. Spatial Correlation in Count Distribution

For analyzing CNFET-based circuits that contain many individual CNFETs, it is necessary to understand how CNT counts of various CNFETs are correlated with one another. We find that the spatial correlation of CNT count for aligned CNT growth is strongly direction-dependent. For the simplicity of analysis, we define "y-direction" as the axial direction of grown CNTs, and "x-direction" as the direction perpendicular to the axial direction of CNTs.

1) Count distribution along the x-direction is highly uncorrelated.

Figure 2.4a plots the correlation coefficient of CNT count as a function of the x distance (shown in the figure) between two boxes with the width of both boxes equal to 1 µm, caculated based on experimental data. When the *x-distance* (Fig.2.4a) between the two boxes is less than 1 µm, the two count distributions exhibits positive correlation since they share some CNTs. However, when the x-distance increases beyond the width of boxes, the correlation coefficient drops to approximately 0.

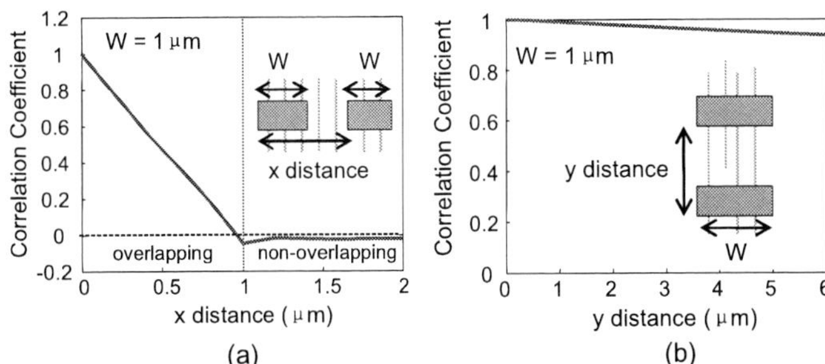

Figure 2.4. CNT correlation in x-direction (a) and y-direction (b). Based on experimentally extracted data.

2) Count distribution along the y-direction is highly correlated

Figure 2.4b shows the correlation coefficient of CNT count as a function of the y-distance between the two boxes. As shown in Fig. 2.4b, the correlation coefficient remains above 0.9 up to a y distance of 6 microns. This is a direct result of well-aligned CNT growth. The gradual decrease in the correlation coefficient is mainly due to CNTs terminating in the middle of the two boxes (Fig 2.4b).

Note that Fig. 2.4 only shows the case for local correlations of CNTs within catalyst stripes (Fig 1.1b). For larger y distances between boxes (CNFETs) that spans across catalyst stripes, e.g. for larger circuits, a change in correlation should be expected. For chip level statistical analysis, such change must be carefully characterized and modeled. In this paper, we focus on CNFET circuits between catalyst stripes.

3. Impact of CNT Density Variation on CNFET Reliability

Reliability of CNFETs has been a particularly important concern due to the presence of fabrication-related non-idealities such as mis-positioned CNTs, metallic CNTs, and CNT density variations [Mitra 09]. An extreme case of CNFET failure occurs when there is no s-CNT left in the CNFET [Zhang 08]. We can derive the probability of such failure using the CNT density distributions derived in Sec. 2. In the case of *ideal removal* of m-CNTs (all m-CNTs are removed without removing any s-CNTs), each CNT has a probability $p_f = p_m$ of not being an s-CNT. In situations when there is inadvertent removal of s-CNTs, this probability increases to

$$p_f = p_m + p_s p_{Rs} \qquad (3.1)$$

where p_{Rs} is the probability that an s-CNT will be inadvertently removed. We generalize the above discussion by considering p_f as the failure probability for each CNT. Then, for a CNFET with N independent CNTs (prior to any removal), the failure probability for the CNFET (p_F) is

$$p_F = p_f^N \qquad (3.2)$$

Equation (3.2) shows that the failure probability of a CNFET exponentially decreases with the number of CNTs. Therefore, wider CNFETs reduce the probability of failure at the cost of increased area, power and performance. When CNT density variations are taken into account, p_F can be derived based on the notion of conditional probability

$$p_F = \sum_{N_i} p_f^{N_i} \text{Prob}\{N(W) = N_i\} \qquad (3.3)$$

where W is the width of the CNFET.

For a given CNFET width, Arithmetic Mean-Geometric Mean Inequality can be applied to (3.3) to show that CNT density variations (or equivalently variation in $N(W)$) always result in higher p_F compared to the cases with uniform CNT density:

$$p_F = \sum_{N_i} p_f^{N_i} \text{Prob}\{N(W) = N_i\}$$

$$\geq \left[\prod_{N_i} \left(p_f^{N_i} \right)^{\text{Prob}[N(W)=N_i]} \right]^{1/\sum_{N_i} \text{Prob}[N(W)=N_i]} = p_f^{\mu[N(W)]} \qquad (3.4)$$

978-1-60558-497-3/09 $25.00 © 2009 ACM

the right hand side of the inequality in (3.4) is the failure probability of a CNFET with uniform CNT density[3]. In other words, to meet a certain requirement of total failure probability p_F, wider CNFETs (corresponding to a larger mean value of $N(W)$ in equation (3.3)) will be needed in the presence of CNT density variations to meet the p_F requirement.

To quantify such effect, we define N_{min} as the minimum average CNT count $\mu(N(W))$ needed to meet a target failure probability p_F. Table 3.1 shows the values of N_{min}, calculated using (3.3) with Gaussian approximation for the CNT count distribution, with varying p_m and p_{Rs} for uniform CNT density as well as in the presence of CNT density variations. As shown in the table, CNT density variations significantly increase N_{min} for all cases compared to uniform CNT density, especially with small values of CNT failure probability (p_f).

Table 3.1. Minimum average number of CNTs per CNFET (N_{min}) for CNFET failure probability (p_F) = 10^{-8} with uniform CNT density and in the presence of CNT density variation.

p_m	33%	33%	10%	10%
p_{Rs}	16%[4]	0%	16%[4]	0%
p_f	43.7%	33%	24.4%	10%
N_{min} (uniform density)	23	17	13	8
N_{min} (with density variation)	29	24	20	18

4. Circuit Level Impact of CNT Density Variation

4.1. CNT Count Correlation between Arbitrary CNFETs

The correlation of CNT counts between two arbitrary CNFETs (shown in Fig. 4.1) is considered in this section. We assume the following two simplifications based on the results in Sec. 2.5:
1) CNT counts from non-overlapping sections of the CNFETs along the x-direction are completely uncorrelated (Fig 2.4a);
2) CNT counts from equally-sized overlapping sections of CNFETs along the x-direction are completely correlated independent of their locations in the y-direction. Figure 2.4b shows that this correlation in fact decreases with increasing y-distance between the 2 CNFETs. This slight decrease in correlation is neglected in this section and more detailed models can be used to extend the results derived below.

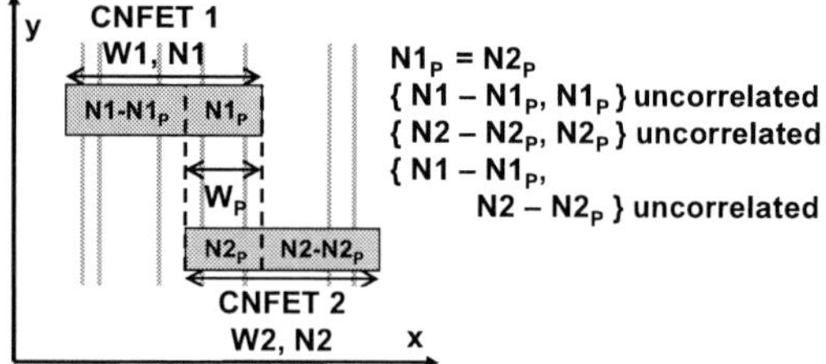

Figure 4.1. CNT count correlation between arbitrary CNFETs.

Let W_1 (W_2), N_1 (N_2) represent the width and CNT count of CNFET 1(2), and W_P, $N_{1P} = N_{2P}$ represent the width and CNT count of the overlapping section of the two CNFETs in the x direction. Then

$$\mu(N_1 N_2) = \mu[(N_{1P} + N_1 - N_{1P})(N_{1P} + N_2 - N_{1P})]$$
$$= \mu(N_{1P}^2) + \mu(N_1 - N_{1P})\mu(N_2 - N_{1P}) + \quad (4.1)$$
$$+ \mu(N_{1P})\mu(N_2 - N_{1P}) + \mu(N_{1P})\mu(N_1 - N_{1P})$$

It can be further shown that the covariance between N_1 and N_2 is given by:

$$Cov(N_1, N_2) = \mu(N_1 N_2) - \mu(N_1)\mu(N_2) \quad (4.2)$$
$$= \sigma^2(N_{1P}) = \sigma^2(N_{2P})$$

where $\sigma^2 (N_{1P})$ or $\sigma^2 (N_{2P})$ can be calculated using (2.8).

4.2. Utilizing CNT Correlation to Optimize the Layout of Cross-Coupled Inverters

In this section, we describe an example to utilize the correction discussed in Sec 4.1 to improve the reliability of CNFET circuits. We consider a non-ideal m-CNT removal process which removes m-CNTs with probability p_{Rm} and inadvertently removes s-CNTs with probability p_{Rs}. Consider a pair of cross-coupled inverters (Fig. 4.2) each with a p-type CNFET (*PFET*) and an n-type CNFET (*NFET*). Static noise margin (*SNM*), defined as the maximum nested square between the normal and mirrored voltage transfer curves (*VTCs*) for the two inverters [Lohstroh 83], is used as a metric for the robustness of the cross-coupled inverters (Fig 4.2b). When m-CNTs are present, the inverters do not produce full rail-to-rail outputs resulting in degraded noise margins (Fig. 4.2c).

Suppose that SNM_R is the required static noise margin that such cross-coupled inverters must satisfy. Given the CNT density distribution, CNT processing parameters (p_{Rm}, p_{Rs}) and the width of the CNFETs (W), we can calculate the probability that a gate fails to satisfy this SNM requirement. We refer to this as *PNMV* or *probability of noise margin violation*. We show that the layout of such cross-coupled inverters can be optimized to reduce the PNMV by up to three-orders of magnitude. Five different layout styles are studied as shown in Fig. 4.3. The different layout styles have different degrees of correlation among the four CNFETs comprising the cross-coupled inverters. The symmetries in the VTCs caused by such correlation are also shown in Fig 4.3. Style 1 (Fig. 4.3a) has perfect correlation among the drive strengths of all four CNFETs. Style 2 (Fig. 4.3b) has perfect correlation between the PFET and NFET of each inverter, while the CNFETs in the different inverters are uncorrelated. Style 3 (Fig. 4.3c) has perfect correlation between the PFET of one inverter and the PFET of the second inverter and similarly for the NFETs. Style 4 (Fig. 4.3d) has perfect correlation between the PFET of one inverter and the NFET the other and vice versa. Style 5 (Fig. 4.3e) contains 4 completely uncorrelated CNFETs. A CNFET SPICE model [Deng 07b] with a 32 nm CNFET technology is simulated to obtain the VTCs. This SPICE model has been calibrated to experimental data with 90% accuracy [Amlani 06]. We assume a uniform CNT diameter of 1.5 nm and $p_m = 1/3$. SNM_R is assumed to be Vdd / 4 and $p_{Rs} = 16\%^4$.

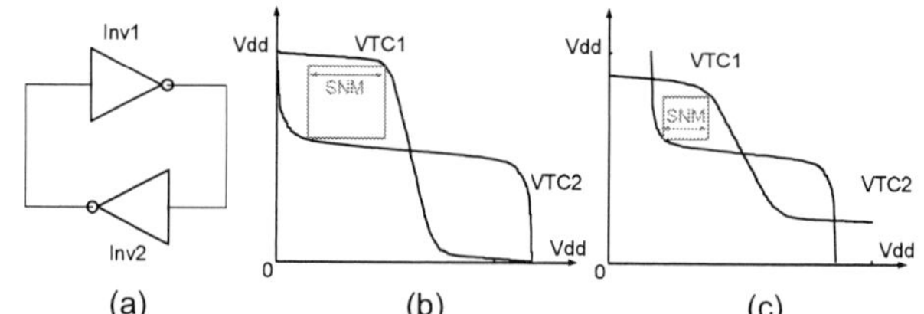

Figure 4.2 Cross-coupled CNFET inverters (a) with Voltage Transfer Curves without m-CNTs (b) and with m-CNTs (c).

Figure 4.4 shows the simulated PNMV for the 5 layout styles as a function of the removal probability of m-CNTs (p_{Rm}). Symmetry in the VTCs of the two inverters, introduced by correlation among CNFETs, plays an important role in determining the PNMV. Layout styles 1, 2 and 3 ensure symmetry between the two "eye-openings" of the VTC curves and therefore have low PNMV values. Style 4 has the highest PNMV for all values of p_{Rm} because of the anti-symmetry in the VTC curves (Fig. 4.3d). Style 5 has no inherent symmetry because all the CNFETs are uncorrelated and has intermediate values of PNMV.

---

[3] Even with uniform CNT density, the CNT count distribution of a CNFET is actually bimodal with possible values of N = [W/s]-1 and [W/s]+1, where s is the uniform inter-CNT spacing. However, the failure probability of this case is very close to the unimodal case and all the discussions are valid.

[4] This value corresponds to the one-sigma cumulative probability point of a Gaussian distribution [Zhang 08].

978-1-60558-497-3/09 $25.00 © 2009 ACM

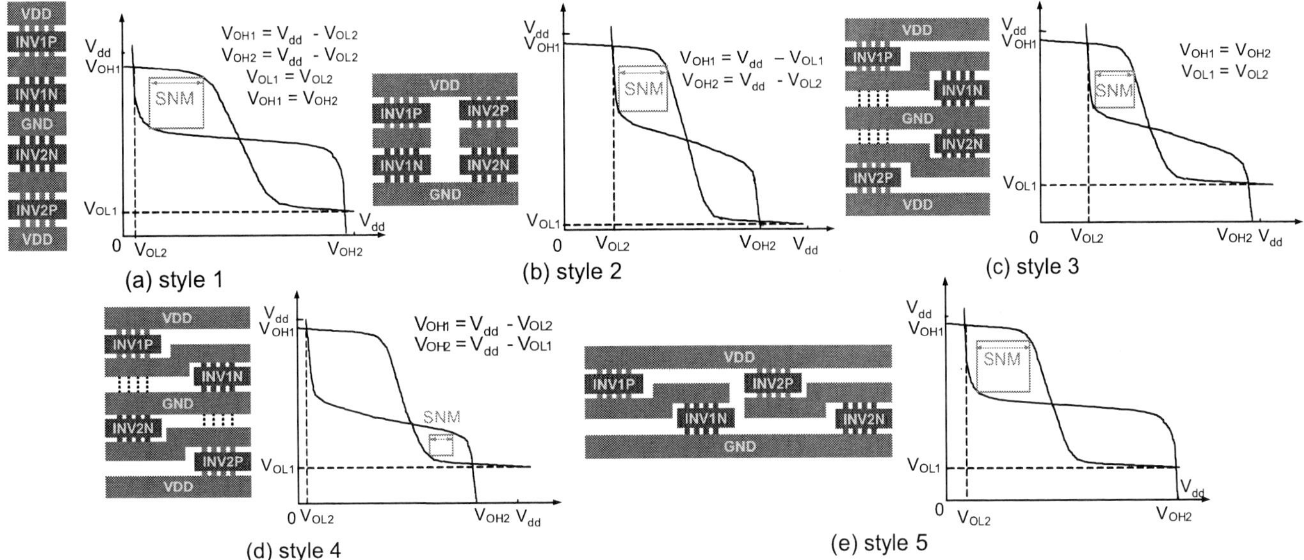

Figure 4.3. CNFET cross-coupled inverter layout styles 1 – 5 (a through e) with varying degrees of CNT correlation.

When the removal probability of m-CNTs is very high ($p_{Rm} > 1 \cdot 10^{-6}$), very few (almost negligible) m-CNTs survive. In this limiting case, failure of this circuit is due to no CNTs left in the CNFETs because of either inadvertent removal of s-CNTs or CNT density variations as discussed in Sec. 3. The PNMV value in this case for layout styles 1, 2, 3 and 5 (style 4 is an exception because of its anti-symmetry) can be given by

$$PNMV = 1 - (1 - p_F)^k \approx k p_F \qquad (4.3)$$

where k is number of CNFETs containing different CNTs in the circuit and p_F is the failure probability for each one of them, calculated using equation (3.3). Style 1 has the lowest PNMV because of perfect correlation between the 4 CNFETs ($k = 1$). Styles 2 and 3 ($k = 2$) and style 5 ($k = 4$) have higher values of PNMV.

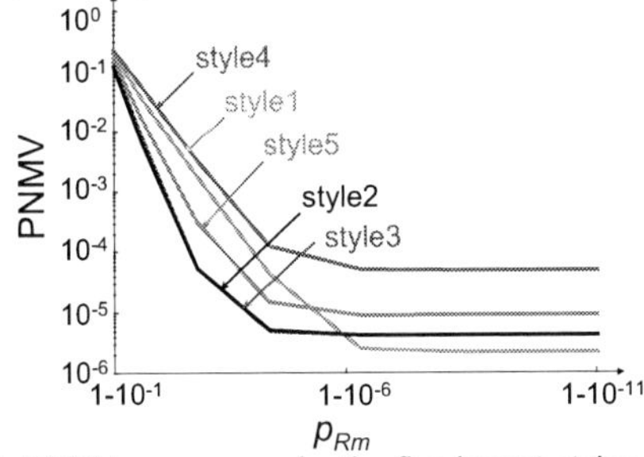

Figure 4.4. PNMV versus p_{Rm} for the five layout styles in Fig. 4.3. Gaussian distribution of CNT count is assumed with $\mu(N) = 20$.

For lower probabilities of m-CNT removal ($p_{Rm} < 1 \cdot 10^{-6}$), the presence of m-CNTs can no longer be ignored, and it becomes the dominant factor in PNMV. In this case, style 1 does not perform as well as styles 2, 3 and 5, since the presence of m-CNTs will cause equal degradation of all four correlated CNFETs in style 1, while such equal degradation is unlikely in the case of uncorrelated CNFETs in styles 2, 3 and 5.

4.3. Delay Variations

In addition to the processes variations introduced by conventional semiconducting processing (e.g. lithography), major sources of variations in CNFET circuits include variations in CNT type (m- or s-), diameter, density and doping concentration. The analysis in [Deng 07a] shows that diameter and doping variations are more tolerable in the cases of multi-CNT CNFETs due to statistical averaging effects

[Borkar 05]. Therefore, in this section, we focus on the delay variations caused by the variations in CNT type and density.

Suppose inverter I_1 has N_1 s-CNTs, driving inverter I_2 with N_2 s-CNTs. We assume that the NFET and PFET of each inverter are perfectly correlated, similar to the layout style 2 in Fig. 4.3b. The first order delay model can be written as:

$$d_{1 \to 2} = \frac{C_{load} V_{dd}}{I_{drive}} = \frac{C_I + N_2 C_0}{N_1 I_0} V_{dd} \qquad (4.4)$$

where C_I is the interconnect capacitance, C_0 is the capacitance per CNT, I_0 is the drive current per CNT. Second order effects such as diameter dependence of CNT capacitance and current drive, and the charge-screening effect are not included [Deng 07b, Kshirsagar 08]. With a first order Taylor expansion ([Chang 03]), (4.4) can be written as

$$d_{1 \to 2} = d_{1 \to 2}^0 + \left(\frac{\partial d_{1 \to 2}}{\partial N_1} \right)_0 \Delta N_1 + \left(\frac{\partial d_{1 \to 2}}{\partial N_2} \right)_0 \Delta N_2 \qquad (4.5)$$

$$= d_{1 \to 2}^0 - \frac{C_I + N_2 C_0}{N_1^2 I_0} \Delta N_1 + \frac{C_0}{N_1 I_0} \Delta N_2$$

The variance in delay caused by CNT density variation is then

$$\sigma^2(d_{1 \to 2}) = \left(\frac{C_I + N_2 C_0}{N_1^2 I_0} \right)^2 \sigma^2(N_1) + \left(\frac{C_0}{N_1 I_0} \right)^2 \sigma^2(N_2) \qquad (4.6)$$

$$- 2 \left(\frac{C_I + N_2 C_0}{N_1^2 I_0} \right) \left(\frac{C_0}{N_1 I_0} \right) Cov(N_1, N_2)$$

The covariance term in (4.6) can be calculated using (4.2). Note that a positive correlation between N_1 and N_2 helps in reducing the overall delay variability. As an example, consider two inverter chains with all inverters equally sized:

1) Correlated inverter chain: CNTs in the PFETs and NFETs of all inverters are perfectly correlated. For example, all the inverters are laid out along in y-axis in Fig. 4.1.

2) Uncorrelated inverter chain: The NFET and the PFET within each inverter are still perfectly correlated but there is no correlation between CNTs in different inverters. For example, all the inverters are laid out along in x-axis in Fig. 4.1.

Using the CNFET SPICE model described in Sec. 4.2, we performed Monte-Carlo simulations, each with 2,000 samples, of these two inverter chains with different number of stages in the presence of CNT density variations and ideal m-CNT removal (defined in Sec. 3). Table 4.1 shows the simulation results using both the experimentally extracted CNT count distribution and the Gaussian approximation to CNT count distribution.

The Gaussian approximation gives less than 5% error when compared to the results using the experimentally extracted CNT count distribution.

Table 4.1 Delay variations for correlated and uncorrelated inverter chains (W = 3 µm for all CNFETs) using 2000 sample Monte Carlo simulations with experimentally extracted CNT count distribution. Results in parenthesis use the Gaussian CNT count distribution.

	Correlated			Uncorrelated		
No. of stages	3	5	10	3	5	10
μ(delay) (ps)	2.93 (2.93)	4.89 (4.88)	9.77 (9.77)	2.98 (2.98)	4.97 (4.97)	9.93 (9.93)
σ(delay) (ps)	0.06 (0.06)	0.10 (0.09)	0.20 (0.19)	0.13 (0.13)	0.15 (0.15)	0.19 (0.18)
σ(delay) / μ(delay) (%)	1.97 (1.90)	2.02 (1.93)	2.04 (1.97)	4.49 (4.31)	3.07 (2.91)	1.88 (1.83)
Average σ(delay) / μ(delay) per stage (%)	1.97 (1.90)	2.02 (1.93)	2.04 (1.97)	8.90 (8.53)	8.72 (8.49)	8.57 (8.50)

The following observations can be made from the simulation results presented in Table 4.1:

1. The correlated inverter chain gives lower per-stage variation. This effect can be shown from (4.6) since the covariance term is positive. In the limiting case when $C_l \rightarrow 0$, the per-stage variation drops down to zero. However, the multiple stages are perfectly correlated, so the σ/μ for the delay per stage is the same as the σ/μ for the total delay.

2. The uncorrelated inverter chain gives higher per-stage variation because the covariance term in (4.6) is zero. However, it can be shown that the delays of successive stages are negatively correlated resulting in faster than $\sqrt{N}$ drop in delay variations as number of stages increases. Therefore, when logic depth is high, the variability is similar to or smaller than what is achieved in the correlated case.

5. Conclusion

Since CNTs are grown using chemical synthesis, it can be extremely difficult to guarantee perfect alignment, positioning and uniform density of CNTs. Hence, circuits must be designed to tolerate CNT imperfections. This paper shows that, in addition to imperfections caused due to mis-positioned CNTs and metallic CNTs, density variations in CNT growth can result in compromised reliability and increased delay variations in CNFET circuits. To analyze the effects of CNT density variations, parameterized models supported by experimental data are needed. This paper presents such a model fitted to data from images of aligned CNTs. This model can potentially be applied to improved future CNT synthesis techniques that produce ultra high density CNTs (>100 CNTs/um).

Correlation in aligned CNT growth has a significant impact on circuit performance metrics such as noise margin and delay variability. The reliability of CNFET circuits can be improved by taking advantage of correlation between CNFETs. With appropriate data input to account for CNT correlation, the CNT density model presented in this paper can be applied to statistical timing and/or yield analysis of VLSI circuits.

6. Acknowledgment

This research was supported in part by the FCRP Center for Circuit and System Solutions (C2S2), the National Science Foundation, and Stanford Graduate Fellowships. The authors would like to thank D. Akinwande, A. Lin, H. Wei, and Prof. H.-S.P. Wong for helpful discussions.

7. References

[Amlani 06] Amlani, I., *et al.*, "First Demonstration of AC Gain From a Single-walled Carbon Nanotube Common-Source Amplifier," *Proc. IEDM*, pp. 1-4, 2006.

[Bobba 09] Bobba, S. *et al.,* "Design of Compact Imperfection-Immune CNFET Layouts for Standard-Cell-Based Logic Synthesis", *Proc.DATE*, 2009.

[Borkar 05] Borkar, S., et al., "Statistical circuit design with carbon nanotubes", U.S. Patent Application 20070155065, 2005.

[Cameron 98] Cameron, A. C. and P. K. Trivedi, *Regression Analysis of Count Data*, Cambridge University Press, 1998.

[Collins 01] Collins, P., S. Arnold, P. Avouris, "Engineering carbon nanotubes and nanotube circuits using electrical breakdown", *Science*, vol. 292, pp. 706 – 709, 2001.

[Chang 03] Chang, H. and S. Sapatnekar, "Statistical timing analysis considering spatial correlation in a PERT-like traversal," *Proc. ICCAD*, pp. 621-625, 2003.

[Cox 62] Cox, D. R., *Renewal Theory*, Methuen & Co, 1962.

[Dai 02] Dai, H., "Carbon Nanotubes: Synthesis, Integration, and Properties," *Acc. Chem. Res.*, 35, 1035-1044, 2002.

[Deng 07a] Deng, J., *et al.*, "Carbon Nanotube Transistor Circuits: Circuit-Level Performance Benchmarking and Design Options for Living with Imperfections", *Proc. ISSCC*, pp. 70-588, 2007.

[Deng 07b] Deng, J., and H. -S. P. Wong, "A Compact SPICE Model for Carbon Nanotube Field Effect Transistors Including Non-Idealities and Its Application — Part I: Model of the Intrinsic Channel Region", *IEEE Trans. Elec. Dev.*, pp.3186-3194, 2007.

[Kang 07] Kang, S. J., et al., "High-Performance Electronics Using Dense, Perfectly Aligned Arrays of Single-Walled Carbon Nanotubes", *Nature Nanotechnology*, vol. 2, pp. 230-236, 2007.

[Kocabas 07] Kocabas, C., et al., "Improved Synthesis of Aligned Arrays of Single-Walled Carbon Nanotubes and Their Implementation in Thin Film Type Transistors", *Journal of Physical Chemistry C* 111(48), 17879-17886 (2007).

[Kshirsagar 08] Kshirsagar, C., H. Li, T. Kopley, and K. Banerjee, "Accurate Intrinsic Gate Capacitance Model for Carbon Nanotube-Array Based FETs Considering Screening Effect", *IEEE Electron Device Letters*, Vol. 29, No. 12, pp. 1408-1411, 2008.

[Lohstroh 83] Lohstroh, J., E. Seevinck, and J.D. Groot, "Worst-Case Static Noise Margin Criteria for Logic Circuits and Their Mathematical Equivalence", *JSSC*, No.6, pp. 803-806, 1983.

[Mitra 09] Mitra, S., *et al.*, "Imperfection-Immune VLSI Logic Circuits using Carbon Nanotube Field Effect Transistors", Proc. *DATE*, 2009.

[Nieuwoudt 07] Nieuwoudt, A., and Y. Massoud, "On the Impact of Process Variations for Carbon Nanotube Bundles for VLSI Interconnect," *IEEE Trans. Elec. Dev.*, Vol. 54, No. 3, 2007

[Patil 08a] Patil, N., *et al.*, "Integrated Wafer-scale Growth and Transfer of Directional Carbon Nanotubes and Misaligned-Carbon-Nanotube-Immune Logic Structures," *Symp. VLSI Tech.*, pp. 205-206, 2008.

[Patil 08b] Patil, N., *et al.*, "Design Methods for Misaligned and Mis-positioned Carbon-Nanotube-Immune Circuits," *IEEE Trans. Computer-Aided Design*, pp. 1725-1736, 2008.

[Patil 09] Patil, N., *et al.*, "Digital VLSI Logic Technology using Carbon Nanotube FETs: Frequently Asked Questions", *Proc. DAC*, 2009.

[Ross 01] Ross, S. M., *A First Course in Probability*, Prentice Hall, 2001.

[Ross 05] Ross, S. M., *Introductory Statistics*, Academic Press, 2005.

[Zhang 06] Zhang, G., *et al.*, "Selective Etching of Metallic Carbon Nanotubes by Gas-Phase Reaction", *Science*, Vol. 314, pp. 974 – 977, 2006.

[Zhang 08] Zhang, J., N. Patil, and S. Mitra, "Design Guidelines for Metallic-Carbon-Nanotube-Tolerant Digital Logic Circuits", *Proc. DATE*, pp. 1009 – 1014, 2008.

Decoding Nanowire Arrays
Fabricated with the Multi-Spacer Patterning Technique

M. Haykel Ben Jamaa, Yusuf Leblebici and Giovanni De Micheli
Swiss Federal Institute of Technology (EPFL)
1015 Lausanne, Switzerland
haykel.benjamaa@epfl.ch

ABSTRACT

Silicon nanowires are a promising solution to address the increasing challenges of fabrication and design at the future nodes of the *Complementary Metal-Oxide-Semiconductor (CMOS)* Technology roadmap. Despite the attractive opportunity that offers their organization onto regular crossbars, the problem of designing the nanowire decoder is still challenging and highly dependent on the nanowire fabrication technology. In this paper, we introduce a novel design style and encoding scheme for decoding nanowires fabricated with the *Multi-Spacer-Patterning Technique (MSPT)*; and we present a method based on Gray codes that reduces the fabrication cost and improves the decoder reliability. We show that by arranging the code in a Gray code fashion, we decrease the fabrication complexity by 17% and the variability by 18% on average. By optimizing the decoder parameters, the simulations showed an improvement of the crossbar yield by 40% and a reduction of the effective bit area by 51% to 169 nm^2.

Categories and Subject Descriptors

B.8.1 [**Performance and Reliability**]: Reliability, Testing, and Fault-Tolerance

General Terms

Design

Keywords

Emerging Technologies, Nanowires, Decoder, Crossbar, Spacer Technique, MSPT

1. INTRODUCTION

Despite the large efforts spent to develop novel solutions in order to reach the future nodes of the technology roadmap, several hurdles are still challenging the evolution of the semiconductor industry. A major issue is the high variability of nanoscale devices, which needs innovative defect tolerance methods at all design levels. In order to go beyond the 22 nm milestone, alternative devices to bulk *Metal-Oxide-Semiconductor Field Effect Transistor (MOSFET)* are currently being investigated, and their design

---

[1]This work is supported by the CCMX/MMNS project, the CSI Center and the Swiss FNS Research Grant 200021-109450/1

Permission to make digital or hard copies of part or all of this work for personal or classroom use is granted without fee provided that copies are not made or distributed for profit or commercial advantage and that copies bear this notice and the full citation on the first page. To copy otherwise, to republish, to post on servers or to redistribute to lists, requires prior specific permission and/or a fee.
DAC'09, July 26-31, 2009, San Francisco, California, USA

methodologies are being optimized. Such devices include quasi one-dimensional silicon nanowires [9, 15]. A promising approach to integrate them is the crossbar architecture [10].

Crossbar circuits are formed by two perpendicular layers of parallel nanowires. At the crossing areas, specific materials are deposited or grafted in order to functionalize the circuit [10]. A major issue has been reported with the integration of crossbar circuits, consisting in interfacing them with a decoder to the outer *Complementary Metal-Oxide-Semiconductor (CMOS)* circuit, which is defined on a much larger pitch [6]. Several types of decoders have been suggested [6, 13, 1, 8]. They are strongly dependent on the nanowire fabrication technology, and some of them bridge the sublitho- to litho-scales in a stochastic way. One of the nanowire fabrication techniques is the *Multi-Spacer Patterning Technique (MSPT)*, which has many advantages in terms of density and compatibility with standard CMOS processing [4]. However, no specific design style for the decoder of MSPT-based nanowires has been proposed so far. And the impact on yield of the encoding scheme, *i.e.*, the code used to identify the nanowires, is not known.

In this paper, we propose a novel design style for decoders of MSPT-based nanowire arrays. The first novelty of this design style is that it is the first specific decoder for this fabrication technology that uniquely addresses every nanowire. The second novelty is that it assigns a deterministic address to every nanowire, unlike other decoders [6, 8]. Furthermore, we suggest an innovative design method based on Gray codes, which reduces the fabrication complexity and improves the fault-tolerance of the decoder.

This paper is organized as follows. The next section surveys different nanowire fabrication technologies and circuits, the decoder design and the nanowire codes. Section 3 presents the MSPT that is currently being studied through experimentation, and the decoder design style that we are suggesting. Section 4 formulates the problem of the decoder design in an abstract way. In Section 5 we derive the optimal code types with respect to different optimization criteria. Section 6 presents the simulation platform and results for nanowire decoders with different encoding schemes. We finally conclude the paper in Section 7.

2. BACKGROUND

This section surveys nanowire fabrication techniques reported in literature. Then, it explains the baseline architecture of nanowire crossbars and highlights the decoder part. Finally, it introduces the code types that will be used in the rest of the paper.

2.1 Fabrication Technologies

Silicon nanowires are quasi one-dimensional long structures fabricated on a sub-lithographic scale; the *lithographic* (or *meso-*) scale being the one defined by the photolithography used to convert the circuit design into physical masks. Many techniques were recently shown to be promising ways to fabricate silicon nanowires. We can distinguish between bottom-up and top-down techniques. Bottom-up techniques are based on the growth of nanowires on a silicon substrate from catalyst seeds. The as-grown nanowires are then collected in a solution and dispersed on the substrate to be functionalized [9, 15]. Top-down approaches define the nanowires

---

978-1-60558-497-3/09 $25.00 © 2009 ACM

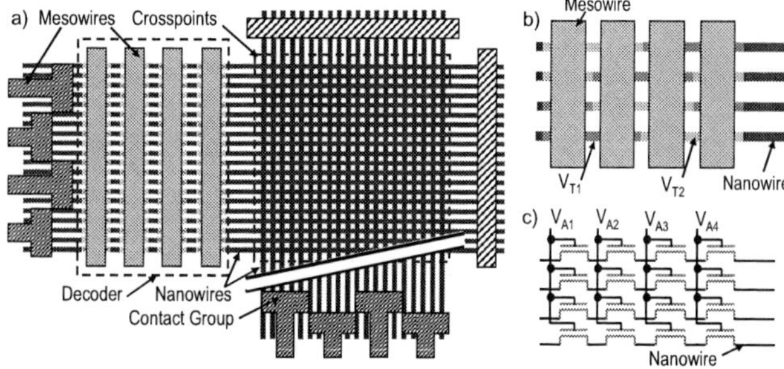

Figure 1: Baseline architecture of a crossbar circuit (a) and highlights of the decoder layout (b) and circuit (c)

directly on the functional substrate by accurately controlling the deposition, oxidation and etching rates [12, 4], or by using nanometer-scale molds [11].

Because of the small dimensions of these devices and the difficulty of placing them accurately, a high variability is expected with respect to the electrical properties of the fabricated circuits. Consequently, a general tendency is in favor of a regular arrangement of these devices in a crossbar architecture where the device placement is easier and the redundancy is higher. Since the nanowires cannot perform computation or information storage, the use of other emerging materials has been proposed to enhance the device functionality. For instance, molecular switches [10] or phase change materials [16] whose on- and off-states can be electrically controlled may be inserted between two layers of crossing nanowires in order to functionalize the circuit.

2.2 Nanowire Circuits and Decoders

The baseline organization of a nanowire crossbar circuit is depicted in Fig 1.a. An arrangement of two orthogonal layers of parallel nanowires defines a regular grid of intersections called crosspoints. The separation between the two layers can be filled with a phase change material or molecular switches at the crosspoints. Information storage, interconnection or computation can be performed with these crosspoints [10, 5]. A set of contact groups is defined on top of the nanowires. Every contact group makes an ohmic contact to a corresponding distinct set of nanowires, which represents the smallest set of nanowires that can be contacted by the lithographically defined lines (mesowires).

This configuration bridges every set of nanowires within a contact group to the outer CMOS circuit. In order to *fully bridge the scales* and make every nanowire within this set uniquely addressable by the outer circuit, a decoder is needed. It is formed by a series of transistors along the nanowire body, controlled by the mesowires and having different threshold voltages V_T [2] (Fig. 1.b). The distributions of V_T's is called the *nanowire pattern*. Depending on this pattern and the pattern of applied voltages in the decoder (V_A's), one single nanowire in the array can be made conductive (Fig. 1.c). In this case, this nanowire is said to be addressed by the applied voltage pattern.

Many decoders have been suggested for nanowire arrays. Their design strongly depends on the nanowire fabrication technology. Axial and radial decoders were suggested for in-situ patterned nanowires [6, 13]. Mask-based decoder [1] and random-contact decoders [8] were proposed for nanowires fabricated with nano-imprint. For other bottom-up fabrication techniques, a gate-all-around decoder was suggested in [12].

2.3 Code Types

We will consider in this paper *hot codes (HC)*, *tree codes (TC)*, *Gray codes (GC)* and *balanced Gray codes (BGC)*. More details about these codes can be found in [2, 7, 3]. In the following, we remind the reader of their basic concepts. A code space means the set of all code words of a given code.

HCs are defined for the multi-valued logic n with the parameters $(M, k) \in \mathbb{N}^2$ such that $M = k \cdot n$. Every code word has M digits, such that every digit takes the value $\in \{0, \ldots, n - 1\}$ and every value exists in the code word exactly k times. For instance, the code words 001122 and 012120 belong to the same HC space with $(M, k) = (6, 2)$ and $n = 3$; but 000121 does not belong to this code space because the values 0 and 2 are not repeated exactly $k = 2$ times.

TCs are defined by the parameter $M \in \mathbb{N}$ for the multi-valued logic n. Every code is represented by M digits such that every digit takes the value in $\in \{0, \ldots, n - 1\}$. For instance, for $n = 3$ and $M = 4$, we obtain the codes: 0000, 0001, 0002, 0010,...,2222. In order to address nanowires with TCs, it was shown that the codes have to be *reflected* [2], i.e., every code word gets its complementary code word appended to it. The complement is obtained by subtracting the code word from the largest code word in the same code space. For instance, the complement of 0010 in this example is $2222 - 0010 = 2212$. Then, 2212 is appended to 0010, giving the reflected code word 00102212. In a similar way, 0000 and 0001 give 00002222 and 00012221 respectively. In the rest of the paper, all TCs are *implicitly* considered to be reflected.

GCs are an arrangement of TCs where successive code words differ in only one digit. For instance, the sequence of code words $0000 \Rightarrow 0001 \Rightarrow 0002 \Rightarrow 0010$ does not belong to the GC because the last two code words differ in two digits. But the sequence $0000 \Rightarrow 0001 \Rightarrow 0002 \Rightarrow 0012$ is eligible.

BGCs are a type of GCs with an additional constraint that every digit is allowed to change at most a given number of times. In the BGCs considered in this paper, the limit on number of changes is set to 2 [3]. In the previously presented GC sequence $0000 \Rightarrow 0001 \Rightarrow 0002 \Rightarrow 0012$, the first and second digits (from the left) do not change, the third digit changes once and the last digit changes twice. Thus, this sequence fulfills the conditions of a BGCs with a limit set to 2.

GCs and BGCs are special arrangements of code words in TC spaces. So, GCs and BGCs will be used *implicitly* in the reflected form by appending their complement, as explained previously for reflected TCs.

3. FABRICATION AND DESIGN OPPORTUNITIES OF NANOWIRE DECODERS

In this section we present the nanowire array fabrication technology that we are currently experimentally investigating. Then, we introduce the decoder fabrication method and we focus on the decoder design opportunities and challenges.

3.1 The Multi-Spacer Patterning Technique

In order to fabricate nanowire arrays with a pitch independent on the photolithography limit, we are currently using the *Multi-Spacer Patterning Technique (MSPT)*, which is a CMOS-compatible fabrication technique using only standard CMOS steps. It yields large arrays of parallel *poly-crystalline Silicon (poly-Si)* spacers separated by *silicon dioxide (SiO_2)* as insulator. The essential fabrication steps are illustrated in Fig. 2 and were described in detail elsewhere [4]. The processing starts with the definition of caves limited by a sacrificial layer. Then, by a series of conformal deposition of thin poly-Si and SiO_2 layers and their anisotropic etching iteratively, it is possible to obtain parallel poly-Si spacers separated by SiO_2 spacers. The technique yields a symmetrical structure with a symmetry axis parallel to the edge of the sacrificial layer.

With the 0.8 μm photolithography available for academic research purposes in our fabrication facilities, we could obtain dense arrays of poly-Si spacers with a sub-lithographic pitch. These spacers represent nanowires having a pitch of a few tens of nanometer, a height of $\sim$ 300 nm and a length of tens of microns. Since the height does not influence the pitch, it was kept as it is; but it can be shrunk by applying the standard chemical and mechanical planarization technique if needed. A *Scanning Electron Microscopy (SEM)* of a cross-section of the obtained nanowire arrays is presented in Fig. 3. The nanowire pitch exclusively depends on the thickness of deposited poly-Si and on the etch, but not on the lithography resolution. The technique yields reliable and reproducible results. It is not intended to compete with bottom-up nanowire fabrication techniques, but it represents a CMOS-compatible and

978-1-60558-497-3/09 $25.00 © 2009 ACM

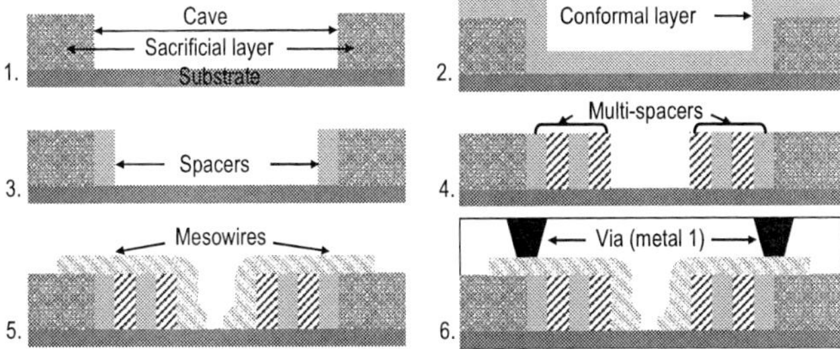

Figure 2: Cave cross-section illustrating the multi-spacer patterning technique. Cave definition (1), conformal poly-Si deposition (2), anisotropic poly-Si etch (3), iteration of 2-3 with poly-Si and SiO$_2$ (4), gate patterning (5), metallization (6)

cost-efficient technique (possible parallelization of spacer definition steps) with an opportunity to investigate the combined design and fabrication aspects of nanowire decoders.

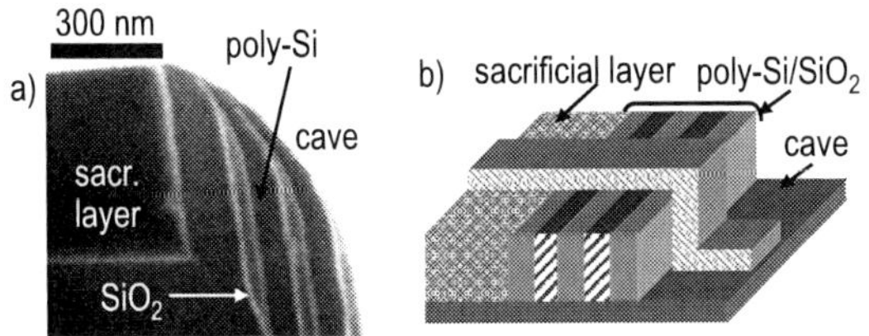

Figure 3: Scanning Electron Microscopy (SEM) of a nanowire array cross-section (a), and a 3-dimensional representation of the full structure (b).

3.2 The Decoder Fabrication Technique

While a pattern can be easily defined during the growth of nanowires in a bottom-up process (Sec. 2.1), it is more difficult to define it with top-down processes. For instance, the MSPT, yields a regular array of undifferentiated nanowires if the bare procedure depicted in Fig. 2 is applied. Once the array is defined on a sublithographic scale, it is difficult to pattern it with standard photolithographic means, unless expensive high-resolution and time-costly methods, such as electron-beam lithography, are applied. Consequently, it is desirable to pattern the nanowires while they are defined: *i.e.*, whenever a new spacer is defined, it has to be patterned before the next spacer is defined.

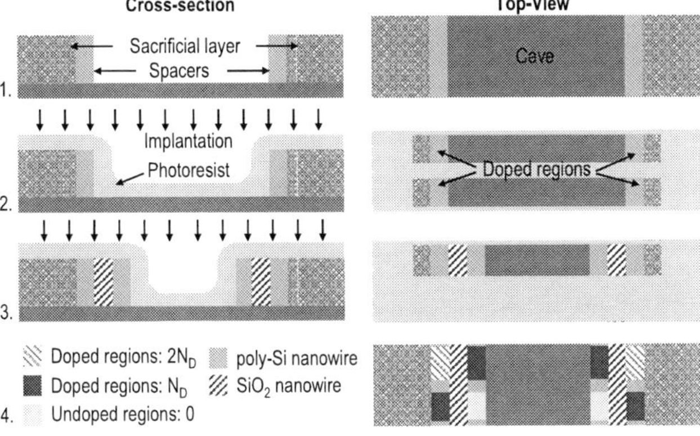

Figure 4: Decoder-aware enhanced fabrication flow: Definition of first poly-Si nanowire (1), photolithography and doping of first poly-Si nanowire (2), photolithography and doping of second poly-Si nanowire (3), final doping patterns (4).

The fabrication flow that includes the decoder is illustrated in Fig. 4 and should be understood as an extension inserted between steps 3 and 4 in Fig. 2. The other steps remain unchanged. The additional steps are lithography patterning and doping after every spacer definition step, using p-type (n-type) doping to increase (decrease) the total doping level. Specific regions from every poly-Si nanowire are defined and doped in this way. Nanowires are fragile and thin structures, and they should be doped carefully with light doses. The total doping level of a specific region is the sum of all (positive and negative) doping levels accumulated in this region throughout the definition of the whole array, as illustrated in step 4 of Fig. 4. An optimized choice of the lithography/doping sequences and the doping doses may result in the desired nanowire pattern.

3.3 Design of the Decoder

The decoder fabrication technique introduced in the previous section yields a decoder operation identical to the description in Sec. 2.2. However, the layout differs in the fact the the nanowires lie within parallel caves having a symmetry axis going through their central axis (Sec. 3.1). The unique addressing of every nanowire in a half cave insures the unique addressing of every nanowire in the whole array. We will therefore consider only half caves in the rest of the paper.

Every half cave contains N nanowires having M doping regions each. Recall that the pattern is the sequence of threshold voltages and the doping profile is the sequence of dopant concentrations along the doping regions of the nanowire. Let $\mathbf{P}_i = \left[P_i^0 \ldots P_i^{M-1}\right]$ and $\mathbf{D}_i = \left[D_i^0 \ldots D_i^{M-1}\right]$ be the pattern and doping profile of the nanowire i respectively. For the considered technique, whenever a nanowire i is patterned by receiving a doping dose, all nanowires $k = 0, \ldots, i-1$ receive the same doping dose simultaneously. Consequently, the doping profile of a nanowire i depends not only on its own doping dose but also on all doping doses received by the nanowires $k = i+1, \ldots, M-1$. We therefore need to determine the analytical multivariable application that links P_i^j and D_i^j for $i = 0, \ldots, N-1$ and $j = 0, \ldots, M-1$, in order to specify whether we can find a set of doping profiles that results in a given set of patterns.

Assuming that a set of doping profiles exists for any set of patterns, it is possible to optimize the choice of patterns according to different cost functions. We consider first the impact of this decoding technique on the fabrication cost. The nanowire profile implies a certain number of lithography/doping steps per nanowire, ϕ_i for $i = 0, \ldots, N-1$. From the fabrication point of view, it is of the highest importance to reduce the total number of lithography/doping steps, *i.e.* $\sum \phi_i$. We therefore need to establish the link between P_i^j and ϕ_i in order to minimize $\sum \phi_i$.

Then, we consider the impact of this decoding technique on the circuit yield by analyzing the variability of the decoder. Every doping region j of the nanowire i, referred to as region (i, j), receives successive doping doses bit by bit. With every additional doping dose, the variability of region (i, j), quantified as the standard deviation of the threshold voltage of this region Σ_i^j, accordingly increases. It is therefore desirable to establish the link between P_i^j and Σ_i^j and to optimize the choice of the patterns in order to minimize the variability.

In the following section, we will derive the analytical mapping between patterns and doping profiles. Then, we will define the cost functions related to the fabrication complexity and to the threshold voltage variability. These cost functions will be minimized by choosing the best code type for the decoder.

4. PROBLEM FORMULATION OF MSPT-BASED NANOWIRE DECODER

In this section we provide an abstract description of the decoder part of the nanowire array in a single half cave. The defined matrices describe the most relevant design and fabrication aspects. Using these definitions, we will derive the cost functions of the fabrication complexity and the circuit variability. Further, we assume a multi-valued logic addressing with n values.

978-1-60558-497-3/09 $25.00 © 2009 ACM

Definition 1. The pattern matrix $\mathbf{P}$ is an $N \times M$ matrix in $\{0, \ldots, n-1\}^{N \times M}$ representing the patterns of N nanowires within a half cave, where every nanowire has M doping regions.

We assume that the N nanowires within every half cave are patterned and have M doping regions each. The pattern corresponds to a set of M V_T's having any one of the n possible values $V_T(0), \ldots, V_T(n-1)$. The patterns are represented by the discrete values $0, \ldots, n-1$, which correspond to the ordered discretization of the threshold voltages $V_T(0), \ldots, V_T(n-1)$. Consequently, the set of patterns on the N nanowires forming one half cave can be represented by an $N \times M$-matrix in $\{0, \ldots, n-1\}^{N \times M}$.

Definition 2. The final doping matrix $\mathbf{D}$ is an $N \times M$ matrix in $\mathbb{R}^{N \times M}$ representing the doping level distribution along the N nanowires within a half cave after the definition of the whole array.

Every V_T needs a unique doping level N_D fixed by the device physics and geometry [14]. Consequently, the pattern matrix, that is uniquely mapped onto a set of V_T's, defines a unique final doping matrix.

Example 1. For $n = 3$, $N = 3$ and $M = 4$, we assume that V_T can have the values 0.1 V, 0.3 V and 0.5 V corresponding to the digits 0, 1 and 2 and to the doping levels 2, 4 and 9×10^{18} cm^{-3}. The patterns are represented with the first N code words of the n-ary tree code. With $\mathbf{V}$ the matrix covering all V_T's, we obtain:

$$\mathbf{P} = \begin{bmatrix} 0 & 1 & 2 & 1 \\ 0 & 2 & 2 & 0 \\ 1 & 0 & 1 & 2 \end{bmatrix} \quad \mathbf{V} = \begin{bmatrix} 1 & 3 & 5 & 3 \\ 1 & 5 & 5 & 1 \\ 3 & 1 & 3 & 5 \end{bmatrix} \cdot 0.1\,\mathrm{V} \quad \mathbf{D} = \begin{bmatrix} 2 & 4 & 9 & 4 \\ 2 & 9 & 9 & 2 \\ 4 & 2 & 4 & 9 \end{bmatrix} \cdot \frac{10^{18}}{\mathrm{cm}^3}$$

PROPOSITION 1. *A non-linear bijective application* h *maps* $\mathbf{P}$ *onto* $\mathbf{D}$ *as follows:* $D_i^j = \mathrm{h}(P_i^j) \, \forall \, i, j$

PROOF. The mapping between digits of the patterns and V_T is a discrete ordering, which is a bijective application g. The mapping between V_T's and N_D's is a monotonic non-linear function f, which is also a a bijection. The interested reader is invited to look into [14] to obtain the exact expression of f. Since h is a composition of f and g, it is bijective as well. $\square$

Definition 3. The step doping matrix $\mathbf{S}$ is an $N \times M$ matrix in $\mathbb{R}^{N \times M}$ representing the additional doping levels after every lithography/doping step.

There is a lithography/doping procedure that follows the definition of every one of the N nanowires. Every procedure $i = 0, \ldots, n-1$ is characterized by M doping levels $\left[N_{D,i}^0, \ldots, N_{D,i}^{M-1}\right]$ along the M doping regions of the nanowires. The set of M doping levels in the N steps can be represented by the matrix $\mathbf{S}$ in $\mathbb{R}^{N \times M}$.

PROPOSITION 2. *A multi-linear application maps the elements of* $\mathbf{S}$ *onto those of* $\mathbf{D}$ *as follows:* $D_i^j = \sum_{k=i}^{N-1} S_k^j$

PROOF. Every nanowire j that is defined, is subsequently patterned by means of doping doses $\left[S_j^0, \ldots, S_j^{M-1}\right]$. Any nanowire i defined before the nanowire j ($i < j$) receives the same dose simultaneously. Thus, the doping level of the nanowire i is the sum of all the levels defined in the steps $i, \ldots, N-1$ following the definition of the nanowire i. $\square$

Example 2. The following step and final doping matrices verify the property stated in Proposition 2. Negative and positive doping levels correspond to the doses with n- and p-type dopants respectively:

$$\mathbf{D} = \begin{bmatrix} 2 & 4 & 9 & 4 \\ 2 & 9 & 9 & 2 \\ 4 & 2 & 4 & 9 \end{bmatrix} \cdot \frac{10^{18}}{\mathrm{cm}^3} \quad \mathbf{S} = \begin{bmatrix} 0 & -5 & 0 & 2 \\ -2 & 7 & 5 & -7 \\ 4 & 2 & 4 & 9 \end{bmatrix} \cdot \frac{10^{18}}{\mathrm{cm}^3}$$

Definition 4. The technology complexity is quantified by Φ representing the total number of additional lithography/doping steps needed to pattern the nanowires.

Every row in $\mathbf{S}$ ($\mathbf{S}_i = \left[S_i^0 \ldots S_i^{M-1}\right]$, $i = 0, \ldots, N-1$) represents the doping doses used in a single step doping procedure. The number of unequal non-zero elements in $\mathbf{S}_i$ represents the number of different doses used at this doping step. The more doping doses, the more lithography steps are needed and the more complex is the fabrication. Let ϕ_i ($i = 0, \ldots, N-1$) be the number of unequal non-zero elements in $\mathbf{S}_i$, then the total number of lithography/doping steps is $\Phi = \sum_i \phi_i$.

Example 3. For $\mathbf{S}$ given in Example 2, we have: $\phi_1 = 2$, $\phi_2 = 4$ and $\phi_3 = 3$. Then, $\Phi = 9$ holds.

Definition 5. The decoder variability is quantified by a $N \times M$ matrix Σ, describing the standard deviation of the threshold voltages in every doping region in the decoder of a half cave.

Every doping operation yields a V_T with a given variability σ_T, measured as the standard deviation. In the proposed technique every doping region is doped at most N times (Proposition 2). We expect the variability to increase with increasing number of doping operations. The number of times a doping regions (i, j) receives a doping dose decreases with increasing i and increasing number of zero-elements in the column j of $\mathbf{S}$. Let ν_i^j be this number, then $\nu_i^j = \sum_{k=i \ldots N-1} (1 - \delta(S_k^j))$, where $\delta(x)$ is the *Kronecker delta function*: $\delta(x) = 1 \Leftrightarrow x = 0$, otherwise $\delta(x) = 0$. Doping operations are assumed to be stochastically independent. The addition of two independent stochastic variables with standard deviations σ_1 and σ_2 respectively yields a stochastic variable with the standard deviation $\sigma_0 = \sqrt{\sigma_1^2 + \sigma_2^2}$. Therefore, if we define Σ as the $N \times M$-matrix describing the variability of the decoder by setting Σ_i^j to the square of the standard deviation of the doping region (i, j), we obtain: $\Sigma_i^j = \sigma_T^2 \cdot \nu_i^j$.

Example 4. For $\mathbf{S}$ given in Example 2, we have:

$$\mathbf{S} = \begin{bmatrix} 0 & -5 & 0 & 2 \\ -2 & 7 & 5 & -7 \\ 4 & 2 & 4 & 9 \end{bmatrix} \cdot \frac{10^{18}}{\mathrm{cm}^3} \quad \Sigma = \begin{bmatrix} 2 & 3 & 2 & 3 \\ 2 & 2 & 2 & 2 \\ 1 & 1 & 1 & 1 \end{bmatrix} \cdot \sigma_T^2$$

PROPOSITION 3. *Optimizing the decoder fabrication complexity consists in finding the best pattern* $\mathbf{P}$ *that minimizes* Φ. *Optimizing the decoder reliability consists in finding the best pattern* $\mathbf{P}$ *that minimizes* $\|\Sigma\|_1$, *where* $\|\Sigma\|_1$ *is the sum of all elements of* Σ, *known as its entrywise 1-norm.*

PROOF. This follows directly from Def. 4 and 5. $\square$

5. OPTIMIZING NANOWIRE CODES

In Sec. 2.3 we reviewed the code types that we are considering to uniquely address the nanowires in any logic with n values: hot codes, tree codes and the arrangement of the tree code words called Gray codes and arranged Gray codes. The last 3 code types are used implicitly in a reflected form. The length of the whole code word–including the reflected part–is M.

In the following, we will prove that the Gray code is the optimal arrangement of the tree code with respect to the defined cost functions. Then, we will investigate the opportunity of arranging the hot codes in a similar fashion to Gray codes in order to optimize the costs of the decoders designed with hot codes.

5.1 The Gray Code

PROPOSITION 4. *Among all arrangements of tree codes, the Gray code minimizes the decoder cost in terms of variability* $\|\Sigma\|_1$.

PROOF. For $i = N-1$, $S_{N-1}^j = D_{N-1}^j = h^{-1}(P_{N-1}^j)$ is fixed by the pattern of the last nanowire, *i.e.* by P_{N-1}^j. Thus, $\nu_{N-1}^j = 1 - \delta(S_{N-1}^j) = 1$, because $S_{N-1}^j \neq 0 \, \forall \, j$, since every region receives a doping dose in order to define the pattern of V_T's of the last nanowire P_{N-1}^j. For $i \neq N-1$, $\nu_i^j - \nu_{i+1}^j = 1 - \delta(S_i^j) = 1 - \delta(D_i^j - D_{i+1}^j)$. This difference is 1 if $D_i^j \neq D_{i+1}^j$, *i.e.* $P_i^j \neq$

978-1-60558-497-3/09 $25.00 © 2009 ACM

P_{i+1}^j, and 0 if $P_i^j = P_{i+1}^j$. Then, ν_i^j can only increase by steps of 1 or remain constant with decreasing i for a fixed j. It remains unchanged if and only if the pattern P_i^j remains unchanged.

Consequently, $\|\Sigma\|_1$ monotonically increases with increasing transitions in the pattern matrix $\mathbf{P}$ between every two successive rows. Given that the rows of $\mathbf{P}$ are the code words in the chosen code space, it is desirable to use the code that minimizes the number of transitions between successive code words in order to minimize $\|\Sigma\|_1$. This condition is fulfilled by the Gray code. $\square$

Example 5. Instead of $\mathbf{P}$ given in Example 1, which includes a tree code sequence with the cost $\|\Sigma\|_1 = 22\cdot\sigma_T^2$ (from Example 4), we use a sequence from the Gray code that avoids the forbidden transition in $\mathbf{P}$ $0220\Rightarrow1012$. Then we obtain $\|\Sigma\|_1 = 18\cdot\sigma_T^2$:

$$\mathbf{P} = \begin{bmatrix} 0 & 1 & 2 & 1 \\ 0 & 2 & 2 & 0 \\ 1 & 2 & 1 & 0 \end{bmatrix}\ \mathbf{S} = \begin{bmatrix} 0 & -5 & 0 & 2 \\ -2 & 0 & 5 & 0 \\ 4 & 9 & 4 & 2 \end{bmatrix}\cdot\frac{10^{18}}{\mathrm{cm}^3}\ \Sigma = \begin{bmatrix} 2 & 2 & 2 & 2 \\ 2 & 1 & 2 & 1 \\ 1 & 1 & 1 & 1 \end{bmatrix}\cdot\sigma_T^2$$

PROPOSITION 5. *Among all arrangements of tree codes, the Gray code minimizes the fabrication cost Φ.*

PROOF. In a similar way to the proof of Proposition 4, we notice that the value of S_i^j is unequal to zero if there is a transition $P_i^j \Rightarrow P_{i+1}^j$ between two successive code words in $\mathbf{P}$ at the digit j. Then, any ϕ_i, and consequently Φ, increase with the number of transitions in $\mathbf{P}$. Since the Gray code minimizes the number of transitions, then it is optimal with respect to Φ. $\square$

Example 6. The Gray code in Example 1 has a fabrication cost $\Phi = 9$ (Example 3). By using the Gray code in Example 5, the fabrication cost was reduced to $\Phi = 7$ ($\phi_1 = 2$, $\phi_2 = 2$ and $\phi_3 = 3$).

5.2 Arranged Hot Codes

The previous section demonstrated that the arrangement of the TC words into a sequence that minimizes the number of transitions between every successive code words, defined a the GC, which minimizes the decoder variability and the fabrication cost. We therefore considered the question whether the code words of HCs can be arranged in a similar way to the GC, such that the number of transitions is minimized, and to assess the possible benefits of such codes, that we called *Arranged Hot Code (AHC)*.

Since the number of values in every HC word (M, k) is fixed (Sec. 2.3), then the minimum number of transitions is 2. We used an exhaustive algorithm for most of the hot codes with a reasonable code space size ($\lesssim 100$) for nanowire arrays, and we found that the arrangement in a *Gray-code-fashion* always exists.

It is possible to show in a very similar way to Proposition 4 and 5 that, when an arrangement of a given hot code exists, in such a way that the number of transitions between every successive code words is minimized, then this arrangement is the optimal hot code with respect to $\|\Sigma\|_1$ and Φ compared to all possible arrangements of the same hot code. In the next section we will therefore assess the performance of the optimized versions of both tree and hot codes in terms of fabrication complexity and circuit costs.

6. SIMULATIONS OF THE DECODER

In this section, the yield and area of crossbar circuits are simulated for different code types and the improvement resulting from the optimized decoder design (defined as its size and the code type) is demonstrated.

6.1 Simulation Platform

In order to assess the impact of the decoder design (meaning the choice of the code type) on the fabrication complexity and the circuit features, we performed a statistical analysis of a crossbar circuit (Fig. 1.a). The function of the crossbar circuit was assumed to be a memory. The defects affecting the molecular switches or the phase change layer were not simulated. Only the defects happening at the decoder part due to the variability of the V_T's in the doping regions were considered. We neglected the probability that

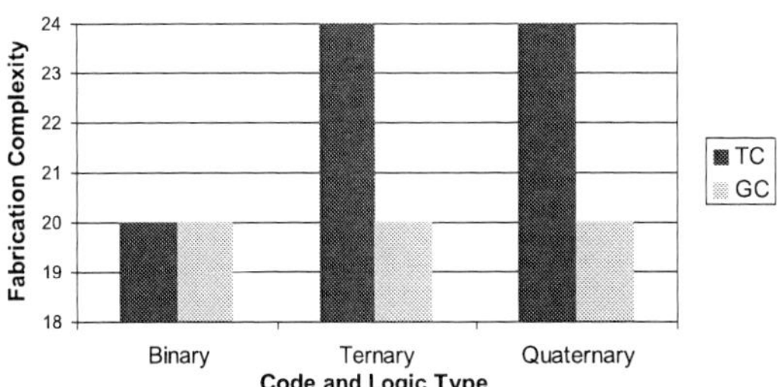

Figure 5: Fabrication complexity in terms of number of additional steps for different code types and logic levels

nanowires can be broken; we actually noticed that the fabricated nanowires (Fig. 3.a) had a yield close to unit.

Both layers forming the crossbar have the same number of caves for a square-shaped crossbar. The cave count and the number of nanowires in every half cave N was fixed according to the raw crosspoint density set to D_{RAW}=16 kB. The number of contact groups per half cave was minimized with respect to the code type (code space size Ω and code length M) and geometry (lithography pitch P_L and nanowire pitch P_N). While Ω and M were used as simulation parameters, P_L was set to 32 nm and P_N to 10 nm. According to standard layout rules, the minimum width of every contact group had to be set to $1.5 \times P_L$. The maximum width of every contact group was limited by the width of Ω nanowires at most that can fit in every contact group.

The threshold voltages V_T's are distributed within the range 0 to 1 V, in order to account for a maximum supply voltage of 1 V. The doping levels were estimated from V_T's by using the assumptions in [14] for the most common materials used in standard CMOS processes. The variability σ_T of V_T was set to 50 mV. A nanowire is addressable if V_T at every doping region varies within a small range as specified in [2]. In this way, the probability that a nanowire is addressed was calculated from the Gaussian distributions of V_T's with the standard deviations given by Σ. We accounted for nanowires that may be addressed by two adjacent contact groups, as explained in [6], and we removed them from the set of addressable nanowires. This gives the estimate for the yield of every cave Y. Consequently, the effective array density that denotes the number of working crosspoints can be estimated as: $D_{\mathrm{EFF}} = D_{\mathrm{RAW}} \cdot Y^2$.

6.2 Simulation Results

We calculated the technology complexity Φ for different code and logic types. The results, plotted in Fig. 5 for $N = 10$ show that Φ is constant for all binary codes and equal to the double of the number of nanowires in a half cave. Higher logic level was suggested as a way to reduce the area overhead of the decoder [2]. However, Fig. 5 shows that the higher logic level comes with some fabrication cost: 20% more steps for the tree code. For ternary and quaternary logic, the Gray code performs better than the tree code (17%) by completely canceling the fabrication complexity overhead.

The variability matrix was calculated for various types of binary codes. N was set to 20 and the plots in Fig. 6 show the variability level at every digit in the $N \times M$-matrix Σ, as square roots of elements of Σ normalized to σ_T. By comparing Fig. 6.a, c and e, we see that the Gray code and its balanced version reduce the variability level at every digit in comparison to the tree code. The balanced Gray code distributes the variability more evenly than the other codes. In this way, the average variability $\|\Sigma\|_1/(N \cdot M)$ could be reduced by 18%. Similar results were obtained for these codes with a higher logic level, as well as for hot codes and their arranged version. Next, we compared the distribution of the elements of Σ for a fixed code type and different code lengths (Fig. 6.a, c and e vs. Fig. 6.b, d and f). We noticed that longer codes have less digit transitions and help reduce the average variability.

The elements of Σ provide the inputs to estimate the crossbar yield, that we quantified as the effective crossbar density normalized by the raw crossbar density of 16 kB. The crossbar yield

978-1-60558-497-3/09 $25.00 © 2009 ACM

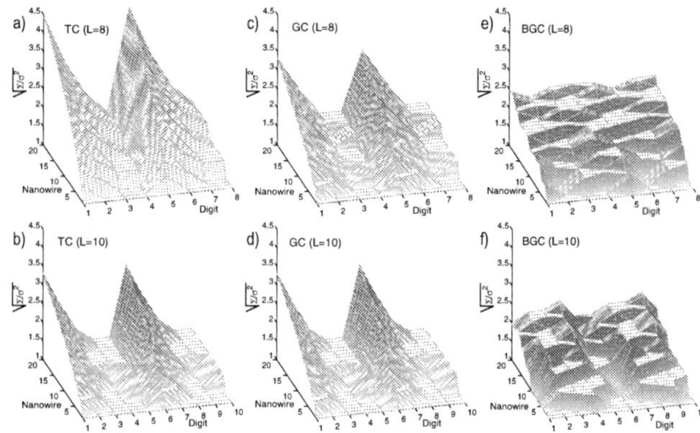

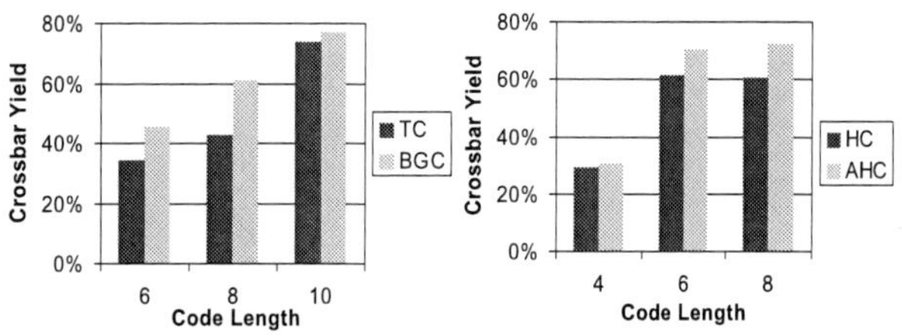

Figure 6: Square root of elements of variability matrix Σ for different binary code types and lengths

Figure 7: Crossbar yield in terms of percentage of addressable crosspoints for different binary code types and lengths

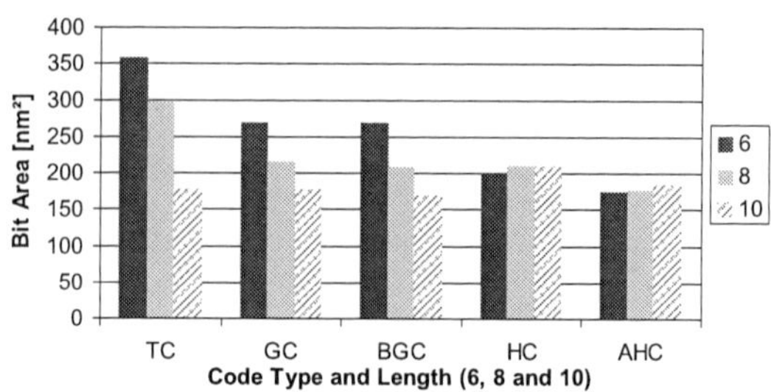

Figure 8: Average area per functional bit for different binary code types and lengths

is plotted in Fig. 7 for various binary code types and lengths. The yield generally increases with increasing code length, until it reaches the maximum ($M \sim 10$ for TC/BGC and $M \sim 6$ for HC/AHC). The yield improvement of the tree code and the arranged hot codes, by increasing the code length from 6 to 10 and 4 to 8 respectively, is $\sim 40\%$. For a fixed code length, the optimized codes (*i.e.* BGC and AHC) perform better than their non-optimized versions (*i.e.* TC and HC respectively). For instance, the balanced Gray code yields 42% more than the tree code, and the arranged hot code 19% better than the hot code with the same length $M = 8$.

The dependency of the yield on the code length is explained by two factors: *i)* the variation with M of the percentage of nanowires at adjacent contact groups, which have to be removed from the set of addressable nanowires; and *ii)* the dependency of the variability on M. On one hand, by increasing M and keeping the code type fixed, the code space size increases and less contact groups are needed, so less nanowires are removed at adjacent contact groups and the yield increases. This effect saturates when the code space size is large enough to neglect the number of nanowires at adjacent contact groups. On the other hand, the average variability $\|\Sigma\|_1 / (N \cdot M)$ decreases with increasing M, because longer codes have less digit transitions, as we showed previously. So, for a fixed code type, the yield, first, increases with increasing M, then it starts decreasing for larger M because the increasing number of digits cancels the benefits of decreasing variability of each digit taken separately. This decrease is just slightly seen for the hot code when M increases beyond 6; and for other code types, it starts appearing from $M \sim 10$.

From the geometrical data, we estimated the crossbar area; then, by considering the effective density, we estimated the bit area for different code types and lengths (Fig. 8). For the tree code and its optimized versions (Gray and balanced Gray codes), the bit area decreases with increasing code length mainly due to the vanishing effect of adjacent contact regions, as explained previously. An area saving by 51% can be achieved by setting M to 10 instead of 6 for the tree code. The balanced Gray code yields a denser crossbar than the Gray code, which in turns yields better than the tree code; for instance crossbars with the balanced Gray code are 30% denser than those with the tree code (for $M = 8$). The hot and arranged hot codes yield the most dense crossbars with $M = 6$.

For larger M, the bit area slightly start to increase, as the yield starts to decrease because of higher variability of longer codes, as explained previously. The arranged hot code performs better than the hot code with 13% less bit area for $M = 6$. Among all these codes, the smallest bit area is 169 nm^2 for the balanced Gray code, followed by the arranged hot code with 175 nm^2.

7. CONCLUSIONS

Silicon nanowires are promising devices that may help tackle many design challenges by means of their organization into regular crossbar circuits. In this paper we presented our validated fabrication technique, the MSPT, that yields very dense parallel arrays of silicon nanowires and we introduced a design style for the nanowire decoder using the same fabrication technique. We identified the design challenges for the MSPT-decoder that result from the interdependence of the nanowire patterns. We quantified these challenges in terms of fabrication cost and decoder variability. Then we proved that the arrangement of the code space in a Gray code fashion minimizes these costs. We performed simulations with different decoder design parameters and code types, including the Gray code, the balanced Gray code and the arranged hot codes. The results showed that the technology cost can be reduced by 17% by using the Gray code. The Gray code and its balanced version help reduce and balance the variability throughout the whole nanowire decoder, resulting in 18% less variability. The decoder design covers not only the code type but also its length: the yield improves by 40% by adding redundancy to the code length and by 19 to 42% by using the optimized code types. The smallest effective area per bit was found to be around 170 nm^2 for the optimized codes: the balanced Gray code and the arranged hot code.

8. REFERENCES

[1] R. Beckman et al. Bridging dimensions: Demultiplexing ultrahigh density nanowire circuits. *Science*, 310(5747):465–468, 2005.

[2] M. H. Ben Jamaa et al. Variability-aware design of multi-level logic decoders for nanoscale crossbar memories. *Trans. CAD*, 27(11):2053–2067, Nov. 2008.

[3] G. S. Bhat and C. D. Savage. Balanced Gray codes. *The Electronic Journal of Combinatorics*, 3(1):R25, 1996.

[4] G. Cerofolini. Realistic limits to computation. II. The technological side. *Applied Physics A*, 86(1):31–42, 2007.

[5] A. DeHon. Design of programmable interconnect for sublithographic programmable logic arrays. In *Proc. Int. Symp. on FPGA*, pages 127–137, 2005.

[6] A. DeHon et al. Stochastic assembly of sublithographic nanoscale interfaces. *IEEE Trans. on Nanotechnology*, 2(3):165–174, 2003.

[7] F. Gray. Pulse code communication. *US Patent No. 2632058*, 1953.

[8] T. Hogg et al. Assembling nanoscale circuits with randomized connections. *Trans. on Nanotechnology*, 5(2):110–122, 2006.

[9] J. D. Holmes et al. Control of thickness and orientation of solution-grown silicon nanowires. *Science*, 287(5457):1471–1473, 2000.

[10] Y. Luo et al. Two-dimensional molecular electronics circuits. *ChemPhysChem*, 3:519–525, 2002.

[11] N. A. Melosh et al. Ultrahigh-density nanowire lattices and circuits. *Science*, 300(5616):112–115, 2003.

[12] K. E. Moselund et al. Prospects for logic-on-a-wire. *Microelectronic Engineering*, (85):1406–1409, 2008.

[13] J. E. Savage et al. Radial addressing of nanowires. *Journal on Emerging Technologies in Computing Systems*, 2(2):129–154, 2006.

[14] S. M. Sze and K. K. Ng. *Physics of Semiconductor Devices*. 2007.

[15] D. Whang et al. Large-scale hierarchical organization of nanowire arrays for integrated nanosystems. *Nano Letters*, 3(9):1255–1259, 2003.

[16] Y. Zhang et al. An integrated phase change memory cell with ge nanowire diode for cross-point memory. *VLSI Technology*, pages 98–99, June 2007.

978-1-60558-497-3/09 $25.00 © 2009 ACM

Boolean Logic Function Synthesis for Generalised Threshold Gate Circuits

Marek A. Bawiec
marek.bawiec@pwr.wroc.pl

Maciej Nikodem
maciej.nikodem@pwr.wroc.pl

The Institute of Computer Engineering, Control and Robotics
Wrocław University of Technology, Poland

ABSTRACT

This paper analyses negative differential resistance (NDR) based logic circuits operating in monostable-bistable transition logic element (MOBILE) regime. We formulate theoretical foundation for formal model of the generalised threshold gate (GTG) and prove that GTG can implement any, n-variable Boolean function. Moreover, we propose the algorithmic approach to GTG structure generation problem.

Categories and Subject Descriptors

B.2 [**Arithmetic And Logic Structures**]: Miscellaneous

General Terms

Algorithms, Design, Theory

Keywords

GTG, Nanoscale Devices, Logic Synthesis, NDR

1. INTRODUCTION

Exploration of negative differential resistance (NDR) and its application to logic circuits focuses on theoretical analyses [2, 3, 4, 8, 9] and physical implementations [5, 6]. Circuits proposed by Avedillo [2, 3] and Berezowski [4] are based on the monostable-bistable transition logic element (MOBILE) concept [1]. Basic MOBILE circuit that operates in such regime, consists of two NDR elements: a load (NDR_l) and a driver (NDR_d) connected in series (Fig. 1–B). By applying bias voltage V_{bias}, which varies between $0V$ and V_{DD} (Fig. 1–A) the circuit switches from monostable OUT_m to one of the bistable states: OUT_{b1} – low voltage or OUT_{b2} – high voltage. The actual output voltage depends on the peak currents I_{pl} and I_{pd} relation (Fig. 1–C). Since the NDR element with smaller peak current switches to high resistance state when V_{bias} increases thus circuit reaches OUT_{b1} state if $I_{pl} < I_{pd}$ and OUT_{b2} if $I_{pl} > I_{pd}$.

Permission to make digital or hard copies of part or all of this work for personal or classroom use is granted without fee provided that copies are not made or distributed for profit or commercial advantage and that copies bear this notice and the full citation on the first page. To copy otherwise, to republish, to post on servers or to redistribute to lists, requires prior specific permission and/or a fee.

DAC'09, July 26-31, 2009, San Francisco, California, USA

Operation of the circuit composed of NDR elements is controlled by four-phase V_{bias} that changes similarly to the traditional clock signal (Fig. 1–A): I-st phase (V_{bias} increases) – circuit evaluates the output switching from monostable to one of bistable states; II-nd phase (V_{bias} high) – circuit remains in the state selected in phase I, independently of the actual peaks I_{pl} and I_{pd} relation; III-rd phase (V_{bias} falls) – circuit returns to the initial monostable state; IV-th phase (V_{bias} low) – circuit remains in monostable state and awaits for I_{pl} and I_{pd} relation adjustments.

Adjusting the relation between I_{pl} and I_{pd} allows to control the output voltage in the bistable state thus enabling to implement Boolean functions. This can be achieved if NDR_l and/or NDR_d is paralleled with another NDR and a transistor that operates as a switch. This concept was presented by Avedillo et al. [2, 3] and Berezowski [4]. They proposed to build circuits from the serially connected NDR_l and NDR_d paralleled by branches of a single NDR-transistor pair [2] or a single NDR serially connected with serial–parallel (SP) transistor network [3, 4].

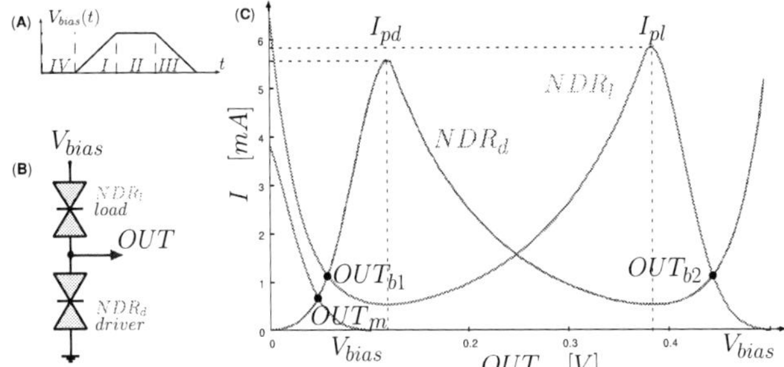

Figure 1: **(A)**–V_{bias} **clocking scheme, (B)–basic MOBILE circuit, (C)–**$I(V)$ **characteristic.**

In this paper we extend the concept presented in [4], propose a modified model of the GTG circuit, prove that GTG enables to implement any n–variable boolean function, and give a GTG synthesis algorithm.

2. MOTIVATION AND PREVIOUS WORK

Avedillo at al. [2] proposed a multi-threshold threshold gate (MTTG) built with NDR devices that have different parameters (i.e. peak currents). Circuit consisting of $2n$ branches, and $2n + 2$ properly adjusted NDR elements is capable of computing weighted sum of n binary inputs and to quantise its result. This allows to implement any 2-variable and small number of $n > 2$-variable boolean functions (e.g. AND, OR). Unfortunately [2] gives no method of circuit synthesis (i.e. selecting parameters of NDR elements) for a given boolean function. Successive paper by Avedillo et al. [3] proposed a modified structure that utilises one or many, serially connected transistors to switch each NDR element. This allows to implement all n-variable boolean functions, however, extends the set of input variables yielding circuit of up to $2^n + 1$ branches. This is infeasible due to two major concerns: (i) inaccuracies in physical design of NDRs accumulate and may result in erroneous quantisation; (ii) power consumption increases with the number of branches in the circuit.

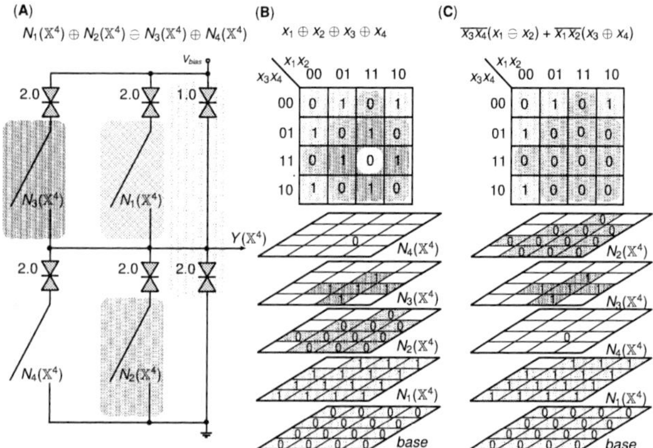

Figure 2: GTG circuit example implementing $Y(\mathbb{X}^4) = x_1 \oplus x_2 \oplus x_3 \oplus x_4$ **(A) circuit structure, (B) correct and (C) incorrect GTG model interpretation.**

Berezowski [4] extended approach of a generalised threshold gate (GTG) and proposed to buit circuit out of series of identical NDR elements, serially connected with SP transistor network without a complementary transistor pair. Such an approach has several advantages: (i) reduced number of branches; (ii) no need to implement NDR elements with different parameters; (iii) no need for complementary transistor pair. Berezowski gave an iterative formula that describes GTG behaviour and used it to show that all 4–input boolean functions can be implemented with at most 6 branches. This was done by exhaustive search and neither proof, that $n + 2$–branch GTG can implement any n–input boolean function, nor the synthesis method was given.

Recent paper by Pettenghi et al. [7] extends the model [4] by using additional inverters and introducing complementary signals. This results in simplified circuit structure, but introduces additional transistors and requires a complementary transistor pair.

3. NEW MODEL

Without loosing generality we can assume that $I_{pd} > I_{pl}$ and thus $Y(0, \ldots, 0) = 0$. If so, then the top branches set '1' to the function output, while the bottom branches reset the output to '0'. Following the [4] function implemented with GTG can be described in a recursive fashion:

$$Y_l(\mathbb{X}^n) = \begin{cases} 0 & l = 0 \\ Y_{l-1}(\mathbb{X}^n) + N_l(\mathbb{X}^n) & l = 2k - 1 \\ Y_{l-1}(\mathbb{X}^n)\overline{N_l(\mathbb{X}^n)} & l = 2k \end{cases} . \quad (1)$$

Unfortunately, (1) is not a GTG general description, since the order of $N_i(\mathbb{X}^n)$ functions (N_i for short) influences the

resulting function $Y(\mathbb{X}^n)$. Consider a 4-input binary logic function (Fig. 2)

$$Y(\mathbb{X}^4) = x_1 \oplus x_2 \oplus x_3 \oplus x_4, \quad (2)$$

which can be implemented with 6–branch GTG where:

$$\begin{aligned} N_1(\mathbb{X}^4) &= x_1 + x_2 + x_3 + x_4, \\ N_2(\mathbb{X}^4) &= x_1x_2 + x_1x_3 + x_1x_4 + x_2x_3 + x_2x_4 + x_3x_4, \\ N_3(\mathbb{X}^4) &= x_1x_2x_3 + x_1x_2x_4 + x_1x_3x_4 + x_2x_3x_4, \\ N_4(\mathbb{X}^4) &= x_1x_2x_3x_4. \end{aligned} \quad (3)$$

It can be easily verified that evaluating (1) for (3) yields (2). However, changing function order by swapping $N_2(\mathbb{X}^4)$ and $N_4(\mathbb{X}^4)$, and evaluating (1) again gives

$$Y_1(\mathbb{X}^4) = \overline{x_3 x_4}(x_1 \oplus x_2) + \overline{x_1 x_2}(x_3 \oplus x_4). \quad (4)$$

This suggests that the related circuit implements different function (Fig. 2–C). Moreover, swapping $N_1(\mathbb{X}^n)$ and $N_3(\mathbb{X}^n)$ gives

$$Y_2(\mathbb{X}^4) = x_2\overline{x_3} + \overline{x_1}x_4 + x_3\overline{x_4} + x_1\overline{x_2}. \quad (5)$$

This obviously cannot be true since the order of branches is not important in physical implementation. This property of (1) model limits its practical application and brings in the need for a new model that will be indifferent to $N_i(\mathbb{X}^n)$ functions order. Proper order of $N_i(\mathbb{X}^n)$ functions for the (1) model can be identified from the analysis of the way a MOBILE circuit operates – assuming $Y(0, \ldots, 0) = 0$, switching on the upper branch turns the output to '1' while subsequent activation of bottom branch turns it back to '0'. The next upper branch switches it again to '1' and so on. Activating upper and bottom branches in turns means that for every two functions $N_i(\mathbb{X}^n)$, $X_j(\mathbb{X}^n)$ $i < j$ the following equation has to be satisfied

$$N_i(\mathbb{X}^n)N_j(\mathbb{X}^n) = N_j(\mathbb{X}^n) \quad (6)$$

Eq.(6) determines the proper order of $N_i(\mathbb{X}^n)$ functions for the (1) model but also brings in additional properties:

$$\begin{aligned} \overline{N_i}N_j &= \left(\overline{N_i} + \overline{N_j}\right)N_j = \overline{N_iN_j}N_j = \overline{N_j}N_j = 0, \\ N_i + N_j &= N_i + N_iN_j = N_i(1 + N_j) = N_i, \\ \overline{N_i}\,\overline{N_j} &= \overline{N_i + N_j} = \overline{N_i}, \\ N_i\overline{N_j} &= N_i\overline{N_j} + \overline{N_i}N_j = N_i \oplus N_j. \end{aligned} \quad (7)$$

Let m denote the total number of $N_i(\mathbb{X}^n)$ functions. Since $m = 2k + \delta$ for some k and $\delta = 0$ if m is even or $\delta = 1$ if m is odd, thus the GTG model (1) can be simplified to

$$Y(\mathbb{X}^n) = \bigcup_{i=1}^{\lfloor \frac{m}{2} \rfloor} N_{2i-1}(\mathbb{X}^n)\overline{N_{2i}(\mathbb{X}^n)} + \delta N_m(\mathbb{X}^n). \quad (8)$$

The following theorem states that (8) can be further transformed to negation–free EXOR sum of $N_i(\mathbb{X}^n)$ functions.

THEOREM 1. *GTG circuit compound of $m + 2$ branches and m unate functions $N_i(\mathbb{X}^n)$ such that $N_i(\mathbb{X}^n)N_j(\mathbb{X}^n) = N_j(\mathbb{X}^n)$ for any $1 \leq i < j \leq m$, implements boolean function*

$$Y(\mathbb{X}^n) = \bigoplus_{i=1}^{m} N_i(\mathbb{X}^n). \quad (9)$$

PROOF. For any $1 \leq i < j < k < l \leq m$ we have:

$$\begin{aligned} N_i\overline{N_j} &+ N_k\overline{N_l} = \\ &= N_i\overline{N_j}\left(\overline{N_k} + N_l\right) + N_k\overline{N_l}\left(\overline{N_i} + N_j\right) \\ &= N_i\overline{N_j}\,\overline{N_k}\,\overline{N_l} + N_k\overline{N_l}\overline{N_i}\,\overline{N_j} = N_i\overline{N_j} \oplus N_k\overline{N_l}. \end{aligned} \quad (10)$$

978-1-60558-497-3/09 $25.00 © 2009 ACM

In a similar way it can be shown that:

$$N_i\overline{N_j} + N_k = N_i\overline{N_j} \oplus N_k \qquad (11)$$

for any $1 \le i < j < k < l \le m$. Considering (8) it follows that $m + 2$–branch GTG circuit implements function $Y(\mathbb{X}^n) = \bigoplus_{i=1}^{m} N_i(\mathbb{X}^n)$. $\square$

The following theorem states that $m = n + 2$ branches are enough to implement any n–variable boolean function.

THEOREM 2. *GTG circuit consisting of $n + 2$ branches can implement any n–variable boolean function.*

PROOF. According to Reed-Muller canonical expression any n-variable boolean function can be represented as:

$$\begin{aligned} Y(\mathbb{X}^n) = a_0 &\oplus (a_1x_1 \oplus a_2x_2 \oplus \cdots \oplus a_nx_n) \\ &\oplus (a_{12}x_1x_2 \oplus a_{13}x_1x_3 \oplus \cdots \oplus a_{n-1\,n}x_{n-1}x_n) \\ &\oplus \cdots \oplus (a_{12\ldots n}x_1x_2\cdots x_n), \end{aligned} \qquad (12)$$

for $a_i \in \{0,1\}$. Reed-Muller formula consist of at most 2^n minterms that are products of at most n variables with no complementary signals. To simplify the notation we denote products from (12) as $I_i(\mathbb{X})$ (I_i for short) and change the indices so they will now correspond to the consecutive numbers rather than variables:

$$\begin{aligned} Y(\mathbb{X}^n) = a_0 &\oplus (a_1I_1 \oplus a_2I_2 \oplus \cdots \oplus a_nI_n) \\ &\oplus (a_{n+1}I_{n+1} \oplus a_{n+2}I_{n+2} \oplus \cdots \oplus a_{\frac{(n-1)n}{2}}I_{\frac{(n-1)n}{2}}) \\ &\oplus \cdots \oplus (a_{2^n-1}I_{2^n-1}). \end{aligned} \qquad (13)$$

In the above equation I_i denotes product of some variables such that for any $i \ne j$, $I_i \ne I_j$. Using this notation any n-variable Boolean function can be represented as EXOR sum of some $I_i(\mathbb{X})$ functions:

$$Y(\mathbb{X}^n) = a_0 \oplus \bigoplus_{i \in \Omega} I_i(\mathbb{X}), \qquad (14)$$

where $\Omega = \{j : a_j = 1\}$. Moreover, for any two terms $I_i(\mathbb{X})$ and $I_j(\mathbb{X})$ from (14), one of two cases occur:

1. $I_iI_j = I_i$ or $I_iI_j = I_j$,

2. $I_iI_j \ne I_i$ and $I_iI_j \ne I_j$.

If 1 holds for every $i \ne j$ then (14) represents the formal model of the GTG and (12) can be directly implemented.

In the other case, i.e. when 2 holds, then:

$$I_i \oplus I_j = (I_i + I_j) \oplus I_iI_j, \qquad (15)$$

$$(I_i + I_j)\,I_iI_j = I_iI_j. \qquad (16)$$

Eq (16) is analogous to (6) which means that by proper grouping of terms I_i and I_j we can transform any n–variable Reed-Muller canonical expression, to the GTG model form (9). This proves that any n–variable boolean function can be implemented with GTG circuit.

To show that at most $n + 2$ branches are required it is enough to realize that there is at most n functions I_i, I_j, being negation free sum-of-products, such that either $I_iI_j = I_i$ or $I_iI_j = I_j$ holds. Additional two branches are required to set the output for all-zero input. $\square$

4. SYNTHESIS

Our circuit synthesis algorithm utilizes two auxiliary functions: Sort and Count. $\text{Sort}(Y(\mathbb{X}^n))$ is a function that given an EXOR sum of $N_i(\mathbb{X}^n)$ terms, outputs $Y(\mathbb{X}^n)$ with EXOR terms ordered according to the smallest number of variables in products and the biggest number of terms in a sum.

Example 1: Given $Y(\mathbb{X}^n) = x_1 \oplus (x_1 + x_3x_4) \oplus x_1x_3x_4 \oplus (x_1x_2 + x_1x_3 + x_2x_3x_4)$ as an input, $\text{Sort}(Y)$ outputs: $(x_1 + x_3x_4) \oplus x_1 \oplus (x_1x_2 + x_1x_3 + x_2x_3x_4) \oplus x_1x_3x_4$.

$\text{Count}(Y(\mathbb{X}^n))$ is a function that outputs the number of EXOR terms in $Y(\mathbb{X}^n)$ expression, e.g. $\text{Count}(Y(\mathbb{X}^n))$ outputs 4 when given $Y(\mathbb{X}^n)$ from the last example.

Algorithm 1 Reed-Muller based GTG circuit synthesis

Require: n–variable Boolean function $Y(\mathbb{X}^n)$
Ensure: NDR_l vs. NDR_d relation, and $N_i(\mathbb{X}^n)$ functions
1: Transform $Y(\mathbb{X}^n)$ to Reed-Muller canonical form, i.e.

$$Y(\mathbb{X}^n) = a_0 \oplus \bigoplus_i N_i(\mathbb{X}^n),$$

2: **if** $Y(0^n) = 0$ **then** $\text{NDR}_l > \text{NDR}_d$
3: **else** $\text{NDR}_l < \text{NDR}_d$,
4: $Y(\mathbb{X}^n) = \text{Sort}(Y(\mathbb{X}^n))$,
5: set $i = 1$, $j = 2$,
6: **if** $N_i(\mathbb{X}^n)N_j(\mathbb{X}^n) \ne N_k(\mathbb{X}^n)$ for $k = i, j$ **then**
7:    set $N_i(\mathbb{X}^n) \leftarrow N_i(\mathbb{X}^n) + N_j(\mathbb{X}^n)$,
8:    set $N_j(\mathbb{X}^n) \leftarrow N_i(\mathbb{X}^n)N_j(\mathbb{X}^n)$,
9:    $Y(\mathbb{X}^n) = \text{Sort}(Y(\mathbb{X}^n))$,
10: **else** set $j = j + 1$,
11: **if** $j > \text{Count}(Y(\mathbb{X}^n))$ **then** $i = i + 1$, $j = i + 1$,
12: **if** $i < \text{Count}(Y(\mathbb{X}^n))$ **then** goto 6-th step,

---

Example 2: Synthesis of a 3-variable Boolean function $Y(x_1, x_2, x_3) = x_1x_2 + x_2\bar{x}_3 + \bar{x}_1\bar{x}_2x_3$.
1. Reed-Muller expression for a given function equals $Y_{RM}(x_1, x_2, x_3) = x_2 \oplus x_3 \oplus x_1x_3$.
2. since $Y(0^3) = 0$ thus $\text{NDR}_l > \text{NDR}_d$, set $i = 1$, $j = 2$,
3. since $N_1N_2 \ne N_k$ for $k = 1, 2$ thus

- $N_1 \leftarrow N_1 + N_2 = x_2 + x_3$ and $N_2 \leftarrow N_1N_2 = x_2x_3$,
- $Y = (x_2 + x_3) \oplus x_2x_3 \oplus x_1x_3$,

4. since $i, j \le \text{Count}(Y)$ thus keep i, j and go to step 6,
5. since $N_1(\mathbb{X}^3)N_2(\mathbb{X}^3) = N_2(\mathbb{X}^3)$ thus $j = j + 1 = 3$,
6. since $i, j \le \text{Count}(Y)$ thus keep i, j and go to step 6,
7. since $N_1(\mathbb{X}^3)N_3(\mathbb{X}^3) = N_3(\mathbb{X}^3)$ thus $j = j + 1 = 4$,
8. since $j > \text{Count}(Y)$ thus $i = i + 1 = 2$, $j = i + 1 = 3$ and go to step 6,
9. since $N_2(\mathbb{X}^3)I_3(\mathbb{X}^3) \ne N_k(\mathbb{X}^3)$ for $k = 2, 3$ thus

- $N_2 \leftarrow x_2x_3 + x_1x_3$, $N_3 \leftarrow x_1x_2x_3$,
- $Y = (x_2 + x_3) \oplus (x_1x_3 + x_2x_3) \oplus x_1x_2x_3$,

10. since $i, j \le \text{Count}(Y)$ thus go to step 6,
11. since $N_2(\mathbb{X}^3)N_3(\mathbb{X}^3) = N_3(\mathbb{X}^3)$ thus $j = j + 1 = 4$,
12. since $j > \text{Count}(Y)$ thus $i = i + 1 = 3$, $j = i + 1 = 4$,
13. since $i = \text{Count}(Y)$ thus finish and output:

$$\begin{aligned} &\text{NDR}_l > \text{NDR}_d, \qquad N_2(\mathbb{X}^3) = x_1x_3 + x_2x_3, \\ &N_1(\mathbb{X}^3) = x_2 + x_3, \qquad N_3(\mathbb{X}^3) = x_1x_2x_3. \end{aligned}$$

We have verified our synthesis method for different functions with different number of variables. Verification was based on selecting random functions, describing with Reed-Muller canonical expression and transforming into the GTG expression. Later on we simulate the circuit using PSpice software. The next example presents a synthesis of 5-variable boolean function, and its simulation results. In order to make the synthesis compact, we simplify the notation and write i instead of x_i, and ij instead of x_ix_j. Moreover, we underline

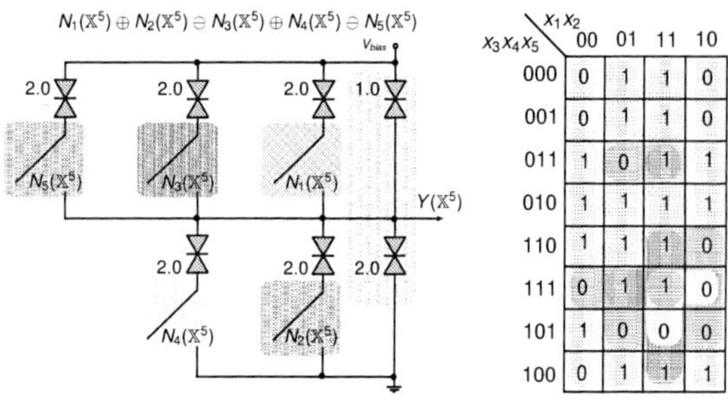
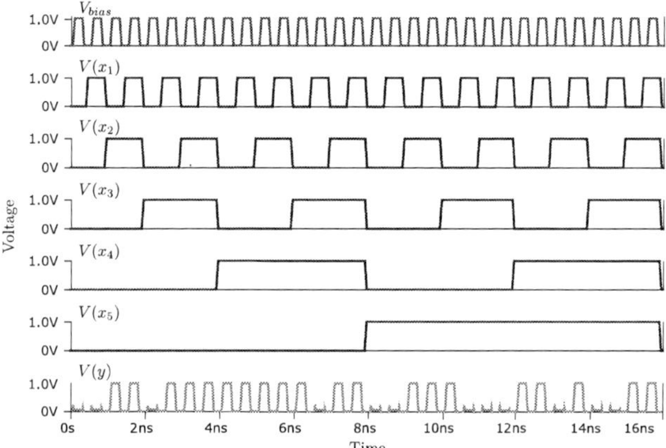

Figure 3: Carnough map, circuit structure and simulation results for 5-variables example.

terms of the EXOR sum that are transformed in each iteration. Figure 3 presents a circuit structure, its GTG model as well as simulation results.

Example 3: Synthesis of 5 variable Boolean function given in Reed-Muller form:

$$Y_{RM} = 2 \oplus 4 \oplus 13 \oplus 35 \oplus 24 \oplus 123 \oplus 245 \oplus 1245 \oplus 1345.$$

1. since $Y(0^5) = 0$ thus $\mathrm{NDR}_l > \mathrm{NDR}_d$, and:

$$
\begin{aligned}
Y &= \underline{2 \oplus 4} \oplus 13 \oplus 35 \oplus 24 \oplus 123 \oplus 245 \oplus 1245 \oplus 1345 \\
  &= \overline{(2 + 4)} \oplus 13 \oplus 35 \oplus 123 \oplus 245 \oplus 1245 \oplus 1345 \\
  &= \overline{(2 + 4 + 13)} \oplus 35 \oplus (123 + 134) \oplus 123 \\
  &\quad \oplus 245 \oplus 1245 \oplus 1345 \\
  &= N_1 \oplus (235 + 345 + 135) \oplus (123 + 134) \\
  &\quad \oplus 123 \oplus 245 \oplus 1245 \oplus 1345 \\
  &= N_1 \oplus (235 + 345 + 135 + 123 + 134) \oplus 123 \\
  &\quad \oplus \underline{245} \oplus (1235 + 1345) \oplus 1245 \oplus 1345 \\
  &= N_1 \oplus N_2 \oplus \underline{123} \oplus (1235 + 1345) \oplus \underline{2345} \oplus 1245 \oplus 1345 \\
  &= N_1 \oplus N_2 \oplus (123 + 2345) \oplus (1235 + 1345) \\
  &\quad \oplus 1245 \oplus 1345 \oplus 12345 \\
  &= N_1 \oplus N_2 \oplus \overline{(123 + 2345 + 1345)} \oplus 1235 \\
  &\quad \oplus \underline{1245} \oplus 1345 \oplus 12345 \\
  &= N_1 \oplus N_2 \oplus N_3 \oplus 1235 \oplus 1345 \\
  &= N_1 \oplus N_2 \oplus N_3 \oplus \overline{(1235 + 1345)} \oplus 12345.
\end{aligned}
$$

2. for given function synthesis algorithm outputs

$$
\begin{aligned}
\mathrm{NDR}_l &> \mathrm{NDR}_d, \\
N_1(\mathbb{X}^5) &= x_2 + x_4 + x_1 x_3 + x_3 x_5, \\
N_2(\mathbb{X}^5) &= x_2 x_3 x_5 + x_3 x_4 x_5 + x_1 x_3 x_5 + x_1 x_2 x_3 \\
                 &\quad + x_1 x_3 x_4 + x_2 x_4 x_5, \\
N_3(\mathbb{X}^5) &= x_1 x_2 x_3 + x_2 x_3 x_4 x_5 + x_1 x_3 x_4 x_5 + x_1 x_2 x_4 x_5, \\
N_4(\mathbb{X}^5) &= x_1 x_2 x_3 x_5 + x_1 x_3 x_4 x_5, \\
N_5(\mathbb{X}^5) &= x_1 x_2 x_3 x_4 x_5.
\end{aligned}
$$

5. CONCLUSIONS

While it is possible to use technology-independent synthesis with AND, OR, NOT gates and perform a technology mapping to NDR devices, the challenge in GTG design is to generate individual gate topologies. As shown in the paper, GTGs of fan-in n can implement all Boolean functions of n variables. Even for a very limited fan-in circuits (like typical library cell) only on-fly synthesis is viable approach because the size of full library of 2^{2^n} cells would be way beyond the excessive. Moreover, some Boolean functions can be implemented in many alternative variants that feature different speed, power and area trade-offs.

Synthesizing GTG circuit for a given Boolean function and exploring different viable alternatives considering the underlying combinatorial structure of the circuit topology as well as timing, power, and area constrains, is not a straightforward problem that can be solved manually. The algorithm proposed in this paper, that generates a GTG circuit consisting of at most $n + 2$ branches, given an n variable Reed-Muller canonical expression is the first synthesis method designed for NDR-based circuits.

6. REFERENCES

[1] T. Akeyoshi, K. Maezawa, and T. Mizutani. Weighted sum threshold logic operation of mobile using resonant-tunneling transistors. *Electron Device Letters, IEEE*, vol.14(10), pp.475–477, October 1993.

[2] M. Avedillo, J. Quintana, H. Pettenghi, P. Kelly, and C. Thompson. Multi-threshold threshold logic circuit design using resonant tunnelling devices. *Electronics Letters*, vol.39(21), pp.1502–1504, October 2003.

[3] M. Avedillo, J. Quintana, and H. Pettenghi. Logic models supporting the design of mobile-based rtd circuits. pp.254–259, July 2005.

[4] K. Berezowski. Compact binary logic circuits design using negative differential resistance devices. *Electronics Letters*, vol.42(16), pp.902–903, 2006.

[5] T. Kim, Y. Jeong, and K. Yang. Low-power high-speed performance of current-mode logic d flip-flop topology using negative-differential-resistance devices. *Circuits, Devices & Systems, IET*, vol.2(2), pp.281–287, April 2008.

[6] H. Kim and K. Seo. Noninverted/inverted monostabel-to-bistable transition logic element circuits using three resonant tunneling diodes and their application to a static binary frequency divider. *The Japan Society of Applied Physics*, vol.47, pp.2854–2857, 2008.

[7] H. Pettenghi, M. Avedillo, and J. Quintana. A novel contribution to the rtd-based threshold logic family. *IEEE International Symposium on Circuits and Systems, 2008*, pp.2350–2353, May 2008.

[8] H. Pettenghi, M. J. Avedillo, and J. M. Quintana. Using multi-threshold threshold gates in rtd-based logic design: A case study. *Microelectronics Journal*, vol.39(2), pp.241 – 247, 2008.

[9] Y. Zheng and C. Huang. Reconfigurable rtd-based circuit elements of complete logic functionality. *Asia and South Pacific Design Automation Conference, 2008*, pp.71–76, March 2008.

978-1-60558-497-3/09 $25.00 © 2009 ACM

Improving STT MRAM Storage Density through Smaller-Than-Worst-Case Transistor Sizing

Wei Xu
Rensselaer Polytechnic Institute
Troy, NY 12180
xuw@rpi.edu

Yiran Chen, Xiaobin Wang
Seagate Technology
Bloomington, MN 55435
{yiran.chen, xiaobin.wang}@seagate.com

Tong Zhang
Rensselaer Polytechnic Institute
Troy, NY 12180
tzhang@ecse.rpi.edu

ABSTRACT

This paper presents a technique to improve the storage density of spin-torque transfer (STT) magnetoresistive random access memory (MRAM) in the presence of significant magnetic tunneling junction (MTJ) write current threshold variability. In conventional design practice, the nMOS transistor within each memory cell is sized to be large enough to carry a current larger than the worst-case MTJ write current threshold, leading to an increasing storage density penalty as the technology scales down. To mitigate such variability-induced storage density penalty, this paper presents a smaller-than-worst-case transistor sizing approach with the underlying theme of jointly considering memory cell transistor sizing and defect tolerance. Its effectiveness is demonstrated using 256Mb STT MRAM design at 45nm node as a test vehicle. Results show that, under a normalized write current threshold deviation of 20%, the overall memory die size can be reduced by more than 20% compared with the conventional worst-case transistor sizing design practice.

Categories and Subject Descriptors

B.7.1 [**Integrated Circuits**]: Types and Design Styles—*Advanced technologies, Memory technologies*

General Terms

Design, Performance

Keywords

STT MRAM, transistor sizing, defect tolerance

1. INTRODUCTION

Recently, spin-torque transfer (STT) magnetoresistive random access memory (MRAM) has received a growing interest due to its great scalability potential [3–5, 9]. The basic building block in MRAM is magnetic tunneling junction (MTJ), which has two ferromagnetic layers separated by one oxide barrier layer. The resistance of MTJ depends on the relative magnetization directions of

Permission to make digital or hard copies of part or all of this work for personal or classroom use is granted without fee provided that copies are not made or distributed for profit or commercial advantage and that copies bear this notice and the full citation on the first page. To copy otherwise, to republish, to post on servers or to redistribute to lists, requires prior specific permission and/or a fee.
DAC'09, July 26-31, 2009, San Francisco, California, USA

the two ferromagnetic layers, i.e., when the magnetization is parallel (or anti-parallel), MTJ is in low (or high) resistance state, as illustrated in Fig. 1 (a) and (b), respectively.

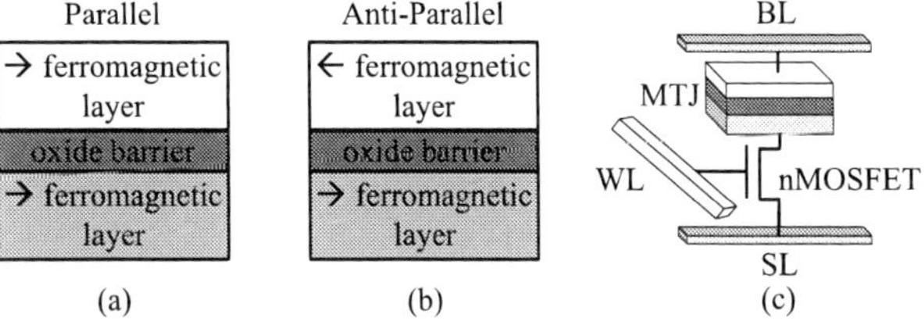

Figure 1: (a) Parallel (low resistance) and (b) anti-parallel (high resistance) of one MTJ, and (c) structure of a 1T1MTJ STT MRAM cell.

One important MTJ device parameter is its write current threshold: To successfully switch the resistance state, the current steered through MTJ must be higher than the write current threshold. Each STT MRAM cell contains one MTJ as the storage element and one nMOS transistor as the access control device as shown in Fig.1(c). To make one memory cell functional, the size of its nMOS transistor should be large enough to support a write current higher than the write current threshold of its MTJ, which tends to make the nMOS transistor largely dominate the overall memory cell size [4–6]. As the technology scales down, although the mean value of MTJ write threshold current reduces, its cell-to-cell variability tends to greatly increase due to the process variability, e.g., with the write latency on the order of 10ns, the standard deviation vs. mean ratio (SDMR) of MTJ write current threshold can be 20% and even higher [7]. As a result, if we follow the conventional design practice that sizes the nMOS transistor for the *worst-case* MTJ write current threshold (e.g., 6σ larger than the mean), the ever increasing process variability may greatly degrade the potential of storage density improvement as the technology continues to scale down.

In this work, we are interested in large-capacity stand-alone STT MRAM that potentially competes with flash memories, where the achievable storage density tends to have the highest priority. In this context, we argue that jointly considering within-cell transistor sizing and memory defect tolerance holds a great promise to minimize the impact of significant MTJ write current threshold variability on the achievable STT MRAM storage density. The basic idea is intuitive: we intentionally use smaller-than-worst-case transistor sizing to reduce the memory cell size at the cost of more cell defects that are further compensated by system-level defect tolerance techniques. This clearly involves a trade-off between the reduced individual memory cell size and increased memory defect tolerance redundancy. Moreover, memory defect tolerance may possibly result in memory read latency penalty that must be very carefully taken into account. This paper presents an approach to effectively implement the above idea, where error correcting code (ECC) is used to realize memory cell defect tolerance. The key is to use a new dual-ECC memory defect tolerance architecture that can minimize the memory defect tolerance implementation overhead and hence enable a large improvement of effective storage density. An algorithm is presented to search for the optimal dual-ECC coding construction that can lead to the highest possible storage density. The effectiveness of the developed design solution has been demonstrated using 256Mb STT MRAM design at 45nm node as a test vehicle. Results show that, under the SDMR of 20%, the overall memory chip size can be reduced by 20% while maintaining almost the same average memory read latency compared with the design using the worst-case transistor sizing.

2. PROPOSED DESIGN STRATEGY

2.1 Basic Design Methodology

Fig. 2 illustrates the proposed smaller-than-worst-case memory cell transistor sizing design methodology. The underlying theme is

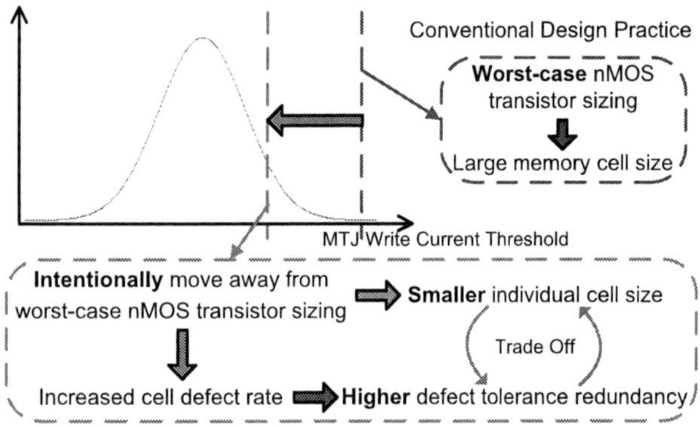

Figure 2: Illustration of the proposed smaller-than-worst-case memory cell transistor sizing design methodology.

to jointly consider the nMOS transistor sizing and memory defect tolerance. In this context, one memory cell is claimed to be defective because the write current threshold of its MTJ is larger than the current driving strength of its nMOS transistor. Conventionally, memory defects are dealt with using redundant row/column repair, which nevertheless may not be adequate to handle a large number of individual defective memory cells induced by smaller-than-worst-case transistor sizing. Hence, we mainly rely on ECC to realize individual memory cell defect tolerance, while the redundant rows/columns are mainly used to handle cluster defects such as open/short word-lines or bit-lines. Notice that, unlike SRAM and DRAM, MRAM is not subject to particle-induced soft errors, hence we only need to consider defect tolerance in this context.

The main design objective in transistor sizing and ECC-based defect tolerance co-design is to minimize the redundancy overhead induced by ECC at minimal impact on memory circuit design and access latency. The first question to be answered is whether we should use a single ECC or multiple ECCs with different error correcting capability vs. coding redundancy trade-offs. Although the use of a single ECC can simplify the memory circuit and architecture design, because the number of defective cells within each memory word may vary from one word to the next, the single ECC solution must employ an ECC that is strong enough to handle the

worst-case scenario. This can result in a significant memory over-protection and hence storage density penalty.

Another intuitive method is to use multiple ECCs with *just enough* error correction capability for each memory word. Different from the random soft errors, the locations of defective MRAM cells can be determined during memory testing, which makes it possible to use different ECC for different memory words. Therefore, with the just enough coding redundancy for each memory word, such an ideal multi-ECC approach has the minimum ECC coding redundancy penalty.

Therefore, this work focuses on the use of multiple ECCs. Fig. 3(a) shows an architecture to implement the above ideal multi-ECC (with n different ECCs) that can realize the *jut enough* defect tolerance for each memory word. Since different ECCs incur different amount of coding redundant bits, it contains n separate coding redundancy storage blocks, as illustrated in Fig 3(a), each block is responsible for storing and accessing coding redundancy for all the memory words being protected by the same ECC. Each coding redundancy storage block consists of one content addressable memory (CAM), which can be efficiently realized using MTJs as presented in [8], and one redundant bits memory. Each entry in CAM contains two portions, including (i) *word_address*: the address of one memory word being protected by the corresponding ECC, and (ii) *redundancy_address*: the address of the corresponding ECC coding redundancy in the associated redundant bits memory. Each CAM is searchable by the *word_address* portion, i.e., given one memory word address input, the CAM will generate a *hit/miss* signal that indicates whether the word is being protected by the corresponding ECC, and if yes, then it will further output the coding redundancy address that is used to fetch the coding redundancy from the associated redundant bits memory.

Although the above straightforward design solution incurs the minimum amount of ECC coding redundancy, it clearly suffers from significant memory read latency degradation because every memory read/write involves the relatively slower CAM search operation. Furthermore, since CAM search operation tends to incur much higher energy consumption than normal memory read operations, it inevitably suffers from high energy consumption.

2.2 Proposed Dual-ECC Architecture

To address above issues, we propose a dual-ECC architecture as shown in Fig. 3(b) that uses only two different ECCs: one is relatively weak (hence incurs a small amount of coding redundancy) and called intrinsic ECC, another one is relatively strong (hence incurs a large amount of coding redundancy) and called extrinsic ECC. It differs from the above straightforward architecture with $n = 2$ at two aspects:

1. The weak intrinsic ECC is *uniformly* applicable to all the memory words, i.e., the intrinsic ECC coding redundancy covers the entire data address space, hence the use of a CAM is obviated.
2. An ECC mode lookup memory is used to store a 2-bit information per memory word to indicate whether each memory word is being protected by the intrinsic ECC, extrinsic ECC, or defect-free.

Accordingly, each memory read operation in the dual-ECC memory may fall into three scenarios, (i) the memory word is defect-free, hence it does not invoke ECC decoding, leading to the minimal read access latency, (ii) the memory word is being protected by the intrinsic ECC, leading to a moderate latency overhead due to intrinsic ECC decoding, and (iii) the memory word is being protected by the extrinsic ECC, leading to the worst-case latency due to the sequential operations of CAM search, redundant bits memory

978-1-60558-497-3/09 $25.00 © 2009 ACM

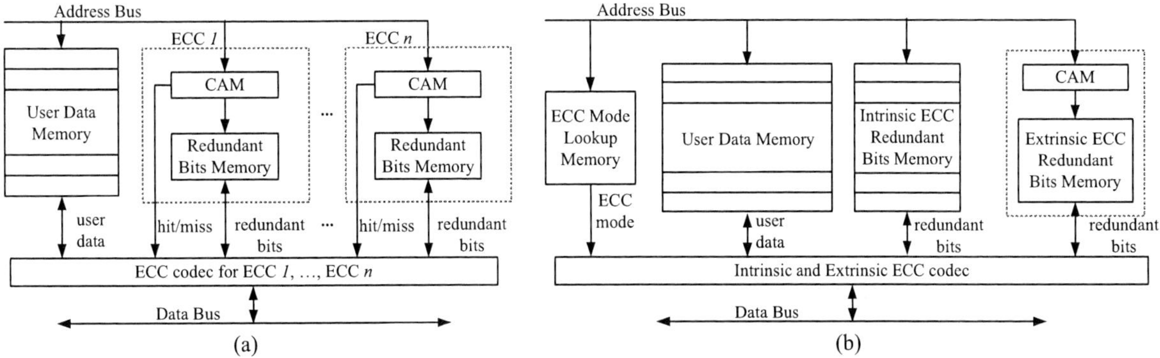

Figure 3: (a) An architecture that implements the ideal multi-ECC approach with n different ECCs and (b) the proposed dual-ECC memory architecture with intrinsic and extrinsic ECCs.

access, and extrinsic ECC decoding. Clearly, compared with the above ideal multi-ECC approach, the average memory read latency can be largely reduced. Meanwhile, we note that the smaller-than-worst-case transistor scaling directly leads to smaller memory array size and hence reduce read access latency, which can compensate the access latency overhead induced by those extra operations to a certain extent.

Under the above dual-ECC memory architecture framework, we further develop a design flow, as shown in Fig. 4, that choose the

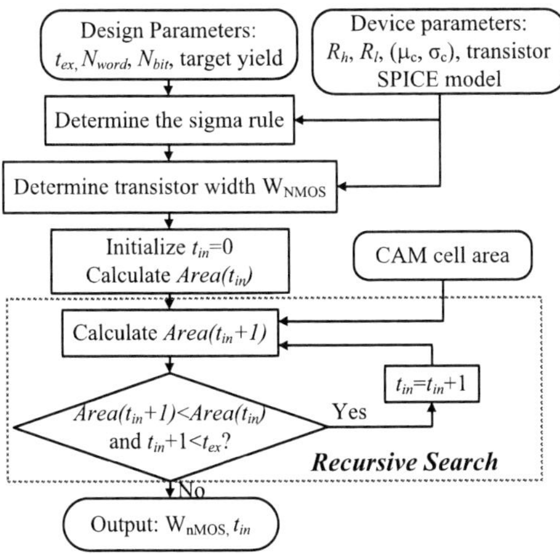

Figure 4: Proposed design flow to choose the optimal intrinsic ECC that can minimize overall memory area.

appropriate error correction capability of the intrinsic ECC in order to maximize the overall storage density. It mainly requires two types of inputs, including

1. *Design parameters* that include the memory size $N_{word} \times N_{bit}$ (N_{word} is the number of words and N_{bit} is the number of user bits per word), target memory yield, and the extrinsic ECC error correction capability (t_{ex}) required to achieve the target memory yield. We note that memory yield only depends on the maximum ECC error correcting capability that is determined by the extrinsic ECC. The use of different intrinsic ECCs only results in different overall area overhead and average memory read latency.

2. *Device parameters* that include MTJ high/low resistance, mean (μ_c) and deviation (σ_c) of MTJ write current threshold, the nMOS transistor SPICE model.

Moreover, we also need to have the estimated CAM cell area. Given the design parameters, we could determine the maximum allowable MRAM cell defect rate P_{cell}, which further determines the minimum allowable sigma rule factor α. Combined with the device parameters, we can determine the MTJ write current and the corresponding nMOS transistor width W_{nMOS} through SPICE simulation. Given t_{ex} and P_{cell} and error correcting capability of the intrinsic ECC, denoted as t_{in}, we can calculate the percentages of the memory words that need to be further protected by the extrinsic ECC. Accordingly, we can calculate the size of the CAM and the redundant bits memory for extrinsic ECC, and then we can finally estimate the overall memory area, denoted as $Area(t_{in})$. To determine which intrinsic ECC can lead to the minimum possible memory area, we can carry out a simple recursive search as shown in Fig. 4: starting from $t_{in} = 0$, we gradually increment t_{in} by choosing a stronger intrinsic ECC and evaluate the corresponding overall memory area $Area(t_{in})$ until we obtain an optimal value of t_{in} that can minimize the overall memory area.

3. MEMORY DESIGN CASE STUDY

This section uses 256Mb memory design at 45nm node as a case study to demonstrate the proposed design method. We assume each memory word contains 256 user bits, hence we have $N_{word}=1M$ and $N_{bit}=256$. We set the high and low MTJ resistance as 2KΩ and 1KΩ, respectively, and the mean of the MTJ write current threshold as 170μA. We use 45nm nMOS transistor PTM model [2]. In terms of intrinsic and extrinsic ECC, we may choose from the single-error-correcting Hamming code and a family of multi-error-correcting BCH codes. Let each ECC is denoted by (n, k, t), where n represents the codeword length, k represents the number of user bits being protected (which is 256 in this case), and t represents the error correction capability. For t ranging from 1 to 5, we construct a Hamming code $(265,256,1)$, and four BCH codes $(274,256,2)$, $(283,256,3)$, $(292,256,4)$, and $(301,256,5)$, which are further denoted as Hamming, BCH-t2, BCH-t2, BCH-t3, BCH-t4, and BCH-t5, respectively. Assuming the MTJ write current threshold has a Gaussian distribution $\mathcal{N}(\mu_c, \sigma_c)$, in order to achieve the same memory yield as the worst-case $6\sigma_c$ rule, we can use $4.6\sigma_c$ rule when using the Hamming code as the extrinsic ECC, and the transistor sizing rule can be reduced to $3.9\sigma_c$, $3.6\sigma_c$, $3.3\sigma_c$, and $3.2\sigma_c$ as we use the BCH codes with t increasing from 2 to 5, respectively.

For each extrinsic ECC option and SDMR, we followed the design flow presented above to search for an optimal intrinsic ECC and the corresponding memory area and read access latency. In this regard, we modified CACTI 5.3 [1], the latest version of the

978-1-60558-497-3/09 $25.00 © 2009 ACM

Table 1: Summary of estimated memory area and read latency results.

SDMR	Extrinsic ECC	Intrinsic ECC	Area (mm^2)					Read latency (ns)			
			User data storage	Extra storage	ECC Codec	Total	Saving	defect-free	Intrinsic ECC	Extrinsic ECC	Average
15%	-	-	60.86	-	-	60.86	-	10.98	-	-	10.98
	Hamming	-	53.98	0.57	0.03	54.58	10.32%	9.86	-	11.36	9.86
	BCH-t2	-	50.55	1.16	0.20	51.91	14.71%	9.31	-	137.31	10.77
	BCH-t3	Hamming	50.40	1.17	0.33	51.90	14.72%	9.30	10.80	205.30	9.60
	BCH-t4	Hamming	48.85	1.45	0.43	50.73	16.64%	9.09	10.59	265.09	11.16
	BCH-t5	Hamming	48.26	2.17	0.53	50.96	16.27%	9.07	10.57	329.07	17.48
20%	-	-	71.78	-	-	71.78	-	12.69	-	-	12.69
	Hamming	-	61.66	0.63	0.03	62.32	13.18%	11.11	-	12.61	11.11
	BCH-t2	-	56.93	1.34	0.20	58.47	18.54%	10.33	-	138.33	11.79
	BCH-t3	Hamming	56.56	1.28	0.33	58.17	18.96%	10.24	11.74	206.24	10.54
	BCH-t4	Hamming	54.43	1.62	0.43	56.48	21.32%	9.91	11.41	265.91	11.98
	BCH-t5	Hamming	53.71	2.49	0.53	56.73	20.97%	9.83	11.33	329.83	18.24
25%	-	-	83.27	-	-	83.27	-	14.37	-	-	14.37
	Hamming	-	69.86	0.85	0.03	70.74	15.05%	12.40	-	13.90	12.40
	BCH-t2	-	63.71	1.66	0.20	65.57	21.26%	11.30	-	139.30	12.76
	BCH-t3	Hamming	62.59	1.56	0.33	64.48	22.57%	10.73	12.23	206.73	11.03
	BCH-t4	Hamming	60.18	1.99	0.43	62.60	24.82%	10.80	12.30	266.80	12.87
	BCH-t5	Hamming	59.24	2.94	0.53	62.71	24.69%	10.66	12.16	330.66	19.07

widely used SRAM/DRAM performance estimation tool CACTI, to estimate the area and read latency of STT MRAM. Furthermore, we design the corresponding ECC codec using Synopsys tool and a 45nm standard cell library to estimate the corresponding ECC codec area and latency cost. Table 1 summarizes the estimated area and read access latency results, in which the extra storage includes ECC mode lookup memory, CAM and extrinsic ECC redundant bits memory. Clearly, the overall memory area will reduce as we start to use stronger extrinsic ECCs. However, this trend tends to become slower and eventually leads to even increased overall memory area (e.g., see the case when increasing from BCH-t4 to BCH-t5 in Table 1), due to the accordingly increased extra storage silicon overhead.

Table 1 further shows the average memory read latency in this case study. Under different extrinsic ECC hence different sigma rule, the percentages of the occurrences of the above three different memory read scenarios will be different, leading to different average memory read latency. This table clearly shows the trade-off between the storage density and memory read access latency. In particular, we note that, when using a relatively weak extrinsic ECC, the average memory read latency may even be less than that of the conventional worst-case design with 6σ rule. This is mainly because the smaller-than-worst-case transistor sizing can reduce the footprint and hence access latency of the primary user data memory, leading to smaller read latency in case of defect-free memory words.

4. CONCLUSIONS

This paper presents a technique that can mitigate the impact of the ever increasing process variability on the achievable STT MRAM storage density. In conventional worst-case design practice, the nMOS transistor within each STT MRAM cell must be sufficiently large to accommodate the worst-case (e.g., 6σ) MTJ write current threshold variation, leading to potentially significant variability-induced storage density degradation. To address this problem, this paper proposes a smaller-than-worst-case nMOS transistor sizing and memory defect tolerance co-design methodology, and develops a practical dual-ECC approach to implement this methodology at minimal memory read latency penalty. Using 256Mb STT MRAM design at 45nm node as a test vehicle, we show that, under 20% of MTJ write current threshold standard deviation vs. mean ratio, the overall memory size can be reduced by more than 20% while maintaining almost the same average read latency compared with the conventional worst-case design.

5. REFERENCES

[1] *CACTI: An integrated cache and memory access time, cycle time, area, leakage, and dynamic power model.* http://www.hpl.hp.com/research/cacti/.

[2] *Predictive Technology Model (PTM).* http://www.eas.asu.edu/~ptm/.

[3] Y. Chen, X. Wang, H. Li, H. Liu, and D. Dimitrov. Design margin exploration of spin-torque transfer RAM (SPRAM). In *Proc. of IEEE International Symposium on Quality Electronic Design (ISQED)*, pages 684–690, 2008.

[4] K.-T. Nam et al. Switching properties in spin transper torque MRAM with sub-50nm MTJ size. In *Proc. of the 7th Annual Non-Volatile Memory Technology Symposium (NVMTS)*, pages 49–51, Nov. 2006.

[5] M. Hosomi et al. A novel nonvolatile memory with spin torque transfer magnetization switching: spin-ram. In *IEEE International Electron Devices Meeting (IEDM)*, pages 459–462, 2005.

[6] T. Kawahara et al. 2Mb spin-transfer torque ram (SPRAM) with bit-by-bit bidirectional current write and parallelizing-direction current read. In *Proc. of 2007 IEEE International Solid-State Circuits Conference*, pages 480–481, Feb. 2007.

[7] X. Wang, Y. Zheng, H. Xi, and D. Dimitrov. Thermal fluctuation effects on spin torque induced switching: Mean and variations. *Journal of Applied Physics*, 103, 2008.

[8] W. Xu, T. Zhang, and Y. Chen. Spin-transfer torque magnetoresistive content addressable memory (CAM) cell structure design with enhanced search noise margin. In *IEEE International Symposium on Circuit and System (ISCAS)*, pages 1898–1901, May 2008.

[9] Z. Diao et al. Spin-transfer torque switching in magnetic tunnel junction and spin-tansfer torque random access memory. *Journal of Physics: Condensed Matter*, 19, 165209(16), April 2007.

978-1-60558-497-3/09 $25.00 © 2009 ACM

EDA in flux – should I stay or should I go?

Eshel Haritan (Organizer)
CoWare Inc.
San Jose, CA, USA

Andreas Kuehlmann (Organizer)
Cadence Design Systems
San Jose, CA, USA

Tina Jones (Chair)
Cadence Design Systems
San Jose, CA, USA

John Epperheimer
Workpath Group LLC
San Jose, CA, USA

Jan Rabaey
UC Berkeley
Berkeley, CA, USA

Rahul Razdan
University of Florida
Gainesville, Florida, USA

Naveen Gupta
Yahoo Inc.
Sunnyvale, CA, USA

PANEL SUMMARY

A crisis is a terrible thing to waste". Quotes like this are often heard by experts in the industry and academia but what does this mean to me? How should I change my professional interests? How should I evolve my career? How is EDA going to evolve? The panel represents multiple points of views on these questions. Four experts will review the current job situation in EDA and provide a historical perspective for previous recessions. The panel will further discuss views on new directions for the Electronics market, how EDA should evolve, and options for career development during the slow down of the industry.

Categories and Subject Descriptors

K7.1 Computing Milieux, The computing Profession, Occupations

General Terms

Design, Economics, Human Factors

Keywords

EDA future, Electronic Markets, EDA research, Career Development, Recession.

PANELIST VIEWPOINTS

Dr. Rahul Razdan: The core value statement of electronic design automation is the ability to provide formalized specifications, enhance verification of function and performance, and in particular situations automatically synthesize an optimal solution. In the last 50 years, three major "mega" markets have driven the electronics marketplace (Information Technology, Internet Backbone, and Consumer),

Permission to make digital or hard copies of part or all of this work for personal or classroom use is granted without fee provided that copies are not made or distributed for profit or commercial advantage and that copies bear this notice and the full citation on the first page. To copy otherwise, to republish, to post on servers or to redistribute to lists, requires prior specific permission and/or a fee.
DAC'09, July 26-31, 2009, San Francisco, California, USA

and in each situation, critical innovations in electronic design automation have been crucial in driving the success of these markets. Today, even before the recent financially driven recession, the existing electronics marketplace was rapidly reaching a point of rapid consolidation and commoditization for all three major markets. In addition, the semi-conductor process platform migration to advanced nodes, a traditional driver of innovation and new growth, had reached a point where the costs of building advanced fabrication facilities was prohibitively expensive. This, in turn, has led to a situation where the cost of new design starts has become prohibitively expensive. The situation seems bleak, not only for EDA, but also for semi-conductors as well as electronics in general. However, there is reason for hope. First, the existing markets are massive and well integrated into the economy, so while a consolidation is occurring, a core base-line business will exist based on just the current markets for some time to come. Second, as in the past, new potential markets have emerged which can significantly change the growth situation. In particular, two new "mega" markets, consumer medical devices and electronics rich infrastructure, offer the hope of adding significant value to the worldwide economy by the significant injection of electronics. These markets have the potential to serve as the next generation engines for the global electronics industry. However, to succeed, the markets require significant innovation in the design practice, and thus can serve as a driver for the EDA marketplace as well.

John Epperheimer: This recession may affect your career more than any other event we have experienced. How should you approach making decisions about your career? What can you learn from today's trends and past downturns? What are the best practices of career self-management you can't afford to ignore in times like these? John Epperheimer, an executive coach and career consultant who has been working with engineers since 1990, will off career tips to guide your thinking and actions.

Jan Rabaey: Never ignore a crisis … there is always some important hidden message. This is most definitely the case for the EDA crisis. In front of our very eyes, the semiconductor industry is undergoing some absolutely fundamental changes, which will change the nature of the design process forever. Moore's law, as we know it, is coming to an end, most probably to be replaced by some other exponential phenomena. Resolving the resulting complexity issues is something that EDA is (or should be) excelling in. Hence, don't whine about the past glory days. There is plenty of exciting and inspiring work ahead if you are willing to open your eyes.

Naveen Gupta: While we're made to believe that this is the worst recession of our lifetime, there's no reason to panic for most of us here. Did you know the unemployment rate for college graduates is 50% lower than the overall average? The employment for engineers is projected to grow by 40% in next decade. Meet Naveen Gupta, Head of Products & Strategy at Yahoo! HotJobs, and find out who is still hiring, where the jobs will be, and what are the tools you need to differentiate yourself for a successful career.

978-1-60558-497-3/09 $25.00 © 2009 ACM

Design Perspectives on 22nm CMOS and Beyond

Shekhar Borkar

Intel Corporation, JF2-04, 2111 NE 25th Ave, Hillsboro, OR 97124.

Shekhar.Y.Borkar@intel.com

ABSTRACT
This paper presents technology and economic challenges posed by 22nm CMOS and beyond, and how they can be addressed by advances in design technology, validation, and testing, to exploit the benefits of scaling we have enjoyed over the decades.

Categories and Subject Descriptors
B.7.1 *Microprocessors and microcomputers, VLSI.*

General Terms
Design, Performance, Reliability.

Keywords
CMOS, Power, Nano, Variability.

1. Introduction
As technology continues to scale beyond 22nm generation, billions of transistors of integration capacity will be available. This scaling will not be easy, and will put constraints on how you design and manufacture. There will be economic challenges too that drive companies to design compelling products that deliver value and performance, and to manufacture in high volumes. *Business as usual* will not be an option, and in this paper we will discuss technology, economic, and design and test challenges to continue to exploit CMOS technology to deliver end user value and performance.

2. CMOS Outlook & Technology Challenges
Table 1 shows CMOS technology outlook extrapolated from the trends that we see today. Notice that today's nano-scale CMOS is at 45nm node, where the transistor channel lengths are even smaller, of the order of 30nm. As the technology scales every two years, the transistor integration capacity doubles (Moore's Law), gate delay reduces by 30%, energy per logic operation reduces by 65%, and power consumption reduces by 50% [1].

High Volume Manufacturing	2006	2008	2010	2012	2014	2016	2018
Technology Node (nm)	65	45	32	22	16	11	8
Integration Capacity (BT)	4	8	16	32	64	128	256
Delay = CV/I scaling	~0.7	>0.7	Delay scaling will slow down				
Energy/Logic Op scaling	>0.5	>0.5	Energy scaling will slow down				
Variability	Medium		High		Very High		

Table 1: Technology Outlook

Transistor threshold voltages have reduced to the extent that subthreshold leakage power has become excessive, and therefore

Permission to make digital or hard copies of part or all of this work for personal or classroom use is granted without fee provided that copies are not made or distributed for profit or commercial advantage and that copies bear this notice and the full citation on the first page. To copy otherwise, to republish, to post on servers or to redistribute to lists, requires prior specific permission and/or a fee.
DAC'09, July 26-31, 2009, San Francisco, California, USA

expect supply voltage and threshold voltage scaling to slow down in the future, resulting in lower energy, power, and delay reduction.

Random dopant fluctuations and sub-wavelength lithography (Figure 1) will result in higher static variations, supply voltage and temperature variations will cause dynamic variations, and the variability in general will continue to become worse [2]. Higher electric fields will worsen transistor reliability and degradation. Total transistors on a chip will continue to double every two years; however, due to variability and degradation, design will be challenging, shifting to designing a system with billions of unreliable components, without reverting to traditional fault tolerant design, but with software playing a critical role as discussed in [3].

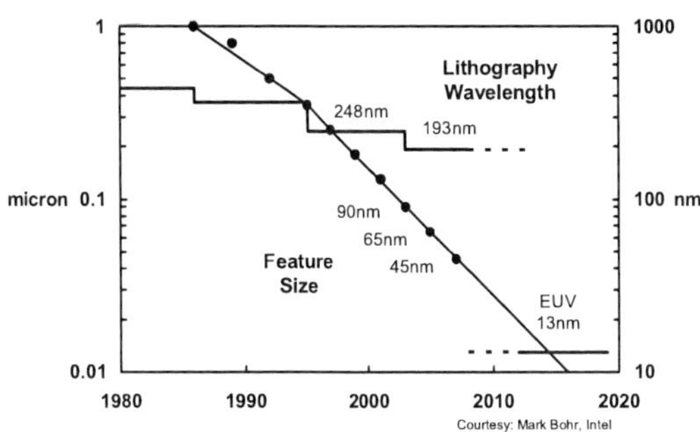

Figure 1: Lithography trends

Design rules become more complex each generation due to sub-wavelength lithography, and more and more restrictions get put on the layout, especially at the lower layers. Moreover, regularity in layout is preferred for ease of patterning and improving yield. Therefore, design technology now faces a new challenge to continue to exploit transistor integration capacity, and continue to deliver performance in the face of these daunting challenges.

3. Economic Challenges
As technology scales, the cost of a transistor goes down by almost half; therefore two times as many transistors can be integrated for the same cost to deliver higher value and performance. However, the cost of fabrication facilities (fab), the cost of a mask set, and turn-around times increase each generation. We expect a mask set to cost more than a million dollars for 22nm generation (Figure 2). Therefore, it is highly desirable that a design is not only functional on the first attempt, but also meets the performance target, and can be put into production.

Although design tools have increased design productivity, the trend has not kept pace with the trend in complexity. Increased transistor integration capacity results in increased design complexity, resulting in higher design and validation costs; we must focus on reducing the design time as well as design efforts.

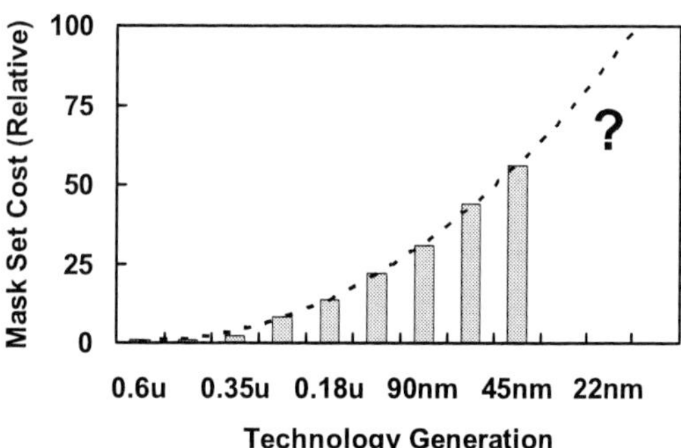

Figure 2: Exponentially increasing mask set cost (estimate).

4. Design Technology & Methodology

Traditionally, a custom design is considered superior because it delivers higher performance and smaller die-size, thus lower cost. But it incurs higher design efforts and potentially longer design cycle (time to market). We can test this tenet in light of challenges described before for 22nm and beyond.

A custom design is higher performance because it allows you to do careful (manual) interconnect engineering to obtain the best frequency of operation. However, in the future, the frequency of operation will be limited by transistor scaling, variability, and power consumption; not so much by the interconnect RC.

Small die-size of a custom design is due to manual physical design to pack transistors and interconnects in the smallest possible area. However, in the future, with increasing design rule complexity and restrictions, imposed regularity, there will be very little room for improvement with manual design. Furthermore, due to increased complexity of the design rules, manual designs may gravitate towards local optimizations rather than global, and thus may even be suboptimal.

This tenet was tested in [4], where the authors took a substantial custom design block of a microprocessor and redesigned it with a new automation methodology. They created a small standard cell library, employed automation, and redesigned the block demonstrating almost 30% improvement in size as well as performance.

We claim that the future of designs in 22nm and beyond is *System Design* with design automation at all levels. And the design automation community needs to research and develop capabilities for this upcoming paradigm shift.

The new design methodology will be mostly SOC-like, yet with some subtle differences. Today's SOC design methodology is built around hard design blocks for a specific technology, and it makes it difficult to migrate a design from one technology to another. The new proposed methodology calls for Soft Design blocks or macros, described at a higher level of abstraction, such as RTL description in Verilog.

The functional blocks are designed once, well characterized, and validated thoroughly for functionality. A system can be built at a higher level of abstraction using these functional blocks, such as cores, special function hardware, an on die network synthesized for the specific purpose, and optimized using system level optimizers [5]. This is where the real design effort will reside—to optimize a system for a given purpose, with validated and well proven functional blocks, rather than to design individual logic blocks from scratch, as it is today.

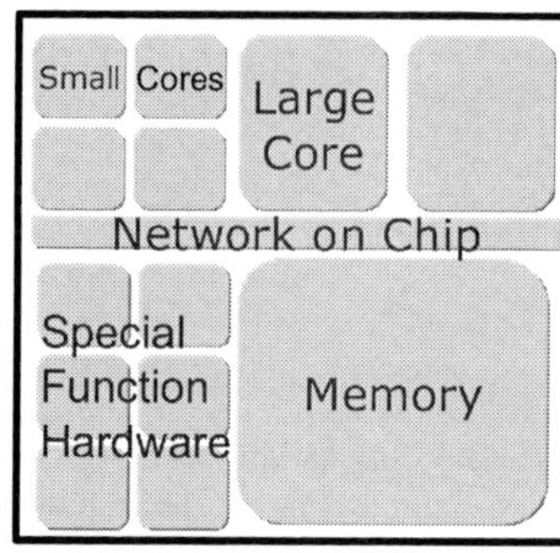

Figure 3: SOC-like design methodology with Soft Macros

Once a system level design optimization and validation is complete, then the physical design too will be automated. Custom design will be limited to memory arrays, register files, and such. Such a system is depicted in **Error! Reference source not found.**.

This new design methodology yields a correct by construction design with minimal validation of the system, details of clocking and interfacing hidden from the user, portable to any technology, reduces design and validation cost by increasing the level of abstraction, and does not compromise die-size (cost) and performance.

5. Test

Needless to say that test should not be an afterthought; it needs to be comprehended in the system design from day one. Each functional block should be built with testing in mind, either for self testing, or to ease testing of the entire system. When the system is put together, the individual test subsystems should work harmoniously towards testing of the entire system.

Test should not be considered as a separate discipline, as it is today; it needs to be merged with design and validation.

6. Conclusion

New technology and economic challenges pose 22nm generation and beyond. The proposed automated design methodology reduces risk and design cost, allows for system optimization, provides flexibility of porting a design to any technology improving longevity—all without compromising cost and performance.

7. References

[1] Shekhar Borkar, "Design Challenges of Technology Scaling, IEEE Micro", July-August 1999.

[2] Shekhar Borkar et al, "Parameter Variations and Impact on Circuits and Microarchitecture", DAC 2003

[3] Shekhar Borkar, "Designing Reliable Systems from Unreliable Components: The Challenges of Transistor Variability and Degradation", IEEE Micro, November-December 2005.

[4] Ueno, et al, "A Design Methodology Realizing an Over GHz Synthesizable Streaming Processing Unit;" VLSI Circuit Symposium 2007

[5] F. Balarin et al, "Metropolis: an Integrated Electronic System Design Environment", IEEE Computer, Vol. 36, No. 4, April, 2003.

978-1-60558-497-3/09 $25.00 © 2009 ACM

Creating an Affordable 22nm Node Using Design-Lithography Co-Optimization

A.J. Strojwas, T. Jhaveri, V. Rovner, L. Pileggi

{andrzej.strojwas, tejas.jhaveri, slava.rovner, larry.pileggi}@pdf.com
PDF Solutions Inc, 333 W. San Carlos Street, Suite 700, San Jose, CA 95110

{ajs, tjhaveri, vrovner, pileggi}@ece.cmu.edu
Carnegie Mellon University, 5000 Forbes Ave, Pittsburgh, PA 15213

Categories and Subject Descriptors

B.7.1 [**Integrated Circuits**]: Types and Design Styles – *Advanced Technologies, VLSI, Standard Cells.*

General Terms

Design, Economics, Experimentation

Keywords

DFM, design technology co-optimization, regular fabric, templates

EXTENDED ABSTRACT

Achieving the required time-to-market with economically acceptable yield levels and maintaining them in volume production has become a daunting task for the advanced technology nodes. These difficulties are primarily attributable to the increase in process variability that is incurred while aggressively scaling technology nodes which are based on the same fundamental device architectures and process solutions. The introduction of a Metal Gate/High-K (MGHK) stack at the 32/28nm technology node will help in addressing the random variations due to random dopant fluctuations (RDF), but its benefit will be exhausted after a single process generation [3]. As a result, for the 22/20nm technology nodes, the only hope to limit RDF will be to adopt novel device architecture, such as FinFET and Ultra Thin Body or Fully Depleted SOI, that would reduce the dopant concentration in the channel.

In addition to the RDF and other random effects, such as line edge roughness, etc., aggressive scaling has been increasing the layout dependent systematic variability largely due to resolution limitations and the application of stressors in modern device architectures [3]. It is important to recognize that if left unchecked, these systematic variations will have a prohibitive impact on the designs targeted towards 22nm technology nodes.

Permission to make digital or hard copies of part or all of this work for personal or classroom use is granted without fee provided that copies are not made or distributed for profit or commercial advantage and that copies bear this notice and the full citation on the first page. To copy otherwise, to republish, to post on servers or to redistribute to lists, requires prior specific permission and/or a fee.
DAC'09, July 26-31, 2009, San Francisco, California, USA

Moreover, the inability to scale the wavelength of the light source used for lithography has led to a rapid increase of process and design costs. In particular, the lack of progress in extreme ultraviolet lithography (EUVL) will result in the need to define 22nm technology node using expensive double patterning technologies (DPT) for critical layers [1,2]. As a result, complex DFM flows have been proposed as an attempt to model various printability and layout dependent effects; however, the increase in design flow complexities, lack of accuracy, and the tremendous expense of maintaining updated models required for these methods has marred their adoption.

We believe a different approach is required to minimize the systematic variations that will plague the future technology nodes. Specifically, the industry should abolish the reactive DFM solutions and adopt a pro-active approach to DFM by increasing the level of abstraction and working towards a co-development of process and design. We propose a novel design methodology based on a set of fully pre-characterized circuit elements, or templates that will ensure a correct by construction integrated

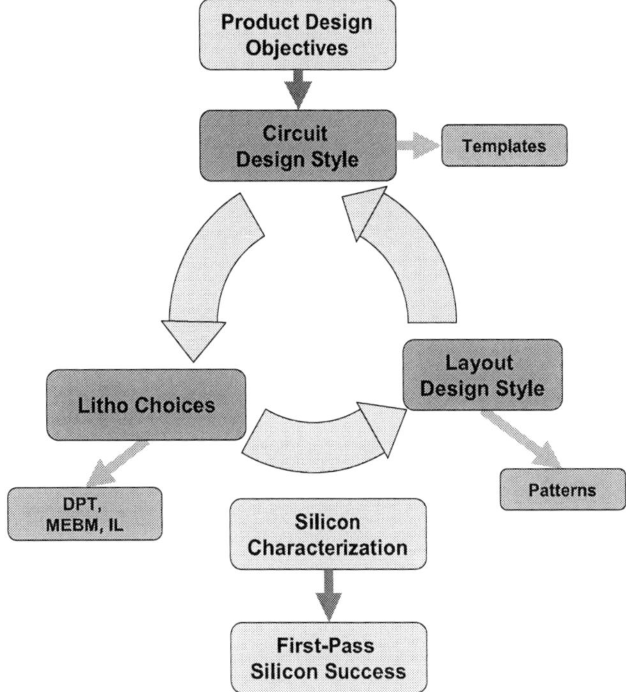

Figure 1: Extremely Regular Layout Methodology

circuit design [2]. The methodology will permit the industry to take the next step in Moore's law with the application of smarter and more efficient circuit, layout, and lithography co-design techniques.

In addition, we will examine the lithographic limitations that will be encountered in future technology nodes, including the 22nm node, and show how the proposed template-based design methodology can be used to overcome these challenges. The key enabler of the methodology is the creation of a regular design fabric onto which one can efficiently map the selected logic-templates using a limited number of printability-friendly layout patterns. The co-optimization of circuit, layout, and process (Figure 1) is achieved by selection of logic functions, circuit styles, layout patterns, and lithography solutions that jointly result in the lowest cost per good die. The resulting set of layout patterns and physical templates are then fully characterized on silicon through the use of specially designed test structures. By ensuring complete coverage, the proposed methodology is able to assure first pass silicon success.

We will show how template-based design methodology can enable future technology nodes that can utilize current generation lithography while minimizing the cost per good die. In particular, we will: discuss the choices of regular design fabrics and their implications on design metrics, such as power, area, and performance, resulting yields, and overall cost; show that the selection of circuit topologies can be mapped efficiently to the choice of regular design fabric; and compare lithography solutions such as DPT, direct write multi-e-beam (MEBM), and interference lithography (IL) for the 22nm technology node and beyond.

ACKNOWLEDGMENTS

The authors acknowledge the support of the Focus Center for Circuit & System Solutions (C2S2), one of five research centers funded under the Focus Center Research Program, a Semiconductor Research Corporation program.

REFERENCES

[1] T. Jhaveri, A.J. Strojwas, L. Pileggi, V. Rover, "Enabling Technology Scaling with In Production Lithography Processes", Proc. SPIE, Vol. 6924 (2008).

[2] L. Liebmann et al, "Simplify to Survive, Prescriptive Layouts Ensure Profitable Scaling to 32nm and Beyond", Proc. SPIE, Vol. 7275 (2009).

[3] A.J. Strojwas, "Taking the next step in Moore's Law: Designs turn to enable next technology node", IEDM (2008).

Device/Circuit Interactions at 22nm Technology Node

Kaushik Roy, Jaydeep P. Kulkarni and Sumeet Kumar Gupta
School of Electrical and Computer Engineering, Purdue University
465 Northwestern Avenue, West Lafayette, IN, USA 765-494-9448
kaushik@ecn.purdue.edu

ABSTRACT

As transition is being made into 22nm node, technology considerations and device architectures suitable for such scaled technologies are being explored. To design circuits and systems at scaled nodes, we believe there is a need for technology aware circuit and system design methodology that considers device architecture, and technology challenges to achieve design optimality. In this paper, we discuss the challenges of device-circuit-system design at the 22 nm node and present techniques at different levels of design abstraction to meet these challenges. In particular, we discuss different device options for multi-gate FETs. Logic and memory design using multi-gate FETs is also considered. Finally, we briefly discuss process variation tolerant system design methodologies for such scaled technologies.

Categories and Subject Descriptors

B.7.1 [Types and Design Styles]: Advanced Technologies

General Terms:

Design

Keywords:

22 nm technology node, DG MOSFETs, FinFETs, scaling, SRAM, transistor sizing

1. INTRODUCTION

The past few decades have seen transistor-scaling driven evolution of semiconductor technology. With every technology generation, devices, circuits and systems are expected to achieve higher performance, lower power consumption and larger integration density. However, the characteristics of highly scaled transistors degrade due to increased short channel effects and larger susceptibility to process variations. Hence, innovative techniques are required to circumvent these problems at all levels of design abstraction. From the process technology point of view, techniques like optical proximity correction (OPC) assume significance for compensating the variations due to light interference effects in lithography. At the device level, short channel effects like DIBL (drain induced barrier lowering) may be mitigated in bulk MOSFETs (Metal Oxide Semiconductor Field Effect Transistors) using halo implants in the channel. Other device structures like UTB SOI (Ultra-thin Body Silicon on

Insulator) MOSFETs and MUGFETs (Multi-gate FETs) such as double gate (DG) MOSFETs [1] and FinFETs [2] are attractive because of larger gate-control of channel and hence, reduced short channel effects. MUGFETs, though suitable as drop-in replacement for bulk-transistors, have different technology options that can best be utilized by a technology-aware circuit design methodology. Example of such technology characteristics and variants include: multi-fin and width quantization, use of the back-gate as an independent gate, gate-underlapping [3], and fin orientation. Bulk devices and MUGFETs are both expected to experience large variations in design parameters. At the circuit level, process variations may be countered using techniques like adaptive body biasing or designing circuits using variation-tolerant logic families. At the system level, techniques like CRISTA [4] and RAZOR [5] may be used.

Semiconductor industry is already producing CMOS digital systems at the 45 nm technology node and considerable amount of research has been carried out for the 32 nm node. The target devices have been bulk CMOS devices or its variants such as floating body SOI. However, to meet 22 nm technology challenges, MUGFETs need to be seriously considered because of their lower short channel effects.

Let us first discuss some of the issues that the 22 nm technology node and the subsequent generations can potentially face. Since the drain will be much closer to the source, the short channel effects (even for bulk devices with halo doping or for MUGFETs), are expected to worsen, giving rise to an exponential increase in the leakage current. In order to increase the gate-control of the channel, the gate oxide needs to be scaled, however that would lead to an increase in the gate leakage current. Use of high-k dielectrics is an option and may be a potential solution to counter short channel effects in sub-32 nm technology nodes.

Due to smaller channel lengths and hence reduced volume of the active region, process variations like random dopant fluctuations (RDF), body thickness variations (for MUGFETs), and line-edge roughness (LER) will have a larger impact on the performance of the devices and circuits. Variation of threshold voltage is expected to increase and may cause timing failures for high performance circuits. A larger design margin may be required leading to lower performance and/or increased power consumption.

Along with scaling the device dimensions, voltage scaling assumes significance in order to keep the electric fields well below their critical level. However, power supply voltage (V_{DD}) scaling is limited by the threshold voltage scaling. Lower threshold voltage leads to an increase in the sub-threshold leakage current while higher threshold voltage leads to the degradation of ON current. Hence, in order to obtain reasonable ON and OFF currents, one may not have much flexibility in scaling the threshold voltage. This results in limited V_{DD} scaling. Due to

Permission to make digital or hard copies of part or all of this work for personal or classroom use is granted without fee provided that copies are not made or distributed for profit or commercial advantage and that copies bear this notice and the full citation on the first page. To copy otherwise, to republish, to post on servers or to redistribute to lists, requires prior specific permission and/or a fee.
DAC'09, July 26-31, 2009, San Francisco, California, USA

978-1-60558-497-3/09 $25.00 © 2009 ACM

increased electric fields, effects like hot carrier injection and negative bias temperature instability (NBTI) are expected to worsen.

Hence, there is a need to analyze such issues in a greater detail and explore potential solutions. This paper discusses the need for close interaction between technology, device, circuit and system level designs to meet the challenges of the 22 nm technology node and beyond.

The organization of the paper is as follows. In Section 2, we discuss the possible device structures that can be suitable for the 22 nm technology node. Section 3 discusses logic and memory design using MUGFETs. In section 4, we briefly discuss technology-independent system level design optimization techniques. Section 5 draws the conclusions.

2. POSSIBLE DEVICE STRUCTURES: FINFETs

As scaling of transistor continues, leakage current increases exponentially every generation. Also with scaling, process imperfections result in significant variations in device characteristics and can have a large impact on the stability of on-chip memory cells. Due to enhanced short channel effects (SCE), scaling single gate bulk devices beyond sub-50nm node is becoming increasingly difficult. Ultra Thin Body Double-Gate MOSFET (UTB DGFET) devices are suitable in sub-50nm technologies due to their higher immunity to SCE, better scalability and increased on-current compared to single gate devices. Furthermore, DGFET has negligible junction capacitance, due to which delay of the circuit is reduced and is mainly limited by the parasitic capacitances. Moreover, the body of DGFET devices are expected to be lightly doped with threshold voltage (V_t) to be principally controlled using metal gate work-function. The lightly doped body eliminates the V_t variations due to random dopant fluctuation (RDF) (note, however, body-thickness variations [6] in ultra-thin devices may lead to threshold variations due to quantum confinement effect). DGFETs, with such attractive properties for scaled technologies, deserve analysis, especially the possibility of having different design options with such devices. Fig. 1 shows some of the technology and device options with UTB DGFETs. We shall discuss them in detail in the subsequent sections.

FinFETs have emerged as one of the most suitable candidates for DGFET structure. FinFETs have quasi-planar structure, the width of the transistor is the height of the fin (H_{fin}). H_{fin} is a technology constraint and depends on the aspect ratio ($H_{fin}:T_{si} \sim$ 4-5). To increase the width of the transistor (in other words, the driving strength of the transistor) the number of fins has to be increased. Hence, the width increases in quanta of fin. In an SRAM cell, the stability depends upon the respective driving strengths of the transistors (β ratio). In conventional planar MOSFET based SRAM cell, β ratio is modified to improve the stability of the cell. However, because of width quantization in FinFETs the range of β ratios is limited. In FinFETs, to improve the stability and to reduce leakage, it has been shown that joint V_{dd}-H_{fin}-V_t design space and/or spacer thickness can be optimized. Multiple fin-orientation (to improve the mobility) can also be used to achieve design optimality.

2.1 Spacer Thickness Optimization

As discussed previously, DG-MOSFETs have negligible junction capacitance, and hence, the drain capacitance is mainly dominated

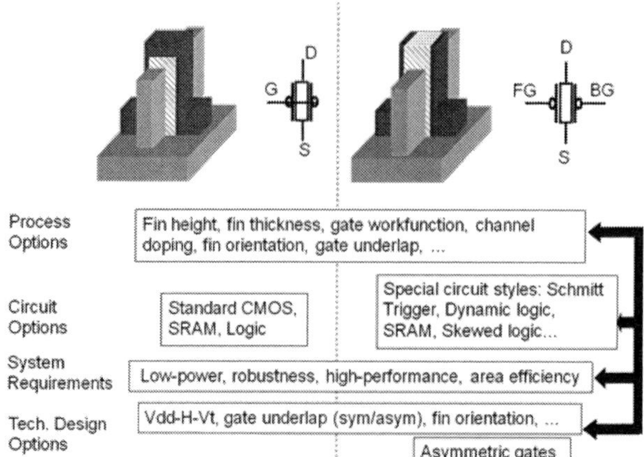

Figure 1. Technology-device-circuit-system co-design options for DG- MOSFETs for the 22 nm technology node and beyond

by the overlap capacitance [7]. In order to reduce the drain capacitance, underlap between the gate and the source/drain (S/D) can be introduced (see Fig. 2). However, this also leads to degradation in the ON current. Since these factors have opposing effects on the delay of a logic gate, there exists an optimal underlap for which the delay is minimum. Introducing S/D underlap has an additional advantage of reduced DIBL due to increase in the channel length, leading to significant reduction in the sub-threshold leakage current.

We obtained optimal source/drain underlap by simulating a DG-MOSFET structure using Taurus device simulator [8] for different source/drain underlap. We first start with a device (L_g=22nm, undoped body, near mid-gap work function, t_{si} =7nm) with no offset spacer (L_{sp}) while meeting a certain sub-threshold leakage target (100nA/um). We then increase L_{sp} to obtain the device with minimum gate delay (CV/I) (corresponding to ΔL_{sp}=6nm on both sides – refer to Fig. 2). Comparison of the optimal device with the conventional DG-MOSFET is shown in Fig. 3. It can be seen that the sub-threshold leakage current (I_{sub}) decreases by 98% for the optimized device. ON current (I_{ON}) degrades by 21% while total capacitance (C) improves by 44% leading to an overall improvement of 29% in the intrinsic delay (CV/I_{ON}) of the device. In addition, an improvement of 67% in DIBL and 27% in sub-threshold swing (SS) is achieved.

To show the feasibility of these devices in circuit applications, we designed two 6-T SRAM cells using conventional DG-MOSFETs and those with optimal S/D underlap, respectively. The results are shown in Fig. 4. It can be seen that there is about 98% reduction in the leakage current (LEAK) for the optimal device. In addition, static noise margin (SNM) and write margin (WM) show an improvement of 22% and 9%, respectively. Access time (T_{Access}) is almost the same for the two devices. Considering tall cell layout for SRAM cell [9], the bit-cell area increases by 14% due to increase in the device footprint.

2.2 Fin Rotation and Fin Orientation

Mobility of PMOS and NMOS devices can be improved by optimizing orientation of crystal surface, resulting in better circuit performance. For quasi-planar multiple-gate devices, such as FinFETs, surface orientation can be changed by modifying the layout of the devices (Fig. 5). This suggests that, orientation

978-1-60558-497-3/09 $25.00 © 2009 ACM

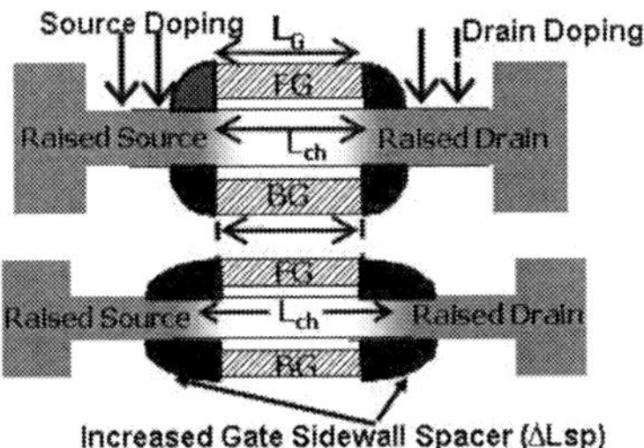

Figure 2. Gate-Source/Drain underlap in FinFET structures

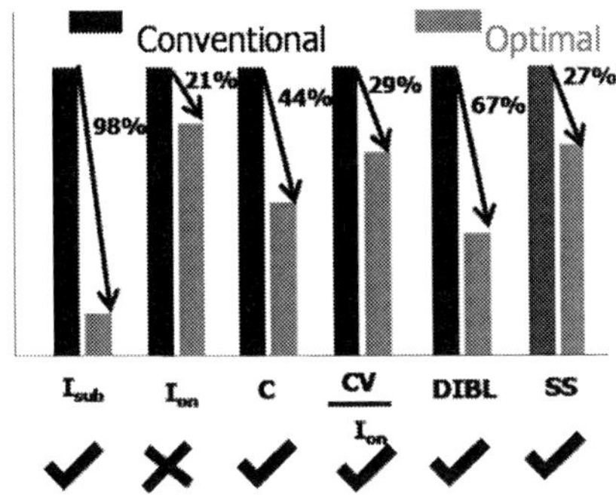

Figure 3. Conventional vs. Optimal Gate underlap structure: Device metrics comparison

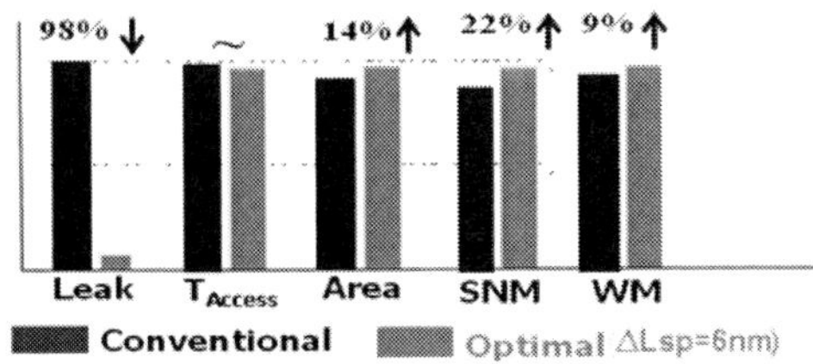

Figure 4. Conventional vs. Optimal Gate underlap structure: SRAM metrics comparison

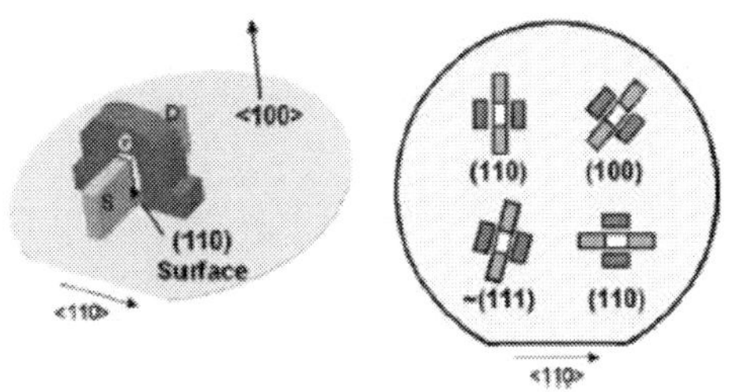

Figure 5. Modification of surface orientation in FinFETs

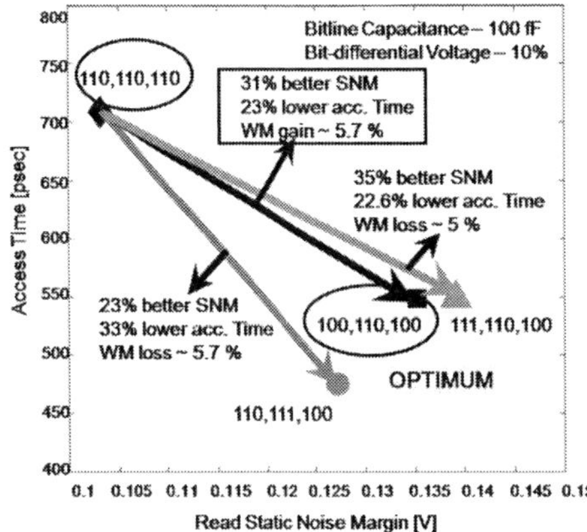

Figure 6. Access time and read SNM variation for different fin orientations

optimization can enhance the performance of FinFET logic circuits. In order to analyze the effect of fin-orientation on cell stability, performance and leakage of SRAM bit-cells, we simulated 6-T and 8-T SRAM cell configurations implemented using FinFETs with different fin orientations. Fig. 6 depicts the comparison of all (110) devices with different device orientations. The results show significant improvement in Read SNM (23-35%) and Access Time (22-33%). It is also observed that the optimized fin-orientations have a marginal impact on Write margin and Hold SNM. It has been shown that achieving (111) orientations on a (110) wafer is more feasible compared to a (100) wafer. Hence, rotated layouts can be avoided in the former, leading to less area penalty. Considering the layout feasibility and area penalty, the orientation (100) PUP (Pull-up), (110) AX (Access), (100) PD (Pull-down) gives a significant improvement in cell stability and performance However, it should be noted that such fin orientations can have detrimental effects on the sensitivity of cell stability and performance with respect to parameter variations.

2.3 Fin Thickness and Fin Ratio Optimization

The impact of width quantization in FinFETs on the area and stability of an SRAM cell has been well studied. It is found that increase in the number of fins reduces the fin-to-fin variability. Hence, increasing the number of fins can reduce the device mismatch in an SRAM cell due to silicon thickness (T_{si}) variation [10]. In FinFET SRAMs, different fin combinations can be realized (under iso-area constraint). Also because of width quantization in FinFETs, SRAM cell area increases in quanta of two fin pitches with increase in the number of fins. We used thin cell layout, fabricated using spacer lithography technique. In this process technology [11], two fins are fabricated in one pitch and to achieve odd number of fins, one fin can be etched away. Fig. 7 shows the layout of an SRAM cell with transistor fin ratios $N_P:N_{AX}:N_N = 1:1:2$ (where N_P, N_{AX} and N_N are the number of fins in the pull-up, access and pull-down transistors respectively). It can be seen that N_{AX} or N_P can be increased to 2 and N_N reduced to 1 without increasing the cell area. Hence, under iso-area, one can have different combinations of fin ratios. For read stability, typically pull down transistor needs to be stronger than the access transistor. Also, for write stability, access transistor should be stronger than pull up transistor. This leaves only one possible combination i.e., $N_P:N_{AX}:N_N = 1:1:2$. Note that the I_{ON} of NMOS is higher than the I_{ON} of PMOS. Hence, even with same number of fins, access transistor is stronger than the pull up transistor, which enhances the write stability. For better read stability, pull-down transistor is stronger than the access transistor. Since, β ratio of the transistors also affects the cell leakage and access time, a suitable combination of fins should be determined for desired SRAM characteristics (leakage/bit, access time, area efficiency, failure probability etc.).

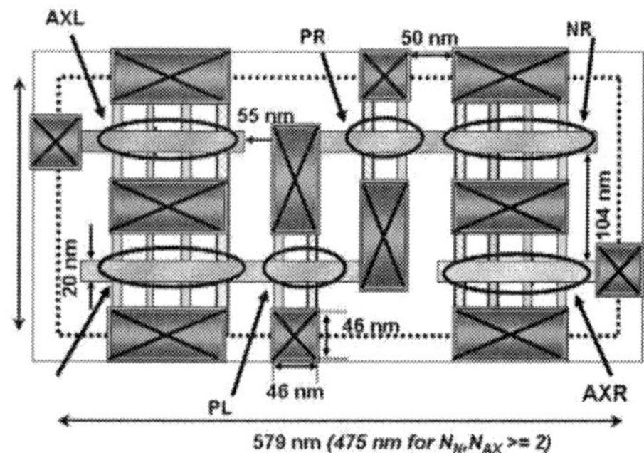

Figure 7. FinFET SRAM thin-cell layout [10]

For better stability, the following design trade-offs exist between T_{si} and fin ratios:

1) increasing T_{si} improves write stability
2) for small N_N:N_{AX}, read stability reduces with increase in T_{si}, however, the difference between read SNM (for increasing T_{si}) diminishes as N_N:N_{AX} increases.
3) increasing N_P:N_N improves read SNM and write stability
4) read stability can be improved by increasing N_N:N_{AX} and write stability can be improved by increasing N_{AX}:N_P.

Considering above trade-offs, we can infer that – under iso-area condition, silicon thickness constraint can be relaxed without affecting the stability of the cell. However, increased Si fin thickness reduces the saturation current (of access transistors) impacting the access time. It has been shown that an increase in the number of fins reduces the variability (σ/μ) of device electrical parameters [10]. Extending the same analysis, one can assume that an increase in the number of fins in an SRAM cell can reduce the variation in T_{si}, resulting in reduced failure probability. Relaxing the silicon thickness constraint in FinFETs improves manufacturability and reduces process variations. Thicker silicon body FinFETs reduce device mismatch among the transistors of an SRAM cell resulting in higher robustness. However, because of increased short-channel-effect in thick silicon body, inter-die variations in gate length and silicon thickness increase.

3. CIRCUIT DESIGN USING FINFETs

The effect of process variations on the system performance may need to be tackled not just at the device level, but also at higher levels of abstraction. In the previous section, we discussed the design and optimization of FinFETs. In this section, we will briefly describe certain circuit design techniques that could be useful for implementing circuits using FinFETs in 22 nm technology.

3.1 A Robust FinFET SRAM

In an SRAM bitcell, the basic element for the data storage is the cross coupled inverter pair. For successful SRAM operation under PVT variations, the stability of the cross coupled inverter is important. Traditionally, device sizing has been adopted to mitigate the effects of process variations. However, device sizing may not be effective in improving the bitcell stability at very low supply voltages. We consider one bit cell configuration suitable for low-voltage FinFET SRAMs.

The design approach is to incorporate built-in feedback mechanism to achieve successful low voltage SRAM operation [12]. Schmitt trigger based SRAM bitcells exhibits robust process variation tolerance. This robust process tolerance can be an essential attribute for SRAM scaling into future nano-scaled technology nodes.

The Schmitt Trigger based (ST) 10 transistor SRAM cell (Fig. 8) focuses on making the basic inverter pair of the memory cell robust. The positive feedback from NFL/NFR adaptively changes the switching threshold of the inverter depending on the direction of input transition ($0 \rightarrow 1$ input transition). The ST bitcell utilizes differential operation and shows better noise immunity. Note that the cell can be used as a drop-in replacement for the 6T cell.

The ST bitcell gives near-ideal inverter characteristics essential for robust memory cell operation (Fig. 9). The cell has 1.56X improvement in read SNM, compared to 6T cell with ~2X larger area (layout of the ST bitcell shown in Fig. 10). Under iso-area condition, the 'minimum area' ST bitcell shows 1.52X read SNM than the 6T cell. Due to the absence of feedback from $1 \rightarrow 0$ input transition and series connected NMOS, the proposed ST bitcell shows higher write-trip-point than 6T cell. Monte-Carlo simulations ($V_{DD} = 400mV$) for read and hold case show that the ST bitcell gives higher mean read/hold SNM compared to the 6T cell

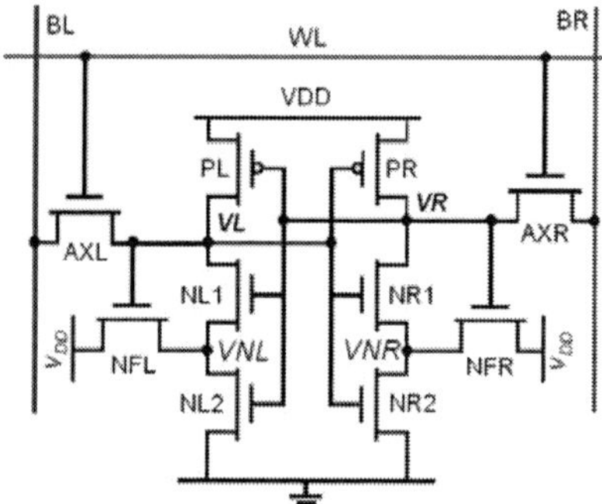

Figure 8. Schmitt Trigger based differential SRAM bitcell [12]

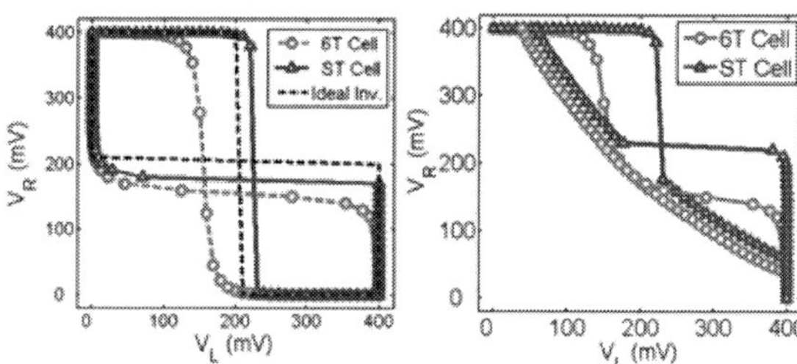

Figure 9. Hold/Read mode characteristics ($V_{DD} = 400mV$) [12]

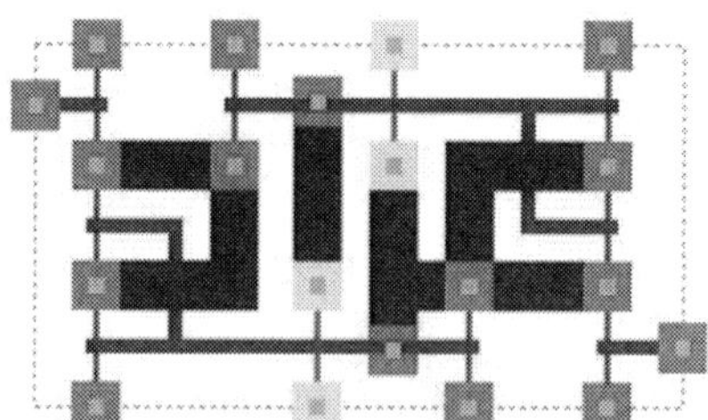

Figure 10. FinFET based Schmitt Trigger SRAM thin cell layout [12]

978-1-60558-497-3/09 $25.00 © 2009 ACM

3.2 Logic Design using FinFETs

Importance of proper transistor sizing in combinational circuits to achieve high circuit speed and/or low power is well understood. As discussed in the previous section, sizing of FinFETs is limited by width quantization. In other words, width of a FinFET, with both the gates tied together (3 Terminal FinFETs or 3-T FinFETs), can be increased in discrete steps of $2H_{fin}$ (where H_{fin} is the height of the fin), thus reducing the design options available to a circuit designer [13]. However, independent control of the front and back gates of FinFETs can potentially provide more design flexibility, since the resolution of discrete sizing step is now H_{fin}, half of what is obtained from the 3-T FinFETs. In this section, we discuss circuit design options using FinFETs with independently controlled front and back gates (4-Terminal FinFETs or 4-T FinFETs).

In order to maximize the frequency of operation, critical path delays are required to be minimized. Since 3-T FinFETs have lower gate delays compared to 4-T FinFETs due to higher I_{ON}, critical paths can be implemented using 3-T FinFETs. However, for non-critical paths, one has more design options.

Let us consider a 2-input NAND gate with input A in the non-critical path and input B in the critical path. Since, the pull down network (PDN) has two transistors in series, 4-T transistors cannot be used in PDN without affecting the critical path delay. However, in the pull-up network (PUN), the PMOS corresponding to signal A can be a 4-T FinFET with the other gate connected to V_{DD} (Independent Gate or IG Cell) as shown in Fig. 11 (a). Now let us consider the case when both the inputs of a logic gate are in non-critical paths. Considering a 2-input NAND gate again, one can implement the PUN using just one 4-T FinFET, with the two input signals driving the independently controlled gates (Merged Gate or MG cell), as shown in Fig. 11(b). Employing 4-T FinFETs in the non-critical paths of a circuit has the following effects on the circuit performance and power: (1) load capacitance of the logic gate driving signal A (refer to Fig. 11(a)) reduces which leads to lower switching power as well as reduced delay (2) the charging time of the output node due to a transition in input A doubles due to the decrease in I_{ON} by half. Hence, the delay of the non-critical path increases by less than 50% with significant power savings. With proper design, it can be ensured that the increase in delay of the non-critical path does not increase the frequency of operation. In fact, if signal A is in the fan-out logic cone of a critical node, the corresponding critical path delay decreases (Fig. 11(c)).

Similarly, one can employ 4-T FinFETs in the implementation of NOR gates as well. Since the β-ratio for a NOR gate is higher compared to a NAND gate, one can achieve more decrease in switching capacitance and hence, higher power savings in NOR gates. It may also be noted that due to the back gate contact overhead in IG cell, there is an area penalty of about 9%. However, for an MC cell, since one less transistor is used, one can achieve ~27% area savings.

We now show some results validating the efficacy of the discussed technique. In order to compare the performance of IG and MC cells with 3-T FinFETs, two libraries were developed: (1) using only 3-T FinFETs and (2) with low power IG cells added to the 3-T FinFET library. A set of ISCAS85 benchmark circuits were synthesized using both the libraries and power and area savings were analyzed. Fig. 12 shows the results. It can be seen

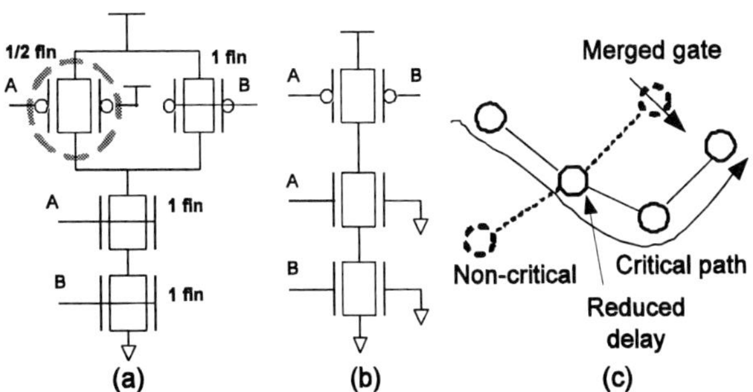

Figure 11. Two low-power options for two-input NAND in IG FinFET technology.(a) Independent gate (IG-Cell) control at A: B is in critical path while A is in noncritical path. (b) Merged gate (MG-Cell): Merging two parallel devices in the pull-up network when both timing arcs are noncritical. (c) Critical path delay reduction due to reduced capacitive loading in the noncritical path [13]

that on an average, 18% power savings and 8% area savings were achieved.

It is also important to consider the impact of process variations on the efficacy of this technique. To analyze the effect of process variations, circuits were synthesized in the worst case corner. Fig. 13 shows the power and area savings for nominal and worst case corners at iso-performance. It can be seen that even in the worst case, one can achieve power savings of about 5% and area savings of 2%.

4. SYSTEM LEVEL DESIGN ISSUES

There has been a lot of interest in recent years to address variations earlier in the design cycle, namely at the architecture and system levels, where it is possible to effectively trade-off variation-tolerance, power, performance, and other metrics such as "quality of results". RAZOR [5], is an approach to Dynamic Voltage Scaling (DVS), based on dynamic detection and correction of circuit timing errors. The key idea of RAZOR is to tune the supply voltage by monitoring the error rate during circuit operation, thereby eliminating the need for large voltage margins. A shadow latch operated with a delayed clock edge is used to detect timing errors. If timing failure is detected, the pipeline is stalled, rolled back, and recomputed with correct values from shadow latches. CRISTA [4] is another approach for low-power, variation-tolerant digital system design, which allows for aggressive voltage over-scaling with a small throughput penalty. The CRISTA design principle (a) isolates and predicts the paths that may become critical under process variations, (b) ensures that they are activated rarely, and (c) avoids possible delay failures in the critical paths by adaptively stretching the clock period to two cycles. In [14], the authors describe a variant of CRISTA called Trifecta, which is an architectural technique that completes common-case sub-critical path operations in a single cycle but uses two cycles when the critical path is exercised. Note that the system level techniques are more or less independent of the underlying technology.

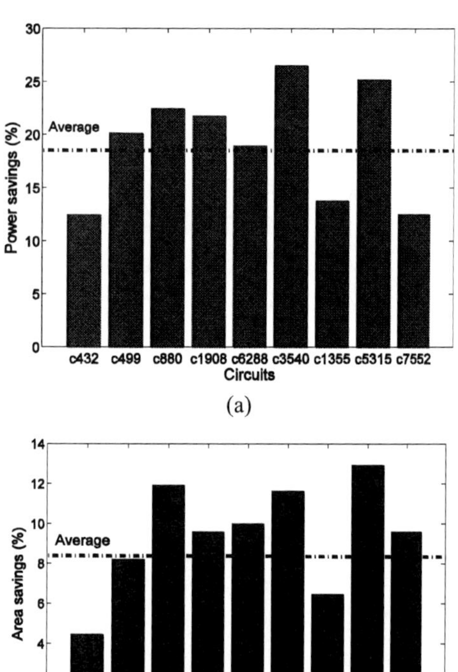

(a)

(b)

Figure 12. (a) Power and (b) area savings for ISCAS85 benchmark circuits using IG FinFET technology [13]

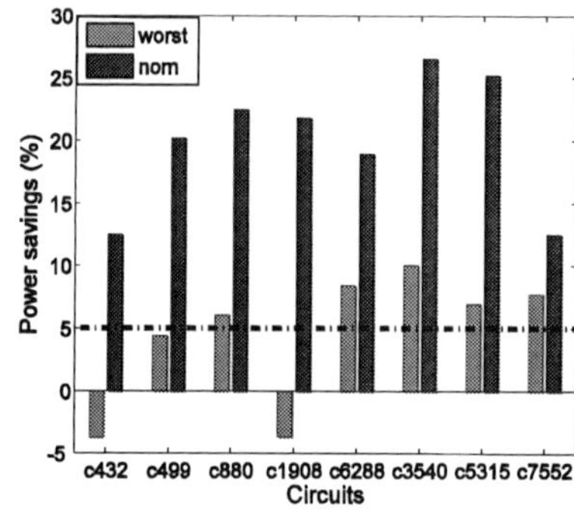

(a)

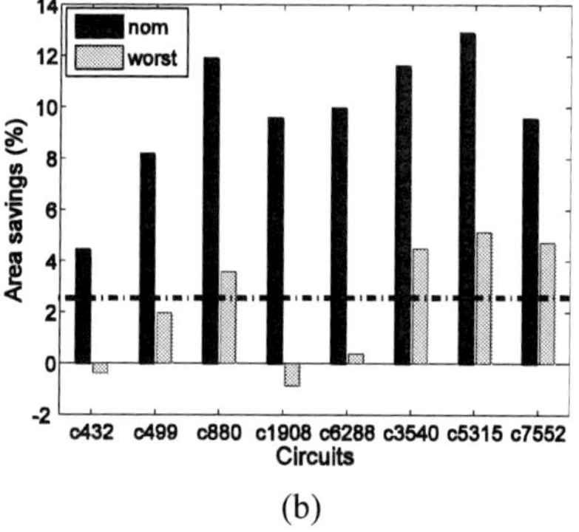

(b)

Figure 13. (a) Power and (b) area savings of ISCAS85 benchmark circuits in nominal and worst case corners at iso-performance [13]

5. CONCLUSIONS

Technology scaling in 22nm node would require closer interaction between process, device, circuit, and architecture. Co-optimization methodologies across different levels would help to mitigate the challenges arising due to changes in transistor topology and increased process variations. Novel device/circuit/system solutions integrated seamlessly with the EDA tools would help meet the desired yield and would help the semiconductor industry to reap the benefits of scaling economics in sub-22nm nodes.

6. ACKNOWLEDGMENTS

Authors would like to thank Semiconductor Research Corporation (SRC), Focused Center Research Program (C2S2), and Intel Corporation for funding this research.

7. REFERENCES

[1] K. Kim et al, "Double-gate CMOS: symmetrical- versus asymmetrical-gate devices", *IEEE Trans. Electron Devices.*, vol. 48, no. 2, pp: 294-299, Feb 2001.

[2] D Hisamoto et al, "FINFET- a self-aligned double-gate MOSFET scalable to 20 nm", *IEEE Trans. Electron Devices*, vol. 47, no. 12, pp: 2320-2325, Dec 2000.

[3] V. Trivedi et al, " Nanoscale FinFET with Gate-Source/Drain Underlap", *IEEE Trans. Electron Devices*, vol. 52, no. 1, pp: 56-62, Jan 2005.

[4] S. Ghosh *et al.*, "CRISTA: A new paradigm for low-power and robust circuit synthesis under parameter variations using critical path isolation," *IEEE Trans. Computer-Aided Design*, vol. 26, no. 11, pp. 1947-1956, Nov 07.

[5] D. Ernst et al "RAZOR: A Low-Power Pipeline Based on Circuit-Level Timing Speculation" *Proc. 36th Annual Int. Symp. Micro-architecture*, Dec 2003.

[6] S. Xiong et al, "Sensitivity of double-gate and FinFET Devices to process variations," *IEEE Trans. Electron Devices*, vol. 50, no. 11, pp: 2255-2261, Nov 2003.

[7] Aditya Bansal et al., "Device Optimization Technique for Robust and Low Power FinFET SRAM Design in Nanoscale Era" *IEEE Trans. Electron Devices*, vol. 54, no. 6, pp:1409-1419 , June 2007

[8] Taurus Device Simulator, Synopsys Inc., v2004.09.

[9] T. Ludwig et al, "FinFET technology for future micro-processors", *Proc. IEEE SOI Conf.*, pp: 33-34. 2003.

[10] D. Lekshmanan et al., "FinFET SRAM: Optimizing Silicon Fin Thickness and Fin Ratio to Improve Stabilty at iso area", *Proc. IEEE Custom Integrated Circuits Conf.*, pp: 623-626, 2007.

[11] Y.-K. Choi et al., "Nanoscale CMOS spacer FinFET for the terabit era", *IEEE Electron Device Letters*, vol. 23, no. 1,pp: 25-27, Jan 2002.

[12] J. P. Kulkarni et al., "160mV Robust Schmitt-Trigger based Sub-threshold SRAM", *IEEE J. Solid State Circuits*, vol. 42, no.10, pp. 2304-2313, Oct 2007.

[13] A Datta et al, "Modeling and Circuit Synthesis for Independently Controlled Double Gate FinFET Devices" *IEEE Trans. Computer-Aided Design.*, vol. 26, No 11 pp: 1957-1966, Nov 2007.

[14] P. Ndai *et al.*, "Trifecta: A non-speculative scheme to exploit common, data-dependent subcritical paths," *IEEE Trans. VLSI Systems* (to appear)

978-1-60558-497-3/09 $25.00 © 2009 ACM

Beyond Innovation: Dealing with the Risks and Complexity of Processor Design in 22nm

Carl J Anderson
IBM Corporation
Austin, TX, USA

ABSTRACT

This talk will describe the challenges of high-performance microprocessor designs in 22nm and beyond. The focus of the talk will be on addressing the risks and complexity of this very demanding design domain and what lies beyond the innovation that has been the driving engine of technology so far.

Categories and Subject Descriptors

B.7.1 [Integrated Circuits]: Advanced technologies, VLSI; B.7.2 [Design Aids];

General Terms

Performance, Design.

Keywords

Microprocessor design, VLSI technology

INTRODUCTION

Technology has been on a relentless path of scaling so far. However, at 45nm and beyond, complex high-performance microprocessor design projects are impacted by numerous factors beyond technology scaling innovation. Innovation plays a part in the design of complex processor chips but does not determine if and when the design gets completed or is successful. The most important aspect of a successful design is its culture and discipline. In today's globally integrated design teams, resources and risk management is crucial in determining the success of large complex high-performance chip projects.

This talk will describe the complexities of high-performance processor design projects. It will address various aspects of this process, such as, its discipline, risk management, resource management and problem identification. It will illustrate solutions that need to be in the culture of a successful design team to conquer 22nm designs.

The 22nm technology will offer chip design teams the use of over a billion transistors. To achieve the most value from these transistors the chip design teams will have to make trade offs between schedule, function, performance, cost and resources. The degree to which design teams have to make these choices will increase significantly at 22nm. It will be very easy and tempting to propose chip designs and goals for 22nm that are unachievable within finite schedules and resources. Trade offs between custom vs synthesized macros, reused vs new custom IP, more features vs power, etc. will have to be made early in the design cycle. The challenge will be to optimize the entire chip and not just a single component. During the execution phase, discipline will be required to ensure that all the components of the design fit together, work together, and are optimized for the targeted technology. A culture of surfacing problems and solving them as early as possible in the design cycle is crucial to keeping the chip design optimized. Problems that are hidden or that only surface near the end of the design cycle may not allow time to implement optimized solutions that keep function and performance. One of the most valued traits of a chip designer is the ability to *innovate inside the box* of schedule, performance, cost and resource constraints. In 22 nm design teams will have embrace the culture of optimizing the chip and not just the components. At 22nm, it is more crucial than ever before for the design teams to have the discipline in bringing together a complex design on schedule, performance, cost and resources.

Permission to make digital or hard copies of part or all of this work for personal or classroom use is granted without fee provided that copies are not made or distributed for profit or commercial advantage and that copies bear this notice and the full citation on the first page. To copy otherwise, to republish, to post on servers or to redistribute to lists, requires prior specific permission and/or a fee.
DAC'09, July 26-31, 2009, San Francisco, California, USA

978-1-60558-497-3/09 $25.00 © 2009 ACM

Physically Justifiable Die-Level Modeling of Spatial Variation in View of Systematic Across Wafer Variability

[†]Lerong Cheng, [†]Puneet Gupta, [§]Costas Spanos, [§]Kun Qian, and [†]Lei He

[†] University of California, Los Angeles

{lerong, puneet, lhe}@ee.ucla.edu

[§] University of California, Berkeley

{spanos, qiankun}@eecs.berkeley.edu

Abstract—**Modeling spatial variation is important for statistical analysis. Most existing works model spatial variation as spatially correlated random variables. We discuss process origins of spatial variability, all of which indicate that spatial variation comes from deterministic across-wafer variation, and purely random spatial variation is not significant. We analytically study the impact of across-wafer variation and show how it gives an appearance of correlation. We have developed a new die-level variation model considering deterministic across-wafer variation and derived the range of conditions under which ignoring spatial variation altogether may be acceptable. Experimental results show that our model is within 1% error from exact simulation result while the error of the existing distance-based spatial variation model is up to 8%. Moreover, our new model is also 10X faster than the spatial variation model for Monte-Carlo analysis.**

Categories and Subject Descriptors: B.7.0 [Integrated Circuits]: General **General Terms:** Performance, Design **Keywords:** Process Variaion, SSTA, Timing, Leakage Analysis

I. INTRODUCTION

With the CMOS technology scaling, process variation has become a major concern for VLSI design. Modeling and analyzing process variation has attracted a lot of attention.

Several works focus on analyzing and modeling of process variation [1]–[8]. The simplest method models process variation as the sum of inter-die (global) variation and independent within-die (local random) variation [8]. Later, people observed that within-die variation is spatially correlated and the correlation depends on the distance between two within-die locations. [1], [2] model spatial variation as correlated random variables, and principle component analysis is applied to perform statistical timing analysis. In this model, a chip is divided into several grids and each grid has its own spatial variation. The spatial variations of different grids are correlated and the correlation coefficient depends on the distance between two grids. [3] focuses on the extraction of spatial correlation. It models the correlation coefficient as a function of distance. Several more complex spatial correlation models have been proposed [9]–[13].

On the other hand, process oriented modeling has concluded that within-die spatial variation is actually caused by deterministic across wafer and across-field variation and purely random within-die spatial variation is not significant [14]–[16]. However, in practice, designers do not know the wafer location or field location of each die; therefore,

we need to analyze the impact of across-wafer variation and across-field variation on die-scale. Moreover, silicon measurements cited in this paper indicate that across-wafer variation is much more significant than the across-field variation, for simplicity, we consider only across-wafer variation in this paper.

In this paper, we first analyze the impact of deterministic across-wafer variation on spatial correlation. We observe that when assuming quadratic across-wafer variation model as in [15], [17], [18]:

1) Different locations of the chip may have different mean and variance. Such differences increase when the chip size increases.

2) When chip size is small, the correlation coefficients for a certain Euclidean distance are within a narrow range. This explains why most existing works find that spatial correlation is a function of distance.

3) Within-die spatial variation is *NOT* spatially correlated when across-wafer systematic variation is removed.

4) Within-die spatial variation is *NOT* independent from inter-die variation.

5) If chip size is small enough, the two-level inter-/within-die decomposition of process variation is still very accurate.

Based on our analysis, we propose an accurate and efficient spatial variation model considering across-wafer variation. Experimental results show that our model is more accurate and efficient compared to the distance-based spatial variation model in [3]. Compared to the exact simulation, the error of our modeled is within 1% and the error of the distance-based spatial correlation model is up to 8%. Moreover, our model is 10X faster than the distance-based spatial correlation model. Finally, we also discuss the case when the across-wafer variation is a non-quadratic function. To the best of our knowledge, this is the first work to mathematically analyze the implication of systematic across-wafer variation at the die-scale.

The rest of this paper is organized as follows: Section II discusses the physical causes for across-wafer variation; Section III analyzes the impact of across-wafer variation on die-scale; Section IV introduces the new variation model and shows experimental results; Section V discusses the case when the across-wafer variation follows a more complicated function; and finally Section VI concludes this paper.

II. PHYSICAL ORIGINS OF SPATIAL VARIATION

In silicon manufacturing, there are many steps that cause non-uniformity of devices across the wafer. Interestingly, most of these processes by the very nature of the equipment follow a radially varying trend across the wafer. Most processes are "center-fed" or "edge-fed" with the boundary conditions at the edge of wafer being substantially different. Moreover, wafers are often rotated to increase process uniformity across them which further leads to radial behavior

Permission to make digital or hard copies of part or all of this work for personal or classroom use is granted without fee provided that copies are not made or distributed for profit or commercial advantage and that copies bear this notice and the full citation on the first page. To copy otherwise, to republish, to post on servers or to redistribute to lists, requires prior specific permission and/or a fee.

DAC'09, July 26-31, 2009, San Francisco, California, USA

of non-uniformity. This is further exacerbated by advent of single-wafer processing for 300mm wafers.

For example, overlay error includes errors in the position and rotation of the wafer stage during exposure, wafer stage vibration, and the distortion of the wafer with respect to the exposure pattern [19]. Magnification and rotation components of overlay error increase from center of the wafer outwards.[1] During the chemical vapor deposition (CVD) step, species depletion and temperature non-uniformity on the wafer at lower temperatures may cause thickness non-uniformity [20], [21]. The redeposition effect in physical vapor deposition (PVD) [22] may cause non-uniformity of etch rate. Moreover, center peak shape of the RF electric field distribution [23] also leads to a center peak shape of etch rate, and chamber wall conditions [24] also cause etch rate non-uniformity. In real processes, the wafers are rotated to improve uniformity. [22], [24] show that the etch rate varies radially across the wafer: the etch rate is high at the center of the wafer and decreases toward the edges. Post-exposure bake (PEB) temperatures are higher at the center of the wafer and decreases outwards [25]. Similarly, other processes ranging from resist coat to wafer deformation due to vacuum chuck holding it follow a bowl-shaped trend across the wafer. All these processes cause a systematic across-wafer variation in physical dimensions. Biggest contributors, by far, are those who impact critical dimension. These are: mask errors and MEEF, lithography (especially PEB) and etch. The rest have minor impact.

Across-wafer variation of gate length observed in several recent silicon measurements [15], [17], [18], [26] validates our arguments. [27] also shows that the ring oscillator frequency and leakage current decrease from the center to the edge of the wafer. Figure 1 shows industry data of ring oscillator frequency for 45nm technology wafers from two different industry processes. In the rest of this paper, all of our simulation and experiments are based on the measurement result of this process. From the figure, we see that ring oscillator frequency decreases from the center to the edge of the wafer.

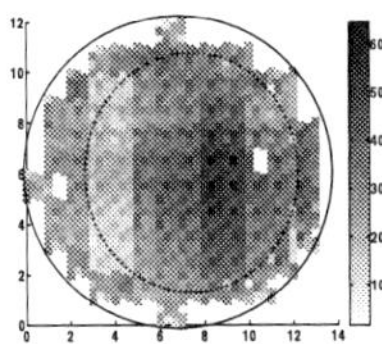

Fig. 1.   Ring oscillator frequency within a wafer.

Besides across-wafer variation, some lithography-induced effects such as lens aberrations can lead to systematic across-field variation and across-die variation. In practice, across-die variation can be modeled as within-die mean shift and will not cause within-die spatial correlation. Moreover, silicon measurements cited in this paper indicate that across-wafer variation is much more significant (probably due to advancements in resolution enhancement and lithographic equipment). Hence, for simplicity, we consider only across-wafer variation in this paper.

III. Analysis of Wafer Level Variation and Spatial Correlation

In this section, we will analyze the impact of across-wafer variation on die-scale. In this paper, we assume the across-wafer variation to be a quadratic function as in [15], [17], [18]:

$$v_p = ax_w^2 + by_w^2 + cx_w + dy_w + m_w + m_f + m_d + r \qquad (1)$$

[1]Overlay error can directly impact critical dimension in double patterning.

where v_p is a variation source, such as L_{eff}, a, b, c, and d are function coefficients, which are obtained from fitting the measurement data from industry process as shown in Figure 1, m_w comprises inter-die random, inter-wafer, inter-lot variation and the fitting error of quadratic fitting as in Equation (1) [2], m_f and m_d are the across-field and across-die variation, respectively. As discussed in Section II, we consider only across-wafer variation and ignore these two types of variations in this paper. r is the random noise, and (x_w, y_w) is across-wafer location. In the rest of this section, we will analyze the spatial variation based on the above model. Table I summarizes the mathematical notations used in this section.

We obtain the coefficients of the above across-wafer variation model by fitting the industrial 45nm process measured ring oscillator delay with 300 wafers from 23 lots. In the rest of this section, all simulations are based-on this extracted model.

Symbols	Description
(x_c, y_c)	Location of the center of the die in the wafer
r_w	wafer radius
m_w	Inter-die variation
r	Within-die random variation
σ_m^2	Variance of m_w
σ_r^2	Variance of r
a, b, c, d	Across-wafer variation coefficients
(l_x, l_y)	x and y dimension die size
(x_w, y_w)	Within-wafer location
(x, y)	Within-die location
ω	Angle between the die and wafer coordinate
$v_p(x, y)$	Variation of within-die location (x, y)
$\dot{x}$	$\dot{x} = x \cos \omega + y \sin \omega$
$\dot{y}$	$\dot{y} = x \sin \omega - y \cos \omega$
$\ddot{x}$	$\ddot{x} = a\dot{x} + c/2$
$\ddot{y}$	$\ddot{y} = b\dot{y} + d/2$
$r_{d\mu}$	$r_{d\mu} = \sqrt{\ddot{x}^2/a + \ddot{y}^2/b}$
$r_{d\sigma}$	$r_{d\sigma} = \sqrt{\ddot{x}^2 + \ddot{y}^2}$
δ	$\delta = \sqrt{(\ddot{x}_1 - \ddot{x}_2)^2 + (\ddot{y}_1 - \ddot{y}_2)^2}$
r_m	$r_m = \sqrt{l_x^2 + l_y^2}$
k_0	$k_0 = r_w^2(a + b)/4 - c^2/4a - d^2/4b$
k_1	$k_1 = r_w^4(a^2 + b^2)/16 - r_w^4 ab/24 + \sigma_m^2$
k_2	$k_2 = k_1/r_w^2$
α	$\alpha = \ddot{x}_1 \ddot{x}_2 + \ddot{y}_1 \ddot{y}_2$
β	$\beta = \sigma_r^2/r_w^2$
s_0	$s_0 = \cos^2 \omega(al_x^2 + bl_y^2) + \sin^2 \omega(bl_x^2 + al_y^2)$
v_g	Inter-die variation
v_s	Within-die spatial variation
v_l	Within-die random variation

TABLE I

Notations.

A. Variation of Mean and Variance with Location

We first consider the variation at a certain location in a die. According to Equation (1), it is easy to find that for a die, whose center lies on (x_c, y_c) wafer coordinates, the variation of location (x, y) in a die (assuming the coordinate of the center of the die to be $(0, 0)$) is:

$$v_p(x, y) = a(x_c + x')^2 + b(y_c + y')^2 + \qquad (2)$$
$$c(x_c + x') + d(y_c + y') + m_w + r(x, y)$$

The definitions of (x', y') are in Table I. In this paper, we assume, a, b, c, and d to be fixed for a process. In practice, these coefficients may vary slightly for wafer-to-wafer or lot-to-lot. In this case, we may either model A, B, C, and D as random variables or lump the fitting residual to inter-die random variation m_w. We will further discuss this in Section V.

[2]We assume the inter-die random variation is induced from the fitting residual, when deterministic wafer level and field level variation are perfectly modeled without residual, m contains only inter-wafer and inter-lot variation.

In practice, (x_c, y_c) are not known to designers. We can convert the wafer-level systematic variation model to a die-level model by noting that the dies are always distributed evenly in the wafer. Therefore, we may model x_c and y_c as random variables which are evenly distributed in the center at $(0, 0)$ with radius r_w (radius of the wafer). It is easy to see that x_c and y_c are identical and their PDF is [3]:

$$PDF(x_c) = 2\sqrt{r_w^2 - x_c^2}/(\pi r_w) \qquad -r_w < x_c < r_w \qquad (3)$$

In this case, the variation at location (x, y), $v_p(x, y)$, is expressed as a function of four random variables: x_c, y_c, m_w, and $R(x, y)$. We may easily calculate the mean and variance of $v_p(x, y)$:

$$\mu_{v_p(x,y)} = k_0 + x''^2/a + y''^2/b = k_0 + r_{d\mu}^2 \qquad (4)$$
$$\sigma_{v_p(x,y)}^2 = k_1 + r_w^2(x''^2 + y''^2) = k_1 + \sigma_r^2 + r_w^2 r_{d\sigma}^2$$

where x'', y'', $r_{d\mu}$, $r_{d\sigma}$, k_0, and k_1 are defined in Table I.

From Equation (4) and (5), it is interesting to note that different die locations may have different means and variances[4]. The location (x_0, y_0) having the smallest mean and variance is given by:

$$x_0 = -c\cos\omega/2a - d\sin\omega/2b$$
$$y_0 = d\cos\omega/2b - c\sin\omega/2a$$

The locations with larger distance to (x_0, y_0) will have larger mean and variance. Figures 2 illustrate the mean and variance ring oscillator delay for different $r_{d\mu}$ (or $r_{d\sigma}$).

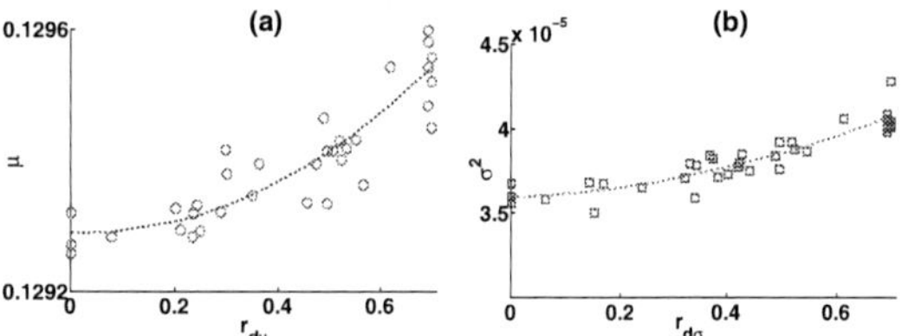

Fig. 2. (a) μ V.S. $r_{d\mu}$. (b)σ^2 V.S. $r_{d\sigma}$.

B. Appearance of Spatial Correlation

Besides mean and variance, we are also interested in the covariance between two locations (x_1, y_1) and (x_2, y_2). From Equation (2), we may calculate the covariance as:

$$Cov = k_1 + r_w^2(x_1'' x_2'' + y_1'' y_2'')$$

With covariance calculated above and variance, we may obtain the correlation coefficient as:

$$\rho = \sqrt{\frac{k_2^2 + 2k_2\alpha + \alpha^2}{(k_2 + \beta)^2 + (r_{d\sigma 1}^2 + r_{d\sigma 2}^2)(k_2 + \beta) + r_{d\sigma 1}^2 r_{d\sigma 2}^2}} \qquad (5)$$

where α, β, and k_2 are defined in Table I. From Equation (5), we obtain the upper bound and lower bound of the correlation coefficient for a certain Euclidean distance:

$$\rho \le \rho_u = \sqrt{1 - \frac{\delta^2 k_2 + \delta u/2 + 2uk_2 + u^2}{(k_2 + \beta)^2 + 2r_m^2(k_2 + \beta) + r_m^4}}$$

$$\rho \ge \rho_l = \sqrt{1 - \frac{\delta^2(k_2 + r_m^2 - \delta^2/4 + \beta) + u(2k_2 + r_m^2)}{(k_2 + \beta)^2 + \delta^2(k_2 + \beta)/2 + \delta^4/16}}$$

[3]x_c and y_c are not independent. In order to generate samples of x_c and y_c, we may assume $x_c = r_s\cos\theta$ and $y_c = r_s\sin\theta$, where r_s and θ are independent. r_s is with triangle distribution ranging from 0 to r_w, θ is with uniform distribution ranging from 0 to 2π

[4]Such difference is caused by the nonlinearity of the across-wafer variation function (we assume quadratic function as in Equation(1)). If the across wafer variation function is a piecewise linear function, the mean and variance will be the same for all locations of a die.

where δ is the Euclidean distance between (x_1'', y_1'') and (x_2'', y_2''), l_x, l_y, and r_m are defined in Table I. From the upper bound and lower bound, we may also calculate the range of correlation coefficient:

$$\rho_u - \rho_l \le \sqrt{r_m^2/(r_m^2 + k_2 + \beta)} \qquad (6)$$

Notice that usually the wafer size is much larger than the die size, that is $k_2 \gg r_m^2$, therefore, the range of correlation coefficient for a certain distance is very narrow. Moreover, from the above equation, we also find that when the variances of the inter-die residual and within-die random variation increase, the range decreases. This explains why most existing works [3], [9] find that spatial correlation is a function of distance. Figure 3(a) illustrates the exact data for 40 locations, the upper bound and the lower bound. From the figure, we find that the range of ρ for a certain distance is very narrow. Although the correlation coefficient is in a narrow range, covariance is not, as shown in Figure 3(b). This is because of the differences of variance across the die.

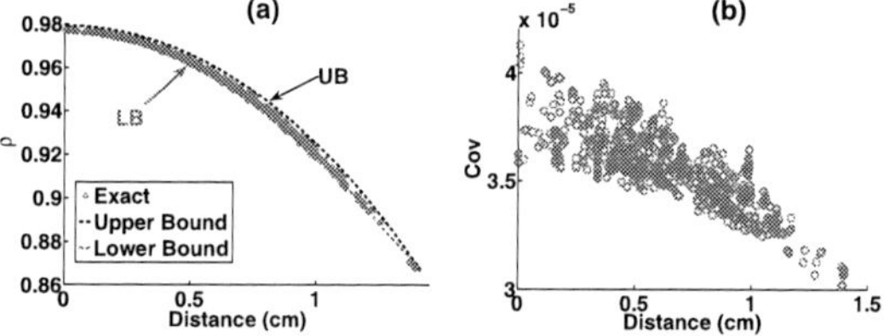

Fig. 3. Apparent spatial correlation and covariance as a function of distance.(a)

Figure 4[5] shows the correlation coefficient for within-die variation after subtracting the mean variation of the die (mainly caused by across wafer variation). We observe that the within-die spatial variation is almost *NOT* spatially correlated, as empirically observed in [3]. This further validates that the spatial variation is caused by systematic across-wafer variation.

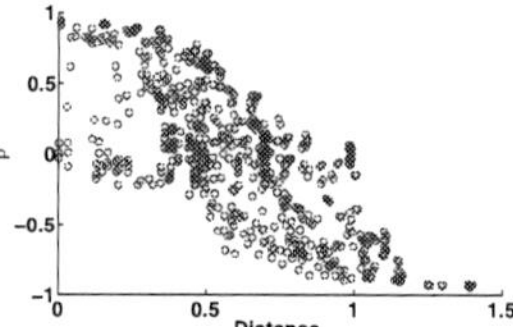

Fig. 4. Correlation coefficient for within-die spatial variation after inter-die variation is removed.

C. Dependence between Inter-die and Within-die Variation

In most existing variation models, process variation is decomposed into inter-die, within-die spatial, and within-die random variation:

$$v_p = v_g + v_s + v_l \qquad (7)$$

where v_g is the inter-die variation, v_s is the within-die spatial variation, and v_l is the within-die variation. Usually v_g is modeled as the variation of the chip mean, v_l is the pure random local variation, and v_s is the residual. v_g, v_s, and v_l are assumed to be independent.

[5]In the figure, the correlation coefficient can is a negative number when distance is large. This is because after subtracting the mean, when the within-die variation of one corner increases, the within-die variation of the opposite corner must decrease. That means, the within-die variations of opposite corners are negative correlated. Moreover even when two locations are very closed, if they lie on the opposite side of the center, their correlation is still near zero.

With the variation model in Equation (2), we may also calculate the inter-die, within-die spatial, and within die random variation. Inter die variation is calculated as the variation of the chip mean:

$$v_g = \frac{1}{l_x l_y} \iint_{\substack{\cdot x \cdot < l_x/2 \\ \cdot y \cdot < l_y/2}} v_p(x,y)dxdy$$
$$= ax_c^2 + by_c^2 + cx_c + dy_c + m_w + s_0$$

where s_0 is defined in Table I.

Within-die random variation is the local random variation: $v_l = R$, and within-die spatial variation is calculated as the remaining variation:

$$v_s(x,y) = v_p(x,y) - v_g - v_l \qquad (8)$$
$$= r_{d\mu}^2 + 2ax'x_c + 2by'y_c - s_1$$

where s_1 is defined in Table I. From the above equations, we find that both inter-die and within-die spatial variations are functions of random variables x_c and y_c. Hence, we may not decompose process variation into independent inter-die and within-die spatial variation.

D. When can Spatial Variation be Ignored?

In the section, we analyze the accuracy of the simple two-level inter-/within-die variation model for different chip sizes. If we only consider inter-/within-die variation, we may lump the across-wafer variation into inter-die variation, that is, approximate the across-wafer variation as a piecewise constant function, as shown in Figure 5(a). To evaluate the impact of the approximation error, we may treat such approximation error as noise and the process variation as signal; and then evaluate the signal to noise ratio. In order to do this, we calculate the mean square approximation error and the total variance of variation. The signal to noise ratio when ignoring the spatial variation is given as:

$$SNR = \sigma_{total}^2/MSE \approx \frac{6abr_w^4 + 6(c+d)r_w^2 + \sigma_M^2 + \sigma_R^2}{abr_w^2(l_x^2 + l_y^2) + 2(c+d)l_x l_y}$$

It can be seen that MSE depends on chip size. When chip size is small, MSE is small. This is because we approximate the across-wafer variation as a piecewise constant function with small steps, hence such approximation is accurate. Figure5(b) illustrates the SNR for different die sizes. It can be seen that the SNR decreases when die size increases as expected. We also observe that when chip size (l_x and l_y) is smaller than 1cm, the SNR is up to 100. That means, two-level inter-/within-die variation model only causes less than 1% error.

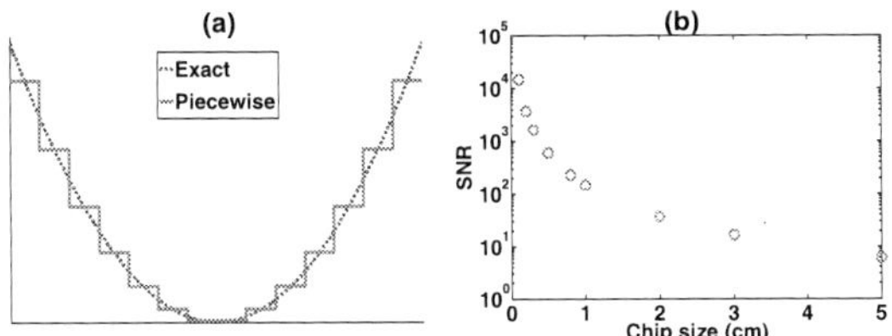

Fig. 5.   Approximating across-wafer variation.

IV. MODELING SPATIAL VARIABILITY

As discussed in Section I, spatial variation largely comes from the deterministic across-wafer variation. Hence, modeling the within-die variation as spatial-correlated random variable is not accurate as discussed in Section III.

Equation (2) calculates the variation for a given location (x,y). In the equation, the die location (x_c, y_c) are modeled as random

variables and their PDF is shown in Equation (3). Equation (2) provides a new spatial variation model. Notice that when we perform Monte-Carlo simulation on this new model, we do not require the across-wafer variation to be a quadratic function. Even the across-wafer variation is with any arbitrary function, we may still apply this model to perform Monte-Carlo simulation. We further discuss this in Section V.

Notice that in this new variation model, there are only four random variables, inter-die random variation M, within-die random variation R, and die location within the wafer x_c and y_c. However, for the traditional distance-based spatial variation model, the number of spatial variation sources depends on the number of grids. Larger chip needs more variables. Therefore, our new model not only models the across-wafer variation accurately, but also is more efficient than the traditional spatial correlation model.

We have implemented the new spatial variation model using Matlab. In order to verify the efficiency and accuracy, we define two comparison cases: 1) the exact deterministic across variation model, 2) distance-based spatial correlation model from [3]. We apply all the above methods to the ISCAS85 suite of benchmarks in PTM 45nm technology [28]. And for each model, we run 10,000-sample Monte-Carlo simulation to obtain the power and delay variation. In the experiment, we consider the gate length variation obtained from minimum square error fitting on the ring oscillator delay from industrial 45nm process measurement from the model as shown in Equation (2).

We first compare the delay variation. We apply SSTA method in [29] to estimate the circuit delay variation. We assume that the chips size is 2cm×2cm and the wafer radius is 15cm. We also assume that the ISCAS85 benchmark circuit is placed in different locations of a chip: center (C), lower left corner (LL), lower right corner (LR), upper left corner (UL), and upper right corner (UR). Table II compares the mean (μ), standard deviation (σ), skewness (γ), and 95% percentile point (95%) for ISCAS85 benchmarks. From the table, we find that our result is closer to the exact model than that of the spatial correlation model from [3]. Moreover, we also find that our method predicts different mean and variance for different location, as discussed in Section III-A. However, the grid-based spatial correlation model can only give the same mean and variance for all locations. The error of our model is within 1% error from the exact simulation and the error of the Spatial correlation model is up to 8%.

Since ISCAS85 benchmarks are very small, the impact of spatial variation on delay is not significant within the circuit. In order to show such impact, we simulate the delay of an inverter chain with 50 inverters. We assume the inverter chain expands from the corner to the center of the chip. For the distance-based spatial correlation model, we divide the chip to $10 \times 10 = 100$ grids. Table III shows the delay variation for the inverter chain expanding from different corners to the center of the chip. From the table, we find that our result is more accurate compared to the grid-based spatial correlation model. Moreover, our model is also more efficient than the spatial correlation model. Our model achieves about 10X speed up compared to the spatial correlation model. This is because there are 100 grids in the spatial correlation model, that means there are 37 spatial random variables[6], our model has only 4 random variables. Moreover, since our model has only four random variables, it is possible for us to apply Fast Quasi-Monte Carlo techniques [30] to improve run time further.

Besides delay, we also apply our model to leakage power analysis.

---

[6]There are 100 correlated spatial random variables, we apply PCA to truncate some insignificant principle components and there remains 37 significant principle components.

bench mark	loca- tion	Exact			Our Model			Spatial correlation model from [3]		
		μ (ns)	σ (ns)	95% (ns)	μ (ns)	σ (ns)	95% (ns)	μ (ns)	σ (ns)	95% (ns)
c7552	C	48.3	6.36	60.0	48.7 (+0.8%)	6.38 (+0.3%)	60.4 (+0.6%)	48.8 (+1.0%)	6.38 (+0.3%)	60.3 (+0.5%)
	LL	47.7	6.10	58.5	47.9 (+0.4%)	6.12 (+0.3%)	58.9 (+0.7%)	48.8 (+2.3%)	6.38 (+4.6%)	60.3 (+3.1%)
	LR	49.2	6.51	62.3	49.1 (-0.2%)	6.53 (+0.3%)	62.5 (+0.3%)	48.8 (-0.8%)	6.38 (-2.0%)	60.3 (-3.2%)
	UL	49.3	6.65	63.1	49.2 (-0.2%)	6.66 (+0.1%)	63.3 (+0.3%)	48.8 (-1.0%)	6.38 (+4.1%)	60.3 (-4.4%)
	UR	49.9	6.91	65.3	49.7 (-0.4%)	6.95 (+0.6%)	65.7 (+0.6%)	48.8 (-1.0%)	6.38 (-7.7%)	60.3 (-7.7%)

TABLE II

DELAY COMPARISON FOR ISCAS85 BENCHMARK IN A 2CM×2CM CHIP.

bench mark	loca- tion	Exact			Our Model				Spatial correlation model from [3]			
		μ (ns)	σ (ns)	95% (ns)	μ (ns)	σ (ns)	95% (ns)	T (s)	μ (ns)	σ (ns)	95% (ns)	T (s)
Inver-	LL-C	1.25	0.17	1.64	1.26 (+0.7%)	0.17 (+1.1%)	1.65 (+0.5%)	15.3	1.28 (+2.4%)	0.20 (+15%)	1.73 (+5.2%)	154 (10.1X)
ter	LR-C	1.28	0.18	1.75	1.28 (+0.2%)	0.18 (+1.1%)	1.75 (-0.2%)	14.7	1.28 (+0.0%)	0.20 (+8.8%)	1.73 (-1.4%)	155 (10.5X)
Chain	UL-C	1.28	0.18	1.73	1.28 (-0.2%)	0.18 (-0.6%)	1.73 (+0.1%)	15.2	1.28 (-0.2%)	0.20 (+7.6%)	1.73 (-0.1%)	153 (10.1X)
	UR-C	1.29	0.19	1.75	1.29 (+0.2%)	0.19 (+0.6%)	1.75 (+0.1%)	14.9	1.28 (-0.7%)	0.20 (+4.8%)	1.73 (-1.3%)	152 (10.2X)

TABLE III

DELAY COMPARISON FOR INVERTER CHAIN.

For the leakage analysis, we assume that 900 copies of ISCAS benchmark circuits are placed in a 2cm×2cm chip. Table IV compares the leakage variation for ISCAS85 benchmarks. Similar to the delay analysis, we observe that our method is more accurate and efficient than the spatial correlation model.

In the above experiment, we only consider big chips. As discussed in Section III-D, when the chip size is small, the traditional inter-/within-die variation model does not introduce much error. In order to verify this, we perform delay estimation of an ISCAS85 bench at different locations in a 3mm×3mm chip and a 6mm×6mm chip. Table V and VI shows the delay variation comparison of our model and the two-level inter-/within-die variation model. From the table, we see that when chip size is small, the traditional inter-/within-die variation model is accurate.

V. GENERAL ACROSS-WAFER VARIATION MODEL

In the previous section, we assumed that the across-wafer variation is a quadratic function as shown in Equation 2. However, in practice, across-wafer variation may not be a perfect parabola. In this section, we will further discuss the case that when the across-wafer variation is with a more general across-wafer variation model.

We assume that the across-wafer variation is with an arbitrary function as follows:

$$v_p = f(x_w, y_w) + m_w + r$$

In this case, the statistical characteristics such as mean, variance, covariance, and correlation coefficient depend on the function f. In most of the cases, we may not have the closed form formulae as in Section III. However, we may still apply similar variation model as in Section IV to perform Monte-Carlo simulation for statistical analysis. Under the general across-wafer variation model, the within-chip variation for location (x, y) are calculated as:

$$v_p(x, y) = f(x_c + x, y_c + y) + m_w + r$$

where (x_c, y_c) are the location of the center of the chip in the wafer, which can be modeled as random variables with PDF shown in Equation (3). In this case, when we know the function f, no matter in close form formula or numerical lookup table, we may generate samples of (x_c, y_c) to perform Monte-Carlo simulation for statistical analysis.

Here we show an example for a non-quadratic across-wafer variation: as shown in Figure 1(b), the across-wafer variation is not a perfect parabola. It can be better modeled as a piece-wise function. It is a parabola in the center region (region in the dotted cycle), and

is almost constant near the edge (region outside the dotted cycle). Similar across-wafer trend is observed in other works [22]. In this case, we model the across-wafer variation as:

$$\begin{aligned} v_p &= x_w^2 + by_w^2 + cx_w + \\ & \quad dy_w + m_w + r \qquad (x_w, y_w) \in Center \\ v_p &= e + m_w + r \qquad\qquad (x_w, y_w) \in Edge \end{aligned} \qquad (9)$$

where $Center$ and $Edge$ are the center region (region in the dotted cycle in Figure 1(b)) and edge region (region outside the dotted cycle), respectively. In practice, we find that the across-wafer variation model in Equation (9) models most processes accurately. In this case, we may perform Monte-Carlo simulation based on Equation (9) to estimate circuit delay and leakage variation.

VI. CONCLUSION

In this paper, we analytically study the impact of systematic across-wafer variation on within-die spatial variation. For simplicity, we assume across-wafer variation is a quadratic function. We first observe that different locations of a chip may have different means and variances and such difference becomes more significant when chip size increases. Secondly, we find that spatial correlation is visible only when the across wafer systematic is not taken into account. When it is taken into account, we show that within die random variability does not exhibit a strong or useful pattern of spatial correlation. We exploited these observations in order to create a much more accurate and efficient model for performance variability prediction. Thirdly, we find that the within-die spatial variation is *NOT* independent of the inter-die variation. However, when chip size is small enough, such dependence is weak and the across-wafer variation can be lumped in to inter-die variation. In this case, the two level inter-/within-die variation model is still accurate. Based on the above analysis, we have proposed an accurate and efficient variation model for deterministic across wafer variation. Experimental result shows that compared to the distance-based spatial variation model, our new model reduces the error from 8% to 1% and improves the run time by 10X.

So far, the proposed variation model in this paper is used for Monte-Carlo analysis. In future, we will further improve our variation model to make it suitable for analytical statistic timing and leakage analysis instead of Monte-Carlo simulation.

REFERENCES

[1] A. Agarwal, D. Blaauw, V. Zolotov, S. Sundareswaran, M. Zhao, K. Gala, and R. Panda, "Statistical delay computation considering spatial correlations," in *ASPDAC*, 2003.

978-1-60558-497-3/09 $25.00 © 2009 ACM

bench mark	Exact			Our Model				Spatial correlation model from [3]			
	μ (mW)	σ(mW)	95% (mW)	μ (mW)	σ(mW)	95% (mW)	T (s)	μ (mW)	σ(mW)	95% (mW)	T (s)
c1908	93.6	19.6	140	94.8 (+1.3%)	20.2 (+3.1%)	145 (+3.6%)	9.7	100 (+7.1%)	24.2 (+23%)	157 (+11%)	121 (12.4X)
c2670	127	22.0	178	129 (+1.6%)	24.5 (+11%)	180 (+1.2%)	9.5	132 (+3.9%)	26.2 (+19%)	190 (+6.7%)	121 (12.7X)
c3540	196	35.6	277	197 (+0.7%)	36.7 (+3.1%)	279 (+1.1%)	10.4	202 (+3.2%)	39.8 (+12%)	287 (+3.6%)	123 (11.8X)

TABLE IV

LEAKAGE COMPARISON FOR ISCAS85 BENCHMARK.

bench mark	loca-tion	Exact			Our Model			Inter-/Within-die		
		μ (ns)	σ (ns)	95% (ns)	μ (ns)	σ (ns)	95% (ns)	μ (ns)	σ (ns)	95% (ns)
c7552	C	48.5	6.37	60.0	48.6 (+0.2%)	6.35 (-0.3%)	60.4 (+0.7%)	48.6 (+0.2%)	6.34 (-0.5%)	60.6 (+1.1%)
	LL	48.4	6.10	60.1	48.3 (-0.2%)	6.12 (+0.3%)	59.9 (-0.3%)	48.6 (+0.4%)	6.34 (+3.9%)	60.6 (+0.9%)
	LR	48.9	6.39	60.7	49.0 (+0.2%)	6.34 (-1.0%)	60.5 (-0.3%)	48.6 (-0.6%)	6.34 (-1.0%)	60.6 (-0.1%)
	UL	49.1	6.34	60.6	49.0 (-0.2%)	6.36 (+0.2%)	60.8 (+0.3%)	48.6 (-1.0%)	6.34 (+0.0%)	60.6 (-0.0%)
	UR	49.1	6.42	60.7	49.0 (-0.2%)	6.37 (-0.5%)	60.9 (+0.3%)	48.6 (-1.0%)	6.34 (-0.8%)	60.6 (-0.1%)

TABLE V

DELAY COMPARISON FOR ISCAS85 BENCHMARK IN 3MM×3MM CHIP.

bench mark	loca-tion	Exact			Our Model			Inter-/Within-die		
		μ (ns)	σ (ns)	95% (ns)	μ (ns)	σ (ns)	95% (ns)	μ (ns)	σ (ns)	95% (ns)
c7552	C	48.8	6.39	59.7	48.6 (-0.4%)	6.34 (-0.8%)	60.0 (+0.5%)	48.5 (-0.6%)	6.31 (-1.3%)	59.4 (-0.5%)
	LL	48.4	6.39	59.8	48.3 (-0.2%)	6.34 (-0.8%)	60.1 (+0.5%)	48.5 (+0.2%)	6.31 (-1.3%)	59.4 (+0.7%)
	LR	48.7	6.39	59.6	48.8 (+0.2%)	6.36 (-0.5%)	60.0 (+0.7%)	48.5 (-0.4%)	6.31 (-1.3%)	59.4 (-0.3%)
	UL	48.8	6.38	59.8	48.7 (-0.2%)	6.36 (-0.3%)	59.6 (-0.3%)	48.5 (-0.6%)	6.31 (-1.1%)	59.4 (-0.6%)
	UR	48.4	6.38	60.1	48.6 (+0.4%)	6.35 (-0.5%)	59.9 (-0.3%)	48.5 (+0.2%)	6.31 (-1.1%)	59.4 (-1.2%)

TABLE VI

DELAY COMPARISON FOR ISCAS85 BENCHMARK IN 6MM×6MM CHIP.

[2] H. Chang and S. S. Sapatnekar, "Statistical timing analysis considering spatial correlations using a single pert-like traversal," in *ICCAD*, 2003.

[3] J. Xiong, V. Zolotov, and L. He, "Robust extraction of spatial correlation," in *ISPD*, Apr 2006.

[4] F. Liu, "A general framework for spatial correlation modeling in vlsi design," in *DAC*, Jun 2007.

[5] J.-H. Liu, M.-F. Tsai, L. Chen, and C. C.-P. Chen, "Accurate and analytical statistical spatial correlation modeling for vlsi dfm applications," in *DAC*, Jun 2008.

[6] T. Sato, H. Ueyama, N. Nakayama, and K. Masu, "Determination of optimal polynomial regression function to decompose on-die systematic and random variations," in *ASPDAC*, Feb 2008.

[7] D. Boning, J. Chung, D. Ouma, and R. Divecha, "Spatial variation in semiconductor processes: Modeling for control," in *Proceedings Process Control Diagnostics and Modeling in Semiconductor Manufacturing II Electrochemical Society Meeting*, May 1997.

[8] S. Borkar, T. Karnik, S. Narendra, J. Tschanz, A. Keshavarzi, and V. De, "Parameter variations and impact on circuits and microarchitecture," in *DAC*, June 2003.

[9] K. Chopra, N. Shenoy, and D. Blaauw, "Variogram based robust extraction of process variation," in *ACM/IEEE International Workshop on Timing Issues*, Feb 2007.

[10] B. Liu, "Spatial correlation extraction via random field simulation and production chip performance regression," in *DATE*, Mar 2008.

[11] W. Z. ect., "An efficient method for chip-level statistical capacitance extractionion via random field simulation and production considering process variations with spatial correlation,". in *DATE*, Mar 2008.

[12] Q. Fu, W.-S. Luk, J. Tao, C. Yan, and X. Zeng, "Characterizing intra-die spatial correlation using spectral density method," in *International Symposium on Quality of Electronic Design*, Mar 2008.

[13] S. ichi OHKAWA, HirooMASUDA, and Y. INOUE, "A novel expression of spatial correlation by a random curved surface model and its application to LSI designs," *IEICE TRANS. FUNDAMENTALS*, 2008.

[14] W. Zhao, Y. Cao, F. Liu, K. Agarwal, D. Acharyya, S. Nassif, and K. Nowka, "Rigorous extraction of process variations for 65nm cmos design," in *Solid State Device Research Conference*, Sep 2007.

[15] P. Friedberg, W. Cheung, and C. J. Spanos, "Spatial modeling of micron-scale gate length variation," in *Proc. SPIE*, 2006.

[16] "Lack of spatial correlation in mosfet threshold voltage variation and implications for voltage scaling," *IEEE Trans. on Semiconductor Manufacturing*, to be appeared.

[17] Q. Zhang, C. Tang, J. Cain, A. Hui, T. Hsieh, N. Maccrae, B. Singh, K. Poolla, and C. J. Spanos, "Across-wafer cd uniformity control through lithography and etch process: experimental verification," in *Proc. SPIE*, 2007.

[18] K. Qian and C. J. Spanos, "A comprehensive model of process variability for statistical timing optimization," in *Proc. SPIE*, 2008.

[19] J. R. Sheats, *Microlithography Science and Technology*. CRC, 1998.

[20] S. J. V., P. S. B., J. S. R., and T. M. G., "Hot-wire cvd growth simulation for thickness uniformity," in *International Conference on Cat-CVD Process*, 2001.

[21] S. Sakai, M. Ogino, R. Shimizu, and Y. Shimogaki, "Deposition uniformity control in a commercial scale HTO-CVD reactor," *MATERIALS RESEARCH SOCIETY SYMPOSIUM PROCEEDINGS*, 2007.

[22] J. Brcka and R. L. Robison, "Wafer redeposition impact on etch rate uniformity in ipvd system," *IEEE Transactions on Plasma Sciences*, Feb 2007.

[23] T. W. Kim and E. S.Aydil, "Investigation of etch rate uniformity of 60 mhz plasma etching equipment," *Japanese Journal of Applied Physics*, Nov 2001.

[24] T. W. Kim and E. S.Aydil, "Effects of chamber wall conditions on cl concentration and si etch rate uniformity in palsma etching reactors," *Journal of The Electrochemical Society*, June 2003.

[25] Q. Zhang, K. Poolla, and C. J. Spanos, "One step forward from run-to-run critical dimension control: Across-wafer level critical dimension control through lithography and etch process," *Journal of Process Control*, vol. 18, no. 10, pp. 937 – 945, 2008. Advanced Process Control for Semiconductor Manufacturing, First IFAC Workshop on Advanced Process Control for Semiconductor Manufacturing.

[26] B. Stine, D. Boning, and J. Chung, "Analysis and decomposition of spatial variation in integrated circuit processes and devices," in *Semiconductor Manufacturing, IEEE Transactions on*, pp. 24 – 41, Feb. 1997.

[27] K. Qian and K. Spanos, "Hierachical Modeling of Spatial Variability of a 45nm Test Chip," in *SPIE*, 2009.

[28] Predictive Technology Model in *http://www.eas.asu.edu/ ptm/*.

[29] L. Cheng, J. Xiong, and L. He, "Non-gaussian statistical timing analysis using Second-Order polynomial fitting," in *ASPDAC*, Feb 2008.

[30] S. Amith and R. R. A., "From finance to flip flops: A study of fast quasi-monte carlo methods from computational finance applied to statistical circuit analysis," in *International Symposium on Quality of Electronic Design*, 2007.

978-1-60558-497-3/09 $25.00 © 2009 ACM

A Gaussian Mixture Model for Statistical Timing Analysis

Shingo Takahashi
NEC Corporation
Kanagawa, 229-1198, Japan
This work was performed in Chuo University.

sint@tsuki.elect.chuo-u.ac.jp

Yuki Yoshida
Chuo University
1-13-27 Kasuga, Bunkyo-ku,
Tokyo, 112-0003, Japan

Shuji Tsukiyama
Chuo University
1-13-27 Kasuga, Bunkyo-ku,
Tokyo, 112-0003, Japan

tsuki@elect.chuo-u.ac.jp

ABSTRACT

This paper introduces a Gaussian mixture model to represent delay and slew distributions in the statistical static timing analysis, and proposes algorithms for propagating them on a given circuit graph. The Gaussian mixture model can represent a non-Gaussian distribution due to the statistical Max operation properly, and any correlation efficiently, since it consists of plural Gaussian distributions. Therefore, not only topological correlations caused by re-convergent paths but also the correlation between each element and the critical delay, which is useful for circuit optimization, are calculated easily. The propagated slews are used to compute delay distributions of circuit elements dynamically so as to improve the accuracy. The proposed Gaussian mixture model is evaluated by comparing with Monte Carlo simulation, and the results show its effectiveness.

Categories and Subject Descriptors

J.6[Computer-Aided Engineering]: Computer-aided design (CAD); B.7.2 [Integrated Circuits]: Design Aids;

General Terms

Algorithms, Design.

Keywords

statistical timing analysis, Gaussian mixture model, delay distribution, slew distribution, variability.

1. INTRODUCTION

Due to the progress of nanometer process technologies, variability of circuit parameters is increasing and timing verification becomes a hard task[1]. To overcome the variability issue in the timing design, statistical static timing analysis (S-STA) has been studied intensively[2]. Existing S-STA algorithms are classified into 2 types: path-based approach and block-based approach[3-7]. Among them, the latter algorithms explore vertices of acyclic graph $G=(N,A)$ representing a given circuit in the topological order, and can be applied to a large circuit.

Since the input slew (transition time) is an important factor of determining delay of a circuit element (logic gate or interconnect), delay d(e) of each edge e must be determined with the use of the slew so as to improve the accuracy of the block-based S-STA. In

[4], a method to compute the output slew of a logic gate in the block-based S-STA is proposed, but since the distribution of slew tends to have more than one peak[8], determining delay d(e) by a single slew value is not appropriate, and a slew must be treated together with an arrival time. Namely, when the delay d(e) of edge e from vertex v to vertex w is determined for computing the arrival time D(w) to w by adding d(e) to arrival time D(v) to v, we must use slew distribution T(v) corresponding to D(v), since D(v) and T(v) represent delay and slew of a signal propagating edge e, respectively. In [9], the slew is treated as a stochastic variable, but it considers only the case of a single path. Hence, we need a more sophisticated method to handle slews.

On the other hand, S-STA algorithms can be classified into 2 types by the shape of delay distribution[2]: one uses Gaussian distribution only[3,4] and the other uses non-Gaussian distribution[5-7]. Among them, the algorithms using Gaussian distribution can represent distributions simply and handle various correlations easily. However, statistical Max/Min operations that are basic and important operations in S-STA do not preserve Gaussian distribution [2]. To solve these problems, various methods[5-7] have been proposed. However, they are more complex and need computational overhead. Moreover, the topological correlations caused by re-convergent path cannot be handled easily, whose effect must be taken into account, if the independent local within die variability becomes large.

In this paper, we propose Gaussian mixture model (GM model) to represent distributions of delays and slews. GM model is a weighted sum of Gaussian distributions, and can re-present correlations easily, since each component of GM model is Gaussian. Moreover, GM model can represent non-Gaussian distributions generated by statistical Max or Min operation properly. For example, the bold dotted line in Fig.1 shows the probability density function (PDF) of $z = $ Max[x, y] for independent variables $x \sim N(140, 23^2)$ and $y \sim N(140, 3.8^2)$, and the thin solid line is the PDF of Gaussian distribution of $N(E[z],V[z])$, where the mean $E[z]$ and variance $V[z]$ of z are calculated by

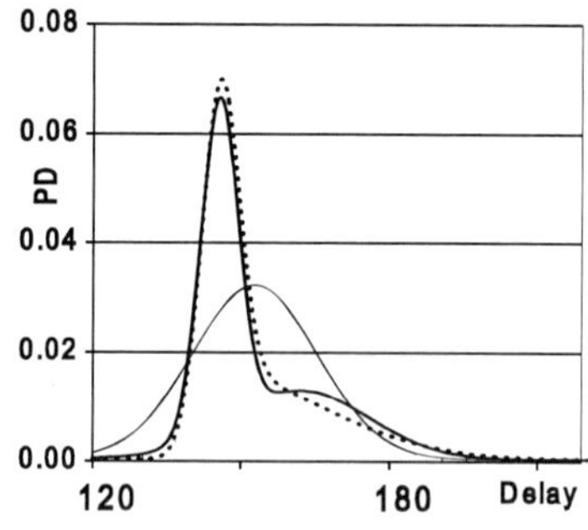

Fig. 1 : PDFs of statistical maximum

Permission to make digital or hard copies of part or all of this work for personal or classroom use is granted without fee provided that copies are not made or distributed for profit or commercial advantage and that copies bear this notice and the full citation on the first page. To copy otherwise, to republish, to post on servers or to redistribute to lists, requires prior specific permission and/or a fee.

DAC'09, July 26-31, 2009, San Francisco, California, USA

978-1-60558-497-3/09 $25.00 © 2009 ACM

Clark's method[10] correctly. The thick solid line in Fig. 1 is the PDF of the proposed GM model.

Moreover, GM model can represent any correlation easily, since each component is Gaussian. Therefore, we can easily compute not only the topological correlation caused by re-convergent paths[2] but also the correlation between the local variation of each logic gate and the critical delay distribution. The latter correlation is useful to identify the most sensitive logic gate such that a variance reduction of the delay distribution of the logic gate decreases the variance of the critical delay distribution. Namely, if we find such a logic gate and can use a cell with a regular layout for the gate, then we have a smaller variance of the critical delay, since the variance of the cell delay with a regular layout is smaller than the one with an irregular layout. Therefore, we may be able to reduce the $\mu+3\sigma$ value of the critical delay with a small area penalty.

The rest of the paper is organized as follows. In Section 2, we describe the outline of the proposed S-STA that propagates delays and slews represented by GM models, and in Sections 3 and 4, we show the basic algorithms at an edge and a vertex, respectively. In Section 5, we show some experimental results which demonstrate the effectiveness of the proposed GM model. Finally, in Section 6, we give conclusions.

2. STATISTICAL TIMING ANALYSIS USING GM MODELS

The circuit to be analyzed is shown by an acyclic graph $G=(N, A)$, where each vertex corresponds to a terminal of the circuit and each edge represents a signal propagation between terminals. For each vertex v of G, we define a pair $(D(v),T(v))$ consisting of delay (the latest or earliest arrival time) $D(v)$ and slew $T(v)$ of the output signal from v, and we assume that such a pair $(D(v),T(v))$ for each source v of G is given. For each edge $e=(v,w)$ of G, we call a pair $(D(v),T(v))$ an input-pair to e, since it corresponds to the input signal to e with delay $D(v)$ and slew $T(v)$. Moreover, we define an output-pair $(D_{out}(e),T_{out}(e))$ of e, which correspond to the output signal from e with delay $D_{out}(e)$ and slew $T_{out}(e)$.

The proposed S-STA algorithm processes each vertex w in the topological order, and computes pair $(D(w),T(w))$ of w from all output-pairs $(D_{out}(e_n),T_{out}(e_n))$ of edges $e_n=(v_n,w)$ coming into w. This process is called *Mix* operation at vertex w. During the Mix operation for w, the algorithm processes each edge $e_n=(v_n,w)$ and computes output-pair $(D_{out}(e_n),T_{out}(e_n))$ from the input-pair $(D(v_n),T(v_n))$. This process is called *Carry* operation at an edge e_n.

Delay D and slew T of a pair (D,T) introduced above are random variables (RVs), and their PDFs are assumed to be mixtures of 2 Gaussian distributions denoted as follows:

$$f_D(D) = P_1 \cdot \frac{1}{\sigma_{D1}} \cdot \phi\left(\frac{D-\mu_{D1}}{\sigma_{D1}}\right) + P_2 \cdot \frac{1}{\sigma_{D2}} \cdot \phi\left(\frac{D-\mu_{D2}}{\sigma_{D2}}\right) \quad (1)$$

$$f_T(T) = P_1 \cdot \frac{1}{\sigma_{T1}} \cdot \phi\left(\frac{T-\mu_{T1}}{\sigma_{T1}}\right) + P_2 \cdot \frac{1}{\sigma_{T2}} \cdot \phi\left(\frac{T-\mu_{T2}}{\sigma_{T2}}\right) \quad (2)$$

Here, P_1 and P_2 are probabilities satisfying $P_1+P_2 = 1$ and $\phi(x)$ is the PDF of the standard Gaussian distribution $N(0,1)$. Henceforth, we denote the cumulative distribution function (CDF) of $N(0,1)$ by $\Phi(x)$.

Equation (1) indicates that the distribution of D consists of 2 Gaussian distributions $N(\mu_{D1},\sigma_{D1}^2)$ and $N(\mu_{D2},\sigma_{D2}^2)$ with *mixture proportions* P_1 and P_2, respectively, and Eq. (2) indicates that T consists of 2 Gaussian distributions $N(\mu_{T1},\sigma_{T1}^2)$ and $N(\mu_{T2},\sigma_{T2}^2)$ with the same mixture proportions. In the following, an RV denoted by Eq. (1) or (2) is called a 2-GMM (Gaussian Mixture Model consisting of 2 Gaussian distributions), and the Gaussian distributions with the mixture proportions P_1 and $P_2=1-P_1$ are called the *first* and the *second components* of the 2-GMM, respectively. The reason why the j-th components ($j \in \{1,2\}$) of D and T have the same mixture proportion is that they represent the delay and slew of the same signal and P_j denotes the probability for the signal to be propagated. In this paper, we denote 2-GMM D represented by Eq. (1) by $D \sim GMM(P_1, \mu_{D1}, \sigma_{D1}^2, \mu_{D2}, \sigma_{D2}^2)$.

As usually done[2], we assume that each distribution of D is represented by a first order canonical form of a local RV x_D and g global RVs r_i ($1 \leq i \leq g$). Namely, let D_j ($j \in \{1,2\}$) be the RV representing the j-th components of D, then we assume that D_j is represented by the following equations.

$$D_j = \mu_{Dj} + s_{x,j}[D] \cdot x_D + \sum_{i=1}^{g} s_{i,j}[D] \cdot r_i \quad (3)$$

Here, $s_{x,j}[D]$ and $s_{i,j}[D]$ are the sensitivities of the j-th components of D to x_D and r_i, respectively. We assume that x_D and r_i ($1 \leq i \leq g$) are all $N(0,1)$ and independent each other. Therefore, there hold the following equation.

$$\sigma_{Dj}^2 = s_{x,j}[D]^2 + \sum_{i=1}^{g} s_{i,j}[D]^2 \quad (4)$$

The local RV x_D is introduced to represent the variation of D which cannot be represented by the global variables only, so as to reduce the number g of global RVs. Henceforth, we call term $s_{x,j}[D] \cdot x_D$ the *local variation* of the j-th components of D.

Each distribution T_j ($j \in \{1,2\}$) of T is also represented by a local RV x_T and the global RVs r_i as follows.

$$T_j = \mu_{Tj} + s_{x,j}[T] \cdot x_T + \sum_{i=1}^{g} s_{i,j}[T] \cdot r_i \quad (5)$$

However, we assume that $s_{x,j}[T] = s_{i,j}[T] = 0$ in this paper, and hence $\sigma_{T1} = \sigma_{T2} = 0$. Although we ignore the variance of slew, we use two means μ_{T1} and μ_{T2} of the first and the second components of T, which are adjunct to the first and the second components of D, respectively, so as to exploit the advantage of mixture of two distributions. It is possible to expand the discussions in this paper to the case of $s_{x,j}[T] \neq 0$ and $s_{i,j}[T] \neq 0$, but we ignore the variance of slew here, so as to simplify the delay calculations. Due to this assumption, we can use ordinary cell libraries used for S-STAs which do not consider slew distributions.

Let D be a 2-GMM $D \sim GMM(P_1, \mu_{D1}, \sigma_{D1}^2, \mu_{D2}, \sigma_{D2}^2)$. Then, μ_{Dj} and σ_{Dj}^2 ($j \in \{1,2\}$) are called the mean and variance of the j-th component of D and denoted by $E_j[D] = \mu_{Dj}$ and $V_j[D] = \sigma_{Dj}^2$, respectively. The 2nd moment of the j-th component of D is denoted by $E_j[D^2] = \sigma_{Dj}^2 + \mu_{Dj}^2$. The mean E[D], the 2nd moment $E[D^2]$, and the variance V[D] of D can be expressed by the following equations, respectively.

$$E[D] = \sum_{j \in \{1,2\}} P_j \cdot E_j[D] \quad (6)$$

$$E[D^2] = \sum_{j \in \{1,2\}} P_j \cdot E_j[D^2] \quad (7)$$

$$V[D] = E[D^2] - E[D]^2 \quad (8)$$

Let D_A and D_B be 2-GMMs such that $D_A \sim GMM(P_{A1}, \mu_{DA1}, \sigma_{DA1}^2, \mu_{DA2}, \sigma_{DA2}^2)$ and $D_B \sim GMM(P_{B1}, \mu_{DB1}, \sigma_{DB1}^2, \mu_{DB2}, \sigma_{DB2}^2)$, respectively, and the j-th component of D_A and the k-th component of D_B are represented as follows.

$$D_{Aj} = \mu_{DAj} + s_{x,j}[D_A] \cdot x_{DA} + \sum_{i=1}^{g} s_{i,j}[D_A] \cdot r_i \qquad (9)$$

$$D_{Bk} = \mu_{DBk} + s_{x,k}[D_B] \cdot x_{DB} + \sum_{i=1}^{g} s_{i,k}[D_B] \cdot r_i \qquad (10)$$

Moreover, let $C_{jk}[D_A, D_B] = C[D_{Aj}, D_{Bk}]$ be the covariance of the j-th component of D_A and the k-th component of D_B, which is defined by regarding each component as a single Gaussian distribution. Then, the covariance of D_A and D_B can be obtained by the following equation.

$$C[D_A, D_B] = \sum_{j,k \in \{1,2\}} P_{Aj} \cdot P_{Bk} \cdot C_{jk}[D_A, D_B] \qquad (11)$$

$$C_{jk}[D_A, D_B] = C_{jk}\big[s_{x,j}[D_A] \cdot x_{DA}, s_{x,k}[D_B] \cdot x_{DB}\big]$$

$$+ \sum_{i=1}^{g} s_{i,j}[D_A] \cdot s_{i,k}[D_B] \qquad (12)$$

Since a single Gaussian distribution can be regarded as a 2-GMM which does not have the second component, we can consider the case of $D_B = r_i$, and we have the following relation.

$$s_{i,j}[D_A] = C_{j1}[D_A, r_i] \qquad (13)$$

3. CARRY OPERATION AT AN EDGE

In this section, we describe the outline of Carry operation at an edge $e=(v,w)$ which computes output-pair $(D_{out}(e), T_{out}(e))$ from a given input-pair $(D(v), T(v))$.

The mixture proportions P_1 and P_2 of the components of output-pair $(D_{out}(e), T_{out}(e))$ are the same as those of input-pair $(D(v), T(v))$, since each component corresponds to a signal with particular slew and delay, and probability of each component does not change through the signal propagation at edge e. Fig. 2 illustrates the outline of Carry operation.

If a signal with delay $D(v)$ and slew $T(v)$ is input to a circuit element corresponding to e, the delay d added to $D(v)$ at edge e and the output-slew t of this signal are determined by $T(v)$, load capacitance $C(e)$, and a set $P(e)$ of circuit parameters associated with e. Let us assume that d is determined by

$$d = E[d] + s_x[d] \cdot x_e + \sum_{i=1}^{g} s_i[d] \cdot r_i \qquad (14)$$

where $x_e \sim N(0,1)$ is the local RV associated with edge e and independent of each r_i ($1 \leq i \leq g$) and the local RV $x_{e'}$ of other edge e'. Since we ignore the variance of slew, this equation does not contain a term concerning the local RV x_{Tin} of $T(v)$.

In order to calculate the mean of slew t and the distribution of delay d, we prepare look-up tables for each $P(e)$, and calculate

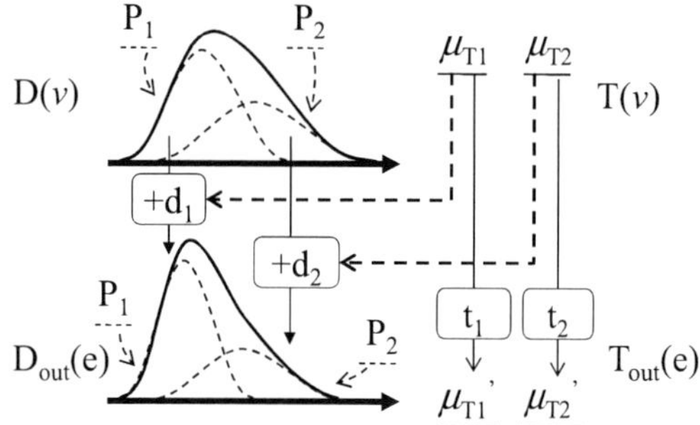

Fig. 2: Carry operation at an edge

them by using $T(v)$ and $C(e)$. By using these tables for the j-th component of $T(v)$, we have the output-slew t_j and the delay d_j for the j-th component. Although the calculation of d_j is the same as the S-STA algorithms which do not treat slews as RVs, our method is different from them in that the delay d_j is calculated for each mean $E_j[T(v)]$ of 2-GMM $T(v)$ propagated dynamically.

Let $E[t_j]$ and $E[d_j]$ be the means of t_j and d_j, respectively, and $s_i[d_j]$ and $s_x[d_j]$ be the sensitivities of d_j to r_i and x_e, respectively. Then the j-th component of $T_{out}(e)$ is

$$E_j[T_{out}(e)] = E[t_j] \qquad (15)$$

and the j-th component of $D_{out}(e)$ is determined by

$$E_j[D_{out}(e)] = E_j[D(v)] + E[d_j] \qquad (16)$$

$$s_{i,j}[D_{out}(e)] = s_{i,j}[D(v)] + s_i[d_j] \qquad (17)$$

$$V_j[D_{out}(e)] = V_j[D(v)] + V[d_j] + 2 \cdot C_{j1}[D(v), d_j]$$

$$= V_j[D(v)] + s_x[d_j]^2$$

$$+ \sum_{i=1}^{g} s_i[d_j]^2 + 2 \cdot \sum_{i=1}^{g} s_{i,j}[D(v)] \cdot s_i[d_j] \qquad (18)$$

From Eqs. (17) and (18), the sensitivity $s_{x,j}[D_{out}(e)]$ to the local RV x_{Dout} of $D_{out}(e)$ is obtained by

$$s_{x,j}[D_{out}(e)]^2 = V_j[D_{out}(e)] - \sum_{i=1}^{g} s_{i,j}[D_{out}(e)]^2 \qquad (19)$$

Let delay Z be $D(u)$ of a vertex u ($\neq w$) or $D_{out}(e')$ of an edge $e'=(v',w)$, for which we must compute covariance $C_{jk}[D_{out}(e), Z]$. If Z is not $D(v)$ of vertex v of $e=(v,w)$, then $C_{jk}[D_{out}(e), Z]$ of the j-th component of $D_{out}(e)$ and the k-th component Z can be calculated by the following equation

$$C_{jk}[D_{out}(e), Z] = C_{jk}[D(v), Z] + C_{1k}[d_j, Z]$$

$$= C_{jk}[s_{x,j}[D(v)] \cdot x_{Din}, s_{x,k}[Z] \cdot x_Z]$$

$$+ \sum_{i=1}^{g} s_{i,j}[D_{out}(e)] \cdot s_{i,k}[Z] \qquad (20)$$

where $C_{jk}[s_{x,j}[D(v)] \cdot x_{Din}, s_{x,k}[Z] \cdot x_Z]$ is the covariance between the local variation of the j-th component of $D(v)$ and the local variation of the k-th component of Z. Since the local RV x_e associated with edge e is independent of all other variables, we have $C_{1k}[d_j, Z] = \sum_{i=1}^{g} s_i[d_j] \cdot s_{i,k}[Z]$, and hence Eq.(20).

From Eqs.(12) and (20), we can see that the covariance between the local variation of the j-th component of $D(v)$ and the local variation of the k-th component of Z does not change in Carry operation. Namely, we have the following equation, if Z is not $D(v)$.

$$C_{jk}[s_{x,j}[D_{out}(e)] \cdot x_{Dout}, s_{x,k}[Z] \cdot x_Z]$$

$$= C_{jk}[s_{x,j}[D(v)] \cdot x_{Din}, s_{x,k}[Z] \cdot x_Z] \qquad (21)$$

If Z is $D(v)$ of vertex v of $e=(v,w)$, then the covariance $C_{jk}[s_{x,j}[D_{out}(e)] \cdot x_{Dout}, s_{x,k}[D(v)] \cdot x_{Din}]$ can be computed by the following equation.

$$C_{jk}[s_{x,j}[D_{out}(e)] \cdot x_{Dout}, s_{x,k}[D(v)] \cdot x_{Din}]$$

$$= C_{jk}[s_{x,j}[D(v)] \cdot x_{Din} + s_x[d_j] \cdot x_e, s_{x,k}[D(v)] \cdot x_{Din}]$$

$$= s_{x,j}[D(v)] \cdot s_{x,k}[D(v)] \qquad (22)$$

This covariance between local variations is necessary for computing the correlation caused by re-convergent paths.

Moreover, the covariance $C_{j1}[D_{out}(e), x_e]$ between the j-th component of $D_{out}(e)$ and the local RV x_e of edge e and the

978-1-60558-497-3/09 $25.00 © 2009 ACM

covariance C[$D_{out}(e)$, x_e] between $D_{out}(e)$ and x_e are obtained by following equations, respectively.

$$C_{j1}[\, D_{out}(e), x_e \,] = s_x[d_j] \tag{23}$$

$$C[D_{out}(e), x_e] = \sum_{j \in \{1,2\}} P_j \cdot s_x[d_j] \tag{24}$$

These covariances are used for computing the correlation coefficient between the local RV of an edge and the critical delay.

4. MIX OPERATION AT A VERTEX

In this section, we propose two methods of representing $D_M =$ Max[D_A, D_B] by a 2-GMM for given 2-GMMs D_A and D_B, and corresponding methods of generating T_M adjunct to D_M for given pairs (D_A, T_A) and (D_B, T_B). These methods are key operations of Mix operation at a vertex w which computes a pair $(D(w), T(w))$ of w from output-pairs $(D_{out}(e_n), T_{out}(e_n))$ of edges $e_n = (v_n, w)$ coming into w in the case where the number of edges coming into w is more than 1. We use these methods according to the probability $\Pr[D_A \geq D_B]$ for D_A to be larger than D_B.

4.1 The Case of $\Pr[D_A \geq D_B] \approx 0.5$

The idea of the algorithm to represent D_M by a 2-GMM is illustrated in Fig. 3. In the figure, ellipses are contour lines of the joint PDF of D_A and D_B, and since both D_A and D_B are 2-GMMs, the joint PDF of D_A and D_B is composed of four components, each of which is a joint PDF of two Gaussian distributions. Since both D_A and D_B are 2-GMMs, the joint PDF of D_A and D_B is composed of four components, each of which is a joint PDF of two Gaussians.

The first and second components of 2-GMM of D_M are obtained from the upper left and the lower right areas of the line $D_A = D_B$, respectively. Namely, the first component of D_M comes from D_A in the case of $D_A \geq D_B$ (upper left area) and the second component from D_B in the case of $D_A \leq D_B$ (lower right area). For given D_A ~GMM(P_{A1}, μ_{DA1}, σ_{DA1}^2, μ_{DA2}, σ_{DA2}^2) and D_B ~GMM(P_{B1}, μ_{DB1}, σ_{DB1}^2, μ_{DB2}, σ_{DB2}^2), we have D_M ~GMM(P_{M1}, μ_{DM1}, σ_{DM1}^2, μ_{DM2}, σ_{DM2}^2) represented by the follow equations, where $q \in \{1,2\}$ indicates a component of D_M. The derivations of these equations can be seen in [11]. In the case of $\alpha_{jk} = 0$, where $V_j[D_A] = V_k[D_B] = C_{jk}[D_A, D_B]$, the j-th component of D_A and the k-th component of D_B are not independent, and hence we must use different equations.

$$\alpha_{jk} = \sqrt{V_j[D_A] + V_k[D_B] - 2 \cdot C_{jk}[D_A, D_B]} \tag{25}$$

$$\beta_{jk} = \frac{\mu_{DAj} - \mu_{DBk}}{\alpha_{jk}} \tag{26}$$

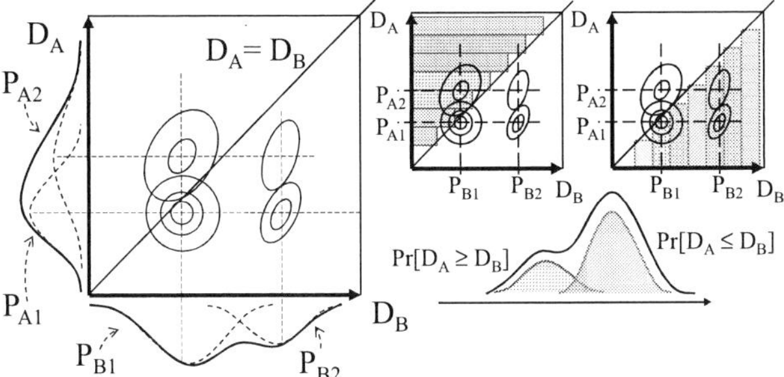

Fig.3: Mix operation for the case of $\Pr[D_A \geq D_B] \approx 0.5$

$$P_{M1} = \Pr[D_A \geq D_B] = \sum_{j,k \in \{1,2\}} P_{Aj} \cdot P_{Bk} \cdot \Phi(\beta_{jk}) \tag{27}$$

$$P_{M2} = \Pr[D_A \leq D_B] = 1 - P_{M1} \tag{28}$$

$$\mu_{DM1} = \sum_{j,k \in \{1,2\}} \frac{P_{Aj} \cdot P_{Bk}}{P_{M1}} \cdot \{\mu_{DAj} \cdot \Phi(\beta_{jk}) + VC_{Ajk} \cdot \phi(\beta_{jk})\} \tag{29}$$

$$\mu_{DM2} = \sum_{j,k \in \{1,2\}} \frac{P_{Aj} \cdot P_{Bk}}{P_{M2}} \cdot \{\mu_{DBk} \cdot \Phi(-\beta_{jk}) + VC_{Bjk} \cdot \phi(\beta_{jk})\} \tag{30}$$

$$VC_{Ajk} = \frac{V_j[D_A] - C_{jk}[D_A, D_B]}{\alpha_{jk}} \tag{31}$$

$$VC_{Bjk} = \frac{V_k[D_B] - C_{jk}[D_A, D_B]}{\alpha_{jk}} \tag{32}$$

$$\sigma_{DMq}^2 = \sum_{j,k \in \{1,2\}} \frac{P_{Aj} \cdot P_{Bk}}{P_{M1}} \cdot m_{2jkq} - \mu_{DMq}^2 \tag{33}$$

$$m_{2jk1} = (\mu_{DAj}^2 + V_j[D_A]) \cdot \Phi(\beta_{jk})$$
$$+ \{2 \cdot \mu_{DAj} - VC_{Ajk} \cdot \beta_{jk}\} \cdot VC_{Ajk} \cdot \phi(\beta_{jk}) \tag{34}$$

$$m_{2jk2} = (\mu_{DBk}^2 + V_k[D_B]) \cdot \Phi(-\beta_{jk})$$
$$+ \{2 \cdot \mu_{DBk} + VC_{Bjk} \cdot \beta_{jk}\} \cdot VC_{Bjk} \cdot \phi(\beta_{jk}) \tag{35}$$

For a 2-GMM Z ~GMM(P_{Z1}, μ_{Z1}, σ_{Z1}^2, μ_{Z2}, σ_{Z2}^2) other than D_A and D_B, the covariance $C_{qk}[D_M, Z]$ between the q-th component of D_M and the h-th component of Z can be written as follows.

$$C_{1h}[D_M, Z] = \sum_{j,k \in \{1,2\}} \frac{P_{Aj} \cdot P_{Bk}}{P_{M1}} \cdot \left\{ C_{jh}[D_A, Z] \cdot \Phi(\beta_{jk}) \right.$$
$$\left. + CC_{jkh} \cdot (\mu_{DAj} - \mu_{DM1} - VC_{Ajk} \cdot \beta_{jk}) \cdot \phi(\beta_{jk}) \right\} \tag{36}$$

$$C_{2h}[D_M, Z] = \sum_{j,k \in \{1,2\}} \frac{P_{Aj} \cdot P_{Bk}}{P_{M2}} \cdot \left\{ C_{kh}[D_B, Z] \cdot \Phi(-\beta_{jk}) \right.$$
$$\left. - CC_{jkh} \cdot (\mu_{DBk} - \mu_{DM2} + VC_{Bjk} \cdot \beta_{jk}) \cdot \phi(\beta_{jk}) \right\} \tag{37}$$

$$CC_{jkh} = \frac{C_{jh}[D_A, Z] - C_{kh}[D_B, Z]}{\alpha_{jk}} \tag{38}$$

The covariance between the q-th component of D_M and the local RV x_e of edge e is computed by setting $x_e = Z$ with only one component.

The sensitivity $s_{x,q}[D_M]$ of the q-th component of D_M to the local RV x_M can be computed by the following equation, where $s_{i,q}[D_M] = C_{q1}[D_M, r_i]$.

$$s_{x,q}[D_M]^2 = \sigma_{DMq}^2 - \sum_{i=1}^{g} s_{i,q}[D_M]^2 \tag{39}$$

In order to devise a method to calculate the slew T_M which constitutes input-pair (D_M, T_M) together with $D_M = $ Max[D_A, D_B], let us consider the case where D_A and D_B are both single Gaussians. In this case, the first component T_{M1} of T_M is adjunct to the first component D_{M1} of D_M with the mixture proportion $P_{M1} = \Pr[D_A \geq D_B]$, and D_{M1} is obtained from D_A. Hence, the mean $E_1[T_M]$ of T_{M1} must be calculated from the slew T_A of output pair (D_A, T_A). Similarly, the second component T_{M2} of T_M is adjunct to the second component D_{M2} of D_M with the mixture proportion $P_{M2} = \Pr[D_A \leq D_B]$, and D_{M2} is obtained from D_B. Hence, the mean $E_2[T_M]$ of T_{M2} must be calculated from the slew T_B of output pair (D_B, T_B).

By expanding this idea to the case of 2-GMM, we compose T_{M1} (T_{M2}) from T_A (T_B) in such a way that the ratio of mixing the components of T_A (T_B) is the same as that of D_A (D_B). Namely, with the use of similar equations to Eqs.(29) and (30), we determine the mean of q-th ($q \in \{1,2\}$) components of T_M as follows.

$$\mu_{TM1} = \sum_{j,k \in \{1,2\}} \frac{P_{Aj}P_{Bk}}{P_{M1}} \cdot E_j[T_A] \cdot \Phi(\beta_{jk}) \qquad (40)$$

$$\mu_{TM2} = \sum_{j,k \in \{1,2\}} \frac{P_{Aj}P_{Bk}}{P_{M2}} \cdot E_k[T_B] \cdot \Phi(-\beta_{jk}) \qquad (41)$$

4.2 The Case of $\Pr[D_A \geq D_B] \approx 0$

The Mix operation described in the previous subsection is effective if $P_{M1} = \Pr[D_A \geq D_B]$ is around 0.5. However, if P_{M1} approaches to 0 or 1, then the advantage of using 2-GMM disappears. For example, in the case where D_B is much greater than D_A, that is, the probability $P_{M1} = \Pr[D_A \geq D_B] \approx 0$, D_M becomes a single Gaussian obtained from the lower right area of the line $D_A = D_B$, and the information of 2-GMM D_B is lost, although $D_M \approx D_B$ in this case. To resolve this problem, we propose another Mix operation which uses the fact that the PDF of D_M must be similar to that of D_B.

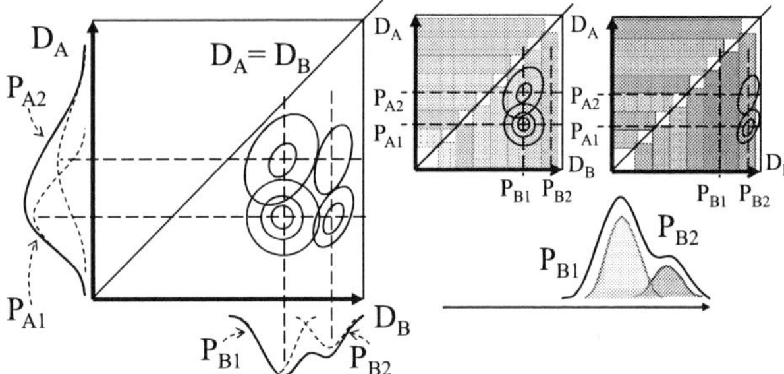

Fig.4: Mix operation for the case of $\Pr[D_A \geq D_B] \approx 0$

The idea of representing D_M by a 2-GMM in the case of $\Pr[D_A \geq D_B] \approx 0$ is shown in Fig.4, where D_M has the same mixture proportions as D_B. In this case, we divide the joint PDF of D_A and D_B into two as shown in the figure, and compute each component of D_M from a divided PDF. Due to the limited space, we omit the details. Each component of the slew T_M which constitutes pair (D_M, T_M) together with D_M is determined by a similar idea shown in the previous subsection.

5. EXPERIMENTAL RESULTS

In order to evaluate the effect of using 2-GMM in S-STA, we compare the following two S-STAs with Monte Carlo simulation (MC): S-STA using canonical (Gaussian) model (C-SSTA) and S-STA using GM model (GM-SSTA). In GM-SSTA, we applied the Mix operation in Subsection 4.1, when $\Pr[D_A \geq D_B]$ is greater than 0.15 and smaller than 0.85, and otherwise, we use the Mix operation in Subsection 4.2.

The experiment was done on the ISCAS'85 bench mark circuits and the circuit shown in Fig.5. The means of gate delays and the interconnect delay used in the experiment are shown in Table 1. The standard deviation of each delay is set to 10% of the mean.

Table 1: The mean of gate delays

BUFF	20	OR,AND	25	XOR	30
NOT	10	NOR,NAND	15	WIRE	5

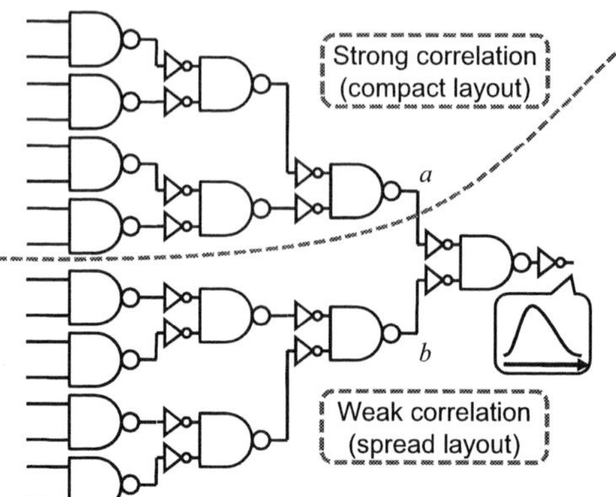

Fig. 5: A test circuit producing a non-Gaussian delay distribution by statistical Max.

In order to represent the spatial correlation of ISCAS'85 bench mark circuits, we divided the circuit area into the quadratic trees[2] and introduced 26 global RVs for the divided subregions. Moreover, we introduced a global RV for each gate so as to represent intra-gate correlation, whose variation is regarded as a local variation. Then, we represent each gate delay by a canonical form consisting of six global RVs, one global RV associated with the gate, and one local RV of an edge. The percentage of local variation in the total variance is 50% and that of local RV of an edge is 40%. Similarly, we represent each interconnect delay by a canonical form consisting of six global RVs and one local RV of an edge. The percentage of local variation in the total variance is 40%.

Since the experimental results of C-SSTA are pretty good, we tested the circuit shown in Fig.5, such that a non-Gaussian delay distribution is produced by a statistical Max operation. The circuit is composed of two subcircuits. The upper subcircuit generating output *a* is assumed to be laid out in a small area (*compact layout*), so that the delays of gates in the subcircuit have strong correlations. The percentage of the local variation in the total variance is 20%. On the other hand, another subcircuit is assumed to be laid out in a large area (*spread layout*), so that the delays of gates in the subcircuit have weak correlations. The percentage of the local variation in the total variance is 90%.

In Table 2, we show the relative errors of $\mu+3\sigma$ values of the critical delay obtained by S-STAs to that obtained by MC, where $\mu+3\sigma$ value means the delay when the value of CDF becomes 99.865%. From the table, we can see that GM-SSTA improved the accuracy from C-SSTA although the difference is small in the case of ISCAS'85 bench mark circuits. However, for the circuit of Fig.5, we see a better improvement. Because the delays to outputs *a* and *b* have similar means and different variances, which generate a non-Gaussian distribution by statistical Max operation.

The computational time of GM-SSTA is 1.48 times larger than that of C-SSTA in average for all ISCAS'85 bench mark circuits, and at most 1.89 times for c6288.

978-1-60558-497-3/09 $25.00 © 2009 ACM   114

Table.2: 99.865% ($\mu+3\sigma$) delay values of CDF

Circuit	MC 10,000 times	C-SSTA		GM-SSTA	
		$\mu+3\sigma$ value	relative error	$\mu+3\sigma$ value	relative error
c5315	1325	1319	−0.47%	1322	−0.25%
c6288	3232	3225	−0.22%	3226	−0.20%
c7552	1119	1116	−0.32%	1117	−0.20%
Fig. 5	121.0	116.1	−4.04%	118.1	−2.43%

One of the merits of using GM model is easiness of handling correlations. To see this merit, we computed the correlation coefficient of the local RV of each edge and the critical delay with the use of Eqs. (36)-(38). If an edge has no possibility to be contained in a critical path (a path whose delay may become the largest), the correlation coefficient becomes 0. On the other hand, if an edge has a large correlation coefficient with the critical delay, then the chance for the edge to be contained in a critical path is large.

In order to check the effect of the correlation coefficient, we found by MC and an exhaustive search the most influenced edge such that the decrement of the variance or the mean of the delay of the edge yields the largest reduction of $\mu+3\sigma$ value of the critical delay, and checked the rank of the edge in the order of the correlation coefficient. Moreover, we checked the rank of the edge in the order of the probability proposed in [3] such that the edge is included in critical paths. This probability is computed by the tightness probability and Gaussian distribution. The results are shown in Table 3, from which we see the effectiveness of the correlation coefficient.

6. CONCLUSION

In this paper, we proposed a new S-STA algorithm propagating delays represented by 2-GMM together with slews. By using 2-GMM, non-Gaussian distribution generated by statistical Max or Min operation can be represented appropriately, and various correlations, such as the topological correlations and the correlation between the critical delay and circuit element delay, can be handled easily. The latter correlation can be used for circuit optimization.

From experimental results, we can see that the proposed S-STA using 2-GMM is effective and attractive as a new S-STA technique. Although we ignore the variance of slew in this paper, the discussions can be extended to the case where slews are also represented by 2-GMMs.

There still remains some future works. Among them, the most important one is to improve the probabilities of the four components of the joint PDF of D_A and D_B shown in Figs. 3 and 4. In this paper, we calculated the probability of the occurrences of the j-th component of D_A and the k-th component of D_B by assuming that they are independent. However, since they are not independent, we must compute a conditional probability. We are tackling this problem and other problems such as to include the effect of multiple input switching and to apply the algorithm for the optimization of circuit.

ACKNOWLEDGMENTS

This research was partially supported by Grant-in-Aid for Scientific Research (C) 19560353.

REFERENCES

[1] A. Srivastava, D. Sylvester, and D. Blaauw, Statistical Analysis and Optimization for VLSI: Timing and Power, Springer, 2005.

[2] D. Blaauw, K. Chopra, A. Srivastave, L. Scheffer, "Statistical timing analysis: From basic principles to state of the art," IEEE Trans. CAD/ICAS, vol.27, no.4, pp.589-607, 2008.

[3] C. Visweswariah, K. Ravindran, K. Kalafala, S. G. Walker, S. Narayan, D. K. Beece, J. Piaget, N. Venkateswaran, and J. G. Hemmett, "First-order incremental block-based statistical timing analysis," IEEE Trans. CAD, Vol. 25, No. 10, pp. 2170-2180, 2006.

[4] H. Chang and S. Sapatnekar, "Statistical timing analysis under spatial correlations," IEEE Trans. CAD, Vol. 24, No. 9, pp. 1467-1482, 2005.

[5] Y. Zhan, A. J. Strojwas, X. Li, L. T. Pileggi, "Correlation-aware statistical timing analysis with non-Gaussian delay distributions," Proc. DA conf. , pp. 77-82, 2005.

[6] K. Chopra, B. Zhai, D. Blaauw, D. Sylvester, "A new statistical max operation for propagating skewness in statistical timing analysis," Proc. ICCAD, pp. 237-243, 2006.

[7] X. Li, J. Le, P. Gopalakrishnan, L. T. Pileggi, "Asymptotic probability extraction for nonnormal performance distributions," IEEE Trans. CAD, Vol. 26, No. 1, pp. 16-37, 2007.

[8] T. Kouno and H. Onodera "Consideration of transition-time variability in statistical timing analysis," IEEE Conf. International SOC Conference, pp.207-210, 2006.

[9] R.Goyal, S.Shrivastava, H.Parameswaran, P.Khurana, "Improved first-order parametarized statistical timing analysis for handling slew and capaciatnce variation," Proc. Int. Conf. VLSI Design, pp.278-282, 2007.

[10] C.E.Clark, "The greatest of a finite set of random variables," Operation Research, vol.9, pp.85-91, 1961.

[11] S. Takahashi and S. Tsukiyama, "A new statistical timing analysis using Gaussian mixture models for delay and slew propagated together," IEICE Trans. on Fundamentals of Electronics, Communications and Computer Sciences, vol.E92-A, no.3, pp.900-911, Mar. 2009.

Table.3: The ranks of the most influenced edge

Circuit	decreasing variance		decreasing mean	
	[3]	Proposed	[3]	Proposed
c1908	36 th	1 st	1 st	2 nd
c2670	7 th	1 st	8 th	1 st
c3540	60 th	2 nd	6 th	4 th
c5315	11 th	1 st	2 nd	2 nd
c6288	1 st	1 st	1 st	1 th
c7552	1 st	1 st	1 st	1 th

978-1-60558-497-3/09 $25.00 © 2009 ACM

A Stochastic Jitter Model for Analyzing Digital Timing-Recovery Circuits

James R. Burnham
High-Q Design, Los Altos, CA
burnham@highqdesign.com

Chih-Kong Ken Yang
UCLA, Los Angeles, CA
yang@ee.ucla.edu

Haitham Hindi
PARC, Palo Alto, CA
haitham.hindi@parc.com

1. ABSTRACT

This paper describes a stochastic jitter model for analyzing the performance and bit error rate (BER) of digital timing recovery circuits. The model uses parallel interconnected Markov chains to simulate the behavior of the system in response to both random and deterministic jitter. Unlike conventional Markov-chain models that require the system to be stationary, the parallel-chain model approximates deterministic changes in conditions with transitions between sub-chains. To verify the accuracy of the model, an analysis was performed on a digital delay-locked loop, and the results compared to measured data. The resulting transition probabilities and BER predicted by the proposed model are more than three orders of magnitude more accurate than those predicted by a conventional Markov-chain model.

Categories and Subject Descriptors

B.7.2 [**Integrated Circuits**]: Design Aids – *Simulation*

General Terms

Algorithms, Measurement, Performance, Design, Reliability

Keywords

Jitter, timing margins, timing recovery circuits, mean-time-between-failures (MTBF), bit-error-rate (BER), stochastic model, Markov chain, delay-locked loop (DLL)

2. INTRODUCTION

Digital timing recovery circuits (DTRCs), such as delay-locked loops (DLLs) and phase-locked loops (PLLs), are widely used feedback circuits that maintain alignment between clock and data signals [1]. Their performance is often measured in terms of bit-error-rate (BER) or mean-time-between failures (MTBF), which provide an estimate of the overall reliability of the system. Both measures are strongly affected by random and deterministic fluctuations in signal phase (jitter), and, both can be difficult to measure directly. This is especially true in systems where the required MTBF is on the order of a year or more, and errors are rarely observed.

There are a number of techniques for estimating the BER (or MTBF) of a system. They generally fall into one of two categories: deterministic and stochastic. Deterministic techniques typically

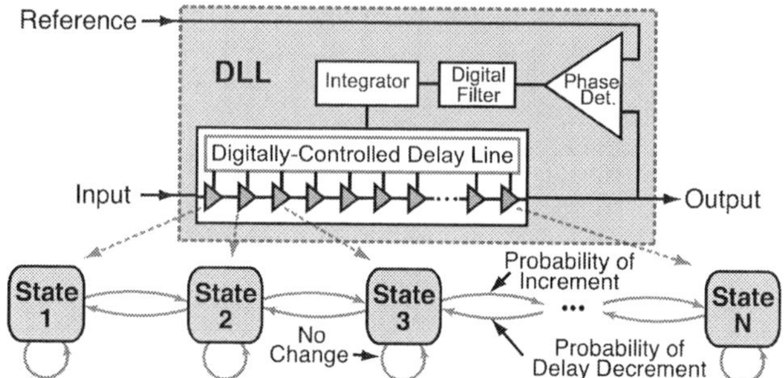

Figure 1: Block diagram and equivalent state-transition diagram of a basic digital DLL.

involve the use of circuit or system-level simulators to model behavior under worst-case conditions [2]. These models can be driven with pseudo-random waveforms to observe the response of the system over an extended period of time. Unfortunately, it is rarely possible to simulate long enough time periods to accurately determine the BER.

Stochastic techniques, such as those based on Markov-chain models [3,4], offer a more efficient method for estimating the BER of a system. They can be used to calculate the probability that the system will occupy a particular state from statistical properties of the input signals and the deterministic behavior of the system. These models are based on the assumption that the system is stationary, and that transitions between states are purely probabilistic. These restrictions exclude many practical systems, including those that operate under changing conditions and those that experience significant levels of deterministic noise and jitter [5].

In this paper, a model that combines stochastic and deterministic elements to predict the behavior of a synchronous system in the presence of noise and jitter is introduced. The model is based on a series of interconnected recurrent Markov chains, with each chain representing a distinct set of operating conditions. A sample analysis of a digital DLL is also presented, including a comparison to a conventional Markov-chain model.

3. CONVENTIONAL MARKOV-CHAIN MODEL

To understand the limitations of a conventional Markov-chain model, it is useful to go through a simple example. A block diagram of a basic digital delay-locked loop (DLL) is shown together with the equivalent state-transition diagram in Figure 1. In this circuit, the digitally-controlled delay line (DCDL) can only occupy one of N states at any given time, and only one unit (gate) delay (UD) can be added or subtracted per clock period. The probability of transitioning to an adjacent state is determined by properties of the DLL and by the phase difference between the output and reference.

Permission to make digital or hard copies of part or all of this work for personal or classroom use is granted without fee provided that copies are not made or distributed for profit or commercial advantage and that copies bear this notice and the full citation on the first page. To copy otherwise, to republish, to post on servers or to redistribute to lists, requires prior specific permission and/or a fee.
DAC'09, July 26-31, 2009, San Francisco, California, USA

978-1-60558-497-3/09 $25.00 © 2009 ACM

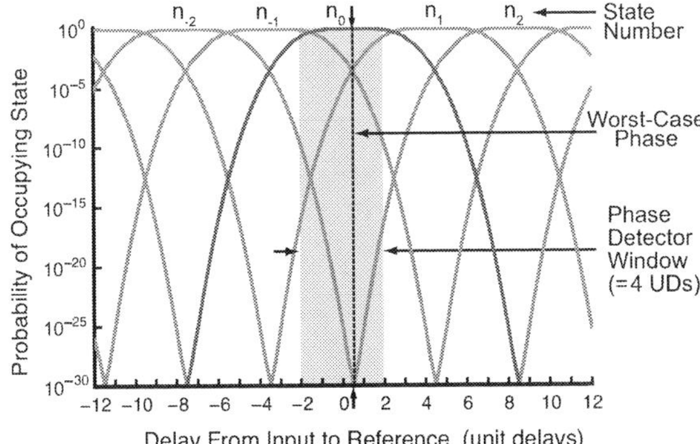

Figure 2: Probability of occupying each state in the DLL as the input signal phase is swept ($\sigma = 1/4$ UD).

Under ideal conditions, the DLL reaches an equilibrium state and remains there indefinitely. However, in the presence of noise and jitter, the output may diverge from the reference far enough to cause a timing error. The probability of an error occurring can be determined by modeling the system as a recurrent Markov chain, and calculating the probability that the phase of the output clock will vary enough to violate the timing requirements of the system.

To perform this calculation, the probability of occupying each state in the system must first be determined. From the basic limit theorem for Markov chains, the equilibrium probability of occupying each state in the system is

$$\Pi = \lim_{i \to \infty} P^i \bullet \Pi_0, \tag{1}$$

where P is a matrix of transition probabilities, Π_0 is some arbitrary initial state vector, and each element π_n in vector Π is the average or equilibrium probability of occupying the n th state in the chain [6].

The transition probabilities are calculated from the reference-clock jitter probability density function (pdf), the timing of the phase detector (PD), the state number, and the spacing between states. For a two-state phase detector, which generates an increment or decrement when the output clock edge transitions before or after the reference-clock edge, the probability of going from state n_m to state n_{m+1} is

$$p(n_m|n_{m+1}) = \int_{-\infty}^{0} f_J(t - m \cdot d_u)dt, \tag{2}$$

where $f_J(t)$ is the jitter pdf between the reference and input clocks, m is the state number, and d_u is the propagation delay through the unit delay element in the delay line. The probability of moving from state n_m to state n_{m-1} is

$$p(n_m|n_{m-1}) = \int_{0}^{\infty} f_J(t - m \cdot d_u)dt. \tag{3}$$

Using (1), (2), and (3), the probability of occupying each state in the DLL is calculated, and the results plotted as a function of input signal phase in Figure 2. As shown in the plot, the worst-case input signal phase, where the probability of occupying the largest number of states reaches a maximum, occurs half a unit delay from the center of the phase detector window. At the worst-case phase there is a 10^{-30} probability that the n_{-2} and n_2 states will be occupied. If a delay variation of four unit delays (a five-state

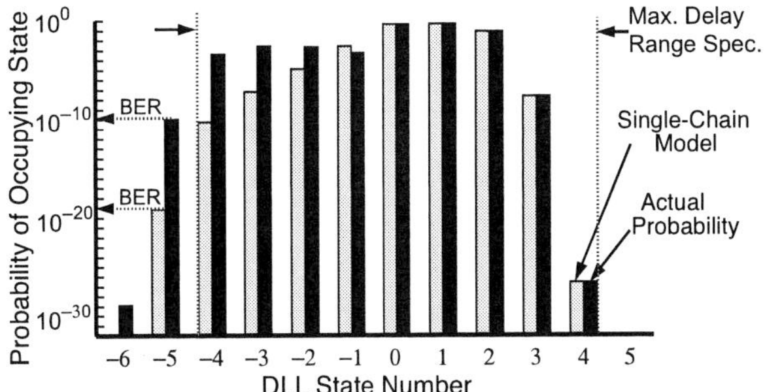

Figure 3: Probability of occupying DLL state when supply levels periodically spike (freq.=$f_s/200k$, 0.5% duty cycle, σ=1/4 UD).

range) is sufficient to cause a timing violation, then the BER of the system is equal to the probability that one of the outer two states will be occupied (2×10^{-30}).

3.1 Limitations of Conventional Model

As mentioned earlier, many practical systems can not be considered strictly stationary, and are difficult to model with a conventional Markov chain. An example is a system where a bank of drivers turns on and off at regular intervals, modulating power supply levels across the chip. As the supply fluctuates, the propagation delay of each gate in the DCDL changes, altering the behavior of the system.

It is possible to approximate a non-stationary system with a stationary Markov chain by modifying the input signal distributions. For example, when supply levels fluctuate, the probability density function (pdf) of the input signal phase can be made bimodal or multimodal to compensate for changes in propagation delay. Unfortunately, this approximation breaks down when the supply levels change at rates low enough to be tracked by the system (which is often the case in practical designs).

To illustrate this point, a conventional Markov-chain model is used to calculate the probability of occupying each state in a digital DLL with periodic spikes on the power supply. The spikes occur at 0.0005% of the DLL sample rate ($f_s/200k$) and last long enough to cause a state change (0.5% duty cycle). The calculations are performed at the worst-case input phase, where the probability of occupying the adjacent states is highest. For comparison, the probabilities are also calculated separately at each supply level, and then normalized and combined into a single histogram. The combined histogram is then used as a reference to evaluate the accuracy of the conventional model. Both histograms are plotted in Figure 3. As shown in the plot, the conventional model underestimates the BER by nine orders of magnitude.

Another limitation of the conventional model is its inaccuracy in modeling the behavior of the system in response to certain types of deterministic jitter. For example, the conventional model can not distinguish between short, frequent bursts of jitter and long, infrequent bursts. Yet the system responds quite differently to these events.

To illustrate the difference, the probability of occupying each state in a digital DLL with long, infrequent bursts of jitter on the reference clock is calculated (frequency=$f_s/100k$, duty cycle=0.1%). In one case the conventional model is used, and in the other the probabilities are calculated separately for each jitter level and then combined in proportion to their overall rate of

978-1-60558-497-3/09 $25.00 © 2009 ACM

117

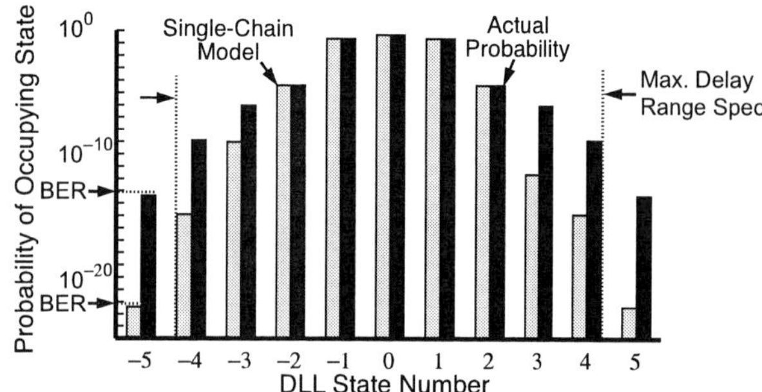

Figure 4: Probability of occupying DLL state when ref. clock jitter slowly toggles ($\sigma1$=1/8 UD, $\sigma2$=1 UD, freq.= $f_s/200k$).

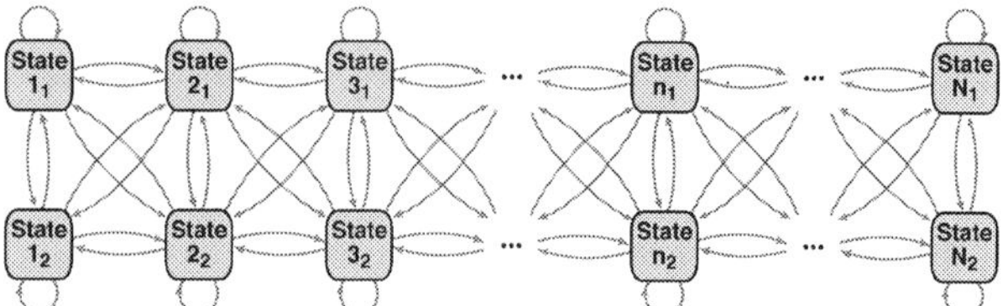

Figure 5: Markov-chain model of bimodal system.

occurrence. In the resulting histograms, plotted together in Figure 4, the conventional model underestimates the BER by six orders of magnitude.

4. PROPOSED MODEL

An alternative to conventional Markov-chain models that more accurately tracks changing conditions and deterministic noise and jitter is a model based on parallel interconnected Markov chains. Each sub-chain in the proposed model is used to represent a distinct set of conditions, and the probability of transitioning into a particular chain is proportional to the amount of time spend under the corresponding set of conditions. Each set of conditions can include different supply levels, temperatures, jitter levels, or other parameters that affect system behavior.

The parallel chain approach is mainly useful for modeling changes that occur at rates slow enough to be tracked by the system. If conditions change at rates too high to be tracked[1], then it may be necessary to either alter the pdf of the input signals, or use a conservative, worst-case model for the system. The first approach can be used when the phase of the input signals is changing rapidly, but the system remains stationary. In this case, the effect can be approximated by distorting the jitter pdf of the input signals to create an equivalent distribution. The latter approach is necessary when the behavior of the system changes in ways that can not be approximated by simply altering the input signal phase.

An example of a parallel two-chain model of an N-state system is shown in Figure 5. For simplicity, the system is limited to transitions between adjacent states. In this model, each chain represents the system under a different set of operating conditions, and states n_1 and n_2 together form the probability of occupying state n in the physical system. The diagonal arrows represent the probability that the system will simultaneously change states and transition to the other chain.

4.1 Transition and State Probabilities

The transition probabilities in the two-chain model are similar to those of the single-chain model, except now there are six non-zero transition probabilities for each state instead of three. If $p(n_0|n_1)$ is the probability of transitioning from state n_0 to state n_1, and $p(c_1|c_2)$ is the probability of transitioning from chain-1 to chain-2, then the complete set of transition probabilities for state n_0 are[2]

$$\{ \, p(n_0|n_{-1}) \cdot p(c_1|c_1), \tag{4}$$
$$p(n_0|n_0) \cdot p(c_1|c_1),$$
$$p(n_0|n_1) \cdot p(c_1|c_1),$$
$$p(n_0|n_{-1}) \cdot p(c_1|c_2),$$
$$p(n_0|n_0) \cdot p(c_1|c_2),$$
$$p(n_0|n_1) \cdot p(c_1|c_2) \, \}.$$

The probability of transitioning between states is calculated as before, by integrating the jitter pdf over the appropriate range for the system. The probability of transitioning between chains is determined by the relative *and* absolute amount of time spent in each chain. The *relative* amount of time spent in chain i is[3]

$$\frac{t_{c_i}}{T}, \tag{5}$$

where t_{c_i} is the average amount of time spent in the i th chain during a time period T. The *absolute* amount of time spent in chain i is proportional to n_{c_i}, the average number of consecutive sample periods spent in chain i. The overall probability of remaining in chain i during any single period is then given by

$$p(c_i|c_i) = (1 - 1/n_{c_i}), \tag{6}$$

and the probability of transitioning out of chain i is

$$p(c_i| \,) = 1/n_{c_i}. \tag{7}$$

The probability of transitioning from chain i to chain j is equal to the probability of transitioning out of chain i multiplied by the proportion of time spent occupying chain j relative to all chains other than i :

$$p(c_i|c_j) = \frac{1}{n_{c_i}} \left(\frac{t_{c_j}}{T - t_{c_i}} \right). \tag{8}$$

The equilibrium probability of occupying each state in the parallel-chain model is calculated as before, using the transition matrix and an arbitrary initial state vector. However, the transition matrix now has $M \times N$ rows and columns, where M is the number of modes (chains) in the model. Also, instead of having three non-zero transition probabilities for each row, there are now $3 \cdot M$ non-zero probabilities. The overall probability of occupying a state can be calculated by summing the individual probabilities for each mode.

4.2 Limitations of Proposed Model

The parallel Markov-chain approach has several important limitations. First, conditions may vary over a continuous range, making it difficult to identify discrete sets. The model is based on the assumption that changes occur in discrete steps, and that on

---

1. This assumes the changes are not correlated with the sample rate of the system, and will not alias to lower frequencies

2. The transition probabilities are slightly different for the end states.
3. Assume T is long enough to capture the average behavior of the system.

978-1-60558-497-3/09 $25.00 © 2009 ACM

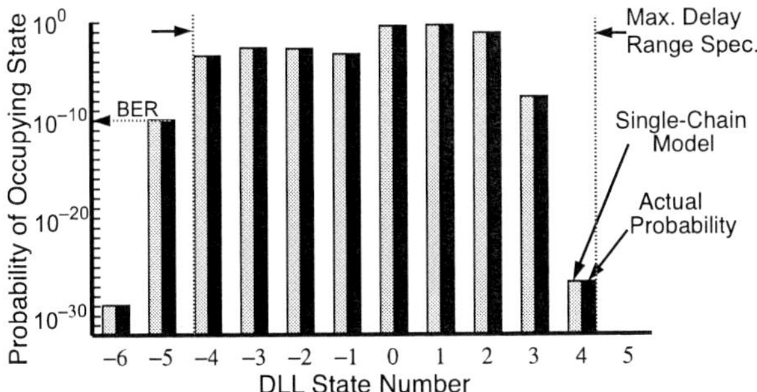

Figure 6: Probability of occupying DLL state when supply levels periodically spike (freq.$=f_s/200k$, 0.5% duty cycle, $\sigma=1/4$ UD).

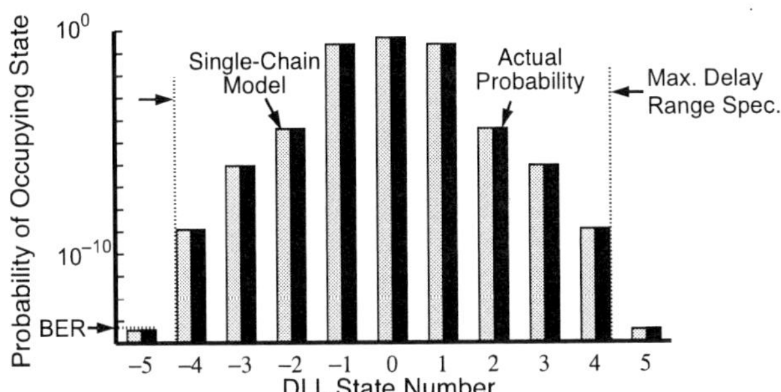

Figure 7: Probability of occupying DLL state when ref. clock jitter slowly toggles ($\sigma1=1/8$ UD, $\sigma2=1$ UD, frequency$=f_s/200k$).

average they repeat over time. If, however, the operating conditions continually change, then it may be difficult to construct an equivalent model. The parallel-chain model is still strictly stationary, and can only approximate non-stationary systems. This limitation can often be overcome by using enough chains to capture the average behavior of the system without reproducing every detail.

Another limitation of both parallel and single-chain models is that conditions are assumed to vary over a finite range. In other words, the system is assumed to be trend-stationary [7]. This limitation is rarely encountered in practical circuits, which tend to operate under conditions that repeat over time.

Another potential source of error comes from modeling the transitions between chains as being stochastic, when in fact they may be deterministic. As a result, only the average behavior is modeled, and some information on the precise sequence of events may be lost. For example, if the power supply voltage in an ideal system is steady for 100 cycles and then drops far enough to cause a state change for the next 100 cycles, there is a zero probability that the system will transition for the first 99 cycles, and a 1.0 probability that it will transition on the 100th cycle. However, in the stochastic model there is a 1/100 probability that the system will transition during *any* cycle.

Despite these limitations, in most cases of practical interest the parallel-chain model is significantly more accurate than a single-chain model in simulating the effects of changing conditions and deterministic sources of noise and jitter.

4.3 Digital DLL Example

To compare the accuracy of the single and parallel-chain models, the examples from section 3.1 are reanalyzed. Figure 6 shows the probability of occupying each state in the DLL when the

power supply toggles slowly between two levels. The probabilities were calculated using a parallel two-chain model, and, for comparison, using the combined method described earlier. As shown in the plot, the probabilities are now much more closely matched to the expected distribution, and the error in the BER has been reduced from more than nine orders of magnitude to approximately 10%.

The calculations were repeated with long, infrequent bursts of jitter on the reference clock. The resulting histogram is plotted together with the reference histogram in Figure 7. As expected, the probabilities match the reference very closely, with the BER now matching to within 5%.

5. MODEL DERIVATION

The proposed parallel-chain model can be derived using simulated results, or it can be fit from measured data after the circuit is fabricated. In both cases the structure of each Markov chain is determined by the circuit architecture, and can be modeled using a procedure similar to that described in Section 3. The procedure for determining the number of chains and the parameters for each chain is more involved, and depends on the noise and jitter characteristics of the system.

5.1 DLL Model Derivation

In DLLs, noise sources contribute to timing uncertainty, which then combines with signal jitter to create an overall jitter distribution. One of the key steps in creating an accurate DLL timing model is to determine the combined worst-case jitter distribution for the input and reference signals. This can be done through simulation, or based on experience with similar circuits.

Once the jitter distribution is known, it can be separated into stochastic and deterministic components. The deterministic components then need to be further broken down into those that are within the tracking bandwidth of the DLL, and those that are not. For those that are within the tracking bandwidth, the dominant components are assigned individual Markov chains. Each component has a mean and variance, as well as a probability of occurring, which together define the parameters for the chain. The remaining components, both deterministic and random, are then combined in a composite jitter distribution that is used to define the reference clock.

As an example, if a DLL with a tracking bandwidth of 20 clock cycles experiences 200-cycle bursts of supply noise every 1000 cycles, then the system can be modeled as two parallel chains. Assuming each burst causes the mean and variance of the reference clock jitter distribution to increase by 20%, chain-1 will have nominal mean and variance and a transition probability of $p(c_1|c_2) = 1/800$, while chain-2 will have 20% higher mean and variance with $p(c_2|c_1) = 1/200$.

5.2 DLL Model Fitting

There are several techniques for fitting parallel-chain models to measured data. The most rigorous is to measure jitter between the input and reference clocks under worst-case conditions, and then follow the procedure above to derive the model. This technique is reliable, but it requires accurate jitter measurements made on internal circuit nodes.

A less accurate but more practical way to fit the model is to use a logic analyzer to measure state transition probabilities (STP) in

978-1-60558-497-3/09 $25.00 © 2009 ACM

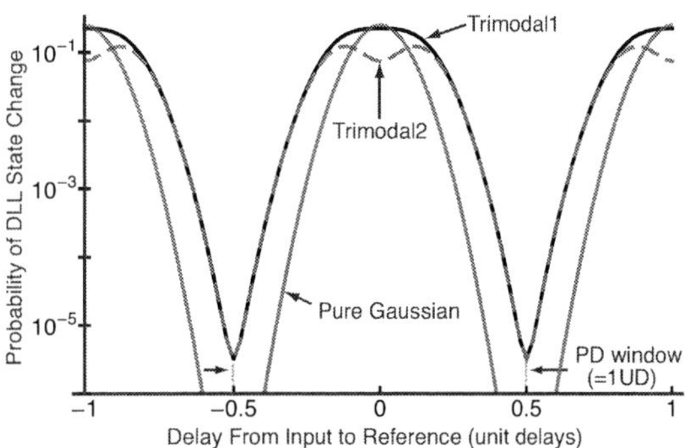

Figure 8: Calculated probability of generating a jitter-induced state change in the DLL.

the DLL while sweeping the phase of the reference clock. The resulting probability curve shows the response of the DLL to noise and jitter on internal circuit nodes without requiring sensitive analog measurements. The STP curve can then be used as a reference to adjust the model until the measured and calculated transition probabilities match. Transition probabilities can be measured over a fairly wide range ($\sim 10^{-13}$-1) in a reasonable amount of time, and the remaining tail regions typically follow a predictable (Gaussian) roll-off and can be extrapolated.

In many cases it is possible to identify the dominant deterministic jitter components in the STP curve and model them as separate chains. Each deterministic component causes the probability curve to deviate from an ideal Gaussian shape, and the distortions can be reproduced by adding parallel chains and adjusting the phase and variance of each until the probability curves match. For example, in Figure 8 the variance of the reference clock jitter in all three curves is identical, but the trimodal curves have additional modes 1/8 UDs before and after the nominal phase. The effect of the additional modes is to widen the main lobe without significantly altering the slope of the tail regions.

Even if individual components are difficult to identify, it is still possible to add parallel chains until the overall shape of the STP curves match. The BER and MTBF are calculated from the probability of occupying outlying states, which in turn depend on the tail portions of the STP curves. Thus, as long as the tail portions of the curves reasonably well matched, the BER and MTBF should also match.

To illustrate this point, three different transition probability curves with different main lobes but identical variance are plotted in Figure 8. The BER of the two trimodal systems are relatively closely matched ($1.2 \cdot 10^{-26}$ and $1.9 \cdot 10^{-26}$), despite significant differences in the main lobes. In contrast, the BER of the purely Gaussian system is more than seven orders of magnitude lower ($8.9 \cdot 10^{-34}$), even though the slope of the tail regions is identical. The key to accurately calculating BER and MTBF is making sure that both the slope *and* the phase of the tail regions of the measured and calculated STP curves match.

6. EXPERIMENTAL RESULTS

To evaluate the accuracy of the parallel-chain model, measurements were performed on a clock-alignment DLL implemented in a large, complex ASIC with numerous sources of noise and jitter. The measured data was used to derive both single

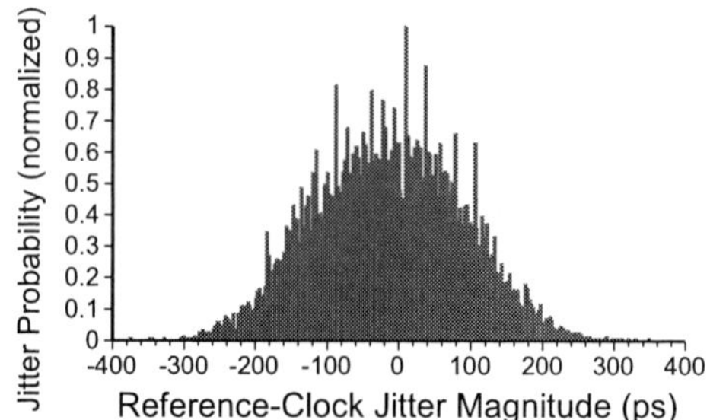

Figure 9: Measured cycle-cycle jitter on the reference clock.

Table 1: Cycle-to-cycle jitter distribution statistics.

Jitter Component	% of Total	$\sqrt{Variance}$ (σ)
Random	97.54%	51 ps
Deterministic	2.46%	291 ps
Total	100%	68 ps

and parallel-chain versions of the model, which were then compared for accuracy.

6.1 Reference-Clock Jitter

A plot of the jitter distribution between the input and reference clocks is shown together with a breakdown of the jitter components in Figure 9 and Table 1. The jitter measurements were made through a system of debug busses, which contributed significantly to the measured noise and jitter. Consequently, the data is not reliable enough by itself to derive a model. However, some useful information can still be gleaned from the data. In particular, the distribution appears random with a Gaussian distribution. But when the data are analyzed[1], significant deterministic components are found. Furthermore, the variance of the deterministic components is significantly larger than that of the random components. It turns out that most of the deterministic jitter is caused by relatively large periodic phase steps on the reference clock, which are infrequent enough that their effect on the overall jitter distribution appears small.

6.2 Transition Probabilities

The transition probabilities in the DLL are plotted as a function of reference-clock phase in Figure 10. The asymmetries visible in the curve were consistently present in the measurements, and are caused by the large deterministic phase steps described earlier.

Five distinct modes were identified in the plot, each resulting in a slight peak in the STP curve and a widening of the main lobe. Separate chains were assigned to each peak, with the means roughly aligned with the center of the peak. The variance of the jitter distribution for each chain was set to 0.3 UDs to fit the slope of the tail portion of the curve on the left. The mean of each distribution was then adjusted together with the probability of transitioning into that chain until the best overall fit was achieved.

For comparison, a conventional single-chain model and a parallel three-chain model were also fit. In all cases the reference-clock jitter distribution was assumed to be Gaussian with between one and five modes. The slope of the tail portion of the single and three-chain unimodal models matches the measured result, but it

---

1. The data was analyzed using Tektronix TDSJIT3 jitter software

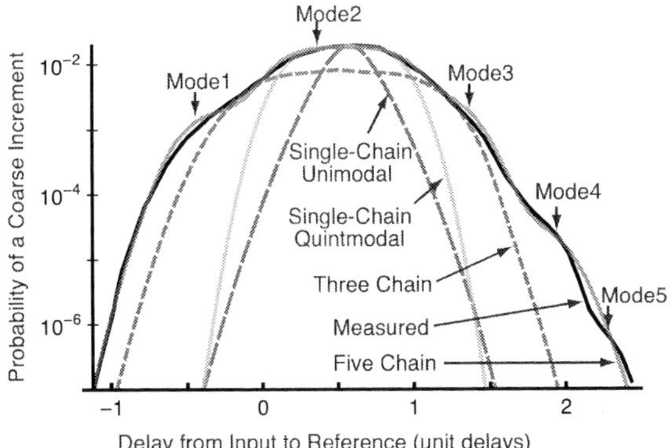

Figure 10: Measured and calculated probability of generating a jitter-induced state change in the DLL (PD width = 4 UDs).

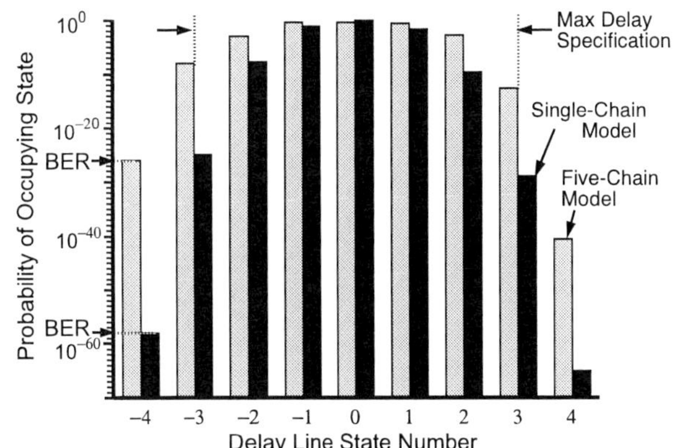

Figure 11: State probabilities and BER using fitted five-chain model and single-chain unimodal model.

wasn't possible to match both the slope and the width of the main lobe with fewer than five chains.

The parameters used to calculate the five-chain probability curve are listed in Table 2. The relatively low probability of switching between chains implies that the transition rates among the various modes in the reference-clock jitter distribution are significantly lower than the sample rate of the DLL. This result is consistent with the jitter measurements listed in Table 1. The relatively low transition rate also explains why the single-chain model is less accurate at reproducing the tail regions of the probability curve.

Table 2: Parameters used in unimodal five-chain model.

Model Type	Chain Number	Probability of Transitioning Into Chain	Mean (UDs)	$\sqrt{Variance}$ (σ)(UDs)
Five Chain Unimodal Gaussian	1	0.00224	-1.18	0.21
	2	0.032	-0.53	0.21
	3	0.04	0.58	0.21
	4	0.004	1.05	0.21
	5	0.00004	1.66	0.21

6.3 State Probabilities and BER

Using the five-chain model, the probability of occupying each state in the DLL was calculated, and the result plotted in Figure 11. For a maximum delay variation of six unit delays (beyond which a timing violation will occur), the BER of the system is $9.6 \cdot 10^{-27}$.

For comparison, the state probabilities were also calculated using the single-chain unimodal model. In the resulting histogram, also shown in Figure 11, the single-chain model significantly underestimates the probability of occupying the outer states. For the same maximum delay variation, the BER is $8.8 \cdot 10^{-59}$, or 32 orders of magnitude lower than that predicted by the five-chain model.

The MTBF calculations described above are based on measurements taken under typical conditions. To calculate the worst-case MTBF, the same model can be used with slight modifications to reflect worst-case operating conditions and jitter levels.

7. CONCLUSION

The stochastic jitter model presented in this paper provides a useful tool for estimating the BER and MTBF of digital timing recovery circuits. The parallel Markov-chain structure overcomes limitations in conventional stochastic models that make it difficult to model non-stationary systems and systems with substantial deterministic noise and jitter.

The accuracy of the parallel-chain model has been verified by calculating the state probabilities of a digital DLL, and comparing them to the expected results. Two systems were modeled, one that is non-stationary, and one with deterministic jitter. In the first case, the BER predicted by the parallel-chain model is approximately twelve orders of magnitude more accurate than the conventional single-chain model, and in the second case the improvement is almost six orders of magnitude.

The proposed model was also fit to measured data, and was found to match the transition probabilities to within approximately 5% over the full range. In contrast, the single-chain model had errors as large as four orders of magnitude in the tail regions of the distribution. The BER predicted by the five-chain fitted model is $9.6 \cdot 10^{-27}$ under typical operating conditions, which is 32 orders of magnitude higher than that of the conventional model.

8. REFERENCES

[1] T. H. Lee and J. F. Bulzacchelli, "*A 155-MHz clock recovery delay- and phase-locked loop*," IEEE Journal of Solid-State Circuits, vol. 27, pp. 1736 - 1746, December 1992.

[2] A. N. Lokanathan and J. B. Brockman, "*Efficient worst case analysis of integrated circuits*," 1995 IEEE Custom Integrated Circuits Conference, pp. 237-240, May 1995.

[3] J. R. Burnham, Design and Analysis of jitter-tolerant digital delay-locked loops and fixed delay lines, Ph.D. Dissertation, Stanford University, June 2007.

[4] A. E. Payzin, "Analysis of a Digital Bit Synchronizer," IEEE Trans. on Comm., Vol. Com-31, No. 4, April 1983.

[5] V. Stojanovic and Mark Horowitz, "*Modeling and Analysis of High-Speed Links*," 2003 IEEE Custom Integrated Circuits Conference, pp. 237 - 240, Sept. 2003.

[6] D. G. Luenberger, *Introduction to Dynamic Systems, Theory, Models, & Applications*: John Wiley & Sons, Inc., pp. 224-245, November 1979.

[7] M. Qi and G. P. Zhang, "*Trend Time-Series Modeling and Forecasting With Neural Networks*," 2008 IEEE Transactions on Neural Networks, vol. 19, no. 5, pp. 808-816, January 2008.

Statistical Ordering of Correlated Timing Quantities and its Application for Path Ranking

Jinjun Xiong, Chandu Visweswariah, Vladimir Zolotov
IBM Thomas J. Watson Research Center, Yorktown Heights, NY, 10598
{jinjun,chandu,zolotov}@us.ibm.com

ABSTRACT

Correct ordering of timing quantities is essential for both timing analysis and design optimization in the presence of process variation, because timing quantities are no longer a deterministic value, but a distribution. This paper proposes a novel metric, called *tiered criticalities*, which guarantees to provide a unique order for a set of correlated timing quantities while properly taking into account full process space coverage. Efficient algorithms are developed to compute this metric, and its effectiveness on path ranking for at-speed testing is also demonstrated.

Categories and Subject Descriptors

B.7.2 [**Integrated Circuits**]: Design Aids

General Terms

Algorithms, Design, Theory

Keywords

Statistical Ordering, Path Ranking, Correlation, Testing

1. INTRODUCTION

One primary use of a static timing analysis program is to provide *feedback* and diagnostics to aid designers in optimizing timing and fixing timing fails. The most common feedback is provided by ordering timing quantities, typically *slacks* such as edge, node, path slacks, in order of criticality so that the designer knows which part of the design to focus on in order to make progress towards timing sign-off. Slack ordering is often required for a number of purposes, including circuit optimization, transistor sizing, timing-driven physical synthesis.

With the advent of increased process variability in deep sub-micron technology, statistical timing is increasingly used for timing verification. Unfortunately, although the need for slack ranking persists, the situation is not as simple as in deterministic timing. Each node, edge and path of the timing

graph has a probability of being critical among all manufactured chips, which is defined as *criticality probability* [1, 2, 3, 4]. Ranking slacks based on criticality probability is useful to identify the most critical slack while taking into account the entire process space. However, it suffers from the problem of *probability masking*. That is, when the top most critical slacks are determined, the ordering of the remaining slacks may not be properly determined. *Dominance masking* and *correlation masking* are the two major masking mechanisms.

The major contributions of this work are as follows. Based on our experience from industrial timing analysis, we share some insights on the difficulty of properly ordering statistical timing quantities resulting from the *probability masking* problem. A new metric, called *tiered criticality*, is proposed to address these masking issues. The new metric has more discriminative capability and it can guarantee to provide a unique ordering of a set of correlated statistical timing quantities, while taking into account full process space coverage. This feature is desirable and may find many potential applications in various CAD problems, such as timing report, path testing, and design optimizations.

2. BACKGROUND

Statistical static timing analysis (SSTA) is a technique to mathematically capture the impact of random process variations on timing, in which all timing quantities in the timing graph, such as AT, RAT, and slacks, are represented as functions of the underlying process parameters

$$S = F(\Delta X_1, \Delta X_2, \ldots, \Delta X_n), \qquad (1)$$

where ΔX_i is a random variable that models the variation of a process parameter. Through different modeling techniques, ΔX_i can capture chip-to-chip, across-chip (spatial), and local random variations. In the literature, the studied functional forms include linear [5, 1] quadratic [6], or general non-linear [7]; and ΔX_i can be either Gaussian or non-Gaussian.

For timing optimization, designers need to constantly compare different timing quantities to make a judgment on the goodness of a design before and after a change. The comparison of deterministic slacks is straightforward, i.e., the smaller the value, the more critical the corresponding slack. But ranking different slacks in the form of (1) is not easy, as S now represents a distribution.

One way to transform a distribution to a value for comparison purposes is via *projection*. For example, the conventional *corner projection* is to project each ΔX_i to one of its

Permission to make digital or hard copies of part or all of this work for personal or classroom use is granted without fee provided that copies are not made or distributed for profit or commercial advantage and that copies bear this notice and the full citation on the first page. To copy otherwise, to republish, to post on servers or to redistribute to lists, requires prior specific permission and/or a fee.
DAC'09, July 26-31, 2009, San Francisco, California, USA

978-1-60558-497-3/09 $25.00 © 2009 ACM

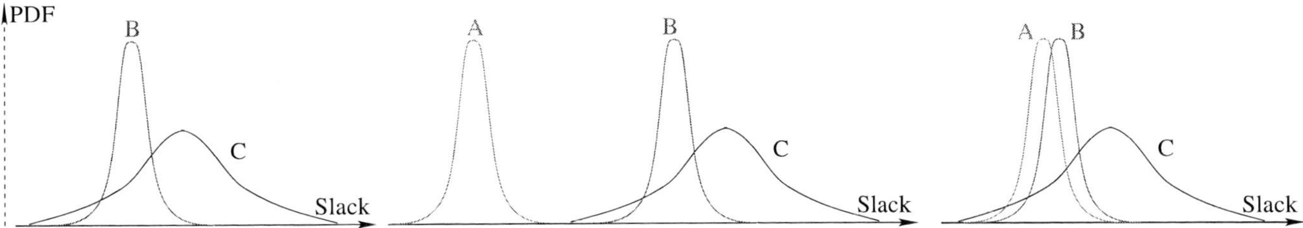

Figure 1: Illustrations of (1) criticality versus projection; (2) dominance masking; and (3) correlation masking.

two boundary (typically ± 3 sigma) values, and obtain one projected value $\bar{s}$ as

$$\bar{s} = F(\delta x_1|_{\pm 3}, \delta x_2|_{\pm 3}, \ldots, \delta x_n|_{\pm 3}). \tag{2}$$

And the *worst corner projection* is the combination of $+3$ sigma or -3 sigma of each ΔX_i that jointly pushes the projected slack $\bar{s}$ to the worst (i.e., smallest).

But using the projected value $\bar{s}$ as a surrogate for the comparison of S may not provide effective guidance for design optimization. For example, the left graph of Fig. 1 shows the distributions of two slacks B and C. If we were to compare B and C based on worst corner projection, then C would appear to be more critical than B. But if we consider all possible (weighted) occurrences of process parameters, it is clear that slack B has a higher probability of being smaller than C. To effectively guide design optimization with limited resources, we would prefer first optimizing B to C. In other words, to correctly rank slack S, we need to consider the multi-dimensional space of process parameters.

This was captured by a concept called *criticality probability* as first defined in [1]. The criticality of S_i is the probability of S_i being the smallest among all other S_j, i.e.,

$$p_i = P\left(S_i \leq min_{j \neq i}(S_j)\right). \tag{3}$$

If we use criticality as a metric to rank B and C as shown in the left graph of Fig. 1, we would determine that B is more critical than C, as B has higher probability of being smaller than C.

3. TIERED CRITICALITY

3.1 Probability Masking

Although the criticality probability as defined in (3) is good at capturing the most critical slack, it suffers from two types of *probability masking* problems: *dominance masking* and *correlation masking*.

Dominance masking happens to dominated slacks with zero criticality and criticality gives us no guidance to order them. For example, the middle graph of Fig. 1 shows the distributions of three slacks A, B and C. Because A is clearly dominating (smaller than) both B and C in the entire space, the criticality probabilities of A, B, and C according to (3) are one, zero, and zero, respectively. Though we are able to rank A as the most critical slack, we cannot properly order B and C based purely on their criticality probabilities. But for most applications, we would like to rank B as more critical than C, as B has a higher probability of being smaller than C.

Correlation masking refers to situations of highly correlated slacks in which criticality probabilities cannot properly order them. If one of them is smaller than the rest

by a small amount, this slack would get all the criticality probability while the rest would get zero criticality probabilities. For example, as shown in the right graph of Fig. 1, assuming that A and B are almost perfectly correlated, and A is only smaller than B by a small amount, the criticality probability of B would be zero, totally dominated by A due to high correlation between them. As a result, B would be ranked as less critical than C. But from most applications' perspectives, it is clear that B should be considered as more critical than C. Also, the small difference between A and B may be caused by a small modeling or numerical error, then the decision of A being more critical than B is not reliable.

One examplar application that suffers from the probability masking problem is path selection for at-speed testing [8, 9]. In the presence of process variation, different paths will become frequency-limiting for chips manufactured under different process conditions. Therefore, it is important to select a set of critical paths to undergo at-speed testing such that every chip, no matter under what process conditions it is manufactured, would have its most critical paths tested. By doing so we can screen out all bad chips and only ship good chips to customers, thus achieving the best yield without hurting the level of shipped-product quality loss.

As the number of paths in a design is exponential, and we only have a limited path budget for at-speed testing, ranking paths is important. The authors of [8, 9] proposed a two-step procedure to address this problem. (1) They first rank all end points in terms of node criticality as computed by [10], and then select all end points with non-zero criticalities. (2) For each chosen critical end point, they select a number of critical paths leading to that end point based on either deterministic slacks, projected slacks [8], or a *test coverage metric* (TQM) [9].

However, because of the probability masking issues, the selected end points with non-zero criticalities in the first step may be limited; while dominance and correlation masking issues may prevent some "useful" end points with zero criticalities from being included. An end point whose slack exhibits distribution same as B in Fig. 1 is valuable to have for at-speed testing. The reasons are multi-fold. First, if the first path A has a negative slack and must be "fixed," then as soon as its slack is improved, the second path B will require attention. Second, due to a small modeling inaccuracy, the second path B may be the one in reality with the high criticality probability. Third, if the first path A is selected for testing purposes and turns out to be either unsensitizable or hard-to-sensitize during the Automatic Test Pattern Generation (ATPG) step, the second path B is extremely valuable for at-speed test to detect process variations.

Therefore, a more sophisticated and discriminating metric is required.

978-1-60558-497-3/09 $25.00 © 2009 ACM

3.2 Definition and Algorithms

To overcome probability masking issues, we propose a new metric called *tiered criticality probability* in this paper.

Without delving into the detailed mathematical definition, we present the intuition behind tiered criticality. Given a set of slacks, we first find the one with the highest probability of being the smallest. In order to avoid masking problems, we then *remove* this slack from the slack set, and repeat the procedure of finding the slack with the highest probability of being the smallest among the remaining slacks. Carrying out this procedure on the examples in Fig. 1 would result in a correct ordering of slacks, i.e., A, B, C. In other words, we organize slacks into a hierarchy of tiers. Each tier t in this hierarchy has exactly one slack S_i, whose tiered criticality probability $p^{(t)}$ is the largest among all slacks excluding those belonging to lower tiers. We have the following theorem.

THEOREM 1. *Ordering slacks based on tiered criticality probabilities is unique, and the lower the tier number of the slack, the more critical that slack.*

The computation of tiered criticality probabilities can be done in a straightforward manner. For each tier t, we compute p_i for all remaining slacks, and find the one with the highest value of p_i, which would be the tiered criticality $p^{(t)}$, and the corresponding slack would be put into the tth-tier as $S^{(t)}$. For each tier, we need to call the algorithm to compute p_i as shown in (3), whose complexity is $O(nm)$ for Monte Carlo simulation, and $O(n)$ for analytical computation [10]. So the total complexity would be either $O(tnm)$ or $O(tn)$.

To speedup the Monte Carlo based computation, we run Monte Carlo only once instead of t times by using a smart book-keeping technique on the Monte Carlo trial data, and then compute the tiered criticality probabilities by repeatedly using the same set of Monte Carlo trial data. Hence reducing the complexity to $O(nm)$.

When (3) is computed by using the statistical min operation as discussed in [1], by leveraging a binary tree data structure, [10] has shown that the p_i via (3) can be computed in linear time complexity $O(n)$ as follows. We build a balanced binary partition tree as shown in Fig. 2, where each leaf node corresponds to one of the n slacks. Through the first bottom-up traversal of the tree as shown in the top plot of Fig. 2, we compute the minimum slack of every node's two direct child nodes, and save it at the node (at the root, we have the minimum slack of all n slacks). Through the second top-down traversal of the tree as shown in the bottom plot of Fig. 2, we compute the complement slack of every node by taking the minimum of its parent node's complement slack and its sibling node's slack (with the root's complement slack initialized as infinity). At the leaf node i, the complement slack would be the minimum slack of the remaining $n-1$ slacks, i.e., $min_{j \neq i}(S_j)$. Then p_i via (3) for each slack S_i can be computed.

Both the bottom-up and top-down traversals have complexity of $O(n)$, so this algorithm has complexity of $O(n)$. For t tiered criticalities, we need to call this binary tree based algorithm t times, thus the total complexity is $O(tn)$.

We propose a speedup technique to compute all tiered criticalities by re-using most of the results saved in the binary tree for the prior tiers' criticality computation without reconstructing the binary tree t times. The idea is better illustrated via a simple example as shown in Fig. 2, where S_1

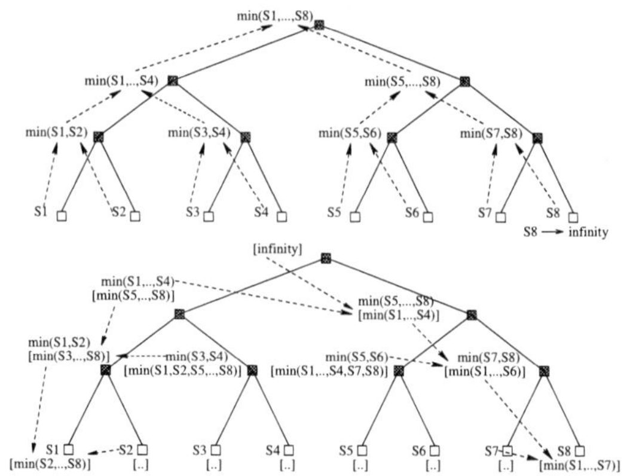

Figure 2: Fast computation of tiered criticalities.

through S_8 represent eight slacks. Assume that for the first tier computation, we find that slack S_8 is the most critical one, and we put S_8 into the first tier and record its tiered criticality. For the second tier, our goal is to compute p_i for the remaining slacks but without reconstructing the binary tree for them. The idea is to replace the slack S_8 with an infinity valued slack, and then do a bottom-up traversal along the path from this leaf node to the root as shown in the top plot of Fig. 2, and update the minimum slack stored at each node along the way. Next, we conduct a top-down traversal from the root to all leaf nodes and update the complement slacks stored at all nodes as before. Then p_i for the remaining slacks can be easily computed at valid leaf nodes (nodes whose slack is not infinity). From this computation, we can determine the second tier criticality easily. We repeat this procedure until all t tiered criticalities are found.

In terms of complexity, this is the same as $O(tn)$, but in practice, the multiplicative constant in the complexity analysis is much lower than reconstruction of the binary tree for each tier.

4. EXPERIMENTAL RESULTS

We have implemented the proposed algorithms to compute the tiered criticality probabilities inside our statistical timing analysis engine, EinsStat [1]. Industrial 90 nm ASIC designs are used for our testing with the process variations determined by foundry data. For comparison purposes, we have also implemented the algorithm of [10] to compute the criticality probability as defined by (3), which we denote as *non-tiered criticalities*.

Table 1 shows the comparison of the order sequences of the top few end points for one design. We report both the ordering, the name of end points, their respective mean and sigma of end point slacks. The first set of rows shows results based on non-tiered criticalities; the second set shows results from tiered criticalities.

We observe that the most critical end point, GRP.1, is the same for both approaches, and this is expected as both take into account the full process space and the most critical slack has the highest probability of being smaller than the rest. However, we can clearly see that the rest of the ordering sequences based on these two metrics are quite different.

For example, by comparing the slack values of RXD.17 with those of GRP.7, we can tell that GRP.7 has higher probability of being smaller than RXD.17. But according

978-1-60558-497-3/09 $25.00 © 2009 ACM

to non-tiered criticalities, RXD.17 is ranked as second while GRP.7 was ranked as seventh. This illustrates the probability masking problem of correlation masking. By computing the correlation between GRP.7 and GRP.1, we find that they have very high correlation ($\rho = 0.98$). In computing their respective non-tiered criticalities, there is almost no chance for GRP.7 being smaller than GRP.1. Hence the non-tiered criticality for GRP.7 is much lower. On the other hand, the slack of RXD.17 is much larger than GRP.7, but because the correlation between RXD.17 and GRP.1 is low, there is still some chance for RXD.17 being smaller than GRP.1. In other words, the non-tiered criticality for RXD.17 is non-zero, and it is thus ranked as more critical than GRP.7.

On the contrary, if we order all slacks based on tiered criticalities, we obtain the proper ordering among different slacks. For example, now GRP.7 is ranked as second while RXD.17 is ranked as 19th. This is true for other end point slacks as well.

Table 1: Comparison on the ordering of end points.

Algorithm	Order	End point	Mean	Sigma
	1	GRP.1	0.315	0.0715
	2	RXD.17	0.361	0.009
Non-tiered criticality	3	RXD.3	0.362	0.009
	4	RXD.0	0.363	0.009
	⋮	⋮	⋮	⋮
	7	GRP.7	0.319	0.072
	1	GRP.1	0.315	0.0715
	2	GRP.7	0.319	0.072
Tiered criticality	3	GRP.11	0.322	0.072
	4	GRP.9	0.322	0.071
	⋮	⋮	⋮	⋮
	19	RXD.17	0.361	0.009

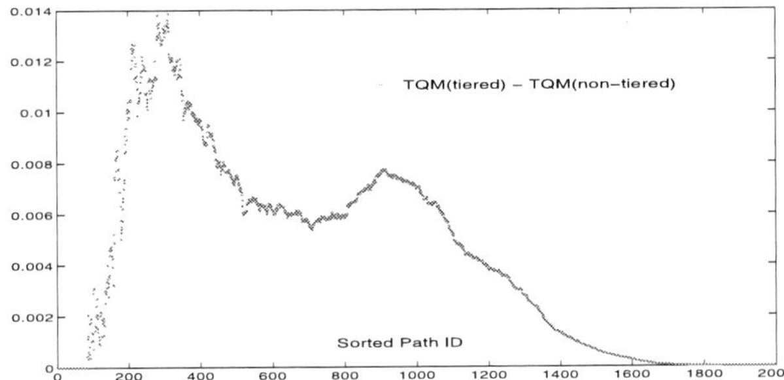

Figure 3: Differences of TQM between paths based on tiered criticalities and non-tiered criticalities.

Next, we show how the tiered criticalities can be used to guide path selection for at-speed testing. We have implemented the same branch-and-bound path selection algorithm as [9]. For comparison, in the first run, we find the top m critical end points based on non-tiered criticalities; while in the second run, we also find m critical end points but based on tiered criticalities. Afterwards, both approaches employ the same test quality metric (TQM) to guide the path selection procedure to select n paths with the best TQM leading to these m end points. To compare the testing quality of these two sets of paths, we sort their individual path TQM, and report the difference between the corresponding sorted path TQM.

Fig. 3 shows the TQM differences between paths selected based on tiered criticalities and paths selected based on non-tiered criticalities. The positive difference in TQM shows that paths selected from tiered criticalities have higher TQM, hence higher testing quality. According to Fig. 3, we observe that the first hundred paths resulting from both approaches have zero differences in TQM, which means that both are able to find the top hundred critical paths. But for the remaining thousands of paths, the differences are always positive, and this convincingly shows that paths selected from tiered criticalities have better TQM. This result is expected, as tiered criticalities provide more accurate ability to distinguish among sub-critical paths, selection of which is also important to protect us against both inaccurate modeling issues and sensitization issues.

5. CONCLUSION AND DISCUSSION

We have shown that statistically ordering a set of correlated timing quantities is difficult due to both dominance and correlation masking problems. A novel metric, called tiered criticalities, have been proposed to solve these masking problems. We have shown that ordering based on tiered criticalities can guarantee the uniqueness of ordering while correctly taking into account the full process space coverage. We envision that this new metric holds the key to solving many potential CAD problems in the presence of process variations.

6. REFERENCES

[1] C. Visweswariah, K. Ravindran, K. Kalafala, S. G. Walker, and S. Narayan. First-order incremental block-based statistical timing analysis. *Proc. 2004 Design Automation Conference*, pages 331–336, June 2004. San Diego, CA.

[2] X. Li, J. Le, M. Celik, and L. T. Pileggi. Defining statistical sensitivity for timing optimization of logic circuits with large-scale process and enviromental variations. *IEEE International Conference on Computer-Aided Design*, pages 844–851, November 2005. San Jose, CA.

[3] K. Chopra, S. Shah, A. Srivastava, D. Blaauw, and D. Sylvester. Parametric yield maximization using gate sizing based on efficient statistical power and delay gradient computation. *IEEE International Conference on Computer-Aided Design*, pages 1023–1028, November 2005. San Jose, CA.

[4] J. Xiong, V. Zolotov, and C. Visweswariah. Incremental criticality and yield gradients. *Design and Test in Europe*, March 2008. Messe Munich, Germany, accepted for publication.

[5] H. Chang and S. S. Sapatnekar. Statistical timing analysis considering spatial correlations using a single PERT-like traversal. *IEEE International Conference on Computer-Aided Design*, pages 621–625, November 2003. San Jose, CA.

[6] L. Zhang, W. Chen, Y. Hu, J. A. Gubner, and C. C. Chen. Correlation-preserved non-gaussian statistical timing analysis with quadratic timing model. In *Proc. Design Automation Conf*, pages 83 – 88, June 2005.

[7] H. Chang, V. Zolotov, C. Visweswariah, and S. Narayan. Parameterized block-based statistical timing analysis with non-Gaussian and nonlinear parameters. *Proc. 2005 Design Automation Conference*, pages 71–76, June 2005. Anaheim, CA.

[8] V. Iyengar, J. Xiong, S. Venkatesan, V. Zolotov, D. Lackey, P. A. Habitz, and C. Visweswariah. Variation-aware performance verification using at-speed structural test and statistical timing. *IEEE International Conference on Computer-Aided Design*, November 2007. San Jose, CA.

[9] V. Zolotov, J. Xiong, H. Fatemi, and C. Visweswariah. Statistical path selection for at-speed test. *IEEE International Conference on Computer-Aided Design*, November 2008. San Jose, CA.

[10] J. Xiong, V. Zolotov, C. Visweswariah, and N. Venkateswaran. Criticality computation in parameterized statistical timing. *Proc. 2006 Design Automation Conference*, pages 63–68, July 2006. San Francisco, CA.

978-1-60558-497-3/09 $25.00 © 2009 ACM

A Parametric Approach for Handling Local Variation Effects in Timing Analysis

Ayhan Mutlu, Jiayong Le, Ruben Molina, Mustafa Celik

Extreme DA Corporation, 3211 Scott Blvd., Santa Clara, CA 95054.

ABSTRACT

In this paper we propose a new methodology, called parametric on chip variation (POCV) analysis, to determine local process variation effects on the timing of designs. The proposed methodology requires relative delay and parasitic variations of cells and interconnects, respectively. Once this information is provided, delays and arrival times are propagated to calculate slacks as a function of these relative variations. A key characteristic of the POCV analysis is that it does not require a statistical library characterization or statistical RC extraction. The POCV method has been implemented in a timing analysis software, and tested on multiple production designs on 65nm and 45nm technology nodes, including multi-million instance designs. Our observation was that compared to the existing methods, POCV removes unrealistical pessimism on the setup paths and captures risks on the hold paths, with no changes to the existing timing sign-off environment.

Categories and Subject Descriptors

B.7.2 [**Integrated Circuits**]: Design aids—*Simulation*

General Terms

Algorithms, performance, design

Keywords

Timing, parametric analysis, On Chip Variation (OCV)

1. INTRODUCTION

Continuous scaling of integrated circuit technologies has made circuits increasingly susceptible to variations in the manufacturing process [1]. These variations can erode timing windows, and eventually contribute to circuit failures. In order to ensure the proper design operation across the variation space, timing verification has to be performed on

Permission to make digital or hard copies of part or all of this work for personal or classroom use is granted without fee provided that copies are not made or distributed for profit or commercial advantage and that copies bear this notice and the full citation on the first page. To copy otherwise, to republish, to post on servers or to redistribute to lists, requires prior specific permission and/or a fee.

DAC'09, July 26-31, 2009, San Francisco, California, USA

all relevant combination of process and environmental parameters. The best way to handle both device and interconnect variations over their entire ranges is to use statistical static timing analysis (SSTA) [2]. However, SSTA requires a variation-aware library characterization as well as interconnect variation extraction effort which may not be available for every technology node. The methodology introduced here does not require any variation-aware cell libraries or interconnect parasitics.

The primary focus of this paper is to analyze the design performance effects of local process variations which can be categorized under on chip variations (OCV), a frequently used term by the designers to describe variations occurring at within-die level. These variations include local process variations, voltage variations such as static and dynamic IR drop, and temperature variations and its long term degradation effects in the device and interconnect parameters. Traditional on-chip-variation (OCV) techniques used in timing analysis use a constant set of derating factors on gate and wire delays and impose unrealistic performance penalties on the designs manufactured at 65nm and below [3]. Clearly, these techniques do not properly model the local process variation effects as they assume perfect positive correlation among the gates of the launch path, and perfect negative correlation between the launch path and capture path [4]. The POCV technique introduced in this paper, reduces the pessimism and increases the design robustness compared to traditional OCV approaches by accounting for the local process variation effects within a statistical timing analysis framework. It inherently handles the path depth and the fact that the delay change of a cell caused by local variations depends on type of the cell.

The rest of the paper is organized as follows. Section 2 gives a brief overview of the local process variation effects on the design timing. Section 3 presents the proposed POCV technique. The findings of our analyses including the one on a test design manufactured at 45-nm process technology are then reported in Section 4. Finally, some concluding remarks regarding the benefits of this new technique and the need for better estimation of local process variation effects in the upcoming technology nodes, are given in Section 5.

2. TIMING EFFECTS OF PROCESS VARIATIONS

From designer's perspective, variations can be divided into two major groups: global variations, and local variations. In order to understand the effects of these variations on a path delay and slack, consider the following circuit given in Fig. 1.

Assume that the logic block has n number of stages. The

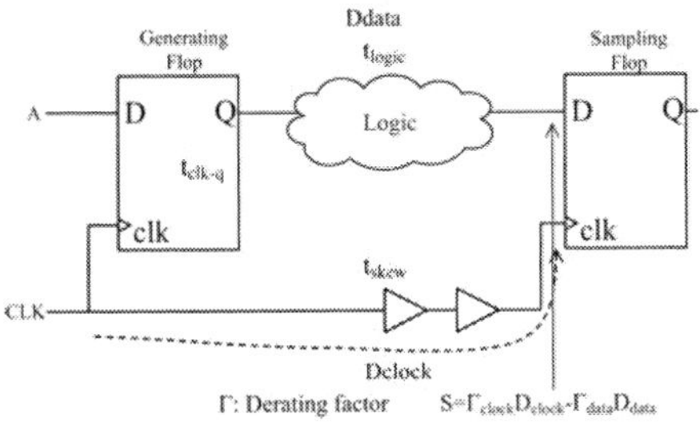

Figure 1: A digital circuit.

mean and standard deviation of the path going from the output of the launch flop to the input of the capture flop can be written as,

$$\mu_{path} = \sum_{i=1}^{n} \mu_i \qquad (1)$$

and

$$\sigma_{path}^2 = \sum_{i=1}^{n} \sum_{j=1}^{n} \rho_{ij} \sigma_i \sigma_j \qquad (2)$$

In equation 2, ρ_{ij} is the correlation between the cells i and j. In case of global variations the correlation coefficient ρ_{ij} is 1 for all i and j. Assuming that the means and standard deviations of all the cells are equal, for global variations, the relative variation of a path can be written as,

$$\frac{\sigma_{path}}{\mu_{path}} = \frac{n\sigma}{n\mu} = \frac{\sigma}{\mu} \qquad (3)$$

In the case of local variations, the correlation coefficient is,

$$\rho_{ij} = 1, \forall i = j$$
$$\rho_{ij} = 0, \forall i \neq j$$

Therefore, for local variations, the relative variation of a path will become,

$$\frac{\sigma_{path}}{\mu_{path}} = \frac{\sqrt{n}\sigma}{n\mu} = \frac{1}{\sqrt{n}} \frac{\sigma}{\mu} \qquad (4)$$

Equation 4 shows the dependency of the relative path variation to the number stages on a path.

In the case of slacks, the effect of the local and global variations is different. When the arrival and required times are statistical quantities with global and local variation components,

$$A = a_0 + a_g p_g + a_l p_{A,l} \qquad (5)$$

$$R = r_0 + r_g p_g + r_l p_{R,l} \qquad (6)$$

where p_g is the global variation parameter, $p_{A,l}$ and $p_{R,l}$ are the local variation parameters associated with the arrival time and the required time, respectively. Assuming that the variation parameters exhibit standard normal behavior, the statistical slack can be expressed as,

$$S = R - A = r_0 - a_0 + (r_g - a_g)p_g + \sqrt{r_l^2 + a_l^2}\, p_l \qquad (7)$$

For slacks, global parameters may get cancelled out due to the correlation between data path and clock path. However, the local variations do not cancel out.

3. POCV METHODOLOGY

POCV analysis has been developed for integrating the local process variation effects into the timing analysis without going through variation-aware library characterization and RC extraction. Unlike the various OCV analyses techniques, POCV does not require any derating factor tables. The required information for POCV analysis is the variations associated with cells and interconnects as shown in Fig. 2, where Δd comes from the variation for intrinsic cell delay which can be further decomposed down to Δd_p and Δd_n to model the mis-correlation between P and N channel transistors. Δr and Δc represents relative variation for resistance and capacitance in parasitics.

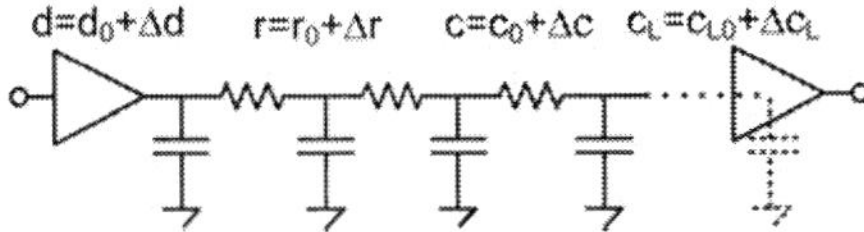

Figure 2: Cell and interconnect variations for POCV analysis.

Most of the information required here is also necessary for calculation of OCV derating factors in traditional corner sign-off methodology. Therefore, input data preparation part is a lot less expensive compared with full SSTA analysis and other advanced STA-based OCV derating techniques. Once all the variation information is available, the parametric delay calculation engine calculates the delay of a stage by integrating the cell and interconnect variations together, shown in basic form below [6],

$$delay = d_0 + \Delta d + \frac{\partial d}{\partial r}\Delta r + \frac{\partial d}{\partial c}\Delta c + \frac{\partial d}{\partial c_L}\Delta c_L \qquad (8)$$

After parametric delay information is computed, the timing propagation is done similar to the one in regular statistical static timing analysis with on-chip variation [2],[5], including both systematic and local random variations.

Parametric OCV method is inherited from SSTA. Therefore it shows superior benefits compared to STA-based advanced OCV methodologies in terms of its required data and realistic handling of the local process variations effects. A widely used OCV method is location-based OCV (LOCV) which builds a derating table as a function of maximum distance and number of stages along the path as shown in Fig. 3.

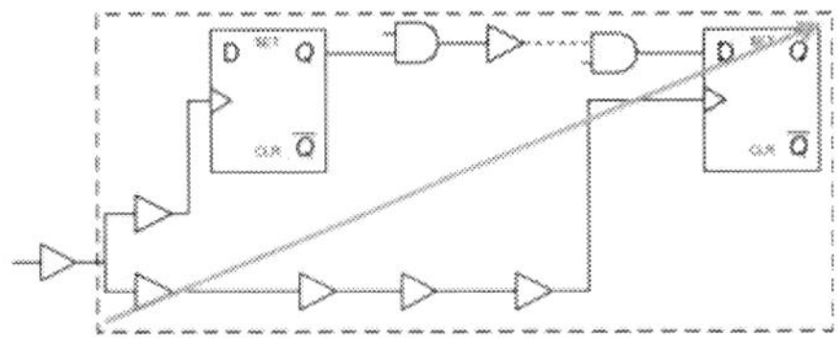

Figure 3: A typical path used to calculate LOCV derating factors.

LOCV relies on accurate tables of timing derating factors, which is a function of cell type, number of stages, and distance, $\Gamma = f\,(cell\_type, number\_of\_stages, distance)$. Building derating tables for each cell in the library is an expensive characterization process.

978-1-60558-497-3/09 $25.00 © 2009 ACM

Table 1 gives and example of LOCV derating factors, where the derating values are characterized for 4 different number of stages and 5 different distance values. For POCV, we only need the value at zero stage and zero distance in the table as input, cutting the characterization cost by 20X for this case.

Table 1: Timing derating table for LOCV analysis.

Distance/Stage	0	10	20	50
0 μm	**1.31**	1.27	1.23	1.18
20 μm	1.29	1.25	1.22	1.17
100 μm	1.28	1.24	1.20	1.16
500 μm	1.27	1.23	1.18	1.15
1000 μm	1.26	1.23	1.17	1.15

Equation 9 gives the formula to construct POCV single stage delay model from an LOCV table, where $\Gamma(0,0)$ represents the first entry value in the table, and p_i is the standard normal random variable.

$$D_i = D_{i,0}\left(1 + (\Gamma(0,0) - 1) \cdot p_i\right) \tag{9}$$

Another problem for LOCV is that it uses distance and number of stages as the only two parameters to describe a path which is not as comprehensive as the way POCV does. It can be proven that the delta delay predicted from POCV is mathematically equivalent to LOCV in simple cases where a path containing the same set of cells over a short distance. For these types of cases, LOCV derating factor for the path is only a function of number of stages, as shown below.

$$\frac{\Delta D}{D} = (\Gamma(0,N) - 1) = \frac{(\Gamma(0,0) - 1)}{\sqrt{N}} \tag{10}$$

In POCV, the path delay is the sum of individual stage delays. When the distance is small, spatial correlation effect can be ignored, and stage delays can be modeled as independent random variables. The relative delay variation can be calculated as in the equation 11 which is equal to equation 10.

$$\frac{\Delta D}{D} = \frac{\sqrt{\sum\left(D_{i,0}(\Gamma(0,0) - 1)\right)^2}}{\sum D_{i,0}} = \frac{(\Gamma(0,0) - 1)}{\sqrt{N}} \tag{11}$$

In more complicated cases, LOCV is less accurate than POCV. For systematic variation, LOCV cannot model exact locations of individual instances inside the path. For local process variations, it cannot model the mis-correlations between P and N channel devices, which are especially important for half-cycle paths that are driven by different clock edges.

4. RESULTS

We have performed various experiments to compare the proposed POCV technique to traditional OCV, LOCV, and finally statistical timing analysis. Results of our studies with multiple designs are provided in the following sections.

4.1 POCV versus traditional OCV

We studied the relative local variation on a 45nm statistical libraries with more than 800 cells, and observed that the 3 sigma local variation effects are mostly within +/- 15%. Based on this observation, we built our parametric

delay model with 15% delay variations as our corner (i.e., 3 sigma) point. Then we performed the POCV analysis to an industry design manufactured with 45-nm process technology. We compared setup and hold slacks obtained from POCV to traditional OCV with 6% and 9% derating factors applied to both early and late paths. Figs 4 and 5 show the results on the hold and setup slacks. As expected, the

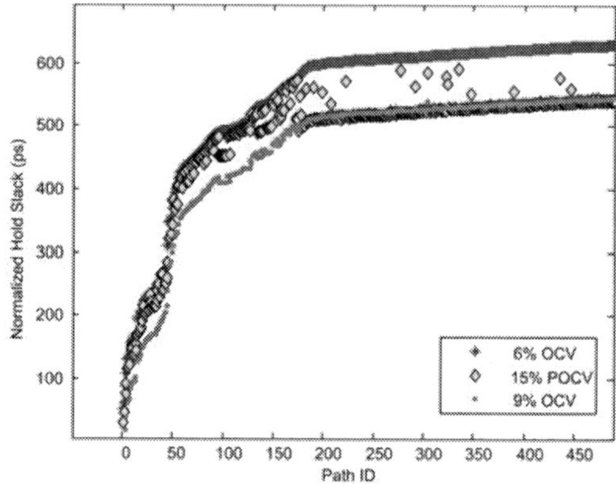

Figure 4: Comparison of POCV to traditional OCV on hold slacks.

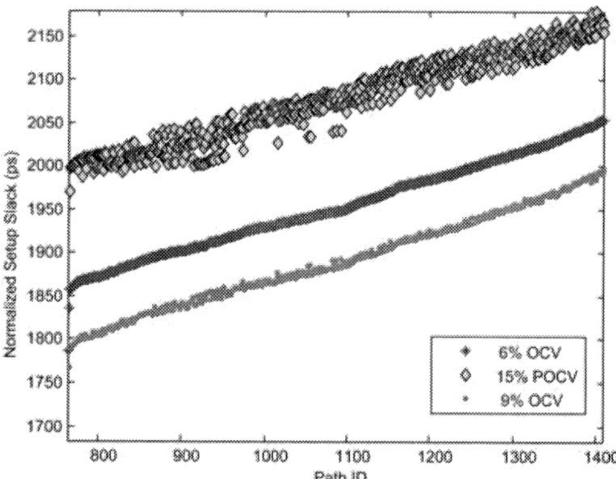

Figure 5: Comparison of POCV to traditional OCV on setup slacks.

hold slack results indicate that single 6% OCV derating factor cannot fully trace the local variation effects. To cover the local variations, 9% OCV derating needs to be applied. However, this causes extra pessimism on some of the hold paths, and further increases the pessimism for the majority of end points on the setup paths. Also, on the setup side, both 6% and 9% deratings are introducing significant pessimism in this design.

4.2 Path based POCV versus LOCV

In our second experiment, we took a 150k block-level 65nm design and selected top 10000 paths and performed both POCV and LOCV analysis. Both POCV delay model and LOCV tables are extracted from selected Monte Carlo spice simulations. The slacks from POCV and LOCV for the paths are compared and the results are presented in Fig. 6.

Fig. 6 shows the histogram of the path-slack difference between two analyses. After some detailed path-level de-

978-1-60558-497-3/09 $25.00 © 2009 ACM

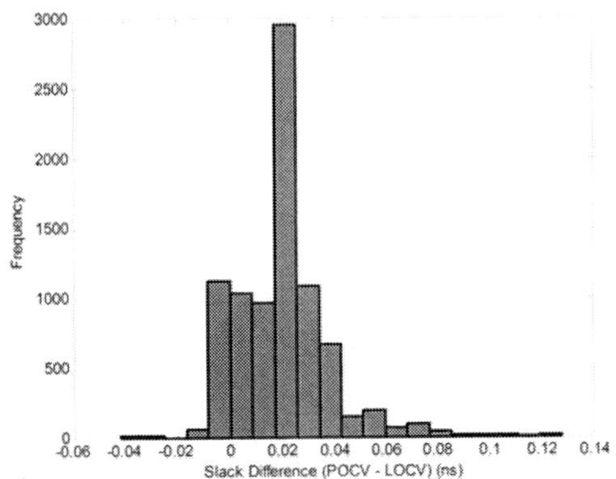

Figure 6: Histogram of slack difference between POCV and LOCV analyses.

bugging, we noticed that there were two main sources that contributed to this difference. The first one was due to P/N transistor mis-correlation from half-cycle paths contributing to those outliers with up to 120ps difference. The second source was due to common-point identification in LOCV analysis, which is more accurately handled by POCV with parametric delay information. The results show that there is still enough margin which would make moving from LOCV to POCV worthwhile.

4.3 Full chip POCV versus SSTA

One advantage for parametric OCV is that it can naturally be applied to block-level and full chip analysis without trading off accuracy. In this last example, we applied POCV analysis on a 100K instance block-level design, and a 6 Million instance full-chip design. For 100k instance block-level design, we compared the slack variation results to full SSTA analysis, shown in Fig. 7. From the plot, we can see that

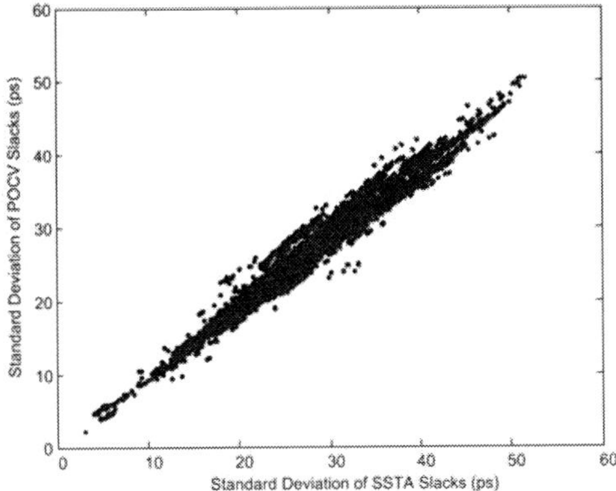

Figure 7: Scatter plot of standard deviations from SSTA and POCV analyses.

the slack sigmas from POCV and SSTA analyses are strongly correlated. The differences are within +/-5ps. The 3 sigma bound will be +/-15ps. Therefore, we can say that as far as the local process variations are concerned, POCV accuracy is within tolerable accuracy range of SSTA.

We studied the performance (speed) and capacity (memory) specifications of our POCV analysis on both designs as well. Since the method has been implemented in a timing analysis tool with multi-threaded computation infrastructure, we also performed studies on the performance scaling with different number of threads. The results of these studies are given in Table 2. Our study shows that the method

Table 2: Performance of POCV analysis.

Design	Block-level, 100K instance		
Number of Threads	1	2	4
Runtime (s)	121	72	40
Memory (MB)	325	327	330
Design	Full-chip, 6M instances		
Number of Threads	1	2	4
Runtime (s)	3425	1750	1117
Memory (GB)	5.8	5.9	5.9

we developed can finish 6 Million instance design in an hour with one thread and 20 minutes with 4 threads. Therefore, it can satisfy the turn-around time (TAT) requirement for production designs in advanced technology nodes.

5. CONCLUSIONS

Local process variability remains to one of the most annoying problems facing the designers particulary at 45-nm node and beyond. Statistical static timing analysis is capable of handling local variations. But the high cost of statistical characterization and parasitic extraction prevents widespread adoption of the technology. In this paper we have proposed a realistic approach which enables designers to tackle down the effects of local variations with minimum amount of required variation data. The POCV analysis proposed in this work provides SSTA-level accuracy, but removes most of the burden to generate statistical libraries and interconnect parasitic variations. This work also demonstrates the benefits of POCV analysis over the traditional derating based OCV and LOCV techniques for handling the local process variation effects in the design's timing.

6. REFERENCES

[1] S. R. Nassif, "Design for variability in DSM technologies," *ISQED Proceedings*, pp. 451 - 454, 2000.

[2] A. Mutlu, K. Le, M. Celik, D. Tsien, G. Shyu, L.C. Yeh, "An exploratory study on statistical timing analysis and parametric yield optimization," *ISQED Proceedings*, pp. 677-684, 2007.

[3] C. Bittlestone, A. Hill, V. Singhal, Arvind N. V., "Architecting ASIC Libraries and Flows in Nanometer Era," *DAC Proceedings*, pp. 776-781, 2003.

[4] A. Nardi, E. Tuncer, S. Naidu, A. Antonau, S. Gradinaru, T. Lin ,and J. Song, "Use of statistical timing analysis on real designs," *DATE Proceedings*, pp. 1605-1610, 2007.

[5] H. Chang, S. Sapatnekar, "Statistical timing analysis under spatial correlations," *IEEE Trans. on CAD of Integrated Circuits and Systems*, pp. 1467-1482, 2005.

[6] J. Le, "Parametric Cell Delay Models and their Applications in 65/45nm Digital Designs," *Electrical Cell Modeling Workshop in IEEE/ACM International Conference on Computer Aided Design (ICCAD)*, San Jose, November, 2008.

978-1-60558-497-3/09 $25.00 © 2009 ACM

Non-Intrusive Dynamic Application Profiling for Multitasked Applications

Karthik Shankar, Roman Lysecky
Department of Electrical and Computer Engineering
University of Arizona, Tucson, AZ
{karthik1, rlysecky}@ece.arizona.edu

ABSTRACT

Application profiling – the process of monitoring an application to determine the frequency of execution within specific regions – is an essential step within the design process for many software and hardware systems. Profiling is often a critical step within hardware/software partitioning utilized to determine the critical kernels of an application. In this paper, we present a non-intrusive dynamic application profiler (*DAProf*) capable of profiling an executing application by monitoring the application's short backwards branches, function calls, function returns, as well as efficiently detecting context switches to provide accurate characterization of the frequently executed loops within multitasked applications. *DAProf* can accurately profile multiple tasks within a software application with 98.5% accuracy using as little as 10% additional area compared to an ARM9 processor.

Categories and Subject Descriptors

C.4 **[Computer Systems Organization]** Performance of Systems – *Measurement Techniques*.

General Terms

Design, Performance.

Keywords

Profiling, multitasking, real-time embedded systems, dynamic optimizations, dynamic hardware/software partitioning.

1. INTRODUCTION

Application profiling – the process of monitoring an application to determine the frequency of execution within specific regions – is an essential step within the design process for many software and hardware systems. Profiling has long been utilized to identify the most frequently executed regions of a software application such that software developers can focus their efforts on optimizing those regions. While static – or offline – profiling is feasible for some applications, dynamic profiling is essential for dynamic optimization techniques. Optimization techniques that leverage dynamic profiling include dynamic binary translation and optimization [3][7], creating multiple specialized software or hardware implementations that can be dynamically selected at runtime [13], storing frequently executed code regions within a low-power loop cache [9][14], and just-in-time compilation.

Dynamic profiling is also an essential task in warp processing [15] – a dynamic hardware/software partitioning approach in which critical kernels within an executing software application are

re-implemented as custom hardware circuits in an on-chip FPGA. As with many dynamic optimization approaches, warp processing relies on accurate, dynamic profiling to determine which software kernels are potential candidates for hardware implementation.

Most previous profiling approaches – intended for desktop computing – introduce runtime overhead, either inserting additional code into the application or interrupting the processor at particular intervals to sample the processor's registers.

A common software-based profiling approach involves instrumenting the application by adding code to count frequencies of the desired code regions [10][12]. For example, if we wish to count the frequency of execution of a subroutine, we can add code to the beginning of the subroutine that increments an associated counter variable. To reduce runtime overhead, other profiling approaches use statistical sampling techniques [1][6][19]. Such methods either interrupt the microprocessor at certain intervals or create an additional software task for profiling and then read the program counter and other internal registers to statistically determine execution behavior. For embedded systems, both instrumentation and statistical profiling approaches potentially change the behavior of the application and incur significant runtime overhead. In the case of real-time systems, which are usually designed with very tight timing constraints, the slightest run time overhead can lead to missed deadlines and potential system failure. For further details and discussion of existing software and hardware based profiling techniques, we refer the interested reader to [16][18].

Due to the limitations of these software profiling methodologies, designers have often resorted to using hardware based profiling approaches. Of notable interest, is the frequent loop detection profiler that non-intrusively monitors the instruction addresses seen on the memory bus and profiles loop iterations by monitoring short backwards branches [8]. A short backwards branch is any branch instruction whose target address has a short negative offset and are typically used to branch at the end of the loop iteration. Whenever a short backwards branch occurs, the frequent loop detection profiler updates a small cache of short backwards branch frequencies and maintains a list of relative branch frequencies. For hardware/software partitioning approaches utilizing profiling to guide the partitioning process, the frequent loop detection profiler can provide a relative ranking of loops to guide the order in which loops are analyzed for hardware implementation. However, without further simulation or application analysis, such limited profiling information may lead to suboptimal hardware/software partitioning results, as performance improvements cannot be accurately estimated using only relative branch execution frequencies.

In [16], we presented an efficient, non-intrusive dynamic application profiler (*DAProf*) capable of profiling an executing application by monitoring the application's short backwards branches and providing detailed profiling statistics for characterizing loop execution behavior, including loop executions, average iterations per execution, and percentage of execution

Permission to make digital or hard copies of part or all of this work for personal or classroom use is granted without fee provided that copies are not made or distributed for profit or commercial advantage and that copies bear this notice and the full citation on the first page. To copy otherwise, to republish, to post on servers or to redistribute to lists, requires prior specific permission and/or a fee.
DAC'09, July 26-31, 2009, San Francisco, California, USA

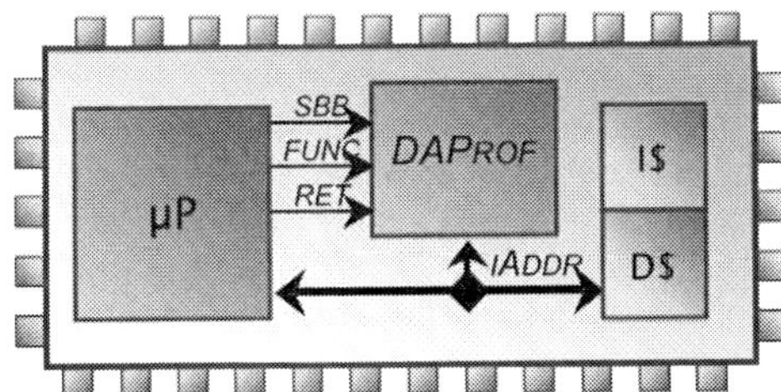

Figure 1. Overview of Dynamic Application Profiler (*DAProf*) integration with microprocessor system utilizing signals for detecting short backwards branches (*sbb*), function calls (*func*), function returns (*ret*).

time. However, *DAProf* was originally designed to profile single task applications. As many embedded systems are multitasked, consisting of several tasks executing within a lightweight kernel or operating system, the original *DAProf* design incorrectly interprets context switches as nonexistent loop execution behavior, thus leading to inaccurate profiling information.

In this paper, we present a non-intrusive, dynamic application profiler with support for profiling multitasked applications by dynamically detecting context switches – in addition to monitoring short backwards branches, function calls, and function returns – to provide detailed and accurate loop execution statistics for embedded applications. The extended *DAProf* design provides an efficient method for non-intrusively detecting context switches to provide accurate profiling results for multitasked applications. In addition, *DAProf* now provides designers with the option to selectively profile specific tasks, functions, library code, or systems calls while filtering out those elements that are not of immediate interest, thereby providing greater profiling accuracy and detail for the profiled elements.

2. MULTITASKED DYNAMIC APPLICATION PROFILER

Figure 1 presents an overview of *DAProf* for multitasked applications. *DAProf* non-intrusively monitors a microprocessor's instruction bus to detect short backwards branches, function calls, function returns, and context switches. *DAProf* considers a short backwards branch as any branch instruction whose target address is a negative offset of less than 1024, which corresponds to small loops containing less than 256 instructions. While the *DAProf* could decode the instructions on the instruction bus, we currently assume the microprocessor provides a one-bit output, *sbb*, indicating a short backwards branch has been executed, a one-bit

output *func* signal indicating a function call is being executed, and a one-bit output *ret* indicating a function has returned. Such support would require minor modification to a microprocessor's decoding logic. To detect function calls and function returns, *DAProf* requires the address from which a function was called and the address to which the function returned.

Figure 2 presents the internal architecture of the extended *DAProf* design with support for multitasked applications. *DAProf* consists of a profiler task filter for specifying the tasks to filter and detecting context switches, a profiler FIFO to synchronize between the microprocessor and profile cache, a profile cache that stores all relevant profile statistics for those loops being profiled, and a profiler controller that analyzes the short backwards branches, function calls, function returns, and context switches to update the profiling statistics within the profile cache.

2.1 Profiler Task Filter

The profiler task filter is primarily utilized to non-intrusively detect context switches between the tasks being profiled. It is implemented as a programmable array storing the starting and ending address of each task, or any region of code, to be profiled. The profiler task filter provides great flexibility in profiling a multitasked application by allowing a designer the option to selectively profile specific tasks, functions, library code, systems calls, etc., while filtering out those elements that are not of immediate interest. To detect context switches, the task filter monitors the processor's instruction bus to determine which task is currently executing or that no profiled tasks are executing. Whenever a change in context is detected – either from one profiled task to another or between a profiled task and a non-profiled task – the profiler task filter will assert the *cs* output along with outputting the current address at which the change was detected. The profiler task filter also filters the *sbb*, *func*, *ret*, and *iAddr* from the processor for all non-profiled regions of code. In other words, if the currently executing instruction does not fall within a profiled task, or code region, then the *sbb*, *func*, and *ret* inputs from the processor will be ignored.

2.2 Profiler FIFO

The profiler FIFO monitors the *sbb*, *func*, *ret*, and *cs* signals from the profiler task filter. Whenever a profile event is detected, the profiler FIFO stores the *Tag* and *Offset* for short backwards branch, the originating address of function calls, the return address of function returns, or the instruction address immediately after a context switch, which are provided by the profiler task filter. The profiler FIFO includes a small FIFO that stores the

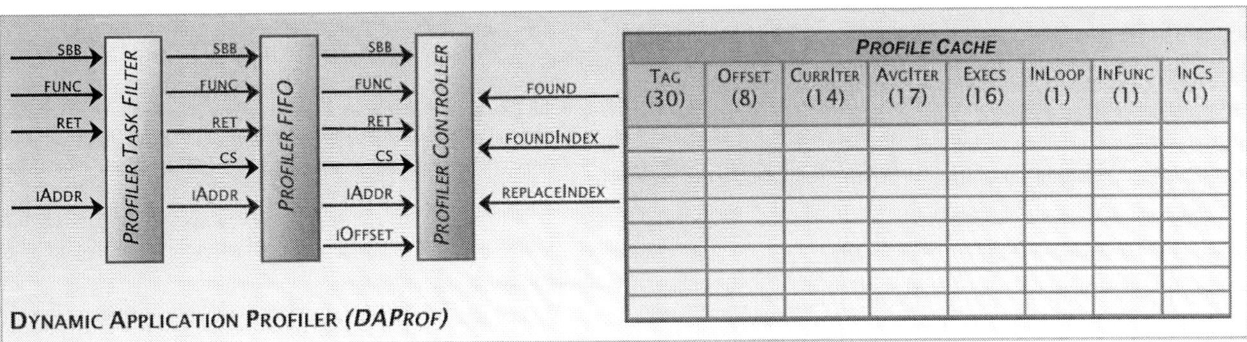

PROFILE CACHE							
TAG (30)	OFFSET (8)	CURRITER (14)	AVGITER (17)	EXECS (16)	INLOOP (1)	INFUNC (1)	INCS (1)

Figure 2. Architectural overview of Dynamic Application Profiler (*DAProf*) supporting multitasked applications consisting of a Profiler Task Filter, Profiler FIFO, Profiler Controller, and Profile Cache (*bit widths for profile cache entries shown in parentheses*).

978-1-60558-497-3/09 $25.00 © 2009 ACM

address of interest, short backwards branch offset when needed, and an encoding indicating if the entry is a short backwards branch, function call, function return, or context switch.

In addition, the profiler FIFO is used to synchronize between the operating frequency of the microprocessor and profiler task filter and the operating frequency of the internal *DAProf* design because the microprocessor and profiler task filter may operate at a higher clock frequency. As short backwards branches do not occur on every clock cycle, the internal *DAProf* profiler design need not operate at the same frequency of the microprocessor. A typical loop of interest within a software application consists of at least two to three instructions in addition to the short backwards branch at the end of the loop. We experimentally determined that the smallest profiled loop within the applications considered consists of 4 instructions. Hence, it should be sufficient to assume that short backwards branches on average occur no more than once every four instructions, implying the internal *DAProf* design can efficiently operate at one fourth the operating frequency of the microprocessor. However, the profiler FIFO should be large enough to accommodate bursts of short backwards branch activity that may occur periodically as an application executes.

In addition, the profiler needs to monitor function call, function return, and context switch events. The combined frequency of all profiling events is not expected to increase the maximum expected frequency of such events. This is evident in the fact that both function calls and function returns require at least several instructions for maintaining the application's execution stack that limits their overall frequency. Similarly, a context switch between tasks requires dozens of instructions to store and restore tasks' contexts.

2.3 Profile Cache

The profile cache is small memory that maintains the current profiling results and intermediate information needed for loop identification, iteration and execution profiling statistics, loop execution monitoring, and determining which profile cache entry should be replaced when new loops are executed. We currently consider a profile cache with 32 entries, which is sufficiently large to profile the embedded software applications considered within this paper – although a larger profile cache may be needed for significantly larger applications.

2.3.1 Loop Identification

Profiled loops are identified within the profile cache by the address of the loop's short backwards branch, which serves as the *Tag* entry for the cache, and by the loop's *Offset* determined by the profiler FIFO. Considering a 32-bit ARM processor and a byte addressable memory, the lower two bits for all instruction addresses will be identical. Hence, the profile cache's *Tag* entry is a 30-bit entry that stores the most significant 30 bits of a loop's short backwards branch address. The profile cache's *Offset* entry is an 8-bit entry that corresponds to the size of the loop in number of instructions. As described earlier, both the *Tag* and *Offset* for a loop are calculated by the Profiler FIFO and provide the mechanism for identifying loop bounds.

2.3.2 Iteration/Execution Statistics

The main profiling information stored within the profile cache includes loop executions, average iterations per loop execution, and loop iterations for the current execution.

Loop *Executions* provide the number of times a loop has been executed throughout the application execution. As the *DAProf* is intended to monitor an application over extended execution periods, regardless of the number of bits used to represent loop executions, the number of loop executions will eventually become saturated. *DAProf* utilizes a 16-bit entry for loop executions that allows 65,536 loop executions to be profiled without saturations. As discussed in the following section, whenever a loop's executions become saturated, *DAProf's* profiler controller will adjust the loop executions for all entries, thereby maintaining a list of relative executions and ensuring all entries do not eventually become saturated

The *Current Iterations* provides the number of times a loop has iterated for the current loop execution and is stored within the profile cache as a 14-bit entry. As a 14-bit entry, *DAProf* can accurately profile loops with a maximum of 16384 iterations per execution, which is very well suited for most applications.

The *Average Iterations* stores the average number of times a loop iterates per execution. As many loops do not iterate a fixed number of times per execution, the average iterations cannot be accurately stored as an integer value. Instead, the profile cache stores the average iterations as a 17-bit fixed point number using 3 bits for the fractional part.

2.3.3 Loop/Function/Context Switch Monitoring

The profile cache contains a 1-bit *InLoop* flag utilized to indicate a loop is currently being executed. The *InLoop* flag is essential in determining if the execution of a short backwards branch corresponds to a new execution or an additional iteration for the current execution. A 1-bit *InFunc* flag is utilized to indicate a loop has called a function that is currently being executed. In addition, a 1-bit *InCS* flag is utilized to indicate a loop's execution has been interrupted due to a context switch. The *InFunc* and *InCS* flags are essential in ensuring that the *InLoop* flag for a loop that has called a function or whose execution has been interrupted due to a context switch is not incorrectly reset.

2.3.4 Associativity

The *Associativity* of the profile cache potentially provides tradeoff between cache size/performance and profiling accuracy. With a fully associative profile cache, the replacement policy must compare all entries within the cache to determine the entry with the smallest total iterations, thereby requiring large hardware resources and reducing the overall performance of the *DAProf*. However, a fully associative profile cache may provide better profiling accuracy as the replacement policy, which is discussed in more detail in the next section, can select from amongst all cache entries. Decreasing the associativity of the profile cache provides increased performance and smaller area requirements by reducing the number of entries the replacement policy must consider, but at the potential cost of reduced accuracy.

2.3.5 Freshness & Replacement Policy

The replacement policy incorporated within the profile cache uses total loop iterations to determine which entry will be replaced when a new loop is executed, where the entry with the lowest total iterations will be replaced. The total loop iterations are calculated as the product of the average iterations and executions. While this policy performs relatively well on its own, newly executed loops may not execute or iterate quickly enough to avoid being immediately replaced.

To solve this problem, the profile cache includes a 3-bit loop *Freshness* value that represents how recently a loop has been executed or iterated, where a larger freshness indicates the loop has been more recently executed. The freshness value is utilized within the replacement policy to only consider loops for replacement if the loops are not fresh – a loop that is not fresh has a freshness value of zero. A 3-bit freshness entry allows up to seven loops per task to be considered fresh and allows newly executed loops to be profiled for an extended duration before their profile cache entry will be considered during replacement.

However, it is necessary to consider the relation between the profile cache's associativity, freshness, and the number of tasks within the application being profiled. For a profile cache with a small associativity, a large maximum freshness, or for an application with a large number of tasks, all entries within the same cache set may be considered fresh and should not be selected for replacement. To avoid this potential problem, the maximum *Freshness* value needs to be determined based on the profile cache associativity and the number of tasks being profiled, as calculated by the following equation:

$$MaxFresh = \left\lfloor \frac{Associativity/2}{\#Tasks} + .5 \right\rfloor - 1$$

2.4 Profiler Controller

Figure 3 provides pseudocode for *DAProf's* profiler controller. The profiler controller interfaces with the profiler FIFO and updates the profiling results for the current loops within the profile cache. The profiler controller either receives the *sbb* signal along with the calculated branch offset, *iOffset*, the *func* signal, *ret* signal, or *cs* signal from the profiler FIFO in addition to a *found*, *foundIndex*, and *replaceIndex* signals from the profile cache. The *found* and *foundIndex* signals indicate if the current short backwards branch is found within the profile cache and at what location. The *replaceIndex* provides the index for the loop entry that will be replaced if the current short backwards branch is not found. In all cases the address of the instruction of interest is provided by the *iAddr* signal from the profiler FIFO.

Whenever a short backwards branch is detected, the profiler controller will determine if the loop is found within the cache. If the loop is found and the loop is currently executing – as indicated by the loop's *InLoop* flag – the short backwards branch execution indicates a loop iteration has been detected and the loop's current iterations are incremented. Otherwise, if the loop is not currently being executed, a new loop execution is detected. For new loop executions, the profile controller increments the loop's executions, sets the *InLoop* flag, sets the current iterations to one, decrements the freshness value for all other loops within the current task, and sets the freshness of the current loop to the maximum freshness. Finally, if the profiler controller detects that the loop's executions have become saturated, the executions for all loops will be divided by two. In addition to ensuring that the executions for all loops never become saturated, this approach provide a mechanism for monitoring the dynamic nature of an application in which loops that were once considered important may no longer be executed as time progresses. Initially, a previously executed loop's high total iterations may ensure the loop is not replaced during profiling. However, after several saturations have been encountered, the reported total iterations will be decreased relative to other loops and can be replaced if the loop is no longer executed.

If a loop's backwards branch is not found within the profile cache, the profiler controller will replace the entry within the cache as indicated by *replaceIndex*. The profiler controller initializes this profile cache entry by setting the *Tag* and *Offset* to those of the newly profiled loop's, setting the executions to one, setting the *InLoop* flag, setting the current iterations to one,

```
DAProf (iAddr, iOffset, sbb, func, ret, cs,
        found, foundIndex, replaceIndex):

1.   if ( cs ) {
2.     for all i, InCS[i] = InLoop[i]
3.     for all i, if ( InLoop[i] && (iAddr <= Tag[i] &&
4.                          iAddr >= Tag[i]-Offset[i])
5.       InCS[i] = 0
6.   }
7.   if ( func ) {
8.     for all i, InFunc[i] = InLoop[i]
9.   }
10.  else if ( ret )
11.    for all i, if ( ( InFunc[i] || InCS[i] ) &&
12.               (iAddr <= Tag[i]  &&
13.                iAddr >= Tag[i]-Offset[i]) ) {
14.      InFunc[i] = 0
15.      InCS[i] = 0
16.    }
17.  }
18.  else if ( sbb ) {
19.    if ( found ) {
20.      if ( InLoop[foundIndex] )
21.        CurrIter[foundIndex] = CurrIter[foundIndex] + 1
22.      else {
23.        for all i, if ( !InCS[i] ) Fresh[i]   = Fresh[i] – 1
24.        Execs[foundIndex]         = Execs[foundIndex] + 1
25.        CurrIter[foundIndex]      = 1
26.        InLoop[foundIndex]        = 1
27.        Fresh[foundIndex]         = MaxFresh
28.        if ( Execs[foundIndex] = MaxExecs )
29.          for all i, Execs[i] = Execs[i] >> 1
30.      }
31.    }
32.    else {
33.      for all i, if ( !InCS[i] ) Fresh[i]   = Fresh[i] – 1
34.      Tag[replaceIndex]          = iAddr
35.      Offset[replaceIndex]       = iOffset
36.      CurrIter[replaceIndex]     = 1
37.      AvgIter[replaceIndex]      = 0
38.      Execs[replaceIndex]        = 1
39.      InLoop[replaceIndex]       = 1
40.      Fresh[replaceIndex]        = MaxFresh
41.      InFunc[replaceIndex]       = 0
42.      InCS[replaceIndex]         = 0
43.    }
44.  }
45.  for all i, if ( InLoop[i] && !InFunc[i] && !InCS[i] &&
46.          !(iAddr <= Tag[i] && iAddr >= Tag[i]-Offset[i]) ) {
47.    InLoop[i] = 0
48.    AvgIter[i] = (AvgIter[i]*7 + CurrIter[i])/8
49.  }
```

Figure 3. Pseudocode for *DAProf* profiler controller with support for multitasked applications.

decrementing the freshness value for all other loops within the current task, and setting the freshness of the newly executed loop to the maximum freshness.

Whenever a context switch is detected, the profiler controller first sets the *InCS* flags for all currently executing loops, i.e., those loops whose *InLoop* flag is still set. This can be efficiently implemented simply by copying all *InLoop* entries to the corresponding *InCS* entries within the profile cache. The profiler controller then determines which loops, if any, will resume execution as the result of the context switch. Thus, if address after a context switch falls within the bounds of any loops whose *InLoop* flag is set, the *InCS* flag for those loops is reset.

978-1-60558-497-3/09 $25.00 © 2009 ACM

Whenever a function call is detected, the profiler controller sets the *InFunc* flags for all currently executing loops, i.e., those loops whose *InLoop* flag is still set. This can be efficiently implemented simply by copying all *InLoop* entries to the corresponding *InFunc* entries within the profile cache. Whenever a function return is detected, the profiler controller resets the *InFunc* and *InCS* flags for those loops that contain the address of the function return's destination, i.e., the loops from which the corresponding function was called. We note that if a function call is executed from the innermost loop of a nested loop, the *InFunc* flag for all loops within the nested loop structure will be set. On return from that function call, the profiler controller must reset the *InFunc* flags for all loops within the nested loops.

For all profiling events, the profiler controller checks all entries of the profile cache whose *InLoop* flag is set to determine if the application is still executing within those loops. The profiler controller also utilizes the *InFunc* and *InCS* flags to ensure that the *InLoop* flag is not incorrectly reset during a function call or context switch. For all detected short backwards branches, function calls, function returns, and context switches, the profiler controller checks all entries of the profile cache whose *InLoop* flag is set and whose *InFunc* and *InCS* flags are not set to determine if the application is still executing within those loops. If a loop is no longer being executed, the profile controller resets the *InLoop* flag and updates the loop's average iterations. *DAProf's* profiler controller utilizes a weighted average in which the previous average iterations accounts for $7/8^{th}$ and the current iterations account for $1/8^{th}$ of the calculated average iterations, as provided by the following equation:

$$AvgIter_i = \frac{7 * AvgIter_i}{8} + CurrIter_i,$$

3. EXPERIMENTAL RESULTS

We consider three alternative profiler implementations including a fully associative, 16-way associative, and 8-way associative *DAProf* designs. *DAProf* was implemented in Verilog and synthesized using Synopsys Design Compiler targeting a UMC 0.18 μm technology. For a fully associative implementation, *DAProf* requires 132,714 gates (2.2 mm^2) and can execute at a maximum operating frequency of 460 MHz. The area required for the fully associative *DAProf* design is approximately 19% of the area of an ARM9 with 32KB cache implemented within a 0.18 μm technology. The 16-way associative *DAProf* design requires 93,194 gates (1.5 mm^2) with a maximum operating frequency of 529 MHz. Finally, the 8-way associative *DAProf* design requires only 71,121 gates (1.2 mm^2) with a maximum operating frequency of 600 MHz. The 8-way associative *DAProf* design requires only 10% of the area of an ARM9 processor. We note that *DAProf's* profile cache is currently implemented using registers. By re-implementing the profile cache using SRAM, we anticipate that a more area efficient design can be created, although we leave this as future work.

To analyze the accuracy of the *DAProf* design, we compare the profiling results of *DAProf* with that of an accurate simulation/instrumentation based profiling method. We created a set of 12 multitasked applications consisting of two to five tasks, where each task corresponds to an application from the MiBench benchmark suite [11]. Table 1 presents an overview of the various multitasked applications considered. All applications were executed within the RTEMS operating system [17].

Table 1. Overview of multitasked applications composed of applications from the MiBench benchmark suite.

	CJPEG	DJPEG	FFT	TIFF2BW	TIFF2RGBA	SUSAN	DIJKSTRA	BIT COUNT	STRINGSEARCH	QSORT	RAWCAUDIO	RAWDAUDIO
MT2.1			✓		✓							
MT2.2	✓									✓		
MT2.3						✓	✓					
MT2.4		✓		✓								
MT2.5								✓	✓			
MT2.6	✓	✓										
MT2.7				✓	✓							
MT3.1								✓	✓			✓
MT3.2			✓		✓					✓		
MT3.3	✓					✓				✓		
MT3.4		✓		✓			✓					
MT4.1	✓					✓	✓			✓		
MT4.2		✓		✓						✓	✓	
MT5.1	✓		✓			✓	✓			✓		

We analyzed the profiling accuracy in terms of percent error in reported average iterations and executions for a fully associative, 16-way associative, and 8-way associative *DAProf* designs for the most frequently executed loops within each multitasked application. For each application, the analysis includes the top ten loops overall within the application, in addition to the top two loops within each task – consisting of 10 to 16 loops combined depending on the application.

Figure 4 presents the percentage error in average iterations of a fully-associative, 16-way associative and 8-way associative *DAProf* design for the multitasked applications. The percent error in average iterations is calculated as the sum of differences between the reported and actual average iterations divided by the sum of the actual average iterations as follows:

$$AvgIter_{\%error} = \frac{\sum_{i=1}^{\#Loops} \left| AvgIter_{i(DAProf)} - AvgIter_{i(actual)} \right|}{\sum_{i=1}^{\#Loops} AvgIter_{i(actual)}}$$

On average, *DAProf* achieves excellent profiling results with an error in reported average iterations of 1.3%, 1.3%, and 1.5% for a fully associative, 16-way associative, and 8-way associative implementations, respectively. In the best case, *DAProf* incurs an error of less than one thousands of one percent for the application *MT3.3*, for all the three implementations. *DAProf* produced a maximum error of 6% for the application *MT3.4* for the 8-way associative design. This error is primarily due to execution behavior of one loop with the application that iterates a single time per execution. In these instances, the backwards branch at the end of the loop is never executed, and our profiling approach is unable to detect that the loop execution, thus leading to reduce profiling accuracy for loops with similar behavior.

Figure 5 presents the percentage error in loops executions of a fully-associative, 16-way associative and 8-way associative *DAProf* design for the multitasked applications. Because of unavoidable execution saturations, the loop executions reported

978-1-60558-497-3/09 $25.00 © 2009 ACM

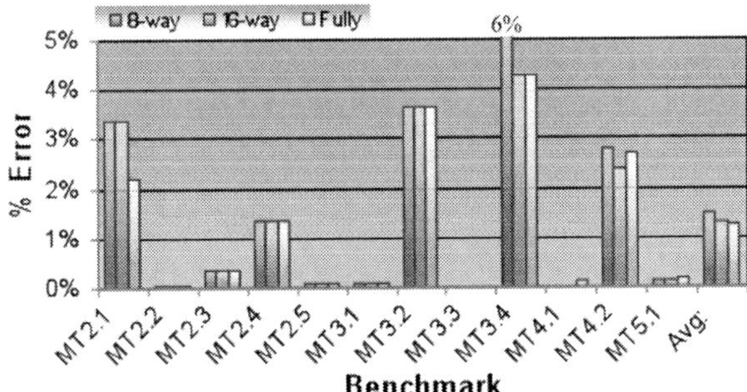

Figure 4. Percentage error in average iterations of fully-associative, 16-way associative and 8-way associative *DAProf* for the multitasked applications presented in Table 1.

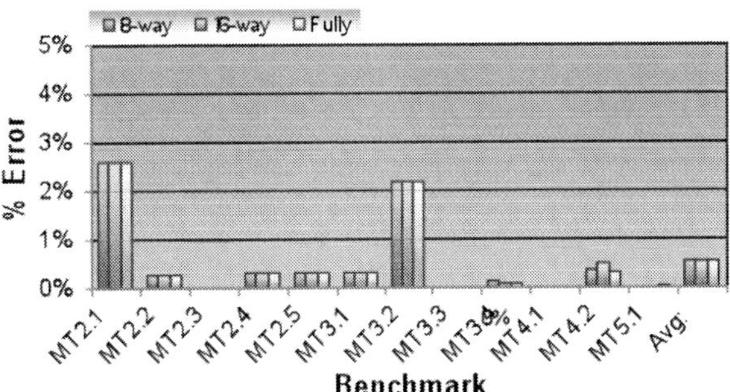

Figure 5. Percentage error in loop executions of fully-associative, 16-way associative and 8-way associative *DAProf* for the multitasked applications presented in Table 1.

by *DAProf* may not directly correspond to the actual total number of loop executions. Thus, the percent error in reported loop executions is calculated as follows:

$$Exec_{\%error} = \frac{\sum_{i=1}^{\#Loops} \left| \frac{Exec_{i(DAProf)}}{\sum_{j=1}^{\#Loops} Exec_{j(DAProf)}} - \frac{Exec_{i(actual)}}{\sum_{j=1}^{\#Loops} Exec_{j(actual)}} \right|}{\#Loops},$$

in which the number of execution for each loop is calculated as that ratio of the reported loop executions of each loop to the total loop executions of the top loops. On average, *DAProf* achieves excellent profiling accuracy for reported loop executions, resulting in an average error of only 0.5% for all associativities. For the applications *MT2.1* and *MT3.2*, the reported loop executions incur an error of 2.6% and 2.2%, respectively. This error can again be attributed to the execution behavior of a few loops that only iterate a single time per execution. For all other applications, a maximum error of 0.5% is achieved.

4. CONCLUSIONS

The dynamic application profiler (*DAProf*) provides an efficient, non-intrusive profiler capable of accurately profiling multitasked applications executing within an operating system. While a fully associative or 16-way associative design provide slightly improved accuracy, an 8-way associative *DAProf* can profile an application executing on a 600 MHz processor with an average profiling accuracy of 98.5% and 99.5% for average iterations and loop executions, respectively, while requiring only 10% area overhead. Thus, an 8-way *DAProf* provides an excellent balance between performance, profiling accuracy, and area.

5. ACKNOWLEDGMENTS

This research was supported in part by Toyota InfoTechnology Center and the National Science Foundation (CNS-0844565).

6. REFERENCES

[1] Anderson, J., L. Berc, J. Dean, S. Ghemawat, M. Henzinger, S.-T. Leung, R. Sites, M. Vandevoorde, C. Waldspurger, W. Weihl. Continuous Profiling: Where Have All the Cycles Gone? ACM Trans. on Computer Systems, Vol. 15, No. 4, 1997.

[2] Arnold, M and B. G. Ryder. A Framework for Reducing the Cost of Instrumented Code. Conf. on Programming Language Design and Implementation (PLDI), 2001.

[3] Bala, V., E. Duesterwald, S. Banerjia. Dynamo: A Ttransparent Runtime Optimization System, Conf. on Programming Language Design and Implementation (PLDI), 2000.

[4] Ball T. and J. Larus. Efficient Path Profiling. Intl. Symp. on Microarchitecture (MICRO), 1996.

[5] Burger, D., T.M. Austin. The SimpleScalar Tool Set, Version 2.0. University of Wisconsin-Madison Computer Sciences Department Technical Report #1342, June 1997.

[6] Dean, J., J. Hicks, C. Waldspurger, G. Chrysos. ProfileMe: Hardware Support for Instruction-Level Profiling on Out-of-Order Processors. Intl. Symp. on Microarchitecture (MICRO), 1997.

[7] Ebcioglu, K., E. Altman, M. Gschwind, S. Sathaye. Dynamic Binary Translation and Optimization. IEEE Trans. on Computers, Vol. 50, 2001.

[8] Gordon-Ross, A., F. Vahid. Frequent Loop Detection using efficient Non-Intrusive On-Chip Hardware. IEEE Trans. on Computers (TC), Vol. 54, 2005.

[9] Gordon-Ross, A., S. Cotterell, F. Vahid. Exploiting Fixed Programs in Embedded Systems: A Loop Cache Example. IEEE Computer Architecture Letters, January 2002.

[10] Graham, S.L., P.B. Kessler, M.K. McKusick. gprof: a Call Graph Execution Profiler. Symp. on Compiler Construction, 1982.

[11] Guthaus, M., J. Ringenberg, D. Ernst, T. Austin, T. Mudge, R. Brown. MiBench: A Free, Commercially Representative Embedded Benchmark Suite. Workshop on Workload Characterization, 2001.

[12] Hazelwood, K., A. Klauser. A Dynamic Binary Instrumentation Engine for the ARM Architecture. Conf. on Compilers, Architectures, and Synthesis for Embedded Systems (CASES), 2006.

[13] Lakshminarayana, G., et al. Common-Case Computation: A High-Level Technique for Power and Performance Optimization. Design Automation Conference (DAC), 1999.

[14] Lee, L.H., Moyer, B., Arends, J. Instruction Fetch Energy Reduction Using Loop Caches for Embedded Applications with Small Tight Loops. Intl. Symp. on Low Power Electronics and Design (ISLPED), 1999.

[15] Lysecky, R., G. Stitt, F. Vahid. Warp Processors. ACM Trans. on Design Automation of Electronic Systems (TODAES), Vol. 11, No. 3, 2006.

[16] Nair, A., R. Lysecky. Non-Intrusive Dynamic Application Profiling for Detailed Loop Execution Characterization. Conf. on Compilers, Architectures, and Synthesis for Embedded Systems (CASES), 2008.

[17] Real-Time Operating System for Multiprocessor Systems (RTEMS), http://www.rtems.org, 2008.

[18] Tony J, Khalid M. Profiling Tools for FPGA-Based Embedded Systems: Survey and Quantitative Comparison. Journal of Computers, Vol. 3, No. 6, June 2008.

[19] Zhang, X., Z. Wang, N, Gloy, J. Chen, M. Smith. System Support for automatic Profiling and Optimization. Intl. Symp. on Operating Systems Principles, 1997.

A Trace-Capable Instruction Cache for Cost Efficient Real-Time Program Trace Compression in SoC

Chun-Hung Lai, Fu-Ching Yang, Chung-Fu Kao, and Ing-Jer Huang
Department of Computer Science and Engineering
National Sun Yat-Sen University, Kaohsiung, Taiwan
{ghlai, fcyang, cfkao}@eslab.cse.nsysu.edu.tw
ijhuang@cse.nsysu.edu.tw

ABSTRACT

This paper presents a novel approach to make the on-chip instruction cache of a SoC to function simultaneously as a regular instruction cache and a real time program trace compressor. This goal is accomplished by exploiting the dictionary feature of the instruction cache with a small support circuit attached to the side of the cache. The trace compression works in both the bypass mode and the on-line mode. Compared with related work, this work has the advantage of utilizing the existing instruction cache, which is indispensable in modern SoCs, and thus saves significant amount of hardware resource. The RTL implementation of a 4KB trace-capable instruction cache, a 4KB data cache and an academic ARM7 processor core has been accomplished. The experiments show that the cache achieves average compression ratio of 90% with a very small hardware overhead of 3652 gates. In addition, the trace support circuit does not impact the global critical path. Therefore, the proposed approach is highly feasible on-chip debugging/monitoring solution for SoCs, even for cost sensitive ones such as consumer electronics.

Categories and Subject Descriptors

B.7.2 [**Integrated Circuits**]: Design Aids - Verification

General Terms

Design, Verification

Keywords

program trace, compression, cache, real time

1. INTRODUCTION

To debug or analyze a software program running in a processor-based system-on-chip (SoC), it is often necessary to collect the program execution trace directly inside the SoC in real time. However, the problem is that the volume of the real time trace grows so rapidly that it is impractical to store the trace on chip or send it outside through limited I/O pins. Therefore, various techniques have been proposed to reduce the trace volume directly in hardware. A straightforward approach is to employ a dedicated hardware compressor, which implements some compression algorithms such as the Lempel-Ziv (LZ) algorithm [1]. However, the hardware cost is very high.

On the other hand, we have observed that the on-chip cache is almost an indispensable component in modern SoC. In this paper, we propose a cost efficient approach to reuse the existing on-chip instruction cache to compress the program execution trace with only minor additional support circuitry. The motivation is that when an instruction address is looked up from the instruction cache, the cache performs a table (or dictionary) lookup operation. When the lookup is successful (called cache hit), it is more economical to record the table index than to record the instruction address. This *trace-capable instruction cache* works as a regular instruction cache and a program trace compressor at the same time, making it more viable than dedicated hardware compressor approaches.

The rest of this paper is organized as follows. Section 2 reviews the related work of hardware-based compression techniques. Section 3 describes the proposed trace-capable instruction cache architecture. Section 4 discusses the data flow of different operation modes of the cache. Section 5 presents the experimental results. Section 6 concludes this paper.

2. RELATED WORK

We characterize the hardware-based compression techniques into two classes. The first class includes the techniques tailored specific to microprocessors. Most works in this class use the filtering mechanisms to select only certain instructions in the trace for recording. The IEEE-ISTO 5001-2003 Nexus Consortium defines a set of specification for on-chip debugging [2]. The Nexus trace module selects only discontinuous instructions for recording. Many modern microprocessors have incorporated Nexus-compliant modules, such as Tensilica's processor cores, Freescale's MPC565, NEC's V850, etc. ARM's Embedded Macrocell (ETM) [3] also uses a similar approach. To further enhance the compression performance, Hopkins and McDonald-Maier proposed a P2G interface [4] which requires the processor core to provide more signals to the trace unit for better filtering.

The second class consists of hardware implementations of lossless data compression algorithms. Among all lossless

Permission to make digital or hard copies of part or all of this work for personal or classroom use is granted without fee provided that copies are not made or distributed for profit or commercial advantage and that copies bear this notice and the full citation on the first page. To copy otherwise, to republish, to post on servers or to redistribute to lists, requires prior specific permission and/or a fee.

DAC'09, July 26-31, 2009, San Francisco, California, USA

978-1-60558-497-3/09 $25.00 © 2009 ACM

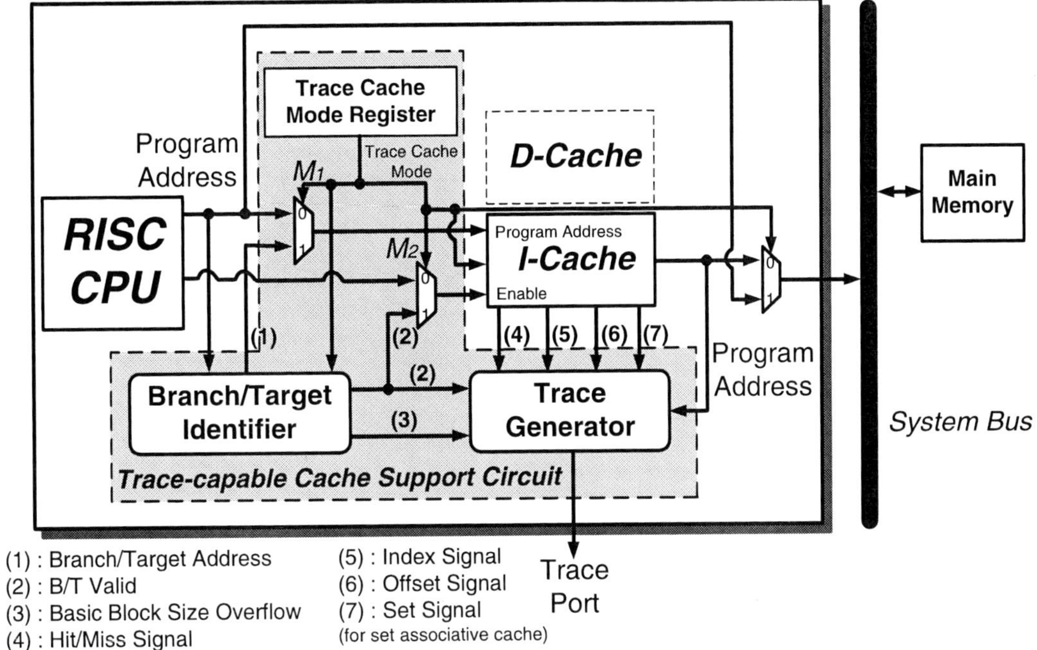

Figure 1: Proposed trace-capable cache architecture

data compression algorithms, dictionary-based approaches such as LZ-related algorithms (e.g., [5] [6]) and other algorithms (e.g., X-MatchPROv4 [7]) are very suitable for hardware implementation [8]. This class of techniques is good for compressing a broad range of data at the cost of higher hardware overhead.

Note that is possible to adopt a hybrid approach by combining techniques from the two classes. An example is proposed by Kao *et al.* [9], in which three phases are adopted to achieve a much higher compression ratio: a branch/target filter, a slicing module, and a LZ-based hardware compressor.

The above techniques focus on maximizing the compression ratio at the cost of higher hardware overhead. On the other hand, the goal of this paper is to develop a more cost effective solution while achieving a considerable compression ratio.

3. TRACE-CAPABLE INSTRUCTION CACHE ARCHITECTURE

The architecture of the trace-capable instruction cache is illustrated in Figure 1. The shaded area depicts the trace-capable cache support circuit, which is integrated with the CPU and the instruction cache. The support circuit consists of a trace cache mode register, a branch/target identifier and a trace generator. The trace cache mode register defines two operation modes for the instruction cache: *bypass mode* and *on-line mode*. Depending on the application need, the instruction cache might be turned on or off while running an application program in the SoC. When the cache is turned off for the program, the cache can then be used solely for the trace compression purpose. This case is called the bypass mode. The bypass mode is used for applications that can not tolerate performance uncertainty caused by the cache behavior. On the other hand, when the cache is turned

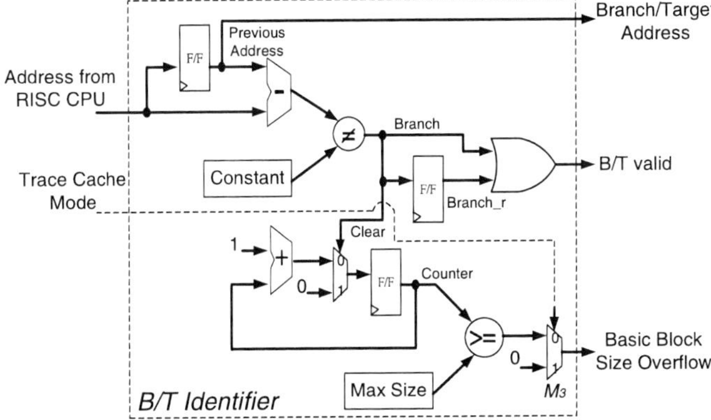

Figure 2: Branch/Target identifier for a RISC CPU

on for the program, the cache is then used simultaneously for the program execution and the trace compression. This case is called the on-line mode. The on-line mode is used for applications that needs the cache to speed up the memory access.

The data cache operates as usual; it does not involve in the program trace compression.

The rest of the components in the figure is described as follows.

3.1 Branch/Target Identifier

The program trace consists of segments of consecutive instruction addresses. It suffices to record only the head (target) address and the tail (branch) address, in order to reduce the trace size. The purpose of the B/T identifier is to select the branch and target addresses. Figure 2 shows the B/T identifier for a RISC CPU. If the offset of previous address and the current address is not equal to a constant offset,

978-1-60558-497-3/09 $25.00 © 2009 ACM

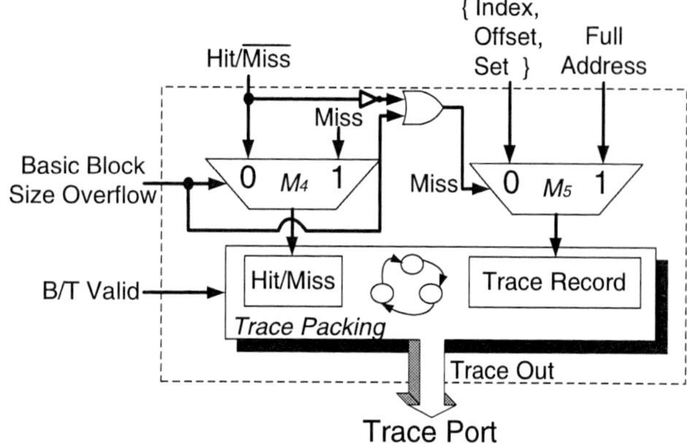

Figure 3: Trace generator

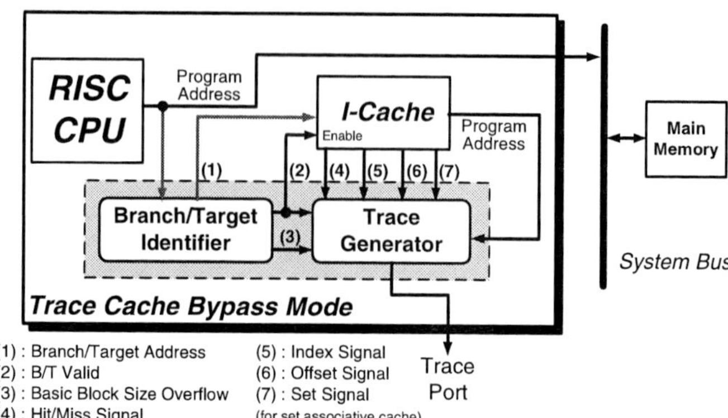

Figure 4: Data path for the bypass mode (M1=1, M2=1)

then a branch has been identified and the branch signal is asserted. This signal is latched to identify the coming target address. For CISC CPU, help from the instruction decoder is needed to identify the branch address.

The basic block size overflow signal in Figure 2 is used to decide whether the consecutive program address references cause cache wrapping around. The counter will increment upon every incoming address, and will be reset when a branch occurs. The max size denotes the maximal length of consecutive address stream that the instruction cache can accommodate without causing the wrapping around. The B/T identifier asserts a basic block size overflow signal when the counter value is greater than or equal to the max size. The basic block size overflow signal is fixed at zero in bypass mode as indicated in M_3 of Figure 2. The use of the basic block size overflow signal will be explained in section 4.2.

3.2   Cache Module

The input of the program address of the instruction cache comes from two possible sources. The address comes from the CPU when in the on-line mode or from the branch/target identifier when in the bypass mode. The selection of the sources of the cache enable signal is the same as the program address. Four internal signals of the instruction cache are exported to the trace generator: hit/miss signal, (cache line) index signal, (cache line) offset signal and set signal (only for set associative cache).

When in the on-line mode, the instruction cache operates normally. Each instruction address is sent to the cache and the CPU reads the corresponding instruction from the cache. The cache performs a refill from the main memory when a read-miss occurs.

When in the bypass mode, the instruction cache operates differently from the normal case. For the CPU, the instruction cache is bypassed. The instructions are directly read from the main memory. On the other hand, Only the addresses generated by the branch/filter identifier are sent to the cache to perform the read operations. In addition, only the tag portion of the cache is operative while the data portion of the cache is idle. When there is a cache hit, the cache hit data is simply ignored. When there is a miss, the tag of the address is written to the tag portion of the corresponding cache line.

3.3   Trace Generator

The trace generator is shown in Figure 3, the cache lookup result is recorded by the trace packing module whenever the B/T valid signal output from the B/T identifier is true. The basic block size overflow signal output from the B/T identifier is used to select the hit/miss signal exported from cache or the forced miss signal. When the basic block size overflow signal is true, the trace generator is forced to record a miss whether the cache lookup result is a cache hit or miss as indicated in M_4. On the other hand, when a cache miss occurs or the basic block size overflow signal is true, the trace generator records the full address; otherwise, it records index, offset, set as indicated in M_5.

Therefore, when a cache hit occurs, the trace generator records a hit and the index, offset, set values. If a cache miss occurs or the basic block size overflow signal is true, the trace generator records a miss and the full address. The trace record length when cache hits depends on the cache configuration, in the basic setting (i.e., direct map cache and cache line size is equal to one), only the index value is recorded. According to this result, the trace file format will be like as follow, where **H** denotes hit and **M** denotes miss: M(address) M(address) H(index) H(index) M(address), etc.

Trace generator will transmit the trace information to off-chip host to save these information as trace files. Trace generator could transmit the data to host via JTAG port or other transmission protocols such as IEEE-ISTO Nexus 5001 AUX port or Ethernet port.

4.   DATA PATH FOR BYPASS MODE AND ON-LINE MODE

4.1   Bypass Mode

The cache operated under the bypass mode is configured by setting the multiplexers M_1 and M_2 to one. The corresponding data path is shown in Figure 4. Therefore, the cache address reference input is the B/T addresses, and the cache enable input is the B/T valid signal output from the B/T identifier. This indicates that the cache is active only whenever a B/T address reference is identified by the B/T identifier. At the same time, the cache lookup result is processed by trace generator only when the B/T valid signal is active.

An example of the compressed trace information in the bypass mode is shown in Figure 5 (a). Column 1 is the

978-1-60558-497-3/09 $25.00 © 2009 ACM

Time	Program Address	After B/T Identifier	After Cache Compression
t	0x0000_0000	(T) 0x0000_0000	M-0x0000_0000
t+1	0x0000_0004		
t+2	0x0000_0008	(B) 0x0000_0008	M-0x0000_0008
t+3	0x0000_0014	(T) 0x0000_0014	M-0x0000_0014
t+4	0x0000_0018		
t+5	0x0000_001C	(B) 0x0000_001C	M-0x0000_001C
t+6	0x0000_0000	(T) 0x0000_0000	H-(0,0)
t+7	0x0000_0004		
t+8	0x0000_0008	(B) 0x0000_0008	H-(0,2)
t+9	0x0000_0014	(T) 0x0000_0014	H-(1,1)
⋮			Index Offset

T : Target Address H : Cache Hit ⤶ : Branch and Target
B : Branch Address M : Cache Miss

(a)

	Word Offset 0	Word Offset 1	Word Offset 2	Word Offset 3
Index 0	0x0000_0000		0x0000_0008	
Index 1		0x0000_0014		0x0000_001C

(b)

Figure 5: (a) Example trace for bypass mode; (b) Cache content of the example trace

instruction sequence. Column 2 is the original program address trace sent by the CPU. Column 3 is the B/T address trace after ignoring the contiguous addresses. Column 4 is the address trace after cache compression. When a cache miss occurs, the trace generator records a miss and the full address, such as the M-0x0000_0000 record at instruction sequence t. If a cache hit occurs, the trace generator records a hit and (index,offset) value, such as the H-(0,0) at $t+6$.

The cache content after executed this address references is shown in Figure 5 (b). The cache configuration in the figure is a 32 bytes direct map cache whose line size is equal to 16 bytes. In **bypass mode**, cache contains the B/T address references only. It implies that the B/T address references will suffer compulsory misses when they are first accessed and then refill into cache as shown in the gray shade of Figure 5 (b).

The trace size in Figure 5 is illustrated as follow, the full address is 32 bits, the Hit/Miss signal is one bit, the index signal is one bit and the offset signal is two bits. Therefore, the original full trace file size will be $(32 \times 10) = 320$ bits, and the compressed trace file size will be $(33 \times 4) + (4 \times 3) = 144$ bits.

4.2 On-line Mode

The cache operated under the on-line mode is configured by setting the multiplexers M_1 and M_2 to zero. The corresponding data path is shown in Figure 6. Therefore, the cache address reference input is the CPU instruction address bus, and the cache enable input is the cache enable bit of the cache control register. This indicates that the memory

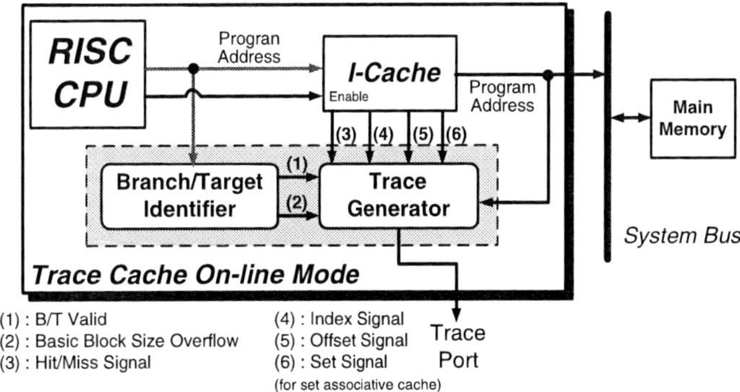

Figure 6: Data path for the on-line mode (M1=0, M2=0)

Time	Program Address	After B/T Identifier	After Cache Compression
t	0x0000_0000	(T) 0x0000_0000	M-0x0000_0000
t+1	0x0000_0004		
t+2	0x0000_0008	(B) 0x0000_0008	H-(0,2)
t+3	0x0000_0014	(T) 0x0000_0014	M-0x0000_0014
t+4	0x0000_0018		
t+5	0x0000_001C	(B) 0x0000_001C	H-(1,3)
t+6	0x0000_0000	(T) 0x0000_0000	H-(0,0)
t+7	0x0000_0004		
t+8	0x0000_0008	(B) 0x0000_0008	H-(0,2)
t+9	0x0000_0014	(T) 0x0000_0014	H-(1,1)
⋮			Index Offset

T : Target Address H : Cache Hit ⤶ : Branch and Target
B : Branch Address M : Cache Miss

(a)

	Word Offset 0	Word Offset 1	Word Offset 2	Word Offset 3
Index 0	0x0000_0000	0x0000_0004	0x0000_0008	0x0000_000C
Index 1	0x0000_0010	0x0000_0014	0x0000_0018	0x0000_001C

(b)

Figure 7: (a) Example trace for on-line mode; (b) Cache content of the example trace

address references are used as the input to B/T identifier and cache at the same time. Whenever the B/T identifier outputs the B/T valid signal as true, the cache lookup result will be processed by the trace generator.

The compressed trace information and the cache content in the on-line mode are shown in Figure 7. The meaning of each column and the cache configuration is the same as illustrated in section 4.1. In **on-line mode**, cache contains the entire address references. It implies that the B/T address references will be refilled into cache in advance due to the cache misses of address references in the same cache line. As shown in the gray shade of Figure 7 (b), the cache misses at instruction sequence t and $t+3$ (please refer to Figure 7 (a)), will refill the entire cache line $index_0$ and $index_1$

978-1-60558-497-3/09 $25.00 © 2009 ACM   139

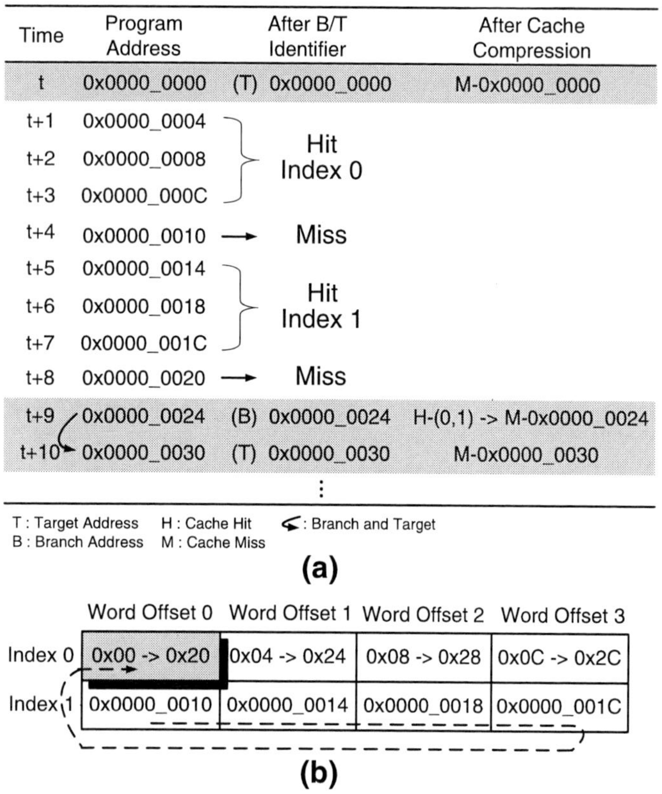

Time	Program Address	After B/T Identifier	After Cache Compression
t	0x0000_0000	(T) 0x0000_0000	M-0x0000_0000
t+1	0x0000_0004		
t+2	0x0000_0008	Hit Index 0	
t+3	0x0000_000C		
t+4	0x0000_0010 →	Miss	
t+5	0x0000_0014		
t+6	0x0000_0018	Hit Index 1	
t+7	0x0000_001C		
t+8	0x0000_0020 →	Miss	
t+9	0x0000_0024	(B) 0x0000_0024	H-(0,1) -> M-0x0000_0024
t+10	0x0000_0030	(T) 0x0000_0030	M-0x0000_0030
		⋮	

T : Target Address   H : Cache Hit   ⤴ : Branch and Target
B : Branch Address   M : Cache Miss

(a)

	Word Offset 0	Word Offset 1	Word Offset 2	Word Offset 3
Index 0	0x00 -> 0x20	0x04 -> 0x24	0x08 -> 0x28	0x0C -> 0x2C
Index 1	0x0000_0010	0x0000_0014	0x0000_0018	0x0000_001C

(b)

Figure 8: (a) Exception in on-line mode; (b) Cache wrap around

respectively, which cause a cache hit when a B/T address is first accessed, such as at instruction sequence $t+2$ and $t+5$. However, in the bypass mode, the cache misses at instruction sequence t and $t+3$ only refill the current address reference as shown in Figure 5 (b), which cause a cache miss at instruction sequence $t+2$ and $t+5$.

The basic block size overflow signal in Figure 2 is only used in the on-line mode, and Figure 8 illustrates an example. As shown in Figure 8 (a), the cache miss at instruction sequence t will cause cache hits at $t+1$, $t+2$, $t+3$ in cache line $index_0$, and the cache miss at instruction sequence $t+4$ will cause cache hits at $t+5$, $t+6$, $t+7$ in cache line $index_1$. And again, the cache miss at instruction sequence $t+8$ will cause a cache hit at $t+9$ in cache line $index_0$. However, the cache refill at instruction sequence $t+8$ will cover the original content in cache line $index_0$ as shown in the gray shade of Figure 8 (b). In this case, the cache has duplicately used the same index and offset value within a basic block, due to a basic block is too large which causes the cache refill wrapping around. And this will cause error in trace reconstruction. Therefore, at instruction sequence $t+9$, the trace generator is forced to record a miss and full address.

5. EXPERIMENTAL RESULTS

The trace-capable cache has been integrated with an academic synthesizable ARM-like CPU N_ARM [10]. The cache module is modified from the caches of LEON2 synthesizable processor [11]. The configurations of the instruction and data caches are the same: 4KB direct map cache with line size of 4 words (16 bytes). The design is synthesized with

Table 1: Gate count of the trace-capable cache (UMC 0.18-μm)

Original CPU + Cache*	N_ARM7[10]	51917
	Cache[11]	261123
	Total (T_1)	321917
Trace-capable Cache Support Circuit	Cache Modification	872
	B/T Identifier	1133
	Trace Generator	1265
	Mode Register	8
	Mux	374
	Total (T_2)	3652 (1.1% w.r.t. T1)

* 4KB direct map instr. cache, 4KB direct map data cache; both with line size of 4 words

Table 2: Timing of the trace-capable cache

	Timing (ns)	
	Global Critical Path	Inst. Cache Path
CPU + Cache	7.3	6
CPU + Trace Capable Cache	7.3	6.3

UMC 0.18-μm standard cell library.

The analysis of the hardware gate counts is shown in Table 1. The trace-capable cache support circuit costs only 3652 gates, which is only 1.1% of the original system. On the other hand, does the support circuit impair the system performance? The circuit delays are shown in Table 2. The support circuit causes the delay of the instruction cache to increase slightly from 6ns to 6.3ns. However, such slight increase does *not* impair the system since it is still less than the global critical path (7.3ns), which is in the CPU.

The compression ratio is measured by executing for 1,000,000 instructions for five applications. The **compression ratio** is defined as follows

$$\left(1 - \frac{Compressed\ Trace\ File\ Size}{Full\ Trace\ File\ Size}\right) \times 100\% \qquad (1)$$

where the **compressed trace file size** can be obtained from

$$\# \ of\ Hit \times (1 + Hit\ Size) + \# \ of\ Miss \times (1 + 32) \qquad (2)$$

The hit size is 9/10/11/12 bits for 2/4/8/16 KB cache respectively.

Table 3 lists the compression ratios for different cache sizes. Due to the page limit, we only show the result of the on-line mode for the direct map cache. (The result of the bypass mode is similar.) The result shows that the compression ratio decreases as the cache size increases. The reason is that although the cache hit rate increases with the cache size, the cache index bitwidth also increases as well, which has more dominate effect on the final trace size. This observation suggests that a small cache such as 2KB is good enough to serve as the trace-capable cache.

Table 4 compares our work with related work. For the average compression ratio, our work (89.8%) is better than the commercial processor-related solutions (ARM7 ETM and generic Nexus module; 80% and 88.5% respectively) and is far better than general purpose data compression solutions (PDLZW+AHDB and X-MatchPROv4; 37.8% and 49% re-

978-1-60558-497-3/09 $25.00 © 2009 ACM

Table 4: Trace compression mechanism comparison

	Average Compression Ratio (R)	Overhead Gate Count (C)	Compression Cost Efficiency $=(R/C) \times 10^4$
Our Work	89.8%*	3652 (UMC 0.18-μm)	245.9
Kao *et al.* [9]	99.7%	51678 (UMC 0.18-μm)	19.3
ARM7 ETM [3]	80.0% [12]	30720 (n/a)	26.0
Generic Nexus	88.5% [4]	61224 (UMC 0.18-μm)	14.5
Hopkins *et al.* [4]	97.0%	31859 (UMC 0.18-μm)	30.4
PDLZW + AHDB [6]	37.8% (exe. file)	133120 (0.35-μm)	3.1
X-MatchPROv4 [7]	49.0%	9039 Tile's 70% of a A500K130-BG45 FPGA	-

* = 1 - geometric mean of trace reduction in 4 KB cache in Table 3

Table 3: Compression ratio vs. cache size (on-line)

Application	Compression Ratio Cache Size (Bytes)			
	2K	4K	8K	16K
DCT	90.8%	90.4%	89.6%	88.9%
FFT	91.2%	92.2%	91.9%	91.8%
JPEG Encoder	88.7%	91.4%	90.7%	89.9%
Fibonacci Sequence	81.6%	79.7%	77.9%	76.1%
Tower of Hanoi	92.0%	91.2%	90.4%	89.6%

* Cache configuration is a direct map cache with line size of 4 words (16 bytes)

spectively). The approaches of Kao *et al.* (99.7%) and Hopkins *et al.* (97%) achieve a higher average compression ratio than our work since their hardware schemes are much more complicated. For the hardware overhead, ours (3652 gates) is far lower than all related approaches (from 30720 to 133120 gates). To further consider the tradeoff between the compression ratio and the hardware cost, the fourth column shows the compression-cost efficiency, which is computed by dividing the average compression ratio with the hardware overhead. The result shows that our work has the most significant efficiency (245.9), which is far better than others (from 3.1 to 30.4). In summary, our work is very cost efficient; it achieves a very good compression ratio with a minimal hardware cost, making it a highly feasible debugging/monitoring solution to SoCs, even for cost sensitive ones such as consumer electronics.

6. CONCLUSIONS

We have presented a novel trace-capable instruction cache which could function simultaneously as a regular instruction cache and as a real time program trace compressor with minor modification to the original processor and cache integration. The proposed cache works in both on-line and bypass modes. Synthesis results have shown that the modification does not slow down the entire system, and the hardware overhead is just 3652 gates while achieving a significant average compression ratio of 90%. Compared with related work, our work has the lowest hardware overhead and the best compression-cost efficiency. Experiments also show that only a small amount of cache, such as 2KB, is sufficient to achieve high compression ratio. Therefore, our approach is a highly feasible debugging/monitoring solu-

tion to SoCs, even for cost sensitive ones such as consumer electronics. Since modern SoCs usually contain very large caches, we would like to investigate the possibility of partitioning the cache into smaller blocks and organizing them in a more intelligent way to support even better versatility/compression/power/performance achievements of caches.

7. REFERENCES

[1] J. Ziv and A. Lempel, "A universal algorithm for sequential data compression," *IEEE Trans. Inform. Theory*, vol. 23, pp. 337–343, May 1977.

[2] *IEEE 5001 Nexus specification*, http://www.nexus5001.org.

[3] *Embedded Trace Macrocell Architecture*, ARM Ltd., http://www.arm.com/products/solutions/ETM.html.

[4] A. Hopkins and K. McDonald-Maier, "Debug support strategy for systems-on-chips with multiple processor cores," *IEEE Trans. Comput.*, vol. 55, pp. 174–184, Feb. 2006.

[5] J.-M. Chen and C.-H. Wei, "VLSI design for high-speed LZ-based data compression," *Proc. IEE Circuits, Devices, Syst.*, vol. 146, pp. 268–278, Oct. 1999.

[6] M.-B. Lin, J.-F. Lee, and G. E. Jan, "A lossless data compression and decompression algorithm and its hardware architecture," *IEEE Trans. VLSI Syst.*, vol. 14, pp. 925–936, Sept. 2006.

[7] J. Nunez and S. Jones, "Gbit/s lossless data compression hardware," *IEEE Trans. VLSI Syst.*, vol. 11, pp. 499–510, June 2003.

[8] S. Kasera and N. Jain, "A survey of lossless data compression techniques," Tech. Rep., 2004.

[9] C.-F. Kao, S.-M. Huang, and I.-J. Huang, "A hardware approach to real-time program trace compression for embedded processors," *IEEE Trans. Circuits Syst. I*, vol. 54, pp. 530–543, Mar. 2007.

[10] Y.-T. Lin and I.-J. Huang, "Enhanced 32-bit microprocessor-based SoC for energy efficient MP3 decoding in portable devices," in *International Conference on Consumer Electronics*, Jan. 2007, pp. 1–2.

[11] *LEON2 Processor User's Manual.*, Gaisler Research.

[12] C. MacNamee and D. Heffernan, "Emerging on-chip debugging techniques for real-time embedded systems," *Proc. IEE Comput. Contr. Eng. J.*, vol. 11, pp. 295–303, Dec. 2000.

Generating Test Programs to Cover Pipeline Interactions

Thanh Nga Dang   Abhik Roychoudhury   Tulika Mitra
National University of Singapore
{dangthit,abhik,tulika}@comp.nus.edu.sg

Prabhat Mishra
Univ. of Florida, Gainesville
prabhat@cise.ufl.edu

ABSTRACT

Functional validation of a processor design through execution of a suite of test programs is common industrial practice. In this paper, we develop a high-level architectural specification driven methodology for systematic test-suite generation. Our primary contribution is an automated test-suite generation methodology that covers all possible processor pipeline interactions. To accomplish this automation, we (1) develop a fully formal processor model based on communicating extended finite state machines, and (2) traverse the processor model for on-the-fly generation of short test programs covering all reachable states and transitions. Our test generation method achieves *several orders of magnitude* reduction in test-suite size compared to the previously proposed formal approaches for test generation, leading to drastic reduction in validation effort.

Categories and Subject Descriptors

B.1.3 [**CONTROL STRUCTURES AND MICROPROGRAM-MING**]: Control Structure Reliability, Testing, and Fault-Tolerance—*Test Generation*

General Terms

Algorithms,Verification

Keywords

Pipelines, Automated test generation, State space exploration.

1. INTRODUCTION

The increasing complexity of embedded systems is driving the dominance of high-performance processors as the design choice. The higher computational requirement of these systems generally translates to the inclusion of complex but performance boosting features, such as caches and pipelines, in the processor architecture. However, the non-trivial interactions among these performance enhancing features are major sources of errors in the processor implementation. A significant portion of the processor design effort is thus spent in validation. A common industrial practice in processor validation is to generate billions of random test programs at the instruction-set architecture (ISA) level. Such a test program generation process is micro-architecture agnostic, and it fails to exercise the subtle micro-architectural artifacts.

Permission to make digital or hard copies of part or all of this work for personal or classroom use is granted without fee provided that copies are not made or distributed for profit or commercial advantage and that copies bear this notice and the full citation on the first page. To copy otherwise, to republish, to post on servers or to redistribute to lists, requires prior specific permission and/or a fee.
DAC'09, July 26-31, 2009, San Francisco, California, USA

In this work, we develop a fully formal framework for micro-architecture aware test program generation. There are three major challenges in developing such a framework. First, an automated test program generator would require a formal specification of the architectural details. We note that architectural components and their interactions are lost at a Register Transfer Level (RTL) model. Instead, we develop a formal processor model based on communicating extended finite state machines. Our model is an extension of the Operation State Machine (OSM) model [8] proposed in the context of MESCAL Architecture Description Language (MADL) [9].

The second and central contribution is the generation of a test suite that covers *all possible pipeline interactions*. In its full generality, a pipeline interaction can be viewed simply as a transition in the global state space of the pipeline. Such a global transition may involve several pipeline resources being acquired or released by multiple in-flight instructions. Existing research on coverage-driven test generation for embedded processor pipelines [5, 6] take a restricted view of pipeline interactions — they typically enumerate and cover interactions among k resources for some small value of k (say $1 \leq k \leq 3$). Clearly, we cannot artificially restrict the number of participating resources in a global transition to be a small number. Moreover, even if the test generation method is scaled to include interactions among large number of resources, *it still fails to cover all pipeline interactions*. This is because a single global transition may also involve interaction among *multiple* instructions in the pipeline for the same resource rather than interaction among multiple resources (e.g., multiple instructions trying to acquire the same functional unit in a single clock cycle). The resource-centric coverage model does not capture such pipeline interactions. To achieve complete coverage of all possible pipeline interactions, both inter-resource interactions and inter-instruction interactions have to be accounted for. We construct an efficient automated method that generates the entire test-suite via a *single on-the-fly* exploration of the global pipeline state space to cover all the transitions and thereby cover all possible pipeline interactions.

Our state space explorer produces paths in the global state space leading to different pipeline interactions. Our last and final task is to generate test programs (sequence of instructions) that actually exercise all these paths. Note that a single test program can potentially cover several pipeline interactions. Suppose transitions t_1 and t_2 are global transitions denoting certain interactions among pipeline components, and t_1 leads to t_2. When we construct/explore the state space, we use the same test program to cover both t_1 and t_2. This reduces the test-suite size by *more than two orders of magnitude* compared to existing works [6].

In summary, we develop a fully formal processor model, traverse the model to cover all possible pipeline interactions, and generate a drastically reduced test suite covering all pipeline interactions.

978-1-60558-497-3/09 $25.00 © 2009 ACM

2. RELATED WORK

Test program generation for processor validation is a well-studied topic. The Genesys tool developed by IBM [1] performs pseudo-random test program generation based on an architecture/testing knowledge base with primary focus on testing the ISA. Subsequently, the focus has been on generating test patterns for micro-architecture. Diep and Shen [4] enumerate the possible pipeline hazards given a low-level processor specification; for each of these hazards a test program is then automatically constructed. Zhu et al. [11] synthesize directed tests from a high-level description to test the bypass paths in a pipelined processor. Mishra and Dutt [7] exploit test program templates for canonical events like pipeline hazards and exceptions. However the template generation is not automated, and this has to be done *manually* by the designer.

Among the formal methods driven approaches to test generation, Ur and Yadin [10] suggest coverage directed test generation. However, they model a very small portion of the processor (only handling of arithmetic instructions) at a low-level (akin to state machine like modeling as in the SMV model checker). It is not at all clear how such an approach will scale up to real-life processors. Moreover, instead of covering all global states/transitions, their test generation covers selected state variables (selected *manually*).

Geist et al. [5] and Koo et al. [6] have used model checkers to generate test programs for processor pipelines. The aim is to achieve a coverage of the state space by covering enough temporal properties. The witness test program for each property is generated automatically by the model checker. However, the designer has to provide the large number of properties required for realistic processors. Apart from automation of the entire process, our key contribution is that we generate the entire test suite through *one* exploration of the global state space. This is in contrast to approaches like [6], which query an external model checker *millions* of times.

3. ARCHITECTURE MODELING

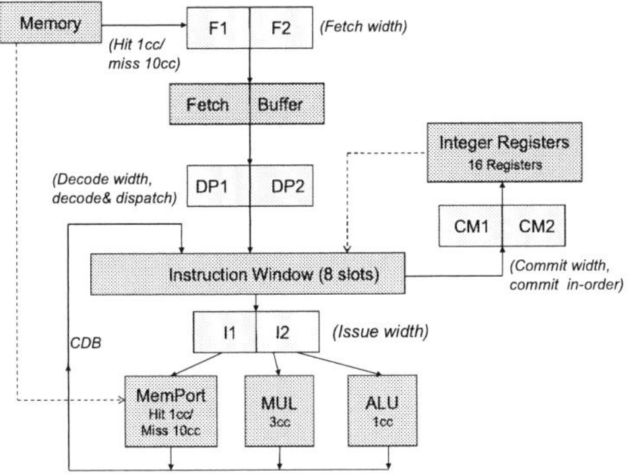

Figure 1: Processor Pipeline Structure

Our formalism models various micro-architectural features for test program generation purposes. For illustration, we will use a 2-*way superscalar, out-of-order execution pipeline* shown in Figure 1. The pipeline consists of the following stages: Instruction Fetch, Instruction Decode and Dispatch, Issue, Execute, Write Back, and Commit. We assume 10 cycle cache miss penalty.

To construct the formal pipeline model, we augment the central ideas in the Operation State Machine (OSM) model [8]. The OSM models a processor at two levels — the operation level and the hardware level. At the operation level, the OSM describes the movement of the instructions across the pipeline stages as Extended Finite State Machines (EFSM) [3]. An EFSM is a finite-state machine with variables; in our pipeline modeling we only consider finite domain variables. Each transition in an EFSM involves (1) a source and a destination state, (2) a guard on the variables (which serves as a pre-condition for the enabledness of the transition), and (3) an action (involving assignments to the variables).

At the hardware level, the OSM models various hardware resources (e.g., instruction window, functional units, etc.) as "token managers". The operation EFSMs progress from one state to another by acquiring/releasing tokens from/to the token managers of the resources. Currently, in the OSM model [8], the communication among operation EFSMs and hardware resources are handled by executing some code corresponding to the token managers. We extend OSMs to develop a fully formal processor model as follows.

The processor pipeline model consists of a collection of EFSMs. This collection can be partitioned into two groups. The first group contains **operation EFSMs** that show the advancement of instructions across the pipeline stages. For a pipeline with maximum N in-flight instructions ($N = 10$ in Figure 1), we have N such EFSMs executing concurrently. In the second group, we have **resource EFSMs** modeling the individual pipeline resources. The instructions progress by acquiring these resources. We also model the instruction/data cache as separate resources that needs to be acquired for an instruction/data to be fetched. This models the interaction between caches and pipeline. In our cache modeling, we ensure that when a memory block is accessed, only the first instruction/data in the block can be cache hit/miss; the remaining accesses in the block result in cache hits. Exceptions are modeled by introducing additional transitions in the corresponding EFSMs.

The pipeline is modeled as a concurrent composition of operation EFSMs and resource EFSMs. These EFSMs communicate among themselves, in particular, an operation EFSM needs to communicate with several resource EFSMs. The communication primitive needs to capture interactions among more than two components as an operation EFSM may need to communicate with *several* pipeline resources to move an instruction from one stage to another. For example, to issue an instruction to the `ALU`, the guard in the `Issue` $\rightarrow$ `ALU` transition (Fig. 2(a)) at the operation level EFSM should check for the availability of the ALU and the source registers. Thus, we cannot use a simple send-receive rendezvous communication involving two components. Moreover, a transition in an EFSM M_i (representing advancement of an instruction in the pipeline) may refer to variables of other EFSMs M_j (the resources).

We propose **communicating EFSMs** for this purpose where the guard and the action in an EFSM transition may involve guards and actions of other EFSMs. Formally, given a collection of communicating EFSMs $M_1, \ldots, M_n$, we define any transition in M_i with: (1) a source state, drawn from the set of states in M_i, (2) a destination state, drawn from the set of states in M_i, (3) a guard, which is a conjunction of *guard methods* (each of these methods is a guard method for some EFSM M_j where j is not necessarily equal to i), and (4) an action, which is a sequence of *action methods* (each of these methods is an action method for some EFSM M_j where j is not necessarily equal to i). Any guard method in our model is a condition that is evaluated without side-effects to true or false. Any action method in our model is a collection of assignments that are executed simultaneously. In other words, the ownership of a variable by an EFSM is for the convenience of the designer. Strictly speaking, these variables function as shared variables where a variable in M_j can be assigned as an effect of a transition in M_i.

Given a collection of communicating EFSMs $M_1, \ldots, M_n$, the execution of a transition t in M_i performs the following atomically.

1. Check that all the guard methods in t are true.

978-1-60558-497-3/09 $25.00 © 2009 ACM

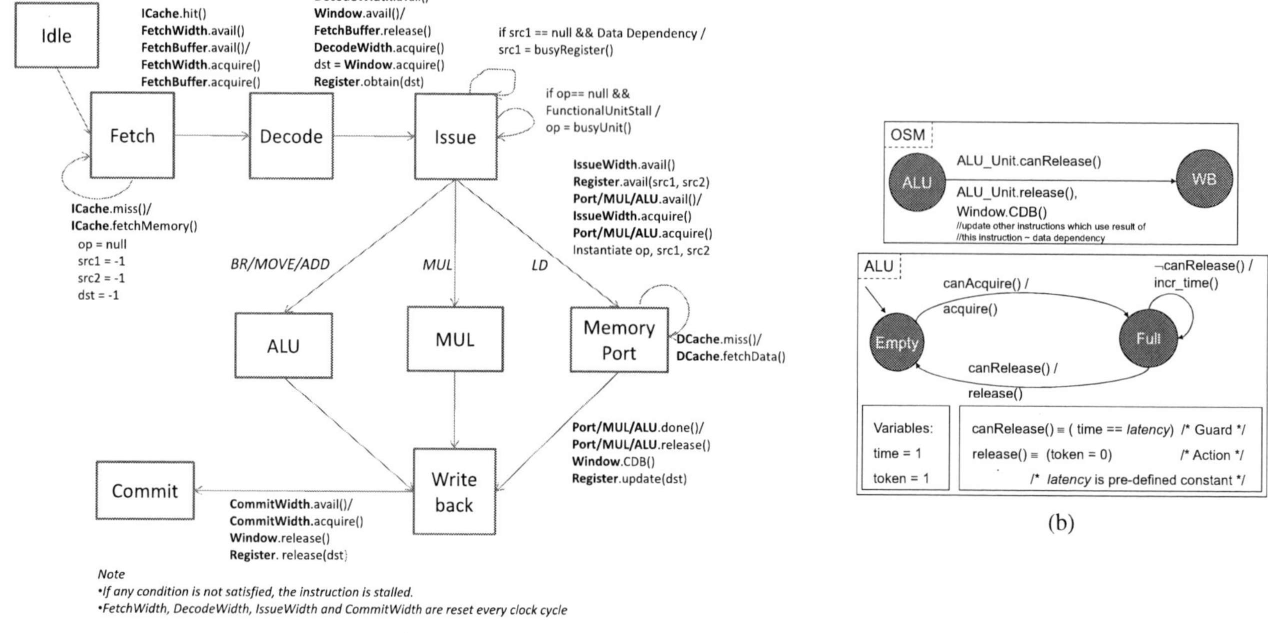

Figure 2: (a) Operation EFSM for the pipeline, (b) A simplified sample transition in the Operation EFSM (shown with guards/actions) and a simplified Resource EFSM for the ALU Unit.

2. Check that all the action methods of t can be executed. If an action method a_t involves assignments to variables in M_i, a_t can be executed. If a_t involves assignments to variables of another EFSM M_j, we check whether in the current state of M_j there exists an enabled transition t' such that (i) the guards of t' are true, and (ii) the action method a_t is the action corresponding to t'.

3. Execute all the action methods of t, changing the state of M_i as well as those of all M_j whose action methods appear in t.

The operation EFSM for our processor pipeline appears in Figure 2(a). In Figure 2(b), we show the guard and action of a sample transition in the operation level EFSM. This transition involves a communication between an operation EFSM and the resource EFSMs for the ALU unit and the instruction Window. Figure 2(b) also shows a simplified version of the ALU EFSM. It has two variables: token (representing whether an instruction is currently executing inside the unit) and time (representing the number of clock cycles since the currently executing instruction started execution). When the operation EFSM makes the transition from ALU state to WB state (signifying end of the EX stage of an arithmetic or logic instruction), it checks whether the ALU unit is ready to finish. This check is captured as a guard method *canRelease*.

4. TEST PROGRAM GENERATION

We now describe our test program generation method.

4.1 Coverage Metric

Previous works on specification driven test generation mostly consider pipeline structure to define the coverage metric. For example, the coverage metric can be defined to ensure that all the pipeline components are exercised. However, given only the structural specification of the pipeline, one cannot define *behavioral* scenarios such as — in one clock cycle, one instruction (say $I1$) is issued to the ALU unit and another store instruction (say $I2$) is stalled in the instruction window as $I2$ is dependent on $I1$. We seek to capture behavioral scenarios in our test suite.

Given the pipeline model as a collection of communicating EFSMs $M_1, \ldots, M_n$, we compose the EFSMs to construct a **global FSM**. Any state of the global FSM consists of a local state and variable valuations for each of the component EFSMs. A transition in the global FSM constitutes a pipeline interaction. Note that a single transition in the global state space may involve progress (or lack thereof) of one or more instructions. In particular, a transition can include (1) the progress of one or more instructions in their operation EFSMs, which in turn requires cooperation from the respective resource EFSMs, and/or (2) progress in one or more resource EFSMs without any change in the operation EFSMs. An example of the latter is when an instruction is executing on the multiplier unit with multi-cycle latency. Here, the number of cycles left to complete execution is a variable associated with the multiplier EFSM. The multiplier EFSM will make a local self-transition and decrease the variable value; the operation EFSM of the corresponding instruction remains in the same state with same variable valuations.

Clearly, if we cover all the reachable states and transitions in the global FSM, we can cover all possible pipeline interactions. This is our coverage metric. That is, our generated test suite must cover all reachable states and transitions of the global FSM.

4.2 An Example

We now describe the test generation algorithm through a simple example. The example illustrates the main features of our method.

Suppose we have reached the global FSM state s illustrated in figure 3. State s has only four active instructions, three of which (I1, I2, I3) are in the instruction window and one is in the fetch queue (I4). We assume that the operation level EFSMs of the inactive instructions are in the idle state. Instruction I1 has already been instantiated as a load instruction and it is now in the "execute" state (operation EFSM state of the instruction I1). This load instruction causes a cache miss; so the data cache (the DCache unit in Fig. 3) is currently in the "miss" state with elapsed time equal to 1. The other two instructions in the instruction window (I2 and I3) are still in the "issue" state; they have not been instantiated to any operation class. Two functional units, ALU and MUL, are in "empty" state. The register resource is in the "partial" state with

978-1-60558-497-3/09 $25.00 © 2009 ACM

144

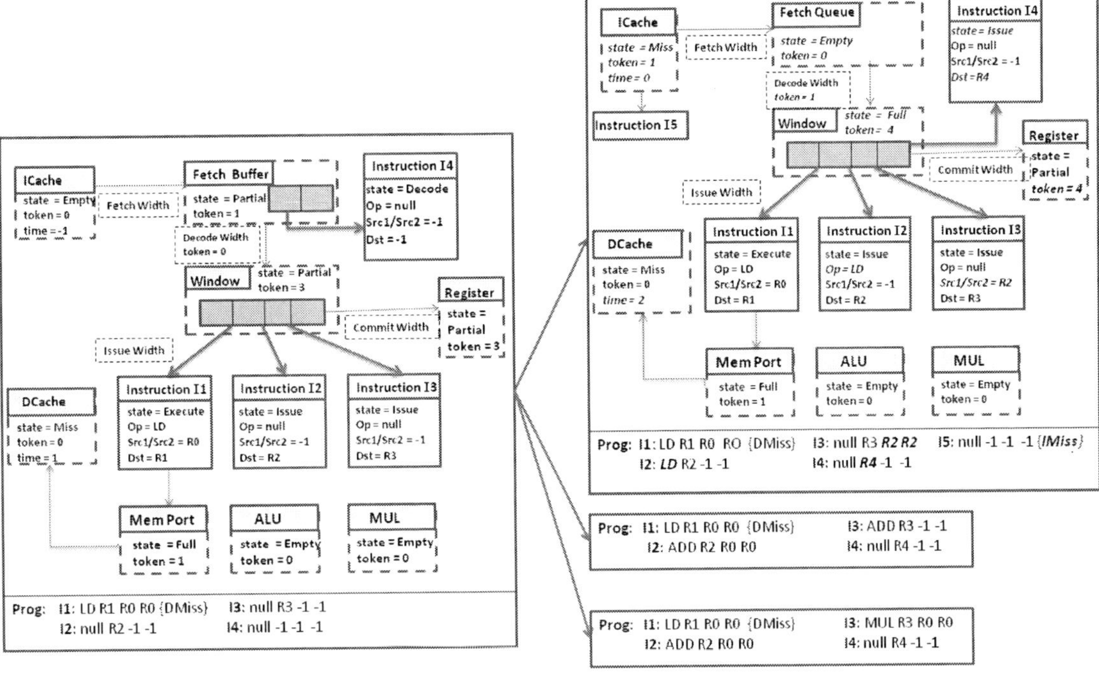

Figure 3: Global FSM states and transitions.

three tokens currently being held by three instructions in the window (register will be in "full" state when all the 16 registers are being used). The last instruction I4 is in the "decode" state and is holding one token from the fetch queue. The fetch queue is in the "partial" state with one free slot available. As the instruction window and the fetch queue preserve the program order, we can deduce program order of I1, I2, I3 and I4. At this point, the partially instantiated test program is as follows. Note that an uninstantiated opcode is written as null and an uninstantiated operand is written as −1.

```
I1: <op>LD   <dst>R1  <src1>R0 <src2>R0 // Dmiss
I2: <op>null <dst>R2  <src1>-1 <src2>-1
I3: <op>null <dst>R3  <src1>-1 <src2>-1
I4: <op>null <dst>-1  <src1>-1 <src2>-1
```

Now we construct all global FSM states reachable from the state s shown in Figure 3. We process the operation level EFSMs in the order of the pipeline stages they are in, i.e., first the operation level EFSMs in the commit state, followed by write-back state, execute state, and so on. So we first process operation EFSM of instruction I1. For I1 to make progress in the operation level EFSM, the resource EFSM of the memory port unit has to move to "completion" state, which in turn requires the data cache EFSM to move to "completion" state. As data cache needs 10 clock cycles to load data from memory, I1 remains in the "execute" state. However, the data cache EFSM updates its elapsed time to 2 units.

Next, we check guard conditions in all possible transitions from the "issue" state in the operation level EFSMs of I2 and I3. Several transitions can be satisfied. For purpose of illustration, let us choose the scenario where I2 is instantiated as a load. I2 will be stalled in the "issue" state due to resource contention for the memory port unit with instruction I1. For operation EFSM of I3, we choose the transition where it is also stalled in the "issue" state due to data dependency with I2. Accordingly, the opcode of I2 and source register of I3 are updated in the test program template.

Now we proceed to the operation level EFSM of instruction I4, which is in the "decode" state. The guard condition for a transition from "decode" to "issue" requires tokens from both the decode width and instruction window. Both of these conditions are satisfied. So I4 returns one token to the fetch buffer, obtains new tokens from decode width and instruction window, and also gets a token

for its destination register. The newly obtained register is recorded in the register resource by increasing it token count.

Finally, we progress the operation level EFSMs currently in the "idle" state. Suppose we fetch one instruction from the instruction cache and this instruction causes a cache miss. Consequently, one operation level EFSM moves from "idle" state to the "fetch" state. A new instruction I5 is appended to our test program. The instruction cache transitions to the "miss" state with zero elapsed time.

This completes one of the global transitions from the global state s and a new global state is reached as illustrated in figure 3. Our test program template is now refined to:

```
I1: <op>LD   <dst>R1  <src1>R0 <src2>R0 // DMiss
I2: <op>LD   <dst>R2  <src1>-1 <src2>-1
I3: <op>null <dst>R3  <src1>R2 <src2>R2
I4: <op>null <dst>R4  <src1>-1 <src2>-1
I5: <op>null <dst>-1  <src1>-1 <src2>-1 // IMiss
```

Clearly we can choose different transitions from the current state of I2, I3, I5. The combination of different local transitions leads to different global states, i.e., different pipeline interactions.

Note that the program with the instructions we generated above uses registers that have not been pre-loaded. So, to make the test program *executable*, we need to introduce instructions in the beginning to load registers from memory. We also make sure that the data for the instruction I1 (above) will not be present in the cache. Furthermore, we need NOP instructions to ensure that the pipeline is empty when our synthesized program starts execution.

4.3 Algorithm

Algorithm 1 is our test generation algorithm. In our approach, (i) construction of the global FSM, (ii) traversal of the paths of the global FSM, and (iii) test program generation of the individual paths — all of these are fused into a single step. *We do not construct and store the global FSM in advance.* Instead, we generate the test suite *on-the-fly*, as we are traversing the FSM.

We start with the initially empty pipeline state S_0. As mentioned before, a global state consists of states and variable valuations for the operation and resource EFSMs. If we can have at most x in-flight instructions, then we have $O_1, \ldots, O_x$ operation EFSMs — all in idle state in the beginning. Similarly, we have $R_1, \ldots, R_y$

978-1-60558-497-3/09 $25.00 © 2009 ACM

resource EFSMs — all in empty state in the beginning. We search the global state space recursively starting with the initial state.

Algorithm 1 Constructing the suite of test programs - Schematic

Require: $O_1, \ldots, O$ operation EFSMs, $R_1, \ldots, R$ resource EFSMs;
Ensure: Test program suite $testSuite$;

$S_0 \leftarrow$ Initial state of $O_1, \ldots, O$ and $R_1, \ldots, R$;
$testPgm(S_0) \leftarrow null; testSuite \leftarrow null; Visited \leftarrow \{S_0\}$;
call $TGen(S_0)$;

function TGen(S: Global state, $testPgm$: Test program){
  for all combinations of enabled local transitions in EFSMs {
    Let $t_1, \ldots, t$ be the chosen transitions in $O_1, \ldots, O$;
    Let $t'_1, \ldots, t'$ be the chosen transitions in $R_1, \ldots, R$;
    /* check feasibility of global transition consisting of local transitions */
    $tempS \leftarrow S$;
    for each operation EFSM O in reverse order of pipeline stages {
      if guard of t is true and action of t is feasible in $tempS$ {
        $tempS \leftarrow update(tempS, O, t)$;
      }
    }
    for each resource EFSM R {
      Let $t' = s \rightarrow s'$;
      if R is not in state s in $tempS$ continue;
      if guard of t' is true and action of t' is feasible in $tempS$ {
        $tempS \leftarrow update(tempS, R, t')$;
      }
    }
    if $(tempS = S)$ continue; /* global transition is not feasible */
    $testPgm(tempS) \leftarrow refine(S, tempS, testPgm(S))$;
    if $(tempS \in Visited)\{$
      $testSuite \leftarrow testSuite \cup \{testPgm(tempS)\}$;
    } else { /* new global state */
      $Visited \leftarrow Visited \cup \{tempS\}$; TGen($tempS$);
    }
  }
}

The recursive routine TGen (see Algorithm 1) identifies all reachable global transitions starting with a global state S. A global transition consists of one or more local transitions in the operation and resource EFSMs. So we choose all feasible combination of local transitions out of state S. Given a combination of local transitions, we give priority to the moves of the operation EFSMs and among them the ones that are in the commit stage first, followed by those in the write-back stage, execute stage and so on. For two operation EFSMs in the same pipeline stage, we give priority to the one corresponding to the earlier instruction in program order. It is *necessary* to consider operation EFSMs in this order. As an operation EFSM makes a transition, it forces some resource EFSMs to make transitions as well. This may enable the next operation EFSM to make a transition. For example, in register bypassing, when an instruction I1 moves from ALU to the write-back stage, another instruction I2 dependent on I1 can move from issue stage to execution. In hardware, these two moves happen in a single clock cycle; we achieve the same effect by advancing the operation EFSM of I1 followed by that of I2 in the *same* global transition.

If the transition t_i of operation EFSM O_i is feasible, then we apply the action method of t_i on the global state to create a temporary global state $tempS$ through the $update$ function. Note that this action method may change the state of some resource EFSMs as well. For the next operation EFSM, the feasibility of the local transition is checked on this temporary global state $tempS$ rather than on the original global state.

Once all the operation EFSMs make their transitions, we move on to the resource EFSMs. The resource EFSMs may make independent transitions without affecting the operation EFSMs. However, the transitions of the operation EFSMs may disable some of the possible transitions of the resource EFSMs. Hence we check that the resource EFSM R_i is in the same local state in both $tempS$

and S. The transition of the resource EFSM may enable transition of the operation EFSMs in the next global state.

While exploring the global FSM corresponding to a processor model, the set of visited states in the global state space is maintained via the $Visited$ set (we implement it as a hash table for efficient access). The path $\pi = S_0 \rightarrow S_1 \ldots \rightarrow S_m$ from the initial global state S_0 to the current state S_m is maintained implicitly via the procedure invocation stack. For each state S_i $(0 \leq i \leq m)$ in π, we maintain a test program $testPgm(S_i)$ which is the sequence of instructions driving execution along the path π from empty state S_0 up to state S_i. The test program at S_i may be "partially instantiated" where the opcodes/operands of some of the program instructions may be un-instantiated (this was shown in our illustrative example). These un-instantiated opcodes/operands are *lazily* instantiated as the instructions proceed along the pipeline stages. Having partially instantiated test programs allows us to maintain *one* test program for each state S_i while exploring a path $S_0 \rightarrow \ldots \rightarrow S_i$. Similarly, when we backtrack during state space exploration — say from state S_{i+1} we backtrack to its parent state S_i and then to another successor S' of S_i — we can readily use the (partially instantiated) test program of S_i to construct the test program for S'.

Given a reachable global transition $S \rightarrow tempS$, the $refine$ function refines the test program at S to create the test program at $tempS$. If the destination state $tempS$ has already been visited, we output the test program corresponding to it; otherwise we continue state exploration from $tempS$. If a generated test program contains any partially instantiated instruction, it means that any un-instantiated opcode/operands can be instantiated arbitrarily.

The following properties hold for our test generation algorithm.

THEOREM 1. *Algorithm 1 visits all global transitions reachable from the initial state S_0.*

Note that for any global state S reachable from S_0, Alg. 1 invokes a recursive call $TGen(S)$. Further, all outgoing transitions from any reachable global state are explored.

THEOREM 2. *The $testSuite$ size from Alg. 1 is bounded by the number of global transitions reachable from the initial state S_0.*

This can be proved by showing that each test program generated covers at least one new reachable global transition. As observed from experiments, our test-suite size is *much smaller* than the number of global transitions (which in turn is much smaller than the test-suite size produced by existing approaches like [6]).

5. EXPERIMENTAL EVALUATION

Our test suite generation framework consists of three main components: *Formal Specification* of ISA and micro architecture for the target processor, *State space exploration*, and target ISA-compatible executable *test program construction*. We have modeled an Alpha 21264 like superscalar processor in the SimpleScalar architectural simulation framework [2]. We specify the ISA in MESCAL Architecture Description Language (MADL) [9]. This includes static properties of the instructions such as operation semantics, assembly syntax, and binary encodings. The current version of MADL can only support the operation level of the OSM model. The hardware level, including the token managers and the structural hardware components, is not part of MADL (as the OSM model itself lacks formalization of the internal behaviors of the token managers). Instead, we rely on Extensible Markup Language (XML) to specify the operation level and hardware level communicating EFSMs. The parser in our test generation framework reads in the ISA (specified in MADL) and the micro-architecture (specified in XML), and creates a mapping between the two. Next, the state space exploration

978-1-60558-497-3/09 $25.00 © 2009 ACM

Processor Configuration	# States $|\mathcal{S}|$	# Transitions $|\mathcal{T}|$	# Test Programs $|\mathcal{TP}|$	Avg. Test Program Length	Test Gen. Time
In-order, superscalarity=1	42,754	95,086	13,232	16	1m30s
In-order, superscalarity=2	43,590	124,256	15,334	17	1m44s
Out-of-order, superscalarity=1	220,039	522,088	84,401	17	8m26s
Out-of-order, superscalarity=2	310,993	851,041	116,260	20	12m24s

Table 1: Statistics for our test generation algorithm for different processor configurations: reachable states, transitions, test programs, length of test programs (in number of instructions), and test generation time.

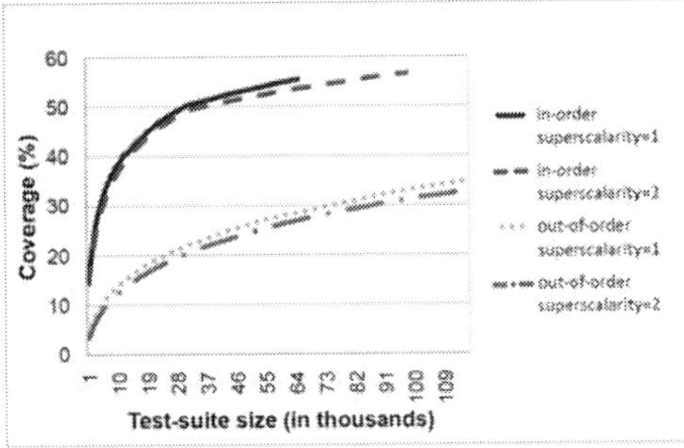

Figure 4: Coverage from random test generation. Our coverage is 100%.

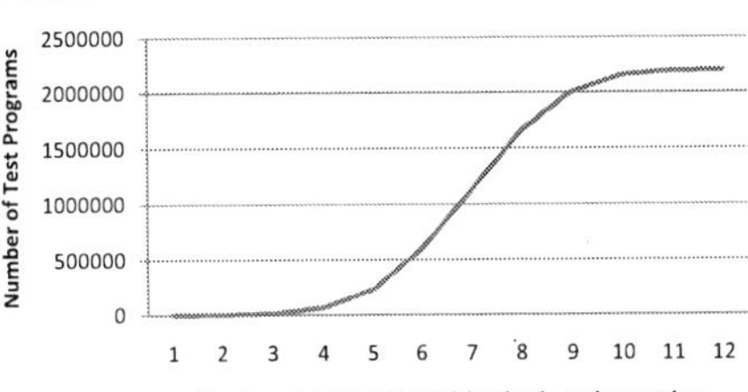

Figure 5: Test-suite size from existing formal approach [6].

module is invoked to traverse and identify all reachable states in the global FSM (obtained by composing the communicating EFSMs). Recall that a partially instantiated test program is maintained as we traverse a path through the global FSM state space. Once a path with at least one previously unexplored global FSM state is formed, our test program construction component creates an *assembly level program* that can exercise this path. Thus, a test-suite is formed. We also validated our test-suite by successfully *executing* the generated test programs.

We consider four different processor pipeline configurations. The most complex pipeline supports out-of-order and 2-way superscalar execution. We restrict superscalarity to one (i.e, scalar pipeline), and also study in-order pipelines (with superscalarity = 1 or 2).

Table 1 presents the various statistics on our test generation algorithm. We report the number of reachable global states/transitions. Note that the number of reachable states and transitions are only a fraction of the total number of global states and transitions (which are in the order of 10^{20}). The number of test programs generated by our method are given in the next column. As discussed earlier, we cover all reachable states and transitions via our test program suite. However, as a generated test program covers several global states/transitions, the test suite size is much smaller than the total number of global states/transitions.

The average length of our test programs vary between 16–20 instructions, which is *pretty small*. The runtime of our test program generation algorithm is very reasonable. Even for the most complex processor, we can generate all the test programs to cover 116K transitions within 12 minute 24 seconds (on a 2.6 GHz Intel Pentium IV machine with 1 GB of main memory).

Next, we evaluate random test generation methods. We identify the set of states $\mathcal{R}$ visited by the random test programs among all reachable states $\mathcal{S}$. Then ($\frac{|\mathcal{R}|}{|\mathcal{S}|} \times 100$) is the coverage percentage. Our method is guaranteed to achieve 100% coverage. The coverage percentage of random method (Y-axis in Figure 4) diminishes quickly to as low as 31% with increasing complexity of processor.

Finally, we compare the quality of our test suite with that of the most recently proposed fault model [6] for processor pipeline interactions (see Figure 5). Assume that there are n resources each with an average of r activities. The work of [6] *estimates* the total number of pipeline interactions as $\sum_{k=1}^{n} {}^{n}C_k \times r^k$, and gener-

ates test programs for these interactions with the help of a model checker. We choose the simplest architecture with in-order execution and superscalarity = 1 for the comparison. The X-axis of Figure 5 shows k, the restriction on the number of functional units participating in a pipeline interaction (this number is imposed by [6]), and the Y-axis shows the number of test programs for varying k. Given k, we find which of the $\sum_{i=1}^{k} {}^{n}C_i \times r^i$ interactions are actually encountered in the global state space; only the number of test programs corresponding to these feasible interactions appears in the Y-axis of Figure 5. We see that [6] generates more than 2 million tests (Fig. 5), whereas we require 13,232 tests (Table 1).

6. CONCLUSION

In this paper, we have presented a fully formal processor modeling framework and used our processor models for systematic test generation. Our test generation method is (i) efficient, (ii) covers *all* pipeline interactions, (iii) produces executable test programs of short length and, (iv) produces a compact test-suite which is less than 1% in size compared to test-suites produced by existing approaches. Thus, our approach greatly reduces validation effort.

Acknowledgments. This work was partially supported by an A*STAR (Singapore) grant R252-000-258-305.

7. REFERENCES

[1] A. Aharon et al. Test program generation for functional verification of PowerPC processors in IBM. In *DAC*, 1995.

[2] T. Austin, E. Larson, and D. Ernst. Simplescalar: An infrastructure for computer system modeling. *IEEE Computer*, 35(2), 2002.

[3] K.T. Cheng and A.S. Krishnakumar. Automatic functional test generation using the extended finite state machine model. In *DAC*, 1993.

[4] T.A. Diep and J.P. Shen. Systematic validation of pipeline interlock for superscalar microarchitectures. In *FTCS*, 1995.

[5] D. Geist, M. Farkas, A. Landver, Y. Lichtenstein, S. Ur, and Y. Wolfsthal. Coverage-directed test generation using symbolic techniques. In *FMCAD*, 1996.

[6] H-M. Koo and P. Mishra. Functional test generation using design and property decomposition techniques. *ACM TECS*, 8(4), 2009.

[7] P. Mishra and N. Dutt. Specification-driven directed test generation for validation of pipelined processors. *ACM TODAES*, 13(2), 2008.

[8] W. Qin and S. Malik. Flexible and formal modeling of microprocessors with application to retargetable simulation. In *DATE*, 2003.

[9] W. Qin, S. Rajagopalan, and S. Malik. A formal concurrency model based architecture description language for synthesis of software development tools. In *LCTES*, 2004.

[10] S. Ur and Y. Yadin. Micro-architecture coverage directed generation of test programs. In *DAC*, 1999.

[11] Q. Zhu, A. Shrivastava, and N. Dutt. Functional and timing validation of partially bypassed processor pipelines. In *DATE*, 2007.

NUDA: A Non-Uniform Debugging Architecture and Non-Intrusive Race Detection for Many-Core

Chi-Neng Wen[1], Shu-Hsuan Chou[1], Tien-Fu Chen[1] and Alan Peisheng Su[2]
[1]National Chung-Cheng University, Taiwan
[1] {wcn93, csh93, chen}@cs.ccu.edu.tw
[2]Global Unichip, Taiwan
[2]alan.su@globalunichip.com

ABSTRACT

Traditional debug methodologies are limited in their ability to provide debugging support for many-core parallel programming. Synchronization problems or bugs due to race conditions are particularly difficult to detect with software debugging tools. Most traditional debugging approaches rely on globally synchronized signals, but these pose problems in terms of scalability. The first contribution of this paper is to propose a novel non-uniform debug architecture (NUDA) based on a ring interconnection schema. Our approach makes debugging both feasible and scalable for many-core processing scenarios. The key idea is to distribute the debugging support structures across a set of hierarchical clusters while avoiding address overlap. This allows the address space to be monitored using non-uniform protocols. Our second contribution is a non-intrusive approach to race detection supported by the NUDA. A non-uniform page-based monitoring cache in each NUDA node is used to watch the access footprints. The union of all the caches can serve as a race detection probe. Using the proposed approach, we show that parallel race bugs can be precisely captured, and that most false-positive alerts can be efficiently eliminated at an average slow-down cost of only 1.4%~3.6%. The net hardware cost is relatively low, so that the NUDA can readily scale increasingly complex many-core systems.

Categories and Subject Descriptors

B.8.1 [PERFORMANCE AND RELIABILITY]: *Reliability, Testing, and Fault-Tolerance*, C.1.3 [Other Architecture Styles]: *Heterogeneous (hybrid) systems*

General Terms

Design, Performance

Keywords

Many-core, Debugging, Architecture, Race detection

1. INTRODUCTION

Multitasking concurrency in the context of parallel programs is essential to the future development of MPSOC. However, nondeterministic parallel programs are hard to debug with tradi-

Permission to make digital or hard copies of part or all of this work for personal or classroom use is granted without fee provided that copies are not made or distributed for profit or commercial advantage and that copies bear this notice and the full citation on the first page. To copy otherwise, to republish, to post on servers or to redistribute to lists, requires prior specific permission and/or a fee.
DAC'09, July 26-31, 2009, San Francisco, California, USA

tional methodologies because subsequent executions with identical inputs are not guaranteed to result in the same behavior. Using a software instrument or debugger to insert probe code into the target program can help to identify potential problems, but may cause a probe effect [1]. Since large-scale many-core systems [2] often feature concurrent execution, concurrent code bases no longer execute in the artificial context of switch-based multi-threading.

Many-core debugging poses several challenges. First, the debugging or monitoring cannot disturb the sequential consistency of the target program. This kind of disturbance is called a "probe effect," and will tend to mask certain synchronization errors after an unknown delay. In fact, it is very easy to cause a probe effect if one fails to adequately separate the debugging data path from the standard data path [3, 4]. To address this issue, OCP-IP [3] recently announced an independent debugging bus for industry many-core debugging. Second, in the context of parallel sequencing, on-chip trace modules are usually employed to record program execution or memory/register state changes and to allow for reliable offline re-execution using this information. This requires a large on-chip trace buffer, and also needs a global timestamp to correlate traces across different cores. Contemporary embedded processor paradigms such as ARM Core-Sight [12] and MIPS On-Chip Instrumentation [3] feature non-intrusive trace modules embedded within the processors. Third, bugs due to unsynchronized access to shared memory in parallel programs (or so-called *race conditions*) are even harder to detect using traditional debuggers; even the on-chip trace modules are supported. A lengthy trace may be needed in order to observe a race condition that results from multiple concurrent accesses. Rather than fully tracing the entire execution, we advocate detecting data races, and alerting the debugger/monitor to record traces for debugging replay.

Software solutions such as Eraser [6] use intrusive software instruments to probe all of the lock/unlock operations and memory access events, in order to detect data races swiftly. However, intrusive instrumentation changes the execution sequence and also contributes to the probe effect. Hence, false positives may be reported. However, such phenomena can be eliminated if a happen-before relation is considered for shared access events [7]. Consequently, such tools [8] need to strike a balance between an impractically long trace and an unacceptably imprecise result. FDR [9] and Rerun [10] are hardware approaches that use a low-overhead hardware recorder in the context of caches or cores, to essentially log the minimum thread ordering information that is necessary to faithfully play back multiprocessor execution after the event. The other hardware approach is HARD [11], which employs additional fields (such as a Bloom filter vector) in the cache and detects the race condition using this Bloom filter.

978-1-60558-497-3/09 $25.00 © 2009 ACM

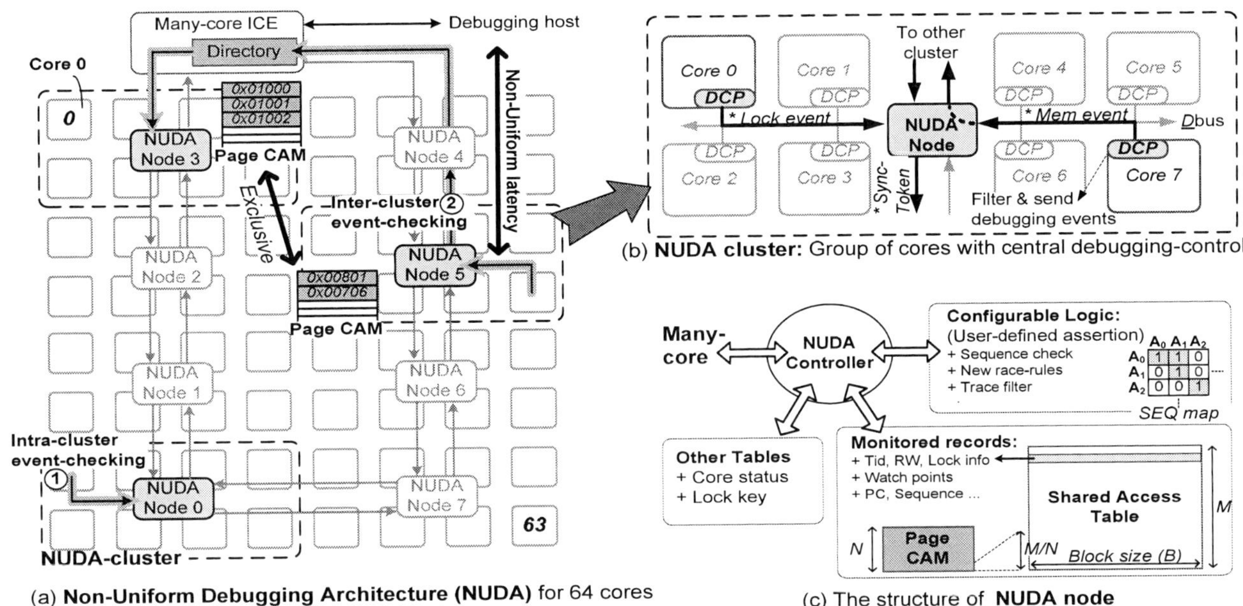

(a) **Non-Uniform Debugging Architecture (NUDA)** for 64 cores

(b) **NUDA cluster:** Group of cores with central debugging-control

(c) The structure of **NUDA node**

Figure 1. The NUDA system architecture.

However, the Bloom filter may cause false positives and the cache flush may empty the BFvector thereby interfering with the filter. Furthermore, it can be challenging to remove locks from the BFvector during unlock operations. The concept of non-uniform cache architecture (NUCA) is commonly used for L2 cache sharing in many-core system [5], because it represents a distributed sharing mechanism with non-uniform memory access latency. The NUCA takes advantages of high data locality to shorten access latency and low cache-miss rates, and without cache-coherency overhead.

Most debugging functions are based on watching instruction and data addresses generated by the processors. The debug engine must constantly check with either breakpoint or watchpoint registers. A many-core system makes the debugging even more complicated. Most traditional debugging facilities rely on globally synchronized signals and are not scalable to many-core systems. Synchronization problems or bugs due to race conditions are particularly difficult to detect. In addition to address comparisons using breakpoints and watchpoints, data race detection involves extra requirements to watch many more lock/unlock and shared addresses at the same time. Hence, this paper proposes a "Non-Uniform Debugging Architecture" (NUDA) to solve the scalability problem in the context of many-core debugging.

The key idea of the NUDA is that we distribute the monitoring data and instruction addresses in a hierarchical manner using a non-uniform address space, caching frequent addresses with distributed CAM tables, and thereby avoiding the need for centralized address comparisons. Once any alert event is detected, the NUDA employs a **synchronization-token** mechanism to globally notify all of the processor cores to stop or to restart, without relying on a global synchronized signal. With NUDA's distributed notification abilities, we can implement various debugging facilities to watch, detect, or even capture the execution trace in order to help develop parallel programs in many-core systems.

Another contribution of this paper is to develop a non-intrusive race detection methodology supported by the NUDA

for use in lock-based programs. A non-uniform page-based monitoring cache in each NUDA node is used to watch the access footprints, and the union of all the caches can serve as a large data-race detection probe. This work also proposes a programmable debugging methodology, by which users specify anchors and execution sequences to ensure that interactions between parallel programs are executed in the correct sequence. The performance slowdown associated with the NUDA is almost negligible because intra-cluster debugging event checks are more frequent than inter-cluster interactions. We demonstrate that the target system can remain non-intrusive when being debugged and probed for synchronization problems using the proposed NUDA mechanism.

This paper is organized as follows. Section 2 details the proposed NUDA for debugging parallel programs in many-core systems. Section 3 presents an example of the use of NUDA for non-intrusive race detection. Section 4 details the proposed software assist and debug methodologies. Section 5 presents our experience of using the proposed solution to analyze a parallel benchmark (SPLAH-2). Finally, Section 6 offers our conclusions.

2. A NON-UNIFORM DEBUGGING ARCHITECTURE

Most of today's many-core debugging is trace-based and relies on off-line verifying of history or execution sequencing. The drawback of this methodology is an unacceptably large storage requirement and slow operating cycles.

2.1 The NUDA for Scalability and Flexibility

Non-Uniform Debugging Architecture (NUDA) is a technique that has been proposed for sequentially consistent parallel programs that must be debugged on many-core system. It has a flexible and configurable infrastructure to handle growing numbers of cores in the future. Figure 1(a) shows a 64-core system with NUDA, which contains a flexible ring interconnection that can be extended as the system grows. The reason for using the ring interconnection is not only to reduce costs but also to take advantage of the ring broadcast mechanism for the synchronization of cores. Figure 1(b) shows a NUDA cluster that features eight neighboring cores and allows communication with the related NUDA node via an independent debugging bus. The NUDA cluster connects to the ring and shared non-uniform memory structure. Eventually, as shown in figure 1(c), the NUDA node will support run-time debugging requirements, such as race detection, thanks to a well-organized and configurable HW structure. In addition, the NUDA node can handle intra-cluster race detection efficiently while still being able to reach the entire system's access record using non-uniform memory structures. There are two kinds of event that occur in the proposed NUDA. One is the transfer of data to a certain NUDA-node as shown in Figure 1(a), and the other involves broadcasting updates or token synchronization to each NUDA-node.

There are three challenges for NUDA. First is the management of distributed sharing pages in figure 1(c), which record the shared memory accesses footprints. This requires reducing the overhead of shared pages access between clusters, when accessing certain remote pages frequently. As with NUCA [4, 5], NUDA's shared pages can be migrated in order to reduce the debugging processing latency in the context of intra-cluster event-checking. Next, to ensure non-intrusive debugging, there are several situations under which all cores should be stopped for proper operation. Therefore, the NUDA needs a synchronized mechanism for starting/stopping all the cores - we call this the "Sync-Token" approach. Finally, the cost is dominated by the memory usage of each NUDA node and the bandwidth of the ring interconnection. Compared to current several MB memory-bit usages, the NUDA only uses a few KB in each cluster. Compared to the bandwidth of current network-on-chip systems (128-bit ↑), the bandwidth of the ring interconnection for debugging only uses 16- or 8-bit operations in the context of infrequent inter-cluster events. The design strategy of CAM involves a page-based fully-associative structure for keeping the monitored shared page index in a shared access table until it is no longer required. Furthermore, the page-based strategy allows for distributed sharing using the non-uniform SRAM associated with every NUDA node.

In Figure 1(b), the NUDA cluster features a NUDA node connecting a group of cores with a local debugging bus (_D_bus). The NUDA node is responsible for run-time debugging, core debugging control, race detection, and communication with other clusters. Debugging events, such as memory accesses, lock/unlock requests, thread activations, check points etc. are all filtered and transmitted by a debugging co-processor (DCP) on each core. Each DCP contains a small number of buffers to ensure robustness to any delays that may be caused by a stream of successive transmissions from the cores to the _D_bus. The NUDA-node can perform flexible and versatile functions, and the NUDA-cluster is scalable and compatible with a wide range of many-core architectures.

2.2 Sync-Tokens for Non-uniform Control

The latency associated with debugging control propagation is challenging. Such latency may interfere with the original execution ordering in a run-time debug scenario. It is essential for programmers to fully understand actual multithreading execution ordering in a system. For example, when a breakpoint is reached by a certain core, all the other cores may stop at different times if no synchronized mechanism exists to bring them to a halt simultaneously. This can lead to a programmer misidentifying an actual break point. Our solution is to stop all the cores

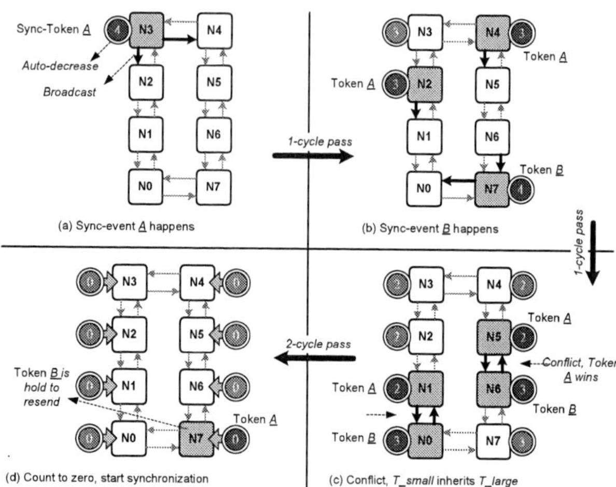

Figure 2. Example of "Sync-Token" for non-intrusive control.

synchronously and to use a fixed latency to maintain non-intrusive operation.

Non-intrusive control plays a role as follows:

- "Start/Stop all cores" → break/watch points, DCP buffer-full, shared access table page-migration, race detected, assertion, etc.

- "Update all cores" → events for checking user-defined assertions (update the SEQ_map as shown in Figure 1(c)).

In this work, we propose a "Sync-Token" mechanism for synchronous debug-control. The key idea behind our "Sync-Token" is to pass a specific and related countdown number to each NUDA node. This countdown number decreases incrementally with time. When the token passes though all of the NUDA nodes, the countdown numbers across the nodes are allowed to reach zero, whereupon they can be synchronized for start/stop/update operations.

Figure 2(a) shows an example of how the sync-token works. Assume Sync-event _A_ happens in NUDA node N3 and is associated with token number 4. One cycle later, Token _A_ broadcasts to NUDA node N2 and NUDA node N4 and transmits token number 3. At the same time, the token number at NUDA node N3 itself decreases to 3. Therefore, NUDA nodes N2, N3, N4 are synchronized as shown in Figure 2(b). However, there is another Sync-event _B_ that is being concurrently processed by NUDA node N7. The tokens broadcast contention, as shown in Figure 2(c). NUDA nodes N1, N2, N3, N4, and N5 are synchro-

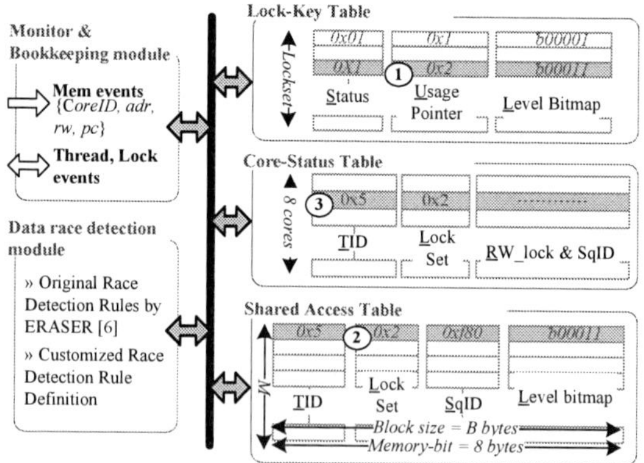

Figure 3. Block diagram of Race Detection.

978-1-60558-497-3/09 $25.00 © 2009 ACM

nized by token number 2 and NUDA nodes N0, N6, N7 are synchronized by token number 3. Under our mechanism, lower-numbered tokens can inherit larger numbers, so that Sync-event *B* will hold and wait to be re-sent at NUDA node N7. While all the tokens across the NUDA nodes count down to zero, all the NUDA nodes can be synchronized for non-intrusive control.

3. NUDA EXAMPLE: NON-INTRUSIVE RUN-TIME RACE DETECTION

Figure 3 shows the block diagram for our proposed real time race detection methodology. The basic idea unites software and hardware approaches to detect races in real time. We use an Eraser-like [6] algorithm to detect races, but we can also address more complicated locking protocols. We insert certain co-coprocessor instructions into the thread library. As the software meets certain conditions, these instructions will invoke the DCP to transmit related events to the NUDA node in the cluster. The hardware consists of a monitor and bookkeeping module, a lock-key table, a core-status table, a shared access table and data race detection logic. The non-uniform property of NUDA allows for both inter-cluster table referral and private use within the cluster.

In order to maintain global sequencing, the "Sequence Identify" (SqID) parameter will be dispatched and incremented with each event. The SqID resets when threads exit. When a thread is created, the DCP sends a new thread packet, which consists of a core identifier (CoreID) and a thread identifier (TID). Once the monitor and bookkeeping module receive this packet, the entry in the core-status table with index CoreID can be initialized and updated.

Following the success of a lock/unlock event, the DCP sends the CoreID, TID and lock key address to the NUDA node. Figure 4 shows a nested write locks example in the context of Thread B. The monitor and bookkeeping module receive this packet. Then, if this is a brand new lock event, the bookkeeping module will rename this event under a new lock set entry and will update this entry index to the corresponding Core-Status table entry's Lock-Set field. We use (*LockSet, Level*) to represent each unique key. In this example, the lock set 0 will be created as write lock S1 is acquired with key (0, 0). In contrast, if a nested lock/unlock event occurs, the bookkeeping module first accesses the Core-Status table to determine the current lock set and update the level bitmap. In this example, the level bit-vector of LockSet0 would be updated from 16'b1 to 16'b11 as the write lock S2 gets acquired with key(0, 1). Our methodology supports a lock renaming mechanism, which allows the process of unlocking to be efficiently conducted for various related variables. This in turn greatly reduces race-detection complexity.

In terms of memory access, DCP allows the access address to filter through the entire range of shared addresses. As above, the monitor and bookkeeping modules receive the memory access packet from the DCP, and they check the page CAM to determine whether this address is within the scope of this cluster monitor. If not, the NUDA node will forward this packet to the destination NUDA node by referencing the directory in the ICE.

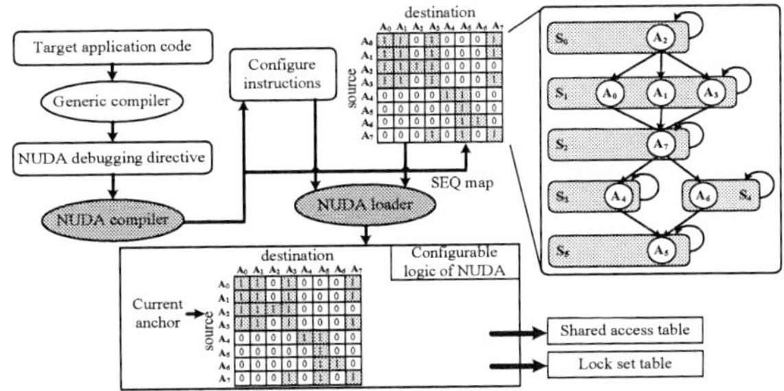
Figure 5. The workflow of NUDA debugging.

Otherwise, the CAM will reveal the start index of the access record table, indexed according to the memory page address.

Each NUDA node features both content-addressable memory (CAM) and SRAM. The monitored capacity is related to the granularity of the smallest monitoring unit for monitoring. The NUDA protocol can scale the granularity between 32 words and 4KB. The selected granularity will influence the number of false positives. For example, a coarse grain resolution means monitoring a larger shared memory scope. Every access event under this scope will be treated as suspect, and this may lead to false identification of race conditions. The shared access table shown in Figure 3 is stored in non-uniform SRAM, which records the footprint of shared memory access events. To more quickly locate the entry in this table, we use a page CAM to address the relationship between shared page addresses and block numbers in the context of a shared access table. The entries (*M*) of the SRAM are partitioned by the entries (*N*) of the page CAM, and the block size (*B* bytes) of the shared access table is consequently the smallest monitoring unit for monitoring. Therefore, the monitored capacity of the shared access table is $M * B$ bytes and the page CAM entries require $(M * B)/4K$ (assuming 1 page = 4KB, $N = 8$). However, as shown in Figure 2, the physical memory-bit usage is only $M * 8$ bytes.

4. SOFTWARE ASSISTANCE

NUDA handles generic debugging tasks through ICE. We propose a debugging programming model that can help the user to configure NUDA. The key idea behind our proposed debugging directives is to place several logic checkpoints in order to ensure that events occur as programmers expect.

4.1 Placing debugging directives in program

Figure 5 illustrates the NUDA debug workflow. We use the #pragma directives in C to command the compiler to appropriately handle NUDA debugging information. The compiler converts the appropriate data to co-processor instructions, which is defined in the DCP. The NUDA-compiler then generates binaries that include both a configuration part and an assertion part. These configurations are then sent to the DCP and NUDA nodes, before the target application execute.

Figure 6 shows some simple sample code for the NUDA de-

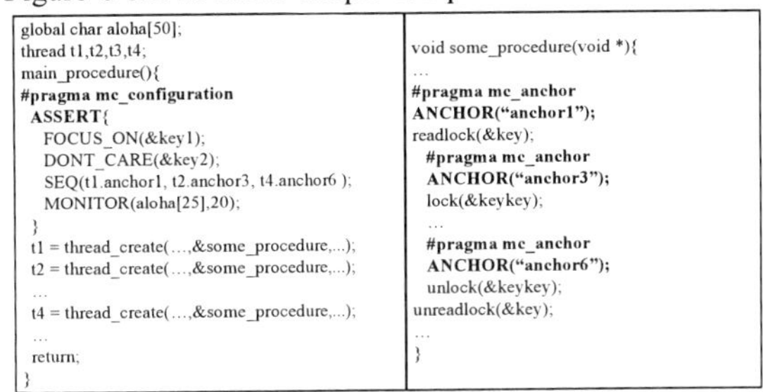

Figure 6. Sample code for NUDA debugging.

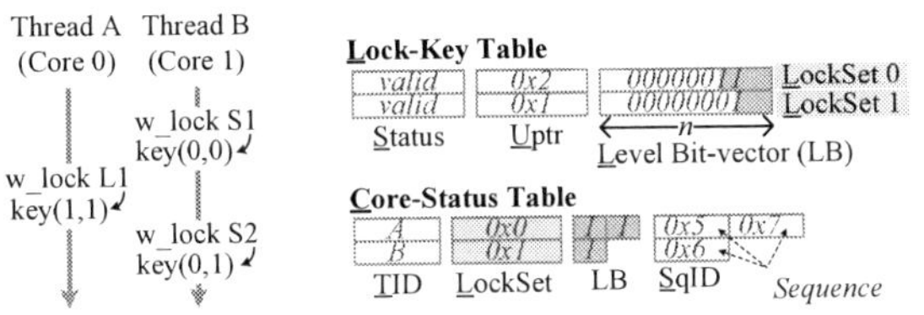
Figure 4. Lock renaming (Lockset & Bitmap) for nested locks.

978-1-60558-497-3/09 $25.00 © 2009 ACM

151

bugging programming. First, a set of generic configuration compounds allow the user to configure debugging commands. The NUDA node can be set to monitor particular shared spaces, validate threading sequences and insert customized watchpoints. Second, anchor point directives can be inserted into any part of the program or into event libraries. These anchors are ultimately translated into co-processor instructions, which in turn are used to notify the DCP that the core is passing a particular anchor. The DCP buffers this information and then transfers it to the appropriate NUDA controller.

We propose a set of primitives to help the programmer better debug multithreaded programs on many-core systems. The directives: **FOCUS_ON(&key)**, **DONT_CARE(&key)**, and **MONITOR(&memory, size)** help the programmer identify any suspect code that requires further analysis. The SEQ syntax allows the developer to identify certain execution sequences. Using this protocol, the system can compare any token's identity with the current SEQ map field in the NUDA configurable logic. The NUDA controller can immediately stop execution if the sequence does not follow the given assertion rules. The cost of implementing the SEQ map is low. Because of the NUDA can handle at most eight steps and only needs an 8x8 SEQ bitmap.

We use Ai to identify anchors and three different symbols to represent the ordering of execution. The statement SEQ{A2,(A3,A1),A0,(A7|A5|A4),A6} denotes that A2 happens at least once. (A3, A1) denotes that A3 should happen after A1 and that this sequence happens at least once and (A7|A5|A4) indicates that each of A7, A5, A4 occur randomly. Hence, the white and black boxes represent the duration of the execution event, and the black box represents a violated sequence caught by NUDA.

4.2 Exception handling (error control)

In many-core systems, especially real time ones, NUDA can serve as a system safety monitor. Certain alternative approaches [12] have attempted to analyze trace records using trace data. However, the number of emergent events can rapidly exceed the available storage capacity. NUDA can throw exceptions when bugs are caught or other situations occur. Exception handlers are used to prevent system crashes or to notify a backend debugger.

5. EXPERIMENTS

The NUDA has been evaluated by a parallel simulator mcore [13] in the context of the SPLASH2 benchmarks [14], as shown in Table I. We used seven race-free SPLASH2 benchmarks to evaluate NUDA. Those benchmarks work with lock-based multi-threading programming model.

Basically race detection by NUDA is real-time and non-intrusive to the original execution. However, in case that the monitoring buffer is full, the whole system has to be stopped. Table II illustrates NUDA overhead in SPLASH-2 for runtime

Table I. Target System Parameters.

Simulator parameters			
simulator	mcore	ISA	x86
#core	64	CPU family	Intel® Atom™
NUDA parameters			
#cores in cluster			4, 8
parallel processing ability in NUDA-node			single, bi-direction
bit-width of ring interconnection (bit)			8, 16
latency of ring interconnection (cycle)			1, 2
DCP buffer size (word)			4,8
NUDA-node monitor block size (words)			8, 16, 32
Shared access table size			16~256KB
CAM mapped page size			0.5~4KB

race detection. We found the overhead to be inversely proportional to buffer size; more DCP buffers can tolerate more concurrent memory access events. Figure 7 shows slowdown effects with respect to buffer size that the average overhead for 4-word to 8-word buffers is small. In addition, the stall caused by shared access table migration is also low. This is because most of the benchmarks use the shared spaces separately. In contrast, we note that benchmarks **ocean**, **mp3d**, and **water-nsquared** all have high density spaces and higher migration stall rates. The page usage column in Table II shows the maximum usage of the shared access table pages across the whole system. Lower page usage rates can reduce inter-cluster traffic and improve performance. Also, the inter-/intra- cluster ratio represents the events relation between inter-cluster and intra-cluster. More inter-cluster events means more chances to do NUDA migration and effects performance.

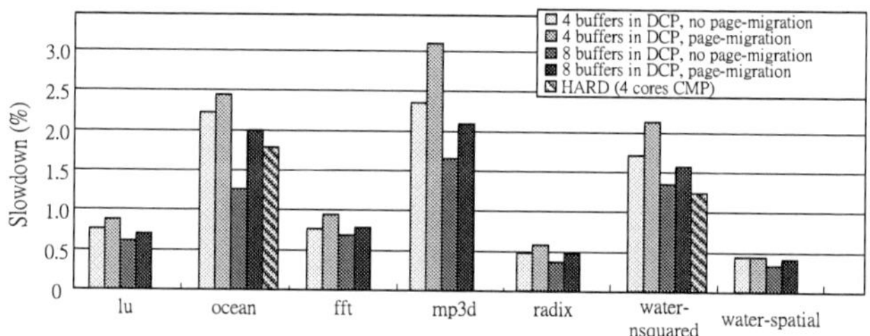

Figure 7. Slowdown effects.

Table III shows the comparison results with other approaches for race detection. The hardware approach, [11], offers fewer false positives than Valgrind and operates as fast as run-time (0.1~2.6% ↓). However, it depends on the cache and cache coherence mechanisms. We believe that this implicit cost of cache coherence is too high and infeasible in most many-core environments. Our proposed NUDA features non-intrusive truly race detection without false positives. The system offers negligible slowdown (0.51~3.06% ↓), and supports user defined assertions. Table III also shows the related hardware cost estimates for several popular many-core processors, interconnection units, and memory by CACTI 5.3 [15]. We assume that a many-core environment is composed of 64 Intel Atom processors, with 16KB I/D cache for each core, and 16MB L2-NUCA for sharing. The

Table II. Benchmarks' statistics, resource requirements, debugging event rate, and performance.

Benchamarks	Data set	Shared Data size	Max. page usage	Shared access events	Locks events (number/10^6cycles)	Inter-/Intra-cluster	Slowdown	Buffer-full stall	Migration stall
lu	512 x 512	132KB	41	10.67%	0.002144	0.04% / 99.6%	0.814%	0.753%	0.061%
ocean	130 x 130	132KB	63	1.8%	432.2	19.3% / 80.7%	2.5%	0.525%	0.975%
fft	256K	396KB	52	3.3%	0.00322	64.9% / 35.1%	0.93%	0.068%	0.25%
mp3d	-	132KB	64	2.66%	0.000133	2.7% / 97.3%	3.06%	2.24%	0.82%
radix	4M keys	396KB	33	9.4%	0.00543	1.2% / 98.8%	0.51%	0.486%	0.024%
water-nsqared	512	132KB	61	4%	0.00771	49.2% / 50.8%	2.2%	1.82%	0.38%
water-spatial	512	132KB	40	1.2%	0.00802	37.56% / 62.44%	0.422%	0.225%	0.197%

* Rate of Shared access events = #shared memory access/#total memory access, Inter-/Intra-cluster, Slowdown, Buffer-full stall, Migration stall = cycles / total execution cycles

Table III. Comparisons of race detection methods (64-core).

	Valgrind[*] [8]	HARD[†] [11]	NUDA[‡] (This Work)
Slowdown	60-400X	0.1-2.6%	0.1-3.06%
Non-intrusive	No	Probably Yes	Yes
False positive/negative	Positive	Both	None
User-defined assertions	No	No	Yes
Cache coherency	No	Required	No
Memory-bit usage[§]	No	64KB extra in L1\$, 1MB extra in L2\$	64KB extra SRAM, 1KB extra CAM
Area estimation[¶] (65nm)	0.0%	5.53mm^2 (0.98%[\\])	2.08mm^2 (0.37%[\\])

* Valgrind works on Intel Core2 Quad CPU 2.5G RAM 2G workstation with Linux kernel 2.6.24

† The SPEC of HARD [11] (normalize to 65nm): 2B BFVector/L1 \$line (1.94mm^2), 2B BFVector/L2 \$line (3.58mm^2), the estimated area: 5.53 mm^2.

‡ The proposed SPEC of NUDA (normalize to 65nm): 8 NUDA clusters, 8KB SRAM (8*0.19mm^2), 32entries CAM (8*0.041mm^2) in each NUDA node (C:32/P:1KB/B:32B), 2B-width local Dbus in each NUDA cluster (0.1mm^2), ring interconnection (9 nodes, 2 rings, 1B-width/per-ring) (0.135mm^2), the estimated area: 2.08 mm^2.

§ Memory ref (cacti 5.3v, 65nm): 32B/line 2048-entry DRAM: 0.34mm^2, 2B/line 2048-entry DRAM: 0.014mm^2, 2B/line 128-entry SRAM: 0.0076mm^2, 8B/line 1024-entry SRAM: 0.19mm^2, 4B/line 32-entry CAM: 0.041mm^2

¶ The assumed SPEC of many-core SOC (normalize to 65nm): 64 Atom cores with 16KB 4-way 32B/line I/D-cache (6.1mm^2 per core), 16MB 8-way 32B/line 256 banks L2-NUCA with Mesh NoC (174.4mm^2), the estimated area: 564 mm^2.

\\ Many-core chip ref: Intel's 80-Core (80 tiles, 65nm): 275mm^2, 0.5mm^2 (per router), The CELL processor (12 tiles, 90nm): 235mm^2, The UltraSPARC T1 (14 tiles, 90nmm,): 378mm^2,The Atom (1 tiles, 45nm): 25mm^2, EIB in CELL (12 nodes, 4 rings, 16B-width/per-ring, 90nm): 5.98mm^2.

total estimated area is **564** mm^2 using 65nm technology. Comparing HARD and NUDA, the main factors in our estimates were memory-usage and interconnections. Other elements (logic, FSM, etc.) do not impact chip area. In HARD, the BFvector (2B/per line) in the L1 cache is estimated from SRAM usage, while the BFvector (2B/per line) in the L2 cache is estimated from DRAM usage. There are 8 NUDA nodes in our proposed system. In addition, the width of the ring interconnection that links all 9 nodes (include the many-core ICE) is 1B, and we also include the ring routers. As shown in Table III, the HARD has **0.98%** (5.53/564) of the area cost compared with the proposed many-core system, and this work has **0.37%** (2.08/564) overhead.

Table V. Race detection by renaming locks.

Benchmarks	Valgrind [8]	NUDA
lu	48688 (errors)	118
ocean	128953(errors)	354
fft	258129(errors)	98
mp3d	-	225
radix	535020(errors)	454
water-nsqared	15375(errors)	55
water-spatial	6159(errors)	34

Table V compares the race detection results. We modified the SPLASH2 macro to create delta locks, which involve assigning different locks every time a lock is acquired. The software, [8], produced too many false positives and the execution speed is slower. This work do not compare with HARD duo to can't precise to model HARD in 64 cores system, but according to the paper of HARD, that reports false positive and false negatives on account of Bloom filter and L1 cache misses. In addition, we emphasize that NUDA is not only a race detector but is also a debugging platform.

6. CONCLUSION

In this paper, we have presented a novel, configurable and efficient non-uniform architecture for many-core debugging. A maximum of four cores can connect with a simple local debugging bus; the debugging bus in turn connects to the NUDA node. Each such node can connect to a maximum of two debugging buses. Each node can handle race detection and user specific tasks, such as event tracing, execution sequential verification, and so on. Our results show that the average overhead of NUDA on a 64-core system is 3.6% when detecting races.

References

[1] McDowell, C. E. and Helmbold, D. P. 1989. Debugging concurrent programs. *ACM Comput. Surv.* vol. 21, no. 4, 593-622. (Dec. 1989),

[2] Borkar, S. 2007. Thousand core chips: a technology perspective. In *Proceedings of the 44th Annual Conference on Design Automation* (San Diego, California, June 04 - 08, 2007).

[3] Standard Debug Interface Socket Requirements For OCP-Compliant SoC. http://www.ocpip.org/

[4] Tang, S. and Xu, Q. 2008. In-band cross-trigger event transmission for transaction-based debug. In *Proceedings of the Conference on Design, Automation and Test in Europe* (Munich, Germany, March 10 - 14, 2008).

[5] Huh, J., Kim, C., Shafi, H., Zhang, L., Burger, D., and Keckler, S. W. 2005. A NUCA substrate for flexible CMP cache sharing. In *Proceedings of the 19th Annual international Conference on Supercomputing* (Cambridge, Massachusetts, June 20 - 22, 2005).

[6] Savage, S., Burrows, M., Nelson, G., Sobalvarro, P., and Anderson, T. 1997. Eraser: a dynamic data race detector for multi-threaded programs. In *Proceedings of the Sixteenth ACM Symposium on Operating Systems Principles* (Saint Malo, France, October 05 - 08, 1997).

[7] Ronsse, M. and De Bosschere, K. 1999. RecPlay: a fully integrated practical record/replay system. *ACM Trans. Comput. Syst.* vol.17, no. 2, 133-152, (May. 1999).

[8] Nethercote, N. and Seward, J. 2007. Valgrind: a framework for heavyweight dynamic binary instrumentation. In *Proceedings of the 2007 ACM SIGPLAN Conference on Programming Language Design and Implementation* (San Diego, California, USA, June 10 - 13, 2007).

[9] Xu, M., Bodik, R., and Hill, M. D. 2003. A "flight data recorder" for enabling full-system multiprocessor deterministic replay. In *Proceedings of the 30th Annual international Symposium on Computer Architecture* (San Diego, California, June 09 - 11, 2003).

[10] Hower, D. R. and Hill, M. D. 2008. Rerun: Exploiting Episodes for Lightweight Memory Race Recording. In *Proceedings of the 35th international Symposium on Computer Architecture* (June 21 - 25, 2008).

[11] Zhou, P., Teodorescu, R., and Zhou, Y. 2007. HARD: Hardware-Assisted Lockset-based Race Detection, In *Proceedings of the International Symposium on High Performance Computer Architecture (HPCA)*, 121-132, 2007.

[12] ARM CoreSight On-chip Debug and Real-time Trace http://www.arm.com/

[13] Nguyen, A.-T., Michael, M., Sharma, A., and Torrellas, J. 1996. The Augmint multiprocessor simulation toolkit for Intel x86 architectures. In *proceedings of Computer Design: VLSI in Computers and Processors*, (Oct 7-9, 1996)

[14] Woo, S. C., Ohara, M., Torrie, E., Singh, J. P., and Gupta, A. 1995. The SPLASH-2 programs: characterization and methodological considerations. In *Proceedings of the 22nd Annual international Symposium on Computer Architecture* (S. Margherita Ligure, Italy, June 22 - 24, 1995).

[15] CACTI: An Integrated Cache Timing, Power, and Area Model http://www.ece.ubc.ca/~stevew/cacti/

Efficient Smart Sampling based Full-Chip Leakage Analysis for Intra-Die Variation Considering State Dependence

Vineeth Veetil, Dennis Sylvester, David Blaauw, Saumil Shah*, Steffen Rochel*

EECS Department, University of Michigan, Ann Arbor, MI

Blaze DFM, Sunnyvale, CA*

{tvvin,dennis,blaauw}@eecs.umich.edu,saumil@umich.edu,s.rochel@ieee.org

Abstract

Leakage power minimization is critical to semiconductor design in nanoscale CMOS. On the other hand increasing variability with scaling adds complexity to the leakage analysis problem. In this work we seek to achieve tractability in Monte Carlo-based statistical leakage analysis. A novel approach for fast and accurate statistical leakage analysis considering inter-die and intra-die components is proposed. We show that the optimal way to select samples, to capture intra-die variation accurately, is according to the probability distribution function of total process variation. Intelligent selection of samples is performed using a Quasi Monte Carlo technique. Results are presented for benchmarks with sizes varying from approximately 5,000 to 200,000 gates. The largest benchmark with 198461 gates is evaluated in 3 minutes with the proposed approach compared to 23 hours for random sampling with comparable accuracy. Compared to a conventional analytical approach using Wilkinson's approximation, the proposed technique offers superior accuracy while maintaining efficiency. State dependence and multiple sources of variation are considered and the approach is scalable with number of process parameter variables for standard cell characterization cost. We also show reduction in sample size to meet target accuracy for computing leakage distribution due to the inter-die component only when compared to random selection of samples.

Categories and Subject Descriptors

J.6 **[Computer Applications]** Computer-Aided Design - *computer-aided design (CAD)*.

General Terms

Algorithms, Verification

Keywords

Monte Carlo, Variance reduction, Statistical leakage

1. Introduction

As circuit design moves to smaller technology nodes the standby power dissipation of devices is of critical concern. According to the 2007 ITRS roadmap circuit leakage control is a challenge for both high performance and low power design. For high performance design increasing design complexity and leakage scaling makes it difficult to control static power while meeting performance requirements. For low power design increase in leakage power is most challenging. Thus accurate and efficient leakage analysis is crucial for the designer. On the other hand increasing process variation with scaling adds complexity to leakage analysis. A promising solution is to perform statistical analysis of leakage and use this to guide leakage optimization and design changes.

Current approaches to calculate full-chip leakage power can be classified into two main categories. The first category of methods are analytical in nature. These attempt to model full chip leakage using a standard distribution, most commonly a lognormal distribution. The moments of this distribution are computed by matching moments with an expression involving summation of leakage distributions at the gate level[1-4]. In [1] a lognormal distribution is used to approximate the leakage current of each gate and the total leakage is obtained by summing the lognormals. A low rank quadratic approximation to capture non-lognormal leakage distributions is proposed in [2]. It is noted that a 20% error is observed when modeling leakage distributions as purely lognormal using a linear approximation. The authors in [3] attempt to capture high level characteristics of a candidate chip design for early mode leakage estimation. In [4] the authors propose a systematic characterization of leakage related parameter variations. A quadratic model of the logarithm of leakage current is also proposed. Traditionally these approaches have provided the desired accuracy. However they make assumptions about either the nature of the statistical distribution of process variation parameters or the nature of the dependence of standard cell leakage on the underlying variables for handling process variation. The process variation parameters are assumed to have a standard distribution, most commonly Gaussian, or the logarithm of standard cell leakage is assumed to be a linear or quadratic sum of the variables modeling process variation. It is not clear that these assumptions will still hold true considering secondary effects in process variation and a growing number of variation sources at technology nodes below 45nm.

The second category of methods fall into the classification of Monte Carlo based techniques involving selection of samples in the process variation space and using these samples to compute leakage distribution. Monte Carlo techniques can handle non-standard distribution of process parameters and lookup tables for dependence of standard cell leakage on process variables. Therefore they do not require simplifying assumptions about the dependence of leakage on process parameters or the nature of process parameter distribution, making them highly scalable. Also the inherent parallelism in evaluating Monte Carlo samples make these techniques amenable to multi-core and Graphics Processing Unit (GPU) computing. However Monte Carlo techniques typically require a large sample size rendering them expensive. There is a need for smart selection of samples to reduce the number of samples that require evaluation without compromising accuracy. In [5] the author describes such techniques, known as variance reduction techniques. These techniques need to be tailored to the system under consideration for efficient reduction in sample size. In the context of integrated circuits it has been shown that a suitable choice of these techniques can lead to significant sample size reduction for statistical timing analysis[6]. We study the applicability of such techniques for leakage analysis in this work.

There are two main contributions in this paper. To the best of our knowledge this work is the first to study sample size reduction for statistical leakage analysis using a Monte Carlo based approach. We consider intra-die variation, state dependence and multiple sources of process variation. Second, we address the issue of standard cell characterization, which is largely ignored in literature. Statistical circuit leakage analysis involves characterization of standard cells at grid points in the process variation space. This is illustrated in the schematic for a traditional flow in Figure 1. Although characterization is only performed once in the design flow for a library the number of grid points grows exponentially with the number of process variation parameters. There is a need to select samples to reduce characterization cost while meeting target accuracy in leakage analysis.

We first consider the problem of leakage analysis for the case of inter-die variation involving multiple process variation parameters. For this we

Permission to make digital or hard copies of part or all of this work for personal or classroom use is granted without fee provided that copies are not made or distributed for profit or commercial advantage and that copies bear this notice and the full citation on the first page. To copy otherwise, to republish, to post on servers or to redistribute to lists, requires prior specific permission and/or a fee.

DAC'09, July 26-31, 2009, San Francisco, California, USA

978-1-60558-497-3/09 $25.00 © 2009 ACM

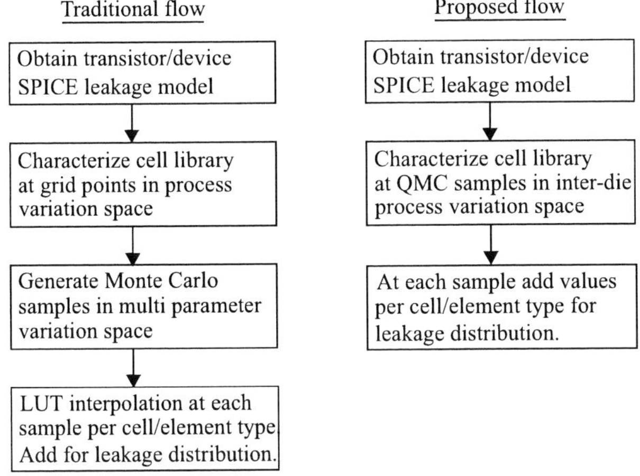

Figure 1. Traditional and proposed leakage analysis flow for global variation with multiple sources.

propose to use a Quasi Monte Carlo technique [7] for selecting samples in the process variation space. We show that for a large benchmark circuit there is significant reduction in sample size to meet target accuracy when compared to a random selection of samples for computing leakage distribution. Standard cell characterization needs to be performed only at these samples which reduces the cost of standard cell leakage characterization. Next we propose a solution for the case of the total leakage distribution considering inter-die and intra-die components which is the major contribution of this paper. We recognize that this problem can be formulated as selecting samples for inter-die variation and computing the local distributions at each of these samples due to intra-die variation. Computation of the moments of the local distribution requires additional samples in the neighborhood of each inter-die sample. The number of these additional samples can be prohibitively high. We propose techniques for efficient selection of the samples. The key ideas are as follows. First we show that the optimal way to select samples to compute local distributions accurately is to select samples according to the probability distribution function of total process variation. Second, the selection of samples is performed intelligently by using the Quasi Monte Carlo technique. Experiments are performed on benchmark circuits synthesized in a 45nm commercial technology. State dependence information is also considered. We compare our technique with 3 approaches 1) random sampling, 2) a technique referred to as *Method1,* and 3) a traditional analytical approach based on [1]. *Method1* involves smart selection of inter-die samples but no intelligence or reuse of samples for intra-die variation. For the largest benchmark considered with 198461 gates, the proposed approach requires 3 minutes whereas random sampling and *Method1* complete the task in 23 hours and 18.4 hours, respectively. We also achieve accurate results for estimation of μ, σ, and the 95th percentile of chip leakage distribution for all benchmarks considered with low runtime.

The paper is organized as follows. Section 2 describes Quasi Monte Carlo approach, which is a standard technique to reduce sample size for Monte Carlo analysis. Section 3 proposes a leakage analysis technique for the case of inter-die variation using a Quasi Monte Carlo technique. Section 4 addresses leakage analysis for total leakage analysis involving inter-die and intra-die variation using smart samples. Results and conclusions are presented in Sections 5 and 6 respectively.

2. Smart Sampling for Leakage Analysis

Monte Carlo-based leakage analysis involves selecting samples in the process variation space to obtain a statistical distribution of circuit leakage. This is mapped to the standard mathematical problem of Monte Carlo (MC), which is to estimate the integral of a function using samples in its domain. There are standard techniques for variance reduction of

MC, including Quasi Monte Carlo techniques. These techniques are detailed in [5].

2.1 Quasi Monte Carlo

The standard Monte Carlo (MC) method addresses the problem of approximating the integral of a function $f(x)$ over the s-dimensional hypercube $C^s = [0, 1)^s$ where x represents a point in an s-dimensional space. The MC estimate of the integral is given by the arithmetic mean of f_i which are values of the function $f(x)$ evaluated at n samples distributed throughout the hypercube. The error bound of a method to numerically estimate an integral using a sequence of samples is mathematically related to a measure of uniformity for the distribution of the points called "discrepancy". A sequence with the smallest possible discrepancy has the property that when used to evaluate the mean it achieves the smallest possible error bound. Sequences constructed to reduce discrepancy are called Low Discrepancy Sequences (LDSs). Quasi Monte Carlo techniques are characterized by their use of LDSs to generate samples. LDSs are deterministic sequences, i.e., there is no randomness in their generation. Intuitively these sequences are well dispersed through the domain of the function, minimizing any gaps or clustering of points. Figure 2 illustrates that quasi random sequences generate samples with lower discrepancy compared to pseudo random sequences (sequences with properties similar to "truly" random sequences). Sobol, Faure, and Niederreiter are LDSs that have been studied extensively. In this work we consider Sobol sequences, which are known to be simple to construct. Interested readers can refer to [7] for a construction of the Sobol sequence. In the context of circuits Quasi Monte Carlo techniques have been studied for statistical timing analysis [6] where results indicate that the techniques are a good fit and are amenable to multi-core and GPU computing. This work is the first to study the application of Quasi Monte Carlo (QMC) techniques for statistical leakage analysis.

3. Leakage Analysis for Inter-Die Variation with Smart Sampling

In this section we first describe the steps in an industrial leakage analysis flow. A typical industrial flow circuit leakage analysis involves characterization of a standard cell library and computation of circuit leakage using the characterized data as explained in Section 3.2. Further we introduce our approach to estimation of statistical leakage due to inter-die parameter variation to achieve tractability for multiple sources of process variation.

3.1 Process Variation Model

Process variation parameters such as critical dimension (CD) and oxide thickness exhibit correlations. To account for correlations between parameters principal component analysis (PCA) is performed. Critical dimension, threshold voltage and oxide thickness are thus expressed as linear combinations of principal components. For process technology nodes 45nm and below some foundries provide such statistical information with principal component analysis. Now process variation models with inter-die and intra-die components are widely used in the literature [1]. Each process variation parameter has a global or inter-die component, which is modeled by a single random variable for a parameter in a die. Intra-die components account for spatial correlation within the die and uncorrelated random variation per device. In this model the die is partitioned into $n * n$ grids and identical parameter variations are assumed within a grid. Therefore, each source of variation is represented by a set of random variables, one for each panel in the grid. For example, transistor gate length variation is represented by a set of random variables for all grids and the set is of multivariate normal distribution with covariance matrix R_{Lg}. As mentioned above the process variation parameters have been resolved into principal components. It follows that each component is represented by a set of random variables for all grids. Principal component analysis (PCA) is again performed on these spatially correlated variables. In addition an independent random variable

978-1-60558-497-3/09 $25.00 © 2009 ACM

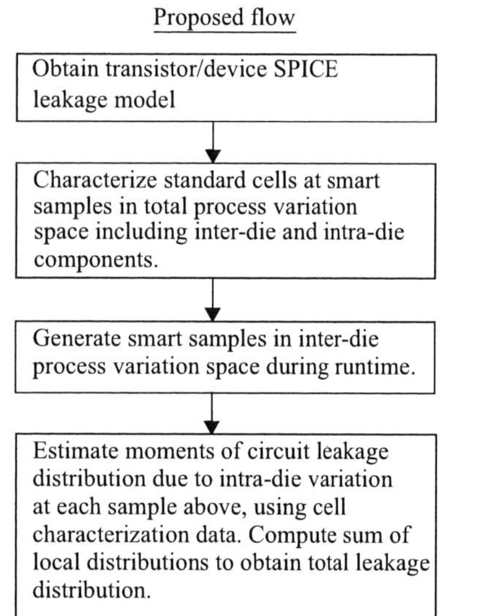

Figure 3. Proposed leakage analysis flow for within-die variation.

accounts for random variation at the device level for components resulting from PCA on process variation parameters.

3.2 Traditional Leakage Analysis Flow for Inter-Die Variation

The standard cell library is characterized for leakage information at grid points in the process variation space. To include state dependence information, standard cells are characterized at the grid points for each input state. If state dependence is not considered then an average of the leakages for all input states is computed.

In a traditional Monte Carlo-based leakage analysis flow (to account for inter-die parameter variation) process parameter variables or their principal components are sampled. As only global variation is considered the same sample set is assigned to every element type in the standard cell library. The leakage value per element type in the library is obtained by interpolation in the leakage lookup table for the element type. The circuit leakage is obtained by adding up the leakage value obtained for each element type after weighting by the number of occurrences of the element type in the circuit.

The above approach does not consider state dependence of standard cell leakage. To enable leakage calculation to account for state dependence, the standard cell characterization data must have leakage information for every cell state as mentioned above. In addition, at the circuit level state probability information is required for every instance of each element type in the circuit. Various approaches exist in the literature to arrive at an estimate of state probability for each instance. For a detailed discussion on this topic refer to [8].

3.3 Proposed Leakage Analysis Flow with Smart Sampling

We propose to use Quasi Monte Carlo based sampling for standard cell library characterization and runtime leakage analysis. In a traditional flow standard cells are characterized at discrete grid points in the space of random variables to model process variation as explained in Section 3.2. In the proposed approach the characterization is performed at samples generated using a Quasi Monte Carlo (QMC) based approach. In

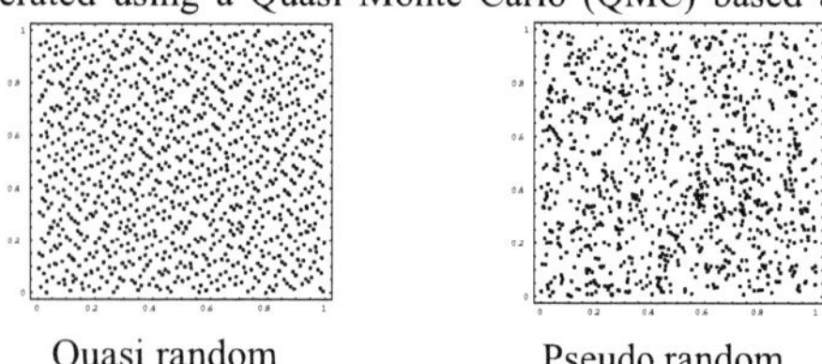

Figure 2. Quasi random and pseudo random sequences.

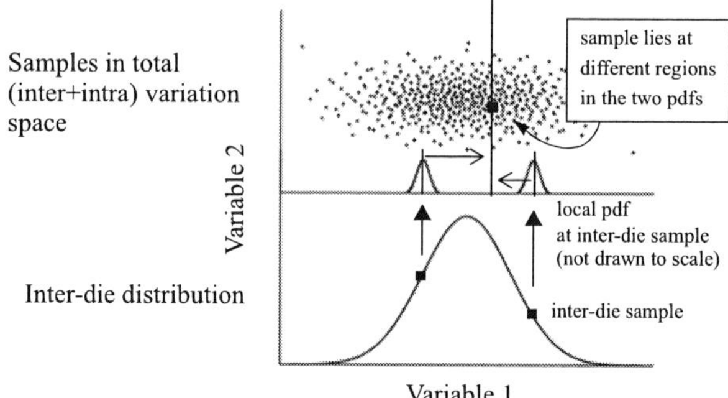

Figure 4. Reusing samples for local distribution computation. Inter-die samples weigh the samples in total variation space according to local probability distribution.

particular we use Sobol sequences in the QMC approach in this work. QMC samples refer to Sobol samples in the rest of the paper. The same process variation samples are used for characterization of all element types in the standard cell library and their states.

The proposed approach differs from a traditional flow during runtime in that samples are not generated at this stage. The inter-die samples are precomputed during cell library characterization. A given inter-die sample is assigned to every element type in the library as before and the circuit leakage is obtained by adding up the leakage values from element types as in the traditional flow. It follows that there is now no need for interpolation in the look-up table from cell characterization. The leakage values are readily available in the tables without need for interpolation. The traditional and proposed flows are illustrated in Figure 1.

4. Leakage Analysis for Total Variation with Smart Sampling

This section proposes an algorithm for estimating full-chip leakage considering inter-die and intra-die components of variation. In sub-45 nm technologies secondary effects in process variation are important and the number of significant sources of process variation is increasing. Existing approaches to calculate full-chip leakage power make simplifying assumptions about either the nature of statistical distribution of process variation parameters or the nature of dependence of the standard cell leakage on these parameters. The parameters are assumed to have a standard distribution or the logarithm of standard cell leakage is assumed to be a linear or quadratic sum of the parameters. Combined with a growing number of process variation sources this is a limitation on the accuracy. Monte Carlo based methods on the other hand are expensive when handling intra-die variation. The proposed approach can efficiently handle any non-standard distribution of variables or dependence of full-chip leakage on these variables.

A schematic of the proposed approach for total variation is illustrated in Figure 3. Process variation consists of inter-die and intra-die components. In Section 3 we discussed generation of samples in the space of inter-die variation distributed according to the joint probability distribution of the variables involved. We apply intra-die variation to such a sample around the nominal and obtain a local leakage distribution for the circuit. The sum of these distributions from all samples should give the total leakage distribution. From a sampling perspective this translates to generating more samples distributed according to the intra-die distribution around each inter-die sample. In this way the problem of total leakage variation can be formulated as a two-level sampling problem, the first level corresponds to inter-die variation and the second level corresponds to intra-die variation at each of the samples in the first level. Using Quasi Monte Carlo sampling, accurate results for inter-die variation can be achieved with few samples as explained in Section 3. However even for a low number of first level, or inter-die, samples the total number of samples in the second level can be prohibitively high. The

978-1-60558-497-3/09 $25.00 © 2009 ACM   156

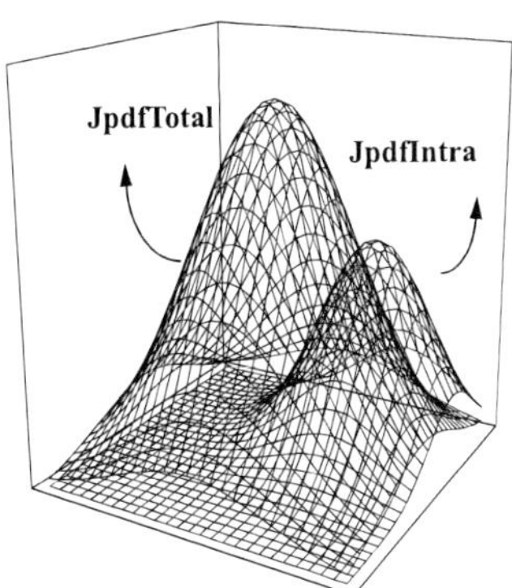

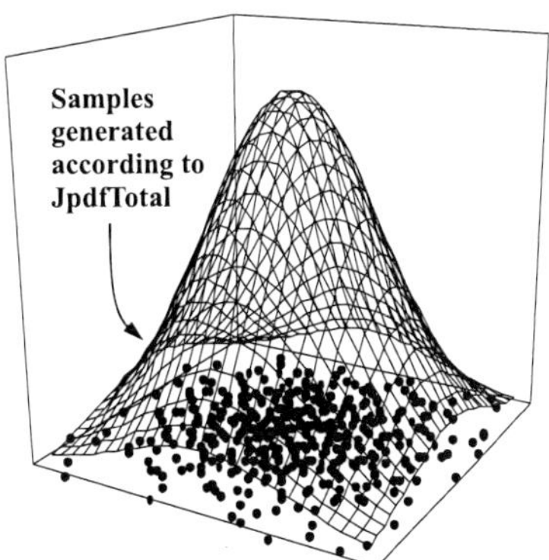

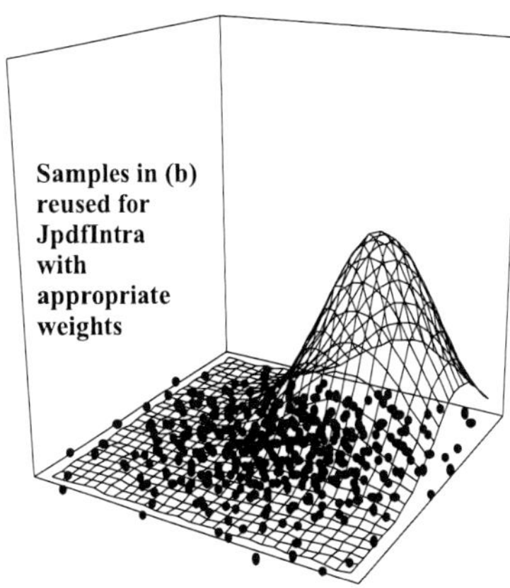

Figure 5. (a) Total (Inter+Intra die) distribution and local pdf at an inter-die sample. (b) QMC based samples are generated according to total variation. (c) For computing mean of local pdf the samples generated in (b) are weighed according to the ratio of the probabilities in the two distribution functions.

idea here is that if the second level samples are chosen optimally such that either the entire set or a subset can be used for computation at every inter-die sample the number of samples can be minimized. A uniform sampling approach in a bounded space enveloping the inter-die samples may be tried. However while this considers outliers in the inter-die distribution this does not weigh samples close to the nominal adequately. The problem is to arrive at a pdf which is optimal for all samples.

Consider the first level or inter-die samples in the process variation space. The problem is to find a pdf for optimality in computation at every inter-die sample. Such a pdf is obtained by summation of the pdfs of local distributions at the inter-die samples. Now we have the surprisingly simple result that if the number of inter-die samples is large enough the summation of the pdfs converges to the pdf for the distribution obtained for total variation with inter-die and intra-die components. The proof has been omitted for brevity. Our experiments indicate that if the inter-die samples are chosen according to a Sobol sequence and the sample size is large enough (typically more than 100) this is indeed true. Therefore we select the second level samples according to the pdf for total variation. To minimize the number of second level samples we use Sobol sequences to sample in this space.

For the case of no spatial correlation the idea is illustrated in Figure 4 where two samples are shown on the inter-die distribution. The second level samples are chosen to be Quasi Monte Carlo based samples in the total process variation space. One such sample in Figure 4 lies in different regions of the pdf for the two inter-die samples. Therefore the first level samples assign different weights to the leakage values obtained at a particular second level sample. The characterization step needs to compute leakages for standard cells at the second level samples only. The procedure to reuse samples is illustrated in Figure 5 for the case of a 2D process variation space. Figure 5a shows the total process variation distribution along with the local distribution at an inter-die sample S. Figure 5b shows the second level samples generated in the total distribution space $x_i : i=1...N$. These samples are reused for computation of moments of local distribution at S as in Figure 5c. In particular the mean of local distribution at S for the circuit, $\overline{L}(S)$ is given by

$$\overline{L}(S) = \sum_{i=1}^{N} \frac{L(x_i) \times JpdfIntra(x_i - S)}{JpdfTotal(x_i)} \quad (1)$$

where *JpdfIntra* is the probability distribution for intra-die variation and *JpdfTotal* is the probability distribution for total variation. Similarly higher moments for the local distribution can be computed. The total leakage distribution is a sum of local leakage distributions and is computed using $\overline{L}(S)$ and the higher moments obtained for all samples. In the case of spatially correlated intra-die variation, the sample for one variable is not a single value but a set of values corresponding to grids in the spatial correlation model. This means that each element of vector x_i in (1)

is not a scalar but a vector with correlated elements. The number of elements in this vector is equal to the number of grids. The functions *JpdfIntra(x_i-S)* and *JpdfTotal(x_i)* are modified to include the spatial correlation. We explore spatial correlation later and show that this level of modeling process variation is not needed for large circuit and full-chip leakage analysis.

Now the local distribution corresponding to one sample can be approximated using Central Limit Theorem abbreviated as CLT [9]. If spatial correlation is not considered then this local distribution has contribution from sum of identical independent random variables from instances of a given element type in the cell library. If there are enough instances the local distribution approaches a normal distribution. Also for a large number of instances the variance of this distribution approaches zero according to CLT. This means that the local distribution approaches a single number which is the mean of the distribution. In the presence of spatial correlation as long as there are sufficient independent regions in a die, i.e., the circuit is large enough the Central Limit Theorem can be applied [10] as if all intra-die variation was uncorrelated. A reduction in variance of the local distribution translates to a reduction in the number of second level or additional samples for a target accuracy. For large circuit blocks and chips the problem essentially is to compute only the mean of the local distribution at each inter-die sample. For circuits where spatial correlation has a significant effect on leakage distribution, the technique can still be applied. The local distribution within a grid panel has contribution from the sum of identical independent random variables from instances of a given element type in the cell library. Therefore the local distribution within each grid panel approaches a normal distribution with number of instances in the panel, which reduces the number of additional samples, with spatial correlation considered, to capture the local distribution.

5. Results

Our simulation results are based on a 45nm commercial technology. Principal component analysis is used to obtain principal components for the correlated process variation parameters including CD, oxide thickness and threshold voltage. Simulations are performed on industrial circuits with sizes ranging from approximately 5000 to 200,000 gates. In our implementation, we only consider inter-die variation and uncorrelated intra-die variation. Spatially correlated intra-die variation is not implemented. In the presence of spatial correlation as long as there are sufficient independent regions in a die, i.e., the circuit is large enough, the Central Limit Theorem can be applied as explained in [10] and therefore the results are accurate for large circuit blocks and chips. This is illustrated in Figure 6 for a benchmark circuit with approximately 43,000 gates. The standard deviation of the leakage distribution without considering spatial correlation is compared to the case where a grid-based spatial correlation model is considered. The total standard deviation for

978-1-60558-497-3/09 $25.00 © 2009 ACM

Table 1. Comparison of proposed approach with Golden (Monte Carlo 20,000 samples) for benchmarks. * indicates that state probability is considered for instances in the circuit.

		Golden (Monte Carlo 20k samples)				Proposed approach				Error (%)			
	Gate count	μ (mW)	σ (mW)	95th percentile (mW)	Runtime	μ (mW)	σ (mW)	95th percentile (mW)	Runtime (s)	μ	σ	95th percentile	Speed up
VD1	5536	0.51	0.18	0.85	1.7 hours	0.51	0.17	0.87	1.77 seconds	0.03	3.55	2.21	3405
VD2	13258	1.21	0.42	1.99	4.8 hours	1.20	0.40	2.04	1.83 seconds	0.30	2.61	2.51	9495
USB	15946	1.11	0.36	1.79	7.4 hours	1.11	0.36	1.85	1.95 seconds	0.01	1.97	3.35	13738
ETHER	23939	1.40	0.46	2.26	10.2 hours	1.40	0.45	2.33	2.09 seconds	0.06	1.99	3.10	17633
VGA	43214	2.85	0.98	4.71	15.6 hours	2.84	0.96	4.85	2.02 seconds	0.49	2.31	2.97	27778
*Chip1	198461	10.63	2.67	15.59	19.2 days	10.64	2.63	15.96	278 seconds	0.10	1.71	2.37	5969

intra-die variation is the same in both cases. The assumption of no spatial correlation accurately estimates standard deviation for number of grid panels above 256, supporting the argument in [10]. Therefore, spatial correlation is not a limitation for circuit blocks and chips with practical sizes for the current implementation, which is our focus in this work. The modification in the algorithm for the case of smaller circuits is discussed in Section 4.

Figure 7 shows the result for our proposed approach for inter-die parameter variation using smart samples. The smart samples are obtained from a Sobol sequence. The error in estimating σ of leakage distribution for inter-die variation using smart samples is compared with a random sampling based approach for a VGA circuit with approximately 43,000 gates. The golden value is obtained from a Monte Carlo simulation with 20,000 samples. We compare the minimum sample size required to achieve target accuracy of 3% error in estimating σ for both methods. The proposed approach requires 9.3X fewer samples compared to random sampling. In a typical industrial flow the standard cells are characterized at grid points in the process variation space. With 7 grid points chosen for each of the three principal components in our implementation the number of points to be characterized is 343 in a traditional flow whereas the proposed approach requires only 150 from Figure 7, a 56% reduction in standard cell characterization overhead.

We now present our results for total process variation considering both inter-die and intra-die components of variation. Table 1 shows results comparing the proposed approach with 20,000 Monte Carlo runs on benchmark circuits. The metrics compared are mean μ, sigma σ, and the 95th percentile of the circuit leakage distribution. The errors in estimating these metrics for the largest benchmark circuit *Chip1* are less than 3%. The errors in estimating the metrics are less than 3.6% for all the benchmark circuits. Note that there is higher accuracy for the largest benchmark studied. The proposed approach has a runtime of less than 3 minutes for the largest benchmark, which illustrates the runtime efficiency. The larger runtime for *Chip1*, even accounting for the larger circuit size, is attributed to the fact that state probability information is only considered for this circuit. State probability consideration for each instance adds significant cost to the computation.

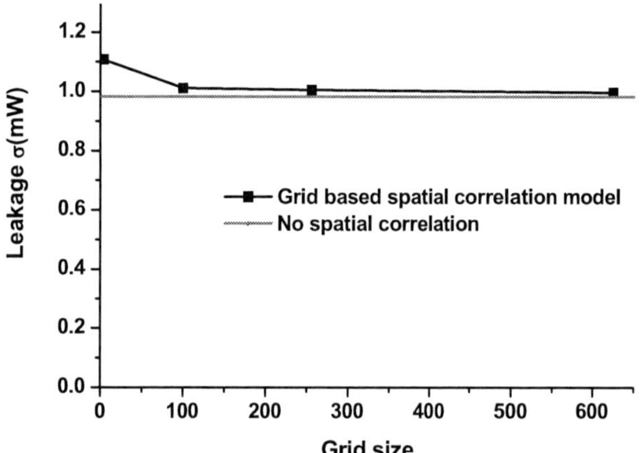

Figure 6. Comparison of sigma of leakage distribution without considering spatial correlation with that of a grid-based spatial correlation model for VGA circuit (43214 gates)

Figure 8 plots the accuracy against runtime of the proposed approach and a random sampling approach. We also compare this with the result

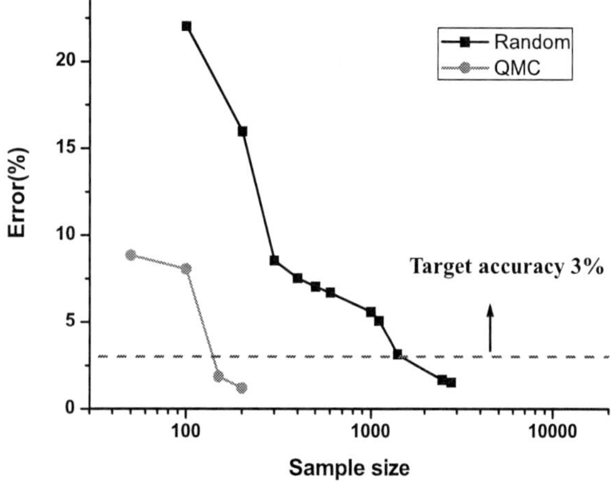

Figure 7. Comparison of error in estimating σ of leakage distribution for inter-die variation using QMC vs random sampling for VGA circuit(43214 gates).

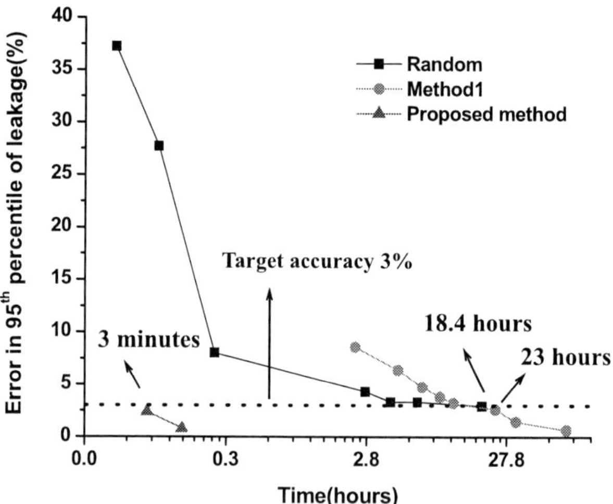

Figure 8. Comparison of accuracy of proposed approach with random sampling based approach vs runtime. The circuit considered is *Chip1* with 200,000 gates.

978-1-60558-497-3/09 $25.00 © 2009 ACM

for another smart sampling based technique called *Method1*. As explained in Section 4 the proposed approach first generates inter-die samples using smart sampling. In the next step a smart selection of samples in the total variation space is coupled with reuse of these samples to compute the mean of local leakage distributions at inter-die samples. In *Method1* inter-die samples are generated using a Sobol sequence as in the proposed approach. However a random sampling based Monte Carlo analysis is performed at each inter-die sample to obtain the local distribution. In other words there is no intelligence or reuse of samples in total variation space, however as inter-die samples are generated using smart samples this method is expected to be faster than random sampling. Figure 8 shows that the proposed approach has a runtime of less than 3 minutes to achieve target accuracy for the largest benchmark whereas *Method1* has a runtime of 18.4 hours. This result illustrates the advantage of smart sampling and reuse of the additional samples in the total variation space. The random sampling approach has a runtime of 23 hours. It may be noted that the slope of the curve for *Method1* is steep in the beginning compared to the rest of the curve. This is because in *Method1* the number of inter-die samples is increased in the beginning till the inter-die component of variation is captured accurately. After that only the number of random samples to capture the local distribution is increased while keeping the number of inter-die samples constant, hence the decrease in slope. The slow convergence of random sampling to capture local distribution is the reason for comparable runtimes of *Method1* and random sampling.

Table 2 compares the proposed approach with an analytical approach to compute leakage distribution based on [1]. In [1] the authors approximate the logarithm of gate leakage as a linear expression involving process variation variables. Wilkinson's approximation is used to compute sum of lognormals to obtain circuit leakage as a lognormal expression. From Table 2 the maximum error in estimating μ is 3.7% for the analytical approach compared to 0.5% for the proposed approach. Similarly the maximum error in estimating σ is 6.1% for the analytical approach compared to 3.6% for the proposed approach. It may also be noted that the proposed approach incurs less error as circuit size increases but no such trend is observed for the analytical approach. For the largest benchmark *Chip1* state dependence has been implemented for both methods. The errors in estimating μ and σ are significantly lower for the proposed approach in this case as illustrated. As mentioned before the runtime for *Chip1* is significantly higher compared to other circuits, even accounting for circuit size because state probability information of instances is considered in this circuit. In the case of the analytical approach the increase in time cost is much higher because the dependence on number of states is quadratic.

Figure 9 compares the total leakage distribution of the largest benchmark circuit with 200,000 gates for the proposed approach with the golden and the analytical approach based on [1]. The leakage variation considering only inter-die variation is also plotted. This analysis considers state probability information for instances in the circuit. The state probability information is extracted using a commercial tool. We see that

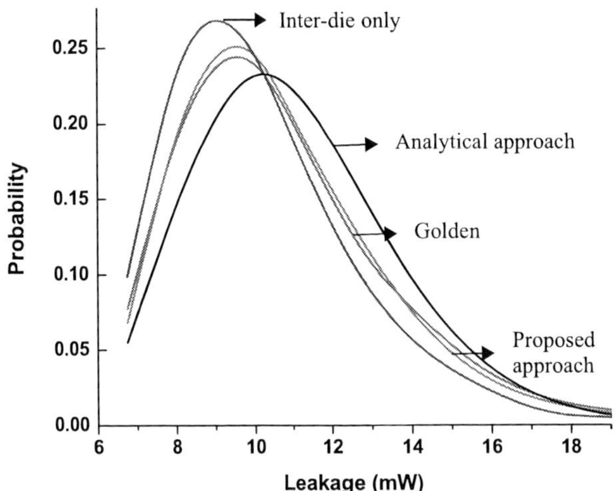

Figure 9. Total leakage distribution considering intra-die variation for *Chip1* (200,000 gates). Proposed approach is compared with the analytical approach based on [1] and the golden. The distribution due to inter-die variation is also plotted. The analysis considers state dependence of leakage for the instances in the circuit.

the distribution curve is captured with accuracy by the proposed approach whereas there is significant error with the analytical approach.

6. Conclusions

Monte Carlo-based techniques are promising for statistical leakage analysis because of the generality and scalability of the approach even when complex relations exist between leakage and process parameters. This work addresses the problem of reducing the sample size for Monte Carlo based leakage analysis. When considering only inter-die variation for a large benchmark circuit the sample size is reduced by 9.3X compared to a random sampling approach to achieve target accuracy. The standard cell characterization cost is also reduced by 56%. We also propose a solution to estimate the total leakage distribution considering inter-die and intra-die components. A novel technique involving smart sampling combined with reuse of samples is introduced to address this issue. The proposed approach is compared with random sampling, *Method1* where samples are not reused, and an analytical approach. For the largest benchmark considered the proposed approach performs the computation in 3 minutes whereas the random sampling approach and *Method1* complete the task in 23 hours and 18.4 hours, respectively. The analytical approach has up to 3.7% and 6.1% in approximating μ and σ compared to 0.5% and 3.6% for the proposed approach. In addition the characterization cost for the total leakage distribution is scalable with respect to the number of process variation variables since Quasi Monte Carlo sample size increases moderately with the number of variables whereas in a traditional grid-based characterization approach the cost grows exponentially with the number of process variation variables.

Table 2. Comparison of proposed approach with Wilkinson's based approach. * indicates that state probability information is considered for instances in the circuit.

Circuit	Gate Count	Proposed approach			Wilkinson's approach		
		% Error		Run-time(s)	% Error		Run-time(s)
		μ	σ		μ	σ	
VD1	5536	0.03	3.55	1.77	3.43	4.81	0.16
VD2	13258	0.30	2.61	1.83	3.16	1.80	0.16
USB	15946	0.01	1.97	1.95	3.62	6.13	0.19
ETHER	23939	0.06	1.99	2.09	3.69	0.53	0.20
VGA	43214	0.49	2.31	2.02	3.03	1.37	0.20
*Chip1	198461	0.10	1.71	278	3.08	5.46	3094

References

[1] H. , S.S.Sapatnekar, "Full-chip analysis of leakage power under process variations including spatial correlations", *Proc. Design Automation Conference*, pp. 523-528, 2005.

[2] X. Li, J.le, L.T.Pileggi, "Projection-based statistical analysis of full-chip leakage power with non-log-normal distributions", *Proc. Design Automation Conference*, pp. 103-108, 2006.

[3] K.R.Heloue, N.Azizi, F.N.Najm, "Modeling and estimation of full-chip leakage current considering within-die correlation", *Proc. Design Automation Conference*, pp. 93-98, 2007.

[4] T.Li, W.Zhang, Z.Yu, "Full-chip Leakage Analysis in Nano-scale Technologies: Mechanisms: Variation Sources, and Modeling", *Proc. Design Automation Conference*, pp. 594-599, 2008.

[5] R.Y.Rubinstein, Simulation and the Monte Carlo Method, John Wiley & Sons Inc., 1981.

[6] V.Veetil, D.Sylvester, D.Blaauw, "Efficient Monte Carlo based Incremental Statistical Timing Analysis", *Proc. Design Automation Conference*, pp. 676-681, 2008.

[7] I.M.Sobol, "The Distribution of Points in a Cube and the Approximate Evaluation of Integrals", *USSR Comp. Math and Math. Phys.*, 7(4), pp. 86-112, 1967.

[8] F.N.Najm, "A Survey of Power Estimation Techniques in VLSI Circuits," *IEEE Transactions on Very Large Scale Integration(VLSI) Systems*, Vol. 2, pp. 446-455, 1994.

[9] A. Papoulis, Probability, Random Variables and Stochastic Processes, McGraw-Hill Inc., New York 1991.

[10] R.Rao, A.Devgan, D.Blaauw, D.Sylvester, "Parametric Yield Estimation Considering Leakage Variability," *Proc. Design Automation Conference*, pp. 442-447, 2004.

Resurrecting Infeasible Clock-Gating Functions

Eli Arbel
IBM Haifa Research
Laboratory
Haifa, 31905, Israel
arbel@il.ibm.com

Cindy Eisner
IBM Haifa Research
Laboratory
Haifa, 31905, Israel
eisner@il.ibm.com

Oleg Rokhlenko
IBM Haifa Research
Laboratory
Haifa, 31905, Israel
olegr@il.ibm.com

ABSTRACT

In this paper we consider the problem of exploiting infeasible clock gating functions. Analysis of industrial designs reveals a large margin of potential for power saving based on clock gating functions that initially appear to be useless due to timing violation or excessive power consumption. We propose two optimization techniques for resurrecting such functions that can be used as a generic post-processing phase in an automatic clock gating tool. The first provides timing-aware approximation and the second aims at generating large gating domains by clustering similar clock gating functions. Our experimental results show that the combination of these two techniques yields an additional power saving of up to 78% in industrial designs.

Categories and Subject Descriptors

B.6.3 [**Design Aids**]: Optimization

General Terms

Algorithms, Design.

Keywords

Clock gating, Approximation, Clustering, Low power

1. INTRODUCTION

Power consumption has become one of the major concerns in modern chip design. While in the past designers were primarily concerned with chip performance, nowadays focus has been shifted towards performance-per-watt and energy efficiency. The motivation for this trend is manifold: from extending battery life of portable devices, to reducing cooling costs of data-centers and to overcoming physical problems (e.g. device variation) due to thermal hot-spots.

Power consumed by a CMOS circuit can be partitioned into two components: (1) static power, primarily caused by sub-threshold and gate leakage, and (2) dynamic power, caused by switching activity and crowbar current. Although in modern sub-micron designs static power is becoming increasingly important, dynamic power is still a major component. One of the most widely used techniques for reducing dynamic power is clock gating, and with the demand for more power-efficient designs and constantly-shrinking time to market of design projects, the EDA industry has witnessed a proliferation of automated tools for clock gating.

Given a logical representation of the design (e.g. RTL), a clock gating tool may infer many opportunities for clock gating, some of which are beneficial in terms of power saving, some of which are not because they consume more power than they save and/or violate timing constraints. One way to deal with such problematical clock gating functions is simply to discard them. However, we argue that in many cases there is still a large margin of potential for power saving based on clock gating functions that are considered problematical at first glance.

In this paper we propose two optimization techniques for exploiting problematical clock gating functions and for maximizing the power saving potential of any type of clock gating: (1) an approximation technique for efficiently reducing the size of clock gating functions, aimed at reducing their leakage as well as improving their delay, and (2) a clustering algorithm for grouping similar clock gating functions that are otherwise discarded, producing large enough hence efficient, clock gating domains. Experimental results show that the combination of these two techniques yields an additional power saving of up to 78% in industrial designs. Also the proposed techniques are not limited to a certain clock gating approach, but rather may be used as a postprocessing procedure for any clock gating algorithm, *e.g.* [7, 6, 9, 10, 13, 14, 16].

2. BACKGROUND AND PRIOR ART

Clock gating decreases dynamic power by reducing the switching activity of certain sequential elements and parts of the clock network. This is achieved by gating the clock signal of sequential elements when new values need not be loaded into them. In this paper, we use the term *activation function* to indicate the function that when high causes the clock to be gated.

An activation function f which has a high on-set probability, i.e. $P(f = 1)$ is high, is considered efficient since it gates the clock signal more often. Obviously the on-set probability of a given activation function is data dependent and can change dramatically based on the inputs and sequential behavior of the design.

Permission to make digital or hard copies of part or all of this work for personal or classroom use is granted without fee provided that copies are not made or distributed for profit or commercial advantage and that copies bear this notice and the full citation on the first page. To copy otherwise, to republish, to post on servers or to redistribute to lists, requires prior specific permission and/or a fee.
DAC'09, July 26-31, 2009, San Francisco, California, USA

Using the on-set probability as the sole metric for deciding whether or not to use an activation function for clock gating is not enough. An activation function can have high on-set probability but at the same time violate timing constraints, or waste more power than it saves due to excessive leakage. Complementary and more predictable metrics for the effectiveness and applicability of activation functions, that can be used in conjunction with the on-set probability metric, are *logic depth* and *gating domain* size.

DEFINITION 1. *The* logic depth *of a single-output combinational logic is defined as the maximal number of gates on any path from logic inputs (either design inputs or sequential element outputs) to the logic's output.*

DEFINITION 2. *A* gating domain *is a set of sequential elements sharing the same activation function.*

Note that while the on-set probability is mainly related to dynamic power, logic depth and gating domain size are related to timing and static power, respectively. In cases where the activation function is problematical, say it violates timing constraints, one would have to trade-off between the efficiency of the function and its logic depth.

An activation function may be impractical to implement for several reasons. First it may have a low on-set probability, thus may not gate the clock signal often enough to justify its circuitry. Second, an activation function whose logic depth is too large will violate timing constraints. In addition, logic depth is highly correlated to the number of gates in the circuit hence such a function will require a substantial amount of gates, increasing the overall leakage and dynamic power. Lastly, the activation function may be applicable to only a small group of sequential elements. Since in most design methodologies clock gating is implemented using a clock buffer which consumes power by itself, clock gating a small gating domain may not be beneficial for power saving.

In case the on-set probability of an activation function is too low, nothing can be done and the function should be discarded. However, functions with excessive logic depth and functions that gate only a small amount of sequential elements can still be optimized and exploited.

Several methods for reducing the size of activation functions are known in the literature. Size reduction is achieved by approximating the original function, hence these methods are referred to as *approximation* methods. The authors of [2] describe an approximation method based on the insight that the less inputs a function has the smaller its implementation. Universal quantification is used for selecting the best subset of inputs of a given activation function. While the method is oriented towards finding the tightest approximation under given constraints, it still works at the coarse granularity of removing function variables, possibly yielding sub-optimal results in terms of the final on-set probability.

In [3], the authors suggest approximating activation functions by pruning minterms with probabilities below a user specified threshold. This approach aims at obtaining high on-set probability of activation functions by making finer approximation steps than in [2]. Nevertheless, since the correlation between the minterm probabilities threshold and timing constraints is low, it is hard to control the delay of the resultant approximated function and its on-set probability using this approach.

Approximation methods based on Binary Decision Diagrams (BDDs) [4] also exist. In [8], an approximation technique whose goal is to reduce the number of BDD nodes below a given threshold is given. However the delay of the circuit is directly related to the depth of the BDD, not to its size [5], thus this approach may find solutions that still exceed the depth limits.

Obtaining large gating domains from a set of smaller ones is another fundamental issue in clock gating, however very little work has been published in this area. A method for grouping similar bits under clock gating is described in [11]. This method exploits special structures like binary counters for the grouping phase, and simulation for quantifying the benefit of clock gating each created group. Unfortunately, this approach cannot be generalized to the more common case of inferred clock gating from arbitrary logic.

A more generalized algorithm for grouping is suggested in [15]. The algorithm starts from a randomly selected function and greedily selects the next function to be merged with the current group so as to improve some optimization criterion, until there are no further unmerged functions. A major disadvantage of this approach is its inability to backtrack while enlarging the current group, making it very sensitive to the random selection of the first function in each group.

In the rest of the paper we describe our approximation and clustering methods. These algorithms were developed in an attempt to solve the aforementioned problems and to exploit the full potential of clock gating. First we describe our approximation method which specifically aims at reducing the delay of a given activation function while trying to preserve its efficiency as high as possible. Next we present our clustering algorithm which is designed to maximize the overall power saving by finding a globally-optimal solution under size constraints.

3. APPROXIMATION

An activation function should not violate timing constraints and/or waste more power than it saves. Therefore, our goal is to reduce the logic depth below a user defined limit, thus reducing its delay and making it more power efficient. The limit is provided by the user either as an integer representing the acceptable depth or as a parameter in the range (0, 1] which indicates the depth of the logic that will be allowed on the clock gate, relative to the logic depth of the original, ungated design. In addition, since it is desirable to maintain high on-set probability for maximizing power saving, we require of our algorithm to give the tightest function approximation that fits within the given depth limit.

Usually logic depth is measured directly on the netlist representation of the circuit. In cases where the activation functions are represented as BDDs, one would have to first synthesize the BDDs in order to measure their logic depth. A commonly used approach for BDD synthesis is based on BDD decomposition algorithms [16, 17]. Using this approach, synthesis of BDDs with long paths will result in a circuit with large delay. Thus, in order to reduce the delay of a given activation function, we should eliminate long paths in the BDD that represents it. Several methods for reducing the depth of a BDD have been suggested in the context of logic synthesis (e.g. [5, 12]), however these methods do not approximate Boolean functions but rather yield optimized representation of the exact functions which may still exceed the depth limit. As a result, these methods may not suffice for making a problematical activation function useful, thus approximation methods must be used.

978-1-60558-497-3/09 $25.00 © 2009 ACM

A fundamental requirement for any approximation algorithm is to maintain correct design functionality. This requirement is translated to obtaining an approximation which does not gate the clock more cycles than the original activation function does. Formally, we define valid approximation of an activation function as follows.

DEFINITION 3. *Let f be an activation function, and f' an approximation of f. We say that f' is a* valid approximation *of f iff $f = 0 \rightarrow f' = 0$.*

Thus, given the truth table of an activation function, a valid approximation can be achieved by setting certain 1 entries in the table to 0. Equivalently, when using BDD representation, a valid approximation can be achieved by eliminating positive BDD paths.

DEFINITION 4. *A* positive BDD path *is a BDD path that corresponds to a true assignment of the Boolean function represented by the BDD.*

Eliminating a positive BDD path is the process of making a BDD path correspond to a false assignment instead of to a true assignment. Note that a BDD path may correspond to several function assignments if one or more variables are not in that path. Thus by eliminating a positive BDD path we set one or more 1 entries in the function's truth table to 0.

Given a BDD, positive paths can be eliminated by first building an auxiliary BDD comprised of a disjunction of smaller BDDs, each of which represents a single path to remove from the original BDD, and then conjuncting the original BDD with the negation of the auxiliary BDD. Hence our approximation algorithm can be formulated as given in Algorithm 1. Note that in our implementation, instead of synthesizing the BDD before measuring the logic depth, we took a heuristic approach which is more efficient.

Conservatively we count one BDD node as 2 logic levels (each BDD node is implemented using a Multiplexer). For example, the corresponding logic depth of the BDD depicted in Figure 1a is 8, as the longest path in the BDD ($\bar{a}b\bar{c}d$) has 4 nodes (BDD terminal nodes are ignored). So in Algorithm 1, BDD path length is defined as the number of nodes comprising the path multiplied by a factor of 2.

Algorithm 1 BDD-paths Elimination Algorithm

Input: Activation function f, depth limit DL
Output: Approximated function f'
1: $aux\_bdd \leftarrow 0$
2: **for each** positive path p in f such that $\text{length}(p) > DL$ **do**
3:    $aux\_bdd \leftarrow aux\_bdd \lor \text{bdd\_of}(p)$
4: **end for**
5: $f' \leftarrow f \land (\neg\, aux\_bdd)$
6: **return** f'

To illustrate the algorithm, consider the activation function f given as a BDD in Figure 1a. In this paper we use the convention in which a dotted line represents the value of the variable being zero, and a solid line represents the value of the variable being one. Assuming depth limit (DL) is 6, the activation function should be approximated since the corresponding logic depth of the function is 8. Thus our approximation algorithm will eliminate the longest path in f ($\bar{a}b\bar{c}d$) by conjuncting f with the BDD $g \equiv \neg(\bar{a}b\bar{c}d)$, resulting in the BDD depicted in Figure 1b. Note that in the

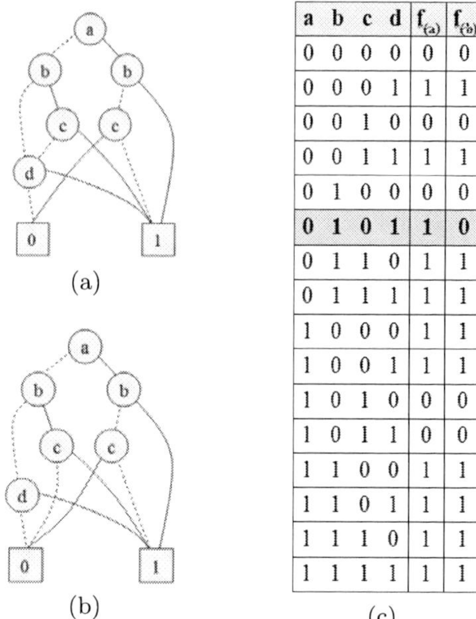

a	b	c	d	$f_{(a)}$	$f_{(b)}$
0	0	0	0	0	0
0	0	0	1	1	1
0	0	1	0	0	0
0	0	1	1	1	1
0	1	0	0	0	0
0	**1**	**0**	**1**	**1**	**0**
0	1	1	0	1	1
0	1	1	1	1	1
1	0	0	0	1	1
1	0	0	1	1	1
1	0	1	0	0	0
1	0	1	1	0	0
1	1	0	0	1	1
1	1	0	1	1	1
1	1	1	0	1	1
1	1	1	1	1	1

(a)   (b)   (c)

Figure 1: An example of approximation using our proposed approach. (a) Original BDD. (b) BDD after approximation. (c) Truth table for both BDDs. Note that the BDD depth in (b) is reduced and that the truth table is changed in only one entry (highlighted).

corresponding truth table (Figure 1c), the approximation eliminated only one row of true assignment (highlighted), meaning this is the tightest approximation possible.

If we assume uniform probabilities of input signals, that is the probability of an input being 0 is equal to it being 1, longest paths in a BDD are the less probable ones. Thus by removing the longest paths first, we are guaranteed to obtain a tight approximation using Algorithm 1 since the on-set probability is defined as the sum of the probabilities of positive BDD paths. In other words, our proposed approximation algorithm is aimed at obtaining activation functions that are highly efficient for clock gating.

To further emphasize the benefit of the proposed approach, consider the algorithm described in [2]. Since the authors suggested to use universal quantification for removing function variables, every variable removed from f given in Figure 1a will result in a less tight approximation than achieved by our proposed approach. Quantification of any variable in f results with a function having an on-set probability of 50%, instead of 62.5% obtained by our approximation. For example, quantifying the a variable from f results in $f' = f|_a \land f|_{\bar{a}} = bc \lor b\bar{c}d \lor \bar{b}\bar{c}d$, which has 8 true assignments.

In the non-uniform case, path probabilities depend on input probabilities, hence the longest positive path is not necessarily the least probable one. However, in order to meet timing constraints one will still have to eliminate the longest BDD paths first.

Another factor which may cause timing violations is late arriving signals. In this work we assume that given prior information regarding signal arrival times, those signals are first quantified from the activation functions before applying our approximation algorithm.

Finally, note that since BDDs are used to store Boolean

978-1-60558-497-3/09 $25.00 © 2009 ACM

functions compactly (due to the subgraph isomorphism reduction rule of BDDs [4]), by removing BDD paths selectively we may increase the total node count of the BDD. This can occur when removal of a path causes a split of two sub-BDDs. However, in all of our experiments this has never been observed and a total reduction of BDD nodes count has been measured.

4. CLUSTERING

In commonly used design methodologies each activation function should be connected to the sequential element via a clock buffer which has its own power demand. Since the feasibility of activation functions is mainly dictated by their power consumption, there is a minimum number of sequential elements that should share an activation function in order to ensure that more power is saved by clock gating than is consumed by it. If the minimum is not met, then different activation functions can be combined by *anding* them together (so that the clock is gated only when gating is allowed for each of the original functions). The primary objective, however, is to combine together sequential elements having similar activation functions. For example, we are not interested in combining sequential elements e_1 and e_2 with activation functions $f_1 = a$ and $f_2 = \neg a$. On the other hand, if sequential elements e_3 and e_4 have activation functions $f_3 = a \wedge b \wedge c$ and $f_4 = b \wedge c$, combining f_3 and f_4 results in a useful activation function $(a \wedge b \wedge c)$.

When combining unique sequential elements into a single gating domain, the activation functions of the sequential elements should be combined as well into a single function, which is eventually connected to the clock buffer of that group. Since activation functions are combined by a logical 'AND' operation, maximization of the overall power saving is obtained by combining sequential elements that share similar activation functions, as otherwise the probability for disabling the clock of the gating domain by the new function is too low. Hence, the requirement of combining similar activation functions along with the above restrictions result in a clustering problem with lower bound constraints on the cluster sizes.

We start by defining an appropriate similarity measure between clock-gating functions f and g:

$$S_1(f, g) = \frac{|f \cdot g| + |\forall_s f \cdot g|}{2|f + g|} \tag{1}$$

where $\cdot$ and $+$ are the logical 'AND' and 'OR' operators respectively, $|f|$ is the size of the on-set of a function f, s (in $\forall_s$) is a set of all the variables not in the support of both f and g, and $\forall_s$ is the universal quantification operator on variables in s. Note, that this similarity measure is strongly oriented towards activation functions rather than general Boolean functions. The reason for this is twofold: first, unlike in Hamming distance where the objective is to count both types of agreements — 0-0 and 1-1 — our measure counts only 1-1 agreement, since we consider functions as similar when they gate the clock of a sequential element at the same cycles. Second, we want to prevent physically distant functions from being clustered together. Thus we use the second term in the numerator to assess the support proximity, keeping in mind that physically distant functions have a smaller chance for common support. Finally, since both terms in the numerator are bounded from above by the size of the on-set in $f + g$ we normalize by $2|f + g|$.

Given the similarities between each pair of activation functions we use a variation of the CAST clustering algorithm [1] to obtain feasible groups of functions. This algorithm receives an $n \times n$ similarity matrix and a cutoff parameter t and partitions the given elements into clusters. For each element e and cluster C the affinity of e to C is defined as an average similarity of e to all the elements in C. The algorithm is iterative and each iteration processes only one cluster. Once the iteration is completed, the cluster becomes final and no more operations are allowed on it. Each iteration has two main phases which repeat until no more changes occur: the ADD phase and the REMOVE phase. In the ADD phase (lines 6-12 in Algorithm 2) an element with the highest affinity to the current cluster is chosen. If its affinity to this cluster is bigger than t times the cluster size, the element is added to the cluster. In the REMOVE phase (lines 14-20 in Algorithm 2), if the affinity of an element with the smallest affinity to the current cluster is lower than t times the cluster size, the element is removed from the cluster.

Algorithm 2 CAST

Input: $\Sigma : n \times n$ similarity matrix, t: cutoff parameter.
Output: C: a set of clusters
1: $C \leftarrow \emptyset$ /* Closed clusters */
2: $C_{open} \leftarrow \emptyset$ /* New cluster */
3: $U \leftarrow \{f_1, \ldots, f_n\}$ /* Functions space */
4: $aff(\cdot) \leftarrow 0$ /* Affinity function */
5: **while** $U \cup C_{open} \neq \emptyset$ **do**
6:     Let u be an element with max. affinity in U
7:     **if** $aff(u) > t|C_{open}|$ **then**
8:        $C_{open} \leftarrow C_{open} \cup \{u\}$
9:        $U \leftarrow U \setminus \{u\}$
10:      **for all** $x \in U \cup C_{open}$ **do**
11:         $aff(x) = aff(x) + \Sigma(x, u)$
12:      **end for**
13:     **else**
14:       Let v be an element with min. affinity in C_{open}
15:       **if** $aff(v) < t|C_{open}|$ **then**
16:          $C_{open} \leftarrow C_{open} \setminus \{v\}$
17:          $U \leftarrow U \cup \{v\}$
18:          **for all** $x \in U \cup C_{open}$ **do**
19:            $aff(x) = aff(x) - \Sigma(x, v)$
20:          **end for**
21:       **else**
22:          $C \leftarrow C \cup C_{open}$
23:          $C_{open} \leftarrow \emptyset$
24:          $aff(\cdot) \leftarrow 0$
25:       **end if**
26:     **end if**
27: **end while**

Since we group activation functions, each obtained cluster corresponds to a new activation function which equals 1 if all the functions in the cluster equal 1. Thus it is computed by conjuncting all the functions in the given cluster. Nevertheless, as mentioned above, not all the resulting clusters meet our cluster size constraints. Thus we enhance the original algorithm with the following heuristics: first we add an external loop iterating over t values in the range [0.1,0.9] with a step of 0.1. Smaller t values encourage clusters to be tighter while t values near 1 allow more remote functions to be combined into one cluster. Then in each iteration we evaluate the merit of the Boolean function corresponding to the cur-

978-1-60558-497-3/09 $25.00 © 2009 ACM

rent cluster. The merit of cluster C is computed as follows:

$$M(C) = \begin{cases} 0 & if |C| < min.size \\ |C| \cdot P(f_C) - COST_{cb} & otherwise \end{cases} \quad (2)$$

where f_C is the gating function corresponding to the cluster C, $P(f_C)$ is the on-set probability of f_C, and $COST_{cb}$ is an expression reflecting the cost of adding a clock buffer for gating the group. $P(f_C)$ is calculated as the sum of products of the variable probabilities on each path in the BDD of f_C from the root to the 1 terminal.

During the algorithm, while the merit is increasing we keep the original t value but when the merit starts to decrease, we decrease t preventing more distant activation functions from joining the cluster and weakening the final activation function. Finally, after each external iteration, each cluster with merit 0 is discarded and its elements are returned to the general pool.

The above procedure produces feasible gating domains that can be efficiently implemented in the design. However, in practice using BDDs for calculating similarities and merit in big designs may results in a significant runtime overhead. To prevent this, we used simulation-based computations both for function similarities as well as for function merits. More formally, given two Boolean vectors V_f and V_g representing activation functions f and g respectively, and their support functions $supp(f)$ and $supp(g)$, the similarity between f and g is computed as:

$$S_2(f,g) = \alpha \frac{|V_f \cdot V_g|}{|V_f + V_g|} + (1 - \alpha) \frac{|supp(f) \cap supp(g)|}{|supp(f) \cup supp(g)|} \quad (3)$$

where $\cdot$ and $+$ are the bitwise 'AND' and 'OR' operators respectively, $|V|$ is the fraction of 1's in V, and $\alpha \in [0,1]$. Note that V_f and V_g are constructed by sampling f and g, and that $supp(f)$ is defined as the variables support set of f. Also, one can see that the two terms of S_2 correspond to the two terms of S_1 from Equation (1) (if we separate the numerator into two parts). In fact, α is used as a balancing factor between the similarity in the simulation vectors and the similarity in support, and it is used to obtain a better correlation between S_2 and S_1 measures. In our settings $\alpha = 0.5$. Using random simulation of 5000 cycles, we observed an error rate of less than 5% in similarity compared to the BDD approach, while reducing run-time by an order of magnitude in lengthy runs. Finally, $P(f)$ in Equation (2) is computed as a fraction of 1's in the simulation vector V_f.

5. RESULTS

We evaluated the proposed methods via power estimation experiments. Our data set contained 93 industrial designs of different sizes. For each design we used the IBM internal clock gating tool (GateAlert) for listing the initial activation functions, which were filtered using logic depth and gating domain size filters meeting industrial bounds for timing and leakage. Implementing the remaining functions in the design, we obtained its power consumption by using a conventional power estimation methodology based on transistor level simulation of technology mapped designs using workloads that represent typical runs. In the following, we term this measurement as *plain* power.

We then applied the approximation and clustering techniques proposed above, setting depth limit to 6 and minimal domain size to 10. First, each of the optimizations was ap-

plied separately and then the combination of the two. When combining the algorithms, we used clustering first and then approximation. The idea is to prevent the approximation algorithm from reducing the similarity level between activation function pairs by eliminating mutual minterms. For example, consider $f_1 = abc \vee \overline{a}$ and $f_2 = bc$ and assume that we restrict the minterms to be of length 2. Using clustering before approximation results in $f_{12} = bc$ which meets the length constraint, while using approximation before clustering results in $f_{12} = \overline{a}bc$ which should be discarded.

In each experiment we measured the resulting power and compared it to the plain power measurement. The 40 top rated results are presented in Table 1. We also show on the last line the average results the remaining designs in the data set. For each design we report the total number of sequential elements, the number of sequential elements filtered by the logic depth and gating domain size filters and the number of sequential elements that have been resurrected by our algorithms. Also we report the power saving of each technique and their combination over the plain measurement.

We observe that using each of the proposed approaches separately may drastically improve the power consumption. However, the best results are obtained by using the combination of both. For example, in *design1* we found no improvement using approximation, and 41% improvement using clustering. However, using the combination of approximation and clustering resulted in 78.3% improvement.

All-in-all, applying the proposed optimization techniques resulted in reducing the power consumption in two-thirds (63) of the designs in our data set, allowing up to almost 80% improvement in power. We observe that for a large fraction of designs there is no power saving at all due to approximation or clustering algorithms, while the combination of the two provides a significant power reduction. Moreover, the power saving obtained due to the combination of the algorithms is greater than sum of each of them when applied separately. This is due to the fact that there are many gating domains that do not meet the size constraint, and when clustered together exceed the depth limit. Therefore, only after approximating these functions, i.e. applying the combination of both, an additional power saving is obtained. Also, as one can see, the additional power saving is poorly correlated with the number of resurrected latches causing a significant variation in the actual power saving between two designs with a similar number of resurrected latches — this is due to the unequal switching factors on the clock gating functions.

The measured runtime of the clock gating algorithm with both optimization techniques was of the same order of magnitude as the original clock gating algorithm (up to several minutes for a single design with up to 1000 sequential elements), and no significant overhead was observed. As one can note, the approximation technique is scalable exactly as other BDD operations and has exactly the same limitations. Clustering bottleneck is in computation of the similarity matrix, and after applying simulation procedure instead of the exact computation it showed quite a good results — up to one minute on the biggest design in our data set.

6. CONCLUSIONS

In this paper we presented two optimization techniques for resurrecting activation functions that violate timing constraints or consume too much power. We showed that using our approximation and clustering techniques an additional

978-1-60558-497-3/09 $25.00 © 2009 ACM

Table 1: Power saving using approximation, clustering and a combination of both.

DESIGN	Number of sequential elements			Power saving over PLAIN		
	total	filtered	resurrected	APP	CLUST	BOTH
design1	239	223	195	0.00%	40.96%	78.30%
design2	225	224	198	57.21%	0.00%	57.21%
design3	984	981	714	45.89%	0.28%	45.89%
design4	46	46	19	12.62%	40.44%	40.44%
design5	402	396	280	2.13%	0.00%	37.86%
design6	296	264	106	23.94%	10.67%	34.60%
design7	266	200	58	14.48%	0.00%	31.11%
design8	51	51	15	10.97%	0.00%	28.87%
design9	115	96	20	0.00%	17.72%	27.27%
design10	496	486	238	0.00%	13.89%	26.96%
design11	129	120	44	0.00%	0.00%	24.99%
design12	143	120	53	0.00%	0.00%	23.54%
design13	466	457	94	0.00%	2.19%	22.74%
design14	904	863	183	0.00%	2.02%	18.22%
design15	309	309	77	0.00%	0.00%	17.20%
design16	618	593	241	0.00%	2.66%	17.00%
design17	795	783	171	0.19%	1.14%	13.99%
design18	223	201	36	3.24%	3.65%	13.41%
design19	288	211	78	5.89%	7.14%	13.03%
design20	542	542	31	0.00%	0.00%	11.56%
design21	174	140	35	9.56%	0.00%	10.24%

DESIGN	Number of sequential elements			Power saving over PLAIN		
	total	filtered	resurrected	APP	CLUST	BOTH
design22	215	192	39	9.32%	0.00%	9.32%
design23	358	358	12	0.00%	0.00%	9.17%
design24	1022	952	348	0.00%	0.00%	8.50%
design25	239	236	73	2.59%	0.00%	8.00%
design26	100	95	17	0.00%	5.50%	7.70%
design27	526	475	52	4.66%	2.84%	7.67%
design28	230	222	38	0.00%	4.73%	7.58%
design29	189	168	35	0.00%	0.01%	7.13%
design30	342	302	80	5.76%	0.00%	7.06%
design31	77	57	4	6.00%	0.00%	6.00%
design32	592	592	58	2.67%	0.00%	5.57%
design33	127	94	8	0.00%	0.00%	5.46%
design34	339	334	27	0.00%	2.16%	5.40%
design35	965	963	84	0.00%	0.00%	4.75%
design36	1060	976	93	0.00%	0.00%	4.66%
design37	75	53	14	4.47%	0.00%	4.47%
design38	133	96	16	0.00%	0.00%	4.25%
design39	505	367	213	0.00%	0.00%	4.21%
design40	35	29	8	0.00%	3.94%	3.94%
The rest	17547	15744	1286	0.24%	0.44%	1.03%

power saving of up to almost 80% can be obtained in typical designs. The proposed techniques are not limited to a certain clock gating algorithm, but rather can be applied as a postprocessing step in any automatic clock gating tool in order to resurrect many of the filtered activation functions. Apart from the presented algorithms, another major contribution of this work is in defining a general framework for utilizing the full power saving potential of automatically inferred clock gating functions in power optimization tools.

7. ACKNOWLEDGEMENTS

We would like to thank our colleagues David Levitt and Oded Fuhrmann for their contribution and fruitful discussion on clustering of activation functions.

8. REFERENCES

[1] A.Ben-Dor, R. Shamir, and Z. Yakhini. Clustering gene expression patterns. *Journal of Computational Biology*, 6(3/4):281–297, 1999.

[2] M. Alidina, J. Monteiro, S. Devadas, A. Ghosh, and M. Papaefthymiou. Precomputation-based sequential logic optimization for low power. *IEEE Transactions on VLSI Systems*, 2(4):426–436, 1994.

[3] L. Benini, G. D. Micheli, E. Macii, M. Poncino, and R. Scarsi. Symbolic synthesis of clock-gating logic for power optimization of synchronous controllers. *ACM Transactions on Design Automation of Electronic Systems*, 4(4):351–375, 1999.

[4] R. E. Bryant. Graph-based algorithms for Boolean function manipulation. *IEEE Transactions on Computers*, 35:677–691, 1986.

[5] L. Cheng, D. Chen, and M. D. F. Wong. DDBDD: delay-driven BDD synthesis for FPGAs. In *Proc. of DAC '07*, pages 910–915, 2007.

[6] D.Banares, J.Cortadella, and M.Kishinevsky. A recursive paradigm to solve boolean relations. In *Proc. of DAC '04*, pages 92–97, 2004.

[7] A. P. Hurst. Automatic synthesis of clock gating logic with controlled netlist perturbation. In *Proc. of DAC '08*, pages 654–657, 2008.

[8] T. Kunjan, U. Hinsberger, and R. Kolla. Approximative representation of Boolean functions by size controllable ROBDD's. Technical report, Inst. fiir Informatik, Univ. Wfirzburg, 1997.

[9] Y. Kuo, S. Weng, and S. Chang. A novel sequential circuit optimization with clock gating logic. In *Proc. of the ICCAD '08*, pages 230–233, 2008.

[10] G. D. Micheli. *Synthesis and Optimization of Digital Circuits*. McGraw-Hill, Inc., 1994.

[11] P. Parra, A. J. Acosta, and M. Valencia. Selective clock-gating for low power/low noise synchronous counters. In *Proc. of the 12th International Workshop on Integrated Circuit Design*, pages 448–457, 2002.

[12] P. Prasad, M. Raseen, A. Assi, and S. Senanayake. BDD path length minimization based on initial variable ordering. *Journal of Computer Sciences*, 1(4):521–529, 2005.

[13] Q.Wu, M.Pedram, and X.Wu. Clock-gating and its application to low power design of sequential circuits. *IEEE Transactions on Computers*, 47(103):415–420, 2000.

[14] R.Wiener, G.Kamhi, and M.Y.Vardi. Intelligate: scalable dynamic invariant learning for power reduction. In *Procõ of PATMOS ,08*, 2008.

[15] F. Theeuwen and E. Seelen. Power reduction through clock gating by symbolic manipulation. In *Proc. of IFIP Int. Workshop on Logic and Architecture Synthesis*, pages 184–191, 1996.

[16] C. Yang and M. Ciesielski. BDS: a BDD-based logic optimization system. In *Proc. of DAC '00*, pages 92–97, 2000.

[17] C. Yang, V. Singhal, and M. Ciesielski. BDD decomposition for efficient logic synthesis. In *Proc. of ICCD '99*, pages 626–631, 1999.

Low Power Gated Bus Synthesis using Shortest-Path Steiner Graph for System-on-Chip Communications

Renshen Wang[†], Nan-Chi Chou[‡], Bill Salefski[‡] and Chung-Kuan Cheng[†]

[†]University of California, San Diego, La Jolla, CA 92093-0404

[‡]Mentor Graphics Corporation, 1001 Ridder Park Drive, San Jose, CA 95131

{rewang, ckcheng}@cs.ucsd.edu, {nanchi_chou, bill_salefski}@mentor.com

ABSTRACT

Power consumption of system-level on-chip communications is becoming more significant in the overall system-on-chip (SoC) power as technology scales down. In this paper, we propose a low power design technique of gated bus which can greatly reduce power consumption on state-of-the-art bus architectures. By adding demultiplxers and adopting a novel shortest-path Steiner graph, we achieve a flexible tradeoff between large power reduction versus small wire-length increment. According to our experiments, using the gated bus we can reduce on average 93.2% of wire capacitance per transaction, nearly half of bus dynamic power and on a scale of 5%~10% of total system power.

Categories and Subject Descriptors

J.6 [**Computer Applications**]: Computer-aided design

General Terms

Algorithms, design, performance

Keywords

Gated bus, Steiner graph, power efficiency

1. INTRODUCTION

System-on-chips (SoC) are nowadays being developed with increasing complexity and on-chip communication demand. However global on-chip wires which are to meet this demand do not scale well towards 35nm feature size [9]. As a result, global interconnect is becoming a bottleneck of improving system performance and power consumption [12]. Bus architectures are therefore regarded as an important aspect in low power SoC design. Current state-of-the-art bus architectures including AMBA [18], CoreConnect [19], Avalon [20], AMBA AHB bus matrix [13], *etc*, provide solutions for SoC on-chip communications. Research in [12] shows that these bus circuits may consume as much power as other major components such as processor, memory controller and

Permission to make digital or hard copies of part or all of this work for personal or classroom use is granted without fee provided that copies are not made or distributed for profit or commercial advantage and that copies bear this notice and the full citation on the first page. To copy otherwise, to republish, to post on servers or to redistribute to lists, requires prior specific permission and/or a fee.

DAC'09, July 26-31, 2009, San Francisco, California, USA

cache. Therefore, reducing power on buses makes significant contribution to the whole system's power consumption.

Techniques of clock gating [6] and power gating [14] have been widely and effectively used to reduce power consumption of electronic systems, among which clock gating reduces dynamic switching power, and power gating reduces static leakage power. Both of the techniques save power by masking off signal/power when/where it is not needed. Since a clock distribution network consumes more than 40% of the total power budget of a CMOS circuit, clock gating has become a necessity in most digital circuit designs. We find that bus connections in current communication architectures are facing a similar (if not the same) situation as clock networks.

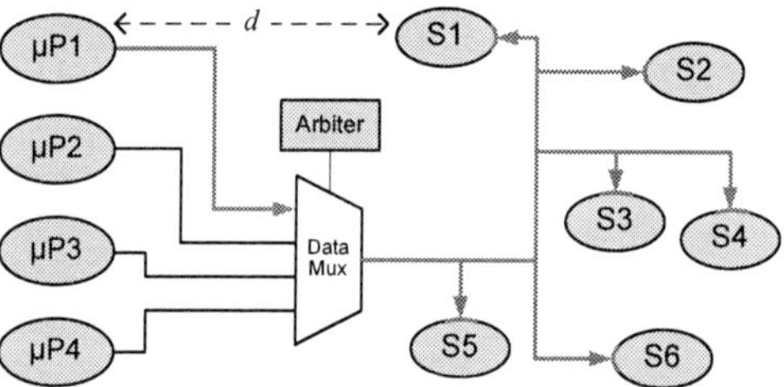

Figure 1: A signal switching on AMBA data bus

Figure 1 shows the wires and fanouts involved in the signal switchings of a data transaction (highlighted arrows) in AMBA bus architecture. Microprocessor 1 sends data to slave 1 through the data multiplexer and a wire net which connects all the slaves. While only the input on slave 1 needs the signals, the entire net and the input stage of other slaves are all driven, resulting in a low power efficiency. Considering the distance d between processor 1 and slave 1, ideally the wire used for the transaction needs only length of d, so most of the switching power on the wire net and fanouts are wasted. If we can mask off the wires not leading to slave 1, like clock gating, the power can then be saved.

In this paper, we propose a physical level on-chip bus gating scheme which can greatly reduce power consumption of current bus architectures. We focus on minimizing the total capacitance of the driven circuit in each transaction which determines its dynamic power. Distributed multiplexers and demultiplexers are used on a shortest-path Steiner graph to ensure minimum wire capacitance, with some increased total wirelength that can be cheaply traded off.

There is a body of work on bus power analysis on system-level and RTL-level such as [3], [12] and [13], which provide valuable data and orientations. Physical-level approaches like bus segmentation [4] and bus splitting [11] reduce bus power by masking off certain bus segments, which has a similar effect as bus gating. However the bus structures in [4] and [11] are limited to tree structures, and the bus type is bi-directional shared bus which is not suitable for on-chip communications. To the knowledge of the authors, this paper is the first attempt to optimize state-of-the-art on-chip bus power consumption by gating techniques applied on a non-tree structure.

2. BACKGROUND

2.1 Current on-chip bus architectures

The AMBA bus architecture [18] is one of the most popular commercial on-chip communication architectures. Figure 2 is a sketch of the components and interconnections in AMBA AHB (Advanced High-performance Bus) protocol. AHB supports up to 16 masters and 16 slaves. A master initiates a bus transaction by providing address to the decoder and control information to the arbiter. Arbiter ensures only one master at a time is allowed to use the bus. Decoder finds the slave corresponding to the address. A slave responds passively. AHB can connect to an APB (Advanced Peripheral Bus) bus on a lower hierarchy via the AHB-APB bridge, which is a slave of AHB and the only master of APB. Another bus architecture CoreConnect [19] contains an OPB (On-chip Peripheral Bus) protocol which has a structure similar to that of AMBA AHB.

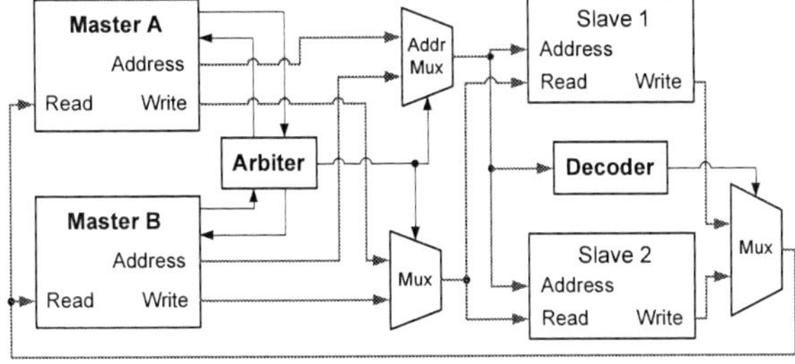

Figure 2: AMBA AHB on-chip bus

In these on-chip buses, the signals coming from a master go though a selection multiplexer, and then sent to all the slave inputs. This applies on the address bus, data bus, and reversely on the slave-to-master data bus. It may be the simplest way of implementing the bus architecture, but as mentioned before (figure 1), the power efficiency of this implementation is quite low. The entire wire net plus all the input stages of connected components are all driven by the multiplexer, while only one component needs the signals. Due to poor scaling of global wires and increasing number of components in SoCs, the limitations of such connections may seriously deteriorate system power consumption.

Gated bus is a solution to eliminate the wasted dynamic power. Like clock gating, bus gating can mask off unnecessary load capacitance on the net, which is especially large on AMBA bus wire nets. Besides, the physical routing of wire nets can change from minimizing total wire length to minimizing individual path lengths, which can further reduce wire capacitance induced by data transactions. Meanwhile, the main arbitration and control mechanism in the original bus architecture does not need to be changed.

2.2 Design flows

Physical-level design of electronic systems starts from gate level netlists and goes through placement, routing, timing/power analysis, *etc.* Since the buses are included in the system as components, bus gating will result in updates on the original netlists of logic design. Figure 3 shows the stage of bus gating fitted in the physical design flow.

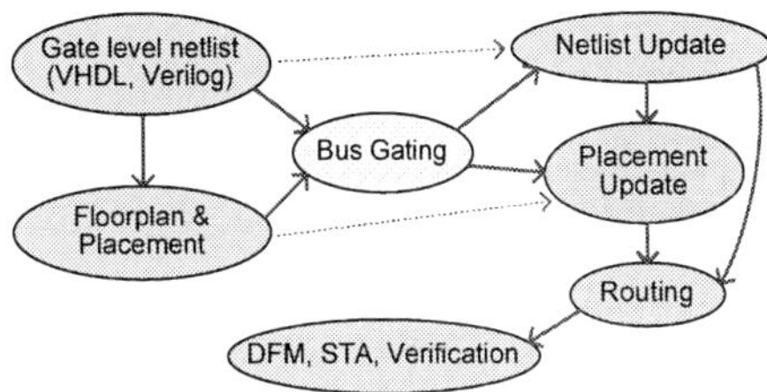

Figure 3: Physical design flow with bus gating

Like clock gating, the added gating stage is dependent on circuit placement, while the placement may also be updated by the gating. This may bring extra complexity on the flow, but the update by bus gating is not large and can be automated with appropriate algorithms.

3. MINIMIZING DYNAMIC POWER

3.1 Distributed multiplexer and demultiplexer

A simple bus gating technique is to add a demultiplexer after the multiplexer, so that only a single path of bus wires is driven at a time. The selection signals are from the arbiter at masters' side and from the decoder at slaves' side. In this way, not only the wire branches to other components are masked off, the connection can also take the shortest path to minimize wire capacitance. However, there are also two problems: single demultiplexer requires more routing resources, and the path length of different data transactions cannot be all minimized.

By distributing the big multiplexer and demultiplexer in to small ones over the nets, the wires can be shared by paths and therefore reduced. The CoreConnect OPB [19] bus supports distributed multiplexers for design flexibility, but no demultiplexers are used. If we add both to AMBA AHB or CoreConnect OPB, as shown in figure 4, the effect is both reduced wire length and masked off wire capacitance. There is an overhead of distributing the control signals, but the number of such wires is small compared to the bus width (at least 32-bit address lines and 32-bit data lines). And

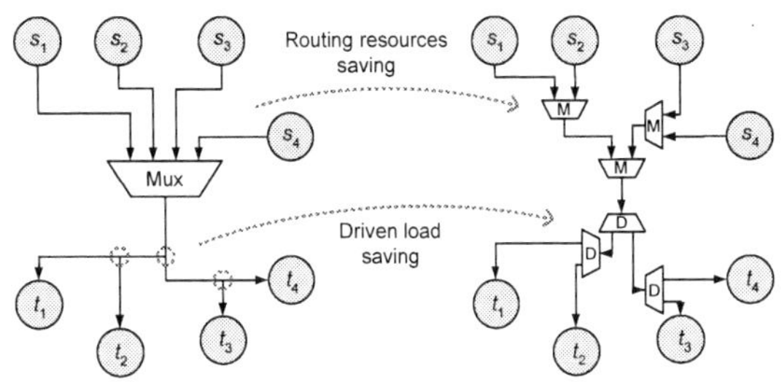

Figure 4: Using distributed multiplexers and demultiplexers on AMBA AHB

978-1-60558-497-3/09 $25.00 © 2009 ACM

since the control signals remain still during a transaction, the dynamic power overhead is negligible.

In current bus architectures, the wire nets connecting all the master (or slave) inputs are best routed on a minimal Steiner tree. With distributed demultiplexers, the best result is a shortest-path Steiner tree/arborescence [15], [5], because on each transaction, signals are delivered on only one path. From the conclusion of [1], switching from RMST (rectilinear minimal Steiner tree) to MRSA (minimal rectilinear Steiner arborescence) only induces 2%~4% of additional wirelength on average. So by the distributed demultiplexer alone, the simple bus gating can effectively eliminate most of the dynamic power on bus wires and input stages of bus components.

3.2 Shortest-path Steiner graph

The modified bus in figure 4 has a tree structure such that every path from its root (set between the last multiplexer and the first demultiplexer) to a leaf is the shortest path, *i.e.* an arborescence structure. However, in figure 4, the path length from source s_1 to terminal t_1 is larger than the Manhattan distance between them. The tree structure has limitations in that there is only one root placed at a certain point, which may not be on the shortest path of every connection. We use Steiner graph replacing the Steiner arborescence to enable overall path optimizations.

As defined in [2], for an unweighted graph $G = (V, E)$, a (possibly weighted) graph $G' = (V', E', \omega)$ is a Steiner graph of G if $V \subseteq V'$, and for any pair of vertices $u, w \in V$, the distance between them in G' (denoted $d_{G'}(u, w)$) is at least the distance between them in G ($d_G(u, w)$). In our bus communication, the original graph G has full connections between a set of masters $s_1, s_2, \cdots, s_m$ and a set of slaves $t_1, t_2, ..., t_n$. Additional vertices are added into G' as Steiner points. If for any s_i, t_j, $d_{G'}(s_i, t_j)$ equals their rectilinear distance, G' is a shortest-path Steiner graph.

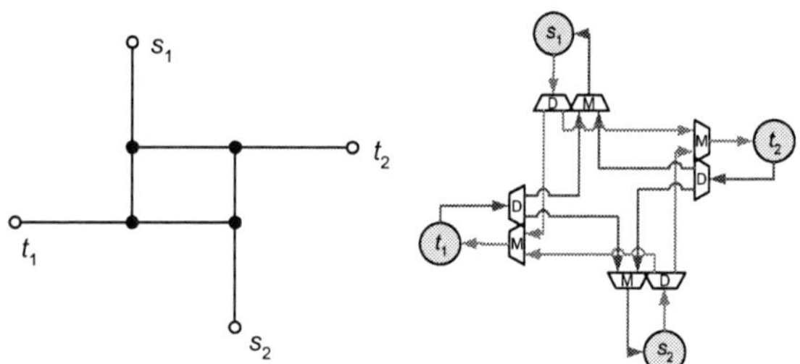

Figure 5: Shortest-path Steiner graph connections

Figure 5 shows a shortest-path Steiner graph of the connections between $\{s_1, s_2\}$ and $\{t_1, t_2\}$, and also its bus implementation on both master-to-slave and slave-to-master directions. This Steiner graph is minimal in terms of total wirelength. The loop in the center is necessary for including all the shortest paths of connections.

A bus architecture consumes minimal dynamic power using wire tracks of a shortest-path Steiner graph, combined with corresponding controls. A possible problem is that the loops require additional routing resources as well as control overhead. But sometimes the wirelength can also be reduced by sharing master-to-slave and slave-to-master connections. As illustrated in figure 6(a), the data bus in current bus architectures needs over $4l$ of length, because on each direc-

tion the multiplexer needs at least length l to connect s_1, s_2 and length l to connect t_1, t_2. We can reduce it to just over $2l$ by combining the nets on the two directions, since only one master device can access the bus at a time, double wires are enough to support all-direction transactions. Plus the address bus, the total wirelength is $4l$ vs $6l$. The signal switch in figure 6(b) acts as a mixed multiplexer and demultiplexer, controlled by 3 or less extra control signals. In average cases with well optimized Steiner graph, the wire length increment is small compared to the power reduction it brings (detailed results in section 5).

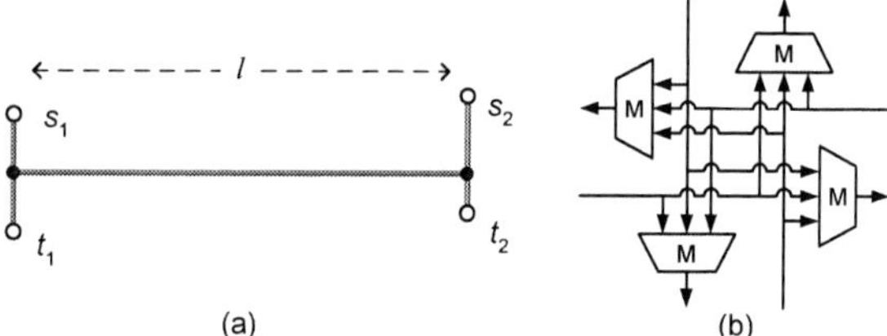

Figure 6: Sharing wires by controlled switches

4. HEURISTICS FOR GENERATING SHORTEST-PATH STEINER GRAPHS

In our problem of minimal shortest-path Steiner graph, the locations of a set of sources $s_1, s_2, \cdots, s_m$ and terminals $t_1, t_2, ..., t_n$ are given, and the objective is to find a rectilinear routing solution containing all the source-to-terminal shortest paths, with total wirelength as small as possible.

In single source cases, this is a rectilinear Steiner arborescence (shortest-path tree) problem which has been studied by [15], [5], *etc.* Finding the exact solution of a mininum rectilinear Steiner tree (MRST) is NP-complete [7], and finding a minimum rectilinear Steiner arborescence (MRSA) is believed to be hard [15] although without hardness proof. For practical use, several efficient heuristic algorithms are introduced and compared in [5], among which the 2-IDeA/G algorithm has the best average performance over runtime. In general cases containing multiple sources, the problem has not been studied before. We adopt the the 2-IDeA/G heuristic as basis and add more heuristics to construct the shortest-path Steiner graph.

4.1 k-IDeA/G heuristic for MRSA

The k-IDeA/G (iterated k-deletion for arborescence) algorithm [5] is based on the RSA heuristic (denoted RSA/G), which is proved to be 2-approximate in [15].

The basic flow of RSA/G is to start with n terminals as n subtrees and iteratively merge a pair of subtree roots v and v' such that the merging point is as far from the source as possible, so that the wires can be shared as much as possible. It terminates when only one subtree remains. For efficient implementation, the RSA/G first sorts all the nodes on the Hanan grid [17] in decreasing distance to the source s, and visits each node maintaining a peer set P of subtree roots. Details are in the pseudo code in table 1. We denote the rectilinear distance from s to v as $\Delta_s(v)$ or $\Delta(s, v)$. Two basic operations are used in RSA/G at: terminal merger opportunity (TMO), when a terminal is added into P as a subtree; and Steiner merger opportunity (SMO), when $|X| \geq 2$ and the subtrees in X are merged.

The heuristic of iterated k-deletion proposed in [5] is to

Second, we construct the MRSA on the set of nodes T' using existing wires. The TMO condition is then changed to $v_i \in T'$. The SMO condition is changed, also for the purpose of wire reusing, from $|X| \geq 2$ to $|X| \geq 2$ or ($|X| = 1$ and $v_i \in G$). Because when v_i is already in the graph, it can share wires with the node in X like the case when $|X| \geq 2$ in RSA/G. As figure 8 shows, when X contains only one node $\{t_2\}$, it should be connected into G when v_i comes to t_3, and half of the connection length can be saved using the existing horizontal wire. The detailed algorithm is described in table 2, where routine 'connect(u, v)' uses existing wires if applicable on shortest connections.

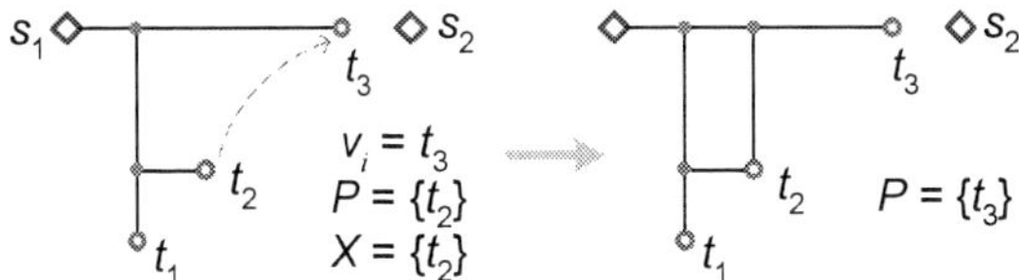

Figure 8: Connecting a node into the Steiner graph

Table 2: Revised RSA/G' algorithm

Given existing Steiner graph G, source s_k, terminals $t_1, \cdots, t_n$, and $v_1, \cdots, v_N$ are same as in RSA/G;

Routine Necessitate(vertex v);
$\quad U \leftarrow \{u \in G$ and exists a wire path from
$\quad\quad v$ to u of length $\Delta_s(v) - \Delta_s(u)\}$;
$\quad T' \leftarrow T' \cup \{u_m \in U$ with minimum $\Delta_s(u)\}$;

$T' \leftarrow \phi$;
for $i = 1$ to n do Necessitate(t_i);
$P \leftarrow \phi$;
for $i = 1$ to N do
$\quad$ if $v_i \in T'$ then $P \leftarrow P \cup \{v_i\}$; $\quad$ **(TMO)**
$\quad X \leftarrow P \cap \{v_j|\Delta_s(v_j) = \Delta_s(v_i) + \Delta(v_i, v_j)\}$;
$\quad$ if ($|X| \geq 1$ and $v_i \in G$) then $\quad$ **(SMO)**
$\quad\quad$ for each $(u \in X)$ connect(v_i, u);
$\quad\quad P \leftarrow P \setminus \overline{X}$;
$\quad\quad$ Necessitate(v_i);
$\quad$ else if ($|X| \geq 2$) then $\quad$ **(SMO)**
$\quad\quad$ merge the nodes in X rooted at v_i
$\quad\quad P \leftarrow (P \setminus \overline{X}) \cup \{v_i\}$;
return; $\quad$ (the MRSA rooted at s_k is added to G)

The k-IDeA iterations remain unchanged. After the shortest path Steiner graph is constructed by applying k-IDeA on the m sources, there are possibly some redundant edges that can be removed. So the final step is to check each edge $(v_i, v_j) \in G$, if G still contains all the master-to-slave shortest paths without (v_i, v_j), remove edge (v_i, v_j).

4.3 Switch control signals and wires

Besides the main part of bus wires, we have control signals for the switches on each Steiner node or terminal of G. When master s_i obtains the bus access from the arbiter and calls slave t_j by giving its address to the decoder, the connection is established through a shortest path in G, and the switches will guide the signals along this path. With pre-computed shortest paths, the control signals of each switch can be saved in a lookup table $C_{dat}(v, s_i, t_j)$, $v \in G$.

The control on data bus is relatively simple. The degree of a Steiner node can be 3 or 4 (s_i or t_j on a 4-degree node can be connected beside the node through a 3-way switch).

Table 1: The RSA/G algorithm

Given a source s and n terminals $t_1, \cdots, t_n$, $v_1, \cdots, v_N$ are the Hanan grid nodes sorted by $\Delta_s(v_1) > \cdots > \Delta_s(v_N)$;

$P \leftarrow \phi$;
for $i = 1$ to N do
$\quad$ if there is t_j at v_i, then $\quad$ **(TMO)**
$\quad\quad P \leftarrow P \cup \{v_i\}$;
$\quad X \leftarrow P \cap \{v_j|\Delta_s(v_j) = \Delta_s(v_i) + \Delta(v_i, v_j)\}$;
$\quad$ if ($|X| \geq 2$) then $\quad$ **(SMO)**
$\quad\quad$ merge the nodes in X rooted at v_i
$\quad\quad P \leftarrow (P \setminus \overline{X}) \cup \{v_i\}$;
return the arborescence rooted at s;

remove up to k nodes from $v_1, \cdots, v_N$ when running the RSA/G algorithm. By removing some nodes, the SMO merges are skipped at those points, which in some cases can result in better overall solution. In each iteration of the k-IDeA algorithm, all the combinations of skipping k or less nodes are tried in the RSA/G and the best set of skipped nodes are marked as permanently deleted. The iterations are then repeated until no further improvement is obtained.

4.2 Multiple MRSA construction with shared wires

With multiple sources $s_1, \cdots, s_m$, our algorithm needs to construct a Steinter graph G which contains all the MRSAs starting form every source. While each MRSA is minimized by k-IDeA, the m arborescences should share as much wire as possible to minimize total wirelength on G. We devise additional heuristics based on k-IDeA to construct multiple MRSAs one by one, explained as follows.

First, on each MRSA (rooted at s_i on ith iteration) construction, the terminals requiring connections can move towards the source s_i along existing edges of G, so that the wires can be reused and shared. As shown in figure 7, with the wires of previous aborescences, we only need to connect 8 nodes instead of the original 16 terminals to form the MRSA rooted at s_2, because all the other terminals can be reached from one of these nodes with shortest path from s_2. This set of nodes (denoted T') can be obtained by checking each terminal t_j, which necessitates a shortest-path connection from s_i. Starting from t_i, we move towards s_i as much as possible along existing wire paths until reaching a vertex v (a terminal or a Steiner node) in G where no vertex closer to s_i can be reached, then add it to T'. When there are multiple paths in the graph, we pick the final vertex closest to s_i so the rest part of the path is short and likely to need less wires. Details are in routine 'Necessitate(v)' in table 2.

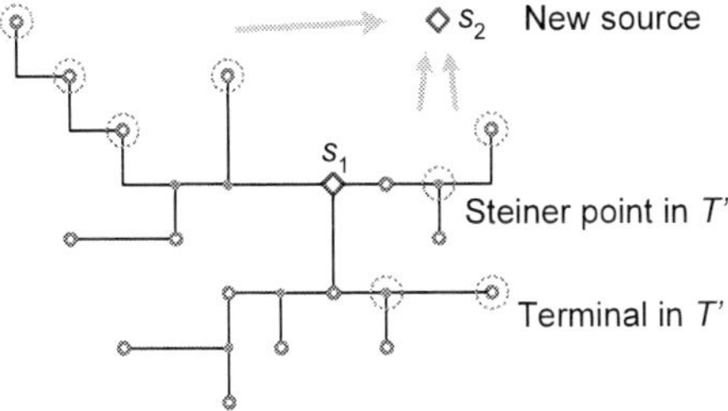

Figure 7: Nodes requiring connections

For a full 4-way switch as in figure 6(b), there are $6 = \binom{4}{2}$ configurations to connect 2 ports out of 4, so 3 wires are enough to send control signals (2^3 configurations) for this switch. And for a 3-way switch with $3 = \binom{3}{2}$ configurations, which is most common on the Steiner graph on random test cases, 2 wires are needed. Note that the connection is always between a master and a slave, so when some ports of a switch lead to only masters (or only slaves), the number of configurations may be reduced. For instance, every Steiner nodes in the tree of figure 7 has degree 3, but only 1 edge leads to the source (tree root). Thus, each switch needs only one control signal to guide the master-to-slave connection.

For address bus, the master-to-slave connection control can be shared with the data bus control. But there is also a master-to-decoder connection for each s_i, so the address signals go through another path to the decoder. All these paths form an MRSA, which is a subgraph of G. Each switch needs 1 or 2 control signals from a separate lookup table $C_{add}(v, s_i)$.

5. EXPERIMENTAL RESULTS

We compare three types of buses in our experiments: a) the original AMBA AHB, b) Steiner arborescence bus (as figure 4), and c) shortest-path Steiner graph bus. The three bus architectures are constructed on a set of cases with fixed placement of components. We list the average wire capacitance driven by data transactions in table 3 and the total wirelength in table 4.

For the original AMBA AHB, the wire net connecting the $s_1, \cdots, s_m$ (or $t_1, \cdots, t_n$) is routed as a Steiner tree. There are heuristics such as BI1S [8] and BOI [10] producing near optimal RMSTs (rectilinear minimum Steiner tree). We implement BOI to minimize the Steiner tree, and the multiplexer (MUX) is placed at the Steiner node where the total wirelength from $t_1, \cdots, t_n$ (or $s_1, \cdots, s_m$) is minimal. The bus control units (arbiter and decoder) in all cases are placed at center of the chip so that the overall wirelength is relatively small for control signals.

Table 3: Average data transaction wire capacitance

Case(m, n)	AHB	MRSA	SPSG	P_{save}
T0 (1,16)	5934	1638	469	72.4%~92.1%
T0 (2,16)	7584	2138	613	71.8%~91.9%
T0 (3,16)	8418	2038	635	75.8%~92.5%
T1 (3,16)	14550	4067	1754	72.0%~87.9%
T2 (2,30)	12305	2074	669	83.1%~94.6%
T3 (3,16)	8230	2019	691	75.5%~91.6%
T4 (5,15)	9674	1862	689	80.8%~92.9%
T5 (6,16)	11210	2012	694	82.1%~93.8%
T6 (8,8)	8952	1945	689	78.3%~92.3%
T7 (12,6)	12107	1933	657	84.0%~94.6%
T8 (16,10)	14377	1905	683	86.7%~95.2%
T9 (8,16)	11652	1883	618	83.8%~94.7%
T10 (8,16)	12899	2103	749	83.7%~94.2%
T11 (6,12)	8970	1897	668	78.9%~92.6%
T12 (12,12)	13242	1936	669	85.4%~94.9%

Table 3 shows the large power saving on wires by bus gating. For each transaction, bus gating can reduce the driven wires from a large net to a single path, by which the wire capacitance on data bus is reduced by over 90% for SPSG bus. Here we only count shortest paths on data bus, because

large, power intensive transactions can be transmitted by burst operation [18] which sends only one value on address bus. By the data in [12], the power on wires presents 14% of the bus power. Another 35% power consumed by bus interfaces (input stage of masters and slaves) is also reduced by about $\frac{m-1}{m}$ or $\frac{n-1}{n}$, because the unselected devices consume power by merely observing the AHB traffic in current AMBA AHB, and by [16] the wasted dynamic power dominates under the 0.15μm technology in [12]. In conclusion, nearly half of the power consumed by the AMBA AHB bus can be saved with bus gating. The percentage may vary as leakage power scales up with technology, but the bus wires are also scaling up and becoming a more important factor [9]. Considering that the bus consumes about 15% of the total system power in [12], bus gating may result in 5%~10% of system power reduction.

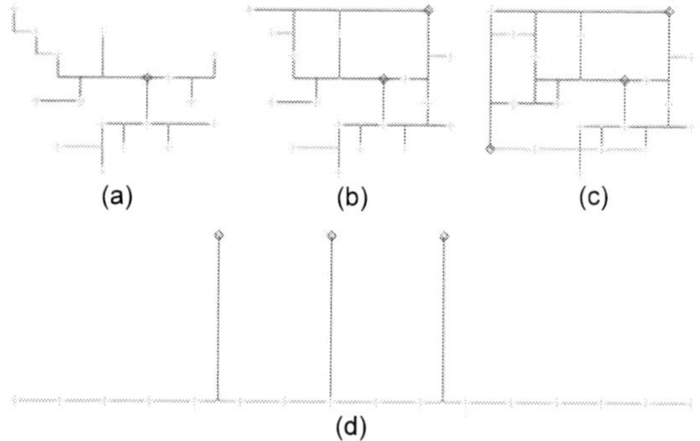

Figure 9: Shortest-path Steiner graphs of T0 & T1

Table 4: Total wirelength

Case(m, n)	AHB	MRSA	SPSG	$L_{increment}$
T0 (1,16)	11200	10316	9656	-7.9%~ −13.8%
T0 (2,16)	14000	11850	14078	−15.4%~0.6%
T0 (3,16)	16000	13350	21697	−16.6%~35.6%
T1 (3,16)	27000	15131	18238	-44.0%~−32.5%
T2 (2,30)	23257	16386	24535	−29.5%~5.5%
T3 (3,16)	15248	11554	14274	-24.2%~−6.4%
T4 (5,15)	17840	13729	22792	−23.0%~27.8%
T5 (6,16)	20988	15778	35287	−24.8%~68.1%
T6 (8,8)	16623	15800	25297	−5.0%~52.2%
T7 (12,6)	22640	13227	27113	−41.6%~19.8%
T8 (16,10)	27316	17018	40844	−37.7%~49.5%
T9 (8,16)	22183	16055	35267	−27.6%~59.0%
T10 (8,16)	24173	15906	34337	−34.2%~42.0%
T11 (6,12)	16920	12566	22070	−25.7%~30.4%
T12 (12,12)	25369	17256	38305	−32.0%~51.0%

The total wirelength (including control wires) of the bus may increase if we ensure all the master-to-slave paths are shortest. With regular placement of components, such as in figure 6 or the artificial case T1 in figure 9(d), the wirelength of SPSG bus can be smaller than AMBA AHB. In random cases (T2~T12), the wirelength is increased because more loops are produced in order to include all the shortest paths, especially when n and m are both large like the cases T5 and T12 in figure 10. T5 has the Steiner graph with most wirelength increment, where we can see some long but thin loops. If we squeeze the loop and combine the two long edges, we reduce wirelength by just slightly increasing the length of some master-to-slave paths. When the routing resource is not abundant, we can reduce wirelength by this operation. Ultimately if we use the MRSA bus which has on average 30% less wirelength than the original AMBA AHB, we still achieve around 80% power reduction on wires and the same mask-off effect on unselected bus interfaces.

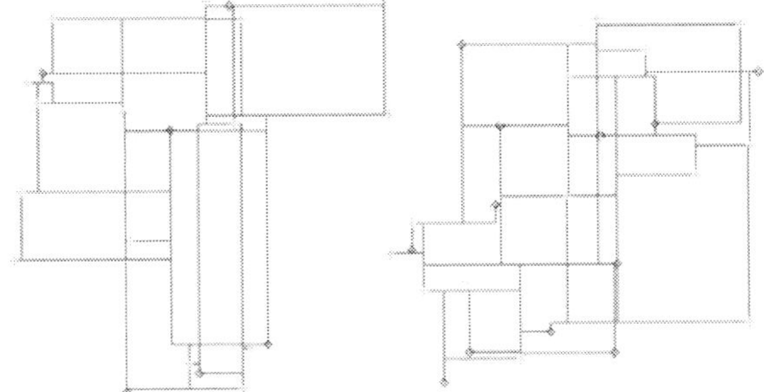

Figure 10: Shortest-path Steiner graphs of T5 & T12

We see a tradeoff between power consumption and wirelength. Figure 11 is drawn by data in the tables above. We have a flexible choice depending on routing resources, while in any case bus gating can always save a large percentage of power consumption on wires. The overhead power on additional controls and switches is low, considering the total number and activity rate of control signals, and that repeaters are usually inserted in bus wires to reduce delay.

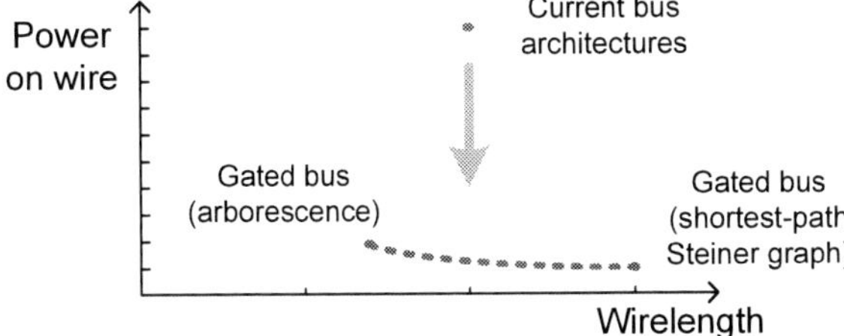

Figure 11: Tradeoff between power and wirelength

6. CONCLUSIONS AND FUTURE WORKS

Bus gating is a technique of reducing power consumption using the same effect as clock gating. Due to that only one pair of master-slave is involved at a time, bus gating can be more effective in terms of the percentage of power reduction. This percentage may become even more significant as technology scales down and the size of global wires relatively scales up.

Moreover, the Steiner graph approach also provides a possibility to allow multiple access as a bus matrix. In fact, if we are not restricted on resources, a full bus matrix in [13] added with demultiplexers can be used as a min-power bus, because the connection between each pair of master-slave is routed independently. In practice where the resources are usually limited, the Steiner graphs in figure 10 with appropriate controls also enable some pairs of connections to be active simultaneously, similar to independent train routes running on streets. Optimization on these multiple connections is a more challenging problem. Based on our shortest-path Steiner graph, power-bandwidth co-optimization on bus matrix will be explored in future works for more efficient on-chip communications.

7. ACKNOWLEDGEMENTS

This research is under the support of NSF CCF-0811794 and California MICRO Program.

8. REFERENCES

[1] C. J. Alpertt, A. B. Kahng, C. N. Szet, and Q. Wang. Timing-driven Steiner trees are (practically) free. *ACM/IEEE Design Automation Conf.*, pages 389–392, 2006.

[2] B. Bollobás, D. Coppersmith, and M. Elkin. Sparse distance preservers and additive spanners. *SIAM Journal on Discrete Math.*, pages 1029–1055, 2005.

[3] M. Caldari, M. Conti, M. Coppola, P. Crippa, S. Orcioni, L. Pieralisi, and C. Turchetti. System-level power analysis methodology applied to the AMBA AHB bus. *Design, Automation and Test in Europe: Designers' Forum*, 2:20032, 2003.

[4] J. Y. Chen, W. B. Jone, J. S. Wang, H. I. Lu, and T. F. Chen. Segmented bus design for low power systems. *IEEE Trans. VLSI Systems*, 7(1):25–29, 1999.

[5] J. Cong, A. B. Kahng, and K.-S. Leung. Efficient algorithms for the minimum shortest path Steiner arborescence problem with applications to VLSI physical design. *IEEE Trans. Computer-Aided Design*, 17(1):24–39, Jan. 1998.

[6] M. Donno, A. Ivaldi, L. Benini, and E. Macii. Clock-tree power optimization based on RTL clock-gating. *ACM/IEEE Design Automation Conf.*, pages 622–627, 2003.

[7] M. R. Garey and D. S. Johnson. The rectilinear Steiner tree problem is NP-complete. *SIAM Journal on Applied Math.*, 32:826–834, 1977.

[8] J. Griffith, G. Robins, J. Salowe, and T. Zhang. Closing the gap: Near-optimal Steiner trees in polynomial time. *IEEE Trans. Computer-Aided Design*, 13:1351–1365, 1994.

[9] R. Ho, K. W. Mai, and M. A. Horowitz. The future of wires. *Proceedings IEEE*, 89:490–504, 2001.

[10] C.-T. Hsieh and M. Pedram. An edge-based heuristic for Steiner routing. *IEEE Trans. Computer-Aided Design*, 13(12):1563–1568, Dec. 1994.

[11] C.-T. Hsieh and M. Pedram. Architectural power optimization by bus splitting. *Design, Automation and Test in Europe*, pages 612–616, 2000.

[12] K. Lahiri and A. Raghunathan. Power analysis of system-level on-chip communication architectures. *Int'l Conf. Hardware-Software Codesign and System Synthesis*, pages 236–241, 2004.

[13] S. Pasricha, Y.-H. Park, F. J. Kurdahi, and N. Dutt. System-level power-performance trade-offs in bus matrix communication architecture synthesis. *Int'l Conf. Hardware-Software Codesign and System Synthesis*, pages 300–305, 2006.

[14] M. Powell, S. H. Yang, B. Falsafi, K. Roy, and T. N. Vijaykumar. Gated-Vdd: a circuit technique to reduce leakage in deep-submicron cache memories. *Int'l Symp. Low Power Electronics and Design*, pages 90–95, 2000.

[15] S. K. Rao, P. Sadayappan, F. K. Hwang, and P. W. Shor. The rectilinear Steiner arborescence problem. *Algorithmica*, 7:277–288, 1992.

[16] S. Rodriguez and B. Jacob. Energy/power breakdown of pipelined nanometer caches (90nm/65nm/45nm/32nm). *Int'l Symp. Low Power Electronics and Design*, pages 25–30, 2006.

[17] M. Zachariasen. A catalog of Hanan grid problems. *Networks*, 38:200–1, 2000.

[18] AMBA 2.0 specification. *http://www.arm.com/products/solutions/AMBA_Spec.html*, 1999.

[19] Coreconnect bus architecture. *IBM White Paper*, 1999.

[20] Avalon interface specifications. *http://www.altera.com/literature*, 2008.

ActivaSC: A Highly Efficient and Non-intrusive Extension for Activity-based Analysis of SystemC Models

Cedric Walravens, Yves Vanderperren, Wim Dehaene
K.U.Leuven - ESAT-MICAS
Kasteelpark Arenberg 10, 3001 Heverlee, Belgium
cedric.walravens@esat.kuleuven.be

ABSTRACT

Today's highly integrated System-on-Chips (SoC) demand innovative architectural modeling means to cope with their energy constraints. In this context, SystemC has become a well-established simulation tool for Transaction Level Modeling (TLM) in the integrated electronics industry but it lacks the support for power modeling. This paper introduces ActivaSC, a flexible, fast, transparent and non-intrusive extension to the SystemC class library which allows capturing the activity information of a digital system being modeled. This information can then be post-processed to estimate power consumption. ActivaSC avoids time consuming design iterations as designers can assess architectural design trade-offs based on system activity early in the design flow. ActivaSC does not require any code alterations nor a specific API, so that it can be used with any modeling style. Finally, several benchmark tests illustrate the superior efficiency of ActivaSC. A speed-up of up to 75% in terms of elaboration overhead and 20% in terms of simulation overhead is realized with respect to prior art.

Categories and Subject Descriptors

B.7.2 [**Integrated Circuits**]: Design Aids; B.8.2 [**Performance and Reliability**]: Performance Analysis and Design Aids

General Terms

Performance, Languages

Keywords

SystemC, Activity monitoring, Power modeling

1. INTRODUCTION

With the advent of battery-powered electronic applications, concerns about the power consumption of such systems and the limited amount of energy available from the

Permission to make digital or hard copies of part or all of this work for personal or classroom use is granted without fee provided that copies are not made or distributed for profit or commercial advantage and that copies bear this notice and the full citation on the first page. To copy otherwise, to republish, to post on servers or to redistribute to lists, requires prior specific permission and/or a fee.
DAC'09, July 26-31, 2009, San Francisco, California, USA

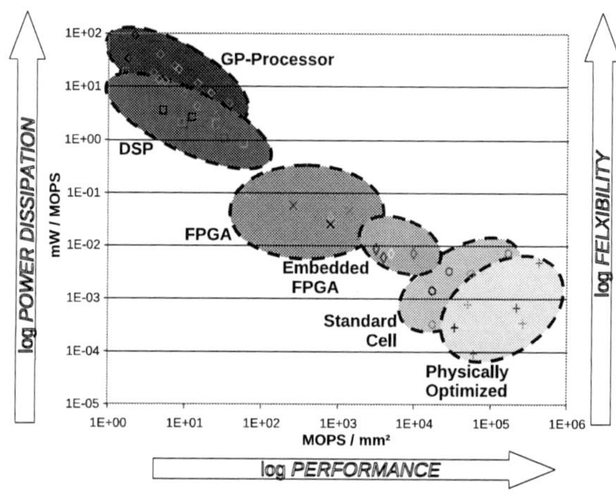

Figure 1: Design tradeoffs
[18]

battery capture an increasing attention. And as performance requirements tend to go *"More-than-Moore"* [22], the selection of a power optimized architecture is vital [12].

In the case of energy-constrained designs, it is particularly crucial to make appropriate architectural decisions, and choose for each block to what degree flexibility must be traded for power consumption (Figure 1).

Numerous power optimization techniques ranging from architectural to circuit level are available [12]. However, the power saving capabilities of these methods are limited –gate tuning at the logic level produces reduction averaging 15%– and concentrate the power optimization effort at a late stage of development. Only optimizations and power management solutions applied sufficiently early in the design cycle have the potential for radical power reduction.

The purpose of our work is therefore to provide designers with means to assess early the impact of their high-level architectural decisions on the power consumption of the system.

Energy-performance can be evaluated using the well-known expression $E = \alpha \cdot f \cdot C \cdot V_{DD}^2$. Frequency f, load-capacitance C and supply voltage V_{DD} can generally be adequately estimated based on simplified circuit-level models or past-experience [16]. The activity α, on the other hand, depends on what both functionality and communication will look like and their interaction. It is this activity which we want to capture.

978-1-60558-497-3/09 $25.00 © 2009 ACM

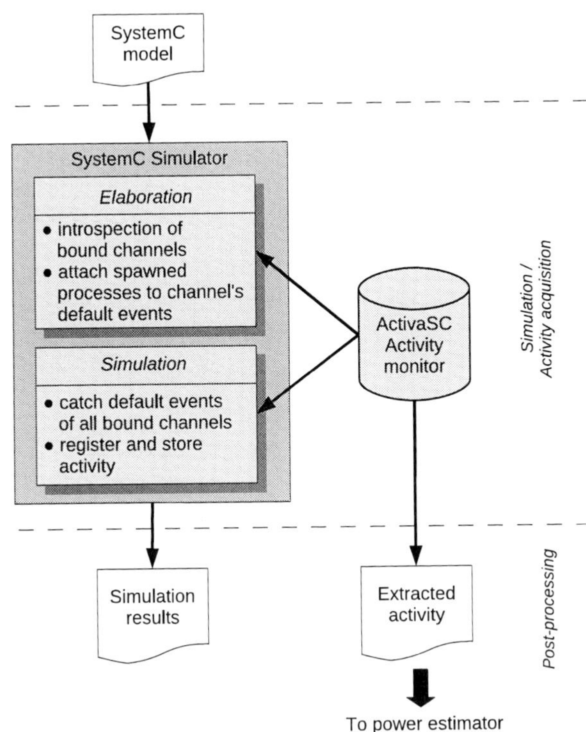

Figure 2: ActivaSC extending SystemC

We chose SystemC [19] as a platform for the basis of our work. Over the past few years, SystemC has become a well-established vehicle around which many model-driven design flows are built [20]. SystemC allows designers to model, simulate and verify the functionality and/or timing at any level of refinement.

Transaction-Level Modeling (TLM) [8] is an abstraction that facilitates an early, efficient and quick exploration of architectures, and it is fully supported by SystemC. A key concept of TLM is the independent development and refinement of communication and computation components within a system. Communication is modeled by channels, while transaction requests take place by making interface method calls of these channel models.

In this paper, we present ActivaSC as an extension of SystemC that monitors and registers the transaction activity within system models and addresses the problem of system-level power estimation with TLMs, focusing on the case where these models originally were not developed with power estimation in mind. ActivaSC allows for the introduction of an early energy-performance evaluation to SoC model-driven digital design flows. Designers now can assess a range of architectural solutions available and modify the systems architecture, if needed. To estimate power consumption, the information gathered by ActivaSC can be post-processed using power modeling tools [7][14] that are based on activity and signal statistics (see Figure 2). We deliberately choose to decouple data acquisition and analysis, in order to reduce simulation run-time and allow different power models to be applied to the same extracted activity data.

This paper is structured as follows: first, Section 2 pro-

vides an overview of related work and positions ActivaSC with respect to current state-of-the-art. Next, Section 3 details on the implementation of ActivaSC and covers all of its features. Finally, Section 4 illustrates the effectiveness of our approach with respect to prior art.

2. RELATED WORK

With the move towards system-level specifications and SoC design methodologies, there has been a significant research interest in power estimation and modeling of system-level, behavioral-level and register transfer (RT) level descriptions.

Power estimation at system-level is based on high-level descriptions using abstract models of capacitance and switching activity [5][1]. At the behavioral-level, power consumption is estimated based on functional descriptions of the hardware components [2]. RT-level estimations model the power consumption of middle-grained components such as multiplexers, adders, multipliers and registers [23][9].

High-level power estimation usually requires a variety of power models. Two types of modeling approaches can be employed: bottom-up and top-down. Bottom-up techniques attempt to capture the power dissipated by modeling the measurements of existing implementations. The resulting power macromodels can be used during high-level power estimation, without the need to perform more computational expensive, lower-level power estimation [14]. The bottom-up approach works best for library and IP based designs. The top-down approach, on the other hand, is best suited for application specific portions of the design that have not been implemented previously. Here, a combinatorial circuit is specified as a Boolean function, without any further information regarding its low-level implementation [7].

Whichever approach is used, they both rely on knowledge regarding input and/or output signal statistics. As SystemC and TLM can model the abstraction levels under consideration, several approaches have been proposed which associate TLM and activity profiling in recent years.

PowerSC [13] registers signal activity by extending the predefined sc_signal SystemC base class. However, this extension is only valid for integers, logical levels and bit values datatypes. As a result, all transactions must be specified at a low abstraction level in terms of data types and user-defined transaction channels which are not derived from sc_signal are prohibited. In contrast, our approach allows activity on any data type to be counted, as will be described in Section 3.3.

DUST [15] registers transactions by extending the predefined sc_port SystemC base class. Again, this prevents the use of ports based on any other class than sc_port, e.g. exports are not supported. DUST uses introspection to analyze automatically the SystemC model hierarchy and retrieve the transaction interfaces at run-time. However, this introspection is implemented using C++ Run-Time Type Information (RTTI) methods to identify the interfaces. This approach is considered as a bad coding practice [10] with a major impact on simulation code speed and efficiency, as confirmed by our benchmark results presented in Section 4. ActivaSC does not use RTTI.

TLM_power [17] presents a power estimation framework for SystemC which makes use of specific API calls to a newly developed power model C++ base class. This approach does not depend on any SystemC base class and is therefore transparent. On the other hand, this requires the introduction

978-1-60558-497-3/09 $25.00 © 2009 ACM

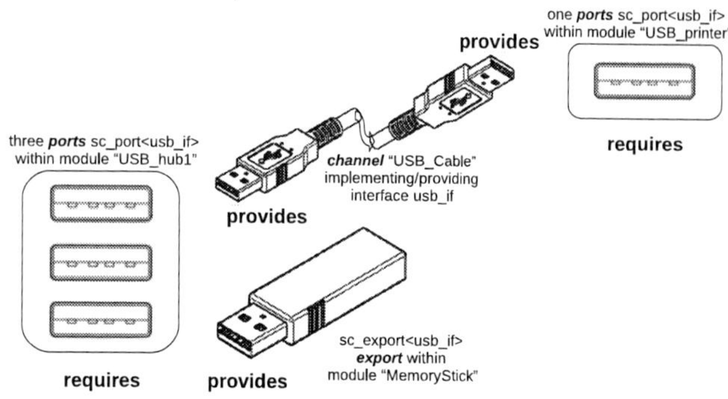

Figure 3: Ports and exports

of several specific power modeling and monitoring API calls into the code, which is fixed, prone to error and also hard to apply to existing models. ActivaSC, however, does not require any additional code change or API to monitor activity.

In conclusion, of all the approaches listed above, none is both fully non-intrusive and fully transparent at the same time, contrary to ActivaSC. Also, apart from TLM_power, they prevent the use of any custom-developed classes to model interfaces, one of the cornerstones of flexible and rapid SystemC modeling. Like DUST, ActivaSC extracts the activity data for post-processing, contrary to PowerSC and TLM_power which include a predefined power modeling approach.

In the following sections, the working principle and performance of ActivaSC will be discussed.

3. EXTENDING SYSTEMC WITH ACTIVITY EXTRACTION

3.1 Monitoring activity using interfaces

A SystemC model consists of several modules that typically separate different functional aspects of the model. Communication across module boundaries is mediated by ports and exports, which serve to forward method calls from the processes within a module to channels to which those ports or exports are bound. SystemC facilitates the development of TLMs and defines primary channels for communicating transactions but gives the user the freedom to define higher-level SystemC channels suited to their needs by using object-orientation, data encapsulation and generic programming concepts inherited from C++.

A channel that implements an interface class 'IF' by inheriting from that class *provides* that interface to the outside world. An sc_port<IF> member of a module *requires* that interface from the outside world. An sc_export<IF> member of a module *provides* that interface to the outside world by exporting a channel within that module (Figure 3). Multiports provide an easier way of binding a large number of ports simultaneously to the same interface, a concept used frequently to expedite code writing.

After investigating these various SystemC communication concepts [11] and their translation into a class hierarchy, we concluded that the optimal place to monitor activity is at the interfaces. The sc_interface class is common to every possible port, export, multiport and channel implementation.

Therefore, using interfaces all means of TLM communication can be monitored.

A further requirement is that the number of changes to existing SystemC code should be kept to a minimum. Creating a subclass of sc_interface with monitoring capabilities would require any reference to the original base class to be replaced by the new subclass in order to extend all interfaces. Furthermore, unless changes are made to the SystemC core library itself, it is impossible to replace the reference to the sc_interface class in the predefined and primitive channel core classes. All this makes such an approach highly impracticable. ActivaSC will therefore employ an alternative way to retrieve the interfaces.

3.2 ActivaSC implementation

ActivaSC uses the hierarchy traversal capability to locate all interfaces in the current model automatically and at runtime. Unlike DUST, this task is now purely accomplished through class-methods provided by SystemC [11] and does not depend on RTTI. As show in detail in Section 4, the result is a 75% faster simulation with respect to DUST, depending on the number of registered ports.

Then, as will be shown below, by retrieving their default events, which are notified when information is passed on across the implementing channel, all TLM activity can be registered. How this is accomplished exactly, will now be covered in further detail. Unless stated otherwise, the term "port" refers to ports, multiports and exports in the remainder.

The execution of a SystemC model consists of *elaboration*, during which the model hierarchy is created and initialized, followed by *simulation*, during which the scheduler runs the model. It is the SystemC kernel that provides core functionality to elaboration and the scheduler. Any instances of ports, channels and modules can only be made during elaboration afterwhich the model hierarchy remains fixed. Hence, performing hierarchy traversal after elaboration, ensures the complete model will be traversed.

The first problem encountered when attempting run-time hierarchy traversal, is that the hierarchical structure of SystemC is defined in terms of sc_object objects, including modules and ports. However, as sc_interface is not derived from sc_object, interfaces are not included. This is an intrinsic property of SystemC and TLM as the goal is to separate communication (i.e. channels, implementations of the sc_interface base class) and functional encapsulation (i.e. modules, derived from the sc_object base class).

SystemC interfaces are implemented by channels which in turn are bound to ports. It is also established that, under the assumption of hierarchical modular system modeling, all communication of interest occurs across module boundaries, passing through ports. Hence, by collecting ports, the bound interfaces can be retrieved. By doing so, every possible way of communicating through ports over channels is mapped. Also, ActivaSC can identify unique communication links across more than one hierarchical levels by supporting exports, as will be discussed at the end of this section.

Next, a mechanism to determine activity on each channel is required. In SystemC, processes can be made sensitive to events to enable synchronized communication. When a process wishes to get notified of new information arriving at its parent module port over the bounded channel, it must be

978-1-60558-497-3/09 $25.00 © 2009 ACM

made sensitive to the default event of that channel. By using the default_event() call provided by the sc_interface base class, it is possible to retrieve this very same default event. After attaching our own dynamically created spawned processes [6] to the default event of every channel, these spawned processes will become notified when the channel to which they became sensitive passes along information.

The following summarizes the novel approach followed by ActivaSC (also see Figure 2):

1. After the SystemC kernel has finished elaborating the model, ActivaSC performs a run-time traversal of hierarchy to collect all port types.

2. The bound channels are determined.

3. Dynamically spawned processes are created and attached to the default events of the channels.

4. Normal simulation starts, and every time a channel notifies its default event because of new information being transferred across the channel, the attached process is called and activity is registered.

ActivaSC is implemented as a header file which needs to be included in the main entry file of the SystemC model (with the #include<activasc.h> statement). During elaboration, ActivaSC instantiates a new EventCollector module. This module holds an implementation of the EventCollector::end_of_elaboration() method, which is called by the kernel right after elaboration, when in fact the model has been completely elaborated and fixed in the simulation context. This method performs all that is required to initialize ActivaSC. All ports in the model are stored in associative multimap arrays [21] that provide efficient methods to browse, search and organize their elements in such a way that is convenient for ActivaSC to identify interfaces and attach spawned processes to all bound channels. All spawned processes remain within the EventCollector module and during simulation each registers the activity on the channel to which it is bound. At the end of the simulation, the extracted activity information is written to a file.

3.3 Discussion

Our solution presents several advantages over prior art.

Thanks to the hierarchy traversal at run-time and the avoidance of class inheritance, ActivaSC becomes extremely *flexible* as it does not require any replacement of existing code. All data types, ports and channels can be employed, even custom-developed ones.

Also, ActivaSC remains completely hidden and is *transparent* to the user when writing the model. This means all SystemC models can be run and monitored with virtually no extra lines of code: required are only the inclusion of the ActivaSC header file and the addition of the `ACTIVASC_START` statement just before simulator invocation.

Further, all concepts employed by ActivaSC have been checked extensively against the IEEE 1666^{TM} SystemC Language Reference Manual Standard specifications [11], and have been found to comply with all requirements, recommendations and suggestions regarding best coding practices (e.g. not using RTTI). This ensures that ActivaSC is fully *portable* and compatible with any C++ compiler and SystemC simulator on the market.

Of course, by performing run-time hierarchy traversal, ActivaSC infers an overhead cost at the start of the simulation. Similarly, the activity monitoring and registration during simulation add to the simulation time. The next section will present benchmark results, comparing the overhead posed by ActivaSC with other works.

Note that ActivaSC requires SystemC release 2.2.0. The impact of deprecated features in this latest release, which already dates from March 2007, can be dealt with [4]. Moreover, with this release, SystemC has fallen in line with the IEEE 1666^{TM} Standard [3], which ensures better portability of the models across simulation platforms.

ActivaSC can handle any depth of hierarchy. Using sc_exports is the only valid way to lay a unique connection across several hierarchical levels in SystemC. ActivaSC supports the use of exports and will identify such a cascade of exports correctly as a single connection, so no multiple registration occurs.

Finally, a limitation of our approach is encountered when communication is modeled merely by issuing events between processes. Here, no port or channel is present and it will remain invisible to ActivaSC. However, such a communication model does not follow the TLM principle of separating functionality from communication.

4. BENCHMARK RESULTS

To confirm the efficiency of ActivaSC and to compare with solutions presented in other works, two benchmarks have been developed. Their goal is to measure and compare the run-time overhead imposed by monitoring activity.

They each use a Producer-Consumer model, consisting of two modules and an interconnecting multiport. Any port within this multiport is bound to a sc_signal channel which implements the required interfaces. Each will form a link that must be monitored for activity. During simulation, the producer generates activity in the model by sending every rising clock edge an unsigned 32 bit integer value (sc_uint<32>) across all links simultaneously.

We will describe the exact content of each benchmark, followed by a summary of the gathered results.

All benchmarks were performed for ActivaSC, PowerSC and DUST. The measured run-times for activity monitoring have been normalized to the run-time of a clean run without any kind of activity monitoring, e.g. 1.0 means no overhead and 2.0 indicates a run-time overhead of 100%. The absolute run-times of the clean benchmarks are provided for reference. All results were collected using the OSCI SystemC 2.2.0 simulator [19] on a dual Quad-Core 2.83 GHz processor workstation and averaged over 20 trials for each benchmark step, with a coefficient of variation of maximally 0,5%. Further, file and screen output was disabled in all tools, so only the activity extraction was measured and not the delays due to these output operations.

4.1 Elaboration overhead

4.1.1 Description

This first benchmark measures the amount of run-time overhead inferred during the elaboration stage. This overhead is a consequence of the amount of time and resources it takes to find and store all links within the model hierarchy.

By ending the simulation before the first rising clock edge, no activity is generated by the Producer and the simulator

978-1-60558-497-3/09 $25.00 © 2009 ACM

Table 1: Normalized Elaboration benchmark

#links	clean (sec)	ActivaSC	PowerSC	DUST
100	<0.01	×1.00	×1.00	×1.00
200	0.01	×1.00	×1.00	×1.00
400	0.01	×1.00	×1.00	×4.00
800	0.03	×1.33	×1.33	×4.00
1600	0.11	×1.18	×1.45	×4.00
3200	0.41	×1.15	×1.46	×4.10
6400	1.62	×1.07	×1.43	×4.07
12800	6.41	×1.03	*	×4.05

*resulted in a segmentation fault.

Table 2: Normalized Simulation benchmark

Sim-time	#links	clean (sec)	ActivaSC	PowerSC	DUST
1 ms	1	1.04	×1.65	×5.11	×2.01
	2	1.14	×1.97	×7.93	×2.55
	5	1.49	×2.63	×13.8	×3.55
	10	2.07	×3.28	×19.0	×4.54
	20	3.26	×3.91	×23.9	×5.25
	50	6.72	×4.78	×28.1	×6.25
	100	11.8	×5.48	×31.4	×6.75
50 ms	1	51.49	×1.65	×5.15	×2.01
	2	56.92	×1.95	×8.44	×2.52
	5	74.21	×2.62	×13.8	×3.54
	10	103.2	×3.28	×18.9	×4.52
	20	163.0	×3.92	×23.6	×5.22
	50	331.4	×4.85	×28.4	×6.30
	100	589.8	×5.49	×31.5	×6.77

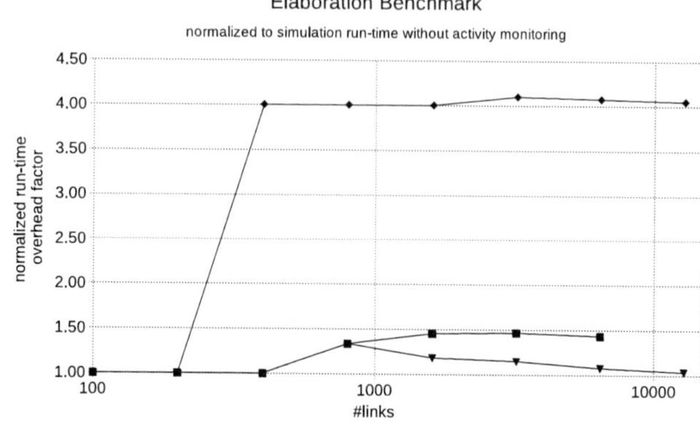

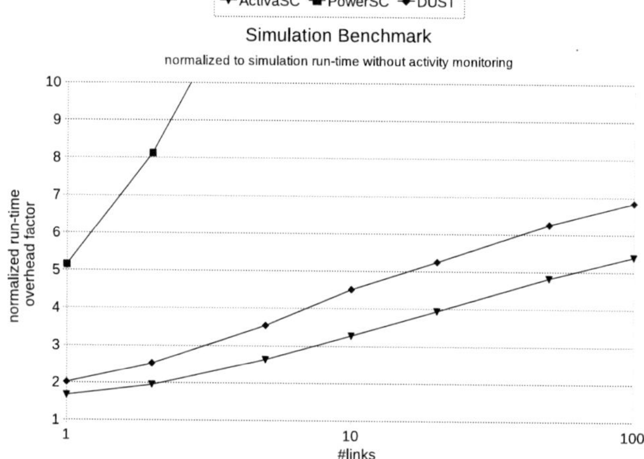

Figure 4: Normalized benchmark results

will exit immediately after elaboration. This allows to effectively measure the contribution of the elaboration phase to the total run-time overhead.

For each benchmark step, the number of links to monitor is increased and the run-time is measured.

Note that the Producer-Consumer model only has 2 modules, which has the for this benchmark desired effect of making the elaboration effort a strong function of the number of links.

4.1.2 Elaboration benchmark results

The absolute clean elaboration run-times together with the overhead factors for ActivaSC, PowerSC and DUST are gathered in Table 1. A graphical representation of the normalized elaboration overhead is shown in the upper plot of Figure 4 on a logarithmic scale.

It can be seen that ActivaSC performs like PowerSC up until 1k links. It should be noted in this context that PowerSC does not employ automated introspection like ActivaSC. Nevertheless, ActivaSC has comparable performance for less than 1k links and, above 1k links, ActivaSC outperforms PowerSC. In addition, PowerSC fails when the model exceeds 7000 links due to a segmentation fault. The inferior performance of DUST in this benchmark can be attributed to the use of RTTI, as mentioned in Section 2. ActivaSC does not suffer from this because it uses native SystemC API calls for introspection.

4.2 Simulation overhead

4.2.1 Description

The second benchmark measures the run-time overhead inferred during actual simulation. Again the same Producer-Consumer model is employed.

The influence of the elaboration phase is kept negligible by running the simulation over a fairly long time (up to 50 ms) with a high clock resolution (1 ns) and a reduced number of links (between 1 and 100 links).

For each benchmark step, the number of links to monitor is increased and the run-time is measured.

4.2.2 Simulation benchmark results

The absolute clean simulation run-times together with the overhead factors for the different monitoring implementations are gathered in Table 2. Only the values for 1 ms and 50 ms of simulation time are presented over the different link increments (between 1 and 100 links). It illustrates that within the margins of accuracy the normalized simulation run-time overhead factors are insensitive to the simulation time. The complete benchmark will take the average of the normalized run-time overhead over the range of 100 us to 50 ms of simulation time for each link incrementation step. The result is shown in the lower plot of Figure 4 on a logarithmic scale.

Here it can be seen that PowerSC presents the poorest

performance. It reaches an normalized overhead factor of 19 for only 10 ports. However, it should be noted that PowerSC performs some run-time analysis of the activity data, which can in part explain this result. DUST, like ActivaSC, does not do any run-time processing of activity, but has more overhead than our approach.

4.3 Summary

In conclusion, ActivaSC has the lowest elaboration overhead thanks to the avoidance of RTTI as well as the lowest simulation overhead thanks to an efficient internal structure to store communication elements and register their activity.

Benchmarks have also shown DUST out-performs PowerSC, as the elaboration does not fail for large amounts of links and as it has a lower simulation run-time overhead.

Comparing ActivaSC to DUST, our work presents a speed-up of up to 75% for elaboration overhead depending on the number of links. With regards to simulation run-time overhead, ActivaSC realizes an average speed-up of 20% compared to DUST.

5. CONCLUSION AND FUTURE WORK

In this paper, we presented an extension of SystemC to extract activity information during simulation, called ActivaSC. Using power modeling tools, designers now can assess a range of architectural alternative solutions and, if required, intervene early in the design process.

Different approaches to monitor activity have been discussed and considered. The implementation of ActivaSC has been covered extensively, as well as its advantages and limitations. ActivaSC has been shown to be a flexible and transparent solution, and great effort has been put in the reduction of additional run-time overhead during both elaboration and simulation. Finally, the efficiency of ActivaSC with respect to prior art has been illustrated using two benchmarks.

In future work, we will investigate how the monitoring overhead can be further decreased by using even more efficient internal data structures. ActivaSC will be used in the development of an ultra low-power controller and bus architecture for wireless sensor nodes. Future research directions might also include comparing several high-level power models and estimators.

6. REFERENCES

[1] A. Abril et al. Energy estimation and optimization in architectural descriptions of complex embedded systems. In *VLSI Circ. and Syst. II*, volume 5837 of *Proc. SPIE*, pages 456–466, Seville, Spain, May 2005.

[2] B. Arts, L. Benini, N. van der Eng, M. Heijligers, A. Kenter, E. Macii, H. Munk, and F. Theeuwen. Enhancing behavioural-level design flows with statistical power estimation capabilities. *Comp. and Dig. Tech., IEE Proc.*, 152(6):731–737, Nov. 2005.

[3] J. Aynsley. *Here's Exactly What You Can Do with the New SystemC Standard!* Doulos Ltd., 2006.

[4] J. Aynsley. *Dealing with Deprecated Features in SystemC 2.2.* Doulos Ltd., 2007.

[5] L. Benini and G. de Micheli. System-level power optimization: techniques and tools. *ACM Trans. Des. Autom. Electron. Syst.*, 5(2):115–192, 2000.

[6] D. C. Black and J. Donovan. *SystemC from the Ground Up*. Springer, 2nd edition, 2004.

[7] K. Buyuksahin and F. Najm. Early power estimation for VLSI circuits. *CAD of Integrated Circ. and Syst., IEEE Trans. on*, 24(7):1076–1088, July 2005.

[8] L. Cai and D. Gajski. Transaction level modeling: an overview. In *CODES+ISSS '03*, pages 19–24. ACM, Oct 1-3 2003.

[9] R. Damasevicius. Estimation of design characteristics at rtl modeling level using SystemC. *Information Tech. and Control*, 35(2):117–123, 2006.

[10] L. Goldthwaite. Technical report on C++ performance. Technical Report N1666, ISO/IEC, July 2004.

[11] IEEE Computer Society. *IEEE Standard SystemC Language Reference Manual*, IEEE Std 1666^{TM}-2005 edition, 31 March 2006.

[12] M. Keating et al. *Low Power Methodology Manual for System-on-Chip Design*. Springer, 2007.

[13] F. Klein, G. Araujo, R. Azevedo, R. Leao, and L. dos Santos. An efficient framework for high-level power exploration. *Circ. and Syst., 2007. MWSCAS 2007. 50th Midwest Symposium on*, pages 1046–1049, Aug. 2007.

[14] F. Klein, G. Araujo, R. Azevedo, R. Leao, and L. C. V. dos Santos. A multi-model power estimation engine for accuracy optimization. In *ISLPED '07: Proc. 2007 Int. Symp. on Low Power Electronics and Design*, pages 280–285, New York, NY, USA, 2007. ACM.

[15] W. Klingauf and M. Geffken. Design structure analysis and transaction recording in systemc designs: A minimal-intrusive approach. In *FDL*, Sep. 2006.

[16] P. Landman and J. Rabaey. Activity-sensitive architectural power analysis. *CAD of Integrated Circ. and Syst., IEEE Trans. on*, 15(6):571–587, Jun 1996.

[17] H. Lebreton and P. Vivet. Power modeling in SystemC at transaction level, application to a DVFS architecture. *Symp. on VLSI, 2008. ISVLSI '08. IEEE Computer Society Annual*, pages 463–466, April 2008.

[18] T. G. Noll. Application specific eFPGAs for SoC platforms. In *International Symposium on IEEE VLSI Design, Automation and Test, VLSI-TSA*, page 28, 2005.

[19] Open SystemC Initiative. http://www.systemc.org.

[20] SOCcentral. SystemC and ESL continue gaining momentum in 2007. http://www.soccentral.com/results.asp?EntryID=22441.

[21] B. Stroustrup. *The C++ Programming Guide*. Addison Wesley, 1997.

[22] G. Zhang et al. The paradigm of "more than Moore". *Electronic Packaging Technology, 2005 6th Int. Conf. on*, pages 17–24, 2005.

[23] L. Zhong. RTL-aware cycle-accurate functional power estimation. *CAD of Integrated Circ. and Syst., IEEE Trans. on*, 25(10):2103–2115, 2006.

GPU friendly Fast Poisson Solver for Structured Power Grid Network Analysis[*]

Jin Shi[1], Yici Cai[1], Wenting Hou[2], Liwei Ma[2], Sheldon X.-D. Tan[3], Pei-Hsin Ho[2], Xiaoyi Wang[1]

[1] EDA Lab, Computer Science Department, Tsinghua University, PRC   [2] Synopsys. Inc

[3] Electrical Engineering Department, University of California, Riverside, CA

shi-j03@mails.tsinghua.edu.cn,   caiyc@tsinghua.edu.cn,   wthou@synopsys.com,   lwma@synopsys.com, stan@ee.ucr.edu,   pho@synopsys.com,   wangxiaoyi00@mails.tsinghua.edu.cn

ABSTRACT

In this paper, we propose a novel simulation algorithm for large scale structured power grid networks. The new method formulates the traditional linear system as a special two-dimension Poisson equation and solves it using an analytical expressions based on FFT technique. The computation complexity of the new algorithm is $O(NlgN)$, which is much smaller than the traditional solver's complexity $O(N^{1.5})$ for sparse matrices, such as the SuperLU solver and the PCG solver. Also, due to the special formulation, graphic process unit (GPU) can be explored to further speed up the algorithm. Experimental results show that the new algorithm is stable and can achieve 100X speed up on GPU over the widely used SuperLU solver with very little memory footprint.

Categories and Subject Descriptors

B.7.2 [**INTEGRATED CIRCUITS**]: Design Aids – *graphics, layout, placement and routing, simulation, verification.*

General Terms

Algorithms, Design, Performance, Verification

Keywords

P/G network, Fast Poisson Solver, GPU

1. INTRODUCTION

Reliable on-chip power delivery is one of the major challenges for 90nm and below silicon technology. As the changes in geometry and physical aspects complicate the on-die power/ground (P/G) grid design, P/G grid simulators have to solve a very large scale linear dynamic system to compute the voltage drop distribution.

Many existing works have been proposed to address P/G grid simulation challenges in the past [1-4]. The mainstream solving algorithms can be classified into two classes: 1) direct solver 2) iterative solver. The direct solvers are robust and the resulting factor matrixes are reusable. The iterative solvers are more memory efficient. However, both types of solvers also have problems. For direct solvers, owing to the fill in the calculation of factor matrix, the memory usage of these algorithms tends to be very large. For iterative solvers, the instable performance of preconditioner has been widely criticized. Because today's P/G grids are highly regular and structured, regularity can be explored

to speed up the simulation. Work [5] gives out a partition method for large grid simulation based on locality property of multi-layer P/G grids under the flip-chip package. The "grid shell" based algorithm in [5] presents a parallel scheme to solve large scale structured P/G grid. And this concept is used successfully to accelerate the speed of power grid synthesis via a macro model technique [6]. The limitation of this technique is that it just can be applied to a grid with flip-chip package. For grids with wire-bond package, the locality characteristic does not exist. Research work [7] introduces the pattern based preconditioned conjugate gradient (PCG) method to use pattern information of structured grids. However, due to its iterative nature, the stability of the preconditioner still remains a problem.

Recently, the parallel computation power of GPU increases rapidly. GPU is suitable for calculating computation intensive problems. Existing works have explored the GPU for solving linear systems, such as [8-10]. However, typically well-optimized classical solvers for P/G simulation are difficult to be parallelized due to communication and synchronization costs. So it is hard to fully unleash power of GPU to accelerate the traditional P/G grid simulation. In this paper, we present a novel simulation algorithm for structured on-chip P/G grids. This algorithm explores the characteristic of structured P/G grids and formulates the linear system as a two dimension Poisson equation. Because two dimension Poisson equation has analtical solution and can be solved efficiently using Fast Fourier Transformation (FFT) technique, the computation complexity can be reduced to $O(NlgN)$, which is compared favorably with the traditional solving techniques, such as SuperLU and the PCG method, whose computation complexity is about $O(N^{1.5})$. Further, as the related 2D FFT calculation can be easily calculated in parallel using GPU, the final speed up ratio over SuperLU can exceed 100X. Also, experimental results show that due to the use of analytical solution, the new solver is as stable as direct solvers and only consumes small memory without suffering the fill-in problem.

Our paper is organized as follows: Section 2 briefly introduces the background of P/G grid simulation; Section 3 introduces the concept of Poisson Block in both flip-chip package and wire-bond package; Section 4 introduces the formulation of Poisson Equation and its analytical solution; in section 5, we introduce how to handle boundary condition using an iterative algorithm; Section 6 introduce how to accelerate the original solver under GPU environment; Section 7 gives out the experimental results and analyzes the performance of the new algorithm. Finally, section 8 concludes the whole paper.

Permission to make digital or hard copies of part or all of this work for personal or classroom use is granted without fee provided that copies are not made or distributed for profit or commercial advantage and that copies bear this notice and the full citation on the first page. To copy otherwise, to republish, to post on servers or to redistribute to lists, requires prior specific permission and/or a fee.

DAC'09, July 26-31, 2009, San Francisco, California, USA

*This work is supported in part by "The National Science and Technology Major Projects No.2008ZX01035-001-05" and in part by "The National Natural Science Foundation of China (NSFC) No.60776026 and No.60828008"

2. P/G grid Simulation Background

For simplification, we only introduce the formulation of resistive P/G grids here. For inductance and capacitance parasitics, the complete RLC model of a P/G grid can be found in [1]. Also, by using the back-Euler or the trapezoidal-Euler method, the inductors and capacitors can be discretized as resistors and equivalent absorbing current sources at each time stamp. Thus the algorithms proposed in this paper can still be applied RLC P/G grids.

For a P/G grid, a topology adjacent matrix A can be obtained. Each column of matrix A represents an edge in the grid. For a specific column, an edge starts from node i to node j (named e_{ij}), will stamp integer number 1 in the ith row and -1 in the jth row of A. Supposing the grid has m rows and n columns, that is total $N = m \times n$ nodes, and E edges, then matrix $A \in R^{N \times E}$. Considering each edge as a resistor which has conductance g_i, then we can use a diagonal matrix $g \in R^{E \times E}$ to represent all resistors in the grid.

Giving any current distribution (using a vector $i \in R^N$ to describe the absorbing current at each node), in order to determine the voltage drop at each node (a vector $x \in R^N$), a linear system which satisfies the Kirchhoff's current law and voltage law can be built as following:

$$Mx' = i', \ M = BgB^T \quad (1)$$

In equation (1), matrix B, vector x' and i' can be obtained by deleting all the pad related rows in matrix A, vector x and vector $\underline{i}$. Thus, if there are total p pads in the grid, then $B \in R^{(N-p) \times E}$, and $M \in R^{(N-p) \times (N-p)}$, $x', i' \in R^{N-p}$. Due to the sparse characteristic of matrix B, matrix M is also a sparse matrix. Furthermore, it is still a definite matrix because of the element fill in pattern [1]. In order to solve the linear system (1), sparse direct solver such as SuperLU [11] has been widely used in industry tools for its efficiency and stability. Also, iterative solver, such as preconditioned conjugate gradient (PCG) method [12] was also applied to solve this linear system. As mentioned in the introduction, both methods have their own problems. To mitigate their problems and further improve the performance of these two methods, we need to explore parallelization schemes and unleash the parallel computing power of GPU.

3. Poisson Block in P/G grid

Research work [5] presents an excellent scheme to divide a large scale P/G grid into small blocks, which can be solved completely and separately without losing any accuracy. It uses a special characteristic of P/G grid with flip-chip package named "locality". "Locality" means that the absorbing current sources only can trigger voltage drop within a local area around it. And beyond this area, the triggered voltage drop will attenuate to zero very fast. However, this property only exists in P/G grids with flip-chip package, because pads are distributed evenly in the grid, and they act as strong current drain points, which prevent the current to go far away from its source point. For P/G grids with wire-bond package, the situation is quite different. Currents have to reach the boundary of the grid to meet the drain points (pads). Thus locality based division cannot be directly applied to solve wire-bond based P/G grids.

Here we extend the locality concept to Poisson Block concept so that we can handle it for both packaging types. A Poisson Block stands for a special area in the grid which satisfies three conditions: 1) the boundary of this area has zero voltage drop. 2)

pad nodes only locate in block boundary 3) the internal nodes only can be attached to absorbing current sources or be floating. Similar to [5], because of the independent nature of Poisson Block, each Poisson Block can be solved completely separately. Further, if we look up industrial designs, we can find that lots of cases can be treated as Poisson Blocks. Blow we list some examples to show practical Poisson Blocks. First, a full P/G grid with wide boundary core-ring and assembled with wire-bond technique can be treated as a Poisson Block. Also, the P/G grid inside an IP block with a wide core-ring can also be treated as a Poisson Block. This is because the wide core-ring makes the voltage drop at boundary nodes to be very close to zero, thus the first rule of Poisson Block can be satisfied. For the second and the third definition of Poisson Block, they can be satisfied naturally under wire-bond package type.

Another case for Poisson Block is under flip-chip package model. As the locality property shows, local current sources just trigger voltage drop within a finite area. Thus, if we enlarge the area to a proper size, then the voltage drop at boundary nodes will be close to zero too. Then the first condition of Poisson Block is met. Furthermore, if we treat the pads inside the block as special boundary nodes instead of the internal nodes by using a special algorithm introduced in Section 5, the last two conditions of Poisson Block can be also met. So the local area in a P/G grid with flip-chip package can also be treated as a Poisson Block.

For an example, we place 5 current sources around the center node of a 50×50 grid with uniformly distributed pads. Figure 1 shows the voltage drop distribution triggered by these current sources. In this figure, we can see that the area contains dominant voltage drop is around the center part of the grid. Also, voltage drop decreases as the distance between the center nodes increases. So if we treat the area bounded by the black dashed rectangle as the local area, we can find that the voltage drop at the boundary is close to zero. Also, beside the nature boundary nodes given by the dashed rectangle, we have to add the pad nodes inside the local area as new boundary nodes. Because the voltage drop on pad node is zero too, adding the pad nodes to the boundary will not break the first condition of Poisson Block while it also satisfies the second condition of Poisson Block. Therefore, an extended local area with internal pad nodes satisfies all the conditions of Poisson Block. In the following section, we can see how we can build our new algorithm under the definition of Poisson Block.

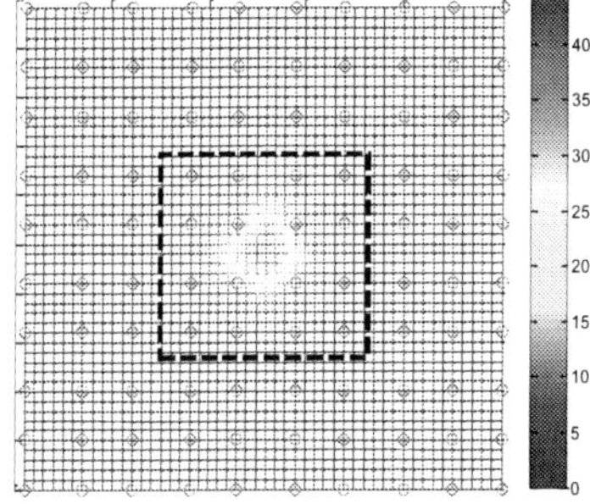

Figure 1. A Poisson Block Case: local area in flip-chip assembled P/G grid (voltage drop unit is milli-volt), red circles stand for pads

4. New Circuit Formulation for Regular Grid and analytical Solution

The classic formulation of P/G grid tries to use a sparse matrix M as in equation (1) to describe the whole linear system. In our new formulation for regular P/G mesh circuits, we use a dense matrix to describe the relationship between current and voltage drop distribution. Before we introduce our detail formulation, we walk through a simple example to show the basic idea. Figure 2 shows a 3×4 grid. Each node in the grid is numbered. The voltage value of each node is marked as u_1 to u_{12}, and independent current sources i_1 to i_{12} are attached to these nodes.

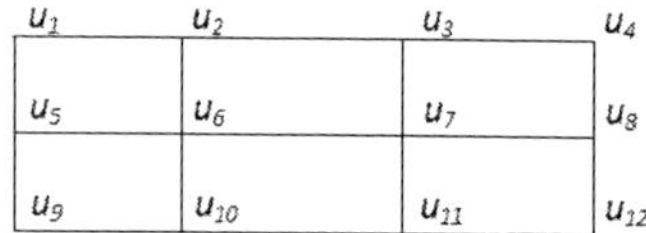

Figure 2. A small grid to show the formulation

We suppose each edge of the grid to have the same resistance r because of the grid regularity. Then we can use a dense matrix U to describe the voltage distribution as shown in equation (2). The relationship of current and voltage distribution which is controlled by Kirchhoff current law can be represented as equation (3).

$$U = \begin{bmatrix} u_1 & u_2 & u_3 & u_4 \\ u_5 & u_6 & u_7 & u_8 \\ u_9 & u_{10} & u_{11} & u_{12} \end{bmatrix} \quad (2)$$

$$\begin{bmatrix} -1 & 1 & \\ 1 & -2 & 1 \\ & 1 & -1 \end{bmatrix}\begin{bmatrix} u_1 & u_2 & u_3 & u_4 \\ u_5 & u_6 & u_7 & u_8 \\ u_9 & u_{10} & u_{11} & u_{12} \end{bmatrix} +$$

$$\begin{bmatrix} u_1 & u_2 & u_3 & u_4 \\ u_5 & u_6 & u_7 & u_8 \\ u_9 & u_{10} & u_{11} & u_{12} \end{bmatrix}\begin{bmatrix} -1 & 1 & & \\ 1 & -2 & 1 & \\ & 1 & -2 & 1 \\ & & 1 & -1 \end{bmatrix} = r\begin{bmatrix} i_1 & i_2 & i_3 & i_4 \\ i_5 & i_6 & i_7 & i_8 \\ i_9 & i_{10} & i_{11} & i_{12} \end{bmatrix} \quad (3)$$

For a more complicated grid, we can get a conclusion that if the grid is regular (metal rails share the same width) and rails are distributed evenly cross the die area (most P/G grid is highly regular), then the Kirchhoff current law can be represented by a matrix equation as shown in (4):

$$T_1U + UT_2 = rI \quad (4)$$

Supposing the P/G grid is an $m \times n$ grid, then in (4), $I \in R^{m \times n}$ represents the current distribution in the grid while $U \in R^{m \times n}$ represents the voltage distribution. $T_1 \in R^{m \times m}$ and $T_2 \in R^{n \times n}$ are tri-diagonal matrices similar to the two matrices used in (3). If we define voltage drop matrix $V = vcc \cdot E - U$ (E is a unit matrix while vcc stands for the supply voltage), then equation (4) can be rewritten as (5):

$$vcc \cdot T_1 \cdot E + vcc \cdot E \cdot T_2 - T_1V - VT_2 = rI \quad (5)$$

From (3) we know that the sum of row of T_1 and T_2 is zero, then we will have $T_1E = ET_2 = 0$. So (5) can be simplified to (6):

$$-T_1V - VT_2 = rI \quad (6)$$

Equation (6) gives a new formulation to describe the voltage distribution under regular P/G grid. To make the formulation practical, we still need to make some changes. First, equation (6) is similar to a two-dimension Poisson Equation which has analytical result [13]. If we can change the form (6) to the form of Poisson Equation, it will make the problem easy to solve. Poisson Equation has coefficient tri-diagonal matrix P which satisfies the following form as shown in (7). By adding diagonal matrix d_1 and d_2 to T_1 and T_2 respectively, we can make the left hand side of

equation (6) to match the form of Poisson Equation, as shown in equation (8):

$$P = \begin{bmatrix} 2 & -1 & & & & \\ -1 & 2 & -1 & & & \\ & -1 & \ddots & & \ddots & \\ & & & \ddots & 2 & -1 \\ & & & & -1 & 2 \end{bmatrix} \quad (7)$$

$$(d_1 - T_1)V + V(d_2 - T_2) = rI + d_1V + Vd_2$$

$$d_1 = d_2 = \begin{bmatrix} 1 & & & \\ & 0 & & \\ & & \ddots & \\ & & & 0 \\ & & & & 1 \end{bmatrix} \quad (8)$$

Because there are only two nonzeros in matrix d_i, if we analyze the matrix d_1V and Vd_2 in the right hand side of equation (8) carefully, we can see that the nonzero elements in d_1V and Vd_2 just represent the boundary elements in matrix V. Remember that the boundary voltage drop of a Poisson Block is equal to zero, so if we apply equation (8) to a Poisson Block, it can be further simplified to (9), which is a real two-dimension Poisson Equation.

$$P_1V + VP_2 = rI \quad (9)$$

Furthermore, we can derive more general equation for the grids having adjacent two metal layers with different routing widths. The derivation of this equation is very similar to (9) as shown in (10). In (10) r_x and r_y represent the resistance of each metal segment in a horizontal routing layer and a vertical routing layer respectively.

$$\frac{P_1V}{r_x} + \frac{VP_2}{r_y} = I \quad (10)$$

The interesting thing about Poisson Equation is that we have the analytical eigen decomposition of tri-diagonal matrix P as shown in (11):

$$P_1 = z_1\Delta_1z_1^T, \quad P_2 = z_2\Delta_2z_2^T \quad (11)$$

In equation (11), z_1 is an $m \times m$ symmetry orthogonal matrix while z_2 is an $n \times n$ symmetry orthogonal matrix given by (12).

$$z_1(i,j) = \sqrt{\frac{2}{m+1}}\sin(\frac{i \cdot j \cdot \pi}{m+1}), z_2(i,j) = \sqrt{\frac{2}{n+1}}\sin(\frac{i \cdot j \cdot \pi}{m+1}) \quad (12)$$

Also, Δ_1 and Δ_2 are two diagonal matrix given by (13)

$$\Delta_1(i,i) = 2(1 - \cos(\frac{i \cdot \pi}{m+1})), \Delta_2(i,i) = 2(1 - \cos(\frac{i \cdot \pi}{n+1})) \quad (13)$$

Let $V = z_1 \cdot X \cdot z_2^T$ (X is a middle result which is very useful to get the final close form expression), substitute (11) into (10) with the definition of V, remember that matrix z_1 and z_2 are orthogonal matrices, then we can get an analytical expression about X as shown in (14):

$$X = (z_1^T \cdot I \cdot z_2) \otimes W \quad (14)$$

In equation (14), operator $\otimes$ means that the result matrix in brackets performs Hadamard matrix multiplication [14] with matrix W, where $W \in R^{m \times n}$ is a dense matrix denoted by the following expression:

$$W(i,j) = \frac{1}{\dfrac{\Delta_1(i,i)}{r_x} + \dfrac{\Delta_2(j,j)}{r_y}}$$

Finally, from (14) we can get the analytical solution of equation (10) as shown in (15).

$$V = z_1 \cdot X \cdot z_2^T = z_1 \cdot ((z_1^T \cdot I \cdot z_2) \otimes W) \cdot z_2^T \quad (15)$$

Further, due to the symmetry property of matrix z_1 and z_2, (15) can be expressed as (16).

$$V = z_1 \cdot ((z_1 \cdot I \cdot z_2) \otimes W) \cdot z_2 \quad (16)$$

So far, once we know the voltage drop distribution, by using (10), we can get the current distribution. On the other hand, if we know the current distribution, using (16) we also can get the voltage drop distribution. In any direction, the solution is absolutely unique and in an analytical form.

5. Boundary Iteration Process

For wire-bond packaging type, all the pad nodes are located at block boundary, thus all boundary elements of matrix I in equation (16) are unknown because they stand for the current drawn by pads. For flip-chip packaging type, the unknown elements of matrix I may be appeared at the internal position of matrix I because there are pads inside each Poisson Block. Because we cannot know the exactly current drained by pads in advance, thus we cannot directly use (16) to get the voltage drop distribution. This problem is named boundary problem. In order to handle this problem, we develop an analytical iterative method. The basic idea here is to use an iteration process to improve the accuracy of initial guess of the current drawn by pad nodes step by step. Once we set up a boundary current distribution drawn by internal pads, then we can merge them with other nodes' current distribution to complete the matrix I. Then by using the equation (16), we can get the overall voltage drop distribution including the boundary voltage drop. The key observation here is that the voltage drop on pads should be zero. If a guessed boundary current distribution causes the calculated result to be nonzero, we know that the boundary current distribution is not accurate. Further, because the solution of equation (10) is unique, if an iterative process can make the boundary voltage drop to converge to zero, then the boundary current distribution will also converge to the accurate value, which will cause the overall voltage drop distribution to converge to the unique solution of equation (10). Thus, we can use a minimization algorithm to compute the accurate solution.

Specifically, by defining the current vector drawn by pad nodes as i_b, other nodes' current distribution vector as i_n, the boundary voltage drop vector as v_b, then the complete simulation process can be formulated as a minimization process as following:

$$\min_{i_b \in R^{np}} \|v_b\| \quad s.t. \begin{cases} v_b = bond(V) \\ V = z_1 \cdot ((z_1 \cdot I \cdot z_2) \otimes W) \cdot z_2 \\ I = i_b \| i_n \end{cases} \quad (17)$$

Here we use a function named "*bond*" to stand for the process that extracts the voltage drop at pad nodes from overall drop matrix V to get vector v_b. Also, the construction process of matrix I using vector i_b and i_n is expressed as $I = i_b \| i_n$. Minimization process (17) tries to find an optimal vector i_b in the geometry space R^{np} (np is the number of pads, m and n are related to the grid size) to minimize the voltage drop on pads nodes. As we known, a Newton gradient method together with a line search strategy can be used to solve the minimization problem given by (17) [15]. Usually, the number of pad nodes is few thousand, so if we can get the gradient direction analytically during the iteration, boundary problem caused by pad nodes can be solved very efficiently.

Suppose v_{ij} to be an element at the ith row and j^{th} column of matrix V, standing for the voltage drop of a pad node. Then the gradient of v_{ij} to any variable i_{st} (at s^{th} row and t^{th} column) in matrix I can be denoted by $\dfrac{\partial v_{ij}}{\partial i_{st}}$. Expand equation (16) and perform derivative operation on both sides, we can get the analytical expression of the gradient in (18), where $z_1(i,:)$, $z_1(j,:)$, $z_2(s,:)$, $z_2(t,:)$ represent the ith row, the jth row of matrix z_1 and the sth row, the tth row of matrix z_2 respectively.

$$\frac{\partial v_{ij}}{\partial i_{st}} = \sum_{a=1}^{m} \sum_{b=1}^{n} Y_{ab} \quad (18)$$

$$Y = [z_1(i,:)^T \cdot z_1(j,:)] \otimes [z_2(s,:)^T \cdot z_2(t,:)] \otimes W$$

From (18) we know that $\dfrac{\partial v_{ij}}{\partial i_{st}}$ is the sum of all elements of matrix Y which can be computed by performing Hadamard matrix multiplication on matrix $[z_1(i,:)^T \cdot z_1(j,:)]$, $[z_2(s,:)^T \cdot z_2(t,:)]$ and W successively. As (18) gives a universal form to calculate the gradient, we can use it to calculate the Jacobian matrix J of vector v_b over vector i_b. After the J matrix is known, we can give out the complete flow of our Fast Poisson Solver algorithm as shown in Figure 3. Due to the known of analytical gradient and the small size of the Jacobian matrix, Fast Poisson Solver converges very fast.

Table 1 shows a convergence ratio comparison between the PCG method and our Fast Poisson Solver. For this test, both the PCG solver and the Fast Poisson Solver's convergence threshold are set to 1e-6. Zero fill-in incomplete LU decomposition (ILU(0)) is used as the precondition matrix for the PCG method. From Table 1, it is clear that as the size of test cases increase, the iteration number of the PCG method increases because the performance degradation of the precondition matrix. However, the iteration number of Fast Poisson Solver remains similar as the size increases. There are two reasons for the fast and stable convergence rate of Fast Poisson Solver: 1) the total pad number does not change obviously from case to case 2) Fast Poisson Solver just searches the result in a low dimension space containing all pad nodes while the PCG method searches for a huge space which contains all nodes of the power grid.

Table 1. Convergence Ratio Comparison between PCG Solver and Fast Poisson Solver

Node Num	PCG Iteration	Max PCG Error	Poisson Solver Iteration	Max Poisson Solver Error
1M	709	9.851000E-07	23	3.356580E-07
1.2M	779	9.661470E-07	34	3.185260E-07
1.4M	851	9.805820E-07	32	4.445130E-07
1.6M	922	9.677930E-07	21	4.100990E-07
1.9M	993	9.915360E-07	29	3.679020E-07
2.3M	1063	9.647550E-07	17	7.098380E-07
2.5M	1142	9.809760E-07	19	1.023180E-06
2.9M	1199	9.981070E-07	18	7.964340E-07
3.2M	1266	9.905290E-07	21	1.065640E-06
3.6M	1347	9.739280E-07	25	1.091320E-06
4M	1421	9.951570E-07	25	1.200610E-06

Also, from Figure 3, we can see that in step 2 of the Fast Poisson Solver, the computation time is mainly spent in sub-step 1, sub-step 4 and sub-step 11 during each iteration. This is because when the grid size is increasing all these sub-steps require large size dense matrix multiplication operations to calculate equation (16) and (18), which are very time consuming. In the following sections, we will introduce a technique to speedup these computation under a GPU environment.

Algorithm Name: Poisson Solver

Input: A P/G grid with current distribution

Output: Voltage drop distribution of the grid

Process:

Step1: Partition the P/G grid as a few Poisson Block P_i together with sub current distribution I_i

Step2: For each Poisson Block pair (P_i , I_i)

1. Calculate the Jacobian matrix J using equation (18)
2. Set initial guess of boundary current distribution i_b to zero
3. Pack I_i and i_b to get I matrix in equation (16)
4. Calculate the boundary element of matrix V using equation (16)
5. Pack the boundary element of matrix V into vector v_b
6. Soving equation $Jf = -v_b$ to get a decent direction vector f
7. Find a scalar α to update i_b in order to get a maximal decrease within an iteration.
8. Update i_b to $i_b = i_b + \alpha f$
9. If not converge, goto 3 else goto 10
10. Pack I_i and i_b to get I matrix in equation (16)
11. Calculate matrix V using equation (16) to get overall voltage drop distribution

End for

Figure 3. Fast Poisson solver algorithm

6. Speedup on GPU

Recently, GUP has been proved to be very efficient for computing intensive tasks due to its inherent parallel computing powers [8-10]. However, the traditional solvers (SuperLU based and PCG based solvers) for P/G grid are difficult to be optimized to leverage the parallelism in GPU. No more than 10X speed up are reported by using GPU architecture in recent research [8-10]. Luckily, in Fast Poisson Solver algorithm, matrix multiplication is the main time consuming operation to calculate equation (16) and (18). Because the matrixes z_1, z_2 and W are only related to the grid size, once the grid size is determined, they keep unchanged during the iterative solving process. Thus, in Fast Poisson Solver, we can load them once and reuse them during iteration process. This property is very friendly for GPU computation because it reduces the communication between CPU and GPU dramatically, which is known as the main bottleneck for performance improvement on GPU architecture.

For equation (16), it can be reduced to 5 steps on GPU: (1) to get $t = z_1 \cdot I$ (2) to get $t = t \cdot z_2$ (3) to get $t = t \otimes W$ (4) to get $t = z_1 \cdot t$ (5) to get $t = t \cdot z_2$. Here all steps are standard matrix multiplication operation, which can be assigned to different core of GPU to achieve extremely high computation performance. Also, for step (3), the Hadamard matrix multiplication can be implemented as scalar vector multiplication which can also be calculated parallel on different GPU cores. For equation (18), it can be implemented on GPU in 4 steps: (1) to get $t_1 = z_1(i,:)^T \cdot z_1(j,:)$ (2) to get $t_2 = z_2(s,:)^T \cdot z_2(t,:)$ (3) to get $t = t_1 \otimes t_2$ (4) to get $t = t \otimes W$. Here steps (3) and (4) can also be implemented using vector scalar multiplication while step (1) and step (2) can be implemented using rank one vector multiplication on GPU. Vector scalar

multiplication, vector rank one multiplication and matrix multiplication are all defined in BLAS (basic linear algebra subroutine) library as level one and level two functions [16]. In this paper we choose NVIDIA CUDA framework [17] to deploy our GPU application. CUDA supports BLAS library and provides a very friendly C interface for programmer. Thus, all sub calculation steps mentioned above can be mapped to a CUDA function directly which are already implemented efficiently on GPU.

CUDA also supports for parallel 2D FFT library which can accelerate FFT computation speed by 15X in recent report [18]. Here due to the special structure of matrix z_1 and z_2, we also can benefit by using CUDA 2D FFT library. The improvement for efficiency by using FFT is caused by the matrix multiplication operation. As denoted by (12), all elements in matrix z_1 and z_2 are results of different sine functions. So $f = z_1 \cdot p$ can be calculated using Discrete Sine Transform method (DFT), which results $f = DFT(p)$. Further, according to basic property of DFT, it can be calculated using FFT on an extended vector with double length [19]. Thus, we can get the following formulation:
$$f = DFT(p) = FFT(q), \quad q = [0, p, 0, -p]$$
Ranging vector p column by column in an object matrix, matrix multiplication with z_1 and z_2 can be calculated using 2D FFT algorithm, i.e, $t = z_1 \cdot I = FFT(I')$, $I' = [0, I, 0, -I]^T$. Because the computation complexity of FFT is just around $O(NlgN)$, and matrix multiplication operation has the highest computation complexity in Fast Poisson Solver algorithm, thus the computation complexity of Fast Poisson Solver is bounded by $O(NlgN)$. Compared with the traditional solver's computation complexity $O(N^{1.5})$, such as SuperLU and PCG, even this algorithm is implemented without any parallel computing resources, the run time efficiency will be increased significantly.

For CUDA implementation, another issue is about its accuracy. All data can only be expressed as float numbers in GPU, which will affect lots of applications. However, due to the iterative nature of Fast Poisson Solver，the problem is partially overcome. One can easily set a convergence threshold to control and minimize the final calculation errors, such that the accuracy is enough for voltage drop analysis.

7. Experimental Results

We implement our Fast Poisson Solver using C++ language and compare the performance of different solving techniques under the same environment. The tested 5 different solvers include: 1) direct sparse solver using SupperLU newest version 3.1 [11]. 2) iterative PCG solver using standard zero fill-in incomplete LU decomposition based preconditioner [12]. 3) Fast Poisson Solver implemented only using CPU 4) Fast Poisson Solver implemented using GPU without FFT acceleration. 5) Fast Poisson Solver implemented using GPU with FFT acceleration. We test these solvers in a 64 bits Linux Server with Intel Xeon CPU running at 3GHz with 8GB memory. For GPU environment, we use a NVIDIA Tesla C870 GPU with 128 cores and 1.5GB display memory. The used CUDA library is in version 2.0. We test 10 P/G grid cases. The smallest one contains one million nodes while the largest one contains four million nodes. Table 2 gives out the run time comparison for the five different solvers.

978-1-60558-497-3/09 $25.00 © 2009 ACM

Table 2: Run Time Performance of different Solvers (unit: sec)

Node Num	SuperLU Solver CPU	ILU(0) PCG Solver CPU	Poisson Solver CPU	Poisson Solver GPU	Poisson Solver GPU+FFT
1M	55.7	138.77	15.64	4.94	3.96
1.2M	72.29	185.98	20.93	4.96	4.33
1.4M	93.12	240.28	27.16	5.3	4.28
1.6M	122.62	307.83	36.33	5.37	3.96
1.9M	142.48	382.09	44.87	6.02	3.96
2.3M	178.92	473.07	54.9	6.75	4.31
2.5M	215.78	572.83	66.27	7.69	4.33
2.9M	307.55	728.83	79.16	9.35	4.45
3.2M	365.06	848.4	94.3	10.9	4.54
3.6M	407.2	960.44	110.51	12.6	4.72
4M	521.43	1159.94	125.63	13	4.94

From Table 2, it is clear that as the grid size is increasing, the run time of the SuperLU method and the PCG method increases super-linearly. The CPU version of the new algorithm is about 5X faster than SuperLU and about 10X faster than the PCG solver. For the new GPU-base Fast Poisson Solver, the run time increases much slowly as the case size increases. When the FFT acceleration strategy is used, the run time almost remains constant for all test cases. For the largest case, the GPU version of the Fast Poisson Solver without FFT acceleration gains about 40X speed up over the SuperLU solver and about 90X speed up than the PCG solver. By using FFT acceleration technique, for the largest test case, the GPU version of Fast Poisson Solver gains 100X speed up over the SuperLU solver and 230X speed up over the PCG solver. Analysis results for the GPU active time show that exchanging data between CPU and GPU costs about 4 seconds for each case. So when matrix multiplication and 2D FFT operations can be calculated very fast on GPU, the run-time is mainly used to exchange data between CPU and GPU memories.

It is well known that the traditional solver is hard to be accelerated by using GPU. Reported speed up of the dense direct solver (LU decomposition based algorithm) under GPU in [8] is limited to about 5X while the reported speed up for PCG based solver in [9] is limited to 4X. Thus, compared with these results, the 100X speed up in our algorithm is a substantial performance improvement.

Further, we compare the memory usage of these different solvers. The results are shown in Table 3. From the table, we can see that the memory usage is the same for both the GPU version and the CPU version of Poisson Solver which is about 1.5X larger than the memory usage of the PCG method. For the FFT version, due to the expanding of object matrices, the memory usage is as 3X as that of the PCG solver. Further, the memory usage of SuperLU increases dramatically due to the fill in elements, when the grid size increases. The GPU FFT version of Fast Poisson Solver only uses as about 1/10 of that of SuperLU. For the largest case, SuperLU solver uses 9GB memory while the Poisson Solver just uses 1.1GB memory. These data proves that the memory performance of Poisson Solver is also excellent.

Table 3. Memory usage comparison among 5 solvers (unit: MB)

Node Num	SuperLU Solver CPU	ILU(0) PCG Solver CPU	Poisson Solver CPU	Poisson Solver GPU	Poisson Solver GPU+FFT
1M	1646.30	80	120.49	120.49	277.12
1.2M	2032.81	96.8	145.74	145.74	335.20
1.4M	2445.93	115.2	173.39	173.39	398.79
1.6M	2941.05	135.2	203.43	203.43	467.90
1.9M	3401.31	156.8	235.88	235.88	542.53
2.3M	3987.59	180	270.73	270.73	622.68
2.5M	4547.00	204.8	307.98	307.98	708.36
2.9M	5729	231.2	347.63	347.63	799.55
3.2M	6829	259.2	389.68	389.68	896.26
3.6M	7486	288.8	434.13	434.13	998.49
4M	9158	320	480.98	480.98	1106.25

8. Conclusion

In this paper, we have proposed a new algorithm to solve large-scale structured P/G grid networks. The new method explores the regular structures of P/G networks by a new formulation, which is very amenable for GPU computing. The CPU version of the new algorithm is about 5X faster than widely used direct solver SuperLU and about 10X faster than iterative PCG solver. The GPU-enabled version with the FFT acceleration technique can achieve 100X speed up over the SuperLU Solver and 200X speed up over the PCG Solver. This speed up ratio is superior to the reported performance of existing GPU based solutions. The memory usage of this new algorithm is only as 3X as that of PCG Solver and is as about 1/10 of the memory usage of SuperLU. Further, the iteration process of the algorithm does not suffer the slower convergence ratio when matrix size increases.

References and Citations

[1]. T. Chen and C. C. Chen, "Efficient large-scale power grid analysis based on preconditioned Krylov-subspace iterative methods", DAC2001 Proceedings, pp. 559-562

[2]. J. N. Kozhaya, S. R. Nassif and F.N. Najm, "A multigrid-like technique for power grid analysis", IEEE Trans. On Computer Aided Design, vol.21, no.10, Oct. 2002, pp. 1148-1160

[3]. H. Qian, S. R. Nassif, S. S. Sapatnekar, "Random walks in a supply network", DAC 2003 Proceedings, pp. 93-98

[4]. M. Zhao, R. V. Panda, S. S. Sapatnekar and D. Blaauw, "Hierarchical analysis of power distribution networks", IEEE Trans. On Computer Aided Design, vol. 21, no. 2, Feb. 2002, pp. 159–168

[5]. E. Chiprout, "Fast flip-chip power grid analysis via locality and grid shells", ICCAD 2004 Proceeding, pp. 485-488

[6]. J. Singh, S. S. Sapatnekar, "Topology optimization of structured power/ground networks", ISPD 2004 Proceedings, pp. 16–123.

[7]. J. Shi, Y. Cai, S. X.-D. Tan, J. Fan, X. Hong, "Pattern-Based Iterative Method for Extreme Large Power/Ground Analysis", IEEE Trans. on CAD of Integrated Circuits and Systems vol.26, no.4, April 2007, pp. 680-692

[8]. Galoppo N, Govindaraju N K, Henson M, et al. "LU-GPU: Efficient Algorithms for Solving Dense Linear System on Graphics Hardware", ACM/IEEE SC 2005 Proceedings, pp.3-8.

[9]. Buatois L, Caumon2 G, Lévy B., "Concurrent Number Cruncher: An Efficient Sparse Linear Solver on the GPU", Proceedings of High Performance Computation Conference, 2007, pp. 358-371.

[10]. Kanupriya Gulati, John F. Croix, Sunil P. Khatr, Rahm Shastry, "Fast Circuit Simulation on Graphics Processing Units", ASP-DAC 2009 Proceedings, pp. 403-408

[11]. http://crd.lbl.gov/ xiaoye/SuperLU

[12]. Saad. Yousef, "Iterative Methods for Sparse Linear Systems", PWS Publishing Company, 1996.

[13]. Strang G. "Introduction to Applied Mathematics", Wellesley-Cambridge Press, 1986.

[14]. http://en.wikipedia.org/wiki/Matrix_multiplication

[15]. Broyden, C.G., "The Convergence of a Class of Double-Rank Minimization Algorithms", Journal Inst. Math. Applic., Vol. 6, pp. 76-90, 1970.

[16]. http://www.netlib.org/blas/

[17]. http://www.nvidia.com/object/cuda_home.html

[18]. http://developer.nvidia.com/object/matlab_cuda.html

[19]. http://heim.ifi.uio.no/~tom/fastpoissonslides.pdf

Fast Vectorless Power Grid Verification Using an Approximate Inverse Technique[*]

Nahi H. Abdul Ghani
Department of ECE
University of Toronto
Toronto, Ontario, Canada
nahi@eecg.utoronto.ca

Farid N. Najm
Department of ECE
University of Toronto
Toronto, Ontario, Canada
f.najm@utoronto.ca

ABSTRACT

Power grid verification in modern integrated circuits is an integral part of early system design where adjustments can be most easily incorporated. In this work, we describe an early verification approach under the framework of current constraints where worst-case node voltage drops are computed via linear programs proportional to the grid size. We propose an efficient method based on a sparse approximate inverse technique to greatly reduce the size of such linear programs while ensuring a user-specified over-estimation margin (in volts) on the exact solution.

Categories and Subject Descriptors

B.7.2 [**Integrated Circuits**]: Design Aids

General Terms

Performance, Algorithms, Verification

Keywords

Power grid, voltage drop, approximate inverse

1. INTRODUCTION

A well-designed power grid in integrated circuits (ICs) should guarantee the proper logic functionality at the intended design speed. As the supply and threshold voltages decrease with technology scaling, full-chip verification is getting more and more challenging. A key concern is the fact that logic circuits slow down under reduced supply voltages thus putting overall circuit timing performance at risk. Hence, power grid analysis and verification has become central to reliable high-speed chip design.

Today, grid verification is typically done by simulation. Such an approach requires full knowledge of the current waveforms drawn by every logic block attached to the grid. These waveforms would then be used to simulate the grid and get the voltage drop at every node. Verifying the grid in this manner proves to be problematic. For one thing, the number of current traces needed to cover the space of voltage drops exhibited on the grid is intractable for modern designs where grids can consist of several million nodes. Another major drawback is that a simulation based flow does not allow for *early* grid verification, when grid modifications can be

[*]This work has been supported in part by the Semiconductor Research Corporation (SRC) (www.src.org).

Permission to make digital or hard copies of part or all of this work for personal or classroom use is granted without fee provided that copies are not made or distributed for profit or commercial advantage and that copies bear this notice and the full citation on the first page. To copy otherwise, to republish, to post on servers or to redistribute to lists, requires prior specific permission and/or a fee.
DAC'09, July 26-31, 2009, San Francisco, California, USA

most easily incorporated. The need, then, is for a verification approach that does not depend on simulation, *i.e.*, a *vectorless* approach. Therefore, we adopt the framework of partial current specifications, in the form of *current constraints* [1], which will be detailed in section 2.2. Under such constraints, power grid verification becomes a problem of computing the maximum voltage drops subject to current constraints.

In [1], the authors adopted a DC grid model and formulated the verification problem as a set of linear programs (LPs). This approach identifies coarse problems with the grid but is oblivious to violations related to the grid dynamics. In [2], the authors considered an RC model of the grid and focused on developing upper bounds on the worst-case voltages using a convergent iterative process that requires the solution of LPs while stepping repeatedly through time. A later work [3] gave a closed form bound which required a single LP per node. For all these formulations, finding the worst-case voltage drop everywhere on the grid would require the application of as many LPs as there are nodes, in a sequence, where every LP is proportional to the size of the grid. This becomes prohibitive for large designs; we often refer to this as the *sequentiality* problem. In this paper, we propose a novel approach to alleviate sequentiality based on an approximate inverse technique. This is a rigorous method that takes advantage of *locality* on the grid. It captures, for any given node, the current sources that are most influential, on that node, in a fast automated way, and then it formulates a reduced-size optimization problem while ensuring a *user-specified* over-estimation margin (in volts) on the solution of the exact LPs. Experimental results in section 5 show that even for small over-estimation values, the improvement in runtime is dramatic.

2. BACKGROUND

2.1 The Power Grid Model

Consider an RC model of the power grid, where each branch is represented by a resistor and where there exists a capacitor from every node to ground. This work is not applicable in its present form to RLC grids, because it takes advantage of special properties of circuit matrices in the RC case. We are working on extending this to the more general case. In a power grid, some nodes have ideal current sources (to ground) representing the currents drawn by the circuits tied to the grid at those nodes, while other nodes may be connected to ideal voltage sources representing the connection to the external voltage supply.

Let the power grid consist of $n+p$ nodes, where nodes $1, 2, \ldots, n$ have no voltage sources attached, and the remaining nodes $(n+1), (n+2), \ldots, (n+p)$ are the nodes where the p voltage sources are connected. Let $i(t)$ be the element-wise non-negative vector of all current sources connected to the grid. We assume that $\forall k = 1, \ldots, n$, $i_k(t)$ is well-defined, so that nodes with no current source attached have $i_k(t) = 0$. With these assumptions, we can write the RC model for the power grid as [2]:

$$C\dot{v}(t) + Gv(t) = i(t) \qquad (1)$$

where $v(t)$ is the $n \times 1$ vector of time varying voltage drops (difference between V_{dd} and node voltages). C is the $n \times n$ *diagonal* capacitance matrix, since all capacitors were assumed to be node-to-ground. G is the $n \times n$ conductance matrix, which is known to be a diagonally-dominant symmetric positive definite $\mathcal{M}$-matrix (so that $G^{-1} \geq 0$). Using a finite difference approximation, a discrete-time version of (1) can be written as:

$$Av(t) = \frac{C}{\Delta t}v(t - \Delta t) + i(t) \qquad (2)$$

where $A = (G + \frac{C}{\Delta t})$ is also a symmetric positive definite $\mathcal{M}$-matrix, so that $A^{-1} \geq 0$. For properties of $\mathcal{M}$-matrices, the reader is refered to [4].

2.2 Current Constraints

In our approach, we perform a verification of the grid in the absence of complete information about currents drawn by the underlying circuit, what may be called a *vectorless* approach. We do this because the currents are typically hard to specify, for at least two reasons: 1) there is a large variety of possible current waveforms, so that the worst case is hard to determine up-front and simulation of a large set of waveforms is prohibitively expensive, and 2) grid design and verification cannot wait until the chip design is nearly-complete, and is typically done early in the design flow, so that the details of the underlying circuitry may not yet be available or complete.

However, while users may not know the exact circuit currents, there is always some knowledge from previous design projects which users can bring to bear. It is such knowledge that one aims to capture with the concept of *current constraints*, originally introduced in [1]. Current constraints capture the uncertainty about the circuit currents arising from both unknown circuit behaviors and the fact that one is uncertain about circuit details early in the design flow. The aim is to verify that the grid is safe (*i.e.*, its voltages remain within certain bounds), under all possible transient current waveforms which satisfy these constraints. Two types of constraints are defined: *local constraints* and *global constraints*. Local constraints are upper bounds on individual current sources, where a current source can represent a single logic gate or cell, but more typically should represent a larger block. They can be expressed as:

$$0 \leq i(t) \leq i_L, \quad \forall t \geq 0 \qquad (3)$$

where i_L represents the peak value of current that this current source can draw - it is a "DC" upper bound on the transient waveform $i(t)$. It is possible to also consider upper bounds that are themselves transient waveforms over time (transient constraints), but in this paper we restrict our work to the case of DC constraints. It is important to remember, however, that it is only the constraints that are DC; the currents themselves are transient. To ensure that these constraints are well-defined for every node of the grid, we enforce the condition that any node with no current source connected would have a zero i_L component.

If *only* local constraints are provided, the problem is much simplified, but the results would be overly pessimistic, because it is never the case that all chip components simultaneously draw their maximum currents, hence the need for global constraints, which are upper bounds on the sums of groups of current sources. They represent the peak total power dissipation of a group of circuit blocks. Assuming we have a total of κ global constraints, they can be expressed in matrix form as:

$$0 \leq Si(t) \leq i_G, \quad \forall t \geq 0 \qquad (4)$$

where S is a $\kappa \times n$ matrix that consists only of 0s and 1s which indicate which current sources are present in each global constraint. As for the case of local constraints, note that i_G is a fixed time-independent upper bound, a DC constraint, but the currents themselves are transient waveforms over time.

How would one obtain/specify these constraints in practice? If a logic block is available and small enough to simulate, then one can generate the constraints by an "offline" process of simulation, which can be viewed as a characterization of that block. If the block is not yet available or is too large to simulate, then one would need to rely on design expertise and engineering judgement (how big it is, what its power needs were in a previous technology and how scaling would affect those needs, etc.). If, early in the design flow, nothing is known about this block, not even its detailed functionality, one typically is able to come up with an area budget for it. From that, and from the projected power density (Watts/μm^2) for the target process technology, one can generate a rough current constraint for it. The bottom line: *something* is typically known about that block, which with good engineering judgement can be formulated into constraints. Chip designers typically use "power budgets" early in the design flow to help verify the grid, often making use of simple spreadsheet applications. The proposed constraint-based approach is a scientific and reliable approach for making use of such budgets, and it depends on good designer expertise and judgement. After all, if truly nothing is known about the circuit currents, then the grid simply cannot be verified.

Another possibility is that the current constraints can be used to implement a "spec-based" design flow; a chip-level designer would simply specify the constraints based on design expertise, and the grid is verified under these constraints. The constraints now become *design guidelines* to be observed in subsequent design activity. If all design teams follow these guidelines and verify their blocks, the end result would be a grid that is *safe by construction*.

Together, the local and global constraints define a *feasible space* of currents, which we denote by $\mathcal{F}$, so that $i(t) \in \mathcal{F}$ if and only if it satisfies (3) and (4). Later in the paper, we will define algorithms that operate on a vector $i(t)$ and which are applicable at any value of time t. In that context, we will use the shorthand $i \in \mathcal{F}$ to denote the fact that $i(t)$ is feasible, for any given t.

3. PROBLEM DEFINITION

In this section, we will show how grid verification boils down to solving linear programs that depend on a certain matrix inverse. The aim is to lay the ground-work for the rest of the paper, in which this inverse will be conservatively approximated using a very efficient approach. The novelty of this paper is in adapting a previous inverse approximation technique to the power grid verification problem. However, we do need to provide some detailed background on grid verification, to motivate the need for finding an approximation to the inverse matrix.

We are interested in the vector of maximum node voltage drops, over all possible currents in $\mathcal{F}$. Using (2), we can write:

$$v(t) = A^{-1}\frac{C}{\Delta t}v(t - \Delta t) + A^{-1}i(t) \qquad (5)$$

Consider the special case where the grid had no stimulus for all $t \leq 0$ so that $v(0) = 0$. At time $t = \Delta t$, we can write:

$$v(\Delta t) = A^{-1}\frac{C}{\Delta t}v(0) + A^{-1}i(\Delta t)$$
$$= A^{-1}i(\Delta t) \qquad (6)$$

At $2\Delta t$ and at $3\Delta t$, we have:

$$v(2\Delta t) = A^{-1}\frac{C}{\Delta t}v(\Delta t) + A^{-1}i(2\Delta t)$$
$$= A^{-1}\frac{C}{\Delta t}A^{-1}i(\Delta t) + A^{-1}i(2\Delta t) \qquad (7)$$
$$v(3\Delta t) = A^{-1}\frac{C}{\Delta t}v(2\Delta t) + A^{-1}i(3\Delta t)$$
$$= \left(A^{-1}\frac{C}{\Delta t}\right)^2 A^{-1}i(\Delta t) + A^{-1}\frac{C}{\Delta t}A^{-1}i(2\Delta t)$$
$$+ A^{-1}i(3\Delta t) \qquad (8)$$

This can be repeated, so that at any future time $p\Delta t$, we have:

$$v(p\Delta t) = \sum_{k=0}^{p-1}\left(A^{-1}\frac{C}{\Delta t}\right)^k A^{-1}i((p-k)\Delta t) \qquad (9)$$

978-1-60558-497-3/09 $25.00 © 2009 ACM

At every point in time $t \in [0, p\Delta t]$, the current vector $i(t)$ must be feasible, *i.e.*, we must have $i(t) \in \mathcal{F}$ and, under that condition, we are interested in the worst-case voltage attained (separately) by each component of $v(p\Delta t)$. In order to compactly capture this notion, we introduce the following notation.

Let $f(x) : \mathbb{R}^n \to \mathbb{R}^n$ be a vector function whose components will be denoted $f_1(x), \ldots, f_n(x)$, and let $\mathcal{A} \subset \mathbb{R}^n$. Now, define a vector $z \in \mathbb{R}^n$, such that z_i is the maximum of $f_i(x)$ over all $x \in \mathcal{A}$. We will denote this by the following operator:

$$z = \underset{\forall x \in \mathcal{A}}{\mathrm{emax}}(f(x)) \tag{10}$$

where the "emax$(\cdot)$" notation is introduced to denote the fact that this is an *element-wise* maximization. Notice that each component z_i may be found separately by solving the optimization problem:

$$\begin{aligned} \text{Maximize: } & f_i(x) \\ \text{Subject to: } & x \in \mathcal{A} \end{aligned} \tag{11}$$

Using this notation, we can write the worst-case voltage drops at all nodes at time $\tau = p\Delta t$ as:

$$v_{max}(\tau) = \underset{\forall i(t) \in \mathcal{F}}{\mathrm{emax}} \left(\sum_{k=0}^{p-1} \left(A^{-1} \frac{C}{\Delta t} \right)^k A^{-1} i((p-k)\Delta t) \right) \tag{12}$$

where the notation "$\forall i(t) \in \mathcal{F}$" means that, for every time point $t \in [0, \tau]$, the current $i(t)$ satisfies all the (local and global) constraints.

In practice, we are interested in the steady state solution where the system becomes independent of the initial condition ($i(t) = 0, \forall t \leq 0$). Since the RC grid model is a dynamical system with a limited memory of its past, then the steady state solution can be obtained by evaluating (12) at points far away from the initial condition, *i.e.*, as $p \to \infty$. Thus, the general solution to the exact voltage drop maximization problem is:

$$v_{max}(t) = \lim_{p \to \infty} \underset{\forall i(t) \in \mathcal{F}}{\mathrm{emax}} \left(\sum_{k=0}^{p-1} \left(A^{-1} \frac{C}{\Delta t} \right)^k A^{-1} i((p-k)\Delta t) \right) \tag{13}$$

Although the RC model is dynamic, *i.e.*, its currents and voltages vary with time, the constraints are DC and do not depend on time. Hence, $\mathcal{F}$ is the same for each time step. With this, the components of (13) can be "decoupled," leading to:

$$v_{max}(t) = \lim_{p \to \infty} \sum_{k=0}^{p-1} \underset{\forall i \in \mathcal{F}}{\mathrm{emax}} \left[\left(A^{-1} \frac{C}{\Delta t} \right)^k A^{-1} i \right] \tag{14}$$

where i is simply an $n \times 1$ vector of variables that satisfies the (local and global) constraints, without reference to any particular point in time. This is a major simplification of the problem, as it has the advantage that the number of constraints for each maximization is fixed and does not span all previous time-points.

Unfortunately, both (13) and (14) are of theoretical interest only. They cannot be directly computed, as they stand, because they have to be evaluated for a large number of time steps until convergence and because the element-wise maximization requires *linear programs* (LPs) proportional to the number of nodes in the grid which for modern designs is in the thousands or even the millions. In previous work, various approaches to simplify (14) have been proposed, two of which will be relevant to our work, and they are as follows.

In [1], a DC model of the grid was verified subject to the current constraints. The approach employed a resistive grid model and offered a major simplification of (14). In this case, the exact maximum voltage drop vector can be computed with a single emax operation as follows:

$$v_{max}(t) = \underset{\forall i \in \mathcal{F}}{\mathrm{emax}} \left(G^{-1} i \right) \tag{15}$$

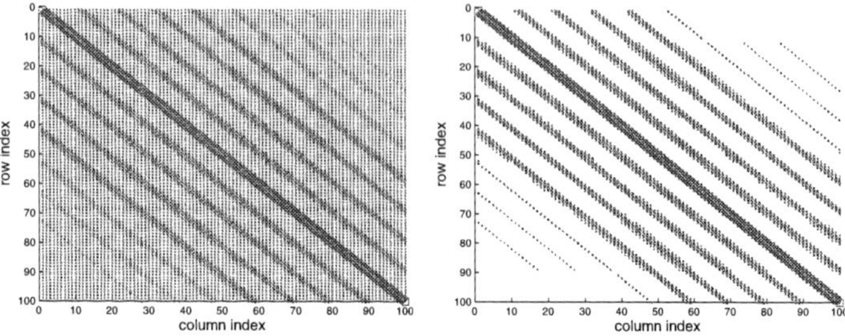

(a) Matrix view of G^{-1} (b) Matrix view of $\hat{M}$

Figure 1: Structure of G^{-1} and $\hat{M}$ for a 100-node grid

In another approach, and given an RC grid, the authors in [3] proposed an upper bound on the exact worst-case voltage drop vector, under transient currents. As $t \to \infty$, the bound was found to be:

$$v_{max}(t) = \left(I + G^{-1} \frac{C}{\Delta t} \right) \underset{\forall i \in \mathcal{F}}{\mathrm{emax}} \left(A^{-1} i \right) \tag{16}$$

where I is the identity matrix. The result in (16) has a run-time advantage over (14) as it requires a single emax operation. Once that is computed, the evaluation of (16) is relatively cheap. It involves a single vector scaling and a single standard linear solve (LU factorization of G).

Thus, we see that both approaches [1] and [3] give a formulation that involves the inverse of a large matrix, G in one case, and A in the other. More generally, both problem formulations involve solving the same/following generic problem:

$$v^\star = \underset{\forall i \in \mathcal{F}}{\mathrm{emax}} \left(E^{-1} i \right) \tag{17}$$

where E is understood to be either A in the case of transient analysis [3] or G in the case of DC analysis [1]. Note that E^{-1} is symmetric, because E (be it A or G) is symmetric. In the mentioned previous work, the matrix inverse E^{-1} was not found explicitly. Instead, the problem was transformed into the voltage domain, so that it makes use of the original sparse matrix E. However, even then, the problem remained quite expensive, limiting the size of grids that may be verified. As an alternative, and in the rest of the paper, we propose an efficient approach to compute a <u>tight</u> conservative approximation of $v^\star$, based on a sparse approximation of the matrix inverse, followed by a solution of (17) using linear programming.

4. PROPOSED SOLUTION

Let M be an $n \times n$ matrix and let the vector m_k denote the kth column of M. Let e_k be the $n \times 1$ vector consisting of all zeros, except for its kth component which is 1. Then, consider the $n \times 1$ vector, called the kth *residual*, defined by:

$$r_k = Em_k - e_k, \qquad \forall k = 1, 2, \ldots, n \tag{18}$$

It is clear that, if we choose $M = E^{-1}$, then $r_k = 0, \forall k$. In general, the norm of the residual $\|r_k\|_2$ (the Euclidean or l-2 norm in this case) is positive, and becomes zero only when $M = E^{-1}$. Thus, the norms of the residuals, for all k, provide a measure of how far M is from being equal to the desired matrix inverse E^{-1}. Let $M^\star \triangleq E^{-1}$, so that when M takes the value $M^\star$, then $\|r_k\|_2 = 0, \forall k$, and we want:

$$v^\star = \underset{\forall i \in \mathcal{F}}{\mathrm{emax}} \left(M^\star i \right) \tag{19}$$

Typical power grids have a mesh structure, where a node has a small number of neighboring nodes. Such a structure gives rise to a matrix E (be it G or A) that is sparse, symmetric, positive definite, and *banded*. For such matrices, it is well known in the literature that their inverse has entries whose values decay

978-1-60558-497-3/09 $25.00 © 2009 ACM

exponentially as one moves away from the diagonal [5]. We illustrate this property in Fig. 1(a), which shows the structure of the G^{-1} matrix for a 100-node grid. In Fig. 1(a), every entry is represented, at the appropriate location, by a square whose area reflects that entry's magnitude, and very small valued entries are represented by dots. It is clear that the majority of the entries are extremely small with the exception of few significant bands in the neighborhood of the main diagonal. As expected, G^{-1} entries decay as one moves away from the diagonal. This is a *key* observation that has been the motivation for published work [6] on finding approximations to the inverse matrix that basically ignore the extremely small values in the inverse. A matrix that has a large number of entries whose values are extremely small is refered to as being *practically* sparse. In this case, $M^\star = E^{-1}$ is practically sparse. Therefore, every component of $v^\star$, which we will denote $v_k^\star$, is mainly the result of the contribution of very few current sources, specifically those that correspond to the significant entries in row k of $M^\star$. This is not surprising, because one expects that the voltage at a node depends mostly on the nearby current sources. This is well-known in power grid research, as in [7] which finds that a current source has a localized contribution on the grid. This is often refered to as the *locality effect*. The difficulty, with locality, is to develop rigorous methods for determining exactly where the neighborhood is, for a given node, and how far it extends. By using methods based on discovering the significant entries of a practically sparse matrix, we believe that we have found a rigorous approach for taking advantage of locality.

Our contribution, described in the next two sub-sections, is as follows. Given a *user-specified* over-estimation margin (in volts) for $v^\star$, we devise a fast automated method to formulate a smaller-size optimization problem, based only on the current sources that are most influential on each node, and we solve it to provide a solution that meets the over-estimation margin specified by the user. We start by briefly describing the existing approach of inverse matrix approximation and then, in the following section, we describe how we have adapted this to solve the power grid verification problem.

4.1 Inverse Approximation Method (SPAI)

A sparse approximate inverse technique (SPAI) technique is given in [6]. Given some user-defined tolerance η, SPAI computes an approximation $\hat{M}$ to the actual inverse matrix $M^\star$. It operates by finding an $\hat{M}$ matrix that gives a very small residual norm, as an approximation to the unknown $M^\star$ (which, as we saw above, would give a zero residual). The technique starts with an arbitrary initial matrix M and iteratively refines its columns by minimizing the Frobenius norm $\|EM - I\|_F$ subject to η. Recall that for any $n \times n$ matrix H, the Frobenius norm is defined as [4]:

$$\|H\|_F = \sqrt{\sum_{i=1}^{n}\sum_{j=1}^{n} |h_{ij}|^2} \qquad (20)$$

where h_{ij} are the entries of H, which can be re-written as:

$$\|H\|_F = \sqrt{\sum_{k=1}^{n} \|He_k\|_2^2} = \sqrt{\sum_{k=1}^{n} \|h_k\|_2^2} \qquad (21)$$

where h_k is the kth column of H. Obviously, the minimum of (21) (which is zero) occurs when $H = 0$. Likewise, the minimum of $\|EM - I\|_F$ (which is zero) occurs when $M = E^{-1} = M^\star$. By reducing $\|EM - I\|_F$ until it is less than η, SPAI finds an $\hat{M}$ that is a good approximation to $M^\star = E^{-1}$. To see how SPAI works, consider that:

$$\|EM - I\|_F^2 = \sum_{k=1}^{n} \|(EM - I)e_k\|_2^2$$

$$= \sum_{k=1}^{n} \|Em_k - e_k\|_2^2 = \sum_{k=1}^{n} \|r_k\|_2^2 \qquad (22)$$

In this way, the problem of finding an approximate inverse $\hat{M}$ separates into n *independent* least-squares problems:

> For each $k = 1, 2, \ldots, n$, vary m_k so as to minimize $\|r_k\|_2^2$ until an m_k is found for which $\|r_k\|_2 \leq \eta$.

While $\|r_k\|_2 > \eta$, the iterative refinement of m_k proposed by [6] attempts to improve on the column by appending the most profitable entries, *i.e.*, the ones that will result in the largest reduction in $\|r_k\|_2$. In this manner, the technique automatically captures the sparsity pattern by including the significant entries of the inverse while avoiding fill-ins.

The numerical study in [6] shows that SPAI captures the main entries of E^{-1} extremely well. Moreover, it is an extremely efficient technique. It is inherently parallel as the columns of $\hat{M}$ can be computed independently of one another. Moreover, the computation of a column is cheap; it requires a matrix-vector product and several QR factorizations of small submatrices of the original sparse matrix E.

4.2 An Upper Bound Vector

The $\hat{M}$ obtained by SPAI can be used to give an approximate solution to the exact optimization problem in (19). However, for power grid verification, we are interested in, not simply *any* approximation to the $v^\star$, but in a *conservative* approximation to $v^\star$, *i.e.*, an upper-bound on it. In other words, we want an $\hat{M}$ which, if used instead of $M^\star$ in (19), would produce a $\hat{v}$ that we can use to construct an upper bound $\bar{v}$ on $v^\star$, so that $v^\star \leq \bar{v}$. In this section, we describe how we adapt SPAI to our needs, by modifying its stopping criterion, and then how we generate our upper bound $\bar{v}$ based on the matrix $\hat{M}$ achieved by SPAI under the new stopping criterion. Crucially, we will show that the over-estimation $\|\bar{v} - v^\star\|$ is under control and can in fact be specified up-front.

4.2.1 *Stopping criterion*

Normally, SPAI terminates when $\|r_k\|_2$ is smaller than a user-supplied η, for every k. However, in order to arrive at predictable over-estimation bounds, as we will see below, a new stopping criterion will now be introduced. If $m_k^\star$ is the kth column of $M^\star = E^{-1}$, then clearly $Em_k^\star = e_k$, so that $r_k = Em_k - e_k = E(m_k - m_k^\star)$. This leads to:

$$\|m_k - m_k^\star\|_\infty = \|E^{-1}r_k\|_\infty \leq \|E^{-1}\|_\infty \|r_k\|_\infty \qquad (23)$$

where $\|\cdot\|_\infty$ is the infinity norm. We define the error tolerance $\epsilon_k > 0$, where $\epsilon_k \in \mathbb{R}$ is a scalar. As we will see below, ϵ_k is derived from another user-specified error-tolerance on voltages. We modify SPAI so that it stops when, for every k, we have:

$$\|r_k\|_\infty \leq \frac{\epsilon_k}{\|E^{-1}\|_\infty} \qquad (24)$$

which achieves the condition that, for every k:

$$\|\hat{m}_k - m_k^\star\|_\infty \leq \epsilon_k \qquad (25)$$

It remains to explain how $\|E^{-1}\|_\infty$ is found, when E^{-1} is unknown! This is easy to do, in fact, because E is an $\mathcal{M}$-matrix, as follows. Let u be the $n \times 1$ vector of all 1s, so that $u^T = [1\ 1\ \cdots\ 1]$. Then, $E^{-1}u$ is a vector whose every component is the sum of all the entries in the corresponding row of E^{-1}. Because $E^{-1} \geq 0$, then each component of $E^{-1}u$ is actually the sum of the absolute values of all the entries in that row of E^{-1}. Thus, the component with the largest value in $E^{-1}u$ is, in fact, the "largest absolute row sum" of E^{-1} which is its infinity norm. In other words, if x is a vector such that $Ex = u$, then it's clear that:

$$\|E^{-1}\|_\infty = \|E^{-1}u\|_\infty = \|x\|_\infty \qquad (26)$$

Finding this x is easy, by performing one LU-factorization of E, which we have to do anyway as part of power grid verification, followed by a backward and a forward solve starting with u.

978-1-60558-497-3/09 $25.00 © 2009 ACM   187

Algorithm 1 UPPER_BOUND

Input: E, δ
Output: $\bar{v}$
1: Compute $\|E^{-1}\|_\infty$ using (26):
    LU-factorize E: $E = L \cdot U$
    Forward solve: $Ly = u$
    Backward solve: $Ux = y$
    Get: $\|E^{-1}\|_\infty = \|x\|_\infty$
2: $(\hat{M}, \epsilon) = $ Modified_SPAI$(E, \|E^{-1}\|_\infty, \delta)$
3: Compute $\hat{v}$ using (27):
    Maximize: $\hat{M}i$
    Subject to: $i \in \mathcal{F}$
4: $\bar{v} = \hat{v} + (u^T i_L)\epsilon$

Algorithm 2 Modified_SPAI$(E, \|E^{-1}\|_\infty, \delta)$

1: $M = I$ {initialize M to the identity matrix}
2: **for** $k = 1 : n$ **do**
3:     $r_k = E m_k - e_k$
4:     Compute ϵ_k in (32)
5:     **while** $(\|r_k\|_\infty > \epsilon_k / \|E^{-1}\|_\infty)$ **do**
6:         Minimize $\|E m_k - e_k\|_2$ and update m_k {this is the result of a QR factorization of a sub-matrix of E }
7:         Update r_k and ϵ_k
8:     **end while**
9: **end for**
10: **return** (M, ϵ)

4.2.2 Upper bound

Once SPAI terminates with its approximation matrix $\hat{M}$, using the new stopping criterion, we run our optimization (19) using $\hat{M}$ instead of $M^\star$, to get:

$$\hat{v} = \operatorname*{emax}_{\forall i \in \mathcal{F}} \left(\hat{M}i \right) \tag{27}$$

and this optimization run is quite fast, as we will see in the results section. In this section, we will show how the $\hat{v}$ resulting from this optimization is used to generate a tight upper bound $\bar{v}$ on $v^\star$, in a way that the over-estimation $\|\bar{v} - v^\star\|$ is under control and predictable up-front. We will do this by first showing how (25) leads to an upper bound on $M^\star$, based on $\hat{M}$ and ϵ_k. We then show how this leads to an upper bound on $v^\star$, based on $\hat{v}$ and ϵ_k.

SPAI starts with an initial $M = I$, and tries to maintain sparsity and reduce fill-ins throughout the process. So, many of the entries of $\hat{m}_k$ will be equal to 0. Let $\mathcal{C}_k$ be the set of non-zero entries in $\hat{m}_k$. The number of such entries, denoted by $|\mathcal{C}_k|$, is often considerably smaller than n. To compute an upper bound on $m^\star_{jk}$, for all j, k, two cases have to be considered, according to whether $\hat{m}_{jk} \in \mathcal{C}_k$ or not. If $\hat{m}_{jk} \notin \mathcal{C}_k$, i.e., $\hat{m}_{jk} = 0$, then (25) yields:

$$|-m^\star_{jk}| \le \epsilon_k \Rightarrow 0 \le m^\star_{jk} \le \epsilon_k \tag{28}$$

so that ϵ_k is an upper-bound on $m^\star_{jk}$, with a maximum over-estimation error of ϵ_k. In the second case, when $m_{jk} \in \mathcal{C}_k$, then, (25) gives:

$$|\hat{m}_{jk} - m^\star_{jk}| \le \epsilon_k \Rightarrow \hat{m}_{jk} - \epsilon_k \le m^\star_{jk} \le \hat{m}_{jk} + \epsilon_k \tag{29}$$

so that $\hat{m}_{jk} + \epsilon_k$ is an upper-bound on $m^\star_{jk}$, with a maximum over-estimation error of $2\epsilon_k$. Let ϵ be the $n \times 1$ vector such that $\epsilon^T = [\epsilon_1 \ \epsilon_2 \ \cdots \ \epsilon_n]$. We will now prove that these upper bounds on the entries of $M^\star$ lead to an upper bound $\bar{v}$ on $v^\star$, as follows:

$$\bar{v} = \hat{v} + (u^T i_L)\epsilon \tag{30}$$

where u is the same vector of all 1s introduced above, so that $u^T = [1 \ 1 \ \cdots \ 1]$. The proof of this result is given in the following claim, which also provides the measure of over-estimation between $v^\star$ and $\bar{v}$.

CLAIM 1. *For every $k = 1, 2, \ldots, n$, let u_k be the $n \times 1$ vector whose jth entry is 1 if $\hat{m}_{jk} \in \mathcal{C}_k$, and otherwise 0. Then, $\bar{v}$ in (30) is an upper-bound on $v^\star$, with a maximum over-estimation error of:*

$$(\bar{v}_k - v^\star_k) \le \delta_k \triangleq \left(u^T i_L + u_k^T i_L \right) \epsilon_k, \quad \forall k = 1, \ldots, n \tag{31}$$

PROOF. In the following, and for any $k \in \{1, 2, \ldots, n\}$, we will prove that $0 \le \bar{v}_k - v^\star_k \le \delta_k$. The proof addresses first the lower bound (of 0) and then the upper bound (of δ_k).

The lower bound: For the kth component of $v^\star$, and using the bounds on the entries of $M^\star$ seen above, and because $M^\star$ and $\hat{M}$ are both symmetric, it is easy to see that:

$$v^\star_k = \max_{\forall i \in \mathcal{F}} \left(\sum_{j=1}^n m^\star_{kj} i_j \right) = \max_{\forall i \in \mathcal{F}} \left(\sum_{j=1}^n m^\star_{jk} i_j \right)$$

$$\le \max_{\forall i \in \mathcal{F}} \left(\sum_{\forall j, \hat{m} \ne 0} (\hat{m}_{jk} + \epsilon_k) i_j + \sum_{\forall j, \hat{m} = 0} \epsilon_k i_j \right)$$

$$= \max_{\forall i \in \mathcal{F}} \left(\sum_{\forall j, \hat{m} \ne 0} \hat{m}_{jk} i_j + \epsilon_k \sum_{j=1}^n i_j \right)$$

$$= \max_{\forall i \in \mathcal{F}} \left(\sum_{j=1}^n \hat{m}_{jk} i_j + \left(u^T i \right) \epsilon_k \right)$$

$$\le \max_{\forall i \in \mathcal{F}} \left(\sum_{j=1}^n \hat{m}_{jk} i_j \right) + \max_{\forall i \in \mathcal{F}} \left(\left(u^T i \right) \epsilon_k \right)$$

$$= \max_{\forall i \in \mathcal{F}} \left(\sum_{j=1}^n \hat{m}_{kj} i_j \right) + \max_{\forall i \in \mathcal{F}} \left(\left(u^T i \right) \epsilon_k \right)$$

$$\le \hat{v}_k + \left(u^T i_L \right) \epsilon_k = \bar{v}_k$$

so that $\bar{v}_k - v^\star_k \ge 0$.

The upper bound: It is easy to see that, if the optimal $\hat{v}_k$ is achieved at some current vector i', then it must also be the case that $\sum_{j=1}^n \hat{m}_{kj} \hat{i}_j = \hat{v}_k$, where $\hat{i}$ is obtained from i' by setting to 0 every jth entry of i' for which $\hat{m}_{kj} = 0$. In the following, we make use of such a vector $\hat{i}$ at which $\hat{v}_k$ is achieved.

If we define $v^\dagger_k \triangleq \sum_{j=1}^n m^\star_{kj} \hat{i}_j$, then obviously $v^\dagger_k \le v^\star_k$, because $v^\star_k$ is optimal and $\hat{i} \in \mathcal{F}$, and $\bar{v}_k - v^\star_k \le \bar{v}_k - v^\dagger_k$. By making use of the fact that $|\hat{m}_{jk} - m^\star_{jk}| \le \epsilon_k$, due to (25), and because $M^\star$ and $\hat{M}$ are both symmetric, this expands as follows:

$$\bar{v}_k - v^\star_k \le \bar{v}_k - v^\dagger_k$$

$$= \sum_{j=1}^n \hat{m}_{kj} \hat{i}_j + \left(u^T i_L \right) \epsilon_k - \sum_{j=1}^n m^\star_{kj} \hat{i}_j$$

$$= \sum_{j=1}^n \hat{m}_{jk} \hat{i}_j + \left(u^T i_L \right) \epsilon_k - \sum_{j=1}^n m^\star_{jk} \hat{i}_j$$

$$= \sum_{j=1}^n \left(\hat{m}_{jk} - m^\star_{jk} \right) \hat{i}_j + \left(u^T i_L \right) \epsilon_k$$

$$\le \sum_{j=1}^n \epsilon_k \hat{i}_j + \left(u^T i_L \right) \epsilon_k = \left(u_k^T \hat{i} \right) \epsilon_k + \left(u^T i_L \right) \epsilon_k$$

$$\le \left(u_k^T i_L + u^T i_L \right) \epsilon_k$$

$\square$

978-1-60558-497-3/09 $25.00 © 2009 ACM

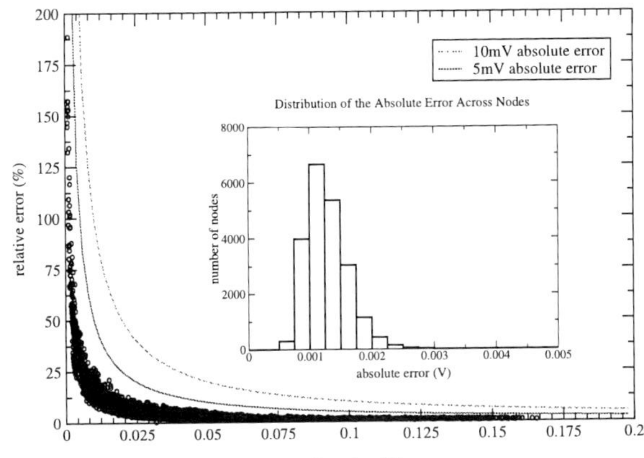

Figure 2: Accuracy of the proposed approach

Our algorithm accepts δ as a user-specified error-tolerance on node voltage drops. During the algorithm, as SPAI is iterating on a column, causing changes in u_k, we repeatedly update ϵ_k based on:

$$\epsilon_k = \frac{\delta}{u_k^T i_L + u^T i_L} \tag{32}$$

until such time when the stopping criterion (24) is met. By claim 1, this achieves the result that:

$$\|\bar{v} - v^\star\|_\infty \leq \delta \tag{33}$$

Algorithm 1 provides a high-level description of our approach which uses our modified version of SPAI (algorithm 2). The algorithm takes a user-defined δ, computes $\|E^{-1}\|_\infty$ and calls Modified_SPAI which returns $\hat{M}$ and ϵ. The algorithm will then compute $\hat{v}$ leading to $\bar{v}$.

5. EXPERIMENTAL RESULTS

The above algorithm, including the modified SPAI component, has been implemented in `C++`. It computes the column approximations of the matrix inverse, generates the reduced optimization problem and solves it using the Mosek optimization package [8]. A number of tests were conducted on a set of randomly-generated power grids and computations were carried on a 64-bit Linux machine with 8 GB of memory. The grids are based on user specifications, including grid dimensions, metal layers (M1-M4), pitch and width per layer, and C4 and current source distribution. Four global constraints were specified for the grids under study.

Fig. 1(b) shows $\hat{M}$ as a result of running Algorithm 2 on the 100-node grid mentioned earlier. By comparing this figure to Fig. 1(a), it is clear that Modified_SPAI captures the significant entries of the inverse extremely well while avoiding fill-ins. The resulting $\hat{M}$ is practically sparse with an average of 20 entries per row/column.

We ran Algorithm 1 on a 21,099-node grid for $\delta = 5$mV and monitored the relative over-estimation error for the kth entry as $(\bar{v}_k - v_k^\star)/v_k^\star$. Fig. 2 shows a scatter plot of such errors, in percent, versus the entries of $v^\star$. The figure also shows the curve corresponding to $\delta = 5$mV where a point on the curve represents a $v^\star$ entry over-estimated by exactly 5mV. On the other hand, a point with an over-estimation greater than 5mV will lie in the region above the curve. From the resulting data distribution, it is clear that our algorithm guarantees the specified over-estimation δ for all entries, as it should, by claim 1. Note that for some entries, especially those with magnitudes less than δ, the relative error can be high. This, however, is of no concern in practice, as the critical $v^\star$ entries are those that are significantly larger than δ. In that case, the relative error incurred is small. The absolute error is always, of course, guaranteed to be less than δ, as we have seen. In fact, the absolute errors can be much less than δ, as seen in the histogram embedded in the above figure, which shows that the actual over-estimation is often less than $\delta/2$.

Before proceeding with the results, we note that computing the exact $v^\star$ in (19), against which we will compare our results, does not have to involve the expensive computation of $M^\star$. The problem can be transformed as shown in [1] from an optimization in the current space to an optimization in the voltage space by defining a vector y, such that $Ey = i$, and then solving $v^\star = $ emax(y), $\forall Ey \in \mathcal{F}$. The exact solution, using this approach, remains expensive and involves solving one LP for every node on the grid. Our algorithm also solves an LP for every node, but each of these LPs is much smaller, because it benefits from locality (we retain only the most influential current sources, as determined by the modified SPAI routine). The result is an extremely fast optimization run.

Table 1 gives runtimes for computing $\bar{v}$ subject to different values of δ. The runtimes take into account the Modified_SPAI algorithm runtime. The table also gives the CPU time required for computing $v^\star$, using the above mentioned approach of converting the problem to the voltage domain. Our method was tested on all grid nodes. Given the size of the test grids, it was impossible to solve all n exact LPs associated with the computation of $v^\star$. For this purpose, we chose 1000 random nodes, solved the 1000 corresponding exact LPs, and estimated the runtime of the exact approach. Results show that even for relatively small δ's, our proposed approach runs several orders of magnitude faster than the exact computation which exhibits impractical runtimes. For instance, computing $\bar{v}$ for a 50,444-node grid subject to $\delta = 5$mV takes around 15.67 hours while computing $v^\star$ requires around 222.7 hours. This is over 14X speedup, and the last test case in the table shows an even higher, 27X, speedup! Thus, our method makes checking a power grid under incomplete current specification a feasible and practical solution. It is also highly parallelizable, both in the modified-SPAI module and in the rest of the algorithm where the separate LPs can be run in parallel.

6. CONCLUSION

We describe an early verification approach under the framework of capturing circuit uncertainty via current constraints and maximizing node voltage drops over the constraint space. Our proposed method takes advantage of locality on the grid and formulates, for any given node, a reduced-size optimization problem while ensuring a user-specified over-estimation margin (in volts) on the exact solution. With this technique, verifying power grids under incomplete current specification becomes scalable and practical. The method shows a drastic improvement in runtime, doing in hours what required days in previous approaches.

7. REFERENCES

[1] D. Kouroussis and F. N. Najm. A static pattern-independent technique for power grid voltage integrity verification. In *ACM/IEEE DAC*, pages 99–104, Anaheim, CA, June 2-6 2003.

[2] M. Nizam, F. N. Najm, and A. Devgan. Power grid voltage integrity verification. In *ACM/IEEE ISLPED*, pages 239–244, San Diego, CA, August 8-10 2005.

[3] I. A. Ferzli, F. N. Najm, and L. Kruze. A geometric approach for early power grid verification using current constraints. In *ACM/IEEE ICCAD*, pages 40–47, San Jose, CA, November 5-8 2007.

[4] Y. Saad. *Iterative Methods for Sparse Linear Systems*. SIAM, Philadelphia, second edition, 2003.

[5] S. Demko, W. F. Moss, and P. W. Smith. Decay rates for inverses of band matrices. *Mathematics of Computation*, 43(168):491–499, October 1984.

[6] M. J. Grote and T. Huckle. Parallel preconditioning with sparse approximate inverses. *SIAM Journal on Scientific Computing*, 18(3):838–853, May 1997.

[7] E. Chiprout. Fast flip-chip power grid analysis via locality and grid shells. In *ACM/IEEE ICCAD*, pages 485–488, San Jose, CA, Novembre 7-11 2004.

[8] Mosek - http://www.mosek.com/.

P e G id	R i e			
	Our Approach			
Nodes	$\delta = 5mV$	$\delta = 10mV$	$\delta = 20mV$	v comp.
8,413	46 min.	38.2 min.	35.65 min.	3.39 h.
21,099	5.42 h.	4.85 h.	4 h.	26.5 h.
39,391	11.48 h.	8.56 h.	8.16 h	103.58 h.
50,444	15.67 h.	11.69 h.	10.8 h.	222.7 h.
72,692	25.7 h.	21.36 h.	19.38 h.	21.38 days
113,304	58.16 h.	47.0 h.	43.8 h.	66.4 days

Table 1: Speed comparisons of the proposed approach with the exact approach

978-1-60558-497-3/09 $25.00 © 2009 ACM

Computing Bounds for Fault Tolerance using Formal Techniques[*]

Görschwin Fey   André Sülflow   Rolf Drechsler

Institute of Computer Science, University of Bremen, 28359 Bremen, Germany
{fey,suelflow,drechsle}@informatik.uni-bremen.de

ABSTRACT

Continuously shrinking feature sizes result in an increasing susceptibility of circuits to transient faults, e.g. due to environmental radiation. Approaches to implement fault tolerance are known. But assessing the fault tolerance of a given circuit is a tough problem.

Here, we propose the use of formal methods to assess the robustness of a digital circuit with respect to transient faults. Our formal model uses a fixed bound in time to cope with the complexity of the underlying sequential equivalence check. The result is a lower and an upper bound on the robustness. The underlying algorithm and techniques to improve the efficiency are presented. In experiments the method is evaluated on circuits with different fault detection mechanisms.

Categories and Subject Descriptors

B.8.1 [**Performance and Reliability**]: Reliability, Testing, and Fault-Tolerance; B.6.3 [**Logic Design**]: Design AidsVerification

General Terms

Verification

Keywords

Fault Tolerance, SAT, Formal Verification

1. INTRODUCTION

According to Moore's Law the number of components per area increases at an exponential rate in integrated circuits. One consequence is an increase of externally induced transient faults [20].

Techniques to cope with transient faults are available on the production level [26] or the design level [1, 10]. Even first tools to improve fault tolerance are available [25].

Proving robustness with respect to transient faults is difficult. Simulation or emulation based methods [6, 18] can only cover a small portion of the states and the input space of a circuit. A formal analysis determines the probability of a fault to propagate to a primary output (see e.g. [16]). But the computational effort is extremely high.

[*]This work has been funded in part by DFG grants DR 287/19-1 and FE 797/5-1.

Permission to make digital or hard copies of part or all of this work for personal or classroom use is granted without fee provided that copies are not made or distributed for profit or commercial advantage and that copies bear this notice and the full citation on the first page. To copy otherwise, to republish, to post on servers or to redistribute to lists, requires prior specific permission and/or a fee.
DAC'09, July 26–31, 2009, San Francisco, California, USA

Methods commonly applied for formal verification can *prove* fault tolerance of an implementation. The approach of [3] proposes to use symbolic methods for the classical analysis of fault trees. But the faults have to be specified manually. Similarly, [13] and [15] rely on symbolic methods. These approaches analyze fault tolerance with respect to mutations of the implementation. As a result, [13] decides whether an implementation is fault tolerant or not, while [15] also provides data about the state space. Both techniques use the original circuit as a specification. The authors of [19] determine fault tolerance with respect to given formal properties. Only faults in state bits are considered. None of the techniques mentioned so far provides insight about circuit structures that are not fault tolerant.

The technique proposed here is similar to [11] and uses the same fault model. A circuit is classified as robust if no fault tampers the output behavior. Detailed feedback about components that are not fault tolerant is returned. Thus, the approach in [11] provides a good basis, but has several limitations when considering practical problems. For example the formal analysis may have to consider a large number of time steps before providing a result and reachability analysis is required to get accurate information.

In contrast, our approach is more efficient and fits practical requirements. The main contributions of our work are:

- Practical model for robustness checking

  We explain why a full formal analysis is an overkill in practice and how a fault detection mechanism helps to prove fault tolerance.

- Fault tolerance within a lower and an upper bound

  While running, our technique delivers bounds on the fault tolerance by determining robust, non-robust and non-classified components. Non-classified components point to potential *Silent Data Corruption* (SDC).

- Restricted observation window for formal analysis

  Restricting the observation time significantly improves the performance.

- Avoiding full reachability analysis

  Full reachability analysis is often too expensive. We show how to use a light-weight reachability analysis when assessing fault tolerance.

Experimental results evaluate the approach and the performance of the algorithm. The formal approach expectedly requires long run times, but proves fault tolerance with respect to any possible input stimulus. Even circuits where full reachability analysis is not feasible are effectively handled by our algorithm. The optimization techniques, i.e. the consideration of structural information and the reuse of learned information, improve the performance by a factor of up to 11.

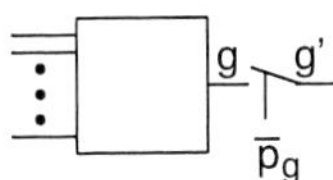

Figure 1: Fault modeling

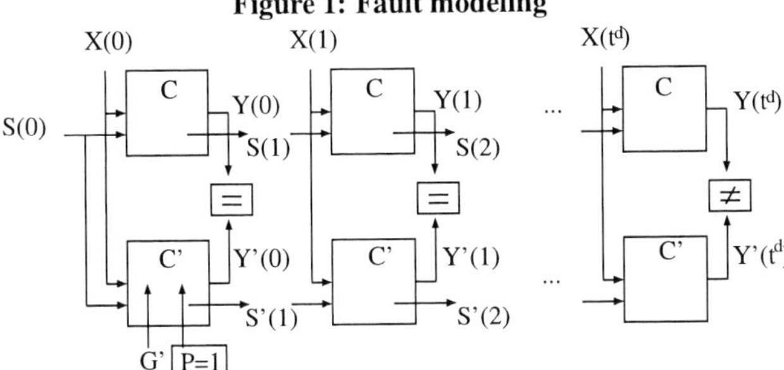

Figure 2: Sequential comparison

This paper is structured as follows: The underlying fault model and a basic approach to determine the robustness are discussed in the next section. Section 3 introduces a model bounded in time that also yields bounds on the robustness of a circuit. The incremental algorithm is explained in Section 4, while Section 5 presents optimization techniques to improve efficiency. Experimental results are given in Section 6. Section 7 concludes the paper.

2. FAULT MODEL AND BASIC APPROACH

We consider a synchronous sequential circuit $\mathbb{C}$ with *Primary Inputs* (PIs) X, *Primary Outputs* (POs) Y and state bits S. The number of components in $\mathbb{C}$ is denoted by $|\mathbb{C}|$. Here, a component may be a gate, a module or a source level expression in the hardware description language. Our fault model assumes that a *faulty* component behaves non-deterministically in one time frame, i.e. the value of the output of the component does not depend on the values of the inputs. We consider single faults only and justify in Section 3 why this is sufficient. A component g is *robust* iff the output behavior of $\mathbb{C}$ cannot change when g is faulty. Let $\mathbb{T}$ be the set of robust components in $\mathbb{C}$, then the robustness of $\mathbb{C}$ is given by $|\mathbb{T}|/|\mathbb{C}|$.

As suggested in [11] we use an instance of *Boolean satisfiability* (SAT) to measure robustness. In the following the output signal of component g is associated to variable g as well. A fault is modeled as follows (the formulation is similar to SAT-based diagnosis [21]): For a component g, a fault predicate p_g and a new variable g' are introduced; then g is replaced by $\overline{p_g} \rightarrow g' = g$ as shown in Figure 1. Consequently, the value of g' is specified by the circuit structure if $p_g = 0$. But if a fault at g is asserted by $p_g = 1$, g' may take any value. Given a circuit $\mathbb{C}$, the circuit $\mathbb{C}'$ is created by replacing each component as explained above. Then, P denotes the set of fault predicates and G' denotes the set of newly introduced variables to replace the outputs of components.

Now, the SAT instance is created as shown in Figure 2: The circuit $\mathbb{C}$ is unrolled for t^d time frames as in bounded model checking [2]; the unrolled circuit is compared to the copy $\mathbb{C}'$ connected to $t^d - 1$ instances of $\mathbb{C}$; the POs in the final time frame t^d are forced to be different; only one variable in P may take the value 1. This SAT instance is satisfied iff the output behavior of the faulty and the original circuit differ in time frame t^d. This may only happen, when a faulty value is injected at component g with $p_g = 1$, i.e. component g is not robust. By finding all satisfying assignments, the non-robust components are retrieved. Once a component g has been found non-robust, further solutions for this component are blocked by inserting the constraint $p_g = 0$ into the SAT instance. All non-robust components are calculated by iteratively incrementing t^d.

Note, that this model already provides an improvement over [11]. Fault injection is only required in the first time step instead of all time steps. If the set of initial states $S(0)$ is equal to the set of

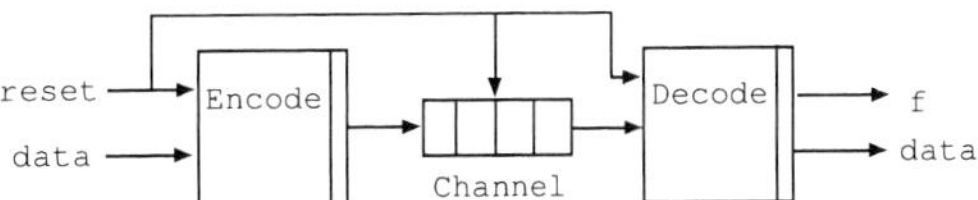

Figure 3: Transmission system

reachable states S^*, the exact value of the robustness is determined when reaching the maximal sequential depth[1] of the correct circuit and a faulty circuit [11]. In the following we assume that $S(0)$ is equal to S^* until relaxing this constraint in Section 3.2.

3. BOUNDS FOR ROBUSTNESS

In this section we adjust the notion of robustness and the model to practical requirements. As a "side effect" the computational effort decreases. First we justify an observation window to restrict the number of time steps considered by the formal analysis. Then we explain how to relax reachability analysis in the context of our approach.

3.1 Observation Window

After detecting an internal malfunction some action has to be taken at the system level in practice. Otherwise the effects of multiple faults may accumulate and may cause a disastrous failure. Therefore, we assume that a fault detection signal f exists. If a malfunction occurs, this is signaled by setting f within a given time bound of no more than t^d time steps. Assuming a very low probability for more than one fault within t^d time steps, this is safe. By this we retrieve exact bounds for the robustness while restricting the formal analysis to an *observation window* of t^d time steps.

We apply a case split to determine the robustness of a component g. Assume component g behaves faulty, then the robustness of g is assessed as follows:

1. Component g is robust, if

   (a) $f = 1$ within t^d time frames before or when a faulty value at the POs occurs or

   (b) $f = 0$, a faulty value at the POs does not occur and after t^d the same state is reached as in the fault free circuit.

2. Component g is non-robust, if
   a faulty value at the POs occurs within t^d time frames and before $f = 1$.

3. Component g is not classified, if
   $f = 0$, a faulty value at the POs does not occur and the state differs from the fault free circuit after t^d time frames.

To guarantee that a circuit is robust, neither non-robust nor non-classified components must remain. A non-robust component is a threat – a fault in this component may cause wrong output values. Knowledge about non-classified components is also essential. A fault in such a component cannot directly influence the output values, but changes the internal state of the circuit, i.e. causes SDC. If this is not detected within the required observation time t^d, an undetected error is immanent in the circuit. Effects of multiple faults may accumulate and eventually cause erroneous output.

The same model also handles circuits that directly correct faults instead of flagging a fault. In this case f is assumed to be constantly 0. As a result case 1(a) of the case split given above does not occur.

EXAMPLE 1. *A (7,4)-Hamming-Code recognizes and repairs single faults [12]. Figure 3 shows a transmission using an encoder for 4 bit* data, *a bit-wise serial channel and a decoder. A failure in the transmitted code word is flagged by setting f. The timing is summarized like this:*

---

[1]The sequential depth is the longest trace without repeating a state.

Table 1: Hamming model

t	$\|\mathbb{T}\|$	$\|\mathbb{S}\|$	$\|\mathbb{U}\|$	R_{lb} %	R_{ub} %
0	5	2	277	1.76	99.30
1	54	28	208	18.72	90.34
2	83	41	170	28.23	86.05
3	112	53	133	37.58	82.21
4	141	65	96	46.69	78.48
5	170	79	57	55.56	74.18
6	217	93	0	70.00	70.00

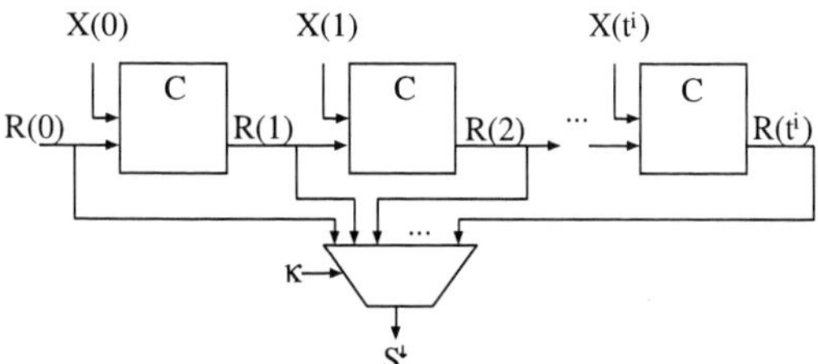

Figure 4: Constraining initial state

- *Encoding and transmission to the channel: 1 time step*

- *Transmission: 4 time steps (registers in the channel)*

- *Decoding, writing to the output, setting f: 1 time step*

The determined robustness depends on the value of t^d:

- $t^d < 6$: *The data from $t = 0$ did not arrive at the POs, yet. Faults in the decoding logic are detected within 1 time frame by setting f. Therefore these components are classified.*

  Faults in the channel change the state, but not all data has been decoded, yet. These faults are undetected, yet, and the components cannot be classified. While incrementing t^d, more and more components are classified.

- $t^d = 6$: *The input data reaches the POs. Faults that can be detected are flagged. All components are classified.*

- $t^d > 6$: *Faults injected at $t = 0$ do not influence the state of the model after more than 6 time frames.*

Let $\mathbb{T}$ be the set of components classified as robust; $\mathbb{S}$ the set of components classified non-robust and $\mathbb{U}$ the set of components not classified, yet. Then, $\mathbb{C} = \mathbb{T} \cup \mathbb{S} \cup \mathbb{U}$. Now, a lower bound R_{lb} and an upper bound R_{ub} for the robustness of the circuit $\mathbb{C}$ are given by:

$$R_{lb} = \frac{|\mathbb{T}|}{|\mathbb{C}|} = 1 - \frac{|\mathbb{S} \cup \mathbb{U}|}{|\mathbb{C}|} \quad \text{and} \quad R_{ub} = \frac{|\mathbb{T} \cup \mathbb{U}|}{|\mathbb{C}|} = 1 - \frac{|\mathbb{S}|}{|\mathbb{C}|}$$

EXAMPLE 2. *The bounds determined for the hamming model of the previous example are shown in Table 1. The bounds approach each other, until all components are classified for $t = 6$.*

3.2 Reachability

Up to now we assumed, that $S(0)$ was the set of reachable states S^*. Consequently, the approach determined exact bounds for the robustness. In practice, reachability analysis is often not feasible due to the computational complexity. For example the computation of S^* with a *Binary Decision Diagram* (BDD) [4] often requires a large amount of main memory or exceeds run time limitations. Therefore we relax this requirement in the following.

In practice the bounds derived above for the robustness depend on the set of states $S(0)$ applied in the initial time frame of the formal analysis. Assume that an overapproximation $S^\uparrow$ and an underapproximation $S^\downarrow$ of the reachable states are available, i.e. $S^\downarrow \subseteq S^* \subseteq S^\uparrow$.

When considering the subset $S^\downarrow$ some states reachable during normal operation are excluded. Therefore some components may not be activated even though relevant during normal operation, e.g.

when the ADD-operation in a CPU is never activated. These components are classified as robust instead of non-robust. The robustness determined when using $S^\downarrow$ is larger than the real value. Therefore, when calculating an upper bound of the robustness using $S^\downarrow$ is safe, i.e. $R_{ub}(S^\downarrow) \geq R_{ub}(S^*)$.

Similarly, considering unreachable states using $S^\uparrow$ decreases the calculated robustness. Consider a circuit with *Triple Modular Redundancy* (TMR). In the fault free case the three redundant modules are in the same state. Consequently, any internal fault of one module is masked and all component are robust. But when also unreachable states are considered where the state of the three modules deviates, a fault may change the output behavior of the overall circuit. Additional components may be classified as non-robust and it is safe to use the overapproximation to determine a lower bound on the robustness, i.e. $R_{lb}(S^\uparrow) \leq R_{lb}(S^*)$.

No full reachability analysis is required. Instead we apply light-weight approximations.

For $S^\downarrow$ we integrate partial reachability analysis into the formal analysis using the structure shown in Figure 4. The original circuit $\mathbb{C}$ is unrolled for t^i time frames and starts from a set of states $R(0)$ known to be reachable, e.g. the reset state. Then, $S^\downarrow$ contains any state reachable from $R(0)$ within t^i time steps as k is left unconstrained. The unrolling depth t^i controls the accuracy: $S^\downarrow$ remains an underapproximation of the reachable states when t^i is smaller than the state space diameter[2].

When leaving $R(0)$ unconstrained, $R(t^i)$ provides an overapproximation $S^\uparrow$ of the reachable states. But for our experiments we use a faster approach that assumes all states are reachable. Alternatively, in case of TMR circuits an invariant can force the states of the three redundant modules to be equal.

Of course, these light-weight approximations may be replaced by more elaborate approaches like the SAT-based procedure in [5]. In this case compactly representing $S(0)$ is crucial.

4. ALGORITHM

This section provides an incremental algorithm to transform the calculation of bounds for the robustness into a sequence of SAT instances [7]. A SAT solver [9] is used to determine the solutions. Parameters for the algorithm are the circuit $\mathbb{C}$, the set of states to be considered S^{use} and the size of the observation window t^d. The algorithm is based on the approach introduced in Section 2: The original circuit $\mathbb{C}$ and a copy $\mathbb{C}'$ are unrolled for an increasing number of time frames $t \in [0 \ldots t^d]$. The initial states of both copies are identical. The POs of time frame t are forced to different values. Circuit $\mathbb{C}'$ contains fault injection logic in time frame 0. If all fault predicates p_g are set to 0, both copies behave identically. The problem is unsatisfiable.

To analyze single faults, fault injection logic in time frame 0 is sufficient which reduces the search space compared to [11]. This is valid, because all states in S^{use} are considered as initial states of the formal analysis, i.e. $S(0) = S^{\mathrm{use}}$. Also at most one fault predicate may take the value 1. The model supports faults in state elements as well as in combinational logic or at primary inputs.

The algorithm in Figure 5 shows the incremental algorithm that determines the lower and upper bound for the robustness. Once a component is classified, this information is used in the following iterations to reduce the run time. Given a circuit $\mathbb{C}$, a copy $\mathbb{C}'$ with fault injection logic is created (Lines 2–5). Both copies are converted into *Conjunctive Normal Form* (CNF) [23] (Line 6). The initial states of both copies are forced to be equal (Line 7). The initial states $S(0)$ for the formal analysis are restricted to S^{use} (Line 8). Whether the algorithm computes exact bounds, a lower bound or an upper bound for the robustness depends on S^{use} as explained in

---

[2]The state space diameter is the largest length of the shortest trace between any two states.

978-1-60558-497-3/09 $25.00 © 2009 ACM

The main routine in Figure 5 proceeds by removing the constraints on POs and f (Line 27). Next, the algorithm determines the remaining non-classified components $\mathbb{U}'$ in a similar way (Lines 29–31). In case of non-classified components the constraints $p_g = 0$ are removed (Line 32) before the next iteration for $t + 1$ starts.

Now, the newly classified set of robust components is available (Line 34). These components do not have to be considered in further iterations and their fault predicates are fixed to 0 (Line 35).

Finally, the sets $\mathbb{T}$, $\mathbb{S}$ and $\mathbb{U}$ are updated by adding or assigning the newly classified components, t is increased, and the additional logic to compare POs and states is removed (Lines 37–43).

If non-classified components remain and t^d has not been reached, the next iteration starts (Line 15). Otherwise the algorithm terminates and returns the three sets $\mathbb{T}$, $\mathbb{S}$ and $\mathbb{U}$. As explained above the parameter S^{use} determines whether exact values or approximations are returned.

5. OPTIMIZATION TECHNIQUES

The algorithm presented so far solves multiple sequential equivalence checking problems and sequential equivalence checking is a hard problem itself. Therefore, optimization is required to improve the performance.

Knowledge about structural dominators is known to be often helpful in CAD algorithms. For example, the output of a fanout free region is a dominator for all nodes within the region. The notion of dominators is more general. A component g is dominated by a component e, if any path from g to a primary output or state bit passes along e. Thus fault effects from g must propagate along e. If component e is robust, component g is robust as well. We determine dominators using the algorithm from [14]. Then, the algorithm to determine robustness runs in two steps. First, faults are only injected into components that dominate others. Second, the dominated components of non-robust and non-classified dominators are considered for a detailed classification. This speeds up the overall run time because the search space is pruned.

Instead of sequential equivalence checking also sequential *Automatic Test Pattern Generation* (ATPG) may be used as the underlying engine. In this case one problem instance is created per fault that has to be considered. As an advantage, the size of the problem instance shrinks by only including those parts of the circuit that may be influenced by the particular fault. Moreover, similar to combinational ATPG, propagation constraints can be used to improve the performance of the engine [22, 8]. For an evaluation we used a SAT-based sequential ATPG engine. However, on the circuits considered the algorithm of Section 4 using equivalence checking was faster than the ATPG approach. This can be explained by the structural similarity of the problem instances. The algorithm of Section 4 creates one problem instance for all faults. This problem instance is kept and extended by further copies of the circuit until t^d is reached. Using the concept of incremental SAT [24], the proof engine keeps learned information for reuse in subsequent calls. Using ATPG, independent problem instances are created for all faults and similar information has to be learned from scratch. Therefore, we choose the algorithm based on equivalence checking for further experiments.

6. EXPERIMENTAL RESULTS

Experimental results are provided in the following. Our benchmark suite contains different types of sequential circuits that allow to explain the results by considering the structure of the circuits:

- without fault tolerance,
- with *Triple Modular Redundancy* (TMR) and
- with fault detection.

The circuits without fault tolerance are taken from the ITC'99 benchmark suite named by their original names b01–b13. Using these circuits, fault tolerant TMR circuits were created. The circuit was

---

```
 1  function robustness (ℂ, S^use, t^d)
 2    create a copy ℂ'_0 of ℂ
 3    foreach component g ∈ ℂ'_0
 4      replace g by g'[g,p_g];
 5    done
 6    convert to SAT instance;
 7    force init states of ℂ'_0 and ℂ_0 to be equal;
 8    force S(0) = S^use
 9    constrain ∑ p_g == 1;
10
11    𝕋 := ∅;
12    𝕊 := ∅;
13    𝕌 := all components g ∈ ℂ'_0;
14    t := 0;
15    while (t ≤ t^d && 𝕌 ≠ ∅)
16      if (t > 0) then
17        create a copy ℂ'_t of ℂ;
18        create ℂ_t and connect to ℂ'_t;
19      fi
20      connect PIs of ℂ'_t and ℂ_t;
21      constrain f = 0 in ℂ_t;
22      cmpPOs := at least one pair of POs differs;
23      cmpFFs := at least one pair of FFs differs;
24
25      add constraint UR := (cmpPOs & !f) = 1;
26      𝕊' := extractAllSolutions();
27      remove constraint UR;
28
29      add constraint UC := (!f & !cmpPOs & cmpFFs) = 1;
30      𝕌' := extractAllSolutions();
31      remove constraint UC;
32      ∀g ∈ 𝕌': remove constraint p_g = 0;
33
34      𝕋' := 𝕌 \ (𝕊' ∪ 𝕌');
35      ∀g ∈ 𝕋': add constraint p_g = 0;
36
37      𝕋 := 𝕋 ∪ 𝕋';
38      𝕌 := 𝕌';
39      𝕊 := 𝕊 ∪ 𝕊';
40
41      t := t + 1;
42      remove cmpPOs;
43      remove cmpFFs;
44    done;
45    return (𝕋, 𝕊, 𝕌);
46  end function;
```

Figure 5: Algorithm

```
 1  function extractAllSolutions ()
 2    M := ∅;
 3    while (satisfiable) do
 4      G = {g | p_g == 1};
 5      M := M ∪ G;
 6      add constraint p_g = 0;
 7    done;
 8    return M;
 9  end function;
```

Figure 6: Retrieving all solutions

Section 3.2. The number of fault predicates with value 1 is limited to one (Line 9).

Then, the sets of robust ($\mathbb{T}$), non-robust ($\mathbb{S}$), and non-classified ($\mathbb{U}$) components are initialized (Lines 11–13). In the beginning all components are non-classified. Next, the sets are incrementally updated for time frame t, starting at $t = 0$ up to $t = t^d$ (Lines 14–44). As soon as all components are classified, i.e. $\mathbb{U} = \emptyset$, the algorithm terminates. Fresh copies of $\mathbb{C}$ are appended to the unrolled circuits for $t > 0$ (Line 16–20). A constraint forces $\mathbb{C}$ to behave fault free (Line 21). Additional logic compares the POs in time frame t (Line 22), where $cmpPOs = 1$ indicates a different value for fault free and faulty copy. Similarly, $cmpFFs$ compares the states (Line 23).

Then the components $\mathbb{S}'$ that can be classified as non-robust in time frame t are determined (Lines 25–27). The POs are forced to different values and the fault detection signal f is forced to 0 (Line 25). Each satisfying solution provides a component that is non-robust. The newly classified non-robust components $\mathbb{S}'$ are returned by the subroutine *extractAllSolutions* shown in Figure 6. The subroutine extracts one non-robust component per satisfying solution (Line 4) and forces the fault predicate of this component to 0 afterwards (Line 6).

978-1-60558-497-3/09 $25.00 © 2009 ACM

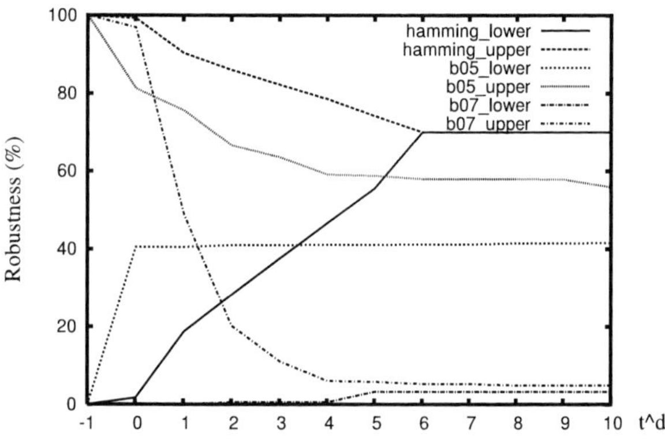

Figure 7: Robustness vs. t^d

replicated three times and a combinational majority voter drives the POs. These TMR circuits have the same sequential depth as the original circuit.

To create circuits with fault detection, the TMR circuits were extended with a signal f. While the states of the three instances are identical, no fault is detected, i.e. $f = 0$. Otherwise a fault is signaled, i.e. $f = 1$. Faults at the PIs do not activate f, because the TMR instances behave equivalently. Additionally, the Hamming model introduced in Example 1, provides fault detection. In all cases PIs, FFs and gates were considered as components.

All experiments were carried out on an AMD Athlon(tm) 64 X2 Dual Core CPU (3.0GHz, 4GB RAM, Linux). The SAT solver Chaff [17] with incremental SAT extension [24] has been used. The consideration of structural dominators improved the performance by a factor of up to 11 (see Section 5).

First, we analyze the influence of the observation window t^d on the robustness. Next the influence of constraints on $S(0)$ and the quality of approximate bounds are evaluated and finally discussed. Note, that a comparison to [11] is not given due to differing models, i.e. the bounds, the fault signal f and handling reachability.

6.1 Influence of t^d

Figure 7 exemplarily visualizes the exact bounds retrieved for three circuits using S^* while extending the observation window. Note, the initial bounds are marked with $t^d = -1$.

As already discussed in Section 3, the exact value for the robustness of the Hamming model is retrieved at $t^d = 6$ where lower and upper bound converge. In case of b05 and b07 the bounds approach each other quite rapidly in the beginning, but do not meet within 10 time frames. While 15% of the components cannot be classified for b05, only 2% remain non-classified for b07. This shows that the convergence behavior of the bounds significantly depends on the design. The incremental algorithm may be stopped as soon as the bounds are close enough or no progress can be observed.

6.2 Influence of t^i

Table 2 summarizes further results. The overapproximation $S \uparrow$ does not constrain $S(0)$. For $S \downarrow$ the approach shown in Figure 4 was used, $R(0)$ denotes the reset state and t^i was set to 0, 1, or 10, respectively. BDD-based exact reachability analysis provides S^*.

Columns $|\mathbb{C}|$ and $|FF|$ give the number of components and state holding elements in the circuit, respectively. Column t^d denotes the length of the observation window required to classify all components. At $t^d = 10$ the algorithm has been stopped. Columns $|U|$ and *time* give the number of non-classified components and the run time in seconds for computation, respectively.

While increasing t^i more reachable states become observable at $S(0)$ and thus the accuracy of the approximate upper bound increases. For example, the results in Table 2 show a significant improvement for b08 and b10 between $t^i = 1$ and $t^i = 10$. But the overhead for computing reachable states "on-the-fly" increases when increasing t^i – more run time is required.

Additionally, for TMR circuits the number of components known to be non-classified increases. Some components classified as robust for small values of t^i become non-classified, e.g. when a fault in one of the three TMR instances does not affect the output behavior but changes the state.

The TMR circuits with fault detection mechanism immediately detect the error in one of the instances and thus an early classification as robust is possible. Already for $t^i = 0$ the upper bound is nearly correct and requires far less computation time in comparison to TMR circuits without fault detection. Only components close to PIs remain non-classified. Therefore t^d has to be increased while the faulty values are not observable at the POs.

6.3 Quality of approximate bounds

For non-TMR circuits, the bounds $R_{lb}(S \uparrow)$ and $R_{ub}(S \downarrow)$, $t^i = 10$ are close to the exact ones $R_{lb}(S^*)$ and $R_{ub}(S^*)$. Only for b05 and b13 the exact bounds differ significantly. Here, increasing t^i or t^d may help to improve the accuracy of the approximate bounds.

For TMR circuits the distance between lower and upper bound is large. The lower bound $R_{lb}(S \uparrow)$ is not tight enough, because the initial states of the TMR instances are allowed to differ. Either an exact analysis or a manual invariant is required to improve the accuracy. Experiments using an invariant to force identical initial states, yielded an almost exact lower bound $R_{lb}(S \uparrow)$. Due to page limitation no details are reported here.

The approximate bounds for circuits with fault detection are close to the exact bounds. Constraining the specification to fault free behavior by setting $f = 0$ in C_t forces the submodules to start in identical states. Consequently, the lower bound becomes more accurate and an exact analysis is not required.

6.4 Discussion

As shown in Table 2, BDD-based reachability analysis often exceeds the run time limitation. Thus, no exact information about fault tolerance can be computed. Our approximation algorithm still provides a partial classification of components giving insight about fault tolerance of the circuits.

For both, the approximate as well as the exact bounds, non-classified components are left for some circuits. Here, a large difference between upper and lower bound due to non-classified components always points to potential immanent undetected errors that do not yet materialize in the output response. Such SDC may lead to faulty output responses in combination with other faults. Thus, the knowledge about the non-classified components is mandatory.

In summary, the main focus of the experiments is on the formal model and the formal algorithm to determine robustness. The proposed model fits practical needs by restricting the observation window. The approximate bounds provide information about fault tolerance even when an exact analysis cannot be applied. For some circuits run time is a bottleneck. Here, providing a set of manual invariants decreases the run time and increases the accuracy.

7. CONCLUSIONS

We presented an approach to formally prove the robustness of a circuit. The algorithm works on a bounded number of time steps and determines a lower bound and an upper bound on the robustness. Approximate sets of reachable states are sufficient to determine these bounds. An incremental algorithm and additional optimization techniques are provided. The results show that even if only a small number of time steps is considered, an exact value of the robustness can often be obtained. Otherwise a subset of non-robust and robust components is provided, that can be used for further design modifications. Especially, for circuits with fault detection mechanism accurate values are determined efficiently.

Future work focuses on improving the effectiveness for very large circuits. That is, using multiple engines like simulation, sequential ATPG and the proposed algorithm within a single framework or by using abstraction and hierarchical information. Moreover, assessing robustness in presence of multiple faults remains an open issue.

978-1-60558-497-3/09 $25.00 © 2009 ACM

Table 2: Influence of constraints on initial states

Part 1 — columns $S^\uparrow$:

circuit	$|\mathbb{C}|$	$|FF|$	t^d	R_{lb} %	$|U|$	time(s)
Without fault tolerance						
b01	64	5	4	**0.00**	-	0.19
b02	34	4	3	**0.00**	-	0.05
b03	199	30	9	**0.00**	-	6.41
b04	832	66	5	**0.00**	-	60.57
b05	1199	34	2	**5.59**	-	57.42
b06	73	9	2	**0.00**	-	0.18
b07	513	49	10	**0.00**	3	22.71
b08	232	21	10	**0.00**	9	5.71
b09	198	28	9	**0.00**	-	4.25
b10	271	17	10	**0.00**	1	6.65
b11	874	31	10	**0.11**	14	123.98
b12	1302	121	10	**0.00**	2	239.98
b13	410	53	10	**0.73**	8	15.60
Tripple modular redundancy						
b01-tmr	212	15	4	**1.89**	-	2.63
b02-tmr	112	12	3	**1.79**	-	0.69
b03-tmr	637	90	9	**1.26**	-	76.56
b04-tmr	2579	198	5	**0.62**	-	697.26
b05-tmr	3922	102	2	**6.96**	-	1137.48
b06-tmr	275	27	2	**3.64**	-	3.00
b07-tmr	1612	147	10	**0.99**	-	2046.31
b08-tmr	741	63	10	**1.08**	35	88.05
b09-tmr	604	84	9	**3.31**	-	51.67
b10-tmr	878	51	10	**1.37**	3	1464.80
b11-tmr	2683	93	10	**0.56**	42	2721.79
b12-tmr	3965	363	[a]6	**0.30**	9	2933.94
b13-tmr	1330	159	10	**2.18**	24	432.77
With fault detection						
b01-tmrflt	215	15	1	**88.84**	-	0.17
b02-tmrflt	123	12	2	**88.62**	-	0.07
b03-tmrflt	648	90	4	**91.36**	-	2.11
b04-tmrflt	2590	198	1	**95.56**	-	45.29
b05-tmrflt	3933	102	1	**45.51**	-	352.61
b06-tmrflt	286	27	1	**71.33**	-	0.56
b07-tmrflt	1623	147	1	**93.53**	-	17.70
b08-tmrflt	752	63	10	**91.89**	8	5.97
b09-tmrflt	615	84	2	**97.72**	-	1.33
b10-tmrflt	889	51	4	**89.99**	-	5.46
b11-tmrflt	2694	93	3	**96.85**	-	23.31
b12-tmrflt	3976	363	1	**97.91**	-	184.09
b13-tmrflt	1341	159	2	**89.56**	-	7.70
hamming	284	7	6	**70.00**	-	20.86

Part 2 — columns $S^\downarrow, t^i=0$ and $S^\downarrow, t^i=1$:

circuit	t^d	R_{ub} %	$|U|$	time(s)	t^d	R_{ub} %	$|U|$	time(s)
b01	5	**10.94**	-	0.26	4	**1.56**	-	0.25
b02	4	**44.12**	-	0.06	4	**20.59**	-	0.06
b03	10	**22.61**	12	9.11	10	**5.53**	3	7.66
b04	3	**86.78**	-	8.32	5	**64.90**	-	29.65
b05	10	**91.33**	124	98.60	10	**90.91**	126	107.26
b06	2	**16.44**	-	0.15	2	**2.74**	-	0.22
b07	10	**87.52**	190	61.06	10	**87.52**	198	68.44
b08	10	**91.38**	81	11.06	10	**90.08**	88	13.29
b09	10	**85.86**	79	10.38	10	**75.76**	88	14.48
b10	10	**62.36**	40	8.88	10	**32.84**	8	7.80
b11	10	**88.44**	133	83.09	10	**83.07**	142	99.76
b12	10	**81.64**	449	470.24	10	**75.88**	386	460.31
b13	10	**66.10**	112	36.31	10	**62.44**	108	38.86
b01-tmr	10	**98.11**	114	15.82	10	**98.11**	132	22.33
b02-tmr	10	**99.11**	42	2.47	10	**98.21**	66	4.39
b03-tmr	10	**98.74**	438	208.95	10	**98.74**	513	286.89
b04-tmr	[a]9	**99.69**	21	19493.90	10	**99.38**	276	9730.20
b05-tmr	10	**99.08**	450	1256.48	10	**99.06**	477	1419.27
b06-tmr	10	**97.45**	81	16.30	10	**97.09**	99	24.63
b07-tmr	10	**99.50**	738	834.42	10	**99.50**	763	925.15
b08-tmr	10	**99.46**	292	142.78	10	**99.46**	330	174.77
b09-tmr	10	**99.83**	315	122.03	10	**99.67**	405	172.53
b10-tmr	10	**99.20**	402	273.23	10	**98.52**	552	484.63
b11-tmr	10	**99.78**	666	1261.08	10	**99.52**	852	1738.78
b12-tmr	10	**99.82**	2031	6657.70	10	**99.82**	2067	7311.42
b13-tmr	10	**99.25**	693	653.31	10	**99.25**	729	766.78
b01-tmrflt	1	**98.14**	-	0.06	1	**98.14**	-	0.10
b02-tmrflt	0	**99.19**	-	0.02	4	**98.37**	-	0.05
b03-tmrflt	4	**98.77**	-	0.77	4	**98.77**	-	0.98
b04-tmrflt	0	**99.69**	-	0.96	3	**99.38**	-	4.91
b05-tmrflt	0	**99.08**	-	4.23	1	**99.06**	-	6.23
b06-tmrflt	1	**97.55**	-	0.09	1	**97.20**	-	0.13
b07-tmrflt	0	**99.51**	-	0.58	10	**99.51**	1	3.31
b08-tmrflt	10	**99.47**	1	1.15	10	**99.47**	9	5.33
b09-tmrflt	0	**99.84**	-	0.10	10	**99.67**	-	0.96
b10-tmrflt	2	**99.21**	-	0.42	3	**98.54**	-	1.25
b11-tmrflt	1	**99.55**	-	1.65	3	**99.52**	-	4.58
b12-tmrflt	5	**99.82**	-	4.59	4	**99.82**	-	6.16
b13-tmrflt	0	**99.25**	-	0.65	0	**99.25**	-	0.90
hamming	6	**70.98**	-	19.57	6	**70.65**	-	21.88

Part 3 — columns $S^\downarrow, t^i=10$ and S^*:

circuit	t^d	R_{ub} %	$|U|$	time(s)	t^d	R_{lb} %	R_{ub} %	$|U|$	time(s)
b01	4	**0.00**	-	0.50	4	**0.00**	**0.00**	-	0.24
b02	3	**0.00**	-	0.13	3	**0.00**	**0.00**	-	0.09
b03	9	**0.50**	-	13.10	9	**0.50**	**0.50**	-	6.23
b04	5	**0.00**	-	131.34					[b]
b05	10	**90.33**	354	477.72	10	**41.54**	**55.88**	172	225.09
b06	2	**0.00**	-	0.50	2	**0.00**	**0.00**	-	0.24
b07	10	**87.33**	268	188.22	10	**3.31**	**4.87**	8	44.21
b08	10	**0.00**	-	35.77	10	**0.00**	**3.88**	9	5.80
b09	10	**47.47**	42	16.34	10	**0.00**	**0.51**	1	5.30
b10	10	**1.85**	4	20.83	10	**0.37**	**1.85**	4	8.73
b11	10	**8.92**	14	216.44	10	**6.18**	**7.78**	14	256.67
b12	10	**52.07**	457	960.20					[b]
b13	10	**52.68**	105	62.04	10	**2.68**	**7.56**	20	1793.98
b01-tmr	10	**98.11**	135	43.08	10	**34.43**	**98.11**	135	26.97
b02-tmr	10	**98.21**	84	9.40	10	**23.21**	**98.21**	84	5.02
b03-tmr	10	**98.74**	534	563.92	10	**14.91**	**98.74**	534	4336.39
b04-tmr	[a]0	**99.69**	2483	31280.70					[b]
b05-tmr	10	**99.06**	1179	5638.39					[b]
b06-tmr	10	**97.09**	102	58.70	10	**60.00**	**97.09**	102	26.22
b07-tmr	10	**99.50**	976	2373.76					[b]
b08-tmr	10	**99.46**	690	634.43	10	**5.94**	**99.33**	692	2180.33
b09-tmr	10	**99.67**	423	302.18	10	**24.17**	**99.67**	456	1572.46
b10-tmr	[a]9	**98.06**	792	17176.30	10	**7.86**	**98.06**	792	2574.79
b11-tmr	10	**99.52**	2412	15212.20					[b]
b12-tmr	10	**99.82**	3210	19253.10					[b]
b13-tmr	10	**99.10**	863	1526.89					[b]
b01-tmrflt	1	**98.14**	-	0.32	1	**98.14**	**98.14**	-	0.26
b02-tmrflt	2	**98.37**	-	0.11	2	**98.37**	**98.37**	-	0.12
b03-tmrflt	4	**98.77**	-	5.85	4	**98.77**	**98.77**	-	4138.53
b04-tmrflt	1	**99.27**	-	5377.38					[b]
b05-tmrflt	1	**99.06**	-	51.10					[b]
b06-tmrflt	1	**97.20**	-	2.38	1	**97.20**	**97.20**	-	0.50
b07-tmrflt	10	**99.51**	1	25.91					[b]
b08-tmrflt	10	**99.47**	9	12.22	10	**98.27**	**99.34**	8	1747.29
b09-tmrflt	10	**99.67**	-	2.36	2	**99.67**	**99.67**	-	1283.33
b10-tmrflt	4	**98.09**	-	29.66	4	**98.09**	**98.09**	-	1172.55
b11-tmrflt	3	**99.52**	-	86.68					[b]
b12-tmrflt	1	**99.82**	-	216.73					[b]
b13-tmrflt	10	**99.11**	8	25.31					[b]
hamming	6	**70.00**	-	38.01	6	**70.00**	**70.00**	-	21.74

[a]Abnormal termination of SAT solver; [b]BDD could not be computed within 5h

8. REFERENCES

[1] T. Austin and V. Bertacco. Deployment of better than worst-case design: Solutions and needs. In *Int'l Conf. on Comp. Design*, pages 550–558, 2005.

[2] A. Biere, A. Cimatti, E. Clarke, and Y. Zhu. Symbolic model checking without BDDs. In *Tools and Algorithms for the Construction and Analysis of Systems*, volume 1579 of *LNCS*, pages 193–207. Springer Verlag, 1999.

[3] M. Bozzano, A. Cimatti, and F. Tapparo. Symbolic fault tree analysis for reactive systems. In *Automated Technology for Verification and Analysis*, volume 4762 of *LNCS*, pages 162–176, 2007.

[4] R. Bryant. Graph-based algorithms for Boolean function manipulation. *IEEE Trans. on Comp.*, 35(8):677–691, 1986.

[5] P. Chauhan, E. Clarke, and D. Kroening. Using SAT based image computation for reachability analysis. Technical Report CMU-CS-03-151, School of Computer Science, Carnegie Mellon University, 2003.

[6] P. Civera, L. Macchiarulo, M. Rebaudengo, M. S. Reorda, and M. Violante. An FPGA-based approach for speeding-up fault injection campaigns on safety-critical circuits. *Jour. of Electronic Testing: Theory and Applications*, 18(3):261–271, 2002.

[7] M. Davis and H. Putnam. A computing procedure for quantification theory. *Journal of the ACM*, 7:506–521, 1960.

[8] R. Drechsler, S. Eggersglüß, G. Fey, A. Glowatz, F. Hapke, J. Schlöffel, and D. Tille. On acceleration of SAT-based ATPG for industrial designs. *IEEE Trans. on CAD*, 27(7):1329–1333, 2008.

[9] N. Eén and N. Sörensson. An extensible SAT solver. In *SAT 2003*, volume 2919 of *LNCS*, pages 502–518, 2004.

[10] D. Ernst, N. S. Kim, S. Das, S. Pant, T. Pham, R. Rao, C. Ziesler, D. Blaauw, T. Austin, and T. Mudge. Razor: A low-power pipeline based on circuit-level timing speculation. In *Micro Conference*, 2003.

[11] G. Fey and R. Drechsler. A basis for formal robustness checking. In *Int'l Symp. on Quality Electronic Design*, pages 784–789, 2008.

[12] R. W. Hamming. Error detecting and error correcting codes. *Bell System Technical Jour.*, 9:147–160, April 1950.

[13] U. Krautz, M. Pflanz, C. Jacobi, H. W. Tast, K. Weber, and H. T. Vierhaus. Evaluating coverage of error detection logic for soft errors using formal methods. In *Design, Automation and Test in Europe*, pages 176–181, 2006.

[14] T. Lengauer and R. E. Tarjan. A fast algorithm for finding dominators in a flowgraph. *ACM Trans. Program. Lang. Syst.*, 1(1):121–141, 1979.

[15] R. Leveugle. A new approach for early dependability evaluation based on formal property checking and controlled mutations. In *IEEE International On-Line Testing Symposium*, pages 260–265, 2005.

[16] M. Miskov-Zivanov and D. Marculescu. Circuit reliability analysis using symbolic techniques. *IEEE Trans. on CAD*, 25(12):2638–2649, 2006.

[17] M. Moskewicz, C. Madigan, Y. Zhao, L. Zhang, and S. Malik. Chaff: Engineering an efficient SAT solver. In *Design Automation Conf.*, pages 530–535, 2001.

[18] A. Pellegrini, K. Constantinides, D. Zhang, S. Sudhakar, V. Bertacco, and T. Austin. CrashTest: A fast high-fidelity FPGA-based resiliency analysis framework. In *Int'l Conf. on Comp. Design*, 2008.

[19] S. A. Seshia, W. Li, and S. Mitra. Verification-guided soft error resilience. In *Design, Automation and Test in Europe*, pages 1442–1447, 2007.

[20] P. Shivakumar, M. Kistler, S. W. Keckler, D. Burger, and L. Alvis. Modeling the effect of technology trends on the soft error rate of combinational logic. In *Int'l Conf on Dependable Systems and Networks*, pages 389–398, 2002.

[21] A. Smith, A. Veneris, M. Fahim Ali, and A.Viglas. Fault diagnosis and logic debugging using boolean satisfiability. *IEEE Trans. on CAD*, 24(10):1606–1621, 2005.

[22] P. Stephan, R. Brayton, and A. Sangiovanni-Vincentelli. Combinational test generation using satisfiability. *IEEE Trans. on CAD*, 15:1167–1176, 1996.

[23] G. Tseitin. On the complexity of derivation in propositional calculus. In *Studies in Constructive Mathematics and Mathematical Logic, Part 2*, pages 115–125, 1968. (Reprinted in: J. Siekmann, G. Wrightson (Ed.), Automation of Reasoning, Vol. 2, Springer, Berlin, 1983, pp. 466-483.).

[24] J. Whittemore, J. Kim, and K. Sakallah. SATIRE: A new incremental satisfiability engine. In *Design Automation Conf.*, pages 542–545, 2001.

[25] C. Zhao and S. Dey. Improving transient error tolerance of digital VLSI circuits using RObustness COmpiler (ROCO). In *Int'l Symp. on Quality Electronic Design*, pages 133–140, 2006.

[26] Q. Zhou and K. Mohanram. Gate sizing to radiation harden combinational logic. *IEEE Trans. on CAD*, 25(1):155–166, 2006.

978-1-60558-497-3/09 $25.00 © 2009 ACM

Clock Skew Optimization Via Wiresizing For Timing Sign-off Covering All Process Corners[*]

Sari Onaissi, Khaled R. Heloue, Farid N. Najm
Department of ECE, University of Toronto, Canada
{sari, khaled, najm}@eecg.utoronto.ca

ABSTRACT

Manufacturing process variability impacts the performance of synchronous logic circuits by means of its effect on both clock network and functional block delays. Typically, variability in clock networks is either handled early in the design flow by assigning margins to clock network delays, or at a later stage through post-processing steps that *only* focus on achieving minimal skew, without regard to functional block variability. In this work, we present a technique that alters clock network lines so that the circuit meets its timing constraints at all process corners. This is done near the end of the design flow while considering delay variability in *both* the clock network and the functional blocks. Our method operates at the *physical level* and provides designers with the required changes in clock network line widths and/or lengths. This can be formulated as a Linear Programming (LP) problem, and thus can be solved efficiently. Empirical results for a set of ISCAS-89 benchmark circuits show that our approach can considerably reduce the effect of process variations on circuit performance.

Categories and Subject Descriptors

B.7.2 [**Hardware**]: Integrated Circuits—*Design Aids*

General Terms

Algorithms, performance, verification, reliability

Keywords

Clock skew optimization, variability, sign-off, parameterized timing analysis, wiresizing

1. INTRODUCTION

The performance of a synchronous logic circuit depends on both circuit delays and clock skews introduced by the clock distribution network. Clock skew is defined as the difference between the arrival times of the clock signal at different clocked circuit elements. Typically, clock networks are designed to minimize skew between the various clocked elements, where the objective in this case is to achieve a *zero-skew clock tree*, as in [1] and [2]. On the other hand, the more aggressive techniques of *skew optimization* or *clock scheduling* (e.g. [3], [4], and [5]) actually assign skews to different clocked elements in order to obtain better performance. These usually formulate optimization problems to find clock network path delays that improve circuit performance. However, in practice, this requires interaction among different stages of the design flow [6] and extends to beyond finding clock network delay values.

With the scaling of VLSI technology, the effect of manufacturing process variations on circuit and clock network delays has increased. Typically the effect of this on clock networks is seen in the form of *unintended skew*, which degrades circuit performance. In the case of zero-skew trees the focus has been on minimizing the maximum skew resulting from process variations. This is typically

---

[*]This project was supported in part by Intel Corp.

Permission to make digital or hard copies of part or all of this work for personal or classroom use is granted without fee provided that copies are not made or distributed for profit or commercial advantage and that copies bear this notice and the full citation on the first page. To copy otherwise, to republish, to post on servers or to redistribute to lists, requires prior specific permission and/or a fee.
DAC'09, July 26-31, 2009, San Francisco, California, USA

done by wire sizing and/or introduction of buffers (e.g. [7] and [8]). Such approaches try to "cap" the maximum skew expected as a result of process variability. However, these approaches deal with variability in the clock network without regard to that in the functional blocks, and are focused on achieving zero-skew rather than timing closure. On the other hand, skew optimization techniques usually deal with variability by incorporating margins [3], or permissible ranges [5], into their optimization formulation. However, this is done at an early stage in the design flow when accurate variability information of the clock network and functional blocks is typically not available and is thus problematic.

In this work we present a linear programming (LP) formulation in order to reduce the effect of process variability on circuit performance. Our approach specifies required changes at the physical level in clock network wire widths or lengths rather than required clock network delay values. It is applied near the end of the design flow as a post-processing step to account for process variability when accurate *variational* clock network and functional path delays are available. As stated earlier, modifying clock line widths or lengths as a post-processing step has been proposed to minimize the unintended skew of zero-skew clock trees. However, in this work we formulate an LP that varies skews to try to meet circuit timing constraints at all process corners rather than try to eliminate skew.

2. PRELIMINARIES

In this section, we first present the process parameter model assumed in our work. Then, we present some definitions and preliminary concepts that are used throughout the paper.

2.1 Variability Model

In our approach, all logic cell and interconnect delays are modeled as linear functions of normalized process parameters, whose values vary between -1 and $+1$. Hence the delay of a logic cell or of an interconnect RC-tree (also referred to as an interconnect structure) can be written as an affine function of these process parameters. Because the delays of individual paths, in both clock networks and functional blocks, are sums of gate and interconnect delays, they also become such affine functions of the process parameters. Let the number of process parameters under consideration be p. Thus, a timing quantity t, representing a timing arc, interconnect, or path delay, can be written as follows:

$$t = \bar{t} + \sum_{l=1}^{p} \delta_l X_l \qquad (1)$$

where $X = (X_1, \ldots, X_p)$ is the set of normalized process parameters, $\bar{t}$ is the *nominal value* of t, and δ_l is its *sensitivity* to process parameter X_l. Such an affine function represents a *hyperplane* in $(p + 1)$-dimensional space, and will often be referred to as, simply, a hyperplane. In this work, it is often required to find the maximum or the minimum value of a hyperplane over the process parameter space. It is trivial to see that the maximum value of t over all process corners is:

$$t_{max} = \bar{t} + \sum_{l=1}^{p} |\delta_l| \qquad (2)$$

which is equivalent to computing t at the process corner $\hat{X} = (\hat{X}_1, \ldots, \hat{X}_p)$, such that $\hat{X}_l = +1$ if $\delta_l \geq 0$ and $\hat{X}_l = -1$ otherwise. Minimizing t involves a very similar operation where the hyperplane is instead computed at $\hat{x} = (\hat{x}_1, \ldots, \hat{x}_p)$, where $\hat{x}_l = -1$ if $\delta_l \geq 0$ and $\hat{x}_l = +1$ otherwise.

2.2 Circuit Model

We focus on synchronous sequential circuits with edge-triggered registers, but the work can be extended to circuits with level-sensitive latches. In such circuits, every combinational logic block receives its inputs from its *input registers* and stores its outputs in its *output registers*. We denote by S the set of combinational blocks in a given circuit. For example, Fig. 1, shows a simple sequential circuit, where in this case $S = \{s_1, s_2\}$. A clock signal is connected to all the registers of a sequential circuit by means of a clock network that consists of buffers and interconnect. Such a clock network can be designed to minimize skew between the clock signals at the various registers, or it can be designed such that it intentionally introduces skew at carefully chosen registers in order to help meet timing constraints. In our approach, we refer to the clock signal received at a register as the register's *clock phase*, and we call the delay of the path from the clock source to the register its *clock phase arrival time*. In general, for a sequential circuit with m registers, the clock phase arrival times are referred to as r_q, $1 \leq q \leq m$. For a combinational logic block $s \in S$, D_s^{ij} refers to the largest delay between the input register controlled by clock phase i and the output register controlled by clock phase j. The term d_s^{ij} is used to refer to the smallest such delay. For example, in Fig. 1, D_1^{24} is the largest delay between reg$_2$ and reg$_4$ through s_1 whereas d_1^{24} is the smallest such delay.

The clock phase arrival times r_q of a sequential circuit are path delays and hence, in our formulation, are hyperplanes in the set of process parameters X. Consider the set P_s^{ij} of all paths through the combinational logic block s between the source register controlled by clock phase i and the sink register controlled by clock phase j. The delays of these paths are, each, a hyperplane in the process parameters, for which D_s^{ij} is their "max" and d_s^{ij} is their "min". Being the maximum and minimum of a set of hyperplanes, it follows that D_s^{ij} and d_s^{ij} are, in general, piecewise planar surfaces and *not* hyperplanes. The notion of piecewise planar timing delays has been used in the context of parameterized static timing analysis (PSTA) and there are existing works (e.g., [9] and [10]) on how these can be found and represented. Typically, such a surface is represented by a set or a "list" of hyperplanes, each of which corresponds to a particular path. Thus D_s^{ij} is the set of hyperplane delays of "potentially longest paths" and d_s^{ij} is the set of hyperplane delays of "potentially shortest paths". However, in our method, it is also possible to approximate D_s^{ij} and d_s^{ij} by single hyperplanes, obtained by finding hyperplane bounds for these two sets, using the approach in [11]. That is, D_s^{ij} would be replaced by a hyperplane approximation of the "max" of the delays of the "potentially longest paths", and similarly d_s^{ij} would be replaced by a hyperplane approximation of the "min" of the "potentially shortest paths" delays. These approximations are conservative, i.e., they never underestimate the true maximum delay or overestimate the true minimum delay for any point in the process space.

3. BACKGROUND

In our approach, we modify the delays of the clock network in order for the circuit to meet its timing constraints in the presence of process variations. This would be applicable near the end of the design flow, as part of physical design while trying to achieve timing sign-off, by modifying the clock network line widths and/or lengths. In this section, we first present a short overview of standard clock tree synthesis and a brief description of how clock skew optimization could be used to improve the performance of a circuit in a typical design flow. We then discuss the effect of process variability on circuit performance and how a circuit's timing constraints can be verified under variability. This will provide some background understanding of the problem and set the stage for the presentation of our proposed clock network optimization method.

3.1 Clock Tree Synthesis

In order to synthesize a clock network, a user can specify certain parameters such as maximum skew, maximum and minimum

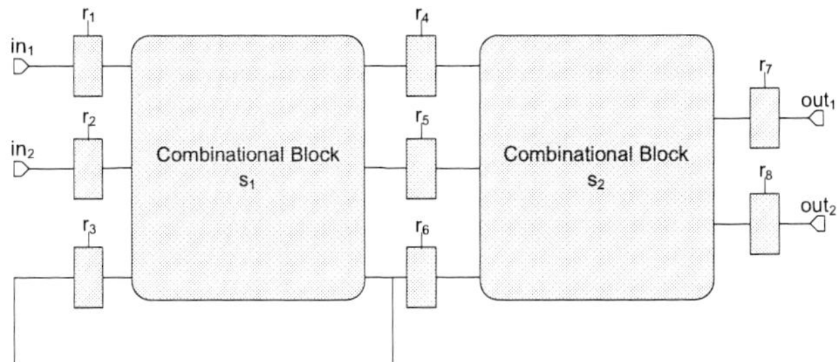

Figure 1: A Simple Sequential Circuit

clock phase arrival times, and a host of other parameters. These are then used to generate a clock network consisting of buffers and interconnect, where the objective would be to have a "balanced" clock tree. That is, designers usually strive to minimize skews between different leaves of the clock tree.

In order to achieve circuits with higher frequencies, or higher slacks, a user can use skew optimization techniques such as those presented in [3] and [4] to find an "optimal" set of clock phase arrival times. These techniques lead to a set of *nominal* clock phase arrival times or a *clock schedule*, which are then translated into a physical structure consisting of buffers and interconnect. The method in [3] is an LP formulation that provides the clock phase delays that maximize the clock frequency for circuits with edge-triggered registers, as follows. Let $\bar{r}_q$, $1 \leq q \leq m$, be the *nominal* clock phase arrival times, and let $\bar{D}_s^{ij}$ and $\bar{d}_s^{ij}$ be, respectively, the largest and smallest nominal delays through the combinational logic block s between the source register with clock phase i and sink register with clock phase j. The minimum acceptable clock period T can be found using the following LP [3]:

$$\text{minimize: } T$$
$$\text{Subject to: } \forall s \in S, \, \forall i, j \in 1, \ldots, m \qquad (3)$$
$$\bar{D}_s^{ij} + \bar{r}_i - \bar{r}_j + t_{setup}^j \leq T$$
$$\bar{d}_s^{ij} + \bar{r}_i - \bar{r}_j - t_{hold}^j \geq 0$$

To allow this general formulation, the convention is adopted that, if logic block s does not contain a combinational path between registers controlled by clock phases i and j, then $\bar{D}_s^{ij} = -\infty$ and $\bar{d}_s^{ij} = +\infty$. For a required clock period of T_c, if $T \leq T_c$ then the clock network which achieves the values of $\bar{r}_q$, $1 \leq q \leq m$ is generated, placed, and routed. However, achieving a clock network that can produce such arrival times is more complex than achieving a minimal skew tree, and clock skew optimization remains an active research area [6]. In any case our method can be used irrespective of whether clock skew optimization is used to generate the clock network or not.

3.2 Verification for Variability

Using the layout of the circuit resulting from either the standard or skew optimization *nominal point* analyses as a starting point, along with characterized cell and interconnect process sensitivities, variational path delays of the circuit can now be extracted. For a register with nominal clock phase arrival time $\bar{r}_q$, its actual r_q under variability is a hyperplane, given by:

$$r_q = \bar{r}_q + \sum_{l=1}^{p} \lambda_l^{(q)} X_l \qquad (4)$$

The dependence of combinational path delays and clock phase arrival times on process parameters may cause some timing constraints to fail for some process settings. In order for the circuit to meet the timing constraints under variability, then, for every logic block s, the following inequalities should be satisfied over the whole process parameter space for all clock phases i and j:

$$D_s^{ij} + r_i - r_j + t_{setup}^j \leq T_c$$
$$d_s^{ij} + r_i - r_j - t_{hold}^j \geq 0 \qquad (5)$$

978-1-60558-497-3/09 $25.00 © 2009 ACM

which is equivalent to saying that:

$$\max_X (D_s^{ij} + r_i - r_j) + t_{setup}^j \leq T_c$$

$$\min_X (d_s^{ij} + r_i - r_j) - t_{hold}^j \geq 0 \qquad (6)$$

Note that here we make the simplifying assumption that t_{setup}^j and t_{hold}^j are constant, rather than dependent on process parameters. Therefore, these are not included in the max and min expressions. However this assumption is by no means necessary, and doesn't affect the applicability of our approach. Performing the verification in (6) can be done as follows. Let us first consider the case where bounding hyperplane approximations are used for D_s^{ij} and d_s^{ij}, as discussed above, so that D_s^{ij} and d_s^{ij} are hyperplanes, given by:

$$D_s^{ij} = \bar{D} + \sum_{l=1}^p \Phi_l X_l, \text{ and } d_s^{ij} = \bar{d} + \sum_{l=1}^p \phi_l X_l \qquad (7)$$

Writing r_i and r_j as in (4), we can now write (6) as:

$$\max_X \left(\bar{D} + \bar{r}_i - \bar{r}_j + \sum_{l=1}^p (\Phi_l + \lambda_l^{(i)} - \lambda_l^{(j)}) X_l \right) + t_{setup}^j \leq T_c$$

$$\min_X \left(\bar{d} + \bar{r}_i - \bar{r}_j + \sum_{l=1}^p (\phi_l + \lambda_l^{(i)} - \lambda_l^{(j)}) X_l \right) - t_{hold}^j \geq 0$$

$$(8)$$

Note that the expressions inside the "max" and the "min" operations are hyperplanes. Finding the maximum or minimum value of a hyperplane over the process space is a straight-forward operation that is performed by computing the value of hyperplane at a carefully chosen corner, as described earlier in (2). For the constraints in (8), we use the term "worst-case" corner to refer to the process corner used in the verification of the constraint. Thus for a setup constraint the "worst-case" corner is the process corner used to maximize its hyperplane, whereas for a hold-time constraint this term refers to the process corner that minimizes its hyperplane.

Now consider the case when D_s^{ij} and d_s^{ij} are written as piecewise planar functions. For example, suppose that D_s^{ij} is the piecewise planar surface formed by taking the "max" of the following hyperplanes:

$$D^{(1)} = \bar{D}^{(1)} + \sum_{l=1}^p \Phi_l^{(1)} X_l$$

$$D^{(2)} = \bar{D}^{(2)} + \sum_{l=1}^p \Phi_l^{(2)} X_l$$

$$\vdots \qquad (9)$$

$$D^{(u)} = \bar{D}^{(u)} + \sum_{l=1}^p \Phi_l^{(u)} X_l$$

Then, in order to verify the setup constraint of D_s^{ij} we have to verify the following constraints:

$$\max_X \left(\bar{D}^{(1)} + \bar{r}_i - \bar{r}_j + \sum_{l=1}^p (\Phi_l^{(1)} + \lambda_l^{(i)} - \lambda_l^{(j)}) X_l \right) + t_{setup}^j \leq T_c$$

$$\max_X \left(\bar{D}^{(2)} + \bar{r}_i - \bar{r}_j + \sum_{l=1}^p (\Phi_l^{(2)} + \lambda_l^{(i)} - \lambda_l^{(j)}) X_l \right) + t_{setup}^j \leq T_c$$

$$\vdots \qquad (10)$$

$$\max_X \left(\bar{D}^{(u)} + \bar{r}_i - \bar{r}_j + \sum_{l=1}^p (\Phi_l^{(u)} + \lambda_l^{(i)} - \lambda_l^{(j)}) X_l \right) + t_{setup}^j \leq T_c$$

These constraints are verified by writing each at its "worst-case" corner as is done for the setup constraint of (8). However, note that "worst-case" corners for these constraints can be different. Let T_v be the smallest clock period for which all the setup constraints of a circuit are satisfied under process variability. This value can be found by evaluating the left-hand side expressions of all setup constraints, such as (10), and choosing the largest value. We call $M_v = T_v - T_c$ the required timing margin; this is the margin that has to be added to T_c so that the circuit can operate correctly in the presence of process variability. The same approach is taken for d_s^{ij}, where its hold-time constraint is also written for each of the hyperplanes that form its piecewise planar surface as in (10) and may be verified in a similar manner.

4. OPTIMIZATION

If the verification process described above reveals that the circuit does not meet its timing constraints for a required T_c under *all* process settings, then an optimization problem can be formulated to modify the clock network so that the required timing margin is minimized, thus possibly allowing the timing constraints to be satisfied at all process corners, as follows.

A set of physical parameters of the clock tree, such as wire lengths and/or widths, are chosen as *variables* to be optimized. By varying these optimization variables, the different clock phase delays can be modified to minimize the required timing margin. However, before we explain how this is done, we will first examine how varying the physical parameters of an interconnect RC-structure of a clock tree affects its clock phase arrival times.

4.1 Physical Parameter to Clock Phase Delay

Consider Fig. 2, which shows a generic segment of the path between a clock source and a register, consisting of the interconnect RC-structure con_k, its "input buffer" buf_k, and one of its "output buffers" buf_{k+1}. We are interested in the effect that varying a physical parameter of con_k would have on the clock phase arrival time r seen in Fig. 2. We will use variation in wire width as the physical parameter of interest in our discussion, however all of the arguments that follow can be made for a variation in wire length as well. Let w_k be the wire width of con_k, and Δw_k be a change in w_k. A variation in w_k would have an immediate effect on the delays of the three "local" circuit elements, con_k, buf_k, and buf_{k+1}. First, the introduction of Δw_k would vary the delay of con_k by Δd_{con_k}. Second, the effective load capacitance seen by buf_k would vary as a result of Δw_k, thus leading to a variation Δd_{buf_k} in its delay, d_{buf_k}. In addition to the variation in the delay of buf_k, the change in its effective load capacitance also leads to a variation in its output signal slew. This in turn varies the delay of buf_{k+1} by $\Delta d_{buf_{k+1}}$. One could argue that this variation in output signal slew of buf_k would also lead to a similar effect in buf_{k+1}, which would then lead to the same effect in buffers further downstream. This would extend the effect of Δw_k to "non-local" circuit elements. However, the impact on the delays of these elements is small in practice, and it can be safely ignored. Thus, the delay variations in the three "local" circuit elements seen in Fig. 2 would account for most of the variation seen in r. This variation, Δr, can now be written as:

$$\Delta r = \Delta d_{buf_k} + \Delta d_{con_k} + \Delta d_{buf_{k+1}} \qquad (11)$$

In our approach, a linear model is used to approximate the dependence of delay variations of circuit elements on Δw_k. Thus, a first order Taylor series approximation of d_{buf_k}, d_{con_k}, and $d_{buf_{k+1}}$, around the original value of w_k is used to write:

$$\Delta d_{buf_k} \approx \frac{\partial d_{buf_k}}{\partial w_k} \Delta w_k, \quad \Delta d_{con_k} \approx \frac{\partial d_{con_k}}{\partial w_k} \Delta w_k \qquad (12)$$

$$\Delta d_{buf_{k+1}} \approx \frac{\partial d_{buf_{k+1}}}{\partial w_k} \Delta w_k$$

and leads to a linear dependence of Δr on Δw_k:

$$\Delta r \approx \left(\frac{\partial d_{buf_k}}{\partial w_k} + \frac{\partial d_{con_k}}{\partial w_k} + \frac{\partial d_{buf_{k+1}}}{\partial w_k} \right) \Delta w_k = \frac{\partial r}{\partial w_k} \Delta w_k \qquad (13)$$

978-1-60558-497-3/09 $25.00 © 2009 ACM

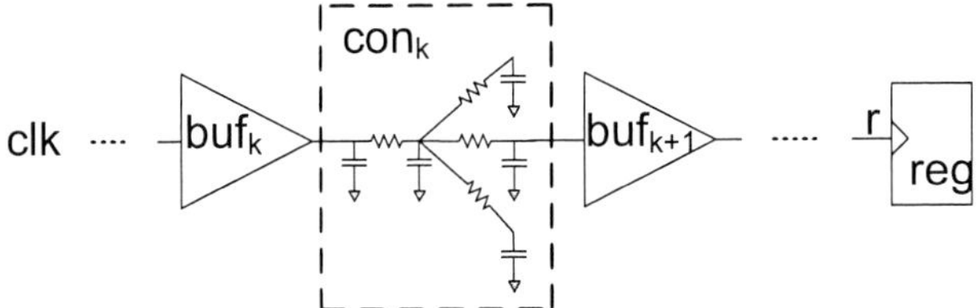

Figure 2: Section of a clock tree path

However, r, d_{buf_k}, d_{con_k}, and $d_{\mathrm{buf}_{k+1}}$ are also hyperplanes in the set of process parameters X. Let r be written as seen in (4), and d_{buf_k}, d_{con_k}, and $d_{\mathrm{buf}_{k+1}}$ as follows:

$$d_{\mathrm{buf}_k} = \alpha_0 + \sum_{l=1}^{p} \alpha_l X_l, \quad d_{\mathrm{con}_k} = \beta_0 + \sum_{l=1}^{p} \beta_l X_l \qquad (14)$$

$$d_{\mathrm{buf}_{k+1}} = \gamma_0 + \sum_{l=1}^{p} \gamma_l X_l$$

Then, using (4), we write:

$$\frac{\partial r}{\partial w_k} = \frac{\partial \bar{r}}{\partial w_k} + \sum_{l=1}^{p} \frac{\partial \lambda_l}{\partial w_k} X_l \qquad (15)$$

and, using (14) and (13), we deduce that:

$$\frac{\partial \bar{r}}{\partial w_k} = \frac{\partial \alpha_0}{\partial w_k} + \frac{\partial \beta_0}{\partial w_k} + \frac{\partial \gamma_0}{\partial w_k}, \quad \text{and} \quad \frac{\partial \lambda_l}{\partial w_k} = \frac{\partial \alpha_l}{\partial w_k} + \frac{\partial \beta_l}{\partial w_k} + \frac{\partial \gamma_l}{\partial w_k}$$

$$(16)$$

where $1 \leq l \leq p$.

4.2 Modeling Multiple Variable Parameters

So far, we have presented an analysis of the effect of varying a single physical parameter on clock phase arrival time. We now describe how one can capture the effect of varying multiple physical parameters on clock phase arrival times. Without loss of generality, we choose to vary only the widths of the various interconnect structures of the clock network. Let W represent the vector of interconnect widths of the clock tree, and let ΔW be the vector of variations in these widths. Let the number of interconnect segments in the clock tree be n, and we write $\Delta W = (\Delta w_1, \Delta w_2, \ldots, \Delta w_n)$, where each component represents the variation in an interconnect segment in the clock tree. For an arbitrary clock phase arrival time r, the introduction of ΔW leads to a change Δr:

$$\Delta r \triangleq r(W + \Delta W) - r(W) \qquad (17)$$

Because Δr is the aggregate result of n "local" delay variations in the clock tree, the result of (13) is generalized to write:

$$\Delta r \approx \sum_{k=1}^{n} \frac{\partial r}{\partial w_k} \Delta w_k \qquad (18)$$

Using (18) and (15) we can write:

$$\Delta r \approx \sum_{k=1}^{n} \frac{\partial \bar{r}}{\partial w_k} \Delta w_k + \sum_{l=1}^{p} \left(\sum_{k=1}^{n} \frac{\partial \lambda_l}{\partial w_k} \Delta w_k \right) X_l \qquad (19)$$

4.3 Formulation of the Optimization Problem

The aim of the optimization step is to determine the required changes in interconnect wire widths, $\Delta W = (\Delta w_1, \Delta w_2, \ldots, \Delta w_n)$, given allowable bounds $\Delta W_{\min} \leq \Delta W \leq \Delta W_{\max}$, so that the required timing margin, M, is minimized, given a required clock period T_c. One possible way to achieve this is to extend the work in [3] by finding an optimal assignment $\Delta W^\star$ such that it minimizes $M = T - T_c$, where T is the clock period for which the timing constraints are met under all process corners. This can be expressed as follows:

minimize: $\quad T - T_c \qquad (20)$

Subject to: $\quad \Delta W_{\min} \leq \Delta W \leq \Delta W_{\max}$

$$\forall s \in S, \forall i, j \in 1, \ldots, m$$

$$\max_{X}(D_s^{ij} + r_i + \Delta r_i - r_j - \Delta r_j) + t_{setup}^j \leq T$$

$$\min_{X}(d_s^{ij} + r_i + \Delta r_i - r_j - \Delta r_j) - t_{hold}^j \geq 0$$

Note that r_i, r_j, D_s^{ij}, and d_s^{ij} are extracted affine functions that depend solely on the process parameters X_l, $1 \leq l \leq p$, whereas the variations in clock phase arrival times, Δr_i and Δr_j, depend on both Δw_k and X_l, as shown in (19), and can be written as:

$$\Delta r_i = \sum_{k=1}^{n} \frac{\partial \bar{r}_i}{\partial w_k} \Delta w_k + \sum_{l=1}^{p} \left(\sum_{k=1}^{n} \frac{\partial \lambda_l^{(i)}}{\partial w_k} \Delta w_k \right) X_l$$

$$\triangleq \sum_{k=1}^{n} g_k^{(0)} \Delta w_k + \sum_{l=1}^{p} \left(\sum_{k=1}^{n} g_k^{(l)} \Delta w_k \right) X_l \qquad (21)$$

$$\Delta r_j = \sum_{k=1}^{n} \frac{\partial \bar{r}_j}{\partial w_k} \Delta w_k + \sum_{l=1}^{p} \left(\sum_{k=1}^{n} \frac{\partial \lambda_l^{(j)}}{\partial w_k} \Delta w_k \right) X_l$$

$$\triangleq \sum_{k=1}^{n} h_k^{(0)} \Delta w_k + \sum_{l=1}^{p} \left(\sum_{k=1}^{n} h_k^{(l)} \Delta w_k \right) X_l \qquad (22)$$

If the minimum value, $M^\star$, of $T - T_c$ is such that $M^\star \leq 0$, then the the timing constraints of the circuit can be met in the presence of process variability. Otherwise, the constraints simply cannot be satisfied given both the required clock T_c and the allowable ranges on ΔW. In that case, achieving sign-off will require different changes in the design, not only *tweaking* the interconnect wire widths, as we propose here. In any case, $M^\star$ corresponds to the smallest required timing margin achieved for $\Delta W^\star \in [\Delta W_{\min}, \Delta W_{\max}]$, in order to meet timing constraints at all process corners.

In the interest of efficiency, we transform the problem (20) into a linear program (LP), by writing each of its constraints at a specific process corner, as follows. Consider the following constraint from (20):

$$\max_{X}(D_s^{ij} + r_i - r_j + \Delta r_i - \Delta r_j) + t_{setup}^j \leq T \qquad (23)$$

Let us start with the case where D_s^{ij} is approximated by a single bounding hyperplane. In this case we can write:

$$D_s^{ij} + r_i - r_j \triangleq t = t_0 + \sum_{l=1}^{p} t_l X_l \qquad (24)$$

Let $\hat{X}_s^{ij} = (\hat{X}_1, \ldots, \hat{X}_p)$ be the process corner that maximizes the expression in (24). Note that this is the *pre-optimization* "worst-case" corner of (23), i.e., when $\Delta r_i = 0$ and $\Delta r_j = 0$. This corner is used to write (23) as:

$$\left. (D_s^{ij} + r_i - r_j + \Delta r_i - \Delta r_j) \right|_{\hat{X}_s^{ij}} + t_{setup}^j \leq T \qquad (25)$$

where writing $D_s^{ij} + r_i - r_j$ at this corner gives a constant value K:

$$\left. (D_s^{ij} + r_i - r_j) \right|_{\hat{X}_s^{ij}} = t_0 + \sum_{l=1}^{p} t_l \hat{X}_l \triangleq K \qquad (26)$$

and using (21), writing Δr_i at $\hat{X}_s^{ij}$ gives:

$$\left. \Delta r_i \right|_{\hat{X}_s^{ij}} = \sum_{k=1}^{n} \left(g_k^{(0)} + g_k^{(1)} \hat{X}_1 + \ldots + g_k^{(p)} \hat{X}_p \right) \Delta w_k$$

$$\triangleq \sum_{k=1}^{n} G_k \Delta w_k \qquad (27)$$

978-1-60558-497-3/09 $25.00 © 2009 ACM

and similarly, Δr_j can be written as:

$$\left.\Delta r_j\right|_{\hat{X}_s^{ij}} = \sum_{k=1}^{n}\left(h_k^{(0)} + h_k^{(1)}\hat{X}_1 + \ldots + h_k^{(p)}\hat{X}_p\right)\Delta w_k$$

$$\triangleq \sum_{k=1}^{n} H_k \Delta w_k \qquad (28)$$

Thus writing (23) at $\hat{X}_s^{ij}$ gives the following constraint:

$$K + \sum_{k=1}^{n}\left(G_k - H_k\right)\Delta w_k + t_{setup}^{j} \leq T \qquad (29)$$

Note that this constraint is now linear in the optimization variables Δw_k, $1 \leq k \leq n$, and no longer has a "max" expression over the process parameters. The same transformation can be done for the rest of the "max" and "min" operations in all of the setup and hold time constraints, respectively, to give the following optimization problem:

$$\text{minimize:} \quad T - T_c \qquad (30)$$

$$\text{Subject to:} \quad \Delta W_{\min} \leq \Delta W \leq \Delta W_{\max}$$

$$\forall s \in S, \ \forall i, j \in 1, \ldots, m$$

$$\left.\left(D_s^{ij} + r_i + \Delta r_i - r_j - \Delta r_j\right)\right|_{\hat{X}_s^{ij}} + t_{setup}^{j} \leq T$$

$$\left.\left(d_s^{ij} + r_i + \Delta r_i - r_j - \Delta r_j\right)\right|_{\hat{x}_s^{ij}} - t_{hold}^{j} \geq 0$$

All the constraints in (30) are now linear in the optimization variables, so that (30) is an LP which can be solved efficiently using commercial LP solvers.

Next, consider the case when D_s^{ij} and d_s^{ij} are expressed as piecewise planar functions. Let D_s^{ij} be the "max" of the set of hyperplanes in (9), then a constraint similar to (29) is written for each of $D^{(1)}, \ldots, D^{(u)}$. This would result in an LP that is very similar to (30), but which has a larger number of constraints. However, since actual path delays are now used in the optimization rather than conservative estimates, this would result in a less pessimistic and more accurate minimum value of the required margin $T - T_c$.

In its present form, (30) performs an optimization where each constraint is written at its own pre-optimization "worst-case" corner, which was found during the verification step. For a particular constraint, say (23), transforming it to (25) and using the latter in (30) is meant to ensure that the post-optimization circuit would satisfy (23) for those process settings X_v such that:

$$\left.\left(D_s^{ij} + r_i + \Delta r_i - r_j - \Delta r_j\right)\right|_{X_v} \leq \left.\left(D_s^{ij} + r_i + \Delta r_i - r_j - \Delta r_j\right)\right|_{\hat{X}_s^{ij}} \qquad (31)$$

However, there is no guarantee that $\hat{X}_s^{ij}$ would be the "worst-case" corner for the post-optimization expression $(D_s^{ij} + r_i + \Delta r_i - r_j - \Delta r_j)$. The reason for this is that the introduction of Δr_i and Δr_j alters the process parameter sensitivities of the expression inside the "max" operation in (23), as can be seen from (21) and (22). Thus, in order for (30) to ensure that the timing margin found allows the circuit to pass timing at *all* process corners, we must make sure that possible changes in "worst-case" corners are accounted for in (30). In the next section, we propose a method that deals with this possibility, in a conservative fashion, while preserving the linearity of the optimization problem.

5. COVERING ALL CORNERS

In this section, we present a method that can be used to determine whether the post-optimization "worst-case" corner of a timing constraint in (20) could possibly be different from its pre-optimization "worst-case" corner. After that, we present our proposed steps to modify (30) if a constraint is found to be as such. The aim of this modification is to ensure that the solution of (30) guarantees that the circuit would pass its timing constraints for all process corners.

5.1 Possible Changes in Worst-Case Corners

Recall from Section 3.2 that the process corner that maximizes or minimizes an affine expression in the space of process parameters can be found by looking at the signs of its sensitivities to each of the these parameters. Therefore, in order to determine whether the "worst-case" corner of an expression might change post-optimization, we need to find which are the sensitivities whose signs might change as a result of the optimization performed. One way of doing this is as follows. Consider the setup constraint in (23) and let $t = D_s^{ij} + r_i - r_j$ as in (24) and let t' be as follows:

$$t' = D_s^{ij} + r_i + \Delta r_i - r_j - \Delta r_j \qquad (32)$$

For the given bounds on ΔW, $\Delta W_{\min}$ and $\Delta W_{\max}$, we want to determine whether the sign of the sensitivity to a process parameter X_l, $1 \leq l \leq p$ in t' could have a sign opposite to its sensitivity in t. Using (21) and (22) we can write the sensitivity to X_l in t', denoted by t'_l, as follows:

$$t'_l = t_l + \sum_{k=1}^{n}(g_k^{(l)} - h_k^{(l)})\Delta w_k \qquad (33)$$

where t_l is the sensitivity of t to parameter X_l, which is known from (26). In addition to determining whether t_l and t'_l are guaranteed to have the same sign or not, we also want to determine the largest absolute value that t'_l can assume, if it is possible for it to have a sign different than that of t_l. This is not necessarily the largest magnitude that t'_l can have when it has a sign opposite to that of t_l, but rather the largest absolute value it can have, irrespective of sign. Assume, without loss of generality, that t_l is negative. Determining whether t'_l can be positive is done as follows. For $1 \leq k \leq n$, if $(g_k^{(l)} - h_k^{(l)}) < 0$, then we set $\Delta w_k = w_k^{\min}$. On the other hand, if $(g_k^{(l)} - h_k^{(l)}) \geq 0$, then we set $\Delta w_k = w_k^{\max}$. The value of t'_l is then computed using these assigned values. If the computed t'_l is positive then it is "at risk" of a sign change, and this value is the largest positive value it can assume. In this case, the smallest negative value of t'_l is also computed in a similar way and is compared to its largest positive value to find its maximum absolute value which we call $t_l^{\max}$.

5.2 Accounting For Changing Corners

For a timing constraint in (20), once all the sensitivities that are "at risk" of having different post- and pre-optimization signs have been determined, we can proceed to modify (30), as follows. Let the constraint under consideration be the setup constraint shown in (23), and let t and t' be as shown in (24) and (33), respectively. Assume, without loss of generality, that for process parameters $X_1, \ldots, X_c$, their sensitivities in t' were found to always have the same signs as their sensitivities in t, while the signs of the sensitivities of process parameters $X_{c+1}, \ldots, X_p$ were found to be possibly different in t' than in t. The largest absolute values $t_{c+1}^{\max}, \ldots, t_p^{\max}$ found for the sensitivities $t'_{c+1}, \ldots, t'_p$ are now summed to find a bound B_s:

$$B_s = \sum_{k=c+1}^{p} t_k^{\max} \qquad (34)$$

Let the pre-optimization "worst-case" corner of (23) be $\hat{X}_s^{ij} = (\hat{X}_1, \ldots, \hat{X}_p)$. The process setting $\hat{X}^* = (\hat{X}_1, \ldots, \hat{X}_c, 0, \ldots, 0)$, and B_s are used to write the following constraint in (30):

$$\left.\left(D_s^{ij} + r_i + \Delta r_i - r_j - \Delta r_j\right)\right|_{\hat{X}^*} + t_{setup}^{j} \leq T - B_s \qquad (35)$$

Note that writing the constraint in (35) is more conservative than the constraint written at $\hat{X}_s^{ij}$ and hence the latter can be removed from (30). For a hold-time constraint, a bound B_h is also computed with the difference that the bound is added to the right-hand side of the inequality and not subtracted from it to write:

$$\left.\left(d_s^{ij} + r_i + \Delta r_i - r_j - \Delta r_j\right)\right|_{\hat{x}^*} - t_{hold}^{j} \geq B_h \qquad (36)$$

Table 1: Exact vs Conservative Approach

ISCAS-89 Circuit	Num Variables m	Pre-optim M_v	Post-optim $M^\star_{\text{exact}}$	Percent Improvement	Post-optim $M^\star_{\text{cons}}$	Percent Improvement
s400	23	9.43%	4.37%	54%	4.69%	50%
s1432	86	11.13%	5.83%	48%	7.60%	31%
s5378	327	13.31%	2.29%	83%	2.89%	78%
s9234	281	10.37%	1.39%	87%	2.58%	75%
s13207	571	8.10%	1.46%	82%	2.28%	72%
s15850	959	10.69%	2.44%	77%	3.12%	71%
s38584	2732	5.12%	-1.88%	136%	-1.00%	120%

Such constraints would ensure that all "potentially worst-case" corners are covered by using the bounds B_s and B_h to tighten the timing constraints of (30). Of course, this comes at the price of introducing some pessimism in the LP.

6. RESULTS

In order to test our approach, we have selected a set of circuits from the ISCAS-89 benchmark suite. These circuits were synthesized and mapped to a 90nm CMOS library, and then placed and routed using available commercial tools. The HSPICE netlists for the logic blocks and clock trees of these circuits were then extracted. Nominal delays and slews were characterized for all cells in the library, and a set of 10 parameters $X_1, \ldots, X_{10}$ was selected to model process variations. The ranges of those parameters were chosen such that their combined effect on the delay or slew of a cell or an interconnect resistance or capacitance value is 15%. The sensitivities of gate delays and interconnect RC-trees to these process parameters were randomly generated, in order to provide difficult test-cases. With the above variational models, the HSPICE netlists of the different test circuits were fed into our STA timing engine, which was implemented in `C++` and extended to handle parameterized static timing analysis (PSTA).

Two PSTA flows were implemented, and consequently two sets of experiments were compared. The first PSTA flow, based on [9], provides an exact analysis as it propagates delays in the circuit using piecewise planar delay models, where all "potentially critical path" delays are preserved at every node. Hence, D_s^{ij} and d_s^{ij} are "lists" of hyperplanes each corresponding to a path delay. The second flow, based on [11], propagates conservative hyperplane bounds to D_s^{ij} and d_s^{ij}. A nominal STA run, without consideration to process variability, is also performed on each of the circuits and the nominal clock period found is considered to be the required clock period T_c for the test circuits.

Using the propagated delays from the exact PSTA, we start by computing the pre-optimization required timing margin, M_v, which guarantees that all constraints are met at their "worst-case" corners. M_v can be easily computed using our verification method in Section 3.2 by maximizing the left-hand side of every setup constraint written as in (10), and recording the largest value achieved over all constraints to find the smallest clock period T_v that will allow correct operation. The required clock period T_c is then subtracted from T_v to find M_v. We then ran our LP formulation of (30) to find the minimum achievable margin $M^\star$. In our LP, the optimization variables are chosen to be variations in interconnect wire widths of the clock tree Δw_k, $1 \leq k \leq n$ where n is the number of interconnect structures in the clock tree. The bounds on variations in wire widths were set such that $0 \leq \Delta w_k \leq w_k$, where w_k is the pre-optimization wire width. In other words, wire widths are allowed, at most, to *double*, as a result of the optimization.

The minimum post-optimization timing margins, $M^\star_{\text{exact}}$ and $M^\star_{\text{cons}}$, achieved based on exact and conservative PSTA flows, respectively, are shown in Table 1, in addition to the pre-optimization margin M_v. Given the above bounds on the allowable wire changes Δw_k, we observed improvements ranging from $50\% - 136\%$. This shows a considerable reduction in the required margin which allows the timing constraints of the circuit to be satisfied for all process corners. Our results also show that although using piecewise planar values guarantees a better post-optimization margin, for most circuits, the difference between the improvements achieved in the two approaches is small. We also recorded and compared the runtimes of the optimization for both the exact and

Table 2: Run-time Comparison

ISCAS-89 Circuit	Exact Approach	Conservative Approach	Speed-up ($\times$)
s400	0.06 s	0.05 s	1.16
s1432	2.45 s	0.98 s	2.5
s5378	0.93 s	0.72 s	1.29
s9234	1.31 s	1.25 s	1.05
s13207	3.56 s	2.3 s	1.55
s15850	13.27 s	9.83 s	1.35
s38584	20.78 s	13.71 s	1.52

conservative delay approaches. The results in Table 2 show that although the hyperplane approach is more conservative in terms of post-optimization margin, it typically produces a speed-up between $1.16\times$ and $2.5\times$. Thus, we see that one is faced with a runtime-accuracy tradeoff, where a good speed-up is achieved at the expense of some pessimism that is introduced to the analysis.

7. CONCLUSION

In this work, we presented a technique that modifies clock network wire widths and/or lengths so that a circuit meets its timing constraints at all process corners. Our method considers delay variability in both the clock network and the functional blocks, and is applicable near the end of the design flow. We showed that the problem of finding the required changes in line widths and/or lengths can be formulated as a Linear Program, which can be solved efficiently. Using our clock skew tuning approach, designers can considerably reduce the effect of process variations on circuit performance, as shown in our results.

8. REFERENCES

[1] R. S. Tsay. Exact zero skew. In *Int. Conf. on Computer Aided Design (ICCAD)*, pages 336–339, 1991.

[2] T. H. Chao, Y. C. Hsu, and J. M. Ho. Zero skew clock net routing. In *DAC*, pages 518–523, June 1992.

[3] J. Fishburn. Clock skew optimization. *IEEE Trans. on Computer-Aided Design*, 39(7):945–951, July 1990.

[4] K. A. Sakallah, T. N. Mudge, and O. A. Olukotun. Analysis and design of latch-controlled synchronous digital circuits. *IEEE TCAD*, 11(3):322–333, March 1992.

[5] J. L. Neves and E. G Friedman. Buffered clock tree synthesis with non-zero clock skew scheduling for increased tolerance to process parameter variations. *Journal of VLSI Signal Processing*, 16:149–161, 1996.

[6] I.M. Liu, T.L. Chou, A. Aziz, and DF Wong. Zero-skew clock tree construction by simultaneous routing, wire sizing and buffer insertion. In *Proceedings of the 2000 international symposium on Physical design*, pages 33–38. ACM New York, NY, USA, 2000.

[7] S. Pullela, N. Menezes, and L. T. Pillage. Reliable non-zero skew clock trees using wire width optimization. In *Design Automation Conference*, pages 165–170, June 1993.

[8] S. Pullela, N. Menezes, and L. T. Pileggi. Post-processing of clock trees via wiresizing and buffering for robust design. *IEEE TCAD*, 15(6):691–701, 1996.

[9] K. R. Heloue, S. Onaissi, and F. N. Najm. Efficient block-based parameterized timing analysis covering all potentially critical paths. In *ICCAD*, pages 173–180, November 2008.

[10] S. V. Kumar, C. V. Kashyap, and S. S. Sapatnekar. A framework for block-based timing sensitivity analysis. In *Design Automation Conference*, pages 688–693, 2008.

[11] S. Onaissi and F.N. Najm. A linear-time approach for static timing analysis covering all process corners. *IEEE Transactions on Computer-Aided Design of Integrated Circuits and Systems*, 27(7):1291–1304, 2008.

978-1-60558-497-3/09 $25.00 © 2009 ACM

Moore's Law:
Another Casualty of the Financial Meltdown?

Jason Cong (Organizer)
University of California, Los Angeles
Los Angeles, CA, USA

Nagaraj NS (Co-organizer)
Texas Instrument
Dallas, TX, USA

Ruchir Puri (Co-organizer)
IBM T J Watson Research Center
Yorktown Heights, NY, USA

William Joyner (Chair)
SRC
Research Triangle Park, NC, USA

Jeff Burns
IBM T J Watson Research Center
Yorktown Heights, NY, USA

Moshe Gavrielov
Xilinx
San Jose, CA, USA

Riko Radojcic
Qualcomm
San Diego, CA, USA

Peter Rickert
Texas Instrument
Dallas, TX, USA

Hans Stork
Applied Materials
Santa Clara, CA, USA

PANEL SUMMARY

Given the exponential increase of fabrication costs, the global recession and credit crunch, one may ask if Moore's Law is financially viable beyond 22nm node. Can we justify the return-of-investment (ROI) for continuous scaling beyond 22nm? Shall we consider other alternatives for integration, such as silicon-in-a-package (SiP) or 3D integrations?

Categories and Subject Descriptors

B.7.1 [Integrated Circuits]: Advanced technologies, VLSI; B.7.2 [Design Aids]; B.7.3 [reliability and Testing].

General Terms

Performance, Design, Economics, Reliability, Experimentation,

Keywords

Keywords are your own designated keywords.

PANELIST VIEWPOINTS

Jeff Burns: Moore's Law 2D density scaling is clearly becoming more challenging, as well as more costly, with each generation. In addition, a major benefit of classical scaling, namely a substantial increase in device-level power/performance with each generation, is diminishing. As a result the return-on-investment prospects for future CMOS technology generations are being energetically debated within the industry, along with related debates on how companies can differentiate their products and continue to deliver increasing cost/performance if the pace of scaling slows significantly or

Permission to make digital or hard copies of part or all of this work for personal or classroom use is granted without fee provided that copies are not made or distributed for profit or commercial advantage and that copies bear this notice and the full citation on the first page. To copy otherwise, to republish, to post on servers or to redistribute to lists, requires prior specific permission and/or a fee.
DAC'09, July 26-31, 2009, San Francisco, California, USA

halts.

The viability of classical scaling, viewed with the expectations of significant generation-over-generation improvements in density and power/performance, is already seriously in question. However, the demand for cost-effective computing is not declining, and new, complementary technologies, especially 3D integration, have the potential to offer new means to deliver substantial increases in value to the end user. Going forward, out-of-the-box innovations in 2D technology along with compatible 3D integration technology can provide the means to continue to enable large improvements in computing cost/performance. Leveraging these capabilities will require innovative thinking across technology, circuits, architectures, and systems.

Moshe Gavrielov: The exponential growth of manufacturing costs as we move to and beyond the 22nm node will indeed provide ample justification for pursuing other integration alternatives such as SIP, 3D integration, etc.... However these approaches will not address the spiraling design complexity and commensurate costs related to developing advanced SOC products for anything other than a handful of the highest volume markets. As both ASSPs and ASICs become increasingly less viable at these advanced process nodes, FPGAs will evolve to service more and more applications that require hardware differentiation.

Riko Radojcic: It is clear that the era of 'happy scaling' has come to an end. Seems like the basic Laws of Physics (finally) trump Moore's Law, and that exponential growth curves really do, sooner or later, taper off.
However this does not mean that the historical rates of increasing the integration levels has come to an end – it just means that relying just on simple shrinking of process dimensions to drive the phenomenal rate of price-performance improvements that we know and love as "Moore's Law" is finished. This is really an opportunity to leverage other, wide open and mostly unexplored, methodologies. Even in the technology domain there are promising possibilities other than

the very exotic 'More Moore' technologies. For example the 3D integration schemes and other 'More than Moore' opportunities are certainly getting some attention now. There are opportunities in the design implementation arenas that could provide additional push – after all the industry is pretty much relying on same design methodologies that have been used for the last 2 to 3 decades, and use of for example, co-design or asynchronous methodologies is pretty limited. And there are certainly possibilities in the architecture design still be exploited. And so on.

The industry should, and will, explore all those – and more – opportunities, with the intent on maintaining Moore's Law. Ongoing improvements in the cost structure should be expected for a while more. History shows that as long as there is an economic motivation a solution is found. And with the convergence of telephony, mobile devices, multimedia, computing, internet and all the other exciting things happening in the market, it feels like we are actually at an inflection point in a renaissance, rather than at a tapering of a growth curve. Therefore, the economic drivers are believed to be there, and hence engineering ingenuity will find a way. After all – Moore's law is an Economic law rather than a Physics law.

Peter Rickert: Our semiconductor business has a long heritage of surviving multiple periodic downturns over the past 20 years. While the current financial downturn is unprecedented, there are no signs that technical innovation will stop! While there are also signs of protracted product introductions as well as reductions in number of product introductions vs. time, we still see multiple product areas which are driving the need for next generation technology to enable their paths to market. Business dynamics will change as well as adapt going forward, but not unlike what we have seen the past 10 years, with the growth of foundries and hybrid manufacturing models. The biggest financial challenges or adaptations going forward may be forced on the backs of the equipment suppliers to find suitable business models for the next decade.

From a product perspective, I see no letting up on key forces which have driven our industry: higher levels of system integration, lower power density, lower system cost, higher throughput, etc. Innovative solutions have already been developed to drive us through the new millennium, most notably in areas now called "More than Moore", pertaining to multiple packaging related 3-D solutions such as SiP, PoP, etc. None of these solutions even existed 10 years ago! I am very confident that the tree of innovation still has plenty of fruit waiting to be harvested in the minds of our engineers across the world. We already have 45/40nm designs ramping production as well as 28nm design kits in the hands of our designers. The path to 22nm and beyond is already starting to be paved with potential solutions like FinFETs, NextGenLitho, subthreshold circuit technology, and even more 3-D solutions in the packaging arena such as TSV. We also see a continual stream of new battery operated applications at the low end and high performance systems at the high end to drive the demand. New applications, such as SmartPhones and Netbooks, and even new techniques such as energy scavenging will drive brand new application spaces. There's still plenty of runway left on this ship!

Hans Stork: Reducing dimensions leads to increased performance and perhaps more importantly reduced chip unit cost, provided the development investment and manufacturing costs do not exceed these benefits. The production costs are strongly affected by the unit capital costs, driving production requirements to very high volumes. The high volume productivity of process equipment allows applications like NAND flash, memory and graphics or micro-processors to continue to benefit from scaling. The seemingly insatiable demand for storage and communication/computation bandwidth provides a pull for ever smaller and faster components until our engineering skills or physics fails. In other words, scaling drives volume and vice versa.

On the other hand, many other applications have left the scaling race. Analog needs the use of voltages that are incompatible with nanometer feature sizes. Consumer logic designs are faced with complexity of architecture, verification and power optimization that are in fact increasingly difficult to solve with scaled dimensions. They are more difficult and more expensive to design and fabricate (e. g. due to rising mask costs) with limited power or performance benefit at the system level. The largest hurdle is the combined development cost of process and product.

The extreme economic downturn that we are experiencing today, is accelerating these trends, and creating a larger gap between the technology leaders and followers. The investment capital required is only available to few, who will want to turn it into a competitive advantage. Moore's Law will continue, but for fewer competitors.

978-1-60558-497-3/09 $25.00 © 2009 ACM

Holistic Verification: Myth or Magic Bullet?

Pradip A. Thaker

Analog Devices Inc., Bangalore, India

pradip.thaker@analog.com

Abstract

Development of multi-million gate SoCs/ICs enabled by advances in submicron technology inherit verification complexity due to feature diversity on a single die and thus convergence of multiple design disciplines on one project with time-to-market pressures unchanged. For example, it is not uncommon for SoCs to integrate mixed-signal components, provide power-management features, contain instances of soft as well as hard 3rd party IPs or large building blocks developed in-house and support compliance/compatibility requirements ranging from functional to electrical aspects for various standard-based protocols. Timely and high-confidence verification sign-off for such SoCs/ICs requires a multi-prong strategy defined in this paper as holistic verification. Holistic verification emphasizes a cocktail of verification methodologies designed to address specific verification challenges of a given SoC as well as advocates aspects of verification planning and design-for-verification techniques that aid verification efforts and enhance verification efficiency. Additionally, this paper describes details of verification challenges associated with mixed-signal integration and power management features of SoCs – two distinctly different design disciplines converging in one project and thus adding growing complexity to verification.

Categories and Subject Descriptors: B.6.3
[**Hardware**]: Logic Design - Design Aids – Verification.

General Terms: Design, Verification.

Keywords: SoC Verification, Emulation, Mixed-signal Verification, Power management Verification

1. Introduction

With advances in submicron technologies over the last decade, multi-million gate SoCs/ICs have become a cliché. With growth in size of the design, the diversity in functionality on a single-chip has proportionally grown while the time-to-market pressures have remained unchanged. On a single-die, it is common to have a variety of combinations of newly developed digital as well as mixed-signal/analog circuits, integration of in-house and/or 3rd party IPs, integration of mega-blocks such as RAMs and ROMs, single or multiple instances of processor core(s), implementation of newly developed algorithms or standards with strict

Permission to make digital or hard copies of part or all of this work for personal or classroom use is granted without fee provided that copies are not made or distributed for profit or commercial advantage and that copies bear this notice and the full citation on the first page. To copy otherwise, to republish, to post on servers or to redistribute to lists, requires prior specific permission and/or a fee.
DAC'09, July 26-31, 2009, San Francisco, California, USA

requirement for logical and electrical compliance, a variety of standard and non-standard interfaces, and integration of building blocks created through orthogonal design flows such as RTL and custom design. With convergence of these multiple disciplines on a single-die, verification of such ICs is beyond the scope of any single verification approach. Brute-force cumulative deployment of all verification techniques, each of which is traditionally used to tackle a respective challenge, is not only cumbersome but also insufficient to produce high-quality robust first silicon. There is clearly a need to assess project-based verification challenges on case-by-case basis and deploy a fitting combination of techniques towards a high-confidence verification sign-off.

The term "Holistic" is defined in the dictionary as "dealing with integrated systems rather than with their parts" or "Multi-pronged wholesome approach to problem solving". This term is used frequently in medical field to emphasize preventive care as well as to refer to deployment of a combination of orthogonal treatment approaches instead of any single method. In both of these contexts, this term fits well for SoC verification as well to describe firstly, design methodologies that enable bug-prevention, simplify verification, as well as increase verification efficiency and secondly, a combo-verification-strategy for project-specific challenges from available options such as, (a) Traditional test-bench based simulation environment, (b) Coverage-driven constrained-random test generation (use of HVLs), (c) Reference modeling (HVLs, SystemC, C/C++), (d) Assertion-based Formal and Semi-formal techniques, (e) Hardware-assisted testing, (f) Mixed-signal simulations, (g) Net-list simulations, (h), Compliance verification with protocol checkers, (i) Asynchronous clock-domain simulations, (j) Power management simulations, (k) simulation v/s silicon gaps verification coverage, etc.

The key components of holistic verification can be broadly classified in three categories:

- Project planning with verification focus

- Design-for-verification techniques

- Verification methodology

These building blocks of holistic verification are described in more detail in subsequent sections.

2. Project Planning with Verification Focus

Verification is a critical step in the SoC design cycle towards silicon quality and project schedule implications. Therefore, technical planning of the project with appreciation for verification as a centerpiece is a must. The technical planning needs to enable bug-prevention, simplify verification and increase verification efficiency. This has implications on both the design team and the verification team and hence, there needs to be clarity on part of both teams as well as the technical leader of the project regarding the respective missions of design and verification teams and mechanisms-of-accountability for their respective deliverables. In this paper, the design team is defined as consisting of architects and engineers involved in micro-architecture definition as well as implementation.

The effort of design creation is prone to errors and inevitably bugs/errors get introduced during this phase of the project. All design errors can be attributed to three bug sources: Error of implementation, Error of specification (or specification interpretation) and Error of omission (of features/behavior).

The effort of verification is to identify bugs/errors that are present in design description with creative stimuli. All design errors exhibit their presence in implementation with appropriate stimuli and these bug manifestations can be classified in three broad categories: Feature malfunction/incorrect behavior, performance (e.g. throughput, latency, etc.) limitation, and compliance and/or compatibility problems.

With the above clarity on three bug sources and three bug manifestations, missions and mechanisms-of-accountability for design and verification teams respectively can be defined as follows:

Design Mission: The main mission of design team is to minimize bug sources in the first, final and all intermediate design releases. The design team needs to operate with passion not only for creativity but also design quality with emphasis on implementation being correct by construction on all three types of bug sources. While there is no design that is absolutely bug-free, the attempts on part of design team to minimize bug sources during design creation phase help improve verification efficiency. Clean design partitioning with simplified inter-block interfaces, well-documented specifications, use of static property-based checks during development as well as use of high-level design description aided by high-level synthesis are some of the techniques that are useful towards this mission. Especially, high level design reduces bug sites since design description is concise and should be given increased consideration for SoC development [1, 2]. Similarly, architecture and performance validation prior to RTL coding completion (preferably prior to RTL coding start) help minimize risk of significant re-design due to architectural and performance-oriented bugs being found late in development cycle with chip-level verification.

Verification Mission: The main mission of the verification team is to expose bug manifestations with efficiency, identify bug sources and drive root-cause analysis. A well thought-out verification platform which inculcates support for a cocktail of verification methodologies targeting project-specific challenges while maintaining homogeneity and portability of environment across blocks and chip-level is a requirement for verification efficiency. Similarly, hierarchical verification is also a must. Critical building blocks of SoC need to be thoroughly verified at block-level as it is easier to set-up stimuli and debug failures with faster run-times compared to verification of the same block with chip-level verification platform. In a well-partitioned SoC, thorough block-level verification of complex blocks allows chip-level verification effort to primarily focus on integration bugs. With appropriate verification architecture and platform definition, hierarchical verification will also promote re-use of Bus Functional Models (BFMs), sequences, constraints, etc. building blocks of verification environment. Deployment of methodologies such as assertion-based formal verification that enables hitting complex and corner-case bugs early in verification cycle also increases efficiency.

While design and verification teams have common goals to meet project schedule and silicon quality, and while they also share responsibility of execution towards these goals, separate accountability metrics that institute awareness for their respective missions is one of the key aspects of successful project planning and execution. Aspects of the bug matrix (such as criticality of the bug, degree of change required in original design for bug fixing, etc.) at the first and the final release of design implementation facilitate as design quality metrics. Functional and code coverage metrics during pre-silicon verification serve as measures of verification team's efforts and accountability. The ultimate measure of design and verification mission success is: (a) Post-silicon bug matrix, (b) Number and nature of re-spins between first silicon and release-to-production (i.e. metal-only spin v/s all-layer, ECO-able changes v/s repetition of whole flow from synthesis through physical design) , as well as (c) Schedule to first tape-out and RTP.

Despite good technical planning, the ultimate question pertaining to verification is "When do we know we are done?" as opposed to "How do we know we are done?". As such, verification is a never ending process and by any measure it only proves presence of bugs but cannot prove absence of bugs in the design description. As a result, there is no set mechanism to define completion of verification effort with binary clarity. But, as part of verification planning, there are some guiding factors that can help make a timely and judicious decision in defining verification cut-off with high confidence in the quality of resulting silicon. The factors that help in ramping down verification effort are: code coverage, functional coverage, number of cycles of random tests and effectiveness trends, bug tracking with bug-matrix and bug-rate along with intuition/gut-feel on part of engineer pertaining to prediction of diminishing returns for the efforts to be spent. While tools and methodologies are certainly great enablers, ultimately it is the engineer who deploys them with efficiency and efficacy. Therefore, while verification planning is important for quality as well as schedule deliverables of a project, the final outcome depends heavily on talent and commitment level of both design and verification engineers.

3. Design-for-Verification Techniques

There are three components to design-for-verification techniques for SoC: Design methodology for the new block, Architecture definition and 3rd Party IP evaluation.

During design creation phase, meeting the functional, performance and compliance requirements is critical. However, it is also critical that the new design meets speed targets as well as design-for-test requirements to enable high overall chip-level coverage. Irrespective of the design flow that each organization may employ, one common fact is that the design changes introduced after verification completion of a block require re-verification of that block once changes are added to the verified design description. This clearly creates overhead on verification team as well as effort. On the other hand, releasing design to verification team late in the design cycle gives verification team a late start and thus less time for a high-quality sign-off.

978-1-60558-497-3/09 $25.00 © 2009 ACM

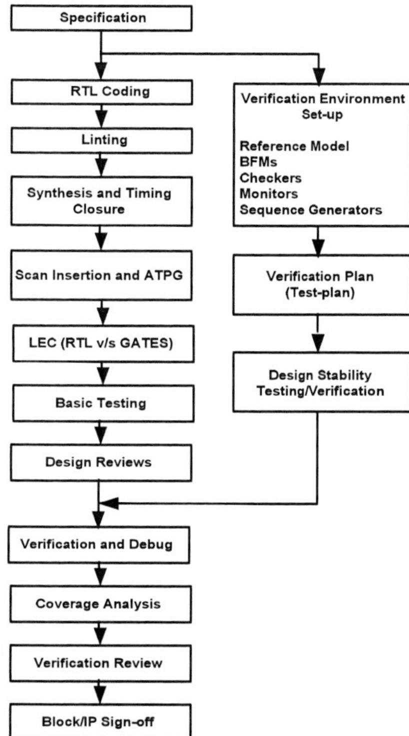

Figure 1: Design Methodology To Aid Verification

Therefore, it becomes important to release design to the verification team at the appropriate time – neither too soon nor not too late. Various organizations attempt to address this question of "When to start the verification effort?". One approach that has been tested across projects as well as across organizations and found to provide a reasonable answer is presented in Figure 1. In the proposed flow, the design and verification teams work in parallel – when the design team is in RTL creation phase and iterating through suitable implementation for architectural trade-offs, pipe-line analysis, timing-closure and design-for-test requirements; the verification team works on setting up the environment and creation of detailed test-plan with an eye towards high functional coverage. Once the design description is deemed stable from not only functional, performance and compliance requirements but also from timing-closure, design-for-test and static verification with rules checking as well as logic equivalency check, the design is released to the verification team. This approach helps shield the verification team from overhead efforts resulting from design iterations. Empirically, it has been observed that this approach introduces about one week delay in releasing design to verification, but it saves significant time by enabling high quality releases from design to verification.

The second component of design-for-verification techniques is architecture definition to help manage verification complexity. Modular architecture is a must for a SoC as that will enable thorough hierarchical verification. Among building blocks of a SoC, it is critical to have standardized block interfaces as that helps reduce the scope of "Error of Specification" and "Error of Omission" as well as promote verification component re-use such as BFMs, sequences, etc. Another key aspect pertaining to block interfaces in a SoC is to minimize the number as well as flavors of physical interconnects. Each interface needs verification and thus

simplifying the connectivity model also helps reduce verification complexity and debug effort. Yet, the most important aspect of architecture definition is to avoid functional snake-paths that traverse through multiple building blocks of a SoC. Verification of functional snake-paths is very complex as it requires integration-level testing as compared to paths contained within blocks which can be tested at a block level. Thorough testing of such functionality that is split across multiple building blocks also increases complexity of sequence generation and debug turn-around times as failure manifestations are much more complex. There are several aspects of architecture decisions such as pull v/s push inter-block data-threads, distributed v/s centralized control processing, etc., that have impact on verification complexity and efficiency. Such decisions need to be made in consultation with verification team.

More often than not, SoC teams make great strides in mitigating risks to schedule and quality for all blocks that are created and verified within the organization/team. However, feature diversity of SoC seldom supports this model of design creation. In most cases, a given SoC relies on re-use of a building block that already exists outside their organization – within the company or an IP vendor. Integrating a complex IP into a design and verifying it at chip-level present challenges of its own. In some cases, due to lack of standardization in verification methodologies across organizations or across IP vendors, IP from selected source does not come with a verification methodology consistent with the native approach. Verifying this integrated IP in the SoC becomes a challenge. In the worst case scenario, the IP has not been previously validated in silicon and that presents a quandary pertaining to confidence in the overall quality of the SoC at first silicon. It is important that IP selection process for a SoC involves the verification team and their key concerns are addressed. A thorough evaluation of a 3rd Party IP must be undertaken among available sources in conjunction with inputs from verification team. It is desirable that the design and EDA industry standardize on one verification methodology to mitigate this problem of lack of uniformity.

In a nutshell, it is strongly recommended against performing design activities in isolation away from the verification team's involvement. Involvement of the verification team from architecture definition onwards is a key requirement.

4. Verification Methodology

With advancement in verification technology, there is a wide range of complementary and supplementary methodologies available. Given the feature diversity and thus complexity of SoCs, a project-specific multi-pronged strategy needs to be devised from available methodologies:

Simulation based Verification (SBV): SBV primarily comprises of coverage-driven constraint random stimulus generation with reference modeling. This class of techniques is the only solution that addresses end-to-end needs for design verification (Concept to tape-out). Simulation-based techniques are not efficient when it comes to run-times since number of simulation cycles loosely correlate to quality of effort and hard-to-reach corner case bugs found after millions of cycles require repetition of effort to validate bug fix [3]. Also, two metrics popularly used in conjunction with simulation-based techniques are incomplete as they measure coverage based on controllability aspects of the stimuli only and completely ignore observability aspects of the response. New observability-based coverage metrics such as VVG [4] provide much better insight into verification quality.

Assertion based Verification (ABV): ABV primarily comprises of formal and semi-formal techniques. These techniques present an orthogonal paradigm to SBV techniques and significantly aid verification efficiency compared to SBV approach since they hit hard-to-reach subtle bugs sooner and bugs are found at the source. However, formal ABV techniques do not provide end-to-end solution for design verification due to technology limitations, but have found niche application in addressing verification challenges of control-oriented IP/block. The assertions written for the formal ABV effort can be used as checks or coverage in the simulation, where they provide benefit of exposing bugs close to their source, in both time and space [5].

Hardware assisted Testing (HAT): This group of techniques brings aspects of hardware implementation in verification. It comprises of FPGA-based testing, hardware acceleration and rapid prototyping. It enables real-life system-level testing, architecture validation, performance validation and hardware-software co-simulation prior to first silicon. Similarly, it offers early platform for software/tools development prior to first silicon. Integrating HAT in SoC verification flow enhances quality of first silicon and allows for earlier software development and bring-up in the flow.

Mixed-signal integration and power-management features add two unique design disciplines to SoC development. Challenges associated with evolving SoC verification flow to address complexity of these features are discussed here.

Mixed-signal simulation: It is not uncommon for the SoC to integrate analog components of varying complexity – e.g. ADC, DAC, PLL, PHY, etc. Analog verification being an entirely different discipline with its unique challenges and methodology deficiencies now contributes to complexity of SoC verification when combined with digital verification techniques. Despite the complexity, mixed-signal verification is important for high-confidence SoC verification sign-off.

Analog verification primarily comprises of circuit simulations (DC, AC, transient, parametric and Monte Carlo analysis across multiple process corners) and characterization. Analog circuit design entails building sufficient margin and this is verified (characterization) with stress testing across different environmental conditions such as power supply, inputs, loads, temperature, etc. Sensitivities of analog circuits are also checked for transistor mismatch, PVT corners, reliability and behavior in presence of package, bond wire, and cable models. For example, the eye-diagram is checked in presence of supply variation to check the jitter effect from data coupling. Similarly, PLL lock period, output clock jitter, etc. parameters are verified across all PVT corners. In yet another example, PCI-Express PHY designs are verified with models based on various cable lengths to be supported for compliance and compatibility requirements.

A simulation tool for analog circuits must have capacity, performance and accuracy. SPICE and FastSPICE are two commonly used simulators for analog circuits. The goal of using SPICE is to achieve highest level of accuracy with transistor device models provided by foundry. The goal of using FastSPICE is to achieve simulation speed up. Irrespective of the simulator used, analog circuit verification still depends on visual analysis of the signal behavior. The designer's knowledge is the single critical factor for analog circuit verification. Much of the advances in digital CAD techniques pertaining to automation for verification are yet to mirror in analog circuit verification. For instance, in analog verification, there are no verification metrics to evaluate coverage of input-output, and parameter space. There are no methods to evaluate the verification progress. There are no methodologies to leverage re-use from one block to another. Techniques of digital verification such as reference modeling, constraint random stimuli generation, auto circuit-response verification, hardware-assisted testing, etc. are virtually non-existent in any useful capacity in domain of analog verification. This makes verification of SoC with mixed-signal components a unique challenge.

The ideal solution for mixed-signal SoC verification requires a high capacity and high performance simulator that can handle transistor-level design description of a multi-million gate SoC while ensuring accuracy for post layout simulation. However, such an ideal solution does not exist. An alternative approach that is widely used for SoC verification is based on (a) chip-level (RTL and Gate-level) simulations with instances of behavioral models for analog components; and (b) limited co-simulation of interfaces between analog and digital logic with analog components being in transistor-level description and digital components being in RTL description. While this approach has provided a workable solution, it is deficient. Firstly, there is no method to establish completeness of behavioral model of an analog component. Any feature/functionality that is not present in behavioral model used in co-simulation leaves opportunity for design error remaining undetected either in the analog design, digital design or interface design between two domains. Secondly, such solution does not simulate multiple levels of circuit abstractions (transistor level, gate-level, RTL level, behavioral, etc.) simultaneously, and is somewhat limited in its capacity to handle layered donuts i.e. nested multiple analog and digital blocks at various levels of design hierarchy. Thirdly, this solution does not support automated response verification and therefore manual/visual analysis of analog behavior within mixed-signal implementation is still by-and-large a common practice.

In summary, while ideal transistor-level simulator with capacity and performance to handle large SoCs with mixed-signal integration is not on horizon; an integrated, automated and re-usable approach with metrics to determine quality of stimuli as well as completeness of behavioral models is urgently needed for mixed-signal SoC verification [6, 7].

Power-management Verification: Advances in sub-micron technologies towards smaller geometries and demand for battery-operated (low-power) devices have driven power management features to become a mainstay on SoCs. Some of the power management implementation techniques are clock-gating, on-chip multi-voltage islands, dynamic voltage scaling, use of multi-threshold libraries, etc. These power management features add another dimension to verification of SoCs. Verification complexity grows with the number of unique power states in design. As design progresses through various stages of development from RTL to physical implementation, power-management features require verification for compliance at each stage and on different design abstractions. Unlike a functional bug where a post-silicon work-around may be possible with software or it may be possible to declare the bug as an errata and release silicon to production, power management bugs are unforgiving and largely unacceptable since power management features are not nice-to-haves but must-haves for a low-power application.

Power management verification deals with unconventional and new classes of bug as well as bug manifestation. These are:

Level-shifting/isolation errors: In order to avoid floating signals driving active logic, power domains should be isolated and voltage domain crossings should be level-shifted accordingly. Any signal being missed from level-shifter insertion could break the entire power management strategy.

Power sequencing errors: With multiple voltage islands on a die and control/data signals interacting among these, there are stringent requirements on power cycling sequence among supply domains for power-up and power-down. Error in implementation leading to dead-lock situations can cause a fatal compliance problem for power-management requirement.

Buffer insertion errors: It is a common practice for logic synthesis or physical optimization tools to perform buffer insertions on nets that need buffers. In power management paradigm, automated buffer insertion on a net with source/destination being on a different supply domain than inserted buffer can be fatal.

Shadow/Retention register errors: Shadow registers may be added on one supply domain to retain logic state from a different supply domain which may be powered down during low-power state. Verification of this seemingly simple feature gets complicated as RTL descriptions do not inherently contain knowledge of supply domains. Often, awareness of power domains is modeled in RTL as "enable" signals which are actively controlled by test-bench for power sequencing simulations.

Power-good signal implementation errors: Design implementation with power management scheme contains several reset trees as registers may have different retention requirements with respect to on-and-off state of voltage islands. It is critical to make sure that all registers receive appropriate reset signal and that no register is on a wrong reset tree.

Delay modeling errors: Delay of a logic cell depends upon various factors such as load, drive-strength, etc. on a given supply domain. However, the same cell instantiated on two different supply domains with different supply voltages will also have different delays for same extrinsic conditions. It is critical that static and dynamic timing analysis is done with correct delay information of the cell with respect to supply voltage assumption for a given voltage island.

Combination of traditional techniques with focus on detection of above errors can be deployed for power-management

verification. These are static checks, assertion-based verification, equivalency checks and simulation-based verification. Traditional RTL simulators should also start building in voltage awareness in order to catch complex bugs in power sequencing.

5. Conclusion

In summary, a combination of verification planning with foresight, design methodologies that aid verification efficiency, insightful leverage from design-for-verification techniques and combo-verification strategy designed for project-specific challenges altogether defined as holistic verification helps to deliver good quality silicon in a timely fashion.

Acknowledgement

I would like to thank Raj S. Mitra of Texas Instruments, Inc. and Badri Gopalan of Synopsys Inc. for their time and efforts for multiple reviews during the progress of this paper and providing valuable feedback.

References

[1] Nikhil, R., "Bluespec System Verilog: efficient, correct RTL from high level specifications", 2nd ACM/IEEE International Conference on Formal Methods and Models for Co-Design, June 2004, pp. 69-70.

[2] Arvind, Nikhil, R. S., Rosenband, D.L, Dave, N., "High-level synthesis: an essential ingredient for designing complex ASICs", ACM/IEEE International Conference on Computer Aided Design, November 2004, pp. 775-782.

[3] J. Bergeron, "Writing Testbenches: Functional Verification of HDL Models, Springer, 2000.

[4] Thaker, P.A., Agrawal, V.D., and Zaghloul, M.E., "Validation Vector Grade (VVG): A New Coverage Metric for Validation and Tests", Proc. 17th IEEE VLSI Test Symposium, April 1999, pp. 182-188.

[5] M. Huth and M. Ryan, "Logic in Computer Science: Modeling and Reasoning about Systems", 2nd edition, Cambridge University Press, 2004.

[6] Ken Kundert. "Principles of top-down mixed-signal design", 2006.

[7] Ken Kundert and Henry Chang, "Verification of complex analog integrated circuits", IEEE Custom Integrated Circuits Conference (CICC), 2006, pp. 177 – 184.

978-1-60558-497-3/09 $25.00 © 2009 ACM

Verification Problems in Reusing Internal Design Components

Warren Stapleton
AMD
7171 Southwest Pkwy
Austin, TX 78735
+01 512-602-3476
Warren.Stapleton@amd.com

Paul Tobin
AMD
90 Central Street
Boxborough, MA 01719
+01 978-795-2545
Paul.Tobin@amd.com

ABSTRACT

Large SOCs provide new verification challenges. This paper discusses how the SOC/IP paradigm has impacted AMD, and highlights some of the design issues that you need to consider if you are going to embark on implementing the SOC/IP approach on large projects in your own company.

Categories and Subject Descriptors:

B.6.3 [**Hardware**]: Logic Design – SOC – Verification.

General Terms: Design, Verification.

Keywords: Verification, SOC, IP, Validation.

1. INTRODUCTION

Even traditionally non-SOC-based companies are faced with the difficulties of figuring out how to take advantage of the SOC/IP paradigm in their in-house design flows. It is a combination of efficiency requirements, and necessity.

In AMDs case, the move towards chips that combine core and graphics processing unit (GPU) functionality, and multiple in-flight projects that take advantage of different core, GPU, and Northbridge combinations to satisfy a broad spectrum of market segments, has practically forced AMD into looking more like an SOC/IP company internally.

The cadence at which projects must be completed is much more rapid than is possible for a single team to develop the chips sequentially. However, the cost for truly independent teams is way too high. Therefore, the only practical approach is to identify common components and treat them like IP delivered to each project with some flexibility for small modifications.

Some of the challenges that AMD faces stem from the fact that we are combining pieces of IP that were never intended to be thought of that way in the first place: there were no reuse considerations when the designs were started. We are sure this is not a unique to AMD, since most high-value pieces of IP have been developed over long periods of time. The typical high-end CPU project takes anywhere from 3-5 years to execute from inception to first production; even then it is likely that test cases -- and possibly some key pieces of design -- were leveraged from previous

Permission to make digital or hard copies of part or all of this work for personal or classroom use is granted without fee provided that copies are not made or distributed for profit or commercial advantage and that copies bear this notice and the full citation on the first page. To copy otherwise, to republish, to post on servers or to redistribute to lists, requires prior specific permission and/or a fee.
DAC'09, July 26-31, 2009, San Francisco, California, USA

generations. After initial production, the design is expected to live on for a decade as it is tweaked for better performance and for different markets.

2. SOCs AT AMD

The SOCs developed at AMD are monsters in terms of complexity, both at the IP level and because of complex system interactions. Typically we are talking about combinations of CPUs, GPUs, and Northbridges. They cannot be thought of as simply three large blocks that are combined using standard buses, since there are many system-level behaviors that are orchestrated among all three of them. For example, the Northbridge -- beyond being heavily involved in supporting memory coherency for the cores -- plays a key role in maintaining the complex memory ordering rules that have been architected for optimal system-level performance. In addition, power management events in the Northbridge are processed by firmware that is implemented in microcode in the CPUs that also responds to control events from the South Bridge. Another example of a system level behavior that requires global consideration is the completely different memory usage patterns generated by the GPU and CPU. Complex pieces of design are implemented for optimizing these memory access patterns, the performance of which must be verified at the system-level.

The IP blocks that go into making up these SOCs are themselves very complicated and require their own large verification efforts. Combined, there are between three and four million lines of RTL combined, with a similar amount of code in the collective verification environments, leading to simulation usage that exceed 50 billion cycles per week.

Traditional simulation is the primary method of verification, although formal verification plays a key role in targeting floating point algorithms and arbitration fairness, and for finding live-lock situations. The simulation farm is truly state of the art, consisting of tens of thousands of cores managed by grid software. Designing large systems under ever-increasing schedule pressure takes large teams; large teams push every aspect of the design process to the limit, since a lot of issues grow exponentially with the size of the team. Some of the obvious ones are disk space, networking requirements, release mechanisms and release qualification procedures. Release systems get complex for a large team because the number of releases performed each day and the amount of time to qualify each one, easily exceed the available time.

Given all this, you would think the problems were the obvious -- simulation speed, formal tool capacity, etc -- but the reality is more complicated. What we see engineers at AMD dealing with regularly is how to combine designs that have been executed by different teams and – in cases of AMD and ATI prior to their merger -- originally from different companies. We see that at every place where there are no existing standards, different solutions -- all good -- have been created by the different engineering teams, and a process needs to get started to reconcile the differences before true progress can be made.

3. REVIEW OF TYPICAL PROBLEMS

To make this short paper generally useful, we thought it might be helpful to provide a list of issues to think about when starting to plan a project like this. Many of these are issues AMD has dealt with when combining designs that originally had been done by ATI with designs from the original CPU side of AMD. We have also added items based on our experiences in other companies.

3.1 Simulators

In the worst case you might be dealing with code that has been developed for simulators from different vendors. Problems could include

- Language construct support, especially related to any *System-Verilog*[1] parts of the language

- Different PLI/VPI [2] support

- Vendor-specific language extensions

- Different compile flows

None of the vendors' supported features are subsets of any of the others today, so there will be a process of reconciling the differences if you are starting with this situation.

3.2 Tool Versions

Even if you are starting with designs that were done with a single vendor's simulator solution, there will be subtle differences between tool releases, especially going from one major release to another. The tool may have become improved in some way, but the changes might highlight incorrect coding, or expose race conditions that have always existed in the design. If some of the IP you are incorporating does not have an active team working on it because it is considered a stable design, then upgrading the design to be compatible with recent releases of the tools can become another unplanned-for exercise.

Tools that are not part of the EDA vendor category can also be a problem. Most complex designs will make extensive use of *UNIX* utilities like *make, sed* and *awk*, scripting languages like *Perl* and *Python*, and different source-control management tools like *CVS, SVN, Perforce* etc. Even something very well standardized like the C++ language has had changed significantly over the time these large projects have survived. In the case of C++, the final executable must be linked with the same version of the C++ library which will force code compilation by versions of the compiler that are at least reasonably close to each other.

3.3 Test Representation

Ask people what they think a test is and you will get as many different answers as people. What is important, though, is that stimulus should be designed to be reusable at different levels of hierarchy. We have found that most purchased IP already had this

in mind when the original developers of the IP implemented their compliance suites, but most in-house designs did not.

Modern verification frameworks address some of this problem with clear separations between stimulus generators and their associated drivers. However, more is required; you should be prepared to add "adaptation" layers that can modify stimulus streams for injection into different drivers. An example of this might be a proprietary internal configuration bus, which -- when accessed at a system level -- becomes memory-mapped. At the lower level the stimulus needs to be connected to a real bus driver, at the higher system level the adapter for the stimulus generator needs to assign addresses and generate transactions for a memory read/write oriented driver.

3.4 Regression Systems

Large projects usually invest quite a bit of effort implementing regressions systems, more extensive than a simple list of command lines to run. These systems will typically use real databases, sometimes holding the test cases themselves, and certainly hold the results from running regressions. The information retained in these databases provides key design health metrics, helps predict the all-important "done" date, and improves subsequent design projects. This is all great until you try to combine two different systems developed by different engineering teams. The question to answer is how to combine the test suites for the IPs and incorporate them into a single regression at the end.

This area is ripe for either a third-party solution or an open source project. In the software industry there are many such commercial solutions; however, they don't appear capable enough to support the scale at which our projects operate. Even if there is not a single solution, some standardization in the area of database schemas could help us move towards combining results from multiple IP-level simulations.

3.5 Design-Related Issues

Beyond tool related problems, the designs themselves will likely have conflicting representations that must be resolved to be able to simulate them together, such as:

- Name collisions: the *Verilog* module name space is flat, so you cannot easily combine designs that use names that collide with each other, without the use of mechanisms like `uselib`, which is often problematic for other tools in the tool chain to deal with. The names for `define` macros are also likely to collide without careful use of prefixes to avoid collisions. Another common source of design-related problems is using common hard macros from a central physical design group, that were done at different times, while the macro has evolved over time.

- Different assertion language standards: *SVA* [1], *PSL* [6], *OVL* [4][5], and some others that are vendor-proprietary.

- Different coverage techniques: it's easier today because of *System-Verilog* [1] coverage groups; however, legacy designs probably used proprietary tools to collect coverage which makes it difficult to collect combined coverage statistics.

3.6 Other More Mundane Things

Other things that get in the way of progress when combining IPs are not as obvious. For example:

- Command line options sound simple, but unfortunately this is not a standardized area, and different teams invent different ways of doing things, which creates issues when these designs are combined into a single simulation.

- Message reporting is a major non-standardized issue. Although modern frameworks try to standardize this, there exist multiple frameworks and most legacy designs will not have been developed using any of them. It is important for message reporting systems to provide callback mechanisms so they can be integrated with a larger framework and report their messages into the global system.

3.7 Division of Labor

An extremely important aspect to making the SOC/IP paradigm work is a clear division of labor between the IP and SOC verification teams. This is more difficult for internally developed IP than for purchased IP because the interfaces between them are not so well defined and are often the result of negotiations between the two design teams that have tried to optimize everything except verification.

Hierarchical verification is rarely 100% complete at lower levels, and is often mistrusted by higher-level teams leading to lots of work being repeated. Also, there are always system-level features to verify that cannot be completed at a lower level; this area needs to be managed carefully to avoid unnecessary work being carried out at the SOC level where the simulation models are naturally going to run more slowly.

Apart from careful management and allocation of test plan items to their appropriate levels, being able to combine coverage data from IP-level testing with data from SOC-level simulations is important in demonstrating that all testing has been completed.

4. RECOMMENDATIONS

AMD has passed through the first phase of this SOC/IP learning process. So, apart from highlighting an endless list of problems to deal with, we will conclude with some suggestions that might help our own company and others doing similar things.

1. Be proactive about keeping IP-level projects, both those in development and considered stable, using recent versions of all tools, including vendor-related, *UNIX*-related, and internally developed tools.

2. Make every attempt to capture information that pertains to building the simulation environment in simple formats that can be processed and repurposed easily such as simple things like file lists that make up the design. AMD has developed a system that solves the problem of combining file lists from different IPs and resolves dependencies amongst them driven by configuration files. This is an area in which the vendor's tools lack any kind of decent solution and we would like to see an industry-driven solution.

3. Where they exist, relentlessly drive the design teams towards standards. For example, the *Spirit* [3] consortium is a great place to start looking for standards related to representing information delivered from the IPs, register descriptions being one of many examples. We would like to see recommendations for standardizing more of the process, down to simple things like representing command line options that control runtime behaviors (especially prevalent in the verification area).

4. Always keep stimulus reuse in mind when designing verification environments. The reuse problem extends way beyond just being able to reuse the monitors and checkers.

5. Develop a delivery layer between the IP design source and the consumer. In this layer, try to eliminate potential issues like name conflicts, any IP configuration, and *Verilog* macros.

5. CONCLUSION

Switching to the SOC/IP paradigm makes a lot of sense for efficiency reasons. However, it is not without its difficulties, especially when it involves converting legacy designs to work within the SOC paradigm. A myriad of language/framework deficiencies, tool flow limitations, and lack of standards, exacerbate the problems, most of which we have outlined in this paper.

6. REFERENCES

[1] IEEE Std 1800-2005 SystemVerilog Language Standard, IEEE 2005.

[2] IEEE 1364 LRM for Verilog.

[3] Spirit Consortium: http://www.spiritconsortium.org.

[4] Foster, H., Larsen, K., Turpin, Mike. Introduction to the new Accellera Open Verification Library. DVCon 2007

[5] Accellera Open Verification Library Standard, Accellera 2007.

[6] Accellera Property Specification Language (PSL) LRM v1.1, June 2004.

978-1-60558-497-3/09 $25.00 © 2009 ACM

Exploiting "Architecture for Verification" to Streamline the Verification Process

Dave Whipp, NVIDIA

[dac09@dave.whipp.name]

Abstract

A typical hardware development flow starts the verification process concurrently with RTL, but the overall schedule becomes limited by the effort required to complete all the necessary verification tasks. Being the limiting factor, verification schedules become unpredictable, often resulting in slippage of the tapeout dates. This paper looks at ways to restructure the flow to complete a significant part of this effort during the architectural phase of the project, prior to the start of RTL. This front-loading of the schedule allows a smaller verification team to complete the process with a tighter schedule.

Categories and Subject Descriptors: B.5.2
[Hardware]: RTL – Design Aids – Verification

General Terms: verification, management

Keywords: verification, executable specification, ESL

1   A Verification Architect

What is the goal of Verification Engineer? Most people faced with this question – I use it for job interviews – will frame the answer in terms of the correct operation of the product or the fidelity of the design to its specification. This is part of the answer but, I believe it misses at least one important element: timeliness. The success or failure of a product can be decided by its launch date. Weeks, and days, do make a difference in consumer products. When I describe my goals as a Verification Architect I focus not on the correctness of the product (which I regard as "a given"), but on the workflow that achieves it. My aim is therefore to construct verification strategies where:

- Verification tasks do not appear on the critical path of the project.
- Estimates of task durations are as accurate as possible.

Permission to make digital or hard copies of part or all of this work for personal or classroom use is granted without fee provided that copies are not made or distributed for profit or commercial advantage and that copies bear this notice and the full citation on the first page. To copy otherwise, to republish, to post on servers or to redistribute to lists, requires prior specific permission and/or a fee.
DAC'09, July 26-31, 2009, San Francisco, California, USA

A task that is on the project's critical path contributes directly to the time between the start of the design and the start of volume manufacturing. The critical path should be dominated by design tasks, not verification. Examples of design tasks are: writing RTL, debugging failures, and optimizing the design to achieve timing, area, and power goals. Verification tasks include: creating and debugging tests and test benches, triaging test failures, and achieving coverage goals.

It is not my intent to disparage the job of verification engineers. I focus on verification efficiency because I am a verifier. If I were a designer then I would seek to minimize the cost of design, and to ensure that design tasks do not appear on the critical path.

A strategy that is used to improve efficiency for both Verifiers and Designers is to attempt to front-load the overall project schedule by moving effort to the product architecture phase.

2   A Hardware Development Flow

The major tasks in chip development flow can be grouped by who is responsible for them. The major groups are "architecture", "design", and "verification". Although different companies may have marginally different flows, I believe that a typical breakdown of the tasks is similar to that shown in table 1.

Table 1: Task Responsibilities in a Traditional Flow

Architecture	Design	Verification
Comprehensive Paper Specification	Write RTL	[UTF Model]
	Debug RTL	Write Testbench(es)
ISS Model	Power, area, and timing closure	Write Test Scenarios
[UTF model]		Debug Tests
	Assertions	Triage RTL failures
		Assertions
		Functional Coverage
		Tune Constraints

The work done by the design group is not the focus of this paper. Their work consists primarily of writing RTL and then optimizing it to achieve the power, area, and timing goals of the module. Where a designer makes assumptions for correct behavior, we ask them to code those assumptions as assertions, and not simply as comments. Finally, we charge the schedule hit for debugging RTL failures as an RTL cost. This is balanced by a "triage" cost for the time spent by Verifiers to figure out the meaning of test failures.

The architecture group is responsible for defining the structure of the design. Major blocks are defined, and their functions described. The interfaces between the blocks are sketched out to some level of detail, typically focused on the performance of the interface (latency, throughput) and whether some standard bus (protocol) is to be used.

The work product of the architecture group almost always includes paper documentation, which will later be read and interpreted by the design and verification teams who are responsible for created the detailed design and affirming that this detailed design is indeed constant with the specification.

It is common for architects to create executable models of the proposed architecture. A common abstraction is the instruction set simulator ("ISS"), which defines the operations while avoiding implementation details. An ISS is not limited to traditional CPU architectures: for example, a hardware block that acts as a network switch may have its routing behavior defined as an ISS.

Less frequently, architects may create a detailed transaction level model of the chip. Such a model will provide a detailed representation of each module in the design, along with its interfaces. A common framework for such models is the untimed functional "UTF" abstraction of SystemC [1]. Ideally for verification, such a model will be interface-accurate and usable as a placeholder in the design for actual RTL units: co-simulating with other units for faster simulation (or before the RTL for the unit is written).

Unfortunately, although detailed executable models are not always created by architects, they are frequently needed. It then falls to the verification team to create the models as an interpretation of the written specification. This has two consequences: first, the model writer may misinterpret the specification; second, the architect doesn't get immediate feedback from running code that is a concrete representation of the specification.

The lesson to learn from this is that an efficient flow requires appropriate architectural representation. I will show that having architects just write transaction-models is insufficient: there are many other forms of executable work-product that we can ask our architects to create.

3   Executable Specifications

3.1   C Models

One thing that both ISS and UTF models have in common is that they are simulation models. The way they are used is to create a stimulus (possibly software) and run the code to see what happens. Although some progress is being made with "sequential equivalence" checking, such tools are not yet ready to replace traditional verification approaches.

Being simulation models, the code is actually an implementation, not a pure specification. Often they are coded in such a way as to improve simulation performance but decrease their usability as a specification. This weakness is a result of the fact that the standard language we have, today, for writing such models is C++.

3.2   Functional Coverage

A C model may be used by the verification team as a reference model. A test is run on both the RTL and the model and the resulting behaviors compared. Depending on the abstraction of the model, the verification effort needed to achieve a reliable comparison can be non-trivial.

When the model is used in this way, the model provides not only a behavioral reference, but also a definition of functional coverage. If a C model defines all the behaviors defined by the architects, then structural coverage of the model is analogous to functional coverage of the design.

When a feature is omitted from an implementation, structural coverage may still achieve 100%. However, if that feature exists in the C model, then structural coverage of that model will show that the test suite did not exercise it. Of course, the feature might have been overlooked even by the architecture group: but if the model is also used for software development then we can have some confidence that its omission will be noticed.

Even if the correlation of model-coverage with functional coverage is not perfect, we can still benefit from the concept. Sometimes we ask: "what 5% of tests should we run for our nightly regression testing?" We have found that a Pareto analysis of structural coverage of the C model is able to answer this question in much less time than it would take to run our test suite on the RTL.

3.3   Leveraging Validation Tests

A simulation model is not the only work-product of architecture that can be directly used for verification. Architects create "validation tests" to check that the behavior of the models matches their expectation. Whilst such tests are often reused for verification, their value can be enhanced if they are able to be retargeted to hit implementation corner cases.

One approach is to create validation tests as self-checking directed tests, and then run the same test multiple times with different "irritators" applied to the design. Examples include context switching between multiple tests, injection of soft errors, and randomizing backpressure from FIFOs. Depending on the design, many other "irritations" may be applied.

Another approach is to construct the validation test suite by generating self-checking directed tests. One way to achieve this

978-1-60558-497-3/09 $25.00 © 2009 ACM

[2] is to define the test space as a graph: random walks of that graph create the tests. The graph itself becomes an executable model of the design, albeit one that is very different from a traditional simulation model. The power of the approach stems from the ability to compose graphs that express difference aspects of the verification space: architects provide a graph for validation; vierifiers add graphs for the implementation corner cases.

3.4 Properties and Assertions

Assertion based verification methodologies often partition assertions into two groups: those written by designers, and those written verifiers [3]. Designers are asked to treat assertions as "executable comments" that document assumptions they are making. Verifiers are asked to create assertions that reflect requirements passed down from the specification.

In the context of an executable specification, it seems absurd to ask verifiers to write architectural assertions. The source of such assertions is the architect, not the verifier, and adding a manual interpretation step into the flow here does not add value.

Today's (temporal) assertion languages (and tools) are not designed to be used with C models. Both SVA and PSL are conceived in terms of sequential regular expressions (SEREs) in the cycle domain. It is possible to map these to transactions and algorithms, but doing so introduces additional complexity that makes it less likely to be adopted. The problem is compounded by the fact that not all architects use Linux as their primary operating system.

Instead, we use an in-house language [4] that enables assertions to be written at higher levels of abstraction, which are then translated to C++, OVL, and SVA for use in different simulation contexts. This is an area that feels ripe for exploration by EDA vendors.

3.5 Interface Definition Language

The UTF C Model, the testbench, and the design all share the same interfaces. It is natural, therefore, to define these interfaces just once, using a formal interface definition language such as IP-XACT [5], and to use code generators to create the RTL and C code that implements them. Indeed, it is only through the use of such a language that it is possible to take architectural assertions and map them to the design.

An interface definition language is the basis of structural models. These define not just the interface signals, but also the modules to which the interfaces connect, and the cycle-level protocols used to mediate the flow of information. From this information it is possible to generate many of the components of the lower layers of a testbench, such as BFMs that translate between the transaction-domain semantics of the higher layers of the testbench (section 3.6) and the cycle-domain semantics of the DUT itself. All that is left, for the testbench author, is to ensure the correct hookup of clocks and resets (plus, perhaps, implementation-specific device ports such as for DFT).

3.6 Test Benches

Comprehensive testing of a transaction level C model demands a testbench capable of interacting with multiple concurrent transaction level interfaces. This is the same requirement as exists for a transaction level testbench for RTL verification. A properly constructed testbench should be reusable from validation to verification.

If the transaction layer of the testbench is written to test the architectural UTF model, and the signal/command layer can be derived from a formal interface definition (section 3.5), then most of the major components of the RTL testbench are derived as work products of architecture.

4 An Improved HW Creation Flow

The task breakdown of table 1 showed that if the work product of the architecture group is limited to just a paper spec (and perhaps a C model or two) then the majority of the downstream tasks fall to the verification group. Table 2 shows how an aggressive embrace of a more general concept of "executable specification" acts to change the distribution of tasks.

Table 2: Task Responsibilities in Enhanced Flow

Architecture	Design	Verification
Minimal Paper Specification	Write RTL	Testbench hookup (clocks and resets)
	Debug RTL	
ISS Model		Write Test Scenarios
	Power, area, and timing closure	
UTF model		Triage RTL Failures
Formal Interface Definitions		Tune Tests
	Assertions	
Assertions and testpoints		
Directed-Test (Graph)		
Transaction Testbench Layer		
Debug Tests		

From the selfish perspective of the verifier it is easy to say that this flow would be preferable. By transferring the majority of the tasks to architecture, the verifier is free to concentrate on the actual goal of verification. But it is not sufficient to improve the workload of the verifier: it must also be advantageous to the architecture group (who are probably not staffed for extra work) and to the efficiency of the overall HW creation flow.

Executable work-products enable architects to better understand the architecture, without depending on the long feedback paths that result from depending upon verification engineers to write models and validate the architectural assumptions. Furthermore, they are creating tangible proof that the architecture works. If the application is video or graphics, then they can see frames (pictures) being created. In other domains the results may be less obvious to outsiders, but no less gratifying to the creators.

The overall flow benefits when bugs can be found earlier in the development process. A model created by an architect is validated as part of its creation, and then stressed by the software group who use it as a platform for their part of the project. The model, validated by both SW and Architecture, can then be trusted by the verification group who then assure that the design is equivalent to the model.

Efficiency is further enhanced when tests and testbenches from architecture are reused for verification. If it is a given that these constructs are needed for architecture, then the additional cost to make them reusable is most likely a cost worth paying.

5   Challenges

When we think of an executable specification, we usually think of a simulation model, written in C++. Such models are useful, but insufficient. Structural coverage of abstract models is a good source of functional coverage, but today's tools do not enable us to fully exploit it. Not only is there limited traceability from lines of code in the model to the equivalent lines in the RTL; but also there is much code in the C++ model that exists only to support its use as a simulator.

When we attempt to stretch a single model between multiple organizations, we find that we need to make tradeoffs. Software groups may want fast models that run on Windows, whereas the Verification group is most likely Linux-centric – and would happily trade performance for a greater fidelity to the micro-architecture.

Initially, the architects reject the idea of adding additional work: instead of just writing a specification document, they are now being asked to write models, tests, and assertions. They feel that these models are not for their benefit: they are for the benefit of verification engineers (who would otherwise be tasked with the work) and for the software group. It is only later that we see attitudes change, and the benefits to both the architects and the project as a whole become apparent.

6   Architecture For Verification

The goals of Verification in the architectural phase of the project are twofold:

- Eliminate unnecessary complexity that would increase verification burden.
- Ensure that the work products of the Architects are directly usable in the verification flow.

Unnecessary complexity is most often found in features that exist to increase performance in areas that are not critical to the product's success. Performance enhancements that are achieved by additional coupling between micro-architectural features lead to a significant increase in the number of corner cases: not only must the features be verified in isolation, but also the optimized interactions must also be verified.

To be most useful, the work products of the architecture group must be more than a set of documents. A multitude of different aspects of executable specification are needed. In addition to ISS and UTF C models, we find value in reusable tests, testbenches, assertions, and various forms of structural definition.

If we understand how the different models will be used by all the stakeholders, then we know how to prioritize the resources to create them. If a model is most useful for verification, then it may make sense to have verification engineers work more closely with architects, and less closely with RTL designers. If the model is to be used by the software authors, then it may make sense to have software engineers work closely with the architects on that model. And, when the model is to be used by all three groups: then all three must work with each other to maximize the overall efficiency of the flow.

7   References

1. SystemC: http://systemc.org
2. Adnan Hamid. "Hope is Not a Verification Strategy – http://www.designcon.com/infovault/paper.asp?PAPER_ID=323 (DesignCon 2008)
3. Harry Foster, Adam Krolnik, David Lacey – "Assertion-Based Design" section 1.4.3 (page 19).
4. Dave Whipp, "Transaction Assertions in an Interface Definition Language" (DesignCon 2008) – http://dave.whipp.name/dv/designcon2008_paper.doc
5. IP-XACT: http://spiritconsortium.org/home

Role of the Verification Team Throughout the ASIC Development Life Cycle

Eric Chesters

Cisco Systems

San Jose

CA, USA

ercheste@cisco.com

ABSTRACT

Traditionally the development of ASICs within a company like Cisco has been tightly coupled with the development of a single end product. However the increasing costs associated with the development of high end ASICs has forced a fundamental shift in the ASIC development process. No longer is it acceptable to develop an ASIC that can only be used in a single product. Instead, the expectation is that ASICs under current development will have a much longer life span and will be used across not just multiple generations of the same product, but in multiple different products across the company.

This has a number of consequences on the ASICs themselves and the manner in which they are developed. In this paper, we discuss the impact on the ASIC development process and the expanded role the verification team now plays throughout the development process in ensuring that ASICs once delivered meet the expectations of the product teams.

Categories and Subject Descriptors: B.6.3 [**Hardware**]: Verification.

General Terms: Verification.

Keywords: Verification, Silicon validation

1. INTRODUCTION

The increasing cost associated with high end ASIC development has prompted us to re-evaluate the very nature of the ASICs we design, their life span and consequently the way in which we develop them. Much has been written elsewhere about the various trends in the industry to optimize the development process through the adoption of techniques such as reuse etc. The focus of this paper, however, is on the broader development life cycle, trends that we observe here, and the role of the verification team in this process.

As previously stated, the very nature of the ASICs we develop today has changed over the last five years. The ASIC development process is no longer tightly coupled with a single endpoint such as the delivery of one product. Instead, the ASICs being developed today are expected to be used in multiple generations of multiple products. In order to support this requirement, the ASICs themselves are becoming far more configurable and in many cases key functions are being implemented via programmable cores on the ASIC.

However, with this flexibility comes a considerable degree of complexity that impacts not just the design verification phase, but also silicon bring-up and the eventual deployment of the ASIC into the hands of the product team.

Based on our experience over the last few years developing high end networking ASICs, we've seen this result in the verification team taking on an expanded role that challenges the view of what a traditional verification team does. As we will see in the following sections, the verification team is naturally placed to mitigate the additional complexity and because of this it now plays a key role throughout the product development process.

2. DEVELOPMENT LIFE CYCLE

The process by which an ASIC is defined, designed, validated and ultimately deployed to the end user can be broken down into a number of discrete phases. For a particular ASIC development, these phases can be viewed as a linear set of events. However for a typical ASIC development organization they represent phases in a cyclic process. As the later phases in the development of the current generation ASIC complete, the team transitions to the early phases of the next generation ASIC and the cycle restarts.

For the purpose of the paper the development cycle is broken down into the following phases:

Permission to make digital or hard copies of part or all of this work for personal or classroom use is granted without fee provided that copies are not made or distributed for profit or commercial advantage and that copies bear this notice and the full citation on the first page. To copy otherwise, to republish, to post on servers or to redistribute to lists, requires prior specific permission and/or a fee.
DAC'09, July 26-31, 2009, San Francisco, California, USA

978-1-60558-497-3/09 $25.00 © 2009 ACM   216

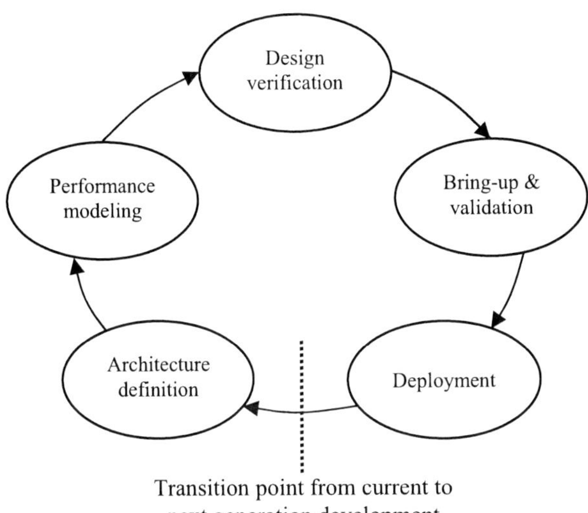

Transition point from current to
next generation development

The role the verification team naturally plays during each of these phases is now considered in more detail.

2.1. Design Verification

The design verification phase is the easiest place to start, since it represents the heart of the verification responsibilities.

A complex ASIC is often decomposed into a number of smaller units or blocks and the design effort is largely focused on the development of these blocks. At the block level, it is relatively easy to fully define how the block should operate, what the expectations are etc. Consequently the goal of the verification task is also simpler to define: ensure that the design meets all the documented expectations for the block.

A smaller subset of the ASIC development team is focused on the ASIC as a single entity and its end to end operation. Typically these people are from the verification team and are tasked with the chip level verification activities. The goals of the chip level verification activities are however significantly harder to define.

The highly configurable nature of the ASIC can result in a very large number of theoretical configurations. However many of the theoretical configurations have no practical use when the ASIC is deployed. For example, consider the case where the ingress and egress interfaces on an ASIC can be independently configured to operate in one of N possible modes. Simply taking the cross product of the ingress and egress configurations would result in N^2 possible combinations that need to be verified. Suppose there exists a system level constraint which dictates that the two interfaces must always be configured in the same mode: this system level constraint would reduce the practical configuration space to only N possible configurations.

One of the first challenges faced by the team working on chip level verification is to reduce the theoretical verification space into one that is both feasible to verify but also covers all the useful configurations. In order to do this, the verification team must first learn about the end systems. What restrictions do those systems place on the way the ASIC is configured and how is the ASIC expected to perform in those systems? Only by

understanding the answers to these questions can the appropriate trade-offs be made.

By the time we reach the end of the design verification phase, the verification team has gained a lot of experience with the ASIC through simulation. It has also learnt a lot about the systems into which the ASIC will be deployed.

2.2. Post Silicon Bring-up and Validation

As stated previously, the ASICs being developed are becoming increasingly configurable and programmable. This introduces a number of complexities during the initial bring-up and validation of the silicon.

First of all, the task of getting the ASIC operational in the bring-up lab often requires a considerable amount of software to configure and operate the ASIC. The verification team is well positioned to assist in this endeavor since they have the requisite practical knowledge and experience with the ASIC, based on the pre-silicon simulation work.

Secondly, during the life span of the ASIC, it is expected that the ASIC will be configured and used in many different ways to meet the needs of the different products. Thus, the first product can no longer be used as the sole platform to validate the silicon since it only represents one of the possible usage models. Instead, a dedicated effort is required to validate and characterize the operation of the ASIC in all of its expected operational modes. This task can be viewed as a natural extension to the pre-silicon verification process and hence is typically driven by the verification team.

Both of these factors result in the role of the verification team being extended well beyond tape-out, and into the post silicon bring-up and validation activities. This extension of the verification process beyond tapeout is an area that needs further investment in the industry. Much of the recent focus has been on optimizing the pre-tapeout activities through reuse, adoption of System Verilog[2] etc. But most of this has been done without considering how this pre-tapeout investment can then be leveraged once silicon is available.

2.3. Deployment

As the ASIC is deployed to the product team, a significant support burden is placed on the ASIC development organization. This burden is greatly increased by the inherent flexibility in the ASIC and the associated configurability and programmability.

Much of this support comes in the form of providing information, answering questions etc., but it also involves close collaboration with the product team to help understand how the ASIC is operating and subsequently to help optimize the configuration of the ASIC to meet the needs of the product.

The same knowledge and experience that is leveraged by the bring-up and post silicon validation activities is vital to providing this support. Hence the verification team again naturally plays a significant role during this phase.

In return, through this support role the members of the verification team gain considerable insight into the end products and their development. A lot of information is gleaned about the product requirements and how the ASIC is being utilized to meet those requirements.

2.4. Next Generation Architecture

One of the biggest challenges faced when starting the next generation development is to define the requirements for the new ASIC. Many requirements are set by external factors such as market forces pushing bandwidths, processing capabilities, new features etc., Others are derived from experience with the current generation, what worked well, what didn't work so well etc.

As discussed in the previous sections, the verification team has gained a lot of experience with the current generation ASIC while driving the post silicon bring-up and validation activities as well as supporting the deployment of the ASIC into the hands of the product teams. This knowledge and experience with the current generation ASIC provides critical input into the requirements definition and early architectural work for the next generation.

Hence, again the verification team finds itself very active during this phase, ensuring that the lessons learned from the current generation are fed back into future generations of the ASIC.

2.5. Architectural Modeling and Validation

The complex nature of the ASICs being developed often requires a certain amount of modeling to be performed to validate the architectural and micro-architectural assumptions. This process can involve a number of different techniques ranging from the more ad-hoc such as the use of spreadsheets etc. to the development of a more refined architectural model written in C++/SystemC[3], System Verilog or some other high level language.

Given the involvement of the verification team during the architectural definition phase and the required software and analytical skills required to drive the modeling activity, such tasks are naturally performed by members of the verification team.

2.6. Return to Verification

And so the development cycle returns to the starting point, verification of the next generation implementation. Due to the integral role of the verification team during the architectural and performance modeling phases, the team embarks on the verification task with a much deeper understanding of the ASIC than could otherwise be gained from simply reading the specifications.

This understanding however is mainly captured in the minds of the engineers. As an industry, we should be looking at better ways to capture this information in a more formal way that can be leveraged during the verification effort, such as, for example, the development of executable specifications, the automated tracing of architectural requirements to the verification test plans etc.,

2.7. Flow of Information

The common theme throughout the previous sections is the flow of information from one phase of the development cycle to the next. The efficient flow of information throughout the development cycle is vital to a successful project and it can be seen that the verification team is naturally positioned to facilitate this communication.

3. IMPACT ON MEHTODOLOGY

As outlined in the previous sections the verification team has a vital role to play throughout the development life cycle by providing continuity and ensuring the efficient flow of information from one phase of the development cycle to the next. However this experience is not just captured in the heads of the engineers but is also captured in the code base that they have been developing throughout the project.

Therefore, in the same way that we want to leverage the knowledge from one phase of the development cycle to the next, we also want to be able to leverage the code that has been developed. For example, we would like to be able to reuse the test scenarios from the architectural modeling phase during the design verification process. Another example would be that we would like to leverage parts of the verification infrastructure that have been proven in simulation during the bring-up process.

For the block level verification efforts, we have adopted a contemporary System Verilog methodology [1]. However, at the chip and system level we augment this with C++ and SystemC.

By following this approach we enable code that was developed in SystemC, as part of the performance modeling phase, to be integrated into our chip level verification environment. The same test scenarios, packet generators and performance monitoring components used to validate the architecture are used to form the basis of the RTL verification environment. Although the use of SystemC enables this potential for reuse it can only be achieved if this end goal is taken into consideration during the initial development of the components. This mindset is achieved by having the same engineers responsible for the modeling activities be responsible for the subsequent chip level verification activities.

Once we enter the post silicon phase of development we typically see the ASIC placed on a board with a number of other custom devices and a generic control processor that is responsible for managing the operation of the ASICs. The control processor runs a generic operating system that is capable of hosting a number of applications written in a variety of languages. However, perhaps the most commonly supported language is C/C++.

By carefully architecting the structure of the verificaiton environment and the associated C++ code, we ensure that at the end of the verification process we can separate out the interesting C++ routines into a pure C++ library.

The resulting C++ library contains routines to initialize and drive the ASICs diagnostic features that have been proven in simulation. This library can then be taken into the bring-up lab where it is highly valuable. Not only does this accelerate the

978-1-60558-497-3/09 $25.00 © 2009 ACM

bring-up activities, but it also acts as a reference implementation during the early deployment activities.

4. CONCLUSIONS

There is a growing trend that sees the role of the verification team extending well beyond the boundaries of the design verification phase. Modern ASICs are becoming highly configurable and programmable and this in turn is increasing the amount of post silicon support that is required from the ASIC development organization. Given the pre-silicon experience with the ASIC, the verification team is often the natural choice to provide this support.

In turn the verification team gains a lot of exposure to the end products, their requirements and the success of current silicon in meeting those needs. It is vital that any future development fully utilizes this knowledge and hence the verification team again has an increased role to play during the early architectural definition phases to ensure that the lessons of the past are applied to the future.

The end result of these trends is that we see the verification team playing a key role throughout the development cycle by enabling the efficient communication of information from one phase to another. Not only does this impact the day to day responsibilities of the team, but it also impacts and shapes the methodologies used by the team.

5. REFERENCES

[1] Tony Tsai, "Techniques for Selective Reuse of Verification Components in Hierarchical Verification of Large Designs.", SNUG, San Jose, 2008

[2] IEEE P1800(™) Standard Language Reference Manual (LRM)

[3] IEEE 1666(™) -2005 Standard SystemC Language Reference Manual (LRM)

An Efficient Approach for System-Level Timing Simulation of Compiler-Optimized Embedded Software

Zhonglei Wang, Andreas Herkersdorf
Lehrstuhl für Integrierte Systeme
Technische Universität München
Arcisstraße 21, 80290 München, Germany
{Zhonglei.Wang, herkersdorf}@tum.de

ABSTRACT

Software accounts for more than 80% of embedded system development efforts, so software performance estimation is a very important issue in system design. Recently, source level simulation (SLS) has become a state-of-the-art approach for software simulation in system level design. However, the simulation accuracy relies on the mapping between source code and binary code, which can be destroyed by compiler optimizations. This drawback strongly limits the usability of this technique in practical system design. We introduce an approach to overcome this limitation by converting source code to a low level representation, called *intermediate source code (ISC)*. ISC has accounted for most compiler optimizations and has a structure close to binary code, so it allows for accurate back-annotation of timing information from the binary level. To show the benefits of our approach, we present a quantitative comparison of the related techniques with the proposed one, using a set of benchmarks.

Categories and Subject Descriptors

C.4 [**Computer Systems Organization**]: Performance of Systems—*Measurement Techniques*; I.6.5 [**Computing Methodologies**]: Simulation and Modeling—*Model Development*

General Terms

Design, Performance

Keywords

Software timing simulation, system level design, iSciSim

1. INTRODUCTION

In modern embedded systems, the importance of software and its impact on the overall system performance is steadily increasing. Applications are often partitioned into several tasks and allocated to multiple processing elements, which are interconnected and cooperate to realize the system functionality. To explore the design space of such a complex system, fast and accurate performance evaluation at a high level of abstraction, i.e., at the system level, is greatly required. The performance evaluation is usually based on simulations to get dynamic statistics of the execution on the system under design. The system architecture will then be adjusted and refined iteratively depending on the simulation results.

For system-level design, software simulation using instruction set simulators (ISSs) is too slow and covers too many unnecessary details. Rather, a technique that allows for fast simulation with enough accuracy for high level design decisions is more desirable. For this purpose, software simulation techniques based on *native execution* have been proposed. The common idea behind them is to generate, for each program, a simulation model that runs directly on the host machine but can produce the performance statistics of target executions. To generate such a simulation model, three issues are to be tackled: functional representation, timing analysis, and coupling of the functional representation and the performance model. The functionality of a program can be represented at the source level, at the intermediate representation (IR) level, or at the binary level. To get accurate timing information, the timing analysis should be performed on binary code, using a performance model that takes the important timing effects of the target processor into account. Coupling of the functional representation and the performance model can be realized by annotating the functional representation with the obtained timing information.

Source level simulation (SLS), using source code as functional representation, is widely used for system level design, because of its high simulation speed and low complexity. However, this technique has a major disadvantage in that it cannot estimate compiler-optimized code accurately, because accurate back-annotation of timing information into source code relies on the mapping between source code and binary code, which might be destroyed by compiler optimizations. Since, in reality, programs are usually compiled with optimizations, this drawback strongly limits the usability of SLS. The major contribution of this paper is to propose an efficient approach, called *iSciSim*, that allows for accurate simulation of compiler-optimized software at a simulation speed as fast as SLS. The idea is to get another representation of source code, which has a structure close to binary code and thus allows for accurate back-annotation of the timing information obtained from the binary level.

The rest of the paper is organized as follows: Section 2 introduces existing software simulation strategies and related work. Following this, Section 3 presents the proposed approach. Experimental results are shown in Section 4. At last, Section 5 concludes this paper.

Permission to make digital or hard copies of part or all of this work for personal or classroom use is granted without fee provided that copies are not made or distributed for profit or commercial advantage and that copies bear this notice and the full citation on the first page. To copy otherwise, to republish, to post on servers or to redistribute to lists, requires prior specific permission and/or a fee.
DAC'09, July 26-31, 2009, San Francisco, California, USA

2. RELATED WORK

The existing techniques related to our approach include ISSs and native execution based techniques, which are further categorized into *source level simulation (SLS)*, *IR level simulation (IRLS)* and *binary (assembly) level simulation (BLS)*, in terms of functional representation levels.

2.1 Instruction Set Simulators

An ISS is a piece of software that imitates the target processor to fetch, decode and execute target instructions one by one, at runtime, similar to a Java interpreter. It provides cycle-accurate estimates of software execution but runs extremely slow due to the on-line interpreting of target instructions. For co-simulation with SystemC, complex timing synchronization between ISSs and the SystemC simulation kernel must be tackled, as shown in [4].

2.2 Source Level Simulation

The major difference among previous works on SLS is the way of getting timing information. Bammi et al. [1] and Giusto et al. [5] associate each instruction with a latency, which is calculated in a statistical way. Such instruction latencies have not taken the timing effects of the target processor into account, and thus, are inaccurate. More accurate timing information can be obtained by means of static analysis [3, 11, 12] or measurement using an ISS [9], for each basic block. In the scope of a basic block, pipeline effects can be analyzed statically, but an approximation must be made on global timing effects, such as the cache effect and the branch prediction effect. These global timing effects can also be analyzed dynamically, using the method proposed in our previous paper [12] and also by Schnerr et al. [11].

Although these recent works on SLS have improved the way of timing analysis, to the best of our knowledge, still no previous work has tackled the problem caused by compiler optimizations, which might lead to inaccurate simulation of compiler-optimized software.

2.3 Binary (Assembly) Level Simulation

For a binary level representation of a program, which is executable on the host machine, the target binary program is translated to either host instructions or a high level language, with the same functionality as the original program. Since both the functional representation and the timing information are generated from the binary level, their coupling is straightforward. Thus, the simulation accuracy depends solely on timing analysis, and in principle, is close to the simulation accuracy of ISSs, if the important timing effects are taken into account [7]. Bammi et al. [1] used inaccurate instruction latencies, and thus, also got inaccurate BLS.

BLS also has disadvantages. Control flows of indirect branches are hard to be constructed statically in a high-level language, because the targets of indirect branches can be resolved only at runtime. Zivojnovic et al. [13] and Lazarescu et al. [7] treat every instruction as a possible target of an indirect branch and thus set a label before the translated code of each instruction. Nakada et al. [10] propose to give control to an ISS when an indirect jump is met. Neither solution is simple and efficient enough. To map the target address space to the memory of the host machine, a memory model is needed. Frequent accesses to such a memory model reduce simulation speed significantly.

2.4 IR Level Simulation

An IR level representation has accounted for processor-independent optimizations and allows for simulation as fast as SLS. Yet, previous work on IRLS has not found a way to describe the mapping between IR and binary code. Without the mapping information, timing information must be estimated also at the IR level for accurate back-annotation. Kempf et al. [6] associate each IR operation with a latency. This latency has not taken into account the timing effects of the compiler back-end and the target processor. Cheung et al. [2] also associate each IR operation with a latency. Then, to take the processor-dependent optimizations into account, they use a SystemC back-end that emulates the back-end of the cross-compiler and optimizes both IR operations and their associated latencies. The output of the SystemC back-end is a simulation model in SystemC, annotated with timing information that has accounted for all the compiler optimizations. A similar approach is proposed by Lee et al. [8]. They modify the back-end of the host compiler to optimize both IR operations and their associated latencies. The output simulation model is a host binary program. Still, the both approaches cannot take into account the timing effects of the target processor. They are compiler-intrusive and need a large effort to write the SystemC back-end or modify the host compiler back-end. Further, neither back-end can behave the same as the back-end of the cross-compiler.

3. THE ISCISIM APPROACH

We propose an approach, called *iSciSim (intermediate source code instrumentation based simulation)*, which allows for fast and accurate simulation of compiler-optimized software. The whole approach consists of three steps: ISC (*intermediate source code*) generation, instrumentation, and simulation, as shown in Figure 1. ISC is a low level representation of source code, generated by converting IR to C code. It is, in principle, an IR level representation, but can be executed as normal source code. When compiled with a C compiler, it shows exactly the same functionality as the original source program. Timing analysis is done on the binary code generated from the ISC, both statically for pipeline effects and dynamically for other timing effects that cannot be resolved statically. Because the ISC has a structure close to the binary code, there is an accurate mapping between the ISC and the binary code, which makes up the basis for accurate back-annotation of the obtained timing information into the ISC. The whole approach can be realized easily, without the need of modifying the compiler. Since source code in different high-level programming languages generate IR in the same form, our approach is applicable for programs written in different languages. The following sub-sections present more details of the respective steps.

3.1 ISC Generation

As already mentioned, ISC is generated by modifying IR, which can be dumped by a cross-compiler during compilation. We use GCC (GNU Compiler Collection) compilers to present our work, but the approach is general enough for other compilers that operate on IR for optimizations and binary code generation. In the optimizer framework of the latest GCC compilers, the processor independent optimizations are performed on IR in GIMPLE form, which is similar to *three-address code*, and then the GIMPLE code is lowered to RTL code for processor-dependent optimizations.

978-1-60558-497-3/09 $25.00 © 2009 ACM

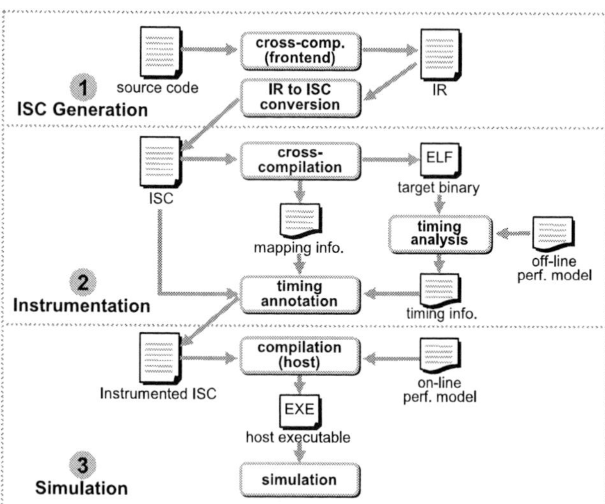

Figure 1: The iSciSim Approach

In Figure 2(b) we use a simple example to show how IR looks like. The shown IR is already the one after all the passes of processor-independent optimizations and has a structure very close to the binary code. The high-level C statements in the original source code (Figure 2(a)) are broken down into 3-address form, using temporary variables to hold intermediate values. For example, the *while* loop is unstructured with an *if-else* statement realizing a conditional branch. As GIMPLE syntax is very close to C syntax, the conversion from IR to C code is quite simple. As shown in the example, the IR is converted to the ISC (Figure 2(c)) after a slight modification of labels, variable names, and the representation of arrays.

ISC has almost the same execution time as its original source code. This has been validated with our experiments (cf. Section 4.1).

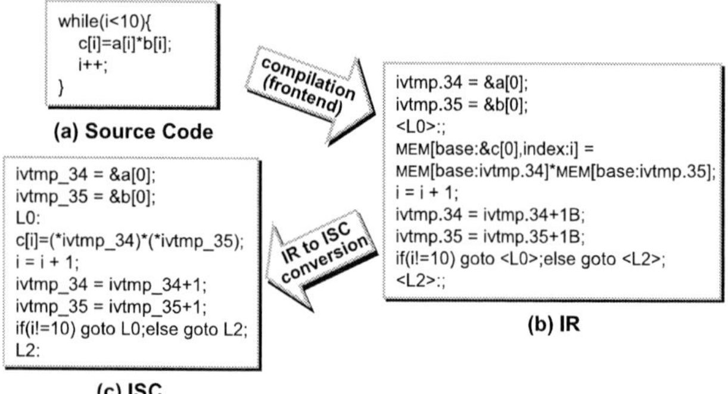

Figure 2: ISC Generation: An Example

3.2 Instrumentation

Instrumentation is regarded as the process of generating a timed simulation model by coupling the functional representation of a program and the performance model. As shown in Figure 1, the work flow of instrumentation contains further three steps: mapping information generation, static timing analysis, and back-annotation of the timing information into ISC according to the mapping information. In Figure 3 the same example is used to illustrate the instrumentation approach. Each step is described as follows.

3.2.1 Mapping Information Generation

For the purpose of accurate instrumentation an efficient way to describe the mapping between the functional representation and the binary code is very important. Most compilers do not provide any information to describe the mapping between IR and binary code. Our solution is very simple. We compile an ISC program again to generate debugging information describing the mapping between the ISC and the generated binary code. As the structure of the ISC has been highly optimized, when compiled again the compiler front-end cannot change it any more.

Figure 3(c) shows a part of mapping information extracted from the debugging information dumped during compilation of the ISC in Figure 3(a). The entry "19→0x18000b4" denotes that the basic block, the last instruction of which corresponds to the memory address "0x18000b4", is generated from a segment of C code, which ends with source line 19.

3.2.2 Static Timing Analysis

Each instruction is associated with a latency. Such instruction latencies are normally specified in the processor's manual. However, the execution time of a sequence of instructions cannot be estimated by simply summing up the instruction latencies, because timing effects of the processor architecture can change the instruction timing.

We use a hybrid approach for timing analysis, namely a mixture of static analysis and dynamic analysis, proposed in our previous paper [12]. Correspondingly, the performance model is also divided into two parts: the off-line performance model and the on-line performance model, which capture *local timing effects* and *global timing effects*, respectively. Pipeline effects are a typical example of local timing effects. When dispatched into a pipeline, only adjacent instructions affect each other. The sequence of instructions in a basic block is known at compile-time, so pipeline effects like data hazards, structural hazards and superscalarity can be analyzed statically in the scope of the basic block. The only approximation is made for the first instructions of the basic block, the timing of which depends on the execution context set by the previous basic block. The advantage of static analysis is that the analysis is needed only once for each instruction, while dynamic analysis must be repeated every time when the same instruction is executed. Figure 3(d) shows the delay of the sample basic block ($cycles = 8$), estimated by analyzing the timing effects of a superscalar pipeline. In addition, we can also extract other useful information from the binary code for other analyses of interest. In the shown example, the number of instructions is also counted.

Global timing effects are highly context-related and different execution paths set different context scenarios for an instruction. They cannot be analyzed statically without complicated control flow analysis. Because the control flow depends on concrete input data during simulation, these global timing effects must be analyzed dynamically. The cache effect and the branch prediction effect are typical global timing effects. In most cases, reasonable approximations can be made for the global timing effects to avoid dynamic timing analysis. For example, we can assume optimistically that all memory accesses hit cache. If an execution of a program has a high enough cache hit rate, this approximation will cause only a small error, which is acceptable in system level design. Otherwise, dynamic analysis is needed. More details about dynamic analysis are presented later.

978-1-60558-497-3/09 $25.00 © 2009 ACM

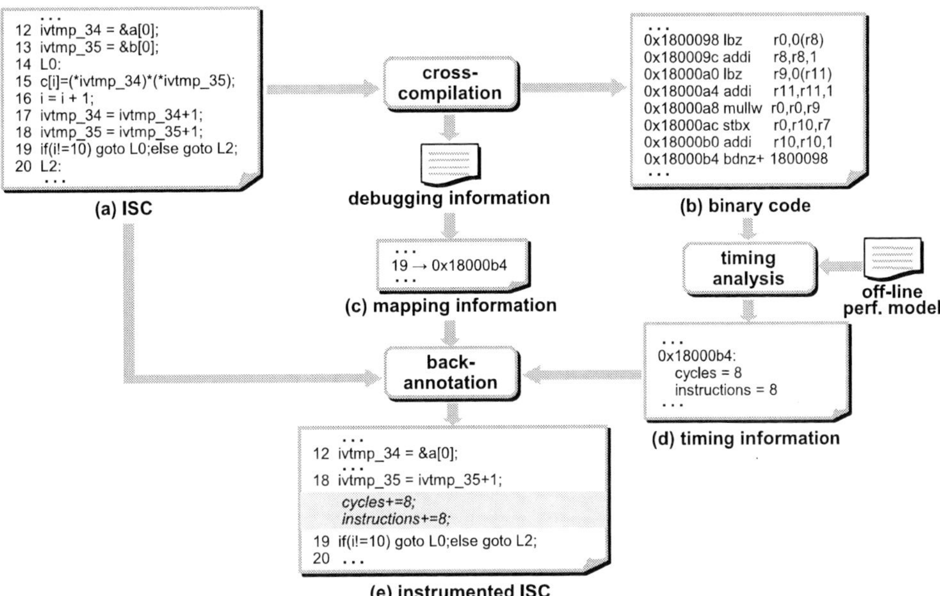

Figure 3: Instrumentation: An Example

3.2.3 Back-Annotation

In this step, we make use of mapping information to insert timing information into ISC. The annotation tool reads ISC line by line until a line that is recorded in the mapping information, e.g., line 19 in the example, is reached. Then, the timing information of the corresponding basic block is inserted before this ISC line. In the example, the timing information of the sample basic block is inserted before line 19. Figure 3(e) shows the instrumented ISC. The numbers of CPU cycles and instructions are counted by two global variables *cycles* and *instructions*, respectively.

Because the control flows of a program at the IR level and the binary level are not exactly the same, the simple algorithm for back-annotation, introduced above, might lead to estimation error. According to our experiments in 4.2, this error is very small. Still, we can use flow analysis of the two control flows to eliminate this error. The flow analysis is not presented here for space reason.

3.3 Simulation

For simulation, the instrumented ISC is compiled and executed on the host machine. During the simulation, the timing information of each basic block is accumulated by the counters to generate the performance statistics of the whole program. In this sub-section, we present dynamic timing analysis and HW/SW co-simulation using SystemC.

3.3.1 Dynamic Timing Analysis

To couple ISC with the on-line performance model, code that triggers dynamic timing analysis is annotated into the ISC. During simulation, the annotated code passes run-time data to the performance model for timing analysis. As an example, we show in Figure 4 how to simulate instruction cache behaviors. At runtime, the addresses of the first instruction and the last instruction of the basic block are sent to an instruction cache simulator by calling *icache(UInt32 first_addr, UInt32 last_addr)*. The cache simulator then decides how many cache hits and how many cache misses are caused by this basic block according to the present state of

the cache. If a cache miss occurs, the cache miss penalty is added to the cycle counter.

Data cache simulation is more complicated, because target data addresses are not known in ISC. A simple solution proposed in [8] is to use a statistical data cache model, which generates cache misses randomly, without the need of data addresses. An alternative is to annotate code, describing register operations, to generate target data addresses at run-time, as proposed in [12]. In this paper, we propose another method, which is more accurate than the former solution and allows for much faster simulation than the latter one. We use data addresses in the host memory space for data cache simulation. It has been validated that data of the same program in the host memory and in the target memory have similar spatial and temporal localities. As shown in Figure 4, *dcache_read(_write)(UInt32 addr, int data_length)* sends host data addresses for data cache simulation.

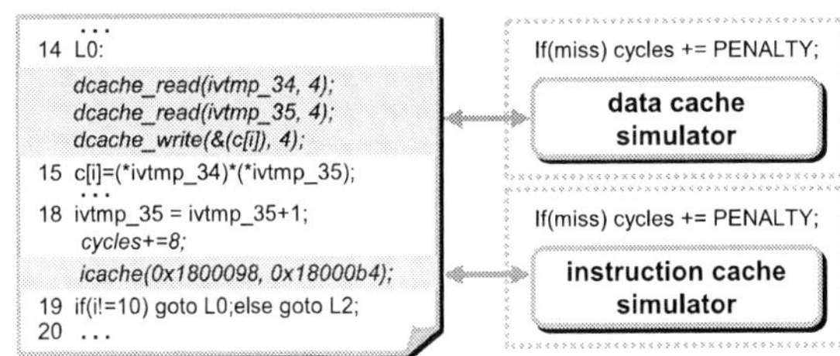

Figure 4: Cache Simulation

In [12], dynamic analysis of the branch prediction effect is also presented. These methods of dynamic timing analysis proposed for SLS are also applicable for our approach. Detailed introduction to all of them is not presented in this paper for space reason. Using dynamic timing analysis, simulation accuracy is gained at the expense of simulation speed. Users should handle this trade-off and decide which timing effects are worth dynamic analysis.

3.3.2 HW/SW Co-Simulation

To model a concurrent system, where multiple processors and hardware components run parallelly, SystemC provides

978-1-60558-497-3/09 $25.00 © 2009 ACM

a desirable simulation framework. To be integrated into a SystemC simulator, the instrumented ISC of each software task is also written in SystemC. There is actually no large difference between the instrumented ISCs in C and in SystemC. The latter has only more code for declaring itself as a SystemC thread. Timing information is still aggregated using variables, instead of being replaced by SystemC *wait()* statements. A wait() statement is generated to advance the simulation time, only when the software task has to interact with external devices. For example, as illustrated in Figure 5, a wait() statement is generated, when there is an access to an external memory in the case of a cache miss. In this way, the number of wait() statements is reduced significantly and the time-consuming context switches between the software thread and the SystemC simulation kernel caused by wait() statements are minimized.

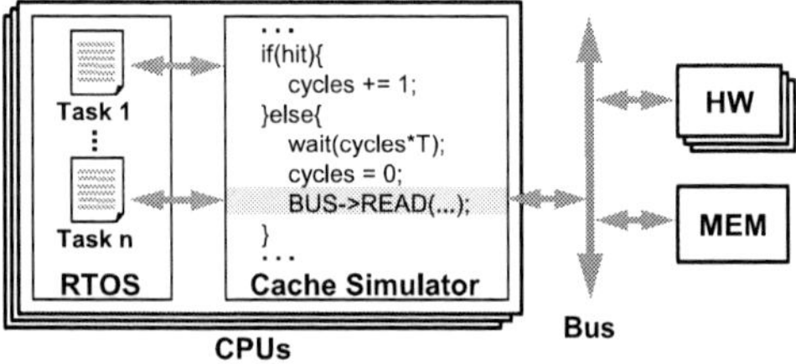

Figure 5: HW/SW Co-Simulation

For simulation with other system components, such as RTOS, peripherals and bus, additional code must be annotated during instrumentation to trigger interactions between software tasks and these system components. Figure 5 gives a highly simplified view of a multiprocessor system and shows an example of modeling accesses to the shared memory. Of course, to simulate such a system, many other complicated issues have to be tackled. We do not go into the details in this paper.

4. EXPERIMENTAL RESULTS

The experiments have been carried out mainly to compare the discussed software simulation strategies and show the benefits of the proposed approach, using 6 benchmark programs with different workloads. In addition, we studied the difference between source code and ISC and also did simulation in SystemC. PowerPC 603e, a superscalar design of the PowerPC architecture, was chosen as the target processor. All the programs were cross-compiled using a GCC compiler with the optimization level -O2. The performance was measured on a 3 GHz Intel CPU based PC with 4 GB memory, running Linux.

4.1 Source Code vs. ISC

As timing analysis is performed on the binary code generated from ISC, a justified concern is whether this binary code is the same as the one generated from the original source code. In the first experiment we have compared the execution times of ISC and source code, using an ISS developed by us. We tested all the 6 benchmarks. As shown in Table 1, the ISCs of 2 benchmarks took exactly the same number of CPU cycles to execute as their source codes. By testing the other 4 programs, a maximal error of -1.92% was found. The average absolute error was 0.63%. This means that the ISC generation might change the temporal behavior of a program, but this change is small enough to be ignored.

4.2 Benchmarking SW Simulation Strategies

In addition to the ISS, we also implemented a BLS approach introduced in [10], which can represent the state-of-the-art. Our SLS tool introduced in [12] can also represent the state-of-the-art of the SLS technique. iSciSim was compared with these related techniques.

We evaluated the four approaches quantitatively, in terms of simulation accuracy and speed. Simulation accuracy was measured by both instruction count accuracy and cycle count accuracy. Instruction count accuracy depends solely on the mapping between functional representation and binary code, while cycle count accuracy depends on both the mapping and the timing analysis.

All the estimates are shown in Table 1. The estimates obtained by the ISS have taken all the timing effects into account and are assumed to be very close to the real execution times, while the other tools performed only static pipeline analysis. Since all the programs are of small size and the target processor has large caches, the timing information obtained by only pipeline analysis without cache simulation was still very accurate. As the mapping between binary-level functional representation and binary code is straightforward, BLS resulted in the instruction count accuracy of 100% for all the programs. It had an average absolute error of 0.83% in cycle counting. SLS was able to count instructions and cycles accurately for data-intensive applications, like *blowfish* and *AES*, where control flows are relatively simple. For the other programs, large errors are seen, because the control flows of these programs were changed by compiler optimizations and the mapping between the binary code and the source code was destroyed. As expected, iSciSim allowed for very accurate simulation with only an average error of 0.99% in instruction counting and an average error of 1.04% in cycle counting.

The errors in the estimates from iSciSim were mainly caused by the ISC generation, the approximations made in timing analysis, and the back-annotation of the timing information. In the last sub-section, it was already found that an average absolute error of 0.63% was caused by the ISC generation. By comparing the cycle counts from the ISS and the BLS, we got the error caused by the simplification of timing analysis, which was 0.83% in average. The mapping error introduced by the back-annotation using the simple algorithm can be measured by comparing the execution times of ISC estimated by the BLS and iSciSim. For the given programs, this error was 0.53% in average.

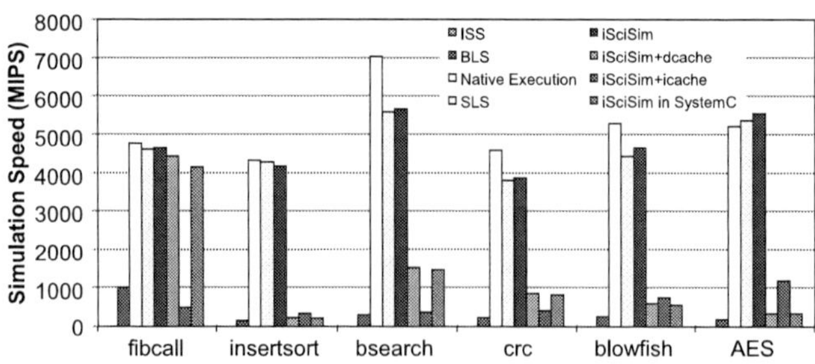

Figure 6: Execution Speeds

Figure 6 shows the speeds of native execution and simulations, measured in MIPS (million instructions per second). The simulation speed of the ISS is too low to be seen in the figure, only 11.9 MIPS in average. the SLS and iSciSim al-

978-1-60558-497-3/09 $25.00 © 2009 ACM

Table 1: Experiment Results

		S ce C de . ISC			Be ch a i g SW Si a i S a egie					
		ISS(SC)	ISS(ISC)		BLS		SLS		iSciSi	
		E i a e	E i a e	E	E i a e	E	E i a e	E	E i a e	E
fibcall	instructions	4384	4384	0.00%	4384	0.00%	12973	195.92%	4356	-0.64%
	cycles	3307	3307	0.00%	3281	-0.79%	12996	292.98%	3263	-1.33%
insertsort	instructions	630	631	0.16%	630	0.00%	975	54.76%	631	0.16%
	cycles	516	508	-1.55%	518	0.39%	672	30.23%	509	-1.36%
bsearch	instructions	59010	59010	0.00%	59010	0.00%	75013	27.12%	58010	-1.69%
	cycles	46057	46057	0.00%	47008	2.06%	51008	10.75%	46008	-0.11%
crc	instructions	17201	17457	1.49%	17201	0.00%	19385	12.70%	17313	0.65%
	cycles	10288	10289	0.01%	10284	-0.04%	15430	49.98%	10313	0.24%
blowfish	instructions	262434	260913	-0.58%	262434	0.00%	265735	1.26%	260899	-0.58%
	cycles	178610	179141	0.30%	175919	-1.51%	187184	4.80%	175975	-1.48%
AES	instructions	3624426960	3544426960	-2.21%	3624426960	0.00%	3624685359	0.01%	3544422940	-2.21%
	cycles	2292275341	2248279321	-1.92%	2296272902	0.17%	2336484895	1.93%	2252276071	-1.74%

low for simulation with average simulation speeds of 4669.0 MIPS and 4765.6 MIPS, respectively, almost as fast as the native execution (5198.2 MIPS in average) and much faster than the BLS (349.4 MIPS in average).

Although cache simulation did not increase the simulation accuracy due to small sizes of the selected programs, we studied how cache simulation affects the simulation performance. As shown in Figure 6, the simulation speeds are slowed down to 1326.4 MIPS and 581.0 MIPS in average, because of data cache simulation and instruction cache simulation, respectively.

4.3 Simulation in SystemC

The last experiment was to study the slowdown of simulation speed caused by wait() statements, when software simulation models are wrapped in SystemC for simulation of a whole system. If one wait() statement was annotated for each basic block, the average simulation speed was slowed down to 4.8 MIPS. If a wait() statement was generated only before an access to the external memory in the case of a data cache miss, the simulation ran at an average speed of 1302.7 MIPS, slightly slower than the simulation in C (1326.4 MIPS in average).

We also simulated a video compression application, which has 12 tasks mapped to four 100 MHz PowerPC cores. Using the iSciSim approach, one SystemC simulation model was generated for each task. These simulation models were combined with transaction level models of RTOS, bus, memory and IO devices. The simulation took 117 seconds to compress 300 video frames in YUV format. The simulated time of the target execution was 243 seconds. Here, we only give information on the speed of such a HW/SW co-simulation. Because HW/SW co-simulation is not the focus of this paper, a detailed description of this case study is out of scope.

5. CONCLUSIONS

This paper presented an approach for fast and accurate timing simulation of compiler-optimized embedded software. We established an automatic process of generating software simulation models. Compared with ISSs and BLS, the proposed approach allows for simulation at much higher speed without compromising accuracy. Compared with SLS, the proposed approach is able to simulate compiler-optimized code with much higher accuracy. Experiments have been presented to validate this. The proposed approach has also a limitation that it requires source code to be available.

6. REFERENCES

[1] J. R. Bammi, W. Kruijtzer, L. Lavagno, E. Harcourt, and M. Lazarescu. Software performance estimation strategies in a system-level design tool. In *Proceedings of the International Workshop on HW/SW Codesign*, 2000.

[2] E. Cheung, H. Hsieh, and F. Balarin. Framework for fast and accurate performance simulation of multiprocessor systems. In *Proceedings of IEEE International Workshop on High Level Design Validation and Test*, 2007.

[3] M.-K. Chung, S. Yang, S.-H. Lee, and C.-M. Kyung. System-level HW/SW co-simulation framework for multiprocessor and multithread SoC. In *Proceedings of IEEE VLSI-TSA international symposium on VLSI Design, Automation and Test*, 2005.

[4] F. Fummi, G. Perbellini, M. Loghi, and M. Poncino. ISS-centric modular HW/SW co-simulation. In *Proceedings of the ACM Great Lakes symposium on VLSI*, 2006.

[5] P. Giusto, G. Martin, and E. Harcourt. Reliable estimation of execution time of embedded software. In *Proceedings of the conference on Design, Automation and Test in Europe*, 2001.

[6] T. Kempf, K. Karuri, S. Wallentowitz, G. Ascheid, R. Leupers, and H. Meyr. A SW performance estimation framework for early system-level-design using fine-grained instrumentation. In *Proceedings of the conference on Design, Automation and Test in Europe*, 2006.

[7] M. Lazarescu, M. Lajolo, J. Bammi, E. Harcourt, and L. Lavagno. Compilation-based software performance estimation for system level design. In *Proceedings of the International Workshop on HW/SW Codesign*, 2000.

[8] J.-Y. Lee and I.-C. Park. Timed compiled-code simulation of embedded software for performance analysis of SOC design. In *Proceedings of the Design Automation Conference*, 2002.

[9] T. Meyerowitz, M. Sauermann, D. Langen, and A. Sangiovanni-Vincentelli. Source-level timing annotation and simulation for a heterogeneous multiprocessor. In *Proceedings of the conference on Design, Automation and Test in Europe*, 2008.

[10] T. Nakada, T. Tsumura, and H. Nakashima. Design and implementation of a workload specific simulator. In *Proceedings of the annual Symposium on Simulation*, 2006.

[11] J. Schnerr, O. Bringmann, A. Viehl, and W. Rosenstiel. High-performance timing simulation of embedded software. In *Proceedings of the Design Automation Conference*, 2008.

[12] Z. Wang, A. Sanchez, and A. Herkersdorf. Scisim: A software performance estimation framework using source code instrumentation. In *Proceedings of the International Workshop on Software and Performance*, 2008.

[13] V. Zivojnovic and H. Meyr. Compiled HW/SW co-simulation. In *Proceedings of the Design Automation Conference*, 1996.

978-1-60558-497-3/09 $25.00 © 2009 ACM

MPTLsim: A Simulator for X86 Multicore Processors

Hui Zeng, Matt Yourst, Kanad Ghose, Dmitry Ponomarev
Department of Computer Science
State University of New York, Binghamton NY 13902-6000

{hzeng, yourst, ghose, dima}@cs.binghamton.edu

ABSTRACT

Current microprocessors are effectively a system-on-a-chip, as they incorporate processing cores, interconnections, shared and private caches and DRAM controllers on a single die. Consequently, it is imperative to have fast and accurate simulation tools for such systems; this paper such a tool for simulating all current and announced variants of multicore processors that use the predominant PC (X86, X86-64) instruction set, as well as external DRAM memory and buses. We discuss the major techniques used for speeding up the simulation and improving the overall accuracy, and the simulation of system-level details such as coherent caches, on-chip interconnections, memory bus and DRAM. We also demonstrate a 8-fold speedup against a widely-used popular tool.

Categories and Subject Descriptors

B.5.2 [Register-transfer-level Implementation]: Design Aids - *Simulation* and B.3.3 [Memory Structures]: Performance Analysis and Design Aids - *Simulation*

General Terms: Measurement, Design

Keywords

Simulator, microprocessor, coherent cache.

1. INTRODUCTION

The microprocessor industry went through a paradigm shift in the recent years from a single core, complex design that pushed the power envelope and clock rates to multicore designs with simpler cores. However, since the introduction of the first desktop/server multicore products (AMD's dual-core Opteron and dual-core Athlon) that essentially put two isolated cores with their private caches into a single die, multicore designs have become increasingly complex. Two imminent product offerings from Intel are good examples of the increasingly complex multicore products. The Intel Nehalem line of CPUs [2] incorporate 6 or more cores, two levels of private caches that support data coherency, on-chip DRAM controllers, interconnections for cache coherence and a fast point-to-point interconnection among the cores and the memory system. The Intel Larabee line of products [15] feature two types of cores - general purpose cores and graphics processing cores (GPUs), with coherent caches and a fast

Permission to make digital or hard copies of part or all of this work for personal or classroom use is granted without fee provided that copies are not made or distributed for profit or commercial advantage and that copies bear this notice and the full citation on the first page. To copy otherwise, to republish, to post on servers or to redistribute to lists, requires prior specific permission and/or a fee.
DAC'09, July 26-31, 2009, San Francisco, California, USA

on-chip ring interconnection.

In effect, these emerging multicore designs are systems on a chip, albeit with components that are at a relatively higher-end than some of the components associated with SoC designs for the embedded market. It has therefore become imperative to have design tools that accurately simulate existing and emerging multicore chips for studying design tradeoffs, new designs, energy estimation and power-performance tradeoffs. We present such a tool, called MPTLsim, targeting the widely used PC instruction set architecture (X86 and X86-64 ISA), that is cycle accurate and also a full-system simulator, in that it simulates application as well as operating system and library codes. MPTLsim also supports native mode execution and this feature is used to provide ultra-fast simulation performance. A preliminary overview of MPTLsim was presented in [19] and this paper focuses on the techniques used for speeding up simulations and for simulating system-level components such as interconnections, coherent caches and DRAM memory systems.

Our goal was to leverage on a cycle-accurate, full-system mode simulator for the X86 and X86-64 ISA that was developed and publicly released [18] as a staring point for developing a cycle accurate, full system simulator for multicore implementations of the X86-64 ISA running on Linux. The use of the X86-64 ISA provides ready access to the full suite of compiler, programming and debugging tool suites that are being constantly developed and upgraded at a time when similar support for other ISAs is becoming unavailable or relatively limited in availability.

2. SIMULATION INFRASTRUCTURES

In this section, we discuss the various infrastructures that are used for our multicore simulation framework.

2.1 The Virtual CPU Mechanism

MPTLsim uses a domain built on a hardware abstraction provided by the Xen hypervisor. Figure 1 shows the relationship of the domains, the virtualized hardware, the hosted kernels and the hypervisor. Domain 0 has a special implication - and provides access to the real drivers and the outside world. All I/O activities from other domains are directed through Domain 0 by forwarding the system calls for I/O to Domain 0. All other domains have a hardware virtualization supported by the core PTLsim components. This virtual hardware supports multiple virtual CPUs (VCPUs) that are implemented using the PTLsim simulation components. When a domain is executing benchmark applications in the native mode, the PTLsim-based VCPUs are effectively bypassed. The VCPUs within a domain effectively implement a process context ("thread" in the industry jargon), that globally share the common address space of the domain. This makes it possible to implement simultaneously multithreaded designs, multiple cores, and multiple multithreaded cores easily using the basic VCPU rubric. Figure 1 also depicts the various components used in implementing MPTLsim.

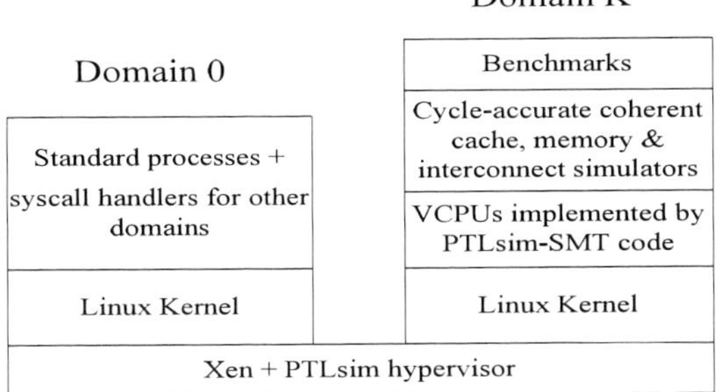

Figure 1: Components of the MPTLsim Framework

MPTLsim builds on PTLsim's approach of using virtual machine technology to host multiple domains on a hypervisor layer. Each domain can run its own kernel on a hardware abstraction provided by the hypervisor; each domain also has an address space exclusive to that domain.

2.2 Cycle-Accurate Simulation

MPTLsim makes use of a detailed uop-accurate, cycle accurate out-of-order (OOO) core design implementing the X86-64 ISA. The simulation loop in this case simulates a processor cycle and round-robins through all the VCPUs to simulate events in the same cycle within each VCPU, the cache and memory, the interconnection and the coherence mechanisms. In spite of this, as we see later, we get about an average speedup of 7-8 times (depending on the number of the simulated cores) against GEMS simulator [13] running in a very similar mode.

2.3 Interrupts and Full System Simulation

To support full system simulation, interrupts need to be correctly simulated in time. As the simulator is running at a slower pace than the real system, interrupts occurring in real-time need to be simulated properly, albeit at the slower simulation pace. To correctly implement interrupt handling by the VCPUs, we use Xen's checkpointing mechanism to checkpoint the state of the VCPUs and modify the device drivers to log the device interrupts and timestamp them with the simulation cycle counter. To process and serve the interrupt, we resume execution from the nearest checkpoint established just prior to the interrupt's time of occurrence and re-execute instructions and the interrupt (which is effectively a *virtual interrupt*!) from the log file - at the *simulation speed*.

3. SPEEDING UP SIMULATIONS

Simulating complex multicore systems in a cycle accurate fashion, as needed for some design tradeoff studies and accurate energy estimation is challenging, as all components need to be simulated and the effect of events in one components can affect one or more of the other components. Several techniques are used in MPTLsim to speed up the simulation:

Native Mode Execution: Since MPTLsim simulates each and every aspect of the X86 and X86-64 ISA (including the decomposition of instructions into micro-operations, uops, within the hardware), the simulator can switch transparently and seamlessly between cycle-accurate full system simulation and execution on the native hardware. Uninteresting portions of the applications being simulated can be skipped in the native mode,

speeding up the overall simulation speed. The availability of fast multicore platforms that simulate the same instruction set is an asset in this respect.

Optimized Simulator Code: Much of the simulation code for the processing cores and caches are well-optimized for caching and execution, including the use of SSE instructions to speed up concurrent operations in the simulation model. Some components that have relatively higher latencies, such as interconnections, DRAM etc are essentially simulated using an event-based infrastructure, but to always maintain the cycle-level accuracy.

Interest-directed Optimization: The simulation of several components in the simulator can leave out details not of direct interest for a specific use of the simulator. MPTLsim is consciously designed to leave these details out where needed. Some examples of this include abilities to turn off instrumentation for performance and energy optimization, to turn off simulation of cache and interconnection buffer contents, to turn off specific cores, to turn off dynamic frequency scaling and so on.

4. SUPPORTING HYPERTHREADING

To support a simultaneously multithreaded (SMT, aka hyperthreaded) core, the OOO single core PTLsim was extensively modified to implement an Intel P4 style hyperthreaded datapath and some variants. Each VCPU maintains a context structure that holds all information about the VCPU/thread context. This includes the values of the X86 architectural registers, X86 machine state registers (MSRs), page tables (as PTLsim implements its own memory management mechanism) and various PTLsim internal state variables. One of these state variables in the context is a variable called running. Suspended contexts - that were suspended on the execution of the X86 hlt instruction, awaiting an external interrupt, for instance or similar blocked on hypercalls - wait on this running flag. When this flag is set, a VCPU resumes its processing. This flag is used in implementing virtual interrupt handling and also during the boot process for the VCPUs.

The basic core design and other datapath components have to be modified as well to support multithreading. A number of core configurations can be simulated, including various combinations of private and shared datapath resources and state-of-the art instructions fetching schemes for SMT, such as I-count [10] or DCRA [6]. The SMT core design also needs to implement atomic operations correctly to preserve the overall semantics of hyperthreading at the simulation and full system level. The X86 ISA allows the uses of the LOCK prefix to execute memory instructions atomically. The xchg and xadd instructions and test-and-set instructions, as well as memory barrier instructions (like sfence) require similar indivisible memory operations.   The original PTLsim design was modified to support these operations correctly in the face of multiple requests by the hardware or the VCPUs and to arbitrate among the atomic accesses in a fair manner across the contenders.

5. MULTIPLE CORES AND COHERENT CACHES

The implementation of the multicore version of PTLsim is based on the SMT datapath design described in the previous section. A maximum of 16 VCPUs can be supported as threads, as various combinations of cores and threads on the cores, ranging from all

978-1-60558-497-3/09 $25.00 © 2009 ACM

16 threads on a single core to a single thread on each of 16 cores, covering all combinations in-between. The main challenge in extending the SMT cores to support multicore designs is centered on the design of the memory hierarchy, including up to two levels of caches and the incorporation of support for cache coherence. We support the models with both private and shared caches at various levels, model support for cache coherency with MESI protocol, and also model various interconnection network designs, such as atomic bus, split-transaction bus or a crossbar switch supporting broadcast.

5.1 Initialization Steps

Two configuration variables are used to specify the multicore configuration and the number of threads running on a core: threads_per_core and number_of_cores. The multithreaded application to be run can specify an affinity of a thread to a particular core, with cores numbered from 0 onwards. When no specific affinity is specified, the threads are simply allocated to cores based on their thread ids. Core affinities and thread ids are specified to the MPTLsim hypervisor through a system call and these are eventually stored as part of the context structure for a thread.

5.2 Caches and Coherence Controller

The original PTLsim design implements caches as an integral part of each core and implements only the tag part of the cache explicitly, with no explicit structures to hold the data part of the caches, as in other simulators, such as [3]. To ease the configuration of caches, including shared caches and the implementation of a cache coherency protocol, all caches are implemented as a separate global structure, distinct from the cores. When two levels of caches are present, a configuration parameter, number_of_core_per_l2 is used to specify how many cores share a L2 cache. Additionally, the data part of the caches is also implemented explicitly to hold both the contents of a cache line and state information in accordance with the coherence protocol.

All of the currently supported cache coherency mechanisms support the sequential consistency model. This requires a global serialization mechanism for memory accesses that needs to be implemented via the local cache controllers, the interconnection and the protocol implementation. A generic structure of the core-local coherence controller for L2 caches, as implemented for MPTLsim is shown in Figure 2. The structure for the coherence controller for L1 caches is similar. Careful attention was paid to the design of the coherence controller to permit them to be adapted to a wide variety of coherence protocols, including directory based protocols.

Each controller maintains a separate set of request queues (in FIFO order) from the core side towards the lower levels – one queue for each of the threads running locally. The use of separate queues permits the implementation of a variety of request scheduling strategies via what is depicted as the "input scheduler" in Figure 2. Once such a request is selected, a "dependence checker" checks if the request is to a cache line with a pending cache miss, stored in a separate pending "request table" within each controller. Each entry in this table consists of a cache tag and index, status flags and a pointer to a list of pending requests queued up on this line. The total number of entries in this table is set to the maximum number of outstanding requests per core, as specified in a configuration parameter. If the dependence checker detects a match with a pending cache line miss already processed off the request queues, the current request is appended to the linked list of pending requests on the same line, as maintained within the request table entry. Once the miss on a line is serviced, locally queued up requests on this line are also revived and serviced in their original request order. The cache controller design for MPTLsim also implements load forwarding and store merging (in a sequentially consistent manner) as part of the logic associated with the various request and response buffers depicted in Figure 2.

Access requests to the local cache can come from a variety of sources – from the local processor, a lookup request initiated by another controller (such as a snoop triggered by bus activity in a bus-based system), a cache lookup required when a response to a previous request comes in from an external source (the lower level or the memory or another cache, depending on the protocol) or dependent requests re-activated on the arrival of a missing line or a previously queued cache read or write initiated by the local core.

These requests are maintained in the "cache access buffer" shown in Figure 2 and serviced using a dedicated port. The read and write requests from the local core use a second dedicated port and if a port happens to be busy in a given cycle, the local read or write is queued up on the "cache access buffer" structure. Requests in this buffer are serviced in a specific order (from highest to lowest) as follows: lookup requests triggered by other controllers, responses to a previous request from an external source, dependent requests that are re-activated dependent requests and previously queued read or write requests from the local core. On cache accesses from either of the ports, state changes are triggered and require the use of protocol engine logic for updates to the state of a cache line. Updates to the data part of a cache line are also made, as necessary.

As in the case of the processor cores, the protocol and coherence controller activities are simulated on a cycle-by-cycle basis. Appropriate statistics are generated and logged for a large variety of events associated with the protocol actions. The nature of the information collected can be pre-selected. Finally, we note that the implementations of the coherence protocol for MPTLsim are sufficiently complex, given the nature of the protocols. Considerable effort was thus directed at verifying the implementations. The details of the verification are beyond the scope of this paper.

5.3 On-Chip Interconnections

At this time, MPTLsim provides three types of built-in interconnection models: (a) a shared bus providing atomic transactions; (b) a split-bus that decouples bus requests and responses using two separate buses – one for addresses and one for data. Both address and data buses have independent arbiters. The specific bus modeled is as described in [7]; (c) a crossbar switch that has separate networks for requests and responses. MPTLsim allows users to specify the duration of interconnection events in terms of an interconnection clock cycle, which in turn is specified as an integer divisor of the clock rate of each core.

978-1-60558-497-3/09 $25.00 © 2009 ACM

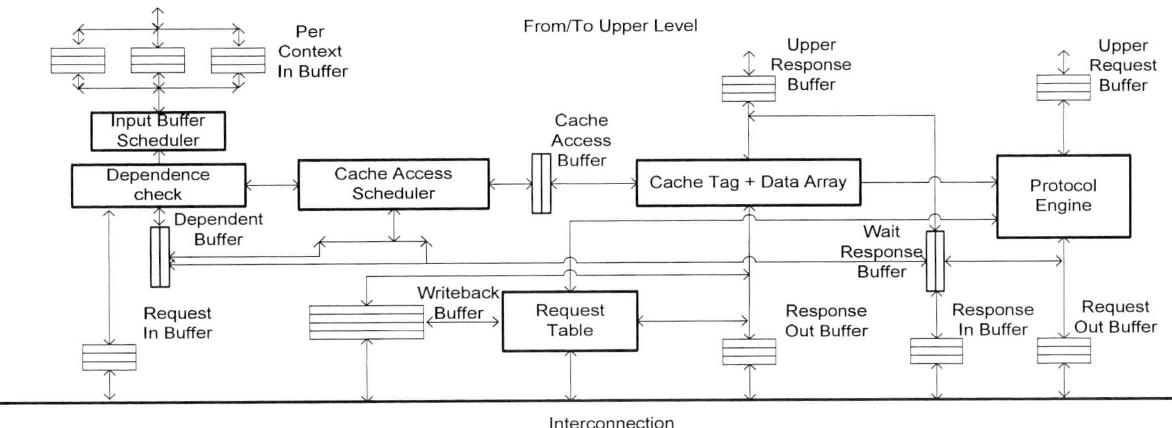

Figure 2: Modeled Per-Core Coherence Controller for L2 Caches

5.4 Memory Simulation

Unlike PTLsim, MPTLsim supports a memory model for external DRAMs. Multiple memory banks and multiple DMA channels are supported. Memory request pipelining on the various DMA channels is also supported. Facilities also exist to specify the characteristics of the interconnection to/from memory and the maximum burst sizes and to reorder the outgoing queue to memory. Additional code to support some memory prefetching schemes is also part of MPTLsim.

6. RELATED WORK

A large number of processor performance simulation tools have been developed in the last decade. One of the most popular cycle-accurate simulators used in the last decade is the Simplescalar toolset [3]. Simplescalar, as well as its variants [16], only supports the simulation of single-threaded workloads, such as SPEC benchmarks, in user-level mode. SESC simulator [9] models a variety of architectures, including dynamic superscalar processors, CMPs, processor-in-memory, and speculative multithreading architectures. However, SESC does not support full-system simulation mode nor does it model contemporary ISAs like X86. A-Sim [8] is a modular and reusable performance modeling framework developed at Intel for Alpha ISA.

While PTLsim is the first open source *cycle-accurate* simulator for X86 instruction sets at the level of uops, numerous *functional* simulators (tools that precisely emulate each instruction in the target instruction set, but do not provide any cycle-accurate timing information) have been developed for x86 [1, 4, 12]. Simics [12] is a commercial functional simulation suite for various processor families (including X86) as well as user-designed plug-in models of real hardware devices. Like QEMU [4], Simics uses X86-to-X86 binary translation to achieve good performance. However, Simics does not include cycle-accurate simulation features below the X86 instruction level (just like any other functional simulator). GEMS [13] is a full-system multiprocessor simulator that leverages the power of Simics [12] as an underlying foundation. In addition to the simulation facilities, GEMS incorporates extensive features to design, specify and validate new cache coherence protocols. This is a feature that all existing multicore simulators, including MPTLsim, lack at this time. For increased simulation speed,

GEMS can be used without the *opal* module by directly feeding the *ruby* module from Simics. For studies where the impact of the out-of-order execution features is secondary, such a configuration can significantly increase the simulation speed. When comparing the simulation speed of the multicore version of PTLsim versus GEMS, we used GEMS with and without *opal* module and report the results for both configurations in the results section. Since it is built on top of the freely available version of Simics, GEMS inherits all of its limitations such as the capability to only simulate the SPARC ISA. Another potential limitation stemming from the reliance on Simics is somewhat restricted flexibility due to the black-box nature of the functional model. Furthermore, licensing limitations can restrict the widespread adoption of the tool, particularly in non-academic environments. SimFlex [17] – another recently designed tool that uses Simics – potentially has similar limitations. Another approach to the design of full-system simulator is to add the full-system simulation capabilities to existing user-level timing simulators – this is exemplified by the M5 simulator [5].

Aside from simulations, another technique to analyze program behavior is through dynamic binary instrumentation. Instrumentation inserts extra code into the application to observe its behavior [11, 14]. Unfortunately, instrumentation does not model the details of the out-of-order processors and is not usable for modeling the non-existing hardware. Ideally, the advantages of instrumentation tools (such as easy prototyping and very fast runtimes) should be combined with the accurate models provided by the cycle-accurate simulators. In this respect, MPTLsim can serve as a perfect complement to dynamic instrumentation tools.

7. RESULTS AND DISCUSSIONS

To measure the simulation speed of MPTLsim, we experimented with parallel benchmarks from Splash-2 suite. In all of our experimental studies, we used configurations of individual cores as shown in Table 1. Unless stated otherwise, both L1 and L2 caches were private to the core and kept coherent using MSI protocol for L1 and MESI protocol for L2 using a split-transaction shared bus. The bus clock cycle was one fourth of the core clock cycle. We estimated the simulation rates of MPTLsim running on an Intel 2.4GHz dual-core 64-bit processor with 4 GBytes of DRAM. We also compared some of

978-1-60558-497-3/09 $25.00 © 2009 ACM

our results to a very similar configuration on the GEMS toolsuite. Figures 3 and 4 depict the number of processor cycles simulated per second for the Splash benchmarks executed on these simulators for 4-core and 8-core configurations. For GEMS, we report the results for two different simulation modes: a faster mode (labeled as *ruby-only* in Figures 3 and 4) that directly uses Simics to drive the GEMS memory module (*ruby*) and a slower mode (labeled as *ruby&opal*) that uses detailed out-of-order timing simulator of the core (*opal* module) to drive *ruby*.

Table 1: Configuration of the Simulated Processor Cores

Parameter	Configuration
Machine width	4-wide fetch, 4-wide issue, 4 wide commit
Window size	Clustered IQ, 16-entry each, 64 entry LSQ, 128–entry ROB
Registers	128 Int and 128 FP
Function Units	2 ALU, 2 LD/ALU, 2 ST/ALU, 2 FP/SSE/MMX
L1 I–cache	64 KB, 4-way, 1 cycle latency.
L1 D–cache	64 KB, 4-way at 1 cycle latency (2 LD, 2 ST)
L2 Cache unified	4 MB, 8-way set-associative, 8 cycles hit latency
BTB and branch predictor	1024 entry, 4–way set–associative. Minimum branch misprediction penalty – 10 cycles. Gshare predictor.
Memory	128-bit wide, 160 cycles first chunk, 2 cycles interchunk
TLB	64 entry (I), 128 entry (D), fully associative

Each benchmark was executed for 200 million committed instructions (in total, counting from all cores). All simulators were configured identically and the same number of instructions was skipped to warm up caches and branch predictors.

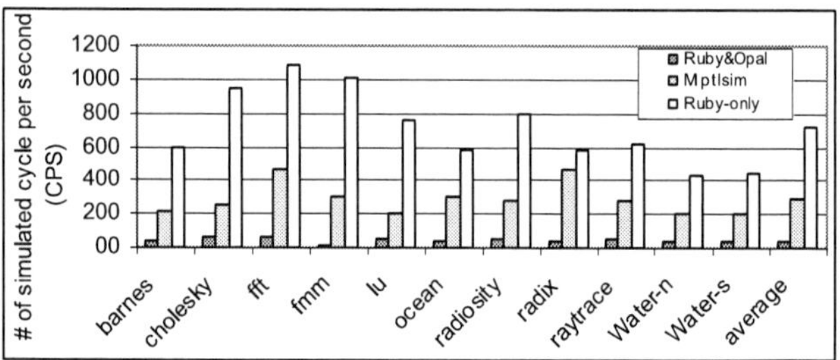

Figure 3: Simulation rates for GEMS and MPTLsim: 4 cores

The results of Figures 3 and 4 clearly demonstrate that the *ruby-only* mode provides the fastest simulation rate. This is no surprise, as this mode provides only the functional simulation of the cores. A more appropriate comparison would be between MPTLsim and *ruby&opal* version of GEMS. Note, however, that MPTLsim provides a cycle-accurate and uop-accurate simulation in contrast to the somewhat inexact timing simulation of opal [13]. In spite of this, MPTLsim provides a significantly faster simulation rate compared to *opal&ruby* variation of GEMS. On the average across all benchmarks, 4400 cycles are simulated in one second in the *ruby+opal* GEMS mode, 29500 cycles are simulated in MPTLsim, and 74200 cycles are simulated by "ruby-only" variation of GEMS. In other words, MPTLsim provides 6.7 times improvement in simulation speeds over GEMS and is about 2.5 times slower than the ruby-only simulations. Interestingly, adding the cycle-accurate out-of-order modeling capabilities to GEMS (through opal module) slows down the simulations almost by a factor of 17.

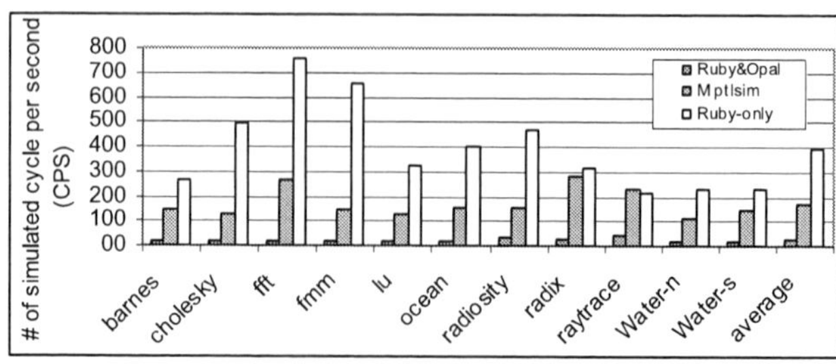

Figure 4: Simulation rates for GEMS and MPTLsim: 8 cores

For the simulated 8-core configuration as shown in Figure 4, *opal&ruby* GEMS simulates about 2200 cycles per second on the average, MPTLsim simulates about 17000 cycles per second and ruby-only variation of GEMS simulates about 41000 cycles per second. With larger number of cores shown in this graph, the relative overhead of cycle-accurate out-of-order simulations increases, compared to the situation shown in Figure 3. Specifically, MPTLsim is now 7.7 faster than the *opal&ruby* GEMS, but still only 2.7 slower than ruby-only model. In other words, the performance gap between MPTLsim and full-fledged GEMS increases much more significantly than the gap between ruby-only model and MPTLsim, as the number of cores increases. These trends are expected to continue as the number of simulated cores increases further, making MPTLsim attractive from the perspective of detailed cycle-accurate simulations and simulation speed. We believe that much of the performance advantages of MPTLsim come from its use of a highly-tuned simulator code and the exploitation of specific features of the X86 ISA in speeding up the MPTLsim code. Some of these reasons are elaborated upon in [18].

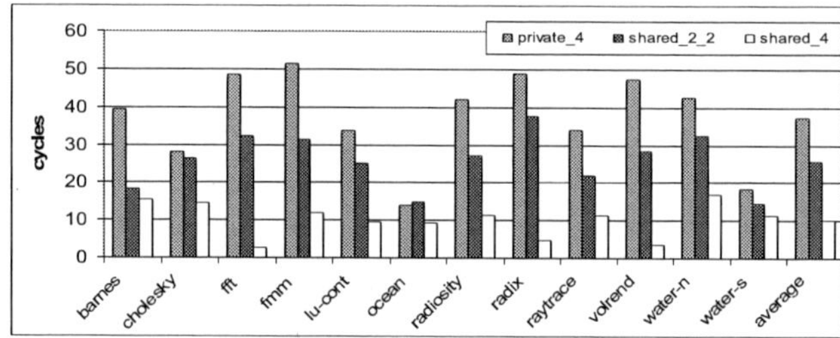

Figure 5: Average L2 read access latency for three L2 cache configurations

We now show some example studies possible with MPTLsim (Figures 5 and 6). This study at the impact of sharing L2 caches across cores. Three configurations of a 4-core design are studied here: a) the "private_4" configuration, where each of the 4 cores have their individual private L2 caches – this is the configuration with no L2 cache sharing; b) the "shared_2_2" configuration, where a pair of cores shares a common L2 cache; and c) the "shared_4" configuration, where all cores share a common L2 cache. The total L2 cache capacity was kept the same across all these configurations. A split-transaction bus was also used in all configurations. Figure 5 shows how the L2 miss handling times decrease as more and more cores share L2 caches. This is due to fewer invalidations and a lower degree of contention on the bus. Figure 6 shows this clearly – the relative number of bus upgrade transactions (labeled as UPGR in Figure 6) which are needed for invalidations, as well as the total number of transactions decrease with an increase in the number of cores that share the L2 cache(s).

978-1-60558-497-3/09 $25.00 © 2009 ACM

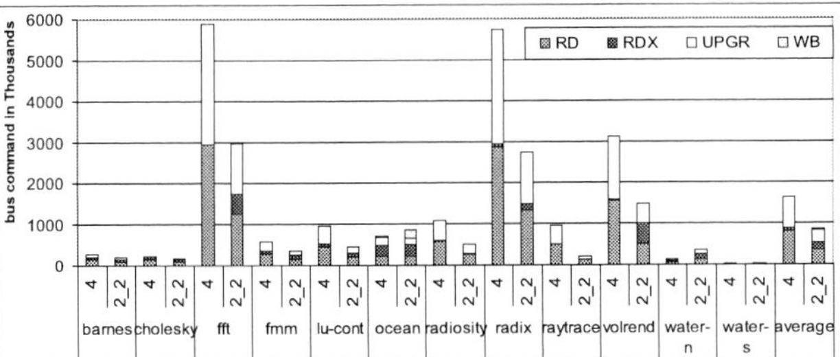

Figure 6: Breakdown of bus transactions for the two of the three L2 cache configurations

8. CONCLUSIONS

We presented the implementation details of a cycle-accurate, uop-accurate full-system simulator for X86 ISA-based multicore designs. MPTLsim models out-of-order cores complete with detailed models for coherent caches, on-chip interconnections and the memory interface. The key advantages to using an X86-based ISA include the ability to access a continuously developed tool chain of programming environments, compilers and debuggers. Another advantage of simulating X86-based ISAs on the widely available X86 hardware platforms include the ability to seamlessly switch between simulation mode and native execution mode. The existence of instrumentation registers on most X86 implementations permit the simulation results to be validated against the results from actual executions.

9. ACKNOWLEDGEMENTS

This work is supported in part by awards CNS-072081 and CNS-0454298 and awards from Intel Corporation.

10. REFERENCES

[1] Bochs IA-32 emulator project and related documentation. *http://bochs.sourceforge.net*.

[2] Multiple presentations on the Intel Nehalem processor line at the Intel developers forum.

[3] AUSTIN, T., LARSON, E., AND ERNST, D. Simplescalar: an infrastructure for computer system modeling. *Computer 35*, 2 (2002), 59–67.

[4] BELLARD, F. QEMU internals. Tech. report at www.lugatgt.org/articles/qemu_internals. 2006.

[5] BINKERT, N. L., DRESLINSKI, R. G., HSU, L. R., LIM, K. T., SAIDI, A. G., AND REINHARDT, S. K. The M5 simulator: Modeling networked systems. *IEEE Micro 26*, 4 (July–Aug. 2006), 52–60.

[6] CAZORLA, F., RAMIREZ, A., VALERO, M., AND FERNANDEZ, E. Dynamically controlled resource allocation in SMT processors. In *Proc. 37th International Symposium on Microarchitecture MICRO-37 2004* (2004), pp. 171–182.

[7] CULLER, D. E., SINGH, J. P., AND GUPTA, A. *Parallel Computer Architecture: A Hardware/Software Approach*. Morgan Kaufmann, 1999.

[8] EMER, J., AHUJA, P., BORCH, E., KLAUSER, A., LUK, C.-K., MANNE, S., MUKHERJEE, S. S., PATIL, H., WALLACE, S., BINKERT, N., ESPASA, R., AND JUAN, T. Asim: a performance model framework. *Computer 35*, 2 (Feb. 2002), 68–76.

[9] J. RENAU, B. FRAGUELA, J. T. W. L. M. P. L. C. S. S. P. S. K. S., AND MONTESINOS, P. SESC simulator. *http://sesc.sourceforge.net* (2006).

[10] LEVY, H., LO, J. L., EMER, J., STAMM, R., EGGERS, S., AND TULLSEN, D. Exploiting choice: Instruction fetch and issue on an implementable simultaneous multithreading processor. In *Proc. 23rd Annual International Symposium on Computer Architecture* (1996), pp. 191–191.

[11] LUK, C.-K., COHN, R., MUTH, R., PATIL, H., KLAUSER, A., LOWNEY, G., WALLACE, S., REDDI, V. J., AND HAZELWOOD, K. Pin: building customized program analysis tools with dynamic instrumentation. In *PLDI '05: Proceedings of the 2005 ACM SIGPLAN conference on Programming language design and implementation* (New York, NY, USA, 2005), ACM, pp. 190–200.

[12] MAGNUSSON, P. S., CHRISTENSSON, M., ESKILSON, J., FORSGREN, D., HALLBERG, G., HOGBERG, J., LARSSON, F., MOESTEDT, A., AND WERNER, B. Simics: A full system simulation platform. *Computer 35*, 2 (Feb. 2002), 50–58.

[13] MARTIN, M. M. K., SORIN, D. J., BECKMANN, B. M., MARTY, M. R., XU, M., ALAMELDEEN, A. R., MOORE, K. E., HILL, M. D., AND WOOD, D. A. Multifacet's general execution-driven multiprocessor simulator (GEMS) toolset. *SIGARCH Comput. Archit. News 33*, 4 (2005), 92–99.

[14] NETHERCOTE, N., AND SEWARD, J. Valgrind: a framework for heavyweight dynamic binary instrumentation. *SIGPLAN Not. 42*, 6 (2007), 89–100.

[15] SEILER, L., CARMEAN, D., SPRANGLE, E., FORSYTH, T., DUBEY, P., JUNKINS, S., LAKE, A., CAVIN, R., ESPASA, R., GROCHOWSKI, E., JUAN, T., ABRASH, M., SUGERMAN, J., AND HANRAHAN, P. Larrabee: A many-core X86 architecture for visual computing. *IEEE Micro 29*, 1 (2009), 10–21.

[16] SHARKEY, J. M-sim: A flexible, multi-threaded simulation environment. Tech. Rep. Tech. Report CS-TR-05-DP1, Department of Computer Science, SUNY Binghamton, 2005.

[17] WENISCH, T. F., WUNDERLICH, R. E., FERDMAN, M., AILAMAKI, A., FALSAFI, B., AND HOE, J. C. Simflex: Statistical sampling of computer system simulation. *IEEE Micro 26*, 4 (July–Aug. 2006), 18–31.

[18] YOURST, M. PTLsim: A cycle accurate full system X86-64 microarchitectural simulator. In *Proc. ISPASS* (2007).

[19] ZENG, H., YOURST, M., GHOSE, K., AND PONOMAREV, D. MPTLsim: a simulator for X86-64 multicore architecture with coherent caches. In *Proc. of the dasCMP Workshop* (2008)

978-1-60558-497-3/09 $25.00 © 2009 ACM

Trace-Driven Workload Simulation Method for Multiprocessor System-On-Chips

Tsuyoshi Isshiki, Dongju Li, Hiroaki Kunieda
Dept. Communications and Integrated Systems
Tokyo Institute of Technology, Tokyo, Japan
isshiki@vlsi.ss.titech.ac.jp

Toshio Isomura, Kazuo Satou
BR Automotive Software Engineering Dept.
TOYOTA MOTOR CORPORATION
Toyota, Japan

ABSTRACT

While Multiprocessor System-On-Chips (MPSoCs) are becoming widely adopted in embedded systems, there is a strong need for methodologies that quickly and accurately estimate performance of such complex systems. In this paper, we present a novel method for accurately estimating the cycle counts of parameterized MPSoC architectures through workload simulation driven by program execution traces encoded in the form of *branch bitstreams*. Experimental results show that the proposed method delivers a speedup factor of 70.15 to 238.58 against the instruction-set simulator based method while achieving high cycle accuracy whose estimation error ranges between 0.016% and 0.459%.

Categories and Subject Descriptors: C.4

[**Computer Systems Organization**]: Performance of Systems – *Modeling techniques*

General Terms: Design, Performance

Keywords: Performance Estimation, MPSoC Architecture Exploration, Simulation, Workload Model

1. INTRODUCTION

While state-of-the-art System-On-Chips are rapidly moving towards multiprocessor architectures (MPSoCs), current design methodologies and tool environment are still insufficient for handling the continuously growing complexities of MPSoC designs under short time-to-market. Competitiveness of the embedded systems market demands a highly optimized design at the system-level, both on MPSoC architecture as well as application software, in order to squeeze out the performance, minimize power consumption, reduce design cost, etc.

Recent commercial ESL (electronic system-level) design tools are targeted to enhance the productivity of these design efforts [1], in particular, to provide a software development platform before the detailed architecture of the MPSoC is fixed, and to drive the architecture optimization process through system-level simulations with the software that have been developed on this *virtual* platform. Nonetheless, these design processes require a great degree of manual designs and iterations, in particular, these software must be written, to a certain extent, *correctly* and *completely* (including device drivers and OS models, however abstract they may be), in order to properly drive the system-level MPSoC simulations. Also,

even with the recent advancements in *instruction-set simulator* (ISS) techniques [2][3], the increasing number of processor cores in current MPSoC results in slow simulation speed which may hinder efficient architecture exploration and optimization.

Another approach to this problem is to apply statistical workload models of the target applications for MPSoC architecture optimization. This approach which does not require a complete software set to drive the optimization process is attractive in this aspect, but deriving a proper workload is highly application dependent, and these models usually apply to data traffic but are difficult to generate accurate computation load models.

Our approach proposed in this paper is a novel *trace-driven workload model* capable of precisely capturing data-dependent behavior of the application to achieve near-cycle-accurate performance estimation. Program execution trace of an application on a given input data set is computed, through source-level instrumentation and native code execution, prior to the trace-driven workload simulation and efficiently encoded as *branch bitstream*. This branch bitstream is then used to steer the workload models in the form of *program trace graphs* generated for the target processor inside our parameterized MPSoC performance estimation framework. Our framework also includes automatic generation of *accurate coarse-grain workloads* directly from the application source code by using the precise execution profile information obtained from the same branch bitstream.

The rest of this paper is organized as follows. Section 2 describes the related works, then the key concepts of our branch bitstreams and program trace graph is described in section 3. Our MPSoC simulation framework is described in section 4. Experimental results are given in section 5, followed by discussions (section 6) and conclusions (section 7).

2. RELATED WORKS

Various simulation-based methods exists in the literature for estimating the performance of single processors, including *sampling-based simulation* [4][5] and *hybrid simulation* [6][7], where these techniques combine the native execution speed with the accuracy of ISS for some parts of the code. Another approach is to apply *source-level timing annotation* of the target processor for generating timing model during software execution [8][9]. Here, the timing information can include both *static timing* (obtained at compile time) and *dynamic timing* (caches and branch predictions) [9]. Our cycle-accurate workload model is derived upon the same static timing model, whereas dynamic timing is not directly addressed in our current framework. Section 6 includes our discussions in this regard. *Trace-path analysis* method in [10] creates task-level timing models through application-specific profiling (MPEG-4 encoding) and manual instrumentation to achieve high cycle accuracy.

At the MPSoC level, *statistical workloads* have been used to model MPSoC subsystems such as bus traffic [11] and NOCs [12], or for certain application domains such as networking [13]. Another approach is to employ system-level simulation on the target

Permission to make digital or hard copies of part or all of this work for personal or classroom use is granted without fee provided that copies are not made or distributed for profit or commercial advantage and that copies bear this notice and the full citation on the first page. To copy otherwise, to republish, to post on servers or to redistribute to lists, requires prior specific permission and/or a fee.
DAC'09, July 26-31, 2009, San Francisco, California, USA

MPSoC architecture driven by application software with *timing annotation* which can evaluate the impacts of interconnect architecture and memory subsystems in detail [14][15]. Our simulation framework includes bus traffic modeling induced by *inter-processor communications*, but currently does not generate memory traffic (this issue is also addressed in section 6).

3. KEY CONCEPTS

3.1 Branch Bitstream

Let us first explain how the program execution trace can be represented by the *branch bitstream*, which is simply a sequence of branch condition bits. Consider a program whose interprocedural control flow graph (interprocedural CFG or *ICFG*) is shown in Figure 1. For clarity, branch operations are explicitly denoted as **bX** as well as calls to function **foo** which are separated from the basic-blocks (**BBX**). Branch edges are labeled as **T** (true) and **F** (false). A *branch bitstream* is generated by simply recording the branch outcome (true or false) with a single bit in the order of program execution. On the program execution trace shown on the right side of Figure 1, its corresponding branch bitstream is **1011001**. The first **1** indicates that **b0** branched in the **T** path (which reaches the call **foo()**), the next **0** indicates that **b2** inside **foo** branched in the **F** path (then returns to **main**), the following **110** indicates that **b1** branched in the **T** path *twice* and then finally to the **F** path, the next **0** indicates that **b0** branched in the **F** path (which reaches the second call **foo()**), and the last **1** indicates that **b2** inside **foo** branched in the **T** path (then returns to **main**).

Rather surprising is the fact that, although no information about the correspondence between the branch condition bits and branch operations is provided, this simple branch bitstream is sufficient for representing the *complete program execution trace*. Decoding the program trace from the branch bitstream is done by simply traversing the ICFG and reading 1 bit at a time upon reaching a branch operation to determine which branch path to traverse. Function calls can be handled by maintaining the call stack during the graph traversal. This capability of encoding the complete program trace further implies that even the *entire instruction trace can be completely reproduced from the branch bitstream* if the CFG reflects the actual target binary code structure.

3.2 Source-Level Instrumentation for Branch Bitstream Generation

In order to obtain the branch bitstream for the application on a given input data set, a simple *source-level instrumentation* is performed automatically on the application using macro call at each branch operation to record the branch condition bits to a file:

```
if(exp)...   ⟹   if(_BC_(exp))...
```

The instrumented code is then compiled and executed on the host machine for fast native code execution. This instrumentation is immune from *any* code optimization during native compilation since the instrumented code section includes global side-effects whose execution order needs to be preserved, thus guaranteeing a correct order of branch condition bit sequence. On the other hand, *the decision of which branch operation to instrument must be carried out in accordance to the actual binary code of the target processor*, since the structure of the program graph must reflect the target binary code structure for precise performance estimation.

There are two cases in C programs that branch bitstreams cannot be directly applied, which are *switch statements* and *calls through function pointers*. Switch statements are currently modeled by a sequence of value-checking branch operations that jump to the

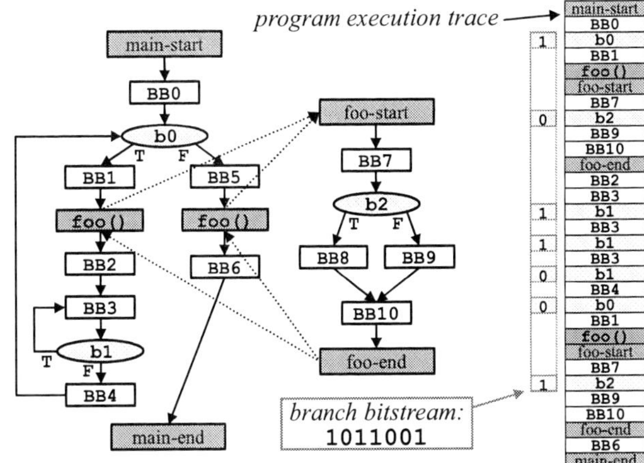

Figure 1: Branch bitstream for program trace encoding

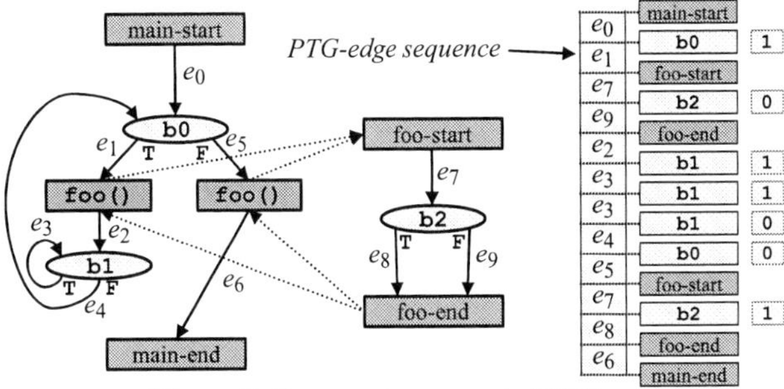

Figure 2: Program trace graph (PTG)

corresponding *case statements*. Calls through function pointers can be modeled in the same manner, but are not yet implemented in our current instrumentation framework.

3.3 Program Trace Graph

For decoding the branch bitstreams more efficiently, we transform the ICFG in Figure 1 to a *program trace graph* (PTG) as shown in Figure 2. PTG is derived by degenerating the ICFG at the basic-blocks while preserving the branch nodes and call nodes. As illustrated in Figure 2, PTG acts as a *translator* between the branch bitstream and the program trace in the form of *PTG-edge sequence*.

A *PTG-node* represents either *function-start*, *function-end*, *branch* or *call*, where each of these PTG-nodes corresponds to a location in the target binary code. A *PTG-edge* carries attributes about the program execution information, including *cycle count*, between the two PTG-nodes it connects to.

Accurate cycle count on the code segment represented by a PTG-edge is obtained by the code generator backend for the target processor in our compiler framework which is equipped with the full knowledge of its pipeline architecture to precisely predict pipeline hazards. This timing annotation can also be accomplished with a back-annotation tool framework as presented in [6]. Operations with *data-dependent latencies* are pre-characterized through additional source-level instrumentation for calculating the *average latencies* during the native code execution that are *back-annotated* in the PTG-edge cycle counts. An example of such data-dependent latency operation in our target processor is an integer division unit whose latency is determined by the *difference in the number of significant bits between the two operands*.

From the PTG-edge sequence (e_0 e_1 e_7 e_9 e_2 e_3 e_3 e_4 e_5 e_7 e_8 e_6) in Figure 2, the *total program execution cycle count* is calculated by accumulating the cycle count on this PTG-edge sequence as:

$$T = c(e_0) + c(e_1) + c(e_7) + c(e_9) + c(e_2) + c(e_3)$$
$$+ c(e_3) + c(e_4) + c(e_5) + c(e_7) + c(e_8) + c(e_6) \quad \text{(eq.1)}$$

where $c(e_i)$ is the cycle count on PTG-edge e_i. Another way of calculating the total cycle count is by counting the *occurrence* of each edge e_i in the PTG-edge sequence (denoted as $n(e_i)$) and accumulating the product of edge cycle count and its occurrence:

$$T = \sum c(e_i) \cdot n(e_i) . \qquad \text{(eq.2)}$$

3.4 Coarse-grain Workload Generation

In this section, we give a simple graph reduction technique on the PTG for obtaining a *coarse-grain workload model* that is *precise* in terms of *average cycle counts*. Here, it is assumed that the branch bitstream obtained from the native execution is *pre-scanned* to obtain the total occurrences $n(e_i)$ on the entire PTG-edge sequence.

3.4.1 PTG Single-entry Single-exit Regions

A *single-entry single-exit* (SESE) region is a distinct CFG sub-structure with entry- and exit-edge pair [16]. SESE regions on the PTG can be defined in the same manner as stated below:

DEFINITION 1: A *PTG-SESE region* $R(e_{in}, e_{out})$ is enclosed by its *entry-edge* e_{in} and *exit-edge* e_{out} where e_{in} dominates e_{out}, e_{out} postdominates e_{in}, and e_{in} and e_{out} are *cycle-equivalent* [16].

DEFINITION 2: An *interior edge set* $E(e_{in}, e_{out})$ of PTG-SESE region $R(e_{in}, e_{out})$ is a set of PTG-edges that are each dominated by e_{in} and postdominated by e_{out}. By convention, an edge dominates and postdominates *itself*, and thus $E(e_{in}, e_{out})$ contains both e_{in} and e_{out}.

One issue arises when dealing with SESE regions on the PTG; that is, SESE region's entry-edge and exit-edge on the CFG may be eliminated in the PTG since *control-merge nodes* are collapsed in the PTG, such as the case of *if-else* block's exit-edge inside function **foo** in Figure 2. This issue is handled by simply restoring these CFG control-merge nodes on the PTG as shown in Figure 3(b). Here, the *out-edge* of such restored control-merge node (such as e_8' in Figure 3(b)) has 0 cycle count since the cycle count on this path is already included in its *in-edges*. In general, the out-edge of *function-start* node and the in-edge of *function-end* node enclose the outermost SESE region of a function.

3.4.2 PTG Reduction on PTG-SESE Regions

Here, we will first consider the PTG-SESE regions which do not contain any call nodes. What follows is that the *precise average workload* of such PTG-SESE region $R(e_{in}, e_{out})$ can be calculated by accumulating the *total workload* of its interior edge set $E(e_{in}, e_{out})$ and dividing it by the *occurrence* of its entry-edge e_{in}:

$$c(R(e_{in}, e_{out})) = \frac{1}{n(e_{in})} \sum_{e_i \in E(e_{in}, e_{out})} c(e_i) \cdot n(e_i) \qquad \text{(eq.3)}$$

In Figure 3(b), the PTG-SESE region $R(e_7, e_8')$ (covering the entire function **foo**) can be replaced with a single PTG-edge e_7' where $c(e_7') = (c(e_7) \cdot n(e_7) + c(e_8) \cdot n(e_8) + c(e_9) \cdot n(e_9)) / n(e_7)$.

The average workload of *non-recursive* call nodes, on the other hand, can be calculated in bottom-up fashion on the *call graph*, starting from the leaf functions where eq.3 can be directly applied, and back-annotating the *function's average workload* at each of its call sites. Figure 4 illustrates this process on the **main**'s PTG in Figure 2, where the *interprocedural edges* to **foo**'s PTG are removed and the out-edges of these call nodes are back-annotated with the average workload of **foo** such that $c(e_2') = c(e_7') + c(e_2)$ and $c(e_6') = c(e_7') + c(e_6)$. Then, the outermost PTG-SESE region $R(e_0, e_6')$ can be reduced to a single PTG-edge e_0' whose cycle count $c(e_0')$ can be calculated by eq.3.

The significance of eq.3 is that this simple formulation is applicable to *any* PTG-SESE region (without recursive calls) with

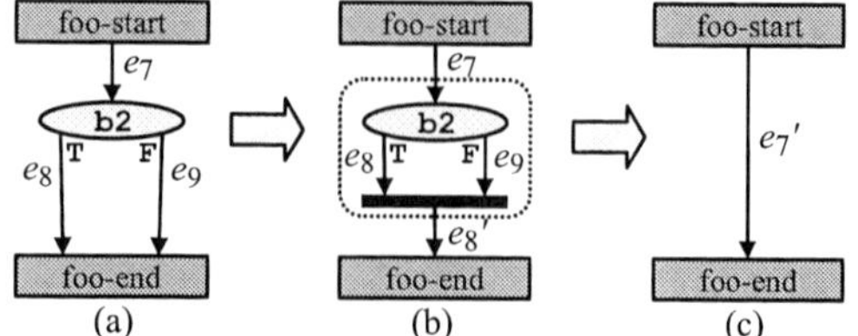

Figure 3: PTG-SESE region reduction without call nodes

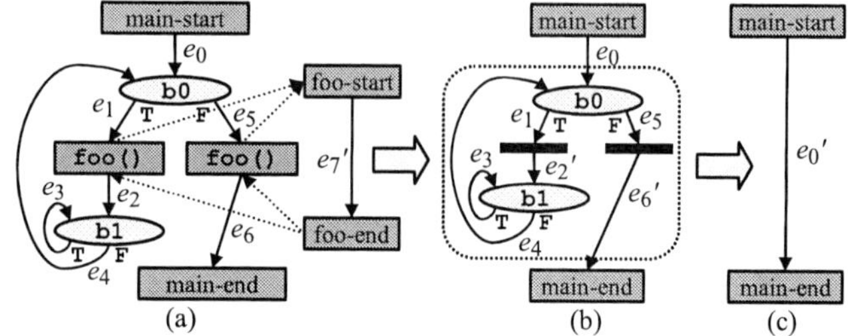

Figure 4: PTG-SESE region reduction with call nodes

arbitrary complex control flow structures. Since every function contains at least one PTG-SESE region, this also implies that the *entire application program can always be reduced to a single PTG-edge*, given that there are no recursive calls and that the PTG-edge cycle counts *alone* are sufficient for estimating the cycle count of the system. The latter assumption holds true for single-processor systems without any dynamic timing effects (such as caches) where *the reduced PTG-edge contains the exact cycle count*, whereas for MPSoC systems, we need to additionally consider the issue of *system synchronization* as discussed next.

3.4.3 Synchronizations Issues on PTG Reduction

In addition to the PTG-edge cycle counts for maintaining the *simulation clocks*, another element required in the system-level simulation is the *synchronization of events* at various MPSoC components such as processors, busses, and channels. Here, we introduce a new PTG-node class called *PTG-sync-node* to represent such events in the application program that must be made visible to the *simulator kernel*, where all such operations in the application program are separated from the basic-blocks in the CFG and appear as *distinct PTG-sync-nodes* in the PTG. These PTG-sync-nodes serve as anchor points for the simulator kernel to update its internal status, and if needed, adjust its simulation clocks by inserting *wait states* on *blocked events*. Details of these PTG-sync-nodes with respect to simulator kernel's behavior are described in section 4.

Consequently, these PTG-sync-nodes must not be removed during the PTG reduction process, as this will alter the sequence of system-level synchronization events. *Therefore, we do not allow reductions to the PTG-SESE regions that contain PTG-sync-nodes.*

3.4.4 Reduced Branch Bitstream Generation

The branch bitstream that corresponds to the reduced PTG can be generated by scanning the original branch bitstream and simply throwing away the branch bits on the eliminated branches.

4. MPSOC TRACE-DRIVEN WORKLOAD SIMULATION FRAMEWORK

Figure 5 illustrates the overall flow of our MPSoC trace-driven workload simulation framework incorporating the methods explained in the previous section. Before we begin to describe the details of our simulation framework, let us first address the issue of MPSoC application development model, since it affects the nature of the application programs executed on the processors, and therefore affecting the overall workload modeling flow of branch bitstream generation and program trace graph generation.

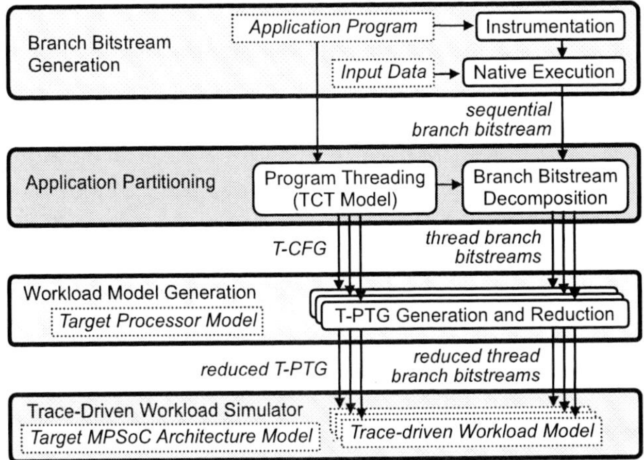

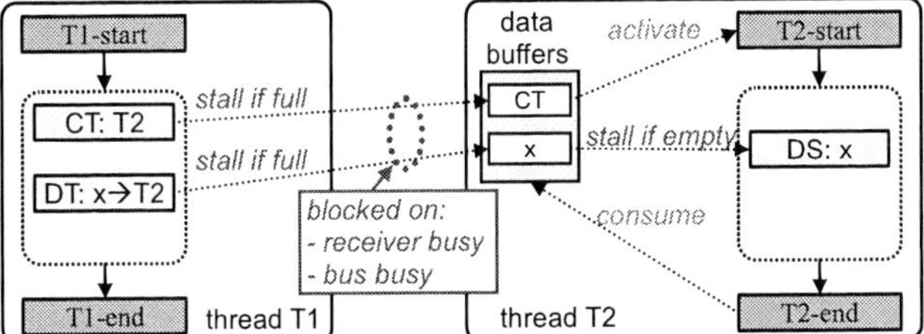

Figure 6: Thread Program Trace Graph (T-PTG)

Figure 5: MPSoC trace-driven workload simulation framework

4.1 MPSoC Application Development Model

The MPSoC application development model employed in our simulation framework is based on the *Tightly-Coupled Thread model* (TCT model) [17][18] that allows a simple coding style on the sequential C code for specifying "threads" without the need to specify any other parallelization directives, such as *concurrency*, *synchronization* and *communication*, since they are all *implicit* in the TCT model. Thread annotation on the C code is done by declaring *thread-scopes* in the form

```
THREAD(name){ ... }
```

where any C statements (including function calls and even *nested* thread-scopes) can be included inside the thread-scope region as long as the thread-scope forms a *SESE region*. Thread annotations can be inserted manually or through the *MAPS framework* [19] which assists the designer with rich program analysis capabilities to emit the thread annotated code semi-automatically. The TCT compiler performs a complete interprocedural dependence analysis to extract *all data dependences between threads* (including globals and pointer dereferences), and inserts *message-passing* instructions automatically (current TCT compiler assumes a *fully distributed memory system without any shared-memory accesses*). These tightly-coupled threads operate in a functional pipeline manner, achieving high degree of parallelism. Also, the TCT compiler guarantees that the behavior of the parallelized code and the original sequential code is identical where race condition and deadlocks are automatically avoided. The TCT framework has been verified on a prototype MPSoC silicon (*TCT-MPSoC*) consisting of a 6-processor array block where each processing element contains a dedicated communication module that executes the TCT communication protocol through a full-crossbar interconnect [18].

4.2 Thread Program Trace Graph

A *thread program trace graph* (*T-PTG*) corresponds to the PTG that is enclosed by the SESE thread-scope region (Figure 6). Each T-PTG is terminated by *thread-start* and *thread-end* nodes instead of *function-start* and *function-end* nodes in the normal PTG. In addition, three types of *inter-processor communication instructions* generated by the TCT compiler also appear in the T-PTG:

- *Control token* (CT) sends an *activation token* to the target thread
- *Data transfer* (DT) sends data to the target thread
- *Data synchronization* (DS) checks if the data is received

Here, DT-node does not require a matching *receive operation* at the receiver end, and thus the data transfer occurs *asynchronously* with the receiver thread state. A finite-size buffer is allocated for individual data at the receiver, and the DT-node *stalls* while the

receiver buffer for that particular data is *full*, where as DS-node stalls while the data buffer is *empty*. CT-node is modeled the same as DT-node in terms of buffer model, where it too stalls if the receiver's *control token buffer* is full. Data buffers and control token buffers are *consumed* at *thread-end* node at the receiver thread. Additionally, DT-nodes and CT-nodes are *blocked* if the receiver is occupied by other transactions or if the bus is occupied.

Above CT, DT and DS nodes along with *thread-start* and *thread-end* nodes are designated as *PTG-sync-nodes* described in section 3.4.3 which need to be preserved during the PTG reduction process. The MPSoC workload simulator kernel performs various tasks associated with these PTG-sync-nodes for synchronizing events among processors as well as updating status of various MPSoC resources such as busses and data buffers.

4.3 MPSoC Workload Model Generation

Below summarizes the basic steps for generating the MPSoC trace-driven workload model in conjuction with the TCT model.

(1) *Branch bitstream generation* is performed on the original sequential code (as described in 3.2), taking full advantage of the fast native code execution speed. In addition, this initial phase is decoupled with the application partitioning, which is very attractive in the sense that a single branch bitstream can drive multiple workload models derived from different application partitioning.

(2) *Application partitioning* corresponds to the task of inserting the thread annotation to the original sequential code. TCT compiler partitions the thread-annotated code into separate CFGs for each thread (T-CFG) with inter-processor communication instructions. At the same time, *the sequential branch bitstream* is decomposed into separate *thread branch bitstreams* which is a straightforward task of scanning the sequential branch bitstream and redirecting it into different *thread branch bitstream files* according to the thread attribute on each branch node.

(3) *Workload model generation* is performed separately on each T-CFG to obtain the *T-PTG* then undergoes the T-PTG reduction process, deriving the *reduced T-PTG* along with the *reduced thread branch bitstreams*.

4.4 MPSoC Architecture Model

MPSoC architecture model in our simulation framework is configurable on parameters explained below.

(1) *Processor type*: only a single processor type is currently supported which is the one implemented in the prototype TCT-MPSoC [18]. The processor (*TCT-PE*) is a simple 4-stage RISC with 32 registers and separate memory ports for instruction and data. In the future, this will be extended to multiple processor types for a *truly heterogenous MPSoC exploration*.

(2) *Thread-to-processor mapping*: each thread is *statically* allocated to one processor, where *multiple threads* can be allocated to each processor (*this feature was not part of the prototype TCT-MPSoC, where TCT-PE could only handle one thread*). The *simulator kernel* directly handles the context switching of threads within the processors using *non-preemptive scheduling* where the *parameterized context switching overhead* is also modeled.

(3) *Bus topology*: MPSoC interconnect is modeled as a parameterized multi-bus. At each processor, its output port connects to a single bus, and its input port connects to all buses. This multi-bus model guarantees that any two processors have direct connectivity, which allows us to model a relatively wide variety of bus topologies from a *single bus* to a *full-crossbar*.

(4) *Communication channel*: communication setup time, data transfer rate and buffer depth can all be configured.

4.5 MPSoC Workload Simulator Kernel

Our *MPSoC trace-driven workload simulator kernel* consists of two major components: *thread workload simulator* and *global scheduler*. *Thread workload simulator* is responsible for updating the *local processor clock* by the method described in section 3.3 (translating the thread branch bitstream into PTG-edge sequence and accumulating the edge cycles). Once activated, the thread workload simulator continues to increment its local processor clock until the generated PTG-edge sequence reaches a *PTG-sync-node* (CT, DT, DS, *thread-start* and *thread-end* nodes) which requires a strict time-ordered simulation by the *global scheduler*. The global scheduler activates the thread trace simulator with the earliest local clock by maintaining a queue of active threads. Mutually exclusive resources (buses and data buffers) also maintain their own clocks to indicate their occupancy status during simulation.

5. EXPERIMENTAL RESULTS

To validate the effectiveness of our proposed trace-driven workload simulation method, we have conducted a series of experiments on single-processor model as well as MPSoC models and compared them against the instruction-set simulator (ISS) for the TCT-PE. All simulations were executed on 3.4GHz Intel Xeon machine with 3.25GB memory.

5.1 TCT-Processor Instruction-Set Simulator

The TCT framework includes a *cycle-accurate ISS* for the TCT-PE, which also can perform cycle-accurate multiprocessor simulations on an arbitrary number of processors that are connected by full-crossbar switch. For processor communication, data transfer setup time is 2 to 6 cycles, and transfer rate is 4 bytes/cycle, and the parallel-ISS accurately simulates all bus contentions due to simultaneous data transfer requests to the same destination processor. As this parallel-ISS was initially developed for the prototype TCT-MPSoC, *this parallel-ISS cannot directly simulate the "multi-threaded" behavior on the processor.*

5.2 Single-Processor Cycle Estimation

Table 1 shows the set of benchmarks used to evaluate the cycle estimation performance for the single-processor model. Here, "# inst" is the number of *TCT-PE instructions* in each program, "# cycles" is the total execution cycles, "ISS (sec)" is the ISS simulation time, "ISS (cycles/sec)" is the ISS simulation speed, and "native (sec)" is the native execution time (all applications were compiled with Microsoft Visual C++ .NET with –O2 optimization).

Table 2 shows our trace-driven workload simulator (*TWS*) performance *without PTG reduction*. Here, "# PTG-edges" is the total number of PTG-edges, "# branch bits" is the length of the branch bitstreams, "TWS (sec)" is the TWS simulation time, "TWS (cycles/sec)" is the *effective* TWS simulation speed, and "TWS speedup" is the speedup over ISS. Even without the PTG reduction, we can observe the *enormous speed advantage of our TWS over ISS*, where the effective TWS simulation speed ranges from 1.7 to 3.0 *billion cycles per second*, while achieving *absolute cycle accuracy*, that is, TWS cycles counts *match exactly with the ISS cycle counts.*

Table 1: Benchmark programs for single-processor cycle estimation

	# inst	# cycles	ISS (sec)	ISS (cycles/sec)	native (sec)
prime500k	100	629,940,586	13.328	47.26M	0.083
string	273	900,232,697	31.266	28.79M	0.172
alphabet	113	716,650,704	25.328	28.29M	0.193
JPEG	1,673	433,010,644	16.829	25.73M	0.076

Table 2: Trace-driven workload simulator (TWS) performance (without PTG reduction)

	# PTG-edges	# branch bits	TWS (sec)	TWS (cycles/sec)	TWS speedup
prime500k	29	53.99M	0.218	2,889M	61.14
string	52	136.18M	0.515	1,748M	60.71
alphabet	21	101.04M	0.375	1,911M	67.54
JPEG	336	27.48M	0.141	3,071M	119.35

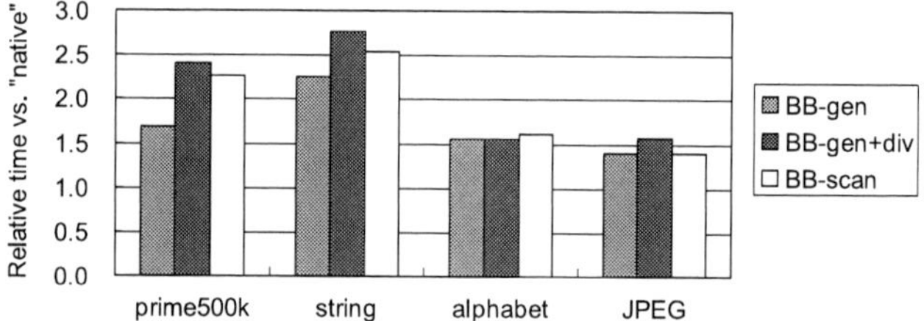

Figure 7: Relative processing times against native execution

Figure 7 shows the processing times inside the TWS workflow that are normalized against the native execution time in Table 1. Here, "BB-gen" is the native execution time of the applications instrumented with branch bitstream generation, and "BB-gen+div" is the execution time with branch bitstream generation and integer division latency profiling. We can observe that the instrumentation overhead for the branch bitstream generation is quite small, ranging from 1.40 to 2.25, and even with the division latency profiling, the overhead increases only up to 2.75. "BB-scan" is the processing time of enumerating the PTG-edge occurrences from the generated branch bitstreams that is performed prior to the PTG reduction. As explained in section 3.4.2, *each of these applications in this single-processor case reduces to a single PTG-edge whose edge cycle match exactly with the ISS cycle count.*

5.3 MPSoC Cycle Estimation

For evaluating the MPSoC cycle estimation performance, we have used the previous two applications, "string" with 9 threads and "JPEG" with 19 threads as shown in Table 4. Here, "# cycles" is the total execution cycles on the parallelized code, "para-speedup" is the speedup ratio from the sequential code, and "para-ISS (sec)" is the parallel-ISS time for the parallelized code. As explained previously, current parallel-ISS can only handle one thread per processor, therefore it models a 9-PE MPSoC for "string" and a 19-PE MPSoC for "JPEG".

Table 5 shows the TWS simulation results for the target MPSoCs on the two applications, with and without PTG reduction. Here, the architecture parameters are set according to the prototype TCT-MPSoC. Unlike the single-processor cases, the MPSoC estimated cycles do not match exactly with the parallel-ISS cycles, although the *estimation errors are well within 1%*, primarily due to the simplistic communication model in our TWS which assumes a fixed setup time for all data transfers, whereas a precise communication hand-shaking is modeled in the parallel-ISS. The effects of the PTG reduction can be observed on the dramatic reductions in branch bitstream length, while the total cycles tend to decrease slightly since the *coarse-grain workloads* obtained by the PTG reduction have the effect of averaging out the temporal fluctuations

978-1-60558-497-3/09 $25.00 © 2009 ACM

of the original *fine-grain workloads.*

Figure 8 shows the simulation results on "JPEG" with PTG reduction *enabled* under different MPSoC parameters. A *19-thread to 15-PE* mapping was derived by selecting 6 threads with low CPU utilization in the previous 19-PE simulation and merging them into two groups (3 threads in each group). In addition to these PE configurations (19-PE and 15-PE), other MPSoC parameters such as bus topologies (full-crossbar and single-bus), data transfer rates (4 bytes/cycle and 2 bytes/cycle), and buffer depth (3 to 9 data elements at each buffer) are altogether *swept* on our TWS framework from the same set of coarse-grain workload models and reduced branch bitstreams. The ability to quickly evaluate these MPSoC parameters (each only taking a fraction of a second) makes our TWS framework ideal for *MPSoC design exploration.*

6. DISCUSSIONS

One important aspect not implemented in our current framework is the ability to incorporate the *runtime behavior* on complex processors which cannot be obtained at compile time. Framework in [9] incorporates *branch prediction* and *instruction cache models* that are triggered during their SystemC simulation. To this end, as discussed in section 3.1, it should again be noted that our branch bitstream (before the PTG reduction) can *reproduce the entire instruction trace of the application execution.* For this reason, it is indeed possible to accurately simulate the *instruction cache* and *branch prediction behaviors* from our branch bitstreams.

Although the current implementation of our TWS framework is customized towards the TCT model and its MPSoC architecture model, we believe that the concept of *PTG-sync-nodes* that bridges the PTG semantics and the TWS kernel can eventually be extended to other MPSoC programming models and MPSoC architecture models. Synchronization primitives used in common parallel programming models and RTOS-based multitask programming can be modeled as PTG-sync-nodes, where the TWS kernel shall be extended to handle the behavior of these additional PTG-sync-node types. Complex MPSoC architectures with memory subsystems and NOCs would require more substantial extensions on the TWS kernel for generating accurate data traffic induced by memory accesses and through the NOC infrastructure.

All these issues shall be addressed in our future works.

7. CONCLUSIONS

In this paper, we have introduced a novel cycle estimation method for single-processors and MPSoCs based on branch bitstream for efficient program execution trace encoding, program trace graph for efficient trace retrieval, and its graph reduction techniques to drastically reduce the branch bitstream lengths. The effectiveness of these methods is confirmed on a set of experiments conducted for single-processor case as well as MPSoCs, showing significant improvement in simulation speed over the ISS, as well as high cycle accuracy. A single sequential branch bitstream generated through code instrumentation and fast native-code execution can drive a variety of workload models, each corresponding to different application partitioning and different MPSoC architectures, thus establishing a practical MPSoC design exploration framework.

REFERENCE

[1] S.K. Shukla, C. Pixley, G. Smith, The True State of the Art of ESL Design, *IEEE Design & Test of Computers,* vol.23, No.5, pp.335-337, 2006.

[2] A. Nohl, G. Braun, O. Schliebusch, R. Leupers, H. Meyr, and A. Hoffmann. A Universal Technique for Fast and Flexible Instruction-set Architecture Simulation. In *Proc. DAC 2002*, pp.22–27, 2002.

[3] M. Reshadi, P. Mishra, and N. Dutt. Instruction Set Compiled Simulation: A Technique for Fast and Flexible Instruction Set Simulation. In *Proc. DAC 2003*, pp.758–763, 2003.

[4] R. Wunderlich, T. Wenisch, B. Falsafi, and J. Hoe. SMARTS: Accelerating

Table 4: Benchmark programs for MPSoC cycle estimation

	# cycles	para- speedup	para-ISS (sec)
string (9 threads)	118,581,455	7.59	51.281
JPEG (19 threads)	41,720,835	10.38	37.219

Table 5: TWS simulation results with and without PTG reduction

	# branch bits	est. # cycles	est. error	TWS (sec)	TWS speedup
string	136.18M	118,893,200	0.263%	0.731	70.15
string (reduction)	41,946	118,562,675	0.016%	0.265	193.51
JPEG	27.48M	41,745,317	0.059%	0.312	119.29
JPEG (reduction)	8,330	41,529,311	0.459%	0.156	238.58

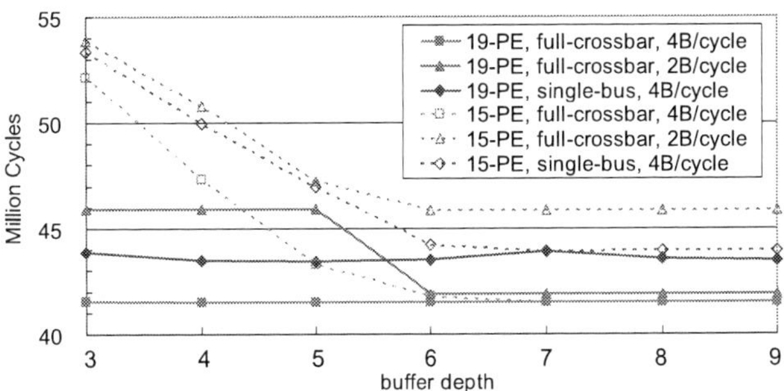

Figure 8: TWS simulations on various MPSoC parameters (JPEG)

Microarchitecture Simulation via Rigorous Statistical Sampling. In *Proc. 30th ISCA,* pp.84–97, 2003.

[5] T. Sherwood, E. Perelman, G. Hamerly, and B. Calder. Automatically Characterizing Large Scale Program Behavior. In *Proc. ASPLOS-X,* pp.45–57, 2002.

[6] P. K. Szwed, D. Marques, R. M. Buels, S. A. McKee, and M. Schulz. SimSnap: Fast-Forwarding via Native Execution and Application-Level Checkpointing. In *Proc. INTERACT-8*, pp.65–74, 2004.

[7] S. Kraemer, L. Gao, J. Weinstock, R. Leupers, G. Archeid, H. Meyr, Hysim: A Fast Simulation Framework for Embedded Software Development, In *Proc. CODES-ISSS*, pp.75–80, 2007

[8] M. T. Lazarescu, J. R. Bammi, E. Harcourt, L. Lavagno, M. Lajolo, Compilation-based Software Performance Estimation for System Level Design, In *Proc. HLDVT'00*, pp.167–172, 2000.

[9] J. Schnerr, O. Bringmann, A. Viehl and W. Rosenstiel. High-Performance Timing Simulation of Embedded Software. In *Proc. DAC 2008,* pp.290–295, 2008.

[10] N. Y. Chang, K. B. Lee and C. W. Jen. Trace-path Analysis and Performance Estimation for Multimedia Application in Embedded Systems. In *Proc. ISCAS '04*, vol.II, pp.129–132, 2004.

[11] R. Giorgi, C. A. Prete, G. Prina, and L. Ricciardi. A Workload Generation Environment for Trace-Driven Simulation of Shared-Bus Multiprocessors. In *Proc. 30th HICSS*, pp. 266–275, 1997.

[12] Jan Madsen et al. Network-on-chip Modeling for System-level Multiprocessor Simulation. In *Proc. 24th RTSS03*, pp. 82–92, 2003.

[13] L. Thiele, S. Chakraborty, M. Gries, S. Kunzli. A Framework for Evaluating Design Tradeoffs in Packet Processing Architectures. In *Proc. DAC 2002*, pp.880–885, 2002.

[14] T. Kempf, K. Karuri, S. Wallentowitz, G. Ascheid, R. Leupers, and H. Meyr. A SW Performance Estimation Framework for Early System-level-design using Fine-grained Instrumentation. In *Proc. DATE '06*, pp. 468–473, 2006.

[15] T. Meyerowitz, A. S. Vincentelli, M. Sauermann, D. Langen, Source-level Timing Annotation and Simulation for a Heterogeneous Multiprocessor, In *Proc. DATE'08*, pp. 276–279, 2008

[16] R. Johnson, D. Pearson, and K. Pingali, The Program Structure Tree: Computing Control Regions in Linear Time. In *Proc. ACM SIGPLAN '94*, pp.171–185, 1994

[17] M. Z. Urfianto, T. Isshiki, A. U. Khan, D. Li, H. Kunieda, Decomposition of Task-Level Concurrency on C Programs Applied to the Design of Multiprocessor SoC, IEICE Trans. 91-A(7), pp.1748-1756, 2008

[18] M. Z. Urfianto, T. Isshiki, A. U. Khan, D. Li, H. Kunieda, A Multiprocessor SoC Architecture with Efficient Communication Infrastructure and Advanced Compiler Support for Easy Application Development, IEICE Trans. 91-A(4), pp.1185-1196, 2008

[19] J. Ceng, J. Castrillon, W. Sheng, H. Scharwächter, R. Leupers, G. Ascheid, H. Meyr, T. Isshiki, H. Kunieda, MAPS: An Integrated Framework for MPSoC Application Parallelization, In *Proc. DAC 2008*, pp.754–759, 2008

978-1-60558-497-3/09 $25.00 © 2009 ACM

Analysis and Mitigation of Process Variation Impacts on Power-Attack Tolerance

Lang Lin

Department of Electrical and Computer Engineering,
University of Massachusetts Amherst, MA

llin@ecs.umass.edu

Wayne Burleson

Department of Electrical and Computer Engineering,
University of Massachusetts Amherst, MA

burleson@ecs.umass.edu

ABSTRACT

Embedded cryptosystems show increased vulnerabilities to implementation attacks such as power analysis. CMOS technology trends are causing increased process variations which impact the data-dependent power of deep submicron cryptosystem designs. In this paper, we use Monte Carlo methods in SPICE circuit simulations to analyze the statistical properties of the data-dependent power with predictive 45nm CMOS device and ITRS process variation models. In addition to the "measurement to disclosure" (MTD) used in [3], we define a lower level metric, Power-Attack Tolerance (PAT), to model both dynamic power and leakage power data-dependence. We show that the PAT of a typical cryptographic component implementation using CMOS standard-cells can significantly deteriorate due to process variations, thus increasing the component's vulnerability to power attacks. Power-attack-resistant logic styles (e.g. SABL [9]) have been developed which increase PAT by an order of magnitude by balancing power consumption at the gate level with considerable overhead. However in the presence of process variations, the degradation probability of MTD is 57%. To mitigate this problem, we demonstrate a transistor sizing optimization method that can reduce such negative impacts to only 18% with minimal power and area overhead.

Categories and Subject Descriptors

B.6 [**Hardware**]: Logic Design; B.7 [**Hardware**]: Integrated Circuits; E.3 [**Data**]: Data encryption.

General Terms

Design, Security, Verification.

Keywords

Process variation, Monte Carlo simulation, differential power analysis, transistor sizing.

1. INTRODUCTION

Embedded system security is realized by an embedded cryptosystem to execute the encryption and decryption

Permission to make digital or hard copies of part or all of this work for personal or classroom use is granted without fee provided that copies are not made or distributed for profit or commercial advantage and that copies bear this notice and the full citation on the first page. To copy otherwise, to republish, to post on servers or to redistribute to lists, requires prior specific permission and/or a fee.

DAC'09, July 26-31, 2009, San Francisco, California, USA

computations with a secret key. By sharing the key with only trusted parties, security can be guaranteed theoretically. However, since every computation process is inevitably accompanied with an amount of power consumption, the logic values involved in the computations can be reflected by specific power patterns. This physical phenomenon enables a typical implementation attack, called "power analysis attack" [1], to extract the secret key by exploiting the correlation between the logic values of the key bits and the measured run-time power traces. The effectiveness of power analysis attacks is determined by two parameters of a cryptosystem implementation [2]:

1). Signal-to-noise ratio (SNR), which is the ratio of the data-dependent power to the data-independent power. For a cryptosystem embedded in an SoC, the on-die power regulators and many non-crypto integrated circuits can lead to large amount of the total data-independent power, which will prevent a simple power analysis attack. To deal with this, differential power analysis (DPA) [1] attack is introduced to statistically collect the subtle data-dependent power and filter out the data-independent power. DPA attacks can efficiently extract the secret key by non-invasive methods and off-the-shelf tools. The attack starts from making a guess of the secret key, and use this key guess to predict the values of some bits (known as the "selection function") involved in the crypto computations. The prediction is used to group the measured power traces into data-dependent and data-independent categories. By accumulating the data-dependent power and neutralizing the data-independent power, a differential power curve (DPC) is generated with respect to a key guess. If the key guess is correct, the corresponding DPC is distinguishable from other DPCs. If no distinguishable DPC can be found, more power traces need to be analyzed to overcome the low SNR. The number of power traces needed to extract the key is named the "measurement to disclosure" (MTD) in [3], which is inversely proportional to the SNR [4].

2). The time t_c, during which the data-dependent power appears on each measured power trace. DPA attacks make use of transient power, instead of time-average power, to extract the secret key. Thus, disarrangement of t_c on each power trace can prevent power analysis. For example, if a random clock-cycle-based delay is inserted to disturb the synchronization of the power traces, DPA attacks based on data-dependent dynamic power become difficult. However, DPA attacks based on data-dependent active leakage power [5] are hardly affected by random delay insertion. The active leakage power can be measured between a clock rising edge and a neighboring clock falling edge, when all the logic gate switching behaviors in the current clock cycle have settled down

978-1-60558-497-3/09 $25.00 © 2009 ACM

before the next switching behavior occurs. The attackers can easily defeat the random delay approach, since the duration of active leakage power is always much longer than the introduced delay uncertainty.

In deep submicron technologies, process variations become a critical concern and most deterministic design features become probabilistic. To model the process variations, the design features are usually assigned with a normal distribution and the probability design functions (PDF) of some circuit performance metrics (such as power and delay) can be derived with Monte Carlo simulations. Although Monte Carlo simulation is time-consuming, it can provide the best theoretical accuracy by modeling both systematic and random process variations.

Process variations can significantly impact both dynamic and active leakage power distributions of fabricated ICs [6]. Intra-die process variations impact every transistor on a single die differently, while inter-die process variations impact all transistors on one die in the same way. In the context of DPA attacks, process variations can also impact the data-dependence of both dynamic and leakage power. Such impacts are not reflected by the measurement noise or SNR of the same chip, but they can lead to a spectrum of SNR across different chips. Process variations have already been predicted to influence the power model during the security evaluation of embedded cryptosystem design [7]. In [8], process variations are modeled as data-independent power that does not impact power attacks. However, in this paper, we use a statistical approach to demonstrate that process variations can increase the inherent SNR of standard-cell CMOS (sCMOS) gates to facilitate power attacks. More importantly, due to the negative impacts of process variations, many existent gate-level countermeasures against DPA attacks can fail. For example, sense-amplifier-based logic (SABL) gates [9] can resist DPA by balancing the power consumption of differential pairs with symmetrically sized transistors to minimize the inherent SNR. In [3] and [10], differential pair routing approaches for ASIC and FPGA are proposed to balance the interconnect capacitance to achieve even lower SNR. However, we demonstrate here that the presence of process variations is an additional factor to impair the power balance and thus can deteriorate the power-attack tolerance (PAT) of all kinds of power-attack-resistant logic styles. As far as we know, no prior work has considered process variation impacts during the early design phase of embedded cryptosystems.

The remainder of this paper is organized as follows. Section 2 defines and analyzes the statistical PAT. Section 3 shows two cases of simulation-based DPA attack to demonstrate the negative impacts of process variations on MTD. Section 4 proposes a transistor sizing method to compensate for the impacts, following by conclusions and suggestions for future works in Section 5.

2. PRELIMINARIES

The power model of sub-90nm CMOS devices considering both dynamic power and active leakage power is given as:

$$P_{total} = P_{dyn} + P_{leak} = f_{0 \to 1} \cdot C_L \cdot V_{dd}^2 + (I_{sub} + I_{gate} + I_{BTBT}) \cdot V_{dd} \quad (1)$$

Dynamic power is consumed when a 0-to-1 logic transition happens at a circuit node to charge the node capacitance. Active leakage power is consumed by each transistor throughout the circuit functioning time. Both power components are highly data-dependent. For a logic gate with N inputs, the data-dependent dynamic power has $d = 2^{2^N}$ values, if we consider all of the logic

transitions. The data-dependent leakage power has $l = 2^N$ values, if we consider all of the logic combinations. The nominal dynamic or leakage power is calculated by the mean of the data-dependent power values.

With the aggressive technology scaling, process variations can significantly impact the power distribution. The most dominant process variations on power distribution are the threshold voltage V_{th} and the effective channel length L_{eff} [11]. First-order design simulations always model them as a normal distribution. Then the dynamic power will meet a normal distribution, and the leakage power will meet a lognormal distribution [12]. As reported from the ITRS [13] and industry sources, the 3σ intra-die variation of V_{th} and L_{eff} can be as large as 42% and 12% in 45nm technology.

2.1 Statistical Power-Attack Tolerance

In the context of power analysis attacks, we are particularly interested in the data-dependence of power consumption instead of mean power. Strong data-dependent power implies large SNR that facilitates power attacks. For an N-input logic gate, the SNR of dynamic power and leakage power can be calculated by:

$$SNR_{dyn} = \frac{\sigma(P_d)}{\mu(P_d)}, SNR_{leak} = \frac{\sigma(P_l)}{\mu(P_l)}. \quad (d = 2^{2^N}, l = 2^N) \quad (2)$$

where $\sigma(P)$ and $\mu(P)$ are the standard deviation and mean of dynamic/leakage power, with respect to all logic transitions/combinations. We define a metric called power-attack tolerance (PAT) to reflect the ability of a circuit to tolerate power attacks. PAT equals to the inverse of SNR. We can respectively use DPAT and LPAT for dynamic power and leakage power. They can be calculated by:

$$DPAT = \frac{\mu(P_d)}{\sigma(P_d)}, LPAT = \frac{\mu(P_l)}{\sigma(P_l)} \quad (3)$$

Each logic gate has the nominal DPAT and LPAT without process variations. To study the statistical PAT under V_{th} and L_{eff} variations, we perform 8000 iterations of Monte Carlo simulations on an NAND2 gate to achieve enough statistical accuracy. We use NAND2 gate because it can generally represent on-chip logic gates. The transistor parameters in our simulation are based on the Predictive Technology Model (PTM) [14]. In Figure 1, we show the PDFs of DPAT and LPAT. We find that the curves have different skews according to the mean value, which cannot be fitted to a normal or lognormal distribution function. The reason is that DPAT and LPAT do not have an analytical dependence on either V_{th} or L_{eff}. Therefore, we use the Weibull distribution function [15] for curve fitting. The PDF of a Weibull random variable x is expressed by:

$$f(x; \alpha, \beta) = \begin{cases} \dfrac{\alpha}{\beta} x^{\alpha - 1} e^{-x^\alpha / \beta}, & x \geq 0 \\ 0, & x < 0 \end{cases}$$

$$\mu = \beta^{1/\alpha} \Gamma\left(\frac{\alpha + 1}{\alpha}\right)$$

$$\sigma^2 = \beta^{2/\alpha} \left[\Gamma\left(\frac{\alpha + 2}{\alpha}\right) - \Gamma^2\left(\frac{\alpha + 1}{\alpha}\right) \right]$$

$$(4)$$

978-1-60558-497-3/09 $25.00 © 2009 ACM

239

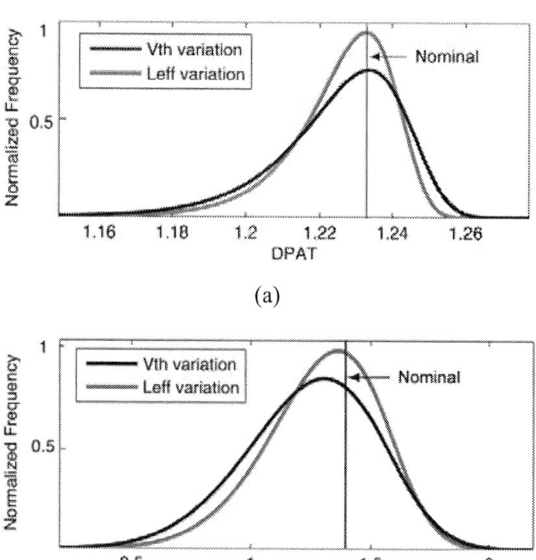

Figure 1. Process variation impacts on: (a) DPAT; (b) LPAT.

In the above expression, α is the shape parameter and β is the scale parameter. The flexibility of the two parameters makes the Weibull distribution a general-purpose function, which can be accommodated to normal distributions, exponential distributions and other distributions.

We show the statistical metrics of the resulting PAT distribution function in Table 1. We can use $\mu(PAT)$ to represent the average PAT under process variations, and $\sigma(PAT)$ to represent the PAT uncertainty under process variations. We find that $\mu(LPAT)$ is negatively biased from the nominal LPAT, which indicates a degraded LPAT due to process variations. The reason behind this is that process variations have significant impacts on leakage power and such impacts vary for different input patterns. To further quantify the degradation of PAT, we calculate the cumulative probability that PAT is less than the nominal value. The degradation probability of DPAT and LPAT are as large as 55% and 66% respectively. This implies that process variations have a negative impact on the PAT of sCMOS gates.

Table 1. Statistical metrics of DPAT and LPAT of sCMOS

	DPAT	LPAT
	V_{th} / L_{eff} variation	V_{th} / L_{eff} variation
Nominal	1.23 / 1.23	1.38 / 1.38
μ (PAT)	1.23 / 1.23	1.25 / 1.30
σ (PAT)	0.017 / 0.014	0.277 / 0.243
P(degradation)	54% / 55%	66% / 60%

2.2 Impacts on SABL

SABL is one of the differential dynamic logic styles that have almost data-independent dynamic power. The schematic of a SABL AND-NAND gate is shown in Figure 2. SABL gates are refreshed during the pre-charge stage to de-correlate the power dependence on logic transition patterns. The pull-down network implements the differential logic that can be sensed by the latch

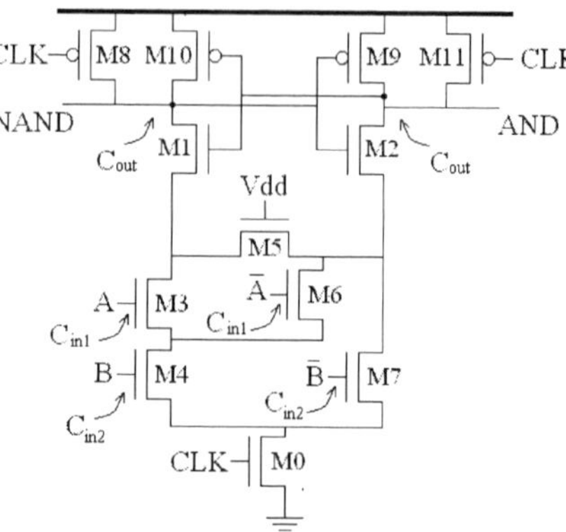

Figure 2. SABL AND-NAND gate without process variations.

on top. The load capacitance (e.g. C_{in1}, C_{in2} and C_{out}) of each differential pair is balanced by full-custom design process, so that the dynamic power for all input patterns is equal.

By introducing the same process variations into the power model, we simulate the PDFs of DPAT and LPAT for a SABL AND-NAND gate. The results are shown in Figure 3, together with the PDFs of the sCMOS NAND for comparison. The PDFs of sCMOS are partly shown, because they are too high compared with those of SABL. The nominal PAT of SABL is inherently 10x larger than that of sCMOS. However, the worst-case process corner can significantly reduce this advantage. Moreover, we can find that LPAT of SABL can deteriorate as low as sCMOS equivalent gate. This indicates that process variations have much more negative impacts on the data-dependent leakage power than the data-dependent dynamic power, especially for large circuits.

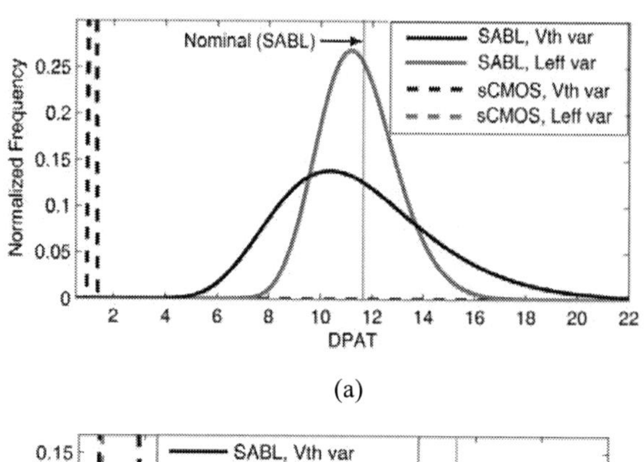

(a)

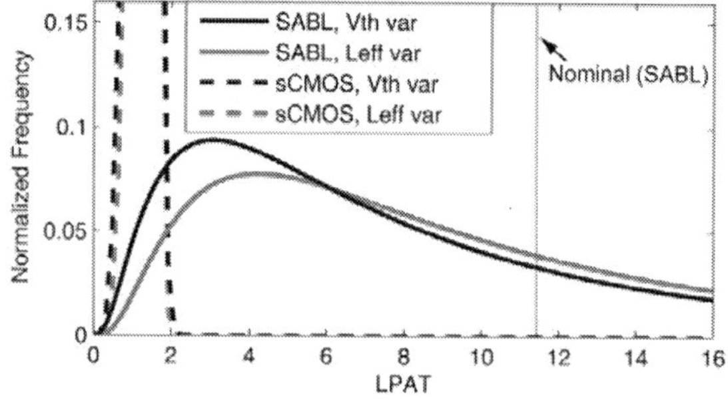

Figure 3. Process variation impacts on SABL gate compared with sCMOS gate: (a) DPAT; (b) LPAT.

We summarize the statistical metrics of DPAT and LPAT in Table 2. Comparing the value of σ, we find LPAT to have a much larger uncertainty due to process variations. The degradation

978-1-60558-497-3/09 $25.00 © 2009 ACM

probability of SABL gates is as large as 59-71%, even more than sCMOS.

Table 2. Statistical metrics of DPAT and LPAT of SABL

	DPAT	LPAT
	V_{th} / L_{eff} variation	V_{th} / L_{eff} variation
Nominal	11.82 / 11.82	11.68 / 11.68
μ (PAT)	11.55 / 11.48	10.75 / 12.45
σ (PAT)	3.15 / 1.52	10.3 / 11.98
P(degradation)	59% / 61%	71% / 64%

2.3 Inter-Die Variation Impacts

Although DPAT of SABL gates is less vulnerable to intra-die variations than LPAT, it can further deteriorate with inter-die variations. Let us assume that every transistor in a SABL AND-NAND has 10% less V_{th} than the nominal value, in addition to the intra-die V_{th} variation used before. In Figure 4, we find that the PDFs with inter-die variations make a left shift to the worse region. In Table 3, we summarize the degradation of PAT due to both intra-die and inter-die variations. The average PAT and the uncertainty of PAT both deteriorate, with the only exception of the DPAT uncertainty. In all, the degradation probability is increased.

Apparently, PAT can be improved by adding positive inter-die V_{th} variation. This indicates that the process variation can be used as a "knob" to calibrate the PAT of an embedded system. Note that the increased V_{th} can lead to performance degradation and power reduction too. Therefore, the optimization of PAT must take other design trade-offs into account.

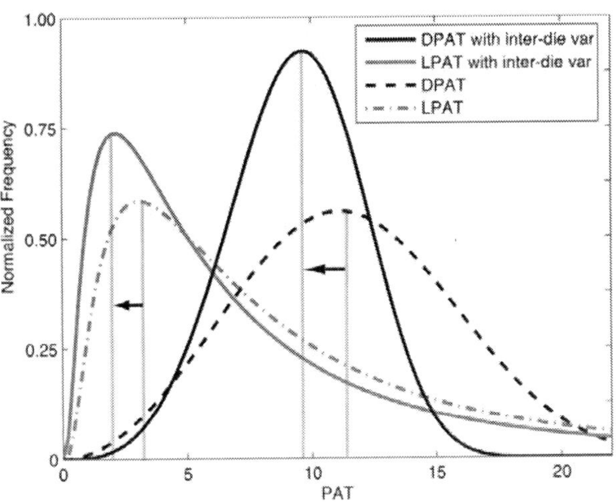

Figure 4. Impacts of additional inter-die process variation on SABL gate

Table 3. Statistical metrics of DPAT and LPAT

	DPAT	LPAT
$\Delta\mu$(PAT)	-2.15	-1.76
$\Delta\sigma$(PAT)	-0.53	1.12
P(degradation)	70%	85%

3. CASE STUDIES

The process variation impacts on the gate-level PAT result in a statistical MTD of DPA attack on a cryptosystem implementation. However, since the impacts on each gate cell of a cryptosystem are not correlated, it is not easy to use the gate-level PAT to derive the statistical MTD. To study this statistical MTD, we need to simulate multiple sets of power traces of a cryptosystem by Monte Carlo method. Then we perform simulation-based DPA attacks on each set of power traces to acquire the corresponding MTD. As an example, we implement a standalone DES (Data Encryption Standard) cryptosystem using 45nm sCMOS library and perform a DPA attack on the fifth substitution-box (Sbox5) during the first round of DES. We use Synopsys Design Compiler to perform the logic synthesis and HSpice to simulate the transient power traces. Piece-Wise-Linear (PWL) voltage source elements are called in HSpice to generate random input plaintexts for the Sbox5 sub-circuit. Then Perl scripts can parse the power trace data and handle the DPA tasks, such as making key guesses, grouping power traces and generating DPCs. The selection function used in our simulations is the Hamming Weight of the Sbox5 ciphertext. The MTD is finally reported, if either a peak dynamic power or a maximum leakage power stands out from the DPCs to indicate a successful key extraction (we set the key as 12 in this example).

Without process variations, our simulation-based DPA attack on the sCMOS DES component results a MTD of 120. Note that the simulated MTD is much smaller than the MTD in a real DPA attack, because a real embedded system has much lower SNR due to the power consumption of non-crypto circuits. By introducing 42% V_{th} intra-die variation, we perform 30 Monte Carlo simulations to get the power traces. It takes 34 hours to finish all these simulations on an Intel 3GHz Pentium 4 processor with 3GB of memory. For each set of power traces, we find out the corresponding MTD. The distribution of the MTDs is shown in Figure 5. The worst-case MTD is as low as 88, which is 27% smaller than the nominal MTD.

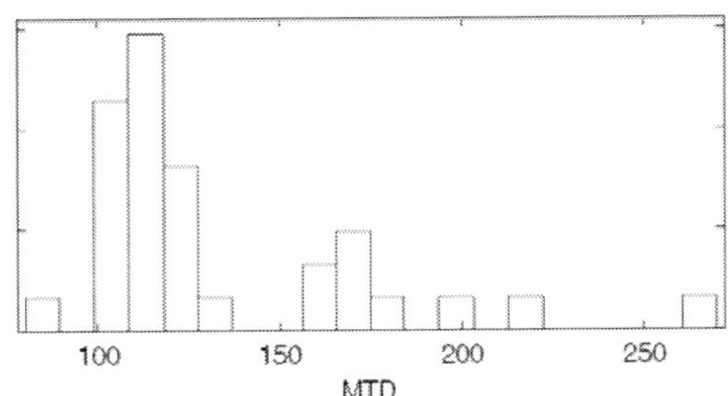

Figure 5. MTD distribution of sCMOS DES

To make a comparison, we design a SABL DES cryptosystem by replacing the sCMOS gates with SABL gate sub-circuits. The nominal MTD is 2300. With the same V_{th} variation, the MTD distribution is shown in Figure 6. The worst-case MTD is 1612, which deteriorates by 30%.

It is difficult to find a fitting curve for the resulting MTD distribution. A high confidence interval to estimate a statistical MTD cannot be achieved, because even thousands of Monte Carlo simulations on a large circuit can take forbidden amount of time. However, the central limit theorem ensures us that the estimation of the mean of MTD can be fairly accurate. Thus, we should use the degradation probability of P(MTD<μ(MTD)) to quantify our results. The degradation probability of MTD is 45% for the sCMOS DES and 57% for the SABL DES. It turns out that process variations have more negative impacts on SABL DES. Note that the above results only consider the intra-die variation. We believe that the addition of inter-die variation can result even higher degradation probability of MTD. Although the process variation impacts cannot be effaced, they can be partially mitigated in secure embedded cryptosystem design.

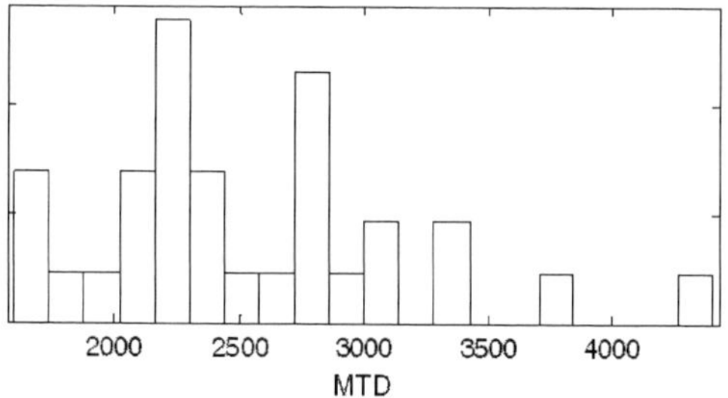

Figure 6. MTD distribution of SABL DES

4. MITIGATIONS

General design methods to deal with process variation impacts on deep submicron circuits are listed in [16]. The increase of transistor channel length can mitigate the L_{eff} variation. For short-channel transistors, the increase of channel length can also increase the threshold voltage and thus mitigate the V_{th} variation [17]. Therefore, we intend to use the transistor sizing method to mitigate process variation impacts on power-attack tolerance. To solve this transistor sizing optimization problem, we can use either the SPICE-based simulation approach or the statistical power estimation approach based on approximate analytical solutions [18]. The former is less efficient than the latter. However, we prefer the SPICE-based approach because power analysis attacks exploit key-dependent power from high-resolution power traces and the evaluation of the power-attack tolerance relies on time-accurate transient power profiles. To design a special-purpose embedded cryptosystem demanding high-level security, the optimization accuracy is more important than the optimization efficiency.

During the optimization procedure, the delay and implementation area penalties caused by sized-up transistors need to be considered. Therefore, we initially set a global transistor sizing constraint to be 10% of the minimum channel length (5nm for 45nm technology), with 1nm sizing resolution [19]. This global constraint can be relaxed if the performance and area are not critical design metrics.

4.1 Gate-level Mitigation

In this work, the ultimate goal is to apply a transistor-level sizing method to decrease the degradation probability of MTD caused by realistic process variations. We assume that each transistor in a logic gate can be individually sized, and the optimized sizing is then applied to the same gate in the design. To optimize the PAT of each gate, two design aims should be considered:

1. Decrease the uncertainty of PAT;
2. Increase the average PAT.

We want to achieve both these two aims, because the failure of one could compromise the success of the other. This can be illustrated in Figure 7. Assume that the original non-optimized PAT distribution of a certain type of logic gate is shown as Design 0. Design 1 is optimized for aim 1, but the degradation probability of PAT is more than 50%. Design 2 is optimized for aim 2, but a small number of chips can have extremely low PAT (less than 5).

Therefore, we should optimize the metric $\mu(PAT)/\sigma(PAT)$ to achieve both the two aims at the gate level.

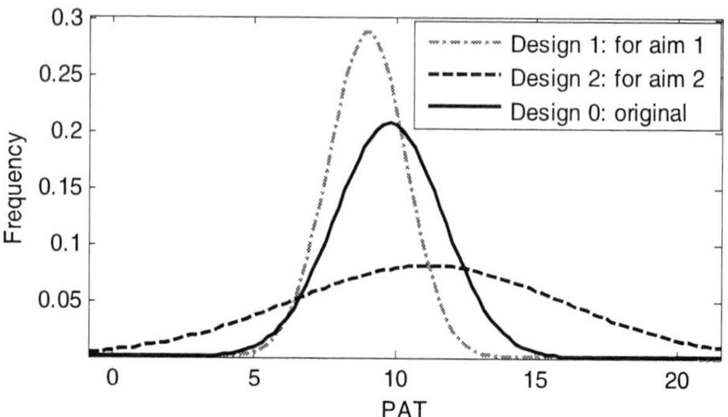

Figure 7. PAT optimization for different design aims.

We demonstrate a PAT optimization of a SABL DES cryptosystem as an example. As a power-attack-resistant logic, the inherent DPAT of SABL gates is ideally infinite (not infinite for a real chip with process variations). The inherent LPAT is not addressed in the original SABL design. As we show before, LPAT is much more sensitive to process variations than DPAT. Therefore, we emphasize more on optimizing the LPAT of SABL gates. To avoid the area overhead of sizing up all the transistors, we focus on the transistors directly connected to the input signals. Thus, we only optimize the length of transistors M3, M4, M6 and M7 in Figure 2. We initialize a local sizing constraint to limit the maximum sizing difference between two transistors to maintain a high DPAT of a power-balanced SABL gate design, and then size up the transistors. Through Monte Carlo simulations, we find that the optimized $\mu(LPAT)/\sigma(LPAT)$ is achieved when the leakage power of 00 pattern and 11 pattern are as similar as possible. As a result, the length of M3 and M6 should be sized up more than M4 and M7 within the global and local sizing constraints.

4.2 System-level Mitigation

At the system level, we aim to mitigate the degradation probability of MTD considering both dynamic power and leakage power. After applying the gate-level optimization, the degradation of MTD can actually become deteriorated. Since the system-level dynamic power has a linear dependence on the total circuit node capacitance, the nodes with sized-up transistors will consume extra amount of dynamic power and the MTD related the dynamic power could be affected. Therefore, we use a heuristic approach to resize some transistors connected to these nodes for a system-level DPAT optimization. If the resizing results still cannot meet the desired MTD, we have to reduce the local sizing constraint at the gate level. The entire design optimization procedure (shown in Figure 8) can take several iterations, and a simulation-based DPA is called to acquire the MTD as long as a new transistor sizing is done. For our DES component with less than thousands of gates, we perform four iterations to reduce the degradation probability of MTD to be 18%, which compensates for the non-optimized design by 40%. The design penalty of our optimization is 0.9% power consumption and 1.5% area overhead. The system performance remains almost the same. For a very large cryptosystem implementation, it is prohibitively expensive to perform too many optimization iterations. Thus, we recommend a divide-and-conquer strategy to optimize only one component of the entire cryptosystem at one time.

978-1-60558-497-3/09 $25.00 © 2009 ACM

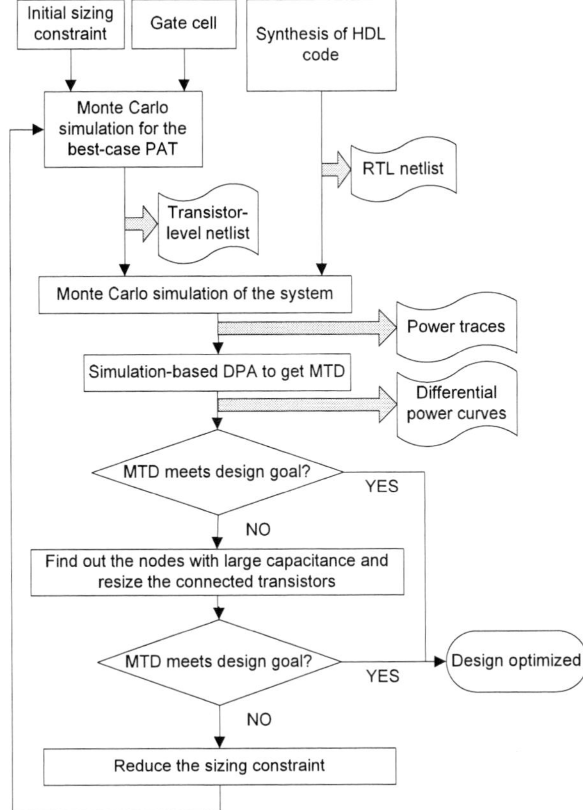

Figure 8. Design optimization procedure of mitigating process variation impacts on power-attack tolerance.

5. CONCLUSIONS

In this work, we analyze the process variation impacts on the power-attack tolerance (PAT) of embedded cryptosystems. We consider a power model with both data-dependent dynamic power and leakage power. Based on Monte Carlo simulations using 45nm CMOS device and process variation models, we demonstrate the negative impacts of process variations on PAT in both standard-cell CMOS and SABL implementation of a DES cryptographic component. Finally, we propose a transistor-level sizing method to mitigate the effect of process variations on PAT and the overall measurement to disclosure (MTD), with minimal design overhead.

Security evaluations of embedded cryptosystems are required at various abstraction levels, with different dominant factors to impact PAT and MTD. For example, the impact of device process variations on PAT could be much less than other impacts such as power supply variations and interconnect variations. However, device process variations are intrinsic parameters of all gates in a cryptosystem, and can be optimized on a netlist prior to the place-and-route design process. Our method for improving PAT can be extended to different design levels with appropriate power models and level-specific optimizations.

ACKNOWLEDGMENTS

We thank Sandip Kundu, Kevin Fu, the members of UMass Amherst VCSG group, and especially the anonymous reviewers for constructive suggestions. We also thank Yan Qi for editing the manuscript. This work was supported in part by the National Science Foundation under Grant CNS-0627529, Semiconductor Research Corporation Task 1595, and Intel Corporation.

6. REFERENCES

[1] P. Kocher, J. Jaffe, B. Jun, "Differential Power Analysis", CRYPTO, LNCS 1666, pp. 388-397, 1999.

[2] S. Mangard, "Hardware Countermeasures Against DPA – A Statistical Analysis of Their Effectiveness", CT-RSA, LNCS 2964, pp. 222-235, 2004.

[3] K. Tiri, I. Verbauwhede, "A digital design flow for secure integrated circuits," IEEE Transaction on CAD, vol. 25(7), pp. 1197-1208, 2006.

[4] K. Tiri, I. Verbauwhede, "Simulation models for side-channel information leaks," ACM/IEEE DAC, pp. 228-233, 2005.

[5] L. Lin, W. Burleson, "Leakage-based differential power analysis (LDPA) on sub-90nm CMOS cryptosystems", IEEE ISCAS, pp. 252-255, 2008.

[6] J. Tschanz, J.Kao, S. Narendra, R. Nair, D. Antoniadis, A. Chandrakasan, V. De, "Adaptive body bias for reducing impacts of die-to-die and within-die parameter variations on microprocessor frequency and leakage," IEEE Journal of Solid-State Circuits, Vol. 37(11), pp. 1396-1402, 2002.

[7] K. Tiri, "Side-channel attack pitfalls," ACM/IEEE DAC, pp. 15-20, 2007.

[8] D. Agrawal, S. Baktir, D. Karakoyunlu, P. Rohatgi, B. Sunar, "Trojan detection using IC fingerprinting," IEEE Symposium on Security and Privacy, pp. 296-310, 2007.

[9] K. Tiri, I. Verbauwhede, "Securing Encryption Algorithms against DPA at the Logic Level: Next Generation Smart Card Technology," CHES, LNCS, vol. 2779, pp. 125-136, 2003.

[10] P. Yu, P. Schaumont, "Secure FPGA circuits using controlled placement and routing," ACM/IEEE CODES+ISSS, pp. 45-50, 2007.

[11] S. Mukhopadhyay, K. Roy, "Modeling and estimation of total leakage current in nano-scaled CMOS devices considering the effect of parameter variation," ACM/IEEE ISLPED, pp. 172-175, 2003.

[12] R. Rao, A. Srivastava, D. Blaauw, D. Sylvester, "Statistical estimation of leakage current considering inter- and intra-die process variation," ACM/IEEE ISLPED, pp. 84-89, 2003.

[13] International Technology Roadmap for Semiconductors, 2006, http://public.itrs.net.

[14] W. Zhao, Y. Cao, "New generation of predictive technology model for sub-45nm design exploration," IEEE ISQED, pp. 585-590, 2006.

[15] W. Mendenhall, and T. Sincich, "Statistics for engineering and the sciences," 5th edition, by Prentice Hall, 2007.

[16] S. Bhunia, S. Mukhopadhyay, K. Roy, "Process variations and process-tolerant design," IEEE VLSI Design, pp. 699-704, 2007.

[17] K. Takeuchi, T. Tatsumi, A. Furukawa, "Channel engineering for the reduction of random-dopant-placement-induced threshold voltage fluctuation," IEEE IEDM, pp.841-844, 1997.

[18] R. Brodersen, M. Horowitz, D. Markovic, B. Nikolic, V. Stojanovic, "Methods for true power minimization", ACM/IEEE ICCAD, pp. 35-42, 2002.

[19] P. Gupta, A. Kahng, P. Sharma, D. Sylvester, "Selective gate-level biasing for cost-effective runtime leakage control," ACM/IEEE DAC, pp. 327-330, 2004.

978-1-60558-497-3/09 $25.00 © 2009 ACM

Evaluating Design Trade-offs in Customizable Processors

Unmesh D. Bordoloi[1]   Huynh Phung Huynh[2]   Samarjit Chakraborty[3]   Tulika Mitra[2]

[1]Verimag Labs, France
[2]Department of Computer Science, National University of Singapore
[3]Institute for Real-Time Computer Systems, TU Munich, Germany
unmesh.bordoloi@verimag.fr, {huynhph1, tulika}@comp.nus.edu.sg, samarjit@tum.de

ABSTRACT

The short time-to-market window for embedded systems demands automation of design methodologies for customizable processors. Recent research advances in this direction have mostly focused on single criteria optimization, e.g., optimizing performance though custom instructions under pre-defined area constraint. From the designer's perspective, however, it would be more interesting if the conflicting trade-offs among multiple objectives (e.g., performance versus area) are exposed enabling an informed decision making. Unfortunately, identifying the optimal trade-off points turns out to be computationally intractable. In this paper, we present a polynomial-time approximation algorithm to systematically evaluate the design trade-offs. In particular, we explore performance-area trade-offs in the context of multi-tasking real-time embedded applications to be implemented on a customizable processor.

Categories and Subject Descriptors

C.3 [**Special-Purpose and Application-Based Systems**]: Real-time and embedded systems

General Terms

Algorithms, Design, Performance

Keywords

ASIP, Processor customization, Multi-objective design space exploration, Pareto-optimal curve.

1. INTRODUCTION

Instruction-set extensible processors that consist of existing processor cores extended with application specific custom instructions are now very popular in the embedded systems domain. Typically, a custom instruction encapsulates the computation of a frequently executed subgraph of the program's dataflow graph. Custom instructions are implemented as hardwired datapath of the existing processor core and this helps improve the performance of the application. Some examples of commercial customizable processors include Lx, ARC[TM] core, Xtensa, Stretch S5 among others.

Permission to make digital or hard copies of part or all of this work for personal or classroom use is granted without fee provided that copies are not made or distributed for profit or commercial advantage and that copies bear this notice and the full citation on the first page. To copy otherwise, to republish, to post on servers or to redistribute to lists, requires prior specific permission and/or a fee.
DAC'09, July 26-31, 2009, San Francisco, California, USA

The past decade has seen a flurry of research activity on automated identification and selection of custom instructions for an application or an application domain. However, most of these design efforts have focused on single-objective optimization, for example, choosing an optimal set of custom instructions either in terms of performance or energy. As the performance/energy improvement offered by custom instructions come at the cost of silicon area, this optimization is typically constrained by a pre-defined silicon area. The designer, on other other hand, can benefit significantly if the automation tools expose all the conflicting trade-offs (e.g., performance versus area) instead of offering a point solution. It is then up to the designer to choose an appropriate trade-off point. More formally, we are interested in generating the *Pareto-optimal curve* in a multi-objective design space (e.g., performance and area) where (a) no point is better than any other point on the curve with respect to both objectives, and (b) no improvement can be made in any objective without trading-off or worsening the other.

Unfortunately, it turns out that computing the Pareto-optimal curve for our design problem is computationally intractable. Therefore, state-of-the-art customization toolchains (such as Tensilica's XPRES compiler [15]) adopt ad-hoc methods that simply compute the best performing design choices at arbitrary silicon area constraints. In this paper, we propose a systematic methodology to explore the performance-area trade-offs in designing customizable processors. We present a polynomial-time approximation algorithm to compute this trade-off. Moreover, as the Pareto curve may potentially contain exponential number of design points, it is impossible to compute this entire set in polynomial time. Hence, our polynomial-time approximation algorithm, by default, has to approximate the (potentially exponential size) set of points on the Pareto curve with only a polynomial number of points. In a typical design cycle of customizable processors, the system designer inspects all the points in the Pareto curve and then selects one, or at most a few implementations. Hence, from a practical perspective, we feel it is more meaningful if the designer is presented with a reasonably few well-distinguishable trade-offs rather than an exponentially large number of solutions, many of which are very similar to each other. Our approximation algorithm is therefore not only attractive in terms of time-complexity, but also returns more meaningful solutions, in terms of the size of the solution set (including the spread/distribution of solutions in this set).

We explore this approximation solution to Pareto curve generation in the context of multi-tasking real-time embed-

978-1-60558-497-3/09 $25.00 © 2009 ACM

ded applications running on customizable processors. Given a multi-tasking application to be implemented on a customizable processor, there are a large number of implementation possibilities with different subsets of custom instructions leading to varying processor utilization versus area trade-offs. We would like to identify *all* schedulable implementations that expose the different performance trade-offs.

Our contributions: Formally, for any schedulable implementation, let (U, A) denote the corresponding utilization of the base processor and the hardware cost arising from the use of custom instructions. We are then interested in identifying all possible Pareto-optimal solutions $\{(U_1, A_1), \ldots, (U_n, A_n)\}$ that capture the different performance trade-offs [7]. Each (U_i, A_i) in this set has the property that there does not exist any schedulable implementation with a performance vector (U, A) such that $U \leq U_i$ and $A \leq A_i$, with at least one of the inequalities being strict. Further, let S be the set of performance vectors (i.e., (utilization, area) tuples) corresponding to all schedulable implementations. Let P be the set of performance vectors $(U_1, A_1), \ldots, (U_n, A_n)$ corresponding to all the Pareto-optimal solutions. Then for any $(U, A) \in S - P$ there exists a $(U_i, A_i) \in P$ such that $U_i \leq U$ and $A_i \leq A$, with at least one of these inequalities being strict (i.e., the set P contains *all* performance trade-offs). The vectors $(U, A) \in S - P$ are referred to as *dominated solutions*, since they are *dominated* by one or more Pareto-optimal solutions.

In this paper we present a polynomial-time approximation scheme for computing the *utilization-area* Pareto curve. Our proposed solution for computing this Pareto curve involves two distinct stages. First, in the *intra*-task stage, *each* individual task is analyzed. Given the library of possible custom instructions for the task, all possible custom instruction configurations are generated exposing the *workload-area* Pareto curve. In the *inter*-task stage, we consider *all* the tasks in the task-set and their *workload-area* configurations, to generate the processor *utilization-area* Pareto curve P for the overall task set. Our framework for approximately computing the trade-offs extends to both of the above stages.

Related Work: Any custom instruction selection problem starts with the identification of a large number of candidate custom instructions from the program's dataflow graph. [2, 5] have proposed different techniques for this problem. Given such a library of valid custom instructions, various approaches have been proposed to select custom instructions to optimize either the performance or the hardware area. For example, different methods that have been proposed include dynamic programming [1], 0-1 Knapsack [6], greedy heuristic [4, 5], and ILP [11] which maximize the performance under different design constraints, e.g., hardware area. However, the above lines of work were aimed at optimizing a single criteria. In this work, we propose a framework for *multi-objective* optimization to expose the performance versus hardware area trade-offs. Moreover, the above approaches do not take into account the real-time constraints. [9, 18] explore processor customization in the context of real-time systems but only perform single-objective (performance) optimization.

The algorithmic techniques presented in this paper have been motivated by [13]. The result that for *any* Pareto curve and *any* ϵ, there *exists* a polynomial-size ϵ-approximate Pareto curve was shown in [13]. However, for many problems, ef-

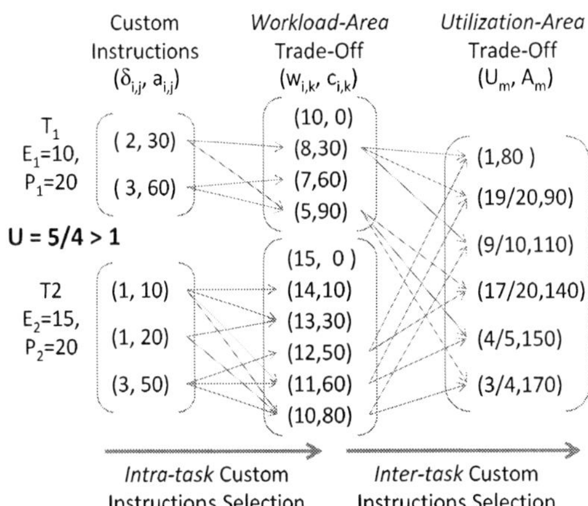

Figure 1: Motivating Example.

ficiently (i.e., in polynomial time) *computing* such approximate Pareto curves might not be possible. Our main technical contribution in this paper is to show that such ϵ-approximate Pareto curves can be efficiently computed in the domain of custom instruction selection. An important consequence of this result is a formal basis for custom instruction selection. It shows that in the quest for efficiency, there is no need to resort to ad-hoc techniques – as currently adopted in state-of-the-art customization toolchains – for identifying best-performing custom instruction choices.

Here, it should be noted that in this paper we have taken a classical approximation algorithms standpoint, where the goal is to provide *formal* guarantees on the quality of the results obtained. Our work differs from the existing large body of work on multiobjective optimization [7] that relies on heuristics and randomized search techniques such as evolutionary algorithms.

2. PROBLEM STATEMENT

2.1 Task Model

We use a very general sporadic task model [3] in a preemptive uniprocessor context. Thus, we are interested in the custom instruction selection for a task set $\tau = \{T_1, T_2, \ldots, T_m\}$ consisting of m hard real-time tasks, with the constraint that the task set is schedulable. Any task T_i can get triggered independently of other tasks in τ. Each task T_i generates a sequence of jobs; each job is characterized by the three parameters – the *minimum separation* (P_i) which is the minimum time interval that must elapse before the successive job of the task T_i is triggered, the *deadline* (D_i) by which each job generated by T_i must complete since its release time, and *workload* (E_i) or the worst case execution requirement of any job generated by T_i.

Throughout this paper, we assume the underlying scheduling policy to be the earliest deadline first (EDF). Assuming that for all tasks T_i, $D_i \geq P_i$, the schedulability of the task set τ can be given by the following condition.

THEOREM 1. *A set of sporadic tasks τ is schedulable under EDF if and only if $(U = \sum_{i=1}^{N} \frac{E_i}{P_i}) \leq 1$ where U is the processor utilization due to τ [3, 12].*

2.2 Intra-Task Custom Instructions Selection

We now state the intra-task custom instruction selection problem for a task T_i. For the task T_i, let there be n_i custom

978-1-60558-497-3/09 $25.00 © 2009 ACM

instruction candidates. Each of these n_i custom instructions is associated with a certain hardware area. Choosing the jth custom instruction will lower the workload of the task T_i by $\delta_{i,j}$. Equivalently, the new workload will be $E_i - \delta_{i,j}$. Hence, for each task T_i we have a set of choices $S_i = \{(\delta_{i,1}, a_{i,1}), \ldots, (\delta_{i,n_i}, a_{i,n_i})\}$, where $a_{i,j}$ is the hardware cost associated with the jth custom instruction. Our objective is to select a set of custom instructions that would minimize the workload on the base processor, as well as use the minimum amount of hardware area for custom instructions. In other words, our goal is to compute all *workload-area* trade-offs in the form of a Pareto curve $\{(w_{i,1}, c_{i,1}), \ldots, (w_{i,N_i}, c_{i,N_i})\}$, where $w_{i,j}$ is the workload of task T_i accelerated with a particular set of custom instructions and $c_{i,j}$ is the corresponding cost in terms of silicon area.

In Figure 1, we illustrate this problem for two different tasks T_1 and T_2. The task characteristics for T_1 are $\{E_1 = 10, P_1 = 20\}$. T_1 has two entries in its library of custom instructions, and thus, $n_1 = 2$. Following our notation, $\delta_{1,1} = 2$ and $\delta_{1,2} = 3$, and the corresponding hardware areas are $a_{1,1} = 30$ and $a_{1,2} = 60$. The goal is to identify the *workload-area* Pareto curve for the task T_1. For example, in this case the Pareto curve consists of four solutions $\{(10, 0), (8, 30), (7, 60), (5, 90)\}$. Note that the solution with zero area does not use any custom instruction. Therefore, the application workload is not reduced and is the highest amongst all the solutions. On the other hand, the solution $(5, 90)$ contains both the custom instructions and has the smallest workload with the largest area. The task characteristics for T_2 and the custom instruction candidates may be read in the same way as described above for T_1.

2.3 Inter-Task Custom Instructions Selection

For each task T_i, let there be N_i hardware implementation choices $\{(w_{i,1}, c_{i,1}), \ldots, (w_{i,N_i}, c_{i,N_i})\}$ in the *workload-area* Pareto curve that is computed by the intra-task custom instruction selection phase. Each of these N_i choices represent a custom instruction configuration for the task. A task may be chosen to run in one these configurations where it will incur certain a hardware cost $c_{i,j}$ and would lower the execution time of the task on the processor from E_i to $w_{i,j}$.

However, in a real-time embedded system there is not one task but a set of tasks running on a processor. Thus, a designer is interested not in the performance of a single task, but rater the *utilization* of the processor by the entire task-set. The goal in the inter-task custom instruction selection phase is to identify one custom instruction configuration for each task, which would minimize the overall processor utilization and minimize the total hardware cost. Therefore, similar to intra-task customization, we generate a *Pareto curve* containing the *Pareto-optimal* set of *utilization-area* vectors $\{(U_1, A_1), \ldots, (U_n, A_n)\}$ with the trade-off between processor utilization U and hardware area A.

In Figure 1, we showed the intra-task custom instruction selection for two different tasks. For T_1 and T_2 we have 4 and 6 elements in the custom instruction configurations respectively. For example, for the task T_1, we have $\{(w_{1,1} = 10, c_{1,1} = 0), (w_{1,2} = 8, c_{1,2} = 30), (w_{1,3} = 7, c_{1,3} = 60), (w_{1,4} = 5, c_{1,4} = 90)\}$. The goal is to identify the *utilization-area* Pareto curve for the task set $\{T_1, T_2\}$. As shown in Figure 1, in this case the Pareto curve consists of six solutions $\{(1, 80), (\frac{19}{20}, 90), \ldots, (\frac{3}{4}, 170)\}$. Without using any custom instructions, the task set $\{T_1, T_2\}$ is not schedulable with $U = \frac{5}{4} > 1$. But the use of custom in-

structions speed up task executions, thereby lowering the utilization. This yields six schedulable solutions with conflicting *utilization-area* trade-offs.

3. EVALUATING DESIGN TRADE-OFFS

We shall now present our algorithms for efficiently computing the Pareto curves in the *intra-task* and the *inter-task* custom instruction selection phases. Note that computing the exact Pareto curves in both these cases is computationally intractable. First, such Pareto curves would typically contain an exponential number of trade-off points (which obviously cannot be computed in polynomial time). Second, it may be shown using a reduction from the classical Knapsack problem, that the problem of computing even one point on the Pareto curve is NP-hard. Hence, our algorithms approximate both – the number of points on the Pareto curve, as well as the "coordinates" of these points on the curve.

3.1 Intra-Task Trade-offs

In what follows, we first present a pseudo-polynomial time dynamic programming algorithm (called DP) to compute the *exact* Pareto curve, which is then used to devise an approximation scheme. Let the maximum cost associated with any custom instruction be M, i.e., $M = max_{(j=1,2,\ldots,n_i)} a_{i,j}$, where n_i is the number of custom instruction candidates for the task T_i and $a_{i,j}$ is the hardware cost associated with the jth custom instruction. Let $\Omega_{k,j}$ be the minimum workload that might be achieved by considering only a subset of custom instructions of task T_i from $\{1, 2, \ldots, k\}$ when the cost is exactly j. The algorithm DP first initializes $\Omega_{0,j} = \infty$ for $j = \{1, 2, \ldots, n_i M\}$ and $\Omega_{j,0} = E_i$ for $j = \{0, 1, \ldots, n_i\}$. Note that $n_i M$ is an upper bound on the total hardware cost that might be incurred. After initialization, the DP computes the values of $\Omega_{k,j}$ ($k = 1$ to n_i) by the recursion:

$$\Omega_{k,j} \leftarrow \min\{\Omega_{k-1,j}, \Omega_{k-1,j-a_{i,k}} - \delta_{i,k}\} \qquad (1)$$

That is, given an area j, we explore all possible configurations and choose the one that results in minimum workload for custom instructions $1, \ldots, k$.

After running DP, we retain the undominated solutions from amongst the solutions in the final iteration ($k = n_i$) to obtain the exact *workload-area* Pareto curve, $\{(w_{i,1}, c_{i,1}), \ldots, (w_{i,N_i}, c_{i,N_i})\}$, where N_i is the size of the Pareto curve. This algorithm runs in pseudo-polynomial time $O(n_i^2 M)$, and will suffer from long running times. Hence, our goal is to *approximately* compute this curve in *polynomial* time.

Our approximation scheme takes an error parameter ϵ as input and returns an ϵ-*approximate Pareto curve* denoted as ϵ-Pareto curve (or $\mathcal{P}_\epsilon$). Given a Pareto curve $\mathcal{P} = \{(x_1, y_1), \ldots, (x_p, y_p)\}$, an ϵ-approximate Pareto curve $\mathcal{P}_\epsilon$ is defined as *any* set $\mathcal{P}_\epsilon = \{(x_1', y_1'), \ldots, (x_q', y_q')\}$ such that for any $(x_i, y_i) \in \mathcal{P}$, there exists a $(x_j', y_j') \in \mathcal{P}_\epsilon$ for which $x_j' \leq (1 + \epsilon)x_i$ and $y_j' \leq (1 + \epsilon)y_i$. In other words, corresponding to any point on the Pareto curve $\mathcal{P}$, there exists a point on $\mathcal{P}_\epsilon$, each of whose coordinates are at most ϵ distance away from the corresponding coordinates of the point on $\mathcal{P}$.

Papadimitriou and Yannakakis [13] has shown that for any multi-objective optimization problem and any ϵ, there *exists* a polynomial-sized ϵ-approximate Pareto curve $\mathcal{P}_\epsilon$. Further, [13] showed that a necessary and sufficient condition for computing such a $\mathcal{P}_\epsilon$ in polynomial time is the existence of a polynomial-time algorithm for solving, what was referred to as the *GAP problem*. In what follows, we state the version

978-1-60558-497-3/09 $25.00 © 2009 ACM

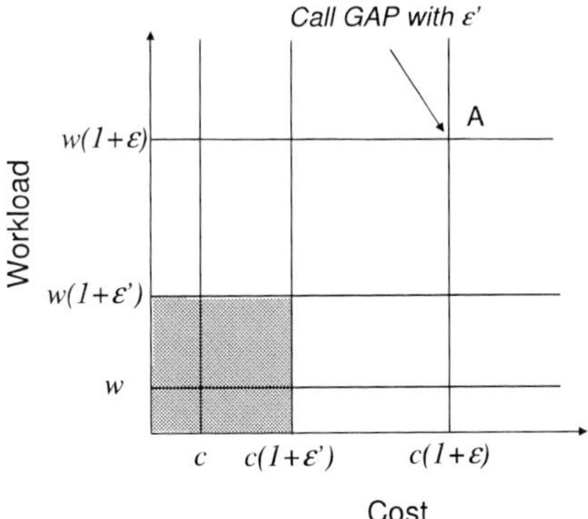

Figure 2: Solving the GAP problem for the corner point A will either return a dominating solution or declare that there is no solution in the shaded area.

of the GAP problem that arises in our setting and show that it can be solved in polynomial time. Finally, we outline our scheme to compute the approximate Pareto curves using the polynomial time GAP subroutine.

3.1.1 The GAP Problem

For a two-dimensional multiobjective optimization problem, the GAP problem can be stated as follows: Given a vector $b = (b_1, b_2)$, either return a solution vector which dominates b, or report that there is no solution better than b by at least a factor of $1 + \epsilon$ in both dimensions. In our setting, the objective is to minimize the workload of a task T_i, $W(S) = E_i - \sum_{j=1}^{n_i} x_{i,j} \delta_{i,j}$ and the cost $C(S) = \sum_{j=1}^{n_i} x_{i,j} a_{i,j}$, where $x_{i,j}$ is a boolean variable which is true if the jth custom instruction is chosen for the solution S. Hence, the corresponding GAP problem can be stated as follows.

Problem Statement: Given a cost c, workload w and an $\epsilon \geq 0$, either return a solution S such that $C(S) \leq c$ and $W(S) \leq w$, or else declare that there is no solution S such that $C(S) \leq \frac{c}{1+\epsilon}$ and $W(S) \leq \frac{w}{1+\epsilon}$.

Solving the GAP Problem: We now present a polynomial-time algorithm to solve this GAP problem. It involves the following two steps:

Step 1: *Transforming Costs*
Let $r = \lceil \frac{n_i}{\epsilon} \rceil$. Modify each cost $a_{i,j}$ of task T_i to $a'_{i,j}$ such that $a'_{i,j} = \lceil \frac{a_{i,j} r}{c} \rceil$. Now, consider the problem of determining whether there exists a solution with the modified costs such that $C'(S) \leq r$. Let us call this problem GAP'. We shall show that solving GAP is equivalent to solving GAP'. Towards this, we enumerate two properties below. These properties can be easily proved with algebraic equations.
  (a) If a solution with the transformed costs satisfies $C'(S) \leq r$, then $C(S) \leq c$.
  (b) If a solution satisfies $C(S) \leq \frac{c}{1+\epsilon}$, then $C'(S) \leq r$.

From property (a), we know that if this problem returns an affirmative answer then the GAP problem would also return a dominating solution. On the other hand, if GAP' returns a negative answer then property (b) leads to the conclusion that there is no solution with cost $\leq c/(1+\epsilon)$. Hence, from the above properties we can infer that solving GAP' is equivalent to solving the original GAP problem.

Step 2: *Solving GAP'*
We present a dynamic programming algorithm to solve the GAP' problem. This algorithm can be constructed with the following adjustments to Algorithm DP.

Algorithm 1 Approximating the Pareto curve.

1: Partition the range of costs from 1 to $n_i M$ geometrically with a ratio $1 + \epsilon' = (1 + \epsilon)^{1/2}$, thus dividing the cost space into $O(log_{1+\epsilon} n_i M)$ coordinates.
2: For each coordinate b, call *Algorithm DP* with transformed costs $a'_{i,j} = \lceil \frac{a_{i,j} r}{b} \rceil$, where $r = \lceil \frac{n_i}{\epsilon} \rceil$.
3: For each run of Step 2, find the solution with the minimum utilization.
4: Retain all the undominated solutions from the solutions found in Step 3. This will represent a ϵ-Pareto curve.

---

1. Run Algorithm DP with the modified costs $a'_{i,j}$.
2. Instead of iterating over all the cost values up to $n_i M$, iterate only up to a cost value of r, where $r = \lceil \frac{n_i}{\epsilon} \rceil$.
3. Finally, if the minimum value in the final array computed by Algorithm DP is such that it is $\leq w$, then return the solution otherwise declare that there is no solution.

Computing each row of the table built by this dynamic programming algorithm requires $O(r)$ running time. Hence, this algorithm runs in time $O(n_i^2/\epsilon)$.

The above polynomial time subroutine for solving the GAP problem proves the existence of a *fully polynomial-time approximation scheme* (FPTAS) for computing the approximate workload-area Pareto curve $\mathcal{P}_\epsilon$ which is polynomial in the input size and in $1/\epsilon$. This is because the following FPTAS can be devised using the algorithm for solving GAP. First, geometrically partition the objective space along all dimensions with a ratio $1 + \epsilon'$, where $\epsilon' = (1 + \epsilon)^{1/2} - 1$. For each corner point of this grid, call the GAP routine (i.e., the algorithm for solving GAP) with the parameter ϵ', and keep all the undominated solutions (see Figure 2 for an illustration of this procedure). This implies that for each rectangle which contains a solution in the exact Pareto curve, there will also be a solution within the same rectangle which belongs to $\mathcal{P}_\epsilon$. The *distance* between these two solutions can be bounded using the dimensions of the rectangle. Hence, for every solution s in the Pareto curve, there exists a solution q in $\mathcal{P}_\epsilon$ such that $\frac{q}{(1+\epsilon)} \leq s$. Moreover, because the number of rectangles is polynomially bounded, it follows that the number of points in $\mathcal{P}_\epsilon$ will also be a polynomial.

Algorithm 1 summarizes the above steps to compute the ϵ-approximate *workload-area* Pareto curve in some more detail. Note, that in step 1 of Algorithm 1 we partition only the area space (and not both workload and area space). This is because if a point (w, c) dominates the corner (w_1, c_1) and $w_1 < w_2$, then (w, c) definitely dominates (w_2, c_2) . In steps 2 and 3, we scale the costs, run Algorithm DP for every co-ordinate in the partitioned cost space and retain the minimum workload at each co-ordinate. The runtime complexity of this algorithm is $O(\frac{n_i^2}{\epsilon} log_{1+\epsilon} n_i M)$.

3.2 Inter-Task Trade-offs

The existence of an inter-task approximation scheme to compute the *utilization-area* Pareto curve may be argued in the same fashion as for the intra-task approximation scheme described above. This scheme takes the set of pareto-optimal solutions $\mathcal{P}_i$ for each task T_i as input (as shown in the previous section), and generates the set of global design trade-offs $\bar{\mathcal{P}}$ for the entire task set. Each global design configuration $S \in \bar{\mathcal{P}}$ contains contains exactly one solution from each $\mathcal{P}_i$ (for each task T_i). If a brute-force approach is

978-1-60558-497-3/09 $25.00 © 2009 ACM

247

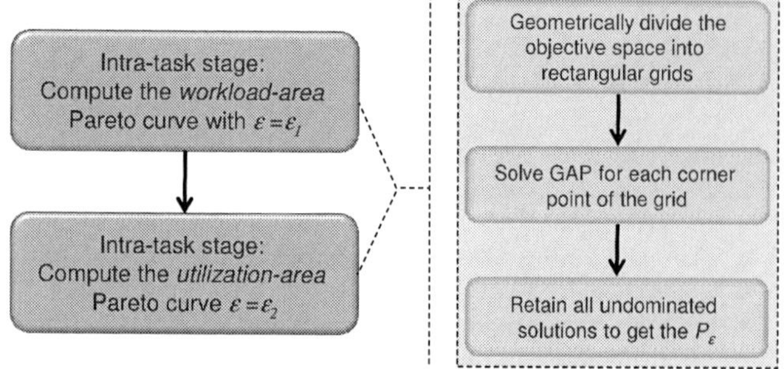

Figure 3: The overall two-stage approximation scheme.

Task Set	Benchmarks
1	cjpeg, adpcm, aes, compress, rijndael ispell
2	djpeg, g721decode, cjpeg, ispell, adpcm jfdctint, aes
3	cjpeg, ispell, edn, sha, g721decode, djpeg compress, ndes
4	adpcm, rijndael, cjpeg, ispell, sha ndes, djpeg, compress, edn
5	aes, djpeg, g721decode, rijndael, jfdctint cjpeg, edn, ispell, sha, ndes

Table 1: Composition of the task sets.

used to examine all possible global design configurations, then $|\mathcal{P}_1| \times \ldots \times |\mathcal{P}_m|$ solutions will have to be examined, where $|\mathcal{P}_i|$ denotes the number of solutions in the set $\mathcal{P}_i$. Hence, the number of solutions grow exponentially with the number of tasks m in the task set, and moreover, not all of these solutions would be optimal (i.e., they would be *dominated* by some Pareto-optimal solution). Our goal is to instead efficiently (but approximately) compute just the Pareto-optimal global design configurations.

Broadly, the procedure follows the same steps as described in Algorithm 1. Due to space constraints, we shall not work out all the details of this stage. However, note that the main difference is the core dynamic programming recursion that is invoked in Step 2 (Algorithm 1), which we present below. Let $U_{i,j}$ be the minimum utilization that might be achieved by considering only a subset of tasks from $\{1, 2, \ldots, i\}$ when the cost is exactly j. Then, $U_{i,j}$ is defined recursively as below, where $\{w_{i,k}, c_{i,k}\} \in \mathcal{P}_i$, (workload-area Pareto curve), and E_i and P_i denote the *execution* time and the *minimum separation* of the task T_i.

$$U_{i,j} \leftarrow \min \left\{ \begin{array}{l} U_{i-1,j}, U_{i-1,j-c_{i,1}} - (E_i - w_{i,1})/P_i \\ \vdots \\ U_{i-1,j}, U_{i-1,j-c_{i,N_i}} - (E_i - w_{i,N_i})/P_i \end{array} \right\} \quad (2)$$

We summarize the overall two-stage approximation scheme in Figure 3. There are two distinct stages: (i) the intra-task stage to compute the *workload-area* Pareto curve for each task, and (ii) the *inter-task* stage which generates the *utilization-area* Pareto curve for the entire task set. The scheme for approximating the Pareto curve follows the three main steps shown on the right hand of Figure 3. Note that at each stage, the approximation scheme takes as an input an error parameter ϵ (chosen by the designer) and returns an ϵ-*approximate* Pareto curve. These parameters might be different for the two stages.

4. EXPERIMENTAL EVALUATION

In this section we report some of the experimental results obtained by running our approximation algorithm on a set of well-known benchmarks. We compare the running times of the optimal algorithm against our approximation scheme, and also illustrate the difference in the sizes of P_ϵ (the approximate Pareto curves) and the exact Pareto curve.

Experimental setup: We use five WCET benchmarks [14] *(compress, jfdctint, ndes, edn, adpcm)*, two benchmarks *(aes, sha)* from MiBench [8], three benchmarks *(g721 encoder, djpeg, cjpeg)* from MediaBench [10] and one *(ispell)* Trimaran benchmark [16] for our experiments.

Given the C code of an application, we use the Trimaran compilation framework to generate optimized intermediate code, as well as profiling information such as basic block execution frequencies. We then construct the data flow graph for each basic block and enumerate all possible custom instructions with at most 4 input operands and 2 output operands [17]. We use Synopsys design tools with 0.18 micron CMOS cell libraries to synthesize and estimate latency/area of custom instructions. Execution cycles of a custom instruction is its latency normalized against a Multiply-Accumulate (MAC) operation, which has 1 cycle latency in the processor running at 120MHz. Hardware area is represented in terms of the number of adders. Further, we assume a single-issue in-order base processor core with a perfect cache. The workload is measured in number of processor cycles required to execute each task.

We create five task sets (see Table 1) with the number of tasks in each set varying from 6–10. We chose a total utilization for the task set (without any custom instructions) and then select individual minimum separation period for each task (P_i) to achieve the corresponding utilization. We also chose five different utilization factors $U = 0.80, 1.00, 1.05, 1.08$ and 1.10. When $U = 0.8$ or 1.0, a task set is schedulable without using any custom instructions. In these cases, we are interested in finding out by how much we can reduce the utilization through custom instructions and what are the hardware trade-offs. For $U > 1.0$, a task set is not schedulable on its own. Here, the goal is to find schedulable solutions by using custom instructions as well as expose the performance-area trade-offs. All CPU times reported below are measured on a desktop with 3.0 GHz Pentium 4 CPU and 1 GB RAM.

Running times: Table 2 shows the running time speedups resulting from our approximation scheme, compared to computing the exact Pareto curve for three different values of ϵ, for each of the five task sets. Computing the exact Pareto curve for task sets 1–5 require 139.78 sec, 514.20 sec, 622.32 sec, 747.17 sec, and 711.52 sec, respectively. Even for small values of ϵ ($\epsilon = 0.44$) our approximation algorithm runs about three orders of magnitude faster than the exact algorithm. For larger values of ϵ (e.g., $\epsilon = 3$), the speedups are even more significant (note that ϵ need not be ≤ 1). The reason behind choosing the values 0.21, 0.44, and 0.69 for ϵ is as follows. Our approximation algorithm involves the computation of $(1 + \epsilon)^{1/2}$. This value might turn out to be an irrational number if ϵ is not carefully chosen. Hence, to avoid any possible rounding-off errors in our implementation, the above values were chosen for ϵ.

Pareto curve size: The *workload-area* Pareto curve (the output of intra-task stage) and the *utilization-area* Pareto curve (the output of intra-task stage) typically contain an exponential number of points. The approximation algorithm generates a polynomial-sized approximate Pareto curve $\mathcal{P}_\epsilon$. We now compare the sizes of the exact Pareto curve and $\mathcal{P}_\epsilon$.

978-1-60558-497-3/09 $25.00 © 2009 ACM

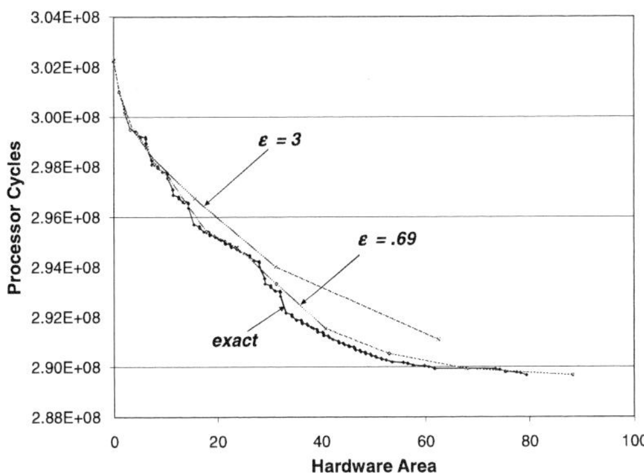

(a) *workload-area* Pareto curve for *g721decode*

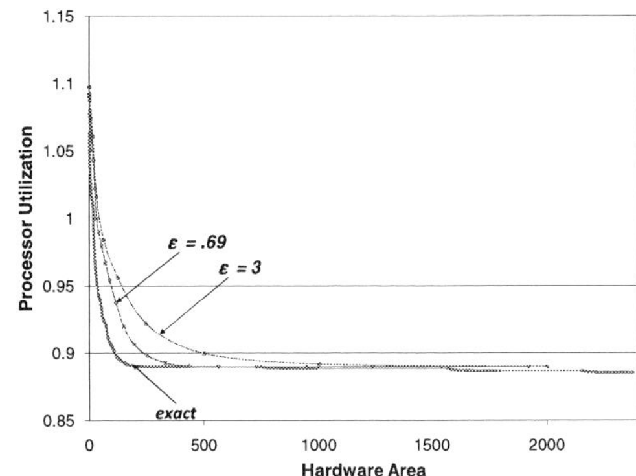

(b) *utilization-area* Pareto curve for task set 1

Figure 4: The exact and approximate Pareto curves for $\epsilon = 0.69, 3$.

Task Sets:	1	2	3	4	5
$\epsilon = 0.21$	643	1075	1037	990	729
$\epsilon = 0.44$	3248	5918	5712	5457	3933
$\epsilon = 0.69$	7106	14587	14389	13922	10208
$\epsilon = 3.0$	29615	72525	89285	69054	77508

Table 2: Speedup obtained from our approximation scheme for the task sets $1 - 5$.

For the intra-task results, Figure 4(a) shows the exact Pareto curve and the P_ϵ generated by our algorithm for the *g721decode* benchmark. For the inter-task case, we show the results for task set 1 in Figure 4(b). For clarity, we have only plotted $\mathcal{P}_\epsilon$ for $\epsilon = 0.69$ and 3. Note that (i) the number of points in P_ϵ decrease as ϵ increases, and (ii) the gap between the exact and approximate curves widen with larger values of ϵ, implying that the relative error indeed increases. We would like to report that even for small values of ϵ (e.g., $\epsilon = 0.21$), P_ϵ contains almost 97% fewer points compared to the exact Pareto curve. Similar trends were seen for all the other benchmarks and task sets, which are not shown here due to space constraints.

Benefits of approximation: Although the running times associated with constructing the exact Pareto curve might seem to be small (10–12 mins), in an interactive design process where the designer repeatedly makes changes and generates new Pareto curves, this might hamper designer productivity. A tool which generates these curves faster (e.g., using our proposed approximation algorithms) would be more usable. Secondly, exact Pareto curves return too many (similar) design trade-offs. Approximate Pareto curves return less, well spread out trade-offs, which might be more manageable for the designer.

5. CONCLUDING REMARKS

In this paper, we proposed a framework for evaluating trade-offs in custom instruction selection for instruction set customizable processors. This framework consists of two stages – in the first custom instruction configurations representing different trade-offs are chosen for each task, and in the second, different configurations from each task are chosen to derive system-level trade-offs. There are a couple of directions in which this framework can be further refined, e.g., by modeling isomorphic instructions across tasks or by accounting for more general scheduling policies for task sets (currently only EDF has been modeled).

6. REFERENCES

[1] M. Arnold and H. Corporaal. Designing domain-specific processors. In *CODES*, 2001.

[2] K. Atasu, L. Pozzi, and P. Ienne. Automatic application-specific instruction-set extensions under microarchitectural constraints. In *DAC*, 2003.

[3] S. Baruah, A.K. Mok, and L.E. Rosier. Preemptively scheduling hard-real-time sporadic tasks on one processor. In *RTSS*, 1990.

[4] N. Cheung, S. Parameswaran, and J. Henkel. INSIDE: INstruction Selection/Identification & Design Exploration for extensible processors. In *ICCAD*, 2002.

[5] N. Clark, H. Zhong, and S. Mahlke. Processor acceleration through automated instruction set customization. In *MICRO*, 2003.

[6] J. Cong, Y. Fan, G. Han, and Z. Zhang. Application-specific instruction generation for configurable processor architectures. In *FPGA*, 2004.

[7] K. Deb. *Multi-Objective Optimization Using Evolutionary Algorithms*. John Wiley & Sons, 2001.

[8] M. R. Guthaus et al. Mibench: A free, commercially representative embedded benchmark suite. In *IEEE Annual Workshop on Workload Characterization*, 2001.

[9] H. P. Huynh and T. Mitra. Instruction-set customization for real-time systems. In *DATE*, 2007.

[10] C. Lee, M. Potkonjak, and W. H. Mangione-Smith. Mediabench: a tool for evaluating and synthesizing multimedia and communicatons systems. In *MICRO*, 1997.

[11] J. Lee, K. Choi, and N. Dutt. Efficient instruction encoding for automatic instruction set design of configurable ASIPs. In *ICCAD*, 2002.

[12] C. L. Liu and James W. Layland. Scheduling algorithms for multiprogramming in a hard-real-time environment. *Journal of the ACM*, 20(1):46–61, 1973.

[13] C. H. Papadimitriou and M. Yannakakis. On the approximability of trade-offs and optimal access of web sources. In *FOCS*, 2000.

[14] F. Stappert. WCET benchmarks. http://www.c-lab.de/home/en/download.html.

[15] Tensilica - XPRES Compiler - Optimized Hardware Directly from C. www.tensilica.com/products/devtools/hw_dev/xpres/.

[16] Trimaran: An infrastructure for research in backend compilation and architecture exploration. http://www.trimaran.org.

[17] P. Yu and T. Mitra. Scalable custom instructions identification for instruction-set extensible processors. In *CASES*, 2004.

[18] P. Yu and T. Mitra. Satisfying real-time constraints with custom instructions. In *CODES+ISSS*, 2005.

978-1-60558-497-3/09 $25.00 © 2009 ACM

A Design Flow for Application Specific Heterogeneous Pipelined Multiprocessor Systems

Haris Javaid   Sri Parameswaran

School of Computer Science and Engineering, University of New South Wales, Sydney, Australia

{harisj, sridevan}@cse.unsw.edu.au

ABSTRACT

This paper describes a rapid design methodology to create a pipeline of processers to execute streaming applications. The methodology is in two separate phases: the first phase, uses a heuristic to rapidly search through a large number of processor configurations (configurations differ by the base processor, the additional instructions and cache sizes) to find the near Pareto front; the second phase, utilizes either the above heuristic or an ILP (Integer Linear Programming) formulation to search a smaller design space to find an appropriate final implementation. By the utilization of the fast heuristic with differing runtime constraints in the first phase, we rapidly find the near Pareto front. The second phase provides either an optimal or a near optimal solution. Both the ILP formulation and the heuristic find a system with the smallest area, within a designer specified runtime constraint. The system has efficiently explored design spaces with over 10^{12} design points.

We integrated this design methodology into a commercial design flow and evaluated our approach with different benchmarks (JPEG Encoder, JPEG Decoder and MP3 Encoder). For each benchmark, the near Pareto front was found in a few hours using the heuristic (took several days for the ILP). The results show that the average area error of the heuristic is within 2.5% of the optimal design points (obtained using ILP) for all benchmarks.

Categories and Subject Descriptors

C.1.3 [**Other Architectural Styles**]: Heterogeneous (hybrid) Systems, Pipeline Processors; C.4 [**Performance of Systems**]: Modeling Techniques, Design Studies

General Terms

Algorithms, Design, Performance

Keywords

MPSoCs, Design Space Exploration, Integer Linear Programming

1. INTRODUCTION

The miniaturization of transistors, coupled with the demand for functionality, performance, and low power has resulted in the emergence of Multi Processor System on Chip (MPSoC) based embedded devices in the market. MPSoCs can be categorized either as homogeneous or heterogeneous. Heterogeneous MPSoCs usually have a smaller footprint than homogenous MPSoCs, and consume less power by mapping an application's tasks onto the most suitable processing elements. The recent emergence of Application Specific Instruction Set Processors (ASIPs) [1, 2, 3] has seen the use of a homogeneous platform to produce heterogeneous processors for seamless use in embedded applications. Both coarse- and fine-grained

Permission to make digital or hard copies of part or all of this work for personal or classroom use is granted without fee provided that copies are not made or distributed for profit or commercial advantage and that copies bear this notice and the full citation on the first page. To copy otherwise, to republish, to post on servers or to redistribute to lists, requires prior specific permission and/or a fee.
DAC'09, July 26-31, 2009, San Francisco, California, USA

parallelism in an application can be exploited using ASIPs in an MP-SoC. An ASIP's architecture and instruction set can be tuned based upon the needs of a specific task, improving performance while minimizing area.

Heterogeneous pipelined multiprocessor architectures are where processing entities are connected in a pipelined fashion via queues. The incoming data stream is processed by each pipeline stage which may contain one or more processing elements. The data stream goes through each stage until the output is finally written by the last pipeline stage. Each of the stages can be customized to suit a particular part of the application which is executed by that pipeline stage processor(s). Pipelined heterogeneous MPSoCs provide a practical implementation platform for streaming applications with high performance gains and reduced area footprint [4].

In this paper, we present a design methodology for implementing streaming applications onto pipelined heterogeneous multiprocessor systems. A partitioned application is taken and the standalone tasks of the application are assigned to processors (ASIPs) in the pipeline. An iterative design flow is used to obtain a design space (consisting of ASIP configurations) which satisfies a set of criteria. Once an approximate design space is selected, either ILP or the heuristic (with design space pruning) can be used to obtain the final design point (set of ASIP configurations) with minimum area under runtime constraint provided by the designer.

The rest of the paper is organized as follows: Section 2 provides the related work and Section 3 provides the background knowledge. The problem addressed in this work is formalized in Section 4 while Section 5 explains the proposed design methodology. Section 6 provides the experimental setup with the results presented in Section 7. Finally, Section 8 summarizes the paper.

2. RELATED WORK

Numerous multiprocessor architectures have been used to implement multimedia applications. For example, a real time video and graphics management system was described in [5] and an HDTV system in [6]. Researchers have also explored different techniques such as loop pipelining and pipelined scheduling of tasks to speed up applications using multiprocessor architectures [7, 8, 9]. In contrast to these works, we focused on mapping streaming applications on ASIP-based pipelined systems to achieve high performance with reduced area footprint owing to heterogeneity of pipelined MPSoCs.

Integer Linear Programming (ILP) [10] is a widely used technique in design space exploration and optimization of multiprocessor architectures [11, 12, 13]. However, none of the previous works specifically targeted ASIP-based pipelined systems which is a viable implementation platform for streaming applications [4]. Since ILP based approaches can be slow for designing complex systems, researchers have proposed heuristics to efficiently explore the design space of multiprocessor systems. Sun et al. [14] examined multi-ASIP systems, simultaneously mapping and scheduling tasks, in addition to custom instruction selection for ASIPs. The authors in [15] represented applications as cyclic directed graphs and explored mapping and partitioning of the application onto homogeneous pipelined multiprocessor architecture.

Implementation of streaming applications onto heterogeneous pipelined multiprocessor systems using ASIPs was explored in [16] and [17]. A heuristic is proposed in [16] to rapidly explore the design space consisting of available ASIP configurations. As compared to [16], this paper focuses on the problem of finding a design with minimum area while a given runtime constraint is satisfied. This is because streaming applications exhibit soft real time constraints and there is no need to add on extra area once the runtime constraint is satisfied. In contrast to [16], Javaid et al. [17] used ILP to find an optimal design for an ASIP-based pipelined multiprocessor system. The ILP formulation in [17] considered single pipeline systems (i.e., only one processor per stage) and only a case study was performed on JPEG. However, this work proposes a complete framework (ILP and a novel heuristic) for design of pipelined multiprocessor systems. To the best of our knowledge, this is the first complete work using both ILP and a heuristic to address the implementation of application specific systems in the context of heterogeneous pipelined multiprocessor systems.

3. BACKGROUND

3.1 Application Model

Sequential applications with the following characteristics are targeted in this work: one, the application contains a kernel which is executed multiple times; and two, the operations in the kernel are independent of each other so that their execution can be overlapped. Note that we refer to the number of times an application kernel is executed as the number of iterations of that application. Standalone tasks of a partitioned application can be executed on different processors communicating with each other via queues. We created four partitioned applications: JPEG Encoder Single Pipeline (JESP); JPEG Encoder Multiple Pipeline (JEMP); JPEG Decoder (JD); and, MP3 Encoder (MP3E). Figure 1 shows the partitioned applications where tasks are connected through queues denoted by the arrows.

3.2 Pipelined Multiprocessor Architecture

A pipelined system consists of processing entities connected in a pipeline using queues (FIFOs) which allow communication at a much higher bandwidth, devoid of the contention typically exhibited by a shared bus architecture. Each stage of the pipeline can contain one or more processors (ASIPs) to execute some part of the application. We assume that the depth of each pipeline stage is one, that is, there cannot be more than one ASIP in series in a pipeline stage. However, there can be more than one ASIP in parallel and we refer to that pipeline stage as a parallel pipeline stage. For example, stage 3 of JEMP in Figure 1 will be a parallel pipeline stage (when mapped to a pipelined system). Further to this assumption, the output of an ASIP in stage i can only be connected to other ASIPs in stage $i + 1$. A pre-partitioned application is taken and the tasks of the application are mapped onto these ASIPs. Each ASIP in the pipelined system has a number of available configurations. ASIP configurations differ by the additional instructions they contain and by the sizes of their instruction and data caches. Additional instructions for an ASIP are generated according to the task mapped on that particular ASIP. A commercial ASIP design tool from Tensilica Inc. [3] is used to automatically generate ASIP configurations from a base processor. Base processors can be identical for each ASIP in the pipelined system or can be different. The processor configuration generation is controlled using a parameter which we refer to as the overhead granularity. The granularity parameter specifies the minimum amount of difference (in terms of gates) between two sets of additional instructions for the same base processor. Permutation of the sets of additional instructions, and instruction and data cache sizes make up the tailored configurations for each of the ASIPs. Area of an ASIP configuration includes base processor, additional instructions (if any) and instruction and data cache sizes.

Given a pipelined multiprocessor system, we used an estimation equation (Equation 1) to calculate system runtime. $R^i(s_i)$, $R^f(s_i)$, $L^1(s_i)$ and $L(s_i)$ refer to initialization time, finalization time, first latency (due to the cache misses on cold start) and averaged latency of pipeline stage i respectively, while s_c stands for critical stage. Critical stage is the one with the worst L form amongst all the stages. I and M refer to the number of iterations of the application (being run on the pipelined system) and the total number of pipeline stages respectively. First latency of a processor is the first iteration's execution time while averaged latency is the average execution time for rest of the $I - 1$ iterations. Initialization and finalization time is the time to execute the non-kernel operations which are executed only once at the start and end of the application. The equation calculates

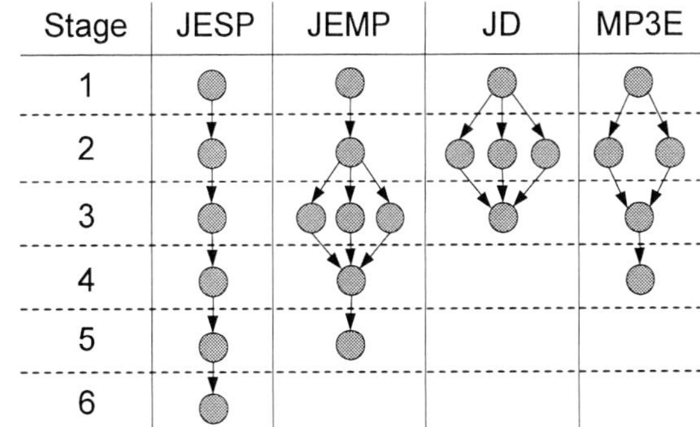

Figure 1: Benchmark Applications

the system runtime by summing up the initialization time of the first stage, time to fill up the empty pipeline, time spent by the critical stage once the pipeline is filled and finalization time of the last stage. For a parallel pipeline stage, the processor with the worst latency is used as the stage latency because it will hide the latency of other processors in that stage. Further details can be found in [18].

$$\mathbb{R} = R^i(s_1) + \sum_{i=1}^{M} L^1(s_i) + (I - 1) \times L(s_c) + R^f(s_M) \quad (1)$$

For an ASIP-based pipelined system where each ASIP can be configured in one of the available configurations, system runtime with a particular set of ASIP configurations can be calculated using Equation 1 given individual configuration latencies. Thus, all the ASIP configurations are simulated separately with their respective tasks mapped on them to record the timings and area information. These recorded values can then be used to calculate system runtime and system area for any combination of ASIP configurations. The area of the pipelined system is the summation of the area cost of all the processors, where area of a processor is measured in terms of gates. Since pre-partitioned applications are used, the size of FIFOs will be constant, thus FIFO areas are not included in the design space exploration. In this work, we also do not consider the area of the processor memories (except for caches).

4. PROBLEM DEFINITION

A partitioned application can be represented as a Directed Acyclic Graph (DAG) where vertices represent tasks and edges represent the data dependencies between the tasks. Thus, an application partitioned for a pipeline design can be represented as graph $G^a = (V, E)$. The set V contains the tasks of the partitioned application. Data dependencies between the stand alone tasks are given by the edges in set E. A pipelined multiprocessor system can also be represented as graph $G^p = (P, F)$ where P is the set of ASIPs in the pipelined system, while F represents the FIFO channels between the ASIPs for communication.

Mapping of a partitioned application on a pipelined system can be defined as a mapping from G^a to G^p. Each task in G^a is mapped to corresponding ASIP in G^p. Similarly, data dependencies from G^a are mapped to corresponding FIFOs in G^p. Once the tasks are mapped to ASIPs and data dependencies to FIFOs in the pipelined system, one configuration of each ASIP is selected to obtain a set of ASIP configurations as the final design. The cost function A to be minimized is defined as summation of the area of all the ASIPs in the pipelined system. The minimization of A is performed while ensuring the system runtime (calculated using Equation 1) remains within the runtime constraint R_c provided by the designer. The problem of selection of configurations (one for each ASIP) with respect to minimization of cost function A and runtime constraint R_c is solved. This selection problem is the design space exploration problem being solved in this paper, and a design flow (refer to Section 5) is proposed for rapid exploration of large design spaces.

5. DESIGN FLOW

The overall design methodology to address the selection problem (described in Section 4) is shown in Figure 2. The methodology consists of two phases: Design Space Generation; and, Design

978-1-60558-497-3/09 $25.00 © 2009 ACM

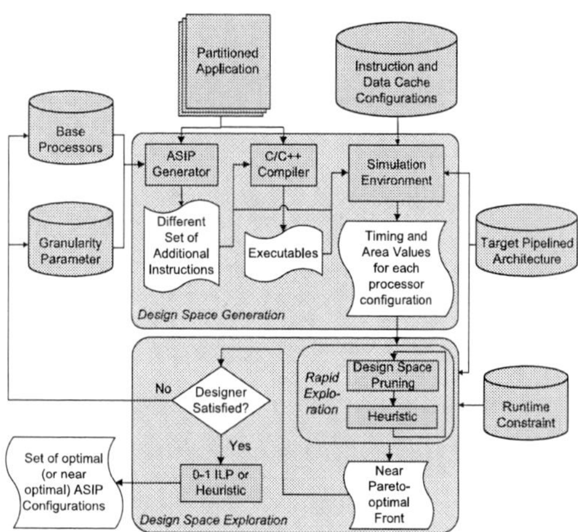

Figure 2: Proposed Design Flow

Space Exploration. The designer provides the partitioned application, pipeline architecture and runtime constraint as input. The partitioned application is provided as separate C code files for each of the ASIPs in the pipelined system.

As shown in Figure 2, the design space generation phase is determined by the base processors, the overhead granularity, and the instruction and data cache configurations. Note that in this design flow the base processors are specified by the designer. While it is possible to include the selection of the base processors in the design flow, experience suggests that the designer prefers to choose the base processors. ASIP configurations are generated using a commercial tool from Tensilica Inc. [3] as described in section 3.2. Using the input parameters, a designer can control the amount of design space to be generated for a particular application. As shown in Figure 2, simulation results are used to record the timing and area values for all the ASIP configurations which will then be used in the exploration phase.

The design space exploration phase shown in Figure 2 is based on a heuristic and/or a 0-1 ILP formulation. At the output of this phase, one configuration of each ASIP is selected so that the resulting pipelined system has minimum area while its runtime is less than runtime constraint R_c. The first step is to perform a rapid design space exploration (shown in Figure 2) with the help of the heuristic and obtain the near Pareto front of the design space. The near Pareto front will reveal the area trend of the design space and will guide the designer to change the input parameters such as base processors, overhead granularity, etc. to obtain a new design space with reduced area footprint. The rapid exploration phase calls the pruning algorithm and the heuristic with all the possible runtime constraints that can be supported by the generated design space in steps of R_s. R_s is the step amount between the runtime constraints and is specified by the designer. This provides the designer with the near Pareto front of the design space in just a few hours for a typical multimedia application.

The Rapid Exploration in Figure 2 can be performed as many times as required until a design space is found which satisfies the designer. This approach provides an opportunity to tweak the design in several hours (excluding simulation time) as compared to ILP which may take days as shown in Section 7. After the approximate design space is finalized, 0-1 ILP (with design space pruning and provided runtime constraint) can be used (only at the end) to obtain the optimal design as shown in Figure 2. If the 0-1 ILP times out, the heuristic can be used instead to obtain the near optimal set of ASIP configurations. The designer can also use a different partitioning of the application and change the instruction and data cache configurations for generation of the new design space (which is not shown in Figure 2 due to space restrictions).

5.1 Design Space Pruning

A number of ASIP configurations can exceed the runtime constraint specified by the designer, thus those ASIP configurations can be removed from the design space. A generalized version of a pruning algorithm in [17] is used in this work. This generalization is applied to pipeline systems with parallel pipeline stages (stages having more than one ASIP in parallel). Readers interested in further details are referred to [18].

5.2 ILP Formulation

The ASIP configuration selection problem can be stated as follows: *Given a pipelined architecture with ASIP configurations and the system runtime constraint R_c, one configuration is selected for each ASIP such that system area is minimized while system runtime satisfies the provided runtime constraint R_c.* The selection problem is directly mapped to a 0-1 ILP problem. The mathematical details of the ILP formulation are omitted but can be found in [18].

5.3 Heuristic

As ILP is exhaustive in the worst case and the focus of this work is to target large design spaces, we propose a novel heuristic which can be used for rapid design space exploration. First of all, an upper bound of the latency of the pipelined system is calculated using the provided runtime constraint R_c as shown in Equation 2. The equation subtracts the maximum possible L^1s of all the stages from R_c and calculates an upper limit of the latency which is applied over all the pipeline stages. Thus, all the ASIP configurations having L greater than L_u are removed. From the remaining configurations, one configuration for each ASIP is selected which has the minimum area. This ensures that the system runtime (calculated using selected configurations) will always be less than the runtime constraint, since the sum of maximum L^1s for all the stages is subtracted before calculating L_u. It should be noted that for some ASIPs all the configurations could have been removed, and thus the heuristic will not be able to find a solution. An error signal is generated where a solution could not be found. This is most likely to happen for runtime constraints that are very close to minimum runtime of the design space.

$$L_u = \frac{R_c - \sum_{i=1}^{M} \max\{L^1(s_i)\}}{(I - 1)} \qquad (2)$$

The heuristic analyzes each configuration of all the ASIPs only once. Hence, its complexity is $O(\sum_{i=1}^{M} N_i \times K_{max})$ where N_i is the number of ASIPs in stage i and K_{max} is the number of configurations for the ASIP with the maximum number of configurations from amongst all the ASIPs in the pipelined system.

6. EXPERIMENTAL SETUP

We integrated the proposed methodology into a commercially available design environment from Tensilica Inc. [3]. The Xtensa LX family of processors with RB-2007.1 toolset is used for our experiments. This toolset includes a C/C++ compiler, Instruction Set Simulator (ISS) and the Xtensa Modeling Protocol (XTMP – used to simulate multiprocessor platforms). The RB-2007.1 toolset also includes the Xtensa PRocessor Extension Synthesis (XPRES), which analyzes the C code and automatically generates new instructions. The generated additional instructions may consist of a combination of fused operations, vector operations, FLIX instructions [19] and specialized operations [20].

We used *lp_solve* [21], a free application, to solve the formulated *0-1 ILP* problem. The design space pruning algorithm and heuristic are implemented in Perl. The whole process of design space generation and exploration as shown in Figure 2 is automated (in Perl) and integrated into the Tensilica's design environment [3].

7. RESULTS & ANALYSIS

The results are presented in two parts: first, we compare the effectiveness of the heuristic with the 0-1 ILP; and second, we show how our methodology can be used to tweak designs in several hours using JPEG decoder as an example. We developed 4 benchmarks to test our methodology as shown in Figure 1. Using different base processors and overhead granularity values, we generated different design spaces for each of the benchmarks. The generated design spaces were in the range of 10^{12} to 10^{16} design points.

Table 1 compares the effectiveness of the proposed heuristic to the ILP. Both the heuristic and ILP (with design space pruning) are executed with runtime constraints spanning the whole design space. Exploring the whole design space provides the Pareto front of the design space that can guide the designer, so that late changes can be

978-1-60558-497-3/09 $25.00 © 2009 ACM

Benchmark	Technique (points)	Total Time	Av. AE	Max. AE
JESP	ILP (1261)	4 hrs	-	-
	Heuristic (971)	17 mins	0.99%	2.72%
JEMP	ILP (1190)	42 days	-	-
	Heuristic (949)	16 mins	0.48%	2.22%
JD	ILP (7000)	28 days	-	-
	Heuristic (7000)	2 hrs	2.22%	10.46%
MP3E	ILP (10700)	18 days	-	-
	Heuristic (10100)	3 hrs	0.25%	5.03%

Table 1: Comparison of Heuristic with ILP

made to the design. In Table 1, Column 3 shows the timing statistics of ILP and the heuristic, while columns 4-5 show the average area error (Av. AE) and maximum area error (Max. AE) respectively for the design points obtained via the heuristic (from the best obtained by the ILP). A timeout of 24 hours is used for the ILP in all the benchmarks. Average Area Error (column 4) is calculated on the basis of number of runtime constraints provided referred to as points in column 2. For example, 1261 different runtime constraints were used for ILP while exploring JESP design space. As explained in Section 5.3, there can be some runtime constraints for which the heuristic is unable to find a solution. This leads to different number of runtime constraints for ILP and the heuristic. The average area error for all the benchmarks is less than 2.5%, and the worst area error in all the benchmarks is 10.46%. Most importantly, the total time to explore the design space (column 3) differs by several magnitudes for ILP and the heuristic. For JEMP, 42 days were spent in obtaining the Pareto front, while the heuristic revealed the near Pareto front in only 16 mins (with 2.22% maximum area error). Thus, the heuristic can be used for rapid design exploration as shown in Figure 2. The number of points (runtime constraints) to be explored depends on the value of runtime step R_s which is specified by the designer. For our experiments, $R_s = 1,000$ is used for JESP, JEMP, and JD and $R_s = 10,000$ is used for MP3E so that there are enough points for average calculation, and thus a reasonable comparison can be done.

To show the effectiveness of our design methodology and how rapid exploration can be used to optimize the area of a pipelined system, we performed a case study on JD. The initial JD implementation was the same as shown in Figure 1 except that we used same base processors to generate and simulate a design space of 5.2×10^{12} design points in 20 hours. The near Pareto front of the design space is shown in Figure 3 marked as 'Same Base Processors' which was obtained in 2 hours using the rapid exploration phase. For sake of simplicity, not all the design points are shown in this graph. We observed that the change in area for runtime range of 1.4×10^7 to 0.9×10^7 clock cycles is miniscule. This means that the processor configurations in the final design are being underutilized for runtimes greater than 0.9×10^7 clock cycles. Thus, we decided to use different base processors. First stage is reading and entropy decoding, second stage is dequantization and DCT, and third stage is color space conversion and writing back to file. A more complex processor is used for stage 1, but a simple processor is used for stage 3 which is not computationally intensive. The second graph annotated as 'Different Base Processors' shows the near Pareto front of the design space with different base processors. As can been seen, there is a significant area reduction for most of the runtimes. This strategy can be repeated again and again until the designer is satisfied. It should be noted that it took only one day (including the simulation time) to generate and explore the new design space. The designer could well choose to only use the ILP solution with the provided runtime constraint, and if the optimal design point obtained is not within the area budget, a new design space can be generated. However, the solution time of the ILP cannot be guaranteed. Thus, for large design spaces the heuristic is more likely to reveal the near Pareto front quickly. Once the designer is satisfied with the design space (after Rapid Exploration in Figure 2), either ILP or the heuristic with the provided runtime constraint can be used to obtain the final optimal or near-optimal design point.

8. CONCLUSION

We presented a framework for efficient implementation of application specific heterogenous pipelined multiprocessor systems. A formal methodology consisting of a 0-1 ILP formulation and a heuristic for design space exploration is presented. A design space pruning algorithm is also used which can enable the use of ILP for large design

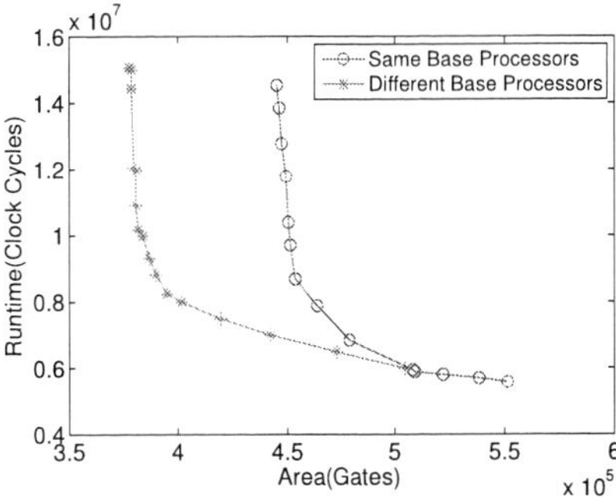

Figure 3: Near Pareto front of JPEG Decoder with same base processors and different base processors

spaces. Our experiments show that the average area error for all the benchmarks when using the heuristic is less than 2.5% and thus, the heuristic can be used for rapid design space exploration. Using the proposed design flow, we were able to explore and obtain optimal or near-optimal designs in one or two days for design spaces in order of 10^{12} design points.[1]

9. REFERENCES

[1] Altera Nios Processor. Altera Corp. (http://www.altera.com).
[2] ARC the leader in configurable processor technology. ARC International (http://www.arc.com).
[3] Xtensa Processor. Tensilica Inc. (http://www.tensilica.com).
[4] S. L. Shee, A. Erdos, and S. Parameswaran. Heterogeneous multiprocessor implementations for jpeg:: a case study. In *CODES+ISSS '06: Proceedings of the 4th international conference on Hardware/software codesign and system synthesis*, pages 217–222, New York, NY, USA, 2006. ACM.
[5] M. Strik, A. Timmer, J. van Meerbergen, and G.-J. van Rootselaar. Heterogeneous multiprocessor for the management of real-time video and graphics streams. *Solid-State Circuits, IEEE Journal of*, 35(11):1722–1731, Nov 2000.
[6] A. Beric, R. Sethuraman, C. Pinto, H. Peters, G. Veldman, P. van de Haar, and M. Duranton. Heterogeneous multiprocessor for high definition video. *Consumer Electronics, 2006. ICCE '06. 2006 Digest of Technical Papers. International Conference on*, pages 401–402, 7-11 Jan. 2006.
[7] T. Kodaka, K. Kimura, and H. Kasahara. Multigrain parallel processing for jpeg encoding on a single chip multiprocessor. In *IWIA '02: Proceedings of the International Workshop on Innovative Architecture for Future Generation High-Performance Processors and Systems (IWIA'02)*, page 57, Washington, DC, USA, 2002. IEEE Computer Society.
[8] S. Banerjee, T. Hamada, P. Chau, and R. Fellman. Macro pipelining based scheduling on high performance heterogeneous multiprocessor systems. *Signal Processing, IEEE Transactions on*, 43(6):1468–1484, 1995.
[9] J. Jeon and K. Choi. Loop pipelining in hardware-software partitioning. In *Asia and South Pacific Design Automation Conference*, pages 361–366, 1998.
[10] T. H. Cormen, C. E. Leiserson, R. L. Rivest, and C. Stien. *Introduction to Algorithms*. MIT Press and MCGraw-Hill, Second edition, 2001.
[11] J. DeSouza-Batista and A. Parker. Optimal synthesis of application specific heterogeneous pipelined multiprocessors. *Application Specific Array Processors, 1994. Proceedings., International Conference on*, pages 99–110, 22-24 Aug 1994.
[12] S.-R. Kuang, C.-Y. Chen, and R.-Z. Liao. Partitioning and pipelined scheduling of embedded system using integer linear programming. In *ICPADS '05: Proceedings of the 11th International Conference on Parallel and Distributed Systems - Workshops (ICPADS'05)*, pages 37–41, Washington, DC, USA, 2005. IEEE Computer Society.
[13] M. Schwiegershausen and P. Pirsch. A formal approach for the optimization of heterogeneous multiprocessors for complex image processing schemes. In *EURO-DAC '95/EURO-VHDL '95: Proceedings of the conference on European design automation*, pages 8–13, Los Alamitos, CA, USA, 1995. IEEE Computer Society Press.
[14] F. Sun, S. Ravi, A. Raghunathan, and N. K. Jha. Synthesis of application-specific heterogeneous multiprocessor architectures using extensible processors. In *VLSID '05: Proceedings of the 18th International Conference on VLSI Design held jointly with 4th International Conference on Embedded Systems Design*, pages 551–556, Washington, DC, USA, 2005. IEEE Computer Society.
[15] J. Cong, G. Han, and W. Jiang. Synthesis of an application-specific soft multiprocessor system. In *FPGA '07: Proceedings of the 2007 ACM/SIGDA 15th international symposium on Field programmable gate arrays*, pages 99–107, New York, NY, USA, 2007. ACM.
[16] S. L. Shee and S. Parameswaran. Design methodology for pipelined heterogeneous multiprocessor system. In *DAC '07: Proceedings of the 44th annual conference on Design automation*, pages 811–816, New York, NY, USA, 2007. ACM.
[17] H. Javaid and S. Parameswaran. Synthesis of heterogeneous pipelined multiprocessor systems using ilp: jpeg case study. In *CODES/ISSS '08: Proceedings of the 6th IEEE/ACM/IFIP international conference on Hardware/Software codesign and system synthesis*, pages 1–6, New York, NY, USA, 2008. ACM.
[18] H. Javaid and S. Parameswaran. Synthesis of application specific heterogeneous multiprocessor systems. Technical Report UNSW-CSE-TR-0911, School of Computer Science and Engineering, The University of New South Wales.
[19] Flix: Fast relief for performance-hungry embedded applications, 2005. Available at: http://www.tensilica.com/pdf/FLIX_White_Paper_v2.pdf.
[20] XPRES Generated Specialized Operations, 2005. Available at: http://tensilica.com/pdf/XPRES%201205.pdf.
[21] lp_solve. Available at: http://lpsolve.sourceforge.net/5.5/.

[1]This research was supported under Australian Research Council's Discovery Projects funding scheme (Project Number DP0986091).

978-1-60558-497-3/09 $25.00 © 2009 ACM

253

Xquasher: A Tool for Efficient Computation of Multiple Linear Expressions

Arash Arfaee[†], Ali Irturk[†], Nikolay Laptev[‡], Farzan Fallah[††], Ryan Kastner[†]

[†] Department of Computer Science and Engineering
University of California, San Diego
{aarfaee, airturk, kastner}@cs.ucsd.edu

[‡] Department of Computer Science
University California, Los Angeles
nlaptev@cs.ucla.edu

[††]Engineering Department
Envis Corporation, CA, 95054
Farzan@envis.com

Abstract— **Digital signal processing applications often require the computation of linear systems. These computations can be considerably expensive and require optimizations for lower power consumption, higher throughput, and faster response time. Unfortunately, system designers do not have the necessary tools to take advantage of the wide flexibility in ways to evaluate these expressions. Therefore, we address the problem of efficiently computing a set of linear systems through a tool, Xquasher, that is developed by us to enable elimination of large common subexpression from expressions with an arbitrary number of terms. Xquasher provides a methodology for efficient computation of both single and multiple linear expressions. We also introduce the concept of power set encoding which helps us to provide an effective optimization method and achieves significant improvement over previously published work. Our tool provides optimized designs with 15% less area with the cost of 3% increase in delay by reducing number of additions on average by 45%.**

Categories and Subject Description

C.3 Special-purpose and application-based systems

General Terms Algorithms, Design

Keywords

Area optimization, common sub-expression elimination, DSP transforms, multiple constant multiplications, linear expression

I. Introduction

Computation of *multiple linear expressions (MLE)* and its sub-computations: *single constant multiplication (SCM)*, *multiple constant multiplications (MCM)* and *linear expression (LE)* are commonplace in linear systems. Many digital signal processing applications such as Finite Impulse Response Filters and Discrete Fourier Transform use linear systems which are expensive and therefore are the dominant factor in the overall performance. It is often advantageous to convert costly constant multiplications into a corresponding set of shifts and additions which can lead to lower power consumption, higher throughput, and faster response time. By careful common sub-expression elimination over these expressions, one can also reduce the total number of additions.

However, current synthesis techniques perform limited transformations, and are unable to do an adequate optimization of these expressions. A major obstacle to the optimization of linear systems is that system designers do not have the necessary tools to take advantage of the wide flexibility in ways to evaluate these expressions. In most cases, designers rely on hand tuned library routines (for software) or Intellectual Property (IP) blocks (for hardware) to implement these computations. The drawback to using these approaches is that the given library routines and IP blocks may not be ideally suited for the platform with respect to the use and/or the constraints specified by the application. Designing a high level tool for the optimization of the linear system computation is crucial.

Therefore, we developed a tool, Xquasher, which provides a general methodology for linear system optimization. Xquasher performs common subexpression elimination to minimize the area of the resulting hardware of the linear system. Our tool is also efficient for optimization of sub-MLE problems: SCM, MCM and LE. Xquasher utilizes a novel powerset encoding format to enable efficient extraction of large common subexpressions which makes it possible to achieve substantial area improvements over previously published works.

In this paper we are mainly focused on reducing the area by reducing the total number of additions. The optimal solution with the minimum possible number of additions is a well known NP-complete problem. By careful analysis of effective factors in this problem and providing novel solutions for each of them, we generate a heuristic algorithm that efficiently reduces number of additions. Our results shows substantial improvement over existing methods [3][4][5][7].

The remainder of this work is organized as follows. The subsequent section formalizes the problem and its definitions. Section III describes our tool, its methodology which uses powerset encoding. Section IV provides experimental analysis of our results and comparisons with previously published works. We conclude in Section V.

II. Problem Formulation and Definitions

This section is devoted to introduction of multiple linear expression (MLE) and its sub-problems: single constant multiplication (SCM), multiple constant multiplication (MCM) and linear expression (LE) where we illustrate the cost of the computation with a three-dimensional space and define some basic terms: *dot*, *line*, *page* and *space* that are used in our methodology for ease of understanding.

A. Definitions

Canonical Signed digits (CSD) is radix-2 signed digit representation with digit set of $\{-1,0,1\}$ where there is no adjacent non-zero digit in the representation. **Bit-Magnitude (BM)** is an integer with the magnitude of $\pm 2^N$. It can be

Permission to make digital or hard copies of part or all of this work for personal or classroom use is granted without fee provided that copies are not made or distributed for profit or commercial advantage and that copies bear this notice and the full citation on the first page. To copy otherwise, to republish, to post on servers or to redistribute to lists, requires prior specific permission and/or a fee.

DAC'09, July 26-31, 2009, San Francisco, California, USA

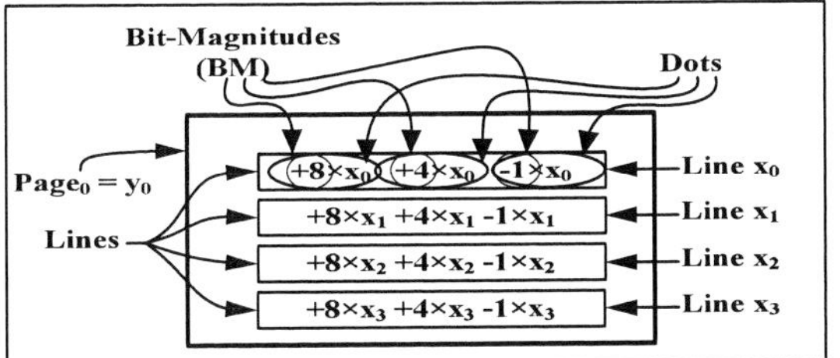

Fig. 1. An example of page and its lines, dots and BMs. Dots are shown with ellipses and in the first line, a circle inside each dot demonstrates the BM part.

described as a CSD number with exactly one non-zero digit. **Dot** is an appearance of an input variable in our expression with a relative magnitude of BM. In hardware, this corresponds to rewiring an input to specific point in the circuit with possible some shifts to the right. **Line** is summation of multiple copies of one variable multiplied by a vector of BMs. Therefore, computation of a line is Single Constant Multiplication (SCM). **Page** is formed by a summation of a set of lines. We present examples for page, lines, dots and BMs in Figure 1. **Space** can be created by union of a set of pages. In our problem, a space is a collection of multiple pages. **Cost of Computation (CoC)** is defined as the number of additions that appears in our object. *Example (Figure 1):* $CoC(Line_{x_0}) = 2$. **Object** refers to *a dot, a line, a page* or *a space.*

B. Problem mapping

In this subsection, we formulize the optimization of a linear system. Our approach is to minimize the number of additions. We first show how to optimize small objects in our space and then apply and improve this methodology for larger objects.

Dot optimization: A dot by itself does not have any cost by definition. They are simply rewiring of copies of input data. We do not perform any optimization over a dot by itself.

Line optimization: Problem of line optimization is known as single constant multiplication (SCM). Modern tools generally use CSD encoding as a solution for this problem. Using a CSD representation with minimum Hamming weight provides a good but not optimal solution in shift and add approach for constant by variable multiplication and therefore is used in our examples as well. Our experiments with Mentor Graphics and Synopsys tools indicate that the area of SCM mirrors the number of additions between non-zero digits in a CSD encoding. Although CSD can be considered as a good solution for optimization of lines, it does not guarantee an optimum solution since further optimization is possible to reduce number of additions inside a line. To see how this is possible, assume a line is given as $Line_1 = 792 \times x_0$ with CSD representation of $(10\bar{1}0010\bar{1}000)_{CSD}$ where a bar in top of a digit means that it is negative. After rewriting corresponding CSD bit values as integer, we have the following vector multiplication:

$$Line_1 = [1024 \quad -256 \quad 32 \quad -8] \times x_0$$
$$= 1024 \times x_0 - 256 \times x_0 + 32 \times x_0 - 8 \times x_0$$

This is the way that modern tools perform this multiplication, and CoC of this multiplication is 3 additions. However, it is possible to optimize this multiplication further by rewriting it:

$$Line_1 = 32 \times (32 \times x_0 - 8 \times x_0) + (32 \times x_0 - 8 \times x_0)$$
$$= (32 + 1) \times (32 \times x_0 - 8 \times x_0)$$

where CoC becomes 2 additions. Elimination of repeated dots by finding common sub-expressions (CS) and Common sub-expression Elimination (CSE) are well known methods for CSM optimization [5][7].

Page optimization: Optimization of a page by applying CSE is similar to the optimization of a line. However searching relatively larger set of dots and applying CSE demand more complex and careful analysis. Therefore, we investigate these issues in more detail and provide a general solution for page optimization. We assume that all lines are encoded with CSD. Assume a page is given as:

$$Page_0 = Line_0 + Line_1 + Line_2 + Line_3$$
$$= +686x_0 + 241x_1 + 991x_2 + 686x_3$$

The un-optimized page results in $4+2+2+4+3 = 15$ additions. We can further optimize this page by rewriting it as:

$$Page_1 = +241x_1 + 991x_2 + 686D_0$$
$$D_0 = x_0 + x_3$$

where we can decrease the number of additions to $2+2+4+2+1 = 11$ by detecting and eliminating an obvious CS: $+686(x_0 + x_3)$. Further optimization is possible by rewriting each line with equivalent CSD encoded version where more CSs become observable:

$$1024,0,-256,0,-64,0,16,0,0,-2,0$$
$$Page_0 = +(-2x_0 + 16x_0 - 64x_0 - 256x_0 + 1024x_0)$$
$$+(+1x_1 - 16x_1 + 256x_1) + (-1x_2 - 32x_2 + 1024x_2)$$
$$+(-2x_3 + 16x_3 - 64x_3 - 256x_3 + 1024x_3)$$

If we optimize $Page_0$ by limiting CSs to have exactly two dots, one of the possible results is:

$$Page_2 = +256x_1 + 1024x_2 - 16D_0 - 256D_0$$
$$+ 1024D_0 + 1D_1 - 1D_2 - 32D_2$$
$$D_0 = +1x_0 + 1x_3, D_1 = +1x_0 - 16x_1, D_2 = +1x_2 - 1D_0$$

where the CoC decreases from 15 to 10. Considering the previous example, one might suggest writing details of lines might lead to a better optimization; however it is more time consuming and hard to determine which CS is the best candidate to eliminate. For example, there are 120 combinations of two dots that could be a CS just in the first step of CSE. In our last attempt over this example, in first and second CSE we ignore details in $Line_0$ and $Line_3$.

$$Page_3 = +256x_1 - 1x_2 - 32x_2 + 1024x_2 + 686D_0 + 1D_1$$
$$D_0 = +1x_0 + 1x_3, D_1 = +1x_1 - 16x_1$$

In the third step we expand $686D_0$ to equivalent CSD encoded version. $Page_4$ is the result of CSE optimization after this step.

$$Page_4 = +256x_1 + 16D_0 - 256D_0 + 1024D_0 + 1D_1$$
$$- 1D_2 - 32D_2$$
$$D_0 = +1x_0 + 1x_3, D_1 = +1x_1 - 16x_1, D_2 = +1x_2 + D_0$$

CoC of $Page_4$ is 9, and we have 40% reduction in CoC from $Page_0$'s. Considering that we just needed 3 CSE and much less complexity in the CSE process to gain the same result, it might be a better approach. Here we just have 28 possibilities for CSs with two terms. It seems that CSE process can get much simpler and faster if we can work with dot representation and constant multiplication form of the line.

Space optimization: We design our tool, Xquasher, for space optimization by introducing two new concepts: **1)** *determination of* the *effective factors in optimization of the run-time* and **2)** *common sub-expression fragmentation (CSF).*

978-1-60558-497-3/09 $25.00 © 2009 ACM

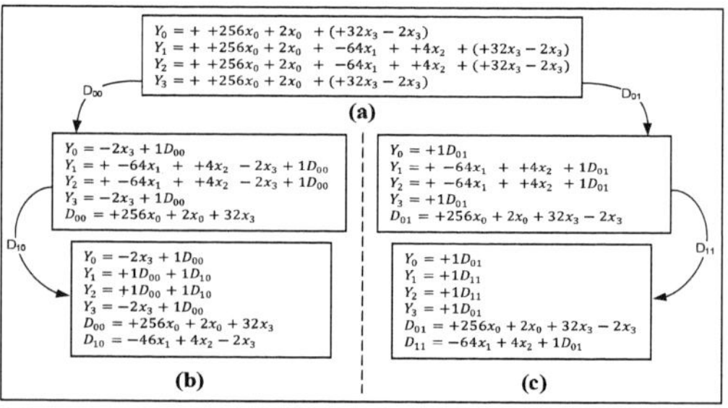

Fig. 2. (a) represents a space with four pages. We show different optimization approaches with (b) and (c). In (b), CS are limited to have exactly three dots. In (c), CSs can have arbitrary large number of dots.

1) *Determination of the effective factors in optimization of the run-time:* Run-time is the main challenge in all CSE algorithms. Finding the optimum solution for an NP-complete problem requires an unacceptable run-time. Heuristic algorithms are well known to be useful in tackling NP-complete problems [3][4][5]. Therefore, Xquasher uses a greedy heuristic algorithm where run-time of the algorithm depends on two factors: **1)** Number of iterations and **2)** Run-time of each iteration. The algorithm that is used in Xquasher can eliminate larger sets of CSs in a single CSE compared to the algorithms that are used in [4-5]. Our algorithm provides a solution with fewer iteration due to the optimization of larger sets of CSs where [4-5] requires multiple iterations for the same optimization. Xquasher is also capable of seeing both dots and sets of dots in a line as a part of CSs which increases the run-time of each iteration but still an effective factor in the first couple of iterations.

Being able to detect and eliminate a CS which contains large sets of dots causes significantly fast reduction of number of dots in the target space. Therefore, the number of remaining dots will be lower for the next iteration. However, finding proper CSs and CSE at run-time is highly dependent on the number of terms inside CSs. The format of CS in our tool is two sets of dots (two lines) instead of two or three dots. Being able to work with sets of dots, allow us to use a lower order algorithm compared to [4-5] and gain lower CoC.

2) *Common sub-expression fragmentation (CSF):* Our optimization approach is to eliminate CSs with large set of dots. CSE of CS with few limited, two or three, dots can easily prevent elimination of larger CSs and result in series of separated dots such that no optimization over them would be possible. Figure 2 (a, b and c) shows examples of effective and ineffective ways of optimization. Given space consists of four pages (a) and CoC for this space is *16* before the optimization.

In (b), we assume that all CSs must contain exactly three dots. We perform CSE for $D_{00} = (-256x_0 + 2x_0 + 32x_3)$ and $D_{10} = (-64x_1 + 4x_2 - 2x_3)$ as the first and second steps respectively. CoC is reduced to *8* after these two optimizations. As mentioned before, elimination of large set of dots may result in better optimization. In (b), we assume that there is no limit for number of dots in a CS. Therefore, we perform CSE for $D_{01} = (-256x_0 + 2x_0 + 32x_3 - 2x_3)$ and $D_{11} = (-64x_1 + 4x_2 + D_{01})$ as the first and second steps respectively. As a result of the CSEs CoC's is reduced to 5 which is a lower cost compared to the first approach of optimization. The reason for this decrease in the cost is that although dot $(-2x_3)$ appeared in all four initial pages (a), in two of these pages dots remain unengaged in all CSEs after the first optimization approach (b). The second approach prevents fragmentation on CSs by considering CSs with more dots.

III. Xquasher's Methodology

The best hand coded optimization for many common linear systems requires the elimination of CSs with large set of dots. Previous works are not able to handle such elimination due to the exponential increase in the cost of considering such CSs. Although our methodology is not able to eliminate all possible large CSs because of the increase in the cost, it is still capable of covering large CSs when CSs' dots are focused inside one or two lines.

A. Power Set Encoding

To convert an integer to PSE, we follow this procedure:

1- Encode an integer, I, to a CSD minimum hamming weight format.

2- Rewrite the CSD encoded number as a series of integer addition where each number has exactly one of the non-zero digits of the original number.

3- Generate power set from step 2's result set.

4- Ignore the null set and generate a new set from the result of step 3 where each element is the sum of numbers inside the related set.

5- Rewrite each number from the result of 4th step as $2^k \times p$, where p is an odd integer, and store these numbers in the following format:

$$[p, k, number\ of\ non - zero\ digit\ in\ CSD\ format]$$

B. Xquasher Algorithm

Here, we present the Xquasher algorithm and describe it below. After encoding all input constants to PSE (1), Xquasher searches for a CS that has the highest number of relevant dots in all appearance of that CS (3-8). Xquasher limits the CS to two PSDs (4-5) and performs CSE for the chosen CS (9), where it updates the related pages to the CS and appends a new page, representing the eliminated CS (10). These steps continue in a loop until there is no more CSE left to be eliminated (2).

```
1   convert all constants to PSE
2   While there is a CS
3       for any page in the space
4           for any PSD1 in any line1 in any page
5               for any PSD2 in any line2 appeared after line1 in any page
6                   find number of occurance of CS =
                        (PSD1, PSD2) in the space
7                   if (number of ocuurance) ×
                    ( number of non − zero digit in PSD1
                      +number of non − zero digit in PSD2 ) is max in the space
8                       best_CF = (PSD1, PSD2)
9           CSE (best_CF)
10          Update PSE valuesS
```

IV. Results

Using the methodology described in Section III along with Synopsys Design Compiler Ultra and TSMC 90nm library, we synthesize several common linear system examples using several techniques. The linear systems that we use as examples are 8-point DCT, 8-point Inverse DCT (IDCT), Discrete Fourier Transform (DFT), Discrete Hartley Transform (DHT), and Discrete Sine Transform (DST) and also three FIR filters: EP24 (6-tap FIR), BT24 (20-tap FIR) and LS24 (41-tap FIR). Xquasher takes a matrix of coefficients as input and then uses our methodology to

978-1-60558-497-3/09 $25.00 © 2009 ACM

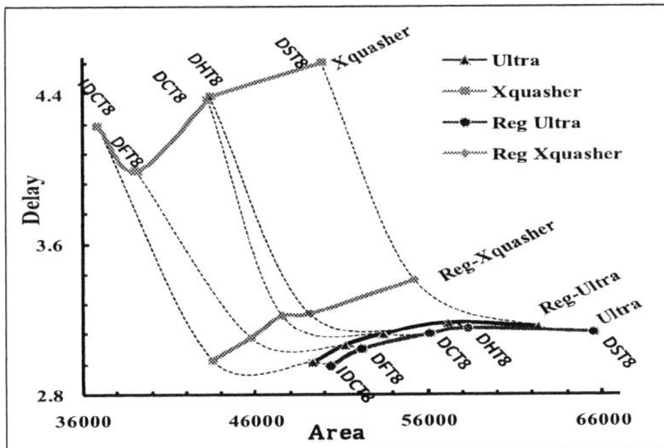

Fig. 3. Synthesis results for benchmarks. Reg Xquasher results shows 15% less area in average with small increase (~3%) in delay compared to the Compile Ultra results [6].

generate a new set of expressions. Xquasher also generates an HDL code based on these expressions. Therefore, we compare our results with the previously published work in terms of number of additions and synthesis results: area and delay.

Table I presents a comparison between the previously published works [1][2][5] and Xquasher in terms of total number of additions and a comparison between [5] and Xquasher in terms of synthesis results: area and delay. As can be seen from Table I, Xquasher provides significant reduction in the number of additions by a careful optimization compared to the previous works. Since the method in [5] is the closest to our results, we take the comparison to another level by presenting the synthesis results of both methods. It can be seen from Table I that Xquasher provides better results in area and as well as delay in our benchmarks.

As can be seen from Table I, previously reported best results [5] are clearly inferior to those provided by Synopsys Compile Ultra since they are significantly worse in both area and delay. To the best of our knowledge and based on our tests, the quality of results for Synopsys' Compile Ultra optimization [6] is the best available. Therefore we compare Xquasher with the results from Compile Ultra in terms of area and delay in Figure 3. Table II also shows this analysis in percentages in terms of area and delay. Xquasher's non-registered results give a significant area reduction (22%) compared to Compile Ultra, though this comes at a price of significant increase in delay (40%).

We also generate a synchronous (Registered) version of our code and compare registered versions of both optimization methods in column 3-4 of Table II. By using registers at the end of each adder tree to save the results by replacing all expressions with registers instead of wires, we compromised partial area saving to gain less delay. Therefore, Xquasher provides 14% decrease in area compared to registered non-optimized version while the average delay penalty is decreased to 3%.

TABLE II
Summary of results of our tool, Xquasher, compared to Synopsys Compiler.

	Xquasher vs. Ultra		Xquasher vs. Ultra (Registered)	
	Delay	**Area**	**Delay**	**Area**
dct8	-41.42%	19.00%	-2.55%	15.23%
dft8	-30.39%	23.49%	-1.97%	12.28%
dht8	-38.49%	23.99%	-3.19%	15.70%
dst8	-46.01%	19.93%	-8.95%	15.76%
idct8	-42.42%	25.21%	-1.02%	13.59%
Average	-39.74%	22.32%	-3.53%	14.51%

V. Conclusion

Linear systems are prevalent in digital signal processing. We developed a tool, Xquasher, that can effectively find and eliminate higher order common subexpression terms. Our tool provides enormous area reductions at the cost of increasing the critical path; 22% reduction in area with the cost of 40% increase in delay. However, we can easily add registers to our design, which yields 15% decrease in area and a minor increase in delay (3%) compared to registered version of the non-optimized design.

VI. References

[1] M. Potkonjak, M.B. Srivastava, and A.P. Chandrakasan, "Multiple Constant Multiplications: Efficient and Versatile Framework and Algorithms for Exploring Common Subexpression Elimination," *IEEE Transactions on Computer Aided Design of Integrated Circuits and Systems*, 1996.

[2] H. T. Nguyen and A. Chattejee, "Number-splitting with shift-and-add decomposition for power and hardware optimization in linear DSP synthesis," *IEEE Transactions on Very Large Scale Integration (Vlsi) Systems*, vol. 8, pp. 419-24, 2000.

[3] H. Safiri, M. Ahmadi, G.A. Jullien, and W.C. Miller, "A New Algorithm for the Elimination of Common Subexpressions in Hardware Implementation of Digital Filters by Using Genetic Programming," *IEEE International Conference on Application-Specific Systems, Architectures and Processors (ASAP)*, 2000.

[4] A. Hosangadi, F. Fallah, and R. Kastner, "Common subexpression elimination involving multiple variables linear DSP synthesis," *IEEE International Conference on Application-Specific Systems, Architectures and Processors*, pp. 202-12, 2004.

[5] A. Hosangadi, F. Farzan, and R. Kastner, "Optimizing high speed arithmetic circuits using three-term extraction," *Design, Automation and Test in Europe*, pp. 6, 2006.

[6] R. Zimmermann, D. Q. Tran, "Optimized Synthesis of Sum-of-Products," *Proceedings 37th Asilomar Conference on Signals, Systems, and Computers*, November 2003

[7] Peter Tummeltshammer, James C. Hoe and Markus Püschel, "Time-Multiplexed Multiple Constant Multiplication", *IEEE Transactions on Computer-Aided Design of Integrated Circuits and Systems*, Vol. 26, No. 9, pp. 1551-1563, 2007.

TABLE I. Comparison of some published work and Xquasher in terms of total of additions. Also synthesis results for [5] and our tool.

Example	Number of Additions					Synthesis Results			
	Original	[1]	[2]	[5]	**XQUASHER**	[5]		**XQUASHER**	
						Delay	Area	Delay	Area
H.264	86	N/A	N/A	63	**53**	4.34	25284.83	**3.50**	**13216.59**
DCT8	274	227	202	188	**161**	5.76	100221.30	**4.37**	**43275.86**
IDCT8	242	222	183	164	**140**	5.33	65135.55	**4.23**	**36907.11**
EP24	26	N/A	N/A	16	**13**	4.49	6394.87	**2.76**	**2272.03**
DST	320	252	238	N/A	**181**	N/A	N/A	**4.57**	**49902.86**
DHT	248	211	209	N/A	**161**	N/A	N/A	**4.39**	**43454.38**
BT24	106	N/A	N/A	48	**48**	6.04	26738.11	**3.12**	**9456.45**
LS24	232	N/A	N/A	112	**99**	5.77	37630.09	**3.17**	**21368.39**

ILP-Based Pin-Count Aware Design Methodology for Microfluidic Biochips *

Cliff Chiung-Yu Lin[1], and Yao-Wen Chang[1,2]
[1]Graduate Institute of Electronics Engineering, National Taiwan University, Taipei 106, Taiwan
[2]Department of Electrical Engineering, National Taiwan University, Taipei 106, Taiwan
chiungyu@eda.ee.ntu.edu.tw; ywchang@cc.ee.ntu.edu.tw

ABSTRACT

Digital microfluidic biochips have emerged as a popular alternative for laboratory experiments. To make the biochip feasible for practical applications, pin-count reduction is a key problem to higher-level integration of reactions on a biochip. Most previous works approach the problem by post-processing the placement and routing solutions to share compatible control signals; however, the quality of such sharing algorithms is inevitably limited by the placement and routing solutions. We present in this paper a comprehensive pin-constrained biochip design flow that addresses the pin-count issue at all design stages. The proposed flow consists of three major stages: (1) pin-count aware stage assignment that partitions the reactions in the given bioassay into execution stages, (2) pin-count aware device assignment that determines a specific device used for each reaction, and (3) guided placement, routing, and pin assignment that utilize the pin-count saving properties from the stage and device assignments to optimize the assay time and pin count. For both the stage and device assignments, exact ILP formulations and effective solution-space reduction schemes are proposed to minimize the assay time and pin count. Experimental results show the efficiency of our algorithms/flow and a 55–57% pin-count reduction over the state-of-the-art algorithms/flow.

Categories and Subject Descriptors

B.7.2 [**Integrated Circuits**]: Design Aids

General Terms

Algorithms, Performance, Design

Keywords

Microfludics, biochip, design methodology, integer linear programming

1. INTRODUCTION

Digital microfluidic biochips, also referred to as lab-on-a-chip or biochips, have emerged as an alternative for conventional laboratory experiments. With lower cost and higher immunity to human errors, the technology is gaining increasing applications including DNA sequencing, immunoassays, environmental toxicity monitoring, and point-of-care diagnosis of diseases [8].

Recently, the second-generation (digital) microfluidic biochips have been proposed [5,12]. Typically, a digital microfluidic biochip consists of a two-dimensional (2D) electrode array and peripheral devices (optical detection sites, dispensing ports, etc.) [5]. On a digital microfluidic biochip, movements of the *droplets* are controlled by the electrohydrodynamic force generated by the electrodes. By assigning time-varying voltage values to turn on/off the electrodes on the digital microfluidic biochip, we can move the droplets around the entire 2D array and perform fundamental microfluidic operations (i.e., mixing reactions) for different bioassays. These operations performed under the control of the electrodes are also called *reconfigurable* operations because of their flexibility in area (electrodes involved) and in execution time. A reconfigurable operation can be carried out anywhere on the 2D plane. And it can be completed, for example, slowly with 4 electrodes, or faster with 8 electrodes.

The reconfigurability of the operations, as it brings about the freedom in design, raises issues both in scheduling and in electrode control. For a given bioassay, different completion time and resource requirements can be achieved by changing the time and location schedule of the involved operations. Therefore, it has raised active design automation discussions in the past few years [10,11,18]. On the other hand, as the chip size grows, it becomes necessary to restructure the electrode control mechanism, or the unlimited number of control signals will be impossible to be implemented.

Originally, the electrodes are addressed and controlled independently, that is, each electrode is assigned a dedicated control pin. This kind of biochips, which are also referred to as *direct-addressing biochips*, provide great flexibility for droplet movement; yet they suffer from the increasing design complexity. Specifically, the routing problem for the large number of control pins has made this architecture only applicable to small-scale biochips [14].

Lately, alternative driving schemes have been proposed to alleviate the growth of the required number of control pins. Pin-constrained digital microfluidic biochips [8], one of the major genres of the pin-count reduction approaches, reduce the number of pins to be routed to the electrodes by assigning each control pin to multiple electrodes; that is, multiple electrodes are controlled by a single control signal, and thus they are turned on/off simultaneously. Regarding this, efforts have been made to cluster the electrodes that can be controlled together without introducing unexpected droplet behaviors [14,16].

However, these currently available pin-count reduction methods focus on the electrode partitioning, control signal merging, and pin assignment as the last step of the design flow, while the feasibility and effectiveness of such methods actually depend on the scheduling, placement, and routing results of the given design. Therefore, it is desirable to consider the pin-count constraint at earlier stages of the pin-constrained biochip design flow. Consequently, we propose in this paper a novel pin-count aware design flow for pin-constrained biochips and efficient algorithms for the corresponding steps in the flow.

*This work was partially supported by ITRI, Springsoft, Synopsys, TSMC, and National Science Council of Taiwan under Grant No's. NSC 97-2221-E-002-237-MY3, NSC 96-2628-E-002-249-MY3, NSC 96-2628-E-002-248-MY3.

Permission to make digital or hard copies of part or all of this work for personal or classroom use is granted without fee provided that copies are not made or distributed for profit or commercial advantage and that copies bear this notice and the full citation on the first page. To copy otherwise, to republish, to post on servers or to redistribute to lists, requires prior specific permission and/or a fee.
DAC'09, July 26-31, 2009, San Francisco, California, USA

978-1-60558-497-3/09 $25.00 © 2009 ACM

1.1 Previous Work

Previous works on the pin-constrained biochip problem generally address the problem after the biochip is placed and routed, and its design flow is illustrated in Figure 1(a). The flow consists of three major stages. The first stage, referred to as scheduling or placement, assigns the time slots and electrodes to each reactions in the bioassay and are resolved in [4,7,11,18]. The second stage, routing, determines the paths that the droplets move around the biochip and are worked out in [2,3,6,13,17]. Note that the solutions for these two stages are designed for direct-addressing biochips and do not address the pin-count constraint. Finally the electrode partitioning or control signal merging, followed by pin assignment, in [14,16], is used to reduce the final pin count. A fundamental problem with the previous flow is that only the last stage of the flow is pin-count aware, while the properties of the scheduling, placement, and routing results do affect the space for pin-count reduction. Currently, due to the lacking of specialized front-end design automation methods for pin-constrained biochips, these properties cannot be maintained/utilized well, and thus the quality of the pin-reduction will inevitably be restricted.

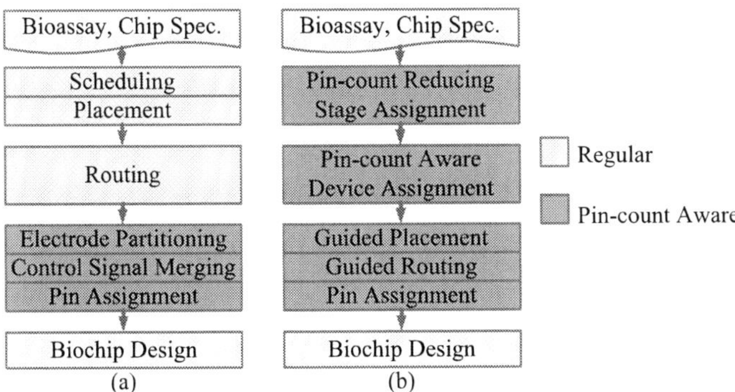

Figure 1: Comparison of the previous design flow and our proposed one: (a) Previous design flow, in which pin count is considered only during sequence merging and pin assignment. (b) Our proposed pin-count aware design flow.

1.2 Our Contributions

Unlike the previous flow that addresses the pin-constrained issue at the last stage, we propose in this paper a novel comprehensive pin-count aware design flow. The flow is summarized in Figure 1(b) and will be detailed in Section 2. It consists of three major stages: (1) pin-count aware stage assignment, (2) pin-count aware device assignment, and (3) guided placement, routing, and pin assignment. Besides adding the pin-count considerations into the flow, we expand the scheduling stage into *stage assignment* and *device assignment* to reflect their criticality in pin-count reduction. For both stages, we propose exact ILP formulations and effective reduction methods to minimize the assay completion time and the pin count. As mentioned earlier, pin-count saving properties should be maintained along the design flow. We also present guidelines for the placement, routing, and pin assignment to maintain the low pin-count properties obtained earlier at the assignment stage.

With the flow and the algorithms for realizing these properties, our contributions can be summarized as follows.

- We propose a dedicated pin-count aware design flow for pin-constrained biochips to consider the pin count throughout all stages of the flow.
- We identify the factors that would affect the pin count along the design and explore the properties that are favorable for pin-count reduction. The properties are universal to the pin-constrained biochips and would be helpful for future development with our and other design flows.
- We derive an exact ILP formulation for stage assignment that models the synchronous control of reactions and minimize the assay completion time. We also provide an effec-

tive scheme to reduce the problem size and an approximation to reduce the assay time to speed up the process.
- We derive an exact ILP formulation for device assignment that minimizes the number of branches from a device fanout, which reduces the corresponding pin-count demand, potential routing complexity, and fault tolerance. A corresponding problem size reduction method is also provided to lower the runtime.
- We present the guidelines for placement, routing, and pin assignment that maintain the minimized pin count derived by the stage and device assignments.

The experimental results show the efficiency of our algorithms for different bioassays (and with different device selections). As our method provides more flexible device count/type choices, we also achieve 55–57% pin-count reductions over the previous works [14,16], which is a very significant improvement and justifies the effectiveness of our design flow/algorithms.

The remainder of this paper is organized as follows. Section 2 analyzes the control-pin demand and gives the overall design flow. Sections 3 and 4 detail the ILP formulation and problem reduction for stage assignment and device assignment, respectively. Section 5 provides the guidelines for follow-up placement, routing, and pin assignment. Finally, the experimental results and conclusions are given in Sections 6 and 7, respectively.

2. PIN DEMAND AND PROPOSED FLOW

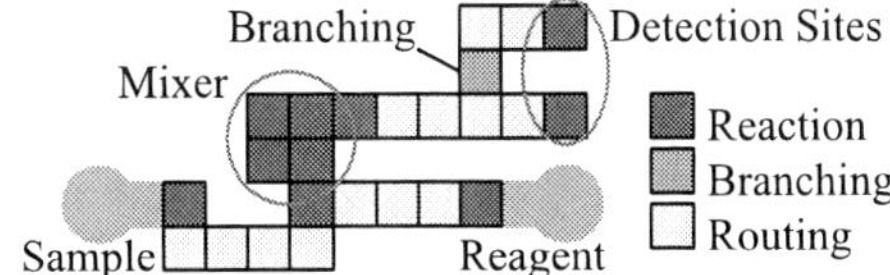

Figure 2: Classification of electrodes.

We classify the demand of pins N_p into three categories:

$$N_p = P_{reaction} + P_{branching} + P_{routing}, \qquad (1)$$

where $P_{reaction}, P_{branching}$, and $P_{routing}$, are the the numbers of pins needed to control the electrodes for reactions, for output branchings from devices, and for routing, respectively. Figure 2 illustrates the classification of electrodes. Although the control signals of these three categories of pins may be further merged, we shall focus on minimizing the three terms before the merging.

Our design flow shown in Figure 1(b) consists of the following three major stages:

1. *Stage assignment:* This stage minimizes $P_{reaction}$ by aligning reactions to execution stages and thus enables the synchronous control of the reactions. For the example bioassay in Figure 3(a), a possible stage assignment result is shown in Figure 3(b). Note that although we represent the generation and optical detection as black boxes in the simplified figure, we also assign corresponding stages to them. In the stage assignment, except for sharing the control pins, we also have to minimize the assay completion time to reduce the execution time overhead introduced during this stage. The constraints and considerations for stage assignment will be detailed in Section 3.

2. *Device assignment:* This stage minimizes $P_{branching}$ by matching the reactions to specific devices. Device assignment is important because different matchings can lead to different feeding relationships between devices and thus affects the number of independent control pins needed for the branchings between devices. A possible device assignment for Figure 3(b) is shown in Figure 3(c), and the device assignment problem will be addressed in Section 4.

3. *Guided placement, routing, and pin assignment:* This stage transforms the stage and device assignment result to actual placement and routing solutions. Following the proposed guideline with careful pin assignment can maintain $P_{routing}$ as a constant.

978-1-60558-497-3/09 $25.00 © 2009 ACM

259

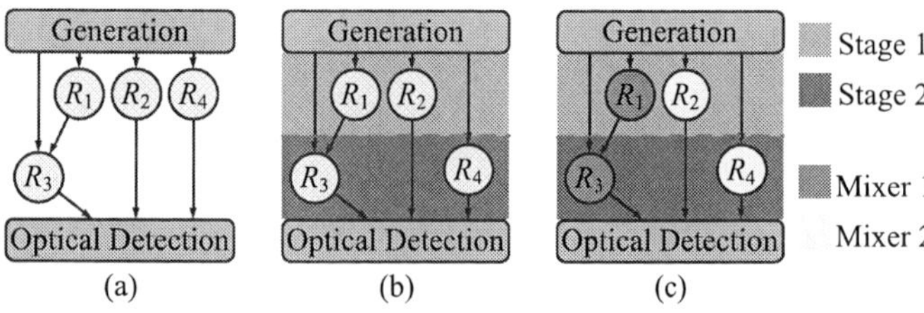

Figure 3: (a) An example bioassay with four mixing reactions. (b) The bioassay after stage assignment. (c) The bioassay after device assignment.

The differences of the proposed flow from the previous one are that the pin-count concerns are considered from the early stages to the end of the flow, and the scheduling problem is expanded to address two major sources of control pin demand.

3. STAGE ASSIGNMENT

This section first introduces the idea of synchronous reactions and the corresponding tradeoff between execution time and pin count. Next, we describe the stage assignment problem that attempts to minimize the assay completion time with the synchronously controlled reactions. Then we present the exact ILP formulation for the stage assignment. Finally, the corresponding solution space reduction schemes and an approximation for execution time are provided to speed up the runtime.

3.1 Advantage of Synchronous Reactions

The motivation for stage assignment is to reduce $P_{reaction}$ by synchronously controlling the reactions assigned to the same stage, while keeping the assay completion time minimized.

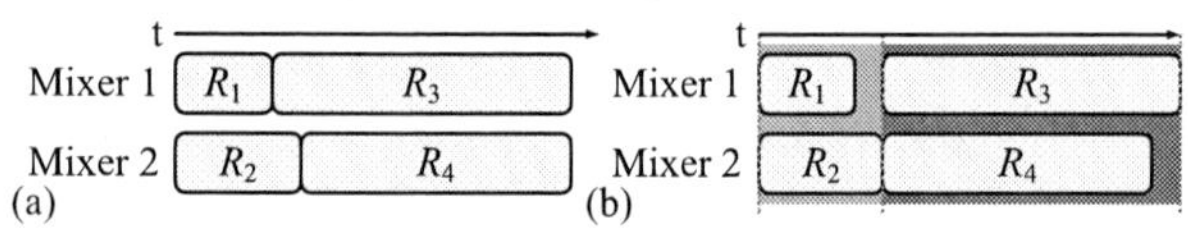

Figure 4: Comparison of asynchronous and synchronous control of two mixers with four reactions R_1, R_2, R_3, R_4: (a) Asynchronous: A shorter completion time is achieved, but the mixers do not share the control signals. (b) Synchronous: Mixers 1 and 2 share their control pins, but the completion time is longer.

The differences between asynchronous and synchronous control can be explained by Figure 4. In Figure 4(a), Mixers 1 and 2 are controlled separately and do not share a control pin; that is, the mixers can decide when to begin and cease their mixing reactions independently. We can see that R_3 can use Mixer 1 right after R_1 is completed, and thus the completion time is shorter. However, Mixers 1 and 2 cannot share a control pin in this way.

On the other hand, in Figure 4(b), Mixers 1 and 2 are controlled together, which means that the mixers must begin and cease their mixing reactions synchronously. Therefore, although R_1 and R_2 take different completion times, the synchronized control signals will cease the reactions only when the slower reaction ends. Thus, R_3 can begin only after R_2 is completed, at the same time as R_4. Despite the execution time overhead, in this case, Mixers 1 and 2 can share their control pins, which is more favorable for pin-constrained biochips.

3.2 The Stage Assignment Problem

Given a bioassay and a chip specification, stage assignment divides the reactions into a set of execution stages, and each stage is dedicated to a single category of reactions (e.g., generation of certain sample/reagent, mixing, optical detection, etc.). Besides, the working time of each stage is also determined in stage assignment. Following are the constraints for the stage assignment.

1. *Capacity constraints:* The number of reactions in a stage is upper-bounded by the number of the devices belonging to the category of the stage.

2. *Uniqueness constraints:* A reaction exists in exactly one stage.

3. *Duration constraints:* The duration of a stage is the duration of the slowest reaction assigned to the stage.

4. *Sequence constraints:* Stages that belong to the same category are sorted and executed sequentially without overlapping.

5. *Precedence constraints:* If there are data dependencies between reactions described in the bioassay such as "reaction R_i must happen before reaction R_j," then the stage that includes R_j can begin only after the stage of R_i ends.

Note that the major differences from the previous scheduling or placement problems in [10, 11, 17] are the duration constraints and sequence constraints superimposed for synchronous control.

Different stage assignment results can lead to different assay completion times. For a given stage assignment, the minimum assay completion time can be calculated with the duration, sequence, and precedence constraints. In our stage assignment, the target is to find the stage assignment with the shortest assay completion time.

3.3 Problem Formulation for Stage Assignment

The stage assignment problem can be formulated as follows.
Given:
1. A set of N reactions $S_r = \{R_1, R_2, \ldots, R_N\}$.
2. A set of M devices categories $S_d = \{D_1, D_2, \ldots, D_M\}$.
3. $V_n \in [1, M], \forall R_n \in S_r$, where $V_n = m$ means that reaction R_n is executed by a device of category D_m.
4. $C_m, \forall D_m \in S_d$, where C_m is the number of the devices of category D_m.
5. $T_{n,m}, \forall R_n \in S_r, D_m \in S_d$, where $T_{n,m}$ is the time that R_n takes with device category m.
6. A set of precedence relationships $S_p \in 2^{S_r \times S_r}$, where $(R_{n_1}, R_{n_2}) \in S_p$ means that R_{n_1} must be completed before R_{n_2} begins.

Find:
1. A partitioning of S_r into independent subsets (stages) $S_{m,1}, S_{m,2}, \ldots, S_{m,I_m}, \forall m \in [1, M]$, where $S_{m,i}$ represents the i-th stage for D_m, and I_m represents the maximum number of stages for D_m.
2. Corresponding start time $B_{m,i}$ and finish time $E_{m,i}$ for these stages.

That minimize: $\max_{\forall m,i} E_{m,i}$
Under the following constraints:
1. *Capacity constraints:* $|S_{m,i}| \leq C_m, \forall(m,i)$,
2. *Uniqueness constraints:*
   (a) $\exists(m,i), R_n \in S_{m,i}, \forall R_n \in S_r$, and
   (b) $S_{m_1,i_1} \cap S_{m_2,i_2} = \emptyset, \forall(m_1,i_1) \neq (m_2,i_2)$,
3. *Duration constraints:* $E_{m,i} - B_{m,i} \geq T_{n,m}, \forall R_n \in S_{m,i}$,
4. *Sequence constraints:* $E_{m,i_1} \leq B_{m,i_2}, \forall m, i_1 < i_2$,
5. *Precedence constraints:* $E_{m_1,i_1} \leq B_{m_2,i_2}, \forall(R_{n_1}, R_{n_2}) \in S_p, R_{n_1} \in S_{m_1,i_1}, R_{n_2} \in S_{m_2,i_2}$.

3.4 ILP Formulation for Stage Assignment

Denote the occurrence of $R_n \in S_{m,i}$ as $g_{n,m,i}$, and follow the notion of the problem formulation, then the stage assignment problem that minimizes assay completion time T_c can be formulated as follows.
Minimize T_c
Subject to
1. *Capacity constraints:*
$$\sum_n g_{n,m,i} \leq C_m, \forall(m,i). \quad (2)$$

2. *Uniqueness constraints:*
$$\sum_{(m,i)} g_{n,m,i} = 1, \forall n. \quad (3)$$

3. *Duration constraints:*
$$E_{m,i} - B_{m,i} - T_{n,m}g_{n,m,i} \geq 0, \forall n, m, i. \quad (4)$$

4. *Sequence constraints:*
$$E_{m,i_1} \leq B_{m,i_2}, \forall m, i_1 < i_2. \quad (5)$$

978-1-60558-497-3/09 $25.00 © 2009 ACM

5. *Precedence constraints:*

$$E_{m_1,i_1} - B_{m_2,i_2} + T_{max}(g_{n_1,m_1,i_1} + g_{n_2,m_2,i_2} - 2) \le 0,$$
$$\forall (R_{n_1}, R_{n_2}) \in S_p, m_1, i_1, m_2, i_2. \quad (6)$$

6. *Assay start and finish:*

$$0 \le B_{m,i} \le E_{m,i} \le T_c, \forall (m, i). \quad (7)$$

In Inequality (6), T_{max} is a large number used to formulate the *AND* logic. It can also be set to an upper bound of the minimum assay completion time obtained by greedy assignment.

3.5 Solution Space Reduction

The naive formulation shown in the previous subsection is complicated and not efficient to be solved without reduction. The number of variables and constraints are $O(NMI_a)$ and $O(|S_p|M^2I_a^2 + NMI_a)$, where I_a is the max number of stages among the device categories, which can be $O(N)$ without careful bounding.

3.5.1 Reaction Category Mapping

To reduce the problem size, we first note that a reaction is mapped to a specific category of devices, and thus we can eliminate redundant variables and constraints. That is, all $g_{n,m,i}$ with $m \ne V_n$ can be removed, and thus all related constraints can be simplified or even abandoned. Therefore, the numbers of variables and constraints become $O(NI_a)$ and $O(|S_p|I_a^2 + NI_a)$, respectively. Note that a pair in S_p actually indicates two corresponding reactions and thus two specific device categories, so we can drop the M^2 from $|S_p|M^2I_a^2$.

3.5.2 Upper Bound for the Stage Number

Then we bound the number of stages used by each device category. The advantage is twofold. First, the effect of the I_a term on the number of variables and constraints can be effectively reduced when the number of stages is tightly bounded. On the other hand, providing a bound for the stage count can restrict the permutations of identical solutions. The exact formulas for the bound will not be presented in this paper due to the page limit.

3.5.3 Lower Bound for Assay Completion Time

Finally, it is helpful to add a lower bound for the assay completion time into the ILP formulation to speed up the runtime of the ILP solver. Sometimes for the case that the assay time is dominated by a few critical paths, the ILP solver may take more time to search for permutations of cells on non-critical paths even when an optimal solution is found. To avoid this situation, we can provide a lower bound by calculating the length values of all paths in advance and use the longest path to bound the assay completion time.

THEOREM 1. *The stage assignment problem can be solved optimally with the ILP formulation in Section 3.4 and the reduction in Section 3.5.*

3.6 Approximation for Assay Completion Time

Even with the reductions in the previous subsection, the formulation is still difficult for ILP solvers. One of the most difficult parts is the precedence constraints around $B_{m,i}$ and $E_{m,i}$ in Equation (6), in which the integer values are conditionally constrained.

To further speed up the ILP solution, we propose to apply ILP only for the stage assignment of mixing reactions, while the stage assignment of the other reactions (i.e., generation, optical detection) are implied by the related mixing reactions called *priors*. By doing so, we can avoid the formulation of the inter-category precedence constraints. Therefore, we only have to formulate the precedence constraints among mixing. To make sure that the minimum assay completion time obtained by the stage assignment of mixing can be used as a good approximate for the minimum execution time for the overall assay, additional constraints must be set on the distribution of the priors. However, the exact constraints for the distribution will not be presented in this paper due to the page limit.

4. DEVICE ASSIGNMENT

This section first explains how device assignment can affect the number of control pins needed for output branchings from devices. Next, the problem description and formulation of device assignment is introduced. Then an exact ILP formulation for the problem is presented. And finally, effective solution-space reduction schemes are proposed to speed up the process.

4.1 Effect of Device Permutation

While $P_{reaction}$ is handled in stage assignment, $P_{branching}$ is to be reduced in device assignment. To illustrate how device assignment can affect the number of branchings, Figure 5(a) shows part of a bioassay, and two possible device assignments for it are shown in Figures 5(b) and (c). In (b), three paths {Mixer 1$\rightarrow$ Mixer 1, Mixer 1$\rightarrow$ Mixer 2, Mixer 2$\rightarrow$ Mixer 1} are used, while in (c), only two paths {Mixer 1$\rightarrow$ Mixer 1, Mixer 2$\rightarrow$ Mixer 1} are used and thus potentially fewer electrodes are needed for controlling the branchings. Generally, device assignment can affect the routing complexity, the potential pin count, and the number of electrodes used.

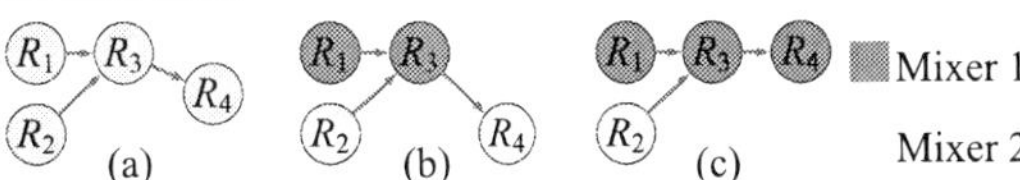

Figure 5: The effect of device assignment. (b)(c) are device assignment results for (a) with different numbers of branchings.

4.2 The Device Assignment Problem

Given a stage assignment result, device assignment decides, for each stage $S_{m,i}$, a 1-1 mapping from the reactions in $S_{m,i}$ to the devices of category D_m, which minimizes the number of branchings introduced by the dependencies in S_p. Note that the sets are obtained by inserting necessary buffers to a direct stage assignment because some droplet may have to wait for a few stages before they are used, and thus the sets may be different from the setup in stage assignment by a buffer category.

4.3 Problem Formulation for Device Assignment

The device assignment problem can be formulated as follows.

Given: $S_r, S_d, V_n, C_m, S_p, S_{m,i}$ same as that from stage assignment after buffer insertion.

Find: $x_{n,z}, z \in [1, C_{V_n}], \forall R_n \in S_r$, where $x_{n,z}$ is the occurrence of the event that R_n is assigned to the z-th device of the category.

That minimize: $\sum_{\forall m_1, z_1, m_2, z_2} p_{m_1, z_1, m_2, z_2}$, where p_{m_1, z_1, m_2, z_2} denotes the existence of a path from the z_1-th device of D_{m_1} to the z_2-th device of D_{m_2}.

Under the following constraints:

1. *1-1 mapping:* For any stage $S_{m,i}$,
$$OR(x_{n,1}, x_{n,2}, \ldots, x_{n,C_m}) = 1, \forall R_n \in S_{m,i}.$$

$$AND(x_{n,z_1}, x_{n,z_2}) = 0, \forall R_n \in S_{m,i}, z_1, z_2 \in [1, C_m], z_1 \ne z_2.$$

$$AND(x_{n_1,z}, x_{n_2,z}) = 0, \forall R_{n_1}, R_{n_2} \in S_{m,i}, n_1 \ne n_2, z \in [1, C_m].$$

2. *Path usage:* For any $m_1 \ne m_2, z_1 \in [1, C_{m_1}], z_2 \in [1, C_{m_2}]$,

$$p_{m_1,z_1,m_2,z_2} = OR_{\substack{\forall (R_{n_1}, R_{n_2}) \in S_p, \\ V_{n_1} = m_1, V_{n_2} = m_2}} (AND(x_{n_1,z_1}, x_{n_2,z_2})).$$

4.4 ILP Formulation for Device Assignment

An exact ILP formulation for the device assignment problem can be written as follows.

Minimize $\displaystyle \sum_{m_1, z_1, m_2, z_2} p_{m_1, z_1, m_2, z_2}.$

Subject to

1. *1-1 mapping:* For any stage $S_{m,i}$,
$$\sum_{z \in [1, C_m]} x_{n,z} = 1, \forall R_n \in S_{m,i}. \quad (8)$$
$$\sum_{R_n \in S_{m,i}} x_{n,z} \le 1, \forall z \in [1, C_m]. \quad (9)$$

2. *Path usage:* For any $m_1 \neq m_2, z_1 \in [1, C_{m_1}], z_2 \in [1, C_{m_2}]$,

$$x_{n_1,z_1} + x_{n_2,z_2} - p_{m_1,z_1,m_2,z_2} \leq 1,$$
$$\forall (R_{n_1}, R_{n_2}) \in S_p, V_{n_1} = m_1, V_{n_2} = m_2. \quad (10)$$

4.5 Solution-Space Reduction

The above formulation involves $O(N+M^2)$ variables and $O(N+M^2+|S_p|)$ constraints. Since the generation reactions can only precede mixing reactions, and the optical reactions can only follow mixing reactions in practical bioassays, we can cut the $O(M^2)$ feeding relationships between devices down to $O(M)$; thus the numbers of variables and constraints become $O(N + M)$ and $O(N + M + |S_p|)$, respectively. We further propose the following solution-space reduction schemes for the ILP formulation.

4.5.1 Permutation Restriction

First, we eliminate unnecessary permutations of the optimum solutions by fixing, for each device category D_m, the device assignment of one of its stages with $|S_{m,i}| = C_m$. Note that this is always feasible because there must be at least one stage with $|S_{m,i}| = C_m$, or the C_m should be reduced because the bioassay never uses all of the devices. By doing so, assume that the original solution space is U_d, the size of the new solution space after such a reduction is

$$|U_d^{\cdot}| = \prod_{\cdot D_m \cdot S_d} \frac{1}{C_m!} |U_d|. \quad (11)$$

4.5.2 Redundancy Pruning

Then we can further reduce the solution space by removing the paths around *universal* peripheral reactions. Here we define a category of peripheral devices D_u to be universal for mixing devices D_m if and only if it is fed by all mixers of some mixing stage. In other words, each mixer must has a path to an optical detector of D_u, and no matter how the priors are permuted, we need exactly C_m paths for them in the best case. Therefore, we can remove the constraints about these paths. Similarly, this method can be applied to generation categories.

Finally, we can iteratively remove *uniform* mixing stages and related paths from the problem. A stage is uniform if and only if it has one of the following properties.

1. All output paths from the stage are removed and the input devices for all reactions in the stage are identical.

2. All input paths to the stage are removed and the output devices for all reactions in the stage are identical.

Note that the process can be iterative because the removal of one stage could make a consecutive stage uniform.

The universal peripherals and uniform mixing stages are removed during the ILP optimization, and their optimal device assignment can be greedily decided in the reverse order of their removal.

THEOREM 2. *The device assignment problem can be solved optimally with the ILP formulation in Subsection 4.4 and the reduction in Subsection 4.5.*

5. PLACEMENT, ROUTING, PIN ASSIGNMENT

This section introduces the general guidelines for pin-count aware placement and routing. Following the proposed guideline with careful pin assignment can keep the low pin-count obtained by the proposed stage assignment and device assignment algorithms and keep $P_{routing}$ as a constant. Due to the page limit, the pin assignment method will not be covered.

5.1 Pin-Count Saving Guidelines

The major reason that the previous placement and routing methods for direct-addressing biochips do not work appropriately for the design of pin-constrained biochips is that the previous methods only preserve segregation (guarding) cells between electrodes that work during overlapping time periods; in contrast, for pin-count reduction, we also have to provide segregation cells between two electrodes that work in separate time spans. For example, electrodes E_1 and E_2 in Figure 6 are turned on for different routing paths in separate time slots, but they still cannot be controlled by the same pin because they are neighbors.

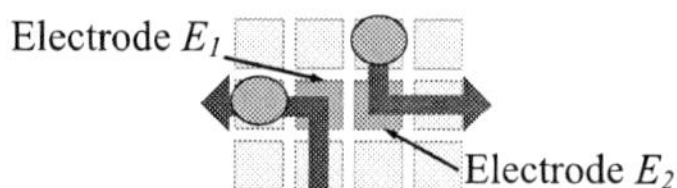

Figure 6: Electrodes used in different time spans that can prevent control pin merging.

Regarding this, we propose the following placement and routing guidelines.

1. The placement and routing for the entire bioassay should be decided simultaneously on a 2D plane.

2. Routing paths should not touch each other, except for necessary crossings or branchings.

3. Segregation (guarding) electrodes should be placed between devices and devices. If we wish to merge the control signals of reactions and routing paths, guarding electrodes are also needed between devices and routing paths.

Note that Items 1 and 2 are not addressed in previous placement and routing works, while similar constraints for Item 3 have been formulated to prevent unexpected merging between droplets [15, 18].

6. EXPERIMENTAL RESULTS

We implemented the proposed design flow in the C++ language with glpk [1] as our ILP solver. All of our experiments were conducted on a Linux machine with two 2.8 GHz AMD-64 CPUs and 8 GB memory.

6.1 Stage Assignment

Two practical bioassays, the *in_vitro* diagnostics used in [10] and the colorimetric protein assay [9], were used to evaluate our stage assignment algorithm. The experimental results for stage assignment are shown in Table 1. Designs "in_vitro 1," "in_vitro 2," and "in_vitro 3" are in_vitro diagnostics with different numbers of samples and reagents and thus different numbers of nodes (reactions) as shown in the table, and Design "protein" is the colorimetric protein assay. For each design, we conducted stage assignment with different mixing device choices to cover both sparse and dense cases, which affect how we approximate the peripheral times. Also, we tested a mini in_vitro diagnostic design "simple" with only 16 nodes just to illustrate the complexity growth of the exact ILP solution.

The runtimes of the ILP solver for both the basic formulation and the formulation after the solution-space reduction and assay time approximation are shown in Table 1. We can see that with the basic formulation, it takes more than 1 day to solve all the cases except for "simple" and "in_vitro 3" with the slowest devices. Note that even for the same design, the choice of devices can greatly affect the runtime. On the other hand, after the solution-space reduction and assay time approximation, the ILP solver took less than one second for all cases. The quality of the assay time approximation is also shown in the table; we can see that the error is smaller than 5% for all cases, with more than half the cases achieving zero errors.

6.2 Device Assignment

We then performed device assignment with the four practical bioassays with the stage assignment results in the previous experiment as reported in Table 2. We can see that the solution-space reduction schemes effectively reduce the number of nodes being formulated by 25% to 84%, and for all cases the optimum device assignment is obtained in less than 0.1 seconds. The results show the effectiveness and efficiency of our device assignment algorithm and the reduction scheme.

6.3 Placement and Routing Quality

With the stage and device assignment results produced by our algorithms, we also generated placement, routing, and pin assignment results for the three designs considered in [14, 16]. Table 3

978-1-60558-497-3/09 $25.00 © 2009 ACM

Table 1: Experimental results for stage assignment.

Design	#nodes	Device Choice	Runtime (sec)		Assay Completion Time		
			Exact	Approx.	Exact	Approx.	Error (%)
simple	16	Fast	0.300	0.032	20	20	0
in_vitro 3	36	Slow	25369.99	0.052	44	46	4.55
		Fast	>1 day	0.036	40	40	0
		Medium	>1 day	0.044	41	43	4.88
in_vitro 2	48	Slow	>1 day	0.044	44	46	4.55
		Fast	>1 day	0.036	40	41	2.50
		Medium	>1 day	0.048	41	43	4.88
in_vitro 1	64	Slow	>1 day	0.540	56	56	0
		Fast	>1 day	0.164	52	52	0
		Medium	>1 day	0.440	53	55	3.77
protein	103	Slow	>1 day	0.116	161	161	0
		Fast	>1 day	0.092	105	105	0
		Medium	>1 day	0.100	121	121	0

Table 2: Experimental results for device assignment.

Design	#nodes		Runtime(sec)
	Original	Reduced	
in_vitro 3	36	27	0.016
in_vitro 2	48	36	0.016
in_vitro 1	64	48	0.020
protein	103	16	0.036

gives the results and comparisons. The results show that our design flow reduces 55–57% of the control pins, which justifies the effectiveness of our design flow and algorithms.

Table 3: Experimental results for placement, routing, and pin assignment.

Design	#control pins		
	[14, 16]	Ours	% improved
multiplexed	25	11	56.00
PCR	14	6	57.14
protein	27	12	55.56

Our result for the 2×2 multiplexed bioassay is shown in Figure 7. In Figure 7(a), a result without control signal merging is illustrated. We can see that even without control signal merging, only 16 pins are required for this bioassay with our algorithms. Pins p, q, r, s, t are shared by the control of the two mixers, while y, z are used to synchronously move the droplets to the two detection sites, and to disposal, respectively. And v, w, x are used to realize the dispensing of the sample/reagent droplets. Pins $1, 2, 3, 4, c$ are used for routing and crossings; b is used to control two branchings.

Figure 7(b) shows the final result that we report in Table 3. The pin count is further reduced from that in Figure 7(a) by merging pins $1, 2, 3, 4$ with p, q, r, s, and t with z. Consequently, only 11 control pins are needed while [14,16] need to use 25 control pins to realize the same bioassay. Note that we also manage to control two mixers simultaneously and lower the time to complete the bioassay for this design.

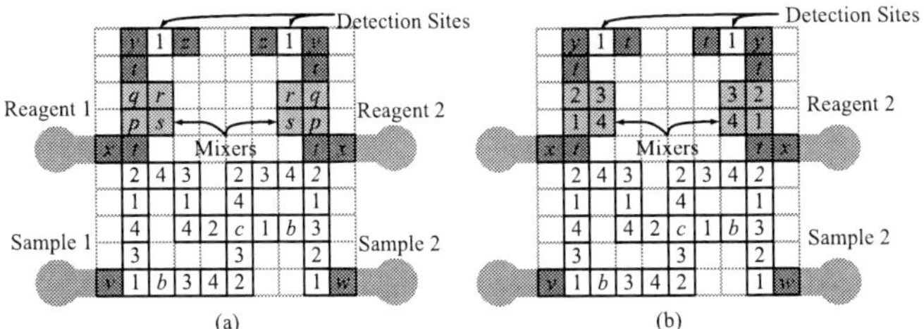

Figure 7: Placement, routing, and pin assignment for 2×2 multiplexed bioassay. (a) Without control signal merging. (b) With control signal merging.

7. CONCLUSIONS

We have presented in this work a pin-count aware design flow for pin-constrained biochips. ILP-based algorithms have been proposed for the stage and device assignments with effective solution-space reductions. Guidelines for placement, routing, and pin-assignment that reduce the pin count have also been discussed.

Experimental results have shown the efficiency of our algorithms and a 55–57% pin-count reduction over the previous flow that only post-processes the placement and routing results.

8. REFERENCES

[1] http://www.gnu.org/software/glpk/.

[2] K. F. Böhringer. Modeling and controlling parallel tasks in droplet-based microfluidic systems. *TCAD*, 25(2):334–344, 2006.

[3] M. Cho and D. Z. Pan. A high-performance droplet router for digital microfluidic biochips. In *ISPD*, pages 200–206, April 2008.

[4] J. Ding, K. Chakrabarty, and R. B. Fair. Scheduling of microfluidic operations for reconfigurable two-dimensional electrowetting arrays. *TCAD*, 20:1463–1468, 2001.

[5] R. B. Fair, V. Srinivasan, H. Ren, P. Paik, V. Pamula, and M. Pollack. Electrowetting-based on-chip sample processing for integrated microfluidics. In *IEDM*, pages 32.5.1–32.5.4, 2003.

[6] E. J. Griffith, S. Akella, and M. K. Goldberg. Performance characterization of a reconfigurable planar-array digital microfluidic system. *TCAD*, 25(2):435–357, 2006.

[7] A. J. Ricketts, K. Irick, N. Vijaykrishnan, and M. J. Irwin. Prtiority scheduling in digital microfluidics-based biochips. In *DATE*, pages 329–334, 2006.

[8] V. Srinivasan, V. Pamula, and R. Fair. An integrated digital microfluidic lab-on-a-chip for clinical diagnostics on human physiological fluids. *LoC*, 4:310–315, 2004.

[9] V. Srinivasan, V. Pamula, P. Paik, and R. Fair. Protein stamping for maldi mass spectrometry using an electrowetting-based microfluidic platform. In *Proceeding of the International Society for Optical Engineering*, pages 26–32, 2004.

[10] F. Su and K. Chakrabarty. Architectural-level synthesis of digital microfluidics-based biochips. In *ICCAD*, pages 223–228, 2004.

[11] F. Su and K. Chakrabarty. Unified high-level synthesis and module placement for defect-tolerant microfluidic biochips. In *DAC*, pages 825–830, June 2005.

[12] F. Su, K. Chakrabarty, and R. B. Fair. Microfluidic-based biochips: Technology issues, implementation platforms, and design-automation challenges. *TCAD*, 25(4):211–223, 2006.

[13] F. Su, W. Hwang, and K. Chakrabarty. Droplet routing in the synthesis of digital microfluidic biochips. In *DATE*, pages 323–328, 2006.

[14] T. Xu and K. Chakrabarty. Droplet-trace-based array partitioning and a pin assignment algorithm for the automated design of digital microfluidic biochips. In *CODES+ISSS*, pages 112–117, Oct. 2006.

[15] T. Xu and K. Chakrabarty. Integrated droplet routing in the synthesis of microfluidic biochips. In *DAC*, pages 948–953, June 2007.

[16] T. Xu and K. Chakrabarty. Broadcast electrode-addressing for pin-constrained multi-functional digital microfluidic biochips. In *DAC*, pages 173–178, June 2008.

[17] P.-H. Yuh, C.-L. Yang, and Y.-W. Chang. BioRoute: A network-flow based routing algorithm for digital microfluidic biochips. In *ICCAD*, pages 752–757, Nov. 2007.

[18] P.-H. Yuh, C.-L. Yang, and Y.-W. Chang. Placement of defect-tolerant digital microfluidic biochips using the T-tree formulation. *JETC*, 3(3):13:1–13:32, Nov. 2007.

O-Router: An Optical Routing Framework for Low Power On-chip Silicon Nano-Photonic Integration

Duo Ding, Yilin Zhang, Haiyu Huang, Ray T. Chen and David Z. Pan
ECE Dept. Univ. of Texas at Austin, Austin, TX 78712
{ ding, yzhang1, dpan }@cerc.utexas.edu, haiyu@mail.utexas.edu, chen@ece.utexas.edu

ABSTRACT

In this work, we present a new optical routing framework, *O-Router* for future low-power on-chip optical interconnect integration utilizing silicon compatible nano-photonic devices. We formulate the optical layer routing problem as the minimization of total on-chip optical modulator cost (laser power consumption) with Integer Linear Programming technique under various detection constraints. Key techniques for variable number reduction and routing speedup are also explored and utilized. *O-Router* is tested on optical netlist benchmarks modified from top global nets of ISPD98/08 routing benchmarks. *O-Router* experimental results are compared with conventional minimum spanning tree algorithm, demonstrating an average of over 50% improvement in terms of total on-chip optical layer power reduction.

Categories and Subject Descriptors

B.7.2 [**Hardware, Integrated Circuit**]: Design Aids

General Terms

Algorithms, Design, Performance

Keywords

Optical Routing, Low Power Nanophotonic Integration, Integer Linear Programming

1. INTRODUCTION

As raised in the International Technology Roadmap for Semiconductors [8], silicon system complexity rockets exponentially due to increasing transistor counts, fueled by smaller feature sizes and increasing demands for higher integration / performances with lower costs. Consequently, interconnect design becomes more and more important for DSM VLSI as technology further scales down, among which on-chip optical interconnect is a potential quantum leap towards next-generation technology. Ever since its first introduction by Goodman in [5], the concept of on-chip optical interconnect has attracted more and more attention over the years in industry (e.g., [9,15]) as well as academia

Permission to make digital or hard copies of part or all of this work for personal or classroom use is granted without fee provided that copies are not made or distributed for profit or commercial advantage and that copies bear this notice and the full citation on the first page. To copy otherwise, to republish, to post on servers or to redistribute to lists, requires prior specific permission and/or a fee.
DAC'09, July 26-31, 2009, San Francisco, California, USA

(e.g., [3,12,14]), with major focus on device fabrication level. As analyzed and projected in [2], on-chip optical interconnect outperforms traditional copper interconnect in power, throughput and delay with apparent gain below 22nm technology node starting from 2016.

As one of the most promising device level break-through for on-chip optical integration, silicon compatible nano photonic devices (e.g., [11,15]) take advantage of optical properties of a signal, characterizing great resilience in terms of small delay, low power and high throughput when compared with traditional copper interconnection. Advances in device level improvements of silicon nano photonics (such as photonic crystal structures in [7,16]) have also been demonstrated. In recent years, low RF power optical modulators operating at a few Gbps speed have been successfully demonstrated [6,7], with compact footprint for potential large scale on-chip integration. Compact photodetecters with up to 50Gbps processing rate have also been demonstrated (such as Germanium-on-Insulator photodetecter in [10]). With a proper collection of current Silicon nano-photonic devices and some extended projections / assumptions based on [2,8], there can be exciting CAD synthesis explorations in interconnect planning for optical on-chip integration.

As a related work, [13] studied timing driven and congestion driven on-chip optical routing CAD algorithms under 3-D system-on-package scenario. Yet the routing geometry in [13] was formulated in a very simple manner: point-to-point straight connection, which also means there is at least 1 optical modulator inserted at each pin and Steiner point in the netlist. There are 3 major issues with such an approach: *First*, it neglects the laser power consumption of optical modulators. Since each modulator requires a laser source for electrical-to-optical data conversion, this approach results in a very power consuming chip; *Second*, it neglects the photon-energy loss constraint on optical interconnect; consequently, there could be pins whose received photon-energy drop below the photo-detectors' detection threshold, leading to inevitable malfunction after optical-to-electrical data conver-

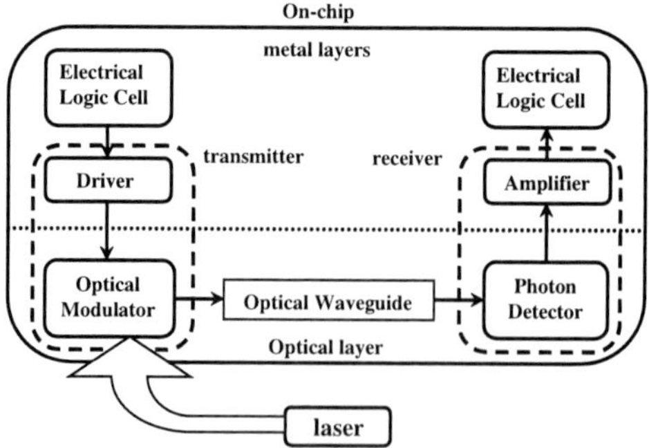

Figure 1: Block diagram for electrical/optical and optical/electrical data conversions

978-1-60558-497-3/09 $25.00 © 2009 ACM

sion. *Third*, optical routing has very different characteristics compared with conventional electrical (copper) interconnect routing, therefore, special routing geometry must be developed to tackle optical interconnect planning problems. In other words, total laser power consumption (proportional to number of modulators inserted) and the constraints for successful optical-to-electrical detection must both be addressed properly for optimized optical routing geometry.

In this work, we present *O-Router*, an optical routing framework that takes into consideration of various constraints and flexibilities that silicon nano-photonics device libraries and optical waveguide models shall impose on the future on-chip optical interconnect. *O-Router* is driven by low power on-chip silicon nano-photonic integration.

The rest of the paper is organized as follows: Section 2 introduces some preliminaries regarding optical and electrical data conversions and silicon photonics, followed by a motivational example and a summary list of key contributions of this paper. Section 3 describes our *Optical Interconnect Library* built for *O-Router*; Section 4 focuses on the optical routing Integer Linear Programming (ILP) problem formulation and speed-up techniques, followed by experimental results in Section 5. Section 6 concludes the paper by a brief summary and some potential future work.

2. PRELIMINARIES AND MOTIVATION

As shown in Fig. 1, on the transmitter's side, the electric signal from the driver (electrical layer) amplitude modulates the light source from the laser inside an optical modulator, and then send the modulated optical signal onto optical interconnect (optical waveguide on optical layer); on the receiver end, a photo-detector detects the photons from the waveguide and converts it into electric signal (back to electrical layer); an amplifier may be needed if this signal drives a high fan-out net on electric layers.

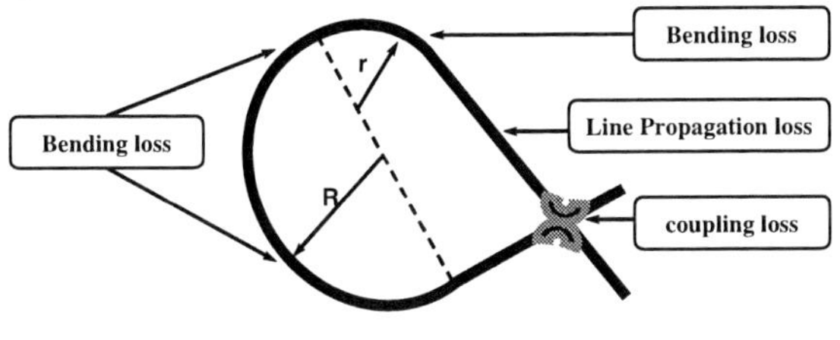

Figure 2: Sources of loss for on-chip optical routing

2.1 Optical Waveguide Routing

As aforementioned, optical routing has unique characteristics when compared with traditional copper routing. Manhattan (X/Y) routing based algorithms are not favored on optical layer because of the huge amount of loss caused by the sharp wire turnings along the data path, unless some special structures be inserted; yet these structures are usually costly in fabrication and/or bulky in footprint size, etc.

O-Router performs gridless optical routing with waveguide couplings/crossings on a single layer. As a result, routing geometry becomes very flexible, with different geometries and penalties according to their respective optical interconnect loss. In order to further explore optical routing geometry, we define the following 3 types of losses (with dB unit) on an optical interconnect path in equations 1- 5.

$$L_{loss} = \alpha \cdot length_{path} \qquad (1)$$

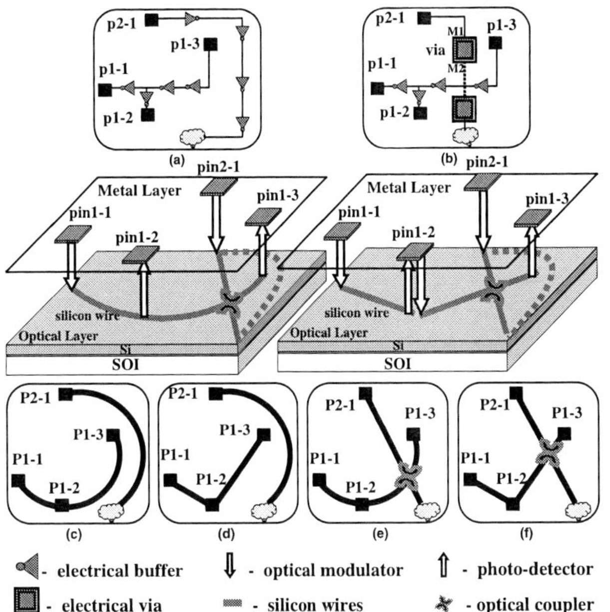

Figure 3: Motivational example for electrical routing v.s optical routing

$$B_{loss} = \beta \cdot \theta \cdot r^{-\eta} \qquad (2)$$
$$C_{loss} = \gamma \cdot Num_{couplers} \qquad (3)$$
$$P_{loss} = L_{loss} + B_{loss} \qquad (4)$$
$$Total_{loss} = P_{loss} + C_{loss} \qquad (5)$$

As shown in Fig. 2, L_{loss} is straight line waveguide loss, it is proportional to the length of optical interconnect, with a coefficient α; B_{loss} is the bending loss, since waveguide cross-section width is negligible compared to the bending radius in *O-Router*, we assume B_{loss} to be proportional to the degree of the optical interconnect (silicon waveguide) arc angle θ, and inversely proportional to the radius r of the interconnect, with an index η; C_{loss} is the coupling loss, proportional to the number of couplers (crossings) on the interconnect, with a coefficient γ. All related coefficients are determined by our *Optical Interconnect Library*, which is built for *O-Router* and will be explained further in Section 3.

2.2 Motivational Example

In this section, we briefly explore the different trade-offs for optical routing. As shown in Figure. 3, there are 2 nets to be routed on a chip, noted as *pini-j*, meaning it is the *jth* member of net *i*; Fig. 3(a) and (b) shows two alternatives for conventional routing on electrical layer with buffers and/or metal via inserted to alleviate the timing penalty caused by the long wires across the chip. Buffers are inserted since RC delay increases quadratically with electrical wire length. Yet buffer insertion is not all-powerful technique. Generally speaking, cross-chip timing critical nets are tough to fix thus impose great difficulty to VLSI design timing closure. As technology further scales down and system integration level rockets, issues with electrical interconnect will get more severe.

Fig. 3(c)-(f) show 4 possible routing geometries for the 2 nets on optical layer, according to our optical routing. Routing geometry (c) requires a total of 2 optical modulators: 1 inserted at P1-1, 1 inserted at P2-1, while for (d), 1 extra modulator will be inserted at P1-2, in order to drive P1-3, since sharp turning at P1-2 is either too lossy or too costly to fix other than using an extra modulator. In (e) and (f), optical coupler is introduced for coupling optical signal across 2 wires, with certain amount of loss. In these 2 cases, couplers can be employed either because doing so results in less amount of loss than taking detours as in (c) and (d), or because taking detours results in more coupling loss with other nets on chip, etc.

We can learn that geometries (c)(e) result in least among of modulating power among (c)-(f), yet optical interconnect bending loss: B_{loss} is also introduced, as well as the coupling loss: C_{loss} (in (e)) so that the constraint for successful detection at P1-3 may be violated due to too much loss on interconnect. To optimally pick the best routing geometry from the (c)-(f) 4 cases is the motivation of *O-Router*.

O-Router targets at finding optimal optical routing geometry to minimize total modulating power, subject to various constraints imposed by the device characterizations.

2.3 Main Contributions of This Paper

Main novelty and contributions of *O-Router* are as follows:

- Based on extensive data collection and road-mapping, we project the technology trends of on-chip silicon nano-photoincs and build *OIL*: an *Optical Interconnect Library* characterized for low-power on-chip integration/synthesis.

- For the first time, we formulate the optical routing problem by taking into considerations of various detection constraints and flexibilities that *OIL* imposes on the future optical interconnect.

- Under gridless single layer optical routing with couplings/crossings, the solution space is theoretically infinite. To reduce solution space without losing optimality, we put a set of constraints on the waveguide routing rules and formulate the optical routing problem with Integer Linear Programming.

- We also propose several key techniques to speed-up the optical routing framework under ILP formulation.

3. OPTICAL INTERCONNECT LIBRARY

To support our *O-Router* framework, we first build an Optical Interconnect Library (OIL), which includes a Mach-Zehnder optical modulator [6], a photo-detector from [10], a fully simulated optical coupler using Rsoft [1], and a set of optical interconnect (silicon waveguide) model. For details regarding OIL, please refer to [4].

3.1 Optical Modulator and Photo-detector

Based on some related research (e.g., [2,8,12]), we project current OIL parameters towards next generation technology, which essentially enables better on-chip integration for nanophotonic devices.

In Table 1, there are 2 sizes of modulators included, one is a normal modulator; the other is ModulatorX: a large modulator with 10X driving power, which will be inserted into a net that suffers greatly from power losses in order to

Table 1: **Major OIL components with high level parameters.**

	throughput	length	width	driving power	loss
modul1	>10Gps	<50um	<10um	1X	-
modulX	>10Gps	<50um	<50um	10X	-
detector	>10Gps	<10um	<10um	-	-
coupler	>10Gps	<50um	<5um	-	<10%

guarantee successful detection. Since the power consumption of ModulatorX is much larger than normal modulator, its usage will be penalized with a constant coefficient $MPow_{penalty}$, details in Section 4.

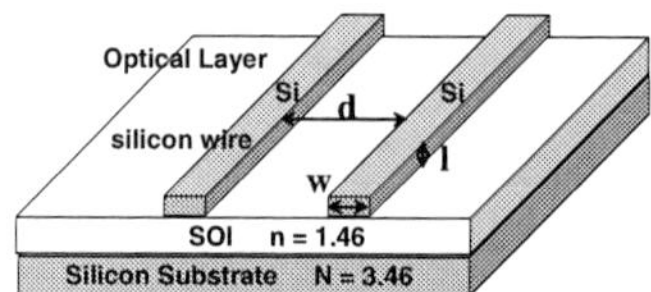

Figure 5: OIL on-chip optical waveguide model

3.2 Optical Coupler and Interconnect Model

The working principle of optical coupler is shown by Fig. 4. There are 4 ports from A to D for each coupler, and the parallel double interconnect region is the arm region. Optical signals will be cross coupled in the arm region. From the 4 simulation cases, we can verify that PortA=PortD and PortB=PortC always satisfy, as if there is wire connection between A-D and B-C. Optical couplers allow us to make full use of the optical layer routing space, making non-planar netlists routable on a single silicon layer. In case (b)(c) in Fig. 4, there is slight loss for high optical logic after the coupler, as is formulated by C_{loss}.

As shown in Fig. 5, the optical waveguide model included in OIL has a reflective index of 3.46, coated on top of a $2um$ thick SOI layer (reflective index < 1.46). The cross-section width of the silicon wire $w=0.5um$, cross-section height $l=0.22um$, wire spacing d between $0.5um$ and $3.0um$. d should be set properly to avoid wire cross-talk.

4. O-ROUTER FORMULATION AND ALGORITHM

Given the pin locations of certain placed netlist for optical routing, *O-Router* seeks optimal routing solution with Integer Linear Programming to minimize total modulating power, meanwhile satisfying various detection constraints according to established OIL parameters. This section is divided into three parts: First is the optical netlist mapping. This is when suitable optical netlist benchmarks for *O-Router* are constructed. Second part is the core ILP formulation, followed by routing speed-up techniques in the third part.

4.1 Optical Netlist Mapping

Given an electrical layer netlist after placement, the goal of this step is to prepare an optical netlist that makes most use of optical layer resource to fix top timing critical nets (i.e., longest) in electrical layer. For our ILP formulation, the resulting optical netlist of this stage consists of only 2, 3 and 4 pin nets. It takes place in 3 phases:

Phase 1: Pre-select top timing critical nets from electrical layer to map onto optical layer. Shown in Fig. 6(a)(b), non-timing critical nets 1/2/3 are not selected.

Phase 2: Cluster within each pre-selected net for routing

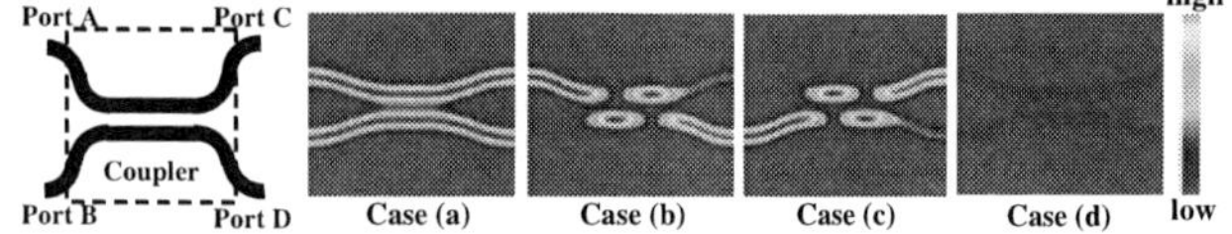

Figure 4: Working mechanism of optical coupler in *O-Router*

978-1-60558-497-3/09 $25.00 © 2009 ACM

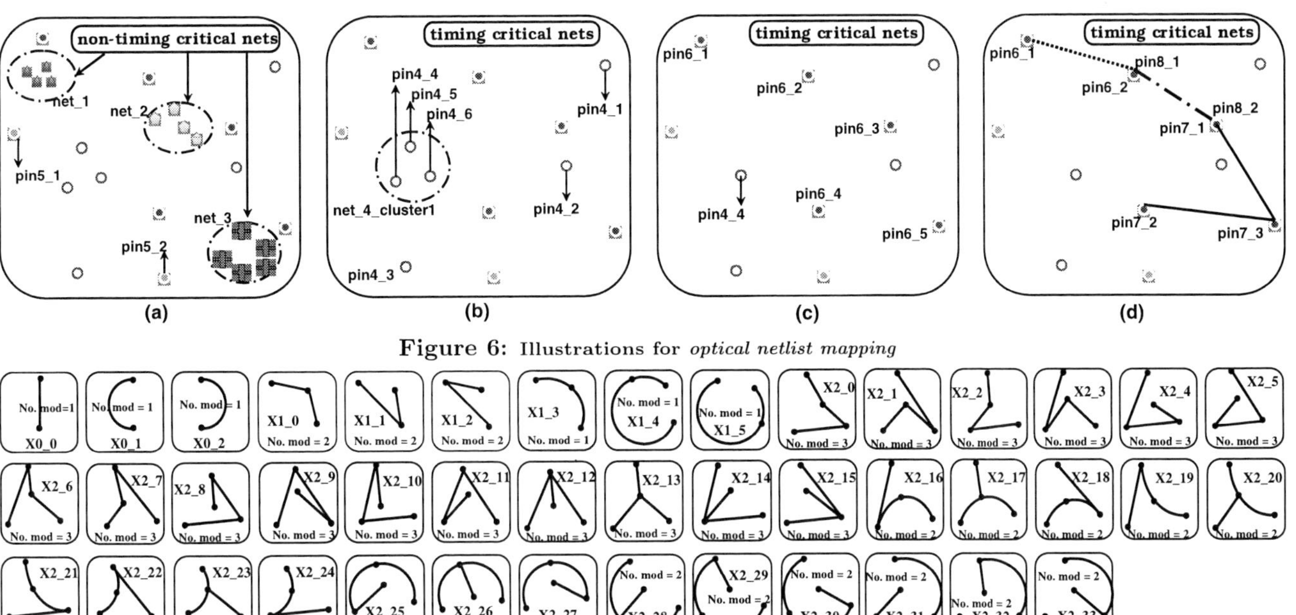

Figure 6: Illustrations for *optical netlist mapping*

Figure 7: A list of optical routing geometries (represented by integer variables) for 2, 3 and 4 pin nets

efficiency enhancement. Since optical routing is most effective dealing with global interconnect, we map a single pin from each local pin cluster onto the optical layer and leave the remaining pins to electrical layer.

As shown from Fig. 6(b) to Fig. 6(c), the *net_4_cluster1* is represented by *pin4_4* on optical layer. With this phase, the pin number for each optical net becomes very small. For *O-Router*, we manage to keep each optical net size to below 5 pins. Practically, nets with more than 4 pins can be decomposed into a set of 2/3/4 pin nets, as illustrated in (c)-(d), where a 5-pin net6 is decomposed into two 2-pin nets and a 3-pin net.

Phase 3: For intersected 2-pin nets in the netlist from Phase2, expand them to have 2 more integer variables if and only if they can avoid crossing each other by taking an arc detour, meanwhile the detour does not cut a third net. This step further expands the feasible solution space for 2-pin nets.

4.2 Integer Linear Programming Formulation

For the original ILP formulation, we enumerate all routing geometries for the 2-pin, 3-pin and 4-pin nets, shown in Fig. 7(a concave shape 4-pin net is shown as an example). Each X_{ij} is an integer variable, where $i \in net\_space$, $j \in sol\_space(net\ i)$. When $X_{ij} = 1$, the corresponding routing geometry from Fig. 7 will be adopted, as part of the final routing solution space. Number of modulators in each X_{ij} is also recorded; OIL will return the actual modulating power based on this number and the ij index.

The ILP formulation is as follows in Equation 6- 15, with all terms and variables explained in Table 2. The objective function is the total power required to drive all the on-chip optical modulators for our optical interconnect framework. The ILP solver will minimize the objective function, subject to constraints imposed from Eq. 7 to Eq. 15. In Eq. 6, the first term $MPow_{Xij}$ is total modulating power consumption for routing geometry X_{ij} using 1X modulators, while the second term $(MPow_{penalty} - P_0) \cdot M_{ij} \cdot N_{ij}$ is for penalizing the usage of 10X driving power ModulatorX: if M_{ij} is

1(hard constraint violation), then ModulatorX will be used to replace all Modulator1's in geometry X_{ij} to meet the constraint (P_0 is the laser power consumption of Modulator1).

$$
min\{ \sum_{\substack{j \in sol\_space(i) \\ i \in net\_space()}} [MPow_{Xij} \cdot X_{ij} + (MPow_{penalty} - P_0) \cdot M_{ij} \cdot N_{ij}]\}
$$

$$
s.t \tag{6}
$$

$$\forall i, m \in net\_space, i \neq m, j \in sol\_space(i), n \in sol\_space(m):$$

$$
P_{loss_{Xij}} \cdot X_{ij} + net\_loss_{Xij} \leq loss\_th_{Xij} + pow \cdot N_{ij} \cdot M_{ij} \tag{7}
$$

$$
P_{loss_{Xij}} = L_{loss_{Xij}} + B_{loss_{Xij}} \tag{8}
$$

$$
net\_loss_{Xij} = \sum_{\substack{m \in net\_space \\ }}^{n \in sol\_space(m)} C_{loss_{Xij\_mn}} \cdot X_{ij\_mn} \tag{9}
$$

$$
C_{loss_{Xij\_mn}} = \gamma_{ij\_mn} \cdot cross\_num < X_{ij}, X_{mn} > \tag{10}
$$

$$
X_{ij} + X_{mn} \leq 1 + X_{ij\_mn} \tag{11}
$$

$$
(1 - X_{ij}) + (1 - X_{mn}) \leq 2 - X_{ij\_mn} \tag{12}
$$

$$
\sum_{j \in sol\_space(i)} X_{ij} = 1, \quad X_{ij} = 0, 1 \tag{13}
$$

$$
X_{ij\_mn} = 0, 1 \tag{14}
$$

$$
M_{ij} = 0, 1 \tag{15}
$$

Constraint Eq. 7 is set for each routing geometry X_{ij}, such that its total loss (propagation loss P_{loss} and coupling loss C_{loss}) is bounded by an upper bound of loss threshold: $loss\_th_{xij}$, once the upper bound of loss is exceeded, it means the photo-detection requirements in routing geometry X_{ij} are violated. If among all feasible X_{ij}, some of such constraint is inevitably violated, then ModulatorX will be inserted into the corresponding geometry X_{ij} and replace existing 1X modulators. Constraint Eq. 10 explicitly maps the crossing number of a net into corresponding coupling loss using OIL.

For ILP formulation of the calculation of optical interconnect coupling number, we introduced the cross-term integer

Algorithm 1 ILP based Optical Routing for low power chip

Require: mapped optical netlist benchmark
  invoke optical netlist parser; **link** OIL
  while $i \in net\_space$ **do**
    while $j \in sol\_space(i)$ **do**
      calculate $(L_{loss_{Xij}}, B_{loss_{Xij}}, MPow_{Xij}, MPow_{penalty}, \text{etc.})$
      while $m \in net\_space, m \neq i$ **do**
        while $n \in sol\_space(m)$ **do**
          calculate $(C_{loss_{Xij\_mn}}, \text{constraint coefficients,etc.})$
        end while
      end while
    end while
  end while
  generate glpk syntax file; **invoke** glpk ILP solver − minimize
  return optical routing for minimum modulating power

Algorithm 2 ILP variable number reduction via trimming

Require: mapped optical netlist benchmark
  while $i \in net\_space$ **do**
    while $j \in sol\_space(i)$ **do**
      calculate $(L_{loss_{Xij}}, B_{loss_{Xij}})$
      if $L_{loss_{Xij}} + B_{loss_{Xij}} \geq threshold_{Xij}$ **then**
        exclude X_{ij}; update data structures
      end if
    end while
  end while
  return trimmed set of routing geometries for each net

Algorithm 3 ILP cross-term variable reduction via merging

Require: mapped optical netlist benchmark
  while $i \in net\_space$ **do**
    while $j \in sol\_space(i)$ **do**
      while $m \in net\_space, m \neq i$ **do**
        while $n \in sol\_space(m)$ **do**
          if $i > m$ **then**
            swap (i,j) with (m,n) in X_{ij_mn}; **calculate** cross-term constraint coefficients; **update** glpk syntax file
          end if
        end while
      end while
    end while
  end while
  return reduced set of cross-terms

Algorithm 4 bounding box for C_{loss} computation speed-up

Require: mapped optical netlist benchmark
  generate $bounding\_box\_matrix[\ ][\ ]$
  while $i \in net\_space$ **do**
    while $m \in net\_space, m \neq i$ **do**
      if $bounding\_box\_matrix[i, m] == 1$ **then**
        calculate $C_{loss_{ij\_mn}}$; update glpk syntax file
      end if
    end while
  end while
  return optical routing for minimum modulating power

---

variables: X_{ij_mn}. Numerically, it is the product of term X_{ij} and X_{mn}. Since variable multiplications are not supported by ILP solver, we add the constraint pair Eq. 11- Eq. 12. Integer constraints Eq. 11 and Eq. 12 bound the X_{ij_mn} term so that it always equals the product of its two corresponding routing geometries. Equality constraint Eq. 13 makes sure that the ILP solver eventually picks only 1 routing geometry out of each net for the final optimal solution. For further details please see Table 2 and Algorithm 1.

Table 2: Descriptions for ILP involved terms and variables.

Name	Description
$net\_space()$	set of nets for an optical netlist
$sol\_space(i)$	set of possible routing geometries for net i
$MPow_{Xij}$	total modulator power consumption of routing geometry X_{ij}
$MPow_{penalty}$	power consumption penalty for using each ModulatorX. Set to 10 times of P_0
P_0	power consumption of Modulator1
N_{ij}	least number of optical modulators used for geometry X_{ij}
$C_{loss_{Xij}}$	coupling loss power between routing geometry X_{ij} and X_{mn}
$P_{loss_{Xij}}$	propagation loss power on silicon wires of X_{ij}
X_{ij}	integer variable. $X_{ij} = 1$ means to accept the jth routing geometry of net i
M_{ij}	integer variable. $M_{ij} = 1$ means to insert modulatorX into jth routing geometry of net i
X_{ij_mn}	integer variable. numerically equals to $X_{ij} \cdot X_{mn}$
$loss\_th_{Xij}$	loss threshold for O-E conversion for X_{ij}
pow	more driving power each ModulatorX brings than Modulator1
γ_{ij_mn}	coupling loss coefficient returned by OIL dependent on geometry X_{ij} and X_{mn}

4.3 Variable Reduction and Speed-up Techniques

Apparently, a direct implementation of Algorithm 1 will result in very large number of variables as well as tremendous among of computations, especially for large optical netlists. Here we propose some useful techniques to speed-up *O-Router*.

4.3.1 Variable Trimming/Merging

Variable trimming procedure first scans through the X_{ij} list and calculate bending loss B_{loss} and line propagation loss L_{loss} for each X_{ij}. If the loss of X_{ij} itself becomes unbearable, then such a routing geometry is dumped before invoking ILP. Variable trimming procedure successfully trims off the infeasible integer variables in Fig. 7 and greatly reduces the variable set. Solution optimality will not be harmed with careful choice of the *threshold* value. Details about this procedure are shown Algorithm 2.

Variable merging procedure runs in parallel with *variable trimming procedure*. As described in Algorithm 3, it cuts the cross-variable set to half of original size, and the idea is amazingly simple:

$$X_{ij} \cdot X_{mn} = X_{ij\_mn} \tag{16}$$

$$X_{mn} \cdot X_{ij} = X_{mn\_ij} \tag{17}$$

$$X_{ij\_mn} = X_{mn\_ij} \tag{18}$$

Essentially, X_{ij} and X_{mn} generate non-zero constraint coefficients only when both of them are adopted, which means the product equality holds in Eq. 16 and 17, consequently, Eq. 18 also holds, so we can rename half of the cross-term variables to the other half, since they are identical. Details about this procedure are shown in Algorithm 3.

4.3.2 Bounding Box Elimination for Speed-up

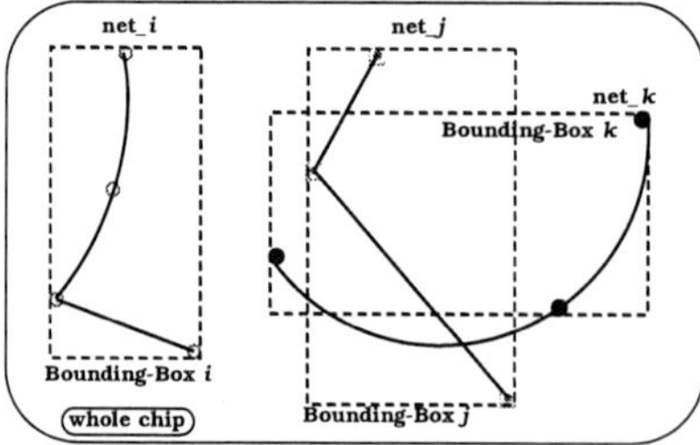

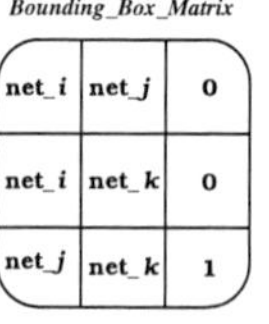

Figure 8: Illustration of Bounding Box Elimination

Table 3: Performance comparisons between *O-Router* and Minimum Spanning Tree algorithm.

	photo-detection threshold : 55%				photo-detection threshold: 75%			
	ibm01	ibm02	ibm03	ibm04	ibm01	ibm02	ibm03	ibm04
Net number	5	20	50	137	5	20	50	137
Pin number	15	50	155	391	15	50	155	391
Pin/net ratio	3	2.5	3.1	2.85	3	2.5	3.1	2.85
MST-routing (normalized power)	3.5	6	35.66	305.13	3.5	12.75	39	306.25
O-Router (normalized power)	1	2.88	10.75	57.75	2.13	5.38	16.5	100.25
Improvement	71.40%	52.00%	69.90%	81.10%	39.10%	57.80%	57.70%	67.30%

The introduction of Bounding Box contributes to computation speed-up of O-Router. *Bounding Box* of net i is defined as the rectangle that bounds all the pins of net i. It is defined by 4 values as in Eq. 19:

$$Bounding\_Box_i = (min(X), max(X), min(Y), max(Y)) \quad (19)$$

$$where \quad X \in x\_axis\{net_i\}, \quad Y \in y\_axis\{net_i\} \quad (20)$$

As illustrated in Fig. 8, a *Bounding_Box_Matrix* will be generated in pre-scanning stage; for any pair of nets with non-overlapping bounding boxes, a 0 is recorded, otherwise, 1 is written; In Fig. 8, *net_j* and *net_k* have potential crossings, thus only them will be processed for constructing coupling loss constraints. This procedure worth the efforts because a general algorithm for calculating C_{loss} is much more complicated than min/max value search. Further details for bounding box elimination procedure are shown in Algorithm 4.

With all 3 speed-up procedures, the original ILP formulation in Section 4.2 is modified, implemented and tested.

5. EXPERIMENTAL RESULTS

Simulations are carried out according to the aforementioned 3 steps in Section 4, and original electrical benchmarks come from ISPD98/08 routing benchmarks. ibm01-04 are the final 4 optical netlists benchmarks, listed as in Table 3. Due to considerations of silicon wire spacing/low coupling noise communication, the sizes of the optical netlists are kept from small to medium, and the optical layer pin density is kept from low to medium. As a baseline for *O-Router*, Minimum Spanning Tree (MST) routing algorithm is implemented on ibm01-04. Both *O-Router* framework and MST algorithm are repeated on ibm01-04 for 2 different photo-detection threshold values: 55% and 75%. Such percentages signify the photo-detection power threshold for received signals at the end of optical interconnect. Therefore, 75% threshold photo-detectors impose stricter detection requirements on *O-Router* framework. In Table 3, the simulated power consumptions are normalized by the amount of power reported by *O-Router* on ibm01, under photo-detection threshold of 55%. For 55% threshold, *O-Router* achieves above 50% of power reduction compared to MST baseline, with a max of 81.1% on ibm04. For the 75% threshold, *O-Router* reports slightly less power reductions due to higher detection requirements; still an average of above 50% reduction, with a max of 67.3% of power reduction on ibm04.

6. CONCLUSION

In this paper, we present the first optical routing framework, *O-Router* for low power on-chip integration of silicon nano-photonics with consideration of various detection constraints. Based on ILP formulation with several variable reduction techniques for routing speed-ups, *O-Router* utilizes Optical Interconnect Library, which is an established collection of some silicon compatible on-chip nano-photonics devices and optical interconnect models, with key parameters projected for future technologies based on optical interconnect roadmap. Experimental results show promising improvements compared with traditional Minimum Spanning Tree routing algorithm. We expect to see a lot of future works along this direction as new nano-photonics devices are introduced for the ultimate global optical and electrical interconnect co-synthesis and planning.

7. ACKNOWLEDGMENT

This work is supported in part by Texas Norman Hackerman Advanced Research Program.

8. REFERENCES

[1] RSoft Photonics CAD Suite version 5.1.7, by RSOFT Inc.

[2] G. Chen et al. Predictions of CMOS Compatible On-Chip Optical Interconnect. In *International Workshop on System Level Interconnect Prediction*, pages 13–20, 2005.

[3] R. T. Chen. Optical Interconnects and VLSI Photonics. In *IEEE LEOS 2004 Summer Topical Meeting, LEOS Newsletters No.5*, pages 8–9, June 2004.

[4] D. Ding and D. Z. Pan. OIL: A Nano-photonics Optical Interconnect Library for a New Photonic Networks-on-Chip Architecture. In *Proc. System-Level Interconnect Prediction*, 2009.

[5] J. W. Goodman. Optical Interconnects for VLSI Systems. In *Proc. of IEEE*, volume 72, pages 850–866, July 1984.

[6] W. M. J. Green et al. Ultra-Compact Low RF Power 10Gb/s Silicon MachZehnder Modulator. In *Proc. of the 20th Annual Meeting of the IEEE Lasers & Electro Optics Society*, 2007.

[7] L. Gu et al. High Speed Silicon Photonic Crystal Waveguide Modulator for Low Voltage Applicatoin. In *Applied Physics Letters, 90, 071105*, 2007.

[8] International Technology Roadmap for Semiconductors. 2008.

[9] M. J. Kobrinsky. On-Chip Optical Interconnects. In *INTEL Technol. J8(2)*, pages 129–141, 2004.

[10] S. J. Koester et al. Germanium-on-Insulator Photodetectors. In *IEEE International Conference on Group VI Photonics*, pages 171–173, 2005.

[11] Y. Massoud et al. Subwavelength Nanophotonics for Future Interconnects and Architectures. In *invited talk*, NRI SWAN Center, Rice University, 2008.

[12] D. A. B. Miller. Device Requirement for Optical Interconnects to Silicon Chips. In *Proc. of IEEE Special Issue on Silicon Photonics*, 2009.

[13] J. R. Minz et al. Optical Routing for 3-D System-on-Package. In *IEEE Transactions on Components and Packaging Technologies*, Dec 2007.

[14] I. O'Conor et al. On-Chip Optical Interconnect for Low-Power. In *E. Macii(Ed), Ultra-Low Power Electronics and Design*. Kluwer Academic Publishers, 2004.

[15] Y. A. Vlasov. Silicon Photonics for Next Generation Computing Systems. In *European Conference and Exhibition on Optical Communication*, Sep 2008.

[16] Y. A. Vlasov et al. Active Control of Slow Light on a Chip with Photonic Crystal Waveguides. In *Nature*, volume 438, Nov 2005.

978-1-60558-497-3/09 $25.00 © 2009 ACM

BDD-based Synthesis
of Reversible Logic for Large Functions

Robert Wille
Institute of Computer Science
University of Bremen
28357 Bremen, Germany
rwille@informatik.uni-bremen.de

Rolf Drechsler
Institute of Computer Science
University of Bremen
28357 Bremen, Germany
drechsler@uni-bremen.de

ABSTRACT

Reversible logic is the basis for several emerging technologies such as quantum computing, optical computing, or DNA computing and has further applications in domains like low-power design and nanotechnologies. However, current methods for the synthesis of reversible logic are limited, i.e. they are applicable to relatively small functions only. In this paper, we propose a synthesis approach, that can cope with Boolean functions containing more than a hundred of variables. We present a technique to derive reversible circuits for a function given by a *Binary Decision Diagram* (BDD). The circuit is obtained using an algorithm with linear worst case behavior regarding run-time and space requirements. Furthermore, the size of the resulting circuit is bounded by the BDD size. This allows to transfer theoretical results known from BDDs to reversible circuits. Experiments show better results (with respect to the circuit cost) and a significantly better scalability in comparison to previous synthesis approaches.

Categories and Subject Descriptors

B.6 [**Hardware**]: Logic Design

General Terms

Algorithms

Keywords

Reversible Logic, Quantum Logic, Synthesis, Decision Diagrams

1. INTRODUCTION

Reversible logic [1, 2, 3] realizes n-input n-output functions that map each possible input vector to a unique output vector (i.e. bijections). Although reversible logic significantly differs from traditional (irreversible) logic (e.g. fan-out and feedback are not allowed), it has become an intensely studied research area in recent years. In particular, this is caused by the fact that reversible logic is the basis for several emerging technologies, while traditional methods suffer from the increasing miniaturization and the exponential growth of the number of transistors in integrated circuits.

In fact, reversible logic can help to face the enormous challenges in the development of future computing machines: While for example the increasing power consumption of electronic devices becomes a serious problem in current technologies, energy dissipation is reduced or even eliminated if computation becomes information-lossless [1]. This holds for reversible logic, since data is bijectivly transformed without loosing any of the original information. Furthermore, reversible computation is the basis for quantum computing [4]. In this domain it has been shown that important problems such as factorization can be solved exponentially faster than by classical methods. Further applications of reversible logic can be found in the domain of optical computing [5], DNA computing [2], and nanotechnologies [6].

However, currently the synthesis of reversible logic or quantum circuits, respectively, is limited. Exact (see e.g. [7, 8]) as well as heuristic (see e.g. [9, 10, 11, 12, 13, 14]) methods have been proposed. But both are applicable only for relatively small functions. Exact approaches reach their limits with functions containing more than 6 variables [8] while heuristic methods are able to synthesize functions with at most 30 variables [13]. Moreover, often a significant amount of run-time is needed to achieve these results.

These limitations are caused by the underlying techniques. The existing synthesis approaches often rely on truth tables (or similar descriptions like permutations) of the function to be synthesized (e.g. in [9, 10]). But even if more compact data-structures like BDDs [11], positive-polarity Reed-Muller expansion [13], or Reed-Muller spectra [14] are used, the same limitations can be observed since all these approaches apply similar strategies (namely selecting reversible gates so that the choosen function representation becomes the identity).

In this work we introduce a synthesis method that can cope with significantly larger functions. The basic idea is as follows: First, for the function to be synthesized a BDD [15] is built. This can be efficiently done for large functions using existing well-developed techniques. Then, each node of the BDD is substituted by a cascade of reversible gates. Since BDDs may include shared nodes causing fan-outs (which are not allowed in reversible logic), this may require additional circuit lines.

As a result, circuits composed of Toffoli [3] or elementary quantum gates [4], respectively, are obtained in linear time and with memory linear to the size of the BDD. Moreover, the size of the resulting circuit is bounded by the BDD, so that theoretical results known from BDDs (see e.g. [16, 17]) can be transferred to reversible circuits. Our experiments show significant improvements (with respect to the resulting circuit cost as well as to the run-time) in comparison to previous approaches. Furthermore, for the first time large functions with more than a hundred of variables can be synthesized at very low run-time.

The remainder of the paper is structured as follows: Section 2 provides the basics of reversible logic and BDDs. Afterwards, in Section 3 the synthesis approach is described in detail. Section 4 briefly reviews some of the already known theoretical results from reversible logic synthesis and introduces bounds which follow from the new synthesis approach. Finally, in Section 5 experimental results are given and the paper is concluded in Section 6.

Permission to make digital or hard copies of part or all of this work for personal or classroom use is granted without fee provided that copies are not made or distributed for profit or commercial advantage and that copies bear this notice and the full citation on the first page. To copy otherwise, to republish, to post on servers or to redistribute to lists, requires prior specific permission and/or a fee.
DAC'09, July 26-31, 2009, San Francisco, California, USA

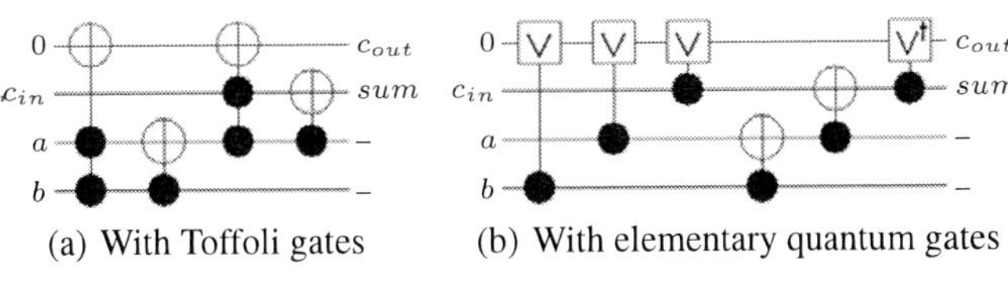

(a) With Toffoli gates   (b) With elementary quantum gates

Figure 1: Two circuits realizing a full adder

2. PRELIMINARIES

To keep the paper self-contained this section briefly reviews the concepts of reversible and quantum logic. We also describe the basics of BDDs which are used as the main data-structure in our synthesis approach.

2.1 Reversible and Quantum Logic

A logic function is reversible if it maps each input assignment to a unique output assignment. Such a function must have the same number of input and output variables $X := \{x_1, \ldots, x_n\}$. Since fanout and feedback are not allowed in reversible logic, a circuit realizing a reversible function is a cascade of reversible gates. A *reversible gate* has the form $g(C, T)$, where $C = \{x_{i_1}, \ldots, x_{i_k}\} \subset X$ is the set of control lines and $T = \{x_{j_1}, \ldots, x_{j_l}\} \subset X$ with $C \cap T = \emptyset$ is the set of target lines. C may be empty. The gate operation is applied to the target lines iff all control lines meet the required control conditions. Control lines and unconnected lines always pass through the gate unaltered.

In the literature, several types of reversible gates have been introduced. Besides the Fredkin gate [18] and the Peres gate [19]), *(multiple controlled) Toffoli* gates [3] are widely used. Each Toffoli gate has one target line x_j, which is inverted iff all control lines are assigned to 1. That is, a multiple controlled Toffoli gate maps $(x_1, \ldots, x_j, \ldots, x_n)$ to $(x_1, \ldots, x_{i_1} x_{i_2} \cdots x_{i_k} \oplus x_j, \ldots, x_n)$.

The *cost* of a reversible circuit is defined either by the number of gates or by so called *quantum cost* [20, 21]. The latter can be derived by substituting the reversible gates of a circuit by a cascade of *elementary quantum* gates [4]. Elementary quantum gates realize quantum circuits that are inherently reversible and manipulate qubits rather than pure logic values. The state of a qubit for two pure logic states can be expressed as $|\Psi\rangle = \alpha|0\rangle + \beta|1\rangle$, where $|0\rangle$ and $|1\rangle$ denote 0 and 1, respectively, and α and β are complex numbers such that $|\alpha|^2 + |\beta|^2 = 1$. The most used elementary quantum gates are the NOT gate (a single qubit is inverted), the controlled-NOT (CNOT) gate (the target qubit is inverted if the single control qubit is 1), the controlled-V gate (also known as a square root of NOT, since two consecutive V operations are equivalent to an inversion), and the controlled-V+ gate (which performs the inverse operation of the V gate and thus is also a square root of NOT).

EXAMPLE 1. *Figure 1(a) shows a Toffoli gate realization of a full adder. A circuit realizing the same function by elementary quantum gates is depicted in Figure 1(b).*

Since quantum circuits are reversible, to realize a non-reversible function (e.g. an n-input m-output function with $n > m$) it must be embedded into a reversible one [22]. Therefore, it is often necessary to add *constant inputs* and *garbage outputs*. The garbage outputs are by definition don't cares and can be left unspecified.

2.2 Binary Decision Diagrams

A Boolean function $f : \mathbb{B}^n \to \mathbb{B}$ can be represented by a *Binary Decision Diagram* (BDD) [15]. A BDD is a directed acyclic graph $G = (V, E)$ where a Shannon decomposition

$$f = \overline{x}_i f_{x_i=0} + x_i f_{x_i=1} \quad (1 \le i \le n)$$

is carried out in each node $v \in V$. In the following the node representing $f_{x_i=0}$ ($f_{x_i=1}$) is denoted by $low(v)$ ($high(v)$) while x_i

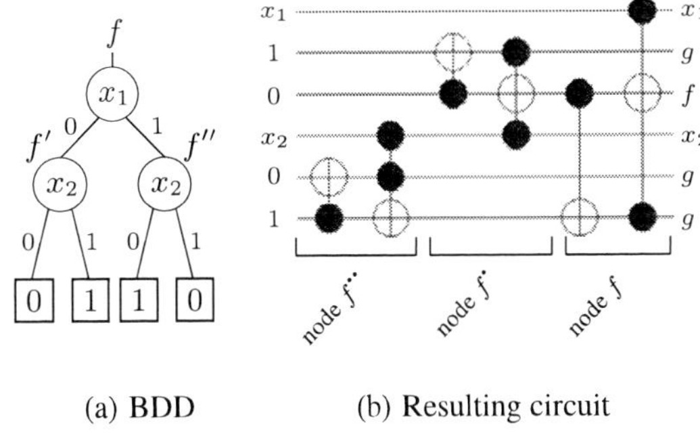

(a) BDD   (b) Resulting circuit

Figure 2: BDD and Toffoli circuit for $f = x_1 \oplus x_2$

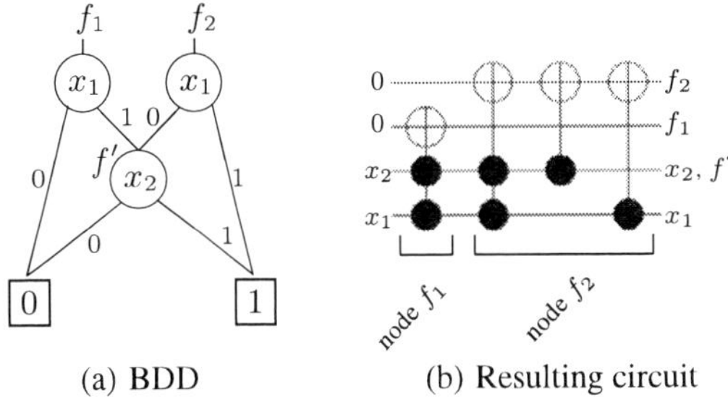

(a) BDD   (b) Resulting circuit

Figure 3: Synthesizing $f_1 = x_1 \wedge x2$ **and** $f_2 = x_1 \vee x_2$

is called the *select variable*. The *size* k of a BDD is defined by the number of non-terminal nodes.

The size of a BDD can be significantly reduced, if *shared nodes* are exploited [15]. That is, if a node v has more than one predecessor. Functions $f : \mathbb{B}^n \to \mathbb{B}^m$ (i.e. functions with more than one output) can be represented more compactly using shared nodes. Further reduction can be achieved if *complement edges* [23] are applied. In particular, this enables the representation of a function as well as its negation by a single node only. To keep the descriptions easier, in the following BDDs without complement edges are used for presentation while they are used in the experiments.

EXAMPLE 2. *Figure 2(a) shows a BDD realizing the function $f = x_1 \oplus x_2$. A BDD representing a function containing shared nodes and including two outputs $f_1 = x_1 \wedge x_2$ and $f_2 = x_1 \vee x_2$ is depicted in Figure 3(a). Edges from a node v to $low(v)$ ($high(v)$) are marked with a small 0 (1).*

3. SYNTHESIS APPROACH

In this section we describe how to derive a reversible circuit from a given BDD representation. First the general idea (i.e. substituting BDD nodes by a cascade of reversible gates) is introduced and discussed. Afterwards the overall synthesis algorithm is described in detail.

3.1 General Idea

Boolean functions can be efficiently represented by BDDs [15]. Having a BDD $G = (V, E)$, a reversible network can be derived by traversing the decision diagram and substituting each node $v \in V$ with a cascade of reversible gates. The respective cascade of gates depends on the concrete type of the node v. For the general case, Figure 4 shows a substitution with two Toffoli gates (and quantum cost of six). This substitution can be applied to derive a complete Toffoli network from a BDD without shared nodes. Thereby, some inputs have to be set to constants, if the respective $low(v)$ or $high(v)$ edge leads to a terminal node.

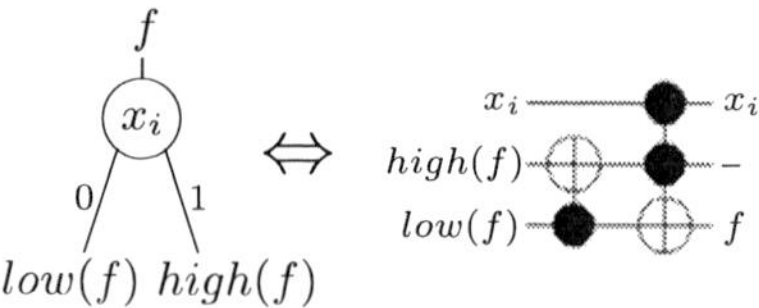

Figure 4: BDD node and equivalent cascade of Toffoli gates

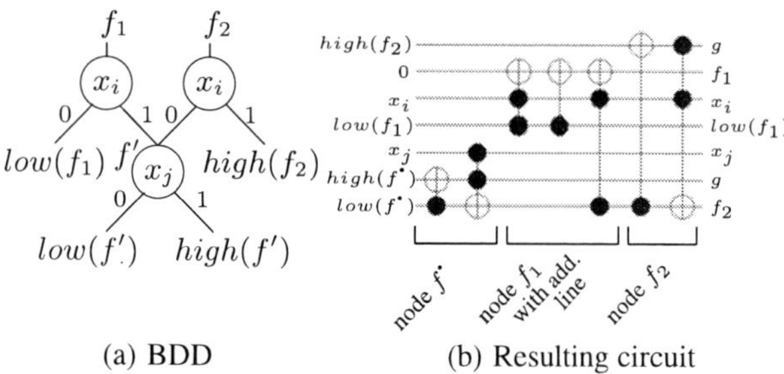

(a) BDD   (b) Resulting circuit

Figure 5: Toffoli circuit for shared BDD

EXAMPLE 3. *Consider the BDD in Figure 2(a). Applying the substitution given in Figure 4 to each node of the BDD, the Toffoli network depicted in Figure 2(b) results.*

However, BDD packages make use of shared nodes. But since shared nodes cause fan-outs (which are not allowed in reversible logic) a modified substitution has to be applied. More precisely, the values on the signals representing the shared node in the circuit to be synthesized must be preserved until they are not needed by any of the remaining nodes. The same holds for signals representing select variables of the nodes since they are also often required more than once. For example, in Figure 3(a) the value of variable x_1 is needed by two nodes. As a result, in some cases the values of all inputs of a BDD node have to be preserved. To represent this in reversible logic, i.e. to "emulate" a fan-out, an additional line and an adjusted cascade of gates as depicted in the first row of Table 1 is needed.

EXAMPLE 4. *In Figure 5(a) a partial BDD including a shared node f' is shown. Since the value of node f' is used twice (by nodes f_1 and f_2), an additional line (the second one in Figure 5(b)) and the cascade of gates as depicted in the first row of Table 1 is applied to substitute node f_1. In doing so, the value of f' is still available such that the substitution of node f_2 can be applied. The resulting circuit is given in Figure 5(b).*

Using these two substitutions (the one from Figure 4 and the one from the first row of Table 1), each BDD can be transformed to a Toffoli network. However, as the remaining rows of Table 1 show, better substitutions are possible, if terminal nodes occur as successors. For example, a node v with $low(v) = 0$ (fourth row) can be synthesized with only one Toffoli gate. Identity nodes, i.e. nodes with $low(v) = 0$, $high(v) = 1$, and a select variable x_i (not depicted in Table 1), can be represented by the same line as the input x_i and thus need no additional gate. Furthermore, due to the additional line, which has to be added if either $low(v)$ or $high(v)$ is a terminal, it is also possible to preserve all inputs of a node. In particular for shared nodes, this allows better substitutions.

REMARK 1. *One might expect, that a circuit line can be saved if one of the edges $low(v)$ or $high(v)$ leads to a terminal node. But this is not possible due to reversibility which has to be ensured when synthesizing reversible logic. As an example consider a node v with $high(v) = 0$ (second row of Table 1). Without loss of generality,*

Table 1: BDD nodes and cascade of Toffoli gates with add. lines

BDD	TOFFOLI CASCADE

Table 2: (Partial) Truth tables for node v with $high(v) = 0$

(a) Without add. line

x_i	$low(x_i)$	f	$-$
0	0	0	0
0	1	1	1
1	0	0	1
1	1	0	?

(b) With additional line

0	x_i	$low(x_i)$	f	x_i	$low(x_i)$
0	0	0	0	0	0
0	0	1	1	0	1
0	1	0	0	1	0
0	1	1	0	1	1

the first three lines of the corresponding truth table can be embedded with respect to reversibility as depicted in Table 2(a). However, since f is 0 in the last line, no reversible embedding for the whole function is possible. Thus, an additional line is required to make the respective substitution reversible (see Table 2(b))[1].

Similar substitutions can also be applied to synthesize circuits containing elementary quantum gates. This even reduces the resulting quantum cost, since the nodes can be substituted at smaller cost than the Toffoli gate cascades shown above. The same holds, if complement edges are applied that may reduce the size of the BDD. Due to page limitation the concrete substitutions are omitted here but can be found in [24] where these aspects are studied in detail. The respective elementary quantum circuit synthesis as well

---

[1] Due to the same reason it is also not possible to preserve the values for $low(v)$ or $high(v)$, respectively, in Figure 4.

as the support of complement edges have been considered in the experiments in Section 5.

3.2 Algorithm

Taking the general idea (substituting BDD nodes by cascades of Toffoli or elementary quantum gates, respectively), a method for synthesizing large functions in reversible logic can be formulated: First, a BDD for the function f to be synthesized is created. This can be done efficiently using state-of-the-art BDD packages (e.g. CUDD [25]).

Next, the resulting BDD $G = (V, E)$ is traversed by a depth-first search. For each node $v \in V$, three checks are performed:

1. *Node v represents the identity of a primary input (i.e. select input)*
   In this case no cascade of gates is added to the circuit since the identity can be represented by the same circuit line as the input itself.

2. *Node v contains at least one edge ($low(v)$ or $high(v)$, respectively) leading to a terminal*
   In this case substitutions as depicted in Table 1 are applied, since they often need a smaller number of gates (or quantum cost) and additionally preserve the values of all input signals.

3. *The successors of node v (i.e. $low(v)$ and $high(v)$) are still needed since they represent either shared nodes or the identity of an input variable*
   In this case the substitution depicted in the first row of Table 1 is applied, since this preserves the values of all input signals.

If none of these cases hold, then the general case substitution from Figure 4 is applied.

EXAMPLE 5. *Consider the BDD shown in Figure 3(a) representing the functions $f_1 = x_1 \wedge x_2$ and $f_2 = x_1 \vee x_2$. First, the synthesis approach traverses node f'. But since f' represents the identity of x_2 no gates are added. Instead, the third line of the circuit in Figure 3(b) is used for storing both, the value of the primary input x_2 and the value of f'. Afterwards, for node f_1 the substitution shown in the fourth row of Table 1 is applied. This not only reduces the number of gates (in comparison to the substitution of Figure 4), but also preserves the value of f' which is still needed by node f_2. In doing so, the remaining node f_2 can be substituted which completes the circuit (see Figure 3(b)).*

As a result, circuits are synthesized which realize the given function f. Since, each node of the BDD is only substituted by a cascade of gates, the proposed method has a linear worst case run-time and linear memory complexity with respect to the number of nodes in the BDD. Furthermore, as discussed in the next section, theoretical results on upper bounds known for BDDs can be transferred to reversible circuits using the proposed approach.

4. THEORETICAL ANALYSIS

In previous work, lower and upper bounds for synthesis of *reversible* functions containing n variables have been determined. In [22], it has been shown that there exists a reversible function, that requires at least $(2^n / \ln 3) + o(2^n)$ gates (lower bound). Furthermore, the authors proved that every reversible function can be realized with no more than $n2^n$ gates (upper bound). For a restricted gate library leading to smaller quantum cost and thus only consisting of NOT, CNOT and two-controlled Toffoli gates (the same as applied for the substitutions in Figure 4 and Table 1), functions can be synthesized with at most n NOT gates, n^2 CNOT gates, and $9n2^n + o(n2^n)$ two-controlled Toffoli gates (according to [9]). A tighter upper bound of n NOT gates, $2n^2 + o(n2^n)$ CNOT gates, and $3n2^n + o(n2^n)$ two-controlled Toffoli gates has been proved

in [14]. In [26] it has been shown, that linear reversible functions are synthesizeable with CNOT gates only. Moreover, their algorithm never needs more than $\Theta(n^2 / \log n)$ CNOT gates for any linear function f with n variables.

Using the synthesis approach proposed in Section 3, reversible networks for a function f with a size dependent on the number of nodes in the BDD can be constructed. More precisely, let f be a function with n primary inputs which is represented by a BDD containing k nodes. Then, the resulting Toffoli circuit consists of *at most*

- $k + n$ circuit lines (since besides the input lines for each node at most one additional line is added) and

- $3 \cdot k$ gates (since for each node cascades of at most 3 gates are added according to Figure 4 and Table 1, respectively).

Asymptotically, the resulting reversible circuits are bounded by the size of the BDD. Since for BDDs many theoretical results exist (see e.g. [16, 17]), using the proposed synthesis approach, these results can be transferred to reversible logic as well. In the following, we sketch some possible results obtained by this observation.

- A BDD representing a single-output function has 2^n nodes in the worst case. Thus, each function can be realized in reversible logic with at most $3 \cdot 2^n$ gates (thereby at most $2 \cdot 2^n$ CNOTs and $2 \cdot 2^n$ Toffoli gates are needed according to the first row of Table 1).

- A BDD representing a symmetric function has n^2 nodes in the worst case. Thus, each symmetric function can be realized in reversible logic with at most $3 \cdot n^2$ gates (thereby at most $2 \cdot n^2$ CNOTs and $2 \cdot n^2$ Toffoli gates are needed according to the first row of Table 1).

- A BDD representing specific symmetric functions, like AND, OR, or EXOR has a linear size. Thus, there exist a reversible circuit realizing these functions in linear size as well.

- A BDD representing an n-bit adder has linear size. Thus, there exist a reversible circuit realizing addition in linear size as well.

Further results (e.g. tighter upper bounds for general function as well as for respective function classes) are also known (see e.g. [16, 17]). Moreover, in a similar way bounds for elementary quantum circuits can be obtained. However, a detailed analysis of the theoretical results that can be obtained by the BDD-based synthesis is left for future work.

5. EXPERIMENTAL RESULTS

We implemented the proposed synthesis approach in C++ on top of the BDD package CUDD [25]. BDDs are constructed with complement edges and optimized using sifting [27, 24]. Both, Toffoli gate circuits and elementary quantum gate circuits have been synthesized. In this section we document experimental results obtained by our approach and compare them to the results generated by (1) the public available RMRLS approach (described in [13]) using version 0.2 in the default settings and (2) the RMS approach (based on the concepts of [14]) in its most recent version including improved handling of don't care conditions at the output.

As benchmarks we used functions provided in RevLib [28] (including most of the functions which have been previously used to evaluate existing reversible synthesis approaches) as well as from the LGSynth package (a benchmark suite for evaluating irreversible synthesis). Since previous approaches (i.e. RMRLS and RMS) require reversible functions as input, non-reversible functions are embedded into reversible ones (based on the concepts of [22]). For BDD-based synthesis, the original function description has been

978-1-60558-497-3/09 $25.00 © 2009 ACM

Table 3: Experimental results

| FUNCTION | | PREVIOUS APPROACHES | | | | | | | BDD-BASED SYNTHESIS | | | | | Δ QC | Δ QC |
| | | RMRLS [13] | | | | RMS [14] | | | | | | | | | |
NAME	PI/PO	L.	GC	QC	TIME	GC	QC	TIME	L.	GC	QC	QC_{EQ}	TIME	(RMRLS)	(RMS)
RevLib Functions															
4mod5_8	4/1	5	9	25	0.86	5	9	<0.01	7	8	24	18	<0.01	**-7**	9
decod24_10	2/4	4	11	55	497.51	7	19	<0.01	6	11	27	23	<0.01	**-32**	4
mini-alu_84	4/2	5	21	173	495.61	36	248	<0.01	10	20	60	43	<0.01	**-130**	**-205**
alu_9	5/1	5	9	49	122.48	9	25	0.01	7	9	29	22	**0.01**	**-27**	**-3**
rd53_68	5/3	7	–	–	>500.00	221	2646	0.14	13	34	98	75	<**0.01**	–	**-2571**
hwb5_13	5/5	5	–	–	>500.00	42	214	0.01	28	88	276	205	**0.01**	–	**-9**
sym6_63	6/1	7	36	777	485.47	15	119	0.13	14	29	93	69	<**0.01**	**-708**	**-50**
mod5adder_66	6/6	6	37	529	494.46	35	151	0.06	32	96	292	213	<**0.01**	**-316**	62
hwb6_14	6/6	6	–	–	>500.00	100	740	0.04	46	159	507	375	<**0.01**	–	**-365**
rd73_69	7/3	9	–	–	>500.00	1344	20779	1.93	13	73	217	162	<**0.01**	–	**-20617**
hwb7_15	7/7	7	–	–	>500.00	375	3378	0.18	73	281	909	653	<**0.01**	–	**-2725**
ham7_29	7/7	7	–	–	>500.00	26	90	0.09	21	61	141	107	<**0.01**	–	17
rd84_70	8/4	11	–	–	>500.00	124	8738	9.92	34	104	304	229	<**0.01**	–	**-8509**
hwb8_64	8/8	8	–	–	>500.00	229	3846	0.90	112	449	1461	1047	**0.01**	–	**-2799**
sym9_71	9/1	10	–	–	>500.00	27	201	3.98	27	62	206	153	<**0.01**	–	**-48**
hwb9_65	9/9	9	–	–	>500.00	2021	23311	1.45	170	699	2275	1620	**0.02**	–	**-21691**
cycle10_2_61	12/12	12	26	1435	491.87	41	1837	26.17	39	78	202	164	**0.09**	**-1271**	**-1673**
plus63mod4096_79	12/12	12	–	–	>500.00	24	4873	17.74	23	49	89	79	**0.08**	–	**-4794**
plus127mod8192_78	13/13	13	–	–	>500.00	25	9131	57.16	25	54	98	86	**0.21**	–	**-9045**
plus63mod8192_80	13/13	13	–	–	>500.00	28	9183	57.19	25	53	97	87	**0.20**	–	**-9096**
ham15_30	15/15	15	–	–	>500.00	–	–	>500.00	45	153	309	246	**1.25**	–	–
LGSynth Functions															
xor5	5/1	6	27	387	484.11	8	68	0.01	6	8	8	8	<**0.01**	**-379**	**-60**
9sym	9/1	10	–	–	>500.00	27	201	4.00	27	62	206	153	<**0.01**	–	**-48**
cordic	23/2	~	~	~	~	~	~	~	52	101	325	247	**0.02**	–	–
bw	5/28	~	~	~	~	~	~	~	87	307	943	693	<**0.01**	–	–
apex2	39/3	~	~	~	~	~	~	~	498	1746	5922	4435	**0.24**	–	–
pdc	16/40	~	~	~	~	~	~	~	619	2080	6500	4781	**0.14**	–	–
seq	41/35	~	~	~	~	~	~	~	1617	5990	19362	14259	**1.14**	–	–
spla	16/46	~	~	~	~	~	~	~	489	1709	5925	4372	**0.10**	–	–
ex5p	8/63	~	~	~	~	~	~	~	206	647	1843	1388	**0.02**	–	–
e64	65/65	~	~	~	~	~	~	~	195	387	907	713	**0.04**	–	–
cps	24/109	~	~	~	~	~	~	~	930	2676	8136	6301	**0.10**	–	–
apex5	117/88	~	~	~	~	~	~	~	1147	3308	11292	8387	**0.14**	–	–
i5	133/66	~	~	~	~	~	~	~	345	530	1738	1382	**0.09**	–	–
i8	133/81	~	~	~	~	~	~	~	955	3550	11478	8212	**0.25**	–	–
i6	138/67	~	~	~	~	~	~	~	280	734	2234	1557	**0.06**	–	–
ex4p	128/28	~	~	~	~	~	~	~	510	1277	4009	3093	**0.03**	–	–
frg2	143/139	~	~	~	~	~	~	~	1411	4472	14944	11323	**0.69**	–	–
i4	192/6	~	~	~	~	~	~	~	729	2115	6827	5158	**0.43**	–	–
i7	199/67	~	~	~	~	~	~	~	403	941	2953	1996	**0.90**	–	–

used which automatically leads to an embedding. All experiments have been carried out on an AMD Athlon 3500+ with 1 GB of memory. The timeout was set to 500 CPU seconds.

The results are summarized in Table 3. The first columns give the name as well as the number of the primary inputs (PI) and primary outputs (PO) of the original function. In the following columns, the number of lines (L.), the gate count (GC), the quantum cost (QC), and the synthesis time (TIME) for the respective approaches (i.e. RMRLS, RMS, and the BDD-BASED SYNTHESIS) are reported[2]. Note that for the BDD-based synthesis two values for quantum cost are given: QC for the cost of the resulting Toffoli gate circuits and QC_{EQ} if elementary gate circuits are synthesized directly. Furthermore, a '~' denotes, that an embedding needed by the previous synthesis approaches could not be created within

the given timeout. Finally, the last two columns (Δ QC) give the absolute difference of the quantum cost for the resultig circuits obtained by the BDD-based elementary quantum circuit synthesis and the RMRLS or RMS approach, respectively.

As a first result, one can conclude, that for large functions to be synthesized it is not always feasible to create a reversible embedding needed by the previous approaches. Moreover, even if this is feasible, both RMRLS and RMS need a significant amount of run-time to synthesize a circuit from the embedding. As a consequence, for most of the LGSynth benchmarks no result can be generated within the given timeout. In contrast, our BDD approach is able to synthesize circuits for *all* given functions within a few CPU seconds.

Furthermore, although the BDD-based synthesis often leads to larger circuits with respect to gate count and number of lines, the resulting quantum cost are significantly lower in most of the cases (except for *4mod5_8*, *decod24_10*, *mod5adder_66*, and *ham7_29*).

---

[2] TIME for BDD-BASED SYNTHESIS includes both, the time to build the BDD as well as to derive the circuit from it.

978-1-60558-497-3/09 $25.00 © 2009 ACM

As an example, for *plus63mod4096_79* the BDD-BASED SYN-THESIS synthesizes a circuit with twice the number of lines but with two orders of magnitude fewer quantum cost in comparison to RMS. In the best cases (e.g. *hwb9_65*) a reduction of several thousands in quantum cost is achieved. Note that quantum cost are more important than gate count since they consider gates with more control lines to be more costly. Furthermore, the total number of circuit lines that have been added by the BDD-BASED SYNTHESIS is moderate considering the obtained quantum cost reductions (in particular since all additional lines have constant inputs).

6. CONCLUSIONS AND FUTURE WORK

In this paper, we introduced a synthesis approach which can cope with large functions. The basic idea is to create a Binary Decision Diagram for the function to be synthesized and afterwards substituting each node by a cascade of Toffoli or elementary quantum gates, respectively. Since BDDs may include shared nodes causing fan-outs (which are not allowed in reversible logic), also substitutions including an additional circuit line are proposed.

While previous approaches are only able to handle functions with up to 30 variables at high run-time, our BDD-based approach can synthesize circuits for functions with more than hundred variables in just a few CPU seconds. Furthermore, in most of the cases reductions in the resulting quantum cost have been observed.

In future work, we will focus on the optimization of the resulting circuits. In particular, the number of additional lines should be reduced. Existing approaches (e.g. [12, 29, 30, 31]) provide a good starting point, but mainly focus on reducing quantum cost. Another idea is to adjust the cost function of exact BDD implementations with respect to quantum cost and to synthesize the circuits from the resulting BDDs. Finally, a detailed analysis of the theoretical results that can be obtained by the proposed approach is left for future work.

7. ACKNOWLEDGEMENTS

The authors thank Daniel Große and D. Michael Miller for helpful discussions.

8. REFERENCES

[1] R. Landauer. Irreversibility and heat generation in the computing process. *IBM J. Res. Dev.*, 5:183, 1961.

[2] C. H. Bennett. Logical reversibility of computation. *IBM J. Res. Dev*, 17(6):525–532, 1973.

[3] T. Toffoli. Reversible computing. In W. de Bakker and J. van Leeuwen, editors, *Automata, Languages and Programming*, page 632. Springer, 1980. Technical Memo MIT/LCS/TM-151, MIT Lab. for Comput. Sci.

[4] M. Nielsen and I. Chuang. *Quantum Computation and Quantum Information*. Cambridge Univ. Press, 2000.

[5] R. Cuykendall and D. R. Andersen. Reversible optical computing circuits. *Optics Letters*, 12(7):542–544, 1987.

[6] R. C. Merkle. Reversible electronic logic using switches. *Nanotechnology*, 4:21–40, 1993.

[7] W.N.N. Hung, X. Song, G. Yang, J. Yang, and M. Perkowski. Optimal synthesis of multiple output Boolean functions using a set of quantum gates by symbolic reachability analysis. *IEEE Trans. on CAD*, 25(9):1652–1663, 2006.

[8] D. Große, R. Wille, G.W. Dueck, and R. Drechsler. Exact multiple control toffoli network synthesis with SAT techniques. *IEEE Trans. on CAD*, 28(5):703–715, 2009.

[9] V. V. Shende, A. K. Prasad, I. L. Markov, and J. P. Hayes. Synthesis of reversible logic circuits. *IEEE Trans. on CAD*, 22(6):710–722, 2003.

[10] D. M. Miller, D. Maslov, and G. W. Dueck. A transformation based algorithm for reversible logic synthesis. In *Design Automation Conf.*, pages 318–323, 2003.

[11] P. Kerntopf. A new heuristic algorithm for reversible logic synthesis. In *Design Automation Conf.*, pages 834–837, 2004.

[12] D. Maslov, G. W. Dueck, and D. Michael Miller. Toffoli network synthesis with templates. *IEEE Trans. on CAD*, 24(6):807–817, 2005.

[13] P. Gupta, A. Agrawal, and N.K. Jha. An algorithm for synthesis of reversible logic circuits. *IEEE Trans. on CAD*, 25(11):2317–2330, 2006.

[14] D. Maslov, G. W. Dueck, and D. M. Miller. Techniques for the synthesis of reversible toffoli networks. *ACM Trans. on Design Automation of Electronic Systems*, 12(4), 2007.

[15] R.E. Bryant. Graph-based algorithms for Boolean function manipulation. *IEEE Trans. on Comp.*, 35(8):677–691, 1986.

[16] I. Wegener. *Branching programs and binary decision diagrams: theory and applications*. Society for Industrial and Applied Mathematics, 2000.

[17] R. Drechsler and D. Sieling. Binary decision diagrams in theory and practice. *Software Tools for Technology Transfer*, 3:112–136, 2001.

[18] E. F. Fredkin and T. Toffoli. Conservative logic. *International Journal of Theoretical Physics*, 21(3/4):219–253, 1982.

[19] A. Peres. Reversible logic and quantum computers. *Phys. Rev. A*, (32):3266–3276, 1985.

[20] A. Barenco, C. H. Bennett, R. Cleve, D. P. DiVinchenzo, N. Margolus, P. Shor, T. Sleator, J. A. Smolin, and H. Weinfurter. Elementary gates for quantum computation. *The American Physical Society*, 52:3457–3467, 1995.

[21] D. Maslov and D. M. Miller. Comparison of the cost metrics through investigation of the relation between optimal ncv and optimal nct 3-qubit reversible circuits. *IET Computers & Digital Techniques*, 1(2):98–104, 2007.

[22] D. Maslov and G. W. Dueck. Reversible cascades with minimal garbage. *IEEE Trans. on CAD*, 23(11):1497–1509, 2004.

[23] K.S. Brace, R.L. Rudell, and R.E. Bryant. Efficient implementation of a BDD package. In *Design Automation Conf.*, pages 40–45, 1990.

[24] R. Wille and R. Drechsler. Effect of BDD optimization on synthesis of reversible and quantum logic. *Workshop on Reversible Computation*, 2009.

[25] F. Somenzi. *CUDD: CU Decision Diagram Package Release 2.3.1*. University of Colorado at Boulder, 2001.

[26] K. Patel, I. Markov, and J. Hayes. Optimal synthesis of linear reversible circuits. *Quantum Information and Computation*, 8(3-4):282–294, 2008.

[27] R. Rudell. Dynamic variable ordering for ordered binary decision diagrams. In *Int'l Conf. on CAD*, pages 42–47, 1993.

[28] R. Wille, D. Große, L. Teuber, G. W. Dueck, and R. Drechsler. RevLib: An online resource for reversible functions and reversible circuits. In *Int'l Symp. on Multi-Valued Logic*, pages 220–225. RevLib is available at http://www.revlib.org.

[29] J. Zhong and J.C. Muzio. Using crosspoint faults in simplifying toffoli networks. In *IEEE North-East Workshop on Circuits and Systems*, pages 129–132, 2006.

[30] A.K. Prasad, V.V. Shende, I.L. Markov, J.P. Hayes, and K.N. Patel. Data structures and algorithms for simplifying reversible circuits. *J. Emerg. Technol. Comput. Syst.*, 2(4):277–293, 2006.

[31] D. Y. Feinstein, M. A. Thornton, and D. M. Miller. Partially redundant logic detection using symbolic equivalence checking in reversible and irreversible logic circuits. In *Design, Automation and Test in Europe*, pages 1378–1381, 2008.

Soft Connections: Addressing the Hardware-Design Modularity Problem

Michael Pellauer[†]   Michael Adler[‡]   Derek Chiou[*]   Joel Emer[††]

[†] Massachusetts Institute of Technology
Computer Science and A.I. Lab
Computation Structures Group
{pellauer, emer}@csail.mit.edu

[‡] Intel Corporation
VSSAD Group
{Michael.Adler,
Joel.Emer}@intel.com

[*] University of Texas at Austin
Electrical and Computer
Engineering Department
derek@ece.utexas.edu

ABSTRACT

Hardware-design languages typically impose a rigid communication hierarchy that follows module instantiation. This leads to an undesirable side-effect where changes to a child's interface result in changes to the parents. Soft connections address this problem by allowing the user to specify connection endpoints that are automatically connected at compilation time, rather than by the user.

Categories and Subject Descriptors

B.5.2 [**Register-Transfer-Level Implementation**]: Design Aids - *hardware description languages*

General Terms

Design, Languages

Keywords

High-Level Communication Description

1. INTRODUCTION

Modularity is a critical feature of high-level hardware description languages (HDLs). Ideally designers should be able to swap alternative modules in a "plug-and-play" manner. Such swapping enables code reuse and design-space exploration, and thus enhances designer productivity.

It is becoming increasingly popular to insert an FPGA into a general-purpose computer using a fast link such as PCIe [8] or Intel Front-Side Bus [7]. In such a setup the FPGA, configured by a standard HDL toolchain, acts as an accelerator to the CPU, running standard software. This usage model is gaining traction in the microprocessor performance modeling community, being used by projects such as Protoflex [6], UT-FAST [4], [5] and our HAsim simulator [2], [9] as part of the umbrella RAMP project [11]. In such an environment FPGA reconfigurations are frequent, so modular refinement and reuse become especially important.

In structural HDLs it can be difficult to swap one module for an alternative in isolation. This is because communication between modules can only follow the instantiation hierarchy. A module can only pass wires to its parent and children. Cross-hierarchical communication goes through the least-common ancestor and every other intervening node. If a new module requires communication with anything other than its direct parent, then we must change the parent module, the parent's parent, and so on.[1]

Permission to make digital or hard copies of part or all of this work for personal or classroom use is granted without fee provided that copies are not made or distributed for profit or commercial advantage and that copies bear this notice and the full citation on the first page. To copy otherwise, to republish, to post on servers or to redistribute to lists, requires prior specific permission and/or a fee.
DAC'09, July 26-31, 2009, San Francisco, California, USA

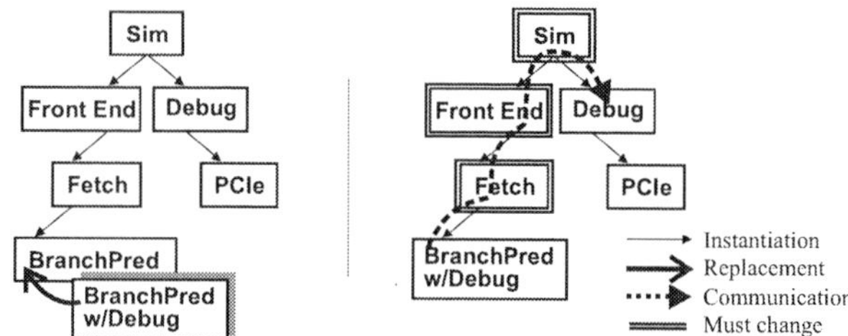

Figure 1. Introducing cross-hierarchical communication.

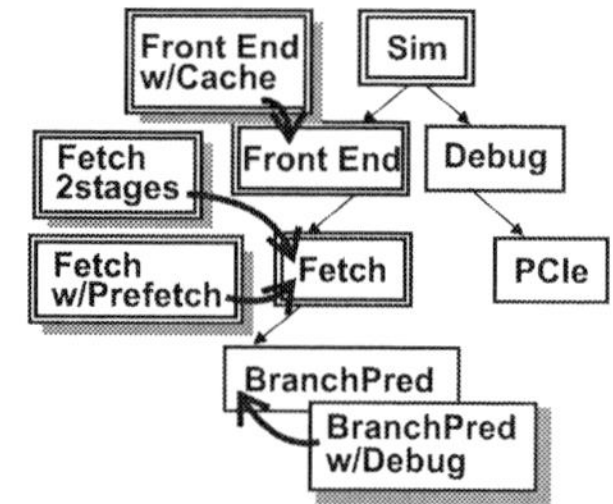

Figure 2. Alternative modules can worsen the problem.

Consider the situation shown in Figure 1. The designer knows that the Branch Predictor on the FPGA has a bug. He wants to swap it for a variant that outputs additional debugging information, that is sent to the host processor using PCIe. In order to do so he must add those wires to the Fetch, Front End, and top-level Simulator modules, then down to the Debug block.

The situation quickly deteriorates as we add more module alternatives to the system. In Figure 2 we have three alternative Fetch units and two Front Ends that the designer is exploring. Each setup uses the branch predictor, and each manifests the bug in different ways. The designer must now produce alternative implementations of these, one of which does not pass debugging wires up, and one which does. In the worst case the number of modules needed grows multiplicatively with the number of alternatives.

In this paper we attack this problem by "softening" the rigid communication hierarchy—thus we name our technique Soft Connections. This scheme restores modularity by allowing the designer to specify a *logical topology* of communication which is separated from its physical implementation. The endpoints are not connected by the user, but rather done automatically using static elaboration. Using Soft Connections restores modularity, allowing individual modules to be swapped in isolation, independent of the instantiation hierarchy.

---

[1] We do not consider Verilog Out-of-Module References (OOMR) to be a satisfactory solution as they break modular abstraction. Languages such as SystemVerilog raise the level of abstraction so that the user works with typed interfaces instead of wires, but the basic problem remains.

This paper deals with whole-design compilation. Discussion of separately-compiled blocks is omitted for space concerns, but is presented in [1]. Although this paper uses the simulation of microprocessors using FPGAs as an ongoing example, the technique is general and could be used for ASIC design.

2. BACKGROUND: STATIC ELABORATION

Implementing Soft Connections in an existing structural HDL such as Verilog would require either modifying the language or using external scripts to transform the source code. Instead, we implement our Soft Connections scheme in Bluespec SystemVerilog [3], an existing hardware description language.

Bluespec provides a powerful *static elaboration* phase which allows users to transform their design arbitrarily without giving up the static safety a hardware-aware language provides. During elaboration statically known values are aggressively propagated in order to resolve polymorphism and "unroll" static loops and function calls. For example, the designer may describe an *n*-bit ripple-carry adder as follows:

```
function bit[n:0] addRC(bit[n:0] x, bit[n:0] y);
  bit[n:0] res = 0;
  bit c = 0;
  for (int k = n; k >= 0; k) begin
      res[k] = x[k] ^ y[k] ^ c;
      c = (x[k] & y[k]) | (x[k] & c) | (y[k] & c);
  end
  return res;
endfunction
```

The designer may then call this `addRC` function multiple times using different types. The HDL compiler will execute the function and its loop, using statically known values of *n* and *k*. If *x* and *y* are known statically than the function itself may result in no hardware, but rather a new static constant. However if *x* and *y* are dynamic inputs to the hardware block then the result is a netlist of AND- and XOR-gates. If for some reason *n* was dynamic, the result would be an error as the loop could not be turned into bounded hardware.

HDLs such as Verilog feature elaboration primarily through the use of *generate* blocks, which allow the user to create static control-flow structures such as loops and if-statements. Bluespec expands this into a Turing-complete software interpreter. This allows the user to work with high-level datatypes such as linked-lists or unbounded integers. These types do not have a hardware representation, but the designer can use them to *influence* the hardware that the compiler generates. For example, here is a Bluespec module that takes as input a list of integers. For each one it instantiates a 32-bit FIFO of that depth (note that < – is the module instantiation operator in Bluespec):

```
module mkFIFOList#(List#(Integer) depths);
  let result_list = nil;
  while (depths != nil)  begin
      Integer d = head(depths);
      FIFO#(bit[31:0]) q <- mkSizedFIFO(d);
      result_list = append(result_list, q);
      depths = pop(depths);
  end
  return result_list;
endmodule
```

This use of static elaboration could be thought of as "embedding a small software program in our hardware description source that the compiler runs to generate hardware." Soft Connections represent a novel use of static elaboration, and help to demonstrate how a more powerful notion of elaboration can benefit hardware designers.

```
method Action train(BPredInfo inf);
  if (inf.branchTaken != table.lookup(inf.pc))
   link_to_debug.send(debugMsgMispredict(inf.pc));
   table.update(inf.pc, inf.branchTaken);
endmethod
```



```
rule debugToPCIe;
  let msg = link_from_sender.receive();
  pcie.transmit(pcieRequest(msg));
  link_from_sender.deq();
endrule
```

Figure 3. This branch predictor sends debug information when it is trained with a misprediction. Separately, the Debug module transmits the debug information to software using PCIe.

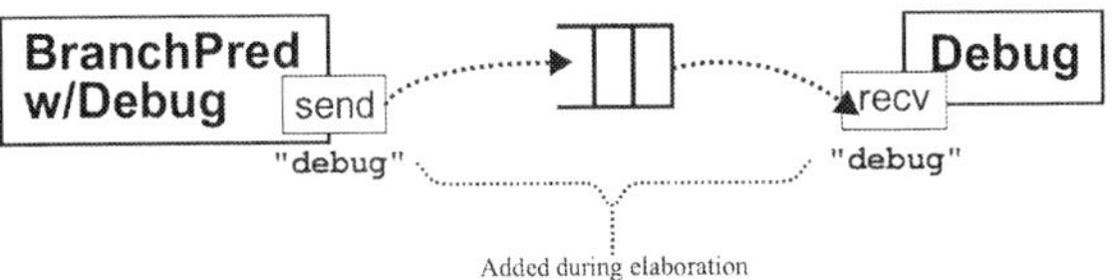

Figure 4. The physical buffering connecting the two endpoints is added during elaboration, as explained in Section 5.

3. SOFT CONNECTIONS
3.1. Point-To-Point Connections

Soft Connections are a library of communication primitives that the designer uses to describe a *logical topology* of communication. The basic Soft Connection is a point-to-point First-In-First-Out channel. This channel is used as if it were a familiar guarded FIFO (Figure 3).

Where Soft Connections differ is the instantiation. Instead of instantiating a single channel module and passing it to both of the users, the communicating modules instantiate the endpoints separately, naming the channel with a unique identifier. For example:

```
let linkToDebug <- mkConnectionSend("debug");
```

Elsewhere, the receiving module instantiates the dual endpoint:

```
let linkFromSender <- mkConnectionRecv("debug");
```

The channel itself is instantiated during elaboration (Figure 4).

As Soft Connections often represent communication between distant modules, we have chosen to implement them using a guarded buffer. Flow-control is handled via Bluespec's standard guarded interface scheme [10], so that the producer's action may not be taken if the buffer is full, nor the consumer's if it is empty.

If our algorithm finds no matching endpoint with the same name, the result is a compilation error. If an error is not desired either endpoint may be specified as optional:

```
let linkFromSender <- mkConnRecvOptional("debug");
```

An optional receiver with no corresponding sender will never receive data. Data can be enqueued to an optional sender with no corresponding receiver but that data will simply disappear. Either are like a wire unterminated on one side - they will have no effect on synthesis results.

978-1-60558-497-3/09 $25.00 © 2009 ACM   277

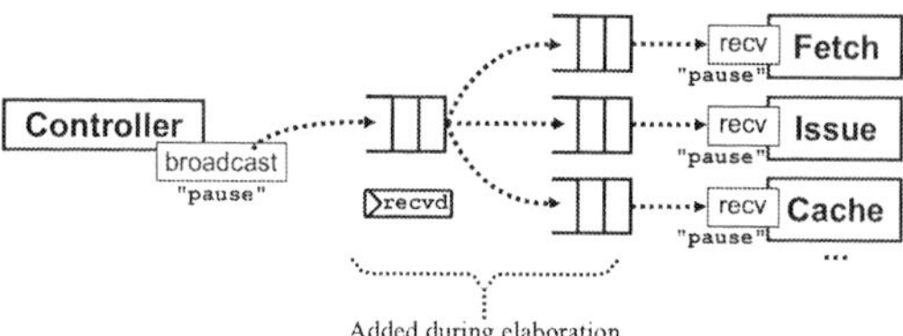

Figure 5. A one-to-many connection. When every receiver has gotten the data the main queue is dequeued. Note that the endpoints of the receivers are standard Point-to-Point receives.

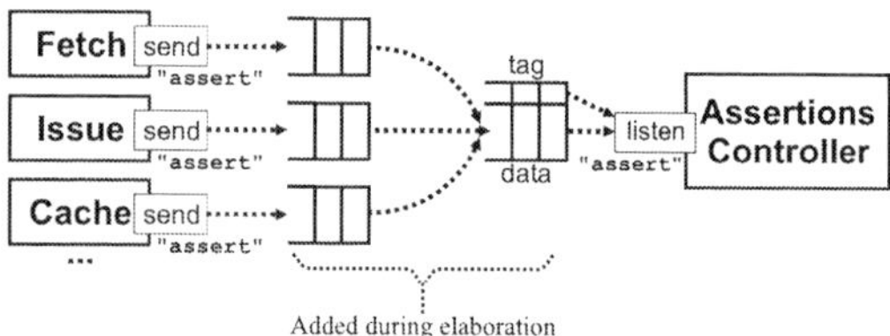

Figure 6. A many-to-one connection. Incoming data is tagged to identify the sender. Note that the senders are standard point-to-point sends.

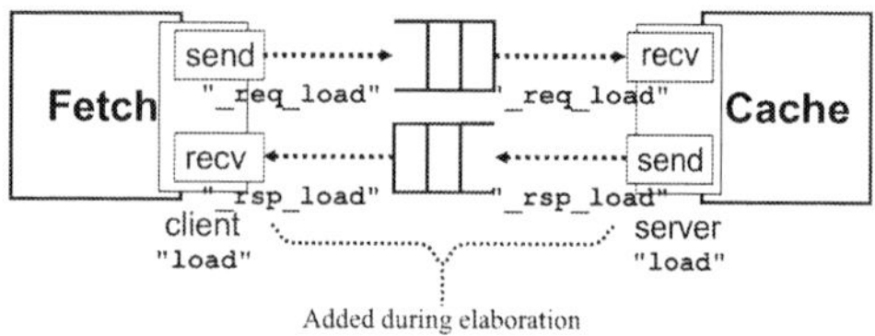

Figure 7. Clients and Servers are abstractions for a bundle of standard send and receive connections.

3.2. One-to-Many and Many-to-One

A one-to-many send is a broadcast that transmits the same data to all listeners (Figure 5). A many-to-one receive is a channel multiplexed by an arbitrator, that also tags the data with a bit field indicating which sender the data comes from (Figure 6). These tags are assigned by our algorithm.

One-to-many connections are useful for relaying control messages from software to many hardware modules—for instance to start, pause, or reset operation. Many-to-one receives are useful for aggregating data such as assertions or debugging information for transmission to software.

3.3 Clients and Servers

The uni-directional channels presented above represent the primitive Soft Connections on which our elaboration algorithm operates. We then use these as building blocks to create useful abstractions for bi-directional communication. The first abstraction is that of a request/response paradigm (Figure 7). The client makes requests and gets responses. The server receives requests and makes responses.

This arrangement is often used to connect functional units to their users. This idea can be combined with one-to-many and many-to-one connections to make multi-user clients and servers. A server with a many-to-one connection can receive requests from multiple clients, and uses many point-to-point connections which deliver responses (Figure 8). The dual of this is a client that is connected to many servers. It broadcasts requests to all of them, then receives the responses in serial. This is a one-to-many send for the requests, and a many-to-one receive for the responses (Figure 9).

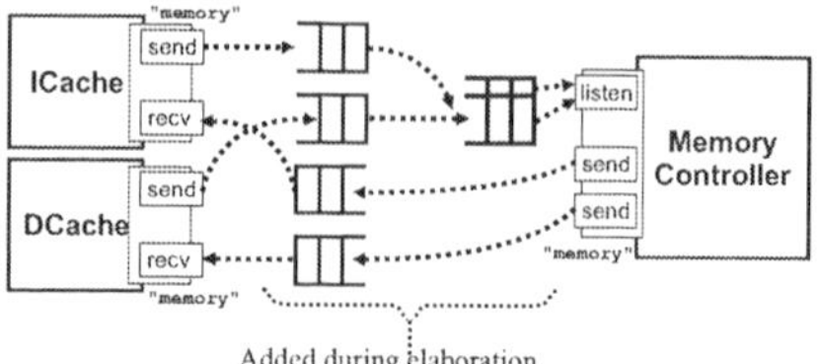

Figure 8. A multi-user server. Note that the clients are standard one-to-one clients.

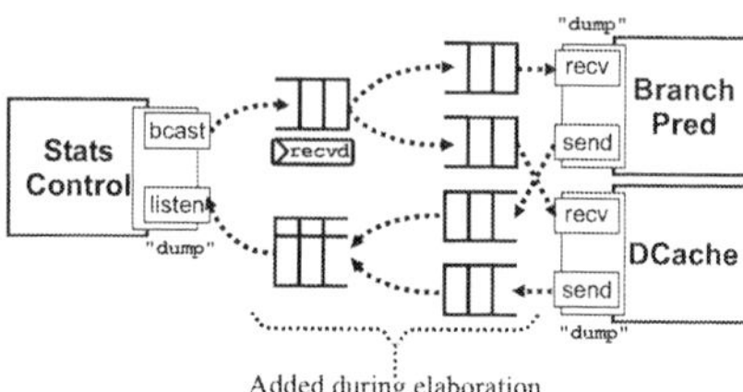

Figure 9. A client connected to multiple servers. Note that the servers are standard one-to-one servers.

3.4. Example: Simulation Controller

The simulation controller presented in Figure 10 represents an example of how Soft Connections can improve designer productivity. The controller is a module that sits on the FPGA and communicates with software on the CPU, mediating interaction with the PCIe link. The controller instantiates six sub-controllers:

- **Commands (Client) :** This receives commands from software such as "start" or "pause" and broadcasts them to listening modules. These modules respond when simulation is finished. Thus this module is a client of many distributed servers.

- **Params (Client):** This receives dynamic parameters set on the command line when the user initiates the software. These parameters are sent to the appropriate listeners. Thus, for example, the cache can be disabled without re-synthesizing the design.

- **Events (Client):** These represent a detailed trace of results from the simulator. Software enables or disable event-dumping dynamically, and these requests are passed on to the modules.

- **Stats (Client):** Periodically the host software can request a dump of statistics. This request is relayed to all listeners, who respond with their current values, which are relayed to the host.

- **Assertions (Listener):** When an assertion fails in a hardware module, it sends a message to this controller, which relays it to software that prints out a message and ends the simulation gracefully.

- **Debug (Listener):** This module listens for debugging messages and relays them to the host software where they are logged.

Using Soft Connections for the communication from these controllers to the simulator modules results in several benefits. First, the designer can fluidly swap modules without rewiring their connection to the controllers. This encourages users to create many variations of their module, without worrying that (for example) a direct-mapped write-through cache contains a smaller set of statistics than an associative write-back cache. Finally, it raises the level of abstraction for the user, who just records stats and assertion failures, without worrying about how this information is communicated to software.

4. PHYSICAL INTERCONNECT SHARING

Soft connections make life easier for the designer by making module communication implicit. The disadvantage of this is that the designer can lose intuition about the implementation cost of their communication network. For example, we have found that the assertions facility is useful for the FPGA in practice. Thus it

978-1-60558-497-3/09 $25.00 © 2009 ACM

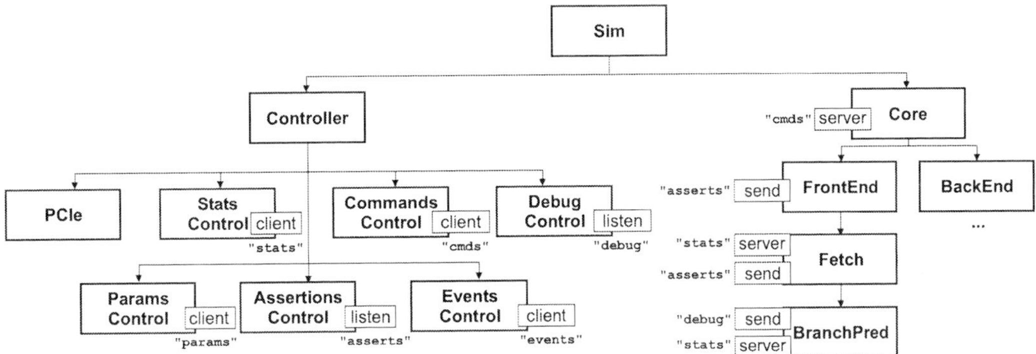

Figure 10. Example: A simulation controller mediates the connection between host software and hardware modules.

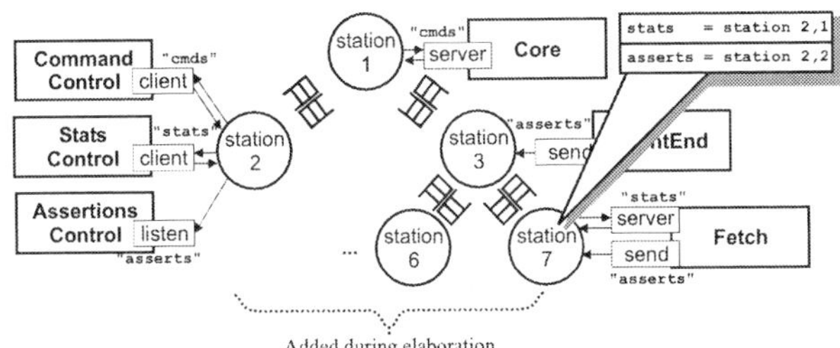

Figure 11. Multiple Soft Connections implemented on a shared physical interconnect. Each station routes logical channels to the appropriate destination using a generated routing table.

becomes frequently used. A typical configuration of our simulator has 42 dynamic assertions, most of them sanity checks relating to correct instruction execution. Implementing these as 42 FIFOs arbitrating directly with the controller is expensive, and places a large burden on the place-and-route tools due to the fan-in. Assertion failures are (hopefully) a rare occurrence, so it makes sense to aggregate them using a multiplexed physical interconnect such as a tree. Such a scheme increases the latency a message takes to reach the endpoint, but can result in more efficient hardware.

Other rarely-used connections such as statistics can also be mapped onto the same interconnect. Thus the user can separate a Soft Connection's *physical representation* (exclusive channel or shared interconnect) from its *logical representation* (point-to-point, one-to-many, etc).

The user creates a shared connection by first instantiating a network station. This station is then passed in to the constructor of the Soft Connection:

```
let fetchStation <- mkStationTree("fetch");
let linkAssert <- mkConnSendShared(fetch_station,
                                   "asserts");
```

Figure 11 shows an example mapping many Soft Connections onto the same shared interconnect. Whether a Soft Connection is implemented as an exclusive or shared interconnect is transparent to the modules which use the endpoints—they use the connection's operations (send, receive, broadcast, etc) as normal. The only difference from the user's point of view is that the latency of communication between sender and receiver has increased, as the messages are in fact being passed over an interconnect which is shared with other endpoints. Our elaboration algorithm connects the stations together into a physical network, and creates a routing table to dynamically guide messages to the appropriate destination. The addressing and routing of messages is handled by the stations themselves.

Algorithm 1. Connecting Soft Connection endpoints directly.

```
 1:   (sends, recvs) = … // Get collected info
 2:   for each s in sends do
 3:       rs = matchByName(s, recvs)
 4:       if rs ={} and not optional(s) then
 5:           error("Unmatched Send " + s)
 6:       else if rs = {r} then
 7:           connect(s, r) // Instantiate buffering
 8:       else
 9:           connectBroadcast(s, rs) // as in Figure 5
10:       recvs = recvs − rs
11:   for each r in recvs do
12:       error("Unmatched Recv" + r)
```

Currently our algorithm connects the stations into a branching tree topology that follows the module instantiation hierarchy. (Layers in the hierarchy with no stations are optimized away.) This topology was chosen because it maximizes spatial locality by keeping the stations near their endpoints, and because it results in a single static route between two given endpoints, which minimizes station routing logic. In the future, support is planned for other physical network topologies such as rings, two-way rings, or grids.

5. CONNECTION ALGORITHM

When a module instantiates a Soft Connection endpoint it is implicitly transforming the interface it presents to the outside world. For a module with interface i its new interface i' is a tuple of i plus linked lists that describe what Soft Connection endpoints the module has instantiated:

$$i' = (i, \{sends\}, \{recvs\})$$

The module's parent (and the parent's parent) see only the original interface i. This, along with collecting all the lists from all of the modules, is accomplished using a standard Bluespec library called `ModuleCollect`.

Algorithm 1 describes our process for connecting Soft Connection endpoints directly. For space reasons we omit many-to-one connections, which work similarly. Connections that are unmatched (and not optional) result in a compilation error via Bluespec's built-in `error` function, which halts elaboration.

The algorithm for instantiating Soft Connections sharing a physical interconnect is most naturally described as a recursive module—it may call itself during elaboration, resulting in a tree-topology of stations connected to each other:

Algorithm 2. Constructing a station's routing table

```
 1:    let (childs, sends, recvs) = ... // Parameters
 2:    // Routing decisions for traffic from local sends.
 3:    for each s in sends do
 4:        if matchByName(s, childs) = {c} then
 5:            // A child (or its descendant) has the recv.
 6:            sendRoute[s] := toChild c
 7:        else // The endpoint is not in this subtree.
 8:            sendRoute[s] := toParent
 9:    // Routing decisions for traffic from children.
10:    for each c in childs do
11:        // Find all sends this child is routing up to us.
12:        for each s in sendsRoutedToParent(c) do
13:            if matchInStation(s, childs) = {c₂} then
14:                // This station is the least-common ancestor.
15:                childRoute[c][s] := toChild c₂
16:            else if matchByName(s, recvs) = {r} then
17:                // The endpoint is local to this station.
18:                childRoute[c][s] := toRecv r
19:            else // The endpoint is not in this subtree.
20:                childRoute[c][s] := toParent
```

```
module mkStationTree#(STATION_INFO info)(STATION);
  List#(STATION) child_stations = nil;
  for (int x = 0; x < length(info.children); x++)
  begin
    let cur_child = info.children[x];
    // Recurse down the tree.
    let c <- mkStationTree(cur_child);
    child_stations = append(child_stations, c);
  end
  let table <- mkRoutingTable(child_stations,
                         info.recvs, info.sends);
  let s <- connectStation(table, child_stations,
                         info.recvs, info.sends);
  return s;
endmodule
```

The routing table is constructed mechanically using Algorithm 2. We have omitted the details of routing one-to-many sends for space concerns. They have the potential to be sent to multiple receivers and children. Additionally, they are always routed up to the parent (which drops the message if none of its other children are receivers). Endpoints that are unmatched at the root station result in an error, as in the unshared case.

6. ASSESSMENT
6.1. Impact on Productivity

In this section we examine a real-world example in order to give some insight into how Soft Connections can improve the process of engineering an FPGA-based accelerator. For the example we have chosen an FPGA-based model of a 5-stage microprocessor pipeline that runs the Alpha instruction set using the HAsim simulator [2].

As shown in Figure 11, the FPGA is configured into a simulator of this target machine. This simulator bears little resemblance to the 5-stage pipeline itself, but accurately computes the performance of the target. This is because the timing properties of the physical implementation—such as the FPGA BlockRAM or the speed of memory through the PCIe—are different from the speeds in the machine we wish to study. Thus we add logic to translate FPGA cycles into model cycles in the target. The simulator is divided into three major partitions: model timing, model functionality, and the simulation controller. The full technique for creating such a simulator is presented in [2].

We synthesized our simulator for a Virtex5 110t part on a PCIe board manufactured by HiTechGlobal [8] using Xilinx ISE 10.1:

Slice LUTs	47214/69120 (68%)
BlockRAM	121/148 (81%)
Critical Path	15.313 ns
Frequency	65 MHZ

It may seem surprising that modeling a simple architecture would use so many FPGA resources. This is because the simulator uses a large number of FPGA resources as on-chip cache. As we have divorced FPGA time from model time, our simulator can devote an arbitrary amount of on-chip memory to cache, even if the target has a smaller cache. This extra cache speeds up simulation, but has no effect on the behavior of the target machine. (In some sense, any unused slice is a wasted resource for an accelerator FPGA.)

Figure 12 gives an overview of how our design uses Soft Connections. We have attempted to quantify the productivity these provide by defining a metric called span. For each connection c between two modules:

$span(c)$ = the number of module instantiation boundaries between the send and receive endpoints.

Span measures the potential work the Soft Connection is saving the designer. Namely, the number of modules that the designer would have to change if she was not using Soft Connections and swapped in a module with a different interface. We acknowledge the limitations of measuring the amount of work that our technique *potentially* can save, but believe that this metric gives valuable insight into the degree that communication between distant endpoints can exist in a hardware design.

Figure 13 shows a histogram of the span of every connection in our simulator—i.e., our simulator contains 74 connections with a span of 7. Spans of 0 represent optional connections which are not being used. We found that the average Soft Connection in our simulator crosses 5.27 module instantiations, and that 50% of them cross 7 or more. This demonstrates that cross-hierarchical communication can be prevalent in real-world situations.

6.2. The Effect of Shared Interconnects

Much of the cross-hierarchical communication—and all of the many-to-one/one-to-many connections—involve communicating data to or from the Simulation Controller (Section 3.4). The cost of multiplexing between these signals can be high, and can result in a burden on the place and route tools. In order to explore this we implemented an alternative version of our simulator where all connections to the controller shared the same interconnect tree.

Overall 100/217 connections were mapped onto this tree, representing the statistics, assertions, commands, parameters, and events facilities. The tree had 14 stations arranged into a depth of 4, with the controller as the root node. All told, this tree spanned 20 module instantiations. We found this version consumes an additional 3076 slice LUTs (4% of total available) because of its extra buffering and routing tables. RAM utilization and clock speed are not affected, as the critical path is elsewhere.

Multiplexing these connections onto the same tree can increase the latency of communication. To measure the impact of this on

978-1-60558-497-3/09 $25.00 © 2009 ACM

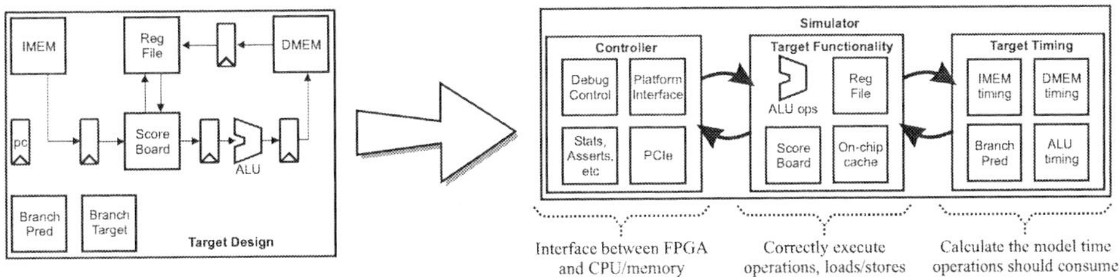

Figure 11. Using the HAsim simulator to model an inorder microprocessor pipeline. The FPGA is not configured into the target pipeline itself, but into three partitions which interoperate to model the pipeline at the fastest rate possible. Only the timing partition must be changed to represent the specifics of the target pipeline. The controller and functional partition can be reused across targets [2].

Category	Number
Intra-Timing	33
Intra-Functional	19
Intra-Controller	20
Timing-Functional	24
Timing-Controller	42
Functional-Controller	76
Unused Optional	3
Total	217

Figure 12. Number and use of Soft Connections in the HAsim inorder pipeline model.

Figure 13. Histogram of Soft Connection span.

Benchmark	Model Cycles	FPGA Cycles: Baseline	FPGA Cycles: Shared	Change
test 164 gzip	7,612,202,736	120,866,746,639	120,848,407,550	-.0002%
test 176 gcc	4,412,926,919	97,284,304,169	96,331,305,044	-.001%
test 181 mcf	515,321,465	13,393,128,486	13,375,809,359	-.001%

Figure 14. Running SPEC benchmarks on the shared interconnect version.

dynamic performance we ran three SPEC benchmarks on each model. The results, shown in Figure 14 demonstrate that over a run which spans billions of model cycles there was no measurable impact on performance—the differences in total FPGA cycles fall within expected run-to-run variation.

7. DISCUSSION

Soft Connections are currently implemented as direct point-to-point connections, or as a shared tree topology. In the future we plan to explore adding support for new physical communication topologies such as rings or grids. We expect grid networks to present a particular challenge as the stations need awareness of the dynamic traffic conditions in order to route messages efficiently. We believe that insights from networks-on-chip—which are traditionally used to connect distinct hardware cores together—may also apply to distributing the connections within the cores themselves.

As FPGA accelerators become more common the barrier to entry becomes a large concern. Traditional tools can force the designer to spend too much effort thinking about on-chip communication and not enough time thinking about the actual logic. Soft Connections are a way to automatically generate a physical implementation of communication from its logical specification. This provides a richer module interface and makes the communication hierarchy less rigid. We believe that these kind of ease-of-use efforts will be critical for FPGAs to gain acceptance as computation accelerators in general-purpose computers.

REFERENCES

[1] M. Pellauer, M. Adler, J. Emer, Modular Soft Connections. Computation Structures Group Technical Report #505, MIT.

[2] M. Pellauer, M. Vijayaraghavan, M. Adler, Arvind and J Emer. Quick Performance Models Quickly: Timing-Directed Simulation on FPGAs. In Proceedings of ISPASS, 2008.

[3] Bluespec Inc. http://www.bluespec.com/, 2008.

[4] D. Chiou, D. Sunwoo, J. Kim, N. A. Patil, W. H. Reinhart, D. E. Johnson, J. Keefe, and H. Angepat. FPGA-accelerated simulation technologies FAST: Fast, full-system, cycle-accurate simulators. In MICRO, 2007.

[5] D. Chiou, D. Sunwoo, J. Kim, N. A. Patil, W. H. Reinhart, D. E. Johnson, and Z. Xu. The FAST methodology for high-speed SoC/Computer simulation. In Proceedings of ICCAD, 2007.

[6] E. Chung, E. Nurvitadhi, J. Hoe K. Mai, and B. Falsafi. Accelerating Architectural-level, Full-System Multiprocessor Simulations using FPGAs. In FPGA '08: 11th International Symposium on Field Programmable Gate Arrays, 2008.

[7] Nallatech, Inc. http://www.nallatech.com/, 2009.

[8] HiTech Global, LLC. http://www.hitechglobal.com/, 2009.

[9] M. Pellauer, M. Vijayaraghavan, M. Adler, Arvind, and J. Emer. APorts: an efficient abstraction for cycle-accurate performance models on FPGAs. In FPGA '08: 11th International Symposium on Field Programmable Gate Arrays, 2008.

[10] D. Rosenband and Arvind. Modular Scheduling of Guarded Atomic Actions. In Proceedings of DAC, San Diego, CA, 2004.

[11] J. Wawrzynek, D. Patterson, M. Oskin, S. L. Lu, C. Kozyrakis, J. C. Hoe, D. Chiou, and K. Asanovic. RAMP: a research accelerator for multiple processors. IEEE Micro, Mar/Apr 2007.

978-1-60558-497-3/09 $25.00 © 2009 ACM

A Computing Origami: Folding Streams in FPGAs

Andrei Hagiescu, Weng-Fai Wong
School of Computing
National University of Singapore
{hagiescu, wongwf}@comp.nus.edu.sg

David F. Bacon, Rodric Rabbah
IBM T.J. Watson
Research Center
{bacon, rabbah}@us.ibm.com

ABSTRACT

Stream processing represents an important class of applications that spans telecommunications, multimedia and the Internet. The implementation of streaming programs in FPGAs has attracted significant attention because of their inherent parallelism and high performance requirements. Languages, tools, and even custom hardware for streaming have been proposed, some of which are commercially available.

There are several significant challenges to realizing streaming applications directly in hardware (FPGAs). Since FPGAs have finite resources, there are often many non-trivial tradeoffs between processing throughput and overall latency. In this paper, we describe an algorithm that computes refinements of stream graphs into designs that optimize processing throughput subject to user-specified area and latency constraints.

Categories and Subject Descriptors

B.6.3 [**Logic design**]: Design aids—*Optimization*

General Terms

Algorithms, Design, Performance

Keywords

FPGA, Streaming, Throughput, Latency

1. INTRODUCTION

There are several existing platforms today that integrate FPGAs with microprocessors, and recent announcements (e.g., [11]) from leading vendors suggest that FPGAs are likely to become widely available as programmable coprocessors. A broad class of applications, including multimedia, networking, graphics, and security codes, provide ample opportunities to exploit FPGA-based acceleration. Sequential parts of a program can be assigned to run on the host processor while the parts of the program with abundant parallelism can be synthesized directly in the FPGA.

Permission to make digital or hard copies of part or all of this work for personal or classroom use is granted without fee provided that copies are not made or distributed for profit or commercial advantage and that copies bear this notice and the full citation on the first page. To copy otherwise, to republish, to post on servers or to redistribute to lists, requires prior specific permission and/or a fee.
DAC'09, July 26-31, 2009, San Francisco, California, USA

In this paper we focus on a class of applications where ample parallelism is available as a result of stream-oriented processing. Stream processing is a data-centric execution model that is dataflow-driven. Streaming codes are naturally expressed as graphs of *filters* that communicate through FIFO data channels. Dependencies between filters are made explicit by the communication channels. Each filter has its own control flow logic and an independent address space, and it executes repeatedly as long as a sufficient number of tokens are available on its input channels.

Our work addresses the following issue: is there a refinement of an input stream graph that can maximize the processing throughput of the overall graph? Furthermore, because FPGA area is finite, and because latency is typically an important consideration in real-time streaming codes, we are interested in maximizing throughput subject to area *and* latency constraints. As far as we know, we are the first in tackling this combined problem.

The intuition behind our proposed algorithm is the following. We inspect the stream graph to identify filters that cause bottlenecks. We observe that if the filters are stateless – they do not maintain a history of their past execution – then we can use data parallelism to increase their throughput. This is achieved by judiciously replicating the bottleneck filters.

Replicating filters has several advantages. The replicated filters do not require resynthesis as they are all instances of the same filter, and the synthesis results are reusable. This is in contrast to prior work on global optimization of loop nests on an FPGA [12] which requires recompilation and evaluation of the recompiled designs based on heuristics. Such an approach will not scale for large designs.

Our algorithm operates on a stream graph, and a set of synthesized filters corresponding to the nodes in the graph, and determines how to assemble the synthesized filters to achieve the best possible throughput. If a filter is replicated, we use a simple hardware template to route dataflow to and from the replicated filters. This approach also makes the issue of filter synthesis orthogonal to design assembly and generation. Hence this work is complementary to a lot of the ongoing research in the community that address high-level synthesis of the filter code itself.

Our algorithm is briefly described as first aggressively replicating candidate filters (graph unfolding), then refolding the graph to reduce the number of replicas if they are not profitable. The next sections provide a motivating example and discuss related work. Next we describe our stream folding algorithm and present the results of our evaluation.

2. EXAMPLE AND BACKGROUND

A stream graph is shown in Figure 1. It consists of three kinds of nodes. A *splitter* node distributes an input stream to other nodes. A *filter* consumes an input stream, performs some computation, and outputs a new stream. A *joiner* aggregates streams, and outputs a single new stream. In the figure, $F1$, $F2$, and $F3$ are filters, the hardware footprint of each filter is correlated to the area of its corresponding rectangle, and the execution latency of the filter is correlated to the length of the rectangle. The splitter and joiner are illustrated using arched double-headed arrows. The edges in the graph describe the dataflow between nodes. Each edge corresponds to a FIFO connecting two nodes together.

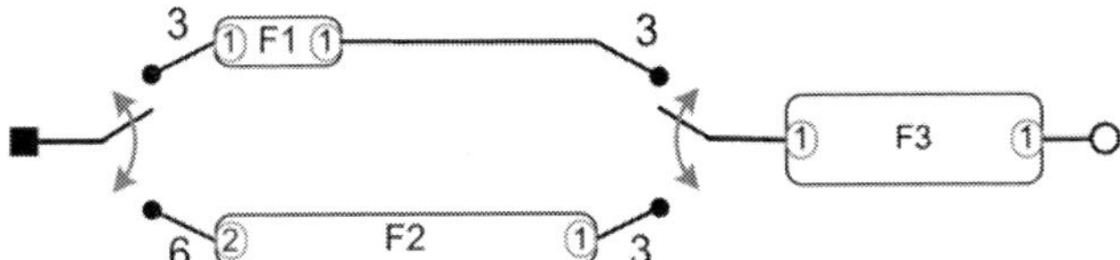

Figure 1: An example stream graph.

Each of the nodes in the stream graph executes when there is a sufficient number of data tokens on its input edge. In the figure, the number of input tokens required to execute each filter is shown in a circle on the left end of each rectangle. The number of data tokens produced in each execution of the filter is shown in a circle on the right side of the rectangle. For example, the filter $F2$ requires 2 data tokens to execute, and when it does, it produces 1 new data token. An execution of a filter is also called a firing.

We use StreamIt [10] to describe stream graphs programmatically and algorithmically. StreamIt is an architecture-independent stream programming language that allows a programmer to focus on describing the algorithm's dataflow (i.e., graph topology) without committing to an implementation or buffering strategy. Each filter in StreamIt declares its data input and output rates per execution. This explicit rate information enables many optimizations that can yield efficient implementations of the stream computation [3, 2, 6]. An example filter declaration is as follows.

```
int->int filter F2(int N) {
  work peek N pop N push N/2 {
    for (int i = 0; i < N/2; i++) {
      int x = pop(); // read/dequeue from input FIFO
      int y = pop();
      push(x+y);      // write/enqueue to output FIFO
    }
  }
}
```

Filters read data from their input FIFO using *pop* statements, and write data to their output FIFO using *push* statements. The filter may have instantiation parameters (N in this case), and always encapsulates its computational logic in a *work* function. Filters may also *peek* at their input data, without altering the state of the FIFO. Peeking is useful for sliding-window computations, and provides an opportunity to optimize filters that otherwise require internal buffering (i.e., state) to preserve previous values.

In Figure 1, the dataflow is split between $F1$ and $F2$ in a periodic and round-robin manner, with 3 tokens dispatched to $F1$ and 6 to $F2$. This information is annotated on the edges that fan-out from the splitter. Similarly, the joiner collects data from the input streams in a round-robin manner. The weight annotations on each edge describe how the data is aggregated from the streams: 3 tokens from $F1$ and 3 from $F2$.

A StreamIt program exposes the communication topology to a compiler or synthesis tool that can then decide on the best implementation choices depending on the target platform. The stream graph in the figure can be described as follows in StreamIt.

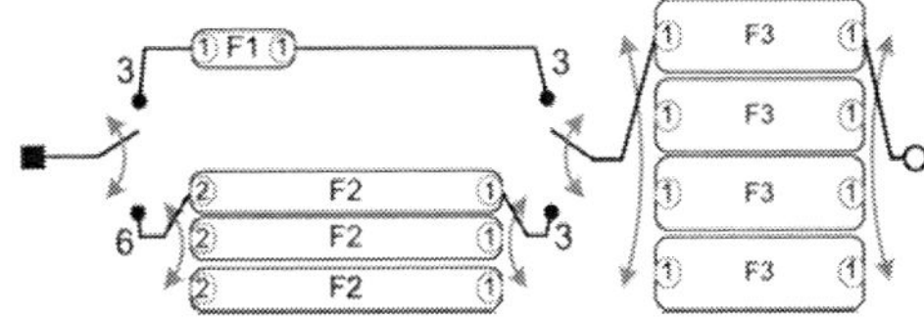

Figure 2: A stream graph with replicated filters that achieves maximum throughput, subject to some area constraint.

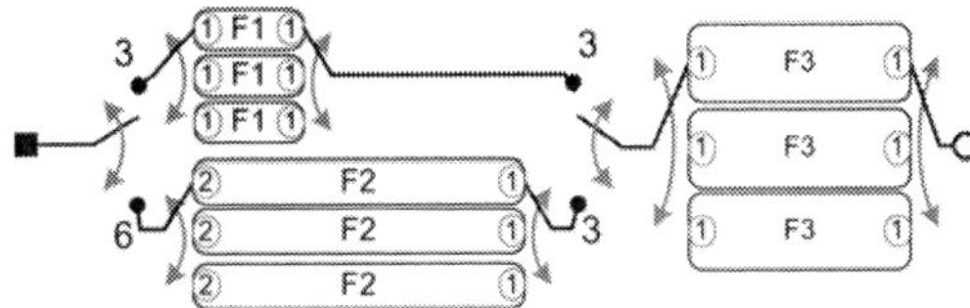

Figure 3: Reducing the latency for the graph in Figure 2 under the same area constraints.

```
int->int pipeline Example() {
  add pipeline {
    add splitjoin {
      split roundrobin(3, 6);
      add F1();
      add F2(2); // instantiating parameterized filter
      join roundrobin(3, 3);
    }
    add F3();
  }
}
```

A *pipeline* in StreamIt is a stream container, connecting a sequence of streams together. Here, the pipeline consists of a *splitjoin* connected to $F3$. The splitjoin is also a stream container, with a splitter at the source, a joiner at the sink, and filters (or other stream containers) between them. The expression *roundrobin(3, 6)* describes the weights of each edge between the splitter and the sibling streams, and similarly for the joiner.

In a stream graph, nodes fire autonomously and concurrently. Since there is no data dependence between $F1$ and $F2$, they can fire in parallel. The firings of filter $F3$ can be pipelined relative to the other filters. It is obvious from the graph that $F1$ and $F2$ execute the same number of times in order for $F3$ to fire (that is, both $F1$ and $F2$ fire 3 times to produce the requisite amount of data for the joiner, which ultimately provides the data to filter $F3$).

A trivial mapping of the stream graph in Figure 1 into hardware does not produce an efficient implementation: the filters $F1$ and $F2$ are not load-balanced. However, if a filter is stateless – that is, it does not maintain any history of its previous executions – we may *replicate* it to achieve a more load-balanced implementation. Replicating a filter creates a new instance of the filter and adds a splitter to distribute data between the filter and its replicas, and a joiner to collect the new data. The replicas effectively increase the firing rate capability of the filter, but also increase hardware costs: each of the replicas incurs a space overhead in hardware that is equal to the original filter, and in addition, there is an added overhead incurred by the splitter-joiner pair that routes the new dataflow. Since we are interested in realizing stream graphs in hardware, and specifically FPGAs with finite resources, we have to judiciously decide which filters to replicate and to what extent. Our algorithm unfolds and refolds a given stream graph to determine where and to what extent replication will be most profitable. We will generally refer to this process of stream graph refinement as *stream folding*.

The stream graph in Figure 2 illustrates the replication of filters $F2$ and $F3$. The graph achieves the best throughput to area ratio: filters fire continuously, making efficient use of the hardware. At steady state, the throughput of the joiner aggregating the outputs of

978-1-60558-497-3/09 $25.00 © 2009 ACM

$F1$ and $F2$ is three times better than the corresponding joiner in the original graph shown in Figure 1. A hardware design and implementation of a stream graph that uses replication increases throughput, but may also affect the latency of the computation. Figure 3 presents an alternate stream folding strategy that tradeoffs throughput to achieve a lower overall latency.

Our stream folding strategy manages the complexity of the design search space by exploiting the hierarchical nature of the stream containers. Peeking filters may be replicated if they appear at the root of a container (e.g., the first filter in a pipeline). The strategy is simple: duplicate the input stream to the filter and its replicas, and then locally discard the parts of the stream that are not relevant to a particular filter. This approach may create a lot of redundant communication, but others have shown that it is possible to design efficient hardware mechanisms to exploit the structured data reuse [4].

3. RELATED WORK

The work that we present is founded on the idea of judiciously replicating filters to increase the throughput of bottleneck filters. In the context of streaming processing, this idea has been explored in the past. For example, [3] describes a greedy strategy to map filters to a multicore architecture. In that work, if the number of filters is less than the number of cores, the compiler replicates the filters (called fission) with the highest computation requirements. If the number of cores is less than the number of filters (which is the common case), the compiler fuses together the filters or stream containers with the least computation until the number of fused filters equals the number of cores. While this strategy works well when mapping to a platform with a finite number of compute engines, it is not clear how to adapt it for an FPGA where the number of compute engines is undefined. Furthermore, the strategy in [3] is not practical if the stream graph consists of filters that are wildly unbalanced since in that case some filters require fission and others require fusion. Our stream folding algorithm is designed to address this issue in the context of FPGAs. The algorithm essentially performs maximal fission of all candidate filters bounded only by the size of the FPGA, then fuses the filters that do not positively effect throughput.

Other relevant stream graph refinements include [2] and [6]. In the former, the cost of the communication introduced by replication is reduced by first fusing filters then replicating the coarsened filters. This reduces the amount of pipeline parallelism that can be effectively exploited in an FPGA. In the latter, an integrated fission and partitioning strategy is offered to replicate filters and assign graph partitions to a finite and predetermined number of cores. The graph partitions are then carefully scheduled using a staging algorithm that attempts to overlap communication and computation. Our algorithm considers and accounts for the latency of the added communication when it determines the extent of replication. Unlike in past work where the communication cost is high and not homogeneous, we can benefit from the FPGA architecture to optimize communication and more accurately account for the added overhead.

This paper does not address the synthesis of the actual computation from StreamIt code (namely the work functions) to a hardware description language. Our emphasis is on the composition of the synthesized modules into an overall space-time efficient design. Recent work [5] specifically addressed the issue of hardware generation from StreamIt, and we believe that work is complementary to our work. Similarly, many of the existing state of the art technologies in this regard can be used to complement our work.

There is also a significant amount of work geared toward improving performance using multiple clocks to drive heterogeneous processors [7], or heterogeneous accelerators synthesized in FPGAs [9]. However, the latter work assumes that the number and type of accelerators in a design is fixed, and the heterogeneous clock assignment finds the optimal set of clock frequencies to assign to each accelerator. The primary contribution of our work compared to these published methodologies is the co-optimization of space and time (latency). Our starting input is a stream graph extracted from a stream program, from which we derive the set of hardware modules to synthesize. In effect, we are simultaneously determining the number and types of "accelerators" to synthesize.

4. STREAM FOLDING

We propose an algorithm that determines which filters to replicate (and by what factor), in order to maximize processing throughput subject to area and latency constraints. Our philosophy is to describe the desired design topology, and instantiate an implementation that simply stitches together the filters and streams as directed by our algorithm. The algorithm assumes that individual filters are already synthesized and both area and behavior (worst-case execution time estimates) information is retrieved from the implementation. If the filters take less time to execute than the worst-case estimate, the correctness of the solution is not affected.

The input to our algorithm is a stream graph, which we derive from StreamIt code. We refer to our graph refinement strategy as stream folding because we first replicate filters to expose data parallelism (graph unfolding) and then refold the graph judiciously to achieve a desired throughput subject to one or more constraints.

We derive a high-throughput design using the steps shown in Algorithm 1. First, we inspect each filter and compute its work factor by multiplying its latency and its firing rate (line 2). The firing rate ($S.runs(f)$) is calculated by the compiler using a Single Appearance Schedule [1]; it equals the number of firings of a filter so that it is rate-matched to its producer and consumer. The computed work factors determine the total area of an initial design point. Lines 5-6 determine the maximum replication factor that matches the area constraint. Stateful filters are not replicated (although past work has shown it may be profitable to do so [6]), and they impose a scaling limit. If a stateful filter dominates the execution, then repli-

---

Algorithm 1: Area / throughput design folding

Input: StreamIt program(S), area constraint (AREA)
Output: Replication coefficients
1 **foreach** *Filter f in S* **do**
2     $workFactor[f] = f.latency \cdot S.runs(f)$;
3     $designPointArea+ = f.area \cdot workFactor[f]$;
4 **end**
5 $scaleLimit = \min\limits_{f.hasState}(\frac{1}{workFactor[f]})$;
6 $scaling = min(AREA/designPointArea, scaleLimit)$;
7 **foreach** *Filter f in S* **do**
8     $replication[f] = \lceil workFactor[f] \cdot scaling \rceil$;
9 **end**
10 **while** $area(replication) > AREA$ **do**
11     $replication = reduceThroughput(replication)$;
12 **end**

Procedure reduceThroughput (*replication*)

$min = \infty$; $t^S_{out} = throughput(replication)$;
foreach *Filter f in S* **do**
    $\delta t = t^S_{out} - throughput(replication.reduce(f))$;
    if $\delta t < min$ **then**
       $min = \delta t$;
       $candidate = f$;
    end
end
return $replication.reduce(candidate)$;

978-1-60558-497-3/09 \$25.00 © 2009 ACM

Algorithm 2: Latency constrained design folding

   Input: Best throughput configuration (baseRepl)
   Output: Latency constrained configuration (latRepl)
1 $latRepl = null$; $T = \infty$;
2 **while** $throughput(baseRepl) \leq T$ **do**
3   **if** $feasibleImprovement(baseRepl)$ **then**
4        $candidates = simAnnealing(baseRepl, T)$;
5   **foreach** $candidate$ in $candidates$ **do**
6   **if** $throughput(candidate) < T$ **then**
7              $latRepl = candidate$;
8              $T = throughput(latRepl)$;
9   **end**
10   **end**
11 **end**
12     $baseRepl = reduceThroughput(baseRepl)$;
13 **end**
14 **return** $latRepl$;

cation of other filters is not likely to be profitable. Line 5 determines which of the stateful filters constrains the replication; it is the stateful filter with the greatest work factor. The first term in the *min* equation on line 6 determines how much of the FPGA resources are available for replication. The resulting scaling factor is used to determine the replication counts for all filters (line 8). Due to rounding, a design that realizes the calculated replicas may have an area slightly larger than the specified area constraint although the design will be on the pareto-optimal frontier with respect to throughput and area. Finally, we refine the design, reducing its throughput while maintaining it on the pareto-optimal front of the design space. Each iteration in lines 10-12 reduces the area of the new design by eliminating filter replicas one at time, starting with filter replicas that only marginally improve throughput. Finally, a maximum throughput design that fits the available area is obtained, and we next determine if it satisfies the latency constraint.

Algorithm 2 generates base designs sorted by throughput, starting with the one achieving the highest throughput as constructed in the previous step. We use simulated annealing to search through the potential candidates while pruning away as much of the infeasible design space as possible. We do this using a custom neighbor visit function that avoids illegal configurations defined by the area constraint, throughput bounds and the maximum synthesizable replication.

The latency of each evaluated design depends on the actual input arrival rates, and the latency constraint may be satisfied only for arrival rates lower than the maximum sustainable by the base design. Finding such a solution adds a lower bound to the throughput of subsequent base designs (line 5-9). Only base designs above this bound are tried (line 2) as they can offer additional replication possibilities (more spare area is available to selectively increase replication) thereby yielding solutions with better throughput than those previously identified.

As long as a candidate is found in one of the steps of the exploration, the search converges easily, being limited to a few tightly constrained simulated annealing steps. However, if no candidate is found, a larger number of possible base designs is explored. We further prune them by checking if an area unconstrained design having the same throughput as the base design can offer the required latency (line 3). Such a design is obtained by replicating all filters except the bottleneck by as much as possible.

4.1 Calculating Throughput

We use the hierarchical nature of the stream graphs derived from StreamIt to efficiently compute the overall throughput of a streaming program. The maximum input throughput t_{in} and output through-

put t_{out} of a filter F_i is defined as follows

$$t_{in}(i) = \frac{pop(F_i)}{latency(F_i)}, t_{out}(i) = \frac{push(F_i)}{latency(F_i)}$$

where $pop(F_i)$ and $push(F_i)$ respectively equal the number of data elements dequeued from and enqueued to the input and output FIFOs of the filter F_i, and $latency(F_i)$ equals the number of cycles spanned by a single firing of F_i. The pop and push values are readily available from StreamIt programs. The throughput of the stream containers follows.

Pipeline and SplitJoin throughput.

The throughput of a pipeline is equal to the lowest throughout of its filters. Furthermore, since individual filters may push and pop at different rates, the rates observed at different points in the pipeline will vary, although filters have to sustain correlated rates. We define the throughput limitation imposed by a filter F_i on the output of a pipeline consisting of the filters $P = \{F_1, \ldots, F_n\}$ as

$$t_{out}^P(i) = t_{out}(i) \cdot \prod_{i<j\leq n} \frac{push(F_j)}{pop(F_j)}$$

and therefore, the actual output throughput of the pipeline is

$$t_{out}^P = \min_{1\leq i\leq n} t_{out}^P(i).$$

For a splitjoin $SJ = \{F_1, \ldots, F_n\}$ where the joiner weights are $(w_1, \ldots, w_n)$, the output throughput is

$$t_{out}^{SJ} = \min_{1\leq n} \left(t_{out}(i) \cdot \frac{\sum\limits_{1\leq j\leq n} w_j}{w_i} \right).$$

Overall throughput.

It is possible to apply these relations to the whole stream graph in a composable manner,

$$t_{out}^S = \min_{i\in S} t_{out}^S(i) = \min_{i\in S} (C_i \cdot t_{out}(i))$$

where C_i is a constant that can be determined by analyzing the unmodified stream graph. To prove this relation, assume a stream can sustain a throughput $t' > t_{out}^S$. Propagating this downwards through the stream hierarchy, for all stream containers $\hat{S}, t_{out}^S(\hat{S}) \geq t'$. Continuing the stream decomposition and applying this relation down to individual filters, results in $\forall i, t_{out}^S(i) \geq t'$, which is a contradiction.

The replication of a filter has the effect of multiplying its throughput by its *replication factor*. If r_i is the replication factor, we can modify the above formula to

$$t_{out}^S = \min_{i\in S} (r_i \cdot C_i \cdot t_{out}(i)).$$

4.2 Calculating Latency

We envision that stream graphs will run on FPGAs that are coupled to host processors. Data transport between the host and FPGA is achieved through a bulk transfer mechanism (e.g., DMA). We call the number of clock cycles between such transfers the *initiation interval* (II). The minimum initiation interval can be computed based on the reciprocal of the highest sustainable throughput in the stream graph. Results are expected to be ready after a time interval called the *latency*.

Data-token reordering and local congestion at a filter's input due to non-periodic data arrival are the major factors for latency variation. While replication improves throughput, it often increases the latency. We believe it is important to obtain exact latency bounds

978-1-60558-497-3/09 $25.00 © 2009 ACM

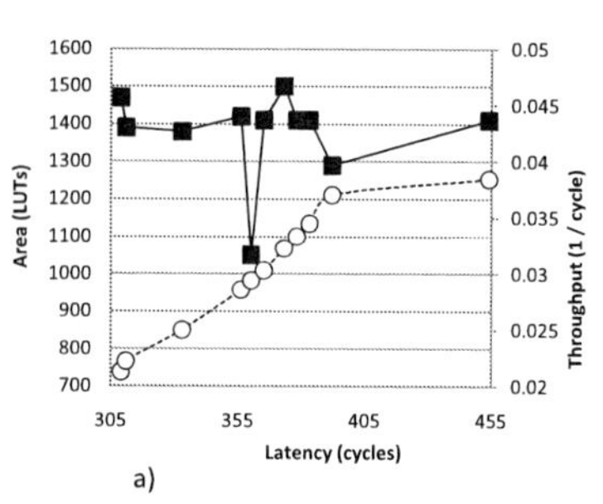
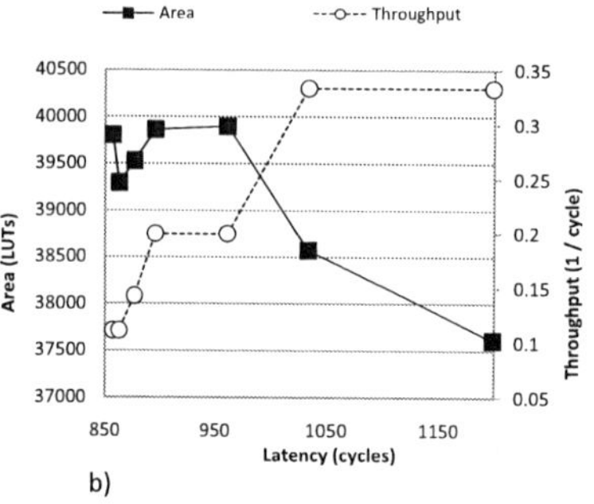
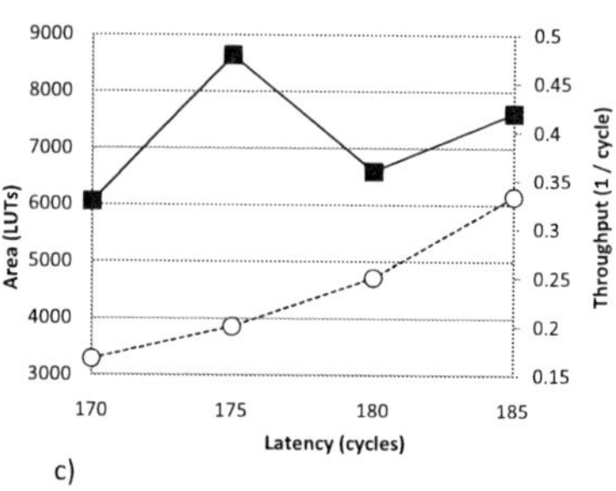

Figure 4: (a) Synthetic example, (b)FFT2, (c) Matrix multiply.

Figure 5: Replication factors for instances of filter *CombineDFT* in FFT. The dotted line separates the replication required to achieve the maximum throughput for a specific design point from the additional replication introduced to decrease latency.

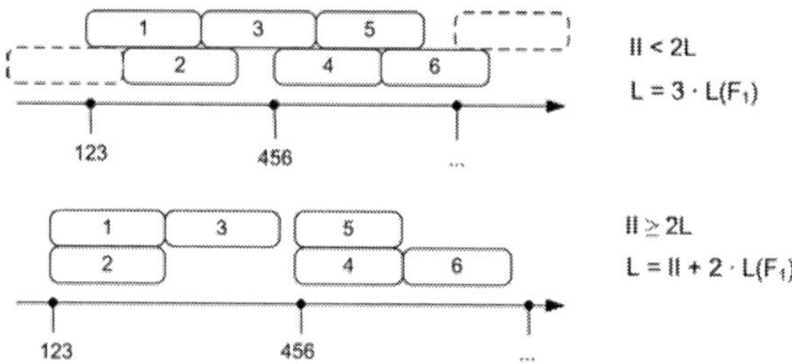

Figure 6: Schedule used to determine latency. Three data tokens arrive every II. With two replicas, computation occurs in parallel.

Table 1: Example latency calculation.

Input	replica 1	replica 2	Constraint	replica 1	replica 2	Constraint
0	$[0, L)$		$II \geq \frac{3L}{2}$	$[0, L)$		$II \geq 2L$
0		$[0, L)$			$[0, L)$	
0	$[L, 2L)$			$[L, 2L)$		
II		$[L, 2L)$			$[II, L)$	
II	$[2L, 3L)$		$II < 2L$	$[II, II + L)$		
II		$[2L, 3L)$			$[II + L, II + 2L)$	
			Interval: $[\frac{3L}{2}, 2L)$			Interval: $[2L, \infty)$

that can offer guarantees especially for real-time stream performance.

Given a stream graph, we determine a valid II where the same set of delays hold. There is a finite set of such intervals and they can be computed starting from the minimum sustainable II. We capture the event arrival time at each filter input as a linear expression $\alpha II + \beta$ and we derive the output time of the result as a linear expression, generating an additional constraint on the upper bound of the II where necessary. We process all filters, until we obtain the overall latency of the stream and a constrained interval where the linear expression holds. We then analyze the adjacent interval, generating a new constraint on the II that will define new intervals recursively.

Table 1 shows the computations necessary in case of a single filter, replicated two times, with 3 input tokens appearing each initiation interval. The corresponding schedule is presented in Figure 6.

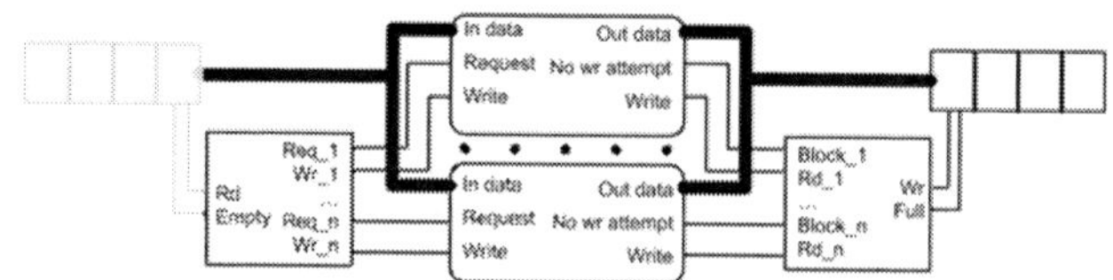

Figure 7: Replication mechanism in hardware.

Our implementation hierarchically iterates over the stream structure, deriving output times based on the input times. In case of replicated filters, it maintains a set of ready times for each replica as linear dependencies on II. The input tokens are already in order and ordering constraints are generated to ensure that the current replica is ready to fire when its data arrive.

4.3 Supporting Hardware

The distribution and gathering of data to and from replicas require hardware resources akin to programmed multiplexers. Our design for such a mechanism is illustrated in Figure 7. The design permits a single data token to be routed per clock cycle. We have automatically synthesized the replication logic for a wide range of replication factors to obtain its associated resource requirements.

5. RESULTS

We evaluated our stream folding algorithm using the streaming benchmarks provided with the StreamIt compiler. These provide us with realistic stream graphs and cover filters with a wide range of area and latency requirements. We also included a synthetic benchmark to check the performance of our implementation beyond the available benchmarks. We implemented our algorithm in Java as a backend for the StreamIt compiler, using Opt4J [8] to perform simulated annealing. For the purpose of achieving low latency designs, we impose a set of constraints on the HDL generation of the filters. These constraints are easily accommodated by current hardware compilers. Namely, we require that a filter reads all of its inputs in consecutive clock cycles. Similarly, we require that

Table 2: Design space solutions under various constraints.

Stream	Minimum area			Best throughput			Constrained design				
	LUTs	Latency	II	LUTs	Latency	II	LUTs	Latency	II	Constraint	Run time
MatrixMult	1498	480	19	7618	185	3	4558	175	7	$Latency \leq 175$	1.14s
Serpent	3028	1027	4	3878	773	2	3053	901	4	$Latency \leq 910$	0.73s
FFT2	37610	1199	3	43370	764	2	39530	868	7	$AREA \leq 40000$ $Latency \leq 880$	34.7s
FMRadio	37458	371	39	87564	371	13	62511	371	20	$Area \leq 65000$	1.01s
DCT	45752	349	3	137256	349	1	91504	349	2	$AREA \leq 120000$	0.73s
BitonicSort	43920	1042	3	131760	1042	1	47400	1282	2	$AREA \leq 50000$	18.3s
Synthetic example	350	309	135	15990	504	2	1490	309	47	$Latency \leq 309$ $Area \leq 1500$	0.43s

a filter writes its output in consecutive cycles. This ensures that arriving data does not block during distribution unless all the filters are currently busy. We perform a one-time hardware synthesis and platform mapping for each filter to empirically determine its resource usage (e.g., area requirements) for the Virtex 4 FPGA architecture, considering LUTs, DSP blocks and Block RAM as part of the area metric.

We explored the throughput achievable under arbitrary area and latency constraints. Our design-space exploration identifies lower latency implementations if it is acceptable to degrade the throughput. As shown in Figure 4, the area of a design is not trivially correlated with either latency or throughput.

We closely inspected FFT2 (a fine grained implementation of FFT) and modified the benchmark so that each floating-point operation is encapsulated in a filter. This allows us to explore the possibility of using stream folding to determine an optimized number of floating-point units to use in a design. We used Xilinx LogiCORE IPs as building blocks for these filters. The total number of filters in this FFT implementation is 118 filters (compared to 22 in the original benchmark). The results are shown in Figure 4(b). The graph shows that there is a significant opportunity in stream folding.

A benchmark that has a tighter range of latency variation is matrix multiply (Figure 4(c)). We found that no solutions were possible if the latency was constrained any tighter than shown. Note that the benchmarks exhibit non-monotonic relationships between throuput and area.

In Figure 5 we represent replication factors of different design solutions of the original FFT. Each group of bars represent the replication factors for instances of the CombineDFT filter (the main computational filter in FFT2) in a particular design point. The dotted line for each group of bars represents the overall replication factor that yields the same maximum throughput for that design point. Designs that are subject to lower latency constraints require greater replication of more filters.

Table 2 shows several design points obtained using our algorithm. For each benchmark, we show the design solution that has (a) the minimum area, (b) the best throughput (area limited only by the device size we considered, a VFX140), and (c) a constrained design between the two. The results are meant to demonstrate the versatility of our algorithm.

Most of our benchmarks are explored in a few seconds on a Core2 Duo 2.33GHz, as long as individual filter replicas monotonically contribute toward lower latencies. In cases with tight latency constraints, joiners might introduce notable adverse latency increases that degrade the convergence of the our algorithm. During our extensive testing, which included additional synthetic benchmarks, the longest runtimes were on the order of minutes.

6. CONCLUDING REMARKS

We have studied the problem of mapping high level descriptions of streaming applications in the form of stream graphs into FPGAs. We proposed replication as a mechanism to increase processing throughput. Our algorithm yields solutions that satisfy area *and* latency constraints. We therefore have a means to automatically and efficiently realize stream applications directly in hardware. Although we have found the scaling of the exploration algorithm to be satisfactory (up to 118 filters), we would like to further investigate and fine-tune the algorithm by introducing more heuristics to guide the design space exploration. Finally, we would like to generalize the technique to decouple filters from each other so as to implement them in different clock domains.

7. REFERENCES

[1] S. Bhattacharyya, P. Murthy, and E. Lee. Kluwer Academic Press, 1996.

[2] M. Gordon, W. Thies, and S. Amarasinghe. Exploiting coarse-grained task, data, and pipeline parallelism in stream programs. In *ASPLOS'06*.

[3] M. Gordon, W. Thies, M. Karczmarek, J. Lin, A. Meli, A. Lamb, C. Leger, J. Wong, H. Hoffmann, D. Maze, and S. Amarasinghe. A stream compiler for communication-exposed architectures. In *ASPLOS'02*.

[4] Z. Guo, B. Buyukkurt, and W. Najjar. Input data reuse in compiling window operations onto reconfigurable hardware. *SIGPLAN Not.*, 39(7):249–256, 2004.

[5] A. Hormati, M. Kudlur, D. Bacon, S. Mahle, and R. Rabbah. Optimus: efficient realization of streaming applications on fpgas. In *CASES'08*.

[6] M. Kudlur, , and S. Mahlke. Orchestrating the execution of stream programs on multicore platforms. In *PLDI '08*.

[7] R. Kumar, K. Farkas, N. Jouppi, P. Ranganathan, and D. Tullsen. Single-isa heterogeneous multi-core architectures: The potential for processor power reduction. In *MICRO '03*.

[8] Optimization framework for java. http://www.opt4j.org/.

[9] S. Sirowy, Y. Wu, S. Lonardi, and F. Vahid. Clock-frequency assignment for multiple clock domain systems-on-a-chip. In *DATE'07*.

[10] W. Thies, M. Karczmarek, and S. Amarasinghe. Streamit: A language for streaming applications. In *CC '02*.

[11] AMD Unveils Torrenza Innovation Socket. http://www.hpcwire.com/hpc/917955.html, 2007.

[12] H. Ziegler and M. Hall. Evaluating heuristics in automatically mapping multi-loop applications to fpgas. In *FPGA'05*.

Retiming and Recycling for Elastic Systems with Early Evaluation

Dmitry E. Bufistov
Universitat Politècnica de
Catalunya
Barcelona, Spain

Jordi Cortadella
Universitat Politècnica de
Catalunya
Barcelona, Spain

Marc Galceran-Oms
Universitat Politècnica de
Catalunya
Barcelona, Spain

Jorge Júlvez
University of Zaragoza
Zaragoza, Spain

Mike Kishinevsky
Strategic CAD Lab, Intel Corp.
Hillsboro, OR, USA

ABSTRACT

Retiming and recycling are two transformations used to optimize the performance of latency-insensitive (a.k.a. synchronous elastic) systems. This paper presents an approach that combines these two transformations for performance optimization of elastic systems with early evaluation. The method is based on Mixed Integer Linear Programming.

On a set of random benchmarks the proposed method achieves, in average, 14.5% performance improvement over min-delay retiming configurations.

Categories and Subject Descriptors: B.5.2 [Register-transfer-level implementation]: Design Aids.
General Terms: Design, Theory, Performance.
Keywords: Elastic systems, early evaluation, optimization.

1. INTRODUCTION

Latency-insensitive (a.k.a. synchronous elastic) systems tolerate changes in communication and computation latencies [5, 8]. The term "elastic system", ES, will be used in this paper.

An ES can be viewed as a composition of combinational blocks and elastic FIFOs connected by channels. A channel is comprised of data wires and a pair of handshake control signals: (valid, stop). The basic case of an elastic FIFO, called elastic buffer, EB, has a latency of one clock cycle and a capacity to store two pieces of information (*tokens*). An EB initially storing one token of information is an elastic equivalent of a synchronous register. An empty EB which contains no tokens is called a *bubble*.

The *valid* and *stop* bits in elastic channels implement a handshake protocol between the sender and the receiver. The valid bit, going in the forward direction, is used by the sender to indicate when useful data is being sent. The stop bit, going in the backward direction, is used for stalling the sender by propagating *backpressure* when the receiver is not ready.

Any synchronous circuit can be transformed into an equivalent ES following a simple automatic flow [6, 8].

A key aspect of ESs is that they accept a set of valid transforma-

Permission to make digital or hard copies of part or all of this work for personal or classroom use is granted without fee provided that copies are not made or distributed for profit or commercial advantage and that copies bear this notice and the full citation on the first page. To copy otherwise, to republish, to post on servers or to redistribute to lists, requires prior specific permission and/or a fee.
DAC'09, July 26-31, 2009, San Francisco, California, USA

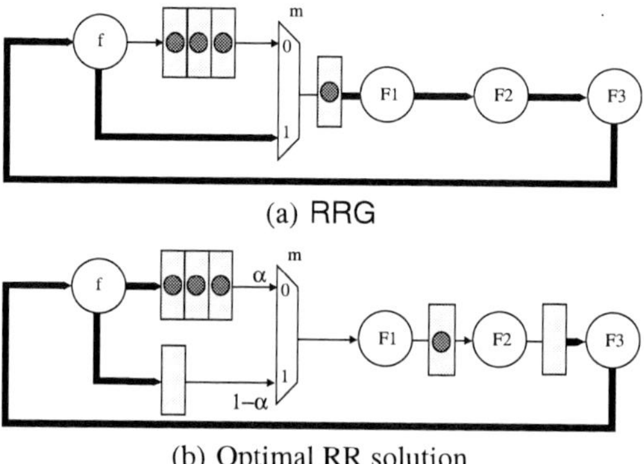

(a) RRG

(b) Optimal RR solution

Figure 1: Retiming and Recycling Graph

tions [10] that preserve the circuit behavior regardless the timing characteristics of its components.

1.1 Retiming and Recycling Graph

Retiming [11] is a well-known technique for sequential optimization. It moves registers across circuit blocks to minimize the clock cycle or area. It preserves the sequential behavior of a circuit. In ESs, retiming moves EBs instead of regular registers.

An ES is modeled by a *Retiming and Recycling Graph* (RRG). Each node of the graph is a combinational block with an associated combinational delay. Each edge represents a connection between blocks labelled with EBs when needed. The RRG can be viewed as an extension of the retiming graph [11].

Figure 1(a) shows an example of an RRG. Only the datapath of the ES is drawn. Each box at the edges represents an EB. If the box is empty, the EB contains no valid information. If the box is marked with a dot, the EB contains one token. E.g., the top edge between nodes f and m has three EBs each labelled with a token. Multiplexors (such as node m) are drawn by using a different symbol than other nodes. Later we will see why.

Assuming that nodes F1, F2, F3 have unit combinational delays while other nodes have zero delays, the *cycle time* of the RRG in Figure 1(a) is equal to three time units, determined by the *critical combinational path* $F1, F2, F3, f, m$.

1.2 Retiming and Recycling (RR)

In ESs it is possible to insert an empty EB at any channel of the system preserving sequential behavior with respect to valid tokens of information. Empty buffer insertion is called *recycling* [6].

The directed cycle $F1, F2, F3, f, m$ in Figure 1(a) with a bottom edge f, m contains only one EB. Retiming preserves the total

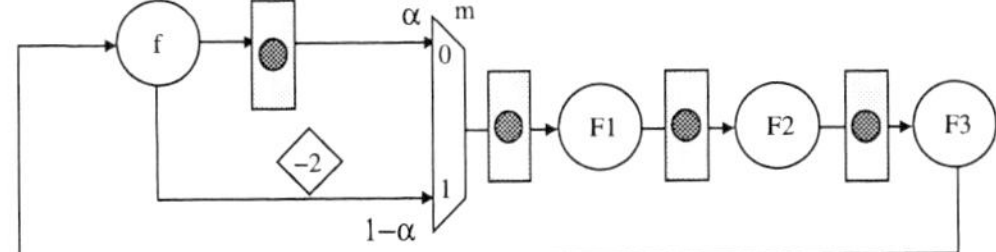

Figure 2: Optimal retiming&recycling with early evaluation

number of EBs at each directed cycle [11]. Thus, the moves of retiming cannot reduce the cycle time in this example: 3 is the minimal cycle time achievable by retiming. Despite, the RRG in Figure 1(b) obtained by applying one retiming move and inserting two bubbles has a smaller cycle time. It is equal to 1 time unit.

This reduction of cycle time is, however, useless, since the actual performance of the ES has not been improved. Indeed, two inserted bubbles reduce the *throughput* of the ES (defined as the amount of useful work done per cycle) to $\frac{1}{3}$. The multiplexor needs to wait for both valid inputs before computing a new token. This is the reason for throughput degradation.

To compare the performance of two ESs the *effective cycle time* metrics is used. The effective cycle time is the ratio of the cycle time to the throughput. Ignoring the delay overhead of inserting extra EBs the effective cycle time of both ESs shown in Figures 1 is the same. It is equal to 3 time units. Minimizing the effective cycle time of an ES by using RR is the main goal of this paper.

1.3 Early evaluation (EE)

Conventional ESs are based on *late evaluation*: the computation is initiated only when all inputs are available. Sometimes this requirement is too strict. For example, once a multiplexor received a select signal, it is sufficient to wait for the selected data channel to produce a token. The other data channel is a "don't care".

Early evaluation (EE) takes advantage of this flexibility to improve the performance of the ES. Care must be taken of the late arriving irrelevant tokens to avoid spurious enabling of functional units. Recently, different schemes to handle EE have been proposed [2, 13, 1, 7]. When EE occurs, a negative token, also called *anti-token*, is generated in the late channels that were not using for enabling the block. When an anti-token and a token meet in the same channel, they cancel each other. Anti-tokens can be *passive*, waiting for the token to arrive, or *active*, traveling backwards through the control until they meet a token. This paper considers only passive anti-tokens.

1.4 Motivational Example

Let us show how RR applied together with EE can improve performance of the RRG from Figure 1. Assume that the select channel of the multiplexor is always valid and it chooses the top data channel with probability $0 < \alpha < 1$, and the bottom channel with probability $(1 - \alpha)$.

The behavior of ESs with EE can be modeled using *Markov chains* [9]. Although this approach does not scale in general, it can be used for analysis of this small example to compute an exact expression for the throughput. Recall that with late evaluation the effective cycle time of the ES in Figure 1(b) is equal to 3. With EE, the throughput is 0.491 for $\alpha = 0.5$. Hence, the effective cycle time is $1/0.491 \approx 2.037$ time units. For $\alpha = 0.9$ the throughput is higher and is equal to 0.719 and the effective cycle time is approximately 1.39 time units.

Using RR it is possible to further improve the performance in the example. The obtained optimal solution is shown in Figure 2. Resolving the Markov chain for the ES in Figure 2, the following expression for the throughput is obtained: $1/(3-2\alpha)$. For $\alpha = 0.9$, the throughput is equal to $\frac{5}{6} \approx 0.833$ that is about 16% better than the throughput for the ES from Figure 1(b) with an EE mux.

The bottom channel coming to the multiplexor contains two anti-tokens (drawn in the rhombus). Note that the total sum of tokens is an invariant and is equal to four for the top cycle and to one $(3-2)$ for the bottom cycle.

The contributions. The first contribution of this paper is the demonstration, as shown in the introductory example, that allowing anti-tokens in initial configurations may help to achieve a better throughput. This is not the case for ESs without EE.

The second contribution is a precise marked graph model for performance estimation of ESs with EE.

The last contribution is a method for minimization of the effective cycle time of ESs with EE. The work is an extension of the paper about performance optimization of ESs with late evaluation [3].

2. BACKGROUND

This section formalizes basic concepts.

A *Retiming and Recycling Graph* (RRG) is a tuple $\langle S, \beta, R_0, R, \gamma \rangle$, where $S = (N, E)$ is the underlying multi-graph of the ES, N is the set of nodes and E is the set of edges. The set N is partitioned into N_1 and N_2: N_1 includes the simple combinational nodes and N_2 the EE nodes. $\beta : N \to \mathbb{R}^+$ assigns combinational delay to each node. $R_0 : E \to \mathbb{Z}$ is the number of the tokens on each edge. If negative, R_0 is the number of anti-tokens. To ensure liveness the sum of tokens on each directed cycle of S must be positive. $R : E \to \mathbb{Z}^+$ is the number of EBs on each edge, condition $R \geq R_0$ must hold. $\gamma : E \to \mathbb{R}^+ \setminus \{0\}$ is the branch selection probability for input edges of EE nodes $n \in N_2$. The sum of the probabilities for all inputs of an EE node $n \in N_2$ is equal to one.

As an example, the values of R_0, R and γ of the top (bottom) edge (f, m) of the RRG in Figure 1(b) are 3, 3 and α (0, 1 and $1 - \alpha$).

Given an RRG, a *combinational path* is a sequence of adjacent edges $e_1, \ldots, e_k$ such that $R(e_i) = 0, 1 \leq i \leq k$. The delay of the combinational path is the sum of the delays of the corresponding nodes. For example, the path formed by the nodes $F1, F2, F3$ in Figure 1(a) is combinational while the path $f, m, F1$ is not.

The *cycle time* of an RRG, $\tau(\text{RRG})$, is the maximum delay of all combinational paths.

Let us assume that combinational delays of nodes $F1, F2, F3$ are equal to 1 time unit while the delays of the rest of the nodes are equal to 0. Then, the cycle time of the RRG in Figure 1(a) is equal to 3. The combinational path $F1, F2, F3, f, m$ is *critical*. Its delay is equal to the cycle time of the RRG.

The *throughput*, $\Theta(n)$, of node $n \in N$ of a RRG is defined as: $\Theta(n) = \lim\limits_{t \to \infty} \frac{\sigma_n(t)}{t}$, where $\sigma_n(t)$ is the number of tokens produced by n till time stamp t. The throughput of every node is the same [9], i.e., $\Theta(n_i) = \Theta(n_j)$ for every $n_i, n_j \in N$. Thus, the throughput of any node can be denoted by $\Theta(\text{RRG})$.

Notice that if an RRG has no bubbles (see Figure 1(a)), one token is produced by each EB each cycle, then $\Theta(\text{RRG}) = 1$. The *effective cycle time* of a RRG, $\xi(\text{RRG})$, is the ratio of the cycle time and the throughput.

A *retiming vector* $r \in \mathbb{Z}^{|N|}$ of a given RRG, is a map $N \to \mathbb{Z}$ that for every edge $e = (u, v)$ transforms R_0 to R_0' as follows: $R_0'(e) = R_0(e) + r(v) - r(u)$.

In contrast to the classical definition in [11] this definition allows negative values for R_0. This is because in ESs EBs can keep anti-tokens [7].

Given an RRG, a RR configuration, RC, is a pair of vectors $R_0' \in \mathbb{Z}^{|E|}, R' \in \mathbb{Z}^{+|E|}$ that satisfies the following constraints:

$$\begin{aligned} R_0'(e) &= R_0(e) + r(v) - r(u), \\ R'(e) &\geq R_0'(e), \quad \text{for each edge } e = (u, v), \end{aligned} \tag{1}$$

978-1-60558-497-3/09 $25.00 © 2009 ACM

where r is a retiming vector.

An RRG has a lot of different RCs. For instance, the retiming vector: $r(m) = -2, r(F1) = -2, r(F2) = -1, r(f) = r(F3) = 0$, transforms the RC in Figure 1(a) to the RC in Figure 2.

Combinational path constraints. In order for an RC to meet a cycle time τ, the delay of every combinational path in the corresponding RRG must be smaller than or equal to τ. There are a set of linear constraints that guaranties this [3].

In the following, these constraints for a given RC and cycle time τ will be refereed as `Path_Constr(RC, \tau)`.

3. THROUGHPUT OF RRG

The performance of an RRG can be estimated by using the result from [9] on performance analysis of *guarded marked graphs*.

A *Guarded Marked Graph* (GMG) is a tuple $\langle N, E, m_0, G \rangle$ where N is the set of nodes which is partitioned into subsets N_1 and N_2: N_1 includes the simple nodes and N_2 - the EE nodes; $E \subset N \times N$ is the set of edges; $m_0 : E \to Z$ assigns an initial number of tokens (possibly negative), $m_0(e)$, to each edge e; $G : N \to 2^{2^E}$ assigns a set of guards to every node, such that the following condition is satisfied. Let us denote the set of input and output edges of a node n_i as ${}^\bullet n_i = \{(n_j, n_i) | (n_j, n_i) \in E\}$ and $n_i^\bullet = \{(n_i, n_j) | (n_i, n_j) \in E\}$, respectively. Then for $n \in N_1$ the guards set $G(n)$ is one element set $\{\{{}^\bullet n\}\}$. This means that all input edges of the node n are in the same guard. For $n \in N_2$ the guards set has $|{}^\bullet n|$ elements, $G(n) = \{{}^\bullet n\}$.

The behavior of an GMG is determined by the following rules:
1. *Guard selection.* A guard $g(n) \in G(n)$ for the next firing of n is selected nondeterministically. The guard selection is trivial for simple nodes, since they only have one guard. For EE nodes any guard in $G(n)$ can be selected.
2. *Enabling.* If the guard $g(n)$ has been selected for the next firing of n, then the node n becomes enabled when corresponding input edge has positive marking.
3. *Firing.* An enabled node n at marking m can fire leading to another marking m' by removing one token from each input edge and adding one token to each output edge.

Timed guarded marked graphs. In order to carry out performance analysis on GMGs a timing interpretation must be added to it. Each guard must be assigned a probability of being selected.

A *Timed Guarded Marked Graph* (TGMG) is a tuple $\langle N, E, m_0, G, \delta, \gamma \rangle$ where $\langle N, E, m_0, G \rangle$ is a GMG; $\delta : N \to \mathbb{R}^+$ assigns a nonnegative delay to every node; $\gamma : G \to \mathbb{R}^+ \backslash \{0\}$ assigns a strictly positive probability to each guard of $G(n)$. It must hold that: $\sum_{e \in G(n)} \gamma(e) = 1$.

For the time evolution of an TGMG it is assumed that the guard selection process has zero duration and that it respects the probabilities (γ) in any infinite execution.

The *throughput*, $\Theta(\mathcal{N})$, of an TGMG is defined as: $\Theta(\mathcal{N}) = \lim_{t \to \infty} \frac{\sigma(t)}{t}$, where t represents the time and $\sigma(t)$ is the firing count vector at time t, i.e., the j's component of $\sigma(t)$ corresponds to the number of times node n_j has fired till the time stamp t.

Notice, $\Theta(\mathcal{N})$ is defined as a vector. In [9] it is shown that all nodes of an TGMG have the same throughput. The throughput is upper bounded by the solution of the following LP problem:

$$\begin{aligned}
Maximize \quad & \phi: \\
& \delta(n) \cdot \phi \leq \widehat{m}(e), & n \in N_1, e \in {}^\bullet n \\
& \delta(n) \cdot \phi \leq \sum_{e \in {}^\bullet n} \gamma(e) \cdot \widehat{m}(e), & n \in N_2 \\
& \widehat{m}(e) = m_0(e) + \sigma(u) - \sigma(v), & e = (u, v).
\end{aligned} \quad (2)$$

The vector σ is real in the constraints.

RRG throughput constraints. There is a simple procedure that

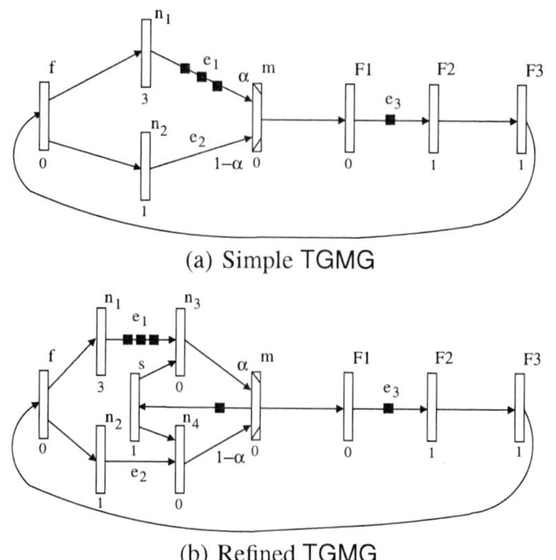

(a) Simple TGMG

(b) Refined TGMG

Figure 3: A TGMG throughput model

constructs a TGMG model for an RRG. Because of the lack of the space the formal description is skipped. It can be found in [4]. For illustration the initial TGMG model for the RRG in Figure 1(b) is shown in Figure 3(a)[1]. Figure 3(b) shows the final version. Basically, a unit delay self loop for each EE node has been added and then TGMG was transformed to preserve the guards set $G(n)$ [4].

Applying (2) to the TGMG model of an RRG it can be guaranteed with the linear constraints that a given RC has throughput upper bound $1/x, x \geq 1$ [4]. Let us denote the set of this constraints as `Thr_Constr(RC, x)`. The throughput upper bound of a given RC can be found as a minimum value of x subject to the `Thr_Constr(RC, x)`. Let us denote this upper bound as $\Theta^{lp}(RC)$ and the corresponding effective cycle time as $\xi^{lp}(RC)$, i.e., $\xi^{lp} = \tau(RC)/\Theta^{lp}(RC)$, where $\tau(RC)$ refers to the cycle time of the RC.

4. RETIMING AND RECYCLING

A method that combines the combinational path and throughput constraints for minimizing the effective cycle time leads to the following non-convex mixed integer quadratic programming problem:

$$\begin{aligned}
Minimize \quad & x \cdot \tau, \\
& R_0'(e) = R_0(e) + r(v) - r(u), \\
& R' \geq R_0, R' \geq 0, \\
& \texttt{Path_Constr(RC, \tau)}, \\
& \texttt{Thr_Constr(RC, x)}, \\
& R' \in INT, r \in INT.
\end{aligned} \quad (3)$$

The exact solution of (3) is not necessarily the one with the minimum effective cycle time, RC_{min}, but it is a very good approximation. On the other hand, (3) represents a big challenge for existing solvers. [4] provides a heuristics based on a MILP to solve (3). This heuristics finds few *non-dominated* RCs and select one with the minimal effective cycle time, RC_{min}^{lp}.

5. EXPERIMENTAL RESULTS

A set of experiments was performed to verify the throughput model and to demonstrate optimization power of the algorithm for ESs with EE.

A set of random RRGs has been generated. The ISCAS89 circuits have been used to extract the underlying graph structures. All

---

[1]Backward arcs modeling backpressure can be eliminated by buffer sizing [12]

978-1-60558-497-3/09 $25.00 © 2009 ACM

parameters have been set randomly, based on simple criterions. For

Table 1: All non-dominated RCs for the test case s526

Name	τ	Θ^{lp}	Θ	err(%)	ξ^{lp}	ξ	$\Delta(\%)$
s526	19.98	0.2500	0.2390	4.6025	79.9200	83.5983	
	24.10	0.3333	0.3050	9.2896	72.3000	79.0164	
	31.74	0.4936	0.4200	17.5219	**64.3041**	75.5714	
	56.54	0.8367	0.7910	5.7787	67.5742	**71.4791**	
	74.52	1.0000	1.0000	0.0000	74.5200	74.5200	5.4

each test case the RC with the minimal effective cycle time, RC^{lp}_{min}, was found. The Verilog representation of elastic controller was generated for each non-dominated RC. The actual throughput was calculated by performing intensive simulations.

Table 1 shows all non-dominated RCs for the test case s526. Rows of the table correspond to different RCs. The column τ provides the cycle time of the RC. The columns Θ^{lp} and Θ provide the throughput upper bound and the actual throughput of the RC (obtained by simulation) respectively. The column $err(\%)$ provides the relative difference between the throughput upper bound Θ^{lp} and Θ. The effective cycle times of RC^{lp}_{min} and RC_{min} are marked in bold in the columns ξ^{lp} and ξ respectively. The last column $\Delta(\%)$ is the relative difference between $\xi(\text{RC}_{min})$ and $\xi(\text{RC}^{lp}_{min})$, e.g., for s526 it is equal to $(75.5714 - 71.4791)/71.4791 \cdot 100\% \approx 5.4\%$. It can be seen that the RC^{lp}_{min} and RC_{min} are different configurations in this case, however RC^{lp}_{min} has only 5.4% worse performance. Also the second best configuration returned by the algorithm does correspond to RC_{min}.

Table 2 shows the obtained results. The first column is the name of the underlying ISCAS89 circuit. The next three columns are the number of simple nodes, EE nodes and edges respectively. The column ξ^* provides the cycle time before the optimization (it is equal to the effective cycle time because originally RRGs have no bubbles). The column ξ_{nee} provides the minimal effective cycle time of the RRG with all nodes being simple (late evaluation). It often coincides with the min-delay retiming cycle time (see [3] for details). In the experiments the ξ_{nee} was always provided by min-delay retiming configuration. The columns ξ^{lp}_{min} and ξ^{sim}_{min} show $\xi(\text{RC}^{lp}_{min})$ and $\xi(\text{RC}_{min})$, respectively. E.g., for s526 the corresponding values are equal to 75.57 and 71.48. The last column $I(\%)$ provides the performance improvement obtained by the proposed method with respect to ES without EE. It is calculated as follows: $I = ((\xi_{nee} - \xi^{sim}_{min})/\xi_{nee}) \cdot 100\%$.

CPLEX was used as an MILP solver. The timeout for integer optimization was set to 20 minutes in all experiments. For all MILPs the optimal solutions were always found.

Table 2: Experimental results.

Name	$|N_1|$	$|N_2|$	$|E|$	ξ^*	ξ_{nee}	ξ^{lp}_{min}	ξ^{sim}_{min}	$I\%$
s641	206	15	270	183.15	109.62	93.72	89.98	17.9
s27	9	5	24	43.73	43.73	32.31	32.31	26.1
s444	45	13	82	174.88	106.75	92.50	92.50	13.3
s386	36	12	131	74.80	74.60	58.55	59.81	21.5
s344	122	13	176	130.63	114.19	90.79	82.89	27.4
s400	37	9	66	149.29	79.50	80.10	77.63	2.3
s526	43	7	71	144.47	74.52	75.57	71.48	4.1
s382	35	7	60	84.65	68.47	66.07	66.07	3.5
s420	7	1	9	76.70	76.70	59.78	59.78	22.1
s832	76	41	462	62.11	50.39	50.39	50.39	0.0
s1488	85	48	572	64.28	59.52	59.52	59.52	0.0
s510	63	40	407	116.63	116.63	73.26	73.26	37.2
s953	232	36	371	354.86	292.28	125.92	119.53	59.1
s713	229	27	341	119.15	96.63	99.13	95.96	0.7
s1494	88	48	572	61.97	55.80	55.80	55.80	0.0
s820	72	38	424	55.64	53.23	46.90	46.90	13.5

Observation 1: The average effective cycle time improvement is equal to 14.5% (the average value of the column $I\%$). The im-

provement strongly depends from the position of EE nodes. The ξ_{nee} was not improved for s832, s1488, s1494. This is because some critical directed cycles (the cycles where bubbles have to be inserted) have no early evaluation nodes. The EE does not affect the performance of such ESs.

Observation 2: The RC^{lp}_{min} coincides with RC^{sim}_{min} in more than half of the examples. In s641, s386, s400, s526, s713, s953 the value of $\Delta(\%)$ is within 5%.

Observation 3: The average error $err(\%)$ of the throughput estimation is equal to 12.5%. The error achieves 35% for some configurations. Usually the error is proportional to the difference between throughputs of an RRG with and without EE nodes.

6. CONCLUSIONS AND FUTURE WORK

A MILP based algorithm for retiming and recycling of elastic systems with early evaluation has been presented. The proposed MILPs are difficult to solve exactly for circuit graphs with more than one thousand edges. However, there are simple and efficient heuristics for solving MILP problems. Exploring such heuristics is a part of the future work.

The proposed model can be extended to handle *telescopic* nodes (i.e., nodes with variable combinational delays).

Acknowledgements. This research has been supported by FPU grant AP2005-4866, FI grant B1 00063, Juan de la Cierva fellowship from the Spanish Ministry of Education and Science, a grant from Intel Corp. CICYT TIN 2004-07925 and research project CICYT TIN2007-66523.

7. REFERENCES

[1] M. Ampalam and M. Singh. Counterflow pipelining: Architectural support for preemption in asynchronous systems using anti-tokens. In *Proc. International Conf. Computer-Aided Design (ICCAD)*, pages 611–618, 2006.

[2] C. F. Brej. *Early Output Logic and Anti-Tokens*. PhD thesis, University of Manchester, 2005.

[3] D. Bufistov, J. Cortadella, M. Kishinevsky, and S. Sapatnekar. A general model for performance optimization of sequential systems. In *Proc. Int. Conf. Computer-Aided Design (ICCAD)*, Nov. 2007.

[4] D. E. Bufistov, J. Cortadella, J. Galceran-Oms, J. Júlvez, and M. Kishinevsky. Retiming and recycling for elastic systems with early evaluation. Technical Report LSI-09-11-R, 2009. http://www.lsi.upc.edu/~techreps/files/R09-11.zip.

[5] L. P. Carloni, K. L. McMillan, and A. L. Sangiovanni-Vincentelli. Theory of latency-insensitive design. *IEEE Transactions on Computer-Aided Design*, 20(9):1059–1076, Sept. 2001.

[6] L. P. Carloni and A. L. Sangiovanni-Vincentelli. Coping with latency in SoC design. *IEEE Micro, Special Issue on Systems on Chip*, 22(5):12, October 2002.

[7] J. Cortadella and M. Kishinevsky. Synchronous elastic circuits with early evaluation and token counterflow. In *Proc. ACM/IEEE Design Automation Conference*, pages 416–419, June 2007.

[8] J. Cortadella, M. Kishinevsky, and B. Grundmann. Synthesis of synchronous elastic architectures. In *Proc. ACM/IEEE Design Automation Conference*, pages 657–662, July 2006.

[9] J. Julvez, J. Cortadella, and M. Kishinevsky. Performance analysis of concurrent systems with early evaluation. In *Proc. Int. Conf. Computer-Aided Design (ICCAD)*, Nov. 2006.

[10] T. Kam, M. Kishinevsky, J. Cortadella, and M. Galceran-Oms. Correct-by-construction microarchitectural pipelining. In *Proc. Int. Conf. Computer-Aided Design (ICCAD)*, Nov. 2008.

[11] C. E. Leiserson and J. B. Saxe. Retiming synchronous circuitry. *Algorithmica*, 6(1):5–35, 1991.

[12] R. Lu and C.-K. Koh. Performance optimization of latency insensitive systems through buffer queue sizing of communication channels. In *Proc. Int. Conf. Computer-Aided Design (ICCAD)*, pages 227–231, Nov. 2003.

[13] R. Reese, M. Thornton, C. Traver, and D. Hemmendinger. Early evaluation for performance enhancement in phased logic. *IEEE Transactions on Computer-Aided Design*, 24(4):532–550, Apr. 2005.

978-1-60558-497-3/09 $25.00 © 2009 ACM

Speculation in Elastic Systems

Marc Galceran-Oms[*]
Universitat Politècnica de
Catalunya
Barcelona, Spain

Jordi Cortadella
Universitat Politècnica de
Catalunya
Barcelona, Spain

Mike Kishinevsky
Strategic CAD Lab, Intel Corp.
Hillsboro, OR, USA

ABSTRACT

Speculation is a well-known technique for increasing parallelism of the microprocessor pipelines and hence their performance. While implementing speculation in modern design practice is error-prone and mostly ad-hoc, this paper proposes a correct-by-construction method for implementing speculation in Elastic Systems. The technique is based on applying provably correct transformations. The benefits of speculation are illustrated with two examples in which these transformations are systematically applied. The method proposed in this paper is amenable for automation in a synthesis flow.

Categories and Subject Descriptors: B.5.2 [Register-transfer-level implementation]: Design Aids.

General Terms: Design, Theory, Verification.

Keywords: Elastic designs, speculation, protocols, synthesis.

1. INTRODUCTION

Speculation is a well-known technique to increase the instruction level parallelism in pipelined microprocessors. When the outcome of an operation is unknown during some cycle, but is required to perform another operation, two schemes can be considered for a correct behavior: (1) stall the pipeline until the first operation has completed, or (2) predict the outcome of the operation and continue the computations without stalling the pipeline. In the second case, the predicted result must be checked for correctness after the first operation has completed and, in case of *misprediction*, the speculated computations must be invalidated. If the predictions are sufficiently accurate, speculation may potentially provide a tangible performance improvement.

Elastic systems, either synchronous or asynchronous, are characterized by their tolerance to the variability of communication and computation latencies or delays [11, 4, 6]. This tolerance enables the exploration of new micro-architectural trade-offs aimed at the optimization for the average case rather that the worst case. Elastic systems use distributed handshake controllers to control the flow of data (*tokens*) along the datapath.

Recently, different schemes to handle **early evaluation** have

---

[*]This work was supported by grants from Intel Corp., CICYT TIN2004-07925 and FI from Generalitat de Catalunya. We thank Verific for permitting us to use their front-end tools.

Permission to make digital or hard copies of part or all of this work for personal or classroom use is granted without fee provided that copies are not made or distributed for profit or commercial advantage and that copies bear this notice and the full citation on the first page. To copy otherwise, to republish, to post on servers or to redistribute to lists, requires prior specific permission and/or a fee.
DAC'09, July 26-31, 2009, San Francisco, California, USA

been proposed [3, 9, 1, 5]. By relaxing the condition that requires *"all inputs to be valid"*, certain operations can be initiated when sufficient information is available to perform the computation. For example, multiplexors only need the select signal and the selected data to be valid. In these cases, the dispensable data must be ignored when arriving at the computational block. *Anti-token* [5] can be used to nullify the dispensable data.

Contribution. This paper presents a novel method to add speculation into elastic systems. Speculative designs are obtained by applying a sequence of provably-correct-by-construction transformations to a non-speculative micro-architecture. Thus, it is guaranteed that the speculative design is functionally equivalent to the original one, regardless the prediction strategy used for speculation. The study of prediction schemes for speculation are out of the scope of this paper, even though they have a crucial impact on the performance of the system. The framework presented can be conceptually applied to any elastic system, either synchronous [4] or asynchronous [11], and customized for any specific elastic protocol. In this paper, we will focus on one specific protocol for synchronous elastic systems for which early evaluation has been incorporated and formally proved to be correct [5].

The paper is structured as follows. Section 2 describes the speculation method by means of a simple example. Section 3 introduces synchronous elastic systems. Section 4 presents the implementation details and verification of the controllers for speculation. Section 5 studies two examples that illustrate the benefits of speculation. Finally, some conclusions are drawn in Section 6.

2. OVERVIEW

The need for speculation arises when there is a decision point in the datapath in which some of the required data arrives late. Fig. 1(a) shows a simple elastic circuit in which speculation can boost up its performance. In this figure, the box is an *Elastic Buffer* (EB), initially containing one valid data item (represented as a token). The circles represent functional blocks. The multiplexor can handle early evaluation when data at the non-selected channel has not arrived yet. Control details are not explicitly displayed, only the data dependencies are drawn.

This scheme could actually be found in a real micro-architecture. For example, the two inputs might be the next PC (Program Counter) and the PC in case a branch instruction is taken. The loop through F and G could represent the computation needed to decide whether the branch is taken. Let us assume that there is a critical path starting at the EB, going through G, the multiplexor, F and arriving at the EB again.

In elastic systems, it is always possible to insert empty EBs. Thus, a possible way to optimize the performance of this design would be to insert an empty EB in the critical path, as shown in Fig. 1(b). While this transformation would improve the cycle time of the design, it would also decrease the throughput, and no real gain would be achieved.

978-1-60558-497-3/09 $25.00 © 2009 ACM

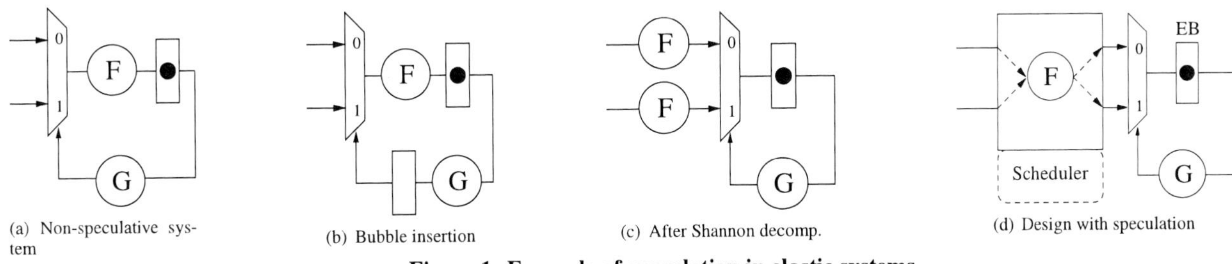

(a) Non-speculative system

(b) Bubble insertion

(c) After Shannon decomp.

(d) Design with speculation

Figure 1: Example of speculation in elastic systems

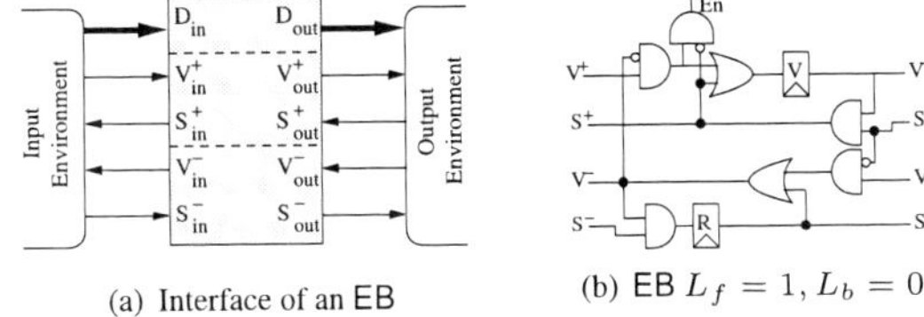

(a) Interface of an EB

(b) EB $L_f = 1, L_b = 0$

Figure 2: EB controllers

Cycle	0	1	2	3	4	5	6
F_{in_0}	A	-	C	-	E	F	F
F_{out_0}	A	-	C	-	E	*	F
F_{in_1}	-	B	D	D	-	G	-
F_{out_1}	-	B	*	D	-	G	-
Sel	0	1	1	1	0	0	0
$Sched$	0	1	0	1	0	1	0
EB_{in}	A	B	*	D	E	*	G

Table 1: Example trace from Figure 1(d). '*' = bubble in the channel, '-' = anti-token in the channel, otherwise token

Given a multiplexor with several inputs, it is possible to move a functional block from the output of the multiplexor to its inputs using Shannon decomposition (viewed also as a multiplexor retiming) [10], as shown in Fig. 1(c). After moving F from the output of the multiplexor to its inputs, F and G are executed in parallel rather than sequentially, achieving a better cycle time. Furthermore, the throughput of the system is optimal as there are no bubbles. However, this speed-up comes at a price: duplication of logic. Here is where speculation can be effectively used. In order to reduce area, all copies of F can be merged into a single one as shown in Fig. 1(d). Thus, the system can now *speculate* which input channel of the multiplexor should first use the *shared functional module*.

When the speculation is correct, the early evaluation multiplexor receives the required data and computes its output. At the same time, an anti-token is propagated backwards through the channel that was not selected, invalidating the unneeded data. When it is not correct, the multiplexor stalls. Then, the correct token must be propagated through F.

The design in Fig. 1(c) is optimal in performance. However, if the prediction strategy in Fig. 1(d) is sufficiently accurate, the cycle clock penalty due to speculation will be rarely paid, thus achieving a similar performance with smaller power and area. A manual implementation of all the stalling/cancelling mechanisms for speculation is complicated and error-prone. We use a set of verified control primitives that can automatically take care of these mechanisms in a distributed control fashion using local handshake protocols. Thus, an automatable and scalable scheme for speculation is provided.

3. SYNCHRONOUS ELASTIC SYSTEMS

An elastic system can be defined as a collection of blocks and FIFOs connected by channels. A channel is a set of data wires with a tuple of control signals associated: (V^+, S^+, V^-, S^-). Synchronous ELastic Flow (SELF) [6, 5] defines a formal protocol and a set of control circuit primitives for creating an elastic system. The *valid* (V^+) and *stop* (S^+) bits implement a handshake protocol between the sender and the receiver of the channel. The valid bit, going in the forward direction, is set by the sender when some piece of data (a *token*) is being sent. The stop bit, going in the backward direction, is used for stalling the sender by propagating *back-pressure* when the receiver is not ready. Analogously, V^- and S^- bits implement the same protocol on the opposite direction. This second pair of handshake bits is used to propagate *anti-tokens*.

Elastic buffers, EB, are the sequential elements in an elastic design. An EB is an unbounded FIFO which stores tokens (data items) and anti-tokens, which cancel each other at the boundaries of the EB. Figure 2(a) shows the interface of an EB with one input channel and one output channel. Two important parameters of an

EB are the forward latency L_f, which is the number of clock cycles needed to propagate tokens through the EB, and the backward latency L_b, which is the number of clock cycles needed to propagate anti-tokens and the stop bit backwards. It is known that the capacity C of an EB must satisfy the following constraint : $C \geq L_f + L_b$. An EB similar to a flip-flop in conventional synchronous designs ($L_f = 1, L_b = 1, C = 2$ initialized with one token) can be efficiently implemented using transparent latches[6]. If an EB does not initially contain any token, it is called a *bubble*. Other implementations can sometimes be useful, Figure 2(b) shows an EB controller with $L_b = 0, C = 1$ for fast propagation of anti-tokens.

Design transformations known from conventional synchronous systems, such as retiming or bypassing, can also be applied in elastic systems. Furthermore, elastic systems support novel correct-by-construction transformations enabling new micro-architectural trade-offs. For example, a method to perform correct-by-construction micro-architectural pipelining was presented in [8].

4. SPECULATION IN ELASTIC SYSTEMS

In this section we will present a method for introducing speculation into an elastic design based on a sequence of provably-correct transformations. This method can be completely automated. As was discussed in section 2, speculation can be achieved following these steps:

1. Find a critical cycle from an output of an early evaluation multiplexor to its select input. If such cycles exist, speculation is the transformation of choice for increasing the performance.
2. Apply Shannon decomposition to move a logic block backward, out of the critical cycle.
3. Apply early evaluation to the moved multiplexor
4. Share the duplicated logic, introducing the speculation control that instantiates some prediction logic.

Table 1 shows a sample trace of the system from Figure 1(d). F_{in_0} and F_{out_0} denote the input and output channel of the shared module F that serve the first input of the multiplexor, while F_{in_1} and F_{out_1} correspond to second channel of the multiplexor. Sel is the select input of the multiplexor connected to the output of G functional block. $Sched$ is the scheduling signal that carries the channel prediction done for the shared unit F. Finally, EB_{in} is the data value at the input channel of the EB connected to the output of the multiplexor. In cycles $0, 1, 3, 4,$ and 6, the correct predictions are made ($Sel = Sched$). During these cycles the early evaluation multiplexor propagates correct value from the selected

978-1-60558-497-3/09 $25.00 © 2009 ACM

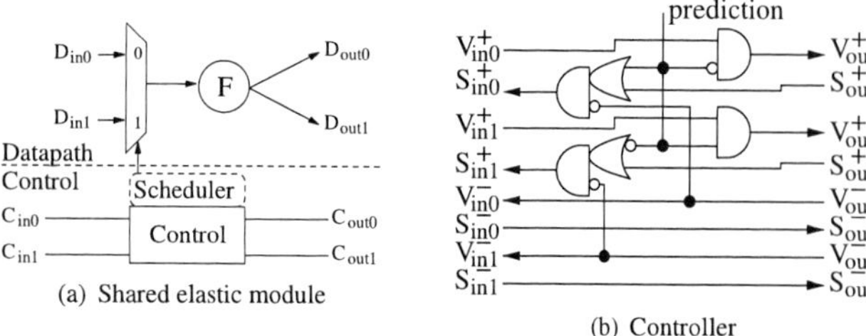

(a) Shared elastic module

(b) Controller

Figure 3: Implementation of a shared module

channel, and the anti-token cancels the token waiting on the unused input channel, since it is not needed. In cycles 2 and 5, however, mispredictions are made ($Sel \neq Sched$). The multiplexor is not enabled, stalling and waiting for the correct token to arrive. The next clock cycle, the scheduler corrects its prediction and selects the other token, that will get propagated through the multiplexor.

4.1 Sharing of Elastic Modules

Speculation is performed for the shared logic that has been re-timed out of the critical cycle. The scheduler selects out of the valid input tokens which one to send for the execution.

Let us assume that sharing occurs between two copies of the block like in the example we have considered so far [1]. There are two input data channels to the shared functional module. Let us assume that elastic buffers are inserted into the channels between the shared module and the multiplexor. Let us also assume that both buffers have identical forward latency L_f and identical backward latency L_b. A special case shown in the example in Figure 1(d) (no buffers inserted), correspond to the case of $L_f = 0$ and $L_b = 0$. Let us also assume (for simplicity of explanation) that there is an EB at each input of the shared module that stores tokens waiting to be served. For better understanding of the behavior of the specu-lation unit let us trace the processing of the i-th token, T_i, arriving at one of the input channels of the shared module. The processing goes through 3 steps:

Propagating to the input of the shared module. Since token or-der is preserved in elastic systems, for the i-th token, T_i, to be available at the input in_1 of the shared module, the $(i - 1)$-th to-ken, T_{i-1}, in this channel must have been processed or canceled by an anti-token. Let us assume that the speculation controller of the shared module predicted channel 2 during the previous transfer. The T_{i-1} (if already arrived) is stalled at in_1 input channel of the shared module, because it needs to be used in case of misprediction. If prediction of channel 2 was correct, an anti-token is generated by the controller of the multiplexor into channel 1 (out_1). This anti-token propagates backwards reaching in_1 in L_b cycles and cancels T_{i-1}. The token T_i will remain stalled, until the previous token, T_{i-1}, is canceled. Thus, the backward latency of EBs can affect the overall system performance and become a bottleneck.

Prediction by the scheduler of the shared module. Once a to-ken T_i gets to the input of the shared module, the scheduler may predict its channel and then T_i will get propagated through the shared module. Otherwise, the token will be stalled until either the scheduler changes its prediction during one of the future cycles or an anti-token generated by the multiplexor arrives and cancels it out.

Early evaluation in the multiplexor. After a token T_i is selected by the scheduler, it is transmitted through the shared module and then, stored by the output EB, reaching the input of the multiplexor in L_f cycles. If T_i was predicted correctly, once the select signal of the multiplexor becomes available, the early evaluation multiplexor will generate a new token at its output. Otherwise, the token will be stalled at the input of the multiplexor, waiting for the correct token

---

[1] This consideration can be easily generalized to k blocks

to arrive in the other channel.

4.1.1 Scheduler

A scheduler predicts at each clock cycle which channel can use the shared resource. The performance gain obtained by applying speculation is based on the assumption that the prediction can be done with a high probability of success. The scheduler can im-plement prediction algorithms of different complexity, from always predicting one of the channels to more advanced algorithms such as the state-of-the-art branch prediction in modern micro-processors.

For better performance, the scheduler should take into account the elastic protocol, since a channel that is not valid, or is stalled, cannot use the shared unit even if selected. Besides, mispredictions can be detected because of back-pressure on the predicted chan-nel. For correctness of behavior a scheduler should avoid potential scheduling deadlocks. To guarantee this, a scheduler should detect and correct all mispredictions. In addition starvation of channels must be avoided, every token that reaches the shared module must eventually be scheduled unless it is cancelled by an anti-token. This property can be formalized as a leads-to constraint: if tokens arrive infinitely often, then they must eventually be served by the shared unit or killed. Formally, for every user of the shared unit, i:

$$\mathsf{G}\ (V_{in_i}^+ \Rightarrow \mathsf{F}\ (V_{out_i}^- \vee (sel = i \wedge \overline{S_{out_i}^+}))) \qquad (1)$$

4.1.2 Design

Figure 3(a) shows the datapath logic for a combinational block shared by two channels, and Figure 3(b) shows its control logic. C_{in_i} and C_{out_i} represent the handshake control bits of the elastic channels; and D_{in_i} and D_{out_i} represent the datapath wires asso-ciated with these control bits. The delay overhead on the datapath is one multiplexor plus the delay in the scheduling decision. One should make sure that the scheduler is out of the critical path.

The controller sets the valid signal of the selected channel as long as its input is valid and keeps the valid signal of the other channel at 0. It also stops the other channel (unless it is killed). The implementation of the controller can be trivially extended to handle more than two channels.

Figure 2(b) depicts a variant of an EB with $L_b = 0$ and $L_f = 1$. This implementation of EB can be used to reduce latency overhead of speculation since stalls and anti-tokens can be detected faster. However, a care must be taken not to chain too many of such con-trollers to avoid potentially long combinational delays in the con-trol.

4.1.3 Verification

The absence of deadlocks has been verified for *any scheduler* that complies with the leads-to property (1). In addition, it has been verified that all controllers comply with the SELF protocol and the interaction between the datapath and the controller is correct. More details about the verification strategy can be found in [7].

5. EXAMPLES

In this section we demonstrate the use of speculation combined with elastic systems on two interesting examples. For performing these experiments we have developed a complete framework for exploring elastic systems. Given an abstract netlist representing an elastic system as a collection of modules and FIFOs connected by elastic channels, our toolkit can apply all of the known correct-by-construction transformations under the user guidance in the form of command scripts within an interactive shell. Since all transfor-mations are local they are very fast to compute.

This environment enables fast exploration of the design space. The user can perform transformations, visualize the modified graph, undo and redo the transformations. At any point, it is pos-sible to generate a Verilog netlist of the elastic controller, a blif

---

978-1-60558-497-3/09 $25.00 © 2009 ACM

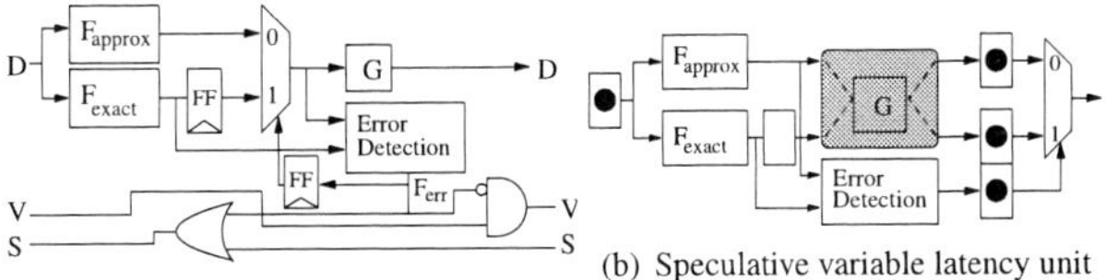

(a) Stalling variable latency unit

(b) Speculative variable latency unit

Figure 4: Speculation used for variable latency

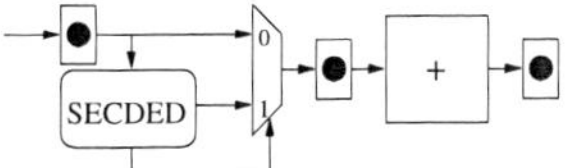

(a) Non-speculative resilient system

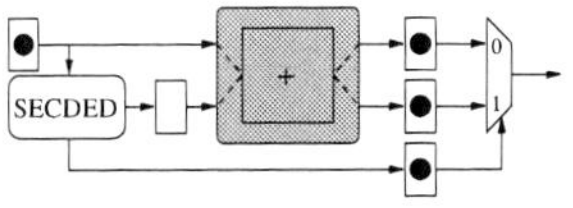

(b) Speculative resilient system

Figure 5: Speculation used for error correction and detection

model for logic synthesis with SIS or a NuSMV model for verification. The elastic controller is built by assembling a set of predefined parameterized control circuit primitives using Verific's front-end tools, and then it is connected to the datapath of the examples. The cycle time is obtained by synthesis, using commercial tools with a 65nm technology library. Finally, the throughput is computed by simulating the whole design.

5.1 Variable Latency Unit

Variable latency units, such as telescopic units [2], optimize the frequent paths of design into a faster single clock cycle, and execute infrequent critical paths in two clock cycles.

Variable latency in elastic systems can be handled in a natural way thanks to the handshake protocols. Figure 4(a) shows a variable latency unit which can take 1 or 2 clock cycles to compute. F_{approx} is an approximation of F_{exact} that can be obtained automatically and it has a shorter critical path. Most of the computation cycles, the approximation is correct ($F_{approx} = F_{exact}$), and thus, $F_{err} = 0$. Therefore, the function can be computed within a single clock cycle with no stalling. However, when the approximation is incorrect, F_{err} inserts a bubble into the receiver channel and stalls the sender. The next cycle F_{exact} can be used to finish the computation. In Figure 4(a), F_{err} is connected directly to the controller, which handles the clock gating mechanism to govern the datapath. Since F_{exact} belongs to the critical path of the original design, it is possible that F_{exact} followed by a few gates of the controller is delay critical.

An alternative implementation that takes the critical path out of the controller can be based on using speculation with replay in case of an error. The system in Figure 4(b) always speculates that the approximate computation is always correct (in which case the computation is finished in one clock cycle). If prediction is incorrect, the next clock cycle, the speculation controller will use the result of the exact computation while the early evaluation multiplexor stalls waiting for the correct data. The shaded box G is shared between the channel coming from F_{approx} and the channel coming from F_{exact} through the bubble.

We have implemented a variable latency ALU using a simple pipeline with an 8-bit datapath. In this pipeline, F_{err} has become critical in the stalling unit like in Figure 4(a), but not in the speculative design. Moreover, the speculative design (Figure 4(b)) improves the effective cycle time by 9% with a 12% area overhead. The area overhead is due to extra EBs storing the results after the shared unit.

5.2 Resilient Designs

Speculation can be used to add soft-error detection and correction in a pipeline without changing the performance of the system in case of error-free behavior. As an example, we have used the single error correction and double error detection mechanism (SECDED)[12]. For each 64 bits of data, 8 extra bits allow to detect and correct any single bit error. Besides, double bit errors are detected as well. Some implementation details of SECDED can be found in [13].

Figure 5(a) shows an adder where soft-error checking is done on each input [2]. SECDED needs a whole pipeline stage, and thus, the pipeline is deeper compared to a design with no error checking.

Speculation can be used to start the addition without waiting for SECDED to finish, as shown if Figure 5(b), after applying Shannon decomposition and sharing. The system always predicts that no errors will be found and the execution of the addition starts normally. At the same time, SECDED is computed on both inputs to detect errors in the input data. Next cycle, if SECDED detected an error, the mispredicted computation is stalled at the input of the multiplexor, and the addition is restarted using the corrected values coming from the SECDED unit.

This design has been synthesized using a 64-bit prefix-adder, and it has been checked that there is no performance penalty during the error-free behaviors. Whenever an error is detected, a single clock cycle is lost in order to correct the data and repeat the computation. This mechanism can also be used for error-protection of memories and register files. The area overhead due to speculation in Figure 5(b) is 36%, caused mainly by the recovery EBs necessary for speculation. Notice that this overhead is paid on a single pipeline stage, and hence, it would be amortized across the whole system when implemented on a real pipeline.

6. CONCLUSIONS

A novel method for applying speculation in elastic systems has been proposed. It is performed by applying Shannon decomposition and module sharing to a non-speculative design. Since both transformations are correct-by-construction, functional equivalence is preserved when applying speculation. It has been shown that speculation can be used to enhance performance of two realistic examples involving variable latency units and resilient designs.

7. REFERENCES

[1] M. Ampalam and M. Singh. Counterflow pipelining: Architectural support for preemption in asynchronous systems using anti-tokens. In *Proc. International Conf. Computer-Aided Design (ICCAD)*, pages 611–618, 2006.

[2] L. Benini, G. De Micheli, A. Lioy, E. Macii, G. Odasso, and M. Poncino. Automatic Synthesis of Large Telescopic Units Based on Near-Minimum Timed Supersetting. *IEEE TRANSACTIONS ON COMPUTERS*, pages 769–779, 1999.

[3] C. Brej. *Early Output Logic and Anti-Tokens*. PhD thesis, University of Manchester, 2005.

[4] L. P. Carloni, K. L. McMillan, and A. L. Sangiovanni-Vincentelli. Theory of latency-insensitive design. *IEEE Transactions on Computer-Aided Design*, 20(9):1059–1076, Sept. 2001.

[5] J. Cortadella and M. Kishinevsky. Synchronous elastic circuits with early evaluation and token counterflow. In *Proc. ACM/IEEE Design Automation Conference*, pages 416–419, June 2007.

[6] J. Cortadella, M. Kishinevsky, and B. Grundmann. Synthesis of synchronous elastic architectures. In *Proc. ACM/IEEE Design Automation Conference*, pages 657–662, July 2006.

[7] M. Galceran-Oms, J. Cortadella, and M. Kishinevsky. Speculation in elastic systems. Technical Report LSI-09-15-R, 2009. www.lsi.upc.edu/~techreps/files/R09-15.zip.

[8] T. Kam, M. Kishinevsky, J. Cortadella, and M. Galceran-Oms. Correct-by-construction microarchitectural pipelining. In *Proc. International Conf. Computer-Aided Design (ICCAD)*, Nov. 2008.

[9] R. Reese, M. Thornton, C. Traver, and D. Hemmendinger. Early evaluation for performance enhancement in phased logic. *IEEE Transactions on Computer-Aided Design*, 24(4):532–550, Apr. 2005.

[10] C. Soviani, O. Tardieu, and S. Edwards. Optimizing sequential cycles through Shannon decomposition and retiming. In *Proceedings of the conference on Design, automation and test in Europe.*, pages 1085–1090. European Design and Automation Association 3001 Leuven, Belgium, Belgium, 2006.

[11] I. Sutherland. Micropipelines. *Communications of the ACM*, 32(6):720–738, 1989.

[12] J. Wakerly, C. Jong, and C. Chang. *Digital Design: Principles and Practices*. Prentice Hall Englewood Cliffs, NJ, 2001.

[13] Xilinx. Single Error Correction and Double Error Detection (SECDED) with CoolRunner-II CPLDs. *Application Note XAPP383*, 1:1–4, 2003.

---

[2]For simplicity, only one of the two inputs of the adder is drawn.

978-1-60558-497-3/09 $25.00 © 2009 ACM

DFM—Don't Care or Competitive Weapon?

Mark Redford (Organizer)
Cambridge Silicon Radio
Cambridge, U.K.

Joseph Sawicki (Moderator)
Mentor Graphics
Wilsonville, OR

Prasad Subramaniam
eSilicon
Murray Hill, N.J.

Cliff Hou
TSMC
Hsinchu, Taiwan

Yervant Zorian
Virage Logic
Fremont, CA

Kimon Michaels
PDF Solutions
San Jose, CA

Categories and Subject Descriptors

B. Hardware; B.7 Integrated Circuits; B.7.2 Design Aids

General Terms: Design, Verification.

Keywords: DFM, design-for-manufacturing, manufacturing variability, sign-off, physical verification, design rules, DRC, RDR, yield.

PANEL SUMMARY

The external specifications of an IC (functions, clock rate, power consumption, etc.) determine the competitiveness of a product. To be successful and profitable in the IC business, designers need to "out-design" their competitors. Usually, Design-For-Manufacturing (DFM) is discussed as a yield improvement strategy, but what is the value of DFM from a competitive point of view? Can DFM gives designers a competitive lever by helping them decide how far to push a design without creating a manufacturing disaster? Can DFM be used to optimize designs rather than just identify hot spots?

This panel will explore these and other questions such as:
o   Signoff as a business decision—avoiding disaster while gaining competitive advantage
o   Rules vs. modeling…style or substance?
o   Do RDRs help or hinder competitive advantage?
o   Can DFM help ensure a parametric yield envelop this is competitive?
o   Design-to-Fab flow as a total quality management process

PANELIST VIEWPOINTS

Mark Redford: The advanced process technologies used for product development is becoming more complex in addition to the continued time to market pressures and the ever increasing chip development costs. The balance of meeting performance specifications, time to market & being able to successfully ramp new products in manufacturing is becoming, and will continue to be, more & more challenging.

There is a requirement for DFM at the most advanced technology nodes. Whilst manufacturing methods can mitigate many systematic issues that can impact yield &/or variability there is still the need to ensure that the integration of complex IP (e.g. digital, analog/RF & memory) blocks, including 3rd party, do not create undetected issues that may manifest themselves when the product ramps in manufacturing.

From a designers perspective the biggest challenge is how such tools integrate into the design flow: used in an efficient & effective way through the flow from cell to macro to full chip where full chip is done in a timely manner & any DFM issues are proactively addressed. For such DFM tools to be of value then the foundries need to provide statistically valid data that covers the layout styles of their product portfolio including analog/RF SoCs, memory, & digital, etc.

The success of DFM relies on a close collaboration between the foundries, tool providers & designers.

a) Foundries: accurately & statistically characterizing their technologies capturing the relevant layout styles used in their designs they will manufacture over the development, early manufacturing, & volume manufacturing phases

b) Tool providers: fast & efficient tools, meaningful information, concurrent capability with, example, P&R. And educate users (foundry/fabless) with regards to tool capabilities and main technical detail behind the tool – e.g. CAA & DLY concepts.

c) Designers: integration of tools into the flow to best captures the tools' value, etc.

Prasad Subramaniam: The relevance of DFM to the general design community is questionable. DFM plays a key role in the initial deployment of a technology by providing foundries with analysis tools and data that help them refine their manufacturing techniques. However, by the time the technology becomes available to the general design community for volume production, the foundries have accumulated enough good experience in scoping and dealing with these DFM issues to make manufacturing yields more predictable. At this stage, the benefit of exercising more DFM optimization for designs is questionable.

Permission to make digital or hard copies of part or all of this work for personal or classroom use is granted without fee provided that copies are not made or distributed for profit or commercial advantage and that copies bear this notice and the full citation on the first page. To copy otherwise, to republish, to post on servers or to redistribute to lists, requires prior specific permission and/or a fee.
DAC'09, July 26-31, 2009, San Francisco, California, USA

From a designer's perspective, the biggest hurdle facing the adoption of DFM is how it fits into the overall design flow. DFM tools are primarily used at the end of the design process today. However, in order to be effective, they need to be used in conjunction with the design process – for example, placement and routing should be DFM aware and the needed DFM optimization should be applied during the place and route flow. Even without DFM, designers face increasing pressures and challenges to get to timely tapeout and it is not realistic to lengthen their schedule by weeks for additional rounds of timing optimization to fix timing issues due to DFM optimization. Secondly, DFM tools have to be practical to be applied efficiently for full chip layout. Many DFM tools are more suitable for smaller cell or hard macro development (like the tools that involve LPC simulation) and should not be advocated for use at the full chip level since their performance and turnaround times are not acceptable. Finally, in order for designers to take a proactive role in mandating DFM optimization, we need more accurate yield measurement tools to determine their effectiveness and predict the yield improvement. Today, these tools do not provide positive results.

The success of DFM requires close cooperation between the foundry and DFM tool providers. This will become more critical as technology shrinks, since it is likely that more DFM related issues will crop up. It is not clear how much process data will be available to DFM providers and the design community. Foundries are reluctant to reveal their manufacturability issues for competitive reasons and may choose to publicly disclose only basic DFM data. In addition, the cost of improving yield may be justified only for high production volumes. As a result mainstream designers are likely to have limited or delayed access to DFM.

In summary, our view is that until the stated issues are resolved, designers are better off dealing with DFM related problems by using sound design practices up-front. Thus, while DFM certainly benefits foundries in bringing up a new technology, its benefit to mainstream designers is marginal and its usage will remain optional.

Cliff Hou: The definition of DFM in a very narrow sense is the final-stage model-based DFM check for hot spots. A general definition of DFM covers a wide spectrum from design rule definition, IP design, chip implementation, to signoff. A good DFM practice takes good balance between design rules and PPA (performance, power, area). This is to say DFM in its very essence can and should be used as competitive edge. With the advent of model-based DFM a few years ago, the focus was on an amount of hot spots that could not be eliminated by design rules alone without area tradeoff. Going forward, to have better simulation and silicon correlation, RDR has been in the picture. Even with RDR, DFM hot spots are still here to stay due to process complexity and the desired packing density, especially for the metal layers. As a result, the hybrid RDR and model-based DFM approach will be needed for many process generations to come.

To use DFM as a competitive weapon to have better simulation and silicon correlation, thorough DFM consideration needs be in place for cell/IP and chip design. Due to the stronger local layout proximity effect on circuit performance in the most advanced process nodes, DFM will play a key role to help

assess layout alternatives and trade-off. Finally, early DFM practice during chip integration instead of waiting until tapeout eliminates the last-minute surprise and the long turnaround time for DFM problem correction. Such an integrated DFM strategy is critical for early yield ramp-up.

Kimon Michaels: The issues incurred by many early designs at 45/40nm highlight that DFM is a band-aid that is losing its ability to cover the growing gap between design and manufacturing. Layout and neighborhood dependent device characteristics, silicon variability, and process complexity have pushed the traditional interface based on design rules and SPICE models beyond its limits.

To successfully and profitably overcome the challenges of advanced nodes, our industry needs to explore redefining the interface between design and manufacturing based on a set of manufacturable patterns. Our industry needs a solution for identifying the manufacturable patterns that are possible at a given node, and then selecting the optimal set of patterns that should be used to meet a given set of product needs. By shifting to a pattern paradigm, DFM considerations would be built into the layout constructs. Logic building blocks can be built by restricting layouts to manufacturable patterns and implemented in standard cell design flows. Such a system reduces variability and eliminates foreign layout patterns thereby enabling full characterization and qualification of the entire layout space. This creates a predictable and profitable path for designers to migrate to leading edge nodes.

Yervant Zorian: Today's continuous process of miniaturization results in manufacturing susceptibility that, in turn, reduces predictability and introduces yield challenges. This lengthens production ramp-up and Time-to-Volume (TTV,) and can negatively impact profitability. These nanometer-era challenges force additional requirements on IP core providers making the role of an IP core not only to be limited to simply enabling design reuse, and thus shortening the time-to-design, but instead, to include in the IP core's function extra capabilities to improve predictability, enhance manufacturability, and help attain adequate yield levels within a shorter ramp up period, thus turning the IP to a Silicon-Aware IP core. Because the Silicon-Aware IP understands the behavior of silicon and is able to address post-silicon issues, it is critical in helping maximize yield, lower test escapes, increase reliability, speed TTV, and improve overall manufacturability.

After the SoC is realized in silicon, there is a critical need for off-chip intelligence to work with the on-chip Silicon Aware IP, in order to boost silicon debug and accelerate TTV. This capability bridges the design and manufacturing disciplines to enable automated test and diagnosis vector generation, analysis of the responses back from silicon, fault isolation, and classification. This off-chip capability is typically used at the critical stages of tape out, bring up and volume manufacturing. Therefore, the main motivation to use such a manufacturing automation is to rapidly, cost effectively and accurately identify, analyze, isolate and classify faults as design are readied for transition from first silicon to volume manufacturing. This is meant to work in concert with the test, diagnosis and repair capabilities of the Silicon Aware IP to speed up SoC TTV and boost the yield percentages of good die per wafer.

The Semiconductor Industry's Nanoelectronics Research Initiative: Motivation and Challenges

Jeff Welser
SRC Nanoelectronics Research Initiative
IBM Almaden Research Center
650 Harry Road, San Jose, CA 95128
1-408-927-1017

jeff.welser@src.org

ABSTRACT

In this presentation, the scaling challenges facing current Complimentary Metal Oxide Semiconductor (CMOS) technology will be discussed, along with the ultimate limits for charge-switching based devices. From this motivation, the current status of the Nanoelectronics Research Initiative (NRI) will be discussed, with an overview of the current research topics being investigated at the NRI centers.

Categories and Subject Descriptors

B.7.1 [**Integrated Circuits**]: Types and Design Styles – *advanced technologies*.

General Terms

Management, Measurement, Experimentation, Theory.

Keywords

Nanoelectronics, logic transistors, research consortium.

1. INTRODUCTION

Since the 1970's, the ability to achieve increased performance per dollar in microprocessor chips by scaling the dimensions of the field-effect transistor (FET) in CMOS has been the driving engine behind the global semiconductor industry. However, in recent generations, exponentially increasing power density due to leakage currents as well as active switching energy of these nanoscale transistors is limiting our ability to reap the historical benefits of continued scaling. We are now forced to trade-off performance and density for reduced power consumption, and hence the fundamental physics of the CMOS transistor operation, rather than fabrication capability, will eventually be the ultimate limit for future scaling.

As the ultimate limits to the scaling of CMOS technology are getting closer, completely new approaches in emerging areas in electronics at the nanoscale need to be explored. Recognizing this critical challenge, the Nanoelectronics Research Initiative (NRI)

Permission to make digital or hard copies of part or all of this work for personal or classroom use is granted without fee provided that copies are not made or distributed for profit or commercial advantage and that copies bear this notice and the full citation on the first page. To copy otherwise, to republish, to post on servers or to redistribute to lists, requires prior specific permission and/or a fee.
DAC'09, July 26-31, 2009, San Francisco, California, USA

was chartered in 2005 by a consortium of Semiconductor Industry Association (SIA) member companies to develop and administer a university-based program to address this issue. The program is managed through the Semiconductor Research Corporation (http://www.src.org).

2. NRI MISSION

NRI Mission: Demonstrate novel computing devices capable of replacing the CMOS FET as a logic switch in the 2020 timeframe.

- These devices should show significant advantage over ultimate FETs in power, performance, density, and/or cost to enable the semiconductor industry to extend the historical cost and performance trends for information technology.

- To meet these goals, NRI is focused primarily on research on devices utilizing new computational state variables beyond electronic charge. In addition, NRI is interested in new interconnect technologies and novel circuits and architectures, including non-equilibrium systems, for exploiting these devices, as well as improved nanoscale thermal management and novel materials and fabrication methods for these structures and circuits.

- Finally, it is desirable that these technologies be capable of integrating with CMOS, to allow exploitation of their potentially complementary functionality in heterogeneous systems and to enable a smooth transition to a new scaling path.

As indicated above, the primary constraint limiting our ability to continue to scale CMOS is power density and hence circuit heat generation. Much of this power is due to the inability to continue to scale operation voltage much below one volt for any switch that relies on a potential barrier to regulate dissipative electron currents at room temperature, due to the finite threshold voltage needed to limit the off-current leakage. While circuit and architecture innovations, such as power-gating, clock-gating, and the use of multiple threshold and oxide thickness devices, can mitigate this increasing power density over the next few CMOS generations, longer term this lack of voltage scaling implies two things:

(1) Simply shrinking an FET to the far nanoscale will not necessarily continue to give the historical benefit of

scaling, as the increasing power density will require trading off switching speed for packing density.

(2) The existing Si FET roadmap is likely to be able to reach the minimum practical dimensions in the next 10-15 years, and while using new FET materials or geometries can yield improved performance, it will not alter the ultimate scaling limits.

A new "switch" for information processing which utilizes an alternative state variable is needed to significantly extend the scaling path. And this switch must be capable of lower power operation or allow energy recovery as part of the switching operation, while still operating at room temperature, to be effective.

3. NRI RESEARCH PROGRAM

The bulk of the NRI research for this new switch takes place in four multi-university, virtual centers, funded by industry, the National Institute of Standards and Technology (NIST), and state and local governments. Each of these centers focuses on a different approach to finding a post-CMOS logic switch:

- **WIN: Western Institute of Nanoelectronics, UCLA, California (Dir: Prof. Kang Wang):** Focuses solely on spintronics and related phenomena, including materials, device structures, and interconnects, for logic applications. [1-6]

- **INDEX: Institute for Nanoelectronics Discovery and Exploration, SUNY-Albany, New York (Dir: Prof. Alain Kaloyeros):** Focuses on a broad range of phenomena for logic devices, organized in centers of competency around excitonic, quantum-dot spin, magnetic, and graphene devices, with emphasis on fabrication and characterization. [7-11]

- **SWAN: SouthWest Academy for Nanoelectronics, UT-Austin, Texas (Dir: Prof. Sanjay Banerjee):** Focuses on a large graphene program, which integrates projects on theory, material fabrication, device structures, and metrology, as well as work on magnetic materials, pseudospintronics, magnetic and multi-ferroic materials, and plasmonics. [12-19]

- **MIND: Midwest Institute for Nanoelectronics Discovery, Notre Dame, Indiana (Dir: Prof. Alan Seabaugh):** Focuses on tunneling and non-equilibrium phenomena for energy efficient devices and architectures, as well as thermal phonon management. [20-25]

In addition, we partner with the National Science Foundation (NSF) to jointly fund NRI-related projects at the existing NSF Nanoscience Centers across the country. We are currently supporting projects ranging from advanced computer simulation of spin-based devices to measurements of non-equilibrium coherent transport in single-layer graphene sheets to directed self-assembly of quantum dot and wire structures for novel devices. More information on all of our research programs, with links to specific projects and research centers, is available on the NRI website: http://nri.src.org

4. STATUS AND FUTURE AREAS

Currently, the NRI research is largely still looking at basic science phenomena, with a focus on studying these new effects to understand which could potentially lead to a new logic device. This means not only investigating the basic science, but also asking the relevant questions – early on – to determine if the effect could ever make a reasonable device: How fast is the switch? How much power is consumed in a switching event? What is the on / off ratio? Can it operate at room temperature? The primary achievement of NRI has been to create a program where physicists are talking with engineers to hopefully accelerate the process of moving from a science discovery to a new device prototype.

While no clear winner has yet emerged, it is possible already to speculate on some attributes of the new device and to consider their implications on future circuits and computation systems. Since power is the primary limitation, it is likely these devices will be slower than CMOS, and hence there needs to be a focus on architectures which can still achieve high computational throughput from slower devices. Note that there is also a large potential market for slower devices, if they are fast enough for a particular product application and significantly cheaper to manufacture. However, that is in some sense the "easier" problem, since it relaxes the constraint of obtaining high performance within an increasingly tight power budget. Similarly, while there is a need for new memory devices as well, there is already a plethora of ideas and solid research efforts in academia and industry that will likely lead to breakthroughs in the future. For more information on the current assessments of many different areas of novel device research, the International Technology Roadmap for Semiconductors (ITRS) Emerging Research Devices (ERD) group has a very comprehensive report available on their website (http://www.itrs.net/Links/2007ITRS/Home2007.htm).

Hence NRI was formed to think about the "harder" problem of how to find a switch that can scale further than CMOS for high-performance logic computation – that is, to find a switch to continue the increase in computations/second/dollar that has driven the industry for half a century. Currently, the industry is moving to multi-core and parallel architectures to continue to increase throughput without increasing frequency (power), but this will only work to a point. Finding new architectures that can achieve high logic throughput with slower, lower power devices – potentially utilizing multi-bit logic devices or individual devices that perform a complex logic function in a single switching event – is extremely important, and an area of research NRI hopes to expand in the future.

Finally, while the ultimate goal is to find a switch capable of replacing (and scaling further than) CMOS, a very likely scenario for the initial adoption of any new device will be in circuits designed for specific applications. If a non-CMOS device / circuit combination was significantly more efficient for an important compute-intensive function (e.g. pattern recognition, encryption, etc.), these could act as accelerators on a chip, with CMOS likely remaining the primary switch for general purpose computation initially. NRI is just starting to do analysis work in this area.

5. ACKNOWLEDGMENTS

Our thanks to the NRI member companies, AMD/GlobalFoundries, IBM, Intel, Micron, and Texas Instruments, as well as to our federal government partners, NSF and NIST, and our state and local government partners, including California, Indiana, New York, Texas, and the City of South Bend.

6. REFERENCES

[1] Y. Acremann, X. W. Yu, A. A. Tulapurkar, A. Scherz, V. Chembrolu, J. A. Katine, M. J. Carey, H. C. Siegmann, and J. Stöhr, "An amplifier concept for spintronics," *Applied Physics Letters*, vol. 93, 102513, 2008.

[2] Y.-H. Chu, L. W. Martin, M. B. Holcomb, M. Gajek, S.-J. Han, Q. He, N. Balke, C.-H. Yang, D. Lee, W. Hu, Q. Zhan, P.-L. Yang, A. Fraile-Rodriguez, A. Scholl, S. X. Wang and R. Ramesh, "Electric-field control of local ferromagnetism using a magnetoelectric multiferroic," *Nature Materials,* vol. 7, June 2008.

[3] K. C. Hall and M. E. Flatté, "Performance of a spin-based insulated gate field effect transistor," *Applied Physics Letters*, vol. 88, 162503, 2006.

[4] D. B. Carlton, N. C. Emley, E. Tuchfeld, and J. Bokor, "Simulation Studies of Nanomagnet-Based Logic Architecture," *Nano Letters*, vol. 8, pp. 4173-4178, Dec 2008.

[5] A. Khitun, M. Bao, and K. L. Wang, "Spin Wave Magnetic NanoFabric: A New Approach to Spin-Based Logic Circuitry," *IEEE Transactions on Magnetics*, vol. 44, no. 9, Sep 2008.

[6] J. D. Koralek, C. P. Weber, J. Orenstein, B. A. Bernevig, Shou-Cheng Zhang, S. Mack, and D. D. Awschalom, "Emergence of the persistent spin helix in semiconductor quantum wells," *Nature*, vol. 458, pp. 610-613, Apr 2009.

[7] S. Salahuddin and S. Datta, "Use of negative capacitance to provide voltage amplification for ultralow power nanoscale devices," *Nanoletters*, vol. 8, p. 405, 2008.

[8] J. R. Williams, L. DiCarlo, and C. M. Marcus, "Quantum Hall Effect in a Gate-Controlled p-n Junction of Graphene," *Science*, vol. 317, p. 638, 2007.

[9] J. Hass, W. A. de Heer, and E. H. Conrad, "The Growth and Morphology of Epitaxial Mulitlayer Graphene," *J. Phys.: Condens. Matter*, vol. 20, 323202, July 2008.

[10] C. A. Ross, F. J. Castano, W. Jung, B. G. Ng, I. A. Colin and D. Morecroft, "Magnetism in multilayer thin film rings," *J. Phys. D: Appl. Phys.*, vol. 41, 113002, 2008.

[11] C. A. F. Vaz, J. Hoffman, A.-B. Posadas, and C. H. Ahn, "Modulation of magnetic anisotropy of magnetite in Fe3O4/BaTiO3(100) epitaxial structures," *Appl. Phys. Lett.*, vol. 94, 022504, 2009.

[12] S. K. Banerjee, L. F. Register, E. Tutuc, D. Reddy and A. H. MacDonald, "Bilayer pseudoSpin Field-Effect Transistor (BiSFET): A proposed new logic device," *IEEE Electron Device Letters*, vol. 30, pp. 158-200, Feb 2009.

[13] B. Sahu, H. Min, A. H. MacDonald, and S. K. Banerjee, "Energy gaps, magnetism, and electric-field effects in bilayer graphene nanoribbons," *Phys. Rev. B*, vol. 78, 045404, 2008.

[14] S. Kim, J. Nah, I. Jo, D. Shahrjerdi, L. Colombo, Z. Yao, E. Tutuc, and S.K. Banerjee, "Realization of a high mobility dual-gated graphene field-effect transistor with Al2O3 dielectric," Applied Physics Letters, vol. 94, 062107, 2009.

[15] X. Li, W. Cai, J. An, S. Kim, J. Nah, D. Yang, R. Piner, A. Velamakanni, I. Jung, E. Tutuc, S. K. Banerjee, L. Colombo, and R. S. Ruoff, "Large-Area Synthesis of High-Quality and Uniform Graphene Films on Copper Foils," *Science*, DOI: 10.1126/science.1171245, May 2009.

[16] H. Min, R. Bistritzer, J.-J. Su, and A. H. MacDonald, "Room-Temperature Superfluidity in Graphene Bilayers," *Phys. Rev. B*, vol. 78, 121401(R), 2008.

[17] J. Wunderlich, T. Jungwirth, A. C. Irvine, J. Zemen, A. W. Rushforth, E. De Ranieri, U. Rana, K. Vyborny, Jairo Sinova, C. T. Foxon, R. P. Campion, D. A. Williams, and B. L. Gallagher, "Local control of magnetocrystalline anisotropy in (Ga,Mn)As: application in spin-transfer-torque microdevices," *Phys. Rev. B*, vol. 76, 054424, 2007.

[18] B. Lee, H.-C. Kim, M. Kim, R. M. Wallace, E. M. Vogel and J. Kim, "Smooth and uniform Al2O3 dielectric layer deposited by atomic layer deposition for graphene-based nanoelectrics," *Applied Physics Letters*, vol. 92, 203102, 2008.

[19] Y. G. Semenov, K. W. Kim, and J. M. Zavada, "Spin Field Effect Transistor with a Graphene Channel," *Applied Physics Letters*, vol. 91, 153105, 2007.

[20] Q. Zhang, T. Fang, H. Xing, A. Seabaugh, and D. Jena, "Graphene nanoribbon tunnel transistors," *IEEE Electron Device Letters*, vol. 29, pp. 1344-1346, 2008.

[21] J. Lee, A. Liao, E. Pop, and W. King, "Electrical and Thermal Coupling to a Single-Wall Carbon Nanotube Device Using an Electrothermal Nanoprobe," *Nano Letters*, vol. 9, p. 1356, 2009.

[22] S. Mookerjea and S. Datta, "Comparative study of Si, Ge, and InAs based steep subthreshold slope tunnel transistors for 0.25 V supply voltage logic applications," *Device Research Conference*, June 2008. Santa Barbara, CA.

[23] T. Fang, A. Konar, H. Xing, and D. Jena, "Mobility in semiconducting graphene nanoribbons: Phonon, impurity, and edge roughness scattering," *Phys. Rev. B*, vol. 78, 205403, 2008.

[24] A. Liao, Y. Zhao, and E. Pop, "Avalanche-induced current enhancement in semiconducting carbon nanotubes," *Phys. Rev. Lett.*, vol. 101, 256804, 2008.

[25] M. Alam, S. Kurtz, M. Niemier, S. Hu, G. Bernstein, and W. Porod, "Magnetic logic based on coupled nanomagnets: Clocking structures and power analysis," *IEEE NANO 2008*, Aug 2008. Arlington, TX.

978-1-60558-497-3/09 $25.00 © 2009 ACM

Single-Electron Devices for
Ubiquitous and Secure Computing Applications

Ken Uchida
Tokyo Institute of Technology
2-12-1-S9-12, O-okayama
Meguro-ku, Tokyo, 152-8552, Japan
+81-3-5734-3591

uchidak@neo.pe.titech.ac.jp

ABSTRACT

Single-electron transistors (SETs) show promise as future functional elements in LSIs, because of their low-power consumption and small size. However, the low driving strength and the oscillating Id-Vg characteristic of SETs make them difficult to use in LSIs as alternatives of conventionally used MOS transistors. As electronics becomes ubiquitous, LSIs are expected to be more widely used in ultra-small low-power electronic devices for various applications such as electronic tags, credit cards, and so on. This strong demand in new application domain leads to increased requirements against LSIs; high functionality, high security and low power-consumption are more strongly required for LSIs than ever. As a result, the change of target applications could offer an opportunity to "new" functional electronic devices such as SETs. This paper firstly gives a strategy for using SETs in conventional logic circuits in order to reduce power consumption. Then, new application domains where SETs could be used by making use of their characteristics, such as high-charge sensitivity and oscillating Id-Vg characteristics are discussed.

B.7.m [**INTEGRATED CIRCUITS MISCELLANEOUS**]

General Terms: Performance, Design

Keywords: Nanotechnology, LSIs, single-electron, true random number generator

1. INTRODUCTION

Single-electron transistors (SETs) are promising as future functional elements in LSIs because of their low-power consumption and small size. Therefore, many attempts have been made to fabricate room-temperature operating SETs in order to replace MOSFETs with SETs. Thanks to the great efforts offered by many researchers, room-temperature operation of SETs is now possible. However, even if the room-temperature operating SETs are successfully fabricated, the low driving strength and the oscillating Id-Vg characteristics of SETs still make it difficult to

Permission to make digital or hard copies of part or all of this work for personal or classroom use is granted without fee provided that copies are not made or distributed for profit or commercial advantage and that copies bear this notice and the full citation on the first page. To copy otherwise, to republish, to post on servers or to redistribute to lists, requires prior specific permission and/or a fee.
DAC'09, July 26-31, 2009, San Francisco, California, USA

utilize them in LSIs as alternative of advanced MOSFETs. Therefore, logic design scheme suitable for SETs has to be developed.

On the other hand, in a ubiquitous computing, LSIs are expected to be widely used not only in high-performance computers, PCs, PDAs and cell phones but also in ultra-small novel electronic devices for various applications such as electronic tags, credit cards, and so on. As a result, the change or the widening of target applications could offer an opportunity to "new" functional electronic devices such as SETs, because SETs have different functionality, such as oscillating characteristics and extremely high charge sensitivity, from those of conventional MOSFETs.

In this paper, we firstly describe the structure and operation of SETs. Then, methodology to design logic circuits with SETs is discussed. Programmable logic circuits utilizing SETs will be introduced. Finally, as an example of an application to a new application domain, true random-number generators (TRNGs) realized with SETs will be presented.

2. STRUCTURE AND OPERATION OF SINGLE-ELECTRON TRANSISTORS

The structure and operation of single-electron transistors are briefly reviewed. Fig. 1 shows the schematic of single-electron transistors (SETs). As shown in the figure, SETs are three-terminal devices. Therefore, the structure of SETs resembles to that of MOSFET. However, SETs have quantum dots and tunnel junctions as counterparts to the channel and pn junctions of MOSFETs. The smaller quantum dots results in higher operation temperature. In other words, the diameter of quantum dots should be less than 5 nm in order to achieve room-temperature operation. The resistance of tunnel junction should be much greater than 25 $k\Omega$; typical resistance of tunnel junctions of Si-based SETs is approximately 1 $M\Omega$. Because of this high resistivity, the driving strength of SETs is intrinsically low.

Furthermore, Id-Vg characteristics of SETs are totally different from those of MOSFETs. SETs show oscillating Id-Vg characteristics as shown in Fig. 2(a), whereas MOSFETs show monotonic increase of drain current with an increase in gate voltage. In Fig. 2(a), two current peaks are shown.

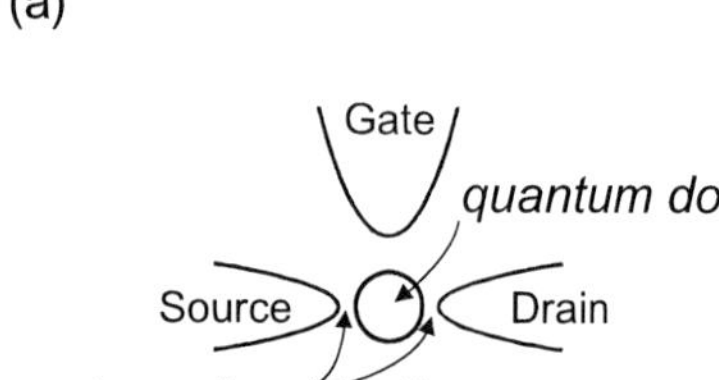

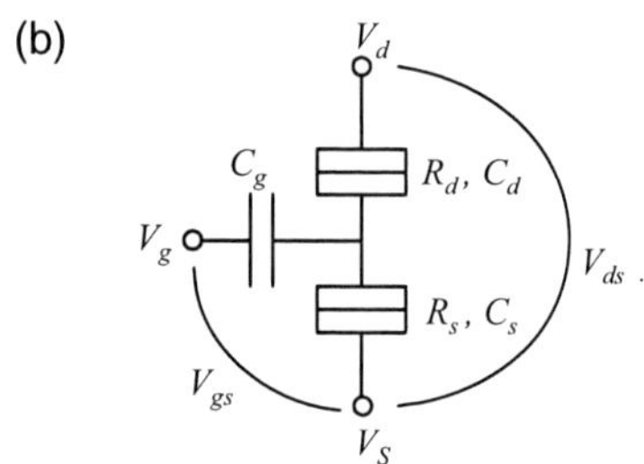

Figure 1. (a) Schematic structure of single-electron transistor. (b) Equivalent circuit model of single-electron transistor. [1]

3. DESIGN SCHEME OF SET LOGIC CIRCUITS

In this section, we introduce a design scheme for SET logic circuits [2]. SETs have a relatively low gain and cannot be used both as pull-up devices and pull-down devices at the same time. In addition, source-to-drain voltage of SETs should be much smaller than the voltage swing applied to gate electrode in order to suppress the leakage current through source-to-drain.

First, logic trees should consist of pull-down SETs only. The SET logic tree implements logic operations. For example, two cascaded switches operate the logic function of AND (NAND) and two parallel SET switches operate the logic function of OR (NOR). By using complementary input, SET trees can implement all the logic operations. Furthermore, the output voltage swing varies from ground to supply voltage, regardless of the number of cascaded SETs, because the logic tree contains no pull-up SETs.

Second, the supply voltage to SET logic trees should be lower than the gate voltage swing of SETs. This is a requirement in order that the SETs can switch the current off, as described above.

Third, one pull-down device controlled by the clock is connected between the ground and the SET logic trees. One pull-up device controlled by the inverse clock is connected between the power supply and the SET logic tree. This guarantees the cyclic operation of the circuit. The pull-up device may be a p-type metal-oxide-semiconductor field-effect-transistor (pMOSFET) or a SET.

Finally, the output voltages of logic trees are amplified to the same voltage as the gate voltage swing of SETs with conventional electronic devices, such as MOSFETs and bipolar transistors, in order to drive the next gates. In any event, these conventional devices are mandatory to compensate for the poor driving capability of SETs.

The proposed SET logic circuit consists of the SET logic tree, a lower supply voltage than the gate voltage swing of SETs, the pull-down switch, the pull-up switch, the clock, and the CMOS amplifier biased by the same supply voltage as the gate voltage swing of SETs. The circuit operation is quite similar to that of the CMOS dynamic logic, such as the P-E logic and CMOS domino logic. In other words, our design scheme of SET logic resembles that of the CMOS dynamic logic, except for the following. First, logic trees consist of SETs. Second, the trees are biased by a lower supply voltage than the input voltage swing of the switching elements. Finally, a voltage amplifier is mandatory. Therefore, the methodology of the CMOS dynamic logic is useful for designing the SET logic.

4. PROGRAMMABLE SET LOGIC

In this section, we propose programmable SET logic consisting of SETs with non-volatile function [3], where the oscillating Id-Vg properties of SETs are fully utilized and the number of switching elements can be greatly reduced.

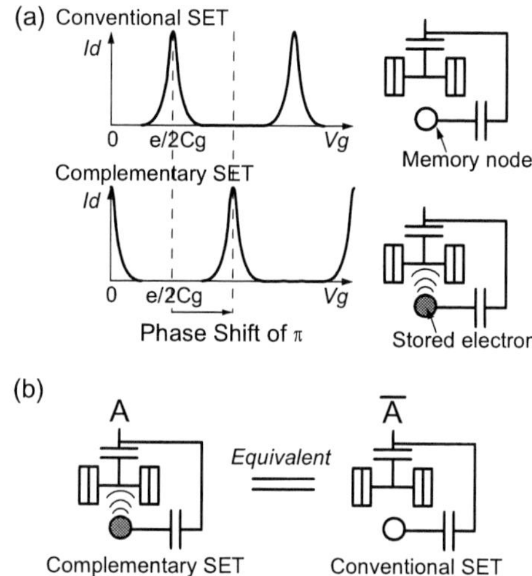

Figure 2. Comparison of conventional SET with complementary SET. SETs have a memory function controlled by the gate voltage. (a) The complementary SET is realized by injecting electrons to the memory node so that the phase shift of Coulomb oscillations becomes π. (b) The operation of the complementary SET is equivalent to that of conventional SET to which inverted signal is inputted. [3] © IEEE 2003.

A SET with a nonvolatile memory function is a key element for the programmable SET logic. The electron injection into the memory nodes makes it possible to tune the phase of Coulomb oscillations, and thus transform the functions of SETs. In fact, the phase shift of π enables realization of SETs complementary to the conventional SETs (Fig. 2). It is therefore suggested that SETs with a nonvolatile memory function have the functionalities of both conventional SETs and complementary SETs. Thus, a logic circuit consisting of SETs with a memory function has multiple functions, from which the desirable one can be selected by programming with the memory function.

5. SINGLE-ELECTRON RANDOM-NUMBER GENERATOR

As we discussed, SETs are very promising as functional elements in logic circuits. However, SETs have much higher charge sensitivity than MOSFETs do. If we can utilize this functionality aggressively, it might be possible for SETs to realize a new function which is important in ubiquitous era.

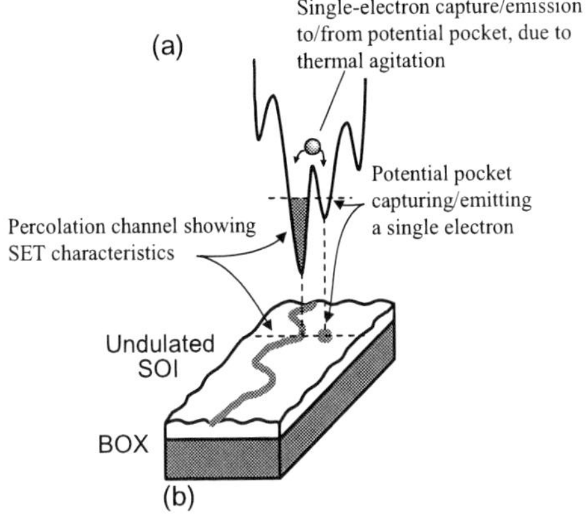

Figure 3. Schematic of proposed single-electron random number generator. [4] © IEEE 2002.

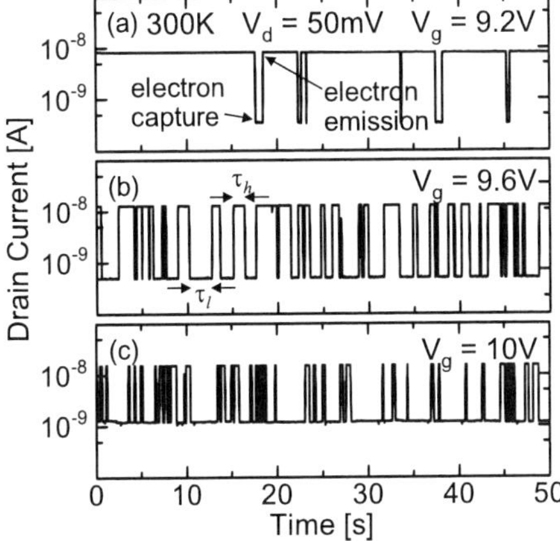

Figure 4. Time dependent Id characteristics ofsingle-electron random number generator.[4] © IEEE 2002.

Here, we propose a new true RNG (TRNG), which consists of a single-electron transistor (SET) and a single-electron trap, called single-electron TRNG [4]. The importance of TRNG in secure application is discussed in several references [5]. The single-electron TRNG is fabricated in an undulated ultrathin SOI film. The nanoscaled undulation in the ultrathin film results in the formation of nanoscale potential fluctuations (Fig. 3a). Thus, both the narrow percolation channel, acting as an SET, and small potential pockets, acting as an electron trap, are formed in the film (Fig. 3b). The trap is designed to stochastically capture or emit a single electron, and thus, the trap makes SET characteristics telegraphic as shown in Fig. 4. Since SET has high charge sensitivity, these telegraphic signals are extremely large (~0.3V@10nA) and can be directly treated as random digital sequences.

The time-dependence of Id clearly depicts RTSs (random-telegraphic signals). It is noteworthy that the RTS ratio (ΔI/I) is greater than one decade, which is the highest room-temperature RTS ratios ever reported. This high value is due to the high charge sensitivity of SETs and the narrow percolation channel.

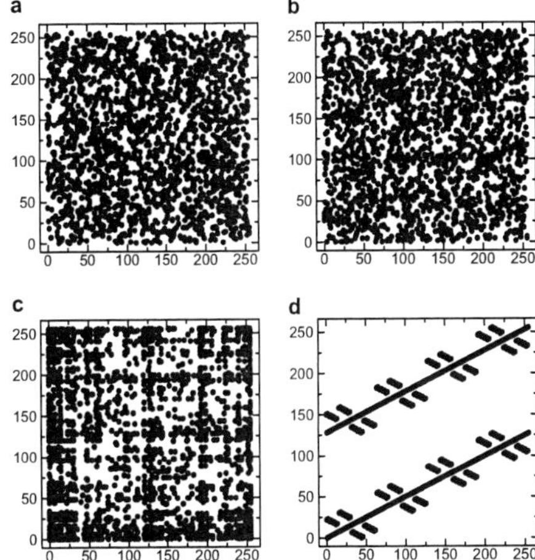

Figure 5. Self-correlation plots for sequential 8-bit random numbers. Randomness is expressed as the randomness of the dot distribution. (a) Noise-based RNG, (b) single-electron RNG, (c) oscillator-based RNG , and (d) pseudo RNG (typical LFSR consisting of 16 shift resistors). The oscillator-based RNG is another common RNG implemented in LSI chips. The oscillator-based RNG utilizes a free-running oscillator, which shows timing jitter caused by thermal noises in transistors. [4] © IEEE 2002.

The quality of random number sequence is evaluated using statistical tests (Fig. 5). It is demonstrated that the single-electron RNG generate high-quality random digital sequence.

6. ACKNOWLEDGMENT

The author would like to thank Prof. A. Toriumi, Drs. S. Fujita, R. Ohba, T. Tanamoto, and A. Ohata for valuable discussions. This work was partly performed under the management of NEDO. The RNG work was partly performed under the management of TAO.

7. REFERENCES

[1] K. Uchida, K. Matsuzawa, J. Koga, R. Ohba, S. Takagi, and A. Toriumi, Jpn. J. Appl. Phys., vol. 39, no. 4B, pp.2321-2324, 2000.

[2] K. Uchida, K. Matsuzawa, and A. Toriumi, Jpn. J. Appl. Phys., vol. 38, no. 7A, pp.4027-4032, 1999.

[3] K. Uchida, J. Koga, R. Ohba, and A. Toriumi, IEEE Trans. Electron Devices, vol. 50, no. 7, pp.1623-1630, 2003.

[4] K. Uchida, T. Tanamoto, R. Ohba, S. Yasuda, and S. Fujita, Technical Digest of International Electron Devices Meeting (IEDM), pp.177-180, 2002.

[5] C. Tokunaga, D. Blaauw, and T. Mudge, "True random number generator with a metastability-based quality control," Technical Digest of International Solid-State Circuit Conference, pp404-405, 2007.

Digital VLSI Logic Technology using Carbon Nanotube FETs:
Frequently Asked Questions

Nishant Patil, Albert Lin, Jie Zhang, H.–S. Philip Wong, Subhasish Mitra
Department of Electrical Engineering and Department of Computer Science
Stanford University, Stanford, CA

Abstract

Carbon Nanotube Field-Effect Transistors (CNFETs) show promise as extensions to silicon-CMOS. Ideal CNFET circuits can potentially provide 20X Energy-Delay-Product benefits over silicon-CMOS at the 16 nm technology node. However, several challenges must be overcome before such performance benefits can be experimentally realized. In this paper, we present a brief overview of CNFET technology, and address commonly raised concerns through a series of Frequently Asked Questions (FAQs). We also provide a CNFET technology outlook which includes a survey of challenges as well as existing and potential solutions to these challenges.

Categories and Subject Descriptors
B.7 [Hardware]: Integrated Circuits

General Terms
Algorithms, Performance, Design, Reliability, Experimentation.

Keywords
Carbon Nanotubes, Carbon Nanotube Transistor, CNFET.

1. Introduction

Carbon Nanotube Field Effect Transistors (CNFETs) are promising candidates as extensions to silicon-CMOS due to excellent device characteristics [Javey 03a, 03b, 05]. Figure 1.1 shows the layout of a CNFET inverter. The gate, source and drain contacts, and interconnects are defined by conventional lithography. The distances between gates and contacts are limited by the lithographic feature size (lithographic pitch in Fig. 1.1, e.g., 64 nm for 32 nm technology node). Since Carbon Nanotubes (CNTs) [1] are grown using chemical synthesis, the inter-CNT distance (sub-lithographic pitch in Fig. 1.1) is not limited by lithography. With CNFETs, a large portion of the existing design and manufacturing infrastructure for FET-based VLSI systems can be utilized. There has been significant progress in CNFETs at the single-CNT level [Javey 03a, 03b, 05]. However, a major gap exists between such CNFETs with single CNTs and the transformation of these results into practical CNFET VLSI technologies competitive with scaled silicon-CMOS.

In this paper, we answer several frequently asked questions about CNFET circuits (Sec. 2). In Sec. 3, we present experimental results from fabricated CNFET circuits. Section 4 concludes the paper and presents a CNFET technology outlook.

2. Frequently Asked Questions about CNFET circuits

Question 1: What are the speed and energy benefits of CNFET circuits?

Simulation results using the Stanford University CNFET Model show that an "ideal" CNFET inverter (Fig. 1.1) using a perfect CNFET technology can be 5.1 times faster, and can

consume 2.6 times lower energy per cycle (i.e., 13X Energy-Delay-Product benefit) compared to a 32 nm silicon CMOS inverter [Deng 07a, Patil 09a]. The model has been calibrated against experimental CNFET data to within 90% accuracy (see Question 3 below). Figure 2.1 show the estimated improvements of CNFET circuits over silicon CMOS for several technology nodes. At the 16 nm and 22 nm technology nodes, the energy-delay product (EDP) advantage over silicon CMOS can improve to 20X [Patil 09a]. These improvements over Silicon-CMOS are due primarily to near-ballistic electrical transport in CNTs [Guo 04, Javey 03a, 03b, 05].

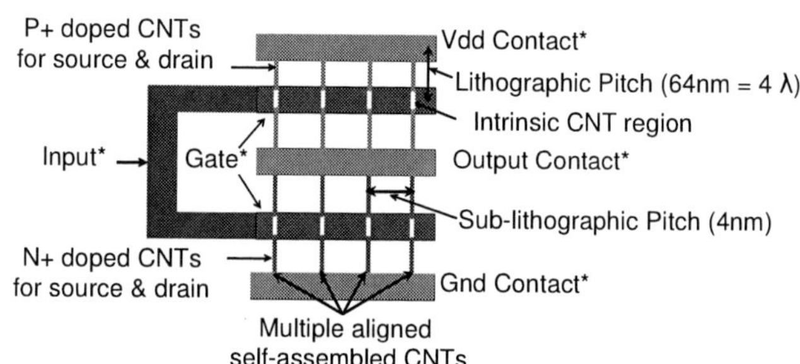

Figure 1.1. CNFET inverter.

Question 2: What are the area benefits of CNFET circuits?

CNFET technologies relying on lithography will not provide significant area benefits compared to silicon CMOS. Although CNTs have diameters between 0.5-3 nm [Kang 07, Patil 08a], CNFET circuit size is not determined only by the diameter of a CNT. As shown in Fig. 1.1, gates, contacts and interconnects are defined using lithography. Hence the spacing between them is limited by the minimum lithographic pitch, similar to today's silicon-CMOS technology.

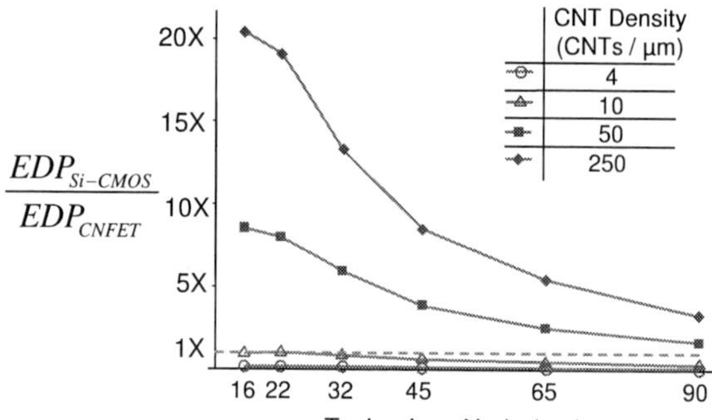

Figure 2.1. CNFET Energy-Delay-Product advantage over silicon-CMOS over several technology nodes.

Question 3: Why does one need multiple CNTs per CNFET?

Multiple CNTs are needed per CNFET to provide sufficient drive current [Deng 07a, Patil 09a]. This situation is similar to conventional multi-gate silicon FETs [Tang 01]. Multiple CNTs are also needed for reliable CNFETs if the CNT electronic properties (e.g. CNT diameter, CNT type (metallic or semiconducting), CNT doping level etc.) are not perfectly controlled (e.g. after metallic CNT removal techniques [Zhang 08,

---

[1] In this paper, all CNTs are Single-Walled Carbon Nanotubes (SWNTs) [Saito 98].

Permission to make digital or hard copies of part or all of this work for personal or classroom use is granted without fee provided that copies are not made or distributed for profit or commercial advantage and that copies bear this notice and the full citation on the first page. To copy otherwise, to republish, to post on servers or to redistribute to lists, requires prior specific permission and/or a fee.
DAC'09, July 26-31, 2009, San Francisco, California, USA

978-1-60558-497-3/09 $25.00 © 2009 ACM

09a] or to control CNT diameter and CNT source and drain doping variation [Deng 07a, Patil 09a]).

Question 4: How does one design CNFET-based circuits? Conventional silicon-CMOS device models are not applicable for CNFETs. Are CNFET device models available?

Several research groups have developed various CNT models to address different aspects of CNTs from device physics to circuit level modeling [Balijepalli 07, Guo 04]. In particular, we have developed the Stanford University CNFET Model[2], a compact model for circuit simulation [Deng 06, 07b, 08]. This model captures the essential physics of the band structure, near-ballistic carrier transport, inter-channel screening, and parasitic capacitances of the device structure. This model has been calibrated against experimental CNFET data to within 90% accuracy [Amlani 06]. Using this compact model, we performed circuit simulations to compare the energy and delay of CNFET circuits with silicon-CMOS in the presence of various sources of variations [Deng 07a, Patil 09a]. In addition, analytical models of the density-of-states, effective mass, carrier density, and CNFET I-V characteristics have also been developed [Akinwande 08b, Liang 08] to aid in the development of analog and digital CNFET circuits [Akinwande 09].

Question 5: How are CNTs synthesized?

CNTs are synthesized (grown) through chemical vapor deposition (CVD) using metal nanoparticles as catalyst (typically iron or cobalt) at high temperatures (700°C – 850°C) and flowing a carbon source (e.g. methane or ethanol) over the substrate [Dai 02]. CNTs can be grown on silicon substrates as well as single-crystal substrates such as quartz [Kang 07] and sapphire [Han 05]. CNTs grown on single-crystal quartz substrates have significantly better alignment as compared to those grown on silicon [Kang 07].

It is often necessary to decouple CNT synthesis from CNFET fabrication. For example, CNT growth is conducted at high temperatures (>850°C); but silicon-CMOS or organic substrates will not be able to tolerate such high temperatures In this case, wafer-scale CNT transfer [Patil 08a, 09b] can be used to transfer CNTs from quartz substrates (for growth) to silicon substrates (for potential integration with silicon-CMOS), while preserving the density and alignment of CNTs [Patil 08a, 09b]. Thus, low temperature CNT transfer (90°C - 120°C) allows one to de-couple the growth substrate from the device fabrication substrate [Kang 07, Patil 08a, 09b].

Question 6: Can CNTs be integrated with silicon-CMOS?

Yes. CNTs can be integrated with silicon CMOS, as CNFETs [Akinwande 08] and as CNT interconnect [Close 08]. Process temperatures must be below 400°C for monolithic integration with silicon CMOS [Wong 07]. Techniques such as dielectrophoresis [Close 08a] and CNT transfer are applicable [Kang 07, Patil 08a, Patil 09b]. In [Akinwande 08a], a CNFET is integrated with silicon-CMOS in a cascode amplifier. In [Close 08b], multi-wall CNTs are integrated into a silicon-CMOS ring oscillator that oscillates rail-to-rail at 1GHz – a speed that is commensurate with digital VLSI applications today.

Question 7: How is the problem of CNT positioning and alignment solved?

A *mis-positioned CNT* is a CNT that passes through a layout region where a CNT was not intended to pass. This may be caused due to misalignment or due to lack of control of correct positioning of CNTs during CNT growth. A large fraction (99.5%) of the CNTs grown on single-crystal quartz substrates grow aligned i.e. straight and parallel to each other [Kang 07]. Even for CNTs on quartz substrates, a non-negligible fraction of CNTs are misaligned (Fig. 2.2). It is very difficult to correctly position and align **all** the CNTs for all CNFETs for VLSI. Mis-positioned CNTs can cause shorts (Fig. 2.2) and incorrect logic functionality in CNFET logic circuits (Fig. 2.3). Using design principles described in [Patil 07, 08b] we experimentally demonstrated in [Patil 08a, 08b] circuits that are immune to a large number of mis-positioned CNTs. We refer to these circuits as *mis-positioned-CNT-immune* circuits. This is accomplished by etching CNTs from regions pre-defined during layout design, so that any mis-positioned CNT cannot result in incorrect logic functionality. This technique is VLSI-compatible since it does not require any die-specific customization or test and reconfiguration. **All** the logic cells have the CNTs removed in the predefined layout regions using oxygen plasma [Patil 08a, 08b]. This is feasible since the penalties associated with this technique are small (see Question 8). Figure 2.4 shows the layout of a complex logic function design using the mispositioned-CNT-immune design. With sufficiently high CNT density, along with the use of mis-positioned-CNT-immune design, CNT alignment and positioning is no longer a problem.

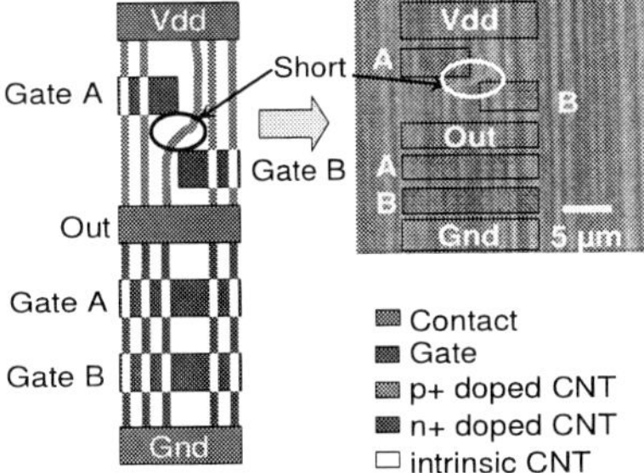

Figure 2.2 Mis-positioned CNT causing a short in NAND logic

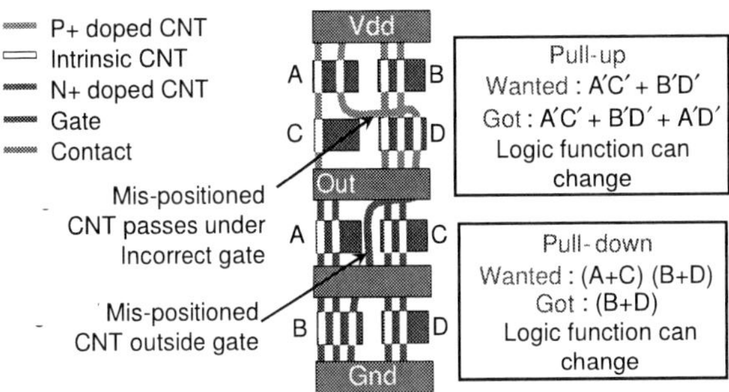

Figure 2.3. Incorrect Logic Functionality caused by mis-positioned CNTs.

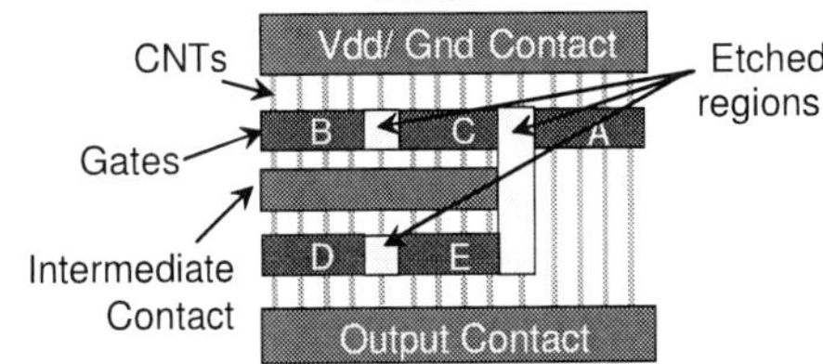

Figure 2.4. Mis-positioned-CNT-immune layout for network represented by the function A+(B+C) (D+E).

Question 8: Do I have to use expensive defect- and fault-tolerance techniques to deal with mis-positioned CNTs?

The worst-case (for minimum sized CNFET logic cells) energy, delay and area penalties of mis-positioned-CNT-immune logic cells are 18%, 13% and 21%, respectively, compared to designs that are not mis-positioned-CNT-immune [Patil 07, 08b].

---

[2] The Stanford University CNFET Model can be downloaded from the Stanford Nano Website: http://nano.stanford.edu/models.php.

These penalties decrease for non-minimum sized standard cells. These penalties are significantly lower than traditional defect-and-fault-tolerance techniques [Dehon 05, Goldstein 01, Rad 06, Rao 05, Tahoori 06]. Furthermore, this design technique is VLSI-compatible since it does not require changes to the existing VLSI design flow [Bobba 09]. There is no need to test or reconfigure for mis-positioned CNTs (logic cells are designed to be immune to mis-positioned CNTs by construction [Patil 07, 08b]).

Question 9: What are metallic CNTs? Why do they cause problems?

Metallic CNTs are also made of carbon, like any other CNT. Depending upon the arrangement of carbon atoms in the CNT (chirality), a CNT can be semiconducting (with a large bandgap, e.g. 0.5 eV for a semiconducting CNT with diameter ~1.5 nm) or metallic (with zero or almost zero bandgap, e.g. ~10 meV) [Saito 98]. Unlike semiconducting CNTs, the conductivity of a metallic CNT is relatively high and cannot be controlled by the gate. This creates source-drain shorts in CNFETs resulting in excessive leakage and severely degraded noise margins in logic gates [Zhang 08, 09a].

Question 10: What can we do about metallic CNTs?

The best way to solve the metallic CNT (m-CNT) problem is to have predominantly semiconducting CNTs (s-CNTs) on the substrate to start with. Typical CNT synthesis techniques yield ~33% metallic CNTs. A preferential CNT growth technique is used by [Li 04] which yields 90% s-CNTs. More recently, a growth technique by [Qu 08] reported 96% s-CNTs (4% m-CNTs). Reduction in percentage of m-CNTs has also been demonstrated by self-sorting when CNTs are deposited onto the wafer ([Engel 08, LeMieux 08]). Such solution-based CNT sorting techniques yield 1-10% m-CNTs [Engel 08, LeMieux 08]. While such improvement in percentage of m-CNTs (i.e., lower percentage of m-CNTs) is necessary, it alone is not enough for VLSI-scale digital circuits.

For VLSI CNFET circuits to meet leakage, noise margins, and delay variation targets, the percentage of m-CNTs must be reduced to less than 0.01% [Zhang 08, 09a]. In the absence of 99.99% s-CNT growth, the other processing option is to remove m-CNTs after CNT growth from an ensemble of m-CNTs and s-CNTs on the substrate. [Collins 01] introduces a current-induced electrical breakdown technique to achieve m-CNT removal. In this technique, s-CNTs are switched off using gate voltage so that current only flows through m-CNTs. At high current levels, oxidation is induced by self-heating of m-CNTs causing them to break down. This approach has the advantage of high removal rate of m-CNTs [Collins 01, Lin 09a]. However, a major drawback for this technique is that it is not scalable for large-scale VLSI systems since it involves individually contacting and breaking down m-CNTs in each CNFET. Compared to electrical breakdown, gas phase chemical reaction-based removal techniques ([Hassanien 05, Yang 06, Zhang 06]) can be more easily applied on a wafer scale. Although high removal rates may be achieved by m-CNT removal techniques, these techniques also inadvertently remove some (useful) s-CNTs, thereby reducing the on-current of a CNFET. Hence, co-optimization of m-CNT removal and circuit design is necessary for enabling robust digital logic using CNFETs [Zhang 08, 09a]. Furthermore, m-CNT tolerant circuits may be used to mitigate the impact of m-CNTs [Lin 09b].

Question 11: Is perfect metallic-CNT removal sufficient for VLSI CNFET technology?

One can define *perfect m-CNT removal* as a process that removes all m-CNTs and preserves all s-CNTs. Since today's CNT synthesis [Dai 02, Kang 07, Qu 08] or CNT sorting techniques [Engel 08, LeMieux 08] do not produce predominantly s-CNTs (>99.99% s-CNTs), m-CNT removal after CNT growth is necessary. However, even with perfect m-CNT removal, since the number of m-CNTs grown is stochastic [Borkar 05, Deng 07a, Patil 09a, Zhang 08, 09a]; there is a non-zero probability that a CNFET will have all m-CNTs. As a result, all the CNTs in such a CNFET will be removed during perfect m-CNT removal creating an open circuit. This probability sets a lower limit on the minimum number of CNTs per CNFET for a required probability of CNFET failure. Further details, including the impact of a realistic non-ideal metallic CNT removal (which does not remove all m- CNTs and removes some s-CNTs), can be found in [Zhang 08, 09a].

Question 12: How well can you control CNT characteristics e.g., CNT diameter, CNT source and drain doping levels, and chirality (metallic or semiconducting)?

Tight control of the CNT characteristics such as CNT diameter and CNT source and drain doping levels will help control the threshold voltage and current variation in a CNFET with multiple CNTs. However, the impact of CNT diameter and CNT source and drain doping variations [Deng 07a, Patil 09a] is not as significant as the variations caused by the presence of metallic CNTs and metallic CNT removal processes [Zhang 08, 09a]. With more than 8 CNTs per CNFETs, the variation in CNFET performance caused due to CNT diameter variation and CNT source and drain doping variations can be reduced to within 5% because of statistical averaging effects [Deng 07a, Patil 09a]. However, design and process co-optimization is required.

Question 13: What is the CNT density needed for high-performance CNFETs?

High CNT density (250 CNTs/µm) is critical to ensure that performance (delay and energy) gains over silicon-CMOS are maintained or improved with shrinking lithographic dimensions [Patil 09a]. This CNT density is about 5-25X higher CNT density than what can be achieved with current CNT synthesis techniques. The average density of CNTs obtained today is 10-50 CNTs/µm [Kang 07, Patil 08a, 09b]. Advances in CNT synthesis are essential to improve this average density from this value to the required density of 250 CNTs/µm. However, mere increase in average CNT density is not enough. CNT density variations, resulting from the lack of precise control of CNT location during synthesis, also pose challenges to circuit design using CNFETs. Modeling and analysis of the impact of CNT density variations on CNFET circuit performance are discussed in [Zhang 09b].

3. Experimental Results

We experimentally demonstrated essential components and their integration for VLSI CNFET technology. We show full-wafer-scale aligned CNT growth on single-crystal quartz wafers (Fig. 3.1) [Patil 08a, 09b]. CNTs grown on single-crystal quartz [Kang 07] are significantly better aligned compared CNTs grown on silicon [Reina 07]. A new technique for full-wafer-scale CNT transfer from quartz to silicon for large-scale silicon integration (Fig. 3.2) has been demonstrated in [Patil 08a, Patil 09b].

Full-wafer-scale aligned CNT growth and transfer, together with m-CNT removal using electrical breakdown [Collins 01], enables mis-positioned-CNT-immune digital logic structures. The circuits are fabricated on full wafers using conventional optical lithography (Fig 3.3, 3.4 and 3.5) [Patil 08a, 08b]. Such logic structures guarantee correct logic functionality in the presence of a large number of mis-positioned CNTs.

4. Conclusions

The first CNFET was reported in 1998 [Martel 98, Tans 98]. Over the past 11 years, remarkable progress has been made in all areas of CNT science and technology from materials, devices, and

978-1-60558-497-3/09 $25.00 © 2009 ACM

circuits [Dai 02, Guo 02, Avouris 07, Cao 08b, Javey 09]. This paper focuses on digital VLSI applications. Other applications such as RF/tera-hertz [Le Louarn 07], sensors [Chen 03, Cho 07, Grüner 06, Kauffman 08a, 08b], large area electronics [Cao 08a] are also very promising, and significant progress has been made over the years.

While the payoff for a new VLSI digital transistor technology is high, the challenges facing the development of a bona fide technology that can rival the incumbent silicon-CMOS technology are daunting. Table 3.1 includes a survey of the challenges faced by CNFET technology and potential solutions to these challenges. Table 3.1 also includes a CNFET technology outlook based on our own experiences. Imperfections in CNT material are major detractors of performance. An **imperfection-immune design paradigm** is required to overcome these barriers. <u>New design techniques must be compatible with VLSI processing and have</u> <u>minimal impact on existing VLSI design flows.</u> For example, techniques that rely on separate customization of every chip can be prohibitively expensive if not designed carefully. Investments made in VLSI design infrastructure are too large to be ignored. With this in mind, a VLSI technology based on an FET-design compatible device is highly desirable as compared to new devices that require a different design paradigm.

Acknowledgment

We thank FCRP GSRC, FENA, and C2S2, NSF and Stanford Graduate Fellowship for support. We thank Dr. I. Amlani, A. Badmaev, Dr. S. Fujita, Dr. J. McVittie, Prof. Y. Nishi, Dr. B. Paul, K. Ryu, S. Yasuda, and Prof. C. Zhou for fruitful collaborations. The work described here includes contributions from several former and current students: D. Akinwande, G. Close, J. Deng, J. Liang, G.C. Wan and H. Wei.

Table 3.1. CNFET Technology Challenges and Outlook

Challenge	Needed For	Current Solutions	Options	Status
High density aligned CNTs	High current density CNFETs Need : 150 – 200 CNTs/µm	Average CNT density: 5-10 CNTs/µm [Kang 07]. Wafer-scale growth demonstrated [Patil 08a, 09b].	Denser CNT Growth; Multiple Transfers of CNTs.	Yellow
CNT alignment and positioning	Correct logic functionality	Mis-positioned-CNT-immune design [Patil 07, 08a, 08b]	Solved	Green
% Metallic CNTs grown or deposited on substrate	Leakage, noise margin Need : < 0.01% metallic CNTs to avoid metallic CNT removal (discussed below)	1. Typical CNT synthesis: 33 % CNTs grown metallic for random chirality [Saito 98]. 2. Plasma Enhanced CNT Synthesis : 10 % CNTs grown metallic [Li 04] 3. Solution Based CNT Sorting : 1% Metallic CNTs [Engel 08], 5% Metallic CNTs [Le Mieux 08]	Tighter control of CNT chirality needed	Red
Metallic CNT Mitigation	Leakage, noise margin Need : > 99.99% m-CNTs removed with > 80% s-CNTs intact	1. Electrical Breakdown of Metallic CNTs [Collins 01] (not VLSI-compatible); 2. Selective Chemical Etching of metallic CNTs [Zhang 06]; 3. Metallic CNT Tolerant Circuits [Lin 09b] (VLSI-compatible).	More VLSI-compatible metallic CNT removal required	Red
CNFET Threshold Voltage Setting	Cascadable logic circuits	Single-CNT Ring Oscillator [Chen 06]. CNT circuits on flexible substrates [Cao 08a].	Air-stable CNT doping with controlled doping level required.	Yellow
CNT Doping	High performance complementary logic circuits	P-type Dopant: Triethyloxonium hexachloroantimonate (OA) [Chen 05]. N-type Dopants: Hydrazine [Chen 05], Polyethylene imine [Shim 01], Potassium (not air-stable) [Javey 05].	Air-stable CNT doping with controlled doping level required.	Yellow
CNFET Source /Drain Contact Metal	High performance complementary logic circuits	p-type : Palladium [Kim 05] n-type : Scandium [Zhang 07], Aluminum [Javey 03b]	Pd, Sc, & Al; Additional, air-stable options would prove useful	Yellow

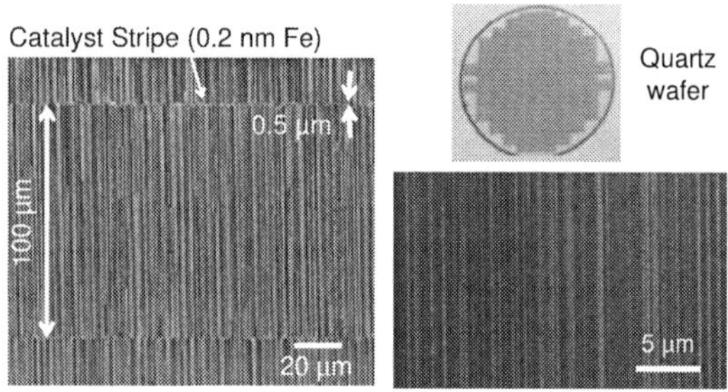

Figure 3.1. Full-wafer-scale aligned CNT growth on single-crystal quartz wafers (details in [Patil 08a, 09b]).

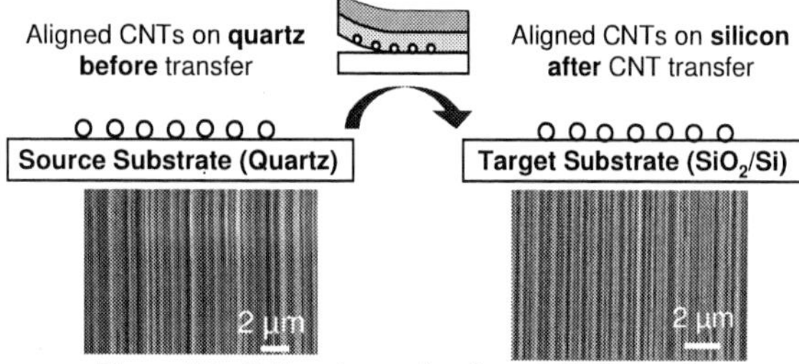

Figure 3.2. Full-wafer-scale aligned CNT transfer from quartz wafers to silicon wafers (details in [Patil 08a, 09b]).

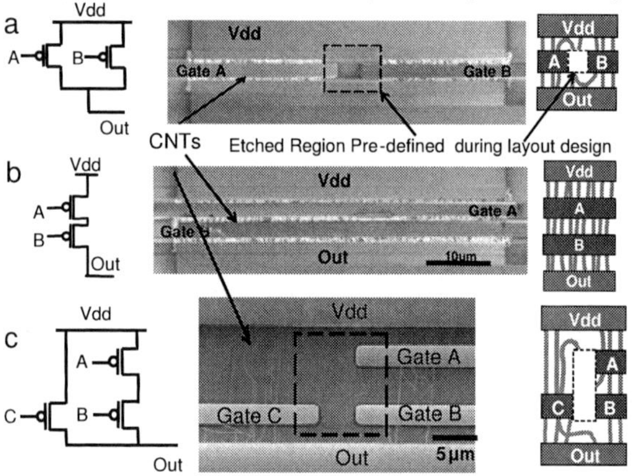

Figure 3.3. Scanning Electron Microscopy (SEM) images and layouts of mis-positioned-CNT-immune logic structures corresponding to (a) NAND pull-up (b) NOR pull-up (c) OR-AND-INVERT pull-up.

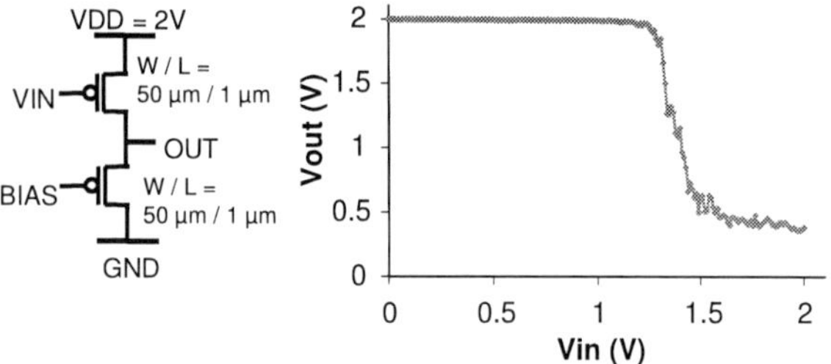

Figure 3.4. Voltage transfer curve of a CNFET inverter.

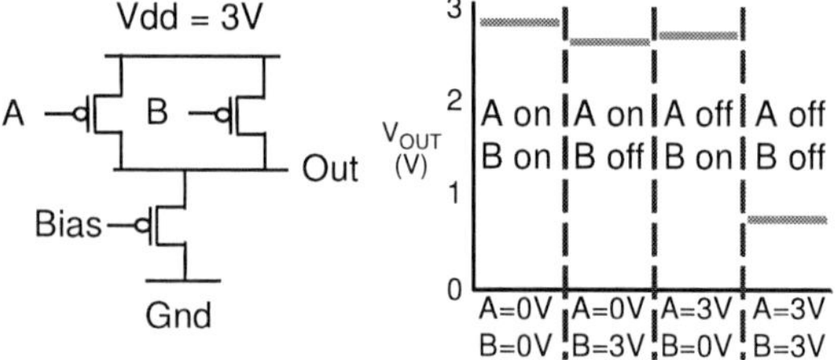

Figure 3.5. Input / Output characteristics of a mis-positioned-CNT-immune CNFET NAND.

References

[Akinwande 08a] Akinwande, D., *et al.*, "Monolithic Integration of CMOS VLSI and Carbon Nanotubes for Hybrid Nanotechnology Application", *IEEE Trans. on Nanotechnology*, vol. 7, issue 5, pp. 636-639, 2008.

[Akinwande 08b] D. Akinwande, *et al.*, "Analytical Ballistic Theory of Carbon Nanotube Transistors: Experimental Validation, Device Physics, Parameter Extraction, and Performance Projection", *Journal of Applied Physics*, vol. 104, pp. 124514-1–124514-7, 2008.

[Akinwande 09] Akinwande, D., Y. Nishi, H.-S.P. Wong, "Carbon Nanotube Quantum Capacitance for Non-Linear Terahertz Circuits", *IEEE Trans. Nanotechnology*, vol. 8, No. 1, pp. 31 – 36, 2009.

[Amlani 06] Amlani, I., *et al.*, "First Demonstration of AC Gain From a Single-walled Carbon Nanotube Common-Source Amplifier", *Proc. IEDM*, pp. 1-4 , 2006.

[Avouris 07] Avouris, P., Z. Chen, V. Perebeinos, "Carbon-based electronics", *Nature Nanotechnology*, 2, p. 605 – 615, 2007.

[Balijepalli 07] Balijepalli, A., S. Sinha, Y. Cao, "Compact modeling of carbon nanotube transistor for early stage process-design exploration", *Proc. International Symposium on Low Power Electronics and Design*, pp. 2-7, 2007.

[Bobba 09] Bobba, S. *et al.*, "Design of Compact Imperfection-Immune CNFET Layouts for Standard-Cell-Based Logic Synthesis", *Proc.DATE*, 2009.

[Borkar 05] Borkar, S., et al., "Statistical circuit design with carbon nanotubes", United States Patent Application 2007015506.

[Cao 08a] Cao, Q., *et al.*, "Medium-Scale Carbon Nanotube Thin-Film Integrated Circuits on Flexible Plastic Substrates", Nature 454, 495-500, 2008.

[Cao 08b] Cao, Q., and J.A. Rogers, "Random Networks and Aligned Arrays of Single-Walled Carbon Nanotubes for Electronic Device Applications", *Nano Res*, 1: 259 272, 2008.

[Chen 03] Chen, R., *et al.*, "Noncovalent functionalization of carbon nanotubes for highly specific electronic biosensors", *Proc. Natl. Acad. Sci.*, vol. 100, pp. 4984–4989, 2003.

[Chen 05] Chen J., C. Klinke, A. Afzali and P. Avouris, "Self-aligned carbon nanotube transistors with charge transfer doping", *Appl. Phys. Lett.*, 86, 123108, 2005.

[Chen 06] Chen Z., *et al.*, "An Integrated Logic Circuit Assembled on a Single Carbon Nanotube", *Science* 311 (5768), 1735.

[Cho 07] Cho, T. S., *et al.*, "A Low Power Carbon Nanotube Chemical Sensor System", *Proc. CICC*, pp.181-184, 2007.

[Close 08a] G.F. Close, H.-S. P. Wong, "Assembly and Characterization of Multi-wall Carbon Nanotubes", *IEEE Trans. Nanotechnology*, vol. 7, issue 5, pp. 596–600, 2008.

[Close 08b] Close G. F., *et al.*, "A 1 GHz Integrated Circuit with Carbon Nanotube Interconnects and Silicon Transistors", *Nano Letters* 2008 8 (2), 706-709.

[Collins 01] Collins, P., S. Arnold, P. Avouris, "Engineering carbon nanotubes and nanotube circuits using electrical breakdown", *Science*, vol. 292, pp. 706 – 709, 2001.

[Dai 02] Dai, H., "Carbon Nanotubes: Synthesis, Integration, and Properties", *Acc. Chem. Res.*, 35, 1035-1044, 2002.

[Dehon 05] DeHon, A., and H. Naeimi, "Seven Strategies for Tolerating Highly Defective Fabrication", *IEEE Design and Test of Computers*, Vol. 22, No. 4, pp. 306-315, 2005.

[Deng 07a] Deng, J., *et al.*, "Carbon Nanotube Transistor Circuits: Circuit-Level Performance Benchmarking and Design Options for Living with Imperfections", *Proc. ISSCC*, pp. 70-588, 2007.

[Deng 07b] Deng, J., and H. -S. P. Wong, "A Compact SPICE Model for Carbon Nanotube Field Effect Transistors Including Non-Idealities and Its Application — Part I: Model of the

Intrinsic Channel Region", *IEEE Trans. Elec. Dev.*, pp.3186-3194, 2007.

[Deng 08] Deng, J., *et al.,* "Carbon nanotube transistor compact model for circuit design and performance optimization", J. *Emerg. Technol. Comput. Syst.* 4, 2, 1-20, 2008.

[Engel 08] Engel, M., *et al.,* "Thin Film Nanotube Transistors Based on Self-Assembled, Aligned, Semiconducting Carbon Nanotube Arrays", *ACS Nano* 2008 2(12), 2445-2452.

[Goldstein 01] Goldstein, S.C., and M. Budiu, "NanoFabrics: spatial computing using molecular electronics", *Proc. Intl. Symp. Computer Architecture,* pp. 178-191, 2001.

[Grüner 06] Grüner, G., "Carbon Nanotube Transistors for Biosensing Applications", *Analytical and Bioanalytical Chemistry*, 384, 322, 2006.

[Guo 02] Guo, J., Mark Lundstrom, and Supriyo Datta, "Performance projections for ballistic carbon nanotube field-effect transistors", Appl. Phys. Lett. 80, 3192, 2002.

[Guo 04] Guo, J., S. Datta, M. Lundstrom, and M. P. Anantram, "Toward multi-scale modeling of carbon nanotube transistors", *Int. J. Multiscale Comput. Eng.*, vol. 2, pp. 257–277, 2004.

[Han 05] Han, S., X. Liu and C. Zhou, "Template-Free Directional Growth of Single-Walled Carbon Nanotubes on a- and r-Plane Sapphire", *J. Am. Chem. Soc.*, vol. 127, pp. 5294–5295, 2005.

[Hassanien 05] Hassanien, A., *et al.,* "Selective etching of metallic single-wall carbon nanotubes with hydrogen plasma", *Nanotechnology*, vol 16, pp. 278 – 281, 2005.

[Javey 03a] Javey, A., *et al.,* "Ballistic Carbon Nanotube Transistors", *Nature,* 424, 654-657, 2003.

[Javey 03b] Javey, A., Q. Wang, W. Kim, H. Dai. "Advancements in Complementary Carbon Nanotube Field-Effect Transistors", *Proc. IEDM*, pp. 31.2.1-31.2.4, 2003.

[Javey 05] Javey, A., *et al.,* "High performance nanotube n-FETs with chemically doped contacts", *Nano Letters*, p. 345-348, 2005.

[Javey 09] Javey, A., and J. Kong (editors), *Carbon Nanotube Electronics*, Springer, 2009.

[Kang 07] Kang, S. J., *et al.,* "High-Performance Electronics Using Dense, Perfectly Aligned Arrays of Single-Walled Carbon Nanotubes", *Nature Nanotechnology*, vol. 2, pp. 230-236, 2007.

[Kauffman 08a] Kauffman, D. R., A. Star, "Carbon nanotube gas and vapor sensors", *Angew. Chem. Int. Ed.*, 47, 6550-6570, 2008.

[Kauffman 08b] Kauffman, D. R., A. Star, "Electronically monitoring biological interactions with carbon nanotube field-effect transistors", *Chem. Soc. Rev.*, 37, 1197-1208, 2008.

[Kim 05] Kim W., *et al.,* "Electrical contacts to carbon nanotubes down to 1 nm in diameter", *Appl. Phys. Lett.* 87, 173101, 2005.

[Le Louarn 07] Le Louarn A., *et al.,* "Intrinsic current gain cutoff frequency of 30 GHz with carbon nanotube transistors", *Appl. Phys. Lett.*, 90, 233108, 2007.

[LeMieux 08] LeMieux, M., *et al.,* "Self-Sorted, Aligned Nanotube Networks for Thin-Film Transistors", *Science*, vol. 321, pp. 101-104, 2008.

[Li 04] Li, Y., *et al.,* "Preferential Growth of Semiconducting Single-Walled Carbon Nanotubes by a Plasma Enhanced CVD Method", *Nano Letters*, vol. 4, pp. 317-321, 2004.

[Liang 08] Liang, J., D. Akinwande, H.-S. P. Wong, "Carrier Density and Quantum Capacitance for Semiconducting Carbon Nanotubes", *J. Applied Physics*, vol.104, pp. 064515-1– 064515-6, September 29, 2008.

[Lin 09a] Lin, A., *et al.,* "Threshold Voltage and On-Off Ratio Tuning for Multiple-tube Carbon Nanotube FETs", *IEEE Trans. Nanotechnology*, vol. 8, No. 1, pp. 4 – 9, 2009.

[Lin 09b] Lin, A., *et al.,* "A Metallic-CNT-Tolerant Carbon Nanotube Technology using Asymmetrically-Correlated CNTs (ACCNT)", to appear in *Proc. VLSI Tech. Symp.*, 2009.

[Martel 98] Martel, R., *et al.,* Single- and multi-wall carbon nanotube field-effect transistors", *App. Phys. Lett.* 73, 2447, 1998.

[Patil 07] Patil, N., *et al.,* "Automated Design of Misaligned-Carbon-Nanotube-Immune Circuits", *Proc. DAC*, pp. 958 - 961, 2007.

[Patil 08a] Patil, N., *et al.,* "Integrated Wafer-scale Growth and Transfer of Directional Carbon Nanotubes and Misaligned-Carbon-Nanotube-Immune Logic Structures," *Proc. Symp. VLSI Technology*, pp. 205-206, 2008.

[Patil 08b] Patil, N., *et al.,* "Design Methods for Misaligned and Mis-positioned Carbon-Nanotube-Immune Circuits", *IEEE Trans. Computer-Aided Design*, pp. 1725-1736, 2008.

[Patil 09a] Patil, N., *et al.,* "Circuit-Level Performance Benchmarking and Scalability of Carbon Nanotube Transistor Circuits", *IEEE Trans. Nanotechnology*, vol.8, no.1, pp.37-45, 2009.

[Patil 09b] Patil, N., *et al.,* "Wafer-Scale Growth and Transfer of Aligned Single-Walled Carbon Nanotubes", *IEEE Trans. Nanotechnology*, 2009.

[Qu 08] Qu, L., D. Feng, and L. Dai, "Preferential Syntheses of Semiconducting Vertically Aligned Single-Walled Carbon Nanotubes for Direct Use in FETs", *Nano Letters*, vol .8(9), pp. 2682-2687, 2008.

[Rad 06] Rad, R.M. and M. Tehranipoor, "A Hybrid FPGA Using Nanoscale Cluster and CMOS Scale Routing", *Proc. Design Automation Conf.*, pp. 727 – 730, 2006.

[Rao 05] Rao, W., A. Orailoglu and R. Karri, "Fault-Tolerant Nanoelectronic Processor Architectures", *Proc. Asia South Pacific Design Automation Conf.*, pp. 311-316, 2005.

[Reina 07] Reina, A., *et al.,* "Growth mechanism of long and horizontally aligned carbon nanotubes by chemical vapor deposition", *J. Phys. Chem. C*, 111, 7292 – 7297, 2007.

[Saito 98] Saito, R., G. Dresselhaus and M. Dresselhaus, *Physical Properties of Carbon Nanotubes*, Imperial College Press, 1998.

[Shim 01] Shim M., *et al.,* "Polymer Functionalization for Air-Stable n-type Carbon Nanotube Field Effect Transistors", *J. Am. Chem. Soc.* 123(46), 11512-11513, 2001.

[Tahoori 06] Tahoori, M.B., "Application-Independent Defect-Tolerance of Reconfigurable Nano-Architectures", *ACM Journal Emerging Technologies in Computing*, 2006.

[Tang 01] Tang, S., *et al.,* "FinFET - A Quasi-Planar Double-Gate MOSFET", *Proc. ISSCC*, pp. 118-119, 2001.

[Tans 98] Tans, S. J., *et al.,* "Room-temperature transistor based on a single carbon nanotube", *Nature* 393, 49, 1998.

[Wong 07] Wong, S., *et al.,* "Monolithic 3D Integrated Circuits", *Proc. VLSI-TSA*, pp.1-4, 2007.

[Yang 06] Yang, C., *et al.,* "Preferential etching of metallic single-walled carbon nanotubes with small diameter by fluorine gas", *Phys Rev B 73*, pp. 75419, 2006.

[Zhang 06] Zhang, G., *et al.,* "Selective Etching of Metallic Carbon Nanotubes by Gas-Phase Reaction", *Science*, Vol. 314, pp. 974 – 977, 2006.

[Zhang 07] Zhang, Z., *et al.,* "Doping-Free Fabrication of Carbon Nanotube Based Ballistic CMOS Devices and Circuits", *Nano Letters* 7 (12), 3603-3607, 2007.

[Zhang 08] Zhang, J., N. Patil, and S. Mitra, "Design Guidelines for Metallic-Carbon-Nanotube-Tolerant Digital Logic Circuits", *Proc. DATE*, pp. 1009 – 1014, 2008.

[Zhang 09a] Zhang, J., N. Patil and S. Mitra, "Probabilistic Analysis and Design of Metallic-Carbon-Nanotube-Tolerant Digital Logic Circuits", to appear in *IEEE Trans. CAD*, 2009.

[Zhang 09b] Zhang, J., N. Patil, A. Hazeghi and S. Mitra, "Carbon Nanotube Circuits in the Presence of Carbon Nanotube Density Variations", *Proc. DAC*, 2009.

CMOS Scaling Beyond 32nm: Challenges and Opportunities

Kelin J. Kuhn

Logic Technology Development, Intel Corporation, Hillsboro, OR, 97124, U.S.A.

kelin.ptd.kuhn@intel.com

ABSTRACT

This paper explores the challenges and opportunities facing CMOS process generations past the 32nm technology node. Planar and multiple-gate devices are compared and contrasted. Resistance and capacitance challenges are reviewed in relation to past history and on-going research. Key enhancers such as high-k metal-gate (HiK-MG), substrate and channel orientation, and NMOS and PMOS strain, are discussed in relation to the challenges of the coming transistor generations.

Categories and Subject Descriptors

B.7.0 [**Hardware**]: Integrated Circuits – *general*

General Terms

Performance.

Keywords

CMOS; high-k; metal-gate; strain; orientation

1. INTRODUCTION

For the past 40 years, relentless focus on Moore's Law transistor scaling has provided ever-increasing transistor performance and density. Interestingly enough, key technologists in each generation of this long history have also looked forward and predicted the "end of scaling" within one or two generations [1-4]. However, each time the technology reached the predicted barriers, scaling did not stop. Instead, imaginative new solutions were developed to further extend Moore's Law.

A decade ago, "CMOS transistor scaling" meant "classic" Dennard scaling [5], where oxide thickness (T_{ox}), transistor length (L_g) and transistor width (W) were scaled by a constant factor ($1/k$) in order to provide a delay improvement of $1/k$ at constant power density. Classic Dennard scaling ended at 130nm (130nm was the last CMOS generation where making the transistor smaller was sufficient to deliver performance enhancement). However (contrary to many predictions [1-4]), CMOS transistor scaling didn't end at the 130nm node. Instead, enhancers were added (strain in the 90nm and 65nm nodes [6,7], and strain + HiK-MG in the 45nm and 32nm nodes [8,9]) to continue to drive the CMOS transistor roadmap forward..

Permission to make digital or hard copies of part or all of this work for personal or classroom use is granted without fee provided that copies are not made or distributed for profit or commercial advantage and that copies bear this notice and the full citation on the first page. To copy otherwise, to republish, to post on servers or to redistribute to lists, requires prior specific permission and/or a fee.
DAC'09, July 26-31, 2009, San Francisco, California, USA

2. TRANSISTOR ARCHITECTURE

2.1 Planar transistors

As we look beyond 32nm CMOS transistors, there are a number of scaling challenges to be addressed in the CMOS planar device (Figure 1). Increased off-state current (I_{off}) from degraded drain-induced barrier lowering (DIBL) and subthreshold slope (SS) caused by poorer short channel effects (SCE) represents a significant limitation for effective gate lengths (L_{eff}) shorter than approximately 15nm. Decreasing T_{ox} to provide better channel control comes with a penalty of increased gate leakage current (I_{gate}) and increased channel doping. Increased channel doping decreases mobility (degrading performance due to impurity scattering) and increases random dopant fluctuations, RDF (degrading the minimum operating voltage, V_{min}). Decreasing gate pitch increases the parasitic capacitance contribution for both contact-to-gate and epi-to-gate thus increasing overall gate capacitance (C_{gs}). Decreasing source/drain opening size increases the source drain resistance (R_{sd}) thus decreasing drive current. Additionally, decreasing gate pitch decreases the volume/quantity of the stressor materials for both NMOS (stress induced by overlayer films) and PMOS (stress induced by embedded-SiGe, e-SiGe) thus decreasing mobility and drive.

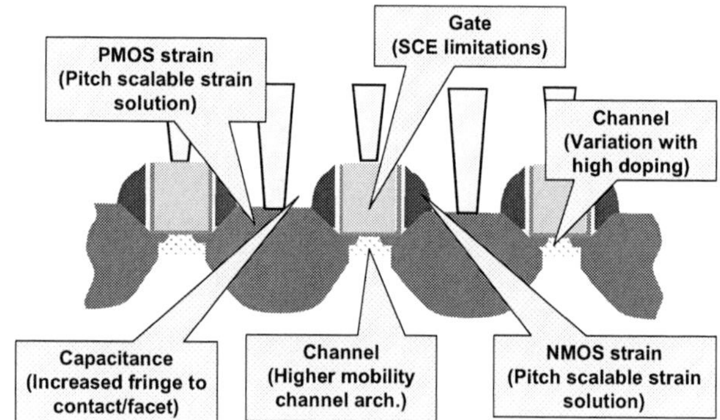

Figure 1. Scaling challenges in planar devices

2.2 Multiple gate or multiple channel devices

Multiple gate or multiple channel FETs (MUGFETS or MUCFETS) have been proposed to help resolve the SCE issues past 32nm (as an example, the ITRS 2007 roadmap [10] predicts the use of multiple gate or channel FETs starting in 22nm).

While MUGFETs provide significant resolution to the DIBL/SS SCE issues, they do have some unique challenges of their own (Figure 2). The most significant of these is the challenge of continuing to maintain a high level of mobility enhancement from stress in an architecture with numerous free surfaces generated by the fins. MuGFETs also require over-scaling the pitch to reduce capacitance and increase drive, (pitch over-scaling will be challenging on nodes still limited to 193nm immersion steppers).

While MUGFETS may provide mitigation for RDF and thus provide V_{min} improvement; they also add new variation sources (fin width W_{si} variation in particular, is similar to L_{eff} variation in modifying the SCE properties of a device). Although C_{gs} and R_{sd} increases are a generic scaling problem for all architectures at tight pitches, the requirements for aggressive fin widths (to maintain SCE) and fin pitches (to meet drive current and capacitance targets) will enhance these problems in MUGFETS. Last, but not least, any 3'D architecture (such as a MUGFET) will require exquisite control of etch and patterning, and introduce topography risks for fill and polish.

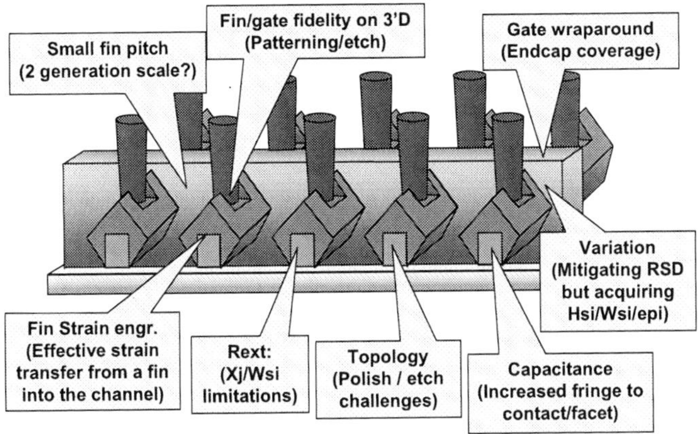

Figure 2. Scaling challenges in MUGFET devices

3. NEXT GENERATION CHALLENGES
3.1 Capacitance challenges

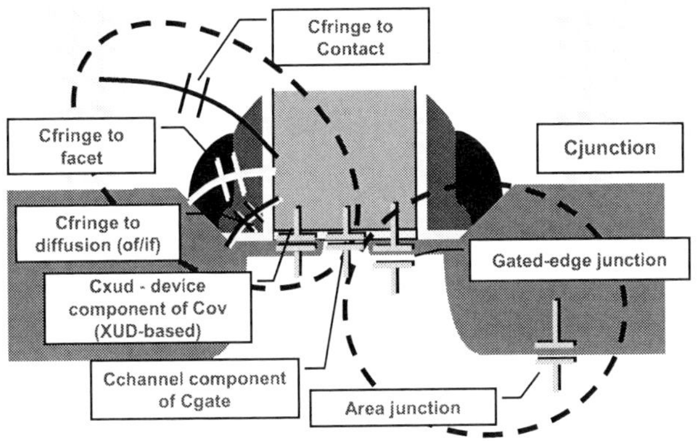

Figure 3. Capacitive elements in planar architectures

The traditional CMOS capacitive elements (Figure 3), such as under-lap capacitance (C_{xud}), channel capacitance, junction capacitances (both gated edge and area) and the inner and outer fringe capacitance; will become more challenging at reduced dimensions of advanced technologies. Furthermore, in recent generations, gate and contact CD dimensions have been scaling slower than contacted gate pitch. This means that parasitic fringe capacitances (for example, contact-to-gate and epi-to-gate) are becoming significant issues. 3'D geometries (such as MUGFET devices) have similar capacitive challenges as planar and also introduce additional "dead space" parasitic capacitance associated with the region between the fins.

Current research to address parasitic capacitance in the front end (for either planar or non-planar devices) is focusing on aggressively reducing the k of the spacer – either by removing it completely (for example, Liow [11]) or changing materials to a lower k alternative (for example, Ko [12]).

3.2 Resistance Challenges

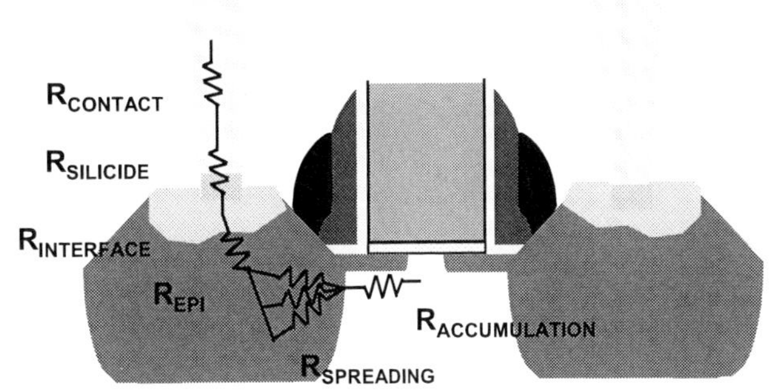

Figure 4. Resistive elements in planar architectures

The traditional CMOS resistive elements (Figure 4), such as the accumulation (R_{acc}), spreading, silicide and contact resistances; will also become more challenging at the reduced dimensions of advanced technologies. Furthermore, resistance elements previously neglected (including interface and epi resistance) are becoming significant issues for planar devices. In addition to the planar challenges, 3'D geometries (FinFET and trigate devices) must also compensate for additional resistance increases with decreasing fin width.

A rich spectrum of techniques are being explored by a variety of groups for improving R_{acc} in a traditional planar architecture. Examples include laser spike anneal (LSA) as explored by Luo [13], Pouydebasque [14], and Yamamoto [15], non-melt LSA as reviewed by Ortolland [16], millisecond anneal of raised source/drain as analyzed by Yako [17], and advanced implantation techniques as discussed by Gelpey [18].

Looking ahead to further resistance improvements, interface resistance improvement through modulation of the Schottky barrier height (SBH) offers significant opportunities. A variety of efforts are in progress to achieve theoretical SBHs through clever modifications to the process. One rich field of research is in alloy modifications to traditional silicides as discussed by Lee [19] and Ouchi [20]. Another area of critical study is implant modifications to traditional silicides, either on single metals or alloys as explored by Zhang [21] and Larrieu [22].

4. PROCESS OPPORTUNITIES
4.1 High-k plus metal gate (HiK-MG)

HiK dielectrics deliver reduced gate leakage and enable further T_{ox} scaling. The use of a metal gate (rather than polysilicon) eliminates poly depletion, resolves the V_T pinning issue seen with poly on HiK, and screens soft optical phonons [23]. However, the challenges with HiK-MG are also significant. There is the requirement to create nearly band-edge workfunctions after thermal processing on both N and PMOS. HiK gate dielectrics contain a high level of traps and charge and thus pose significant challenges for reliability. A variety of scattering mechanisms result in reduced mobility. Finally, HiK-MG flows are complex with significant challenges on thermal budget and on the etch, polish and fill integration.

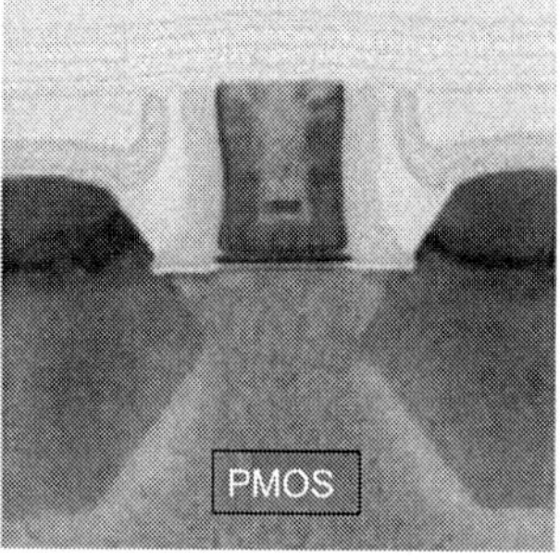

Figure 5. 45nm manufacturable HiK-MG process [8, 24]

In 2007, Mistry showed that the challenges of HiK-MG could be overcome by presenting a manufacturable 45nm HiK-MG process which demonstrated an 0.7X T_{ox} scale while simultaneously reducing gate leakage by 25X for NMOS and 1000X for PMOS with ring oscillator performance 23% better than 65 nm at the same leakage and 100mV lower Vcc [8, Figure 5]. Natarajan continued this work in 2008 showing 2nd generation HiK-MG results with 0.9A equivalent oxide thickness (EOT) and CV/I gate delay improved 22% compared to 45nm at the same leakage and Vcc [9].

4.2 Wafer and channel orientation

Maintaining the scaling roadmap will require continual improvement in channel mobility. While advanced materials such as Ge or III-V materials offer potential long-term options, a shorter term approach for the 22nm or 15nm nodes may be reorient the surface or the channel orientation.

The classic orientation for silicon is a (100)-type surface; with two <110> channels perpendicular to each other and a <100> channel at 45 degrees. The competing orientation is a (110)-type surface with three possible channel directions; <110>, <111> and <100>. If both crystal orientations are possible simultaneously, the best unstrained NMOS is on the (100) surface <110> direction, and the best unstrained PMOS is on the (110 surface) <110> direction (see Figure 6).

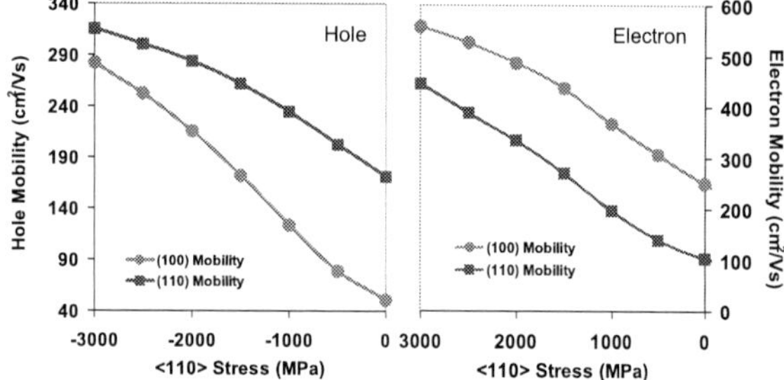

Figure 6. Simulated electron and hole mobility for (100) and (110) substrates as a function of stress [25]

Significant research in the last five years has focused on the challenge of trying to take advantage of the enhanced PMOS mobility on (110) <110> type material, without degrading the NMOS [25, 26]. While it is theoretically possible to create both orientations on one wafer (by using a FinFET or vertical FET device on (100)-type material, and orienting the NMOS at 45 degrees, [27]) the 45 degree orientation poses significant challenges for lithography.

The alternative approach is to simply integrate both orientations on the same wafer. This process (termed HOT) was first reported by Yang in 2003 [28]. In this process, a wafer bonding technique is used to create SOI of the opposite type as the handle wafer for one of the transistor types. Epi from the handle wafer is then used to create a bulk transistor of the opposite type to the SOI. Sung in 2005 [29] introduced a direct silicon bond version of HOT that used bulk transistors for both N and P (rather than one SOI and one bulk). Yang in 2006 [30]; reversed the concept with a SuperHot process that integrates a pure SOI (100)-type and a pseudo SOI (110)-type device on the same wafer.

4.3 Strain

Strain has had tremendous positive impact on maintaining the transistor scaling roadmap after the end of "classic" Dennard scaling at the 130nm node. Because of the high gains provided by strain in today's processes, future transistor architecture solutions (whether on (100) or (110), or with alternative channel materials) will require significant strain enhancement on both N and PMOS.

Reviewing the rich history of strain technology provides insight on what may be possible in subsequent generations. Early work in CMOS strain followed examples from III-V systems, and introduced channel strain by epitaxial growth of lattice mismatched Si/SiGe systems. The seminal work in this area was done by Welser in 1992 [31] when he first quantified the strain enhancement in strained Si on relaxed SiGe. Hoyt in 2002 [32] expanded on Welser's work by exploring the strain enhancement with vertical effective field and doping . Also in 2002, Rim [33] further expanded on the Welser and Hoyt work by quantifying and controlling the lower V_T associated with short channel strained NMOS.

In 2000, Ito explored a different approach to strain engineering in CMOS [34]. He demonstrated a highly-manufacturable NMOS strain process by using the SiN contact etch stop layer (CESL), which introduced channel strain without the need to re-architect the channel. Pidin expanded on this concept in 2004 [35], by demonstrating simultaneous NMOS and PMOS CESL. Mayuzumi extended this to HiK-MG processes in 2007 [36], by illustrating that strain improvement was possible even with dual-cut CESL stressors.

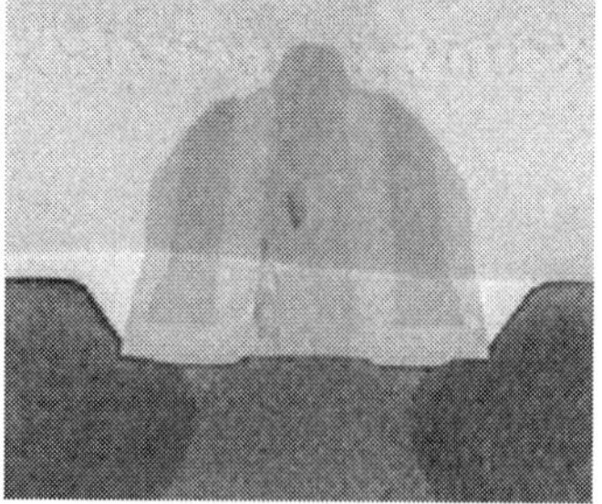

Figure 7. Embedded SiGe (e-SiGe) is a highly-manufacturable uniaxial PMOS strain solution [6, 37]

For PMOS, Thompson in 2002 [6] presented a highly-manufacturable uniaxial PMOS strain solution (again with stressors outside the channel) using embedded SiGe (e-SiGe) material in the source/drain regions. Ghani [37] expanded on this work in 2003, and Chidambaram in 2004 [38].

The elegant e-SiGe technique caught on quickly and by 2005 many researchers were exploring this process, with representative examples including Lee with e-SiGE with SOI [39], Ohta illustrating the impact of e-SiGe profile engineering [40], and Zhang demonstrating e-SiGe on thin body SOI [41]. As an additional enhancement, Wang in 2007 [42] and Auth in 2008 [24] demonstrated that HiK-MG in a replacement gate flow enabled additional PMOS strain enhancement as a consequence of first straining the PMOS with e-SiGe and then removing the gate (see Figure 8).

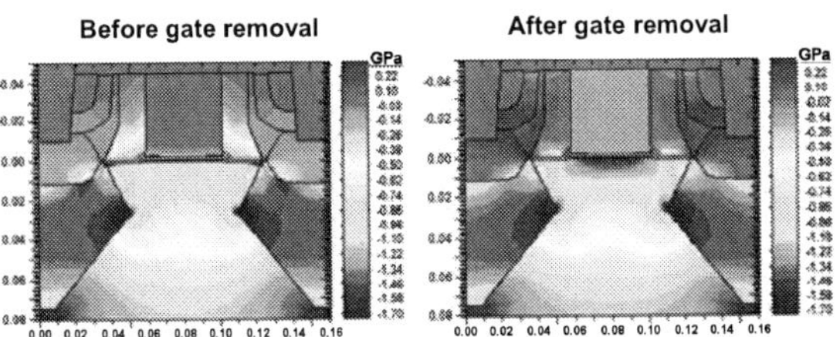

Figure 8. Poly gate removal increases channel stress 50% [24]

Stress memorization (where the gate is implanted, capped, then annealed to introduce channel stress) was an unexpected and highly successful stress enhancement technique. Ota in 2002 [43] first demonstrated the stress memorization technique (SMT) with Chen in 2004 [44] reporting the first large enhancements. More recently, Wei in 2007 [45] demonstrated that the process can be repeated multiple times for additional enhancement and Kubicek in 2008 [46] demonstrated SMT on HiK-MG.

There has been much recent interest in taking the PMOS enhancements with e-SiGe and applying the same techniques to NMOS. Ang in 2004 [47] first reported SiC enhancement of NMOS. Liu in 2007 [48], demonstrated SiC enhancement with implant plus SPE technique. Ren in 2008 [49], demonstrated long channel and Yang in 2008 [50] demonstrated short channel SiC enhancement from in-situ phosphorus doped embedded SiC.

Last, but certainly not least, are the enhancements due to metal stress in the front end. Kang in 2006 [51] first reported enhancement from gate metal stress. Auth in 2008 [24] reported enhancement for gate metal stress and also reported enhancement with tensile contact stress.

5. CONCLUSION

While significant transistor challenges (SCE, resistance, capacitance, mobility, etc.) exist for CMOS technologies past 32nm, a rich selection of potential solutions are being explored. As has been true historically, we do not expect Moore's Law (Figure 9) to be stopped by the latest round of challenges, but anticipate imaginative new solutions which will further extend Moore's Law and the CMOS transistor scaling roadmap.

6. ACKNOWLEDGMENTS

The author gratefully acknowledges the many people in Intel's 45nm and 32nm generations who contributed to this work.

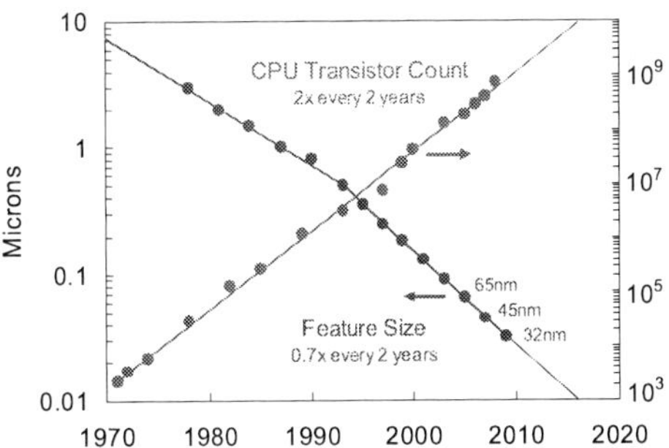

Figure 9. Moore's Law will continue in spite of challenges

7. REFERENCES

1. Broers, A. N.; IEDM Tech. Dig., pp. 2-6, Dec. 1980.
2. Meindl, D.; IEDM Tech. Dig., pp. 8–13, Dec. 1983
3. Wright, P. J. et al.; IEEE TED, Vol. 37, pp. 1884-1892, Aug 1990.
4. Heilmeier, G. H.; IEDM Tech. Dig., pp. 2-5, Dec. 1984
5. Dennard, R.H et al.; IEEE JSSC, Vol .9, No. 5, pp. 256–268, Oct 1974
6. Thompson, S. et al.; IEDM Tech. Dig., pp. 61–64, Dec. 2002
7. Bai, P. et al.; IEDM Tech. Dig., pp. 657-660, Dec. 2004
8. Mistry, K. et al.; IEDM Tech. Dig., pp. 247-250, Dec. 2007
9. Natarajan, S. et al.; IEDM Tech. Dig., 941-944, Dec. 2008
10. A. Allan, ITRS roadmap, 2007 ITRS Conf., Dec. 2007
11. Liow, T.Y. et al.; IEEE EDL, Vol. 29, Issue 1, Jan. 2008 pp.80 - 82
12. Ko, C.H. et al.; 2008 Symp. on VLSI Tech, pp. 108–109, June 2008.
13. Luo, Z. et al.; IEDM Tech. Dig., pp. 489–492, Dec. 2005
14. Pouydebasque, et al.; IEDM Tech. Dig., pp. 663–666, Dec. 2005
15. Yamamoto, T. et al.; IEDM Tech. Dig., pp. 143-146 Dec. 2007
16. Ortolland, et al. ; Symp. on VLSI Tech; pp.186–187, June 2008
17. Yako, K. et al.; IEDM Tech. Dig., pp. 909-912, Dec. 2008
18. Gelpey, J. et al.; IWJT '08. pp. 82–86, 15-16 May 2008
19. Lee, R.T.P. et al.; IEDM Tech. Dig., pp. 851-854, Dec. 2006
20. Ohuchi, K. et al.;. IEDM Tech. Dig., pp. 1029–1031, Dec. 2007
21. Zhang, Z. et al.; IEEE EDL,; Vol. 28, Issue 7, July 2007 pp.565–568
22. Larrieu, G. et al.; IEDM Tech. Dig., pp.147–150, Dec. 2007
23. Chau, R. et al; IEEE EDL, Vol. 25, No. 6, pp. 408– 410, June 2004
24. Auth, C. et al.; 2008 Symp. on VLSI Tech; pp. 128–129, June 2008
25. Packan, P. et al.; IEDM Tech. Dig., pp. 63-66, Dec. 2008
26. Shimizu, K. et al.; IEDM Tech. Dig., pp. 67-70, Dec. 2008
27. Chang, L. et al.; IEEE TED, Vol. 51, No.10, pp. 1621–1627, Oct. 2004
28. Yang, M. et al.; IEDM Tech. Dig., pp. 453-456, Dec. 2003
29. Sung, C.Y. et al.; IEDM Tech. Dig., pp. 225–228, Dec. 2005.
30. Yang, M. et al.; 2006 Symp. on VLSI Tech, pp. 166–167, June 2006.
31. Welser, J. et al.; IEDM Tech. Dig., pp. 1000–1002, Dec. 1992
32. Hoyt, J.L. et al.; IEDM Tech. Dig., pp. 23–26, Dec. 2002
33. Rim, K. et al.; 2002 Symp. on VLSI Tech, pp. 98-99, June 2002
34. Ito, S.; Namba et al.; IEDM Tech. Dig., pp. 247–250, Dec. 2000
35. Pidin, S. et al.; IEDM Tech. Dig., pp. 213–216, Dec. 2004
36. Mayuzumi, S. et al.; IEDM Tech. Dig., pp. 293–296, Dec. 2007.
37. Ghani, T. et al.; IEDM Tech. Dig., pp. 978–980, Dec. 2003
38. Chidambaram, et al.; Symp. on VLSI Tech, pp. 48–49, June 2004
39. Lee, W.-H. et al.; IEDM Tech. Dig., pp. 61-64, Dec. 2005
40. Ohta, H. et al.; IEDM Tech. Dig., pp. 247-250, Dec. 2005
41. Zhang, D. et al.; Symp. on VLSI Tech, pp. 26–27, June 2005
42. Wang, J. et al.; Symp. on VLSI Tech, pp. 46–47, June 2007
43. Ota, K. et al.; IEDM Tech. Dig., pp. 27–30, Dec. 2002
44. Chen, C.H et al.; Symp. on VLSI Tech, pp. 56–57, June 2004
45. Wei, A. et al.; IEEE Symp. on VLSI Tech; pp. 216–217, June 2007
46. Kubicek, S. et al.; IEEE Symp. on VLSI Tech; pp. 130-131,June 2007
47. Ang, K.W. et al.; IEDM Tech. Dig., pp.1069–1071, Dec. 2004
48. Liu, Y. et al.; Symp. on VLSI Tech, pp. 44–45, June 2007
49. Ren, Z. et al.; Symp. on VLSI Tech; pp.172–173, June 2008
50. Yang, B.; et al.; IEDM Tech. Dig., pp. 51-54, Dec. 2008
51. Kang, C.Y.; et al.; Symp. on VLSI Tech, pp. 885-8, Dec. 2006

An O(n log n) Path-Based Obstacle-Avoiding Algorithm for Rectilinear Steiner Tree Construction

Chih-Hung Liu[†], Shih-Yi Yuan[§], and Sy-Yen Kuo[†‡] and Yao-Hsin Chou[‡]

[†] Graduate Institute of Electronics Engineering and [‡]Department of Electrical Engineering, National Taiwan University
[§]Department of Communications Engineering, Feng Chia University
sykuo@cc.ee.ntu.edu.tw

ABSTRACT

For the *obstacle-avoiding rectilinear Steiner minimal tree* problem, this paper presents an $O(n \log n)$-time algorithm with theoretical optimality guarantees on a number of specific cases, which required $O(n^3)$ time in previous works. We propose a new framework to directly generate $O(n)$ critical paths as essential solution components, and prove that those paths guarantee the existence of desirable solutions. The path-based framework neither generates invalid initial solutions nor constructs connected routing graphs, and thus provides a new way to deal with the OARSMT problem. Experimental results show that our algorithm achieves the best speed performance, while the average wirelength of the resulting solutions is only 1.1% longer than that of the best existing solutions.

Categories and Subject Descriptors:
B.7.2 [**Integrated Circuits**]: Design Aids

General Terms: Algorithms, Performance, Design

Keywords: Physical design, Routing, Steiner tree, Spanning tree

1. INTRODUCTION

Given a set of pins and a set of obstacles, an *obstacle-avoiding rectilinear Steiner minimal tree* (OARSMT) connects all the pins possibly through some additional points (Steiner points) using rectilinear edges with minimal total wirelength and without intersecting any obstacles. Since modern IC designs contain more and more rectilinear obstacles, such as macro cells, IP blocks, and pre-routed nets, the OARSMT problem has become more important, and has attracted increasing attentions [1]–[7]. Since the rectilinear Steiner minimal tree problem has been proved to be NP-Complete [8], the presence of obstacles further increases the difficulty.

The large and increasing number of obstacles also raises the requirement of efficiency. According to ITRS [9], the hard IP count per chip will be more than one thousand in the recent future. Therefore, since the Steiner tree construction will be invoked millions of times in floorplanning and placement phases [10], it is necessary to develop an $O(n \log n)$-time method to handle numerous obstacles.

Most recent works fall into two categories. The first class generates an initial solution without considering obstacles, and then legalizes the edges intersecting obstacles. Yang et al. [11] proposed a four-step method to remove invalid edges. This kind of approach may lack the global view of obstacles, and thus the solution quality might be limited. The second class constructs a connected graph embedding a valid solution, and applies graph algorithms. Li and Young [7] proposed a maze-routing based method on extended

Permission to make digital or hard copies of part or all of this work for personal or classroom use is granted without fee provided that copies are not made or distributed for profit or commercial advantage and that copies bear this notice and the full citation on the first page. To copy otherwise, to republish, to post on servers or to redistribute to lists, requires prior specific permission and/or a fee.
DAC'09, July 26-31, 2009, San Francisco, California, USA

Hanan grid, and their method outperforms all previous works in terms of wirelength. Since the second class contains more global information of obstacles, the solution quality would be better.

Three recent works [1, 5, 6] in the second category share a common structure. Shen et al. [1] proposed the structure: (1) *obstacle-avoiding spanning graph* (OASG) construction, (2) *minimal terminal spanning tree* (MTST) construction, and (3) rectilinear Steiner tree construction. Lin et al. [5] proposed another OASG with more essential edges, and thus Lin's method guarantees an optimal solution for any two-pin nets and specific multiple-pin nets (not all cases). However, since Lin's OASG has $O(n^2)$ edges, the worst-case time complexity is $O(n^3)$. Long et al. [6] proposed a faster MTST algorithm with time complexity of $O(|E| \log |V|)$. Since Long's OASG has $O(n)$ edges, Long's method takes $O(n \log n)$ time. Nevertheless, Long's method cannot guarantee the same optimality in [5], and the solution quality is still worse than [5].

To conclude, it is very desirable to develop an $O(n \log n)$-time algorithm with the same theoretical optimality guarantee as [5]. However, under the common structure, it requires a routing graph which has only $O(n)$ edges but guarantees the existence of a rectilinear shortest path between any two pins. Therefore, a new framework should be employed to resolve the bottleneck.

In this paper, for the OARSMT problem, we propose a new framework to develop an $O(n \log n)$-time algorithm with the same theoretical optimality guarantee as [5]. Unlike the two major categories, the new framework directly generates essential solution components without constructing a routing graph or generating invalid initial solutions. We first analyze the geometry mapping of previous works [5,6], and derives a critical path generation method (Section 3.1) to generate $O(n)$ critical paths, which guarantees the existence of an optimal solution for a number of specific multiple-pin nets (not all cases). Our algorithm also increases the overlapping between different paths for improving the wirelength, and performs an $O(n \log n)$-time dynamic local refinement scheme.

We believe that our work has the following contributions:

- Provides an effective and efficient path-based framework to deal with the OARSMT problem.

- Brings about an $O(n \log n)$-time algorithm with theoretical optimality guarantees on a number of specific cases.

Experimental results on 22 benchmarks show that our algorithm is very effective and efficient. Compared with [6], the most effective $O(n \log n)$ method, our algorithm achieves the same speed performance and further improves by 2.7% in wirelength on average, which is significant considering the routability for millions of signal nets. Compared with [7], which achieves the best solution quality, our algorithm achieves 50.1 times speedup on average, while the resulting wirelengths are only 1.1% longer on average. Considering the large and increasing number of obstacles and signal nets, the improvement in run time is very significant.

The rest of this paper is organized as follows. Section 2 formulates the OARSMT problem. Section 3 describes the algorithm and the corresponding theoretical foundations. Section 4 shows the experimental results, and Section 5 makes the conclusion.

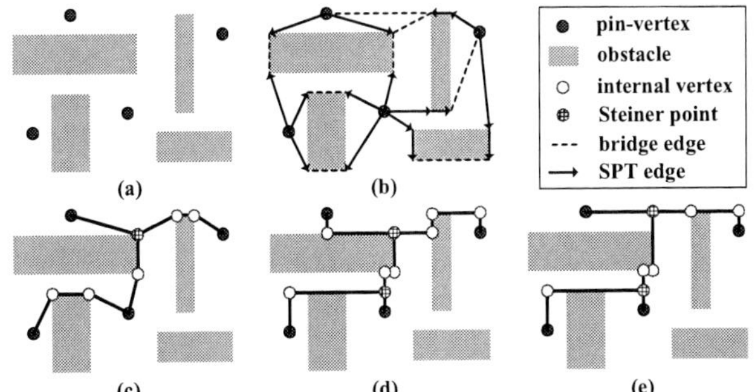

Figure 1: (b)–(e) four phases of our algorithm.

2. PROBLEM FORMULATION

DEFINITION 1. *An **obstacle** is a rectilinear polygon on a plane. Any two obstacles cannot overlap with each other, but they could be line-touched at the boundary and point-touched at the corner.*

Let $P = \{p_1, p_2, \cdots, p_m\}$ be a set of m pin-vertices, $O = \{o_1, o_2, \cdots o_k\}$ be a set of k obstacles, C be the set of N_c corners of obstacles in O, and n be the size of $P \cup C$ as the input size. When all the obstacles are rectangles, we have $n \leq m + 4k$.

All edges/paths/distances are measured by rectilinear distance (L_1 metric), e.g., the length of an edge (v_i, v_j), denoted as $|(v_i, v_j)|$, is $|x_i - x_j| + |y_i - y_j|$. For simplification, we assume that neither an edge nor a path runs over any obstacles except their boundaries. For two vertices v and u, $SP(v, u)$ represents a shortest path between v and u, and the length of $SP(v, u)$ is $|SP(v, u)|$.

- **The Obstacle-Avoiding Rectilinear Steiner Minimal Tree Problem:**

  Given a set P of pins and a set O of obstacles on a plane, construct a tree connecting all the pins in P possibly through some additional points (Steiner points) using vertical or horizontal edges, such that no tree edges intersect the interior of any obstacles in O and the total wirelength is minimized.

Hereafter, we denote an *obstacle-avoiding rectilinear Steiner tree* (OARST) as a solution and an OARSMT as an optimal solution.

3. ALGORITHM

Our path-based algorithm consists of the following four phases. The first two phases generate a solution using critical paths without constructing a routing graph or generating an invalid solution.

1. Critical Path Generation: In this phase, critical paths are generated as solution components [See Fig. 1(b)]. Those critical paths guarantee the existence of desirable solutions.

2. *Obstacle-Avoiding Steiner Tree* (OAST) Construction: An OAST connecting all pin-vertices is constructed by selecting those critical paths in Phase 1 [See Fig. 1(c)]. A greedy path-based method is employed to reduce the wirelength.

3. OARST Construction: An OARST is constructed from the OAST in Phase 2 by transforming slant edges into rectilinear ones [See Fig. 1(d)].

4. Local Refinement: The wirelength of the OARST in Phase 3 is be reduce by an $O(n \log n)$-time dynamic refinement scheme [See Fig. 1(e)].

3.1 Critical Path Generation

We generate $O(n)$ critical paths in $O(n \log n)$ time, and claim that those critical paths guarantee the existence of optimal solutions on a number of specific cases. Section 3.1.1 and Section 3.1.2 analyze Lin's OASG construction [5] and Long's MTST algorithm [6]

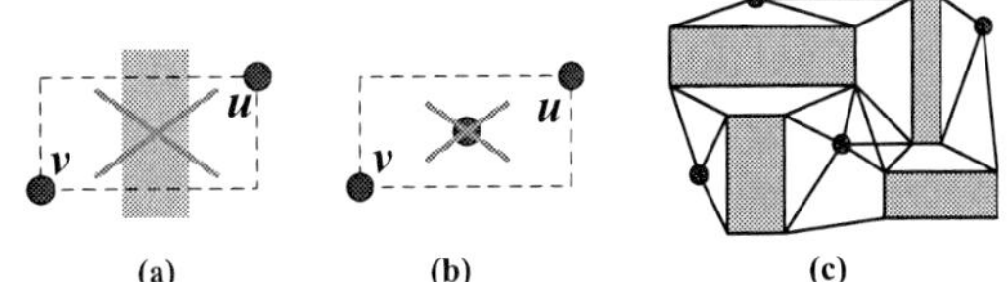

Figure 2: Example Definition 2 and Lin's OASG. (a) no path between u and v exists in the bounding box. (b) one vertex locates inside the bounding box. (c) Lin's OASG for Fig. 1(a).

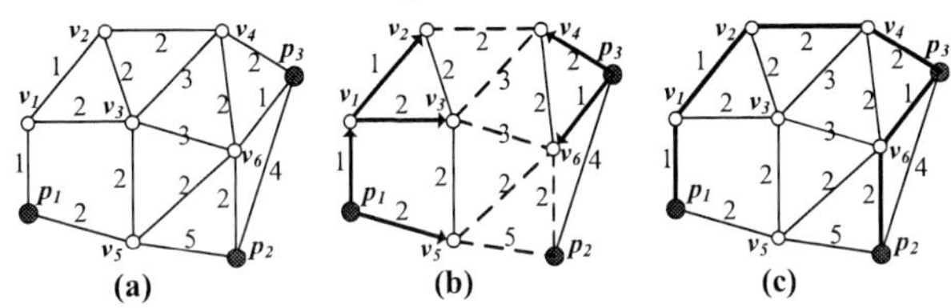

Figure 3: Examples Definition 3 and Definition 4. (a) an instance, where $S = \{p_1, p_2, p_3\}$. (b) a terminal forest and bridge edges of (a), where arrow segments are forest edges and dash segments are bridge edges. (c) an MTST of (a).

from our viewpoints. Section 3.1.3 simulates Long's MTST algorithm on Lin's OASG to analyze the corresponding geometry mapping and conclude the bottleneck of the time complexity. Section 3.1.4 applies shortest path map to generate critical paths, and justifies our claim using geometry information and graph theory.

3.1.1 Lin's OASG Construction

DEFINITION 2. *For two vertices u and v, u is **immediately visible** from v, if (1) there is a path between u and v in their bounding box and (2) no other vertex in $P \cup C$ locates inside or on the boundary of the bounding box [12]. If u is immediately visible from v, u is a **neighbor** of v [5]. For example, u is not a neighbor of v in Fig. 2 (a)–(b). An edge connecting two neighbors is a **visible** edge.*

Lin et al. [5] constructed an OASG by connecting each vertex in $P \cup C$ to all its neighbors in $P \cup C$. Fig. 2(c) shows Lin's OASG for Fig. 1(a). They proved that their OASG guarantees the existence of a shortest path for any two vertices in $P \cup C$, implying that an MTST (Definition 3) of Lin's OASG is an OARSMT for any two-pin net or multiple-pin nets where an OARSMT contains only simple paths between pin-vertices. However, since a vertex has $O(n)$ neighbors, Lin's OASG has $O(n^2)$ edges.

3.1.2 Long's MTST algorithm

DEFINITION 3. *For a graph $G(V, E)$ and a terminal set $S \subseteq V$, a **minimum terminal spanning tree** (MTST) connects all vertices in S using a set of terminal paths with the minimum sum of the lengths of those terminal paths, where a **terminal path** is a path between two vertices in S without other internal vertices in S [6].*

Fig. 3(c) shows an MTST of Fig. 3(a) which consists of two terminal paths, $(p_1 \leftrightarrow v_1 \leftrightarrow v_2 \leftrightarrow v_4 \leftrightarrow p_3)$ and $(p_2 \leftrightarrow v_6 \leftrightarrow p_3)$.

DEFINITION 4. *For a graph $G(V, E)$ and a terminal set $S \subseteq V$, a **terminal forest** F consists of $|S|$ disjoint shortest path trees where each tree T of F is rooted at a terminal $s \in S$ and for each vertex v of T, s is the nearest terminal of v in G (s is called the **root terminal** of v) [6]. An edge $(v, u) \in E$ is a **bridge edge** if v and u belong to different trees in F.*

Fig. 3(b) shows a terminal forest and bridge edges of Fig. 3(a). v_1 belongs to the shortest path tree rooted at p_1 since p_1 is the nearest terminal of v_1; the dash segment connecting v_2 and v_4 is a bridge edge since v_2 and v_4 belong to different trees in the forest.

978-1-60558-497-3/09 $25.00 © 2009 ACM

In [6], Long et al proposed a two-step MTST construction with time complexity of $O(|E|\log|V|)$. At the first step, they extended Dijkstra algorithm [15] to construct a terminal forest. At the second step, they first generated terminal paths via all the bridge edges, e.g., a terminal path via a bridge edge (v_2, v_4) consists of $SP(p_1, v_2)$, (v_2, v_4), and $SP(v_4, p_3)$, i.e., $(p_1 \leftrightarrow v_1 \leftrightarrow v_2 \leftrightarrow v_4 \leftrightarrow p_3)$. In other words, a bridge edge represents a terminal path. Then, they constructed an MTST by applying Kruskal algorithm [15] on a graph whose vertices are terminals and whose edges are those terminal paths generated via bridge edges.

3.1.3 Shortest Path Trees

Since an MTST of Lin's OASG is an OARSMT in many cases [5], we attempt to construct an equivalent solution. However, since Lin's OASG has $O(n^2)$ edges, for $O(n\log n)$ time complexity, we should avoid constructing Lin's OASG. Toward this end, we simulate Long's MTST algorithm on Lin's OASG as shown in Fig. 4 to analyze the corresponding geometry mapping. Since Lin's OASG [5] guarantees a shortest path for any two vertices in $P \cup C$, we study the literature about L_1 shortest paths, and conclude that it is feasible to map a terminal forest of Lin's OASG to Definition 5.

DEFINITION 5. *Given a set P of pin-vertices and a set O of obstacles with a set C of corners, **multi-source shortest path trees** (multi-source SPTs) connect all the vertices in $P \cup C$ such that (1) for each vertex v in $P \cup C$, v belongs to a tree rooted a pin-vertex p in P, (2) p is the nearest pin-vertex of v, and (3) $SP(v, p)$ of the tree is also $SP(v, p)$ in the plane.*

Mitchell [12] proposed a wavefront-based method to construct multi-source SPTs in $O(n\log^2 n)$ time, and later improved the time complexity to $O(n\log n)$ in [13]. For each edge (v, u) of multi-source SPTs in [12], v and u are immediately visible to each other as the same as that of Lin's OASG, implying that the multi-source SPTs are equivalent to a terminal forest of Lin's OASG. Therefore, a terminal forest of Lin's OASG can be constructed in $O(n\log n)$ time without constructing Lin's OASG, and the bottleneck of the time complexity has become the handling of those bridge edges in Fig. 4(b), each of which represents a terminal path.

Although the number of those bridge edges could be $\Omega(n^2)$, in Fig. 4, we observe many redundant bridge edges which must not be in an MTST of Lin's OASG. Therefore, we conjecture that to obtain the equivalent solution, only $O(n)$ bridge edges need to be considered and can be constructed in $O(n\log n)$ time.

3.1.4 Shortest Path Map

To resolve the bottleneck of the time complexity and prove our conjecture, we should reduce redundant bridge edges. In [14], for constructing a *minimum spanning tree* (MST) in a plane, a Voronoi diagram can be applied to divide a plane and thus reduce the redundant edges. Similarly, we use a shortest path map to divide a plane, consider the obstacles, and reduce the redundant bridge edges.

DEFINITION 6. *Given a set P of pin-vertices and a set O of obstacles with a set C of corners, a **shortest path map** (SPM) is a subdivision of plane where (1) each region belongs to a vertex $v \in P \cup C$ (called the site of region), (2) all points in the region of v share the same nearest pin-vertex $p \in P$, and (3) those points have the same predecessor v (the site of the region) along their shortest path to p. Each vertex in $P \cup C$ has only one region.*

Fig. 5(b) shows an SPM for Fig. 1(a). For example, all points in the region R_6, whose site is v_6, share the same nearest pin-vertex p_3, and all those points have at least one shortest path to p_3 passing through v_6. Noticeably, a region may have no area, e.g., the region R_5, which is represented by a red bold line segment and referred

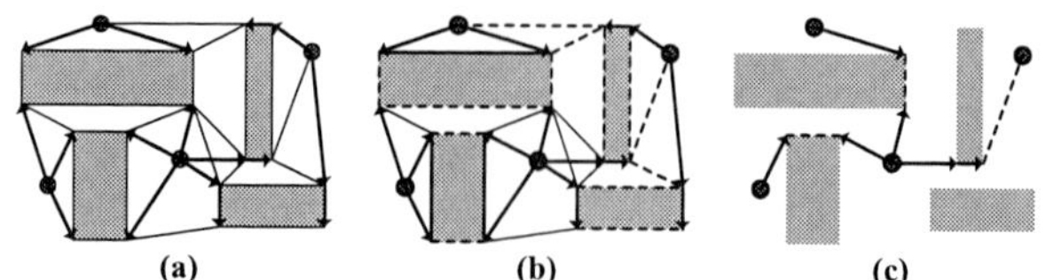

Figure 4: Long's MTST algorithm on Lin's OASG. (a) a terminal forest of Fig. 2(c). (b) terminal paths via bridge edges of (a). (c) an MTST generated by those terminal paths in (b).

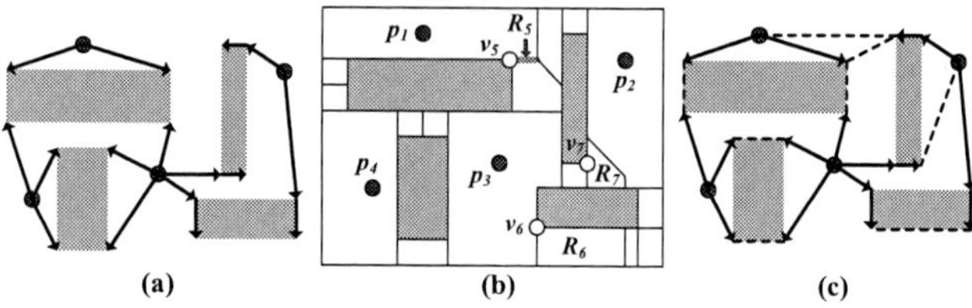

Figure 5: The critical path generation for the instance in Fig. 1(a). (a) multi-source shortest path trees. (b) a shortest path map. (c) critical paths represented by bridge edges (dash segments) and multi-source SPTs.

by a blue arrow, has no area. The slopes of those slant boundaries are 1 or -1 since in L_1 metric, the slope of a bisector must be one of $\{-1, 1, 0 \ (horizontal), \infty \ (vertical)\}$

We propose the critical path generation as follows:

1. Multi-source SPTs are constructed using the wavefront-based method in [12, 13] as shown in Fig. 5(a).

2. An SPM is constructed using the information of the multi-source SPTs as shown in Fig. 5(b).

3. For each two adjacent regions, if their sites have different nearest pin-vertices and are immediately visible to each other, a bridge edge between the two sites is constructed to generate a critical path. Fig. 5(c) shows those generated critical paths and represents them using bridges edges and multi-source SPTs. For a bridge edge (v_1, v_2), assuming the nearest pin-vertices of v_1 and v_2 to be p_1 and p_2 respectively, the critical path is $SP(p_1, v_1) \leftrightarrow (v_1, v_2) \leftrightarrow SP(v_2, p_2)$.

We will prove that those critical paths guarantee the existence of an equivalent solution to an MTST of Lin's OASG. Since a terminal forest of Lin's OASG is multi-source SPTs (discussed in Section 3.1.3), according to Long's MTST algorithm, we only need to discuss those bridge edges. If the critical path generation fails the guarantee, at least one essential bridge edge is not constructed. Without loss of generality, we assume the essential bridge edge to be (v_1, v_2). Since (v_1, v_2) is not constructed by the critical path generation, (v_1, v_2) should be broken by a region belonging to other vertex v_3. Assume the nearest pin-vertices of v_1, v_2, and v_3 to be p_1, p_2, and p_3 respectively, and v_4 to be a point on (v_1, v_2) in the region of v_3. According to Definition 6, $|SP(p_3, v_3)| + |(v_3, v_4)|$ is smaller than $|SP(p_1, v_1)| + |(v_1, v_4)|$ and $|SP(p_2, v_2)| + |(v_2, v_4)|$, implying that $|SP(p_3, v_3)| + |(v_3, v_4)| + |SP(p_1, v_1)| + |(v_1, v_4)|$ and $|SP(p_3, v_3)| + |(v_3, v_4)| + |SP(p_2, v_2)| + |(v_2, v_4)|$ is smaller than $|SP(p_1, v_1)| + |(v_1, v_4)| + |SP(p_2, v_2)| + |(v_2, v_4)|$, which is the length of the terminal path via (v_1, v_2). That is, at least two terminal paths for (p_1, p_3) and (p_2, p_3) respectively are shorter than the terminal path for (p_1, p_2) via (v_1, v_2). According to the cycle property of an MST [15], the terminal path via (v_1, v_2) can be safely deleted, i.e., (v_1, v_2) is not essential. There exists a contradiction.

Since an MTST of Lin's OASG is an OARSMT for any two-pin nets and multiple-pin nets where an OARSMT consists of only simple paths between pin-vertices, we conclude Theorem 1.

978-1-60558-497-3/09 $25.00 © 2009 ACM

316

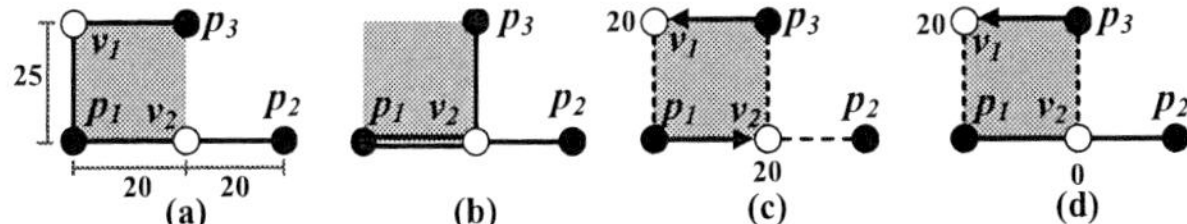

Figure 6: The overlapping of different paths. (a) no overlapping between the two paths. (b) overlapping at (v_2, p_3). (c) the critical paths generated by Section 3.1. (d) $|SP(p_1, p_3)|$ via (v_2, p_3) has become 20 after update operation.

THEOREM 1. *Our critical path generation guarantees the existence of an OARSMT for any two-pin net or multiple-pin nets where an OARSMT consists of only simple paths between pin-vertices.*

We analyze the time complexity of the critical path generation and the number of those generated critical paths below. Since the multi-source SPTs construction takes $O(n \log n)$ time in [12, 13], we only analyze step 2–3. In [12], Mitchell also claimed that an SPM, i.e. our step 2, can be extended from multi-source SPTs in $O(n \log n)$ time. By Euler's formula, the number of boundaries must be linear to the number of faces for a planar subdivision. Since there are $O(n)$ regions in an SPM (by Definition 6), there are $O(n)$ pairs of adjacent regions, implying that our critical path generation constructs $O(n)$ bridge edges (critical paths). To check if a bridge edge is visible can be done by searching other vertices locating inside the bounding box of the two endpoints. The search operation takes $O(\log n)$ time after constructing a range search tree with fractional cascading [14] among $P \cup C$, which takes $O(n \log n)$ time. Therefore, we can conclude Theorem 2. Theorem 1 and Theorem 2 strongly support our conjecture in Section 3.1.3.

THEOREM 2. *Our critical path generation provides $O(n)$ critical paths in $O(n \log n)$ time.*

3.2 OAST Construction

In this Section, we propose a greedy method to construct an *obstacle-avoiding Steiner tree* (OAST) based on those critical paths in Section 3.1. Unlike most previous works, for better solution quality, our greedy method attempts to increase the overlapping between different paths, which is critical but often neglected.

Most MTST-based algorithms [1, 5, 6] select paths (edges) according to their lengths without considering the overlapping between different paths, and thus may lose much potential improvement of wirelength. For example, as shown in Fig. 6, $|SP(p_1, p_2)|$, $|SP(p_2, p_3)|$, and $|SP(p_1, p_3)|$ are 40, 45, and 45 respectively, implying that an MTST can consist of $SP(p_1, p_2)$ and $SP(p_1, p_3)$. However, there are two possibilities to select $SP(p_1, p_3)$, $p_1 \leftrightarrow v_1 \leftrightarrow p_3$ and $p_1 \leftrightarrow v_2 \leftrightarrow p_3$, as shown in Fig. 6(a) and Fig. 6(b) respectively. In the former case, $SP(p_1, p_2)$ and $SP(p_2, p_3)$ do not overlap, while in the latter case, $SP(p_1, p_2)$ and $SP(p_2, p_3)$ overlap at (p_1, v_2) and result in an obviously redundant edge, which will be removed easily. After removing the redundant edge (p_1, v_2), the wirelength of Fig. 6(b) is 65, while that of Fig. 6(a) is 85. Thus, from our viewpoint, Fig. 6(b) contains more potential improvement of wirelength than Fig. 6(a). In short, increasing the overlapping between different paths will lower the wirelength.

Our path-based framework provides a potential way to increase the overlapping of different paths. For example, Fig. 6(c) shows those critical paths generated by Section 3.1 and represents them using three shortest path trees and three bridge edges (dash segments). In detail, bridge edges, (p_1, v_1), (v_2, p_2), and (v_2, p_3), represent $p_1 \leftrightarrow v_1 \leftrightarrow p_3$, $p_1 \leftrightarrow v_2 \leftrightarrow p_2$, and $p_1 \leftrightarrow v_2 \leftrightarrow p_3$, whose lengths are 45, 40, and 45, respectively. The values adjacent to v_1 and v_2 are the distances from them to their nearest pin-vertices, e.g., $|SP(p_1, v_2)|$ is 20. Since $p_1 \leftrightarrow v_2 \leftrightarrow p_2$ has the lightest length, we will first select this critical path. After the se-

```
Algorithm: Greedy_OAST_Construction(P, C, BE, TE, E)
Input:  P /* the set of pin-vertices */
        C /* the set of obstacle corners */
        BE /* the set of bridge edges */
        TE /* the set of directed tree edges
              of multi-source SPTs*/
Output: E /* the set of edges of the OAST */
1   for each directed tree edge e = (v, u) ∈ TE
2       p(u) ← v
3   for each vertex v in P ∪ C
4       w(v) ← |SP(v, np(v))|
        /* np(v) is the nearest pin-vertex of v */
5   Heap H_cp ← φ
6   for each bridge edge e = (v, u) ∈ BE
7       w(cp(e)) ← w(v) + |(v, u)| + w(u)
        /* cp(e) denotes the critical path via e */
8       H_cp.insert(e, w(cp(e)))
9   while H_cp is nonempty
10      e(v, u) ← H_cp.extractMin()
11      if Find-Set(np(v)) ≠ Find-Set(np(u))
12          Set-Union( Find-Set(np(v)) , Find-Set(np(u)) )
13          E ← E ∪ {(v, u)}
14          for each vertex v' ∈ {v, u}
15              while v'.makred = false and p(v') ≠ null
16                  w(v') ← 0
17                  for each bridge edge e' = (v', u') ∈ BE
18                      w(cp(e')) ← w(v') + |(v', u')| + w(u')
19                      H_cp.decreaseKey(e', w(cp(e')))
20                  v'.marked ← true
21                  E ← E ∪ {(v', p(v'))}
22                  v' ← p(v')
```

Figure 7: The greedy OAST construction algorithm.

lection, as shown in Fig. 6(d), the distance between v_2 and p_1 can be viewed as 0 since any path to p_1 through v_2 can share the path between p_1 and v_2. Then, we update bridge edges connected to v_2, i.e. (v_2, p_3), to reduce $|p_1 \leftrightarrow v_2 \leftrightarrow p_3|$ to be 25. Therefore, we will connect $p_1 \leftrightarrow v_2 \leftrightarrow p_3$ instead of $p_1 \leftrightarrow v_1 \leftrightarrow p_3$ to obtain the same result with Fig. 6(b). To conclude, by updating bridge edges connected to selected critical paths, our path-based framework shares the edges of selected paths with unselected ones, and thus increases the overlapping between different paths.

The greedy OAST construction algorithm is summarized in Fig. 7. Lines 1–8 initialize the information representing those critical paths. Lines 1–2 assign the predecessor to each vertex, Lines 3–4 assign weight to each vertex, and Lines 5–8 initial H_{cp} by assigning weight to each critical path and inserting the corresponding bridge edge to H_{cp}. Lines 9–22 construct an OAST, and increase the overlapping of different paths by updating bridge edges. Line 11 determines if a critical path results in a cycle in P; Lines 13–22 recursively connect edges in the critical path and update the corresponding bridge edges. In details, Lines 16–19 update the corresponding bridge edges; Lines 20–22 connect the edges leading to the nearest pin-vertex. It is clear that the recursive updates of bridge edges will further take benefit of the path overlapping. In fact, the OAST construction does not connect those obviously redundant edges.

We prove the time complexity to be $O(n \log n)$. Since multi-source SPTs are a forest connecting all vertices in $P \cup C$, $|TE|$ is $O(n)$, and thus Lines 1–2 take $O(n)$ time. Lines 3–4 can be done by traversing tree edges in TE from each root, which takes $O(n)$ time. Lines 5–8 take time loglinear to the size of BE, which is $O(n)$ by Theorem 2. Since Lines 14–22 trace a vertex at most once and update a bridge edge at most twice, Lines 9–22 perform $O(n)$ times find-set, trace-vertex, decrease-key operations. Since each operation takes $O(\log n)$ time, Lines 9–22 take $O(n \log n)$ time.

In not considering the overlapping between different paths, we will obtain an OAST as shown in Fig. 4(c), while our greedy OAST construction obtains a better OAST as shown in Fig. 1(c).

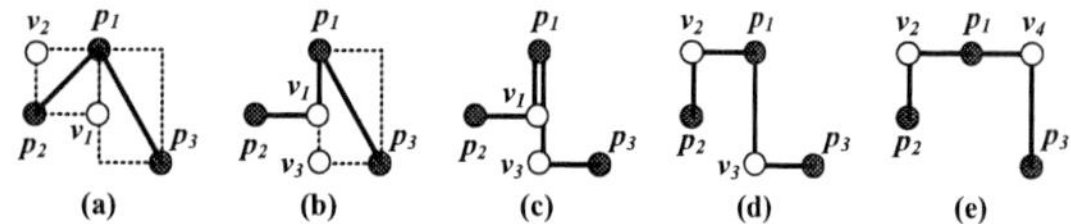

Figure 8: The rectilinear transformation for slant edges. (a) an instance. (b)–(c) transformation with edge overlapping. (d)–(e) other cases without edge overlapping.

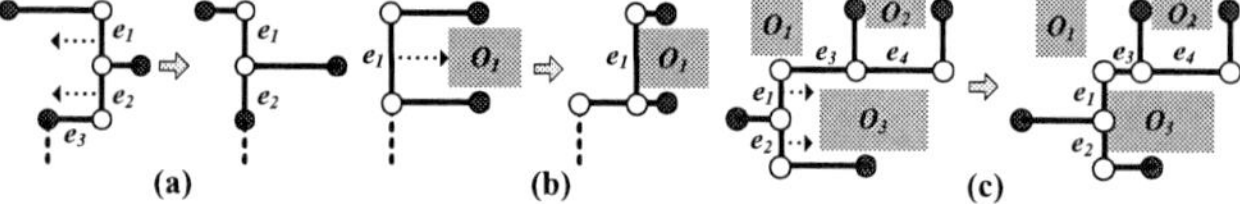

Figure 9: Refinement Patterns. (a)-(b) two cases of U-shaped patterns. (c) the up nearest obstacle of a segment, $\{e_3, e_4\}$, has been changed to O_2 from O_1 after moving $\{e_1, e_2\}$ right.

3.3 OARST Construction

We construct an *obstacle-avoiding rectilinear Steiner tree* (OARST) from the OAST in Section 3.2 by transforming all slant edges into rectilinear ones. Since all edges of the OAST are visible we can directly transform an edge of the OAST into L-shaped rectilinear edges. For example, as shown in Fig. 8(a), since no obstacle intersects the bounding box of p_1 and p_2 (otherwise, (p_1, p_2) must not be visible), (p_1, p_2) can be transformed to L-shaped rectilinear edges, (p_1, v_1) and (v_1, p_2) (or (p_1, v_2) and (v_2, p_2)).

The OARST construction also considers the overlapping between different edges. For example, if (p_1, p_2) has been transformed to (p_1, v_1) and (v_1, p_2) as shown in Fig. 8(b), (p_1, p_3) will be transformed to (p_1, v_3) and (v_3, p_3) to result in the edge overlapping $(p1, v1)$ as shown in Fig. 8(c), which can be directly removed to improve the wirelength. On the other hand, if (p_1, p_2) has been transformed to (p_1, v_2) and (v_2, p_2), Fig. 8(d) or Fig. 8(e) will occur. Nevertheless, since no obstacle intersects the bounding boxes of (p_1, p_2) and (p_1, p_3), Fig. 8(d) and Fig. 8(e) can be refined to Fig. 8(c) by the local refinement in Section 3.4. It is clear that the whole OARST construction takes $O(n \log n)$ time.

3.4 Efficient Local Refinement

DEFINITION 7. *A **segment** is a connected graph whose edges are arranged in the same line, e.g., $\{e_1, e_2\}$ in Fig. 9(a) is a segment. For a segment s, an **adjacent** segment is a segment connected to and perpendicular to s, e.g., $\{e_1, e_2\}$ in the left part of Fig. 9(a) has three adjacent segments.*

We attempt to perform dynamic local refinements to reduce the wirelength in $O(n \log n)$ time. Lin et al. [5] introduced a U-shaped refinement handling the two cases in Fig. 9(a)–(b). Based on the U-shaped refinement, we conclude that if a segment has more adjacent segments on one side than the other side, moving this segment to the former side may reduce the wirelength. We called a segment as *movable* if it can be moved to reduce the wirelength.

As shown in Fig. 9(b), the moving offset of a segment may depend on the nearest obstacle. However, the nearest obstacle of a segment cannot be pre-computed before dynamic local refinements since the nearest obstacle may be changed. For example, in Fig. 9(c), the up nearest obstacle of $\{e_3, e_4\}$ originally is O_1, while the up nearest obstacle has become O_2 after moving $\{e_1, e_2\}$ right. For $O(n \log n)$ time complexity, we should find a way to compute the nearest obstacle for a movable segment in $O(\log n)$ time.

The right nearest obstacle of a segment is also the first touched obstacle when moving the segment right. As shown in Fig. 10(a)–(b), moving a segment to touch an obstacle can be divided into two cases, (1) touch the obstacle corners and (2) just touch the obstacle boundary. In the latter case, if the nearest obstacle influences

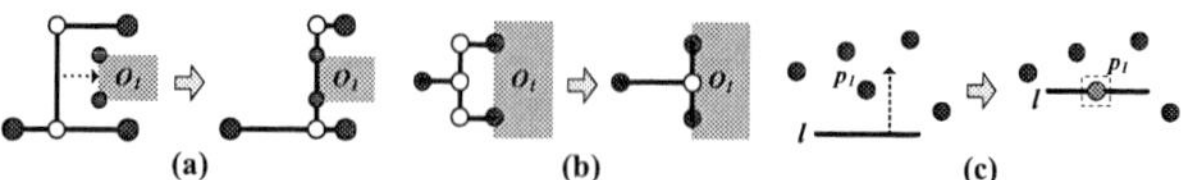

Figure 10: Computation of the nearest obstacles. (a) touching the corners of an obstacle. (b) just touching the boundary of an obstacle. (c) the segment dragging problem.

the moving offset, the movable segment should have an adjacent segment with one endpoint locating at the boundary of the obstacle. Therefore, the moving offset in the latter case can be computed using those adjacent segments, and we only discuss the first case.

Chazelle [16] studied *the segment dragging query problem*: Given a set of n points, pick a horizontal (vertical) segment and answer the first hit point when dragging the segment vertically (horizontally). As shown in Fig. 10(c), dragging l up will first hit p_1, and p_1 is the answer of this segment dragging query. Chazelle proposed a method to answer the segment dragging query in $O(\log n)$ time after an $O(n \log n)$-time preprocessing. Therefore, by applying Chazelle's method on the corner set C, we can compute the nearest obstacle for the first case in $O(\log n)$ time.

Considering the tradeoff between efficiency and solution quality, our refinement scheme operates each movable segment with at most 8 adjacent segments until no such segment exists. Moving a movable segment may generate a new movable segment (e.g. merging two segments), but in the case, the movement will remove at least one edge such as e_3 in Fig. 9(a). Since the OARST has $O(n)$ edges, the refinement scheme will operate $O(n)$ movable segments. For each segment s, since s has at most 8 adjacent segments and the nearest obstacle is computed in $O(\log n)$ time, the moving offset can be computed in $O(\log n)$ time. Therefore, the refinement scheme takes $O(n \log n)$ time.

3.5 Discussion

Since each phase takes $O(n \log n)$ time, the time complexity of the whole construction is still $O(n \log n)$. By Theorem 1 and the details in Section 3.2 to 3.4, our algorithm guarantees all the optimality in Section 3.5 of [5]. We can conclude the following.

THEOREM 3. *Our algorithm constructs an OARST in $O(n \log n)$ time, and guarantees to provide an OARSMT for any two-pin nets and multiple-pin nets where the topology of an OARSMT contains only simple paths between pin-vertices.*

The output of Phase 1–2 is similar to [5] and [6], while [5] takes $O(n^3)$ time and [6] cannot guarantee the same optimality. Long et al. [6] reduced the redundant terminal paths in a graph, but for $O(n \log n)$ time, they omitted some essential edges to construct an $O(n)$-space graph. Our framework directly reduces the redundant terminal paths in geometry domain without constructing a graph. This is why our algorithm can satisfy Theorem 1.

To conclude, compared with recent works [1]– [6], our path-based framework has a global view of obstacles, guarantees the existence of desirable solutions, and keeps only $O(n)$-space solution components. It really gives a new efficient and effective way to deal with the OARSMT problem.

4. EXPERIMENTAL RESULTS

We have implemented our algorithm in C. There are 22 benchmark circuits, five industrial test cases (ind01–ind05) from Synopsys, 12 test cases used in [2] (rc01–rc12), and five random test cases used in [5] (rt01–rt05). Our experiments were conducted on a Linux server with two 2.4-GHz Intel processors and 8-GB memory.

We compare our algorithm with those presented in [5], [6] and [7]. To the best of our knowledge, the approach in [6] is the most

978-1-60558-497-3/09 $25.00 © 2009 ACM

Table 1: Comparison on the Total Wirelength and the CPU Time.

Test Cases	m / k	Total Wirelength				Imp.(%) ($\frac{X-D}{X}$)			Time (sec.)				speedup ($\frac{E}{F}$)
		[6] (A)	[5] (B)	[7](C)	ours (D)	(X=A)	(X=B)	(X=C)	[6]	[5]	[7] (E)	ours (F)	
IND01	10 / 32	639	632	619	626	2.0	0.9	-1.1	0.01	< 0.01	0.007	0.001	7.00 x
IND02	10 / 43	10,000	9,600	9,500	9,700	3.0	-1.0	-2.1	0.01	< 0.01	0.042	0.001	42.00 x
IND03	10 / 50	623	613	600	600	3.7	2.1	0.0	0.01	< 0.01	0.009	0.001	9.00 x
IND04	25 / 79	1,126	1,121	1,096	1,095	2.8	2.3	0.1	0.02	< 0.01	0.033	0.002	16.50 x
IND05	33 / 71	1,379	1,364	1,360	1,364	1.1	0.0	-0.3	0.02	0.01	0.008	0.002	4.00 x
RC01	10 / 10	27,540	26,900	25,980	26,740	2.9	0.6	-2.9	0.01	< 0.01	0.044	< 0.001	-
RC02	30 / 10	41,930	42,210	42,010	42,070	-0.3	0.3	-0.1	0.01	< 0.01	0.103	0.001	103.00 x
RC03	50 / 10	54,180	55,750	54,390	54,550	-0.7	2.2	-0.3	0.01	< 0.01	0.097	0.001	97.00 x
RC04	70 / 10	59,050	60,350	59,740	59,390	-0.6	1.6	0.6	0.02	< 0.01	0.121	0.001	121.00 x
RC05	100 / 10	75,630	76,330	74,650	75,430	0.3	1.2	-1.0	0.02	0.01	0.092	0.001	92.00 x
RC06	100 / 500	86,381	83,365	81,607	81,903	5.2	1.8	-0.4	0.13	0.24	0.577	0.017	33.94 x
RC07	200 / 500	117,093	113,260	111,542	111,752	4.6	1.3	-0.2	0.15	0.43	0.714	0.026	27.46 x
RC08	200 / 800	122,306	118,747	115,931	118,349	3.2	0.3	-2.1	0.27	0.83	1.194	0.043	27.77 x
RC09	200 / 1,000	119,308	116,168	113,460	114,928	3.7	1.1	-1.3	0.36	0.91	1.502	0.049	30.65 x
RC10	500 / 100	167,978	170,690	167,620	167,540	0.3	1.8	0.0	0.08	0.62	0.291	0.016	18.19 x
RC11	1,000 / 100	232,381	236,615	235,283	234,097	-0.7	1.1	0.5	0.14	3.15	0.844	0.021	40.19 x
RC12	1,000 / 10,000	842,689	789,097	761,606	780,528	7.4	1.1	-2.5	5.88	118.52	192.920	0.681	283.29 x
RT01	10 / 500	2,362	2,267	2,231	2,259	4.4	0.4	-1.3	0.12	0.06	0.171	0.017	10.06 x
RT02	50 / 500	52,218	48,441	47,297	48,684	6.8	-0.5	-2.9	0.11	0.11	0.612	0.018	34.00 x
RT03	100 / 500	8,645	8,368	8,187	8,347	3.4	0.3	-2.0	0.13	0.47	0.211	0.020	10.55 x
RT04	100 / 1,000	10,580	10,306	9,914	10,221	3.4	0.8	-3.1	0.42	0.95	0.364	0.040	9.10 x
RT05	200 / 2,000	55,286	53,993	52,473	53,745	2.8	0.5	-2.4	1.43	2.06	2.759	0.078	35.37 x
Average						2.7	0.9	-1.1	Average				50.10 x

effective $O(n \log n)$-time method, and the method in [7] achieves the best solution quality. The results of [6] are quoted from the paper, where the algorithm was conducted on a Redhat Linux server with two 2.1-GHz AMD processors and 2-GB memory. The results of [5] are quoted from the paper, where the algorithm was conducted on a Ubuntu 6.06 server with one 2-GHz AMD-64 CPU and 8-GB memory. We have obtained the program in [7], and execute their program on our platform.

Table 1 lists the total wirelengths and the CPU times of these algorithms. "speedup" compares the run time of [7] and ours.

Compared with [6], the most effective $O(n \log n)$-time method, our algorithm achieves the same speed performance and further improves by 2.7% in wirelength on average, which is significant considering the routability for millions of signal nets. The wirelength of our algorithm is at most 7.4 % shorter than that of [6] (rc12), but at most only 0.7% longer (rc03). Besides, our algorithm guarantees the same theoretical optimality as that in [5], but [6] cannot. Therefore, the solution quality of our algorithm could be more stable.

Compared with [7], which achieves the best solution quality, our algorithm runs much faster, while the wirelengths of the resulting solutions are only 1.1% longer on average. As shown in Table 1, across all the 22 benchmarks, our algorithm achieves 50.1 times speedup on average compared with [7]. For rc12, which contains 10 000 obstacles, our algorithm terminates within only one second, while that in [7] takes 192 seconds. Considering the large and increasing number of obstacles and signal nets in a modern IC design, the advantage in the run time is very significant.

To conclude, the above analysis shows that our algorithm achieves the best speed performance, and the solution quality is still comparable to the best previous work [7]. Our algorithm meets the requirements of the OARST construction in a modern IC design.

5. CONCLUSIONS

In this paper, we propose a path-based framework to deal with the OARSMT problem. We prove that if using the new framework, the theoretical optimality guarantee in [5] can be obtained in $O(n \log n)$ time rather than the original $O(n^3)$. Compared with previous works, our framework has a global view of obstacles, guarantees the existence of desirable solutions, and keeps only $O(n)$ solution components in $O(n)$ space. Therefore, our framework gives key insights into the OARSMT problem, and provides a new efficient and effective way to develop more desirable OARSMT algorithms. Experimental results have shown the high efficiency and effectiveness of our algorithm, and thus our algorithm can meet the requirements of the OARST construction in a modern IC design.

6. ACKNOWLEDGEMENTS

This research is supported by the National Science Council, Taiwan under Grant NSC 97-2221-E-002-216-MY3 and Excellent Research Projects of National Taiwan University, 95R0062-AE00-05.

7. REFERENCES

[1] Z. Shen, C. Chu, Y. Li, "Efficient rectilinear steiner tree construction with rectilinear blockages," *in Proc. ICCD*, pp. 38–44, 2005.

[2] Z. Feng, Y. Hu, T. Jing, X. Hong, X. Hu and G. Yan, "An $O(n \log n)$ algorithm for obstacle-avoiding routing tree construction in the lambda-geometry plane," *in Proc. ISPD*, pp 48–55, 2006.

[3] Y. Shi, P. Mesa, H. Yu and L. He, "Circuit simulation based obstacle-aware Steiner routing," *in Proc. DAC*, pp. 385–338, 2006.

[4] P. C. Wu, J. R. Gao, and T. W. Wang, "A fast and stable algorithm for obstacle-avoiding rectilinear Steiner minimal tree," *in Proc. ASP-DAC*, pp. 262–267, 2007.

[5] C. W. Lin, S. Y. Chen, C. F. Li, Y. W. Chang and C. L. Yang, "Obstacle-avoiding rectilinear Steiner Tree construction based on spanning graphs," *IEEE Trans. Computer-Aided Design*, Vol. 27, No.4, pp. 643–653, 2008.

[6] J. Long, H. Zhou, and S. Memik, "EBOARST: an efficient edge-based obstacle-avoiding rectilinear steiner tree construction algorithm ," *IEEE Trans. Computer-Aided Design*, Vol. 27, No.12, pp. 2169–2182, 2008.

[7] L. Li and Evangeline F. Y. Young, "Obstacle-avoiding rectilinear steiner tree construction," *in Proc. ICCAD*, 2008.

[8] M. Garey and D. Johnson, "The Steiner tree problem is NP-Complete," *SIAM J.APPL.MATH.*, Vol. 32, No.4, pp. 826–834 ,June 1977.

[9] *International Technology Roadmap for Semiconductors* (ITRS), 2007, http://www.itrs.net.

[10] M. Pan and C. Chu, "FastRoute: a step to integrate global routing into placement," *in Proc. ICCAD*, pp. 464–471, 2006.

[11] Y. Yang, Q. Zhu, T. Jing, X.Hong, and Y. Wang, "Rectilinear steiner minimal trees among obstacles," *in Proc. ASIC*, pp. 630–635, 2003.

[12] J. S. B. Mitchell, "L_1 shortest paths among polygonal obstacles in the plane," *Algorithmica*, Vol. 8, pp. 55–88, 1992.

[13] J. S. B. Mitchell, "An optimal algorithm for shortest rectilinear paths among obstacles," *in Proc. CCCG*, 1989.

[14] M. de Berg, M. van Kreveld, M. Overmars, and O. Schwarzkopf, *Computational Geometry: Algorithms and Applications*, 2nd edition, Springer, 2000.

[15] T. Cormen, C. Leiserson, R. Rivest, and C. Stein, *Introduction to Algorithms*, 2nd edition, The MIT Press, 2001.

[16] B. Chazelle, "An algorithm for segment dragging and its implementation," *Algorithmica*, Vol. 3, pp 205–221, 1988.

978-1-60558-497-3/09 $25.00 © 2009 ACM

GRIP: Scalable 3D Global Routing Using Integer Programming

Tai-Hsuan Wu, Azadeh Davoodi
Department of Electrical and Computer
Engineering

Jeffrey T. Linderoth
Department of Industrial and Systems
Engineering

University of Wisconsin - Madison WI 53706
{twu3,adavoodi,linderoth}@wisc.edu

ABSTRACT

We propose GRIP, a scalable global routing technique via Integer Programming (IP). GRIP optimizes wirelength and via cost without going through a layer assignment phase. GRIP selects the route for each net from a set of candidate routes that are generated based on an estimate of congestion generated by a linear programming *pricing* phase. To achieve scalability, the original IP is decomposed into smaller ones corresponding to balanced rectangular subregions on the chip. We introduce the concept of a *floating terminal* for a net, which allows flexibility to route long nets going through multiple subregions. We also use the IP to plan the routing of long nets, detouring them from congested subregions. For ISPD 2007 benchmarks, we obtain 3.9% and 11.3% average improvement in wirelength and via cost for the 2D and 3D versions respectively, compared to the best results reported in the open literature.

Categories and Subject Descriptors

B.7.2 [**Integrated Circuits**]: Design Aids

General Terms

Algorithms, Design

Keywords

Global Routing, Integer Programming

1. INTRODUCTION

Design of Integrated Circuits in nanometer regime is subject to many obstacles such as manufacturability, variability, yield-loss and timing failures. With increasing design sizes and shrinking device geometries, the severity of many of these issues is impacted by the routing of interconnects. Global routing (GR) is the primary step of routing during which the net regions will be planned. It has increasingly gained significance in recent years due to its larger role on the above-mentioned issues. It is crucial that the GR generates a high-quality routing solution in a manageable runtime that scales well with the design size.

Permission to make digital or hard copies of part or all of this work for personal or classroom use is granted without fee provided that copies are not made or distributed for profit or commercial advantage and that copies bear this notice and the full citation on the first page. To copy otherwise, to republish, to post on servers or to redistribute to lists, requires prior specific permission and/or a fee.
DAC'09, July 26-31, 2009, San Francisco, California, USA

Among the two-categories of *concurrent* [8], [20], [21], [15], [4], [3], [5] and *sequential* [19], [22], [6], [18], [17] global routers, the latter has been more successful in terms of the tradeoff between solution quality and execution runtime. Sequential approaches have much smaller runtime but rely on an ordering of the nets and applying rip-up and re-route.

Much attention has been given to sequential approaches because the concurrent ones are inherently more time consuming. The most recent concurrent approach is the IP-based BoxRouter [8]. BoxRouter is extremely fast, but it only considers L-shaped routes in the IP. Recently [15] proposes the use of a few more basic patterns for routing each net in a progressive congestion-driven IP formulation. Similarly, [15] also has the downside of only considering a limited number of pre-determined patterns in the IP formulation. This in turn requires applying complicated pre- and post-processing steps to generate a final solution [8]. Also recently [21] proposes a hierarchical IP formulation for GR. However, as we discuss, the major downside of any hierarchical GR is failure to effectively account for the impact of short nets. In this paper, we make the following contributions to overcome some of these challenges:

1. We propose an IP formulation that simultaneously minimizes wirelength and via cost, thereby skipping the traditional layer assignment phase. The IP works with 3D Steiner routes and heavily penalizes overflow.

2. Promising routes for each net are generated by a linear programming *pricing* phase that takes into account a measure of current congestion at each iteration. The IP decides among these many promising routes for each net while considering capacity constraints.

3. To achieve scalability, we decompose the chip area into rectangular subregions to achieve *balanced* and smaller-sized IPs and then effectively integrates their solutions. The execution runtime depends on the number of subregions, some of which can be processed in parallel.

4. We introduce the concept of *floating terminals* for a net. Floating terminals allow flexibility in routing long nets through subregions while remaining compatible with our pricing phase for candidate route generation. We also discuss a pricing procedure for planning the regions through which long nets will travel during the decomposition of the IP into subregions.

Compared to [3], our candidate routes are generated by varying a base Steiner route, considering congestion as weights in a grid-graph as well as the other candidate routes generated so far at each iteration of column generation.

In our simulation results, we achieve an average 11.3% improvement in total wirelength and via cost of 3D ISPD 2007 benchmarks, compared to the best result reported for each benchmark. This is due to the concurrent nature of our approach, the pricing phase for candidate route generation, and directly working with the 3D model of the problem.

The organization of the paper is as follow. In Section 2, we discuss the IP formulation and customized column generation procedure. In Section 3, we discuss IP decomposition, subregion extraction, long net planning, and subregion solution integration. Simulation results are in Section 4.

2. PRICE AND BRANCH FOR GR

The algorithm proposed for global routing is based on the (approximate) solution of a large-scale integer program (IP). The solution procedure begins with a column generation (pricing) phase, followed by branch-and-bound.

2.1 An Integer Program for the GR Problem

In a mathematical description of the global routing problem, we are given a grid-graph $G = (V, E)$ describing the network topology, a set of (multi-terminal) nets given by $\mathcal{N} = \{T_1, T_2, \ldots, T_N\}$, (with $T_i \subset V$), and edge capacities u_e and weights c_e $\forall e \in E$. Denote by $\mathcal{T}(T_i)$ the collection of *all* Steiner trees (routes) connecting the terminals in T_i, and let the parameter $a_{te} = 1$ if Steiner tree t contains edge $e \in E$, $a_{te} = 0$ otherwise. Define the binary decision variable x_{it} that is equal to 1 if and only if net T_i is routed with route $t \in \mathcal{T}(T_i)$. An integer program for the global routing problem can be written as

$$\min_{x,s} \sum_{i=1}^{N} \sum_{t \in \mathcal{T}(T_i)} c_{it} x_{it} + \sum_{i=1}^{N} M s_i \qquad \text{(ILP-GR)}$$

$$\begin{cases} \sum_{t \in \mathcal{T}(T_i)} x_{it} + s_i = 1 & \forall i = 1, \ldots, N \\ \sum_{i=1}^{N} \sum_{t \in \mathcal{T}(T_i)} a_{te} x_{it} \leq u_e & \forall e \in E \\ x_{it} = \{0, 1\} & \forall i = 1, \ldots, N, \forall t \in \mathcal{T}(T_i), \\ s_i \geq 0 & \forall i = 1, \ldots, N. \end{cases}$$

The parameter c_{it} is the cost of route t for net T_i which is computed as the total length of the 3D route, $c_{it} = \sum_{e \ni t} c_e$, where the notation $e \ni t$ denotes that edge $e \in E$ is contained in route $t \in \mathcal{T}(T_i)$. The first set of equations in the model enforces the routing of each net. The decision variable s_i will be positive if net T_i cannot be routed, and the objective function trades off the total routing length with the number of nets that are routed. Typically M is chosen sufficiently large to ensure that all nets are routed. The second set of equations in the model ensure that the given edge capacities are not exceeded. The formulation (ILP-GR) has a number of appealing properties.

1. The exact properties of the route, such as topology and metal layer can be incorporated into the "cost" of a route. The objective is to minimize this cost. The formulation can thus handle the 3D GR problem to include both wirelength and via cost as the cost of a route. It then avoids a traditional layer-assignment phase which can be a source of sub-optimality.

2. The formulation does not require that the nets be *a priori* broken into two-terminal segments. Breaking nets before doing routing can be a significant source of sub-optimality in the resulting final routing [19]. We note that the final version of our proposed algorithm has some "net-breaking" to define subproblems for scalability. (See Section 3).

3. The slack variables s_i and the corresponding objective penalty factor M push the optimization to generate a *no-overflow* routing solution. The model is quite flexible, as with minor modifications, the integer program can be set to minimize the total overflow.

A significant disadvantage of the formulation (ILP-GR) is its size. First, for a given net T_i, the number of decision variables for this net is equal to $|\mathcal{T}(T_i)|$—the number of possible Steiner trees connecting the terminals in T_i. Second, the number of nets N may also be very large. Nevertheless, we use (ILP-GR) as the basis of our GR algorithm. In the subsequent discussion, we outline the manner in which we deal with the issues posed by large formulation size.

2.2 Column Generation

The first step in an IP-based approach to global routing is to solve the linear-programming (LP) relaxation of (ILP-GR), a relaxation obtained by replacing the binary requirement on the variable $x_{it} \in \{0, 1\}$ with a nonnegativity restriction $0 \leq x_{it} \leq 1$. The linear program is solved by a *column-generation* (CG) procedure [11, 12].

To describe the column generation procedure it is helpful to consider the dual (LPD-GR) of the linear programming relaxation of (ILP-GR):

$$\max_{\lambda \leq M, \pi \leq 0} \sum_{i \in N} \lambda_i + \sum_{e \in E} \pi_e u_e \qquad \text{(LPD-GR)}$$

$$\text{s.t.} \quad \lambda_i + \sum_{e \ni t} \pi_e \leq c_{it} \ \forall i = 1, \ldots, N, \forall t \in \mathcal{T}(T_i). \quad (1)$$

In a column generation procedure, only a small subset of all possible routes is explicitly included in the LP relaxation of (ILP-GR). Let $\mathcal{S}(T_i) \subset \mathcal{T}(T_i)$ be the set of routes considered for net T_i. The restricted master problem for (ILP-GR) is

$$\min_{x \geq 0, s \geq 0} \sum_{i=1}^{N} \sum_{t \in \mathcal{S}(T_i)} c_{it} x_{it} + \sum_{i=1}^{N} M s_i \qquad \text{(RMLP-GR)}$$

$$\begin{cases} \sum_{t \in \mathcal{S}(T_i)} x_{it} + s_i = 1 & \forall i = 1, \ldots, N \\ \sum_{i=1}^{N} \sum_{t \in \mathcal{S}(T_i)} a_{te} x_{it} \leq u_e & \forall e \in E. \end{cases}$$

Solving (RMLP-GR) yields a (primal) solution $(\hat{x}, \hat{s})$ as well as values $\hat{\lambda} \leq M$ and $\hat{\pi} \leq 0$ for the dual variables in (LPD-GR). By linear programming duality, if the solution $(\hat{\lambda}, \hat{\pi})$ satisfies all the dual constraints (1), then $(\hat{x}, \hat{s})$ is an optimal solution to the LP relaxation of (ILP-GR). If not, then the violated dual constraint suggests a column (variable) that may be added to (RMLP-GR) to reduce its objective value.

To determine if the dual solution $(\hat{\lambda}, \hat{\pi})$ is feasible, we must determine if there exists a route $t \in \mathcal{T}(T_i)$ with $\hat{\lambda}_i + \sum_{e \ni t} \hat{\pi}_e > c_{it}$. This is itself an optimization problem, known as the *pricing problem*, that can be decomposed into independent problems for each individual net $i = 1, \ldots, N$. Specifically, given net T_i, for each edge $e \in E$ define the binary decision variables t_e, taking value 1 if and only if edge e is used in a route for net T_i. The pricing problem for net T_i is then

$$\min_t \{ \sum_{e \in E} (c_e - \hat{\pi}_e) t_e \mid t \in \mathcal{T}(T_i) \}. \qquad \text{(PP}(T_i))$$

978-1-60558-497-3/09 $25.00 © 2009 ACM

Let t^* be an optimal solution to $(\mathrm{PP}(T_i))$. If $\sum_{e \in E} \hat{\pi}_e t_e^* + \sum_{e \in E} c_e t_e^* < \hat{\lambda}_i$, then t^* identifies a violated constraint (1) in (LPD-GR), and the current solution to (RMLP-GR) can be improved. The CG procedure is summarized as follows:

0. For each $i = 1, \ldots, N$, initialize $\mathcal{S}(T_i)$ with at least one route. (In our implementation, we use the route generated for net T_i by the package Flute [9]).

1. Solve (RMLP-GR), yielding primal solution $(\hat{x}, \hat{s})$ and dual values $(\hat{\lambda}, \hat{\pi})$.

2. For each $i=1,\ldots, N$, solve $(\mathrm{PP}(T_i))$, yielding a route t^*. If $\hat{\lambda}_i + \sum_{e \in E} \hat{\pi}_e t_e^* > \sum_{e \in E} c_e t_e$, then $\mathcal{S}_i = \mathcal{S}_i \cup \{t^*\}$.

3. If improving routes for some net T_i were found, return to step 1. Otherwise, stop—the solution $(\hat{x}, \hat{s})$ is an optimal solution to the LP relaxation of (ILP-GR).

In order to speed solution time, we typically stop the procedure once the solution value has "tailed off." Specifically, if the objective value of (RMLP-GR) has made little or no improvement in the last 10 iterations, the CG procedure is terminated.

2.3 Solving the Pricing Problem

In the pricing phase (step 2) of the CG procedure, small-weight Steiner trees with respect to the weights $\hat{w}_e = c_e - \hat{\pi}_e$ must be identified. Finding a minimum-weight Steiner tree is in general NP-Hard [14], so our approach for finding columns that reduce the optimal value of (RMLP-GR) is based on local search. Given a dual solution $(\hat{\lambda}, \hat{\pi})$, the *reduced cost* of route t of net T_i is $\bar{c}_{it} = c_{it} - \hat{\lambda}_i - \sum_{e \ni t} \hat{\pi}_e$. Note that the pricing problem $(\mathrm{PP}(T_i))$ can be viewed as a procedure for identifying a Steiner tree t for net T_i whose reduced cost $\bar{c}_{it} < 0$. By the complementary slackness conditions of linear programming, for any optimal solution $(\hat{x}, \hat{s})$ to (RMLP-GR) and corresponding dual solution $(\hat{\lambda}, \hat{\pi})$, the reduced cost $\bar{c}_{it} = 0$ if $\hat{x}_{it} > 0$.

Our local improvement procedure for solving $(\mathrm{PP}(T_i))$ uses this fact as well as the following simple observation. Given a route $t \in \mathcal{S}(T_i)$, let $V(t)$ be the set of vertices in t. If the variable $\hat{x}_{it} > 0$, and if there exists a path P' from some terminal $u \in T_i$ to a vertex $v \in V(t)$ such that the weight of P' (with respect to weights $\hat{w}$) is less than the weight of the path P from u to v using edges in t, the reduced cost of tree $t' = t \cup P' \setminus P$ is negative. Thus, adding the variable corresponding to route t' to (RMLP-GR) may reduce its objective value. Figure 1 demonstrates how new routes can be constructed by finding short u-v paths from $u \in T_i$ to a vertex v on the base Steiner tree. An interesting feature of this pricing algorithm is that the new routes can use different Steiner points than the original routes.

To approximately solve $(\mathrm{PP}(T_i))$ for a net T_i, our procedure starts with the tree $t \in \mathcal{S}(T_i)$ with largest value of $\hat{x}_{ti}$. Using edge weights $\hat{w}_e = c_e - \hat{\pi}_e$, a single-source shortest path problem from some terminal $u \in T_i$ to each vertex $v \in V(t)$ is solved. If the uv path length is smaller than the existing path length, a new route has been identified. Dijkstra's single-source shortest path algorithm [13] generates an entire tree of shortest path weights, thus possibly identifying *many* routes that would reduce the objective value of (RMLP-GR). In our implementation, we add a pre-specified maximum number of routes selected uniformly from the set of all identified negative cost routes per net.

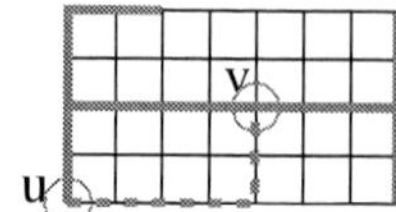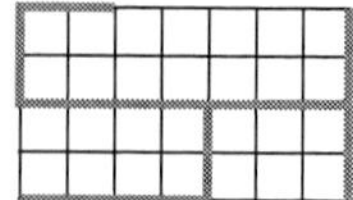

Figure 1: Improving routes via a shortest path algorithm on a weighted grid-graph

An important component of our pricing algorithm for a given net T_i is the selection of the starting terminals from which Dijkstra's algorithm to identify improving routes is run. In our implementation, Dijkstra's algorithm is run using the most congested terminals as starting points.

We identify these congested terminals as follows. For each terminal we compute the weight of the path P' that connects it to the base Steiner tree. The most congested terminals are those for which the corresponding P' has maximum weight.

2.4 Selecting Nets to Price

For large instances of (ILP-GR), the CG procedure can be significantly accelerated by only solving the pricing problem $(\mathrm{PP}(T_i))$ for a subset of all the nets. To select the nets $T_i \in \mathcal{N}$ for which $(\mathrm{PP}(T_i))$ is solved, our procedure takes advantage of information provided by the solution of the restricted master problem. Specifically, if $\hat{s}_i > 0$, then the net T_i is not completely routed using the existing routes in $\mathcal{S}_i$, so net T_i is priced by step (2) of the CG procedure.

Nets for pricing are also selected based on measures of congestion in the current LP solution to (RMLP-GR). Congestion may be identified in one of two ways. First, congested edges are those edges e that have the most negative value of $\hat{\pi}_e$. The intuition behind this choice is that $\hat{\pi}_e$ provides the rate of change in the objective function of (RMLP-GR) per unit additional capacity on edge e. A second way to identify a congested edge is to let $r_i \in \arg\max_{t \in \mathcal{S}(T_i)} \hat{x}_{ti}$ be the route for net T_i with the largest solution value in (RMLP-GR). The value $\eta_e = \sum_{i=1}^{N} a_{r_i e}$ is the number of units of capacity on edge e that would be used if the routes r_i were used for each net $T_i \in \mathcal{N}$. If the value $(\eta_e - u_e)$ is large, then edge e is highly-congested. In our algorithm, a bounding box around a congested edge e (identified by either of the two measures) is created, and all nets T_i that contain a terminal inside the bounding box are also candidates to be priced by $(\mathrm{PP}(T_i))$ in step 2 of the CG procedure.

2.5 Branch and Bound

Once the CG procedure for the solution of the LP relaxation of (ILP-GR) is complete, either because no improving routes were found in the pricing phase, or because an iteration limit was reached, a promising candidate subset of routes $\mathcal{S}(T_i) \subset \mathcal{T}(T_i)$ has been identified for each net T_i. Using only these route variables, the integer program (ILP-GR) is formulated and solved by the commercial integer programming solver CPLEX (v9) [10]. The solution returned by CPLEX is a feasible solution to the problem.

The proposed approach, based on the direct solution of (ILP-GR), has significant promise to improve the solution quality of existing GRs. For example, using this approach, we solved the 2D IBM01 circuit of the ISPD1998 suite [1] and were able to improve the wirelength by approximately 5% compared to the best solution found by FGR [19], without any overflows. However, the runtime to achieve this high-quality solution was prohibitively long—a few hours. Thus, in the following section, we discuss mechanisms for decomposing the full global routing (ILP-GR) into smaller instances in order to accelerate the overall runtime.

978-1-60558-497-3/09 $25.00 © 2009 ACM

3. DECOMPOSITION FOR SCALABILITY

Many existing global routing algorithms define reasonably-sized subproblems and create a full global routing out of solutions to these subproblems. For example, to achieve a good runtime, BoxRouter [8] starts by solving an IP over a small rectangular box on the chip and then progressively increases the size of the box to generate new IPs, fixing the solution to the previous IP. Fixing the solution of previous IP when increasing the box size may lead to a degradation in solution quality. SideWinder [15] solves an IP over the entire chip by gradually introducing more base patterns for the nets in the congested areas at each iteration. However, Sidewinder only works with three simple-shaped patterns which are defined *a priori*. The work [21] proposes a hierarchical IP approach that first solves a small IP to plan the routing of the longest nets. However, the impact of the shorter nets is neglected.

As demonstrated in Section 2, our proposed algorithm for global routing has potential to find high-quality solutions, but also requires a mechanism to accelerate the procedure. In this section, we first discuss a decomposition of the integer program (ILP-GR) into smaller ones that correspond to non-overlapping rectangular "subregions" on the chip. We introduce the concept of "floating-terminals" to define the IP of each subregion, providing significant flexibility for routing nets that might enter or exit that subregion. We then discuss effective integration of the subregion solutions to generate a valid and high quality final solution. Finally, we discuss a technique to plan long nets that pass multiple subregions.

3.1 Subregion Extraction / IP Decomposition

The goal of our decomposition procedure is to define non-overlapping rectangular subregions on the chip. Each subregion defines the boundaries of a smaller-sized GR problem which we solve using the IP-based procedure outlined in Section 2. The objective of the subregion definition is to define *balanced* subregions, resulting in "equally-difficult" optimization problems that take approximately the same time to solve. We first tried a coarse, uniform grid to define the subregions. However, we noticed that the IPs corresponding to the congested subregions were taking significantly longer time to be solved by our procedure (e.g., hours for congested subregion and minutes for the less congested ones).

To decrease the gap in solution times, our procedure attempts to create subregions having the same average edge utilization (AEU). To define the AEU, we first assume that all nets are routed using the Steiner route generated by [9] in the 2D-projected problem. For each edge, we define a utilization factor as the ratio of the number of routes that cross the edge to the edge capacity. Many edges might have a utilization factor higher than one, indicating an overflow. For a subregion, the AEU is the the average utilization factor of all edges contained in the subregion.

The subregions are defined using a partitioning-based strategy, depicted graphically in Figure 3. We recursively apply bi-partitioning to subregions to obtain two new smaller-sized subregions, at each step ensuring that the generated subregions have similar AEU. During one bi-partitioning step, to decide between a vertical or horizontal partitioning, we choose the one that results in the smaller aspect ratio of the generated subregions. The partitioning of a subregion is stopped when any of its sides reaches 32 units of the routing grid, a size empirically set to generate an IP that can be typically solved by the procedure outlined in Section 2 in an acceptable runtime.

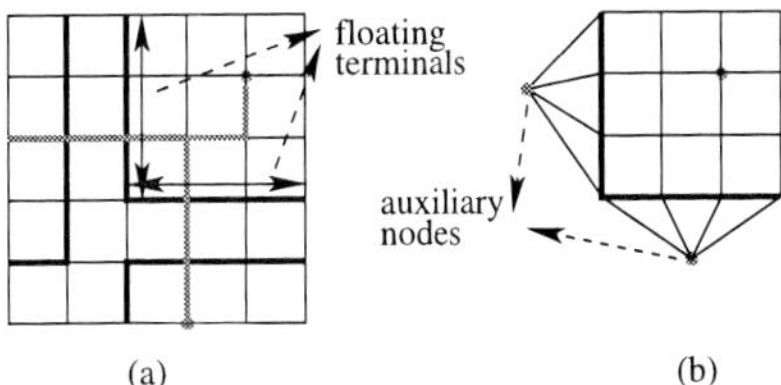

Figure 2: Modifying grid-graph of a subregion to handle floating terminals within our IP procedure.

Each subregion defines a new grid-graph $G'(V', E')$ and set of nets $\mathcal{N}' \subset \mathcal{N}$. The set $\mathcal{N}'$ is composed of two types of nets, nets that have all terminals inside the subregion ($T_i \subseteq V'$), and nets that have at least one terminal outside the subregion ($T_i \not\subseteq V'$). Figure 2(a) shows the latter type of these nets. The net in the figure belongs to three different subregions. The common boundaries of these subregions are shown in bold. Considering the top-right subregion, we can think of having a net with one fixed and two "floating" terminals. Each floating terminal represents a portion of a subregion boundary through which the net will connect to another subregion.

To specify such nets in our IP, we represent each floating terminal using an auxiliary node. The auxiliary node is added to the set of nodes V' in the grid-graph. Edges connecting the nodes that are on the subregion boundary to their corresponding auxiliary node are added to the set E'. The added edges have infinite capacity and zero cost in the definition of the integer program (ILP-GR). Figure 2(b) illustrates the addition of auxiliary nodes and edges. After applying this simple construction, the integer program (ILP-GR) is well-defined, and can be solved by the procedure outlined in Section 2.

The example of Figure 2 is for 2D routing, but in the general 3D case, each boundary of a subregion is a plane and graph G' extends to the third dimension. The nodes on this vertical boundary plane are connected to their corresponding auxiliary node.

3.2 Handling Long Nets

In our subregion extraction procedure, the regions through which net T_i are routed, and hence the locations of floating terminals for T_i are taken from a given Steiner topology (e.g., the route generated by Flute for T_i). Even though the subregion IP has significant flexibility in implementing the routes that connect floating terminals, the entire procedure relies on knowing the assignment of each net to one or more subregions. For long nets that are assigned to more than two subregions, the subregion assignment issue becomes particularly important. (This is in spite of the fact that the number of subregions is typically much smaller than the number of bins of the routing grid). Figure 3 illustrates this point. The two long nets are routed using their Steiner routes, both of which pass from subregion A. If A is congested, it is better to detour these nets from A.

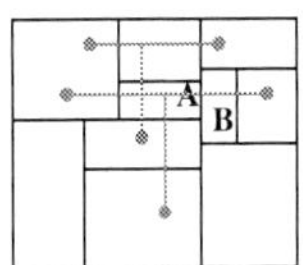

Figure 3: Assigning long nets to subregions using an initial Steiner route can cause unresolvable overflow.

978-1-60558-497-3/09 $25.00 © 2009 ACM

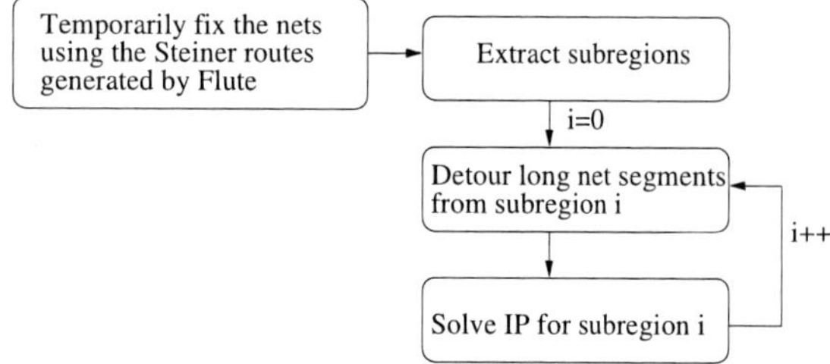

Figure 4: Dynamic planning of long nets

Many procedures for assigning the subregions of long nets were investigated. First, similar to [21], we tried a standard hierarchical IP formulation in terms of long nets. This approach helped considerably with the removal of overflow, but its failure to accurately consider the impact of short nets falling completely inside a subregion led to poor quality solutions in terms of wirelength. Instead, we use the procedure depicted in Figure 4, which is explained below:

1. Before extracting the subregions, all nets are broken into two-terminal segments using the Steiner trees generated by Flute.

2. Next, the subregion extraction begins. The routes used to compute the AEU during subregion extraction are those generated by step 1.

3. Once the subregions are created, we begin solving the IPs for each subregion in a congestion-based order that is discussed in the next subsection. Before solving a subregion IP, we use a procedure to detour as many long nets that pass from the subregion as possible. Once a subregion is solved, its solution remains fixed. Before solving the IP for the next subregion, we apply the same procedure to detour as many passing nets as possible to the remaining subregions.

To detour long net segments outside a subregion, as required by step 3, we apply a shortest path algorithm on the grid-graph. We set the edge weights inside the target subregion to a large number to avoid getting re-routed inside that subregion. Also, outside the target subregion, the edges that have overflow will also have a large weight. The weights of the grid-graph gets updated every time a long net is detoured.

Our procedure provides two significant benefits. First, it detours the long nets dynamically, every time a new subregion is processed. Second, it considers an estimate of the current congestions based on continually updating the edge weights to detour long nets.

3.3 Subregion Integration

So far we explained how the subregions are extracted and the long nets are dynamically planned. We then finalize the routing solution using a two-phase approach. In the first phase, we fix the locations of the floating terminals and generate a routing solution for all the routes that completely fall within a subregion. For the routes that cover more than one subregion, we fix a "backbone" inside each of its subregions. In the second phase, we connect the backbones of these longer nets in adjacent subregions. This two-phase approach is entirely based on integer programming and solved using formulation (ILP-GR) as we elaborate next.

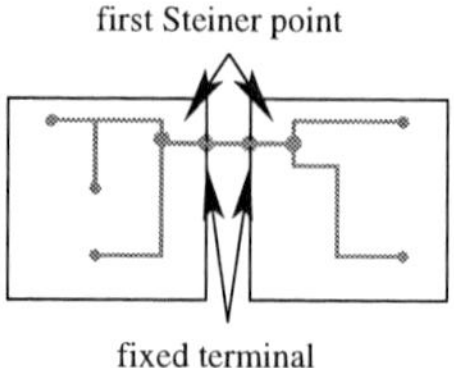

Figure 5: We remove the segment connecting the fixed boundary terminal (previously floating) to the first Steiner point in the route backbone. We reroute these connecting segments given the fixed boundary terminals using our IP procedure.

In phase 1, we visit and process the subregions in the following order. We first compute the total edge overflow (TEO) based on the initial Flute Steiner-routes in each subregion. Subregions are then processed in decreasing order of their TEOs. Every time a subregion is solved, the floating terminals for a net T_i will get fixed at a boundary of the region. Specifically, the net T_i is partially routed, and subsequent subregion IPs must respect this partial routing. If two consecutive subregions (in terms of TEO) are not physically adjacent, we process them in parallel.

Phase 1 fixes the locations of the floating terminals on the subregion boundaries and also generates an initial routing solution. For this routing solution, we fix all the (short) nets that completely fall within a subregion. For the (long) nets covering more than one subregion, we fix a "backbone" inside each of its subregions as follows. For a long net we visit each of its subregions. Inside each subregion, we remove the "branch" that connects the fixed terminal on the subregion boundary to the route backbone. Figure 5 illustrates. Considering the route that goes within two subregions, for each subregion, we remove the segment that connects the identified fixed terminal on the boundary to the backbone of the route. Specifically, the removed segment is one that connects the boundary terminal to the first Steiner point of the route in the subregion.

In phase 2, once these "connecting segments" are removed, routing resources will be freed and we reconnect these segments using the formulation (ILP-GR) while fixing the routes of short nets and backbones of long nets from phase 1.

In summary, our procedure uses the IP formulation of Section 2 as a basic component throughout the routing process. Subregion extraction aims to generate equally-difficult optimization problems. Subregions are initially solved while using the flexibility of floating terminals on their boundaries as they are visited in the order of their TEO. After the initial phase, all the short nets and the backbone of long nets within each subregion are routed. In addition, the locations of the floating terminals on the subregion boundaries will get fixed to ones that ensure obtaining a feasible solution. The final phase effectively connects the backbones of the long nets in different subregions. In our simulation results, we observed significant improvement in solution quality from phase 2 for connecting the subregions using integer programming.

The result is a scalable, effective global router, called GRIP (Global Routing via Integer Programming). GRIP is a *robust* tool and does not rely on any design-dependent tuning. Everything is based on integer programming. The defined parameters AEU (for defining subregions) and TEO (for ordering them) both depend on our congestion estimate which is only based on the Flute-Steiner routes.

978-1-60558-497-3/09 $25.00 © 2009 ACM

Table 1: Results for ISPD 2007 benchmarks.

Benchmark	Best Approach			GRIP		
	tool	TOF	WL	TOF	WL	%Impr
adaptec1.2D	FGR	0	54.7	0	52.8	3.5
adaptec2.2D	FGR	0	52.4	0	50.1	4.4
adaptec3.2D	FGR	0	131.5	0	125.9	4.3
adaptec4.2D	FGR	0	125	0	122.1	2.3
adaptec5.2D	FGR	0	153.2	0	144.5	5.7
newblue1.2D	FGR	400	46.3	0	44.9	3.1
newblue2.2D	FastRoute	0	76.4	0	73.2	4.1
newblue3.2D	NTHU-R	31454	110.8	35573	107.9	N/A
adaptec1.3D	FGR	0	88.6	0	78.9	10.94
adaptec2.3D	FGR	0	90.1	0	80.7	10.41
adaptec3.3D	FGR	0	200.6	0	182.2	9.17
adaptec4.3D	FGR	0	183.0	0	167.7	8.36
adaptec5.3D	NTHU-R	0	260.2	0	227.8	12.45
newblue1.3D	NTHU-R	0	91.0	0	78.1	14.14
newblue2.3D	FGR	0	132.5	0	114.7	13.46
newblue3.3D	NTUgr	31454	167.0	33158	162.7	N/A

Table 2: Our grid size, subregion count and runtime

benchmark	#nets	grid	#subregions	2D	3D
adaptec1	176715	324x324	91	290	440
adaptec2	207972	424x424	169	229	366
adaptec3	368494	774x779	562	262	387
adaptec4	401060	774x779	558	240	398
adaptec5	548073	465x468	199	410	688
newblue1	270713	399x399	137	319	513
newblue2	373790	557x463	242	230	296
newblue3	442005	973x1256	1162	701	1389

4. SIMULATION RESULTS

We implemented GRIP using C++. For solving individual LPs and IPs we used CPLEX (v9) [10]. We demonstrate the performance of GRIP on the ISPD 2007 benchmarks [2]. Table 2 reports the total number of routed nets and the grid size for each benchmark. Each benchmark has a 3D as well as a projected-2D version, and the grid size is the same in both versions. The 2D and 3D benchmarks have two and six metal layers, respectively.

We compare the summation of wirelength and via cost (denoted by WL) in Table 1[1]. The comparison is made against the best reported solution for each benchmark, found by either FGR 1.1 [19], NTHU-Route 2.0 [6], FastRoute 3.0 [22] or NTUgr [7]. For the 2D and 3D benchmarks, we obtained an average improvement of 3.9% and 11.3%. The improvement in the 3D benchmarks were more significant than in the 2D case, because GRIP considers explicit 3D Steiner routes, skipping the layer assignment phase.

The total overflow (denoted by TOF) is also given in Table 1. The GRIP solutions had zero overflow for all the benchmarks except newblue3 which is known to be unroutable. For newblue3 we report NTUgr as the best tool only because it generates the smallest overflow (and not the smallest WL).

Table 2 reports the running time (wall clock time) of both 2D (column 5) and 3D (column 6) benchmarks. The runtime unit is minutes. The number of subregions created by the subregion extraction procedure described in Section 3.1 for each benchmark is given in column 4. In reporting the runtimes we process independent subregions in parallel, which is why wall time is the appropriate measure. GRIP was run on a heterogenous grid of CPUs, shared by many users, and controlled by the Condor grid computing toolkit [16]. When solving the IP formulation, the majority of runtime was spent on the linear program (for column generation) rather than solving the integer program using branch-and-bound. This helped us to effectively identify candidate routes; the number of candidate routes reached up to a hundred for some nets, while for some nets only a few routes were generated in the linear program. Overall, our runtimes are scalable and adjustable, since they depend on the number of subregions we chose to create. Continuing work is aimed at further exploiting parallelism to obtain similar high-quality solutions in a smaller run time.

---

[1]Benchmark solutions are available for download at: http://wiscad.ece.wisc.edu/gr/

5. CONCLUSIONS

We presented GRIP, a tool for global routing using integer programming. We introduced a novel IP formulation to select candidate routes for each net based on a continually-updated congestion metric, while directly working with the 3D model of the routing problem. To achieve reasonable runtime, we discussed subregion extraction and IP decomposition as well as a method for planning long nets and integrating the subregion solutions.

6. REFERENCES

[1] ISPD 1998 global routing benchmark suite, 1998.

[2] ISPD global routing contest and benchmark suite, 2007.

[3] C. Albrecht. Global routing by new approximation algorithms for multicommodityflow. *IEEE TCAD*, 20(5):622–632, 2001.

[4] L. Behjat and A. Chiang. Fast integer linear programming based models for VLSI global routing. In *ISCAS (6)*, pages 6238–6243, 2005.

[5] M. Burstein and R. Pelavin. Hierarchical wire routing. *IEEE TCAD*, 2(4):223–234, 1983.

[6] Y.-J. Chang, Y.-T. Lee, and T.-C. Wang. Nthu - route 2.0: A fast and stable global router. In *ICCAD*, pages 338–343, 2008.

[7] H.-Y. Chen, C.-H. Hsu, and Y.-W. Chang. High-performance global routing with fast overflow reduction. In *ASPDAC*, pages 582–587, 2009.

[8] M. Cho, K. Lu, K. Yuan, and D. Z. Pan. Boxrouter 2.0: architecture and implementation of a hybrid and robust global router. In *ICCAD*, pages 503–508, 2007.

[9] C. C. N. Chu and Y.-C. Wong. Flute: Fast lookup table based rectilinear Steiner minimal tree algorithm for VLSI design. *IEEE TCAD*, 27(1):70–83, 2008.

[10] CPLEX Optimization, Inc., Incline Village, NV. *Using the CPLEX Callable Library, Version 9*, 2005.

[11] G. Dantzig and P. Wolfe. Decomposition principle for linear programs. *Operations Research*, 8:101–111, 1960.

[12] J. Desrosiers and M. E. Lübbecke. A primer in column generation. In G. Desaulniers, J. Desrosiers, and M. M. Solomon, editors, *Column Generation*, chapter 1. Springer, 2005.

[13] E. W. Dijkstra. A note on two problems in connetion with graphs. *Numerische Mathematik*, 1:269–271.

[14] M. R. Garey and D. S. Johnson. The rectilinear Steiner tree problem is NP-complete. *SIAM Journal of Applied Math*, 32:826–834, 1977.

[15] J. Hu, J. A. Roy, and I. L. Markov. Sidewinder: a scalable ILP-based router. In *SLIP*, pages 73–80, 2008.

[16] M. J. Litzkow, M. Livny, and M. W. Mutka. Condor—A hunter of idle workstations. In *Proceedings of the 8th International Conference on Distributed Computing Systems*, pages 104–111, 1998.

[17] M. D. Moffitt. Maizerouter: Engineering an effective global router. In *ASPDAC*, pages 226–231, 2008.

[18] M. M. Ozdal and M. D. F. Wong. Archer: a history-driven global routing algorithm. In *ICCAD*, pages 488–495, 2007.

[19] J. A. Roy and I. L. Markov. High-performance routing at the nanometer scale. In *ICCAD*, pages 496–502, 2007.

[20] T. Terlaky, A. Vannelli, and H. Zhang. On routing in VLSI design and communication networks. *Discrete Applied Mathematics*, 156(11):2178–2194, 2008.

[21] Z. Yang, S. Areibi, and A. Vannelli. An ILP based hierarchical global routing approach for VLSI ASIC design. *Optimization Letters*, pages 281–297, 2007.

[22] Y. Zhang, Y. Xu, and C. Chu. Fastroute3: a fast and high quality global router based on virtual capacity. In *ICCAD*, pages 344–349, 2008.

978-1-60558-497-3/09 $25.00 © 2009 ACM

Automatic Bus Planner for Dense PCBs [*]

Hui Kong, Tan Yan and Martin D.F. Wong

Department of Electrical and Computer Engineering, University of Illinois at Urbana-Champaign

{huikong2,tanyan2,mdfwong}@illinois.edu

ABSTRACT

Since no commercial PCB routing tools can solve the routing problem for today's complex PCBs, these circuit boards have to be routed manually, taking about 2 months of time per board. Bus planning is one of the most time-consuming steps of PCB routing. It consists of assigning buses to multiple layers of the PCB and routing them in a planar fashion on each layer. Routing congestion between on-board components and the min-max length bounds of the buses must also be considered during routing. In this paper, we present the first automatic bus planner. We tested our system on a state-of-the-art industrial circuit board with over 7000 nets and 12 signal layers. All the nets on this board were already manually routed. Our bus planner is able to achieve 100% routing completion using the layer assignment extracted from manual design. For simultaneous layer assignment and bus routing, we are able to successfully route 98.5% of the nets. The remaining 1.5% can be routed either manually or by using vias. The runtime of our bus planner is less than 3 hours on a 3 Ghz workstation.

Categories and Subject Descriptors: B.7.2 [Integrated Circuits] Design Aids - Placement and Routing
General Terms: Algorithms, Performance
Keywords: Topological Routing, Layer Assignment, Bus Planning, PCB Routing

1. INTRODUCTION

As IC technology advances rapidly, the dimensions of packages and PCBs are decreasing while the pin counts and routing layers keep increasing. Today, a high-performance PCB usually contains thousands of pins and more than ten signal layers [10, 14, 15]. Such a problem scale makes manual design extremely time-consuming. Moreover, due to the high clock frequencies on modern PCBs, the routed nets must be

[*]This work was partially supported by the National Science Foundation under grant CCF-0701821 and a grant from IBM.

Permission to make digital or hard copies of part or all of this work for personal or classroom use is granted without fee provided that copies are not made or distributed for profit or commercial advantage and that copies bear this notice and the full citation on the first page. To copy otherwise, to republish, to post on servers or to redistribute to lists, requires prior specific permission and/or a fee.
DAC'09, July 26-31, 2009, San Francisco, California, USA

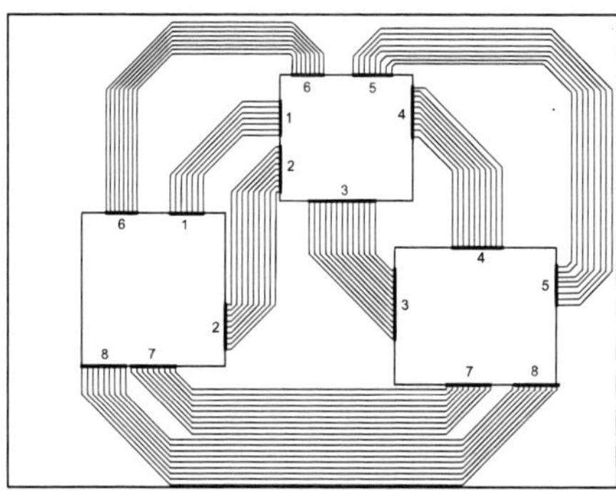

Figure 1: One layer of bus planning solution.

subject to very stringent min-max length constraints [13], which make the design even more difficult. Unfortunately, since current auto-routers are not able to produce acceptable solutions, many industrial designs are still completed by manual efforts, which typically take 2 months per board. In the future, PCB designs will be substantially more complex, making manual routing approaches infeasible.

Designers always prefer the nets in a PCB to be bundled together as buses. All the nets in a bus are expected to be routed together. Planning the buses, which consists of assigning the buses to the routing layers and topologically routing them on each layer in a planar fashion, is one of the most time-consuming steps in PCB routing. Figure 1 shows an example of the bus planning result on one layer. Practical boards may contain many such layers. During the planning, we have to consider crossings of the bus topologies and the routing congestion between the on-board components as well as the min-max length bounds of the nets. As far as we know, existing commercial tools that help bus planning are interactive tools [2, 3, 9].

Notice that the bus planning problem is different from the multi-layer global routing problem for ICs [4, 11, 12] because the X-Y style for IC routing breaks a net into horizontal and vertical segments on different layers, while in PCBs, all the nets in a bus are expected to be routed on a single layer because any via introduced to the net can reduce the reliability and performance of circuits, damage signal integrity and increase the manufacturing cost. This fundamental difference makes IC routers incapable of performing bus planning.

We are developing a complete PCB auto-routing system, and this paper reports the bus planning part of the system. To the best of our knowledge, this is the first automatic bus planner for complex PCB design. Our bus planner consists of three main stages:

1. Global routing, which routes all buses on a single layer with all routing resources mapped onto one layer.
2. Layer assignment of the routed bus.
3. Iterative improvement by reassignment and rerouting.

Our bus planner has the following features:

- We introduce a novel dynamic routing graph to guide the search of topological bus routes. The dynamic structure of the graph enables us to find planar routes for the buses more efficiently.
- We introduce a bin-packing-based congestion estimation. It provides a more accurate estimation than traditional congestion models. Owing to its accuracy, our planner effectively avoids violations of routing capacity constraints.
- We introduce an iterative improvement technique that resembles the bubble-sort. It helps further improve the routing solution.

We tested our bus planner on a state-of-the-art industrial circuit board with over 7000 nets and 12 signal layers. It has been completely routed by manual efforts. We perform two experiments. First, we take the layer assignment of the buses from the manual solution and apply our bus planner to route each layer. We are able to achieve 100% routing completion while satisfying the length constraints within 40 minutes. Then, we use our bus planner to perform simultaneous layer assignment and topological bus routing. We are able to successfully route 98.5% of the nets while satisfying length constraints. The remaining 1.5% can be routed either manually or by using vias. The runtime of our planner is less than 3 hours.

The rest of this paper is organized as follows: Section 2 introduces the bus planning problem. The three stages of our planner—global routing, layer assignment and iterative improvement—are introduced in Section 3, Section 4 and Section 5 respectively. Experimental results are given in Section 6, and Section 7 concludes our paper.

2. BUS PLANNING PROBLEM

A typical high-end PCB board contains a number of components, such as MCM, memory, and I/O modules, and a large number of bus structures between these components. Each bus is a group of two-pin nets between components. Inside the component, the nets of a bus will be escaped to the component boundaries to generate bus escape intervals, which are the starting points of our bus planner. Each bus can be represented by a pair of escape intervals on the component boundaries. We note that for any pair of buses, their internal escape routes may conflict with each other, as displayed in Figure 2, which means they cannot be assigned to the same layer.

The input of our bus planning problem is the number of routing layers and a set of bus intervals as well as a bus internal conflict graph, whose vertices correspond to buses and whose edges correspond to internal route conflicts between buses. The expected output is the layer assignment of the topological routes of buses, where each bus is routed in planar fashion considering crossings, routing congestions and min-max length bounds. Figure 3 gives an example of a bus planning problem and its two-layer planning solution. In this paper, we define the bus net count as its number of nets and the bus route width as the width of the space its net routes need to occupy, which is proportional to the bus

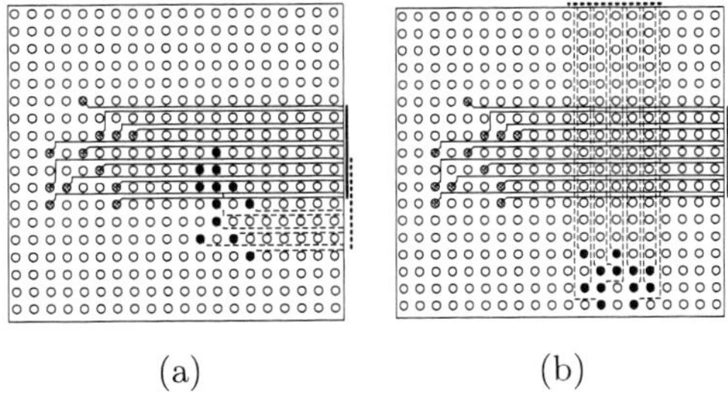

Figure 2: Internal route conflicts.

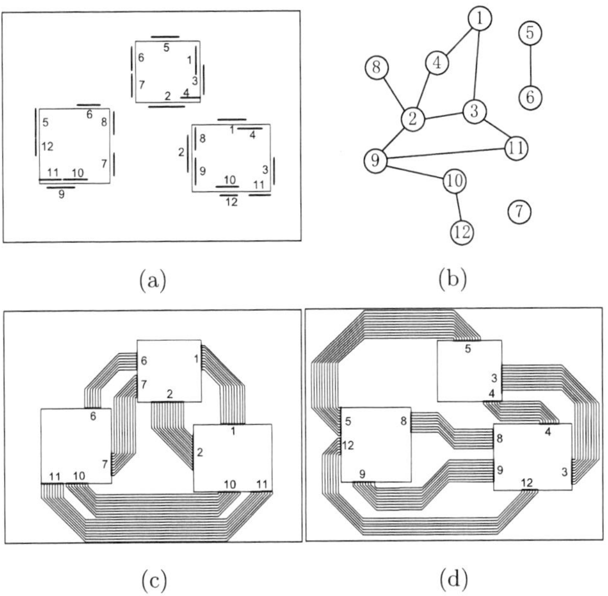

Figure 3: Given bus escape intervals (a) and bus internal conflict graph (b) as input, the expected output of planning buses to 2 layers is (c) and (d).

net count.

Our bus planning consists of 3 stages: (1) bus global routing, (2) bus layer assignment and (3) iterative improvement by rerouting and reassigning. We will explain them in detail one by one.

3. GLOBAL ROUTING

Global routing is the first stage of our bus planner. It generates the initial topological bus routes by routing buses on a single layer which includes the routing resources of all routing layers. A negotiated-congestion router is applied in this stage. The generated initial routes are then assigned to given routing layers in the layer assignment stage and then the planning solution will be improved in the iterative improvement stage.

In this section, we first introduce the Hanan grid and the dynamic routing graph data structure, then present the bin-packing-based congestion estimation approach and finally describe the negotiated-congestion router.

3.1 Hanan Grid

The routing graph of our global routing is based on the Hanan grid, which is constructed from the components by extending their boundaries until they touch other components or the board boundaries. The board and its compo-

978-1-60558-497-3/09 $25.00 © 2009 ACM

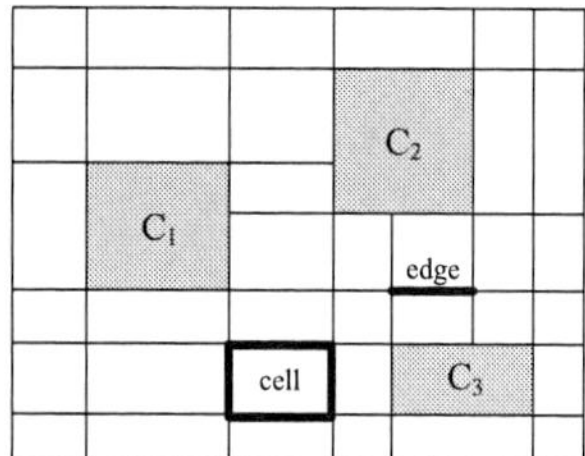

Figure 4: Hanan grid.

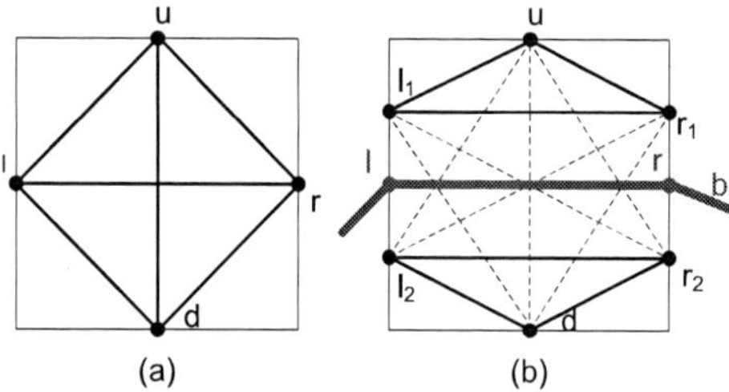

Figure 5: The dynamic routing graphs in a Hanan grid cell. (a) and (b) give the routing graphs before and after a bus is routed. The thin dashed lines in (b) represent edges intersecting with the routed bus and they may be given high penalties.

nents are all represented by their bounding boxes. Figure 4 shows a Hanan grid for 3 components. The grid cells and edges are refined to as Hanan grid cells and Hanan grid edges, respectively.

3.2 Dynamic Routing Graph

In the global routing stage, we route all buses on a single layer in which the routing resources of all routing layers are lumped together. On one hand, we want to avoid crossings between buses without internal conflicts since we do not know if they will be assigned to the same layer in later stages, and on the other hand, we allow crossings between buses with internal conflicts since we know they will not be assigned to the same layer. Thus we need a flexible data structure to handle both of these situations.

For this purpose, we introduce a *dynamic routing graph*, whose vertices are the middle points of all Hanan grid edges and whose edges are connections of vertices in the same cell. When an edge is occupied by a bus, the cell is split into two parts and new vertices and new edges are created. Figure 5 gives an example. Figure 5(a) shows the initial routing graph in a Hanan grid cell, and Figure 5(b) shows the new routing graph after a bus b passes through this cell, where two vertices (l and r) representing the left and right Hanan grid edges are eliminated and four vertices (l_1, l_2, r_1 and r_2) representing four smaller grid edges are added. In Figure 5(b), the thick solid lines represent the route of bus b, which cuts the left and right Hanan grid edges into two smaller ones. The thin solid lines represent the new routing graph edges that do not cross with bus b. The dashed lines represent the new routing graph edges that cross b. The edges represented by dashed lines will be given high penalties to avoid bus crossings with b. Figure 6 gives a global view of how the routing graph change happens after a bus is routed.

By using this dynamic routing graph, our global router can handle all possible bus topologies, and each topological

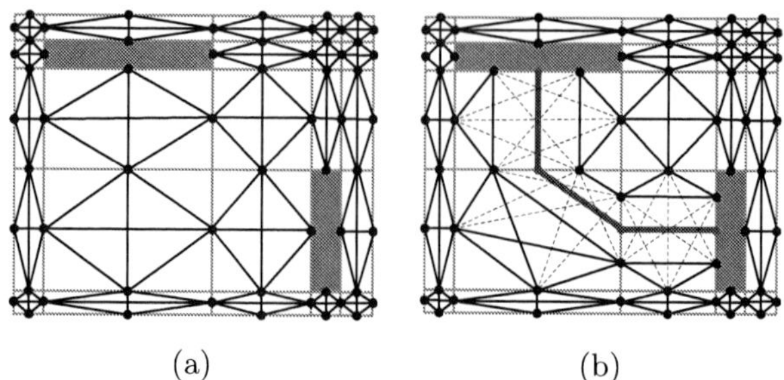

Figure 6: The dynamic routing graphs of a PCB before (a) and after (b) a bus is routed. The thin dashed lines in (b) represent routing graph edges that are given high penalties.

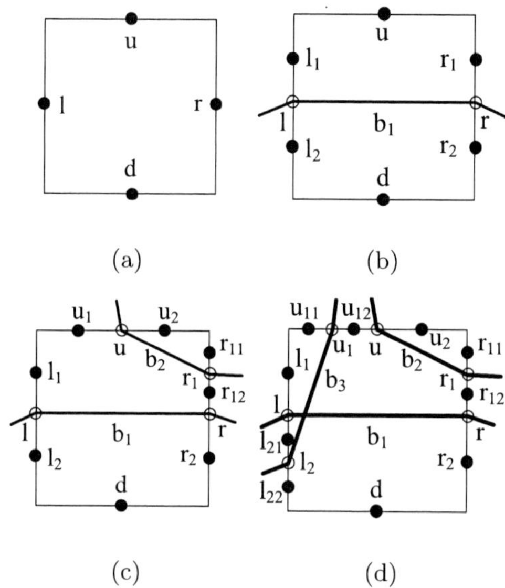

Figure 7: The dynamic routing graph in a Hanan grid cell as 3 buses b_1, b_2 and b_3 pass through it consecutively. Here b_1 has internal conflict with b_3. (a) is the initial routing graph. (b), (c) and (d) are routing graphs when b_1, b_2 and b_3 are routed consecutively.

bus route is a path on this graph. Figure 7 gives an example, where 3 buses b_1, b_2 and b_3 pass through a Hanan grid cell consecutively and b_1 and b_3 have internal conflict. In Figure 7, to simplify the demonstration, we only show the graph vertices. Figure 7(a) gives the initial routing graph. As shown in Figure 7(b), when b_1 is routed, four new graph vertices are added and two old vertices are deleted. When routing b_2, the routing graph edges crossing with b_1 are given high penalties. As a result, b_2 passes through this cell without crossing b_1 and the routing graph changes again, as shown in Figure 7(c). When routing b_3, since b_3 has internal conflict with b_1, the two nets will be routed on different layers. Therefore, we do not need to assign high penalty cost to the edges crossing with b_1. To avoid bus crossing between b_1 and b_2, edges crossing with b_2 are given high penalties. Finally, as shown in Figure 7(d), b_3 is routed to pass through this cell intersecting b_1 but not intersecting b_2.

Figure 8 gives an example of all the routing paths on one layer. It can be seen that each routing path goes through

978-1-60558-497-3/09 \$25.00 © 2009 ACM

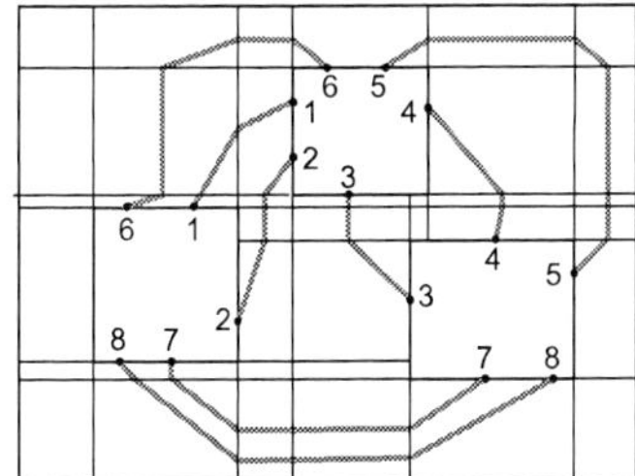

Figure 8: An example of all the routing paths on one layer.

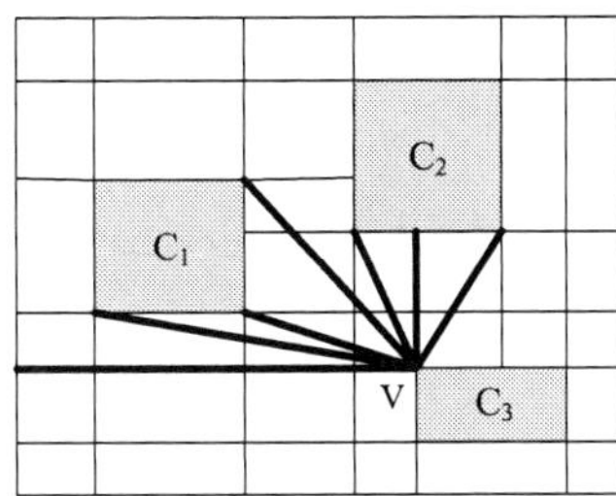

Figure 9: The critical cuts from corner v

either the middle points of a Hanan edge or newly created middle points after splitting a Hanan cell.

3.3 Congestion Estimation

In the global routing stage, to capture the routing space outside the components, we introduce the *critical cuts*, which are the line segments connecting each component corner to all its visible component corners or component or board boundaries. A corner is visible to another corner if the line segment between the two corners does not intersect any other component. Similarly, for a component corner, if the line segment that starts from this corner and is normal to a component or board boundary does not intersect any component, then the component or board boundary is visible to the corner. The thick lines in Figure 9 gives all critical cuts starting from the component corner v.

Since the critical cuts capture the routing space outside the components, congestions may happen on them and our global router should be able to estimate their congestions. All critical cuts can be captured by our routing graph, so their congestions can be estimated. As shown in Figure 9, each orthogonal critical cut consists of a sequence of consecutive Hanan grid edges, so it can be captured by routing graph vertices on these edges. Each diagonal critical cut intersects some Hanan grid cells, so it can be captured by routing graph edges crossing this cut inside these cells. Since our routing graph is dynamic, whenever the graph changes, the graph vertices and edges capturing critical cuts will be updated.

Given a critical cut t and L routing layers, the traditional approach estimates its congestion by comparing the total number of nets passing through it and the maximum number of nets it can accommodate, as listed in Equation 1. In the equation, $c(t)$ is the maximum number of nets t can accommodate on a single layer, b_i $(1 \leq i \leq m)$ are buses passing through t and $s(b_i)$ is the net count of b_i.

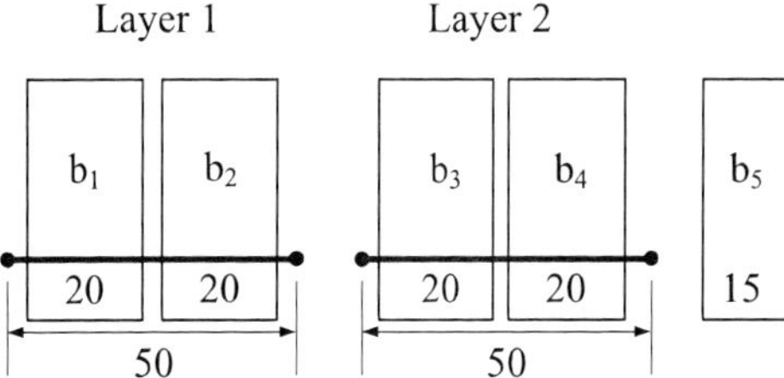

Figure 10: An example of bin-packing-based congestion estimation.

$$Congestion(t) = max(0, \sum_{i=1}^{m} s(b_i) - L \cdot c(t)) \qquad (1)$$

But this traditional approach estimates congestion at the net level instead of at the bus level, so it is not accurate. Figure 10 gives an example, where there are 5 buses b_1, b_2, b_3, b_4 and b_5 with 20, 20, 20, 20 and 15 nets respectively passing through a critical cut with single-layer capacity 50 in 2 layers. By using capacity-based estimation, this cut is not congested, since the total net count, 95, is less than its overall capacity, 100. But as shown in Figure 10, we can not assign all 5 buses onto 2 layers.

To estimate congestion at the bus level, we propose the bin-packing-based estimation. Here we regard critical cut t as a bin with size $c(t)$ and regard a bus b_i $(1 \leq i \leq m)$ as an item with size $s(b_i)$. The bin number b is the minimum number of bins to accommodate all these m items. The congestion it estimates is:

$$Congestion(t) = max(0, b - L) \qquad (2)$$

For the example shown in Figure 10, if we apply the bin-packing-based estimation to it, the bin number is 3, but the layer number is 2, so the critical cut must be congested on some layer.

Although the bin packing problem is NP-hard [7], there exist many effective and fast heuristic algorithms [5].

3.4 Negotiated-Congestion Routing

In this section, we will present our negotiated-congestion router. Our negotiated-congestion algorithm is similar to the Pathfinder negotiated-congestion algorithm [1,6,8], which was originally proposed for FPGA routing. It rips up and reroutes buses iteratively. In each iteration, each bus is ripped up and rerouted once by following the same ordering. When routing a bus b, the router is to find a minimum cost path on the routing graph. The cost function consists of 4 terms: length cost, crossing cost, congestion cost and length-bound violation cost. The length cost is the primary objective and it is proportional to the route length of b. The other three terms are penalty costs. The crossing cost is proportional to total net count of buses b crosses. The congestion cost is proportional to the total congestion of congested critical cuts b passes through. The length-bound violation cost is proportional to the amount b's route length exceeds its maximum length bound. Note that the minimum length bound of nets can be satisfied by extending net routes.

4. LAYER ASSIGNMENT

978-1-60558-497-3/09 $25.00 © 2009 ACM

Simulated Annealing (SA) [16] is used to assign the global routes of the buses obtained from the previous stage onto the given layers. The initial assignment is a random assignment of all the buses. In each step of the annealing process, we attempt to move one bus to another layer and evaluate the cost of the new assignment. The cost of an assignment is a weighted sum of two parts: (1) intersection cost, which is the total number of nets that have crossings with each other, and (2) congestion cost, which is the total congestion cost of all critical cuts by using bin-packing estimation.

The output of this SA procedure is a layer assignment of all the topological bus routes. Later, we will iteratively improve this layer assignment as well as the individual routes by using a bubble-sort-like bus recycling technique.

5. ITERATIVE IMPROVEMENT

After the initial bus topologies are assigned to multiple layers, there might still be conflicts between the bus routes on some layers, such as bus intersections and routing congestions. This is because when we perform global routing, we did not know the subsequent layer assignment result and thus may have produced suboptimal routes for some buses. To address this issue, we perform another negotiated-congestion routing to the existing buses on each layer after the layer assignment. In this routing step, we assign very high cost to bus intersections so that they are minimized.

If routing conflicts still exist, we perform a bubble-sort-like iterative improvement. First we collect the conflicting buses on all layers together, reassign them to the first layer and reroute all buses on this layer. Then we choose and fix the maximum set of non-conflicting buses on the first layer. The remaining conflicting buses are then moved to the second layer and rerouted together with the buses already assigned on it. The maximum set of non-conflicting buses are then selected to be fixed on the second layer and the remaining conflicting ones are moved to the third layer. Such procedure is performed iteratively until all buses are routed without conflicts or the result cannot be further improved.

Figure 11 gives an example of bubble-sort-like iterative improvement, where the numbers in the dark shaded boxes and in the light shaded boxes represent the number of non-conflicting and conflicting nets assigned to the layers, respectively. Figure 11(a) gives the initial layer assignment result after rerouting nets layer by layer, resulting 40, 30 and 30 conflicting nets on the three layers. All these 100 conflicting nets need to be recycled. Figure 11(b) shows that all 100 conflicting nets are reassigned to layer 1. In Figure 11(c), 20 of 100 conflicting nets are added to layer 1 without conflicts after rerouting all nets on layer 1, then the remaining 80 conflicting nets are reassigned to layer 2. In Figure 11(d), 50 of 80 conflicting nets are added to layer 2 after net rerouting and the remaining 30 nets are reassigned to layer 3. Finallly, these 30 nets are all successfully added to layer 3, as shown in Figure 11(e).

6. EXPERIMENTAL RESULTS

We test our bus planner on a state-of-the-art PCB from industry that has 7000+ nets and 12 routing layers and has been manually routed. The manual solution is able to route all buses without any intersection or congestion. We run our experiments on a workstation with two 3.0 Ghz Intel Xeon CPUs and 4 GB memory.

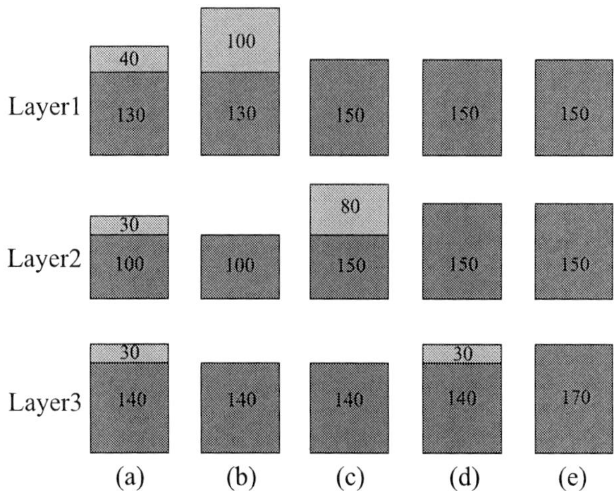

Figure 11: An example of bubble-sort-like improvement.

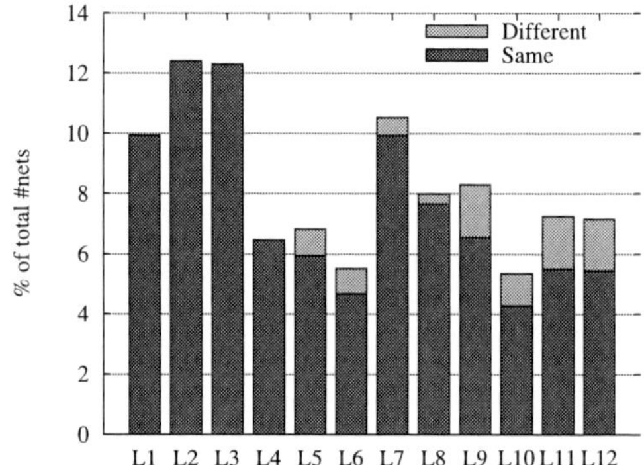

Figure 12: Layer-by-layer comparison between our routing solution and manual routing solution. y-axis is the percentage over the total number of nets in the design.

We did two kinds of experiments. The first one was to test the effectiveness of our single-layer bus router. We directly used the layer assignment extracted from the manual solution and run our negotiated-congestion router on each layer to obtain the topological routing of each bus. We were able to route *all the buses* with no conflicts or congestions within 40 minutes. The length constraints of all the buses were also satisfied. We compared our routing with the manual routing and found that most of our routing is the same as the manual solution. Figure 12 shows the distribution of the nets among the 12 layers in the manual solution and how many of the nets have the same routing as our solution. It can be seen that on the first four layers, the bus routes are exactly the same while on the other layers, the majority of the routes are the same.

We performed another experiment to test our bus planner including layer assignment and topological routing. In this experiment, we were only given the intervals of all the buses and output the layer assignment as well as topological routing. After the layer assignment stage, 95.9% of the nets were successfully assigned and routed. The iterative improvement procedure increased the completion rate to 98.5%. This completion rate is very close to 100%. Only 1.5% of the nets are left unrouted. These nets can be routed

978-1-60558-497-3/09 $25.00 © 2009 ACM

Layer	net#	L1	L2	L3	L4	L5	L6	L7	L8	L9	L10	L11	L12
1	573	81	183	269	0	0	26	0	0	0	0	0	14
2	486	0	0	0	0	0	91	0	95	103	60	13	124
3	698	56	0	82	84	31	0	117	14	128	0	94	92
4	611	78	14	25	0	31	31	0	0	0	0	280	152
5	741	0	243	42	84	0	0	84	124	41	0	62	61
6	646	173	82	0	28	0	0	84	155	43	81	0	0
7	498	0	81	140	170	62	31	0	0	0	14	0	0
8	507	0	60	0	60	59	113	83	118	0	14	0	0
9	622	78	0	0	0	81	0	155	0	84	102	61	61
10	437	0	159	0	0	185	0	0	0	93	0	0	0
11	557	143	0	225	14	31	62	82	0	0	0	0	0
12	549	89	51	82	14	0	34	81	30	92	76	0	0

Table 1: Comparison between our layer assignment and the manual layer assignment.

by manual effort or by using vias. The total CPU time was less than 3 hours.

Table 1 compares our layer assignment result with the manual layer assignment. The first two columns give the layer as well as the number of nets assigned to each layer by our planner. The next 12 columns show the number of nets that are assigned to each of the 12 layers in the manual solution. For example, among the 573 nets that our planner assigned to layer 1, only 81 nets are still assigned to layer 1 in the manual routing. It can be concluded from this table that the correlation between our layer assignment and the manual assignment is very small, which means our SA-based approach obtained a very different solution.

7. CONCLUSION

In this paper, we reported an automatic bus planner, which is one of the key parts of a complete PCB auto-routing system we are developing. In this bus planner, we use a dynamic routing graph to guide the efficient search of planar topologies of the bus routes and a bin-packing-based congestion estimation to effectively avoid violations of routing capacity constraints. We also use a bubble-sort-like iterative improvement to further increase the number of routed nets. We tested our bus planner on a state-of-the-art industrial circuit board with over 7000 nets and 12 signal layers. Our bus planner is able to achieve 100% routing completion using the layer assignment extracted from manual design. For simultaneous layer assignment and bus routing, we are able to successfully route 98.5% of the nets while satisfying length bounds. The remaining 1.5% can be routed either manually or by using vias. The runtime of our bus planner is less than 3 hours.

8. REFERENCES

[1] V. Betz and J. Rose. VPR: A new packing, placement and routing tool for FPGA research. In *FPL '97: Proceedings of the 7th International Workshop on Field-Programmable Logic and Applications*, pages 213–222, 1997.

[2] Cadence Allegro PCB Design. *(http://www.cadence.com/products/pcb/pcb_design)*.

[3] Cadence OrCAD PCB Designer. *(http://www.cadence.com/products/pcb/pcb_designer)*.

[4] M. Cho, K. Lu, K. Yuan, and D. Z. Pan. BoxRouter 2.0: architecture and implementation of a hybrid and robust global router. In *ICCAD '07: Proceedings of the 2007 IEEE/ACM international conference on Computer-aided design*, pages 503–508, 2007.

[5] J. E. G. Coffman, M. R. Garey, and D. S. Johnson. Approximation algorithms for bin packing: a survey. pages 46–93, 1997.

[6] C. Ebeling, L. McMurchie, S. A. Hauck, and S. Burns. Placement and routing tools for the Triptych FPGA. *IEEE Trans. Very Large Scale Integr. Syst.*, 3(4):473–482, 1995.

[7] M. R. Garey and D. S. Johnson. *Computers and Intractability; A Guide to the Theory of NP-Completeness*. W. H. Freeman & Co., 1990.

[8] L. McMurchie and C. Ebeling. PathFinder: A Negotiation-Based Performance-Driven Router for FPGAs. In *Field-Programmable Gate Arrays, 1995. FPGA '95. Proceedings of the Third International ACM Symposium on*, pages 111–117, 1995.

[9] Mentor Graphics Topology Router. *http://www.mentor.com/products/pcb-system-design/layout-routing/topology-router*.

[10] M. M. Ozdal. *Routing algorithms for high-performance VLSI packaging*. PhD thesis, University of Illinois at Urbana-Champaign, 2005.

[11] M. M. Ozdal and M. D. F. Wong. Archer: a history-driven global routing algorithm. In *ICCAD '07: Proceedings of the 2007 IEEE/ACM international conference on Computer-aided design*, pages 488–495, 2007.

[12] M. Pan and C. Chu. FastRoute 2.0: A High-quality and Efficient Global Router. In *ASP-DAC '07: Proceedings of the 2007 conference on Asia South Pacific design automation*, pages 250–255, 2007.

[13] L. Ritchey. Busses: What are they and how do they work. In *Printed Circuit Design Magazine*, 2000.

[14] J. C. Whitaker. *The Electronics Handbook (2nd Edition)*. CRC Press, 2005.

[15] D. Wiens. Printed circuit board routing at the threshold. *White Paper, Mentor Graphics*, 2000.

[16] D. F. Wong, H. W. Leong, and C. L. Liu. *Simulated Annealing for VLSI design*. Kluwer Academic Publishers, 1988.

978-1-60558-497-3/09 $25.00 © 2009 ACM

A Correct Network Flow Model for Escape Routing[*]

Tan Yan
tanyan2@illinois.edu

Martin D. F. Wong
mdfwong@illinois.edu

Department of Electrical and Computer Engineering
University of Illinois at Urbana-Champaign

ABSTRACT

Escape routing for packages and PCBs has been studied extensively in the past. Network flow is pervasively used to model this problem. However, none of the previous works correctly models the diagonal capacity, which is essential for 45° routing in most packages and PCBs. As a result, existing algorithms may either produce routing solutions that violate the diagonal capacity or fail to output a legal routing even though there exists one. In this paper, we propose a new network flow model that guarantees the correctness when diagonal capacity is taken into consideration. This model leads to the first optimal algorithm for escape routing. We also extend our model to handle missing pins.

Categories and Subject Descriptors: B.7.2 [Integrated Circuits]: Design Aids – Placement and Routing

General Terms: Algorithm, Theory

Keywords: Escape routing, diagonal capacity, missing pin, network flow, package routing, PCB routing

1. INTRODUCTION

Escape routing is an important problem in package and PCB design. Traditionally, the escape routing problem is to route from specified pins in a pin array to the boundary of the array (see Figure 1) [2, 4, 6, 14–18]. Some other works also study variants of this problem: the escaped nets around the array boundary are required to follow some ordering constraints [5, 7, 8, 10, 13]. Both the traditional version and its variants have many applications. In this paper, we focus on the traditional escape routing problem.

In the pin array, the design rules limit the number of wires between two orthogonally or diagonally adjacent pins. We call such constraints *orthogonal capacity* and *diagonal capacity* respectively, or O-cap and D-cap for short (see Figure 1). If we consider a *tile* of the pin array, which is the square formed by four adjacent pins, we can see that O-cap limits the number of wires that go through its four sides while D-cap limits the number of wires that go through its two diagonals.

Network flow is pervasively used to model this problem [2, 4, 6, 14, 15, 18]. However, none of the previous network

[*]This work was partially supported by the National Science Foundation under grant CCF-0701821 and a grant from the Fujitsu Laboratories.

Permission to make digital or hard copies of part or all of this work for personal or classroom use is granted without fee provided that copies are not made or distributed for profit or commercial advantage and that copies bear this notice and the full citation on the first page. To copy otherwise, to republish, to post on servers or to redistribute to lists, requires prior specific permission and/or a fee.
DAC'09, July 26-31, 2009, San Francisco, California, USA

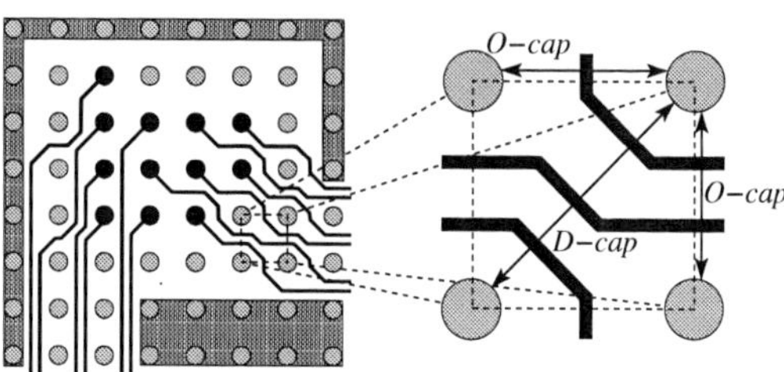

Figure 1: An example of the escape routing problem (left) and the enlarged view of a tile (right). Shaded areas denote blockages. Specified pins (black) are escaped to the boundary of the pin grid. In this example, O-$cap = 2$ and D-$cap = 3$.

flow models is correct. The models in [2, 14] simply ignore the diagonal capacity and may therefore give results violating design rules. On the other hand, the models [4, 6] set a limit on the number of wires inside a tile. Since tile capacity does not correctly reflect diagonal capacity, their models may miss the optimal solution. Detailed discussion about these models can be found in the next section. The other two models are not correct either. [18] uses triangulation, which captures *only one* of the two diagonals in a tile, and therefore its solution may still contain capacity violations on the other diagonal. [15]'s network only gives an upper bound estimation on the number of routable nets. Therefore, none of the previous network flow models is able to correctly model the diagonal capacity.

Non-flow solutions were also proposed in some early works [16, 17]. Those works assume that all the pins in the array must be escaped on a single layer and the routing is symmetrical. This symmetry assumption is the foundation of their algorithms. However, due to the high pin density in modern packages and PCB designs, not all pins in the array can be escaped on one layer and the escape routing is very likely to be asymmetrical. Therefore, their algorithm is not applicable to the more general escape routing problem we discuss here (we do not require all the pins to be escaped on one layer or the routing to be symmetrical).

In this work, we propose a network flow model that correctly models the diagonal capacity. Our model guarantees to give a legal routing if one exists. We then build an algorithm based on this model. As far as we know, this is the first algorithm that guarantees optimality. We also extend our model to handle missing pins, which are unused pins removed to increase the routing resource. Experimental results show that our algorithm has very short runtime.

The rest of this paper is organized as follows: Section 2 introduces some background information. Section 3 presents our network flow model and escape routing algorithm, and Section 4 extends our model to deal with missing pins. Fi-

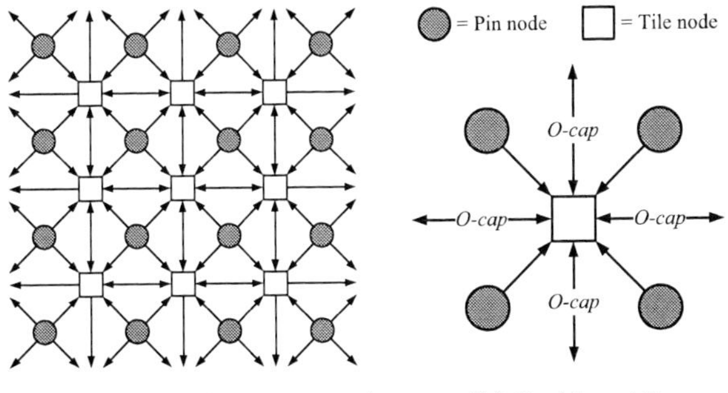

(a) Flow model for a pin grid   (b) Inside a tile

Figure 2: The traditional network flow model used in previous works.

nally, experimental results and some concluding remarks are given in Section 5.

2. BACKGROUND

In this section, we will first formulate the escape routing problem and then introduce the network flow model used by some previous works.

2.1 Problem Formulation

The input to the escape routing problem is an $m \times n$ pin array with p pins specified as to-be-escaped pins. Certain areas of the pin array are marked as blockages, which do not allow any routing inside. Such blockages are used to model pre-routed nets or to guide the routing to escape through certain preferred boundaries. *O-cap* and *D-cap* are also given to specify the orthogonal capacity and diagonal capacity in a tile. We can safely assume that $O\text{-}cap \leq D\text{-}cap \leq 2 \cdot O\text{-}cap$ for all our inputs due to the following two facts:

- The diagonal of a square tile is longer than the side. Therefore, $D\text{-}cap \geq O\text{-}cap$.
- The *O-cap* constraint already implies that at most $2 \cdot O\text{-}cap$ wires can pass the diagonal. So setting *D-cap* to be larger than $2 \cdot O\text{-}cap$ is meaningless.

The expected output of the problem is an octilinear routing from the to-be-escaped pins to the boundary of the pin array satisfying the capacity constraints and avoiding the blockages. We would also like the total length of the routing to be minimized.

2.2 Previous works

In the traditional network flow model used by $[2,4,6,14]^1$, each pin is represented by a pin node and each tile is represented by a tile node. Edges are added between horizontally and vertically adjacent tile nodes and from pin nodes to their adjacent tile nodes (see Figure 2). Edges extending out of the pin grid boundary are connected to a super sink. There are also edges from a super source to the pin nodes that are expected to be escaped. This model works fine if we only consider orthogonal capacity *O-cap* because we can add capacity constraints on the orthogonal edges in the network to realize *O-cap*. However, diagonal capacity *D-cap* is not captured by this model. Consider the case where $O\text{-}cap = 2$ and $D\text{-}cap = 3$, which is very common for PCB routing. Since

---

^{1}Some of these works assume monotonic and/or symmetric routing. Therefore, some orthogonal edges are unidirectional in their models. However, the network structure is the same as what we show here.

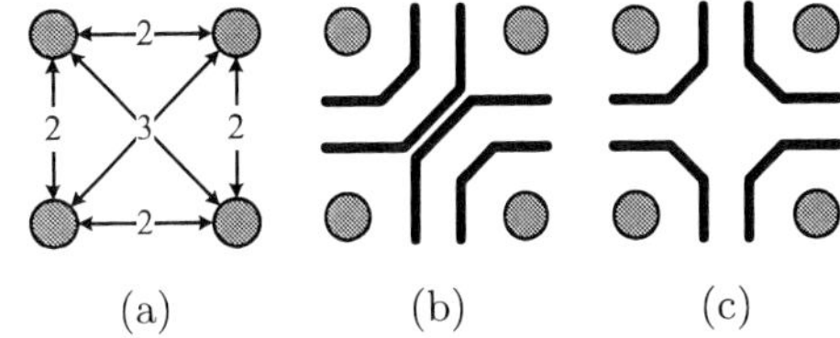

(a)   (b)   (c)

Figure 3: Previous network flow models cannot handle diagonal capacity *D-cap* correctly. For the case $O\text{-}cap = 2$, $D\text{-}cap = 3$ in (a), models in $[2, 14]$ may produce illegal routing in (b) and models in $[4, 6]$ cannot capture the valid routing in (c).

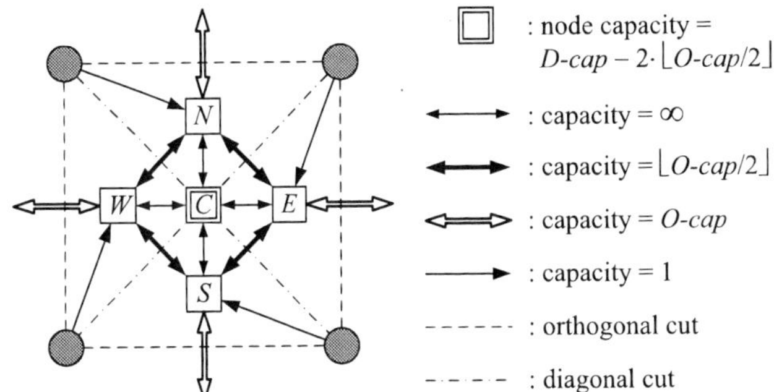

Figure 4: Our network flow model inside a tile.

the model has no control over the number of wires passing through the diagonals of the tile, it may produce routing like Figure 3 (b) which violates *D-cap*. In $[4, 6]$, the number of wires inside a tile is also limited by adding node capacities to tile nodes. However, such node capacity does not correctly reflect the diagonal capacity. No matter how we set the tile node capacity, there are always counter-examples:

- Tile node capacity ≥ 4: This is the same as having no tile capacity because $O\text{-}cap = 2$ already implies at most 4 wires in a cell. The network may produce the routing in Figure 3 (b), which violates *D-cap*.
- Tile node capacity ≤ 3: The network cannot model the routing in Figure 3 (c) because there are 4 wires inside the tile. However, the routing itself is legal because only 2 wires pass each diagonal. As a result, the network model may fail to capture a legal routing solution even when one exists.

From the discussion above, it can be seen that such a simple network cannot model diagonal capacity correctly. A more sophisticated network model is needed to capture the diagonal capacity.

3. OUR NETWORK FLOW MODEL

In our proposed network flow model, each tile contains five nodes, namely *N*-node on the north, *E*-node on the east, *S*-node on the south, *W*-node on the west and *C*-node in the center (see Figure 4). The first four nodes are called *peripheral nodes* and the last node is called the *center node*. We give the center node a capacity $D\text{-}cap - 2 \cdot \lfloor O\text{-}cap/2 \rfloor$. Node capacity can be realized by splitting the node into two and adding an edge with the same capacity between them.

We create bidirectional edges (which are realized by two directed edges: a forward edge and a backward edge) between every peripheral node and the center node and give these edges infinite capacity. We also introduce bidirectional

978-1-60558-497-3/09 $25.00 © 2009 ACM

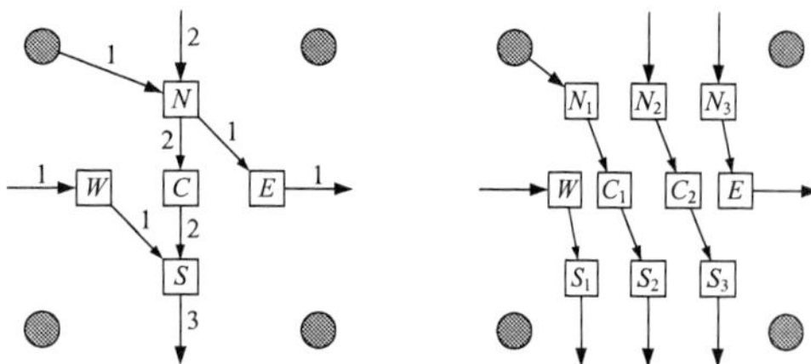 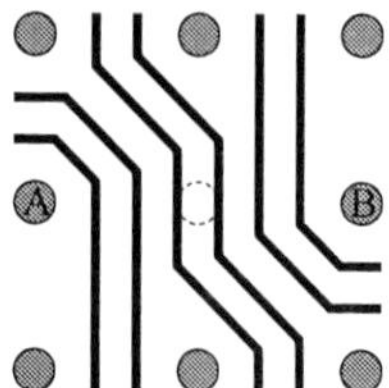

Figure 5: Split the nodes and edges of a flow (left) to obtain the topology (right).

edges between peripheral node pairs (N, E), (E, S), (S, W) and (W, N). We call such edges *peripheral edges* and give each of them capacity $\lfloor O\text{-}cap/2 \rfloor$. Connections between tiles are also necessary. We use bidirectional edges to connect the N-node of a tile to the S-node of the tile above it as well as the E-node of a tile to the W-node of the tile to the right of it. Such edges are called *inter-tile edges* and have capacity $O\text{-}cap$. In order to escape the pins, we also create a pin node for each pin and create unidirectional edges from the four pins at the corners of a tile to the four peripheral nodes in the tile. The edges are from the pin at the NW corner to N-node, from the pin at the NE corner to E-node, from the pin at the SE corner to S-node and from the pin at the SW corner to W-node. All edges have capacity 1. Of course, if any nodes or edges lie in the blockage, we will not create them.

We also introduce a super source s and a super sink t. All edges from the boundary tiles to the outside of the pin array are connected to t. Finally, we add edges with capacity 1 from s to the pin nodes of all the to-be-escaped pins.

The following theorem guarantees the correctness of our model. The proof is omitted due to the page limit.

THEOREM 1. *The max-flow value of our network is the same as the number of to-be-escaped pins if and only if there exists a legal escape routing for all to-be-escaped pins.*

The max-flow solution of the network model can then be converted into escape routing by node and edge splitting (see Figure 5): for a node v, if the flow through it $flow(v) > 1$, we split the node into $flow(v)$ copies. Similarly, we also split each edge with flow larger than 1. The split nodes and edges are then connected in a planar fashion.

The result of node and edge splitting is essentially a planar topology of the escape routing. This topology can then be converted into an octilinear routing by existing algorithms [11, 12].

In order to minimize the wirelength, we can assign cost 1 to the inter-tile edges (the hollow edges in Figure 4) and zero cost to all other edges. If we compute the min-cost max-flow of the network, we can minimize the number of tiles each wire traverses and thus the total wirelength can be minimized.

4. MODELING THE MISSING PINS

In practical PCB designs, designer may remove some unused pins in the array to increase the routing resource. Figure 6 gives an example in which $O\text{-}cap = 2$ and $D\text{-}cap = 4$. It can be seen that by removing the pin at the center, the maximum number of wires allowed between A and B increases from 4 to 6. The difference, 2, is called the *extra horizontal capacity* of the missing pin. Similarly, we can define *extra*

Figure 6: Missing pins increase the routing resource.

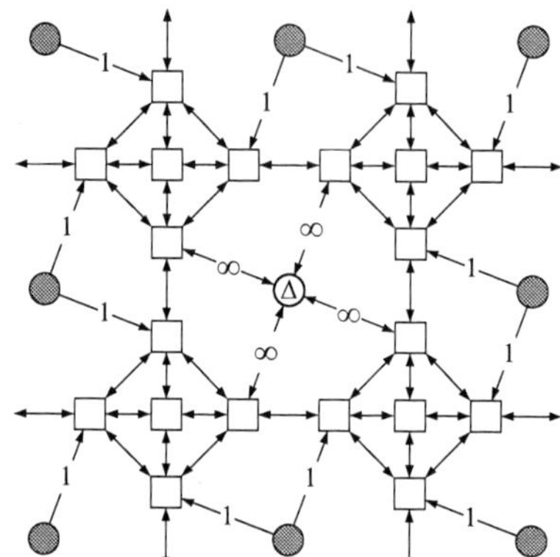

Figure 7: Modeling missing pins in our network.

vertical capacity and *extra diagonal capacity*. Since usually the pin is round, the three types of extra capacities are the same. So we do not distinguish them but call them *extra capacity* and denote it as Δ. This extra capacity depends on the design rules and is usually given as input.

To model this extra capacity, we use a resource node to replace the pin node (see Figure 7). The resource node has node capacity same as Δ. The unidirectional edges from the pin node to the peripheral nodes are now replaced with bidirectional edges between the resource node and the peripheral nodes. We give such edges infinite capacity.

The resource node essentially increases the horizontal and vertical capacity of the network by Δ. So it is able to capture the extra capacity introduced by the missing pin.

5. EXPERIMENTAL RESULTS AND CONCLUDING REMARKS

We implement our network flow-based escape routing algorithm in C++ and test it on several industrial data sets. We use the min-cost flow solver CS2 [1] to obtain the min-cost flow solution of our network model. All experiments are carried out on a workstation with two 3.0GHz Intel Xeon processors and 4GB memory. The operating system is Red-Hat Linux 2.6.9.

We test our router on eight data sets and the result is reported in Table 1. Among the eight data sets, *industrial_1* to *industrial_7* are actual industrial data and *modified_8* is derived from industrial data with some modification. The left five columns of the table gives the information on the data including the name, the pin array size, the number of to-be-escaped pins, the number of missing pins, and the capacity rules ($O\text{-}cap$, $D\text{-}cap$ and extra capacity Δ). The two following columns show the number of $D\text{-}cap$ constraint violations in our result as well as the runtime of our router. The last two columns show the number of $D\text{-}cap$ violations

978-1-60558-497-3/09 $25.00 © 2009 ACM

Table 1: Experimental results

	array $w \times h$	escape pin no.	missing pin no.	capacities			our model		traditional model	
				O	D	Δ	D-cap vio.	runtime	D-cap vio.	runtime
industrial_1	14×16	78	13	2	3	3	0	0.16 s	0	0.14 s
industrial_2	29×11	66	20	2	3	3	0	0.22 s	10	0.17 s
industrial_3	33×14	120	46	2	3	3	0	0.33 s	6	0.28 s
industrial_4	35×17	160	30	2	3	3	0	0.61 s	9	0.28 s
industrial_5	35×35	108	105	2	3	3	0	0.86 s	1	0.68 s
industrial_6	35×17	143	38	2	3	3	0	0.39 s	0	0.30 s
industrial_7	35×35	220	106	2	3	3	0	1.01 s	53	0.79 s
modified_8	35×35	225	33	2	3	3	0	0.99 s	53	0.79 s

and the runtime of the traditional model used in [2, 14].

It can be seen that our model gives *zero D-cap* violations while the traditional model leads to as many as 53 *D-cap* violations because it ignores the diagonal capacity. Since the total runtime is only one second or less, the runtime difference of the two methods is insignificant.

Figure 8 shows our routing solution of *modified_8*. We can see that there are several missing pins on the north, west and east side of the array (highlighted by the gray zones), and their spaces are fully utilized in our result. To show that our router can handle dense designs, we draw a dashed polygon on top of the routing result. It can be seen that almost all the *O-cap* and *D-cap* along the polygon is used up by our routing, which indicates that the routing is very dense. Notice that if the max-flow solution of our network has value less than the number of to-be-escaped pins, then not all these pins can be escaped on a single layer. We have to use multiple layers to escape all of them. It is also possible to extend our model to find out the minimum number of layers to escape all the pins.

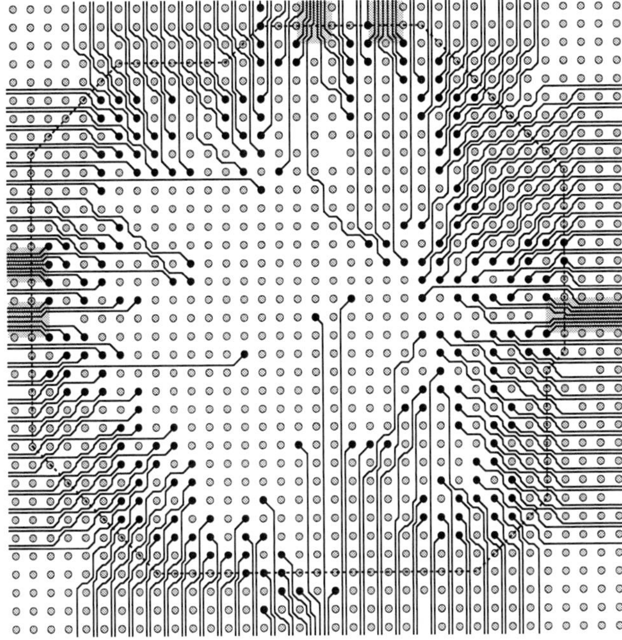

Figure 8: Routing solution of *modified_8*. The shaded zones highlight the spaces of the missing pins that are fully utilized by our router. The dashed polygon is drawn on top of the result to show that the routing uses up almost all routing resources.

6. REFERENCES

[1] CS2: min-cost flow solver. http://www.igsystems.com/cs2/index.html.

[2] W.-T. Chan, F. Y. L. Chin, and H.-F. Ting. A faster algorithm for finding disjoint paths in grids. In *Proc. Int. Symp. on Algorithms and Computation*, pages 393–402, 1999.

[3] W. W.-M. Dai, R. Kong, and M. Sato. Routability of a rubber-band sketch. In *Proc. Design Automation Conf.*, pages 45–48, 1991.

[4] J.-W. Fang and Y.-W. Chang. Area-I/O flip-chip routing for chip-package co-design. In *Proc. Int. Conf. on Computer-Aided Design*, pages 518–522, 2008.

[5] J.-W. Fang, C.-H. Hsu, and Y.-W. Chang. An integer linear programming based routing algorithm for flip-chip design. In *Proc. Design Automation Conf.*, pages 606–611, 2007.

[6] J.-W. Fang, I.-J. Lin, Y.-W. Chang, and J.-H. Wang. A network-flow-based RDL routing algorithmz for flip-chip design. *IEEE Trans. Computer-Aided Design*, 26(8), Aug. 2007.

[7] Y. Kubo and A. Takahashi. A global routing method for 2-layer ball grid array packages. In *Proc. Int. Symp. on Physical Design*, pages 36–43, 2005.

[8] Y. Kubo and A. Takahashi. Global routing by iterative improvements for two-layer ball grid array packages. *IEEE Trans. Computer-Aided Design*, 25(4), Apr. 2006.

[9] C. E. Leiserson and F. M. Maley. Algorithms for routing and testing routability of planar VLSI layouts. In *Proc. Ann. Symp. on Theory of Computing*, pages 69–78, 1985.

[10] L. Luo and M. D. F. Wong. Ordered escape routing based on boolean satisfiability. In *Proc. Asia and South Pacific Design Automation Conf.*, pages 244–249, 2008.

[11] D. Staepelaere, J. Jue, T. Dayan, and W. W.-M. Dai. SURF: Rubber-band routing system for multichip modules. *IEEE Des. Test. Comput.*, 10(4):18–26, Dec. 1993.

[12] D. J. Staepelaere. Geometric transformations for a rubber-band sketch. Master's thesis, University of California at Santa Cruz, Santa Cruz, CA, USA, Sept. 1992.

[13] Y. Tomioka and A. Takahashi. Monotonic parallel and orthogonal routing for single-layer ball grid array packages. In *Proc. Asia and South Pacific Design Automation Conf.*, pages 642–647, 2006.

[14] D. Wang, P. Zhang, C.-K. Cheng, and A. Sen. A performance-driven I/O pin routing algorithm. In *Proc. Asia and South Pacific Design Automation Conf.*, pages 129–132, 1999.

[15] R. Wang, R. Shi, and C.-K. Cheng. Layer minimization of escape routing in area array packaging. In *Proc. Int. Conf. on Computer-Aided Design*, pages 815–819, 2006.

[16] M.-F. Yu and W. W.-M. Dai. Pin assignment and routing on a single-layer pin grid array. In *Proc. Asia and South Pacific Design Automation Conf.*, pages 203–208, 1995.

[17] M.-F. Yu and W. W.-M. Dai. Single-layer fanout routing and routability analysis for ball grid arrays. In *Proc. Int. Conf. on Computer-Aided Design*, pages 581–586, 1995.

[18] M.-F. Yu, J. Darnauer, and W. W.-M. Dai. Interchangeable pin routing with application to package layout. In *Proc. Int. Conf. on Computer-Aided Design*, pages 668–673, 1996.

978-1-60558-497-3/09 $25.00 © 2009 ACM

Flip-Chip Routing with Unified Area-I/O Pad Assignments for Package-Board Co-Design *

Jia-Wei Fang[1,2], Martin D. F. Wong[2], and Yao-Wen Chang[1,3]
[1]Graduate Institute of Electronics Engineering, National Taiwan University, Taipei, Taiwan
[2]Department of Electrical and Computer Engineering, University of Illinois at Urbana-Champaign, USA
[3]Department of Electrical Engineering, National Taiwan University, Taipei, Taiwan

ABSTRACT

In this paper, we present a novel flip-chip routing algorithm for package-board co-design. Unlike the previous works that can consider only either free- or pre-assignment routing, our router is the first work in the literature that can handle both the free- and pre-assignment routing. Based on the computational geometry techniques (e.g., the Delaunay triangulation and the Voronoi diagram), the router applies a unified network-flow formulation to perform congestion estimation for the pre-assignment routing. According to the congestion map, the network-flow formulation can also consider the free-assignment nets during the routing for the pre-assignment ones. Then, the router modifies the network-flow formulation to optimally assign and route the free-assignment nets, considering the routed pre-assignment nets. With the package and board co-design flow, we can achieve 100% routing completion. Experimental results based on industry designs demonstrate the high-quality of our algorithm.

Categories and Subject Descriptors: B.7.2 [Integrated Circuits]: Design Aids - Layout, Placement and Routing
General Terms: Algorithms, Design
Keywords: Detailed Routing, Global Routing, Physical Design

1. INTRODUCTION

In VLSI designs, the increasing complexity and decreasing feature size have made the demand of more I/Os a significant problem to packaging technologies. An advanced packaging technology, the *flip-chip package*, as shown in Figure 1(a), is created for higher integration density and larger I/O counts. In recent IC's, designers place the I/O pads (buffers) in the whole area of a die, instead of just along the die boundary, to result in shorter wirelength, higher chip density, and better signal and power integrity.

For the flip-chip applications, typically the top metal or an extra metal layer, called a *re-distribution layer* (*RDL*) as illustrated in Figure 1(b), is used to redistribute/connect the *I/O pads* to the *bump pads* without changing the placement of the I/O pads [5].

There are two kinds of the RDL routing problems for the flip-chip design. The first one is the *free-assignment* (*FA* for short) routing problem. In this problem, a router has the freedom to assign each I/O pad with a signal to any bump pad with no signal during routing [3]. Therefore, it ignores the routing constraints from a printed circuit board (PCB). The second kind of RDL routing is the *pre-assignment* (*PA* for short) routing problem where the mapping among the I/O pads and the bump pads is pre-defined before routing and cannot be changed [2]. Further, the pre-defined netlist can provide the routing constraints

*This work was partially supported by ITRT, SpringSoft, Synopsys, TSMC, National Science Foundation of the US under Grant No's. CCF-0701821, and National Science Council of Taiwan under Grant No's. NSC 97-2221-E-002-237-MY3, NSC 96-2628-E-002-248-MY3, NSC 96-2628-E-002-249-MY3, and NSC 096-2917-I-002-120.

Permission to make digital or hard copies of part or all of this work for personal or classroom use is granted without fee provided that copies are not made or distributed for profit or commercial advantage and that copies bear this notice and the full citation on the first page. To copy otherwise, to republish, to post on servers or to redistribute to lists, requires prior specific permission and/or a fee.
DAC'09, July 26-31, 2009, San Francisco, California, USA

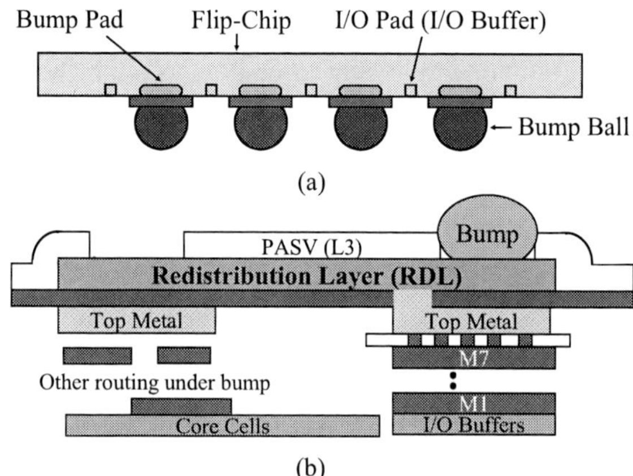

Figure 1: (a) Flip-Chip Package. (b) Section of RDL.

from the PCB. However, if the netlist generation does not consider the package routing constraints, the RDL routing might not be completed. Hence, for package-board (PB) co-design, the two routing problems shall be considered simultaneously to achieve greater design flexibility and better design performance. Figure 2 shows two routing examples. An I/O pad and a bump pad with the same label i form a PA net i, e.g., nets 1–4. Other I/O pads which are not paired with any bump pad are the FA nets, e.g., nets 5–8. In Figure 2(a), since the routing does not consider PB co-design, net 2 might block net 7, and net 8 forces net 1 to detour. In contrast, in Figure 2(b), considering PB co-design gives a feasible routing for net 7 and further reduces the total wirelength. Therefore, it is desirable to develop an RDL routing algorithm to consider both the PA and FA nets for PB co-design.

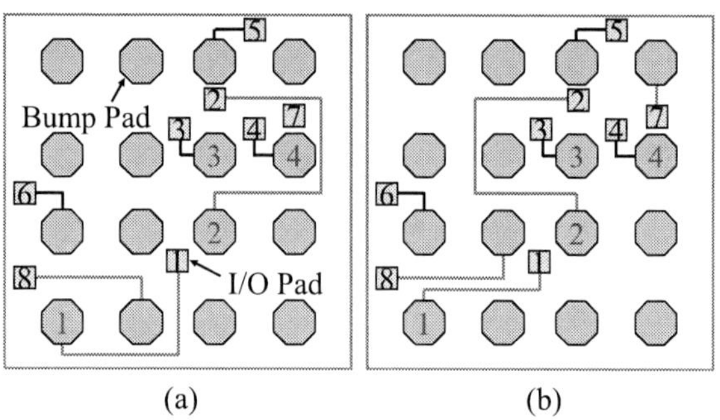

Figure 2: (a) Routing without Package-Board Co-Design. (b) Routing with Package-Board Co-Design.

1.1 Previous Work

To the best of our knowledge, there is no existing work in the literature specially for the area-I/O flip-chip routing that consider both the PA and FA nets for PB co-design. Two related works on flip-chip routing only consider the FA routing problem [3] or the PA routing problem [2], but not both together. Furthermore, the work [2] can only handle the flip-chip structure with the I/O pads placed along the die boundary. Although the work [3] is for the area-I/O flip-chip structure, its *tile-based* flow-network construction did not consider the PA nets and the positions of I/O pads which can avoid over design. Another related work is the routing for Pin-Grid-Array (PGA) packages [6]. The work [6] used triangulation and its dual to assign pins in the PGA packages. Since the work represented pre-routed nets by edges (obstacles) in the triangulation (dual), the topology of the nets cannot be changed during routing. For PB co-design, however, FA and PA nets affect each other, and thus the changes of the topology are inevitable. In addition to the aforementioned differences, a major deficiency of the previous works is that they cannot handle the unified constraints induced by both the FA and PA routing.

1.2 Our Contributions

We present in this paper a novel flip-chip routing algorithm for package and board co-design. Unlike the previous works that can consider only either FA or PA routing, our router is the first work in the literature that can handle both the FA and PA routing. Based on the computational geometry techniques (e.g., the Delaunay triangulation [DT] and the Voronoi diagram [VD] [4]), the global routing applies a unified network-flow formulation to perform congestion estimation for the PA routing. With the DT and VD techniques, the positions of I/O pads can be captured more precisely to avoid wire congestion. According to the congestion map, the network-flow formulation can also consider the FA nets during the routing for the PA ones. Then, the router modifies the network-flow formulation to optimally assign and route the FA nets, considering the routed PA nets. With the package and board co-design flow, we can achieve 100% routing completion. Experimental results based on industry designs demonstrate that our router can achieve 100% routability and shorter routed wirelength, compared with the related works.

The rest of this paper is organized as follows. Section 2 gives the formulation of the routing problem. Section 3 details our routing algorithm. Section 4 reports the experimental results. Finally, our conclusion is given in Section 5.

2. PROBLEM FORMULATION

We introduce the notation used in this paper and formally define the RDL routing problem with both the FA and PA routing constraints for the PB co-design. Figure 3 shows the modelling of the routing structure of the area-I/O flip-chip package. Let $N = N_f \cup N_p$ be the set of FA nets (N_f) and PA nets (N_p) for routing. $Q = Q_f \cup Q_p$ is the set of I/O pads where Q_f and Q_p are the I/O pads of the FA and PA nets, respectively. B is the set of bump pads. For practical applications, each I/O pad is paired with only one bump pad, and several I/O pads can be routed to the same bump pad to form a multi-pin net. Let l_i be the bump-pad column i, and let w_j be the bump-pad row j. Each bump pad column/row (l_i/w_j) consists of a set of m bump pads, and each bump pad is represented by $b_{i,j}$. A tile is constructed by four adjacent bump pads $(b_{i,j}, b_{i+1,j}, b_{i+1,j+1}, b_{i,j+1})$. Since the RDL routing is typically on a single layer, it does not allow *wire crossings*, for which two wires intersect each other in the RDL.

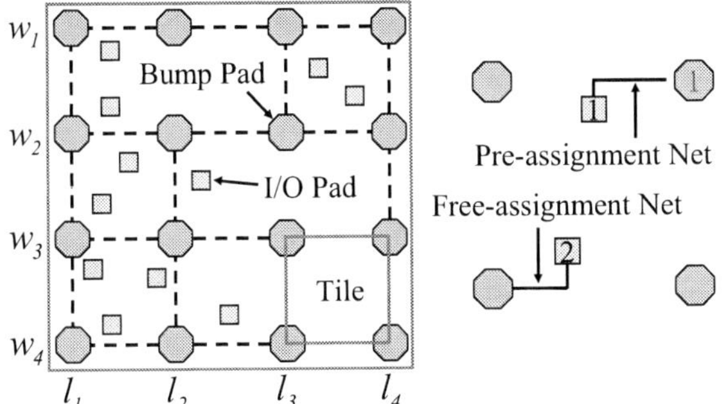

Figure 3: Routing Structure in a Flip-Chip Package.

Now we formally define the addressed problem as follows:

PROBLEM 1. *Given a netlist containing both free- and pre-assignment nets, and a placement of I/O pads and bump pads, the RDL routing problem for PB co-design is to connect a set of I/O pads and a set of bump pads so that there is no wire crossing in RDL and the total wirelength is minimized.*

3. THE ROUTING ALGORITHM

3.1 Algorithm Overview

In the routing flow of Figure 4, our algorithm consists of two major phases: (1) global routing based on DT, VD, and the *minimum-cost maximum-flow algorithm (MCMF)* [1], and (2) detailed routing based on routing-path refinement, net-ordering determination, and maze routing.

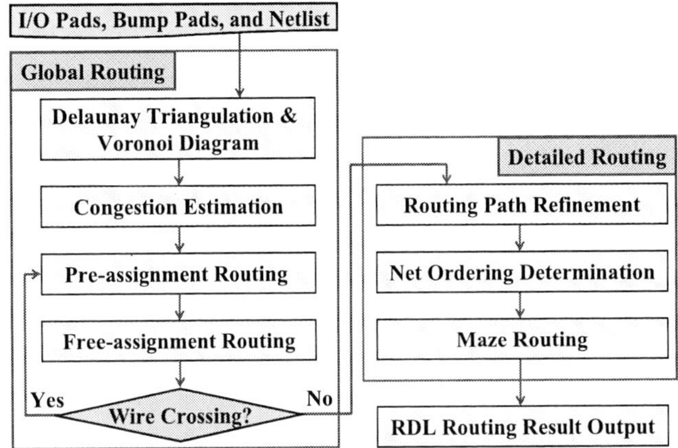

Figure 4: Routing Flow of Package-Board Co-design.

In the first phase, for PA nets, we construct the flow network G based on DT and VD to perform congestion estimation by MCMF. By the congestion map, we can apply maze routing to route the PA nets on G considering FA nets. Then we modify G to optimally assign and route the FA nets by MCMF based on the routed PA nets. To get better routing results, we allow the FA nets to cross the PA nets. If there are wire crossings, the crossed PA nets will be ripped up and rerouted or the crossed FA nets will be forbidden to choose the same routing paths.

In the second phase, we first separate every global-routing path from VD. Then we create a routing sequence in each tile that can guarantee to route all nets. Finally, we use maze routing to route each net based on the routing sequence.

3.2 Global Routing

3.2.1 Congestion Estimation

We construct the flow network G to perform the *concurrent assignment and routing* for the FA nets without considering the PA ones. Since this network-flow based routing can achieve the optimal results for the FA nets, the results can provide an accurate estimation of the congestion for routing the PA ones. Figure 5 shows an example where nets 1–4 are PA nets, and nets 5–8 are FA ones. In Figure 5(a), we first insert a set D of DT nodes on the die boundary and at the middle of each side of the bump pads. Based on the I/O pads and the DT nodes, we can obtain the corresponding DT (see the dashed edges). By the DT nodes, we can avoid edge crossings with the bump pads and use the routing space between the bump pads and the die boundary. Based on the DT, we can get VD and its corresponding edges (see the solid edges in Figure 5(a); for better illustration, only a partial VD is shown in the figure). The white nodes give the set M of VD nodes. They are inserted at each crossing of VD edges and the center of every FA bump pad. Figure 5(b) shows part of G. According to the VD, we construct $G = (V, E)$, where E denotes the edge set, V denotes the vertex set: $Q \cup M \cup \{s, t\}$, s is the source node, and t is the sink node. There are five types of edges:
1. Directed edge from an I/O pad to a VD node,
2. Bi-directional edge between a VD node and another one,
3. Directed edge from a VD node to a VD node in a bump pad,
4. Directed edge from the source node to an I/O pad,
5. Directed edge from a VD node in a bump pad to the sink node.
The last two types of edges are not shown in Figure 5. Type-1 edges are constructed from an I/O pad to its surrounding VD nodes. Type-2 edges are the VD edges. Type-3 edges are connected to a VD node in a bump pad from its surrounding VD nodes. Each edge is associated with a (*cost, capacity*) ordered-pair. The cost is the length l_e of the edge e. The capacity is used to avoid wire congestion and equals the maximum number of wires allowed to pass through the edge. For each Type-2 edge (VD edge), its capacity is n_{max} which can be measured by the

978-1-60558-497-3/09 $25.00 © 2009 ACM

length of the DT edge crossing the VD edge. The capacity of the other edges is 1 because only one signal can be transmitted from an I/O pad and to a bump pad. Since we also want to avoid wire congestion in every triangle, the capacity of each VD node is set to be the maximum capacity of all the edges connecting the VD node. Figure 5(c) shows an example routing result. Recall that we do not allow crossings for all wires. Since E represents the potential global-routing paths of all nets, we can guarantee that no wire crossing will occur if there exists no crossing among edges. As a result, we construct all edges and avoid their crossings at the same time. After applying MCMF, we can use the routing results for the FA nets to accurately estimate the congestion (see the arrows in Figure 5(a)) because MCMF gives the optimal routes (with the minimum wirelength).

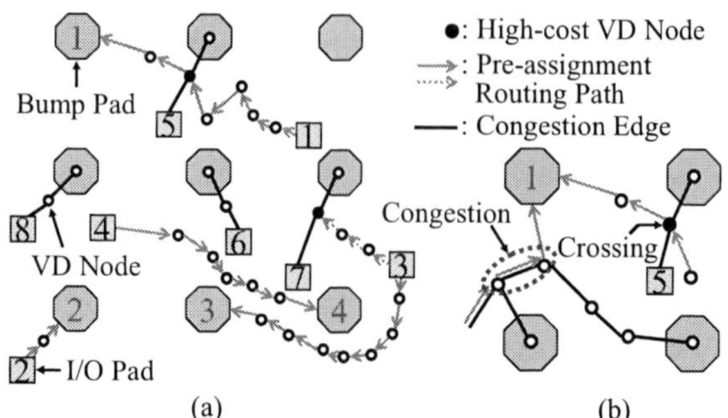

Figure 6: (a) PA Routing based on VD. (b) Additional Costs of Wire Crossings and Wire Congestion.

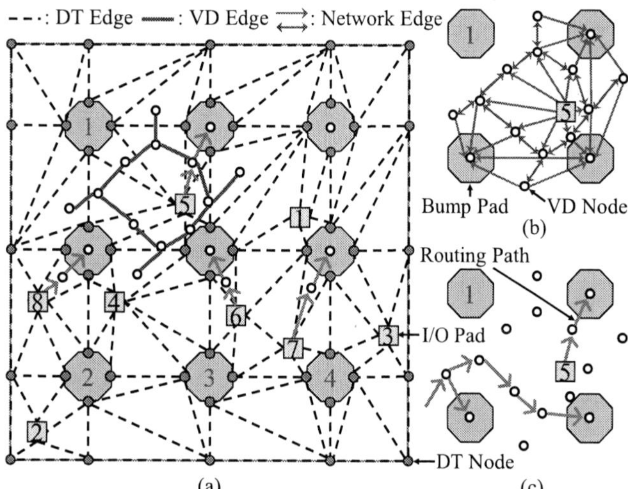

Figure 5: (a) DT and VD. (b) Flow Network. (c) Global-routing Paths of FA Nets.

Figure 7: (a) FA Routing based on Modified Flow Network. (b) Modified Flow Network.

3.2.2 Pre-Assignment Routing

According to the congestion estimation, some edges of G might be identified as congestion edges with routed FA wires (see Figure 6(a)). We first divide each multi-pin PA net into a set of 2-pin nets. Then we apply maze routing to route the PA nets on G. The new cost of each vertex v in G is $\alpha \times d_b$, where d_b is the central distance between two adjacent bump pads, and α is a user-defined positive constant if the vertex is a high-cost VD node which represents a crossing between two nets; $\alpha = 0$, otherwise. The new cost function of each edge e in G is defined by

$$Cost(e) = l_e + \beta \times d_b \times (n_e - n_{max} + 1), \qquad (1)$$

where n_e is the number of routed wires through e, including the FA and PA ones. If $n_e - n_{max} + 1 > 0$, β is also a user-defined positive constant. Otherwise, $\beta = 0$. In Figure 6(b), if the PA routing path crosses the wires of I/O pad 5, there is an additional cost $\alpha \times d_b$ for passing the high-cost VD node. Once a cross is generated, we make the cost of the high-cost VD node zero in order not to count the cross again. The reason is that a crossed FA net can be removed to avoid the crossings. If the PA routing path is routed through the congestion edges (see the dotted ellipse), the cost of each edge will be computed by Equation 1. Figure 6(a) shows the routing result of the PA nets, where net 1 crosses net 5, and net 3 detours to avoid crossing net 7. Since the routing results for the FA nets are just used for congestion estimation, we allow wire crossings at this step. Note that no wire crossing is allowed between the PA nets.

3.2.3 Free-Assignment Routing

After the PA routing, we modify G to optimally route all FA nets based on the routed PA wires. In Figure 7(a), the routed PA wires become obstacles during routing the FA nets. Then we add extra VD nodes into G (see Figure 7(b)) to route FA nets along the PA wires and thus to avoid wire crossings. The extra VD nodes are inserted around the VD nodes on the routed PA wires. The number of extra VD nodes around a VD node equals the number of edges connecting the VD node. For example, there are three edges of the VD node connecting to extra VD node 1. Thus, we insert an extra VD node between two adjacent edges.

After inserting the extra VD nodes, we remove all the dashed edges and the edges connecting the routed PA wires from G to avoid wire crossings. We will construct additional edges later to replace the removed ones. The thick edges will be kept because the FA nets can be routed between nets i and j (extra VD nodes 1 and 4). Then in Figure 7(b), we have additional five types of edges in G (over the previous types in Section 3.2.1) as follows:

6. Directed edge from an I/O pad to an extra VD node,
7. Bi-directional edge between an extra VD node and another one along a VD edge,
8. Bi-directional edge between an extra VD node and another one around the same VD node,
9. Bi-directional edge between an extra VD node and a VD node,
10. Directed edge from an extra VD node to a VD node in a bump pad.

Type-6 edges are constructed from an I/O pad to its surrounding extra VD nodes. Every Type-7 edge (e.g., the edge between extra VD nodes 3 and 5) is constructed without crossing the PA wires (VD edges). Type-8 edges are constructed between two adjacent extra VD nodes around the same VD node to cross the PA wires. Type-9 edges consist of two categories. In the first category, if only one terminal v of a VD edge is on the routed PA wires (e.g., VD node 7), a Type-9 edge is constructed between the other terminal and the closest one of the extra VD nodes surrounding v (e.g., VD node 6 and extra VD node 3). In the second category, the edges are constructed between a VD node and its surrounding extra VD nodes. Type-10 edges are connected to a VD node in a bump pad from its surrounding extra VD nodes. Each edge is also associated with a (cost, capacity) ordered-pair.

The cost of the Type-8 edge is $\gamma \times d_b \times n_w$, where γ is a user-defined positive constant and n_w is the number of the crossed PA wires. In the second category of the Type-9 edges, the cost is $\gamma \times d_b \times n_r$. If an edge is like the one connecting extra VD node 2 and the thick edges, n_r equals the n_w between extra VD nodes 1 and 2. Otherwise, $n_r = 0$ like the one on the edge between extra VD node 1 and the connected VD node. That is because there is no wire crossing to route into nets i and j. The cost of any other edge is the edge length l_e. The capacity of the Type-7 edge is defined as follows:

$$Capacity(e) = \left\lfloor \frac{n_{max} - n_w}{\nu} \right\rfloor. \qquad (2)$$

Recall that each Type-7 edge is along a VD edge. On the VD

978-1-60558-497-3/09 $25.00 © 2009 ACM

edge, n_w is the number of routed PA wires and n_{max} is the maximum capacity. $\nu = 3$ if the VD edge is a thick one. That is because we can route wires into the middle of the thick edges except at the two sides. Otherwise, $\nu = 2$. The capacity δ of the Type-8 edge is infinity. In Section 3.2.4, we will discuss how to change δ to improve the routing results. For the edge in the first category of the Type-9 edges, its capacity equals n_{max} of the deleted VD edge between one VD node and the other VD node surrounded by extra VD nodes (e.g. VD nodes 6 and 7). In the second category of the Type-9 edges, the capacity is given by Equation 2 where $\nu = 3$. The capacity of the remaining edges is 1. After applying MCMF, Figure 7(a) shows the routing result of the FA nets. Net 5 crosses net 1 without detouring bump pad 1 since we only set high costs on the Type-8 edges and the edges in the second category of the Type-9 edges to avoid wire crossings. The reason is that ripping up and rerouting some PA nets may further improve the routing results.

3.2.4 *Iterative Improvement*

After the FA routing, there may be some wire crossings. The reasons are that there is no solution without wire crossings or the detours of the FA nets are too long. Then we rip up and reroute the PA nets crossing the routed FA wires. In Figure 7(a), the total additional cost of net 5 to cross net 1 on G is γ. In Figure 8, we rip up and reroute net 1 if and only if the increase of the total wirelength is less than γ, or there is no other routing path of net 5. If we choose not to rip up and reroute net 1, we will set the capacity δ of the edge which makes net 1 crossed to be 0. By doing so, we can forbid net 5 to cross net 1 again. Finally, the iterative improvement is converged.

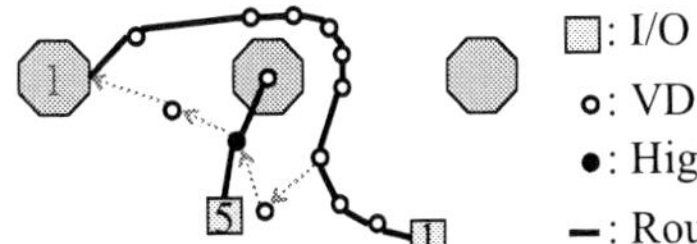

Figure 8: Improvement of Routing Results.

3.3 Detailed Routing

3.3.1 *Routing-Path Refinement*

After the global routing, only the edge with wires is left. In Figure 9(a), there are two FA nets and two PA nets. According to the number of wires (#Wires) on each edge, we can remove the VD nodes and refine the routing paths without generating any wire crossing as in [3]. Then, in Figure 9(b), we can split the edges of the VD nodes into independent wires with wire nodes.

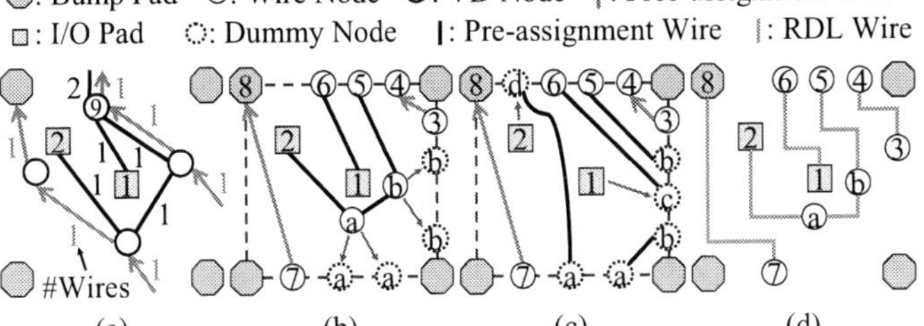

Figure 9: (a) Global-routing Paths. (b) Path Refinement. (c) Net-ordering Determination. (d) Maze Routing.

3.3.2 *Ordering Determination and Maze Routing*

In each tile, we can treat the routing as a channel routing. So we can modify the net-ordering determination algorithm presented in [3] to generate a routing sequence of the wires in a tile. Note that the net-ordering determination algorithm restricts each terminal to be on the boundary of a tile and to have no detours, but in the detailed routing problem, there may be I/O pads with detours inside a tile. Hence, in Figure 9(b) and (c), we insert dummy nodes instead of each I/O pad or wire node inside the tile. For example, since wire node a shares the same VD node with wire node 7, two dummy nodes a are inserted at the right side of wire node 7 to avoid crossing net 8. Since all terminals are on the boundary now, we can apply the determination algorithm. The resulting routing sequence $S = <
w(8,7), w(a,b), w(3,4), w(b,5), w(1,6), w(2,a) >$ and then we can route each wire $w(i,j)$ in a tile without intersecting each other by maze routing [3]. Figure 9(d) shows the result of maze routing.

4. EXPERIMENTAL RESULTS

We implemented our algorithm in the C++ programming language on a 1.2GHz SUN Blade 2000 workstation with 8 GB memory. The benchmark circuits, listed in Table 1, are real industry designs. In Table 1, "Circuits" is the names of circuits, "#I/O Pads (Free-/Pre-assignment)" is the number of I/O pads of FA/PA nets, and "#Bump Pads" is the number of bump pads.

Table 1: Benchmarks for the Package-Board Co-design.

Circuits	#I/O Pads (Free-/Pre-assignment)	#Bump Pads
fcpb1	162 (68/94)	289
fcpb2	301 (117/184)	400
fcpb3	346 (153/193)	441
fcpb4	333 (137/196)	441
fcpb5	1059 (394/665)	1764

Table 2: Comparison between FA-maze, PF-maze, and Ours. (N/A: Not Available.)

Routers / Circuits	Routability (%)			Total Wirelength (um)			CPU Time (s)		
	FA-maze	PF-maze	Ours	FA-maze	PF-maze	Ours	FA-maze	PF-maze	Ours
fcpb1	86.42	90.74	100	N/A	N/A	4539121	11	19	33
fcpb2	76.74	85.38	100	N/A	N/A	9920878	28	50	145
fcpb3	80.06	88.15	100	N/A	N/A	15431763	40	84	163
fcpb4	76.88	93.99	100	N/A	N/A	16584677	35	82	164
fcpb5	77.81	93.39	100	N/A	N/A	192887686	1124	1655	3438
Average	79.58	90.33	100						

We compared our algorithm with two heuristics, namely FA-maze and PF-maze. FA-maze integrates maze routing with some techniques presented in [3]. In FA-maze, it first tried to optimize the routing of FA nets by [3]. Then, the PA nets were routed sequentially by maze routing. PF-maze is our algorithm without performing the congestion estimation and the iterative improvement. Therefore, we set the values of α, β, and γ in PF-maze to be 0, 0, and ∞, respectively. In our proposed algorithm, in contrast, the values of α, β, and γ are set to be 3, 1, and 5, respectively. To fairly compare the three algorithms, we applied the same routing sequence of the PA nets. The routing sequence was decided by the non-decreasing order of the Manhattan lengths of the nets. The experimental results are shown in Table 2. We report the routability, the total wirelength, and the CPU times. Compared with FA-maze, our algorithm improves the routability by 20.42%. Note that for all circuits, FA-maze fails to find a routing solution while ours can achieve 100% routability. Compared with PF-maze, our algorithm improves the routability by 9.67%, which reveals the effects of the congestion estimation and the iterative improvement. The results show that our unified FA and PA routing algorithm is effective and efficient for the PB co-design.

5. CONCLUSIONS

We have developed a unified FA and PA area-I/O flip-chip router for PB co-design. Our DT- and VD-based network-flow algorithm guarantees to find the optimal solution with the minimum wirelength for the free-assignment nets. Our PB co-design routing flow can facilitate the interactions among free- and pre-assignment nets and lead to superior routing solutions with 100% routability and the shorter routed wirelength.

6. REFERENCES

[1] R. K. Ahuja, T. L. Magnanti, and J. B. Orlin, *Network Flows: Theory, Algorithms, and Application*, Prentice Hall, 1993.
[2] J.-W. Fang, C.-H. Hsu, and Y.-W. Chang, "An integer linear programming based routing algorithm for flip-chip design," *Proc. of DAC*, pp. 606–611, 2007.
[3] J.-W. Fang and Y.-W. Chang, "Area I/O flip-chip routing algorithm for chip-package co-design," *Proc. of ICCAD*, pp. 518–522, 2008.
[4] F. P. Preparata and M. I. Shamos, *Computational Geometry: An Introduction*, Springer, 1985.
[5] UMC, "0.13μm flip-chip layout guideline," p. 6, 2004.
[6] M.-F. Yu, J. Darnauer and W.-M. Dai, "Interchangeable pin routing with application to package layout," *Proc. of ICCAD*, pp. 668–673, 1996.

Statistical Multilayer Process Space Coverage for At-Speed Test

Jinjun Xiong, Yiyu Shi[*], Vladimir Zolotov, Chandu Visweswariah
IBM Thomas J. Watson Research Center, Yorktown Heights, NY, 10598
[*]Electrical Engineering, UCLA, CA, 90095

ABSTRACT

Increasingly large process variations make selection of a set of critical paths for at-speed testing essential yet challenging. This paper proposes a novel *multilayer process space coverage metric* to quantitatively gauge the quality of path selection. To overcome the exponential complexity in computing such a metric, this paper reveals its relationship to a concept called *order statistics* for a set of correlated random variables, efficient computation of which is a hitherto open problem in the literature. This paper then develops an elegant recursive algorithm to compute the order statistics (or the metric) in provable linear time and space. With a novel data structure, the order statistics can also be incrementally updated. By employing a branch-and-bound path selection algorithm with above techniques, this paper shows that selecting an optimal set of paths for a multi-million-gate design can be performed efficiently. Compared to the state-of-the-art, experimental results show both the efficiency of our algorithms and better quality of our path selection.

Categories and Subject Descriptors

B.7.2 [**Integrated Circuits**]: Design Aids

General Terms

Algorithms, Design, Theory

Keywords

Order Statistics, Path Selection, Process Space Coverage

1. INTRODUCTION

Modern high performance chips coming off the manufacturing line are tested *at-speed*, i.e., a set of paths is selected to undergo at-speed tests to identify "bad" chips in which one or more of the selected paths fail timing requirements [1]. Selection of these paths is complicated by the presence of process variations, as different paths can be critical in different chips manufactured under different process conditions

Permission to make digital or hard copies of part or all of this work for personal or classroom use is granted without fee provided that copies are not made or distributed for profit or commercial advantage and that copies bear this notice and the full citation on the first page. To copy otherwise, to republish, to post on servers or to redistribute to lists, requires prior specific permission and/or a fee.
DAC'09, July 26-31, 2009, San Francisco, California, USA

[2]. Selection of paths to cover various process conditions is important to ensure high quality of testing [3].

Most existing work targets transition delay faults that are large and localized changes of delay [4], and they tend to find short paths through those fault sites rather than critical paths. Delay defects resulting from process variations are, however, small and distributed, and it is the accumulated delay variation along the signal propagation path that causes critical paths to fail. The authors of [5] proposed finding the longest paths through each fault site, but it was based on a deterministic delay model. The authors of [6, 7, 2, 3] tried to take process variation into account, but their approaches either suffer from high algorithmic complexity, or are inaccurate due to simplified statistical timing models. Thus existing path selection techniques are not effective in detecting delay faults caused by process variations.

Recently, the authors of [8] proposed a metric called *Test Quality Metric* (TQM) for at-speed testing. Such a metric provides a measure of process space coverage. Guided by such a metric, the authors proposed a branch-and-bound path selection algorithm to find a set of paths to maximize TQM. But there is a major drawback, i.e., the TQM metric is a *single layer* process space coverage metric, which gives rise to two main objections. First, the statistical timing model is not perfect, and any model inaccuracy can easily cause the loss of process space coverage. Second, not all selected paths are sensitizable. Because a path's sensitizability is known only after an expensive procedure called ATPG (Automatic Test Pattern Generation), so even if the TQM of the set of selected paths is high, after ATPG many paths may turn out to be unsensitizable, and the post-ATPG TQM may be significantly reduced.

In this work, we address these issues by enforcing that every point in the multi-dimensional space of process variations is covered multiple times. The major contributions of this paper are multi-fold. We first define a new metric, called multilayer test quality metric (mTQM), to quantitatively measure this multilayer process space coverage property for any given set of paths. We show that direct computation of such a metric has exponential complexity. A novel observation is made to relate the computation of mTQM to the *order statistics* of a set of correlated random variables. An elegant recursive formula is developed to compute the order statistics, a hitherto open problem. We develop an efficient algorithm to compute mTQM with provable linear complexity in time and space. Under the guidance of mTQM, we extend the branch-and-bound path selection algorithm of [8] to perform path selection. A novel data structure is developed to help efficient and incremental updating of mTQM (or order statistics) during path selection.

All the above techniques enable us to optimally and efficiently select a set of paths to achieve the best multilayer process space coverage. Experimental results show that our algorithm can improve the post-ATPG TQM by more than 10% compared to [8].

The remainder of the paper is organized as follows: section 2 reviews existing work; section 3 discusses the multi-layer coverage metric and its computation; the metric is then used to guide a brand-and-bound path selection algorithm in section 4; experimental results are provided in section 5, and concluding remarks are given in section 6.

2. PRELIMINARIES

Given a set of paths $\Pi = \{\pi_1, \pi_2, \ldots, \pi_n\}$, the *test quality metric* (TQM) of this set of paths [8] is defined as the probability that a tested chip has no timing violation conditional upon all paths in Π passing at-speed testing, i.e.,

$$Q(\Pi) = P(S_c \geq 0 | S_\Pi \geq 0), \tag{1}$$

where S_c is the statistical chip slack, and S_Π is the statistical slack of the corresponding path set Π, i.e.,

$$S_\Pi = min(s_1, \ldots, s_n), \tag{2}$$

where s_i is the statistical slack of the corresponding path π_i. Clearly, the larger the $Q(\Pi)$, the better the testing quality, hence the better the path selection.

It has also been shown in [8] that TQM is related to PCM (Process space Coverage Metric) $q(\Pi)$ through

$$Q(\Pi) = \frac{P(S_c \geq 0)}{1 - q(\Pi)}, \tag{3}$$

with the PCM being defined as

$$q(\Pi) = P(S_\Pi \leq 0) = P(min(s_1, \ldots, s_n) \leq 0). \tag{4}$$

For a given design, chip slack S_c does not depend on the paths selected for testing. Hence maximizing TQM $Q(\Pi)$ in (3) is equivalent to maximizing the PCM $q(\Pi)$ in (4).

Under the guidance of PCM (or equivalently TQM), the authors of [8] proposed a branch-and-bound path selection algorithm to select the top n paths that gives the largest (best) TQM among billions of paths from the design.

The algorithm traverses the timing graph in a depth-first manner, and the efficiency is achieved through an effective pruning strategy. It is proved in [8] that there exists an *upper bound* for PCM as

$$q(\Pi) \leq P(min(s_1, \ldots, s_{i-1}, s_{\tilde{\Pi}}, s_{i+1} \ldots, s_n) \leq 0), \tag{5}$$

where slack s_i of path π_i is replaced with slack $s_{\tilde{\Pi}}$ of a set of paths $\tilde{\Pi}$ with $\pi_i \subseteq \tilde{\Pi}$. In the context of path selection, the set of paths $\tilde{\Pi}$ corresponds to the partial path segment. During traversal of the partial path, any of its branches can be pruned immediately if its upper bound metric is worse than the current metric, as there is no way the metric can be improved by continuing path search through that branch.

3. MULTILAYER PROCESS SPACE COVERAGE METRIC

3.1 Metric Definition

In statistical timing, all timing quantities such as slack s are represented as functions (e.g., linear [9] or quadratic [10]) of the underlying process parameters ΔX, i.e., $s = F(\Delta X)$, where ΔX is a $k \times 1$ vector containing normalized Gaussian random variables to model the variation of process parameters, including chip-to-chip, within-chip, and local random variations. The entire k-dimensional space spanned by ΔX is also called the *process space* and is denoted by $\Omega = \{\Delta X | \Delta X \in \mathbf{R}^k\}$.

The meaning of process space coverage and its metric PCM (4) is better explained by defining a mapping from the path slack s_i to a subspace $\omega_i \subseteq \Omega$ as

$$\omega_i = \{\Delta X | s_i = F_i(\Delta X) \leq 0\}. \tag{6}$$

For a set of n paths, its corresponding subspace $\omega_n^{(1)} \subseteq \Omega$ is the union of subspaces defined by each individual path, i.e.,

$$\omega_n^{(1)} = \omega_1 \cup \ldots \cup \omega_n. \tag{7}$$

In other words, a set of paths defines (or covers) a subspace of the entire process space such that by testing this path set at-speed we would sort out all bad chips manufactured under those process conditions. The corresponding PCM as defined in (4) can thus be equally interpreted as

$$q(\Pi) = \frac{|\omega_n^{(1)}|}{|\Omega|} \tag{8}$$

where $|\cdot|$ is a Lebesgue measure (i.e., probability-weighted area) [11] of the process space.

Fig. 1 shows a small example with three path slacks s_1, s_2 and s_3 in a two-dimensional process space. Within the 3-sigma circular region, the subspace covered by slack s_1 is $C \cup D \cup E$, the subspace covered by s_2 is $B \cup C \cup D \cup F$, and the subspace covered by s_3 is $A \cup B \cup C$.

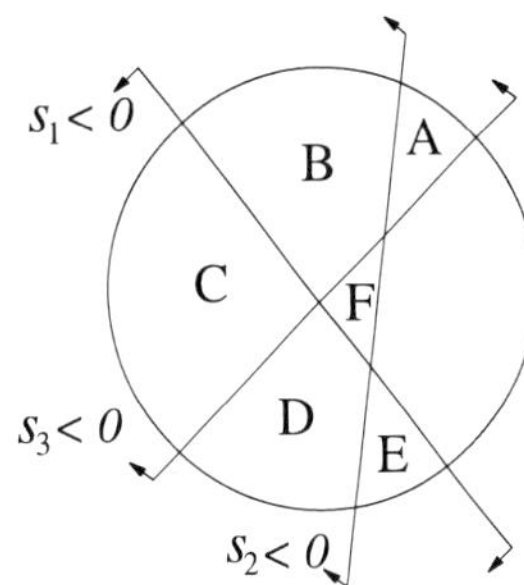

Figure 1: Example of process space coverage.

We call the PCM from [8] as given by (7) and (8) as *single layer process space coverage metric* (sPCM) and the corresponding TQM as *single layer TQM* (sTQM), because every point in the space will be counted towards $\omega_n^{(1)}$ as long as there is one path covers that point. In other words, sPCM will not distinguish the subspaces that are covered by only one path (such as A in Fig. 1) or by multiple paths (such as C in Fig. 1).

The so-defined sPCM is not desirable for path selection for at-speed testing. For example, if one of the paths cannot be sensitized, the subspace that is covered only by that path will be left untested. On the other hand, if we guarantee that all subspaces are covered multiple times (e.g., at least twice as B, C, D in Fig. 1), then even if some paths turn

978-1-60558-497-3/09 $25.00 © 2009 ACM

out to be unsensitizable after ATPG, their subspaces will likely still be covered by other remaining sensitizable paths. In other words, we want to find paths to maximize the multiple coverage area as much as possible. To capture this property, we propose a new metric called *multilayer process space coverage metric* (mPCM) (or mTQM for *multilayer TQM*) in the following.

We first note that the subspace covered m times by a set of m paths is given by

$$\omega_m^{(m)} = \omega_1 \cap \ldots \cap \omega_m. \qquad (9)$$

For a set of n paths $\mathbf{\Pi} = \{\pi_1, \pi_2, \ldots, \pi_n\}$ with the corresponding coverage subspace as $\{\omega_1, \omega_2, \ldots, \omega_n\}$, the subspace covered by this set of paths at least m times, $\omega_n^{(m)} \subseteq \Omega$, would be the union of $\omega_{m,i}^{(m)}$ with each $\omega_{m,i}^{(m)}$ covered m times by a set of m paths, which is further a subset of $\mathbf{\Pi}$. There are total $\binom{n}{m}$ number of $\omega_{m,i}^{(m)}$ (or equivalently, subsets of size m from $\mathbf{\Pi}$). Mathematically, this is given by

$$\omega_n^{(m)} = \cup_{\forall i \in \binom{n}{m} \text{combinations}} \left(\omega_{m,i}^{(m)} \right). \qquad (10)$$

Similar to (8), the multilayer process space coverage metric (mPCM) is thus defined as

$$q^{(m)}(\mathbf{\Pi}) = \frac{|\omega_n^{(m)}|}{|\mathbf{\Omega}|} \qquad (11)$$

Because for $m=1$, (10) and (11) would give back the same results as (7) and (8), sPCM (or sTQM) can be interpreted as a special case of mPCM (or mTQM).

For the example as shown in Fig. 1, to find the subspace covered by $n=3$ path slacks at least $m=2$ times, we can find the union of the subspace covered by both s_1 and s_2 (i.e., $C \cup D$), the subspace covered by both s_1 and s_3 (i.e., C), and the subspace covered by both s_2 and s_3 (i.e., $C \cup B$). The result is $B \cup C \cup D$.

3.2 Metric Computation and Order Statistics

From the process space coverage point of view, we have defined mPCM as shown in (11). This definition is convenient for conceptual understanding, but not for computation. Therefore, we have the following lemma to help us perform the computation.

LEMMA 1. *For a set of slacks* $\{s_1, s_2, \ldots, s_m\}$ *and its corresponding coverage subspaces* $\{\omega_1, \omega_2, \ldots, \omega_m\}$, *the* union *of these subspaces can be represented by the statistical minimum of all slacks, i.e.,*

$$\omega_1 \cup \ldots \cup \omega_m = \{\Delta X | min(s_1, \ldots, s_m) \leq 0\}. \qquad (12)$$

And the intersection *of these subspaces can be represented by the statistical maximum of all slacks, i.e.,*

$$\omega_1 \cap \ldots \cap \omega_m = \{\Delta X | max(s_1, \ldots, s_m) \leq 0\}. \qquad (13)$$

In other words, the union operation on subspaces corresponds to the statistical minimum operation on slacks; while the intersection operation on subspaces corresponds to the statistical maximum operation on slacks.

Since we know how to perform the statistical minimum and maximum operations as in any statistical timer[1], ac-

---

[1] All derivations as shown in this work are exact as long as the statistical min/max operations are exact. Only when the approximated min/max operations are used (e.g., [9, 10]) will our computation incur some approximation error.

cording to (12) and (13), the subspace in (10) can be obtained as

$$\omega_n^{(m)} = \{\Delta X | f_{n,m}(s_1, \ldots, s_n) \leq 0\}, \qquad (14)$$

with slack $f_{n,m}$ as a function of $s_1, \ldots, s_n$ given by

$$f_{n,m} = min_{\forall i \in \binom{n}{m} \text{combinations}} \left(max(s_{1,i}, \ldots, s_{m,i}) \right), \qquad (15)$$

where $\{s_{1,i}, \ldots, s_{m,i}\}$ is one of the $\binom{n}{m}$ combinations from $\{s_1, s_2, \ldots, s_n\}$. Therefore, rather than using (11), we compute mPCM as follows

$$q^{(m)}(\mathbf{\Pi}) = P(f_{n,m}(s_1, \ldots, s_n) \leq 0). \qquad (16)$$

Computing sPCM according to (4) is straightforward, but computation of mPCM according to (16) is not, because the latter requires enumeration of all $\binom{n}{m}$ combinations, which is on the order of exponential complexity $O((\frac{n}{m})^m)$.

Before we propose our efficient solution to this problem, we introduce a well-known concept called *order statistics* [12]. Given n random variables $s_1, \ldots, s_n$, the mth-order statistic for $m \leq n$ is a random variable equal to the mth-smallest value of a statistical sample of $s_1, \ldots, s_n$. For simplicity, we denote the mth-order statistic of n random variables as $h_{n,m}$. We can prove the following theorem:

THEOREM 1. *The random variable* $f_{n,m}$ *as defined in (15) is the same as the* m^{th}*-order statistic of the* n *input random variables* $s_1, \ldots, s_n$, *i.e.,*

$$f_{n,m}(s_1, \ldots, s_n) = h_{n,m}(s_1, \ldots, s_n). \qquad (17)$$

For example, when $m=1$ or $m=$n, we have

$$f_{n,1} = min(s_1, s_2, \ldots, s_n) = min(f_{n-1,1}, s_n), \qquad (18)$$
$$f_{n,n} = max(s_1, s_2, \ldots, s_n) = max(f_{n-1,n-1}, s_n). \qquad (19)$$

The detailed proof through induction is omitted in the interest of space. We give an intuitive explanation via a simple example with $n=3$. Assume that we want to find mPCM for $m=2$, then $f_{n,m}$ in (15) is given by

$$f_{3,2} = min\left(max(s_1, s_2), max(s_1, s_3), max(s_2, s_3) \right). \qquad (20)$$

When s_1, s_2 and s_3 are deterministic values, it is easily seen that $f_{3,2}$ would give the second smallest number among the inputs. When s_1, s_2 and s_3 are random variables, then $f_{3,2}$ would be the second-order statistic of the input set.

If we knew how to compute the order statistics $h_{n,m}$ for any n and m, computing mPCM according to (16) would be simple. There are abundant research efforts in the literature that have showed how to compute the order statistics for a set of independent random variables (e.g., [13]). But when random variables are correlated (such as in our case), computing the order statistics efficiently is an open problem. Monte Carlo based approach would work, but the computation cost is high, and it is not feasible for repeated evaluation in most applications.

In this work, we propose an efficient way to compute $f_{n,m}$ (or equivalent $h_{n,m}$) by leveraging our understanding of the mapping between slacks s_i and process space coverage w_i, i.e., $f_{n,m}$ corresponds to the space covered by w_i at least m times as shown in (10). Then instead of enumerating all $\binom{n}{m}$ combinations, the same m-layer subspace is obtained by the union of two subspaces: (1) the subspace covered by the first n-1 slacks m times; and (2) the intersection of the subspace covered by the last slack s_n and the subspace covered by the first n-1 slacks m-1 times. Mathematically, this can be formalized as

$$\omega_n^{(m)} = \omega_{n-1}^{(m)}(s_1, \ldots, s_{n-1}) \cup \left(\omega_n \cap \omega_{n-1}^{(m-1)}(s_1, \ldots, s_{n-1}) \right).$$

For example, as shown in Fig. 1, the first subspace is $C \cup D$ covered by both s_1 and s_2, the second subspace is $B \cup C$, and the result would be same as before, i.e., $B \cup C \cup D$. Hence we have the following theorem.

THEOREM 2. *Given n slacks $s_1, s_2, \ldots, s_n$, the mth-order statistic $f_{n,m}$ (or equivalent $h_{n,m}$) is given by*

$$f_{n,m} = min\left(f_{n-1,m}, max(s_n, f_{n-1,m-1})\right). \quad (21)$$

Based on this recursive formula (21), we can develop an efficient algorithm to compute all order-statistics $f_{n,1}, \ldots, f_{n,m}$.

The idea behind the algorithm can be understood with a simple example as shown in Fig. 2 with $n{=}6$ and $m{=}3$. To compute $f_{6,3}$ according to (21), we need two inputs $f_{5,3}$ and $f_{5,2}$ and two min/max operations; and this holds similarly for $f_{5,3}$ and $f_{5,2}$. By organizing the data needed for each computation into a directed graph and sharing intermediate results among different computations, we obtain the pattern in Fig. 2 with each node's two input nodes and edges corresponding to the two inputs and two min/max operations, respectively, needed to compute the results stored at that node according to (21) (or (18) and (19) for those nodes with one input edge). By performing the computation along the dashed lines bottom-up and only keeping those results necessarily for the next level of computation, we can compute all required order statistics ($f_{6,3}$, $f_{6,2}$ and $f_{6,1}$) in linear time $O(nm)$ and space $O(n+m)$.

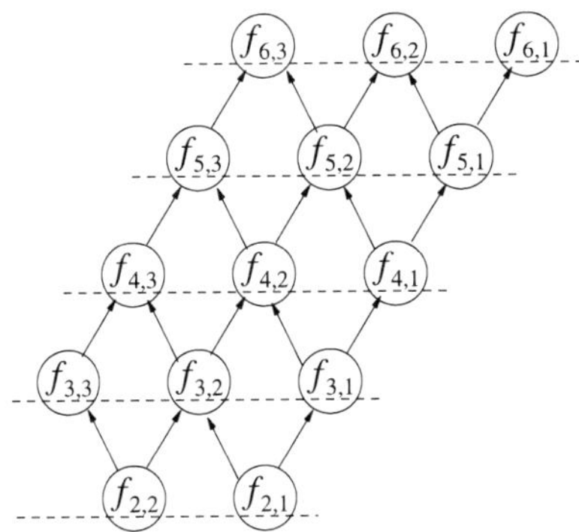

Figure 2: Computation of order-statistics.

4. PATH SELECTION FOR MULTILAYER PROCESS SPACE COVERAGE

In this section, we extend the work of [8] to select the best n paths such that mTQM (or equivalently mPCM) is maximized. The key to the branch-and-bound algorithm is an effective pruning strategy that hinges on the computation of a metric bound. For sPCM as defined in (4), its *upper bound metric* is given by (5). In the case of mPCM as defined in (16), we can also show that there exists a similar upper bound metric

$$q^{(m)}(\mathbf{\Pi}) \leq P(f_{n,m}(s_1, \ldots, s_{i-1}, s_{\tilde{\Pi}}, s_{i+1}, \ldots, s_n) \leq 0), \quad (22)$$

where slack s_i of path π_i is replaced with slack $s_{\tilde{\Pi}}$ of a set of paths $\tilde{\Pi}$ with $\pi_i \subseteq \tilde{\Pi}$.

Once we know how to compute the upper bound for mPCM, we can utilize the same branch-and-bound algorithm framework as discussed in [8] by (1) replacing the optimization metric from sPCM to mPCM; and (2) during path traversal, replacing the sPCM bound with the mPCM bound for pruning. Hence the entire branch-and-bound algorithm can be utilized to find the top n paths with maximum mPCM. In the interest of space, we refer readers to [8] for the detailed discussion on the overall branch-and-bound algorithm framework.

During path selection, for every potential branch to grow the current partial path of interest, we obtain a new slack $\tilde{s}$ of the growing partial path, and we need to decide (1) whether we should continue to search through this branch for a new path (i.e., pruning) ; and (2) when a new path is found, which existing path should be replaced by this new one. All these decisions involve the updating of mPCM with new slack $\tilde{s}$, and since we always keep the best so-far n slacks, we need to temporarily update mPCM n times. Hence the complexity is $O(n^2m)$ for every branch, as we can compute mPCM in $O(nm)$ time. For a large design, such complexity is too high. In the following, we propose a more efficient way for incremental mPCM updating.

We utilize a similar binary-tree data-structure as discussed in [8] with the following major modifications. For the retained n slacks, we organize them into a balanced binary-tree with each leaf node corresponding to one slack s_i. At every node D in the tree, we store two sets of slacks: the *node slack set* $\mathbf{S}_D$, and the *complement node slack set* $\overline{\mathbf{S}_D}$. The former contains the 1st- to m^{th}-order statistics of all its downstream leaf node slacks; while the latter contains the 1st- to m^{th}-order statistics of all but its downstream leaf node slacks.

We denote $g_m(s_1, \ldots, s_r)$ as the set of 1st- to m^{th}-order statistics from the algorithm as illustrated in Fig. 2, i.e.,

$$g_m(s_1, \ldots, s_r) = \begin{cases} f_{r,1}, \ldots, f_{r,m} & \text{if } r \geq m \\ f_{r,1}, \ldots, f_{r,r} & \text{otherwise.} \end{cases} \quad (23)$$

Then we can compute node slack set $\mathbf{S}_D$ and complement node slack set $\overline{\mathbf{S}_D}$ as follows

$$\mathbf{S}_D = g_m(\mathbf{S}_{D_{left}} \cup \mathbf{S}_{D_{right}}), \quad (24)$$

$$\overline{\mathbf{S}_D} = g_m(\overline{\mathbf{S}_{D_{parent}}} \cup \mathbf{S}_{D_{sibling}}), \quad (25)$$

where D_{left}, D_{right}, D_{parent} and $D_{sibling}$ are node D's left child, right child, parent, and sibling nodes, respectively.

To fill-up the binary tree with proper node information, we need two traversals of this tree. The first traversal is bottom-up from leaves to the root by computing all node slack sets via (24); and the second traversal is top-down from the root to leaves by computing all complement node slack sets via (25). As the input to $g_m(\cdot)$ is limited to $2m$ number of slacks, the complexity of constructing the binary tree is $O(nm^2)$.

Once the binary-tree has been constructed, the new mPCM after replacing one leaf slack s_i with the new slack $\tilde{s}$ can be updated in $O(m^2)$ time by calling $g_m(\overline{\mathbf{S}_i} \cup \tilde{s})$ with the complement leaf slack set $\overline{\mathbf{S}_i}$ already known. The mth-order statistic of $g_m(\overline{\mathbf{S}_i} \cup \tilde{s})$ gives the updated $f_{n,m}$ for mPCM computation. To decide which slack should be replaced to achieve the best improvement on mPCM, we only need to loop through all leaf nodes once, hence the complexity is $O(nm^2)$. Once the leaf node has been replaced with the new slack $\tilde{s}$, we need to do a bottom-up traversal from this

978-1-60558-497-3/09 $25.00 © 2009 ACM

leaf node to the root by updating the node slack set via (24); and a top-down traversal from the root to all leaves to update the complement node slack set via (25). Again, the complexity is $O(nm^2)$.

To summarize, by utilizing the binary-tree type of data-structure and maintaining proper node slack sets and complement node slack sets, we have reduced the complexity of updating mPCM from $O(n^2m)$ to $O(nm^2)$. Typically, the number of required paths n is on the order of thousands, while the number of required layers m for coverage is less than 10. Therefore, we have reduced the complexity from quadratic $O(n^2)$ to linear $O(n)$.

5. EXPERIMENTAL RESULTS

We have implemented the algorithm for mTQM (or equivalently mPCM) computation and the branch-and-bound based path selection algorithm inside an in-house statistical timing tool [9]. Three industrial 90 nm and 65 nm designs are used for experiments, and we denote them as D_1, D_2 and D_3, and the size of each design is $5K$, $170K$, and $3.2M$ gates, respectively. The amount of process variation is obtained according to foundry rules for both front-end and back-end process parameters.

5.1 mTQM Comparison

For a given set of n path slacks, we compute its mTQM via (3) for different required layers m; and the mPCM is computed by (16) with $f_{n,m}$ given by (21). To check the accuracy of our algorithm, we also compute mTQM through Monte Carlo simulation ($100K$ trials). We report the mTQM comparison based on D_2 in Table 1 for different n and m. We observe that our approach computes mTQM with almost the same accuracy as Monte Carlo (less than 0.2% error) while achieving two orders of magnitude speedup. This conclusion holds for different path number n and layer number m; and the runtime grows almost linearly.

Table 1: Comparison of mTQM computation.

Path n,	mTQM		Runtime (sec)	
Layer m	MC	Ours (diff)	MC	Ours (speedup)
$n=8K$, $m=1$	0.9868	0.9847 (-0.2%)	577.3	0.621 (930×)
$n=8K$, $m=2$	0.8349	0.8359 (-0.1%)	589.9	0.648 (910×)
$n=8K$, $m=3$	0.7902	0.7919 (0.2%)	595.2	0.706 (843×)
$n=8K$, $m=4$	0.7493	0.7494 (0.0%)	612.6	0.732 (837×)
$n=8K$, $m=5$	0.7343	0.7329 (-0.2%)	637.4	0.768 (830×)
$n=1K$, $m=2$	0.7530	0.7523 (-0.1%)	157.3	0.116 (1356×)
$n=3K$, $m=2$	0.8344	0.8359 (0.2%)	241.6	0.172 (1405×)
$n=5K$, $m=2$	0.8359	0.8359 (0.0%)	421.5	0.316 (1334×)
$n=7K$, $m=2$	0.8359	0.8359 (0.0%)	583.0	0.457 (1276×)
$n=9K$, $m=2$	0.8358	0.8359 (0.0%)	765.6	0.598 (1280×)

5.2 Quality of Path Selection

We run the branch-and-bound path selection algorithm as discussed in Section 4 to select the top n paths with the maximum mTQM, and we denote it as mTQM-BnB. To illustrate the value of a multilayer coverage based path selection, we compare the quality of path selection with [8], which targets single layer process space coverage and is denoted as sTQM-BnB.

We observe that (1) when $m=1$ our path results are exactly the same as [8], and this is expected as our mTQM metric is the same as sTQM when $m=1$; (2) when m is more than one, i.e., we require more number of layers for process space coverage, our algorithm selects different set of

paths. To compare the quality of these two sets of paths, ideally we could run ATPG on both sets of paths and find out what paths are sensitizable and what are not, and then recompute the sTQM for only sensitizable paths. Obviously, the higher the post-ATPG sTQM, the better the results for testing. Such experiment should be run for multiple designs, and the algorithm with more likelihood of getting higher post-ATPG sTQM wins.

Table 2: Post-ATPG sTQM comparison.

sTQM	Post-ATPG Avg		Post-ATPG Min	
Design	sTQM-BnB	mTQM-BnB	sTQM-BnB	mTQM-BnB
D_1	0.772	0.898	0.515	0.637
D_2	0.827	0.905	0.551	0.696
D_3	0.716	0.830	0.511	0.640

But the ATPG process is typically a time-consuming and cumbersome step, instead, we use a randomized approach to mimic the ideal experiment but with much less effort. For a given design, we run both algorithms to obtain two sets of n paths; then we arbitrarily choose the same number of paths from each set and mark those paths as unsensitizable, and compute post-ATPG sTQM for the remaining sensitizable paths. We repeat this arbitrary choice of paths many times ($10K$ in our experiments), and report the comparison of the post-ATPG sTQM from both algorithms. Table 2 shows some statistics of the comparison, i.e., the average and minimum post-ATPG sTQM, from both sTQM-BnB and mTQM-BnB (with $m=2$). We observe that mTQM-BnB always achieves both higher average and minimum post-ATPG sTQMs. This observation is expected because, according to mTQM-BnB, every point in the process space has been covered multiple times, and the loss of space coverage only happens if all those paths are unsensitizable. Apparently, such a likelihood is lower than paths selected with only one layer coverage.

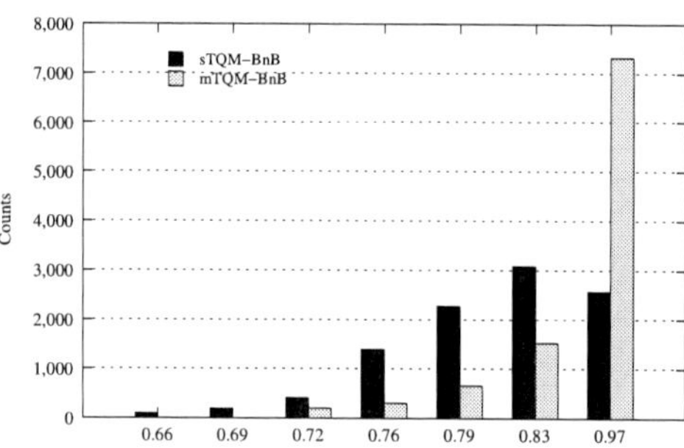

Figure 3: Histogram comparison of the final sTQM.

We also show the histogram comparison for D_2 in Fig. 3. We observe that, among the 10 K experiments, the post-ATPG TQM varies significantly, ranging from 0.66 to 0.97. There is about 25% chance for sTQM-BnB to retain the same sTQM as before (around 0.97), while mTQM-BnB has more than 70% chance of retaining the same sTQM as before. In other words, mTQM-BnB has much more likelihood to achieve higher post-ATPG sTQM than sTQM-BnB. Hence the quality of path selection based on mTQM-BnB is better; and this improvement would be even more significant when mTQM-BnB targets mTQM with larger m.

For the required m-layer mTQM maximization, mTQM-BnB guarantees to find the best set of n paths among all possible choices of paths. This can be confirmed by running

978-1-60558-497-3/09 $25.00 © 2009 ACM

mTQM-BnB targeting different m to obtain different sets of paths, and re-evaluating these sets of paths in terms of mTQM for $m=1, 2, \ldots$ -layer coverage. One such experiment for D_3 is shown in Table 3. We see that the set of paths achieving the best mTQM only comes from the mTQM-BnB algorithm targeting the same required layer m (diagonal cell in the table). For example, the mTQM-BnB targeting 4-layer coverage has the best 4-layer coverage mTQM (0.558) compared to others.

Table 3: Evaluation of mTQM for different m.

mTQM evaluation	mTQM-BnB targeting different m			
	$m=1$	$m=2$	$m=3$	$m=4$
$m=1$	**0.881**	0.881	0.881	0.871
$m=2$	0.650	**0.763**	0.738	0.728
$m=3$	0.562	0.553	**0.591**	0.578
$m=4$	0.513	0.503	0.533	**0.558**

5.3 Efficiency of Path Selection

We compare the runtime of our mTQM-BnB algorithm with or without employing the proposed incremental mTQM updating techniques. Results based on D_2 are presented in Fig. 4. The top plot shows the comparison for selecting 8000 paths targeting mTQM with different m; and the lower plot shows the comparison for selecting different number of paths targeting 2-layer mTQM. We see that mTQM-BnB without using the incremental mTQM updating technique has about quadratic complexity; while mTQM-BnB with the speedup technique scales almost linearly with both the path number and the layer number, confirming our prior complexity analysis.

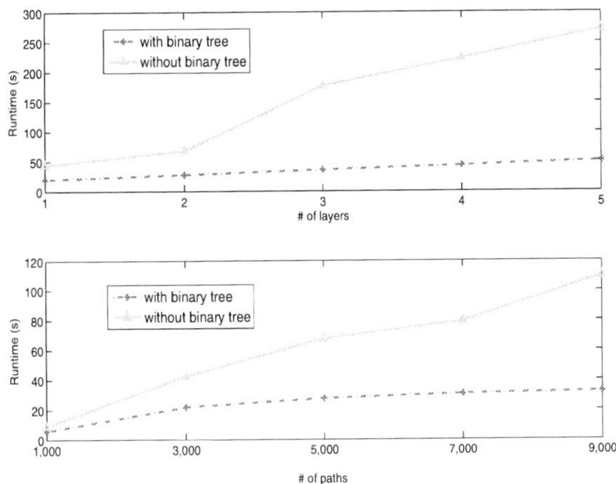

Figure 4: Runtime scalability for mTQM-BnB.

We also compare the runtime of our 2-layer mTQM-BnB algorithm with the single sTQM-BnB algorithm. For design D_1, D_2, and D_3, we notice that sTQM-BnB has runtime of 5.39, 19.76, and 22.85 CPU seconds, respectively; while the runtime for mTQM-BnB in second is 8.87, 27.66, and 31.64, respectively. Therefore, by selecting paths to target more layers of process space coverage, mTQM-BnB incurs less than 50% runtime overhead compared to the existing path selection targeting single layer coverage, and leads to superior path selection.

6. CONCLUSION AND DISCUSSION

We have proposed a novel *multilayer process space coverage* metric to quantitatively evaluate the testing quality of a set of paths in detecting bad chips manufactured under all process conditions. By leveraging the mapping between statistical path slacks and process coverage, we have developed an elegant recursive algorithm to compute the metric in linear time and space. We have also extended the branch-and-bound based path selection algorithm in [8] to target multilayer process space coverage. A novel data structure has been developed to make incremental updating of the metric during path selection to run in linear time. Experimental results have shown consistent improvement on testing quality over the state-of-the-art [8].

We have also demonstrated an analogy of the metric with order statistics for a set of correlated random variables, thus providing an efficient algorithm to compute order statistics analytically, a hitherto open problem in the literature. Since order statistics have found use in different fields, we envision this algorithm to be of value to other domains for many other applications as well. Our future work will explore this possibility.

7. REFERENCES

[1] V. Iyengar, T. Yokota, K. Yamada, T. Anemikos, R. Bassett, M. Degregorio, R. Farmer, G. Grise, M. Johnson, D. Milton, M. Taylor, and F. Woytowich. At-speed structural test for high-performance ASICs. *Proc. International Test Conference*, pages 2.4:1–10, October 2006. Santa Clara, CA.

[2] L-C. Wang, J-J. Liou, and K-T. Cheng. Critical path selection for delay fault testing based upon a statistical timing model. *IEEE Transactions on Computer-Aided Design of ICs and Systems*, 23(11):1550–1565, November 2004.

[3] V. Iyengar, J. Xiong, S. Venkatesan, V. Zolotov, D. Lackey, P. A. Habitz, and C. Visweswariah. Variation-aware performance verification using at-speed structural test and statistical timing. *IEEE International Conference on Computer-Aided Design*, November 2007. San Jose, CA.

[4] M. L. Bushnell and V. D. Agrawal. *Essentials of electronic testing for digital, memory and mixed-signal VLSI circuits*. Kluwer Academic Publishers, 2000.

[5] M. Sharma and J. H. Patel. Finding a small set of longest testable paths that cover every gate. *International Test Conference*, pages 974–982, October 2002. Baltimore, MD.

[6] J-J. Liou, A. Krstic, L-C. Wang, and K-T. Cheng. False-path-aware statistical timing analysis and efficient path selection for delay testing and timing validation. *Proc. 2002 Design Automation Conference*, pages 566–569, June 2002. New Orleans, LA.

[7] W. Qiu and D. M. H. Walker. An efficient algorithm for finding k longest testable paths through each gate in a combinational circuit. *International Test Conference*, pages 592–601, Setpember 2003. Charlotte, NC.

[8] V. Zolotov, J. Xiong, H. Fatemi, and C. Visweswariah. Statistical path selection for at-speed test. *IEEE International Conference on Computer-Aided Design*, November 2008. San Jose, CA.

[9] C. Visweswariah, K. Ravindran, K. Kalafala, S. G. Walker, and S. Narayan. First-order incremental block-based statistical timing analysis. *Proc. 2004 Design Automation Conference*, pages 331–336, June 2004. San Diego, CA.

[10] L. Zhang, W. Chen, Y. Hu, J. A. Gubner, and C. C. Chen. Correlation-preserved non-gaussian statistical timing analysis with quadratic timing model. In *Proc. Design Automation Conf*, pages 83 – 88, June 2005.

[11] R. M. Dudley. *Real Analysis and Probability*. Cambridge University Press, 2002.

[12] H. A. David and H. N. Nagaraja. *Order Statistics*. John Wiley and Sons, 2003.

[13] H. M. Barakat and Y. H. Abdelkader. Computing the moments of order statistics from nonidentical random variables. *Statistical Methods and Applications*, 13:13–24, 2004.

978-1-60558-497-3/09 $25.00 © 2009 ACM

Speedpath Analysis Based on Hypothesis Pruning and Ranking *

Nicholas Callegari[1], Li-C. Wang[1], Pouria Bastani[2][†]
[1]University of California - Santa Barbara
[2]Intel Corporation

ABSTRACT

In optimizing high-performance designs, speed limiting paths (*speedpaths*) impact the performance and power trade-off. Timing tools attempt to model and capture all such paths on a chip. Due to the high performance nature of these designs, critical paths predicted by the timing tools often do not match the actual speedpaths found on silicon chips. Early silicon data therefore is used to identify the speedpaths, and further performance optimization is carried out by pushing the delays on these paths. In this context, the paper presents a novel data mining approach that analyzes a small number of identified speedpaths against a large number of non-speedpaths. The result of this analysis for each speedpath is a set of hypotheses explaining why the path is special. These hypotheses can be used in guiding the search for the root causes, or in predicting additional paths as potential speedpaths. We demonstrate the feasibility of this approach and summarize our findings based on analysis of silicon speedpaths collected from a 65nm microprocessor.

Categories and Subject Descriptors

B.8.2 [**Hardware**]: Performance Analysis and Design Aids

General Terms

Algorithm, Performance, Reliability

Keywords

Speedpath, Timing Analysis, Data Mining

1. INTRODUCTION

For high-performance design, timing analysis and performance optimization does not stop at the first silicon. Silicon information is analyzed carefully to drive further speed and power optimization. This process is called *silicon steppings*, that involves detection and improvement of speed limiting paths (*speedpaths*). A speedpath is a path that limits the performance of a chip where the performance can be measured by observing the result of applying (functional) legacy tests. Because speedpaths can be observed at different cycles of such a test, a single chip can have multiple speedpaths [1].

Due to the nature of high performance design, speedpaths are not well-predicted by typical timing flows. The deficiency comes due to a multitude of process, design and environmental effects that are either unknown or too difficult to model and simulate accurately in

---

*This work supported in part by CA Micro/Intel project No. 07-079.
†This work began while Dr. Bastani was a UCSB graduate student.

Permission to make digital or hard copies of part or all of this work for personal or classroom use is granted without fee provided that copies are not made or distributed for profit or commercial advantage and that copies bear this notice and the full citation on the first page. To copy otherwise, to republish, to post on servers or to redistribute to lists, requires prior specific permission and/or a fee.
DAC'09, July 26-31, 2009, San Francisco, California, USA

the design process. Therefore, for high-performance high-volume microprocessors, it is often more (cost) effective to uncover speedpaths using actual silicon samples. In each silicon stepping, a set of speedpaths are identified where performance is further improved by pushing the delays on these paths. If causes can be established to explain the speedpaths, this information can be feedback to the design and more paths can be identified as potential speedpaths which can be fixed before they show up as speedpaths in the subsequent silicon stepping. These iterations continue until the performance is pushed to a satisfactory level, at which time design changes are frozen and mass production begins.

Each silicon stepping has significant associated manufacturing cost. Additionally, identifying and verifying silicon speedpaths demands a huge amount of engineering effort. Because of the high cost, the number of speedpaths found at each silicon steppings is typically small, for example in the 10's or 100's. Hence, once this small set of speedpaths are collected, they are treated as precious resources. One desires to uncover and utilize the information presented with these paths as much as possible for guiding performance optimization before the next silicon stepping. However, it is not always obvious what the most effective approach is to best uncover and utilize the information on a few identified speedpaths.

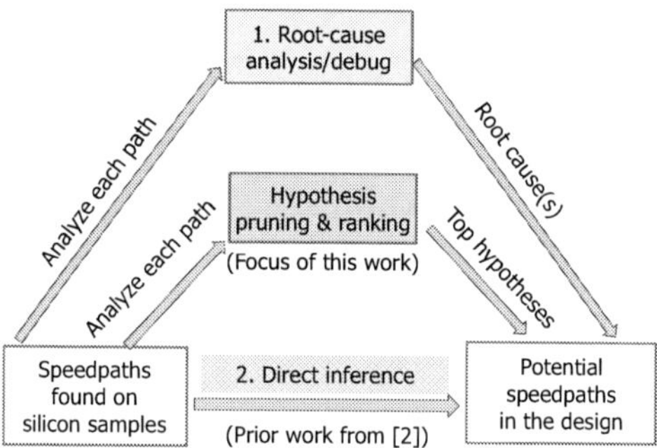

Figure 1: Three approaches to analyze speedpaths

There exists two ways by which one may use the information to find additional speedpaths. Figure 1 illustrates the two approaches: *root-cause analysis*, and *direct inference* [2]. In root-cause analysis, one tries to obtain the root cause(s) as why a path is a speedpath. The root cause(s) can then be used to identify potential speedpaths by finding all paths containing the same cause(s). While root-cause analysis is intuitive, it may not be the most desirable approach because of the usual high cost associated with it [3]. *Direct inference* offers a short-cut for finding potential speedpaths. If finding potential speedpaths is our ultimate goal, then one can simply collect all paths whose characteristics are very similar to the identified speedpaths. This is a search-by-similarity strategy and in this approach, one does not need to know the reasons behind speedpaths.

In [2], the authors present a direct inference method that utilizes one-class *support vector machine* (SVM) [4] to implement the search-by-similarity strategy. The result of this learning is a one-class SVM model that characterizes the speedpaths. This model can then be used to scan all other paths and rank them according to their similarities to the speedpaths. From this ranking and a proper selection criterion, potential speedpaths can be derived.

The direct inference approach is attractive because it avoids the difficulty of finding the root causes. However, it does not offer the advantage for finding guides to improve timing models and tools. Furthermore, it may not be as effective as the root-causing approach for finding all potential speedpaths. For example, a speedpath and a potential speedpath may share the same root cause and hence, they are similar if one focuses the comparison of their characteristics only on the cause. However, they may be deemed dissimilar by the direct inference because direct inference has no idea about the specific cause and utilizes the overall characteristics of the two paths as the comparison basis for deciding similarity. From this perspective, we see that the direct inference, although much more efficient, may only find a subset of the potential speedpaths.

The limitation of the direct inference approach and the high cost of the root-cause analysis approach, motivate the development of a third approach, the *hypothesis pruning and ranking* (HPR) as shown in Figure 1. In the HPR approach, we assume two sets of paths are given, a small set of speedpaths and a large set of non-speedpaths. Because the speedpath set is small, we further assume that there is a significant degree of diversity in the characteristics of those paths. Note that if this assumption is not true, one can use the approach in [2] to group paths because one-class SVM in essence produces clusters where similar paths are grouped together. The set of non-speedpaths can be found by finding all paths sensitized by the tests and do not show up as speedpaths in testing. While a speedpath set may contain tens of paths, a non-speedpath set may contain millions.

The strategy of the HPR approach is to analyze each speedpath individually against the information presented in the non-speedpath set. Conceptually, this is achieved by two steps: (1) forming a hypothesis space that contains all the hypotheses of interest, (2) pruning the impossible or unlikely hypotheses based on information presented with the non-speedpaths, and if possible further ranking the remaining hypotheses. The result of HPR approach is therefore a rank of hypotheses and this rank may be based on a partial ordering. Top hypotheses (that describe special characteristics of the speedpaths) from the rank can be used for two purposes, for guiding the search in root-cause analysis and, as that in the direct inference approach, for finding potential speedpaths.

In this paper, we present two methods to implement the HPR approach. For forming the hypothesis space, the two methods use the same idea of incorporating *features* to describe path characteristics as that introduced in prior works [2, 5]. However, for pruning and ranking, the two methods employ different techniques. The first method utilizes techniques derived from Association Rule Mining [6, 7] that is often applied to mining large business database. The second method utilizes the kernel learning concept in SVM [4] in conjunction with the rule analysis. While the first method is more intuitive to understand, we will demonstrate that the second method is much more flexible and easier to implement. Through experiments, we will demonstrate that the second method can find all top hypotheses found by the first method. The experiments are based on silicon speedpaths from a 65nm microprocessor.

The rest of the paper is organized as the following. Section 2 discusses the use of *features* to form the hypotheses space and points out the fundamental challenge in the HPR approach. Section 3 presents the basic concepts in rule mining. Section 4 discusses the concept of *confidence* and presents a way to evaluate the confidence of a hypothesis in the context of proposed hypothesis pruning and ranking. Section 5 presents the first method and Section 6 presents the second method. Section 7 presents the experimental results and Section 8 concludes.

2. FEATURES AND HYPOTHESIS SPACE

A hypothesis space is formed based on *features*. A feature can be seen as an indicator of potential concern on a path. There can be two types of features, *occurrence based* and *description based*. An occurrence based feature has a binary value. For example, a cell by itself can be a feature. Then, given a path, if the path contains the cell, the value of that feature is 1. Otherwise, the value is 0. A description based feature has a numerical value. For example, the capacitive load associated with a specific cell on a path can be a description based feature.

Given a set of n occurrence based features $F = \{f_1, \ldots, f_n\}$, the initial hypothesis space is the power set 2^F. In other words, a hypothesis is simply a combination of features. However, on a speedpath, usually not all features would appear. Suppose m features (denoted set Q) appear on the path, then the hypothesis space for the speedpath is the power set 2^Q. This is because if a feature does not appear on the path, it cannot be used to explain why the path is special. From this perspective, we see that even though one may use a large set of n features to describe all paths, on analyzing a single speedpath, the search space can be much smaller, i.e. $m \ll n$. However, even for m in the order of tens, this search space can still be too large to enumerate explicitly. Hence, in the analysis, the hypothesis space exists implicitly.

If F contains n description based features, then a hypothesis space essentially contains an infinite number of hypotheses. The intuition to see this is by linking description based features back to occurrence based features. To form a hypothesis space as the power set of the features, one needs to convert n description based features into n corresponding occurrence based features. However, there are infinite many ways for this conversion.

For example, suppose we have two description features, A and B. Suppose on a speedpath p the characteristic is described as $\{A = 0.2001, B = 0.4\}$. Suppose we have two non-speedpaths with the feature vectors described as $\{A = 0.2, B = 0.5\}$ and $\{A = 0.3, B = 0.3\}$. To explain p, one can convert A, B into two occurrence based features A_0, B_0 where A_0 has value 1 if $A = 0.2001$, $A_0 = 0$ if $A \neq 0.2001$, $B_0 = 1$ if $B = 0.3$ and $B_0 = 0$ if $B \neq 0.3$. Observe that A_0, B_0 are very specific to the path p.

Alternatively, one may think that A_0, B_0 are too specific and replace them with two new features: A_1, B_1 where $A_1 = 1$ if $A_1 \in \{A > 0.2001 \cap A < 0.2002\}$ and $B_1 = 1$ if $B_1 \in \{B > 0.35 \cap B < 0.45\}$. In the second conversion method, two ranges are used for the two features. As we can see, there are infinite many ways to decide on these two ranges to create the two occurrence based features for deriving the power set where each way would interpret the hypotheses in the hypothesis space differently. Hence the hypothesis space becomes infinite.

This simple example illustrates that in the continuous space, there are infinite many ways to partition the space to form a set of hypotheses. Hence, the fundamental challenge in hypothesis ranking is how to perform effective search in a continuous hypothesis space.

2.1 Selecting features

Usually, a speedpath is due to effects that are not accounted for in the timing flow. This can be because their rarity of occurrence along with the large additional cost to model them. Selecting a proper set of features to begin undoubtedly requires some knowledge about these unaccounted effects. Although this is an important step for the

978-1-60558-497-3/09 $25.00 © 2009 ACM

	Feature				
Path	F1	F2	F3	F4	F5
P1	**1**	**1**	**1**	**1**	**0**
P2	1	1	1	0	1
P3	0	1	1	1	0
P4	0	1	1	1	1
P5	1	1	1	0	1
P6	0	1	1	1	0

Table 1: Sample Data

proposed approach to be effective, it should not be seen as a severe limitation for the following reasons. First, the approach does not require to select a minimal set and hence, if one is not sure about some features, the solution is simply to include them. Second, a feature is a single-order effect that can be part of the concern. Usually, it is not too difficult for an experienced designer (or a process engineer) to point out such single-order effects. For example, a design can list a set of cells that she/he feel less confident on their timing modeling. Third, a missing feature does not necessarily degrade the effectiveness of the approach significantly. For example, suppose the root cause is a combination of three features $\{A, B, C\}$ and suppose the feature set only includes A and B. This does not imply that the hypothesis $\{A, B\}$ will not be ranked high. From the viewpoint of fixing the potential speedpaths, fixing all paths with $\{A, B\}$ would have included the paths with $\{A, B, C\}$. This may lead to over-fixing, but the effectiveness with respect to the performance improvement does not suffer.

The authors in [2, 5] group features into five categories of effects: **Topological effects** that are layout dependent, **Dynamic effects** that are test patterns dependent, **Static effects** that are independent of test patterns, **Statistical effects** that are process dependent, and **Random effects** that do not belong to the previous four. From the set of effects described above, an engineer develops thus a set of features that can potentially be used to differentiate between non-speedpaths and speedpaths. In this work, based on our domain knowledge and application scenario we chose the following features:

1. Active device type in a path due to logic sensitization. These are based on four different flavors of N and P MOS transistors.
2. The predicted arc or stage delay for a particular cell in the path. This is a function of input transition time, output load and net resistance and capacitance.
3. Percent of total path delay due to the nets as opposed to cells. $\%Net = NetDelay/(TotalPathDelay)$
4. Dynamic switching activity in the region.
5. Cross coupling aggressor impact.
6. Temperature variation, using Infrared Emission Microscopy to obtain a temperature map during functional operation [8].

Note that most of the features are description based. With these features every path can now be described as a feature vector.

3. HYPOTHESIS PRUNING

To illustrate the process of hypothesis pruning, Table 1 shows a simple example with a single speedpath, P1, and five other non-speedpaths, paths P2,...,P6, where each path is descried with with five occurrence based features, F1-F5. The hypothesis space, i.e. the power set, is illustrated as a *concept lattice* as shown in Figure 2. Notice that in the lattice, feature F5 does not appear because it does not appear on the speedpath. In this lattice, each node represents a hypothesis which can also be called a *monomial*.

For example, $\{2\}$ denotes the single-order hypothesis $\{F2\}$. We call it "single-order" because it consists of one feature. Note that since F2 appears on all five non-speedpaths, its *coverage* by the non-speedpath set is 5. Similarly, the coverage of $\{F2,F3\}$ is also 5. This is why a label "5" is on the left of the node $\{2, 3\}$.

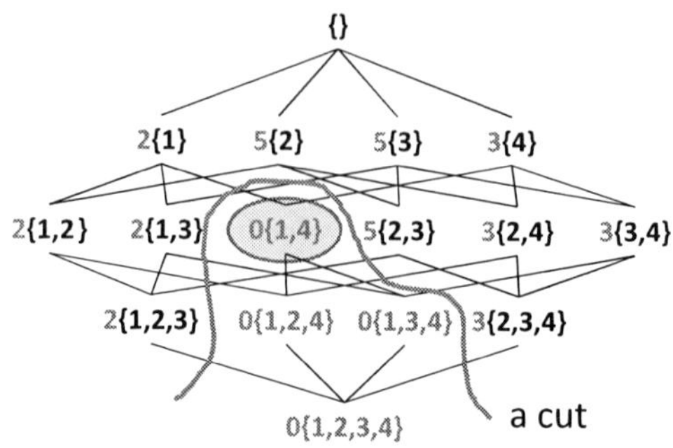

Figure 2: The lattice view of the hypothesis pruning

Because the monomial $\{1,4\}$ does not exist on any of the non-speedpaths, its coverage is zero. Since $\{1, 4\}$ is not covered, all its decedents in the lattice are not covered. The four hypotheses with coverage zero are therefore the ones left after the pruning process. They are considered *consistent* with the dataset presented, i.e. not appearing on any of the non-speedpaths. The monomial $\{1, 4\}$ is a *2-order* hypothesis because it consists of two features. In general, one would like to find the lowest-order hypotheses that are consistent with the dataset and are not the decedents of any other hypothesis. This is because (1) it is easier to process the information presented in lower-order hypotheses than higher-order hypotheses, and (2) any higher-order hypothesss consistent with the dataset would have been included in its corresponding lower-order hypothesis.

Finding all the lowest-order hypotheses consistent with the dataset translates to finding a *cut* on the concept lattice. Above this cut are all nodes with non-zero coverage by the non-speedpaths. In Figure 2, the cut would separate the four nodes $\{1, 4\}$, $\{1, 2, 4\}$, $\{1, 3, 4\}$, $\{1, 2, 3, 4\}$ from the rest of the nodes.

Finding a cut in such a lattice can be challenging if the lattice size is very large. Various algorithms were proposed in association rule mining [6] for efficient finding of such a cut. For example the very first Apriori algorithm [9] is based on a breadth-first search strategy. If one expects the cut deep in the lattice (from the top), the maximum itemset algorithm MAFIA [10] that follows a depth-first search strategy, would be more efficient. The difference between association rule mining and the hypothesis pruning in our work, lies in the ways to decide *consistency*. In hypothesis pruning, this consistency is decided by checking if a hypothesis appears on any one of the non-speedpaths. In rule mining, the consistency of an itemset (an itemset corresponds to a node in the lattice) is decided by a *support-confidence* evaluation [6]. The support calculation corresponds to the consistency checking in our formulation. Several methods have been proposed for efficient calculation of the support [11]. For hypothesis pruning, the bit-set based representation proposed in [10] provides good efficiency for implementation.

4. CONFIDENCE AND ITS EVALUATION

The next step is to evaluate the confidence for hypothesis $\{F1,F4\}$ (which will be same for all three higher-order hypotheses below it). The simplest way to evaluate the confidence is to calculate the ratio of the lattice space that is covered. Figure 3-(I) illustrates this concept. By assuming that the empty hypothesis is always covered, we see from Figure 2, 12 out of the 16 nodes are covered. Hence, the confidence is 75%. In evaluating all speedpaths individually, all decedents of $\{F_1, F_4\}$ would also have the same confidence. Hence, with this method, we cannot differentiate in between the hypotheses that are consistent with the dataset. We can, however, compare the confidence obtained based on one speedpath path to another.

The idea to relate confidence to the size of the non-covered subspace H can be justified with a simple probability argument as the following. For a hypothesis $h \in H$, denote the probability that h

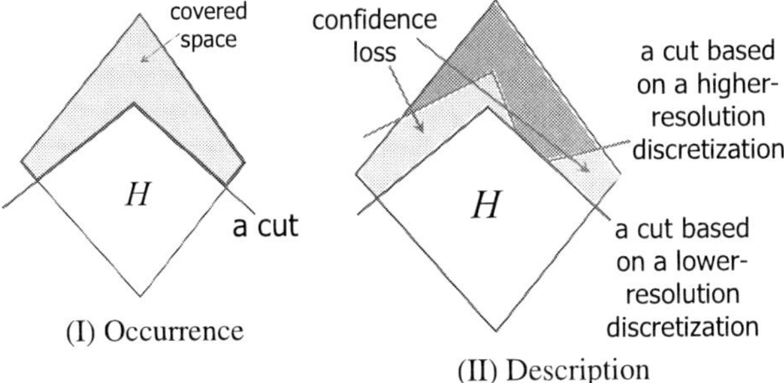

Figure 3: Confidence measured as coverage in hypothesis space

is not true as $error(h)$. We want $error(h) < \epsilon$. The probability that h is not true and h does not appear on m randomly selected non-speedpaths is $\leq (1 - \epsilon)^m$. For all hypotheses in H, any one of them is not true and does not appear on m randomly selected non-speedpaths is therefore $\leq |H|(1 - \epsilon)^m \leq |H|e^{-m\epsilon}$. Suppose we want this probability to be bounded by δ. Then, we can set $\delta = \frac{1}{m}ln(\frac{|H|}{\epsilon})$. In this equation $(1 - \delta)$ is the confidence that is a function of $|H|$ and the error probability ϵ. We see that for a given ϵ, the larger the $|H|$ is, the smaller the $(1 - \delta)$ is.

4.1 Dealing with description based features

As discussed in Section 2, for description based features, a concept lattice can be formed by converting the features into occurrence based features. Once such a conversion is decided, the resulting lattice for the speedpath is determined. Then, a cut can be found on this lattice based on the non-speedpaths. Below we discuss how the conversion method can influence the location of such a cut.

Suppose we are given with a set of features whose values fall into the range $[0, 10]$. A simple way to convert these features into occurrence based features is to divide the range into two sub-ranges of equal length: $[0, 5), [5, 10]$. Note that given a speedpath, any one of its features will fall into either one of these two sub-ranges but not both. This is also true if we, instead of dividing the range into two, divided it into four sub-ranges of equal length. Suppose in a process, we continue to divide $[0, 10]$ into into $2, 4, \ldots, 2^i$ sub-ranges of equal length to build a sequence of hypothesis lattices for a speedpath. Notice that in this lattice, the name(s) of the feature(s) labeling each node do not change. They are the features appearing on the speedpath. Hence, all the lattices have the same size. The only thing that changes is the sub-range each feature falls into. In the process, we call each way to divide the range a *discretization* method. We say that the jth discretization has a *lower resolution* than the $(j + 1)$th discretization.

Given such a sequence of lattices and a non-speedpath set, we can obtain a sequence of cuts, one on each lattice. Figure 3-(II) illustrates that the cut from a *higher resolution* discretization cannot be lower than a cut from a *lower resolution* discretization. The intuition behind this is that a higher resolution discretization means a smaller sub-range specified for each feature on the speedpath. A smaller sub-range means it would be harder for feature values from the non-speedpaths to fall into that sub-range, and therefore some nodes covered before with a bigger sub-range, could become uncovered. Intuitively, what this is showing is that, a higher resolution discretization scheme would result in hypotheses that are "more specific" to the speedpath, which may reduce their confidence because of reduced coverage on its corresponding lattice.

Following the example discussed in Section 2 before, for example, a hypothesis $h_1 : \{A \in [0.2, 0.25], B \in [0.35, 0.45]\}$ would be considered more specific than $h_2 : \{A \in [0.2, 0.3], f_2 \in [0.3, 0.5]\}$ because the sub-ranges associated with the two features in h_1 are smaller. Based on our confidence evaluation scheme, the confidence

of h_1 is $\leq$ the confidence of h_2.

It is clear from the discussion that if we intend to find the hypotheses with the highest confidence, then we would use the lowest-resolution discretization scheme. However, this may correspond to a cut lower in the lattice and return hypotheses with very high-orders. High-order hypotheses are less preferred because they can also be too specific to the speedpath (in the extreme case, one can say that all features together on the speedpath is the reason but this tells us little information regarding why the path is special). Hence, in searching for good hypotheses, we will try to find the ones in the low orders (such as 2nd, 3th) with the lowest discretization resolution such that in the resulting lattice, the hypotheses are consistent with the given dataset and never appear on the non-speedpaths.

5. DISCRETIZATION METHOD

The discretization strategy is commonly utilized most of the association rule mining research where items are with numerical values. The most reasonable method for discretizing continuous features is based on one's intuition, i.e. design knowledge on the features. For those features that are not so intuitive, systematic methods can be used. The three basic approaches are, equal frequency (EF), equal width (EW), and clustering which includes methods such as fuzzy clustering, K-means, and Self Organizing Map [12].

Equal width divides a range into k equal sized intervals. Equal frequency divides a range into k intervals where in each interval, the number of instances within is the same. Clustering on the other hand, tries to find best boundaries to partition a range so that most of the values would be away from the boundaries.

Figure 4 shows a histogram of the feature values based on a particular feature on both the speedpaths and the non-speedpaths. The feature is the output load of a specific gate type. In the figure, x axis shows the number of paths, and the y axis shows the feature values, both are normalized to 1. When deciding how to divide the feature values into sub-ranges, there is no easy way to determine the best number of bins to be partitioned and the location of those bins.

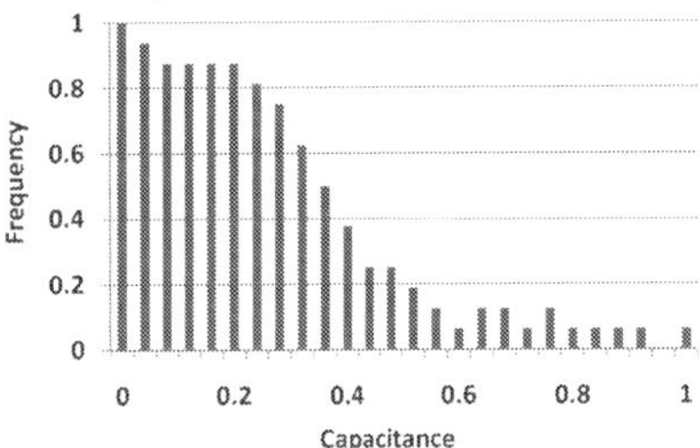

Figure 4: Normalized Capacitance Distribution

We began by identifying which features could be divided using intuition based on domain knowledge. For example, if the load of a gate is a feature, and design engineers believe that gates with a load within 5% of the maximum specified load are more likely to contain delay mismatch and should be treated separately from all other gates, it could be binned accordingly.

For the remaining features we experimented with equal width, equal frequency as well as fuzzy clustering algorithm. We discovered that because most of the features have a distribution similar to the one shown above, equal frequency is not effective. When experimented with equal width and clustering, we discovered that results are similar. Since equal width is more intuitive, we chose equal width as the main discretization method for the experiments.

Given a hypothesis order i, we tried to find the lowest resolution discretization such that there exists one or more ith-order hypotheses consistent with the dataset. This can be achieved by finding the lowest resolution discretization for each of the i feature combination such that with the discretization, the resulting hypothesis is consistent with the dataset. To do this, we implemented a greedy search approach. For each i feature combination $f_1 \ldots f_i$ with corresponding value ranges $R_1 \ldots R_i$, in each step of the greedy search, one of the ith range is selected and the partitioning on the range is increased by one. Take each range R_j. Let the hypothesis based on the current discretization be h_1. Let the hypotheses after increasing its partitioning by one, be h_2. Let n_1 be the number of non-speedpaths on which h_1 appear, and n_2 be the number of non-speedpaths on which h_2 appear. The greedy is to select R_j such that its $n_1 - n_2$ is the largest. This tries to make a consistent hypothesis out of the i features by minimizing the number of partitions across all ranges.

6. MAXIMUM MARGIN METHOD

Given a fixed order, for finding the highest confidence hypotheses in that order, the discretization method tries to find the lowest resolution discretization such that one or more resulting hypotheses in that order do not appear on any non-speedpaths. Although discretization is conceptually easy to understand, we also observe that the greedy method is only a heuristic that conceptually tries to achieve the "lowest resolution."

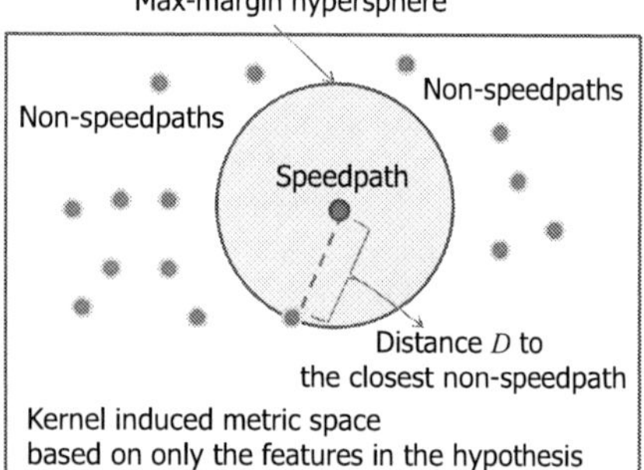

Figure 5: Find the hypothesis with maximum margin

Figure 5 presents an alternative method that is easier to implement and more robust. The method is based on finding a hypersphere with *maximum separation margin* in a *kernel* induced metric space. This concept is borrowed from a Support Vector Machine (SVM) classification algorithm [4] where a kernel function is used to map the input space onto a high dimensional kernel space (formally as Reproducing Kernel Hilbert Space (RKHS)) [4] and find a maximum margin separation hyperplane in that space.

Given a hypothesis, we use the Gaussian kernel function $k(\vec{v}, \vec{v_i})$ to evaluate the similarity between the speedpath $\vec{v}$ and each non-speedpath $\vec{v_i}$ based on *only* the features in that hypothesis. Then, for the hypothesis, we find the maximum margin hypersphere, centered at the speedpath path, to separate the speedpath from all good paths in the kernel induced metric space. Finding the hypersphere is equivalent to find the good path with the closest distance. Given a kernel $k()$, the distance between two feature vectors x, z can be calculated as $k(x, x) - 2k(x, z) + k(z, z)$ [4].

The important thing to note is that the distance is calculated based on only the feature values presented in the hypothesis. If a non-speedpath does not contain any one of the features in the hypothesis, the distance becomes infinity. Furthermore, the distance is calculated in a space induced by the kernel. Hence, if we desire to change the space (i.e. change how feature values are interpreted), we only need to change the kernel function. The work in [5] discusses the flexibility of using such a kernel. Because different feature values can meaning different things, proper interpretation of their values can be built into such a kernel. This provides the flexibility for a user to change how feature values are interpreted, for example weighting one subset of features more than another subset.

With the maximum margin method, the radius (distance) D of

the hypersphere is output as the new confidence measure for the hypothesis. With this new confidence measure, on a given hypothesis order, we can rank all hypotheses in that order and output the top N hypotheses. The intuition behind using D as a confidence measure is illustrated in the simple one-dimensional picture in Figure 6.

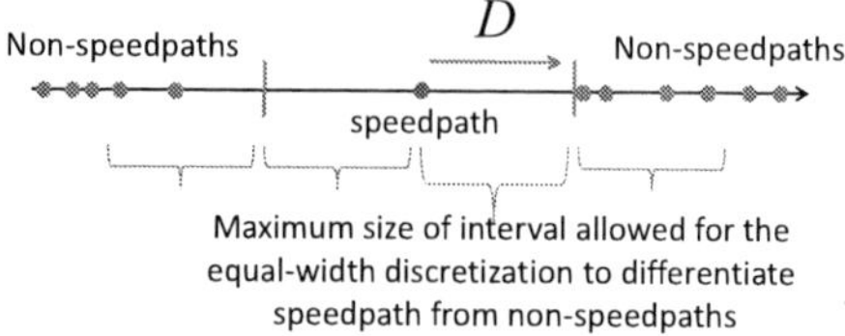

Figure 6: Linking maximum distance to the smallest number of bins in equal width discretization method

The example illustrates what maximum margin separation means in a one-dimensional Euclidean space. We see that the maximum margin D is the distance to the closest non-speedpath. Using D, one can think that it translates into the width in equal width method. This figure shows that the larger the D is, the less the number of bins can be with the equal width method, i.e. lower resolution and hence higher confidence. We see that intuitively, using D as a confidence measure concurs with the confidence definition from the perspective of a concept lattice discussed before. Because of this, we expect the discretization method and the maximum margin methods to give similar results. However, it is also noted that the one-dimensional Euclidean view does not imply the two methods are the same. When considering a multi-dimensional space and a kernel, the maximum margin method may rank hypotheses differently from the less flexible equal width discretization method.

7. EXPERIMENTAL RESULTS

For the experiments, we use the same speedpath dataset used in [2]. The design is a high performance microprocessor fabricated in a 65nm node. The dataset contains 56 speedpaths and 19000 non-speedpaths. These paths are encoded with 74 description features. In each analysis, hypotheses are ranked and derived from a speedpath. For example, by applying the maximum margin method to one speedpath and ranking all its 2nd-order hypotheses, we obtained the result in Figure 7. We see that two hypotheses stand out.

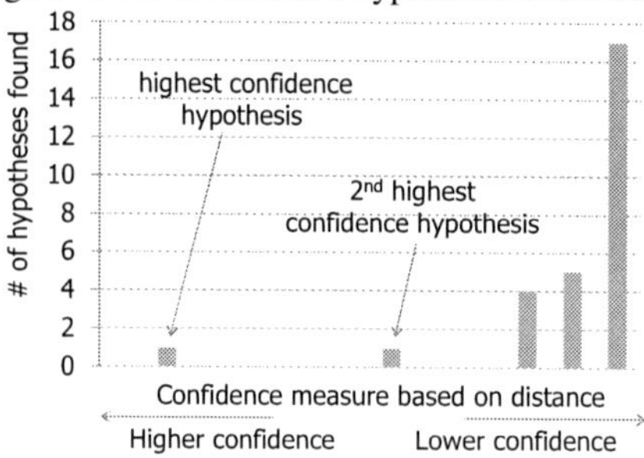

Figure 7: An example of hypothesis ranking (2nd-order)

We applied both methods, the equal width discretization (DS) method and the maximum margin (MM) method, to analyze the 56 speedpaths individually against the 19K non-speedpaths. We tried to find hypotheses in the order from 1 to 4. With the DS method, the lowest resolution discretization for each order returns one or more hypotheses. With the MM method, in each order the top N hypotheses based on their confidence ranking were selected, for $N = 1 \ldots 5$. To compare the MS method and the DS method, all hypotheses found across the 56 paths were considered together. Figure 8 shows the comparison results. The % number shown indicates in each case, how many hypotheses collected by the DS method are also in the set of top N hypotheses found by the MM method. We see that for the 1st-order, all hypotheses found by the DS method

978-1-60558-497-3/09 $25.00 © 2009 ACM

are also found by the the MM method for $N = 1$ (and for $N = 2\text{-}5$). For the 2nd-order, this is true for $N = 3$ and for the 3rd and 4th orders, this is true for $N = 5$. In general, we see that the MM method find all hypotheses found by the DS method with a small N.

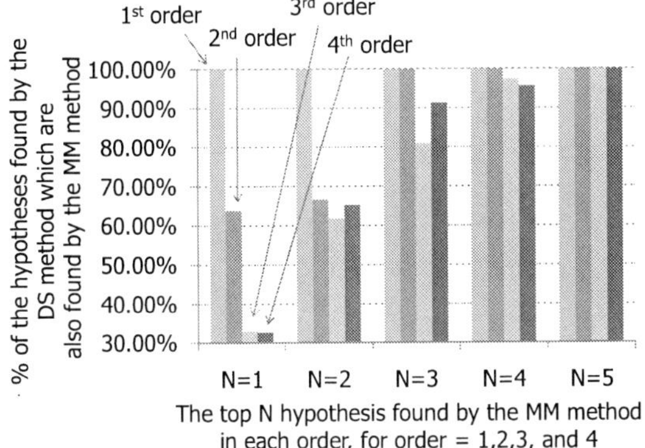

Figure 8: Discretization method vs Maximum margin method

To find potential speedpaths, we devise a voting scheme where top hypotheses found based on individual speedpaths are used to vote for the possibility that a non-speedpath is a potential speedpath. Recall that for a given hypothesis on a speedpath, its confidence is calculated as the distance to the closest non-speedpath. We normalize this distance by the order of the hypothesis and call it the *normalized confidence*. The reason for doing this is that a higher order hypothesis tends to have a larger distance because of the curse of dimensionality [4]. Normalizing the distance avoids giving preference to the higher order hypotheses. Based on normalized confidence, we can then rank all 1st-4th order hypotheses together on a speedpath. From each path, we select the top X hypotheses, and each hypothesis is the closest path in the maximum margin calculation. We add the normalized confidence of the hypothesis to that closest path as a confidence vote for the path to be a potential speedpath.

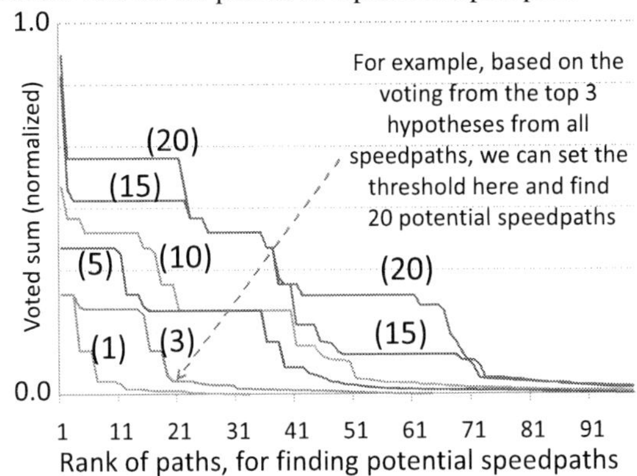

Figure 9: Finding potential speedpaths based on the voted sum from the top (X) hypotheses from all 56 speedpaths, where X is shown in the figure for 1,3,5,10,15, and 20 (using the maximum margin method).

Figure 9 shows the results for X=1,3,5,10,15, and 20. If one fix a threshold on the y-axis and says that all paths above that horizontal line are potential speedpaths, then it is intuitive to see that if more top hypotheses are used from each speedpath in the voting, more potential speedpaths will be found. In this figure, one can also select potential speedpaths by the trend of the curve, for example, if the voted sum from the voting drops to close to zero, then it is a good point to discard the paths behind. Note that by using such a voting scheme, a path is considered as a potential speedpath with higher confidence if it receives votes from more hypotheses that are also with higher confidence. This schemes also reduce the chance of letting noise in the dataset to influence the result.

7.1 Manual analysis of top hypotheses

We manually analyzed several top hypotheses (MM method). The first hypothesis was a top 2nd-order hypothesis shared by 20 speedpaths. The hypothesis consisted of a significantly high delay push-out due to coupling noise in combination with a slower than average cell that had a specific active transistor device type. It is believed the combination of these two features causes a speedpath due to inaccuracy in timing analysis for this device type when accounting for coupling noise. The second hypothesis was a top 3rd-order hypothesis shared by 10 speedpaths. The hypothesis consisted of two slower than average cells, in combination with a significantly high amount of switching activity. The third hypothesis was a top 2nd-order hypothesis shared by 3 speedpaths. This hypothesis consisted of a combination of two slower than average cells unique to the speedpaths. Additionally, unique 2nd-order and 3rd-order hypotheses were found on 21 speedpaths (no sharing with other speedpaths). Similar manual analysis on them found them also reasonable.

8. CONCLUSION

This paper presents a novel hypothesis pruning and ranking approach that analyzes a small number of identified speedpaths against a large number of non-speedpaths for explaining why the speedpaths are special. In our methodology, a hypothesis space is implicitly encoded using a set of features that represents single-order of concerns in modeling, design methodology, and/or process. When using occurrence based features, the hypothesis space is a concept lattice. Hypothesis pruning is to find a cut on this lattice. Confidence is evaluated based on lattice coverage. When using description based features, we present two methods to prune and rank hypotheses. The first method is based on value discretization and the second method is based on finding maximum separation hypersphere. We explain that the two methods are consistent and experimental results also show that they can find common hypotheses. To find potential speedpaths, we develop a voting scheme where a path is considered as a potential speedpath with high confidence if it receives more votes from high-confidence hypotheses found across all speedpaths.

9. ACKNOWLEDGMENTS

We acknowledge the following people at Intel for their contributions to our speedpath identification and root cause analysis in the form of guidance, discussions and implementation: Eli Chiprout, Kip Killpack, and Chandramouli Kashyap.

10. REFERENCES

[1] L.Lee, et al. "On Silicon-Based Speed Path Identification," *Proc. VTS*, 2005.

[2] P. Bastani, K. Killpack, L.-C. Wang, and E. Chiprout. "Speedpath prediction based on learning from a small set of examples," *Proc. DAC*, 2008.

[3] K. Killpack, C. Kashyap, E. Chiprout, "Silicon Speedpath Measurement and Feedback into EDA flows," *Proc. DAC*, 2007.

[4] N. Cristianini, and J. Shawe-Taylor, "An Introduction to Support Vector Machine," *Cambridge University Press*, 2002.

[5] P. Bastani, N. Callegari, L. Wang, and M. Abadir. "Statistical diagnosis of unmodeled systematic timing effects," *Proc. DAC*, 2008.

[6] Chengqi Zhang and Shichao Zhang. "Association Rule Mining, Models and Algorithms," *Lecture Notes in Computer Science* Vol. 2307, Springer 2002.

[7] S. Brin, R. Motwani, et al. "Dynamic itemset counting and implication rules for market basket data," *Proc. ACM SIGMOD*, 1997, pp. 255-264.

[8] D.Barton, P. Tangyunyong, J. Soden, A. Liang, and e. a. F. Low. "Infrared light emission from semiconductor devices," *ISTFA Proc.*, 1996.

[9] R. Agrawal, T. Imielinksi, and A. Swami. "Mining association rules between sets of items in large databases," *Proc. ACM SIGMOID*, 1993, pp. 207-216.

[10] D. Burdick, M.Calimlim, and J. E. Gehrke. "MAFIA: A Maximal Frequent Itemset Algorithm for Transactional Databases," *International Conference on Data Engineering*, 2001, pp. 443-452.

[11] H. Mannila, et al. Discovery of Frequent Episodes in Event Sequences," *Data Mining and Knowledge Discovery*, Vol 1, (3), 1997, pp. 259-289.

[12] M. Vannucci and V. Colla. "Meaningful discretization of continuous features for association rules mining by means of a som," *ESANN*, 2004, pp. 489-494.

Interconnection Fabric Design for Tracing Signals in Post-Silicon Validation

Xiao Liu and Qiang Xu
CUhk REliable computing laboratory (CURE)
Department of Computer Science & Engineering
The Chinese University of Hong Kong, Shatin, N.T., Hong Kong
Email: {xliu,qxu}@cse.cuhk.edu.hk

ABSTRACT

Post-silicon validation has become an essential step in the design flow of today's complex integrated circuits. One effective technique that provides real-time visibility to the circuit under debug (CUD) is to monitor and trace internal signals during its normal operation. Typically, a large number of signals are tapped and a subset of them are selected to be observed in each debug process. These trace signals need to be transferred to on-chip buffers and/or off-chip trace ports for analysis. Existing solutions use multiplexer trees or specific access networks to conduct the above duty, which, however, either provide less visibility to the CUD or result in high hardware cost. In this paper, we propose a novel trace signal interconnection fabric design to tackle the above problem. Experimental results on benchmark circuits show the efficacy of the proposed solution.

Categories and Subject Descriptors

B.7.3 [**Integrated Circuits**]: Reliability and Testing

General Terms

Verification, Design.

Keywords

Post-Silicon Validation, Trace-Based Debug

1. INTRODUCTION

With the ever-increasing design complexity and the ever-shrinking market window for today's integrated circuit (IC) products, it is increasingly difficult to guarantee the correctness of the design solely through pre-silicon verification. Post-silicon validation has thus become an essential step in the design flow of complex ICs, which has a significant impact on the profitability of these products because of its associated time-to-market delay and high re-spin cost [2].

Since the circuit under debug (CUD) is a piece of silicon that has already been fabricated, there is only limited visibility of its internal signals, which makes post-silicon validation an extremely difficult task. One widely-used technique to mitigate this problem is to reuse the CUD's existing test structure (e.g., JTAG and scan chains) to run/stop its operation and observe whether the values in

the circuit's storage elements are the expected values [9]. This debug methodology facilitates to identify those easy-to-find bugs that leave "evidences" when the circuit halts at very low hardware cost. Many tricky bugs, however, only manifest themselves after a long period of operations, and it is quite difficult to identify them with the above post-mortem debug strategy. Moreover, the behavior of many bugs is not repeatable, making diagnosis with this run/stop debug methodology even more difficult.

In order to reason the root cause of the design's erroneous behavior effectively, it is invaluable to have real-time visibility to the CUD during its normal operation, which can be obtained by monitoring and tracing the CUD's internal signals [7, 18]. As shown in [1], for million-gate industrial designs, it is common to tap thousands of signals in the circuit and select a subset of them (say, 32 signals) to trace concurrently in each debug process[1] [10, 12]. These trace signals are then transferred to on-chip trace buffers and/or off-chip trace ports for later analysis.

Interconnecting the large number of tapped signals to the trace buffers/ports is not an easy task and involves non-trivial design-for-debug (DfD) overhead. Existing solutions typically use pipelined multiplexer (MUX) trees to conduct the above duty (e.g., [1, 11, 20]). As any signals going through the same multiplexer cannot be observed concurrently, this ad-hoc technique limits the visibility to the CUD. At the same time, since bugs often occur in unexpected scenarios, designers are typically not knowledgeable about exactly which signals should be traced together at the design stage and it is a rather cumbersome process for them to manually build the MUX network that satisfies debug needs.

In this paper, we propose a novel interconnection fabric design to tackle the above problem, which contains two main parts: (i). a *MUX network* that connects those mutually-exclusive tapped signals, which can be designated by designers and/or extracted automatically based on structural analysis; (ii). a *non-blocking concentration network* that is able to transfer any m signals out of n inputs ($m \leq n$) to the trace buffers/ports. With the proposed method, designers are able to flexibly select any trace signal combinations (so long as they are not mutually-exclusive and do not exceed the provided trace bandwidth) in each debug process, which significantly enhances the CUD's debuggability at low DfD hardware cost.

The remainder of this paper is organized as follows. Section 2 reviews related prior work and motivates this paper. The proposed interconnection fabric for tracing signals in post-silicon validation is detailed in Section 3. Section 4 then presents our experimental results on benchmark circuits. Finally, Section 5 concludes this work.

Permission to make digital or hard copies of part or all of this work for personal or classroom use is granted without fee provided that copies are not made or distributed for profit or commercial advantage and that copies bear this notice and the full citation on the first page. To copy otherwise, to republish, to post on servers or to redistribute to lists, requires prior specific permission and/or a fee.
DAC'09, July 26-31, 2009, San Francisco, California, USA

---

[1] Due to the limited trace bandwidth, it is impossible to trace all the tapped signals at the same time.

978-1-60558-497-3/09 $25.00 © 2009 ACM

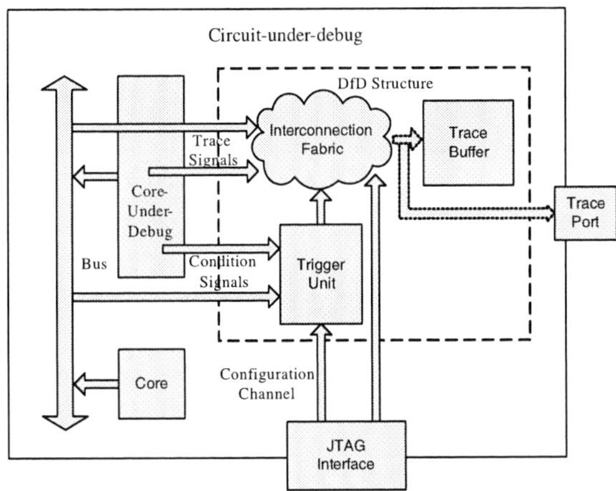

Figure 1: Trace-Based Debug Architecture.

2. PRELIMINARIES AND MOTIVATION

2.1 Trace-Based Debug Methodology

Introducing DfD circuitries to trace the CUD's internal signals provides real-time observabilities into the circuit. This debug methodology facilitates designers to conduct root cause analysis for the abnormal behavior of the CUD and has gained wide acceptance in the industry (e.g., [1, 3, 4, 13, 21]). As designers are not knowledgeable about which part of the design may contain bugs, a large number of signals need to be tapped in the system, typically in the thousand range for million-gate designs [1]. Due to the associated DfD area cost and debug bandwidth requirement, however, it is impossible to *concurrently* trace all the tapped signals. Instead, only a small number of internal signals (say, 32) can be real-time observed together, and it is up to the designers to determine which signals to trace at a specific debug phase, according to the system's erroneous behavior. These signals are then transferred to on-chip trace buffers and/or off-chip trace ports for diagnosis, through an *interconnection fabric* (see Fig. 1). Typically, a trigger unit controls the start and stop of the tracing, in which the triggering mechanism can be configured through JTAG interface through the debug configuration channel [19]. This so-called *trace qualification* process is very useful to reduce the large volume of debug data that are to be analyzed.

Interconnecting a large number of tapped signals to the trace buffers and/or trace ports involves non-trivial design-for-debug (DfD) overhead. In the following, we overview the possible design options for this interconnection fabric in prior work.

2.2 Related Work and Motivation

Industrial designs typically use MUX trees to select a subset of the tapped signals to trace in each debug process, in which the control signals to the multiplexers can be configured through the JTAG interface (e.g., [1, 20]). To satisfying the timing constraint for the tracing logic, the MUX trees can be pipelined. In addition, when the tapped signals are coming from different clock domains, first-in first-out (FIFO) buffers and/or flip-flop chains can be used to ensure data safety [1].

To conduct root cause analysis for design bugs effectively, it is desired to have the flexibility to concurrently observe certain related tapped signals (e.g., state elements and those control signals that activate state transitions for finite state machine) under the trace bandwidth limit. MUX-based interconnection fabric, however, limits this flexibility and reduces the visibility to the CUD, as any signals going through the same multiplexer cannot be traced concur-

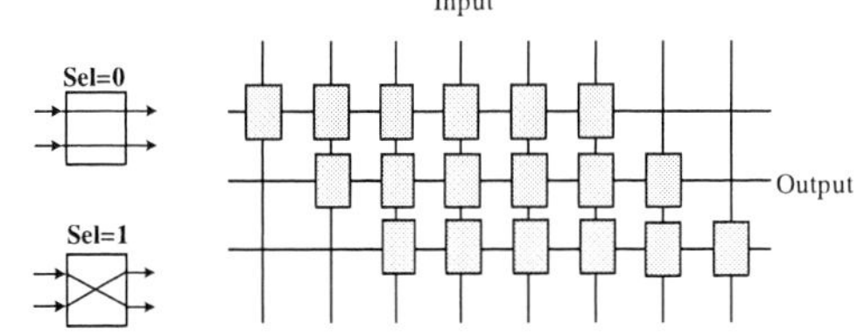

Figure 2: Sparse Crossbar Concentrator.

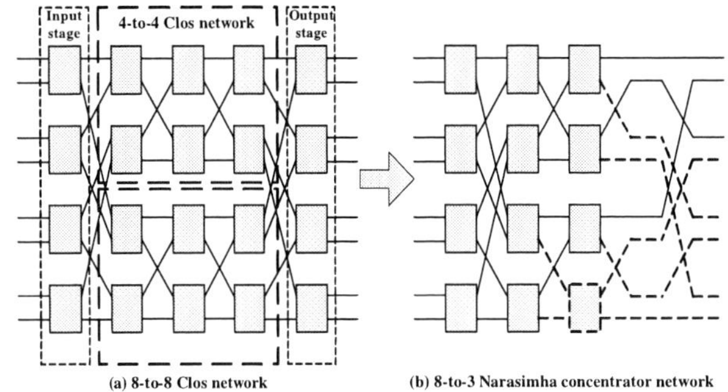

Figure 3: Narasimha Network.

rently. It is hence up to the designers to manually build the MUX network based on their design knowledge, which is a rather cumbersome process for them. More importantly, since bugs often occur in unexpected scenarios, designers are typically not knowledgeable about exactly which signals should be traced together at the design stage to satisfy their debug needs.

To overcome the above limitation, designers can resort to *non-blocking concentration networks*, which have been extensively studied in the context of communication theory [16]. A *n*-to-*m* concentrator is able to transfer *any m* out of *n* ($m \leq n$) input signals to the output side, which naturally satisfies our interconnection fabric design needs that tries to select a subset of trace signals from a large number of tapped signals. A well-known concentrator, namely *sparse crossbar concentrator*, is a direct-connected network constructed using crossbar switches. [14] proved that the number of switches required to build such *single-stage* concentrators is at least $(n - m + 1) \times m$ and and an example 8-to-3 concentrator is shown in Fig. 2.

Narasimha [15] proposed a multi-stage concentrator that is able to dramatically reduce the required number of crossbar switches. This *Narasimha concentrator* is developed on top of the well-known 3-stage *Clos network* [5]. Clos network is able to provide all possible permutations of the *n* ($n = 2^k$) inputs at the output side of the network, and it is constructed in a recursive manner. An example 8-to-8 Clos network is depicted in Fig. 3 (a). The 8 input signals are firstly assigned to four 2×2 crossbar switches, and their outputs are connected *evenly* to the upper and lower sub-networks, which are 4-to-4 Clos networks. The output stage is constructed in a reverse manner, and the whole process ends when the sub-network are all 2×2 crossbar switches. More details about the Clos network can be found in [8]. As it is not necessary to provide the "permutation" capability in a concentrator, Narasimha concentrator eliminates the output stage of the Clos network. It also removes those crossbar switches that do not drive outputs. An example 8-to-3 Narasimha concentrator is shown in 3 (b).

Recently, Quinton and Wilton [17] proposed to use the Narasimha concentrator to connect a programmable logic block to the other cores in SoC designs. While the original Narasimha concentrator requires the number of inputs to be 2^k (similar to the Clos network), it can be easily revised to take arbitrary number of inputs, as shown

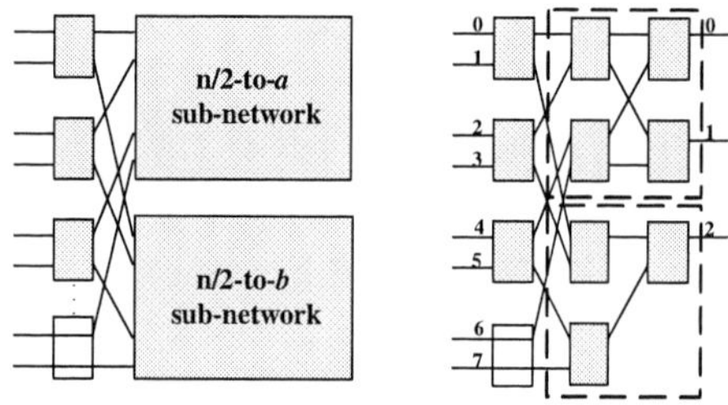

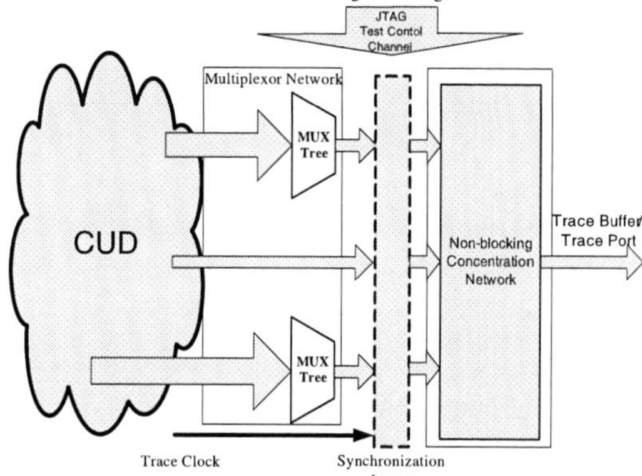

Figure 4: Revised Narasimha Concentrator in [15].

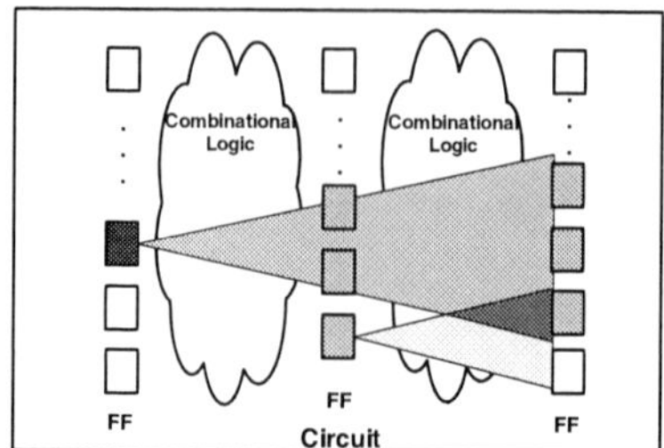

Figure 6: Forward Propagation for Correlation Extraction.

Figure 5: Proposed Interconnection Fabric.

in Fig. 4. [17] also did minor modifications to the Narasimha concentrator design to further reduce its hardware cost. To be specific, [17] proposed to replace the last crossbar unit with two direct wires at the input stage, as shown in Fig. 4. This replacement, however, cannot always provide the desired functionality. A counter example for a 8-to-3 network is shown in Fig. 4. If the last crossbar at input stage is reduced and we would like to trace inputs {0, 1, 7} at the same time, apparently, input 7 should be transferred to output 2 with the lower sub-network. Then, one of the other two inputs cannot find a path to the output. Consequently, the combination of inputs {0, 1, 7} cannot be traced together, which violates the functionality for non-blocking concentrator design.

Apparently, using concentrators to construct trace signal interconnection fabric provides better visibility to the CUD when compared to MUX trees, but at the cost of more DfD area. In practice, some tapped signals are not highly correlated with each other, and hence it is not necessary to observe them concurrently. This observation motivates us to combine MUX trees and concentrators to have a flexible yet low-cost tracing network. In addition, even for the revised Narasimha concentrator, it still contains lots of redundant elements. Take a special case for 8-to-8 network as an example, the simplified concentrator requires 9 crossbar switches to construct it, but these switches actually can all be eliminated. This motivates us to propose a new concentrator design with much lower hardware cost.

3. PROPOSED INTERCONNECTION FABRIC DESIGN

The problem investigated in this paper can be formulated as:

Given N_{tp} tapped signals in the CUD, those highly-correlated signals should be able to be traced concurrently, while others not. We are to design an interconnection fabric to transfer a subset of N_{bf} (typically, $N_{bf} \ll N_{tp}$) tapped signals to the trace buffers/ports at runtime, which satisfies the above requirement at the minimum hardware cost.

To tackle the above problem, we propose to design the interconnection fabric as shown in Fig. 5, which contains two main parts: (i). a *multiplexer network* that connects those mutually-exclusive tapped signals, which can be designated by designers and/or extracted automatically based on structural analysis. This stage outputs N_{pc} potentially-correlated signals. (ii). a *non-blocking concentration network* that is able to transfer any N_{bf} signals out of N_{pc} inputs to the trace buffers and/or trace ports. One of the main objectives of our design is to minimize N_{pc}. The reason behind this is that, accessing the same signals with MUX tree always results in smaller DfD cost when compared to using nonblocking concentrator, since the latter one requires more resources to achieve the "any combination" objective.

3.1 Multiplexer Network for Mutually-Exclusive Signals

Firstly, we need to determine which tapped signals are highly-correlated and hence may need to be traced together in post-silicon validation. As discussed earlier, it is a rather cumbersome process for the designers to manually conduct this duty. We introduce a simple yet effective method to facilitate this process through circuit structural analysis. For a tapped signal, its *related logic elements* are those that are either in the logic cone that drives this signal starting from inputs or in the logic cone that this signal drives until outputs. For two tapped signals, if there is no overlap between their respective related logic elements, they are not correlated at all. The above constraint is quite stringent and it does not reflect how high the correlation is among tapped signals. We therefore first levelize the sequential elements of the circuit. Obviously, the closer a logic element is to the tapped signal, the more related it is with the signal. Then, instead of checking the overlapping of all the related logic elements of two tapped signals, we only check whether there is any overlap within the neighboring N_l levels, in which N_l is a user-defined value. If there is, we call these tapped signals *highly-correlated* and should be able to be observed together. Otherwise, they are *mutually-exclusive*.

According to the above, we build an *uncorrelation graph* among tapped signals. In this graph, each vertex represents a tapped signal and we add an edge between two vertices if they are not highly correlated. The graph is initialized as a complete graph. The edges are then gradually removed from the graph by conducting forward propagation analysis for the circuit from the first sequential logic level, as shown in Fig. 6. Note, there is no need to conduct backward analysis for a tapped signal, since the correlations are already obtained by those tapped signals in previous logic level through forward propagation, if any.

Our MUX network is composed of a number of MUX trees (see Fig. 5), which are used to connect mutually-exclusive signals and only one of the signals is required to be observed from each MUX tree at a time. In our uncorrelation graph, each MUX tree corresponds to a *clique*. To minimize the number of outputs N_{pc} from the MUX network, it is equivalent to use the minimum number of

978-1-60558-497-3/09 $25.00 © 2009 ACM

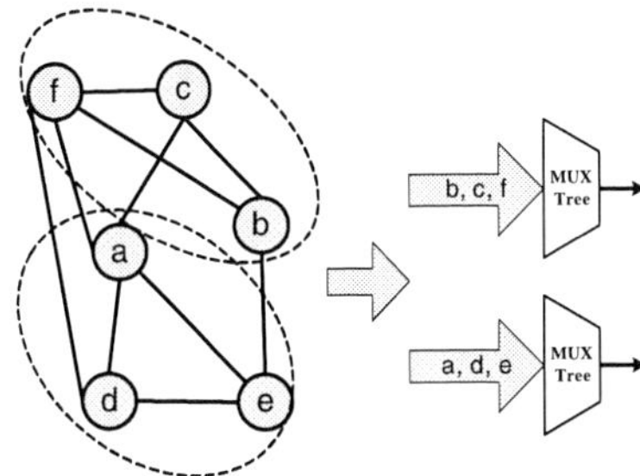

Figure 7: An Example of Uncorrelation Graph.

cliques to cover all the vertices in the uncorrelation graph. This "minimum clique cover problem" is a well-known NP-hard problem in graph theory, and we can resort to a classical greedy heuristic to solve it [6]. As shown in Fig. 7, at least two cliques are required to cover all 6 nodes, and two MUX trees are introduced accordingly to merge these signals and $N_{pc} = 2$.

The above discussions are applicable when all tapped signals are from the same clock domain. If, however, we have tapped signals from different clock domains to transfer to the same trace buffer and/or trace port, we need to build the uncorrelation graph separately for signals from each domain and we need to add a synchronization layer before these signals going into the non-blocking concentrator (see Fig. 5). At the same time, in order not to miss any traced data, it is important to trace signals using the fastest clock among all domains.

3.2 Non-Blocking Concentration Network for Concurrently-Accessible Signals

Starting from the revised Narasimha concentrator that is able to take arbitrary number of inputs as shown in [17] (without using their simplification method as it violates our desired concentration capability, see Section 2.2), we propose several simplification rules to reduce the DfD cost of the concentration network, detailed in the following.

- **Rule 1: Replacement of crossbars that provide redundant path**

As mentioned before, the structure presented in existing work provides redundant paths where signals will never go through. We first introduce a theorem for the output port assignment of the revised Narasimha network, which is missed in [17]. This theorem guarantees that any m out of n input signals are accessible by this concentrator.

Theorem 1 For any n-to-m Narasimha-based concentrator, if the m output nodes are evenly distributed into the top half and the bottom half of the concentrator, it is able to provide n to m accessibility.

Proof: Any two input signals of the crossbar can be connected to the top half and the bottom half of the concentrator, respectively. Therefore, any m out of n input signals can evenly flow into two sub-fabrics with no larger than one difference (i.e., $\{k+1, k\}$ or $\{k, k\}, k = \lfloor n/2 \rfloor$). Recursively, every sub-fabric also has *non-blocking* feature. As a result, if the output nodes are assigned accordingly, the accessibility is guaranteed. □

To propagate the simplification effect, we firstly assign the input and output signal effect on each switch. As depicted in Fig. 8, every switch has a (Input Effect, Output Effect) initialized as (0, 0). Then Input/Output Effect of switches in input/output stage is updated as the number of assigned input/output signal. After that, the Input/Output Effect are propagated forwardly/backwardly. The forward propagation is as follows. Consider one switch at middle

stage, if its Input Effect is 2, two back-end connected switches will add 1 for their own Input Effect, since both paths are required to transfer signals. Similarly, when its Input Effect is 1, then only one connected switch will add 1 for its Input Effect (here we choose the upper one). The backward propagation is conducted in the same way but with the opposite direction.

After both propagations, the redundant units in the original concentrator can be identified and simplified as shown in Fig. 8. If both Input Effect and Output Effect are 2 for a switch, it remains to be a crossbar unit. If they are 2 and 1 respectively, the switch is simplified to be a MUX. Otherwise, it is replaced by a wire. With the above simplification rule, for a 9-to-4 concentration network, its cost will be reduced from 16 crossbars to 8 crossbars and 5 MUXes.

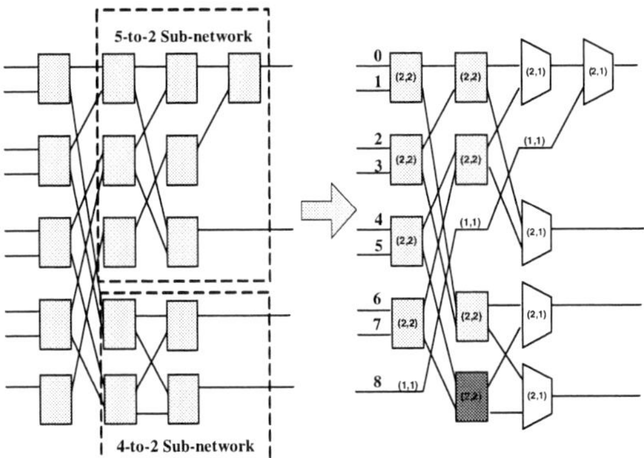

Figure 8: 9-to-4 Network Example of Rule 1.

- **Rule 2: Reduction of crossbar at input stage**

As discussed earlier, [17] proposed to replace the last crossbar unit at the input stage with two direct wires, but it cannot always hold the desired property for concentration network. The following theorem presents the condition for such replacement.

Theorem 2 For any n-to-m network, a crossbar switch at the input stage switch can be replaced by two direct wires, provided both n and m are even.

Proof: We start from the case that n is even. By Theorem 1, at the input stage, m out of n signals can be evenly distributed into two sub-networks (i.e., $n/2$-to-a and $n/2$-to-b, where $|a - b| \leq 1$). We consider the to-be-reduced unit as the one whose two inputs are connected directly to two sub-networks, given m is even. If neither signals from two sub-networks are traced, the replacement does not affect the original transfer capability. If one of them is traced, other units at this stage is able to evenly distribute $m - 1$ signals. If both are traced, since they are separated into two sub-networks, the functionality is reserved.

However, if m is odd (say, $a = (m-1)/2$ and $b = (m+1)/2$), suppose the signal connected to the sub-block with a outputs is traced, the other $m - 1$ signals should be connected to the sub-networks with $a - 1$ and b outputs respectively. Since $b - (a - 1) = 2$, this requirement cannot be satisfied.

Similarly, when n is odd, if we trace the last signal (e.g., Input 8 in Fig. 8) and an input that also connects to the top half sub-network (Input 6), the replacement is not applicable. □

We apply this rule to further simplify the fabric at the input stages. For the 9-to-4 network depicted in Fig. 8, only one switch can be reduced in a 4-to-2 sub-network. To note, this rule does not work for the special case that the input stage contains only one switch.

- **Rule 3: Replacement of crossbars that provide unnecessary permutation**

Even after applying previous simplification rules, the structure still unnecessarily cost large amount of crossbar units for n-to-m concentrator, when m is large. Let us use a 8-to-7 network shown in Fig. 9 to demonstrate our simplification rule. Starting from the last stage, it is obvious that the switches here can only change the order of two signals to provide unnecessary "permutation". Consequently, these crossbars can be replaced with two wires as depicted in Fig. 9 (b). Next, from these direct wires, the frontend crossbar switch may be directly connected to output ports with two wires too. Based on the same principle, it can be further replaced and we continue to process the frontend crossbar switches until the input side is reached. The result for this example is shown in Fig. 9 (c) and we are able to remove four crossbar switches.

In particular, for the special case with n-to-n structure, after applying this rule, no crossbar units is required.

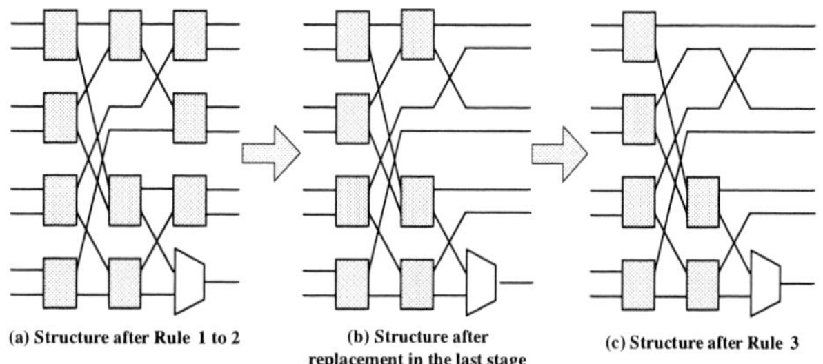

(a) Structure after Rule 1 to 2   (b) Structure after replacement in the last stage   (c) Structure after Rule 3

Figure 9: 8-to-7 Network Example of Rule 3.

4. EXPERIMENTAL RESULTS

We conduct experiments on three large ISCAS'89 benchmark circuits to evaluate the DfD cost of the proposed interconnection fabric and compare against existing structures. The correlation constraints are extracted by using the methods presented in Section 4.1 with $N_l = 3$. We randomly select 50, 100, 150, and 200 tapped signals and automatically generate interconnection fabric to connect those signals to trace buffers with various widths. The fabrics are inserted into the benchmark circuits and the DfD cost is obtained by using a commercial synthesis tool.

Table 1 shows the DfD cost introduced by the proposed approach and three existing methods. Column 1 presents the number of tapped signals; Column 2 is the proportion of the edges among all possible connections in the uncorrelation graph; Column 3 is the number of signals output from our multiplexer network; Column 4 is the trace buffer width; Column 5-8 are the DfD area cost in terms of 2-input NAND equivalent gates; and Column 9 shows the DfD area reduction resulted from the proposed method, when compared with that in [17] (again, without applying their simplification method). In this table, *MUX* refers to the MUX tree that does not consider signal correlations. It indicates the lower bound for the DfD area cost to connect N_{tp} signals to N_{bf} outputs. *Sparse* corresponds to the sparse crossbar concentrator design in [12].

As expected, MUX tree and sparse crossbar concentrator costs the least and the most DfD area, respectively. With the growth of the tapped signal count, generally speaking, more DfD area is required for the interconnection fabrics (e.g., s38417 case shown in Fig. 10). When the number of tapped signal is fixed, the cost of MUX tree slightly decreases with the increase of buffer width, because less signals are required to be concentrated. By contrast, we observe proportional increase of the sparse concentrator DfD area with various buffer widths in almost all cases. Similarly, the DfD cost using [17] grows steadily. For the proposed interconnection fabric, however, its area cost depends on not only the signal count

Table 1: Experimental Results for DfD Area Cost

Input #	Edge %	N_{pc}	Buffer width	DfD Cost (2-input NAND gates)				
				MUX	Sparse	[17]	Prop.	Δ
s35932								
50	94.1%	13	8	384	7224	2571	651	74.7%
			16	339	11760	3177	339	89.3%
			32	339	12768	3177	339	89.3%
100	95.2%	13	8	828	15624	5280	1101	79.2%
			16	789	28560	6192	789	87.3%
			32	789	46484	7380	789	89.3%
150	95.2%	28	8	1275	24024	8157	2055	74.8%
			16	1215	45476	9909	2067	79.1%
			32	1110	80084	11223	1110	90.1%
200	95.0%	32	8	1728	32424	10740	2589	75.9%
			16	1662	62276	12660	2685	78.8%
			32	1524	113692	14484	1524	89.5%
s38584								
50	32%	40	8	1533	8399	3746	3210	14.3%
			16	1461	12935	4334	3474	19.8%
			32	1352	13943	4352	3330	23.5%
100	44%	68	8	4444	19234	8871	6456	27.2%
			16	4372	32170	9783	6888	29.6%
			32	4236	49980	10971	7080	35.5%
150	39%	104	8	4301	27106	11215	8213	26.8%
			16	4245	48429	12967	9149	29.4%
			32	4096	83014	14281	9677	32.2%
200	40%	143	8	6466	37197	15514	10840	30.1%
			16	6402	66907	17433	12091	30.6%
			32	6263	118335	19257	13249	31.2%
s38417								
50	69.8%	27	8	528	7374	2721	1233	54.7%
			16	462	11910	3309	1266	61.7%
			32	363	12918	3327	363	89.1%
100	64.6%	60	8	1280	16047	5239	3036	42.1%
			16	1208	28983	6119	3384	44.7%
			32	1026	46794	7307	3492	52.2%
150	67.6%	74	8	1923	24649	8782	4049	53.9%
			16	1851	46039	10534	4607	56.3%
			32	1707	80647	11848	4883	58.8%
200	64.6%	109	8	2314	33025	11341	5622	50.4%
			16	2242	62784	13261	6495	51.0%
			32	2098	114215	15085	7233	52.1%

$$\text{area overhead reduction } \Delta = \left(1 - \frac{\text{Area of proposed fabric}}{\text{Area of [17]}}\right) \times 100\%$$

and buffer width but also the correlation constraints. For the case that signals have low correlations (e.g., circuit s35932), the number of signals feeding into the non-blocking concentrator is usually quite small, as they have been processed in the MUX network stage. In this case, the cost of the proposed interconnection fabric can be significantly reduced, without sacrificing the debug flexibility. Taking s35932 as an example, as there are very few correlations among singals, the DfD cost of our interconnection fabric is the same as or slightly higher than that of MUX tree. In average, the DfD cost is reduced up to 83% when compared with [17]. For the case that traced signals are highly related (e.g., s38584), the DfD cost for the proposed design reduces for roughly 27% by applying the simplification rules on non-blocking concentrators. One thing to be noted is that that DfD area cost reported in this table seems to be relatively high when compared to the original circuit size, we attribute this to the large percentage of tapped signals (in practical designs, the percentage is smaller [1]).

To investigate the impact of the number of propagation levels N_l during correlation extraction, we conduct a case study for s35932 when tapping 200 signals and transferring to 16-bit trace buffer. When we set N_l to be 6, N_{pc} increases to 68 and the DfD area cost grows to 4197 2-input NAND gates. If $N_l = 12$, the DfD area cost further grows to 7809 gates. From the above, we can see that building correlation constraints among signals has a significant impact on the cost of the interconnection fabric. In practice, designers should carefully select N_l based on their design knowledge.

We also study the effectiveness of each simplification rule on non-blocking concentrator. In this experiment, we trace 100 sig-

978-1-60558-497-3/09 $25.00 © 2009 ACM

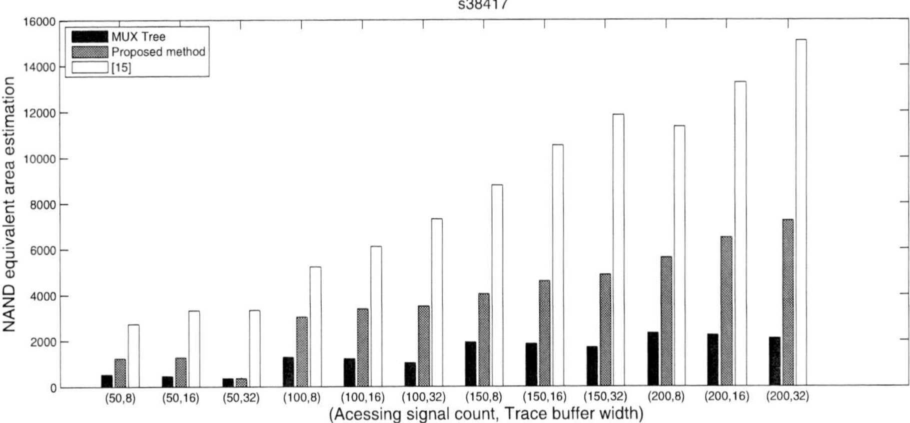

Figure 10: Experimental results for DfD area cost of s38417.

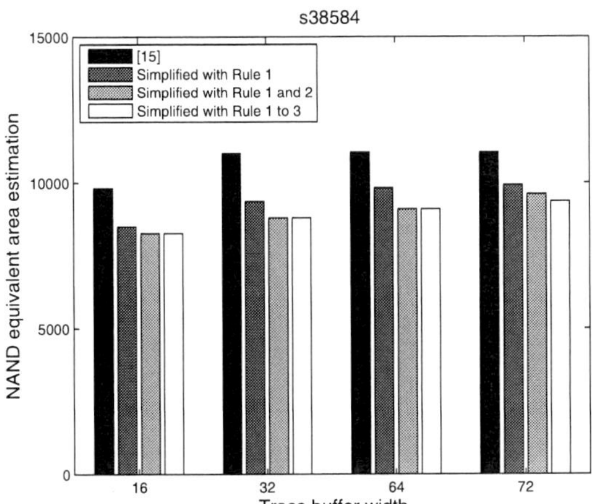

Figure 11: Experimental results for DfD area cost of s38584 for simplification rule evaluation.

nals in s38584 and build the interconnection fabric using the concentrator only. As depicted in Fig. 11, Rule 1 is the most effective one when the fabric contains 32 outputs, because the original fabric contains most redundant paths at this moment. Rule 2 saves most area for 64-bit buffer. It is more effective than 16-bit and 32-bit cases because more crossbar switches at input stages that have not been replaced by MUXes can be reduced by this rule. It is more effective than 72-bit because the rule does not work for the subnetwork containing an odd number of outputs. The impact of Rule 3 is observable only when the output number is 72, since only under such circumstances the switch at the last stage of the fabric is connected with two outputs.

Finally, to address the timing issue, we can pipeline the proposed interconnection fabric by inserting flip-flops into it. Consider the case that tracing 200 signals with 8-bit buffer in s38417. The original proposed fabric introduces a few critical paths. After pipelining, these critical paths can be eliminated, with additional 3199 gates.

5. CONCLUSION

In this paper, we propose a novel trace signal interconnection fabric design that contains a multiplexer network that connects those mutually-exclusive tapped signals and a novel simplified non-blocking concentrator. Experimental results on benchmark circuits show that the proposed solution is able to significantly reduce DfD area cost while satisfying designers' debug flexibility requirement.

6. ACKNOWLEDGEMENTS

This work was supported in part by the General Research Fund CUHK417406, CUHK417807, and CUHK418708 from Hong Kong SAR Research Grants Council, and in part by a grant N_CUHK417/08 from the NSFC/RGC Joint Research Scheme.

7. REFERENCES

[1] M. Abramovici. In-System Silicon Validation and Debug. *IEEE Design & Test of Computers*, 25(3):216–223, May-June 2008.

[2] M. Abramovici, et al. A Reconfigurable Design-for-Debug Infrastructure for SoCs. In *Proc. ACM/IEEE Design Automation Conference (DAC)*, pp. 7–12, 2006.

[3] Altera Inc. Design Debugging Using the SignalTap II Embedded Logic Analyzer.

[4] ARM. Embedded Trace Macrocell Architecture Specification.

[5] C. Clos. A Study of Non-Blocking Switching Networks. *Bell System Technical Journal*, 32:406–424, 1953.

[6] J. Gramm, J. Guo, F. Huffner, and R. Niedermeier. Data Reduction, Exact, and Heuristic Algorithms for Clique Cover. In *Workshop on Algorithm Engineering and Experiments*, pp. 8–94, 2006.

[7] A. B. Hopkins and K. D. McDonald-Maier. Debug Support for Complex Systems On-chip: A Review. In *IEE Proc., Computers and Digital Techniques*, pp. 197–207, July 2006.

[8] F. Hwang. *The Mathematical Theory of Nonblocking Switching Networks*. World Scientific Publishing Company, 1999.

[9] D. Josephson and B. Gottlieb. Debug Methodology for the McKinley Processor. In *Proc. IEEE International Test Conference (ITC)*, pp. 451–460, 2001.

[10] H. F. Ko and N. Nicolici. Automated Trace Signals Identification and State Restoration for Improving Observability in Post-Silicon Validation. In *Proc. IEEE/ACM Design, Automation, and Test in Europe*, pp. 1298–1303, 2008.

[11] H. F. Ko, A. B. Kinsman, and N. Nicolici. Distributed Embedded Logic Analysis for Post-Silicon Validation of SOCs. In *Proc. IEEE International Test Conference (ITC)*, paper 16.3, 2008.

[12] X. Liu and Q. Xu. Trace Signal Selection for Visibility Enhancement in Post-Silicon Validation. In *Proc. IEEE/ACM Design, Automation, and Test in Europe (DATE)*, pp. 1338–1343, 2009.

[13] MIPS Technologies Inc. EJTAG Trace Control Block Specification.

[14] S. Nakamura and G. M. Masson. Lower Bounds on Crosspoints in Concentrators. *IEEE Tran. on Computers*, C-31(12):1173–1179, Dec. 1982.

[15] M. J. Narasimha. A Recursive Concentrator Structure with Applications to Self-Routing Switching Networks. *IEEE Tran. on Communications*, 42(234):896–898, Feb.-Apr. 1994.

[16] N. Pippenger. On the Complexity of Strictly Nonblocking Concentration Networks. *IEEE Tran. on Communications*, 22(11):1890–1892, Nov. 1974.

[17] B. R. Quinton and S. J. E. Wilton. Concentrator Access Networks for Programmable Logic Cores on SoCs. In *Proc. International Symposium on Circuits and Systems (ISCAS)*, pp. 45–48, 2005.

[18] S. Tang and Q. Xu. A Multi-Core Debug Platform for NoC-Based Systems. In *Proc. IEEE/ACM Design, Automation, and Test in Europe (DATE)*, pp. 45–48, 2007.

[19] B. Vermeulen and S. K. Goel. Design for Debug: Catching Design Errors in Digital Chips. *IEEE Design & Test of Computers*, 19(3):37–45, May 2002.

[20] B. Vermeulen, S. Oostdijk, and F. Bouwman. Test and Debug Strategy of the PNX8525 Nexperia™ Digital Video Platform System Chip. In *Proc. IEEE International Test Conference (ITC)*, pp. 121–130, 2001.

[21] Xilinx Inc. Chipscope Pro Software and Cores User Guide.

978-1-60558-497-3/09 $25.00 © 2009 ACM

357

Online Cache State Dumping for Processor Debug

Anant Vishnoi, Preeti Ranjan Panda, and M. Balakrishnan
Department of Computer Science and Engineering
Indian Institute of Technology Delhi
Hauz Khas, New Delhi 110016, India
{anant,panda,mbala}@cse.iitd.ac.in

ABSTRACT

Post-silicon processor debugging is frequently carried out in a loop consisting of several iterations of the following two key steps: (i) processor execution for some duration, followed by (ii) dumping out of the processor's internal state into an external logic analyzer for further offline processing. Internal state of the processor is dominated by the L2 cache. During the process of dumping the cache content, the processor's execution is halted so that the state can be faithfully reproduced offline. In order to reduce the duration for which the processor is halted, and indirectly reduce debug time, we propose two *Online Cache Dumping* strategies, *Retransmit Non-dumped Line (RNL)* and *Dump History Table (DHT)*, with the objective of transferring the cache contents while the processor is executing, and yet maintaining fidelity of the dumped data. For typical experimental debug scenarios, we observe that the effective dump times are reduced to between 0.01% and 3.5% of the original times. We also employ compression to reduce the cache content transfer time and logic analyzer space. Our experiments indicate an average compression ratio of 59.2%.

Categories and Subject Descriptors

B.7.2 [**Integrated Circuits**]: Design Aids—*Verification*; B.8.1 [**Performance and Reliability**]: Reliability, Testing, and Fault-Tolerance

General Terms

Reliability, Verification

Keywords

Post-silicon Validation, Processor Debug, Cache Compression, Design for Debug, Silicon Debug

1. INTRODUCTION

Post-silicon processor debugging is frequently carried out in a loop consisting of several iterations of the following two

key steps: (i) processor execution for some duration, followed by (ii) dumping out of the processor's internal state into an external logic analyzer for further offline processing. Internal state of the processor is dominated by the L2 cache (and L3 cache, if on-chip; we use the term L2 to include both L2 and L3 in this paper). During the process of dumping the cache content, the processor's execution is halted so that the state can be faithfully reproduced offline [16]. Freezing the processor execution for the entire cache dump duration leads to loss of precious time in the debug infrastructure, particularly since the action is repeated a large number of times [15]. One way to reduce the debug cycle time is to attempt *Online Cache Dumping* – the L2 cache dump proceeds simultaneously with processor execution – while maintaining valid data in the dump. The obvious challenge here is to ensure that the L2 cache state being dumped out is not corrupted by the executing processor.

We present two strategies for online cache dumping, with different cost-performance trade-offs. When the signal for starting the cache dump is received, we start the dumping, maintaining a logical division between the *dumped* and *non-dumped* areas of the cache. In *Retransmit Non-dumped Line (RNL)*, when the processor attempts to update a cache line that is not yet dumped, we first dump the existing line so that it can be faithfully reproduced offline. In the *Dump History Table (DHT)* mechanism, we ensure that dumping occurs only once per line in the non-dumped area by maintaining a table that keeps track of whether each line has been already dumped. The dumped cache content is then taken through a compression unit in order to reduce the transfer time and expensive logic analyzer space (Figure 1).

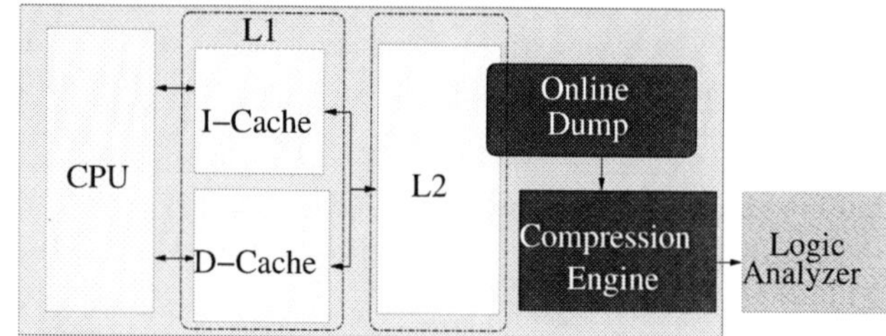

Figure 1: Online Dumping with Cache Data Compression

2. RELATED WORK

A large body of work already exists in the area of test compression. Balakrishnan et al. [5] address the problem

Permission to make digital or hard copies of part or all of this work for personal or classroom use is granted without fee provided that copies are not made or distributed for profit or commercial advantage and that copies bear this notice and the full citation on the first page. To copy otherwise, to republish, to post on servers or to redistribute to lists, requires prior specific permission and/or a fee.
DAC'09, July 26-31, 2009, San Francisco, California, USA

978-1-60558-497-3/09 $25.00 © 2009 ACM

of faster loading of test vectors. Their proposed improvement uses a software compression mechanism to improve the loading time. Anis et al. [4], propose a post silicon debug process where they capture signatures at intervals through a lossy compression mechanism, which are later offloaded to the debugging software. Cheng et al. [8] propose a test data compression technique, CacheCompress, to compress scan chain data. The above line of research is orthogonal to our context of processor debug.

Cache partitioning, which is relevant to our work, has been attempted in the past with a different objective: improving performance and reducing power. Albonesi [3] proposed a software controlled on-demand cache partitioning architecture. Zhang et al. [19] proposed a way halting mechanism to reduce cache energy consumption. They activate only those cache ways whose initial few tag bits match the corresponding bits of the requested address. A configurable cache architecture has also been proposed by Zhang et al. [20], offering flexibility in parameters such as cache associativity, which can be adjusted according to the requirement of the application.

The techniques used in [1, 11, 12] attempt to fit multiple cache line data into a single cache line. The FPC cache compression algorithm [1] attempts to locate frequently occurring patterns in a cache line and compresses the line using a static coding technique. Lee et al. [11] proposed a compressed memory hierarchy model that selectively compresses L2 cache and memory blocks. Their selective compressed memory system (SCMS) uses a hardware implementation of the X-RL compression algorithm [10]. Lekatsas et al. [12] use SAMC to compress the cache.

Most previous work on the cache partitioning and compression have focused on performance and power improvement, which is orthogonal to our targeted scenario, and could be applied in conjunction. Recent work on post-silicon validation has addressed the topic of debug hardware mechanisms in processors [13, 17, 9], but none has directly examined the cache state dump scenario. In this paper, we extend our previous work [16], where the cache state was compressed before being dumped out for debug purposes, but in a way that required the CPU to be frozen for the entire duration of the cache dump.

3. ONLINE CACHE DUMPING

The objective of online cache dumping is to reduce the total debug cycle time by overlapping the the dumping of L2 cache state with processor execution. The key requirement here is that, if we need the state at time $t = T_1$ dumped, the state should not be corrupted by the processor when it resumes execution. Stalling is necessary to dump out the contents of the register file, pipeline registers, L1 caches, etc. However, the bulk of the processor state consists of L2 cache which is not so frequently accessed, and it is possible to overlap the dumping of L2 cache with the processor's execution when it resumes after T_1, with minor impact on execution time.

The central idea in achieving overlapping execution with dumping is as follows. We carry out the cache dump one line at a time in increasing order of the line index number. A counter is maintained to keep track of what part of the cache has already been dumped; this partitions the cache logically into two areas: *dumped area* and *non-dumped area*. If, during dumping, the processor requests an update to line

B_1 in the dumped area, then execution proceeds normally since the state for B_1 has already been captured. If an update is requested to line B_2 in the non-dumped area, then line B_2 (consisting of data, tag, control bits, and the identifying line number) is first dumped out before updation. Processor execution and dumping proceed simultaneously in normal sequence after this. The line number and data is used later by the offline state reconstruction program to retain the right state for B_2 and ignore the (corrupted) state dumped out later when the dump sequence reaches B_2.

3.1 Cache Partitioning

Caches can be naturally partitioned at different levels of granularity in different ways. Way-wise partitioning suggested in [3] could possibly be adapted for our purpose, where we can dump the cache one way at a time, with the corresponding cache way/bank becoming unavailable for processor execution during this time. However, this interferes with the cache behavior of the executing program and may obstruct the debugging process. Other problems such as data duplication may occur in the different cache ways. We use a set-wise partitioning approach using a simple counter to partition the cache. The counter keeps track of the current index being dumped, and is incremented after dumping the lines in the set corresponding to each index. Determining whether an update request to an address occurs in the dumped or non-dumped area is done by comparing the index with the counter.

3.2 Handling Cache Updates

A key issue in overlapping execution with cache dump is the proper handling of cache updates while the dumping is in progress. We describe two mechanisms for achieving this, representing a trade-off between the volume of additional dumped data and area overhead.

3.2.1 Retransmit Non-dumped Line (RNL)

In Retransmit Non-dumped Line (RNL), on a cache update during processor execution, we first check whether the update being requested is to the dumped or non-dumped area. No additional action is necessary if the line has already been dumped. If it belongs to the non-dumped area, we first dump the specific line before permitting the cache to be updated. Along with the cache line data, we also dump the cache line and way number to help identify the line/way during offline cache state reconstruction. The important relevant steps in the cache controller algorithm is explained in Algorithm 1

Algorithm 1 takes as inputs the memory address, write request (w_req), dump request (dump_s) and counter and returns the cache line data. When a cache set dump is requested by setting dump_s to TRUE, we extract the cache set indicated by the counter (lines 1 to 5 of Algorithm 1). Function dump-cachline1(), used for dumping the cache line, takes the index and way number as input and dumps the corresponding cache line data, consisting of data, tag, and control bits (including any state bits maintained by the replacement policy algorithm, such as LRU bits) to the compression engine. After dumping the cache set the counter is incremented.

If dump_s is FALSE (i.e., the processor has requested the cache update), we search the TAG in all ways of the cache set addressed by the index bits. If TAG is found, then the way

978-1-60558-497-3/09 $25.00 © 2009 ACM

Algorithm 1 RNL Procedure

Input: Address, w_req, dump_s, Counter
Output: Cache line data
```
 1: if dump_s == TRUE then
 2:     for i = 0 to num_ways do
 3:         dump-cacheline1 (Counter,i)
 4:     end for
 5:     Counter ⇐ Counter + 1
 6: else
 7:     TAG ⇐ Address.TAG
 8:     index ⇐ Address.index
 9:     miss ⇐ TRUE
10:     for  i = 0 to num_ways do
11:         if TAG == cache.TAG (index,i) then
12:             miss ⇐ FALSE
13:             way ⇐ i
14:         end if
15:     end for
16:     if miss == TRUE then
17:         // if miss, the way number is determined through
            replacement policy
18:         way ⇐ replacement-way(index)
19:     end if
20:     if index > Counter then
21:         if miss == TRUE OR w_req == TRUE then
22:             dump-cachline2 (index,way)
23:         else
24:             dump-control-bits (index,way)
25:         end if
26:     end if
27: end if
```

---

number is noted. If not, then the cache way to be replaced is decided by the replacement policy of the cache (function replacement-way()).

Line 16 of Algorithm 1, checks whether the requested cache update will affect the non-dumped area, by comparing the index with counter. If the index is greater, then we are in the non-dumped area and the existing state needs to be dumped before being updated by the cache write or miss action. We use function dump-cacheline2() for this purpose, which is similar to dump-cachline1(), except that, along with the line data, it also dumps out the index and way number, which are required to reconstruct the cache state offline. If we don't have a cache update (e.g, due to a cache READ hit operation) then the cache access can still affect some control bits (e.g., state bits maintained by the LRU algorithm). Hence, to accurately reproduce the cache contents, we capture the control bits with dump-control-bits() function.

3.2.2 Dump History Table (DHT)

Typical application behaviour may cause the processor to update the same cache line multiple times due to locality of reference. When such updates happen in the non-dumped area, the existing cache line is dumped each time the line is updated. Clearly, only the first such dump is useful for our purpose. We propose to maintain a Dump History Table (DHT) in which we maintain one bit corresponding to each cache line. We set this bit the first time the corresponding line is dumped due to an update in the non-dumped area. Future updates refer to this bit, and on finding this set, avoid the dumping. This leads to reduction in the total dump size

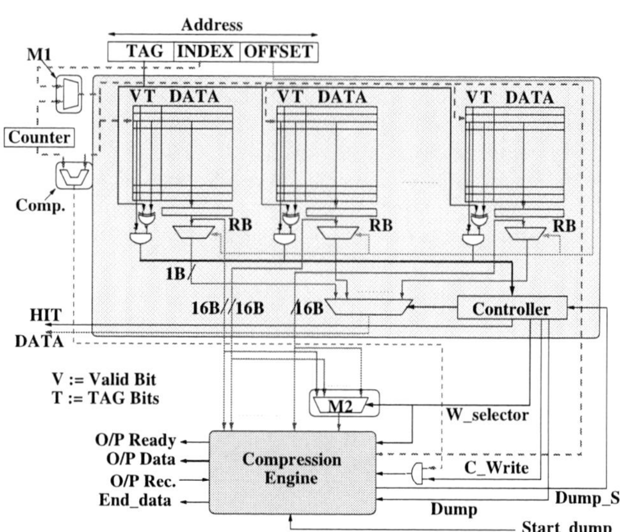

Figure 2: Retransmit Non-dumped Line (RNL) architecture

at the expense of the area overhad of maintaining the table.

DHT can be handled with a simple modification in Algorithm 1. The history table entries are initialised to FALSE. The condition (line 17) for calling function dump-cacheline2() is extended by also including a check for whether the corresponding DHT bit is FALSE. If yes, then dump-cacheline2() is called and the DHT bit is set to TRUE.

3.3 Compressor

Cache state data is amenable to compression because of repeated patterns occurring in the cache fields. A hardware compression engine that compresses the cache dump on its way out to the logic analyzer helps bring down both the dump time as well as logic analyzer memory. The compressor implementation is discussed in Section 4.

4. ARCHITECTURE DETAILS

We discuss here the architecture and implementation issues involved in realizing the above logical alterations in caches.

4.1 Modification in Cache Architecture

We attempt to reuse the hardware in the read path of the cache to the extent possible. A counter and comparator shown in Figure 2 are used to distinguish between dumped and non-dumped areas. The comparator result is AND'ed with a *C_Write* signal generated by the cache controller (representing a cache update requested by the processor) to enable the compression engine to accept a new cache line, due to cache update. A selector MUX is used to route the read address either from the counter or the index bits.

Figure 3 shows a simplified version of the modified cache controller FSM with S_0 being the initial/idle state. On Read or Write request, we proceed to and return from S_1 if there is a hit. Read and Write misses are serviced from states S_2 and S_4. The FSM jumps to a new dump state S_5 when it receives the *Dump_S* (D) HIGH from the compression engine. In S_5, cache lines of the set pointed to by the counter are read into their respective buffers (Figure 2) for transfer

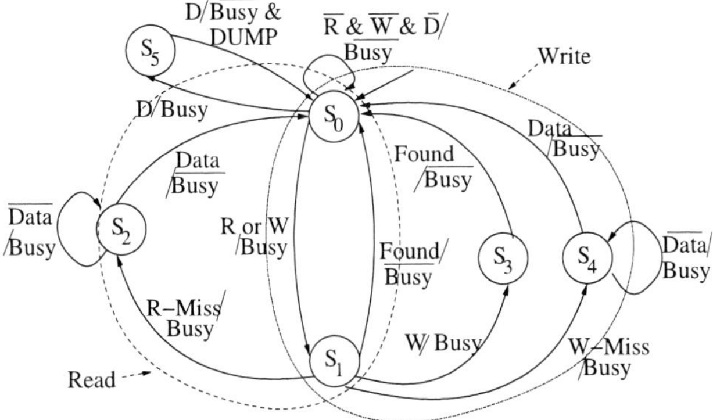

Figure 3: Modified Cache Controller FSM

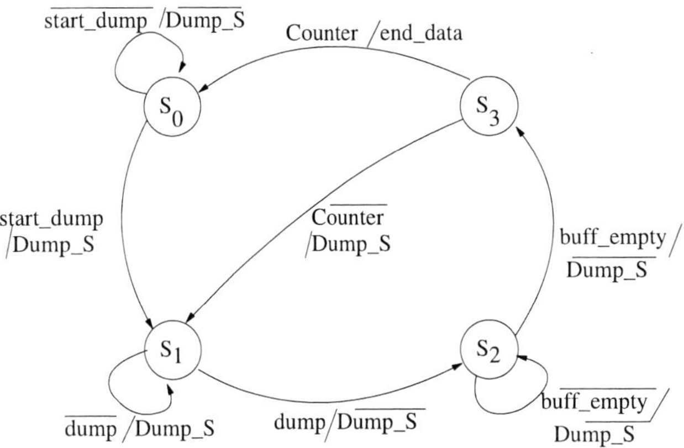

Figure 4: Compression Engine Communication FSM

to the compression engine. After the transfer, the counter is incremented.

The controller generates the following additional signals: (i) *C_Write*, informing the compression engine about updates to the cache that would result in data transfer from the non-dumped area; and (ii) *W_selector*, for selecting the way from where data needs to be dumped. Other details, omitted here, are as per the algorithms described in Section 3.

4.2 Compression Engine

The compression engine plays a significant role in reducing the cache dump time by compressing the dumped data. Different fields (data, tag, control bits, and ECC) of the cache exhibit different types of behavior, which can be exploited by using separate compressors, resulting in a more aggressive compression than when treating the entire cache data as just a bit stream. Additional dumped data due to cache updates are treated as a separate field. We experimented with two compression algorithms: X-Match [10] and LZW-SLU, a hardware implementation of an LRU-based dictionary compression scheme. Details of the compressor architecture are described in [16].

The compression engine activates the *Dump_S* signal to receive the next cache set, and is also responsible for communication with the logic analyzer (Figure 2). Figure 4 shows a 4-state communication FSM of the compressor. S_0 is the initial state. On receiving *start_dump* signal, it moves to S_1 raising the *Dump_S* signal leading the cache controller to initiate data transfer from the set pointed to by the counter. On receiving the *dump* signal from the controller (indicating that data is ready in the buffers), a transfer is initiated of this data into the compressor's internal buffers in state S_2. If the engine's buffer has available space, it moves to state S_3 checking the counter. If we are at the last line, then the FSM returns to S_0, otherwise, dumping is resumed at S_1.

4.3 RNL Architecture

Figure 2 shows the RNL hardware architecture. On a cache update, the controller sets *C_Write* signal HIGH. Simultaneously, the comparator checks if the index field is greater than counter. If yes, the compression engine fetches the cache line of the addressed way indicated by the *W_selector*. The *W_selector* and index values are also dumped.

4.4 DHT Architecture

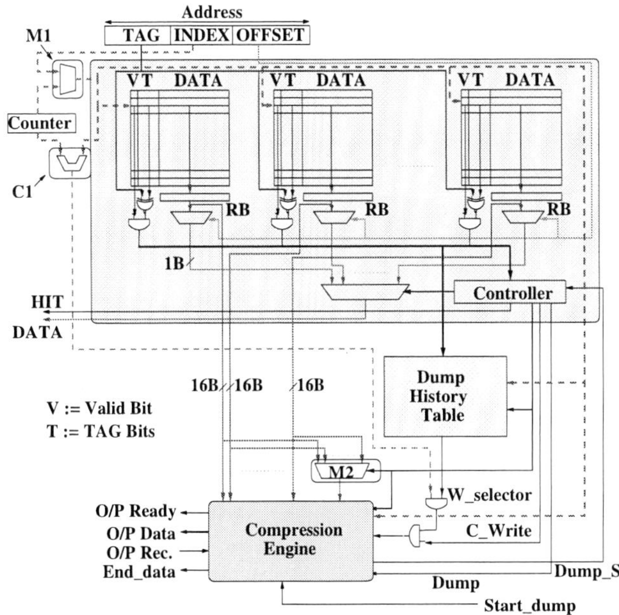

Figure 5: Dump History Table (DHT) Architecture

The DHT hardware is similar to the RNL hardware, except for the History Table, as shown Figure 5. This history table is checked before a line scheduled for updating is dumped to the compression engine. If the corresponding bit is set, then the dump is suppressed. For an A-way associative cache with N sets, the size of the history table is $N \times A$ bits.

5. EXPERIMENTS

We evaluated our proposed debug mechanism by running applications from the Mediabench and CPU-SPEC 2000 benchmarks on our framework consisting of processor and cache simulators Simplescalar [6] and Dinero [7] modified to support halting the simulation and dumping cache contents. To create realistic debug scenarios where one application runs after another, we also invoked the benchmarks sequentially as subroutines. Our experiments were carried out assuming a 2.0 GHz processor in 180 nm technology with the following L2 unified cache configuration: 512 KB, 4-way, 16-byte

978-1-60558-497-3/09 $25.00 © 2009 ACM   361

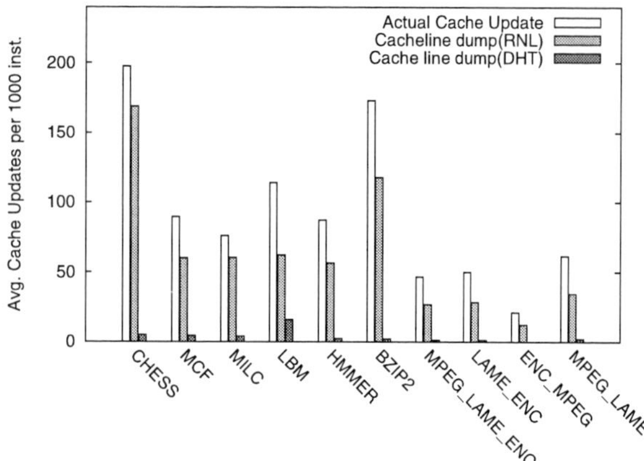

Figure 6: Average #cache line dumps due to cache update)

line, 3 control bits, 7 ECC bits. The cycle time and compression data shown here is the average of readings for each benchmark for dumps taken at intervals of 10 million cycles.

5.1 Saved Debug Time

Since L2 cache misses, which lead to traffic over the memory bus, are relatively rare (average of 0.6% for our benchmarks), the overlapping with the cache dump leads to a relatively small number of wasted cycles. We call this the effective dump time, since this is the overhead incurred by the program execution due to the dumping. The effective. Column 3 in Table 1 shows the effective dump time for the benchmark. Column 4 shows this information as a percentage of the cache dump time if the processor was stalled for dumping; the cache configuration used results in a dump time of 6.12×10^5 cycles. We note that the effective dump time is very small – 0.01% to 3.5%, with the average being 0.67% – compared to stalling the processor for the entire cache dump duration.

5.2 Additional cache lines dumped

Additional cache lines are dumped in our mechanism that overlaps dumping with processor execution. The number of additional lines dumped is an important metric for comparison of the cache update handling strategies: RNL and DHT. Figure 6 shows the number of actual cache updates (per 1000 instructions) and the additional cache lines dumped due to updates in the non-dumped area for both strategies RNL and DHT.

On an average, RNL dumps 31% less number of additional cache lines than the actual number of updates, and DHT dumps a further 96% fewer lines. The *Bzip2* and *Chess* examples exhibit the largest differences between RNL and DHT (66% and 86% respectively). For the remaining benchmarks, DHT always dumps significantly fewer number of additional cache lines compared to RNL.

5.3 Compression Results

Columns 5 and 6 of Table 1 shows the compression ratios (computed as $(1 - O/I) \times 100\%$, where O is the number of output bytes of the compressor and I is the cache dump size) obtained by two hardware implementations of compression algorithms LZW_SLU (hardware implemention of LZW [16]) and X-Match [10], of the cache data dumped by RNL

and DHT. We observe that, DHT always outperforms RNL on an average with 8.25% and 10.23% more compression with LZW-SLU and X-Match respectively. The maximum gain of 17% in compression ratio is noticed in *Bzip2*. Also, LZW-SLU gives 15% and 13% more compression than X-Match for RNL and DHT respectively.

Benchmarks	Comp. Algo.	Eff. Dump Time		Compression Ratio	
		N $\times 10^3$ CPU Cycles	Perc. (%)	RNL	DHT
Chess	LZW-SLU	1.44	0.22	62.29	77.24
	X-Match	2.25	0.37	40.22	56.32
MCF	LZW-SLU	19.02	3.11	47.77	56.49
	X-Match	21.20	3.46	30.31	44.88
MILC	LZW-SLU	2.50	0.41	27.59	35.47
	X-Match	8.75	1.43	21.10	33.67
LBM	LZW-SLU	4.95	0.81	86.36	92.11
	X-MATCH	12.08	1.97	55.70	69.95
HMMER	LZW-SLU	6.66	1.09	36.32	43.02
	X-Match	6.63	1.08	34.75	47.86
BZIP2	LZW-SLU	5.80	0.95	11.76	29.26
	X-Match	5.41	0.88	8.47	10.32
MPEG_LAME_ENC	LZW-SLU	0.09	0.01	53.95	59.69
	X-Match	0.12	0.02	38.75	47.66
LAME_ENC	LZW-SLU	0.09	0.01	55.56	61.93
	X-MATCH	0.17	0.03	38.50	48.95
ENC_MPEG	LZW-SLU	0.12	0.02	75.31	76.84
	X-Match	0.08	0.01	56.97	59.25
MPEG_LAME	LZW-SLU	0.04	0.01	52.14	59.44
	X-Match	0.12	0.02	32.86	47.15
Average	LZW-SLU	4.07	0.67	50.90	59.15
	X-Match	5.68	0.93	35.77	46.00

Table 1: Dump Time and Compression Results

5.4 Implementation and Synthesis Results

We synthesized VHDL descriptions of the new cache controller and compression engine with Synopsys Design Compiler with a 180nm library, and used estimations from CACTI [14] for cache and SRAM components.

For our targeted cache of 512KB, 16 byte cache line and 4 way associativity in 180 nm, the CACTI estimates are: 2.63 ns for access time and 38.9 mm^2 for area. The area difference for the synthesized cache controller with and without our proposed modification is 0.0002 mm^2, which is negligible in comparison to the cache area. The critical path of both synthesised modules are almost identical.

The area impact on the cache with the modification required for the RNL, is negligible in comparison to the cache area, as only a few muxes, a counter, and controller gates are needed, as shown in Figure 2. DHT requires an SRAM module for implementing the history table along with the controller modification. The estimated area of the history table is 0.15 mm^2 and access time 1.47 ns. Since the table is accessed in parallel with the cache access and the latter dominates the delays, its access does not cause any delay overhead.

For the LZW-SLU implementation we used 4 parallel compressors for the data field of the cache, 1 compressor for the additional cache lines dumped due to cache update and 1 compressor for TAG and ECC fields. The compression engines have two input buffers with size equal to the cache line data size and two output buffers of 256 bytes.

Table 2 shows a comparison of the area, critical path delay,

978-1-60558-497-3/09 $25.00 © 2009 ACM

Compression Algorithm	Critical Path(ns)	Bytes Per Cycle	Compressor Area (mm^2)
LZW-SLU	5.11	1	10.206
X-Match	15.09	4	2.760

Table 2: Area, Critical Path and Processing speed of Compression Algorithms

bytes processed per cycle, and hardware area for the two compression engines. The comparison shows X-Match to be superior in area and delay (its latency is longer, but it processes 4 bytes per cycle). However, as seen in Table 1, its compression ratio is worse.

5.5 Limitations

The following limitations still exist with our proposed strategy. The overlapping of dumping and execution places an upper bound on the interval between successive dumps; it cannot be smaller than the dump duration. If a smaller duration is needed, we can either use multiple compression engines, or, in the worst case, revert to the original of execution and dumping in sequence.

There are some classes of errors that cannot be caught with the proposed debug procedure. Problems involving interaction over the memory bus may not be reproduced since new traffic is introduced by the dumping procedure. Sequentialisation may be needed in this case also.

Finally, dumping the state of other components carrying state information closer to the CPU, such as pipeline registers, register file, L1 cache, etc., still require the CPU to be halted. Nevertheless, these components represent a relatively small fraction ($< 7\%$ in our example) of the total processor state. In general, even though online dumping results in significantly lesser duration for which the processor is frozen, the mechanism still cannot be used to reproduce accurately scenarios that occur due to external events.

6. CONCLUSION

We proposed and evaluated a mechanism for overlapping processor execution with transfer of cache contents off-chip for debug purposes with the objective of reducing the total debug cycle time which includes both processor execution and internal state transfer time. In order to prevent corruption of the cache state to be dumped by the executing processor, we proposed two alternatives based on the overall plan of explicitly transferring a cache line not already dumped before a regular cache update is permitted to occur. Experimental evaluation shows that the mechanism reduces the effective L2 cache dump time to between 0.01% and 3.5% of the original duration. The limitations identified above are the subject of future work.

7. ACKNOWLEDGMENT

The authors wish to thank G. Jaya Krishna and Shiva Sadashivaiah from Intel for their contributions to this research.

8. REFERENCES

[1] A. R. Alameldeen and D. A. Wood, "Adaptive cache compression for high-performance processors," in *ISCA*, June 2004.

[2] A. R. Alameldeen and D. A. Wood, "Frequent Pattern Compression: A Significance-Based Compression Scheme for L2 Caches," Tech. Rep. CS-TR-2004-1500, UW Madison, April 2004.

[3] D. H. Albonesi, "Selective cache ways: On-demand cache resource allocation," in *J. Instruction-Level Parallelism*, vol. 2, 2000.

[4] E. Anis and N. Nicolici, "Low cost debug architecture using lossy compression for silicon debug," in *DATE*, April 2007.

[5] K. J. Balakrishnan, N. A. Touba, and S. Patil, "Compressing functional tests for microprocessors," in *ATS*, December 2005.

[6] D. Burger and T. Austin, "The Simplescalar tool set, version 2.0," Technical Report cs-tr-97-1342, June 1997.

[7] J. Edler and M. Hill, "Dinero IV Trace-Driven Uniprocessor Cache Simulator." http://www.cs.wisc.edu/ markhill/DineroIV/.

[8] H. Fang, C. Tong, B. Yao, X. Song, and X. Cheng, "CacheCompress: a novel approach for test data compression with cache for IP embedded cores," in *ICCAD*, November 2007.

[9] D. Josephson, "The good, the bad, and the ugly of silicon debug", in *DAC*, June 2008.

[10] M. Kjelso, M. Gooch, and S. Jones, "Design and Performance of a Main Memory Hardware Data Compressor," in *22nd EUROMICRO*, September 1996.

[11] J.-S. Lee, W.-K. Hong, and S.-D. Kim, "Design and evaluation of a selective compressed memory system," in *ICCD*, October 1999.

[12] H. Lekatsas, J. Henkel, and W. Wolf, "A decompression architecture for low power embedded systems," in *ICCD*, September 2000.

[13] S.-B. Park and S. Mitra, "IFRA: instruction footprint recording and analysis for post-silicon bug localization in processors", in *DAC*, June 2008.

[14] P. Shivakumar and N. Jouppi, "CACTI 3.0: An integrated cache timing, power and area model," WRL Research Rep., August 2001.

[15] B. Vermeulen, M. Z. Urfianto, and S. K. Goel, "Automatic generation of breakpoint hardware for silicon debug," in *DAC*, June 2004.

[16] A. Vishnoi, P. R. Panda, and M. Balakrishnan, "Cache aware compression for processor debug support", in *DATE*, April 2009.

[17] I. Wagner and V. Bertacco, "Reversi: Post-silicon validation system for modern microprocessors", in *ICCD*, October 2008.

[18] T. A. Welch, "A technique for high-performance data compression," *IEEE Computer*, vol. 17, June 1984.

[19] C. Zhang, F. Vahid, J. Yang, and W. A. Najjar, "A way-halting cache for low-energy high-performance systems," *TACO*, 2(1), March 2005.

[20] C. Zhang, F. Vahid, and W. A. Najjar, "A highly-configurable cache architecture for embedded systems," in *ISCA*, June 2003.

Finding Deterministic Solution from Underdetermined Equation: Large-Scale Performance Modeling by Least Angle Regression

Xin Li

ECE Department, Carnegie Mellon University
5000 Forbs Avenue, Pittsburgh, PA 15213
xinli@ece.cmu.edu

ABSTRACT

The aggressive scaling of IC technology results in high-dimensional, strongly-nonlinear performance variability that cannot be efficiently captured by traditional modeling techniques. In this paper, we adapt a novel L_1-norm regularization method to address this modeling challenge. Our goal is to solve a large number of (e.g., $10^4 \sim 10^6$) model coefficients from a small set of (e.g., $10^2 \sim 10^3$) sampling points without over-fitting. This is facilitated by exploiting the underlying sparsity of model coefficients. Namely, although numerous basis functions are needed to span the high-dimensional, strongly-nonlinear variation space, only a few of them play an important role for a given performance of interest. An efficient algorithm of least angle regression (LAR) is applied to automatically select these important basis functions based on a limited number of simulation samples. Several circuit examples designed in a commercial 65nm process demonstrate that LAR achieves up to 25× speedup compared with the traditional least-squares fitting.

Categories and Subject Descriptors

B.7.2 [**Integrated Circuits**]: Design Aids – Verification

General Terms

Algorithms

Keywords

Process Variation, Response Surface Modeling

1. INTRODUCTION

As IC technologies scale to 65nm and beyond, process variation becomes increasingly critical and makes it continually more challenging to create a reliable, robust design with high yield [1]. For analog/mixed-signal circuits designed for sub-65nm technology nodes, parametric yield loss due to manufacturing variation becomes a significant or even dominant portion of the total yield loss. Hence, process variation must be carefully considered within today's IC design flow.

Unlike most digital circuits that can be efficiently analyzed at gate level (e.g., by statistical timing analysis), analog/mixed-signal circuits must be modeled and simulated at transistor level. To estimate the performance variability of these circuits, response surface modeling (RSM) has been widely applied [2]-[6]. The objective of RSM is to approximate the circuit performance (e.g., delay, gain, etc.) as an analytical (either linear or nonlinear) function of device parameters (e.g., V_{TH}, T_{OX}, etc.). Once response surface models are created, they can be used for various purposes, e.g., efficiently predicting performance distributions [7].

Permission to make digital or hard copies of part or all of this work for personal or classroom use is granted without fee provided that copies are not made or distributed for profit or commercial advantage and that copies bear this notice and the full citation on the first page. To copy otherwise, to republish, to post on servers or to redistribute to lists, requires prior specific permission and/or a fee.

DAC'09, July 26-31, 2009, San Francisco, California, USA

While RSM was extensively studied in the past, the following two trends in advanced IC technologies suggest a need to revisit this area.

- **Strong nonlinearity**: As process variation becomes relatively large, simple linear RSM is not sufficiently accurate [7]. Instead, nonlinear (e.g., quadratic) models are required to accurately predict performance variability.

- **High dimensionality**: Random device mismatch becomes increasingly important due to technology scaling [1]. To accurately model this effect, a large number of random variables must be utilized, rendering a high-dimensional variation space [6].

The combination of these two recent trends results in a large-scale RSM problem that is difficult to solve. For instance, as will be demonstrated in Section 5, more than 10^4 independent random variables must be used to model the device-level variation of a simplified SRAM critical path designed in a commercial 65nm CMOS process. To create a quadratic model for the critical path delay, we must determine a $10^4 \times 10^4$ quadratic coefficient matrix including 10^8 coefficients!

Most existing RSM techniques [2]-[5] rely on least-squares (LS) fitting. They solve model coefficients from an over-determined equation and, hence, the number of sampling points must be equal to or greater than the number of model coefficients. Since each sampling point is created by expensive transistor-level simulation, such high simulation cost prevents us from fitting high-dimensional, nonlinear models where a great number of sampling points are required. While the existing RSM techniques have been successfully applied to small-size or medium-size problems (e.g., 10~1000 model coefficients), they are ill-equipped to address the modeling needs of today's analog/mixed-signal system where $10^4 \sim 10^6$ model coefficients must be solved. The challenging issue is how to make RSM affordable for such a *large* problem size.

In this paper, we propose a novel RSM technique that aims to solve a large number of (e.g., $10^4 \sim 10^6$) model coefficients from a small set of (e.g., $10^2 \sim 10^3$) sampling points without over-fitting. While numerous basis functions must be used to span the high-dimensional, strongly-nonlinear variation space, not all these functions play an important role for a given performance of interest. In other words, although there are a large number of unknown model coefficients, many of these coefficients are close to zero, rendering a unique *sparse* structure. Taking the 65nm SRAM in Section 5 as an example, the delay variation of its critical path can be accurately approximated by around 50 basis functions, even though the SRAM circuit contains 21310 independent random variables! However, we do not know the right basis functions in advance; these important basis functions must be automatically selected by a "smart" algorithm based on a limited number of simulation samples.

Our proposed RSM algorithm borrows the recent advance of statistics [8] to explore the underlying sparsity of model coefficients. It applies *L_1-norm regularization* [8] to find the unique sparse solution (i.e., the model coefficients) of an underdetermined equation. Importantly, the proposed L_1-norm regularization is formulated as a convex optimization problem and, therefore, can be solved both robustly and efficiently.

An important contribution of this paper is to apply an efficient algorithm of *Least Angle Regression* (LAR [8]) to solve the L_1-norm regularization problem. For our RSM application, LAR is substantially more efficient than the well-known interior-point method [13] that was developed for general-purpose convex optimization. In addition, LAR results in more accurate response surface models than the statistical regression (STAR) algorithm proposed in [6], which can be proven by both theoretical analyses [10] and numerical experiments. Compared with STAR, LAR reduces modeling error by 1.5~3× with negligible computational overhead, as will be demonstrated by the numerical examples in Section 5.

The remainder of this paper is organized as follows. In Section 2, we review the background on principal component analysis and response surface modeling, and propose our L_1-norm regularization in Section 3. The LAR algorithm is used to efficiently solve all model coefficients in Section 4. The efficacy of LAR is demonstrated by several numerical examples in Section 5, followed by the conclusions in Section 6.

2. BACKGROUND

2.1 Principal Component Analysis

Given N process parameters $X = [x_1\ x_2\ ...\ x_N]^T$, the process variation $\Delta X = X - X_0$, where X_0 denotes the mean value of X, is often modeled by multiple zero-mean, correlated Normal distributions [2]-[7]. Principal component analysis (PCA) [11] is a statistical method that finds a set of independent factors to represent the correlated Normal distributions. Assume that the correlation of ΔX is represented by a symmetric, positive semi-definite covariance matrix R. PCA decomposes R as [11]:

$$R = U \cdot \Sigma \cdot U^T \qquad (1)$$

where $\Sigma = diag(\lambda_1, \lambda_2, ..., \lambda_N)$ contains the eigenvalues of R, and $U = [U_1\ U_2\ ...\ U_N]$ contains the corresponding eigenvectors that are orthonormal, i.e., $U^T U = I$. (I is the identity matrix.) PCA defines a set of new random variables $\Delta Y = [\Delta y_1\ \Delta y_2\ ...\ \Delta y_N]^T$:

$$\Delta Y = \Sigma^{-0.5} \cdot U^T \cdot \Delta X . \qquad (2)$$

The new random variables in ΔY are called the principal components. It is easy to verify that all principal components in ΔY are independent and standard Normal (i.e., zero mean and unit variance). More details on PCA can be found in [11].

2.2 Response Surface Modeling

Given a circuit design, the circuit performance f (e.g., delay, gain, etc.) is a function of the process variation ΔY defined in (2). RSM approximates the performance function $f(\Delta Y)$ as the linear combination of M basis functions [2]-[6]:

$$f(\Delta Y) \approx \sum_{i=1}^{M} \alpha_i \cdot g_i(\Delta Y) \qquad (3)$$

where $\{\alpha_i;\ i = 1,2,...,M\}$ are the model coefficients, and $\{g_i(\Delta Y);\ i = 1,2,...,M\}$ are the basis functions (e.g., linear, quadratic, etc.). The unknown model coefficients in (3) can be determined by solving the following linear equation at K sampling points:

$$G \cdot \alpha = F \qquad (4)$$

where

$$G = \begin{bmatrix} g_1(\Delta Y^{(1)}) & g_2(\Delta Y^{(1)}) & \cdots & g_M(\Delta Y^{(1)}) \\ g_1(\Delta Y^{(2)}) & g_2(\Delta Y^{(2)}) & \cdots & g_M(\Delta Y^{(2)}) \\ \vdots & \vdots & \vdots & \vdots \\ g_1(\Delta Y^{(K)}) & g_2(\Delta Y^{(K)}) & & g_M(\Delta Y^{(K)}) \end{bmatrix} \qquad (5)$$

$$\alpha = [\alpha_1\ \ \alpha_2\ \ \cdots\ \ \alpha_M]^T \qquad (6)$$

$$F = [f^{(1)}\ \ f^{(2)}\ \ \cdots\ \ f^{(K)}]^T . \qquad (7)$$

In (5)-(7), $\Delta Y^{(k)}$ and $f^{(k)}$ are the values of ΔY and $f(\Delta Y)$ at the k-th sampling point respectively. Without loss of generality, we assume that all basis functions are normalized:

$$G_i^T G_i = 1 \quad (i = 1,2,\cdots,M) \qquad (8)$$

where

$$G_i = [g_i(\Delta Y^{(1)})\ \ g_i(\Delta Y^{(2)})\ \ \cdots\ \ g_i(\Delta Y^{(K)})]^T . \qquad (9)$$

This assumption simplifies the notation of our discussion in the following sections.

Most existing RSM techniques [2]-[5] attempt to solve the least-squares (LS) solution for (4). Hence, the number of samples (K) must be equal to or greater than the number of coefficients (M). It, in turn, becomes intractable, if M is large (e.g., 10^4~10^6). For this reason, the traditional RSM techniques are limited to small-size or medium-size problems (e.g., 10~1000 model coefficients). In this paper, we propose a novel RSM algorithm that aims to create high-dimensional, strongly-nonlinear response surface models (e.g., 10^4~10^6 model coefficients) from a small set of (e.g., 10^2~10^3) simulation samples without over-fitting.

3. L_1-NORM REGULARIZATION

Our proposed RSM technique utilizes a novel L_1-norm regularization scheme that is derived from advanced statistics theories [8]. In this section, we describe its mathematical formulation and highlight the novelties.

3.1 Mathematical Formulation

Unlike the traditional RSM techniques that solve model coefficients from an over-determined equation, we focus on the nontrivial case where the number of samples (K) is less than the number of coefficients (M). Namely, there are fewer equations than unknowns, and the linear system in (4) is underdetermined. In this case, the solution α (i.e., the model coefficients) is not unique, unless additional constraints are added.

In this paper, we will explore the sparsity of α to uniquely determine its value. Our approach is motivated by the observation that while a large number of basis functions must be used to span the high-dimensional, nonlinear variation space, only a few of them are required to approximate a specific performance function. In other words, the vector α in (6) only contains a small number of non-zeros. However, we do not know the exact locations of these non-zeros. We will propose a novel L_1-norm regularization scheme to find the non-zeros so that the solution α of the underdetermined equation (4) can be uniquely solved.

To derive the proposed L_1-norm regularization, we first show the idea of L_0-norm regularization. To this end, we formulate the following optimization to solve the sparse solution α for (4):

$$\begin{aligned} &\underset{\alpha}{\text{minimize}} && \|G \cdot \alpha - F\|_2^2 \\ &\text{subject to} && \|\alpha\|_0 \le \lambda \end{aligned} \qquad (10)$$

where $\|\bullet\|_2$ and $\|\bullet\|_0$ stand for the L_2-norm and L_0-norm of a vector, respectively. The L_0-norm $\|\alpha\|_0$ equals the number of non-zeros in

978-1-60558-497-3/09 $25.00 © 2009 ACM

the vector α. It measures the sparsity of α. Therefore, by directly constraining the L_0-norm, the optimization in (10) attempts to find a sparse solution α that minimizes the least-squares error.

The parameter λ in (10) explores the tradeoff between the sparsity of the solution α and the minimal value of the cost function $\|G \cdot \alpha - F\|_2^2$. For instance, a large λ will result in a small cost function, but meanwhile it will increase the number of non-zeros in α. It is important to note that a small cost function does not necessarily mean a small modeling error. Even though the minimal cost function value can be reduced by increasing λ, such a strategy may result in over-fitting especially because Eq. (4) is underdetermined. In the extreme case, if λ is sufficiently large and the constraint in (10) is not active, we can always find a solution α to make the cost function exactly zero. However, such a solution is likely to be useless, since it over-fits the given sampling points. In practice, the optimal value of λ can be automatically determined by cross-validation, as will be discussed in detail in Section 4.

While the L_0-norm regularization can effectively guarantee a sparse solution α, the optimization in (10) is NP hard [8] and, hence, is extremely difficult to solve. A more efficient technique to find sparse solution is based on L_1-norm regularization – a relaxed version of L_0-norm:

$$\begin{aligned} \underset{\alpha}{\text{minimize}} \quad & \|G \cdot \alpha - F\|_2^2 \\ \text{subject to} \quad & \|\alpha\|_1 \leq \lambda \end{aligned} \qquad (11)$$

where $\|\alpha\|_1$ denotes the L_1-norm of the vector α, i.e., the summation of the absolute values of all elements in α:

$$\|\alpha\|_1 = |\alpha_1| + |\alpha_2| + \cdots + |\alpha_M|. \qquad (12)$$

The L_1-norm regularization in (11) can be re-formulated as a convex optimization problem. Introduce a set of slack variables $\{\beta_i; i = 0,1,...,M\}$ and re-write (11) into the following equivalent form [13]:

$$\begin{aligned} \underset{\alpha, \beta}{\text{minimize}} \quad & \|G \cdot \alpha - F\|_2^2 \\ \text{subject to} \quad & \beta_1 + \beta_2 + \cdots + \beta_M \leq \lambda \\ & -\beta_i \leq \alpha_i \leq \beta_i \qquad (i = 1,2,\cdots,M) \end{aligned} \qquad (13)$$

In (13), the cost function is quadratic and positive semi-definite. Hence, it is convex. All constraints are linear and, therefore, the resulting constraint set is a convex polytope. For these reasons, the L_1-norm regularization in (13) is a convex optimization problem and it can be solved by various efficient and robust algorithms, e.g., the interior-point method [13].

The aforementioned L_1-norm regularization is much more computationally efficient than the L_0-norm regularization that is NP hard. This is the major motivation to replace L_0-norm by L_1-norm. In the next sub-section, we will use a two-dimensional example to intuitively explain why solving the L_1-norm regularization in (11) yields a sparse solution α.

3.2 Geometrical Explanation

To understand the connection between L_1-norm regularization and sparse solution, we consider the two-dimensional example (i.e., $\alpha = [\alpha_1 \; \alpha_2]^T$) in Fig. 1. Since the cost function $\|G \cdot \alpha - F\|_2^2$ is quadratic and positive semi-definite, its contour lines can be represented by multiple ellipsoids. On the other hand, the constraint $\|\alpha\|_1 \leq \lambda$ corresponds to a number of rotated squares, associated with different values of λ. For example, two of such squares are shown in Fig. 1, where $\lambda_1 \leq \lambda_2$.

Studying Fig. 1, we would notice that if λ is large (e.g., $\lambda = \lambda_2$), both α_1 and α_2 are not zero. However, as λ decreases (e.g., $\lambda =$

λ_1), the contour of $\|G \cdot \alpha - F\|_2^2$ eventually intersects the polytope $\|\alpha\|_1 \leq \lambda$ at one of its vertex. It, in turn, implies that one of the coefficients (i.e., α_1 in this case) becomes exactly zero. From this point of view, by decreasing λ of the L_1-norm regularization in (11), we can pose a strong constraint for sparsity and force a sparse solution. This intuitively explains why L_1-norm regularization guarantees sparsity, as is the case for L_0-norm regularization.

In addition, various theoretical studies from the statistics community demonstrate that under some general assumptions, both L_1-norm regularization and L_0-norm regularization result in the same solution [9]. Roughly speaking, if the M-dimensional vector α contains L non-zeros and the linear equation $G \cdot \alpha = F$ is well-conditioned, the solution α can be uniquely determined by L_1-norm regularization from K sampling points, where K is in the order of $O(L \cdot \log M)$ [9]. Note that K (the number of sampling points) is a logarithm function of M (the number of unknown coefficients). It, in turn, provides the theoretical foundation that by solving the sparse solution of an underdetermined equation, a large number of model coefficients can be uniquely determined from a small number of sampling points.

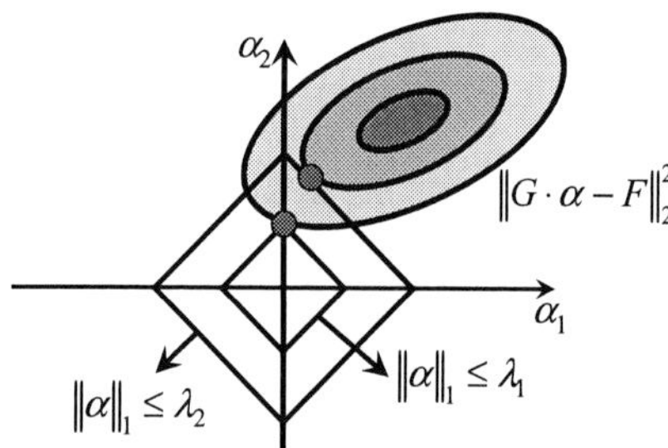

Fig. 1. The proposed L_1-norm regularization $\|\alpha\|_1 \leq \lambda$ results in a sparse solution (i.e., $\alpha_1 = 0$) if λ is sufficiently small (i.e., $\lambda = \lambda_1$).

4. LEAST ANGLE REGRESSION

While Eq. (11) gives the mathematical formulation of L_1-norm regularization, a number of implementation issues must be carefully considered to make it of practical utility. Most importantly, an efficient algorithm is required to automatically determine the optimal value of λ. Towards this goal, a two-step approach can be used: (a) solve the optimization in (11) for a set of different λ's, and (b) select the optimal λ by cross-validation. In this section, we will introduce an efficient algorithm of least angle regression (LAR [8]) to accomplish these two steps with minimal computational cost.

4.1 Piece-wise Linear Solution Trajectory

To solve the optimization in (11) for different λ's, one straightforward approach is to repeatedly apply the interior-point method [13] to solve the convex programming problem in (13). This approach, however, is computationally expensive, as we must run a convex solver for many times in order to visit a sufficient number of possible values of λ. Instead of applying the interior-point method, we propose to first explore the unique property of the L_1-norm regularization in (11) and minimize the number of the possible λ's that we must visit.

As discussed in Section 3.2, the sparsity of the solution α depends on the value of λ. In the extreme case, if λ is zero, all coefficients in α are equal to zero. As λ gradually increases, more and more coefficients in α become non-zero. In fact, it can be proven that the solution α of (11) is a piece-wise linear function of

978-1-60558-497-3/09 $25.00 © 2009 ACM

366

λ [8]. To intuitively illustrate this concept, we consider the following simple example:

$$f(\Delta Y) = -0.43 \cdot \Delta y_1 - 1.66 \cdot \Delta y_2 + 0.12 \cdot \Delta y_3 \\ + 0.28 \cdot \Delta y_4 - 1.14 \cdot \Delta y_5 \quad (14)$$

We collected 50 random sampling points for this function and solved the L_1-norm regularization in (11) to calculate the values of α associated with different λ's. Fig. 2 shows the solution trajectory $\alpha(\lambda)$, i.e., α as a function of λ, which is piece-wise linear. The details of the mathematical proof for this piece-wise linear property can be found in [8].

The aforementioned piece-wise linear property allows us to find the entire solution trajectory $\alpha(\lambda)$ with low computational cost. We do not have to repeatedly solve the L_1-norm regularization at many different λ's. Instead, we only need to estimate the local linear function in each interval $[\lambda_i, \lambda_{i+1}]$. Next, we will show an iterative algorithm to efficiently find the solution trajectory $\alpha(\lambda)$.

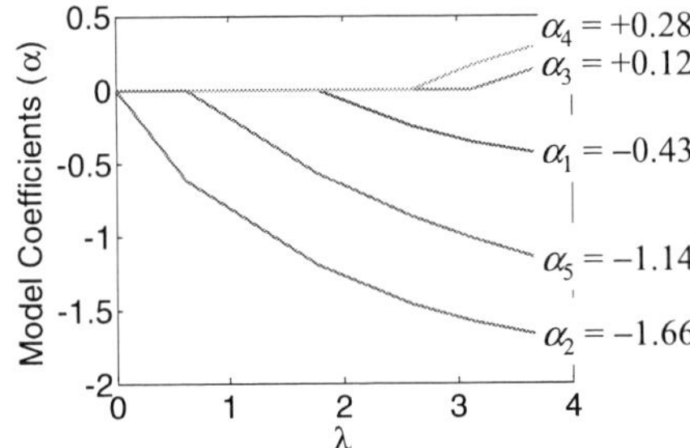

Fig. 2. The solution trajectory $\alpha(\lambda)$ of the L_1-norm regularization in (11) is a piece-wise linear function of λ.

4.2 Iterative Algorithm

We start from the extreme case where λ is zero for the L_1-norm regularization in (11). In this case, the solution of (11) is trivial: $\alpha = 0$. Our focus of this sub-section is to present an efficient algorithm of least angle regression (LAR) to calculate the solution trajectory $\alpha(\lambda)$, as λ increases from zero. To this end, we re-write the linear equation $G \cdot \alpha = F$ in (4)-(9) as:

$$F = \alpha_1 G_1 + \alpha_2 G_2 + \cdots + \alpha_M G_M . \quad (15)$$

Eq. (15) represents the vector F (i.e., the performance values) as the linear combination of the vectors $\{G_i; i = 1,2,...,M\}$ (i.e., the basis function values). Each G_i corresponds to a basis function $g_i(\Delta Y)$. As λ increases from zero, LAR [8] first calculates the correlation between F and every G_i:

$$r_i = \left| G_i^T F \right| \quad (i = 1,2,\cdots,M) \quad (16)$$

where G_i is a unit-length vector (i.e., $G_i^T G_i = 1$) as shown in (8). Next, LAR finds the vector G_{s1} that is most correlated with F, i.e., r_{s1} takes the largest value. Once G_{s1} is identified, LAR approximates F in the direction of G_{s1}:

$$F \approx \gamma_1 G_{s1} . \quad (17)$$

At this first iteration step, since we only use the basis function $g_{s1}(\Delta Y)$ to approximate the performance function $f(\Delta Y)$, the coefficients for all other basis functions (i.e., $\{\alpha_i; i \neq s1\}$) are zero. The residual of the approximation is:

$$Res = F - \gamma_1 G_{s1} . \quad (18)$$

To intuitively understand the LAR algorithm, we consider the two-dimensional example shown in Fig. 3. In this example, the vector G_2 has a higher correlation with F than the vector G_1. Hence, G_2 is selected to approximate F, i.e., $F \approx \gamma_1 G_2$. From the geometrical point of view, finding the largest correlation is

equivalent to finding the least angle between the vectors $\{G_i; i = 1,2,...,M\}$ and the performance F. Therefore, the aforementioned algorithm is referred to as least angle regression in [8].

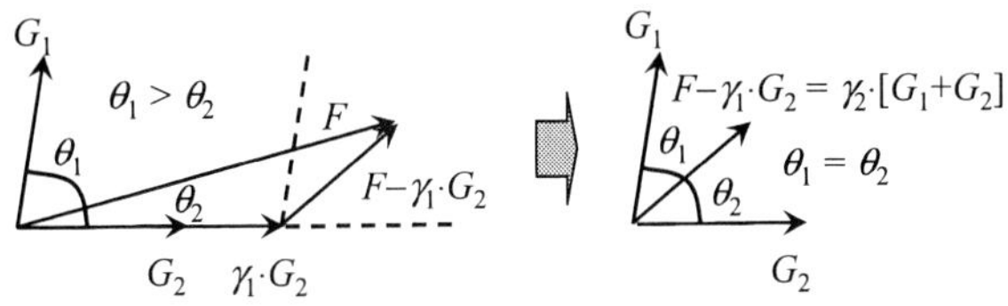

Iteration 1: $F \approx \alpha_2 G_2$ where $\alpha_2 = \gamma_1$

Iteration 2: $F \approx \alpha_1 G_1 + \alpha_2 G_2$ where $\alpha_1 = \gamma_2$ & $\alpha_2 = \gamma_1 + \gamma_2$

Fig. 3. LAR calculates the solution trajectory $\alpha(\lambda)$ of a two-dimensional example $F = \alpha_1 G_1 + \alpha_2 G_2$.

As $|\gamma_1|$ increases, the correlation between the vector G_{s1} and the residual $Res = F - \gamma_1 G_{s1}$ decreases. LAR uses an efficient algorithm to compute the maximal value of $|\gamma_1|$ at which the correlation between G_{s1} and $F - \gamma_1 G_{s1}$ is no longer dominant. In other words, there is another vector G_{s2} that has the same correlation with the residual:

$$\left| G_{s1}^T \cdot (F - \gamma_1 G_{s1}) \right| = \left| G_{s2}^T \cdot (F - \gamma_1 G_{s1}) \right| . \quad (19)$$

At this point, instead of continuing along G_{s1}, LAR proceeds in a direction equiangular between G_{s1} and G_{s2}. Namely, it approximates F by the linear combination of G_{s1} and G_{s2}:

$$F \approx \gamma_1 G_{s1} + \gamma_2 \cdot (G_{s1} + G_{s2}) \quad (20)$$

where the coefficient γ_1 is fixed at this second iteration step.

Taking Fig. 3 as an example, the residual $F - \gamma_1 G_2$ is approximated by $\gamma_2 \cdot (G_1 + G_2)$. If $|\gamma_2|$ is sufficiently large, F is exactly equal to $F = \gamma_2 G_1 + (\gamma_1 + \gamma_2) \cdot G_2$. In this example, because only two basis functions $g_1(\Delta Y)$ and $g_2(\Delta Y)$ are used, LAR stops at the second iteration step. If more than two basis functions are involved, LAR will keep increasing $|\gamma_2|$ until a third vector G_{s3} earns its way into the "most correlated" set, and so on. Algorithm 1 summarizes the major iteration steps of LAR.

Algorithm 1: Least Angle Regression (LAR)
1. Start from the vector F defined in (7) and the normalized vectors $\{G_i; i = 1,2,...,M\}$ defined in (8)-(9).
2. Apply (16) to calculate the correlation $\{r_i; i = 1,2,...,M\}$.
3. Select the vector G_s that has the largest correlation r_s.
4. Let the set $Q = \{G_s\}$ and the iteration index $p = 1$.
5. Approximate F by:

$$F \approx \gamma_p \cdot \sum_{G_i \in Q} G_i . \quad (21)$$

6. Calculate the residual:

$$Res = F - \gamma_p \cdot \sum_{G_i \in Q} G_i . \quad (22)$$

7. Use the algorithm in [8] to determine the maximal $|\gamma_p|$ such that either the residual in (22) equals 0 or another vector G_{new} ($G_{new} \notin Q$) has as much correlation with the residual:

$$\left| G_{new}^T \cdot Res \right| = \left| G_i^T \cdot Res \right| \quad (\forall G_i \in Q). \quad (23)$$

8. If $Res = 0$, stop. Otherwise, $Q = Q \cup \{G_{new}\}$, $F = Res$, $p = p+1$, and go to Step 5.

It can be proven that with several small modifications, LAR will generate the entire piece-wise linear solution trajectory $\alpha(\lambda)$ for the L_1-norm regularization in (11) [8]. The computational cost of LAR is similar to that of applying the interior-point method to solve a *single* convex optimization in (11) with a fixed λ value.

Therefore, compared to the simple approach that repeatedly solves (11) for multiple λ's, LAR typically achieves orders of magnitude more efficiency, as is demonstrated in [8].

4.3 Cross-Validation

Once the solution trajectory $\alpha(\lambda)$ is extracted, we need to further find the optimal λ that minimizes the modeling error. To avoid over-fitting, we cannot simply measure the modeling error from the same sampling data that are used to calculate the model coefficients. Instead, modeling error must be measured from an independent data set. Cross-validation is an efficient method for model validation that has been widely used in the statistics community [12]. An S-fold cross-validation partitions the entire data set into S groups, as shown by the example in Fig. 4. Modeling error is estimated from S independent runs. In each run, one of the S groups is used to estimate the modeling error and all other groups are used to calculate the model coefficients. Different groups should be selected for error estimation in different runs. As such, each run results in an error value ε_i ($i = 1,2,...,S$) that is measured from a unique group of sampling points. In addition, when a model is trained and tested in each run, non-overlapped data sets are used so that over-fitting can be easily detected. The final modeling error is computed as the average of $\{\varepsilon_i; i = 1,2,...,S\}$, i.e., $\varepsilon = (\varepsilon_1+\varepsilon_2+...+\varepsilon_S)/S$.

For our application, LAR is used to calculate the solution trajectory during each cross-validation run. Next, the modeling error associated with each run is estimated, resulting in $\{\varepsilon_i(\lambda); i = 1,2,...,S\}$. Note that ε_i is not simply a value, but a one-dimensional function of λ. Once all cross-validation runs are complete, the final modeling error is calculated as $\varepsilon(\lambda) = (\varepsilon_1(\lambda)+\varepsilon_2(\lambda)+...+\varepsilon_S(\lambda))/S$, which is again a one-dimensional function of λ. The optimal λ is then determined by finding the minimal value of $\varepsilon(\lambda)$.

The major drawback of cross-validation is the need to repeatedly extract the model coefficients for S times. However, for our circuit modeling application, the overall computational cost is dominated by the transistor-level simulation that is required to generate sampling data. Hence, the computational overhead by cross-validation is almost negligible, as will be demonstrated by our numerical examples in Section 5.

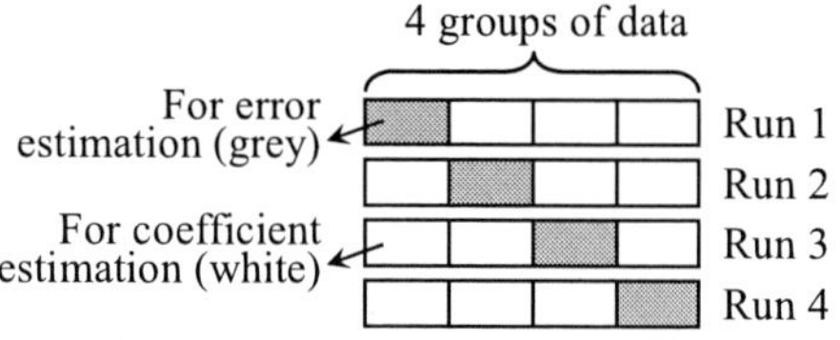

Fig. 4. A 4-fold cross-validation partitions the data set into 4 groups and modeling error is estimated from 4 independent runs.

5. NUMERICAL EXAMPLES

In this section we demonstrate the efficacy of LAR using several circuit examples. For each example, two independent random sampling sets, called training set and testing set respectively, are generated using Cadence Spectre. The training set is used for coefficient fitting (including cross-validation), while the testing set is used for model validation. All numerical experiments are performed on a 2.8GHz Linux server.

5.1 Two-Stage Operational Amplifier

Fig. 5 shows the simplified circuit schematic of a two-stage operational amplifier (OpAmp) designed in a commercial 65nm process. In this example, we aim to model four performance metrics: gain, bandwidth, offset and power. The inter-die/intra-die variations of both MOS transistors and layout parasitics are considered. After PCA based on foundry data, 630 independent random variables are extracted to model these variations.

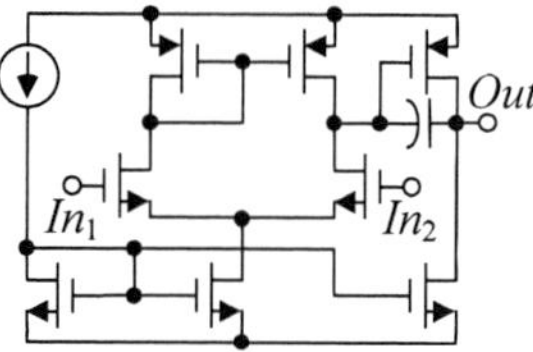

Fig. 5. Simplified circuit schematic of an operational amplifier.

A. Linear Performance Modeling

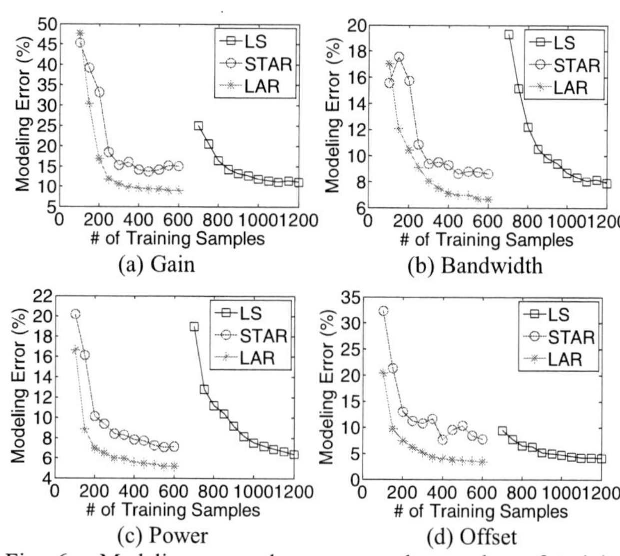

(a) Gain (b) Bandwidth (c) Power (d) Offset

Fig. 6. Modeling error decreases as the number of training samples increases.

Table 1. Linear performance modeling cost for OpAmp

	LS	STAR	LAR
# of Samples	1200	600	600
Spectre (Sec.)	16140	8070	8070
Fitting (Sec.)	2.6	1.2	44.2
Total (Sec.)	16142	8071	8114

Fig. 6 shows the modeling error for three different techniques: least-squares fitting (LS), STAR [6] and LAR. To achieve the same accuracy, both STAR and LAR require much less training samples than LS, because they do not solve the unknown model coefficients from an over-determined equation. On the other hand, given the same number of training samples, LAR yields better accuracy (up to 2~3× error reduction) than STAR. STAR is similar to the orthogonal matching pursuit algorithm developed for signal processing [10]. It has been theoretically proven that the L_1-norm regularization used by LAR is more accurate, but also more expensive, than the orthogonal matching pursuit used by STAR [10]. However, for our circuit modeling application, the overall modeling cost is dominated by the Spectre simulation time that is required to generate sampling points. Therefore, the computational overhead of LAR is negligible, as shown in Table 1. LAR achieves 2× speedup compared with LS in this example.

978-1-60558-497-3/09 $25.00 © 2009 ACM

B. Quadratic Performance Modeling

To further improve accuracy, we select 200 most important process parameters based on the magnitude of the linear model coefficients. Next, we create quadratic performance models for these critical process parameters. In this example, the 200-dimensional quadratic model contains 20301 unknown coefficients. Compared with STAR, LAR reduces the modeling error by 1.5~3×, as shown in Table 2. In addition, compared with LS, LAR reduces the modeling time from 4 days to 4 hours (24× speedup) while achieving similar accuracy, as shown in Table 3.

Table 2. Quadratic performance modeling error for OpAmp

	LS	STAR	LAR
Gain	4.21%	8.03%	5.77%
Bandwidth	3.84%	5.36%	4.11%
Power	1.52%	4.37%	1.69%
Offset	3.69%	9.15%	2.94%

Table 3. Quadratic performance modeling cost for OpAmp

	LS	STAR	LAR
# of Samples	25000	1000	1000
Spectre (Sec.)	336250	13450	13450
Fitting (Sec.)	51562	92	1449
Total (Sec.)	387812	13542	14899

5.2 Simplified SRAM Read Path

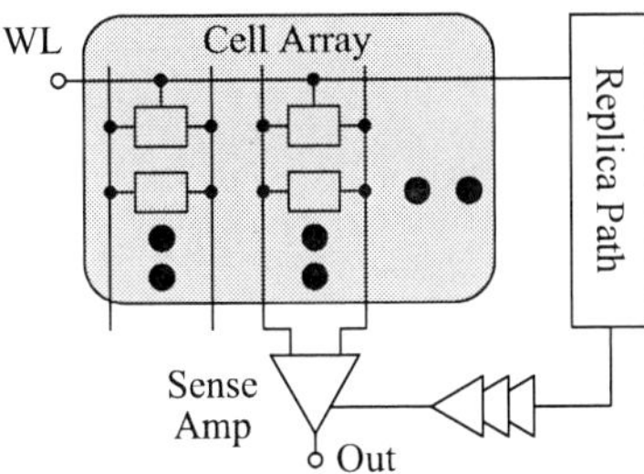

Fig. 7. Simplified circuit schematic of an SRAM read path.

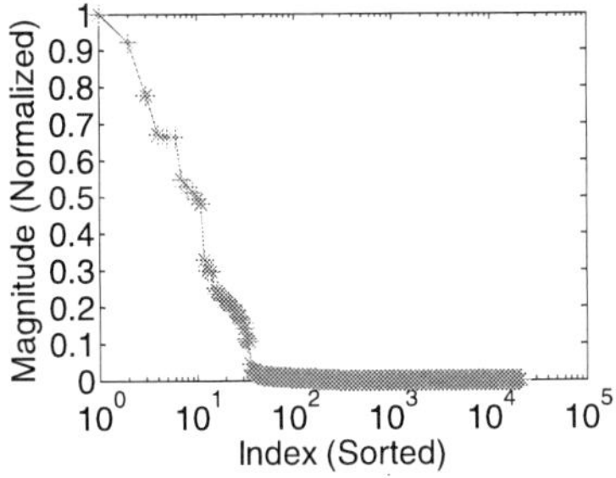

Fig. 8. Magnitude of the model coefficients estimated by LAR.

Table 4. Linear performance modeling error and cost for SRAM

	LS	STAR	LAR
Modeling Error	9.78%	6.34%	4.94%
# of Samples	25000	1000	1000
Spectre (Sec.)	728250	29130	29130
Fitting (Sec.)	13856.1	26.5	338.3
Total (Sec.)	742106	29156	29468

Shown in Fig. 7 is the simplified circuit schematic of an SRAM read path designed in a commercial 65nm process. The read path contains cell array, replica path for self-timing and sense amplifier. In this example, the performance of interest is the delay from the word line (WL) to the sense amplifier output (Out). Both inter-die and intra-die variations are considered. After PCA based on foundry data, 21310 independent random variables are extracted to model these variations.

Three different techniques are implemented for linear performance modeling: least-squares fitting (LS), STAR [6] and LAR. As shown in Table 4, LAR is most accurate among these three methods. Compared with LS, LAR reduces the modeling time from 8.6 days to 8.2 hours (25× speedup). Fig. 8 shows the magnitude of the linear model coefficients estimated by LAR. Even though there are 21311 basis functions in total, only 50 model coefficients are not close to zero. These 50 basis functions are automatically selected by LAR to accurately approximate the performance of interest in this example.

6. CONCLUSIONS

In this paper, we propose a novel L_1-norm regularization to efficiently create high-dimensional linear/nonlinear performance models for nanoscale circuits. The proposed method is facilitated by exploring the unique sparse structure of model coefficients. An efficient algorithm of least angle regression (LAR) is used to solve the proposed L_1-norm regularization problem. Our numerical examples demonstrate that compared with least-square fitting, LAR achieves up to 25× runtime speedup without surrendering any accuracy. LAR can be incorporated into a robust circuit design flow for efficient yield prediction and optimization.

7. ACKNOWLEDGEMENTS

This work has been supported by the National Science Foundation under contract CCF–0811023.

8. REFERENCES

[1] Semiconductor Industry Associate, *International Technology Roadmap for Semiconductors*, 2007.

[2] X. Li, J. Le, L. Pileggi and A. Strojwas, "Projection-based performance modeling for inter/intra-die variations," *IEEE ICCAD*, pp. 721-727, 2005.

[3] Z. Feng and P. Li, "Performance-oriented statistical parameter reduction of parameterized systems via reduced rank regression," *IEEE ICCAD*, pp. 868-875, 2006.

[4] A. Singhee and R. Rutenbar, "Beyond low-order statistical response surfaces: latent variable regression for efficient, highly nonlinear fitting," *IEEE DAC*, pp. 256-261, 2007.

[5] A. Mitev, M. Marefat, D. Ma and J. Wang, "Principle Hessian direction based parameter reduction for interconnect networks with process variation," *IEEE ICCAD*, pp. 632-637, 2007.

[6] X. Li and H. Liu, "Statistical regression for efficient high-dimensional modeling of analog and mixed-signal performance variations," *IEEE DAC*, pp. 38-43, 2008.

[7] X. Li, J. Le, P. Gopalakrishnan and L. Pileggi, "Asymptotic probability extraction for non-normal distributions of circuit performance," *IEEE ICCAD*, pp. 2-9, 2004.

[8] B. Efron, T. Hastie and I. Johnstone, "Least angle regression," *The Annals of Statistics*, vol. 32, no. 2, pp. 407-499, 2004.

[9] E. Candes, "Compressive sampling," *International Congress of Mathematicians*, 2006.

[10] J. Tropp and A. Gilbert, "Signal recovery from random measurements via orthogonal matching pursuit," *IEEE Trans. Information Theory*, vol. 53, no. 12, pp. 4655-4666, 2007.

[11] G. Seber, *Multivariate Observations*, Wiley Series, 1984.

[12] T. Hastie, R. Tibshirani and J. Friedman, *The Elements of Statistical Learning*, Springer, 2003.

[13] S. Boyd and L. Vandenberghe, *Convex Optimization*, Cambridge University Press, 2004.

A Robust and Efficient Harmonic Balance (HB) using Direct Solution of HB Jacobian

Amit Mehrotra
Berkeley Design Automation
amit.mehrotra@berkeley-da.com

Abhishek Somani
Berkeley Design Automation
asomani@berkeley-da.com

ABSTRACT

In this paper we introduce a new method of performing direct solution of the harmonic balance Jacobian. For examples with moderate number of harmonics and moderate to strong nonlinearities, we demonstrate that the direct solver has far superior performance with a moderate increase in memory compared to the best preconditioned iterative solvers. This solver is especially suited for Fourier envelope analysis where the number of harmonics is small, circuits are nonlinear and Jacobian bypass can be used for additional speed. For examples with large number of harmonics and moderate to strong nonlinearities, the performance advantage is maintained but the memory requirements increase. We propose efficient preconditioners based on direct solution of harmonic balance matrices which provide the user with a memory-speed trade-off.

Categories and Subject Descriptors

B.7.2 [**Integrated Circuits**]: Design Aids—*simulation*

General Terms

Algorithms, Performance

Keywords

Simulation, Harmonic Balance, Preconditioning

1. INTRODUCTION

Harmonic balance analysis (HB) has been the preferred simulation method for RF and microwave applications for a long time [18, 9, 7]. For circuits where the nonlinearities are moderate and the signal transitions are not very sharp, harmonic balance is the preferred method of simulation over time-domain techniques such as shooting Newton because it gives unmatched performance and accuracy [17, 5, 13]. Furthermore, RF and microwave applications rely heavily on measured frequency domain data such as s-parameters of linear networks. These can be directly simulated in frequency domain without requiring expensive convolutions or fitting. The drawback of traditional HB implementations is that the underlying Jacobian is large and not very sparse and it is very inefficient to factor and solve directly. Practical HB implementations tried to circumvent the problem by separating the linear and nonlinear portions of the circuit and formulating the problem so as to achieve balance between the harmonics of the waveforms of the two portions (hence the name harmonic balance). Since there is no coupling between harmonics in the linear portion, it can be arbitrarily large and

Permission to make digital or hard copies of part or all of this work for personal or classroom use is granted without fee provided that copies are not made or distributed for profit or commercial advantage and that copies bear this notice and the full citation on the first page. To copy otherwise, to republish, to post on servers or to redistribute to lists, requires prior specific permission and/or a fee.
DAC'09, July 26-31, 2009, San Francisco, California, USA

the limitation of this approach was the size and nonlinearity of the portion which contained all the nonlinear elements [18].

Krylov subspace methods addressed these limitations of HB [5, 12]. These iterative methods do not require the matrix to be explicitly formed, let alone factored. Instead they require the matrix-vector product to be formed at each iteration. As shown in Section 2, matrix-vector product with the HB Jacobian can be formed very efficiently using a series of FFTs and sparse matrix-vector multiplications with circuit matrices. The relative success of iterative methods with increasing circuit sizes resulted in restricting the use of traditional direct solution method to small circuits only.

However, the effectiveness of these iterative methods critically depends on selecting appropriate preconditioners, otherwise the number of iterations becomes prohibitively large. In addition, as the complexity and nonlinearity of circuits increases, Krylov subspace based HB engines require more advanced preconditioners [11] and still require a large number of iterations to reliably solve the system of linear equations, and in many cases, are unable to solve the system (See Section 3 for a brief description of existing preconditioners for HB Jacobian).

In this paper, we develop a direct method for solving linear systems that have the HB Jacobian as the coefficient matrix. Unlike previous such attempts which suffer from the problems mentioned earlier, we exploit the unique structure of the HB Jacobian and develop a highly specialized direct solver specifically for the problem (Section 4). We show that compared to the best preconditioned Krylov subspace methods, this direct solver gives competitive performance on circuits with mild nonlinearities and far superior performance and robustness for circuits with moderate to strong nonlinearities. We also show that this specialized direct solver gives far superior performance compared to a generic sparse direct solver. Moreover, for Fourier envelope applications, the direct solver lends itself very naturally to Jacobian bypass and therefore gives additional performance advantage as demonstrated in Section 5.

As the number of harmonics increases, as in three or more tone HB problems or users request a larger number of harmonics, the memory requirement of direct method based HB solvers increases rapidly. To address this issue, we developer new preconditioners for iterative methods based on the direct solver for HB Jacobian for problems which have a large number of harmonics (Section 6). These preconditioners provide a direct trade-off between performance and memory. In some cases they provide both speed and memory improvements.

2. HB PRELIMINARIES

We first review the basic theory of Harmonic Balance and introduce the notations we will use in this paper. Consider a non-autonomous circuit whose equations are given by

$$\int_{-\infty}^{t} y(t-s)x(s)\mathrm{d}s + \frac{\mathrm{d}q(x(t))}{\mathrm{d}t} + f(x(t)) + b(t) = 0 \quad (1)$$

where $x(t) \in \mathbb{R}^n$ represents the state variables in the circuit, n is the circuit size, y represents the matrix-valued impulse response

978-1-60558-497-3/09 $25.00 © 2009 ACM

function of frequency-domain linear elements (such as s parameters), $q : \mathbb{R}^n \to \mathbb{R}^n$ represents the nonlinear charge and flux storage in the circuit, $f : \mathbb{R}^n \to \mathbb{R}^n$ represents the memoryless nonlinearities and $b : \mathbb{R} \to \mathbb{R}^n$ represents the time-dependent excitations in the circuit which are assumed to be periodic with period T. Since the circuit is nonautonomous, the circuit steady-state response $x(t)$ will also be periodic with period T and its functions $q(x)$ and $f(x)$ are also T-periodic. Therefore all these waveforms can be expanded in Fourier series as follows:

$$b(t) = \sum_{i=-\infty}^{\infty} B_i \exp(j2\pi i f_0 t), \; x(t) = \sum_{i=-\infty}^{\infty} X_i \exp(j2\pi i f_0 t)$$

$$f(t) = \sum_{i=-\infty}^{\infty} F_i \exp(j2\pi i f_0 t), \text{ and } q(t) = \sum_{i=-\infty}^{\infty} Q_i \exp(j2\pi i f_0 t)$$

where $f_0 = \frac{1}{T}$.

(1) can be written in frequency-domain as

$$\sum_{i=-\infty}^{\infty} [Y_i X_i + j2\pi i f Q_i + F_i + B_i] \exp(j2\pi i f t) = 0$$

where $Y(f) = \int_{-\infty}^{\infty} y(t) \exp(-j2\pi f t) \mathrm{d}t$ is the Fourier transform of $y(t)$ and $Y_i = Y(if_0)$. Since $\exp(j2\pi i f_0 t)$ are orthogonal, it follows that

$$Y_i X_i + j2\pi i f_0 Q_i + F_i + B_i = 0$$

$\forall i$. In practice the infinite summations are truncated to a finite number of harmonics k. Collocating the above equation at $i \in [-k, k]$, we have

$$Y_{-k} X_{-k} + j2\pi(-k)f_0 Q_{-k} + F_{-k} + B_{-k} = 0$$
$$\cdots$$
$$Y_0 X_0 + j2\pi 0 f_0 Q_0 + F_0 + B_0 = 0$$
$$\cdots$$
$$Y_k X_k + j2\pi k f_0 Q_k + F_k + B_k = 0$$

In matrix form

$$F_{hb} = \mathcal{Y}\mathcal{X} + j2\pi f_0 \Omega \mathcal{Q} + \mathcal{F} + \mathcal{B} = 0 \qquad (2)$$

where, $\Omega = \mathrm{diag}([-kI \; \ldots \; 0 \; \ldots \; kI])$ and

$$\mathcal{Q} = [Q_{-k}, \ldots, Q_0, \ldots, Q_k]^T$$

$\mathcal{B}$, $\mathcal{F}$ and $\mathcal{X}$ are similarly defined and

$$\mathcal{Y} = \mathrm{diag}[Y_{-k} \; \ldots \; 0 \; \ldots \; Y_k]$$

(2) represents a system of $m(2k+1)$ nonlinear equations in $m(2k+1)$ unknowns $\mathcal{X}$ which can be solved using Newton's method. The Jacobian for (2) is given by

$$J_{hb} = \mathcal{Y} + j2\pi f_0 \Omega \frac{\partial \mathcal{Q}}{\partial \mathcal{X}} + \frac{\partial \mathcal{F}}{\partial \mathcal{X}}$$

It can be easily shown that

$$J_{hb} = \mathcal{Y} + j2\pi f_0 \Omega \Gamma \mathcal{C} \Gamma^{-1} + \Gamma \mathcal{G} \Gamma^{-1} \qquad (3)$$

where

$$\mathcal{C} = \begin{bmatrix} C(t_1) & 0 & 0 \\ 0 & \ddots & 0 \\ 0 & 0 & C(t_{2k+1}) \end{bmatrix}$$

where $C(t_i) = \frac{\mathrm{d}q}{\mathrm{d}x}\big|_{x(t_i)}$. $\mathcal{G}$ is also similarly defined. Γ represents time to frequency translation and consists of a series of permutations and Fourier transforms. This Jacobian is large and not very sparse and therefore a generic dense or sparse solver will not be able to solve this system very efficiently. Note that

$$\Gamma \mathcal{C} \Gamma^{-1} = \begin{bmatrix} C_0 & C_1 & \ldots & & C_{-k} \\ C_{-k} & C_0 & C_1 & \ldots & \\ & & \ddots & & \\ C_1 & C_2 & \ldots & C_{-k} & C_0 \end{bmatrix} \qquad (4)$$

where C_i is the ith Fourier coefficient of $C(t)$. I.e., the matrix block-circulant.

However, Krylov subspace methods only require a subroutine to perform matrix vector multiplication with J_{hb} which can be achieved using permutations (no cost), Fourier transforms ($\mathcal{O}\left(n(2k+1)\log(2k+1)\right)$) and sparse matrix vector multiplications $\mathcal{O}\left(n(2k+1)\right)$. Therefore *if the problem is properly preconditioned*, Krylov subspace methods converge quickly to the solution.

3. EXISTING PRECONDITIONERS FOR HB JACOBIAN

In this section we briefly review various preconditioners we will be using to compare with our approach. A more elaborate and detailed review can be found in [4].

3.1 Averaging preconditioner

If $G(t_i)$s and $C(t_i)$s can be assumed constant then (3) reduces to

$$J_{hb_{avg}} = \mathcal{Y} + \begin{bmatrix} j2\pi(-k)f_0 C_{avg}+G_{avg} & \ldots & 0 \\ & \ddots & \\ & \ldots & j2\pi k f_0 C_{avg}+G_{avg} \end{bmatrix}$$

which is a block diagonal matrix and each block has the sparsity structure of the circuit transient matrix, i.e., $J_{hb_{avg}}$ can be very easily inverted. For problems with mild nonlinearities, this preconditioner works extremely well and is probably the best preconditioner for such problems.

3.2 Averaging preconditioner with one step correction

As the circuit becomes more nonlinear, off-diagonal entries in J_{hb} become more dominant and $J_{hb_{avg}}$ becomes less effective. A one-step correction scheme includes them in the following manner [4]

$$J_{hb} = J_{hb_{avg}} + \left(j2\pi f_0 \Omega \Gamma \mathcal{C}_{diff} \Gamma^{-1} + \Gamma \mathcal{G}_{diff} \Gamma^{-1}\right)$$

where $\mathcal{G}_{diff} = \mathcal{G} - \mathcal{G}_{avg}$. J_{hb} is expanded in Taylor series to find a better approximation of J_{hb}^{-1} as follows:

$$J^{-1} = \left[I + J_{hb_{avg}}^{-1}\left(j2\pi f_0 \Omega \Gamma \mathcal{C}_{diff} \Gamma^{-1} + \Gamma \mathcal{G}_{diff} \Gamma^{-1}\right)\right]^{-1} J_{hb_{avg}}^{-1}$$

$$\approx \left[I - J_{hb_{avg}}^{-1}\left(j2\pi f_0 \Omega \Gamma \mathcal{C}_{diff} \Gamma^{-1} - \Gamma \mathcal{G}_{diff} \Gamma^{-1}\right)\right] J_{hb_{avg}}^{-1}$$

3.3 Averaging preconditioner with super diagonals

Due to nonlinearities, if a few of the harmonics are large, then they can also be included in $J_{hb_{avg}}$. However, inspecting the form of the circulant matrix in (4) it is obvious that if all entries of a given harmonic are included, the matrix becomes difficult to invert. Therefore, we only include entries in the super-diagonal and discard the entries in the subdiagonal. The resulting matrix is block upper triangular and therefore easier to invert. However, if the harmonic index is large, then the corresponding entries are further away from the block-diagonal and more entries of that harmonic are discarded from the lower diagonal part of the matrix and the preconditioner becomes less effective. This is specially true for multi-tone problems, where the artificial frequency map may place waveforms with significant harmonic content, far away from the diagonal.

3.4 One-step correction on averaging preconditioner with super diagonals

This is similar in spirit to averaging preconditioner with one-step correction where the difference matrix is generated by subtracting the averaging preconditioner along with the superdiagonals from the full harmonic balance preconditioner. One-step correction is performed using this modified difference matrix.

978-1-60558-497-3/09 $25.00 © 2009 ACM

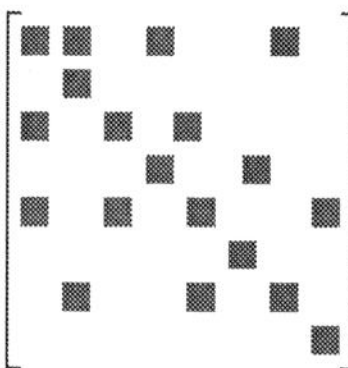

Figure 1: Sparsity structure of HB Jacobian

3.5 Finite-difference Jacobian

If the circuit is highly nonlinear then the preconditioner of the finite difference Jacobian can be used. The form of this preconditioner is

$$\begin{bmatrix} \frac{C(t_1)}{h_1} + G(t_1) & 0 & \cdots & & 0 \\ -\frac{C(t_1)}{h_2} & \frac{C(t_2)}{h_2} + G(t_2) & & & \\ & & \ddots & & \\ & \cdots & 0 & -\frac{C(t_{n-1})}{h_n} & \frac{C(t_n)}{h_n} + G(t_n) \end{bmatrix}$$

where $h_i = t_i - t_{i-1}$. The limitation of this method is that it cannot be used for harmonic balance problems with more than one tones. Another option is to use the full finite-difference Jacobian as a preconditioner.

$$\begin{bmatrix} \frac{C(t_1)}{h_1} + G(t_1) & 0 & \cdots & & -\frac{C(t_n)}{h_1} \\ -\frac{C(t_1)}{h_2} & \frac{C(t_2)}{h_2} + G(t_2) & & & \\ & & \ddots & & \\ & \cdots & 0 & -\frac{C(t_{n-1})}{h_n} & \frac{C(t_n)}{h_n} + G(t_n) \end{bmatrix}$$

This works better for more nonlinear problems compared to the preconditioner of finite-difference but is more expensive to apply since every solve with this requires another set of Krylov subspace iterations.

3.6 Schur-complement preconditioner

This preconditioner is based on the assumption that the matrix can be permuted such that the number of columns (c) containing the nonlinear entries are small $c \ll n$ and these columns can be permuted to the right of the matrix. The first $n - c$ columns are factored using a block diagonal solver, though a rectangular LU decomposition is required. The resulting Schur complement matrix can be factored either exactly or using incomplete LU [11]. This preconditioner relies on the fact that the number of columns with nonlinear elements is very small compared to the entire circuit and also that permuting these columns to the right of the matrix does not cause significant fills. Both these assumptions are not generally true.

4. DIRECT METHOD FOR HB JACOBIAN

4.1 Basic procedure

Note that the structure of J_{hb} is similar to (4) except that it is not block circulant as $\Gamma C \Gamma^{-1}$. In this formulation, all variables of a single harmonic are grouped together. If we apply a permutation to J_{hb} such that all the harmonics of a single variable are grouped together then the matrix structure will look as in Figure 1 [5]. At the top level, the matrix has the same sparsity structure as the transient matrix but the difference is that in the transient matrix, each entry is a number whereas in this matrix, each "entry" is a block matrix of size $(2k + 1) \times (2k + 1)$.

This matrix can be efficiently factored in the following manner. We first form a test matrix which has the same sparsity structure as the transient matrix and place *representative* entries and factor it using a modern sparse LU solver. The solver, along with factoring the matrix will determine a row and column permutation and a row and column scaling scheme which will minimize the number of fills in factoring the test matrix. Moreover, it will also provide the locations of the fills in the L and U matrices.

Given this information, it is trivial to write a matrix factor routine for J_{hb} as follows:

For each column j

1. identify the permuted column, perform row permutation and row and column scaling

2. for each row i

(a) if there is a structural nonzero entry in U

$$U_{ij} = A_{ij} - \sum_{k=1}^{i-1} L_{ik} U_{kj} \text{ if } i \leq j$$

(b) if there is a structural nonzero entry in L

$$L_{ij} = \left(A_{ij} - \sum_{k=1}^{j-1} L_{ik} U_{kj} \right) U_{jj}^{-1} \text{ if } i > j$$

Here A_{ij}, U_{ij}, L_{ij} are the dense matrices of size $(2k+1) \times (2k+1)$ located at the $(i, j)^{th}$ entry of J_{hb}, it's upper (U) and lower (L) triangular factors respectively. These operations can be performed very efficiently using aggressively optimized BLAS [3] and LAPACK [1] implementations for the target CPU architecture. The test matrix should be constructed so that it mimics the properties of J_{hb}. For instance, if there is an entry in the transient matrix at $(i, j)^{th}$ location where only C matrix entry is present, then the corresponding block in J_{hb} will be $j2\pi f_0 \Omega C_{i,j}$ which is singular. This entry should not be a choice of pivot for the test matrix otherwise the matrix factor of J_{hb} will fail. This can be avoided by introducing numerical zeroes in such locations which will prevent these entries to be used as pivots when factoring the test matrix.

Furthermore, these BLAS and LAPACK routines are also available in parallel form and multi-threaded versions of these routines can be used to further improve performance as the performance of these basic routines scales almost linearly with the number of processors. However, in this paper all the results are reported with the sequential implementations of BLAS and LAPACK since we are comparing performance with sequential implementation of preconditioners in Krylov subspace methods.

4.2 Complexity analysis

We now calculate the storage and computational complexity of direct solve and compare it with iterative solvers. Note that J_{hb} need not be constructed explicitly and only the factors L and U are required. During the factorization, the block matrices of J_{hb} can be constructed on the fly and each block is only needed once. Therefore the storage requirement of the direct solver is

$$(nnzl + nnzu)(2k + 1)^2$$

where $nnzu$ and $nnzl$ are the nonzeroes in U and L respectively. For preconditioned Krylov subspace methods, typically C and G matrices are stored in time domain for efficient matrix vector multiplication with J_{hb} and the storage requirement is $2nn_t$ where n_t is the number of time points. For multitone problems the artificial frequency map can result in large n_t. However, for the direct solver, the storage requirement is quadratic in k and therefore can be quite steep for multitone problems.

The computation complexity of the direct solver is

$$\mathcal{O}\left((nnzl^{\alpha_l} + nnzu^{\alpha_u}) (2k + 1)^3 \right)$$

where $\alpha_l, \alpha_u > 1$ but are close to 1. The complexity of a Krylov subspace based method is somewhat less straight-forward to express. If a Krylov subspace method converges in l iterations, then the complexity of HB matrix vector multiply is

$$l\mathcal{O}\left((nnzc + nnzg)(2k + 1) \log(2k + 1) \right)$$

where $nnzc$ and $nnzg$ are the number of nonzeroes in C and G respectively. For an Arnoldi based Krylov subspace method such as GMRES [16] (preferred over Lanczos based methods such as

978-1-60558-497-3/09 $25.00 © 2009 ACM

Table 1: Run time comparison of HB using direct solver and best preconditioned Krylov subspace solver

Example	circuit size	number of tones	number of harmonics	problem size	direct solver		best Krylov solver		
					time (s)	memory (GB)	time (s)	memory (GB)	iterations
mixer	58	1	40	4698	0.45	0.086	0.068	0.063	15
LNA + mixer	1135	1	30	69235	23.4	0.45	15.4	0.16	52
		3	125	284885	991	6.5	602	1.31	92
receiver 1	2158	1	15	66898	28.5	0.28	23.5	0.16	33
		3	125	541658	3297	12.7	962	1.66	58
receiver 2	8254	3	140	2319374	20917	64.4	DNF	—	—
LNA + mixer + filter	8444	1	20	346204	104.6	1.42	242.1	0.56	124
		3	80	1359484	3661	20	DNF	—	—
PA	341592	1	7	5123880	46582	25.6	DNF	—	—

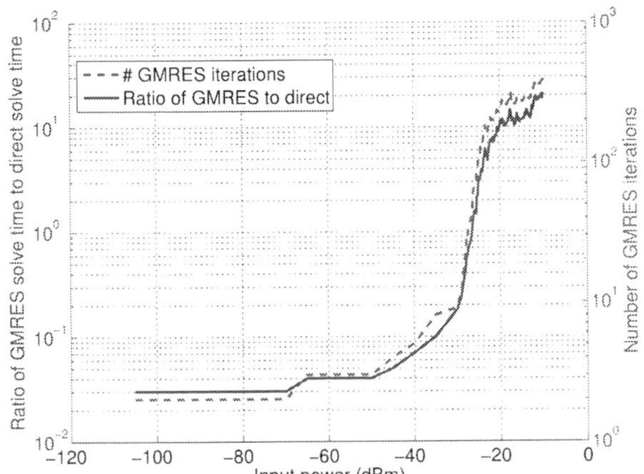

Figure 2: Ratio of GMRES solve time to the direct solve time as a function of input power for an LNA.

TFQMR [6] due to superior convergence and numerical properties), the complexity of running l iterations is $\mathcal{O}\left(l^2 n(2k+1)\right)$. The complexity of applying the preconditioner per iteration is typically $\mathcal{O}\left(nnz^\alpha(2k+1)\right)$.

The key observation is that if the number of Krylov subspace iterations l required to achieve convergence is greater than $2k+1$, then the computational complexity of a Krylov subspace based method is similar to the direct method! I.e., for such problems, the direct method can be competitive or significantly superior in performance compared to Krylov subspace methods.

Specifically for problems with moderate to strong nonlinearities, the number of Krylov subspace iterations can be very large even though the number of harmonics requested is moderate and it is *precisely* these situations where the direct method outperforms its more decorated Krylov subspace siblings.

4.3   A representative example

Before we present a comprehensive set of numerical results, it is instructive to compare the direct and the preconditioner Krylov solver performance on a simple example. The circuit is a low noise amplifier with two tones of equal power applied at the input. The frequency separation of these tones (Δf) is much smaller than their respective frequencies (i.e., $f_1, f_2 \gg \Delta f$). Two-tone harmonic balance is run with all harmonic mixes which add up to ten or less. The input power is swept from -105 dBm to -10 dBm.

Figure 2 shows the relative time for solving the harmonic balance Jacobian using the best preconditioned GMRES to the CPU for the direct solver, along with the number of GMRES iterations. For low powers, we used the averaging preconditioner but for high powers, we needed to switch to the 1-step correction preconditioner. It is obvious from the figure that for small input power levels, the LNA operates almost linearly and GMRES is much faster than direct solver. However, as the input power increases which causes the circuit to become more nonlinear, the number of GMRES iterations and therefore the CPU time

of GMRES become much worse compared to the direct solver. Note that since the number of harmonics required over the entire power sweep is constant, direct solver CPU time is constant.

4.4   Numerical results

The above procedure was implemented in our experimental HB simulator along with most of the above mentioned preconditioners. In subsequent tables, all examples which required less than 2.4 GB memory for direct solver were run on an Intel 32 bit machine and used Intel Math Kernel Library (MKL) [8] for LAPACK and BLAS. All examples requiring more than 2.4 GB were run on an AMD Opteron machine and use Automatically Tuned Linear Algebra Subroutines (ATLAS) [2]. Both direct solver and best preconditioned Krylov subspace methods were run on the same unloaded machine.

Table 1 shows the CPU time and memory comparisons between the direct solver and the best preconditioned Krylov subspace solver (in terms of CPU time), for various circuits with increasing size and harmonics along with the number of Krylov subspace iterations for the iterative methods. It is clear that direct solver is slower (but not excessively slow) for mildly nonlinear problems compared to Krylov subspace methods but becomes markedly superior as the circuit complexity and nonlinearity increases. In many examples, Krylov subspace methods exceeded the generous iteration limit of 500 and are reported as DNFs. The mixer example is a four quadrant Gilbert's cell mixer. Averaging preconditioner was the fastest on this example. For the LNA + mixer example, the averaging preconditioner with 1-step correction (Section 3.4) was the fastest. For the receiver 1 example, averaging preconditioner with super-diagonals corresponding to the first and second harmonic (Section 3.3) was the fastest. The LNA + mixer + filter example converged in 1-tone HB using only the finite difference preconditioner and this could not be used in the 3-tone HB and no other preconditioner could be used for this circuit for 3-tone HB.

Table 1 also points out a very desirable characteristic of the direct solver: robustness. If the process fits in the CPU memory and the test matrix is formed correctly, HB Jacobian solve will succeed for all the Newton iterations. The same cannot be claimed for Krylov subspace solvers which may exceed the iteration limit as Newton is progressing and may require a preconditioner change in the middle or may abort all together.

4.5   Comparison with general purpose sparse LU solvers (GLU)

It is instructive to compare the factor time of the proposed specialized solver for HB Jacobian with the factor time of a GLU for this matrix. Table 2 shows the ratio of the factor time of the general purpose sparse solver with the factor time of our proposed direct solver. It is clear that the proposed solver is much more efficient, and this efficiency increases with the example size. Further, for many larger examples in Tables 1 and 3, GLU ran out of memory even on a machine with 64 GB RAM.

5.   FOURIER ENVELOPE ANALYSIS

We will first briefly review the Fourier envelope analysis in the

978-1-60558-497-3/09 $25.00 © 2009 ACM

Table 3: Run time comparison of Fourier envelope using direct solver and best preconditioned Krylov subspace solver

Examples	circuit size	number of harmonics	problem size	direct solver			best Krylov solver		
				time (sec)	memory (GB)	bypass %	time (sec)	memory (GB)	iterations
LNA + mixer	1135	15	35185	4835	0.22	31	9522	0.20	60
receiver 1	2158	15	66898	1951	0.62	63	3628	0.35	34
LNA + mixer + filter	8444	20	346204	5547	1.45	49	27467	0.68	102
receiver 3	14685	10	308385	17467	1.0	47	90734	0.756	86
GSM transmitter	29053	13	784431	16570	2.57	76	109540	1.42	52
extracted receiver	32466	10	681786	6590	1.84	58	59024	1.34	42
LNA + mixer + filter	44150	20	1810150	360250	15.36	51	311040	3.63	102

Table 2: Ratio of factor time of a general purpose sparse solver to the factor time of the proposed direct solver

Example	circuit size	problem size	ratio
LNA + mixer	1135	69235	3.5
receiver 1	2158	66898	7.0

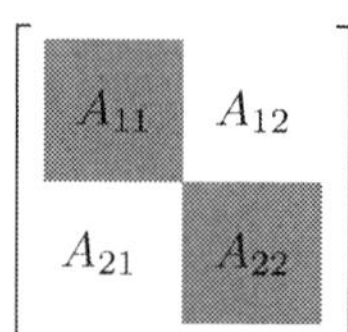

Figure 3: Splitting HB Jacobian into two segments

framework of multi-rate signals. When circuits are driven by or respond with signals of widely separated time scales, it is advantageous to view the circuit equations in more than one time scales [14, 15]. The resulting partial differential equation can be efficiently solved and the circuit response can be easily derived from the multi-rate solution. The multi-rate PDE version of (1) is given by

$$\frac{\partial q\left(\hat{x}(t_1, t_2)\right)}{\partial t_1} + \frac{\partial q\left(\hat{x}(t_1, t_2)\right)}{\partial t_2} + f\left(\hat{x}(t_1, t_2)\right) + \hat{b}(t_1, t_2) = 0$$

where the convolution term has been omitted. If $x(t_1, t_2)$ is the solution of the above equation then it can be shown that $x(t, t)$ is the solution of the original DAE [15]. Fourier envelope assumes that the signals are periodic in t_2 and therefore can be expanded in Fourier series as before. After following essentially the same steps outlined above we get

$$\frac{\partial \mathcal{Q}(t_1)}{\partial t_1} + 2\pi j f_2 \Omega \mathcal{Q}(t_1) + \mathcal{F}(t_1) + \mathcal{B}(t_1) = 0$$

The above DAE is discretized using an appropriate linear multi-step method (such as Backward Euler) and integrated in t_1

$$\frac{\mathcal{Q}_i - \mathcal{Q}_{i-1}}{h_i} + 2\pi j f_2 \Omega \mathcal{Q}(t_{1_i}) + \mathcal{F}(t_{1_i}) + \mathcal{B}(t_{1_i}) = 0$$

The Jacobian for this is of the form

$$\frac{1}{h_i}\mathcal{C}_i + \mathcal{G}_i + 2\pi j f_2 \Omega \mathcal{C}_i$$

Note that the Fourier envelope Jacobian is almost identical to HB Jacobian except for an additional term $\frac{1}{h_i}\mathcal{C}_i$. Therefore direct solve can be used for Fourier envelope as well.

In fact, direct solve is ideally suited for this analysis because it is typically performed with one large periodic tone (usually the LO in an RF transceiver), the nonlinearities are strong and at every envelope step, it usually takes 2-3 Newton iterations to converge.

Since direct solve lends itself very naturally to Jacobian bypass, we can take advantage of this. The Jacobian matrix is not factored (bypassed) and the previous factors are used to perform triangle solves with L and U. As the number of bypassed Jacobians increases, the total number of Newton iterations increases but the time taken per Newton iteration reduces because of time savings due to Jacobian bypass. Jacobian bypass can result in 2-3× performance improvement over non-bypassed implementation of the direct solver. Table 3 compares the CPU time and memory requirement for various examples. Many of these examples are also present in Table 1. Note that even for those

examples where 1- or 3-tone HB with direct solver was slower than the best preconditioned Krylov subspace method, envelope analysis is faster with the direct solver. This is entirely due to the use of Jacobian bypass with the direct solver which does not give any speed improvement in the case of Krylov subspace methods.

6. PRECONDITIONERS BASED ON DIRECT SOLVE

Li and Pillage [10, 4] introduced a new preconditioner of HB Jacobian. They segmented the HB matrix into two parts of approximately equal number of harmonics, as shown in Figure 3, solved the two portions independently, and used the combined solution as the result of a preconditioner solve. By recursively applying this method on each sub-problem they developed efficient methods for solving the hierarchical problem. Our proposed preconditioner is inspired by their approach with two crucial modifications

1. we use the direct solver for solving each of the segments

2. we also include A_{21} in the preconditioner, i.e., for two segments we are including 75% of the matrix in the preconditioner.

I.e., if we need to solve for $[b_1, b_2]^T$ using this preconditioner, we first solve $A_{11}x_1 = b_1$, and then solve $A_{22}x_2 = b_2 - A_{21}x_1$. This requires an extra matrix vector multiply with J_{hb}. $A_{21}x_1$ can be obtained by multiplying J_{hb} with $[x_1, 0]^T$ and taking the second block vector of the product. To generalize this, if the matrix is divided into l segments then for the ith block row, we need to form $\sum_{j=1}^{i-1} A_{ij}x_j$, which can be formed by multiplying J_{hb} with $[x_1, \ldots, x_{i-1}, 0, \ldots]^T$ and taking the result of the ith block row. Note that the inclusion or non-inclusion of the A_{ij} blocks applies only to the preconditioner matrix, and the accuracy of the solution of HB Jacobian is *unaffected* in both the cases.

The advantage of using direct solve of A_{11} and A_{22} is that these matrices need to be factored only *once* and in each subsequent Krylov subspace iteration, we only need to perform *triangle solves* with L and U factors of these matrices which gives significant speed advantage over the preconditioner where A_{11} and A_{22} are also solved using Krylov subspace methods.

6.1 Complexity analysis

Assuming that the matrix is divided into l segments of approximately equal number of harmonics, the storage requirement of each segment is reduced by the factor l^2 but we need

978-1-60558-497-3/09 $25.00 © 2009 ACM

Table 4: Memory and CPU time for HB Jacobian using various values of segments

| | direct | | Preconditioned by direct | | | | | |
| | | | segment=2 | | | segment=3 | | |
Example	time(s)	mem(GB)	time(s)	mem(GB)	iterations	time(s)	mem(GB)	iterations
LNA + mixer	990	6.5	1161	3.7	32	1325	2.8	67
receiver 1	3297	12.7	1905	7.0	15	2202	5.2	26
LNA + mixer + filter	3361	20	8052	15.7	50	12450	12.6	80

Table 5: Memory and CPU time for HB Jacobian using various values of segments without using A_{21}

| | direct | | Preconditioned by direct | | | | | |
| | | | segment=2 | | | segment=3 | | |
Example	time(s)	mem(GB)	time(s)	mem(GB)	iterations	time(s)	mem(GB)	iterations
LNA + mixer	990	6.5	1487	3.8	61	1479	2.8	94
receiver 1	3297	12.7	2328	7.1	28	2340	5.3	44
LNA + mixer + filter	3361	20	9513	16.0	99	10551	12.6	151

to store l of these. Hence the overall memory requirement reduces by a factor of l. Similarly the matrix factor time is reduced by l^2 but we need to perform l triangle solves at each Krylov iteration. If m iterations are required for Krylov subspace method to converge to the solution, then the total solve time is $\mathcal{O}\left(ml\left(nnzu_u^\alpha + nnzl_l^\alpha\right)\left(\frac{2k+1}{l}\right)^2\right)$. We need $m-1$ matrix vector multiplies with the HB Jacobian. Therefore the overall cost of this is

$$\mathcal{O}\left(\left(nnzu_u^\alpha + nnzl_l^\alpha\right)\frac{(2k+1)^3}{l^2}\right)$$

$$+\mathcal{O}\left(m\left(nnzu_u^\alpha + nnzl_l^\alpha\right)\frac{(2k+1)^2}{l}\right)$$

$$+\mathcal{O}\left(ml(nnzc + nnzu)(2k+1)\log(2k+1)\right) + \mathcal{O}\left(m^2 n(2k+1)\right)$$

This provides a direct trade-off between memory and speed. As l is increased, the memory consumption goes down but m increases which may increase the HB solve time depending how steep the increase in m is.

6.2 Results

Table 4 shows the memory and CPU time for various segment values for 3-tone HB examples from Table 1. For comparison, the same examples are run for various segment values but without including A_{21} (Table 5). This saves the matrix vector multiply with J_{hb} but increases the number of Krylov subspace iterations. It is clear that whether CPU time increases or decreases with increasing number of segments is heavily dependent on the problem.

7. CONCLUSION

In this paper we introduced a new method of performing direct solve of the harmonic balance Jacobian. The proposed block LU decomposition is novel as it takes advantage of the unique structure of the HB Jacobian which has the same block-sparsity pattern as the Jacobian formed in transient analysis. The resulting matrix and LU data-structures are more compact than is possible with a general purpose sparse matrix solver and allow handling much larger problems. This solver also results in far superior performance as the operations on the dense blocks are performed efficiently using LAPACK/BLAS routines.

For examples with moderate number of harmonics and moderate to strong nonlinearities, we demonstrated that the proposed direct solver is significantly faster than the best preconditioned iterative solvers with a moderate increase in the memory. We also showed that this solver is especially suited for Fourier envelope analysis where the number of harmonics is small, circuits are nonlinear and the direct solver is very well suited for using Jacobian bypass. For examples with large number of harmonics and moderate to strong nonlinearities, the performance advantage of the new solver is maintained but the memory requirements increase. To address this we proposed preconditioners based on direct solution of harmonic balance matrices which provide the user with a memory-speed trade-off. Above all, the proposed method robustly solves problems with strong nonlinearities, which were often found to be beyond the reach of traditional iterative methods with comparable computing resources.

8. REFERENCES

[1] E. Anderson, Z. Bai, C. Bischof, S. Blackford, J. Demmel, J. Dongarra, J. D. Croz, A. Greenbaum, S. Hammarling, A. McKenney, and D. Sorensen. *LAPACK Users' Guide, 3rd ed.* Society for Industrial and Applied Mathematics, Philadelphia, PA, 1999.

[2] ATLAS. http://math-atlas.sourceforge.net/.

[3] L. S. Blackford, J. Demmel, J. Dongarra, I. Duff, S. Hammarling, G. Henry, M. Heroux, L. Kaufman, A. Lumsdaine, A. Petitet, R. Pozo, K. Remington, and R. C. Whaley. An updated set of Basic Linear Algebra Subprograms (BLAS). *ACM Transactions on Mathematical Software*, 28:135–151, 2002.

[4] W. Dong and P. Li. Hierarchical harmonic-balance methods for frequency-domain analog-circuit analysis. *IEEE Trans. Computer-Aided Design*, 26:2089–2101, Dec. 2007.

[5] P. Feldmann, B. Melville, and D. Long. Efficient frequency domain analysis of large nonlinear analog circuits. In *Proceedings of the IEEE 1996 Custom Integrated Circuits Conference*, pages 461–464, 1996.

[6] R. Freund, G. H. Golub, and N. M. Nachtigal. Iterative solutions of linear systems. *Acta Numerica*, pages 57–100, 1991.

[7] R. J. Gilmore and M. B. Steer. Nonlinear circuit analysis using the method of harmonic balance, a review of the art. Part I. Introductory concepts. *International Journal on Microwave and Millimeter Wave Computer Aided Engineering*, 1, Jan. 1991.

[8] Intel. http://www.intel.com/cd/software/products/asmo-na/eng/307757.htm.

[9] K. S. Kundert, J. K. White, and A. L. Sangiovanni-Vincentelli. *Steadystate methods for simulating analog and microwave circuits*. Kluwer, Boston, MA, 1990.

[10] P. Li and L. T. Pillegi. Efficient harmonic balance simulation using multi-level frequency decomposition. In *IEEE/ACM International Conference on Computer Aided Design*, pages 677–682, 2004.

[11] D. Long, R. Melville, K. Ashby, and B. Horton. Full-chip harmonic balance. In *Proceedings of the IEEE 1997 Custom Integrated Circuits Conference*, pages 379–382, 1997.

[12] R. C. Melville, P. Feldmann, and J. Roychowdhury. Efficient multi-tone distortion analysis of analog integrated circuits. In *Proceedings of the IEEE 1995 Custom Integrated Circuits Conference*, pages 241–244, 1995.

[13] O. Nastov, R. Telichevesky, K. Kundert, and J. White. Fundamentals of fast simulation algorithms for RF circuits. *Proceedings of the IEEE*, 93:600–621, Mar. 2007.

[14] J. Roychowdhury. Efficient methods for simulating highly nonlinear multi-rate circuits. In *Proceedings 34th Design Automation Conference*, pages 269–274, 1997.

[15] J. Roychowdhury. Analyzing circuits with widely separated time scales using numerical PDE methods. *IEEE Trans. Circuits Sys. 1*, 48:578–594, May 2001.

[16] Y. Saad. *Iterative methods for sparse linear systems*. Society for Industrial and Applied Mathematics, 1996.

[17] R. Telichevesky, K. Kundert, I. Elfadel, and J. K. White. Fast simulation algorithms for RF circuits. In *Proceedings of the IEEE 1996 Custom Integrated Circuits Conference*, pages 437–444, 1996.

[18] V.Rizzoli, C.Cecchetti, A. Lipparini, and F. Mastri. General-purpose harmonic balance analysis of nonlinear microwave circuits under multitone excitation. *IEEE Trans. Microwave Theory Tech.*, 1, Jan. 1991.

978-1-60558-497-3/09 $25.00 © 2009 ACM

Stochastic Steady-State and AC Analyses of Mixed-Signal Systems

Jaeha Kim[1,2], Jihong Ren[3], and Mark A. Horowitz[1]

[1]Stanford University, Stanford, CA, [2]Kenosys Research, Inc., Los Altos, CA, [3]Rambus, Inc., Los Altos, CA.

ABSTRACT

This paper demonstrates that the steady-state and adjoint sensitivity analyses can be extended to stochastic mixed-signal systems based on Markov chain models. The examples of such systems include digital phase-locked loops and delta-sigma data converters, of which steady-state response is statistical in nature, consisting of an ensemble of waveforms with probability distribution. For efficient Markov-chain analysis, the paper describes three methods that can limit the number of states: a state discretization scheme based on Gaussian decomposition, a state exploration algorithm that discovers the recurrent states, and a state truncation algorithm that eliminates the states with negligible stationary probabilities. The stochastic AC analysis is performed by deriving a first-order ordinary differential equation governing the perturbations in the stationary probabilities and solving it via phasor analysis. In the digital PLL and first-order $\Delta\Sigma$ ADC examples, the number of states was reduced by a factor of 35 and the frequency-domain phase and noise transfer functions were simulated with a $57\sim22,000\times$ speed-up compared to using transient, Monte-Carlo simulations.

Categories and Subject Descriptors
B.7.2 [**Integrated Circuits**]: Design Aids – *Simulation*

General Terms
Design, Algorithms

Keywords
Markov chains, steady-state analysis, adjoint sensitivity analysis, stochastic systems.

I. INTRODUCTION
Linear system analysis has long been an indispensable tool for analog circuit designers. In hand analysis, circuits are typically analyzed in their small-signal models from which various linear system parameters can be derived: e.g., poles, zeros, gain, and bandwidth [1]. In simulation, the adjoint linear analyses such as the AC and periodic AC (PAC) analyses [2-5] have been the most effective tools to simulate the linearized response of the circuits. Due to such prevalence of linear analysis tools, it is not surprising that most analog circuits strive to achieve certain linear system behavior while minimizing all nonlinearities such as offset and distortion. This paper aims to extend the adjoint linear analysis to

Permission to make digital or hard copies of part or all of this work for personal or classroom use is granted without fee provided that copies are not made or distributed for profit or commercial advantage and that copies bear this notice and the full citation on the first page. To copy otherwise, to republish, to post on servers or to redistribute to lists, requires prior specific permission and/or a fee.
DAC'09, July 26-31, 2009, San Francisco, California, USA

mixed-signal circuits and simulate their linearized responses even when the circuits do not have a steady-state response that is either time-invariant or periodically time-varying (PTV).

It is a steady trend in IC design that increasingly more analog circuits are being implemented as mixed-signal systems with the digital circuits either replacing or assisting the analog circuits [7]. It is mainly because in advanced CMOS technologies, transistors are poor gain elements while they are still good, fast switches. For example, by leveraging the fast speed of digital logic, delta-sigma ($\Delta\Sigma$) data converters can achieve high resolutions without requiring stringent accuracy for analog circuits [8,9]. Also, many phase-locked loops (PLLs) are now implemented with time-to-digital converters and digital-controlled oscillators so that the loop filters can be realized entirely in digital, reducing the large filter area and enabling more sophisticated control of the loop dynamics [10-12]. Furthermore, many analog front-end circuits in wireless and wireline communication systems are equipped with digital calibration loops that monitor and correct the undesired properties of the circuits such as offset, skew, or duty-cycle error.

A common characteristic of these emerging mixed-signal systems is that they exhibit random behaviors. For example, most digitally-controlled feedback loops including digital PLL/DLLs and calibration loops show aperiodic dithering as they approach to the steady states rather than exponential convergence in linear circuits. Sometimes, this randomness is even intentional. $\Delta\Sigma$ modulators rely on this randomness to shape the deterministic quantization errors to out-of-band random noise. Furthermore, injecting pseudo-random dithering into $\Delta\Sigma$ data converters has been found to improve linearity and suppress periodic idle tones. Dynamic element matching is another technique that randomizes the selection of the sub-elements so that any mismatch among them appears as random noise. Also in digital calibration loops, it has been reported that randomizing the calibration period can help reduce undesired harmonic tones that arise due to periodicity [13].

Despite the fact that many mixed-signal systems are stochastic, their design intents are still linear and simulating their adjoint linear system responses at the steady-states (e.g. the frequency-domain transfer function) is of high interest to the designers. For example, the $\Delta\Sigma$ data converters are designed with specific signal transfer function (STF) and noise transfer function (NTF) in mind that maximize the signal-to-noise ratio (SNR). Also, phase transfer function is the most common way of describing the characteristics of a digital PLL such as bandwidth and jitter peaking. While AC or PAC analyses available from today's circuit simulators can efficiently simulate the adjoint linear responses for the circuits that have DC or periodic steady-states, they cannot be applied to the above-mentioned mixed-signal systems because they have *stochastic* steady-states [6].

This paper demonstrates that the steady-state and AC analyses can be extended to these stochastic mixed-signal systems based on Markov chain models. The steady-state response of such a system is statistical in nature, representing an ensemble of waveforms rather than a single one; the probability density function (PDF) and the power spectral density (PSD) are the examples. While computing the steady-state distribution of a Markov chain model is a well-established art [14,15], the main challenge lies in constructing a finite Markov chain model that has a feasible number of states [16]. We describe three methods to limit the number of states: a state discretization scheme that uses Gaussian decomposition to represent the probability distributions across continuous state space, a state exploration algorithm that includes only the states that are part of the steady-state response, and a state truncation algorithm that eliminates the states whose stationary probabilities are negligible.

The adjoint sensitivity analysis and phasor analysis used in the AC/PAC analysis can also be extended to the Markov chain models of the mixed-signal systems and efficiently simulate their linear, frequency-domain responses. In case of a digital PLL and a $\Delta\Sigma$ analog-to-digital converter (ADC), the linear responses of interest are the jitter transfer function and the signal or noise transfer function, respectively. The proposed stochastic AC (SAC) analysis achieves a large speed-up of 57~22,000× compared to using the alternative Monte-Carlo transient simulations. This is a similar speed-up that the AC/PAC analyses have over the equivalent transient simulations.

This paper is organized as follows. Section II describes the stochastic steady-state (SSS) analysis algorithm based on Markov chain models and ways to reduce the number of states. Section III describes the SAC analysis. Section IV discusses the experimental results with a digital PLL and a first-order $\Delta\Sigma$ ADC example.

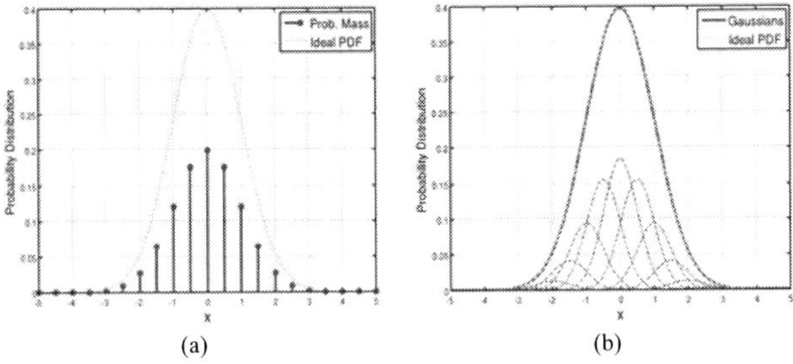

FIGURE 1. Representation of probability distributions over continuous state space: (a) with a set of probability masses, (b) with a set of Gaussian basis functions.

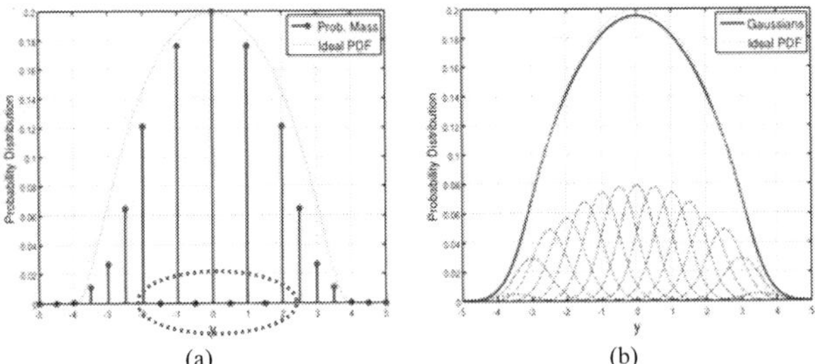

FIGURE 2. Probability distribution after the state projection $y=4\cdot\tanh(x/2)$: (a) with a set of probability masses, (b) with a set of Gaussian basis functions.

II. STOCHASTIC STEADY-STATE (SSS) ANALYSIS

A finite, discrete-time Markov chain is characterized by its transition probability matrix $\mathbf{T} = \{T_{ij}\}$'s whose elements describe the probabilistic progression of the states from time n to $n+1$:

$$T_{ij} = P(\mathbf{s}[n+1] = \mathbf{s}_i \mid \mathbf{s}[n] = \mathbf{s}_j) \text{ for } i, j = 1, 2, ..., N, \qquad (1)$$

where $\mathbf{s}[n]$ is the system state vector at time n and the Markov chain is assumed to have N discrete states $\mathbf{s}_1$, $\mathbf{s}_2$, ..., $\mathbf{s}_N$. The Markov chain property states that the state probability distribution at time $n+1$, $\mathbf{p}[n+1]$, depends solely on $\mathbf{p}[n]$ via the expression:

$$\mathbf{p}[n+1] = \mathbf{T} \cdot \mathbf{p}[n], \qquad (2)$$

where $\mathbf{p}[n]$ is an $N\times1$ vector $\{p_i[n]\}$'s and $p_i[n]$ is $P(\mathbf{s}[n]=\mathbf{s}_i)$. If the matrix $\mathbf{T}$ is constant regardless of n (i.e. time-homogeneous), then Perron-Frobenius theorem states that the stationary distribution vector $\boldsymbol{\pi}$ of the Markov chain is found by solving:

$$\lim_{n\to\infty} \mathbf{p}[n] = \boldsymbol{\pi} = \mathbf{T} \cdot \boldsymbol{\pi} . \qquad (3)$$

There are a number of ways of solving this equation numerically and their efficiencies have been well studied. As these computational methods are relatively mature, this paper does not discuss them in further details and refers the readers to [14,15].

The rest of this section focuses on the problem of constructing a Markov chain model from a circuit model with a feasible number of states. With continuous state variables in mixed-signal systems, the number of states in the Markov chain model can grow quickly with the number of variables. A related work by Demir and Feldman [16] also recognized this problem and used a decompositional graph representation in order to efficiently store and compute the large transition probability matrices with millions of states. As an orthogonal approach, we try to reduce the number of states in the Markov chain itself.

II.A. Discretization of Continuous PDFs via Gaussian Decomposition

Since a Markov chain model describes how the state probability distribution evolves over time, the first consideration should be on how to represent the probability distribution over continuous state space. In theory, the distribution is best represented with continuous probability density function (PDF). But in practice, it must be discretized both in time and value to enable numerical computation. Since most mixed-signal systems are sampled-data systems, the time axis is already discretized and the remaining problem is to find an efficient way of expressing the probability distribution and projecting it across a unit time step.

A simple way of approximating a PDF is to use a probability mass function defined for a set of discrete points covering the continuous state space, as illustrated in Fig. 1(a). If a system is governed by a state-space equation shown below:

$$\mathbf{s}[n+1] = f(\mathbf{s}[n]), \qquad (4)$$

then the probabilities at time $n+1$, $\mathbf{p}[n+1]$, can be computed based on $\mathbf{p}[n]$ via the following expression:

$$P(\mathbf{s}[n+1] = \mathbf{s}_i) = \sum_j \delta_{ij} \cdot P(\mathbf{s}[n] = \mathbf{s}_j) , \qquad (5)$$

where δ_{ij} is 1 if the quantized $f(\mathbf{s}_j)$ is equal to $\mathbf{s}_i$ and 0 otherwise.

One problem with this approach is that the probability masses from time n may project non-uniformly onto the state grids at time $n+1$, resulting in a probability distribution that differs from

978-1-60558-497-3/09 $25.00 © 2009 ACM

the true results. Fig. 2(a) illustrates this problem. When the state-projection function $f(\cdot)$ has high sensitivity to $\mathbf{s}[n]$, the adjacent points at time n may map to distant points at time $n+1$, leaving the points in-between with zero probabilities. A brute-force solution is to use the finer grids. Alternative solutions may include using adaptive grids or interpolation schemes.

To mitigate this problem, we propose representing a PDF with a set of Gaussian basis functions rather than with point-mass probabilities. That is, the state probability distribution $f_{PDF}(\mathbf{s}[n])$ is approximated as a weighted mixture of Gaussian PDFs centered at each discrete state point with fixed variance:

$$f_{PDF}(\mathbf{s}[n]) \cong \sum_i w_i[n] \cdot g(\mathbf{s}[n] - \mathbf{s_i}), \qquad (6)$$

where w_i's are non-negative real numbers with their sum equal to 1 and $g(\cdot)$ is the normal distribution function with mean of 0 and variance of σ_G^2. It has been proven that with σ_G equal to half the grid spacing, any function which is smooth on scales below σ_G can be decomposed into a set of Gaussians [17]. For brevity, we will limit our discussion to a one-dimensional case, but it can be extended to multi-dimensional cases as well.

The main advantage of Gaussian decomposition is the simplicity and accuracy in computing the projected probability distribution at the next time step, $f_{PDF}(\mathbf{s}[n+1])$, even when the grids are coarse. Assuming the same state-projection function $f(\cdot)$ defined as before, $f_{PDF}(\mathbf{s}[n+1])$ can be computed via:

$$f_{PDF}(\mathbf{s}[n+1]) \cong \sum_i w_i[n] \cdot g\!\left(\frac{\mathbf{s}[n+1] - f(\mathbf{s_i})}{d\mathbf{f}/d\mathbf{s}_i} \right). \qquad (7)$$

That is, each of the sub-Gaussian component $g(\mathbf{s}[n]-\mathbf{s_i})$ in (6) takes a new mean $f(\mathbf{s_i})$ and a new variance $\sigma_G^2 \cdot (d\mathbf{f}/d\mathbf{s_i})^2$ where $d\mathbf{f}/d\mathbf{s_i}$ is the derivative of f with respect to the state $\mathbf{s_i}$. If $g((\mathbf{s}[n+1]-f(\mathbf{s_i}))/(d\mathbf{f}/\mathbf{s_i}))$ can be decomposed to the original set of Gaussians $g(\mathbf{s}[n+1]-\mathbf{s_j})$ with a new set of weighting coefficients w'_{ij}, for $j=1,$ ... N, the expression for $f_{PDF}(\mathbf{s}[n+1])$ becomes:

$$f_{PDF}(\mathbf{s}[n+1]) \cong \sum_i \sum_j w'_{ij} \cdot g(\mathbf{s}[n+1] - \mathbf{s}_j)$$
$$= \sum_j w_j[n+1] \cdot g(\mathbf{s}[n+1] - \mathbf{s}_j) \qquad (8)$$

Fig. 2(b) plots the projected PDF based on the Gaussian decomposition method with the same set of grid points as in the previous case. Scaling the variance of each Gaussian component with the derivative of the projection function essentially performs the optimal interpolation between the discrete points.

Decomposition of the projected Gaussian PDF into the original set of Gaussian bases can be carried out by a simplified EM algorithm which performs the maximum likelihood fitting of the weighting coefficients [18]. The initial weights computed based on the adjusted Gaussian PDF with variance equal to $\sigma^2 - \sigma_G^2$ give very close estimates that limit the number of iterations to less than 20 with 0.1% tolerance.

Based on our Gaussian mixture representation, the probability distribution vector $\mathbf{p}[n]=\{p_i[n]\}$'s corresponds to the weights $\{w_i[n]\}$'s and the transition probability T_{ji} can be computed as:

$$T_{ji} = w_j[n+1] \text{ of } f_{PDF}(\mathbf{s}[n+1]) \text{ in (8) given } \mathbf{s}[n] \sim N(\mathbf{s_j}, \sigma_G^2). \qquad (9)$$

Note that this new transition probability matrix $\mathbf{T}$ remains stochastic (i.e. the elements are non-negative and the sum of each row is equal to 1) and all the existing methods to compute the stationary distribution $\boldsymbol{\pi}$ of a Markov chain still apply.

Once $\boldsymbol{\pi}$ is computed, one can derive various statistics of the system such as mean E[X], variance Var[X], auto-correlation $R_{XX}[n]$, and power spectral density PSD(f) as the followings:

$$E[X] = \mathbf{v_X}^T \cdot \boldsymbol{\pi}, \qquad (10)$$
$$Var[X] = (\mathbf{v_X}-E[X])^T \cdot \text{diag}(\boldsymbol{\pi}) \cdot (\mathbf{v_X}-E[X]) + \sigma_G^2, \qquad (11)$$
$$R_{XX}[n] = \mathbf{v_X}^T \cdot \mathbf{T}^n \cdot \text{diag}(\boldsymbol{\pi}) \cdot \mathbf{v_X} + \sigma_G^2, \qquad (12)$$
$$PSD(f) = \text{Discrete Fourier Transform of } Rxx[n] \qquad (13)$$

where $\mathbf{v_X}$ is an Nx1 vector of $\{v_{x,i}$'s$\}$ and $v_{x,i}$ is the value of the signal X at state $\mathbf{s_i}$. We note that for discrete states whose probabilities are represented with point masses, the equations (11) and (12) still remain valid except that σ_G is set to zero.

II.B. State Exploration Algorithm

Once we decide on a set of discrete points to cover the continuous state space, the next step is to determine which state points should be included in the Markov chain model. As discussed earlier, including all the points can be infeasible as the number increases exponentially with the dimension of the state space. Since we are interested only in the steady-state probability distribution, we need not include the transient states but only the recurrent states, i.e., the states with positive stationary probabilities.

Algorithm 1 outlines the pseudo-algorithm for building the transition probability matrix $\mathbf{T}$ consisting only of the states that are reachable from a seed state, $\mathbf{s_1}$. It is assumed that the seed state is a recurrent state obtained either by running an initial transient simulation or by inspection (in many situations the desired steady state is known). Then all the states reachable by the recurrent seed state are also recurrent [14-15]. The algorithm incrementally updates the list of reachable states as well as the matrix $\mathbf{T}$ whenever it encounters a new reachable state. The algorithm terminates when no more reachable state is found and the transition probabilities for all the known states are computed.

We note several observations regarding the state exploration algorithm. First, the number of states no longer scales with the total state space dimension, but rather with the intrinsic dimension of the stationary distribution. For example, if a system has a deterministic periodic steady state, the Markov chain model would contain the states that span a one-dimensional space (i.e. a loop) even if the system state has >100 dimensions. Second, provided that the seed state is a positively recurrent state, it is

---

ALGORITHM 1. State exploration algorithm for finding all reachable states from a seed state s_1.

1. Initialize two sets S_e and S_u:
    $S_e \leftarrow \{\}$: set of evaluated states
    $S_u \leftarrow \{s_1\}$: set of unevaluated states

2. Take one state s_i from S_u and move it to S_e.

3. Simulate the stochastic system response starting from s_i:
    $S_t \leftarrow \{s_j\}$'s   : a set of reachable states from s_i after one step
    $P_t \leftarrow \{P_j\}$'s   : a set of transition probabilities from s_i to s_j

4. For each pair (s_j, P_j) from the sets S_t and P_t:
    $T(s_i \rightarrow s_j) = P_j$
    If $s_j \notin S_e$, then add s_j to S_u.

5. Repeat steps 2, 3, and 4 until S_u is empty.

---

guaranteed that the constructed transition probability matrix $\mathbf{T}$ is positive and irreducible, which has a unique eigenvector for the eigenvalue of 1 along with many desired properties for the sound computation of π [15]. However, the algorithm is worthwhile only when the recurrent state space occupies a small portion of the entire state space.

II.C. Low-Probability State Truncation Algorithm

Although all the reachable states found by the state exploration algorithm are recurrent and have positive stationary probabilities, a sizable portion of them have the probabilities that are very close to zero. In fact, these states can pose ill-conditioned problems when computing π (e.g. the sub-dominant eigenvalues can be very close to 1). In a digitally PLL example discussed later, about 43.4% of the recurrent states occupy an aggregate probability less than 0.1%. Since they contribute so little, these states can be truncated to further reduce the number of states in the Markov chain and improve the numerical soundness of the solution.

Algorithm 2 lists the revised state exploration algorithm that terminates early if it determines that the remaining states have negligible probabilities. The original state exploration algorithm computed the transition probabilities for the reachable states in no particular order. In contrast, the revised algorithm tries to evaluate the states with large stationary probabilities first (but it does not guarantee to). Every time it chooses the next state to evaluate the transition probabilities for, it computes the probability leakage from the set of evaluated states (S_e) to the set of unevaluated states. If the total probability leakage is below a certain threshold (e.g., 0.001%), the algorithm terminates. Otherwise, the unevaluated state through which the probability leaks the most is selected as the next state.

This so-called *probability leakage* is computed by solving an intermediate Markov chain with all the transition probabilities originating from the unevaluated states and arriving at the evaluated states set to $1/N_e$, where N_e is the number of the currently evaluated states. In other words, the probability flux that leaves the evaluated set immediately returns to the evaluated set uniformly. It can be shown that the stationary probability π_i associated with an unevaluated state $\mathbf{s_i}$ corresponds to the probability leakage from the evaluated states via the state $\mathbf{s_i}$:

$$\pi_{\mathbf{i}} = \sum_{\mathbf{s_j} \in S_e} T_{ij}\, \pi_{\mathbf{j}} \tag{14}$$

The unevaluated state with large probability leakage is likely to

---

ALGORITHM 2. State truncation algorithm for finding the reachable states with relevant stationary probabilities.

1. Initialize two sets S_e and S_u:
   $S_e \leftarrow \{\}$: set of evaluated states
   $S_u \leftarrow \{s_1\}$: set of unevaluated states

2. Compute the probability leakage from S_e to S_u.

3. Take the state s_i from S_u with the largest leakage and move it to S_e.

4. Simulate the stochastic system response starting from s_i:
   $S_t \leftarrow \{s_j\}$'s   : a set of reachable states from s_i after one step
   $P_t \leftarrow \{P_j\}$'s   : a set of transition probabilities from s_i to s_j

5. For each pair (s_j, P_j) from the sets S_t and P_t:
   $\mathbf{T}(s_i \rightarrow s_j) = P_j$
   If $s_j \notin S_e$, then add s_j to S_u.

6. Repeat steps 2, 3, 4 and 5 until S_u is empty.

---

have significant contribution by itself or lead to other unvisited states that have large contributions.

The computational cost of repeatedly solving the intermediate Markov chain can be kept low by iteratively refining the solution found in the previous step. Using the power method, the solution π is refined by the following iteration:

$$\pi \leftarrow \mathbf{T}_{1:N_e} \cdot \pi_{1:N_e}, \tag{15}$$

$$\pi_{\mathbf{j}} \leftarrow \pi_{\mathbf{j}} + \frac{1}{N_e} \sum_{s_i \in S_u} \pi_i \text{ for } \mathbf{s_j} \in S_e. \tag{16}$$

The iteration is a sparse matrix multiplication and converges to a solution within 0.1% in less than 10 iterations. These intermediate solutions need not be accurate as they are only to evaluate the termination criteria and to select the next state for evaluation.

III. STOCHASTIC AC ANALYSIS

Stochastic AC (SAC) analysis measures the steady-state, sinusoidal perturbations in the stationary distribution π in response to a small-signal sinusoidal input u. As the AC and PAC analyses do for time-invariant and PTV circuits, respectively, the SAC analysis provides the frequency-domain transfer function of a stochastic mixed-signal system linearized at its steady states. While the perturbation analysis for Markov chains has been studied, it was done mostly in the context of investigating the numerical conditions of Markov chains [19].

To simulate the SAC transfer function, we first consider the perturbation matrix $\Delta \mathbf{T}$, whose elements $\{\Delta T_{ij}\}$'s describe the changes in the transition probabilities due to a change in u:

$$\Delta T_{ij} = \partial T_{ij} / \partial u. \tag{17}$$

The adjoint linear system can be derived by substituting $\mathbf{T}$ with $\mathbf{T} + \Delta \mathbf{T} \cdot \delta u$ and $\mathbf{p}[n]$ with $\pi + \delta\pi[n]$ in the Markov chain model (2):

$$\pi + \delta\pi[n+1] = (\mathbf{T} + \Delta\mathbf{T} \cdot \delta u)\cdot(\pi + \delta\pi[n]) \cong \mathbf{T} \cdot \pi + \mathbf{T} \cdot \delta\pi[n] + \Delta\mathbf{T} \cdot \pi \delta u. \tag{18}$$

Since $\pi = \mathbf{T} \cdot \pi$, we obtain a linear, first-order ODE that governs the time progression of the perturbation in the probability vector $\delta\pi$:

$$\delta\pi[n+1] = \mathbf{T} \cdot \delta\pi[n] + \Delta\mathbf{T} \cdot \pi \delta u. \tag{19}$$

Now the steady-state sinusoidal response can be computed by solving this adjoint linear ODE via phasor analysis. The phasor analysis starts by substituting the input and output variables in the ODE with their respective phasor expressions:

$$\delta u[n] = \mathbf{U} \cdot \exp(jn\omega T_s), \tag{20}$$

$$\delta\pi[n] = \mathbf{P} \cdot \exp(jn\omega T_s), \tag{21}$$

where $\mathbf{U}$ is a complex scalar, $\mathbf{P}$ is a $N \times 1$ complex vector, and T_s is the sampling period of the system. Substituting these expressions in (19) yields:

$$\mathbf{P} \cdot \exp(j(n+1)\omega T_s) = \mathbf{T} \cdot \mathbf{P} \cdot \exp(jn\omega T_s) + \Delta\mathbf{T} \cdot \pi \cdot \mathbf{U} \cdot \exp(jn\omega T_s) \tag{22}$$

$$(\exp(j\omega T_s) \cdot \mathbf{I} - \mathbf{T}) \cdot \mathbf{P} = \Delta\mathbf{T} \cdot \pi \cdot \mathbf{U} \tag{23}$$

Therefore, the SAC response $\mathbf{P}$ can be obtained by solving (23).

From the SAC response phasor $\mathbf{P}$, the frequency-domain transfer function for various statistical parameters of the system can be computed using the similar equations to (10)~(13). For example, the transfer function from the input u to the mean value of the signal X, E[X], can be computed by:

$$H_X(j\omega) = E[X(j\omega)]/U(j\omega) = \mathbf{v_X}^{\mathbf{T}} \cdot \mathbf{P}(\omega). \tag{24}$$

978-1-60558-497-3/09 $25.00 © 2009 ACM

IV. STOCHASTIC MIXED-SIGNAL SYSTEM EXAMPLES

IV.A. Digitally-Controlled Phase-Locked Loop

Fig. 3 depicts an example of a digitally-controlled PLL with a binary phase detector (PD) and its discrete-time system model. The binary phase detector measures the polarity of the difference between the input and output clock phases, Φ_{in} and Φ_{out}, and a digital loop filter updates the frequency of a digitally-controlled oscillator (DCO) accordingly. The loop filter may incur a delay of D cycles. The output phase variable Φ_{out} is continuous while other variables within the loop filter (u, u_d, and u_{int}) are discrete. We also model two noise sources in this system: the phase noise of the input clock (N_{in}) and that of the oscillator (N_{osc}).

First, the stochastic steady-state (SSS) analysis is performed in various noise settings. The results are summarized in Table I and the jitter histograms are shown in Fig. 4. The values of Φ_{out} is discretized in 101 steps between -0.5~0.5 UI and u_{int} has a range of -150~150. With the loop filter delay (D) of 4, the total number of states is 121,606. However, the results show that our state exploration/truncation algorithms reduced the number of states to well below 10,000 with the probability leakage threshold of 0.001%. The results from a 100,000-cycle Monte-Carlo transient simulation are also shown. The jitter histograms show good match between the two cases. Both simulations were run in Matlab on a 64-bit Linux machine with 2.53-GHz Intel Core 2 Duo processor and 4-GB memory. While the simulation times for the transient simulations stay constant regardless of the noise conditions, the SSS simulations take longer with the increasing number of states.

Next, the stochastic AC (SAC) analysis is performed to measure the jitter transfer function of the PLL. The results are plotted in Fig. 5 and summarized in Table II. To measure the AC transfer function with transient analysis, a series of simulations with different sample noise waveforms was run with and without a small sinusoidal perturbation on the input phase. That is, an ensemble of the resulting perturbations in the system responses was collected. Then the steady-state sinusoidal response in the mean perturbation response was estimated and the procedure was repeated for each excitation frequency. In our simulation, we simulated an ensemble of 100 perturbation responses and each response spanned 10,000 cycles. Apparently, it was very time-consuming for the transient analysis to measure the AC transfer functions and the proposed SAC analysis achieved large speed-up factors ranging from 57 to 170.

IV.B. Delta-Sigma Analog-to-Digital Converter

Fig. 6 illustrates the schematic of a first-order low-pass $\Delta\Sigma$ ADC with 4-level quantizer and its discrete-time system model. The $\Delta\Sigma$ modulator accumulates all the quantization errors in the past and adds its value to the input signal when performing the next data conversion. The quantizer output D_{out} is in 4 discrete levels while the integrator output V_{int} is continuous. We also model the dither noise (N_{dither}) being injected into the quantizer input.

The SSS and SAC analyses are performed for various levels of dither noise and the simulated quantization noise PSD and noise transfer function (NTF) are plotted in Fig. 7 and 8, respectively. Also, Table III and IV summarize the results along with the transient simulation results. The Markov chain model for this first-order $\Delta\Sigma$ ADC was rather small with only 804 states in total and both SSS and SAC analyses showed very short simulation

times. Nonetheless, the largest speed-up of 22263 was achieved by the state exploration/truncation algorithms which reduced to the number of states down to 61.

V. REFERENCES

[1]   P. R. Gray and R. G. Meyer, *Analysis and Design of Analog Integrated Circuits*, 3rd Ed., J. Wiley, New York, 1993.

[2]   L. T. Pillage, et al., *Electronic Circuit and System Simulation Methods*, McGraw-Hill, New York, 1995.

[3]   S. W. Director and R. A. Rohrer, "The Generalized Adjoint Network and Network Sensitivities," *IEEE Trans. Ckt. Theory*, vol 16, pp. 318-323, Aug. 1969.

[4]   J. W. Bandler, et al., "A Unified Theory for Frequency-Domain Simulation and Sensitivity Analysis of Linear and Nonlinear Circuits," *IEEE Trans. Microwave Theory and Tech.*, vol 36. pp. 1661-1669, Dec. 1988.

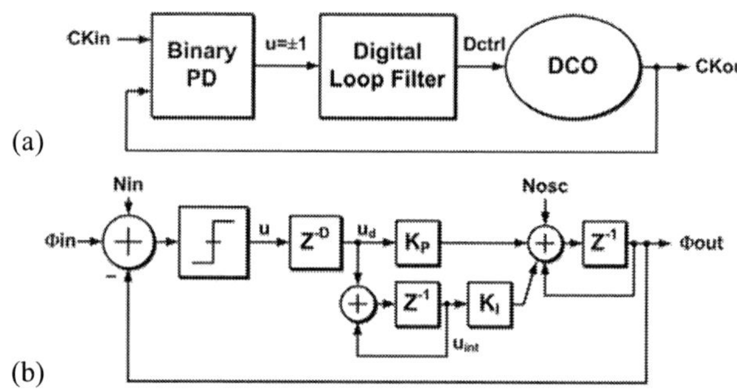

FIGURE 3. (a) A digitally-controlled phase-locked loop and (b) its discrete-time system model.

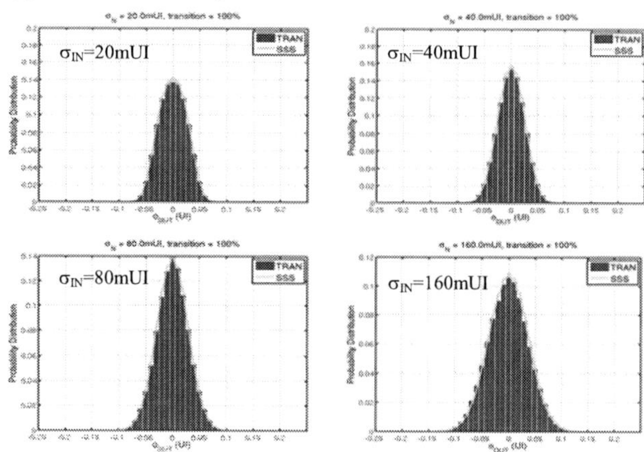

FIGURE 4. Simulated jitter histograms (σ_{osc}=0.002).

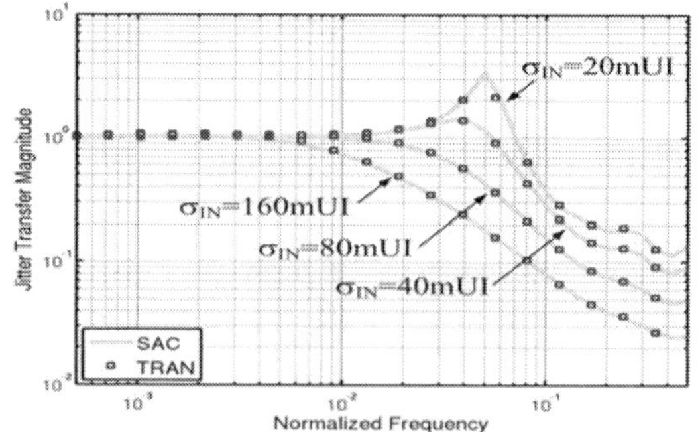

FIGURE 5. Simulated jitter transfer functions.

978-1-60558-497-3/09 $25.00 © 2009 ACM   380

[5]   K. S. Kundert and A. Sangiovanni-Vincentelli, "Simulation of Nonlinear Circuits in the Frequency Domain," *IEEE Trans. Computer-Aided Design of Integrated Circuits and Systems*, vol. CAD-5, pp. 521-535, OCt. 1986.

[6]   K. S. Kundert, "Challenges in RF Simulation," *IEEE Radio Frequency Integerated Circuits Symp.*, pp. 105-108, Jun. 2005.

[7]   B. Murmann, "Digitally Assisted Analog Circuits," IEEE *Micro*, vol. 26, pp. 38-47, Mar.-Apr. 2006.

[8]   J. C. Candy and G. C. Temes, "Oversampling Methods for A/D and D/A Conversion," *Oversampling Delta-Sigma Data Converters*, IEEE Press, New York, 1992, pp. 1-275.

[9]   B. E. Boser, and B. A. Wooley, "The Design of Sigma-Delta Modulation Analog-to-Digital Converters," *IEEE J. Solid-State Circuits*, pp. 1298-1308, Dec. 1988.

[10] P. K. Hanumolu, et al., "Digitally-Enhanced Phase-Locking Circuits," *IEEE Custom Integrated Circuits Conf.*, pp. 361-368, Sep. 2007.

[11] I. Hwang, et al., "A Digitally Controlled Phase-Locked Loop with a Digital Phase-Frequency Detector for Fast Acquisition," *IEEE J. Solid-State Circuits*, pp. 1574–1581, Oct. 2001.

[12] A. Rylyakov, et al., "A Wide Power-Supply Range (0.5V-to-1.3V), Wide Tuning Range (500MHz-to-8GHz) All-Static CMOS AD PLL in 65nm SOI," *IEEE Int'l Soild-State Circuits Conf.*, pp. 172-173, Feb. 2007.

[13] M. Clara, et al., "A 1.5V 200MS/s 13b 25mW DAC with Randomized Nested Background Calibration in 0.13um CMOS," *IEEE Int'l Solid-State Circuits Conf.*, pp. 250-251, Feb. 2007.

[14] W. J. Stewart, *An Introduction to the Numerical Solution of Markov Chains*, Princeton University Press, Princeton, 1994.

[15] W. J. Stewart, "Numerical Methods for Computing Stationary Distributions of Irreducible Markov Chains," *Advances in Computational Probability*, Ch. 3, pp. 81-111, Kluwer Academic Publishers, 1997.

[16] A. Demir and P. Feldmann, "Modeling and Analysis of Communication Circuit Performance Using Markov Chains and Efficient Graph Representations," *IEEE/ACM Int'l Conf. on Computer-Aided Design*, pp. 290-295, Nov. 2000.

[17] K. Kuijken and M. R. Merrifield, "A New Method for Obtaining Stellar Velocity Distribution from Absorption-Line Spectra: Unresolved Gaussian Decomposition," *Monthly Notices Royal Astronomical Soc.*, vol. 264, pp. 712-720, Oct. 1993.

[18] T. Hastie, R. Tibshirani, J. Friedman, *The Elements of Statistical Learning: Data Mining, Inference, and Prediction*, Springer Series in Statistics, 2001.

[19] C. D. Meyer, "Sensitivity of the Stationary Distribution of a Markov Chain," *SIAM J. Matrix Anal. Appl.*, vol. 15, pp. 715-728, Jul. 1994.

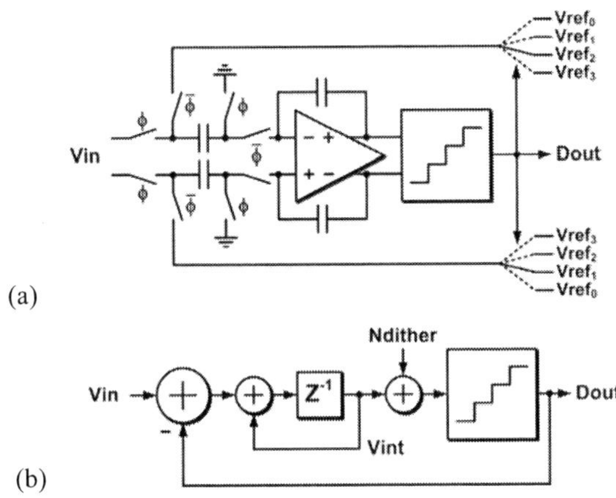

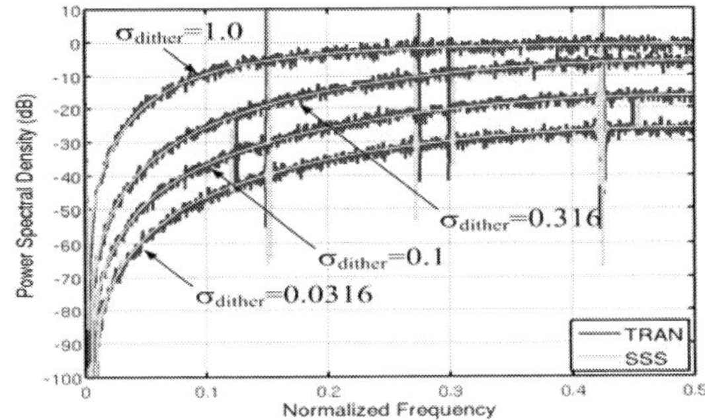

FIGURE 6. (a) A first-order $\Delta\Sigma$ ADC and (b) its discrete-time system model.

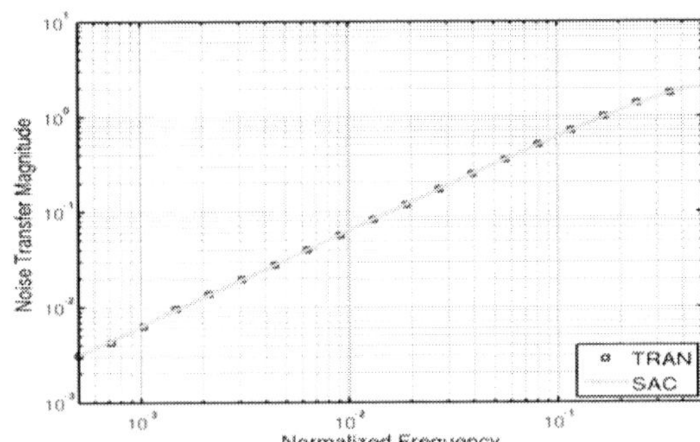

FIGURE 7. Simulated quantization noise spectrum.

FIGURE 8. Simulated noise transfer function (σ_{dither}=0.1).

Table I. SSS simulation results for the digital PLL.

σ_{IN} (mUI)	# of Evaluated States / # of Non-Zeros	Simulation Time SSS / TRAN (sec.)	σ_{OUT} SSS / TRAN (mUI)
20	3496 / 20916	31.73 / 0.84	25.44 / 25.29
40	4301 / 25806	40.79 / 0.80	25.10 / 24.95
80	5555 / 33330	56.11 / 0.82	28.92 / 29.02
160	7915 / 47490	89.03 / 0.85	37.03 / 37.73

Table II. SAC simulation results for the digital PLL.

σ_{IN} (mUI)	Simulation Time SAC / TRAN (sec.)	Speed-Up Factor
20	1.85 / 315.51	170.5×
40	2.55 / 314.61	123.4×
80	3.39 / 316.69	93.42×
160	5.53 / 315.12	56.98×

Table III. SSS simulation results for $\Delta\Sigma$ ADC.

σ_{dither}	# of Evaluated States / # of Non-Zeros	Simulation Time SSS / TRAN (sec.)	σ_{OUT} SSS / TRAN
0.0316	61 / 109	0.14 / 0.74	0.1111 / 0.1111
0.1	158 / 504	0.17 / 0.74	0.1129 / 0.1131
0.316	436 / 1741	0.37 / 0.73	0.2628 / 0.2627
1.0	795 / 3180	0.57 / 0.74	0.6377 / 0.6380

Table IV. SAC simulation results for $\Delta\Sigma$ ADC.

σ_{dither}	Simulation Time SAC / TRAN (sec.)	Speed-Up Factor
0.0316	0.013 / 289.42	22263.1×
0.1	0.034 / 291.25	8566.2×
0.316	0.14 / 290.99	2078.5×
1.0	0.35 / 289.89	828.3×

Parallelizable Stable Explicit Numerical Integration for Efficient Circuit Simulation *

Wei Dong and Peng Li
Department of ECE, Texas A&M University, College Station, TX 77843
{weidong, pli}@neo.tamu.edu

ABSTRACT

This work exploits the recently developed telescopic projective numerical integration method for efficient parallel circuit simulation. Stable explicit numerical integration is achieved by adopting an explicit inner integrator (e.g. forward Euler) in a multi-level telescopic projective framework, thereby addressing the well-known stability limitation of many explicit numerical integration methods. In the presented approach, the *effective* time step of the entire multi-level integration is no longer limited by the smallest time constant of the circuit so as to safeguard stability. Rather, it is controlled solely by the accuracy requirement. This makes it possible to explore the natural parallelizability of such explicit integration method for parallel circuit simulation. We demonstrate the potential of the presented approach and its parallel implementation on multi-core machines with encouraging initial results.

Categories and Subject Descriptors

B.7.2 [**Integrated Circuits**]: Design Aids—*simulation*

General Terms

Algorithms, Design, Performance

Keywords

Transient Simulation, Explicit Numerical Integration, Parallel Computing

1. INTRODUCTION

SPICE-like transistor-level transient simulation is a critical means of pre-silicon verification and forms the basis for simulation-based design tuning and optimization [1,2]. However, the simulation runtime bottleneck presents a significant

---

*This work was supported in part by SRC and Texas Analog Center for Excellence under contract 2008-HC-1836, and NSF Career Award #CCF-0747423.

The authors thank Chandramouli Kashyap and Chirayu Amin at Intel Strategic CAD Lab for helpful discussions.

Permission to make digital or hard copies of part or all of this work for personal or classroom use is granted without fee provided that copies are not made or distributed for profit or commercial advantage and that copies bear this notice and the full citation on the first page. To copy otherwise, to republish, to post on servers or to redistribute to lists, requires prior specific permission and/or a fee.
DAC'09, July 26-31, 2009, San Francisco, California, USA

challenge and prevents efficient analysis of medium-to-large IC designs. A body of research has been devoted to improve the performance and runtime efficiency of transient analysis via algorithmic innovation and parallelization [3–9]. Given the recent industry's shift to multi-core parallel computing, it is particularly meaningful to develop efficient simulation algorithms that lend themselves naturally to parallelization.

Numerical integration plays a critical role in transient analysis. It allows the numerical solution of a system of algebraic-differential equations (ADEs) by converting it into a sequence of algebraic equations at discretized time points. Unlike implicit integration methods, an explicit method can efficiently extrapolate the transient response based on the responses at previous time points, circumventing the need for solving the coupled system of equations. However, this potential computational advantage comes with a severe limitation: instability. The largest time step of an explicit method, say forward Euler, is constrained by the smallest time constant in the circuit, presenting a severe limitation to many practical circuits with widely separated time constants. The stability limitation of explicit integration methods may offset the potential computational benefit and therefore limits their wide application to circuit simulation. Nevertheless, efforts have been made to address the stability issue of explicit methods [8, 9]. Particularly, in the ACES simulator developed at IBM [8], a number of techniques were adopted to alleviate the stability constraint. However, such techniques are heuristics in nature and are only applicable to digital timing simulation.

This work leverages on the recent development on numerical solution of ordinary differential equations (ODEs) from the numerical analysis community [10, 11]. The so-called *telescopic projective* integration framework has been shown to be efficient for certain physical simulation problems with widespread time constants. In this paper, we show how the same principle can be applied to general circuit simulation. The use of an explicit integrator at the inner loop of the telescopic projective framework helps maintain the explicit nature while relaxing the stability constraint on the *effective* time step of the entire scheme. By addressing the known stability limitation of explicit numerical integration with a theoretically sound foundation, this allows explicit numerical integration be applied in a very general way. Equally importantly, the explicit nature of the approach breaks the entire simulation task into independent sub-tasks of device model evaluation, small-scale node (or device) based system solves, facilitating natural parallelization.

2. TELESCOPIC PROJECTIVE INTEGRATION

The telescopic projective framework in essence is a multi-level numerical integration method for solving initial value ODE problems [10, 11]. In principle, the telescopic method recursively employs projective integration on multiple levels as in Fig. 1.

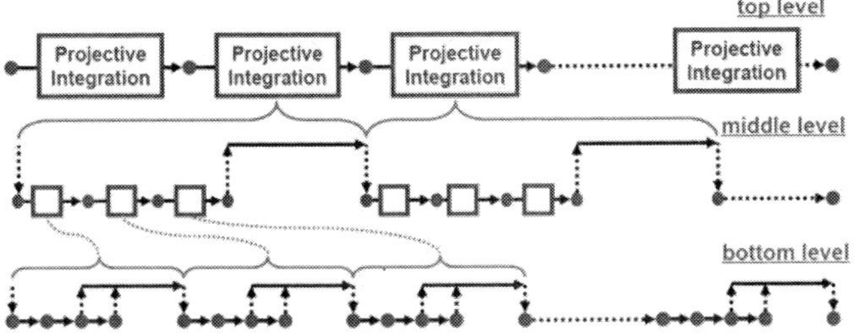

Figure 1: telescopic projective framework.

To better understand the key idea of the telescopic projective integration method, let us first consider the case of projective integration under the context of circuit simulation. Practical circuits may posses time constants with varying magnitudes. While the long-time circuit transient response is mainly determined by the slow components in the circuit, fast components exist for a short period of time and dissipate quickly. For practical purposes, it is often sufficient to only track the slow components in the transient response. Hence, it is desirable to use time steps with a size comparable to large time constants of the system to gain simulation efficiency. Unfortunately, the use of explicit integration methods is severely limited by stability concerns; the largest time step must be set to be comparable to the smallest time constants in the system to ensure stability. It is especially inefficient where there exists a large gap between the time constants of fast and slow components. Projective integration is specifically designed to accelerate the solving of ODE problems under such a situation.

In projective integration, a combination of an *inner* integrator and *outer* projective integrator is employed to achieve efficiency and stability at the same time. As shown in Fig. 1, in the first few inner (explicit) integration steps (integrate for $k + 1$ steps from t_n to t_{n+k+1}), a small time step is used to heavily decay the fast components in the system such that the accumulated integration error is exponentially damped. In other words, the purposely chosen small time step in inner integration steps alleviates the stability concern. Then, we may perform an outer projection using a much larger step without violating the stability constraint. The time step size of the outer projection step is chosen to track the slow components and set solely by (local) accuracy control. Because in the projection step, the solution at $t_{n+k+1+M}$ is extrapolated based on the solutions at t_{n+k} and t_{n+k+1} as

$$x_{n+k+1+M} = (M + 1)x_{n+k+1} - Mx_{n+k}, \qquad (1)$$

the accumulated error is only linearly amplified after it is exponentially damped in the preceding inner integration steps. As a result, the stability of projective integration method is maintained and the overall efficiency of projective integration is boosted by the outer projective step.

However, in more general cases, the eigenvalues or time constants of the system may be widely distributed and there

may not be any obvious isolations between eigenvalue clusters. In practice, neither is it trivial to know the distribution of the eigenvalues in advance. Under these cases, the step size of the outer projective step is significantly constrained ($M < 3k$) to ensure stability [10], which heavily deteriorates the efficiency of projective integration. To remedy this problem, a multi-level projective integration approach is suggested in [11]. The basic idea is that although at each level a limited speedup of $M + k + 1/k + 1$ is obtained when M is relatively small, a significant overall runtime speedup $(M + k + 1/k + 1)^q$ can be obtained in a q-level telescopic framework. Therefore, we can still maintain a good simulation efficiency without loss of stability.

3. STABLE EXPLICIT INTEGRATION FOR CIRCUIT SIMULATION

An electronic circuit can be described using differential equations in time domain as follows

$$F(X(t)) + \frac{d}{dt}Q(X(t)) + U(t) = 0, \qquad (2)$$

where $X(t)$ is the vector of nodal voltages and branch currents, $U(t)$ is the input, $F(\cdot)$ and $Q(\cdot)$ are functions describing static and dynamic nonlinearities. In our proposed explicit telescopic projective integration, we adopt forward Euler as the inner integrator at the bottom level of the hierarchical projective framework, as shown in Fig. 2. Since the inner forward Euler integrator and the outer projective integrator are both explicit, the entire integration scheme is explicit in nature. As will be seen in the following stability analysis, the presented integration scheme has good stability properties, which is crucial for practical circuit simulation applications.

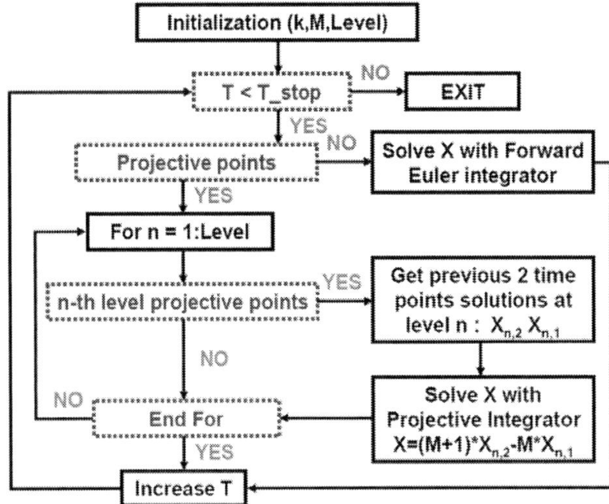

Figure 2: Stable explicit numerical integration for circuit simulation.

Different from the analysis in [10, 11], we analyze the stability property based on the modified nodal analysis and focus on practical issues in circuit simulation. As a standard practice, we analyze the stability of the adopted numerical integration method using a homogenous linear (test) circuit

978-1-60558-497-3/09 $25.00 © 2009 ACM

with a nonzero initial condition:

$$G \cdot X(t) + C \cdot \frac{dX(t)}{dt} = 0, \tag{3}$$

where G and C are the linear conductance and capacitance matrices. Without loss of generality, we assume C is invertible and $C^{-1}G$ is diagonalizable, which be represented as $C^{-1}G = P\Lambda P^{-1}$. Since we first employ $k+1$ inner forward Euler integration steps within each projective integration loop, we represent the solutions after k and $k+1$ inner steps as

$$\begin{cases} X_{n+k} = P\left(I - h\Lambda\right)^k P^{-1} \cdot X_n \\ X_{n+k+1} = P\left(I - h\Lambda\right)^{k+1} P^{-1} \cdot X_n \end{cases}, \tag{4}$$

After $k+1$ inner steps, one outer projective step would directly extrapolate the solution at time point $t_{n+k+1+M}$ based on the known solutions at time points t_{n+k} and t_{n+k+1}. We substitute (4) into (1) and reach the following relationship:

$$X_{n+k+1+M} = P\left((M+1)\left(I - h\Lambda\right)^{k+1} - M(I - h\Lambda)^k\right) P^{-1} X_n. \tag{5}$$

The telescopic projective method can be understood as a multi-level projective method in essence. Without loss of generality, a two-level telescopic projective integration framework is considered. Rewritting (5) as $X_{n+(k+1+M)} = P\Phi P^{-1} X_n$, we have

$$X_{n+(k+1+M)^2} = (M+1) X_{n+(k+1)(k+1+M)} - M X_{n+k(k+1+M)}$$
$$\overset{Y = P^{-1}X}{\Rightarrow} Y_{n+(k+1+M)^2} = \left((M+1)\Phi^{k+1} - M\Phi^k\right) Y_n \tag{6}$$

Note that Φ is a diagonal matrix. Denote its i-th diagonal entry as ϕ_i, the corresponding components of $Y_{n+(k+1+M)^2}$ and Y_n as $y_{n+(k+1+M)^2,i}$ and $y_{n,i}$, respectively. Then we have

$$\phi_{telescopic,i} = \frac{y_{n+(k+1+M)^2,i}}{y_{n,i}} = (M+1)\phi_i^{k+1} - M\phi_i^k, \tag{7}$$

where $\phi_i = (M+1)\left(1 - h\lambda_i\right)^{k+1} - M\left(1 - h\lambda_i\right)^k$ and λ_i is the i-th diagonal entry of Λ correspondingly. Till now, we have derived the equations for stability evaluation under the context of circuit analysis. The stability region is defined as $|\phi_{telescopic,i}| \le 1$. We can use the same procedure as in [11] to prove that there exists a $[0, 1]$ stability region for telescopic projective method. Since the inner integrator is forward Euler in our approach, it means that for any real $\lambda_i \in \left[0, \frac{1}{h}\right]$, the telescopic projective method can guarantee the stability with certain choices of parameters k and M.

4. PARALLEL IMPLEMENTATION

The presented technique addresses the known stability limitation of existing explicit numerical integration methods and makes it possible to explore the resulting computational advantages. With the use of explicit numerical integration, it is desirable to integrate the transient circuit response on a per-node or per-device basis without solving any coupled large systems of equations. This leads to natural parallelization. As shown in Fig. 3 (a), while this goal is straightforward to achieve when there exists no coupling between different circuit nodes, complication arises if coupling does exists, as illustrated in Fig. 3 (b). In the latter case, capacitance currents i_{c1} and i_{c2}, which are needed in explicit numerical integration, can no longer be determined

individually at each circuit node. Instead, a coupled system involving both nodes needs to be solved. To remedy this

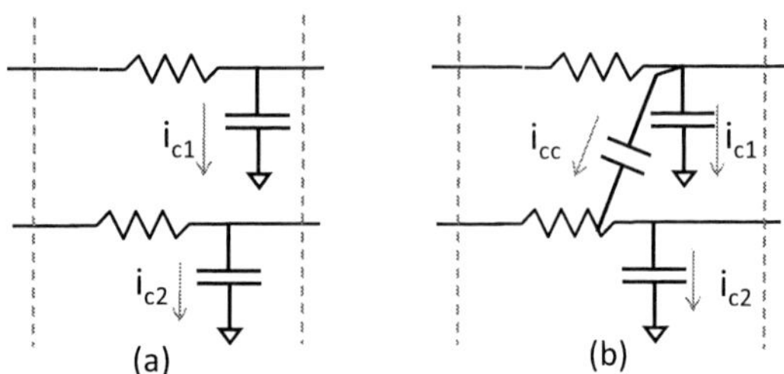

Figure 3: Two circuit nodes: (a) without coupling, and (b) with coupling.

problem, a handling similar to what is in [8] is taken. The forward Euler based integration can be formally described in the matrix form as:

$$G_n V_n + C_n \frac{V_{n+1} - V_n}{h} + U_n = 0, \tag{8}$$

where symbols with subscript n are known variables at the present time point n, and V_{n+1} is the unknown nodal voltage vector at the next time point $n+1$. Note that (8) is in general a coupled system of equations. We may split the capacitance matrix C_n into a diagonal part Λ_n and an off-diagonal part N_n and approximate (8) as:

$$G_n V_n + (\Lambda_n + N_n)\frac{V_{n+1} - V_n}{h} + U_n = 0 \Rightarrow$$
$$V_{n+1} \approx \Lambda_n^{-1}\left[\left(-hG_n + \Lambda_n - N_n\right)V_n + N_n V_{n-1} - hU_n\right], \tag{9}$$

where V_{n-1} is the known nodal voltage vector at the preceding time point $n-1$. Note that Λ_n is diagonal, implying that each node voltage now can be computed on an individual node basis, leading to straightforward parallelization. With proper handling, the error introduced in the above approximation can be well controlled through time step control and the overall stability probability of telescopic projective integration is maintained. An in-depth discussion on these issues is out the scope of this paper due to space limitation.

Care shall be taken to handle more complex devices. For example, a MOSFET is considered as a four-terminal device and the nonlinear gate capacitance is modeled by specifying the coupled charge equations at the four terminals. The capacitive terminal currents must be decided by solving the four coupled equations together. This implies that multiple small nonlinear systems of equations need to be solved as a whole, which by no means presents any practical difficulty to parallel implementation.

As shown in Fig. 6, a properly handling of all the above issues leads to a relatively straightforward parallelization of the presented stable explicit integration method.

5. EXPERIMENTAL RESULTS

The explicit telescopic projective integration method was incorporated into a SPICE-like simulator. Parallelization was realized using C/C++ with Pthreads on a multi-core Linux server.

First, a simple stiff RC circuit, as shown in Fig.5, is used to demonstrate the favorable characteristics of telescopic projective integration. The smallest capacitance in the circuit has a value of value $1fF$, implying that the largest time step

978-1-60558-497-3/09 $25.00 © 2009 ACM

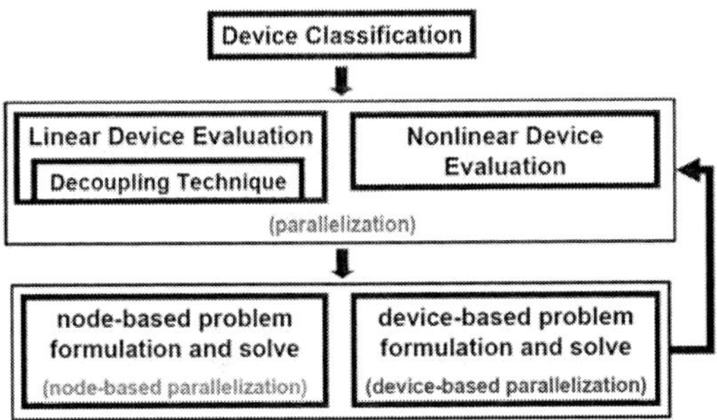

Figure 4: Parallel telescopic projective integration.

that can be used with the standard forward Euler method while ensuring stability is on the order of 10^{-15} seconds.

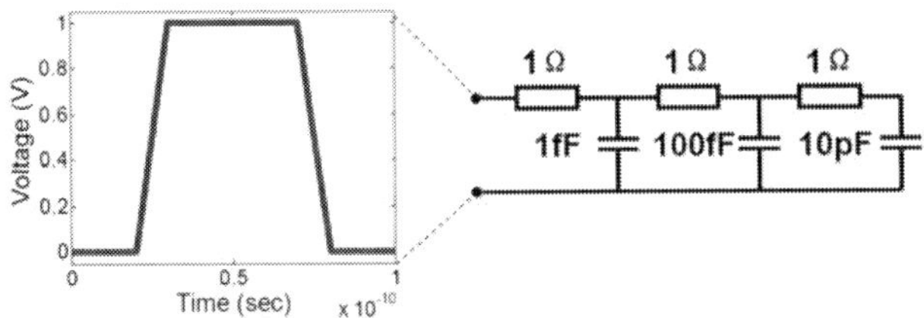

Figure 5: Test circuit and test input waveform.

Now apply the two-level telescopic projective method to the same circuit. If $k = 9$ and $M = 20$, the theoretical time-step boost achieved through the new integration scheme is $(M + k + 1/k + 1)^2 = 9$; similarly, the boost is 16 when $k = 9$ and $M = 30$. In Fig. 5, we show the output waveforms for the node connected to the $10p$F capacitor. In the figure, BE, FE, $TP1$ and $TP2$ denote the simulated waveforms using backward Euler, forward Euler and the two versions of the two-level telescopic projective methods, respectively. The time steps of BE, FE and that of the inner integrator in telescopic projective integration are set to be the same. The result of the telescopic projective methods agrees well with the other two methods.

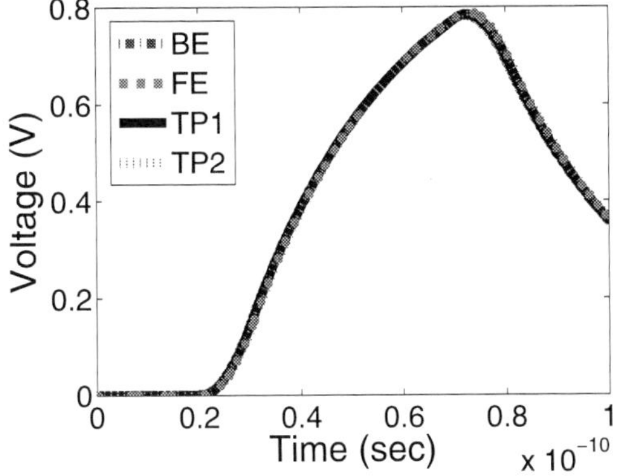

Figure 6: Transient simulation of the test circuit.

Next, we apply backward Euler, two-level serial explicit telescopic projective integration, and its two-thread and four-

thread parallel versions to a number of test circuits. For the two-level telescopic method, we set $k = 3$ and $M = 1$. The runtime statistics are collected in Table 1. In the table, $Time$ is the total runtime and $Speedup$ represents the speedup over backward Euler method. When the number of threads is low (one or two), telescopic integration can be actually slower than BE. This is not very surprising since in telescopic integration multiple inner forward Euler integration steps are needed to ensure stability. However, because of the explicit nature of the method, parallelisms can be easily exploited by adding more threads to gain runtime benefits. This would be particularly meaningful for large circuits, where potentially a large number of threads can be executed currently to process the large workload.

Table 1: Statistics of the transient simulations on serial and parallel platforms

| Circuit | Serial | | 2-thread | 4-thread |
| | BE | Proposed | Proposed | Proposed |
	Time(s)	Speedup	Speedup	Speedup
Buffer chain	36	0.54	0.72	1.17
DB mixer	33	0.61	0.87	1.52
4-bit adder	207	0.73	1.18	1.79
RC mesh 1 w/drivers	582	1.06	1.79	2.84
RC mesh 2 w/drivers	2,611	2.15	3.60	5.67

6. CONCLUSIONS

In this paper, we have demonstrated the application of stable explicit telescopic integration to transient simulation, especially for some stiff circuit problems. While adopting forward Euler as the inner integrator of the multi-level telescopic projective framework, the presented approach guarantees the stability of the overall integration scheme. At the same time, the explicit nature of the proposed method can be exploited to speed up transient simulation via efficient parallelization.

7. REFERENCES

[1] L.W.Nagel. *SPICE2: A Computer Program to Simulate Semiconductor Circuits.* ERL-M520, University of California, Berkeley, CA, USA, 1975.

[2] K.S.Kundert. *The Designer's Guide to Spice and Spectre.* Kluwer Academic Publishers, Massachusetts, USA, 1995.

[3] E. Lelarasmee, A. E. Ruehli, and A. L. Sangiovanni-Vincentelli. The waveform relaxation method for time-domain analysis of large scale integrated circuits. *TCAD*, 1(3):131–145, July 1982.

[4] M.W.Reichelt, A.Lumsdaine, and J.K.White. Accerlerated waveform methods for parallel transient simulation of semiconductor devices. In *ICCAD*, pages 270–274. IEEE/ACM, November 1993.

[5] U.Wever and Q.Zheng. Parallel transient analysis for circuit simulation. In *HICSS*, pages 442–447. IEEE, January 1996.

[6] W.Dong, P.Li, and X.Ye. Wavepipe: parallel transient simulation of analog and digital circuits on multi-core shared-memory machines. In *DAC*, pages 238–243, 2008.

[7] X.Ye, W.Dong, P.Li, and S.Nassif. Maps: Multi-algorithm parallel circuit simulation. In *ICCAD*, pages 73–78, 2008.

[8] A.Devgan and R.A.Rohrer. Adaptively controlled explicit simulation. *TCAD*, 13(6):746–762, June 1994.

[9] R.Griffith and M.Nakhla. A new high-order absolutely-stable explicit numerical integration algorithm for the time-domain simulation of nonlinear circuits. In *ICCAD*, pages 276–280, 1997.

[10] C.W.Gear and I.G.Kevrekidis. Projective methods for stiff differential equations: Problems with gaps in their eigenvalue spectrum. *SIAM J. Sci. Comput.*, 24(4):1091–1106, 2002.

[11] C.W.Gear and I.G.Kevrekidis. Telescopic projective methods for parabolic differential equations. *J. Comput. Phys.*, 187(1):95–109, 2003.

Efficient Design-Specific Worst-Case Corner Extraction for Integrated Circuits

Hong Zhang
Mentor Graphics Corporation
1001 Ridder Park Drive
San Jose, CA 95131
hongzhang@umail.ucsb.edu

Tsung-Hao Chen and Ming-Yuan Ting
Mentor Graphics Corporation
1001 Ridder Park Drive
San Jose, CA 95131
{howard_chen, ming_ting}@mentor.com

Xin Li
Carnegie Mellon University
5000 Forbs Avenue
Pittsburgh, PA 15213
xinli@ece.cmu.edu

ABSTRACT

While statistical analysis has been considered as an important tool for nanoscale integrated circuit design, many IC designers would like to know the design-specific worst-case corners for circuit debugging and failure diagnosis. In this paper, we propose a novel algorithm to efficiently extract the worst-case corners for nanoscale ICs. Our proposed approach mathematically formulates a quadratically constrained quadratic programming (QCQP) problem for corner extraction. Next, it applies the Lagrange duality theory to convert the non-convex QCQP problem to a convex semi-definite programming (SDP) problem that is easier to solve. Our circuit example designed in a commercial CMOS process demonstrates that the proposed SDP formulation can find the worst-case corners both efficiently and robustly, while the traditional QCQP fails to achieve global convergence.

Categories and Subject Descriptors

B.7.2 [**Integrated Circuits**]: Design Aids – Verification

General Terms

Algorithms

Keywords

Process Variation, Worst-Case Corner, Integrated Circuit

1. INTRODUCTION

As IC technologies scale to 45nm and below, integrated circuits are significantly impacted by large-scale process variations, including both inter-die and intra-die variations [1]. Inter-die variations model the common/average variations across the die, while intra-die variations model the individual, but spatially correlated, local variations (e.g., random device mismatches) within the same die. To achieve high parametric yield and, hence, low manufacturing cost, it is of great importance to statistically model and analyze inter-die/intra-die variations for integrated circuits.

To address this need, extensive researches have been performed during the past two decades. As one of the most important techniques, response surface modeling (RSM) was widely applied to statistical circuit analysis and optimization [2]-[11]. Recently, it has been demonstrated that linear RSM is not sufficiently accurate to capture the large-scale variations that are observed for nanoscale technologies; instead, quadratic or even

Permission to make digital or hard copies of part or all of this work for personal or classroom use is granted without fee provided that copies are not made or distributed for profit or commercial advantage and that copies bear this notice and the full citation on the first page. To copy otherwise, to republish, to post on servers or to redistribute to lists, requires prior specific permission and/or a fee.

DAC'09, July 26-31, 2009, San Francisco, California, USA

strongly nonlinear RSM is required to improve modeling accuracy [2]-[6]. Once RSM is created, it can be used to efficiently predict performance distribution and/or parametric yield [7]-[9].

While yield prediction is a necessary step for design sign-off, calculating the yield value only does not meet the needs of circuit designers. If a circuit fails the yield specification, it is important for the CAD tool to provide additional information that helps circuit designers to improve yield. Towards this goal, worst-case corner extraction aims to identify the unique process condition at which a given circuit fails to work [10]-[11]. Once the worst-case corners are determined, designers can simulate their circuit at these corners, find the reason for its performance failure, and eventually come up with the appropriate solution to improve robustness. From this point of view, worst-case corner extraction enables IC designers to easily debug the circuit so that they can efficiently use their design knowledge for yield enhancement.

Extracting the worst-case corners, however, is not trivial, primarily due to the following two reasons. First, realistic worst-case corners cannot be accurately predicted by IC foundries. Instead, they are topology-dependent and performance-dependent [10]-[11]. An efficient algorithm is required to find the worst-case corner for a particular circuit topology and a particular performance metric. Second but more importantly, quadratic RSM results in a non-convex quadratically constrained quadratic programming (QCQP) problem for corner extraction [10]-[11]. As will be demonstrated by our numerical example in Section 4, directly solving such a non-convex optimization using local search methods (e.g., gradient-based search) can easily get stuck at a local minimum. The challenge here is how to find the worst-case corners both *efficiently* and *robustly*.

In this paper, we propose a new mathematical formulation that converts the non-convex QCQP problem to a convex semi-definite programming (SDP) problem that is easier to solve. The proposed approach is derived from the Lagrange duality theory of nonlinear optimization [15]. It explores the unique property that the QCQP formulated for worst-case corner extraction only contains a single quadratic constraint. In such a special case, the dual form of QCQP is a convex SDP. In addition, under some general assumptions, there is no duality gap between the primal problem (i.e., QCQP) and the dual problem (i.e., SDP). Namely, once the dual problem is solved by convex SDP, the solution of the primal problem can be easily determined. By converting the non-convex QCQP to a convex SDP, we can efficiently and robustly find the worst-case corners with global convergence.

The remainder of this paper is organized as follows. In Section 2, we review the background on principal component analysis, response surface modeling, and performance distribution estimation. We propose our mathematical formulation for worst-case corner extraction in Section 3. The efficacy of the proposed corner extraction is demonstrated by the numerical example in Section 4. Finally, we conclude in Section 5.

978-1-60558-497-3/09 $25.00 © 2009 ACM

2. BACKGROUND

2.1 Principal Component Analysis

Principal component analysis (PCA) [13] is a statistical method that finds a set of independent factors to represent a set of correlated Normal random variables. Given N process parameters $X = [x_1\ x_2\ ...\ x_N]^T$, the process variations $\Delta X = X - X_0$, where X_0 contains the mean values of X, is often modeled by multiple zero-mean Normal distributions [2]-[11], and the correlation of ΔX can be represented by a symmetric, positive semi-definite covariance matrix R [13]. PCA decomposes R as [13]:

$$R = V \cdot \Sigma \cdot V^T \tag{1}$$

where $\Sigma = diag(\lambda_1, \lambda_2, ..., \lambda_N)$ contains the eigenvalues of R, and $V = [V_1\ V_2\ ...\ V_N]$ contains the corresponding eigenvectors that are orthonormal, i.e., $V^T V = I$ (I is the identity matrix). Based on Σ and V, PCA defines a set of new random variables:

$$\Delta Y = \Sigma^{-0.5} \cdot V^T \cdot \Delta X . \tag{2}$$

These new random variables in ΔY are called the principal components or factors. It is easy to verify that all principal components in ΔY are independent and standard Normal (i.e., zero mean and unit variance).

2.2 Response Surface Modeling

Given a circuit design, the circuit performance (e.g., delay, gain, etc.) is a function of process parameters (e.g., V_{TH}, T_{OX}, etc.). A circuit performance f can be approximated as a quadratic response surface model (RSM) of the process variations [2]-[11]:

$$f(\Delta Y) = \Delta Y^T \cdot A \cdot \Delta Y + B^T \cdot \Delta Y + C \tag{3}$$

where $\Delta Y = [\Delta y_1\ \Delta y_2\ ...\ \Delta y_N]^T$ represents the principal components extracted by PCA, and $C \in R$ and $B \in R^N$ and $A \in R^{N \times N}$ stand for the model coefficients. The quadratic model in (3) provides superior accuracy over a simple linear model, when applied to capture the large-scale manufacturing variations observed in today's IC technologies.

2.3 Performance Distribution Estimation

Once the quadratic response surface model in (3) is available, it can be used to predict the probability distribution of the given performance metric. Although no analytical form exists to represent the probability distribution of the performance $f(\Delta Y)$, the probability density function (PDF) and the cumulative distribution function (CDF) can be numerically calculated by a number of efficient algorithms, e.g., APEX [9]. The performance distribution can be further used to estimate a number of robustness metrics (e.g., worst-case performance, parametric yield, etc.) of the design. However, knowing the worst-case performance and/or parametric yield only is not sufficient. Circuit designers are particularly interested in the unique process condition (i.e., process corner) at which their circuit fails to work. If the worst-case corners are identified, circuit designers can simulate their circuit at these corners and find the appropriate solution to improve its robustness. Motivated by this observation, we aim to develop an efficient algorithm to find worst-case corners in this paper.

3. EFFICIENT CORNER EXTRACTION

In this section, we describe the proposed corner extraction algorithm in detail and highlight its novelties. We first mathematically formulate the corner extraction problem and then develop a convex semi-definite programming (SDP) method to solve it.

3.1 Mathematical Formulation

Once the response surface model $f(\Delta Y)$ is extracted for a performance metric f, it can be used to estimate the probability density function $pdf(f)$, the cumulative distribution function $cdf(f)$, and finally the worst-case performance f_{WC}. Here, the worst-case performance f_{WC} is defined by a given percentile point of $cdf(f)$. Taking the delay of a digital path as an example, the worst-case delay can be defined by the 99% point of $cdf(f)$ [9].

Our goal for worst-case corner extraction is to find the unique process condition (i.e., the value of ΔY^*) at which $f(\Delta Y^*)$ is equal to the worst-case performance value f_{WC}, i.e.:

$$f(\Delta Y^*) = f_{WC} . \tag{4}$$

This problem, however, is not mathematically well-defined. Studying (4), we would notice that there are N problem unknowns (i.e., $\Delta Y^* \in R^N$) but only one equation. In other words, Eq. (4) is underdetermined. Mathematically, we can find an infinite number of solutions (i.e., process corners) that satisfy (4).

On the other hand, not all these process corners are useful from the viewpoint of a circuit designer. Some of these corners will "never" occur, because the probability density function $pdf(f)$ is almost zero at these locations. To find the useful process corners, we must take the probability into account. In particular, we want to find the process corner that is most likely to occur, i.e., the maximum likelihood solution of (4) [10].

Recall that the random variables in ΔY are independent and standard Normal after PCA. To find the maximum likelihood solution of (4), we formulate the following optimization problem:

$$\max_{\Delta Y} \left(\frac{1}{\sqrt{2\pi}} \right)^N \cdot \exp\left(-\|\Delta Y\|_2^2 / 2\right) \tag{5}$$
$$S.T. \quad f(\Delta Y) = f_{WC}$$

where $\|\bullet\|_2$ denotes the L_2-norm of a vector. Since $\exp(\bullet)$ monotonically increases, Eq. (5) can be re-written as:

$$\min_{\Delta Y} \|\Delta Y\|_2^2 \tag{6}$$
$$S.T. \quad f(\Delta Y) = f_{WC}$$

Substituting (3) into (6) yields the following quadratically constrained quadratic programming (QCQP) problem:

$$\min_{\Delta Y} \|\Delta Y\|_2^2 \tag{7}$$
$$S.T. \quad \Delta Y^T \cdot A \cdot \Delta Y + B^T \cdot \Delta Y + C = f_{WC}$$

While the cost function in (7) is convex, the constraint set is not convex. In this case, directly solving (7) using local search methods (e.g., gradient-based search) can easily get stuck at a local minimum. In what follows, we will propose a novel technique to convert the non-convex QCQP problem in (7) to a convex semi-definite programming (SDP) problem that is easier to solve.

3.2 Lagrange Dual Formulation

Our proposed SDP formulation is derived via three steps: (a) relax the equality constraint in (7) to an inequality constraint; (b) find the Lagrange dual formulation that turns out to be an SDP problem; (c) demonstrate the strong duality so that solving the dual problem yields the solution of the primal problem. In this sub-section, we describe the mathematical details of all these three steps.

First, we relax the equality constraint in (7) to an inequality constraint that depends on the definition of the worst-case performance. If the worst case is defined as the upper bound of the performance variation (e.g., the worst-case delay of a digital path), we write the constraint as:

$$\Delta Y^T \cdot A \cdot \Delta Y + B^T \cdot \Delta Y + C \geq f_{WC} . \tag{8}$$

Otherwise, if the worst case is defined as the lower bound of the performance variation (e.g., the worst-case bandwidth of an analog amplifier), the constraint is represented as:

$$\Delta Y^T \cdot A \cdot \Delta Y + B^T \cdot \Delta Y + C \leq f_{WC} . \tag{9}$$

Without loss of generality, we represent both (8) and (9) by the following standard form:

$$\Delta Y^T \cdot \widetilde{A} \cdot \Delta Y + \widetilde{B}^T \cdot \Delta Y + \widetilde{C} \leq 0 . \tag{10}$$

For example, to convert (8) to (10), we simply have: $\widetilde{A} = -A$, $\widetilde{B} = -B$ and $\widetilde{C} = f_{WC} - C$.

Based on (10), we obtain the following relaxed formulation of the QCQP problem in (7):

$$\begin{aligned} \min_{\Delta Y} \quad & \|\Delta Y\|_2^2 \\ S.T. \quad & \Delta Y^T \cdot \widetilde{A} \cdot \Delta Y + \widetilde{B}^T \cdot \Delta Y + \widetilde{C} \leq 0 \end{aligned} . \tag{11}$$

It is important to note that the two optimization problems in (7) and (11) are exactly equivalent for our corner extraction application. Namely, solving (11) results in the solution of (7). Intuitively, if the constraint in (11) is not active, removing the constraint does not change its solution. In this case, the optimization becomes unconstrained and its solution is simply $\Delta Y = 0$. This, however, should never happen in our application, since the worst-case corner cannot be at the nominal condition. For this reason, the constraint in (11) must be active. Mathematically, this conclusion can be formally proven by using the Karush-Kuhn-Tucker condition from the optimization theory [15].

The relaxed optimization in (11), however, may not be convex either. As will be demonstrated by the numerical example in Section 4, the quadratic coefficient matrix $\widetilde{A}$ in (11) can be neither positive semi-definite nor negative semi-definite. In this case, the constraint set in (11) is not convex. To efficiently solve (11), we write the corresponding Lagrange dual problem [15]:

$$\begin{aligned} \max_{\lambda,\gamma} \quad & \gamma \\ S.T. \quad & \lambda \geq 0 \\ & \begin{bmatrix} I + \lambda\widetilde{A} & 0.5\widetilde{B} \\ 0.5\widetilde{B}^T & \lambda\widetilde{C} - \gamma \end{bmatrix} \geq 0 \end{aligned} . \tag{12}$$

Eq. (12) is a semi-definite programming (SDP) problem [15]. It can be proven that SDP is convex and, hence, it can be solved both robustly and efficiently [15].

While the dual problem in (12) is convex and easy to solve, we need to further demonstrate that solving the dual problem in (12) yields the solution of the primal problem in (11). Denote the cost function value at the optimal solution as $p*$ and $d*$ for (11) and (12), respectively. The Lagrange duality theorem guarantees [15]:

$$p* \geq d* . \tag{13}$$

Namely, $d*$ is a lower bound of $p*$. In addition, the following theorem gives a sufficient condition for strong duality [15].

Theorem 1: Strong duality holds for (11) and (12), i.e., $p* = d*$, if there exists an ΔY with:

$$\Delta Y^T \cdot \widetilde{A} \cdot \Delta Y + \widetilde{B}^T \cdot \Delta Y + \widetilde{C} < 0 . \tag{14}$$

In other words, the constraint set in (11) is strictly feasible.

Eq. (14) is referred to as the Slater's constraint qualification. It is obvious that this condition is typically satisfied for our corner extraction application. Namely, we can find a process corner ΔY at which the performance $f(\Delta Y)$ is worse than f_{WC}.

Solving the SDP in (12) yields the optimal value $\lambda = \lambda*$ that maximizes the cost function γ. Once $\lambda*$ is known, the solution ΔY of (11) is given by [15]:

$$\Delta Y^* = -\frac{\lambda^*}{2} \cdot \left(I + \lambda^* \cdot \widetilde{A}\right)^{-1} \cdot \widetilde{B} . \tag{15}$$

Eq. (15) gives the unique process corner where the performance $f(\Delta Y)$ reaches the worst-case value f_{WC}.

3.3 Summary

Algorithm 1: Worst-case corner extraction
1. Given a performance function $f(\Delta Y)$ of interest, generate a number of sampling points $\{(\Delta Y_{(i)}, f_{(i)}); i = 1,2,...,K\}$ based on the design of experiments (DOE) [14].
2. Fit the quadratic performance model in (3) to approximate $f(\Delta Y)$.
3. Based on the fitted quadratic performance model, predict the cumulative distribution function $cdf(f)$ by either Monte Carlo analysis or APEX [9].
4. Estimate the worst-case performance value f_{WC} that is defined by the 99% (or 1%) point of $cdf(f)$.
5. Based on the quadratic model coefficients A, B and C in (3) and the worst-case performance value f_{WC}, calculate the coefficients $\widetilde{A}$, $\widetilde{B}$ and $\widetilde{C}$ in the standard form (10) by using (8)-(9).
6. Solve the semi-definite programming (SDP) in (12) for $\lambda*$.
7. Calculate the worst-case corner $\Delta Y*$ using (15).

Algorithm 1 summarizes the major steps of the proposed worst-case corner extraction: (a) quadratic performance modeling, (b) worst-case performance estimation, and (c) worst-case corner extraction. Once the corners are extracted, designers can simulate their circuit at these corners, find the reason for its performance failure, and eventually come up with the appropriate solution to improve robustness.

4. NUMERICAL EXAMPLE

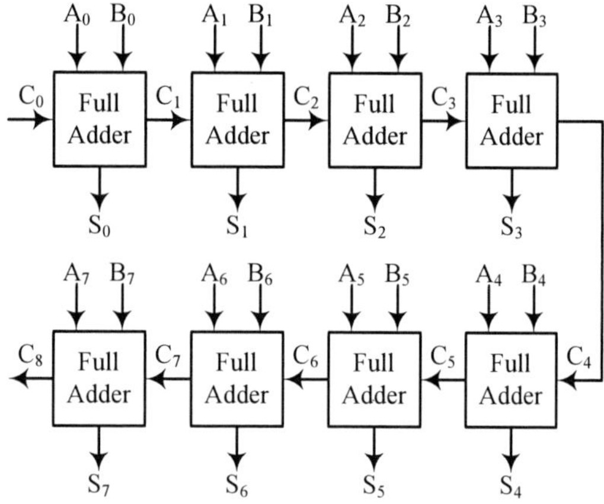

Fig 1. Simplified circuit schematic of an 8-bit ripple carry adder.

In this section, we demonstrate the efficacy of the proposed corner extraction algorithm using an 8-bit ripple carry adder, as shown in Fig 1. The circuit is implemented in a commercial CMOS process. It consists of a chain of 1-bit adders. In this example, only inter-die variations are considered for MOS transistors. After PCA based on foundry data, 34 independent random variables are extracted to model these variations. It should

978-1-60558-497-3/09 $25.00 © 2009 ACM

be noted, however, that nothing would preclude us from handling intra-die variations by using the proposed methodology.

In this example, we are interested in the variability of the propagation delay from "B_0" to "S_7". We first apply quadratic performance modeling [14] to approximate the delay (i.e., f) as a function of the random variations (i.e., ΔY). The relative modeling error is 0.39%. Fig 2 shows the eigenvalues of the quadratic coefficient matrix A for the performance model $f(\Delta Y)$. Similar to the previous example, the quadratic function $f(\Delta Y)$ is neither positive semi-definite nor negative semi-definite in this example.

Next, we apply Monte Carlo analysis to the quadratic performance model $f(\Delta Y)$, and calculate the worst-case delay that is defined as the 99% point of $cdf(f)$. The semi-definite programming (SDP) problem in (12) is then solved by CVX [12], and the worst-case corner ΔY^* is calculated by using (15). In this example, the runtime for SDP is 1.5 seconds on a LINUX 2.8GHz server with 2GB memory.

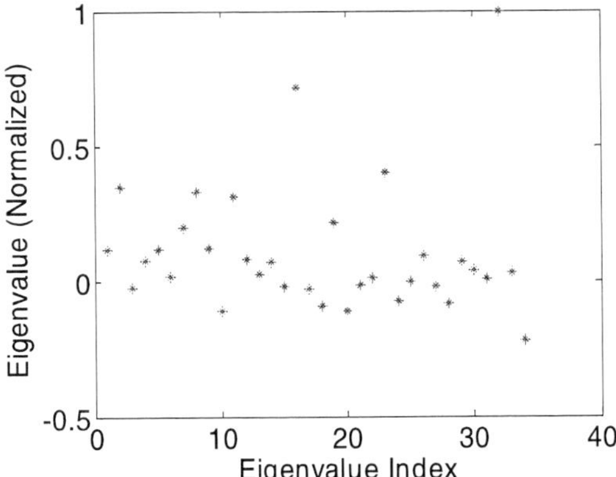

Fig 2. Eigvalues of the quadratic coefficient matrix A for the performance model $f(\Delta Y)$.

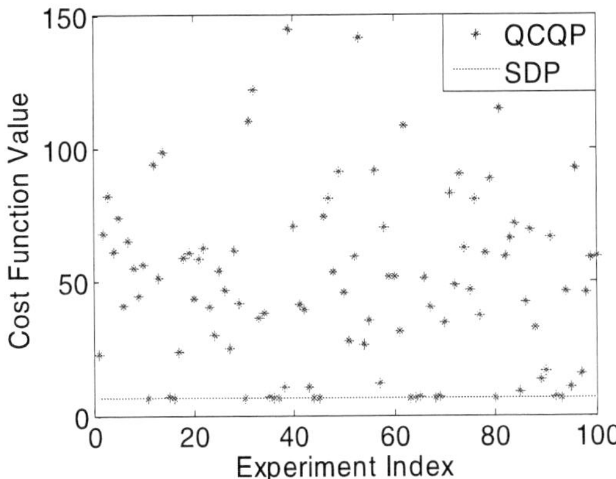

Fig 3. The minimized cost function $\|\Delta Y\|_2^2$ by semi-definite programming (SDP) and 100 runs of quadratically constrained quadratic programming (OCQP) with random initial guess.

For testing and comparison, we apply sequential quadratic programming (SQP) [15] to solve the quadratically constrained quadratic programming (QCQP) problem in (11) to extract the corner ΔY^*. One hundred independent QCQP runs are performed with random initial guess in order to check the convergence property. As shown in Fig 3, a large number of (83 out of 100) QCQP runs get stuck at a local minimum. It, in turn, demonstrates that QCQP fails to robustly extract the worst-case corner ΔY^*

from a non-convex quadratic model.

5. CONCLUSIONS

In this paper, we propose a novel semi-definition programming (SDP) method to efficiently extract design-specific worst-case corners for nanoscale circuits. The proposed SDP approach is facilitated by the Lagrange duality theory of nonlinear optimization [15]. It converts a non-convex quadratically constrained quadratic programming (QCQP) problem to a convex SDP problem that is easier to solve. Our numerical experiments demonstrate that while the traditional QCQP fails to achieve global convergence, our proposed SDP formulation can find the worst-case corners both robustly and efficiently. The proposed SDP method can be further incorporated into a statistical circuit analysis flow to provide design insights and assist failure diagnosis.

6. REFERENCES

[1] Semiconductor Industry Associate, *International Technology Roadmap for Semiconductors*, 2007.

[2] X. Li, J. Le, L. Pileggi and A. Strojwas, "Projection-based performance modeling for inter/intra-die variations," *IEEE ICCAD*, pp. 721-727, 2005.

[3] Z. Feng and P. Li, "Performance-oriented statistical parameter reduction of parameterized systems via reduced rank regression," *IEEE ICCAD*, pp. 868-875, 2006.

[4] A. Singhee and R. Rutenbar, "Beyond low-order statistical response surfaces: latent variable regression for efficient, highly nonlinear fitting," *IEEE DAC*, pp. 256-261, 2007.

[5] A. Mitev, M. Marefat, D. Ma and J. Wang, "Principle Hessian direction based parameter reduction for interconnect networks with process variation," *IEEE ICCAD*, pp. 632-637, 2007.

[6] X. Li and H. Liu, "Statistical regression for efficient high-dimensional modeling of analog and mixed-signal performance variations," *IEEE DAC*, pp. 38-43, 2008.

[7] G. Debyser and G. Gielen, "Efficient analog circuit synthesis with simultaneous yield and robustness optimization," *IEEE ICCAD*, pp. 308-311, 1998.

[8] F. Schenkel, M. Pronath, S. Zizala, R. Schwencker, H. Graeb and K. Antreich, "Mismatch analysis and direct yield optimization by spec-wise linearization and feasibility-guided search," *IEEE DAC*, pp. 858-863, 2001.

[9] X. Li, J. Le, P. Gopalakrishnan and L. Pileggi, "Asymptotic probability extraction for non-normal distributions of circuit performance," *IEEE ICCAD*, pp. 2-9, 2004.

[10] A. Dharchoudhury and S. Kang, "Worst-case analysis and optimization of VLSI circuit performances," *IEEE Trans. CAD*, vol. 14, no. 4, pp. 481-492, Apr. 1995.

[11] M. Sengupta, S. Saxena, L. Daldoss, G. Kramer, S. Minehane and J. Cheng, "Application-specific worst case corners using response surfaces and statistical models," *IEEE Trans. CAD*, vol. 24, no. 9, pp. 1372-1380, 2005.

[12] M. Grant and S. Boyd, *CVX Users' Guide*, 2008. (www.stanford.edu/~boyd/cvx/)

[13] G. Seber, *Multivariate Observations*, Wiley Series, 1984.

[14] R. Myers and D. Montgomery, *Response Surface Methodology: Process and Product Optimization Using Designed Experiments*, Wiley-Interscience, 2002.

[15] S. Boyd and L. Vandenberghe, *Convex Optimization*, Cambridge University Press, 2004.

Timing-driven optimization using lookahead logic circuits

Mihir Choudhury and Kartik Mohanram

Department of Electrical and Computer Engineering, Rice University, Houston

{mihir,kmram}@rice.edu

Abstract

This paper describes a timing-driven optimization technique for the synthesis of multi-level logic circuits. Motivated by the parallel prefix problem, the proposed timing-driven optimization produces logic circuits with "lookahead" properties due to the inherent parallelism among the synthesized sub-circuits. Lookahead logic circuits are synthesized using global critical path sensitization information to decompose and reduce the Boolean functions of the nodes in the technology-independent representation of the logic circuit. Unlike prior timing-driven optimization techniques, where synthesis of the decomposition functions is potentially expensive, the proposed technique has the advantage that the decomposition functions are discovered in the synthesized form. On average, the proposed technique reduces the number of logic levels (mapped delay) of 15 benchmark circuits by 40%, 56%, and 22% (21%, 56% and 10%) over the best results of SIS, ABC, and an industry-standard synthesizer, respectively.

Categories and Subject Descriptors: B.6.3 [LOGIC DESIGN]: Design Aids—Automatic synthesis
General Terms: Algorithms, Design, Performance
Keywords: Logic synthesis, timing optimization, lookahead

1. Introduction

Timing-driven optimization during multi-level logic synthesis is a well-researched area, and several solutions have been proposed in literature [1–9]. These techniques either restructure the critical paths or perform decomposition-based resynthesis of the circuit. Restructuring techniques, such as [1–4], are computationally efficient but the improvements from these techniques are limited because restructuring is restricted to cutsets of nodes on the critical path. On the other hand, decomposition-based resynthesis techniques, such as [5–7], have immense scope for optimization because the space of possible transformations is vast. However, the algorithms proposed in literature are computationally intensive and the improvements achieved are limited by available computational resources.

Motivated by the parallel prefix problem [10, 11], this paper describes a timing-driven optimization technique for the synthesis of multi-level logic circuits. The prefix problem is one of the funda-

mental approaches to build parallel algorithms and has been extensively studied in literature, with successful applications to problems including sorting, parallelizing compilers, task scheduling, etc. The classic example of the application of the prefix problem to logic circuits is in the design of the tree-structured carry lookahead adder (CLA), where it is used to reduce the delay of carry propagation in an n-bit ripple carry adder from $O(n)$ to $O(\log(n))$.

The basic property that allows the application of prefix theory to these problems is the identification of intermediate computation that can be performed in parallel. For instance, in an n-bit binary adder, the generate bit (g_i) and propagate bit (p_i) for each bit-slice can be computed in parallel and the carry bit (c_i) is a prefix computation defined over the pair (p_j, g_j), $1 \leq j \leq i$. The regular modular structure of the adder makes it easy to identify parallel computation of the pairs (p_j, g_j) for each bit-slice in the adder. Once the pairs are computed, parallel prefix theory is used to synthesize a fast implementation for the carry, such as the CLA. For an adder, parallel intermediate computation are simple functions with disjoint support, i.e., (g_i, p_i) and (g_j, p_j), $i \neq j$ do not have any common inputs. However, in general multi-level logic circuits such as control logic in microprocessors, intermediate computations are complex functions with non-disjoint support due to logic sharing. Hence, identifying parallel computation to apply the principles of the prefix problem to general multi-level logic circuits is significantly more challenging.

In this paper, we propose a timing-driven optimization technique to identify intermediate computation that can be parallelized in general multi-level logic circuits. The optimized logic circuits exhibit "lookahead" properties due to the inherent parallelism among the synthesized sub-circuits, and are hence called lookahead logic circuits in this paper. When applied to a ripple carry adder, our technique can systematically derive different realizations of high-performance adders including the carry lookahead, carry skip, and carry select adders. The main advantage of our technique is that it synthesizes decomposed circuits with smaller delays in the form of lookahead logic circuits, instead of searching for decomposition functions. Lookahead logic circuits are synthesized by decomposing and reducing the Boolean functions corresponding to the internal nodes in the technology-independent representation of the original logic circuit. The lookahead logic circuits are then combined using Shannon's decomposition and its implication-based simplifications to reconstruct the original logic circuit. Unlike prior timing-driven optimization techniques, where the synthesis of the decomposition functions is potentially expensive, our technique has the advantage that the decomposition function is discovered in the synthesized form of the lookahead logic circuit.

The performance of lookahead logic circuits is compared to the state-of-the-art academic tools SIS and ABC, and an industry-standard synthesizer. First, a case study of the n-bit ripple carry adder is used to compare our technique with these tools. Results indicate that our technique discovers several interesting decompositions with fewer levels of logic. Next, 15 circuits from the MCNC

The authors would like to acknowledge Prof. Peter Varman at Rice University and Prof. Adnan Aziz at the University of Texas, Austin for helpful discussions and suggestions. This research was supported in part by NSF CAREER Award CCF-0746850 and in part by a gift from the Fujitsu Laboratories of America.

Permission to make digital or hard copies of part or all of this work for personal or classroom use is granted without fee provided that copies are not made or distributed for profit or commercial advantage and that copies bear this notice and the full citation on the first page. To copy otherwise, to republish, to post on servers or to redistribute to lists, requires prior specific permission and/or a fee.
DAC'09, July 26-31, 2009, San Francisco, California, USA

978-1-60558-497-3/09 $25.00 © 2009 ACM

and ISCAS benchmark suites and the OpenSPARC T1 processor are used to compare our technique to the best results obtained using these tools. On average, our technique reduces the number of logic levels in the final circuit by 40%, 56% and 22% over the best results of SIS, ABC, and the industry-standard synthesizer, respectively. When mapped delays are evaluated, our technique achieves an average reduction of 21%, 56% and 10% over the best results of SIS, ABC, and the industry-standard synthesizer, respectively. Our approach is computationally efficient, with a runtime of 100 seconds on the largest circuit considered in this paper.

This paper is organized as follows. Sec. 2 provides a background on existing timing-driven optimization techniques. Sec. 3 introduces lookahead logic circuits and describes the proposed synthesis algorithm. Sec. 4 presents a case study for n-bit adders. Sec. 5 presents results. Sec. 6 presents conclusions.

2. Timing-driven decomposition

The timing-driven optimizations proposed in literature can be broadly divided into two classes: (i) structure-based and (ii) decomposition-based. The earliest techniques for timing-driven optimization were based on restructuring critical paths to reduce circuit delay [1–4]. Most structure-based techniques have used the transformation of a ripple carry adder into a fast implementation like the CLA, carry select adder or carry bypass adder as motivation for their techniques. The technique proposed in [1], called tree height reduction, uses a CLA as motivation to reduce the delay of the circuit by rescheduling computation along critical paths. The technique presented in [2], called the generalized select transform, uses a carry select adder as a motivating example and proposes a technique that identifies late arriving signals, performs computation using both 0 and 1 as the value for the signal, and then uses that signal to select the correct output through a multiplexer. In [3], the carry bypass adder is used as motivation to propose the generalized bypass transform that reduces the critical path delay by adding redundant bypass paths and turning the critical paths into false paths. The false paths can then be eliminated without increasing the delay of the circuit using a technique presented in [4].

Decomposition-based techniques fundamentally differ from structure-based techniques in that they do not directly restructure the circuit. Instead, the circuit structure is changed as a result of changing the functionality of the internal nodes, while maintaining functional equivalence at the primary outputs. Decomposition-based techniques are capable of exploring a much richer design synthesis space, at higher computational cost, as compared to structure-based techniques. A decomposition-based technique using partial collapsing and simplification of nodes to reduce the delay is proposed in [5]. The technique proposed in [6] uses permissible functions to resynthesize sets of nodes that lie on the critical path to reduce the delay. In [7], additional redundant circuitry is added to compute the output on input patterns that sensitize the critical paths. This approach includes features of structure-based techniques, but suffers the following drawbacks. Since redundant logic is added in the form of bypass paths to the original circuit, the technique leads to a circuit with a high area and/or power footprint. The improvements in delay are limited because the additional redundant logic is restricted to only implications (0-approximation or 1-approximation) of the original function. The scalability of this approach is also limited due to a bottom-up synthesis approach for the additional redundant logic starting from an incompletely specified Boolean function with a large don't care space. Finally, although not directly related to the present work, BDD-based decomposition techniques for timing optimization have also been proposed [9, 12–15] and to this day are an active area of research.

In this paper, we propose a decomposition-based timing-driven optimization technique using lookahead logic circuits. Unlike prior techniques, where the synthesis of the decomposition functions is potentially expensive, our technique has the advantage that the decomposition functions are discovered in the synthesized form. It can explain conversion of a ripple carry adder into several fast implementations including the carry lookahead, carry select, and carry bypass adders. Like most other timing-driven optimization techniques, it also complements existing logic optimization algorithms. The next section develops the theory of lookahead logic circuits and describes the synthesis algorithm for lookahead logic circuits.

3. Lookahead logic circuits

With the background on timing-driven optimization, we use binary addition to introduce the basic principles of prefix computation and then develop the theory of lookahead logic circuits. The most common approach to speed up carry computation in adders with large operand sizes is to exploit the observation that carry propagation in binary addition is a *prefix* problem [16].

Prefix problem: Given n values $z_1, z_2, \ldots, z_n$ and an associative binary operator $\otimes$, the prefix computation problem, or simply the prefix problem, is to compute the n values $z_i \otimes z_{i-1} \otimes \ldots \otimes z_1$, $1 \leq i \leq n$. In the context of binary addition of two n-bit numbers a and b, the carry for the ith bit can be expressed as

$$\text{Prefix element}: z_i = (g_i, p_i)$$
$$\text{Operator}: (u_1, v_1) \otimes (u_2, v_2) = (u_1 + v_1 u_2, v_1 v_2)$$
$$(G_{i:j}, P_{i:j}) = (g_i, p_i) \otimes (g_{i-1}, p_{i-1}) \otimes \ldots \otimes (g_j, p_j), i > j$$
$$c_i = G_{i:1} + P_{i:1} c_{\text{in}}, 1 \leq i \leq n$$
$$c_{\text{out}} = g_n + p_n g_{n-1} + \cdots + p_n p_{n-1} \cdots p_1 c_{\text{in}} \tag{1}$$

where $g_i = a_i b_i$ and $p_i = a_i \oplus b_i$ represent the generate and propagate bits. Since the prefixes g_i and p_i can be computed in parallel, the prefix problem reduces to efficient prefix computation and several tree structures, with size and depth trade-offs, have been proposed in literature to realize parallel-prefix adders [11].

We make the important observation that the parallel-prefix CLA can be thought of as an optimal timing-driven decomposition for carry computation and we generalize this as follows. Consider a Boolean function $f(x_1, x_2, \ldots, x_n)$ of n inputs $x_1, x_2, \ldots, x_n$. Consider the decomposition for the Boolean function f given by the identity

$$f = \overline{\Sigma}_l f_l + \Sigma_l \overline{\Sigma}_{l-1} f_{l-1} + \cdots + \\ \Sigma_l \Sigma_{l-1} \cdots \overline{\Sigma}_1 f_1 + \Sigma_l \Sigma_{l-1} \cdots \Sigma_1 f_0 \tag{2}$$

where Σ_i (called the window function) and f_i are all functions of $x_1, x_2, \ldots, x_n$. By drawing an analogy to the CLA representation from equation 1, we can interpret the CLA representation from equation 2 as a *lookahead* decomposition for the Boolean function f. Here, $\overline{\Sigma}_i f_i$ corresponds to the generate bit g_i and Σ_i corresponds to the propagate function p_i, $1 \leq i \leq l$. The interesting connection between the CLA representation and the timing-driven decomposition lies in the expressions for Σ_i and f_i. Let us look at the timing-critical computation for the carry bit, c_i, of each stage of the n-bit adder. Note that c_i can be computed without the carry, c_{i-1}, of the previous stage when $a_i = b_i = 0$ ($c_i = 0$) and when $a_i = b_i = 1$ ($c_i = 1$). Thus, the case $a_i = b_i$ is not a timing-critical computation at the ith bit-slice. However, when $a_i \neq b_i$ ($a_i \oplus b_i = 1$), the carry of the previous stage is necessary to compute c_i. Hence, $a_i \neq b_i$ is a timing-critical computation at the ith bit-slice. When Σ_i is set to $a_i \oplus b_i$ and f_i is set to the value of c_i for

978-1-60558-497-3/09 $25.00 © 2009 ACM

$\overline{\Sigma}_i = 1$, i.e., $f_i = a_i$ or $f_i = b_i$, the timing-driven decomposition for c_{out} for an n-bit adder is given by

$$
\begin{aligned}
c_{\text{out}} &= (a_n\overline{\oplus}b_n)a_n + \ldots + (a_n \oplus b_n)\ldots(a_2 \oplus b_2)(a_1\overline{\oplus}b_1)a_1+ \\
&\quad + (a_n \oplus b_n)\ldots(a_1 \oplus b_1)c_{\text{in}} \\
&= a_nb_n + \ldots + (a_n \oplus b_n)\ldots(a_2 \oplus b_2)a_1b_1+ \\
&\quad + (a_n \oplus b_n)\ldots(a_1 \oplus b_1)c_{\text{in}} \\
&= g_n + p_ng_{n-1} + \ldots + p_np_{n-1}\ldots p_1c_{\text{in}},
\end{aligned} \qquad (3)
$$

which is equivalent to the expression for c_{out} obtained using the prefix problem in equation 1.

Thus, the key contribution of this paper for timing-driven optimization is the use of information about timing critical computation to identify window functions Σ_i that produce lookahead logic circuits f_i with fewer levels of logic. The regular modular structure of a binary adder makes it easy to identify a good timing-driven decomposition, Σ_i and f_i. However, applying this technique to the synthesis of multi-level control logic circuits is challenging for the following reasons:

1. Control logic is irregular with multiple critical paths. Due to logic sharing, control logic defies the easy modularity that makes it possible to write a CLA-like representation for the Boolean expression of the critical paths.

2. The Boolean expression for the critical-path in control logic is significantly more complex, i.e., it cannot be expressed as a simple expression such as that for the carry in adders.

3. Both Σ_i and f_i for an adder have a disjoint support set for $1 \leq i \leq n$, i.e., Σ_i and Σ_j as well as f_i and f_j ($i \neq j$) do not have common inputs in their support. However, for multi-level logic circuits, Σ_i and f_i may not have disjoint support sets. Hence, realizing them independently using separate logic circuits can be very expensive, and a good tradeoff would be to share logic between these functions.

4. Unlike an adder where the delay of each p_i and g_i term is equivalent to a single level of logic, the functions Σ_i and f_i may have different levels of logic and delays and hence combining them optimally is a challenge.

In the rest of this section, we will describe a synthesis technique for lookahead logic circuits (circuits for implementing Σ_i and f_i) that addresses these challenges.

Definitions

Decomposed logic circuit: A decomposed logic circuit is a directed acyclic graph (DAG) with nodes representing AND gates. The edge connecting a node i to another node j can be of two types: (i) complemented, when there is an inverter between the output of node i and input of node j and (ii) uncomplemented, when there is no inverter. Thus, a decomposed circuit uses AND and NOT gates as building blocks, and is referred to as an and-invert-graph (AIG).

Technology-independent network: A technology-independent network is an intermediate DAG representation of a circuit in which the internal nodes are arbitrary Boolean functions. An AIG can be converted into a technology-independent representation using clustering algorithms ("renode" command in the tool ABC [17]).

3.1 Synthesis of lookahead logic circuits

Given a decomposed circuit $\mathcal{C}$ with n inputs, $x_1, x_2, ..., x_n$, and m outputs, let l_C denote the number of levels of logic in $\mathcal{C}$. Although our implementation considers all outputs simultaneously, for ease of notation and without loss of generality, we refer to a primary output y containing at least one critical path, i.e., at least one

path with l_C levels of logic for the rest of this discussion. Consider the problem of obtaining a single level of timing-driven decomposition for the Boolean function, y, given by

$$
y = \overline{\Sigma}_1y_1 + \Sigma_1y_0 \qquad (4)
$$

as proposed in equation 2 to improve the performance of the circuit by reducing the number of logic levels.

Attempting a function-based decomposition of the Boolean function y as shown in equation 4 has two major disadvantages. First, there is no knowledge of the circuit implementation of Σ_1, y_0, and y_1. Hence, a function-based decomposition may result in a bad choice of Σ_1, y_0, or y_1 that may lead to a higher number of logic levels than the original circuit. Since the space of decompositions is vast, finding a good function-based decomposition based on equation 4 with lesser levels of logic than the original circuit is challenging. Second, even with the knowledge of the functions Σ_1, y_0, and y_1 that can potentially produce a good decomposition, directly synthesizing AIGs for these Boolean functions is a challenge and does not scale as the complexity of the function increases.

In this paper, we propose a novel approach to address the issues of finding the functions Σ_1, y_0, and y_1 and synthesizing their AIGs to have fewer logic levels than the original circuit. Our technique is based on two key ideas. First, we use transformations on the technology-independent network, $\mathcal{T}$, of the original decomposed circuit, $\mathcal{C}$, to synthesize the technology-independent networks for Σ_1, y_0, and y_1. The transformations are made by simplifying the Boolean functions of the internal nodes in the technology-independent network to reduce the logic levels of the circuit. In this process, the functions Σ_1, y_0, and y_1 are derived dynamically during simplification. Second, we use global path sensitization information, extracted from the given decomposed circuit, $\mathcal{C}$, as a metric to guide the simplification. This ensures that the simplifications transform the functionality of the internal nodes significantly to reduce delay while preserving the functionality at the primary outputs. Our technique has two stages: (i) extracting global critical path sensitization information from $\mathcal{C}$ and (ii) simplifying the technology-independent network $\mathcal{T}$ to obtain the technology-independent representations of Σ_1, y_0, and y_1. We will now describe each step in greater detail.

Path sensitization information: The aim of obtaining path sensitization information for output y is to identify minterms in the input space of y that are responsible for exercising *all* the speed-paths (critical or near-critical paths) in the decomposed circuit. These minterms are referred to as the timing-critical minterms or speed-path minterms in the input space of y. We shall refer to this set of minterms as the speed-path characteristic function (SPCF) for y. Thus, for a given delay Δ, the SPCF for y contains all minterms that sensitize paths of length greater than or equal to Δ. To compute the SPCF for a decomposed circuit, in which the delay is given by the number of levels of logic, Δ may be set to an integer value greater than 0. In that case, the SPCF will contain all minterms that sensitize paths with greater than or equal to Δ levels of logic.

Several algorithms have been proposed for the exact computation of the SPCF [7, 18]. These algorithms compute the exact set of minterms that sensitize paths with a delay greater than or equal to a desired value. These algorithms are path-based and require traversal of each critical path. Other algorithms that compute an approximation of the SPCF have also been proposed [19, 20]. These algorithms compute an over-approximation of the SPCF, i.e., minterms that do not sensitize critical paths may be included in the SPCF. The over-approximation algorithms are computationally more efficient than path-based algorithms because they are node-based and

978-1-60558-497-3/09 $25.00 © 2009 ACM

require computation only at nodes that lie on the critical path. Note that the SPCF is used only as a metric to guide the synthesis of the lookahead logic circuit. Although our implementation computes the SPCF exactly, it is possible to use the over-approximation techniques to compute the SPCF for computational efficiency.

After the SPCF for output y is computed, simplifications are made to the technology-independent network $\mathcal{T}$. The simplification of $\mathcal{T}$ is performed in two stages. The first simplification, referred to as the primary simplification, is used to synthesize the technology-independent networks for Σ_1 and y_0 and the second simplification, referred to as the secondary simplification, is used to synthesize the technology-independent network for y_1. Both primary and secondary simplifications involve simplifying the Boolean expressions of the internal nodes in $\mathcal{T}$. As a result of the simplifications, the Boolean function for output y is transformed to y_0 in the primary simplification and to y_1 in the secondary simplification. In the primary simplification, additional logic for the technology-independent network of Σ_1 is also added to $\mathcal{T}$.

Primary simplification of $\mathcal{T}$: The pseudo-code for the primary simplification algorithm is shown in algorithm 2. The main goal of the primary simplification of $\mathcal{T}$ is to reduce the number of logic levels by simplifying the Boolean function of the internal nodes in $\mathcal{T}$. When an internal node in $\mathcal{T}$ is simplified, the original Boolean function at y is changed to y_0. By adding additional logic to $\mathcal{T}$, the algorithm ensures that the window function, Σ_1, is altered suitably (as described in algorithm 1) so that $y_0 = y$ when $\Sigma_1 = 1$. The algorithm ensures that the additional logic does not cancel the reduction in logic levels obtained as a result of the simplification of the internal node. Another goal of the primary simplification is to obtain a good window function Σ_1. As we have seen in the carry lookahead adder example in equation 3, functions containing timing-critical minterms or speed-path minterms form good window functions. Thus, the SPCF for the output y is used as a metric to guide the simplification of the internal nodes as explained below.

Using the SPCF: Consider an internal node j in the fanin cone of output y in $\mathcal{T}$. Let b_j denote the Boolean function of this node. Thus, b_j is a typical Boolean function with 10–15 inputs. The SPCF contains the global critical-path sensitization minterms for output y. Let $\tilde{b}_j$ denote the Boolean function obtained after simplification of b_j. In order to use the SPCF information for simplification of b_j, we assign a weight $w(c)$ to each prime implicant cube c in the off-set and on-set of b_j. The weight $w(c)$ is the fraction of minterms in the SPCF that will be covered in the window function Σ_1 if $\tilde{b}_j(c) = b_j(c)$. Thus, $w(c)$ is the metric based on which the Boolean function of the internal node is simplified. The cube weights can be easily computed for each node using the global Boolean functions of each node and the SPCF. Note that the cube weights for a node are computed only if the node is chosen for simplification. The function reduce in algorithm 2 describes the procedure for choosing nodes for simplification. The function simplify in algorithm 1 describes the procedure for simplifying the Boolean function of a node using the SPCF. At the end of the primary simplification, a technology-independent network for Σ_1 and y_0 is obtained for every output y with l_c levels of logic.

Secondary simplification of $\mathcal{T}$: The primary simplification determines the window function Σ_1. In the secondary simplification, $\mathcal{T}$ is reduced to generate the technology-independent network for y_1. Thus, in the secondary simplification, the complement of the window function, $\overline{\Sigma}_1$, is used to assign cube weights for the internal nodes. However, unlike the primary simplification, where the nodes had to be carefully chosen for simplification in order

to obtain a good window function Σ_1, the only objective of the secondary simplification is to generate the technology-independent network for y_1. Hence, the objective is to reduce the levels of logic in $\mathcal{T}$ as much as possible. This is done by replacing all cubes with zero weight by don't cares to simplify the Boolean function of every node. After the secondary simplification, the technology-independent network for y_1 is obtained for every output y with l_c levels of logic.

Reconstructing y: In general, equation 4 can be used to reconstruct y from Σ_1, y_1, and y_0. However, there are several simplifications that can be applied when Σ_1, y_1, and y_0 satisfy implication properties with y. For example, consider $\overline{y_0} \Rightarrow \overline{y}$ and $y_1 \Rightarrow y$. This means that y_1 is a 1-approximation for y and y_0 is a 0-approximation for y. This can be used to reduce y to $\Sigma_1 y_0 + y_1$. In this manner, 28 unique implication-based rules can be identified for the simplification of the Shannon decomposition in equation 4. We do not list them in the paper for brevity. In our optimization runs, we have observed that the implication-based rules are frequently used to reduce the number of levels of logic while reconstructing y. Finally, the technology-independent network for the reconstructed y is converted into a decomposed circuit by converting each node in the technology-independent network into an AIG. Area recovery is then performed using standard redundancy elimination algorithms.

Algorithm 1: simplify(j)

---

    input   : j is a node in $\mathcal{T}$ with Boolean function b_j and logic level l_j

    output   : $\tilde{b}_j$, the simplified Boolean function for node j

    $S_0(S_1)$ is the minimum 0(1)-SOP of b_j

    $w(c)$ is the weight of cube c, $c \in S_0$ or $c \in S_1$

    if $w(c) = 0 \; \forall c \in S_0(S_1)$ **then**

        $\tilde{b}_j = 0(1)$

        $\mathcal{L}$ – Cubes of S_1 (S_0) in decreasing order of weight

        foreach $c \in \mathcal{L}$ **do**

            $\tilde{b}_j(c) = 1(0)$

            Compute level (j) assuming $\tilde{b}_j$ is the Boolean function of j

            if level (j) $\geq l_j$ **then**

                $\tilde{b}_j(c) = 0(1)$

        window(j) = $\tilde{b}_j$ $(\overline{\overline{\tilde{b}_j}})$

    else

        Both 0-SOP and 1-SOP for j have non-zero weights

        Initialize $\tilde{b}_j = x$ /* don't care */

        $\mathcal{L}$ – Cubes of S_0 and S_1 in decreasing order of weight

        foreach $c \in \mathcal{L}$ **do**

            Set $\tilde{b}_j(c) = b_j(c)$

            Compute level (j) assuming $\tilde{b}_j$ is the Boolean function of j

            if level (j) $\geq l_j$ **then**

                $\tilde{b}_j(c) = x$ /* don't care */

        window(j) = $\tilde{b}_j \oplus b_j$

    mark (j)

Quantifying logic levels in $\mathcal{T}$: The logic levels for the nodes in a technology-independent network is used during the simplification of the technology-independent network in the proposed algorithm and is also used to keep track of the progress in the reduction of the logic levels. The logic level for a node j, level(j), is computed using the minimum sum-of-products (SOP) representation of the off-set and on-set for the Boolean function of node j. The minimum logic level is computed for the Huffman AND tree of each prime-implicant cube in the off-set and on-set. The minimum logic level for the Huffman OR tree is then computed using the minimum logic level of each cube. The smaller logic level value, between the off-set and the on-set, is defined as the logic level for node j. In addition, to computing the level of each node, the critical inputs

978-1-60558-497-3/09 $25.00 © 2009 ACM

Algorithm 2: reduce($\mathcal{C}$, $\mathcal{T}$, SPCF($l_\mathcal{C}$))

input	: Decomposed circuit $\mathcal{C}$ with l. levels of logic
input	: SPCF(l.) $\forall$ output $y \in \mathcal{C}$
input	: Technology-independent network $\mathcal{T}$ for $\mathcal{C}$ with l. levels of logic
output	: Modified $\mathcal{T}$ with y_0 and Σ_1 $\forall$ output $y \in \mathcal{C}$

```
foreach output y of T do
    if SPCF(y) = 0 then
      └ continue   /* output does not contain critical path */
    repeat
      │ j = Unmarked node with highest logic level in fanin (y)
      │ b_j = simplify (j)
      │ Recompute logic level of nodes in T
    until level (y) < l.
    y_0 = y   /* output of the reduced network */
    Σ_1 = ⋀_marked nodes j (window(j))
  └ Unmark all nodes in T
```

can also be identified for each node. An input to a node is critical if the reduction of its level is a necessary condition for reducing the level of the node. The critical inputs to a node are also used in the the function `reduce` to explore candidate nodes for the function `simplify`.

4. Case study: n-bit adder

Historically, the adder has been an excellent example for evaluating various timing-driven optimization techniques primarily because of its regular prefix structure. Fast implementations of an n-bit adder include the (i) carry lookahead adder (CLA), (ii) carry select or conditional carry adder, and (iii) carry bypass or carry skip adder. In Sec. 2, we have described how existing timing-driven optimization techniques have used one of these adders as a motivating example to develop timing-driving optimizations for general multi-level logic circuits. In contrast, our timing-driven optimization technique can be used to derive *all* these fast adders from a ripple carry adder. Let a and b be two 2-bit binary numbers and c_{in} be the carry-in bit. Let y denote the two bit sum and c_{out} denote the carry. Let $g_i = a_i b_i$ denote the generate bit and $p_i = a_i + b_i$ denote the propagate bit.

The simplest implementation of an n-bit adder is a ripple carry adder that can be realized by linearly cascading n full adders. Although the ripple carry-adder has a small area, the critical path delay of the ripple carry adder is $O(n)$. The carry-propagation logic is the most delay-intensive operation in a ripple carry adder. In a 2-bit ripple carry adder, $c_{\text{out}} = g_2 + p_2(g_1 + p_1 c_{\text{in}})$ with 5 levels of logic. We will now explain how our timing-driven decomposition can transform a ripple carry adder into all these fast adders.

CLA (4 levels, disjoint): Based on the discussion in Sec. 3, two levels of timing-driven decomposition, i.e., (Σ_2, y_2) and (Σ_1, y_1) can be used to convert a ripple carry adder into a CLA. The window functions at the two levels are disjoint.

$$\Sigma_1 = (a_1 \oplus b_1) \text{ and } \Sigma_2 = (a_2 \oplus b_2)$$
$$y_0 = c_{\text{in}}, y_1 = a_1, \text{ and } y_2 = a_2$$
$$c_{\text{out}} = \overline{\Sigma}_2 y_2 + \Sigma_2 \overline{\Sigma}_1 y_1 + \Sigma_2 \Sigma_1 y_0$$

Carry select and carry bypass adders (4 levels, overlapping): For the carry select and carry bypass adders, a single-level of decomposition is sufficient to realize the final implementation. However, it is important to note that 2-bit carry select and carry bypass adders have 4 levels of logic if a multiplexer is considered as a single level of logic. Both decompositions are overlapping because y_1 and y_0 have common inputs in their support. For the carry select

adder, we have:

$$\Sigma_1 = c_{\text{in}}, y_0 = g_2 + p_2 p_1, \text{ and } y_1 = g_2 + p_1 g_1$$
$$c_{\text{out}} = \overline{\Sigma}_1 y_1 + \Sigma_1 y_0$$

For the carry bypass adder, we have:

$$\Sigma_1 = p_2 p_1, y_0 = c_{\text{in}}, \text{ and } y_1 = g_2 + p_2 g_1$$
$$c_{\text{out}} = \overline{\Sigma}_1 y_1 + \Sigma_1 y_0$$

New decomposition (4 levels, overlapping): The proposed technique also reveals another decomposition of the 2-bit adder with 4 logic levels. This decomposition also falls under the category of a single-level overlapping decomposition.

$$\Sigma_1 = c_{\text{in}} + g_2 + p_2 g_1, y_0 = g_2 + p_2 p_1, \text{ and } y_1 = 0$$
$$c_{\text{out}} = \overline{\Sigma}_1 y_1 + \Sigma_1 y_0$$

From these examples, it is clear that even a simple circuit like a 2-bit adder has four different decompositions with the optimal number of logic levels. This illustrates the expressive power of overlapping timing-driven decomposition techniques to extract equivalent descriptions with area-delay tradeoffs.

For a 2-bit adder, it is easy to identify many different fast implementations. In general, for an n-bit adder ($n \geq 4$), identifying the adder implementation with the optimal number of logic levels is non-trivial. To illustrate this, we present the best results from SIS, ABC, an industry-standard synthesizer, and our technique to optimize an n-bit ($n = 2, 4, 8, 16, 32$) ripple carry adder (details of the scripts used are given in the next section). We compare the results of synthesis to the theoretical number of logic levels required to generate the carry in a tree-structured CLA for each value of n in table 1. Note that in the optimum tree-structured CLA, the critical path terminates in the output computing the most significant bit (MSB) of the sum. Hence, the optimum number of logic levels for a 2-bit tree-structured CLA is 5, even though c_{out} has 4 logic levels. The number of logic levels obtained using existing techniques is higher than the theoretical optimum for the tree-structured CLA. In contrast, our technique provides the optimum solution for $n = 2$ and returns a circuit with one level of logic less than the optimum for $n \geq 4$. This is because our approach is able to identify a Boolean factoring for the MSB of the sum and c_{out} simultaneously.

Table 1: Comparison of best AIG levels after timing optimization of an n-bit adder, $n = 2, 4, 8, 16, 32$.

n	Tree-structured CLA	SIS [21]	ABC [17]	Industry-standard synthesizer	Lookahead logic circuits
2	5	6	6	5	5
4	7	11	9	8	6
8	9	17	18	11	8
16	11	28	34	15	10
32	13	51	66	18	12

5. Results

Our timing-driven optimization technique for synthesis of lookahead logic circuits is implemented within ABC [17]. All experiments were run on a 64-bit 2.4 GHz Opteron-based system with 6 GB memory. The performance of lookahead logic circuits is compared to state-of-the-art academic tools SIS and ABC, and an industry-standard synthesizer. Fifteen circuits from the MCNC and ISCAS benchmark suites and the OpenSPARC T1 processor are used to compare our technique to the best results obtained using

978-1-60558-497-3/09 $25.00 © 2009 ACM

Table 2: Comparison of the proposed technique with the best algorithms in SIS, ABC, and the industry-standard synthesizer

Name	PI/POs	SIS [21]				ABC [17]				Industry-standard synthesizer				Lookahead logic circuits			
		Gates	Levels	Delay	Power	Gates	Levels	Delay	Power	Gates	Levels	Delay	Power	Gates	Levels	Delay	Power
rot	135/107	621	14	167.5	4.4	453	21	216	3.6	591	15	174.7	5.5	624	11	172.2	5.8
dalu	75/16	1604	20	227.4	2.8	1046	31	355.6	5	703	14	138.5	3.5	966	11	123.4	5.4
i10	257/224	2454	29	436.2	12.6	1784	32	407.1	10.1	1935	26	284.4	13.9	1931	22	262	13.5
C432	36/7	273	22	260.4	2	136	23	248.7	1.2	197	19	205.4	2.4	250	15	174.4	3
C880	60/26	376	16	177.8	3	310	21	198.9	2.4	402	17	171	3.7	280	13	141.5	2.6
C2670	233/140	765	18	224.3	6.2	555	17	188.6	4.8	599	15	151.6	6.5	973	14	144.4	9.8
C5315	178/123	1784	21	245.4	13.3	1295	32	315.9	10.7	1464	26	261.8	14.1	1655	19	222	17.7
sparc_exu_ecl_flat	572/634	2409	13	184.8	14.2	2108	13	161.4	14.5	2422	12	144.6	19.5	2191	11	150.1	20.1
lsu_stb_ctl_flat	182/169	838	16	191.3	4.8	712	20	213.3	4.3	896	16	160.2	7.3	909	12	134.4	7.3
sparc_ifu_dcl_flat	136/94	487	13	146.1	3.1	414	19	192.5	2.6	474	15	151.7	3.5	517	12	153.2	4
sparc_ifu_dec_flat	131/146	881	14	153.3	4.4	797	14	158	4.4	923	13	186.4	6.2	828	13	151.8	6.6
lsu_excpctl_flat	251/179	670	12	130.9	4.9	567	13	137.6	4.3	685	12	133.7	5.8	743	12	127.7	6.6
sparc_tlu_intctl_flat	82/80	227	8	96.5	1.2	174	11	115.4	1	304	7	77.7	2.6	266	6	76.9	2.5
sparc_ifu_fcl_flat	465/522	2254	14	247.4	15.8	2043	17	256.1	13.8	2387	14	176.6	19.1	2459	11	184.4	22.4
tlu_hyperv_flat	449/464	2397	17	198.9	14.5	2278	11	309.7	17	2573	11	128.6	20.3	2424	10	102.3	17
Relative average	–	1.15	1.09	1.22	0.80	**0.88**	1.28	1.40	**0.70**	1	1	1	1	0.98	**0.78**	**0.90**	1.10

[†] SIS: `delay`, `rugged`, `algebraic` and `speed_up`; ABC: `resyn2rs`;
Industry-standard synthesizer: `-map-effort high -area-effort high`

these tools. Each benchmark circuit is optimized with each tool and mapped to a library of gates for the 65nm CMOS technology. For each circuit, an equivalence check is performed after optimization to ensure that the original and optimized circuits are equivalent. Our approach is computationally efficient, with a runtime of 100 seconds on the largest circuit considered in this paper.

The first two columns in table 2 give the circuit information. Subsequent columns report the number of gates in the AIG, logic levels in the AIG, technology-mapped delay, and the power consumption at 1GHz for the best results obtained with each optimization tool. Within SIS, the scripts `delay`, `rugged`, `algebraic`, and `speed_up` were used. For each benchmark circuit, the *best* results with the lowest technology-mapped delay are reported in the table. Within ABC, script `resyn2rs` was used. Within the industry-standard synthesizer, each design was compiled with the options `-map-effort high` and `-area-effort high`. The last row in the table compares the tools, on average and normalized to the industry-standard tool. On average, our technique shows a 40%, 56%, and 22% reduction in the number of logic levels in the optimized circuit over SIS, ABC, and the industry-standard synthesizer, respectively. Note that, on average, the size of the decomposed circuit obtained using our technique and the industry-standard tool are comparable. When mapped delays are evaluated, our technique achieves an average reduction of 21%, 56% and 10% over the best results of SIS, ABC, and the industry-standard synthesizer, respectively. For our technique, the trade-off for a 10% improvement in mapped delay over the industry-standard synthesizer is a 10% increase in the total power consumption.

6. Conclusions

This paper described a timing-driven optimization technique based on lookahead logic circuits. Lookahead logic circuits are synthesized by simplifying the technology-independent network of the original circuit using path sensitization information. The original logic circuit is then reconstructed from the lookahead logic circuits using Shannon's decomposition and its implication-based simplifications. The use of a technology-independent network for simplifications provides a computationally efficient means for searching a rich space of circuit decompositions to enhance the performance of the original circuit.

References

[1] K. Singh *et al.*, "Timing optimization of combinational logic," in *Proc. Intl. Conference Computer-aided Design*, pp. 282–285, 1988.

[2] C. Berman *et al.*, "Efficient techniques for timing correction," in *Proc. Intl. Symposium on Circuits and Systems*, pp. 415–419, 1990.

[3] P. McGeer *et al.*, "Performance enhancement through the generalized bypass transform," in *Proc. Intl. Conference Computer-aided Design*, pp. 184–187, 1991.

[4] K. Keutzer *et al.*, "Is redundancy necessary to reduce delay?," *IEEE Trans. Computer-aided Design*, vol. 10, no. 4, pp. 427–435, 1991.

[5] H. Touati *et al.*, "Delay optimization of combinational logic circuits by clustering and partial collapsing," in *Proc. Intl. Conference Computer-aided Design*, pp. 188–191, 1991.

[6] K. Chen *et al.*, "Timing optimization for multi-level combinational networks," in *Proc. Design Automation Conference*, pp. 339–344, 1991.

[7] A. Saldanha *et al.*, "Performance optimization using exact sensitization," in *Proc. Design Automation Conference*, pp. 425–429, 1994.

[8] G. de Micheli, *Synthesis and Optimization of Digital Circuits*. McGraw-Hill, 1994.

[9] M. Fujita *et al.*, "Multi-level logic optimization," in *Logic synthesis and verification* (S. Hassoun, T. Sasao, and R. K. Brayton, eds.), ch. 2, Kluwer Academic Publishers, Boston, MA, 2002.

[10] F. Leighton, *Introduction to Parallel Algorithms and Architectures: Arrays, Trees, Hypercubes*. Morgan Kaufmann Publishers, 1991.

[11] S. Lakshmivarahan and S. K. Dhall, *Parallel Computing Using the Prefix Problem*. Oxford University Press, 1994.

[12] Y.-T. Lai *et al.*, "OBDD-based function decomposition: Algorithms and implementation," *IEEE Trans. Computer-aided Design*, vol. 15, no. 8, pp. 977–990, 1996.

[13] B. Becker *et al.*, "On the expressive power of OKFDDs," *Formal Methods in System Design*, vol. 11, no. 1, pp. 5–21, 1997.

[14] C. Yang *et al.*, "BDS: A BDD-based logic optimization system," *IEEE Trans. Computer-aided Design*, vol. 21, no. 7, pp. 866–876, 2000.

[15] D. Wu *et al.*, "FBDD: A folded logic synthesis system," in *Intl. Conference on ASIC*, pp. 746–751, 2005.

[16] R. Ladner and M. Fischer, "Parallel prefix computation," *Journal of the ACM*, vol. 27, no. 4, pp. 831–838, 1980.

[17] "ABC Logic synthesis tool." Please visit the URL `http://www.eecs.berkeley.edu/~alanmi/abc/` for further details.

[18] L. Benini *et al.*, "Telescopic units: A new paradigm for performance optimization of vlsi designs," *IEEE Trans. Computer-aided Design*, vol. 17, no. 3, pp. 220–232, 1998.

[19] L. Benini *et al.*, "Automatic synthesis of large telescopic units based on near-minimum timed supersetting," *IEEE Trans. Computers*, vol. 48, no. 8, pp. 769–779, 1999.

[20] Y.-S. Su *et al.*, "An efficient mechanism for performance optimization of variable-latency designs," in *Proc. Design Automation Conference*, pp. 976–981, 2007.

[21] E. Sentovich *et al.*, "SIS: A system for sequential circuit synthesis," Tech. Rep. UCB/ERL M92/41, EECS Department, University of California, Berkeley, 1992.

978-1-60558-497-3/09 $25.00 © 2009 ACM

Simulation and SAT-Based Boolean Matching for Large Boolean Networks [*]

Kuo-Hua Wang, Chung-Ming Chan, and Jung-Chang Liu
Dept. of Computer Science and Information Engineering
Fu Jen Catholic University
Hsinchuang City, Taipei County 24205, Taiwan, R.O.C.
khwang@csie.fju.edu.tw

ABSTRACT

Boolean matching is to check the equivalence of two target functions under input permutation and input/output phase assignment. This paper addresses the permutation independent (P-equivalent) Boolean matching problem. We will propose a matching algorithm seamlessly integrating Simulation and Boolean Satisfiability (S&S) techniques. Our proposed algorithm will first utilize functional properties like unateness and symmetry to reduce the searching space. In the followed simulation phase, three types of input vector generation and checking method will be used to match the inputs of two target functions. Experimental results on large benchmarking circuits demonstrate that our matching algorithm is indeed very effective and efficient to solve Boolean matching for large Boolean networks.

Categories and Subject Descriptors: B.6.3 [**Hardware**] Logic Design: Design Aids - Verification.

General Terms: Algorithms, Design, Verification.

Keywords: Boolean Matching, Simulation and SAT.

1. INTRODUCTION

Logic simulation technique had been widely used in design verification and debugging over a long period of time. The major disadvantage of using simulation comes from the fact that it is very time consuming and almost impossible to catch completely the functionality of a very large Boolean network. To solve this issue, Boolean satisfiability (SAT) technique was proposed and exploited in many industrial formal verification tools. In recent years, the technique of combining simulation and SAT was popular and successfully applied in many verification and synthesis problems like equivalence checking [1] and logic minimization [2]-[4].

Boolean matching is to check whether two functions are equivalent or not under input permutation and input/output phase assignment (so-called *NPN-class*). The important applications of Boolean matching involve the verification of two circuits under unknown input correspondences, cell-library

binding, and table look-up based FPGA's technology mapping. In the past decades, various Boolean matching techniques had been proposed and some of these approaches were discussed in the survey paper [5]. Among those previously proposed approaches, computing *signatures* [5][6] and transforming into *canonical form* [7][8] of Boolean functions were the most successful techniques to solve Boolean matching. Recently, SAT technique was also applied for Boolean matching with don't cares [9]. Most of these techniques were proposed to handle completely specified functions, comparatively little research had focused on dealing with Boolean functions with don't cares [6][9].

The first and foremost issue for Boolean matching is the data structure for representing Boolean functions. As we know that many prior techniques used truth table, sum of products (SOPs), and Binary Decision Diagrams (BDDs) to represent the target functions during the matching process, they suffer from the same memory explosion problem. By our knowledge, many types of Boolean functions can not be represented by SOPs for large input set and the memory space will explode while constructing their BDD's. Therefore, these Boolean matching techniques were constrained to apply for small to moderate Boolean networks (functions). To address the above issues, And-Inverter Graphs (AIGs) [10] had been utilized and successfully applied in verification and synthesis problems [4][10]. In this paper, we will propose a permutation independent (P-equivalent) Boolean matching algorithm for large Boolean functions.

This paper is organized as follows. Section 2 gives a brief research background on our work. Section 3 shows our procedure of detecting functional properties. Some definitions and notations are given in Section 4. Section 5 presents our simulation strategy for distinguishing the input variables of Boolean functions. Section 6 and Section 7 show our S&S-based matching algorithm with implementation issues and the experimental results, respectively. Section 8 concludes this paper and suggests some directions for the future work.

2. BACKGROUND

2.1 Boolean Matching

Boolean matching is to check the equivalence of two target functions $f(X)$ and $g(Y)$ under input permutation and input/output phase assignment. To solve this problem, we have to search a *feasible* mapping ψ such that $f(\psi(X)) = g(Y)$ (or $\bar{g}(Y)$). It is impractical to search all possible mappings because the time complexity is $O(n! \cdot 2^{n+1})$, where n is the number of input variables. Among those previously proposed techniques for Boolean matching, *signature*

---

[*]This work was supported in part by the National Science Council of Taiwan under Grant NSC 96-2221-E-030-014-MY2.

Permission to make digital or hard copies of part or all of this work for personal or classroom use is granted without fee provided that copies are not made or distributed for profit or commercial advantage and that copies bear this notice and the full citation on the first page. To copy otherwise, to republish, to post on servers or to redistribute to lists, requires prior specific permission and/or a fee.
DAC'09, July 26-31, 2009, San Francisco, California, USA

978-1-60558-497-3/09 $25.00 © 2009 ACM

Table 1: Definition and S&S-Based Checking of Functional Properties.

Name	Property		S&S Checking		
	Definition	Notation	Disjoint	Removal_Condition	SAT_Check
Positive Unate	$f_{\bar{x}_i} \subseteq f_{x_i}$	$PU(x_i)$	x_i	$f(v_1)=1,\ val(v_1,x_i)=0$	$f_{\bar{x}_i} \cdot \bar{f}_{x_i} = 0$
Negative Unate	$f_{x_i} \subseteq f_{\bar{x}_i}$	$NU(x_i)$	x_i	$f(v_1)=1,\ val(v_1,x_i)=1$	$\bar{f}_{\bar{x}_i} \cdot f_{x_i} = 0$
NE Symmetry	$f_{\bar{x}_i x_j} = f_{x_i \bar{x}_j}$	$NE(x_i,x_j)$	x_i, x_j	$val(v_1,x_i) \neq val(v_1,x_j)$	$f_{x_i x_j} \oplus f_{x_i \bar{x}_j} = 0$
E Symmetry	$f_{\bar{x}_i \bar{x}_j} = f_{x_i x_j}$	$E(x_i,x_j)$	x_i, x_j	$val(v_1,x_i) = val(v_1,x_j)$	$f_{\bar{x}_i \bar{x}_j} \oplus f_{x_i x_j} = 0$
Single Variable Symmetry	$f_{\bar{x}_i \bar{x}_j} = f_{x_i \bar{x}_j}$	$SV(x_i,\bar{x}_j)$	x_i	$val(v_1,x_j)=0, \forall x_j \in X - \{x_i\}$	$f_{\bar{x}_i \bar{x}_j} \oplus f_{x_i \bar{x}_j} = 0$
	$f_{\bar{x}_i x_j} = f_{x_i x_j}$	$SV(x_i,x_j)$	x_i	$val(v_1,x_j)=1, \forall x_j \in X - \{x_i\}$	$f_{\bar{x}_i x_j} \oplus f_{x_i x_j} = 0$
	$f_{\bar{x}_i \bar{x}_j} = f_{\bar{x}_i x_j}$	$SV(x_j,\bar{x}_i)$	x_j	$val(v_1,x_i)=0, \forall x_i \in X - \{x_j\}$	$f_{\bar{x}_i \bar{x}_j} \oplus f_{\bar{x}_i x_j} = 0$
	$f_{x_i \bar{x}_j} = f_{x_i x_j}$	$SV(x_j,x_i)$	x_j	$val(v_1,x_i)=1, \forall x_i \in X - \{x_j\}$	$f_{x_i \bar{x}_j} \oplus f_{x_i x_j} = 0$

is one of the most effective approaches. Various signatures were defined to characterize input variables of Boolean functions [5]. Since these signatures are invariant under the permutation or complementation of input variables, the input variables with different signatures can be distinguished with each other and many infeasible mappings can be pruned quickly. However, it had been proved that signatures have the inherent limitation to distinguish all input variables for those functions with $\mathcal{G}$-symmetry [11].

2.2 Boolean Satisfiability

The Boolean Satisfiability (SAT) problem is to find a variable assignment to satisfy a given conjunctive normal form (CNF) or prove it is equal to the constant 0. Despite the fact that SAT problem is NP-complete, many advanced techniques like *non-chronological backtracking*, *conflict driven clause learning*, and *watch literals* have been proposed and implemented in the state-of-art SAT solvers [13][14]. Consequently, SAT technique has been successfully applied on solving many EDA problems [15] over the past decades. Among these applications, combinational equivalence checking (CEC) is an important one of utilizing SAT solver to check the equivalence of two combinational circuits. The following briefly describes the concept of SAT-based equivalence checking. Consider two functions (circuits) f and g to be verified. A *miter* circuit with functionality $f \oplus g$ is constructed first and then transformed into a SAT instance (circuit CNF) by simple gate transformation rules. If this circuit CNF can not be satisfied by the SAT solver, then f and g are equivalent; otherwise, they are not equivalent.

2.3 And-Inverter Graph

And-Invert Graphs (AIGs) is a directed acyclic graph which can be used as the structural representation of Boolean functions. It consists of three types of nodes: primary input, 2-input AND, and constant 0(1). The edges with IN-VERTER attribute denote the Boolean complementation. It is easy to transform a Boolean network (function) into AIGs by simple gate transformation rules. Moreover, it is very fast to perform simulation on AIGs with respect to (w.r.t.) a large set of input vectors at one time. However, it is unlike to the BDDs which is a canonical form of Boolean functions w.r.t. a given input variable ordering. It may also have many functionally equivalent nodes in the graph. In the paper [10], SAT sweeping and structural hashing techniques were applied to reduce the graph size. More recently, Mishchenko et al. exploited SAT-based equivalent checking techniques to remove equivalent nodes while constructing an AIG, i.e., Functionally Reduced AIGs (FRAIGs) [12]. By our experimental observation, FRAIGs can represent many large Boolean functions that can not be constructed as BDDs due to the memory explosion problem.

3. DETECTING FUNCTIONAL PROPERTY

Consider a function $f(X)$ and an input $x_i \in X$. The *cofactor* of f w.r.t. x_i is $f_{x_i} = f(x_1, \cdots, x_i = 1, \cdots, x_n)$. The cofactor of f w.r.t. $\bar{x}_i$ is $f_{\bar{x}_i} = f(x_1, \cdots, x_i = 0, \cdots, x_n)$. A function f is *positive (negative) unate* in variable x_i if $f_{\bar{x}_i} \subseteq f_{x_i}$ ($f_{x_i} \subseteq f_{\bar{x}_i}$). Otherwise, it is *binate* in that variable. Given two input variables $x_i, x_j \in X$, the definitions of *non-equivalence symmetry* (NE), *equivalence symmetry* (E), and *single variable symmetry* (SV) of f w.r.t. x_i and x_j are summarized in Table 1.

S&S approach is also applied to check functional symmetry and unateness of target functions in our matching algorithm. Instead of enumerating all items (possible functional properties) and checking them directly, we exploit simulation to quickly remove impossible items. For those items that can not be removed by the simulation results, SAT technique is exploited to verify them. Consider a function f and some functional property p to be checked. Our detecting procedure starts by using random simulation to remove impossible items as many as possible. If there still exist some unchecked items, it repeats taking an item and checking it with SAT technique until the taken item is a true functional property of f. Guided simulation will then be used to filter out the remaining impossible items. Rather than generating pure random vectors, guided simulation will generate simulation vectors based on counter examples by SAT solving, i.e., solutions of the SAT instance.

In order to remove impossible items, we generate many pairs of random vectors (v_1, v_2)'s with Hamming distance 1 or 2 for simulation. The vector pairs with distance 1 are used to remove functional unateness and single variable symmetries, while the pairs with distance 2 are used to remove non-equivalence and equivalence symmetries. Suppose that v_1 and v_2 are disjoint on input x_i (and x_j) if their distance is 1 (2). Without loss of generality, let $f(v_1) \neq f(v_2)$ and $f(v_1) = 1$. The conditions for removing impossible functional properties and SAT-based equivalence checking are briefly summarized in Table 1, where the notation $val(v_1, x_i)$ denotes the value of x_i in the vector v_1.

4. DEFINITIONS AND NOTATIONS

Let $P = \{X_1, X_2, \cdots, X_k\}$ be a *partition* of input set X, where $\bigcup_{i=1}^{k} X_i = X$ and $X_i \cap X_j = \emptyset$ for $i \neq j$. Each X_i is an *input group* w.r.t. P. The *partition size* of P is the number of subsets X_i's in P, denoted as $|P|$. The *group size* of X_i is the number of input variables in X_i, denoted as $|X_i|$.

DEFINITION 4.1. *Given two input sets X and Y with the same number of input variables, let $P_X = \{X_1, X_2, \cdots, X_k\}$ and $P_Y = \{Y_1, Y_2, \cdots, Y_k\}$ be two ordered input partitions of X and Y, respectively. A **mapping relation** $R = \{G_1, G_2, \cdots, G_k\}$ is a set of mappings between the in-*

978-1-60558-497-3/09 $25.00 © 2009 ACM

put groups of P_X and P_Y, where $G_i = X_i{}^{Y_i}$ and $|X_i| = |Y_i|$. Each element $G_i \in R$ is a **mapping group** which maps X_i to Y_i. ∎

DEFINITION 4.2. *Consider a mapping relation R and a mapping group $G_i = X_i{}^{Y_i}$ in R. The **mapping relation size** is the number of mapping groups in R, denoted as $|R|$. The **mapping group size** of G_i, denoted as $|G_i|$, is the group size of X_i (or Y_i), i.e., $|G_i| = |X_i| = |Y_i|$.* ∎

DEFINITION 4.3. *Consider two functions $f(X)$ and $g(Y)$. Let R be a mapping relation and $G_i = X_i{}^{Y_i}$ be a mapping group in R. G_i is **unique** if and only if $|G_i| = 1$ or $X_i(Y_i)$ is a NE-symmetric set of f (g). The mapping relation R is **unique** if and only if all the mapping groups in R are unique.* ∎

DEFINITION 4.4. *Let v_i be an input vector w.r.t. the input set X. The **input weight** of v_i is the number of inputs with binary value 1. It is denoted as $\rho(v_i, X)$.* ∎

DEFINITION 4.5. *Consider a function $f(X)$ and a vector set V involving m distinct input vectors. The **output weight** of f w.r.t. V is the number of vectors v_i's in V such that $f(v_i) = 1$. It is denoted as $\sigma(f, V)$ and $0 \le \sigma(f, V) \le m$.* ∎

5. SIMULATION APPROACH FOR DISTINGUISHING INPUTS

The idea behind our simulation approach is the same as the concept of exploiting signatures to quickly remove impossible input correspondences as many as possible. Consider two target functions $f(X)$ and $g(Y)$. Let $R = \{G_1, \cdots, G_i, \cdots, G_c\}$ be the current mapping relation. Without loss of generality, groups $G_1, \cdots, G_i$ are assumed non-unique while the remaining groups are unique. To partition a non-unique mapping group $G_i = X_i{}^{Y_i}$, for each input $x_j \in X_i$ ($y_j \in Y_i$), we generate a vector v or a set of vectors V for simulation and use the simulation results as the signature of x_j (and y_j) w.r.t. f (and g). It can further partition G_i into two or more smaller mapping groups by means of these signatures. Suppose the mapping size of G_i is m, i.e., $|G_i| = |X_i| = |Y_i| = m$. In the following, we will propose three types of input vectors and show how to distinguish input variables of X_i (and Y_i) in terms of the simulation results.

5.1 Type-1

We first generate c subvectors $v_1, \cdots, v_i, \cdots, v_c$, where v_i is a random vector with input weight 0 or m, i.e., $\rho(v_i, X_i) = \rho(v_i, Y_i) = 0$ or m. For each input variable $x_j \in X_i$ (and $y_j \in Y_i$), the subvector $\tilde{v}_i$ with input weight 1 or $m-1$ can be obtained by complementing the value of x_j (and y_j) in v_i. The concatenated vector $v^j = v_1 | \cdots | \tilde{v}_i | \cdots | v_m$ will then be used as input values for simulating on f (and g) and its corresponding output value $f(v^j)$ ($g(v^j)$) can be viewed as the signature of x_j (y_j). Fig. 1 demonstrates the vector set V_i used to partition G_i, where A_i is the set X_i or Y_i. Each vector (row) in V_i is dedicated to an input variable in X_i (Y_i). Using such a set V_i for simulation, in most cases, X_i (Y_i) can be partitioned into two subsets X_{i0} (Y_{i0}) and X_{i1} (Y_{i1}), where the signatures of input variables in these two sets are output value 0 and 1, respectively. Therefore, G_i can be divided into two mapping groups $G_{i0} = X_{i0}{}^{Y_{i0}}$ and $G_{i1} = X_{i1}{}^{Y_{i1}}$. Moreover, in our matching algorithm all non-unique mapping groups can be partitioned simultaneously.

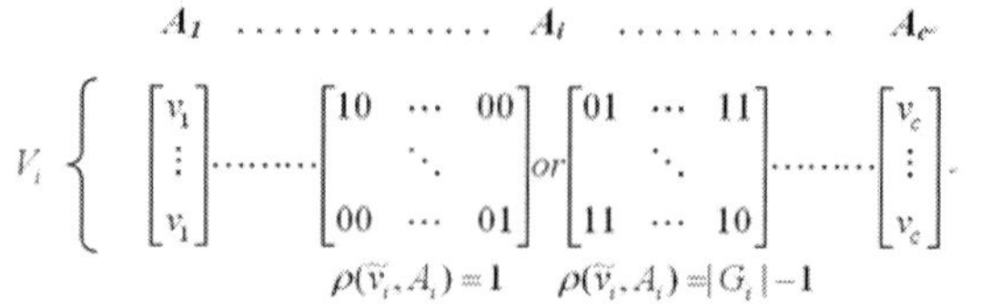

Figure 1: Type-1 Simulation Vectors V_i of G_i.

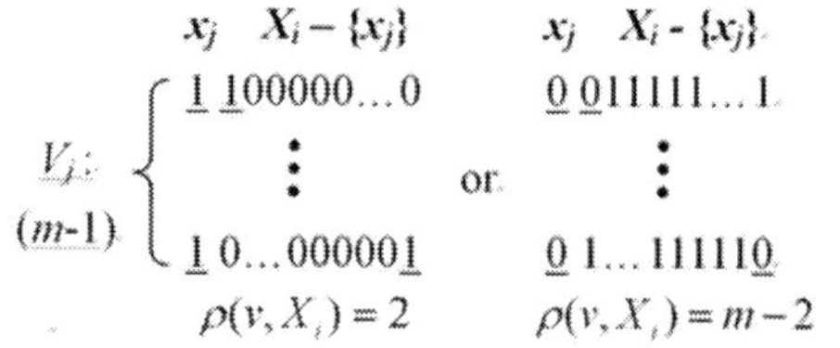

Figure 2: Type-2 Simulation Vectors V_j.

5.2 Type-2

For each input $x_j \in X_i$ ($y_j \in Y_i$), a vector set V_j involving $|G_i| - 1$ vectors with weight 2 or $m - 2$ will be generated. For simplicity, Fig. 2 only shows out the subvectors w.r.t. the input set X_i while the subvectors w.r.t. the remaining inputs sets can be generated like the initial subvectors v_i's of Type-1. Consider a vector in V_j. We assign 1 (or 0) to the input variable x_j and one of the remaining inputs, while the other inputs are assigned 0 (or 1). After the simulation, the output weight $\sigma(f, V_j)$ ($\sigma(g, V_j)$) will be used as the signature of x_j (y_j). The input variables with the same output weight can match to each other. Consequently, we can partition G_i into at most m groups because of $0 \le \sigma(f, V_j) \le m - 1$. Moreover, if there exists some input $x_j \in X_i$ which can uniquely map to an input $y_j \in Y_i$, we can apply Type-1 checking to further partition the set $X_i - \{x_j\}$ ($Y_i - \{y_j\}$) using the simulation results for the vector set V_j.

5.3 Type-3

The third type of vectors can only be used for the mapping group $G_i = X_i{}^{Y_i}$, where X_i and Y_i involves several *NE-symmetric* sets. The idea is mainly based on functional symmetry that function f is invariant under the permutation of inputs in its NE-symmetric set. Suppose X_i consists of e symmetric set $S_1, S_2, \cdots, S_e$ each with k input variables. To partition the group G_i, two random vectors a_1 and a_2 with different input weight w_1 and w_2 will be generated, where $0 \le w_1, w_2 \le k$. For each symmetric set S_i, we then generate a vector v_i by assigning a_2 and a_1's to S_i and the remaining sets, respectively. Fig. 3 shows only the weight distribution of vectors v_i's, where v_i is dedicated to S_i for simulation. As to the other mapping groups, the weights of their subvectors must be 0 or $|G_i|$ like we used in Type-1. Using such a vector set for simulation, we can partition $S_1, S_2, \cdots, S_e$ into two groups of symmetric sets which have output simulation value 0 and 1, respectively. It is easy to show that at most $k \times (k+1)$ combinations of w_1 and w_2 are required for this type of checking.

Our matching algorithm checks that $f(X)$ and $g(Y)$ can not match to each other using the following observation.

OBSERVATION 5.1. *Consider a non-unique mapping group $G_i = X_i{}^{Y_i}$. Through the simulation and checking steps as we described above, let P_{X_i} and P_{Y_i} be the resultant partitions w.r.t. to X_j and Y_j, respectively. For each group $A \in P_{X_i}$, there is a corresponding group $B \in P_{Y_i}$. The following shows two situations that $f(X)$ and $g(Y)$ can not match to each other:*

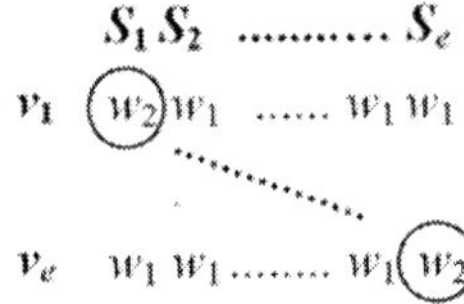

Figure 3: Type-3 Simulation Vectors of G_i.

- $|P_{X_j}| \neq |P_{Y_j}|$, i.e., their partition sizes are different.
- $|A| \neq |B|$, i.e., their set sizes are different. ∎

6. S&S-BASED MATCHING ALGORITHM

6.1 Our Matching Algorithm

Our S&S-based Boolean matching algorithm shown in Fig. 4 can match two target functions $f(X)$ and $g(Y)$ under a *threshold* bound, i.e., the maximum number of simulation rounds. It is mainly divided into three phases: *Initialization*, *Simulation*, and *Recursion*. In the *Initialization* phase, it exploits functional symmetries and unateness to initialize the mapping relation R followed by computing the maximum number of mapping groups in R, denoted as $MaxSize$. Since the input variables in an NE-symmetric set can be permuted without affecting the functionality, $MaxSize$ is equal to the number of NE-symmetric sets adding the number of non-symmetric inputs. It is the upper bound used to check the first terminating condition of the second phase. In the *Simulation* phase, it improves the mapping relation R by the *Simulate-and-Update* procedure implementing the simulation approach described in Section 5. This step is repeated until it can find a unique mapping relation R with $|R| = MaxSize$ or no improvement can be made to R under the *threshold* bound. If this phase is ended by the second terminating condition, it calls the *Recursive-Matching* procedure and enters into the *Recursion* phase to search the final mapping relation. Otherwise, the *SAT-Verify* procedure exploiting SAT technique is called to verify if two target functions are matched under the unique mapping relation R searched by the *Simulation* phase.

6.2 Recursive Matching Algorithm

Fig. 5 shows our recursive matching algorithm. Consider target functions f, g, a mapping relation R, and the size bound $MazSize$ of R. It starts by checking if R is unique and so to verify R using SAT technique. For the case that R is not unique, the smallest non-unique mapping group $G_i = X_i{}^{Y_i}$ in R will be selected to be partitioned. Consider the mapping of an input $x_j \in X_i$ to an input $y_k \in Y_i$. It will partition G_i into two mapping groups $A = \{x_j\}^{\{y_k\}}$ and $B = (X_i - \{x_j\})^{Y_i - \{y_k\}}$. With such a partitioning, we can derive a new mapping relation $tmpR = R \cup T - \{G_i\}$ with mapping relation size $|R| + 1$, where $T = \{A, B\}$ is derived from G_i. *Simulate-and-Update* procedure is then called to further improve $tmpR$ and return a new mapping relation $NewR$. If $NewR$ is not empty, it will call itself again; otherwise, it indicates a wrong selection of input mapping. In addition, this algorithm can be easily modified to find all feasible mapping relations as shown in Fig. 5.

6.3 Implementation Issues

6.3.1 Control of Random Vector Generation

By our experimental results, most of the runtime was consumed by the simulation phase for some benchmarking circuits. The reason is that too many random vectors gener-

---

```
Algorithm S&S-Boolean-Matching(f(X), g(Y), threshold)
Input: f and g are target functions;
       threshold: the maximum # of simulation rounds;
Output: ∅ or ψ, i.e., the feasible mapping relation of f and g;
Begin
   ψ = ∅;    cnt = 0;
   R = Initial-Mapping(f, g);    // Phase 1
   MaxSize = #NE_Classes + #Non_Symm_Inputs;
   while (|R| < MaxSize and cnt < threshold) do
      NewR = Simulate-And-Update(f, g, R);    // Phase 2
      if (NewR = ∅) return ∅;
      if (|R| == |NewR|) then
         cnt = cnt + 1;       // no improvement made on R
      else
         R = NewR;    cnt = 0;
      endif
   endwhile
   if (|R| < MaxSize) then
      ψ = Recursive-Matching(f, g, R, MaxSize);    // Phase 3
   else
      if (SAT-Verify(f, g, R) is TRUE) ψ = R;
   endif
   return ψ;
End
```

Figure 4: Our Boolean Matching Algorithm.

ated for simulation are useless to improve current mapping relation. Thus it will incur a large amount of iterations on the simulating and updating steps in this phase. Instead of generating random vectors without using any criterion, we propose a simple heuristic to control the generation of two adjacent random vectors. Let v_1 be the first random vector. The second vector v_2 can then be generated by randomly complementing $n/2$ inputs in v_1, where n is the number of input variables of target functions. We expect it can evenly distribute the random vectors in the Boolean space and so that it can quickly converge to find feasible mapping relation. Our experimental result shows the runtime can be greatly reduced for some circuits.

6.3.2 Reduction of Simulation Time

Our matching algorithm can be easily extended to deal with Boolean functions with multiple outputs. While matching two target functions with multiple outputs, we can reduce the simulation time by utilizing the mapping relation found so far. Clearly, the more the unique mapping groups we find, the less the number of outputs with indistinguishable inputs is. So, rather than simulating the whole Boolean network, simulating the subnetwork involving these outputs and their transitive fanin nodes is enough. Our experimental result reveals that our matching algorithm can reduce the runtime significantly as it approaches the end of matching process.

6.3.3 Analysis of Space Complexity

During the simulation process, we need to store the simulation vectors for all nodes in the Boolean network. Let the number of inputs and number of nodes in the Boolean network be I and N, respectively. The memory space used by our matching algorithm is $M \times I \times N$ words (4 bytes), where M is an adjustable parameter, i.e., the number of sets used in each simulation round. The smaller M means that it can reduce the storage space and simulation time in a simulation round. On the contrary, it may need more simulation rounds to improve the mapping relation. Besides,

978-1-60558-497-3/09 $25.00 © 2009 ACM

Algorithm *Recursive-Matching*$(f(X), g(Y), R, MaxSize)$
Input: f and g are target functions;
    R: the current mapping relation;
    $MaxSize$: the maximum size of mapping relation R;
Output: $\emptyset$ or ψ, i.e., the feasible mapping relation of f and g;
Begin
    if $(|R| == MaxSize)$ **then**   // the terminating condition
        if $(SAT\text{-}Verify(f, g, R)$ is **TRUE**$)$ **return** R;
        return $\emptyset$;
    endif
    G_i = the smallest *non-unique mapping group* in R;
    $\psi = \emptyset$; Choose an input $x_j \in X_i$;
    for each possible mapping relation T w.r.t. G_i **do**
        $tmpR = R \cup T - \{G_i\}$;
        $NewR = Simulate\text{-}And\text{-}Update(f, g, tmpR)$;
        if $(NewR \neq \emptyset)$
            $\psi = Recursive\text{-}Matching(f, g, NewR, MaxSize)$;
        // comment the next line if to find all solutions
        if $(\psi \neq \emptyset)$ **return** ψ;
    endfor
    return ψ;
End

Figure 5: Recursive Matching Algorithm.

Table 2: Matching Results for Threshold 1000

#Input	#Circuit	#Solved	CPU Time (sec.)		
			Min	Avg	Max
4~10	31	31	0.00	0.04	0.30
11~20	21	21	0.01	0.55	8.04
21~30	14	14	0.03	0.21	1.29
31~40	10	9	0.07	1.16	4.79
41~50	8	8	0.22	2.94	5.83
51~257	28	26	0.14	2.57	17.56

it may stop the simulation phase early, and thus enter into the recursive matching phase. If there exist many large non-unique mapping groups, then the runtime will increase significantly because it may incur a large amount of simulation and SAT verification on infeasible mapping relations.

7. EXPERIMENTAL RESULTS

The proposed S&S-based Boolean matching algorithm had been implemented into Berkeley's ABC system on the Linux platform with dual Intel Xeon 3.0 GHz CPU's. To demonstrate the efficiency of our algorithm, MCNC and LGSyn benchmarking sets were tested in our experiments. For each tested circuit, we randomly permuted its input variables to generate a new circuit for being matched. In addition, to make our experimental results more convincible, we restructured this new circuit by executing a simple script file including some synthesis commands offered by ABC. Two sets of experiments were conducted to test our matching algorithm.

The first experiment was conducted to search all feasible mapping relations on 112 circuits with input number ranging from 4 to 257. The experimental results showed that three circuits C6288, i3, and o64 can not be solved within 5000 seconds. To dissect these circuits, we found that one of the two mapping relations of C6288 can not be verified by SAT technique while the other two circuits have a large amount of feasible mapping relations because they owns a great many G-symmetries [11]. If only to search one feasible mapping, the execution times for i3 and o64 are 0.49 sec. and 8.67 sec., respectively. The experimental results were summarized in Table 2 w.r.t. the circuit input size. In this table, the first two columns show the input ranges of benchmarking circuits and number of circuits in different input ranges, respectively.

Table 3: Comparison on the Effects of Three Phases

	Functional Property(1)			+Sim.(2)	+Rec.(3)
	+Unate	+Symm	+SVS		
#Circuit	19	49	71	94	109
#Inc	19	30	22	23	15
ratio (%)	17.4	44.9	65.1	86.2	100

Table 4: The Results of Circuits with Inputs > 50

Circuit	#I	#O	#Sol		CPU Time (sec.)		
			O	S	Orig	Unate	+Symm
apex3	54	50	1	1	0.10	0.10	0.38
apex5	117	88	144	1	7.11	2.96	0.68
apex6	135	99	2	1	1.86	0.42	0.33
C2670	233	140	-	2	*	*	7.96
C5315	178	123	4	1	6.31	2.86	3.29
C7552⋆	207	108	-	1	*	*	14.56
C880	60	26	8	1	0.28	0.20	0.25
dalu	75	16	2	1	1.20	3.36	5.47
des	256	245	1	1	10.21	0.25	2.33
e64	65	65	1	1	0.01	0.79	0.32
ex4p	128	28	-	4096	*	*	6.08
example2	85	66	1	1	0.05	0.02	0.23
frg2	143	139	1	1	0.45	0.10	0.72
i10	257	224	48	2	25.63	15.16	17.56
i2	201	1	-	1	*	*	1.02
i4	192	6	-	1	*	*	0.22
i5	133	66	1	1	0.18	0.03	0.35
i6	138	67	1	1	0.50	0.02	0.14
i7	199	67	1	1	0.82	0.04	0.19
i8	133	81	1	1	0.57	0.06	0.40
i9	88	63	1	1	0.18	0.03	0.16
pair	173	137	1	1	0.84	0.64	2.44
rot	135	107	72	1	3.79	1.69	1.25
x1	51	35	2	1	0.17	0.13	0.14
x3	135	99	2	1	2.05	0.28	0.32
x4	94	71	2	1	0.62	0.37	0.15
Total					> 25063	> 25029	66.94
Avg					> 964	> 963	2.57

-: unknown   *: CPU time > 5000 sec.   ⋆: memory explosion

The third column labeled **#Solved** shows the number of circuits solved by our matching algorithm. The next three columns **Min**, **Avg**, and **Max** show the minimum, average and maximum runtime for the solved circuits, respectively. It shows that our algorithm is very efficient for the circuits with moderate to large input sets.

For those solved circuits, we also compared the effectiveness of three phases and the comparison results were shown in Table 3. The rows labeled **#Circuit** and **#Inc** show the number of circuits that have been matched successfully and the number of increased matched circuits in each individual step, respectively. In the first phase, we compared the effect of incrementally applying different functional properties. The columns named as **+Unate**, **+Symm**, and **+SVS** show the results of only using unateness, adding E&NE symmetry, and adding SV symmetry, respectively. It is clear that the more functional properties are used, the more circuits can be solved in the first phase. It shows that 71 and 94 circuits can be solved after the first phase and the second (simulation) phase, respectively. All the remaining circuits can be solved by the third (recursion) phase.

Table 4 shows the experimental results for those circuits with input size greater than 50. The first three columns labeled **Circuit**, **#I**, and **#O** show the circuit name, number of input variables, and number of outputs in this benchmarking circuit, respectively. The next two columns **O** and **S** are the numbers of mapping relations found by our matching algorithm without using and using functional properties. It should be noted that the solutions induced by NE-symmetry is not taken into account on the numbers shown in the **S** column. Moreover, the numbers greater than one indicate that these benchmarking circuits own G-symmetry. The last three columns named as **Orig**, **Unate**, and **+Symm** compare the execution times of without using functional property, using only functional unateness, and adding functional symmetries, respectively. The result shows that there are

978-1-60558-497-3/09 $25.00 © 2009 ACM

Table 5: Benchmarking Results for s-Series Circuits

Circuit	#I	#O	#N	#BDD	Symmetry	#Sol	CPU Time (sec.)						
							Orig		Unate		+Symm		
							First	All	First	All	First	All	SAT
s4863 †	153	120	3324	56691	1(8),1(9)	$4 \cdot 8! \cdot 9!$	*	*	2.56	*	1.87	1.88	0.01
s3384	264	209	2720	882	22(2)	2^{22}	4.79	*	2.14	*	4.02	4.02	0.00
s5378	199	213	2850	⋆	4(2),1(5),1(7)	$(2!)^4 \cdot 5! \cdot 7!$	1.31	*	3.38	*	2.42	2.42	0.00
s6669 †	322	294	4978	22957	32(2), 1(17)	$4 \cdot 2^{32} \cdot 17!$	6.30	*	2.83	*	4.08	50.54	46.50
s9234.1	247	250	4023	4545	-	1	3.41	3.41	5.84	5.84	7.82	7.82	0.00
s38584.1	1464	1730	26702	22232	1(3),1(9)	$3! \cdot 9!$	76.31	*	210.13	*	457.82	457.82	0.03
s38417	1664	1742	23308	55832099	2(2),1(3)	$(2!)^2 \cdot 3!$	91.81	*	324.57	*	998.53	998.55	0.13
Total							183.93		551.45		1476.56	1523.05	46.67
Avg							30.66		78.78		210.94	217.58	6.67
Ratio							0.15		0.37		1.00	1.03	0.03

†: circuits own $\mathcal{G}$-*symmetry*   -: no symmetry $m(n)$: m NE-symmetric sets with n inputs   *: CPU time > 5000 sec.   ⋆: memory explosion

5 out of 26 circuits can not be solved by our matching algorithm without using full functional property within 5000 seconds. The reason why this situation occurs is that these circuits have a great many NE-symmetries. However, it can resolve all cases if functional symmetries are utilized to reduce the searching space. The average runtime of using full functional property is 2.57 second. It clearly reveals that our S&S-based matching algorithm is indeed effective and efficient for solving the Boolean matching problem. In this experiment, the BDD's of these circuits were also built for comparison with AIGs. It shows the circuit C7752 had the memory explosion problem while constructing BDD without using dynamic ordering.

In order to test our matching algorithm on very large Boolean networks, the second experiment was conducted to test the circuits in ISCAS89 benchmarking set. Since these circuits are sequential, we executed the *comb* command in ABC to transform them into combinational circuits. Table 5 shows the partial experimental results. For each circuit, the columns **#N**, **#BDD**, **Symmetry**, and **#Sol** show the number of nodes in the Boolean network (AIG), number of nodes in the constructed BDD, NE-symmetry, and number of feasible mappings. The CPU times of finding the first feasible mapping, finding all feasible mapping relations, and performing SAT verification of two target circuits are shown in the **First**, **All**, and **SAT** columns, respectively. The experimental results show that our algorithm can not find all feasible mappings for those circuits with a great many NE-symmetries unless we detect them in advance. It also reveals that only searching the first feasible mapping relation without using symmetry is faster than the one using symmetry in some cases. The reason why this situation occurs is that it took too much time on detecting symmetries for the circuits. The result also shows only a very small amount of runtime was consumed by SAT verification for all circuits except the s6669 circuit. Besides, the AIG size was far less than the BDD size in many tested circuits and the s5378 circuit had the memory explosion problem. In summary, our S&S-based Boolean matching algorithm can be easily adjusted to fulfill different requirements for large Boolean networks.

8. CONCLUSIONS

We had proposed a S&S-based Boolean matching algorithm in this paper. Three types of input vectors were generated for simulation and their simulation results were used to distinguish the inputs of two target functions. Our matching algorithm had been tested on a set of large benchmarking circuits. The experimental results reveal that our algorithm is indeed effective and efficient for solving the Boolean

matching problem on very large functions. The future work will be dynamically adjusting threshold value in the simulation process and extending our matching algorithm to deal with input/output phase assignment.

9. REFERENCES

[1] A. Mishchenko, S. Chatterjee, R. Brayton, and N. En, "Improvements to Combinational Equivalence Checking," in *Proc. of Internatioanl Conference on Computer-Aided Design*, pp. 836-843, Nov. 2006.

[2] S. Plaza, K. Chang, I. Markov, and V. Bertacco, "Node Mergers in the Presence of Don't Cares," in *Proc. of Asia and South Pacific Design Automation Conference*, pp. 414-419, Jan. 2007.

[3] Qi Zhu, N. Kitchen, A. Kuehlmann, A. Sangiovanni-Vincentelli, "SAT Sweeping with Local Observability Don't-Cares," in *Proc. of Design Automation Conference*, pp. 229-234, July 2006.

[4] A. Mishchenko, et. al, "Using Simulation and Satisfiability to Compute Flexibilities in Boolean Networks," *IEEE Transaction on Computer-Aided-Design of Integrated Circuits and Systems*, Vol. 25, No. 5, pp. 743-755, May 2006.

[5] L. Benini and G. De Micheli, "A Survey of Boolean Matching Techniques for Library Binding," *ACM Trans. on Design Automation of Electronic Systems*, Vol. 2, No. 3, pp. 193-226, July 1997.

[6] Afshin Abdollahi, "Signature Based Boolean Matching in the Presence of Don't Cares," in *Proc. of Design Automation Conference*, pp. 642-647, June 2008.

[7] G. Agosta, et. al, "A Unified Approach to Canonical Form-based Boolean Matching," in *Proc. of Design Automation Conference*, pp. 841-846, June 2007.

[8] A. Abdollahi and M. Pedram, "A New Canonical Form for Fast Boolean Matching in Logic Synthesis and Verification," in *Proc. of Design Automation Conference*, pp. 379-384, June 2005.

[9] K.H. Wang and Chung-Ming Chan, "Incremental Learning Approach and SAT Model for Boolean Matching with Don't Cares," in *Proc. of International Conference on Computer-Aided Design*, pp. 234-239, November 2007.

[10] A. Kuehlmann, V. Paruthi, F. Krohm, and M. K. Ranai, "Robust Boolean Reasoning for Equivalence Checking and Functioanl Property Verification," *IEEE Transaction on Computer-Aided-Design of Integrated Circuits and Systems*, Vol. 21, No. 12., pp. 1377-1394, Dec. 2002.

[11] J. Mohnke, P. Molitor, and S. Malik, "Limits of Using Signatures for Permutation Independent Boolean Comparison," in *Proc. of ASP Deisgn Automaiton Conference*, pp. 459-464, 1995.

[12] A. Mishchenko, S. Chatterjee, R. Jiang, and R. K. Brayton, "FRAIGS: A Unifying Representation for Logic Synthesis and Verification," *ERL Technical Report*, EECS Dept., UC Berkeley, March 2005.

[13] J. Marques-Silva and K. A. Sakallah, "GRASP: A Search Algorithm for Propositional Satisfiability," *IEEE Transactions on Computer-Aided-Design of Integrated Circuits and Systems*, Vol. 48, No. 5, pp. 506-521, May 1999.

[14] M. W. Moskewicz, C. F. Madigan, Y. Zhao, L. Zhang, and S. Malik, "Chaff: Engineering an Efficient SAT Solver," in *Proc. of Design Automation Conference*, pp. 530-535, June 2001.

[15] J. P. Marques-Silva and K. A. Sakallah, "Boolean Satisfiability in Electronic Design Automation," in *Proc. of Design Automation Conference*, pp. 675-680, June 2000.

978-1-60558-497-3/09 $25.00 © 2009 ACM

New Spare Cell Design for IR Drop Minimization in Engineering Change Order

Hsien-Te Chen
Department of Computer Science, National Tsing Hua University
HsinChu, Taiwan 300

Chieh-Chun Chang
Department of Computer Science, National Tsing Hua University
HsinChu, Taiwan 300

TingTing Hwang
Department of Computer Science, National Tsing Hua University
HsinChu, Taiwan 300

ABSTRACT

Unused spare cells occur inevitably in traditional ECO design flow. It results in inefficient area usage, more leakage, and more IR drop impacts. To tackle these problems, a reconfigurable cell is proposed which serves the dual purposes of decoupling capacitance and spare cell in this paper. Before Engineering Change Order (ECO) is applied, these cells are pre-placed as decoupling capacitors. When ECO is applied, these cells are configured as functional cells. To demonstrate the efficiency of our configurable cell, we propose an algorithm for timing closure and IR drop minimization. Compared with traditional ECO flow, our method shows 16% reduction in maximum IR drop and 56% reduction in leakage before applying ECO, and 8% reduction in maximum IR drop after applying ECO, with 10% area of spare cells. In addition, we show that there are less unsolved ECO timing paths left after applying our ECO timing optimization algorithm due to free selection of ECO gate type.

Categories and Subject Descriptors

B.7.2 [**Hardware**]: Integrated Circuits—*Layout, Design Aids*

General Terms

Design, Algorithms

Keywords

IR Drop, Decoupling Capacitor, Spare Cell, ECO

1. INTRODUCTION

Synthesis and physical design methodologies from leading suppliers have tremendously relied on ECO (Engineering Change Order) flow to solve timing closure issues and to accommodate incremental change of chip function. In conventional ECO flow, redundant standard cells, known as spare cells, are pre-placed uniformly in CORE area for ECO changes. Rather than building a new layout, ECO flow using spare cells is employed so as to minimize impacts on original chip layout. Traditionally, the ratio of spare cells to total number of cells is about 5% to 20%. Those spare cells are very effective in solving timing closure [6, 7] and accommodating incremental change of chip function [8, 10], however, at cost of extra area and leakage overhead.

On the other hand, as feature size of MOS technology continues to shrink, supply voltage is reduced. Significant IR-drop (voltage-drop) on power lines affects not only chip noise margin but also chip speed [3]. In order to reduce IR-drop, on-chip decoupling capacitor (DECAP) [9, 11, 12] was introduced as one of the most effective methods to reduce current fluctuation in power lines.

In traditional chip implementation flow, spare cells are pre-placed to meet ECO requirements and decoupling-capacitor cells are filled after cell placement whenever there is space. Both spare and decoupling-capacitor cells may appear in the same layout and serve different goals.

To serve the dual purposes of spare cell and decoupling-capacitor cell, we propose a reconfigurable (RECON) cell where a base cell is first designed. Based on this base cell, decoupling cell and spare cells such as inverter, nand, nor gates, etc. can be configured by changing layer of metal-1. Moreover, to demonstrate the effectiveness of our RECON cell, an ECO algorithm to solve timing closure and IR drop minimization problem by selecting a set of decoupling cells and transforming them to spare cells is proposed. Our specific contributions in this paper are:

- to propose a new reconfigurable (RECON) cell structure to serve the dual roles of spare cells and decoupling-capacitor cells. This new RECON cell achieves the following goals:

  - elimination the necessity of designing two types of cells (i.e., spare cells and decoupling-capacitor cells).

  - reduction of leakage resulting from the elimination of tie-high and tie-low cells for spare cells.

  - full utilization of spare cells and decoupling-capacitor cells; In other words, there are no idle spare cells.

  - free selection for ECO gate type because a base cell can be configured to any type of gate.

  - easy reconfiguration of RECON cell to other functionality by changing one layer of metal.

- based on our RECON cell, to propose an algorithm to demonstrate the effectiveness of RECON cell for timing closure and IR drop minimization.

Permission to make digital or hard copies of part or all of this work for personal or classroom use is granted without fee provided that copies are not made or distributed for profit or commercial advantage and that copies bear this notice and the full citation on the first page. To copy otherwise, to republish, to post on servers or to redistribute to lists, requires prior specific permission and/or a fee.

DAC'09, July 26-31, 2009, San Francisco, California, USA

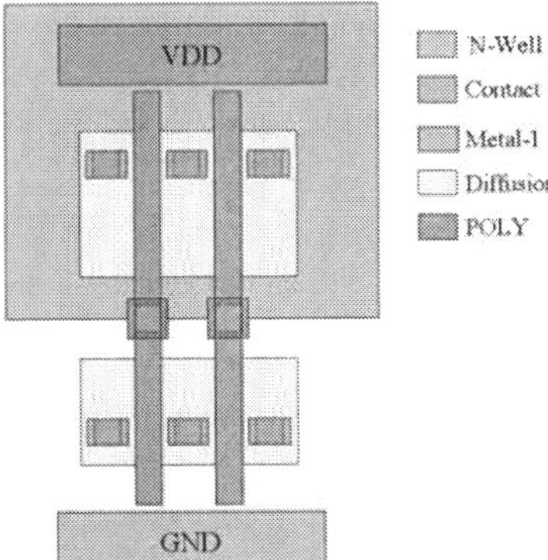

Figure 1: RECON Base Cell

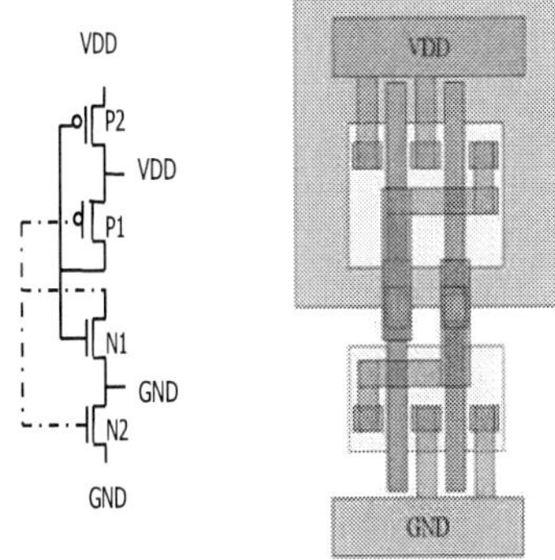

Figure 2: Reconfigured Decoupling Capacitor Using RECON

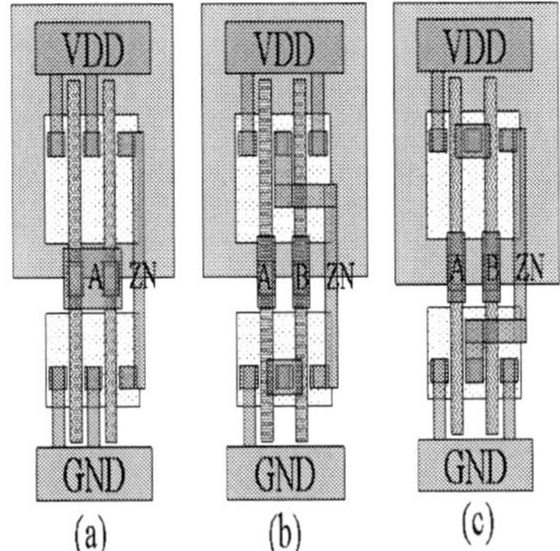

Figure 3: Reconfigured Cells: (a) INVERTER; (b) 2-Input NAND; (c) 2-Input NOR;

The rest of the paper is organized as follows. Section 2 proposes the architecture of our reconfigurable cells and compares them to traditional standard cells. Section 3 formulates the problem of timing closure and IR drop minimization in ECO flow using RECON cell. Section 4 shows the experimental results. Conclusions are put forth in Section 5.

2. NEW DESIGN STYLE

In this section, we will discuss basic characteristics of reconfigurable (RECON) cells and compare them with traditional standard cells, namely spare cell and decoupling-capacitor cell. In addition, chip-level implementation in an ECO design flow using reconfigurable cells is discussed.

2.1 Structure and Layout of Base Cell

Fig. 1 illustrates the base cell layout of RECON cell structure. The basic characteristics of base cell are:

- Minimum transistor length;

- Two PMOS gates with the same transistor width;

- Two NMOS gates with the same transistor width;

- Eight CONTACT's;

- VDD and GND implemented by layer of metal-1;

- Same cell height as standard cell;

- No well pick-up created in base cell to go with deep sub-micron standard cells;

A base cell is able to be configured to decoupling capacitor and different functional cells with different driving capabilities by metal-1 connections. The most frequently used cells, such as inverter, 2-input nand, 2-input nor and decoupling-capacitor, are easily implemented by one base cell. Cells of more complex function and different driving capabilities are implemented by connecting two or more base cells.

First, a decoupling-capacitor (RECON DECAP) cell configured from the base cell of RECON is shown in Fig. 2. Compared with RECON base cell, RECON DECAP has two Z-shaped metal-1 polygons connecting gates and drains in the cell. The left side of Fig. 2 is its schematic capture that shows connections among PMOS gates ($P1$ and $P2$) and NMOS gates ($N1$ and $N2$). When RECON DECAP cell becomes stable, $P1$ and $N1$ are on and RECON cell acts as decoupling capacitance. This design is the same as

that in [2]. Furthermore, $P2$ and $N2$ in serial with $P1$ and $N2$, respectively, are added to the original layout [2] to minimize the leakage of RECON DECAP cell. When the cell is in stable state, $P2$ and $N2$ are turned off.

Similarly, other functional cells are implemented in the same fashion. With metal-1 connection, three basic functional cells configured from RECON base cell are represented in Fig. 3. From the above examples, we know that it is easy to reconfigure from base cell to RECON DECAP cells and to functional cells by changing only metal-1 layer.

2.2 Comparisons between RECON Cells and Standard Cells

In this section, performance comparison between RECON cells and their corresponding standard cells is conducted. For functional cells, standard cells in $TSMC\ 0.13um$ library are used while RECON cells are generated by the following steps:

1. Cell layouts in GDSII are created with $TSMC\ 0.13um$ process in complying with layout style of traditional standard cells. Area data is extracted accordingly.

2. SPICE net-lists are extracted from GDSII files by RC extractor.

3. Frame views of automatic placement-and-route kits are manually constructed based on cell layouts.

4. SPICE simulations are applied to generate delay, leakage, internal power and input pin capacitance.

978-1-60558-497-3/09 $25.00 © 2009 ACM

Table 1: Performance Ratio of RECON DECAP Cell to Traditional Decoupling Capacitor

Cell	Area	Capacitance	Leakage
DECAP4	1.00	0.39	0.34
DECAP8	1.00	0.18	0.12
DECAP16	1.00	0.16	0.10
DECAP32	1.00	0.16	0.09
DECAP64	1.00	0.18	0.09

For decoupling capacitor, steps to generate RECON DE-CAP cells are as follows:

1. The first three steps are the same as those of RECON functional cells.

2. Leakage of RECON DECAP cells is extracted by SPICE DC simulations.

3. Resistance and capacitance of RECON DECAP cells are generated by SPICE AC simulations based on method described in [1].

Since there is no decoupling capacitor in $TSMC$ $0.13um$ library, we apply the same circuit structure described in Fig. 2 to generate a standard decoupling capacitor. To emulate commercial decoupling capacitor, the design is tailored to have long channel devices so as to store more charge.

First, we compare our RECON cell when it is configured as a decoupling-capacitor cell (RECON DECAP) with conventional standard cell. The result is shown in Table 1. The comparison shows that RECON DECAP cell can store only about one third to one seventh capacitance of traditional decoupling capacitor using the same area. The reason is that RECON cell uses a standard channel device for all types of gates and thus has no flexibility to fine tune the layout while conventional decoupling capacitor uses long channel devices to store large capacitance. On the other hand, the comparison of leakage shows that RECON DECAP cell has smaller leakage and is only 9% to 34% leakage of conventional decoupling capacitor as a result of smaller channel area.

Next, performance ratio of selected RECON cells to their corresponding spare cells in $TSMC$ $0.13um$ standard cell library is shown in Table 2. In this experiment, RECON spare cells are reconfigured from RECON DECAP cells. For standard spare cells, it is necessary to connect tie cells to their inputs and outputs to avoid gate floating. Henceforth, area of tie cell is added to that of standard spare cell in this table. On the same ground, leakage of conventional spare cell includes the leakage of tie cell.

The columns labeled w/o tie-$cells$ and $RECON$ shows the area/leakage ratios of standard cell without tie cell and RE-CON cell to that of standard cell with tie cell labeled w tie-$cells$, respectively. Let us take INVX4 in the 3rd row as an example. It shows that tie cells take 33% of the total area (standard cell + tie cells). Area ratio of 89% shows that RECON INVX4 is 11% smaller than conventional INVX4 standard cell with tie-high or tie-low cells. For delay, RE-CON INVX4 is 1% faster. For leakage, tie cell takes 3% of the total leakage (standard cell + tie cells) and RECON is 7% less than the standard cell with tie cells. For internal dynamic power and pin capacitance, RECON is 2% less than the standard cell.

However, due to less flexibility to fine tune layout and to size transistor, area of RECON cells with large driving capabilities (e.g., INVX16, BUFX8) is about 33% larger than their corresponding standard cells with tie-high or tie-low cells.

In average, RECON cells occupy 14% (4th column) smaller area and consume 10% (8th column) less leakage than conventional standard cells with tie cells. For delay (5th column) and pin capacitance (10th column), RECON cells are almost the same as the standard ones. Lastly, without tie high and tie low cells, the area of RECON cell is, in average, 20% (4th column to 2nd column) more than that of standard cell, which explains why internal dynamic power of RECON cells is about 9% (9th column) larger.

In summary, RECON spare cells outperform traditional ECO spare cells (reconfigured from RECON DECAP cell) in area and leakage while have the same performance level in delay and pin capacitance. However, in terms of decoupling capacitance, RECON is less effective than traditional one.

2.3 Chip-Level Implementation

2.3.1 Modeling of Power Supply Network

First, a layout is partitioned to rows and power network is also distributed in rows. Next, to analyze IR drop, three assumptions in [9] are assumed in our RECON ECO design.

- metal layer of VDD and GND are modeled as a power-grid resistance network.

- standard cells are modeled as time-varying current sources.

- RECON DECAP cells are modeled as capacitors connected between VDD and GND.

With those assumptions, supply voltage variation of each cell in cycle-based timing frame will be derived from switching activities temporally and power-grid network spatially. The procedure to calculate IR drop for each row is elaborated as follows. First of all, clock cycle is divided into a specified number of time slots (e.g., 100 time slots in one clock cycle). Secondly, active (switching) gates in one time slot are derived from static timing analysis for all ECO paths. Thirdly, maximum current consumption in each time slot is computed by counting the number of active gates sharing the same power line in that time slot. Fourthly, maximum current is fed to power-grid resistance network to calculate supply voltage variations for each time slot.

2.3.2 RECON and Traditional ECO Flow

The major differences between RECON ECO flow and traditional ECO flow are:

- Instead of spare (functional and tie) cells, RECON cells configured as DECAP cells are pre-placed.

- When RECON ECO flow is not applied, these RECON DECAP cells acting as decoupling-capacitor are able to reduce IR drop. The role is the same as traditional decoupling capacitor cells.

- When RECON ECO design flow is applied, some RE-CON DECAP cells are selected and reconfigured to specific types of RECON functional cells by only modifying metal-1 layer. These RECON functional cells are used in ECO flow as spare cells to accommodate

978-1-60558-497-3/09 $25.00 © 2009 ACM

Table 2: Performance Ratio of RECON Spare Cell to Traditional Spare Cell

Cell	Area			Delay	Leakage			Power	Pin Cap
	w/o tie-cells	w tie-cells	RECON		w/o tie-cells	w tie-cells	RECON		
INVX1	0.50	1	0.67	1.02	0.88	1	0.81	0.89	0.93
INVX2	0.57	1	0.57	1.01	0.94	1	0.93	0.94	0.94
INVX4	0.67	1	0.89	0.99	0.97	1	0.93	0.98	0.98
INVX8	0.79	1	1.14	0.98	0.98	1	0.89	1.02	0.98
INVX16	0.88	1	1.33	0.98	0.99	1	0.87	1.00	0.99
ND2X1	0.40	1	0.40	1.05	0.8	1	0.80	1.19	0.98
ND2X2	0.50	1	0.67	1.05	0.89	1	0.84	1.30	0.98
NR2X1	0.40	1	0.67	1.09	0.81	1	0.79	1.33	0.98
NR2X2	0.50	1	0.94	1.11	0.89	1	0.86	1.55	1.04
BUFX1	0.57	1	0.57	0.84	0.92	1	1.21	1.07	1.50
BUFX2	0.63	1	1.00	1.00	0.96	1	0.93	0.97	0.95
BUFX4	0.75	1	1.00	0.99	0.98	1	0.90	0.98	0.92
BUFX8	0.83	1	1.33	0.99	0.99	1	0.88	0.99	0.96
Avg.	0.61	1	0.86	1.01	0.92	1	0.90	1.09	1.01

incremental change of functions or to solve timing closure problem.

- Those unselected RECON DECAP cells keep acting as decoupling capacitors.

One major advantage of RECON ECO flow is that no redundant RECON spare cell exists when ECO design flow is not applied. All cells are configured as decoupling-capacitor cells so as to reduce both leakage and IR drop.

3. ECO USING NEW DESIGN STYLE

To demonstrate the effectiveness of our RECON ECO flow, we will develop a simple algorithm to solve timing closure problem.

3.1 Problem formulation

Before we formally define the problem, the following definitions are given:

- An ECO path is a path that violates the timing constraint.

- Gate sizing operation is a technique to change driving capability of cell on ECO path with partial reroute.

- Buffer insertion operation is a technique to insert a buffer along a path with partial reroute.

Now, our problem is defined as follows: Given a set of placed gate level net-list, ECO paths and timing constraint, our objective is to perform gate sizing or buffer insertion on ECO paths so that timing constraint is met and IR drop is minimized.

3.2 RECON ECO Algorithm

In ECO timing optimization, one key observation in [6] is applied in our $RECON\_ECO\_Algorithm$.

- A gate with large output capacitance load is the most critical one in ECO timing optimization. Therefore gates are prioritized by their output capacitance load in $RECON\_ECO\_Algorithm$.

Our algorithm, $RECON\_ECO\_Algorithm$, is shown in Fig.4. Before we proceed our algorithm, IR drop analysis and static timing analysis are performed. The procedure starts by checking if there is any path that violates the timing constraint. If yes, we select the most critical one for

```
Algorithm RECON_ECO_Algorithm
Input:a set of ECO paths to be optimized
   While(there exits an ECO path)
     Select the most critical path p_i;
     ECO_gate_list = gates on p_i are sorted by output loading
       in decreasing order;
     While (timing(p_i) not satisfied and ECO_gate_list ≠ φ )
       Select the first gate, g_j, in ECO_gate_list;
       region_Sizing = Search_region_GateSizing(g_j);
       region_Buffer = Search_region_BufferInsertion(g_j);
       For all configurable cell rc_k in region_Sizing
         If IRdrop(rc_k)>threshold
           Remove rc_k for region_Sizing;
       For all configurable cell rc_k in region_Buffer
         If IRdrop(rc_k)>threshold
           Remove rc_k for region_Buffer;
       candidate_list = configuration(region_Sizing)
               +configuration(region_Buffer);
       Compute path delay_gain for each rc_k in candidate_list;
       best_gain =   MAX   delay_gain(rc_k);
                  for all rc_k
       Select best configuration corresponding to best_gain;
       Update the delay of p_i;
       Delete g_j from ECO_gate_list;
     End while
   End while
```

Figure 4: $RECON\_ECO\_Algorithm$

ECO optimization. Then, for an ECO path to be processed, we sort the gates on the path by output capacitance loading in decreasing order.

Next, we select the first gate, g_j in the sorted list. To improve the delay, we can either size up g_j or insert a buffer at the output of g_j. In either cases, it requires a search region where DECAP cells can be found for reconfiguration.

For gate sizing, a search region is a bounding box defined by Eq. (1). Let g_j be the active gate, g_{j_fanin} be fanin gates of g_j and g_{j_fanout} be fanout gates of g_j. Then search region for gate sizing is defined as:

$$Search\_Region_{GateSizing}(g_j) =$$

$$Bounding\_Box(g_j \cup g_{j\_fanin} \cup g_{j\_fanout}) \qquad (1)$$

For buffer insertion, a search region is defined by Eq. (2).

$$Search\_Region_{BufferInsertion}(g_j) =$$

$$Bounding\_Box(g_j \cup g_{j\_fanout}) \qquad (2)$$

In either cases, we collect all configurable cells in the regions. Before we select a DECAP cell for reconfiguration, we check if IR drop of its neighboring cell is larger than a threshold value. If yes, the cell will remain to be a decoupling cell and is deleted from configurable cell list. Next, for all configurable cells in the list, we connect them in all possible ways to form gates with different driving capabilities. For example, three abutted RECON DECAP cells are able to construct one NR2X4 (NOR gate with driving capability of 4) or three NR2X1 (NOR gate with driving capability of 1). Then, these candidate configurations are put into a list, denoted as *candidate_list*. Next, for each candidate configuration in the list, we compute the path *delay_gain*. The path *delay_gain* is computed taking into consideration the location of configurable cell and IR drop on the path.

Finally, the configuration which results in the maximum gain is selected. The next iteration starts by selecting the next gate in the ECO path for process. The procedure does not stop until the timing constraint is satisfied or there is no more candidate. The whole procedure repeats till all ECO paths are processed.

4. EXPERIMENTS

4.1 Experiment Setup

Five benchmark examples are selected from ITC99 benchmarks [4]. The benchmark examples are first synthesized by Synopsys synthesis tool to gate-level net-lists in VERILOG format using $TSMC$ 0.13um standard cells. Secondly, automatic placement-and-route tool by Cadence [5] is applied to place the standard cells and our RECON cells. Then, IR drop and static timing analysis are performed. Finally, given a set of ECO paths, $RECON\_ECO\_Algorithm$ is called. Fig. 5 illustrates our experimental flow. For comparison, a traditional ECO flow is also performed. Before ECO flow, spare cells including Inverter, 2-input NAND, 2-input NOR and Buffer are used. The same ECO flow is applied except that the candidate configurations for timing improvement are these fixed spare cells.

In our experiment, 10% area are used to place our RECON DECAP cells or traditional spare cells. Either spare cells or our reconfigurable cells are evenly distributed in the layout.

4.2 Experiment Results

The statistics of circuits in benchmark set are listed in Table 3. Column labeled $\#Gates$ is the number of gates after synthesis. Column labeled $\#Paths$ is the number of ECO paths which violates timing constraint. Timing constraint is set to be 90% of critical path delay in the original circuit after synthesis.

The statistics of circuits in benchmark set are listed in Table 3. Column labeled $\#Gates$ is the number of gates after synthesis. Timing constraint is set to be 90% of critical path delay in the circuit with all spare cell configured as RECON DECAP. Column labeled $\#Paths$ is the number of ECO paths which violates timing constraint. Each column consists of two values. Columns labeled $Trad.$ and $RECON$ are the number of violated paths for traditional ECO flow and for RECON ECO flow respectively. Note that the number of critical paths under traditional and RECON ECO flows are different. This is because the timing calculation takes IR drop into consideration. In RECON ECO flow, spare cells are acting as decoupling capacitance, and thus IR drop in the circuit is less serious.

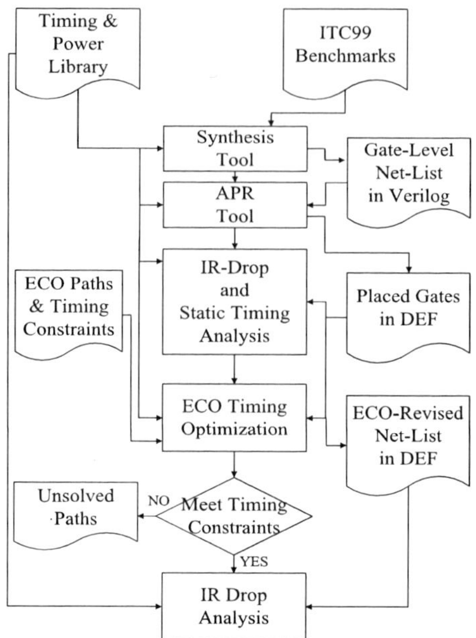

Figure 5: Experimental ECO flow

Table 3: Statistics of Circuits in Benchmark Set

Ben.	#Gates	#Paths	
		Trad.	RECON
b14	13203	323	250
b15	11148	62	52
b20	27823	1476	1237
b21	27571	1061	783
b22	41445	1862	1246

Table 4 summarizes the performance of benchmark circuits before ECO flow is applied. Column labeled *Leakage* is the leakages induced from spare cells in milliampere. Column labeled *Max. IRDrop* is maximum IR-drop in millivolt. Column labeled *Avg. IR−Drop* is average IR-drop. The last column reports power noises as defined in [9]. Each column consists of three values. Columns labeled *Trad.* and *RECON* are for traditional ECO flows and for RECON ECO flow respectively. Column labeled *Ratio* is the ratio of RECON ECO flow to traditional one. From this table, RECON ECO flow outperforms traditional ECO flow 56% in leakage reduction, 27% in maximum IR-drop reduction, 41% in average IR-drop reduction and 46% in power noise. The same leakage ratio of traditional ones to RECON for all benchmark cases is because the same 20% area is applied for both flows.

Next, performance summary of benchmark circuits after ECO flow is performed is illustrated in Table 5. In this table, RECON ECO flow compared with traditional flow is 8% and 13% better in maximum IR-drop reduction and in average IR-drop reduction, respectively.

Intuitively IR-drop in RECON ECO flow will become worse after ECO flow is applied because some of RECON DECAP cells are reconfigured to functional cells for ECO timing optimization. However, in some cases (e.g., *b20* with 5% area), we observe that IR-drop is improved after ECO flow. This is because automatic placement-and-route tool does not take IR-drop into account and timing-critical cells may be placed in the same row. After ECO flow taking IR-drop into consideration, those timing-critical cells are re-

Table 4: Performance Comparison before ECO

Ben.	Leakage(mA)			Max. IR-Drop(mV)			Avg. IR-Drop(mV)		
	Trad.	RECON	Ratio	Trad.	RECON	Ratio	Trad.	RECON	Ratio
b14	8315	3658	0.44	97.7	82.5	0.84	49.6	42.7	0.86
b15	8370	3689	0.44	62.9	50.8	0.81	28.1	26.1	0.93
b20	17399	7648	0.44	122.5	110.4	0.90	70.2	53.7	0.76
b21	17396	7652	0.44	160.9	126.7	0.79	87.0	71.2	0.82
b22	26094	11479	0.44	204.5	179.7	0.88	127.0	104.4	0.82
Avg.	15515	6825	0.44	129.7	110.0	0.84	72.4	59.6	0.84

Table 5: Performance Comparison after ECO

Ben.	Max. IR-Drop(mV)			Avg. IR-Drop(mV)		
	Trad.	RECON	Ratio	Trad.	RECON	Ratio
b14	102.5	99.3	0.97	54.7	49.8	0.91
b15	62.9	50.0	0.79	32.7	20.7	0.63
b20	114.3	103.1	0.90	69.5	63.0	0.91
b21	143.2	138.4	0.97	89.4	85.1	0.95
b22	196.3	193.1	0.98	126.0	121.5	0.96
Avg.	123.8	116.8	0.92	74.5	68.0	0.87

Table 6: Unsolved Paths Comparison

Ben.	Trad.	RECON	Ratio
b14	0	0	-
b15	0	0	-
b20	459	10	0.02
b21	30	0	-
b22	377	0	-

placed by cells in different rows to reduce current induced from simultaneous switching. This replacement reduces delay and IR-drop at the same time.

Lastly, Table 6 shows the number of unsolved ECO paths after ECO flow. In this table, it shows that RECON ECO flow has less number of unsolved paths. This is because RECON cell is able to provide the flexibility of configuring a spare cell to different gate type. In summary, RECON ECO flow outperforms conventional ECO flow in all performance indexes for five ITC99 benchmarks.

5. CONCLUSION

In this paper, reconfigurable cells and their corresponding ECO flow have been presented. Reconfigurable ECO flow has been shown to be effective as compared to traditional ECO flow in all aspects including leakage, IR drop and timing.

6. REFERENCES

[1] Patrik Larsson, "Parasitic Resistance in an MOS Transistor Used as On-Chip Decoupling Capacitance," *IEEE Journal of Solid-State Circuits*, vol. 32, no. 4, April 1997, pp. 574-576.

[2] Neil Weste and David Harris, "A circuits and Systems Perspective(3rd Edition)," *Addison Wesley*

[3] H. B. Bkoglu, "Circuits, Interconnections, and Packaging for VLSI," *Addison Wesley*, 1990.

[4] http://www.cerc.utexas.edu/itc99-benchmarks/bench.html, Benchmark Suite for ITC 1999.

[5] http://www.cadence.com/, Manual of SOC Encounter 2007.

[6] Yen-Pin Chen, Jia-Wei Fang, and Yao-Wen Chang, "ECO Timing Optimization Using Spare Cells," *International Conference on Computer Aided Design*, 2007, pp. 530-535.

[7] S. Pant, D. Blaauw, V. Zolotov, S. Sundareswaran,and R. Panda, "Vectorless Analysis of Supply Noise Induced Delay Variation," *International Conference on Computer-Aided Design*, 2003, pp. 184-190.

[8] Yu-Min Kuo, Ya-Ting Chang, Shih-Chieh Chang and Malgorzata Marek-Sadowska, "Engineering Change Using Spare Cells with Constant Insertion," *International Conference on Computer-Aided Design*, 2007, pp. 544-547.

[9] Haihua Su, Sachin S. Sapatnekar and Sani R. Nassif, "An Algorithm for Optimal Decoupling Capacitor Sizing and Placement for Standard Cell Layouts," *Proceedings of the International Symposium on Physical Design*, 2002, pp.68-73.

[10] Chih-chang Lin, Kuang-Chien Chen, Shih-Chieh Chang , Malgorzata Marek-Sadowska and Kwang-Ting Cheng, "Logic Synthesis for Engineering Change," *Conference on Design Automation Conference*, 1995, pp. 647-652.

[11] Sanjay Pant and David Blaauw, Timing-Aware Decoupling Capacitance Allocation in Power Distribution Networks," *Conference on Design Automation Conference*, 2007, pp. 757-762.

[12] Hang Li, Zhenyu Qi, Tan, S.X.D., Lifeng Wu, Yici Cai and Xianlong Hong "Partitioning-Based Approach to Fast On-Chip Decap Budgeting and Minimization," *Conference on Design Automation Conference*, 2005, pp. 170-175.

Matching-Based Minimum-Cost Spare Cell Selection for Design Changes

Iris Hui-Ru Jiang
Dept. of Electronics Engineering
National Chiao Tung University
Hsinchu 30010, Taiwan
huiru.jiang@gmail.com

Hua-Yu Chang
Freelance

Taipei, Taiwan
huayu.chang@gmail.com

Liang-Gi Chang and Huang-Bi Hung
Institute of Electronics
National Chiao Tung University
Hsinchu 30010, Taiwan
{lgchang, penn}.ee96g@nctu.edu.tw

ABSTRACT

Metal-only ECO realizes the last-minute design changes by revising the photomasks of metal layers only. This task is challenging because the pre-injected spare cells are limited both in number and in cell types. This paper proposes a matching-based ECO synthesizer, named ECOS, that correctly implements the incremental design changes using the available spare cells as well as tries to reduce the prohibitive photomask cost at the same time. The experiments are conducted on five industrial testcases. ECOS uses less photomask costs to complete design changes for all cases than the direct method that transforms the widely-used hand-editing procedure into an automatic one.

Categories and Subject Descriptors

B.7.2 [**INTEGRATED CIRCUITS**]: Design Aids – *Placement and Routing*

General Terms

Algorithms, Design

Keywords

ECO, spare cells, resynthesis, physical synthesis, matching

1. INTRODUCTION

Engineering change order (ECO) is a process that implements incremental design changes [1]. These modifications may result from functionality debugging, timing improvement, and specification revision. The later stage where ECO is performed, the less resources are available, and the greater challenges can be met. After the base layers (placement) are frozen during the design cycle, ECO not only can shorten design time by avoiding rebuilding the design from scratch but also can reduce fabrication time by manufacturing the base-layer photomasks in advance. After the first silicon chips are produced, ECO can save the prohibitive photomask cost by reusing the base-layer part in the next tape-out. Modifying only a few photomasks of metal layers to realize design changes is referred to as metal-only ECO. To

Permission to make digital or hard copies of part or all of this work for personal or classroom use is granted without fee provided that copies are not made or distributed for profit or commercial advantage and that copies bear this notice and the full citation on the first page. To copy otherwise, to republish, to post on servers or to redistribute to lists, requires prior specific permission and/or a fee.

DAC'09, July 26-31, 2009, San Francisco, California, USA

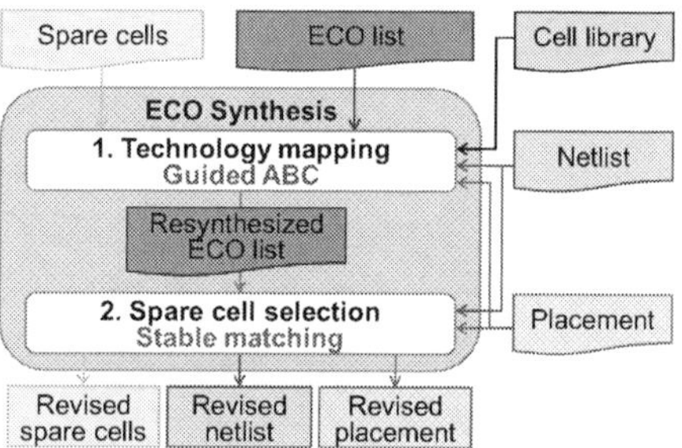

Figure 1. The overview of ECOS.

facilitate metal-only ECO, a design is sprinkled with unused (spare) cells at placement, and their inputs are tied to either logic high or low to prevent floating signal. ECO is then performed by rewiring the inputs and outputs of spare cells.

Good metal-only ECO relies on the following three techniques:
1. Sufficient and evenly sprinkled spare cells can accommodate design changes at all possible locations [2][3][4].
2. A good incremental eco-router can handle tremendous obstacles & design rules and complete routing with minimum changes [2][3][4][5].
3. A powerful ECO synthesizer can fulfill the design changes on functionality and/or timing by including the physical information into logic synthesis wisely and modeling the impact on the photomask cost when the selected spare cells deviate from the ideal locations [2][3].

For timing ECO, [6] proposed a technology-remapping technique based on dynamic programming. [7] fixed the input slew, output loading, and delay violations by inserting buffers (spare cells). However, these methods might not easily be extended for functional ECO. On the other hand, for functional ECO, [8] remapped the expected cells with available spare ones whose inputs could be inserted with constant values (logic high/low). However, this method did not consider the issues of non-tree type cells (e.g., multiplexors, full-adders), freeing up unused cells, and minimizing the photomask cost. It can hardly handle timing ECO.

To overcome the aforementioned drawbacks, this paper proposes a matching-based ECO synthesizer, named ECOS, to complete the functional changes with the minimum photomask cost. As shown in Figure 1, given the list of functional changes, the original netlist and placement, and the available spare cells, ECOS first resynthesizes the given ECO list using affordable spare cell types with geometry proximity consideration. Then, each instance in the resynthesized list is replaced by an adequate spare cell based on stable matching, as well as the related nets are reconnected. Moreover, the induced unobservable cells can be freed up for later

ECO runs. The objective of spare cell selection is to minimize the photomask cost of metal layers. Without loss of generality, this cost is modeled by the summation of the half-perimeter bounding box (HPBB) of each net in the revised design. This cost model benefits short interconnect delay, thus readily extending to timing ECO. Afterwards, formal equivalence checking can be performed to verify whether the revised design matches the revised functionality. ECOS has the following distinguished features:

1. It handles non-tree type spare cells and ECO functions.
2. It considers constant value insertion for spare cells.
3. It recycles freed-up cells for subsequent ECO runs.
4. It integrates physical information into resynthesis.
5. It solves the competition among spare cells.
6. It minimizes the photomask cost (also benefits timing).
7. It can readily extend to timing ECO.
8. It easily collaborates with existing synthesizers.

To demonstrate the effectiveness, we automate the common hand-editing ECO flow and then conduct the experiments on five industrial testcases. The results show that ECOS is promising.

2. PRELIMINARIES AND PROBLEM FORMULATION

2.1 Common ECO Synthesis Flow

Metal-only ECO is commonly performed by hand-editing the netlist [1]. However, this ad hoc method is time-consuming and resource intensive because the design related files have to be searched and edited many times during the whole process. Figure 2(a) shows an example design with two inputs, two outputs, and four logic cells. Spare cells include two AND and one NOT cells. The placement is also illustrated. For simplicity, the area of each cell in this example is 0, and all pins are located at the same point.

The photomask cost of metal layers depends on how complicated the whole routing could be. Hence, the photomask cost of each net is modeled by the half-perimeter bounding box (HPBB) over its related pins; the total cost of a design is the sum of HPBBs over all nets. For the design in Figure 2(a), we have its total cost:

HPBB(net1) + HPBB(net2) + HPBB(net3) + HPBB(in1) + HPBB(in2) + HPBB(out1) = 6,000 + 1,000 + 3,000 + 5,000 + 5,000 + 2,000 = 22,000.

Assume the functional ECO is to replace the functionality of cell U3 by AND. We have two options: spare cells SPARE1 and SPARE2. We prefer SPARE2 because of the better proximity to cells U1, U2, U4 and output out2. Moreover, in the revised design, the output of cell U3 becomes unused. Hence, the inputs of U3 can be connected to constant values instead. This kind of freed-up cells can be gathered and reused at subsequent ECO runs. The revised design can be improved as Figure 2(b). The total cost becomes 20,000.

Although not shown here, a spare cell can be used to implement more than one function, e.g., a two-input NAND cell can be a NOT cell by inserting a logic high to one input. The constant insertion can maximize the capability of each spare cell. Please note that the cell types of spare cells in most of cases do not directly match the ECO functionality. We need to translate the ECO functionality into pieces, and realize each piece by a spare cell. When the size of the ECO list is large and the resource of spare cells is limited, the ad hoc method would be time-consuming and may fail due to the competition among ECOs.

2.2 Problem Formulation

This paper solves the following metal-only ECO problem.

> **Minimum-Cost Spare Cell Selection for Functional ECO:**
> Given the original netlist, placement, a set of spare cells, and a list of functional changes, complete the ECO list using the available spare cells, create the revised netlist with the minimum cost, and generate the revised set of spare cells.

3. MATCHING-BASED ECO SYNTHESIS

This paper develops a matching-based ECO synthesizer, named ECOS, to solve the problem formulated in Section 2.2. Figure 1 details ECOS' inputs/outputs and summarizes its two steps.

3.1 Technology Mapping: Guided ABC

The first step of ECOS performs technology mapping to resynthesize the given ECO list using the available spare cell types. After this step, a resynthesized ECO list is produced. We build our synthesizer based on the well-established environment, ABC [9]. Basically, ABC performs optimal-delay DAG-based technology mapping, while guided ABC uses the physical information of spare cells to direct technology mapping. The cell library is modified for each ECO. Each spare cell corresponds to one library cell; its cell area reflects the cost when it is selected for this ECO, while its cell delay is set to zero. Doing so can trigger ABC to perform area-recovery since each possible mapping has the same delay.

As mentioned in Section 2.1, the cells that become unobservable after ECO are freed up and included in the revised set of spare cells for later ECO runs. They cannot affect the costs of their input and output nets after being recycled. In Figure 2(a), cell U3 does not affect the costs of net1, net2, and net3 after ECO1 is applied. After cell U3 in Figure 2(b) is freed up, the HPBBs of touched nets will be:

HPBB(net1) = 3,000; HPBB(net2) = 0; HPBB(net3) = 0.

Moreover, we can compute the HPBB of each ECO by finding the bounding box over all active pins in its related nets, e.g., we have

HPBB(ECO1)
= HPBB(net1, net2, net3) = HPBB(U1, out2, U2, U4) = 6,000.

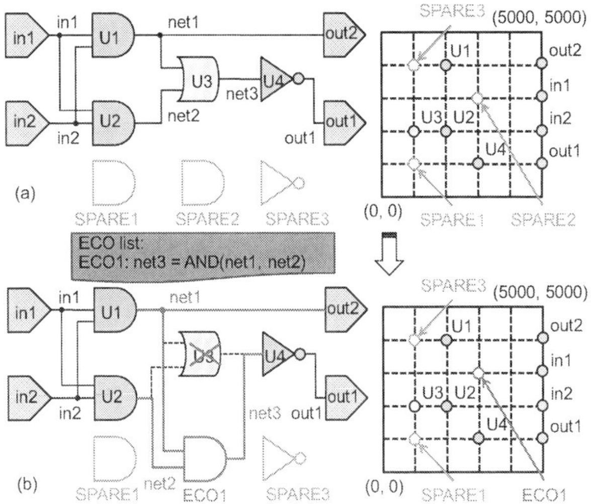

Figure 2. (a) The original design. (b) The revised design.

StableMatching(*M*, *W*)
1. Initialize all $m \in M$ and $w \in W$ to free
2. **while** $\exists$ free man *m* with a woman *w* to propose to **do**
3.   *w* = *m*'s highest ranked such woman
4.   **if** *w* is free **then**
5.   (*m*, *w*) become engaged
6.   **else** // some pair (*m'*, *w*) already exists
7.   **if** *w* prefers *m* to *m'* **then**
8.   (*m*, *w*) become engaged
8.   *m'* becomes free

Figure 3. The stable matching algorithm [10].

The value can be viewed as the lower bound of the total cost induced by an ECO. The cost of selecting a spare cell for a given ECO can be computed accordingly. For ECO1, the costs of SPARE1, SPARE2, and SPARE3 are 7,000, 6,000, and 7,000, respectively. Hence, the cell library for ECO1 contains one SPARE1 cell of function/area/delay AND/7,000/0, one SPARE2 cell of function/area/delay AND/6,000/0, and one SPARE3 cell of function/area/delay NOT/7,000/0. Based on the modified cell library, guided ABC can generate the best choice for each ECO. For example, for ECO1, SPARE1 and SPARE2 have the same delay (zero), so the cell of smaller area is selected, i.e., ECO1 is resynthesized as a SPARE2 cell. Then, the resynthesized ECO list contains only the cell types of available spare cells. Sometimes, an ECO in the original ECO list may be converted into several cells. Moreover, DAG-based technology mapping cannot handle non-tree type spare cells and ECO functions. We resort this problem to ROBDDs. If the spare cell types are a mixture of only multiplexors (MUX) and inverters (NOT), the ECO list will be transformed to ROBDDs first; these ROBDDs are then simplified and converted to MUX/NOT cells. In addition, constant value insertion can naturally be implemented in technology mapping, thus maximizing the capability of spare cells. It can be seen that the guidance made for ABC indeed can also easily be built in other existing logic synthesizers provided by EDA vendors.

3.2 Spare Cell Selection: Stable Matching

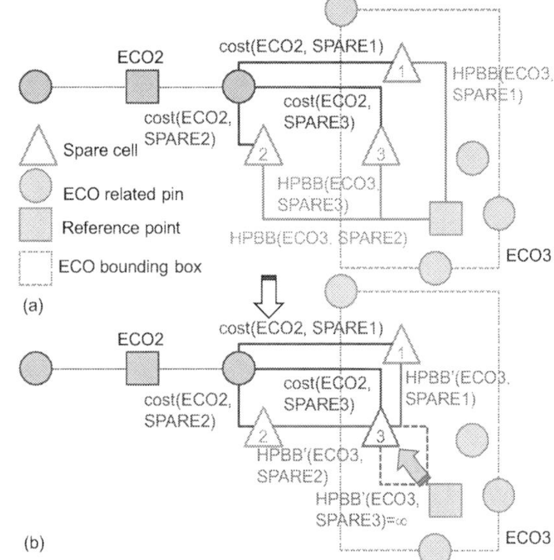

Figure 4. The reference points and bounding boxes:
(a) Before engagement. (b) After engagement.

Although guided ABC considers physical information of spare cells, it cannot handle the competition among ECOs. If a spare cell is the best choice of several ECOs, guided ABC duplicates it for these ECOs. To solve this problem, we do not select spare cells directly in guided ABC, but defer the decision making to step 2. With the global view of costs, deferred decision making at step 2 may lead to good results.

3.2.1 The Stable Marriage Problem

Stable Marriage:
Given a set of *n* men and *m* women, marry them off in pairs after each man has ranked the women in order of preference and each women has done likewise such that no pair of man and woman who would both rather have each other than their current partners. If there are no such pairs, all the marriages are stable.

This paper reduces spare cell selection to the stable marriage problem, handling the competition among candidates in nature. Gale and Shapley proposed a stable matching algorithm listed in Figure 3, which is male-optimal [10].

3.2.2 The Reduction

Due to male-optimality, each ECO in the resynthesized list is modeled as a man, while each spare cell is modeled as a woman. The preference is defined by the added cost resulting from assigning a spare cell to an ECO. The less added cost, the more preference. The added cost contains the difference made on the existing nets and the induced cost on the newly created nets. (Please note that the added cost is different from the cost used in guided ABC.) If there are no newly created nets among the ECOs in the resynthesized list, the preference order can be determined directly, e.g., ECO1 in Figure 2(a) has the following preferences:

pref(ECO1, SPARE1) = cost(ECO1, SPARE1) = 1,000;
pref(ECO1, SPARE2) = cost(ECO1, SPARE2) = 0.

Thus, ECO1 prefers SPARE2 and proposes to SPARE2. SPARE2 then accepts, and the stable matching is found. If no ECOs in the resynthesized list are connected each other, the stable matching algorithm leads to good results for all ECOs. However, when there are internal connections among some ECOs, the spare cell selection for ECO would affect each other. This case may occur when an ECO is resynthesized into several ECOs; some newly created nets connect among ECO cells only. Because the designated spare cells for ECOs have not been decided yet, the cost cannot be determined. To estimate the induced cost on the newly created nets, the reference point of each ECO is introduced to represent a good location for spare cell selection. It is initially set as the average *x*- and *y*-coordinates over all pins on its related nets. For example, the reference point of ECO1 in Figure 2(a) is at:

$x(\text{ECO1}) = 3,000; y(\text{ECO1}) = 2,750.$

With setting the reference points, we can compute the induced cost on newly created nets and then rank the preference between ECOs and spare cells. For example, Figure 4(a) depicts the reference points and bounding boxes of ECO2 and ECO3. Assume a newly created net connecting ECO2 and ECO3. The preference of ECO2 proposing to SPARE1 consists of the added cost between SPARE1 and the existing nets of ECO2 and the estimated cost of the net between ECO2 and ECO3.

pref(ECO2, SPARE1) =
cost(ECO2, SPARE1) + HPBB(ECO3, SPARE1).

978-1-60558-497-3/09 $25.00 © 2009 ACM   410

As the stable matching algorithm progresses, once an ECO is engaged to a spare cell, its reference point is updated to the real location of the engaged spare cell. As shown in Figure 4(b), after ECO3 proposes to SPARE3 and gets accepted, the estimated cost of the net between ECO2 and ECO3 is updated as follows.

HPBB'(ECO3, SPARE1) = HPBB(SPARE3, SPARE1).
HPBB'(ECO3, SPARE2) = HPBB(SPARE3, SPARE2).
HPBB'(ECO3, SPARE3) = ∞.

The estimated cost between ECO3 and SPARE3 is set to a large value rather than 0 to prevent ECO2 from proposing to SPARE3. Doing so can guarantee that the method is stable and always has a solution. Spare cell selection follows the stable matching algorithm, except lines 5 and 8 in Figure 3. The reference point of an ECO is updated when it is engaged. The preference ranking related to this ECO is updated accordingly; this update would not affect the processed proposals to maintain stability.

3.2.3 The Complexity of Spare Cell Selection

The preference can be computed during step 1. Ranking n ECOs and m spare cells can be done in $O(nm\lg m + mn\lg n)$ time. The worst case of the stable matching algorithm is $O(nm\lg n)$. Hence, the overall time complexity of spare cell selection is $O(nm(\lg m + \lg n)) = O(nm\lg m)$ since n is less than or equal to m. Because only the spare cells of the same type of an ECO in the resynthesized list can be matched, the expected number of possible proposals is far less than nm, thus the practical complexity is quite lower than the worse case.

4. EXTENSION TO TIMING ECO

For timing ECO, we may insert buffers into timing critical paths. These buffers can be viewed as functional ECO, and ECOS naturally fixes timing by minimizing the related HPBBs.

5. EXPERIMENTAL RESULTS

We implemented our algorithm in C++ language and executed the program on a PC with an Intel® Core™2 CPU T7400 of 2.16 GHz frequency and 2 GB memory. Totally five industrial testcases are used. The first three use a general cell library with basic and complex logic cells, while the last two use a cell library with only multiplexors and inverters. The left part of Table 1 lists the number of pins, the number of cells, the number of nets, and the number of spare cells of testcases. The netlist and placement of the original design are described in DEF format, while the ECO list is specified in VERILOG format. We automated the common ECO method as the direct method. It searches for a spare cell with the required functionality within the bounding box of each ECO. If none, blind ABC resynthesizes each ECO with the available spare cell types (the area & delay of each spare cell are 0). Then, each ECO in the resynthesized list is directly mapped to one spare cell of the same type inside its bounding box. If failed, it is

Table 1. Statistics on testcases and on ECO.

Case	Statistics				#ECOs			
	#Pin	#Cell	#Net	#Spare	#Req	#FR	Blind ABC	Guided ABC
testcase1	483	28,591	28,705	350	7	7	16	11
testcase2	483	28,591	28,705	2,300	49	51	295	153
testcase3	483	28,591	28,705	2,300	94	121	313	162
testcase4	33	198	181	40	3	3	5	4
testcase5	30	938	850	100	4	4	6	5

#Req: number of required ECOs; #FR: number of freed up cells

alternatively mapped to someone else outside.

The right part of Table 1 lists the number of the given ECO list, the number of freed-up cells and the sizes of resynthesized ECO lists for blind & guided ABC. The number of freed-up cells could be greater than the size of the ECO list when the freed up cell has multiple outputs. Guided ABC generates a much fewer ECOs in the resynthesized list than blind ABC. The smaller resynthesized ECO list may result in the smaller overlapped bounding boxes and then lead to the lower cost. Table 2 compares the total cost, the ECO related HPBB before & after ECO, and CPU times. The original values are listed here for reference. Ratio means the cost normalized to the direct method. The direct method incurs average 7.66% degradation on the total cost. Considering the HPBB of ECO (which can be viewed as the lower bound of the cost induced by ECO), ECOS on average outperforms the direct method by 47.09% on the total cost and always produces the better results. In addition, compared with the timing-consuming manual method, the automatic methods are efficient.

6. REFERENCES

[1] S. Golson. The human eco compiler. In *Proc. SNUG*, pages 1-57, 2004.
[2] Synopsys, Inc. http://www.synopsys.com/.
[3] Cadence Design System. http://www.cadence.com/.
[4] Magma Design Automation. http://www.magma-da.com/.
[5] J.-Y. Li and Y.-L. Li. An efficient tile-based ECO router with routing graph reduction and enhanced global routing flow. In *Proc. ISPD*, pages 7-13, 2005.
[6] Y.-P. Chen, J.-W. Fang and Y.-W. Chang. ECO timing optimization using spare cells and technology remapping. In *Proc. ICCAD*, pages 530-535, 2007.
[7] C.-P. Lu, M. C.-T. Chao, C.-H. Lo, and C.-W. Chang. A metal-only-ECO solver for input slew and and output loading violations. In *Proc. ISPD*, pages 191-198, 2009.
[8] Y.-M. Kuo, Y.-T. Chang, S.-C. Chang and M. Marek-Sadowska. Engineering change using spare cells with constant insertion. In *Proc. ICCAD*, pages 544-547, 2007.
[9] ABC: A system for sequential synthesis and verification. http://www.eecs.berkeley.edu/~alanmi/abc/.
[10] D. Gale and L. S. Shapley. College admissions and the stability of marriage. *American Mathematical Monthly*, vol. 69, pages 9-14, 1962.

Table 2. Comparison on total cost, HPBB of ECO, and CPU time.

	Total cost			HPBB of ECO			CPU time (sec)	
	Before ECO	Direct	ECOS	Before ECO	Direct	ECOS	Direct	ECOS
testcase1	4,049,536,290	4,062,343,750	**4,051,305,350**	8,124,160	24,336,840	**12,621,060**	0.35	**0.51**
testcase2	4,142,631,960	4,414,703,560	**4,336,177,520**	78,648,240	301,202,720	**194,842,960**	1.20	**5.41**
testcase3	4,142,631,960	4,640,619,480	**4,553,033,540**	166,895,860	572,380,280	**466,332,680**	1.63	**9.38**
testcase4	2,310,270	2,652,430	**2,323,150**	114,800	501,200	**164,080**	0.10	**0.14**
testcase5	12,945,900	14,098,100	**12,989,860**	413,250	1,677,700	**567,220**	0.15	**0.17**
Ratio	0.9234	1.0000	**0.9516**	0.2724	1.0000	**0.5291**	-	-

Handling Don't-Care Conditions in High-Level Synthesis and Application for Reducing Initialized Registers

Hong-Zu Chou[†], Kai-Hui Chang[‡], and Sy-Yen Kuo[†]
[†]Electrical Engineering Department, National Taiwan University, Taipei, Taiwan
[‡]Avery Design Systems, Inc., Andover, MA, USA
sykuo@cc.ee.ntu.edu.tw

ABSTRACT

Don't-care conditions provide additional flexibility in logic synthesis and optimization. However, most work only focuses on the gate level because it is difficult to handle such conditions accurately at the behavior and register transfer levels, which is problematic since the trend is to move toward high-level synthesis. In this work we propose innovative methods to handle such conditions accurately at high-level designs. In addition, we propose two novel algorithms based on our new methods to minimize the number of registers that need to be initialized at the architecture level, which can reduce the routing resources used by the reset signals and alleviate the routing problem. Our results show that we can identify 53% of the registers that can be uninitialized in a 5-stage pipelined processor within 5 minutes, demonstrating the effectiveness of our approach.

Categories and Subject Descriptors

B.5.2 [**Register-Transfer-Level Implementation**]: Design Aids—*Optimization*

General Terms

Design, Verification

Keywords

Don't-Care (DC), RTL symbolic simulation, Synthesis

1. INTRODUCTION

Don't-care conditions have been widely exploited by synthesis tools to improve netlist quality. In order to enhance the efficiency and accuracy of such synthesis techniques, a great deal of effort has been invested [3,6,9,11,12,15]. However, most existing techniques focus on optimizing an existing gate-level netlist only. Not being able to utilize such DCs at a higher-level of abstraction, such as the Register Transfer Level (RTL), greatly reduces the optimization power of high-level synthesis tools [11]. In particular, once a netlist has been generated, DCs often provide local optimization opportunities only and cannot change the netlist structure significantly. On the other hand, if DCs are known at the RTL, the synthesis tool can potentially generate quite different netlists for better optimizations. Not being able to utilize don't-cares in the RTL will undoubtedly become a serious limitation of such synthesis tools in the future.

One major reason why DCs are rarely utilized at the RTL for logic synthesis is that it is difficult to handle such DCs

Permission to make digital or hard copies of part or all of this work for personal or classroom use is granted without fee provided that copies are not made or distributed for profit or commercial advantage and that copies bear this notice and the full citation on the first page. To copy otherwise, to republish, to post on servers or to redistribute to lists, requires prior specific permission and/or a fee.
DAC'09, July 26-31, 2009, San Francisco, California, USA

correctly at this level. As Haufe and Rogin suggested [7], DC (or X) is one major source of RTL and gate-level mismatch. One previous research suggests to eliminate X-values throughout the design using a two-state methodology [2]. However, this methodology can significantly reduce synthesis quality because DCs are no longer available for optimization. Our first contribution in this work is to use high-level[1] symbolic simulation to accurately handle the propagation of X values. In addition, we propose a SAT formulation that can prove whether or not the X values will be propagated to any observation points, such as primary outputs or registers. Our techniques not only provide more optimization opportunities in high-level synthesis but also solve many verification problems [1, 14]. In this way, designers can use X values whenever appropriate and verify whether those Xs are indeed don't cares using our methods, producing more optimization opportunities for synthesis tools.

Being able to handle X conditions accurately at higher levels enables many logic optimizations. For instance, with the miniaturization of transistors, routing has become a serious problem that attracts much attention [10]. By reducing the number of registers which need to be initialized, the routing problem can be alleviated because routing resources used by the reset signal can be reduced. Our second contribution is two algorithms for finding registers that can be uninitialized in a design. The first algorithm is optimal and can find the maximum set of such registers; however, it is computation intensive. The second algorithm is a faster heuristic that finds approximate solutions. Our empirical results show that the heuristic algorithm can produce good results within significantly shorter time compared with the optimal one. Moreover, when the algorithm is applied to a 5-stage pipelined processor, it found that approximately 53% of the control registers do not need to be initialized for a reset period that is 5 cycles long. These results show that our techniques can improve synthesis quality dramatically and help alleviate routing problems.

The rest of this paper is organized as follows. Our symbolic-based X-value checking algorithm is presented in Section 2. In Section 3, we propose new methodologies to find the minimum number of registers that need to be initialized. Empirical results are shown in Section 4, and Section 5 concludes this paper.

2. HANDLING X-VALUES AT THE RTL

In this section, we describe how symbolic simulation can be used to accurately handle X-propagation at the RTL and check the observability of those X-values. The reason why we chose symbolic simulation is because it is a combination of hardware and software verification methods, thus it can natively handle high-level code without the synthesis step [5]. Furthermore, handling X-values at higher levels can also provide additional benefits for synthesis and verification [11].

---

[1]We only use RTL in the text for simplicity, but our techniques can also be applied to ESL as long as the design can be symbolically simulated.

978-1-60558-497-3/09 $25.00 © 2009 ACM

2.1 Handling X-Propagation at the RTL

Although many logic simulators support 3-value (0/1/X) simulation, the handling of X-values is often inaccurate. Consider a simple example shown in 1(a), the RTL code includes an uninitialized register i which controls the value of output o. In this example, if i has X-value, output o can be either 0 or 1. However, logic simulation can only take one branch and makes $o = 1$ according to the Verilog standard. Consequently, simulation mismatch between RTL code and the synthesized netlist could occur. Unlike logic simulation, symbolic simulation treats the X-value in i as a symbol and produces output expressions in terms of the symbol. Since a symbol represents both 0 and 1, symbolic simulation can handle all possible values of i simultaneously, as shown in Figure 1(c). Therefore, we can employ symbolic simulation to accurately handle X-propagation at the RTL.

(a)	module X (o);	(b)	Logic simulation results:
	output reg o; reg i;		$o = 1$
	always @(i)	(c)	Symbolic trace:
	if ($i == 1$) $o = 0$;		$o = \text{multiplex}(i) -$
	else $o = 1$;		1: 1'b0;
	endmodule;		0: 1'b1;

Figure 1: Simulating X using logic and symbolic simulation. Assume i=1'bx, symbolic simulation produces more accurate results.

2.2 Checking the Observability of X-Values

By using symbolic simulation to handle the propagation of X-values, those values can be handled accurately at the RTL. However, even if the symbols representing X propagated to primary outputs, it does not necessarily mean that those X-values will affect primary outputs. In this section we formulate the X-value checking problem as a SAT problem. In particular, we use symbolic simulation to generate logic expressions to form a SAT instance, and rely on SAT engines to solve the problem. Note that although similar SAT constructions have been proposed for error diagnosis and equivalence checking [4, 13], the SAT instance formulation proposed in this paper is still different.

2.2.1 Problem Formulation

The main objective of X-value checking is to determine whether any X will propagate to any observation points (e.g., primary outputs at each cycle, register values at the end of the reset period, etc.), and the problem is formulated as follows. Given a design containing N primary inputs PI_1, PI_2, ..., PI_N, M registers R_1, R_2, ..., R_M and K observation points O_1, O_2, ..., O_K, check whether any O_i can be affected by an uninitialized R_j within a given number of cycles C, where $1 \leq i \leq K$ and $1 \leq j \leq M$. C is the number of reset cycles for the design, and it is often defined in the specification. Alternatively, our experiments suggest that one can incrementally prolong the reset period until the number of uninitialized registers no longer increases. Note that although in our current formulation we only consider X-values in registers, our methods can be applied easily to any signal in the design.

2.2.2 Converting the Problem to a SAT Instance

To solve the problem formulated in Section 2.2.1, we propose a new method that converts the problem to a SAT instance. The built instance is shown in Figure 2, and the method works as follows.

1) Duplicate design A to create an identical copy A'. Use symbol X_{Ai} to represent the initial state of register R_i in design A. Similarly, use symbol $X_{A' \cdot i}$ for design A'.
2) Inject a symbol $PI_{i@c}$ to both design's primary inputs PI_i at cycle c, perform symbolic simulation for C cycles.
3) Observation points $OA_{1@c}$, ..., $OA_{K@c}$ and $OA'_{1@c}$, ..., $OA'_{K@c}$ are connected to a miter. When the miter's out-

put is 1, it means there exists a pair of observation points, $OA_{i@c}$ and $OA'_{i@c}$, whose values are different; otherwise, the equivalency of these observation points is proved.

Note that Figure 2 only shows a snapshot at cycle c for PI and O. The SAT instance should include PI and O from cycle 1 to cycle C. Finally, a SAT solver is called to find if a solution exists to make miter $= 1$. If the solver can find a solution, then "X"s can propagate to one or more observation points. The reason is that if we have a circuit and apply the SAT solutions to the primary inputs, then we will have two different initial states that can make the values at observation points different, which means the "X" can be propagated out. On the other hand, if the solver proves that the SAT instance is unsatisfiable, then we know that none of the "X" can be propagated to the observation points.

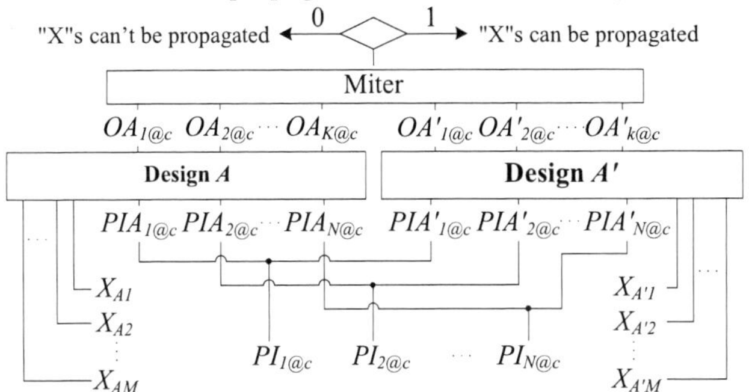

Figure 2: Illustration of symbolic X-propagation checking ($PI_{i@c}$, $PIA_{i@c}$ and $PIA'_{i@c}$ denote primary inputs, X_{Ai} and $X_{A' \cdot i}$ denote registers' initial states, and $OA_{i@c}$ and $OA'_{i@c}$ denote observation points)

3. REDUCING INITIALIZED REGISTERS

In this section we propose two algorithms to minimize the number of registers based on the techniques described in Section 2. The first one can find the minimum set of registers that need to be initialized, while the second one can only get an approximate result but requires lower computation cost.

3.1 Finding Optimal Solution

Our first method that finds the minimal number of initialized registers is based on the SAT-instance construction described in Section 2.2.2. To serve the new purpose, however, two components are added to the SAT-instance formulation, including 1) multiplexers and 2) cardinality constraints [13]. A multiplexer is added to every register to model the initial state in the SAT formulation. More specifically, when the select line S_i is 1, the initial states of X_{Ai} and $X_{A' \cdot i}$ are both 0, which means register R_i is initialized to 0. When S_i is 0, X_{Ai} and $X_{A' \cdot i}$ are regarded as free variables and become the primary inputs of the SAT instance; i.e., they are uninitialized. An example of the multiplexer is given in Figure 3(a), where symbol "X" denotes a free variable that can take any value. Cardinality constraints, as shown in Figure 3(b), restrict the number of select lines that can be set to 1 simultaneously to m. This number also represents the number of registers that need to be initialized. The cardinality constraints are implemented by an adder that performs a bitwise addition of the select lines and outputs the sum, m.

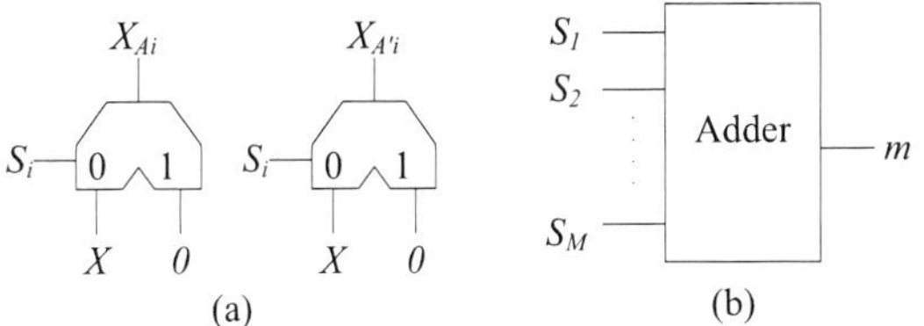

Figure 3: Additional constructs added to Figure 2 for finding the minimum set of initialized registers

The algorithm is presented in Figure 4. Variable ss denotes a set of registers to be initialized, and SS_m denotes the solution space that contains different combinations of exactly m registers that should be initialized, where $0 \leq m \leq M$ (M=*total number of registers*). Function $check\_X(miter, m, SS)$ denotes the procedure to check the observability of X-values in solution space SS under the constraints $miter$ and m using the constructs shown in Figure 2 and Figure 3. Initially, m is set to 0, and a SAT solver is called to find if a solution exists to make $miter = 1$ under constraint m. If the problem is satisfiable and solution ss is returned, it means that initializing the combination of m registers returned by $check\_X$ in the solution ss will propagate X-values to observation points, and ss should be excluded from the solution space SS_m. The procedure is repeatedly performed until no solution is returned. At this point, two situations should be considered: 1) SS_m is empty, which means that if only m registers are initialized, no matter which combination, X values in one or more of the uninitialized registers will be propagated to observation points. When this happens, we increment m to look for solutions with one more register to initialize. 2) SS_m is not empty, which means initializing m registers using any combination of registers that remain in SS_m can make sure no X in the uninitialized registers can be propagated to observation points, and the solution is returned. Note that observation points typically include two types of signals: primary outputs at every cycle and registers at the end of the reset period. In practice, users can decide which signals to observe based on their requirements. For instance, if the Xs at primary outputs during reset can be ignored, users do not have to make them observation points.

```
01  for (m = 0; m <= M; m + +)
02      ss = check_X(miter = 1, m, SS_m);
03      while (ss)
04          SS_m = SS_m \ {ss};
05          ss = check_X(miter = 1, m, SS_m);
06      if (SS_m is not empty) return SS_m;
```

Figure 4: Optimal method for finding a minimum set of registers that need to be initialized.

The above formulation uses the constructs shown in Figure 2 and 3 to find and exclude all combinations of m registers that when only those registers are initialized, the X values in the uninitialized variables will be propagated to observation points. When $check\_X$ can no longer find a solution under constraint m, it means choosing any combination of m registers that still exists in SS_m will not propagate X-values to the observation points. Since we start our search from $m = 0$, we can find the minimal number of registers that should be initialized. This problem is at least NP-complete (the proof is omitted due to space limitations).

3.2 Fast Heuristic Algorithm

In order to reduce the computation cost, we design a heuristic algorithm that finds a large set of registers that can be uninitialized. The pseudo-code of the algorithm is presented in Figure 5. Initially, $check\_X$ is used to find sets of registers to initialize so that no X can be propagated to the observation points under constraint $m = M - 1$ (i.e., only one register is uninitialized), and these solutions are added to a set $seed_{M-1}$. After that, solution space SS_m is created by spanning the solutions comprised in $seed_{m+1}$, where $0 \leq m \leq M - 2$. This is achieved in the *span* function by adding each register in $seed_{M-1}$ to each combination of registers in $seed_{m+1}$ in line 4 to form a new set of candidate combinations of registers for $check\_X$ in SS_m. To avoid significant cost for checking all possible solutions in SS_m, the algorithm randomly selects only T solutions, $ss_1, ss_2, \ldots, ss_T$, for X-propagation checking, where T is a pre-defined threshold determined empirically. If ss_i cannot

propagate X-values, it will be added to $seed_m$. The same procedure is performed repeatedly with a smaller m until $seed_m$ is empty. At this point, $seed_{m+1}$ includes the solutions that $M - (m + 1)$ registers can be uninitialized.

```
01  seed_{M-1} = {ss_i | ss_i ∈ SS_{M-1} &&
02          check_X(miter = 1, m = M - 1, ss_i) == null};
03  for (m = M - 2; m >= 0; m − −)
04      SS_m = span(seed_{m+1}, seed_{M-1});
05      {ss_1, ss_2, . . . ss_T} = randomly_select(SS_m);
06      foreach (ss_i)
07          if (check_X(miter = 1, m, ss_i) == null)
08              seed_m = seed_m ∪ ss_i;
09      if (seed_m is empty)
10          return seed_{m+1};
```

Figure 5: Heuristic method for finding a large set of registers that can be uninitialized.

Two efficient methods can be employed to improve the performance of the heuristic algorithm. The first one is that for the registers that should be initialized, we can actually make them stay at 0/1 for all cycles, not just the first cycle. The reason is that since they are going to have the reset signal, they can be forced to 0/1 during the reset period. The second one is that the primary inputs can be constant scalar values instead of symbols. Since people usually have direct access to primary inputs, and we are only concerned about the final state after the reset period, we can apply a known sequence to the primary inputs during the reset period. These scenarios can reduce the solution space and make SAT solver run faster. Obviously, the sequence used may affect the number of registers that need to be initialized. Finding a good sequence is our future work.

4. EXPERIMENTAL RESULTS

Designs PIPE and DLX are used in our experiments to evaluate the performance of optimal and heuristic algorithms. Since the heuristic algorithm is based on a random scenario, we ran each configuration of experiment 30 times to obtain their average. Only register values at the end of the reset period are considered as the observation points, and the experiments are conducted with a commercial symbolic simulator called Insight [16] that supports high-level code.

Comparing Optimal and Heuristic Algorithms Using PIPE: To compare the performance of the optimal and the heuristic algorithms, we developed a simple design called PIPE containing 12 word-level registers, and each register can be uninitialized after certain cycles. The results are shown in Table 1. The optimal algorithm can always find the minimum set of registers that need to be initialized as expected; however, its runtime is long due to its optimal nature (more than 1000 seconds when the number of cycles is smaller than 5). Note that the runtime of optimal algorithm increases rapidly as the number of registers increases, making it difficult to be applied to practical designs. Comparatively, the heuristic algorithm is a greedy approach that checks T possible combinations per iteration only. As shown in Table 1, when $T = 2$, the heuristic algorithm can get the same results as the optimal algorithm. We also observe that as the number of cycle increases, SAT solver also requires more time to produce the result. However, the maximum runtime is still shorter than 4 seconds, suggesting that our heuristic algorithm can produce good results within a significantly shorter time.

Evaluating Heuristic Algorithm on DLX: DLX is a 32-bit RISC microprocessor with 5-stage pipeline designed by Hensessy and Patterson [8]. The BugUnder Ground project from Michigan [17] provides a DLX implementation, and it is used to evaluate the performance of our heuristic approach. The DLX implementation has 75 control registers, 32 general

978-1-60558-497-3/09 $25.00 © 2009 ACM

Table 2: The performance of heuristic algorithm on DLX (C=1 to 5, T=2 to 6)

	$T = 2$		$T = 3$		$T = 4$		$T = 5$		$T = 6$	
Cycles C	Uninit. Reg.	Runtime(s)	Uninit. Reg.	Runtime(s)	Uninit. Reg.	Runtime(s)	Uninit. Reg.	Runtime(s)	Uninit. Reg.	Runtime(s)
1	18.3	101.98	21.07	117.8	24.9	154.01	25.37	196.04	26.33	207.88
2	30.6	126.22	32.93	156.12	32.83	209.34	34.47	226.83	34.5	230.55
3	35.87	139.78	36.93	170.43	37.47	235.02	37.67	246.44	37.43	251.45
4	37	141.91	37.67	171.65	37.83	241.89	37.9	250.43	37.97	254.56
5	40	144.5	40	177.26	40	250.82	40	249.99	40	260.38

Table 1: Comparison of optimal and heuristic algorithms using PIPE

	Optimal		Heuristic ($T = 2$)	
Cycles C	Init. Reg.	Runtime(s)	Init. Reg.	Runtime(s)
1	11	2414.69	11	1.91
2	10	2401.49	10	1.82
3	8	2326.83	8	1.8
4	6	1531.38	6	1.98
5	5	841.76	5	2.15
6	3	82.23	3	2.07
7	3	85.3	3	2.06
8	2	17.46	2	2.22
9	1	2.39	1	2.48
10	1	2.47	1	2.55
11	1	2.39	1	2.54
12	0	0.01	0	3.48

registers and 3 dummy registers (RDaddr_reg, Wrt_en and Control_state in cpu.v). Note that the number of registers in the synthesized netlist is 1652. It is obvious that utilizing don't-cares at the RTL is much more efficient than the gate level because only 75 word-level registers need to be handled instead of 1652 gate-level registers.

We initialize general and dummy registers by default because their X-values are known to be propagated. In addition, we applied the methods mentioned in Section 3.2 to improve the performance of SAT solving. Specifically, primary inputs Iin and Din are constrained to 0, and registers that need be initialized are forced to 0 for all cycles.

We examined the effect of varying the number of cycles C and the threshold T on the results. As shown in Table 2, as the number of cycles increases, there is a gradual increase of registers that can be uninitialized because the Xs may be masked with additional cycles. Therefore, in practice one can gradually increase C until the number of uninitialized registers stops increasing.

Since DLX is a 5-stage processor, many registers that can be uninitialized will be identified within 5 cycles. Therefore, when $C = 5$ the proposed approach found that 40 registers do not propagate their X-values. We also observe that when the threshold T becomes larger, more combinations of registers will be checked, and better results can be produced. It is interesting to note that the results of $C = 5$ are the same regardless of the value of T. The reason is that all registers contained in $seed_{74}$ can be uninitialized after 5 cycles, and the SAT solver cannot find any satisfiable solution from the constrained solution space. We also analyzed the reason why some registers must be initialized and found that they are intermediate registers to store temporary values for specific instructions. Since we bind primary inputs to 0, those instructions cannot be generated to initialize those registers.

Runtime of the above experiments are listed in Table 2. Apparently, the required runtime increases with the number of cycles C and the threshold T because of the complexity in SAT translation and satisfiability checks. However, the average maximum runtime is still smaller than 5 minutes. The empirical results show that the heuristic approach can find approximately $40/75 = 53\%$ registers that can be uninitialized; therefore, we do not need to use a large threshold which may cause significant increase in runtime.

5. CONCLUSION

Don't-care (X) conditions provide additional flexibility for synthesis optimizations. However, existing techniques typically focus on optimizing gate-level netlists only, mostly due to the difficulty to handle X accurately at the RTL. Not being able to use don't-care conditions at those levels hurts the quality of synthesized netlists, and this limitation will become more problematic in the future since the trend is to move even more toward high-level synthesis. To address this problem, we propose innovative techniques to accurately handle X-values using high-level symbolic simulation and check observability of X-values using SAT-solvers. In addition, we utilize the don't-cares for a novel optimization: identifying the registers that can be uninitialized in high-level designs for synthesis optimizations. This new optimization reduces the number of registers that need reset signals and can alleviate routing problems. Our proposed methods found that 53% of control registers in a DLX processor can be uninitialized for a reset period that is 5 cycles long, and the runtime is smaller than 5 minutes, suggesting that our methods provide a practical building block for new high-level synthesis optimizations.

Acknowledgments

This research was supported by the National Science Council, Taiwan under Grant NSC 97-224-E-002-216-MY3; Excellent Research Projects of National Taiwan University, 95R0062-AE00-05; and Small Business Innovation Project of Ministry of Economic Affairs, Taiwan, 1Z960401.

6. REFERENCES

[1] D. Brand, R. A. Bergamaschi and L. Stok, "Be Careful with Don't Cares," *DAC'95*, pp. 83-86.

[2] L. Bening, "A Two-State Methodology for RTL Logic Simulation," *DAC'99*, pp. 672-677.

[3] R. A. Bergamaschi, D. Brand, L. Stok, M. Berkelaar, and S. Prakash, "Efficient Use of Large Don't Cares in High-Level and Logic Synthesis," *ICCAD'95*, pp. 272-278.

[4] K. H. Chang, I. L. Markov, and V. Bertacco, "Functional Design Errors in Digital Circuits: Diagnosis, Correction and Repair", Springer 2008.

[5] D. W. Currie, A. J. Hu, and S. Rajan, "Automatic Formal Verification of DSP Software," *DAC'00*, pp. 130-135.

[6] M. Damiani and G. De Micheli, "OObservability Don't Care Sets and Boolean Relations," *ICCAD'90*, pp. 502-505.

[7] C. Haufe and F. Rogin, "Ad-Hoc Translations to Close Verilog Semantics Gap," *workshop on IEEE DDECS'08*, pp. 1-6.

[8] J. L. Hensessy, and D. J. Patterson, "Computer Architecture: A Quantitative Approach, 2nd edition," Morgan Kaufman, 1996.

[9] A. Mishchenko and R. K. Brayton, "SAT-Based Complete Don't-Care Computation for Network Optimization," *DATE'05*, pp. 412-417.

[10] M. D. Moffitt, J. A. Roy, and I. L. Markov, "The Coming of Age of (Academic) Global Routing," *ISPD'08*, pp. 148-155.

[11] R. Ranjan, Y. Antonioli, A. Hunter, and O. Petlin, "Formal Verification Enables Safe X Handling," Dec. 2008. http://www.scdsource.com/article.php?id=324

[12] H. Savoj and R. K. Brayton, "Observability Relations and Observability Don't Cares," *ICCAD'91*, pp. 518-521.

[13] A. Smith, A. Veneris and A. Viglas, "Design Diagnosis Using Boolean Satisfiability," *ASPDAC'04*, pp.218-223.

[14] M. Turpin, "The Dangers of Living with an X," SNUG Boston, 2003.

[15] Q. Zhu, N. Kitchen, A. Kuehlmann, and A. Sangiovanni-Vincentelli, "SAT Sweeping with Local Observability Don't-Cares," *DAC'06*, pp. 229-234.

[16] Avery Design Systems Inc., http://www.avery-design.com

[17] Bug UnderGround, http://bug.eecs.umich.edu

978-1-60558-497-3/09 $25.00 © 2009 ACM

Oil Fields, Hedge Funds, and Drugs

Patrick Groeneveld (Organizer and Chair)
Magma Design Automation, Inc.

Rob A. Rutenbar
Carnegie Mellon University
Pittsburgh, PA, USA

Erik Carlson
Armored Wolf
Aliso Viejo, CA, USA

Jed Pitera
IBM Almaden Research Lab
San Jose, CA, USA

Jinsong Chen
Lawrence Berkeley Labs
Berkeley, CA, USA

Categories and Subject Descriptors

G Mathematics of Computing, G.3 Probability and Statistics

General Terms: Algorithms

Keywords: Monte Carlo methods, Oil field discovery, Financial market analysis, Drug discovery

PANEL SUMMARY

Statistical analysis is a fundamental method in analysis, design, and optimization of large systems with uncertainties. It is being applied in drug development, analyzing financial markets, search for new oil fields, and many more areas. As the silicon process technology scales to its limits, uncertainty plays an increasing role and has made its way into tools such as yield analysis, statistical time analysis, etc. In this educational panel, we will start with a short tutorial on Monte Carlo methods including their use in EDA. Three experts from distinct application fields will then follow and discuss their experience in using Monte Carlo methods for solving large-scale problems in their domain. It may come as a surprise to some attendees of DAC to learn how much commonality there is between methods used in EDA and other field that seem far-fetched.

PANELIST PRESENTATIONS

Rob A. Rutenbar: Moore's law device scaling dramatically increases the statistical variability with which tomorrow's chips must contend. Devices with atomic dimensions don't have deterministic parameters: every behavior we want to model is a messy smear of probability. How do we attack such problems? For a small handful of tasks, we have robust, analytical techniques; statistical static timing analysis (SSTA) is the obvious example. But in the general case, we really only one

Permission to make digital or hard copies of part or all of this work for personal or classroom use is granted without fee provided that copies are not made or distributed for profit or commercial advantage and that copies bear this notice and the full citation on the first page. To copy otherwise, to republish, to post on servers or to redistribute to lists, requires prior specific permission and/or a fee.
DAC'09, July 26-31, 2009, San Francisco, California, USA

class of solutions: Monte Carlo (MC) analysis. Monte Carlo techniques have the virtue that they are widely applicable and as accurate as we like; their downside is their reputation for agonizingly slow execution time.

In this talk, I will discuss the following question: *Is the silicon community unique in facing such statistical problems?* As it turns out, the answer is: *No.* Problems in computational finance and risk analysis share many of the characteristics that challenge us in statistical circuit analysis: high dimensionality, profound nonlinearity, stringent accuracy requirements, and expensive analysis (i.e., circuit simulation). In this talk I'll show examples of adapting computational MC ideas from finance and risk analysis for use in the silicon world. I'll show how the same methods used to price complex securities can be adapted to compute silicon yields, giving speedups of 2x - 50x. I'll show how methods used to analyze the statistics of rare events (like the size of the biggest wave in a hurricane like Katrina) can be used to analyze failures in SRAM, giving speedups of 20,000x.

Recent events on Wall Street have given "computational finance" a rather bad name. I will argue that the mathematics is not the culprit here – it's the set of data and assumptions you start with, and how you use the results. A lovely attribute of the silicon world is that our statistics derive from real (and measurable) physics, which lets us adapt and exploit these methods with great success.

Erik Carlson: Monte Carlo methods have been used for security analysis since the 1970s. My presentation will focus on their use and shortcomings for mortgage analysis and risk management.

In finance, the underlying random variables are usually assumed to follow a path that is a function of a Brownian motion. Estimates for the sensitivity with respect to input parameters can be obtained by numerical differentiation. This can be a time-consuming process.

Monte Carlo methods can be used for portfolio evaluation. The value of each component instrument of the portfolio is computed for each simulation, and the portfolio value is observed factoring in the correlation among the instruments. The resulting portfolio values for each scenario create a probability distribution. The statistical characteristics of the portfolio such as Value at Risk [VaR] (the amount you can lose given a threshold and time horizon), Volatility (the standard deviation of the returns) and Expected Shortfall (tail risk) are then observed.

Proper use of Monte Carlo methods has allowed some firms to better weather the current financial crisis. For example one of the better performers, Bank of America, reported two exceedences during 2007 of a popular risk measure VaR, which is inline with the 99% confidence interval of the analysis. Alternatively, (UBS) reported 50 exceedences of the same 99% confidence interval during the same period.

Jed Pitera: Monte Carlo methods, particularly Markov Chain Monte Carlo (MCMC) methods, are a useful tool in several different areas of the drug discovery process. The goal of drug discovery is the development of a novel molecule with targeted biological activity, in order to cure or mitigate a particular disease. These applications of MCMC in the pharmaceutical industry span length scales from atoms to meters (molecules to patients) and time scales from picoseconds (10^{-12} s) to years. At a molecular level, MCMC simulations are used to predict how potential drug molecules will bind to their biological target. On intermediate scales, a Markovian view of physical systems is a useful approach for connecting molecular dynamics to slower processes, such as protein folding, membrane fusion, or crystallization. At an even coarser level, these methods are used to model processes at the organism (pharmacokinetics/ dynamics), population (disease progression) or global (pandemic) level. The intrinsically sequential nature of genomic information (DNA, RNA or protein sequences) provides another fertile application area for Markov Chain methods, but the rich field of bioinformatics is outside the scope of this panel.

There are two primary reasons for selecting MCMC methods over simpler MC approaches in these applications. In some cases, such as molecular simulations of drug-receptor binding, the integrand to be sampled is "stiff" – small coordinate changes rapidly move the system from stable (statistically important) configurations to unstable regions with negligible contributions to the integral. In other cases, there is some temporal behavior of the system of interest that needs to be modeled. If this temporal behavior can be captured as a series of memoryless transitions between well-defined states (healthy; sick; dead), MCMC is an ideal tool. Regardless of the specific application, the main challenge in this domain is the development of accurate, complete and predictive models of the biological process.

Jinsong Cheng: In my part I will start with an introduction into geophysical techniques for discovery of oil/gas fields, including

a historical perspective and main methods for deepwater exploration (e.g. Gulf of Mexico). The goal of geophysical exploration is to find economic oil/gas reservoirs, specifically estimating oil/gas saturation in a spatial domain, using surface, crosswell, and borehole based geophysical methods. The most widely used technique is the seismic method, in which seismic sources send out waves into the subsurface; reflected waves are then recorded at the surface by a series of receivers. Another recently introduced technique is the controlled-source electromagnetic (CSEM) methods, in which electric transmitters are towed by boats and keep sending EM signals into the subsurface; reflected signals are then recorded by many receivers deployed on the seafloor. Both seismic and CSEM methods provide an enormous amount of information about the subsurface that need to be analyzed.

Conventional methods for integrating those data follow a two-step approach: (1) the data sets are first inverted individually using traditional optimization (or Gauss-Newton iterative) based methods for extracting geophysical attributes (e.g., seismic velocity and electrical conductivity) along a two-dimensional (2D) cross-section or in a three-dimensional (3D) domain; (2) the inverted 2D or 3D images and borehole data are jointly interpreted to identify possible oil or gas reservoirs. The above approach suffers from several limitations. The first is that the conventional methods for inverting geophysical data are local searching methods, which are sensitive to the choice of initial values; different initial values yield different solutions. In addition, those methods provide less or no information about the uncertainty associated with the solutions, which is critical for oil companies to make decisions. The second limitation is that the integration methods used for combining multiple data sets do not permit information sharing across multiple sources.

Bayesian methodology provides a systematical, coherent approach for combining multiple sources of information and Markov chain Monte Carlo (MCMC) sampling methods provide a practical way to obtain solutions from the complex joint posterior distribution function. However, the use of MCMC methods for oil and gas discovery is relatively new (starting in the late 90s). The main reason for this is that forward simulation of geophysical data is time-consuming while MCMC methods need a large amount of forward modeling. This situation has changed due to the fast growth of computing resources. As a result, MCMC based Bayesian methods for inverting geophysical data are becoming a popular approach in geophysical exploration of oil and gas reservoirs. In this talk, I will demonstrate the benefits of jointly inverting the two types of data sets for oil and gas exploration.

978-1-60558-497-3/09 $25.00 © 2009 ACM

Human Computation

Luis von Ahn
Computer Science Department
Carnegie Mellon University
biglou@cs.cmu.edu

ABSTRACT

This talk is about harnessing human brainpower to solve problems that computers cannot. Although computers have advanced dramatically over the last 50 years, they still do not possess basic conceptual intelligence or perceptual capabilities that most humans take for granted. By leveraging human abilities in a novel way, I solve large-scale computational problems and collect data to teach computers basic human talents. To this end, I treat human brains as processors in a distributed system, each performing a small part of a massive computation.

Categories and Subject Descriptors

I.2.m Artificial Intelligence – *miscellaneous*.

General Terms

Design, Security, Human Factors, Theory.

Keywords

Distributed Knowledge Acquisition, Games With A Purpose, Human Computation, Image Labeling, Online Games, World Wide Web, CAPTCHAs.

1. INTRODUCTION

At the height of its construction, 44,733 people worked on the Panama Canal. The Great Pyramid of Giza required 50,000 workers and the Apollo Project 400,000. No matter what you put on this list, humanity's largest achievements have been accomplished with less than a few hundred thousand workers because it has been impossible to assemble (let alone pay!) more people to work together—until now. With the Internet, we can coordinate the efforts of billions of humans. If 400,000 people put a man on the moon, what can we do with 100 million? My research aims to develop theories and build computer systems that enable massive collaborations between humans and computers for the benefit of humanity. I am developing a new area of computer science called human computation, which studies how to harness the combined power of humans and computers to solve problems that would be impossible for either to solve alone.

Permission to make digital or hard copies of part or all of this work for personal or classroom use is granted without fee provided that copies are not made or distributed for profit or commercial advantage and that copies bear this notice and the full citation on the first page. To copy otherwise, to republish, to post on servers or to redistribute to lists, requires prior specific permission and/or a fee.
DAC'09, July 26-31, 2009, San Francisco, California, USA

2. EXAMPLES

An example of this is my reCAPTCHA project [1], which has enlisted hundreds of millions of people to help digitize books by solving CAPTCHAs on the Internet. CAPTCHAs are widespread security measures used throughout the Web. You've seen them: images of squiggly characters that you must type to obtain free email accounts and access to other sites. By asking humans to do a task that computers cannot, CAPTCHAs prevent automated programs from abusing online services. It is estimated that over 200 million CAPTCHAs are typed every day, each taking roughly ten seconds of human effort—that's 500,000 hours per day. My new reCAPTCHA project channels this effort into a dual purpose: transcribing books.

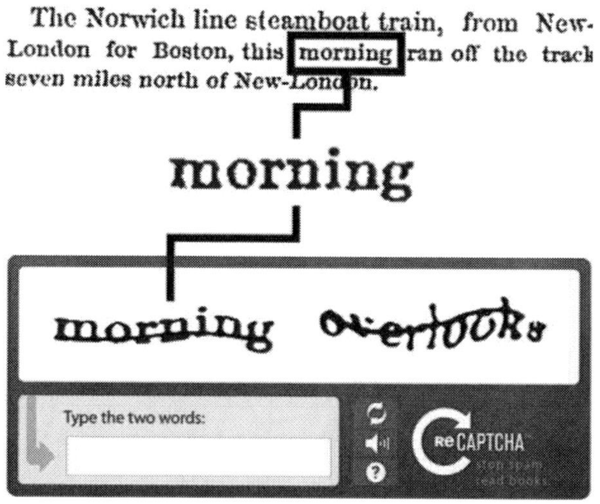

The Norwich line steamboat train, from New-London for Boston, this morning ran off the track seven miles north of New-London.

morning

morning overlooks

Type the two words:

reCAPTCHA

Figure 1. The reCAPTCHA system displays words from scanned texts to humans on the World Wide Web. In this example, the word "morning" was unrecognizable by OCR. ReCAPTCHA isolated the word, distorted it using random transformations including adding a line through it, and then presented it as a challenge to a user. Since the original word ("morning") was not recognized by OCR, another word for which the answer was known ("overlooks") was also presented to determine if the user entered the correct answer.

Physical books and other texts written before the computer age are currently being digitized en masse (e.g., by Google Books and the Internet Archive) to preserve human knowledge and make information more accessible. The pages are photographically scanned and the resulting bitmap images are transformed into text files using optical character recognition (OCR) software. This

978-1-60558-497-3/09 $25.00 © 2009 ACM

transformation into text enables the books to be indexed and searched. Unfortunately, OCR is far from perfect. In older prints where the ink has faded, OCR cannot recognize about 30% of the words. On the other hand, humans are far more accurate at transcribing such print.

ReCAPTCHA demonstrates that old print material can be transcribed, one word at a time, by people typing CAPTCHAs on the Internet. Whereas the original CAPTCHAs displayed images of random characters rendered by a computer, reCAPTCHA displays words taken from scanned texts that OCR could not decipher. The solutions entered by humans are then used to improve the digitization process. However, to meet the goal of a CAPTCHA (differentiating humans from computers) the system must be able to verify the user's answer. To do this, reCAPTCHA gives the user two words, one for which the answer is not known and a second "control" word for which the answer is known. If the user correctly types the control word, the system assumes they are human and gains confidence that they also typed the unknown word correctly. *To date, over 400 million people—6% of humanity!—have helped digitize at least one word through reCAPTCHA, making it perhaps the largest example of massive collaboration.* The words come partly from the New York Times Archive, which we will fully transcribe in 2009.

Another example of my research on human computation is the ESP Game [2]. Many people play this compelling online game over 40 hours a week, and in the process they help determine the contents of images on the Web by providing meaningful labels for them (e.g., a picture of a white dog gets the labels "white" and "dog"). In this game, randomly paired players are shown an image and challenged to "think alike" while typing possible one-word descriptions. When partners match on a word, they receive points and that word becomes a label for the image. This approach is unusual but effective: rather than using computer vision techniques that do not yet work well enough, the game encourages people to do the work in exchange for entertainment. Because partners are randomly chosen and because the game randomly

tests players with images whose correct tags are already known, the labels are guaranteed to be accurate even if the players do not want them to be. So far, the ESP Game has collected over 50 million labels. Since computers cannot yet do this, attaching proper labels to images on the Web allows for more accurate image search and improves accessibility of Web sites to visually impaired individuals. For these reasons, Google licensed the ESP Game and uses it to improve its popular image search engine.

Based on the ESP Game, my students and I have created many other Games With A Purpose (GWAPs), each transforming a task into an enjoyable game. Our latest gaming site, GWAP.com, draws thousands of players to solve problems that computers cannot yet solve. Many other researchers are now taking up this technique. For example, scientists at the University of Washington have created a game in which players determine the optimal structure of proteins that computers cannot analyze. Will somebody find a cure for cancer while playing a game?

3. ACKNOWLEDGMENTS

This work was partially supported by generous gifts from the Heinz Endowment and the Fine Foundation, and by an equipment grant from Intel Corporation. Luis von Ahn was partially supported by a MacArthur Fellowship, a Sloan Fellowship, and a Microsoft Research New Faculty Fellowship.

4. REFERENCES

[1] Luis von Ahn, Benjamin Maurer, Colin McMillen, David Abraham, and Manuel Blum, 2008. reCAPTCHA: Human-Based Character Recognition via Web Security Measures. *Science,* 321(5895):1465-1468.

[2] Luis von Ahn and Laura Dabbish, 2004. Labeling Images with a Computer Game. *Proceedings of the 22nd International Conference on Human Factors in Computing Systems (CHI 2004),* 319-326.

How to Make Computers that Work Like the Brain

Dileep George
Numenta Inc.
811 Hamilton St.
Redwood City, CA, USA
+1-650-369-8282

dgeorge@numenta.com

ABSTRACT

By using neuroanatomy and neurophysiology as a set of constraints, we believe that we have started to uncover how the brain uses hierarchy and time to create a model of the world, and to recognize novel patterns as part of that model. Hierarchically organized memory is fundamentally different than the linear memory used in current computers, and therefore offers the potential for new computer architectures. Today, we are exploring and advancing this technology by using traditional computer architectures (benefited by multiple CPU cores) to emulate the hierarchical structure of the neocortex. Exploiting the hierarchical temporal structure of the neocortex to build intelligent machines could open up many opportunities to rethink how integrated circuits and systems can play a leading role.

Categories and Subject Descriptors

I.2.0 [**Computing Methodologies**]: Artificial Intelligence– *philosophical foundations.*

General Terms

Algorithms, Theory

Keywords

hierarchical learning, neocortex, spatio-temporal learning, brain

1. INTRODUCTION

Suppose you are traveling to India for the first time. You see a new contraption that runs on three wheels. From your tourist guide you learn that it is called an auto rickshaw. After this incident, you notice that roads in India are filled with auto rickshaws. They come in different sizes, shapes, colors and adornments. After seeing a few examples of auto rickshaws, you are able to correctly recognize all the other auto rickshaws.

It is unlikely that evolution programmed the knowledge of auto rickshaws into our brains. You had to learn about them by getting exposed to patterns. After seeing a few auto rickshaws, you are able to recognize novel ones although the patterns that correspond to those novel ones are not identical or close to the patterns of the ones you first saw. This capability of learning systems to recognize new patterns is termed *generalization*. The learning and generalization capabilities of human brains are yet to be matched by our algorithms.

Permission to make digital or hard copies of part or all of this work for personal or classroom use is granted without fee provided that copies are not made or distributed for profit or commercial advantage and that copies bear this notice and the full citation on the first page. To copy otherwise, to republish, to post on servers or to redistribute to lists, requires prior specific permission and/or a fee.
DAC'09, July 26-31, 2009, San Francisco, California, USA

The main goal of this paper is to outline some principles required for the building of machines that learn, generalize and exhibit some form of 'intelligence'. In the following sections, I will describe how the principles for building intelligent machines can be understood by systematically studying the organization of the neocortex.

2. WHY STUDY THE NEOCORTEX?

The necessity of studying the brain can be argued from the point of view of machine learning. In machine learning, a researcher first constructs a parameterized adaptive model for the data to be modeled. This initial model, constructed based on domain knowledge, will have several parameters that can be tuned based on the training data. The researcher then devises algorithms that can adapt these parameters. Typically, the parameters are tuned by adapting them to minimize an error criterion. A model whose parameters are tuned can then be used to classify novel patterns, predict future patterns, compress, reconstruct etc. However, despite its mathematical elegance, this method hides some fundamental problems associated with learning.

One aspect of learning is the sample complexity – the number of examples required for training a model to a reasonable level of performance. If the hypothesis space for a learning problem is very large, then the construction of a learned model can take a large number of training examples and long training times. The No Free Lunch (NFL) theorem, a fundamental result on search and optimization, addresses aspects about the complexity of learning [7]. Intuitively, the NFL theorem for learning argues that no learning algorithm has an inherent superiority over another algorithm for all learning problems. If an algorithm is superior for a particular problem, it is only because that algorithm exploited assumptions that are suitable for that problem. The same algorithm will not do well on a problem that has assumptions different from that of the original one. This means that, to make the algorithms effective, the machine-learning researcher has to encode knowledge about the problem domain into the initial model structure. The more prior knowledge that is put in, the easier it is to learn. Does that mean we need to create a new model for each new problem that we try to solve using machine learning? Clearly, such an approach would be very labor intensive.

On the other hand, humans and other mammals seem to solve these problems in a different way. The fact that humans can learn and adapt to problems that did not exist when the initial model (the neocortex) was created is proof of the generic nature of the mechanisms used by the human brain. Moreover, a large number of researchers now conjecture that the neocortex is using fundamentally the same algorithm for learning different modalities. This means that the same algorithm is in operation for learning models for audition, vision, somato-sensory perception and language. Although not proven conclusively, there is increasing amount of evidence for a *common cortical algorithm*. The existence of a common cortical algorithm was conjectured first by a neuroscientist who studied cortical tissues of different mammals [6]. The remarkable similarity of the cortical connectivity across different modalities and even across different species was the reason for this conjecture.

The combination of the common cortical algorithm and the NFL theorem produces important implications for machine learning and for our quest to build intelligent machines. On the surface, the NFL theorem seems to create problems for the idea of a common cortical algorithm. How can one mechanism/algorithm do very well on tasks as different as vision, audition and language? The common cortical algorithm and the NFL theorem can be consistent with each other provided we can exploit the same basic assumptions to learn vision, audition, language etc. If the cortex is good at learning a wide variety of tasks using a common mechanism, then there must be something common about these seemingly different tasks. Evolution must have discovered this commonality and neocortex must be exploiting that.

From the NFL theorem we conclude that a universal theory of learning is, in essence, a universal theory of assumptions. The set of assumptions that a learner uses to predict outputs, given inputs that it has not encountered, is known as the inductive bias of that learning algorithm. The more assumptions we make, the easier it becomes to learn. However, the more assumptions we make, the fewer the number of problems we can solve. If we need to design an algorithm that can be applied to a large class of problems, the question we need to ask is: What is the basic set of assumptions that are specific enough to make learning feasible in a reasonable amount of time while being general enough to be applicable to a large class of problems? We are in luck if the same set of assumptions works for a large class of problems.

2.1 How should we study the neocortex?

The problem of reverse-engineering the neocortex is a daunting one. How should we search for these universal assumptions? There are many anatomical and physiological details in the brain; how is one to know what is important and what is not? What is a mere implementation detail that biology has to employ because it has only neurons to work with, and what is an important computational principle that cannot be missed? A good strategy would be to study the neocortex and the world at the same time. Studying the anatomy and physiology of the neocortex should give us important clues about the nature of the assumptions made by the neocortex. While studying the organization of the neocortex, we will have to look for general principles that can be relevant from the point of view of computational learning. We should select only those principles for which we can find organizational counterparts in the world. If we look for structural properties of the neocortex that are matched to the spatial/temporal statistics of signals in the world, we can be reasonably certain that we will find principles that are relevant from a learning point of view.

3. HIERARCHICAL ORGANIZATION

The neocortex, indeed the entire nervous system, has a hierarchical organization. Knowledge about the world, whether it is visual, auditory, or tactile information, is stored in this hierarchically organized memory. The hierarchical organization of the neocortex could be a fundamental property that mirrors the hierarchical organization of the world.

3.1 Hierarchies in the brain

It is well established that the visual is organized as an anatomical hierarchy [1]. The structural hierarchy of the primate visual cortex has been well studied. In the ventral stream that is generally accepted as responsible for object recognition, information gets passed successively to visual areas V2, V4 and IT. The neurons in region V1 see only a small portion of the visual field. The neurons in region V2 receive their inputs from several neurons in area V1. Therefore, the effective receptive field of a V2 neuron is several times bigger than that of a V1 neuron. In general the receptive field sizes increase as you go up in the hierarchy. A neuron in area IT has a receptive field that is almost equal to the size of the whole visual field. If we present an object as input to the visual cortex, the neurons in area V1 will respond to small local features like oriented lines in that object. The pattern of neuron firings in area IT will correspond to the identity of the object itself. This kind of organization of the neurons gives rise to a structural hierarchy.

Interspersed with the structural hierarchy of the visual cortex is a temporal hierarchy. To see this, imagine that we slowly move the object presented to the visual cortex, while still keeping the whole object within its receptive field. The neurons in area V1 will see local features moving in and out of their receptive fields. That is, their responses will change quickly as we move the object. Since the neurons in area IT code for the identity of the object, their responses will remain largely stable as the object is moved. Therefore the lower levels of the hierarchy change states at a faster rate compared to the higher levels. The idea of a temporal hierarchy is consistent with the idea of a structural hierarchy where the neurons specialize for object identities as you ascend in the hierarchy.

3.2 Hierarchies in the world

Natural and man-made dynamic systems tend to have a nested multi-scale organization. In such systems, there are large-scale system-level variables and small-scale sub-system-level variables. Often, the larger system level variables are slower compared to the smaller sub-system level variables. If we consider weather systems, for example, winter is a high-level system variable that affects the whole country. During winter, there will be local system variables like snow storms that affect the weather in, say, Minnesota. While the winter lasts for several months, the local variations usually last only for days. The weather system can be thought of as organized as a hierarchy. This hierarchy is spatial as well as temporal. The higher-level variables like winter correspond to large spatial area and slow temporal variations. Local variations like snow storms are faster compared to the large-scale variables but influence smaller spatial areas.

It is important to note that the hierarchical organization is in both time and space. The hierarchy in space can be thought of as the construction of a complex system using sub-systems. The sub-

978-1-60558-497-3/09 $25.00 © 2009 ACM

systems provide stable re-usable components that can be assembled together to build larger systems. Herbert Simon was among the first to point out this arrangement [9]. According to Simon, the hierarchy in time can be thought of as an ordering of characteristic frequencies of these sub-systems. Different sub-systems will have different characteristic frequencies associated with their dynamics. In general there is mapping between the sub-system ordering and the frequency ordering. It is generally observed that smaller subsystems (small spatial scales) are associated with faster frequencies of vibrations and larger subsystems are associated with slower frequencies of vibrations.

4. HIERARCHICAL TEMPORAL MEMORY

Hierarchical Temporal Memory (HTM) is a theory of the neocortex that postulates that the neocortex builds a model of the world using a spatio-temporal hierarchy [2,3,4]. According to this theory, the operation of the neocortex can be approximated by replicating a basic computational unit - called a node - in a tree structured hierarchy. Each node in the hierarchy uses the same learning and inference algorithm. Each node stores spatial patterns and then sequences of those spatial patterns. The feed-forward output of a node is in terms of the sequences that it has stored. The spatial patterns stored in a higher-level node records co-occurrences of sequences of its child nodes. The HTM hierarchy is organized in such a way that the higher levels of the hierarchy incorporate larger amounts of space and longer durations of time. The states at the higher levels of the hierarchy vary at a slower rate compared to the lower levels. It is speculated that this kind of organization leads to efficient learning and generalization because it mirrors the organization of the world.

HTMs originated from a model called the *Memory-Prediction Framework*, first described by Jeff Hawkins in a book titled On Intelligence [5]. The Memory-Prediction Framework proposed the neocortex used memory of sequences in a hierarchy to model and infer causes in the world. The Memory-Prediction Framework proposed several novel learning mechanisms and included a detailed mapping onto large scale cortical-thalamic architecture as well onto the microcircuits of cortical columns.

In our research, HTMs have been used in invariant pattern recognition on gray-scale images, in the identification of speakers in the auditory domain and in learning a model for motion-capture data in an unsupervised manner. Other researchers have reported success in using HTMs in content-based image retrieval and object categorization.

HTM can be specified mathematically using a generative model. Each node in the hierarchy contains a set of *coincidence patterns* and a set of Markov chains where each Markov chain is defined over a subset of the set coincidence patterns in that node. A coincidence pattern in a node represents a co-activation of the Markov chains of its child nodes. A coincidence pattern that is generated by sampling a Markov chain in a higher-level node concurrently activates its constituent Markov chains in the lower level nodes. For a particular coincidence-pattern and Markov chain that is active at a higher-level node, sequences of coincidence patterns are generated by concurrently sampling from the activated Markov chains of the child nodes. This kind of generative model is a spatio-temporal hierarchy where the higher levels of the hierarchy affect more output streams and remain stable for longer durations to influence longer temporal scales.

The process of constructing an HTM model for spatio-temporal data is the process of learning the coincidence patterns and Markov-chains in each node at every level of the hierarchy. Although algorithms of varying levels of sophistication can be used to learn the states of an HTM node, the basic process can be understood using two operations (1) memorization of coincidence patterns and (2) learning of a mixture of Markov chains over the space of coincidence patterns. In the case of a simplified generative model, an HTM node remembers all the coincidence patterns that are generated by the generative model. In real world cases, we have found that storing a fixed number of a random selection of the coincidence patterns suffice if we allow multiple coincidence patterns to be active at the same time.

Each Markov chain in a node represents a set of coincidence patterns that are likely to occur close-by in time. The learning of the mixture of Markov chains is simplified considerably because of this constraint. We have found that a simple way to learn the mixture of Markov chains for real world cases is to learn a large transition matrix that is then partitioned using a graph-partitioning algorithm.

HTM networks use Bayesian belief propagation for inference. Bayesian belief propagation was originally derived for inference in Bayesian networks [8]. Inference mechanism in an HTM network deals with the propagation of evidence from anywhere in the network to all other parts of the network. The presentation of a new image to the first level of an HTM network is an example of new evidence. Propagation of this evidence to other parts of the network results in the network adjusting their belief states given this evidence. For example, a new image can lead to a different belief in the top level of the network regarding the identity of the object in that image. Information can also be propagated down in the hierarchy for attention, segmentation and fill-in.

5. COMPUTATIONAL CONSIDERATIONS

We are beginning to understand how hierarchical representations solve many problems in both neuroscience and machine intelligence, and how to build computer systems that perform in a similar way. Such machines are naturally parallel. At a time when the semiconductor industry is looking for ways to utilize multi-core CPUs, a new hierarchical computing paradigm is emerging, one that holds the promise of productively utilizing large amounts of parallel computing power.

Today, our HTM systems are implemented completely in software. Because HTMs comprise many nodes performing a similar function, they are naturally parallel and easily divided among multiple CPUs. Our current implementation will run on single-core CPUs, multi-core CPUs, and large computer clusters. HTM software readily takes advantage of multi-core-processor architectures, and adding more cores to CPUs will benefit our work in HTMs. As useful as this is, most of the performance improvements we have made to date are a result of software changes. For example, through software tuning, alone, we have significantly improved the time it takes to train an HTM and to do inference.

HTMs consist of a hierarchy of memory nodes. The computational burden of each node is not great, but in order to train an HTM, it can require exposure to many training patterns. Thus, although each iteration is quick, overall training time can be an obstacle. As our systems get larger, this becomes more critical.

978-1-60558-497-3/09 $25.00 © 2009 ACM

Each node has a chunk of memory associated with it. The data stored in a node consists primarily of large sparse matrices. The computations performed by a node are typically vector and matrix operations on these large sparse matrices. Many of the software performance improvements that we have made to date have come from taking advantage of the sparsity and other unique attributes of the data. The same idea could apply to custom silicon.

It might be possible to embed some of the matrix and vector operations into the memory chips themselves. Thus, instead of retrieving a matrix from memory, and multiplying it by a vector, we could instead pass the vector directly to the memory chip and the memory chip itself would perform the operation, and return the result.

Numenta has now built and experimented with a few generations of algorithms for its Hierarchical Temporal Memory (HTM) technology. From the very beginning, it has been very clear to us that hardware implementations of HTM algorithms will not only be possible but also be required for several applications. However, we decided to stick to a software-only strategy during the initial stages of building HTM technology. We knew that our algorithms would need to go through several generations before we can be confident about their correctness and applicability to real-world problems. For this reason, we built a very flexible software platform that let us quickly prototype algorithms and test them on a large scale using a computer cluster. We have done this for the last 4 years and we are at a stage where we think it is advantageous to take some of our algorithms to hardware for the following reasons:

1) Algorithms are becoming mature: We are now at the third generation of HTM algorithms and our experience with them on real world problems indicate that these algorithms are going to remain stable for some time.

2) We have mathematical models for cortical circuits: We now have mathematical models for cortical circuits that include feed forward, feed back and temporal inference. The mathematical abstraction helps us anticipate and parameterize future algorithm iterations. This model also gives us some design metrics for hardware implementations.

3) Real-world applications are being built: We have tested our algorithms on real world image recognition and recognition of people and objects in videos. These real-world applications have given us several data points on the amount of memory, processing and interconnections required for building hardware for real applications.

We anticipate that real world hardware implementations will first appear for inference only applications where the HTM was trained offline. In those cases, the trained network files are downloaded on to an embedded hardware chip. These applications could be built on existing silicon technologies. More exciting hardware developments will come from implementing the learning algorithms in hardware in such a way as to exploit the properties of novel device materials.

6. REFERENCES

[1] Felleman, D. J. and Van Essen, D. C., "Distributed hierarchical processing in the primate cerebral cortex", *Cerebral Cortex*, 1(1) , Jan/Feb 1991.

[2] George, D. and Hawkins, J., "A hierarchical Bayesian model of invariant pattern recognition in the visual cortex", in *Proceedings of the International Joint Conference on Neural Networks*, volume 3, 2005.

[3] George. D., "How the brain might work: a hierarchical and temporal model for learning and recognition", *PhD thesis, Stanford University*, June 2008.

[4] Hawkins, J. and George, D., "Hierarchical temporal memory: Concepts, theory and terminology", http://www.numenta.com/Numenta_HTM_Concepts.Pdf

[5] Hawkins, J. and Blakeslee, S., *On Intelligence*, Henry Holt and Company, 2004.

[6] Mountcastle, V. B., "An organizing principle for cerebral function: the unit model and the distributed system", in The Mindful Brain, MIT Press, 1978.

[7] Y. C. Ho and D. L. Pepyne, "Simple explanation of the no-free-lunch theorem and its implications", *Journal of Optimization Theory and Applications*, 115(3), December 2002.

[8] Pearl, J., *Probabilistic reasoning in intelligent systems: networks of plausible inference.*, Morgan Kaufmann, 1988.

[9] Simon, H.A., *The Sciences of the Artificial.* MIT Press, 1981.

978-1-60558-497-3/09 $25.00 © 2009 ACM

A Fully Polynomial Time Approximation Scheme for Timing Driven Minimum Cost Buffer Insertion

Shiyan Hu
Dept. of Electrical and Computer Engineering
Michigan Technological University
Houghton, Michigan 49931
shiyan@mtu.edu

Zhuo Li and Charles J. Alpert
IBM Austin Research Laboratory
11501 Burnet Road
Austin, Texas 78758
{lizhuo, alpert}@us.ibm.com

ABSTRACT

As VLSI technology enters the nanoscale regime, interconnect delay has become the bottleneck of the circuit timing. As one of the most powerful techniques for interconnect optimization, buffer insertion is indispensable in the physical synthesis flow. Buffering is known to be NP-complete and existing works either explore dynamic programming to compute optimal solution in the worst-case exponential time or design efficient heuristics without performance guarantee. Even if buffer insertion is one of the most studied problems in physical design, whether there is an efficient algorithm with provably good performance still remains unknown.

This work settles this open problem. In the paper, the first fully polynomial time approximation scheme for the timing driven minimum cost buffer insertion problem is designed. The new algorithm can approximate the optimal buffering solution within a factor of $1 + \epsilon$ running in $O(m^2 n^2 b/\epsilon^3 + n^3 b^2/\epsilon)$ time for any $0 < \epsilon < 1$, where n is the number of candidate buffer locations, m is the number of sinks in the tree, and b is the number of buffers in the buffer library. In addition to its theoretical guarantee, our experiments on 1000 industrial nets demonstrate that compared to the commonly-used dynamic programming algorithm, the new algorithm well approximates the optimal solution, with only 0.57% additional buffers and 4.6× speedup. This clearly demonstrates the practical value of the new algorithm.

Categories and Subject Descriptors

B.7.2 [**Integrated Circuits**]: Design Aids - Placement and Routing; J.6 [**Computer-aided Engineering**]: Computer-aided Design

General Terms

Algorithms, Performance, Design

Keywords

Buffer Insertion, Fully Polynomial Time Approximation Scheme, NP-complete, Cost Minimization, Dynamic Programming

Permission to make digital or hard copies of part or all of this work for personal or classroom use is granted without fee provided that copies are not made or distributed for profit or commercial advantage and that copies bear this notice and the full citation on the first page. To copy otherwise, to republish, to post on servers or to redistribute to lists, requires prior specific permission and/or a fee.
DAC'09, July 26-31, 2009, San Francisco, California, USA

1. INTRODUCTION

As VLSI technology enters the nanoscale regime, interconnect delay has become the bottleneck of the circuit timing since devices scale much faster than interconnects. As one of the most effective interconnect timing optimization engines, buffer insertion is indispensable in the physical synthesis flow [1, 2, 3]. It is demonstrated in [4] that in two recent IBM ASIC designs, over one-fourth gates are buffers.

Buffer insertion is one of the most studied problems in physical design. Existing works include [5, 6, 7, 8] on exploring dynamic programming techniques with advanced data structures to compute optimal timing driven buffering solutions. Since buffers themselves are a drain on power, it is highly desirable to use as little buffering resources as possible in buffer insertion. Buffering with cost (power or area) minimization has been considered in [6]. It proposes a dynamic programming algorithm which runs in pseudo-polynomial time. This surprises no one as the minimum cost timing driven buffering problem is NP-complete [9]. Derived from this classic problem, buffering techniques have been developed for various scenarios. For example, there are works [6, 10, 11, 12] handling slew, noise and/or variations, and works exploring the interaction of buffering with routing [13], placement [14], and floorplanning [15, 16].

Despite the fact that many buffering techniques have been developed, the underlying buffering problem is still less studied especially in theory. The dynamic programming can compute the optimal solution but not runs in polynomial time, while heuristic algorithms can run fast but without any performance guarantee. Whether there is any efficient algorithm with provably good performance still remains unknown. This work aims to settle this open problem and advance the understanding of minimum cost timing driven buffering problem from a theoretical point of view. Yet, our new algorithm is highly practical.

In this paper, we propose a fully polynomial time approximation scheme (FPTAS) for the NP-complete timing driven minimum cost buffering problem. In our context, a fully polynomial time approximation scheme refers to a buffering algorithm which is able to compute a solution with the cost at most $1 + \epsilon$ times worse than the cost of the optimal buffering solution for any $\epsilon > 0$. It runs in time polynomial in the input size of the problem instance and $1/\epsilon$. Given an NP-complete problem, an FPTAS is generally regarded as an ultimate solution in theory. The main contribution of this paper is summarized as follows.

- An FPTAS algorithm is proposed to approximate the optimal buffering solution within a factor of $1 + \epsilon$ in

$O(m^2 n^2 b/\epsilon^3 + n^3 b^2/\epsilon)$ time for any $0 < \epsilon < 1$ and in $O(m^2 n^2 b/\epsilon + mn^2 b + n^3 b^2)$ time for any $\epsilon \geq 1$, where n is the number of candidate buffer locations in the tree, m is the number of sinks in the tree, and b is the number of buffers in the buffer library.

- This work presents the first provably good approximation algorithm on the timing-driven minimum cost buffering problem.
- The proposed FPTAS is motivated from the algorithms in [17, 18] for the layer assignment problem. Nevertheless, our FPTAS features novel techniques such as double-ϵ oracle based solution search and timing-cost approximate dynamic programming algorithm. The latter runs in $O(\frac{m^2 n}{\epsilon_1 \epsilon_2} + \frac{mn^2 b}{\epsilon_1} + \frac{m^2 n^2}{\epsilon_1^2 \epsilon_2} + \frac{mn^2 b}{\epsilon_1 \epsilon_2} + \frac{n^3 b^2}{\epsilon_1})$ time to compute a buffering solution with the cost at most $(1 + \epsilon_1)W^*$ and the timing at most $(1 + \epsilon_2)T$ where W^* refers to the optimal cost and T refers to the timing constraint.
- The new FPTAS algorithm is highly practical. Experimental results on 1000 industrial nets using a buffer library consisting of 48 buffer types demonstrate that the FPTAS approximates the optimal solution by only 0.57% additional buffers with $4.6\times$ speedup compared to the dynamic programming algorithm.

2. PRELIMINARIES

A routing tree $\mathcal{T} = (V, E)$ is given as an input to the minimum cost timing buffering problem, where $V = \{s_0\} \cup V_s \cup V_n$, and $E \subseteq V \times V$. As in many previous works [9, 11], the routing tree is assumed to be binary in this paper. Trees in other topologies can be easily converted to a binary tree [7]. Vertex s_0 is the root/driver of $\mathcal{T}$, V_s is the set of sink vertices, and V_n is the set of candidate buffer locations. For each sink $s \in V_s$, there is a sink capacitance $C(s)$ and a required arrival time (RAT). The driver s_0 has an arrival time, denoted by $AT(s_0)$. A net satisfies the timing constraint if the arrival time is no greater than the required arrival time at driver. By subtracting $AT(s_0)$ and each sink RAT by $AT(s_0)$, one can modify the net such that $AT(s_0) = 0$ and all RAT are positive. Note that this will not impact buffering solutions. Define the timing constraint T to be the maximum RAT after the above modification. A buffer library B containing all buffer types which can be assigned to candidate buffer locations is also given. Note that B includes both non-inverting buffers and inverting buffers.

As is commonly used in physical synthesis, the Elmore delay model is adopted. The Elmore delay on an edge $e = (v_i, v_j)$ is computed by $D(e) = R(e)\left(\frac{C(e)}{2} + C(v_j)\right)$, where $C(e), R(e), C(v_j)$ refer to the edge capacitance, the edge resistance and the downstream capacitance viewing at v_j, respectively. For a buffer b placed at vertex v_j, its buffer delay is computed by $D(b) = R(b) \cdot C(v_j) + K(b)$, where $R(b)$ and $K(b)$ refer to the driving resistance and the intrinsic delay of buffer b, respectively. Each buffer b has also an input capacitance $C(b)$ and a cost $w(b)$. In this paper, the buffer area is used as buffer cost to illustrate our new algorithm. A buffer assignment γ is a mapping $\gamma : V_n \to B \cup \{\bar{b}\}$ where $\bar{b}$ denotes the case with no buffer inserted. The total cost of a buffering solution γ for the tree $\mathcal{T}$ is defined as the sum of the costs over all inserted buffers. The timing driven minimum cost buffering problem, known as NP-complete [9], can be formulated as follows.

Timing Constrained Minimum Cost Buffering: Given a binary routing tree with n candidate buffer locations and a buffer library with b buffer types, to compute a buffer assignment solution such that the timing constraint is satisfied and the total buffer cost is minimized.

3. ALGORITHMIC FLOW

Our FPTAS algorithm for timing constrained minimum cost buffering problem is motivated from [17, 18] for a layer assignment problem. Let W^* denote the cost of the optimal buffering solution. At a high level, the FPTAS works in the following intuitive framework. It first makes a guess x on W^* and uses a procedure called *oracle* to check whether such a guess is good, i.e., sufficiently close to W^* or not. If this is the case, return x. Otherwise, make a new guess on W^*. This procedure is iterated until a good guess is made.

Certainly, there are two algorithmic design challenges in the above framework. First, one does not know the optimal cost W^*, which is our target, then how to decide whether $x \geq W^*$ for any x? Moreover, one needs to perform this check efficiently. Second, if the current guess is not good, how to find a possibly better guess? The first difficulty needs to a salient design of the oracle and the second difficulty needs the usage of an efficient oracle based solution search. These two key components will be described in this paper.

4. DOUBLE-ϵ ORACLE SEARCH

4.1 The Oracle

Given any positive number x, the oracle can efficiently decide whether $W^* \geq x$, where W^* is the total buffer cost of the optimal buffering solution. In fact, the decision is answered approximately depending on ϵ, meaning that the answer is either $W^* \geq x$ or $W^* < (1 + \epsilon)x$.

A key component in the oracle is a *polynomial time timing-cost approximate dynamic programming* algorithm which will be described in Section 5. This algorithm has three critical properties. First, the algorithm can be performed efficiently, in time polynomial in $\overline{W} = n/\epsilon$. Second, it will either return a buffering solution with the cost no greater than a cost budget value $\overline{W}$, or conclude that this cost budget $\overline{W}$ is too low to find any buffering solution (approximately) satisfying the timing constraint. Third, the dynamic programming algorithm will return a solution with timing slightly larger than the timing constraint T with controlled error. Precisely, the timing of the solution is bounded by $(1 + \epsilon)T$ where ϵ is the target approximation ratio. This is acceptable especially considering that in early stage of physical synthesis flow (or when chip is in the prototype stage), the timing constraint is often set according to the designer' experience and thus in general the timing constraint is not stringent. In addition, varying ϵ, our algorithm can provide more design flexibility (in terms of timing and cost tradeoff) to the circuit designers. Even in the late physical design stage where timing constraint is stringent, in practice, one can still rip-up and rebuffer the nets with timing violations using [6] to compute the optimal buffering solutions. We call this procedure *timing recovery*. Even if FPTAS with timing recovery is not guaranteed to run in polynomial time, in practice it would still run much faster than using [6] alone. This is the case since there are very few nets which violate the timing constraints by our FPTAS as indicated in the experiments.

Given a timing-cost approximate dynamic programming algorithm, we are ready to present the oracle. First, for any

978-1-60558-497-3/09 $25.00 © 2009 ACM

positive number ϵ, each buffer cost w is scaled by the factor of $\frac{x\epsilon}{n}$ followed by down-rounding, i.e., w becomes $\lfloor \frac{wn}{x\epsilon} \rfloor$. The dynamic programming is performed to the scaled and rounded buffering problem with the cost budget set to n/ϵ. There are two possible decision results in an oracle query.

1. By dynamic programming, a buffering solution with timing no greater than $(1+\epsilon)T$ is found for the total buffer cost $\leq n/\epsilon$. This means that using the unscaled and unrounded costs, the cost of the obtained buffering solution will be smaller than $\frac{n}{\epsilon} \cdot \frac{x\epsilon}{n} + x\epsilon = (1+\epsilon)x$. This is the case since the rounding error in cost is at most $x\epsilon$ by noting that the rounding error is at most $\frac{x\epsilon}{n}$ at each buffer and there are only n candidate buffer locations. Therefore, there is a buffering solution with cost smaller than $(1+\epsilon)x$ and the timing at most $(1+\epsilon)T$ in the original buffering problem. We conclude that $W^* < (1+\epsilon)x$.

2. By dynamic programming, there is no buffering solution with cost $w = n/\epsilon$ which can satisfy even the relaxed timing constraint $(1+\epsilon)T$. This means that the original unscaled buffering problem does not have a buffering solution within the cost $\frac{n}{\epsilon} \cdot \frac{x\epsilon}{n} = x$ satisfying $(1+\epsilon)T$ and thus T. We conclude that $W^* \geq x$ in the original buffering problem.

Since both timing and cost are rounded, our FPTAS actually computes the $(1+\epsilon)$ approximation to the buffering problem with the longest path delay bounded by $(1+\epsilon)T$. According to Lemma 2 in Section 5, the proposed dynamic programming algorithm runs in $O(\frac{m^2 n}{\epsilon_1 \epsilon_2} + \frac{mn^2 b}{\epsilon_1} + \frac{m^2 n^2}{\epsilon_1^2 \epsilon_2} + \frac{mn^2 b}{\epsilon_1 \epsilon_2} + \frac{n^3 b^2}{\epsilon_1})$ time which gives the time for an oracle query.

4.2 The Double-ϵ Oracle Based Solution Search

After obtaining the oracle, a $(1+\epsilon)$ approximation could be computed as follows. Start with an initial lower bound W_l^* and an initial upper bound W_u^* on the optimal buffering cost W^*. For example, W_u^* could correspond to always using largest buffer at every candidate buffer location and W_l^* could be set to the cost of the single buffer with smallest cost in the buffer library. One just needs to compare the timing of the given unbuffered net to the timing constraint to account for the case where no buffer needs to be inserted. Subsequently, a binary search is performed within these bounds. That is, each time $x = \frac{W_l^* + W_u^*}{2}$ is used to query the oracle and depending on the decision results, W_l^* and W_u^* will be updated accordingly, i.e., update $W_u^* = (1+\epsilon)x$ in Case 1 and update $W_l^* = x$ in Case 2. The complexity of this algorithm certainly depends on the gap between the initial upper and lower bounds.

One can make the time complexity independent of the initial bound values as in [18]. This is accomplished by utilizing the fact that an oracle query takes the time inversely proportional to ϵ. Precisely, a larger ϵ leads to a coarser approximation but runs faster while a smaller ϵ leads to finer approximation but runs slower. This provides the opportunity in varying approximation ratio ϵ adaptively to accelerate the whole procedure of oracle based solution search. Instead of sticking to the target approximation ratio ϵ, one can use larger ϵ initially and gradually reduce it to the target ϵ. When these ϵ form a decreasing geometric sequence (e.g., $\ldots, 27, 9, 3, 1, 1/3, \ldots, \epsilon$), the total asymptotic runtime will be bounded by the last query since the an oracle query takes the time proportional to $1/\epsilon$.

Our oracle based solution search is similar to the one in [18] with an important difference as follows. Since their algorithm only rounds one parameter while our approach rounds two parameters (cost W and timing Q), applying the technique in [18] leads to adaptively setting both rounding parameters. This is not desired for rounding Q since the timing error will not be controlled by ϵ. That is, in first few iterations during oracle based search, the timing of the solutions may be significantly larger than $(1+\epsilon)T$ since approximation ratios there would be big (e.g., the first few approximation ratios in $\ldots, 27, 9, 3, 1, 1/3, \ldots, \epsilon$ are much bigger than the target ϵ). This means that by dynamic programming, one only knows whether the cost budget n/ϵ is sufficient for computing a solution with much larger timing, which may lead to the wrong decision in narrowing down the gap between upper and lower bounds. It motivates us to propose to only change the ϵ corresponding to cost but not the ϵ corresponding to timing. We call it *double-ϵ oracle based solution search*. This is why there are two ϵ_1 and ϵ_2 in the dynamic programming algorithm in Section 5. Namely, ϵ_1 corresponds to the approximation ratio on cost W and ϵ_2 corresponds to the approximation ratio on timing Q. If we fix both of them to ϵ, one cannot reduce the total runtime to be independent of the initial bounds. Our idea is to fix ϵ_2 at ϵ while adaptively changing ϵ_1.

In details, let $W_{u,i}^*$ denote the upper bound and $W_{l,i}^*$ denote the lower bound after i-th iteration in the oracle based solution search. Initially, $W_{u,0}^* = W_u^*$ and $W_{l,0}^* = W_l^*$. A geometric sequence of approximation ratios ϵ_i' converging to ϵ will be used in oracle based solution search for ϵ_1 (but not ϵ_2 since it is fixed at ϵ). Following [18], set

$$\epsilon_i' = \sqrt{\frac{W_{u,i}^*}{W_{l,i}^*}} - 1. \tag{1}$$

In each iteration,

$$x = \sqrt{\frac{W_{u,i}^* \cdot W_{l,i}^*}{1 + \epsilon_i'}} \tag{2}$$

is used to query the oracle. It can be proved that after i-th oracle query,

$$\frac{W_{u,i+1}^*}{W_{l,i+1}^*} = \left(\frac{W_{u,i}^*}{W_{l,i}^*}\right)^{3/4}, \tag{3}$$

since for Case 1 $W_{u,i+1}^* = (1+\epsilon_i')x = W_{u,i}^{*\,3/4} W_{l,i}^{*\,1/4}$, $W_{l,i+1}^* = W_{l,i}^*$, and for Case 2 $W_{u,i+1}^* = W_{u,i}^*$, $W_{l,i+1}^* = x = W_{u,i}^{*\,1/4} W_{l,i}^{*\,3/4}$. This process is iterated until the ratio between upper and lower bound is no greater than 2. Let i_θ be the first i such that $\frac{W_{u,i}^*}{W_{l,i}^*} \leq 2$. According to Lemma 2 in Section 5, an oracle query takes $O(\frac{m^2 n}{\epsilon_1 \epsilon_2} + \frac{mn^2 b}{\epsilon_1} + \frac{m^2 n^2}{\epsilon_1^2 \epsilon_2} + \frac{mn^2 b}{\epsilon_1 \epsilon_2} + \frac{n^3 b^2}{\epsilon_1})$ time. We first bound the time until i_θ, which is $O(\frac{m^2 n}{\epsilon_2} \sum_{1 \leq i \leq i_\theta} \frac{1}{\epsilon_i'} + mn^2 b \sum_{1 \leq i \leq i_\theta} \frac{1}{\epsilon_i'} + \frac{m^2 n^2}{\epsilon_2} \sum_{1 \leq i \leq i_\theta} \frac{1}{\epsilon_i'^2} + \frac{mn^2 b}{\epsilon_2} \sum_{1 \leq i \leq i_\theta} \frac{1}{\epsilon_i'} + n^3 b^2 \sum_{1 \leq i \leq i_\theta} \frac{1}{\epsilon_i'})$. Since $i \leq i_\theta$ and $\frac{W_{u,i}^*}{W_{l,i}^*} > 2$, $\frac{1}{\epsilon_i'} = \frac{1}{\sqrt{\frac{W_{u,i}^*}{W_{l,i}^*}} - 1}$ means $\frac{1}{\epsilon_i'^2} \leq (2 + \sqrt{2})^2 \cdot \frac{W_{l,i}^*}{W_{u,i}^*}$. Thus,

$$O\left(\frac{m^2 n^2}{\epsilon_2} \sum_{1 \leq i \leq i_\theta} \frac{1}{\epsilon_i'^2}\right) = O\left(\frac{m^2 n^2}{\epsilon_2} \sum_{1 \leq i \leq i_\theta} \frac{W_{l,i}^*}{W_{u,i}^*}\right). \tag{4}$$

978-1-60558-497-3/09 $25.00 © 2009 ACM

It is shown in [18] that

$$\sum_{1 \le i \le i_\theta} \frac{W_{l,i}^*}{W_{u,i}^*} = \sum_{1 \le i \le i_\theta} (\frac{W_{l,i_\theta}^*}{W_{u,i_\theta}^*})^{(\frac{4}{3})^{i_\theta \cdot i}} = \sum_{0 \le j < i_\theta} (\frac{W_{l,i_\theta}^*}{W_{u,i_\theta}^*})^{(\frac{4}{3})^j}, \quad (5)$$

(note that j starts from 0), and $\frac{W_{l,t}^*}{W_{u,t}^*} < \frac{1}{2^{3/4}}$ by noting that i_θ is the first i such that $\frac{W_{u,i}^*}{W_{l,i}^*} \le 2$. Subsequently,

$$\sum_{1 \le i \le i_\theta} \frac{W_{l,i}^*}{W_{u,i}^*} = \sum_{0 \le j < i_\theta} (\frac{W_{l,i_\theta}^*}{W_{u,i_\theta}^*})^{(4/3)^j} < \sum_{0 \le j < i_\theta} \frac{1}{2^{3/4}}^{(4/3)^j}. \quad (6)$$

The last term is the sum of a monotonically decreasing geometric sequence which is certainly bounded by $O(1)$. Thus,

$$O(\frac{m^2 n^2}{\epsilon_2} \sum_{1 \le i \le i_\theta} \frac{1}{c_i'^2}) = O(\frac{m^2 n^2}{\epsilon_2}). \quad (7)$$

Similarly, since $\sum_{1 \le i \le i_\theta} \sqrt{\frac{W_{l,i}^*}{W_{u,i}^*}} < \sum_{0 \le j < i_\theta} 0.59^{1/2 \cdot (4/3)^j} = O(1)$ [18], we have $O(\frac{mn^2 b}{\epsilon_2} \sum_{1 \le i \le i_\theta} \frac{1}{c_i}) = O(\frac{mn^2 b}{\epsilon_2})$ and $O(n^3 b^2 \sum_{1 \le i \le i_\theta} \frac{1}{c_i}) = O(n^3 b^2)$. The total runtime for oracle based solution search is bounded by

$$O(\frac{m^2 n^2 b}{\epsilon} + mn^2 b + n^3 b^2), \quad (8)$$

since ϵ_2 is fixed at ϵ.

After i_θ oracle queries, the ratio between upper and lower bounds is at most 2. With this better starting point, the timing-cost approximate dynamic programming can be efficiently performed with both ϵ_1 and ϵ_2 set to ϵ. First set x to the current lower bound W_{l,i_θ}^*. Subsequently, scale and round the cost w of each buffer to $\lfloor \frac{wn}{x\epsilon} \rfloor$ where ϵ is the target ϵ. Note that W^* is no greater than $W_{u,i_\theta}^* \le 2W_{l,i_\theta}^*$, which means that there is at least one solution with scaled cost no greater than $2n/\epsilon$. Otherwise, similar to Case 2, there is no buffering solution within cost $\frac{2n}{\epsilon} \cdot \frac{x\epsilon}{n} = 2x \ge W_{u,i_\theta}^*$ which is a contradiction. This solution is the $(1 + \epsilon)$ approximation which will be returned by our FPTAS. Similar to the argument in Case (1), scaling the result back by the factor of $\frac{x\epsilon}{n}$ forms a lower bound on W^* and the maximum rounding error is $\frac{x\epsilon}{n} \cdot n = x\epsilon = W_{l,t}^*\epsilon \le W^*\epsilon$. Thus, the cost of the obtained buffering solution is at most $(1+\epsilon)W^*$. This last step uses the dynamic programming which takes $O(\frac{m^2 n}{\epsilon^2} + \frac{mn^2 b}{\epsilon} + \frac{m^2 n^2}{\epsilon^3} + \frac{mn^2 b}{\epsilon^2} + \frac{n^3 b^2}{\epsilon})$ time by noting that both ϵ_1 and ϵ_2 are set to ϵ. The efficient computation is due to the fact that the ratio between the current upper and lower bounds has been reduced to ≤ 2. Together with the time for oracle based solution search in Eqn. (8), the total time is $O(\frac{m^2 n}{\epsilon^2} + \frac{mn^2 b}{\epsilon} + \frac{m^2 n^2}{\epsilon^3} + \frac{mn^2 b}{\epsilon^2} + \frac{n^3 b^2}{\epsilon} + \frac{m^2 n^2 b}{\epsilon} + mn^2 b + n^3 b^2)$ which can be simplified to $O(m^2 n^2 b/\epsilon^3 + n^3 b^2/\epsilon)$ for $0 < \epsilon < 1$ and to $O(m^2 n^2 b/\epsilon + mn^2 b + n^3 b^2)$ for $\epsilon \ge 1$. We reach the following theorem.

Theorem 1: A $(1 + \epsilon)$ approximation to the timing constrained minimum cost buffering problem can be computed in $O(m^2 n^2 b/\epsilon^3 + n^3 b^2/\epsilon)$ time for any $0 < \epsilon < 1$ and in $O(m^2 n^2 b/\epsilon + mn^2 b + n^3 b^2)$ time for $\epsilon \ge 1$, where n is the number of nodes in the tree, m is the number of sinks in the tree, and b is the number of buffers in the buffer library.

5. POLYNOMIAL TIME TIMING-COST APPROXIMATE DYNAMIC PROGRAMMING

5.1 Bound Distinct Cost W and RAT Q

A careful investigation in Lillis' algorithm [6] would reveal that the number of solutions is not polynomially bounded which is why Lillis' algorithm is a pseudo-polynomial algorithm. To design an efficient algorithm, we certainly need to bound the number of solutions during solution propagation. A major innovation in the proposed FPTAS algorithm is a dynamic programming algorithm with polynomially bounded W and Q. As a result, the number of solutions will also be polynomially bounded (since there is only one possible non-dominated solution with each pair of W and Q, namely, the one with the smallest C). First note that we have two ϵ in the algorithm, one being ϵ_1 which refers to the approximation in cost and the other being ϵ_2 which refers to the approximation in timing. ϵ_1 is varying while ϵ_2 is fixed to ϵ in the fast double-ϵ oracle based solution search.

To bound W, recall that in Section 4, one first scales and rounds each buffer cost w to an *integer* as $w = \lfloor \frac{wn}{x\epsilon_1} \rfloor$. After that, the oracle only wants to know whether there is a solution (approximately) satisfying the timing constraint with cost up to n/ϵ_1. Let $\overline{W} = n/\epsilon_1$. The oracle uses this cost bound to perform the dynamic programming. Thus, whenever there is a solution with cost greater than $\overline{W}$, it will be eliminated from the solution set. Consequently, there are at most $\overline{W} + 1$ distinct W $(0, 1, \dots, \overline{W})$ at any location during solution propagation.

Our new dynamic programming algorithm is as follows. First, it always works with cost bins (W-bin) since the oracle scales and rounds each buffer cost before performing the dynamic programming algorithm. In dynamic programming, *right before a branch merge*, all the solutions are also discretized into timing bins (Q-bin) and then the solution pruning is performed. This allows us to bound Q as well. Consequently, the number of non-dominated solutions during solution propagation is bounded.

To bound the number of distinct Q, right before each branch merge, for all $Q \ge 0$, round up Q of each branch to the nearest value in $\{0, \epsilon_2 T/m, 2\epsilon_2 T/m, \dots, T\}$, where m is the number of the sinks. We underestimate the delay by rounding. For example, when $\epsilon_2 = 0.5$ and $m = 2$, $Q = 0.7T$ and will be up-rounded to $3\epsilon_2 T/m$. The solutions with $Q < 0$ will be pruned since the arrival time at driver is 0. Thus, there are at most $\frac{T}{\epsilon_2 T/m} + 1 = m/\epsilon_2 + 1$ distinct Q after rounding. Together with the fact that there are at most $O(\overline{W})$ distinct W at any point, at most $O(\overline{W}m/\epsilon_2) = O(\frac{mn}{\epsilon_1 \epsilon_2})$ non-dominated solutions can be obtained after any branch merge (the number of non-dominated solutions before branch merge will be discussed soon). This is due to the fact that there is only one solution for each pair of Q, W, namely, the one with minimum C. Note that the rounding error in timing at each branching point is at most $\epsilon_2 T/m$ and thus at most $\epsilon_2 T$ for the whole tree with $m - 1$ branching points. Note that the timing of the obtained solution (in the scaled problem) is at most T but it is with the rounded Q. This means that after rounding Q back, we obtain a buffering solution with timing at most $(1 + \epsilon_2)T$ for the original buffering problem.

The time complexity analysis is as follows. After performing a branch merge, there are at most $O(\frac{mn}{\epsilon_1 \epsilon_2})$ non-dominated solutions. After that, solutions are propagated

978-1-60558-497-3/09 $25.00 © 2009 ACM

427

in its upstream branch. An add wire operation does not introduce new solution. After performing buffer insertion operation at a node v, for solutions with new buffers inserted at v, there are only b distinct C (where b is the number of buffer types). Since the number of W is always bounded by $O(\overline{W})$, there are at most $O(\overline{W}b) = O(nb/\epsilon_1)$ non-dominated buffered solutions. This is due to the fact that there is only one solution for each pair of C, W, i.e., the one with maximum Q.

We are to bound the time for computing these $O(\overline{W}b) = O(nb/\epsilon_1)$ non-dominated buffered solutions. Given a solution, buffer insertion at v leads to b possible new solutions. Since there are at most $O(mn/(\epsilon_1\epsilon_2) + n^2b/\epsilon_1)$ non-dominated solutions anywhere (see below), computing non-dominated buffered solutions at node v takes $O(mnb/(\epsilon_1\epsilon_2) + n^2b^2/\epsilon_1)$ time. This is due to that within a single W bin, (Q, C) based pruning takes linear time in the number of generated solutions which is the same as the pruning without considering W in [5]. Note that cross W-bin pruning is not performed since this will not improve the asymptotic complexity. Together with those $O(\overline{W}m/\epsilon_2)$ unbuffered solutions (which are propagated by add wire from the last branch merge), there are at most $O(\overline{W}m/\epsilon_2 + \overline{W}b)$ non-dominated solutions. When these solutions are propagated along this branch, there are at most $O(\overline{W}m/\epsilon_2 + n\overline{W}b) = O(mn/(\epsilon_1\epsilon_2)+n^2b/\epsilon_1)$ non-dominated solutions at any point before next branch merge since there are at most n candidate buffer locations and Q is not rounded until the branch merge.

Before branch merge, all solutions are first put into cost bins (W-bin) and timing bins (Q-bin). That is, they are placed into the bins with integer costs from 0 to n/ϵ_1. In each cost bin, there are m/ϵ_2+1 timing bins with the timing as $0, \epsilon_2T/m, 2\epsilon_2T/m, ..., T$ by up-rounding each Q. By a linear traversal of all bins, dominated solutions will be pruned. The whole process certainly takes time linear in the number of solutions, i.e., $O(mn/(\epsilon_1\epsilon_2) + n^2b/\epsilon_1)$ time. In solution pruning, since W is always an integer, there are $W+1$ possible merging results of left and right branch solutions for each W. For example, to obtain merged cost $W = 3$, the four possibly combinations of left-branch solution γ_1 and right-branch solution γ_2 are (1) $W(\gamma_1) = 0, W(\gamma_2) = 3$, (2) $W(\gamma_1) = 1, W(\gamma_2) = 2$, (3) $W(\gamma_1) = 2, W(\gamma_2) = 1$, and (4) $W(\gamma_1) = 3, W(\gamma_2) = 0$. For a fixed combination on W, there are a list of solutions with distinct Q along each branch. For each Q-bin, the time complexity can be easily bounded since all the solutions in each branch are non-dominated and thus for the same W, C are increasingly sorted. One just needs to correspondingly merge two solutions with the same Q.

Table 1: An example for branch merge.

Left branch (Q, W, C)				
C	$W = 0$	$W = 1$	$W = 2 \leftarrow$	$W = 3$
$Q = 0$	10	7	5	2
$Q = \epsilon_2T/m$	20	17	15	12
$Q = 2\epsilon_2T/m$	50	45	40	35
Right branch (Q, W, C)				
C	$W = 0$	$W = 1 \leftarrow$	$W = 2$	$W = 3$
$Q = 0$	15	12	10	5
$Q = \epsilon_2T/m$	25	22	15	10
$Q = 2\epsilon_2T/m$	55	50	45	20

It is helpful to look at an example to illustrate the above analysis. Refer to Table 1. To obtain the merged cost of 3, suppose that we are merging solutions with $W = 2$ in the left branch and solutions with $W = 1$ in the right branch.

The corresponding W are shown with arrows. For $Q = 0$ after branch merge, the minimum C is $C = 5 + 12 = 17$. For $Q = \epsilon_2T/m$ after branch merge, the minimum C is $C = 15 + 22 = 37$. For $Q = 2\epsilon_2T/m$ after branch merge, the minimum C is $C = 40 + 50 = 90$. One then also needs to consider the other three merging possibilities for the merged cost to be 3, i.e., (1) $W(\gamma_1) = 0, W(\gamma_2) = 3$, (2) $W(\gamma_1) = 1, W(\gamma_2) = 2$, and (4) $W(\gamma_1) = 3, W(\gamma_2) = 0$. Subsequently, the solution with the minimum C for each pair of W, Q is picked. This process takes $O(m/\epsilon_2 \cdot W)$ time for each merged W since there are only $O(m/\epsilon_2)$ distinct Q. Summing over all W till $\overline{W}$, the time complexity for branch merge and solution pruning in branch merge is $O(m/\epsilon_2 \cdot (\overline{W})^2) = O(mn^2/(\epsilon_1^2\epsilon_2))$. Together with the time for putting the solutions into bins before branch merge, the total runtime is $O(mn/(\epsilon_1\epsilon_2) + n^2b/\epsilon_1 + mn^2/(\epsilon_1^2\epsilon_2))$ for a single branch merge.

5.2 Time Complexity

As mentioned above, a single buffer insertion together with pruning takes $O(mnb/(\epsilon_1\epsilon_2) + n^2b^2/\epsilon_1)$ time. Summing over n candidate buffer locations, total buffer insertion takes $O(mn^2b/(\epsilon_1\epsilon_2) + n^3b^2/\epsilon_1)$ time. This certainly upper bounds the time for add wire. A single branch merge takes $O(mn/(\epsilon_1\epsilon_2) + n^2b/\epsilon_1 + mn^2/(\epsilon_1^2\epsilon_2))$ time. Summing over all $m - 1$ branch merges, $O(m^2n/(\epsilon_1\epsilon_2) + mn^2b/\epsilon_1 + m^2n^2/(\epsilon_1^2\epsilon_2))$ time is needed. Thus, the dynamic programming will run in

$$O\left(\frac{m^2n}{\epsilon_1\epsilon_2} + \frac{mn^2b}{\epsilon_1} + \frac{m^2n^2}{\epsilon_1^2\epsilon_2} + \frac{mn^2b}{\epsilon_1\epsilon_2} + \frac{n^3b^2}{\epsilon_1}\right) \quad (9)$$

time to compute a solution with the cost at most $(1 + \epsilon_1)$ optimal cost and with timing at most $(1 + \epsilon_2)T$. We reach the following lemma.

Lemma 2: The timing-cost approximate dynamic programming algorithm can compute a timing driven buffering solution with the cost at most $(1+\epsilon_1)$ optimal cost and with timing at most $(1+\epsilon_2)T$ in $O(\frac{m^2n}{\epsilon_1\epsilon_2} + \frac{mn^2b}{\epsilon_1} + \frac{m^2n^2}{\epsilon_1^2\epsilon_2} + \frac{mn^2b}{\epsilon_1\epsilon_2} + \frac{n^3b^2}{\epsilon_1})$ time for any $\epsilon_1, \epsilon_2 > 0$, where n is the number of candidate buffer locations, m is the number of sinks, and b is the number of buffers in the buffer library.

6. EXPERIMENTAL RESULTS

We compare the proposed FPTAS for the timing driven minimum cost buffering problem to the dynamic programming algorithm [6] which computes optimal buffering solution. The experiments are performed on a set of 1000 nets at various scales extracted from an industrial ASIC chip. The buffer library consists of 48 buffer types including buffers and inverters. The buffer cost is measured by buffer area in this paper. However, other metric can be easily handled in FPTAS.

Refer to Table 2 for the comparison. Cost Ratio and Speedup are computed by comparing to the total buffer cost and the runtime of dynamic programming algorithm. # Vio. specifies the number of nets with timing violations. The results with small $\epsilon < 1$ are shown. This range of ϵ is desired in practice since one always wishes to compute solutions close to the optima. We make the following observations.

- The dynamic programming in [6] computes the optimal solution. Total buffer cost is 3304.09, no net has timing violation, and CPU time is 625.2 seconds.

978-1-60558-497-3/09 $25.00 © 2009 ACM

Table 2: Comparison of the dynamic programming [6] and the FPTAS algorithm on 1000 industrial nets. In dynamic programming solution, total buffer cost is 3304.09, and CPU is 625.2 seconds. # Vio. specifies the number of nets with timing violations. Cost Ratio and Speedup are computed by comparing to dynamic programming.

	FPTAS				
ϵ	# Vio.	Total Cost	CPU(s)	Cost Ratio	Speedup
0.01	3	3322.91	135.5	0.57%	4.6×
0.05	9	3390.64	127.6	2.6%	4.9×
0.10	12	3499.78	122.1	5.9%	5.1×
0.20	30	3541.17	117.2	7.2%	5.3×
0.30	53	3578.80	114.5	8.3%	5.5×
0.40	51	3669.12	112.9	11.1%	5.5×
0.50	53	3766.96	110.5	14.0%	5.7×
Average					5.2×

Table 3: The obtained timing on the nets violating the timing constraints for FPTAS with $\epsilon = 0.01$.

	FPTAS with $\epsilon = 0.01$		
Net	Timing Constraint	Actual Delay	Timing Violation
1	18.043	18.2140	0.95%
2	4.239	4.2504	0.27%
3	4.054	4.0555	0.04%

- Our FPTAS works very well in practice. Compared to the dynamic programming solutions, there are only slight solution degradations in total buffer costs while on average over 5× speedup is obtained. For example, when the target approximation ratio is set to $\epsilon = 0.01$, the actual approximation ratio (cost ratio) is only 0.57% while 4.6× speedup is achieved. The speedup is so significant since the total number of solutions at driver over 1000 nets is 166,676 for FPTAS with $\epsilon = 0.01$ (for all iterations in performing double-ϵ oracle based solution search) while it is 1,221,604 in the dynamic programming algorithm [6]. It is important to note that the cost ratio is theoretically guaranteed to be no greater than ϵ. In practice, it is much smaller as is shown in Table 2. This clearly demonstrates the effectiveness of our FPTAS algorithm.
- Larger ϵ leads to more speedup while smaller ϵ leads to less solution quality degradation. This is again as guaranteed theoretically.
- Since timing is rounded in our timing-cost approximate dynamic programming, there are timing violations in the obtained buffering solutions. However, it is clear that this happens with very small probability in practice as indicated by our experimental results. When $\epsilon = 0.01$, only 3 out of 1000 nets have timing violations. In addition, the obtained timing is theoretically guaranteed to be within $(1 + \epsilon)T$ where T is the timing constraint. For example, the nets with timing violations for FPTAS with $\epsilon = 0.01$ are shown in Table 3. Their actual delays are clearly bounded by the $(1 + \epsilon)T = 1.01T$.
- When the timing constraints are stringent, one may need every net to satisfy the timing constraint. For this, the following timing recovery procedure could be performed. Those nets with timing violations can be ripped up and rebuffered using optimal dynamic programming [6]. The overall runtime would still be much better than [6] since few nets need to be rebuffered. For the nets without timing violations, the approxi-

mation ratio is bounded by $1 + \epsilon$. For the nets with timing violations, the optimal solutions are computed in rebuffering. Thus, the approximation ratio is still bounded by $1 + \epsilon$ after timing recovery. Refer to Table 4 for the results. The cost is increased compared to FPTAS without timing recovery since rounding on timing in FPTAS underestimates delay and may make the cost of the obtained buffering solution smaller than the optimal cost (esp. for many of the nets with timing violations) even if rounding on cost increases it. Empirically, FPTAS with $\epsilon = 0.01$ gives the best performance since it has fewest nets which need rebuffering.

Table 4: The results of FPTAS with timing recovery.

	FPTAS with Timing Recovery				
ϵ	# Vio.	Total Cost	CPU(s)	Cost Ratio	Speedup
0.01	0	3331.57	143.2	0.83%	4.4×
0.05	0	3415.21	157.7	3.4%	4.0×
0.10	0	3592.76	209.5	8.7%	3.0×
0.20	0	3654.91	223.0	10.6%	2.8×
0.30	0	3702.88	262.2	12.1%	2.4×
0.40	0	3867.12	261.8	17.0%	2.4×
0.50	0	4028.25	265.7	21.9%	2.4×
Average					3.0×

7. REFERENCES

[1] P. Saxena and N. Menezes and P. Cocchini and D.A. Kirkpatrick, "Repeater scaling and its impact on CAD," *TCAD*, vol. 23, no. 4, pp. 451–463, 2004.

[2] J. Cong, "An interconnect-centric design flow for nanometer technologies," *Proceedings of the IEEE*, vol. 89, no. 4, pp. 505–528, 2001.

[3] Z. Li, C. Alpert, S. Hu, T. Muhmud, S. Quay, and P. Villarrubia, "Fast interconnect synthesis with layer assignment," *ISPD*, 2008.

[4] P.J. Osler, "Placement driven synthesis case studies on two sets of two chips: hierarchical and flat," *ISPD*, pp. 190–197, 2004.

[5] L.P.P.P. van Ginneken, "Buffer placement in distributed RC-tree networks for minimal Elmore delay," *in Proceedings of the IEEE International Symposium on Circuits and Systems*, pp. 865–868, 1990.

[6] J. Lillis and C.-K. Cheng and T.-T.Y. Lin, "Optimal wire sizing and buffer insertion for low power and a generalized delay model," *IEEE Journal of Solid State Circuits*, vol. 31, no. 3, pp. 437–447, 1996.

[7] W. Shi and Z. Li, "A fast algorithm for optimal buffer insertion," *TCAD*, vol. 24, no. 6, pp. 879–891, 2005.

[8] R. Chen and H. Zhou, "A flexible data structure for efficient buffer insertion," *ICCD*, 2004.

[9] W. Shi and Z. Li and C. Alpert, "Complexity analysis and speedup techniques for optimal buffer insertion with minimum cost," *ASPDAC*, pp. 609–614, 2004.

[10] C.J. Alpert and A. Devgan and S.T. Quay, "Buffer insertion for noise and delay optimization," *DAC*, pp. 362–367, 1998.

[11] S. Hu, C. J. Alpert, J. Hu, S. Karandikar, Z. Li, W. Shi, and C. N. Sze, "Fast algorithms for slew constrained minimum cost buffering," *DAC*, 2006.

[12] R. Chen and H. Zhou, "Fast min-cost buffer insertion under process variations," *DAC*, 2007.

[13] H. Zhou, D.F. Wong, I.-M. Liu, and A. Aziz, "Simultaneous routing and buffer insertion with restrictions on buffer locations," *DAC*, 1999.

[14] T.-C. Chen, A. Chakraborty, and D. Z. Pan, "An integrated nonlinear placement framework with congestion and porosity aware buffer planning," *DAC*, 2008.

[15] J. Cong, T. Kong, and D. Z. Pan, "Buffer block planning for interconnect-driven floorplanning," *ICCAD*, 1999.

[16] C.J. Alpert, J. Hu, S.S. Sapatnekar and P. Villarrubia, "A practical methodology for early buffer and wire resource allocation," *DAC*, 2001.

[17] S. Hu, Z. Li, and C.J. Alpert, "A polynomial time approximation scheme for timing constrained minimum cost layer assignment," *ICCAD*, 2008.

[18] S. Hu, Z. Li, and C.J. Alpert, "A faster approximation scheme for timing constrained minimum cost layer assignment," *ISPD*, 2009.

978-1-60558-497-3/09 $25.00 © 2009 ACM

Spare-Cell-Aware Multilevel Analytical Placement *

Zhe-Wei Jiang[1], Meng-Kai Hsu[1], Yao-Wen Chang[12], and Kai-Yuan Chao[3]
[1]Graduate Institute of Electronics Engineering, National Taiwan University, Taipei, TW
[2]Department of Electrical Engineering, National Taiwan University, Taipei, TW
[3]Enterprise Microprocessor Group, Intel Corporation, Hillsboro, OR
{crazying, kaie}@eda.ee.ntu.edu.tw; ywchang@cc.ee.ntu.edu.tw; kaiyuan.chao@intel.com

ABSTRACT

Post-silicon validation has recently drawn designers' attention due to its increasing impacts on the VLSI design cycle and cost. One key feature of the post-silicon validation is the use of spare cells. In the literature, most existing works focus on developing new delicate spare cell structures. On the other hand, the placement of spare cells has a crucial impact on the design cycle and cost of the post-silicon debugging; however, there exists not much work on this placement problem. In this paper, we propose *the first* spare-cell-aware analytical placement framework which predicts the spare cell requirement and considers spare cell insertion during global placement. We also propose a multilevel spare cell insertion technique which provides a more efficient spare cell planning and a better control of quality impact due to spare cell insertion. To guide the selection of available spare cell positions during insertion, we propose a mixed-integer-linear-programming formulation to determine the optimal spare cell positions. Experimental results show that our algorithm can averagely achieve 17–33% and 1.77–2.61X better quality of spare cell insertion than that of the existing spare cell insertion algorithms, UniSpare [10] and PostSpare [22, 26], on the tested real designs with 1–5% spare cell insertion rates.

Categories and Subject Descriptors

B.7.2 [**Integrated Circuits**]: Design Aids [Placement and Routing]; J.6 [**Computer-Aided Engineering**]: Computer-Aided Design

General Terms

Algorithms, Performance

Keywords

Physical Design, Spare Cells, Placement

1. INTRODUCTION

Due to the increasing complexity of modern VLSI designs, more and more errors escape pre-silicon validation. Those errors can only be found after a chip is manufactured, and thus *post-silicon debugging* is required to make the manufactured design work correctly. According to [3], the post-silicon debugging now contributes averagely 35% of the development cycle. It is also pointed out in a recent EE Times article [14] that "post-silicon debugging can cost $15 to $20 million and take six months to complete." These phenomenons have clearly indicated the urgent need of new methodologies to reduce the design cycle and cost of post-silicon debugging.

There are two different ways of performing post-silicon debugging. The first way is to reuse the transistor masks and change only the metal layer masks, which is called *metal fix* [10,15]. Since the transistor masks are usually much more costly to be respun than the metal layer masks, changing only the metal masks can effectively save the post-silicon debugging cost. The other way to perform post-silicon debugging is through *Focus Ion Beam (FIB)* to modify individual chips. The FIB technique has the capability to cut unwanted electrical connections, or to deposit conductive material in order to make a connection after the chip is manufactured. Both of the above two techniques require *spare cells* to be placed before manufacturing in order to perform the desired debugging steps.

Up to present, not much work discusses the spare cell placement problem. The spare cell placement usually follows the intuitive guidelines of placing spare cells close to potentially buggy regions or modules. In the literature, most existing works focus on developing new delicate spare cell structures [5–7,13,19,21–26]. Among these works, one strategy, called *PostSpare*, of inserting spare cells after design placement was proposed by Yee [26] and Payne [22]. However, the PostSpare strategy might not be able to insert spare cells among blocks[1], while blocks are sometimes packed due to performance optimization and buffer insertion. Thus the spare cells will be inserted around all blocks and might be far from the blocks which also require the spare cell. Recently, Chang *et al.* provided a comprehensive survey on existing spare cell insertion techniques in [10]. They also proposed a new spare cell insertion strategy, called *UniSpare*, to place spare cells among the chip uniformly before circuit placement. Such a strategy is expected to reduce the average distance from blocks to spare cells (one most commonly used metric for spare cell insertion quality). However, according to our observation, since the blocks that require the identical spare cell type may not be uniformly distributed in the chip, and the distribution cannot be predicted before placement, the UniSpare strategy may insert spare cells far from those blocks that require them. Figure 1 illustrates such difficulties faced by the PostSpare and UniSpare strategies.

To overcome the limitations and drawbacks of the previous spare cell insertion strategies, we propose *the first* spare-cell-aware analytical placement framework which predicts the spare cell requirement and preserves whitespace for later insertion during global placement. Such a framework enables the opportunities of the optimization process to search for an optimal placement solution with the consideration of spare cell insertion. Afterwards, the spare cells are inserted after the legalization and detailed placement stages, which eases the potential problems with unknown block distributions, and makes the quality impact analysis and control easier. We summarize our major contributions as follows:

*This work was partially supported by ITRI, Springsoft, Synopsys, TSMC, and National Science Council of Taiwan under Grant No's. NSC 97-2221-E-002-237-MY3, NSC 96-2628-E-002-249-MY3, NSC 96-2628-E-002-248-MY3.

Permission to make digital or hard copies of part or all of this work for personal or classroom use is granted without fee provided that copies are not made or distributed for profit or commercial advantage and that copies bear this notice and the full citation on the first page. To copy otherwise, to republish, to post on servers or to redistribute to lists, requires prior specific permission and/or a fee.
DAC'09, July 26-31, 2009, San Francisco, California, USA

---

[1]Note that to distinguish from spare cells, we use "blocks" to represent the regular cells within a design in this paper.

978-1-60558-497-3/09 $25.00 © 2009 ACM

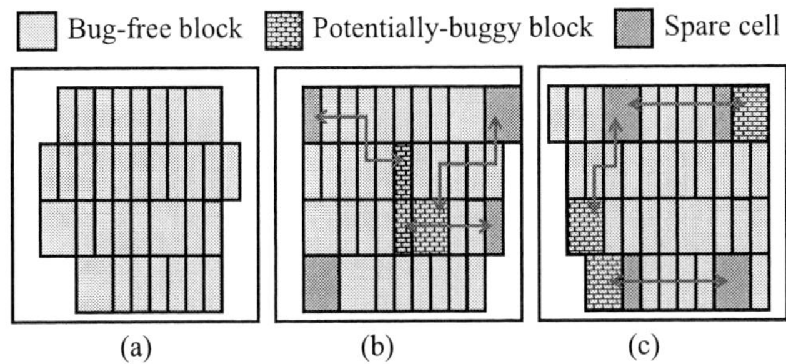

Figure 1: Illustration of existing spare cell insertion strategies. The blocks and spare cells with the identical size are of the identical type. (a) The original placement without spare cell insertion. (b) The PostSpare strategy places spare cells around the packed blocks, but the inserted spare cells are far from the central blocks within the placement region. (c) The UniSpare strategy inserts spare cells between blocks, but the inserted spare cells may be far from those blocks which require them, due to the lack of block distribution information.

- This is *the first* work in the literature to consider the spare cell insertion within the analytical placement framework. The proposed *cluster expansion* and *density constraint determination* can effectively preserve whitespace during global placement for later spare cell insertion.

- This paper proposes a multilevel spare cell insertion technique, which provides a more efficient spare cell planning according to known block distribution and gives a better control of quality impact due to spare cell insertion.

- We propose a mixed-integer-linear-programming formulation to determine the optimal spare cell positions with known block positions. Those optimal positions can provide a good guidance for the selection of available spare cell positions during spare cell insertion.

The remainder of this paper is organized as follows. Section 2 reviews the analytical placement framework used in this paper. The proposed spare-cell-aware global placement techniques and multilevel spare cell insertion are explained in Section 3. Section 4 reports the experimental results. Finally, the conclusions are given in Section 5.

2. REVIEW ON THE ANALYTICAL PLACEMENT FRAMEWORK

The circuit placement problem can be formulated as a hypergraph $H = (V, E)$ placement problem. Let vertices $V = \{v_1, v_2, ..., v_n\}$ represent blocks and hyperedges $E = \{e_1, e_2, ..., e_m\}$ represent nets. Let x_i and y_i be the respective x and y coordinates of the center of block v_i. The circuit may contain some *preplaced blocks* which have fixed x and y coordinates and cannot be moved. We intend to determine the optimal positions of movable blocks so that the total wirelength is minimized, and there is no overlap among blocks. The placement problem is usually solved in three stages, (1) global placement, (2) legalization, and (3) detailed placement. Global placement evenly distributes the blocks and finds the best position for each block to minimize the target cost (e.g., wirelength). Then, legalization removes all overlaps. Finally, detailed placement refines the solution.

We summarize the notations used to explain this framework in Figure 2.

x_i, y_i	center coordinate of block v_i
w_b, h_b	width and height of bin b
P_b	base potential (area of preplaced blocks) in bin b
D_b	potential (area of movable blocks) in bin b
M_b	the maximum potential in bin b
$t_{density}$	target placement density

Figure 2: Notation used in this paper.

To evenly distribute the blocks, we divide the placement region into uniform non-overlapping bin grids. Then, the global placement problem can be formulated as a constrained minimization problem as follows:

$$\begin{aligned} \min \quad & W(\mathbf{x}, \mathbf{y}) \\ \text{s.t.} \quad & D_b(\mathbf{x}, \mathbf{y}) \le M_b, \quad \text{for each bin b,} \end{aligned} \quad (1)$$

where $W(\mathbf{x}, \mathbf{y})$ is the wirelength function, $D_b(\mathbf{x}, \mathbf{y})$ is the potential function that is the total area of movable blocks in bin b, and M_b is the maximum allowable area of movable blocks in bin b. M_b can be computed by $M_b = t_{density}(w_b h_b - P_b)$, where $t_{density}$ is a user-specified target density value for each bin, w_b (h_b) is the width (height) of bin b, and P_b is the *base potential* that equals the preplaced block area in bin b. Note that M_b is a fixed value as long as all preplaced block positions are given and the bin size is determined.

The wirelength $W(\mathbf{x}, \mathbf{y})$ is defined as the total half-perimeter wirelength (HPWL). Since $W(\mathbf{x}, \mathbf{y})$ is not smooth and non-convex, it is hard to minimize it directly. Thus, several smooth wirelength approximation functions are proposed, such as quadratic wirelength [12,18], L_p-norm wirelength [9, 17], and log-sum-exp wirelength [8, 16, 20]. The log-sum-exp wirelength model,

$$\gamma \sum_{e \in E} (\log \sum_{v_k \in e} \exp(x_k/\gamma) + \log \sum_{v_k \in e} \exp(-x_k/\gamma) +$$
$$\log \sum_{v_k \in e} \exp(y_k/\gamma) + \log \sum_{v_k \in e} \exp(-y_k/\gamma)), \quad (2)$$

proposed in [20], achieves the best result among these three models [9]. When γ is small, log-sum-exp wirelength is close to the HPWL [20].

Since density $D_b(\mathbf{x}, \mathbf{y})$ is neither smooth nor differentiable, mPL [9] uses inverse Laplace transformation to smooth the density, while APlace [16] and NTUplace3 [11] use the bell-shaped function for each block to smooth the density. We express the function $D_b(\mathbf{x}, \mathbf{y})$ as $D_b(\mathbf{x}, \mathbf{y}) = \sum_{v \in V} P_x(b, v) P_y(b, v)$, where P_x and P_y are the overlap functions of bin b and block v along the x and y directions. In this paper, we adopt the bell-shaped potential function [16] p_x to smooth P_x. By doing so, the non-smooth function $D_b(\mathbf{x}, \mathbf{y})$ can be replaced by a smooth one, $\hat{D}_b(\mathbf{x}, \mathbf{y}) = \sum_{v \in V}^n c_v p_x(b, v) p_y(b, v)$, where c_v is a normalization factor so that the total potential of a block equals its area. Besides, in order to reduce the "valleys" generated by the bell-shaped function, we use the Gaussian function to further smooth the base potential [11].

The quadratic penalty method is used to solve Equation (1), implying that we solve a sequence of unconstrained minimization problems of the form

$$\min \quad W(\mathbf{x}, \mathbf{y}) + \lambda \sum_b (\hat{D}_b(\mathbf{x}, \mathbf{y}) - M_b)^2 \quad (3)$$

with increasing λ's. The solution of the previous problem is used as the initial solution for the next one. We solve the unconstrained problem in Equation (3) by the conjugate

978-1-60558-497-3/09 $25.00 © 2009 ACM

gradient (CG) method. Further, we apply the dynamic step size approach in [11] to speed up the process of minimizing Equation (3).

3. MULTILEVEL SPARE-CELL-AWARE PLACEMENT

In this paper, we propose the spare-cell-aware multilevel analytical placement. Figure 3 summarizes the flow chart of the proposed algorithm extended from the analytical placement framework introduced in Section 2. Note that due to the increasing complexity of modern circuit designs, the multilevel framework is usually applied on the analytical placement to improve the scalability. Since in the upper level of global placement, blocks are grouped into clusters, we propose the *cluster expansion* to provide a rougher prediction of spare cell insertion to each cluster before solving the analytical placement formulation. Then in the finest level of global placement, with known block distribution, we propose the *density constraint determination* to transform the spare cell requirement into density constraints, which achieves a more accurate whitespace preservation for spare cell insertion. Finally, after legalization and detailed placement, the *multilevel spare cell insertion* is proposed to insert spare cells to near-block and less-quality-impact positions. In this section, we will introduce the details of all proposed techniques.

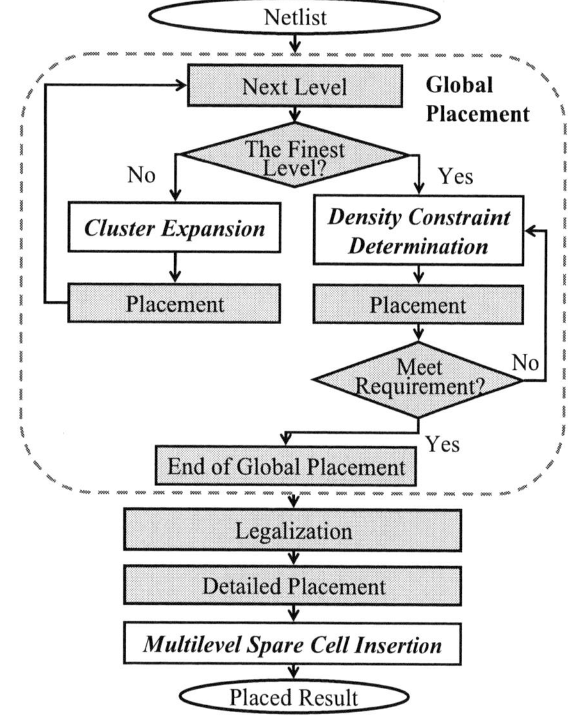

Figure 3: The proposed flow of our spare-cell-aware multilevel analytical placement.

3.1 Cluster Expansion

The multilevel framework applies a two-stage technique of bottom-up coarsening followed by top-down uncoarsening. During the coarsening stage, the blocks are clustered level by level to reduce the number of movable blocks. There have been many famous clustering techniques proposed in the literature, e.g., the best-choice clustering [4] and the first-choice clustering [9]. The clustering process continues until the number of blocks is reduced significantly.

After clustering, the analytical placement problem is solved by the conjugate gradient method at each level of the uncoarsening stage. However, since in the uncoarsening stage, each cluster contains multiple movable blocks, and the exact placement within the cluster remains unknown, it is relatively hard to preserve whitespace for individual blocks within each cluster directly. Therefore, in the upper level of uncoarsening, we evaluate the total area of the spare cell requirement for each cluster, and attach the area to the corresponding cluster when solving the analytical placement formulation.

In most cases, the spare cell requirement is determined according to designers' experience. The requirement can be specified by the number or the insertion rate of each type of spare cells. Without loss of generality, in this paper, we adopt the insertion rate to represent the spare cell requirement. Let r_k represents the given required insertion rate of spare cells of type t_k. For a block v_i, let $T(v_i)$ represent its required spare cell type, and $R(v_i)$ represent the required insertion rate of $T(v_i)$; in other words, if the required spare cell type of v_i is t_k, then $T(v_i) = t_k$ and $R(v_i) = r_k$. Let $Area(t_k)$ represent the area of the spare cell of type t_k. For a cluster C, its required area $Area_s(C)$ for spare cell insertion can be computed by

$$Area_s(C) = \sum_{v_i \in C} R(v_i) \times Area(T(v_i)). \qquad (4)$$

Then when solving the analytical placement formulation (Equation (1)) for the current level, $Area_s(C)$ is attached to cluster C when computing the potential function $D_b(\mathbf{x}, \mathbf{y})$. By doing so, we can preserve the whitespace for spare cell insertion during the upper levels of uncoarsening, and thus give the global placement engine more chance to optimize its objectives with the consideration of spare cell insertion.

3.2 Density-Constraint Determination

When the uncoarsening stage reaches the finest level, the optimization process of the analytical placement framework is directly applied on blocks of the circuit instead of clusters. Therefore, with known block positions, we can now provide a more accurate spare cell prediction to the global placement engine. Since the density constraints give a better control of whitespace than expanding block sizes, in the finest level, we propose to transform the spare cell requirement into the density constraints in Equation (1) to further plan the whitespace during global placement.

In the analytical placement framework reviewed in Section 2, the potential function $D_b(\mathbf{x}, \mathbf{y})$ and the maximum allowable potential M_b are determined according to pre-divided uniform non-overlapping bins. To preserve whitespace for spare cell insertion through density constraints, we need to modify M_b for each bin according to its spare cell requirement. Similar to the cluster expansion technique, each block will contribute the required area for spare cell insertion to the corresponding bins, which is proportional to the overlapping area between the block and the bin. For a density bin b, its required area $Area_s(b)$ for spare cell insertion can be written as

$$Area_s(b) = \sum_{v_i \in V} \left(R(v_i) \times Area(T(v_i)) \times \frac{O_x(v_i, b) \times O_y(v_i, b)}{Area(v_i)} \right), \qquad (5)$$

where $O_x(v_i, b)$ and $O_y(v_i, b)$ are the amounts of the overlaps between block v_i and bin b along the x and y directions, respectively, and $Area(v_i)$ is the area of block v_i. Then the maximum allowable potential $\hat{M}_b$ with the spare cell consid-

978-1-60558-497-3/09 $25.00 © 2009 ACM

eration can be updated by $\hat{M}_b = M_b - Area_s(b)$, and the analytical placement formulation will be transformed to

$$\begin{aligned}
\min \quad & W(\mathbf{x}, \mathbf{y}) \\
\text{s.t.} \quad & D_b(\mathbf{x}, \mathbf{y}) \leq \hat{M}_b \quad \text{for each bin b.}
\end{aligned} \qquad (6)$$

Solving this formulation can thus preserve whitespace for spare cell insertion. Note that this update for the maximum allowable potential will be performed several times during the conjugate gradient optimization process to monitor the block position change and provide a more accurate prediction of spare cell insertion.

3.3 Multilevel Spare Cell Insertion

Since the spare cell insertion always desires as less impact on the placement quality as possible, the spare cell insertion should therefore be performed after the placement is legalized and fully optimized, which eases the potential problems with unknown block distributions, and makes the quality impact analysis and control easier. Further, because the spare cell requirement may not uniformly distribute within the chip boundary, a whole-chip analysis is required to determine an efficient spare cell planning. Therefore, in this paper, we propose a multilevel spare cell insertion technique which provides a more efficient spare cell planning according to known block distribution and a better control of quality impact due to spare cell insertion.

For each given insertion rate r_k of spare cell type t_k, the total number of required spare cells of type t_k can be easily computed by the total number of blocks that require spare cells of type t_k multiplied by r_k. After obtaining the required number of spare cells, we recursively partition the placement region into uniform sub-regions. Each time the partitioning is performed, the required number of spare cells is also divided and assigned to two sub-regions in proportion to the number of blocks belonging to each sub-region that require spare cells of type t_k. Such an allocation strategy can make sure that the spare cell distribution will be similar to the block distribution. A slicing tree is constructed at the same time to record the cut directions and the number of spare cells assigned to each sub-region. The recursive partitioning process continues until the sub-region is small enough or the number of assigned spare cells to this sub-region equals 0 or 1.

After the construction of the partitions and the slicing tree, we insert the spare cells in a bottom-up fashion. For each sub-region located in the leaf of the slicing tree, we first compute the optimal position for each inserted spare cell according to the blocks that require spare cells of type t_k. Then each row within this sub-region is explored to find available spare cell positions with their quality impacts under a given threshold. Then to achieve a lower average distance from blocks to spare cells, those positions are chosen in an increasing order of the minimum distance to the optimal positions. If some spare cells fail to be inserted within this sub-region, they are reserved and will be inserted again in the upper level of the slicing tree within the union region of this sub-region and those corresponding to its siblings.

Figure 4 gives an example of the multilevel spare cell insertion. Figure 4(a) shows the initial partitions and the block distribution within each partition. Figure 4(b) illustrates the allocation process along the slicing tree. Assume that there are 10 spare cells of type t_k to be inserted in this region. In the root of the slicing tree, the number of allocated spare cells is divided to 7 and 3 because the two sub-regions contain 35 and 15 blocks that require spare cells of type t_k respectively. The partitioning process continues to allocate spare cells into all sub-regions. Note that the recursive partitioning stops at sub-region G since only one spare cell is allocated in G. Figure 4(c) illustrates the bottom-up spare

cell insertion process. All spare cells are inserted into their assigned sub-regions. However, in this example, one spare cell failed to be inserted in sub-region A, and another spare cell failed to be inserted in sub-region D. They are reserved and will be inserted into the union of sub-regions A and B and the union of sub-regions C and D respectively, as shown in Figure 4(d).

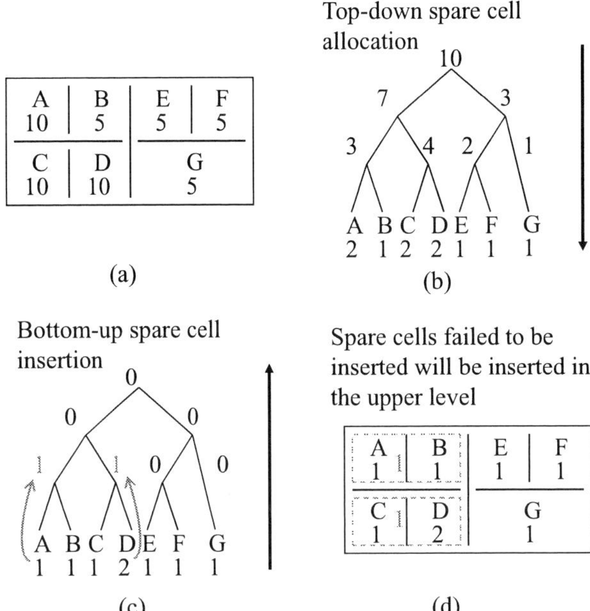

Figure 4: (a) Number of blocks that require spare cells of type t_k in each sub-region. (b) Illustration of the top-down spare cell allocation. The numbers on the tree nodes indicate the amounts of assigned spare cells. (c)(d) Bottom-up spare cell insertion. Spare cells will be inserted in an upper level on the slicing tree if there is no available position in their corresponding sub-region.

3.4 Determination of Optimal Spare Cell Positions

The optimal positions that minimize the average distance from blocks to spare cells can provide a good guidance for the selection of available spare cell positions in our multilevel spare cell insertion. For a sub-region that contains exactly one spare cell, its optimal position is at the gravity center of all blocks that require it. However, when a sub-region contains more than one spare cells, the determination of the optimal spare cell positions becomes much more complex. For a given sub-region, let V_P represent the set of blocks in this sub-region that require spare cells of type t_k, and S represent the set of to-be-inserted spare cells of the same type. For a block $v_p \in V_P$ and a spare cell $s_q \in S$, their distance is represented by $d(v_p, s_q)$. Then, with the objective of minimizing the average distance from placed blocks to spare cells, the determination of optimal spare cell positions can be formulated as

$$\begin{aligned}
\min \quad & (\textstyle\sum_{V_P} d_p^{min})/|V_P| \\
\text{s.t.} \quad & d_p^{min} = \min_S d(v_p, s_q), \forall v_p \in V_P,
\end{aligned}$$

where d_p^{min} represents the distance from block v_p to its clos-

978-1-60558-497-3/09 $25.00 © 2009 ACM

est spare cell, and $|V_P|$ denotes the size of set V_P. However, it is hard to solve such a formulation directly, since it tries to minimize the minimum of functions.

To cope with such difficulty, we introduce a binary variable $\gamma_{p,q}$ for each pair of (v_p, s_q). If $\gamma_{p,q}$ is set to 1, it means that the closest spare cell for block v_p is s_q. We can then rewrite the formulation as follows:

$$\min \quad (\textstyle\sum_{V_P} d_p^{min})/|V_P|$$

$$\text{s.t.} \quad d_p^{min} \geqslant d(v_p, s_q) + (\gamma_{p,q} - 1) \cdot D_{max}, \forall p, q \quad (7)$$

$$\textstyle\sum_q \gamma_{p,q} = 1, \forall p \quad (8)$$

$$\gamma_{p,q} \in \{0, 1\}, \forall p, q,$$

where D_{max} is a constant representing the maximum possible distance from blocks to spare cells, and is set to the summation of the width and height of the given sub-region. For each block v_p, the inequality (8) constrains that exactly one $\gamma_{p,q}$ will be set to 1. Then in inequality (7), the chosen $\gamma_{p,q}$ will constrain that d_p^{min} is greater or equal to the chosen $d(v_p, s_q)$, while others with value 0 make no influence on d_p^{min}, because D_{max} is always larger or equal to $d(v_p, s_q)$. Since our objective tends to minimize d_p^{min}, the assignment of $\gamma_{p,q}$'s will tend to select the closet spare cell s_p for each block v_p, which makes d_p^{min} equal to the exact minimum distance from block v_p to all spare cells. Such a formulation can be solved by *mixed integer linear programming (MILP)*, and has the capability to find the optimal spare cell positions and the closest spare cell assignment at the same time. Since the sub-regions are usually small and contain few spare cells, say less than 5 in practice, at the leaf nodes of the constructed slicing tree by the multilevel spare cell insertion, the optimal spare cell positions can thus be determined within reasonable running time.

4. EXPERIMENTAL RESULTS

We conducted several experiments to justify the effectiveness of the proposed method. Our algorithm has been integrated into NTUplace3 [11], which is a leading academic placer for the large-scale mixed-size designs and is available to the public. It should be noted that the proposed algorithm is very flexible and can also be integrated into other placers that are based on a similar framework introduced in Section 2 with slight modifications. We compare our proposed spare cell insertion algorithm with two previously proposed ones, PostSpare [22, 26] and UniSpare [10]. We also integrated both algorithms into NTUplace3. All the experiments were performed on the same PC workstation with eight Intel Xeon 2.5 GHz CPUs and 26 GB memory. To demonstrate the effects of the spare cell insertion on real designs, we picked nine largest OpenCores [2] circuits in the IWLS 2005 benchmark suite [1]. The benchmark statistics are given in Table 1.

Table 1: Statistics of the OpenCores circuits used in this paper.

Circuit Name	#Movs	#Nets	Util (%)
ac97_ctrl	11855	11947	70.01
aes_core	20795	21055	70.01
des_perf	98341	98576	70.01
ethernet	46771	46889	70.01
mem_ctrl	11440	11560	70.03
pci_bridge32	16816	16989	70.02
usb_funct	12808	12967	70.01
vga_lcd	124031	124133	70.01
wb_conmax	29034	30165	70.01

4.1 Quality Comparison of Spare Cell Insertion

To demonstrate the effectiveness of our proposed algorithm, we compare the quality of spare cell insertion in this experiment. One most commonly used metric to measure the quality of spare cell insertion is the *average spare distance*. Let $d_{min}(v_i, t_k)$ represent the minimum distance from block v_i to spare cells of type t_k. Then the average spare distance d_{avg} for a placement solution can be computed by

$$d_{avg} = \frac{\sum_{v_i \in V} d_{min}(v_i, T(v_i))}{|V|}, \quad (9)$$

where $|V|$ indicates the total number of blocks in this circuit. Table 2 reports the average spare distances of all three compared spare cell insertion algorithms for 1–5% spare cell insertion rates. We list the exact values of d_{avg} of our proposed algorithm, and for the UniSpare and PostSpare algorithms, we list the ratios of their d_{avg} to that of our algorithm. Among all tested insertion rates, our algorithm can consistently achieve 17–33% better average d_{avg} than UniSpare due to its lack of block distribution information before placement, while our algorithm applies the multilevel spare cell insertion to place spare cells close to blocks that require them. Compared with PostSpare, since it can hardly insert spare cells between packed blocks without pre-allocated whitespace, the average spare distance d_{avg} of PostSpare is about 1.77–2.61X worse than that of our algorithm.

4.2 HPWL Comparison after Spare Cell Insertion

In this experiment, we compare the HPWL degradation due to spare cell insertion for each algorithm. Table 3 lists the HPWLs for NTUplace3 without spare cell insertion, and the ratios of the HPWLs for our algorithm and UniSpare to those of NTUplace3, among 1–5% spare cell insertion rates. Note that the results of PostSpare are not listed here since its HPWLs are all the same with those of NTUplace3, and its average spare distance d_{avg} is much worse than that of our algorithm and UniSpare. Since the spare cell insertion methods for our algorithm and UniSpare add extra cells between the placed blocks, they often induce HPWL overheads in order to improve the quality of spare cell insertion[2]. Under the identical spare cell insertion rate, since UniSpare distributes spare cells uniformly into the chip, part of the spare cells will be placed far from the placed blocks. Therefore, the number of spare cells inserted between blocks of UniSpare should be less than that of our algorithm, which inserts all spare cells between placed blocks. Consequently, the HPWL overhead of UniSpare should be less than that of our algorithm. However, according to Table 3, the average HPWL overhead of our algorithm (ranges from 2% to 4%) is always less or equal to that of UniSpare (ranges from 3% to 5%) among all tested insertion rates. Such a phenomenon implies that our proposed spare-cell-aware analytical placement framework provides accurate spare cell prediction, and thus enables the opportunities of the global placement engine to optimize its original objectives with the additional consideration of spare cell insertion.

5. CONCLUSIONS

In this paper, we have proposed the first spare-cell-aware analytical placement framework. A multilevel spare cell in-

---

[2]Due to the unstability of the nonlinear analytical placement formulation, NTUplace3 may sometimes find a slightly better HPWL placement solution with the existence of pre-placed spare cells from UniSpare, or with the use of cluster expansion and density constraint determination for our algorithm.

978-1-60558-497-3/09 $25.00 © 2009 ACM

Table 2: The comparison of the average spare distance (d_{avg}).

Circuit Name	1% Insertion Rate			2% Insertion Rate			3% Insertion Rate			4% Insertion Rate			5% Insertion Rate		
	Ours	Uni-Spare	Post-Spare	Ours	Uni-Spare	Post-Spare	Ours	Uni-Spare	Post-Spare	Ours	Uni-Spare	Post-Spare	Ours	Uni-Spare	Post-Spare
	d_{avg} ($\times 10^5$)	Ratio	Ratio	d_{avg} ($\times 10^5$)	Ratio	Ratio	d_{avg} ($\times 10^5$)	Ratio	Ratio	d_{avg} ($\times 10^5$)	Ratio	Ratio	d_{avg} ($\times 10^5$)	Ratio	Ratio
ac97_ctrl	2.94	1.33	1.87	2.05	1.30	2.23	1.70	1.53	2.43	1.48	1.45	2.60	1.34	1.46	2.81
aes_core	2.50	1.32	2.23	1.75	1.30	2.71	1.47	1.31	2.80	1.27	1.28	2.88	1.17	1.26	3.16
des_perf	3.26	1.17	3.29	2.22	1.23	4.15	1.93	1.16	4.47	1.75	1.10	4.88	1.73	0.95	4.69
ethernet	2.91	1.30	3.19	2.18	1.30	3.54	1.89	1.30	3.61	1.59	1.35	3.85	1.51	1.28	3.80
mem_ctrl	2.93	1.31	1.90	2.24	1.31	2.35	1.96	1.20	2.40	1.68	1.21	2.92	1.58	1.18	2.87
pci_bridge32	3.06	1.26	2.27	2.14	1.32	2.61	1.87	1.24	2.51	1.67	1.28	2.57	1.49	1.26	2.71
usb_funct	3.01	1.28	2.06	2.17	1.19	2.50	1.86	1.19	2.75	1.72	1.16	2.92	1.46	1.23	3.31
vga_lcd	3.40	1.07	4.82	2.74	0.96	5.33	2.00	1.13	6.77	2.32	0.80	5.49	2.37	0.68	5.05
wb_conmax	2.16	1.88	3.29	1.53	1.37	3.78	1.35	1.30	4.37	1.23	1.24	4.03	1.19	1.21	4.08
average	-	1.33	2.77	-	1.25	3.24	-	1.26	3.57	-	1.21	3.57	-	1.17	3.61

Table 3: The comparison of HPWL ratios.

Circuit Name	NTUplace3 HPWL ($\times 10^9$)	1% Insertion Rate		2% Insertion Rate		3% Insertion Rate		4% Insertion Rate		5% Insertion Rate	
		Ours	UniSpare	Ours	UniSpare	Ours	UniSpare	Ours	UniSpare	Ours	UniSpare
ac97_ctrl	1.16	1.02	1.02	1.01	1.01	1.02	0.98	1.02	1.02	1.04	1.03
aes_core	1.98	1.02	1.05	1.04	1.05	1.03	1.06	1.05	1.06	1.03	1.07
des_perf	14.08	1.07	1.04	1.13	1.05	1.10	1.23	1.11	1.14	1.14	1.05
ethernet	8.62	1.01	1.04	1.04	1.03	1.03	0.96	0.97	0.98	0.98	0.98
mem_ctrl	1.30	1.03	1.09	1.03	1.10	1.02	1.11	1.05	1.11	1.04	1.13
pci_bridge32	1.86	1.00	0.94	0.98	0.96	1.02	0.97	1.00	0.96	1.01	0.95
usb_funct	1.35	1.03	1.05	1.05	1.06	1.04	1.04	1.06	1.04	1.05	1.03
vga_lcd	29.71	0.98	1.00	0.97	1.06	0.97	1.02	0.99	1.02	1.03	1.03
wb_conmax	3.21	1.01	1.02	0.98	1.02	1.02	1.04	1.04	1.03	1.03	1.05
average	-	1.02	1.03	1.03	1.04	1.03	1.05	1.03	1.04	1.04	1.04

sertion technique has also been presented with proven capability of providing an efficient spare cell planning and an excellent control of quality impact due to spare cell insertion. To guide the selection of available spare cell positions during insertion, we have also proposed a mixed-integer-linear-programming formulation to determine the optimal spare cell positions. Experimental results have shown that our algorithm can achieve a much better efficiency of spare cell insertion than existing algorithms with only slight quality overhead.

6. REFERENCES

[1] IWLS 2005 Benchmarks. http://iwls.org/iwls2005/benchmarks.html.

[2] OpenCores. http://www.opencores.org.

[3] M. Abramovici, P. Bradley, K. Dwarakanath, P. Levin, G. Memmi, and D. Miller. A reconfigurable design-for-debug infrastructure for SoCs. In Porc. of DAC, 2006.

[4] C. Alpert, A. Kahng, G.-J. Nam, S. Reda, and P. Villarrubia. A semi-persistent clustering technique for VLSI circuit placement. In Proc. of ISPD, pages 200–207, 2005.

[5] C. Bingert, C. D. Gorsuch, O. G. Mercado, A. K. Myers, J. A. Schadt, and B. W. Yeager. US patent 6,600,341: Integrated circuit and associated design method using spare gate islands. 2003.

[6] M. Brazell and A. Essbaum. US patent 6,993,738: Method for allocating spare cells in auto-place-route blocks. 2006.

[7] P. Chaisemartin. US patent 6,586,961: Structure and method of repair of integrated circuits. 2003.

[8] T. Chan, J. Cong, J. Shinnerl, K. Sze, and M. Xie. mPL6: Enhanced multilevel mixed-size placement. In Proc. of ISPD, 2006.

[9] T. Chan, J. Cong, and K. Sze. Multilevel generalized force-directed method for circuit placement. In Proc. of ISPD, 2005.

[10] K.-H. Chang, I. L. Markov, and V. Bertacco. Reap what you sow: Spare cells for post-silicon metal fix. In Porc. of ISPD, 2008.

[11] T.-C. Chen, Z.-W. Jiang, T.-C. Hsu, H.-C. Chen, and Y.-W. Chang. NTUplace3: A high-quality mixed-size analytical placer considering preplaced blocks and density constraints. In Proc. of ICCAD, 2006.

[12] H. Eisenmann and F. M. Johannes. Generic global placement and floorplanning. In Proc. of DAC, 1998.

[13] C. M. Giles. US patent 6,650,139: Modular collection of spare gates for use in hierarchical integrated circuit design process. 2003.

[14] R. Goering. Post-silicon debugging worth a second look. EE Times, February 5, 2007.

[15] D. Josephson. The good, the bad, and the ugly of silicon debug. In Porc. of DAC, 2006.

[16] A. B. Kahng and Q. Wang. Implementation and extensibility of an analytic placer. IEEE Trans. on CAD, 24(5), 2005.

[17] A. B. Kahng and Q. Wang. A faster implementation of APlace. In Proc. of ISPD, 2006.

[18] M. Kleinhans, G. Sigl, F. M. Johannes, and K. J. Antreich. Gordian: VLSI placement by quadratic programming and slicing optimization. IEEE Trans. on CAD, 10(3), 1991.

[19] D. Lee. US patent 5,696,943: Method and apparatus for quick and reliable design modification on silicon. 1997.

[20] W. C. Naylor, R. Donelly, and L. Sha. US patent 6,301,693: Non-linear optimization system and method for wire length and dealy optimization for an automatic electric circuit placer. 2001.

[21] Z. Or-Bach. US patent 6,756,811 B2: Customizable and programmable cell array. 2004.

[22] R. L. Payne. US patent 5,959,905: Cell-based integrated circuit design repair using gate array repair cells. 1999.

[23] J. A. Schadt. US patent 6,404,226 B1: Integrated circuit with standard cell logic and spare gates. 2002.

[24] A. Vergnes. US patent 6,791,355 B2: Spare cell architecture for fixing design errors in manufactured integrated circuits. 2004.

[25] J. Wong, D. Chiang, and J. Tolentino. US patent 6,255,845 B1: Efficient use of spare gates for post-silicon debug and enhancements. 2001.

[26] C. L. Yee, S. Aji, and S. Rusu. US patent 5,623,420: Method and apparatus to distribute spare cells within a standard cell region of an integrated circuit. 1997.

978-1-60558-497-3/09 $25.00 © 2009 ACM

Handling Complexities in Modern Large-Scale Mixed-Size Placement[*]

Jackey Z. Yan   Natarajan Viswanathan   Chris Chu

Department of Electrical and Computer Engineering
Iowa State University, Ames, IA 50010
{zijunyan, nataraj, cnchu}@iastate.edu

ABSTRACT

In this paper, we propose an effective algorithm flow to handle large-scale mixed-size placement. The basic idea is to use floorplanning to guide the placement of objects at the global level. The flow consists of four steps: 1) The objects in the original netlist are clustered into blocks; 2) Floorplanning is performed on the blocks; 3) The blocks are shifted within the chip region to further optimize the wirelength; 4) With big macro locations fixed, incremental placement is applied to place the remaining objects. There are several advantages of handling placement at the global level with a floorplanning technique. First, the problem size can be significantly reduced. Second, exact HPWL can be minimized. Third, precise object distribution can be achieved so that legalization only needs to handle minor overlaps among small objects in a block. Fourth, rotation and various placement constraints on macros can be handled. To demonstrate the effectiveness of this new flow, we implement a high-quality floorplan-guided placer called *FLOP*. We also construct the Modern Mixed-Size (MMS) placement benchmarks which can effectively represent the complexities of modern mixed-size designs and the challenges faced by modern mixed-size placers. Compared with state-of-the-art mixed-size placers and leading macro placers, experimental results show that *FLOP* achieves the best wirelength, and easily obtains legal solutions on all circuits.

Categories and Subject Descriptors

B.7.2 [**Hardware, Integrated Circuits, Design Aids**]: Placement and routing

General Terms

Algorithms, Design, Performance

Keywords

Floorplanning, Incremental Placement, Mixed-size Design

1. INTRODUCTION

[*]This work was partially supported by IBM Faculty Award and NSF under grant CCF-0540998.

Permission to make digital or hard copies of part or all of this work for personal or classroom use is granted without fee provided that copies are not made or distributed for profit or commercial advantage and that copies bear this notice and the full citation on the first page. To copy otherwise, to republish, to post on servers or to redistribute to lists, requires prior specific permission and/or a fee.
DAC'09, July 26-31, 2009, San Francisco, California, USA

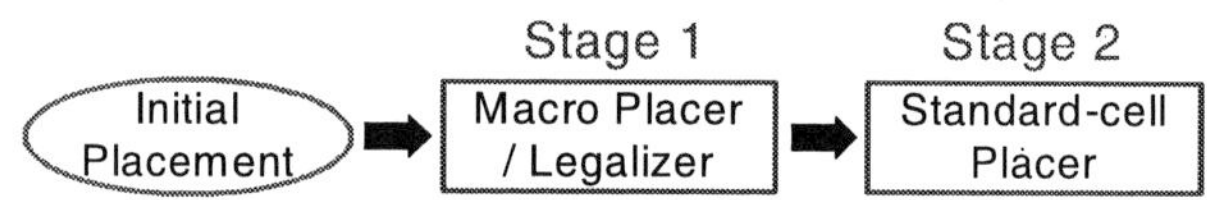

Figure 1: Previous two-stage approach.

In the nanometer scale era, placement has become an extremely challenging stage in modern VLSI designs. Millions of objects need to be placed legally within a chip region, while both the interconnection and object distribution have to be optimized simultaneously. As an early step of VLSI physical design flow, the quality of the placement solution has significant impacts on both routing and manufacturing. In modern System-on-Chip (SoC) designs, the usage of Intellectual Property (IP) and embedded memory blocks becomes more and more popular. As a result, a design usually contains tens or even hundreds of big macros. A design with big movable macros and numerous standard cells is known as mixed-size design, where the placement of big macros plays a key role. Due to the big size difference between big macros and standard cells, the placement of mixed-size designs is much more difficult than the standard-cell placement. Existing placement algorithms usually cannot generate a legal solution by themselves. They have to rely on a post-placement legalization process. However, legalizing big macros with wirelength minimization has been considered very hard to solve for a long time.

1.1 Previous Work

Most mixed-size placement algorithms place both the macros and the standard cells simultaneously. Examples are the annealing-based placer Dragon [1], the partitioning-based placer Capo [2], and the analytical placers FastPlace3 [3], APlace2 [4], Kraftwerk [5], mPL6 [6], and NTUplace3 [7]. The analytical placers are the state-of-the-art placement algorithms. They can produce the best result in the best runtime. But, the analytical approach has two problems. First, only an approximation (e.g., by log-sum-exp or quadratic function) of the Half-Perimeter Wirelength (HPWL) is minimized. Second, the distribution of objects is also approximated and that usually results in a large amount of overlaps. They have to rely on a legalization step to resolve the overlaps. For mixed-size designs, such legalization process is very difficult and is likely to significantly increase the HPWL.

Other researchers apply a two-stage approach as shown in Figure 1 to handle the mixed-size placement. An initial wirelength-driven placement is first generated. Then a macro placement or legalization algorithm is used to place only the macros, without considering the standard cells. After that, the macros are fixed, and the standard cells are re-placed in the remaining whitespace from scratch. As the macro placement is a crucial stage in this flow, people propose

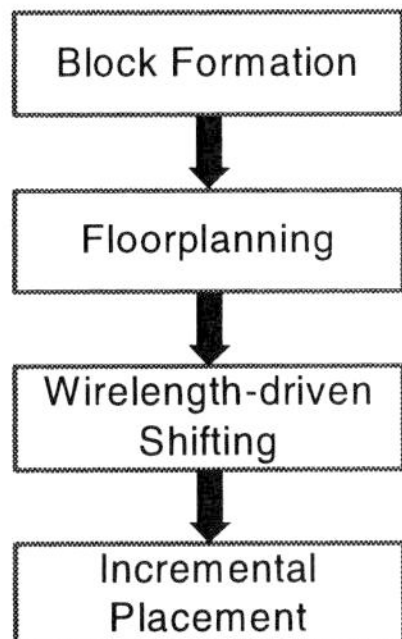

Figure 2: New algorithm flow for mixed-size placement.

different techniques to improve the quality of result (QoR). Based on the MP-tree representation, Chen et al. [8] used a packing-based algorithm to place the macros around the four corners of the chip region. In [9], a transitive closure graph (TCG) based technique was applied to enhance the quality of macro placement. One main problem with the above two approaches is that the initial placement is produced with large amount of overlaps. Thus, the initial solution may not provide good indications on the locations of objects. However, the following macro-placement stage still determines the macro locations by minimizing the displacement from the low-quality initial placement. Alternatively, Adya et al. [10] used an annealing-based floorplanner to directly minimize the HPWL among the macros and clustered standard cells at the macro-placement stage. But, they still have to rely on the illegal placement to determine the initial locations of macros and clusters. For all of the above two-stage approaches, after fixing the macros, the initial positions of standard cells have to be discarded to reduce the overlaps.

1.2 Our Contributions

To effectively handle the complexities of mixed-size placement, we present a new algorithm flow which efficiently integrates floorplanning and incremental placement algorithms. As floorplanners have a good capability of handling a small number of objects [2], we apply floorplanning on the clustered circuit to generate a global overlap-free layout, and use it to guide the subsequent placement algorithm. This new flow is as follows (see Fig. 2).

1. **Block Formation**: The purpose of the first step is to cut down the problem size. We define "small objects" as small macros and standard cells. The small objects are clustered into soft blocks, while each big macro is treated as a single hard block.

2. **Floorplanning**: In this step, a floorplanner is applied on the blocks to directly minimize the *exact* HPWL. Simultaneously, the objects are precisely distributed across the chip region to guarantee an overlap-free layout.

3. **Wirelength-driven Shifting**: In order to further optimize the HPWL, the blocks are shifted at the floorplan level. After shifting, big macros are fixed. The remaining movable objects are assumed to be at the center of the corresponding soft block.

4. **Incremental Placement**: Lastly, the placement algorithm will place the remaining objects. The initial positions of such objects provided by the previous step are used to guide the incremental placement.

Comparing this new methodology with the state-of-the-art analytical placers, we can see that it is superior in several aspects: 1) The exact HPWL is optimized in Steps 1–3; 2) The objects are more precisely

distributed in Step 2; 3) Placement constraints and macro orientation optimization can be handled in Step 2. Compared with the previous two-stage approach, instead of starting from an illegal initial placement, we use the floorplanner to directly generate a global overlap-free layout among the big macros, *as well as between big macros and small objects*. In addition, the problem size has been significantly reduced by clustering. A good floorplanner should be able to produce a high-quality global layout for the subsequent incremental placer. Furthermore, the initial positions of the small objects are not discarded. We keep such information as a starting point of incremental placement. Since the big macros have already been fixed, the placer avoids the difficulty of legalizing the big macros.

Based on the new algorithm flow, we implement a robust, efficient and high-quality floorplan-guided placer called *FLOP*. It can effectively handle mixed-size placement with all movable objects including both macros and standard cells. *FLOP* can also optimize the macro orientation respecting to packing and wirelength optimization.

To show the effectiveness of *FLOP*, we derive the Modern Mixed-Size (MMS) placement benchmarks from the original ISPD05/06 Placement Benchmarks. These new circuits can represent the challenges of modern large-scale mixed-size placement.

The rest of this paper is organized as follows. Section 2 describes the overview of *FLOP*. Section 3 introduces the block formation and floorplanning algorithms. Section 4 presents the wirelength-driven shifting technique. Section 5 describes the incremental placement algorithm. Section 6 describes the MMS benchmarks. Section 7 presents the experimental results. Finally this paper ends with the conclusion and future work.

2. OVERVIEW OF FLOP

FLOP follows the same algorithm flow as shown in Figure 2.

The block formation is based on the result of recursive partitioning of the original circuit. After partitioning, small objects in each partition are clustered into a soft block and each big macro becomes a single hard block.

In the floorplanning step, *FLOP* adopts a min-cut based fixed-outline floorplanner similar to *DeFer* [11]. In *DeFer*, a hierarchy of the blocks needs to be derived using recursive partitioning. Because such a hierarchy has already been generated during the block formation step, it will be passed down and will not be generated again. Another way to look at the flow of *FLOP* is that the block formation step is merged into the floorplanning step as the first stage of *DeFer*.

We formulate the wirelength-driven shifting problem as a linear programming (LP) problem. Therefore, we can find the *optimal* block position in terms of the HPWL minimization among the blocks. In the LP-based shifting we only ignore the local netlist among small objects within each soft block.

Because analytical placers have the best capability in placing a large number of small objects, we use an analytical placer as the engine in the incremental placement step.

3. BLOCK FORMATION AND FLOORPLAN-NING

A high-quality and non-stochastic fixed-outline floorplanner *DeFer* was presented in [11]. It has been shown that, compared with other fixed-outline floorplanners, *DeFer* achieves the best success rate, the best wirelength and the best runtime on average.

Here is a brief description of the algorithm flow of *DeFer*: Firstly the original circuit is partitioned into several subcircuits, each of which contains at most 10 objects. After that, a high-level slicing tree structure is built up. Secondly, for each subcircuit an associated shape curve is generated to represent all possible slicing layouts within the subcircuit. Thirdly, the shape curves are combined from bottom-up following the high-level slicing tree. In the final shape curve at the root the points within the fixed outline are chosen for further HPWL optimization. At the end *DeFer* outputs a final layout.

In *FLOP*, we use *DeFer* in the floorplanning step. To make it more robust and efficient for mixed-size placement, we propose some new techniques and strategies, which are described in Sections 3.1–3.3.

3.1 Usage of Exact Net Model

We use the exact net model in [12] to improve the HPWL in partitioning. By applying this net model in partitioning, the cut value becomes exactly the same as the placed HPWL, so that the partitioner can directly minimize the HPWL instead of interconnections between two partitions. In *FLOP* at the first β levels of the high-level slicing tree ($\beta = 3$ by default), we apply two cuts on the original partition. One is horizontal cut, and another is vertical cut. We compare these two cuts and pick the one with less cost, i.e. HPWL.

However, for a vertical/horizontal cut, the cut value returned by the net model is only equal the horizontal/vertical component of HPWL. So for two cuts with different directions, it is incorrect to decide a better cut direction based on the two cut values generated by these two cuts. The authors in [12] avoided such comparison by fixing the cut direction based on the dimension of the partition region. Nevertheless, this may potentially lose the better cut direction. Here we propose a simple heuristic to solve the cut value comparison between the cuts from two different directions.

Suppose K is the total number of nets in one partition that we are going to cut. For the horizontal cut (H-cut), $L^x_{H_i}/L^y_{H_i}$ is the horizontal/vertical component of the HPWL of net i, the same as $L^x_{V_i}$ and $L^y_{V_i}$ for the vertical cut (V-cut). So the total HPWL of the K nets in this partition are:

$$\text{For H-cut}: \quad L_H = \sum_{i=1}^{K} L^x_{H_i} + \sum_{i=1}^{K} L^y_{H_i}$$
$$\text{For V-cut}: \quad L_V = \sum_{i=1}^{K} L^x_{V_i} + \sum_{i=1}^{K} L^y_{V_i}$$

Thus, the correct way to make the comparison between H-cut and V-cut should be:

$$\text{if } L_H \geq L_V \Rightarrow \text{V-cut is better}$$
$$\text{if } L_H < L_V \Rightarrow \text{H-cut is better}$$

As the net model only returns $\sum_{i=1}^{K} L^y_{H_i}$ for H-cut, and $\sum_{i=1}^{K} L^x_{V_i}$ for V-cut, we need find a way to estimate $\sum_{i=1}^{K} L^x_{H_i}$ and $\sum_{i=1}^{K} L^y_{V_i}$. Let the aspect ratio (i.e. height/width) of the partition region be γ. When K is very big, based on statistics we can have:

$$\frac{\sum_{i=1}^{K} L^y_{H_i}}{\sum_{i=1}^{K} L^x_{H_i}} \approx \frac{\sum_{i=1}^{K} L^y_{V_i}}{\sum_{i=1}^{K} L^x_{V_i}} \approx \gamma$$

Thus,

$$\text{if } L^y_H \geq L^x_V \cdot \gamma \Rightarrow \text{V-cut is better}$$
$$\text{if } L^y_H < L^x_V \cdot \gamma \Rightarrow \text{H-cut is better}$$

Two reasons prevent us from applying the net model in lower levels ($> \beta$): 1) As partitioning goes on, K becomes smaller and smaller, which makes the approximation of $\sum_{i=1}^{K} L^x_{H_i}$ and $\sum_{i=1}^{K} L^y_{V_i}$ inaccurate; 2) Using the net model, we restrict the combine direction in the Generalized Slicing Tree [11], which hurts the packing quality. To make a trade-off we only apply the net model in the first β levels.

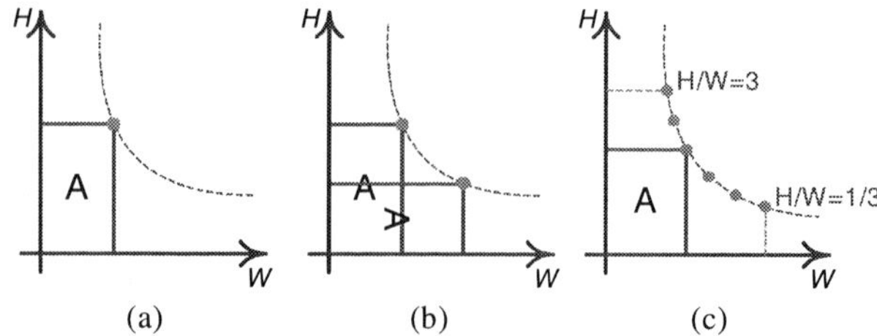

Figure 3: Generation of shape curves for blocks.

3.2 Block Formation

As mentioned earlier, since *DeFer* starts with a min-cut partitioning, *FLOP* merges the block formation step into the floorplanning step. After the original circuit is partitioned into multiple subcircuits, in each subcircuit we treat a big macro as a hard block, and cluster all small objects into a soft block.

However, in *DeFer* the partitioning will not stop until each subcircuit contains less than or equal to 10 objects. If the same stopping criteria is used in *FLOP*, then most subcircuits will contain at most 10 standard cells, which means by clustering we can only cut down the problem size by at most 90%. Nevertheless, for a typical placement problem with millions of objects, the resulted circuit size is still too big for the floorplanning algorithm. So here we propose a more suitable stopping criteria. Let A_o be the total area of all objects in the design. In one partition there are N_p objects of which the total area is A_p, α is the area bound ($\alpha = 0.15\%$ by default). We will stop cutting this partition, if *either* one of the following conditions is satisfied: 1) $\frac{A_p}{A_o} \leq \alpha$; 2) $N_p \leq 10$.

3.3 Generation of Shape Curve for Blocks

To capture the shape of the blocks, we generate an associated shape curve for each block. For the hard block if a macro cannot be rotated, only one point representing the user-specified rotation is generated (see Fig. 3 (a)). Otherwise two points representing two different rotations are generated (see Fig. 3 (b)). For the soft block we bound its aspect ratio from $1/3$ to 3, and sample multiple points on the shape curve to represent its shape (see Fig. 3 (c)). Considering the target density constraint in the placement, we add some white space in each soft block. In some sense, we "inflate" the soft block based on the target density.

$$A'_{s_i} = \frac{A_{s_i}}{TD} \times (max((TD - 0.93), 0) \times 0.5 + 1) \tag{1}$$

In Equation 1, for soft block i, A'_{s_i} is the "inflated area", A_{s_i} is the total area of objects within soft block i, and TD is the target density. Based on this formula, if the target density is more than 93%, we add some white space into the soft block. The purpose is to leave some space for the analytical placer to place the small objects.

4. WIRELENGTH-DRIVEN SHIFTING

In *FLOP* the wirelength-driven shifting process is formulated as a linear programming (LP) problem, which is the same as in [13]. We use the contour structure [14] to derive the horizontal and vertical non-overlapping constraints among the blocks.

The LP-based shifting is an essential part in *FLOP*. In terms of the HPWL minimization it can find the optimal position for each block, and basically provides a globally optimized layout for the analytical placer. Since the LP-based shifting optimizes the HPWL at the floorplan level, it only ignores the local nets among the small objects within each soft block. The smaller the soft block is, the less nets it ignores, and the better the HPWL we will get at last. However, if the soft blocks become too small, numerous nets will be considered in the shifting. This would slow down the whole algorithm. Because of this, in the partition stopping criteria we set an area bound α, so that the soft blocks would not become too small. On the other hand, we only need the shifting step to generate a globally good layout. Regarding the local nets within the soft blocks, the following analytical placer can handle them very efficiently and effectively.

5. INCREMENTAL PLACEMENT

As mentioned before, the output of the wirelength-driven shifting step is a layout with legal, non-overlapping locations for the big macros. These big macros are then fixed in place to prevent further movement during any subsequent steps. But, there are multiple "soft blocks" in the layout, each containing numerous "small objects" (i.e., small macros and standard cells). The shifting step assigns these small objects to the center of the corresponding soft block. In this respect, the placement step has two key tasks: 1) Spread the small objects over the placement region and obtain a final overlap free placement among all objects; 2) Use the initial locations of the small objects as obtained by the shifting step.

To satisfy these two tasks, we use an efficient analytical incremental placement algorithm (see Algorithm 1).

Algorithm 1 Analytical Incremental Placement

1: **Phase 0: Physical and Netlist based clustering**
2:     *initial_objects* ← *number_of_small_objects*
3:     set locations of small objects to center of their soft blocks
4:     **while** *number_of_clusters* > *target_number_of_clusters* **do**
5:         cluster netlist using Best-choice clustering [15]
6:         use physical locations of small objects in clustering score
7:         set *cluster_location* ← center of gravity of the objects within cluster
8:     **end while**
9: **end**
10: **Phase 1: Coarse global placement**
11:     generate "fixing forces" for clusters based on their initial locations
12:     solve initial quadratic program (QP)
13:     **repeat**
14:         perform *Cell Shifting* [3] on coarse-grain clusters
15:         add spreading forces to QP formulation
16:         solve the quadratic program
17:     **until** placement is roughly even
18:     **repeat**
19:         perform *Iterative Local Refinement* [3] on coarse-grain clusters
20:     **until** placement is quite even
21:     uncluster movable macro-blocks
22:     legalize and fix movable macro-blocks
23: **end**
24: **Phase 2: Refinement of fine-grain clusters**
25:     **while** *number_of_clusters* < 0.5*number_of_small_objects* **do**
26:         uncluster netlist
27:     **end while**
28:     perform *Iterative Local Refinement* on fine-grain clusters
29: **end**
30: **Phase 3: Refinement of flat netlist**
31:     **while** *number_of_clusters* < *number_of_small_objects* **do**
32:         uncluster netlist
33:     **end while**
34:     perform *Iterative Local Refinement* on flat netlist
35: **end**
36: **Phase 4: Legalization and detailed placement**
37:     Legalize the standard cells in the presence of fixed macros
38:     Perform detailed placement [16] to further improve wirelength
39: **end**

---

6. MMS BENCHMARKS

The only publicly available benchmarks for mixed-size designs are ISPD02 and ICCAD04 IBM-MS [10, 17] that are derived from ISPD98 Placement Benchmarks. As pointed out in [18], these circuits can no longer be representative of modern VLSI physical design. To continue driving the progress of physical design for the academic community, two suites of placement benchmarks [18, 19] have been released recently. They are directly derived from modern industrial ASICs design. Unfortunately, however, in the original circuits most macros have been fixed due to the difficulty of handling movable macros for the existing placers. The authors in [8,9] freed all fixed objects in ISPD06 benchmarks and created new mixed-size placement circuits. But seven out of eight circuits *do not* have any fixed I/O objects, which is not realistic in the real designs. In order to recover the complexities of modern mixed-size designs, we modify the original ISPD05/06 benchmarks and derive the Modern Mixed-Size (MMS) placement benchmarks (see Table 1). Essentially, we make the following changes on the original circuits.

I. Macros are freed from the original positions. In the GSRC Bookshelf format that the original benchmarks use, both fixed macros and fixed I/O objects are treated as fixed objects. There is no extra specification to differentiate them. So we have to distinguish them only based on the size differences. Basically, if the area of one fixed object is more than $\lambda \times$ the average area of the whole circuit, we will recognize it as a macro. Otherwise, it is a fixed I/O object. Because for each circuit the average area is different, we need to use a different λ (see the last column in Table 1) to decide a reasonable number and suitable threshold size for the macros. There is one exception: in both circuits *bigblue2* and *bigblue4*, there is one macro that does not connect with any other objects. If this macro is freed, it may cause some trouble for quadratic-based analytical placers. So we keep it fixed. Since this macro is also very small compared with other macros, it would not affect the circuit property.

II. The sizes of all I/O objects are set to zero. In MMS benchmarks there are two types of I/Os: *perimeter I/Os* around the chip boundary and *area-array I/Os* spreading across the chip region. Generally, the area-array I/Os are allowed to be overlapped with other movable objects in the design. But existing placers treat all fixed I/Os as fixed objects, so that their algorithms *internally* do not allow such overlaps during the legalization. Since the macros have already been freed in MMS benchmarks, the placers should ignore the overlaps between fixed I/O objects and movable objects, and concentrate on the legalization of movable objects. As we cannot change the code of other placers, one simple way to enforce this is to set the sizes of all I/O objects to zero.

The target density constraints are the same as the original circuits. The same scoring function [1] is used to calculate the scaled HPWL. However, since the macros are movable in the MMS circuits, we need to modify the script used in [19] to get the correct "scaled_overflow_factor". The modification being: Any movable macro that has a width or height greater than the bin dimension used for scaled overflow calculation, is now treated as a fixed macro during scaled overflow calculation. Note that, this was the method employed by the original script on *newblue1*, which is the only design that has big movable macros in the original circuits. It is required to treat big movable macros as fixed, otherwise we will get an incorrect picture of the placement density.

We have discussed the MMS benchmarks setup with the authors in [18, 19]. To keep the original circuit properties as much as possible, the above changes are the best we can do without accessing the original industrial data of the circuits. The MMS benchmarks are publicly available at [20].

7. EXPERIMENTAL RESULTS

All experiments were performed on a Linux machine with AMD Opteron 2.6 GHz CPU and 8GB memory. We use *hMetis2.0* [21] as the partitioner and *QSopt* [22] as the LP solver. The seed of *hMetis2.0* is set to 5. Essentially, we set up four experiments.

---

[1]scaled_HPWL = HPWL * (1 + scaled_overflow_factor)

Circuit	#Objects	#Movable Objects	#Standard Cells	#Macros	#Fixed I/O Objects	#Net	#Net Pins	Target Density%	λ
adaptec1	211447	210967	210904	63	480	221142	944053	100	70
adaptec2	255023	254584	254457	127	439	266009	1069482	100	160
adaptec3	451650	450985	450927	58	665	466758	1875039	100	650
adaptec4	496054	494785	494716	69	1260	515951	1912420	100	460
bigblue1	278164	277636	277604	32	528	284479	1144691	100	120
bigblue2	557866	535741	534782	959	22125	577235	2122282	100	30
bigblue3	1096812	1095583	1093034	2549	1229	1123170	3833218	100	470
bigblue4	2177353	2169382	2169183	199	7970	2229886	8900078	100	550
adaptec5	843128	842558	842482	76	570	867798	3493147	50	440
newblue1	330474	330137	330073	64	337	338901	1244342	80	2000
newblue2	441516	440264	436516	3748	1252	465219	1773855	90	190
newblue3	494011	482884	482833	51	11127	552199	1929892	80	170
newblue4	646139	642798	642717	81	3341	637051	2499178	50	400
newblue5	1233058	1228268	1228177	91	4790	1284251	4957843	50	570
newblue6	1255039	1248224	1248150	74	6815	1288443	5307594	80	650
newblue7	2507954	2481533	2481372	161	26421	2636820	10104920	80	650

Table 1: Statistics of the Modern Mixed-Size placement benchmarks.

I. To test the capability of handling the large-scale mixed-size placement, we compare *FLOP* with five state-of-the-art mixed-size placers *APlace2*, *NTUplace3*, *mPL6*, *Capo10.5* and *Kraftwerk* on MMS benchmarks. *Before the experiments, we have contacted the authors of each placer above, and they provided us their best-available binary for MMS circuits.* In Table 2, for the ISPD06 circuits (*adaptec5 – newblue7*) the reported HPWL is the scaled HPWL. *FLOP* is the default mode of *FLOP* with all macros rotatable, and *FLOP-NR* restricts the rotation on all macros. *APlace2* crashed on every circuit, so we do not report its results. For the default mode, *FLOP* generates 8%, 2%, 44% and 26% better HPWL compared with *NTUplace3*, *mPL6*, *Capo10.5* and *Kraftwerk*, respectively. About the runtime, *FLOP* is 7× and 3× faster than *Capo10.5* and *mPL6*. Also *FLOP* achieves legal solution on all circuits. Compared with *FLOP-NR*, *FLOP* generates 4% better HPWL by rotating the macros.

II. To show the importance of the initial positions of small objects in the incremental placement step, we generate the results of *FLOP-NI* that *discards* such information and places all small objects from scratch. As shown in Table 2, *FLOP-NI* produces 5% worse HPWL and 17% slower than *FLOP*.

III. We compare *FLOP* with leading macro placers *CG*, *MPT* and *XDP*. Due to the *IP* issues, their binaries are not available. But the authors sent us the benchmarks used in [9]. So in Table 3 the other placers' results are cited from [9]. These benchmarks allow the rotation of macros and do not consider the target density. As shown in Table 3, *FLOP* achieves 1%, 12%, 7% and 14% better HPWL compared with *CG*, *MPT*, *XDP* and *NTUplace3*, respectively. To show which algorithm provides the best macro location, we use *NTUplace3* to substitute the incremental placer inside *FLOP* (*NTUplace3* does not support incremental placement). The results show that *FLOP+NTUplace3* generates 9% worse HPWL than *CG*. But this does not mean *FLOP* is weaker than *CG* in terms of handling the macros. We observe that *FLOP+NTUplace3* produces significantly worse HPWL on *newblue7*. However, using the *same* macro locations generated by *FLOP*, the incremental placer inside *FLOP* achieves the *best* HPWL on *newblue7*. We believe this is because *NTUplace3* is not an incremental placer. As shown earlier, non-incremental placement will significantly degrade *FLOP's* result.

IV. The runtime breakdown of *FLOP* is shown in Figure 4. We can see that the LP-based shifting takes almost $1/3$ of the total runtime. This is the main bottle neck of the runtime in *FLOP*.

8. CONCLUSION

This paper presents a new algorithm flow for large-scale mixed-size placement. To show the effectiveness of such flow, a high-quality mixed-size placer *FLOP* is proposed. Compared with state-

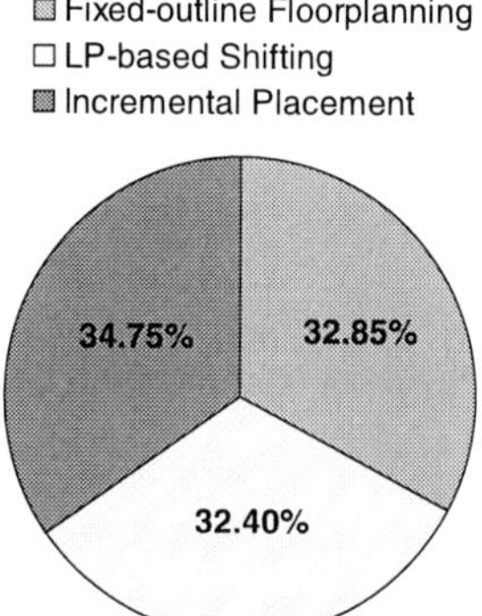

Figure 4: Runtime breakdown of *FLOP*.

of-the-art mixed-size placers and leading macro placers, *FLOP* achieves the best HPWL, and easily produces the legal layout for every circuit.

We believe there is much room to further improve the QoR of *FLOP*. For example, we can use the min-cost flow algorithm to substitute the linear programming formulation in order to speed up the LP-based shifting step. We also observe that the partitioning takes around 80% of the total runtime in the floorplanning step. Thus a stand-alone clustering algorithm is needed in the block formation step to cut down the problem size before partitioning. This will definitely improve both the runtime and HPWL. In the future, different floorplanners and placers can be incorporated into this flow to handle other problems, e.g. placement with geometry constraints.

Acknowledgment

The authors would like to thank Dr. George Karypis, Dr. Gi-Joon Nam, Guojie Luo, Tung-Chieh Chen and Peter Spindler for the help with hMetis2.0, ISPD05/06 Placement Benchmarks, mPL6, NTUplace3 and Kraftwerk, respectively. Also thanks goes to Yi-Lin Chuang and Dr. Yao-Wen Chang for providing us their benchmarks.

9. REFERENCES

[1] T. Taghavi, X. Yang, B.-K. Choi, M. Yang, and M. Sarrafzadeh. Dragon2006: Blockage-aware congestion-controlling mixed-size placer. In *Proc. ISPD*, pages 209–211, 2006.

[2] J. A. Roy, S. N. Adya, D. A. Papa, and I. L. Markov. Min-cut floorplacement. *IEEE Trans. on Computer-Aided Design*, 25(7):1313–1326, July 2006.

Circuit	NTUplace3 [7]		mPL6 [6]		Capo10.5 [2]		Kraftwerk [5]		FLOP-NR		FLOP-NI		FLOP	
	HPWL	Time(s)	HPWL	Time(s)	HPWL	Time(s)	HPWL	Time(s)	HPWL	Time(s)	HPWL	Time(s)	HPWL	Time(s)
adaptec1	80.45	630	77.84	2404	84.77	5567	86.73	285	77.18	722	85.27	931	76.83	824
adaptec2	136.46	1960	88.40	2870	92.61	7373	108.61	408	87.17	1219	86.72	1371	84.14	1192
adaptec3	*illegal*	–	180.64	5983	202.37	16973	272.76	773	182.21	1943	173.07	2352	175.99	2116
adaptec4	177.23	1896	162.02	5971	202.38	17469	260.96	962	166.55	2398	175.67	2689	161.68	2427
bigblue1	95.11	955	99.36	2829	112.58	8458	130.78	348	95.45	1776	98.91	2174	94.92	2007
bigblue2	144.42	1661	144.37	12202	149.54	17647	*aborted*	–	150.66	2373	162.40	3622	153.02	3110
bigblue3	*illegal*	–	319.63	9548	583.37	69921	417.73	2369	372.79	7037	394.75	6531	346.24	5437
bigblue4	764.72	8182	804.00	23868	915.73	109785	969.03	6000	807.53	13816	839.53	21927	777.84	19671
adaptec5*	*aborted*	–	376.30	22636	565.88	23957	402.21	1653	381.83	5048	385.07	3882	357.83	3293
newblue1*	60.73	875	66.93	3171	110.54	3133	76.22	481	73.36	1368	71.69	1280	67.97	1012
newblue2*	*aborted*	–	179.18	6044	303.25	8156	272.67	615	231.94	2646	190.50	2830	187.40	2414
newblue3*	*aborted*	–	415.86	17623	1282.19	73339	374.54	578	344.71	2336	355.07	3237	345.99	2757
newblue4*	*aborted*	–	277.69	9732	300.69	6589	291.45	1266	256.91	2575	268.46	3282	256.54	2455
newblue5*	*aborted*	–	515.49	24806	570.32	16548	503.13	2639	516.71	8801	536.38	10546	510.83	9163
newblue6*	*aborted*	–	482.44	13112	609.16	18076	651.34	2726	502.24	9450	506.99	10740	493.64	9563
newblue7*	*aborted*	–	1038.66	31680	1481.45	43386	*illegal*	–	1113.07	18765	1101.07	27029	1078.18	25104
Norm	**1.08**	**0.78**	**1.02**	**2.95**	**1.44**	**6.56**	**1.26**	**0.35**	**1.04**	**1.00**	**1.05**	**1.17**	**1**	**1**

Table 2: Comparison with mix-size placers on MMS Benchmarks (* comparison of scaled HPWL), HPWL($\times$10e6).

Circuit	CG [9]+NTUplace3	MPT [8]+NTUplace3	XDP [23]+NTUplace3	NTUplace3	FLOP+NTUplace3	FLOP
	HPWL	HPWL	HPWL	HPWL	HPWL	HPWL
adaptec5	29.46	31.01	31.08	29.03	27.94	27.36
newblue1	6.23	6.50	6.32	6.06	7.41	7.32
newblue2	18.89	22.60	18.90	28.09	25.38	23.79
newblue3	30.18	37.57	37.64	53.48	31.20	33.61
newblue4	21.38	23.77	22.01	22.83	21.03	19.72
newblue5	42.92	43.71	45.41	39.91	37.21	36.66
newblue6	44.93	50.50	46.43	44.24	45.80	41.87
newblue7	99.03	108.06	102.21	100.06	170.00	86.96
Norm	**1.01**	**1.12**	**1.07**	**1.14**	**1.10**	**1**

Table 3: Comparison with macro placers on modified ISPD06 benchmarks [9] with default chip utilization, HPWL($\times$10e7).

[3] N. Viswanathan, M. Pan, and C. Chu. Fastplace 3.0: A fast multilevel quadratic placement algorithm with placement congestion control. In *Proc. ASP-DAC*, pages 135–140, 2007.

[4] A. B. Kahng and Q. Wang. A faster implementation of APlace. In *Proc. ISPD*, pages 218–220, 2006.

[5] P. Spindler and F. M. Johannes. Fast and robust quadratic placement combined with an exact linear net model. In *Proc. ICCAD*, pages 179–186, 2006.

[6] T. Chan, J. Cong, J. Shinnerl, K. Sze, and M. Xie. mPL6: Enhanced multilevel mixed-sized placement. In *Proc. ISPD*, pages 212–214, 2006.

[7] T.-C. Chen, Z.-W. Jiang, T.-C. Hsu, H.-C. Chen, and Y.-W. Chang. A high-quality mixed-size analytical placer considering preplaced blocks and density constraints. In *Proc. ICCAD*, pages 187–192, 2006.

[8] T.-C. Chen, P.-H. Yuh, Y.-W. Chang, F.-J. Huang, and D. Liu. MP-tree: A packing-based macro placement algorithm for mixed-size designs. In *Proc. DAC*, pages 447–452, 2007.

[9] H.-C. Chen, Y.-L. Chuang, Y.-W. Chang, and Y.-C. Chang. Constraint graph-based macro placement for modern mixed-size circuit designs. In *Proc. ICCAD*, pages 218–223, 2008.

[10] S. Adya and I. Markov. Consistent placement of macro-blocks using floorplanning and standard-cell placement. In *Proc. ISPD*, pages 12–17, 2002.

[11] J. Z. Yan and C. Chu. DeFer: Deferred decision making enabled fixed-outline floorplanner. In *Proc. DAC*, pages 161–166, 2008.

[12] T.-C. Chen, Y.-W. Chang, and S.-C. Lin. IMF: Interconnect-driven multilevel floorplanning for large-scale building-module designs. In *Proc. ICCAD*, pages 159–164, 2005.

[13] X. Tang, R. Tian, and M. D. F. Wong. Minimizing wire length in floorplanning. *IEEE Trans. on Computer-Aided Design*, 25(9):1744–1753, September 2006.

[14] P.-N. Guo, C.-K. Cheng, and T. Yoshimura. An O-tree representation of non-slicing floorplan and its applications. In *Proc. DAC*, pages 268–273, 1999.

[15] G.-J. Nam, S. Reda, C. J. Alpert, P. G. Villarrubia, and A. B. Kahng. A fast hierarchical quadratic placement algorithm. *IEEE Trans. on Computer-Aided Design*, 25(4):678–691, April 2006.

[16] M. Pan, N. Viswanathan, and C. Chu. An efficient and effective detailed placement algorithm. In *Proc. ICCAD*, pages 48–55, 2005.

[17] S. Adya, S. Chaturvedi, J. Roy, D. Papa, and I. Markov. Unification of partitioning, placement and floorplanning. In *Proc. ICCAD*, pages 550–557, 2004.

[18] G.-J. Nam, C. J. Alpert, P. Villarrubia, B. Winter, and M. Yildiz. The ISPD2005 placement contest and benchmarks suite. In *Proc. ISPD*, pages 216–220, 2005.

[19] G.-J. Nam. ISPD 2006 placement contest: Benchmark suite and results. In *Proc. ISPD*, pages 167–167, 2006.

[20] MMS Placement Benchmarks. http://www.public.iastate.edu/~zijunyan/.

[21] hMetis2.0. http://glaros.dtc.umn.edu/gkhome/.

[22] QSopt LP Solver. http://www.isye.gatech.edu/~wcook/qsopt/.

[23] J. Cong and M. Xie. A robust detailed placement for mixed-size ic designs. In *Proc. ASP-DAC*, pages 188–194, 2006.

978-1-60558-497-3/09 $25.00 © 2009 ACM 441

RegPlace: A High Quality Open-source Placement Framework for Structured ASICs

Ashutosh Chakraborty, Anurag Kumar, David Z. Pan
ECE Dept. Univ. of Texas at Austin, Austin, TX 78712
{ashutosh, anurag}@cerc.utexas.edu, dpan@ece.utexas.edu

ABSTRACT

Structured ASICs have recently emerged as an exciting alternative to ASIC or FPGA design style as they provide a new trade-off between the high performance of ASIC design and low non-recurring engineering (NRE) costs of FPGA design. To fully utilize the benefits of structured ASICs, key physical design stage like placement should be made aware of modularity of their architecture. In this work, we propose a novel solution to placement for structured ASICs. In particular, we propose creation of intermediate virtual platform to exploit the regularity of structured ASIC and Integer Linear Program and network flow formulations for satisfying constraints associated with typical structured ASIC clock architectures. A preliminary version of this work recently won the structured ASIC placement contest by eASIC [1]. Our placer achieves 35% and 5% wirelength improvement over other placers and can place a design of 1 million cells in approximately 4 hours.

Categories and Subject Descriptors

B.7.2 [**Hardware, Integrated Circuit**]: Design Aids

General Terms

Algorithms, Design

Keywords

Placement, Structured ASIC, FPGA, Regular ASIC, Global Placement, legalization

1. INTRODUCTION

Skyrocketing costs and increasing variability associated with an ASIC design flow and unacceptable power and delay penalty associated with FPGA design flow have forced semiconductor companies to look for alternatives. One viable alternative that has emerged over time is the use of *structured* ASICs. Structured ASICs provide an exciting middle-ground between high performance of ASIC designs and short time-to-market FPGA designs. They exploit the fact that not all mask-layers provide equal value for the customers and these layers can be pre-fabricated amortizing their cost

Permission to make digital or hard copies of part or all of this work for personal or classroom use is granted without fee provided that copies are not made or distributed for profit or commercial advantage and that copies bear this notice and the full citation on the first page. To copy otherwise, to republish, to post on servers or to redistribute to lists, requires prior specific permission and/or a fee.

DAC'09, July 26-31, 2009, San Francisco, California, USA

over multiple designs [2]. Structured ASIC flow is much simpler than that for traditional ASICs because majority of deep submicron issues such as signal integrity, power grid optimization, low skew clock tree distribution are already taken care of by the structured ASIC vendors. Structured ASICs can be used to implement designs consisting of millions of gates in contrast to FPGAs which can implement designs with much lesser number of gates. There are a wide variety of structured ASIC architectures, but all of them have a fundamental repeated logic element called *tile* which may contain pre-defined combinational logic, small RAM, and registers [3].

To fully utilize the benefits of structure ASICs, tools for high quality placement and routing need to be developed. Placement for structured ASIC requires cells to be mapped exactly on a compatible site, similar to the case for FPGA. However, since the problem size of structured ASIC can be an order of magnitude bigger than FPGA [2], the existing FPGA tools cannot be used for structured ASICs. Not only does the placer have to handle millions of cells along with the site compatibility requirement, the task is made much more difficult due to the clocking schemes in structured ASICs. The clocks are built with pre-allocated resources to provide low skew clocking and simplified design flow. This restricts the number of clocked elements which can be placed in proximity of each other. In this work we present RegPlace [4], a high quality open source placement tool for structured ASICs which can deal with the above mentioned legality and clock constraints.

The rest of the paper is organized as follows: We describe the intricacies of placement for structured ASIC in Section 2. Section 3 discusses the previous work done and our key innovations. We present our placement solution in Section 4. The efficacy of our solution is demonstrated by experimental results in Section 5. Section 6 concludes our paper.

2. PROBLEM DESCRIPTION

As discussed earlier, large problem size, site compatible requirements, whitespace requirement for routing tracks and strict clock constraints make the task of placement in structured ASICs very challenging. In this work, we propose a solution to the placement problem of structure ASICs. We base our work on the architecture of the popular Nextreme line of structured ASICs from eASIC Corporation [1]. Nextreme is the most *structured* of structured ASIC solutions with only one via layer available for customization. However, our formulations are very generic and can be applied to any structured ASIC.

There are four types of cells in Nextreme line of struc-

978-1-60558-497-3/09 $25.00 © 2009 ACM

tured ASICs. They are SRAM programmable Logic cells (LCELLS), flip flops (DFF), registers (REG) and memories (BRAM). In the rest of the paper, we will refer to these types of cells with their short names and all these types will collectively referred to as *cells*. The placement problem requires that a cell can only be placed on a location which is reserved for that type of cell. In the basic repeating *tile* of Nextreme architecture, the space reserved for LCELL, DFF, REG and BRAM are as shown in Figure 1 (figure is only for illustration and not drawn to scale). There are 36 LCELL and 24 DFF columns in each tile. Each of these columns accommodate 64 LCELLs and DFFs respectively. A total of 4 REG and 1 BRAM can be accommodated in each tile. This tile repeats all over the chip with some horizontal and vertical inter-tile whitespace between adjacent tiles. Depending on the size of the netlist to be placed, the number of times these tiles need to be repeated can be configured. Table 1 shows the details of two such configured platforms along with the maximum cells of each type that can be accommodated in them.

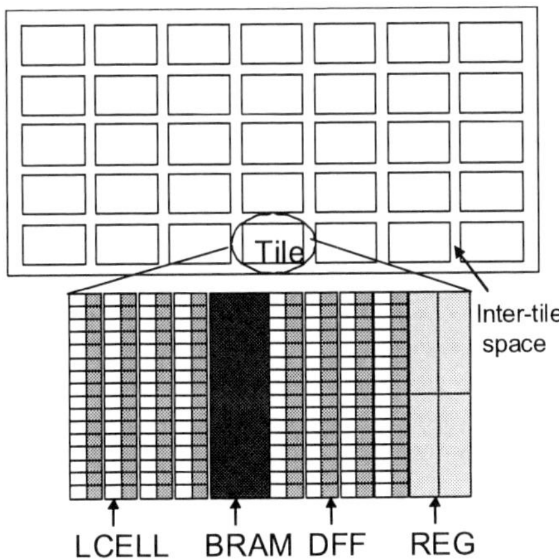

Figure 1: Structured ASIC platform.

Table 1: Platform Characteristics

Plat.	Max LCELL	Max DFF	Max REG	Max BRAM	Intertile Space Horz	Vert
A	1M	675k	1760	440	8.00	6.45
B	1.2M	811k	2112	528	8.00	8.92

The sizes of each LCELL, DFF, REG and BRAM in the mapped design netlist are 1x1, 1x1, 8x32 and 16x64. These cells are to be placed according to the reserved column locations described above. The presence of these four different types of cells require that the global placement ensure that the density requirement for each type of cells is met. The presence of whitespaces and interleaving of space for each of the cell types makes the physical location available for each type of cell very discrete.

Because of architectural constraints, only a certain number of different clocks can go inside a tile. For example, in Figure 1, each tile can have $N(=4$ in our benchmarks) different clocks. Moreover, all the cells in one column of DFF can have only one clock type. The clock constraint on placement has big impact on final placement because clock violations may require that the violating cells be reallocated to another region far away from their optimal location.

3. PREVIOUS WORKS AND OUR CONTRIBUTION

There are several commercial and academic placers for ASIC designs e.g. [5] [6] [7] [8] [9]. These placers implement sophisticated mathematical or min-cut formulations to generate very good quality results for ASIC designs. However, due to site compatibility and hard clock constraints they cannot be directly used for placement of structured ASIC designs. Nevertheless, we note that if the key algorithms in these placers can be somehow used to solve the structured ASIC placement problem, the solutions would be of good quality. Another class of placement tools which are designed for FPGAs can take care of site constraints. However, due to smaller sizes of FPGA based designs, the existing FPGA placement approaches such as [10] [11] rely on slow algorithms like simulated annealing which cannot scale to problem sizes frequently encountered in structured ASIC designs.

There is limited research in the domain of placement for structured ASICs. The work in [12] only addresses incremental placement issues in structured ASICs. Industries are currently using existing row-based placers with product specific legalizers or heuristics [13]. However, there is a dearth of tools which can handle the clock constraints or exploit the modularity of structured ASIC platforms.

Our main technical contributions in this work are

- We propose row-based placer friendly virtual platform generation. This method achieves even density distribution and much faster placement.

- An integer linear program (ILP) formulation is proposed to satisfy clock constraints on the number and type of clocks that can appear in a tile as well as each column of a tile.

- A detail placement flow is proposed specifically for tile based structured ASIC architectures.

4. OUR PLACEMENT FLOW

Our complete placement flow is depicted in Figure 2. We outline the major steps here while the highlights of each step are discussed in next sections. In the first phase, given a design netlist and physical platform specification, we transform them into virtual netlist and virtual platform. This key step mitigates the problems arising due to severely discretized space available to the placer. A high quality row-based placer is run to place the virtual netlist on the virtual platform which generates an initial solution for our problem. In the second phase, we transform the placement solution back to the real platform while minimizing the impact on solution quality due to this transformation. The third stage performs the key step of satisfying clock and density requirement at the platform level using efficient mathematical formulation. This is followed by intra-tile clock assignment and perform aggressive wirelength reduction while maintaining the site legality of the solution. The key highlights of the above stages are presented below.

4.1 Virtual Platform Generation

The physical space available for placement of each type of cell is very discrete in our placement problem. One obvious method to take care of this is through the use of blockages. Though some existing placement tools do have the capability to consider blockages, we cannot specify our placement

978-1-60558-497-3/09 $25.00 © 2009 ACM

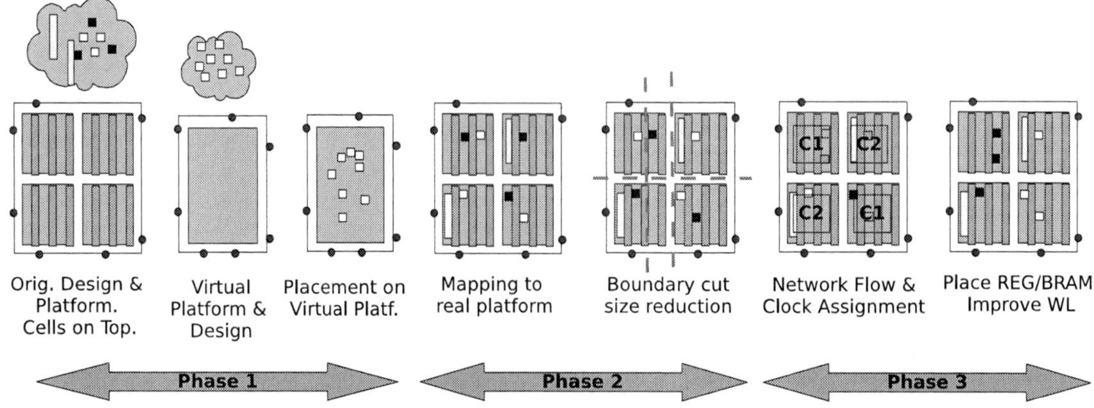

Figure 2: Flow of RegPlace. Different colored columns on platform represent different compatible sites.

problem with blockages because blockages for one type of cell can be valid site for another type of cell. Further, the solution quality and the time required by placers degrade significantly when considering several thousands of blockages. To overcome this challenge, we generated a virtual platform and corresponding virtual netlist. The virtual platform is a physically shrunk copy of the real platform with only the reserved space for one type of cell adjacent to each other. Since majority of the cells are of the type LCELLs, we made the virtual platform by stacking together all spaces reserved for LCELLs. In other words, the inter-tile spacing and the space reserved for DFFs, REGs, and BRAMs is removed leaving a contiguous virtual platform which has nearly the same height as the real platform but nearly 25% the width of real platform. The virtual nelist is generated by forcing the cell size of each type of cell to be equal to that of an LCELL. A standard row-based placer can now be exploited to place the virtual netlist onto the virtual platform. In the general case, the space available in the virtual platform may not be sufficient to contain the virtual netlist. In such a case, we expand the virtual platform horizontally until its size is at least 10% larger than the virtual netlist's space requirement to allow sufficient whitespace for the placer.

4.2 Transforming Virtual Placement Solution

The results of global placer on virtual platform needs to be mapped back to the real platform. We perform this step by inserting whitespace corresponding to the blockages for LCELLs. This step, which is depicted in Figure 3, can cause large increase in wirelength (WL) because several nets which were earlier small would be elongated by the amount of whitespace inserted. Note that in Figure 3, we assume only 4 columns of LCELL instead of 24 to avoid cluttering the diagram. To reduce this impact, we perform cut minimization on cells immediately lying on the boundary of the inserted whitespace.

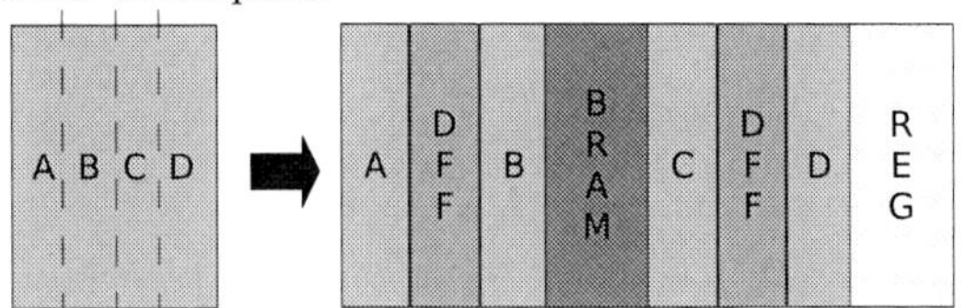

Figure 3: Example of shifting of LCELL columns to accommodate other cell types in between. Columns A, B, C and D spread to reform the tile structure

Since the initial placement is performed on a continuous virtual platform, all the cells get placed in a unique tile as a result of mapping the virtual placement solution on to the real platform. In other words, no cell occupies the spaces reserved for DFF, REGs and BRAMs as well as the spacing between individual tiles.

4.3 Chip Level Density & Clock Legalization

In the placement obtained from previous step, under the assumption that all cells are of the same type, each tile has density less than one i.e. they can be placed without overlap. However, for all the cells to be placed in their matching cell type, each tile's density for *each* type of cell should be less than one. It is easy to see that if, after all the cells have been assumed to be of type LCELL, there is no density violation, then, there can never be any density violation if some of the cells are turned back into their own type (thus leaving empty space in the space for LCELLs). Due to this reason, we never need to consider density of LCELLs in our approach and only the DFFs, BRAM and REG file need to be considered.

Satisfying DFF density: The first step for DFF density and clock constraint is to determine which set of clocks will be present in each tile under the constraint that no more than N different clocks exist per tile (N=4 in our implementation). Our placer formulates this problem as an ILP as follows:

In a design with T tiles and total of C clocks, let boolean variable E_{ti} denote the existence of clock type i in tile t. The effort of removing k_i cells of clock type i from a tile t is $(1 - E_{ti}) \times k_i$. The contribution of tile t to the cost function is the total effort to clean up the cells of a clock type which cannot be placed in the tile which is $\sum_{i=1}^{C}(1 - E_{ti}) \times k_i$. The complete cost function is simply addition of the above cost function for each tile. To guarantee enough space for all the cells of each clock type, few more constraints need to be put.

$$\text{Minimize} \quad \sum_{t=1}^{T}\left(\sum_{i=1}^{C}(1 - E_{ti}) \times k_i\right)$$

$$\text{Subject to :}$$

$$\sum_{i=1}^{C} E_{ti} \leq N \quad \forall t \in (1, T)$$

$$\sum_{t=1}^{C} E_{ti} \geq Min_i \;\; \forall i \in (1, C)$$

$$E_{ti} \quad \in \; 0, 1$$

978-1-60558-497-3/09 $25.00 © 2009 ACM

where Min_i is the minimum number of tiles required for the DFF with clock i for the legalization to be possible. The above ILP formulation has O(T×C) boolean variables, O(T) constraints for guaranteeing no more than N clocks in each tiles, and O(TC) constraints to guarantee enough space for all cells. Considering that T is in hundreds and N is in tens, the above ILP formulation can be solved within seconds.

Once the clock assignment is fixed, we formulate the problem of determining how many DFFs to move among various tiles as a min-cost network flow problem with capacity constraints. The network flow is constructed by connecting a tile congested w.r.t. DFFs with the four tiles around it. The maximum capacity of any edge is taken as the minimum of the number of DFFs that need to exit a congested tile and the number of DFFs which can move into neighboring tile without congesting it. In our implementation, the cost of moving a cell out of its current tile is 1. Thus the network flow solution minimizes the *number of DFF* that need to move for satisfying density constraints. It is possible to extend our work by modeling the cost as change in WL due to moving a DFF which corresponds to a solving density constraints with least movement. At this stage, each tile satisfies density constraints w.r.t. DFFs.

Satisfying REG Density: Since each tile can have at most 4 REGs, and the number of REG cells to be placed are usually very small (several hundreds), it is possible to solve their placement globally. For T tiles, there are 4T valid places for REGs. We cast this selection as an assignment problem which can be solved efficiently using Munkres algorithm [14] whose efficient implementation is O(n^3) complexity. The cost of placing a REG into a valid position is modeled as the wirelength(WL) of all the nets incident to the REG cell in that position. The solution to the assignment problem not only identifies which REG goes to which tile, but also fixes its position to a valid location in the tile. Even for the biggest benchmark and platform, this step takes less than 30 seconds CPU time.

Satisfying BRAM Density: The case for BRAM is exactly the same as that of REGs and can be solved by assignment problem readily. Notice that owing to very small number of BRAMs and only one space per tile, the size of assignment problem for BRAM is an order of magnitude smaller than that for REGs and is solved within 10 seconds for biggest benchmark.

4.4 Tile Level Site Legalization

Once all of the above steps are run, each tile's density for each type of cell is under control. As pointed out earlier, the BRAMs and REGs are already placed at their best legal location by virtue of the solving assignment problem which maps them to legal locations. However, the LCELLs and DFFs need to be put in their respective columns (see Figure 1). There are two main steps to perform this function:

Column's clock type determination: In the first step for tile level site legalization, we determine which *column* should be reserved for which clock (recall that each column can have cells of only one type of clocks). Our experiments show that this step is very critical since if the column reservation is not done carefully, wirelength can degrade significantly when moving the DFFs to column which can accommodate its clock type. We solve this problem using ILP in a manner similar to solving the problem of deciding which clocks can come in which tile (see Section 4.3). Each col-

umn c is assigned as many binary variables as their are clock types. The cost function is the sum of efforts required to clean up a column by removing the DFFs whose clock type is not the same as that returned by ILP solution. A column which initially does not have any DFF in it is assigned a clock number -1 indicating that it can be reserved at a later time while wirelength minimization. Consider a case where upto M DFFs can be accommodated in a column and let each tile have P DFF columns and currently C clocks in it with N_i ($i \in [1, C]$) DFFs for each of these clock types. Binary variable p_i, when true, expresses that column p is reserved for clock i. Let the number of cells before DFF site legalization in the column p of clock type i be given by n_{pi}.

The complete ILP can be written as:

$$\text{Minimize} \quad \sum_{p=1}^{P} \left(\sum_{i=1}^{C} p_i \times \left(\sum_{j=1, j \neq i}^{C} n_{pj} \right) \right)$$

$$\text{Subject To} : \quad \sum_{i=1}^{C} p_i \leq 1 \qquad \forall p \in [1, P]$$

$$\sum_{p=1}^{P} p_i \geq \lceil n_{pi}/M \rceil \quad \forall i \in [1, C]$$

$$p_i \quad \in \quad 0, 1$$

The ILP above has O(PC) binary variables and O(P + C) constraints. The first set of constraints make sure only one clock can occupy a column. The second set of constraints guarantee enough columns reserved within the tile for each clock type so that its cells can be placed. The values of C and P is determined by the structured ASIC's architecture and is generally a small constant number (e.g. $C = 4$ and $P = 24$ in our benchmarks). In our experiments we observed that the above ILP takes less than 0.2 seconds to solve.

Site Legality of LCELLs and DFFs: With the assignment of all columns to a clock type or unreserved type, we iterate over the LCELLs and DFFs to legalize them. These cells are sorted in non-decreasing horizontal coordinate and are assigned to their closest unoccupied and clock compatible (in case of DFFs) possible location. A procedure *rotateAndPlace* has been implemented which takes as input a given point and a cell and rotates in circle with increasing radius around the point until the given cell can be placed legally depending on the type of the cell.

4.5 Wirelength Recovery

The chip and tile level density legalization outlined in the Section 4.3 and Section 4.4 produce placement results which are completely legal w.r.t. clock, overlap and site constraints. However, due to movement of several cells away from the initial relative order suggested by the placer results on virtual platform, significant increase in the wirelength may occur. Our wirelength optimization procedure depicted in Algorithm 1 recovers wirelength. Our philosophy during wirelength minimization is to never break the clock, site or overlap legality. Algorithm 1 shows the three main phases of our wirelength recovery. In the first phase, large distance inter-tile movement of cells takes place. Using the median idea [15], the best bounding box (BB) of each cell is computed. The cells are then sorted in the non-increasing order of their distance from their best BB and processed in that order. For the cell being currently optimized, a procedure returns the list of all the tiles which have non-empty geometrical intersection with its best BB. This list is iterated while trying to place the cell near the center of the intersection of best BB and the tile currently being tried using routine *rotateAndPlace*. After this stage, all inter-tile movements have been done and we focus on intra tile wirelength minimization. The second phase improves adjacent columns in the tile to identify any horizontal movement of cells. This is achieved by looking at each consecutive pair of columns such that both are of the same type (DFF or LCELL). Further, if both the columns are DFF columns, their clock type should be same. For such a pair of row, horizontal swapping of a cell in first with the neighboring cell or space in the second column is tried. In the third phase, we look at three consecutive cells/spaces in each column starting from lowest vertical position and enumerate all six possible combinations and pick the best. In all the three phases, we check to see if a movement has caused WL increase and if so, we revert back the change. Our third phase is similar to the method outlined in [16] with the major difference being that we consider white-space also during explicit combination enumeration while FastPlace does not.

Algorithm 1 Wirelength Minimization

Ensure: Placement Is Legal

    { **Large Distance Movement** of Cells}

1: **while** WL Improvement $\geq \delta$ **do**
2:     Compute Best BB b_c for each cell c
3:     Sort cells according to distance from b_c
4:     **for all** cell c in sorted list **do**
5:         List lst = All tiles intersecting b_c
6:         **for all** tile t in list lst **do**
7:             rotateAndPlace(c, center of t's intersection with b_c)
8:             **if** could place in t **then**
9:                 **if** WL Improved **then**
10:                   break
11:                 **else**
12:                   continue
    { **Intra Tile Inter Column** WL Improvement}
13: **for all** tile t in platform **do**
14:     **while** WL Improvement $\geq \delta$ **do**
15:         **for all** adjacent columns $p1$ & $p2$ **do**
16:             **if** $p1$ and $p2$ are not same type or $p1$ & $p2$ have different clocks **then**
17:                 continue
18:             **for all** vertical positions in columns **do**
19:                 Swap cell-cell/cell-space pair adjacent in $p1, p2$
20:                 **if** WL NOT Improved **then**
21:                   Swap back to original position
    { **Intra Tile Intra Column** WL Improvement}
22: **for all** Tile t in platform **do**
23:     **while** WL Improvement $\geq \delta$ **do**
24:         **for all** Column p in t **do**
25:             Sort cells and spaces in p by vertical coordinate
26:             Try all 6 combination of 3 consecutive cells/space
27:             Pick the best configuration

---

5. EXPERIMENTAL SETUP AND RESULTS

All the above algorithms are implemented in C++ in our placer tool RegPlace (REGular PLACEr). All our experiments were run on a dual-core 3.3GHz 64-bit AMD linux workstation with 4GB RAM and 8GB swap. We used GLPK [17] as the ILP and network flow solver. The benchmarks were provided by eASIC as part of the placement challenge. The initial placement on the virtual platform were generated using two global placers: mPL [6] and CAPO [5].

Table 2: Benchmark Characteristics

Bench	#LCELL	#DFF	#REG	#BRAM	#CLK
easic1	832,824	87,052	110	172	18
easic2	812,200	45,478	175	686	7
easic3	961,063	52,780	192	0	3
easic4	102,038	23,330	0	44	7
easic5	913,853	84,505	145	262	26

5.1 Advantage of VP Generation

To demonstrate the efficacy of generation of virtual platform (VP) instead of solving the placement problem by blockages, we did the following experiment: Existing row-based placers were run on two benchmarks which were identical to each other except that in one case the placer was run on virtual platform and then mapped back to real platform whereas in the second case, the placer was run by considering blockages. In both the cases, all the cell types were converted to LCELLS, and only the space for LCELLS were available to the placers. Table 3 shows the WL results and time taken to complete the placement for several benchmarks.

Table 3: Advantage of using Virtual Platform (VP) compared to placement with blockages (WB) when using CAPO. Wirelength (WL) are in millions of units. All CPU time are in minutes.

	WL			CPU		
	VP	WB	Δ	VP	WB	Δ
easic1	11.1	11.1	0%	230	464	2.0X
easic2	21.3	21.2	-0.47%	231	476	2.1X
easic3	17.1	17.3	1.15%	374	639	1.7X
easic4	2.0	2.1	5%	29	133	4.5X
easic5	14.55	15.2	4.2%	247	516	2.1X
Average			2.6%			2.5X

Table 3 shows the advantage of using VP to migrate the discrete placement region into a contiguous region compared to the approach of specifying unplacable area as blockages. Overall, by using VP method, the wirelength and runtime improves by 2.6% and 2.5X respectively. On further analysis, we found that most of the extra time is spent by CAPO in overlap removal due to lots of overlaps of cells with the blockages.

While mapping the solution of placement on VP back to real platform, we re-introduce all the blockages. As pointed out in Section 4.2, the wirelength increase due to this step can be reduced by reducing the cut-size by swapping cells lying immediately on the both sides of the introduced blockage. To quantify this benefit, we performed cut-minimization during mapping step on our placement results of Table 3. Only the cells upto two unit distance away on both sides of the blockage re-insertion point were considered during cut minimization to reduce the performance overhead. On an average, we are able to further reduce the WL by 2% over the numbers in previous table.

5.2 Overall Wirelength

A preliminary version of RegPlace competed and won the eASIC placement contest [1]. In its current form, we have further improved the wirelength and runtime of RegPlace by 9% and 10% respectively. Table 4 shows the comparative data for our placer vs. other participants Team1 [18] and Team2 [19]. The WL numbers for Team1, Team2 are as

978-1-60558-497-3/09 $25.00 © 2009 ACM

Table 4: Comparison of our placer with other contestants. CPU numbers for Team 2 unavailable.

Bench	Team1		Team2		RegPlace (M)		RegPlace (C)	
	WL	CPU	WL	CPU	WL	CPU	WL	CPU
easic1	16.9M	3499	10.6M	N.A.	11.3M	13849	12.8M	14538
easic2	32.3M	2444	25.8M	N.A.	23.4M	9529	25.1M	9720
easic3	20.9M	2907	20.5M	N.A.	18.8M	6627	19.6M	7045
easic4	2.3M	216	1.8M	N.A.	2.0M	695	2.1M	735
easic5	20.4M	3055	14.4M	N.A.	13.9M	10197	16.6M	10479
Total	92.8M	12122	73.1M	N.A.	69.4M	40897	76.2	42517

reported in the contest results. However, since we have upgraded our placer, the runtime shown for Team1 and Team2 teams have been scaled by the proportion of runtime our initial version of placer takes on our workstation compared to the contest benchmarking workstations. To understand the impact of using different global placers to generate an initial solution on the virtual platform, we repeated our placer flow with two state-of-the-art placer: mPL and CAPO. The initials (M, C) in last four column names depict use of the above placers respectively.

From Table 4, several observations can be made. Our wirelength results improve significantly (10%) when the initial placement on virtual platform is generated using mPL rather than CAPO. Similarly, the runtime improves on using mPL. Compared to other teams, RegPlace beats Team1 in all the benchmarks irrespective of the initial placer solution being generated with CAPO or mPL. The wirelength reduction w.r.t. Team1 is as much as 33%. However, Team1 does have the advantage of being very fast (3.3X faster than RegPlace). RegPlace when used with mPL for initial placement, is better than Team2 in 3 out of 5 benchmarks as well as the total wirelength. Due to absence of runtime information of Team2, we cannot compare the runtime. The reduction in total wirelength compared to Team2 is 5%. These results show that RegPlace is a high quality solution for structured ASIC placement problem. Figure 4 shows the output of our placer for the benchmark easic2 where each cell type has been plotted with a different color.

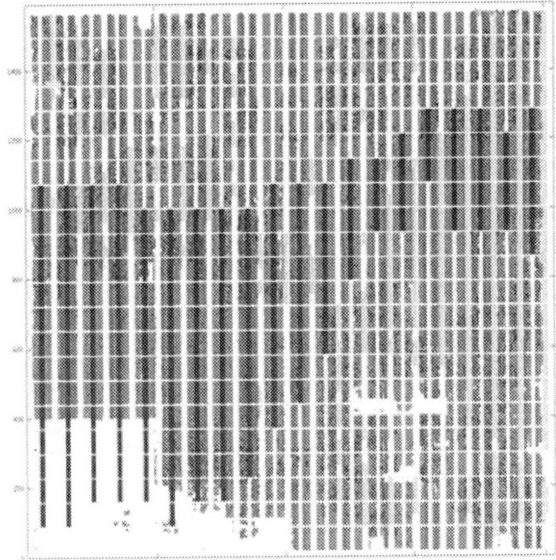

Figure 4: Final layout of benchmark easic2. Cells are shown as LCELL(red), DFF(black), REG(green), and BRAM(blue).

6. CONCLUSIONS

In this work, we proposed a new flow for efficient solution of placement problem for structure ASICs. The key novelty of this work are: a) concept of virtual platform for obtaining better initial placement solution, b) network-flow based inter-tile density correction, c) very fast ILP based clock constraint correction at both chip and tile level. Powered by these techniques, our placer won the eASIC placement challenge. In future, we propose to incorporate multiple density constraints in global placer which can significantly reduce the wirelength penalty associated with site and clock legalization.

7. REFERENCES

[1] "eASIC Corporate Website." http://www.easic.com/.

[2] H. Schmit, A. Gupta, and R. Ciobanu, "Placement Challenges for Structured ASICs," in *ISPD '08: Proceedings of the 2008 International Symposium on Physical design*, pp. 84–86, ACM, 2008.

[3] "SOC Central." http://www.soccentral.com/.

[4] A. Chakraborty, A. Kumar, and D. Pan, "Regplace," *http://www.cerc.utexas.edu/utda/download/RegPlace.htm*.

[5] J. A. Roy, D. A. Papa, S. N. Adya, H. H. Chan, A. N. Ng, J. F. Lu, and I. L. Markov, "Capo: Robust and Scalable Open-source min-cut Floorplacer," in *ISPD '05: Proceedings of the 2005 International Symposium on Physical design*, pp. 224–226, ACM, 2005.

[6] T. F. Chan, J. Cong, J. R. Shinnerl, K. Sze, and M. Xie, "mpl6: Enhanced Multilevel Mixed-size Placement," in *ISPD '06: Proceedings of the 2006 International Symposium on Physical design*, pp. 212–214, ACM, 2006.

[7] N. Viswanathan, M. Pan, and C. Chu, "FastPlace 3.0: A Fast Multilevel Quadratic Placement Algorithm with Placement Congestion control," in *ASP-DAC '07: Proceedings of the 2007 conference on Asia South Pacific design automation*, pp. 135–140, IEEE Computer Society, 2007.

[8] T. Luo and D. Z. Pan, "DPlace2.0: A Stable and Efficient Analytical Placement Based on Diffusion," in *ASP-DAC '08: Proceedings of the 2008 conference on Asia and South Pacific design automation*, pp. 346–351, IEEE Computer Society Press, 2008.

[9] "SOC Encounter tool." http://www.cadence.com/products/di/soc_encounter.

[10] V. Betz and J. Rose, "VPR: A New Packing, Placement and Routing Tool for FPGA Research," in *FPL '97: Proceedings of the 7th International Workshop on Field-Programmable Logic and Applications*, pp. 213–222, Springer-Verlag, 1997.

[11] K. Eguro, S. Hauck, and A. Sharma, "Architecture-adaptive Range Limit Windowing for Simulated Annealing FPGA Placement," in *DAC '05: Proceedings of the 42nd annual conference on Design automation*, pp. 439–444, ACM, 2005.

[12] A. C. Ling, D. P. Singh, and S. D. Brown, "Incremental Placement for Structured ASICs Using the Transportation Problem," *Very Large Scale Integration, 2007. VLSI - SoC 2007. IFIP International Conference on*, pp. 172–177, Oct. 2007.

[13] T. Okamoto, T. Kimoto, and N. Maeda, "Design Methodology and Tools for NEC Electronics' Structured ASIC ISSP," in *ISPD '04: Proceedings of the 2004 International Symposium on Physical design*, pp. 90–96, ACM, 2004.

[14] H. W. Kuhn, "The Hungarian Method for the Assignment Problem," *Naval Research Logistics Quarterly*, vol. 2, pp. 83–97, 1955.

[15] S. Goto, "An Efficient Algorithm for the Two-Dimensional Placement Problem in Electrical Circuit layout," *Circuits and Systems, IEEE Transactions on*, vol. 28, pp. 12–18, Jan 1981.

[16] M. Pan, N. Viswanathan, and C. Chu, "An Efficient and Effective Detailed Placement Algorithm," *Computer-Aided Design, International Conference on*, vol. 0, pp. 48–55, 2005.

[17] "GNU Linear Programming Kit." http://www.gnu.org/software/glpk/.

[18] "Michigan Wolverine Placer by Dongjin Lee and Igor Markov."

[19] "The Esteem Placer, written by Bob Erickson, an independent software consultant." http://www.linkedin.com/in/boberickson.

978-1-60558-497-3/09 $25.00 © 2009 ACM

A Novel Verification Technique to Uncover Out-of-Order DUV Behaviors

Gabriel Marcilio, Luiz C. V. Santos
Computer Sciences Department
Federal University of Santa Catarina, UFSC
Florianopolis, Brazil
{gaheris, santos}@inf.ufsc.br

Bruno Albertini, Sandro Rigo
Institute of Computing
University of Campinas, UNICAMP
Campinas, Brazil
{bruno.albertini, srigo}@ic.unicamp.br

ABSTRACT

Post-partitioning verification has to deal with abstract data, implementation artifacts, and the order of events may not be preserved in the DUV due to the concurrency treatment in the golden model. Existing techniques are limited either by the use of greedy heuristics (jeopardizing verification guarantees) or by black-box approaches (impairing observability). This work proposes a novel white-box technique that overcomes those limitations by casting the problem as an extended bipartite graph matching. By relying on proven properties, solid verification guarantees are provided. Experimental validation was performed upon platforms built around contemporary real-life applications.

Categories and Subject Descriptors

B.5.2 [**Design Aids**]: Simulation, Verification.

General Terms

Design, Verification

Keywords

White-Box Verification, Bipartite Graphs

1. INTRODUCTION

With the rise of *Electronic System Level* (ESL), the scope of verification has been extended and post-partitioning verification has come into play [7]. In such a scope, verification has to deal with distinct resolutions for data and timing (partially ordered events, cycle-approximate/accurate, etc). Timing and concurrency imposes challenges for post-partitioning design verification by allowing behaviors in the *device under verification* (DUV) that do not preserve the order specified in the *reference golden model* (RGM). Dealing with such out-of-order behaviors is probably one of the most complex aspects of ESL verification [7].

Permission to make digital or hard copies of part or all of this work for personal or classroom use is granted without fee provided that copies are not made or distributed for profit or commercial advantage and that copies bear this notice and the full citation on the first page. To copy otherwise, to republish, to post on servers or to redistribute to lists, requires prior specific permission and/or a fee.
DAC'09, July 26-31, 2009, San Francisco, California, USA

One of the first steps in an ESL flow is the creation of a behavioral (functional) design representation that obeys the *specification* (SPEC). This model should abstract from timing aspects, because the higher computational complexity of timed models makes HW/SW co-design prohibitive [3]. Besides, timed representations may lead to non-reusable testbenches [7] when design refinement changes timing. The need for an untimed behavioral model has two consequences: 1) Post-partititioning verification must check the order of events but not their timing; 2) The order of events may not be preserved in the DUV due to the concurrency treatment adopted in the behavioral model.

Unfortunately, the existing techniques are limited either by the use of greedy heuristics or by black-box approaches. That is why this work proposes a technique that *simultaneously* handles out-of-order DUV behaviors while monitoring DUV internals, without requiring timed models, RTL preview, or greedy heuristics to harmonize order mismatches. The problem is cast as an extended bipartite graph matching that takes event order into account and distinguishes implementation artifacts from actual errors.

This paper is organized as follows. Section 2 points out limitations of existing approaches on handling out-of-order behaviors. Section 3 shows an illustrative example. Sections 4 and 5 formalize the target problem and a few fundamental properties. Section 6 proposes algorithms and analyzes their verification guarantees. Experimental results are shown in Section 7 and our conclusions in Section 8.

2. RELATED WORK

The long history of verification and its crucial role in EDA resulted in a large amount of techniques for checking the functionality of an implementation with respect to its pre-specified behavior. Thorough surveys may be found in [2] [11], for instance. Despite the achievements of formal verification of out-of-order pipelines [6] [10], they are unlikely to replace simulation in ESL flows, since theorem-proving requires the manual construction of complex formulas and model-checking leads to large search spaces. Therefore, let us focus on ESL-compliant approaches to overcome the non-preserving order of DUV behaviors:

Use of cycle-accurate RGM [3]: A timed RGM induces the preservation of order but penalizes model performance and testbench reuse.

Ordering based upon RTL preview [12]: Whenever a decision is needed to address concurrent events, a sample ("trace") of the RTL implementation is synthesized on demand to support such decision so that the proper order is se-

a) SPEC	b) RGM	c) Latencies		d) DUV	e) BVG

Inputs: a, b, c, d
Outputs: x, y, z
Algorithm:
x = a + b;
y = c * x;
z = d - x;
	add r1, r2, r3
	// Event 0: r1, x
	
	mul r4, r5, r1
	// Event 1: r4, y
	
	sub r6, r7, r1
	// Event 2: r6, z
		Instruction / add / sub / mul	**Latency** / 1 cycle / 1 cycle / 4 cycles	add r1, r2, r3
				// Event 0: r1, x
				
				sub r6, r7, r1
				// Event 2: r6, z
				
				mul r4, r5, r1
				// Event 1: r4, y
					v_0^+ (r1, x) — (r1, x) v_0^-
					v_1^+ (r4, y) (r6, z) v_1^-
					v_2^+ (r6, z) (r4, y) v_2^-

Figure 1: The induction of out-of-order DUV behaviors

lected. The output differences between the functional and the RTL model can guide designers to harmonize behavior order. Since heuristics are needed to help distinguishing real defects from implementation artifacts [12], this approach cannot provide precise verification guarantees.

Use of scoreboard [7]: A scoreboard is a data structure that stores expected results (or injected stimuli) and compares the result obtained from the DUV to the stored result (or the result transformed from the stored stimuli). A scoreboard can address the problem of non-preserving order within its comparator. A serious limitation of scoreboards is their inability to verify DUV internals [7].

Although still under-exploited for verification, a recent technique [1] may potentially overcome such inability through computational reflection. That technique was judiciously crafted as a minimal-intrusive white-box enabler that can be used as implementation infrastructure.

Those limitations motivated us to develop an alternative approach that *simultaneously* handles out-of-order DUV behaviors while monitoring DUV internals, without requiring timed models, RTL preview, or greedy heuristics to harmonize order mismatches.

3. AN ILLUSTRATIVE EXAMPLE

Figure 1a shows a SPEC describing three partiallly ordered events. Assume that the adopted RGM is an untimed model that addresses concurrent events by simply serializing them in instruction fetching order (i.e. it is a simple instruction-set simulator). Figure 1b illustrates the effect of this untimed model (r, v represents the writing of a value v to a destination register r).

Suppose that the specified behaviors are implemented by an arithmetic co-processor (DUV) whose micro-architecture is based on Tomasulo's Algorithm [9]. Assume that two instances of a *monitor* m are attached to the RGM (m^+) and DUV (m^-) register-file write ports. Figure 1d shows the instruction order in the DUV. Event 0 occurs at the end of the first cycle, when the other two instructions are launched simultaneously. Due to the different latencies (Figure 1c), Event 2 occurs at the end of the second cycle, whereas Event 1 occurs at the end of the fifth cycle.

Every change in value observed by a monitor is recorded as an entry of a *log*. To check if every specified behavior in the RGM is implemented by the DUV, let us find a correspondence between compatible values in the logs of m^+ and m^-. To reason about them, let us build a bipartite graph, where each partition, say V^+ (V^-), represents the log of a monitor m^+ (m^-). An edge (v^+, v^-) means that the values associated with vertices v^+ and v^- are compatible. Figure 1e shows the resulting *bipartite verification graph* (BVG). To find classes of compatible values such that all specified

behaviors are exhibited by the DUV, we can search for a matching M.

Since the vertices v_i^+ (v_i^-) in partition V^+ (V^-) are ordered by the time t_i when a new value is monitored, the crossing between edges (v_1^+, v_2^-) and (v_2^+, v_1^-) represents an inversion in the order of events (with respect to RGM's) due to the distinct treatment of concurrency. To decide whether that inversion violates or not the SPEC, we rely on a relation R $=\{(v_0^+, v_1^+), (v_0^+, v_2^+)\}$, which captures the SPEC's partial order of events. Since $(v_1^+, v_2^+) \notin R$ and $(v_2^+, v_1^+) \notin R$, we can include edges (v_1^+, v_2^-) and (v_2^+, v_1^-) in matching M, since their choice does not violate the specified order, i.e. it preserves the *causality* of events.

Classical matching algorithms cannot directly tackle out-of-order DUV behaviors, since they do not assume a precedence relation R between the vertices of partition V^+, but only a compatibility relation between vertices of distinct partitions. That is why we conceived a suitable problem modeling and proposed new algorithms to tackle it.

4. PROBLEM MODELING

Suppose that a module has two representations, I^+ (RGM) and I^- (DUV), whose functional compatibility is to be verified for a given set of stimuli. Assume that k points of such module are observed through *monitors* $m_1, m_2, ..., m_k$.

Let t_0 denote the time when a monitor captures its initial value. Every time a monitor detects a change to a new value v at instant t_i ($i \neq 0$), we say that an event *event* v_i has happened. The value observed by a monitor is sampled each time it changes, within a given observation interval.

Definition 1: The *log of a monitor* m_k is a series of events $\langle v_{k0}, v_{k1}, ..., v_{ki}, ..., v_{kn} \rangle$ observed at times $\langle t_0, t_1, ..., t_i, ..., t_n \rangle$.

Definition 2: We say that two values u and v are *compatible*, written $u \approx v$, iff they are equal or their difference is bounded by a predefined range of values.

Definition 3: Let V^+ be the set of events monitored in the RGM. The *event precedence relation* is the set $R = \{(v_i^+, v_j^+) \in V^+ \times V^+ : v_i^+$ precedes v_j^+ in the SPEC$\}$.

The correlation between RGM and DUV is acceptable when every specified behavior is preserved and exhibited only once, while keeping value compatibility ($\approx$) and satisfying the specified order of events (R), as formalized below:

Definition 4: Given a monitor m, two representations I^+ and I^- of a module, and their respective sequences of events $\langle v_0^+, v_1^+, ..., v_n^+ \rangle$ and $\langle v_0^-, v_1^-, ..., v_p^- \rangle$, we say that there is a *proper mapping* of events, if there is a mapping $\mu\colon V^+ \rightarrow V^-$, where $V^+ = \{v_0^+, v_1^+, ..., v_n^+\}$ and $V^- = \{v_0^-, v_1^-, ..., v_p^-\}$, such that all the following holds:

- μ is an injection (completeness and uniqueness);
- $\mu(v_i^+) = v_j^- \Rightarrow v_i^+ \approx v_j^-$ (compatibility);

978-1-60558-497-3/09 $25.00 © 2009 ACM

- $\forall v_i^+, v_x^+ : ((v_i^+, v_x^+) \in R) \wedge (\mu(v_i^+) = v_j^-) \wedge (\mu(v_x^+) = v_k^-) \Rightarrow j < k$ (causality)

As implementation artifacts do introduce unspecified behavior ($|V^+| \leq |V^-|$), μ is not required to be a bijection.

Definition 5: I^+ is $\{m_k\}$-*compatible* with I^- iff there is a proper mapping for every monitor m_k.

Target Problem: Given two representations I^+ and I^- of a module and a set of stimuli, verify if they are $\{m_k\}$-compatible.

Definition 6: Let m^+ and m^- be instances of a monitor m whose logs are $\langle v_0^+, v_1^+, ..., v_n^+ \rangle$ and $\langle v_0^-, v_1^-, ..., v_p^- \rangle$. A *bipartite verification graph of a monitor* m, written $\mathrm{BVG}(m)$, is a bipartite graph $G(V, E)$, where:

- $V^+ = \{v_0^+, v_1^+, ..., v_n^+\}$, $V^- = \{v_0^-, v_1^-, ..., v_p^-\}$;
- There is an edge $(v_i^+, v_j^-) \in E$ iff $v_i^+ \approx v_j^-$.

Remind that a matching $\mathrm{M} \subseteq E$ is such that no vertex is incident to more than one edge in M. If an edge of M is incident to every vertex of partition V^+, the matching is *complete*. Recall also that a *connected component* (CC) is a subgraph $C_j(V_j, E_j)$ of G such that, for each $u, v \in V_j$, there is a path between u and v. Let us now formalize the key notions to solving the target problem.

Definition 7: A matching $\mathcal{M}$ of a $\mathrm{BVG}(m)$ is *proper* iff it induces a proper mapping μ.

Definition 8: Given two edges (v_i^+, v_j^-) and (v_x^+, v_k^-), with $i \neq x$, their *crossing function*, written $\chi((v_i^+, v_j^-), (v_x^+, v_k^-))$, returns true iff: $(((v_i^+, v_x^+) \in R) \wedge (j > k) \vee ((v_x^+, v_i^+) \in R) \wedge (k > j)))$.

Definition 9: We say that there is an *improper crossing* between edges (v_i^+, v_j^-) and (v_x^+, v_k^-), with $i \neq x$, iff $\chi((v_i^+, v_j^-), (v_x^+, v_k^-))$ is true.

Definition 10: Given an edge (v_i^+, v_j^-), we say that it is *causal*, written $\mathtt{causal}((v_i^+, v_j^-))$, iff $\forall (v_i^+, v_x^+) \in R, (v_x^+, v_k^-) \in \mathrm{M} : \chi((v_i^+, v_j^-), (v_x^+, v_k^-)) = \mathrm{false}$.

Note that the adopted notation distinguishes a subset of edges with a well-know property (a general matching M) from a subset of edges with additional properties (a more restrictive proper matching $\mathcal{M}$). The next section enumerates a few properties that allows us to tell whether an edge belongs or not to a proper matching.

5. PROPER MAPPING PROPERTIES

The notions formalized in Section 4 lead to important properties to be enforced while building a proper matching. Since their proofs are either simple or can be found in the literature, they are omitted due to lack of space.

Lemma 1: Identification of improper edge (IE) through improper crossing. Let $\mathcal{M}$ be a proper matching such that the edge $(v_i^+, v_j^-) \in \mathcal{M}$. If there is a vertex v_x^+ such that $\chi((v_i^+, v_j^-), (v_x^+, v_k^-)) = \mathrm{true}$, then $(v_x^+, v_k^-) \in E$ is improper.

Lemma 2: Identification of compulsory edge (CE). If there exists a proper matching $\mathcal{M}$ and there is a single edge $(v^+, v^-) \in E$ incident to v^+, then $(v^+, v^-) \in \mathcal{M}$ and this edge is said to be *compulsory*.

Lemma 3: Identification of IE induced by CE. If $(v^+, v^-) \in E$ is a compulsory edge, then all the remaining edges incident to v^- are improper.

Lemma 4: (Hall's Theorem [4]). Let $\mathrm{S} \subseteq V^+$ and $\mathrm{Adj(S)} = \{v^- \in V^- | ((v^+, v^-) \in E) \wedge (v^+ \in S) \}$. There exists a complete matching M iff $|S| \leq |Adj(S)|$ for all $\mathrm{S} \subseteq V^+$.

Definition 11: A CC, say $C_j(V_j, E_j)$, of a bipartite graph $G(V, E)$ is said to be *improper* iff the graph $C_j(V_j, E_j)$ does not satisfy Hall's Theorem.

Lemma 5: Identification of improper CC. Let $C_j(V_j, E_j)$ be a CC of a bipartite graph $G(V, E)$. If $|V_j^+| > |V_j^-|$ then there exists no proper matching for G.

Lemma 6: Identification of IE through improper CC. Let E' be the set of edges resulting in improper crossings with some $(v^+, v^-) \in E$. If a reduced graph $G'(V, E - E')$ has a CC, say $C_j(V_j, E_j)$, such that $|V_j^+| > |V_j^-|$, then (v^+, v^-) is improper.

Lemma 7: Identification of IE through uniqueness of behavior. Let E'' be the set of edges incident to v_k^- and different from a given edge $(v_i^+, v_k^-) \in E$. If the reduced graph $G'(V, E - E'')$ has a CC, say $C_j(V_j, E_j)$, such that $|V_j^+| > |V_j^-|$, then (v_i^+, v_k^-) is improper.

Lemma 8: Hall's Theorem corollary for equivalence relations. Let E be induced by an equivalence relation $\equiv$. There exists a complete matching M for $G(V, E)$ iff $|V_j^+| \leq |V_j^-|$ for each CC, say $C_j(V_j, E_j)$, of G.

6. THE PROPOSED TECHNIQUE

This section describes our algorithms and discusses verification guarantees by relying on Section 5's background. For simplicity, we omit a few auxiliary functions and assume a global scope for sets V, E, R, and M.

The overall key idea underlying the proposed algorithms is the following: assuming that a proper matching $\mathcal{M}$ might exist, build a matching M iteratively by selecting compulsory edges (CEs) and discarding improper edges (IEs) until M is complete (M = $\mathcal{M}$) or a complete matching cannot be found ($\mathcal{M}$ does not exist).

6.1 The underlying algorithms

Algorithm 1 $\mathtt{match\text{-}CE}((v^+, v^-))$

1: **if** $\mathtt{causal}((v^+, v^-))$ **then**
2:     $\mathrm{M} \leftarrow \mathrm{M} \cup \{(v^+, v^-)\}$
3:     $\mathtt{remove\text{-}IEs\text{-}crossing}((v^+, v^-))$ [Lemma 1]
4:     $\mathtt{remove\text{-}IEs\text{-}incident\text{-}to}(v^-)$ [Lemma 3]
5:     **return** true
6: **else**
7:     **return** false
8: **end if**

Algorithm 2 $\mathtt{remove\text{-}IE}((v^+, v^-))$

1: $E \leftarrow E - \{(v^+, v^-)\}$
2: **if** $|Adj(v^+)| = 1$ [Lemma 2] **then**
3:     **let** $\{u^-\} = Adj(v^+)$
4:     $\mathtt{match\text{-}CE}((v^+, u^-))$
5: **end if**

Algorithm 1 includes a compulsory edge (CE) in the matching, if it is causal. If a CE cannot be matched, M cannot be proper, and the algorithm returns false. When matched, a CE may induce improper edges (IEs), which are pruned according to Lemmas 1 and 3. For such removals, the functions at lines 3 and 4 invoke a primitive function $\mathtt{remove\text{-}IE}$, which is described by Algorithm 2. Since a removal can turn another edge into a CE, Algorithm 1 is recursively invoked through Algorithm 2 (at line 4).

978-1-60558-497-3/09 $25.00 © 2009 ACM

Algorithm 3 prunes IEs according to Lemmas 1 and 3 (line 3) and Lemmas 6 and 7 (line 9). It is recursively invoked (at line 10), since the removal of an edge (v^+, v^-) may turn other edges into CEs or IEs. Algorithm 3 returns false if at least one CE resulted unmatched (lines 4 or 11). When it returns true, the unpruned edges are either matched or eligible for a proper matching.

Algorithm 3 prune-IEs()

```
1: CEs ← find-CEs() [Lemma 2]
2: for all (v⁺, v⁻) ∈ CEs do
3:     if ¬ match-CE((v⁺, v⁻) [Lemmas 1 and 3] then
4:         return false
5:     end if
6: end for
7: for all (v⁺, v⁻) ∈ E do
8:     if implies-IE((v⁺, v⁻)) [Lemma 5] then
9:         E ← E - {(v⁺, v⁻)} [Lemmas 6 and 7]
10:        if ¬ prune-IEs() then
11:            return false
12:        end if
13:    end if
14: end for
15: return true
```

The function implies-IE, described by Algorithm 4, decides if (v_i^+, v_j^-) is an IE according to Lemmas 6 and 7. It first finds the edges that would not preserve causality (E') or uniqueness of behaviors (E''). Then it obtains the set of reduced edges E^* (line 3) resulting from speculatively matching (v_i^+, v_j^-). Next, it checks if a CC becomes improper as a result of such speculation (line 5). If so, the algorithm returns true to indicate that (v_i^+, v_j^-) is an IE. (The description of implies-improper-CC is omitted for simplicity, as it merely applies Lemma 5).

Algorithm 4 implies-IE((v_i^+, v_j^-))

```
1: E' ← {(v_x⁺, v_k⁻) ∈ E|χ((v_i⁺, v_j⁻), (v_x⁺, v_k⁻))}
2: E'' ← {(v_y⁺, v_j⁻) ∈ E|y ≠ i}
3: E* ← E - (E' ∪ E'')
4: for all (v_x⁺, v_k⁻) ∈ (E' ∪ E'') do
5:     if implies-improper-CC(v_x⁺, E*) [Lemma 5] then
6:         return true
7:     end if
8: end for
9: return false
```

Algorithm 5 match(v_i^+)

```
1: if Adj(v_i⁺) = ∅ then
2:     return false
3: end if
4: let v_j⁻ be an arbitrary member of Adj(v_i⁺)
5: M ← M ∪ {(v_i⁺, v_j⁻)}
6: E' ← {(v_x⁺, v_k⁻) ∈ E|χ((v_i⁺, v_j⁻), (v_x⁺, v_k⁻))}
7: E'' ← {(v_y⁺, v_j⁻) ∈ E|y ≠ i}
8: E ← E - (E' ∪ E'')
9: if (E' ∪ E'' ≠ ∅) ∧ (¬ prune-IEs()) then
10:    return false
11: end if
12: return true
```

Algorithm 5 matches a vertex v_i^+, performs the necessary bookkeeping, and re-launches pruning. An arbitrary

edge (v_i^+, v_j^-) is selected for the matching (line 5), possibly inducing non-causal edges (line 6) and edges leading to non-uniqueness of behavior (line 7), which are removed (line 8). (As Algorithm 4 is always invoked before Algorithm 5, such removal does not jeopardize the proper matching of any CC). Since edge removal may expose IEs or CEs, Algorithm 3 is invoked (line 9). Algorithm 5 returns false either if all the edges incident to v_i^+ were pruned (line 2) or if a CE cannot be matched (line 10).

Algorithm 6 proper-matching()

```
1: M ← ∅
2: if ¬ prune-IEs() then
3:     return (false, M)
4: end if
5: for all v_i⁺ ∈ V⁺ such that v_i⁺ is unmatched do
6:     if ¬match(v_i⁺) then
7:         return (false, M)
8:     end if
9: end for
10: return (true, M)
```

Algorithm 6 is the entry point of the proposed technique. Given a BVG, it builds a proper matching M = $\mathcal{M}$ (if any) or otherwise returns false to indicate that M $\neq \mathcal{M}$. First, it removes as many IEs and matches as many CEs as possible (line 2). Then it attempts to match every vertex left unmatched (line 6). As soon as a vertex cannot be matched (lines 2 and 6), the algorithm returns false.

6.2 Verification guarantees

Let us refer to a verification outcome as a false negative when no fault is detected, but actually exists. Conversely, an outcome is a false positive if an apparent fault is detected, but none exists. The following theorems provide grounds for false positive and false negative analysis.

Theorem 1: If Algorithm 6 returns true for every monitor m_k, then I^+ and I^- are $\{m_k\}$-compatible.

Proof: Suppose that Algorithm 6 is applied to BVG(m_k). Since it returns true only if match(v_i^+) = true for every unmatched v_i^+, then we have M = $|V^+|$ and, therefore, the first clause of Definition 4 holds. Since M $\subseteq E$ and E captures by construction the relation $\approx$, the second clause of Definition 4 also holds. Thus, to prove that Algorithm 6 induces a proper mapping, we have to show that the third clause of Definition 4 holds, i.e., that a non-causal edge is never matched. Only two functions match edges: match-CE(v^+, v^-) matches each edge under the precondition causal(v^+, v^-) = true; match(v^+) matches each edge under the precondition prune-IEs() = true, which holds when $\forall (v^+, v^-) \in$ M: match-CE(v^+, v^-) = true, which in its turn holds when $\forall (v^+, v^-) \in$ M: causal(v^+, v^-) = true. Thus, M = $\mathcal{M}$. Since, by hypothesis, this holds for every monitor, we conclude that I^+ and I^- are $\{m_k\}$-compatible. □

Theorem 1 guarantees that, within the range of monitored values, our technique never results in false negative.

Theorem 2: If Algorithm 6 returns false for some monitor m_k, then I^+ and I^- are not $\{m_k\}$-equivalent.

Proof: Assume, by absurd, that a proper mapping actually exists for a given monitor m_k, but was overlooked by Algorithm 6. This could only happen if at least one proper edge was excluded from M. Only two functions exclude edges: match(v_i^+) only excludes edges provenly improper by Lemmas 1 and 3 (Algorithm 5, lines 6 and 7);

978-1-60558-497-3/09 $25.00 © 2009 ACM

Table 1: Benchmark and platform characterization

Application	Components	Function
Blowfish [8]	PPC+MEM+IP	f-function
SHA-0 [8]	PPC+MEM+IP	sha-transform
JPEG [8]	PPC+MEM+IP	dct
Tomasulo [9]	PPC+MEM+IP	coprocessor
MP3 [5]	2xPPC+3xMEM+2xIP	idct

Table 2: Fault characterization

Application	Monitor	Location	Fault Description
Blowfish [8]	m_1	output	An S-box entry was assigned the same value as a neighbor's
SHA-0 [8]	m_2	output	Number of required invocations was decremented by one
JPEG [8]	m_3	output	Value read from a register was shifted 1 bit to the left
Tomasulo [9]	m_4 / m_5	RFWP / IU	Instructions issued without checking for operand availability
MP3 [5]	m_6	output	Multiplier-accumulator performing multiplication only
	m_7	output	Output multiplexer inverting Layer 2 and Layer 3 frames
	m_8	output	Values sampled one clock cycle later than expected

Table 3: Verification results

R	Monitor	No fault in DUV						Fault induced in DUV					
		$|E|$	$|V^+|$	$|V^-|$	$|E_p|$	$|M|$	time [s]	$|E|$	$|V^+|$	$|V^-|$	$|E_p|$	$|M|$	time [s]
none	m_1	8398	8367	25104	0	8367	0.2	80	8367	25104	0	80	0.02
	m_2	4873	4873	4873	0	4873	0.1	4872	4873	4872	0	4872	0.1
	m_3	685	37	37	0	37	0.02	361	37	37	0	19	0.01
total	m_1	8398	8367	25104	31	8367	0.2	80	8367	25104	0	80	0.02
	m_2	4873	4873	4873	0	4873	0.1	4872	4873	4872	0	4872	0.1
	m_3	685	37	37	648	37	0.02	361	37	37	342	19	0.02
partial	m_4	4136	440	440	3696	440	8.5	3550	440	440	3483	67	8.3
	m_5	15200	440	440	14760	440	29	15200	440	440	14824	376	30
	m_6	2666	2632	2632	34	2632	0.04	52	2632	1813	48	4	0.01
	m_7	2666	2632	2632	34	2632	0.04	308	2632	1819	290	18	0.03
	m_8	2666	2632	2632	34	2632	0.04	2869	2632	2628	310	2559	0.05

prune-IEs() excludes edges not only under the same Lemmas 1 and 3 (Algorithm 3, line 3), but also under Lemmas 6 and 7 (Algorithm 3, line 9). Since, through Lemma 8, Lemmas 6 and 7 become necessary and sufficient conditions under an equivalence relation $\equiv$ (due to transitivity), no proper edge is ever excluded. Thus, if Algorithm 6 returns false, there is no proper mapping $\mathcal{M}$ and we conclude that I^+ and I^- are actually not $\{m_k\}$-equivalent. $\square$

Theorem 2 guarantees that, within the range of monitored values, our technique never results in false positive when verifying an equivalence relation between RGM and DUV (for instance, when monitoring integer type attributes). Note that, since Lemmas 6 and 7 establish sufficient (but not necessary) conditions, proper edges may be overlooked under a more general compatibility relation $\approx$. To overcome this limitation, Algorithm 3 would have to use Lemma 4 at line 8, which would be computationally inefficient, as it requires the enumeration of all subsets of partition V^+.

7. EXPERIMENTAL RESULTS

Tables 1 and 2 summarize the experimental set-up. Table 3 shows the results under distinct scenarios: unordered, totally, and partially ordered events (the first two scenarios are unlikely, but they grant wider-range validation). IPs to be verified were cast as modules of SystemC platform representations. Platforms contains one or more PowerPC (PPC) processors, buses and memory (MEM) modules. Two distinct models were used for each IP: the DUVs were func-

tional, cycle-approximate representations, while their RGM counterparts were purely functional, untimed models. Run-times were measured at 2.66 GHz on a Quad-Core workstation with 4GB of memory (DDR-2, 667 MHz).

Observe that, before faults were induced, all specified behaviors in the RGM were reproduced by the DUV ($|M| = |V^+|$) for every monitor. Note that, for monitor m_1, the number of unspecified behaviors ($|V^-| - |V^+|$) in the DUV is larger than the number of behaviors specified in the RGM ($|V^+|$). This indicates that implementation artifacts were introduced as a result of refinement. Note also that, as a consequence of the induced faults, no DUV exhibited all specified behaviors ($|M| < |V^+|$). Observe that the more severe the fault, the lower the matching cardinality. The fact that edges not in M were pruned ($|E_p| = |E|$ - $|M|$) is an evidence that the technique has found values that, although compatible, would subvert the specified order of events.

In the first scenario, no edges could be removed ($|E_p|$=0), since our pruning relies on causality violation induced by event ordering. In the second scenario, on the contrary, pruning had some effect for monitor m_1 (under no fault) and a substantial impact for monitor m_3 (with and without faults). Obviously, pruning had no effect when all edges were compulsory (as in the corner case captured by monitor m_2) or when the severity of the fault trivialized the compatibility relation (monitor m_1 under fault).

Let us now focus on the third scenario. To induce out-of-order execution in the DUV, we selected a dynamic single-

978-1-60558-497-3/09 $25.00 © 2009 ACM

issue representation for an arithmetic coprocessor that uses a simplified version of Tomasulo's Algorithm [9]. Note that, in this case, the order of events is dictated by data and control dependencies. Two monitors were inserted inside the IP: m_4 at the register-file write port (RFWP) and m_5 at the issue unit (IU) output. As a consequence of the induced fault, although all instructions were launched ($|V^-| = |V^+|$), the execution order of some instructions in the DUV did not satisfy the partial order and this was reflected into sheer pruning (14760/14824 out of 15200 edges) for the monitor m_5. Most pruned edges represent RAW or WAR hazards induced by the fault. As a result of hazards, wrong values were written in the DUV's register file, which was also detected by our technique, giving rise to a matching of low cardinality ($|M| = 67$) for monitor m_4.

For MP3, faults of distinct severity levels were induced. Under the most severe fault (m_6), only 2% of the compatible values ($|E|$) remain in the verification space; the fault is easily detected since most of them are pruned. Under a milder fault (m_7), 11% of the verification space is kept. On the contrary, under the most subtle fault (m_8), induced by lack of synchronism, the verification space slightly increases (2869/2666) and only 11% of it can be pruned, making the fault harder to uncover. In all MP3 experiments exhibiting out-of-order DUV behaviors, the technique was able to distinguish them from actual errors.

8. CONCLUSIONS AND PERSPECTIVES

Our technique rules out heuristics that limit the verification space. Its pruning relies solely on proven properties. Validation combined theoretical proofs and experimental results upon real-life applications. The worst rates for the current implementation are 34ms/vertex or 2ms/edge, although there are opportunities for further improvement. As opposed to scoreboards, which make local decisions, our technique analyzes logged information globally. As future work, we intend to elaborate a more detailed (quantitative) comparison with scoreboards.

9. ACKNOWLEDGMENTS

The authors would like to thank the Brazilian Research Agencies that partially supported this work either by funding the co-operative projects CAPES/PROCAD 0326054 and 0018058 or by individual grants CNPq/PQ 300977/2006-9 (2nd author) and FAPESP 07/58129-8 (3rd author).

10. REFERENCES

[1] B. Albertini, S. Rigo, G. Araujo, C. Araujo, E. Barros, and W. Azevedo. A Computational Reflection Mechanism to Support Platform Debugging in SystemC. In *Proc. of the International Conference on Hardware/Software Codesign and System Synthesis (CODES+ISSS)*, pages 81–86, 2007.

[2] B. Bailey. *The Functional Verification of Electronic Systems: An Overview from Various Points of View.* International Engineering Consortium, 2005.

[3] F. Ghenassia. *Transaction-Level Modeling with SystemC: TLM Concepts and Applications for Embedded Systems.* Springer, 2005.

[4] M. Hall. An Algorithm for Distinct Representatives. *American Mathematical Monthly, No. 10,* 63:716–717, 1956.

[5] ISO IEC 11172-3 Standard Implementation. ftp://ftp.tnt.uni-hanover.de/pub/mpeg/audio/mpeg2/software.

[6] C. Jacobi. Formal Verification of Complex Out-of-Order Pipelines by Combining Model-Checking and Theorem-Proving. *Lecture Notes in Computer Science,* 2404:211–226, 2002.

[7] G. Martin, B. Bailey, and A. Piziali. *ESL Design and Verification: A Prescription for Electronic System Level Methodology.* Morgan Kaufmann, 2007.

[8] Mibench suite. http://www.eecs.umich.edu/mibench.

[9] D. A. Patterson and J. L. Hennessy. *Computer Organization and Design: The Hardware/Software Interface.* Morgan Kaufmann, 2007.

[10] J. U. Skakkebaek, R. B. Jones, and D. L. Dill. Formal Verification of Out-of-Order Execution with Incremental Flushing. *Lecture Notes in Computer Science,* 1427:98–109, 1998.

[11] B. Wile, J. Goss, and W. Roesner. *Comprehensive Functional Verification: The Complete Industry Cycle.* Morgan Kaufmann, 2003.

[12] J.-S. Yim, C.-J. Park, W.-S. Yang, H.-S. Oh, H.-C. Lee, H. Choi, T.-H. Kim, S.-J. Lee, N. Won, Y.-H. Lee, I.-C. Park, and C.-M. Kyun. Verification Methodology of Compatible Microprocessors. In *Proc. of the Asia and South Pacific Desing Automation Conference*, pages 173–180, 1997.

978-1-60558-497-3/09 $25.00 © 2009 ACM

Shortening the Verification Cycle with Synthesizable Abstract Models

Alon Gluska
Intel MMG
MATAM, Haifa, Israel
+972-4-8656088
alon.gluska@intel.com

Lior Libis
Intel MMG
MATAM, Haifa, Israel
lior.libis@intel.com

ABSTRACT

Abstract modeling has been widely used, albeit independently, for both formal verification and high-level modeling of SoC designs. In this paper we show that proper selection of modeling language and abstraction level can make the same code useful for both formal and simulation-based techniques. The abstract model enables architecture exploration and the development of verification collateral pre-RTL, and can be used as a behavioral checker in simulation against the RTL and in hardware emulation. In parallel, it enables applying formal verification techniques to verify the specification and implementation of the design. We provide examples of the successful application of abstract models developed in SystemVerilog in the course of the verification of the newest Intel® Core™ microprocessor.

Categories and subject descriptors

B.5.2 [**Verification**]: Automatic synthesis, Hardware description languages, Verification.

General terms: Design. Verification.

Keywords: Logic design. Verification. Abstract Modeling.

1. INTRODUCTION

The use of abstract modeling for verification is a well-known technique in verification. The abstract model acts as an executable specification of the hardware to be designed, and thus is used, pre-RTL, for architecture exploration and software development, typically for large SoC designs. Independently, removal of the irrelevant details of the RTL creates an abstract model which enables proving properties using formal verification techniques.

In this paper we present a novel approach that seeks to leverage simultaneous use of the abstract models for both formal and simulation-based verification. This approach advocates using the very same models for early development of simulation testbenches and verification collateral, and later on as checkers in RTL simulation and emulation, while in parallel using them as formal reference against the RTL in order to formally prove properties of the RTL. Simultaneous use of the same abstract models for both simulation and FV requires a proper selection of abstraction level and language that may compromise some aspects that are optimized for only one of those areas. In the course of the

Permission to make digital or hard copies of part or all of this work for personal or classroom use is granted without fee provided that copies are not made or distributed for profit or commercial advantage and that copies bear this notice and the full citation on the first page. To copy otherwise, to republish, to post on servers or to redistribute to lists, requires prior specific permission and/or a fee.
DAC'09, July 26-31, 2009, San Francisco, California, USA

verification of the newest Intel Core microprocessor, we chose the SystemVerilog language, primarily because of its ability to be easily synthesized, a fact that makes it applicable both for formal verification and for emulation [5]. This compensates for its weaker abstraction capabilities as compared to C++ based languages, which are very difficult to synthesize [6]. We chose a relatively low level of abstraction to improve the effectiveness of the abstract models for FV and as behavioral checkers for post-RTL simulation and emulation. However, the models were still abstract enough to identify holes and flaws in the specification. We will discuss our considerations for the selection of the modeling language and abstraction level later.

As we will show, this approach proved to be very successful. In the pre-RTL stage, we used the abstract executable models to expedite the development of the verification testbench and simulation collateral. This included in particular the development of large set of assertions on model inputs (assumptions) and internal behaviors (properties) that were to be tested in simulation and enabled the cleanup of the abstract model. In addition, we applied formal property verification (FPV) to verify the properties, given the assumptions. This enabled further cleanup of the abstract models and the creation of a mature verification platform pre-RTL. In the post-RTL stage, we used the abstract models as behavioral checkers during simulation. Some of the abstract models were used as reference for Formal Verification as well. This enabled the RTL designers to find a large number of RTL bugs prior to the release of the code to the verification engineers, and had a critical role in in-depth verification post-release. Finally, the synthesizable abstract models were easily integrated into the emulation of the design, making for minimal impact on performance.

This paper is structured as follows. We first briefly summarize the use of abstraction for high-level modeling in SoC verification and for FV. We then present our approach, which combines the use of such verification models for both simulation and FV. We discuss the selection of the abstraction language, in particular comparing the use of the standard languages SystemVerilog, SystemC, and Specman/e for abstraction. We briefly discuss alternatives for abstraction levels, with their advantages and drawbacks. Finally we present a number of examples and results from the verification of the microprocessor.

2. BACKGROUND

The following section briefly describes the verification gaps that abstraction copes with, how abstraction is applied to simulation (primarily in the scope of SoC designs in the pre-RTL stage), and how it is applied for formal verification of RTL modules.

978-1-60558-497-3/09 $25.00 © 2009 ACM

2.1 Verification crisis and gaps

Verification already lies on the critical path towards the completion of many designs, and its complexity limits the content of products and impacts tape-out schedules [1][2]. While the size, and hence complexity, of designs are increasing, market time windows are decreasing; the result is a rapid increase in the cost of failures [2]. To cope with growing complexity, verification engineer headcount growth has overtaken that of design engineers. In spite of this, the rate of first silicon success is steadily declining [1]. This trend is expected to continue and further increase the verification crisis and its impact on design cost, content, and schedule.

In current industrial practice, verification consists of generating and simulating large numbers of random tests on an RTL representation of the design. Formal verification of properties can also be applied to selected RTL modules in order to provide superior verification for these modules. However, in advanced designs such as microprocessors, the RTL includes, in addition to functional specifications, many non-functional implementation details which support meeting aggressive circuit timing, power, and area requirements. This means that for the purposes of functional verification, an RTL model simply holds too much information. This slows down simulation, complicates FV, and adds complexity, which is reflected by the late discovery of a large number of functional bugs. The slow pace of simulation prevents adequate testing of some complex scenarios, and their verification is targeted late in emulation or even delayed until silicon.

Moreover, current RTL-based verification methodology depends on the availability of the RTL model. As a result, development and cleanup of verification testbenches and collateral, as well as of formal proofs, are delayed until the RTL code is released to verification. The RTL code cannot be adequately tested by the RTL designers because of the incomplete verification environment. The result is that verification is delayed, and functional bugs that might have been found by designers are delivered to verification, slowing down the progress of the project even further. In addition, the sharing of collateral between FV and simulation is hard to achieve. [3]

Abstraction copes first and foremost with the verification gaps created by the dependence on the completeness of the RTL. Appropriate abstractions increase the comprehensibility of the design and enable the development of verification collateral and the application of verification techniques over the abstract model in the pre-RTL stages. They also serve as reference models for the RTL in post-RTL verification. [4]

2.2 The role of abstraction in simulation

Using abstract models in pre-RTL has become an established approach at most of the world's leading System-on-a-Chip (SoC) design companies, and is increasingly used in system design. Models are used for architecture exploration, early system verification development, and development of software. Enriched with implementation details, abstract models can serve in post-RTL development stages as reference against the RTL; verification collateral from the pre-RTL stage can frequently be used as-is.

The level of abstraction varies from transaction-level modeling (TLM), which describes the data flow, through cycle-approximate modeling and all the way down to complete cycle-accurate modeling. Even at this latter level many of the specific implementation details frequently included in the RTL (such as timing, power, and area details) are omitted, a fact which enables simulation to run between 20x to 1000x faster than RTL simulation, depending on the level of abstraction.

The level of abstraction used for architecture exploration is usually too high to be effectively reused for the verification of the derived logic. Architectural exploration and software development require compact, fast, latency-accurate high-level un-timed models. Verification of the logic design, on the other hand, requires functional completeness and the ability to correlate the abstract models with the RTL implementation (for formal verification or for use as checkers in simulation), requirements that make such models too slow for software development and arch exploration. These conflicting requirements inhibit the reuse of highly abstract models for logic verification.

Correctly selecting the level of abstraction to be used for the verification collateral in simulation can simplify the development of the verification collateral itself, increase the debug efficiency of failures, reduce the number of released bugs, and significantly reduce the testing space at the system level.

2.3 The role of abstraction role in formal verification

Formal Verification (FV) also employs an abstract mathematical model of the logic, primarily in model checking techniques. Model checking consists of a systematic exhaustive exploration of all states and transitions in the model using domain-specific abstraction techniques to consider whole groups of states in a single operation, thereby reducing computing time. The abstraction of the model reduces the number of states necessary to perform the formal proof while maintaining the functionality of the original model with respect to the specifications to be verified. The meaning of this is that abstraction makes it practical to apply FV on significantly larger designs. The abstract model can be extracted automatically by the FV tools, or be developed independently.

Most formal verification tools used in the hardware industry still operate on a low-level net-list representation, where the design is first synthesized and flattened into a net-list. This requires that the source design representation be synthesizable. To date, synthesis tools are tuned to operate on standard RTL languages, but are not yet mature with respect to languages with higher abstraction capabilities such as SystemC, although some preliminary work has recently been presented (e.g. [8] [10]).

In [7] a different approach is presented. There, FV is applied pre-RTL on a TLC abstract model in order to verify the correctness and completeness of the specification. Although our approach does not target the complete cleanup of the specification, applying FV techniques on the abstract model pre-RTL makes it possible to find flaws in the specification that otherwise would only be found much later.

978-1-60558-497-3/09 $25.00 © 2009 ACM

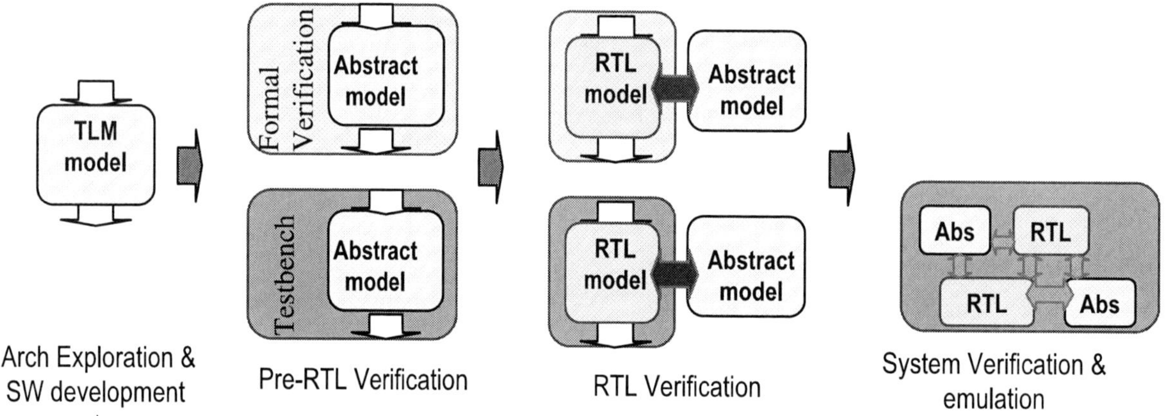

Arch Exploration &
SW development

Pre-RTL Verification

RTL Verification

System Verification &
emulation

Figure 1: Proposed flow for sharing abstract models along the design flow

For our purposes, the abstract model can serve as a formal reference for model checking against the RTL. In addition, a formal proof of properties given assumptions can be developed on the early abstract model and later applied to the RTL model.

3. OUR PROPOSED FLOW: REUSE OF ABSTRACT MODELS

We believe, and demonstrate by example, that carefully selecting the abstraction level used for verification enables leveraging most of the benefits of both simulation and formal verification using the same verification collateral. The proposed flow is presented herein.

A schematic flow of the use of abstract models for simulation is illustrated in Figure 1. The un-timed TLM models are used for architectural exploration and software development. These models include implementation hooks that are added and are applicable for RTL verification only. During RTL verification, these abstract models serve as checkers against the RTL, using the hooks as references. Finally, in system verification and in emulation of the design, the abstract models can be used either as checkers or instead of the corresponding RTL components in order to achieve faster integration and speed.

The proposed flow places the following requirements on the implementation of the abstract models:

- **Partition** of the abstract model should conform to the partition of the design, in order to enable substituting the RTL module with its corresponding abstract model. This also enables leveraging FV proofs from the abstract model for the RTL model.

- **Level of abstraction** should be relatively low in order to facilitate root causing bugs, both earlier and closer to the place where they occur. However, it should be high enough to enable the fast simulation runs required for software development and architecture exploration. This can be achieved through compilation "ifdef" sections that replace un-timed transaction-level blocks with concrete implementation-dependent logic.

- **Language** should be synthesizable in order to enable the use of the abstract model for FV and for fast emulation.

4. CONSIDERATIONS IN ABSTRACTION AND OUR DECISIONS

In this section we discuss our considerations regarding the application of abstract models.

Partition matching with RTL was a mandatory requirement. It enables reuse of the TLM models for verification. It also soon became evident that the abstract models must be **synthesizable**. This makes it possible to apply FV techniques to verify them and later on to use them against the RTL. It also makes it easier to integrate them into the emulation model with virtually no impact on performance.

Abstraction level and **language**, however, were subject to different considerations.

4.1 Use of synthesizable models

First for foremost, synthesizable models enable applying FV and Assertion Formal Verification (FV based solely on assertions as assumptions and properties, hereinafter referred to as AFV) in order to prove properties on the abstract models during *Pre-RTL Verification*. Standard FV techniques such as model checking or symbolic simulation can be applied. Such FV proofs consist of assumptions on the inputs to the logic for formal proofs of properties. These assumptions and properties are frequently developed anyway as assertions for RTL verification. Their early use for the verification of the abstract models improves their accuracy and pulls in the maturity of the verification collateral. Abstract models are typically far smaller than the corresponding RTL, thereby enabling FV tools to cope with large pieces of code effectively. Our experience shows that FV can be effectively applied to cycle-accurate abstract models that cover 3-5x more functionality than RTL. Naturally, during *RTL Verification*, abstract models can be reused as reference for formal proofs.

Functional and program-based coverage can also be very effective during *Pre-RTL Verification* for measuring the effectiveness of the testbench in providing adequate stimulation of the micro-architectural characteristics of the module. This can reveal bugs, holes, or imbalances in the implementation of the testbench or in the coverage monitors. During *RTL Verification*, the same coverage monitors can be reused to ensure adequate stimuli in RTL simulation. Most standard functional coverage tools can

978-1-60558-497-3/09 $25.00 © 2009 ACM

refer simultaneously and seamlessly to events from both RTL and the abstract models.

In the post-RTL stage, abstract models function as checkers against the RTL implementation. Being synthesizable, such checkers can be seamlessly integrated within the emulation models with a drastically lower impact on performance than checkers that run externally and sample RTL nodes via an API.

Finally, synthesizable abstract models are easier to debug because they are usually natively supported by RTL debuggers.

4.2 Selection of Abstraction Level

The abstract model acts as an executable specification of the hardware and can be implemented at different levels of abstraction. Models of higher levels of abstraction, such as the transaction-level, are smaller, easier to develop, run faster in simulation, and are less sensitive to changes in implementation. Such models are very useful for architecture exploration, software development, and early system development. In many cases, the high-level abstract models do not need to be complete or accurate, as long as they enable measuring the important and frequent characteristics of the design. Low-level models provide better correlation with the implementation and can therefore serve as effective checkers against the RTL. For the latter purpose, completeness and accuracy of the models are critical. Because RTL simulation is slow anyway, the impact on simulation is usually not a primary area of concern.

Because the abstract models targeted a range of purposes, we chose a medium level of abstraction. The models developed were abstract enough to facilitate exploration of the specification of the design, and detailed enough to adequately serve as a reference for FV simulation against the RTL implementation. Our goal was to achieve a high degree of abstraction while maintaining low-level cycle accuracy for pre-defined key states, as follows:

- Algorithms were implemented in high-level procedural code using modules and functions to emulate classes. This high level of abstraction was transaction based, functionally reuseable, devoid of timing and staging constraints.

- Code for staging was done using flipflops and latches, utilizing SV's ability to stage not only simple constructs such as signals but also complex constructs such as structs. This low-level staging targets cycle accuracy, matching the low-level implementation of the RTL to the verification hooks.

- Checking consists of a set of hooks, that is, key states identified as being of known logical behavior, either internally or on logic boundaries. SVA assertions were used to compare the RTL hooks and the checker implementation of the same set of hooks.

4.3 Selection of Abstraction Language

In the selection process we analyzed and compared the following languages:

- **SystemC** - the de-facto industry standard language for high-level abstraction

- **Specman/e** - an established standard verification language

- The synthesizable portion of **SystemVerilog** (SV/RTL) - an RTL modeling language that supports abstraction constructs

- **SVTB** – SystemVerilog Testbench, the verification subset of the SystemVerilog language

Our findings were as follows:

- **Abstraction expressiveness** - SV/RTL was found to be inferior to all the other languages. Synthesizable SV only provides structs and modules, and lacks inheritance and polymorphism available in SystemC and Specman/e.

- **Cycle accuracy modeling** - Specman/e and SVTB were weaker. They are procedural languages and do not support even simple constructs for synchronous logic such as flops and latches.

- **Footprint** - SystemC and SV/RTL provide a smaller footprint, which means fewer bugs, lower development costs, and greater readability, due to the fact that synchronous constructs make it easier to focus on the algorithmic parts only. Slightly modified instances are better supported in SystemC.

- **Event based vs. procedural writing** – Languages which support event-based writing, such as SV, extract the order of statements from the context of the code. Specman/e and SVTB are procedural, and therefore the topological order has to be maintained by the code developer. In SystemC, order can be derived from a manually-written sensitivity list. For example, SV interprets the following code:

```
always_comb begin
      c = a | d;
end
always_comb begin
      a = b | f;
end
```

and executes the assignment to **a** before the assignment to **c**, in spite of the reversed topological order. In Specman/e (and similarly in SystemC) topological order must be kept for the algorithm to work, i.e.:

```
algorithm() : is {
      a = b | f;
      c = a | d;
};
```

This is significant when it comes to breaking large and complex algorithms into small and simple reusable sections. Manually maintaining the topological order of the constituents while at the same time maintaining the high level of abstraction can become very difficult. It should be noted however, that debug of event-based code is usually harder, and the code is harder to read.

- **Synthesis** - Of all these languages, SV/RTL alone can be easily synthesized into a net-list as required for FV or emulation. As pointed out in [6][9][10], adapting software formal verification techniques to SystemC is a formidable task, mainly because of its object-oriented nature and its support for both synchronous and asynchronous semantics with a notion of time. Specman/e and SVTB cannot be synthesized.

In conclusion, all the languages enable the development of abstract models to various levels of effectiveness, compaction and footprint. However, we came to the conclusion that the ability to synthesize SV/RTL was the most important consideration as an enabler for FV and emulation. As a result, we chose SV/RTL as our modeling language.

5. APPLICATION OF ABSTRACT SYNTHESIZABLE MODELING

5.1 Background

We successfully used synthesizable abstract models in the course of the verification of a new Intel® Core™ microprocessor. We developed abstract models in SV for seven modules in the design that were considered complex and risky. These models were developed in parallel with architectural exploration, and were used at early stages for the development of testbenches and FV verification collateral. Later on they were turned into checkers that run in simulation and emulation against the RTL.

The examples below demonstrate some of the advantages of using synthesizable abstract modeling. In particular:

- Early cleanup of the micro-architecture specification.

- Pull-in of the verification flow by early development and cleanup of verification collateral using the abstract models as the DUT.

- Early cleanup of the abstract models using simulation, functional coverage, and FV.

- Prevention of many bugs from being released to verification through early discovery by the RTL designers pre-release.

- Reuse as behavioral model checkers for simulation and emulation, and as formal specs for bounded model checking against the RTL.

5.2 Example 1: Development of RC testbench using RC abstract models and use of abstract models for bug prevention and early bug finding

RC is a new, large, and complex cache cluster designed to improve power and performance in the newest Core™ microprocessor design. It consists of four functional units: a cache Data, a Fill mechanism to fill it up, and Control (RC-C) and Queue (RC-Q) modules to extract and manage the stored data.

The cluster level testbench of the RC was first built around abstract models of the RC-C and the RC-Q as the DUT (see Figure 2). The abstract models were used to develop and clean up the testbench and verification collateral (such as tests and basic coverage monitors) several months before the corresponding RTL was released to verification. It also enabled the early development of assertions for the

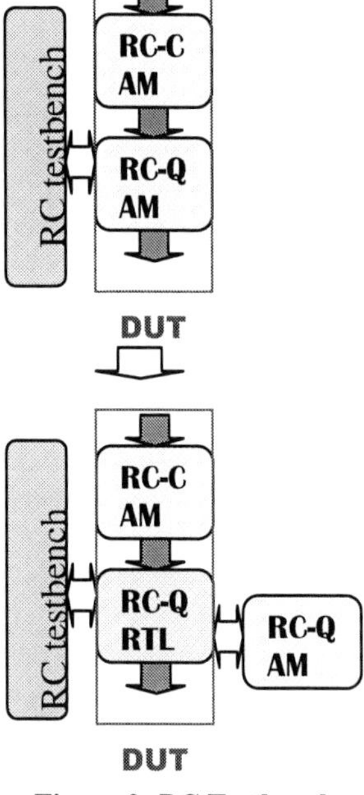

Figure 2: RC Testbench development stages

interfaces. These assertions were used to verify the correctness of the RC-C abstract model both by simulation and AFV. This activity uncovered several specification bugs before the RTL was released. These assertions were later reused to verify the RTL itself.

Since the abstract models were ready before the RTL, they were used to verify the behavior of the RTL during the RTL coding itself. This enabled a bug prevention methodology where the abstract models were used as checkers to gate new code insertion into the verification environment. This resulted in a much cleaner model and facilitated the fast convergence of the RTL.

The RTL of the RC-Q was released to verification after intensive FV and AFV testing (see Example 2) before the RTL of the RC-C was available. We then plugged the RC-Q RTL into the DUT and ran the abstract model of the RC-Q as a checker against the RTL. This enabled bug finding in RC-Q RTL before RC-C RTL was available. Currently, all abstract models are being used as checkers against the RTL.

The results of using abstract models for testbench development for the RC cluster were:

- Pull-in of development and cleanup of the verification testbench.

- Pull-in of RC-Q RTL verification, and its decoupling from the design of the RC-C.

- Improved understanding of the complex specification and identification of specification bugs

- Bug prevention and earlier bug finding due to a robust testbench being available to designers and verification engineers as soon as RTL was developed.

5.3 Example 2: Spec cleanup and deployment of FV techniques using the RC-Q abstract model

The abstract model of the RC-Q is cycle accurate within the unit boundaries. In particular, all its computation is done in cycle zero (namely, when the request is accepted), and its outputs are delayed until they are expected to be issued. The abstract model was written in synthesizable portions of SV, and contains only 29% as many lines as the RC-Q RTL. The abstract model was developed within 6 weeks, as opposed to 3.5 months for the RTL, and it was available to verification about 3 months before first RTL. Both the RTL and the abstract model used assertions intensively, and interface assertions were reused in both the RTL and the abstract model.

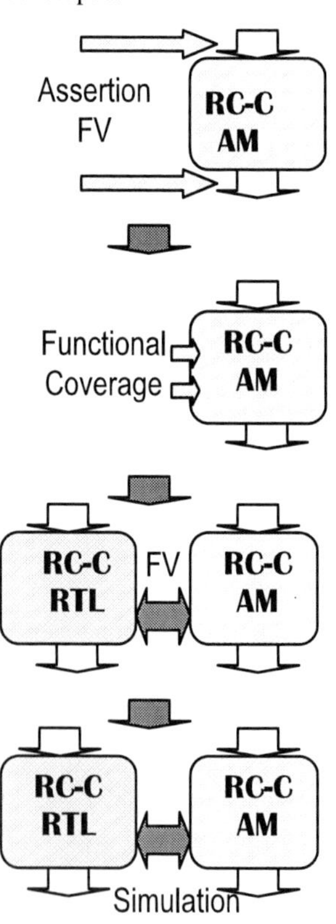

Figure 3: Stages in RC-Q testing

978-1-60558-497-3/09 $25.00 © 2009 ACM

Testing of the RC-Q was carried out in the following stages (see Figure 3):

1. We first applied AFV on the abstract model using the unit interface assertions. This uncovered 2 microarchitecture specification holes and bugs and 9 implementation bugs.

2. When the RC testbench became available pre-RTL, the abstract model was simulated and debugged using the RC testbench and interface assertions. This uncovered 3 specification bugs and over 12 implementation bugs in the abstract model.

3. Consequently, we used functional coverage to improve the quality of the stimuli. This enabled improvements to the testbench and guided the development of new tests. We tracked key micro-architecture-level events which were easily portable to the RTL.

4. When the first RTL became available, and before its integration into the simulation environment, we applied FV on the RTL using the abstract model as a spec. This uncovered 21 bugs in the RTL and 4 modeling bugs all in internal logic that was not covered by assertions.

5. In post-RTL simulation, the abstract model was turned into a checker against the RTL. In the few months that have followed, over 40 RTL bugs have been found in simulation by failures of the abstract model, most of them discovered by the designer prior to the release of the code to verification.

To summarize the results:

The results of using abstract models for the RC-Q were:

- Early convergence of the uArch specification prior to RTL.

- Early cleanup of verification collateral before RTL.

- Bug prevention by providing the RTL designer with an improved means of finding bugs before release to verification.

- Pull-in of RTL bug finding.

- Reuse as a checker for FV and for simulation.

5.4 Summary of examples by numbers

	RC-Q	RC-C
Number of spec bugs	5	4
Modeling bugs pre-RTL	21	32
Modeling bugs post-RTL	6	22
RTL bugs pre-release	41	66
RTL bugs post-release	11	34
% code lines w.r.t. RTL	29%	36%
% RTL bugs pre/post release	79%	66%

Note that:

- Numbers refer to the original code that was released to verification. Both units were modified in later stages.

- The RC-C model was too large and enabled only partial FV between the RC-C model and RTL. This reduced the percentage of bugs found pre-release.

6. SUMMARY

In this paper we have presented the application of synthesizable abstract modeling to the verification of a microprocessor. Abstract modeling starts at the un-timed TLM level and can be used for architecture exploration. It is then refined with some implementation details to enable early verification pre-RTL. Finally, it turns into a reference model checker post-RTL and in hardware emulation.

The use of abstract models enables the pull-in of specification cleanup, development and testing of verification collateral, and early verification. It accelerates bug finding and prevents bugs from being released to verification. It reduces the dependence of verification on the availability of RTL. Furthermore, the use of synthesizable code enables RTL-based technologies, such as functional coverage or FV, to be used to further clean up the abstract models. There are, however, issues that still need to be resolved in order to make this strategy more universally applicable; in particular, the need to support the conflicting requirements of architecture exploration and functional verification, and the need to bring to maturity FV and functional coverage for SystemC modeling.

We successfully applied the technique in the course of the verification of the newest Intel® Core^{TM} microprocessor design. The examples shown demonstrate the potential abstraction has for pulling in and shortening the verification process and for enhancing the effective use of coverage and FV techniques. In all these examples, early verification enabled a significant cleanup of the models, and provided the verification and design engineers with a better means of cleaning up the RTL. This in turn prevented many bugs from being released and reduced the total effort invested in verification.

7. REFERENCES

[1] J. Yuan, C. Pixley (et al.). Constraint-Based Verification. Springer, 2006.

[2] H. Carter, S. Hemmady. Metric-Driven Design Verification. Springer, 2007.

[3] A. Flaisher, A. Gluska and E. Singerman. Case study: Integrating FV and DV within the Verification of Intel® Core™2 Microprocessor. FMCAD, 2007.

[4] B. Bailey, G. Martin, A. Piziali, ESL Design and Verification: A Prescription for Electronic System Level Methodology. Morgan Kaufmann/Elsevier, 2007.

[5] J. Bergeron, Verification Methodology Manual for SystemVerilog, Springer 2006.

[6] S. Kundu, M. Ganai, R. Gupta, Partial Order Reduction for Scalable Testing of SystemC TLM Designs. DAC 2008.

[7] R. Beers, Pre-RTL Formal Verification: An Intel Experience, DAC 2008.

[8] M. Vardi, Formal Techniques for SystemC Verification. In DAC 2007.

[9] D. Kroening and S. Seshia, Formal Verification at Higher Levels of Abstraction, ICCAD 2007.

[10] D. Karlsson, P. Eles, Z. Peng, Formal verification of SystemC designs using a petri-net based representation, DATE 2006

Non-cycle-accurate Sequential Equivalence Checking

Pankaj Chauhan, Deepak Goyal, Gagan Hasteer, Anmol Mathur, Nikhil Sharma
Calypto Design Systems, Inc., Santa Clara, CA 95054.
{pchauhan, dgoyal, ghasteer, amathur, nsharma}@calypto.com

ABSTRACT

We present a novel technique for Sequential Equivalence Checking (SEC) between non-cycle-accurate designs. The problem is routinely encountered in verifying the correctness of a system-level model versus an RTL design which has been derived from the former either manually or through high-level synthesis. The existing state-of-the-art in formal verification/SEC does not provide an efficient mechanism to perform such an equivalence check. Our technique reduces the SEC problem to a cycle-accurate equivalence-checking problem by constructing a pair of normalized cycle-accurate designs from the original designs, on which standard equivalence-checking techniques can then be deployed. We report the results of deploying our techniques on several industrial examples.

Categories and Subject Descriptors

B.2.2 [**Performance Analysis and Design Aids**]: Verification; B.5.2, B.6.3, B.7.2 [**Design Aids**]: Verification

General Terms

Verification, Algorithms

Keywords

Sequential Equivalence Checking, Model Checking, Formal Verification, Unit Product Machine, High Level Synthesis

1. INTRODUCTION

The problem of Sequential Equivalence Checking (SEC) [10, 18, 20] between a pair of designs with different cycle behavior is of growing importance because of the increased popularity of High Level Synthesis (HLS, also known as Behavioral synthesis) tools which can transform untimed or partially-timed high-level specifications into timed RTL models [14]. For example, an HLS tool may take as input, a partially-timed hardware model of an image processor specified in SystemC, which, takes in an array of 256 inputs, processes them, and produces an array of 256 outputs, all in a single time step. The output RTL produced by HLS may schedule

the inputs to be read serially over a period of 256 clock cycles, may process the computation on these inputs over a period of 512 cycles and produce the outputs serially over another period of 256 cycles. Furthermore, the RTL model may be able to start reading the next set of 256 inputs as soon as the previous computation is finished, *i.e.* in parallel with the outputs of the previous computation. Thus, the first set of inputs would be read during cycles 0 to 255, the second set of inputs during cycles 768 to 1023, and so on. The first set of outputs would be produced during cycles 768 to 1023, the next set of outputs during cycles 1536 to 1791, and so on. In steady state, the original design completes a computation and produces 256 new outputs every cycle, whereas the synthesized RTL model does the same every 768 cycles. We refer to such pairs of designs as non-cycle-accurate designs. In contrast, cycle-accurate designs have inputs and outputs matching every cycle.

In this paper, we present the concept of input, output and state maps with latency which can be used to precisely formulate an equivalence checking problem between non-cycle-accurate designs. We present an algorithm for doing SEC between such designs by transforming them into a product machine of a cycle-accurate pair of designs (called the normalized product machine) such that the transformed designs are equivalent *iff* the original designs are equivalent. We also give an algorithm for converting any counterexample on the normalized product machine into a counterexample on the original pair of designs. Thus,' the problem of SEC between non-cycle-accurate designs is reduced to the problem of model checking [5] on the normalized product machine.

Our method is not just relevant in the context of HLS but also in the context of cycle-accurate RTL-RTL verification. If the RTL designs perform a full computation (called a transaction) over a period of a fixed number of cycles, it may be easier to verify the one-transaction normalized machines against each other. This is because the process of creating the normalized machine can sometimes eliminate the entire state machine that schedules the computation and may reduce a sequential verification problem to a combinational verification problem.

2. DEFINING NON-CYCLE-ACCURATE SEQUENTIAL EQUIVALENCE

In this section we define non-cycle-accurate sequential equivalence, and some terminology used in the rest of the paper.

DEFINITION 1. *A sequential machine is a tuple* $M = (I, S, R, Y, T)$, *where* $I = \{i_0, i_1, \ldots, i_{n-1}\}$ *is the set of* inputs, $S = \{s_0, s_1, \ldots, s_{m-1}\}$ *is the set of* state variables (flops), $R = \{r_0, r_1, \ldots, r_{m-1}\}$ *is the set of* reset values *for the state variables*, $Y = \{y_0, y_1, \ldots, y_{k-1}\}$ *is the set of* outputs, *and* $T = \{t_0, t_1, \ldots, t_{m-1}\}$ *is the set of* transition functions. *A transition function* $t_i(I, S)$ *de-*

Permission to make digital or hard copies of part or all of this work for personal or classroom use is granted without fee provided that copies are not made or distributed for profit or commercial advantage and that copies bear this notice and the full citation on the first page. To copy otherwise, to republish, to post on servers or to redistribute to lists, requires prior specific permission and/or a fee.
DAC'09, July 26-31, 2009, San Francisco, California, USA

978-1-60558-497-3/09 $25.00 © 2009 ACM

scribes the next state value of the state variable s_i. Each output is a function of the present state and the inputs. We use y_i to denote both the ith output variable, as well as the function that computes that output.

The reset value for a state variable may either be a constant or unspecified (symbolic). A state is an initial state, *iff* the values of the state variables are consistent with the partially specified reset values.

The concept of *period* of a sequential machine is central to our notion of equivalence. This denotes the period in terms of number of cycles, or clock-ticks, after which the equivalence assumptions and checks are repeated. The period naturally corresponds to the notion of the length of a *transaction*, in which a finite state system finishes a computation. For the image processor example mentioned earlier, 768 is the period. The algorithm we describe in the next section accelerates a pair of sequential machines by their period, to get a unit-period product sequential machine. In this paper, we only consider sequential machines with a fixed period. Note that in many cases, SEC between two variable period designs can be cast into a fixed period SEC problem.

Fig. 1 shows two examples of sequential machines, one with period 1, and the other with period 4, both of which compute the sum of 4 inputs. The output of the parallel and serial machines are comparable at cycles $0, 1, 2, \ldots$ and $3, 7, 11, \ldots$ respectively which correspond to transactions $1, 2, 3 \ldots$ for each machine. Note that we always start counting cycles from 0 and transactions are counted starting from 1.

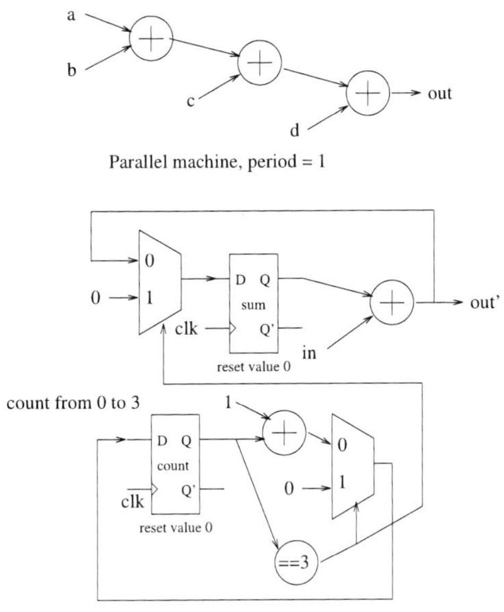

Figure 1: Two example sequential machines, one sums up four inputs in a single cycle, while the other takes four cycles.

Our model of a sequential machine is similar to a Mealy machine, except that it does not have a generalized transition relation, but a set of transition functions for state variables. Most synchronous digital hardware system descriptions fit this model.

We consider equivalence between two sequential machines, M_S and M_I (*specification* and *implementation*), along with their periods P_S and P_I. The variables corresponding to the two machines have the subscripts S or I, e.g., i_S and y_I. We use the @t notation to indicate the cycle (or clock-tick), starting with cycle 0, e.g., y_S@3 denotes the output y_S at cycle 3. The @t notation is also

used in the specification of mapping (input/output/state) relations, described next.

Our system allows the specification of a variety of input, output, and state maps which together capture the assumptions to be made and the equalities to be proven. The maps are meant to be repeatable every *period* number of cycles. In our system, every equality (assumed or derived) is effected by a process called *mitering*, in which one signal in the equality (miter winner) is used to drive the fan out of the other signal (miter loser). The term *speculative reduction* [2] has also been used to refer to mitering when applied to assumptions.

Input Maps: Input maps describe assumptions to be made about a pair of inputs being equal at certain times. An input map is a tuple $(i_S@l_1, i_I@l_2)$. Here, the specification input i_S at cycles $l_1, l_1 + P_S, l_1 + 2 \cdot P_S, \ldots$ is to be assumed equal to the implementation input i_I at cycles $l_2, l_2 + P_I, l_2 + 2 \cdot P_I, \ldots$. Typically, $0 \leq l_1 < P_S$ and $0 \leq l_2 < P_I$. We also use the term *latency* for the cycle offsets l_1 and l_2. After all the input assumptions are specified, timed inputs that are not constrained by any assumption are assumed to be independent free symbolic variables. For the example in Fig. 1, $(a@0, in@0)$, $(b@0, in@1)$, $(c@0, in@2)$, and $(d@0, in@3)$ are reasonable input maps.

Our system also allows for the specification of constants, partial constants, ranges of constants, symbolic constants, etc. for input assumptions, but since these are not central to the main ideas in this paper, we do not discuss them further. Moreover, any such constraints can be easily modeled by adding suitable functionality to the designs themselves.

Output Maps: Output maps describe proof obligations about a pair of outputs being equal at certain times. An output map is a tuple $(y_S@l_1, y_I@l_2)$. Here, the specification output y_S at cycles $l_1, l_1 + P_S, l_1 + 2 \cdot P_S, \ldots$ is to be proven equal to the implementation output y_I at cycles $l_2, l_2 + P_I, l_2 + 2 \cdot P_I, \ldots$. For the running example, $(out@0, out'@3)$ is a reasonable output map. Note that the latencies l_1 and l_2 may be different, and in some cases, much larger, than the periods. E.g., a deep pipeline may take a long time to fill up, but once it does, an output is produced every cycle.

State Maps (Flop Maps): Maps between state variables of the designs help reduce the complexity of verification by allowing a divide and conquer approach. A state map takes the form of a tuple $(s_S@l_1, s_I@l_2)$. A state map generates an *assume-guarantee* pair. The *assume* part of the state map corresponds to the assumption that $s_S@(l_1 + (j-1) \cdot P_S)$ is equal to $s_I@(l_2 + (j-1) \cdot P_I)$ for transaction j. The *guarantee* part of the state map corresponds to the proof obligation that $s_S@(l_1 + j \cdot P_S)$ and $s_I@(l_2 + j \cdot P_I)$ are equal for transaction j. The state maps also introduce a one-time initial check $s_S@(l_1) = s_I@(l_2)$ starting from the reset sets of states R_S and R_I. This *assume-guarantee* setup ensures the soundness of state maps by ensuring that the *assume* part of transaction j has been proven through the *guarantee* part of transaction $j - 1$ for $j > 1$. For transaction 1, the *assume* part has been proven by the one-time initial check. By implementing equality assumptions through mitering, a state map may help replace two state variables in the product machine by one state variable. An extension of the state maps is that of *cutting* the state maps, in which we remove both state variables altogether, and drive the fanout of the present state value of both state variables by a fresh primary input. Note that the proof obligation still remains the same irrespective of whether the state map is cut or not. The *cut* is a form of safe abstraction, under which a proof remains a valid proof, but a falsification may be spurious. Combinational equivalence checking corresponds to the special case of SEC in which all state elements are required to be mapped and cut. However, for general SEC, a partial state map or no map at all still allows one to make progress.

Sequential machines, their periods, input maps, output maps, and state maps thus form the core concepts that allow one to precisely formulate a non-cycle-accurate SEC problem between designs with arbitrary timing differences. Except for state maps, all other components need to be provided by the user to formulate the problem. State maps are hints that, if available, simplify the verification task. Techniques for finding state maps automatically [21, 19] are not further discussed in this paper. In the next section, we describe how we take such a problem setup, and create an idealized, period-1, cycle-accurate product sequential machine (or a unit product machine), with latency 0 on all the outputs, on which further analysis, *e.g.*, bounded or unbounded model checking, can be done.

3. CONSTRUCTING UNIT PRODUCT MACHINE

Our system provides an on the fly, lazy unrolling scheme for sequential netlists, where the fanin of a signal x at cycle t ($x@t$) is created only when needed. Sequential machines are stored as word-level (bit-vector) netlists containing combinational operators (such as arithmetic), signals, flops, and input/output ports. The outputs and next state functions of flops are computed by combinational netlists, that realize the respective functions.

The unrolled netlists are stored in a database, where each netlist entity has a cycle counter and attributes correlating the entity to the one in the original netlist for book-keeping purposes. Unrolling through combinational operators keeps the same cycle but when unrolling across a flop (from the output Q-port to the next state input D-port), we decrement the cycle counter, as long as the cycle counter does not become negative. Otherwise, the unrolling stops at the initial state value of the flop, and a primary input is created in the unrolled netlist. These inputs may be replaced by the reset values, as and when needed. Facilities exist to interpret unrolled netlists as regular netlists, with enough book-keeping to allow for finding the original signal, and the cycle. Our system also provides combinational solvers that can provide answers to equality queries at the level of abstraction (word-level or bit-level) that the original sequential machines are in.

We are given two sequential machines, M_S and M_I, their periods P_S and P_I, a set of input maps Φ, a set of output maps Ψ, and a set of flop maps Θ. Thus, the equivalence check problem is $\langle M_S, P_S, M_I, P_I, \Phi, \Psi, \Theta \rangle$. The output of the algorithm is a single sequential machine M_U, along with a set of output maps Ψ_U. All the signals in Ψ_U are drawn from M_U, and have their latencies as zero. The set of output maps Ψ_U may be seen as the union of two disjoint sets of maps Ψ_{1_U} and Ψ_{2_U}, where the former corresponds to the original output maps Ψ and the latter corresponds to the *guarantee* part of the state maps Θ. The period P_U of M_U is 1. The *assume* part of the state maps Θ are factored into the construction of $M_U = (I_U, S_U, R_U, Y_U, T_U)$ through mitering.

In the following algorithm, the period of a flop refers to the period of the design that the flop belongs to. The high-level algorithm is described below in two phases.

Preprocessing:

1. Process the input maps Φ to obtain equivalence classes of timed inputs. Unconstrained timed inputs end up in an equivalence class of size 1. Recall that the maps repeat every period number of cycles of the two sequential machines. For each equivalence class, choose one as the *leader*. Miter all members of each equivalence class choosing the *leader* as the miter-winner.

2. **Initial State Checks:** For each state map $(s_S@l_1, s_I@l_2) \in \Theta$, add $s_S@l_1$ and $s_I@l_2$ to a joint unrolling frontier F_0. Unroll F_0, and apply reset values whenever unrolling reaches a flop output at cycle 0. Use combinational solvers, to compare $s_S@l_1$ and $s_I@l_2$ (if both latencies are 0, and both reset values are constants, then it is merely a comparison of constants). If they are not equal, report a counterexample and quit.

3. In order to reduce the number of flops in the product machine, we use the procedure OPTIMIZESTATEVARS (Fig. 2) on both M_S and M_I independently. It is a 1-step induction to find the set S_{ind} of state variables that are equal to their reset values every period cycles, in other words a fixed-point of the constant state variables. The term phase abstraction is used in [3, 16], for a similar process. For each flop $s \in S_{\text{ind}}$ the flop output at cycle 0 obtained during the unrolling process is replaced by the respective constant reset value. This step is very critical to our algorithm as it can help reduce the number of flops in the product machine significantly. For example, this step followed by constant propagation converts the serial machine of Fig. 1 to one with no flops, and 3 adders. In the absence of this optimization, the serial machine would have resulted in 2 flops, 8 multiplexors, and 8 adders.

Algorithm CONSTRUCTUNITPRODUCTMACHINE

1. For each state map $(s_S@l_1, s_I@l_2) \in \Theta$, add $s_S@l_1$ and $s_I@l_2$ to a joint unrolling frontier F_U. Since the initial state assumptions are proven, miter $s_S@l_1$ and $s_I@l_2$. This corresponds to the *assume* part of the flop maps. If the miter winner flop is part of S_{ind} and has a latency 0, then replace it by the constant reset value, otherwise leave it as a primary input to the unrolled netlist. If the state map is specified to be cut, then remember the miter winner state variable in a set S_c.

2. For each output map $(y_S@l_1, y_I@l_2) \in \Psi$, add $y_S@l_1$ and $y_I@l_2$ to F_U, and to Y_U. Also, add $((y_S@l_1)@0, (y_I@l_2)@0)$ to Ψ_U, the new set of output maps. Note the overloaded @ notation, the first referring to the realized output in M_U, and the second @ for the actual latency specification.

3. For each flop map $(s_S@l_1, s_I@l_2) \in \Theta$, add $s_S@(l_1 + P_S)$ and $s_I@(l_2 + P_I)$ to F_U, and to Y_U. Also, add $((s_S@(l_1 + P_S))@0, (s_I@(l_2+P_I))@0)$ to Ψ_U. This corresponds to the *guarantee* part of state maps.

4. Unroll from frontier F_U until we reach either primary inputs or flop outputs at cycle 0. Let S_r be the set of flops whose cycle-0 outputs are hit during unrolling. For each flop $s \in S_r$ with a period P, add $s@P$ to the frontier F_U and unroll again. The unrolling from the new frontier may find more flops whose cycle-0 outputs are hit. Add these flops to S_r, and keep repeating this step until convergence.

5. The next state function of each $s \in S_r$ is just $s@P$. Replace the primary input corresponding to $s@0$ with the output of a new flop s'. Add s' to S_U. Connect the input D-port of flop s' by the signal corresponding to $s@P$. The set of reset values R_U is made up of the reset values in the original sequential machines.

6. Optionally, apply combinational optimizations to the unrolled netlist, such as constant propagation, structural hashing, functional sweeps [12, 13], etc.

Fig. 3 shows the unit product machine obtained from the above algorithm for the example in Fig. 1. The unit machine contains the

978-1-60558-497-3/09 $25.00 © 2009 ACM

1   Let S_{ind} be the set of all relevant flops, e.g., the ones in the fanin of outputs being compared.

2   Remove from S_{ind} all the flops whose reset values are not constants.

3   Repeat the following 3 steps until S_{ind} does not change.

4   For each flop $s \in S_{\text{ind}}$, add $s@0$ and $s@P$ (where P is its period) to a frontier F, and unroll.

5   Replace cycle-0 values of the flops in S_{ind} in the unrolled netlist by the respective reset values.

6   Use combinational solver to find out for each flop $s \in S_{\text{ind}}$, if $s@P$ is equal to its reset value. If not, remove s from S_{ind}.

7   Return S_{ind}.

Figure 2: Procedure to optimize the number of state variables in the unit machine.

output at time 0 of the parallel machine, and the output at time 3 of the serial machine. The *count* and the *sum* state variables are always equal to their reset values (0) every period number of cycles, so they get eliminated by the OPTIMIZESTATEVARS procedure. A simple structural hashing on the unrolled machine causes the two outputs to be trivially mitered in the unit product machine, and hence, a trivial SEC proof is obtained.

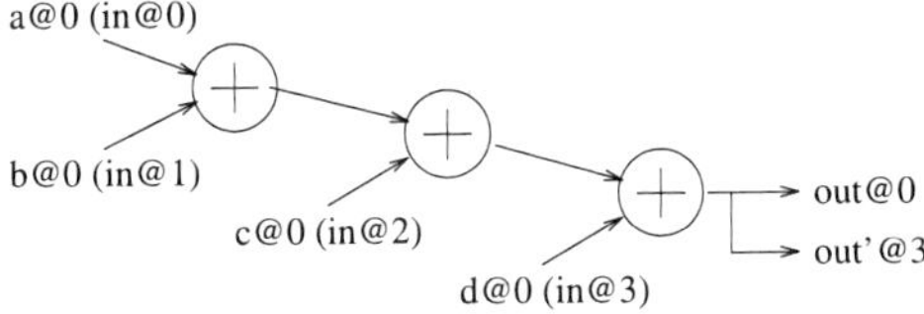

Figure 3: Unit product machine for the example in Fig. 1.

4. TIME CORRELATIONS

The joint unrolling based algorithm described in the previous section misses important time correlations between related input signals. Consider just a single machine in Fig. 4(a) with a latency of 1 on the output signal *out*. On applying the above algorithm to produce a unit machine with latency 0, we get the machine in Fig. 4(b). The output sequence of the unit machine $out'@0, out'@1, \ldots$ and of the original machine $out@1, out@2, \ldots$ are supposed to match. However, the unit machine clearly does not capture the computation of the original machine. To start with, it consumes two inputs every clock cycle, whereas the original machine consumes just one. In essence, the temporal correlation that the input x (corr. to $in@1$) is really just input y (corr. to $in@0$) one cycle later is lost. This lost correlation can be captured by inserting a flop between x and y as shown in Fig. 4(c). In general, if the unrolled inputs corresponding to two inputs on the unit machine are m−multiples of their period apart, there need to be m flops between them. Moreover, the reset values of the flops should be symbolic, to allow for fully symbolic inputs in the earlier cycles.

The situation is exacerbated with two designs unrolled together, even more so when the input maps are such that the first few values of some input are unconstrained, and they get constrained in future. Consider the example of Fig. 5. To construct the unrolled machines, we start with the frontier $out1@0$ and $out2@1$, giving us $G1@0$, z (corr. to $b@0$) and $F2@0$ as the inputs of the unrolled netlist. Based on the above algorithm, we add $G1@2$ and $F2@2$ to the

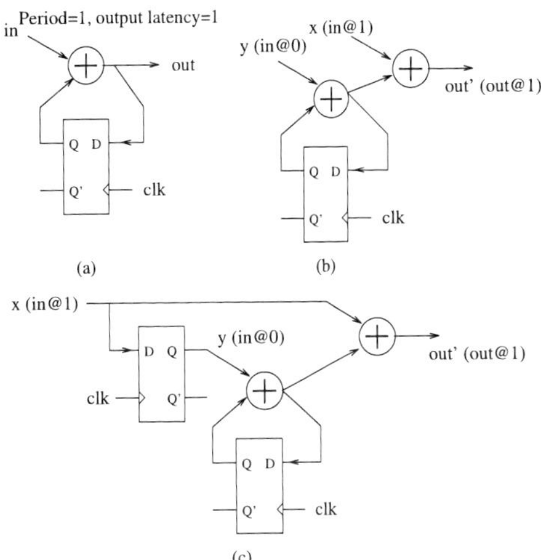

Figure 4: (a) A simple sequential machine with output latency 1, (b) corresponding unit machine, with output latency 0, and (c) Correct unit machine with correlated inputs.

frontier and unroll again, giving us y (corr. to $a@0$ and $b@1$), and x (corr. to $a@1$) as more inputs. After inserting flops $G1'$ and $F2'$ corresponding to the original flops $G1$ and $F2$, we get the unit machine in Fig. 6.

The outputs of the unit machine at cycle 0 i.e. $out1'@0$, and $out2'@0$, are $G1'@0$ and $F2'@0 \times z@0$, both of which evaluate to 0 since the reset values of both flops $G1'$ and $F2'$ are zero. However, at cycle 1 we get a falsification. The outputs at cycle 1, i.e. $out1'@1$, and $out2'@1$ (corr. to $out1@2$ and $out2@3$ on the original machines) evaluate to $x@0 \times y@0$ and $y@0 \times z@1$ respectively which results in a falsification. On the original machine, $x@0 \times y@0$ corresponds to $a@1 \times a@0$ and $y@0 \times z@1$ corresponds to $a@0 \times b@2$. Moreover, from the input constraints (as seen in Fig. 5), $b@2$ is equal to $a@1$, and hence, the counterexample is invalid on the original machines. The problem is that we are missing the correlation between $x@0$ and $z@1$ on the unit product machine in Fig. 6. The fix is to have input z (corr. to $b@0$) of the unit product machine be fed by a flop whose next state input is fed by input x (corr to $a@1$ and $b@2$). It is also important to leave the reset value of this inserted flop unspecified, as it allows a fully unconstrained symbolic value for $b@0$. Also note that $a@0$ and $a@1$ are not 2 cycles (i.e. period) apart, so they remain independent. Fig. 7 shows the fixed unit machine on which we get a proof of equality on the outputs $out1'$ and $out2'$.

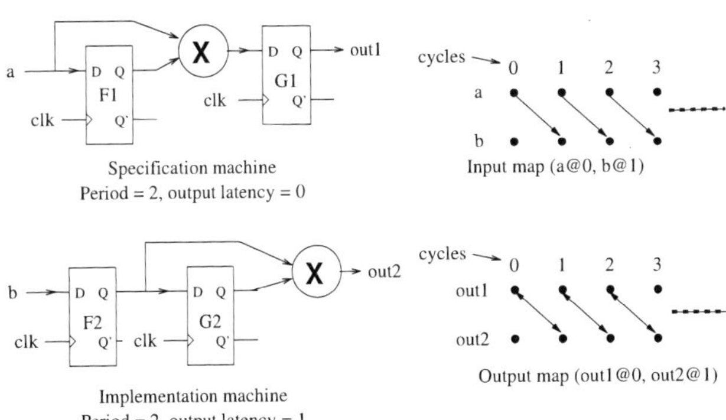

Figure 5: Two period 2 sequential machines, with input and output maps having a non-zero latency. The reset value of all flops is 0.

The lost correlations can be fixed by analyzing the inputs of the unit machine, and if they are a m-multiples of their respective period apart, inserting m flops between them. The reset values of these flops are left symbolic. For a set of correlated inputs, the only one that remains after flop insertion is the one with the highest transaction counter (furthest in the future). We also need to properly account for signals such as $b@0$, which become miter losers in the future. Due to the lack of space, we will skip the detailed algorithm for correlating unit machine inputs.

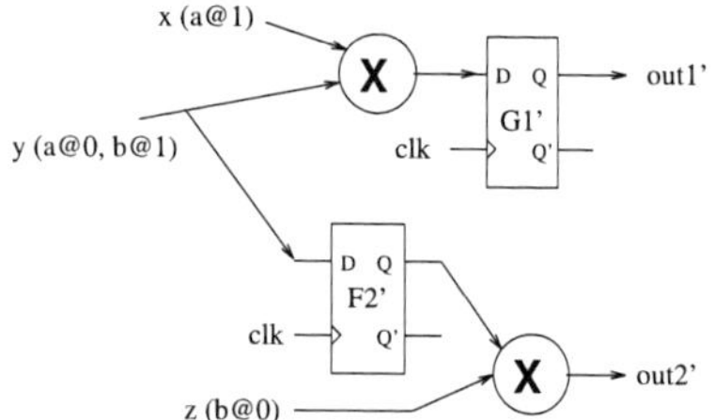

Figure 6: Unit product machine for the example in Fig. 5 with uncorrelated inputs.

Constructing the unit product machine, and applying the time correlations captures all the input/state assumptions, and proof obligations. Moreover, each primary input and state variable of the unit machine has enough book-keeping information to provide the cycle and the original signal/state variable that it came from. This book-keeping enables us to translate any trace on the unit machine to the pair of traces on the original sequential machines. For example, if a signal $c@l$ (input, output, state) is assigned value v in cycle k of the unit machine, it corresponds to the assignment of value v to the original signal c at cycle $l + k \cdot period(c)$. Moreover, if $d@m$ is a miter loser to $c@l$, it corresponds to the value assignment of v to signal d at cycle $m + k \cdot period(d)$.

Note that the cutting at state-maps is a user specified abstraction, so a counterexample on the unit machine may not necessarily reproduce on the original sequential machines. However, it faithfully reproduces on the original machines, if the state variables are abstracted correspondingly in the original machines.

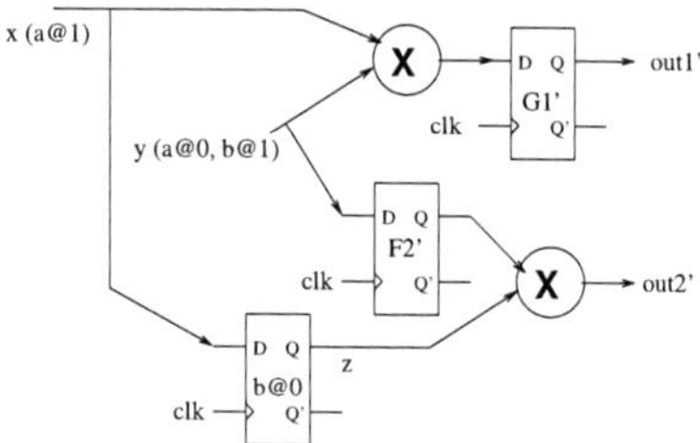

Figure 7: Correct unit machine for the example in Fig. 5 with correlated inputs.

We thus state the following theorem (proof omitted for lack of space) to capture the correctness of the unit machine.

THEOREM 1. *Given a sequential equivalence checking problem* $\langle M_S, P_S, M_I, P_I, \Phi, \Psi, \Theta \rangle$, *let* M_U, *with proof obligations* Ψ_U *be constructed using the algorithm* CONSTRUCTUNITPRODUCTMACHINE *followed by input correlation capture. Then the original proof obligations* Ψ *and* Θ *are valid if and only if* Ψ_U *on* M_U *are. Moreover, any counterexample on* M_U *can be translated to an equivalent counterexample on the pair of machines* M_S *and* M_U.

5. EXPERIMENTS

We present experiments to show the variety and scale of non-cycle-accurate SEC problems to which our technique has been successfully applied. Since we are not aware of any other system which does non-cycle-accurate SEC by the way of constructing cycle accurate unit machine, we cannot compare our approach to other approaches. Other systems that deal with non-cycle-accurate SEC are either proprietory, or limited in their applicability. Table 1 shows characteristics of a small sample of real industrial examples, on which SEC was successfully done using our system SLEC [4]. These designs come from a variety of problem domains, such as video/audio processing, wireless, encryption and were obtained using various approaches such as HLS, manual C v/s RTL, C v/s C, and RTL-RTL.

It should be obvious to the reader that the unit machine construction is only a part of a solution to SEC. Our system provides cycle accurate bounded/unbounded equivalence checking using word-level state-abstraction, induction, reachability, model checking, etc. Word-level solvers, word-level and bit-level netlist optimizations, efficient unrolling, are critical components of our system.

In Table 1, for each SEC problem, we report the sizes of the input, output and state map sets. We report the period, gate-count and number of word-level flops in each design M_S and M_I. Multiplying design sizes by their periods provides an upper bound on the size of the netlists for pure unrolling based bounded proofs, such as [6]. Next, we report the size of the unit machine relevant to each design. The size of the combined unit machine may be smaller than the sum of the individual sizes because of the sharing induced by mitering. Finally, we report the verification times and memory usage.

It is noteworthy that the size of the unit product machine is much smaller than the upper bound given by pure unrolling for most cases. The dramatic reductions are largely attributed to state variable optimization, and aggressive combinational optimizations. Many problems in the table are comparisons between untimed, period-1 specifications to implementations with large periods and a large number of flops. There are also some cases where periods of both machines are similar. In the case of FIR, even though a cycle-accurate, period-1 problem setup was available, lack of an effective state mapping made the verification infeasible. For the setup with period 350 reported here, state variable optimization eliminated the scheduling state machines which helped simplify the problem significantly.

6. RELATED WORK

Sequential equivalence is an important, and well studied problem. Theoretically, bisimulation equivalence [17] of transition systems is a close representation of sequential equivalence in the presence of resets. Pixley [18] provided a comprehensive theoretical framework for hardware equivalence, and introduced the concept of *alignability* for bringing two state machines into corresponding states. Singhal et al. [19] explored the notion of *safe-replaceability* and *delayed safe-replaceability* for non-reset based sequential equivalence. Authors in [9, 10] further formalized various approaches to equivalence, including compositional equivalence checks, and presented a SAT-based algorithm for alignability. van Eijk et al. [20, 21] presented methods for detecting equivalent state variables, and structural similarities for sequential equivalence. In addition to these, various forms of state mappings in different contexts have been described in the past, e.g., [1, 11].

The state machine reduction algorithm we proposed is similar to phase abstraction introduced in [3]. Lu et al. [15], Mishchenko et al. [16] use inductive techniques to determine state variables that follow repeatable patterns. However, their techniques are demonstrated for boolean representations only, which we are not restricted to. Finally, [2, 8, 16] present sequential equivalence checking sys-

Name	$\|\Phi\|$	$\|\Psi\|$	$\|\Theta\|$	M_S			M_I			unit M_S	unit M_I	verification	
				P_S	GC	$\|S\|$	P_I	GC	$\|S\|$	GC	GC	time (s)	mem (MB)
video1	73	64	192	1	1.623 M	256	147	83 K	1121	26 K	33 K	17988	7750
video2	53	6	19	34	25 K	56	36	26 K	60	77 K	80 K	12381	988
video3*	195	64	64	1	5.638 M	128	36	18 K	260	186 K	187 K	10693	644
video4*	50	1	1	1	217 K	33	30	29 K	459	35 K	65 K	1025	435
video5*	3	1	1	181	24 K	1324	182	21 K	260	6 K	6 K	433	405
FIR	740	2162	2156	350	593 K	2198	350	606 K	2198	42 K	41 K	9375	9966
FFT256	31	82	0	1091	12 K	263	1075	12 K	195	3.358 M	3.358 M	11272	1175
wireless1	3	2	0	1	25 K	3	128	10 K	791	16 K	18 K	8811	325
wireless2	8	2	0	1	168 K	2	132	163 K	2639	158 K	185 K	3306	1216
DES	6	1	0	1	171 K	2	129	112 K	461	97 K	74 K	3658	923
AES	19	1	8	1	1.164 M	8	424	19 K	87	73 K	135 K	6494	994

Table 1: SEC for large industrial designs. $\|\Phi\|/\|\Psi\|/\|\Theta\|$-number of input/output/state maps, GC-optimized gate count, $\|S\|$-number of word-level flops. A falsification was found for starred designs.

tems, that incorporate various techniques. In [6], bounded model checking was used for SEC between C programs and RTL in Verilog. Extraction of verification models from high-level descriptions, such as C++, is separately described in [7].

None of the approaches above address the problem of non-cycle-accurate SEC, especially while preserving the abstraction-level of the netlists. Our normalization method can be used get a cycle-accurate SEC problem, to which any of the above approaches can be applied.

7. CONCLUSIONS AND FUTURE WORK

We have presented a definition of non-cycle-accurate SEC, and a novel method for reducing a non-cycle-accurate SEC problem to a cycle-accurate SEC problem. Our normalization method provides many benefits. All state variables that go through a set pattern in every transaction completely disappear from sequential analysis. Unit machine construction puts the computations in a normal form, enabling meaningful intermediate equivalences, crucial for divide and conquer approaches, such as cuts and heaps, or structural similarities [13, 20]. Moreover, our approach does not restrict us to just bit-level representations. We have successfully demonstrated our method on a variety of industrial designs.

There are a few limitations of our approach, which we are working towards lifting, such as variable periods, and more complex mappings.

8. REFERENCES

[1] M. Aagaard, B. Cook, N. A. Day, and R. B. Jones. A framework for microprocessor correctness statements. In *CHARME*, pp. 433–448, 2001.

[2] J. Baumgartner, H. Mony, V. Paruthi, R. Kanzelman, and G. Janssen. Scalable sequential equivalence checking across arbitrary design transformations. In *ICCD*, pp. 259–266, 2006.

[3] P. Bjesse and J. Kukula. Automatic generalized phase abstraction for formal verification. In *ICCAD*, pp. 1076–1082, 2005.

[4] Calypto Design Systems. Sequential Equivalence Checking: A new approach to functional verification of datapath and control logic changes. http://www.calypto.com/wp_request.php?paper=sequential, 2007.

[5] E. M. Clarke, O. Grumberg, and D. Peled. *Model Checking*. MIT Press, 2000.

[6] E. M. Clarke, D. Kroening, and K. Yorav. Behavioral consistency of C and Verilog programs using bounded model checking. In *DAC*, pp. 368–371, 2003.

[7] M. Haldar, G. Singh, S. Prabhakar, B. Dwivedi, and A. Ghosh. Construction of concrete verification models from C++. In *DAC*, pp. 942–947, 2008.

[8] S.-Y. Huang, K.-T. Cheng, K.-C. Chen, C.-Y. Huang, and F. Brewer. AQUILA: An equivalence checking system for large sequential designs. *IEEE TOC*, 49(5):443–464, 2000.

[9] D. Kaiss, M. Skaba, Z. Hanna, and Z. Khasidashvili. Industrial strength SAT-based alignability algorithm for hardware equivalence verification. In *FMCAD*, pp. 20–26, 2007.

[10] Z. Khasidashvili, M. Skaba, D. Kaiss, and Z. Hanna. Post-reboot equivalence and compositional verification of hardware. In *FMCAD*, pp. 11–18, 2006.

[11] A. Koelbl, J. R. Burch, and C. Pixley. Memory modeling in ESL-RTL equivalence checking. In *DAC*, pp. 205–209, 2007.

[12] A. Kuehlmann. Dynamic transition relation simplification for bounded property checking. In *ICCAD*, pp. 50–57, 2004.

[13] A. Kuehlmann, V. Paruthi, F. Krohm, and M. K. Ganai. Robust Boolean reasoning for equivalence checking and functional property verification. *IEEE TCAD*, 21:1377–1394, 2002.

[14] Y.-L. Lin. Recent developments in high-level synthesis. *ACM TODAES*, 2(1):2–21, 1997.

[15] F. Lu and K.-T. Cheng. Sequential equivalence checking based on k^{th} invariants and circuit sat solving. In *HLDVT*, pp. 45–51, 2005.

[16] A. Mishchenko, M. L. Case, R. K. Brayton, and S. Jang. Scalable and scalably-verifiable sequential synthesis. In *ICCAD*, 2008.

[17] D. Park. Concurrency and automata on infinite sequences. In P. Deussen, editor, 5th GI-Conference, volume 104 of *Theoretical Computer Science*, pp. 167–183, Karlsruhe, 1981. Springer-Verlag.

[18] C. Pixley. A theory and implementation of sequential hardware equivalence. *IEEE TCAD*, 11(12):1469–1478, 1992.

[19] V. Singhal, C. Pixley, A. Aziz, S. Qadeer, and R. K. Brayton. Sequential optimization in the absence of global reset. *ACM TODAES*, 8(2):222–251, 2003.

[20] C. A. J. van Eijk. Sequential equivalence checking based on structural similarities. *IEEE TCAD*, 19(7):814–819, 2000.

[21] C. A. J. van Eijk and J. A. G. Jess. Exploiting functional dependencies in finite state machine verification. In *DATE*, pp. 266–271, 1996.

978-1-60558-497-3/09 $25.00 © 2009 ACM

Regression Verification

Benny Godlin
CS, Technion, Haifa, Israel.
bgodlin@cs.technion.ac.il

Ofer Strichman
IE, Technion, Haifa, Israel.
ofers@ie.technion.ac.il

ABSTRACT

Proving the equivalence of successive, closely related versions of a program has the potential of being easier in practice than functional verification, although both problems are undecidable. There are two main reasons for this claim: it circumvents the problem of specifying what the program should do, and in many cases it is computationally easier. We study theoretical and practical aspects of this problem, which we call *regression verification*.

Categories and Subject Descriptors

D2.4 [**Software/program verification**]

General Terms

Verification

Keywords

Software verification, Equivalence checking

1. INTRODUCTION

Proving the equivalence of successive, closely related versions of a program has the potential of being easier in practice than applying functional verification to the newer version against a user-defined, high-level specification. There are two reasons for this claim. First, it mostly circumvents the problem of specifying what the program should do. The user can take a no-action 'default specification' by which the outputs of the program (or even only its return value) should remain unchanged if the two programs are run with the same inputs. Second, as we show in this article, there are various opportunities for abstraction and decomposition that are only relevant to the problem of proving equivalence between similar programs, and these techniques make the computational burden less of a problem.

Both functional verification and program equivalence of general programs are undecidable. Coping with the former

was declared in 2003 by Tony Hoare as a "grand challenge" to the computer science community [8]. Program equivalence can be thought of as a grand challenge in its own right, but there are reasons to believe, as indicated above, that it is a 'lower hanging fruit'. The observation that equivalence is easier to establish than functional correctness is supported by past experience with two prominent technologies: *regression testing* – the most popular automated testing technique for software – and *equivalence checking* – the most popular formal verification technique for hardware. In both cases the reference is a previous version of the system. Equivalence checking also demonstrates the difference in the computational effort: it is computationally easier than model-checking, at least under the same assumption that we make here, namely that the two compared systems are mostly similar. One may argue, however, that the notion of correctness is weaker: rather than proving that a model is 'correct', we prove that it is 'as correct' as the previous version. In contrast one may argue that it can still expose functional errors since failing to comply with the equivalence specification indicates that something is wrong with the assumptions of the user. In any case, it might be a feasible venue even in cases where the alternative of functional verification is not.

We call the problem of proving the equivalence of closely related programs *regression verification*. It can be useful wherever regression testing is useful, and in particular for guaranteeing backward compatibility. This statement holds even when the programs are *not* equivalent. Our system allows the user to define an 'equivalence specification' in which the compared values (e.g., the outputs) are checked only if a user-defined condition is met. For example, if a new feature – activated by a flag – is added to the program and we wish to verify that all previous features are unaffected, we condition the equivalence requirement with this flag being turned off. Backward compatibility can also be useful when introducing new performance optimizations or applying *refactoring*[1].

The idea of proving equivalence between programs is not new, and in fact preceded the idea of functional verification.[2] We delay a detailed survey of earlier work to later in this section, but let us just mention that as far as we know no one has focused so far on coping with this problem for gen-

Permission to make digital or hard copies of part or all of this work for personal or classroom use is granted without fee provided that copies are not made or distributed for profit or commercial advantage and that copies bear this notice and the full citation on the first page. To copy otherwise, to republish, to post on servers or to redistribute to lists, requires prior specific permission and/or a fee.
DAC'09, July 26-31, 2009, San Francisco, California, USA

---

[1]Refactoring is a popular set of techniques for rewriting existing code for various purposes.
[2]In his 1969 paper about axiomatic basis for computer programming [6], Hoare points to previous works from the late 50's on axiomatic treatment of the problem of proving equivalence between programs.

978-1-60558-497-3/09 $25.00 © 2009 ACM

eral programs in real programming languages nor on how to exploit the fact that large parts of the code in the two compared programs is identical. Ideally the complexity of the solution should be dominated by the (semantic) difference between the two compared programs rather than on their size.

There are many different ways to define the notion of Input/Output equivalent programs (six different definitions appear in [3]). Here we focus on the following definition:

Definition 1. Partial equivalence Two programs P_1 and P_2 are said to be *partially equivalent* if any two terminating executions of P_1 and P_2 starting from the same inputs, return the same value.

The problem of program equivalence according to this definition can be reduced to one of functional verification of a single program P rather easily: P should simply call the two programs (after possible renaming of the global variables) consecutively with nondeterministic but equal inputs, and assert that they return the same output. The problem with this direct approach is that it makes no use of the expected similarity of the code. Rather, it solves a monolithic functional verification problem without decomposition. As we show in this article, the similar code structure can be beneficial exactly for this reason.

Sect. 2 below describes briefly an inference rule that we introduced in [3] for proving the partial equivalence of recursive programs. This rule is obviously not complete, but turns out to be strong enough for proving partial equivalence in many realistic cases. In Sect. 3 we will present an algorithm for decomposing the equivalence proof – ideally to the granularity of pairs of functions. We report on some experiments in Sect. 4. Due to lack of space many details about our system are left out, as well as more references to earlier works. The interested reader may find them in the first author's thesis [2]. The theoretical background on the inference rule that we use can be found also in [3].

Related work As mentioned earlier, the idea of proving equivalence between programs is not new. It is a rather old challenge in the theorem-proving community. A lot of attention has been given to this problem in the ACL2 community (see, e.g., [11, 12]). These works are mostly concerned with program equivalence as a case study for using proof techniques that are generic (i.e., not specific for proving equivalence). We are not aware of such works that are targeted at programs that are mostly syntactically equivalent, which is the target of regression verification.

Attempts to build fully automatic proof engines for industrial programs concentrated so far, to the best of our knowledge, on very restricted cases. Arons et al. [1] developed a tool in Intel for proving the equivalence of two versions of microcode, with the goal of proving backwards compatibility, but the programs were assumed to be loop-free.

Another relevant line of research is concerned with *translation validation* [14, 13], the process of proving equivalence between a source and a target of a compiler or a code generator. The fact that the translation is mechanical allows the verification methodology to rely on various patterns and restrictions on the generated code. A recent example of translation validation, from the synchronous language SDL to C, is by Haroud and Biere [5].

2. PARTIAL EQUIVALENCE

To prove partial equivalence we suggested in [3] to use an inference rule in the style of Hoare's rule for recursive invocation [7]. Hoare's rule is

$$\frac{\{p\} \text{ call } proc \{q\} \vdash_H \{p\} \, proc\text{-body} \, \{q\}}{\{p\} \text{ call } proc \{q\}} \quad (\text{REC}) , \quad (1)$$

where $proc$-body is the body of the procedure $proc$, in which the recursive call is ignored. The only effect of the recursive call on the proof is that we assume that it maintains the (p, q) relation. This unintuitive rule was described by Hoare in [7] as follows: *The solution... is simple and dramatic: to permit the use of the desired conclusion as a hypothesis in the proof of the body itself.* The correctness of rule (REC) is proved by induction, where the base case corresponds to the base of the recursion.

Our rule (PROC-P-EQ) (for 'Procedures Partial Equivalence') for partial equivalence between functions[3] P_1 and P_2 has the same flavor. For the case of P_1, P_2 being recursive functions without calls to other functions, it can be stated as follows:

$$\frac{\begin{array}{c} in[\text{call } P_1] = in[\text{call } P_2] \to out[\text{call } P_1] = out[\text{call } P_2] \vdash \\ in[P_1 \text{ body}] = in[P_2 \text{ body}] \to out[P_1 \text{ body}] = out[P_2 \text{ body}] \end{array}}{in[\text{call } P_1] = in[\text{call } P_2] \to out[\text{call } P_1] = out[\text{call } P_2]}$$
$$(2)$$

Informally, this means that if assuming that the recursive calls maintain the *congruence* condition (i.e., equal inputs lead to equal outputs) enables us to prove this condition over the bodies of P_1 and P_2 (i.e., P_1, P_2 without the recursive calls), then the congruence relation holds for P_1, P_2. In [3] we proved that this rule is sound (in fact the rule in [3] generalizes the rule presented here to the case of mutual recursion). Although the soundness proof refers to an artificial language, it has most of the features of an imperative language such as C. In Sect. 3.2 we will elaborate further on this point.

A convenient method for checking the premise of rule (PROC-P-EQ) is to replace the recursive call with an *uninterpreted function* (see, e.g., chapter 3 in [10]), because by definition it maintains the congruence condition. After performing this replacement we say that the calling function is *isolated*. Denote by P^{UF} the isolated version of a function P. Rule (PROC-P-EQ) can be reformulated accordingly:

$$\frac{\vdash_{\mathcal{UF}} in[P_1^{UF}] = in[P_2^{UF}] \to out[P_1^{UF}] = out[P_2^{UF}]}{in[\text{call } P_1] = in[\text{call } P_2] \to out[\text{call } P_1] = out[\text{call } P_2]} ,$$
$$(3)$$

where $\vdash_{\mathcal{UF}}$ is some sound inference system that can also reason about uninterpreted functions. The key observation in using this rule is that its premise is decidable for a language with finite domains such as C because, recall, it contains no loops or recursive calls. The following example demonstrates the use of this rule and shows how partial equivalence is proven.

---

[3]We use the term 'function' here although the rule refers to procedures, i.e., functions that can have multiple outputs. In a languages such as C such multiple outputs can be returned by a function if they are gathered first into a single structure. In addition, global variables that are written to by a function can be modeled as part of its list of outputs.

978-1-60558-497-3/09 $25.00 © 2009 ACM

Example 1. Consider the two functions in Fig. 1. Let H be the uninterpreted function to which we map `gcd1` and `gcd2`. Figure 2 presents the isolated functions.

```
gcd1(int a, int b)        gcd2(int x, int y)
{ int g;                  { int z;
  if (!b) g = a;            z = x;
  else
     a = a%b;               if (y > 0)
     g = gcd1(b, a);           z = gcd2(y, z%y);
  return g;                 return z;
}                         }
```

Figure 1: Two functions to calculate GCD of two positive integers.

```
gcd1(int a, int b)        gcd2(int x, int y)
{ int g;                  { int z;
  if (!b) g = a;            z = x;
  else
     a = a%b;               if (y > 0)
     g = H(b, a);              z = H(y, z%y);
  return g;                 return z;
}                         }
```

Figure 2: After isolation of the functions, i.e., replacing their function calls with calls to the uninterpreted function H.

To prove the partial equivalence of the two functions, we need to first translate them to formulas expressing their respective transition relations. A convenient way to do so is to use Static Single Assignment (SSA) (see, e.g., [10]). Briefly, this means that in each assignment of the form `x = exp;` the left-hand side variable `x` is replaced with a new variable, say x_1. Any reference to `x` after this line and before `x` is assigned again is replaced with the new variable x_1 (this is done in a context of a program without unbounded loops). In addition, assignments are guarded according to the control flow. After this transformation, the statements are conjoined: the resulting equation represents the computations of the original program.

The SSA form of `gcd1`, denoted T_{gcd_1}, is

$$\begin{pmatrix} a_0 = a & \wedge \\ b_0 = b & \wedge \\ b_0 = 0 \to g_0 = a_0 & \wedge \\ (b_0 \neq 0 \to a_1 = (a_0 \% b_0)) \wedge (b_0 = 0 \to a_1 = a_0) & \wedge \\ (b_0 \neq 0 \to g_1 = H(b_0, a_1)) \wedge (b_0 = 0 \to g_1 = g_0) & \wedge \\ g = g_1 & \end{pmatrix} \tag{4}$$

The SSA form of `gcd2`, denoted T_{gcd_2}, is

$$\begin{pmatrix} x_0 = x & \wedge \\ y_0 = y & \wedge \\ z_0 = x_0 & \wedge \\ y_0 > 0 \to z_1 = H(y_0, (z_0 \% y_0)) & \wedge \\ y_0 \leq 0 \to z_1 = z_0 & \wedge \\ z = z_1 & \end{pmatrix} \tag{5}$$

The premise of rule (PROC-P-EQ) requires proving the validity of the following formula over nonnegative integers:

$$(a = x \wedge b = y \wedge T_{gcd_1} \wedge T_{gcd_2}) \quad \to \quad g = z. \tag{6}$$

Many theorem provers can prove such formulas fully automatically, and hence establish the partial equivalence of `gcd1` and `gcd2`. $\square$

Now suppose that the two compared functions P_1, P_2 call other functions P_1^c, P_2^c, respectively. Rule (PROC-P-EQ) can still be used if one of the following holds:

1. If P_1^c, P_2^c were already proven to be equivalent then they can be replaced with uninterpreted functions. Such a replacement imposes an overapproximating abstraction. The soundness of the rule is maintained.

2. Otherwise, if P_1^c, P_2^c and their descendants are not recursive then they can be inlined in their callers. The premise of rule (PROC-P-EQ) is then checked as before.

3. Otherwise, if some of the descendants of P_1^c, P_2^c are recursive but were proven partially equivalent then these descendants can be abstracted with uninterpreted functions. As in the previous case P_1^c, P_2^c can then be inlined into their callers.

In the next section we describe an algorithm that attempts to prove partial equivalence of general programs by traversing the call graphs bottom-up and replacing functions with their uninterpreted versions when possible, based on these generalizations.

3. REGRESSION VERIFICATION

Our Regression Verification Tool (RVT) is geared towards C programs and hence we begin with a brief description of CBMC [9], the underlying decision procedure that we use for checking the premise of rule (PROC-P-EQ) and its generalizations as described in the previous section. CBMC, developed by D. Kroening, is a bounded model checker for C programs that supports almost all of the features of ANSI-C. It requires from the user to define a bound k on the number of iterations that each loop in a given ANSI-C program is taken, and a similar bound on the depth of each recursion. This enables CBMC to symbolically characterize the full set of possible executions restricted by these bounds, by a decidable formula f. The existence of a solution to $f \wedge \neg a$, where a is a user defined assertion, implies the existence of a path in the program that violates a. Otherwise, we say that CBMC established the k-correctness of the checked assertions. We use CBMC in a very restricted way, however: recall that the premise of rule (PROC-P-EQ) is over nonrecursive functions without loops (hence in our case $k = 1$).

RVT generates small loop-free and recursion-free C programs – each corresponds to a pair of functions that it attempts to prove equal – which it sends to CBMC for decision. Before this iterative process begins, RVT makes two preliminary steps.

Loops All loops in P_1 and P_2 are replaced with recursive functions. This process is described in [2].

Pairing A pairing *pair* is built by pairing functions and global variables between the two compared programs.

Pairing is done recursively, in a manner reminiscent of computing congruence closure. The algorithm works on the parse trees of both programs and pairs nodes, where a node can be either a variable, a function, or a type. Note that wrong pairing does not affect soundness: pairing is used for generating the verification conditions, and hence wrong pairing can only fail a proof. The pairing algorithm works top-down: it initially pairs global variables with the same name and type. It then pairs functions with the same name, return type, and prototype. Then, within paired functions that are also syntactically equivalent up to variable names, it attempts to pair elements that appear in isomorphic locations. If these elements were already paired it just checks that the pairing according to this function agrees with the previous one, and otherwise it issues a warning. This process is repeated until no new pairing is discovered.

We assume here that as a minimum this process results in a bijective pairing between all *recursive* functions. Without this condition fulfilled, RVT can only attempt to prove partial equivalence of subprograms rooted at paired functions that fulfill this condition.[4]

The input to the main algorithm is thus two recursive programs without loops and a pairing $pair$. We denote by $pair_f$ the pairs of functions in $pair$.

3.1 A bottom-up decomposition algorithm

The equivalence check in RVT, in its basic form, is presented in Algorithm 1. It is based on traversing bottom-up the call graphs of the two programs to be compared. This algorithm can be applied directly to two call graphs without loops of length larger than 1 (i.e., no mutual recursion). The more general case is considered in [2].

The algorithm progresses bottom-up on the call graphs, and updates the labels on the nodes to "Equivalent" or "Failed". The progress on the graph is made by a function NEXT-UNMARKED-PAIR() (not presented) which returns the next unmarked pair in $pair_f$, according to a BFS order on the reversed call graph of one of the two compared programs. This function aborts if either all pairs are already marked or it finds that the call-graphs are inconsistent (inconsistency means that there are two pairs of functions $\langle f, f' \rangle \in pair_f$ and $\langle g, g' \rangle \in pair_f$ such that f is a descendant of g but f' is an *ancestor* of g').

The equivalence check in line 5 is conducted by verifying, with CBMC, that various assertions hold in a single loop-free and recursion-free C program $check\text{-}block(f, g)$ that RVT constructs (see below). CBMC returns TRUE if the assertions hold and FALSE otherwise. We call these checks 'semantic checks' to distinguish them from the syntactic checks in line 2. We now proceed by describing $check\text{-}block(f, g)$.

Check-blocks Consider the maximal connected subDAG rooted at f that contains only functions that are unpaired or marked "Failed". Let S_f denote this set of functions (excluding f). S_g is defined similarly with respect to g. The program $check\text{-}block(f, g)$ consists of the following elements:

1. The functions f, g and all functions in S_f, S_g, such that

- Name collisions in global identifiers of the two programs are solved by renaming;
- All calls to f, g are replaced with calls to $UF(f)$, $UF(g)$, respectively;
- For all $\langle h_1, h_2 \rangle \in pair_f$ such that $h_1, h_2 \notin \{S_f \cup S_g\}$, calls to h_1, h_2 are replaced, respectively, with calls to $UF(h_1)$ and $UF(h_2)$. (Observe that the pair $\langle h_1, h_2 \rangle$ is marked "Equivalent").

2. The `main()` function, which consists of:

- Assignment of nondeterministic but equal values to inputs of f and g;
- Calls to f, g; and
- Assertion that the outputs of f and g are equal.

Following are several notes on the definition of *check-block(f, g)*:

- The check-block is nonrecursive. This is because when a recursive pair $\langle f, g \rangle \in pair_f$ is labeled "Failed" the algorithm aborts in line 8, and hence the code of f, g will not be part of future check-blocks.

- The code of each nonrecursive pair $\langle f, g \rangle \in pair_f$ that is labeled "Failed" is included when checking the equivalence of their parents, and possibly more ancestors, until reaching a provably equivalent pair or reaching the roots. We call this process *logical inlining*, since it is equivalent to inlining but is more faithful to the program's original structure. This enables RVT to prove equivalence in case, for example, that some code was moved from the parent to the child, but together they still perform the original computation.

- The code of a pair $\langle f, g \rangle \in pair_f$ that is proven to be equivalent does not participate in any subsequent check-block. It is replaced with uninterpreted functions in all subsequent semantic checks, or disappears altogether if some ancestor pair is also marked "Equivalent" in each of its paths to the roots of the related subprograms.

- The replacement of recursive calls of paired functions with uninterpreted functions corresponds to isolation (see Sect. 2). Recall that proving equivalence of paired isolated functions also proves their partial equivalence by rule (PROC-P-EQ).

Equivalence specification Since just proving the equivalence of the return value is insufficient in practice, RVT allows the user to specify the equivalence criterion with pairs of tuples of the form $\langle label, condition, expression \rangle$. It then adds assertions to the check-blocks that check that at the locations specified by the labels the expressions are equivalent every time the conditions hold.

Complexity Each pair in $pair_f$ is labeled at most once by either "Failed" or "Equivalent". Thus, if $n = |pair_f|$, which, in turn, cannot be larger than the number of functions, then the algorithm performs not more than n syntactic and n semantic checks.

Example 2. Consider the call graphs in Fig. 3. Assume that for $i = 1, \ldots, 6$ we have $\langle f_i, f_i' \rangle \in pair_f$, and that the functions marked by grey nodes in Fig. 3 are syntactically equivalent to their counterparts. We describe step by step the execution of Algorithm 1:

---

[4]RVT can also attempt to prove k-equivalence in this case, i.e., prove that there is no trace contradicting the equivalence which requires a recursion depth larger than k. The description of this feature is beyond the scope of this paper.

978-1-60558-497-3/09 $25.00 © 2009 ACM

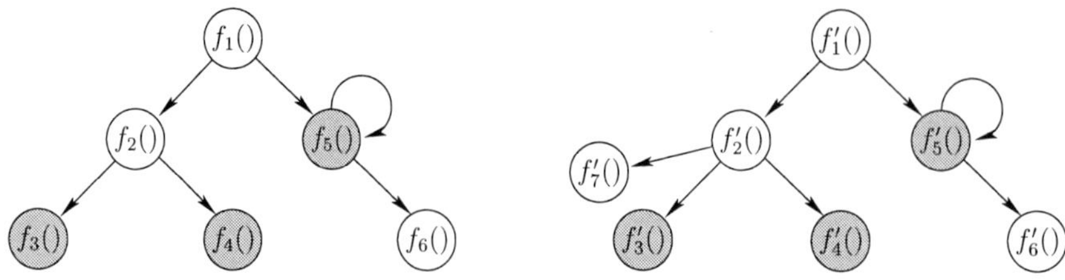

Figure 3: Two call graphs for Example 2. A node is grey if it is syntactically equivalent to its counterpart.

Algorithm 1 A basic call-graph based algorithm for attempting to prove the partial equivalence of pairs of functions.

Procedure COMPARE()
input: Call graphs CG_1 and CG_2 and a pairing $pair_f$.
output: Marking of pairs in $pair_f$ with "Equivalent" or "Failed".

1: $\langle f, g \rangle$ = NEXT-UNMARKED-PAIR() ▷ Bottom-up. Aborts if none.
2: **if** $\langle f, g \rangle$ are syntactically equivalent and all their children (not including recursive calls) are marked by "Equivalent" **then**
3:     Mark $\langle f, g \rangle$ by "Equivalent".
4: **else**
5:     **if** CBMC($check\text{-}block(f, g)$) **then** ▷ Semantic check
6:         Mark $\langle f, g \rangle$ "Equivalent".
7:     **else**
8:         **if** f, g are recursive **then** abort.
9:         Mark $\langle f, g \rangle$ "Failed".
10: goto line 1.

---

1. In line 3 pairs $\langle f_3, f_3' \rangle$, $\langle f_4, f_4' \rangle$ are marked "Equivalent".

2. The program $check\text{-}block(f_2, f_2')$ is sent to CBMC. Now $S_{f_2'} = \{f_7'\}$ and hence this program contains also the code of f_7', whereas f_3, f_3', f_4, f_4' are replaced by uninterpreted functions. Assume that this semantic check fails. Then the pair $\langle f_2, f_2' \rangle$ is marked "Failed".

3. The program $check\text{-}block(f_6, f_6')$ is sent to CBMC. Assume that the check fails and hence the pair is marked "Failed".

4. The program $check\text{-}block(f_5, f_5')$ is sent to CBMC. Since $S_{f_5} = \{f_6\}$ and $S_{f_5'} = \{f_6'\}$, this program contains also f_6, f_6'. The recursive calls are replaced with uninterpreted functions(based on rule (PROC-P-EQ)). Assume that this time the check succeeds. Then $\langle f_5, f_5' \rangle$ is marked "Equivalent" in line 6.

5. The program $check\text{-}block(f_1, f_1')$ is sent to CBMC. Now $S_{f_1} = \{f_2\}$ and $S_{f_1'} = \{f_2', f_7'\}$. Hence, the respective call subgraphs contain also f_2, f_2' and f_7', whereas $f_3, f_3', f_4, f_4', f_5, f_5'$ are replaced with uninterpreted functions. Assume this check succeeds. The algorithm marks $\langle f_1, f_1' \rangle$ with "Equivalent".

At this stage all function pairs are marked by either "Equivalent" or "Failed" and the algorithm terminates.

The soundness proof of algorithm 1 appears in [G08] and is omitted here due to space limitations.

3.2 Specific issues with C programs

RVT works on C (reference ANSI C99) programs, although not all features are supported. A major issue in applying rule (PROC-P-EQ) to C programs is that of dynamic data structures. Recall that deciding formulas with uninterpreted functions requires the comparison pair-wise of the arguments with which such functions are called, and a similar comparison of their outputs. If some of these arguments are pointers, such a comparison is meaningless. In this section we briefly describe RVT's method of treating pointer arguments of functions and dynamic data structures.

Whereas in nonpointer variables the comparison is between values, in the case of pointer variables the comparison should be between the data structures that they point to. A dynamic data structure can be represented as a graph whose vertices are structs and edges are the pointers that point from one struct to the other. We call such graph a *pointer-element graph*. We make a simplifying assumption that all dynamic structures that are passed to a function through pointer arguments or globals are in the form of trees, i.e., aliasing within dynamic structures and between function arguments is not allowed. We define equality of structures as follows:

Definition 2. (Iso-equal structures) Two structures are *iso-equal* if their pointer-element graphs are isomorphic and the values at structs related by the isomorphism are equal.

Let p_1, p_2 be paired pointer variables that are arguments to the functions that we wish to compare. RVT generates a tree-like data structure with a bounded depth (see below) and with nondeterministic values. It then makes both p_1 and p_2 point to this tree. This guarantees that the input structure is arbitrary but equivalent up to a bound, and under the assumption that on both sides it is a tree. A similar strategy is activated when we compare p_1 and p_2 that point to an output of the compared functions.

What should be the bound on this tree? Recall that the code of the related subprograms that we check (the check-block) does not contain loops or recursion, and hence there is a bound on the maximal depth of the items this code can access in any dynamic data structure that is passed to the roots of the related subprograms. It is possible, then, to compute this bound, or at least overestimate it, by syntactic analysis. For example, searching for code that progresses on the structure such as `n = n -> next` for a pointer `n`. Such a mechanism is not implemented yet in RVT, however, and it relies instead on a user-defined bound.

4. EXPERIMENTS

We tested RVT on several synthetic and limited-size industrial programs and attempted to prove equivalent dif-

978-1-60558-497-3/09 $25.00 © 2009 ACM

ferent versions of these programs. In each case we specified simple equivalence criteria in the form of conditional expressions as explained before. We tested RVT with the following sets of programs.

Random programs We used a random program generator to create several dozen recursive programs of different sizes. The user specifies the probability to generate each type of variable, block, or operator. Variables can be global, local or formal arguments of functions. Types can be basic C types, structures or pointers to such types. Small differences in versions are introduced in random places. We used this program to generate random yet executable C programs with up to 20 functions and thousands of lines of code. When the random versions are equivalent, RVT proves them to be partially-equivalent relatively fast, ranging from few seconds to 30 minutes. On non-equivalent versions, on the other hand, attempts to prove partial equivalence may run for many hours or run out of memory.

Industrial programs The small industrial programs that we tried are:

TCAS (Traffic Alert and Collision Avoidance System) is an aircraft conflict detection and resolution system used by all US commercial aircraft. We used a 300-line fragment of this program that was also used in [4].

MicroC/OS The core of MicroC/OS which is a low-cost priority-based preemptive real time multitasking operating system kernel for microprocessors, written mainly in C. The kernel is mainly used in embedded systems. The program is about 3000 lines long.

Matlab examples Parts of engine-fuel-injection simulation in Matlab which was generated in C from engine controller models. The tested parts contain several hundreds lines of code and use read-only arrays.

All these tests exhibit the same behavior as the random programs above. For equivalent programs, semantic-checks are very fast, proving equivalence in minutes. We did not encounter a case in which partially equivalent programs cannot be proven to be so due to the incompleteness of (PROC-P-EQ).

Recall that in the process of semantic checks, paired functions that cannot be proven equivalent are (logically) inlined. Our experience was that in such cases the proof becomes too hard: the decision procedure runs for hours or even fails to reach a decision at all. In some examples the bottleneck is the use of operators that burden the SAT solver, such as multiplication (*), division (/) and modulo (%) over integers. A simple solution in such cases is to *outline* these operators (i.e., take them out to a separate function). The reason is that RVT proves the equivalence of these separate functions syntactically and then replaces them with uninterpreted functions, which reduces the computation time dramatically.

5. SUMMARY

We started the introduction by mentioning Tony Hoare's grand challenge, namely that of functional verification, and by mentioning that proving equivalence is a grand challenge in its own right, although an easier one. In this work we started exploring this direction in the context of C programs, and reported on our prototype tool RVT with which we were able to prove the equivalence of several small industrial programs. Our technique can be improved in several dimensions, such as strengthening rule (PROC-P-EQ) with automatically generated invariants and finding more opportunities for making the verification conditions easier to decide. Investigating such opportunities for object-oriented code is another big challenge.

To summarize, the main contribution of this article is a method for an automatic, incremental proof, based on isolating functions from their callees and abstracting them with uninterpreted functions. This method keeps the verification conditions decidable and small relative to the size of the input programs. The initial syntactic checks and the decomposition mechanism helps meeting our goal of keeping the complexity sensitive to the changes rather than to the original size of the compared programs.

6. REFERENCES

[1] T. Arons, E. Elster, L. Fix, S. MadorHaim, M. Mishaeli, J. Shalev, E. Singerman, A. Tiemeyer, M. Y. Vardi, , and L. D. Zuck. Formal verification of backward compatibility of microcode. In *CAV05*, volume 3576 of *LNCS*. Springer, 2005.

[2] B. Godlin. Regression verification: Theoretical and implementation aspects. Master's thesis, Technion, Israel Institute of Technology, 2008.

[3] B. Godlin and O. Strichman. Inference rules for proving the equivalence of recursive procedures. *Acta Informatica*, 45(6):403–439, 2008.

[4] A. Groce, D. Kroening, and F. Lerda. Understanding counterexamples with explain. In *CAV*, pages 453–456, 2004.

[5] M. Haroud and A. Biere. SDL versus C equivalence checking. In *SDL Forum*, pages 323–338, 2005.

[6] C. Hoare. An axiomatic basis for computer programming. *Comm. ACM*, 12(10):576–580, 1969.

[7] C. Hoare. Procedures and parameters: an axiomatic approach. *In Proc. Sym. on semantics of algorithmic languages*, (188), 1971.

[8] T. Hoare. The verifying compiler: A grand challenge for computing research. *J. ACM*, 50(1):63–69, 2003.

[9] D. Kroening, E. Clarke, and K. Yorav. Behavioral consistency of C and Verilog programs using bounded model checking. In *Proceedings of DAC 2003*, pages 368–371. ACM Press, 2003.

[10] D. Kroening and O. Strichman. *Decision procedures – an algorithmic point of view*. Theoretical computer science. Springer, May 2008.

[11] P. Manolios and M. Kaufmann. Adding a total order to acl2. In *The Third International Workshop on the ACL2 Theorem Prover*, 2002.

[12] P. Manolios and D. Vroon. Ordinal arithmetic: Algorithms and mechanization. *Journal of Automated Reasoning*, 2006. to appear.

[13] A. Pnueli, M. Siegel, and O. Shtrichman. Translation validation for synchronous languages. In *Proc. 25th Int. Colloq. on Automata, Languages and Programming (ICALP'98)*, volume 1443 of *LNCS*, pages 235–246. Springer, 1998.

[14] A. Pnueli, M. Siegel, and E. Singerman. Translation validation. In *TACAS'08*, volume 1384 of *LNCS*, pages 151–166. Springer, 1998.

2009 46th ACM/IEEE Design Automation Conference

(DAC 2009)

San Francisco, CA, USA
26 – 31 July 2009

Pages 472-968

IEEE Catalog Number: CFP09DAC-PRT
ISBN: 978-1-60558-497-3

Copyright © 2009, ACM
All Rights Reserved

****This publication is a representation of what appears in the IEEE Digital Libraries. Some format issues inherent in the e-media version may also appear in this print version.*

IEEE Catalog Number: CFP09DAC-PRT
ISBN 13:   978-1-60558-497-3
Library of Congress No.:   85-644924
ISSN: 0738-100X

Additional Copies of This Publication Are Available From:

Curran Associates, Inc
57 Morehouse Lane
Red Hook, NY 12571 USA
Phone: (845) 758-0400
Fax: (845) 758-2633
E-mail: curran@proceedings.com

TABLE OF CONTENTS

SESSION 12 – DESIGN INTEGRITY CHALLENGES

SESSION 13 – PANEL

SESSION 14 – SPECIAL SESSION:
VERIFYING AN SOC MONSTER: WHOSE JOB IS IT ANYWAY?

SESSION 15 – TIMING SIMULATION: OPTIMIZED EMBEDDED SOFTWARE AND MPSOCS

SESSION 16 – ADVANCES IN EMBEDDED SYSTEM MODELING AND OPTIMIZATION

SESSION 23 – ANALOG/RF SIMULATION AND STATISTICAL MODELING

SESSION 24 – RECENT ADVANCES IN TIMING, ECO AND LOGIC OPTIMIZATION

SESSION 25 – PANEL

SESSION 26 – SPECIAL SESSION: COMPUTATION IN THE POST-TURING ERA

SESSION 27 – ADVANCES IN PHYSICAL SYNTHESIS

SESSION 39 – EMBEDDED SYSTEM DESIGN FOR LOW-POWER

SESSION 40 – HARDWARE AUTHENTICATION, CHARACTERIZATION AND TRUSTED DESIGN

SESSION 41 – TARGETED TEST AND DIAGNOSIS

SESSION 42 – CHALLENGES OF MEMORY-AWARE DESIGN FOR EMBEDDED SYSTEMS

SESSION 54 – MODEL ORDER REDUCTION TECHNIQUES AND APPLICATIONS

Accurate Temperature Estimation Using Noisy Thermal Sensors

Yufu Zhang and Ankur Srivastava
Department of ECE, University of Maryland, College Park, MD, 20742
yufuzh, ankurs@umd.edu

ABSTRACT

Multicore SOCs rely on runtime thermal measurements using on-chip sensors for DTM. In this paper we address the problem of estimating the actual temperature of on-chip thermal sensor when the sensor reading has been corrupted by noise. Thermal sensors are prone to noise due to fabrication randomness, V_{DD} fluctuations etc. This causes discrepancy between actual temperature and the one predicted by thermal sensor. Our experiments estimate this variation to be around 30%. In this paper we present a statistical methodology for predicting the actual temperature for a given sensor reading. We present two techniques: single sensor prediction and multi-sensor prediction. The latter tries to estimate the actual temperature for each sensor (of the many on-chip sensors) simultaneously while exploiting the correlations between temperature and noise of different sensors. When the underlying randomness follows a Gaussian characteristic, we present optimal schemes of estimating the expected temperature. We also present heuristic schemes for the case where the Gaussian assumption fails to hold. The experiments showed that using our estimation schemes the RMS error can be reduce as much as 67% as compared to blindly trusting the sensors to be noise free.

Categories and Subject Descriptors

B.7.2 [**Integrated Circuit**]: Design Aids – *verification*

General Terms

Algorithms, Measurement, Verification

Keywords

Temperature, On-chip Sensor, Estimation, Multicore, DTM

1. INTRODUCTION

Dynamic thermal management (DTM) is the technique of controlling the temperature surges on chip usually caused due to heavy task execution. Typically, this techniques uses a set of on-chip sensors to estimate the silicon temperature. If the temperature is above a given threshold then corrective action is taken to bring the chip back within acceptable temperature levels [18,19]. Accurate thermal sensing using on-chip sensors during the runtime is essential for effective DTM. Several multicore processors have been equipped with thermal sensors (AMD Opteron has 38 thermal sensors).

Although several existing approaches have addressed the thermal sensor design and placement problem, the noise associated with thermal sensor readings has not received much attention. The state of

Permission to make digital or hard copies of part or all of this work for personal or classroom use is granted without fee provided that copies are not made or distributed for profit or commercial advantage and that copies bear this notice and the full citation on the first page. To copy otherwise, to republish, to post on servers or to redistribute to lists, requires prior specific permission and/or a fee.
DAC'09, July 26-31, 2009, San Francisco, California, USA

the art assumes that once placed and calibrated, the on-chip thermal sensors provide accurate thermal information during runtime. The entire dynamic thermal management process assumes that the sensed thermal data is accurate and noise free. In reality, on chip thermal sensors are affected by a range of noise sources. Some of these noise sources include fabrication randomness, power grid noise, cross coupling and non-linear dependence between temperature and circuit parameters. Assuming ideal thermal sensor operation can either lead to failure to detect overheating or false alarms that result in costly and un-necessary response from the thermal management unit. Accounting for the noisy behavior in thermal sensor operation is therefore crucial.

In this paper, we present a statistical methodology for estimating the real temperature given noisy thermal sensor observations. For any given sensor observation, we model the actual sensor temperature as a random variable with an associated probability density function (PDF). The nature of this PDF depends upon the noise sources such as fabrication randomness, power grid noise and others. The problem is then formulated as estimating the expected silicon temperature for a given sensor observation. Such a methodology is developed for two cases: single sensor and multi-sensor. When we have only one thermal sensor, the expected temperature is calculated by applying the Bayes rule on the joint-PDF (JPDF) of actual temperature and sensor reading. When this JPDF is Gaussian, we present an optimal scheme for estimating the temperature. When we have multiple sensors, the real temperatures of different sensors and sensor noise will in general exhibit correlation because of the correlated fabrication randomness and correlated switching activities. Due to such correlations the sensor readings contain temperature information about each other as well. In this paper we develop a scheme for simultaneously estimating the temperatures for different sensors while exploiting these correlations. Even in the multi-sensor case, if the underlying joint PDF between sensor readings and actual temperatures is Gaussian, we present an optimal scheme for estimation. When the underlying joint PDF is non-Gaussian we use a moment matching based scheme for approximating the sensor temperatures.

Our experimental results indicate that our methods can significantly increase the estimation accuracy of sensor temperature by up to 67% as compared to ignoring the error in sensor readings. We also found that exploiting the correlations of different sensors results in better thermal estimates than ignoring them and estimating each sensor temperature individually.

Our statistical scheme for reading noisy sensors can be implemented with very low overhead either as a hardware module or as a software kernel in the operating system. Our MATLAB implementations took only about one millisecond to execute (on a 2.0 GHz Pentium Dual Core machine). Thermal profiles of chips take several hundred milliseconds to change [17], therefore a few millisecond overhead for reading noisy thermal sensors will not impact the effectiveness of DTM.

2. MOTIVATION

2.1 Unpredictable On Chip Thermal Profile

Unpredictability in workload overhead and variability in transistor & interconnect parameters causes randomness in the chip thermal profile. If we divide the entire chip into small grids and approximate the temperature in each grid cell as a constant (not an unreasonable assumption for large number of grids), then the grid temperature values form a correlated vector of random variables. The joint-PDF of this random thermal vector could be computed by exploiting the dependence of temperature on random on-chip power dissipation values (due to unpredictable workload) and circuit parameters. One such approach could be as follows. Let each grid have an associated power density. The thermal distribution on chip is guided by the Poisson equation which can be solved using the method of Green Function [1]. The solution suggests that the relationship between the power density and temperatures in each grid can be approximated by the following linear transformation [1].

$$\vec{T} = T_0 + A\vec{P} \tag{1}$$

where $T = \{T(1),\ T(2),...T(i),...\}$ is the vector that represents the average temperature at all grid locations and $P = \{P_d(1), P_d(2),...P_d(j),...\}$ is the vector of all average power density values at the grid locations. The coefficient matrix A can be computed as in [1]. If P is random (due to random workload characteristics) with an associated PDF, we can use the above relationship to compute the PDF of T. Hence in this paper we assume that the randomness associated with T is known and/or computable.

2.2 On-chip Thermal Sensors

Thermal sensors provide real time measurements pertaining to the thermal state of the chip. Since thermal profile exhibits unpredictable behavior, thermal sensors have become an integral part of future Multicore systems and SOCs. For example IBM Power5 employs 24 thermal sensors, IBM Cell processor has 10 digital thermal sensors, AMD Opteron quad-core processor has about 38 sensors. Consider a popular implementations of on-chip temperature sensor --- Ring Oscillator (abbr. RO).

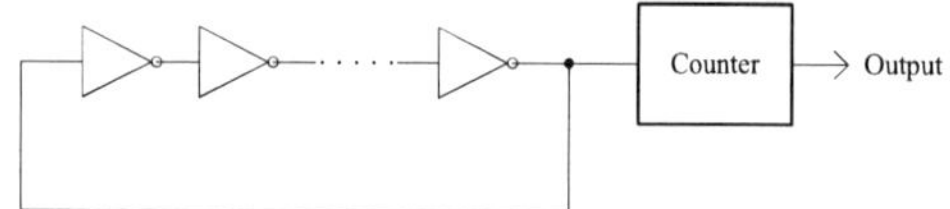

Fig.1 Typical ring oscillator consists of
N stages of inverters, where $N = 2k+1$ is an odd number.

The above figure highlights the structure of a typical ring oscillator. The oscillating frequency is determined by the propagation delay of each individual inverter and will be observed at the output by the counter.

$$t_{PHL} = \frac{2C}{\mu_n C_{ox}(W/L)_n(V_{DD}-V_t)}\left[\frac{V_t}{V_{DD}-V_t} + \frac{1}{2}\ln\left(\frac{3V_{DD}-4V_t}{V_{DD}}\right)\right] \tag{2}$$

The above equation shows the transition time for the inverter to switch from high voltage level to low voltage level. μ_n is the electron mobility in silicon for NMOS, C_{ox} is the capacitance per unit gate area, $(W/L)_n$ is the width/length ratio for NMOS, C is the effective load capacitance that the inverter is driving. V_{DD} is the supply voltage and V_t is the threshold voltage. The expression for t_{PLH} is identical to equation (2) except that μ_n and $(W/L)_n$ are replaced with μ_p and $(W/L)_p$, which are the corresponding parameter values for PMOS. The frequency of the ring oscillator is given by the following equation

$$f = \frac{1}{P} = \frac{1}{N(t_{PHL}+t_{PLH})} \tag{3}$$

where P is the period of the RO and N is the total number of inverter stages.

Temperature effects: In equation (2), both V_t and μ_n (or μ_p) are temperature sensitive. The magnitude of V_t decreases by about $2mV$ for every $1\,°C$ increase in the temperature [6]. μ_n (or μ_p) also decreases with temperature rise with a more complex relationship. Because the effect of the latter is a more dominant one, the overall observed effect of the temperature rise is a decrease in the RO frequency [6, 7, 8]. Hence RO frequency observation can be used as a predictor of chip temperature. Though on-chip temperature sensors are powerful, if not used properly, however, the information collected by them can be highly erroneous and misleading. This is because the accuracy of the sensors themselves is also affected significantly by fabrication randomness and the environment uncertainty (like V_{DD} noise, cross coupling etc).

2.3 Noisy Sensor Behavior

As we can see from equation (2) and (3), the frequency of RO is a function of many parameters which are subject to fabrication randomness and environmental uncertainty.

$$f = F(W, L, V_{DD}, V_t(T), \mu_{n/p}(T), C) \tag{4}$$

For example, W, L etc are subject to fab-variability while V_{DD} fluctuations are common during execution time. Assuming all the noisy parameters in the above expression are random variables with a certain probability distribution, the observed frequency will be a random variable as well whose probability density function (PDF) is determined by its local temperature. For any given temperature value, we can get many potential sensor readings based upon the nature of the fabrication and environmental uncertainty. Hence the temperature reading provided by such a sensor cannot be blindly relied upon.

We performed the following Monte Carlo simulation based on equations (2) and (3) with 100000 samples for each different temperature value (ranging from $20\,°C$ to $100\,°C$ with $20\,°C$ increments). We assume all parameters to be Gaussian random variables with mean and variance specified as below.

Table 1 Characteristics of the random Parameters used in Monte Carlo simulation

Parameters	W_n/W_p (nm)	L_n/L_p (nm)	T_{ox} (nm)	V_{DD} (V)	V_t (V)	μ_n/μ_p ($m^2/V{\cdot}s$)
Mean	270	180	4.1	3	0.45	0.034
Std_dev	5%	6%	3%	5%	4%	3%

As shown in the above table, we use 180nm technology process parameters, the mean of the parameters are set to their nominal values and the standard deviations from the mean are set in accordance with the predictions from [9]. The sole purpose of this simulation is to demonstrate how the sensor observations (RO frequency) exhibits randomness under the influence of fabrication variability and environmental uncertainties. Hence the values used in table 1 are only representative rather than precise. To account for the temperature effects we use the following empirical equations relating V_t and $\mu_{n/p}$ with temperature T [6,10].

$$V_t = V_{t0} + 0.002(T-T_0) \tag{5}$$

$$\mu_{n/p} = \mu_0(T/T_0)^{-1.5} \tag{6}$$

where V_{t0} and μ_0 are the nominal values of V_t and $\mu_{n/p}$ evaluated at the room temperature T_0. Figure 2 has the result of this MC simulation.

978-1-60558-497-3/09 $25.00 © 2009 ACM

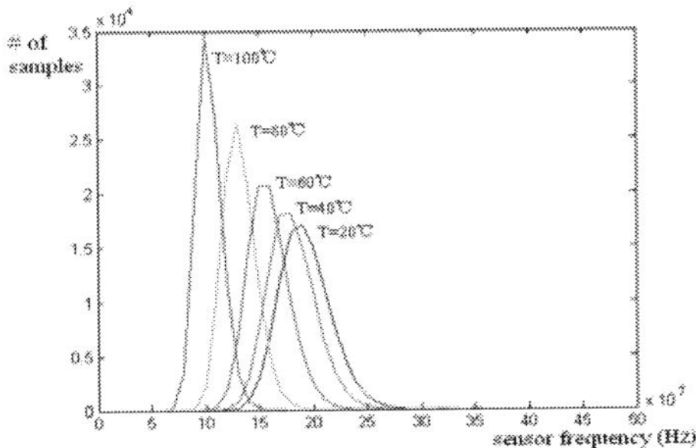

Fig.2 Simulated RO frequency PDF under fabrication randomness for each different temperatures ranging from 20℃ to 100℃ with 20℃ increment (5 curves with different colors shown together for comparison).

Essentially figure 2 highlights the potential noise in the RO frequency readings. The horizontal axis shows the frequency values and the vertical axis shows the number of samples fallen into each small range around a certain frequency value. Thus each curve in the figure shows the potential PDF of the sensor frequency when its underlying temperature is fixed at the value as shown in the figure. One important observation is that these distribution curves heavily overlap with each other and all of them have significant amount of variation. Hence if the sensors are blindly trusted the possible error could be significant. This clearly demonstrates the need for temperature estimation methodologies to predict the accurate temperature value from noisy sensor readings.

3. SINGLE SENSOR ESTIMATION

3.1 Randomness Modeling:

Consider the case where we have a single sensor with f being the sensor reading. Under nominal circumstances where all other sensor parameters (except temperature) are nominal and known, the sensor reading w.r.t. sensor temperature can be approximated by the following linear relationship [7,8].

$$T = a_0 f + b_0 \qquad (7)$$

The one-to-one correspondence between temperature T and sensor frequency f can be used to accurately estimate T when a certain f is observed. However in reality, as mentioned earlier, each sensor parameter like V_{DD}, V_t, W, L etc are random owing to fabrication imperfections and environmental uncertainties. Hence the constant a_0 and b_0 become random as well. The PDF associated with these random parameters could be characterized using some of the current statistical analysis techniques that are being developed for countering the effects of fabrication randomness [11,12]. The exact schemes and details for generating these PDFs are ignored here because the focus of this paper is different.

The real temperature is a random variable with a known PDF (to be determined using sensors). As illustrated in section 2.1, one possible way of generating the thermal PDF at a point could be as follows. If we know the PDF of the power density vector P then using equation (1) we can determine the PDF of the temperature vector T as well. For example, figure 3 highlights the PDF of the power density at a specific location in an out-of-order processor chip. We simulated a high performance aggressive out-of-order processor with pipeline width of 8 instructions and an instruction window of 128 instructions. Level 1 caches (both instruction cache and data cache) are 32KB 4-way set associative. All the caches in the hierarchy are using LRU replacement policy and a block size of 64 bytes. For benchmarks, we simulated all the SPEC 2000 CPU benchmark suite [13] compiled

with the default parameters provided with the suite. We bypassed the startup part, based on simpoint [14], and simulated a representative 250M instructions for each of the benchmarks. Figure 3 below shows the power density PDF for the instruction window module.

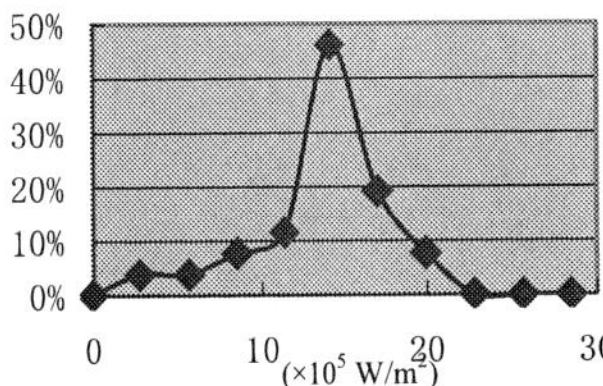

Fig.3 Power density distribution for instruction window module.

Even though we may not accurately know the exact value of the power density of any part of the chip, using statistical technique on representative benchmarks, we can generate the power density PDF for those parts. Based on this randomness associated with the power density vector P, we can characterize the randomness associated with the temperature vector T across various representative benchmarks based on the linear transformation highlighted in equation (1). In this way, we can account for the un-predictable workload in our formulation.

Given the random characteristics of each parameter on which sensor frequency depends we can evaluate the joint probability density function JPDF for different sensor parameters like Temperature T, V_{DD}, V_t, W, L, T_{ox} etc. There is already a significant effort towards statistical analysis of timing and leakage given random fabrication parameters [11,12,15]. Similar approaches could be used here to generate the JPDF also.

Once we generated the JPDF for various random sensor parameters like T, V_{DD}, V_t, W, L etc, we could evaluate their accumulative effect on the frequency of the RO based on equation (2) and (3). Using this relationship we can compute the joint-PDF between the frequency and the sensor parameters T, V_{DD}, V_t, W, L etc. Let us suppose that JPDF (f, T, V_{DD}, V_t, W, L..) represents this joint-PDF, then the joint PDF between the sensor reading f and temperature T at the sensor location could be evaluated as follows

$$p(f,T) = \int \int ... \int jpdf(f,T,V_{DD},V_t,W,L...)dV_{DD}dV_t dWdL...$$

Essentially we are integrating over all other random parameters besides f and T to generate the marginal joint probability density function for f (our observation) and T (the value that we want to estimate). Note that this joint PDF between f and T accounts for any correlation (spatial or logical) between different parameters for this sensor. Given this JPDF and a sensor reading f, we would like to find the expected value of the temperature that might have caused that specific reading.

$$E(T \mid f) \qquad (8)$$

This strategy of estimating the expected temperature for a given sensor reading exploits the statistical information that could be generated a-priori.

3.2 Optimal Solution for Gaussian Distribution

Let us assume that the function $p(f,T)$ is jointly Gaussian. In realistic scenarios this may not be a very inaccurate assumption. Consider figure 2 which illustrates the PDF of the sensor frequency for each value of sensor temperature. It can be inferred that a Gaussian approximation of the illustrated PDFs is a reasonably accurate approximation.

978-1-60558-497-3/09 $25.00 © 2009 ACM

474

When $p(f,T)$ is jointly Gaussian, the following equation could be used to estimate the expected T for the given observation f [16].

$$E(T|f) = \mu_T + \frac{cov(T,f)}{\sigma_f^2}(f - \mu_f) \qquad (9)$$

where μ_T and μ_f is the mean of T and f respectively, $cov(T,f)$ is their covariance and σ_f^2 is the variance of f (all of them are known). Under the joint Gaussian assumption, the above formula provides the optimal estimate of the temperature for a given observation f which has been corrupted by noise. Proof is omitted for brevity. The interested readers can refer to [16] for details.

3.3 Non-Gaussian Case

In the previous sub-section we presented a technique for estimating the $E(T|f)$ when $p(f,T)$ was jointly Gaussian. From our experience, although $p(f,T)$ may not be exactly Gaussian, a Gaussian approximation usually provides reasonable accuracy towards the estimation of T. Essentially we can approximate the non-Gaussian $p(f,T)$ with a Gaussian JPDF. Then the result above could be used to calculate the approximate value for T. As will be shown in the experiment section later, for our application of estimating temperature from noisy sensor readings, this method does provide highly accurate estimates of expected temperature.

We use moment matching to fit a Gaussian JPDF $p_G(f,T)$ over a non-Gaussian $p(f,T)$. The approximation is done by matching the characteristic functions of $p(f,T))$ and $p_G(f,T)$. The characteristic function is defined as the Fourier transform of any JPDF function. For the two variable case the characteristic function of JPDF $f_{X1,X2}(x_1,x_2)$ is

$$\Phi(y_1, y_2) = \int_{-\infty}^{\infty} \int_{-\infty}^{\infty} e^{i(y_1 x_1 + y_2 x_2)} f_{X_1,X_2}(x_1,x_2) dx_1 dx_2$$

Expanding the exponential term in the above equation gives a series representation of $\Phi(y_1, y_2)$:

$$\Phi(y_1, y_2) = 1 + iy_1 \iint x_1 f_{X_1,X_2}(x_1,x_2) dx_1 dx_2 + iy_2 \iint x_2 f_{X_1,X_2}(x_1,x_2) dx_1 dx_2$$
$$- \frac{y_1^2}{2} \iint x_1^2 f_{X_1,X_2}(x_1,x_2) dx_1 dx_2 - \frac{y_2^2}{2} \iint x_2^2 f_{X_1,X_2}(x_1,x_2) dx_1 dx_2$$
$$- y_1 y_2 \iint x_1 x_2 f_{X_1,X_2}(x_1,x_2) dx_1 dx_2 + \cdots$$

In the above equation, the coefficients of y_1 and y_2 are defined as the moments of $f_{X1,X2}$, which are formalized below:

$$m_{ij} = \int_{-\infty}^{\infty} \int_{-\infty}^{\infty} x_1^i x_2^j f_{X_1,X_2}(x_1,x_2) dx_1 dx_2 = E[X_1^i X_2^j]$$

Therefore the characteristic function of $f_{X1,X2}(x_1,x_2)$ can be represented as a infinite series in terms of the moments:

$$\Phi(y_1, y_2) = 1 + iy_1 m_{10} + iy_2 m_{01} - \frac{y_1^2}{2} m_{20} - \frac{y_2^2}{2} m_{02} - y_1 y_2 m_{11} + \cdots$$

Note that m_{10} is simply μ_{x1}, m_{01} is μ_{x2}, $m_{11} = E[X_1 X_2] = $ Covariance$(X_1, X_2) + \mu_{x1}\mu_{x2}$. $m_{20} = E[X_1^2] = $ Variance$(X_1) + \mu_{x1}^2$ and so on. Now let us suppose that we want to approximate $f_{X1,X2}$ with $f_{X1,X2}^G$ where $f_{X1,X2}^G$ is a Gaussian distribution. For this we match the moments of the two distributions so as to evaluate the unknowns of $f_{X1,X2}^G$. Since $f_{X1,X2}^G$ is bivariate Gaussian density function, it has 5 unknowns $(\mu_{x1}, \mu_{x2}, \sigma_{x1}, \sigma_{x2}, \rho_{x1x2})$ which can be obtained by matching the first 5 moments of $f_{X1,X2}$, i.e.

$$m_{10} = m_{10}^G, m_{01} = m_{01}^G, m_{20} = m_{20}^G, m_{02} = m_{02}^G, m_{11} = m_{11}^G$$

Thus in order to approximate $p(f,T)$ using a simplified Gaussian JPDF $p_G(f,T)$, the moments of the two distributions are matched, or equivalently the characteristic functions are matched. Matching these moments essentially gives us information pertaining to the mean, variance and covariance of the approximating Gaussian JPDF $p_G(f,T)$ which is enough to determine this function. Following this moment matching based Gaussian approximation, we can now use equation (9) to estimate the approximate value of T.

4. MULTI-SENSOR ESTIMATION
4.1 Exploiting Correlations

Typical VLSI chips use multiple sensors to estimate the thermal state of different hotspot areas. The previous section described ways of reading these noisy sensors and estimating the expected temperature at each sensor that might have caused the specific reading. Each sensor was treated independently.

One important observation is that both the temperature and the sources of noise at different sensor locations can be correlated. This correlation can be exploited to improve the quality of temperature estimation. In this section we describe ways of estimating the expected temperature at each sensor location while effectively exploiting the correlation between sensor readings caused by correlated fabrication randomness and correlated thermal behavior. Exploiting this correlation helps us perform more accurate estimation of the sensor temperatures as compared to treating each sensor independently.

Before we begin the description of our scheme, we illustrate the degree of correlation that temperature and noise of different sensors can have and how it impacts the degree of correlation in observed sensor frequencies.

Table2. Power Density Characteristics of Different Modules.

Correlation	branch predictor	rename	Instruction Window	load/store queue	register file	ALU	Icache	Dcache
Dcache	0.42	0.54	0.69	0.98	0.42	0.47	0.56	
Icache	0.66	0.97	0.89	0.61	0.58	0.68		
ALU	0.36	0.74	0.90	0.51	0.93			
register file	0.13	0.63	0.84	0.47				
load/store queue	0.45	0.58	0.74					
Instruction Window	0.54	0.91						
rename	0.61							
Mean ($\times 10^5$ W/m^2)	0.34	4.33	12.48	1.09	3.88	3.30	4.71	1.03
Std_Dev ($\times 10^5$ W/m^2)	0.18	1.68	4.21	0.42	1.49	1.33	1.87	0.41

We simulated a high performance aggressive out-of-order processor with the same experimental settings as described in section 3. We calculated the distribution in power dissipation in different blocks of the CPU for representative benchmarks and used the data to generate the correlations between the power dissipation in different modules of the CPU. Table 2 presents this correlation data. It can be seen that the correlation between some modules can indeed be as high as 0.9. Hence the power density vector P in equation (1) represents a correlated random vector. The temperature vector at a set of sensors T (vector composed of individual sensor temperatures) can be computed using equation (1). Let us assume that using various statistical characterization schemes we can compute the mean, variance, covariance etc of the power density vector P (as illustrated

978-1-60558-497-3/09 $25.00 © 2009 ACM

in table 2), the randomness associated with different sensor temperatures can be computed as follows [16].

$$\overrightarrow{\mu_T} = T_0 + A\overrightarrow{\mu_P}, \qquad \Sigma_{TT} = A\Sigma_{PP}A^T \qquad (10)$$

where μ_p and Σ_{pp} are the mean vector and the covariance matrix of the power density vector (just like the table above) and A is the coefficient matrix relating vector T and P as described in section 2 whereas T is the vector of *sensor* temperatures.

The noise sources in sensor parameters can be correlated as well. One source of noise is fabrication randomness. Spatially correlated fabrication randomness (correlation in channel length, width, oxide thickness etc.) causes correlated sensor noisy behavior. Such spatial correlation can be captured by statistical analysis techniques such as the model described in [11]. Essentially if two sensors are placed close together certain process parameters such as W, L, T_{ox} will have higher correlation due to the vicinity in their location. On the other hand if they are placed far apart the correlation will be low. Such spatial correlation information is also exploited in our methodology to increase the accuracy of the estimated temperature.

4.2 Modeling the randomness

From above, we can conclude that exploiting the correlation between different sensor readings would result in more accurate estimates of the expected sensor temperature. Let us suppose we have n sensors and therefore n temperature values of interest and n sensor readings. Just like section 3, let us assume that we can characterize the joint probability density function between all the transistor parameters, temperatures and sensor readings using statistical schemes JPDF(T_1, T_2,…, T_n, f_1, f_2,…, f_n, V_{DD1}, V_{DD2}, … W_1, W_2, …). The following equation can be used to generate the marginal joint PDF between sensor temperatures and sensor readings.

$$p(\overrightarrow{f},\overrightarrow{T}) = \int\int...\int jpdf(\overrightarrow{f},\overrightarrow{T},\overrightarrow{V_{DD}},\overrightarrow{V_t},\overrightarrow{W},\overrightarrow{L}...)d\overrightarrow{V_{DD}}d\overrightarrow{V_t}d\overrightarrow{W}d\overrightarrow{L}...$$

where f and T are vectors representing n sensors readings and n real temperatures. Given a set of n sensor readings f, we would like to find the expected value of the n temperature values.

$$E(\overrightarrow{T} \mid \overrightarrow{f}) \qquad (11)$$

This strategy of estimating the expected values of n sensor temperatures given n sensor readings exploits both the statistical characteristics of the random variables themselves (T, V_{DD}, V_t, W, L etc.) on which the frequency depends and the correlation of these variables between different sensors. This information is generated a-priori to improve the accuracy of our temperature estimation at multiple sensor locations.

4.3 Optimal Solution for Gaussian Distribution

As in single sensor's case, let us first assume that $p(f,T)$ has a joint Gaussian distribution, then the following expression can be used to estimate the temperature vector T when given a vector of frequency observations f [16]

$$E(\overrightarrow{T}\mid\overrightarrow{f}) = \overrightarrow{\mu_T} + \Sigma_{Tf}\Sigma_{ff}^{-1}(\overrightarrow{f} - \overrightarrow{\mu_f}) \qquad (12)$$

where μ_T μ_f is the mean of vectors T and f respectively, Σ_{Tf} is the covariance matrix for T and f and Σ_{ff} is the covariance matrix of f itself. When $p(f,T)$ is jointly Gaussian, the above formula provides the optimal estimator for the temperature vector T when given a set of observations f which has been corrupted by noise. For brevity we omit the proof of this result. The interested readers can refer to [16] for the detailed proof for the multi-sensor case.

4.4 Non-Gaussian Case

It is very hard to prove that indeed the JPDF $p(f,T)$ is Gaussian for our case. In general we can expect that the nature of this JPDF is clearly non-Gaussian. When this is true an efficient closed form expression of the optimal estimator for the temperature is very hard to obtain.

For the case of measuring noisy on-chip temperature values, empirically we found the JPDF $p(f,T)$ not to be very far from Gaussian. Hence in order to extend the result of equation (12) to the non-Gaussian case, we fit a Gaussian density function over the real JPDF $p(f,T)$ using the method of moment matching (just like the single sensor case described above). Essentially our method is to approximate the actual joint probability distribution with a Gaussian distribution. This enables us to use the estimator of equation (12) to approximate the result of (11).

As in the single sensor case, the method of moment matching matches the mean, variance and covariance of the real JPDF $p(f,T)$ and approximating Gaussian JPDF $p_G(f,T)$. Knowledge of mean, variance and covariance is enough to characterize the approximating Gaussian JPDF. In this way, we can take the prior knowledge of the non-Gaussian joint distribution and fit a Gaussian model to it. Then the estimator shown in equation (12) can be applied to estimate the temperatures at multi-sensor locations. The accuracy of this heuristic strategy will be demonstrated in the next section.

5. IMPLEMENTATION OVERHEAD

Our statistical scheme for reading noisy sensors can be implemented with very low overhead either as a hardware module or as a software kernel in operating system. Equations (9) and (12) are simple linear equations and their implementation complexity will be very low. For example they could be implemented using look up tables in hardware without incurring much area/power overhead. In software, their evaluation would be reasonably fast also. Our MATLAB implementations (which are generally slower) took only about one millisecond. The experiments were done on a dual-core 2.0GHz Pentium machine. Thermal profiles of chips take several hundred milliseconds to change [17], therefore a few millisecond overhead for reading noisy thermal sensors will not impact the effectiveness of DTM. In fact reducing the thermal estimation error would actually improve the DTM efficiency.

6. EXPERIMENTAL RESULTS

In this section we demonstrate the effectiveness of our methodology through Monte Carlo simulation. In the following experiments we consider the case where two thermal sensors are placed on chip and we are trying to estimate their temperatures from the two available sensor readings (note that for more than two sensors the approach remains the same). We assume the temperature dependency of sensor frequency is governed by equation (5) (6) according to [6,10]. The temperatures of these two sensors are assumed to be jointly Gaussian with mean, variance and covariance information obtained from power density characteristics and equation (10) (see table 3). The source of the sensor noise is assumed to be supply voltage fluctuation. We assumed it to have a Gaussian PDF with mean and variance illustrated in the table below. The supply voltage randomness was uncorrelated. Based on all the information above, we can fit an approximating joint-Gaussian distribution on $p(T,f)$ and use equation (9)/(12) to calculate the expected values of sensor temperatures conditioned on the sensor observations.

978-1-60558-497-3/09 $25.00 © 2009 ACM

Table 3. Results comparison for 3 different schemes (* Values shown in terms of percentage of nominal values.)

V_{DD} Mean/Std_dev*	Temperature Mean/Variance*	Correlation in T between two sensors	Relative RMS error			Accuracy improvement	
			Scheme 1	Scheme 2	Scheme 3	Scheme 3 over 1	Scheme 3 over 2
3V / 2%	300K/30%	0.8	5.9%	2.9%	2.7%	54.5%	6.9%
3V / 3%	300K/30%	0.8	8.9%	3.1%	2.9%	66.8%	6.5%
3V / 4%	300K/30%	0.8	11.3%	3.2%	3.0%	72.7%	6.2%

We show results for the following 3 schemes. The accuracy for each are compared in table 3.

1) Scheme 1: The sensor readings are trusted blindly and the corresponding temperatures are estimated using equation (7). In this scheme all sensor parameters are assumed to be fixed at their nominal values with zero noise;

2) Scheme 2: We use equation (9) to estimate the sensor temperatures. Each sensor is treated independently without considering the correlation information between them.

3) Scheme 3: Equation (12) is used to estimate sensor temperatures. The correlation information is exploited for more accuracy.

The RMS error shown in table 3 is the average value for two sensors and is computed based on 10000 iterations of Monte Carlo simulation. As can be seen from the table, when the variation in random sensor parameters (V_{DD} in this case) increases so does the RMS error. However scheme 2 and 3 always give better estimates as compared to scheme 1 because they take into account the realistic situation of noisy sensors. The improvements become more significant (see second-last column) as noise variance increases. Scheme 3 performs better than scheme 2 since it exploits correlations. As the noise variance increases, the degree of correlation decreases and the effectiveness of scheme 3 reduces as well (see last column of table 3).

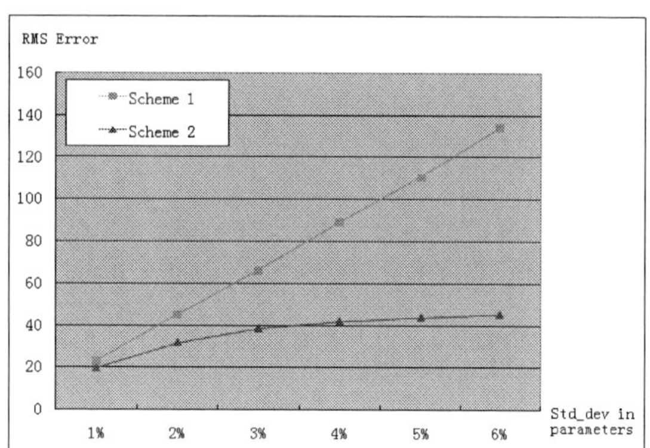

Fig.4 RMS error for different amount of std_dev in sensor parameters

Figure 4 above highlights the improvement of scheme 2(blue) over scheme 1(magenta) when we increase the variation in random sensor parameters (we did not show scheme 3 here because the trend is similar).

7. Conclusion

We addressed the problem of estimating the accurate sensor temperatures when given noisy sensor readings. Both single sensor case and multi-sensor case are investigated with different strategies of exploiting the correlation information. Optimal and heuristic estimation schemes are proposed to address the problem when the underlying nature of the sensor noise is Gaussian and non-Gaussian.

* This work is partly supported by NSF grant CCF-0728969 and CCF-0634321

8. REFERENCES

[1]   Yong Zhan and Sachin S.Sapatnekar, "High-Effciency Green Function-Based Thermal Simulation Algorithms", IEEE Trans. Comput.-Aided Design Integr. Circuits Syst., vol. 26, pp. 1661, Sept. 2007.

[2]   R. Gharpurey and R. G. Meyer, "Modeling and analysis of substrate coupling in integrated circuits", IEEE J. Solid-State Circuits, vol. 31, pp. 344, Mar. 1996.

[3]   A. M. Niknejad, R. Gharpurey and R. G. Meyer, "Numerically Stable Green Function for Modeling and Analysis of Substrate Coupling in Integrated Circuits", IEEE Trans. Comput.-Aided Design Integr. Circuits Syst., vol. 17, pp. 305, Apr. 1998.

[4]   T. Y. Wang and C. P. Chen, "3-D Thermal-ADI: A Linear-time Chip Level Transient Thermal Simulator", IEEE Trans. Comput.-Aided Design Integr. Circuits Syst., vol. 21, pp. 1434, Dec. 2002.

[5]   P. Li , L. T. Pileggi , M. Asheghi and R. Chandra, "Efficient Full-chip Thermal Modeling and Analysis", Proc. IEEE/ACM Int. Conf. Comput.-Aided Des., Dig. Tech. Papers, Nov. 2004, p. 319.

[6]   Adel S. Sedra and Kenneth C. Smith, "Microelectronic Circuits" 5th Edition, Oxford University Press, 2004.

[7]   K.Arabi, B.Kaminska, "Built-in temperature sensors for on-line thermal monitoring of VLSI circuits", Proc. of Int. Conf. on Comp. Design, 1997.

[8]   Basab Datta and Wayne Burleson, "Low-Power and Robust On-Chip Thermal Sensing Using Differential Ring Oscillators" Proc. of IEEE Midwest Symposium on Circuits and Systems, 2007.

[9]   S. Nassif, "Delay Variability: Sources, Impact and Trends", Proc. ISSCC, pp. 368-369, 2000.

[10] Sze, Simon M, "Physics of Semiconductor Devices", Second Edition, John Wiley and Sons, New York, 1981.

[11] Brian Cline, Kaviraj Chopra et al. "Analysis and Modeling of CD Variation for Statistical Static Timing", ICCAD'06

[12] Ying-Yen Chen, Jing-Jia Liou, "Extraction of Statistical Timing Profiles Using Test Data" DAC, 2007

[13] Greg Hamerly, Erez Perelman, Jeremy Lau, and Brad Calder, "SimPoint 3.0: Faster and More Flexible Program Analysis", Workshop on Modeling, Benchmarking and Simulation, June 2005.

[14] http://www.spec.org.

[15] Lerong Cheng, Jinjun Xiong and Lei He, "Non-Linear Statistical Static Timing Analysis for Non-Gaussian Variation Sources", DAC 2007.

[16] Steven M. Kay, "Fundamentals of Statistical Signal Processing: Estimation Theory", Prentice Hall, 1993

[17] J. Choi, C. Cher et al, "Thermal-aware Task Scheduling at the System Software Level", Int. Symp. on Low Power Electronics and Design, 2007

[18] D. Brooks, M. Martonosi, "Dynamic Thermal Management for High Perf. Microprocessors", Int. Sym. on High Perf. Comp. Arch. January 2001.

[19] A.K. Coskun, T. Simunic Rosing, K.C. Gross, "Proactive Temperature Management in MPSoCs" International Symposium on Low Power Electronics and Design, pp 213-218, August 2008.

Spectral Techniques for High-Resolution Thermal Characterization with Limited Sensor Data

Ryan Cochran
Division of Engineering
Brown University
Providence, RI 02912
Email: ryan_cochran@brown.edu

Sherief Reda
Division of Engineering
Brown University
Providence, RI 02912
Email: sherief_reda@brown.edu

ABSTRACT

Elevated chip temperatures are true limiters to the scalability of computing systems. Excessive runtime thermal variations compromise the performance and reliability of integrated circuits. To address these thermal issues, state-of-the-art chips have integrated thermal sensors that monitor temperatures at a few selected die locations. These temperature measurements are then used by thermal management techniques to appropriately manage chip performance. Thermal sensors and their support circuitry incur design overheads, die area, and manufacturing costs. In this paper, we propose a new direction for full thermal characterization of integrated circuits based on spectral Fourier analysis techniques. Application of these techniques to temperature sensing is based on the observation that die temperature is simply a space-varying signal, and that space-varying signals are treated identically to time-varying signals in signal analysis. We utilize Nyquist-Shannon sampling theory to devise methods that can almost fully reconstruct the thermal status of an integrated circuit during runtime using a minimal number of thermal sensors. We propose methods that can handle uniform and non-uniform thermal sensor placements. We develop an extensive experimental setup and demonstrate the effectiveness of our methods by thermally characterizing a 16-core processor. Our method produces full thermal characterization with an average absolute error of 0.6% using a limited number of sensors.

ACM Categories & Subject Descriptors
B.7.1 [Integrated Circuits]: Types and Design Styles.
General Terms: Design, Performance, Algorithms.
Keywords: Thermal management, spectral methods, spatial estimation, thermal sensors.

1. INTRODUCTION

Runtime thermal variations arise due to the spatially and temporally non-uniform power density distributions that occur during chip operation. These variations lead to thermal gradients and hot spots that degrade performance and shorten the mean time to circuit failure. State-of-the-art chips typically employ thermal sensor(s) in order to capture a number of on-chip temperature measurements during runtime. Thermal management techniques use these measurements to control cooling fan speeds, adjust workload scheduling, and assign appropriate voltages and frequencies [12, 16, 19, 4]. Thermal sensors (e.g., thermal diodes) and their support circuitry utilize silicon real estate and increase design complexity. These sensors are typically placed during design time at locations where thermal hot spots are expected to occur. However, unpredictable workloads and within-die process variations alter the locations of runtime hot spots from their expected locations.

By using Fourier analysis techniques, this paper proposes a new direction for full thermal characterization of integrated circuits using a limited number of thermal sensors. We formally ground our ideas in the classical foundations of Fourier signal analysis, which leads to superior results when compared to existing methods. Our contributions are as follows.

- We discuss the implications of using Nyquist-Shannon sampling theory to determine the fewest number of thermal sensors that can fully characterize the runtime thermal status of a processor.

- Given a limited number of thermal sensors, we propose signal reconstruction techniques that generate a full-resolution thermal characterization of a processor. We propose methods that handle uniform and non-uniform sensor placements, the latter of which may arise due to design and layout constraints.

- Using a comprehensive tool chain of thermal simulators, power estimators, and workloads, we demonstrate that our proposed methods can provide an accurate high-resolution thermal characterization for a 16-core processor. We also demonstrate the trade-off between the number of thermal sensors and the attained accuracy of the thermal characterizations.

- In addition to full thermal characterization, we quantify the effectiveness of our methods in estimating hot spot magnitudes. We demonstrate the superiority of our methods over other methods (e.g., grid-based interpolation [8] and geostatistical Kriging estimators [7]).

The organization of this paper is as follows. Section 2 provides the necessary motivation and background for this work. In Section 3, we discuss Nyquist-Shannon theory and its implications on thermal sensing. We also discuss uniform and non-uniform sampling cases. Section 4 develops a number of signal reconstruction methods. Using a 16-core processor model, we provide experimental results in Section 5 that demonstrate the effectiveness of our methods. Finally, Section 6 summarizes the main conclusions of this work and indicates directions for future work.

Permission to make digital or hard copies of part or all of this work for personal or classroom use is granted without fee provided that copies are not made or distributed for profit or commercial advantage and that copies bear this notice and the full citation on the first page. To copy otherwise, to republish, to post on servers or to redistribute to lists, requires prior specific permission and/or a fee.
DAC'09, July 26-31, 2009, San Francisco, California, USA

978-1-60558-497-3/09 $25.00 © 2009 ACM

2. BACKGROUND

High-end chips, such as multi-core processors and graphical processor units, are equipped with integrated thermal sensors that monitor the on-chip temperatures during runtime. Thermal diodes, which translate temperature variations into voltage variations, are a popular choice for temperature sensing. The diode voltage signal is routed to and measured by an analog to digital (A/D) converter. In older processor technologies, the A/D conversion takes place on the board, while in newer processors (e.g., Intel's Core i7), the A/D conversion takes place on-chip. State-of-the-art chips are typically equipped with more than one thermal sensor. There are a number of reasons for this: (1) complex chips with large die area require more thermal sensors to capture temperatures at a wide range of locations; (2) the unpredictability of a processor's workload can lead to continuous migration of hot spots; and (3) within-die manufacturing variations lead to leakage variability that can further conceal the locations of the thermal hot spots. The thermal sensors, together with their support circuitry and wiring, complicate the design process and increase the total die area and manufacturing costs. Thus, there is an inherent tradeoff in thermal monitoring. On the one hand, designers would like to reduce costs by using the fewest number of thermal sensors, while on the other hand, the gravity of runtime thermal problems require higher thermal resolution.

Given the limitations on the number of thermal sensors, it is necessary to optimally place them near potential hot spot locations. Mukherjee and Memik [10, 9] describe a clustering algorithm that computes the thermal sensor positions that best serve clusters of potential hot spot locations. The locations of these hot spots are identified via extensive workload thermal simulation. Even with optimized placement, it is likely that sensors will fail to detect hot spots, especially when there are a large number of cores. Thus, Long *et al.* [8] advocate using a grid-based interpolation scheme that identifies the hot spot around each sensor by interpolating the measurements at its immediate neighbors. Another entirely different potential method for full thermal characterization is based on geostatistical techniques. Liu [7] proposes using *Kriging* estimation as a general framework for estimating variability (whether manufacturing, thermal or IR drop) at various chip locations during design time. A Kriging temperature estimator computes the unknown temperature at a particular location using a weighted combination of the known measurements at other locations. In choosing the optimal weights, temperature is modeled as a random field. The Kriging estimator then chooses a set of weights such that the variance of the error values is minimized. In order to do this, Kriging estimators utilize *variograms*, which capture the spatial correlations between the temperatures at various locations on the die [7]. While Liu [7] does not specifically advocate the use of Kriging estimators for runtime thermal characterization, we implemented a Kriging adaptation for the sake of comparison.

Our objective is to achieve full characterization of on-chip runtime temperature using the measurements of a few thermal sensors. Our approach is formally grounded in spectral Fourier analysis techniques. Application of these techniques to temperature sensing is based on recognition that die temperature is simply a space-varying signal, and that space-varying signals are treated identically to time-varying signals in Fourier signal analysis.

While temperature is a continuous variable, any representation in computer memory must be discretized. Thus, the temperature $t(m, n)$ is a discrete function that is defined over a finite region $0 \leq m \leq M - 1$ and $0 \leq n \leq N - 1$, where M and N are the *resolutions* required for thermal characterization. For example, if a die has dimensions $1\ cm \times 1\ cm$, then with resolutions $M = N = 128$, temperatures are evaluated for every $78\ \mu m \times 78\ \mu m$

square. The two-dimensional Discrete Fourier Transform (DFT) is given by

$$T(p, q) \;=\; \sum_{m=0}^{M-1} \sum_{n=0}^{N-1} t(m, n) e^{-j2\pi pm/M} e^{-j2\pi qn/N} \quad (1)$$

for all $p = 0, 1, \ldots M - 1$ and $q = 0, 1, \ldots N - 1$, and the inverse DFT is given by

$$t(m, n) \;=\; \frac{1}{MN} \sum_{p=0}^{M-1} \sum_{q=0}^{N-1} T(p, q) e^{j2\pi pm/M} e^{j2\pi qn/N} \quad (2)$$

for all $m = 0, 1, \ldots M - 1$ and $n = 0, 1, \ldots N - 1$. To compute the DFT and the inverse DFT efficiently, the Fast Fourier Transform (FFT) and the inverse FFT are used, achieving an $O(MN \log MN)$ runtime as opposed to $O((MN)^2)$ [13].

There are two main aspects to full thermal characterization:

1. **Thermal Sensing or Sampling.** Thermal sensors are treated as discrete temperature samples at various die locations. With a handful of thermal sensors in even state-of-the-art chips, thermal measurements are very sparse relative to the objective of full thermal resolution. An interesting aspect of Fourier analysis is that it is capable of determining the exact number of thermal sensors required for full thermal reconstruction.

2. **Signal Reconstruction.** Given thermal sensor values, signal reconstruction techniques based on Fourier analysis can be used to reconstruct the temperature at all die locations, thus achieving full-resolution thermal characterization. Given appropriately spaced samples, thermal reconstruction can theoretically be performed with no information loss. In reality, a number of constraints hinder this theoretical possibility, and one can only hope for minimal information loss. An attractive feature of thermal characterization is that the continuous nature of heat flow gives relatively smooth on-chip temperature gradients that do not change abruptly. This feature limits the bandwidth of the thermal signal in the spectral domain, which permits near perfect signal reconstruction.

In the next two sections we will describe our methods for signal sampling and signal reconstruction. Methods for signal sampling are given in Section 3, and methods for signal reconstruction are given in Section 4.

3. THERMAL SAMPLING

With thermal sensors treated as discrete samples of a continuous phenomenon, the following important question arises: *what is the minimum number of sensors required to fully characterize the thermal status of a chip?* According to the Nyquist-Shannon sampling theorem, a continuous signal can be perfectly reconstructed from its samples if the sampling frequency is at least twice the highest frequency contained in the signal. This result is remarkable because it shows that sampling entails no loss of information for an appropriately band-limited signal. Another classical result in signal processing is that time-limited signals are not band-limited, and that band-limited signals are not time-limited; i.e., a signal cannot be simultaneously time-limited and band-limited [14]. Temperature, a space-varying signal, is space-limited by the edges of the chip. Thus, the spectral representation of the temperature is not band-limited. As a result, perfect reconstruction is not possible in the case of on-chip temperatures; however, near-perfect reconstruction with negligible loss of information is possible if these *edge effects* are appropriately handled during reconstruction such that the higher frequency magnitudes in the spectral domain are minimized.

978-1-60558-497-3/09 $25.00 © 2009 ACM

Techniques for handling edge effects are mentioned in Section 4. We now consider two possible sampling schemes.

Uniform Sampling. We first consider the case in which the locations of the sensors are aligned on a lattice such that they are equally spaced from each other. There are a number of lattices for which such uniformity can be achieved, including hexagonal lattices, diamond lattices, and rectangular grids. We focus on rectangular grids as they are more suitable for the tiled layouts typically encountered in multi-core processors and GPUs. To sample the temperature $t(m, n)$, we assume that the thermal sensors have been placed with a horizontal/vertical spacing of $P \in \mathcal{Z}^+$. Thus, the sampling signal $s(m, n)$ can be described with

$$s(m, n) = \sum_{u=0}^{\lfloor \frac{M-1}{P} \rfloor} \sum_{v=0}^{\lfloor \frac{N-1}{P} \rfloor} \delta(m - uP, n - vP), \qquad (3)$$

where $\delta(\cdot, \cdot)$ is the Dirac delta function. To obtain the sampled temperature signal $t_s(m, n)$, we multiply the temperature with the sampling function to get

$$
\begin{aligned}
t_s(m, n) &= t(m, n)s(m, n) \\
&= t(m, n) \sum_{u=0}^{\lfloor \frac{M-1}{P} \rfloor} \sum_{v=0}^{\lfloor \frac{N-1}{P} \rfloor} \delta(m - uP, n - vP) \\
&= \sum_{u=0}^{\lfloor \frac{M-1}{P} \rfloor} \sum_{v=0}^{\lfloor \frac{N-1}{P} \rfloor} t(uP, vP)\delta(m - uP, n - vP) \quad (4)
\end{aligned}
$$

Consider the DFT of $t_s(m, n)$. Since $t_s(m, n)$ is the product of $t(m, n)$ and $s(m, n)$, then the DFT $T_s(p, q)$ of the sampled signal $t_s(m, n)$ is equal to the convolution of the $T(p, q)$ and $S(p, q)$, which are the DFTs of $t(m, n)$ and $s(m, n)$ respectively. The Fourier transform of a periodic impulse train is a periodic impulse train as well; i.e.,

$$\sum_{u=0}^{\lfloor \frac{M-1}{P} \rfloor} \sum_{v=0}^{\lfloor \frac{N-1}{P} \rfloor} \delta(m-uP, n-vP) \leftrightarrow \frac{1}{P^2} \sum_{u=0}^{P-1} \sum_{v=0}^{P-1} \delta(p-u\frac{M}{P}, q-v\frac{N}{P})$$

$$(5)$$

Thus, the convolution of $T(p, q)$ and $S(p, q)$ gives

$$T_s(p, q) = \frac{1}{P^2} \sum_{u=0}^{P-1} \sum_{v=0}^{P-1} T(p - u\frac{M}{P}, q - v\frac{N}{P}). \qquad (6)$$

Temperature sampling by the thermal sensors in the space-domain has led to periodic copies of the Fourier transform of the temperature signal in the spectral domain. Figure 1 visually illustrates the impact of sampling, where Figure 1.a gives a thermal map of a 16-core processor in both the space and spectral-domain. After sampling, the spectral-domain representation of Figure 1.b shows the repetition of the spectral map of Figure 1.a. According to the Nyquist-Shannon theorem, if copies of the spectral-domain temperature signals are spread far "enough" apart, then overlap or *aliasing* between the copies will not occur. The necessary spreading, which is controlled by the sampling frequency, depends on the highest frequency seen in the temperature signal. If the temperature signal is band-limited with frequency B, then sampling at a rate higher than $2B$ guarantees full reconstruction of the original signal. Sampling at a higher rate is achieved by decreasing the spacing P between the thermal sensors. Thus, thermal sensors should be spaced such that $1/P \geq 2B$, or equivalently $P \leq 1/(2B)$. As mentioned earlier, the temperature signal is space-limited by the size of the chip,

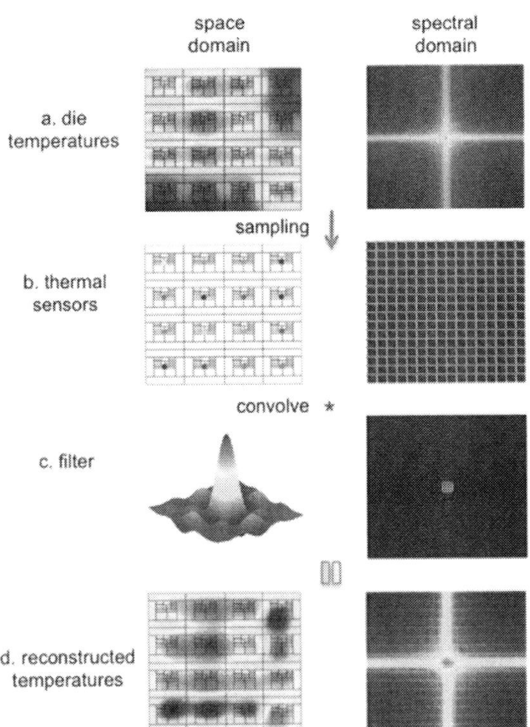

Figure 1: Main steps used for signal reconstruction. The log of the magnitude of the 2D DFT is plotted.

and consequently, the spectral-domain representation of the temperature is not band-limited. To minimize information loss due to sampling, one must pick a bandwidth B below which most of the signal's energy is concentrated. The impact of sensor spacing on the thermal characterization accuracy will be quantified in our experimental results section (Section 5).

Non-uniform Sampling. In many cases, it may be necessary to place sensors in a non-uniform pattern for the following reasons:

1. There could be regions on the die that have greater utilization and higher thermal gradients than other regions with uniform utilization. Thus, it may be advantageous to allocate more sensors to locations that have higher gradients or unpredictability.

2. Design constraints could force designers to place the sensors in a non-uniform pattern. Even in the case of multi-core processors where one can expect regular layouts, on-chip switches and cache structures can create irregular design layouts. Automatic layout and placement tools can also move blocks from their intended locations if it leads to timing or die area improvements.

Full thermal characterization consists of two problems: optimal sensor placement, and maximal use of thermal information given a set of temperature samples. The first problem is not our focus in this work. In the next section we provide solutions for the second problem for both uniform and non-uniform sample patterns.

4. THERMAL RECONSTRUCTION

We have seen in the previous section that sampling a signal in the space-domain leads to periodic copies of the Fourier transform, $T(p, q)$, of the original signal $t(m, n)$, in the spectral-domain. If the temperature signal is band-limited (or has negligible energy beyond a certain frequency) then these periodic copies are well separated from each other and aliasing is minimal. The Whittaker-Shannon-Kotelnikov (WSK) classical theorem states that to recover the original signal from the samples, it is sufficient to extract only one copy of the signal in the spectral-domain [14]. This extraction can be achieved using a low-pass box filter as shown in Figure 1.c. In the frequency domain, this box can be expressed as

$$F(p,q) = 1 \quad \text{if} \quad |p| \leq B \quad \text{and} \quad |q| \leq B$$
$$= 0 \quad \text{otherwise.} \tag{7}$$

Taking the inverse DFT of the box filter gives the spatial-domain representation of the filter $f(m,n)$ which is equal to

$$f(m,n) = \text{sinc}(\frac{m}{B})\text{sinc}(\frac{n}{B}). \tag{8}$$

Reconstruction is achieved by *convolving* the space-domain samples with the space-domain filter representation. That is, the reconstructed temperature of a chip $t_r(m,n)$ can be found using

$$t_r(m,n) = \sum_{u=0}^{M-1} \sum_{v=0}^{N-1} t_s(u,v)\text{sinc}(\frac{m}{B} - u)\text{sinc}(\frac{n}{B} - v). \tag{9}$$

This result is illustrated in Figure 1.d. One of the practical problems that arises when using the sinc function is that it is not space-limited. In any implementation, the sinc function must be truncated, which has the effect of smearing its spectral-domain representation, leading to a less than sharp box filter edge. This smearing effect can be minimized by *windowing* the sinc function. In our implementations, we multiply the sinc function by a *Hamming window* of the same size. The severity of the edge effects depends on the size of the sinc function used. Edge effects can be minimized by extending the temperature data by a distance larger than half the size of the filter function. The values in the extended region can be copies of the edge values, periodic repetitions of the temperature signal, or even a mirror image of the temperature signal. It is also useful to investigate other filter functions that approximate a low-pass box filter in the spectral domain without windowing [11]. We investigate three other functions.

- **Nearest neighbor:** The simplest interpolation function is nearest neighbor, in which each location is given a temperature equal to the value measured by the sensor closest to it. This is achieved by convolving the sampled temperature signal with a rectangular function expressed as follows:

$$f(x,y) = 1, \quad \text{for } x \in [-0.5, 0.5] \text{ and } y \in [-0.5, 0.5]$$
$$f(x,y) = 0, \quad \text{otherwise}$$

- **Linear function:** Linear interpolation amounts to convolving the temperature samples with a round cone function. This function corresponds to a modestly good low-pass filter in the spectral-domain. However, it attenuates frequencies near the cut-off frequency, resulting in smoothing of the thermal characterization results. It also passes a good amount of energy above the cut-off frequency. The linear function is expressed as follows:

$$f(x,y) = g(x)g(y), \quad \text{where}$$
$$g(u) = (1-u) \quad \text{for } u \in [0,1]$$

- **Cubic B-spline function:** Cubic B-spline functions are reasonably good low-pass filters. They are positive in the whole interval from 0 to 2, so they smooth somewhat more than necessary below the cut-off frequency. These filters are symmetric, so they only need to be expressed on the interval $[0,2]$. The cubic B-spline function we use is expressed as follows:

$$f(x,y) = g(x)g(y), \quad \text{where}$$
$$g(u) = \frac{u^3}{2} - u^2 + \frac{4}{6} \quad \text{for } u \in [0,1]$$
$$g(u) = \frac{-u^3}{6} + u^2 - 2u + \frac{8}{6} \quad \text{for } u \in [1,2]$$

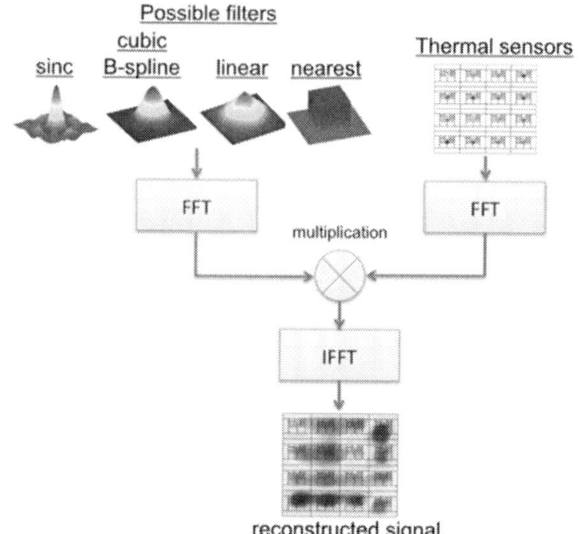

Figure 2: Flow of the proposed runtime thermal characterization technique.

All three interpolation functions are contrasted with the sinc function in Figure 2. To achieve a computationally efficient thermal characterization, we propose the flow in Figure 2. Instead of directly convolving in the space-domain, we first take the Fast Fourier Transform (FFT) of the temperature samples and the space-domain representation of the interpolation functions. We then multiply the resultant spectral-domain representations to get the FFT of the reconstructed 2D thermal signal. We then apply the inverse FFT (IFFT) to get the full-resolution thermal characterization in the space-domain. If the required thermal resolution is $R = MN$, then application of the proposed flow of Figure 2 has a runtime of $O(R \log R)$, which gives a significant advantage over the $O(R^2)$ runtime achieved by straightforward convolution. Such speedup boost is necessary for runtime thermal characterization.

Reconstruction from Non-Uniform Samples. If thermal sensors are placed non-uniformly, then directly applying the proposed flow of Figure 2 could lead to large errors. Instead, it is necessary to apply signal reconstruction algorithms that are devised to handle non-uniform samples. One such algorithm is proposed by Sauer and Allebach [15]. This iterative algorithm consists of two steps:

1. Given the locations of the thermal sensors, construct a Voronoi diagram for the die. All die locations that belong to the Voronoi cell of a thermal sensor are assigned the same temperature as the thermal sensor at center of the cell.

2. The resultant 2D temperature map obtained from the first step is low-pass filtered B (by convolving the temperature with the sinc function) using the flow of Figure 2.

Sauer and Allebach [15] prove that iterating these two steps leads to convergence and reconstruction of the original signal if it is band-limited.

5. EXPERIMENTAL RESULTS

Simulation Infrastructure. To evaluate the effectiveness of our methodology, we set up a *tool chain* that simulates temperatures for a 32 nm 16-core processor. Our tool chain takes as inputs the processor's floor-plan and the workload that will run on each core and produces as output the steady-state temperatures at various

978-1-60558-497-3/09 $25.00 © 2009 ACM

grid locations. Using workload instruction traces, dynamic power traces for each micro-architectural unit are calculated and then fed together with the floor-plan into the thermal simulator. Once the steady-state temperatures are calculated, they are fed into a leakage calculator which outputs the corresponding leakage power of each unit. The leakage power values are then added to the original dynamic power values, and a new thermal simulation is performed. This process is iterated until the temperatures converge to stable values. Power and thermal simulations are performed using the following tools.

- For dynamic power estimation, we use a Wattch-like [1] power simulator, with the power consumption appropriately scaled to 32 nm technology based on ITRS predictions [3]. For the leakage consumption of the processor core units, we construct a leakage model using the expressions for leakage power from PTScalar [6]. To accurately model cache leakage power, we use CACTI 5.0 [18], which has accurate cache leakage values at current and future technology nodes.

- We utilize HotSpot (version 4.0) [17] for thermal simulation. HotSpot takes as inputs the processor floor-plan and workload power traces and produces as output the steady-state temperatures for a set of grid locations.

- We use the Alpha 21264 processor as our baseline core [5]. The 21264 is an out-of-order speculative execution core that is commonly used as a test-bench core in thermal management research [6, 8]. We create a 16-core processor based on the Alpha processor. The die area of the processor is $1.1\ cm \times 1.1\ cm$, and we discretize the temperature by defining a grid with resolution $M = 64 \times N = 64$, where each grid location represents the temperature for an area of size $172\ \mu m \times 172\ \mu m$.

- For workloads, we use eight benchmarks from the SPEC2000 suite [2]. We use four integer benchmarks: gcc, bzip, mcf, and twolf, and four floating point benchmarks: ammp, equake, lucas, and mesa.

In each simulation, we assign each core in the 16-core processor a workload from our SPEC2000 benchmark selection such that each core gets a random workload from the available eight. We then assign each core a random frequency in the range $1.5 - 3$ GHz. We then execute the tool chain to find the true temperatures at all grid locations. Thermal sampling is accomplished by zeroing temperatures at all grid locations except those corresponding to thermal sensors. The number of samples and the sample locations are varied, and each proposed method is evaluated for each sensor configuration using the *average* and *maximum* absolute post-reconstruction error across a set of 32 simulations.

Experiments. In our first set of experiments, we determine the average and maximum absolute error for each reconstruction method for uniformly spaced samples while varying the total number of sensors on a 16-core processor. We present results for the following cases: 1 sensor per core, 4 sensors per core arranged on a 2×2 grid, and 9 sensors per core arranged on a 3×3 grid. The first case corresponds to 16 total sensors, while the second and third cases correspond to 64 and 144 total sensors respectively. The bar-plot of Figure 3 summarizes the average absolute error calculated for each proposed reconstruction method. We include error values for a geostatistical-based Kriging estimator for sake of comparison. The results show that as the number of sensors increases, the thermal characterization error decreases. Furthermore, the results show that our proposed reconstruction methods are capable of achieving full thermal characterization with minimum average absolute error

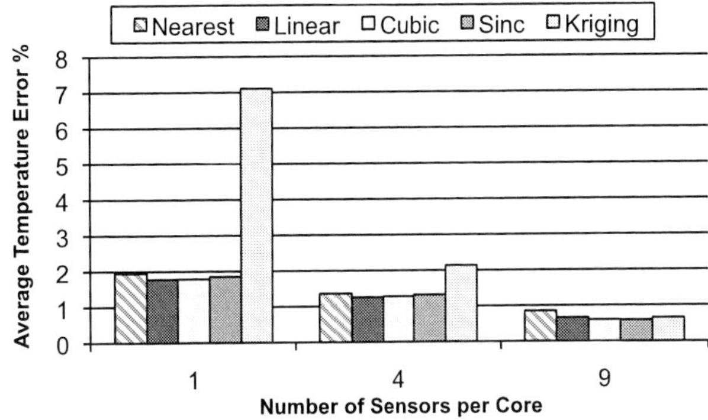

Figure 3: Results of full thermal characterization. We report the average error percentage for temperature estimation at all grid locations.

ranging from 1.8% to 0.6%, depending on the number of the sensors. In this experimental setting, our convolution filters deliver near identical performance, and they significantly outperform the Kriging estimator, especially when the number of sensors are few. Our second set of experiments will reveal a differentiation in the performance of our proposed convolution filters. Using a single representative workload and frequency assignment, Figure 4 compares the thermal resolution attained by using 1, 4, and 9 sensors per core to the true thermal characterization. We include the latter case, despite a possibly unrealistic number of total sensors for a 16-core processor, to illustrate that increasing the number of sensors eventually reconstructs the original signal to near perfection.

Full thermal characterization is particularly useful for advanced thermal management techniques, examples of which include workload scheduling and per-core frequency and voltage assignment. In simpler thermal management techniques (e.g., fan speed control), only the magnitude and the location of the maximum hot spot are relevant. Thus, in our second set of experiments, we consider the hypothetical performance of our methods in such applications. We first identify the location and magnitude of the maximum hot spot in the true thermal characterization. We then evaluate the temperature at that location in our reconstructed results and report the average absolute error. We report error values for each method explored in our first set of experiments, and in addition we implement and compare to the neighborhood interpolation scheme in [8]. Our results in Figure 5 show that the sinc filter function is consistently better at interpolating the maximum hot spot (0.4% errors for the case of 9 sensors per core), and that all of our reconstruction tech-

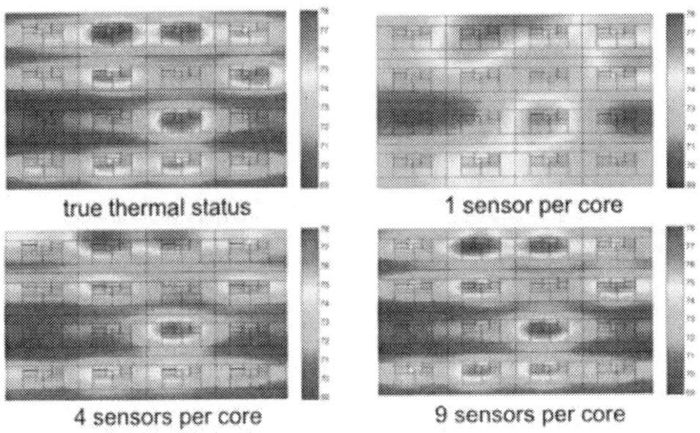

Figure 4: Impact of increasing the number of thermal sensors on the full thermal characterization.

978-1-60558-497-3/09 $25.00 © 2009 ACM

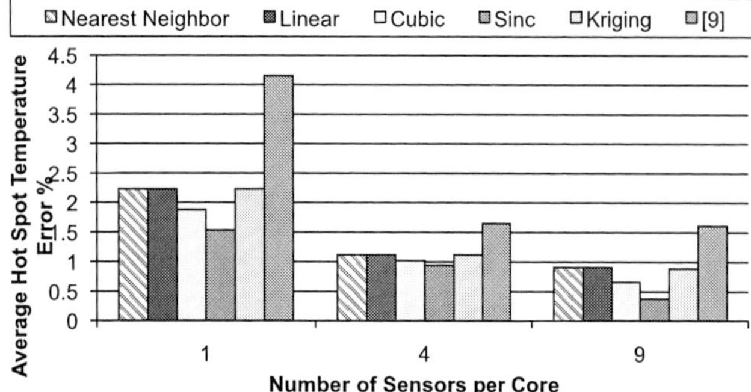

Figure 5: Hot spot estimation. We report the average error percentage for temperature estimation at the hottest die location.

a. Voronoi diagram   b. Final characterization

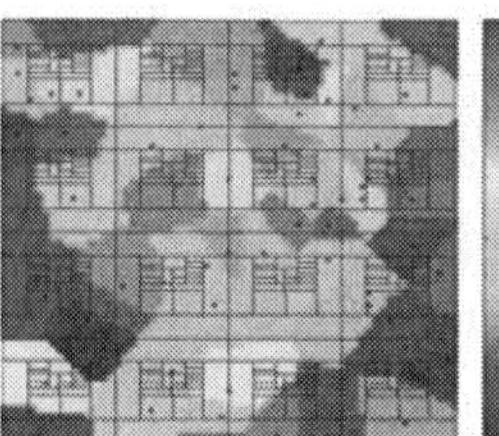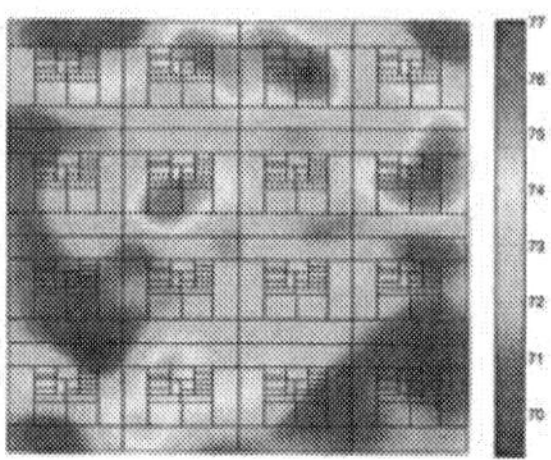

Figure 6: Results of thermal characterization using non-uniform sampling.

niques outperform the Kriging estimator and the method proposed in [8].

In our third set of experiments, we consider non-uniform sensor placements. Figure 6 shows a representative workload and marks the locations of the thermal sensors in the processor's floor-plan with '*'. We use the same workload and frequency assignment as in Figure 4. Figure 6.a gives the output of the first step of the iterative algorithm which constructs the Voronoi diagram, and 6.b gives the results after the algorithm converges using a sinc filter function. The average absolute error for full thermal characterization using non-uniform samples is found to be 2.39%. This value confirms that our methods are capable of handling signal reconstruction for both uniform and non-uniform sensor placements.

6. CONCLUSIONS

In this paper we have presented a new direction for runtime thermal characterization in integrated circuits based on Fourier analysis techniques. The application of Fourier analysis to thermal characterization creates a theoretical foundation for this field of research. We have analyzed the implications of the Nyquist-Shannon sampling theory in determining the optimal number of thermal sensors. We have also analyzed the trade-off between thermal estimation error and the number of thermal sensors. We have proposed the use of signal reconstruction methods for full thermal characterization with limited sensor measurements. We have analyzed the case in which the thermal sensors are non-uniformly spaced, and we have proposed and implemented an algorithm that is capable of handling such a case. The effectiveness of our techniques have been evaluated on a 16-core processor model, and we have demonstrated the superiority of these techniques to existing techniques.

Our proposed work is the first to consider full thermal characterization with limited sensor data. The success of our proposed methods will allow thermal management techniques to take advantage of a full-resolution thermal map for a processor. Our future work will focus on building a software prototype that utilizes actual thermal sensor measurements obtained from Intel Core i7 to achieve full thermal characterization. We will also investigate the impact of calibration errors in the thermal sensor measurements on the results of our proposed methods.

7. REFERENCES

[1] D. Brooks, V. Tiwari, and M. Martonosi, "Wattch: A Framework for Architectural-Level Power Analysis and Optimizations," in *International Symposium on Computer Architecture*, 2000, pp. 83–94.

[2] J. Henning, "SPEC CPU2000: Measuring CPU Performance in the New Millennium," *IEEE Computer*, vol. 33(7), pp. 28–35, 2000.

[3] ITRS, "International Technology Roadmap for Semiconductors," http://public.itrs.net, 2007.

[4] M. Kadin and S. Reda, "Frequency and Voltage Planning for Multi-Core Processors Under Thermal Constraints," in *International Conference on Computer Design*, 2008, pp. 463–470.

[5] R. Kessler, E. McLellan, and D. Webb, "The Alpha 21264 Microprocessor Architecture," in *Proc. International Conference on Computer Aided Design*, 1998, pp. 90–95.

[6] W. Liao, L. He, and K. Lepak, "Temperature and Supply Voltage Aware Performance and Power Modeling at Microarchitecture Level," *Transactions on Computer-Aided Design of Integrated Circuits and Systems*, vol. 24(7), pp. 1042–1053, 2005.

[7] F. Liu, "A General Framework for Spatial Correlation Modeling in VLSI Design," in *Proc. Design Automation Conference*, 2007, pp. 817–822.

[8] J. Long, S. Memik, G. Memik, and R. Mukherjee, "Thermal Monitoring Mechanisms for Chip Multiprocessors," in *ACM Transactions on Architecture and Code Optimization*, vol. 5(2), 2008, pp. 9:1–9:23.

[9] S. O. Memik, R. Mukherjee, M. Ni, and J. Long, "Optimizing Thermal Sensor Allocation for Microprocessors," *IEEE Transactions on Computer-Aided Design of Integrated Circuits and Systems*, vol. 27(3), pp. 516–527, 2008.

[10] R. Mukherjee and S. Memik, "Systematic Temperature Sensor Allocation and Placement for Microprocessors," *Design Automation Conference*, pp. 542 – 547, 2006.

[11] J. Parker, R. Kenyon, and D. Troxel, "Comparison of Interpolating Methods for Image Resampling," *IEEE Transactions on Medical Imaging*, vol. 2(1), pp. 31–39, 1983.

[12] M. D. Powell, M. Gomaa, and T. N. Vijaykumar, "Heat-and-Run: Leveraging SMT and CMP to Manage Power Density Trhough the Operating System," in *International Conference on Architectural Support for Programming Languages and Operating Systems*, 2004, pp. 260–270.

[13] W. H. Press, S. A. Teukolsky, W. T. Vetterling, and B. P. Flannery, *Numerical Recipes in C*, second, Ed.   Cambridge University Press, 1996.

[14] M. J. Roberts, *Signals and Systems*, First, Ed.   McGraw Hill, 2004.

[15] K. D. Sauer and J. P. Allebach, "Iterative Reconstruction of Band-Limited Images from Nonuniformly Spaced Samples," *IEEE Trans. Circuits and Systems*, vol. 34, no. 12, pp. 1497–1506, 1987.

[16] K. Skadron, "Hybrid Architectural Dynamic Thermal Management," in *Design, Automation and Test in Europe*, 2004, pp. 10–15.

[17] K. Skadron, S. Ghosh, S. Velusamy, K. Sankaranarayanan, and M. Stan, "HotSpot: A Compact Thermal Modeling Methodology for Early-Stage VLSI Design," *Transactions on VLSI Systems*, vol. 15(5), pp. 501–513, 2006.

[18] S. Wilton and N. P. Jouppi, "CACTI: An Enhanced Cache Access and Cycle Time Model," *IEEE Journal Solid-State Circuits*, vol. 31(5), pp. 677–688, 1996.

[19] S. Zhang and K. S. Chatha, "Approximation Algorithm for the Temperature-Aware Scheduling Problem," in *Proc. International Conference on Computer Aided Design*, 2007, pp. 281–288.

978-1-60558-497-3/09 $25.00 © 2009 ACM   483

Dynamic Thermal Management via Architectural Adaptation

Ramkumar Jayaseelan, Tulika Mitra
School of Computing
National University of Singapore
{ramkumar,tulika}@comp.nus.edu.sg

ABSTRACT

Exponentially rising cooling/packaging costs due to high power density call for architectural and software-level thermal management. Dynamic thermal management (DTM) techniques continuously monitor the on-chip processor temperature. Appropriate mechanisms (e.g., dynamic voltage or frequency scaling (DVFS), clock gating, fetch gating, etc.) are engaged to lower the temperature if it exceeds a threshold. However, all these mechanisms incur significant performance penalty. We argue that runtime adaptation of micro-architectural parameters, such as instruction window size and issue width, is a more effective mechanism for DTM. If the architectural parameters can be tailored to track the available instruction-level parallelism of the program, the temperature is reduced with minimal performance degradation. Moreover, synergistically combining architectural adaptation with DVFS and fetch gating can achieve the best performance under thermal constraints. The key difficulty in using multiple mechanisms is to select the optimal configuration at runtime for time varying workloads. We present a novel software-level thermal management framework that searches through the configuration space at regular intervals to find the best performing design point that is thermally safe. The central components of our framework are (1) a neural-network based classifier that filters the thermally unsafe configurations, (2) a fast performance prediction model for any configuration, and (3) an efficient configuration space search algorithm. Experimental results indicate that our adaptive scheme achieves 59% reduction in performance overhead compared to DVFS and 39% reduction in overhead compared to DVFS combined with fetch gating.

Categories and Subject Descriptors

C.1.0 [**Processor Architectures**]: General

General Terms

Design, Performance, Reliability

Keywords

Dynamic Thermal Management, Architecture Adaptation

Permission to make digital or hard copies of part or all of this work for personal or classroom use is granted without fee provided that copies are not made or distributed for profit or commercial advantage and that copies bear this notice and the full citation on the first page. To copy otherwise, to republish, to post on servers or to redistribute to lists, requires prior specific permission and/or a fee.
DAC'09, July 26-31, 2009, San Francisco, California, USA

1. INTRODUCTION

Exponentially increasing power density due to technology scaling has made thermal management an important aspect of computer systems design. Traditionally, the problem of high on-chip temperature has been solved by employing more advanced packaging and cooling solutions. But modern high-performance processors are already pushing the limits of what the cooling solutions can offer. This has led to widespread interest in thermal-aware design at all levels of the computer systems. Recently, researchers have explored architectural and software-based techniques for thermal management with the aim to maximize performance while maintaining the on-chip temperature below a specified threshold. The on-chip temperature is continuously monitored and when it exceeds a predefined threshold, appropriate mechanisms are engaged to lower the temperature.

The most popular choice of mechanisms for thermal management include dynamic voltage/frequency scaling (DVFS), clock gating, and fetch gating. Unfortunately, each of these mechanisms, once engaged, is associated with significant performance degradation as well as invocation overhead. For instance, a system employing DVFS for thermal management incurs performance loss due to lower operating frequency plus non-negligible overhead in scaling voltage/frequency [17]. Hence, thermal management techniques must be judicious in choosing the severity of the response mechanism in proportion to the severity of the thermal stress.

We show that runtime micro-architectural adaptivity, such as scaling instruction window size and issue width, is more effective at managing thermal stress with minimal performance impact. These mechanisms are easy to configure at runtime and have significant impact on temperature. As most applications exhibit only limited instruction-level parallelism (ILP), they cannot exploit the wide issue width and large instruction window available in high performance processors. If these micro-architectural parameters can be scaled appropriately to track the available ILP of the program, we get significant reduction in on-chip temperature (due to reduced power dissipation) with hardly any impact on performance. On the other hand, for an application with higher ILP, we have no choice but to reduce the operating frequency of the processor to maintain on-chip temperature below the threshold. Therefore, we argue that a combination of multiple mechanisms (architectural adaptation with DVFS) can provide significantly better performance that current techniques while still satisfying the thermal constraints.

The motivation behind exploiting an adaptive micro-architecture (DVFS, fetch gating, issue width scaling and issue window scaling) for DTM is illustrated in Figure 1. The figure shows the peak temperature and throughput (in billion instructions per second or BIPS) at different processor configurations for benchmark `crafty`. Clearly, *there exist multiple configuration points with varying per-*

978-1-60558-497-3/09 $25.00 © 2009 ACM

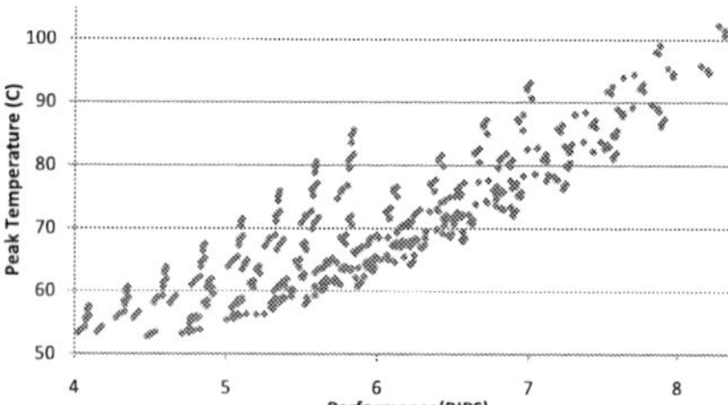

Figure 1: *Performance versus temperature for different configuration points with benchmark crafty.*

formance that result in the same peak temperature. In other words, given a thermal constraint, appropriate choice of configuration parameters can lead to significantly higher performance.

The main challenge in using multiple mechanisms is to chose at runtime an appropriate configuration for a time varying workload. Conventional single response DTM techniques employ feedback control to determine the best performing operating point (frequency level) for a given workload/thermal stress [16]. Architecture adaptation such as window scaling and issue width scaling have not been exploited for thermal management but have been used for power/performance tradeoffs. Techniques such as [4, 12] opportunistically exploit the workload characteristics to scale micro architecture structures for power savings. However, like single response DTM, these techniques are single point optimizations and do not employ multiple adaptations synergistically.

To control multiple response mechanisms, we rephrase the thermal management problem as a configuration space exploration problem. We design a software-based framework that identifies the optimal configuration under thermal constraints. The configuration management routine wakes up once every adaptation interval (in the order of milliseconds) to collect workload statistics. These statistics are used to choose the optimal configuration that is thermally safe for the next interval.

As our DTM framework is prediction based (relies on neural network classifier to determine if a configuration is thermally safe), we assume the presence of a simple hardware DTM mechanism such as clock gating as a backup technique. In case we choose a configuration that exceeds thermal threshold (which is very rare), clock gating will be engaged to lower the temperature.

2. RELATED WORK

With temperature constraints becoming one of the key limiters of performance in computer systems, there has been widespread interest in the design of efficient thermal management techniques [16]. The goal is to maximize performance of the system while maintaining the temperature below a critical point. The on-chip temperature is monitored continuously and when it exceeds a threshold, appropriate mechanisms are engaged to lower the temperature. Commonly employed mechanisms to control temperature include fetch gating [15], activity migration [18], and DVFS [7]. There is a performance loss associated with all of these mechanisms and the design of thermal management schemes involve adjusting the extent of response (performance loss) to the severity of thermal stress. Feedback controllers have been used for this purpose and it has been shown that they achieve near optimal results [6].

We show that employing multiple mechanisms synergetically for thermal management can provide better performance. One possible approach to engage more than one response for thermal management is to determine a crossover point between the thermal management techniques as in [15] where fetch gating and dynamic voltage scaling are combined. Similarly Jung et al. [8] employ exhaustive simulation and stochastic modeling to determine the

best power management policy with two configuration parameters, namely, cache size and frequency. It is unclear how to determine the cross-over point when multiple mechanisms are employed. Similarly, the large configuration space for multiple mechanisms makes exhaustive simulation like [8] infeasible. In our approach, we view the thermal management problem as a configuration space exploration problem and design an efficient online technique to determine the configuration that results in maximum performance for the workload under a given temperature constraint.

Orthogonal to existing hardware based schemes, software based thermal management have also been explored. These schemes exploit the variation in the thermal behavior of different tasks in a multitasking scenario and perform scheduling to maintain thermal constraints. Common approaches involve adjusting time slices between hot/cold tasks [10, 5] and migration [13]. Instead of changing the workload executing on the processor in response to a thermal stress, our DTM technique alters the hardware configuration for a given workload.

Adaptive hardware components that can change complexity at runtime in terms of width and size have been used previously to provide power/performance tradeoffs. Existing hardware adaptation techniques are single point optimizations that rely on local information about the workload [4] to reduce power. We use multiple adaptations synergistically along with frequency scaling for thermal management. [17] is the only other work that uses architectural adaptivity for thermal management. However, it uses off-line profiling to guide adaptation and hence is only applicable for multimedia applications. In contrast, ours is an online technique that is applicable to any workload (including multimedia applications).

3. ADAPTIVE MICRO-ARCHITECTURE

The micro-architectural parameters that we control at runtime for effective thermal management are (1) instruction window size, (2) issue width, and (3) fetch gating. These structures have been chosen as (a) it is easy to reconfigure them at runtime, and (b) they contribute (either directly or indirectly) to the thermal hotspots of the processor. In addition, we also scale the operating frequency/voltage.

We model an adaptive instruction window similar to the design in [4] that contains four partitions each of which can be enabled or disabled at runtime. The issue width can be altered between two and six instructions per cycle. When the issue width is reduced, the additional functional units are disabled and the corresponding register file ports are not precharged. The fetch unit is controlled by setting the appropriate gating level. Fetch gating level T implies instruction fetch is halted once after every T cycles. We assume special instructions to resize the adaptive structures in software [17].

Moving along each axis in our configuration space has different impact on performance and temperature. Fetch gating lowers the active power dissipation by reducing the number of instructions delivered to the back-end. Scaling the window size and issue width changes the power dissipation per operation (adaptive structures consume less power when scaled down) in addition to altering the activity factor (less number of instructions are issued per cycle at smaller issue width and window size). Finally, power dissipation reduces with reduced operating voltage/frequency.

4. DYNAMIC THERMAL MANAGEMENT

We now present our software-based DTM framework. Figure 2 presents the components of our software-based dynamic thermal management framework that exploits the adaptive micro-architecture presented in Section 3. The configuration management routine (on the right in Figure 2) runs in software. It collects the performance

978-1-60558-497-3/09 $25.00 © 2009 ACM

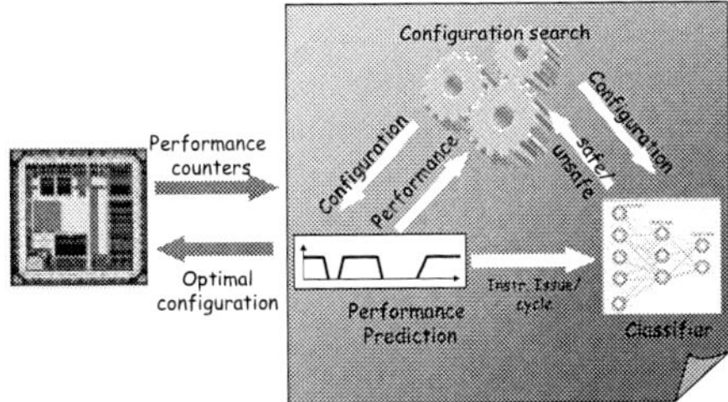

Figure 2: *Components of the Adaptive DTM Framework.*

counters from the processor once every adaptation interval (10^7 cycles or 2.8 ms at 3.6 GHz). As temperature change occurs slowly [16], the adaptation interval is set in the order of milliseconds, which is the period for timer interrupts in many systems. These workload statistics are used to guide the choice of configuration parameters for the next interval. The goal of the configuration search routine is to find the configuration with the maximal performance that satisfies the thermal constraints. At a particular configuration C' in the search space, we need to answer two questions.

1. What is the expected performance of this configuration? For this purpose, we develop a model (Section 4.2) that predicts the performance of configuration C' given the counter values for the currently running configuration C.
2. Is the configuration C' thermally safe? We design a neural network classifier (Section 4.1) that takes in configuration C' plus the number of instructions of each class (integer, floating point, branch, and load/store) issued per cycle as input and predicts if C' is thermally safe.

Note that the classifier requires the number of instructions issued per cycle as input as the temperature depends on the issued instructions. The performance, on the other hand, is determined only by the committed instructions. To bridge this gap, our performance model also estimates the number of instructions of each class issued per cycle corresponding to configuration C'.

The configuration space of our adaptive micro-architecture consists of 1,280 points (8 fetch gating levels × 4 window sizes × 5 issue widths × 8 frequency levels). Clearly, it is not feasible to evaluate all the configurations and find the optimal one. Instead, we design an efficient search strategy (Section 4.3) that (a) reduces the four-dimensional configuration space (fetch gating levels, windows sizes, issue widths, frequency levels) to two dimensions (IPC and frequency levels) based on insights gained from the performance model, and (b) further prunes the two-dimensional configuration space based on some properties of the space. Due to these optimizations, our search strategy evaluates only a small subset of the configuration space (32 points in the worst case).

4.1 Neural Network Classifier

While searching for the optimal configuration, we need to determine if a particular configuration is thermally safe. The thermal profile of a processor typically shows large variations among the different components of the processor (up to $15^{o}C$ difference) [16]. In our adaptive micro-architecture, the temperature of a processor component depends both on the configuration parameters as well as the usage pattern of the component (workload). However, an analytical framework to determine the temperature of the different components is too computationally expensive to be employed in an online DTM framework like ours. Instead, we model the problem of determining if a particular configuration is thermally safe for the current workload as a classification problem. A classification problem consists of a set of input features, output classes and a trained classifier. When an input is given (i.e., the input features are assigned values), the classifier predicts the class to which the input

belongs. In our framework, we design a neural network classifier that partitions the {configuration, workload} pairs into thermally safe and thermally unsafe classes.

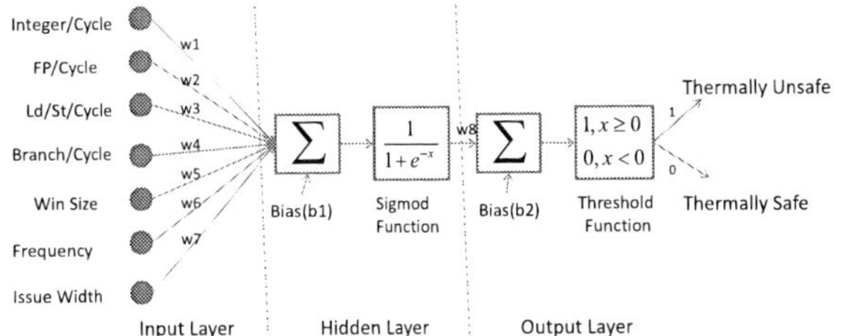

Figure 3: *Neural network classifier architecture.*

Classifier Architecture. Figure 3 shows the structure of our neural network classifier. The input features consist of three configuration parameters: (1) instruction window size, (2) issue width, (3) operating frequency plus four workload parameters: number of integer, branch, load/store and floating point instructions issued per cycle. We choose the workload features that correlate well with the usage pattern of the branch predictor and the execution core (instruction window, register file, and execution units) — the hottest components of the processor [16]. We verify this by employing principal component analysis [1], and observe that these features account for 98% of the variance in the observed temperature.

The values of the workload features vary according to the configuration as well as the workload. For example, the number of integer instructions issued per cycle depends on the issue width, instruction window size as well as fetch gating level. Given a configuration, the workload parameters are obtained from the performance prediction model (see Section 4.2). Fetch gating level is excluded from input features of the classifier as it only alters the usage patterns of the processor components, which is reflected sufficiently in the four workload features. The rest of the configuration parameters, on the other hand, also impact power consumption per usage.

We use a neural classifier with a single hidden layer and one neuron in the hidden layer as shown in Figure 3. The hidden layer neuron uses a sigmod transfer function and the output layer neuron uses a threshold transfer function. Our classifier architecture results in high prediction accuracy with minimal classification time.

Training the Classifier. We use the Levenberg-Marquardt training algorithm [1] for training our classifier. The training algorithm is an iterative off-line procedure that adjusts the weights and bias values in the neural classifier to minimize the classification error. We train the classifier during system installation and/or when the system conditions (heat sinks, ambient conditions, etc.) change.

The training set is generated by running a set of micro-benchmarks under different configurations and checking if the resulting execution hits the thermal threshold. Each micro-benchmark consists of a loop body with 100 instructions. The number of loop iterations is large enough to ensure that the loop execution time is longer than the thermal time constant of the different processor components. The loop body of each micro-benchmark contains a mix of integer, floating point, load/store, and branch instructions. The shares of the instruction classes in this mix are generated randomly for each micro-benchmark.

Accuracy of the Classifier. We first train the classifier with 30 micro-benchmarks each running on 10 randomly chosen configurations from the configuration space. After training, we test the accuracy of our classifier for real programs by comparing the classification results against actual benchmark runs. We simulate each

978-1-60558-497-3/09 $25.00 © 2009 ACM

benchmark at 160 configuration points and determine if the execution hits threshold temperature and compare it with the corresponding classifier prediction. The results show that our neural network is highly accurate at predicting if a {workload,configuration} pair is thermally safe. The average classification error is less than 3%. Wrong classifications are of two types: (a) the classifier predicts a thermally safe configuration as unsafe (false positive), (b) the classifier predicts a thermally unsafe configuration as thermally safe (false negatives). In the case of false negative errors, the fail safe hardware DTM (clock gating) is engaged if the corresponding configuration is chosen by our search algorithm. False negative errors are observed only for 1.14% of the configurations and hence the backup hardware mechanism is rarely engaged in our scheme.

4.2 Performance Prediction Model

The performance prediction model is used as part of the configuration search process to predict the performance at a given configuration. It uses the performance counter values collected in the previous interval as input to characterize the workload and predict the performance for this workload under any given configuration.

First, we present the input and output parameters of the performance model. Let $C = \langle T, IW, W, F \rangle$ denote the configuration in the current adaptation interval, where T is the fetch gating level, IW is the issue width, W is the instruction window size, and F is the frequency level. We collect the following statistics from performance counters: Number of committed instructions $N_{useful}(C)$, Number of cycles in the interval $Cycles$, Number of committed instructions of type X: $N_{useful}^X(C)$ where X can be of type integer, floating point, branch, or load/store, Instruction cache misses IC_{miss}, Data cache misses DC_{miss} and Branch mispredictions Br_{miss}.

The model is used to produce two outputs. First, it estimates performance for each configuration $C' = \langle T', IW', W', F' \rangle$ that is visited; the performance is expressed as number of useful instructions committed per second (to include the effect of frequency scaling). Second, given a configuration C', the neural network classifier needs the number of instructions issued per cycle for each instruction class to predict if C' is thermally safe. Therefore, the performance prediction model has to estimate number of issued integer, floating point, branch, and load/store instructions per cycle. Note that number of issued instructions is typically more than the number of committed instructions due to branch misprediction.

Our performance prediction model is an extension of the interval analysis [9] by Karkhanis and Smith. Interval analysis is based on the notion that any superscalar processor has a sustained background level of performance that is interrupted by miss events such as cache misses and branch misprediction. Based on this assumption, the CPI (cycles per instruction) of a processor is

$$CPI = CPI_{steady} + CPI_{miss}$$

where CPI_{steady} is the background sustainable performance when there are no miss events and CPI_{miss} is the loss in performance due to branch mispredictions, instruction cache misses and data cache misses. For the current configuration C, we get

$$CPI_{steady}(C) = CPI(C) - CPI_{miss}(C) \text{ where } CPI(C) = \frac{Cycles}{N_{useful}(C)}$$

CPI_{miss} can be computed from the number of miss events and their corresponding latencies and is largely independent of the configuration [9] because changing window size, fetch gating level or issue width has minimal impact on miss ratios and their penalties.

$$CPI_{miss} = CPI_{miss}(C') = CPI_{miss}(C)$$

Now the CPI of configuration C' for which we are estimating performance can be expressed as

$$CPI(C') = CPI_{steady}(C') + CPI_{miss}(C') = CPI_{steady}(C') + CPI_{miss}$$

Thus for C', we only need to compute the steady background performance $CPI_{steady}(C')$ or $IPC_{steady}(C') = \frac{1}{CPI_{steady}(C')}$.

Karkhanis and Smith [9] observe that the IPC in the absence of miss-events and unbounded issue width is approximately $\sqrt{(W)}$. Under limited issue width, the IPC follows the unbounded characteristics and saturates at the issue width. Thus

$$IPC_{steady}(IW, W) = min(IW, \sqrt{W})$$

The fetch gating level affects steady IPC by changing the number of instructions delivered to the window. When the throttling level is T, the fetch unit is inactive one cycle after every T cycles of activity and $\frac{T}{T+1} \times FW$ instructions are delivered per cycle, where FW is the fetch width. At steady state, the number of instructions fetched per cycle should be equal to the number of instructions issued per cycle. Therefore, the steady IPC at configuration $C' = \langle T', IW', W', F' \rangle$ can be expressed as

$$IPC_{steady}^{ideal}(C') = min(\frac{T'}{T'+1} \times FW', IW', \sqrt{W'})$$

We call this ideal IPC as the characterization does not account for non-unit latency instructions, limited number of functional units of different types, and commit of multi-cycle operations [9]. To factor in these effects on the steady IPC, we compute a ratio η between *ideal* steady IPC and *observed* steady IPC for configuration C.

$$\eta = \frac{IPC_{steady}(C)}{IPC_{steady}^{ideal}(C)} = \frac{1}{CPI_{steady}(C) \times IPC_{steady}^{ideal}(C)}$$

As we do not adapt the latency of the functional units etc., the factor η remains constant across different configurations. So

$$IPC_{steady}(C') = \eta \times IPC_{steady}^{ideal}(C') = \eta \times min(\frac{T'}{T'+1} \times FW', IW', \sqrt{W'})$$
$$(1)$$

$$CPI(C') = \frac{1}{IPC_{steady}(C')} + CPI_{miss};$$

Finally, the performance for C' is estimated in terms of the number of instructions committed per second

$$Performance(C') = IPC(C') \times F' = \frac{1}{CPI(C')} \times F'$$

Estimating Issued Instructions. We also estimate the number of instructions issued per cycle at configuration C' ($IPC_{issue}(C')$) by extending our model to count both correct path and wrong path instructions [9]. The instruction mix at issue needed by the neural classifier can be computed as

$$IPC_{issue}^X(C') = IPC_{issue}(C') \times \frac{N_{useful}^X(C)}{N_{useful}(C)}$$

where $IPC_{issue}^X(C')$ is the number of instructions of type X issued per cycle and X can be of type integer, floating point, branch, or load/store. $N_{useful}^X(C)$ are input to the performance model.

Accuracy of the Performance Prediction Model. We evaluated the accuracy of our performance model for 64 randomly selected configuration points. The error is less than 5% for any benchmark and the average error for all the benchmarks is only 3.8%.

4.3 Configuration Search Strategy

We perform an intelligent search of the configuration space to determine the best performing configuration that is within the thermal limit. The search process explores the configuration space employing the neural classifier and the performance model.

Reducing Search Space. An exhaustive evaluation of all the 1,280 points in the configuration space is infeasible. We exploit insights derived from the performance prediction model in Section 4.2 to reduce the search space. From Equation 1 it is clear that that the steady IPC is constrained either by the fetch gating level (T), the window size (W) or the issue width (IW). Therefore, the performance of a processor cannot be improved by over-designing along one of the configuration parameters while restricting the other parameters — a balanced architecture provides the best performance. In other words, given a target steady IPC, we can compute appropriate values of T, W, and IW from Equation 1 as follows.

$$W = \left(\frac{IPC_{steady}}{\eta}\right)^2 ; \quad IW = \frac{IPC_{steady}}{\eta} ;$$

$$T = \frac{K}{1-K} \quad \text{where} \quad K = \frac{IPC_{steady}}{\eta \times FW}$$

So we can reduce the four dimensional configuration space (T, W, IW and frequency F) into a two dimensional search space consisting of only frequency and steady IPC (see Figure 4).

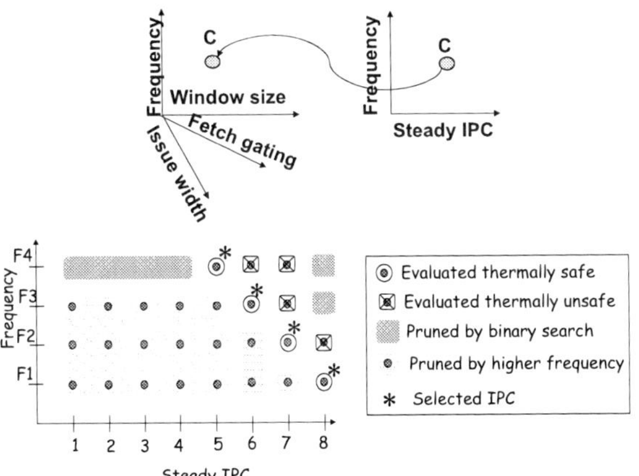

Figure 4: *Pruning of the configuration search space.*

Pruning Search Space. In the reduced search space {steady IPC, Frequency}, any increase in either steady IPC or Frequency would result in both higher temperature and higher performance. We exploits this to further prune the search space.

1. If a configuration $\langle F, IPC \rangle$ is thermally unsafe, then all the configurations $\langle F, X \rangle$ where $X > IPC$ are also thermally unsafe and can be pruned.

2. If $\langle F, IPC \rangle$ is thermally safe, then all the configurations $\langle F, X \rangle$ where $X < IPC$ have lower performance than the known thermally safe point $\langle F, IPC \rangle$ and can be pruned.

3. If $\langle F, IPC \rangle$ is the highest performing, thermally safe configuration at frequency level F, then we can prune all the configurations $\langle F', IPC' \rangle$ where $F' < F$ and $IPC' \leq IPC$ as they are guaranteed to have lower performance than $\langle F, IPC \rangle$.

We perform a linear search along the frequency dimension and binary search along the steady IPC dimension. Our search strategy for a hypothetical search space with four frequency levels and eight IPC values is shown in Figure 4. The search starts at the highest frequency $F4$. It evaluates $\langle F4, 5 \rangle$, which is the midpoint of the IPC space and determines that it is thermally safe. Therefore, points with lower values along the IPC axis are pruned. The search proceeds to the remaining points along the IPC axis and evaluates the midpoint $\langle F4, 7 \rangle$ as thermally unsafe. Now the higher IPC part ($\langle F4, 8 \rangle$) of the search space is pruned because they will be thermally unsafe. The search returns $\langle F4, 5 \rangle$ as the feasible point with

highest IPC at frequency $F4$. At the next frequency level, the configuration space $\langle F3, 1 \rangle \dots \langle F3, 5 \rangle$ is pruned as the points within this space are guaranteed to have lower performance than the point $\langle F4, 5 \rangle$. In this fashion, the search selects the best performing thermally safe point at each frequency level and finally chooses the highest performing point among them.

Complexity of Search Algorithm. Our algorithm has a worst case complexity of $O(L_f \times \ln(L_{IPC}))$ where L_f and L_{IPC} are the number of frequency levels and steady IPC levels. In our implementation, $L_f = 8$ and $L_{IPC} = 9$ (between 2 and 6 in increments of 0.5) resulting in 32 configuration points in the worst case. Our optimized search routine (with key optimizations such as using constants and pre-computations wherever possible, fast exponentiation [14], etc.) takes around 8,000 cycles on our simulated architecture in the worst case (32 search points). This represents an overhead of about 0.3% for a configuration interval of 1 ms.

5. EXPERIMENTAL RESULTS

We now present our experimental methodology and an evaluation of our software-based DTM management scheme against state-of-the-art DTM management techniques.

5.1 Processor Model and Workloads

We use Simple Scalar-3.0 simulator with Wattch power models [2, 3] for our experimental evaluation. We model an out-of-order superscalar processor with an issue width of 6 instructions per cycle, 128 entry active list (reorder buffer), 64 entry issue window and 64 KB instruction and data caches. The model also includes a 128 entry fully associative TLB, 2MB unified L2-cache, and 4KB entry bimod branch predictor. As mentioned earlier, our adaptive architecture has four possible window sizes (16,32,48,64), five possible issue widths (2–6), and eight fetch gating levels.

We use linear scaling in Wattch to obtain the power consumption with a supply voltage of 1.4 Volt and a frequency of 3.6 GHz at 100 nm, which corresponds to the supply voltage and frequency of the Pentium 4 processor. The power consumption at different window sizes are obtained based on [17]. When the issue width is altered, we assume that the additional functional units are switched off (no leakage power) and the corresponding lines of the wake-up logic and register ports are not driven [17]. For dynamic voltage/frequency scaling, we consider eight different levels between 3.6 GHz and 2.5 GHz. We obtain the corresponding supply voltages following the methodology proposed in [16]. We assume a penalty of $10\mu s$ per frequency/voltage transition [16].

We use HotSpot-3.0 [16] for thermal simulation with a floor-plan similar to Alpha floor-plan scaled to 100 nm and a convection resistance of 1.0 K/W. The power values are collected once every $2.8\mu s$ (10^4 cycles at 3.6 GHz). We feed the power trace to the thermal model and calculate the temperatures. The leakage power is obtained based on a simple model that computes the ratio between the active power and leakage power as a function of temperature [16]. The maximum allowed temperature is 85^oC and adjusting for sensor placement and reading errors, we get a threshold of 82^oC.

We use 14 benchmarks from the SPEC 2000 benchmark suite. For each of these benchmarks, we fast forward to the simulation point specified by [11] and simulate a total of 500 million instructions. Our simulation consists of an architectural warmup phase and a thermal warmup phase after which the statistics are collected.

5.2 Dynamic Thermal Managements Schemes

We compare our software-based thermal management scheme exploiting architectural adaptation (called *adaptive DTM*) against

978-1-60558-497-3/09 $25.00 © 2009 ACM

two state-of-the-art hardware based schemes namely *DVS* and *hybrid DTM* (DVS + fetch gating). We use a PI-control based DVS scheme and use Matlab to design a PI controller with a set point of 81.8^oC which includes a low-pass filter to prevent frequent voltage transitions [16]. The hybrid DTM scheme [15] combines fetch gating and DVS for thermal management.

As discussed earlier, the configuration search algorithm (implemented in software) of *adaptive DTM* has worst-case overhead of 8,000 cycles. In our experiments, we assume this worst-case overhead for each invocation of the search routine. Another important parameter for *adaptive DTM* is the configuration interval or the interval at which the search routine is invoked. We set the configuration interval to 2.8ms (10^7 cycles at 3.6GHz).

5.3 Performance Comparison

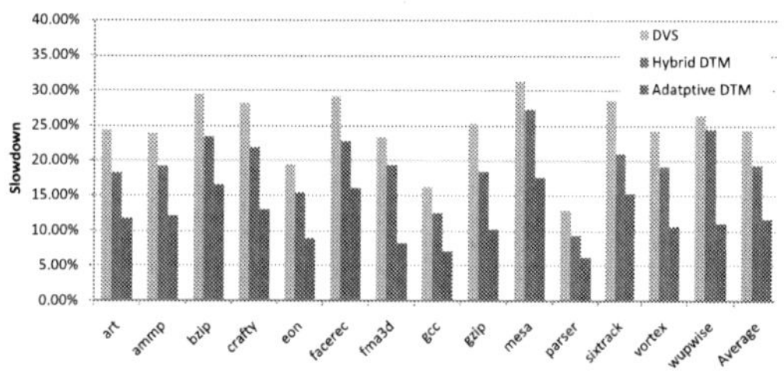

Figure 5: *Performance comparison of different DTM schemes.*

Figure 5 plots the slowdown of the three DTM schemes compared to the baseline architecture operating at the maximum frequency (i.e., without any thermal constraints). Any DTM schemes incurs some slowdown when the temperature of the processor exceeds the threshold. It is clear that *adaptive DTM* has significant performance benefit (lower slowdown) compared to the existing DTM schemes. On an average, *adaptive DTM* has 11.68% slowdown while *DVS* and *hybrid DTM* have 24.4% and 19.37% slowdown, respectively. In other words, *adaptive DTM* has 52% reduction in slowdown compared to *DVS* and 39% reduction in slowdown compared to *hybrid DTM*. Next, we try to explain where this performance benefit of *adaptive DTM* is coming from.

5.4 Temperature Profiles and Throughput

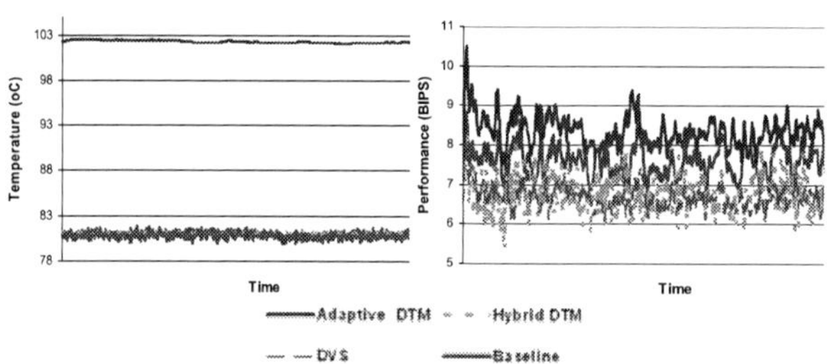

Figure 6: *Temperature profile and throughput for crafty.*

Figure 6 plots the time varying temperature profiles and throughput for benchmark `crafty`. The temperature of the baseline configuration remains above the thermal threshold for the entire duration of execution. The DTM mechanisms keep the temperature below the threshold. The corresponding performance plots show that keeping the temperature below the threshold results in loss of performance (billion instructions per second or BIPS). The performance of all three DTM schemes are lower than the baseline. However, the performance of *adaptive DTM* is higher than *DVS* and *hybrid DTM* for most points in the plot.

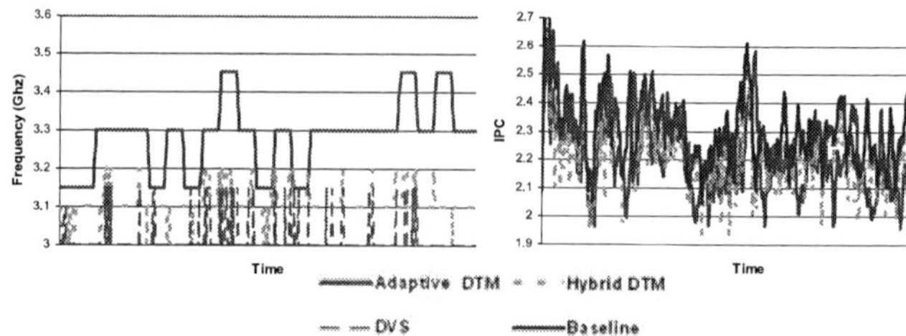

Figure 7: *Operating Frequency and IPC plots for crafty.*

We further analyzed the loss in performance in terms of frequency and IPC components. Figure 7 shows the IPC and frequency plots for crafty. Clearly, *adaptive DTM* resulted in a higher operating frequency than *DVS* and *hybrid DTM* for the same thermal constraint and this results in an improved performance. This in because unlike the other DTM schemes, *adaptive DTM* scales the micro-architecture structures in conjunction with frequency scaling. However, this impacts IPC. Our configuration search strategy optimizes along both the frequency and IPC axes and hence achieves better performance than existing techniques.

6. CONCLUSIONS

In this paper, we explore micro-architectural adaptivity, such as scaling instruction window size and issue width, for dynamic thermal management (DTM). We formulate the thermal management issue as a configuration space exploration problem and present a software-based framework that determines the best performing configuration under thermal constraints. Our method results in 59% reduction in performance overhead in comparison to DVS based DTM scheme and 39% reduction in comparison to a hybrid scheme that combines fetch gating and DVS.

7. REFERENCES

[1] Matlab Neural Network Toolbox. www.mathworks.com/access/helpdesk/help/pdf_doc/nnet/nnet.pdf.

[2] T. Austin, E. Larson, and D. Ernst. SimpleScalar: An Infrastructure for Computer System Modeling. *IEEE Computer*, 35(2), 2002.

[3] D. Brooks, V. Tiwari, and M. Martonosi. Wattch: A Framework for Architectural-level Power Analysis and Optimizations. In *ISCA 2000*.

[4] A. Buyuktosunoglu et al. A Circuit Level Implementation of an Adaptive Issue Queue for Power-Aware Microprocessors. In *GLSVLSI*, 2001.

[5] J. Choi et al. Thermal-aware task scheduling at the system software level. In *ISLPED 2007*.

[6] A. Cohen et al. On Estimating Optimal Performance of CPU Dynamic Thermal Management. *IEEE Computer Architecture Letters*, 2, 2003.

[7] H. Hanson et al. Thermal response to DVFS: Analysis with an Intel Pentium M. In *ISLPED 2007*.

[8] H. Jung, P. Rong, and M. Pedram. Stochastic modeling of a thermally-managed multi-core system. In *DAC 2008*.

[9] T. S. Karkhanis and J. E. Smith. A First-Order Superscalar Processor Model. In *ISCA 2004*.

[10] A. Kumar et al. HybDTM: A Coordinated Hardware-Software Approach for Dynamic Thermal Management. In *DAC 2006*.

[11] E. Perelman, G. Hamerly, and B. Calder. Picking Statistically Valid and Early Simulation Points. In *PACT 2003*.

[12] D. Ponomarev, G. Kucuk, and K. Ghose. Reducing Power Requirements of Instruction Scheduling through Dynamic Allocation of Multiple Datapath Resources. In *MICRO 2001*.

[13] R. Rao, S. Vrudhula, and K. Berezowski. Analytical results for design space exploration of multi-core processors employing thread migration. In *ISLPED 2008*.

[14] N. N. Schraudolph. A Fast, Compact Approximation of the Exponential Function. In *Technical Report INDISA-07-98*.

[15] K. Skadron. Hybrid Architectural Dynamic Thermal Management. In *DATE 2004*.

[16] K. Skadron et al. Temperature-aware Microarchitecture: Modeling and Implementation. *ACM TACO*, 1(1), 2004.

[17] J. Srinivasan and S. V. Adve. Predictive Dynamic Thermal Management for Multimedia Applications. In *ICS 2003*.

[18] X. Zhou, C. Yu, and P. Petrov. Compiler-driven register re-assignment for register file power-density and temperature reduction. In *DAC 2008*.

On-line Thermal Aware Dynamic Voltage Scaling for Energy Optimization with Frequency/Temperature Dependency Consideration

Min Bao
IDA, Linkoping
University SE-58183
Linkoping, Sweden
minba@ida.liu.se

Alexandru Andrei
Ericsson, Linkoping
SE-58330, Sweden
alexandru.andrei
@ericsson.com

Petru Eles
IDA, Linkoping
University SE-58183
Linkoping, Sweden
petel@ida.liu.se

Zebo Peng
IDA, Linkoping
University SE-58183
Linkoping, Sweden
zpe@ida.liu.se

ABSTRACT

With new technologies, temperature has become a major issue to be considered at system level design. Without taking temperature aspects into consideration, no approach to energy or/and performance optimization will be sufficiently accurate and efficient. In this paper we propose an on-line temperature aware dynamic voltage and frequency scaling (DVFS) technique which is able to exploit both static and dynamic slack. The approach implies an off-line temperature aware optimization step and on-line voltage/frequency settings based on temperature sensor readings. Most importantly, the presented approach is aware of the frequency/temperature dependency, by which important additional energy savings are obtained.

Categories and Subject Descriptors

C.3 [**Special-Purpose and Application-Based Systems**]: *Microprocessor/microcomputer applications, real-time and embedded systems*; D.4.1 [**Operating Systems**]: Process Management— *scheduling*; J.6 [**Computer-Aided Engineering**]: *computer-aided design*; J.7 [**Computers in Other Systems**]: *real time*

General Terms

Algorithms, Design, Performance, Theory

Keywords

temperature dependency, energy, voltage/frequency scaling

1. INTRODUCTION

Technology scaling and ever increasing demand for performance have resulted in high power densities in current circuits, which not only results in huge energy consumption but also leads to increased chip temperature. Temperature has become a major issue to be considered at system level design. Of particular importance in this context is the development of adequate temperature modeling, analysis, and measuring tools. HotSpot [24] is an architecture and system-level temperature model and simulator, based on elaborating an equivalent circuit of thermal resistances and capacitances that correspond to the architecture blocks and to the elements of the thermal package. A similar approach is proposed in [27] where dynamic adaptation of the resolution is performed, in order to speed up the thermal analysis. Simpler, analytical temperature models, which are much less accurate, have been also proposed [23]. For on-line temperature monitoring, sensors have been used [22] together with techniques for collecting and analyzing their values with adequate accuracy [9].

Permission to make digital or hard copies of part or all of this work for personal or classroom use is granted without fee provided that copies are not made or distributed for profit or commercial advantage and that copies bear this notice and the full citation on the first page. To copy otherwise, to republish, to post on servers or to redistribute to lists, requires prior specific permission and/or a fee.
DAC'09, July 26-31, 2009, San Francisco, California, USA

Several approaches to thermal aware system-level design aiming at energy optimization or temperature control have been proposed. Static approaches are exclusively based on temperature models used at design time. For example, in [21], a simulated-annealing based thermal aware floorplanning approach has been proposed. Thermal aware task allocation and scheduling have been addressed in [26] while in [23] an approach to task scheduling under peak temperature constraints is presented. An approach to design space exploration for multiprocessor SoCs under area and thermal constraints is presented in [14].

Various techniques have been proposed in which decisions are taken based on temperature measurements on the chip at execution time. Jung [12] proposed a temperature management approach based on a Markovian decision process aiming at minimizing energy under temperature constraints. The abstract model of the managed system is constructed at design time, based on a very simple temperature model. Using this off-line generated abstract model, at run time, decisions are taken based on temperature sensor reading. No strict timing constraints are considered. The techniques proposed in [8] for dynamic OS-level workload scheduling are exclusively based on execution time temperature sensor reading. The goal is to avoid thermal hot spots and large temperature variations which have a negative impact on system reliability.

One of the preferred approaches for reducing the overall energy consumption is dynamic voltage/frequency scaling (DVFS). This technique exploits the available slack times by reducing the voltage and frequency at which the processors operate and, thus, achieves energy efficiency. There are two types of slacks: (1) static slack, which is due to the fact that, when executing at the highest (nominal) voltage level, tasks finish before their deadline even when executing their worst number of cycles (WNC); (2) dynamic slack, due to the fact that most of the time tasks execute less cycles than their WNC. Offline DVFS techniques [11], [13] can only exploit static slack while online approaches [4], [3], [25] are able to further reduce energy consumption by exploiting the variation of the workload generated by the tasks. None of the above approaches considers any implication of the chip temperature.

As mentioned earlier, high power densities on current microprocessors result in high energy consumption and increased chip temperature. Growing temperature leads to an increase in leakage power and, consequently, energy, which, again, produces higher temperature. Thus, temperature is an important parameter to be taken into consideration at voltage selection. However, very few of the proposed DVFS techniques are considering the temperature issue. In [16] an offline, static DVFS scheme is presented, aimed at reducing peak temperature. In [5] we have proposed a temperature and leakage aware offline DVFS approach for energy minimization. The approach proposed in [19] is based on a design time optimization procedure which is performed considering various start time temperatures and workloads. At run-time, frequency setting is based on actual temperatures received from sensors. The approach ignores the dependency of leakage (and, consequently, energy) on temperature and assumes (as in offline DVFS techniques) that the number of cycles executed by a given task is fixed and known at design time.

The basic idea of DVFS approaches is to achieve energy minimization by reducing the supply voltage. Since frequency depends on the voltage, every change in supply voltage has to be followed, in principle, by a frequency adjustment. In order to provide performance, the frequency is usually set to the maximum value allowed by the current supply voltage. However, temperature has an important impact on circuit delay and, implicitly, on frequency,

978-1-60558-497-3/09 $25.00 © 2009 ACM

mainly through its influence on carrier mobility and threshold voltage [7]. Thus, the frequency does not only depend on the voltage but also on the temperature. This aspect has been completely ignored even by temperature aware approaches to DVFS. In fact, when calculating the allowed frequency for a given supply voltage V_{dd}, it is implicitly assumed that this is the frequency f corresponding to a maximum temperature T_{max} at which the chip is allowed to run. While this is a safe assumption, it is far from efficient. If we are aware that the chip is running at a temperature $T < T_{max}$, the frequency could be fixed at $f' > f$ and, thus, performance is increased for the same energy consumption. Or, maybe more important, the same frequency f could be achieved with a supply voltage $V'_{dd} < V_{dd}$ and, thus, further energy is saved. As we will show in this paper, such a temperature aware dynamic frequency setting can be an important contributor to energy efficiency.

The main contribution of this paper is an online temperature aware DVFS approach which can exploit both static and dynamic slack. The approach implies an offline temperature aware optimization step and online voltage/frequency settings based on temperature sensor reading. Most importantly, the presented approach is aware of the frequency/temperature dependency by which important additional energy savings are obtained.

The paper is organized as follows: In section 2 we introduce the power, delay and application model as well as the voltage selection techniques used in the paper. Section 3 gives a motivational example. In section 4 we present our DVFS approaches. Finally, experimental results and conclusions are presented in sections 5 and 6 respectively.

2. PRELIMINARIES
2.1 Power and Delay Model

For dynamic power we use the following equation [18], [6]:

$$P_{dyn} = C_{eff} * f * V_{dd}^2 \qquad (1)$$

where C_{eff}, V_{dd}, and f denote the effective switched capacitance, supply voltage, and frequency, respectively.

The leakage power is expressed as follows [15], [18], [16]:

$$P_{leak} = I_{sr} * T^2 * e^{\frac{\alpha * V_{dd} + \beta * V_{bs} + \gamma}{T}} * V_{dd} + |V_{bs}| * I_{ju} \qquad (2)$$

where I_{sr} is the reference leakage current at reference temperature. T is the current temperature, V_{bs} is the body bias voltage, and I_{ju} is the junction leakage current. α, β and γ are curve fitting circuit technology dependent coefficients.

The frequency at a reference temperature is calculated as follows [18]:

$$f = \frac{1}{d} = \frac{((1 + k_1) * V_{dd} + K_2 * V_{bs} - v_{th1})^\alpha}{K_6 * Ld * V_{dd}} \qquad (3)$$

where Ld is the logic depth. K_1, K_2, K_6, and v_{th1} are technology dependent coefficients. α reflects the velocity saturation imposed by the used technology (common values $1.4 < \alpha < 2$). The scaling of frequency with temperature is given by equ.4 [15] :

$$f \propto \frac{(V_{dd} - (v_{th1} + k * (T - T_{ref})))^\xi}{V_{dd} * T^\mu} \qquad (4)$$

T is the temperature while k, ξ, and μ are empirical technology dependent constants.

2.2 Application and System Model

The functionality of the application is captured as a set of task graphs. In a task graph $G(\Pi, \Gamma)$, nodes $\tau \in \Pi$ represent computational tasks, while edges $\eta \in \Gamma$ indicate data dependencies between tasks (communication). Each task is characterized by the following parameters: the worse case (WNC), best case (BNC), and expected (ENC) number of clock cycles to be executed, a deadline, and the average switched capacitance. ENC is the arithmetic mean value of the probability density function $p(NC)$ of the task execution cycles NC, i.e., $ENC = \sum_{j=BNC}^{WNC} j * p_j(j)$

The application is mapped and scheduled on an embedded system with a voltage scalable processor which can operate at several discrete supply voltage levels. The processor has internal temperature sensors that can be accessed during execution. The application and a set of look up tables (LUT), one for each task, are stored in memory. The LUTs are generated offline and are used during execution whenever the scheduler has to adjust the processor's performance to the appropriate level, via voltage/frequency scaling. An appropriate performance level allows the tasks to meet their deadlines while maximizing the energy savings.

2.3 Temperature-Aware DVFS

In [2] we have presented a DVFS approach which given a mapped and scheduled application, calculates the appropriate voltage levels for each task, such that the total energy consumption is minimized. Another input to the algorithm is the dynamic power profile of the application, which is captured by the average switched capacitance of each task. This information will be used for calculating the dynamic energy consumed by the task executed at certain supply voltage levels according to equ.1. The leakage energy is calculated using equ.2. However, since leakage strongly depends on temperature, an obvious question is which temperature to use for leakage calculations. Ideally, it should be the temperature at which the chip will work when executing the application. This temperature, however, is not known, since the algorithm is just calculating the voltages at which to run the system and these voltages are influencing the energy dissipation which, again, is determining the temperature.

The algorithm in [2] requires the designer to introduce an assumed temperature which is used at energy optimization. This, of course, leads to suboptimal results, since the temperature used for energy calculation during voltage selection is different from the actual temperature at which the chip works. In order to overcome the above problem, in [5] we have proposed a temperature aware DVFS technique which is based on an iterative approach as illustrated in Fig.1. The approach starts from an initial "assumed" temperature, at which the processor is supposed to run. The voltage selection algorithm will determine, for each task, the voltage levels such that energy consumption is minimized. Based on the determined voltage (and the switched capacitances known for each task) the dynamic power profiles are calculated, the thermal analysis is performed, and the processor temperature profile is determined in steady state. This new temperature information is now used again for voltage selection and the process is repeated until the temperature converges. Convergence means that the actual temperature values used at voltage selection correspond to the temperature at which the chip will function when running with the calculated voltages. As shown in [5], in most of the cases, convergence is reached in less than 5 iterations.

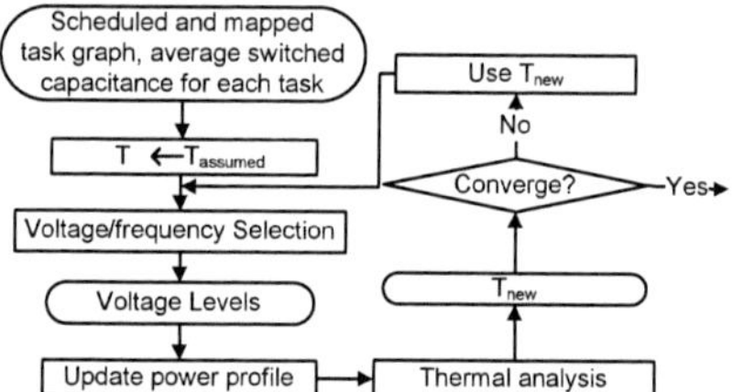

Figure 1: Temperature-aware Voltage Selection

Thermal analysis in our DVFS technique is based on the HotSpot environment [24]. Our modifications to HotSpot, in order to capture the dependence of leakage on temperature, are presented in [5].

Two limitations of the above approach are important in the context of this paper:

1. It ignores frequency/temperature dependency and, thus, produces solutions which are excessively conservative.

2. It is a static technique, assuming that tasks always execute their WNC and, thus, cannot exploit dynamic slack.

3. MOTIVATIONAL EXAMPLE

In this section, we consider an application consisting of three tasks as shown Fig.2. The WNC of τ_1, τ_2, and τ_3 is $2.85 * 10^6$, $1.0 * 10^6$, and $4.30 * 10^6$, respectively, and their average switched capacitance (in F) is $1.0 * 10^{-9}$, $0.9 * 10^{-10}$, and $1.5 * 10^{-8}$, respectively. The application has a global deadline of 0.0128s. We assume that the three tasks are executed on a processor which has 9 discrete V_{dd} levels from 1.0V to 1.8V with a step of 0.1V. The chip size is 0.007m*0.007m with a peak working temperature of $T_{max} = 125°C$.

For the above example we perform energy minimization using the temperature-aware DVFS method outlined in section 2.3. As mentioned, this DVFS approach (like all other in literature) ignores the frequency/temperature dependency and, when calculating the maximum allowed frequency for a certain supply voltage,

978-1-60558-497-3/09 $25.00 © 2009 ACM

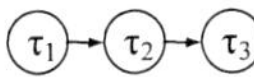

Figure 2: Motivational Example

the maximum allowed working temperature for the chip $T_{max}=125°C$ is considered. In Table 1 we show the actual voltages and frequencies for each task, as calculated by the DVFS algorithm, and the consumed energy. We also show the peak temperature for each task, obtained with dynamic thermal analysis. As can be observed, this peak temperature is far below the T_{max} of the chip.

Table 1: DVFS without Frequency/Temperature Dependency

Task	Peak Temp($°C$)	Voltage(V)	Freq(MHz)	Energy(J)	Total
τ_1	74.6	1.8	717.8	0.063	
τ_2	73.3	1.7	658.8	0.017	
τ_3	74.7	1.6	600.1	0.228	0.308

From equ. 3 and 4 it is obvious that by taking into consideration the actual temperature at frequency calculation there is a large margin for reducing the supply voltage without compromising on performance. We have performed a DVFS based energy optimization, similar to the one above, but with the difference that frequencies corresponding to the different voltage settings are calculated taking into consideration the peak temperature at which the actual task runs. Table 2 shows the results. We can see that a substantial energy reduction of 33% has been obtained.

Table 2: DVFS with Frequency/Temperature Dependency

Task	Peak Temp($°C$)	Voltage(V)	Freq(MHz)	Energy(J)	Total(J)
τ_1	61.1	1.8	836.7	0.051	
τ_2	59.9	1.7	765.1	0.013	
τ_3	61.1	1.3	483.9	0.142	0.206

The DVFS approach used above is an off-line, static one which assumes that tasks execute their WNC and, thus, can only exploit the static slack. However, in reality, there are huge variations in the number of cycles executed by a task, from one activation to the other, which leads to an important amount of dynamic slack. Imagine an activation scenario for which each of the three tasks in Fig. 2 execute a number of cycles equal to 60% of their WNC. If we use the offline DVFS approach above and run at the voltages calculated as in Table 2 the total energy consumption would be 0.122J. However, much more can be done by also exploiting the dynamic slack. This implies that, at run-time, whenever a task terminates, the voltage level and the frequency for the next task is calculated by taking into consideration the current time and the current chip temperature. Table 3 shows the voltage and frequency levels determined in this way as well as the corresponding energy consumption. The total energy consumed is 0.106J, which means a reduction of 13.1% compared to the off-line DVFS approach.

The examples presented in this section demonstrate that (1) considering the frequency/temperature dependency at DVFS can lead to substantial energy savings and (2) an on-line temperature aware approach is needed in order to make use of the dynamic slack created due to variable number of clock cycles executed at different activations.

4. DVFS WITH FREQUENCY/TEMPERATURE DEPENDENCY

4.1 Static Approach

The static approach is based on the technique outlined in section 2.3 and described in detail in [5]. As can be observed in Fig. 1, the successive iterations are leading, after convergence, to a temperature profile which corresponds to the one at which the chip will work. This temperature profile is used for energy calculation by the voltage/frequency selection algorithm.

For each task τ_i the voltage/frequency selection algorithm calculates a certain supply voltage V_{dd_i} such that energy consumption is minimized and deadlines are satisfied. When calculating the frequency setting for τ_i, as opposed to [5] and all other previous approaches, we now consider the thermal profile of the task and determine the maximum temperature T_{peak_i} at which that task runs. At voltage/frequency selection, the frequency is calculated

Table 3: Dynamic DVFS

Task	Peak Temp($°C$)	Voltage(V)	Freq(MHz)	Energy(J)	Total(J)
τ_1	50.5	1.5	625.2	0.018	
τ_2	50.4	1.5	625.2	0.005	
τ_3	51.4	1.3	481.2	0.083	0.106

based on equ. 3 and 4 (instead of being fixed, in a conservative way, considering the worst case temperature T_{max} for which the chip is designed).

4.2 Dynamic Approach

The above approach determines start times for tasks and their voltage/frequency levels assuming that they execute their WNC. By this, only static slack is considered for energy minimization [1].

In order to exploit the dynamic slack, at the termination of each task and before starting the next one, voltage and frequency settings have to be determined based on the values of the current time and temperature. In principle, calculating the appropriate voltage/frequency settings implies the execution of the temperature aware DVFS algorithm from section 4.1. Running this algorithm on-line, after each task execution, implies a huge time and energy overhead which can be even higher than the execution time and energy consumption of the actual application.

To overcome the above problem, we have divided our dynamic DVFS approach into two phases. In the first phase, performed off-line, voltage/frequency settings for all tasks are pre-computed, based on possible start times of the tasks and the possible temperatures at that start time. The resulting voltage/frequency settings are stored in look-up tables (LUTs), one for each task. In Fig. 3 we show two such tables. They contain voltage and frequency settings for combinations of possible start time t_s and start temperature T_s of a task. For example, the line with start time 1.3 and start temperature 55 stores the voltage and frequency setting for the situation when τ_2 starts in the time interval (1.2s,1.3s] and the start temperature is in the interval ($45°C,55°C$]. In section 4.2.1 we will present the generation procedure of the LUTs.

The second phase is performed on-line and is illustrated in Fig. 3. Each time a task terminates and a new voltage/frequency level has to be fixed for the next one, the on-line scheme looks up the appropriate setting from the LUT, depending on the actual time and temperature reading. If there is no exact entry in the LUT corresponding to the actual time/temperature, the entry corresponding to the immediately higher time/temperature is selected. For example, in Fig.3, τ_1 finishes at time 1.25s with a temperature $49°C$. To determine the appropriate voltage and frequency for τ_2, LUT_2 is accessed based on these time and temperature values. There is no exact entry for 1.25s and $49°C$, so the entry corresponding to start time 1.3s and start temperature $55°C$ is chosen. This on-line phase indicated with VS at Fig. 3 is of very low, constant time complexity O(1) and, thus, very efficient.

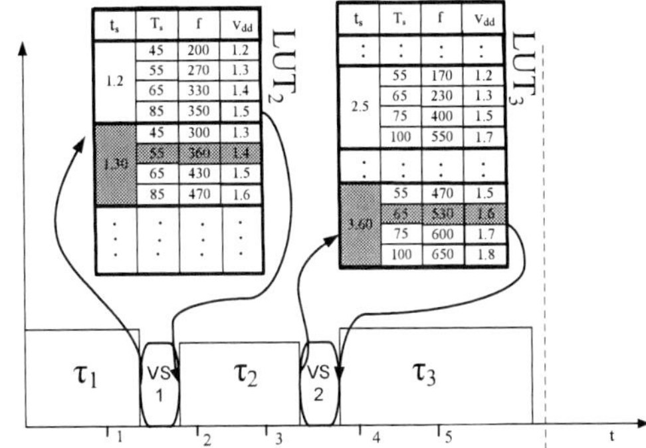

Figure 3: On-line Phase

4.2.1 LUT Generation

Considering an application and a system as described in section 2.2, the goal is to generate a LUT for each task τ_i, such that the

---

[1]It should be mentioned that, as opposed to the dynamic one, the static approach can be used even in the case that no temperature sensors are available on the chip.

energy consumption during execution is minimized. The task execution order is fixed according to a scheduling policy (e.g. EDF). According to this order, task τ_i has to be executed after τ_{i-1} and before τ_{i+1}. It is important to notice that the voltage levels and frequencies are calculated so that the energy consumption is optimal in the case that the tasks execute their expected number of cycles ENC (which, in reality, happens with a much higher probability than e.g. the WNC). Nevertheless, voltages and frequencies are fixed such that, even in the worst case (tasks execute WNC), deadlines are satisfied.

The LUT generation algorithm is presented in Fig.4. The outermost loop iterates over the set of tasks and successively constructs the table LUT_i for each task τ_i. The next loop generates the entries of LUT_i corresponding to the various start temperatures T_{s_i} of τ_i. Finally, the innermost loop iterates, for each possible start temperature, over all considered start times t_{s_i} of task τ_i. The algorithm starts by computing the earliest and latest possible start times for each task. The earliest start time EST_i is calculated based on the situation that all tasks execute with their best case number of cycles BNC at the highest voltage setting and lowest temperature (the ambient temperature). The latest start time LST_i is calculated as the latest start time of τ_i that still allows to satisfy the deadlines for all tasks τ_j, $j \geq i$, when executed with the worst case number of cycles WNC at the highest voltage and the maximum temperature T_{max} allowed for the chip.

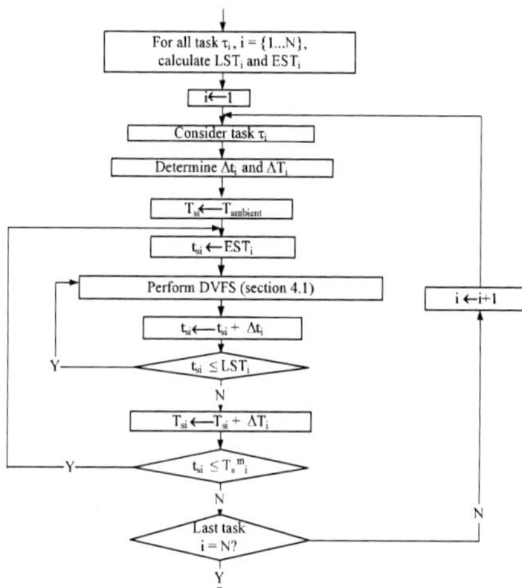

Figure 4: LUT generation

Considering the intended granularity of the LUT, the time and temperature quanta $\triangle t_i$ and $\triangle T_i$ are determined. Thus, for task τ_i, the number of time entries (the number of different start times considered) will be $(LST_i - EST_i)/\triangle t_i$, while, for each possible start time, the number of temperature entries is $(T_{s\ i}^m - T_{ambient})/\triangle T_i$, where $T_{s\ i}^m$ is the maximum possible temperature at the start time of τ_i. In sections 4.2.2 and 4.2.3 we will further elaborate on the granularity and size of the LUTs.

When calculating the actual LUT entries for a task τ_i, the calculation of the voltage and frequency setting is performed by running the DVFS algorithm outlined in section 4.1, for all tasks τ_j, $j \geq i$, considering t_{s_i} and T_{s_i} as start time and starting temperature, respectively, for τ_i.

4.2.2 Temperature Bounds and Granularity

As discussed before, the number of entries generated in LUT_i along the temperature dimension is $(T_{s\ i}^m - T_{ambient})/\triangle T_i$. The basic idea is that the lowest possible temperature is the temperature of the ambient, while $T_{s\ i}^m$ is the highest possible temperature, in the worst case, at the start time of task τ_i. But what is the value of $T_{s\ i}^m$? One solution is to consider for $T_{s\ i}^m$ the maximum temperature T_{max} at which the chip is allowed to work. While this assumption is safe, it leads to unnecessarily large tables since, during the execution of most of the tasks, the chip will never reach temperatures close to T_{max}. In order to avoid unnecessarily large tables, we need a safe but tighter upper bound on the temperature $T_{s\ i}^m$. In order to achieve this goal, our LUT generation algorithm in Fig.4 is executed several times in successive iterations before the final LUT tables are obtained.

We start by considering that for the first task the maximal starting temperature is the ambient temperature ($T_{s\ 1}^m = T_{ambient}$).

The two inner loops in Fig. 4 will generate LUT_1. As part of the DVFS procedure executed during generation of LUT_1 we obtain the possible temperature profiles of τ_1 and, thus, also the peak temperature T_{peak_1} reached during execution of this task. The worst case starting temperature of task τ_2 is $T_{s\ 2}^m = T_{peak_1}$. Considering this value for $T_{s\ 2}^m$, table LUT_2 is generated and the procedure is continued for all tasks τ_i. After the algorithm in Fig. 4 has been executed once, we have all LUT tables, based on the assumption that the maximal possible temperature at the start of τ_1 is equal to $T_{ambient}$. This, however, is not the case, since the application is executed periodically and τ_1 is started again after the last task τ_N. Thus, in fact, the maximal staring temperature of τ_1 is, in the worst case, equal to the worst case peak temperature of τ_N. Therefore, we repeat the LUT generation algorithm, this time considering that $T_{s\ 1}^m = T_{peak_N}$. This will lead to a higher T_{peak_1} than in the previous iteration and, thus, a new larger $T_{s\ 2}^m = T_{peak_1}$. Thus, new lines will be generated in the LUTs. The procedure is continued iteratively, until, for a certain task, the peak temperature over two successive iterations does not change, which means that no new entries into the LUT tables will be generated. Our experiments have shown that convergence is reached after not more then 3 iterations. This procedure also allows to detect if there exists a possibility for the design to reach, in the worst case, a thermal runaway situation (in which case the iterations do not converge) or if the maximum allowed temperature can be violated (there is convergence but there are peak temperatures which are beyond T_{max}).

The above technique leads to a tightening of the range of temperatures in the LUT. There are two more questions to be answered regarding the number of temperature entries (1) What should be the granularity of the temperature investigation and (2) how to reduce the number of entries if only a limited amount of memory is available?

It is obvious that a finer granularity and larger number of entries will, potentially, produce better energy savings at the cost, however, of increased memory consumption. With regard to the granularity $\triangle T_i$, our experiments have shown that values around $15°C$ are optimal, in the sense that finer granularities will only marginally improve energy efficiency. If, due to memory limitations, we only can afford a certain number of temperature entries NT_i to be stored for a task τ_i, we have to decide which lines of LUT_i to preserve and which to eliminate. One straightforward approach would be to maintain an even distribution of the selected NT_i lines over the range $[T_{ambient}, T_{s\ i}^m]$. However, start temperatures of tasks, during execution, do not spread evenly over this range. Thus, it is more efficient to have the NT_i lines more dense around the temperature values that are more likely to happen, and sparse towards the extremes. This means that less pessimistic voltage/frequency settings will be used for the most likely cases, while cases that are much less likely to happen are handled in a more pessimistic way. Thus, after the LUT tables have been generated, in order to select the appropriate NT_i lines along the temperature dimension for each task τ_i, we run a temperature analysis session in which all tasks are executed for their expected number of cycles ENC. From this analysis, we can observe which is the most likely starting temperature for each task and we select the NT_i lines among those close to this most likely temperature.

4.2.3 LUT Granularity Along the Time Dimension

A straightforward approach would be to allocate the same number of entries, along the time dimension, to each task (Nt_i is the same for all tasks τ_i, $i = 1..N$). However, the start time interval sizes $LST_i - EST_i$ can differ very much between tasks, which should be taken into consideration when deciding on the number of time entries. Therefore, given a total number of entries along the time dimension NL_t, we determine the number of time entries in each LUT_i, as follows [2]:

$$Nt_i = NL_t * \frac{(LST_i - EST_i)}{\sum\limits_{i=1}^{N}(LST_i - EST_i)} \qquad (5)$$

4.2.4 Accounting for Analysis Accuracy and Ambient

The solutions produced by our techniques presented in section 4.1 and 4.2 are safe. By this we mean that:

---

[2]Let us mention that while the start time intervals are very different from task to task, this is much less the case with temperature interval. Therefore, the number of entries along the temperature dimension (NT_i, see section 4.2.2 has been kept identical for all tasks in our experiments.

1. It is guaranteed that deadlines are satisfied;

2. If, at run time, a certain frequency setting is selected for a task τ_i, it is guaranteed that the temperature during execution of τ_i will not exceed the limit allowed for the chip to run at the selected frequency.

There are two aspects which have to be discussed with respect to the second of the two statements above. First is the issue of ambient temperature. If a task τ_i is starting its execution at a certain temperature T, the temperature profile during task execution depends on the actual ambient temperature. Thus, a safe frequency selection has to also take into consideration the current ambient temperature. Two possible solutions can be considered:

1. Generate the voltage/frequency settings considering the highest ambient temperature under which the system is supposed to function. This is a safe but pessimistic solution with, potentially, smaller energy savings.

2. Generate alternative voltage/frequency settings for a set of ambient temperatures in the range assumed for the system to function. During run time, using sensors for the ambient temperature, the system will switch to those tables corresponding to that ambient temperature that is immediately higher than the actual measured one. This solution requires additional memory for storing a larger amount of tables but could lead to better energy efficiency.

The second aspect to be considered is the accuracy of the temperature analysis. The fact that a certain frequency setting is safe, with regard to the peak temperature reached during execution of a task, is based on the temperature analysis performed as part of the DVFS procedure. Thus, the results can be safe only to the extent to which this analysis provides safe temperatures. Of course, system level thermal analysis tools are not provably accurate. Nevertheless, relative precisions are reported for the various analysis tools and we are using this information in order to account for the inaccuracy of the thermal analysis. More precisely, given a certain relative precision of the temperature analysis tool that we use, we account for this precision in a conservative way when determining the peak temperatures used for frequency calculation.

In section 5, we will evaluate the impact of both ambient temperature and potential analysis inaccuracy on the energy optimization results.

5. EXPERIMENTAL RESULTS

Our experiments have been performed on both generated applications and a real-life example.

We have randomly generated applications consisting of 2 to 50 tasks. The WNC of the tasks are in the range $[10^6, 10^7]$. The applications are executed on a processor which can run at 9 different supply voltage levels in the range [1.0, 1.8]. The temperature model related coefficients are the same as in [15], while the power model related coefficients are as in [18]. Similar to [15] and [20], in equ.4 we use the coefficients $\mu = 1.19$, $\xi = 1.2$, and k $= -1.0V/°C$. If not mentioned differently we assume an ambient temperature of 40°C.

It is important to mention that in all our experiments, we have accounted for the time and energy overhead produced by the on-line component of our dynamic approach. Similarly, we have also taken into consideration the energy overhead due to the memories. This overhead has been calculated based on the energy values given in [10] and [17].

The first set of experiments is aimed at evaluating the efficiency of taking into consideration the frequency/temperature dependency. We first compare the static DVFS approach in [5], which ignores the frequency/temperature dependency, to our static approach proposed in section 4.1. Considering the average over all 25 applications the energy consumption obtained by considering frequency/temperature dependency is 22% smaller than when ignoring this dependency.

For the dynamic approach, described in section 4.2, we have run the same set of experiments, first ignoring frequency/temperature dependency and than considering it. The energy consumption has been reduced, on average, by 17% in the latter case.

The next set of experiments compares the energy consumption with the static DVFS approach in section 4.1 and the dynamic one in section 4.2 (both considering frequency/temperature dependency). As the ratio BCN/WCN has a strong influence on the potential efficiency of a dynamic approach, we run the experiments considering three different ratios: 20%, 50%, and 70%. Also, we assume that the workload distribution of each task conforms

to a normal distribution $N(ENC, \sigma^2)$, where ENC is the mean value, and σ is the standard deviation. For our energy evaluations we have generated actual numbers of executed clock cycles for each task considering standard deviations of $(WNC - BNC)/3$, $(WNC - BNC)/5$, $(WNC - BNC)/10$, and $(WNC - BNC)/100$. Fig.5 shows the energy savings with the dynamic approach relative to the static one. As can be observed, the efficiency of the dynamic approach, compared to the static one, increases as the ratio between BNC and WNC becomes smaller. Remember that our DVFS algorithm is targeted torwards optimizing the energy consumption for the case that tasks execute the expected nunmber of cycles ENC. Therefore, energy savings are larger, compared to the static approach, when the standard deviation σ is smaller (more of the actual executed number of clock cycles are clustering around the ENC).

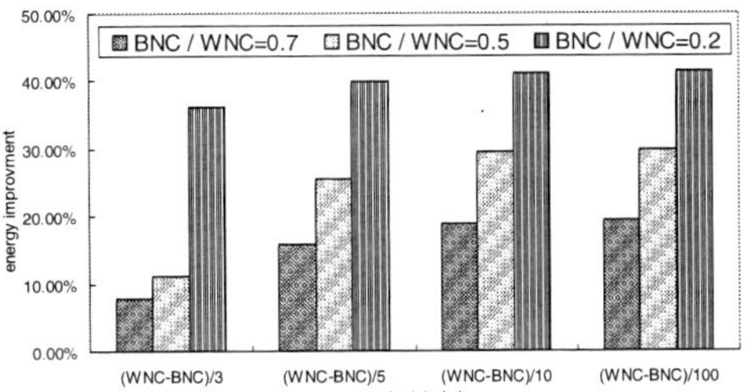

Figure 5: Dynamic vs Static Approach

The third set of experiments is aimed at exploring the impact of the LUT sizes. In particular, for this paper, we are interested in the impact of entries along the temperature dimension. The number of lines along the time dimension has been kept constant for these experiments and is distributed according to the discussion in section 4.2.3. First we run, for all applications, our dynamic DVFS approach considering a granularity $\triangle T = 10°$. We evaluate the average energy reduction with the obtained LUTs, compared to the static approach. Then we impose a certain limitation on the number of lines along the temperature dimension and we construct the corresponding LUTs as discussed in section 4.2.2. We again evaluate the energy consumption considering these reduced LUTs. The diagram in Fig. 6 shows the average results for different number of lines and two different standard deviations of the actual number of clock cycles executed by tasks. Having one single temperature entry will produce energy reductions compared to the static case which are 37% smaller (for $\sigma = (WNC - BMC)/3$) than with an unreduced LUT. However, with 2 entries the results are already very close to those obtained with an unreduced LUT and with 3 entries they are, in practice, identical. This is good news, since it shows that significant energy savings can be obtained with relatively small memory overhead. It should be mentioned that all other experiments presented in this section have been performed with 2 entries along the temperature dimension.

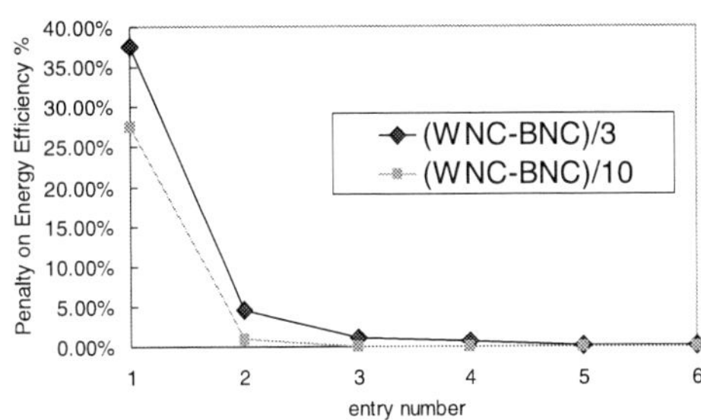

Figure 6: Impact of Temperature Line Number

Our final experiments have been performed in order to explore the impact of ambient temperature and temperature analysis accuracy. For all experiments above, we have assumed that $T_{ambient}$ is 40°C and is known at design time. In order to evaluate the impact of the ambient, we considered all the generated applications and constructed LUTs for values of $T_{ambient}$ in the range [-10°C, 40°C]. For each (application, LUTs) pair corresponding to a certain $T_{ambient}$ we evaluated the energy consumption considering that the $T_{ambient}$ is identical with the one assumed at LUT generation. Then we run the simulations for the same (application, LUTs) pair, but considering that $T_{ambient}$ deviates with $10°$, $20°$,...,$50°$ from the value assumed at design time. The results are shown in Fig.7. We can see that if $T_{ambient}$ is different by, for example, $20°$ from the one assumed at design time, the energy consumption increases by only 7% on average. This shows that, if the predicted range of ambient temperature is, for example, $40°$, generating two sets of LUTs (granularity of $20°$) will lead to energy losses, on average, less than 7%.

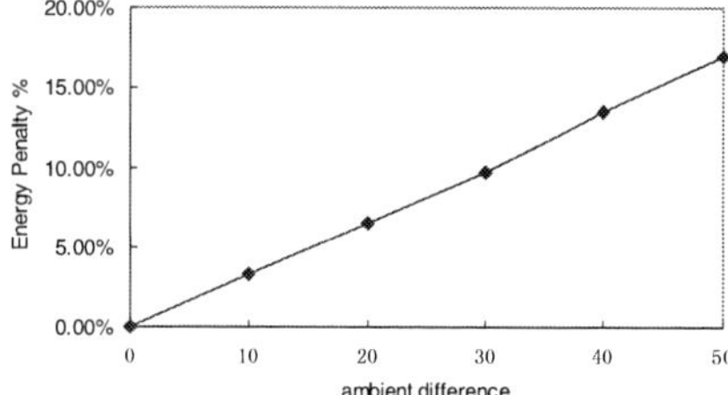

Figure 7: Impact of the Ambient Temperature

All the presented experiments have been performed considering that the temperature modeling and analysis is accurate. We have repeated the experiments considering a relative accuracy of 85%. When calculating frequency settings we accounted, in a conservative way, for this degree of accuracy. Our experiments have shown that the energy degradation due to the 85% relative accuracy is less than 3%.

Finally, we apply our static and dynamic approach described in section 4 to a real life case, namely an MPEG2 decoder which consists of 34 tasks and is described in more detail in [1]. The energy consumption using the static approach is reduced by 22% when considering frequency/tempeature dependency. For the dynamic approach the reduction is 19%. When considering frequency/temperature dependency, the dynanmic approach gives an energy reduction of 39% compared to the static one.

6. CONCLUSION

We have proposed an on-line temperature aware dynamic voltage and frequency scaling (DVFS) technique which is able to exploit both static and dynamic slack and takes into consideration the frequency/temperature dependency. The approach consists of two parts: an off-line temperature aware optimization step and on-line voltage/frequency setting based on temperature sensor readings. Experiments show that significant energy improvements can be achieved using the proposed techniques.

7. REFERENCES

[1] http://ffmpeg.mplayerhq.hu/.

[2] A. Andrei, P. Eles, Z. Peng, M. Schmitz, and . . B. M. Al-Hashimi. , energy optimization of multiprocessor systems on chip by voltage selection. *IEEE Transactions on Very Large Scale Integration Systems*, 15((3)):pp.262–275.

[3] A. Andrei, M. Schmitz, P. Eles, Z. Peng, and B. Al-Hashimi. Quasi-static voltage scaling for energy minimization with time constraints. *Design Automation and Test (DATE)*, April 2005.

[4] H. Aydin, R. Melhem, D. Moss, and P. Alvarez. Dynamic and aggressive scheduling techniques for power-aware real-time systems. *22nd IEEE Real-Time Systems Symposium (RTSS'01)*, pages pp. 95– 105, Dec. 2001.

[5] M. Bao, A. Andrei, P. Eles, and Z. Peng. Temperature-aware voltage selection for energy optimization. *Design Automation and Test (DATE)*, April 2008.

[6] A. P. Chandrakasan and R. W. Brodersen. *Low Power Digital CMOS Design*. Norwell, MA: Kluwer, 1995.

[7] R. Cobbold. Temperature effects on mos transistors. *Electronic Letters*, 2:pp. 190ĺC192, 1966.

[8] A. Coskun, T. Rosing, and K. Whisnant. Temperature aware task scheduling in mpsocs. *Design, Automation & Test in Europe Conference & Exhibition, 2007. DATE '07*, (No.7):pp. 1–6, April 2007.

[9] K. Gross, K. Whisnant, and A. Urmanov. Electronic prognostics through continuous system telemetry. *In 60th Meeting of the Society for Machine Failure Prevention Technology (MFPT)*, (1):pp. 53ĺC62, Apr. 2006.

[10] S. Hsu and A. A. et al. A 4.5-ghz 130-nm 32-kb l0 cache with a leakage-tolerant self reverse-bias bitline scheme. *IEEE JOURNAL of Solid-State Circuits*, May 2003.

[11] T. Ishihara and H. Yasuura. Voltage scheduling problem for dynamically variable voltage processors. *Low Power Electronics and Design, 1998. Proceedings. 1998 International Symposium on*, pages pp. 197– 202, Aug. 1998.

[12] H. Jung, P. Rong, and M. Pedram. Stochastic modeling of a thermally-managed multi-core system. *Design Automation Conference, 2008*, pages pp. 728–733, June 2008.

[13] W. Kwon and T. Kim. Optimal voltage allocation techniques for dynamically variable voltage processors. *ACM TECS*, 4(1):pp. 211–230, 2005.

[14] Y. Li, B. C. Lee, D. Brooks, Z. Hu, and K. Skadron. Cmp design space exploration subject to physical constraints. *HPCA06*, pages pp. 15–26, 2006.

[15] W. P. Liao, L. He, and K. M. Lepak. Temperature and supply voltage aware performance and power modeling at micro-architecture level. *IEEE TonCAD*, 24(No.7):pp. 1042ĺC1053, July 2005.

[16] Y. Liu, H. Yang, R. Dick, H. Wang, and L. Shang. Thermal vs energy optimization for dvfs-enabled processors in embedded systems. *Symp. on Quality Electronic Design (ISQED07)*, (International Symposium on Quality Electronic Design, 2007. ISQED '07. 8th):pp. 204–209, Mar. 2007.

[17] A. Macii, E. Macii, and M. Poncino. Improving the efficiency of memory partitioning by address clustering. *Design, Automation and Test in Europe Conference and Exhibition,*, 2003.

[18] S. Martin, K. Flautner, T. Mudge, and D. Blaauw. Combined dynamic voltage scaling and adaptive body biasing for lower power microprocessors. *ICCAD, .*, pages pp. 721–725, 2002.

[19] S. Murali, A. Mutapcic, and D. A. et al. Temperature control of high-performance multi-core platforms using convex optimization. *JOURNAL of VLSI Signal Processing Systems*, (Design, Automation and Test in Europe, 2008. DATE '08):pp. 110–115, Mar. 2008.

[20] B. Razavi. *Design of Analog CMOS Integrated Circuits*. McGraw-Hill Science Engineering, Erewhon, NC, August 2000.

[21] K. Sankaranarayanan, S. Velusamy, and K. S. M.R. Stan. A casefor thermal-aware floorplanning at the microarchitectural level. *the JOURNAL of Instruction-Level Parallelism*, 7:pp. 1–16, Oct 2005.

[22] M. Sasaki, M. Ikeda, and K. Asada. -1/+0.8°c error, accurate temperature sensor using 90nm 1v cmos for on-line thermal monitoring of vlsi circuits. *Microelectronic Test Structures, 2006 IEEE International Conference*, pages pp.9–12, March 2006.

[23] S. Wang and R. Bettatin. Delay analysis in temp.-constrained hard real-time systems with general task arrivals. *RTSS06*, pages pp. 323–334, 2006.

[24] W.Huang, S.Ghosh, S.Velusamy, K. Sankaranarayanan, K. Skadron, and M. Stan. Hotspot: A compact thermal modeling methodology for early-stage vlsi design. *IEEE on VLSI Systems*, 14(5):pp. 501–513, May 2006.

[25] C. Xian, Y.-H. Lu, and Z. Li. Dynamic voltage scaling for multitasking real-time systems with uncertain execution time. *IEEE Transactions on Computer-Aided Design of Integrated Circuits and Systems*, 27(8):pp. 1467–1478, Aug. 2008.

[26] Y. Xie and W.-L. Hung. Temperature-aware task allocation and scheduling for embedded multiprocessor systems-on-chip (mpsoc) design. *JOURNAL of VLSI Signal Processing Systems*, 45(3):pp. 177–189, Dec. 2006.

[27] Y. Yang and Z. G. et al. Isac: Integrated space and time adaptive chip-package thermal analysis. *IEEE Trans. Computer-Aided Design of Integrated Circuits and Systems*, Jan. 2007.

978-1-60558-497-3/09 $25.00 © 2009 ACM

SRAM Parametric Failure Analysis

Jian Wang[1], Soner Yaldiz[2], Xin Li[2], Lawrence T. Pileggi[2]

[1] PDF Solutions, Inc., San Jose, CA 95110
[2] Dept. of Electrical and Computer Engineering, Carnegie Mellon University, Pittsburgh, PA 15213
[1] jian.wang@pdf.com, [2] {syaldiz, xinli, pileggi}@ece.cmu.edu

ABSTRACT

With aggressive technology scaling, SRAM design has been seriously challenged by the difficulties in analyzing rare failure events. In this paper we propose to create statistical performance models with accuracy sufficient to facilitate probability extraction for SRAM parametric failures. A piecewise modeling technique is first proposed to capture the performance metrics over the large variation space. A controlled sampling scheme and a nested Monte Carlo analysis method are then applied for the failure probability extraction at cell-level and array-level respectively. Our 65nm SRAM example demonstrates that by combining the piecewise model and the fast probability extraction methods, we have significantly accelerated the SRAM failure analysis.

Categories and Subject Descriptors

B.8.2 [**Performance and Reliability**]: Performance Analysis and Design Aids

General Terms: Algorithms, Reliability

Keywords: SRAM, Parametric Failure, Failure Probability Estimation, Response Surface Model

1. INTRODUCTION

As the most commonly used embedded memory of modern on-chip systems, SRAM plays a crucial role in defining the system performance [1]. For maximum storage density, SRAM bit-cells have always been designed to use near-minimal devices in a given technology node. Such tight layout footprints make SRAM cells extremely vulnerable to the performance degradation and failures caused by random dopant fluctuation (RDF), which is inversely proportional to the layout area [2,3].

The RDF induced variations are spatially independent in nature. It follows that a rare cell-level failure caused by such variations can become quite significant for a system with many replicated cells. With the aggressive technology scaling, such local random variations are becoming more dominant. Therefore SRAM cells have to be carefully designed to provide an exceedingly low failure rate, such that the functionality is maintained for the system which contains thousands, or even millions of such

Permission to make digital or hard copies of part or all of this work for personal or classroom use is granted without fee provided that copies are not made or distributed for profit or commercial advantage and that copies bear this notice and the full citation on the first page. To copy otherwise, to republish, to post on servers or to redistribute to lists, requires prior specific permission and/or a fee.
DAC'09, July 26-31, 2009, San Francisco, California, USA

identical cells [4]. How to probe such rare failure events has become a serious challenge for SRAM design and analysis.

A widely used approach to study such a probability problem is the Monte Carlo method [5], which provides statistical estimations based on experiments performed at randomly selected samples in the variation space. To estimate the probability of the rare failure events in SRAM circuits, however, the Monte Carlo method typically requires millions, or even billions of samples to reach reasonable accuracy [6,7]. Such prohibitive cost severely limits the application of the Monte Carlo method in SRAM analysis. To alleviate this problem, several improvements have been proposed by integrating some controlling scheme in the sampling process, such as Latin hypercube sampling [8] or low-discrepancy sampling [7]. These methods claim to offer comparable accuracy to Monte Carlo methods while running up to a hundred times faster. Nevertheless, as demonstrated later in this paper, such an enhancement is still insufficient to enable SRAM failure analysis.

An alternative avenue to attack the problem is by modeling the stability metric as a known distribution and calculating the failure probability explicitly. For example, Gaussian distributions, non-central F distributions, or generalized Pareto distributions are among the commonly-used forms for approximating the stability metrics [9-12]. The accuracy of these methods, however, heavily relies on the validity of their assumptions, which unfortunately, are often questionable for nanoscale SRAM circuits. As pointed out by [6] and [13], even when the center portion of the stability metric distribution closely matches the presumed type, the tail part, where the failures usually happen, often deviates from the assumed form due to the increased nonlinearity in those regions.

In this paper we address the problem of SRAM parametric failure analysis with two novel techniques. Firstly, a response surface modeling (RSM) approach is applied to reduce the cost of transistor-level simulations. The critical problem here is how to accurately model the performance metric over a large variation space. We designed a piecewise approach, which by adaptively partitioning the variation space, renders a set of models covering the entire space while providing superb accuracy in regions critical for failure classification. Secondly, based on the statistical performance models, we explicitly identify the failure regions and apply a controlled sampling scheme to better probe those areas. Our approach effectively reduces the sample size without any assumptions on the performance metric distribution. By combining the model-based evaluation and the efficient sample-allocation, we significantly accelerate SRAM failure analysis, as compared to the traditional Monte Carlo approach.

The remainder of the paper is organized as follows. In Section 2, we define the SRAM failure analysis problem and review some background techniques. Section 3 presents our methods for fast

978-1-60558-497-3/09 $25.00 © 2009 ACM

probability extraction, i.e. the piecewise modeling and the controlled sampling techniques. We then escalate the problem in Section 4 and discuss how to estimate the failure probability for a SRAM array. Some results from a 65nm design are then shown in the next section, followed by our conclusions in Section 6.

2. BACKGROUND

2.1 SRAM Parametric Failure Analysis

To ensure proper functionality over the process variations, we focus on *stability margins* that are defined for SRAM cells as measures of robustness under different operating scenarios. Violation of a stability margin specification at certain process point is referred to as a *parametric failure* (or in short, a *failure*) in this paper. It should be noted, however, nothing would preclude us from considering other SRAM performance metrics for failure analysis by using the proposed techniques.

Assume that the process variations are described by the vector $x \in \Re^k$ of k independent random variables, which include both the global process variables and the local process variables from all transistors in the SRAM cell. The distribution of x is defined by its probability density function (PDF) $p(x)$. We choose a metric $S(x)$ and a specification S_{SPEC} to test the stability of the SRAM cell. $S(x) \geq S_{SPEC}$ signifies that the cell is stable at the process point x, and vice versa. Thereby we define the indicator function:

$$\mathcal{J}(x) \equiv \begin{cases} 1 & S(x) < S_{SPEC} \\ 0 & S(x) \geq S_{SPEC} \end{cases} \qquad (1)$$

where $\mathcal{J}(x) = 1$ indicates the failure event. One very important problem in the SRAM stability analysis is to find the parametric yield loss, i.e. the failure probability due to the process variations:

$$P \equiv \mathcal{P}(\mathcal{J}(x) = 1) \qquad (2)$$

The most commonly used approach for probability estimation is the Monte Carlo (MC) method [5]. Traditional Monte Carlo approach consists of three steps: (a) a total number of N samples, $x^{\{i\}}$ $(i = 1, \cdots, N)$, are randomly selected in the process variation space according to the distribution $p(x)$; (b) the interested stability metric is tested at all the sample points; (c) the results, $S(x^{\{i\}})$ and $\mathcal{J}(x^{\{i\}})$, are aggregated for estimation.

The failure probability can be estimated by averaging the indicator function:

$$P_{MC} = \frac{1}{N} \sum_{i=1}^{N} \mathcal{J}(x^{\{i\}}) \qquad (3)$$

Due to the randomness in the sampling, the estimated probability varies with different set of samples with a variance as [5]:

$$\sigma^2[P_{MC}] = \frac{P(1-P)}{N} \cong \frac{P_{MC}(1-P_{MC})}{N-1} \qquad (4)$$

where the operator "$\cong$" means "estimated by" (since the actual probability P is not known a priori).

As previously discussed, when analyzing the failure events in SRAM cells, we expect the probability to be extremely small. In such case, a meaningful estimation requires $\sigma[P_{MC}]$ to be sufficiently smaller than the actual probability. We define the *confidence ratio* to qualify the "accuracy" of the estimation:

$$R_C \equiv \frac{\sigma[P_{MC}]}{P} = \sqrt{\frac{1-P}{NP}} \approx \sqrt{\frac{1}{NP}} \qquad (5)$$

Clearly, if we want to retain a low R_C while P is very small, the sample count N has to be extremely large. This point is illustrated with a simple example in Fig. 1, where we use the Monte Carlo method to study the tail of a standard normal distribution and obtain the probability $\mathcal{P}(x > n_\sigma)$ (for convenience we often map an extremely small probability to a normal distribution and represent its location in the unit of σ). The plot shows the number of samples needed to reach a R_C of 0.05. Evidently, the sample count increases intractably when the event being analyzed moves towards the tail of its distribution.

For SRAM cell failure analysis, it is very usual to have failure probabilities beyond 5σ [4,6]. As demonstrated in Fig. 1, the excessive samples required by direct Monte Carlo method are often beyond practical means.

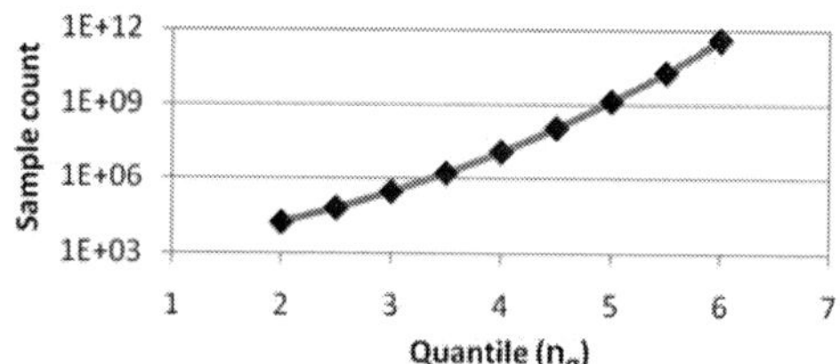

Figure 1. Samples needed by Monte Carlo method

2.2 Importance Sampling

Intuitively, with random sampling and a limited sample size, it is very unlikely that many samples are placed into the region that causes rare failures. This, in turn, renders large estimation error (confidence ratio) in the direct Monte Carlo method. To alleviate this problem, importance sampling (IS) is applied to the SRAM applications [6].

In importance sampling, a biased sample distribution $g(x)$ is introduced to intentionally place more samples into the failure region. After the controlled sampling and experiments, the failure probability is estimated as [14]:

$$P_{IS} = \frac{1}{N} \sum_{i=1}^{N} \left[\mathcal{J}(x^{\{i\}}) \frac{p(x^{\{i\}})}{g(x^{\{i\}})} \right] \qquad (6)$$

Note that compared to (3), the averaging here is weighted to unbias the estimation. The estimation variance can be derived as:

$$\sigma^2[P_{IS}] \cong \frac{\sum_{i=1}^{N} \left[\mathcal{L}(x_{\{i\}})^2 - P_{IS}^2 \right]}{N(N-1)} \qquad (7)$$

where $\mathcal{L}(x^{\{i\}}) \equiv \mathcal{J}(x^{\{i\}}) p(x^{\{i\}}) / g(x^{\{i\}})$. If $g(x) = p(x)$, Equ. (7) converges to Equ. (4). But as $g(x)$ shifts more emphasis onto the failure region, the dispersion in $\mathcal{L}(x^{\{i\}})$ decreases, as well as the variance $\sigma^2[P_{IS}]$.

By placing more samples in the failure region, importance sampling is expected to provide a much tighter estimation, compared to direct Monte Carlo method with the same number of random samples. One key issue, however, is how to design the new distribution $g(x)$. The authors of [6] attempt to first locate

978-1-60558-497-3/09 $25.00 © 2009 ACM

the region of failures by uniformly sampling the variation space, and then construct $g(x)$ based on such information. This causes two problems, however: (a) searching the process space with uniform samples can be very ineffective, especially when the failure events are rare or the space dimension is high [14]; (b) even the sample number is effectively reduced by importance sampling, the simulation cost for evaluating the samples could still be very expensive, as demonstrated later in this paper.

3. FAST ESTIMATION OF SRAM CELL FAILURE PROBABILITY

In view of the challenges in the SRAM stability analysis, we propose to build accurate statistical performance models that facilitate the probability extraction. A controlled sampling scheme is then performed for better failure region coverage and sample reduction. This section describes these two techniques.

3.1 Performance Modeling of Stability Metrics

There are two motivations for introducing response surface model (RSM) into the SRAM stability analysis problem. First, with the model it is possible to accurately locate the region of failures and apply a precisely-designed controlled sampling scheme. Second, the RSM hides the internal workings of the circuit and therefore can greatly accelerate the sample evaluation process.

With the increasing process variations, however, the circuit performance exhibits stronger nonlinear effects and is very difficult to be accurately captured. Furthermore, particularly in the SRAM applications, the cell stability failures have to be controlled at an exceedingly low level, which demands the scope to be extended further into the tails of the parameter distributions.

Facing these challenges, we propose to partition the variation space and create piecewise response surface models for the stability metrics. The authors of [15] developed a systematic approach to partition a parameter space and create piecewise models. Although initially proposed for analog macromodeling, with proper modifications this method is applicable to SRAM failure analysis as well.

In summary, we use the upper and lower bounds of the process parameters to define the initial region as a hypercube. This space is then adaptively divided into smaller pieces: (a) the inscribed ellipsoid of the current polytope (a hypercube in the first step) is first found by convex optimization; (b) the local space is sampled by a Design-of-Experiment (DoE) approach, and at each sample point the SRAM cell is simulated for its stability margin; (c) with the samples, a linear local model is created and the modeling error is evaluated to decide if further partitioning is necessary; (d) partitions with large errors will be further divided and the above steps are recursively applied. After the partitioning, the stability margin is modeled as a collection of the local models in all final partitions. More details of the partitioning formulation can be found in [15] and are neglected due to space limitations here.

Unlike [15], where a simple error-based criterion is applied to decide the partitioning direction, here we employ a set of criteria to better suit the particular probability extraction problem. Specifically, since our eventual goal is to estimate the failure probability, the modeling error does not need to be uniformly controlled over the entire variation space. For regions where the value of the stability metric is close to the predefined

specification, high accuracy is required because it directly affects the correctness of the failure identification. For the remaining regions, the model does not need to be as accurate, as long as the modeling error does not reverse the specification pass/fail status at that process point.

For this reason, we selectively apply a response-based criterion and an error-based criterion in different stages of the partitioning process. Both methods are illustrated in Fig. 2. The error criterion first finds the direction with the maximum modeling error. A hyperplane passing the ellipsoid center is then selected to be orthogonal to that direction and is used to divide the current partition into two pieces. The purpose of such partitioning is to reduce the size the local space and to increase the model accuracy. Alternatively, the response criterion constructs the contour of the specification, which is a hyperplane since the local model is linear. With some guard band added to both sides of the contour, the current polytope is divided into three parts. The purpose of this is to isolate the region requiring higher accuracy.

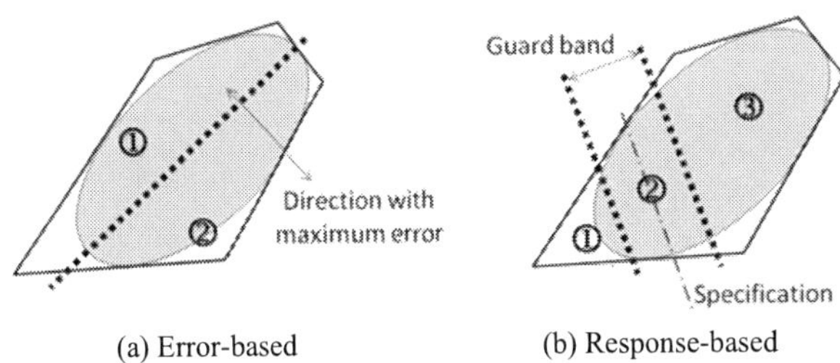

(a) Error-based   (b) Response-based

Figure 2. Partitioning criteria

Before the partitioning, we define a larger error tolerance ε_H and a smaller one ε_L. Intuitively, the partitioning always starts with the error criterion approach, until the higher tolerance ε_H is met. Then, the response-based approach is applied to divide the current polytope into three pieces. For the two pieces away from the specification, we stop further partitioning. While for the center piece, the partitioning carries on using again the error-based approach but with the lower tolerance ε_L as target. By this flow, we identify the regions where the stability metric is close to the specification, and create local models with higher accuracy in those regions. For the rest of the parameter space, the accuracy requirement is relaxed to reduce the modeling cost.

3.2 Model-based Probability Extraction

After the piecewise modeling, we have the stability metric represented as a linear model in each partition. We can then estimate the failure probability of the SRAM cell in three steps. Firstly, we isolate the critical partitions that contain failure regions. This is conveniently done by solving a linear programming problem to find the worst-case stability metric value within each polytope. Secondly, in each critical partition, we construct a new sample distribution to provide better coverage in the failure region and estimate the failure probability with importance sampling. Finally, we aggregate the results from all partitions and gradually append more samples until a target estimation confidence is achieved.

In each critical partition, we construct a new sample distribution as illustrated in Fig. 3. In such a partition, the specification contour $S(x) = S_{SPEC}$ defines a hyperplane. Therefore, very similar to that in the space partitioning process, we can solve a convex optimization problem to obtain an ellipsoid $\Phi = \left\{ Ex + d \,\big|\, \|x\|_2 \leq 1 \right\} \subseteq \Re^k$ that approximates the shape of the failure region [16]. As the sample distribution for the importance sampling, we construct a standard multivariate Gaussian distribution residing in the space spanned by the ellipsoid axes. If observed from the original parameter space, the new distribution is centered at the ellipsoid center d, and along each ellipsoid axis it presents an independent normal distribution whose standard deviation is proportional to the corresponding axis length. The PDF of the new sample distribution is given as:

$$g(x) = \frac{1}{\left(\sqrt{2\pi}\right)^k \sqrt{|\Lambda|}} \exp\left(-\frac{1}{2}(x-d)^T E^{-1}(x-d)\right) \quad (8)$$

where Λ is a diagonal matrix of the eigenvalues of E.

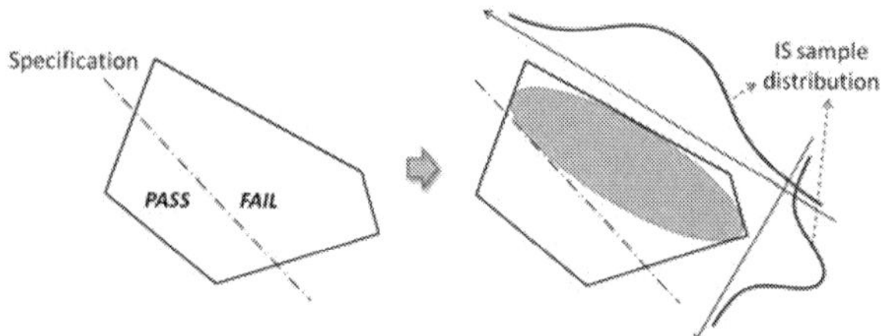

Figure 3. Defining biased sample distribution

We now can apply the importance sampling technique to estimate the failure probability. The samples are generated according to the new distribution $g(x)$ and are evaluated by the RSM. From the samples, we can estimate the failure probability by Equ. (6), and the estimation variance by Equ. (7). When all the partitions are processed, we aggregate the results and obtain the overall failure probability and estimation variance as:

$$P_{cell} = \sum_i P_{IS}^{(i)}, \quad \sigma^2[P_{cell}] = \sum_i \sigma^2[P_{IS}^{(i)}] \quad (9)$$

where the bracketed superscripts indicate the corresponding model partition.

In addition, due to the *sample superimposability* of the Monte Carlo method [5], it is unnecessary to apply all the samples at once. Instead, a better strategy is to first use a small amount of samples and then gradually append more until a predefined objective is met. In our implementation, we define a target R_C and first use a small number of samples in each partition. Next we select a partition which potentially provides the maximum estimation variance drop. Additional samples are appended to the chosen partition to improve the estimation accuracy there. This process is repeated until the pre-defined target is reached.

In summary, the proposed algorithm accelerates the analysis of SRAM cell failure events from three aspects: (a) With the RSM, we can easily identify the failure region and apply importance sampling with a controlled sample distribution. (b) For sample evaluation, response surface model is used in place of the expensive transistor-level simulation. (c) The samples are added in small chunks and the supplement stops as soon as the predefined estimation confidence is reached.

4. ARRAY-LEVEL ESTIMATION

In addition to the cell-level failure probability described in the previous section, it is often of even more interest to know such probability at array-level or system-level. This analysis problem is not trivial since the failures of individual cells are affected by both independent factors (such as local variations) and correlated factors (such as global variations). To solve this problem, we propose a nested Monte Carlo (NMC) method, where the outer level handles the global variations and the inner level handles the local variations.

The process variation information currently provided by IC foundries is usually defined in two levels. The global variations, defined as vector x_G, affect all the cells uniformly, while the local variations, defined as vector x_L, apply to each cell independently. For simplicity we assume that the SRAM array consists of M identical bit-cells without any redundancy (the handling of redundancy will be discussed at the end of this section). The failure event at the array-level, $\mathcal{I}_{sys}(x_G, x_{L[1]}, \ldots, x_{L[M]})$, is defined as the situation that at least one cell fails (variables connected to an individual cell are marked with square brackets in subscript). Our goal here is to find the failure probability P_{sys} of such array failures.

The nested Monte Carlo method can be described as the following steps: (a) We create the piecewise RSM for the stability metric of an individual cell with both the global and local variations as parameters, i.e. $S(x_G, x_{L[i]})$. (b) K samples are generated in the x_G space, according to the global variation distributions. (c) At each sample $x_G^{\{i\}}$, we apply the technique described in the previous section and estimate the cell failure probability $P_{cell}\big|_{x_G = x_G^{\{i\}}}$. (d) Given point $x_G^{\{i\}}$, the cell failures are only affected by local variations and are independent. Therefore, the array-level failure probability is obtained as:

$$P_{sys}\big|_{x_G = x_G^{\{i\}}} = 1 - \left(1 - P_{cell}\big|_{x_G = x_G^{\{i\}}}\right)^M \quad (10)$$

(e) When all the samples are processed, we estimate the array-level failure probability and the estimation variance as:

$$P_{NMC} = \frac{1}{K} \sum_{i=1}^{K} \left(P_{sys}\big|_{x_G = x_G^{\{i\}}} \right) \quad (11)$$

$$\sigma^2[P_{NMC}] \cong \frac{\sum_{i=1}^{K} \left(M^2 \left(P_{cell}\big|_{x_G = x_G^{\{i\}}} \right)^2 + M^2 \sigma^2 \left[P_{cell}\big|_{x_G = x_G^{\{i\}}} \right] \right) - K P_{NMC}^2}{K(K-1)} \quad (12)$$

The nested Monte Carlo method expedites the analysis of SRAM array failures in several ways. Firstly, the RSM is created only once with both the global and local variations as parameters, and there is no need to create the model over again at different global variation points. Secondly, at the inner-level we estimate the cell failure probability with the proposed importance sampling technique, then the failure probability of the array is analytically calculated by Equ. (10). Thirdly, by analyzing the estimator variance in (12), we notice that it is minimized when P_{cell} does not fluctuate much at different x_G points, i.e. the failure is dominated by the local variations. Fortunately, this is what we expect from the trends for the latest technology nodes [4].

As an alternative approach, we can also estimate the upper bound of the array-level failure probability by assuming all cell failures are independent, i.e.,

$$P_{sys} \leq 1 - \left(1 - P_{cell}\right)^M \quad (13)$$

978-1-60558-497-3/09 $25.00 © 2009 ACM

Since the local variations are becoming dominant, we expect such an upper bound to be fairly close to the actual value such that can be used as a fast approximation when accuracy is not very critical.

For simplicity we did not include redundancy in the above description. It should be noted, however, by formulating (10) and (13) accordingly, the proposed methods can be easily applied to systems with redundancy and/or ECC.

5. NUMERICAL EXPERIMENTS

In this section we demonstrate the efficiency of the proposed algorithms with a 6T SRAM cell designed in IBM 65nm CMOS process. Three stability metrics are selected for experiments: static noise margin (SNM) [17], read noise margin (RNM) [17] and write margin (WM) [18] (since we do not require any assumption on the metric distribution, other definitions can be used as well). The proposed methodology is implemented in MATLAB, using Spectre as transistor-level simulation engine.

5.1 Model Creation

For each stability metric, we apply the piecewise modeling methodology to create a response surface model, capturing the variation space up to $\pm 6\sigma$ for the local variations. For comparison, we also create a linear model over the entire region. To evaluate the model accuracy, we select 10,000 points uniformly distributed in the process space and choose those points that are in the tail part of the stability metric distribution (close to the specification) to compute the modeling error. The average errors and other relevant results are summarized in Table 1.

Evidently, linear template yields significant errors for all three metrics, including SNM and RNM which many believe to be normally distributed. In contrast, the proposed methodology effectively captures the large variation space and provides superior accuracy for all three metrics.

Table 1. Modeling results

Stability metric	Model template	Partition #	Total simulation #	Runtime (min)	Avg. err. (%)
SNM	Piecewise	13	0.6 k	7.0	1.9
	Linear	1	25	0.1	8.1
RNM	Piecewise	16	0.8 k	10.0	3.4
	Linear	1	27	0.1	12.1
WM	Piecewise	23	0.8 k	11.3	3.4
	Linear	1	19	0.1	8.2

5.2 Cell-level Failure Probability Extraction

When the piecewise models are available, the RSM-based importance sampling technique is applied to estimate the probability of the selected stability failure for an individual SRAM cell. We set the target confidence ratio R_C to 0.1 and run the proposed estimation flow (IS+RSM). The estimation results are listed in Table 2, which also includes the results from RSM-based and simulation-based Monte Carlo runs (MC+RSM, MC+Sim) with similar estimation confidence. Other methods, such as those in [11] or [12], are not included here since the estimation accuracy information is not attainable in those methods. It should be noted, however, that since the sample sizes required by direct Monte Carlo method are in general beyond our

computational capacity, some data (in shaded cells) are extrapolated from actual runs with fewer samples. Data that cannot be extrapolated (such as the failure probability from simulation-based Monte Carlo analysis) are left blank in the table.

The results demonstrate that RSM evaluation is over 400 times faster than transistor-level simulation in general, while the importance sampling technique, as compared to direct Monte Carlo analysis, is able to achieve a sample size reduction of over 2,000 fold. By combining these two features, the proposed method accelerates the SRAM failure analysis by 10^5 to 10^8 times, thereby enabling such difficult analyses.

Table 2. Cell-level estimation results

Metric	Method	Sim./eval. #	Runtime	P_{cell}	$\sigma[P_{cell}]$
SNM	IS+RSM	175 k	1.3 min	5.31e-8	5.3e-9
	MC+RSM	1 B	5.1 day	——	5.4e-9
	MC+Sim	1 B	6.6 yr	——	——
RNM	IS+RSM	105 k	2.5 min	1.17e-7	1.2e-8
	MC+RSM	0.7 B	4.0 day	——	1.2e-8
	MC+Sim	0.7 B	4.9 yr	——	——
WM	IS+RSM	60 k	1.1 min	5.48e-9	4.7e-10
	MC+RSM	40 B	0.8 yr	——	5.0e-10
	MC+Sim	40 B	386 yr	——	——

5.3 Array-level Failure Probability Extraction

Next we demonstrate the nested Monte Carlo (NMC) method on a 16 Kb SRAM array. For the inner-level estimation of the cell failure probability, we again set the R_C as 0.1. At the outer-level, we fix the number of the global variation (x_G) samples to be 100, which renders sufficiently accurate results in our experiments. For comparison, we calculate the sample number needed by direct Monte Carlo method to achieve similar accuracy and project its runtime (based on RSM evaluations). The results are summarized in Table 3. Again extrapolated data are in shaded cells.

Table 3. Array-level estimation results

Metric	Method	x_G #	RSM eval. #	Runtime	P_{sys}	$\sigma[P_{sys}]$
SNM	NMC	100	9.1 M	6.5 min	7.35e-4	8.0e-5
	MC	115 k	1.9 B	9.7 day	——	8.0e-5
	UpB	——	175 k	1.3 min	8.70e-4	8.6e-5
RNM	NMC	100	9.9 M	7.5 min	1.92e-3	1.9e-4
	MC	58 k	0.95 B	5.4 day	——	1.9e-4
	UpB	——	105 k	2.5 min	2.18e-3	1.9e-4
WM	NMC	100	3.1 M	7.8 min	7.86e-5	8.1e-6
	MC	1.2 M	20 B	0.4 yr	——	8.1e-6
	UpB	——	60 k	1.1 min	8.98e-5	7.7e-6

By handling the global variations and the local variations at two different levels, the nested Monte Carlo method is able to utilize the benefits of the importance sampling technique for estimating the cell-level probability. The probability obtained is then analytically translated from cell-level to array-level at the global variation point. Compared with direct Monte Carlo method, the nested approach provides over 1000× speed-up in our experiments (both methods use RSM for sample evaluation).

978-1-60558-497-3/09 $25.00 © 2009 ACM

Table 3 also lists the results of the upper bound (UpB) estimation. As expected, such method overestimates the failure probability by certain amount for all three examples. This is once more confirmed in Figure 4, where the WM failure probabilities for various array sizes are estimated with both the NMC and the UpB methods. For this 65nm process, however, the local variations dominant the overall effects. Therefore the overestimation by the UpB method is in fact only marginal. This suggests the UpB method as a candidate where runtime is more critical than accuracy (e.g. early-stage design decisions).

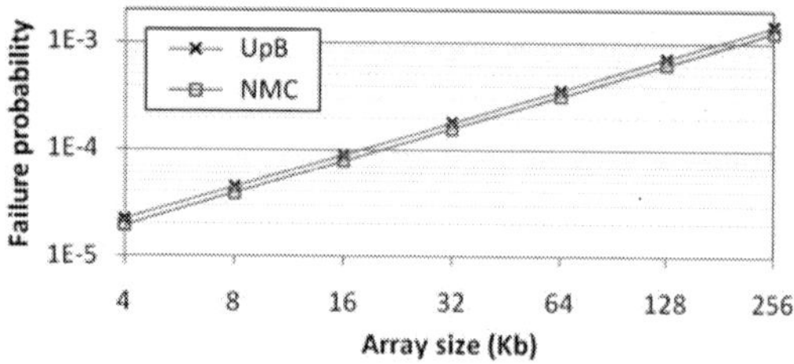

Figure 4. WM failure probabilities for different array sizes

6. CONCLUSIONS

With aggressive technology scaling, the design and analysis of SRAM circuits have become increasingly challenging. The most critical issues stem from the growing process uncertainties and the stringent functionality requirements. In view of the challenges, we propose a model-based importance sampling approach to facilitate the analysis of the rare failure events in SRAM circuits. Our proposed methodology extracts the SRAM failure probability at both the cell-level and the array-level. At cell-level, a piecewise modeling framework is applied to accurately model the cell stability metric over the large process variation space. A controlled sampling is then performed for better failure region coverage and sample reduction. At the array-level, we propose a nested Monte Carlo method that handles both the global and the local variations, as well as fully incorporates the benefits of the above cell-level failure probability extraction method.

In our experiments for a 65nm SRAM design, the piecewise model accurately captures the variation space up to $\pm6\sigma$ for the local variations. Our proposed methods for failure probability estimation also clearly outperform the conventional Monte Carlo approach at both the cell-level and the array-level. With the piecewise model and the fast probability extraction method, we accelerate the SRAM failure analysis by magnitudes and enable such difficult analyses.

As an extension to the proposed modeling and analysis flow, we are currently working to include design parameters (transistor sizes, array configuration parameters, etc.). The eventual goal is to help building up the complete methodology and tool-set to support the variation-aware design flow for SRAM systems.

7. ACKNOWLEDGEMENTS

The authors acknowledge the support of the Focus Center for Circuit & System Solutions (C2S2), one of five research centers funded under the Focus Center Research Program, a Semiconductor Research Corporation program, and the National Science Foundation under contract CCF-0702278.

8. REFERENCES

[1] B. H. Calhoun, et al., "Digital circuit design challenges and opportunities in the era of nanoscale CMOS," *Proceedings of the IEEE*, vol. 96, no. 2, pp. 343-365, Feb. 2008.

[2] T. Mizuno, J. Okamura, and A. Toriumi, "Experimental study of threshold voltage fluctuation due to statistical variation of channel dopant number in MOSFET's," *IEEE Trans. on Electron Devices*, vol. 41, no. 11, pp. 2216-2221, Nov. 1994.

[3] A. J. Bhavanagarwala, X. Tang, and J. D. Meindl, "The impact of intrinsic device fluctuations on CMOS SRAM cell stability," *IEEE JSSC*, vol. 36, no. 4, pp. 658-665, Apr. 2001.

[4] R. Heald, and P. Wang, "Variability in sub-100nm SRAM designs," *Proc. of ICCAD '04*, pp. 347-352, 2004.

[5] I. M. Sobol, *A Primer for the Monte Carlo Method*. CRC, Boca Raton, FL, 1994.

[6] R. Kanj, R. Joshi, and S. Nassif, "Mixture importance sampling and its application to the analysis of SRAM designs in the presence of rare failure events," *Proc. of DAC '06*, pp. 69-72, 2006.

[7] A. Singhee, and R. A. Rutenbar, "From finance to flip flops: a study of fast quasi-Monte Carlo methods from computational finance applied to statistical circuit analysis," *Proc. of ISQED '07*, pp. 685-692, 2007.

[8] R. L. Iman, J. C. Helton, and J. E. Campbell, "An approach to sensitivity analysis of computer models," *Journal of Quality Technology*, vol. 13, no. 3, pp. 174–183, Jul. 1981.

[9] C. Wann, et al., "SRAM cell design for stability methodology," *Proc. of IEEE VLSI-TSA '05*, pp. 21-22, 2005.

[10] K. Agarwal, and S. Nassif, "Statistical analysis of SRAM cell stability," *Proc. of DAC '06*, pp. 57-62, 2006.

[11] S. Mukhopadhyay, H. Mahmoodi, and K. Roy, "Modeling of failure probability and statistical design of SRAM array for yield enhancement in nanoscaled CMOS," *IEEE Trans. on CAD*, vol. 24, no. 12, pp. 1859-1880, Dec. 2005.

[12] A. Singhee, and R. A. Rutenbar, "Statistical blockade: a novel method for very fast Monte Carlo simulation of rare circuit events, and its application," *Proc. of DATE '07*, pp. 16-20, 2007.

[13] K. Takeda, et. al., "Redefinition of write margin for next-generation SRAM and write-margin monitoring circuit," *Proc. of ISSCC '06*, pp. 2602-2611, 2006.

[14] T. C. Hesterberg, "Advances in importance sampling," PhD Dissertation, Stanford University, Palo Alto, CA, 1988.

[15] J. Wang, X. Li, and L. T. Pileggi, "Parameterized macromodeling for analog system-level design exploration," *Proc. of DAC '07*, pp. 940-943, 2007.

[16] S. Boyd, and L. Vandenberghe, *Convex Optimization*. Cambridge University Press, Cambridge, United Kingdom, 2004.

[17] E. Seevinck, F. J. List, and J. Lohstroh, "Static-noise margin analysis of MOS SRAM cells," *IEEE JSSC*, vol. 22, no. 5, pp. 748-754, Oct. 1987.

[18] J. M. Rabaey, A. Chandrakasan, and B. Nikolić, *Digital Integrated Circuits: A Design Perspective*, 2nd ed. Prentice Hall, Englewood Cliffs, NJ, 2003.

978-1-60558-497-3/09 $25.00 © 2009 ACM

Soft Error Optimization of Standard Cell Circuits Based on Gate Sizing and Multi-objective Genetic Algorithm

Weiguang Sheng
Microelectronic Center
Harbin Institute of Technology
Harbin, China
wgsheng@hit.edu.cn

Liyi Xiao
Microelectronic Center
Harbin Institute of Technology
Harbin, China
xiaoly@hit.edu.cn

Zhigang Mao
Microelectronic Center
Harbin Institute of Technology
Harbin, China
mao@hit.edu.cn

ABSTRACT

A radiation harden technique based on gate sizing and multi-objective genetic algorithm (MOGA) is developed to optimize the soft error tolerance of standard cell circuits. Soft error rate (SER), chip area and longest path delay are selected as the optimization goals and fast fitness evaluation algorithms for the three goals are developed and embedded into the MOGA. All the three goals are optimized simultaneously by optimally sizing the gates in the circuit, which is a complex NP-Complete problem and resolved by MOGA through exploring the global design space of the circuit. Syntax analysis technique is also employed to make the proposed framework can optimize not only pure combinational logic circuit but also the combinational parts of sequential logic circuit.Optimizing experiments carried out on ISCAS'85 and ISCAS'89 standard benchmark circuits show that the proposed optimization algorithm can decrease the SER 74.25% with very limited delay overhead (0.28%). Furthermore, the algorithm can also reduce the area for most of the circuit under test by average 5.23%. The proposed technique is proved to be better than other works in delay and area overhead and suitable to direct the design of soft error tolerance integrated circuits in high reliability realms.

Categories and Subject Descriptors

B.8.1 [**Performance and Reliability**]: Reliability, Testing, and Fault-Tolerance; G.1.6 [**Optimization**]: Global optimization

General Terms

Reliability

Keywords

multi-objective, genetic algorithm, soft error, optimization

1. INTRODUCTION

Radiation induced soft error has been the main reliability issue for the deep sub-micron VLSI ICs in recent years[4]. Intensive works have been done to rapidly characterize the soft error rate (SER) of the circuit in design stage and try to optimize the reliability with minimal area and delay costs.

In the SER evaluation techniques, field testing and hardware-based fault injection are not suitable for design stage[21, 22]; Monte Carlo simulation and 3-D device simulation are too expensive for large circuits[16, 20, 8]; software-based fault injection and simulated fault injection are limited by the low accuracy or the slow speed[2, 3, 11].

Recently many approaches based on analytical model are developed to characterize the SER rapidly with minimal accuracy loss compared to Spice. The ASERTA in [7] inject glitches into combinational circuit and express the unreliability with the glitches widths other than the commonly accepted FITs (*failure-in-time, defined as 1 failure in 10^9 hours*). The work presented in [19] use parameterized descriptors to model the soft error rate and propagate the descriptors to the output to evaluate the SER of the circuit. However, techniques in [7, 19] can only analyze pure combinational logic circuits. The MARS-C and MARS-S in [13, 14] use binary decision diagrams (BDD) and algebraic decision diagrams (ADD) to model the circuit's structure and the fault events. However, the methods based on BDD&ADD are expensive for large circuit because they are inherently limited by the memory blowup problems. The SERA in [23] combines probability theory, circuit simulation and fault simulation to characterize the SER of small circuits, which is accurate but much slower than other tools.

The radiation harden techniques based on hardware redundancy or error detection/correction codes (EDC/ECC) will introduce excess area and power penalties. Hence, many approaches other than redundancy and EDC/ECC have been introduced to lower the soft error sensitivity. The author of [13, 14] develops a method to identify the weak part of the circuit and then harden the weak part by resizing the gate. SERTOPT of [7]combines capacity adding, gate sizing and voltage variation to optimize the circuit with 79.30% SER gain and 20.50% area overhead, 6.20% delay penalty. The heuristic method of [24] can reduce SER by 90.00% with 19.30% area overhead and 1.24% delay penalty. The harden technique based on shadow gates in [1] can introduce 30.00% area overhead and 4.00% delay penalty.

The main contributions of this paper are as follows.

1) Developing a fast SER evaluation method that can be used for not only pure combinational logic circuit but also combinational part of sequential logic circuit;

2) A multi-objective optimization algorithm (MOGA) is introduced into the SER tolerance realm, which has same SER optimization effect but much smaller area overhead and delay penalty than other works.

The remainder of this paper are organized as follows. In Section

Permission to make digital or hard copies of part or all of this work for personal or classroom use is granted without fee provided that copies are not made or distributed for profit or commercial advantage and that copies bear this notice and the full citation on the first page. To copy otherwise, to republish, to post on servers or to redistribute to lists, requires prior specific permission and/or a fee.
DAC'09, July 26-31, 2009, San Francisco, California, USA

978-1-60558-497-3/09 $25.00 © 2009 ACM

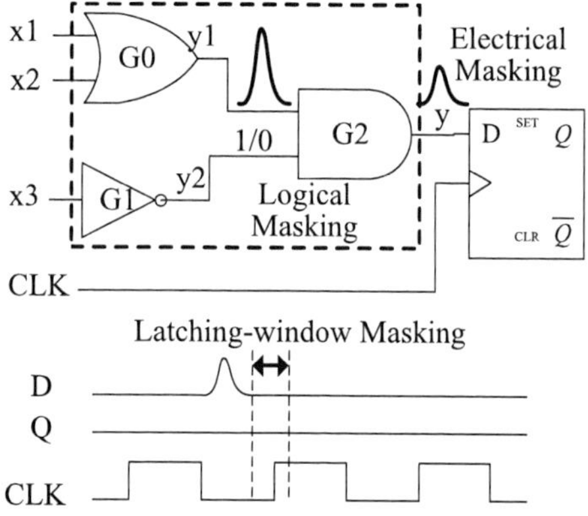

Figure 1: Soft error masking in the circuit

2, we describe the fitness evaluation algorithms for the SER, chip area and delay that used in the MOGA. The multi-objective optimization framework and the configuration, variation to the MOGA are provided in Section 3. The experiments and optimized results are given in Section 4. Section 5 summarizes the work and suggests future considerations.

2. FITNESS EVALUATION ALGORITHMS

The gate's generated width and propagated width of soft error glitch are affected by it's size (W/L)[7]. Hence, we can optimize the SER tolerance of the circuit by altering the size internal gates, which may introduce excess area and delay penalty[7, 24]. In this paper, MOGA is employed to simultaneously optimize the SER, area and delay of the circuit. In order to minimize the time overheads, MOGA demands fast fitness evaluation algorithms for the SER, area and delay goal, which are described in this section.

2.1 SER and area evaluation

When cosmic ray or heavy ions strike at the drain of sensitive transistor (Off State), numerous electron/hole pairs will be produced and the drift and diffusion of the electron/hole pairs will generate a current pulse between the drain and substrate of the transistor and a observable transient glitch will appear at the output of the transistor.

The glitch generating is affected by the gate size, power supply voltage, collected charge (depends on the particle's energy) and driving loads. In this paper, the glitch widths are stored in a Spice look up table (LUT) with 4 columns: {Cell Name (Type&Size), Charge, Load, Glitch Width}. When the LUT has been constructed, the width of the glitch generated can be retrieved from the LUT by specifying the cell name (including size data), charge value and loads.

The glitch propagating is affected by three masking factors[15, 10]: **i) logical masking** — to be latched, the glitch must be on a sensitive path from it's originated place to the latch; **ii) electrical masking** — the glitch may be attenuated or completely masked due to the electrical properties of the gates which propagates; **iii) latching-window masking** — to be latched, the glitches reach the latch must be in the latching-window of the latch. The three masking mechanisms are depicted in Fig. 1.

The SER can be computed by injecting various charge Q into each gate and propagating the glitches generated along different pathes. The glitches will be attenuated by the gates along the propagation path depends on the three masking factors. Electrical mask-

ing is computed as follows[7].

$$w_o = \begin{cases} 0, & w_i < d \\ 2 \cdot (w_i - d), & d < w_i < 2 \cdot d \\ w_i, & w_i > 2 \cdot d. \end{cases} \quad (1)$$

In (1), w_i, w_o corresponds to the input glitch width and the output glitch width; d is the delay of the specified gate affected by the gate type and the loads.

Logical masking is investigated by propagating the glitches with input vectors or node state probability (the probability that a node in '1' or '0' state). For example, if a glitch originate from gate i is expressed as $\{w_i, lm_i\}$ where $\{w_i, lm_i\}$ is a 2-tuple, w_i is the initial glitch width and lm_i is the logical masking probability initialized to $lm = 1$. If the glitch $\{w_i, lm_i\}$ entering pin A of an AND gate (e.g., gate G2 in Fig. 1), the output glitch at pin Y will be $\{w_o, lm_i * prob(B = '1')\}$. $prob(B = '1')$ can be computed in two modes: the input vector propagating mode and the state probability propagating mode. In the input vector propagating mode, $prob(B = '1') \in \{0, 1\}$ and the state of pin B depends on the current input vector; in the state probability propagating mode, $prob(B = '1')$ is a real number in $[0, 1]$ that was pre-calculated by applying 10,000 random input vectors to the circuit and extracting the node state probability through logic simulation. For the other type cells, the logical masking formula can also be inferred similarly. The glitches of one gate i are propagated to the fanout gate j and then to the fanout gate of j, until to the primary output of the circuit. For the glitches with $lm = 0$, they will be erased from the glitches list of the gate. Because the glitches in node state probability mode can be propagated in 1 pass traversing, it has much faster speed with some accuracy lost. When all the glitches have reached to the primary output, the latching-window masking probability that the glitch tuple be latched by the flip-flop is given as follows for input vector propagating mode and state probability propagating mode separately.

$$P(\varepsilon) = \begin{cases} 0, & T_{pw} < T_{dly} \\ \dfrac{T_{pw} - T_{dly}}{T_C}, & T_{pw} \geq T_{dly} \end{cases} \quad (2)$$

$$P(\varepsilon) = \begin{cases} 0, & T_{pw} < T_{dly} \\ \dfrac{(T_{pw} - T_{dly}) \cdot P_{lm}}{T_C}, & T_{pw} \geq T_{dly} \end{cases} \quad (3)$$

In (2) and (3), T_{pw} is the width w of the glitch tuple propagated to the output of the circuit; P_{lm} is the logical masking probability lm of the glitch tuple; T_{dly} is the sum of the setup time and hold time of the flip-flop and T_C is the circuit clock period.

Hence, the total failure rate of the circuit for the input vector propagating mode and the state probability propagating mode can be expressed by (4) and (5) separately.

$$P(F) = \sum_{u=1}^{O} \sum_{i=1}^{N} \sum_{j=1}^{G} \sum_{k=1}^{K} P_{u,i,j,k}(\varepsilon) \Big/ G \cdot N \cdot K \quad (4)$$

$$P(F) = \sum_{u=1}^{O} \sum_{j=1}^{G} \sum_{k=1}^{K} P_{u,j,k}(\varepsilon) \Big/ G \cdot K \quad (5)$$

In (4) and (5), O, G, N, K are the number of outputs, gates, input vectors and different charge values; $P_{u,i,j,k}(\varepsilon)$ and $P_{u,j,k}(\varepsilon)$ are the latching probability for one hit. The $P(F)$ is the expectation of the latching probability (failure rate) for one effective particle strike; $R_{PH} = 56.5 \, m^{-2} S^{-1}$ is the sea level neutron flux[13, 19]; $R_{eff} =$

978-1-60558-497-3/09 $25.00 © 2009 ACM

2.2×10^{-5} is the fraction of effective particle hits[25, 9]. We can now derive the expression for SER (FITs) as

$$SER = P(F) \cdot R_{PH} \cdot R_{eff} \cdot A_{circuit} \cdot 10^9 \cdot 3600 \qquad (6)$$

In (6), $A_{circuit}$ is the total silicon area of the circuit and ($R_{PH} \cdot R_{eff} \cdot A_{circuit} \cdot 10^9 \cdot 3600$) represents the effective particle strikes at the circuit in 10^9 hours (for computing SER in FITs).

In order to apply above technique to sequential logic circuit, a verilog netlist parser is developed with C/C++ and ANTLR[18] to extract the auxiliary information for the analysis. It extracts the input ports, output ports to build the scripts for the analyzing; it also extracts gate type, gate name, gate size and the interconnect data for building the graph representation of the circuit. For sequential circuits, the parser automatically excludes the flip-flops from the netlist, makes the input of flip-flops as the circuit's primary output and makes the output of flip-flops as the primary input of the circuit, and then the sequential circuit is converted to combinational circuit that can be analyzed by the program.

From (6), $A_{circuit}$ must be computed before the calculation of SER, which can be used as the approximation of chip area fitness and be accumulated by traversing the graph representation of the circuit in analyzing.

2.2 Delay evaluation

The delay fitness is approximated by the longest path delay (LPD) of the circuit. The LPD algorithm may get conservative delay results because the fault path problem have not been considered, but it is sufficient to be used as the delay fitness evaluation algorithm in multi-objective optimization. The LPD algorithm is listed in algorithm 1 [17], which is embedded into the SER evaluation algorithm to reuse the data structures such as the graph representation of the circuit. In timing analysis algorithms, gates are usually described by vertexes and nets is described by edges, hence the whole circuit can represented by a graph data structure. Algorithm 1 reads the graph representation of the circuit, the cell library information and computing the circuit delay by traversing the graph.

```
    Input  : Circuit graph representation, Cell library
             information, etc.
    Output : Max delay of the circuit
 1  foreach V in Graph do          /* V is a vertex */
 2      V.MaxDly ← 0;              /* Max delay of V */
 3      V.Visited ← FALSE;         /* Visited flag */
 4  end
 5  QUEUE ←{PrimaryInput};         /* Input ports */
 6  while QUEUE.NotEmpty do
 7      V ←Dequeue(QUEUE);         /* Fetch vertex */
 8      if V.Visited=FALSE then
 9          foreach U∈Fanins(V) do
10              if U.MaxDly+V.Dly>V.MaxDly then
11                  V.MaxDly ← U.MaxDly + V.Dly;
12              end
13          end
14          foreach W∈Fanouts(V) do
15              Enqueue(QUEUE, W);  /* Save W */
16          end
17          V.Visited ← TRUE;
18      end
19  end
20  return Max(PrimaryOutput.MaxDly);
```

Algorithm 1: The LPD algorithm

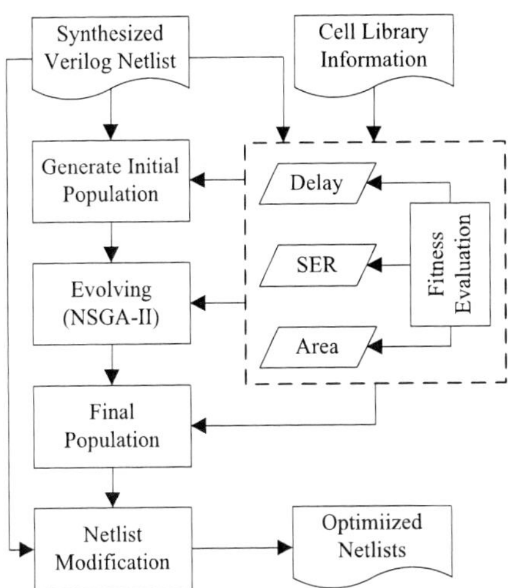

Figure 2: Structure of the framework

3. PROPOSED OPTIMIZATION FRAMEWORK

3.1 Overview of the framework

The structure of the optimization framework can be figured as Fig. 2. The framework read verilog synthesized netlist and cell library parameters as the input data. In Fig. 2, the module surrounded by dashed line is the fitness evaluation module to calculate the SER, area and delay of the circuit. Firstly, the framework make a initial population and then use MOGA to evolve the initial population. When the evolution finished, the optimized results are stored in the final population and can be rewrote to optimized verilog netlist by a conversion program.

3.2 The multi-objective optimization algorithm

The presence of multiple objectives in a problem, in principle, gives rise to a set of optimal solutions (largely known as Pareto-optimal solutions), instead of a single optimal solution[6]. MOGA has the ability to find multiple Pareto-optimal solutions in one single simulation run. In this paper, we use a algorithm derived from the classical NSGA-II of [6] to find the Pareto-optimal solutions that have better SER, area and delay simultaneously than the Non-Pareto-optimal results. The main structure of the NSGA-II algorithm is figured in algorithm 2.

The detailed description of algorithm 2 and the content of the function **FastNondominatedSort**, **CrowdingDistanceAssignment**, **PartialSort** in algorithm 2 can be found from [6]. The genetic operators (selection, crossover and mutation) are implemented in function **MakeNewPop** and described in algorithm 3. Algorithm 3 use binary tournament selection to select two individuals from the old population and do crossover and mutation on these two individuals to create two new individuals. After calculating SER, area and delay fitness (see Section 2) for the new individuals, they are pushed into the new population. The design of genetic operators are described in next subsection.

3.3 Chromosome coding and genetic operator

The chromosome coding scheme in this paper is real coding, which is figured in Fig. 3. For a typical gate in TSMC cell library, e.g. AND2 has four sizes: "XL", "X1", "X2" and "X4". We can use '0', '1', '2' and '4' to represent the four sizes and encode them into the chromosome, where each gene represents one gate size. However, illegal size may be generated by crossover and mutation operator in this encoding scheme, e.g. size '3' for AND2

978-1-60558-497-3/09 $25.00 © 2009 ACM

```
Input  : Circuit graph representation, Cell library
          information, Paremeters of MOGA, etc.
Output : Optimum gate size configuration schemes
          with better SER, area and delay
1  P_0 =InitPop(), Q_0 =MakeNewPop(P_0);
2  for t ← 0 to T do        /* For each generation */
3      R_t ← P_t ∪ Q_t;
4      F_{1...r} =FastNondominatedSort(R_t);
5      P_{t+1} ← ∅, i ← 1;
6      while |P_{t+1}| + |F_i| ≤ N do
7          CrowdingDistanceAssignment(F_i);
8          P_{t+1} ← P_{t+1} ∪ F_i;
9          i ← i + 1;
10     end
11     PartialSort(F_i,>);
12     P_t ← P_{t+1} ← P_{t+1} ∪ F_i[1 : N − |P_{t+1}|];
13     Q_t ← Q_{t+1} =MakeNewPop(P_{t+1});
14 end
15 F =FastNondominatedSort(P_t);
16 return F_1;              /* Return optimum result */
```

Algorithm 2: The NSGA-II algorithm

Gate1.Size OR SizeIndex	Gate2.Size OR SizeIndex	...	GateN.Size OR SizeIndex

Figure 3: The chromosome coding scheme

gate. The solution is to use size index to replace the real size in the chromosome, e.g. using indices '0', '1', '2' and '3' to represent the sizes "XL", "X1", "X2" and "X4", which will not generate illegal gate sizes.

"Whole Arithmetic Crossover" is employed in this work as the crossover operator. Given two chromosomes A and B, A' and B' are the new chromosomes, suppose that α is the cross vector ($\alpha_i \in [0, 1]$), then A' and B' are expressed as $A' = \alpha \times A + (1 − \alpha) \times B$, $B' = \alpha \times B + (1 − \alpha) \times A$.

The mutation operator employed is the inconsistent mutation operator derived from [12], which introduce annealing mechanism to the mutation operator to speed the convergence of the genetic algorithm. Given a parent chromosome $V = (v_1, v_2, \cdots, v_k, \cdots, v_n)$, suppose gene $v_k \in [a_k, b_k]$ will be mutated where a_k and b_k are the lower and upper bounds of v_k, the mutated chromosome is $V' = (v_1, v_2, \cdots, v'_k, \cdots, v_n)$ and v'_k is given by the following equation.

$$v'_k = \begin{cases} v_k + \varphi(t, b_k − v_k), & \text{random()}\%2 == 0 \\ v_k − \varphi(t, v_k − a_k), & \text{random()}\%2 == 1 \end{cases} \quad (7)$$

In equation (7), $\varphi()$ is a function described by (8). In equation (8), $r \in [0, 1]$ is a random number; t is the current temperature (current generation No.) and T is the maximum temperature (the maximum generation); λ is a inconsistent factor between 2 and 5. As the MOGA evolving, φ will be decreasing along with t increasing, which can stabilize the solutions and speed up the convergence of the genetic algorithm.

$$\varphi(t, y) = y \times (1 − r^{(1−t/T)^{\lambda}}) \quad (8)$$

4. EXPERIMENTS AND RESULTS

In this section, by using the proposed framework, optimization experiments are carried out, experiment results are analyzed and the performance of the algorithm are also been investigated.

```
Input  : OldPop              /* Old population */
Output : NewPop              /* New population */
1  NewPop ← ∅;
2  while NewPop.Size<=PopSize do
3      Parent1 =BinaryTournament(OldPop);
4      Parent2 =BinaryTournament(OldPop);
5      Child1, Child2 =CrossOver(Parent1,Parent2);
6      Mutate(Child1);
7      Mutate(Child2);
8      CalFitness(Child1);   /* SER,area,delay */
9      CalFitness(Child2);
10     NewPop.Push(Child1);     /* Save child */
11     NewPop.Push(Child2);
12 end
13 return NewPop
```

Algorithm 3: MakeNewPop algorithm

4.1 Experiments setup

We implement the framework in a Pentium-M machine with a 1.6-GHz processor and 1-GB RAM running Microsoft Windows XP. TSMC $0.18\,\mu m$ standard cell library is used as the target library. In (2) and (3), the T_C is set to be 4 ns for all the circuits except S1196 and S1238 (8 ns for these two circuits) and T_{dly} is set to be 120ps[23]. In (6), we set the neutron flux R_{PH} as $56.5\,m^{-2}S^{-1}$ and set the fraction of effective particle hits R_{eff} to be 2.2×10^{-5}[25, 9]. ISCAS'85 and ISCAS'89 benchmark circuit is selected as the research prototype. In algorithm 3, node state propagating mode is used to calculate the SER fitness to reduce the time of optimization. In the MOGA of this paper, the population size N is set to 100 and maximum generation T is set to 200. The crossover probability is set to 0.9 and mutation probability is set to 0.1. When multi-objective optimization is finished, all the SER results are calibrated in input vector propagating mode (10,000 randomly generated vectors) to get the accurate result.

4.2 Optimization results

The SER, area and delay of the ISCAS'85 and ISCAS'89 circuits before and after the optimization are listed in Table 1. Notes that the results of ISCAS'89 circuit is only for the combinational part, not for the whole circuit. Though MOGA can produce many solutions (Pareto-optimization solutions), solution with smaller delay and area is selected to do the comparison, which are described in Table 1.

In Table 1, the column header with "Opt" indicates the optimized result. From Table 1 we can figure out that MOGA has dramatic optimization ability to find the solutions with smaller SER, area and delay. In order to do comparison with other works, the ratio of SER decrease, area overhead and delay penalty are listed in Table 2. The results in Table 2 show that MOGA based optimization algorithm can reduce the SER average 74.25 with very small area overhead (−5.23%) and delay penalty (0.28%).

The average results compared to the other works are listed in Table 3. The data in Table 3 show that the proposed framework has similar optimization ability in SER with much smaller area and delay penalties than the others. Our method can reduce much of the benchmark circuits' area by average 5.23% during the optimization. As comparison, Ref [7] reports that only C3540's area is reduced by 9.6% and average 20.50% area overhead is introduced in the optimization; the area overhead of Ref [24] and [1] are 19.30% and 30.00% separately. The delay penalty of our work is also the smallest. Given relaxed delay restriction, better SER and area optimization results will be obtained.

978-1-60558-497-3/09 $25.00 © 2009 ACM

Table 1: Optimization result for ISCAS benchmark circuits

Circuit	SER (#FITs)	OptSER (#FITs)	Area (um²)	OptArea (um²)	Dly (ns)	OptDly (ns)
C432	0.265	0.219	5030	4100	1.29	1.29
C499	0.839	0.153	9010	7030	1.43	1.47
C880	0.586	0.078	7680	7390	1.20	1.23
C1355	0.915	0.006	12700	9270	1.35	1.44
C1908	0.650	0.126	8430	6540	1.88	2.00
C2670	3.180	0.428	16900	19600	1.81	1.81
C3540	1.770	0.029	15800	14400	2.17	2.25
C5315	3.590	0.321	26700	28400	1.79	1.86
C7552	3.310	0.239	29800	29700	1.74	1.83
S298	0.125	0.051	1160	1210	0.91	0.83
S444	0.140	0.019	1450	1570	0.88	0.84
S526	0.215	0.188	1850	2050	0.87	0.81
S1196	0.242	0.057	8360	7740	1.01	0.97
S1238	0.209	0.062	7730	7060	1.02	0.99

Table 2: Ratio of optimization and the overheads

Circuit	SER Decrease (%)	Area Overheads (%)	Delay Penalty (%)
C432	17.36	−18.49	0.00
C499	81.76	−21.98	2.80
C880	86.69	−3.78	2.50
C1355	99.34	−27.01	6.67
C1908	80.62	−22.42	6.38
C2670	86.54	15.98	0.00
C3540	98.36	−8.86	3.69
C5315	91.06	6.37	3.91
C7552	92.78	−0.34	5.17
S298	59.20	4.31	−8.79
S444	86.43	8.28	−4.55
S526	12.56	10.81	−6.90
S1196	76.45	−7.42	−3.96
S1238	70.33	−8.68	−2.94
Average	**74.25**	**−5.23**	**0.28**

Table 3: Comparison to other works

Circuit	SER Decrease (%)	Area Overheads (%)	Delay Penalty (%)
This work	74.25	−5.23	0.28
Ref [7]	79.30	20.50	6.20
Ref [24]	≈90.00	19.30	1.24
Ref [1]	−	30.00	4.00

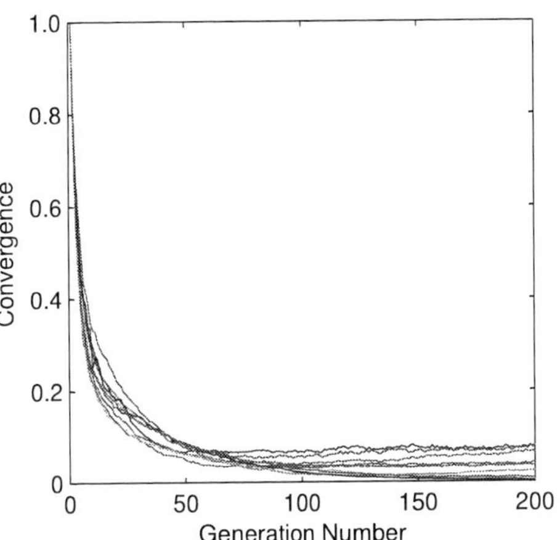

Figure 4: Convergence curve for ISCAS'85 circuits

4.3 Performance of the algorithm

This section give some description on the convergence and solution diversity of the proposed optimization method. The convergence metric employed is derived from [5], which is normalized to a real number $C \in [0,1]$. The smaller C indicates better convergence and bigger C represent bad convergence. The convergence curve for the ISCAS'85 circuits is showed in Fig 4. We can see that the convergence is well and most of the simulation run of the circuits converge after 100 generations.

The quantitative measuring of the diversity is complex for the MOGA with more than three optimization goals. Hence, we optimize the ISCAS'85 C432 circuit with a bigger population size (400) and plot the Pareto-optimization solutions (Pareto front) in a 3-D surface. In order to get smooth surface, interpolation technique is employed in Matlab and the surface is ploted in Fig. 5. The surface of Fig. 5 is very complex with many peaks and valleys, which is hard for the ordinary optimization algorithms to find optimum solutions.

5. CONCLUSIONS

In this paper, a versatile SER tolerance optimization framework is developed and employed to optimize the circuit's SER, area and delay simultaneously. The main achievements of this paper can be categorized as follows.

1) Versatile SER evaluation algorithm is developed, which can analyze verilog synthesized netlist automatically and be applicable for not only pure combinational logic circuit but also the combinational part of sequential logic circuit.

2) Multi-objective genetic algorithm is introduced to the SER optimization realm. The results show that the proposed algorithm can reduce the SER by average 74.25%; it can also optimize the area of the circuit with average 5.23% area reduction. The delay penalty of the proposed algorithm is only 0.28%. The comparison to the other works show that Multi-objective genetic algorithm is a better algorithm for SER tolerance, which can optimize the SER, area and delay simultaneously and get better results than the heuristic methods.

3) The framework is developed for industry standard cell library which make it more applicable and can be integrated with the standard digital design flow for SER tolerance design.

978-1-60558-497-3/09 $25.00 © 2009 ACM

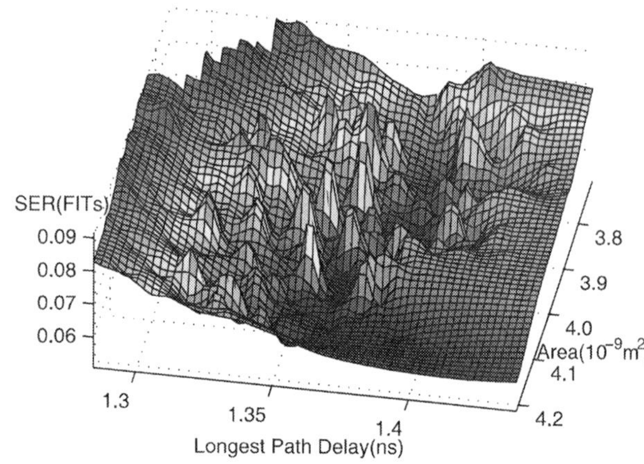

Figure 5: Pareto front surface for C432

6. REFERENCES

[1] A design approach for radiation-hard digital electronics. In *Proceedings of the 43rd annual conference on Design automation*, pages 773 –778, San Francisco, CA, USA, 2006. ACM.

[2] J. Arlat, Y. Crouzet, J. Karlsson, et al. Comparison of physical and software-implemented fault injection techniques. *IEEE Transactions on Computers*, 52(9):1115 – 1133, 2003.

[3] J.-C. Baraza, J. Gracia, S. Blanc, et al. Enhancement of fault injection techniques based on the modification of vhdl code. *IEEE Transactions on Very Large Scale Integration (VLSI) Systems*, 16(6):693 – 706, 2008.

[4] R. C. Baumann. Radiation-induced soft errors in advanced semiconductor technologies. *IEEE Transactions on Device and Materials Reliability*, 5(3):305 – 316, 2005.

[5] K. Deb and S. Jain. Running performance metrics for evolutionary multi-objective optimization. Technical report, Kanpur Genetic Algorithms Laboratory, Indian Institute of Technology, 2002.

[6] K. Deb, A. Pratap, S. Agarwal, et al. A fast and elitist multiobjective genetic algorithm: Nsga-ii. *IEEE Transactions on Evolutionary Computation*, 6(2):182–197, 2002.

[7] Y. S. Dhillon, A. U. Diril, A. Chatterjee, et al. Analysis and optimization of nanometer cmos circuits for soft-error tolerance. *IEEE Transactions on Very Large Scale Integration Systems*, 14(5):514–524, 2006.

[8] P. E. Dodd. Physics-based simulation of single-event effects. *IEEE Transactions on Device and Materials Reliability*, 5(3):343 – 357, 2005.

[9] P. Hazucha and C. Svensson. Impact of cmos technology scaling on the atmospheric neutron soft error rate. *IEEE Transactions on Nuclear Science*, 47(6):2586 – 2594, 2000.

[10] T. Karnik, P. Hazucha, and J. Patel. Characterization of soft errors caused by single event upsets in cmos processes. *IEEE Transactions on Dependable and Secure Computing*, 1(2):128 – 143, 2004.

[11] R. Leveugle and A. Ammari. Early seu fault injection in digital, analog and mixed signal circuits:a global flow. In *Proceedings of the Design, Automation and Test in Europe Conference and Exhibition (DATE'04)*, volume 1, pages 590 – 595, Paris, France, 2004.

[12] Z. Michalewicz. *Genetic Algorithms + Data Structures = Evolution Programs*. Springer, 1998.

[13] N. Miskov-Zivanov and D. Marculescu. Circuit reliability analysis using symbolic techniques. *IEEE Transactions on Computer-Aided Design of Integrated Circuits and Systems*, 25(12):2638 – 2649, 2006.

[14] N. Miskov-Zivanov and D. Marculescu. Modeling and optimization for soft-error reliability of sequential circuits. *IEEE Transactions on Computer-Aided Design of Integrated Circuits and Systems*, 27(5):803 – 816, 2008.

[15] K. Mohanram and N. A. Touba. Cost-effective approach for reducing soft error failure rate in logic circuits. In *International Test Conference 2003 (ITC'03)*, pages 893 – 901, 2003.

[16] P. C. Murley and G. R. Srinivasan. Soft-error monte carlo modeling program, semm. *IBM Journal of Research and Development*, 40(1):109 – 118, 1996.

[17] C. Oh and M. R. Mercer. Efficient logic-level timing analysis using constraint-guided critical path search. *IEEE Transactions on Very Large Scale Integration (VLSI) Systems*, 4(3):346–355, 1996.

[18] T. J. Parr. *The Definitive ANTLR Reference: Building Domain-Specific Languages*. Pragmatic Bookshelf, 2007.

[19] R. R. Rao, K. Chopra, D. T. Blaauw, et al. Computing the soft error rate of a combinational logic circuit using parameterized descriptors. *IEEE Transactions on Computer-Aided Design of Integrated Circuits and Systems*, 26(3):468 – 479, 2007.

[20] H. H. K. Tang. Semm-2: A new generation of single-event-effect modeling tools. *IBM Journal of Research and Development*, 52(3):233 – 244, 2008.

[21] Y. Tosaka, R. Takasu, T. Uemura, et al. Simultaneous measurement of soft error rate of 90 nm cmos sram and cosmic ray neutron spectra at the summit of mauna kea. In *46th Annual International Reliability Physics Symposium*, pages 727 – 728, Phoenix, 2008.

[22] M. Violante, L. Sterpone, A. Manuzzato, et al. A new hardware/software platform and a new 1/e neutron source for soft error studies: Testing fpgas at the isis facility. *IEEE Transactions on Nuclear Science*, 54(4):1184 – 1189, 2007.

[23] M. Zhang and N. R. Shanbhag. Soft-error-rate-analysis (sera) methodology. *IEEE Transactions on Computer-Aided Design of Integrated Circuits and Systems*, 25(10):2140 – 2155, 2006.

[24] Q. Zhou and K. Mohanram. Gate sizing to radiation harden combinational logic. *IEEE Transactions on Computer-Aided Design of Integrated Circuits and Systems*, 25(1):155 – 166, 2006.

[25] J. F. Ziegler. Terrestrial cosmic rays. *IBM Journal of Research and Development*, 40(1):19 – 39, 1996.

Improving Testability and Soft-Error Resilience through Retiming [*]

Smita Krishnaswamy [†], Igor L. Markov [♯], John P. Hayes [♯]
[†] IBM T.J. Watson Research Center, Rt. 134, Yorktown Heights, NY 10598
[♯] University of Michigan, EECS Department, Ann Arbor, MI 41809
{smita, imarkov, jhayes}@eecs.umich.edu

Abstract

State elements are increasingly vulnerable to soft errors due to their decreasing size, and the fact that latched errors cannot be completely eliminated by electrical or timing masking. Most prior methods of reducing the soft-error rate (SER) involve combinational redesign, which tends to add area and decrease testability, the latter a concern due to the prevalence of manufacturing defects. Our work explores the fundamental relations between the SER of sequential circuits and their testability in scan mode, and appears to be the first to improve both through retiming. Our retiming methodology relocates registers so that 1) registers become less observable with respect to primary outputs, thereby decreasing overall SER, and 2) combinational nodes become more observable with respect to registers (but not with respect to primary outputs), thereby increasing scan-testability. We present experimental results which show an average decrease of 42% in the SER of latches, and an average improvement of 31% random-pattern testability.

Categories and Subject Descriptors B.6.2 [**Logic Design**]
Reliability and Testing— *Redundant Design, Testability*
General Terms Algorithms, Design, Reliability
Keywords Testability, Soft Errors, Retiming

1 Introduction

Single-event upsets (SEUs) caused by α-particles, high-energy neutrons, and cosmic rays are of concern in CMOS logic circuits. As the energy threshold for causing an error decreases, the number of particles with sufficient energy increases rapidly [18]. For instance, at lower energy thresholds, even trace amounts of radioactive contaminants in solder can affect CMOS circuits [8]. Additionally, transient errors in ICs may occur through a variety of hard-to-model phenomena, including capacitive and inductive noise, as well as thermal and power supply fluctuations.

At the same time, circuit test is becoming more important because fabrication process parameters are harder to control in sub-wavelength lithography. Fluctuations in dopant concentrations, transistor gate length, wire shapes, and via alignments can lead to hard errors in chips [2]. Therefore, methods that increase testability without compromising SER are necessary to identify incorrectly manufactured chips.

Heidel et al. [8] observe that registers are a major contributor to fail rates in high-performance ICs since latched errors are often not subject to electrical or timing masking. Several papers propose partial replication to improve the reliability of combinational logic [11, 15]. However, these

methods only account for the combinational portion of the logic circuit and often incur significant area overhead. Further, these methods have to be used sparingly in order to maintain testability. Other techniques [21, 16] utilize electrical and timing masking to prevent the latching of errors originating in combinational logic. These techniques do not affect previously latched errors.

Errors in combinational logic are only becoming problematic now, while errors in registers are already a problem for critical applications [8]. In this paper, we specifically aim to reduce the soft-error susceptibility of registers, while simultaneously improving circuit testability. The main idea is to design circuits such that, even if a register experiences an SEU, its chances of propagating to a primary output are small. We account for logic masking in both combinational logic and registers during sequential operation and use this information to improve both the overall SER and testability of the design through retiming. Since we focus on logic masking, our solution is applicable to the various sources of errors mentioned above.

Retiming is the process of relocating registers to improve an objective (usually area or clock period) such that the functionality of the circuit remains unchanged.[1] Our retiming method utilizes the relationship between signal observability, soft-error propagation, and random-pattern testability. We derive linear programs (LPs) that relocate registers so as to minimize their sequential observability. Our main contributions are:

- An observability-based method for computing the error-susceptibility, and random-pattern testability for potential register locations in a sequential circuit.

- Linear programs for retiming that reduces circuit vulnerability to soft errors, and improves circuit testability simultaneously.

The remainder of the paper is organized as follows. Section 2 reviews relevant previous work on register retiming and functional simulation. Section 3 analyzes the effects of register relocation on error propagation and testability. Section 4 presents our retiming formulation and various extensions. Empirical validation is given in Section 5, followed by conclusions in Section 6.

2 Background

We now summarize the necessary background in retiming, soft-error mitigation, and testability.

2.1 Previous Work on Retiming

Leiserson and Saxe [10] first developed algorithms for minimum-period and -area retiming of edge-triggered circuits. For minimum-area retiming, a sequential circuit is represented by a graph $G(V, E)$, where each vertex $v \in V$ represents a combinational gate, and each edge $(u, v) \in E$

Permission to make digital or hard copies of part or all of this work for personal or classroom use is granted without fee provided that copies are not made or distributed for profit or commercial advantage and that copies bear this notice and the full citation on the first page. To copy otherwise, to republish, to post on servers or to redistribute to lists, requires prior specific permission and/or a fee.
DAC'09, July 26-31, 2009, San Francisco, California, USA

---

[1] In the past, the verification of retiming has been a problem, but solutions available in the past 4-5 years facilitate the practical use of retiming [14].

978-1-60558-497-3/09 $25.00 © 2009 ACM

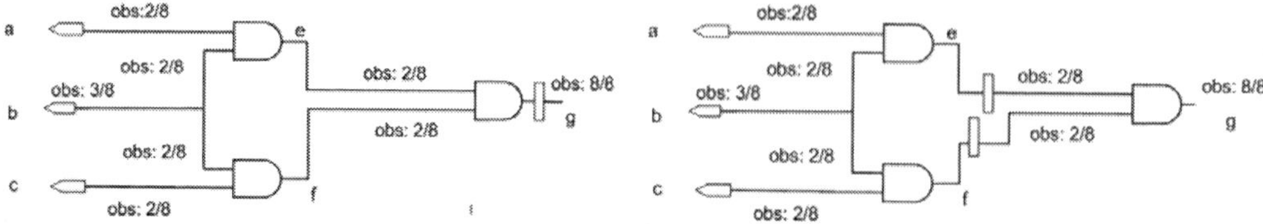

Figure 1: An illustration of observability reduction through retiming.

represents a connection between a driver u and sink v. An edge is labeled by a weight $w(u,v)$, indicating the number of registers (flip-flops) between u and v. The objective of minimum-area retiming is to determine labels $r(v)$ for each vertex v such that the total sum of edge weights is minimized. Here, $r(v)$ denotes the number of registers that are moved from the outputs to the inputs of v. The weight of an edge after retiming is given by:

$$w_r(u,v) = w(u,v) - r(u) + r(v)$$

Therefore, the total number of registers in the retimed circuit can be minimized by the following expression.

$$\sum_{(u,v)\in E} w(u,v) - r(u) + r(v)$$

Additionally, the retiming labels have to meet *legality* constraints, $w(u,v) \geq r(u) - r(v)$, for each edge to enforce the fact that edges cannot have negative weights. A linear program for the minimum-area retiming problem is given in Figure 2. Leiserson and Saxe [10] observe that this problem is the dual of a min-cost network flow problem and can therefore be solved in polynomial time.

```
Minimize
    Σ(u,v)∈E w(u,v) − r(u) + r(v)

subject to
    ∀(u,v) ∈ E, r(u) − r(v) ≤ w(u,v)
```

Figure 2: An LP for minimum-area retiming.

It is also possible to constrain the period in minimum-area retiming by ensuring that every path between two vertices with greater delay than the target period P has weight ≥ 1. In minimum-period retiming a binary search is conducted for the target clock period P and the feasibility of each period according to the legality constraints is checked using the Bellman-Ford algorithm [10].

Aspects of circuit testing have also been improved using retiming. Dey and Chakradhar [5] aim to reduce the lengths of partial scan chains in order to decrease testing time. Das and Bhattacharya [4] observe that combinational redundancies can be converted into sequential redundancies (unobservable changes in the state diagram) to improve the scan-based testability of circuits. In [20], the authors use the reverse process to convert sequential redundancies to combinational ones and then remove this redundancy using combinational optimization techniques. We note that these works are significantly different in their focus from ours. We aim to reduce the average observability – in effect the random-pattern observability of registers during normal operation. However, given that many registers are scanned, this retiming improves their observability during testing.

2.2 Techniques for SER Mitigation

Several techniques have been developed for improving the SER of logic circuits. These can be categorized by their error-mitigation mechanism. Soft errors in combinational logic are affected by three sources of masking [19]: 1) logic masking, where errors stop propagation due to the lack of a sensitized path to primary outputs or latches; 2) electrical masking, where soft errors are attenuated before being latched because of insufficient glitch duration or amplitude; 3) timing masking, when soft errors arrive at a register prior to a latching clock edge. Of these, only logic masking affects latched soft errors since latches can retain and drive erroneous values. Soft errors in combinational logic, however, are merely glitches that are likely to disappear due to the above mentioned sources of masking.

Techniques that increase logic masking include triple-modular redundancy, partial logic replication [15], guided rewiring [1], and signature-based partial redundancy addition [11]. These techniques mask errors on state elements, but often adversely affect testability and generally involve significant area overhead. Error-correction techniques such as Reed-Solomon and Hamming codes can also directly be used for state encoding, but these techniques incur so much overhead that they are not considered practical.

Techniques such as BISER [21, 16] use repeated sampling to determine the value of a signal before latching, relying on the assumption that erroneous glitches have shorter duration than the difference between sampling times. This assumption does not hold for errors already present in latches. Gate hardening [3] increases the energy threshold for error propagation such that only high-energy particles cause soft errors. While gate hardening can be used for state elements, this technique becomes less effective and requires proportionally more overhead as device technologies shrink. In this paper, we specifically focus on mitigating errors in registers through logic masking.

2.3 Logic Masking and Testability

A soft error in a logic circuit is propagated to a primary output only if there is a sensitized and observable path from the error to the output. This occurs only when a suitable test vector is applied at the primary input. Therefore, estimating the fraction of test vectors or *testability* is equivalent to estimating the probability that the error propagates to at least one primary output. This testability measure can be computed through functional simulation signatures in linear time. Note that this is testability with respect to the primary outputs. An alternative testability measure can be computed for scan-mode, i.e., assuming that the latches are scanned. We utilize these differing notions of testability—one applicable in sequential operation, and one applicable in scan-mode—to improve reliability and testability.

Two major parameters affecting testability are signal probability (controllability) and observability. Here, we review the use of functional simulation signatures for estimating the sequential observability of nodes in a circuit in linear time. Signal probability is computed in [11] by simulating random input vectors through logic in topological order. The collection of output responses at logic gates with respect to a collection of input vectors is known as a *signature*. More

978-1-60558-497-3/09 $25.00 © 2009 ACM

formally, for input vectors $\{v_1, v_2, v_3, \ldots v_k\}$, the signature at a node f with respect to the input vectors is given by $Sig(f) = \{f(v_1), f(v_2), f(v_3), \ldots f(v_k)\}$.

To evaluate signal observability, observability don't-care masks (ODC masks) are computed from the signatures [17]. These correspond to input vectors for which the value of the signal affects a primary output. Observability is computed in reverse topological order by flipping bits in the signature and checking if the change is propagated to a primary output. For greater efficiency, this computation can be done in two steps. The first step is to check whether the change locally propagates through neighboring gates. The second step is to check for further propagation through the circuit by examining pre-computed ODC masks of the outputs of the neighboring gates. Corresponding to the K-bit signature $sig(f)$, we define $ODC(f)$ as the K-bit sequence whose i-th bit is 0 if input vector X_i is in the don't-care set of f; otherwise the ith bit is 1. Formally, $ODC(g) = (X_1 \in care(F_g), X_2 \in care(F_g), \ldots, X_K \in care(F_g))$. Figure 3 shows an example of signature and ODC mask computation on a small circuit. Figure 4 summarizes the ODC computation algorithm of [17] for reference.

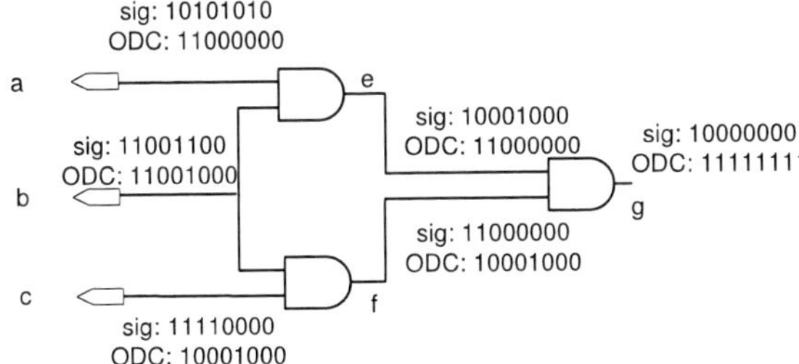

Figure 3: Signature and ODC illustration.

```
compute_odc_approx (Circuit C, size K)
{
  compute_sigs(C,K)
  sort_reverse_topological(C)
  for(all nodes g ∈ C)
    newsig(g) =  sig(g)
    for(each fan-out branch f ∈ fanout(g))
      sig(f) = Op < f > (inputsigs(f))
      localodc(g,f) = newsig(f) ⊕ sig(f)
      globalodc(g,f) = localodc(g,f) & odc(f)
      odc(g)| = globalodc(g,f)
}
```

Figure 4: An approximate ODC computation algorithm.

Signatures and ODC masks are both stored as bit-vectors and computed using bitwise operations for increased scalability. Together, the signature and the ODC mask give information about testability. For instance, all bit positions where the signature and ODC mask are 1 correspond to test vectors for the 0-1 fault or glitch at the node. Formally, for a node f the testability of a transient 1-to-0, and 0-to-1 bit-flip errors are:

$$test_1(f) = \text{num\_ones}\big(sig(f) \& ODC(f)\big)/K$$

$$test_0(f) = \text{num\_ones}\big(\sim sig(f) \& ODC(f)\big)/K$$

This measure for testability was validated by comparison to ATPG software in [11]. The SER, incorporating logic masking, is simply the sum of testabilities at each node f,

weighted by gate error rates $gerr0(f)$ and $gerr1(f)$ (given in units of FITs) for 1-to-0 and 0-to-1 errors respectively.

$$SER(C) = \sum_{f \in C} test_1(f)gerr0(f) + test_1(f)gerr1(f)$$

The logical SER is generally proportional to the observability, and only different by a multiplicative factor if the probabilities of 0-to-1 and 1-to-0 errors are the same. This assumption can be made due to lack of information during technology-independent optimization.

$$obs(f) = \text{num\_ones}\big(ODC(f)\big)/K$$

$$SER(C) = \sum_{f \in C} obs(f)gerr(f)$$

Note that logic design cannot change the actual probabilities of error occurrence ($gerr(0)$ and $gerr(1)$). However, design decisions can decrease the testability of registers or signals. Since we are performing technology-independent logic optimization for reliability, and do not have information about the relative probabilities of 0-to-1 vs 1-to-0 errors, we use the observability to improve the estimated SER.

3 Circuit Analysis

In this section, we analyze the effect of register relocation on the SER and testability of sequential circuits.

3.1 Retiming and Sequential SER

Retiming can improve circuit reliability by relocating registers such that soft errors are more likely to be masked. In order to account for error propagation through multiple stages, we modify the circuit using time-frame expansion. In an n-frame expansion, n copies of the circuit are made and each register is simply replaced by a wire. The outputs of the kth stage are fed back into the inputs of the $k + 1$th stage (as appropriate). Register inputs in the 0th frame are treated as primary inputs and register outputs of the n frame are treated as primary outputs.

The sequential observability is similar to the measure introduced in the previous section. However, we consider multiple cycles of operation through time-frame expansion. Only the primary outputs of each frame are considered completely observable. This accounts for errors that propagate past a single cycle before appearing at a primary output. It is true that some errors in registers at the n-th frame may appear at primary outputs, but this is unlikely since the probability of errors in the n-th frame are small. Experiments reported in several publications [6, 13] show that the majority of errors are flushed out in 2-3 cycles.

Additionally, in order to compute observability from a reasonable set of start states we obtain a sample of reachable states by simulating the sequential circuit for 20 cycles starting from a reset state. Experiments have shown that 10-15 cycles of simulation suffice to reach steady state on most ISCAS89 benchmarks [7, 13]. We denote this measure $seqobs(f,n)$ where f is the name of the signal, n is the number of frames of expansion. The sequential observability is given by the fraction of ones in the ODC mask of f in the n-frame expanded circuit.

$$seqobs(f,n) = \text{ones}\big(ODC(f,n)\big)/K$$

The SER of a sequential circuit is therefore:

$$SER(C,n) = \sum_{f \in C} gerr0(f)seqobs(f,n)$$

978-1-60558-497-3/09 $25.00 © 2009 ACM

If we separate the contributions of the registers from the contributions of the combinational logic to the SER of the sequential circuit, then we obtain:

$$SER\_Comb(C,n) = \sum_{f \in Comb(C)} gerr0(f)seqobs(f,n)$$

$$SER\_Reg(C,n) = \sum_{r \in Reg(C)} gerr0(r)seqobs(r,n)$$

$$SER(C) = SER\_Reg(C,n) + SER\_Comb(C,n)$$

The $SER\_Comb(C,n)$ portion of the error does not change under register relocation, since functionally, registers simply transmit the input signal to the output unchanged (after some delay). Therefore, registers do not logically mask errors when considering multi-cycle operation, and in turn, do not affect the observability of other nodes in the circuit. However, $SER\_Reg(C,n)$, does change with the movement of registers. For instance, if registers move from the output of a node f to its inputs then, errors on the registers can be additionally logically masked by f. The sequential observability of a register r is generally (with some exceptions) the same as that of its driving gate, therefore if a register is moved, its sequential observability changes. This suggests that registers should be placed in locations such that their sequential observability is low.

3.2 Retiming and Random-Pattern Testability

Since the signature-based framework computes the testability of a circuit based on random simulation vectors, $test_1(f)$ computes the random-pattern testability of node f for 0-1 errors, and $test_0(f)$ computes the same for 1-0 errors. In the absence of registers, the random-pattern testability of the entire circuit can be computed by:

$$Rand\_Test(C,n) = \frac{1}{|Comb(C)|} \sum_{f \in Comb(C)} test_1(f,n) + test_0(f,n)$$

In modern logic circuits, most registers are directly scanned out and read in test mode, i.e., registers can be treated as primary outputs when considering testability. Therefore, if registers were added to nodes f with low testability, then $Rand\_Test(C)$ would increase. This again leads to the conclusion that registers should be placed in regions of low observability.

The random-pattern testability of a circuit can be improved iteratively. At iteration 0, we assume that there are no fixed registers in the design. Therefore, $Rand\_Test(C,n)$ is improved by placing registers in locations of low observability—in our case, through retiming. In iteration 1, the testability is analyzed with respect to the current register locations. Additional test points are placed at locations that have low observability with respect to the circuit of iteration 0, and so on. Therefore, iteration 0 of this process involves the same goal as minimizing SER, i.e., decreasing the total sequential observability of registers.

Example 1 *For the circuit in Figure 1, the testabilities of nodes e, f and g are as follows.*

$$test_0(e,1) = 1/8, test_1(e,0) = 1/8$$
$$test_0(f,1) = 1/8, test_1(e,0) = 1/8$$
$$test_0(g,1) = 7/8, test_1(g,0) = 1/8$$

The random-pattern testability of this circuit is

$$Rand\_Test(C,n) = (2/8 + 2/8 + 1)(1/3) = (1/2)$$

In this circuit registers should be placed at nodes e and f due to their low observability.

4 Capturing Retiming by Linear Programs

We now derive LPs for retiming, accounting for the sequential observability of each register location. First, we present the basic retiming formulation assuming no register sharing, i.e, if a latch driven by a node u has fanouts v, w, then we model this as though there was a latch both at (u,v) and (u,w). In other words $w(u,v) = w(u,w) = 1$. Then, we account for register sharing at fanout branches.

4.1 Minimum-Observability Retiming

The sequential observability of each edge (u,v) is the same as the output of a buffer that is placed on edge (u,v). We denote the observability of edge (u,v), $seqobs((u,v),n)$ which is computed using signatures as described in Section 2. In the case where u only has one fanout $seqobs((u,v),n) = seqobs(u,n)$, in other words, the observability is the same as that of its driver. Since registers logically act as buffers, a register output has the same observability as a register input, and edges with registers still have the same observability after the registers are moved. The objective function accounting for total register observability is given by:

$$\sum_{(u,v) \in E} w_r(u,v)seqobs((u,v),n)$$

Additionally, if u is a primary input then $r(u)$ is necessarily 0 and similarly for v. This ensures that no peripheral retiming is done, and that the overall period of the circuit does not increase beyond the longest combinational path in the module being optimized. The modified LP is shown in Figure 5.

Minimize
$\sum_{(\mathbf{u,v}) \in \mathbf{E}}(\mathbf{w(u,v)} - \mathbf{r(u)} + \mathbf{r(v)})\mathbf{seqobs((u,v),n)}$

subject to
$\forall (\mathbf{u,v}) \in \mathbf{E}, \mathbf{r(u)} - \mathbf{r(v)} \leq \mathbf{w(u,v)}$

Figure 5: Minimum-observability retiming formulation.

Example 2 *For the circuit shown in Figure 1 the edges include $(a,e), (b,e), (b,f), (c,f), (e,g), (f,g), (g,o)$. Note that input and output wires are also considered valid edges. However, we only derive retiming labels for the intermediate nodes e, f, g. The objective function is:*
$$w_r(a,e)(2/8) + (w_r(b,e) + w_r(b,f))(3/8) + w_r(c,f)(w/8)$$
$$+ w_r(e,g)(2/8) + w_r(f,g)(2/8) + w_r(g,o)(8/8)$$
The retimed weight, for instance, of edge (e,g) is $w_r(e,g) = w(e,g) - r(e) + r(g)$.

Once the circuit is retimed, registers can be shared again during post-processing. For instance, if edges (u,v) and (u,w) both have registers after retiming then these are simply shared. In general, the number of registers required at the output of u is $max(w_r(u,f_1), w_r(u,f_2) \ldots w_r(u,f_n))$ where $f_1, f_2, \ldots f_n$ are fanouts of u.

The formulation in Figure 5 can be modified to constrain the area and period of the circuit. For area constraints, we can perform a binary search for the smallest feasible area M by including the constraint $(\sum_{(u,v) \in E} w(u,v) - r(u) + r(v)) < M$. The period can be constrained to a target P by the method of [10]. Here, the D matrix stores the delay of longest path between the vertices (u,v) in $D(u,v)$ and the W matrix stores the weight of the said path. These additional constraints are shown in Figure 6.

$$\begin{aligned}
&\texttt{Minimize}\\
&\quad \sum_{(u,v)\in E}(w(u,v)-r(u)+r(v))\,seqobs((u,v),n)\\[4pt]
&\texttt{subject to}\\
&\quad \forall (u,v)\in E, r(u)-r(v)\le w(u,v)\\
&\quad \left(\sum_{(u,v)\in E} w(u,v)-r(u)+r(v)\right) < M\\
&\quad \forall u,v\in V \quad \text{s.t.}\quad D(u,v)>P, r(u)-r(v)\le W(u,v)-1
\end{aligned}$$

Figure 6: Area- and period- constrained retiming for minimum observability.

4.2 Incorporating Register Sharing

In the previous section, we derived a minimum-observability retiming formulation that does not account for register sharing. Hence, the optimization takes place on a version of the circuit with registers cloned at each fanout branch. The difficulty in incorporating sharing is that observability is a non-linear property of edges.

Example 3 *Consider again the circuit C of Figure 1. Suppose the retimed weights of edges (b,e) and (b,f) are $w_r(b,e)=2$ and $w_r(b,f)=1$. According to the formulation of Figure 5, the objective function for this portion of the circuit equals $(2/8)w_r(b,e)+(2/8)w_r(e,g)=(1/3)$. However, the two registers at (b,f) and (b,e), can be replaced by a single register with fanouts to both e and f. This register does not have observability $seqobs((b,e),n)+seqobs((b,e),n)=4/8$. Instead, the observability is computed by counting the fraction of $1's$ in its ODC mask. The ODC mask, in turn, is computed as the bitwise OR of the ODC masks through each fanout, as shown in Figure 4. Thus, $ODC(b)=globalODC(b,e)\mid globalODC(b,f)=3/8$.*

For each register with driver u and fanouts $S=\{s_1,s_2\ldots s_m\}$, the correct sequential observability must be computed using the method of Figure 4. This observability is equivalent to the $seqobs$ of a buffer with input u and outputs S, denoted $seqobs((u,S),n)$. Here, S is a subset of all of the fanout branches of u, $F_u=\{f_1,f_2,\ldots f_n\}$. For any node u with fanout branches $F_u=\{f_1,f_2,\ldots f_n\}$, we can compute the total number of registers that can be shared by any subset of these branches, using the edge weights introduced in the previous section as follows. The number of registers that can be shared by *all* the fanout branches is given by:

$$w_r(u,F_u)=min(w_r(u,f_1),w_r(u,f_2)\ldots w_r(u,f_n))$$

The number of registers shared by a subset of fanouts, $S=\{f_1,f_2,\ldots f_{n-1}\}\subset F_u$ of size $|F-u|-1$ is $min(w_r(u,f_1),w_r(u,f_2)\ldots w_r(u,f_{n-1}))-w_r(u,F_u)$. In general, the number of registers shared by a subset $S\subset F_u$ is the minimum weight of any edge of the form $(u,s_i),s_i\in S$, minus the registers that are shared by any larger subset S' of F_u. Hence, the total weight of a subset of fanout branches S, using the principle of inclusion and exclusion, is given by:

$$w_r(u,S)=\sum_{S':S\subset S'\subseteq F}(-1)^{(|S'|-|S|)}min(w_r(u,s_1'),w_r(u,s_2'),\ldots),s_i'\in S'$$

Then, these register counts, $w_r(u,S)$, are weighted by their sequential observability $seqobs((u,S),n)$. The sum of such quantities over all possible subsets of F_u, gives us the correct total observability of registers driven by u, assuming maximal sharing. Maximal sharing is desired because a shared register always has observability less than or equal to that of its cloned registers combined.

$$totobs(u)=\sum_{S:S\subseteq F_u} w_r(u,S)*seqobs((u,S),n) \tag{1}$$

The function $totobs(u)$ has to be linearized in order to be incorporated into the LP. This requires linearizing the min function—the only non-linear element of the $totobs$ function. Generally, the function $min(a_1,a_2,\ldots a_n)$, for any real values a_i can be linearized by introducing a new variable MIN, along with the constraints $MIN\le a_1, MIN\le a_2,\ldots MIN\le a_n$. Then, the objective function has to maximize the value of MIN as LP converges to a solution so that $MIN=min(a_1,a_2\ldots a_n)$.

We introduce a variable $MIN_{u,S}$ for each function $min(w_r(u,s_1),w_r(u,s_2),w_r(u,s_3)\ldots)$ in $totobs(u,F_u)$. The associated constraints are of the form $MIN_{u,S}\le w_r(u,s_1)$, $MIN_{u,S}\le w_r(u,s_2)$, etc. Finally, we append $-c(MIN_{u,S})$ to the end of the objective function for each variable $MIN_{u,S}$ introduced. Here, c is any sufficiently large constant, i.e., $\forall S, c >> \sum seqobs((u,S),n)$. Since the retiming linear program has a minimization objective, the additional terms ensure that the $MIN_{u,S}$ variables are set to their highest (correct value) when the objective is optimized. The remaining retiming variables will be optimized for low observability as before. This altered LP, incorporating register sharing, is given in Figure 7.

$$\begin{aligned}
&\texttt{Minimize}\\
&\quad \sum_{(u)\in V}totobs(u)-\left(c\sum_{S\subset F_u} MIN_{u,S}\right)\\[4pt]
&\texttt{subject to}\\
&\quad \forall u\in V, S\in F_u, \forall(s_j\in S) Min_{u,S}\le w_r(u,s_1)\\
&\quad \forall(u,v)\in E, r(u)-r(v)\le w(u,v)
\end{aligned}$$

Figure 7: Minimum-observability retiming formulation with register sharing.

While the formulation in Figure 7 correctly captures register sharing, it can become intractable for nodes with many fanouts. By recollecting the coefficients next to the MIN variables, we can write the $totobs$ function as follows:

$$totobs(u)=\sum_{S\subseteq F_u} C_{u,S} Min(u,S),$$

$$C_{u,S}=\sum_{S':S'\subset S\subseteq F_u}(-1)^{|S|-|S'|}seqobs(u,S',n)$$

From this formulation, it is clear that the number of additional terms generated in the objective function for each node u is on the order of $2^{|F_u|}$. However, many practical circuits have low maximum fanout, due to drive-strength limitations of available standard cells.

5 Empirical Validation

We now describe experiments to validate our proposed retiming formulation. Our signature and observability computations are implemented in C++, while the linear programs are solved using CPLEX v.10.1 [9]. Note that an integer optimal solution is guaranteed without explicitly enforcing integer constraints [10].

Figure 8 summarizes the propagation of errors through sequential circuits, as estimated by bit-parallel functional simulation [11] extended to sequential circuits. The figure indicates that most errors are apparent at the outputs in immediate cycles after their occurrence. The error probability in later cycles diminishes rapidly. Therefore, we compute sequential observability from a ten-frame expansion of the circuit.

Table 1 shows results on ISCAS-89 benchmark circuits with the minimum-observability retiming formulation where each edge (u,v) is weighted by a 10-frame sequential observability measure. We use the formulation shown in Fig-

978-1-60558-497-3/09 $25.00 © 2009 ACM

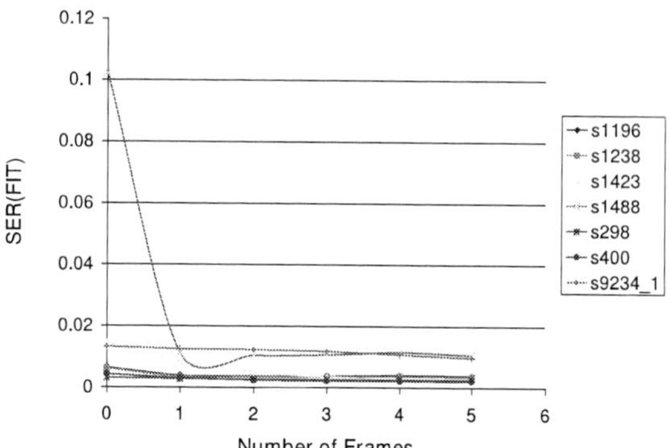

Figure 8: Error propagation in sequential circuits through multiple cycles of operation.

Circuit	Before		After		Change	
	No. FFs	Total Obs	No. FFs	Total Obs	% Area Inc.	% Obs. Dec.
s208	8	9.570	10	9.173	1.785	4.147
s298	14	15.579	34	5.059	15.037	67.527
s344	15	12.770	11	8.539	-2.29	33.126
s386	6	4.206	4	3.8256	-1.121	9.035
s444	21	14.752	34	4.146	6.435	71.895
s526	21	15.906	94	3.567	34.112	77.568
s832	5	25.311	30	2.669	8.561	89.3365
s1238	18	2.848	27	2.147	1.711	24.631
s1196	18	2.911	29	2.115	2.010	27.373
s1494	6	5.746	7	3.248	0.153	43.47
s1488	6	5.744	8	3.8256	0.303	42.787
s1423	73	25.273	115	20.115	5.753	20.110
Avg.					6.53	42.61

Table 1: Decrease in register observability through retiming.

Circuit	No. Gates	Scan Testability		
		Before	After	% Improved
s208	104	0.494	0.499	1.03
s298	119	0.523	0.699	33.72
s386	159	0.360	0.389	7.77
s444	181	0.474	0.689	45.42
s526	193	0.445	0.614	37.75
s832	287	0.222	0.368	65.28
s1238	508	0.300	0.317	5.78
s1196	529	0.302	0.321	6.45
s1494	647	0.269	0.413	53.69
s1488	653	0.267	0.410	53.57
s1423	657	0.424	0.552	30.08
Avg.				30.96

Table 2: Improvements in random-pattern scan-testability.

ure 5. The LP was solved in all cases in less than 0.1 seconds by *CPLEX*. The observability, which is proportional to the SER susceptibility is decreased by an average of 42% with only a 7% overall area increase. Table 2 shows an average improvement of 31% in the random-pattern testability of nodes in the combinational portion of the circuit. Our results indicate an interesting feature of error tolerance—that it is possible to decrease the overall sequential SER while increasing what is traditionally computed as the SER of a combinational logic circuit, i.e., with registers treated as primary outputs [12, 11]. Therefore, it is insufficient to solely consider combinational behavior for SER analysis. While combinational circuit optimization for SER remains important and generally decreases overall circuit SER, it is necessary to account for error propagation behavior at the sequential level. For instance, combinational circuits can be designed such that they reduce error propagation to registers with high observability.

6 Conclusions

We have developed a novel retiming approach to improve both the soft-error tolerance and testability of sequential circuits. The proposed techniques exploit certain relationships between observability, testability and SER, and incorporate them into linear programs for retiming optimization. We also extended these programs to handle area constraints and fanout sharing. Area-unconstrained minimization was shown to reduce the average error susceptibility of registers by 40%, and improve testability by 31% on average.

Acknowledgment. This work was sponsored in part by the Air Force Research Laboratory under Agreement No. FA8750-05-1-0282.

References

[1] S. Almukhaizim, Y. Makris, et al., "Seamless Integration of SER in Rewiring-based Design Space Exploration," *ITC*, 2006, pp. 1-9.

[2] S. Borkar, et al., "Parameter Variations and Impact on Circuits and Microarchitecture,"*DAC*, 2003, pp. 328-342.

[3] T. Calin, M. Nicolaidis, R. Velaco, "Upset Hardened Memory Design for Submicron CMOS Technology," *IEEE Trans. Nucl. Sci.*, Dec. 1996, vol. 43, pp. 2874-2878.

[4] D. K. Das, B.B. Bhattacharya, "Does Retiming Affect Redundancy in Sequential Circuits?," *VLSID*, 1996, pp. 260-263.

[5] S. Dey, S.T. Chakradhar,"Retiming Sequential Circuits to Enhance Testability," *VTS*, 1994, 25-28.

[6] J. P. Hayes, I. Polian, B. Becker, "An Analysis Framework for Transient-Error Tolerance," *VTS*, 2007, pp. 249-255.

[7] G.D. Hachtel, E. Macii, A. Pardo, F. Somenzi, "Markovian Analysis of Large Finite State Machines," *TCAD*, Dec. 1996, vol. 15, no. 12, pp. 1479-1493.

[8] D. F. Heidel et al., "Alpha-particle Induced Upsets in Advanced CMOS Circuits and Technology," *IBM Journal of Research and Dev.*, May 2008, vol. 52, no. 3, pp. 225-232.

[9] ILOG CPLEX: http://www.ilog.com/products/cplex

[10] C.E. Leiserson, J.B. Saxe,"Retiming Synchronous Circuitry," *Algorithmica*, 1991, vol. 6,pp. 5-35.

[11] S. Krishnaswamy, S. M. Plaza, I. L. Markov, J.P. Hayes, "Enhancing Design Robustness with Reliability-aware Resynthesis and Logic Simulation," *ICCAD*, 2007, pp. 149-154.

[12] N. Miskov-Zivanov, D. Marculescu, "MARS-C: Modeling and Reduction of Soft Errors in Combinational Circuits," *DAC*, 2006, pp.767-772.

[13] N. Miskov-Zivanov, D. Marculescu, "Modeling and Optimization for Soft-Error Reliability of Sequential Circuits," *IEEE TCAD*, 2008, vol 27. no.5 pp. 803-816.

[14] M. N. Mneimneh, K. A. Sakallah, "Principles of Sequential-Equivalence Verification," *IEEE DT*, 2005, vol. 22, no. 3, pp. 248-257.

[15] K. Mohanram, N. A. Touba, "Partial Error Masking to Reduce Soft Error Failure Rate in Logic Circuits," *DFT*, 2003, pp. 433-440.

[16] M. Nicolaidis,"Time Redundancy Based Soft-Error Tolerance to Rescue Nanometer Technologies," *VTS*, 1999, pp. 86-94.

[17] S. Plaza, K-H. Chang, I. Markov, V. Bertacco,"Node Mergers in the Presence of Don't Cares," *ASPDAC*, 2007, pp. 414-419.

[18] K. Rodbell, D. F. Heidel, et al., "Low-Energy Proton-Induced Single-Event-Upsets in 65 nm Node, Silicon-on-Insulator, Latches and Memory Cells," *IEEE trans. on Nuclear Science*, vol. 54, no. 6, Dec. 2007, pp. 2474-2479.

[19] P. Shivakumar, M. Kistler, et al., "Modeling the Effect of Technology Trends on Soft Error Rate of Combinational Logic" *DSN 2002*, pp. 389-398.

[20] H. Yotsuyanagi, S. Kajihara, K. Kinoshita, "Synthesis for Testability by Sequential Redundancy Removal using Retiming," *Fault-Tolerant Computing*, 1995, pp. 33-40.

[21] M. Zhang, S. Mitra, et al.,"Sequential Element Design with Built-in Soft Error-Resilience" *TVLSI*, Dec. 2006, vol. 14, no.12, pp. 1368-1378.

978-1-60558-497-3/09 $25.00 © 2009 ACM

513

Statistical Reliability Analysis Under Process Variation and Aging Effects

Yinghai Lu[1], Li Shang[2], Hai Zhou[1,3], Hengliang Zhu[1], Fan Yang[1], Xuan Zeng[1*]
[1]State Key Lab of ASIC & System, Microelectronics Dept., Fudan University, China
[2]ECEE, University of Colorado, Boulder, U.S.A., [3]EECS, Northwestern University, U.S.A.

Abstract—Circuit reliability is affected by various fabrication-time and run-time effects. Fabrication-induced process variation has significant impact on circuit performance and reliability. Various aging effects, such as negative bias temperature instability, cause continuous performance and reliability degradation during circuit run-time usage. In this work, we present a statistical analysis framework that characterizes the lifetime reliability of nanometer-scale integrated circuits by jointly considering the impact of fabrication-induced process variation and run-time aging effects. More specifically, our work focuses on characterizing circuit threshold voltage lifetime variation and its impact on circuit timing due to process variation and the negative bias temperature instability effect, a primary aging effect in nanometer-scale integrated circuits. The proposed work is capable of characterizing the overall circuit lifetime reliability, as well as efficiently quantifying the vulnerabilities of individual circuit elements. This analysis framework has been carefully validated and integrated into an iterative design flow for circuit lifetime reliability analysis and optimization.

Categories and Subject Descriptors:
J.6 [Computer-Aided Engineering]: Computer-Aided Design
B.8.1 [Performance and Reliability]: Reliability, Testing, and Fault-Tolerance
General Terms: Design, Algorithms, Performance
Keywords: NBTI, Yield, Process variations

I. Introduction

Aggressive scaling of CMOS process technology poses serious challenges on the lifetime reliability of integrated circuits (ICs). IC lifetime reliability is affected by fabrication-induced process variation and run-time aging effects [3]. Feature size reduction increases the difficulty of precise fabrication process control. Fabrication induced geometric and electrical parameter variations, e.g., changes in device effective channel length and threshold voltage, have significant impact on IC performance and reliability. Meanwhile, run-time aging effects, such as electromigration, thermal cycling, and negative bias temperature instability (NBTI), have become another fast-growing concern of IC lifetime reliability. NBTI is known to be the dominating circuit lifetime aging effect [12], [7]. The occurrence of NBTI is due to the generation of traps at $Si\text{-}SiO_2$ interface when PMOS devices are negatively stressed, e.g., $V_{gs} = -V_{dd}$. This effect causes temporal increase of PMOS threshold voltage (V_{th}) and long-term performance degradation.

Process variation [16], [15] and NBTI effect [13], [6], [14], [21], [26], [20] have both drawn significant attention in the recent past. Most of the past work treats them as two independent issues, and addresses the impact of each effect on IC reliability and performance

*Corresponding author. E-mail: xzeng@fudan.edu.cn

Permission to make digital or hard copies of part or all of this work for personal or classroom use is granted without fee provided that copies are not made or distributed for profit or commercial advantage and that copies bear this notice and the full citation on the first page. To copy otherwise, to republish, to post on servers or to redistribute to lists, requires prior specific permission and/or a fee.
DAC'09, July 26-31, 2009, San Francisco, California, USA

separately. However, IC reliability is jointly affected by both effects. In addition, process variations and NBTI effect have strong influence on each other. As reported by Bhardway et al., NBTI-induced threshold voltage shift of PMOS transistor depends not only on its working condition (such as input duty cycle and temperature), but also on the underlying process parameters such as original threshold voltage and dioxide thickness [6]. Due to fabrication-induced process variations, NBTI effect shall be modeled as a random process. On the other hand, the circuit timing statistics will be affected by the NBTI effect as well as the process variation as time evolves.

Recently, work starting to consider both effects has been reported. In [11], NBTI effect (ΔV_{th}) is modeled as a random process. This work, however, ignores the variation of other process parameters. In [4], the authors consider both process variation and NBTI effect for standard cell modeling and optimization. Process parameters are treated as random variables and modeled with response surface method. In this work, however, NBTI effect is incorporated with the worst-case delay model. Since NBTI effect is a strong function of circuit run-time condition, the worst-case approximation is pessimistic (up to $30\times$ increase of ΔV_{th} versus the nominal has been reported in [4]). In [22], the authors present a thorough analysis of circuit aging under the variation of threshold voltage. This work's main focus is on single path circuit modeling. A comprehensive analysis of IC performance and reliability yet requires statistical techniques to address correlated paths and other process parametric variations. We believe this work starts an important direction, and our study will expand and build up on this foundational work.

In this article, we present an analysis framework to evaluate the IC lifetime reliability by jointly considering process variation and NBTI effect. This work makes the following contributions:

- We present a nonlinear scalable statistical gate delay aging model, which considers both run-time gate working condition and fabrication-induced process variation.
- We present a statistical timing analysis framework using the proposed gate delay model, which is capable of characterizing the performance and reliability degradation under process variation and run-time aging. A fast pruning algorithm is proposed to improve the analysis efficiency.
- We present a criticality and sensitivity analysis method to quantify the reliability impact of each individual circuit element. Such quantification enables efficient iterative IC reliability optimization flow.

The rest of the article is organized as follows. Section II introduces the proposed analysis framework. Sections III, IV, and V describe in detail the proposed modeling, analysis, and optimization methods. Section VI reports the experimental results. The paper is concluded in Section VII.

II. Overview of Aging-Aware Statistical Framework

The proposed aging-aware statistical timing analysis framework is shown in Figure 1. The analysis framework consists of three key components:
Gate-Level Aging-Aware Statistical Timing Model: Given a technology library in the SPICE netlist form, it characterizes the process

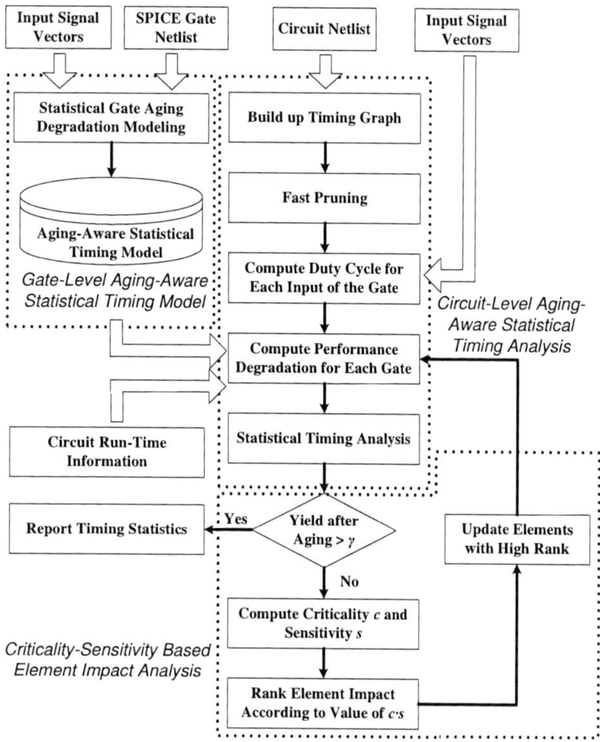

Fig. 1. Aging-aware statistical framework.

variation and NBTI aging effect of each gate element, and generates a statistical model in polynomial chaos expansion (PCE) form incorporating NBTI-induced aging degradation and process parameter variation. The NBTI aging model considers various run-time factors, such as temperature, input signal probability, and duty cycle.

Circuit-Level Aging-Aware Statistical Timing Analysis Method: It conducts statistical timing analysis on the entire circuit using the proposed gate-level model. Logic simulation is first conducted to collect the signal duty cycle information of each logic element. Given process variations and run-time circuit conditions, such as voltage and temperature, gate performance degradation is expressed in a unified PCE form, which is used by statistical static timing analysis (SSTA) to compute the performance degradation of the entire circuit. A circuit pruning algorithm is developed, which incrementally removes gates and edges with sufficiently large time slacks during analysis, thereby improving analysis efficiency without sacrificing accuracy.

Criticality-Sensitivity Based Element Vulnerability Analysis Method: It conducts circuit criticality and sensitivity analysis in order to quantify the impact of individual elements on circuit lifetime timing improvements. This analysis provides guidance to IC aging-aware reliability optimization, and enables an efficient iterative IC reliability analysis and optimization framework.

The following sections discuss the details of the proposed modeling and analysis methods.

III. Statistical Gate Aging Modeling

A. Parametric NBTI Aging Modeling

This section describes a parametric method to model PMOS NBTI effect, which extracts and formulates NBTI run-time dependencies, e.g., temperature and signal probability, in a compact form that allows rapid estimation of NBTI-induced time degradation under arbitrary run-time conditions. This proposed method can also be extended to model other lifetime aging effects. The NBTI effect manifests itself as increase of PMOS threshold voltage and degradation of circuit timing. NBTI physical mechanism has been studied in [7], [19]. A NBTI

model under arbitrary dynamic temperature variation is proposed in [26]. In [6], the authors propose a long-term NBTI model, which provides an analytical upper bound estimation of the NBTI impact over time, described as follows.

$$\Delta V_{th}(t) = \left(\frac{\sqrt{K_v^2 \alpha T_{clk}}}{1 - \beta_t^{1/2n}} \right)^{2n} \tag{1}$$

where

$$K_v = \left(\frac{q t_{ox}}{\epsilon_{ox}} \right)^3 K^2 C_{ox}(V_{gs} - V_{th})\sqrt{C} \exp\left(\frac{2E_{ox}}{E_o} \right) \tag{2}$$

$$C = T_o^{-1} \exp\left(-\frac{E_\alpha}{kT} \right) \tag{3}$$

$$\beta_t = 1 - \frac{2\xi_1 t_e + \sqrt{\xi_2 C (1 - \alpha) T_{clk}}}{2 t_{ox} - \sqrt{Ct}} \tag{4}$$

where α is the average signal duty cycle, T is the average temperature, V_{th} is the initial threshold voltage and t_{ox} is thickness of gate dielectric. For the sake of simplicity, please refer to [6] for detailed explanation of other parameters.

This model shows that the NBTI effect is a strong function of the run-time temperature T and signal probability (duty-cycle) α of the logic gate. In this work, we extract the dependency on T and α from the long-term model described in Equation (1). We further assume T_{clk} is small based on the fact that modern high-speed IC designs are typically clocked at the multi-gigahertz range. Using the similar reduction technique as in [20] and considering the exponential temperature dependency shown in Equation (3), the NBTI-induced threshold shift model can be approximately reduced to

$$\Delta V_{th}(T, \alpha, t) = b e^{-\frac{nE_\alpha}{kT}} \left(\frac{\alpha}{1 - \alpha} \right)^n t^n \tag{5}$$

where k is Boltzmann constant, $E_\alpha = 0.49eV$ and b is a fitting constant. Figure 2 demonstrates the relative error of our simplified model in Equation 5 under varying working conditions against the original long-term model in Equation 1, using the 65nm CMOS technology. The temperature ranges from $320K$ to $380K$ and the average duty cycle ranges from 0.1 to 0.95. As is displayed in the figure, the simplified model achieves very good accuracy within normal work conditions.

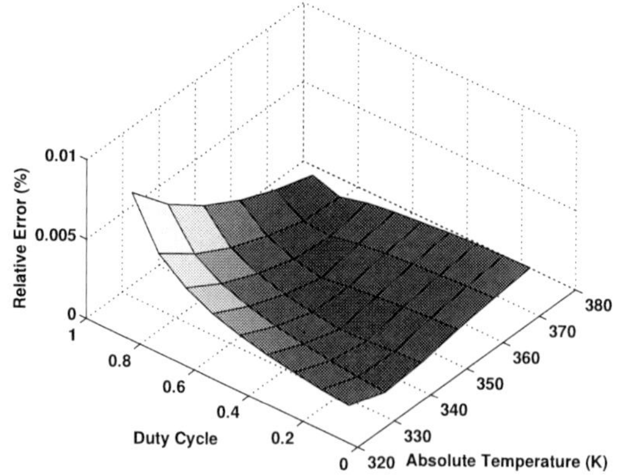

Fig. 2. Relative Error of Model in Equation 5 Against the Long-term Model (Equation 1).

Following the alpha-power law [18], the first-order gate delay can be approximated as a linear function of the threshold voltage. The gate delay can then be expressed as follows [20], [23]:

$$\Delta d(T, \alpha, t) = \tilde{b} e^{-\frac{nE_\alpha}{kT}} \left(\frac{\alpha}{1 - \alpha} \right)^n t^n \tag{6}$$

978-1-60558-497-3/09 $25.00 © 2009 ACM

where constant $\tilde{b}$ can be fitted from SPICE characterization. A primary advantage of using Equation (6) to characterize gate-level NBTI effect is that, given a reference model pre-characterized at T_{ref} and α_{ref}, the aging effect of a logic gate under any arbitrary T and α can be efficiently calculated using parameter scaling, which will be discussed in detail in Section IV-A.

B. Considering Process Variations in NBTI Aging Modeling

The NBTI-induced gate delay degradation depends not only on circuit run-time conditions but also on fabrication-determined process parameters, such as the initial threshold voltage V_{th} and oxide thickness t_{ox}. Due to fabrication-induced process variations, the NBTI aging process and its impact on circuit timing become a random process. To model NBTI-induced aging under process variation, we apply the stochastic collocation method, which was originally proposed to model the gate delay under uncertainty [15]. Considering a set of process parameters as a random variable vector with normal distribution

$$\vec{\xi} = [L, V_{th}, t_{ox}, \ldots], \tag{7}$$

the variation of gate delay degradation can be modeled using the polynomial chaos expansion on the set of random variables.

$$\Delta d(\vec{\xi}, T, \alpha, t) \approx \sum_{j}^{P} c_j \Phi_j(\vec{\xi}) e^{-\frac{nE_\alpha}{kT}} \left(\frac{\alpha}{1-\alpha} \right)^n t^n \tag{8}$$

where $\{\Phi_j\}_{j=1}^{P}$ is the complete set of the d-dimension Hermite polynomials up to the l-th order, and c_j is the unknown coefficient. Hermite polynomials form a set of orthogonal basis of Hilbert space under Gaussian measure, and thus is the best if $\vec{\xi}$ is Gaussian. Second order Hermite polynomials are sufficient in practice. The coefficients c_j can be determined using the stochastic collocation method [5], [15]. Using Equation (8), the delay of a logic gate at time t can then be expressed as follows.

$$d(\vec{\xi}) = d_0(\vec{\xi}) + \Delta d(\vec{\xi}, T, \alpha, t) \tag{9}$$

where $d_0(\vec{\xi})$ is the initial gate delay after chip fabrication. $\Delta d(\vec{\xi}, T, \alpha, t)$ represents the time-dependent gate aging effect. $d_0(\vec{\xi}) = \sum_{j}^{P} d_j \Phi_j(\vec{\xi})$ is expressed in polynomial chaos form with respect to the same set of random variables as Δd, and the coefficients are computed using the same stochastic collocation method. For gates with multiple inputs, this model is applied for each input of the gate.

IV. AGING-AWARE STATISTICAL TIMING ANALYSIS

Given a circuit netlist and the aging-aware variational gate delay model (Equation 9) as inputs, the proposed aging-aware statistical timing analysis method computes the aging effect of each logic gate based on its run-time condition, and carries out circuit-level statistical timing analysis.

A. Computation of Gate NBTI Aging Effect

For each type of logical gate provided by the technology library, the NBTI aging effect is characterized once under a reference working condition T_{ref} and α_{ref}, and is expressed as $\Delta d(\vec{\xi}, T_{ref}, \alpha_{ref}, t)$. During circuit-level logic analysis, for each logic gate i, the duty cycle α_i of its input signal is estimated by circuit simulation using user-provided input signal vectors, and the gate temperature T_i is provided from chip thermal profile. The aging effect of gate i can then be calculated by scaling the referenced model as follows.

$$\Delta d(\vec{\xi}, T_i, \alpha_i, t) = \Delta d(\vec{\xi}, T_{ref}, \alpha_{ref}, t) \cdot R_T \cdot R_\alpha \tag{10}$$

where the scaling factors for temperature T_i and signal duty cycle α_i are

$$R_T(T_{ref}, T_i) = exp(\frac{nE_\alpha}{k} \cdot \frac{T_i - T_{ref}}{T_{ref} T_i}) \tag{11}$$

and

$$R_\alpha(\alpha_{ref}, \alpha_i) = \left[\frac{\alpha_i(1 - \alpha_{ref})}{\alpha_{ref}(1 - \alpha_i)} \right]^n \tag{12}$$

Due to the exponential dependence of the temperature and error introduced during model simplification, a maximum of $\pm 25K$ temperature difference is allowed in order to achieve good accuracy of the scaled aging effect for a gate (This setting is used in the experimental result section.). In order to cover the complete circuit operation range, we develop piece-wise gate aging model. In addition, the above scaling model is not applicable when $\alpha = 1$, which indicates that the gate is under static stress. Using the static NBTI model, a scaling aging model can also be developed for $\alpha = 1$ using the same method described above.

B. Equivalent Aging Time Analysis

Circuit workload may vary over time, so is the aging process. For example, as shown in Figure 3, a gate experiences three different working conditions, (T_1, α_1), (T_2, α_2) and (T_3, α_3), with a duration of t_1, t_2 and t_3, respectively. In this work, we introduce *equivalent aging time* to facilitate characterization of the aging effect under such varying conditions. For the sake of clarity, the dependency of the aging effect on $\vec{\xi}$ is omitted. Given the aging effect of a logic gate under condition (T_2, α_2), the equivalent aging effect under condition (T_1, α_1) is described as follows.

$$\Delta d(T_2, \alpha_2, t_{eqv1}) = \Delta d(T_1, \alpha_1, t_1) \tag{13}$$

Using Equation 6, the equivalent aging time t_{eqv1} can be computed as

$$t_{eqv1} = t_1 \cdot [R_T(T_2, T_1) R_\alpha(\alpha_2, \alpha_1)]^{1/n} \tag{14}$$

where R_T and R_α are defined in Equation (11) and (12). Then, the aging effect of the gate at t_2 equals that of the gate working under (T_2, α_2) during time $(0, t_{eqv1} + t_2)$, and can be computed as $\Delta d(T_2, \alpha_2, t_{eqv1} + t_2)$. This procedure can be carried out inductively whenever the working condition changes.

Figure 3 demonstrates the use of equivalent aging time to estimate the overall aging process under three different working conditions. The dotted lines are the equivalent aging times computed at each transition using Equation (14). The solid lines are actual aging durations.

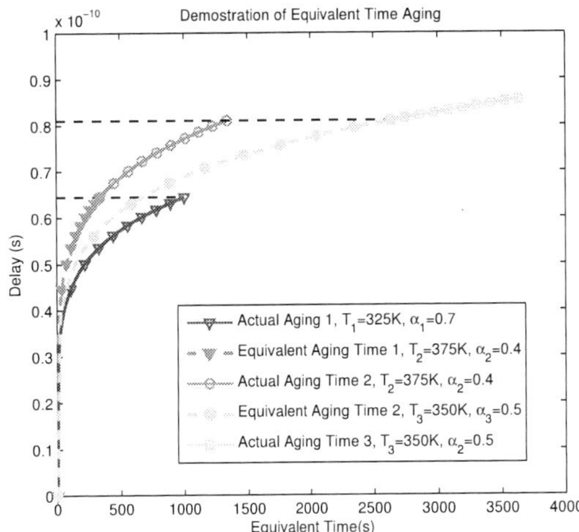

Fig. 3. Demo of equivalent aging time. A NAND2 gate experiences three different conditions consecutively. Each working condition lasts for 1000s.

978-1-60558-497-3/09 $25.00 © 2009 ACM

C. Aging-Aware SSTA and Pruning Algorithm for Repeated Analysis

Once the aging effect of a gate i is computed, the delay of each gate is expressed in PCE form (Equation 9) of its underlying process parameter $\vec{\xi}_i$. To handle the correlation between the process parameters $\{\vec{\xi}_i\}_1^q$ from q different gates, principle component analysis [2] is carried out to extract a set of independent variables $\vec{z}$ according to the correlation matrix of $\{\vec{\xi}_i\}_1^q$. And each set of process parameters $\vec{\xi}_i$ is represented by a linear combination of $\vec{z}$. After this transformation, a PCE delay model based timing analysis method [5] is adopted to compute the arrival time (AT) and required time (RT) of each node on the graph. The arrival time and required time are also represented in the PCE form.

In practice, the circuit needs to be tested under different combinations of working conditions and it is time-consuming and thus undesirable if aging-ware SSTA has to be re-run each time the working condition of the circuit changes. Inspired by the work of [24], we propose a fast pruning algorithm which prunes redundant timing nodes and edges and thus improves the subsequent repeated analysis.

It is observed in [24] that the distributions of the arrival time at the different inputs of a node may be distantly separated. The statistical maximum of them, which is the distribution of arrival time at the output of the node, will also be separated from some of the inputs. Such input can be identified using the following equation:

$$\mu_{out} - \mu_{in} \geq \sigma_{in} + \sigma_{out} \tag{15}$$

where μ and σ are the means and variances of the input and output arrival times. Inputs of this type do not contribute to the timing statistics of the later stages and therefore the incident edges corresponding to this type of inputs can be removed from the timing graph of the circuit. Note that as the circuit ages, it is possible that inputs not contributing to circuit performance at time zero become relevant due to the NBTI aging effect. However, as studied in [21] and [20], the maximum delay degradation along a timing path is limited by 20%. Considering this factor, we add a safe margin to the edge pruning condition to prevent the removal of any potentially important edge due to aging effect:

$$\mu_{out} - \mu_{in}(1 + \epsilon) \geq (\sigma_{in} + \sigma_{out}) \times (1 + \epsilon) \tag{16}$$

where ϵ is set to be 20%.

Furthermore, a node with all its fan-out edges pruned can also be removed from the timing graph, along with its fan-in edges. In summary, the pruning algorithm consists of two phases. Firstly, it does a forward topological search to prune unimportant timing edges using Equation (16). Next, it carries out a backward topological search to prune nodes with no remaining fan-out edges, and their fan-in edges. The running time of the pruning algorithm is dominated by SSTA in the first phase of the algorithm. The payoff of this algorithm is that the aging-aware SSTA under different working conditions can be sped up on the pruned graph. In addition, optimization needs only to consider the pruned graph.

V. Criticality-Sensitivity Based Element Impact Analysis

Statistical analysis fulfills two purposes: first, it calculates the reliability after several years' aging; second, when the target reliability is not achieved, it provides guidance for circuit optimization. In this section, we propose a criticality-sensitivity based element impact analysis, which can be embedded in an iterative optimization framework to improve reliability of the circuit after years of aging. The whole flow is shown in the lower right part of Figure 1. Our aging optimization procedure is similar to TILOS [9] and the work in [10] in the sense that it chooses a group of most *effective* gates for sizing at each iteration. However, we propose a criticality-sensitivity based analysis to measure, in the probability space, the impact of sizing this gate on the reduction of aging effect of the whole circuit. At each

iteration of the optimization flow, the aging-aware statistical timing analysis proposed in Section VI is used to update the aging effect and timing information of the circuit. Then, criticality-sensitivity based element impact analysis discussed below is carried out to select a group of gates Φ, whose change affects the timing of the circuit most effectively. At the end of the iteration, gates in Φ are optimized to improve the circuit yield under process variation and aging effect. The optimization iteration repeats until it meets the expected reliability or violates the constraints of area or power.

Because of circuit structure, the effects of updating different elements are different and interdependent. Updating one element in each iteration solves the interdependence problem but is expensive. It is also unnecessary to update an element on a non-critical path. Therefore, it is important to select a small group of most effective elements to optimize in each iteration. As a study, we focus on gate sizing where the designed width μ_1 of each gate will be selected to optimize the circuit reliability at the end of a given aging period. Notice that the fabricated actual width ξ_1 of a gate is still a random variable, but its mean value and variance are decided by μ_1.

The effect of sizing gate i on the reliability y of the whole circuit, which is defined as the circuit yield after a given period of aging, is just the reliability gradient over μ_1, which can be expressed as

$$\frac{\partial y}{\partial \mu_1} = \frac{\partial y}{\partial A_i} \cdot \frac{\partial A_i}{\partial d_i} \cdot \frac{\partial d_i}{\partial \xi_1} \cdot \frac{\partial \xi_1}{\partial \mu_1} \tag{17}$$

where d_i is the gate delay, and A_i is the slack given by

$$A_i(\vec{\xi}) = AT_{i,in}(\vec{\xi}) + d_i(\vec{\xi}) + RT_{i,out}(\vec{\xi}) \tag{18}$$

where $AT_{i,in}$ and $RT_{i,out}$ are the arrival time at the input and the required time at the output of gate i. From the above relationship, we know $\partial A_i/\partial d_i = 1$. The product of the last two terms in Equation (17) gives the **sensitivity** $\partial d_i/\partial \mu_1$, which is the derivative of the gate delay with respect to the designed width. Since gate delay is a distribution, the sensitivity, as its derivative over a constant variable, is also a distribution. We will use its mean value to guide the optimization. For this, we have

$$E(\partial d_i/\partial \mu_1) = E\left(\frac{\partial d_i}{\partial \xi_1}\right) \cdot E\left(\frac{\partial \xi_1}{\partial \mu_1}\right) \tag{19}$$

where the first term is a constant and can be obtained from the coefficient of the ξ_1 term in Equation (9), and the second term is 1 if μ_1 gives the mean of ξ_1.

The term $\partial y/\partial A_i$ captures the complex relationship between gate slack and circuit reliability. When the slack is positive, it will be zero. Even when the slack is negative, if the gate is not on all critical paths, the term will still be zero. Furthermore, with process variation, A_i becomes a distribution instead of a deterministic figure. Therefore, we use criticality [25] to evaluate how important a gate is to improving the reliability. Considering the aging effect, the **criticality** c_i is defined as the probability that gate i lies on the critical path of the circuit after a period of aging time, due to the process variation. To compute the criticality of each gate in the circuit, we adopt the cutset-based method proposed in [25], [17]. Since our timing distribution is expressed in PCE form instead of first-order canonical form, the tightness probability, a basic building block of the criticality computation algorithm, is computed using either numerical integration [8] or the APEX method [16].

After the criticality c_i and sensitivity s_i of every gate in the circuit are computed, we rank the gates according to the product $c_i \times s_i$. Although $c_i \times s_i$ is not the same as $\partial y/\partial \mu_1$, it provides an approximate ranking of the gates by their impact on the improvement of the circuit reliability after aging. Using this criterion, a group of n highest-ranking gates are chosen for optimization. After optimization, the aging-aware statistical timing analysis is carried out to update the aging effect, timing, and reliability of the circuit. Notice that in the analysis, not all the gates in the circuit need to be updated. Only the

978-1-60558-497-3/09 $25.00 © 2009 ACM

fan-in gates of the n modified gates and those on their fan-out cones need to be reexamined.

VI. Experimental Results

The proposed statistical reliability analysis framework is implemented in C++, including a modeling engine, a timing analysis engine, and a criticality-sensitivity based element impact analysis engine. The effectiveness of the impact analysis is tested in an iterative optimization approach. Circuits from ISCAS85 are used as the testbench. The experiments are run on a 64-bit Linux server with 3.0GHz Xeon CPU and 2G memory.

A. Verification of the Gate Aging Model

This section evaluates the proposed statistical gate aging model shown in Equation 8. Various types of logic gates, such as BUF, NAND, and NOR, are considered with 65nm PTM [1] technology. For each type of gate, the channel width, gate length, and threshold voltage of PMOS and NMOS are modeled using Gaussian random variables. The variance of each random parameter is set as 10% of its mean. The reference aging model is characterized once using the proposed modeling method under normal working condition with $T = 325K$ and $\alpha = 0.5$. To test the accuracy of our aging model in different working conditions, the aging effect of the gates under a different working condition is computed by scaling from the reference model using Equation 10. We select $T = 350K$ and $\alpha = 0.75$ as the testing condition, which achieves the maximum temperature difference against the reference condition allowed in our model, as is discussed in Subsection IV-A. The aging effects of each gate are compared against the results of corresponding 5000-point Monte-Carlo simulation. In each of the Monte-Carlo instances, the random ΔV_{th} is computed using long-term model in Equation 1 according to the sampled process parameters.

The relative error results against Monte-Carlo are given in Table I. The aging effect is modeled accurately at the reference working condition. As the working condition deviates from the reference one, the modeling error begins to grow, which is not unexpected since the modeling error stems not only from the coefficient regression but also from the fact that the scaling relationship in α and T does not exactly satisfy the long-term model. Still, the aging model has sufficient accuracy to be used in the aging-aware statistical analysis framework under allowed fluctuation of working conditions.

TABLE I

Error of Gate Aging Model against Monte-Carlo

Gate	Ref.		Scaled	
	μ	σ	μ	σ
NOT	0.01%	0.15%	2.45%	4.87%
BUF	0.02%	1.12%	1.02%	2.53%
NAND2	0.05%	0.86%	2.19%	5.32%
NOR2	0.11%	1.38%	2.15%	5.34%
XOR2	0.05%	2.67%	2.78%	4.11%
NAND4	0.06%	2.42%	3.98%	4.96%
NOR4	0.21%	1.83%	2.87%	4.25%

B. Aging Effects of Circuit Delay Distribution

Next, using the proposed aging-aware statistical timing analysis method, We characterize circuit delay distributions under aging effects. The proposed analysis method uses the SSTA method which has been validated against Monte-Carlo simulations in the past work [5]. To evaluate temperature-dependent aging effect, we constructed a run-time temperature profile gathered from a computer server. The chip temperature varies from $36°C$ (low workload) to $75°C$ (high workload). The temperature profile is then repeated in a 5-year time span. In addition, during the first two years, the ratio between high

workload and low workload is set to 0.5, and for the next three years, the ratio is increased to 2. The aging history for ISCAS85 circuits under the changing temperature profile is computed using the proposed equivalent aging time technique. The delay distributions of each circuit are computed at years 0, 2 and 5. The evolution of the means and the variances at years 2 and 5 are shown with respect to year zero in Figure 4.

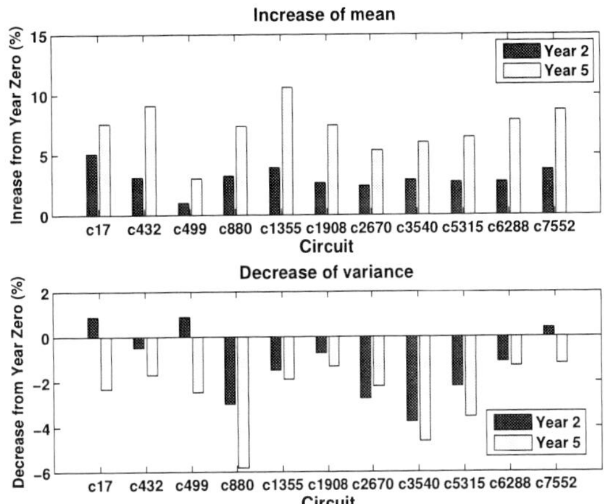

Fig. 4. Degradation of circuit delay distribution due to aging effect.

These results demonstrate that, due to process variation and NBTI aging effect, mean delay increases while delay variance decreases in general, as was first discovered by Wang et al [22]. However, in some circuits, such as c17, c449 and c7552, delay variance increases over time. This is due to the fact that, unlike the single path case [22], the distributions of the paths converged at a gate inputs may approach or depart from each other, resulting in a possible increase of circuit delay variance.

C. Pruning Effects

We next evaluate the effectiveness of the proposed aging-aware pruning algorithm (Section IV-C). We choose $\epsilon = 20\%$ as the delay degradation of the paths in the circuit will not exceed 20% under normal operation, which is also suggested by Figure 4. Table II lists the pruning statistics, accuracy and speedup against the results of using SSTA without pruning. With the maximum difference of mean and variance under 1% and 4%, the pruning algorithm introduces little error for the subsequent timing analysis. From Table II, it is observed that the analysis speedup is proportional to the number of pruned edges and nodes, which depends on the structure of the circuit and the different aging effect of each gate in the circuit. For ISCAS85 benchmarks, speedups from 10% to 70% are achieved. However, if a circuit is well-balanced, that is, the arrival time distributions at the inputs of every node are close to each other, the pruning algorithm is not effective, as shown in the result of c7552.

D. Effectiveness of Element Impact Analysis

Finally, we designed and implemented an iterative gate sizing optimization framework to test the effectiveness of the proposed criticality-sensitivity based element impact analysis (Section V). The four largest benchmarks shown in Table II are used in this experiment. For each benchmark, we consider the high workload condition with temperature of $75°C$. The optimization flow is set to target a 5-year time span. During each optimization iteration, the proposed aging-aware timing analysis is used to estimate the circuit timing information. Gates are then ranked by the product of criticality and sensitivity in a non-increasing order. The first Φ gates are then chosen,

978-1-60558-497-3/09 $25.00 © 2009 ACM

518

TABLE II
PRUNED SSTA WITH AGING EFFECT

Circuit	Pruned Nodes	Pruned Edges	Error μ	Error σ	Speedup
c17	33.3%	35.7%	0.25%	1.48%	1.29X
c432	33.5%	40.8%	0.04%	2.16%	1.24X
c499	22.4%	43.6%	0.06%	2.29%	1.31X
c880	50.6%	58.6%	0.20%	3.21%	1.75X
c1355	19.2%	30.6%	0.03%	2.90%	1.22X
c1908	36.4%	41.4%	0.02%	2.52%	1.46X
c2670	47.4%	51.3%	0.20%	1.98%	1.76X
c3540	42.3%	49.9%	0.08%	3.74%	1.52X
c5315	10.1%	17.3%	0.03%	1.87%	1.06X
c6288	11.9%	22.0%	0.08%	1.31%	1.10X
c7552	6.31%	5.45%	0.02%	2.60%	1.01X

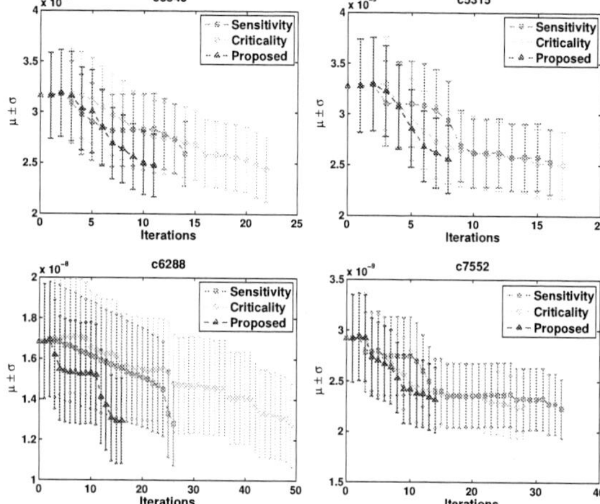

Fig. 5. Error bars of the mean and variance during optimization.

and the size of each gate is increased by ϵ. $\Phi = 50$ provides the best performance-time trade-off in our experiments. This procedure is repeated until the $\mu + \sigma$ of the circuit delay is improved by 25%. For comparison purposes, we also consider another two gate impact analysis methods which rank the gates using either sensitivity or criticality alone. The traces of the mean and variance of the delays of these circuits after each iteration are plotted in Figure 5. It can be observed that the progress of the sensitivity-guided method is unstable while that of the criticality-guided method is not fast enough. On the other hand, the progress of the proposed method using the product of sensitivity and criticality is both stable and fast, showing the effectiveness of criticality-sensitivity analysis.

VII. CONCLUSION

In this work, we have studied the impact of process variation and run-time aging effect on IC lifetime reliability. This work yields a comprehensive IC reliability analysis framework, which considers the joint impact of NBTI aging effect and parameter variations. Techniques are proposed to optimize the accuracy and efficiency of IC reliability analysis, including scalable NBTI aging model, equivalent aging time and aging-aware timing graph pruning. A novel criticality-sensitivity based analysis method is proposed to allow rapid estimation of the impact individual circuit element on the overall circuit lifetime reliability. Leveraging the proposed analysis framework, we have designed and implemented a statistical reliability optimization flow for nanometer-scale IC design.

ACKNOWLEDGMENT

This research is supported partially by NSFC research project 60676018 and 60806013, China National Basic Research Program under the grant 2005CB321701, China National Major Science and Technology special project 2008ZX01035-001-06 during the 11th five-year plan period, the doctoral program foundation of Ministry of Education of China under 200802460068, the International Science and Technology Cooperation program foundation of Shanghai under 08510700100, the program for Outstanding Academic Leader of Shanghai, NSF under CCF-0811270, and SRC under awards 2007-HJ-1593 and 2007-TJ-1589.

REFERENCES

[1] *Predictive technology model, http://www.eas.asu.edu/ptm.*
[2] A. Agarawal, D. Blaauw, and V. Zolotov. Statistical timing analysis for intro-die process variations with spatial correlations. In *ICCAD*, 2003.
[3] M. Alam, K. Kang, B. C. Paul, and K. Roy. Reliability- and process-variation aware design of VLSI design. In *IPFA*, India, 2007.
[4] S. Basu and R. Vemuri. Process variation and nbti tolerant standard cells to improve parametric yield and lifetime of ICs. In *ISVLSI*, 2007.
[5] S. Bhardwaj, P. Ghanta, and S. Vrudula. A framwork for statistical timing analysis using non-linear delay and slew models. In *ICCAD*, Nov. 2006.
[6] S. Bhardwaj, W. Wang, R. Vattikonda, Y. Cao, and S. Vrudhula. Predictive modeling of the NBTI effect for reliable design. In *CICC*, pages 189–192, Sep. 2006.
[7] S. Chakravarithi, A. T. Krishman, V. Reddy, C. F. Machala, and S. Krishnan. A comprehensive framework for predictive modeling of negative bias temperature instability. In *Annual International Reliabilty Physics Symposium*, Phoenix, 2004.
[8] H. Chang, V. Zolotov, S. Narayan, and C. Visweswariah. Parameterized block-based statistical timing analysis with non-gaussian parameters, nonlinear delay functions. In *DAC*, 2005.
[9] J. P. Fishburn and A. E. Dunlop. TILOS: a posynomial programming approach to transistor sizing. In *ICCAD*, 1985.
[10] M. R. Guthaus, N. Venkateswarann, C. Visweswariah, and V. Zolotov. Gate sizing using incremental parameterized statistical timing analysis.
[11] K. Kang, S. P. Park, K. Roy, and M. A. Alam. Estimation of statistical variation in temporal NBTI degradation and its impact on lifetime circuit performace. In *ICCAD*, 2007.
[12] H. Kufluoglu and M. Alam. A generalized reaction-diffusion model with explicit $H - H_2$ dynamics for negative-bias temperature-instability (NBTI) degradation. *IEEE Trans. on Electron Devices*, 54(5):1101–1107, May 2007.
[13] S. Kumar, C. H. Kim, and S. Sapatnekar. An analytical model for negative bias temperature instability. In *ICCAD*, Nov. 2006.
[14] S. Kumar, C. H. Kim, and S. Sapatnekar. NBTI-aware synthesis of digital circuits. In *DAC*, June 2007.
[15] S. Kumar, J. Li, C. Talarico, and J. Wang. A probabilistic collocation method based statistical gate delay model considering process variations and multiple input switching. In *DATE*, 2005.
[16] X. Li. Asymptotic probability extraction for non-normal distribution of circuit. In *ICCAD*, 2004.
[17] D. Mogal, H.Qian, S. Sapatnekar, and K. Bazargan. Clustering based prining for statistical criticality computation under process variation. In *ICCAD*, 2007.
[18] T. Sakurai and A. R. Newton. Alpha-power law mosfet model and its application to cmos invert delay and other formulars. *IEEE Journal of Solid-State Circuits*, 25(2):584–594, Apr. 1990.
[19] D. Shroder and J. A. Babcock. Negative bias temperature instability: Road to cross in deep submicron silicon semiconductor manufacturing. *Journal of Applied Physics*, 94(1), July 2003.
[20] W. Wang, Z. Wei, S. Yang, and Y. Cao. An efficient method to identify critical gates under circuit aging. In *ICCAD*, 2007.
[21] W. Wang, S. Yang, S. Bhaedwaj, R. Vattikonda, S. Vrudhula, F. Liu, and Y. Cao. The impact of NBTI on the performance of combinational and sequential circuits. In *DAC*, June 2007.
[22] W.-P. Wang, V. Reddy, B. Yang, V. Balakrishnan, S. Krishnan, and Y. Cao. Statistical predction of circuit aging under process variation. In *CICC*, 2008.
[23] W.-P. Wang, S.-Q. Yang, and Y. Cao. Node criticality computation for circuit timing analysis and optimization under NBTI effect. In *ISQED*, 2008.
[24] Y. Wang, X. Zeng, J. Tao, H. Zhu, X. Luo, C. Yan, and W. Cai. Adaptive stochastic collocation method (ASCM) for parameterized statistical timing analysis with quadratic delay model. In *ISQED*, 2008.
[25] J. Xiong, V. Zolotov, C. Visweswariah, and N. Venkateswara. Criticality compuatation in parameterized statistical timing. In *DAC*, 2006.
[26] B. Zhang and M. Orshansky. Modeling of nbti-induced pmos degradation under arbitrary dynamic temperature variation. In *ISQED*, 2008.

978-1-60558-497-3/09 $25.00 © 2009 ACM

Guess, Solder, Measure, Repeat -
How do I get my mixed-signal chip right?

Geoffrey Ying (Organizer)
Synopsys Inc.
Mountain View, CA, USA

Andreas Kuehlmann (Organizer)
Cadence Design Systems
San Jose, CA, USA

Ken Kundert (Chair)
Designer's Guide Consulting
Los Altos, CA, USA

George Gielen
Katholieke Universiteit Leuven
Leuven, Belgium

Eric Grimme
Intel Corp.
Hillsboro, OR, USA

Martin O'Leary
Cadence Design Systems
San Jose, CA, USA

Sandeep Tare
Texas Instruments
Dallas TX, USA

Warren Wong
Synopsys Inc.
Maintain View, CA, USA

Categories and Subject Descriptors

B Hardware, B.7 Integrated Circuit, B.7.2 Design Aids

General Terms: Performance, Design, Verification.

Keywords: Mixed-signal Verification, Functional Verification, Performance Verification, Low Power Verification, Analog Behavioral Modeling, SPICE, Verilog, Verilog-AMS, VHDL

PANEL SUMMARY

Over the past 20 years, EDA has developed a solid digital implementation methodology that combines some restrictions on the design style with a set of comprehensive tools leading to predictable design flows. The recent increased use of analog components in complex SOC designs triggered a set of verification challenges ranging from simple connectivity problems to complex interferences between analog and digital data blocks. This panel discusses the state of the affairs in analog-mixed signal verification and draws a picture of future directions in terms of new approaches and tools.

PANELIST PRESENTATIONS

Sandeep Tare: Verification of complex mixed-signal ICs is a daunting and expensive proposition. Using behavioral models for analog circuits is widely advocated and accepted means to speedup mixed-signal simulations. Since behavioral models are used for verification at sub-IP, IP, subsystem and SOC levels, they have to satisfy widely varying (and often conflicting) requirements. There is no modeling language or standard that

Permission to make digital or hard copies of part or all of this work for personal or classroom use is granted without fee provided that copies are not made or distributed for profit or commercial advantage and that copies bear this notice and the full citation on the first page. To copy otherwise, to republish, to post on servers or to redistribute to lists, requires prior specific permission and/or a fee.
DAC'09, July 26-31, 2009, San Francisco, California, USA

renders itself suitable for all purposes. A concerted effort by EDA industry and their customers to address this gap will be very helpful. Creating these models and keeping them in sync with the implementation (schematic) can be quite challenging. Any EDA technology that automatically creates behavioral models from a language neutral format will help address both the above problems. The validity of simulations done using behavioral models is closely tied to the quality of these models. Therefore, it is imperative that these models be rigorously verified against the spec or schematic (preferably both) depending on the context in which they will be used. There is a need to define standard methods to tackle this challenge systematically.

Digital IC verification has made huge advances in simulation-based verification (E.g. assertions, checkers, constrained random stimulus, transaction models, coverage driven verification etc). Since A/MS verification primarily relies on exhaustive simulation, there is great benefit in exploiting these techniques to fit A/MS needs. Specifically, application of these techniques to analog behavioral models and circuits must be enabled in such a way that the entire DV environment can be systematically reused for MSV. The digital world has also made huge advances in novel non simulation-based verification techniques (structural, static, formal, equivalency checks). These techniques must be explored for A/MS especially because A/MS simulations tend to be much more expensive than digital. In addition to tackling verification challenges, some thought, research and energy must also be directed towards defining standards to improve testability of A/MS designs.

Eric Grimme: The last five years have seen a rapid rise in mixed-signal verification at Intel. Efforts changed from experimental tinkering by a handful of early adopters to dedicated mixed-signal validation teams in nearly every Intel project. The need is fueled by a drastic increase in analog content and the continued growth of SOC design methodologies. The mixed-signal solutions are greatly aided by the establishment and proliferation of standardized languages such as Verilog-AMS.

978-1-60558-497-3/09 $25.00 © 2009 ACM

One can question though if after this period of rapid adoption, a somewhat static and not entirely satisfactory state remains. Even with the wide adoption of mixed-signal verification techniques, estimates indicate that the effort for the design and validation of analog blocks can still reach an order of magnitude more than for comparably sized digital blocks. Throughput times for the validation of large mixed-signal components are still bounded by either the effort to construct satisfactory behavioral models and/or the maturity/capacity of fast-Spice simulators for transistor level modeling. Formal analog methods to verify the equivalency of the behavioral and transistor representations are still in their infancy. Tool complexities arise, particularly for SOC, as one crosses the boundaries of languages or becomes overly dependent on tool features that fall outside the scope of language standardization. Adoption of mixed-signal verification practices in early design work at Intel has been slowed due to the limited patchwork of mixed-signal oriented enhancements in the centralized design environment. All of these limitations present opportunities for extending the much needed impact of mixed-signal verification techniques.

Warren Wong: The convergence of computing, consumer and mobile applications necessitates integrating complex mixed-signal functions with sophisticated digitally adjusted analog circuits, multiple supply domains, and thousands of interface signals between the digital and analog blocks. While the digital and analog verification engineers are well versed in the art of verification in their respective domains, when these worlds collide, even these veteran verification engineers are often left perplexed trying to verify the digital and analog blocks together to meet the tight tape-out deadline.

Mixed signal verification methodology ranges from bottom-up approach where behavioral models are created from transistors to facilitate better simulation performance, to top-down approach where behavioral models are validated before transistor implementations. On both ends, model creations and validations are challenged by engineering expertise because few engineers are knowledgeable in both digital and analog. Automatic tools are at infancy handicapped by the fact that there is no standard on analog behaviors similar to those in digital domain, hence often requiring custom interventions. Simulation tools are more mature with capability for auto-insertion of interface elements. Pragmatic verification methodologies leveraging current tool capabilities need to be used in the short term while waiting for the new tools to mature. Some of these methodologies include using more digital verification techniques, replacing AMS behavioral models which are more error prone with digital models and real numbers, and running faster circuit simulations while keeping more components at SPICE level.

Martin O'Leary: There is a strong industry trend toward more mixed-signal capabilities in both digital and analog designs. Reasons for this trend are various: SOCs are having increasingly

higher levels of integration which is pulling more analog, mixed-signal and RF blocks into them; at lower process nodes analog designers are adding digital controllers to compensate for process effects; analog part vendors are tending to add digital controllers and programmability to make their products more attractive and differentiated. However this has lead to mixed-signal issues becoming an increasing cause of silicon failures and as a result, loss of productivity because traditional digital-centric approaches to verification and traditional analog-centric approaches to verification are not reliably detecting mixed signal failure modes.

The solution to this challenge will be a combination of things such as infusing analog verification with digital approaches (e.g. power specification, assertions and coverage), infusing digital verification with analog approaches (e.g. behavioral modeling) and as well as some novel concepts specific to mixed signal. This requires a unification of the tools and flows used in analog and digital verification. A mixed-signal simulator is required which can assemble and simulate circuits containing digital, analog and mixed signal blocks represented in a variety of HDLs (Verilog, VHDL, SystemVerilog, SystemC, SPICE) and seamlessly supporting verification methodologies such as assertions, coverage, randomization, low power specification across these blocks while completing the required simulations in a reasonable time by using highest performance digital and analog simulation engines in this mixed-signal simulator. Such a simulator must support mixed signal languages such as Verilog-AMS and VHDL-AMS to enable the creation of mixed-signal test-benches, mixed-signal behavioral models and automatic cross-HDL connectivity. The simulator should enable the creation of behavioral models with different performance and accuracy tradeoffs e.g. real-valued models for analog blocks to enable high performance mixed-signal SOC simulation using the Verilog-AMS wreal concept. A mixed-signal flow is also needed that bridges the gap between analog and digital flows in an evolutionary way rather than a revolutionary way to enable easier adoption. Much of this technology is already in place for the industry to adopt.

Georges Gielen: Verifying complex integrated mixed-signal systems requires higher levels of analog modeling. One of the biggest bottlenecks for its widespread use in industry are the insufficient skills and experience of most designers in model writing. Fortunately, in recent years large progress is being made towards computer-automated model generation, based on techniques such as model order reduction among others. At the same time, future designs tend to become even more complicated to verify. Digital-assisted analog circuits with lots of local and global tuning and reconfiguration loops just multiply the verification effort. Also, increasing variability and reliability problems (e.g. EMC) require efficient yet accurate solutions. Finally, the trend towards heterogeneous 3D systems adds another level of complexity to the verification problem because of all the parasitics involved.

978-1-60558-497-3/09 $25.00 © 2009 ACM

The Cilk++ Concurrency Platform

Charles E. Leiserson

MIT CSAIL and *Cilk Arts, Inc.*

Abstract

The availability of multicore processors across a wide range of computing platforms has created a strong demand for software frameworks that can harness these resources. This paper overviews the Cilk++ programming environment, which incorporates a compiler, a runtime system, and a race-detection tool. The Cilk++ runtime system guarantees to load-balance computations effectively. To cope with legacy codes containing global variables, Cilk++ provides a "hyperobject" library which allows races on nonlocal variables to be mitigated without lock contention or substantial code restructuring.

Categories and Subject Descriptors

D.1.3 [**Software**]: Programming Techniques—*Concurrent Programming*

General Terms

Algorithms, Performance, Design, Reliability, Languages.

Keywords

Amdahl's Law, dag model, hyperobject, multicore programming, multithreading, parallelism, parallel programming, race detection, reducer, span, speedup, work.

1   Introduction

Although the software community has extensive experience in serial programming using the C [18] and C++ [30] programming languages, they have found it hard to adapt C/C++ applications to run in parallel on multicore systems. In earlier work, the MIT Cilk system [14, 32] extended the C programming language with parallel computing constructs. The Cilk++ solution similarly extends C++, offering a gentle and reliable path to enable the estimated three million C++ programmers [31] to write parallel programs for multicore systems. Cilk++ is available for the Windows Visual Studio and the Linux/gcc compilers.

This work was supported in part by the National Science Foundation under Grants 0615215, 0712243, and 0822896 and in part by the Defense Advanced Research Projects Agency under Contract W31P4Q-08-C-0156. Cilk, Cilk++, and Cilkscreen are registered trademarks of Cilk Arts, Inc.

Permission to make digital or hard copies of part or all of this work for personal or classroom use is granted without fee provided that copies are not made or distributed for profit or commercial advantage and that copies bear this notice and the full citation on the first page. To copy otherwise, to republish, to post on servers or to redistribute to lists, requires prior specific permission and/or a fee.
DAC'09, July 26-31, 2009, San Francisco, California, USA

```
1    // Parallel quicksort
2    using namespace std;
3
4    #include <algorithm>
5    #include <iterator>
6    #include <functional>
7
8    template <typename T>
9    void qsort(T begin, T end) {
10     if (begin != end) {
11       T middle = partition(begin, end, bind2nd(
              less<typename iterator_traits<T>::
              value_type>(),*begin));
12       cilk_spawn qsort(begin, middle);
13       qsort(max(begin + 1, middle), end);
14       cilk_sync;
15     }
16   }
17
18   // Simple test code:
19   #include <iostream>
20   #include <cmath>
21
22   int main() {
23     int n = 100;
24     double a[n];
25
26     cilk_for (int i=0; i<n; ++i) {
27       a[i] = sin((double) i);
28     }
29
30     qsort(a, a + n);
31     copy(a, a + n, ostream_iterator<double>(cout
            , "\n"));
32
33     return 0;
34   }
```

Figure 1: Parallel quicksort implemented in Cilk++.

Like the MIT Cilk system [14, 32], Cilk++ is a *faithful* linguistic extension of C++, which means that parallel code retains its serial semantics when run on one processor. The Cilk++ extensions to C++ consist of just three keywords, which can be understood from an example. Figure 1 shows a Cilk++ program adapted from http://www.cvgpr.uni-mannheim.de/heiler/qsort.html, which implements the quicksort algorithm [7, Chapter 7]. Observe that the program would be an ordinary C++ program if the three keywords cilk_spawn, cilk_sync, and cilk_for were elided.

Parallel work is created when the keyword cilk_spawn precedes the invocation of a function. The semantics of spawning differ from a C++ function (or method) call only in that the parent can continue to execute in parallel with the child, instead of waiting for the child to complete as is done in C++. The scheduler in the Cilk++ runtime system takes the responsibility of scheduling the spawned functions on the individual processor cores of the multicore computer.

A function cannot safely use the values returned by its children until it executes a cilk_sync statement. The cilk_sync statement is a local "barrier," not a global one as, for example, is used in message-passing programming [23, 24]. In the quicksort example, a cilk_sync statement occurs on line 14 before the function returns to avoid the anomaly that would occur if the preceding calls to qsort were scheduled to run in parallel and did not complete before the return, thus leaving the vector to be sorted in an intermediate and inconsistent state.

In addition to explicit synchronization provided by the cilk_sync statement, every Cilk function syncs implicitly before it returns, thus ensuring that all of its children terminate before it does. Thus, for this example, the cilk_sync before the return is technically unnecessary.

Cilk++ improves upon the original MIT Cilk in several ways. It provides full support for C++ exceptions. Loops can be parallelized by simply replacing the keyword for with the keyword cilk_for keyword, which allows all iterations of the loop to operate in parallel. Within the main routine, for example, the loop starting on line 26 fills the array in parallel with random numbers. In the MIT Cilk system, such loops had to be rewritten by the programmer as divide-and-conquer recursion, but Cilk++ provides the cilk_for syntax for automatically parallelizing such loops. In addition, Cilk++ includes a library for mutual-exclusion (mutex) locks. Locking tends to be used much less frequently than in other parallel environments, such as Pthreads [17], because all protocols for control synchronization are handled by the Cilk++ runtime system.

The remainder of this paper is organized as follows. Section 2 provides a brief tutorial on the theory of parallelism. Section 3 describes the performance guarantees of Cilk++'s "work-stealing" scheduler and overviews how it operates. Section 4 briefly describes the Cilkscreen race-detection tool which guarantees to find race bugs in ostensibly deterministic code. Section 5 explains Cilk++'s "hyperobject" technology, which allows races on nonlocal variables to be mitigated without lock contention or restructuring of code. Finally, Section 6 provides some concluding remarks.

2 An overview of parallelism

The Cilk++ runtime system contains a provably efficient work-stealing scheduler [4, 14], which scales application performance linearly with processor cores, as long as the application exhibits sufficient parallelism (and the processor architecture provides sufficient memory bandwidth). Thus, to obtain good performance, the programmer needs to know what it means for his or her application to exhibit sufficient parallelism. Before describing the Cilk++ runtime system, it is helpful to understand something about the theory of parallelism.

Many discussions of parallelism begin with Amdahl's Law [1], originally proffered by Gene Amdahl in 1967. Amdahl made what amounts to the following observation. Suppose that 50% of a computation can be parallelized and 50% cannot. Then, even if the 50% that is parallel were run on an infinite number of processors, the total time is cut at most in half, leaving a speedup of at most 2. In general, if a fraction p of a computation can be run in parallel and the rest must run serially, Amdahl's Law upper-bounds the speedup by $1/(1-p)$.

Although Amdahl's Law provides some insight into parallelism, it does not *quantify* parallelism, and thus it does not provide a good understanding of what a concurrency platform such as Cilk++ should offer for multicore application performance. Fortunately, there is a simple theoretical model for parallel computing which provides a more general and precise quantification of parallelism that subsumes Amdahl's Law. The *dag (directed acyclic graph) model of multithreading* [3] views the execution of a multithreaded program as a set of instructions (the vertices of the dag) with graph edges indicating dependencies between instructions. (See Figure 2.) We say that an instruction x **precedes** an instruction y,

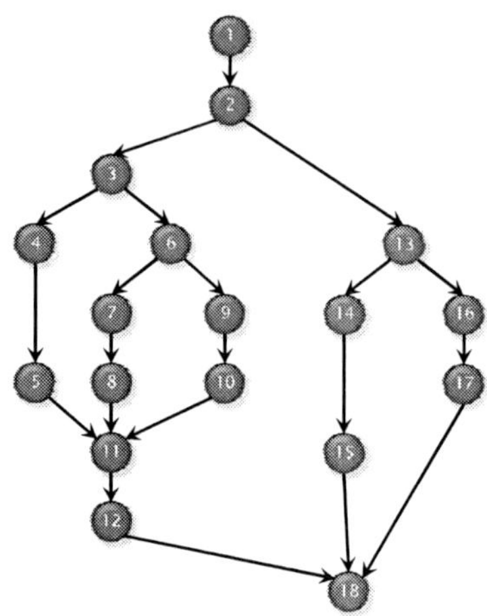

Figure 2: A directed acyclic graph representation of a multithreaded execution. Each vertex is an instruction. Edges represent ordering dependencies between instructions.

sometimes denoted $x \prec y$, if x must complete before y can begin. If neither $x \prec y$ nor $y \prec x$, we say that the instructions are in *parallel*, denoted $x \parallel y$. In Figure 2, for example, we have $1 \prec 2$, $6 \prec 12$, and $4 \parallel 9$.

The dag model of multithreading can be interpreted in the context of the Cilk++ programming model. A cilk_spawn of a function creates two dependency edges emanating from the instruction immediately before the cilk_spawn: one edge goes to the first instruction of the spawned function, and the other goes to the first instruction after the spawned function. A cilk_sync creates dependency edges from the final instruction of each spawned function to the instruction immediately after the cilk_sync. A cilk_for can be viewed as divide-and-conquer parallel recursion using cilk_spawn and cilk_sync over the iteration space.

The dag model admits two natural measures that allow us to define parallelism precisely, as well as to provide important bounds on performance and speedup.

The Work Law

The first important measure is *work*, which is the total amount of time spent in all the instructions. Assuming for simplicity that it takes unit time to execute an instruction, the work for the example dag in Figure 2 is 18.

We can adopt a simple notation to be more precise. Let T_P be the fastest possible execution time of the application on P processors. Since the work corresponds to the execution time on 1 processor, we denote it by T_1. Among the reasons that work is an important measure is because it provides a lower bound on P-processor execution time:

$$T_P \geq T_1/P .\tag{1}$$

This **Work Law** holds, because in our simple theoretical model, each processor executes at most 1 instruction per unit time, and hence P processors can execute at most P instructions per unit time. Thus, with P processors, to do all the work, it must take at least T_1/P time.

We can interpret the Work Law (1) in terms of the *speedup* on P processors, which using our notation, is just T_1/T_P. The speedup tells us how much faster the application runs on P processors than on 1 processor. Rewriting the Work Law, we obtain $T_1/T_P \leq P$, which is to say that the speedup on P processors can be at most P. If the application obtains speedup proportional to P, we say that the application exhibits *linear speedup*. If it obtains speedup exactly

978-1-60558-497-3/09 $25.00 © 2009 ACM

P (which is the best we can do in our model), we say that the application exhibits *perfect linear speedup*. If the application obtains speedup greater than P (which cannot happen in our model due to the Work Law, but can happen in models that incorporate caching and other processor effects), we say that the application exhibits *superlinear speedup*.

The Span Law

The second important measure is *span*, which is the longest path of dependencies in the dag. The span of the dag in our example is 9, which corresponds to the path $1 \prec 2 \prec 3 \prec 6 \prec 7 \prec 8 \prec 11 \prec 12 \prec 18$. This path is sometimes called the **critical path** of the dag, and span is sometimes referred to in the literature as critical-path length. Since the span is the theoretically fastest time the dag could be executed on a computer with an infinite number of processors (assuming no overheads for communication, scheduling, etc.), we denote it by T_∞. Like work, span also provides a bound on P-processor execution time:

$$T_P \geq T_\infty \, . \tag{2}$$

This **Span Law** arises for the simple reason that a finite number of processors cannot outperform an infinite number of processors, because the infinite-processor machine could just ignore all but P of its processors and mimic a P-processor machine exactly.

Parallelism

We define *parallelism* as the ratio of work to span, or T_1/T_∞. Parallelism can be viewed as the average amount of work along each step of the critical path. Moreover, perfect linear speedup cannot be obtained for any number of processors greater than the parallelism T_1/T_∞. To see why, suppose that $P > T_1/T_\infty$, in which case the Span Law (2) implies that the speedup satisfies $T_1/T_P \leq T_1/T_\infty < P$. Since the speedup is strictly less than P, it cannot be perfect linear speedup. Another way to see that the parallelism bounds the speedup is to observe that, in the best case, the work is distributed evenly along the critical path, in which case the amount of work at each step is the parallelism. But, if the parallelism is less than P, there isn't enough work to keep P processors busy at every step.

As an example, the parallelism of the dag in Figure 2 is $18/9 = 2$. That means that there's little point in executing it with more than 2 processors, since additional processors will be surely starved for work.

As a practical matter, many problems admit considerable parallelism. For example, matrix multiplication of 1000×1000 matrices is highly parallel, with a parallelism in the milliions. Many problems on large irregular graphs, such as breadth-first search, generally exhibit parallelism on the order of thousands. Sparse matrix algorithms can often exhibit parallelism in the hundreds.

3 Runtime system

Although optimal multiprocessor scheduling is known to be NP-complete [15], Cilk++'s runtime system employs a "work-stealing" scheduler [4, 14] that achieves provably tight bounds. An application with sufficient parallelism can rely on the Cilk++ runtime system to dynamically and automatically exploit an arbitrary number of available processor cores near optimally. Moreover, on a single core, typical programs run with negligible overhead (less than 2%).

Performance bounds

Specifically, for an application with T_1 work and T_∞ span running on a computer with P processors, the Cilk++ works-stealing sched-

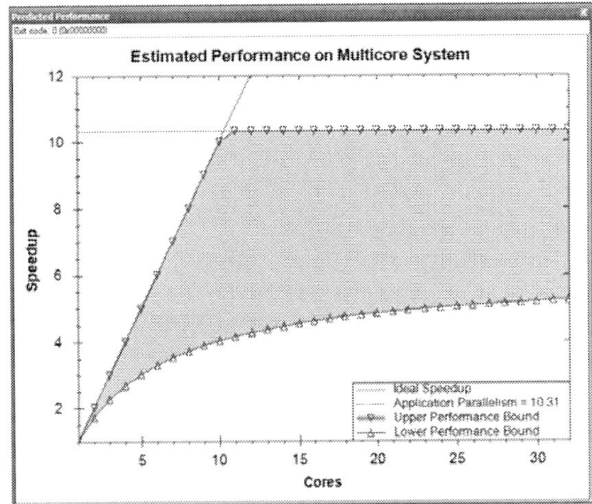

Figure 3: Parallelism profile of quicksort produced by the Cilk++ performance analyzer.

uler achieves expected running time

$$T_P \leq T_1/P + O(T_\infty) \, . \tag{3}$$

If the parallelism T_1/T_∞ exceeds the number P of processors by a sufficient margin, this bound (proved in [4]), guarantees near-perfect linear speedup. To see why, assume that $T_1/T_\infty \gg P$. Equivalently, we have $T_\infty \ll T_1/P$. Thus, in Inequality (3), the T_1/P term dominates the $O(T_\infty)$ term, and thus the running time is $T_P \approx T_1/P$, leading to a speedup of $T_1/T_P \approx P$.

The Cilk++ development environment contains a performance-analysis tool that allows a programmer to analyze the work and span of an application. Figure 3 shows the output of this tool running the quicksort program from Figure 1 on 100 million numbers. The upper bound on speedup provided by the Work Law corresponds to the line of slope 1, and the upper bound provided by the Span Law corresponds to the horizontal line at 10.31. The performance analysis tool also provides an estimated lower bound on speedup — the lower curve in the figure — based on *burdened parallelism*, which takes into account the estimated cost of scheduling. Although quicksort seems naturally parallel, one can show that the expected parallelism for sorting n numbers is only $O(\lg n)$. Practical sorts with more parallelism exist, however. See [7, Chapter 27] for more details.

In addition to guaranteeing performance bounds, the Cilk++ run-time system also provides bounds on stack space. Specifically, on P processors, a Cilk++ program consumes at most P times the stack space of a single-processor execution. Consider the following simple code fragment:

```
for (int i=0; i<1000000000; ++i) {
    cilk_spawn foo(i);
}
cilk_sync;
```

This code conceptually creates one billion invocations of foo that operate logically in parallel. Executing on one processor, however, this Cilk++ code uses no more stack space than a serial C++ execution, that is, the call depth is of whichever invocation of foo requires the deepest stack. On two processors, it requires at most twice this space, and so on. This guarantee contrasts with that of more naive schedulers, which may create a work-queue of one billion tasks, one for each iteration of the subroutine foo, before executing even the first iteration, thus blowing out physical memory.

978-1-60558-497-3/09 $25.00 © 2009 ACM

Work stealing

Cilk++'s work-stealing scheduler operates as follows. When the runtime system starts up, it allocates as many operating-system threads, called **workers**, as there are processors (although the programmer can override this default decision). Each worker's stack operates like a work queue. When a subroutine is spawned, the subroutine's activation frame containing its local variables is pushed onto the bottom of the stack. When it returns, the frame is popped off the bottom. Thus, in the common case, Cilk++ operates just like C++ and imposes little overhead.

When a worker runs out of work, however, it becomes a **thief** and "steals" the top frame from another **victim** worker's stack. Thus, the stack is in fact a double-ended queue, with the worker operating on the bottom and thieves stealing from the top. This strategy has the great advantage that all communication and synchronization is incurred only when a worker runs out of work. If an application exhibits sufficient parallelism, one can prove mathematically [4,14] that stealing is infrequent, and thus the cost of communication and synchronization to effect a steal is negligible.

The dynamic load-balancing capability provided by the Cilk++ runtime system adapts well in real-world multiprogrammed computing environments. If a worker becomes descheduled by the operating system (for example, because another application starts to run), the work of that worker can be stolen away by other workers. Thus, Cilk++ programs tend to "play nicely" with other jobs on the system.

Cilk++'s runtime system also makes Cilk++ programs **performance-composable**. Suppose that a programmer develops a parallel library in Cilk++. That library can be called not only from a serial program or the serial portion of a parallel program, it can be invoked multiple times in parallel and continue to exhibit good speedup. In contrast, some concurrency platforms constrain library code to run on a given number of processors, and if multiple instances of the library execute simultaneously, they end up thrashing as they compete for processor resources.

4 Race detection

The Cilk++ development environment includes a **race detector**, called Cilkscreen, a powerful debugging tool that greatly simplifies the task of ensuring that a parallel application is correct. We define a **strand** to be a sequence of serially executed instructions containing no parallel control, that is, a path in the multithreaded dag, where each vertex except the first in the path has at most one incoming edge and every vertex except the last in the path has at most one outgoing edge. A **data race** [26] exists if logically parallel strands access the same shared location, the two strands hold no locks in common, and at least one of the strands writes to the location. A data race is usually a bug, because the program may exhibit unexpected, nondeterministic behavior depending on how the strands are scheduled. Serial code containing nonlocal variables is particularly prone to the introduction of data races when the code is parallelized.

As an example of a race bug, suppose that line 13 in Figure 1 is replaced with the following line:

```
qsort(max(begin + 1, middle-1), end);
```

The resulting serial code is still correct, but the parallel code now contains a race bug, because the two subproblems overlap, which could cause an error during execution.

Race conditions have been studied extensively [6, 8–12, 16, 20–22, 25, 27–29]. They are pernicious and occur nondeterministically. A program with a race bug may execute successfully millions of times during testing, only to raise its head after the application is

```
 1   bool has_property(Node *);
 2   std::list<Node *> output_list;
 3   // ...
 4   void walk(Node *x)
 5   {
 6     if (x)
 7     {
 8       if (has_property(x))
 9       {
10         output_list.push_back(x);
11       }
12       walk(x->left);
13       walk(x->right);
14     }
15   }
```

Figure 4: C++ code to create a list of all the nodes in a binary tree that satisfy a given property.

shipped. Even after detecting a race bug, writing regression tests to ensure its continued absence is difficult.

The Cilkscreen race detector is based on provably good algorithms [2,6,11] developed originally for MIT Cilk. In a single serial execution on a test input for a deterministic program, Cilkscreen guarantees to report a race bug if the race bug is **exposed**: that is, if two different schedulings of the parallel code would produce different results. Cilkscreen uses efficient data structures to track the series-parallel relationships of the executing application during a serial execution of the parallel code. As the application executes, Cilkscreen uses dynamic instrumentation [5, 19] to intercept every load and store executed at user level. Metadata in the Cilk++ binaries allows Cilkscreen to identify the parallel control constructs in the executing application precisely, track the series-parallel relationships of strands, and report races precisely. Additional metadata allows the race to be localized in the application source code.

5 Reducer hyperobjects

Many serial programs use **nonlocal variables**, which are variables that are bound outside of the scope of the function, method, or class in which they are used. If a variable is bound outside of all local scopes, it is a **global variable**. Nonlocal variables have long been considered a problematic programming practice [33], but programmers often find them convenient to use, because they can be accessed at the leaves of a computation without the overhead and complexity of passing them as parameters through all the internal nodes. Thus, nonlocal variables have persisted in serial programming. In the world of parallel computing, nonlocal variables may inhibit otherwise independent parts of a multithreaded program from operating in parallel, because they introduce races. This section describes Cilk++ reducer hyperobjects [13], which can mitigate races on nonlocal variables without creating lock contention or requiring code restructuring.

As an example of how a nonlocal variable can introduce a data race, consider the problem of walking a binary tree to make a list of those nodes that nodes satisfy a given property. A C++ code to solve the problem is abstracted in Figure 4. If the node x being visited is nonnull, the code checks whether x has the desired property in line 8, and if so, it appends x to the list stored in the global variable output_list in line 10. Then, it recursively visits the left and right children of x in lines 12 and 13.

Figure 5 illustrates a straightforward parallelization of this code in Cilk++. In line 12 of the figure, the walk function is spawned recursively on the left child, while the parent continues on to execute an ordinary recursive call of walk in line 13. As the recursion unfolds, the running program generates a tree of parallel execution

```
1   bool has_property(Node *);
2   std::list<Node *> output_list;
3   // ...
4   void walk(Node *x)
5   {
6     if (x)
7     {
8       if (has_property(x))
9       {
10        output_list.push_back(x);
11      }
12      cilk_spawn walk(x->left);
13      walk(x->right);
14      cilk_sync;
15    }
16  }
```

Figure 5: A naive Cilk++ parallelization of the code in Figure 4. This code has a data race in line 10.

```
1   bool has_property(Node *);
2   std::list<Node *> output_list;
3   mutex L;
4   // ...
5   void walk(Node *x)
6   {
7     if (x)
8     {
9       if (has_property(x))
10      {
11        L.lock();
12        output_list.push_back(x);
13        L.unlock();
14      }
15      cilk_spawn walk(x->left);
16      walk(x->right);
17      cilk_sync;
18    }
19  }
```

Figure 6: Cilk++ code that solves the race condition using a mutex.

that follows the structure of the binary tree. Unfortunately, this naive parallelization contains a data race. Specifically, two parallel instantiations of walk may attempt to update the shared global variable output_list in parallel at line 10.

The traditional solution to fixing this kind of data race is to associate a mutual-exclusion lock (mutex) L with output_list, as is shown in Figure 6. Before updating output_list, the mutex L is acquired in line 11, and after the update, it is released in line 13. Although this code is now correct, the mutex may create a bottleneck in the computation. If there are many nodes that have the desired property, the contention on the mutex can destroy all the parallelism. For example, on one set of test inputs for a real-world tree-walking code that performs collision-detection of mechanical assemblies, lock contention actually degraded performance on 4 processors so that it was worse than running on a single processor. In addition, the locking solution has the problem that it jumbles up the order of list elements. That might be okay for some applications, but other programs may depend on the order produced by the serial execution.

An alternative to locking is to restructure the code to accumulate the output lists in each subcomputation and concatenate them when the computations return. If one is careful, it is also possible to keep the order of elements in the list the same as in the serial execution. For the simple tree-walking code, code restructuring may suffice, but for many larger codes, disrupting the original logic can be time-consuming and tedious undertaking, and it may require expert skill,

```
1   #include <reducer_list.h>
2   bool has_property(Node *);
3   cilk::hyperobject<cilk::reducer_list_append<
        Node *> > output_list;
4   // ...
5   void walk(Node *x)
6   {
7     if (x)
8     {
9       if (has_property(x))
10      {
11        output_list().push_back(x);
12      }
13      cilk_spawn walk(x->left);
14      walk(x->right);
15      cilk_sync;
16    }
17  }
```

Figure 7: A Cilk++ parallelization of the code in Figure 4, which uses a reducer hyperobject to avoid data races.

making it impractical for parallelizing large legacy codes.

Cilk++ provides a novel approach [13] to avoiding data races in code with nonlocal variables. A Cilk++ *reducer hyperobject* is a linguistic construct that allows many strands to coordinate in updating a shared variable or data structure independently by providing them different but coordinated views of the same object. The state of a hyperobject as seen by a strand of an execution is called the strand's "view" of the object at the time the strand is executing. A strand can access and change any of its view's state independently, without synchronizing with other strands. Throughout the execution of a strand, the strand's view of the reducer is private, thereby providing isolation from other strands. When two or more strands join, their different views are combined according to a system- or user-defined reduce() method. Thus, reducers preserve the advantages of parallelism without forcing the programmer to restructure the logic of his or her program.

As an example, Figure 7 shows how the tree-walking code from Figure 4 can be parallelized using a reducer. Line 3 declares output_list to be a reducer hyperobject for list appending. The reducer_list_append class implements a reduce function that concatenates two lists, but the programmer of the tree-walking code need not be aware of how this class is implemented. All the programmer does is identify the global variables as the appropriate type of reducer when they are declared. No logic needs to be restructured, and if the programmer fails to catch all the use instances, the compiler reports a type error.

This parallelization takes advantage of the fact that list appending is associative. That is, if we append a list L_1 to a list L_2 and append the result to L_3, it is the same as if we appended list L_1 to the result of appending L_2 to L_3. As the Cilk++ runtime system load-balances this computation over the available processors, it ensures that each branch of the recursive computation has access to a private view of the variable output_list, eliminating races on this global variable without requiring locks. When the branches synchronize, the private views are reduced (combined) by concatenating the lists, and Cilk++ carefully maintains the proper ordering so that the resulting list contains the identical elements in the same order as in a serial execution.

6 Conclusion

Multicore microprocessors are now commonplace, and Moore's Law is steadily increasing the pressure on software developers to multicore-enable their codebases. Cilk++ provides a simple but

978-1-60558-497-3/09 $25.00 © 2009 ACM

effective concurrency platform for multicore programming which leverages almost two decades of research on multithreaded programming. The Cilk++ model builds upon the sound theoretical framework of multithreaded dags, allowing parallelism to be quantified in terms of work and span. The Cilkscreen race detector allows race bugs to be detected and localized. Cilk++'s hyperobject library mitigates races on nonlocal variables. Although parallel programming will surely continue to evolve, Cilk++ today provides a full-featured suite of technology for multicore-enabling any compute-intensive application.

Acknowledgments

Many thanks to the great team at Cilk Arts and to our many customers who have helped us refine the Cilk++ system. Thanks to Patrick Madden of SUNY Binghamton for proposing extensive revisions to the original manuscript.

References

[1] Gene Amdahl. The validity of the single processor approach to achieving large-scale computing capabilities. In *Proceedings of the AFIPS Spring Joint Computer Conference*, pages 483–485, April 1967.

[2] Michael A. Bender, Jeremy T. Fineman, Seth Gilbert, and Charles E. Leiserson. On-the-fly maintenance of series-parallel relationships in fork-join multithreaded programs. In *Proceedings of the Sixteenth Annual ACM Symposium on Parallel Algorithms and Architectures (SPAA 2004)*, pages 133–144, Barcelona, Spain, June 2004.

[3] Robert D. Blumofe and Charles E. Leiserson. Space-efficient scheduling of multithreaded computations. In *Proceedings of the Twenty Fifth Annual ACM Symposium on Theory of Computing*, pages 362–371, San Diego, California, May 1993.

[4] Robert D. Blumofe and Charles E. Leiserson. Scheduling multithreaded computations by work stealing. *Journal of the ACM*, 46(5):720–748, September 1999.

[5] Derek Bruening. *Efficient, Transparent, and Comprehensive Runtime Code Manipulation*. PhD thesis, Department of Electrical Engineering and Computer Science, Massachusetts Institute of Technology, 2004.

[6] Guang-Ien Cheng, Mingdong Feng, Charles E. Leiserson, Keith H. Randall, and Andrew F. Stark. Detecting data races in cilk programs that use locks. In *Proceedings of the Tenth Annual ACM Symposium on Parallel Algorithms and Architectures (SPAA '98)*, pages 298–309, Puerto Vallarta, Mexico, June 28–July 2 1998.

[7] Thomas H. Cormen, Charles E. Leiserson, Ronald L. Rivest, and Clifford Stein. *Introduction to Algorithms*. The MIT Press, third edition, 2009.

[8] Anne Dinning and Edith Schonberg. An empirical comparison of monitoring algorithms for access anomaly detection. In *Proceedings of the Second ACM SIGPLAN Symposium on Principles & Practice of Parallel Programming (PPoPP)*, pages 1–10. ACM Press, 1990.

[9] Anne Dinning and Edith Schonberg. Detecting access anomalies in programs with critical sections. In *Proceedings of the ACM/ONR Workshop on Parallel and Distributed Debugging*, pages 85–96. ACM Press, May 1991.

[10] Perry A. Emrath, Sanjoy Ghosh, and David A. Padua. Event synchronization analysis for debugging parallel programs. In *Supercomputing '91*, pages 580–588, November 1991.

[11] Mingdong Feng and Charles E. Leiserson. Efficient detection of determinacy races in Cilk programs. In *Proceedings of the Ninth Annual ACM Symposium on Parallel Algorithms and Architectures (SPAA)*, pages 1–11, Newport, Rhode Island, June22–25 1997.

[12] Yaacov Fenster. Detecting parallel access anomalies. Master's thesis, Hebrew University, March 1998.

[13] Matteo Frigo, Pablo Halpern, Charles E. Leiserson, and Stephen Lewin-Berlin. Reducers and other Cilk++ hyperobjects. In *Proceedings of the Twenty-First Annual ACM Symposium on Parallel Algorithms and Architectures (SPAA '09)*, Calgary, Canada, August 2009. To appear.

[14] Matteo Frigo, Charles E. Leiserson, and Keith H. Randall. The implementation of the Cilk-5 multithreaded language. In *Proceedings of the ACM SIGPLAN '98 Conference on Programming Language Design and Implementation*, pages 212–223, Montreal, Quebec, Canada, June 1998. Proceedings published ACM SIGPLAN Notices, Vol. 33, No. 5, May, 1998.

[15] Michael R. Garey and David S. Johnson. *Computers and Intractability*. W.H. Freeman and Company, 1979.

[16] David P. Helmbold, Charles E. McDowell, and Jian-Zhong Wang. Analyzing traces with anonymous synchronization. In *Proceedings of the 1990 International Conference on Parallel Processing*, pages II70–II77, August 1990.

[17] Institute of Electrical and Electronic Engineers. Information technology — Portable Operating System Interface (POSIX) — Part 1: System application program interface (API) [C language]. IEEE Standard 1003.1, 1996 Edition.

[18] Brian W. Kernighan and Dennis M. Ritchie. *The C Programming Language*. Prentice Hall, Inc., second edition, 1988.

[19] Chi-Keung Luk, Robert Cohn, Robert Muth, Harish Patil, Artur Klauser, Geoff Lowney, Steven Wallace, Vijay Janapa Reddi, and Kim Hazelwood. Pin: building customized program analysis tools with dynamic instrumentation. In *PLDI '05: Proceedings of the 2005 ACM SIGPLAN Conference on Programming Language Design and Implementation*, pages 190–200, New York, NY, USA, 2005. ACM Press.

[20] John Mellor-Crummey. On-the-fly detection of data races for programs with nested fork-join parallelism. In *Proceedings of Supercomputing'91*, pages 24–33. IEEE Computer Society Press, 1991.

[21] Barton P. Miller and Jong-Deok Choi. A mechanism for efficient debugging of parallel programs. In *Proceedings of the 1988 ACM SIGPLAN Conference on Programming Language Design and Implementation (PLDI)*, pages 135–144, Atlanta, Georgia, June 1988.

[22] Sang Lyul Min and Jong-Deok Choi. An efficient cache-based access anomaly detection scheme. In *Proceedings of the Fourth International Conference on Architectural Support for Programming Languages and Operating Systems (ASPLOS)*, pages 235–244, Palo Alto, California, April 1991.

[23] The MPI Forum. MPI: A message passing interface. In *Supercomputing '93*, pages 878–883, Portland, Oregon, November 1993.

[24] The MPI Forum. MPI-2: Extensions to the Message-Passing Interface. Technical Report, University of Tennessee, Knoxville, 1996. Available from: citeseer.ist.psu.edu/517818.html.

[25] Robert H. B. Netzer and Sanjoy Ghosh. Efficient race condition detection for shared-memory programs with post/wait synchronization. In *Proceedings of the 1992 International Conference on Parallel Processing*, St. Charles, Illinois, August 1992.

[26] Robert H. B. Netzer and Barton P. Miller. What are race conditions? *ACM Letters on Programming Languages and Systems*, 1(1):74–88, March 1992.

[27] Itzhak Nudler and Larry Rudolph. Tools for the efficient development of efficient parallel programs. In *Proceedings of the First Israeli Conference on Computer Systems Engineering*, May 1986.

[28] Dejan Perković and Peter Keleher. Online data-race detection via coherency guarantees. In *Proceedings of the Second USENIX Symposium on Operating Systems Design and Implementation (OSDI)*, Seattle, Washington, October 1996.

[29] Stefan Savage, Michael Burrows, Greg Nelson, Patric Sobalvarro, and Thomas Anderson. Eraser: A dynamic race detector for multithreaded programs. In *Proceedings of the Sixteenth ACM Symposium on Operating Systems Principles (SOSP)*, October 1997.

[30] Bjarne Stroustrup. *The C++ Programming Language*. Addison-Wesley, third edition, 2000.

[31] Bjarne Stroustrup. *C++ in 2005*. Addison-Wesley, 2005. Preface to the Japanese translation.

[32] Supercomputing Technologies Group, Massachusetts Institute of Technology Laboratory for Computer Science. *Cilk 5.4.2.3 Reference Manual*, April 2006. Available from: http://supertech.csail.mit.edu/cilk/home/software.html.

[33] William Wulf and Mary Shaw. Global variable considered harmful. *SIGPLAN Notices*, 8(2):28–34, 1973.

978-1-60558-497-3/09 $25.00 © 2009 ACM

Misleading Performance Claims in Parallel Computations

David H Bailey[*]
Lawrence Berkeley National Laboratory
Berkeley, CA 94720
dhbailey@lbl.gov

ABSTRACT

In a previous humorous note entitled "Twelve Ways to Fool the Masses ... ," I outlined twelve common ways in which performance figures for technical computer systems can be distorted. In this paper and accompanying conference talk, I give a reprise of these twelve "methods" and give some actual examples that have appeared in peer-reviewed literature in years past. I then propose guidelines for reporting performance, the adoption of which would raise the level of professionalism and reduce the level of confusion, not only in the world of device simulation but also in the larger arena of technical computing.

Categories and Subject Descriptors

D.2.8 [**Software Engineering**]: Metrics—*performance measures*

General Terms

Performance

Keywords

Parallel computing

1. INTRODUCTION

Some readers may have read (or heard about) a tongue-in-cheek article I wrote entitled "Twelve Ways to Fool the Masses When Giving Performance Reports on Parallel Computers" [1]. This article attracted an astonishing amount of attention at the time, including mention in the *New York Times* [5]. Evidently it struck a responsive chord among many professionals in the field of technical computing who shared my concerns. The following is a very brief summary of the "Twelve Ways":

1. Quote only 32-bit performance results, not 64-bit results, and compare your 32-bit results with others' 64-bit results.

2. Present inner kernel performance figures as the performance of the entire application.

3. Quietly employ assembly code and other low-level language constructs, and compare your assembly-coded results with others' Fortran or C implementations.

4. Scale up the problem size with the number of processors, but don't clearly disclose this fact.

5. Quote performance results linearly projected to a full system.

6. Compare your results against scalar, unoptimized, single processor code on Crays [prominent vector computer systems at the time].

7. Compare with an old code on an obsolete system.

8. Base megaflops operation counts on the parallel implementation instead of on the best sequential implementation.

9. Quote performance in terms of processor utilization, parallel speedups or megaflops per dollar (peak megaflops, not sustained).

10. Mutilate the algorithm used in the parallel implementation to match the architecture. In other words, employ algorithms that are numerically inefficient in order to exhibit artificially high megaflops rates.

11. Measure parallel run times on a dedicated system, but measure conventional run times in a busy environment.

12. If all else fails, show pretty pictures and animated videos, and don't talk about performance.

Both at the time this article appears, and in the intervening years, some have suggested that my intent in publishing the "Twelve Ways" article was to criticize computer vendors for their sometimes exuberant claims of performance. There have been some instances of dubious claims of performance by computer vendors, but my primary targets were scientists and engineers themselves, when presenting performance results of their applications. This is also the main focus of this

---

[*]Supported in part by the Director, Office of Computational and Technology Research, Division of Mathematical, Information, and Computational Sciences of the U.S. Department of Energy, under contract number DE-AC02-05CH11231.

©2009 Association for Computing Machinery. ACM acknowledges that this contribution was authored or co-authored by an employee, contractor or affiliate of the U.S. Government. As such, the Government retains a nonexclusive, royalty-free right to publish or reproduce this article, or to allow others to do so, for Government purposes only."
DAC'09, July 26-31, 2009, San Francisco, California, USA

978-1-60558-497-3/09 $25.00 © 2009 ACM

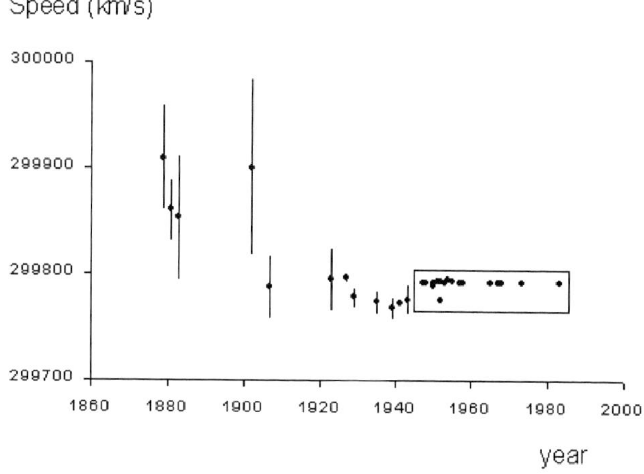

Figure 1: History of measurements of the speed of light

article. Here I present a brief summary of some of the most common abuses, and then close by presenting some principles that will help reduce confusion in the field. Portions of this material are condensed from a previous paper [2].

2. THE NEED FOR PROFESSIONAL DISCIPLINE

We do not need to look very far to see examples in other fields where well-meaning but sloppy practices in reporting experimental results have led to distortion and major embarrassment. One amusing example is the history of measurements of the speed of light [4]. Accurate measurements performed in the late 1930s and 1940s seemed to be converging on a value of roughly 299,760 km/s. After World War II, some researchers made more careful measurements, and thereafter the best value converged to its present-day value of 299,792.458 km/s (which is now taken as a standard, indirectly defining the meter and the second). Certainly few researchers seriously believe that the speed of light changed over a few years. Thus we are left with the conclusion that sloppy experimental practices and possibly some form of self-deluding "group-think" led researchers to converge on the earlier (incorrect) value.

Difficulties with experimental methods and dispassionate assessment of results have also been seen in the social sciences. In his book *The Blank Slate: The Modern Denial of Human Nature*, Harvard psychologist Steven Pinker chronicles the fall of the "blank slate" paradigm of the social sciences, namely the assertion that heredity and biology play no significant role in human psychology—all personality and behavioral traits are socially constructed. This paradigm prevailed in the field until the 1980s, when indisputable empirical evidence forced a change. The current consensus, based on latest research, is that humans at birth universally possess sophisticated facilities for language acquisition, pattern recognition and social life, and that heredity, evolution and biology are major factors in human personality—some personality traits are as much as 70% heritable.

Along this line, anthropologists, beginning with Margaret Mead in the 1930s, painted an idyllic picture of primitive societies such as South Sea Islanders, claiming that they had

little of the violence, jealousy, warfare or social hangups that afflict Western societies. Beginning in the 1980s, a new breed of anthropologists began to re-examine these findings. They found, contrary to earlier results, that these societies typically had murder rates several times higher than large U.S. cities, and death rates from inter-tribe warfare exceeding those of warfare among Western nations by factors of 10 to 100. What's more, complex, jealous taboos surrounded courtship and marriage—some even condoned violent reprisals if a bride were found not to be a virgin on her wedding night.

How did these scientists get it so wrong? Pinker and others have concluded that the principal culprit was sloppy experimental methodology and wishful-thinking analysis of results [6].

3. THE PRESSURE FOR REPORTING HIGH PERFORMANCE

For many years, parallel computers were almost exclusively the province of universities and government laboratories. More recently, processor manufactures such as Intel and Advanced Micro Devices, recognizing that basic physical laws no longer permit them to continue increasing the clock frequency, are instead offering increased performance by incorporating multiple processing cores on a single chip. In other words, like it or not even single-user personal computers are now parallel computers, and engineering workstations often incorporate 16, 32 or more processing cores. The message to both end users and to third-party technical software firms is clear: adopt your codes to run on parallel systems or risk being left behind.

The field of semiconductor device modeling and system engineering is no exception. In fact, the exploding complexity of designs by itself is requiring engineers to utilize the most powerful systems available for all stages of design, ranging from basic device physics to chip layout to full-system simulation. What's more, the increasing emphasis on reducing time to market by itself is a strong incentive to utilize highly parallel systems wherever possible. The price of failing to utilize this technology is shown by such unfortunate episodes as the Pentium divide fiasco, which in the end cost Intel over $500 million.

When this external pressure is added to the natural human tendency of scientists and engineers to be exuberant about their own work, it should come as little surprise that some have presented sloppy and potentially misleading performance claims in papers and conference presentations. And since the reviewers of these papers are themselves in many cases caught up in the excitement of this new technology, it should not be surprising that they have tended to be relatively permissive with questionable aspects of these papers.

Clearly the field of technical computing does not do itself a favor by condoning inflated performance reports, whatever are the motives of those involved. In addition to fundamental issues of ethics and scientific accuracy, the best way to insure that computer systems are effective for engineering design or other applications is to provide early feedback to manufacturers regarding their weaknesses. Once the reasons for less-than-expected performance rates on certain problems are identified, improvements can be made in the next generation.

In the next section, I will present several examples of ques-

978-1-60558-497-3/09 $25.00 © 2009 ACM

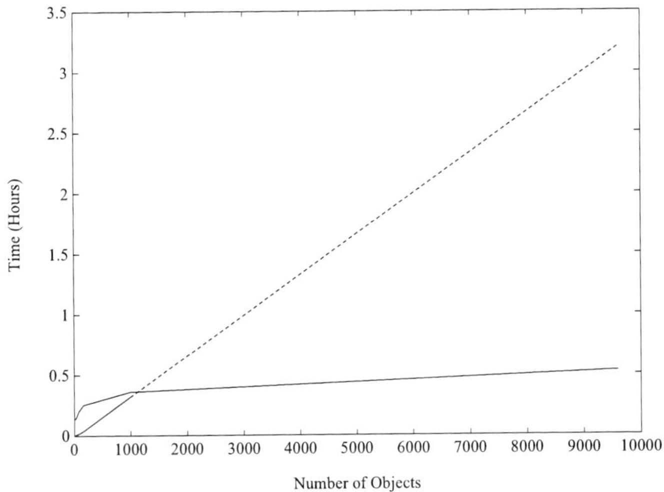

Figure 2: Timings of parallel (lower) and single-processor vector (upper) on a defense application

Total Objects	Parallel Run Time	Vector Run Time
20	8:18	0:16
40	9:11	0:26
80	11:59	0:57
160	15:07	2:11
990	21:32	19:00
9600	31:36	*3:11:50

Table 1: Raw data for plot in Figure 2. * denotes estimate.

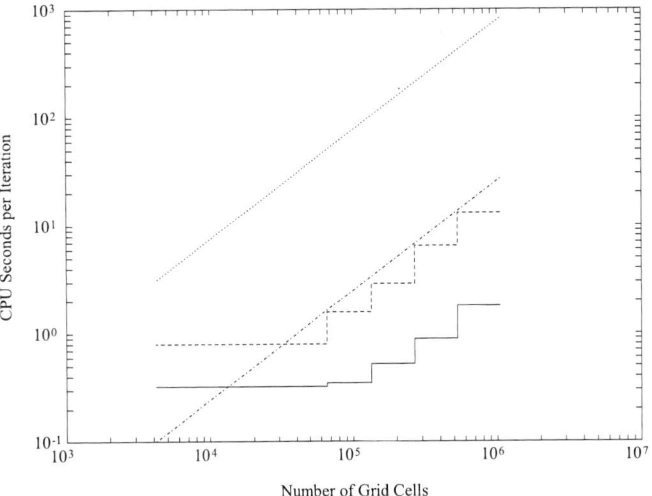

Figure 3: Timings of parallel system (solid and dashes) and single-processor vector system (dash-dots and dots) on a fluid dynamics application

tionable performance reporting that have appeared in peer-reviewed papers. I confess that these examples are a bit "dated" at this point in time, but they are nonetheless instructive in pointing out practices that should be avoided. My only purpose in citing these examples is to provide concrete instances of the performance issues in question. I do not wish for these citations to be misconstrued as criticisms of individual scientists, laboratories or supercomputer manufacturers. This is because I personally do not believe this problem is restricted to a handful of authors, laboratories and vendors, but that many of us in the field must share blame. For this reason I do not include references for these papers.

4. PLOTS

Plots can be effective vehicles to present technical information, particularly in verbal presentations. Further, plots are in many cases all that high-level managers (such as those with authority for computer acquisitions) have time to digest. Unfortunately, plots can also mislead an audience, especially if prepared carelessly or if presented without important qualifying information.

Figure 2 is a reconstruction of the final performance plot from a paper describing a defense application. The plot compares timings of the authors' parallel system code with timings of a comparable code running on a single-processor vector system. The plot appears to indicate an impressive performance advantage for the parallel system, on all problem sizes except perhaps a small region at the far left.

However, examination of the raw data used for this plot, which is shown in Table 1, gives a different picture. First of all, except for the largest problem size, all data points lie in the small region at the far left. In other words, most of the two curves shown are merely the linear connections of the next-to-last data points with the final points. Further, the single-processor vector system is actually faster than the parallel system for all sizes except for the largest problem size. At the least, it is clear that a logarithmic scale would have been far more appropriate for this data.

Other difficulties are encountered when one reads the text

accompanying this graph and table. First of all, the authors concede that the runs on the vector system used a code that "has not been optimized" for that system. Secondly, for the largest problem listed, the only one where the vector fails to out-perform the parallel system, the vector timing is, by the authors' admission, an estimate, an extrapolation based on a smaller run. In the paper, as in Figure 2, the lower curve leading out to the last point is dashed, possibly intending to indicate that this is an estimate, but this feature is not explained in either the caption or the text.

Figure 3 is a reconstruction of another performance plot from a paper describing a fluid dynamics application. This plot compares timings of the author's codes running on a parallel system with those of comparable codes running on a single-processor vector system. The two curves shown for each computer system represent a structured and an unstructured grid version of the code, respectively. As before, the plot appears to indicate a substantial performance advantage for the parallel system for all problem sizes and both types of grids.

Once again, careful examination of the text accompanying this plot places these results in a different light. First of all, the author admits that his parallel results have been linearly extrapolated to a full-sized system system from a smaller system. The author also acknowledges explains that the vector version of the unstructured code has not been tuned for vector computation. An additional difficulty with

978-1-60558-497-3/09 $25.00 © 2009 ACM

this plot can be seen by carefully examining the two vector system curves. In the original, as in Figure 2, these are precisely straight lines. Needless to say, it is exceedingly unlikely that a vector system code, scaled over nearly three orders magnitude in problem size, exhibits precisely linear timings. Thus one has to suspect that the two vector system "curves" are simply linear extrapolations from single data points. In summary, it appears that of all points on four curves in this plot, at most two points are clearly real timings. Also, it appears that the author is comparing 32-bit floating-point performance on the parallel system with 64-bit floating-point performance on the vector system.

5. PROJECTIONS AND EXTRAPOLATIONS

The practice of citing estimated and extrapolated performance results is, unfortunately, fairly widespread in the field. This may in part be an unintended consequence of limited budgets at many research institutions and private firms, where scientists and engineers often have to settle for scaled-down versions of highly parallel systems. As a result, authors frequently cite performance results that are merely linear projections from much smaller systems, often without the slightest justification.

The practice of linearly extrapolating one's performance results to a larger system is doubly perplexing because the question of whether various computer designs and applications will "scale" is in fact an important topic of current research in the field of parallel computing. It seems that many scientists and engineers using parallel computers are willing to assume as an established fact one of the most fundamental questions in the field!

We have already seen one instance of citing extrapolated results. In another paper in my files, the authors compare their application running on one moderately parallel system with comparable codes running on a single-processor vector system and a massively parallel system. Fortunately, all of the moderately parallel timings in the three tables are real timings. But out of a total of 33 figures listed for the vector system and the massively parallel system, more than half (17) are merely projections or estimates. There does not appear to be any attempt to mislead the reader, since the authors indicate which figures in each table are projections by means of asterisks. Nonetheless, one is left to wonder about how reliable these comparisons are, and whether they will always be quoted with the appropriate disclaimer.

Some authors have taken the practice of citing projections one step further. In one peer-reviewed paper in my files, the author states in his abstract that his code runs "at the speed of a single processor of a Cray-2 [a 1986 vintage 4-processor vector computer] on 1/4 of a CM-2 [a 1990 vintage massively parallel computer]". Some thirteen pages later, the author cites a timing on a Convex C210 [a 1990 vintage minicomputer] and then states "experience indicates that for a wide range of problems, a C210 is about 1/4 the speed of a single processor Cray-2." No further mention is made of the Cray-2.

It is well known that for any computer system, timings and megaflops rates can vary dramatically depending on how much effort has been expended in tuning for the particularly architecture being used. Thus any blanket performance ratio such as 1/4 is dubious. But the most troubling item here is the fact that the author, in the abstract of his paper, clearly implies a performance comparison with a Cray-2, even though he evidently has never run his code on a Cray-2.

6. COUNTING FLOPS

A common practice in the field of scientific computing is to cite performance rates in terms of millions of floating point operations per second (megaflops) or in billions of floating-point operations per second (gigaflops). For various reasons, some in the field have suggested that the practice of citing megaflops or gigaflops be abandoned. However, I am of the opinion that while direct timing comparisons are always preferred, megaflops or gigaflops rates may be cited if calculated and reported consistently.

Part of the confusion derives from the method used to determine the number of floating point operations (flops) performed. Many authors count the number of flops actually performed in their parallel implementations, a number usually obtained by analyzing the parallel source code. However, parallel implementations almost always perform significantly more flops than serial implementations. For example, some calculations are merely repeated in each processor. Using the actual number of flops performed on a parallel computer thus results in megaflops rates that are inflated when compared to rates obtained from corresponding single-processor implementations.

Another difficulty with basing megaflops or gigaflops rates on the actual parallel flop count is that this practice tacitly encourages researchers to employ numerically inefficient algorithms in their applications, algorithms often chosen mainly for the convenience of the particular architecture being used. It is easy to understand how such choices can be made, since it is widely accepted in the field that algorithmic changes are often necessary when porting a code to a parallel computer. But when this practice is carried too far, both the audience and the scientist may be misled.

Because of the potential for misleading comparisons, it is clear that a single standard flop count should be used when comparing rates for a given application. In my view, the most sensible flop count for this purpose is the minimal flop count — the value based on an efficient implementation of the best practical serial algorithms. In this way, one is free to use an implementation with a higher flop count on a particular architecture if desired, but no extra credit is given for these extra operations when megaflops rates are computed. This standard also acts as a deterrent to the usage of numerically inefficient algorithms.

7. OTHER ISSUES

Many authors report "speedup" figures for their parallel applications. Such figures indicate the degree to which the given application "scales" on a particular architecture. However, here also there is potential for the audience to be mislead, especially when speedup figures are based on inflated single processor timings.

For example, users of message passing parallel systems often base speedup figures on a single node timing of the multiple node version of the program. When running on a single node, the multiple node program needlessly synchronizes with itself and passes messages to itself. These "messages" are handled quite rapidly, since the operating system recognizes that these are local transmissions. Nonetheless, a significant amount of overhead is still required, and it is not unusual for the single node run time to increase by 20

978-1-60558-497-3/09 $25.00 © 2009 ACM

percent with the addition of message passing code.

Some authors present "scaled speedup" figures, where the problem size is scaled up with the number of processors. Such figures may be informative, but it is essential that authors who quote such figures clearly disclose the fact that they have scaled their problem size to match the processor count. It is also important that authors provide details of exactly how this scaling was done.

Another aspect of performance reporting that needs to be carefully analyzed is how the authors measure run time. Most of the scientists I have queried about this issue feel that elapsed wall clock time is the most reliable measure of run time, and that if possible it should be measured in a dedicated environment. By contrast, CPU time figures may mask extra elapsed time required for input and output.

One final aspect of performance reporting is the source of untold confusion in the field: are the results for 32-bit or 64-bit floating point arithmetic? Since on many systems, 32-bit computational performance rates are nearly twice as high as 64-bit rates, there is a temptation for authors to quote only 32-bit results, to fail to disclose that rates are for 32-bit data, and to compare their 32-bit results with others' 64-bit results. It is clear that 32-bit/64-bit confusion is widespread in performance reporting, since we have seen several examples already.

In my view, quoting 32-bit performance rates is permissible so long as: (1) this data type is clearly disclosed and (2) a brief statement is included explaining why this precision is sufficient. Along this line, it should be kept in mind that with new much more powerful computer systems, it is now possible to attempt much larger problems than before. Inevitably, these much larger calculations greatly exacerbate any numerical sensitivities that may exist in the code. As a result, numerical issues that previously were not significant now are significant, and some programmers are discovering to their dismay that higher precision is necessary to obtain meaningful results. In fact, there are many applications where even 64-bit floating-point arithmetic is not sufficient, and where much higher precision is required [3].

I suspect that in the majority of cases where the authors do not clearly state the data type, the results are indeed for 32-bit data. One example is an award-winning paper, where the authors never state whether their impressive performance rates are for 32-bit or 64-bit calculations, at least not in any place where a reader would normally look for such information. That their results are indeed for 32-bit data can however be deduced by a careful reading of their section on memory bandwidth, where we read that operands are four bytes long.

8. PROPOSED GUIDELINES

Clearly this field needs a detailed set of guidelines for reporting supercomputer performance, guidelines which are formally adopted and widely disseminated to authors and reviewers. Virtually every field of science has found it necessary at some point to establish rigorous standards for the reporting of experimental results, and ours should be no exception. To that end, I propose the following. These guidelines focus on computational performance, since that is the topic of this paper and apparently the most frequent arena of confusion. However, it is hoped that the spirit of these guidelines will be followed by researchers reporting performance in other areas of technical computing, such as in mass storage and local area networks.

1. If results are presented for a well-known benchmark, comparative figures should be truly comparable, and the rules for the particular benchmark should be followed.

2. Only actual performance rates should be presented, not projections or extrapolations. For example, performance rates should not be extrapolated to a full system from a scaled-down system. Comparing extrapolated figures with actual performance figures, such as by including both in the same table, is particularly inappropriate.

3. Comparative performance figures should be based on comparable levels of tuning.

4. Direct comparisons of run times are preferred to comparisons of megaflops rates or the like. Whenever possible, timings should be true elapsed time-of-day measurements (this might not be possible in some "production" environments).

5. Megaflops or gigaflops rates should be computed from consistent flop counts, preferably flop counts based on efficient implementations of the best practical serial algorithms. One intent here is to discourage the usage of numerically inefficient algorithms, which may exhibit artificially high performance rates on a particular parallel system.

6. If speedup figures are presented, the single processor rate should be based on a reasonably well tuned program without multiprocessing constructs. If the problem size is scaled up with the number of processors, then the results should be clearly cited as "scaled speedup" figures, and details should be given explaining how the problem was scaled up in size.

7. Any ancillary information that would significantly affect the interpretation of the performance results should be fully disclosed. For example, if the results are for 32-bit rather than for 64-bit data, or if assembly-level coding was employed, or if only one processor of a conventional system is being used for comparison, these facts should be clearly stated.

8. Due to the natural prominence of abstracts, figures and tables, special care should be taken to insure that these items are not misleading, even if presented alone. For example, if significant performance claims are made in the abstract of the paper, any important qualifying information should also be included in the abstract.

9. Whenever possible, the following should be included in the text of the paper: the hardware, software and system environment; the language, algorithms, the datatypes and programming techniques employed; the nature and extent of tuning performed; and the basis for timings, flop counts and speedup figures. The goal here is to enable other scientists and engineers to accurately reproduce the performance results presented in the paper.

978-1-60558-497-3/09 $25.00 © 2009 ACM

9. CONCLUSIONS

The examples I have cited above are somewhat isolated in the literature, and I see no evidence that the problem of inflated performance reporting is out of control. However, clearly those of us in the technical computing field would be wise to arrest any tendency in this direction before we are faced with a significant credibility problem. As was mentioned above, scientists in many other disciplines have found it necessary to adopt rigorous standards for reporting experimental results, and ours should be no exception. It is my hope that this article, with the proposed guidelines above, will stimulate awareness and dialogue on the subject and lead to standards in the field.

10. REFERENCES

[1] D. H. Bailey. Twelve ways to fool the masses when giving performance results on parallel computers. *Supercomputing Review*, pages 54–55, August 1991.

[2] D. H. Bailey. Misleading performance reporting in the supercomputing field. *Scientific Programming*, 1:141–151, 1992.

[3] D. H. Bailey. Resolving numerical anomalies in scientific computation. `http://crd.lbl.gov/~dhbailey/dhbpapers/numerical-bugs.pdf`, 2008.

[4] A. N. Cutler. A history of the speed of light. `http://www.sigma-engineering.co.uk/light/lightindex.shtml`, 2001.

[5] J. Markoff. Measuring how fast computers really are. *New York Times*, page 14F, September 1991.

[6] S. Pinker. *The Blank Slate: The Modern Denial of Human Nature*. Viking, New York, 2002.

Massively Parallel Processing: It's Déjà Vu All Over Again

Steven P. Levitan
Electrical and Computer Engineering
University of Pittsburgh
Pittsburgh, PA 15261
+1-412-648-9663

levitan@pitt.edu

Donald M. Chiarulli
Department of Computer Science
University of Pittsburgh
Pittsburgh, PA 15260
+1-412-624-8839

don@cs.pitt.edu

ABSTRACT

In this paper we will identify those aspects of the concurrent computing landscape that have changed since the 1980's and how those changes might impact the efficacy of parallel computing as we move from single- to multi- to many- and to massive numbers of processing cores.

Categories and Subject Descriptors

C.1.4 [Parallel Architectures]. D.1.3 [Concurrent Programming]: Distributed programming, parallel programming. B.7.1 [Integrated Circuits]: Microprocessors, Very Large Scale Integration (VLSI).

General Terms

Algorithms, Performance, Design.

Keywords

Multicore, Massively Parallel Processing, Parallel Architectures and Algorithms.

1. INTRODUCTION

Based on the ideas of Amdahl's law [1], Gustafson's rule [9], Stone's performance metrics [16], Synder's corollary of modest potential [15], and Finnegan's approach [5], the promise of massively parallel processing (MPP) is only superseded by the hyperbole of the discussions to date. What is clear is that while some applications naturally benefit from parallel processing others simply do not. And, simple extrapolation of increased performance with increasing hardware is naïve and perhaps dangerous [2].

This means that, from the user perspective, we will not see significant qualitative changes in system performance by applying parallelism to current applications. Rather, all is not lost. What have changed over the 30 years of parallel computing history are the user's expectations. Rather than single algorithms, there are significant new application domains such as conversational interfaces, personal sensor networks, and various forms of intelligent assistants, that can be enabled by highly integrated and massively parallel hardware.

Permission to make digital or hard copies of part or all of this work for personal or classroom use is granted without fee provided that copies are not made or distributed for profit or commercial advantage and that copies bear this notice and the full citation on the first page. To copy otherwise, to republish, to post on servers or to redistribute to lists, requires prior specific permission and/or a fee.

DAC'09, July 26-31, 2009, San Francisco, California, USA

Over the next decade we must go beyond the use of threads for small scale parallelism, and process migration for medium scale systems to concurrent implementations of computationally intensive applications.

There are three obvious differences between the integrated multicore environments proposed for the next decade and the parallel supercomputers of the previous century. First, the hardware environment on-chip is different than the supercomputer environment in terms of communications latency, memory bandwidth, and I/O overhead relative to CPU processing power. Second, the applications themselves will not be the scientific supercomputer applications that were the focus of much research in the past. While many applications will be based on the signal, image, and graphic processing algorithms developed for parallel super-computers these new applications will be focused on meeting individual user's needs. Finally, these new systems will be general purpose personal computers. They will be running many applications in a multiprocessing environment that is likely to be on a mobile platform versus the dedicated applications that in the past ran on special purpose processors in "batch mode."

In the rest of this paper we introduce the rational for multicore processors, using concurrency rather than machine complexity or clock speed to increase performance. Then we present some of the problems with simple models for the performance gains predictions that have been used in the past and describe the software and hardware issues that are key to performance gains. Finally, we discuss the differences between multicomputer based parallel processing and the multicore environment.

2. THE DRIVE TO MULTICORE

The drive to multicore is propelled by the goal of increased performance and the "walls" single CPU architectures have encountered in terms of instruction level parallelism (ILP), clock frequency, and ultimately power. Architectural techniques to provide more performance have reached a point of diminishing returns and simply turning up the clock speed is no longer an option. Therefore, it makes sense to back-off from the high ILP performance, high clock speed, and deeply pipelined designs to more moderate, lower speed, lower voltage, and lower power CPU designs and utilize parallelism to achieve the performance gains we are seeking.

However, the performance gains of a multicore processor are predicated on the effective utilization of the cores to perform useful work. Therefore, it is interesting to examine the effectiveness of the potential parallelism for given applications and the factors that limit the efficiency of multiprocessors. Over

978-1-60558-497-3/09 $25.00 © 2009 ACM

the last several decades, the question of "how much speedup is possible?" has been examined for a large variety of architectures, algorithms, and fundamental assumptions about the kinds of programs that would be run in a parallel computing environment. As we show below, the differences in the assumptions made by researchers has been quite varied.

3. SPEEDUP ANALYSES

One way to analyze advantages of parallel execution is to measure the speedup over an equivalent serial implementation of the same algorithm: $Speedup = Time_{serial}/Time_{parallel}$. In 1967 Gene Amdahl noted that, if any part of the algorithm was "intrinsically serial" then that fraction would limit the overall performance of any parallel implementation [1]. To see this effect, we use the time to do the work on a serial machine with S as the fraction of that work that is intrinsically serial and P as the fraction that is parallelizable, and hypothesize that the speedup for a parallel machine would be: $Speedup = (S + P)/(S + P/N)$, with $P = 1 - S$. This gives: $Speedup = 1/(S + (1 - S)/N)$. This function is plotted in Figure 1 for values of S ranging from 10% to 90% and N from 1 to 1024. Even in the best case shown, for S = 10% and N = 1024 the speedup is only 10x. This pessimistic result comes from the fact that the serial fraction of the code cannot be sped up, no matter how many processors you can recruit to solve the problem.

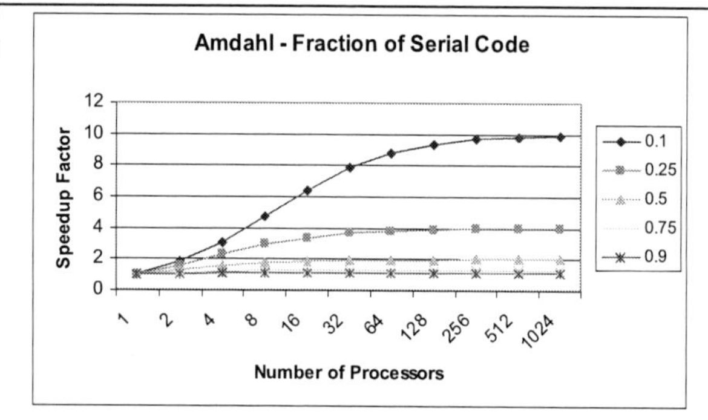

Figure 1: Speedup as a function of the number of processors, for a fixed problem size and various fractions of intrinsically serial code.

A different view has been taken by many researchers who claim that the real win for parallel processing is that it allows you to scale the problem size to larger instances, and then solve the larger problem quickly with a large number of processors. This interpretation was taken to extreme by D. Cohen in 1981 who reported "…in the 70s … scientists discovered… [that] problems with [high] time complexity can be solved in [little] time… using a number of processors which is a function of the problem size..." And further that "…the cost of VLSI processors decreases exponentially. … Hence, the application of an exponential number of processors does not cause any cost increase…" He attributed this result to Finnegan [5].

For our interpretation of Finnegan's rule we let the work grow with the number of processors while keeping the amount of serial work constant, this gives: $Speedup = (c + P)/(c + P/N)$ which is shown in Figure 2

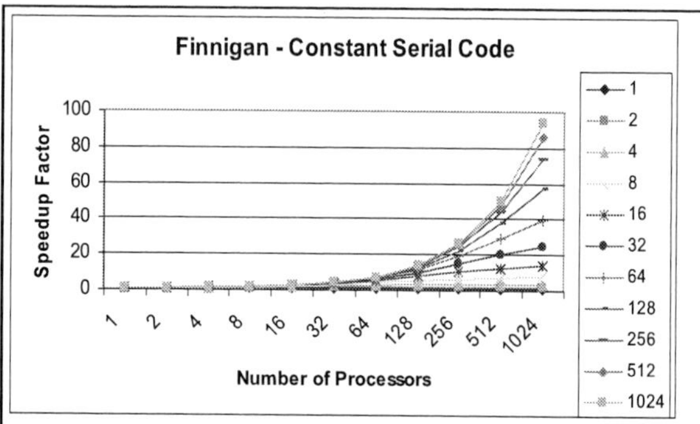

Figure 2: Speedup as function of the number of processors, for various data-set sizes given a fixed quantity of intrinsically serial code.

for a value of C=10. This gives a much more optimistic speedup of almost 100 for a 1024 processor architecture.

Another optimistic analysis was performed by Gustafson and Barsis [9]. In that work, they use the time to solve a large problem in a parallel environment, and then examine what fraction could not be run in parallel. From that information, they hypothesize the time the same problem would take in a serial environment. $Speedup = (S' + P' \cdot N)/(S' + P')$ where the prime indicates the fact that these are the relative serial and parallel fractions of the parallel code implementation. Thus, the hypothesized serial time is computed based on the data scale factor N. This gives a very optimistic: $Speedup = N + (1 - N) \cdot S'$, which is to say nearly linear speedup with problem and architecture size. To understand this relation, shown in Figure 3, one can consider an application running on a 1024 node parallel processor and taking 2 seconds. If half of that code ran as serial code and half as parallel code, then one could assume that a uniprocessor would take 1+1024 seconds to solve the same problem. Thus giving a speedup of 512x.

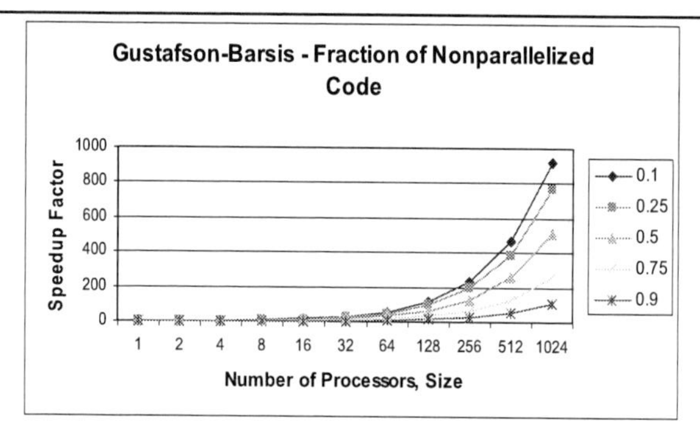

Figure 3: Speedup, in terms of the relative slowdown of a serial processor, as a function of the number of processors and data set size, for various fractions of intrinsically serial code.

Looking beyond the runtime for a fixed size problem, Snyder and others note that what matters is the amount of useful work that can be done in a fixed time as the number of processors increase [15]. Figure 4 shows that for algorithms with

978-1-60558-497-3/09 $25.00 © 2009 ACM

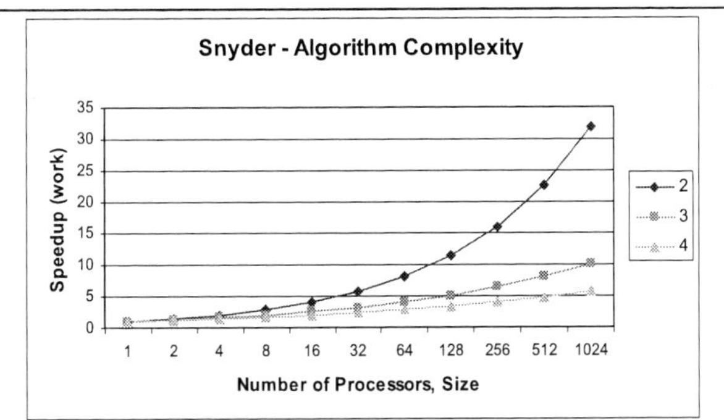

Figure 4: Speedup, in terms of work completed, as a function of the number of processors for various exponents, x, for polynomial time algorithms $O(N^x)$.

polynomial complexity, $O(N^x)$, the increase in the amount of work done grows as the inverse power ($1/x$). As shown in Figure 4, a 1024 node multicore machine solving an $O(N^2)$ problem will only be able to complete 35x more work in the same time as a uniprocessor.

A further complexity in predicting runtime performance is due to the communications overhead of the cooperating processes or threads in many parallel applications. Stone looks at runtime in terms of both computation and communication costs [16]. He defines the ratio R/C as the ratio of runtime, R, to communication time, C, for each block of code that could be run in parallel.

Stone notes that using massive parallelism where short blocks of code communicate frequently can lead to very poor speedups if the communications take finite time. Given M tasks and N processors, and for simplicity it is assumed that all tasks need to communicate with each other, the speedup can be characterized as:

$$\frac{RM}{(\frac{RM}{N}+\frac{CM^2}{2}-\frac{CM^2}{2N})} = \frac{N\,R/C}{R/C + \frac{M(N-1)}{2}}$$

Where we can see, on the left, the numerator is the total runtime of all the tasks on a serial processor and the denominator has the run time for the parallel tasks and the communications time for all tasks except those tasks that share processors. The re-ordered equation, on the right, shows that only if R/C is much larger

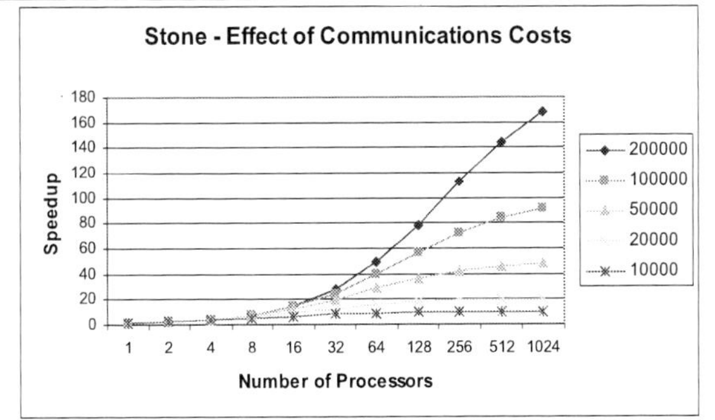

Figure 5: Speedup as a function of number of processors for various values of the ratio of runtime to communication time, R/C, for 2000 processes.

than MN will the speedup grow linearly with N. This effect is shown in Figure 5, where for all the plots, M, the number of processes is kept constant at 2000.

This ratio of computation to communication time is critical to achieving any significant speedups in parallel processors and therefore it has been the subject of research from both a hardware and a software perspective. First, in terms of computer architecture, the question is: How can we build interconnect systems that minimize the latency of communications between processors? And second, in terms of parallel algorithm design, the question is: How can we partition our problems such that we can minimize the communications between processes?

4. COMMUNICATIONS
4.1 Interconnection Networks

The design of system level interconnection (SLI) networks is one of the classic problems in computer architecture. The space of possible networks can be defined in several ways. One classification scheme is based on the four axes: operation mode (synchronous vs. asynchronous), control strategy (centralized vs. distributed), switching method (circuit vs. packet), and network topology [7]. Topologies can be classified as static or dynamic, single or multi-hop, directed graph, hierarchical, or multi-stage [14].

One key metric for the design of interconnection networks is their scaling behavior. For the desirable case of single hop fully-connected networks the space of network topologies span the range from pair-wise direct connections, which are limited in scale by their need for $O(N^2)$ wires, to many-to-many bus-style interconnections, which are limited by their fan-in/fan-out loading that scales as $O(N^2)$. Corresponding to those choices, the control and arbitration of these networks range from simple pair-wise handshaking to global arbitration of the shared resource.

Traditionally network designers have addressed these scaling issues by choosing to use multi-hop or hierarchical switched networks. For example, in a crossbar switch interconnection network $O(N^2)$ switches replace the $O(N^2)$ wires of a fully interconnected network, with bandwidth, routing, and control latency managed at the switch in the network core. To make this approach scalable, the switching architecture is broken down in various ways to create a network of smaller switches such that each message makes multiple hops within the network core. This is the most common architecture of modern network-on-chip (NOC) designs for providing interconnection between components at the system level [3].

Both the earlier work and newer analysis, such as [10] have shown that nearest-neighbor, mesh interconnection networks do not scale well for large systems due to the large diameter of nearest neighbor topologies. Rather, networks that have higher node degrees but lower diameters such as hyper-cubes, multi-grids, pyramids, or meshes-of-trees topologies have better performance for moderate increase in costs. In fact, even random networks turn out to be better than meshes for many applications [6].

A good deal of research has been done on mapping specific parallel algorithms onto specific interconnection networks [11][12]. While this exercise has highlighted the power and

978-1-60558-497-3/09 $25.00 © 2009 ACM

weaknesses of particular networks it has not converged on a single "universally best" network for all applications. Rather the communication graph of each application often determines the optimal embedding. These graphs come from the method used to decompose the original problem.

4.2 Parallel Programming Paradigms

From a high level perspective there are four classic methods for decomposing a problem for solution on a parallel computer. These are:

- Decomposition in "space" – Map some aspect of the data representation of the problem to individual processors. This can be straightforward such as putting a 2D weather map onto a mesh of processors or mapping nodes of a distributed data-structure such as a graph onto subsets of processors.

- Decomposition in "time" – Pipeline partial results either at a fine granularity, as in the vector unit of a single CPU, or across processing units in signal processing pipelines or systolic arrays.

- Decomposition by "function" – Assign tasks to processors based on specific capabilities of the heterogeneous processors, or to balance workloads.

- Decomposition by "instance" – Run multiple instances of problem, with different input conditions, on different processors.

While the last technique is trivial, it is probably the most common and effective use of parallel computing clusters. On the other hand, the first technique encompasses most of the research on parallel algorithm (and language) design.

One thing that is clear is that the same problem can be solved with algorithms that might or might not lend themselves to parallel speedups. The amount of "intrinsically serial code" in an implementation is not the same as any intrinsically serial part of the original problem.

What we have learned is that finding good parallel algorithms is not easy, and once those are found, mapping the algorithms onto an interconnection network to minimize the amount and cost of communications is also difficult. However, it is not clear that the success of multicore systems will simply be based on how well they run particular parallel algorithms.

We now turn to looking at three key differences between parallel processing in the past and the prospects for multicore architectures of the future.

5. WHAT IS DIFFERENT NOW?

5.1 The Hardware

The first question we consider is: "Does the change in size from macro-scale to chip-scale parallel computing really change anything?"

The obvious advantage of multicore platforms is the high level of integration. Processor cores are smaller, logic is faster, and individual threads run faster. But, as we showed above, none of this means anything if the interconnection networks cannot efficiently move the data between computational cores. Decreased size and increased scale give a two edged sword, on

the one edge interconnection length is, on average, a function of the die size, while on the other edge increased logic density puts N^2 more resources at each incremental unit of distance away. This problem is exacerbated by the trend to keep clock rates fixed even as technology scales down.

What sort of interconnection networks can we design in this technology regime that will support efficient, low latency communications? While design paradigms have shifted in multicore environments, the same impediments that have historically blocked a general-purpose parallel architecture still exist. We are likely to continue to see more application driven architectures that attempt to balance the market size for the application with development costs. Algorithm development will precede these architectures, as new user applications emerge.

On the positive side, there is a significant paradigm shift that we can exploit in on-chip interconnection networks for multicore processors. Looking at the interconnection networks operating at the board and rack level of modern supercomputers, we typically find hierarchical switched networks that store and forward packets of data over shared physical links. These networks provide the low diameter requirements to keep latency low, but often are bandwidth limited due to cost constraints.

Many academic proposals for large-scale multicore chips are based on Network-on-Chip architectures (NOC's) that scale-up the switched hierarchical networks of supercomputers to chip-scale solutions. For these switched networks, one of the primary motivations is to efficiently share the physical links in the networks, which are a conserved resource in traditional supercomputer designs. This efficiency is gained at the price of end-to-end message latency associated with multiple store-and-forward operations as each message traverses the network.

On the other hand, slotted rings are common in current commercial multicore processors such as the Intel Larrabee [13] and IBM Cell Broadband Engine [8] where bandwidth is not so costly and network diameters are still moderate.

It is not clear that either design paradigm holds in highly integrated systems. In this environment, the number of physical interconnection is only related to the number of metal layers and shorter distances reduce the cost of delivering data across any single interconnection. This is not to say that multicore interconnection network would not include shared links and hierarchy. The key is to find a network management paradigm that can manage this sharing in a way that exploits the technology to minimize latency.

There are two design paradigms that seem capable of meeting this requirement. The first is a traditional switched approach paired with some sort of look-ahead mechanism to reduce the network setup time. The second is to make the network purely passive, for example a hierarchy of multiple fan-in/fan-out structures with all control operations moved to edges and reduced to a combination of bus/route selection and access arbitration [4].

5.2 The Software

The second question is really: "Have the user's needs changed?"

There is in computing a synergistic relationship between available technology and the user's needs and expectations. The

978-1-60558-497-3/09 $25.00 © 2009 ACM

classic modality is that new technology created new markets. However, today it's just as likely that it is user demands and expectations that are pushing the development of new technologies to meet these needs.

To understand how the answer to this question can drive the development of multicore parallel architectures, we need to make the connection between these new demands and the potential for new parallel system solutions that pair these applications with multicore parallel computing engines.

Without suggesting that we have a crystal ball about the "next big thing," there are certainly applications that seem to have intense computation requirements, whatever algorithms ultimately emerge as a solution, that are close and personal to end user. Spoken interfaces are an obvious application. However, more generally there is a great variety of end-user oriented applications that may emerge which range from cognitive prosthesis and quality-of-life enhancement for older individuals, to data-fusion and processing for personal "sensor networks", to wearable or embedded applications in our homes and transportation systems, and personal data miners that watch what we do and then suggest what we want.

The answer to the role of multicore architectures in the future lies in the relationship between the system architecture and the solutions that meet these new user demands.

5.3 The Application Domain

Rather than consider "what algorithms will we be running?" the last question is "What will these systems really be doing?"

The answer to that, we think, is that they will be interacting with individual users in their day-to-day lives. To that end, we should not be thinking about individual algorithms but rather higher level computational metaphors that reflect the need to work in the user's problem domain with constructs that directly support computational intensive applications. These domains include:

- Interpretation – reacting to commands by delegating the execution of abstract protocols, controlling physical and computational systems.

- Transformation – partitioning and translating different representations, performing data analysis, signal processing, and communication.

- Simulation – using cooperating processes to predict the behavior of physical, artificial, and social systems.

- Optimization and search – exploring state-spaces representations and processes with distributed resources.

Each of these domains, as well as others, provides a rich field of opportunities for using concurrency to solve problems more effectively, consuming less power, than a uniprocessor solution.

The difference between the problems that were solved by parallel processing systems of the past decades and the multicore processors of the future will be in their value to individual consumers.

6. ACKNOWLEDGMENTS

This work was supported, in part, by the National Science Foundation under grant CCF-0541150.

7. REFERENCES

[1] Amdahl, Gene, "Validity of the Single Processor Approach to Achieving Large-Scale Computing Capabilities", AFIPS Conference Proceedings, (30), pp. 483-485, 1967.

[2] Bailey, David H., "Misleading Performance Reporting in the Supercomputing Field," RNR Technical Report RNR-92-005, December 1, 1992. also in *Scientific Programming, vol. 1.*, no. 2 (Winter 1992), pg. 141–151.

[3] Benini, L. and De Micheli, G. ,"Networks on Chips: A New SoC Paradigm," *IEEE Computer*, Jan. 2002, pp. 70-78.

[4] Chiarulli, D.M., Levitan, S.P., Melhem, R.G., Teza, J.P., Gravenstreter, G., "Partitioned Optical Passive Star (POPS) Multiprocessor Interconnection Networks with Distributed Control, *IEEE Journal on Lightwave Technology, Vol. 14*, No. 7, pp. 1601-1612, July 1996.

[5] Cohen, D, "The VLSI approach to computational complexity," in *CMU Conference on VLSI Systems and Computations*, Kung, Sproul Steele ed. Computer Science Press, Rockville, MD., pp. 124-125 1981

[6] Falman, S.E., "The hashnet interconnection scheme," Carnegie-Mellon University, Dept. of Computer Science (CMU-CS-80-125) 1980.

[7] Feng, T.Y, "A Survey of Interconnection Networks," *IEEE Computer*, December, 1981, pp 12-27.

[8] Gschwind, M., Hofstee, H. P., Flachs, B., Hopkins, M., Watanabe, Y., and Yamazaki, T. 2006, "Synergistic Processing in Cell's Multicore Architecture," *IEEE Micro* 26, 2 (Mar. 2006), 10-24.

[9] Gustafson, John L., "Reevaluating Amdahl's Law," *Communications of the ACM 31*(5), 1988. pp. 532-533.

[10] Kim, J., Balfour, J., and Dally, W., "Flattened butterfly topology for on-chip networks," *IEEE Computer Architecture Letters,* vol. 6, 2007.

[11] Leighton, F.T., *Introduction to Parallel Algorithms and Architectures: Arrays, Trees, and Hypercubes*, Morgan Kaufman, San Mateo, CA, 1992.

[12] Levitan, S.P., "Measuring Communication Structures in Parallel Architectures and Algorithms," (in) *The Characteristics of Parallel Algorithms*, L. Jamieson, D. Gannon, R. Douglass, Eds., Cambridge, MA, MIT Press, 1987, pp. 101-137.

[13] Seiler, L., Carmean D., Sprangle, E., Forsyth, T., Abrash, M., Dubey, P., Junkins, S., Lake, A., Sugerman, J., Cavin, R., Espasa, R., Grochowski, E., Juan, T., Hanrahan, P., "Larrabee: A Many-Core x86 Architecture for Visual Computing," *ACM Transactions on Graphics*, 27, 3, 2008.

[14] Siegel, H.J. *Interconnection Networks for Large-scale Parallel Processing: Theory and Case Studies*, McGraw-Hill, 1990.

[15] Snyder, Lawrence, "Type Architectures Shared Memory and the Corollary of Modest Potential," Ann. Rev. Comput. Sci. 1986. 1:289-317.

[16] Stone, Harold, *High-Performance Computer Architecture*, Addison-Wesley, MA, 1987. (see chapters 1 and 6).

Provably Good and Practically Efficient Algorithms for CMP Dummy Fill

Chunyang Feng[1], Hai Zhou[1,2], Changhao Yan[1], Jun Tao[1], Xuan Zeng[1*]
[1]State Key Lab of ASIC & System, Microelectronics Dept., Fudan University, China
[2]EECS, Northwestern University, U.S.A.

Abstract—To reduce chip-scale topography variation in Chemical Mechanical Polishing (CMP) process, dummy fill is widely used to improve the layout density uniformity. Previous researches formulated the dummy fill problem as a standard Linear Program (LP). However, solving the huge linear program formed by real-life designs is very expensive and has become the hurdle in deploying the technology. Even though there exist efficient heuristics, their performance cannot be guaranteed. In this paper, we develop a dummy fill algorithm that is both efficient and with provably good performance. It is based on a fully polynomial time approximation scheme by Fleischer [4] for covering LP problems. Furthermore, based on the approximation algorithm, we also propose a new greedy iterative algorithm to achieve high quality solutions more efficiently than previous Monte-Carlo based heuristic methods. Experimental results demonstrate the effectiveness and efficiency of our algorithms.

Categories and Subject Descriptors:

J.6 [Computer-Aided Engineering]: Computer-Aided Design

General Terms: Design, Algorithms

Keywords: Design for Manufacturability, Dummy Fill Problem, Covering Linear Programming

I. INTRODUCTION

Chemical Mechanical Polishing (CMP) is widely used as the primary planarizing technique in the fabrication of integrated circuits. Despite being a predominant planarizing technique, CMP is known to suffer from undesired pattern dependent problems. Previous studies show that post-CMP topography is strongly dependent on the underlying feature density [11]. To achieve layout density uniformity, dummy fill is a highly recommended technique by foundries to increase the density of sparse regions.

In general, layout density control consists of two phases: density analysis and fill synthesis. Density analysis determines the area available for filling, while fill synthesis computes the amount of dummy fills for each density tile of the layout [3]. In this paper, we address the main problem of the dummy fill synthesis.

The existing work in the area of fill synthesis can be classified into two categories [10]: *linear-programming* (LP) based approaches and Monte-Carlo (or greedy) based heuristic approaches. Two objectives, the Min-Var objective and Min-Fill objective are proposed. The Min-Var objective seeks the most uniform density distribution possible, and the Min-Fill objective seeks to minimize the dummy feature

*Corresponding author. E-mail: xzeng@fudan.edu.cn

Permission to make digital or hard copies of part or all of this work for personal or classroom use is granted without fee provided that copies are not made or distributed for profit or commercial advantage and that copies bear this notice and the full citation on the first page. To copy otherwise, to republish, to post on servers or to redistribute to lists, requires prior specific permission and/or a fee.
DAC'09, July 26-31, 2009, San Francisco, California, USA

insertion cost while satisfies a density uniformity constraint [8]. Kahng et al. [7] proposed the first LP formulation for the Min-Var objective. Tian et al. [13] gave the first LP formulation for the Min-Fill objective. Although LP solvers produce optimal solution for these formulations, the runtime is too expensive, in the order of n^3, where n is the number of variables in the LP. As stated in [8], in the fixed-dissection paradigm, if the window size $w = 200\mu m$, and each window is divided into $r = 4$ steps, then for the problem with a chip of $20mm$ on each side, there would be 160000 variables. This renders the LP-based method infeasible. This problem will get even worse with the density window size getting smaller. Recent experiments found that a window of about $50\mu m$ to $60\mu m$ is necessary to obtain the pattern density for copper CMP process [9]. Therefore, solving the linear program has become the computational bottleneck in the fill synthesis problem.

To alleviate this difficulty, several Monte-Carlo or greedy based heuristic approaches were proposed [1], [2], [14]. These methods select a tile according to certain criteria in each iteration, and fill it with a predetermined amount of dummy fills. These methods have been shown to take less runtime. However, due to their heuristic nature, their performance cannot be guaranteed or even bounded. Moreover, there is no clear guidance for the determination of the predefined filled amount during each iteration. If a large amount of fill is added into a tile during each iteration as in the greedy method [2], an excessive amount of total fill could be inserted in the fill. If only a single filling geometry is added per iteration as in the Monte Carlo approach [1], the runtime might become too long.

To address these issues in the existing approaches, the present work

- proposes, for the first time, a *covering linear program* (CLP) formulation for the Min-Fill objective of the dummy fill problem, and presents a Fully Polynomial Time Approximation Scheme (FPTAS) for the Min-Fill CLP problem based on the work of Fleischer [4]. Experimental results show that the new FPTAS algorithm is highly practical.
- develops a new greedy iterative method based on the concept of the new FPTAS algorithm for solving the Min-Fill problem, which improves the existing heuristic methods both in performance and runtime.

The rest of this paper is organized as follows. In Section II, we describe the Min-Fill LP formulation of the dummy fill problem. In Section III, we first give the formal formulation of the Min-Fill CLP problem, and then present the provably good and efficient approximation algorithm. In Section IV, we propose a new greedy iterative method based on the approximation algorithm. The experimental results are presented in Section V, and Section VI concludes the paper.

II. PROBLEM FORMULATION

According to several widely accepted chip-scale CMP models [11], [6], the post-CMP topography is proportional to the feature density

978-1-60558-497-3/09 $25.00 © 2009 ACM

within a given window. Thus, to improve the CMP quality, dummy fill is adopted to reduce the layout density variation. However, dummy fill also introduces undesirable side effects. Therefore, it is desirable to uniform the layout density and at the same time minimize the total insertion amount of dummy fill.

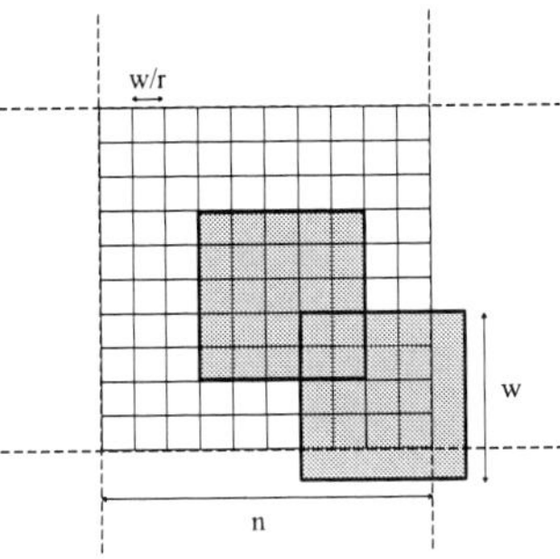

Fig. 1. The layout is discretized into $\frac{nr}{w} \times \frac{nr}{w}$ tiles ($r = 5$), and each $w \times w$ window contains r^2 tiles.

As in [3], [13], the Min-Fill objective of dummy fill problem can be described as: Given a design rule-correct layout in an $n \times n$ layout region, along with a window size $w < n$, and upper (U) and lower (L) bounds on the feature density in any window, add dummy fill features to create a filled layout such that the insertion cost is minimized while the density of any window remains in the give range (L, U).

Due to the uncountable number of windows in the layout, to make the filling problem tractable, the standard practice is to consider only a finite set of overlapping $w \times w$ windows of a fixed r-dissection [3], where r determines the window shift step w/r, as illustrated in Fig. 1. The $n \times n$ layout is partitioned into tiles $T_{ij}, i, j = 0, 1, \cdots, (nr/w) - 1$, with tile size w/r, such that each window W_{ij} centered at T_{ij}, $i, j = 0, 1, \cdots, (nr/w) - 1$, consists of r^2 tiles. Note that windows are "wrapped around" the layout, that is, a window that overlaps with the upper edge of the layout also contains tiles on the bottom of the layout. This is used to model the wafer with consecutive chips.

The density of the window W_{ij} in the "fixed-dissection" can be calculated using the following equation

$$\rho_w(i, j) = \sum_{k=i-r/2}^{i+r/2} \sum_{l=j-r/2}^{j+r/2} [d(k, l) \times f(k - i, l - j)], \quad (1)$$

where $f()$ is the CMP filter function and $d(i, j)$ is the feature density in tile T_{ij}. The shape of the filter function depends on the used CMP models. In [7], the square function is used, which means the window density is simply equal to the average of tile densities in the window, while a low-pass filter function is incorporated in [13].

Note that Eq. (1) does not consider the multi-layer cumulative effect of topography thickness [13]. Even though only the single-layer dummy fill problem is discussed in this paper, our formulations and algorithms can be easily extended to multi-layer case.

By now, if we arrange both the tile density d and the window density ρ_w in vector forms, then the Min-Fill LP formulation proposed in [13] is given as follows:

$$\begin{aligned} \text{minimize} \quad & c^T x \\ \text{subject to} \quad & l \leq \rho_w \leq u \\ & 0 \leq x \leq slack \end{aligned} \quad (2)$$

In the above formulation, x is a vector of size $\left(\frac{nr}{w}\right)^2$ representing the dummy fill amounts in tiles. The term c is referred to as the insertion cost factor. If all the elements in c are equal, then the objective becomes to minimize the total amount of dummy fills. Terms l and u are also both vectors with size of $\left(\frac{nr}{w}\right)^2$. All the elements of l equals to L, and all the elements of u equals to U, representing the window density constraints. The window density ρ_w can be expressed as a system of linear equations based on Eq. (1)

$$\rho_w = Ad = A(x + x^0), \quad (3)$$

where x^0 is the original tile density before filling. A is the window matrix, giving the mapping from the tile density to the window density.

The term *slack* is the density upper bound of dummy fills for each tile, which is determined in the density analysis phase. For example, a coupling-constrained density analysis algorithm was proposed in [15] which identifies feasible locations and amounts of dummy fills such that the fill-induced coupling capacitance is bounded. Based on [15], *slack* can be easily derived and the final dummy fill solution is guaranteed to satisfy both the density rules and coupling constraints.

The Min-Fill LP in Eq. (2) imposes density constraints only on windows in the "fixed-dissection". Therefore, density constraints can still be violated by other windows within the layout. Fortunately, for the window density calculated using the square filter function, the discrepancy is exactly characterized in the following two theorems [8].

Theorem 1: Suppose all $w \times w$ windows in the "fixed-dissection" of the layout have area density at least L and at most U. Then any $w \times w$ window has density at least $L - \frac{1}{r} + \frac{1}{4r^2}$ and at most $U + \frac{1}{r} - \frac{1}{4r^2}$, and these bounds are tight.

Theorem 2: Suppose all $\frac{w}{r} \times \frac{w}{r}$ tiles in the "fixed-dissection" of the layout have area density at least L and at most U. Then the exact lower bound on the area density of any $w \times w$ window equals

$$\frac{(r-1)^2}{r^2} L + \frac{4(r-1)}{r^2} \max\{L - 0.5, 0\} + \frac{4}{r^2} \max\{L - 0.75, 0\}$$

and the exact upper bound equals

$$\frac{(r+1)^2}{r^2} U - \frac{4(r-1)}{r^2} \max\{U - 0.5, 0\} - \frac{4}{r^2} \max\{U - 0.25, 0\}$$

Although Eq. (2) can be solved exactly by standard LP solvers, these solvers are often too time-consuming for large problems. The number of variables and the number of constraints in Eq. (2) are both $O\left(\left(\frac{nr}{w}\right)^2\right)$. As discussed in previous section, the window size of $50\mu m - 100\mu m$ is often used to calculate the window density. If the shift step $r = 4$ or 5, even for a medium-sized chip with $n = 1mm$, the total number of variables will be $O(10^4)$, making the standard LP method too expensive to finish in reasonable time.

III. Fast Approximation Scheme

A. The Min-Fill CLP Formulation

If we change the Min-Fill LP in Eq. (2) slightly by moving the upper bound u on the window density to the tile density, we will get the following formulation:

$$\begin{aligned} \text{minimize} \quad & P(x) = c^T x \\ \text{subject to} \quad & Ax \geq b \\ & x \leq s \\ & x \geq 0 \end{aligned} \quad (4)$$

where $b = \max\{l - Ax_0, \mathbf{0}\}$. It is the density lower bound of dummy fills for each window. The dummy density upper bound for each tile is redefined as

$$s = \min\{u - x_0, slack\}. \quad (5)$$

978-1-60558-497-3/09 $25.00 © 2009 ACM

The formulation in Eq. (4) is a standard *covering linear program* (CLP) due to the nonnegative nature of matrix A and vectors b, c and s (assuming the original layout density does not exceed U).

The reason that we can simply apply density upper bound on tiles instead of windows is that the Min-Fill CLP formulation can still guarantee similar density bounds on any $w \times w$ windows in the layout as the Min-Fill LP formulation. According to Theorem 1 and 2, the lower and upper density bounds of any $w \times w$ window in the layout of the Min-Fill LP formulation is $L - \frac{1}{r} + \frac{1}{4r^2}$ and $U + \frac{1}{r} - \frac{1}{4r^2}$, while the Min-Fill CLP formulation can ensure that the density of any $w \times w$ window in the layout can be bounded between $L - \frac{1}{r} + \frac{1}{4r^2}$ and $\frac{(r+1)^2}{r^2}U - \frac{4(r-1)}{r^2}\max\{U - 0.5, 0\} - \frac{4}{r^2}\max\{U - 0.25, 0\}$.

B. Provably Good Algorithm for Dummy Fill

The advantage of formulating the Min-Fill problem as a CLP problem is to leverage the efficient Fully Polynomial Time Approximation Scheme (FPTAS) for such a problem. Comparing with standard LP solvers, the scheme is much more efficient and produces better results if given longer running time. Our algorithm is mainly based on Fleischer [4]; the pseudo-code is given in Alg. 1.

Algorithm 1 FPTAS for Dummy Fill

Input: window matrix A, density bound b, cost factor c, tile bound s, precision ε

Output: filling density x

1: **Initialize** $x = \delta/C$
   find window p with min relative density $\rho(p)/b(p)$
2: **while** $P(x) < \theta$ **do**
3:    $\alpha = (1 + \varepsilon)\rho(p)/b(p)$
4:    **while** $\rho(p)/b(p) < \alpha$ and $P(x) < \theta$ **do**
5:       **if** tile $j \in$ window p and $x(j) < \alpha s(j)$ **then**
6:          put j in set $Q(p)$
7:       **end if**
8:       compute min weighted density price η among $Q(p)$
9:       **for** $j \in Q(p)$ **do**
10:         $x(j) = x(j)(1 + \varepsilon\eta/price(j,p))$
11:       **end for**
12:       find window p with min relative density $\rho(p)/b(p)$
13:    **end while**
14:    **if** $P(x/\alpha) < P(x^*/\alpha^*)$ **then**
15:       $x^* = x$, $\alpha^* = \alpha$
16:    **end if**
17: **end while**
   return x^*/α^*

The general idea of the algorithm is very simple. It will iteratively find a window whose density is relatively low, and increase the window density by increasing the tile densities in the window. Increasing the density of a tile may also cause other window density to increase, giving an unexpected side-effect. A key feature of the approach is to compensate such a wrong decision not by removing the fill, but by scaling down its impact. In other words, the x value will keep increasing, but at the end of day, only x/α will be used as the solution, for a scaling factor of α.

In more detail, the relative density of a window i is given by $\rho(i)/b(i)$, where $\rho = Ax$. The minimal relative density among all the windows will be iteratively increased. The increases are grouped into phases, each of which will ensure that the minimal relative density is at least α, and α will be increased by at least a factor of $1 + \varepsilon$ from phase to phase. This is given by the outer loop of Alg. 1. The inner loop will boost the minimal relative density by increasing x in the window p of the minimal relative density. Since each tile j has a density upper bound $s(j)$, we will only increase $x(j)$ satisfying $x(j)/\alpha < s(j)$. All such tiles are collected in Q. The price of increasing unit density of window p by tile j is given by $price(j, p) = c(j)/A(p, j)$. Obviously, we should increase more on tiles with lower prices. However, since we have upper bounds on tiles, the price should be weighted by $\frac{\alpha s(j) - x(j)}{\varepsilon x(j)}$. Denoting by η the minimal weighted price among all tile in $Q(p)$, each $x(j)$ in $Q(p)$ will be increased by a factor of $\varepsilon\eta/price(j,p)$. With such an update rule, tiles with lower prices will be rewarded with larger factors, benefiting $P(x)/\alpha$.

The iterations will continue till $P(x)/\alpha$ is within ε-optimal or, equivalently, $P(x) \geq \theta$ with a properly chosen θ. It should be noticed that during iterations x/α is always a feasible solution. Therefore, the best solution will be kept for the final answer.

C. Algorithm Analysis

We are going to show that the algorithm described in the previous subsection approximates the optimal solution within a constant factor. Due to space limitation, we will only be able to give a qualitative analysis of the algorithm.

The approximation algorithm simultaneously solves the Min-Fill CLP and its dual LP. The dual solution is used in proving the approximation guarantee of the algorithm. The dual of the Min-Fill CLP is:

$$\text{maximize} \quad D(y, z) = b^T y - s^T z$$

$$\text{subject to} \quad Ay - z \leq c \tag{6}$$
$$y \geq 0$$
$$z \geq 0$$

where y, z are both variables of the dual problem.

According to the duality theory, the strong duality holds for linear programs. Our approximation algorithm obtains exactly feasible solutions x and y, z to both primal and dual problems with $P(x)/D(y, z) \leq 1 + \varepsilon$. This means that we will get both feasible solutions and the solution is ε-optimal from the exact solution.

The algorithm maintains dual variables (y, z) during α-phases. The increase of $P(x)$ each time by the increase of the primal variable x is balanced in $D(y, z)$ by the increase of the dual variable y. However, if the variable $x(j)$ has already reached its upper bound, the dual variable $z(j)$ will be increased by a proper amount, so that the increases of both dual variables $y(j)$ and $z(j)$ cancel each other and make $D(y, z)$ unchanged. This update procedure guarantees that the ratio of $P(x)$ and $D(y, z)$ is within a $1 + \varepsilon$ factor, so that the ε-optimality of the algorithm can be proved via LP duality theory. The whole update procedure of the dual variables is shown as follows

$$y(p) = y(p) + \eta$$
$$\textbf{for } j \notin Q(p) \textbf{ do}$$
$$z(j) = z(j) + \eta A(p, j) \tag{7}$$
$$\textbf{end for}$$

The above segment of instructions could be put between line (12) and line (13) of Alg. 1. Since they are not necessary in finding the optimal solution to the Min-Fill problem, they are omitted in the implementation.

If $Q(p) = \emptyset$ during the iteration, the algorithm will get stuck. However, it also means that the primal problem is infeasible. That is because if the problem is feasible, then $As \geq b$. Thus in each iteration the following inequality will hold

$$b(p)\alpha \; > \; \sum_{j=1}^{m} A(p,j)x(j) \geq \sum_{j \notin Q(p)} A(p,j)x(j)$$
$$\geq \; \alpha \sum_{j \notin Q(p)} s(j)A(p,j), \tag{8}$$

978-1-60558-497-3/09 $25.00 © 2009 ACM

where m is the number of tiles, which equals to $(\frac{nr}{w})^2$. Therefore, we have

$$b(p) > \sum_{j \notin Q(p)} A(p,j)s(j),$$

which implies that the problem is infeasible if $Q(p) = \emptyset$.

In practice, if we find $Q(p) = \emptyset$ in the execution, we know that the density lower bound L cannot be achieved even when all the available area in window p is filled. We can thus fill all the tiles in window p, and reduce the problem size by resetting all the other bounds as $b(i) - \sum_{j:A(p,j)>0} A(i,j)s(j)$.

Theorem 3: The FPTAS for Dummy Fill in Alg. 1 finds a $(1+\omega)$-approximation solution for the Min-Fill CLP in $O(\varepsilon^{-2} m \log(mC))$ iterations by choosing

$$\begin{aligned} \delta &= ((C(1+\varepsilon))^{1-\varepsilon} m)^{-1/\varepsilon} \\ \varepsilon &< \min(0.15,\ \omega/4) \\ \theta &= 1 \end{aligned}$$

where

$$C = \frac{||c||_\infty}{\min_{j:c(j)>0} c(j)}$$

and $m = (\frac{nr}{w})^2$ is the number of tiles.

The detailed proof of the above theorem can be found in [5], [4].

IV. A New Greedy Iterative Approach

In this section, a new greedy iterative method will be developed based on the approximation algorithm in the previous section. The new heuristic method builds on the concept of window density increase of the approximation algorithm. The main difference of two algorithms is that the greedy method abandons the α-phases procedure, making it more efficient and simpler to implement. The greedy iterative method is described in Alg. 2.

Similar to the approximation algorithm, in each iteration, it picks the most underfilled window, and inserts dummy features into the tiles in that window. In more detail, the method maintains three sets T, W and $Q(p)$. A tile will be put in T if and only if either it belongs to a window which has already achieved the density upper bound U, or all its available area has been filled. In each iteration, the algorithm only increases the density of tiles in $Q(p)$, which collects all the tiles of window p that does not belong to set T. Set W is used to ensure the method contains no infinite loop. Initially, all windows are collected in W. If $Q(p) = \emptyset$ during the iteration, which means the density of the window p cannot be improved any more, then window p will be removed from set W.

If the user decides to apply a pre-defined dummy fill pattern to the layout, the final CLP solution may need to be rounded so that an integer number of dummy fills can be filled. The new greedy method can easily handle this rounding issue by applying randomized rounding [12] during each iteration, and its final solution is still close to the optimal. Note that in the implementation of the method, the rounding step should ensure that at least one single fill pattern be filled.

Compared with the previous Monte Carlo and Greedy methods [2], which either insert a single filling geometry or insert the maximal possible amount during each iteration, the new greedy method uses the weighed price concept of the approximation algorithm to decide the actual filled amount in each tile, making it closer to optimal, since the dynamic information of the layout is considered and used in each insertion, instead of a static and brute-force insertion.

V. Experimental Results

We have implemented both the fast approximation scheme and the greedy algorithm in C++. All simulations given in this section were

---

Algorithm 2 The New Greedy Iterative Method for Dummy Fill

Input: window matrix A, density bounds L and U, cost factor c, tile bound $slack$, precision ε
Output: filling amount x

```
 1: Initialize x = 0, T = ∅, put all windows in W
    find window p with the minimum density in W
 2: while ρ_w(p) < L do
 3:    if tile j ∈ window p and tile j ∉ T then
 4:       put j in set Q(p)
 5:    end if
 6:    if Q(p) = ∅ then
 7:       remove p from W
 8:    else
 9:       compute min weighted density price η among Q(p)
10:       for j ∈ Q(p) do
11:          γ = η/price(j, p)
12:          δ = rounding(εγx(j))
13:          x(j) = x(j) + δ
14:          if x(j) ≥ slack(j) then
15:             put the tile j in set T
16:          end if
17:       end for
18:    end if
19:    for i ∈ {1, ..., m} do
20:       if ρ_w(i) ≥ U then
21:          put all the tiles in window i in set T
22:       end if
23:    end for
24:    find the window p with the minimum density in W
25: end while
    return x
```

---

performed on a Unix workstation with 3.0 GHz processor and 2 GB memory. For comparison, we use the GNU Linear Programming Kit as our standard LP solver.

To test the performance of the proposed methods, we use both randomly generated layout cases and a real design case. For the randomly generated layout, similar to the procedure described in [14], we assume the maximal signal metal density is 50% in each tile, and use a random number uniformly distributed between 0 and 0.5 to represent the original tile density before filling. Further we assume that the dummy fills are square blocks separated by required spacing, and the maximal dummy fill density is also assumed to be 50%. So the density slack in each tile can be calculated as

$$slack(i) = 0.5 \times \max(0, 1 - x_0(i) - slack\_reduction). \quad (9)$$

As discussed before, when calculating the slack areas, the foundry design rules or other considerations such as the coupling capacitance constrains [15] have to be satisfied. This makes the allowable areas smaller than the empty areas. We model this by using a parameter $slack\_reduction$ that reduces the allowable area density. In the following experiments, $slack\_reduction$ is a random number between 0 and 0.2.

Throughout the experiments, for minimizing the total amount of dummy fill, we set each element of the cost factor c to 1. We also adopt the filter function used in [13] to calculate the density of each window:

$$f(x,y) = c_0 exp(c_1(x^2 + y^2)^{c_2}), \quad (10)$$

where $c_0 = 0.1$, $c_1 = -0.1$ and $c_2 = 1$.

A. Approximation Precision

In order to test the approximation precision of the approximation algorithm with respect to the parameter ε, we compare the solution of our algorithm with the solution of the LP method. In the test case, we assume the layout size $n = 800\mu m$, the window size $w = 100\mu m$, the step number $r = 5$, and the lower and upper density bounds are 50% and 80%, respectively. We solve the fractional CLP problem to get a fair comparison with the optimal LP solution. Here, two approximation ratios are defined for comparison. The first approximation ratio is used to compare the total inserted amount of dummy fill of these two methods, and it is defined as:

$$fill\_ratio = abs(CLP\_fills - LP\_fills)/LP\_fills, \qquad (11)$$

where $abs()$ is the absolute value function, and CLP_fills and LP_fills represent the obtained total inserted amount of the CLP and LP approaches, respectively. The second one is used to compare the window density distribution, and is defined as:

$$std\_ratio = abs(CLP\_std - LP\_std)/noFill\_std, \qquad (12)$$

where CLP_std, LP_std and $noFill\_std$ represent the standard deviations of the window density distributions of the CLP solution, the LP solution, and the original layout, respectively.

Fig. 2 shows these approximation ratios with respect to the parameter ε for the test case. It is clear that the accuracy of the

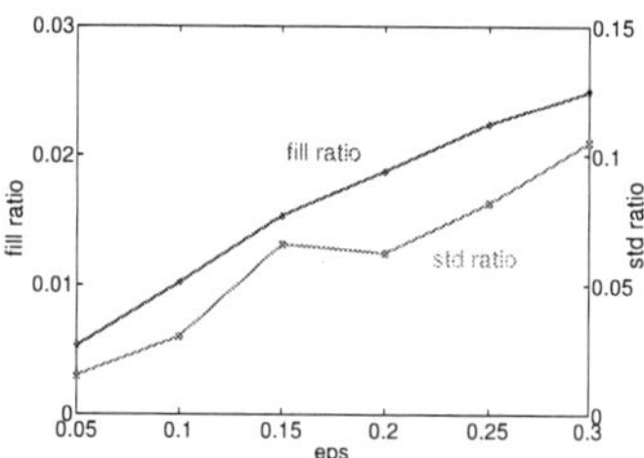

Fig. 2.   The approximation ratios for different values of ε.

approximation algorithm is very high. For the total inserted amount, when $\varepsilon = 0.3$, the fill ratio is only 2.5%. As for the window density distribution, it is important to note that smaller ε leads to smaller window density variation, and the solution of our method is very close to the optimal LP solution when ε is small.

Fig. 3 illustrates the ratio of the primal solution $P(x)$ and the dual solution $D(y, z)$ of our approximation algorithm for the test case. The approximation bound $1 + 4\varepsilon$ is also drawn for comparison. It is important to note that the ratio of the primal and dual solutions is theoretically guaranteed to be no greater than $1 + 4\varepsilon$. In practice, it is much smaller as is shown in Fig. 3. This clearly demonstrates the effectiveness of our approximation algorithm.

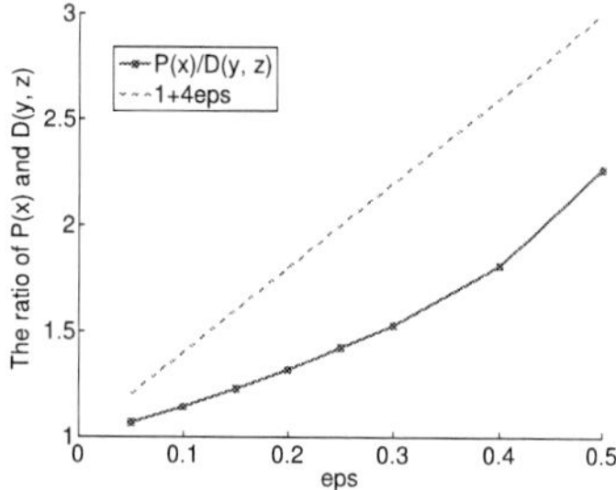

Fig. 3.   The ratio of the primal and dual solutions.

B. Scalability

The fast speed is the significant advantage of our approximation method over the standard LP method. In the test cases, we keep w and r fixed as $100\mu m$ and 5, while the layout size n changes from $600\mu m$ to $1.6mm$. The CPU time comparison of these two methods is shown in Fig. 4. The horizontal coordinate is the number of unknowns, which is equal to $\left(\frac{nr}{w}\right)^2$. The values of ε used in this experiment are 0.1, 0.2 and 0.3. The scalability of our approximation algorithm is clearly demonstrated in Fig. 4.

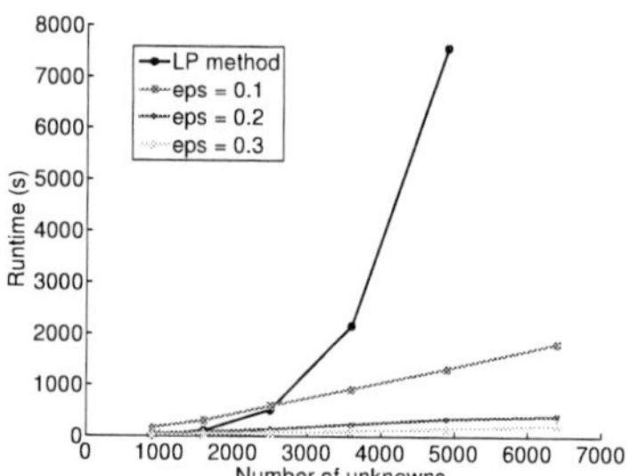

Fig. 4.   The runtime comparison of our approximation method with LP method.

C. Rounding Issue

In this section, we use pre-defined squared dummy patterns to test the performance of our methods. The size of the dummy blocks used in this experiment is $0.5\mu m \times 0.5\mu m$.

The first experiment in this section is intended to make a comparison between the new FPTAS algorithm and the new greedy algorithm. We keep w and r fixed as $100\mu m$ and 5, while the layout size n changes from $200\mu m$ to $1000\mu m$. Speedup of the greedy method is computed by comparing to the runtime of the FPTAS algorithm, which is illustrated in Fig. 5. The values of ε used in this experiment are 0.1, 0.25 and 0.5. It can be seen that the greedy algorithm is more efficient than the approximation algorithm in practice, especially with smaller ε. Furthermore, there are only slight solution degradations in both total inserted amount and window density distribution for the greedy method compared to the FPTAS algorithm. The approximation ratios (*fill_ratio* and *std_ratio*) are no more than 5% for the test cases.

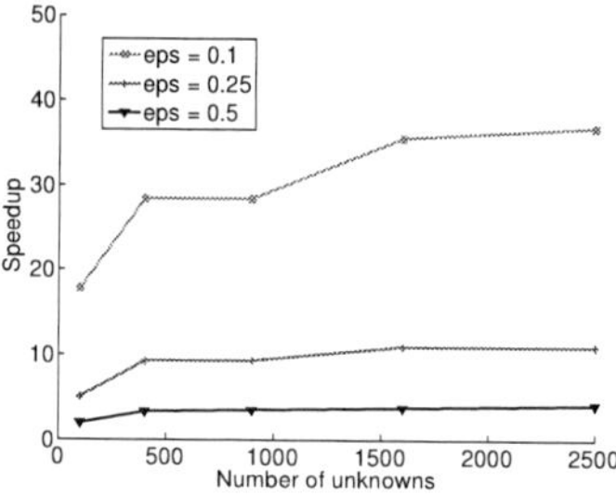

Fig. 5.   The speedup of the greedy algorithm compared to the approximation algorithm.

In the second experiment, a comparison of our greedy algorithm with our implementation of the Monte-Carlo method [1] is conducted and reported in Table I. "MaxDens" is the maximum density of all the windows. "#Dummy" shows the number of inserted dummy fills. In the Monte-Carlo method, the priority of a tile p is chosen to be proportional to $U - MaxDens(p)$, where $MaxDens(p)$ is the maximum density over windows containing the tile p. In each iteration of the

978-1-60558-497-3/09 $25.00 © 2009 ACM

TABLE I

COMPARISON OF OUR GREEDY METHOD WITH MONTE-CARLO APPROACH

Test case (n/w/r)	Monte-Carlo				Greedy			
	MaxDens	MinDens	#Dummy	Time(s)	MaxDens	MinDens	#Dummy	Time(s)
800/200/4	0.5000	0.4434	951186	44.01	0.4946	0.4720	950365	0.71
1000/100/5	0.5000	0.4437	969389	154.82	0.4954	0.4719	947426	15.32
800/100/5	0.5000	0.4388	594805	79.37	0.4946	0.4701	574469	9.45
600/100/5	0.5000	0.4536	341082	39.77	0.4992	0.4773	337867	5.23
400/50/5	0.5000	0.4399	149358	20.10	0.4952	0.4705	145084	6.38
Average	0.5000	0.4439	601164	67.62	0.4958	0.4724	591040	7.42

Monte-Carlo method, a single filling geometry is inserted into a tile. The approximation parameter ε used in the greedy method is 0.1.

It should be noted that there are certain differences between the objectives of our methods and the Monte-Carlo method [1]. In the Monte-Carlo method, the window density upper bound is the input of the algorithm, and it tries to maximize the window density lower bound. Our methods, however, with the given density lower bounds, try to minimize the total number of dummy fills. In order to make a fair comparison, we use the medium value of the window density solution of the Monte-Carlo method as the density lower bound input to our greedy method.

We can observe that our greedy method performs better than the Monte-Carlo method under all known metrics. Fig. 6 shows the window density solutions of our greedy method and the Monte-Carlo method for one of the test cases. Our method achieved roughly 50% reduction on the density variation.

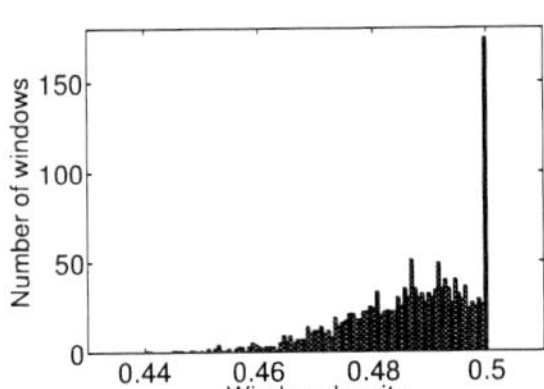
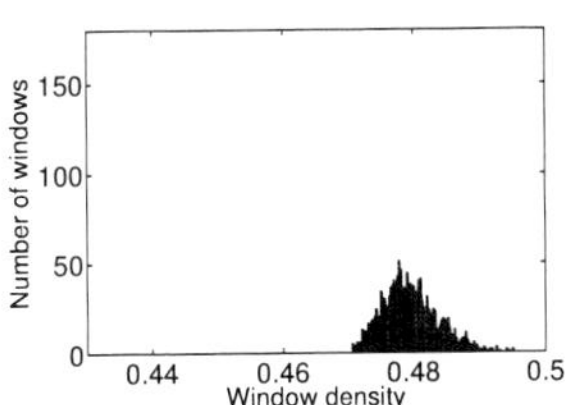

(a) The solution of the Monte-Carlo method. (b) The solution of the greedy method.

Fig. 6. Comparison of the window density distributions.

D. Test on Real Layout Design

We also perform experiments on a real design case. The layout size is $854\mu m \times 243\mu m$, and it has 202238 rectangles. We duplicate the layout 3 times, and assume the window size is $100\mu m$ and step size $r = 5$. So there are totally 1849 variables. It takes the LP solver 375 seconds to get the optimal solution, while our approximation algorithm with $\varepsilon = 0.25$ only takes 3 seconds. The standard deviation of the window density distribution of the original layout is 0.0567. After filling by our approximation algorithm, it is reduced to 0.0065.

VI. CONCLUSIONS

In this paper, we present the first covering linear program formulation for the dummy fill problem with Min-Fill objective, and propose a probably good and efficient algorithm based on the recent fast approximation scheme [4]. The efficiency and the scalability of the algorithm are clearly demonstrated in the experimental results. Moreover, based on the approximation algorithm, a new greedy iterative method is proposed. Simulation results show that our greedy method outperforms the Monte-Carlo method [1] both in performance and runtime. We are currently developing multicore parallel dummy fill algorithms based on the approximation scheme.

ACKNOWLEDGMENTS

This research is supported partially by NSFC research project 60676018 and 60806013, China National Basic Research Program under the grant 2005CB321701, China National Major Science and Technology special project 2008ZX01035-001-06 during the 11th five-year plan period, the doctoral program foundation of Ministry of Education of China under 200802460068, the International Science and Technology Cooperation program foundation of Shanghai under 08510700100, the program for Outstanding Academic Leader of Shanghai, and NSF under CCF-0238484 and CCF-0811270.

REFERENCES

[1] Y. Chen, A. B. Kahng, G. Robins, and A. Zelikovsky. Monte-Carlo algorithms for layout density control. In *Proceedings of ASP-DAC*, pages 523–528, 2000.

[2] Y. Chen, A. B. Kahng, G. Robins, and A. Zelikovsky. Practical iterated fill synthesis for CMP uniformity. In *Proceedings of ACM/IEEE Design Automation Conference*, pages 671–674, 2000.

[3] Y. Chen, A. B. Kahng, G. Robins, and A. Zelikovsky. Area fill synthesis for uniform layout density. *IEEE Trans. on CAD*, 21(10):1132–1147, 2002.

[4] L. Fleischer. A fast approximation scheme for fractional covering problems with variable upper bounds. In *Proceedings of ACM-SIAM symposium on Discrete algorithms*, pages 1001–1010, 2004.

[5] N. Garg and J. Könemann. Faster and simpler algorithms for multicommodity flow and other fractional packing problems. In *Proceedings of IEEE Symposium on Foundations of Computer Science*, pages 300–309, 1998.

[6] T. E. Gbondo-Tugbawa. *Chip-Scale Modeling of Pattern Dependencies in Copper Chemical Mechanical Polishing Processes*. PhD thesis, Massachusetts Institute of Technology, 2002.

[7] A. Kahng, G. Robins, A. Singh, and A. Zelikovsky. Filling algorithms and analyses for layout density control. *IEEE Trans. on CAD*, 18(4):445–462, 1999.

[8] A. B. Kahng and K. Samadi. CMP fill synthesis: A survey of recent studies. *IEEE Trans. on CAD*, 27(1):3–19, 2008.

[9] S. Lakshminarayanan, P. J. Wright, and J. Pallinti. Electrical characterization of the copper CMP process and derivation of metal layout rules. *IEEE Trans. on Semiconductor Manufacturing*, 16(4):668–676, 2003.

[10] M. Mukherjee and K. Chakraborty. A randomized greedy method for rectangular-pattern fill problems. *IEEE Trans. on CAD*, 27(8):1376–1384, 2008.

[11] D. O. Ouma. *Modeling of Chemical Mechanical Polishing for Dielectric Planarization*. PhD thesis, Massachusetts Institute of Technology, 1998.

[12] P. Raghavan and C. D. Thompson. Randomized rounding: A technique for provably good algorithms and algorithmic proofs. *Combinatorica*, 7(4):365–374, 1978.

[13] R. Tian, D. F. Wong, and R. Boone. Model-based dummy feature placement for oxide chemical mechanical polishing manufacturability. In *Proceedings of ACM/IEEE Design Automation Conference*, pages 667–670, 2000.

[14] X. Wang, C. C. Chiang, J. Kawa, and Q. Su. A min-variance iterative method for fast smart dummy feature density assignment in chemical-mechanical polishing. In *Proceedings of ISQED*, pages 258–263, 2005.

[15] H. Xiang, L. Deng, R. Puri, K.-Y. Chao, and M. D. F. Wong. Fast dummy-fill density analysis with coupling constraints. *IEEE Trans. on CAD*, 27(4):633–642, 2008.

978-1-60558-497-3/09 $25.00 © 2009 ACM

544

Predicting Variability in Nanoscale Lithography Processes

Dragoljub Gagi Drmanac
University of California,
Santa Barbara

Frank Liu
IBM
Austin Research Lab

Li-C. Wang
University of California,
Santa Barbara

ABSTRACT

As lithography process nodes shrink to sub-wavelength levels generating acceptable layout patterns becomes a challenging problem. Traditionally, complex convolution based lithography simulations are used to estimate areas of high variability. These methods are slow and infeasible for large scale full chip analysis. This work proposes a solution to this problem by using machine learning techniques to identify layout areas that are more prone to variability. A novel target layout representation is proposed, and the latest support vector machine (SVM) algorithms are used to detect variability within standard cells and between cells in a simulated full chip layout.

Categories and Subject Descriptors

B.7.2 [**Integrated Circuits**]: Design Aids; I.4.7 [**Feature Measurement**]: Feature representation

General Terms

Algorithms, Performance, Reliability, Verification

Keywords

Photo Lithography, Process Variation, Modeling Variability, Machine Learning, Kernel Methods

1. INTRODUCTION

As IC process nodes continue to shrink from 65 nm to 45nm and below, generating acceptable layout patterns becomes an increasingly difficult problem. Industry leading foundries continue to depend on 193nm lithography systems to print the latest 45nm designs. Sub-wavelength photolithography results in unavoidable manufacturability issues due to process variations and unforeseen cell-to-cell interactions. To combat these problems, several resolution enhancement techniques (RET) have been employed such as optical proximity correction (OPC) and phase shift masks

This work is supported in part by National Science Foundation, Grant No. 0541192 and Semiconductor Research Corporation, project task 1585.001.

Permission to make digital or hard copies of part or all of this work for personal or classroom use is granted without fee provided that copies are not made or distributed for profit or commercial advantage and that copies bear this notice and the full citation on the first page. To copy otherwise, to republish, to post on servers or to redistribute to lists, requires prior specific permission and/or a fee.
DAC'09, July 26-31, 2009, San Francisco, California, USA

(PSM); however, these methods often rely on complex simulation models and are usually applied late in the design cycle when only minor layout changes can be made. Lithography simulation is the current golden standard for understanding variability of new process nodes and cell designs, yet it is prohibitively time consuming for full chip analysis. A first order layout analysis approach is needed to quickly detect areas of high variability within standard cells, and between adjacent cells in full chip placements.

Dealing with variability early in the design process helps prevent manufacturability issues, yet very few tools exist that can choose between acceptable designs to minimize variability. Several model-based design for manufacturability (DFM) methodologies use variability analysis to manage complex process-design interactions, however none of them directly predict variability. For example, simulating standard cells in various configurations to measure context based variability, or performing litho-aware electrical current calculations to extract transistor parameters for accurate timing analysis [9]. The aforementioned DFM methods could all benefit from a direct variability prediction tool, to avoid the complications of lithography simulation and process parameter selection.

Understanding and predicting sources of variability is a ubiquitous problem that has applications in various forms of DFM. Most work in this area has attempted to combine variability metrics or models with existing DFM frameworks. For example, pattern matching techniques have been used in conjunction with standard design rule checking (DRC) software to identify layout configurations that are difficult to manufacture in the presence of lithographic process variation [4]. While this helps DRC software capture some problematic 2D geometries, it relies on manual pattern selection based on slow lithography simulation and human judgment.

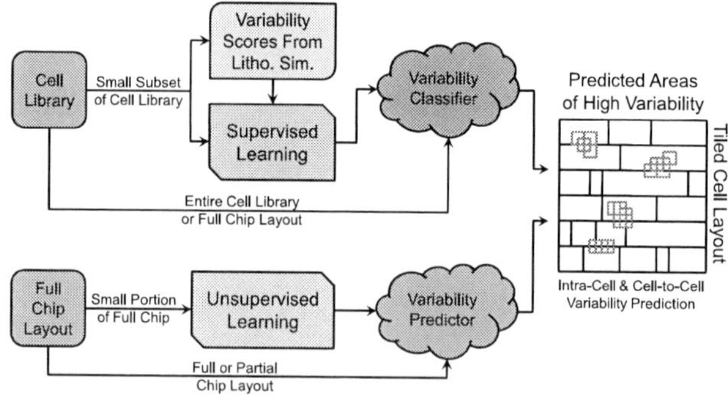

Figure 1: Variability Prediction Overview

978-1-60558-497-3/09 $25.00 © 2009 ACM

Improvements to standard OPC techniques have also been made by incorporating analytical process variation models that account for fluctuations in dosage and focus [10]. While this improves OPC's awareness to variability issues, it does so at a cost of two to three times traditional OPC runtime. Most existing DFM tools work to improve printability of placed and routed layouts, however many key problems are generated early in the design cycle when details about the manufacturing process are unavailable [6].

This paper proposes a novel machine learning framework for quickly predicting areas of high variability within standard cells, and between cells in full chip layouts. Figure 1 shows an overview of two proposed machine learning approaches for predicting variability. The first method builds a variability classifier using training samples labeled with lithography simulation scores from a subset of cells. The second approach learns to predict variability by analyzing a small layout portion without simulation. Both methods can predict variability in standard cell libraries or across full chip layouts. The rest of the paper is organized as follows. Section 2 will go over measuring variability using lithography simulation. Section 3 will describe our novel layout representation that makes variability prediction possible. Section 4 will outline the two variability prediction methods in detail, and section 5 will discuss the experiments and results. Section 6 will conclude the paper followed by references in section 7.

2. MEASURING VARIABILITY

In this work we focus on predicting variability in printed 45nm metal layers with respect to changes in dosage and focus. To quantify variability, an industry leading lithography simulator is used in conjunction with a defined variability metric to guide and validate our predictions.

2.1 Lithography Simulation

Lithography simulation is currently the most common approach for predicting intra-cell and cell-to-cell variability. We use a lithography simulator to create process variation (PV) bands showing the possible areas within which a given metal layer will print, as dosage and focus conditions vary. Figure 2(a) shows a metal target layer and its corresponding PV bands. Notice how the width of the bands changes depending on different neighboring geometries; the thicker the band, the more uncertainty there is in the final edge placement.

2.2 Variability Metric

Lithography simulators commonly scan PV bands and assign variability scores to layout areas according to a simple metric. Usually, the metric is a measure of PV band area normalized by the window under which the variability calculation is performed. In this work we introduce an analogous variability measure, the PV band area normalized by the empty target area. Figure 2(b) shows how this calculation is performed. First a square window is selected to perform the calculation under, then the PV band area within the window is calculated and normalized by the area within the window not covered by metal layer polygons. We normalize by the empty target area to ensure that variability in denser layout regions is treated as more critical. For a fair comparison of variability we purposely chose a metric similar to existing industry standards.

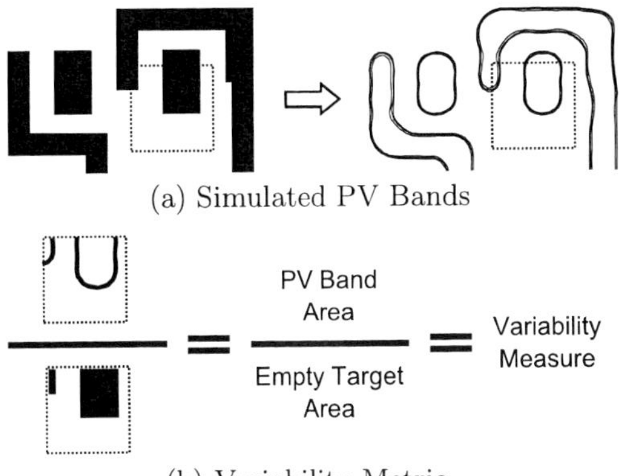

(a) Simulated PV Bands

(b) Variability Metric

Figure 2: Calculating the Variability Measure

3. NOVEL LAYOUT REPRESENTATION

The graphic data system (GDSII) file format is the industry standard for storing and transferring large layout designs, however its representation is limited since entire layouts are saved as a set of planar geometric shapes. To perform variability prediction we propose an intermediate layout representation that captures relative shape and position information, called the histogram distance transform (HDT). Concepts from image processing and feature encoding are discussed to introduce this novel representation.

3.1 Image Processing

Since variability changes based on configurations in a two dimensional space, image processing techniques can help decompose target layouts, and encode them such that efficient variability prediction is possible. In this section we discuss two bitmap image formats, image histograms, the distance transform, and raster scanning.

3.1.1 Bitmap Image Representations

All target layouts, PV bands, and intermediate transformations are represented using simple bitmap images. The simplest image format we use is called the portable bitmap (PBM) which is just an array of 1's and 0's representing black and white pixels. This image format was selected to represent our target layouts because of its simplicity and software compatibility. To store results of image transforms we used a grayscale image format called portable gray map (PGM) which is similar to PBM except each pixel is an integer from 0-255 representing the 256 grayscale levels. Since we do not store intermediate images during processing our methodology does not require a large memory or compressed image formats.

3.1.2 Image Histogram

An image histogram is a compact representation that captures the distribution of intensity levels present within a grayscale image. In PGM images there are 256 grayscale levels so the image histogram has 256 bins on the x-axis and displays the number of pixels per bin on the y-axis. An example grayscale image and its corresponding histogram are shown in Figure 3. Usually, one grayscale value dominates the image so to get a more normalized result we take the log of the number of pixels for each intensity value. This is an abstract translation and rotation invariant representation that captures a great deal of information content [3].

978-1-60558-497-3/09 $25.00 © 2009 ACM

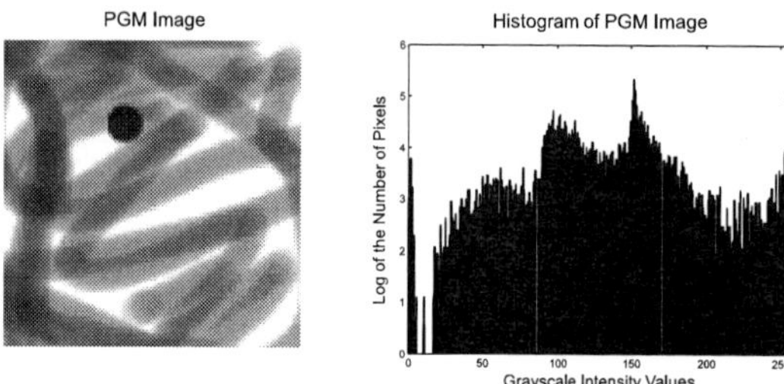

Figure 3: Image Histogram

3.1.3 Distance Transform

The distance transform is an image processing technique that converts a black and white bitmap image into a grayscale image by replacing each white pixel with the distance to the nearest black/white pixel boundary. Since computed distances can be large, each pixel is renormalized to fit in the 256 grayscale range. An example binary image and its distance transform are shown in Figure 4. The distance transform has converted the white rectangle into a grayscale image where each pixel value represents its distance to the nearest boundary. In this simple example the boundary is the perimeter of the white rectangle.

Performing the distance transform can be seen as an encoding of shape information for a given binary image. After the transform is performed a more descriptive image results that captures overall shape, length, and width [5]. The distance transform also approximates the skeleton of a binary shape, which shows up as the bright pixel areas similar to an x-ray. We use an efficient Euclidian distance transform algorithm proposed in [5].

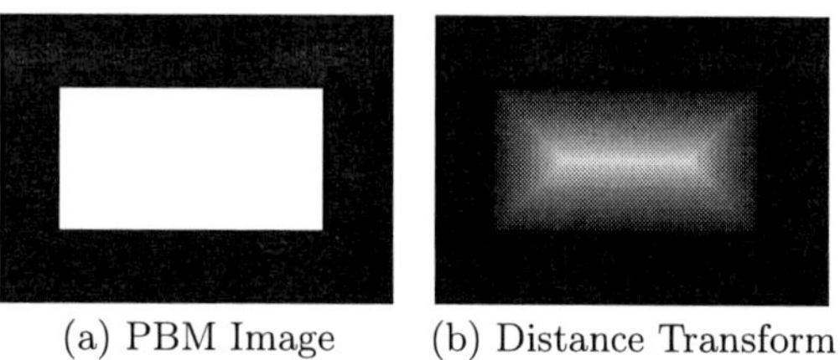

(a) PBM Image   (b) Distance Transform

Figure 4: Distance Transformed Rectangle

3.1.4 Raster Scanning

Since layouts are often large compared to the local features contributing to variability, it is important to raster scan layouts to extract local features. Our raster scanning procedure is shown in Figure 5. We start with a 100x100 pixel window and move it around the target layout capturing samples with a step size of 50 pixels. This gives us a 50% overlap between raster windows and ensures good coverage of each cell. The scanned images are scaled so each pixel corresponds to $3nm^2$ on the actual target layout. This resolution is fine enough to capture detailed spatial information without slowing the system down with unnecessary data processing.

3.2 Encoding Layout Features

Figure 6 shows the procedure for transforming a GDSII layout into our novel HDT representation. We extract the target outline from a GDSII file and save it as a binary PBM image, next we compute the distance transform of the

Figure 5: Raster Scanning a Cell

target outline and obtain a grayscale PGM image, finally we take the histogram of a small raster scanned portion of the resulting image as our target layout representation for that small area. The distance transformed layout is fully raster scanned; with each step of the raster window a histogram is computed and stored as a 256 dimensional vector. After the transformation is complete the target layout is represented as a set of histogram vectors.

The distance transform of a target layout captures the distribution of spacing between polygons. Bright areas correspond to spread out polygons and dark areas correspond to tightly packed polygons. Notice that polygon shape is also captured by the transform computed within the target outlines. Fine skeleton structures appear encoding the shape and size of the original layout. Although this is an abstract representation of relative shape and spacing, small layout areas can accurately be represented in this way. The distance transform captures local shape and spacing information, while the histogram gives us an efficient means of storing and working with that information.

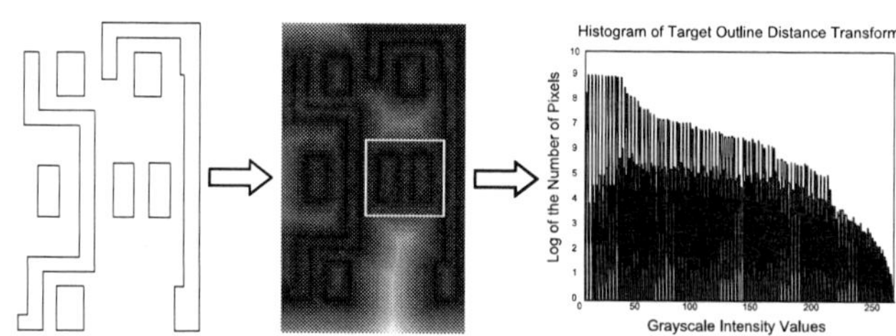

Figure 6: Novel Target Layout Representation

4. PREDICTING VARIABILITY

Variability prediction is performed using SVM learning algorithms. In machine learning theory there are two fundamental learning approaches, supervised and unsupervised. Supervised learning estimates a function from a set of labeled training examples, to solve regression or classification problems, while unsupervised learning estimates the support of the training data without labels, to solve clustering or outlier detection problems. In this work we use two-class and one-class nu support vector classifier (ν-svc) algorithms proposed by [7] and [8], and a modified version of the open source LibSVM software package implemented by [2].

4.1 Layout Similarity Measure

To use our HDT layout representation in a machine learning framework we define a function to measure similarity between two raster windows. A kernel function $K(\vec{x}, \vec{y})$ computes the similarity between two sample vectors. Since our target layout representation is histogram based we use the

978-1-60558-497-3/09 $25.00 © 2009 ACM   547

histogram intersection kernel shown below, where $\vec{x}$ and $\vec{y}$ are 256 dimensional histogram vectors.

$$K(\vec{x}, \vec{y}) = \sum_{i=1}^{256} min(x_i, y_i)$$

This kernel is efficient because it calculates the intersection of each histogram bin without time consuming multiplication or exponentiation. Intuitively, we can see that the larger the intersection between two histograms the more similar they are. The histogram intersection kernel has been shown to perform well in image recognition tasks by [1].

4.2 Two-Class Supervised Learning

Two-class SVM analysis generates a predictive classifier that detects areas of high variability based on labeled training samples picked from a standard cell library. A training sample is a 100×100 pixel layout window encoded using our HDT layout representation. Each window is assigned a variability score calculated from PV bands using the variability metric previously defined. Since ν-SVC performs pass/fail classification, a variability score threshold is established to label high variability areas as failing and low variability areas as passing. Each sample is constructed as a training-label histogram-vector pair $(l_i, \vec{s_i})$, where $\vec{s_i} = (h_1, h_2, ..., h_{256})$ contains 256 histogram components $h_1 - h_{256}$, encoded using our HDT representation, and a pass/fail classification label l_i, determined by variability score thresholding.

To train a variability prediction model the ν-SVC soft margin classifier is used to separate high and low variability samples with maximum margin in a 256 dimensional feature space. Since the classification space is large many separating hyperplanes exist, so optimization is used to find one that maximally separates the training data. A maximally separating hyperplane ensures that our classifier will be general enough to correctly label unseen samples as having low or high variability. The problem is formulated as the following quadratic programming optimization.

$$Maximize \quad -\frac{1}{2} \sum_{i,j=1}^{m} \alpha_i \alpha_j l_i l_j K(\vec{s_i}, \vec{s_j})$$

$$Subject\ to\ \ 0 \leq \alpha_i \leq \frac{1}{m},\ \sum_{i=1}^{m} \alpha_i l_i = 0,\ \sum_{i=1}^{m} \alpha_i \geq \nu.$$

Where $\alpha_{i,j}$ are Lagrangian multipliers, $\vec{s_{i,j}}$ are training samples, $l_{i,j} \in \{\pm 1\}$ are pass/fail training labels, ν is the fraction of accepted training errors, and $K(\vec{s_i}, \vec{s_j})$ is the histogram intersection kernel. In general, training samples may not be linearly separable so ν is used to control the fraction of misclassified samples allowed when establishing a boundary, namely how soft the margin will be. When optimization is complete, some Lagrangian multipliers are zero, indicating that their corresponding training samples play no role in establishing the classification boundary. Samples with nonzero Lagrangian multipliers influence the boundary and are called support vectors. The resulting two-class SVM classifier consists of a decision function $f(\vec{x})$ used to classify new samples, and set of support vectors defining the classification boundary. The decision function is shown below.

$$f(\vec{x}) = sgn\left(\sum_{i=1}^{m} \alpha_i l_i K(\vec{x}, \vec{s_i}) + b \right)$$

$f(\vec{x})$ is a weighted sum of similarity between all support vectors and the current sample being classified. The histogram intersection kernel $K(\vec{x}, \vec{s_i})$ compares new samples to the set of support vectors, so that a pass/fail decision can be made.

4.3 One-Class Unsupervised Learning

One-class SVM can be viewed as an outlier analysis algorithm that places the ultimate outlier at the origin. Samples far from the origin are labeled as normal and samples near the origin as abnormal. In the context of our problem and the HDT layout representation, the origin represents layouts that have uniform shape and spacing. When raster scanning metal layers, uniform shape and spacing most often occurs in areas with tightly packed parallel wires. In these areas, the critical dimension of processing is being pushed to the limit, and surrounding geometries substantially influence parallel wire widths, so we expect these areas to exhibit variability.

Outlier analysis is performed using the one-class ν-SVC algorithm. Each raster scanned window is constructed as a sample vector $\vec{s_i} = (h_1, h_2, ..., h_{256})$. In this case our sample has no training label attached to it since the algorithm assumes all samples are of the same class, and the origin is the only sample of the outlier class. The one-class algorithm maximally separate samples from the origin using the following quadratic programming optimization.

$$Minimize \quad \frac{1}{2} \sum_{i,j=1}^{m} \alpha_i \alpha_j K(\vec{s_i}, \vec{s_j})$$

$$Subject\ to\ \ 0 \leq \alpha_i \leq \frac{1}{\nu m},\ \sum_{i=1}^{m} \alpha_i = 1.$$

Where $\alpha_{i,j}$ are Lagrangian multipliers, $\vec{s_{i,j}}$ are training samples, ν is the fraction of samples we expect to have high variability, and $K(\vec{s_i}, \vec{s_j})$ is the histogram intersection kernel. This is similar to the previous optimization problem, except it does not depend on any training labels. After optimization is complete, samples with nonzero Lagrangian multipliers are support vectors and define the classification boundary. Once the boundary is established the outlier measure for each sample can be computed using the function $g(\vec{x})$ shown below.

$$g(\vec{x}) = \sum_{i=1}^{m} \alpha_i K(\vec{x}, \vec{s_i}) - \rho$$

In general, there will be many outliers of various degrees, so it is important to perform an outlier ranking to examine the worst outliers. $g(\vec{x})$ returns positive and negative numbers, where positive numbers correspond to normal samples and negative numbers correspond to outliers. Using $g(\vec{x})$ it is possible to rank-select the top 500 outlier windows and draw them on the target layout giving a clear picture of where variability occurs.

5. EXPERIMENTS AND RESULTS

Variability analysis was performed on 45nm metal layers extracted from 22 standard cells, and a large random tiling of those cells. Analysis performed on individual cells, from which more complicated layouts can be constructed, is called cell-based, while analysis performed on a tiling of cells is called chip-based. Both cell-based and chip-based analysis was performed using two-class and one-class algorithms.

978-1-60558-497-3/09 $25.00 © 2009 ACM

First, a 22 cell library was analyzed using both methods to ensure prediction accuracy, then a large layout based on a random cell tiling was analyzed to measure variability prediction accuracy on a realistic scale.

5.1 Cell Library Analysis

The first cell based experiment examined how accurately a classifier trained with 11 of 22 simulated cells could correctly predict variability in the unseen half. To verify prediction accuracy all 22 cells were simulated and pass/fail variability labels were calculated. Overall prediction accuracy was determined by comparing simulation and classification results. To train the classifier 11 large cells were scanned and generated 3960 training samples. To validate the classifier 11 smaller cells were scanned and generated 1584 testing samples. Larger cells generating more samples were used for training so that a generally representative variability classifier could be built. The second cell-based approach directly applied one-class outlier analysis on all 22 cells to compare how accurately it could classify samples relative to simulation results. In this case, training labels were not needed for analysis but only used to compute the classification accuracy.

Table 1 shows the classification accuracy for cell-based analysis using both one-class and two-class methods. Two-class SVM analysis achieved a 92.4% classification accuracy on unseen cells, and one-class outlier analysis correctly classified 85.1% of all cells without simulating to generate variability scores. The one-class method classified low and high variability samples with the same 85% accuracy, while the two-class method more accurately classified low variability samples than high variability samples. This is expected, since the two-class method was trained with more low variability samples, it learned a better concept of low variability. In one-class analysis no training labels were used, so the learned model was not biased toward predicting low or high variability.

Learning Method	Variability Prediction Accuracy		
	Total	Low Var.	High Var.
2-Class	92.4%	95.0%	86.5%
1-Class	85.1%	85.2%	84.8%

Table 1: Cell-Based Prediction Accuracy

5.2 Full Chip Analysis

Full chip analysis was performed in two ways; either by binary classification, or by one-class SVM outlier analysis. The advantage of one-class analysis is that it is an unsupervised method capable of classifying samples without depending on lithography simulation for training labels. The disadvantage is that the only way to introduce prior knowledge about variability is through your data representation and kernel function. According to features encoded by our HDT layout representation, the one-class outlier analysis method will identify abnormal layout areas that are likely to exhibit high variability, while the two-class method will identify layout areas that are similar to previously seen high variability samples.

The tiled layout shown in Figure 7 was analyzed using a two-class variability predictor trained with the full 22 cell library, as well as a one-class outlier model trained with sam-

ples from the scanned tiled cell layout. Raster scanning the layout generated 50,000 samples that needed to be analyzed and validated. Since the constructed tiling was very large it was time consuming to fully simulate, so three predicted clusters of high variability were selected for validation by lithography simulation. These cell clusters are thickly outlined in Figure 7 and labeled Tiled Simulations 1-3.

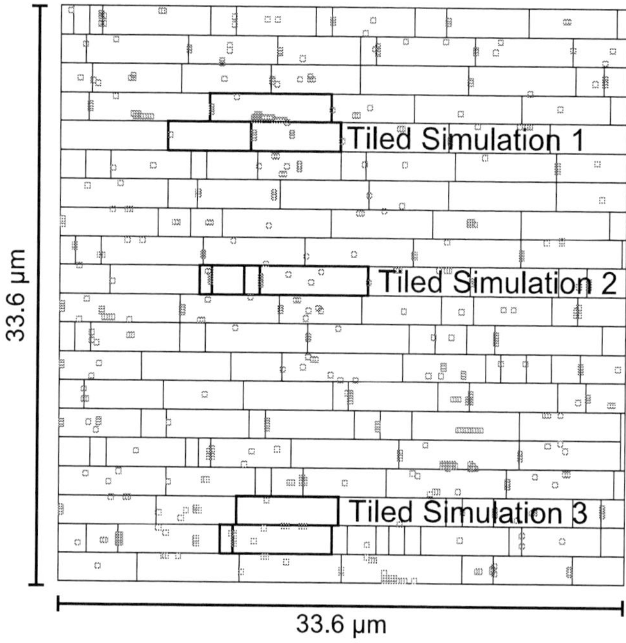

Figure 7: Predicted Variability on a Chip Layout

Two-class SVM analysis trained a high/low variability predictor based on lithography simulation of a cell library, and subsequently used that predictor to perform full layout analysis. Figure 7 show the top 500 areas of high variability detected by the binary classifier marked with small red squares. High variability areas are detected within and between standard cells showing that the model can generalize well and detect variability caused by unforeseen cell-to-cell interactions. One-class outlier analysis was performed on the same tiled layout and the top 500 areas of high variability were found to be very similar to those predicted in Figure 7 indicating that our two variability prediction approaches are consistent.

To show full-chip variability prediction is accurate, we chose three highlighted groups of adjacent cells in Figure 7, to verify by lithography simulation. After simulation, a grayscale map corresponding to the variability scores was printed across each set of adjacent cells. Darker areas correspond to higher variability, and lighter areas indicate lower variability. Figure 8 shows high variability areas predicted by lithographic simulation as well as areas of high variability predicted by our two-class full chip method. Across all three simulations it is clear that the darkest areas corresponding to high variability are well predicted by our SVM models. Moreover, we see that accurate variability prediction is possible between adjacent cells, without explicit enumeration of cell-to-cell interaction contexts.

978-1-60558-497-3/09 $25.00 © 2009 ACM

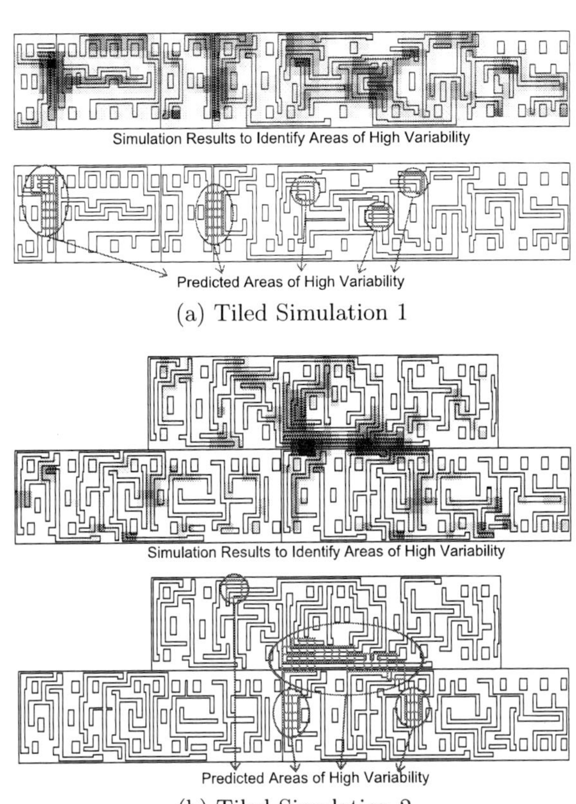

Simulation Results to Identify Areas of High Variability

Predicted Areas of High Variability

(a) Tiled Simulation 1

Simulation Results to Identify Areas of High Variability

Predicted Areas of High Variability

(b) Tiled Simulation 2

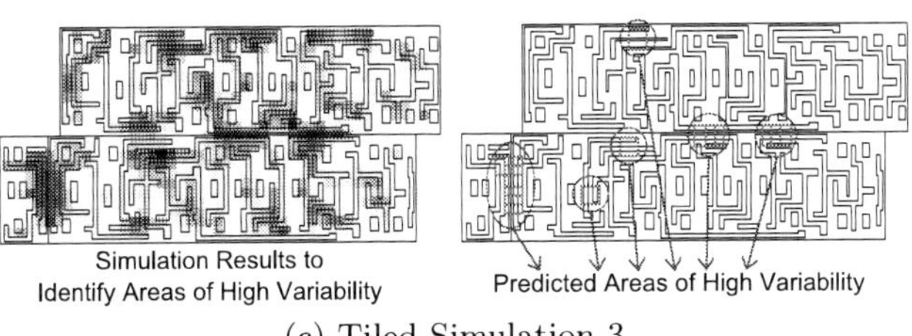

Simulation Results to Identify Areas of High Variability

Predicted Areas of High Variability

(c) Tiled Simulation 3

Figure 8: Simulated vs. Predicted Variability

Method	Raster	Train	Predict	Total
2-Class Cell	18 s	48 s	5 s	71 s
1-Class Cell	18 s	11 s	6 s	35 s
Litho. Cell	-	-	1.5 h	1.5 h
2-Class Chip	165 s	48 s	36 s	249 s
1-Class Chip	165 s	865 s	278 s	1308 s
Litho. Chip	-	-	33 h*	33 h*

*Based on 10 minute per-cell average runtime.

Table 2: Runtime Comparison

6. CONCLUSION

This paper introduced a new machine learning methodology for detecting areas of high variability within and between standard cells in full chip layouts. A novel HDT layout representation was used in conjunction with the latest SVM learning techniques to correctly predict areas of high variability according to a defined variability metric. Several experiments were performed and validated by an industry leading lithography simulator, showing that intra-cell and cell-to-cell variability could be predicted with high accuracy. Runtime of variability prediction was shown to be much faster than that of conventional lithography simulation, allowing high-speed first-order variability prediction.

7. REFERENCES

[1] S. Boughorbel, J. Tarel, and N. Boujemaa. Generalized histogram intersection kernel for image recognition. In *International Conference on Image Processing*, 2005.

[2] C. Chang and C. Lin. *LIBSVM: a library for support vector machines*, 2001. Software available at http://www.csie.ntu.edu.tw/ cjlin/libsvm.

[3] O. Chapelle, P. Haffner, and V. N. Vapnik. Support vector machines for histogram-based image classification. *Transactions on Neural Networks*, 1999.

[4] V. Dai, J. Yang, N. Rodriguez, and L. Capodieci. DRC Plus: Augmenting Standard DRC with Pattern Matching on 2D Geometries. In *SPIE*, 2007.

[5] P. F. Felzenswalb and D. P. Huttenlocher. Distance transforms of sampled functions. *Cornell Computing and Information Science Technical Report*, 2004.

[6] J. C. Rey, N. Nagaraj, A. Kahng, R. Aitken, L. Capodieci, F. Klass, C. Hou, and V. Singh. DFM in Practice: Hit or Hype? In *Design Automation Conference*, 2008.

[7] B. Schölkopf, J. Platt, J. Shawe-Taylor, A. J. Smola, and R. C. Williamson. Estimating the support of a high-dimensional distribution. *Neural Computation*, 2001.

[8] B. Schölkopf, A. J. Smola, R. C. Williamson, and P. L. Bartlett. New support vector algorithms. *Neural Computation*, 2000.

[9] N. Verghese, R. Rouse, and P. Hurat. Predictive Models and CAD Methodology for Pattern Dependent Variability. In *Design Automation Conference*, 2008.

[10] P. Yu, S. X. Shi, and D. Z. Pan. Process Variation Aware OPC with Variational Lithography Modeling. In *Design Automation Conference*, 2006.

5.3 Performance

Table 2 summarizes the runtime of SVM variability prediction compared to conventional lithography simulation. Both cell-based models ran quickly, taking on average 30 seconds to train, and 6 seconds to classify the full cell library. This speed can be attributed to the small number of support vector comparisons required to determine the class of a sample. Chip-based analysis was slower due to the larger area being scanned and analyzed. Two-Class chip-based analysis took 249 seconds to complete, since training only required scanning the small cell library. One-class chip-based analysis took 1308 seconds to analyze all 50,000 layout samples. This was a much slower analysis approach since it trained with more samples and classification depended on more support vectors. Classification can be accelerated by splitting test data into small subsets and using multiple processors to classify each subset in parallel. In comparison, lithography simulation took 1.5 hours to analyze the cell library, and approximately 33 hours to analyze the tiled cell array, which is 70-90x slower than SVM classification. All analysis was performed on a 3 GHz Core 2 Quad running Linux.

978-1-60558-497-3/09 $25.00 © 2009 ACM

Variability Analysis under Layout Pattern-Dependent Rapid-Thermal Annealing Process

Yun Ye[1], Frank Liu[2], Min Chen[1], Yu Cao[1]

[1]*Department of Electrical Engineering, Arizona State University, Tempe, AZ 85287*
[2]*IBM Austin Research Laboratory, Austin, TX 78758*
{yun.ye, min.chen, ycao}@asu.edu, frankliu@us.ibm.com

ABSTRACT

Rapid-Thermal Annealing (RTA) with radiation heating is recently adopted in nanoscale CMOS fabrication in order to achieve ultra-shallow junction with maximum dopant activation rate. However, recent results report the systematic shift of threshold voltage (V_{th}) and increased V_{th} variation due to RTA process [1-2]. The exact amount of variations depends on layout pattern density, RTA heating temperature (T) and effective annealing time. In this work, we develop joint thermal/TCAD simulation and compact modeling tools to analyze performance variability under various layout pattern densities and RTA conditions. With the new simulation capability, we recognize two major variation mechanisms under RTA: the change of effective channel length (L_{eff}) induced by lateral dopant diffusion, and the fluctuation of equivalent oxide thickness (EOT) due to incomplete dopant activation. We perform device simulations to quantify transistor performance shift due to L_{eff} and EOT variations. Moreover, we propose a suite of compact models that bridge the underlying RTA process with device parameter change for efficient design optimization. The new tools are validated with published silicon data at 45nm and 65nm nodes. They will facilitate physical designers to predict and mitigate circuit performance variability due to the layout-dependent RTA process.

Categories and Subject Descriptors

B.7.2 [**Integrated Circuits**]: Design Aids – *layout, simulation*;
B.8.2 [**Performance and Reliability**]: Performance Analysis and Design Aids.

General Terms

Design, Experimentation, Performance, Reliability

Keywords: Rapid-Thermal Annealing, Layout Pattern, Dopant Activation, Threshold Voltage Variation, Physical Design.

1. INTRODUCTION

The aggressive scaling of CMOS devices significantly improves the performance, but also leads to many undesirable effects. One of the most profound impacts is short-channel effects, such as Drain Induced Barrier Lowering (DIBL) that sharply

Permission to make digital or hard copies of part or all of this work for personal or classroom use is granted without fee provided that copies are not made or distributed for profit or commercial advantage and that copies bear this notice and the full citation on the first page. To copy otherwise, to republish, to post on servers or to redistribute to lists, requires prior specific permission and/or a fee.
DAC'09, July 26-31, 2009, San Francisco, California, USA

reduces threshold voltage at shorter channel length and leads to a dramatic increase in the leakage. To mitigate short-channel effects in scaled CMOS devices, advanced fabrication technology has to adopt rapid-thermal annealing (RTA) process in order to achieve ultra-shallow junction in the source/drain region. Different from traditional thermal annealing, the RTA process applies a much shorter pulse (e.g., Lamp RTA [3] or Laser Annealing [5][7]) to heat the silicon substrate to a much higher temperature (e.g., 4ms annealing at 1250°C) [4], such that dopants in the source/drain and gate regions receive sufficient energy to be activated, but only have the minimal period of time for the diffusion.

The RTA process is a must for nanoscale CMOS fabrication, achieving ultra-shallow junction depth and low source/drain/gate resistance. On the other hand, one distinct property of RTA is that the entire silicon substrate does not reach thermal equilibrium due to the extremely short heating period. During this period, the exact amount of energy and thus, the annealing temperature, depends on the reflectivity of the silicon substrate: the reflectivity of the gate is

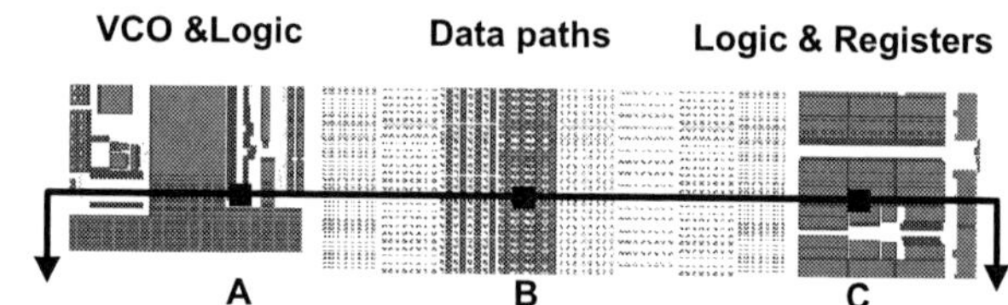

(a) Partial layout of a 45nm test chip.

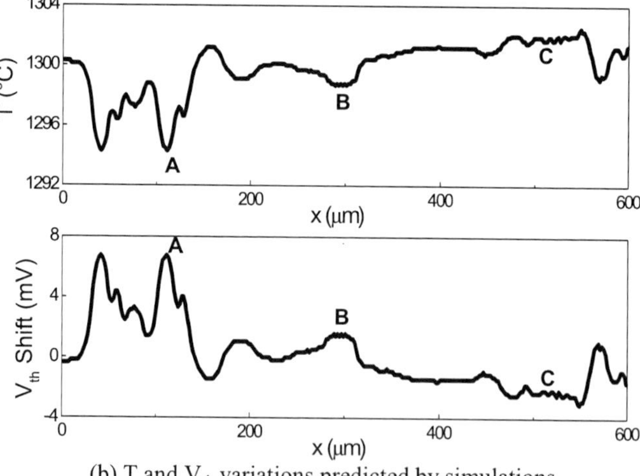

(b) T and V_{th} variations predicted by simulations.

Figure 1. RTA induced annealing temperature and threshold voltage variations under various pattern densities (45nm test chip, 10ms annealing at 1300°C).

usually lower than that of the source/drain region, while the isolation region has the lowest reflectivity due to the STI structure. As a result, different layout pattern densities lead to different annealing temperature, transistor definition and performance. This phenomenon has been observed in threshold and delay shift in test circuits [3]. The length scale of such variations is determined by the thermal diffusion distance in the silicon substrate, which is proportional to $(Dt)^{1/2}$, where D is the thermal conductivity of silicon, t is the annealing time. In the ms RTA process, this length is typically around hundreds of μm [3][6].

With such a thermal diffusion length, different silicon regions will have their own annealing temperatures, depending on local layout pattern. The fluctuation in T significantly affects transistor performance, such as I_{on}/I_{off} ratio and V_{th} [1][2][5]. As an example, Fig. 1 illustrates the layout of a realistic 45nm test chip. Due to the layout style, various components have a pronounced difference in pattern density (Fig. 1a). With the proposed thermal simulation tool, Fig. 1b illustrates the fluctuation in the annealing temperature, which directly induces V_{th} variation by more than 30mV. The largest difference is observed close to the boundary of different components, such Points A and C, where the non-uniformity of circuit layout reaches the maximum.

To minimize the variations, improvements in both process technology and physical design are necessary. In this work we develop a set of simulation and modeling capability, as shown in Fig. 2, to bridge the understanding of the underlying physics and circuit analysis under RTA, including (1) Transient thermal simulation to predict the dependence of the annealing temperature on layout pattern density; (2) Process simulation and modeling to analyze the primary impacts on equivalent gate oxide thickness and effective channel length; and (3) Device simulation and compact models to predict the change of device and circuit performance.

Using thermal simulation codes, we first obtain an appropriate simulation window to define the pattern density. This helps us efficiently track the temperature profile in a large scale layout. Then we investigate two independent variation mechanisms under RTA, i.e., dopant activation and lateral source/drain diffusion. These two effects further influence several important device parameters, including EOT, V_{th}, and L_{eff}. Based on TCAD simulation, we propose a set of analytical models that directly predict these parameter changes from RTA conditions and PD. The new models will enable efficient layout optimization in order to reduce the systematic variation due to RTA.

We systematically validate the method with TCAD tools and published silicon data under various conditions, including different annealing time, annealing temperature, and doping in devices. This paper is organized as follows: Section 2 presents the theoretical background and the development of thermal simulation capability. It defines an appropriate window size to extract pattern density under different RTA conditions. Section 3 integrates process and device simulations to analyze dopant activation lateral junction diffusion. Base on TCAD simulation results, we further develop compact models to predict the change of threshold voltage and other device parameters. Finally we apply the new models to benchmark the variability of circuit delay and leakage due to pattern-dependent RTA at 45nm node.

2. THERMAL SIMULATION OF PATTERN-DEPENDENT RTA PROCESS

In this section, we present the theoretical background and the code development of RTA thermal simulation. The study is performed with a representative 45nm technology. We use the equivalent emissivity in our thermal simulation in order to simplify the simulation. To further improve the simulation efficiency, we study the maximum window size that can be used to define local pattern density, as well as its dependence on RTA conditions.

2.1 Thermal Simulation

In a typical RTA process, the energy source, such as the lamp or laser, emits energy to the surface of the substrate. On the chip surface, different structures, e.g., gate, source/drain, and STI, have different emissivity [1][2] and thus, they absorb different amount of energy during the annealing. This is the origin that induces non-uniform temperatures within a die.

In order to investigate the interaction between pattern density and the annealing temperature, we develop transient thermal simulation on the cross section of the chip. Figure 3a illustrates the structure in thermal simulation. The initial condition of the full wafer is usually at a constant preheat temperature, e.g., 650°C [3]. A typical power profile of the annealing process is shown in Fig. 3b. Finite Difference Method (FDM) is applied to solve the initial value problem (IVP). Such a method is accurate, but not efficient enough if the chip size is large. To speed up the simulation in mm scale, we introduce the concept of the equivalent emissivity for the surface of the chip. The equivalent emissivity (ε_{eq}) is defined as the weighted average of the emissivity of each layout pattern within a simulation window, depending on the area density. The equivalent

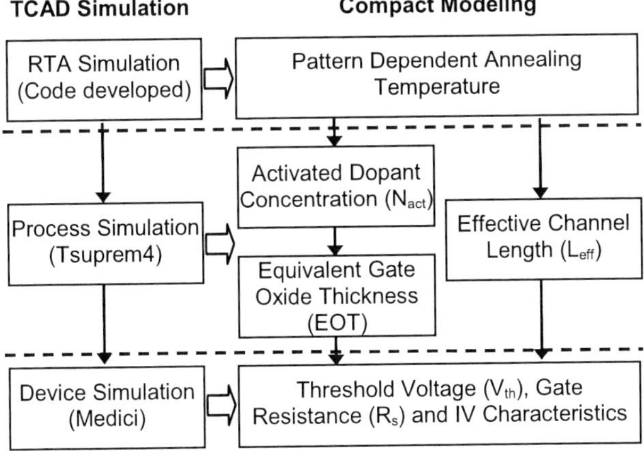

Figure 2. The flow of joint thermal/TCAD simulation and compact modeling to investigate the RTA process.

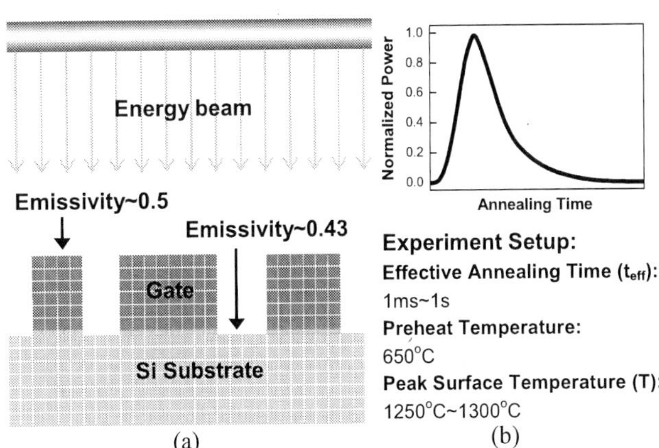

Figure 3. Layout thermal simulation. (a) simulation structure; (b) power profile in the experiment.

978-1-60558-497-3/09 $25.00 © 2009 ACM

emissivity is given as the following:

$$\varepsilon_{eq} = \sum_{i=1}^{n} \varepsilon_i \cdot \left(A_i / A_{window} \right) \qquad (1)$$

where ε_i refers to the emissivity of a pattern, A_i is the area of the pattern in the window, and A_{grid} is the total area of the simulation window. When the simulation window is much smaller than the thermal diffusion distance, e.g. 1μm, a single value of ε_{eq} is a sufficient representation of the thermal characteristics within the simulation grid. For the vertical direction, since the active regions of study are much thinner than the wafer thickness (usually around hundreds of μm), we treat the cross section of wafer as a homogenous material. Overall, the process of radiation heating based RTA has a much shorter ramp rate than traditional conduction based annealing process. The entire chip area does not reach thermal equilibrium during the annealing period [3].

2.2 Simulation Window Size

Using the newly developed thermal simulation tool, we are able to investigate the thermal conduction and temperature distribution inside the substrate. However, there is still a limitation on the window size we select to define the equivalent emissivity (Eq. 1) and calculate the pattern density. The window size is preferred to be large for fast simulation, but it should be small enough to track the change of the temperature profile for an accurate performance prediction.

To identify the appropriate window size, we simulate a sample 45nm design with various sizes of the simulation window. Within each window size, the emissivity of different patterns is averaged to the equivalent emissivity. Figure 4 shows the results from several representative window sizes using 20 ms RTA at 1300°C. The maximum temperature error is defined as the maximum difference as compared to the result from the minimum window size (100nm). Under a larger window size, the components with higher spatial frequency are filtered out and therefore, the simulation error increases. To guarantee sufficient accuracy in device performance prediction, we define the threshold window size when the maximum T error reaches 0.6°C, which corresponds to 0.8mV V_{th} shift in this 45nm technology. In the sample case, the appropriate window size is 22.4μm. The exact value depends on the RTA conditions, such as the annealing time and T.

Figure 5 illustrates the window size dependence on annealing temperature and time through thermal simulations. The criteria of post-annealing V_{th} shift is 1.1mV. The window size can be approximated as a linear function of T. It is also proportional to $t^{1/2}$ due to the thermal diffusion process [7]. Therefore, we express the overall dependence as:

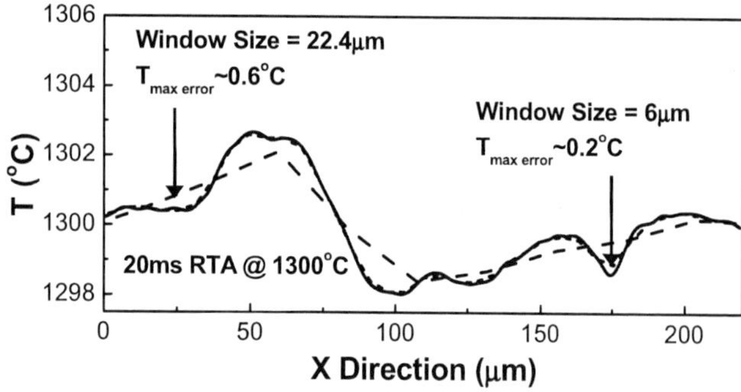

Figure 4. The search of the maximum window size during thermal simulation of the RTA process.

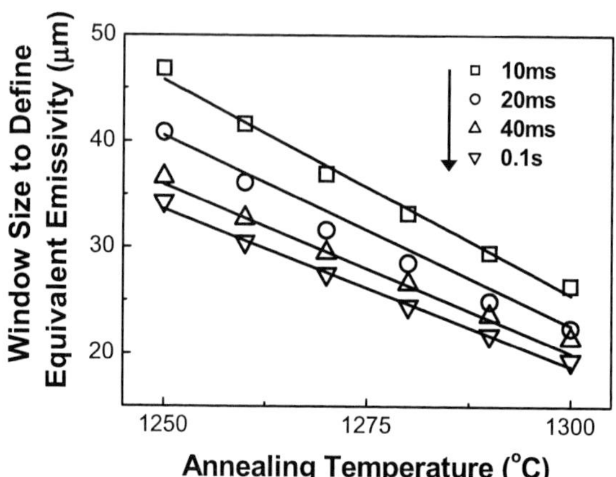

Figure 5. Simulation window sizes under different RTA conditions at 45nm node.

$$L_{window} = \left(A_1 \cdot T + A_2 \right)\left(B_1 \cdot t^{-1/2} + B_2 \right) \qquad (2)$$

where t refers to the annealing time, and A_1, A_2, B_1, B_2 are fitting parameters. Within the simulation window, the pattern density and the equivalent emissivity are averaged for fast thermal simulation, in order to predict the value of the annealing temperature for further model calculations. Through this method, the simulation efficiency is high enough to support chip-scale thermal simulations.

3. COMPACT MODELING OF PERFORMANCE VARIABILITY

There are two primary mechanisms that affect the threshold voltage in the RTA process (Fig. 6). The first one is dopant activation in the gate. We propose compact models to connect the annealing condition with dopant activation rate, equivalent oxide thickness and threshold voltage. The second factor is effective channel length defined by lateral thermal diffusion in the source/drain region. Due to the DIBL effect, V_{th} is highly sensitive to the change of L_{eff}. We describe the impact of the two mechanisms in compact models and validate them against TCAD simulations and published silicon data.

3.1 Dopant Activation

One major purpose of the RTA process is to electrically activate the dopants in the gate and source/drain regions. Depending of the activation rate, the polysilicon gate will have a finite doping level. When a suitable gate voltage is applied, it leads to the depletion close to the interface between the gate and the dielectric. This depletion is equivalent to the increase in oxide thickness and results in threshold voltage change. The concept of equivalent oxide thickness (EOT) is usually used to describe the

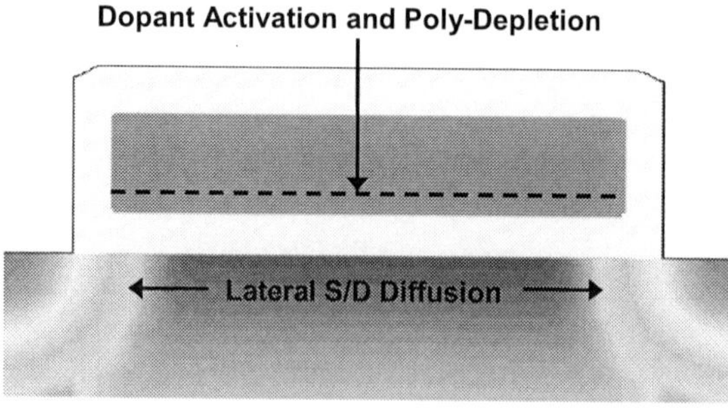

Figure 6. Two mechanisms affect device parameters in RTA process: dopant activation and lateral diffusion.

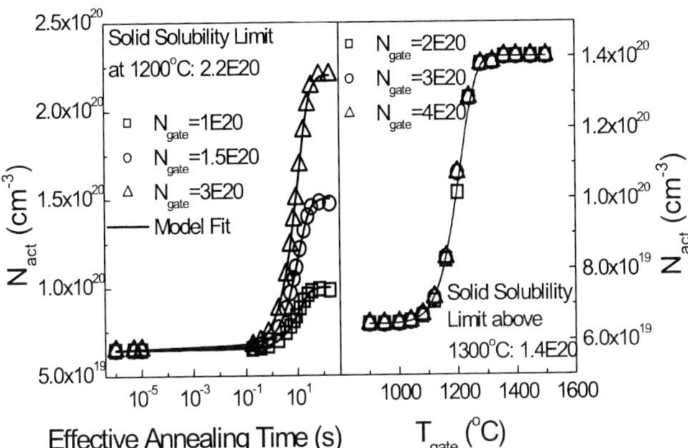

Figure 7. Dopant activation rate depends on RTA conditions and is limited by solid solubility (2.2E20 at 1200°C and 1.4E20 at 1300°C).

phenomenon. The EOT is given as the following:

$$EOT = W_{poly} \cdot \varepsilon_{Si} / \varepsilon_{ox} + t_{ox} \qquad (3)$$

where the W_{poly} is the depletion width in polysilicon gate, t_{ox} is the gate oxide thickness, ε_{Si} and ε_{ox} are dielectric constant for silicon and oxide respectively. If t_{ox} is large enough, the impact of poly-depletion can be ignored. However as technology scaling continues, the oxide thickness is as thin as 1nm and thus, the variation in poly-depletion can no longer be neglected.

In the RTA process, the dopants may not be completely activated due to the short time, even though T is high. Therefore the depletion width, which is inversely proportional to the square root of activated dopant concentration, becomes larger. The increase in EOT further leads to larger V_{th} after the annealing.

Here we employ a simple model to connect the activated dopant concentration (N_{act}) to the RTA process [8]:

$$N_{act} = N_{max} + (N_{min} - N_{max}) \cdot e^{-t_{eff}/\tau} \quad (4)$$

$$\tau = \tau_0 \cdot e^{E_a / k \cdot T^{-1}} \qquad (5)$$

where N_{max} refers to the maximum concentration of activated dopants; N_{min} is the minimum activated doping concentration, which refers to the activated doping concentration before the annealing; τ refers to the activation time constant, which is defined at the time that 50% dopants activated; t_{eff} is the effective annealing time such that the activation rate is equivalent to that of the simulated temperature profile [8]. Their values are usually available from RTA process parameters. Figure 7 shows the matching between analytical models and TCAD simulation results.

With active doping concentration, we are able to compute the change of EOT. In order to achieve the same I-V characteristics, the electric field in the oxide-channel surface should be constant, i.e., the electric field in the interface of the gate and the dielectric is a constant. Since the electrical field is proportional to $W_{poly} \cdot N_{act}$, we obtain $W_{poly} \sim 1/N_{act}$. We can express the EOT dependence as:

$$EOT = t_{ox} + t_{poly} = t_{ox} + a / N_{act} \qquad (6)$$

where a is a fitting parameter. Figure 8 validates the equation as compared to TCAD simulations, which are extracted from the C-V characteristics [8]. By integrating Eqs. 4-6, we are able to analytically predict the change of EOT from a given RTA.

3.2 Effective Channel Length

A side effect of the annealing process is the lateral diffusion that changes the value of effective channel length (Fig. 6). In RTA process the variation in temperature may result in channel length variation in nm scale. Although the diffusion of source/drain is relatively small and has a marginal impact on the junction depth, L_{eff}, which has a nominal value around 30nm at 45nm node, is very sensitive to the lateral junction change. Even with the change of several nanometers, threshold voltage is dramatically different, due to the exponential dependence of L_{eff} through the DIBL effect. We investigated the sensitivity of the junction change on annealing conditions by performing Tsuprem4 conditions and extracting compact models, as shown in Fig. 9. As a characteristic of the diffusion process, the junction change is dependent on $(Dt)^{1/2}$, where D is the diffusion coefficient, and t is the annealing time. We apply a polynomial equation to fit the dependence on the annealing temperature as the following:

$$\Delta X_j = a(T - T_0)^b \qquad (7)$$

where ΔX_j is the junction change after the annealing, T_0 is a reference temperature where the junction move is zero in a typical RTA process and a, b are a fitting parameters.

The TCAD simulations further confirms that the junction change is relatively insensitive to the doping level, as shown in the right figure in Fig. 9. Based on Fick's Law, the junction move has a square root dependence on the annealing time. Extracted from our simulation, the dependence is described as:

$$\Delta X_j = \sqrt{D(t + t_0)} - \sqrt{Dt_0} \qquad (8)$$

The parameter t_0 is the equivalent time before RTA to account for the preheat and other conditions, as well as to approximate the junction as an ideal abrupt shape for model derivation. Fig. 9 evaluates the model with Tsuprem4 results under various temperatures and the annealing time.

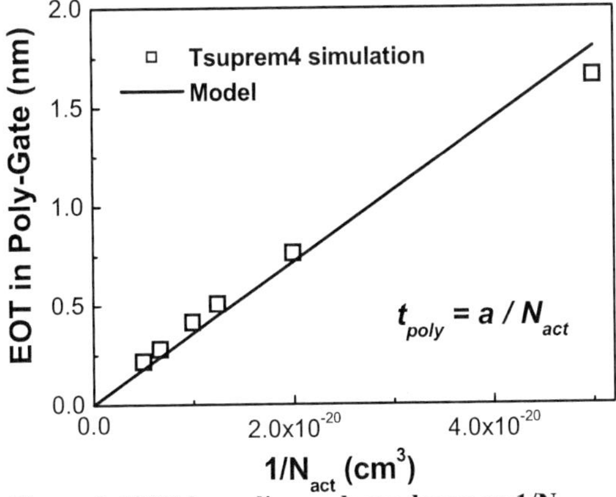

Figure 8. EOT has a linear dependence on $1/N_{act}$.

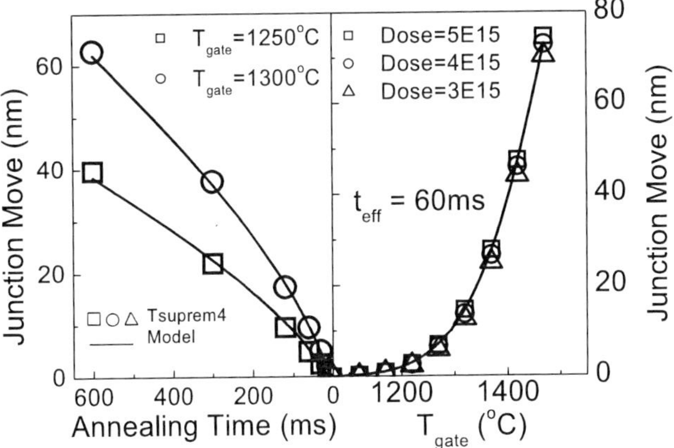

Figure 9. The dependence of ΔX_j on the RTA process.

978-1-60558-497-3/09 $25.00 © 2009 ACM

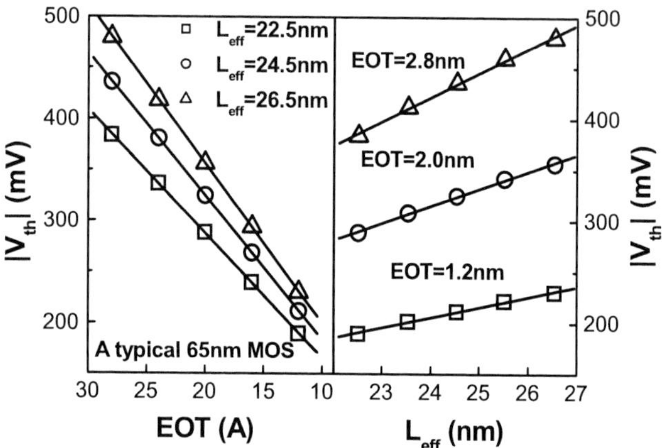

Figure 10. The shift of V_{th} due to RTA, as a compound of changes in EOT and effect channel length.

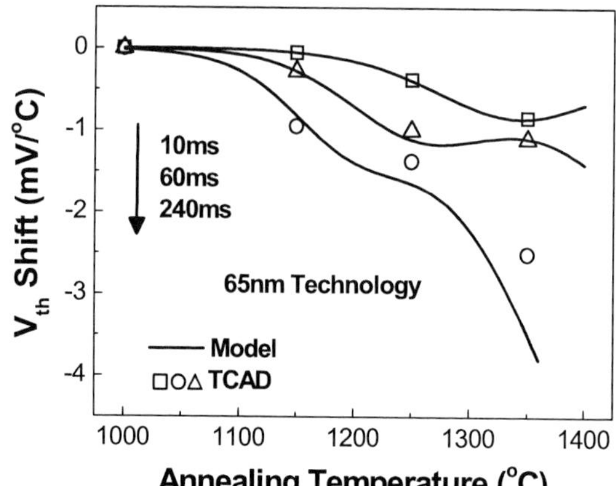

Figure 11. The dependence of threshold shift on the annealing temperature and time.

3.3 Impact on Threshold Voltage

With the models of L_{eff} and EOT variations, we further investigate the impact on device parameter by performing device simulations. The typical variation of the RTA annealing temperature in a 45nm design ranges from several °C to tens of °C. Such a change results in the junction move within 2nm and the variation in L_{eff} smaller than 4nm. In this small range of variations, the shift of V_{th} is approximately linear to L_{eff}, even though the DIBL effect is an exponential function of L_{eff}. Figure 10 illustrates the matching between models and TCAD simulations. Within the reasonable range of EOT, we are also able to expand V_{th} as a linear function of EOT. We propose the following model for the threshold dependence on L_{eff} and EOT:

$$V_{th} = V_{ref} + \left(A + B \cdot L_{eff}\right) \cdot t_{ox} \qquad (9)$$

where the A, B, V_{ref} are fitting parameters.

Table I summarizes the entire set of models to calculate Vth variation from the effective annealing temperature and time, which are predicted from thermal simulation on a given layout with the

appropriate window size. Based on these results, a physical designer will be able to efficiently diagnose and optimize layout pattern density to reduce performance variability. Figure 11 shows an example of V_{th} variation induced by the RTA for a 65nm technology. As shown in Fig. 11, the curve is not monotonic. This is because the threshold change is induced by both L_{eff} change and EOT change. While the L_{eff} change is proportional to $t^{1/2}$ and T^b, the shift of EOT shift rate is only pronounced when the activation rate is larger than 50%, as shown in Fig. 7. Such differences lead to the behavior in Fig. 11 that is well predicted by the new models.

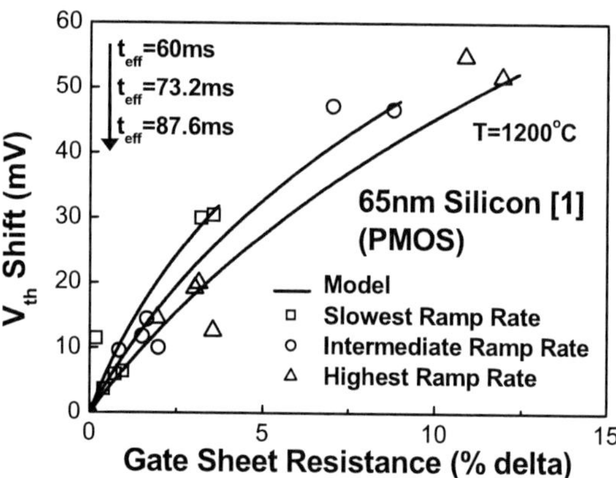

Figure 12. Compact models accurately predict the change of V_{th} and gate resistance under various RTA conditions.

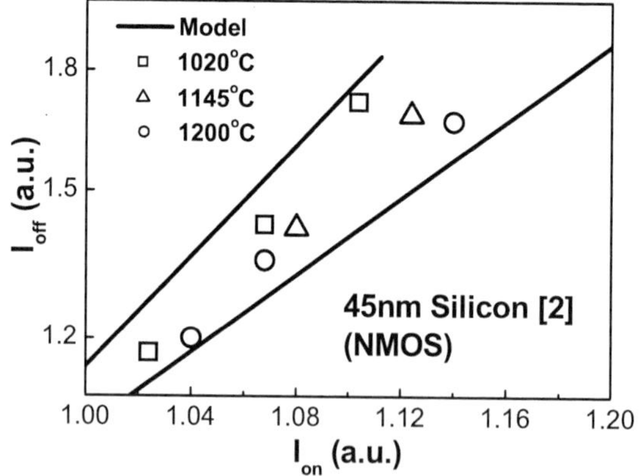

Figure 13. Higher annealing temperature improves I_{on}/I_{off}.

Table I. Compact models to predict V_{th} variation of a device under the RTA process. T is predicted by thermal simulations with a given layout pattern.

Thermal Annealing
Dopant Activation
$N_{act}(T,t) = N_{max} + (N_{min} - N_{max}) \cdot \exp(-t_{eff}(T,t)/\tau)$
$t_{eff} = \int_0^t \exp\left[E_a/k \cdot \left(T^{-1} - T'(t)^{-1}\right)\right] dt$
S/D Lateral Diffusion (to define L_{eff})
$\Delta X_j(T,t) = a(T - T_0)^b \left(\sqrt{D(t + t_0)} - \sqrt{Dt_0}\right)$
$L_{eff}(T,t) = L_{eff_0} - 2 \cdot \Delta X_j(T,t)$
Device Parameters
$EOT(T,t) = T_{ox} + T_{poly} = T_{ox} + a / N_{act}(T,t)$
$V_{th}(T,t) = V_{ref} + \left(a + b \cdot L_{eff}(T,t)\right) \cdot EOT(T,t)$
$\Delta V_{th}(T,t) = \partial V_{th}(T,t)/\partial T \cdot \Delta T$

978-1-60558-497-3/09 $25.00 © 2009 ACM

555

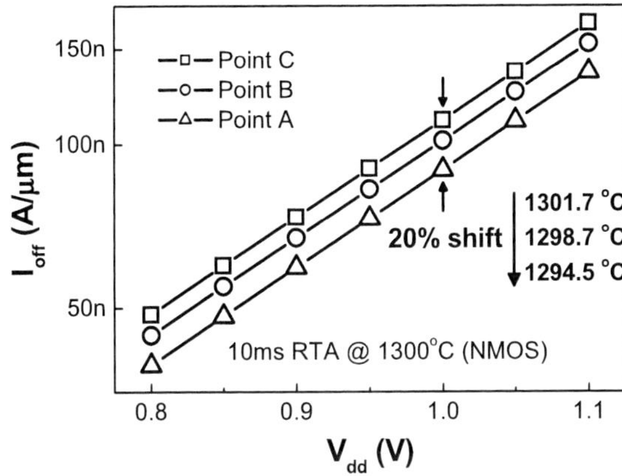

Figure 14. Within-die variation of the leakage at different sampling points in Fig. 1.

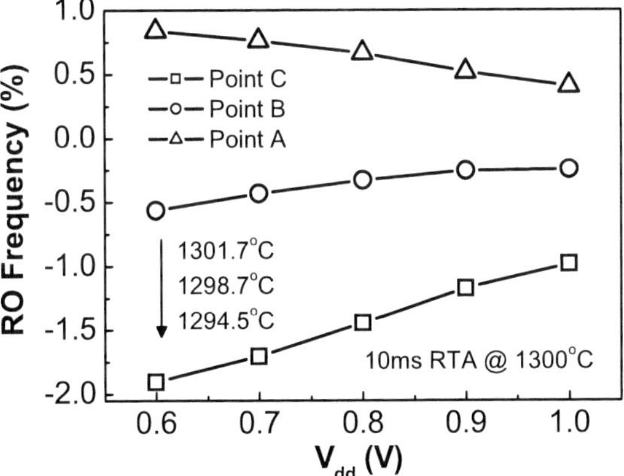

Figure 15. Within-die variation of RO frequency in the 45nm design in Fig. 1.

3.4 Validation with Silicon Data

We implement the models in SPICE simulator and validate its prediction with available published silicon data. Figure 12 shows the V_{th} shift vs. the sheet resistance of the gate, which is an index of the activation rate in polysilicon [1]. The silicon data are under different RTA annealing ramp rate, which is equivalent to different effective annealing time (Eq. 4). Furthermore, we evaluate the newly developed models the other set of 45nm silicon data. Figure 13 shows that at higher annealing temperature, I_{on}/I_{off} can be improved by ~10%, benefiting from higher dopant activation rate and therefore, thinner poly-depletion thickness in EOT. In both cases, our model well matches the published data using the same RTA conditions.

4. IMPACT ON CIRCUIT PERFORMANCE VARIABILITY

With the capabilities of compact modeling and circuit analysis, we benchmark the change of circuit performance change under different layout pattern densities. The patterns are taken from the 45nm layout as shown in Fig. 1. We simulate the leakage current and the frequency of identical 11-stage ring oscillators at three representative locations, Point A, B, and C in Fig. 1. Depending on their unique pattern density, the annealing temperature is generated from thermal simulation and used for model calculation of V_{th} shift. Other technology specifications are from PTM 45nm technology [10]. Figs. 14 and 15 highlight the simulation results. About 20% variation in the leakage and 3%

variation in the frequency are observed due to the non-uniformity in the layout. The leakage current becomes larger at the position with higher emissivity and thus, lower T. Within the increasing tight budget in power and timing, such an amount of variations need to be effectively reduced by joint process and design efforts.

5. CONCLUSION

In this work, we develop the capabilities of thermal simulation and compact models to analyze within-die variability due to pattern-dependent RTA process. The thermal simulation tool predicts the annealing temperature from the layout pattern. The new compact models further capture two major variations sources, EOT and L_{eff}, during the RTA, and calculate the shift of device parameters. The results are validated with TCAD simulations and silicon data at 45nm and 65nm generations. They effectively close the gap between the process knowledge and circuit simulation in order to minimize transistor and circuit performance variability due to systematic RTA effects.

6. ACKNOWLEDGEMENT

The authors acknowledge the support of the Materials, Structures, and Device Center (MSD), and the Center for Circuit and System Solutions (C2S2), two of five research centers funded under the Focus Center Research Program, a Semiconductor Research Corporation program.

REFERENCES

[1] I. Ahsan, et al., "RTA-Driven Intra-Die Variations in Stage Delay, and Parametric Sensitivities for 65nm Technology," *VLSI Symposium on Technology*, pp. 170-171, 2006.

[2] E. Granneman, X. Pages, H. Terhorst, K. Verheyden, K. Vanormelingen, E. Rosseel, "Pattern-Dependent Heating of 3D Structures," *Advanced Thermal Processing of Semiconductors*, pp. 131-138, 2007.

[3] T. Gebel, L. Rebohle, R. Fendler, W. Hentsch, W. Skorupa, M. Voelskow, W. Anwand, R. A. Yankov, "Millisecond Annealing with Flashlamps: Tool and Process Challenges," *RTP*, pp. 47-55, 2006.

[4] P. Timans, J. Gelpey, S. McCoy, W. Lerch, S. Paul, "Millisecond Annealing: Past, Present and Future," *Mater. Res. Soc. Symp. Proc.* vol. 912, 2006.

[5] M. Bidaud, "High-Activation Laser Anneal Process for the 45nm CMOS Technology Platform," *RTP*, pp. 251-256, 2007.

[6] T. Kubo, T. Sukegawa, E. Takii, T. Yamamoto, S. Satoh and M. Kase, "First Quantitative Observation of Local Temperature Fluctuation in Millisecond Annealing," *RTP*, pp. 321-326, 2007.

[7] L. M. Feng, Y. Wang, D. A. Markle, "Minimizing Pattern Dependency in Millisecond Annealing," *International Workshop on Junction Technology*, pp. 25-30, 2006.

[8] A. Mokhberi, P. B. Griffin, J. D. Plummer, E. Paton, S. McCoy, K. Elliott, "A Comparative Study of Dopant Activation in Boron, BF$_2$, Arsenic, and Phosphorus Implanted Silicon," *IEEE TED*, vol. 49, no. 7, pp. 1183-1191, July 2002.

[9] F. Ootsuka, "An Engineering Method to Extract Equivalent Oxide Thickness and its Extension to Channel Mobility Evaluation," *IEEE TED*, vol. 49, no. 12, pp. 2345-2348, Dec. 2002.

[10] W. Zhao, Y. Cao, "New generation of predictive technology model for sub-45nm design exploration," *IEEE TED*, vol. 53, no. 11, pp. 2816-2823, Nov. 2006

Event-Driven Gate-Level Simulation with GP-GPUs

Debapriya Chatterjee, Andrew DeOrio and Valeria Bertacco

Department of Computer Science and Engineering, University of Michigan
{dchatt, awdeorio, valeria}@umich.edu

ABSTRACT

Logic simulation is a critical component of the design tool flow in modern hardware development efforts. It is used widely – from high-level descriptions down to gate-level ones – to validate several aspects of the design, particularly functional correctness. Despite development houses investing vast resources in the simulation task, particularly at the gate-level, it is still far from achieving the performance demands required to validate complex modern designs.

In this work, we propose the first event-driven logic simulator accelerated by a parallel, general purpose graphics processor (GP-GPU). Our simulator leverages a gate-level event-driven design to exploit the benefits of the low switching activity that is typical of large hardware designs. We developed novel algorithms for circuit netlist partitioning and optimized for a highly-parallel GP-GPU host. Moreover, our flow is structured to extract the best simulation performance from the target hardware platform. We found that our experimental prototype could handle large, industrial scale designs comprised of millions of gates and deliver a 13x speedup on average over current commercial event-driven simulators.

Categories and Subject Descriptors. B.6.3 [**Logic Design**]: Design Aids—*Simulation*; C.1.2 [**Processor Architectures**]: Multiple Data Stream Architectures (Multiprocessors)—*Parallel Processors*

General Terms. Verification, Performance

Keywords. Gate-level simulation, High-performance simulation, General Purpose Graphics Processing Unit(GP-GPU)

1. INTRODUCTION

Logic simulation is the validation workhorse of modern digital designs. It is used to verify designs at the behavioral level, as well as the structural level, ensuring that a synthesized circuit's netlist matches the functionality and timing of the behavioral model. Structural netlists are particularly cumbersome for simulation because of their low-level specification and the fine granularity of the structural definition, which consists of gate primitives in the target technology library. It is typical for design houses to invest the computational power of large simulation "farms" to complete as many simulation cycles as possible before final design tapeout. However, even with such investment in today's development efforts, large portions of a design go unverified. The result is unforeseen bugs that are released into silicon, which may have drastic impacts, ranging from silicon respins to market recalls.

The root cause of this situation lies in the vast complexity of modern designs (several million gates) and the fact that the performance of commercial logic simulators is inversely proportional to their size. In addition, the technology in commercial logic simulators today is fairly mature, thus their performance improvement between subsequent releases largely relies on the performance trends of the underlying simulating hardware host.

Modern gate-level simulators proceed in two phases: during the first phase, the circuit netlist to be simulated is restructured and optimized (compilation phase); in the second phase, the netlist is simulated ("executed") using the input stimuli specified in the testbench. The performance of the simulator is driven by this second phase, since the compilation step is only required once per netlist. In this work, we propose a novel simulator design that leverages the high-performance of general purpose graphics processing units (GP-GPUs) for the execution phase of the simulation, leading to a major improvement in simulation performance. During the compilation phase, a netlist is "levelized", that is, gates are organized into levels so that all the gates in one level depend only on simulation values generated in previous levels. Thus, gates in a same level are not directly connected and can be simulated in parallel, an advantage that can be leveraged when many parallel processing units are available, as in GPUs. During the execution phase, gates are simulated by level; however, in an *event-driven* simulation, a gate is simulated only if at least one of its input values had changed, while in an *oblivious* simulator all gates are evaluated with each cycle. While oblivious simulation has the advantage of simple, efficient static gate scheduling, event-driven simulation has been noted to perform better in practice. This is because it is typical for large designs to only simulate a small fraction of the gates (1 to 10%) during any given cycle. Thus, even in face of a more complex dynamic scheduling architecture, most commercial simulators rely on an event-driven approach for performance reasons.

The recent availability of general purpose computing programming models for high-performance and highly parallel GPUs led us to explore a new simulation architecture targeting these hardware platforms, with the hope of delivering a conspicuous performance advantage at a small hardware cost (that of a GPU peripheral). Specifically, the NVIDIA's CUDA architecture provides a programming interface that enables users to develop software applications for their vastly parallel co-processor GPU. However, CUDA exposes its parallel architecture directly to the programmer, with the result that applications must be designed specifically for this structure in order to derive benefit from it.

1.1 Contributions

In this work, we present the first event-driven GPU-based logic simulator, which leverages GPUs' massive parallelism to achieve large performance speedups over commercial logic simulators. Our solution leverages a novel *macro-gate* segmentation algorithm, designed specifically to benefit from the CUDA architecture. A macro-gate comprises several gates of the original netlist connected to each other. The macro-gates generated cover the entire circuit's netlist; they are compiled into a suitable data structure and transferred to the GPU's memory. During simulation, those macro-gates that require simulation because their input values have changed, are tagged for execution and handed over to the CUDA's low-level scheduler. We developed a prototype of our simulator and applied it to a range of designs, including a SPARC multiprocessor of more

Permission to make digital or hard copies of part or all of this work for personal or classroom use is granted without fee provided that copies are not made or distributed for profit or commercial advantage and that copies bear this notice and the full citation on the first page. To copy otherwise, to republish, to post on servers or to redistribute to lists, requires prior specific permission and/or a fee.
DAC'09, July 26-31, 2009, San Francisco, California, USA

than a million gates. We developed several testbench infrastructures, from random generators running on the GPU, to assembly programs for processor designs; and simulated the designs for millions of cycles. We found that our GPU-based simulator delivers performance speedups from 4 to 44 times over the performance of a top-end commercial simulator, with 13 times being the average.

2. RELATED WORK

Research on logic simulators bloomed in the 1980s, when the concepts of circuit netlist compilation, oblivious and event-driven simulation were first explored [6, 3, 14, 2]. In particular, [2] provides a comparative analysis of early attempts to parallelize event-driven simulation by dividing the processing of individual events across multiple machines with fine granularity. This fine granularity would generate a high communication overhead and, depending on the solution, the issue of deadlock avoidance could require specialized event handling. Parallel logic simulation algorithms were also proposed for distributed systems [16, 15] and multiprocessors [12]. In these solutions, individual execution threads would operate on distinct netlist clusters and communicate in an event-driven fashion, with a thread being activated if switching activity was observed at the inputs of its netlist cluster. Both conservative [7, 17, 10] and speculative techniques, such as time warp [5, 4], were proposed to handle synchronization in these discrete event algorithms. Today, several commercial simulators building on these concepts are available: they execute on a single CPU and adopt aggressive compiled-code optimization techniques to boost their performance.

In addition, specialized hardware solutions (*emulation systems*) have also been explored to boost simulation performance. These systems typically consist of several identical hardware units connected together, with units optimized for the simulation of small logic blocks. To emulate a circuit netlist, a "compiler" partitions the netlist into blocks and then loads each block into separate units [9, 1, 13]. Modern emulators can deliver 3-4 orders of magnitude speedup and they can handle very large designs. However, their cost is prohibitively large and the process of successfully mapping a netlist to an emulator can take up to few months.

Most recently, a few research solutions have been proposed to run simulations on GPUs: a first attempt by Perinkulam [20] did not provide performance benefits due to lack of general purpose programming primitives for their platform and the high communication overhead generated by their solution. An oblivious simulator solution was proposed in [8]: their software design is simpler, and can be optimized statically, but simulating all gates in each cycle limits the performance of this approach. Moreover, the size of the circuits that can be simulated with the solution in [8] is severely limited by the size of the shared memory in the GPU platform. Another recent solution in this space [11] introduces parallel fault simulation on a CUDA GPU target. It derives its parallelism by simulating distinct fault patterns on distinct processing units, with no partitioning within individual simulations or the design. In contrast, we target fast simulation of complex designs, thus we must explore circuit partitioning and optimizations techniques in order to leverage the parallelism of the target platform. Moreover, we optimize the performance of individual simulation runs, in contrast with [11], which optimizes over all faults simulations.

3. INTRODUCTION TO CUDA

The architecture of modern graphic processing units (GPUs) comprises a large number of data streaming processing units. They are commonly fairly simple, programmable and together can execute an astonishing amount of floating point (or integer) instructions in parallel. Typically, GPUs can be programmed via a graphics

library interface, however NVIDIA has made available a general purpose programming interface for their CUDA platform (Compute Unified Device Architecture [18]), enabling the development of a broader range of applications. A CUDA GPU consists of a set of multiprocessors (Figure 1, each comprising several functional units (FUs), which can execute multiple program threads concurrently (up to 512). Threads are organized into blocks; with one or more blocks in concurrent execution on individual multiprocessors. All threads running in a multiprocessor have fast access (1 cycle) to a small shared local memory (16KB), and also to a much larger device memory (up to 1GB - 300-400 cycles latency for access). The CUDA architecture can be programmed using C language extension in a SIMD (single-instruction-multiple-data) fashion: all FUs across the entire GPU must be executing the same code, operating on different data. Finally, data placement to shared or device memory must be handled explicitly by the programmer. When executing a program on CUDA, also called a *kernel*, the host computer uploads the data and compiled program (in our case the netlist and simulation code) to the GPU's device memory, and then relinquishes control to the GPU scheduler, which executes all required threads autonomously until simulation completes.

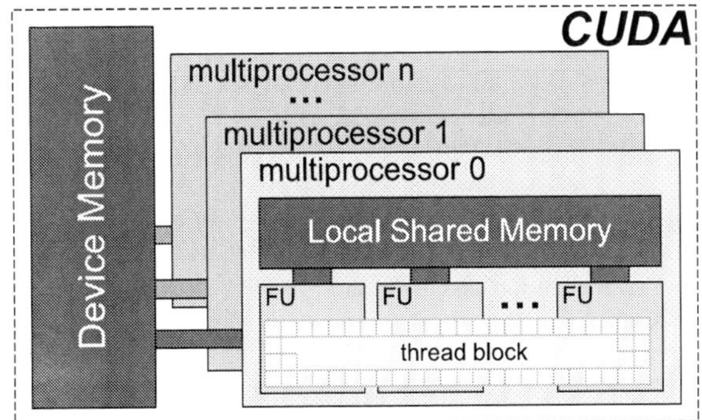

Figure 1: NVIDIA CUDA architecture. A GPU includes a number of multiprocessors, each comprising 8 execution units. Several threads (up to 512) may execute concurrently within a multiprocessor and communicate through a small shared memory bank (16KB). The larger device memory has much higher access latency.

4. OVERVIEW

Our event-driven CUDA-accelerated simulator first applies a *compilation phase*, during which it transforms the netlist to leverage the raw performance of the target architecture. This is followed by a *simulation phase* where the compiled netlist is uploaded to the GPU co-processor and one or more simulations may be executed with different input testbenches.

The compilation phase is responsible for segmenting a large monolithic netlist into blocks amenable to simulation by individual execution units within the GPU. This requires segmenting the netlist into *macro-gates*: a set of several connected gates within the netlist of ideal size, optimizing the logic within each macro-gate, and finally producing the data structures and the CUDA programs necessary to carry out the simulation. During simulation, both program and data reside on the GPU. Testbenches can be implemented using many different solutions; if they are encoded in a CUDA program (possibly with associated stimuli data), then the simulation can be completely offloaded from the host with direct performance benefits. If the testbench resides on the host, control alternates between host and GPU to simulate and generate stimuli.

4.1 Netlist generation

The first step of compilation considers a digital design and synthesizes it to a flattened netlist using a target technology library (we

978-1-60558-497-3/09 $25.00 © 2009 ACM

used the GTECH library by Synopsys for our experiments). If the design is a gate-level description (as in the case of synthesis validation), the synthesis step may be unnecessary. Finally, the combinational portion of the netlist is extracted for further processing, while the storage elements will be mapped to memory during simulation. Note that in our implementation, we excluded tri-state buffer and latches from the synthesis library to obtain a simple synchronous netlist. Latches could be easily included by adapting our simulator to operate at a finer granularity, that is, time units instead of clock cycles. Tri-state elements can be included by using 4-valued logic instead of binary. Both of these are straightforward extensions to the simulator. The combinational netlist is finally *levelized*, that is, logic gates are organized into levels, so that the fanin of all gates in one level is computed in previous levels. With this organization, it is possible to simulate the entire netlist one level at a time, from inputs to outputs, with no backward dependency. In our prototype implementation, we used an ALAP (as-late-as-possible) levelization, though other solutions are also possible.

4.2 Segmentation into macro-gates

To exploit the parallelism available in the GPU, we must segment the gate-level netlist into several logic blocks (called *macro-gates*), and assign the simulation of each macro-gate to a distinct CUDA multiprocessor. During simulation, we maintain a *sensitivity list* of nets at the inputs of each macro-gate: if any net in a sensitivity list changes value, then the corresponding macro-gate will be affected by the change and must be simulated (*i.e.activated*). Otherwise, the macro-gate can be skipped during the current cycle.

In determining how to partition the netlist into macro-gates, we took into consideration several factors: (i) the time required to simulate a macro-gate should be greater than overhead of determining which macro-gates to simulate; (ii) CUDA's multiprocessors can only communicate through device memory, thus macro-gates should not share data. To this end, we occasionally duplicate small portions of logic, so that each macro-gate can compute the value of its outputs independent of other concurrent macro-gates. Finally, (iii) we want to avoid cyclic dependencies between macro-gates, so to simulate each macro-gate at most once per cycle.

To address the constraint list, we segment the netlist by partitioning the netlist into *layers*: each layer encompasses a fixed number of the netlist's levels. Macro-gates are then defined by selecting a set of nets at the top boundary of a layer, and including its cone of influence back to the input nets of the layer. The number of levels

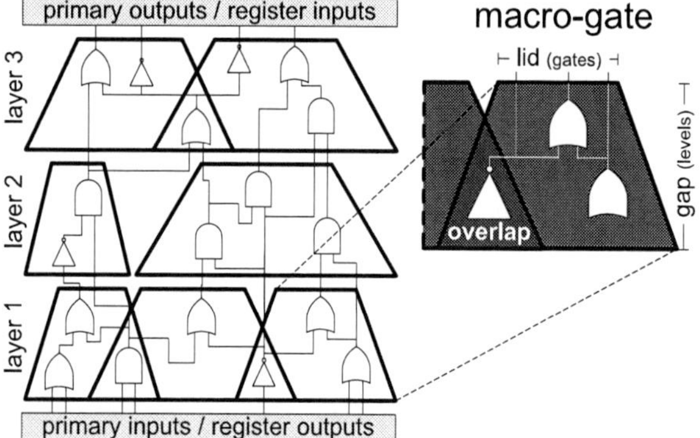

Figure 2: Segmentation topology. The levelized netlist is partitioned into layers, each encompassing a fixed number of levels (gap). Macro-gates are then carved out by extracting the transitive fanin from a set of nets (lid) at the output of a layer, back to the layer's input. If an overlap occurs, the gates involved are duplicated to all associated macro-gates.

within each layer is called the *gap* and corresponds to the height of the macro-gate. By using this procedure, it is possible that a given logic gate is assigned to two or more macro-gates. In this case, we duplicate it, so that each macro-gate can compute the value of its output nets without sharing any data with other macro-gates (second requirement). Finally the number of output nets used to generate each macro-gate is a variable parameter (called *lid*), whose value is selected so that the number of logic gates in all macro-gates is approximately the same. Figure 2 shows a schematic of the segmentation technique, while figure 3 presents the pseudo-code of the algorithm. The set of nets that cross the boundary between each pair of layers is monitored during simulation to determine which macro-gates should be activated.

Section 5 discusses the process that we used to select optimal values for gap and lid, so as to achieve a high-level of parallelism during simulation with little macro-gate overlap and low activation rates. Note that, in our prototype implementation, we fixed gap and lid across the entire netlist: additional performance could be achieved if each layer had its own associated gap and each macro-gate had an associated lid.

```
segmentation (netlist, gap, lid) {
    levelized_netlist = ALAP_schedule(netlist)
    layers = gap_partition(levelized_netlist)
    for (layer in layers) {
        macro-gates = lid_partition(layer)
        macro-gates_pool = append(macro-gates);
        compute_monitored_nets (layer);
    }
    return macro-gates_pool }
```

Figure 3: Macro-gate segmentation algorithm. The levelized netlist is partitioned into layers: several macro-gates are carved from each layer and appended to the macro-gates pool to be simulated. The nets to be monitored are also tagged at this stage.

4.3 Macro-gate balancing

Each macro-gate is designed to be simulated in a single CUDA multiprocessor. Because our lowest-level primitives are basic logic gates, we designed our CUDA simulation program so that the execution threads simulate all the gates in the same level, then move on to the next level, and so on, until an entire macro-gate has been simulated. Thus the gap is directly proportional to layer simulation performance. However, the segmentation procedure tends to generate macro-gates with a large base (many gates) and a narrow tip. Correspondingly, we have many active threads in the lower levels, and just a few in the top levels.

To maximize concurrency throughout the simulation, we optimize each macro-gate individually with a *balancing* step, as outlined in the schematic of Figure 4. This is the last step of the compilation phase: it exploits the slack available in the levelization within each macro-gate and restructures macro-gates to have approximately the same number of logic gates in each level. As a result, a smaller number of threads will be required to simulate the base of the macro-gate. Note that it is always possible to "shrink" the size of the base, at the price of an increased gap.

4.4 Simulation phase

As mentioned earlier in this section, simulation is carried out directly on the GPU co-processor. Each multiprocessor is responsible for the simulation of one or more macro-gates. Each macro-gate corresponds to one thread block. In determining the number of macro-gates that should be simulated concurrently on a multiprocessor, the number of concurrent thread blocks allowed in a multiprocessor (3), was the limiting factor. A single allocation would enable larger macro-gates, however, mapping several smaller ones

978-1-60558-497-3/09 $25.00 © 2009 ACM

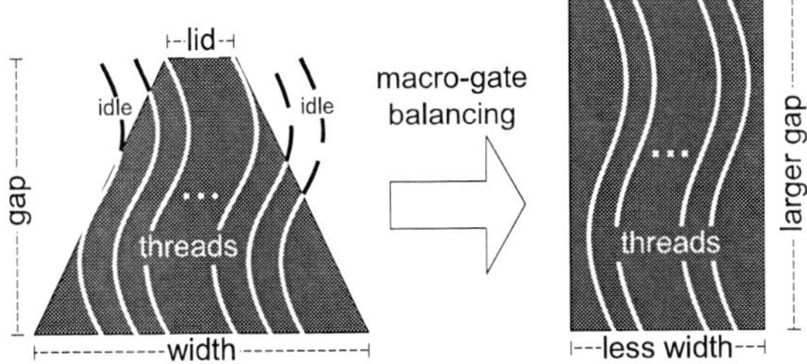

Figure 4: Macro-gate balancing. The balancing algorithm exploits the levelization slack within a macro-gate to restructure it so that fewer execution threads are required to simulate the lower levels, and idle threads are minimized at the top levels.

concurrently allows us to hide the memory latency in retrieving structural netlist data from device memory. We found experimentally that the latter solution provides better performance.

The overall simulation alternates executing all active macro-gates in a layer, with analyzing the corresponding monitored nets to determine which macro-gates should be activated for the next layer. The CUDA scheduler is responsible for assigning activated macro-gates to individual multiprocessors. Figure 5 illustrates the layered structure of macro-gates and monitored nets. It also shows how activated macro-gates are transferred from the pool to a multiprocessor for execution. Within a macro-gate simulation, multiple concurrent threads simulate all the gates in same level, then synchronize, and finally advance to the next level, until completion.

Data placement is organized as follows: primary inputs, outputs, register values and monitored nets are mapped to device memory, since they must be shared among several macro-gates (multiprocessors). Truth tables for the gates in the technology library are mapped to shared memory because of their frequent access. In addition, intermediate net values generated within a macro-gate are also placed in shared memory. Finally, the netlist structure is stored in device memory and accessed during each macro-gate simulation.

5. OPTIMIZATIONS

5.1 Macro-gate sizing and activation

In segmenting a netlist into macro-gates, the selection of *gap* and *lid* values have critical impact on the simulation performance (see also Section 4.2). During the compilation phase, we select these values by evaluating a range of solution points; for each candidate value we collect several metrics: number of macro-gates, number of monitored nets, size of macro-gates and activation rate. The activation rate is obtained by a mock-up of the simulation on a micro testbench. We then select the locally optimal values and perform

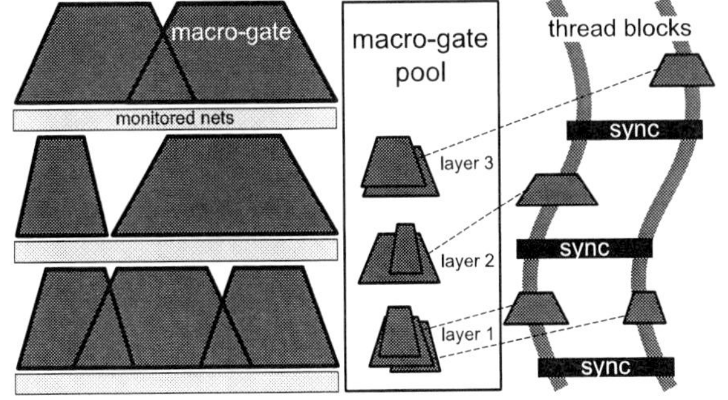

Figure 5: The event-driven simulation operates by layer. Within each layer, it simulates activated macro-gates and then analyzes the monitored nets to tag additional macro-gates for activation. Activated macro-gates are transferred by the CUDA scheduler to an available multiprocessor for simulation.

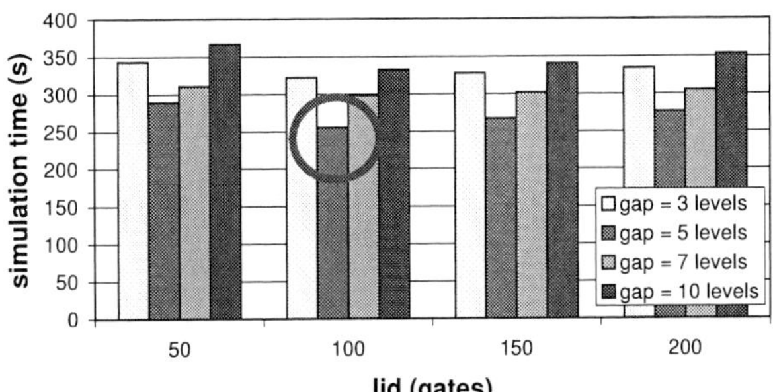

Figure 6: Estimation of optimal gap and lid for the LDPC testbench design. We run a mock simulation with a micro testbench using a range of gap and lid values and found that optimal performance is achieved for gap=5 and lid=100.

detailed segmentation. Figure 6 shows an example of this selection. The chart reports the simulation times for the LDPC benchmark design when running the micro-testbench: each bar corresponds to a unique <gap,lid> value pair. In this example the best estimates are 5 for gap and 100 for lid. The boundaries for the range of gap values considered are derived from the number of monitored nets generated: we only consider gap values for which no more than 50% of the total nets are monitored. In practice, small gap values tend to generate many monitored nets, while large gap values trigger high activation rates. For lid determination, we bound the analysis by estimating how many macro-gates will be created at each layer: We strive to run all the macro-gates concurrently. The GPU used for our evaluation included 14 multiprocessors and the CUDA scheduler allows at most three thread blocks in concurrent execution on a same multiprocessor. Thus we only consider lid values that generate no more than $14 \cdot 3 = 42$ macro-gates per layer. Note that this analysis is performed only once per compilation.

After the simulation of all active macro-gates in a layer is completed, the GPU executes a *scheduling kernel* that evaluates the array of monitored nets to determine which macro-gates should be activated in the next layer. This array is organized as a bit vector, with each monitored net being implicitly mapped to a unique location in the array. If a macro-gate simulation modifies the value of any of these nets, its corresponding location is tagged. Each macro-gate has a corresponding *sensitivity list* where all the input nets triggering its activation are tagged. With this structure, a simple bit-wise AND operation between the monitored nets array and a macro-gate's sensitivity list determines if any input change has occurred and the macro-gate should be activated. The alternative of maintaining the sensitivity lists as linked lists within the monitored nets array would require variable size data structures, which are extremely cumbersome to manage in a GP-GPU architecture.

5.2 CUDA-specific optimizations

We also explored a few optimizations that are specific of the CUDA architecture. For instance, CUDA has an additional memory block, called *texture memory* that can be used as an intermediary to access device memory. The texture memory controller operates by conglomerating adjacent memory accesses and sending block requests to device memory. We leveraged this memory when retrieving the netlist structure of a macro-gate during simulation: since gates in a same level are placed in contiguous locations in device memory, the access through texture memory could bypass most of the latency for these data.

6. EXPERIMENTAL RESULTS

We evaluated the performance of our simulator on a broad set of designs ranging from purely combinational circuits such as an

Design	Testbench	# Gates	# Flops
Alpha no pipeline	recursive Fibonacci program	17546	2795
Alpha pipeline	recursive Fibonacci program	18222	2804
LDPC encoder	random stimulus	62515	0
JPEG decompressor	1920x1080 image	93278	20741
3x3 NoC routers	random legal traffic	64432	13698
4x4 NoC routers	random legal traffic	144098	23875
OpenSPARC core	OpenSPARC regression suite	262201	62001
OpenSPARC-2 cores	OpenSPARC regression suite	610670	124002
OpenSPARC-4 cores	OpenSPARC regression suite	1221340	248004

Table 1: Testbench designs for evaluation of the simulator.

LDPC encoder, to a multicore SPARC design containing over 1 million logic gates. Designs were obtained from OpenCores [19] and from the Sun OpenSPARC project [21]; the Alpha processors and NoC designs were developed in advanced digital design courses by student teams at the University of Michigan.

We report in Table 1 the key aspects of these designs: number of gates, flip-flops and type of stimulus that was used during simulation. The first two designs are processors implementing the Alpha instruction set, the first can execute one instruction at a time, while the second has a 5-stage pipelined architecture. Both were simulated executing a binary program that computed Fibonacci series recursively. The LDPC encoder outputs an encoded version of its input; for this design we developed a random stimulus generator that run directly on the GPU platform. The JPEG decompressor would decode an input image. The NoC designs consist of a network of 5-channel routers connected in a torus network and simulated with a random stimulus generator sending legal packets through the network. Finally, the OpenSPARC designs use processors from the OpenSPARC T1 multi-core chip (excluding caches) and run a conglomeration of assembly regressions provided with Sun's open source distribution. We built several versions of this processor: single-core, two cores, and four cores and we simulated local cache activity by using playback of pre-recorded signal traces from processor-crossbar and processor-cache interactions.

6.1 Macro-gates

We studied several aspects of the compilation phase of our simulators and report here our findings. Figure 7 shows the total number of macro-gates generated for each design when using the gap and lid values determined in Section 5.1. On average each macro-gate includes 400 logic gates. In addition, we indicate the number of layers used in the segmentation of each design. Note that the largest design include many more macro-gates in each layer that could be simulated concurrently(42 as per section 5.1).

As mentioned in Section 4.2, gate duplication is a necessary consequence of the high communication latency between multiprocessors. However, we strive to keep duplication low, so not to inflate the number of simulated gates during each cycle. Figure 8 plots the fraction of gates that were duplicated, averaged over all our experimental designs: more than 80% incurred no duplication, less than

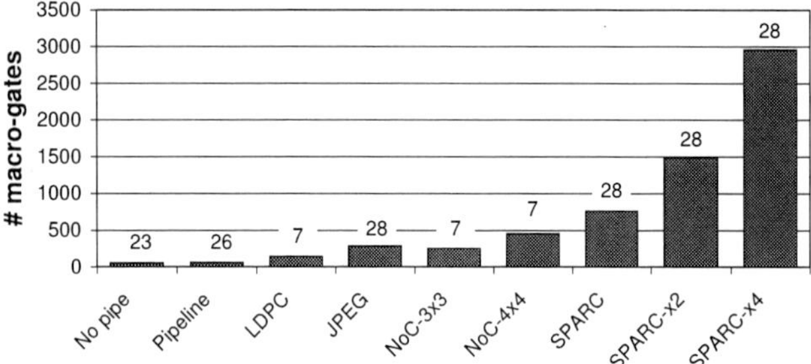

Figure 7: Macro-gates and layers. The plot shows the total number of macro-gates for each design. The value above each bar indicates the number of layers in the segmentation.

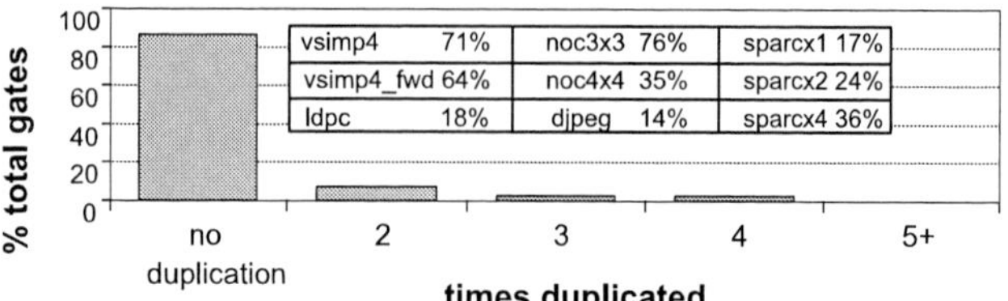

Figure 8: Gate duplication due to macro-gate overlap. The graph reports the number of times that gates were duplicated. The overlapping table indicates the gate inflation that each design incurred as a result of duplication.

10% were duplicated once, very few incurred more than one duplication. The table reports the overall rate of "gate inflation" in each design, resulting in an overall average of 39%.

6.2 Monitored nets

The number of monitored nets has a high impact on the simulator performance thus segmentation strives to keep the fraction of nets that are monitored low. As an example, in Figure 9 we plot the structure of the LDPC encoder design after segmentation: for each layer, we plot the corresponding number of macro-gates and monitored nets. Note how middle layers have more macro gates and how lower layers tend to generate the most monitored nets. Finally, we analyzed the fraction of total nets in the design that require monitoring because they cross layer boundaries. The compilation phase should strive to keep this fraction low, since it is directly related to the size of the sensitivity list that must be checked when evaluating a macro-gate for possible activation. Figure 10 reports our findings for experimental testbench designs after the segmentation phase.

6.3 Macro-gate activity

The activation rate of macro-gates is an important metric for event-driven simulation (an oblivious simulator has an activation rate of 100% on any design). The goal of an event-driven simulator is to keep this rate as low as possible, thus leveraging the fact that not all gates in a netlist switch on every cycle. Figure 11 reports the macro-gate activation rates for a number of our designs. Plots show the cumulative distribution of activation rates among all the macro-gates for distinct designs. Note how, for most designs, the majority of the macro-gates have an activation rate between 10 and 30% only. However, for LDPC, most macro-gates experience a high activation rate ($> 80\%$): this is due to the inherent nature of this design. The designs that are not reported had a cumulative distribution similar to that of the OpenSPARC and NoC designs. Note that activation rate in our solution does not directly relate to performance gain over oblivious simulation. As an example, the JPEG decoder has an average activation rate of 40%. This does not mean that, on average, the JPEG decoder is simulated 2.5 times faster when compared to an oblivious simulation. Indeed, even if

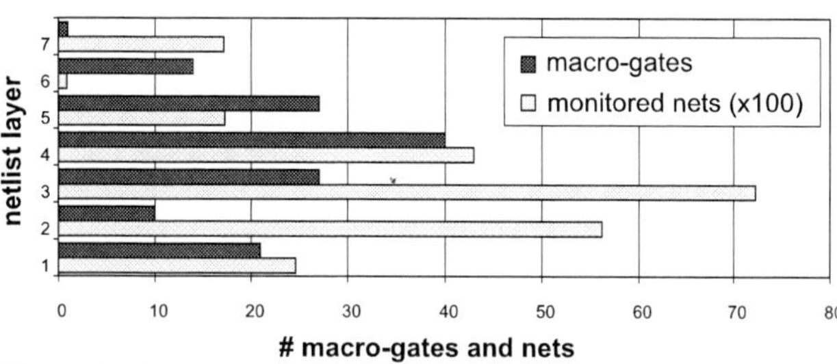

Figure 9: Segmented structure for LDPC encoder design. The plot shows the geometry of the LDPC encoder after segmentation. For each layer we report the number of macro-gates and of monitored nets in hundreds.

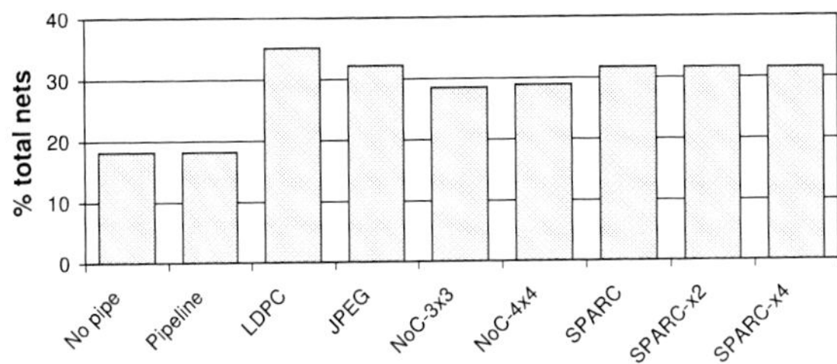

Figure 10: Fraction of monitored nets. Percentage of all nets that are monitored for each testbench design

very few macro-gates are activated, one for each layer, the performance would be just the same as if several macro-gates were simulated in each layer. This is because the parallel processing units can hide much of the additional computation required when activating many macro-gates, while the synchronization barriers force macro-gates to be simulated in layer-order. The overall performance of the JPEG design in event-driven simulation is 1.55 times faster than in oblivious simulation.

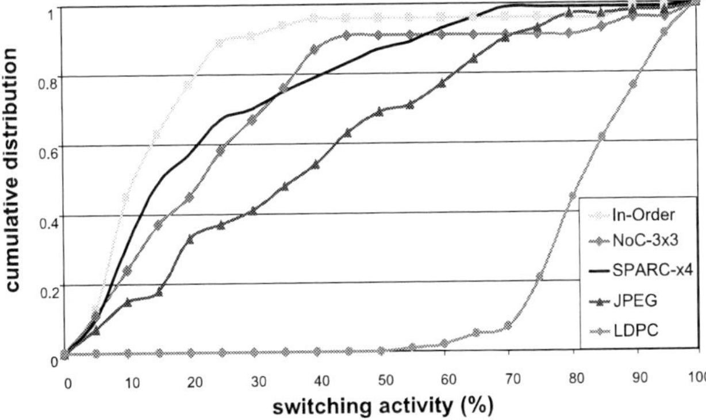

Figure 11: Cumulative distribution of macro-gates w.r.t. to activation rate. The plots show which fraction of macro-gates have an activation rate below a threshold indicated on the x axis. Most designs have very low activation rate (<30%).

6.4 Performance Evaluation

Finally, we evaluated the performance of our prototype event-driven simulator against that of a commercial, event-driven sequential simulator. Our graphics coprocessor was a CUDA-enabled 8800GT GPU with 14 multiprocessors and 512MB of device memory, operating at 600 MHz for the cores and 900MHz for the memory. The current implementation has 83% occupancy and achieves a bandwidth of 20.4 GB/s. The commercial simulator was run on a 2.4 GHz Intel Core 2 Quad running RH-EL5, enabling 4 parallel simulation threads. For each design, Table 2 reports the number of cycles simulated, the runtimes in seconds for both the GPU-based simulator and the commercial simulator (compilation times are excluded), and the relative speedup. Note that our prototype simulator outperforms the commercial simulator by 4 to 44 times. Despite the LDPC encoder having a very high activation rate, we report the best speedup for this design. As mentioned before, most gates in this design are switching in each cycle: this affects our activation rates, but hampers the sequential simulator performance. Thus, the speedup obtained is due to sheer parallelism of our architecture.

7. CONCLUSIONS

In this work, we have presented a novel event-driven simulator architecture that leverages the high-level of parallelism of general purpose GPUs. By extracting parallelism in the simulation of gate-level netlists, we are able to realize a 13 times speedup over traditional sequential simulators, on average. Our simulator carves out

design	sim cycles	seq sim(s)	GPU sim(s)	speed up
Alpha no pipeline	12,889,495	31,678	2,567	**12.15x**
Alpha pipeline	13,423,608	54,789	7,781	**7.04x**
LDPC encoder	1,000,000	115,671	2,578	**44.87x**
	10,000,000	>48h	25,973	**43.49x**
JPEG decompressor	2,983,674	12,146	599	**20.28x**
3x3 NoC routers	1,967,155	3,532	397	**8.90x**
4x4 NoC routers	10,000,001	28,867	3,935	**7.34x**
sparc core x1	1,074,702	27,894	6,077	**4.59x**
sparc core x2	1,074,702	40,378	8,229	**4.91x**
sparc core x4	1,074,702	61,678	10,983	**5.62x**

Table 2: GP-GPU simulator performance. Performance comparison between our CUDA-based event-driven simulator and a commercial event-driven simulator. Our prototype simulator outperforms the commercial simulator by 13 times on average.

macro-gates from the structural netlist of a design and schedules them for simulation on the multiprocessors of the NVIDIA CUDA architecture, only if they are activated by switching events at their inputs. We show in our experimental results that the simulator is capable of delivering a remarkable performance speedup on large, industrial-scale designs of over a million gates, thus bringing about new validation frontiers for the digital design industry. In the future, we plan to explore further optimizations in the segmentation algorithm to deliver even higher simulation performance.

8. REFERENCES

[1] J. Babb, R. Tessier, M. Dahl, S. Hanono, D. Hoki, and A. Agarwal. Logic emulation with virtual wires. *IEEE Trans. on CAD*, 1997.

[2] W. Baker, A. Mahmood, and B. Carlson. Parallel event-driven logic simulation algorithms: Tutorial and comparative evaluation. *IEEE Journal on Circuits, Devices and Systems*, 1996.

[3] Z. Barzilai, J. Carter, B. Rosen, and J. Rutledge. HSS–a high-speed simulator. *IEEE Trans. on CAD*, 1987.

[4] H. Bauer and C. Sporrer. Reducing rollback overhead in time-warp based distributed simulation with optimized incremental state saving. *Proc. ANSS*, 1993.

[5] O. Berry and G. Lomow. Speeding up distributed simulation using the time warp mechanism. In *Proc. of workshop on Making distributed systems work*, 1986.

[6] R. Bryant, D. Beatty, K. Brace, K. Cho, and T. Sheffler. COSMOS: a compiled simulator for MOS circuits. In *Proc. DAC*, 1987.

[7] K. Chandy and J. Misra. Asynchronous distributed simulation via a sequence of parallel computations. *Comm. ACM*, 1981.

[8] D. Chatterjee, A. DeOrio, and V. Bertacco. High-performance gate-level simulation with GP-GPUs. In *Proc. DATE*, 2009.

[9] M. Denneau. The Yorktown simulation engine. *Proc. DAC*, 1982.

[10] R. Fujimoto. Parallel discrete event simulation. *Comm. ACM*, 1990.

[11] K. Gulati and S. Khatri. Towards acceleration of fault simulation using graphics processing units. *Proc. DAC*, 2008.

[12] H. Kim and S. Chung. Parallel logic simulation using time warp on shared-memory multiprocessors. *Proc. IPPS*, 1994.

[13] Y.-I. Kim, W. Yang, Y.-S. Kwon, and C.-M. Kyung. Communication-efficient hardware acceleration for fast functional simulation. *Proc. DAC*, 2004.

[14] D. Lewis. A hierarchical compiled code event-driven logic simulator. *IEEE Trans. on CAD*, 1991.

[15] N. Manjikian and W. Loucks. High performance parallel logic simulations on a network of workstations. *Proc. of workshop on Parallel and distributed simulation*, 1993.

[16] Y. Matsumoto and K. Taki. Parallel logic simulation on a distributed memory machine. *Proc. EDAC*, 1992.

[17] J. Misra. Distributed discrete-event simulation. *ACM Computing Surveys*, 1986.

[18] NVIDIA. *CUDA Compute Unified Device Architecture*, 2007.

[19] Opencores. http://www.opencores.org/.

[20] A. Perinkulam and S. Kundu. Logic simulation using graphics processors. In *Proc. ITSW*, 2007.

[21] Sun microsystems OpenSPARC. http://opensparc.net/.

978-1-60558-497-3/09 $25.00 © 2009 ACM

Efficient SAT Solving for Non-Clausal Formulas Using DPLL, Graphs, and Watched Cuts[*]

Himanshu Jain[†]
Verification Group
Synopsys, Inc.

Edmund M. Clarke
School of Computer Science
Carnegie Mellon University

ABSTRACT

Boolean satisfiability (SAT) solvers are used heavily in hardware and software verification tools for checking satisfiability of Boolean formulas. Most state-of-the-art SAT solvers are based on the Davis-Putnam-Logemann-Loveland (DPLL) algorithm and require the input formula to be in conjunctive normal form (CNF). We present a new SAT solver that operates on the negation normal form (NNF) of the given Boolean formulas/circuits. The NNF of a formula is usually more succinct than the CNF of the formula in terms of the number of variables. Our algorithm applies the DPLL algorithm to the graph-based representations of NNF formulas. We adapt the idea of the two-watched-literal scheme from CNF SAT solvers in order to efficiently carry out Boolean Constraint Propagation (BCP), a key task in the DPLL algorithm. We evaluate the new solver on a large collection of Boolean circuit benchmarks obtained from formal verification problems. The new solver outperforms the top solvers of the SAT 2007 competition and SAT-Race 2008 in terms of run time on a large majority of the benchmarks.

Categories and Subject Descriptors: J.6 [Computer Aided Engineering]: [Computer-Aided Design]

General Terms: Algorithms, Design, Verification

Keywords: Boolean Satisfiability, Verification, DPLL, NNF

1. INTRODUCTION

The problem of propositional (Boolean) satisfiability (SAT) is to decide whether a given propositional formula is satisfiable. This problem is of central importance in hardware and software verification, logic synthesis, automatic test generation, and artificial intelligence. Most state-of-the-art SAT procedures [3, 4, 5, 17, 18, 13, 11, 19, 8] are variations of the Davis-Putnam-Logemann-Loveland (DPLL) algorithm and require the input formula to be in conjunc-

---

[*]This research was sponsored by the Semiconductor Research Corporation (SRC), the Gigascale Systems Research Center (GSRC), the Office of Naval Research (ONR), the Naval Research Laboratory (NRL), the Army Research Office (ARO) and General Motors Lab at CMU.

[†]The author did this research as a graduate student at CMU.

Permission to make digital or hard copies of part or all of this work for personal or classroom use is granted without fee provided that copies are not made or distributed for profit or commercial advantage and that copies bear this notice and the full citation on the first page. To copy otherwise, to republish, to post on servers or to redistribute to lists, requires prior specific permission and/or a fee.
DAC'09, July 26-31, 2009, San Francisco, California, USA

tive normal form (CNF). Typical formulas arising in practice are not necessarily in CNF. We refer to propositional formulas not in CNF form as *non-clausal* formulas (Boolean circuits). As argued by Thiffault et al. [20] converting a non-clausal formula to CNF introduces overhead in form of a large number of new variables and may destroy the initial structure of the formula, which can be crucial in efficient satisfiability checking.

Suppose we are given a non-clausal formula ϕ. In order to check the satisfiability of ϕ using a CNF based SAT solver, ϕ needs to be converted to CNF. Let us assume for now that ϕ is in *negation normal form (NNF)*. This means that ϕ contains only "$\wedge$" ("AND"), "$\vee$" ("OR"), "$\neg$" ("NOT") operators and the scope of each "$\neg$" is a propositional variable. There are two ways of converting ϕ to a CNF formula. The first way of converting ϕ to a CNF formula is by introduction of new variables [21]. This produces a CNF formula ϕ' that is equi-satisfiable to ϕ and is linear in the size of ϕ. This is most common and practical way of converting ϕ to a CNF formula.

The second method is to expand ϕ using the distributive properties of $\wedge, \vee$ in order to obtain a CNF formula. Let us denote the CNF formula obtained by expansion of ϕ as $CNF(\phi)$. For example let ϕ be $(a \wedge \neg b) \vee (c \wedge (d \vee \neg f))$. Then $CNF(\phi)$ is $(a \vee c) \wedge (a \vee d \vee \neg f) \wedge (\neg b \vee c) \wedge (\neg b \vee d \vee \neg f)$. Note that $CNF(\phi)$ and ϕ contain the same set of variables and are logically equivalent. This method of obtaining $CNF(\phi)$ from ϕ is impractical because the size of $CNF(\phi)$ can be exponential in the size of ϕ.

In this paper we present a new SAT solver that checks the satisfiability of ϕ by applying the DPLL algorithm to the *hpgraph* [7, 15] of ϕ. Each path in the hpgraph of ϕ, starting from a root node and ending at a leaf node, corresponds to a clause in $CNF(\phi)$. That is, the hpgraph of ϕ implicitly encodes $CNF(\phi)$. Fig. 1(a) shows the hpgraph of the formula $(a \wedge \neg b) \vee (c \wedge (d \vee \neg f))$. The size of the hpgraph of ϕ is linear in the size of ϕ. Thus, the hpgraph compactly represents $CNF(\phi)$. By using the hpgraph we avoid explicitly listing out the clauses in $CNF(\phi)$, which can require exponential time and space in the size of ϕ.

Let us denote the disjunctive normal form (DNF) formula obtained by converting ϕ to DNF by expanding out ϕ as $DNF(\phi)$. Let ϕ be $(a \wedge \neg b) \vee (c \wedge (d \vee \neg f))$. Then $DNF(\phi)$ is $(a \wedge \neg b) \vee (c \wedge d) \vee (c \wedge \neg f)$. Once again explicitly listing out $DNF(\phi)$ can be exponential in the size of ϕ. Our solver utilizes the *vpgraph* [7, 15] of ϕ. The vpgraph of ϕ implicitly encodes $DNF(\phi)$ and is linear in the size of ϕ. The vpgraph for our example is shown in Fig. 1(b).

1.1 Our Contributions

- We present a new SAT solver that checks the satisfiability of a NNF formula ϕ by applying the DPLL algorithm to the hpgraph of ϕ. Our solver also utilizes the vpgraph of ϕ in certain steps of SAT solving. If the input formula/circuit is not in NNF it is

---

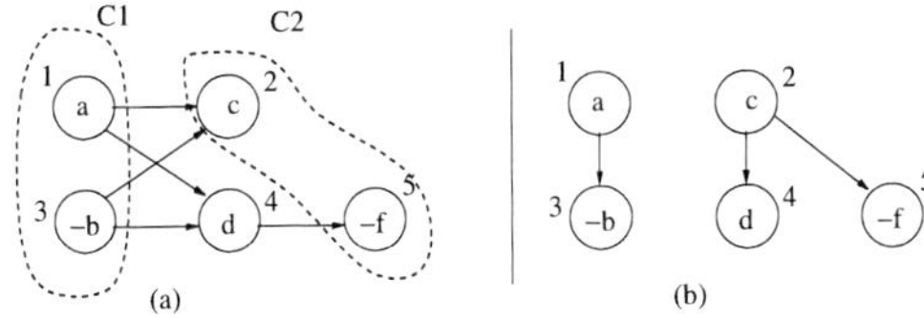

Figure 1: Let ϕ be $(a \wedge \neg b) \vee (c \wedge (d \vee \neg f))$. **(a) The hpgraph of ϕ. Two node cuts C_1, C_2 are shown. (b) The vpgraph of ϕ. We show the negation of a variable by using a minus sign.**

first converted to an equi-satisfiable NNF formula as explained in Section 2.

- The most crucial component of our DPLL SAT solver is an efficient Boolean Constraint Propagation (BCP) algorithm on the hpgraph. We describe an algorithm for performing BCP on hpgraph that *adapts* the *two-watched-literal scheme* [18] found in CNF SAT solvers. In particular, a "watch" in an hpgraph corresponds to a *node cut* in the hpgraph. By maintaining two node cuts for each connected component in the hpgraph we achieve the same effect as the two watched-literal scheme found in the CNF SAT solvers. Fig. 1(a) shows two node cuts C_1, C_2 (possible watches) for a hpgraph. Two node cuts allow watching two nodes (literals) on each path (clause) in a hpgraph component. The two-watched-literal scheme used in CNF SAT solvers is a special case of our algorithm (when the hpgraph represents a CNF formula). As in CNF SAT solvers *non-chronological backtracking* is cheap as the node cuts are not updated when backtracking. (Section 3)

- We show how to update the node cuts (watches) in the hpgraph efficiently by using the vpgraph of the given formula. We show that a *minimal* cut in a hpgraph corresponds to a path in the corresponding *vpgraph*. Thus, finding a small node cut in an hpgraph corresponds to finding a path in the corresponding vpgraph. For example, notice that paths $\langle 1,3 \rangle, \langle 2,5 \rangle$ in the vpgraph shown in Fig 1(b) correspond to cuts C_1, C_2, respectively, in the hpgraph shown in Fig 1(a). (Section 4)

- We have carefully implemented these ideas in a non-clausal SAT solver. We evaluate the solver on 2541 non-clausal benchmarks obtained from publicly available sources. Our solver outperforms the top SAT solvers of the SAT 2007 competition and SAT-Race 2008 in terms of runtime. (Section 5)

Related Work: Most state-of-the-art SAT procedures are based on DPLL search and require the input formula to be in CNF [3, 4, 5, 17, 18, 13, 11, 19, 8]. There has been work on applying DPLL directly to circuit [12, 16, 20] representations. In [12] a hybrid SAT solver is described where the original formula is processed in circuit form, and learned clauses are processed separately in CNF. The circuit-based BCP is implemented by means of a lookup table. In [20] a watched literal scheme is proposed for efficient BCP on a given circuit. Unlike existing circuit SAT solvers our SAT solver does not operate on the circuit representation directly. In our approach a given formula/circuit is converted to an equi-satisfiable NNF formula. The NNF formula is then represented in the form of two graphs called the vpgraph and hpgraph. These graphs are used in our SAT algorithms. Jain et al. [15] use a vpgraph/hpgraph in a *General Matings* based SAT solver. This work uses the vpgraph/hpgraph in a DPLL-based SAT solver. In our unreported experiments we found the solver in [15] to have poor performance on the benchmarks we consider in this paper.

A technical report version of this paper with proofs can be found in the chapters 2 and 4 of [14].

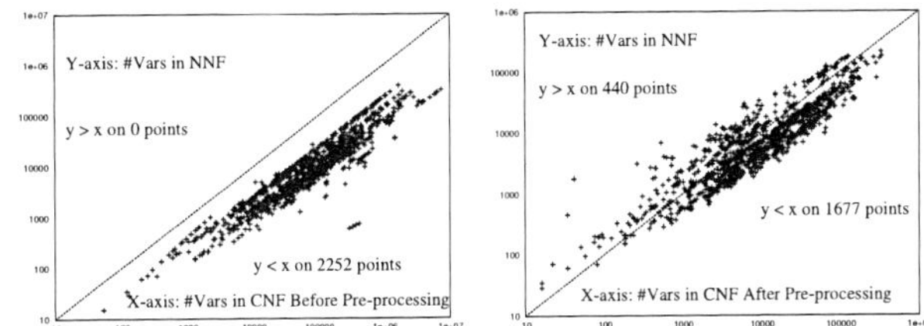

Figure 2: Number of variables comparison.

2. PRELIMINARIES

A Boolean formula is in *negation normal form (NNF)* iff it contains only the Boolean connectives "$\wedge$" ("AND"), "$\vee$" ("OR") and "$\neg$" ("NOT"), and the scope of each occurrence of "$\neg$" is a Boolean variable. We also require that there is no *structure sharing* in a NNF formula, that is, output from a gate acts as input to at most one gate. A NNF formula is tree-like, while a circuit can be DAG-like.

Conversion of Boolean Circuits to NNF: Most formulas obtained in practice are Boolean circuits. In our work Boolean circuits are converted to NNF formulas in two stages. The first stage re-writes other operators (such as xor, iff, implies, if-then-else) in terms of $\wedge, \vee, \neg$ operators. In order to avoid a blowup in the size of the resulting formula we allow sharing of sub-formulas. Thus, the first stage produces a formula containing $\wedge, \vee, \neg$ gates, possibly with structure sharing. The second stage gets rid of the structure sharing in order to obtain a NNF formula. This is done by introduction of new variables. We introduce a new variable for each gate that has a fanout *greater* than one. Once the sharing is removed the negations can be pushed to the variables using DeMorgans laws. Observe that the conversion of a Boolean circuit to a NNF formula can be done in linear time in the size of the Boolean circuit. We compare the number of variables in the NNF and CNF representations of a collection of industrial Boolean circuits in Fig. 2. The NNF forms have $5 - 10$ times fewer variables than CNF. Modern CNF SAT solvers use pre-processing techniques in order to eliminate variables and clauses from the input CNF formula [9]. Fewer variables reduce overhead during BCP and can make the decision heuristics more effective. Even after pre-processing of CNF formulas the NNF forms have fewer variables than CNF on 80% of our benchmarks. The fewer variables in the NNF form (without any pre-processing) motivates the need for exploring SAT solving techniques that operate on NNF directly. We provide experimental evidence that shows the utility of the NNF forms in SAT solvers.

In the subsequent sections we assume that the input Boolean circuit has been converted to a NNF formula. Given a NNF formula ϕ our SAT algorithms check the satisfiability of ϕ without introducing any more new variables. We use the work by Andrews [7] and Jain et al. [15] in order to represent the NNF formula in form of two graphs the *hpgraph* and *vpgraph*.

Hpgraph: Given a NNF formula ϕ, the hpgraph $G_h(\phi)$ is defined as a tuple (V, R, L, E, Lit), where V is the set of nodes corresponding to all occurrences of literals in ϕ, $R \subseteq V$ is a set of root nodes, $L \subseteq V$ is a set of leaf nodes, $E \subseteq V \times V$ is the set of edges, and $Lit(n)$ denotes the literal associated with node $n \in V$. A root node $n \in R$ has no incoming edges and a leaf node $n \in L$ has no outgoing edges.

The hpgraph for a NNF formula $\phi := (a \wedge \neg b) \vee (c \wedge (d \vee \neg f))$ is shown in Fig. 1(a). We have $V = \{1,2,3,4,5\}$, $R = \{1,3\}$, $L = \{2,5\}$, $E = \{(1,2),(1,4),(3,2),(3,4),(4,5)\}$ and for each $n \in V$, $Lit(n)$ is shown inside the node labeled n.

Vpgraph: Given a NNF formula ϕ, the vpgraph $G_v(\phi)$ is also defined as a tuple (V', R', L', E', Lit'). The vpgraph for the formula ϕ is shown in Fig. 1(b). We have $V' = \{1,2,3,4,5\}$, $R' = \{1,2\}$,

978-1-60558-497-3/09 $25.00 © 2009 ACM

$L' = \{3,4,5\}$, $E' = \{(1,3),(2,4),(2,5)\}$ and for each $n \in V$, $Lit(n)$ is shown inside the node labeled n.

Jain et al. [14, 15] present a recursive (linear time) procedure for obtaining the hpgraph and vpgraph from a given NNF formula. We briefly review the recursive steps involved in the creation of an hpgraph: 1) the hpgraph for a literal l is a new node n with $Lit(n) = l$, 2) the hpgraph for $\phi_1 \wedge \phi_2$ is the union of the hpgraphs of ϕ_1 and ϕ_2, and 3) the hpgraph for $\phi_1 \vee \phi_2$ is obtained by connecting leaves in the hpgraph of ϕ_1 to the roots in the hpgraph of ϕ_2.

For this paper we only need to understand the key properties of the hpgraph and vpgraph. A path $\pi = \langle n_0, \ldots, n_k \rangle$ in $G_h(\phi)$ or $G_v(\phi)$ is said to be a **rl-path** iff it starts at a root node and ends at a leaf node. In Fig. 1(a), $\langle 1,2 \rangle$, $\langle 1,4,5 \rangle$, $\langle 3,2 \rangle$, $\langle 3,4,5 \rangle$ are rl-paths.

Key Property of the Hpgraph and Vpgraph: Each rl-path π in the hpgraph $G_h(\phi)$ corresponds to a clause obtained by collecting the literals occurring on π. For example, the rl-path $\langle 1,4,5 \rangle$ in Fig. 1(a) corresponds to the clause $a \vee d \vee \neg f$. The collection of all clauses occurring in the hpgraph gives a CNF form for the given formula. Each rl-path in the vpgraph $G_v(\phi)$ corresponds to a cube (term). For example, the rl-path $\langle 2,4 \rangle$ in Fig. 1(b) corresponds to the term $c \wedge d$. The collection of all terms occurring in the vpgraph gives a DNF form of the given formula.

THEOREM 1. *Let $G_h(\phi)$ be the hpgraph and $G_v(\phi)$ be the vpgraph of ϕ. Let π denote a rl-path and n denote a node on π.*
(a) ϕ is equivalent to the CNF formula $\bigwedge_{\pi \in G_h(\phi)} \bigvee_{n \in \pi} Lit(n)$.
(b) ϕ is equivalent to the DNF formula $\bigvee_{\pi \in G_v(\phi)} \bigwedge_{n \in \pi} Lit(n)$.

The clauses (terms) corresponding to rl-paths in hpgraph (vpgraph) can be redundant (that is, contain a literal and its negation).
Using Hpgraph inside a SAT Solver: Given an assignment σ to a subset of variables in ϕ, we say that there is a **conflict** iff σ falsifies ϕ. A literal l is an **implied (unit)** literal iff l must be set to true in order to obtain a satisfying assignment under σ. We say an assignment *falsifies* a node n in $G_h(\phi)$ or $G_v(\phi)$ iff the assignment falsifies $Lit(n)$. We use the hpgraph of ϕ in order to detect conflicts and implied literals. Recall that each rl-path in the hpgraph corresponds to a clause. The following corollary states that if each node of a rl-path is false then there is a conflict.

COROLLARY 1. *Given an assignment σ to variables in ϕ the following are equivalent: 1) σ falsifies ϕ. 2) there exists a rl-path π in $G_h(\phi)$ such that σ falsifies every node on π.*

Consider the hpgraph in Fig. 1(a). The assignment $\sigma := \{a = 0, c = 0\}$ falsifies every node on rl-path $\langle 1,2 \rangle$. Thus, σ falsifies ϕ.

COROLLARY 2. *Let σ be an assignment to variables in ϕ. If there is a rl-path π in $G_h(\phi)$ and a node $m \in \pi$ such that σ falsifies every node in π except m, and $Lit(m)$ is not assigned in σ, then $Lit(m)$ is an implied literal.*

The implied literal detected using the above corollary will be termed as an *h-implied* literal. We will refer to the node m in the above corollary as an *h-implied* node. Consider the hpgraph shown in Fig. 1(a) and an assignment $\sigma := \{a = 0, d = 0\}$. σ falsifies all but node 5 on the rl-path $\langle 1,4,5 \rangle$ in the hpgraph. It follows that $Lit(5) = \neg f$ is an implied literal (due to clause $a \vee d \vee \neg f$). That is, f must be set to zero under the current assignment.

A main difference between the existing DPLL SAT solvers and our solver is in the Boolean constraint propagation (BCP). In our solver the BCP algorithm uses the hpgraph of a given formula.

3. BCP ON THE HPGRAPH

Let V denote the set of variables in a given formula ϕ. Given an assignment σ of truth values to a set of variables $W \subseteq V$, the Boolean constraint propagation (BCP) algorithm detects two cases. (1) It reports if σ falsifies ϕ (conflict). (b) If there is no conflict, the BCP algorithm provides a set of implied (unit) literals. Before we describe the BCP algorithm on a hpgraph, we briefly review the BCP algorithm used in modern CNF SAT solvers.

3.1 Review of BCP in CNF SAT solvers

Most modern CNF SAT solvers use the *two-watched-literal scheme* [18] in order to obtain an efficient BCP algorithm. Suppose we are given a CNF formula ϕ. Let D be a clause in ϕ. In the two-watched-literal scheme two *watches* are associated with D. A *watch* is simply a literal l occurring in D. Before the search (DPLL algorithm) starts any two literals in D can be designated as its watches. Let l_1, l_2 be the watches corresponding to D. Four cases arise depending upon the status of l_1, l_2 given the current assignment σ.
Case A: Watches l_1, l_2 are either true or unassigned. In this case there cannot be any conflict or an implied literal due to D. The clause D is not even examined during BCP.

The clause D is examined only when one of its watches becomes false. Without loss of generality assume that l_2 becomes false in the remaining three cases.
Case B: If l_1 is already true, then D is already satisfied. In this case nothing needs to be done even though l_2 is false.

If l_1 is not true, the solver tries to replace the falsified watch (l_2) by another watch that is not false. If there is a literal l_3 in D that is not false and $l_3 \neq l_1$, then l_2 is replaced by l_3. However, such a literal (l_3) may not exist in the remaining two cases.
Case C (conflict): All the literals in D are false. In this case D is false under the current assignment.
Case D (implied literal): l_1 is unassigned but all other literals in D are false. In this case, l_1 is reported as an implied literal.

The main benefit of the two-watched-literal scheme is that it reduces the number of times the solver examines the clauses in a given CNF formula. This is crucial for obtaining efficient solvers that can handle CNF formulas with millions of clauses. Another advantage is that the *non-chronological backtracking* is cheap. This is because the watched literals do not need to be updated during backtracking. We now describe how the two-watched-literal scheme found in CNF SAT solvers can be adapted to obtain an efficient algorithm for BCP on a hpgraph.

3.2 Generalizing Two-Watched-Literal Scheme to Two-Watched-Cut Scheme for Hpgraph

Let ϕ be a NNF formula. We are given an assignment σ to a subset of variables occurring in ϕ. The BCP algorithm uses the hpgraph $G_h(\phi)$ of ϕ to detect conflicts and implied literals. Given $G = (V, E, R, L, Lit)$ we say that $C \subseteq V$ is a **rl-cut** in G iff removal of all nodes in C from G disconnects all rl-paths in G. For example, $\{1,3,4\}, \{2,3,4\}, \{5,6,8\}, \{5,7,8\}$ are some of the rl-cuts in the hpgraph shown in Fig. 3. The node set $\{2,7,8\}$ is not an rl-cut as it does not disconnect the rl-paths $\langle 3,5 \rangle, \langle 4,5 \rangle$. An rl-cut contains at least one node from each rl-path. More precisely let C be a rl-cut in $G_h(\phi)$. For every rl-path π in $G_h(\phi)$ there exists a node n such that $n \in \pi$ and $n \in C$. Two rl-cuts C_1, C_2 are said to be **node-disjoint** if $C_1 \cap C_2 = \emptyset$. For example the rl-cuts $\{1,3,4\}, \{5,7,8\}$ in Fig. 3 are node-disjoint.
Watches in a Hpgraph: Each rl-path in a hpgraph corresponds to a clause. Let the clause corresponding to an rl-path π be D. In order to apply the two-watched-literal scheme found in CNF SAT solvers we want to *watch* two nodes n_1, n_2 on π. This in turn amounts

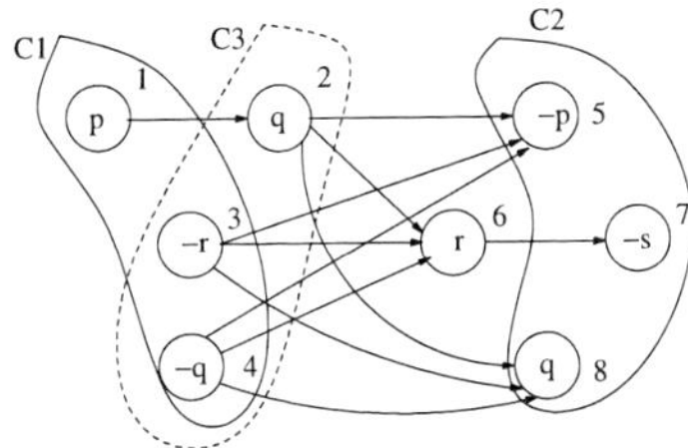

Figure 3: An hpgraph. Three rl-cuts $C_1 := \{1,3,4\}, C_2 := \{5,7,8\}, C_3 := \{2,3,4\}$ are shown.

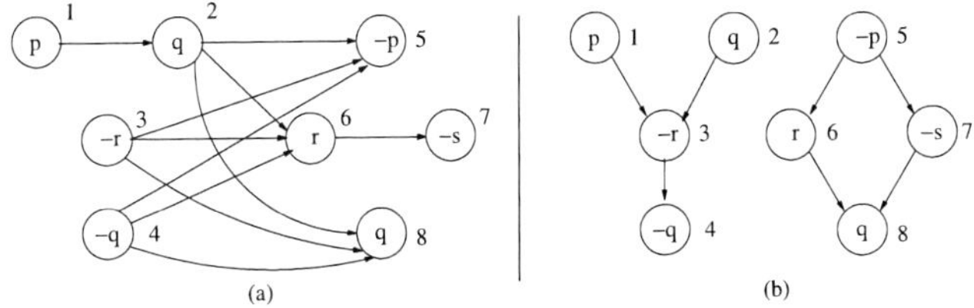

Figure 4: (a) Hpgraph for formula $(((p \lor q) \land \neg r \land \neg q) \lor (\neg p \land (r \lor \neg s) \land q))$. (b) The corresponding vpgraph.

to watching two literals $Lit(n_1), Lit(n_2)$ in D. However, there are usually exponentially many paths (clauses) in a hpgraph. So it is expensive to maintain watches for each rl-path (clause) explicitly.

This intuition leads us to define a *watch* in a hpgraph as a *rl-cut* in the hpgraph. By taking a rl-cut as a watch we make sure that at least one node on every rl-path is present in our watch. This in turn corresponds to watching a literal on each clause in the hpgraph. For example, the rl-cut $C := \{1,3,4\}$ is a possible watch for the hpgraph in Fig. 3. Note that C contains at least one node from each rl-path in Fig. 3. Watching node 3 on rl-paths $\langle 3,5\rangle, \langle 3,6,7\rangle, \langle 3,8\rangle$, amounts to watching literal $Lit(3) = \neg r$ on clauses $\neg r \lor \neg p, \neg r \lor r \lor \neg s, \neg r \lor q$, respectively.

In order to get the effect of the two-watched-literal scheme we watch two rl-cuts in the hpgraph. By maintaining two rl-cuts for a hpgraph we are able to watch two nodes (literals) on each rl-path (clause) in the hpgraph. For example the rl-cuts $C_1 := \{1,3,4\}$ and $C_2 := \{5,7,8\}$ are two possible watched cuts for the hpgraph in Fig. 3. For the rl-path $\langle 1,2,6,7\rangle$, the rl-cuts C_1, C_2 allow us to watch nodes 1, 7 (literals $p, \neg s$).

Suppose we are given $G = (V, E, R, L, Lit)$, a partial assignment σ, and a rl-cut $W \subseteq V$. We say that W is **acceptable** iff there is no node $m \in W$ such that $Lit(m)$ is false in σ. We say that W is **satisfied** iff for all $m \in W$ $Lit(m)$ is true in σ. For example, given the hpgraph in Fig. 3 and $\sigma := \{q = 1\}$ the rl-cuts $\{5,6,8\}, \{5,7,8\}$ are acceptable. The rl-cuts $\{2,3,4\}, \{1,3,4\}$ are not acceptable given σ (due to node 4).

3.3 BCP on Hpgraph Using Two Watched Cuts

For a given hpgraph $G_h(\phi) = (V, E, R, L, Lit)$ we maintain two rl-cuts C_1, C_2 (watches). Before the DPLL algorithm starts C_1, C_2 can be initialized to any two rl-cuts in $G_h(\phi)$ that are node-disjoint. As the search progresses the algorithm tries to maintain the invariant that at least one of C_1, C_2 is acceptable. The algorithm also tries to maintain C_1, C_2 as node-disjoint as possible. This is useful for detecting implied literals. Recall, that implied literals detected using the hpgraph (Corollary 2) are called h-implied literals. We describe the various cases that may arise during the BCP on a hpgraph below. Each of the cases below generalize the cases that occur in the two-watched-literal scheme for CNF SAT solvers.

Case A: Both rl-cuts (watches) C_1, C_2 are node-disjoint and acceptable. Then there can be no conflict or h-implied literals due to the current assignment. This is because each clause in the hpgraph contains two literals that are not false. In this case there is no need to look at any other part of the hpgraph. In Fig. 3 let $C_1 := \{1,3,4\}, C_2 := \{5,7,8\}$ and $\sigma := \{r = 0\}$. Observe that both C_1, C_2 are acceptable rl-cuts and are node-disjoint.

Suppose one of the rl-cuts say C_2 is no longer acceptable. Then we have the following cases.

Case B: For each node $n \in C_1$, $Lit(n)$ is already true, that is, C_1 is *satisfied*. In this case there cannot be any conflict or an h-implied literal in the hpgraph. Intuitively, every clause present in the hpgraph is satisfied. The algorithm leaves C_2 unchanged in this case.

In Fig. 3 let $C_1 := \{1,3,4\}, C_2 := \{5,7,8\}$ and $\sigma := \{p = 1, r = 0, q = 0\}$. Observe that C_2 is not acceptable as $Lit(5), Lit(8)$ are false under σ. However, C_1 is satisfied. So C_2 is not updated.

If the previous cases do not apply, then the algorithm tries to find a replacement rl-cut for C_2. When searching for a replacement to C_2, the algorithm tries to find a rl-cut that is as different from C_1 as possible. Intuitively, this is similar to why we keep two *distinct* watched literals in a clause in the CNF two-watched-literal scheme. If a replacement cut C_3 is found such that C_3 is acceptable and $C_3 \cap C_1 = \emptyset$, then C_2 is replaced by C_3. Otherwise, we have the following two cases.

Case C (conflict): There is no acceptable rl-cut in the hpgraph. In this case the current assignment σ falsifies the given formula. In Fig. 3 let $C_1 := \{1,3,4\}, C_2 := \{5,7,8\}$ and $\sigma := \{p = 1, r = 1\}$. In this case neither C_1 nor C_2 is acceptable and there is no possible replacement for them. This is expected as the clause $(\neg r \lor \neg p)$ corresponding to the rl-path $\langle 3,5\rangle$ is false.

Case D (implications): There is an acceptable rl-cut C_3 but C_3 is not node-disjoint from C_1. In this case, for every $n \in C_1 \cap C_3$ the corresponding literal $Lit(n)$ is an h-implied literal (assuming $Lit(n)$ is not already true). If $C_3 \neq C_1$, then C_2 is replaced by rl-cut C_3.

In Fig. 3 let $C_1 := \{1,3,4\}, C_2 := \{5,7,8\}$ and $\sigma := \{p = 1\}$. Observe that C_2 is not acceptable as $Lit(5)$ is false under σ. Also note that case B does not hold as C_1 is not satisfied. Thus, we seek a replacement for C_2. Note that any new acceptable rl-cut must include nodes 3,4 since they are the only possible nodes that can be watched on the paths $\langle 3,5\rangle, \langle 4,5\rangle$, respectively. Thus, a possible rl-cut C_3 is $\{2,3,4\}$. Both C_1, C_3 contain nodes 3,4. It can be seen that $Lit(3), Lit(4)$ are precisely the h-implied literals. $Lit(3) = \neg r$ is h-implied due to the rl-path $\langle 3,5\rangle$, which corresponds to the clause $\neg r \lor \neg p$. Similarly, $Lit(4) = \neg q$ is h-implied due to the rl-path $\langle 4,5\rangle$, which corresponds to the clause $\neg q \lor \neg p$. Since h-implied literals are also implied literals it follows that $\neg r, \neg q$ are implied literals given the current assignment.

4. MINIMAL RL-CUTS IN HPGRAPH

In section 3.3 we adapted the CNF two-watched-literal scheme to hpgraphs by using two rl-cuts in the hpgraph $G_h(\phi)$. We now describe how rl-cuts are obtained and updated efficiently during BCP. The key idea is to make use of the vpgraph $G_v(\phi)$. For example, consider the hpgraph in Figure 4 (a) and the corresponding vpgraph in Figure 4 (b). Observe that any rl-path in the vpgraph corresponds to a rl-cut in the hpgraph. The rl-paths $\langle 1,3,4\rangle, \langle 2,3,4\rangle, \langle 5,6,8\rangle, \langle 5,7,8\rangle$ in the vpgraph corresponds to rl-cuts $\{1,3,4\}, \{2,3,4\}, \{5,6,8\}, \{5,7,8\}$, respectively, in the hpgraph.

Given $G = (V, E, R, L, Lit)$ we say that $C \subseteq V$ is a **minimal rl-cut** in G iff C is a rl-cut in G and no proper subset of C is a rl-cut in G. Let $nodes(\pi)$ denote the set of nodes occurring on a rl-path π. A surprising fact is that the rl-paths in the vpgraph correspond to minimal rl-cuts in the hpgraph. One can also prove that every minimal rl-cut in the hpgraph corresponds to a rl-path in the vpgraph.

978-1-60558-497-3/09 $25.00 © 2009 ACM

566

THEOREM 2. *Let $G_h(\phi)$ be a hpgraph and $G_v(\phi)$ be a vpgraph for a NNF formula ϕ. (a) Let π be a rl-path in $G_v(\phi)$. Then $nodes(\pi)$ form a minimal rl-cut in $G_h(\phi)$. (b) Let C be a minimal rl-cut in $G_h(\phi)$. Then there exists a rl-path π in $G_v(\phi)$ such that $C = nodes(\pi)$.*

4.1 Updating Minimal RL-cuts in Hpgraph

Our algorithm always maintains two minimal rl-cuts in the hpgraph as the watched cuts. These cuts are obtained and updated by finding rl-paths in the corresponding vpgraph by using a depth-first search like routine. The BCP algorithm relies on the ability to find acceptable rl-cuts in the hpgraph. This is done by searching for acceptable rl-paths (rl-paths with no falsified nodes) in the vpgraph. The BCP routine also requires that we find disjoint rl-cuts in the hpgraph (if possible). This is done by searching for disjoint rl-paths in the vpgraph. More precisely, suppose we are trying to replace rl-cut C_2 in the hpgraph. Let the other rl-cut in the hpgraph be C_1 and let π_1 denote the rl-path corresponding to C_1 in the vpgraph. Then we search for a rl-path in the vpgraph that is completely disjoint from π_1. If we succeed in finding a path π_3 in the vpgraph that is completely disjoint from π_1, then we obtain a replacement C_3 for C_2 in the hpgraph such that $C_1 \cap C_3 = \emptyset$. If there is no rl-path in the vpgraph that is completely disjoint from π_1, we find the set of all nodes N on π_1 that must be shared by any acceptable rl-path in the vpgraph (this can be done in linear time). Intuitively, the nodes in N give us the precise set of h-implied literals.

THEOREM 3. *Let $G_h(\phi)$ be a hpgraph and $G_v(\phi)$ be a vpgraph for a NNF formula ϕ. Let π_1 be an acceptable rl-path in $G_v(\phi)$. Suppose there is no other acceptable rl-path in $G_v(\phi)$ that is completely node disjoint from π_1. Let N denote the set of nodes that must be shared in any acceptable rl-path in $G_v(\phi)$. (a) Then $N \neq \emptyset$. For every $m \in N$ either $Lit(m)$ is true or $Lit(m)$ is implied under the current assignment. (b) If m is an h-implied node, then $m \in N$.*

Optimizations: BCP based on *only* two-watched rl-cuts can be inefficient when the hpgraph has millions of nodes. This is because even the minimal rl-cuts for the entire hpgraph can be large and will be updated frequently during BCP. In practice, there are usually many hpgraph components (due to top level conjunctions and structure sharing). Each component is small as compared to the entire hpgraph in terms of number of nodes. For efficiency we maintain two watched rl-cuts for each hpgraph component. The algorithms above apply to each individual hpgraph/vpgraph component. See [14] for other important optimizations.

The hpgraph for a CNF formula is a disjoint union of various line graphs (hpgraph components) where each line graph represents a clause. A minimal rl-cut in a line graph is simply a cut of size one. Thus, the two-watched rl-cuts for each hpgraph component reduces to two-watched-literal scheme when the input is a CNF formula.

The other important components of our SAT solver such as decision heuristics, conflict driven learning, non-chronological backtracking, and restarts are implemented in a similar manner as other state-of-the-art SAT solvers. The conflict driven learning [17, 18] generates new clauses, which are added to a CNF clause database. The BCP routine takes into account both the hpgraph and the clause database in order to detect conflicts and implied literals.

5. EXPERIMENTAL RESULTS

The experiments are performed on a 1.86 GHz Intel Xeon (R) machine with 4 GB of memory running Linux. The techniques described in the paper have been implemented in a SAT solver called NFLSAT (Non-clausal FormuLas SATisfiability checker). The input formula is given in AIG (And Inverter Graph) [1], or ISCAS format. We evaluate the solver on a collection of 2541 Boolean circuits obtained from publicly available sources. These benchmarks consist of 1) 839 bounded model checking problems and 857 k-induction problems obtained from all sequential circuits used in the 2007 hardware model checking competition [2], 2) all 341 benchmarks that were used in the AIG track in the SAT competition 2007 [6]. The remaining benchmarks are obtained from microprocessor verification and equivalence checking domains. Out of 2541 Boolean circuits, 2192 circuits are in the AIG format. We use a timeout of 10 minutes per problem per solver.

We compare NFLSAT against three state-of-the-art CNF solvers RSAT [5], PicoSAT [4], and MiniSAT [3]. In SAT 2007 competition the solvers RSAT, PicoSAT, and MiniSAT were ranked first, second, third, respectively in the industrial category. We use the SAT 2007 competition version of RSAT and PicoSAT. We use the current public version of MiniSAT (minisat2-070721). The CNF versions of the above circuits were obtained by means of the standard Tseitin transformation [21]. We use `aigtocnf` [1] to convert the benchmarks in AIG format to CNF. Note that RSAT and MiniSAT use pre-processing [9] to simplify the input CNF formulas. We include the time required to obtain hpgraph/vpgraph from a Boolean circuit in NFLSAT's runtime.

We also compare NFLSAT with MiniSAT++ 1.0 which was ranked first in the AIG track of SAT-Race 2008. MiniSAT++ simplifies the given AIG circuit using DAG-aware rewriting and then converts the simplified circuit to CNF by using an improved Tseitin translation [10]. The resulting CNF is then passed to MiniSAT 2.1, which was ranked first in the CNF track in SAT-Race 2008. We also compare NFLSAT with PicoaigerSAT which was ranked second in the AIG track of SAT-Race 2008. MiniSAT++ and PicoaigerSAT directly accept AIG inputs. We compare NFLSAT with MiniSAT++ and PicoaigerSAT on 2192 AIG benchmarks.

Figure 5 gives scatter plots comparing NFLSAT with other solvers. NFLSAT has better performance on points below the line $y = x$. NFLSAT has better runtimes than RSAT, PicoSAT, MiniSAT, MiniSAT++, and PicoaigerSAT, on respectively 89%, 91%, 86%, 83%, and 86% of the benchmarks.

The experimental results are shown in Table 1. The first four rows summarize the results for all 2541 benchmarks, while the last three rows summarize the results for 2192 AIG benchmarks. For each solver we report: 1) Number of problems solved within timeout in the "Solved" column. 2) Number of problems where a timeout occurred in the "Timeout" column. 3) The total time spent in seconds on the problems that were solved in the "STime" column. 4) Sum of "Solved Time" and timeouts in the "TTime" column.

NFLSAT solves more problems than RSAT, PicoSAT, MiniSAT, PicoaigerSAT, and it is also faster in terms of run time. The main competition to NFLSAT is given by MiniSAT++ which solves 14 more problems within timeout. These 14 benchmarks are from SAT competition 2007 AIG benchmark suite and were obtained from CNF formulas by reverse engineering. The extraction of circuit structure from CNF is not perfect and many of the extracted circuits are simply a conjunction of clauses. On such CNF-like benchmarks NFLSAT is not able to match MiniSAT++ performance.

The main conclusion is that NFLSAT is competitive to the existing state-of-the-art SAT solvers on a majority of the Boolean circuit benchmarks in terms of runtime. The NNF form of Boolean circuits has fewer variables than (pre-processed) CNF in the majority of the cases. Fewer variables in turn reduce the overhead during the BCP and can make the decision heuristics more effective. The two-watched-cut scheme carries more overhead than the two-watched-literal scheme. However, the two-watched-cut scheme can potentially update the watches for a large number of clauses with-

978-1-60558-497-3/09 $25.00 © 2009 ACM

Solver	Solved	Timeout	STime	TTime
NFLSAT	**2364**	**177**	**29753**	**135953**
RSAT	2310	231	45794	184394
PicoSAT	2281	260	43297	199297
MiniSAT	2270	271	39489	202089
NFLSAT	**2060**	**132**	**26585**	**105785**
MiniSAT++	2074	118	32457	103257
PicoaigerSAT	2033	159	35892	131292

Table 1: Summary of each SAT solvers performance.

out having to look at each clause individually.

6. SUMMARY

We presented a DPLL-based SAT solver that operates on the graph-based representations of non-clausal formulas (Boolean circuits). The input formula is converted to a NNF formula that is represented using two graphs. The hpgraph encodes the CNF form of the given NNF formula, while the vpgraph encodes the DNF form of the given NNF formula. The key step in the DPLL algorithm is Boolean constraint propagation (BCP). We adapt the idea of the two-watched-literal scheme from CNF SAT solvers in order to efficiently carry out BCP on hpgraph. In our algorithm two cuts are watched for each hpgraph component. The watched cuts are used to detect conflicts and implied literals. We use the duality between the cuts in a hpgraph component and the paths in the corresponding vpgraph component for efficiently updating the cuts. Experimental results show that the new SAT solver is faster than the state-of-the-art solvers on a majority of the benchmarks.

Acknowledgment. We thank Per Bjesse, Sicun Gao and Will Klieber for their valuable comments. The first author is grateful to his thesis committee members for their feedback on this work.

7. REFERENCES

[1] AIGER, http://fmv.jku.at/aiger.
[2] Hardware model checking competition, http://fmv.jku.at/hwmcc07/.
[3] Minisat sat solver. http://minisat.se/.
[4] Picosat sat solver. http://fmv.jku.at/picosat/.
[5] Rsat sat solver. http://reasoning.cs.ucla.edu/rsat/.
[6] SAT competition 2007, www.satcompetition.org/2007/.
[7] P. B. Andrews. *An Introduction to Mathematical Logic and Type Theory: to Truth through Proof.* Kluwer Academic Publishers, second edition, 2002.
[8] A. Biere. Picosat essentials. *Journal on Boolean Satisfiability, Boolean Modeling and Computation (JSAT)*, 4:75–97, 2008.
[9] N. Eén and A. Biere. Effective Preprocessing in SAT Through Variable and Clause Elimination. In *SAT*, pages 61–75, 2005.
[10] N. Eén, A. Mishchenko, and N. Sörensson. Applying Logic Synthesis for Speeding Up SAT. In *SAT*, pages 272–286, 2007.
[11] N. Eén and N. Sörensson. An Extensible SAT-solver. In *SAT*, pages 502–518, 2003.
[12] M. K. Ganai, P. Ashar, A. Gupta, L. Zhang, and S. Malik. Combining Strengths of Circuit-based and CNF-based Algorithms for a High-performance SAT solver. In *DAC*, 2002.
[13] E. Goldberg and Y. Novikov. BerkMin: A Fast and Robust Sat-Solver. In *DATE*, 2002.
[14] H. Jain. Verification using satisfiability checking, predicate abstraction, and craig interpolation. Technical Report CMU-CS-08-146, SCS, CMU, 2008.
[15] H. Jain, C. Bartzis, and E. M. Clarke. Satisfiability checking of non-clausal formulas using general matings. In *SAT*, pages 75–89, 2006.
[16] Feng Lu, Li-C. Wang, Kwang-Ting Cheng, and Ric C.-Y. Huang. A Circuit SAT Solver With Signal Correlation Guided Learning. In *DATE*, 2003.
[17] J. P. Marques-Silva and K. A. Sakallah. GRASP - A New Search Algorithm for Satisfiability. In *ICCAD*, pages 220–227, November 1996.
[18] M. W. Moskewicz, C. F. Madigan, Y. Zhao, L. Zhang, and S. Malik. Chaff: Engineering an efficient SAT solver. In *DAC*, pages 530–535, June 2001.
[19] K. Pipatsrisawat and A. Darwiche. A lightweight component caching scheme for satisfiability solvers. In *SAT*, pages 294–299, 2007.
[20] C. Thiffault, F. Bacchus, and T. Walsh. Solving Non-clausal Formulas with DPLL Search. In *SAT*, 2004.
[21] G.S. Tseitin. On the complexity of derivation in propositional calculus. In *Studies in Constructive Maths and Mathematical Logic*, pages 115–125, 1968.

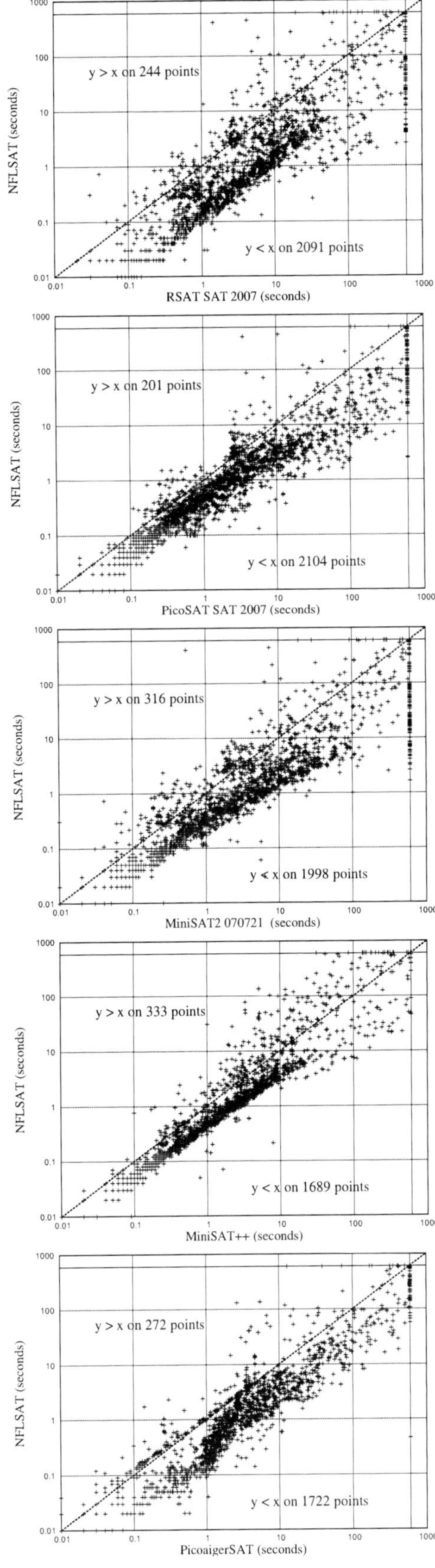

Figure 5: Scatter plots comparing run times of NFLSAT and other solvers.

978-1-60558-497-3/09 $25.00 © 2009 ACM

Constraints in One-to-Many Concretization for Abstraction Refinement*

Kuntal Nanshi, Fabio Somenzi

University of Colorado at Boulder

Abstract

In one-to-many concretization for model checking based on abstraction refinement, constraints on input vectors that are pseudo-randomly generated are often essential to the success of the procedure. These constraints have to do with both primary inputs and invisible state variables. We discuss algorithms that address both types and we show their effectiveness through experiments.

Categories and Subject Descriptors
   B.6.3 [**Logic design**]: Design aids—*Verification*
General Terms: Verification, Algorithms
Keywords: model checking, abstraction refinement, simulation

1.   Introduction

The name of *semi-formal* verification is sometimes given to the collection of techniques that combine aspects of model checking or other formal approaches with simulation of input vectors. The combination aims at jointly overcoming the capacity limitations of formal methods and the lack of completeness of simulation.

When a model fails a property in its specification in such a way that the shortest error traces are hundreds or thousands of cycles long, a combination of systematic search, as employed in model checking, and guided random simulation has advantages over other techniques. On the one hand, both BDD-based and SAT-based algorithms usually run into capacity limits for long traces. On the other hand, simulation is weak at detecting bugs that require long and complex sequences of events to be exposed.

In recent years, ways to enhance state exploration with simulation have been proposed. In this paper we are concerned with the *simulAbs* approach of [7, 8], which is based on the notion of *one-to-many mapping* in counterexample concretization. In accordance with the *abstraction refinement* paradigm, an abstract model is analyzed by model checking for satisfaction of a universal property. Since the abstract model simulates [6] the concrete one, if it satisfies the property at hand, then so does the concrete model. If, however, an *abstract counterexample* is found, it is subjected to a *concretization check*. If the check passes, the property fails; otherwise, the abstract model is refined to remove the spurious error trace, and the process is repeated until a conclusive answer is found or resources are exhausted.

While in standard abstraction refinement, concretization check consists in deciding whether the abstract counterexample corresponds to an execution of the concrete model of the same length, in the *simulAbs* approach one abstract state may be mapped to many concrete states. Thus, one may avoid the excessive refinement necessary to stretch the abstract counterexample to the length of the concrete one.

The one-to-many mapping that characterizes the concretization step of [7] derives from each abstract state a *milestone* to be reached and makes use of *pseudo-random simulation* to find sequences of states that connect those milestones. Pseudo-random vectors, however, are often not effective, especially when only a few input combinations allow extension of a trace toward a target state. For that reason, [8] introduced the use of SAT checks to connect states, and the addition of certain variables to the abstract model that enhance its *visibility*. These techniques improve the ability to concretize segments of the error trace that can be mapped one-to-one to concrete segments. While this is important, there are cases when only one-to-many mapping will avoid recourse to refinement.

To increase the chance of success of simulation-based, one-to-many mapping, one has to constrain certain inputs of the model to specific values during the pseudo-random generation. It is also important to be able to force the value of certain state variables that are *invisible*, that is, missing from the abstract model. Failure to do so results in *mis-alignment* of the visible state variables, which hinders concretization. The constraining of input and state variables is therefore the focus of this paper.

We propose an algorithm that computes all the constraints on the primary inputs of the abstract model and a heuristic algorithm for state variable alignment. Experiments show that these techniques allow the efficient verification of properties that were problematic for simulation-based concretization.

The rest of this paper is organized as follows. Sections 2 and 3 describe prior art and provide background. Section 4 discusses input constraint computation and Section 5 deals with the variable alignment problem. Our algorithm is detailed in Section 6 and experimental results are presented in Section 7.

2.   Related Work

Abstraction refinement techniques are popular in both hardware and software verification. An abstract model over-approximates the concrete model. Therefore an abstract counterexample (ACE) may not exist in the concrete model. SAT-based abstraction refinement procedures have been discussed in [5, 13]. These methods insist on finding a concrete counterexample of the same length as the abstract one and are therefore not suitable for long counterexamples.

Other SAT-based techniques have been proposed to deal with long counterexamples. In [1], the authors proposed a technique that does not limit the search of concrete counterexamples to the length of the abstract one, while using the abstract counterexample as *beacons*. The "short-sighted" approach limits the guidance offered by the abstract counterexample as discussed in [8].

The authors of [4] and [12] have proposed SAT-based techniques that deal with repeating patterns in counterexamples. While the technique of [4] heuristically identifies loops and then uses SAT to

---

*This work was supported in part by SRC contract 2008-TJ-1859.

Permission to make digital or hard copies of part or all of this work for personal or classroom use is granted without fee provided that copies are not made or distributed for profit or commercial advantage and that copies bear this notice and the full citation on the first page. To copy otherwise, to republish, to post on servers or to redistribute to lists, requires prior specific permission and/or a fee.
DAC'09, July 26-31, 2009, San Francisco, California, USA

978-1-60558-497-3/09 $25.00 © 2009 ACM

iterate through the loop, the technique of [12] attempts to identify a repeating pattern and use induction to prove the existence of counterexample. Both these techniques work well only when dealing with a regular repeating pattern.

A major alternative to SAT-based techniques is simulation-based concretization. The approaches of [14, 3, 10] describe various implementations to guide simulation using the state distance in the abstract model as a metric. These methods are not complete as they do not refine the abstract model, and therefore need a larger model to refute a property.

3. Preliminaries

We are concerned with functional verification of digital systems that are modeled as finite state machines. We assume that a (concrete) model has a finite set of state variables. Each valuation of these state variables is a *state*.

Let $x = \{x_1, \ldots, x_n\}$ and $y = \{y_1, \ldots, y_n\}$ be the present-state and next-state variables. Let $w = \{w_1, \ldots, w_m\}$ be the primary inputs. A concrete model is a 4-tuple $\langle x, w, I, T \rangle$, where $I(x)$ is the initial state predicate and $T(x, w, y)$ is the transition relation. We assume that the transition relation is given by

$$T(x, w, y) = \bigwedge_{1 \leq i \leq n} y_i \leftrightarrow \delta_i(x, w) \ , \tag{1}$$

that is, the next value of each state variable is a function of the current state and the primary inputs. *Invariant checking* consists in deciding whether a predicate on the state variables is true for all reachable states. An error trace is a path connecting an initial state to a state violating the candidate invariant.

We assume a form of abstraction known as *localization reduction*, which is obtained by dropping from (1) some of the conjuncts. The corresponding present-state variables are turned into primary inputs. Localization reduction exploits the fact that many properties are *local*: They can be verified by considering only the variables in the property and those in their vicinity. Let the abstract model contain a subset $\hat{x} \subseteq x$ of state variables with $|\hat{x}| = k \leq n$, and let $\hat{y} \subseteq y$ be the corresponding subset of next-state variables. The abstract transition relation is given by

$$\hat{T}(\hat{x}, \hat{w}, \hat{y}) = \bigwedge_{1 \leq i \leq k} y_i \leftrightarrow \delta_i(\{\hat{x}, \check{x}\}, w) \ . \tag{2}$$

The variables in $\hat{x}$ are *visible* and those in $\check{x} = x \setminus \hat{x}$ are *invisible*.

Localization reduction can be seen as a form of predicate abstraction in which the abstraction function α maps a concrete state to an abstract state if and only if it agrees with it on the visible variables. That is, state abstraction amounts to projection of the state on the visible variables. Though the techniques defined in this paper use localization reduction as abstraction technique, they can be extended to other abstraction schemes as well.

The abstraction refinement approach to invariant checking used in this paper is the one of [13], called GRAB. Reachability analysis of an abstract model allows the GRAB algorithm to compute its *synchronous onion rings* (SORs), which collectively contain all abstract states along all shortest paths from the initial states to states violating the given invariant. Each ring contains states at the same distance from the initial states. The SORs drive the refinement process: They are analyzed to decide which invisible variables should be made visible. The SORs are also checked for concretization. If a path exists in the concrete model such that the i-th state of the trace is mapped by α to the i-th SOR, then the path is an error trace. The existence of such a path is decided by a SAT solver. The abstract error trace provides *guidance* in the search of the concrete trace. By capturing multiple ACEs, the SORs improve the chance of success of concretization.

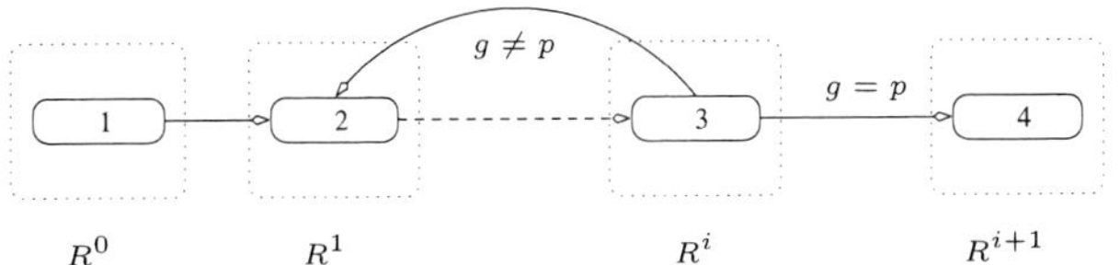

Figure 1: Example of input constraint

4. Constraints

Hardware designs are usually equipped with reset signals. In normal operation, these signals are asserted infrequently. In most semi-formal techniques, constraints for reset signals are given by the user; therefore, they do not need a dedicated algorithm. However, other inputs of the design that are asserted infrequently may not be constrained externally in a meaningful manner. A dedicated algorithm is needed to constrain their values during simulation.

Consider a communication circuit, in which an input signal indicates an end-of-frame. Being part of the input control, this signal may not be constrained like a reset signal. The end-of-frame is asserted infrequently, and if prematurely asserted, it may derail concretization of the counterexamples. This is hard to achieve with pseudo-random simulation; hence the need to constrain the input.

Consider the example shown in Fig 1. It represents a portion of abstract counterexamples. Each dotted box represents an SOR and g is a primary input variable. Assume that the transition from R^i to R^{i+1} does not match any length-1 concrete transition, else the abstract transition could be mapped using a SAT solver. Progress from SOR R^i to SOR R^{i+1} requires that value p be applied to g. Suppose that instead of $g = p$, the pseudo-random vector generation procedure chooses $g \neq p$; as a consequence the concrete trace returns to a previous SOR, say R^1. As discussed in [7], a concrete trace that goes back to a previous SOR does not necessarily move away from the target. Concretization algorithms in general do not stop at such an event and continue to extend the concrete trace.

Assume that the backward transition to SOR R^1 enabled by $g \neq p$ is in fact a transition away from the target in the concrete trace. Pseudo-random vectors therefore may cause the concrete trace to "reset," thus making concretization difficult.

4.1 Computing Constraints

Computing constraints on the input variables w is cast as the problem of finding satisfying assignments for transitions from the current abstract state to all states in the next SOR reachable from it. The transition relation for the abstract model $\hat{T}(\hat{x}, \hat{w}, \hat{y})$ is conjoined with the current abstract state $S_i(\hat{x})$ and next SOR $R^{i+1}(\hat{y})$. The characteristic function of all the transitions from the current abstract state to the next SOR is therefore given by

$$S_i(\hat{x}) \wedge \hat{T}(\hat{x}, \hat{w}, \hat{y}) \wedge R^{i+1}(\hat{y}) \ . \tag{3}$$

In order to compute constraints on primary inputs, (3) is cofactored with every value v_{ij} of each primary input w_i of the abstract model $\hat{M}$. If the cofactor results in the empty set, no transitions are possible with $w_i = v_{ij}$; therefore, $w_i \neq v_{ij}$ is a constraint on w_i.

The constraint on variable w_i is therefore a disjunction of all values v_{ij} which result in a non empty cofactor. For example, in case of a binary primary input, if the cofactor for value 0 results in the empty set, the remaining value 1 is the constraint. Let ψ_{ij} be the result of the cofactor of (3) with respect to value v_{ij} of input variable w_i, then the constraint for variable w_i is given by the equation,

$$C_i = \bigvee_{j | \psi_{ij} \neq 0} w_i = v_{ij} \ .$$

A constraint on an input is therefore a subset of all possible val-

978-1-60558-497-3/09 $25.00 © 2009 ACM

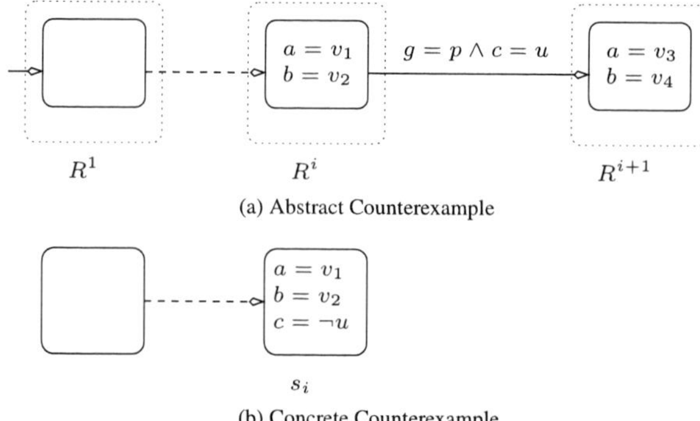

(a) Abstract Counterexample

(b) Concrete Counterexample

Figure 2: Example of mis-alignment

ues of the input such that all transitions to the next SOR are still enabled. Notice that the constraints are only computed on the real primary inputs of the abstract model and not the invisible variables that are inputs to the abstract model or the real primary inputs that are not in the abstract model. The constraints are computed before every attempt to map a transition to the next SOR. They are then used to generate vectors such that an input is pseudo-randomly assigned a value from the set that satisfies the constraint.

The constraint on each primary input is independent of the constraints on the other primary inputs. The set of generated vectors can therefore be characterized by the equation

$$\bigwedge_i \bigvee_{j\,|\,\psi_{ij}\neq 0} w_i = v_{ij} \ .$$

Our implementation computes constraints with Binary Decision Diagrams (BDD), although it could also use satisfiabilty checks.

5. Alignment

Section 4 deals with generating values on the primary inputs that enable transition to the next SOR. However, invisible variables figure more prominently than primary inputs in preventing concretization of abstract counterexamples. They cannot be constrained to specific values like primary inputs; we need a more sophisticated approach to achieve correct values on invisible variables.

Consider the example shown in Fig 2. It represents a portion of the abstract counterexamples in form of SORs. Let a and b be the visible variables of the abstract model, g a primary input variable and c an invisible variable. Assume that next state functions for a and b contain the predicates

$$\pi_a = (a = v_1) \wedge (b = v_2) \wedge (g = p)$$
$$\pi_b = (b = v_2) \wedge (c = u) \ .$$

Assume that the concretization procedure has traced a concrete path to state $s_i \in \alpha^{-1}(R^i)$. Also assume that for state s_i, $c \neq u$ holds. Therefore the abstract transition to SOR R^{i+1} cannot be mapped unless the concrete trace moves to a concrete state s_i' where $c = u$ holds. Further assume that to trace a concrete path from s_i to s_i' certain primary inputs must be constrained. Computing constraints on inputs as explained in Section 4 is not enough, since the constraints are computed only on the abstract model and invisible variable c would be ignored. The only constraint for the abstract transition from SOR R^i to SOR R^{i+1} is $g = p$. If we simulate the concrete model from s_i with constraint $g = p$, the result is a state in the concrete model where $a = v_3$ holds, but $b = v4$ does not. While variable a can change to a value that it takes in SOR R^{i+1}, variable b cannot transit to its value in SOR R^{i+1}. The rest of the counterexample therefore may not be concretized, leading to refinement

of the abstract model. Subsequent concretization attempts may also fail until invisible variable c is added to the abstract model.

For concrete state s_i, variable a has all the necessary conditions to move to a value in the next SOR, whereas variable b has to wait until c changes value. We call this condition "mis-alignment."

Definition 1 *For concrete state $s_i \in \alpha^{-1}(R^i)$, let x_s and x_t be such that,*

- $\hat{x} = x_s \cup x_t$
- $\forall x_j \in x_s, |s_i(x) \wedge y_j \leftrightarrow \delta_j \wedge R^{i+1}(\hat{x})| \neq 0$
- $\forall x_j \in x_t, |s_i(x) \wedge y_j \leftrightarrow \delta_j \wedge R^{i+1}(\hat{x})| = 0$

SOR R^i is mis-aligned if and only if both x_s and x_t are not empty.

Mis-alignments may be gradually removed by subsequent refinements of the abstract model. The refinements may get expensive since the invisible variables that cause mis-alignments may be many and to remove mis-alignment all such invisible variables may need to be added to the abstract model.

As an alternative to refinement, we propose to identify invisible variables causing mis-alignment and compute constraints that allow them to change their value so that the transition to the next SOR may take place. In order to remove the mis-alignment, the concrete trace has to reach a state in the concrete model where x_t is an empty set, i.e., no invisible variable has a value that prevents the transition to the next SOR. To find the values on invisible variables that would make the transition possible, we need a *stepping stone* between current state in R^i and next SOR R^{i+1}, which still lies in SOR R^i, but is closer to R^{i+1} in the concrete model. We do this by computing a pre-image of SOR R^{i+1} over an *interim* abstract model that includes invisible variables like c. Since, the variables that cause mis-alignment are difficult to find without refining the abstract model, we add all closest invisible variables—state variables that appear in the support set of bit relations of the visible variables—for the pre-image computation. The pre-image is conjoined with the current SOR R^i to yield a series of candidates for the *stepping stone*. The candidates are given by

$$R^i(\hat{x}) \wedge \exists \hat{y} \,.\, \exists \hat{w} \,.\, \hat{T}(\hat{x}, \hat{w}, \hat{y}) \wedge R^{i+1}(\hat{y}) \ .$$

Invisible variables that cause mis-alignment are not known in advance. We therefore define heuristics to select a representative variable, called a *guide* variable, that is used to compute constraints to change values on invisible variables causing mis-alignment. Other schemes may be derived that use multiple guide variables in place of a single guide variable. However, our use of a single *guide* variable is motivated by the observation that invisible variables causing mis-alignment tend to have the same Cone of Influence (COI) and therefore inputs constraints to change values of all such variables is the same.

5.1 Computing Constraints with Alignment

Assume that the abstract counterexamples have been concretized up to SOR R^i. Concrete state s_i, whose corresponding abstract state is in R^i, is the current state from which a transition to the next SOR R^{i+1} will be attempted.

The heuristics for selecting a guide variable has two components. The first selects invisible variables whose values in s_i make forward progress to next SOR impossible. To do so, each invisible variable x_k is assigned a score based on θ_{x_k}, which is given by

$$\theta_{x_k} = (\exists x_E \,.\, s_i(x)) \wedge \exists \hat{y} \,.\, R^i(\hat{x}) \wedge \hat{T}(\hat{x}, \breve{x}, \hat{y}) \wedge R^{i+1}(\hat{y}) \ ,$$

where $x_E = \breve{x} \setminus x_k$ (set of invisible variables without x_k). If $\theta_{x_k} = 0$ then the score of x_k is 0; otherwise it is 1. The invisible variables

978-1-60558-497-3/09 $25.00 © 2009 ACM

571

assigned a score of 0 need to change their values for the next SOR to be reached.

The second component of the heuristic is taken from GRAB [13]; it selects invisible variables that have the best chance to kill all transitions from R^i to R^{i+1}. It does so by assigning each invisible variable a score given by the equation

$$\gamma_{x_k} = \frac{|\exists w_E \,.\, \forall x_k \,.\, \exists \hat{y}.R^i(\hat{x}) \wedge \hat{T}(\hat{x}, \breve{x}, \hat{y}) \wedge R^{i+1}(\hat{y})|}{|R^i(\hat{x})|} \;,$$

where $x_E = \breve{x} \setminus x_k$ as before. The two scores θ and γ are added to get a combined score for each invisible variable. The combined score is between 0 and 2. The invisible variable with the lowest score is selected as "guide" to compute *guide constraints*. The procedure to compute the guide constraints is similar to that described in Section 4.1, with the exception that only the transition relation for guide variable x_k is involved in the computation. The characteristic function for guide variable from current state to the candidate *stepping stones* is given by

$$R_i(\hat{x}) \wedge y_k \leftrightarrow \delta_{y_k}(\hat{x}, \hat{w}) \wedge R^{i+1}(\hat{y}) \;. \tag{4}$$

In order to compute constraints on primary inputs, (4) is cofactored with respect to every value v_{ij} of each primary input w_i. As in Section 4.1, if the cofactor is the empty set, no transitions are possible with $w_i = v_{ij}$; therefore, $w_i \neq v_{ij}$ is a constraint on w_i. The constraint on variable w_i is therefore a disjunction of all values v_{ij} which result in a non-empty cofactor.

In addition to the guide constraints, constraints are also computed for transitions from R^i to *stepping stones* in the abstract model as described in Section 4.1. The two constraints are combined to derive the overall constraints to transit from R^i to *stepping stones*. If the two constraints do not contradict each other, i.e., their conjunction is satisfiable, the constraint for the input variable is the conjunction. If, however, the constraints do contradict each other, we stop the current simulation attempt and call for remedial measures like backtracking and refinements.

6. Concretization Algorithm

The pseudo-code for our concretization algorithm is shown in Fig. 3. The procedure takes as inputs the concrete model M, current abstract counterexamples in form of SORs R, and a set of abstract states that lie on a potential counterexample path, (*care set Z* [7]). Our concretization procedure is adopted from [8] with modifications described in this Section.

Let $R = \{R^0, \ldots, R^L\}$ be the SORs representing length-L abstract counterexamples, where R^0 is a set of abstract initial states and R^L is a set of abstract states where the given property ϕ does not hold. With $r_i \in R^i$, let $r_0, r_1, \ldots, r_L$ represent an abstract counter example, such that for $0 \leq i \leq L$, there is an abstract transition between r_i and r_{i+1}. A counterexample segment (CES) as defined in [7] is a sequence of states on a path from concrete state $s_i \in \alpha^{-1}(R^i)$ to state $s_{i+1} \in \alpha^{-1}(R^{i+1})$. The concrete states s_i on the trace that map to abstract states are called *milestones*.

The procedure initially assumes that no SOR is mis-aligned. If it cannot map a transition to the next SOR within the given resource bounds, it assumes that the current SOR is mis-aligned and proceeds to align it. The initial set of aligned SORs is therefore the set of current SORs. The procedure to align the SOR is combined with the *backtrack* procedure and is explained in Section 6.1.

The algorithm begins by selecting an initial state in the concrete model $s_0 \in \alpha^{-1}(R^0)$. It then attempts to find a longest segment of concrete counterexample that maps one-to-one to an abstract one beginning from s_0. This search is set up as a set of satisfiability

problems [8]. The CHECKABSTRACTSEGMENT procedure returns a counter example segment (CES) which is appended to the current concrete counterexample. If CHECKABSTRACTSEGMENT reaches the last SOR, the process is complete and the CCE is returned.

Otherwise, the procedure attempts to map transitions to the next SOR using simulation. The simulation phase of our procedure differs from that in [8] in two respects. First, it uses constrained vectors; second, simulation is stopped as soon as the concrete trace reaches a state s_t whose corresponding abstract state lies in an SOR different from the current one. The latter change makes the procedure more effective when a portion of concrete counterexample matches a repeating pattern in an abstract counterexamples.

The simulation phase of our procedure begins by generating n constrained vectors. The procedure to generate constrained vectors is explained in Section 6.2. The concrete model is then sequentially simulated with the generated vectors starting from s_i, the current *milestone* state. A successful simulation procedure returns the CES, which is added to the CCE. If simulation mapped a transition to the last SOR, concretization is successful and the CCE is returned. Otherwise, it repeats the concretization process starting from the new *milestone*. The procedure continues until the abstract counterexample is completely mapped or the simulation phase exhausts resources without mapping a transition to the next SOR.

If the simulation attempt fails, the procedure *backtracks* and attempts to align the current SOR as explained in Section 6.1. After a successful alignment, the concretization procedure attempts to map a transition to the next SOR again, this time with the help of *stepping stones*. The simulation procedure searches for a trace from the current milestone to a state in the *stepping stones*. Once such a trace is found, the gap to the next SOR can be bridged with a simple SAT search. If simulation cannot find a trace to a *stepping stone*, the procedure *backtracks* further and attempts to align the previous SOR. If alignment fails or the number of backtrack attempts exceeds a bound, the concretization procedure terminates and calls for refinement of the abstract model.

6.1 Aligning SORs

The pseudo-code for the alignment procedure is shown in Fig 4. It takes as inputs the set of currently aligned SORs $\hat{R}$, a set of visible variables $\hat{x}$ and a set of invisible variables $\breve{x}$. The procedure returns a set of newly aligned SORs.

The procedure begins by adding all invisible variables $\breve{x}$ to the set of visible variables $\hat{x}$. It then temporarily builds a new abstract

```
1   CONCRETIZATION(M, R, Z) {
2       R̂ = DUPLICATE(R)
3       while (TRUE) {
4           if (i == 0) s_i = GETINITIALSTATE(M, R^0)
5           CES = CHECKABSTRACTSEGMENT(M, R̂, i)
6           add CES to CCE
7           if (i == |R|-1) return CCE
8           V = GENCONSTRAINEDVECTORS(M, R, R̂, s_i, n)
9           CES = SIMULATEANDCHECK(V, s_i, R, R̂, Z, M, i)
10          if (CES ≠ NULL) {
11              add CES to CCE
12              if (i == |R|-1) return (CCE)
13          } else
14              if (!ALIGNSOR(R̂, x̂, x̆, i, CCE)) return NULL
15      }
16  }
```

Figure 3: Concretization algorithm

978-1-60558-497-3/09 $25.00 © 2009 ACM

```
1    ALIGNSOR($\hat{R}, \hat{x}, \check{x}, i$,CCE) {
2        $\hat{x} = \hat{x} \cup \check{x}$
3        $\bar{M}$ = BUILDABSTRACTMODEL($\hat{x}$)
4        if (CHECKALIGN($\hat{R}^i$)) {
5            $i--$
6            ADJUST(CCE,$i$) }
7        $\hat{R}^i = \hat{R}^i \wedge$ preimg($\hat{R}^{i+1}$)
8        if ($\hat{R}^i$ is empty) return NULL
9        else return $\hat{R}$
10   }
```

Figure 4: Aligning synchronous onion ring

```
1    GENCONSTRAINEDVECTORS($M, R, \hat{R}, s_i, n$) {
2        if (!CHECKALIGN($\hat{R}^i$)) {
3            for each $w_j \in w$
4                $c_{w_j}$ = FINDSATASSIGNS($M, R^i, R^{i+1}, \hat{x}$)
5        } else {
6            $x_k$ = SELECTGUIDEVARIABLE($M, R, s_i$)
7            for each $w_j \in w$ {
8                $c_{w_j}^a$ = FINDSATASSIGNS($M, R^i, \hat{R}^i, \hat{x}$)
9                $c_{w_j}^b$ = FINDSATASSIGNS($M, R^i, \hat{R}^i, x_k$)
10           }
11           $c$ = COMBINECONSTRAINTS($c, c_a$)
12       }
13       $V$ = GENERATEVECTORS($w, c, n$)
14       return $V$
15   }
```

Figure 5: Generating constrained vectors

model $\bar{M}$. The procedure then aligns the current SOR by computing the pre-image of the next SOR with the newly-built abstract model. The pre-image is conjoined with the current SOR to get the aligned SOR. The aligned SOR is therefore a set of *stepping stones* as described in Section 5. If pre-image computation results in the empty set, the alignment process is aborted, else the SOR in the set is replaced by the aligned SOR. If the current SOR was previously aligned, the procedure *backtracks* to the previously recognized SOR and attempts to align that SOR. Whenever the procedure *backtracks*, the segment of CCE starting from the previous milestone is discarded.

6.2 Generating Constrained Vectors

The pseudo-code for generating constrained vectors is shown in Fig 5. The procedure takes as inputs the concrete model M, the original SORs R and the aligned SORs $\hat{R}$, the current state s_i and n, the number of vectors to be generated.

If the current SOR is not aligned, the procedure computes constraints on all primary inputs of the abstract model for the transition from the current abstract state to the next SOR as discussed in Section 4.1. On the other hand, if the current SOR is aligned, the constraints need to be refined to enable certain invisible variables to change values. (See Section 5.1.) The procedure performs a localized refinement by adding the selected invisible variable to the abstract model. This addition is temporary and is reversed before the next refinement iteration in case concretization fails. The procedure then computes the guide constraints as explained in Section 5.1. In addition the procedure computes constraints on all inputs for transition from the current abstract state to *stepping stones* over the visible variables.

The two constraints are then combined into refined constraints as described in Section 5.1. The procedure generates input vectors by pseudo-random assigning to each primary input values from its set of constraints. An unconstrained input may take any value.

7. Experimental Results

We implemented our algorithm in VIS-2.1 [2, 11]. The experiments were run on a Linux workstation with a 2.06 GHz Intel Core 2 Duo processor and 2GB of RAM. We compare our algorithm CO-LIGN to GRAB [13], *simulAbs* [7] and IVHAC [7], which are also implemented in VIS. In addition to the set of test cases used in [8], we also use circuits from OPENCORES [9].

For CO-LIGN heuristic the bounds are same as those for IVHAC [8], i.e., the lower bound on number of simulation vectors is 2500 and the upper bound is 50000; the number of backtrack attempts is set to 5. The heuristic bounds for *simulAbs* are set as described in [7]. All experiments are run with the same values of heuristic parameters and the heuristic parameters are not tuned. Further improvements may be achieved by tuning them. All experiments are run five times, each time with a different random seed. The results presented are the median value of each experiment. Each run is limited to 2 hours, with BDD dynamic variable ordering enabled.

Table 1 presents the comparison of various methods. Column 1 refers to the example and the property under study. Column 2 lists the number of binary variables in the cone of influence (COI) of the property. In Column 3, we list the length of the shortest counterexample if known. Passing properties are shown as 'T' in Column 3. For each method being compared we list **CPU**, the total run time in seconds and **ratio**, the ratio of binary state variables in the last abstract model to the binary state variables in the concrete model. In addition, we also list **cex**, the length of concrete counterexample found by *simulAbs*, IVHAC and CO-LIGN. A lower ratio demonstrates the ability of the approach to find a concrete counterexample with a smaller abstract model. For experiments that did not finish in the alloted time, the ratio is reported in parentheses, and refers to the last abstract model computed by the approach.

Our experiments use a mixture of passing and failing properties with counterexamples of varying lengths. Our experiments demonstrate that CO-LIGN fared significantly better then previous approaches when counterexamples are long. CO-LIGN was able to find counterexamples for examples like eth-p1, dma-p1, spi-p1 and i2c-p1, where none of the compared approaches was successful. For these examples the progress of the concrete trace requires specific values on inputs, which pseudo-random vectors cannot provide; hence the failure of previous simulation-based approaches. The run times for CO-LIGN are significantly better than GRAB for long counterexamples as SAT runs out of steam. For passing properties, concretization procedures are an overhead, and GRAB performs better since abstract counterexamples being small, the overhead is less than for the simulation-based approaches.

Our approach runs SAT checks as part of CHECKABSTRACTSEGMENT (Section 6), every time it detects a repeating pattern in the concrete counterexample. Sometimes these checks are futile. In case of b12-p2 and ar-p1, where the repeating pattern runs into thousands, the futile SAT checks add a significant overhead. Even so, our approach is significantly better than GRAB for those two examples. This overhead, may significantly be reduced with a different implementation scheme for CHECKABSTRACTSEGMENT.

The results demonstrate that our approach finds counterexamples for a wide variety of problems which were not solved by previous simulation-based techniques or by GRAB, and it does so with little overhead.

Table 1: Performance comparison of CO-LIGN with *simulAbs*, IVHAC and GRAB

circuit	COI regs	cex len	GRAB[13] CPU	ratio	*simulAbs*[7] CPU	ratio	cex	IVHAC[8] CPU	ratio	cex	CO-LIGN CPU	ratio	cex
d1-p2	101	13	6.3	0.22	1.9	0.03	100	5.4	0.03	678	4	0.03	711
d1-p4	101	T	1.3	0.03	1.3	0.03	-	1.2	0.03	-	1.2	0.03	-
d2-p1	94	14	22.1	0.53	295.0	0.47	1149	58.9	0.47	777	29.3	0.47	555
d4-p2	230	T	103.5	0.16	>2 h	(0.05)	-	1086.2	0.16	-	393.7	0.16	-
d24-p1	179	9	4.2	0.02	4.6	0.02	9	4.5	0.02	9	4.2	0.02	9
d24-p4	179	T	4.5	0.05	298.6	0.05	-	76.0	0.05	-	27.8	0.05	-
rcu1-p2	49	T	0.8	0.10	0.8	0.10	-	0.7	0.10	-	0.7	0.10	-
rcu1-p3	44	48	58.0	0.25	90.6	0.09	111	34.9	0.09	104	7.7	0.09	49
ba1-p1	44	-	>2 h	(0.60)	1.4	0.18	564	2.6	0.18	1854	2.6	0.18	610
ba2-p1	120	-	>2 h	(1.00)	76.9	0.07	5126	78.6	0.07	5145	65.4	0.07	389
b12-p2	110	-	>2 h	(0.15)	301.9	0.09	32296	116.1	0.09	32163	595.0	0.09	31898
rcu2-p3	137	-	>2 h	(0.24)	38.5	0.03	1628	32.9	0.03	1470	28.2	0.03	369
ar-p1	111	-	>2 h	(0.28)	17.8	0.02	3107	81.1	0.08	3132	322.2	0.08	3081
daio-p2	27	-	>2 h	(0.70)	4.6	0.33	12796	18.1	0.33	12796	17.6	0.33	12796
b13-p1	34	-	>2 h	(0.85)	33.3	0.05	40036	73.7	0.59	40033	43.7	0.59	40033
am2901-p1	68	16	47.5	0.25	>2 h	(0.50)	(16)	24.2	0.25	16	22.5	0.25	16
blkjk-p1	103	13	12.8	0.08	1709.4	1.00	(13)	16.1	0.04	15	9.5	0.04	111
lkysven-p1	30	63	34.7	0.86	177.6	1.00	(63)	24.2	0.80	2108	14.6	0.80	1038
two-p1	30	29	3.0	0.83	80.4	1.00	(29)	16.1	0.83	29	8.5	0.83	29
iftch-p1	28	8	0.4	0.39	194.3	1.00	(8)	19.6	0.32	8	9.3	0.32	8
vga-p1	62	-	>2 h	(0.12)	>2 h	(0.50)	-	57.6	0.08	2420	4.7	0.08	1678
eth-p1	29	-	>2 h	(0.58)	>2 h	(0.58)	-	>2 h	(0.58)	-	495.5	0.17	8197
dma-p1	20	2050	34.0	0.90	355.7	1.00	(2050)	738.0	0.90	(2050)	22.8	0.52	3071
spi-p1	28	-	>2 h	(0.62)	>2 h	(0.62)	-	>2 h	(0.62)	-	94.7	0.18	57348
i2c-p1	33	-	>2 h	(0.70)	>2 h	(0.70)	-	>2 h	(0.70)	-	87.6	0.27	12294

8. Conclusions

We presented an algorithm that enhances concretization checks for abstraction refinement, where an abstract counterexample is allowed to map a concrete one of possibly different length. The main contributions of this paper are techniques that constrain the input vectors in the pseudo-random simulation phase of the algorithm. We have identified the alignment problem that arises from invisible variables that take on values that prevent extension of the error trace and we have proposed a heuristic algorithm to re-align variables. Our experiments show that examples that are hard for concretization procedures based on simulation are verified efficiently by our new algorithm.

References

[1] P. Bjesse and J. Kukula. Using counter example guided abstraction refinement to find complex bugs. In *Proceedings of the Conference on Design, Automation and Test in Europe*, pages 156–161, Feb. 2004.

[2] R. K. Brayton et al. VIS: A system for verification and synthesis. In T. Henzinger and R. Alur, editors, *Eighth Conference on Computer Aided Verification (CAV'96)*, pages 428–432. Springer-Verlag, Rutgers University, 1996. LNCS 1102.

[3] F. M. de Paula and A. Hu. An effective guidance strategy for abstraction-guided simulation. In *Proceedings of the Design Automation Conference*, pages 63–68, San Diego, CA, June 2007.

[4] D. Kroening and G. Weissenbacher. Counterexamples with loops for predicate abstraction. In *Computer-Aided Verification*, pages 152–165, Aug. 2006.

[5] R. P. Kurshan. *Computer-Aided Verification of Coordinating Processes*. Princeton University Press, Princeton, NJ, 1994.

[6] R. Milner. An algebraic definition of simulation between programs. *Proc. 2nd Int. Joint Conf. on Artificial Intelligence*, pages 481–489, 1971.

[7] K. Nanshi and F. Somenzi. Guiding simulation with increasingly refined abstract traces. In *Proceedings of the Design Automation Conference*, pages 737–742, San Francisco, CA, July 2006.

[8] K. Nanshi and F. Somenzi. Improved visibility in one-to-many trace concretization. In *Design Automation and Test in Europe*, pages 819–824, Munich, Germany, Mar. 2008.

[9] URL: http://www.opencores.org.

[10] S. Shyam and V. Bertacco. Distance-guided hybrid verification with GUIDO. In *Proceedings of the Conference on Design, Automation and Test in Europe*, pages 1211–1216, Munich, Germany, 2006.

[11] URL: http://vlsi.colorado.edu/~vis.

[12] C. Wang, A. Gupta, and F. Ivancic. Induction in CEGAR for detecting counterexamples. In *Formal Methods in Computer Aided Design*, pages 77–84, Nov. 2007.

[13] C. Wang, B. Li, H. Jin, G. D. Hachtel, and F. Somenzi. Improving Ariadne's bundle by following multiple threads in abstraction refinement. In *Proceedings of the International Conference on Computer-Aided Design*, pages 408–415, Nov. 2003.

[14] W. Wu and M. S. Hsiao. Efficient design validation based on cultural algorithms. In *Proceedings of the Conference on Design, Automation and Test in Europe*, pages 402–407, Mar. 2008.

978-1-60558-497-3/09 $25.00 © 2009 ACM

Spectrum: A Hybrid Nanophotonic–Electric On-Chip Network

Zheng Li*, Dan Fay[†], Alan Mickelson[†], Li Shang[†], Manish Vachharajani[†], Dejan Filipovic[†], Wounjhang Park[†], and Yihe Sun*

* Tsinghua National Laboratory for Information Science and Technology, Inst. of Microelectronics, Tsinghua University, Beijing 100084, China
[†] Department of Electrical, Computer, and Energy Engineering, University of Colorado, Boulder, CO 80309, U.S.A
zheng-li@mails.tsinghua.edu.cn {daniel.fay, mickel, li.shang, manish.vachharajani, dejan, wpark}@colorado.edu sunyh@tsinghua.edu.cn

ABSTRACT

On many-core chip designs, short, often-multicast, latency-critical messages, used extensively in high-level coherence and synchronization protocols, often become the bottleneck of parallel performance scaling. This paper presents Spectrum, a hybrid nanophotonic-electric on-chip network that optimizes both throughput and latency. Spectrum's novel planar nanophotonic subnetwork broadcasts latency-critical messages through a wavelength-division multiplexed (WDM) two-dimensional waveguide. Spectrum's throughput-optimized packet-switching electrical subnetwork handles high bandwidth traffic. Overall, Spectrum delivers an almost ideal CMOS-compatible interconnection network for many-core systems.

Categories and Subject Descriptors: C.1.2 [Processor Architectures]: Multiple Data Stream Architectures (Multiprocessors) – Interconnect Architectures.

General Terms: Design, Performance.

Keywords: Networks-on-Chip, silicon photonics, optical communication.

I. MOTIVATIONS

While fabrication technology scaling has steadily improved the performance of transistors, the performance of global wires has not scaled correspondingly, and has become the primary performance bottleneck in microprocessors. Moreover, the power consumption of global wires and repeaters are consuming an increasingly significant portion of the overall system power budget. As a result, core-to-core communication is now the major bottleneck in scaling *parallel* program performance on many-core chip multiprocessors.

Recently introduced electrical packet-switched on-chip interconnect fabrics promise to dramatically improve the throughput and scalability of global interconnect over today's common approaches (e.g., shared buses and dedicated point-to-point link circuitry), and are gradually becoming the *de facto* on-chip communication backplane in the many-core era. Many multi-core chips default to this solution (e.g., TRIPS [1], TILE64 [2], and Intel's 80-core Teraflops prototype chip [3]). These interconnects are able to provide highly-scalable throughput at large processor counts; for example, the on-chip network in Intel's 80-core chip offers 320 GB/s bisection bandwidth.

While electrical packet-switched on-chip networks provide excellent throughput, they introduce excess and variable latency because of run-time multi-hop data buffering, resource arbitration and link contention. As the number of on-chip processor cores increases, worst-case latency suffers as the average number of hops per packet increases. Thus, in future many-core systems, short, latency-critical packets, like synchronization and coherence messages, are likely to become the main performance bottleneck.

Compounding the problem is that a significant portion of latency-critical messages are short messages employed in synchronization and coherence protocols that must be multicast or broadcast [4]. Broadcast or multicast messages are translated into a series of independent unicast messages and relayed across the electrical packet-switched network. The latency is thus determined by the slowest unicast packet. Worse still, recent studies show that even if only a small portion of the total on-chip network traffic is multicast traffic, the overall network latency and throughput can degrade substantially, negating the point-to-point efficiency and performance advantages of the electrical packet-switched on-chip network [5].

Optics, as the ultimate on-chip communication technology, have been explored to address the on-chip interconnection problem [6]. Various solutions have been proposed to replace global wires with one-dimensional waveguide-based optical interconnect. Dainesi et al. [7] fabricated 3-D integrable optoelectonic switches and filters on silicon-on-insulator (SOI) for telecommunication ICs. Haurylau et al. [8] study the roadmap of on-chip optical interconnect design, and identify major design and fabrication challenges of on-chip optical solutions. Kirman et al. [9] investigate the possible use of on-chip optical interconnect in high-performance chip-multiprocessors. They propose a hierarchical snoopy bus design consisting of global optical loops and local electrical interconnect. Shachama et al. [10] present a hybrid on-chip network for future 22 nm technology. It combines a high bandwidth circuit-switched torus photonic network and a packet-switched electrical control network. Vantrease et al. [11] present a nanophotonic interconnect targeting throughput optimization in many-core systems for future 16nm technology. The proposed design leverages WDM technology and consists of a novel all-optical arbitration logic. Petracca et al. [12] explore topologies of photonic network design using one real application. Most of these works focus on the bandwidth and power benefits of on-chip optics. Assumptions are also made that future nanophotonic technologies will make these designs compatible with CMOS electronics.

In this work, we observe that both interconnect throughput *and latency* will quickly become performance bottlenecks in many-core microprocessors. As a result, it is essential for interconnect technology to efficiently provide both low latency and high throughput. Moreover, the fabrication of interconnect design shall be compatible with existing CMOS process. It should not interfere with the fabrication of the processor logic in a state-of-the-art CMOS process.

We propose and develop Spectrum, a hybrid nanophotonic-electric on-chip interconnect. The nanophotonic and electrical subnetworks operate in parallel. Spectrum makes the following contributions:

1) A simplified, throughput-optimized, packet-switched on-chip electrical network that supports 3D die-stacking integration. Large, latency-tolerant data packets are relayed via this network.

2) A broadcast-based nanophotonic network based on Wave Division Multiplexing (WDM) over a 2D SOI waveguide. This network efficiently supports low-latency broadcasting of short messages, such as those used in coherence and synchronization protocols.

3) A realizable nanophotonic network. We have successfully fabricated key structures in the nanophotonic network, in a CMOS-compatible manner.

II. NANOPHOTONIC SUBNETWORK

In nanophotonic interconnect, delay is not constrained by the RC constant as is electrical interconnect. Thus, it can provide light speed communication – i.e., low latency. This section discusses the specifics of the broadcast-based nanophotonic network and explains how the fabrication challenges have been addressed.

This paper was supported in part by the NSF under awards CCF-0829950, in part by the National Natural Science Foundation of China (NSFC) under grant #60236020 and the Specialized Research Fund for the Doctoral Program of Higher Education (SRFDP) #20050003083.

Permission to make digital or hard copies of part or all of this work for personal or classroom use is granted without fee provided that copies are not made or distributed for profit or commercial advantage and that copies bear this notice and the full citation on the first page. To copy otherwise, to republish, to post on servers or to redistribute to lists, requires prior specific permission and/or a fee.
DAC'09, July 26-31, 2009, San Francisco, California, USA

978-1-60558-497-3/09 $25.00 © 2009 ACM

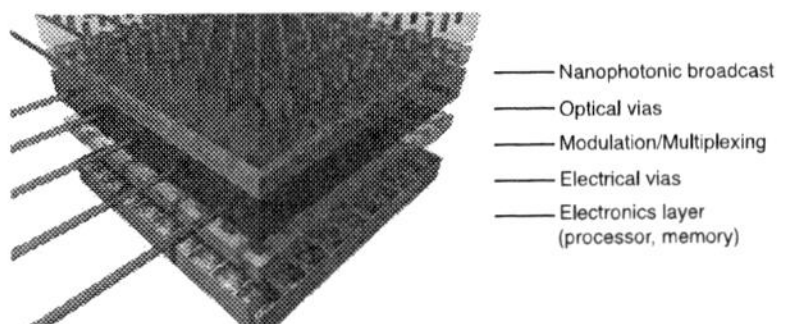

(a) Schematic depiction.

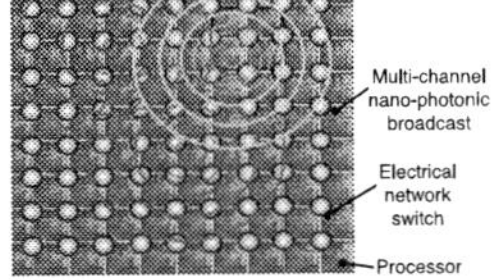

(b) Top-down view.

(c) Components of the photonic network and their frequency responses.

Figure 1.   Spectrum.

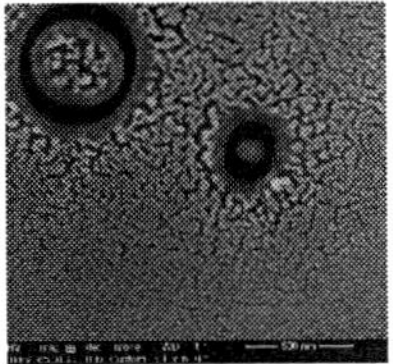

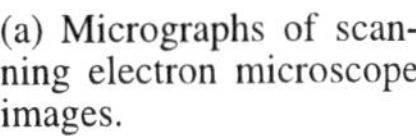

(a) Micrographs of scanning electron microscope images.

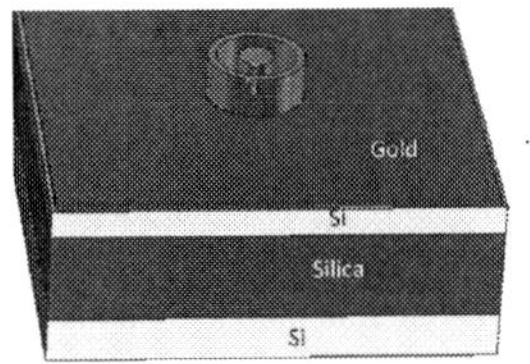

(b) Plasmonic antenna model.

Figure 2.   Antenna

As shown in Figure 1(c), like other general optical interconnects, Spectrum's nanophotonic subnetwork consists of transmitters, channels and receivers. These components are stacked layer by layer, as depicted in Figure 1(a).

To provide a CMOS-compatible transmission channel for broadcasting latency-critical messages, all nanophotonic components are fabricated in separate layers, with available technology, and stacked on a standard CMOS die. The waveguide is a planar SOI waveguide, the top layer in Figure 1(a). Within this layer, an array of nanophotonic antennas, nanoscopically printed on the surface of the silicon, broadcast and receive the optical signal. These antennas are unique to Spectrum, and thus are the key unproven nanophotonic component – however we discuss how we have successfully fabricated these antennas.

The antenna feeds extend through several layers of polymer-encased polaritonic nanophotonic devices that relay messages between the send/receive electronics and the planar waveguide. The devices in these layers are mainly ring resonators, which can serve as filters [13], modulators [14] and WDM demultiplexers [15]. Germanium-doped silicon photon detectors and amplifiers sit in the bottom layer together with other electrical components. Since each component rests in its own layer, each layer can be optimized in the horizontal dimension independently, without excessive constraints on area or the need for waveguide crossovers.

II.A. 2D Broadcast Transmission Channels

In Spectrum, we choose SOI for the 2D waveguide. SOI waveguide technology has progressed from a barely demonstrable idea to a viable technology [16] in the last 15 years. Small losses, e.g., $0.7\,dB/cm$, are now obtainable in single mode structures that achieve the maximum confinement (minimum spot) possible.

SOI waveguide processing generally starts from a silicon substrate and proceeds by epitaxial growth of silicon layers. These silicon layers are then processed by indiffusion followed by metallization and/or more epitaxial growth. One indiffusion process that is commonly applied is oxygen indiffusion to produce silicon dioxide. The index of refraction contrast between silicon ($n_{Si} \approx 3.475@1550$ nm) and silicon dioxide ($n_{SiO_2} \approx 1.45@1550$ nm) is one of the largest that can be achieved without lattice straining. This large index differential can be used to obtain tight confinement in silicon layers with a minimum of scattering off the indiffused boundaries. In our

fabrication sample, the Si layer is $310\,nm$ thick on top of a $1\,\mu m$ layer of SiO_2 on top of a $1\,mm$ layer of silicon.

The antennas are fabricated by vacuum depositing a thin layer ($10 - 40nm$) of Au on the top silicon layer of an SOI wafer. The Au layer is then photolithographically processed to produce a set of square Au islands in a two-dimensional square pattern. The lithography process performs a reasonable replication of the mask detail apart from some loss of edge sharpness. The sharpness may not be so desirable for antenna operation, but a square is the easiest shape to draw on a mask. The optical antenna structures are then formed by using a focused ion beam (FIB) that focuses gallium ions onto the gold surface, thereby etching through the deposited layer and down to the surface of the silicon. A scanning electron micrograph of the antenna structures written in $10\,nm$ Au on top of the SOI layer is shown in Figure 2(a). The shape that we were attempting to produce is the ideal structure of the *Pistolkors* [17] diffraction antenna as is depicted in Figure 2(b).

The bit error rate for end-to-end transmission is the criterion to determine the optical power supply for our broadcast network. Given the need for low latency for short messages, frequent error detection and correction are not options. It is generally assumed that a bit error rate (BER) of 10^{-12} is required for multi-gigabit transmission. The loss budgeting method [18] requires one to determine the minimum required signal power at the detector for a given error rate and system noise. Assuming that we will operate with the lowest received power possible (the thermal noise limit), the required output power of the link can be calculated (Gaussian model) as follows:

$$P_{detector} = 4 \times Q^2 \times P_{NA} \approx -44dBm \qquad (1)$$

where thermal noise power $P_{NA} = kT\Delta f$, and Q, the signal to noise ratio, is calculated through $BER = \frac{1}{2}erfc(\frac{Q}{\sqrt{2}})$. Thus, we can safely assume a $-30dBm$ sensitivity of the receiver, which is $1\mu W$. In the proposed design, where signal dispersion is negligible at the targeted several Gbps rate and centimeter propagation distance, the system components are treated as a series of transmission coefficients such that the system can be modeled as an attenuator applied to the input power. The fundamental and major loss in the proposed photonic network is the broadcast loss. The two-dimensional Friis antenna equation then gives

$$T = \frac{P_r}{P_t} = G_t G_r \frac{1}{2\pi} \frac{\lambda}{2\pi nr} \qquad (2)$$

where T is the power transmission, G_t and G_r are the gains for the transmitter and receiver antenna respectively, P_r is the received power, P_t is the transmitted power, λ is the operating wavelength, r is the distance from source to receiver, and n is the index of refraction. Assuming that our antennas are both omnidirectional diffraction antennas with gains of unity, the factor is around 58 dB, for a maximum $0.7\,cm$ propagation radius. There are some obvious corrections that make the situation much less dire. For one, the outer ring of antennas should not be omnidirectional and the antennas in the center have a half radius. Thus, we can gain around $6\,dB$ for each antenna's directivity. The above analysis shows that if the laser source provides over $16\,dBm$ power per wavelength ($16\,dBm = 58$ dB

978-1-60558-497-3/09 $25.00 © 2009 ACM

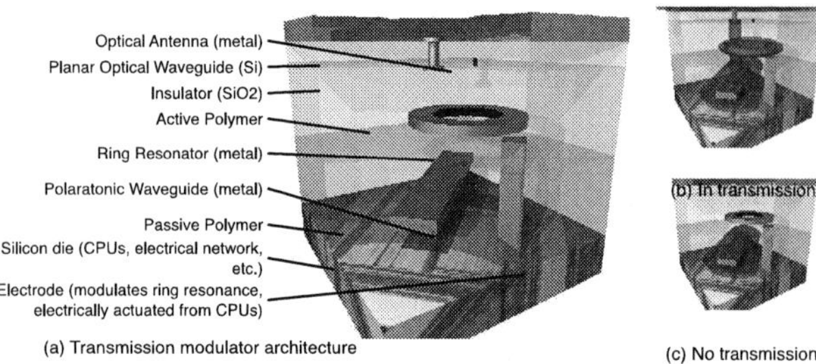

Figure 3.   A single modulator and its operation.

- 6 dB transmission antenna - 6 dB receiving antenna - 30 dBm), the proposed design will be able to achieve the required bit error rate. Detailed power calculations will be provided in Section IV.

The proposed design operates in the C+L bands (1530 nm–1625 nm). All the components of the proposed nanophotonic network can operate throughout the C+L bands within achievable fabrication tolerance and electrical tuning. At 1500 nm, a 0.1 nm band, including a sufficient protection region on either side of its center, can carry up to a 10 GHz signal if the center wavelength of the source is fixed and stabilized. Our working band can easily hold hundreds of carrier wavelengths of 10 Gb/s capacity.

II.B. Transmitter

The transmitter provides a means to take an electrical signal and add an optical carrier. The electrical signal is a baseband digital signal from standard CMOS drivers and the optical carrier is generated by the off-chip light source.

II.B.1) Optical sources: Past work shows that locating the laser light source off chip is an optimal solution today and will remain so in the immediate future [6]. We will multiplex multiple erbium-doped fiber lasers onto the on-chip waveguides, one for each wavelength in the prototype. These lasers can generate more than 100 W per wavelength, which is more than enough for our purposes.

II.B.2) Source waveguide: This layer of polymer waveguide spreads the wide band light to on-chip nodes. The loss of the source waveguide is approximated 1.5 dB/cm based on current technology.

II.B.3) Filter and modulator: Depicted in Figure 1(c), the polaritonic waveguide carries a variety of wavelengths that may be selected, modulated and ultimately transmitted via the antenna. Wavelength selection is performed using high finesse ring resonators [19] as a filter. The voltage on the electrode is capable of tuning the ring resonator to a desired frequency by changing the reflection index and thus the free spectrum range.

After wavelength selection, intensive modulation is performed using a higher layer ring resonator, depicted in Figure 3. Figure 3(b) and Figure 3(c) show how the modulator operates. When the ring resonator is tuned to match a particular wavelength on the waveguide, the antenna, waveguide and ring resonator are all coupled, resulting in a transmission in the 2D planar waveguide. We choose electro-optic polymers (dendrimers) to load plasmonic modulating structures. Recent advances in electro-optic polymers have resulted in electro-optic coefficients that exceed the commonly-used lithium niobate by factors of 5 or more [20]. Small variations in the applied voltage modulate the signal. For electrical drivers, a ring resonator is seen as a capacity load of around 10 fF.

II.C. Receiver Design

Next, we describe receiver design. Past work has shown that compact and complete WDM demultiplexers can be implemented in multiple layers [21]. Ge doped silicon detectors and transamplifers can then translate the optical signal into electrical signals to feed the CMOS digital circuits.

II.C.1) Demultiplexer: We use a three-dimensional implementation of a ring demultiplexer, and couple all of the wavelengths that the antenna can receive from the antenna feedline into a full band

ring resonator. The work of Xu et al. [19] has demonstrated that the full-band ring may have a radius of 1.5 μm. By coupling that ring to four quarter-band rings with offset center passbands in a lower layer, we can divide the original band into four sub-bands. By recursively stacking such structures as in 1(c), we can divide the incoming signal into an arbitrary number of bands, i.e., one band per WDM channel of the proposed design. The area of such a decoder mapped into a single plane is approximately 6×6 μm^2. A 64-channel demultiplexer then would have an area of 48×48 μm^2. The loss of this multilayer demultiplexer is estimated to be 5 dB.

II.C.2) Ge doped silicon PIN detector and transimpedance amplifier: The Ge doped silicon detector [22], which serves as the receiver end, resides in the bottom CMOS layer, together with amplifiers and electrical logic. The photocurrent from the photodetector is amplified by a transimpedance amplifier (TIA), which provides digital voltage signals to the CMOS gates that follow. The TIA consists of a high-speed inverter, and a feedback resistor using a PMOS transistor.

III. Electrical Subnetwork

Spectrum consists of a compact electrical packet-switched subnetwork. The proposed design aims to deliver scalable on-chip throughput. It provides the interface to the nanophotonic subnetwork and conducts nanophotonic-electric resource arbitration. In addition, the electrical subnetwork also supports 3D die-stacking integration, providing high-throughput and low-latency connections to vertically adjacent processor cores and memory layers, thereby minimizing memory access latency.

III.A. Router Design

Figure 4(a) shows the router microarchitecture of the proposed electrical subnetwork, which consists of electrical interconnects, buffers, crossbar switches, and routing and arbitration logic. The electrical interconnects connect processor cores (e.g., L2 cache interfaces), memory blocks, adjacent routers, and the nanophotonic interface. Unlike modern low-latency on-chip networks, the design of Spectrum can be greatly simplified, since the design of the electrical subnetwork in Spectrum focuses on scalable throughput instead of latency optimization. Recent studies in electrical on-chip network design have focused on reducing latency at the expense of logic complexity and/or throughput, including high radix networks [23], express virtual channels [24], low latency routers [25], etc. However, even if these techniques eliminate the arbitration and router pipeline delay, the nanophotonic waveguide still offers a significant latency advantage over an electrical interconnect.

Spectrum's router is unique in that it arbitrates broadcast nanophotonic network resources among the processor cores and network blocks. Figure 4(b) shows the router pipelines. Incoming latency-critical short protocol messages and missed data word packets are directly forwarded to the nanophotonic–electric arbiter to compete for the broadcast nanophotonic channel. Latency-tolerant data traffic, e.g., eviction and writeback messages, goes through a five-stage pipeline path. Another feature of Spectrum's router network is its support for 3D integration. Recently-developed 3D IC integration technology has the potential to overcome the limitations of 2D IC technology and provide efficient support for many-core system integration [26], [27], [28]. One interesting feature of 3D integration is the short inter-layer connections (in the range of tens of micrometers) [26], [28]. As shown in Figure 4(a), the router consists of multiple input and output ports to both vertically and laterally adjacent processor cores and memory blocks.

III.B. Nanophotonic-Electric Arbitration

A key factor in the latency of packets offloaded onto the nanophotonic network is the arbitration latency for a particular WDM channel. This section describes the efficient single-cycle arbitration scheme implemented in Spectrum's router.

Spectrum is equipped with a distributed global arbitration scheme to enable one-cycle arbitration for the nanophotonic network. This

978-1-60558-497-3/09 $25.00 © 2009 ACM

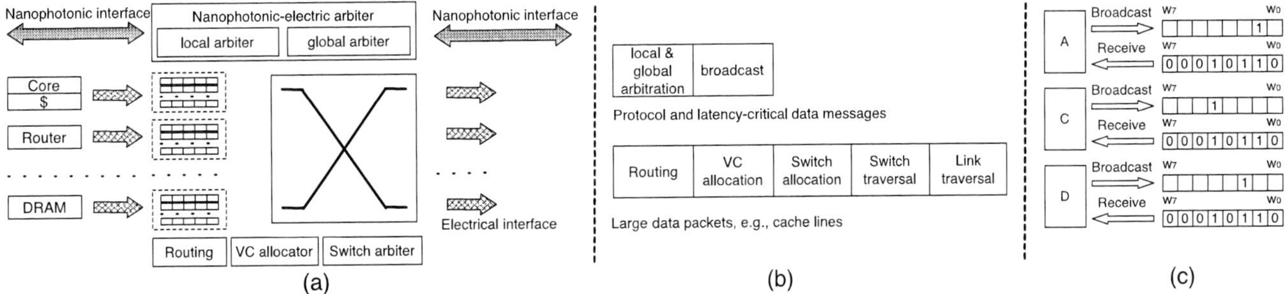

Figure 4.   (a) Spectrum router microarchitecture. (b) Router pipelines. (c) An arbitration example.

scheme not only provides low-latency arbitration, but also enforces global communication ordering, which is critical for correctly implementing many coherence and consistency models. By leveraging WDM technology, the nanophotonic 2D waveguide is capable of transmitting data bits in parallel through multiple wavelengths. Spectrum's nanophotonic network design divides the total number of wavelengths N into K channel groups. Thus each transmission channel consists of M wavelengths, with $M = N/K$, and can transmit rM data bits in an electrical clock cycle, where r is the frequency ratio of the optical network to the electrical network.

Since the goal of the optical network is to provide the lowest latency possible for short messages, arbitration should be as fast as possible. Figure 4(c) shows an example of the proposed arbitration scheme. The arbitration scheme starts with each node being assigned a unique dynamic priority. This dynamic priority corresponds to one of the wavelengths in a channel group. After each transmission, each router changes its dynamic priority using a deterministic random number generator. Since all routers share the same seed, they always agree on a global priority.

To arbitrate for access to the nanophotonic network, a node broadcasts a 1 value on the wavelength corresponding to its dynamic priority. For example, for an eight-node network, Node A could broadcast W_1 ([− − − − − − 1−]), Node C broadcasts W_4 ([− − −1− − − −]), and Node D broadcasts W_2 ([− − − − −1− −]). As the node sends its dynamic priority, it also listens on the nanophotonic subnetwork for the resulting combined priority. If one or more other transmitters also try to arbitrate for the channel group, the routers will all see a bit vector that is the bitwise OR of all of the dynamic priorities sent. For example, if Node, A, C, and D arbitrate for access at the same time, they will all see a combined priority of [− − −1 − 11−], which is the bitwise OR of Router A, C, and D's dynamic priorities.

This dynamic priority scheme can be used to implement two different forms of channel assignment. The first scheme is dynamic, where, for K channel groups, the K arbitrating routers with the highest priorities will each be allowed to transmit on one of the K channel groups. This scheme requires each router to use a hierarchical, parallel adder to perform a population count on the bit vector to see if it was one of the K highest dynamic priorities. The goal of the nanophotonic network is to provide extremely low latency for short messages. An adder circuitry, however, may have too much delay to allow for single-cycle arbitration. To deal with this potential problem, we also propose another scheme in which channels are pre-assigned to routers based on their dynamic priorities. Each router then arbitrates with other routers that have the same channel assignment in a manner similar to the previous channel assignment technique with $K = 1$. Since there is only one channel to arbitrate for, the bit vector checking logic can be vastly simplified to an OR operation, where if any of the bits in the bit vector with a higher priority than the router's are 1, the router does not gain access to the channel.

In addition to time-critical traffic, Spectrum also allows noncritical traffic to be sent over the nanophotonic network. The noncritical traffic, however, is assigned a lower priority, thus is only allowed to use the nanophotonic network when it is idle.

As the broadcast nanophotonic subnetwork provides a global ordering of transactions, coherence and consistency protocols are efficiently supported. While the proposed Spectrum design uses a classical snoopy MESI cache coherence protocol, Spectrum's low-latency nanophotonic broadcast subnetwork can also benefit other cache coherence protocols, such as directory-based coherence or token coherence [4].

IV. EVALUATION

This section evaluates Spectrum, the proposed hybrid nanophotonic–electric on-chip network. We compare Spectrum's performance and power consumption against electrical on-chip alternatives using scientific and commercial workloads, and then explore the design space of Spectrum.

IV.A. Spectrum Parameters

The design and fabrication plan targets mainstream, i.e. 65 nm, CMOS technology. We consider the following configuration, including a 16-core chip-multiprocessor with four die-stacked DRAM layers. A 16-way tiled CMP running at 2 GHz is assumed to be integrated into a 1 cm square silicon die. Each tile is equipped with an Alpha 21264-like processor, a private 64KB L1 cache, a private 256KB L2 cache, and a Spectrum router.

The Spectrum electrical router is implemented using TSMC 65 nm low-power technology library. Each router is approximately $0.6 \times 0.6mm^2$. Each Ge doped silicon PIN detector will take approximately $10 \times 10\mu m^2$. The overall area router with photodetectors will be less than $0.7 \times 0.7mm^2$, which is large enough for stacking the antennas and ring resonators (filters, modulators, and demultiplexers) on top. Since antennas can be placed at each corner of the tiles towards the center of the chip, each antenna only needs to cover an area of $0.514 \times 0.514cm^2$ for signal broadcast.

The frequency of the nanophotonic subnetwork is determined by the propagation latency, as follows.

- $T_{CLK \to Q}$: Clock to Q propagation latency of the driving latch storing the information to be transferred. Assuming a load of 12fF, $T_{CLK \to Q} \approx 62ps$, based on the technology library.
- T_m: Optical transmission of the modulator, $T_m \approx 40ps$ ([29]).
- T_p: Time of flight. $T_p = \frac{0.514\sqrt{2} \times 3.5}{3} \times 10^{-10}s \approx 85ps$.
- T_r: Receiver delay, $T_r \approx 14.6ps$ ([29]).
- T_{setup}: Setup time of the latch receiving the signal, $T_{setup} = 38.1ps$, based on the technology library.

The total delay is approximately 240ps. The nanophotonic network thus operates at 4 GHz, twice as fast as the electrical network. In each clock cycle, $2M$ bits per channel can be sent in parallel. The only fundamental delay is T_p, and with future technology scaling, Spectrum can potentially enjoy even higher frequencies.

Spectrum leverages 3D integration to provide low latency access to die-stacked DRAM memory. Each of the four DRAM layers is partitioned into 4×4 DRAM blocks, each with a 64-bit interface.

978-1-60558-497-3/09 $25.00 © 2009 ACM

TABLE I
CONFIGURATION

Nanophotonic subnetwork		Electrical subnetwork		Memory Hierarchy	
Throughput per wavelength	4Gb/s	Router freq.	2GHz	L1 cache per core	64KB, 2-way 64-byte line
Number of wavelengths	N	Topology	4-ary 2-mesh	L2 cache per core	256KB, 16-way 128-byte line
Broadcast channels	K	Routing	Dimension-ordered	L2 access latency	8 cycles
Coherence protocol	Snoopy protocol MESI	Flow control	Credit-based	Memory access latency	14 cycles
Protocol msg size	64 bits	Buffers per phy. channel	25 (5 per virtual channel)	Flit size	64 bits

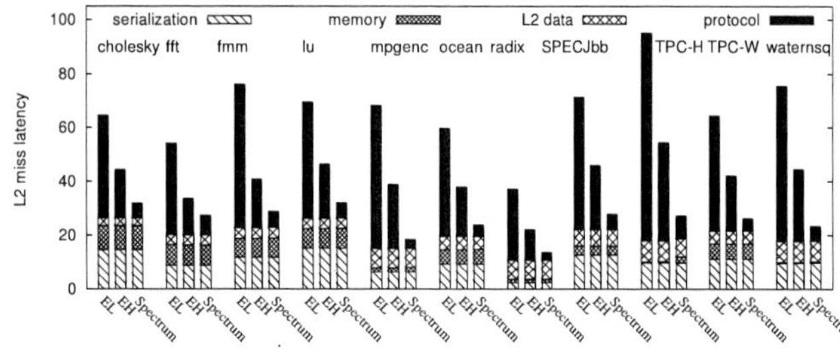

(a) L2 miss latency and decomposition.

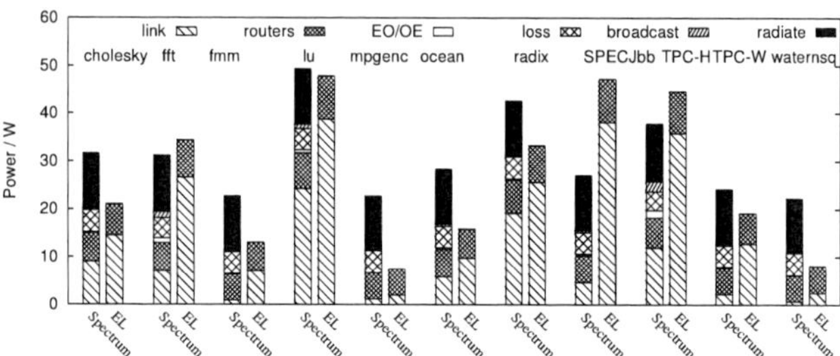

(b) Power consumption (Note: off-chip radiation does not cause on-chip heating).

Figure 5.   Comparison between electrical network and Spectrum.

Using CACTI [30] version 5.3, which supports modeling DRAM memories, it was determined that at the 65nm process node, we can have 2MB memory per block totaling 128MB memory with 25.8 Gb/s bandwidth per block. Memory access latency is estimated to be 14 cycles considering DRAM access, layer traversal, and router delay.

We have implemented a trace-driven cycle-accurate cache-network simulator, which simulates the activities of the memory hierarchy and the nanophotonic-electric network, as well as all coherence protocol transactions. Table I shows the configuration of the modelled network and memory hierarchy. The network consists of a broadcast WDM nanophotonic subnetwork supporting a snoopy MESI protocol and a pipelined virtual channel input-buffered electrical subnetwork.

Network traffic traces are gathered using the M5 full-system simulator [31] running several SPLASH2 [32] and ALPBench [33] multithreaded benchmarks that we have, including `ocean`, `radix`, `cholesky`, `waternsq`, `lu`, `fft`, `fmm`, and `mpgenc`. We also consider memory access traces collected from three commercial server workloads: `TPC-H`, `TPC-W` [34], and `SpecJBB` [35]. Each benchmark is spawned into 16 threads, which are distributed amongst the processor cores and executed concurrently.

IV.B. Comparison of Spectrum with Electrical Alternatives

This experiment evaluates the performance of Spectrum using parameters $N = 64$ and $K = 2$, which provide the best performance tradeoffs under the current experimental setup (see Section IV-C).

We compare Spectrum against two electrical on-chip network alternatives. The first electrical network, named EL, is a 2D mesh design using a five-stage pipeline router (routing, virtual channel allocation, switch arbitration, switch traversal, and link traversal). EL forms the baseline electrical solution. The second electrical network reference, named EH, is equipped with recently-proposed aggressive latency minimization techniques, which are capable of bypassing control, arbitration, and buffering stages, and only consist of communication-essential stages, including switch and link traversal plus serialization. EH models the culmination of years of on-chip interconnect research and represents a nearly ideal minimal-latency bandwidth-abundant circuit-switched electrical interconnect. Both EL and EH support a distributed directory-based MESI protocol, where the home nodes are distributed across the 4×4 system with an evenly-partitioned address space.

Figure 5(a) shows the average L2 cache miss latency (read and write) of Spectrum compared with the two electrical alternatives, also with 3D support. Latency is decomposed into four parts: serialization latency of the cache line, memory access latency, L2 cache data array access latency, and protocol transaction latency. The first three parts are the intrinsic delay in L2 miss latency, and the last part is

determined by the protocol and on-chip network design. This study shows that Spectrum can significantly reduce the protocol transaction latency (i.e., 88.0% improvement over EL and 72.7% improvement over EH), which will lead to an L2 cache miss latency decrease (i.e 61.0% improvement over EL and 31.0% improvement over EH).

These results demonstrate that Spectrum can effectively speed up on-chip cache transactions compared to the best possible on-chip electrical alternatives. These results also demonstrate that, as on-chip data sharing increases, Spectrum offers more performance improvement, as the broadcast nanophotonic subnetwork can effectively handle latency-critical cache coherency protocol messages. (Note: on-chip data sharing can be roughly estimated by comparing the total local L2 cache misses and the actual off-chip memory accesses.) On the other hand, it is challenging to efficiently handle on-chip broadcast and multicast messages using electrical alternatives.

Figure 5(b) shows the power consumption breakdown of Spectrum. The power consumption of the electrical subnetwork is contributed by routers (labeled as "routers") and link circuitry (labeled as "link"). The electrical power also needs to consider EO/OE devices (labeled as "EO/OE"), including modulators and amplifiers. The power consumption of the nanophotonic subnetwork is contributed by the overhead of the source waveguide which spreads the wide band light to on-chip nodes (labeled as "loss", please see Section II-B2 and Figure 1(a)), the proposed broadcast nanophotonic network (labeled as "broadcast"), and off-chip laser radiation into the ambient (labeled as "radiate"). Only the on-chip "loss" and "broadcast" contribute to the chip power dissipation.

Figure 5(b) shows that, since electrical network offloads latency-critical protocol and data traffic as well as a portion of noncritical traffic to the nanophotonic subnetwork, the power consumption of Spectrum's electrical subnetwork decreases significantly compared to that of EL. In addition, without considering the off-chip radiation, the on-chip power efficiency of Spectrum ("link" + "routers" + "EO/OE" + "loss" + "broadcast") is higher than EL for most cases. As technology scales further, the power efficiency of nanophotonic network is expected to improve, but the power consumption of electrical network is expected to increase. Therefore, using nanophotonic on-chip interconnect will become more power beneficial in future technologies. This study also shows that, the power efficiency of the nanophotonic network is mainly constrained by the overhead of the source waveguide which feeds the laser signal across the chip, shown as the "loss" bar in Figure 5(b). The loss is approximately 1.5 dB/cm based on current technology. To further improve the power efficiency of nanophotonic on-chip interconnect, it is important to optimize the source waveguide technology and minimize the loss.

978-1-60558-497-3/09 $25.00 © 2009 ACM

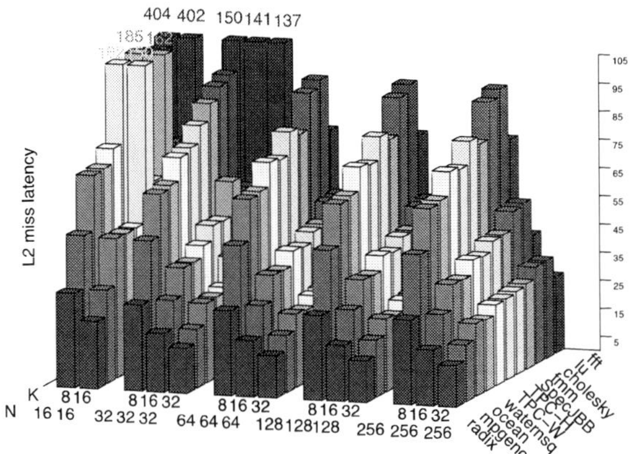

Figure 6.   Design exploration of the nanophotonic network.

IV.C. Design Exploration of Spectrum

Next, we explore the design space of the 2D WDM nanophotonic broadcast network and evaluate the design under various configurations of the following three parameters: the total number of wavelengths (N), the number of channels (K), and the number of wavelengths per channel (M). Recall that $N = K \times M$. The results are shown in Figure 6. This study yields the following observations. First, cache transaction latency increases as N decreases. However, because the optical power linearly depends on N, for a given power budget, N is fixed and K is inversely proportional to M.

K represents the maximum number of concurrent transactions, which affects channel contention. M determines the number of clock cycles required for relaying each message, which is the serialization latency. The relationship of K and M is a tradeoff between contention delay and serialization delay. As shown in this figure, increasing M reduces the latency of each on-chip cache transaction, which further reduces the total number of pending transactions, thereby reducing the need for more parallel channels. On the other hand, increasing K reduces the contention among pending transactions, thereby reducing the overall latency of each transaction. Given the total number of supported wavelengths, this study provides the best-performance tradeoffs and settings of the nanophotonic subnetwork. Given 64 wavelengths, the best-performance setting for the 4×4 network is 2 channels and 32 wavelengths per channel. This is what we use in Section IV-B.

V. CONCLUSION

While on-chip electrical packet-switched networks promise scalable on-chip throughput in many-core systems, they cause long latency as core count increases. Thus, in many-core systems, this type of electrical interconnect will cause latency-critical messages such as synchronization and coherence messages to become the performance scaling bottleneck.

Spectrum, the proposed hybrid nanophotonic-electric network, addresses the challenge of latency and throughput scalability by combining a throughput-optimized packet-switched electrical network with a latency-and-broadcast-optimized nanophotonic network to handle latency-critical communication. By removing latency-critical and multicast traffic from the packet-switched electrical network, we can simplify the design of the electrical network while improving its throughput. By leveraging a 2D waveguide and photonic nanostructures, we can build a nanophotonic network that has low latency and can be fabricated today in a CMOS-compatible manner. Experimental results show that a conservative implementation of Spectrum can significantly improve on-chip cache transaction performance compared to an aggressive electric-only on-chip interconnects, and has the potential to scale to future technology.

REFERENCES

[1] P. Gratz, et al., "On-chip interconnection networks of the TRIPS chip," *IEEE Micro*, vol. 27, no. 5, pp. 41–50, Sept.-Oct. 2007.

[2] "Tilera TILE64 chip-multiprocessor," http://www.tilera.com.

[3] S. Vangal, et al., "An 80-tile 1.28TFLOPS networks-on-chip in 65nm CMOS," in *Proc. Int. Solid-State Circuits Conf.*, Feb. 2007.

[4] M. M. K. Martin, "Token coherence," Ph.D. dissertation, Univerity of Wisconsin–Madison, 2003.

[5] N. E. Jerger, L.-S. Peh, and M. Lipasti, "Virtual circuit tree multicasting: A case for on-chip hardware multicast support," in *Proc. Int. Symp. Computer Architecture*, June 2008, pp. 229–240.

[6] R. Beausoleil, et al., "Nanoelectronic and nanophotonic interconnect," *Proceedings of the IEEE*, vol. 96, no. 2, pp. 230–247, Feb. 2008.

[7] P. Dainesi, et al., "3-D integrable optoelectronic devices for telecommunications ICs," in *Proc. Int. Solid-State Circuits Conf.*, vol. 1, 2002, pp. 360–473.

[8] M. Haurylau, et al., "On-chip optical interconnect roadmap: challenges and critical directions," *Group IV Photonics, 2005. 2nd IEEE International Conference on*, pp. 17–19, 21-23 Sept. 2005.

[9] N. Kirman, et al., "On-chip optical technology in future bus based multicore designs," *IEEE Micro*, vol. 27, no. 1, pp. 56–66, 2007.

[10] A. Shacham, K. Bergman, and L. Carloni, "Photonic networks-on-chip for future generations of chip multiprocessors," *Computers, IEEE Transactions on*, vol. 57, no. 9, pp. 1246–1260, Sept. 2008.

[11] D. Vantrease, et al., "Corona: System implications of emerging nanophotonic technology," in *Proc. Int. Symp. Computer Architecture*, 2008, pp. 153–164.

[12] M. Petracca, et al., "Design exploration of optical interconnection networks for chip multiprocessors," in *Proc. of Symp. on High Performance Interconnects*, 2008, pp. 31–40.

[13] J. K. S. Poon, et al., "Wide-range tuning of polymer microring resonators by the photobleaching of the cld-1 chromophores," *Optics Letters*, vol. 29(22), pp. 2584–2586, 2004.

[14] Q. Xu, et al., "12.5 Gbit/s carrier-injection-based silicon micro-ring silicon modulators," *Optics Express*, vol. 15, no. 2, pp. 430–436, 2007.

[15] B. E. Little, et al., "Wavelength switching and routing using absorption and resonance," *IEEE Photonics Technology Letters*, vol. 10, no. 6, p. 816=818, 1998.

[16] Y. A. Vlasov and S. J. McNab, "Losses in single-mode silicon-on-insulator strip waveguides and bends," *Optics Express*, vol. 12, no. 8, pp. 622–631, 2004.

[17] A. A. Pistolkors, "Theory of the circular diffraction antenna," *Proceedings of the IRE*, vol. 36, no. 1, pp. 56–60, 1948.

[18] C. H. Cox, *Analog Optical Links: Theory and Practice.* Cambridge University Press, 2006.

[19] Q. Xu, D. Fattal, and R. G. Beausoleil, "Silicon micro-ring resonator with 1.5 μm radius," *Optics Express*, vol. 16, no. 6, pp. 4309–4315, 2008.

[20] E. M. McKenna, et al., "Comparison of r_{33} values for aj404 films prepared with parallel plate and corona poling," *Journal of the Optical Society B*, vol. 24, no. 11, pp. 2888–2892, 2007.

[21] M. Sumetsky, "Vertically-stacked multi-ring resonator," *Optics Express*, vol. 13, no. 17, pp. 6354–6375, 2005.

[22] T. K. Woodwqard and A. V. Krishnamoorthy, "1Gb/s integrated optical detectors and receivers in commercial CMOS technologies," *IEEE Journal of Selected Topics in Quantum Electronics*, vol. 5, no. 2, pp. 146–156, 1999.

[23] J. Kim, J. Balfour, and W. Dally, "Flattened butterfly topology for on-chip networks," *Computer Architecture Letters*, vol. 6, no. 2, pp. 37–40, Feb. 2007.

[24] A. Kumar, et al., "Express virtual channels: towards the ideal interconnection fabric," in *Proc. Int. Symp. Computer Architecture*, 2007, pp. 150–161.

[25] R. Mullins, A. West, and S. Moore, "Low-latency virtual-channel routers for on-chip networks," in *Proc. Int. Symp. Computer Architecture*, 2004, pp. 188–197.

[26] K. Banerjee, et al., "3-D ICs: a novel chip design for improving deep-submicrometer interconnect performance and systems," *Proc. IEEE*, vol. 89, no. 5, 2001.

[27] A. W. Topol, et al., "Three-dimensional integrated circuits," *IBM J. Research and Development*, vol. 4, pp. 491–506, 2006.

[28] J. Kim, et al., "A novel dimensionally-decomposed router for on-chip communication in 3D architectures," in *Proc. Int. Symp. Computer Architecture*, June 2007.

[29] G. Chen, et al., "Predictions of CMOS compatible on-chip optical interconnect," *Integration, the VLSI Journal*, vol. 40, no. 4, pp. 434–446, 2007.

[30] S. Thoziyoor, N. Muralimanohar, and N. Jouppi, "Cacti 5.0," Hewlett-Packard, Tech. Rep., Oct. 2007.

[31] N. L. Binkert, et al., "The M5 simulator: Modeling networked systems," *Proc. Int. Symp. Microarchitecture*, vol. 26, no. 4, pp. 52–60, 2006.

[32] "SPLASH2 website," http://www-flash.stanford.edu/apps/SPLASH/.

[33] M.-L. Li, et al., "The ALPbench benchmark suite for complex multimedia applications," in *Int. Symp. Workload Characterization*, Oct. 2005, pp. 34–35.

[34] "Transaction processing and database benchmarks," http://www.tpc.org.

[35] "Java business benchmark," http://www.spec.org.

978-1-60558-497-3/09 $25.00 © 2009 ACM

Exploring Serial Vertical Interconnects for 3D ICs

Sudeep Pasricha

Department of Electrical and Computer Engineering
Colorado State University, Fort Collins, CO
sudeep@engr.colostate.edu

ABSTRACT

Three-dimensional integrated circuits (3D ICs) offer a promising solution to overcome the on-chip communication bottleneck and improve performance over traditional two-dimensional (2D) ICs. Long interconnects can be replaced by much shorter vertical through silicon via (TSV) interconnects in 3D ICs. This enables faster and more power efficient inter-core communication across multiple silicon layers. However, 3D IC technology also faces challenges due to higher power densities and routing congestion due to TSV pads distributed on each layer. In this paper, serialization of vertical TSV interconnects in 3D ICs is proposed as one way to address these challenges. Such serialization reduces the interconnect TSV footprint on each layer. This can lead to a better thermal TSV distribution resulting in lower peak temperatures, as well as more efficient core layout across multiple layers due to the reduced congestion. Experiments with several 3D multi-core benchmarks indicate clear benefits of serialization. For instance, a 4:1 serialization of TSV interconnects can save more than 70% of TSV area footprint at a negligible performance and power overhead at the 65nm technology node.

Categories and Subject Descriptors:

B.7.1 [**Integrated Circuits**]: Types and Design Styles—*Advanced technologies, VLSI*;

General Terms: Performance, Design

Keywords: 3D ICs, Serial Interconnect, Networks on Chip, VLSI

1. MOTIVATION

In recent years, the rapid scaling of semiconductor technology has led to more and more processing cores and memories being integrated on a single chip. These highly integrated chip multiprocessors (CMPs) have provided the high performance needed for supporting complex emerging applications particularly in the multimedia and networking domains. However, these planar CMP architectures are now facing fundamental challenges due to on-chip interconnects not scaling well with technology [1][9]. Interconnect delay has increased significantly compared to gate delay in ultra deep submicron (UDSM) technologies as a result of increased crosstalk coupling noise and parasitic resistivity [2]. According to the International Roadmap for Semiconductors (ITRS) [3], delay on global interconnects has become a major source of performance bottlenecks and is one of the semiconductor industry's topmost challenges.

Permission to make digital or hard copies of part or all of this work for personal or classroom use is granted without fee provided that copies are not made or distributed for profit or commercial advantage and that copies bear this notice and the full citation on the first page. To copy otherwise, to republish, to post on servers or to redistribute to lists, requires prior specific permission and/or a fee.
DAC'09, July 26-31, 2009, San Francisco, California, USA

One promising solution to overcome the interconnect bottleneck and continue the pace of growth of CMP systems is the use of three-dimensional (3D) integration, in which multiple active device layers are vertically stacked and interconnected [4]-[7]. Such 3D ICs not only allow more cores to be integrated on a chip, but also provide potential performance advances, as each core can access a greater number of nearest neighbors, and thus has a higher supportable communication bandwidth. Most importantly, since the inter-layer distance is small, there is an opportunity to replace long global interconnects (~several mm) between communicating cores in a horizontal plane, by stacking the cores on adjacent layers and connecting them with shorter vertical interconnects. Since wire delay depends on the square of the wire length (or has a linear dependence if repeaters are used), this results in a reduction in inter-core latency. Wire length reduction in 3D ICs also translates into lower power dissipation in interconnects and repeaters.

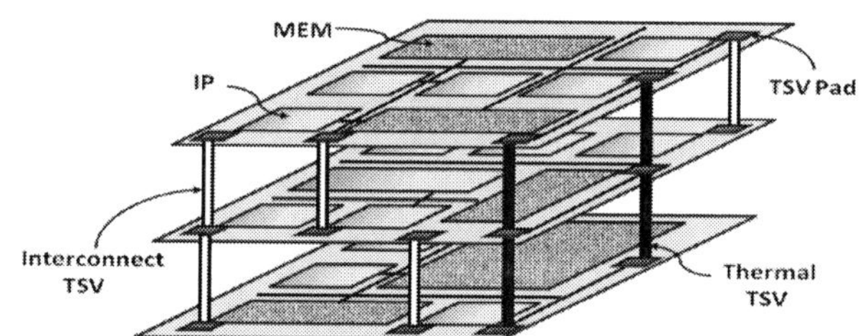

Figure 1. 3D IC with three layers

A number of different 3D IC manufacturing technologies have been explored in recent years, including transistor stacking, die-on-wafer stacking, chip stacking, and wafer stacking. Out of these, wafer stacking is one of most promising high-performance yet inexpensive implementation technology for 3D ICs [7], and the focus of this paper. Figure 1 shows an example of such a 3D IC with three active silicon layers. Each layer can have potentially multiple processing cores (IPs) and memories (MEM). Inter-layer communication is facilitated by vertical through silicon via (TSV) interconnects. A major concern in the adoption of 3D ICs is the increased power densities that can result from placing a core on top of another core. Since high peak temperatures due to increasing power densities can cause catastrophic IC failure and this is already a major concern in 2D architectures, the move to 3D will accentuate the thermal problem. Strategically placed thermal TSVs (Figure 1) can help establish a thermal path from the core of a chip to the heat sink and are one possible solution for cooling 3D ICs. Traditional 2D network-on-chip (NoC) topologies such as mesh, torus and butterfly will most likely be extended to create scalable networks that can handle intra- and inter-layer communication requirements in 3D ICs [8]-[10]. As the number of cores increases in each layer to support rising application complexity, the amount of communication between layers is also expected to increase. This will lead to an increase in the number of interconnect TSVs. Since each interconnect TSV requires a pad for bonding to a wafer layer, this will lead to an interesting scenario where the area footprint of interconnect TSVs in each layer can no longer be ignored.

As an example, consider a 3D NoC with a hundred 64-bit vertical TSV links between layers (for CMPs with hundreds of cores, such a large number of vertical links is to be expected). Assuming TSV pad dimensions of 10μm×10μm and a pitch of 16μm, the TSVs will take up an area of approximately 1.6mm^2 in each layer, which is equivalent to the size of a computation core! Unlike a computation core however, these interconnect TSVs are spread out (uniformly or non-uniformly) in each layer, which will make floorplanning and routing extremely challenging. Furthermore, the need to use thermal TSVs (which can take 10-20% of total chip area [11][12]) to create thermal-efficient 3D ICs will lead to an even greater TSV footprint on each layer. This will further complicate efficient chip layout, and overall performance.

In this paper, serialization of TSV interconnects is proposed to overcome the abovementioned challenge for 3D ICs. Serialization of TSV interconnects will have the benefit of reducing the number of TSV interconnects and line drivers, which in turn will reduce the TSV interconnect area footprint in each layer. This will make it easier to obtain a more efficient chip layout. Additionally, since TSV density is limited by fabrication cost factors, fewer interconnect TSVs can make way for more thermal TSVs, which will lead to more thermal-efficient IC designs. To the best of the authors' knowledge, this is the first work to explore the impact of using serial vertical interconnects in 3D ICs. Experimental results with several CMP applications indicate that using serial TSV interconnects can significantly reduce TSV area footprint, at a negligible performance and power overhead.

2. RELATED WORK

In the last few years, there has been a growing interest in 3D ICs from academia and industry as a means to alleviate the interconnect bottleneck problem currently facing 2D ICs. IBM [4][5] and Tezzaron [6] have recently presented promising preliminary results and test chips with 3D IC technology. Here, research on 3D ICs and on-chip serialization is briefly reviewed.

Several researchers have proposed thermal-aware floorplanning techniques for 3D ICs [12]-[16]. In particular, [14] proposed inserting thermal vias to reduce temperature hotspots during floorplanning. A few researchers have explored interconnect architecture design for 3D ICs [8][10][17]. In [8], 2D mesh and 2D folded torus topologies were compared with 3D mesh and 3D stacked mesh topologies. It was shown that 3D NoCs have more complex switches but offer better performance and lower energy for communication. In [10] a hybrid bus-NoC 3D interconnect architecture was proposed. In [17] circuit level models for TSVs were explored. Some recent work has looked at decomposing cores (processors [18][19], NoC routers [20], and on-chip cache [21]) into the third dimension which allows reducing wire latency at the intra-core level, as opposed to the inter-core level.

Serialization has been explored for long parallel on-chip global interconnects as a way to overcome UDSM artifacts such as crosstalk, wiring congestion, skew and high power dissipation [22]-[30]. In [22][23] it was shown that serial links can reduce communication area and power dissipation for not only long global links but also for shorter links in future technologies. In [24][25] a serial bus architecture was proposed to reduce on-chip bus energy. In [26]-[28] high speed ring oscillators and shift registers for communication serialization were described. In [29][30] fast asynchronous serial links were explored. None of the above works have explored the impact of using serial vertical interconnects in 3D ICs. As will be shown later, vertical TSV serialization is a powerful means of reducing TSV area footprint, and thus improving 3D IC cost, routability, and thermal efficiency.

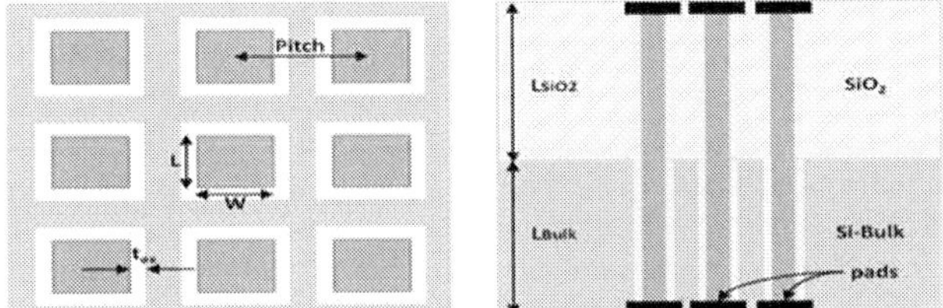

Figure 2. 3D TSV bundle schematic

3. VERTICAL TSV OVERVIEW

3.1 Interconnect TSVs

Vertical interconnects implemented as through silicon vias (TSVs) provide the highest interconnect bandwidth in 3D ICs, compared to wire bonding, peripheral vertical interconnects, and solder ball arrays. To be useful in 3D CMP communication (in NoCs for instance), TSV interconnects should not be used in isolation, but rather as a group or bundle. Figure 2 shows a bundle of 9 TSVs placed in a 3×3 grid structure, in the bulk-silicon manufacturing technology [17]. Each TSV has a length L, width W, and an oxide coating of thickness t_{ox} around it in the bulk-silicon. Pads on the wafer surfaces are needed to bond to the vertical TSVs, using mechanical thermo-compression [31]. Typically, the pads tend to be larger than via cross-section to account for the oxide coating. For instance, for a via cross section of 4μm×4μm, and an oxide thickness of 1μm, the pad thickness has to be at least 5μm×5μm.

Interconnect TSVs have inherent resistance, capacitance and inductance, and can be modeled as RLC interconnects. As discussed in [7], the impact of inductance on delay and power dissipation for frequencies of a few GHz can be ignored for vertical TSV interconnects. In such a scenario, TSV interconnect resistance can be described as a function of via length l, cross-section σ and resistivity ρ:

$$R = \frac{\rho \times l}{\sigma} \qquad (1)$$

For example, a copper TSV with a 4μm×4μm cross section has a resistance of about 1.18mΩ/μm in the 130nm technology node. The skin effect for TSVs at frequencies of a few GHz is negligible. Since TSVs are interconnected using metal bonding, an appropriate contact resistance must also be considered. In this work, a contact resistance of 100mΩ per layer is considered [31].

The capacitance of TSVs must account for coupling between adjacent TSVs in a bundle. The following capacitance matrix is used for this purpose:

$$\tilde{C} = \begin{pmatrix} C_{1,1} & -C_{1,1} & ... & -C_{1,n} \\ -C_{2,1} & C_{1,1} & ... & -C_{1,n} \\ ... & ... & ... & ... \\ -C_{n,1} & -C_{n,2} & ... & C_{n,n} \end{pmatrix} \qquad (2)$$

where the elements outside of the diagonal represent inter-via coupling (with inverted signs), while the elements along the diagonal are the sum of the capacitances towards the ground plane ($C_{i,0}$ not explicitly shown in the matrix) plus the coupling capacitances:

$$C_{ii} = C_{i,0} + C_{i,1} + ... + C_{i,i-1} + C_{i,i+1} + ... + C_{i,n} \qquad (3)$$

The capacitance for TSVs in bulk-silicon was extrapolated from extraction results in [17] for TSV bundle densities corresponding to commonly used on-chip bus/link sizes (32, 64, 128 bits).

3.2 Thermal TSVs

In addition to acting as vertical interconnects, TSVs can also be

978-1-60558-497-3/09 $25.00 © 2009 ACM

used in a non-electrical capacity to conduct heat and alleviate hot spots in 3D ICs [12][14][33]. The idea of using thermal TSVs to overcome thermal problems was first utilized in the design of packaging and printed circuit boards (PCBs). Lee et al. [34] studied arrangements of thermal vias in the packaging of multichip modules (MCMs) and found that as the size of thermal via islands increased, more heat removal was achieved but less space was available for routing. This observation holds for 3D ICs as well, where thermal problems are greater than in 2D ICs because of the many dielectric layers. Thermal TSVs create efficient thermal conduits and greatly reduce chip temperatures, but have been shown to take up to 10-20% of total chip area to achieve a reduction in maximal chip temperature of up to 47% [11][12].

3.3 TSV Reliability Issues

Since TSV fabrication technology is not yet mature, the reliability of TSV interconnects is expected to be a limiting factor for 3D IC performance and yield for the near future [7]. Unsuccessful wafer alignment prior to and during the wafer bonding process is one of the primary mechanisms of failure for TSVs. To improve yield, hardware redundancy is often used. A simple and effective way to add redundancy and improve yield is to use double pads [35]. Since misalignments are caused by the unavoidable shift of bonding pads with respect to their nominal position, using large square pads twice as wide as standard pads can improve misalignment tolerance by an order of magnitude [31][36].

4. TSV SERIALIZATION SCHEME

To reduce the number of interconnect TSVs in 3D ICs, a shift-register based serialization scheme is proposed, similar to [26]-[28]. A single serial line is used to communicate both data and control signals between the source and destination nodes. A frame of data transmitted on the serial line using this scheme consists of $n+2$ bits, which includes a start bit ('1'), n bits of data, and a stop bit ('0').

Figure 3(a) shows the block diagram of the transmitter (or serializer) at the source. When there is no transmission, the output of the flip-flop is zero, the ring oscillator and shift registers are disabled, and the $n+2$ bit counter is in the reset state, with a '1' in its least significant bit output ($r0$). When a word becomes available for transfer in the transmission buffer, the R-S flip-flop is enabled, thereby enabling the ring oscillator, which generates a local clock signal and can oscillate above 2 GHz to provide high transmission bandwidth. At the first positive edge of this clock, an $n+2$ bit data frame is loaded in the shift register. In the next $n+1$ cycles, the shift register shifts out the data frame bit by bit. The stop bit is eventually transferred on the serial line after $n+2$ cycles, and $r0$ becomes '1'. At this time, if the transmission buffer is empty, the ring oscillator and shift registers are disabled, and the serial line goes into its idle state. Otherwise, the next data word is loaded into the shift register from the transmission buffer on the next positive clock edge. Thus data transmission continues without interruption.

Figure 3(b) shows the block diagram of the receiver (or de-serializer) at the destination. The R-S flip-flop in the receiver is activated when a low-to-high transition is detected on the input serial line (the 'low' corresponds to the stop bit of the previous frame, while the 'high' corresponds to the start bit of the current frame). After being activated, the flip-flop enables the receiver ring oscillator (which has a circuit similar to the transmitter ring oscillator) and the ring counter. The n-bit data word is read bit by bit from the serial line into a shift register, in the next n clock cycles. Thus, after n clock cycles, the n bit data will be available on the parallel output lines, while the least significant bit output of the ring counter ($r0$) becomes '1' to indicate data word availability at the output. With the assertion of $r0$, the R-S flip-flop is also

reset, disabling the ring oscillator. At this point the receiver is ready to start receiving the next data frame. Note that in case of a slight mismatch between the frequencies of the transmitter and receiver ring oscillators, correct operation can be ensured by adding a small delay in the clock path of the receiver shift register. The preceding discussion assumed $n{:}1$ serialization, where n data bits are transmitted on one serial line (i.e., a serialization degree of n). If wider links are used, this scheme can be easily extended. For instance, consider the scenario where $4n$ data bits need to be transmitted on four serial lines. In such a case, the number of shift registers in the transmitter must be increased from 1 to 4. However the control circuitry (flip-flop, ring oscillator, ring counter) can be reused among the multiple shift registers and remains unchanged. At the destination, every serial line must have a separate receiver to eliminate jitter and mismatch between parallel lines.

Note that it is possible to modify the proposed scheme by using an additional strobe line for synchronization, similar to [26]. This would reduce the number of bits transferred on the serial data line, and the sizes of the shift register and ring counter (which must be rewired to disable the oscillator after the desired number of clock cycles have been generated) from $n+2$ to n. However, for the case of TSV serialization, the overhead of the extra TSV line and pads are prohibitive enough to overshadow the very slight improvement in energy and performance that may be obtained for this case.

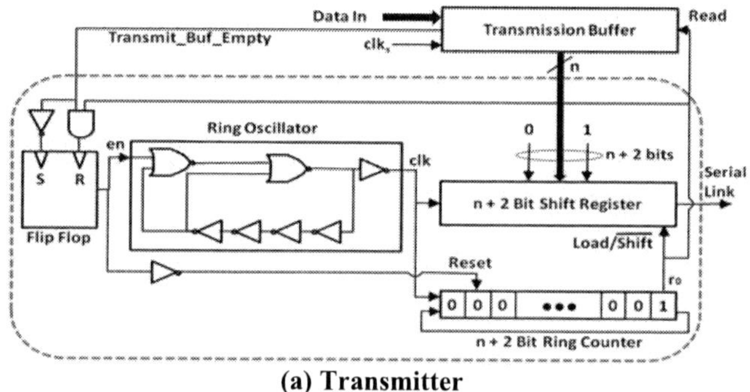

(a) Transmitter

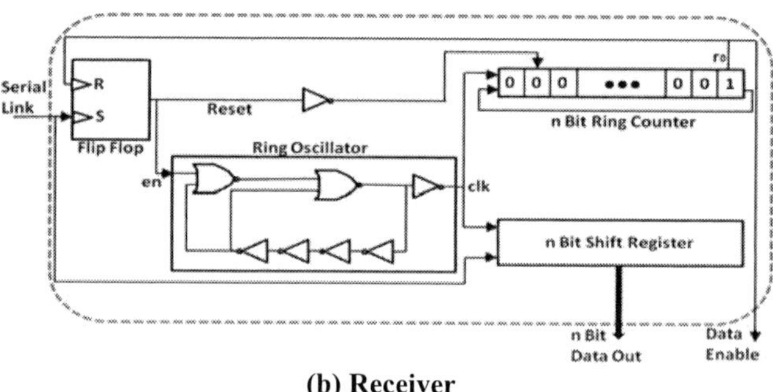

(b) Receiver

Figure 3. TSV Serialization scheme

5. EXPERIMENTS

To explore the impact of using the proposed serialization scheme for vertical TSV interconnects in 3D ICs, several experiments were conducted. First the area overhead of using the proposed scheme was explored. Subsequently, the power and performance impact of using the serialization scheme was explored, in the context of several CMP applications.

5.1 Impact on Area

The goal of the first experiment was to quantify the impact of using serialization on TSV area footprint. The serialization scheme was implemented at the RTL level and synthesized down to a gate-level net-list using Synopsys Design Compiler [38]. The synthesis was performed for the 130, 90, and 65nm TSMC standard cell libraries. This enabled a determination of the area overhead of the transmitter and receiver logic used in the serialization scheme.

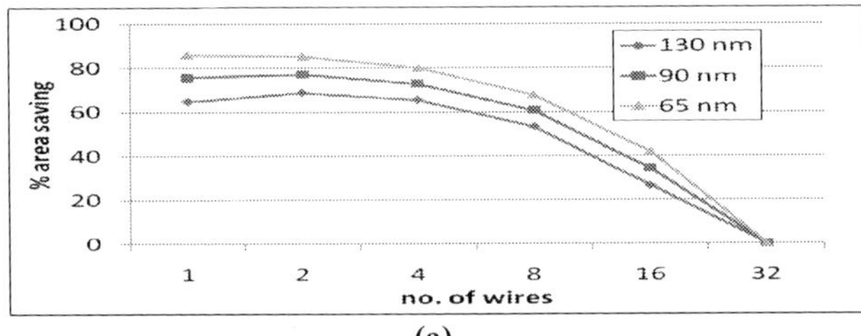

(a)

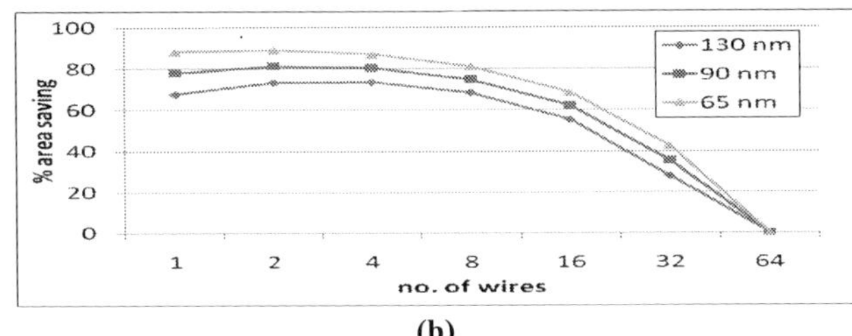

(b)

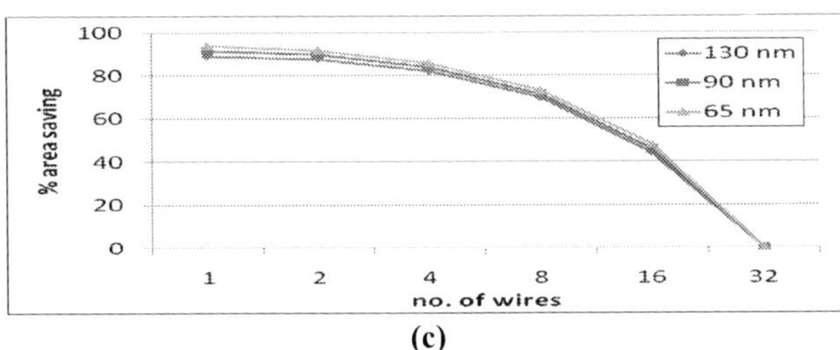

(c)

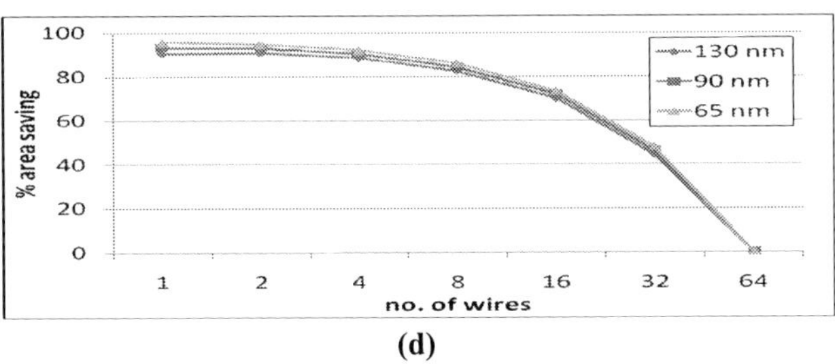

(d)

Figure 4. % area saving with serialization for 130–65 nm (a) 8μm TSV pitch, 32-bit link, (b) 8μm TSV pitch, 64-bit link, (c) 16μm TSV pitch, 32-bit link, (d) 16μm TSV pitch 64-bit link

Figure 4(a) and (b) show the area savings when using the proposed serialization scheme for a 4μm×4μm interconnect TSV cross section with 1μm oxide thickness (i.e., 5μm×5μm TSV pads) and an 8μm pitch. Figure 4(a) shows the area savings for a 32-bit link. The x-axis shows the number of wires in the link with varying degrees of serialization and the y-axis shows the percentage savings in area compared to the base case (32 wires) with no serialization used. It can be seen that the proposed serialization scheme can significantly reduce the area footprint of TSVs in 3D ICs. Even a 4:1 serialization can result in as much as a 55-70% area savings. As technology scales, the savings in area increase, as can be seen with the curves for the 130, 90, and 65nm implementations. This is a result of lower area footprint of the transmitter and receiver logic with shrinking sizes of gates with technology. A similar analysis for area saving is presented for a 64-bit link in Figure 4(b).

An interesting observation from both Figure 4(a) and (b) is that the

saving in area reaches a critical point beyond which serialization does not lead to further area savings. For instance, in Figure 4(b), the most area saving for a 130nm implementation is achieved with four wires. Any further serialization (i.e., reduction to 2 or 1 wires) causes a reduction in area savings. Similarly, for 90 and 65nm implementations, the most area saving is obtained for a two wire solution. This phenomenon is a result of the overhead of the transmission and receiver circuitry taking up more area as the level of serialization is increased. Serialization beyond a critical point is thus not beneficial for area savings.

As discussed in Section 3.3, often the TSV pad area is increased to ensure fault tolerance, since misalignment of wafers is not uncommon in 3D ICs. To explore area savings due to serialization under this phenomenon, TSV pad dimensions and pitch were increased to 10μm×10μm and 16μm respectively, while TSV dimensions were kept the same. Figure 4(c) and (d) show the area savings for 32-bit and 64-bit links when the proposed serialization scheme. The savings in area are more in this case, as compared to the previous non fault-tolerant case. For instance with a 4:1 serialization, the area savings are greater than 70% for implementations across 130-65nm libraries. For a TSV footprint of 1.6 mm^2 in each layer of a 3D CMP design (as explained in Section 1), the results show an area saving ranging from 45-95% depending upon the level of serialization. Such a reduction in area has significant benefits for reducing routing congestion and fabrication cost, and potentially improving thermal TSV distribution that can lead to lower peak chip temperatures.

5.2 Impact on CMP Power and Performance

Of course the area savings due to serialization come at a cost: a reduction in performance. A serialization degree of n theoretically reduces link throughput by a factor of n. There is additionally also an impact on the power dissipation of the communication fabric due to the additional serialization transmitter and receiver circuitry used. The next set of experiments attempt to quantify the power and performance impact of using the proposed serialization scheme for multi-core CMP designs implemented in 3D ICs.

5.2.1 Experimental Setup

Six applications from the well known SPLASH-2 benchmark suite (*Barnes, Water-NSq, FFT, Cholesky, Ocean, Raytrace*) [39] were selected, then parallelized and implemented on multiple cores that were mapped to a 3D IC. The die size was assumed to be 2cm×2cm. The cores were connected with a packet switched 2D mesh NoC communication fabric with 64-bit wide vertical and planar links, clocked at a frequency of 1 GHz. Table 1 summarizes the implementation details of the CMP applications, such as number of cores (including processors and on-chip memories), the number of (64-bit) vertical links and the number of layers on which the cores were mapped in the 3D IC implementation.

Table 1. 3D CMP Implementation Details

CMP applications	Description	Cores	Vert. links	Layers
Barnes	Galaxy evolution	32	16	2
Water-NSq	Forces/potentials of H$_2$O molecules	38	18	2
FFT	FFT kernel	44	20	2
Cholesky	Cholesky factorization kernel	76	54	4
Ocean	Ocean movements	88	66	4
Raytrace	3-D ray tracing	112	82	4

The 64-bit vertical TSV links were implemented as vertical buses with an interface at each router. There are two reasons for using

978-1-60558-497-3/09 $25.00 © 2009 ACM

vertical buses instead of vertical packet switched links. Firstly, vertical packet switched links would require the addition of two more ports and links (up and down) to each router, which would increase its complexity. In contrast, a vertical bus requires the addition of only a single new port to a router. Secondly, the distance between layers is relatively small (~20-100μm) in 3D ICs compared to inter-router distance in 2D ICs (~1000μm or more). As a consequence, multi-hop and router delay for vertical packet switched links would dominate the vertical propagation time and reduce performance [10]. Therefore a vertical bus with a dynamic TDMA arbitration to support programmable quality of service (QoS) is considered instead of vertical packet switched links.

The CMP applications were modeled in SystemC [40] using a fast and accurate transaction-based bus cycle accurate (T-BCA) modeling abstraction [41]. The cores were modeled at the behavioral level granularity, while the inter-core communication was modeled at a cycle accurate granularity. Each of the applications was simulated with testbench traffic (~few hours) to estimate performance of the implementations. A high level simulated annealing floorplanner for 2D ICs [32] was used to create a thermal-aware layout of the CMP application on the 3D-IC, and Manhattan distance based wire routing estimates were used to determine wire lengths. The wire length information, together with application traffic profiles obtained from simulation were plugged into an on-chip communication architecture power estimation framework [37] to determine link and communication-centric logic component (routers, NoC interfaces) power dissipation for the target technology library.

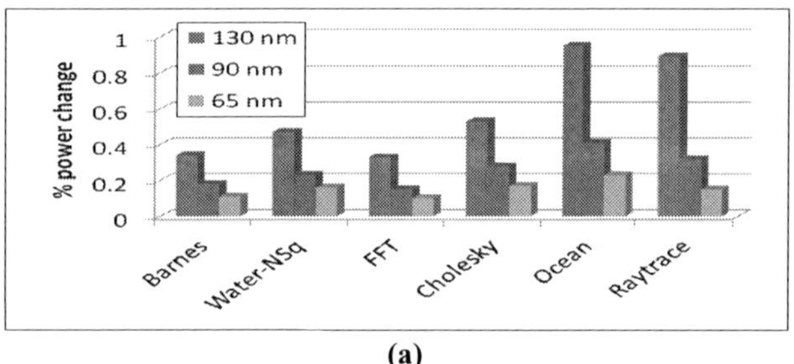

(a)

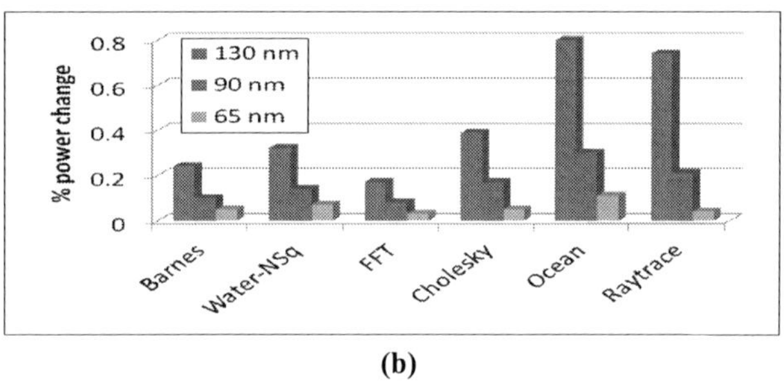

(b)

Figure 5. % power change for CMP applications for 130–65nm nodes when using (a) 2:1 serialization (64 to 32 bit links), and (b) 4:1 serialization (64 to 16 bit links)

5.2.2 Impact on Power Dissipation

Figure 5 shows the percentage change in power dissipation of the on-chip communication architecture fabric when the proposed serialization scheme is used, for the CMP applications implemented in the 130, 90, and 65 nm technology libraries. Figure 5(a) shows the case of 2:1 serialization (64-bit links reduced to 32-bit links) while figure 4(b) shows the case of 4:1 serialization (64-bit links reduced to 16-bit links). Higher degrees of serialization are not as important because they can cause a notable

degradation in application performance (as discussed in the next subsection). It can be seen that serialization causes a slight increase in power dissipation. This is an interesting result because a lot of previous work on serialization for planar interconnects has indicated the potential for power savings with serialization [22][24][25]. The reason for the reduction in power dissipation in those schemes is because of two reasons: *(i)* the schemes allow aggressive reduction in crosstalk capacitance due to greater freedom with wire spacing and sizing after serialization reduces the number of wires, and *(ii)* the length of the wires considered are long enough for the saving in crosstalk to mitigate power dissipation in the serialization circuitry. However, for the proposed vertical TSV serialization case, while there is a reduction in crosstalk capacitance with a reduction in number of TSVs, the width and spacing (pitch) of the remaining TSVs is not altered. In addition, the length of the TSVs is much smaller, which limits the savings due to reduction in crosstalk capacitance. Thus, the power dissipation overhead of the serialization transmitter and receiver circuitry dominates, leading to an increase in power dissipation.

Note that the power dissipation for the 4:1 serialization case is less that for the 2:1 case, because of the greater sharing of resources in the transmitter and receiver circuitry (Section 4), as well as the reduction in switching power on the fewer wires for the 4:1 serialization case. The power dissipation overhead of serialization decreases with technology scaling, and is almost negligible for the 65 nm library. Further scaling below 65 nm will potentially lead to the wire power dissipation dominating the power dissipation of the serialization circuits. In such a scenario, serialization of interconnect TSVs will lead to a reduction in power dissipation.

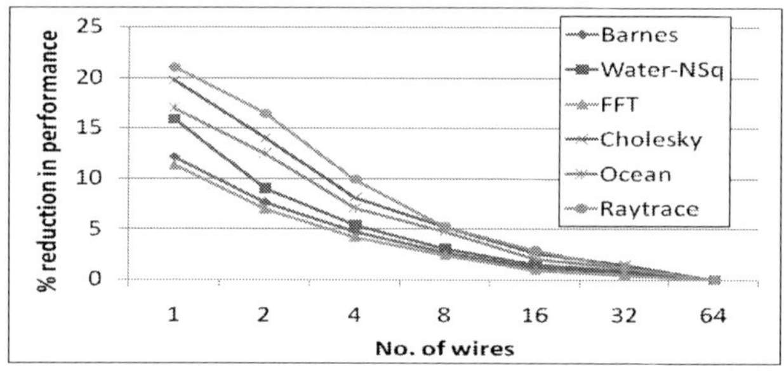

Figure 6. % performance change with varying degrees of serialization for 64-bit links in a 65nm technology implementation for CMP applications (1 GHz clk frequency)

5.2.3 Impact on Performance

Serialization of links can save area, but only at the cost of performance. Figure 6 shows the percentage reduction in overall application performance (y-axis; measured in number of cycles to finish application execution) for the CMP applications, with varying degrees of serialization (represented using number of wires; x-axis). It can be seen that as the degree of serialization increases, and the number of wires in the links are reduced from the original value of 64, the performance degrades. This performance degradation is low for low degrees of serialization, but can become high for higher degrees of serialization. For instance, the performance degradation is about 1.7% on average for a 4:1 serialization (16 wires) but reaches around 16.1% for a 64:1 serialization (1 wire). The exact value of performance degradation depends on the frequency of vertical transfers and the number of TSV interconnects, which varies across the CMP applications. It is clear however that lower degrees of serialization such as 2:1 and 4:1 are more practical because of their low performance and power overhead and an appreciable area saving.

978-1-60558-497-3/09 $25.00 © 2009 ACM

One advantageous consequence of serialization is that it reduces crosstalk capacitance, which in turn reduces propagation delay. This can enable higher clock frequencies on the wires. Figure 7 shows the percentage performance overhead for the 4:1 serialization case, when TSV interconnects are clocked at higher clock frequencies than the baseline 1GHz, for the 65 nm technology node. The higher clock frequencies can improve performance, but only to a limited degree. Even without such frequency increase however, serialization has clear benefits for 3D IC technology. For instance, 4:1 serialization of 64-bit links can save more than 70% of the TSV area footprint on each layer, at the minimal cost of 0.06% power and 1.86% performance overhead on average for CMP applications. This is a strong motivation for considering serialization of TSV links in emerging 3D CMP architectures.

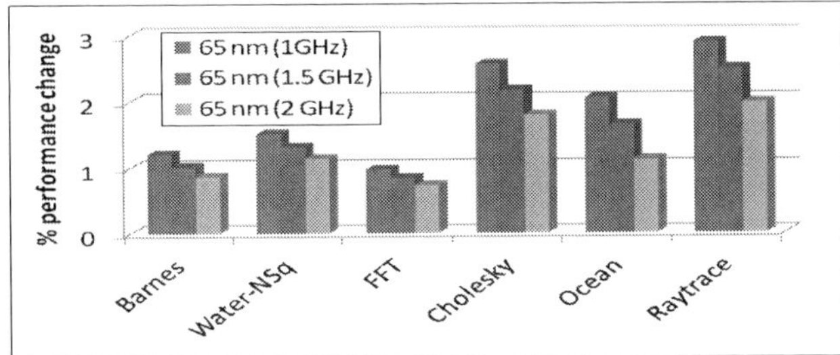

Figure 7. % performance change with 4:1 serialization (64 to 16 bit links) for CMP applications for 65 nm technology node, with scaled TSV clock frequencies

6. CONCLUSION

Vertical interconnect TSVs in 3D ICs take up significant chip area, and can cause routing congestion because of their typically spread-out distribution. In this paper, the use of serialized vertical interconnects in 3D ICs was explored to reduce the area footprint of interconnect TSVs. A shift register based serialization scheme was proposed and it was shown how varying the degree of serialization can result in significant savings in TSV area on each layer of a 3D IC. It was also demonstrated through experiments on several 3D CMP architectures that the proposed serialization scheme can significantly reduce interconnect TSV area footprint, at a nominal power and performance overhead. The extra space made available on each layer due to serialization can be used for better core placement and routing, as well as more efficient thermal TSV insertion for temperature management. Future work will quantify the temperature reduction due to serialization, as well as explore other serialization schemes for 3D ICs.

7. REFERENCES

[1] S. Pasricha and N. Dutt, "On-Chip Communication Architectures", Morgan Kauffman, ISBN 978-0-12-373892-9, Apr 2008

[2] S. M. Rossnagel, T. S. Kuan, "Alteration of Cu Conductivity in the Size Effect Regime," JVST, Vol. 22, Iss. 1, pp. 240-247, Jan. 2004.

[3] International Technology Roadmap for Semiconductors (ITRS), System Drivers, 2007.

[4] A. W. Topol et al., "Three-dimensional integrated circuits," IBM J. Res. & Dev. Vol. 50 No. 4/5 Jul/Sep 2006.

[5] K. Bernstein, et al., "Interconnects in the Third Dimension: Design Challenges for 3D ICs," Proc. DAC 2007, pp.562-567.

[6] R. S. Patti, "Three-Dimensional Integrated Circuits and the Future of System-on-Chip Designs", Proc IEEE, Vol 94, No. 6, Jun 2006.

[7] V. F. Pavlidis, E. G. Friedman, "Three-dimensional Integrated Circuit Design", Morgan Kaufmann, Sep 2008.

[8] B. Feero, P.P. Pande, "Performance Evaluation for Three-Dimensional Networks-On-Chip", Proc. ISVLSI 2007.

[9] S. Pasricha, N. Dutt, "Trends in Emerging On-Chip Interconnect Technologies", IPSJ Transactions on System LSI Design Methodology, Vol. 1, Sep 2008

[10] F. Li et al., "Design and Management of 3D Chip Multiprocessors Using Network-in-Memory", Proc. ISCA 2006, pp. 130-141.

[11] B. Goplen, S. Sapatnekar, "Thermal via placement in 3D ICs", Proc. ISPD 2005.

[12] Z. Li, et al., "Efficient thermal-oriented 3D floorplanning and thermal via planning for two-stacked-die integration", ACM TODAES 11:2, Apr 2006, pp. 325-345.

[13] C. Addo-Quaye, "Thermal-aware mapping and placement for 3-D NoC designs," Proc. IEEE Int. Syst.-on-Chip Conf., 2005, pp. 25–28.

[14] E. Wong, S. K. Lim, "3D Floorplanning with Thermal Vias", Proc. DATE 2006, pp. 1-6.

[15] J. Cong, Jie Wei and Yan Zhang, "A thermal-driven floorplanning algorithm for 3D ICs", Proc. ICCAD 2004, pp. 306-313.

[16] P. Zhou et al., "3D-STAF: scalable temperature and leakage aware floorplanning for three-dimensional integrated circuits", Proc ICCAD 2007.

[17] I. Loi et al., "Supporting vertical links for 3D networks on chip: toward an automated design and analysis flow", Proc. NanoNet 2007.

[18] K. Puttaswamy and G.H.Loh, "Thermal Herding: Microarchitecture Techniques for Controlling Hotspots in High-Performance 3D-Integrated Processors", Proc. HPCA 2007, pp. 193-204.

[19] Y. Liu, et al., "Fine Grain 3D Integration for Microarchitecture Design Through Cube Packing Exploration", Proc. ICCD, 2007.

[20] D. Park et al. "MIRA: A Multi-layered On-Chip Interconnect Router Architecture", Proc. ISCA 2008, pp. 251-261.

[21] K. Puttaswamy, G. H. Loh, "Implementing caches in a 3D technology for high performance processors" Proc. ICCD 2005, pp. 525-532.

[22] A. Morgenshtein et al., "Comparative Analysis of Serial vs Parallel Links In NoC", Proc. SSOC, 2004.

[23] R. Dobkin, et al., "Parallel vs. Serial On-Chip Communication", Proc. SLIP 2008.

[24] M. Ghoneima et al., "Serial-Link Bus: A Low-Power On-Chip Bus Architecture", Proc. ICCAD 2005.

[25] N. Hatta et al., "Bus Serialization for Reducing Power Consumption", IPSJ TACS 47:3, Mar 2006.

[26] S. Kimura et al., "An On-Chip High Speed Serial Communication Method Based on Independent Ring Oscillators", Proc. ISSCC 2003.

[27] I-Chyn Wey et al., "A 2Gb/s High-Speed Scalable Shift-Register Based On-Chip Serial Communication Design for SoC Applications", Proc. ISCAS 2005.

[28] M. Saneei, A. Afzali-Kusha1, M. Pedram, "Two High Performance and Low Power Serial Communication Interfaces for On-chip Interconnects", Proc. CJECE 2008.

[29] R. Dobkin, et al., "Fast Asynchronous Shift Register for Bit-Serial Communication," Proc. ASYNC, 117-126, 2006.

[30] S. Ogg et al., "Serialized Asynchronous Links for NoC", DATE 2008.

[31] K.-N. Chen, A. Fan, and R. Reif, "Microstructure examination of copper wafer bonding," in http://www-mtl.mit.edu/ reif/papers/2001-knchen-JEM-manuscript.pdf.

[32] S. N. Adya, I. L. Markov, "Fixed-outline Floorplanning: Enabling Hierarchical Design", IEEE TVLSI, Dec. 2003

[33] T. Zhang, Y. Zhan and S. Sapatnekar, "Temperature-aware routing in 3D ICs", Proc. ASPDAC 2006.

[34] S. Lee, et al, "Analysis of Thermal Vias in High Density Interconnect Technology," Proc. IEEE Semi-Therm Symposium, pp. 55-61, Feb. 1992.

[35] I. Loi, et al., "A Low-overhead Fault Tolerance Scheme for TSV-based 3D Network on Chip Links", Proc. ICCAD 2008.

[36] K. N. Chen, A. Fan, and R. Reif, "Interfacial morphologies and possible mechanisms of copper wafer bonding," in http://www-mtl.mit.edu/users/reif/papers/2002-knchen-JMS-manuscript.pdf.

[37] S. Pasricha, Y. Park, F. Kurdahi, N. Dutt, "System-Level Power-Performance Trade-Offs in Bus Matrix Communication Architecture Synthesis", IEEE/ACM CODES+ISSS 2006

[38] Synopsys Design Compiler, PrimeTime PX, www.synopsys.com.

[39] S.C. Woo et al."The SPLASH-2 programs: Characterization and methodological considerations", Proc. ISCAS, 1995.

[40] SystemC initiative. www.systemc.org.

[41] S. Pasricha, N. Dutt, M. Ben-Romdhane, "Extending the Transaction Level Modeling Approach for Fast Communication Architecture Exploration", IEEE/ACM DAC 2004.

No Cache-Coherence: A Single-Cycle Ring Interconnection for Multi-Core L1-NUCA Sharing on 3D Chips

Shu-Hsuan Chou, Chien-Chih Chen, Chi-Neng Wen, Yi-Chao Chan,
Tien-Fu Chen, Chao-Ching Wang and Jinn-Shyan Wang

Dept. of CSIE and EE, National Chung Cheng University, Taiwan, R.O.C.
{csh93, ccchi96m, wcn93, cyc95m, chen}@cs.ccu.edu.tw,
92ccwang@vlsi.ee.ccu.edu.tw, ieegsw@ccu.edu.tw

ABSTRACT

Consistent with the trend towards the use of many cores in SOC and 3D Chip techniques, this paper proposes a "single-cycle ring" interconnection (SC_Ring) with ultra-low latency and minimal complexity. The proposed SC_Ring allows multiple single-cycle transactions in parallel. The main features of the circuit-switched design include a set of 3-ported circuit-switched routers (4~16) and a performance/timing effective arbiter. The arbiter, called "BTPC", features single-cycle arbitration and routing-control by means of the novel Binary-Tree paths convergence and path-prediction mechanisms, to provide a highly reduced time complexity. By combining this with the integration of 3D chips, the proposed ring-based interconnection offers several advantages for hierarchical clustering in future many-core systems, in terms of cost, latency, and power reductions. Moreover, based on the proposed SC_Ring, this work realizes a "level-1 non-uniform cache architecture" (L1-NUCA) for fast data communication without cache-coherency in facilitating multithreading/multi-core as a case study. Finally, experimental results show that our approach yields promising performance.

Categories and Subject Descriptors

B.4.3 [Interconnections (Subsystems)]: *Topology*, C.1.2 [Multiple Data Stream Architectures (Multiprocessors)]: *Interconnection architectures*, C.1.4 [Parallel Architectures]: *Distributed architectures*

General Terms

Design, Management, Performance

Keywords

Ring interconnection, Single-cycle transactions, Arbitration, Level-1 non-uniform cache architecture, Memory structure, Multi-core, NOC, SOC

1. INTRODUCTION

The SOC design trend has recently experienced a movement from multi-core (4~16) to many-core (16~256) design. The concepts of communication-centric interconnections and memory structure designs greatly impact the success of many-core systems. A scalable and efficient interconnection can be characte-

Permission to make digital or hard copies of part or all of this work for personal or classroom use is granted without fee provided that copies are not made or distributed for profit or commercial advantage and that copies bear this notice and the full citation on the first page. To copy otherwise, to republish, to post on servers or to redistribute to lists, requires prior specific permission and/or a fee.

DAC'09, July 26-31, 2009, San Francisco, California, USA

rized by its topology, routing strategy, highly regular wiring strategy, and method of flow-control [1, 2, 3]. Resource limitations, in terms of area and power, are the major constraints on network on-chip (NoC) designs [1]. Packet and circuit-switched NoCs offer significant differences in memory latency, area/power, frequency, and design complexity [1, 3]. Currently, issues of reliability, hot spots, circuit variation, and low power problems are more important than frequency issues. Hence, several recently proposed works still stick on circuit-switching, but the disadvantages of these approaches, including unacceptable wire delay, requiring an efficient arbiter, and so on, [3] become more significant as the systems are scaled up.

Compared to mesh, the ring topology requires fewer connections, simpler router design, and easier routing-path control [2, 4]. A ring interconnection with a circuit-switched strategy may be sufficiently performance and power effective to be competitive, although it is comparably simpler. However, problems of wire delay and efficient arbitration limit the maximum number of transactions. Fortunately, the development of 3D die-stacking techniques (4~5 layers) has the potential to offer significant reductions in the number of wires, through 3D placement and routing [5, 6, 7]. The distance between two device layers is 10~20 μm, and the wire delay is negligible (~8 ps) [5]. Moreover, IBM reduces the pitch to a state-of-the-art $0.2 \times 0.2\ \mu m^2$ [6], which may greatly increase the capacity for inter-wave wires. Furthermore, several architecture designs have already been discussed for integrating all the components on a 3D chip [5, 6, 7], including NoCs, cache, memory, critical paths in the processor, and so on.

The non-uniform cache architecture (NUCA) is a well-known option for the multi-core memory structure. In general, L2 caches are organized as a L2-NUCA with dynamic data-migration and private/shared mechanisms [8, 9] for improving the cache utilization and lower the miss rate of the multi-core sharing. Moreover, the L2-NUCA also stands to benefit from the no cache-coherency property [8, 9], because the many-core cache-coherency can cost much latency and power. L1 caches, which are sensitive to the processor clock rate, are usually organized by a centralized coherent bus, instead of a L1-NUCA. However, a bus is not scalable to support a large-scale multi-core.

In order to achieve ultra-low latency and minimal complexity for on-chip networks, this work proposes a "single-cycle" ring (SC_Ring) interconnection, using the circuit-switched strategy (no buffers) and the ring topology (minimum wires). By single-cycle latency, we mean that the proposed SC_Ring allows data transmission between two nodes to be performed in one cycle, and it allows multiple single-cycle transactions in parallel. The

978-1-60558-497-3/09 $25.00 © 2009 ACM

SC_Ring is simply constructed by a set of 3-port circuit-switched routers and a performance/timing effective arbiter. A variable number of routers (4~16) are connected by one clockwise and one counterclockwise ring-paths, for various applications to extend flexibility and scalability. Moreover, the number of ring-paths can be determined application-specific. However, several critical design challenges require further examination, including the wire delay, effective arbitration for maximum parallelism, and the timing complexity of arbitration. Section 2 illustrates some approaches to the design challenges by means of binary-tree paths convergence (BTPC) arbitration flow, path-prediction, and 3D placement.

Based on the proposed SC_Ring interconnection, with low latency and high bandwidth, this paper also proposes a "L1 non-uniform cache architecture" (L1-NUCA) for fast data communication and no cache-coherency in facilitating multithreading/multi-core (especially in streaming processing [4]), as a case study in Section 3. In our experiments, synthesis results (technology node 90 nm) of the proposed SC_Ring interconnection by different wire loads model show benefits to placing SC_Ring on the 3D chip with the performance/timing effective arbiter (900MHz/8-router ring, 600 MHz/16-router ring). In the performance evaluation of the SC_Ring, the average parallel transactions, average memory latency, and power consumption are estimated by random and multithreading benchmarks. Moreover, some performance and power results are used to compare with the proposed L1-NUCA and the original cache-coherent model.

2. PROPOSED SINGLE-CYCLE RING

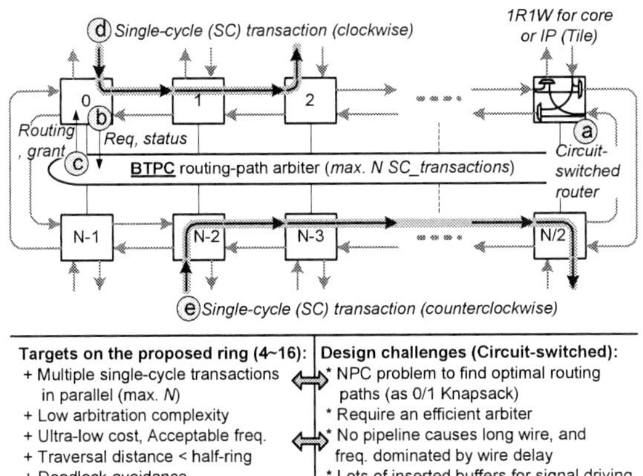

Figure 1. Conceptual view of the single-cycle ring (SC_Ring).

Figure 1 illustrates the organization of the proposed ring (SC_Ring) interconnection and its design challenges. Due to the circuit-switched design, the wire delay problem seems almost intractable. Fortunately, with the development of the 3D chip [5, 6, 7], the wire length (delay) and lots of buffer insertion (cost/power) are greatly reduced shown in Subsection 2.3. The routing path arbitration on rings is critical, and finding the optimal solution can be proved to be an instance of the NP-complete problem of 0/1 Knapsack. This work proposes an efficient "Binary-Tree Paths Convergence" (BTPC) arbitration flow with only $O(\log_2 N)$ time complexity (N = number of routers), and still giving good arbitration performance. Figure 1(a) shows the 3-ported circuit-switched router, which is configured by the BTPC routing-path arbiter. Each router supports a set of read/write ports for the processor core or IP (namely the tile) connection. In Figure 1(b)(c), the request/status signals from tiles are provided for arbitration, and the BTPC arbiter delivers the routing configuration and grant signals after the computation. Figure 1(d)(e) presents a single-cycle (SC) transaction on the clockwise ring, and another SC transaction on the counterclockwise ring. In the best case, it allows N SC transactions to take place concurrently on the proposed ring interconnection.

2.1 BTPC Arbitration for Routing-Paths

2.1.1 Transform into vector-based ring-paths

A transaction from Tile 0 to Tile 2 (Tile 0→2 on 8 routers ring) can generally be represented in binary-encode notation as "3'b000→3'b010", but we employ the bit-vector notation instead as 8'b00000**110**, as shown in Figure 2. Although the cost of bit-vector notation is more than the binary-encode, the bit-vector not only implies several transactions on the ring-path, but the operation of arbitration also becomes quite simple. For example, we can see that Tile 0→2 (8'b00000**110**) and Tile 1→3 (8'b0000**1100**) conflict in the ring-path, by applying the operation "AND" to their bit-vectors[2]. In another example, Tile 0→2 (8'b00000**110**) and Tile 4→6 (8'b0**1100**000) can be merged as Tile 0→2, 4→6 (8'b0**1100110**) by the operation "OR". As the proposed SC_Ring interconnection has both clockwise and counterclockwise ring-paths, a pair of N-bit (N = number of routers) bit-vectors can be represented as the two reversed ring-paths, and the transactions transformed to either bit-vector. In Figure 2, the transaction (Tile 0→2) is transformed into 8'b00000**110** as the clockwise ring-paths, and the transaction (Tile 3→1) is transformed into 8'b000000**110** as the counterclockwise ring-paths.

Simple rules for keeping traversal distance < half-ring:
- "Clockwise", if dst_id > src_id & distance < $N/2$.
- "Counterclockwise, Cross", if dst_id>src_id & distance>$N/2$.
- "Counterclockwise", if dst_id < src_id & distance < $N/2$.
- "Clockwise, Cross", if dst_id < src_id & distance > $N/2$.
(If dst_id(D) > src_id(S), distance= D-S; else, distance= S-D;)

Under the half-ring constraint on the transaction traversal distance, simple rules are provided for the vector-based ring-path transformation. Comparing the source's tile_id (src_id) and the destination's tile_id (dst_id) is the primary way to decide between clockwise and counterclockwise. However, if the distance between two nodes is larger than $N/2$, it should be transformed into the reversed ring-path (namely "cross", as Tile 6→0) because the distance on the reversed ring-path must be smaller than N/2. Therefore, the half-ring constraint is guaranteed. The proposed transformation rules show several compare/minus operations, but they can be simplified to only a few levels of gates to improve the timing.

Ring-path (Tile_id)	0	1	2	3	4	5	6	7
Clockwise Ring								
Tile 0 → 2 (T)	0	1	1	0	0	0	0	0
Tile 6 → 0 (L)	1	0	0	0	0	0	0	1
CounterClockwise Ring								"cross"
Tile 3 → 1 (T)	0	1	1	0	0	0	0	0

T: true request, L: no request but inherit the last

Figure 2. Example of vector-based ring-paths (8-router ring).

Another important design strategy is to reserve the previous routing-path, so the next transaction will have a good opportunity to access the same destination. In Figure 2, we define the true request (T) and no request but inherit the last (L) and transform them both into bit-vector. If there is no request, the bit-vector will be inherited by the last-request (ex. Tile 6→0 (L)), but the fake transaction will have lower priority to not affect true transactions.

2.1.2 Binary-tree paths convergence (BTPC)

The problem of arbitration of ring routing-paths can be identified with the NP-Complete problem of 0/1 Knapsack, because the best solution is to maximize the transferring transactions, subject to limited routing resources. The number of transactions transferred is regarded as the value in 0/1 Knapsack, and the limited routing resources corresponds to the constraint of total weight. Thus, an optimal arbitration result is unfeasible, but our approach provides an approximated solution that is efficient and feasible in the hardware implementation. Two approximate solutions, "Shortest-path first" and "Priority first", are discussed as comparisons for the ring routing-path arbitration. At first, the "Shortest-path first" always picks up the shortest transferring distance transaction and marks it as a permitted routing-paths if it does not encounter any conflicts (step-by-step). It is a greedy algorithm to take the minimum routing resources to permit as many transactions as possible. However, it is unfeasible to find the shortest path for the hardware in run-time (time complexity $O(N^2)$, where N is the number of routers in the ring-path). It also has a serious deadlock problem. In the "Priority first" strategy, the transactions are picked up according to their priority and marked as permitted routing-paths if they do not encounter conflicts (step-by-step). The algorithm is deadlock-free by the round-robin priorities, but it is also hardware unfeasible (time complexity $O(N^2)$).

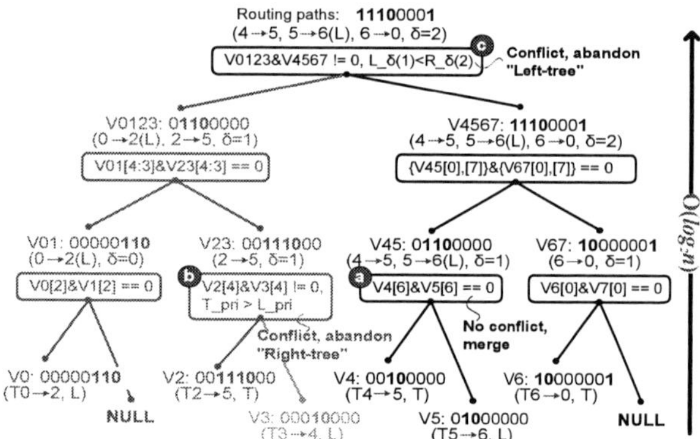

Figure 3. The $O(\log_2 N)$ arbiter by BTPC mechanism. (BTPC arbitration example for 8 routers clockwise ring)

We propose a "binary-tree paths convergence" (BTPC) algorithm, which provides efficient arbitration with low time complexity $O(\log_2 N)$. In Figure 3, showing the arbitration for the clockwise ring, the main idea is that two transactions (represented as bit-vectors for ring-path) in two neighboring routers are compared to see if they conflict, and merged. After the first level convergence, there are only 4 bit-vectors (8/2) for next-level convergence, and some transactions are abandoned. Finally, it only costs three levels of convergence, and the last ring-path bit-vector indicates the permitted transactions. The algorithm uses a comparison-based binary-tree, and substantially reduces the time complexity to $O(\log_2 N)$. However, it will abandon a sub-tree while the routing-paths have conflicts, and it will also sacrifice some un-conflicting routing-paths, especially in abandoning sub-trees at higher levels. For this reason, two design strategies are proposed to minimize the negative impact.

In the first design strategy, comparing/merging two neighboring bit-vectors will find conflicts of routing-paths in lower convergence level, because there is a greater opportunity for conflict between neighboring routers on the single direction ring (e.g., the clockwise ring). In the second design strategy, the

variable δ is defined for the number of true transactions (T) in the bit-vector as shown in Figure 3, and the arbiter will abandon the sub-tree with smaller δ for more permitted transactions, the so-called "local optimization" strategy. Figure 3(a) shows two neighboring bit-vectors detecting the routing-path with no conflict and merging, and the operations are quite simple. The bit[6] of two neighboring bit-vectors checks whether they conflict ("V4[6] & V5[6] == 0"), and the "OR" operand defines the merging of two bit-vectors. Figure 3(b) shows how the conflict is detected by checking "V2[4] & V3[4] != 0", and the arbiter abandons the right sub-tree. Because the bit-vector (L) of router3 is not truly a request, but inherited the last request in router3 (the fake transaction) that is mentioned in Subsection 2.1.1. Therefore, the fake transaction will have a low priority for arbitration. Therefore, it not only enables the arbitration of true transactions, but also helps with the prediction of routing-paths for most single-cycle transactions. Figure 3(c) shows how the conflict in the final convergence level is detected, and the algorithm abandons the left sub-tree, which has smaller δ (δ of left sub-tree (1) < δ of right sub-tree (2)) according to the "local optimization" strategy.

However, the "local optimization" strategy may also lead to deadlock. Hence, we also support the starvation detection feature, and then the BTPC arbiter will set a larger value of δ to prevent transactions with higher priority from starvation. In figure 4, our experiment shows an average of 6.3 transactions transferred by successive random patterns on the proposed 16 routers of the SC_Ring interconnection (clockwise and counter-clockwise) if a tile only communicates with its 8 neighbors (Locality-8). However, it shows only an average of 3.9 transactions transferred if each tile communicates with all tiles randomly (Locality-16). In addition, it also shows the arbitration efficiency, the permitted transaction count of BTPC is near to the permitted count of two approximation solutions whose can't be hardware implementation. The BTPC algorithm plays an important role in reducing the time complexity of ring routing-paths arbitration to $O(\log_2 N)$, and some synthesis results demonstrate the promised performance in Figure 9.

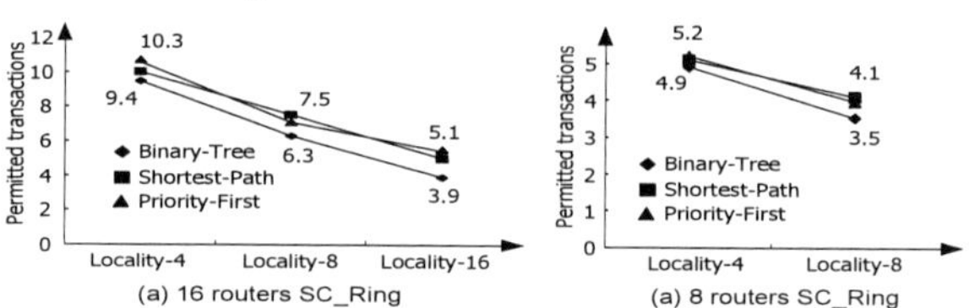

Figure 4. Arbitration efficiency v.s. communication locality.

2.1.3 Remove conflicts between rings

After the binary-tree paths convergence (BTPC), the result is a number of bit-vectors for routing configuration by a number of ring-paths. The BTPC removes ring routing-path conflicts and intra-ring destination conflicts. However, destination conflicts between rings (inter-ring) should also be resolved. Therefore, the final step of the BTPC arbitration flow checks destination conflicts between rings and selects one by priority. Finally, the routing configuration is generated.

2.2 Path-Prediction Mechanism

The time complexity of the BTPC arbitration flow is shown, including transforming the vector-paths (O(1)), BTPC arbitration ($O(\log_2 N)$), and removing conflicts (O(1)). In general, one data transaction can be divided into two processing stages, namely the arbitration and routing stages. Usually, the two stages would be pipelined to maintain frequency without sacrificing performance. Fortunately, many applications sequentially access just one memory segment at a time. Accordingly, the prediction strategy speculates that the subsequent transaction(s) may be similar to these previous transactions. Our proposed approach provides a prediction mechanism to combine these steps into a single stage to achieve efficient single-cycle transaction performance. Figure 5 shows a single-cycle transaction block diagram that maintains frequency. In brief, the path-prediction predicts the subsequent-cycle routing paths by assessing the status of current or previous requests. The predicted routing paths will be prepared for the subsequent cycle, and these will then be used to configure the routers by the "pipe". Of course, the prediction is not always correct, and it should be compared with the real requests and with the status of tiles in the context of destination fulfillment procedures (the "Check" module) during the routing stage. In summary, the path-prediction and routing stages can process in parallel to reduce the critical path.

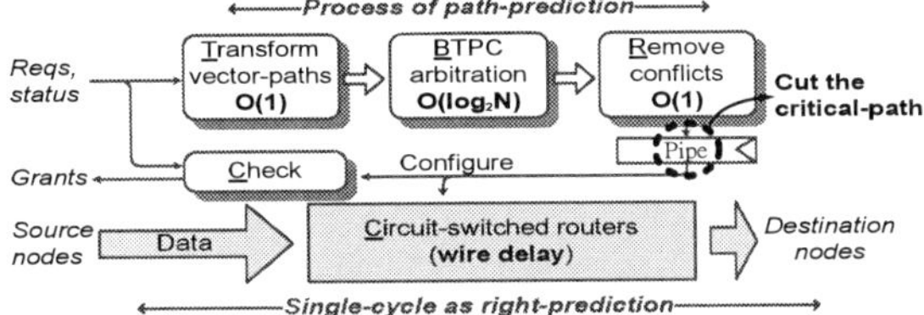

Figure 5. A pipelined arbitration for single-cycle transaction.

Figure 6 and 7 show experiments of average transaction latency on 16-router SC_Ring by random and multimedia workloads, including 1 cycle memory access latency. By increasing injection rate, the memory latency increases rapidly if each tile communicates with all tiles randomly (Locality-16). Thus, the communication locality is an important factor for average latency reduction. A small case study on 16-router SC_Ring interconnection indicates that due to a low injection rate, the data communication is fast, excepting H.264. However, an overall better results can be obtained if we perform optimized placement of threads onto processors by considering communication locality (5.77→4.86).

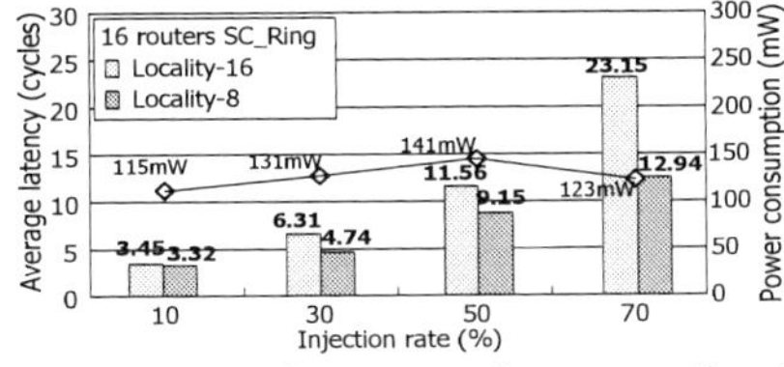

Figure 6. Average latency and power estimations.

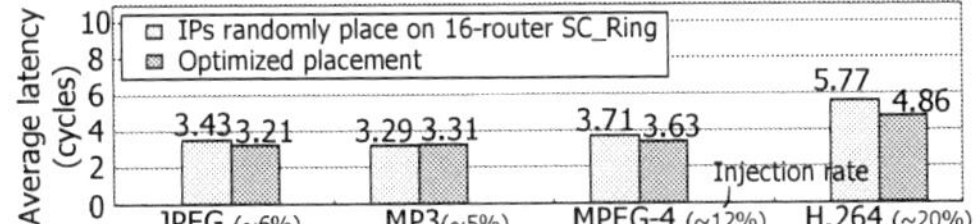

Figure 7. Average latency on multimedia workloads.

2.3 Placement on 3D Chips

To further overcome long wire delays, we may employ three-dimensional (3D) designs to support the proposed SC_Ring interconnection, where multiple device layers are stacked together. Given the thermal and process limitations of 3D chips [5, 6, 7], we only propose 4 device layers for a maximum of 16 routers in the SC_Ring interconnection. Figure 8(a) shows the 8-router SC_Ring placement on a 4-layer stack, and each layer has 2 routers connecting 2 tiles. For reference, in the 3D stacking technique in technique node 70 nm [5], the distance

between 2 layers is 10~20 μm, and the wire delay is about 8 ps. The wire delay is negligible, and the maximum transferring distance on the 8-router SC_Ring is estimated to be less than 100 μm with less than a 100 ps delay. As regards the inter-wave wire area, there are 172b signals (128b data, adr, control) between two routers by one ring. Figure 8(b) shows the view of the BTPC arbiter placement, with 22b signals between the arbiter and each router. Hence, there are total (172*4+22*4) d2d paths between two layers. From a current reference [6], it only costs about 2,850 μm² by via pitch-1 μm. Therefore, this shows that the inter-wave wire area cost is acceptably small.

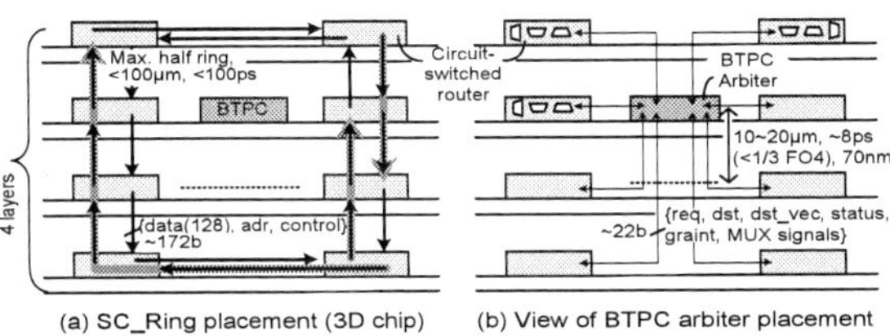

Figure 8. Side view of the 3D chip (4 layers) with the SC_Ring.

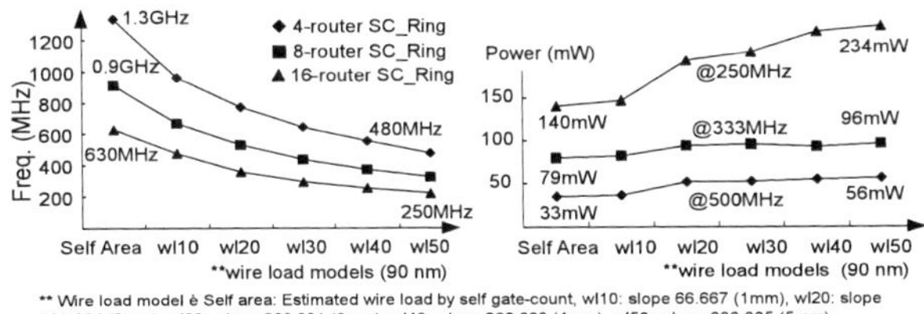

Figure 9. Timing/Power by different **wire load models (90 nm).

While the wire delay is negligible in a 3D chip, the frequency of the proposed SC_Ring interconnection is dominated by arbitration and MUXs in each router. Figure 9 shows timing/power evaluations of the proposed SC_Ring by different wire load models simulating different wire delay models, and the results show that larger wire delays will cause lower frequencies, larger area 20~25% ↑ (buffer insertion), and larger power consumption 20~70% ↑ (signal driven). Moreover, it also shows that the BTPC routing-path arbiter is timing efficient, and the proposed SC_Ring interconnection is quite suitable to integrate into a 3D chip. The experiment shows that a circuit-switched 16-router SC_Ring can achieve up to a frequency of 630 MHz (90 nm), which still include the design features of multiple single-cycle transactions.

3. L1-NUCA FOR NO COHERENCY

Given a single-cycle ring, we further propose the "L1 Non-Uniform Cache Architecture" (L1-NUCA) to realize the no cache-coherency system. The key point of the approach is to distribute the shared address space to different L1 data caches, depending on the locations of the data. As shown in Figure 10, each processor has its own data cache, and these caches are connected via a SC_Ring, constructing the proposed L1-NUCA. The proposed MMU (to add fields for identifying memory location) helps to differentiate private (such as stack accesses), shared, and non-cacheable address regions without extra latency (gray is a private cache-line, and black is a shared cache-line). Figure 10(a)(b) shows the core accessing private and shared address space in its L1 D$. Figure 10(c) illustrates how a remote shared cache is accessed by searching the memory location in the MMU stage, and the cases of non-cacheable in 10(d) and cache-miss will be passed to L2-NUCA.

978-1-60558-497-3/09 $25.00 © 2009 ACM

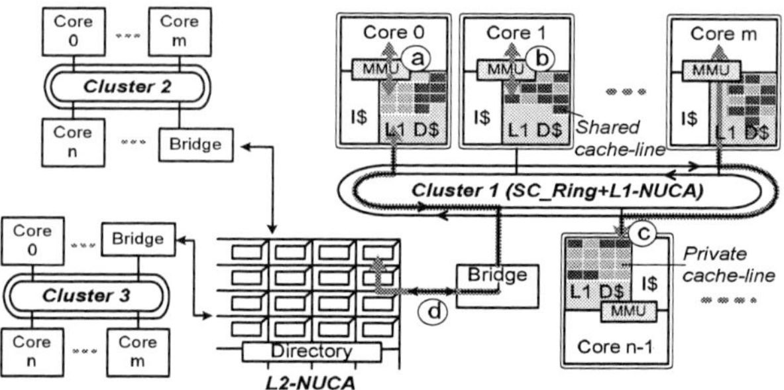

Figure 10. Conceptual view of the no cache-coherency system.
(a) Private cache access, (b) Local shared cache access,
(c) Remote shared cache access, (d) Non-cacheable or cache-miss.

In our proposed no cache-coherency system, multi-cores share the L1-NUCA in the same cluster, and multiple clusters share the L2-NUCA. Hence, by the low latency/high bandwidth SC_Ring interconnection, the multi-cores in a given cluster feature fast data communication and no cache-coherency, especially for applications with streaming processing. In addition, the inter-cluster communication will go though the L2-NUCA via the non-cacheable address space, and the address space partition will be set by multithreading programs.

3.1 Private, Shared, and Non-Cacheable

In order to search distributed shared memory locations in L1 D$ more quickly, this work proposes a page-based hardware mechanism by adding fields to the entries in the TLB and the smart address space mapping controller. Therefore, we can insert the information of the distributed shared memory locations into each page entry of MMU by looking ahead along the shard mapping, and the searching process is hidden by processing the virtual address translation in parallel without extra latency. Each page consists of four sub-pages, which can be located at four different cores. Upon a new page, the mapping controller will check if the page is located in a private, shared, or non-cacheable address region, and determine to which L1 D$ the location maps by handshaking with other cores if it is shared. Afterwards, the page will be updated to the MMU of this core with the physical address and address partition information. The state field in Figure 11 indicates the state of the corresponding memory location as private (01), shared (10), or non-cacheable (11). The other four fields identify where four sub-pages are locate in which L1 D$. The example in Figure 11 illustrates that the second entry of MMU is a shared page, which is divided into four sub-pages, located at the 7th, 7th, 10th, and 0th L1 D$ respectively. In general, the granularity of the sub-pages determines the amount of shared data per processor, and will be the basic unit in cache migration. A large unit of sub-pages may lead to unnecessary remote shared accesses due to false sharing. However, a lower granularity of sub-pages will require more sub-page fields entered in each MMU (4 KB per page and 1 KB per sub-page in Figure 11). Fortunately, the cost is negligible (each memory location identifier is 4b).

The mapping controller (Mapping Ctrl.) of each core is responsible for the management of distributed shared memory locations, and it contains private/shared region registers for address space partition and records the access frequencies of shared pages for data-migration. In particular, a distributed control strategy is provided for the distributed shared management, in which the mapping controller of each core handshakes with other cores to update its location information in case of new page allocation, data-migration, and memory location changes.

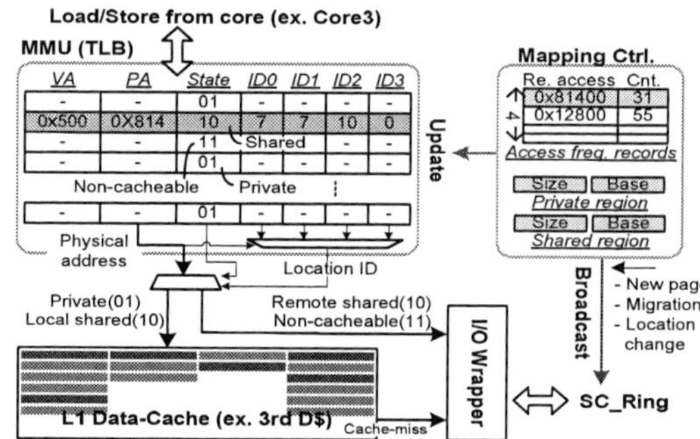

**Figure 11. Page-based hardware mechanism for
fast searching memory locations with no extra latency.**

3.2 Distributed Shared Strategy

By the proposed distributed control strategy for shared memory management, several distributed control behaviors complete the efficient L1-NUCA. In Figure 11, the mapping controller of each core operates distributed control by broadcasting commands on the SC_Ring. Figure 12(a) describes a shared memory location change. Core0 receives the location change command from SC_Ring, and updates the sub-page (0x81400) located in the 7th L1 D$, if the sub-page exists in Core0's MMU. When a core (Core3) allocates a new page, it will broadcast to inquire about the new page's location on SC_Ring. If none of the cores responds, the page's location remains in Core3, as shown in Figure 12(b). On the other hand, in this case when there are replies giving the page location in 1st, 7th, 12th, 12th L1 D$, the core will update them to MMU.

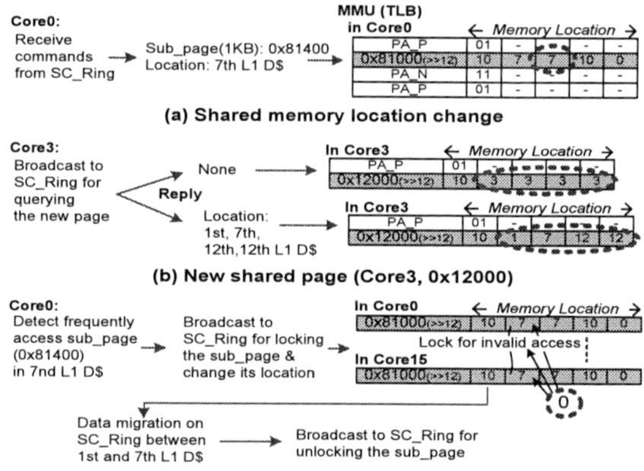

Figure 12. The management of distributed shared mapping.

3.3 Attracting Sub-pages by Migration

Although the SC_Ring is supported, the remote shared access still pays some overhead (minimum 2~3 cycles). The behavior of data migration is necessary to keep data localized within L1 D$ for performance/power effectiveness, and the migration range is 1 KB (sub-page) in our proposed L1-NUCA prototype. In the example in Figure 12(c), when Core0 detects frequent access to a remote shared sub-page (ex. 0x81400, in 7th L1 D$) from the record of access frequency, it will determine to migrate the sub-page into its own L1 D$. At first, Core0 broadcasts to lock the sub-page and change its location commands to all cores in the cluster, and then the sub-page data migrates from the 7th to 0th L1 D$. When the migration finishes, Core0 will broadcast to unlock the sub-page to complete the migration process. When a task is done by multiple threads, the address partition mapping may change for the new task. Thus, the multithreading scheduling handler should flush all L1 D$ in the cluster and reset the MMU by new private/shared regions before restarting.

4. EXPERIMENTS

In the proposed single cycle ring (SC_Ring) experiment, we implemented a cycle-accurate model connected with many-core simulator **mcore** similar to [11] and a configurable RTL model for synthesis and power evaluation in technology node 90 nm. Table I shows the target system parameters. This work evaluated by a parallel simulator in the context of the SPLASH-2, Parsec benchmarks. This experiment constructs 16 cores with intra-cluster L1 cache system as MESI snooping cache-coherency protocol, static and dynamic L1-NUCA, and there are some latency/power results for comparison. In Figure 4, 6, 7, 9, there are several SC_Ring experimental results, which including the efficiency of BTPC arbiter, average transfer latency by the injection rate, small case studies for multimedia workloads, and timing/power evaluations by different wire load models to show assumed results in 3D chips. In overall, the SC_Ring performs ultra-low latency and minimal complexity for on-chip networks.

We focus on the shared memory access events. Table I gives those parameters, where the migration size is 1KB (sub-page) on frequently remote accessing and the L2 cache latency is average 20 cycles. The power model is composed of cache bank (CACTI 5.3 [10]), SC_Ring interconnection, and snooping bus by hspice tools.

Table I. Target System Parameters.

Many-core simulator parameters			
Simulator	mcore	ISA	x86
#core	16	CPU family	Intel® Atom™
L1 data-cache	32B/cache line, 4-way, 16KB		
Data sync	MESI snooping coherency protocol, L1 S-NUCA and L1 D-NUCA		
Data migration	Size: 1KB (sub-page), Event trigger: 100 remote access		
L2 cache	L2-NUCA, 1-port, average latency: 20 cycles		
Power model (90 nm)	Cache Bank (CACTI 5.3), SC_Ring interconnection, snooping bus by hspice tools		
Multithreading benchmarks		SPLASH-2, Parsec	

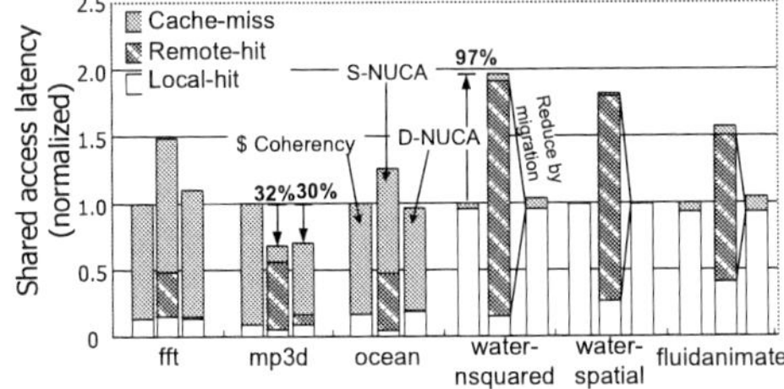

Figure 13. Comparisons of shared access latency.

Figure 13 shows the shared access latency (normalized by $ coherency) of three L1 $ system (Static-NUCA, Dynamic-NUCA), and the latency is primarily contributed by cache-miss, remote hit, and local hit. In mp3d, the result shows access latency reducing (35% ↓), because there are lots of data overlapping and communication between threads (low cache-miss rate and fast data communication). However, the other benchmarks are lacking of the streaming property, and their shared data address space is less overlapped. Hence, the L1-NUCA will not perform better and pay for the latency of remote shared access contrarily. Moreover, the S-NUCA with no data-migration has high ratio of remote shared access to cause high latency, especially in **water-nsquared**, **water-spatial**, and **fludanimate**. The D-NUCA with proposed data-migration mechanism eliminates the penalty greatly shown in Figure 13. Comparing average power consumption, the cache snooping protocol sends commands to snooping bus and broadcasts to all tag banks in each core for checking. Our power model count is about 680mW (40mW for tag banks per core and 40mW when snooping bus activating). In

L1-NUCA, the remote shared access will cost extra power by transactions on SC_Ring without broadcasting. One transaction on SC_Ring costs only 20mW. Figure shows significant power saving in mp3d (43%). However, the S-NUCA consumes large power in water-spatial because of too many remote shared access. Therefore, the unnecessary remote shared access can be avoided.

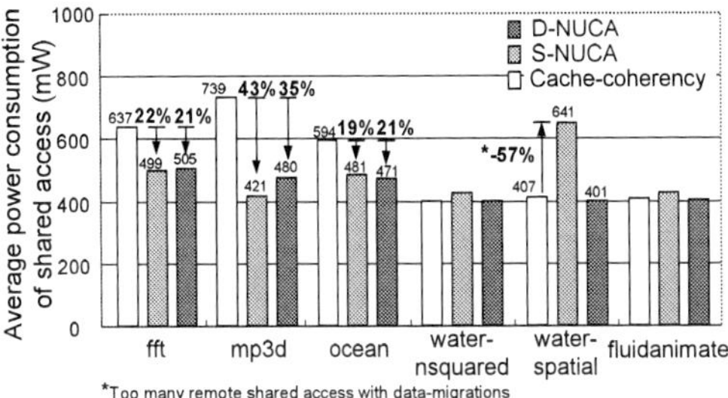

Figure 14. Comparisons of shared access power.

5. CONCLUSION

We propose a SC_Ring interconnection with ultra-low latency and minimal complexity. By combining this with the integration of 3D chips, the proposed ring-based interconnection offers several advantages for hierarchical clustering in future many-core systems, in terms of cost, latency, and power reductions. Moreover, by SC_Ring supporting, this work realizes L1-NUCA in facilitating multithreading/multi-core but also shows great reduction of shared access latency and power consumption in coherent multithreading programs.

References

[1] Kim, J., Park, D., Theocharides, T., Vijaykrishnan, N., and Das, C. R. A low latency router supporting adaptivity for on-chip interconnects. In *Proceedings of Annual Conference on Design Automation.* 2005.

[2] Bourduas, S. and Zilic, Z. A Hybrid Ring/Mesh Interconnect for Network-on-Chip Using Hierarchical Rings for Global Routing. In *Procs of the 1st international Symposium on Networks-on-Chip.* 2007.

[3] Chang, K., Shen, J., and Chen, T. Tailoring circuit-switched network-on-chip to application-specific system-on-chip by two optimization schemes. *ACM Trans. Des. Autom. Electron. Syst.* vol. 13, no. 1. 2008.

[4] Kistler, M., Perrone, M., and Petrini, F. Cell Multiprocessor Communication Network: Built for Speed. *Micro* vol. 26, no. 3. 2006.

[5] Loh, G. H., Xie, Y., and Black, B. Processor Design in 3D Die-Stacking Technologies. *IEEE Micro* vol. 27, no. 3. 2007.

[6] Li, F., Nicopoulos, C., Richardson, T., Xie, Y., Narayanan, V., and Kandemir, M. Design and Management of 3D Chip Multiprocessors Using Network-in-Memory. In *Procs of the ISCA.* 2006.

[7] Pavlidis, V. F. and Friedman, E. G. 3-D topologies for net-works-on-chip. *IEEE Trans. Very Large Scale Integr. Syst.* vol. 15, no. 10. 2007.

[8] Huh, J., Kim, C., Shafi, H., Zhang, L., Burger, D., and Keck-ler, S. W. A NUCA substrate for flexible CMP cache sharing. In *Procs. of Intl. Conference on Supercomputing.* 2005.

[9] Dybdahl, H. and Stenstrom, P.. An Adaptive Shared/Private NUCA Cache Partitioning Scheme for Chip Multiprocessors. In *Proceedings of the Intl. Symp. on High Performance Computer Architecture.* 2007.

[10] CACTI: An Integrated Cache Timing, Power, and Area Model http://www.ece.ubc.ca/~stevew/cacti/

[11] Nguyen, A.-T.; Michael, M.; Sharma, A.; Torrellas, J. The Augmint multiprocessor simulation toolkit for Intel x86 architectures. In *procs of Computer Design: VLSI in Computers and Processors,* 1996.

Thermal-driven Analog Placement Considering Device Matching

Po-Hung Lin[1], Hongbo Zhang[2], Martin D. F. Wong[2], Yao-Wen Chang[1,3]

Graduate Institute of Electronics Engineering, National Taiwan University, Taipei, Taiwan[1]

Dept. of Electrical and Computer Engineering, University of Illinois at Urbana-Champaign, IL, USA[2]

Dept. of Electrical Engineering, National Taiwan University, Taipei, Taiwan[3]

marklin@eda.ee.ntu.edu.tw; {hzhang27, mdfwong}@illinois.edu; ywchang@cc.ee.ntu.edu.tw

ABSTRACT

With the thermal effect, improper analog placements may degrade circuit performance because the thermal impact from power devices can affect electrical characteristics of the thermally-sensitive devices. There is not much previous work that considers the desired placement configuration between power and thermally-sensitive devices for a better thermal profile to reduce the thermally-induced mismatches. In this paper, we first introduce the properties of a desired thermal profile for better thermal matching of the matched devices. We then propose a thermal-driven analog placement methodology to achieve the desired thermal profile and to consider the best device matching under the thermal profile while satisfying the symmetry and the common-centroid constraints. Experimental results based on real analog circuits show that our approach can achieve the best analog circuit performance/accuracy with the least impact due to the thermal gradient, among existing works.

Categories and Subject Descriptors: B.7.2 [Integrated Circuits]: Design Aids - Layout, Placement and Routing

General Terms: Algorithms, Design, Reliability

Keywords: Analog placement, thermal matching

1. INTRODUCTION

In modern RF or analog and mixed-signal IC design, the thermal issue becomes more and more important during device placement, especially when integrating power amplifiers and other analog or mixed signal circuits into the same chip, such as the RF system [13] shown in Figure 1. The RF system contains power devices in the power amplifiers and thermally-sensitive matched devices which appear in the mixer, the low-pass filter, and other sub-circuits. Generally, the power devices consume much more power than all the other devices and may generate significant heat which may affect the electrical properties of the thermally-sensitive matched devices, such as the saturation current, I_{dsat}, of a MOS transistor. Consequently, it may degrade the circuit performance or even change the whole circuit behavior.

According to [4, 5], the matched devices should be in *symmetric* and/or *common-centroid* placements. Ideally, if the heat of the whole chip is evenly distributed, the devices can be thermally

Permission to make digital or hard copies of part or all of this work for personal or classroom use is granted without fee provided that copies are not made or distributed for profit or commercial advantage and that copies bear this notice and the full citation on the first page. To copy otherwise, to republish, to post on servers or to redistribute to lists, requires prior specific permission and/or a fee.
DAC'09, July 26-31, 2009, San Francisco, California, USA

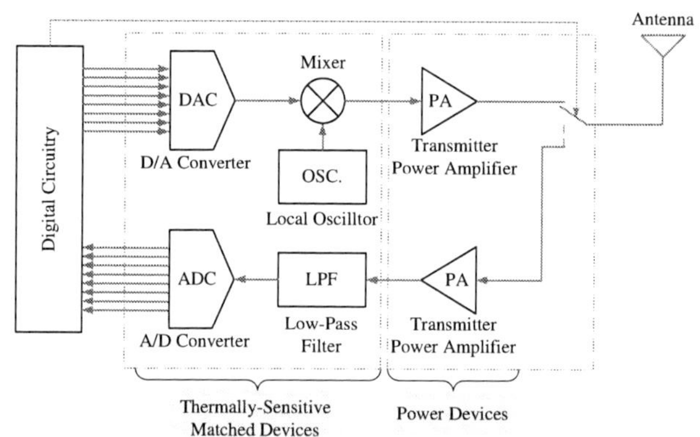

Figure 1: The block diagram of a generic RF system.

matched very well by these techniques. However, the power devices always generate certain thermal gradients on the chip which cause the devices with symmetric and/or common-centroid placements to become mismatched. To consider the thermally induced mismatch, the *thermal profile* of the chip induced by the arrangements of power devices should further be considered together with symmetric and common-centroid placements of thermally-sensitive matched devices in analog layouts. Since the devices other than the power devices in Figure 1 consume much less power, we simply consider them as non-power devices. Figure 2 shows two different thermal profiles based on different arrangements of the power devices in the power amplifiers. For better thermal matching of all thermally-sensitive matched devices, the thermal profile in Figure 2(b) is superior to that in Figure 2(a), which will be further discussed in Section 2.

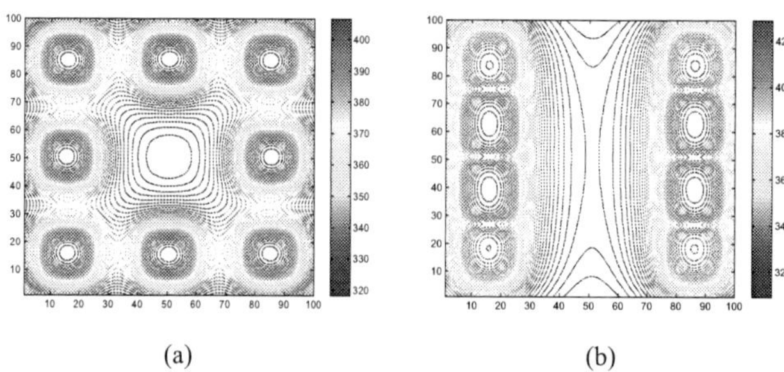

(a) (b)

Figure 2: Thermal profiles based on two kinds of power device arrangements. (a) The thermal profile where power devices are evenly distributed at four sides of the chip. (b) The thermal profile where power devices are evenly distributed at two opposite sides of the chip.

1.1 Previous Work

Analog placement considering device matching constraints has been extensively studied based on various floorplan representations, such as the absolute floorplan representation [4, 7], B*-tree [2, 17], hierarchical B*-tree (HB*-tree) [8], sequence pair (SP) [1, 18], transitive closure graphs (TCG) [10, 22], and corner block list (CBL) [11] for symmetry constraints, and CBL and grid-based approaches [14] for common-centroid constraints. Among these works, only [4, 7, 11] addressed thermally constrained symmetric placement.

Cohn et al. [4] introduced a basic placement configuration for thermal device matching, which is to position the power devices along a thermal symmetry line bisecting the chip such that the isothermal contours are symmetric across the symmetry line. The thermally-sensitive matched devices are then placed symmetrically about the power devices to have the same ambient temperature. Consequently, the thermal mismatch between the matched devices is reduced. Although such an approach is very effective, it limits the layout design with only one symmetry line on the chip. Such a configuration for thermal device matching is not applicable to modern RF or analog and mixed-signal design as seen in Figure 1, which contains multiple symmetry groups with different symmetry lines in different sub-circuits.

Both Lampaert et al. [7] and Liu et al. [11, 12] presented their thermally constrained analog placement by the thermal profile computation. During placement iterations, the temperature of all matched devices are calculated based on certain thermal models. The thermally-induced mismatch is then optimized by minimizing the temperature differences between the symmetric devices. Although their approaches do not limit the layout with only one symmetry line, it is time-consuming to calculate the temperature of all matched devices during placement iterations when the number of the matched devices is large. In addition, it is difficult to guarantee that all devices are thermally matched by summing up the temperature differences between symmetric devices in each symmetry group and other placement objectives such as reducing placement area and thermal hot spots [11]. None of the previous works directly optimizes the thermal profile based on the power device arrangement to achieve better thermal matching of the devices.

1.2 Our Contributions

In this paper, we propose the *first* thermal-driven analog placement considering thermal device matching by directly optimizing the thermal profile of analog layouts. We introduce the desired thermal profile and the corresponding placement configuration for better device matching, especially when placing multiple symmetry groups with different symmetry lines. We then present our placement methodology to simultaneously place all devices, including power devices and thermally-sensitive matched devices with either the symmetry or the common-centroid constraint. We adopt a table-lookup approach to speed up the thermal profile computation. The thermal profile is optimized based on coarse-grid and fine-grid thermal tables at different placement stages. Since the objective based on our approach is to generate the desired thermal profile, instead of to minimize the temperature differences between matched devices, the time complexity is only dependent on the number of power devices, but is independent of that of matched devices. Therefore, our approach is more efficient and scalable, which significantly improves the runtime when placing a large number of thermally-sensitive matched devices in modern analog designs. Finally, we propose the *first* thermal-driven common-centroid placement (TCCP) algorithm that considers the best device matching under the desired thermal profile. Experimental results show that our approach can achieve better runtime and the best analog circuit performance/accuracy in the presence of thermal gradients, compared with the previous works.

The remainder of this paper is organized as follows. Section 2 introduces the desired thermal profile for thermal device matching and the corresponding placement configuration. Section 3 presents our thermal-driven analog placement to generate the desired thermal profile based on the placement configuration while considering both symmetry and common-centroid constraints. Section 4 reports the experimental results, and finally Section 5 concludes this paper.

2. THE DESIRED THERMAL PROFILE

Before introducing our thermal-driven analog placement, we shall first consider the desired thermal profiles and the corresponding placement configuration. Inspired by manual layouts, the desired thermal profile should have the following properties:

- Lower temperature at thermal hot spots.
- Smoother thermal gradients at the non-power device areas.
- More separation between power and thermally-sensitive devices.
- More regular isothermal contours in either the horizontal or the vertical direction such that the matched devices can easily be placed along the contours.
- Larger accommodation areas for multiple thermally-sensitive device groups

By comparing both thermal profiles at the non-power device areas in Figure 2, although the one that the power devices are evenly distributed at four sides of the chip has lower temperature at the thermal hot spots, its thermal gradient, isothermal contours, and accommodation area are not as good as the other one that the power devices are evenly distributed at two opposite sides of the chip. Therefore, the thermal profile in Figure 2(b) is more desirable than that in Figure 2(a) when considering thermal matching in analog layouts. Since the isothermal contours in Figure 2(b) are very regular in the vertical direction, the thermally-sensitive matched devices can be placed along the isothermal contours anywhere in the placement area to have the same ambient temperature so that thermally induced mismatches between the matched devices are reduced. In addition, when placing different symmetry groups in different sub-circuits, they are not necessary to share the common symmetry line bisecting the power devices. Consequently, the area utilization and the interconnecting wire length of the whole analog layout can further be optimized.

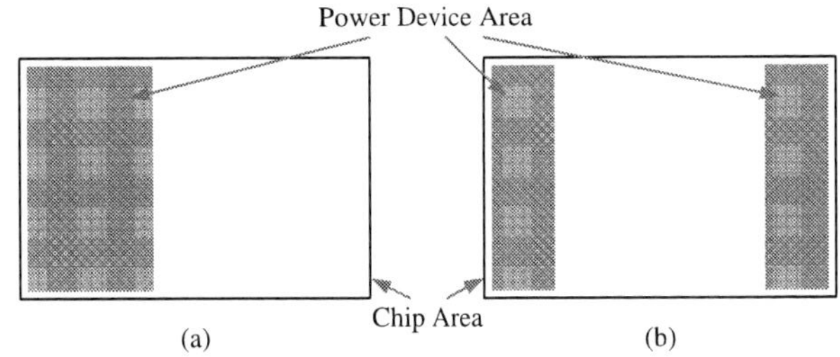

Figure 3: Placement configurations of power device area arrangements. (a) The power device area is arranged at one short side of the chip. (b) The power device areas are arranged at both short sides of the chip.

Based on the desired thermal profile in Figure 2(b), the corresponding placement configuration, especially the arrangement of power device areas, should be considered. According to [5], it is always recommended to place non-power, thermally-sensitive devices as far away from power devices as possible to alleviate thermal impacts from power devices. To allow more separation between power and thermally-sensitive devices on the same chip, the power devices are preferred to be arranged in the rectangular areas located at either one or both short sides of the rectangular chip as shown in Figure 3. The rest of the chip area is reserved for the placement of non-power devices, including the thermally-sensitive matched devices with either the symmetry or the common-centroid constraint. As both arrangements in Figure 3 are preferable, choosing the better arrangement further depends on other factors, such as the reduction of interconnections among devices and/or I/O pins, and the alleviation of thermal hot spots.

978-1-60558-497-3/09 $25.00 © 2009 ACM

594

3. THERMAL-DRIVEN ANALOG PLACEMENT

We propose our thermal-driven analog placement to fulfill the desired thermal profile and the placement configuration introduced in the previous section by applying the simulated annealing algorithm [6] based on the hierarchical B*-tree (HB*-tree) and automatically symmetric-feasible B*-tree (ASF-B*-tree) floorplan representations [8] due to its efficiency and effectiveness to handle symmetry constraints based on the symmetry-island formulation. Figure 4 shows a symmetric placement and its corresponding HB*-tree and ASF-B*-tree. Each module node, n_i, corresponds to a module b_i, and the hierarchy node n_{S0} corresponds to the symmetry island of the symmetry group S_0 containing a self-symmetric module, b_3^s, and a symmetry pair, (b_4, b_4'). The ASF-B*-tree represents the symmetric placement of S_0.

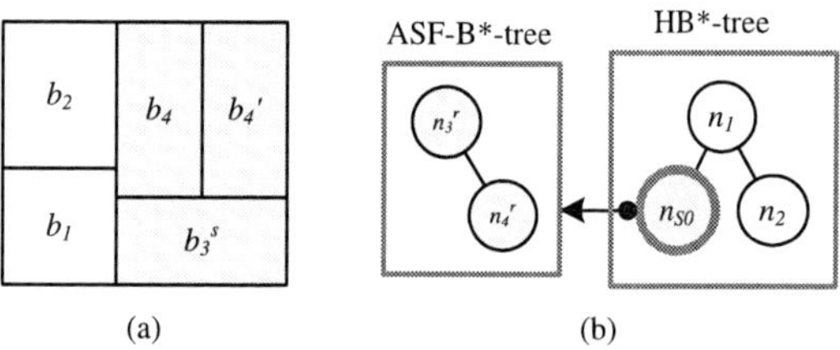

Figure 4: (a) A symmetric placement containing a symmetry group $S_0 = \{b_3^s, (b_4, b_4')\}$, and two non-symmetric modules, b_1 and b_2. (b) The corresponding HB*-tree and ASF-B*-tree of the placement in (a).

In addition to handling symmetry constraints, the HB*-trees can also be hierarchically constructed based on the hierarchical circuit clustering [9] so that the close proximity of devices in the same sub-circuit is preserved during placement. For example, we can use different HB*-trees to model the device placements in different sub-circuits such as those in Figure 1. Each HB*-tree modelling the placements of a sub-circuit is further linked by a hierarchy node in the top-level HB*-tree which models the top-level placement considering the topology among different sub-circuits.

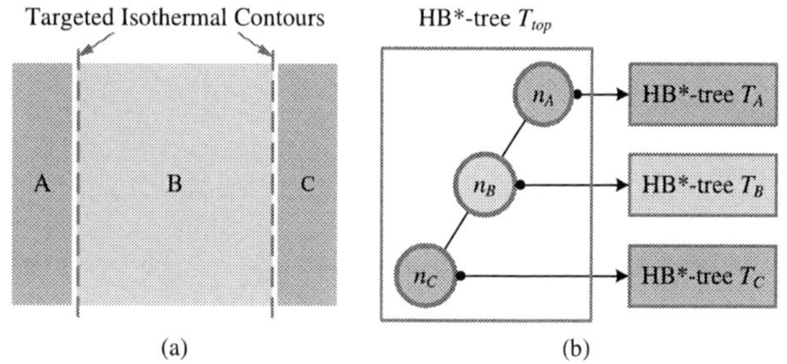

Figure 5: The placement configuration and its corresponding HB*-trees. (a) The placement configuration based on the power area arrangement in Figure 3. (b) The HB*-trees representing the topology among the three regions in (a).

We further extend the HB*-trees to handle the problem of thermal-driven analog placement. Figure 5(a) shows three regions A, B, and C in the whole placement area based on the desired placement configuration in Figure 3(b). The regions A and C are arranged to place power devices, while the region B is arranged to place non-power devices, including all thermally-sensitive matched devices. To represent the placement configuration in Figure 5(a), we consider the fixed structure of the top-level HB*-tree, T_{top}, as shown in Figure 5(b). The placements of power devices in the regions A and C are modelled by the HB*-trees T_A and T_C which are linked by the hierarchy nodes n_A and n_C respectively, while the placement of non-power devices in the region B is modelled by the HB*-tree T_B which is linked by the hierarchy node n_B. During the simulated

annealing, a node can be moved from T_A to T_C, or vice versa, to optimize the interconnection wire length. If one of T_A and T_C becomes null, the placement configuration will be automatically reduced to that in Figure 3(a).

Given the following inputs and constraints:

- a set of device modules including power and non-power devices,
- power densities of all power devices,
- the targeted aspect ratio of the placement area,
- symmetry and common-centroid constraints for all matching device groups,

the objective of our thermal-driven analog placement is to obtain a placement P that minimizes the cost function, $\Phi(P)$, defined in Equation (1). In this equation, α, β, γ, and δ are user-specified parameters, A_P is the area of the bounding rectangle for the placement, W_P is the half-perimeter wire length (HPWL), R_P is the difference between the aspect ratio of P and the targeted aspect ratio, and T_P is the thermal cost of P based on the targeted placement configuration, which is further defined in Equation (2).

$$\Phi(P) = \alpha A_P + \beta W_P + \gamma R_P + \delta T_P. \tag{1}$$

$$T_P = (T_{l,max} - T_{l,min}) + (T_{r,max} - T_{r,min}). \tag{2}$$

Based on the desired thermal profile, we consider two targeted straight isothermal contours near the boundaries between the power and non-power device regions as seen in Figure 5(a). In Equation (2), $T_{l,max}$ and $T_{l,min}$ denote the maximum and minimum temperatures at the left targeted isothermal contour in Figure 5(a), while $T_{r,max}$ and $T_{r,min}$ denote the maximum and minimum temperatures at the right targeted isothermal contour. Since the power consumption between power and non-power devices is large in the typical RF system as seen in Figure 1, the heat generated by the non-power devices can hardly affect the thermal profile contributed by the power devices. Therefore, by minimizing the differences between the maximum and minimum temperatures at the same targeted isothermal contour, the desired thermal profiles in Figure 2(b) can be obtained.

3.1 Thermal Profile Computation

To obtain the temperature at each point on the targeted isothermal contours in Figure 5(a), it is required to compute the thermal profile based on a certain thermal model. The previous works [7, 11] compute the thermal profile by calculating approximated thermal equations based on different thermal models. Although it is fast to compute the thermal profile of a certain placement, it becomes inefficient when calculating those equations more than hundreds of thousands times to evaluate the thermal profiles of different placements during the simulated annealing process.

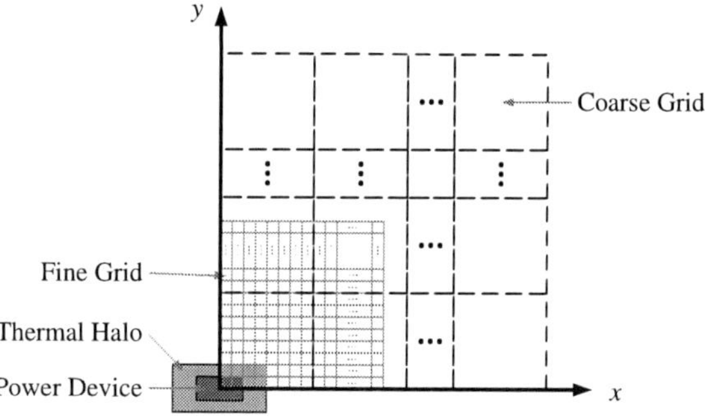

Figure 6: The coarse-grid and fine-grid thermal tables indicating the thermal profile of the power device with different precisions and scales.

Since the temperature at each point in the placement area can be calculated by the superposition of the thermal profiles contributed by all power devices placed at different locations according to [7, 11, 19], we adopt a table-lookup approach by constructing thermal tables that store the thermal profile of each power device to facilitate the thermal profile computation. The thermal profiles of all power devices are pre-simulated using a thermal simulation tool, such as 3D-Thermal-ADI [19, 20] which is available in the public domain. Given the device area, device location, device power densities, targeted chip area, and other thermal coefficients, it will compute the thermal profile of the corresponding device. After the thermal simulation of all power devices, a *coarse-grid* and a *fine-grid* thermal tables are then constructed for each device to represent its thermal profile with different precisions at different scales as shown in Figure 6. Each grid (i, j) in the thermal table records a certain temperature $T(i, j)$ contributed by the corresponding power device. The coarse-grid thermal table indicates a global thermal profile covering the whole placement area, while the fine-grid thermal table shows the detail thermal profile near the placement of the corresponding power device. The sizes of the thermal tables depend on the trade-off between the memory usage and the precision we need for the thermal profile optimization. By assuming the isomorphic thermal profile of each device in four quadrants and sharing common thermal tables of some identical power devices, the size and the number of the thermal tables can effectively be reduced.

3.2 Thermal Profile Optimization

Based on the thermal tables illustrated in Figure 6, we optimize the thermal profile at three different placement stages. Before the placement process, the thermal halo of each power device is allocated. The global thermal profile optimization is then performed during the simultaneous placement of power and non-power devices. Finally, the detailed thermal profile optimization is processed for local placement refinements of power devices.

3.2.1 Thermal Halo Allocation

Since most of the power devices are arranged in the same power device area, the area is prone to have thermal hot spots. To effectively reduce the temperature at the thermal hot spots, a thermal halo should be added to each power device as shown in Figure 6. The thermal halo covers an area above a certain temperature in the fine-grid thermal table of the power device.

3.2.2 Global Thermal Profile Optimization

During the simulated annealing based on the HB*-trees, the placement of power and non-power devices are simultaneously optimized by minimizing the cost functions in Equations (1) and (2). At this stage, we only consider the coarse-grid thermal tables for the thermal cost in Equation (2) which is obtained by calculating the difference between the maximum and minimum temperatures at the coarse grids passed by the targeted isothermal contour as shown in Figure 7(a).

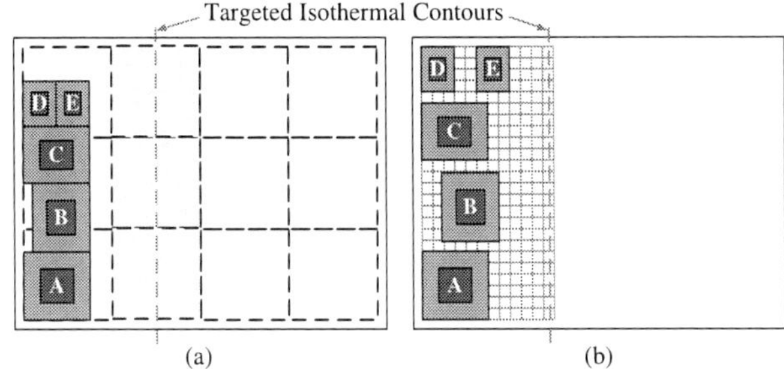

Figure 7: The placement of power devices are optimized based on (a) global and (b) detailed thermal profile optimization.

3.2.3 Detailed Thermal Profile Optimization

Once the placement of the power devices are optimized based on the global thermal profile optimization as seen in Figure 7(a), the detailed thermal profile optimization is further performed to obtain more desirable isothermal contours. We apply both vertical and horizontal local movements for the power devices on the fine grids as shown in Figure 7(b). The thermal cost in Equation (2) is calculated based on the fine-grid thermal tables to minimize the temperature difference among the fine grids passed by the targeted isothermal contour in Figure 7(b).

Since the vertical movement of the power devices does not affect the shape of the isothermal contour very much, we simply evenly distribute the power devices vertically by traversing the vertical constraint graph (VCG) representing the vertical relationship among the power devices, which can be converted from a B*-tree as described in [17].

To minimize the number of power devices that need to be moved during the horizontal local refinement, only those adjacent to the right boundary of the power device area should be considered, which are devices A, B, C, and E in Figure 7(a). These boundary devices can be identified by the contour data structure during packing a B*-tree [3]. By the iterative horizontal movement of the boundary devices, the thermal cost T_P in Equation (2) can further be minimized.

3.3 Thermal-driven Matching Device Placement

We consider the desired thermal profile in Figure 2(b) to place the thermally-sensitive matched devices with either the symmetry or the common-centroid constraint. For a matching device group with the symmetry constraint, the matched devices should be placed on the same isothermal contours to have the same ambient temperature so that the thermally-induced mismatch is minimized. Since the desired thermal profile has regular isothermal contours in either the horizontal or the vertical direction, all the symmetry device groups can simply be placed with their symmetry lines being perpendicular to the isothermal contours. The symmetric placements of all symmetry device groups with different symmetry lines can simultaneously be optimized during the simulated annealing based on the HB*-trees.

For a matching device group with the common-centroid constraint as shown in Figure 8, none of the previous works considers the thermal profile during the common-centroid placement. We propose our algorithm to generate a common-centroid placement for a matching device group while considering the desired thermal profile. Based on our approach, all possible common-centroid placements of each matching group with different aspect ratios are pre-generated, which is the same as the approach in [14]. When integrating the placement with other devices or device groups, a candidate of the pre-generated common-centroid placements is randomly selected during the simulated annealing based on the HB*-trees. The final candidates of all matching groups are simultaneously optimized based on the cost function in Equation (1).

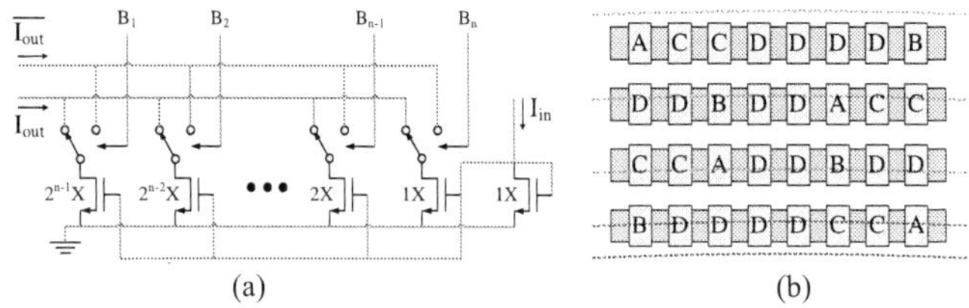

Figure 8: A matching device group with the common-centroid constraint. (a) A matching device group in the binary weighted current network. (b) A common-centroid placement of the matching device group in (a) containing four MOS devices, A, B, C, and D, having 4, 4, 8, and 16 sub-devices respectively. The dotted lines denote the isothermal contours.

Given a common-centroid device group G_{cc} containing q devices, i.e. $G_{cc} = \{b_1, b_2, \ldots, b_q\}$, and each device b_j has n_{b_j} sub-devices, to better match the device layouts, the size of all the sub-devices should be identical. Besides, in G_{cc}, the relationship between any two sub-device numbers n_{b_j} and n_{b_k} of devices b_j

and b_k is usually the ratio of power of two, i.e. $n_{b_j} = 2^l \times n_{b_k}$, where l is an integer. Furthermore, the sub-devices are preferred to be regularly placed in a two-dimensional array as shown in Figure 8(b). To minimize the thermal mismatch among the q devices in G_{cc}, we need to evenly distribute the sub-devices of each device along the direction of the thermal gradient.

Figure 8(b) shows the common-centroid placement of the 3-bit binary weighted current network illustrated in Figure 8(a). Device A denotes the MOS transistor connected to I_{in}, and devices B, C, D denote the MOS transistors connected to the control signals of B_3, B_2, and B_1, which have 4, 4, 8, and 16 sub-devices respectively. The isothermal contours, i.e. the dotted lines, are in the horizontal direction, implying that the direction of the thermal gradient is vertical. In the following, we simply consider the direction of the row (column) of the 2D array to be the same as that of the thermal gradient (contours).

To assign the sub-devices into a k-row 2D array while minimizing the thermal mismatch, the sub-devices of each device should be equally divided by k and assigned into one of the rows. For some cases, if the sub-device number of a device is not dividable with respect to the row number, we allow the sub-device number in each row with ± 1 tolerance. It should be noted that even the sub-device numbers of a device assigned to different rows are not equal, the sub-device number in the ith row should be the same as that in the $(k - i + 1)^{th}$ row, or the symmetric row. Therefore, the common-centroid placement must be feasible. In Figure 8(b), the respective sub-device numbers of devices A, B, C, and D in each row are 1, 1, 2, 4 after the even assignment.

Once the sub-devices of each device are evenly assigned into rows, we should consider the diffusion-sharing for MOS transistors. We construct the diffusion graph of the sub-circuit in each row, and then find the Eulerian trail on the diffusion graph [15]. After the Eulerian trails are found, the sub-devices on the same Eulerian trail are merged. Considering the first row in Figure 8(b), there are three Eulerian trails, $C - C$, $D - D$, and $D - D$. Consequently, the six devices are merged into three sub-device groups, and the sub-devices in the symmetric row are also merged accordingly.

After considering the diffusion sharing, the column position of each sub-device or sub-device group in each row is assigned in a random order while keeping the symmetric row in the reverse order. Figure 8(b) shows the final common-centroid placement that minimizes the mismatch between the four devices due to the thermal gradient. The algorithm of the k-row thermal-driven common-centroid placement (namely, k-row TCCP) is summarized in Algorithm 1.

Algorithm 1 k-row TCCP

1: **for all** q devices in G_{cc} **do**
2:     Evenly assign the sub-devices into k rows with the same number in $row[i]$ and $row[k - i + 1]$;
3: **end for**
4: **for** $i = 1$ to $\lceil k/2 \rceil$ **do**
5:     Merge the diffusion of the sub-devices in $row[i]$ and $row[k - i + 1]$ based on the same Eulerian trails;
6:     Place the (merged) sub-devices in $row[i]$ in a random order;
7:     Place the (merged) sub-devices in $row[k - i + 1]$ in the reverse order of that in $row[i]$;
8: **end for**

---

4. EXPERIMENTAL RESULTS

We implemented our placement algorithm in the C++ programming language on a Dual 2.8GHz Intel Pentium4 PC under the Linux operation system. We performed two sets of experiments: (1) one is based on the analog placement benchmarks in [2, 8, 17] consisting of analog designs, biasynth_2p4g and lnamixbias_2p4g, with different numbers of symmetry groups, and (2) the other is based on the real analog circuit, the binary weighted current network shown in Figure 8(a), containing a large common-centroid device group in which each device has different numbers of sub-devices of uniform sizes.

In the first set of the experiments, we compared our approach that optimizes the desired thermal profile with the other one that minimizes the temperature differences between devices of each symmetry pair. Both approaches applied the simulated annealing algorithm based on the HB*-trees and the cost function in Equation (1), while the later one applied a different thermal cost function T_P of placement P defined in Equation (3), which is the summation of the temperature difference of m symmetry pairs. In Equation (3), T_{b_i} denotes the temperature of the device b_i, and $T_{b_{i,sym}}$ corresponds the temperature of the symmetric device of b_i.

$$T_P = \sum_{i=1}^{m} |T_{b_i} - T_{b_{i,sym}}|. \tag{3}$$

Table 1 lists the names of the benchmark circuits ("Circuit"), the numbers of modules ("# of Mod."), the numbers of symmetry modules ("# of Sym. Mod."), the numbers of power device modules ("# of Power Mod."), the total module areas ("Mod. Area"), and the maximum temperature of the whole chip ("T_{Max}"), the maximum temperature difference of each symmetry pair ("Max ΔT_{sym}"), the total areas ("Area") and the runtimes ("Time") for both approaches, the temperature difference optimization ("Temperature Diff. Opt.") and the thermal profile optimization ("Thermal Profile Opt.").

Since there was no power device specified in the original benchmarks, we simply selected 11 devices in biasynth_2p4g and 10 devices in lnamixbias_2p4g as the power devices, and simulated the thermal profile of each device to obtain its thermal table. Compared with the approach based on the temperature difference optimization, our proposed approach results in less than one quarter temperature difference of the matched devices in a symmetry pair and 5.28X faster running time with comparable maximum chip temperature and total chip area. Figure 9 shows the resulting placement of lnamixbias_2p4g and its corresponding thermal profile. The devices in red color are the power devices, and those in other colors denote the symmetric devices.

In the second set of the experiments, we evaluated the thermally-induced mismatch within a common-centroid placement under the desired thermal profile by performing HSPICE simulation with pre-assigned temperature for each sub-device. The temperature of each sub-device in the common-centroid group can be extracted according to its location in the thermal profile once the whole common-centroid device group is placed at a certain position in region B in Figure 5(a). After performing HSPICE simulation, the temperature-dependent electrical parameters of each device were measured. Since our experiment is based on the binary weighted current network in Figure 8(a), which is commonly used in data converter systems, we measured the drain current I_D of each MOS transistor. For an n-bit data converter system, the I_D linearity may change maximally by $\pm \frac{1}{4}$LSB over the full temperature range to maintain monotonicity of the system, where LSB stands for the least significant bit in data converters [16]. If the difference between the ideal I_D and the real one, I_D', is larger than $\pm \frac{1}{4}$LSB, the D/A converter will fail. If the thermally-induced mismatch value, σ, of a common-centroid placement shown in Equation (4) is less than one, the circuit is within the tolerance of the accuracy; otherwise, the circuit will fail.

$$\sigma = \frac{|I_D' - I_D|}{\frac{1}{4}LSB}. \tag{4}$$

Therefore, we compared the σ value of the resulting common-centroid placement based on our k-row TCCP algorithm with that based on the grid-based approach in [14] under the same ambient temperature or the same location in the thermal profile.

Table 2 lists the names of the benchmark circuits ("Circuit"), the numbers of devices ("# of Dev."), the number of sub-devices in each device ("# of Sub-devices"), and the σ values based on the grid-based approach [14] and our k-row TCCP algorithm. Methods 1–3 give three different assignments of the positions of the sub-devices based on the grid-based approach [14]. For Method 1 (Method 2), the devices containing the most (least) sub-devices were assigned first, so they were placed close to the centroid. For Method 3, the sub-devices were assigned in a random order. The results show

Table 1: Comparisons of the maximum temperature difference of each symmetry pair, area utilization, and CPU times for the approaches based on the temperature difference optimization and our thermal profile optimization.

Circuit	# of Mod.	# of Sym. Mod.	# of Power Mod.	Mod. Area $(10^3 \mu m^2)$	Temperature Diff. Opt.				Thermal Profile Opt. (This Work)			
					T_{Max} (K)	Max ΔT_{sym} (K)	Area $(10^3 \mu m^2)$	Time (s)	T_{Max} (K)	Max ΔT_{sym} (K)	Area $(10^3 \mu m^2)$	Time (s)
biasynth_2p4g	65	8+12+5	11	4.70	318.18	0.91	5.42	1524	321.10	0.22	5.47	385
lnamixbias_2p4g	110	16+6+6+12+4	10	46.00	322.58	2.66	56.71	5340	322.09	0.62	55.87	809
Comparison					1.00	4.21	1.00	5.28	1.00	1.00	1.00	1.00

Table 2: Comparisons of the circuit accuracy due to thermally-induced mismatches for common-centroid placements based on the grid-based approach and our k-row TCCP algorithm. The numbers in bold font mean that the circuits are within the tolerance of the accuracy.

Circuit	# of Dev.	# of Sub-devices	σ value			
			The grid-based approach [14]			TCCP
			Method 1	Method 2	Method 3	
bwcn_4bit	5	{4, 4, 8, 16, 32}	**0.06784**	**0.07040**	**0.04480**	**0.00128**
bwcn_5bit	6	{4, 4, 8, 16, 32, 64}	**0.26624**	**0.28416**	**0.00512**	**0.00512**
bwcn_6bit	7	{4, 4, 8, 16, 32, 64, 128}	1.06496	1.13664	**0.73728**	**0.04096**
bwcn_7bit	8	{4, 4, 8, 16, 32, 64, 128, 256}	4.24960	4.50560	**0.06144**	**0.20480**
bwcn_8bit	9	{4, 4, 8, 16, 32, 64, 128, 256, 512}	16.9984	18.14528	11.81696	**0.43008**

that our k-row TCCP algorithm obtained accurate results for all the bwcn circuits. The results based on "Method 3" are not as good as ours when the device/sub-device number in the common centroid group becomes larger. Both Methods 1 and 2 have very poor performance against the impact from the thermal gradient since only small circuits behave accurately. Therefore, our approach is the most effective one that considers the thermal gradient. The runtime of each approach is less than one second on a Dual 2.8GHz P4 PC.

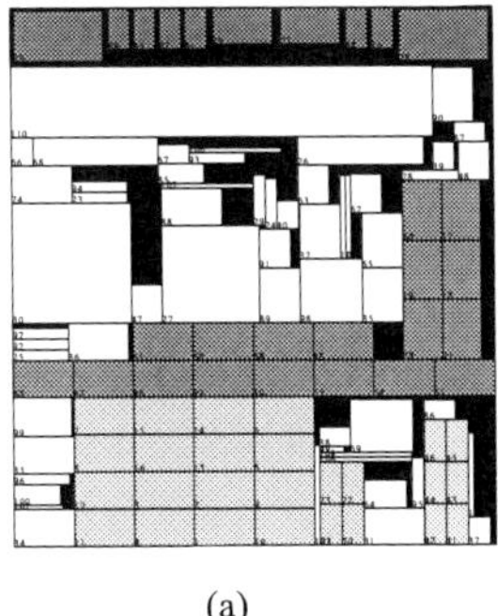

(a)

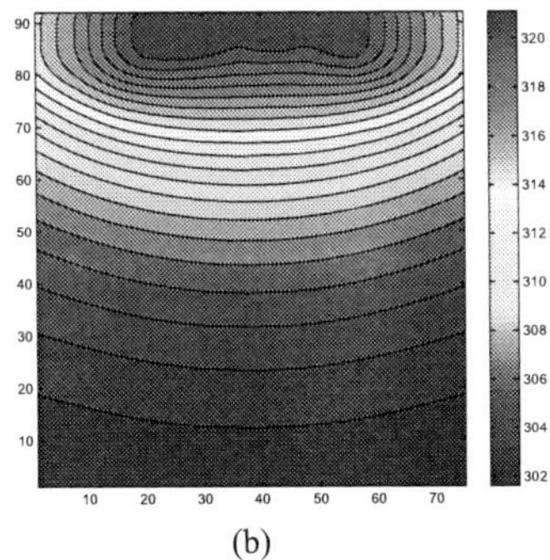

(b)

Figure 9: (a) The resulting placement of lnamixbias_2p4g. (b) The corresponding thermal profile.

5. CONCLUSIONS

In this paper, we have addressed the thermal issue in analog placement and studied the thermal-driven analog placement problem. We have proposed our algorithms to simultaneously optimize the placements of power and non-power devices to generate a desired thermal profile for thermally-sensitive matched devices. We have also proposed our analog placement methodology that considers the best device matching under the thermal profile while satisfying the symmetry and the common-centroid constraints. Experimental results based on the analog benchmark circuits and the real analog circuit show that our approach can achieve the best analog circuit performance/accuracy with the least impact due to the thermal gradient.

6. ACKNOWLEDGMENTS

We would like to thank Prof. Yun Chiu, Mr. Dae Hyun Kwon, and Mr. Wenbo Liu of University of Illinois at Urbana-Champaign for many very helpful discussions on analog layout design. This work was partially supported by ITRI, Springsoft, Synopsys, TSMC, National Science Council of Taiwan under Grant No's. NSC 96-2917-I-002-121, NSC 97-2221-E-002-237-MY3, NSC 96-2628-E-002-249-MY3, and NSC 96-2628-E-002-248-MY3, and National Science Foundation of the US under Grant CCF-0701821.

7. REFERENCES

[1] F. Balasa and K. Lampaert, "Symmetry within the sequence-pair representation in the context of placement for analog design" *IEEE TCAD*, vol. 19, no. 3, pp. 721–731, July 2000.

[2] F. Balasa, S. C. Maruvada, and K. Krishnamoorthy, "On the exploration of the solution space in analog placement with symmetry constraints," *IEEE TCAD*, vol. 23, no. 2, pp. 177–191, Feb. 2004.

[3] Y.-C. Chang, Y.-W. Chang, G.-M. Wu, and S.-W. Wu, "B*-trees: a new representation for non-slicing floorplans," *Proc. DAC*, pp. 458–463, June 2000.

[4] J. M. Cohn, D. J. Garrod, R. A. Rutenbar, and L. R. Charley, *Analog Device-level Layout Automation*, Kluwer Academic Publishers, 1994.

[5] A. Hastings, *The Art of Analog Layout*, 2nd Ed., Prentice Hall, 2006.

[6] S. Kirkpatrick, C. D. Gelatt, and M. P. Vecchi, "Optimization by simulated annealing," *Science*, vol. 220, no. 4598, pp. 671–680, May 1983.

[7] K. Lampaert, G. Gielen, and W. Sansen, *Analog Layout Generation for Performance and Manufacturability*, Kluwer Academic Publishers, 1999.

[8] P.-H. Lin and S.-C. Lin, "Analog placement based on novel symmetry-island formulation," *Proc. DAC*, pp. 465–470, June 2007.

[9] P.-H. Lin and S.-C. Lin, "Analog placement based on hierarchical module clustering," *Proc. DAC*, June 2008.

[10] J.-M. Lin, G.-M. Wu, Y.-W. Chang, and J.-H. Chuang, "Placement with symmetry constraints for analog layout design using TCG-S," *Proc. ASP-DAC*, pp. 1135–1138, Jan. 2005.

[11] J. Liu, S. Dong, Y. Ma, D. Long, and X. Hong, "Thermal-driven symmetry constraint for analog layout with CBL representation," *Proc. ASP-DAC*, pp. 191–196, Jan. 2007.

[12] J. Liu, S. Dong, X. Hong, Y. Wang, O. He, and S. Goto, "Symmetry constraint based on mismatch analysis for analog layout in SOI technology," *Proc. ASP-DAC*, pp. 772–775, Jan. 2008.

[13] R. Ludwig and G. Bogdanov, *RF Circuit Design: Theory and Applications*, 2nd Ed., Prentice Hall, 2009.

[14] Q. Ma and F.-Y. Young, "Analog placement with common centroid constraints," *Proc. ICCAD*, pp. 579–585, Nov. 2007.

[15] R. Naiknaware and T. S. Fiez, "Automated hierarchical CMOS analog circuit stack generation with intramodule connectivity and matching considerations," *IEEE JSSC*, vol. 34, no. 3, pp. 304–317, March 1999.

[16] R. Plassche, *CMOS Integrated Analog-to-Digital and Digital-to-Analog Converters*, 2nd Ed., Kluwer Academic Publishers, 2003.

[17] M. Strasser, M. Eick, H. Graeb, U. Schlichtmann, and F. M. Johannes, " Deterministic analog circuit placement using hierarchically bounded enumeration and enhanced shape functions," *Prof. ICCAD*, Nov. 2008.

[18] Y.-C. Tam, E. F.-Y. Young and C. Chu, "Analog placement with symmetry and other placement constraints," *Proc. ICCAD*, pp. 349–354, Nov. 2006.

[19] T.-Y. Wang and C. C.-P. Chen, "3D thermal-ADI: a linear time chip-level transient thermal simulator," [Online tool] 2003, http://cc.ee.ntu.edu.tw/~cchen/3D_Thermal_ADI.htm

[20] T.-Y. Wang, Y.-M. Lee, and C. C.-P. Chen, "3D thermal-ADI: an efficient chip-level transient thermal simulator," *Proc. ISPD*, pp. 10–17, Apr. 2003.

[21] N. H.E. Weste and D. Harris, *CMOS VLSI design: A Circuits and System Perspective*, 3rd Ed., Addison Wesley, 2006.

[22] L. Zhang, C.-J. R. Shi, and Y. Jiang, "Symmetry-aware placement with transitive closure graphs for analog layout design," *Proc. ASP-DAC*, pp. 180–185, Jan. 2008.

978-1-60558-497-3/09 $25.00 © 2009 ACM

Yield-driven Iterative Robust Circuit Optimization Algorithm

Yan Li
Department of EECS,
Massachusetts Institute of Technology,
77 Mass Ave., Cambridge, MA 02139
liyan@mit.edu

Vladimir Stojanović
Department of EECS,
Massachusetts Institute of Technology,
77 Mass Ave., Cambridge, MA 02139
vlada@mit.edu

ABSTRACT

This paper proposes an equation-based multi-scenario iterative robust optimization methodology for analog/mixed-signal circuits. We show that due to local circuit performance monotonicity in random variations constraint maximization can be used to efficiently find critical constraints and worst-case scenarios of random process variations and populate them into a multi-scenario optimization. This algorithm scales gracefully with circuit size and is tested on both two-stage and fully differential folded-cascode operational amplifiers with a 90 nm predictive model. The improving yield-trends are confirmed across process and random variations with Hspice Monte-Carlo simulations.

Categories and Subject Descriptors

B.7.2 [**Integrated Circuits**]: Design Aids

General Terms

Algorithms

Keywords

Robust Circuit Optimization, Variability, Yield, Analog Circuits

1. INTRODUCTION

Traditional robust circuit design routine involves design iterations over process, voltage, and temperature (PVT) corners coupled with Monte-Carlo simulations for yield estimation. It is up to designers to interpret these estimation results to find design sensitivities, which is often complicated by the large dimensionality of design space, and results in long design time, over-design, and wasted power and area.

Equation-based circuit optimization enables easier global design-space exploration in a large-dimensional space and could potentially be used in design of robust circuits. Geometric programming (GP) has been used extensively in circuit design examples, from operational-amplifiers [1], [2] to phase-locked loops [3]. A straightforward approach to extend this methodology to robust design would be reformulating the optimization problem into a robust

Permission to make digital or hard copies of part or all of this work for personal or classroom use is granted without fee provided that copies are not made or distributed for profit or commercial advantage and that copies bear this notice and the full citation on the first page. To copy otherwise, to republish, to post on servers or to redistribute to lists, requires prior specific permission and/or a fee.
DAC'09, July 26-31, 2009, San Francisco, California, USA

optimization problem. However, this robust optimization problem usually loses the tractability. To make the robust GP more tractable, [4] and [5] consider a special case of ellipsoidal uncertainty and use piecewise-linear convex approximation techniques. These techniques, however, do not scale well due to the number of piecewise-linear approximation terms, and are limited to circuits that are strictly GP compliant.

To circumvent the intractability in the robust optimization problem, and make use of the circuit optimization framework, in this paper, we propose an iterative yield-driven robust optimization algorithm. It inherits the iterative design fashion of a traditional robust design routine, but also takes advantage of the optimization engine's global design space exploration power. The methodology gracefully extends the multi-scenario optimization used for design over PVT corners, by adding only the most critical random variation corners discovered through an efficient search step (worst-case analysis).

Other similar 'worst-case analysis' ideas have appeared in [6], [7] and [8]. To compute the 'worst-case distance', [6] uses local sensitivity analysis assuming performance linearization. In [7], the worst-case analysis is based on response surface modeling (RSM), which trades-off accuracy for size of design space and can become computationally expensive for large circuits. In [8], a robust taper is designed in a similar iterative fashion, but uses entirely different search technique due to a different problem structure. Without the dependence on RSM or local sensitivities, our algorithm is able to search globally for a robust design. Another advantage is that the size of the multi-scenario robustifying optimization problem scales linearly with the number of performance constraint, which in most cases, is much smaller than the number of variation variables. This makes our algorithm readily scalable to large circuits.

The remainder of the paper is organized as follows. Section 2 elaborates the details of the algorithm. Section 3 presents a two-stage and a folded-cascode operational amplifiers (op-amp) examples to show monotonic yield convergence and applicability of the algorithm for fast analog/mixed-signal robust circuit optimization.

2. ALGORITHM

2.1 Sources of Variability

In this paper, we focus on with-in die mismatch induced by random dopant fluctuations and line-edge roughness. Pelgrom's model [9] and derivative work, to the first order show that the impact of process variations on deviations of threshold voltage V_T and the relative current factor β are inversely dependent on the area of the transistor, as

$$\sigma_{V_T} = \frac{A_{V_T}}{\sqrt{W \cdot L}}, \text{ and } \frac{\sigma_\beta}{\beta} = \frac{A_\beta}{\sqrt{W \cdot L}}, \tag{1}$$

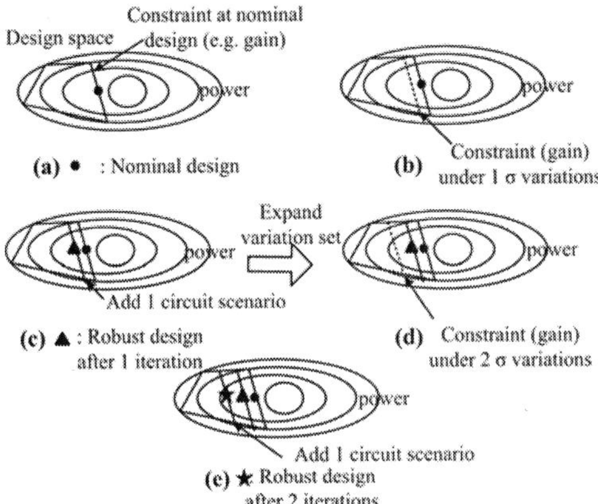

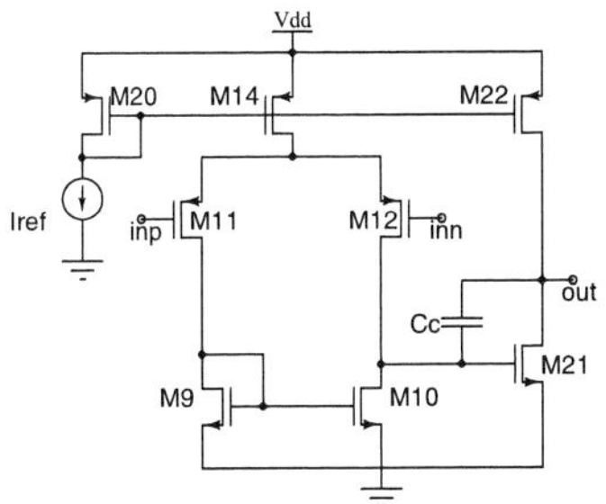

Figure 3: The two-stage Op-amp schematic.

Table 1: Specifications for nominal design

open loop gain	300
unity-gain bandwidth (ω_u)	160MHz
phase margin	60°
CMRR	100
slew rate	100MV/s
input-referred spot noise @ 1MHz (vnoise)	$64\mathrm{nV}/\sqrt{\mathrm{Hz}}$

Figure 1: An intuitive illustration of the two iterations of the algorithm.

where W and L are the width and length of a transistor; A_{V_T} and A_β are mismatch parameters.

2.2 Proposed robust circuit optimization framework

The intuition behind this algorithm is illustrated in Fig. 1. In Fig. 1, (a)-(e) depict the impact of variations on feasible design space, the nominal design point, and how the algorithm pushes the nominal design point to be more robust in two iterations. Here, we assume a circuit design problem with underlying design variables as circuit sizing parameters(x-y plane, not shown here) and objective to minimize the power consumption, represented by different level contours. The feasible design space is constrained by a set of biasing and performance constraints simplified to a polygon in the figures. In (a), a nominal design point is marked with a dot and one of the constraints (e.g. gain) is active. Under variations, the gain constraint shifts with respect to the x-y plane (the dash line) and nominal design point now violates the design specification with some probability. Although the objective can also change under variations, it is not shown here for simplicity. In (c), the algorithm

generates a variation-aware circuit scenario (adds constraints) to exclude the design space that would fail the gain specification under variations and push the design to a more robust design point, the triangle marker. In (d) and (e), the algorithm iterates once again, expanding the variation set.

While this type of iterative algorithm has been considered previously in other applications, e.g. [8], here we outline the necessary steps to map the algorithm to a circuit optimization scenario, by solving the inherent circuit biasing and large variation issues. The following subsections elaborate the algorithm flow as shown in Fig. 2 and map the algorithm steps on a two-stage op-amp circuit example in Fig. 3, to illustrate the specific circuit-related refinements.

2.2.1 Initial design

The initial design is a nominal circuit optimization problem without consideration of variations. It corresponds to the Robustified circuit optimization block without any added circuit scenarios in Fig. 2. The formulation is as the following:

$$\begin{aligned}
\underset{\mathbf{x}}{\text{minimize}} \quad & f_0(\mathbf{x},\boldsymbol{\mu}^*) \\
\text{subject to} \quad & f_i(\mathbf{x},\boldsymbol{\mu}^*) \leq 1, i = 1,2,\ldots,n, \\
& g_j(\mathbf{x},\boldsymbol{\mu}^*) = 1, j = 1,2,\ldots,m, \\
& h_k(\mathbf{x},\boldsymbol{\mu}^*) \leq 1, k = 1,2,\ldots,l, \\
& \boldsymbol{\mu}^* = 0, \quad (2)
\end{aligned}$$

where $f_i(\mathbf{x},\boldsymbol{\mu}^*)$ and $g_j(\mathbf{x},\boldsymbol{\mu}^*)$ are biasing constraints and $h_k(\mathbf{x},\boldsymbol{\mu}^*)$ are performance constraints. $\boldsymbol{\mu}^* = 0$ means that no variation is considered here. In the two-stage op-amp example, we minimize a weighted sum of power and area, subject to the performance constraints illustrated in Table 1, and biasing constraints (KVL, KCL, transistor regions).

After the initial design, we make the optimization variation-aware in the following iterations by adding the critical variation direction information. To obtain the critical variation directions, the next subsection Constraint maximization performs a smart search and adds the information in the form of new circuit scenarios to the Robustified circuit optimization block.

2.2.2 Constraint maximization

Constraint maximization identifies the set of critical performance constraints that would violate the corresponding specifications under process variations. It involves solving the optimization problem defined in (3) for each **performance** constraint. This makes the al-

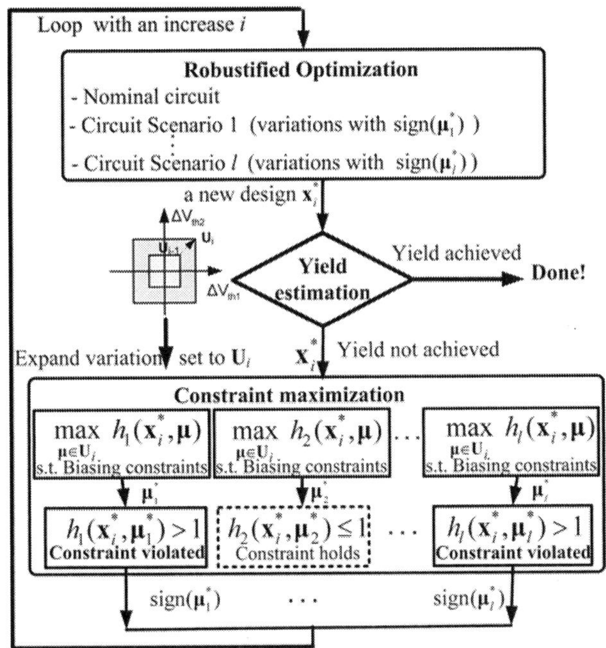

Figure 2: Flow of the iterative robust algorithm.

978-1-60558-497-3/09 $25.00 © 2009 ACM

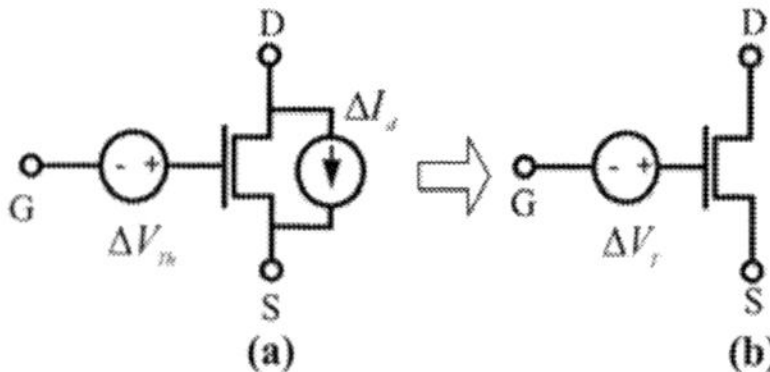

Figure 4: Transistor Macro model.

gorithm scale well, since the number of performance constraints does not necessarily grow as the size of the circuit grows.

The problem can be formulated as

$$\underset{\boldsymbol{\mu}}{\text{maximize}} \quad h_k(\mathbf{x}^*, \boldsymbol{\mu}), k = 1, 2, \ldots l,$$

$$\text{subject to} \quad \boldsymbol{\mu} \in \mathbb{U},$$

$$\text{Biasing constraints}, \qquad (3)$$

where $h_i(\mathbf{x}^*, \boldsymbol{\mu})$ is the performance constraint evaluated at design point $\mathbf{x}^*$ from previous circuit optimization step, with variation variable vector $\boldsymbol{\mu}$. The constraints are bounding and biasing constraints. No probability distribution is assumed. The range for the ith variation variable can be written to be proportional to its standard deviation as $\mu_i = [-k\sigma_i, k\sigma_i]$, where σ_i is the standard deviation of the ith variation and k is a constant. In each iteration, the variation set can be expanded by increasing k. In the two-stage op-amp example, for the bandwidth constraint maximization, the objective in (3) becomes $h_k(\mathbf{x}^*, \boldsymbol{\mu}) = \text{spec}.\omega_u/\omega_u(\mathbf{x}^*, \boldsymbol{\mu})$.

The purpose of expanding the variation set in each iteration is to take more variations into consideration and search for more critical variation directions that would fail the design. Therefore, as iterations go on, the robustified circuit optimization is aware of more critical directions and pushes the design to become more and more robust. By expanding the variation set iteratively, we achieve a design just meeting the yield target, avoiding over-design.

The key circuit-specific refinement in this step is to realize that during the maximization, circuit biasing constraints have to be satisfied even in the presence of random variations for the circuit to be realizable. Therefore, all circuit descriptions are defined with the macro transistor model in Fig. 4. Fig. 4(a) reflects the effect of threshold and current factor variations through the voltage source ΔV_{Th} and the current source ΔI_d, defined in (1). We can further simplify the model to Fig. 4(b), which has an aggregate standard deviation accounting for both threshold and current factor variations, as shown in (4),

$$\sigma_{V_T}^2 = \frac{A_\beta^2}{W \cdot L} \left(\frac{I_d}{gm}\right)^2 + \frac{A_{V_{Th}}^2}{W \cdot L}. \qquad (4)$$

Although (3) seems to be a general optimization problem, a close examination on the dependence of circuit performances on process variations reveals that circuit performances can be assumed to be locally monotonic on process variations (but not circuit design parameters). This assumption holds reasonably well according to the work in RSM, since linear regression model is a traditional way to fit the circuit performance on variation variables [10] [11] and linear model is a monotonic function on fitting variables, the process variations. Furthermore, the variation set is always set to be small initially, i.e. $[-0.2\sigma_{V_T}, 0.2\sigma_{V_T}]$, and often does not need to be expanded to as large as $[-3\sigma_{V_T}, 3\sigma_{V_T}]$. In the two examples we show in the next section, k only increases to around 1 with yield already approaching 100%. Under this small variation range, the accuracy of the linear model is satisfactory, hence our assumption on monotonicity is also reasonable.

To validate the assumption on this op-amp example, we let $k = [1, 2]$, with each transistor's ΔV_T ranging from $-k\sigma_{V_T}$ to $k\sigma_{V_T}$. For

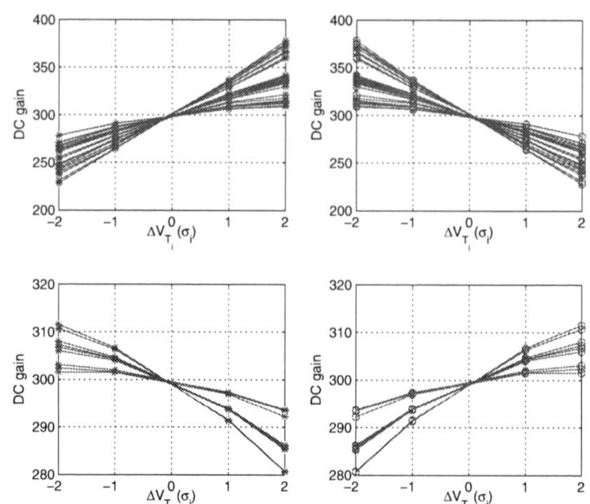

Figure 5: Monotonicity of circuit performances on variation variables in a two-stage op-amp example.

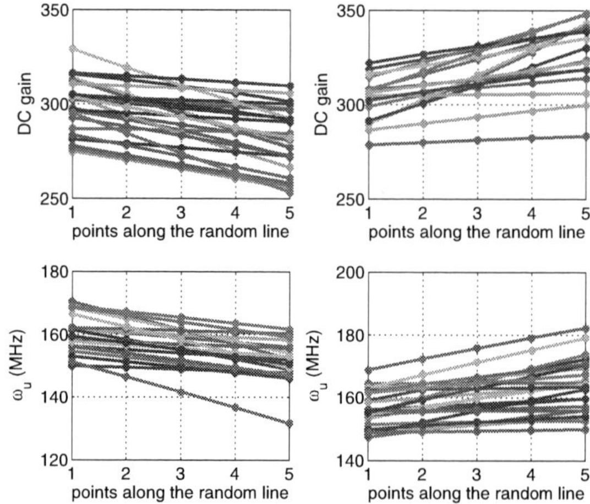

Figure 6: Monotonicity of circuit performances along random lines in variation variable space in a two-stage op-amp example.

each k, we select random variation corners out of the total 2^8 corners and record the DC gain and unity-gain bandwidth of the design under these variations. The performances show nicely monotonic functions as we increase the k, as shown in Fig. 5, where the x-axes denotes equally spaced points along the line. A more general test is to randomly select a line in the variation variable space, i.e., $\mathbf{s}, \mathbf{v} \in [-2\sigma_{V_T}, 2\sigma_{V_T}]$ and let $\mathbf{s} + t \cdot \mathbf{v}$ be the variation vector with t as a varying scalar. Fig. 6 shows monotonic performance variations along different random lines. Because of the monotonicity of performances on variation variable, the objective function in (3) becomes monotonic together with bounding constraints, and the optimal solution of the maximization problem can be found easily with general gradient-based optimization solvers.

Although there is no theoretical proof showing the monotonicity property on variations, it is confirmed in the two examples presented here. In the worst situation where the monotonicity does not hold, the algorithm should still work, with the price of adding sub-worst circuit scenarios during iterations. Those sub-worst circuit scenarios can still help push the design to be more robust.

We also notice that since the optimization solver is inherently a GP-based solver, it requires all design variables to be positive. However, ΔV_T range obviously includes negative values. A brute-force way to overcome this problem is to prefix the sign of ΔV_T

978-1-60558-497-3/09 $25.00 © 2009 ACM

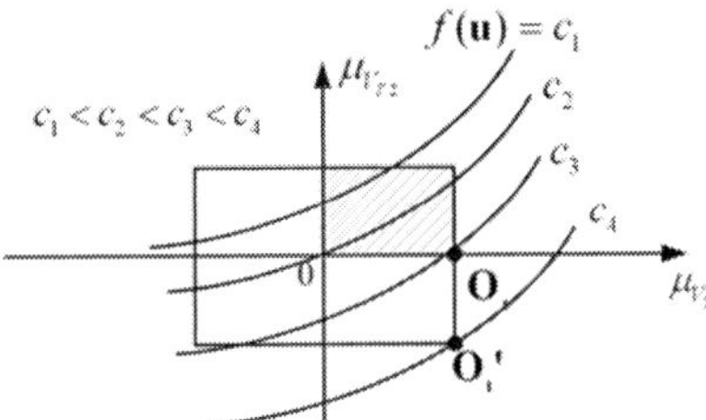

Figure 7: Constraint maximization with two variation variables.

and maximize over all sign combinations, which makes the number of maximization grow exponentially with the number of variation variables. However, with the objective function being monotonic over variation range, it is easy to map the solution from the maximization done under the positive variations to a solution under the whole variation range. This is illustrated in Fig. 7, in which case two transistor variations are considered. Suppose we maximize spec.ω_u / ω_u over the shadow area. Then, because of monotonicity, the optimal solution should be on one of the four vertices of the shadow area. If we obtain the optimal solution at O_1 and know that the objective function increases as $\mu_{V_{T2}}$ decreases, then if not restricted to positive range, the solution O_1 should shift to O_1'. This simple mapping can help obtain the optimal solution to the maximization problem efficiently.

2.2.3 Robustified circuit optimization

With the critical variation directions obtained in constraint maximization, the next step is the Robustified circuit optimization block in Fig. 2, to achieve a new robust design. If the design has not been robust over the whole variation range, we should be able to identify a nonempty set of constraints whose right-hand sides are violated, i.e. $h_k(\mathbf{x}^*, \boldsymbol{\mu}^*) > 1$. Since each of the worst-case variation vectors $\boldsymbol{\mu}_i^*$ can cause a different biasing condition, a circuit scenario has to be instantiated for each of them. The new circuit scenario is setup such that it has the same topology and shares the optimization variables $\mathbf{x}$ with the nominal circuit, except that it uses the macro model in Fig. 4(b) for transistors, with variability vector having the sign of $\boldsymbol{\mu}_i^*$ and magnitudes parameterized by optimization variable $\mathbf{x}$.

The key poit in the context of the circuit optimization, e.g. two-stage op-amp, is that the maximization solution $\boldsymbol{\mu}^* = \Delta \boldsymbol{V_T}^*$ is used to determine only the polarity of $\Delta \boldsymbol{V_T}$, and the value of $\Delta \boldsymbol{V_T}$ in the macro model is a function of design variables as $k\sigma_{V_T}$. This enables the optimizer to recognize that resizing the circuit will help decrease the variation and improve the worst performance. Here, k is the constant used to determine the range of the variation set in maximization, and σ_{V_T} is defined in (4). Therefore, the new scenario reflects the real situation when the nominal circuit is under variation with degraded performance. By doing multi-scenario optimization, we ensure that the degraded performance meets the specification, thus giving the nominal performance more margin to fail.

From the perspective of computation cost, adding clone circuits does create a larger optimization problem to solve. However, notice that the number of clone circuits is equal or less than the number of performance constraints, which in usual case is on the order of ten or less. Besides, the interior-point method used in GP-based solvers scales gracefully with increase in the number of constraints [12], since the number of optimization variables remains the same as in the nominal optimization problem.

2.2.4 Yield estimation

Yield estimation closes the loop in Fig 2. Although many fast yield estimation methods exist, for instance importance sampling

Table 2: Robust two-stage op-amp designs in iterations from optimization and Hspice simulation

	optimization		simulation	
k	DC gain	ω_u (MHz)	DC gain	ω_u (MHz)
N/A	300	160	263	125
0.2	311	163	268	130
0.4	320	166	273	133
0.6	328	168	275	135
0.8	337	170	281	137
1	344	172	282	139
1.2	351	173	285	141
1.4	360	174	288	142

[13], pseudo-noise analysis [14], here we use the direct Monte-Carlo sampling method performed in optimization domain. For each Monte-Carlo variability sampling point the optimizer solves a feasibility problem with all design variables fixed, checking the yield with proper biasing and performance constraints. In the result section we will see that the yield estimated in optimization domain follows the yield obtained from Hspice simulation, making these other simulation-based yield estimation methods possible to use in this context as well.

3. EXPERIMENTAL RESULTS

Two circuit examples are presented in this section: a two-stage and a folded-cascode op-amp. In the first example, we consider two different starting design points and show that in both cases, the algorithm is able to add more robustness to the circuit and increase the design yield. The second example has four times the number of variation variables as in the first example and the algorithm still behaves nicely. In both examples, the optimizer uses signomial transistor models derived from a 90 nm predictive technology model.

3.1 A two-stage op-amp example

This is the example discussed in Section 2, with the schematic shown in Fig. 3. The specifications of the op-amp are listed in Table 1 with 2 pF load capacitance and design objective to minimize a weighted sum of gain and area. The variation sources under consideration are threshold voltage and current factor of total 8 transistors.

3.1.1 One-corner initial design

The initial nominal design is optimized under **tt** corner with 1 V supply voltage, 10 μA reference current and temperature at 298 K. The nominal design just meets the gain and bandwidth (ω_u) constraints. In the first iteration, only these two constraints are found to violate the specifications after maximizations. Based on optimal solutions from maximization, we instantiate two circuit scenarios where the polarity of the ΔV_T voltage sources are determined from the sign of maximization solutions. Then a 3-scenario circuit optimization is solved and we reach a robust design with gain and ω_u improved to 311 and 163 MHz (measured without variability). As iterations go on, the design margins increase at the same time, shown in the two columns under optimization in Table 2. Table 2 also shows the simulation results for verification. Because of the transistor model's signomial fitting error, there is around 20% mismatch between Hspice simulations and optimization results. However, the simulation results do show the same trend of the improved performance and yields, with yields shown in the top two figures in Fig. 8 (the Hspice yields are calculated with respect to the *simulated* nominal design performance without variations). The cost of robustified design are increased power and area shown in the bottom two figures in Fig. 8.

978-1-60558-497-3/09 $25.00 © 2009 ACM

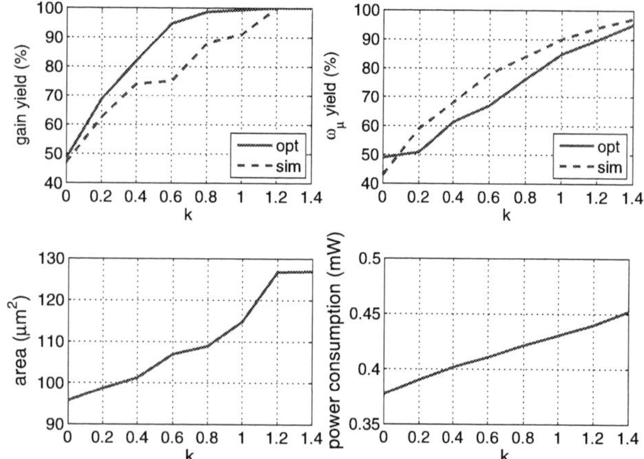

Figure 8: Two-stage op-amp with 1-corner initial optimization design: yield improvement of gain and ω_u, power and area consumptions.

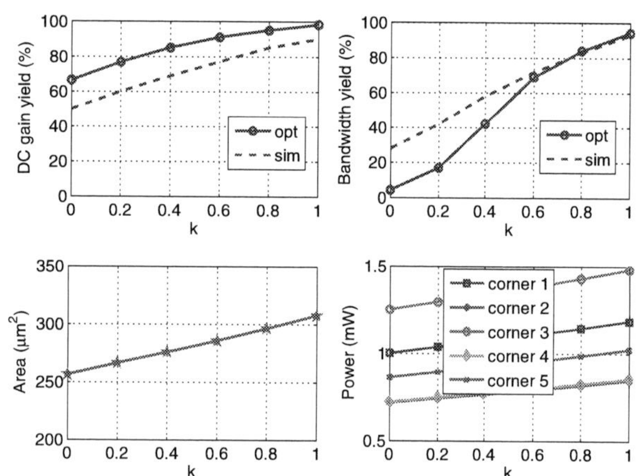

Figure 9: Two-stage op-amp five-corner optimization design: yield improvement on gain, ω_u and power, area consumptions.

3.1.2 Five-corner initial design

Here, we consider the situation where designers start from a fairly robust design. The initial design is a multi-scenario optimized design for the 5-corner listed in Table 3. After applying random variations to the optimized design, gain and bandwidth at the second corner are found to violate the specifications and the corresponding yields for that corner drop to around 60% and 20% respectively, as shown in the top two figures in Fig. 9. This implies that in a design optimized over multiple PVT corners, even though some PVT corners show performances well above specifications even under random variations, the yield of some corners can be very low under random variations.

After adding two circuit-scenarios representing gain and bandwidth with variation vectors found in maximization at the second corner (ss), and optimizing the multi-scenario circuit, the yields for

Table 3: Five-corner of the two-stage op-amp initial design.

number	corner	temp (K)	vdd (V)	Iref (μA)
1	tt	298	1	10
2	ss	398	0.9	8
3	ff	233	1.1	12
4	fs	398	0.9	8
5	sf	233	1.1	8

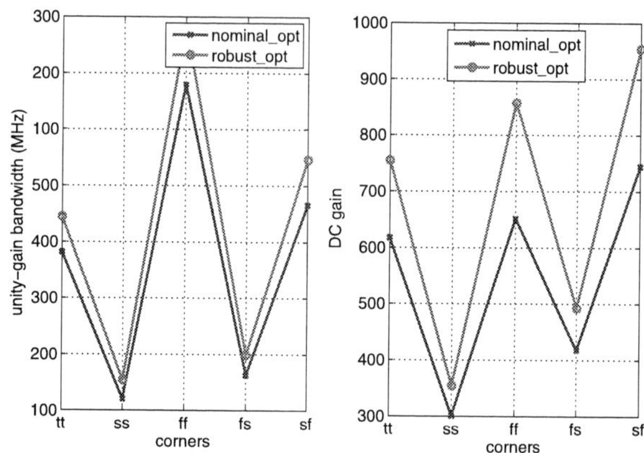

Figure 10: Two-stage op-amp five-corner optimization design: DC gain and ω_u comparison of initial and final robust designs.

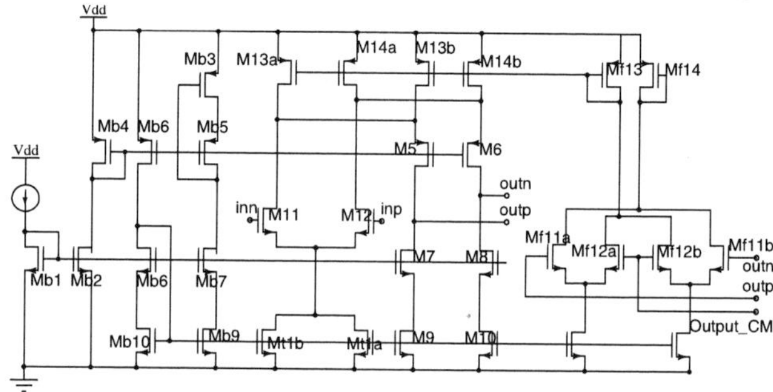

Figure 11: The folded-cascode op-amp schematic.

the second corner gradually increase to around 100%, shown in Fig. 9. The final robust design achieved in optimization is shown in Fig. 10, compared with the initial design. The cost to achieve the improvement is shown in the bottom two figures in Fig. 9. The simulations in Hspice of the second corner again show the same trend in yields in Fig. 9.

3.2 A fully differential folded-cascode op-amp with common-mode feedback

Fig. 11 shows the schematic of a fully differential folded-cascode op-amp with common-mode feedback (CMFB). The specifications considered in this example are listed in Table 4, with the objective to minimize a weighted sum of power and area. The variation sources under consideration are the threshold and current factor variations from total 32 transistors.

Table 5 shows the robust design evolution during iterations. The initial design has a transconductance value just on the lower bound of the specification and this critical constraint is found in the first maximization. To prevent this from happening, a circuit scenario is instantiated and along with the nominal circuit, it pushes the new robust design to a high transconductance value close to the upper bound. Therefore, in the second iteration, maximization finds another critical constraint, the transconductance upper bound con-

Table 4: The folded-cascode op-amp example: specifications for nominal design

DC gain	50dB
phase margin	60°
Gm	[0.5 mS, 0.6 mS]
CMFB DC gain	60 dB
CMFB unity-gain bandwidth ($\omega_{u_{CMFB}}$)	5.9 MHz

978-1-60558-497-3/09 $25.00 © 2009 ACM

Table 5: Folded-cascode op-amp: iterations of the robust designs from optimization: gm $\in [0.5\ \mathbf{mS}, 0.6\ \mathbf{mS}]$

k	violated constraints	Gm (mS)	yield (%)	area (μm^2)	power (mW)
0	gm ≥ 0.5	0.5	49	539	0.27
0.2	gm ≤ 0.6m	0.59	69	544	0.30
0.4	None	0.58	84	627	0.31
0.6	None	0.57	92	659	0.32
0.8	None	0.55	94	668	0.33

Table 6: The computational time breakdown per iteration step for the two-stage and folded-cascode robust design examples

	Two-stage op-amp	Folded-cascode op-amp
Nominal design	10 s	17 s
Maximization	6 specs: 20 s	5 specs: 1 min
Redesign with 1-scenario	11 s	30 s
Redesign with 2-scenarios	20 s	40 s

straint, and instantiates another circuit scenario to guard the upper bound. The next robust design that comes out of a three-scenario circuit optimization gives a lower *Gm* value and a higher yield. As iterations continue, no more constraint violations are found and the variation set keeps expanding, until the yield is satisfactory. As expected, the transconductance *Gm* from the optimizer finally converges to the middle point of the specification range, i.e. 5.5 mS. Fig. 12 shows the Monte-Carlo performed in optimization domain of the initial and final robust design's *Gm* and the area and power consumption are shown in Table 5 . Again, we can see the tradeoff between the circuit performance and area, power cost.

3.3 Computational efficiency

Optimizations ran on a server with 3.16 GHz Intel® Xeon® processor and 16 GB of memory running Linux. The time cost in one iteration consists of maximization time and multi-scenario redesign time. The approximate solving time for these two examples in one iteration is shown in Table 6. It takes $6-7$ iterations ($10-15$ minutes) to achieve a robust design under random variations. The problem size of the two-stage op-amp redesign with 2-scenarios is around 200 variables and 600 constraints while for the folded-cascode op-amp redesign with 2-scenarios, it has around 800 variables and 2500 constraints.

4. CONCLUSION

In this paper we show that complicated and often intractable yield-driven robust circuit optimization can be performed in a scalable and tractable manner in a few iterative steps. The proposed algorithm is iterative and first decomposes the yield-driven robust optimization into minmax optimization followed by yield-estimation. We further show that the maximization step can be performed efficiently, due to the circuit-related monotonic properties on variations of the performance functions. We find that the key to success in this stage is to maintain the circuit biasing constraints during the maximization step. The multi-scenario optimization is used in the last step, where the key ingredient is to introduce only the relative values of the variability corners and let the solver figure out the updated absolute values based on the state of optimization variables. Lastly, the variability space bound is increased and algorithm iterates until the satisfactory yield has been achieved. Coupled with these key circuit optimization related insights, this iterative minmax optimization presents an efficient and scalable solution for yield-driven circuit optimization.

5. ACKNOWLEDGMENTS

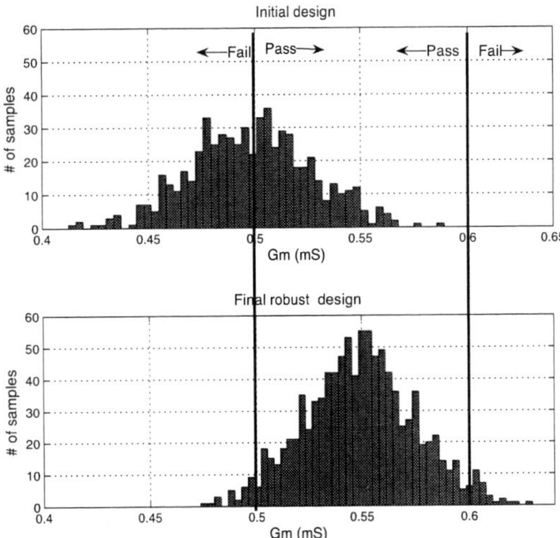

Figure 12: Monte-Carlo check with the initial and final robust designs: Gm of the folded-cascode op-amp.

The authors would like to thank Mar Hershenson, Sunderarajan Mohan, Dave Colleran, Almir Mutapčić and Shyne Tseng at Magma Design Automation, Inc for all their technical discussions and support. This work is funded by the Center for Integrated Circuits and Systems at MIT .

6. REFERENCES

[1] M. Hershenson, "Design of pipeline analog-to-digital converters via geometric programming," in *Proc. of ICCAD*, 2002, pp. 317–324.

[2] M. Hershenson, S. Boyd, and T. Lee, "Optimal deisgn of a cmos op-amp via geometric programming," *IEEE Trans. Computer-Aided Design*, vol. 20, no. 1, pp. 1–21, 2001.

[3] D. Colleran, C. Portmann, A. Hassibi, C. Crusius, S. Mohan, S. Boyd, T. Lee, and M. del Mar Hershenson, "Optimization of phase-locked loop circuits via geometric programming," *Custom Integrated Circuits Conference, 2003. Proceedings of the IEEE 2003*, pp. 377–380, Sept. 2003.

[4] Y. Xu, K.-L. Hsiung, X. Li, I. Nausieda, S. Boyd, and L. Pileggi, "Opera: optimization with ellipsoidal uncertainty for robust analog ic design," in *IEEE/ACM Design Automation Conference (DAC)*, 2005, pp. 632–637.

[5] K. Hsiung, S. Kim, and S. Boyd, "Tractable approximate robust geometric programming," *Optimization and Engineering*, vol. 9, no. 2, pp. 1389–4420, 2008.

[6] K. Antreich and H. Graeb, "Circuit optimization driven by worst-case distances," *Computer-Aided Design, 1991. ICCAD-91. Digest of Technical Papers., 1991 IEEE International Conference on*, no. SN -, pp. 166–169, 1991.

[7] A. Dharchoudhury and S. Kang, "Worst-case analysis and optimization of vlsi circuit performances," *Computer-Aided Design of Integrated Circuits and Systems, IEEE Transactions on*, vol. 14, no. 4, pp. 481–492, Apr 1995.

[8] A. Mutapcic, S. Boyd, A. Farjadpour, S. Johnson, and Y. Avniel, "Robust design of slow-light tapers in periodic waveguides," *Engineering Optimization*, vol. 41, no. 4, pp. 365–384, 2009.

[9] M. Pelgrom, A. Duinmaijer, and A. Welbers, "Matching properties of mos transistors," *IEEE JSSC*, vol. 24, no. 5, pp. 1433–1440, 1989.

[10] S. Nassif, "Modeling and analysis of manufacturing variations," *Custom Integrated Circuits, 2001, IEEE Conference on.*, pp. 223–228, 2001.

[11] X. Li, J. Le, P. Gopalakrishnan, and L. T. Pileggi, "Asymptotic probability extraction for nonnormal performance distributions," *Computer-Aided Design of Integrated Circuits and Systems, IEEE Transactions on*, vol. 26, no. 1, pp. 16–37, Jan. 2007.

[12] S. Boyd and L. Vandenberghe, *Convex Optimization*. New York: Cambridge Univeristy Press, 2004.

[13] R. Kanj, R. Joshi, and S. Nassif, "Mixture importance sampling and its application to the analysis of sram designs in the presence of rare failure events," in *DAC '06: Proceedings of the 43rd annual conference on Design automation*. New York, NY, USA: ACM, 2006, pp. 69–72.

[14] J. Kim, K. D. Jones, and M. A. Horowitz, "Fast, non-monte-carlo estimation of transient performance variation due to device mismatch," in *DAC '07: Proceedings of the 44th annual conference on Design automation*. New York, NY, USA: ACM, 2007, pp. 440–443.

978-1-60558-497-3/09 $25.00 © 2009 ACM

Contract-Based System-Level Composition of Analog Circuits

Xuening Sun   Pierluigi Nuzzo   Chang-Ching Wu   Alberto Sangiovanni-Vincentelli

Department of Electrical Engineering and Computer Science
University of California, Berkeley
Email: {xuening, nuzzo, jameswu, alberto}@eecs.berkeley.edu

ABSTRACT

Efficient system-level design is increasingly relying on hierarchical design-space exploration, as well as compositional methods, to shorten time-to-market, leverage design re-use, and achieve optimal performances. However, in analog electronic systems, circuit behaviors are so tightly dependent on their interface conditions that accurate system performance estimations based on characterizations of individual stand-alone circuits is a hard task. Since there is no general solution to this problem, analog system integration has traditionally used *ad-hoc* solutions heavily dependent on designers' experience. In this paper, we build upon the *analog platform-based design* methodology by exploiting *contracts* to enforce *correct-by-construction* system-level composition. Contracts intuitively capture the thought process of a designer, who aims at *guaranteeing* circuit performance only under specific *assumptions* (e.g. loading and dynamic range) on the interface properties. Our approach allows *automatic* detection and composition of compatible components in a given library. We apply our methodology to an ultra-wide band receiver front-end to show that contracts allow pre-designed IP components to be smoothly integrated and design decisions to be reliably made at a higher abstraction level, both key factors to improve designer productivity.

Categories and Subject Descriptors

B.7.2 [**Integrated Circuits**]: Design Aids

General Terms

Design, Reliability, Theory

Keywords

analog, platform, composition, contract, assume-guarantee, platform-based design, UWB, radio-frequency, system, integration

1. INTRODUCTION

The increasing complexity of today's electronic systems seriously hinders designers' productivity. At the same time, nonrecurring engineering (NRE) costs in modern chips continue to increase, making design re-spins extremely expensive. To lower design cost,

Permission to make digital or hard copies of part or all of this work for personal or classroom use is granted without fee provided that copies are not made or distributed for profit or commercial advantage and that copies bear this notice and the full citation on the first page. To copy otherwise, to republish, to post on servers or to redistribute to lists, requires prior specific permission and/or a fee.
DAC'09, July 26-31, 2009, San Francisco, California, USA

a disciplined design methodology is needed which reduces iterations in the flow and enables design reuse. While productivity for digital circuits greatly benefited from RTL synthesis and standard cell libraries, their analog domain counterparts remain elusive. Most efforts in analog design automation have traditionally focused on *circuit-level* design. However, in recent years *system-level* analog design has become a research focus.

The earliest approaches to system-level analog design were hierarchical. In fact, mixed-signal system designs, typically optimizing several competing continuous-valued performance specifications on a set of a few hundred assembled circuits, were no longer tractable as a "flat" problem. The top-down constraint-driven design methodology [1] propagated constraints down the hierarchy for complex mixed-signal systems. Very recently, multi-objective bottom-up methodologies, as in [2], have been proposed, where systems are built by evaluating Pareto-optimal performance points for components at a lower level.

However, hierarchical methodologies tend to lose in accuracy when propagating information across design levels. Consequently, reuse and smooth integration of pre-designed and pre-characterized IP blocks is very difficult. One major problem stems from the fact that performances of analog components are not only a function of design parameters, such as transistor sizings and voltage biasing, but also a function of the interface conditions at the input and output ports. Even in the simple case of two cascaded analog circuits, performances of the composition cannot be generally obtained by directly cascading behavioral and performance models of the stand-alone components, since the behavior of the loading block substantially affects that of the driver.

In both top-down and bottom-up paradigms, the problem of interface conditions, under which hierarchical system compositions are legal, has not been rigorously addressed. In this paper we approach this problem in the context of Platform-Based Design (PBD) that has been recently extended to the analog domain [3] and proved to be, because of its "meet-in-the-middle" approach, a natural framework for system-level analog design. In particular, we tackle the problem of correct composition of analog elements in the context of APBD. For each platform element, we generate a *contract* that must be satisfied to preserve accuracy in performance estimations and guarantee feasibility of composition with other elements, thus allowing smooth hierarchical design of mixed-signal integrated systems. Contracts intuitively capture the thought process of a designer, who aims at *guaranteeing* circuit performance only under specific *assumptions* on the interfaces (e.g. loading and dynamic range). By formalizing the interface problem, our approach allows *automatic* detection and composition of compatible components in a given library. In addition, we broaden the design search space by eliminating the need for standardized "hard" inter-

978-1-60558-497-3/09 $25.00 © 2009 ACM

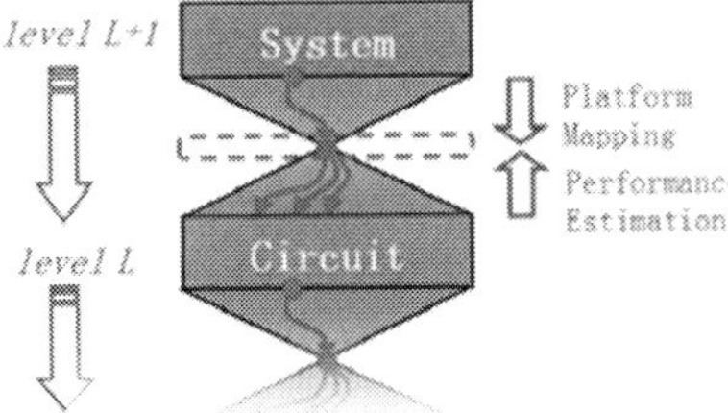

Figure 1: Each platform stack of APBD features a meeting-in-the-middle of system constraints and performance characterizations.

face constraints and by developing general performance models to be used in numerous systems.

The rest of the paper is organized as follows. Section 2 gives an overview of APBD and contract-based design. Section 3 provides the mathematical framework to describe system level composition of analog components. We illustrate our methodology on the design of an UWB receiver front-end in Section 4, and present the results and conclusion in Section 5 and Section 6.

2. BACKGROUND

2.1 Analog Platform-based Design

Analog platform-based design (APBD) is naturally amenable to design reuse. A platform is a collection of components and composition rules. Each component is characterized by a behavioral model that represents the functionality of the component and a performance model that represent a set of feasible performances of the component. A design is obtained by composing components of the platform into a platform instance. The design is successively refined using "meet-in-the-middle" (Figure 1) mappings of higher-level design decisions onto the feasible performance regions of the lower-level platforms.

In this framework, new designs can be assembled quickly from a library of pre-designed and pre-characterized components, giving the highest priority to design reuse, correct assembly of components, and an efficient flow from specification to implementation. Since design decisions are made at the system level, APBD can locate globally optimal design solutions by evaluating system trade-off across all components, instead of just using locally optimized designs of each individual component. In addition, platforms permits hierarchical design space exploration, which progressively reduce the number of design variables and separate design concerns.

Support-vector-machines (SVM) have been proposed for circuit-level performance characterizations in [4]. Based on the set of discrete performance points from transistor-level electrical simulations, an m-dimensional SVM is used to provide a continuous region that tightly estimates the feasible performance space of interest for a circuit block, where m is the number of performance parameters for the circuit.

APBD has been successfully applied to several case studies [5, 3] . However, the problem of defining correct composition rules for components of a platform library has not been rigorously addressed. Correct composition guarantees implementation feasibility and performance estimation accuracy. Indeed, performances of analog components are strongly dependent on interfaces with other components and loading conditions. In previous APBD case studies, the interface conditions were hand-tuned and fixed to constants during performance characterization. In [6] the introduction of ad-hoc interconnection blocks is suggested to accurately model interface effects in analog circuits. However, the provided guide-

lines can be hardly generalized, basically consisting in case-by-case topology-related heuristics which require non negligible effort for the designer to build the models. To formulate a more rigorous composition approach, we utilize contract-based composition.

2.2 Contract-based Design

Contracts are fundamentally rooted in compositional reasoning. As advocated in [7],[8], and [9], compositional reasoning reduces design and verification complexity by decomposing system-level tasks into manageable problems at the component-level. System properties are inferred or proved based on component properties. More recently, contracts have been exploited in component-based software engineering (CBSE) [10] to facilitate correct reuse and composition of software components. The core of contract-based design (CBD) is a set of assume-guarantee relationships between the environment (e.g. interfaces) and the component (e.g. performances). During composition, a legal connection must satisfy the assumptions of all components involved to guarantee the correct system behavior. We borrow the assume-guarantee concept for analog component composition, where each analog component assumes that interface conditions (e.g. load impedance, input voltage level) are within a specified range and guarantees feasibility for a set of corresponding performance outputs. Composition is only allowed if the contract assumptions of all interconnected components are simultaneously satisfied.

In fact, assume-guarantee relationships have always been intuitively used by analog designers. However, in most cases, only a few fixed interface assumptions are considered in an ad-hoc manner. In contrast, contracts allow rigorous definitions and analysis of the validity of compositions. Furthermore, contract models of each component may be reused in multiple application domains. More details are provided in Section 3 and 4.

3. MATHEMATICAL FORMULATION

3.1 Definitions

An Analog Platform (AP) is a library of components [3], each denoted by:

- a set of input variables $u \in \mathcal{U}$, a set of output (and performance) variables $y \in \mathcal{Y}$, a set of internal variables $x \in \mathcal{X}$ (including state variables), a set of configuration parameters $\kappa \in \mathcal{K}$;
- a behavioral model $\mathcal{F}(u, y, x, \kappa) = 0$, which implicitly represents the behavior of the component; in general, $\mathcal{F}(.) = 0$ is a set of integro- differential equations uniquely determining y and x given u and κ;
- a *feasible performance model*. Let $\phi_y(u, \kappa)$ denote the map that computes the performance y corresponding to particular values of u and κ by solving the behavioral model. The feasible performance set is then the set described by the relation $\mathcal{P}(y(u)) = 1 \Leftrightarrow \exists \kappa', y(u) = \phi_y(\kappa', u)$.
- *validity laws* $\mathcal{L}(u, y, x, \kappa, \delta) \leq 0$, i.e. constraints (or *assumptions*) on the variables and parameters of the component that define the range of the variables for which the behavioral and performance models are *guaranteed* as being valid.

In what follows, $\zeta \in \mathcal{Z} \subseteq \mathcal{Y}$ denote the subset of performances in y to distinguish them from the output variables. To formalize composition, we extend the validity laws for each component by introducing the *contract* concept through the following definitions:

Definition 1. A contract **C** is a set of assume/guarantee tuples c: (a, g) where for each c_i,

978-1-60558-497-3/09 $25.00 © 2009 ACM

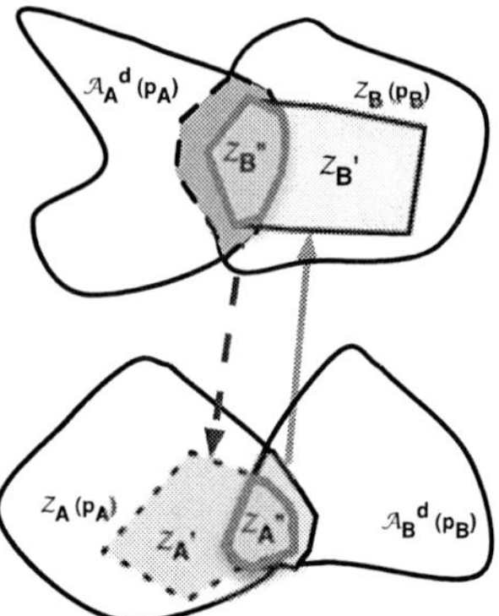

Figure 2: Projections of the contracts of A and B onto assumptions and performance guarantees of p_A and p_B during composition. Composition can only occur between design instances that simultaneously satisfy assumptions of both components involved, which is confined to the *consistent regions* $\mathcal{Z}_A''$ and $\mathcal{Z}_B''$.

1. $a_i = \{a_i^s, a_i^d\}$, where $a_i^s = \mathcal{A}^s$, the set of static assumptions such as supply voltage value, process technology, DC input voltage values; $a_i^d \subseteq \mathcal{A}^d$, the set of dynamic assumptions, which can vary with component configurations and external environment (e.g. load impedance $Z_L \geq 50\Omega$); a_i^s is, in general, a set of equality relations, whereas a_i^d is a set of inequalities, or the solution set to those inequalities; both $\mathcal{A}^s$ and $\mathcal{A}^d$ are subsets of the component validity laws;

2. $g_i \subseteq \mathcal{Z}$ is a set of guaranteed feasible performances of the component under the assumption a_i; $\bigcup_{i=1}^{n} g_i = \mathcal{Z}$, n being the total number of assume/guarantee tuples in the contract **C**.

Definition 2. Components A and B are consistent *iff* there exists an output port $p_A \in \mathcal{Y}_A$ and an input port $p_B \in \mathcal{U}_B$, such that

1. For all static assumptions $\mathcal{A}_A^s(p_A)$ of A on p_A and $\mathcal{A}_B^s(p_B)$ of B on p_B, $\mathcal{A}_A^s(p_A) = \mathcal{A}_B^s(p_B)$.
2. For all contracts $\mathbf{C}_A(p_A)$ on p_A and $\mathbf{C}_B(p_B)$ on p_B, $\exists c_{Ai} \in \mathbf{C}_A(p_A)$ and $c_{Bi} \in \mathbf{C}_B(p_B)$ such that $g_{Ai} \cap a_{Bi} \neq \emptyset$ and $g_{Bi} \cap a_{Ai} \neq \emptyset$, i.e. both a_{Ai} and a_{Bi} can be satisfied simultaneously, where a's and g's are assume and guarantee parts of the c tuples.
3. The union of all performance guarantees, g_{Ai} and g_{Bi}, for all consistent tuples form the *consistent regions* $\mathcal{Z}_A''$ and $\mathcal{Z}_B''$ for A and B, respectively. $\mathcal{Z}_A''$ and $\mathcal{Z}_B''$ are the *legal* search spaces of performance parameter for the composed system, in which assumptions of both components are satisfied simultaneously.

A composition is valid only when the design search spaces of its components are bound inside the consistent regions. Figure 2 shows that, in general, only a subset of $\mathcal{A}_A^d(p_A)$ is satisfied by the performance space, $\mathcal{Z}_B$, related to the input port p_B. The reduced assumption space that is satisfied upon composition directly corresponds to a reduced performance space of A, denoted with $\mathcal{Z}_A'$. Similarly for B, $\mathcal{Z}_B$ and $\mathcal{A}_A^d(p_A)$ are recursively reduced. Thus, in general, $\mathcal{A}_B^d(p_B) \cap \mathcal{Z}_B$, and vice versa, do not necessarily form the consistent regions. In the example of Figure 2, $\mathcal{Z}_B'' = \mathcal{A}_A^d(p_A) \cap \mathcal{Z}_B'$ and $\mathcal{Z}_A'' = \mathcal{A}_B^d(p_B) \cap \mathcal{Z}_A'$.

3.2 Composition

Contracts allow AP components to be reliably composed, while preserving correct system behavior and performance estimation accuracy. Without loss of generality, in this section, we formalize composition of consistent components A and B at platform level l,

to generate $C = A \| B$ as an AP component at level $l+1$. C can be denoted with:

- a set of internal variables $x_C \in \mathcal{X}_C$,
- a set of input variables $u_C \in \mathcal{U}_C$,
- a set of output (and performances) $y_C \in \mathcal{Y}_C$,
- a set of configuration parameters $\kappa_C \in \mathcal{K}_C = \{\mathcal{Z}_A; \mathcal{Z}_B\}$.

Some level l component variables may change roles in composition, e.g. an input variable may become an internal variable, depending on the application. Composition is allowed when components are *consistent*. A composition is characterized by the interconnect equations which specify which variables are shared when composing components and by the contracts that are defined when the composition is indeed possible. Formally, a connection is establishing a pairwise equality between variables and performances (e.g. $p_A = p_B$) of two components. Interconnect relations are generally a set of linear equalities. In this paper, we call λ the set of interconnected (shared) variables between A and B.

Similar to process behavior composition in [11], the behavioral model, $\mathcal{F}_C$ is $\mathcal{F}_A \times \mathcal{F}_B$ conjoined with the interconnect relations. The performances $\zeta_C \in \mathcal{Z}_C$ are obtained through a map $\zeta_C = \phi_C(\zeta_A, \zeta_B)$, where $\zeta_A \in \mathcal{Z}_A, \zeta_B \in \mathcal{Z}_A$, and ϕ_C is a performance map from platform l to platform $l+1$ (e.g. cascade equations). The feasible performance model, $\mathcal{P}_C$, is defined afresh based on ϕ_C. To provide the validity laws for the new platform instance, we construct the new contract $\mathbf{C}_C$ as follows:

$$\forall (a_{Ai}, g_{Ai}) \in \mathbf{C}_A, \forall (a_{Bj}, g_{Bj}) \in \mathbf{C}_B \quad i = 1 \ldots n, j = 1 \ldots m$$
$$a_{Cij} = \{a_{Ai}; a_{Bj}\} \text{ if} \tag{1}$$
$$\{a_{Ai}(\lambda) \cap g_{Bj}(\lambda)\} \neq \emptyset \text{ and } \{a_{Bj}(\lambda) \cap g_{Ai}(\lambda)\} \neq \emptyset$$
$$g_{Cij} = \{\zeta_C = \phi_C(\zeta_A, \zeta_B)\} \,\forall \zeta_A, \zeta_B \text{ s.t.} \tag{2}$$
$$\zeta_A(\lambda) \in \{a_{Bj}(\lambda) \cap g_{Ai}(\lambda)\} \text{ and } \zeta_B(\lambda) \in \{a_{Ai}(\lambda) \cap g_{Bj}(\lambda)\}$$
$$\mathcal{Z}_C = \bigcup g_{Cij} \quad \mathcal{A}_C = \bigcup a_{Cij}$$
$$\mathbf{C}_C = (\mathcal{A}_C, \mathcal{Z}_C)$$

where n and m are the total number of contract tuples for A and B, respectively. $\zeta_A(\lambda)$ and $\zeta_B(\lambda)$ are performances of A and B, respectively, on λ. $\mathcal{Z}_C$ is the feasible performance space for C. Essentially, (1) and (2) state that design instances of A and B can compose if and only if the performances of A satisfies the assumptions of B on the shared variables λ, and vice versa. Note that (2) finds only *consistent* component performances of A and B.

Contracts allow design decisions to be made accurately at a raised level of abstraction since feasibility and performance estimation accuracy are preserved by composition. The accuracy of the performance model of the composed system is only dependent on the accuracy of ϕ_C and the performance estimations of the components (e.g. ζ_A and ζ_B). No additional estimation errors are introduced by composition as long as all components operate within consistent regions.

3.3 Properties of Composition

To deal with composition of complex systems, we list here some properties that composition exhibits. First of all we notice that composition is not *commutative*. In fact, besides the trivial cases, where A and B perform different functions, which are not interchangeable, in [12], it is demonstrated that, even in cascaded filter designs, the order of cells is key in accurately estimating the total system noise figure and linearity. On the other hand, in this section, we prove that composition is *associative*, which enables the hierarchical exploration of the design space.

978-1-60558-497-3/09 $25.00 © 2009 ACM

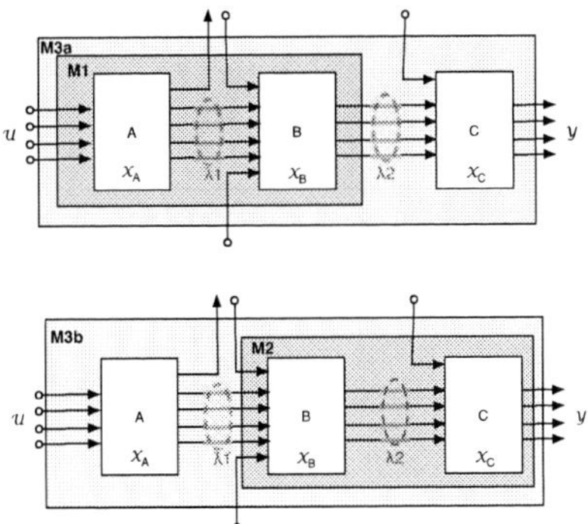

Figure 3: Composition Associativity

Interconnection equations are independent of the order in which composition is evaluated, as shown in Figure 3 for the case of three components. Accordingly,

THEOREM 1. *The contract assumptions of composed systems obey the associative property.*

PROOF. Given components A, B, and C, as in Figure 3, contract assumptions of $M1 = A\|B$ is given as:

$$\mathcal{A}_{M1}^s = \{\mathcal{A}_A^s \times \mathcal{A}_B^s\} \setminus \mathcal{A}^s(\lambda_1) \tag{3}$$

$$\mathcal{A}_{M1}^d = \{\mathcal{A}_A^d \times \mathcal{A}_B^d\} \mid \mathcal{A}_A^d(\lambda_1) = \mathcal{Z}_{B\lambda_1}'', \mathcal{A}_B^d(\lambda_1) = \mathcal{Z}_{A\lambda_1}'' \tag{4}$$

where λ_1 is set of consistent connections between A and B, and $\mathcal{Z}_{A\lambda_1}''$ and $\mathcal{Z}_{B\lambda_1}''$ are, respectively, *consistent regions* of A and B on λ_1. Contract assumptions of $M2 = B\|C$ is given as

$$\mathcal{A}_{M2}^s = \{\mathcal{A}_B^s \times \mathcal{A}_C^s\} \setminus \mathcal{A}^s(\lambda_2) \tag{5}$$

$$\mathcal{A}_{M2}^d = \{\mathcal{A}_B^d \times \mathcal{A}_C^d\} \mid \mathcal{A}_B^d(\lambda_2) = \mathcal{Z}_{C\lambda_2}'', \mathcal{A}_C^d(\lambda_2) = \mathcal{Z}_{B\lambda_2}'' \tag{6}$$

where λ_2 is set of consistent connections between B and C. Therefore, contract assumptions of $M3a = M1\|C$ is

$$\begin{aligned}
\mathcal{A}_{M3a}^s &= \{\mathcal{A}_{M1}^s \times \mathcal{A}_C^s\} \setminus \mathcal{A}^s(\lambda_2) \\
&= \{\mathcal{A}_A^s \times \mathcal{A}_B^s \times \mathcal{A}_C^s\} \setminus \{\mathcal{A}^s(\lambda_1); \mathcal{A}^s(\lambda_2)\}
\end{aligned} \tag{7}$$

$$\begin{aligned}
\mathcal{A}_{M3a}^d &= \{\mathcal{A}_{M1}^d \times \mathcal{A}_C^d\} \mid \mathcal{A}_{M1}^d(\lambda_2) = \mathcal{Z}_{C\lambda_2}'', \mathcal{A}_C^d(\lambda_2) = \mathcal{Z}_{M1\lambda_2}'' \\
&= \{\mathcal{A}_A^d \times \mathcal{A}_B^d \times \mathcal{A}_C^d\} \mid \\
&\quad \mathcal{A}_A^d(\lambda_1) = \mathcal{Z}_{B\lambda_1}'', \mathcal{A}_B^d(\lambda_1) = \mathcal{Z}_{A\lambda_1}'', \\
&\quad \mathcal{A}_B^d(\lambda_2) = \mathcal{Z}_{C\lambda_2}'', \mathcal{A}_C^d(\lambda_2) = \mathcal{Z}_{B\lambda_2}''.
\end{aligned} \tag{8}$$

Similarly, contract assumptions of $M3b = A\| M2$ is

$$\mathcal{A}_{M3b}^s = \{\mathcal{A}_A^s \times \mathcal{A}_{M2}^s\} \setminus \mathcal{A}^s(\lambda_1) = \mathcal{A}_{M3a}^s \tag{9}$$

$$\begin{aligned}
\mathcal{A}_{M3b}^d &= \{\mathcal{A}_A^d \times \mathcal{A}_{M2}^d\} \mid \mathcal{A}_A^d(\lambda_1) = \mathcal{Z}_{M2\lambda_1}'', \mathcal{A}_{M2}^d(\lambda_1) = \mathcal{Z}_{A\lambda_1}'' \\
&= \mathcal{A}_{M3a}^d
\end{aligned} \tag{10}$$

Therefore, contract assumptions of $(A\|B)\|C$ and $A\|(B\|C)$ is identical. $\square$

It should be noted that we only consider cases where $\lambda_1 \cap \lambda_2 = \emptyset$. Since assumptions are directly associated with the feasible performance spaces of A, B and C, associative properties of the performance models of composed systems are dependent only on the associativity of the performance maps, $\phi_{M1/2}$. Similarly, associativity of the behavioral models depend only on the properties of $\mathcal{F}_{A/B/C}$. Given that $\phi_{M1/2}$ and $\mathcal{F}_{A/B/C}$ are associative, then $(A\|B)\|C$ is the same as $A\|(B\|C)$. Therefore, the composition operation is associative.

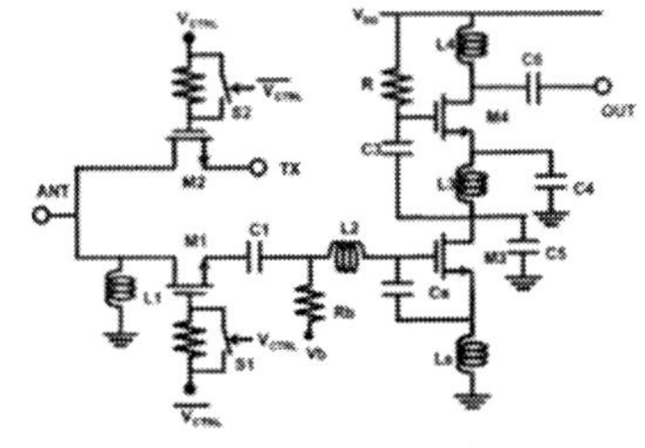

(a) LNA Schematic

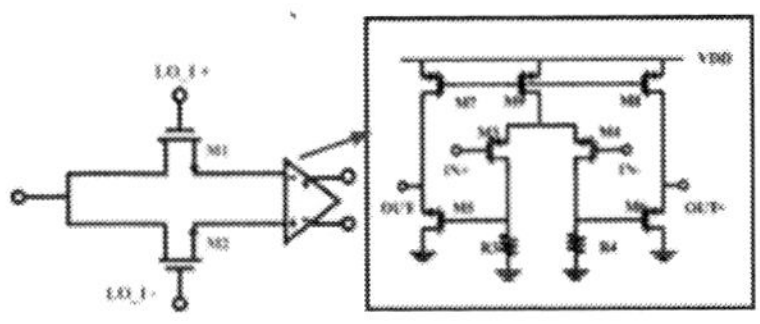

(b) Mixer Schematic

Figure 4: Circuit Schematic

4. CASE STUDY

We apply our methodology to optimize the radio-frequency (RF) front-end for an ultra-wideband (UWB) receiver based on the architecture in [5]. We aim to minimize power consumption while satisfying a set of given constraints using pre-characterized components in our RF platform library. The authors of [5] showed that APBD can be used to automatically optimize RF system design; however, fixed interface conditions were enforced via an interface buffer to produce accurate and reliable results. By relying on contracts, we can remove the interface buffer between the LNA and mixer. The RF front-end can then be automatically composed without incurring design overhead from the interface circuitry.

4.1 UWB RF Components

As shown in Figure 4a, the RF front-end includes two main building blocks. The first block (Figure 4a) consists of the T/R switch (M_1 and M_2), the wideband (3.1 - 4.8 GHz) input matching network (L_1, L_2, L_S, and C_S), and the LNA, which features a stagger tuning technique to achieve gain flatness over the wide band. The second block(Figure 4b) includes a passive mixer (M_1 and M_2) and a low-noise buffer amplifier (M_3-M_8) to boost mixer gain.

4.2 Contracts

Both LNA and mixer blocks were characterized using a 0.13μm 1.2V technology and assuming an input signal bandwidth of 500MHz centered at 3.96 GHz, so that the static assumptions $\mathcal{A}_{LNA}^s$ and $\mathcal{A}_{mixer}^s$ were inherently equal. Being mainly concerned with the interface effects between the LNA and the mixer, we defined contracts for the output of the LNA and the input of the mixer.

The dynamic assumptions, $\mathcal{A}_{LNA}^d$, of the LNA include conditions on the equivalent load impedance (specified through an RC network) as follows:

$$\begin{aligned}
\mathcal{A}_{LNA}^d &= \{a_{LNA1}^d(x) = r, a_{LNA2}^d(x) = c\}; \\
&\quad r \in R_{load}, c \in C_{load}
\end{aligned} \tag{11}$$

where x is the output port of the LNA, $R_{load} \in [85, 520]\Omega$ and $C_{load} \in [0.03, 0.25]$pF. The mixer dynamic assumption on the source impedance is:

$$\mathcal{A}_{mixer}^d = \{a_{mixer1}^d(y) \geq 20\Omega, a_{mixer2}^d(y) \leq 1000\Omega\} \tag{12}$$

where y is the input port of the mixer.

978-1-60558-497-3/09 $25.00 © 2009 ACM

	1	2	3	4	5
ζ_{LNA}	NF	gain	S11	power	IIP3
ζ_{Mixer}	Rin	Cin	gain	power	NF
	6	7	8	9	10
ζ_{LNA}	IIP2	P1dB	Zout	Rload	Cload
ζ_{Mixer}	IIP3	IIP2	P1dB		

Table 1: Elements of performance vectors ζ_{LNA} and ζ_{Mixer}

The contracts of the two components are:

$$\mathbf{C_{LNA}} = \bigcup_{\alpha_{iLNA}^d \in \mathcal{A}_{LNA}^d} (\{\mathcal{A}_{LNA}^s, \alpha_{iLNA}^d\}, g_{iLNA}) \tag{13}$$

$$\mathbf{C_{Mixer}} = \bigcup_{\alpha_{imixer}^d \in \mathcal{A}_{mixer}^d} (\{\mathcal{A}_{mixer}^s, \alpha_{imixer}^d\}, g_{imixer}) \tag{14}$$

where $g \subseteq \mathcal{Z}$ are the subsets of the characterized performance space that are guaranteed under the given assumptions. The elements of performance vectors, $\zeta \in g$, are shown in Table 1 .

To characterize the feasible performance regions of each component, we first generate a sample set by running a series of batch simulation. Then, SVM is used to characterize the sample points into a continuous region, which forms the feasible performance space, $\mathcal{P}$, used in system-level optimization. 1338 configurations of the LNA were simulated, using uniformly randomized RC loads within the assumed range. 2132 configurations of the mixer were simulated, using a source impedance of 50Ω. Since the mixer block just had one set of dynamic assumptions, we do not need to vary the source impedance during characterization. On the contrary, dynamic assumptions of the LNA directly correspond to the guaranteed performance, that is varied loading conditions causes varied performances. During performance characterization, this phenomenon can be reflected by including dynamic assumptions as part of the performance vector. Appending dynamic assumptions increase the dimensions of the SVM by d, where d is the number of columns of the dynamic assumption. In the case of the LNA, d is 2 for *Rload* and *Cload*.

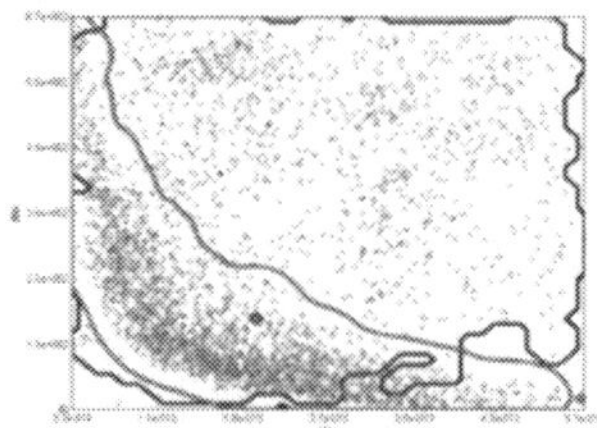

Figure 5: Intersection of LNA RC loading assumptions (blue) and mixer input RC performances (red).

Before composition, we must check for component consistency conditions:

$$\mathcal{A}_{LNA}^s = \mathcal{A}_{mixer}^s \tag{15}$$

$$\exists\, (\alpha_{LNAi}, g_{LNAi}) \in \mathbf{C}_{LNA}, (\alpha_{mixeri}, g_{mixeri}) \in \mathbf{C}_{mixer} \text{ s.t.}$$

$$\{\alpha_{LNAi}^d \cap g_{mixeri}(\lambda)\} \neq \emptyset \tag{16}$$

$$\{\alpha_{mixeri}^d \cap g_{LNAi}(\lambda)\} \neq \emptyset \tag{17}$$

where λ is the interconnection between LNA and mixer, and $g(\lambda)$ are the set of guaranteed performances on λ. Specifically, $g_{LNA}(\lambda)$ is the Z_{out} value of the LNA, and $g_{mixer}(\lambda)$ are R_{in} and C_{in} values of the mixer. Automatic construction of the consistent region is, in general, non-trivial and computationally expensive. Thus,

Perf. @3.96 GHz	LNA Opt.	Mixer Opt.	LNA NN	Mixer NN	System Estimated	System Simulation	Ref. [5]
NF(dB)	3.49	13.3	3.49	13.2	3.67	3.89	4.36
Power Gain(dB)	23.5	-3.57	23.5	-3.34	19.9	18.74	17.2
IIP3(dBm)	-13.7	-7.37	-13.7	-7.26	-30.9	-27.96	-22.5
P1dB(dBm)	-21.2	-14.7	-21.2	-14.8	-38.3	-38.85	-34.6
Power(mW)	5.39	1.62	5.39	1.67	7.07	**7.07**	**10.8**

Table 2: Optimization and simulation results

we first estimate its existence by checking for the bounding-box intersection between the assumptions of the LNA and the guarantees of the mixer, and vice versa. If this preliminary consistency check passes, we frame the contract conditions as constraints in the system-level optimization problem, which will bound the search space to within the consistent regions. By inspection, Figure 5 shows that (16) indeed is true, similarly (17) also holds. In fact, all of the characterized LNA performance space satisfied (12), so (17) is always true in this case.

4.3   Composition and Optimization

We aim to minimize power consumption of the RF front-end while meeting system constraints on IIP3, P1dB, Gain, and NF. The system level optimization problem is:

$$min_{\zeta_{LNA}, \zeta_{Mixer}} \omega_1 \cdot power + \omega_2 \cdot \Theta_1(NF) \tag{18}$$

$$\text{s.t.} \begin{cases} \left.\begin{array}{l} IIP3 \geq -35dBm \\ Gain \geq 18dB \\ NF \leq 5dB \end{array}\right] & \text{System Specifications} \\ \left.\begin{array}{l} \mathcal{P}_{LNA}(\zeta_{LNA}) = 1 \\ \mathcal{P}_{mixer}(\zeta_{mixer}) = 1 \end{array}\right] & \text{Feasibility Constraints} \\ \zeta_{LNA}(9, 10) = \zeta_{mixer}(1, 2) & \text{Contract Constraints} \end{cases}$$

where ω_i are weight coefficients and Θ is a penalty function. System performances (power, NF, gain, IIP3, and P1dB) are calculated from the ζ_{LNA} and ζ_{mixer} using simple cascade equations. $\mathcal{P}_{LNA}$ and $\mathcal{P}_{mixer}$ are SVM classifiers to bound ζ_{LNA} and ζ_{mixer} in the feasible performance space. Furthermore, a legal composition must satisfy contract constraints between elements $\zeta_{LNA}(9, 10)$, representing a_{LNA}^d, and $\zeta_{Mixer}(1, 2)$, representing $g_{mixer}(\lambda)$. As in [5],we employ adaptive simulated annealing [13] as the optimization engine, modified such that the stochastic search is within the feasibility regions given by $\mathcal{P}$, as well as consistent regions of composition.

5.   RESULTS

Optimization finished after evaluating 20730 design instances of the LNA and mixer, 7186 of which satisfied contract assumptions and formed *legal* compositions. The total optimization time took 21 minutes on a 3.16 GHz Intel Core2 Duo Workstation. Since we used equality constraints for the LNA assumptions, extra optimization time were required for the simulated annealing engine to find satisfactory points. Less stringent inequality constraints may also be used for assumptions to speed up optimization, but at the cost of less accurate performance estimations.

We verify optimization results by comparing between estimated system performances and transistor-level simulations from Spectre-eRF. We use a nearest-neighbor (NN) search to map the optimization results onto a set of configuration vectors saved from the performance characterization phase.

Table 2 shows the optimized results, NN performance parameters, estimated system performance, and simulated results of the RF front-end, based on NN configurations. The estimated system performance, based on simple cascade equations, closely matched simulation results without any function fitting or adjustments, showing that contracts preserved model accuracy at the system level after

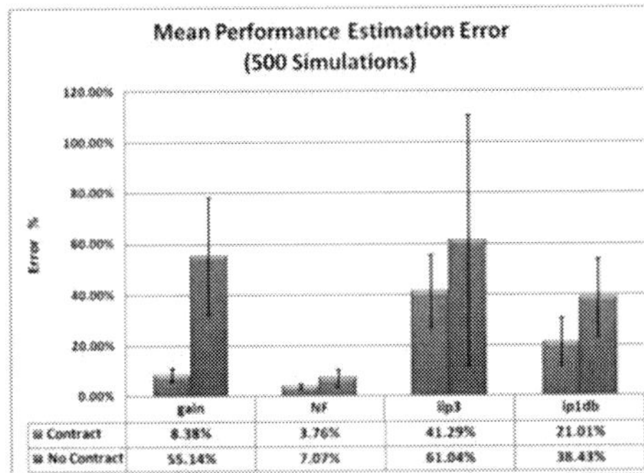

	gain	NF	iip3	ip1db
Contract	6.38%	3.76%	41.29%	21.01%
No Contract	55.14%	7.07%	61.04%	38.43%

Figure 6: Contract-based composition yielded lower average estimation error for all performances, as well as a lower standard deviation, as represented by the error bars.

composition. The quality of the optimized system is investigated by comparing with the reference system [5], which was optimized without using contracts and required an intermediate buffer to fix interface conditions. By removing the interface buffer, the system power consumption reduced by 34.5%, while improving system noise figure and gain performances. The linearity of our optimized system is slightly degraded in comparison. The main reason is because by removing the interface buffer, certain combinations of the LNA and mixer designs were not contained in the consistent regions. Indeed, a higher level tradeoff exists for the usage of interface buffers, which our methodology now allows. It should be noted that the reported voltage gain from [5] was 29.2 dB, based on a high-impedance output loading ($1\text{M}\Omega$ in parallel with 1pF) on the mixer during characterization. For a fair comparison, we normalized the mixer output load in [5] to 50Ω, which is the mixer output loading used in our characterization environment, and found that the equivalent power gain to be 17.2 dB.

To further examine improvements in accuracy, we compare the errors between systems composed with and without contract conditions. Figure 6 shows the errors between estimated performances and simulation results, averaged over 500 individual system compositions, for both *contract* and *no contract* cases. For all performance parameters, contract-based composition resulted in significantly lower average estimation error, as well as lower standard deviation on the error. Errors on the linearity performance parameters remained relatively high because the interface models considered in this case study were simplified RC models and may not represent additional non-linear loading effects (e.g. parasitic capacitance) introduced at the mixer input. However, even this simple approximation of the interface shows that contracts indeed help in maintain model accuracy at high abstraction levels.

6. CONCLUSIONS

We proposed a *contract* extension to APBD to enable system-level composition of analog circuits. Contract-based composition extends the set of validity laws in the current APBD definition and allows automatic detection and integration of consistent components in a given AP library. As an example, we applied contract-based composition to the design of an UWB receiver RF front-end system, consisting of a T/R switch, LNA, and a mixer. Results show *automatic* system integration with a 34.5% power reduction from the reference design by removing extra circuitry needed to fix interface conditions. Furthermore, contracts improve upon the system-level performance estimation accuracy. Without contracts, direct composition of pre-characterized components is highly unreliable, forcing designers to rely on either manual integration or extra circuitry/design cost to "match" components. Thus, contracts improve design productivity by enabling design reuse and automatic composition of analog circuits at the system level, while preserving the accuracies of the underlying performance models.

7. ACKNOWLEDGEMENT

The authors wish to thank Fernando De Bernardinis and Yanmei Li for helpful discussions. They also thank the Gigascale Systems Research Center, United Microelectronics Corporation, and the sponsors of the Berkeley Wireless Research Center (BWRC) for supporting this work.

8. REFERENCES

[1] H. Chang, A. Sangiovanlli-Vincentelli, F. Balarin, E. Charbon, U. Choudhury, G. Jusuf, E. Liu, E. Malavasi, R. Neff, and P. Gray, "A top-down, constraint-driven design methodology for analog integrated circuits," *Custom Integrated Circuits Conference, 1992., Proceedings of the IEEE 1992*, pp. 8.4.1–8.4.6, May 1992.

[2] T. Eeckelaert, T. McConaghy, and G. Gielen, "Efficient multiobjective synthesis of analog circuits using hierarchical pareto-optimal performance hypersurfaces," *DATE '05 Proceedings*, pp. 1070–1075 Vol. 2, March 2005.

[3] F. De Bernardinis, P. Nuzzo, and A. Sangiovanni-Vincentelli, "Mixed signal design space exploration through analog platforms," in *DAC 2005 Proceedings*, 2005, pp. 1390–1393.

[4] F. De Bernardinis, M. I. Jordan, and A. Sangiovanni-Vincentelli, "Support vector machines for analog circuit performance representation," in *DAC '03 Proceedings*, 2003, pp. 964–969.

[5] Y. Li, C.-C. Wu, A. Sangiovanni-Vincentelli, and J. Rabaey, "Design and optimization of an mb-ofdm ultra-wideband receiver front-end," in *ICCSC '08 Proceedings*, April 2008.

[6] F. De Bernardinis, S. Gambini, R. Vincis, F. Svelto, A. Sangiovanni-Vincentelli, and R. Castello, "Design space exploration for a umts front-end exploiting analog platforms," in *ICCAD '04 Proceedings*, 2004, pp. 923–930.

[7] J. Misra and K. Chandy, "Proofs of networks of processes," *Software Engineering, IEEE Transactions on*, vol. SE-7, no. 4, pp. 417–426, July 1981.

[8] K. L. McMillan, "A compositional rule for hardware design refinement," in *CAV '97: Proceedings of the 9th International Conference on Computer Aided Verification*. London, UK: Springer-Verlag, 1997, pp. 24–35.

[9] L. Benvenuti, A. Ferrari, E. Mazzi, and A. Sangiovanni-Vincentelli, "Contract-based design for computation and verification of a closed-loop hybrid system," in *HSCC '08 Proceedings.*, 2008, pp. 58–71.

[10] H. Giese, "Contract-based component system design," in *HICSS-33 Proceedings*, 2000.

[11] E. Lee and A. Sangiovanni-Vincentelli, "A framework for comparing models of computation," *Computer-Aided Design of Integrated Circuits and Systems, IEEE Transactions on*, vol. 17, no. 12, pp. 1217–1229, Dec 1998.

[12] V. Giannini, P. Nuzzo, F. De Bernardinis, J. Craninckx, B. Come, S. D'Amico, and A. Baschirotto, "A synthesis tool for power-efficient base-band filter design," in *DATE '06 Proceedings*, 2006, pp. 162–163.

[13] L. Ingber, "Adaptive simulated annealing (asa): Lessons learned," *Control and Cybernetics*, vol. 25, pp. 33–54, 1996.

Serial Reconfigurable Mismatch-tolerant Clock Distribution

Atanu Chattopadhyay   Zeljko Zilic

McGill University, Department of Electrical and Computer Engineering
3480 University, Montreal, Quebec, Canada, H3A 2A7
atanu.chattopadhyay@mail.mcgill.ca, zeljko.zilic@mcgill.ca

ABSTRACT

We present an unconventional clock distribution that emphasizes flexibility and layout independence. It suits a variety of applications, clock domain shapes and sizes using a modular standard cell approach that compensates intra-die temperature and process variances. Our clock distribution provides control over regional clock skew, permits use in beneficial skew applications and facilitates silicon-debug. By adding routing to the serial clock network, we permit post-silicon resizing and reshaping of clock domains. Defective sections of the clock network can be bypassed, providing post silicon repair capability to the network.

Categories and Subject Descriptors

B.7.3 Reliability and Testing [**Integrated Circuits**].

General Terms

Design, Reliability.

Keywords

Clock networks, process variation, clock skew.

1. INTRODUCTION

In deep sub-micron technologies, device and interconnect variance is leading to an ever-increasing amount of uncertainty that must be addressed [1], particularly with clock distribution networks (CDNs).

We present a CDN that differs radically from standard designs, by using a serial approach tolerant to clock buffer mismatches and capable of post-silicon re-shaping of clock domains. The system provides all the benefits of closed loop CDNs, using an active synchronization stage to eliminate clock skew between regional clocks, while avoiding many of their pitfalls. The all-digital circuitry uses an open loop approach at run-time to provide a simple to implement low-power operating mode.

2. SERIAL CLOCK NETWORKS

Our serial clock distribution network aligns each local clock to half the phase difference between two reference clocks traveling in opposite directions. This averaging technique was first proposed by Grover et al. [2] and has been used in [3,4]. Our

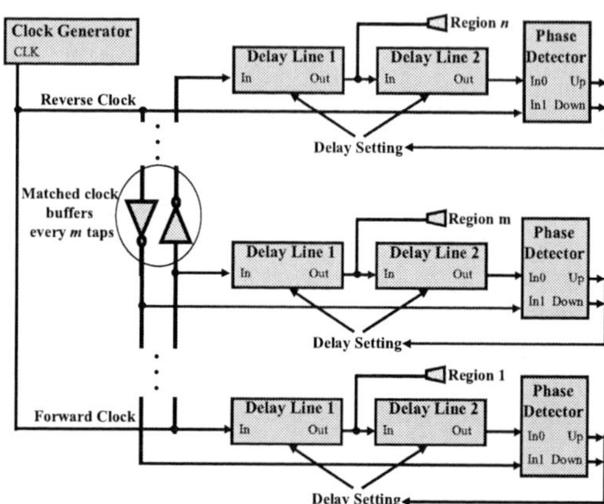

Figure 1. Dual reference line clock network.

clock network is the first to use it to mitigate mismatch effects in a clock network.

2.1 Concept

The underlying concept of our clock network is shown in Figure 1 for *n* taps. All the taps are connected together as a *thread* using a pair of wires to propagate forward and reverse reference clock signals, creating a clock domain with the required shape and size. While there is more than one method to perform the required averaging, we employ a technique that delays the forward clock through two identical delay lines. The signal between these delay lines is used as the local clock. A simple phase detector is required to determine which reference signal edge occurs first.

Our dual reference signal averaging clock network simplifies clock network design since there are no constraints placed on the location of clock regions and the clock path taken between regions. Our network can be implemented with standard cell components and allows modification of portions of the clock network without complete reconstruction. The technique is easily ported to other technologies since the characteristics of devices and interconnect do not matter as much as how they match.

2.2 Reconfigurability

Reconfiguring clock domains post-silicon is not easy for typical IC clock networks. The clock threads in our serial network can be reconfigured using routing switches between local clock regions. This functionality is impossible when clock signals are broadcast through an integrated circuit, as is the case with clock trees. The extent of flexibility is variable; it can be as small as connecting a shared resource synchronously between two domains to a full fledged multi-clock mesh where local taps can be arbitrarily connected to any clock in the system. Devadas et al. [5] has used a bidirectional mesh similar to ours for data networks, but our application of this approach for clock networks is unique.

Permission to make digital or hard copies of part or all of this work for personal or classroom use is granted without fee provided that copies are not made or distributed for profit or commercial advantage and that copies bear this notice and the full citation on the first page. To copy otherwise, to republish, to post on servers or to redistribute to lists, requires prior specific permission and/or a fee.

DAC'09, July 26-31, 2009, San Francisco, California, USA

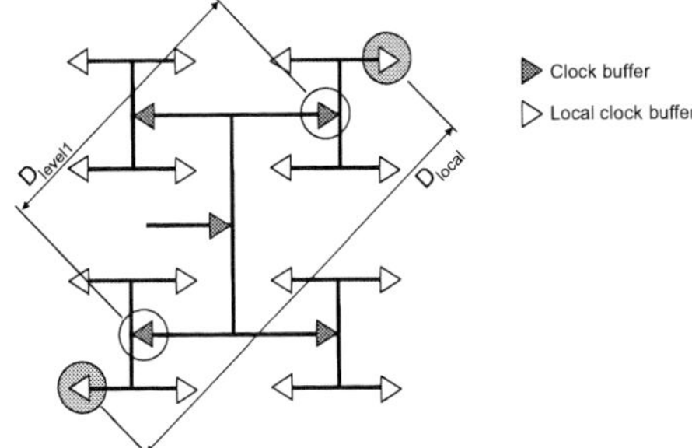

Figure 2. Increasing distance between clock buffers.

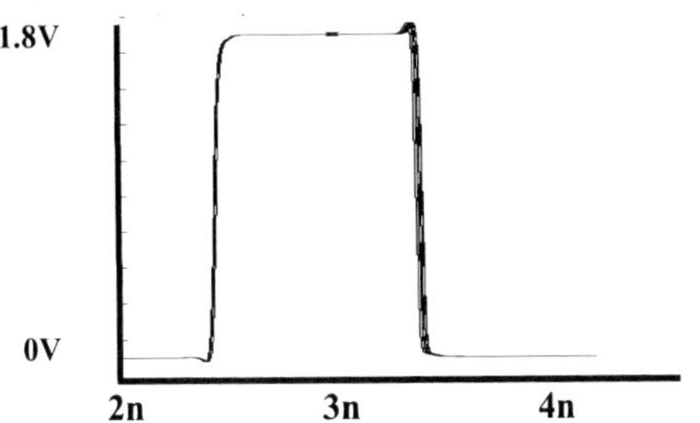

Figure 3. Six clock outputs for a serial clock thread.

3. VARIABILITY IN CLOCK NETWORKS

Process variation results in two kinds of mismatch in an integrated circuit (IC). Inter-die mismatch affects all devices on a die equally and does not alter the matching of components on a single die. Intra-die mismatch has been modeled by Pelgrom et al.'s relation for variance due to parameter (P) deviation [6]:

$$\sigma^2(\Delta P) = \frac{A_p^2}{WL} + S_p^2 D^2 \qquad (1).$$

W and L represent transistor width and length, respectively, and D represents the distance between devices. The A_P^2 term models the distance variance and the S_P^2 term models the discrete variance. Minimizing process variance requires using sufficiently large transistors to decrease discrete variance and locating the centers of devices as close together as possible using centroid layouts to minimize the distance variance. Equation 1 can be extended to these centroid layouts:

$$\sigma^2_{centroid}(\Delta P) = \frac{A_p^2}{WL} + \frac{S_p^2 D_x^2 D_y^2}{12 D_w^2} \qquad (2)$$

where D_x and D_y are the horizontal and vertical distances between devices and D_w is the wafer diameter [7]. Even though process gradients are never perfect planes, Equation 2 shows that placing clock buffers close to each other will result in much better matching than the dispersed clock buffers typical of current CDNs that do not allow them to be co-located. The clock buffers requiring matching in our clock network are adjacent, Figure 1.

As clock drivers get further apart, the potential mismatch increases. The total distance related skew accumulates through every level of clock buffers. In a symmetric tree structure, the worst-case skew will occur between diagonally opposite local buffers since driver pairs here are furthest apart at every level, Figure 2. The skew performance of our serial clock network depends on the matching of the forward and reverse reference signal segments between adjacent clock regions. Since co-located clock drivers are inherently well-matched, they are tolerant to distance related skew. By the same argument, distance related interconnect variance is also suppressed by our system.

Increased power density in ICs can cause significant cross-die temperature fluctuation, or so called "hot spots" that alter transistor and interconnect behavior. Power supply variation can also modify the delay of clock buffers, creating clock skew. Placing devices requiring matching close together will expose them to the same power supply and temperature environment, so our system can be synchronized to correct these conditions locally, but traditional distributed buffers in clock trees cannot be.

4. SIMULATION RESULTS

Our clock network has been designed using TSMC's 180 nm standard process using Cadence Virtuoso. Extracted layout simulations show that our proof of concept design can operate with clock signals between 500 MHz and 2.5 GHz and provides a sub-15 ps skew bound for 6 clock regions, Figure 3.

5. CONCLUSION

The system provides multi-point active skew compensation and a power-saving open-loop operating mode. Our cell based approach to clock distribution allows components to be designed independently and to be moved around conveniently since the clock network can be modified with a simple change in the number or location of the clock taps. The presence of the digitally programmable delay lines allows the system to accommodate blocks with different tree depths and latencies.

Using a dual reference signal averaging technique allows designers to delay clock tuning and provides additional debug and repair capability to the clock network. Programmable repeater stages allow us to redirect clocks post-silicon. Using a serial approach minimizes the total clock line length, reducing the total clock load and potentially clock power. By placing clock buffers close together and using centroid layout techniques, it is possible to practically eliminate all distance induced variation. Clock buffers and delay lines will exhibit similar temperature and power supply characteristics allowing compensation of temperature and long term power supply fluctuation in our clock network.

6. REFERENCES

[1] J. Rosenfeld and E.G. Friedman, "Design methodology for global resonant H-tree clock distribution networks," *Proc. ISCAS* 2006.

[2] W.D. Grover, J. Brown, T. Friesen and S. Marsh, "All-digital multipoint adaptive delay compensation circuit for low skew clock distribution," *Electronics Letters*, vol. 31, issue 23 (9, Nov. 1995), 1996-1998.

[3] A. Kapoor, N. Jayakumar and S.P. Khatri, "A novel clock distribution and dynamic de-skewing methodology," *Proc. ICCAD* 2004, 626-631.

[4] A. Chattopadhyay and Z. Zilic, "Reconfigurable clock distribution circuitry," *Proc. ISCAS* 2007, 877-880.

[5] M. H. Cho, M. Lis, M. Kinsy, K. S. Shim, T. Wen, and S. Devadas, "Oblivious Routing in On-Chip Bandwidth-Adaptive Networks," CSAIL Technical Report TR-2009-011, March 2009.

[6] M.J.M Pelgrom, A.C.J. Duinmaijer and A.P.G. Welbers, "Matching properties of MOS transistors," *IEEE Journal of Solid-State Circuits*, 24, 5 (Oct 1989), 1433-1439.

[7] B. Linares-Barranco and T. Serrano-Gotarredona, "Cheap and easy systematic CMOS transistor mismatch characterization," *Proc. ISCAS* 1998, pp. 466-469.

Thermal-Aware Data Flow Analysis

José L. Ayala[1]   David Atienza[2]   Philip Brisk[2]

[1]DACYA, Complutense University of Madrid – 28040 Madrid (Spain)
E-mail: jayala@fdi.ucm.es
[2]ESL and LAP, EPFL – 1015 Lausanne (Switzerland)
E-mail: {david.atienza, philip.brisk}@epfl.ch

ABSTRACT

This paper suggests that the thermal state of a processor can be approximated using data flow analysis. The results of this analysis can be used to evaluate the efficacy of thermal-aware compilation strategies, or as input to thermal-aware optimizations that occur in the early stages of back-end compilation. We propose different ways how the exploitation of thermal behavior knowledge can be included in the different compilation phases.

Categories and Subject Descriptors

C.1 [**Computer Systems Organization**]: Processor Architectures

General Terms

Algorithms, Theory

Keywords

Thermal Management, Compiler.

1. INTRODUCTION

Thermal management has become an increasingly important issue in modern semiconductor devices. In particular, steep thermal gradients have been shown to significantly reduce the reliability of silicon systems. In processors, one particular area of concern is the *register file (RF)*, which has high power density and is accessed every cycle [1]. In response, several research groups have recently proposed thermal-aware register assignment techniques for the RF [2, 3], or thermal-aware instruction binding in VLIW processors [4]. One of the key challenges for thermal-aware compilation in the near future is to model the effects of program transformations on power density and thermal gradients. State-of-the-art thermal emulation tools require compiled programs in order to characterize the thermal state of the processor [5]; this limits their usage, in practice, to feedback-driven optimization frameworks.

This paper suggests a more radical approach, namely, a compiler may be able to predict, with reasonable accuracy, the thermal state of the processor at every point in the program. Although our idea is to consider the benefits and possibilities of thermal-aware data flow analysis in general, the motivating example and following discussion will emphasize optimizations relating to the RF as an exploratory case study. The key point of the analysis—and the reason that we consider the idea to be "wild and crazy"—is that, as the analysis is applied prior to register allocation and assignment, and at this stage there is no information about the layout of the RF and the placement of registers, some accuracy could be lost due to the taken assumptions; but, this is the only alternative to feedback-driven thermal optimization.

Permission to make digital or hard copies of part or all of this work for personal or classroom use is granted without fee provided that copies are not made or distributed for profit or commercial advantage and that copies bear this notice and the full citation on the first page. To copy otherwise, to republish, to post on servers or to redistribute to lists, requires prior specific permission and/or a fee.
DAC'09, July 26-31, 2009, San Francisco, California, USA

2. MOTIVATING EXAMPLE

Two variables *interfere* in a program if their lifetimes overlap. Interfering variables cannot be assigned to the same register or memory location in order to avoid any overwriting. As a result, when a compiler assigns a register to variable v, it can choose from any register *not* assigned to a variable that interferes with v. In general, the compiler maintains an ordered list of registers and selects the first one in the list that is free. As the list is always traversed in order, the same small set of registers is chosen again and again. As far as performance is concerned, this is generally permissible, as registers are assumed to be interchangeable; however, repeatedly choosing the same registers for assignment is not the best choice from a thermal management point of view.

Fig. 1 shows thermal maps for three register assignment policies: choosing the first free register from an ordered list (Fig. 1(a)); randomly choosing a free register (Fig. 1(b)); and a "chessboard" pattern (Fig. 1(c)) [2]. The first two policies result in clear RF hot spots with steep thermal gradients; the chessboard policy, however, shows a homogenized temperature map, which occurs because the accesses are distributed uniformly across a large surface. The chessboard policy, however, only works if the program only uses half of the registers in the RF. Indeed, if register pressure is high, then all registers will be used, and may be accessed repeatedly. If certain registers are accessed more than others, then thermal gradients may still appear and reliability can suffer even trying to apply the chessboard pattern of Fig. 1(c).

3. BASICS OF DATA FLOW ANALYSIS

A compiler uses data flow analysis to determine properties about a program that are either required for correctness or useful for optimization. Depending on the needs of the analysis, different types of information must be propagated through the data flow solver. For example, *liveness analysis* asks whether a variable is *live* at each point in the program—if so, then a storage location must be allocated for that variable. In this case, a single bit of information per variable is required. *Bitwidth analysis* [7], which is more complex, propagates an interval for each variable, which represents the range of possible values that it can take; the bitwidth can be derived from this interval. The thermal analysis, proposed in the next section, must propagate a floorplan-aware estimate of the thermal state of the processor, which is considerably more complex than a bit or an interval for each variable. The thermal state is a continuous function that can only be approximated, typically as a discrete set of points. The fidelity of the analysis will depend on the granularity of the approximation—increasing the number of points would increase accuracy, but at the cost of increased computation time.

4. THERMAL DATA FLOW ANALYSIS

The proposed data flow analysis would be a forward analysis [6]. For simplicity, we describe it in the context of a single procedure. A pseudocode description is shown in Fig. 2. The analysis repeatedly computes the thermal state of the RF following each instruction, which relates the technology coefficients of logic activity and peak

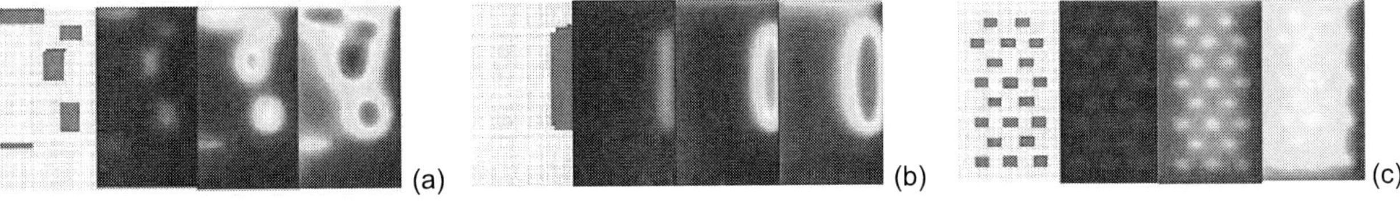

Figure 1. Thermal maps for register assignment policies: (a) deterministic order; (b) random; and (c) chessboard.

power found in the thermal models [1, 5]. The coefficients are linked in an analytical way to the high-level information of instruction execution and variables assignment found in the early compilation stages. If the change in thermal state prediction for at least one instruction exceeds a user-supplied parameter (δ), then another iteration is required; otherwise, the analysis terminates and the thermal state following each instruction is output.

```
Do
      Boolean: stop ← True
      For each basic block B
            For each instruction I∈ B, taken in forward order
                  Estimate thermal state after I
                  If the change in I's thermal state exceeds δ
                        stop ← False
                  EndIf
            EndFor
      EndFor
While( stop = False )
Output the thermal state of each instruction
```

Figure 2. Pseudocode of proposed data flow analysis.

Unlike traditional data flow analyses, in the proposed thermal-aware analysis does not appear to be a way to guarantee convergence; however, if the analysis does not converge after a reasonable number of iterations ("reasonable" must be determined empirically or user-defined), this suggests that the thermal state of the program may be too difficult to predict at compile time due to a very irregular data usage. Thus, to ensure reliability, the program could be re-optimized so that its thermal state becomes more predictable. Also, the result of the analysis phase can be used to conduct the compilation process achieving a temperature-aware compilation at different stages.

The proposed thermal analysis makes the most sense if applied after register assignment, as the precise registers that are accessed by each instruction are known, but then no aggressive thermal-aware high-level optimizations are possible. However, the more ambitious possibility that we propose in this paper, which has never been considered before, would be to develop predictive analyses that would be performed at earlier stages of compilation, i.e., before register allocation and assignment, or perhaps even before instruction scheduling. While it is not easy to promote the thermal information to the early compilation stages, the development of a set of rules that qualify the impact of the compiler decisions on the thermal profile will allow the envisioning of later thermal-aware compilation without the feedback of temperature information, which involves a time-consuming thermal simulation phase of the target processor [2, 5].

The goal of these analyses would not be to accurately predict the thermal state of the process at each point; instead, the goal would be to determine precisely which parts of the program are likely to exacerbate power density and thermal problems in the RFs, and to determine which variables are most likely to be involved. In particular, if just two variables are involved, they can easily be assigned to registers in disparate regions of the RF; however, when more variables are likely to create hot spots, it becomes increasingly

difficult to assign them to registers in different regions of the RF, especially when register pressure is high.

For the purposes of thermal management, the greatest benefit will be achieved by spilling these "critical" variables to memory, or splitting them (via copy insertion) to spread their accesses across a multitude of registers. As it has been seen in the presented thermal maps, spreading (in space) the accessed registers reduces the appearance of hotspots and thermal gradients. Also, the thermal diffusion between the accessed registers homogenizes the temperature on the device and improves its reliability by decreasing leakage. However, power reduction techniques based on switching off register banks could not theoretically be applied after the spread register assignment, and a compromise between these types of techniques for different optimization metrics can be explored at the compiler level.

Other possible optimizations that could be driven by this analysis refer to spreading accesses to registers in time, either using instruction scheduling, to avoid consecutive accesses to already hot registers, or using register promotion (i.e., promoting some memory-resident variables into registers), which would help on avoiding the thermal gradients between hot and cold registers, by making more uniform the use of registers in time. Finally, the insertion of NOP instructions gives the RF a chance to cool down between accesses in extremely hot situations, although it can affect overall system performance and should be applied only if no other option to cool down the system is feasible.

5. CONCLUSIONS

This paper proposes that compilers can estimate the thermal state of a processor in early stages of compilation using data flow analysis; in particular, we have suggested how different compiler transformations can guide thermal-aware optimization methods for the RF, which is the first step to develop even more general thermal-aware high-level optimization in the future. In the long-term, our goal is to develop comprehensive data flow thermal analyses and rules relating to all parts of the processor and to use them to better understand the impact of high-level transformations on thermal gradients and power density at the same time.

REFERENCES

[1] Srinivasan, J., et al. Predictive dynamic thermal management for multimedia applications. In *Proc. 17th ICS*, pp. 109-120, 2003.

[2] Atienza, D., et al. Reliability-aware design for nanometer-scale devices. In *Proc. ASPDAC*, pp. 549-554, 2008.

[3] Zhou, X., et al. Compiler-driven register re-assignment for register file power-density and temperature reduction. In *Proc. DAC*, pp. 750-753, 2008.

[4] Schafer, B. C., et al. Temperature-Aware Compilation for VLIW Processors. In *Proc. RTCSA*, pp. 426-431, 2007.

[5] Atienza, D., et al. HW-SW emulation framework for temperature-aware design in MPSoCs. *ACM TODAES*, 12(3), pp. 1-26, August, 2007.

[6] Cooper, K. D., and Torczon, L. Engineering a Compiler, *Morgan-Kaufmann*, San Francisco, CA, USA, 2003.

[7] Stephenson, M., et al. Bitwidth analysis with application to silicon compilation, In *Proc. of PLDI*, pp. 108-120, 2000.

Nanoscale Digital Computation Through Percolation

Mustafa Altun, Marc D. Riedel, and Claudia Neuhauser

University of Minnesota, USA

E-mail: {altu0006, mriedel, neuha001} @umn.edu

ABSTRACT

In this study, we apply a novel synthesis technique for implementing robust digital computation in nanoscale lattices with random interconnects: *percolation theory on random graphs*. We exploit the non-linearity that occurs through percolation to produce Boolean functionality. We show that the error margins, defined in terms of the steepness of the non-linearity, translate into the degree of defect tolerance. We study the problem of mapping Boolean functions onto lattices with good error margins.

Categories and Subject Descriptors

B.7.1 **[integrated circuits]**: Types and Design Styles – *advanced technologies*.

General Terms: Design, Reliability.

Keywords: Percolation, Nanoscale Digital Computation, Logic Synthesis.

1. PERCOLATION THEORY

Percolation theory is a rich mathematical topic that forms the basis of explanations of physical phenomena such as diffusion and phase changes in materials. It tells us that in media with random local connectivity, there is a critical threshold for global connectivity: below the threshold, the probability of global connectivity quickly drops to zero; above it, the probability quickly rises to one [2]. This is illustrated in Figure 1.

Consider the lattice shown in Figure 2(a). Suppose that each square in the lattice is colored black with independent probability p_1. Let p_2 be the probability that a connected path exists between the top and bottom plates. Figure 2(b) shows the relationship between p_1 and p_2 for different square lattice sizes.

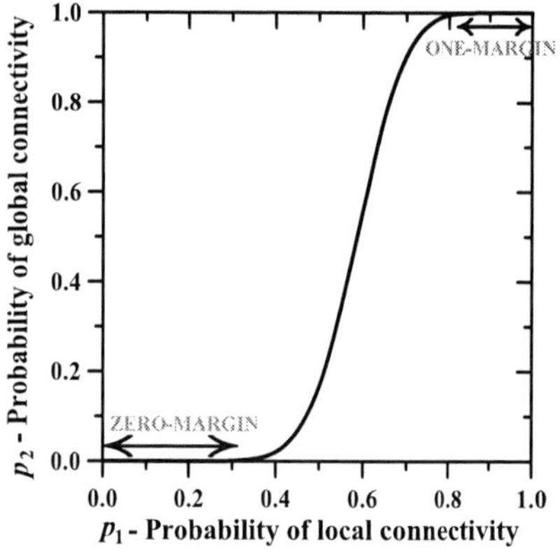

Figure 1. Non-linearity through percolation in random media.

Permission to make digital or hard copies of part or all of this work for personal or classroom use is granted without fee provided that copies are not made or distributed for profit or commercial advantage and that copies bear this notice and the full citation on the first page. To copy otherwise, to republish, to post on servers or to redistribute to lists, requires prior specific permission and/or a fee.

DAC'09, July 26-31, 2009, San Francisco, California, USA

Percolation theory tells us that with increasing lattice size, the curve steepness increases. (In the limit, an infinite lattice produces a perfect step function.) Here p_c is defined as a critical probability below which p_2 is approximately 0 and above which p_2 is approximately 1. We define the **one margin** and the **zero margin** as the corresponding ranges of p_1, i.e., values of p_1 that produce values of p_2 that we interpret as *logical* one and zero, respectively. We exploit the theory in a novel way: we use the nonlinearity produced by percolation to implement digital computation.

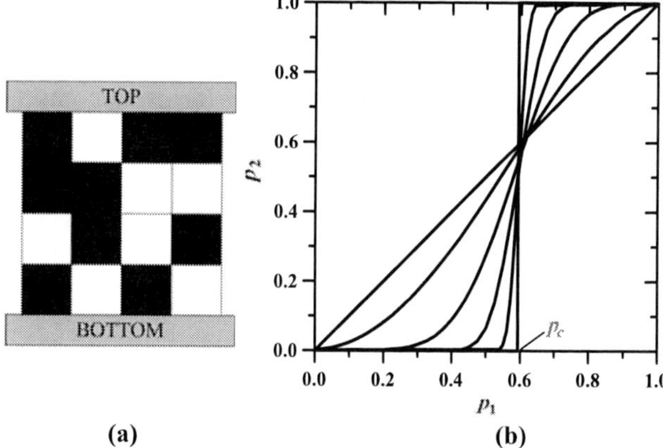

Figure 2. (a) Percolation lattice; (b) p_2 versus p_1 for 1×1, 2×2, 6×6, 24×24, 120×120 and infinite size lattices.

2. NANOWIRE CROSSBAR ARRAYS

Although not tied to any specific technology, we frame our discussion in terms of a conceptual model of nanowire arrays. Figure 3 illustrates a nanowire crossbar array with four plates: left, right, top, and bottom. Figure 4 illustrates connectivity in terms of squares: black squares represent crosspoints that are ON and white squares crosspoints that are OFF. Suppose that in this technology crosspoints between the horizontal and vertical nanowires are FET-like junctions [1]. When a high or low voltage is applied, these develop low or high impedances, respectively. Ideally, if the applied voltage is 0 (corresponding to logic zero), then all the crosspoints are OFF and so there is no connection between any of the plates. If the applied voltage is V_{DD} (corresponding to logic one), then all the crosspoints are ON and so the plates are connected. However, with defects in the lattice, not all crosspoints will respond this way [3]. This is illustrated in Figure 5.

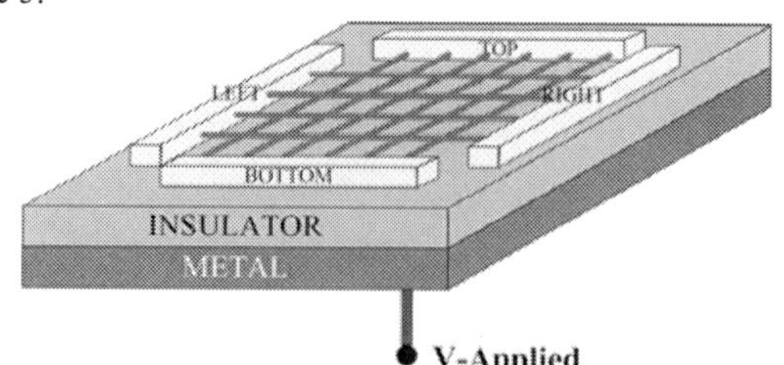

Figure 3. 3-D representation of a nanowire crossbar array.

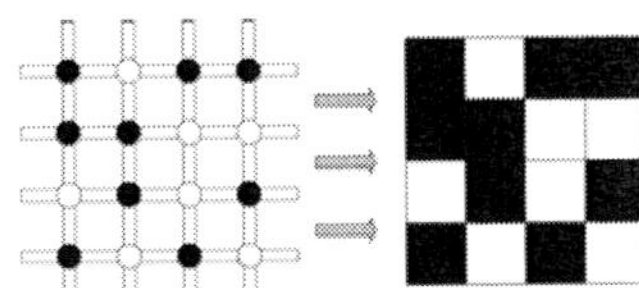

Figure 4. Nanowire crossbar array with random connections.

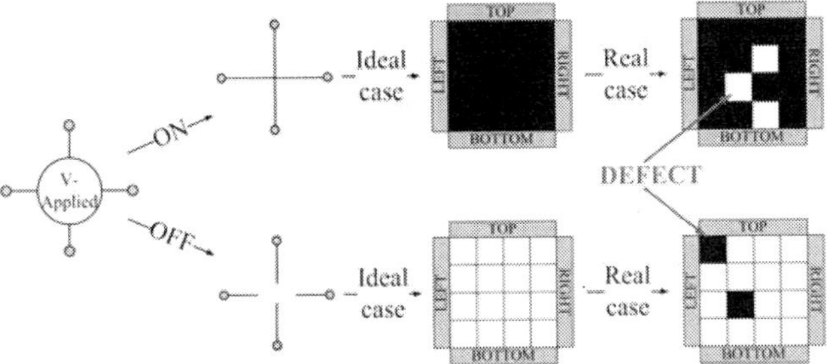

Figure 5. Schematic of a nanowire crossbar based switch.

3. IMPLEMENTING BOOLEAN FUNCTIONS THROUGH PERCOLATION

Our approach for implementing Boolean functions is illustrated in Figure 6. The input values, $X_{11}, ..., X_{rc}$, are applied to distinct regions in the array. In each region, the probability of local connectivity p_1 is ideally only dependent on the corresponding Boolean input value. Connectivity through the entire array is now a Boolean function of the input values via the relationship between the input values and the probability of local connectivity. Call the Boolean functions that are implemented according to the top-to-bottom and left-to-right plate connectivities f and g, respectively. As shown in Figure 7, each Boolean function evaluates to one if there exists a path between corresponding plates, and evaluates to zero otherwise. In this way, digital computation is achieved without switches or logic elements; it occurs due to the random connectivity of the fabric.

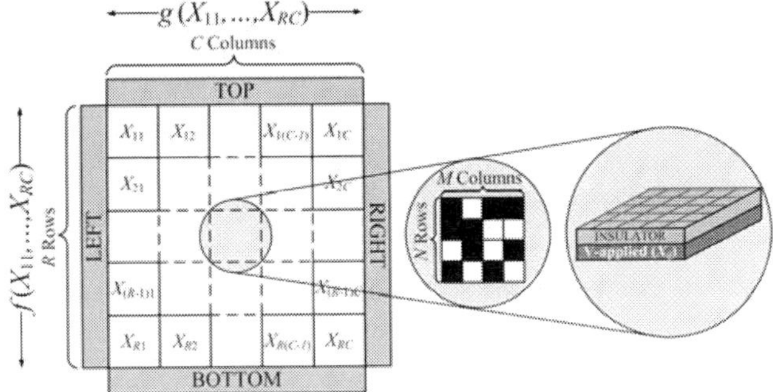

Figure 6. Boolean computation in a random network.

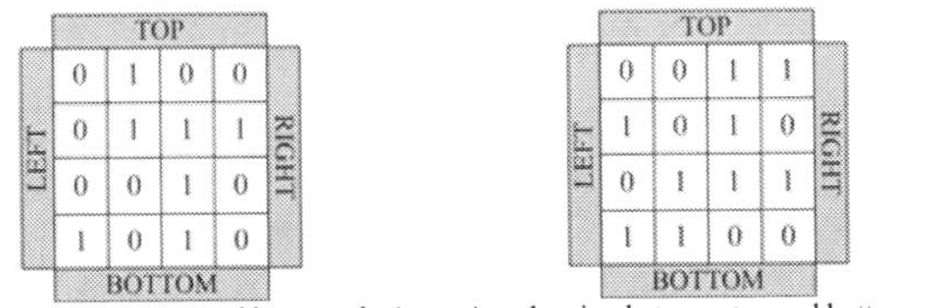

Figure 7. Relation between Boolean functionality and paths.

An important consideration in the synthesis methodology is the quality of the margins. Suppose that the zero and one margins are the range of values for p_1 for which p_2 is always below ε and above $1 - \varepsilon$, respectively, where ε is a very small number. We selected $\varepsilon = 0.001$ in this study. Accordingly, these margins

correlate with the degree of **defect tolerance**. For instance a 10% one margin means that even though we expect that one out of ten crosspoints will not operate correctly – because these are defective, or because of transient errors, or because of noise – the circuit still evaluates to one with high probability ($p_2 > 0.999$). The higher the margins, the higher the defect tolerance that we achieve.

Different assignments of input variables to the regions of the network affect the margins. A 4-input 2×2 lattice is analyzed in Figure 8. As can be seen, the row highlighted in grey has very low margins – indeed, these are nearly zero –so the circuit is likely to produce erroneous values for this input computation. Let's examine why.

X_{11}	X_{21}	X_{12}	X_{22}	f	Margin	g	Margin
0	0	0	0	0	40%	0	40%
0	0	0	1	0	25%	0	25%
0	0	1	1	1	14%	0	23%
0	1	0	1	0	23%	1	14%
0	1	1	0	0	0%	0	0%
0	1	1	1	1	14%	1	14%
1	1	1	1	1	25%	1	25%

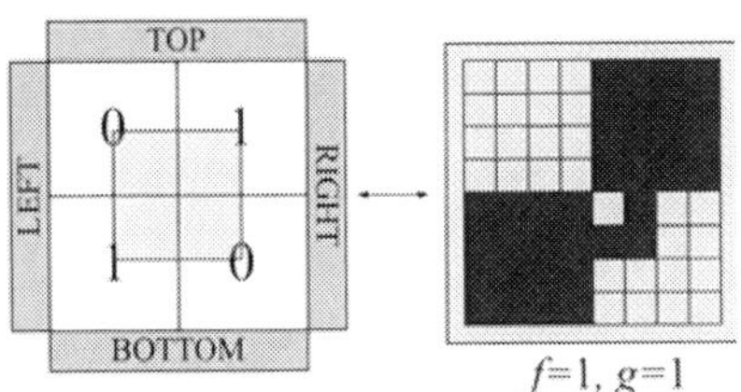

$f = X_{11}X_{21} + X_{11}X_{21}$

$g = X_{11}X_{12} + X_{21}X_{22}$

Figure 8. Four-input lattice and possible 0/1 assignments to the inputs (up to symmetries) and the corresponding margins.

Assignments that evaluate to zero but have diagonally adjacent assignments of blocks of one's result in poor zero margins because there is a good chance that a weak connection from top to bottom will form through stray, random connections across the diagonal. This is illustrated in Figure 9. Note that each 4-connected path corresponds to the AND of the inputs; the paths taken together correspond to the OR of these AND terms, so implement a sum-of-products expression. This interpretation gives rise to an interesting problem in logic optimization: given a Boolean sum-of-products expression as the target function, how should one assign the variables to the regions of the fabric such that there are no diagonally adjacent variables that are one in any assignment that evaluates to zero? This question takes us to the concept of **lattice duality**. A necessary and sufficient condition for good error margins is that the Boolean functions corresponding to the top-to-bottom and left-to-right plate connectivities f and g are dual functions.

Figure 9. An input assignment with a low zero margin.

4. REFERENCES

[1] A. DeHon, "Nanowire-based programmable architectures," *ACM Journal on Emerging Technologies in Computing Systems*, Vol. 1, No. 2, pp. 109–162, 2005.

[2] G. Grimmet, *Percolation*, Springer, New York, NY, 1989.

[3] J. Huang, M. Tahoori, and F. Lombardi, "On the defect tolerance of nano-scale two-dimensional crossbars," *Proc. IEEE Int'l Symp. on Defect and Fault Tolerance of VLSI systems*, pp. 96 - 104, 2004.

A *Learning* Digital Computer

Bo Marr, Arindam Basu, Stephen Brink, Paul Hasler

Georgia Institute of Technology

Atlanta, GA

{hbmarr, arindamb, phasler} *@ece.gatech.edu*

ABSTRACT

The concept of learning digital hardware is presented here. A proof of concept of a circuit that can arbitrarily control the current, and thus the switching speed and power consumption, of a digital circuit is given. This control of current is directly tuned by the feedback from the digital circuit itself, thus a learning digital computer. An argument for a completely new paradigm in digital computing follows whereby an entire system of learning digital circuits is proposed.

Categories and Subject Descriptors

B.7.1 [**Integrated Circuits**]: Types and Design Styles— *Advanced Technologies*

General Terms

Design

Keywords

Floating Gates, Learning

1. INTRODUCTION

What if processors could learn? With all of the myriad applications that our embedded systems, general purpose processors, and reconfigurable arrays of hardware are required to run, we could benefit greatly if our processors could learn exactly what it was we wanted them to do and how we wanted them to do it. Better software is not the solution for this adaptibility problem; afterall, the ultimate performance of software is limited by the hardware itself. For low power processors, software only complicates the matter – the more software, the more instructions, and the more power is burned.

We propose to create a processor where the hardware itself *learns*. The hardware will learn which application it is running and adapt to create stronger circuits in the critical path of the application and will learn which paths are not critical

Permission to make digital or hard copies of part or all of this work for personal or classroom use is granted without fee provided that copies are not made or distributed for profit or commercial advantage and that copies bear this notice and the full citation on the first page. To copy otherwise, to republish, to post on servers or to redistribute to lists, requires prior specific permission and/or a fee.

DAC'09, July 26-31, 2009, San Francisco, California, USA

and thus tune down the power in those areas. The processor will remember what it has learned so that even when the hardware is powered down and the application reloaded at some later time, the processor will go back to the optimal state it learned for that application. Hardware designers or synthesis algorithms would no longer have to spend hours tweaking designs when the designer doesn't even know for which application the circuit will ultimately be tweaked. Microcontrollers and complicated dynamic voltage scaling (DVS) algorithms would become unnecessary. Software or firmware will not be what tweaks the processor, but the fabric of the circuits themselves.

Alas, many have proposed neuronal models of learning and implemented these in analog hardware [4], and some have even proposed this neuronal process in digital by implementing equations that model learning in FPGAs [2]. However, these methods all depend on spike-based neuron models using the spike time dependent plascitiy (STDP) algorithm and cannot be of use to us for a general theory on a learning digital hardware.

2. A KEY CIRCUIT ELEMENT

In order for a processor, or digital circuits, to *learn* a novel circuit element would be introduced with a couple of key features. It would need to be able to

- Be dynamically programmable (during run-time).

- Control current flow arbitrarily in digital circuits.

- Remember or have a memory capacity.

- Be implemented with insignificant overhead to performance or power.

Since the flow of current is what ultimately determines the speed at which a digital circuit switches and its power consumption, a circuit element with the above characteristics would allow for digital circuits to tune their own performance and power. Such a circuit element is given in Figure 1.

Represented in Figure 1 is a floating gate transistor used to control the speed and power of a digital circuit. The gate of the pFET is floating as it has no DC connection and is only capacitively coupled to other nodes, which means it can hold an arbitrary charge on the node. For a faster digital circuit, more charge is allowed onto the floating node, opening up the FET allowing more current to flow. Charge can be taken off for a more power efficient digital circuit.

978-1-60558-497-3/09 $25.00 © 2009 ACM

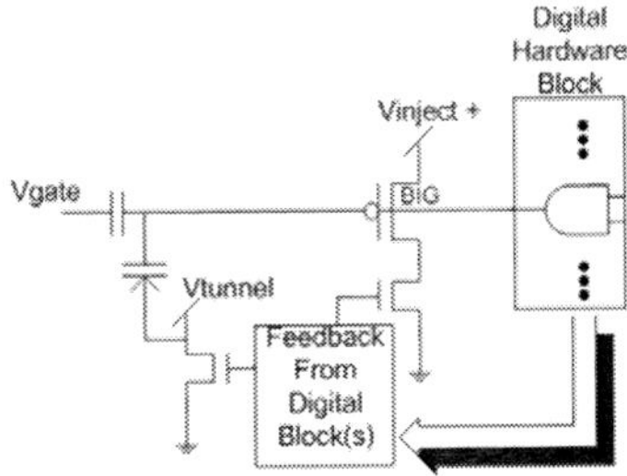

Figure 1: A floating gate transistor with digital feedback to control charge injection onto the pFET's floating node and to control electron tunneling off of the pFET's floating node. This node is also connected to a FET in the digital circuit allowing for a bias current of an arbitrary value (digital circuit of arbitrary speed).

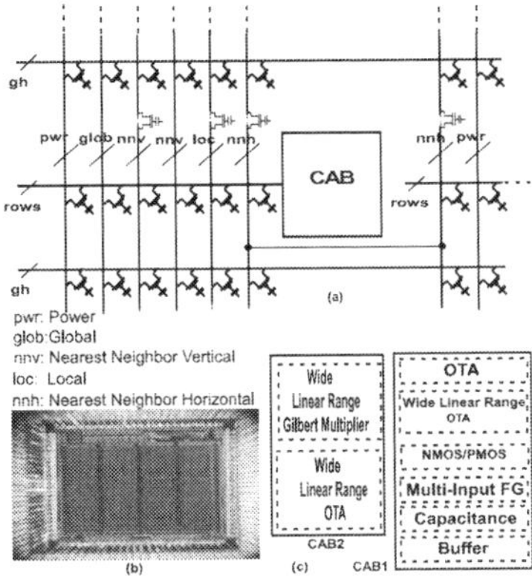

Figure 2: Die photo a reconfirgurable chip used to show proof of concept of this work. It is being redesigned and updated for testing of a large scale version of the digital study shown herein.

The voltage on the gate, V_{fg}, is reduced by charge injection (putting electrons onto the gate node) and is increased through *Fowler-Nordheim* electron tunneling [3], but otherwise does not change and *remembers* the current charge.

Recall that current through a transistor above threshold neglecting velocity saturation is

$$I_{ds} = \mu C_{ox} \frac{W}{L} \frac{(V_{gs} - Vt)^2}{2} \qquad (1)$$

$$I_{transistor} \propto V_{fg}^2 \qquad (2)$$

Thus changing the charge on the floating node would change the current, and thus the speed, in the digital circuit quadratically. In subthreshold digital, of which there has been much interest of late, the relationship is $I \propto e^{V_{fg}}$, which allows for control in ulta-low power circuits as well [3].

An experimental chip has been fabricated showing this concept which will be released in early 2009, a prototype shown in Figure 2.

3. DATAPATH: A CASE STUDY

Now that we have our key circuit element, a case study of how a processor's datapath would benefit is given. Take for example the image processing path of a digital signal processor (DSP). In many compression algorithms, video clips, and movie sequences the pixel data only changes for a very

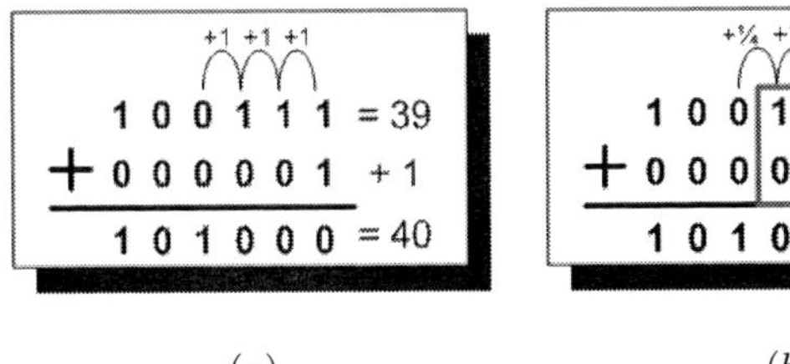

Figure 3: Example showing number of carry operations in the addition of 39 and 1, where 4 carry overs are needed for the critical path. (a) Each 1-bit addition takes unit time, the critical path is equal to 4 time units here. (b) Our learning adder speeds up the first 3 1-bit additions to $\frac{1}{2}$, $\frac{1}{4}$, and $\frac{1}{4}$ time units respectively so the critical path only takes 2 time units total.

few pixels from frame to frame. Typical image data for an H.264 decoder yields repeated inputs to an FIR filter, made up of strictly of adders and multipliers, due to the repeated pixel values being generated by the movie.

Using this H.264 movie decoder example, Figure 3 shows there are 4 carry-overs needed in the addition of a pixel value, 39, being incremented by 1, and this is the critical path using a standard ripple-carry adder. If each 1-bit addition takes unit time, the critical path takes 4 time units to complete. Now, since pixel values are repeated 100s if not 1000s of times in a typical movie scene, it is likely this exact addition, or one close to it, would be repeated 1000s of times. Our datapath *learns* and *strengthens* the critical path (Figure 3b); the first 3 1-bit additions are sped up, and the critical path now takes only 2 time units total for a 2X speed increase. It has been shown that a 2X current increase, and thus this scenario, is quite plausible with floating gate technology [1].

4. SYSTEM OF THE FUTURE

In summary, the techniques presented here could be used to create an entirely new paradigm of a learning digital computer. The next steps to be taken are to determine the best feedback mechanism (asynchronous completion cells are one method), the range of current, speedups, and power gains that can be realized, and to fabricate a datapath with this technology.

5. REFERENCES

[1] A. Basu, C. M. Twigg, S. Brink, P. Hasler, C. Petre, S. Ramakrishnan, S. Koziol, and C. Schlottmann. Rasp 2.8: A new generation of floating-gate based field programmable analog array. *Proceedings, Custom Integrated Circuits Conference CICC*, Sept. 2008.

[2] A. Cassidy and A. Andreou. Dynamical digital silicon neurons. In *IEEE Symposium on Biological Circuits and Systems (BioCAS)*, Nov. 2008.

[3] P. Hasler and J. Dugar. Correlation learning rule in floating-gate pfet synapses. *IEEE Transactions on Circuits and Systems II: Analog and Digital Signal Processing*, Jan. 2001.

[4] S.-C. Liu and R. Mockel. Temporarily learning floating gate vlsi synapses. In *IEEE International Symposium on Circuits and Systems (ISCAS)*, May 2008.

978-1-60558-497-3/09 $25.00 © 2009 ACM

Programmable Neural Processing on a Smartdust

Shimeng Huang[1], Joseph Oresko[1], Yuwen Sun[1], and Allen C. Cheng[1234]
Departments of Electrical and Computer Engineering[1], BioEngineering[2], Neurological Surgery[3], and Computer Science[4]
University of Pittsburgh
Pittsburgh, PA 15260

{shh61, jjo5, yus25, acc33}@pitt.edu

ABSTRACT

To date, spike sorting for neural processing in brain-computer interfaces has been performed using off-chip analysis or custom ASIC designs. In this paper, we propose and test the feasibility of performing on-chip, real-time spike sorting on a commercially available and programmable smartdust wireless sensor network mote since such a platform is currently the closest to an implantable chip due to its small size and similar physical and environmental constraints.

Categories and Subject Descriptors

C.3 [**Computer Systems Organization**]: Special-purpose and Application-based Systems – *real-time and embedded systems*. J.3 [**Computer Applications**]: Life and Medical Sciences – *biology and genetics*.

General Terms

Performance, Design, Experimentation.

Keywords

Brain-computer interface, bio-implantable computing, smartdust, programmable spike sorting.

1. INTRODUCTION

Brain-computer interfaces (BCIs) offer tremendous promise for improving the quality of life for disabled people, such as those who suffer from spinal cord injury, stroke, neuromuscular disorders, and amputees. In order to decode the intensions of users, one core function of the BCI is to map the measured neural action potentials to their source neurons; this process is known as spike sorting. In order to make these systems practical and portable for everyday use, they need to be implantable and free of wires connecting to external computers. Therefore, the processing of the neural signals via spike sorting needs to be performed in real-time and on-chip.

To date, there are many spike sorting hardware implementations, but none provide a complete on-chip solution. First, there are implementations which only perform bio-signal amplification [1] or data acquisition and transmission [2]. Most recently, some

Permission to make digital or hard copies of part or all of this work for personal or classroom use is granted without fee provided that copies are not made or distributed for profit or commercial advantage and that copies bear this notice and the full citation on the first page. To copy otherwise, to republish, to post on servers or to redistribute to lists, requires prior specific permission and/or a fee.
DAC'09, July 26-31, 2009, San Francisco, California, USA

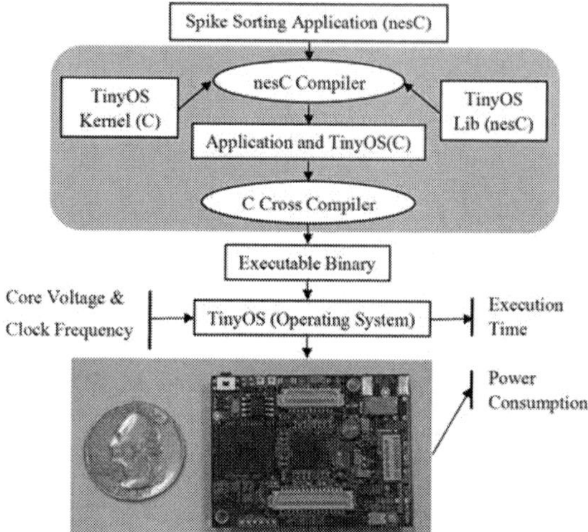

Figure 1. Experimental workflow diagram.

platforms with more functions have been developed, including data filtering [3][4] and feature extraction [5]. Besides these custom VLSI designs, the embedded system TelosB [6] also performs signal acquisition, processing and communications. However, neither the custom VLSI designs nor the embedded design perform the whole spike sorting process and lack the ability to perform on-chip clustering.

In this research we investigate the ability of a smartdust wireless sensor network mote to perform spike sorting, since it is currently the smallest, most lightweight platform available. Both the smartdust and implantable chips require a small size, low power dissipation, an event-driven architecture and wireless communication. Programmability is another key feature that a smartdust has versus a traditional ASIC. With the wide range of spike sorting algorithms that already exist and the continual design and improvement of algorithms, the ability to easily change the spike sorting code running on the smartdust without changing the architecture of the chip is a key benefit since changes in software are much easier, cheaper and more risk-free than changes in hardware, especially when the chips are implanted inside the body.

2. EXPERIMENTAL PROTOTYPE

For the experiment, the Imote2 was chosen as the embedded spike sorting hardware platform. It is a commercially available prototype board for wireless sensor networks (WSNs). The operating system used for the Imote2 board is TinyOS. It is an extremely memory and power efficient OS designed for smartdust applications, thus making it attractive for an implantable chip. Figure 1 describes the workflow of the proposed design. In this

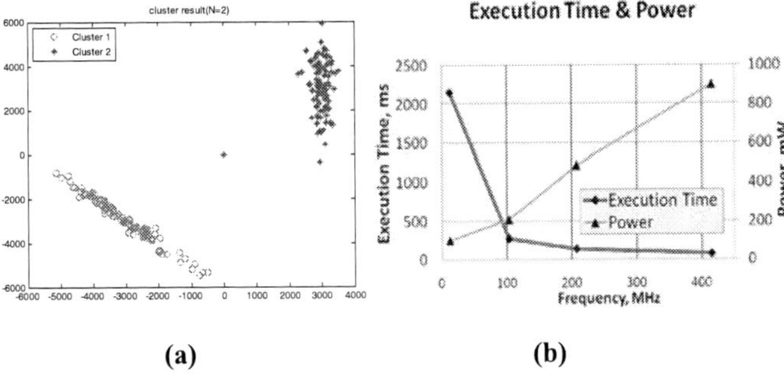

(a)   (b)

Figure 2. (a) Clustered spikes (b) Execution Time and Power.

case study, we implemented the classification stage of spike sorting on the Imote2 board. Feature morphology is the input to the Imote2 and the clustered spikes are the output. TinyOS manages the light-weight k-means clustering algorithm developed to perform the sorting. In order to determine the performance over the range of operating conditions of the Imote2, dynamic voltage and frequency scaling is used. Execution time, code size, and power are measured as performance indicators.

The executable binary running on the Imote2 is generated on the PC. The workflow of the compilation process is shown in Figure 1. For the application, the spike sorting algorithm is first implemented in C, and then packaged by nesC in order to be compatible with TinyOS. Application code and TinyOS code are compiled by a nesC complier (the language of TinyOS) into C, which is then cross-compiled by a C compiler to get the executable binaries for the Imote2's Intel PXA271 processor.

3. PRELIMINARY RESULTS

For input neural data, we used the principal components of 200 spikes. Figure 2 (a) shows the distribution of the input data as well as the two sorted clusters of neural data.

For the code size, required are 179,180 bytes of ROM and 16,964 bytes of RAM, which utilizes only 0.5% of the ROM and less than 0.1% of the available RAM. Since the clustering algorithm is compiled directly with the TinyOS source tree, the code size includes both the clustering contribution and the contribution of the TinyOS kernel and modules being used.

Next, the frequency of the Imote2 was iterated across the four allowed frequencies (13, 104, 208, and 416 MHz), while the full clustering is performed under each condition. The relationship between the core frequency and the execution time as well as the frequency and power was measured. The results are plotted in Figure 2 (b) and behave as expected – with increasing clock frequency the power increases and the execution time decreases.

In order to perform on-the-fly on-chip spike sorting in real time, the processing speed must meet with the firing rate recorded from one electrode, the maximum of which is 50Hz, observed from motor cortex recording of primates [7].

For our test of clustering 200 spikes, 100 from each of the two neurons were clustered. This corresponds to a worst-case acquisition time of roughly 4 seconds of continuous bursts of neural activities from one electrode. Therefore, in order to be able to keep meeting the real-time deadlines, the execution of the spike

sorting code must complete every 4 seconds in order to be ready for the next set of samples. The spike sorting deadline can be met when running at all four frequencies. However, since there is a trade-off between power and execution time, it would be more reasonable to run at the frequency of 13 MHz, since it uses less power. In conclusion, at 13 MHz it takes the program 2.1 seconds to complete, which meets the 4-second real-time requirement.

4. CONCLUSIONS & FUTURE WORK

Based on the preliminary experimental results, real-time on-chip spike sorting on the Imote2 using the k-means clustering algorithm presented is feasible. This proof of concept establishes the ground for the further exploration of spike sorting on a programmable smartdust. With the programmable ability of the smartdust, additional features can be implemented in software, as opposed to designing a new custom ASIC, allowing for a more flexible, multi-use implantable computing platform with rapid development.

Looking forward, now that the clustering algorithm can run on the platform the next step is to implement a PCA feature extraction algorithm to obtain a more complete spike sorting result. Next, real-time acquisition of neural signals will be implemented on the smartdust, allowing for their complete processing. Finally, the wireless communication features of the smartdust will be utilized to send the finished spike sorting data to a neuroprosthetic limb or similar device to complete the brain-computer interface.

5. REFERENCES

[1] R.Harrison, "A low-power low-noise CMOS amplifier for neural recording applications," IEEE Journal of Solid-State Circuits, vol. 38, pp. 958-965, 2003.

[2] P. Mohseni and K. Najafi, "A battery-powered 8-channel wireless FM IC for biopotential recording applications," Solid-State Circuits Conference 2005, vol. 1, pp. 560-617, 2005.

[3] M. Chae, K. Chen, W. Liu, J. Kim and M. Sivaprakasam, "A 4-channel wearable wireless neural recording system," Circuits and Systems 2008, pp. 1760-1763, 2008.

[4] G. Santhanam, M. D. Linderman, V. Gilja, A. Afshar, S. I. Ryu, T. Meng, and K. Shenoy, "HermesB: A Continuous Neural Recording System for Freely Behaving Primates," IEEE Transactions on Biomedical Engineering, vol. 54, pp. 2037-2050, 2007.

[5] M. Chae, W. Liu, Z. Yang, T. Chen, J. Kim, M. Sivaprakasam and M. Yuce, "A 128-Channel 6mW Wireless Neural Recording IC with On-the-Fly Spike Sorting and UWB Tansmitter," Solid-State Circuits Conference 2008, pp. 146-603, 2008.

[6] S. Farshchi, A. Pesterev, E. Guenterberg, I. Mody and W. J., "An Embedded System Architecture for Wireless Neural Recording," Neural Engineering 2007 3rd International IEEE/EMBS Conference on, pp. 327-332, 2007.

[7]]Z. Zumsteg, et al., "Power Feasibility of Implantable Digital Spike Sorting Circuits for Neural Prosthetic Systems," IEEE Transactions on Neural Systems and Rehabilitation Engineering, vol. 13, pp272-279, 2005.

978-1-60558-497-3/09 $25.00 © 2009 ACM

Human Computing for EDA

Andrew DeOrio and Valeria Bertacco

Department of Computer Science and Engineering, University of Michigan
{awdeorio, valeria}@umich.edu

ABSTRACT

Electronic design automation is a field replete with challenging –
and often intractable – problems to be solved over very large instances. As a result, the field of design automation has developed a
staggering expertise in approximations, abstractions and heuristics
as a means to side-step the NP-hard nature of these problems. Approximations and heuristics are at heart a natural application of human reasoning. In this work we propose to harness human potential
to solve some of these problems. Specifically, we propose FunSAT,
a massively multi-player puzzle game for SAT solving. FunSAT
leverages visual pattern recognition skills, abstract perception and
intuitive strategy skills of humans to solve complex SAT instances.
Players are motivated by the puzzle-solving challenges of the game
and by its social interaction aspects.

Categories and Subject Descriptors

B.6.3 [**Logic Design**]: Design Aids—*Verification*; H.1.2 [**Information
Systems**]: User/Machine Systems—*Human information processing*

General Terms

Verification, Human Factors

Keywords

Satisfiability, Human Computing

1. INTRODUCTION

Despite great advances in computer hardware and algorithm design, many Electronic Design Automation (EDA) problems are still
beyond the reach of available computing systems, because of their
intractable complexity and large scale. In this work, we propose to
use human-powered reasoning to tackle these complex problems.

Compared to computers, humans rely on very different approaches
to problem solving. There are a number of activities in which humans are more skilled than computers, and they tend to attack problem solving with the skills in which they are most comfortable. As
an example, humans can easily recognize visual images and patterns and, in general, they are skilled in visual reasoning. People
can identify musical melodies and use creativity to generate new
ones; they can cleverly apply adaptive abstraction techniques to
problems, often without even being aware of the specific technique
that they are using. All of these are tasks in which computers perform very poorly. In the game of Go for instance, human players,
with their intuition and propensity for pattern recognition, perform
consistently better than software agents.

Not only do humans excel in a number of unique areas of intelligence, but they frequently expend this cognitive potential playing
computer games: a recent study estimates that 34% of internet-
connected adults play online games regularly [1]. Massively Mul-
tiplayer Online (MMO) games offer the additional attraction of a

Permission to make digital or hard copies of part or all of this work for
personal or classroom use is granted without fee provided that copies are
not made or distributed for profit or commercial advantage and that copies
bear this notice and the full citation on the first page. To copy otherwise,
to republish, to post on servers or to redistribute to lists, requires prior
specific permission and/or a fee.
DAC'09, July 26-31, 2009, San Francisco, California, USA

social component to gaming: the number of players registered with
"World of Warcraft", a popular role-playing game, recently reached
10 million [3]. Some games naturally resemble problem solving
tasks in EDA problems. For instance, "Pipe Dream", a popular
game from the 1980s, requires players to arrange pipe segments
on a grid. Pipe Dream is clearly reminiscent of routing wires in
physical synthesis. The vast gaming population provides a rich opportunity to harness human intelligence to perform useful tasks,
provided that they are presented as games.

In this work, we propose to leverage human skills such as visualization, abstraction and strategy to solve complex EDA problems.
In order to achieve this goal, we need to present the problem in a
format that is intuitive and amenable to human skills. Moreover,
we need to cast the problem as a game, where players are rewarded
for solving a puzzle. Typically, EDA problem instances are complex, often containing millions of primitives or solution paths. To
address this complexity, we devise two strategies: (i) problem partitioning and abstraction, and (ii) our games should be multi-player.
In a multi-player game, players coordinate their activity as a team
to achieve a goal that is beyond the reach of any individual. The
reward for solving a puzzle (a sub-problem) could range from advancing levels in the game, to the opportunity for socialization by
interaction with other players.

Several CAD problems are suitable for games. Beside "Pipe
Dream", discussed earlier, placement could be cast as a packing
puzzle, possibly similar to the game of "Tetris". Both of these
tasks are naturally visual and have potential for providing more
effective solutions than machines. Indeed, engineers today routinely intervene in the place and route processes to correct and solve
sub-problems that machines cannot tackle. In this work, we focus
on SATisfiability solving, a central problem in EDA known to be
NP-complete. While available SAT solvers can solve many complex problem instances, other instances often arise that are beyond
the reach of software solvers, due to either the sheer complexity
of the instance or the narrow solution space. Typical solvers employ a wide range of heuristic approaches, strategy, randomization,
etc. We propose presenting SAT instances in visual and abstract
ways, so that human players have the potential of finding solutions
where software solvers fail.

2. HUMAN COMPUTING TODAY

Few applications have been recently proposed for human computing. Von Ahn [7] leverages human computing to index web
images: he has created a game in which a player's goal is to tag
pictures. Images are shown to users who suggest associated words,
which are then tabulated for popularity and later used as search
terms. A more nefarious example of human computing is in solving captchas, a popular security measure on the web that strives to
ensure that a user is indeed human. Captchas leverage computers'
lack of skill in visual recognition to block software applications
from accessing a site. To counter this security measure, spammers
have developed infrastructure to have other humans solve captchas
on other sites by enticing them with incremental rewards [2]. Other
applications of human computing are set forth in [6], including language translation and text summarization, among others.

Complementary to these efforts, recent research has attempted to infer the computational model of the human brain [5], and has found that this model can deliver notable advantages for certain tasks (such as image recognition) over traditional computation [7]. Finally, Amazon's Mechanical Turk [4] operates in the space of crowdsourcing by also leveraging a pool of human users to solve problems. However, in this case problems are not camouflaged as games and there is no explicit attempt to cast them so that humans have a direct advantage over a machine. The Mechanical Turk consists of task listings, ranging from categorizing products to writing articles, and offering a wage to anyone who completes a task: it appears to be an effective way to motivate a large group of people to perform menial, yet essentially human tasks. Absent from these applications are the intractable problems extremely common in EDA, some of which have great potential for human computing.

3. A HUMAN-POWERED SAT SOLVER

FunSAT is a visual puzzle game. It transforms SAT instances into puzzles and presents them to human players who solve them. In developing FunSAT, we strove to present the problem in a way that leverages the unique visual reasoning and pattern recognition abilities of humans. SAT instances are presented visually and hierarchically, similarly to a map, where a player can zoom in and out to approach the problem at different levels. We envision the possibility of structuring the game into levels so that a player at first is only allowed to zoom out a certain amount; once players have solved local portions of the puzzle, they are promoted to the next level, where they can access a larger region of the problem.

Figure 1 shows a screenshot of the game. Boolean variables are mapped to buttons, which can be assigned to true (purple), false (yellow) or unassigned (grey) with a click of the mouse. Clauses are represented by circles whose size is proportional to the number of literals in the clause, so that players can intuitively rank the difficulty of satisfying each clause. At the beginning of the game, all clauses are undetermined (grey); they become green when a partial assignment satisfies them, and red if they are falsified. To leverage the human ability of spatial perception and area, we lay out clauses in a grid. Variables surround the "clause grid" and the relationship of variables to clauses is shown by hovering the mouse over a variable, highlighting the affected clauses.

We implemented a prototype of FunSAT in LabVIEW. As the game progresses, players click on different variables, changing their values, while observing the color impact in the grid of clauses. The final goal is to light the entire grid with green, thus satisfying all clauses. In attempting different assignments, humans learn color and shape patterns that are generated by different selections and develop gaming strategies based on this. FunSAT has some similarity with a handheld electronic game called "Lights Out", a logic puzzle where players must switch off all the lights in the game by manipulating a pool of buttons.

3.1 Visual Aids

To help humans reach the goal of satisfying all the clauses, we implemented a number of visual aids, including an intuitive layout scheme, informative feature sizing and smart reminders of recent activity. A clever clause layout could suggest which clauses are affected by each variable, thus helping the player develop a winning strategy. To this end, during the construction of the visual problem interface for an instance, we arrange clauses sharing the most variables in close physical proximity.

Puzzle solving is bound to encounter dead ends: logging of past selection activity provides users with backtracking capabilities. This enables users to revisit the past few variable assignments, as well as providing an "undo" function.

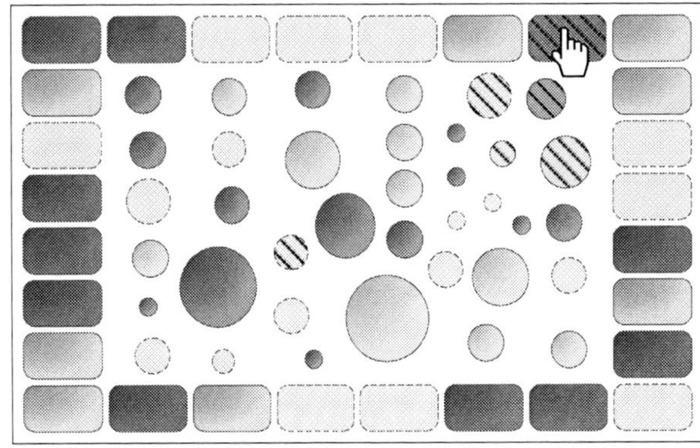

Figure 1: Snapshot of FunSAT. Circles represent clauses and the surrounding buttons correspond to variables. Assignments are applied by clicking on a button, which switches from unassigned (grey), to purple (true), to yellow (false).

3.2 Scaling

Scaling FunSAT to large instances presents a challenge, as the game limited by screen real estate and human patience. We propose an approach inspired by navigable maps: users can work on the instance at different levels of detail by zooming in and out. At small scale levels, an inset abstract view of the entire SAT "globe" provides information on the system-level impact of selections. This hierarchical approach enables us to enhance the game in the multi-player direction, where multiple people can operate in different areas of a same "map" and coordinate their efforts. In order to minimize the impact of selections made in one part of the map to other regions, instances could be partitioned with a min-cut procedure.

4. CONCLUSIONS AND FUTURE WORK

We propose bringing the art of human computing to the science of SAT solving. Leveraging the human passion for games, along with its propensity for visual reasoning and pattern recognition, FunSAT presents SAT instances as a visual challenge amenable to these strengths. In addition, it has the scaling potential to realize a large multi-player online game, where players collaborate in solving the same instance. In the future, we hope to further strengthen player motivation by providing social networking benefits for successful players. Finally, casting EDA problems as games will allow unskilled players, to solve complex engineering problems with a low barrier to entry. Hopefully, a byproduct of this area of research will be an increased appreciation for engineering tasks in the general population.

5. REFERENCES

[1] Online game playing is still the most popular online activity. *Associated Content*, 17 August 2007. www.associatedcontent.com.

[2] PC stripper helps spam to spread. *BBC News*, 30 October 2007. www.news.bbc.co.uk.

[3] WoW surpasses 10 million subscribers, now half the size of australia. *Joystiq*, 22 January 2008. www.joystiq.com.

[4] Amazon. Mechanical turk, 2008. www.mturk.com.

[5] J. Hawkins and S. Blakeslee. *On Intelligence*. Owl Books, 2005.

[6] L. von Ahn. Games with a purpose. *IEEE Computer*, 39(6):92–94, June 2006.

[7] L. von Ahn and L. Dabbish. Labeling images with a computer game. In *ACM SIGCHI*, pages 319–326, 2004.

978-1-60558-497-3/09 $25.00 © 2009 ACM

Synthesizing Hardware from Sketches

Andreas Raabe
International Computer Science Institute (ICSI)
1947 Center St., Suite 600
Berkeley, CA 94704, USA
Raabe@ICSI.Berkeley.edu

Rastislav Bodik
Computer Science Division, #1776
University of California, Berkeley
Berkeley, CA 94720-1776
Bodik@CS.Berkeley.edu

Categories and Subject Descriptors

B.6.3 [**Design Aids**]: Hardware description languages

General Terms

Design, Languages

Keywords

Sketching

ABSTRACT

This paper proposes to adapt *sketching*, a software synthesis technique, to hardware development. In sketching, the designer develops an incomplete hardware description, providing the "insight" into the design. The synthesizer completes the design to match an executable specification. This style of synthesis liberates the designer from tedious and error-prone details—such as timing delays, wiring in combinational circuits and initialization of lookup tables—while allowing him to control low-level aspects of the design. The main benefit will be a reduction of the time-to-market without impairing system performance.

1. INTRODUCTION

Due to the on-going micro-miniaturization in chip production, hardware development faces new challenges. First, the designer productivity grows slower than the number of transistors per chip. Second, the rate of innovation has increased to a point where only the first to the market makes a profit. Hardware synthesis has been successfully used to improve designer productivity, but in general it results in designs considerably slower and bigger than hand-coded hardware descriptions.

In high-level synthesis, the synthesizer acts as a smart compiler, translating high-level specifications into low-level designs. We recognize that the designer is best equipped to create smart low-level designs. Therefore, we allow the designer to produce intricate low-level designs. However, low-level designs are tedious to write. To this end, sketching will allow the designer to leave tricky details unspecified. The incomplete design description is called a *sketch*.

The sketch will be completed by the synthesizer to meet a separate executable specification that might contain timing constraints. The correctness of the synthesized design will be guaranteed by an existing verifier or model checker, delivering guarantees as strong as the designers are currently used to. If the sketch contains a bug, the synthesizer will reject it, optionally producing diagnostics as to why the sketch cannot be completed. The sketch may have multiple completions, each with different characteristics. Synthesizing these alternatives enables automatic design space exploration.

Sketching was originally developed for software synthesis [2], as a form of constraint programming [3]. We believe that applying sketching in hardware will decrease design space exploration time as well as implementation time.

At the technical level, the designer omits details by leaving in the code a wildcard keyword. On compilation the sketch compiler figures out how this wildcard has to be replaced in order to meet the specification. Sketch synthesis makes intense use of verification techniques to generate candidate implementations as well as checking them for correct behavior with respect to the specification. Thus it is enabled by the recent improvements to SAT solvers [1].

Sketching may be more suitable for hardware than it is for software. First, executable specifications are not common in software. Additionally, in general software developers will not tolerate compilation times longer than a few minutes.

None of these problems applies to hardware design. Design space exploration in hardware development proceeds in several refinement steps, each producing a specification for the next step. Sketching will enable the designer to express insight into how a hardware description should be refined in each step, without having to implement all details himself.

2. APPLYING SKETCHING

The simplest software sketch one can think of might be:

```
int spec(int x) {return (x+20);}
int sketch(int x) implements spec
    {return (x+10+??);}
```

Here the sketch synthesizer will substitute the wildcard keyword ?? with a suitable constant value that makes the sketch meet the specification (here, the value is 10). This primitive wildcard is always replaced with a constant, which might seem limiting, but even complex functionality like allowing the compiler to reorder statements has been demonstrated for software as syntactic sugar.

Nevertheless, the primitive (constant) wildcard will be useful in hardware implementations, e.g., , for wiring components. Consider as an example the well known Wallace Tree for fast integer multiplication. This tree is notoriously hard to implement due to its irregular structure. Using only the constant wildcard and a bound on the circuit's depth enables a very compact description of a multiplier with the depth of Wallace. Figure 1 shows how a multiplier sketch could look like in a hypothetical sketching-extended hardware description language. The sketch proceeds in three phases. First n^2 partial prod-

Permission to make digital or hard copies of part or all of this work for personal or classroom use is granted without fee provided that copies are not made or distributed for profit or commercial advantage and that copies bear this notice and the full citation on the first page. To copy otherwise, to republish, to post on servers or to redistribute to lists, requires prior specific permission and/or a fee.

DAC'09, July 26-31, 2009, San Francisco, California, USA

```
A = fold(new Bit[W,W], lambda(i,j).x[i]*y[j])
bit[depth(W)+1][W*W] C;
    --inputs and outputs to each stage
    --assume depth calculates log_3/2(W/2)
C[0] = flatten(A) ;
    --turn 2-dim into a 1-dim array of size W*W
for d in 0 to depth(W) do
    --generate adders for all levels
  int i=0;
  for ?? do(C[d+1][i++],C[d+1][i++]) =
          adder(C[d][??],C[d][??],C[d][??]); od
  for ?? do (C[d+1][i++],C[d+1][i++]) =
          halfAdder(C[d][??],C[d][??]); od
  for ?? do C[d+1][i++] = C[d][??]; od
od
int[W*W] P = ??;
C[depth(W)]=C[depth(W)][P];
    --P permutes the bits in the array C,
    --C[P] expands to [C[P[0]], ...C[P[n-1]]]
Bit[2*W] out;
out=trunc(C[depth(W)],W-1)+trunc(C[depth(W)],W-1);
    --trunc truncates argument to lowest W-1 bits
return out
```

Listing 1: A multiplier sketch in a hypothetical sketching-extended HDL.

ucts are calculated; next an adder tree is constructed that reduces the problem to adding two numbers of length $2n - 1$, which is executed in the last phase. The adder tree has a very irregular structure due to the varying numbers of bit-wise identity, half and full adders, used intermixed on all layers of the tree. Note that the sketch does not constrain the number of adders, allowing automated search of the design space spanned by the sketch for the best implementation. For $n = 4$, an equivalent description is synthesized in a few seconds.

Sketches could be used in other applications, too. To simplify the designer tasks of suitably wiring components, one can write a sketch that synthesizes all possible permutations; the synthesizer selects a permutation that leads to a correct design. Such permutations are useful, for example, in butterfly networks in FFT components or for building bigger multipliers from smaller ones in FPGA development.

Using wildcards is not limited to wiring. Other interesting applications of wildcards include

- computing the depth of a FIFO, synchronizing its output to some concurrent pipelined datapath (e.g., a pipelined blockram).
- deciding if a certain functionality of a module needs to be used or not (e.g., FIFO fall-through, signed or unsigned multiplication).
- simplifying implementation of indexing (e.g., of bit-vectors, arrays, or memory cells), avoiding the notorious off-by-one error.
- implementing the succession of states in a finite state machine.

3. THE SKETCH SYNTHESIS LOOP

To enable synthesis of a candidate implementation, the sketch is transformed into a Boolean circuit S with input x. Therefore all loops are unrolled. Wildcards are transformed into control variables c and interpreted as input variables. The resulting circuit is represented as a function $S(x,c)$. Initially arbitrary values are assigned to the control-variables c forming a candidate. This candidate is then verified against the specification using a SAT solver [2]. If the candidate is correct, the synthesis is complete. Otherwise, the verifier provides an input for which the candidate implementation misbehaves. The correct output for this input is obtained using the specification. This input/output pair (i, o) is added to a set of clauses of the form $S(i, c) = o$ called observations. Another SAT solver is used to generate the next candidate c that satisfies all clauses obtained so far.

These two interacting SAT solvers guarantee that either an implementation is obtained which adheres to the specification or a set of

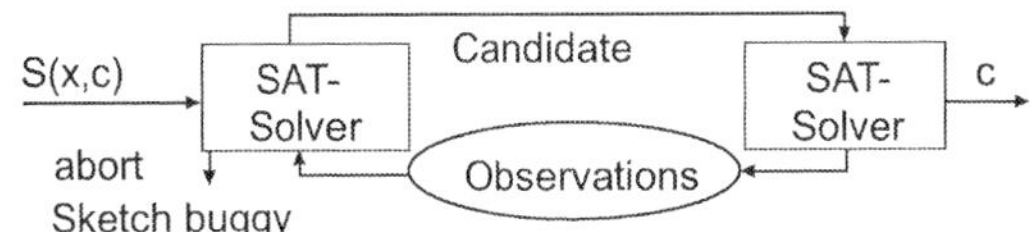

contradictory clauses results. In the latter case the sketch is buggy and cannot be made equal to the specification.

4. CHALLENGES IN APPLYING SKETCHING TO HARDWARE DEVELOPMENT

Although sketching is a promising technique, challenges unique to hardware design present themselves. First, it may be desirable to describe the specification at a different level of abstraction than the sketch. The specification might be functional, while the sketch allows a pipelined implementation. In this case it is not necessarily obvious what is to be considered a correct solution. E.g., the timing behavior of a pipelined implementation of the sketch will differ considerably from that of a functional specification. Thus it will be necessary to provide the designer with means to express whether a sketch is allowed to behave differently in that respect. Similar issues arise when the specification and the sketch are given with different types, e.g., , floating-point vs. fixed-point. Here, the sketch implementation will yield different results than the spec. The designer will want to specify an acceptable deviation or ask for the implementation with the minimal deviation. This will in general need some domain knowledge about floating-point algorithms. In software sketching resolution time grows exponentially with the size of c. Thus, scalability of sketch resolution will be an important future research topic.

5. CONCLUSION

Sketching has the potential to change the way hardware is developed. On the one hand it may render many steps in a refinement hierarchy unnecessary. A sketch where many details are left open is in itself an abstract representation of the implementation and can be refined stepwise by filling in those details by hand or constraining the search performed by the sketch synthesizer. On the other hand it provides a new way to stepwise improve a component's performance, size or power consumption. Co-simulation of components on different abstraction levels enables refining the components individually. Sketching will enable micro-refinement of the individual component, by implementing additional detail, while still having a simulable (or depending on the level of abstraction synthesizable) hardware description. This will be an enormous benefit since it avoids design errors and unproductive over-optimizing. Moreover, a sketch implementation will be obtained by verification against its specification, it will be correct by design. All this applies even in areas were high-level synthesis cannot be applied, since low-level interfacing capabilities are required (e.g., if complex timing behavior of peripheral components needs to be respected).

Therefore, sketching will liberate the designer from implementing tedious details and thus enable an extraordinary increase in productivity. This will effectively decrease time-to-market and improve performance and correctness of the product at the same time.

6. REFERENCES

[1] M. J. Heule. *SmArT solving: Tools and techniques for satisfiability solvers*. PhD thesis, TU Delft, 2008.
[2] A. Solar-Lezama, L. Tancau, R. Bodik, V. Saraswat, and S. A. Seshia. Combinatorial Sketching for Finite Programs. In *Proc. of ASPLOS 2006*, 2006.
[3] P. Van Roy and S. Haridi. *Concepts, Techniques, and Models of Computer Programming*. The MIT Press, 2004.

978-1-60558-497-3/09 $25.00 © 2009 ACM

Endosymbiotic Computing:
Enabling Surrogate GUI and Cyber-Physical Connectivity

Pai H. Chou
University of California, Irvine, USA, and
National Tsing Hua University, Hsinchu, Taiwan
phchou@uci.edu

ABSTRACT

Endosymbiotic Computing entails attaching an RF-enabled micro-controller module (endomodule) to an appliance such that it appears as a networked device in the cyber world. It enables a smart phone to work as not only a universal remote control but also a surrogate GUI for inspecting the attributes of these appliances, without modifications to legacy circuits. To minimize the cost and resource requirements of the endomodules, we propose a generalized active message programming method that executes dynamically-loaded threaded code on-demand without requiring parsing.

Categories and Subject Descriptors

J.7 [**Computers in Other Systems**]: Consumer products; C.3 [**Special-Purpose and Application-Based Systems**]: Real-time and embedded systems

General Terms

Design

Keywords

Endosymbiotic computing, wireless interface, surrogate GUI

1. INTRODUCTION

Nearly 30 years ago, graphical user interface (GUI) helped users make sense of computing by borrowing metaphors from real-life objects. Thirty years later, GUI operations such as cut-and-paste and drag-and-drop are as natural to modern users as any real-life ones. Unfortunately, many every-day appliances such as microwave ovens, TVs, air conditioners (A/C), and telephone/answering machines (TAM) still rely on simple buttons, LEDs, and remote controls as their primary user interface. Although PCs and cell phones may be programmed as universal remote controls [1], they merely turn hardware buttons into software ones. One trend with networked devices is *surrogate GUI*, where devices such as printers and routers contain embedded web servers that provide the GUI as HTML. The Surface Computer [2] provides a GUI for devices in

Permission to make digital or hard copies of part or all of this work for personal or classroom use is granted without fee provided that copies are not made or distributed for profit or commercial advantage and that copies bear this notice and the full citation on the first page. To copy otherwise, to republish, to post on servers or to redistribute to lists, requires prior specific permission and/or a fee.
DAC'09, July 26-31, 2009, San Francisco, California, USA

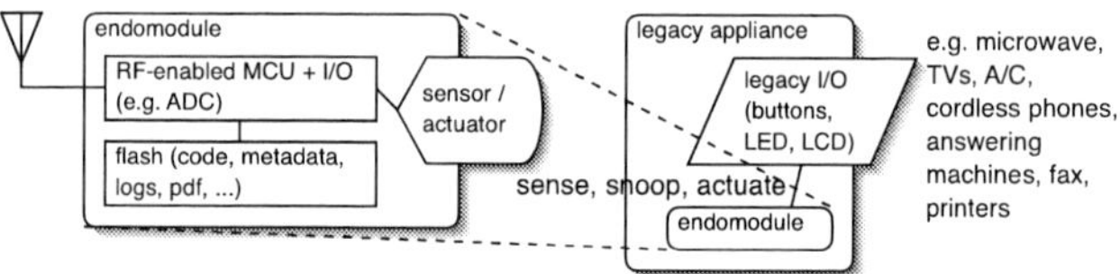

Figure 1: Endosymbiotic module.

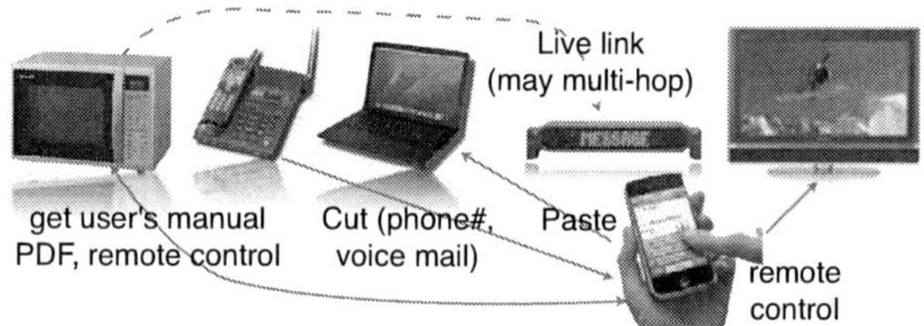

Figure 2: Wireless data grabbing and live linking.

proximity, but it assumes large displays with no mobility. We propose to generalize surrogate GUI with low-cost, short-range wireless technologies available as integrated, RF-enabled, 8-bit MCU (microcontroller unit) system-on-chips (SoCs).

Our vision is called *Endosymbiotic Computing*, where *endo-* is a Greek prefix for "internal, within," and *symbiosis* means a mutually beneficial relationship. An *endomodule* containing an RF-enabled MCU is to be embedded inside an appliance (Fig. 1), draws power from it and controls it programmatically.

2. ILLUSTRATIVE EXAMPLES

Scenario 1: Smart Phone as Universal GUI

A compatible phone discovers a list of endosymbiotically-enabled appliances in proximity, including a microwave, an A/C, a TV, a fax, a TAM, and a marquee. The user taps on the menu for the microwave, whose endomodule replies to offer more choices, including the serial number, model number, and a script for a software remote control to be run on the smart phone. Moreover, the endomodule keeps a usage log for the user to track energy usage for the past year. This particular endomodule contains flash memory with plenty of room to provide a PDF file for the user's manual, a file for the instruction video, and voice files in several languages to help the visually-impaired, all without having to decode these files.

Scenario 2: Inter-Device Cut-and-Paste

The user now inspects the TAM and wants to copy the caller-ID from this TAM to her computer. The smart phone would act as not only a remote control but more generally a *data grabber*. The endomodule may snoop the pins of the LCD to reconstruct the content

978-1-60558-497-3/09 $25.00 © 2009 ACM

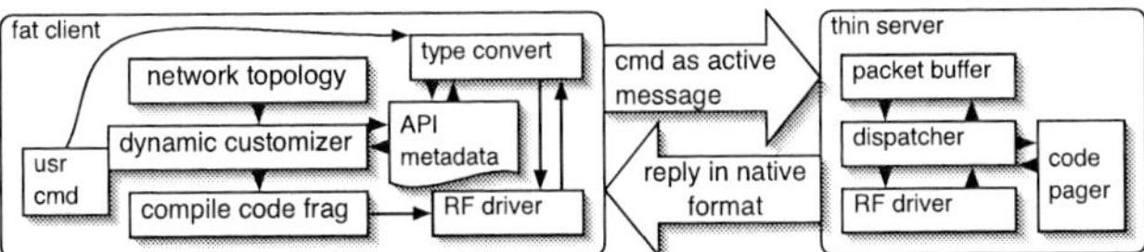

Figure 3: Software architecture with fat client and thin server.

being displayed on the LCD. The data grabber can first wirelessly grab the text from the endomodule, connect to a PC, and paste the text possibly by a gesture. The voice message can also be digitized by the endomodule's built-in ADC and grabbed similarly.

Scenario 3: Inter-Appliance Live Linking

The user discovers that she can not only grab a snapshot of the data but also establish a *live link* from the source to the destination. She uses her cell phone to grab the TAM's caller-ID as a virtual "outlet," walks outside the RF range of the TAM, and virtually connects it to the "inlet" on the marquee. Even though the TAM cannot connect directly to the marquee, a multi-hop route is discovered automatically to realize this end-to-end connection. Moreover, additional agent processes are synthesized and injected into the network to actively pull and push data on behalf of the two passive endpoints.

3. MECHANISMS

Hardware

The endomodule hardware must be simple and low cost such that appliance manufacturers will be willing to incorporate them and add sufficient value at a very small cost. In the extreme case, the endomodule may be as simple as a passive RFID, and the smart phone would look up all attributes from the Internet or a database. RFID may certainly co-exist with active RF devices, though here we focus the discussion on active RF interfaces.

Short-range wireless protocols include Bluetooth[3], Wibree[4], ZigBee[5], Z-Wave[6], and others. Bluetooth is popular for computer peripherals and cell phones. ZigBee is positioned for wireless sensor networks, though it is common to use just the 802.15.4 MAC layer and replace the ZigBee stack with a simpler networking layer. Z-Wave supports home control and is simpler than ZigBee but at a lower data rate. Wibree, or Bluetooth Low Energy Technology, will be a subset of Bluetooth and requires only a few KB of memory, compared to 30-100 KB with others. Wibree, 802.15.4, and Z-Wave are available in integrated 8051 MCU+RF chips, making them cheaper (around US$1 in large quantities) and simpler to implement. Power consumption is less of an issue here than cost and complexity. We believe that, similar to the Internet, the physical and MAC layers do not matter so much, as they can be easily integrated by wireless routers that can bridge these standards. What is important is how to support the execution of software on highly resource-constrained hardware.

Software Architecture

Endosoftware, namely that running on the endomodule to be attached to the appliance, should be kept as simple as possible. Java's hardware demand may be too high [7], as we assume something closer to an 8-bit, 8051-class MCU. Our approach is to move most of the complexity to the *fat clients*, a collective term for general-purpose computers, smart phones, gateways, and routers. Each endomodule works as a *thin server* that passively responds to commands. The software architecture is shown in Fig. 3.

To eliminate the parsing overhead, we borrow the idea from *active messages* [8], where a packet contains the address of the han-

dling routine (in the receiver's address space) followed by the parameters. This way, the receiving endomodule just needs to jump to the address stored at the beginning of the packet without parsing. Active messages can be viewed as a special form of *threaded code* [9], a classical technique for efficiently implementing interpreters by representing virtual instructions as the addresses of the corresponding routines. Such a "command interpreter" does not understand the meaning of the command, as it obliviously jumps to the target address. With this active message layer, an endomodule can bootstrap wirelessly and execute any function that it loads. Security may be a concern, and our solution is to exploit hardware-supported encryption and decryption engines that are found as part of most transceivers of this class.

For "standard I/O" with minimal overhead, an endomodule would read and write data in the most convenient, native format. For instance, for date and time, general-purpose computers commonly use the "seconds since the epoch" representation, but it is not convenient for most 8-bit MCUs. The ordering, size, and endian of the day, month, year, hour, minute, and second are described as metadata in a form such as XML. The fat client is responsible for looking up the metadata associated with each command and formatting the data in the packets accordingly. This way, the replying endomodule does not have to know anything about the data types on the fat client side. More generally, to support the live linking in Scenario 3 above, the fat client would actively poll the source, perform type conversion, and actively push the data to the sink, or delegate any of these tasks to a thin server by dynamic compilation.

We propose to optimize the code for endomodules by dynamic compilation on the host side and wireless code update. The fat client may synthesize and dynamically compile a customized version of a function by exploiting run-time constants. For instance, the fat client may use its knowledge about the network topology to create a specialized version of the multi-hop routing function for each thin client with its neighbors' IDs hardwired.

4. CONCLUSIONS

Users today spend possibly more time in the cyber world than in the physical world. To truly bridge the gap, we propose the idea of Endosymbiotic Computing. It enables the user to (1) grab snapshots of data from appliances that previously had no connectivity to general-purpose computers, (2) query and control these appliances including accessing their manuals stored on the endomodules, and (3) customize the way these appliances should work with each other by inferring the necessary conversions and helper processes to inject into the network. It does not just dumb down the interface for the lowest common denominator users; it enables them to perform powerful distributed programming at their fingertips.

5. REFERENCES

[1] Brian X. Chen. Update turns iPhone into an Apple TV remote. http://blog.wired.com/gadgets/2008/07/update-turns-ip.html, 2008.

[2] Microsoft Surface. http://www.microsoft.com/surface/.

[3] The official Bluetooth® technology info site. http://www.bluetooth.com/.

[4] Bluetooth low energy technology. http://www.bluetooth.com/Bluetooth/Products/low_energy.htm.

[5] ZigBee Alliance – Home Page. http://www.zigbee.org/.

[6] Z-Wave. http://www.z-wave.com/.

[7] CASCADAS Consortium. CASCADAS Project. http://www.cascadas-project.org/, 2009.

[8] D.E. Culler, J. Hill, P. Buonadonna, R. Szewczyk, and A. Woo. Active message communication for tiny networked sensors. In *EMSOFT*, 2001.

[9] James R. Bell. Threaded code. *Communications of the ACM*, 16(6):370–372, 1973.

978-1-60558-497-3/09 $25.00 © 2009 ACM

Debugging from High Level down to Gate Level

Masahiro Fujita   Yoshihisa Kojima   Amir Masoud Gharehbaghi

University of Tokyo and CREST, Japan Science and Technology (JST)

Tokyo, Japan

fujita@ee.t.u-tokyo.ac.jp   kojima@cad.t.u-tokyo.ac.jp   amir@cad.t.u-tokyo.ac.jp

ABSTRACT

C-based hardware designs are now accepted as means to increase design productivity. Starting with rather algorithmic design descriptions, incremental refinements are applied to generate high-level synhesizable descriptions which are further processed by high-level and logic synthesis tools. C-based system level design descriptions, such as in SpecC [?] and SystemC [?], can give concise and global views on the behaviors of the designs as well as structures, and various types of dependencies, such as control, data, concurrency, and others, can be extracted quickly. These dependencies can be the bases for efficient and effective debugging for all levels of design descriptions. In this paper, graph representations for various dependencies which are extracted from C-based descriptions are introduced. Then techniques on their uses for debugging in various design levels are discussed. We present static and dynamic tracing methods for dependence analysis as well as techniques that try to establish mapping between implementations and C-based design descriptions.

Categories and Subject Descriptors

B.7.2 [**Hardware**]: Integrated Circuits—*Design Aids*

General Terms

Algorithms, Verification, Design

Keywords

High-Level Design, System Level Design, Dependence Analysis, Equivalence Checking, Post-Silicon Debug

1. INTRODUCTION

Due to increased complexity and capacity of target designs, it takes more and more time to establish their logical correctness through verification and debugging of designs. One way to realize shorter design cycles is to start design with more abstracted representations, such as system level designs where entire designs are represented as hardware/software combined systems. Because

Permission to make digital or hard copies of part or all of this work for personal or classroom use is granted without fee provided that copies are not made or distributed for profit or commercial advantage and that copies bear this notice and the full citation on the first page. To copy otherwise, to republish, to post on servers or to redistribute to lists, requires prior specific permission and/or a fee.

DAC'09, July 26-31, 2009, San Francisco, California, USA

design descriptions are more abstracted than the lower level ones, such as the ones in register transfer level (RTL) and gates, the numbers of lines in design descriptions becomes much smaller, which makes verification and debugging efforts much simpler. Currently C-based languages, such as SpecC [?] and SystemC [?], are commonly used to describe designs in system level. Starting from system level, design descriptions are gradually refined either manually or automatically all the way down to the ones in RTL and gates. Design models which are more abstracted than RTL, such as algorithmic and behavior ones, are called system level models (SLM). In this paper we discuss the use of such SLMs for debugging in all levels of design descriptions.

Since SLMs are more abstracted than RTL or gate level, global views on design descriptions including various dependencies, such as data, control, concurrent interactions, etc., which form the bases for analyzing and debugging design descriptions in all levels of designs, can be captured concisely. These dependencies have been used in analysis of design descriptions through the use of static and model checking based techniques. With these approaches, cause-effect analysis can be efficiently made, which is a central issue for debugging design descriptions not only for high-level descriptions but also lower ones such as RTL designs and their implementations. This paper presents three techniques that utilize C-based design descriptions for debugging designs of various levels. In section 2, debugging methods that statically utilize graph based representations for various dependencies, which are extracted from C-based descriptions, are presented. In the following section, techniques that dynamically analyze various dependencies for debugging are introduced. In section 4, methods to utilize C-based design descriptions for post-silicon debugging are presented.

2. GRAPH REPRESENTATIONS OF VARIOUS DEPENDENCIES

There have been developed several tools and associated methodologies to use C or C++ languages and their extensions to describe hardware parts of the target designs as well as their software parts. Although there are differences in their details, the ways to describe hardware parts in C and C++ languages share common features, which are the followings:

- Parallel (concurrent) descriptions with synchronization statements, such as $par, notify, wait$

- Introduction of modules and ports for structural hierarchy

- Various statements to describe hardware oriented controls

The above features are realized either extending the languages or introducing some fixed types of descriptions styles.

978-1-60558-497-3/09 $25.00 © 2009 ACM

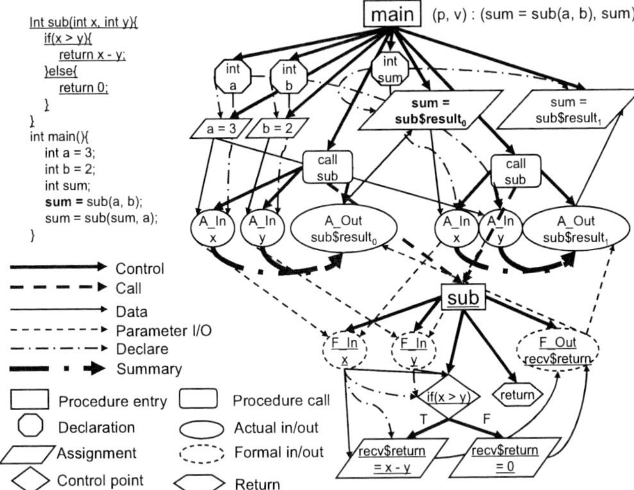

Figure 1: A simple C description and its system dependence graph

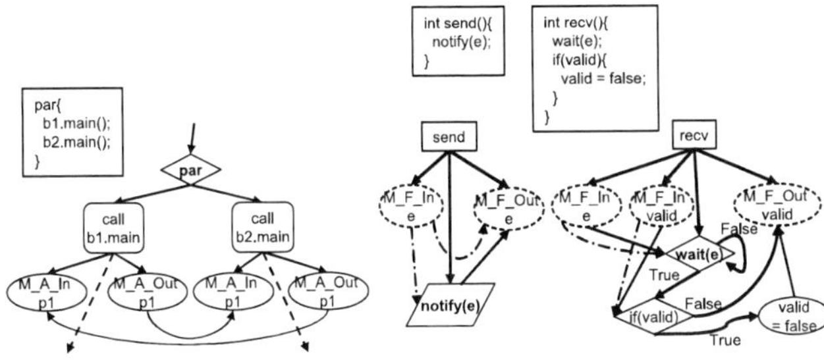

Figure 2: Extended nodes and edges for par, notify and wait statements

In software engineering fields, program slicing is a technique to extract portions of the original programs which are relevant to the variables at some statements specified by users. Slicing can be one of the most basic operations for debugging the descriptions. It is computed given two parameters, program point p and the set of variables v which appear in p. Program slicing first computes various dependence graphs. Horwitz [?] et al defines System Dependence Graphs (SDGs), which contains multiple Procedure Dependence Graphs (PDGs) and expresses dependencies between procedures. An example system dependence graph generated from a simple C description is shown in Figure ??. As for the details of various edges and nodes defined in system dependence graphs, please refer to [?]. SDGs or their extensions for C-based hardware descriptions can capture various dependencies which are required for debugging designs of various levels.

In SpecC language, there are hierarchical structures such as the *behavior*, *channel* and *interface*, concurrent parallel execution syntax as *par*, and synchronization syntax as *wait* and *notify*. To address these SpecC language's features, an SDG for SpecC with new nodes and edges is used [?]. The new nodes and edges are designed and inserted in such a way that various dependence analysis techniques developed for the original SDG can also be used for SpecC with slight modifications. To deal with concurrent executions realized by using *par* statements, a node for *par* as a control point

node, similar to that of *if*, *while* and *for*, is defined. From *par* node, control dependence edges labeled **true** are drawn to every node corresponding to the statements that are executed concurrently under the semantic *par*. For example, in Figure ??, since *b1.main()* and *b2.main()* are executed concurrently, there must be control edges from the *par* node to each of them. Extra data dependence edges for representing the shared ports and parameters are also required to represent structural hierarchies. In the figure, *b1* and *b2* are running in parallel and there is a shared variable *p1*, hence, the data edge for *M_A_Out p1* to *M_A_In p1* are constructed. In Figure ?? the prefix M_ is used to represent member variables.

Wait statement is also defined as a control point node in a dependence graph, and control dependence edges labeled **true** are constructed to every node executed after it. This is because whether *wait* is passed or not affects the executions of those statements, just like *if* or *for* statements. In addition, a data dependence edge of an event variable, e, used in *wait(e)* statement is constructed. This is because an event, e, is a sort of virtual data communicated between *notify(e)* and *wait(e)* statements. In Figure ??, there is a *control edge* from the *wait* node to the control point node for *if(valid)*. The *notify(e)* statement is defined as an assignment node to an event variable, e. In the figure, a data dependence edge is constructed from *notify(e)* to the formal out node corresponding to the event variable e. Typically an event variable is communicated through *channel* connected to *behavior*'s port. Therefore we can traverse data dependence edges from *notify* node to *wait* node across *behaviors* and *channels* in the SDG.

With the above extensions over SDGs for C languages, various dependencies in SpecC descriptions can be represented as Extended System Dependence Graphs. By traversing the dependence graphs, various dependencies among variables and statements can be obtained which help debugging design descriptions. For example, starting from the buggy outputs (outputs which generate wrong values), by traversing the dependence graphs in a backward way, we can reach potential buggy statements. If there are multiple error traces available, the intersections of those potential buggy statements are most suspicious portions of design descriptions. Concurrent statements are one of the sources of bugs relating to synchronization of executions. By establishing the mappings between error traces and dependence graphs including synchronization statements such as *notify* and *wait*, debugging processes can become much simpler. Various traversal techniques used in static checking methods [?, ?] can provide means to reason about the descriptions with respect to given error traces. In nature static checking must reason about things in conservative ways. Therefore dynamic concurrency cannot be analyzed accurately. Instead concurrent executions under co-begin and co-end-like statements, which is the semantics of SpecC *par* and *channel*, can be correctly traced. Moreover, once the portions to be analyzed are fixed with various dependence analysis, similar formulation to formal debugging methods, such as [?], can be used to reason about C-based descriptions completely. Also, SDG can be utilized under a software debugging method called Delta Debugging [?] to deal with hardware oriented features of C-based design languages.

3. CAUSE-EFFECT ANALYSIS THROUGH COMBINED USE OF CONCRETE AND SYMBOLIC SIMULATION

In order to trace design descriptions dynamically for debugging, we need to "simulate" the design descriptions. In traditional simulation, the state space is explored by multiple runs of concrete simulation with the different input vectors. Clearly exercising the

978-1-60558-497-3/09 $25.00 © 2009 ACM

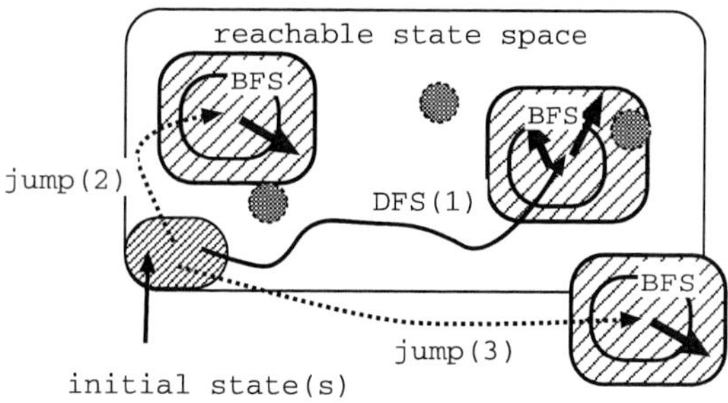

Figure 3: The user-driven BFS approach.

simulation for all possible inputs exhaustively is not feasible. So the problem is how to prepare the *good* input vectors, and the critical issue is how to activate corner cases. Traditional simulation can be said to exercise a depth-first search in the state space. On the other hand, symbolic algorithms, such as symbolic simulation, can traverse the state space in breadth-first ways. They do not manipulate the states one by one explicitly, but manipulate the set of reachable states symbolically and can handle all the possible input values implicitly at once. It may, however, suffer from blow-up of symbolic expressions when the target designs become large, and so only the limited cycles from the given states can be inspected.

To overcome the limitation of the concrete simulation in width and that of the symbolic simulation in depth, there have been proposed approaches shown in Figure ?? [?, ?, ?]. First, users run concrete simulation (DFS) only to gather the reachable states randomly. Then, the user picks up some of the reachable states (e.g., (1) in Figure ??), or make up some states even if they are not known as feasible ((2) and (3) in Figure ??), and try symbolic simulation (BFS) from them. Weighted random and/or user-specified constraints-based generations could be used to make up states to further proceed symbolic simulations.

This approach realizes the short-range exhaustive search from the states which are far from the initial states in flexible ways under user control. Instead of just simulating designs either in concrete and symbolic ways, we can introduce control mechanisms which analyze the simulations results and modify the input vectors based on the results. Here our goal is to find the situations or system states which are indicated by the designers. A set of control commands on concrete and symbolic simulations can automatically generate the desired input vectors for concrete and symbolic simulations that reach the target states. This can be an effective approach not only for verification but also for debugging. For example, based on the symbolic simulation results, input vectors for concrete simulations can be adjusted and vice versa. When debugging, typically multiple error traces are given. Those traces can be considered as the concrete simulation results. By utilizing symbolic simulation partially on the traces, we may be able to collect more general conditions under which buggy behaviors appear. Also, if designers know suspicious portions of the descriptions corresponding to the given error traces, we can use the control mechanisms to let concrete simulations to go through those suspicious portions.

To demonstrate the effectiveness of the above debugging approach, a tool, which has various commands to control concrete and symbolic simulations as well as input vector generations, has been implemented [?] and is applied to a finite impulse response filter design. This filter reads two 8 bit input signals and writes one 8 bit output signal through 7 stage pipelines in every cycle. This design contains 24 conditional branches in 200 lines of SpecC codes. Suppose a designer wants to find an input sequence which

goes through a suspicious execution path in his/her mind. Given the design description, the declarations of the input signals for pattern generation, the assumed conditions for input values, and the specification of the execution path for each branch over 30 cycles, the tool has successfully generated an input sequence that goes through the specified path, within one minute. The generated input sequence can be used to inspect the suspicious portion for debugging.

4. POST-SILICON DEBUGGING WITH SYSTEM LEVEL MODELS

There have been a number of researches in post-silicon debugging utilizing SLMs. One common way is to try to establish mappings between implementations, such as RTL descriptions, and SLMs [?]. Also, a technique that tries to generate transaction level descriptions, which are sorts of SLMs, from given RTL descriptions has been proposed [?]. These help efficient analyses of chip behaviors significantly. Current designs, however, may have many more features than SLMs, and mappings between SLMs and implementations can be just too complicated to be used as they are for debugging. For example, various low power mechanisms, such as sleeping modes and power-off modes, are generally not explicit in SLMs.

Here we introduce a method to raise the debug abstraction level of communications to transaction level in post-silicon debugging. The method is based on on-chip instrumentation through the on-chip bus communication protocols. Furthermore, we present a method for automatic analysis of extracted transactions to find potential erroneous transaction sequences. The post analysis method searches for certain patterns to find potential problems such as deadlock and race from transaction sequence that is previously extracted.

Transaction level modeling (TLM) concept has been introduced to help the designers to cope with the complexity of systems consisting of both hardware and software parts with complicated communications. Using the TLM methodology, systems are partitioned into computation parts and communication parts. During the system refinement and synthesis, computation parts become functional units/cores or programs running on processors. Communication parts become on-chip communication lines, buses or networks. There are two main problems with the RTL models of communication parts that are generated from their corresponding transaction level models. The first problem is that currently the RTL generation is mostly done manually. This means that there may be bugs in this process. Being able to extract transaction level communication behavior from actual system behavior on the chips can help to find these bugs. The second problem is that due to the complexity of system, there may be unexpected behaviors of the system that are not handled in transaction level design. For example interactions of the system with environment such as interrupts, are main sources of nondeterministic behavior of the system that are not usually considered in transaction level. Extracting transaction level information from system runs can reveal the actual behavior of system in an abstracted way that is better understandable and traceable by designers than pure signals on real chips. Here we introduce a method to raise the abstraction level of debugging to transaction level for communication parts of systems. First, we extract the basic events such as requests and responses from control signals following the communication protocol using on-chip monitor circuits. Then, according to the definition of different transactions in TLM, we merge the basic events to build up the transactions.

To handle all kinds of transactions uniformly we model the transactions as follows. Each transaction consists of the following four

978-1-60558-497-3/09 $25.00 © 2009 ACM

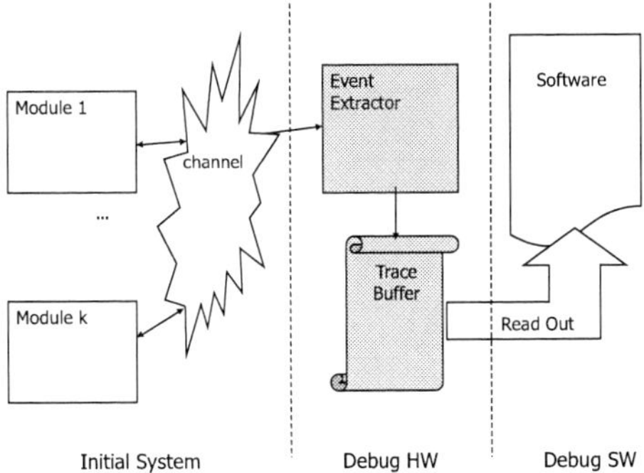

Figure 4: Post-silicon debug infrastructure

basic phases and two ones for error handling:

1. Start of Request (SoRq): in TLM corresponds to putting the request in the channel by the initiator. In RTL it is usually when the master tries to take the bus ownership, but may be different depending on on-chip bus protocol.

2. End of Request (EoRq): in TLM corresponds to getting the request by target. In RTL it is the time when slave is selected to start the operation. It may be after address decoding or by explicit signal activation from bus, depending on on-chip bus protocol.

3. Start of Response (SoRp): in TLM corresponds to putting the response in the channel by the target. In RTL it is usually when the slave acknowledges end of transfer, but may be different depending on on-chip bus protocol.

4. End of Response (EoRp): in TLM corresponds to getting the response from the channel by initiator. In RTL it is usually when the master is acknowledged, but may be different depending on on-chip bus protocol.

5. Request Error (RqErr): corresponds to any kind of error that causes a request to be ended unsuccessfully after SoRq.

6. Response Error (RpErr): corresponds to any kind of error that causes a response to be ended unsuccessfully after SoRp.

Given formal definitions of on-chip bus protocol such as the one used in [?], the above basic transactions can be automatically extracted, which become the bases of our mapping between SLMs and implementations. Transaction level debug requires both on-chip hardware support and off-chip software support. The on-chip part mainly deals with extraction of basic events to build up transactions and also storing them on an on-chip trace buffer. The software is responsible to read the trace buffer, build the transactions and provide transaction analysis. As shown in Figure **??**, the system consists of k modules (Module 1 to Module k) communicating with each other through the channel (e.g. on-chip bus). Debug hardware, as shown in the figure, consists of two main parts: event extractor and trace buffer. Event extractor is a module that is connected to communication channel and is responsible to extract the basic events that have been explained above. The data that is extracted from the event extractor is stored in a trace buffer to enable transaction extraction and also further analysis by software. There is also a mechanism (like scan) to read out the trace buffer contents.

Sizes of extra hardware in our method has been evaluated through an FPGA implementation with OPB and PLB on-chip buses that are parts of IBM CoreConnect bus system. The on-chip extractors for OPB and PLB buses are coded in VHDL hardware description language (around 150 lines each), which are synthesized on Xilinx Vitex4 devices assuming 4 master and 4 slave devices are attached to buses. It takes only 1,966 equivalent gates for OPB and 2,206 equivalent gates for PLB bus which are very low area overhead for a typical design. Assuming that there are 4 masters and 4 slaves, each entry of the trace buffer becomes 10 bits (2 bits for master ID, 2 bits for slave ID, 2 bits for slave address, 1 bit for rd/wr and 3 bits for command). Therefore by allocating a 16 KB memory block, we can save more than 1,600 transaction elements in a trace buffer, which helps post-silicon debugging with references to C-based design descriptions.

5. CONCLUSIONS

Various techniques which utilize SLMs for debugging have been presented. As shown in the paper, formal analysis of SLMs helps a lot even for post-silicon debugging along with on-chip instrumentation. As designs become more and more complicated, uses of SLMs are essential for efficient and effective debugging. Many more researches which integrate various formal analysis of SLMs with debugging processes in various design levels are expected.

6. REFERENCES

[1] D. Gajski, J. Zhu, R. Doemer, A. Gerstlauer, and S. Zhao, *SpecC: Specification Language and Methodology*, Kluwer Academic Publisher, Mar. 2000.

[2] SystemC: http://www.systemc.org/

[3] S. Horwitz, T. Reps, and D. Binkley, "Interprocedural Slicing Using Dependence Graphs," *ACM Transactions on Programming Languages and Systems*, Vol.12, No.1, 1990.

[4] S. Sasaki, T. Nishihara, D. Ando, M. Fujita, "Hardware/Software Co-design and Verification Methodology from System Level Based on System Dependence Graph," *Journal of Universal Computer Science*, Vol.13, No.13, pp.1972-2001, 2007.

[5] S.Safarpour, A.Veneris, "Abstraction and Refinement Techniques in Automated Design Debugging," *DATE*, 2007.

[6] M. Fujita, "Trends in Formal Verification Techniques for C-based Hardware Designs," *IPSJ Transactions on System LSI Design Methodology (TSLDM)*, Vol. 2, February 2009.

[7] Y. Kojima, T. Nishihara, T. Matsumoto, M. Fujita, "Interactive Verification and Debugging Environment by Concrete/Symbolic Simulations for System-level Designs," *ATS*, 2008.

[8] K. Sen, D. Marinov, G. Agha, CUTE: a concolic unit testing engine for C *Foundations of Software Engineering*, 2005.

[9] S. Shyam, V. Bertacco, "Distance-Guided Hybrid Verification with GUIDO", *DATE*, 2006.

[10] C.R. Ho, M. Theobald, B. Batson, J.P. Grossman, S.C. Wang, J. Gagliardo, M.M. Deneroff, R.O. Dror, D.E. Shaw, "Post-Silicon Debug Using Formal Verification Waypoints," *DVCon*, 2009.

[11] N. Bombieri, F. Fummi, G. Pravadelli, "A methodology for abstracting RTL designs into TL descriptions," *Formal Methods and Models for Co-Design*, 2006.

[12] A. Zeller, "Isolating Cause-Effect Chains from Computer Programs," *ACM FSE-10*, 2002.

[13] S. Watanabe, K. Seto, Y. Ishikawa, S. Komatsu, M. Fujita, "Protocol Transducer Synthesis using Divide and Conquer Approach," *ASPDAC2007*, 2007.

978-1-60558-497-3/09 $25.00 © 2009 ACM

The Day Sherlock Holmes Decided to do EDA

Andreas Veneris
University of Toronto, ECE Dept. & CS Dept.
Toronto, ON, Canada
veneris@eecg.toronto.edu

Sean Safarpour
Vennsa Technologies, Inc.
Toronto, ON, Canada
sean@vennsa.com

ABSTRACT

Semiconductor design companies are in a continuous search for design tools that address the ever increasing chip design complexity coupled with strict time-to-market schedules and budgetary constraints. A fundamental aspect of the design process that remains primitive is that of debugging. It takes months to close, it introduces costs and it may jeopardize the release date of the chip. This paper reviews the debugging problem and the research behind it over the past 20 years. The case for automated RTL debug tools and methodologies is also made to help ease the manual burden and complement current industrial verification practices.

Categories and Subject Descriptors

B.7.2 [**Integrated Circuits**]: Design Aids—*Simulation, Verification*

General Terms

Design, Verification

Keywords

Debugging, Error Localization, Verification

1. INTRODUCTION

In the past decade, there has been an exponential increase in the cost and time required for verification and debugging of VLSI systems. Verification checks the correctness of a design and if faulty, debugging identifies the root-cause of the problem. Although debugging manifests itself in every step of the design cycle, in this study, we are interest in functional Register Transfer Level (RTL) debugging.

It is a well-accepted fact that debugging and verification take up to 70% of the chip design time. With debugging contributing to as much as half of this time, it directly results in millions of dollars in non-recurring costs and may jeopardize the release date of the end product. To make things worst, silicon prototypes today are rarely bug-free. Functional bugs may escape pre-silicon verification only to be discovered during in-system silicon validation. It comes as no surprise that more than 60% of design tape-outs require at least one re-spin and more than half of the failures are not due to power, timing or manufacturing defects but due to logical or functional errors not discovered or properly fixed during verification [15].

Permission to make digital or hard copies of part or all of this work for personal or classroom use is granted without fee provided that copies are not made or distributed for profit or commercial advantage and that copies bear this notice and the full citation on the first page. To copy otherwise, to republish, to post on servers or to redistribute to lists, requires prior specific permission and/or a fee.
DAC'09, July 26-31, 2009, San Francisco, California, USA

Without a doubt, verification and debug are major bottlenecks. This burden is expected to increase 675% by 2015 as reported by EETimes [10]. There are a number of reasons to justify this trend. Firstly, the modern semiconductor design flows and verification methodologies are complex in nature. The processes of design and verification are comprised of heterogeneous components implemented at multiple levels of abstractions (procedural, behavioral, synthesizable, etc.) using different languages (Verilog, System Verilog, VHDL, PSL, etc) and ever changing standards and protocols. Interacting with such components adds layers of complexity and overhead while hindering transparency and efficiency. The lack of a unified and centralized verification environment makes the debugging pain a growing challenge for the end engineer.

Further, design specifications often described in abstract models, may not directly correspond to signals and transactions at the design level. For example, the specification can be described in a plain document, a Matlab model or in a software language such as C/C++, whereas the design is implemented in cycle-accurate Verilog or in VHDL. The separation between these two layers can result in misinterpretations and usually complicates verification/debugging efforts. Additionally, the ever increasing size of modern devices poses challenges both to Electronic Design Automation (EDA) tools and engineers alike. Typical design blocks grow beyond the 400,000 synthesized gate mark and error-traces extend past thousands of cycles. Task outsourcing and geographical dispersed teams only add extra layers of communication overhead to these processes.

In 2006, the International Technology Roadmap for Semiconductors (ITRS), issued its new set of needs for the current and next generation design semiconductor processes. Although most topics saw minor numeric revisions, the roadmap contains a major fourteen-page update in design verification with a strong emphasis in debugging. The report [11] states that *"technological progress depends on the development of rigorous and efficient methods to achieve high-quality verification results ... and techniques to ease the burden of debugging a design once a bug is found ... without major breakthroughs, verification will be a non-scalable, show-stopping barrier to further semiconductor progress"*. Without a doubt, the roadmap depicts a grim yet realistic picture that establishes an urgent need for scalable automated debugging tools and methodologies.

To a certain extent, the tremendous growth of the semiconductor industry over the past decades can be partially attributed to the amount of automation provided by the EDA community. Most manual steps of the design flow (synthesis, placement, routing, test, verification, etc.) have been automated to help close designs faster and cheaper. Unlike these processes, debugging remains a time-consuming and resource intensive manual task where graphical navigators and waveform viewers allow engineers to perform simple "what-if" analysis. With no form of exaggeration, the engineer today resembles the famous detective Sherlock Holmes who searches for needles in a haystack and relies on a "hunch" or "gut feel" to localize the culprit bug when verification fails.

In past years, VLSI design companies have in part alleviated this debugging pain by allocating more verification engineers to the problem. As a net effect, it has been reported, there are two to three times more verification engineers than designers in design teams [3]. It is clear that adding verification engineers cannot provide a sustainable solution as the pain continues to clime. Automated RTL debugging techniques to localize the source of an error have become an urgent necessity if we

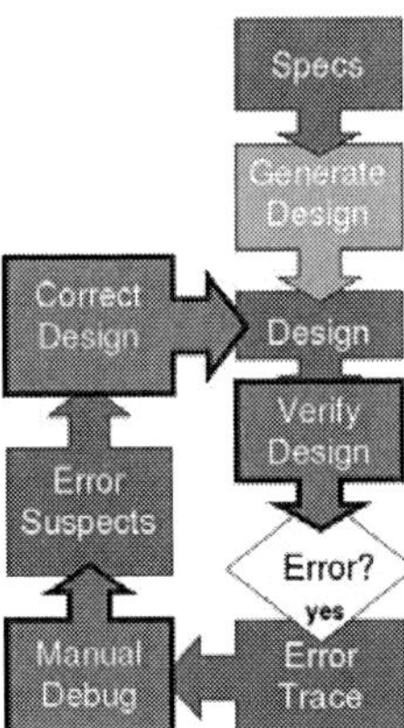

Figure 1: Typical verification and debugging design flow

desire to circumvent the manual problem and drastically improve the verification flow.

This paper builds the case for robust automated RTL debugging tools and methodologies to aid the engineer. We first define the problem and outline its multiple facets in the design cycle. Then we give a historical review of the research in debugging from the early days to state-of-the-art advances and we present industrial case-studies. The paper concludes by re-iterating the need for further research to provide the platform for cost-effective automated debugging tools and scalable methodologies to complement current industrial practices.

2. WHAT IS DEBUGGING?

Every time a design or a silicon prototype fails to adhere to a set of specifications, a debugging problem usually follows. As such, debugging manifests itself in virtually every step of the design cycle. When the design does not meet its power requirements, the engineer has to debug the problem and fix it by optimizing certain portions. When a place and route tool cannot meet timing closure, the designer does it manually by exploiting flexibilities not seen by the tool. When a silicon prototype fails test, silicon debug identifies the error root-cause to fix it so that the re-spun prototype passes test.

In this study we are interested in the problem of RTL debug for functional failures. Once verification fails, it returns with an *error-trace* or a *counter-example* that exhibits the erroneous behavior at some observation points. The input to debugging is the actual design, the set of error-traces V and the correct responses to those error-traces as shown in Fig. 1. A debugging tool localizes the error source or *suspects* with references to the RTL files, gate-level netlists or design schematics.

Note that a debugging tool utilizes a golden or reference model that provides the expected logic values for the erroneous design. In our context, there is a fundamental assumption that this model acts as a "black box", *i.e.,* there is no structural correspondence between internal lines of the model with this of the design. For example, the golden model can be a Matlab program while the design is in Verilog. This complicates the debugging effort dramatically because the solution space explodes exponentially to the number of errors in the design [21]:

$$solution\,space = (circuit\,lines)^{\#\,errors} \qquad (1)$$

Another implication is that the only observation points to the debugger are the primary output design signals or the embedded assertions/properties. These may be cycle- or no-cycle- accurate values captured by interface monitors, checkers or assertions. In other words, we are interested in debugging that follows simulation-based verification, formal verification and emulation flows [12]. We do not include combinational equivalence checking in this category since it utilizes structural equivalences to solve the problem [8]. To that end, the problem of functional RTL debug resembles this of fault diagnosis (or silicon debug) [20].

Once verification identifies that a design contains an error(s), debugging usually involves the following questions:

- Is there a bug in the design or is the bug in the testbench?

- Which block in the design and which RTL line(s) should we focus on?

- What is the root-cause of the bug?

- Who should fix the bug and how should it be fixed?

The process that answers these questions today involves an arduous manual task with many iterations to close it. It delays the subsequent steps of the design cycle and introduces significant non-recurring costs.

3. DEBUGGING: THE EARLY DAYS

There is a consensus that the term "design error" is attributed to the paper by Abadir et al. from 1988 [1]. That paper outlines a set of typical errors found in the design flow also known as *design error models*. Essentially, this is a dictionary of possible simple error types such as gate replacement errors, missing or misplaced input gate line errors, etc. In the same work, the authors prove theorems for the test set V to guarantee 100% verification coverage using previous results for stuck-at faults.

Following that work, in the 1990s, a great deal of automated algorithms were developed to tackle the problem using the design error model in [1]. A comprehensive review of those methods is found in [14]. Depending on the underline engine used to drive the algorithm, those techniques can be classified as *symbolic-based* and *simulation-based*.

3.1 Symbolic-based Debugging

Symbolic approaches typically perform diagnosis by generating and solving an *error equation* using Binary Decision Diagrams (BDDs) based on the functionality of both the correct and erroneous circuits. In [9], algorithms for single and multiple design error diagnosis and correction are presented. For single errors, an error equation is generated in turn for each line l in the netlist. For a netlist with inputs X, the error equation for line l is noted $E^l(X, z(X))$, where $z(X)$ is an unspecified function over the circuit inputs. By construction, if there exists some function $z(X)$ that satisfies the equation $E^l = 0$, then replacing the line l with a circuit implementing $z(X)$ will "correct" the behavior of the netlist according to its specification. Moreover, $z^*(X) \in [E^l(X,0), \overline{E^l(X,1)}]$ specifies the family of all solutions $z^*(X)$ that correct the circuit at l.

Essentially, E^l is an algebraic representation of a miter for the two circuits. Although effective for single errors, it exhibits memory problems as it uses BDDs [5] to build the error equation. Furthermore, its applicability to multiple errors is limited to cases depending on their structural proximity [9].

3.2 Simulation-based Debugging

To overcome the excessive memory requirements of BDD-based approaches, debugging with simulation has been extensively investigated. Those methods provide a time/space trade-off as they remain polynomial in the input size but they may require more time to give an answer.

Simulation-based techniques [14] [21] [22] simulate an error-trace and trace backwards from primary output to primary input marking suspect lines using different criteria. For each error-trace, they collect the suspect lines and since the error is present in each one of them, they intersect the results for all runs. Although their memory requirements remain linear to the size of the circuit, the complexity mitigates to the time domain. As the number of errors increases, their performance degrades. For this reason, their applicability to sequential circuit debugging has been rather limited [14].

To overcome these obstacles, the concept of simulation-based debugging has been enriched with simulation of unknown values [4] to alleviate the need for an error model. Although practical in some cases, unknowns can decrease the resolution of the solution for large designs or for sequential designs. In the work of Liu et al. [16], an incremental debugging method is proposed. That method outperforms conventional techniques for multiple errors but since it is not exhaustive in the solution space, it may miss finding solutions.

978-1-60558-497-3/09 $25.00 © 2009 ACM

4. DEBUGGING WITH SATISFIABILITY

Recently, the introduction of Boolean Satisfiability (SAT) in the field opened new opportunities for cost effective automated debugging tools.

4.1 SAT-based Debugging

A *SAT-based debugging* technique was proposed in SAT [20] where the problem is formulated as a SAT instance for a conventional solver to return solutions corresponding to suspects. A variety of SAT-based debugging formulations have been proposed building on the initial work [6] [13] [18]. Experiments show that SAT-based debugging outperforms traditional simulation- and BDD-based techniques, in both time and space, sometimes by orders of magnitude.

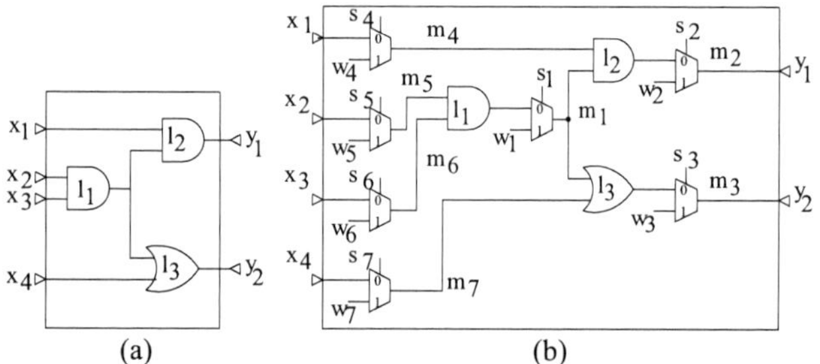

(a)   (b)

Figure 2: SAT-based debugging

To model debugging [20], a multiplexer m_i is added for every gate (and primary input) l_i. The output of this multiplexer, m_i, is connected to the fanouts of l_i while l_i is disconnected from its fanouts. This construction has the following effect: when the select line s_i of a multiplexer is inactive ($s_i = 0$), the original gate l_i is connected to m_i, otherwise, when $s_i = 1$ a new unconstrained primary input w_i is connected. Figure 2 illustrates the above transformation for a combinational circuit.

This construction is later constrained with the input/output values of the expected primary output responses for the particular error-trace. A potential correction on line l_i is indicated when the select line s_i is assigned to 1 under which condition the correction value is stored in w_i. The SAT solver can assign any value $\{0,1\}$ to the s_i and w_i variables such that the resulting Conjunctive Normal Form (CNF) satisfies the constraints applied by the vectors V. To force the SAT solver to find a specific number N of error locations, further logic is added to activate at most N select lines [20]. Thus for $N = 1$, a single s_i is set to 1 which *corresponds* to candidate error location l_i, etc. Finally, the construction is repeated and constrained for each error-trace before given to a SAT solver.

Following the initial formulation, a variety of advances have been proposed to improve performance. More notably, the work in [18] borrows the concepts of abstraction and refinement from formal verification to ease the debugging effort. The authors in [2] use Quantified Boolean Formula (QBF) Satisfiability to "compress" the memory required by replicating the CNF for the different error-traces. Finally, orthogonal to [20], the research in [7] and [19] introduces SAT-based techniques to reduce the length of the error-traces and further reduce the memory requirements for existing debugging methods.

4.2 QBF-based Debugging

The backbone of the SAT-based formulation proposed by [20] when applied on sequential circuits is the repetition of the combinational circuitry for a number of cycles equal to this of the error-trace. This is also known as the Iterative Logic Array representation or time-frame expansion [17]. Clearly, replicating a half million gate block for possibly thousands of cycles of industrial-size error-traces, it may require prohibitively excessive memory resources. Evidently, more compact representations of sequential debugging problems are required to ensure scalability with no sacrifice in performance.

To that end, [17] presents a parameterizable encoding for debugging using QBF Satisfiability. In Boolean SAT all variables in the CNF are existentially quantified. QBF is a generalization of SAT that also allows for universal ($\forall$) quantification of the variables. A QBF formula in *prenex normal form* is written as:

$$Q_1 \mathcal{V}_1 \quad Q_2 \mathcal{V}_2 \quad \cdots \quad Q_r \mathcal{V}_r \quad | \quad \Phi \qquad (2)$$

The design debugging formulation using QBF is given by the following equation:

$$\exists e, s^0, s^1, \ldots, s^k, X, Y \;\; \forall t \;\; \exists s, s', x, w, y \;|$$

$$\bigwedge_{j=1}^{k} t^k(j) \rightarrow [(s = s^{j-1}) \wedge (s' = s^j)] \;\wedge$$

$$\bigwedge_{j=1}^{k} t^k(j) \rightarrow [(x = x^j) \wedge (y = y^j)] \;\wedge$$

$$T_{en}(s, s', \langle x, w, e \rangle, y) \wedge \Phi_C(s^0, X, Y) \wedge \Phi_N(e) \qquad (3)$$

where e are the error location select lines, $s^0 \ldots s^k$ are state elements for the k-cycle error-trace, and X (Y) is the set of design primary input (output). Although the intricate details of the encoding are beyond the context of this paper [17], pictorially the hardware construction that corresponds to Eq. 3 is shown in Figure 3. In that figure, T_{EN} is a single copy of the combinational circuitry (*i.e.,* transition relation) from Figure 2(b). Intuitively, the QBF formulation mitigates the space expansion of the circuit into time using universal quantification. Experiments shows the favorable nature of this encoding as it achieves a dramatic 92% reduction in space when compared to SAT and sometimes outperforms it in terms of run-time.

5. INDUSTRIAL CASE STUDIES

With the help of recent advances, automated debugging tools for industrial problems are within reach. In this section, we present three case studies representative of common bugs found in the RTL. For each of the cases, we show a code snippet in Verilog containing the bug as well as the correct implementation shown in comments (*i.e.* //). We present how a state-of-the-art industrial automated debugger using the methodologies from Sections 4 and 4.2 efficiently tackles the problem in Table 1.

In the first example, shown in Fig. 4, the 0 and 1 input of the multiplexer are incorrectly connected. This type of mistake is very common, as it is easy to mix-up the signals or forget the polarity of the condition. In this case, the bug is detected by a self-checking testbench when the output of the memory controller is found to be different than the expected value.

In the second example, shown in Fig. 5, a single clock fifo module, vga_fifo, is erroneously instantiated instead of dual clock fifo module, vga_fifo_dc. This type of error can be caused by missing details in specifications that do not explicitly require a dual clock fifo. This bug is caught by an end checker when the pixels generated by the vga controller do not match those of a golden C model. In this case, it can be very hard to narrow down the problem from the controller output all the way to the fifo module.

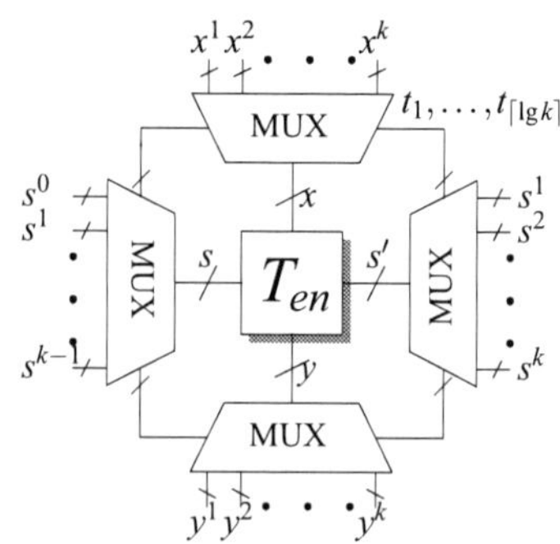

Figure 3: Design debugging construction

978-1-60558-497-3/09 $25.00 © 2009 ACM

```
always @(posedge clk)
   // correct line:
   //wb_data <= #1 `MC_MEM_SEL ? mem : rf;
   // bug below:
   wb_data <= #1 `MC_MEM_SEL ? rf : mem;
```

Figure 4: Case study 1: misinterpretation of condition in memory controller

```
// correct block:
//    vga_fifo_dc line_fifo(
//       .rclk  ( clk_p_i           ),
//       .wclk  ( wb_clk_i          ),
//       .rclr  ( 1'b0              ),
//       .wclr  ( ctrl_ven_not      ),
//       .wreq  ( line_fifo_wreq    ),
//       .d     ( line_fifo_d       ),
//       .rreq  ( line_fifo_rreq    ),
//       .q     ( line_fifo_q       ),
//       .empty ( line_fifo_empty_rd ),
//       .full  ( line_fifo_full_wr )
//    );
//
// bug below:
   vga_fifo  line_fifo (
      .clk   ( clk_p_i           ),
      .aclr  ( 1'b0              ),
      .sclr  ( ctrl_ven_not      ),
      .wreq  ( line_fifo_wreq    ),
      .d     ( line_fifo_d       ),
      .rreq  ( line_fifo_rreq    ),
      .q     ( line_fifo_q       ),
      .empty ( line_fifo_empty_rd ),
      .full  ( line_fifo_full_wr )
   );
```

Figure 5: Case study 2: wrong fifo instantiation in vga controller

The third example, shown in Fig. 6, contains multiple errors where the `if` conditions for signals RX_W and RC_W are misinterpreted. In this large communication block, the errors affect control circuitry which triggers an assertion to fail dozens of clock cycles after the bug has been excited.

```
always@(posedge CLK)
   // correct line:
   //if(RX_W == 1'h1)begin
   // bug below:
   if(RX_W == 1'h0)begin
      NTY_CTL <= RX_CTL;
      NTY_DATA <= RX_DATA;
   end
   else begin
      // correct line:
      //if(RC_W == 1'h1)begin
      // bug below:
      if(RC_W == 1'h0)begin
         NTY_CTL[63:48]<= RC_CTL[63:48];
         NTY_CTL[7:5]<= REG_RES_FRAME[2:0];
      end
   end
```

Figure 6: Case study 3: multiple wrong conditions in communication block

For the three case studies, Table 1 shows the design type and size of the circuit in primitive gates in columns one and two respectively. Columns three and four present the time required by the debugger to localize the suspects, and the total number of suspects returned to the engineer. In all cases, the debugger eliminates more than 99% of the code in a few seconds or minutes. The suspects returned point the engineer to a small set of Verilog lines where the design can be rectified and the bug removed. The relatively small run-time of the tool makes the automated debugger a powerful and practical tool to aid engineers in the daunting debugging task.

6. CONCLUSION

Debugging of RTL designs remains a manual and resource-intensive task today. This paper outlines the rich research in debugging over the past 20 years. It also builds the case for novel automated debugging methodologies for industrial applications. In the near future, these tools can help reduce the manual debugging pain as well as the overall verification effort.

Table 1: Case study statistics

Design type	# gates	Debug time	# Suspects
memory controller	46K	26 sec.	5
vga controller	150K	32 sec.	12
communication block	800K	672 sec.	10

7. REFERENCES

[1] M. S. Abadir, J. Ferguson and T. E. Kirkland, "Logic Verification Via Test Generation", in *IEEE Trans. on CAD*, vol. 7, pp. 138–148, Jan. 1988.

[2] M. Fahim Ali and S. Safarpour and A. Veneris and M. Abadir and R. Drechsler, "Post-verification debugging of hierarchical designs," in *IEEE Int. Conf. on Computer-Aided Design*, pp. 871-876, 2005.

[3] J. Bergeron, *Writing Testbenches: Functional Verification of HDL Models*, Kluwer Academic Publishers, 2003.

[4] V. Boppana and M. Fujita, "Modeling the unknown! Towards model-independent fault and error diagnosis," in *IEEE Intern. Test Conference*, pp. 678-687, 1998.

[5] R. E. Bryant, "Graph–based algorithms for Boolean function manipulation," in *IEEE Trans. on Computers*, vol. C–35, no. 8, pp. 677-691, 1986.

[6] K.H-. Chang, I. Markov and V. Bertacco, "Automating post-silicon debugging and repair," in *IEEE Int. Conf. on Computer-Aided Design*, pp. 91-98, 2007.

[7] K.H-. Chang, I. Markov and V. Bertacco, "Simulation-based Bug trace minimization with BMC-based Refinement," in *IEEE Trans. on Computer-Aided Design*, vol. 26, no. 1, Jan. 2007.

[8] S. Chatterjee, A. Mishchenko, R. Brayton and A. Kuehlmann, "On resolution proofs for combinational equivalence," in IEEE/ACM Design Automation Conference, pp. 600-605, 2007.

[9] P. Y. Chung and I. N. Hajj, "Diagnosis and correction of multiple design errors in digital circuits," in *IEEE Trans. on VLSI Systems*, vol. 5, no. 2, pp. 233-237, June 1997.

[10] D. Geus, *ICs will rebound*, EETimes, Mar. 3, 2009.

[11] International Technonology Roadmap for Semiconductors, 2006.

[12] R. Drechsler, *Advanced Formal Verification*, Kluwer Academic Publishers, 2004.

[13] G. Fey, S. Staber, R. Bloem and R. Drechsler, "Automatic Fault Localization for Property Checking," in *IEEE Trans. on Computer-Aided Design,*, vol. 27, no. 6, June 2008.

[14] S.-Y. Huang and K.-T. Cheng, *Formal Equivalence Checking and Design Debugging*, Kluwer Academic Publishers, 1998.

[15] J. Jaeger, *"Virtually every ASIC ends up an FPGA*, EETimes, Dec. 7, 2007.

[16] J. B. Liu and A. Veneris, "Incremental Diagnosis," in *IEEE Trans. on Computer-Aided Design, vol. 24, no. 2, pp. 240-251*, Febr. 2005.

[17] H. Mangassarian, A. Veneris, S. Safarpour, M. Benedetti and D. Smith, "A performance-driven QBF-based iterative logic array representation with applications to verification, debug and test," in *IEEE Int. Conf. on Computer-Aided Design*, pp. 240-245, 2007.

[18] S. Safarpour and A. Veneris, "Abstraction and Refinement techniques in Automated Design Debugging," in *IEEE/ACM Design and Test in Europe*, pp. 1-6, 2007.

[19] S. Safarpour, A. Veneris and H. Mangassarian, "Trace Compaction using SAT-based Reachability Analysis," in *IEEE Asian-South Pacific Design Automation Conference*, pp. 23-26, 2007.

[20] A. Smith, A. Veneris, M. Fahim Ali and A. Viglas, "Fault Diagnosis and Logic Debugging Using Boolean Satisfiability," in *IEEE Trans on Computer-Aided Design*, vol. 24, no. 10, pp. 1606-1621, 2005.

[21] M. Tomita, T. Yamamoto, F. Sumikawa and K. Hirano, "Rectification of multiple logic design errors in multiple output circuits," in *Proc. of the Design Automation Conf.*, pp. 212–217, 1994.

[22] A. Veneris, and I. N. Hajj, "Design Error Diagnosis and Correction Via Test Vector Simulation," in *IEEE Trans. on Computer-Aided Design*, vol. 18, no. 12, pp. 1803–1816, Dec. 1999.

978-1-60558-497-3/09 $25.00 © 2009 ACM

Debugging Strategies for Mere Mortals

Valeria Bertacco

Department of Computer Science and Engineering, University of Michigan
valeria@umich.edu

ABSTRACT

Recent improvements in design verification strive to automate error detection and greatly enhance engineers' ability to detect functional errors. However, the process of diagnosing the cause of these errors, and subsequently fixing them, remains one of the most difficult tasks of verification. The complexity of design descriptions, paired with the scarcity of software tools supporting this task lead to an activity that is mostly ad-hoc, labor intensive and accessible only to a few debugging specialists within a design house.

This paper discusses some recent research solutions that support the debugging effort by simplifying and automating bug diagnosis. These novel techniques demonstrate that, through the support of structured methodologies, debugging can become a task pursued by the average design engineer. We also outline some of the upcoming trends in design verification, postponing some the verification effort to runtime, and discuss how debugging could leverage these trends to achieve better quality of results.

Categories and Subject Descriptors

B.6.2 [**Logic Design**]: Reliability and Testing—*Error-checking*;
B.6.3 [**Logic Design**]: Design Aids—*Verification*

General Terms

Verification, Algorithms

Keywords

Design verification, Validation, Error diagnosis, Error correction.

1. INTRODUCTION

Digital integrated circuit design has reached unparalleled levels of complexity. In this context, verification has become a pivotal aspect of electronic design automation. In fact, various estimates indicate that functional errors are still responsible for 40% of failures at first tape-out, and that verification accounts for two thirds of the design cycle and effort [4, 18].

Resolving design bugs in the early development stages is, at the same time, a sophisticated and time-consuming activity, as well as a crucial task for the project development and for the success of a design team. In the past few decades, much research has been dedicated to improving the quality and the effectiveness of verification, however, much less effort has been devoted to supporting a design team in resolving a functional bug, that is, finding the root cause of the bug and devising a modification to the design that corrects it. A few commercial software applications are available that provide minimal debugging support, for the most part in the form of visualization tools that can connect a signal transition observed

Permission to make digital or hard copies of part or all of this work for personal or classroom use is granted without fee provided that copies are not made or distributed for profit or commercial advantage and that copies bear this notice and the full citation on the first page. To copy otherwise, to republish, to post on servers or to redistribute to lists, requires prior specific permission and/or a fee.
DAC'09, July 26-31, 2009, San Francisco, California, USA

in simulation to a specific location in the source Register-Transfer Level (RTL) description [22]. While these aids are valuable when investigating a bug, they are far from solving the problem, particularly when the problem manifests itself through a bug trace, several millions cycles long, producing an erroneous outcome at the end. As a result, bug diagnosis and correction is an extremely time-consuming challenge, with some bugs imposing delays of several days, or even weeks, to the development schedule. Occasionally, the correction of a bug may affect so many components of a circuit, that the design team may choose not to pursue it. This is particularly prone to occur in the late development stages of a system, or if the effects of the bug under evaluation may be countered through other means (such as microcode or compiler patches). Figure 1 shows a schematic of the design flow, highlighting how *debugging, that is, bug diagnosis and correction*, is an integral part of the design/verification loop, often disregarded in high-level flow diagrams and when planning a development schedule, but almost always the most time-consuming component.

The debugging methodologies available today rely, for the most part, on the skill and creativity of individual designers, making it more of an art for a few gifted people, than a science that can be

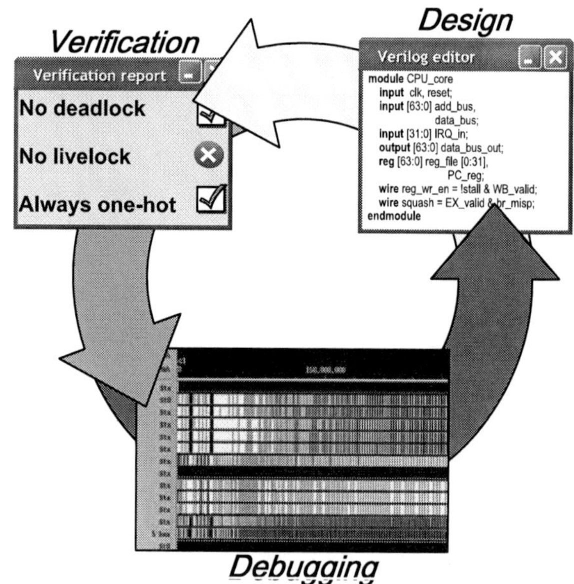

Figure 1: The design, verification and debugging cycle. During integrated circuit development, the system is designed, usually by means of a hardware description language, then verified by a combination of simulation-based and formal techniques. Each time a new bug is exposed in verification, it must be diagnosed and a design fix must be developed for it. This constitutes debugging. Once the fix is deployed, the new version of the design must undergo verification again to detect additional bugs, and/or bugs introduced by the fix. Overall debugging constitutes a major component, in terms of time and effort, of a digital system development cycle.

taught to the average engineer. While this is a high risk proposition as it stands, the growing complexity of design, verification, and even of simulation traces is quickly making this approach crumble. What is critically needed to overcome this situation are tools and techniques to support engineering in debugging, so that the complexity of the task can be reduced and constructive methodologies, accessible by people of ordinary skill, can be developed.

This paper outlines some of the work that is ongoing in our research group at the University of Michigan to address the problem of resolving functional bugs exposed through verification. Specifically, I will present two key projects that support and automate both diagnosis and correction of functional bugs. Both projects share a very low barrier to entrance, that is, they complement the current debugging methodology, without requiring any change to it. Since most verification today still occurs by means of logic simulation, the first solution, called *Butramin* tackles the complexity of debugging based on simulation traces. Such traces are often tens or hundreds millions of cycles long and may take hours or days just to be replayed so to re-create the bug condition. Butramin leverages a number of simulation-based analyses to reduce their size and length (in simulation cycles) by three to six orders of magnitude.

The second solution, called *REDIR*, is an automatic diagnosis technique that considers a set of bug traces and the corresponding correct system responses. It then leverages the RTL description of the design to isolate the root cause of a bug. The diagnosis is presented to the user as a set of signals (wire, registers, etc.) in the RTL that are responsible for the incorrect computation. Note that, "correctness" for REDIR is simply defined with respect to the traces and responses provided by the user. The correct behavior of the system in response to other stimuli is unknown, since no golden model is used, and the RTL contains functional bugs.

Finally, the industry today is becoming aware that design and verification complexities are such that digital systems are bound to be released with latent functional bugs. Thus, researchers are starting to develop correction solutions to be deployed in silicon and that operate at system runtime, being activated only if and when a bug is manifested. While this trend by no means diminishes the importance of debugging, it does allow for trade offs. For instance, if a bug entails such a widespread set of modifications to endanger design stability, it may be wiser to rely on runtime correction. Or, when detailed diagnosis becomes extremely time consuming, a better option may be to simply derive a system-level condition that may trigger the bug, and use that in runtime verification.

2. FOCUSING TRACES ON A BUG

Among the techniques and methodologies available for functional verification, simulation-based verification is prevalent in the industry because of its linear and predictable complexity. A common methodology in this context is *random simulation*, where stimuli are provided by a constraint-based random generator. Such generators can automatically produce random legal input for the design at a very high rate, based on a set of rules (or constraints) derived from the specification document. In random simulation bugs are detected by means of assertion statements, or checkers, embedded in the design. When a bug is detected, the simulation trace leading to it is stored aside and can be replayed at later times to analyze the conditions that led to the failure. Because of the randomized nature of this methodology, and because it is usually applied in late design stages (when simple bugs have already been flushed out), it is very common for the bug traces generated to be lengthy and complex. Another family of techniques, attracting increasing attention from industry, is that of semi-formal verification. These tools combine a mix of formal and simulation-based techniques with the goal of

producing high-coverage verification results on complex designs [1, 14, 12]. While semi-formal tools are a promising direction in terms of high-quality verification for industrial size designs, little concern has been given to the reduction in complexity for the bug traces generated. As a result, once a bug is found, a copious amount of effort goes into tracking it back to its root cause: either an incorrect design implementation or an erroneous property definition.

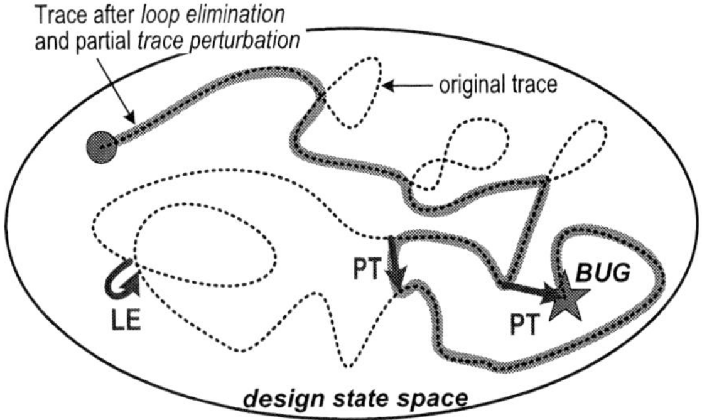

Figure 2: Two trace minimization techniques used by Butramin. The schematic shows a bug trace within the design state space (dashed line), starting from an initial state (circle) and ending at a bug state (star). Butramin attempts to remove "loops" within the trace using its *loop elimination* (LE) technique, it then explores possible *trace perturbations* (PT) by eliminating cycles and input events. These perturbations may lead to a shortcut in the trace. The solid line in the picture represents a trace obtained after cycle elimination and after discovering one of the PT shortcuts indicated.

The solution we set forth to address diagnosis of complex bug traces is called Butramin ("BUg TRAce MINimization") [7]. The objective of Butramin is to consider a bug trace and the corresponding checker (or property) that triggers the bug, and seek a shorter and simpler trace to falsify the same property. Previous work in this area has been mostly centered on using formal techniques to simplify a property's counterexample [20, 9]. In a separate context, the problem of trace minimization has also been addressed in software verification [11, 13]. We instead place most of the effort on simulation-based techniques so that we can apply our solution to complex modules, as those developed in the industry. Moreover, we do not strive to obtain a minimum-length trace, but simply one of manageable size for debugging purposes. However, in our experimental evaluation, we find that in practice our minimized traces are extremely compact, of similar size than those obtained through formal minimization techniques.

Butramin simplifies a trace by iteratively eliminating redundant portions of the trace. For instance, it checks if there are redundant sequential steps, or sequential loops that can be removed. It also checks for possibly redundant combinational input events. In addition, it attempts to "perturb" a trace by eliminating one full simulation cycle, and/or one input event. The perturbed trace obtained is re-simulated to check if it still exposes the original bug (the original user-provided assertion is used for this purpose). If the test is successful the new trace replaces the original one, otherwise it is discarded. Often, perturbation leads to fairly different traces that may expose the bug much earlier than the original ones. When these mechanisms are exhausted, Butramin further simplifies a trace by using X-value simulation to determine which input signals are essential in exposing a bug. In the final stage, a SATisfiability (SAT)-based, fixed-window bounded model checker seeks additional "shortcuts" in the trace obtained so far, typically already much smaller than the original one.

978-1-60558-497-3/09 $25.00 © 2009 ACM

As an example of some of the techniques Butramin uses, Figure 2 shows schematically how *loop elimination* (LE) and *trace perturbation* (PT) work. Loop elimination removes sequential loops in a trace by hashing the states encountered during re-simulation and detecting when a trace enters the same state twice. Trace perturbation uses a trial-and-error approach, whereby a trace is modified by either removing one of more simulation cycles or eliminating input events. Then Butramin checks if the new trace still reaches a bug state through re-simulation.

We found experimentally that Butramin can reduce the size of a bug trace by three to six orders of magnitude in terms of cycles in the trace and, consequently, size of the trace. We note that traces generated by constrained-random simulation are more susceptive to benefit from Butramin, and also that traces derived from more complex designs (which usually entail more events and longer traces) present more opportunities for reduction. The impact of Butramin appears to be uncorrelated with the frequency of occurrence of the bug configuration targeted by the trace, that is, the number of distinct design states that expose the bug. In many cases we could reduce traces of several million cycles down to a few tens or a few hundreds cycles, that is, a trace that is much simpler to analyze and to re-simulate. In terms of execution time, we did not focus on optimizing the performance of this solution, but gave top consideration to the quality of the results, since the engineering time saved by the latter well outweighs the execution time of the software. We envision a deployment scenario where Butramin is run overnight to prepare simplified traces to be analyzed, and we found that all of our execution times are well within this limit: most commonly just one, or a few, hours. Butramin can be deployed in practically any simulation-based and semi-formal verification methodology with no effort: it is simply applied to bug traces generated in verification to greatly reduce them before they are analyzed for diagnosis.

3. AUTOMATIC DIAGNOSIS

One of the most difficult aspects of debugging is diagnosis, that is locating the error source within a design. *REDIR* (RTL Error DIagnosis and Repair) [8] is a scalable and powerful RTL error diagnosis and correction system, which adopts some of the hardware analysis techniques prevalent at the gate-level into the more designer-friendly and succinct RTL descriptions. The approach is significantly more accurate than previous software-based solutions in that it can analyze designs rigorously using formal verification techniques. At the same time, it is also considerably faster and more scalable than gate-level diagnosis because it models errors at a higher abstraction level (RTL), and thus there is a smaller number of candidate error sites to be evaluated.

The inputs of REDIR include a design containing one or more bugs, a set of simulation vectors exposing them, and the correct responses for the primary outputs over the given test vectors (usually generated by a high-level behavioral model). Note that REDIR only requires the correct responses at the primary outputs of the high-level model, not at any internal node. The correct output responses could be the primary outputs of the design, or the outputs of a set of checkers in the context of assertion-based verification. REDIR can then output a minimum cardinality set of RTL signals that should be corrected in order to eliminate the erroneous behavior. We call this set the *symptom core*. When multiple cores exist, REDIR provides all of the possible minimal cardinality sets. In addition, the framework suggests several possible ways of modifying the signals in the symptom core to help in the correction of the design. Our empirical evaluation shows that REDIR can diagnose and correct multiple errors in design descriptions with thousands of lines of Verilog code (corresponding approximately to 100K cells

after synthesis), a typical block size developed by individual engineers. As a result, REDIR can assist in everyday debugging tasks and fundamentally accelerate the design development.

The objective of error diagnosis is to identify a minimal number of variables in the RTL description that are responsible for the design's erroneous behavior. Moreover, errors can be corrected by modifying the statements related to those variables. Each signal affecting the design's correctness is called a *symptom variable*. Correcting all the symptom variables that contribute to a bug would eliminate it. A key idea in REDIR is error modeling: we embed additional constructs in the RTL design to evaluate a number of possible variants of the design and determine which of these variants would produce the expected output responses for each input test vector. Gate-level solutions for automatic diagnosis used a similar concept applied at the gate-level.

```
module half_adder(a, b, s, c);
  input a, b; output s, c;
  assign s = a ^ b;
  assign c = a | b;
endmodule
module half_adder_MUX_enriched(a, b, s_n, c_n,
s_sel, c_sel, s_f, c_f);
  input a, b, s_sel, c_sel, s_f, c_f;
  output s_n, c_n;
  assign s = a ^ b;
  assign c = a | b;
  assign s_n = s_sel ? s_f : s;
  assign c_n = c_sel ? c_f : c;
endmodule
```

Figure 3: An RTL error-modeling example. Module half_adder shows the original code, where c is erroneously driven by "$a \mid b$" instead of "a & b". Module half_adder_MUX_enriched shows the corresponding MUX-enriched version. Differences are marked in boldface. (*Figure reproduced from [8]*)

To model errors in the design, we introduce conditional assignments for each RTL variable, as shown in the example in Figure 3. Note that we insert only one conditional assignment even if the variable contains multiple bits. These assignments allow the REDIR framework to locate sites of erroneous behavior in RTL. Suppose that the output responses of the design are incorrect because c should be driven by "a & b" instead of "$a \mid b$". Obviously, to produce the correct outputs, the behavior of c must be modified. To model this situation, we insert a conditional assignment, "assign $c_n = c_{sel}$? c_f : c", into the code. Next, we replace all uses of c in the code with c_n (but not the assignments to c). Variable c_{sel} allows simulation of the design using c_f instead of c; moreover c_f is what we call a *free variable*, that is, we can assign it as deemed necessary to achieve the correct output response. If we can identify the $_{sel}$ variables that should be asserted, and the correct signals that should drive the corresponding free variables to produce the desired circuit behavior, we can diagnose and fix the errors.

Once errors are modeled as described, we rely on a Pseudo-Boolean solver or RTL symbolic simulation to perform the diagnosis and infer which design signals are responsible for incorrect output behavior. By forcing the Pseudo-Boolean solver to find a set of assignments that satisfy all the given <test vector, output response> pairs while minimizing the number of $_{sel}$ variables assigned to true, we obtain the complete set of signals that concur to the bug and that must be thus corrected.

The diagnosis capability of REDIR has been evaluated on a number of microprocessor modules and designs. The bugs injected in the design ranged from using an incorrect operator or complemented operand or simply wrong operand, to incorrect data forwarding and to incorrect state transition or execution of an instruction. For one simple microprocessor design, we had available a number of buggy versions [5] with bugs that were present in the design since its development and had been fixed in validation. Thus, for this testbench, no artificial bug injection occurred. The experimental evaluation indicates that REDIR could isolate the bug sources in each case, and that, because it addresses the problem at a higher abstraction level (that is, RTL) it can cope with much more complex designs than gate-level diagnosis solutions. The computation time of this diagnosis solution is also very practical, always less than one hour, even for systems sufficiently complex that a gate-level analysis would take more than two days, possibly running out of memory. One of the limitations of REDIR, however, is that in performing diagnosis it may detect several distinct sets of signals that can independently be modified to correct the bug. Thus, a user would have to manually determine which of these sets entails the minimum amount of source code modification.

Most techniques that have been proposed so far in this space target the gate-level description of a design [6, 17, 24, 25, 27] or even the transistor-level [16]. However, most debugging effort occurs in the early development phases of a design, when the system is described by an RTL model. The lack of powerful and automatic debugging tools at this level greatly reduces designers' productivity when fixing even very simple errors. Recently, a few techniques that work directly at the RTL have been developed. Some of them [15, 19, 21] employ a software analysis approach that implicitly makes use of multiplexers to identify statements in the RTL code potentially responsible for the errors. These techniques can suffer from a large number of false positives, and return too many candidate error sites. To address this problem, recent work by Staber *et al.* [23] inserts multiplexers explicitly into the RTL code. This enables the use of hardware analysis techniques and greatly improves the accuracy of diagnosis. Other techniques, such as [10], analyze an RTL description and failed properties using state-transition diagrams and model checking. REDIR is similar to several successful gate-level methods [2, 3, 6, 24, 27] in that it only requires test vectors and output responses to diagnose a functional error.

4. CONCLUSIONS

Looking forward, bug diagnosis can leverage and benefit from current trends in verification. Until recently, the goal for verification was to achieve complete functional correctness before tapeout. However, today, due to the unattainable complexity of verification, this task is becoming more selective. In recent trends in academia and, in lesser measure, in industry, engineering teams strive to verify the most common system's behaviors, and then they complement this effort with runtime detection and correction techniques for functional correctness. These techniques provide the advantage of shortening the digital system development cycle, but also come at a price in chip area and performance [26]. Thus, verification is striking a new trade-off between development effort/time and runtime performance, and development teams can choose to halt verification, once they can guarantee than any residual bug would occur with less than a specified frequency on average.

Bug diagnosis and correction can also benefit from this trend by choosing to only fix bugs for which a correction is known that does not run the risk to jeopardize the stability of a design close to tapeout. Or, by budgeting the time that can be spent in finding a bug, or a class of bugs, based on their criticality, frequency of occurrence, *etc.* All the techniques discussed above and the trade-offs enabled by novel runtime solutions contribute to taming the complexity of bug diagnosis and make it a task that can be approached by mere engineering mortals.

5. REFERENCES

[1] M. Aagaard, R. Jones, and C.-J. Seger, "Combining theorem proving and trajectory evaluation in an industrial environment", in *Proc. DAC*, 1998, pp. 538–541.

[2] M. F. Ali, S. Safarpour, A. Veneris, M. Abadir and R. Drechsler, "Post-verification debugging of hierarchical designs", *Proc. ICCAD*, 2005, pp. 871–876.

[3] M. F. Ali, A. Veneris, S. Safarpour, R. Drechsler, A. Smith and M. Abadir, "Debugging sequential circuits using Boolean satisfiability", *Proc. ICCAD*, 2004, pp. 44–49.

[4] J. Bergeron, *Writing testbenches: functional verification of HDL models*, Kluwer Academic Publishers, 2nd edition, 2003.

[5] "Bug UnderGround", http://bug.eecs.umich.edu/

[6] K.-H. Chang, I. L. Markov and V. Bertacco, "Fixing design errors with counterexamples and resynthesis", *Proc. ASPDAC*, 2007, pp. 944–949.

[7] K.-H. Chang, V. Bertacco and I. L. Markov, "Simulation-based bug trace minimization with BMC-based refinement", *IEEE Transactions on CAD*, Jan. 2007, pp. 152–165.

[8] K.-H. Chang, I. Wagner, V. Bertacco and I. L. Markov, "Automatic error diagnosis and correction for RTL designs", *Proc. HLDVT*, 2007, pp. 65–72.

[9] Y. A. Chen and F. S. Chen, "Algorithms for compacting error traces", in *Proc. ASPDAC*, 2003, pp. 99–103.

[10] G. Fey, S. Staber, R. Bloem and R. Drechsler, "Automatic fault localization for property checking", *IEEE Transactions on CAD*, Jun. 2008, pp. 1138–1149.

[11] P. Gastin, P. Moro and M. Zeitoun, "Minimization of counterexamples in SPIN", in *Proc. SPIN*, 2004, pp. 92–108.

[12] S. Hazelhurst, O. Weissberg, G. Kamhi and L. Fix, "A hybrid verification approach: getting deep into the design", in *Proc. DAC*, 2002, pp. 111–116.

[13] R. Hildebrandt and A. Zeller, "Simplifying failure-inducing input," in *Proc. ISSTA*, 2000, pp. 134–145.

[14] P.-H. Ho, T. Shiple, K. Harer, J. Kukula, R. Damiano, V. Bertacco, J. Taylor and J. Long, "Smart simulation using collaborative formal and simulation engines," in *Proc. ICCAD*, 2000, pp. 120–126.

[15] T.-Y. Jiang, C.-N. J. Liu and J.-Y. Jou, "Estimating likelihood of correctness for error candidates to assist debugging faulty HDL designs," *Proc. ISCAS*, 2005, pp. 5682–5685.

[16] A. Kuehlmann, D. I. Cheng, A. Srinivasan and D. P. Lapotin, "Error diagnosis for transistor-level verification", *Proc. DAC*, 1994, pp. 218–224.

[17] J. C. Madre, O. Coudert and J. P. Billon, "Automating the diagnosis and the rectification of design errors with PRIAM", in *Proc. ICCAD*, 1989, pp. 30–33.

[18] P. Rashinkar, P. Paterson and L. Singh, *System-on-a-chip verification: methodology and techniques*, Kluwer Academic Publishers, 2002.

[19] J.-C. Rau, Y.-Y. Chang and C.-H. Lin, "An efficient mechanism for debugging RTL description", *Proc. IWSOC*, 2003, pp. 370–373.

[20] K. Ravi and F. Somenzi, "Minimal satisfying assignments for bounded model checking," *Proc. TACAS*, 2004, pp. 31–45.

[21] C.-H. Shi and J.-Y. Jou, "An efficient approach for error diagnosis in HDL design", in *Proc. ISCAS*, 2003, pp. 732–735.

[22] SpringSoft, http://www.springsoft.com/

[23] S. Staber, G. Fey, R. Bloem and R. Drechsler, "Automatic fault localization for property checking", *Springer-Verlag LNCS 4383*, 2007, pp. 50–64.

[24] A. Smith, A. Veneris and A. Viglas, "Design diagnosis using Boolean satisfiability", *Proc. ASPDAC*, 2004, pp. 218–223.

[25] A. Veneris and I. N. Hajj, "Design error diagnosis and correction via test vector simulation", *IEEE Transactions on CAD*, Dec. 1999, pp. 1803–1816.

[26] I. Wagner and V. Bertacco, "Engineering trust with semantic guardians", *Proc. DATE*, 2007, pp. 743–748.

[27] Y.-S. Yang, S. Sinha, A. Veneris and R. Brayton, "Automating logic rectification by approximate SPFDs", *Proc. ASPDAC*, 2007, pp. 402–407.

978-1-60558-497-3/09 $25.00 © 2009 ACM

MAGENTA: Transaction-based Statistical Micro-Architectural Root-Cause Analysis

Gila Kamhi, Alexander Novakovsky, Andreas Tiemeyer, Adriana Wolffberg

Intel Corporation, Haifa, Israel

ABSTRACT

Adopting an ESL based design and validation methodology, we introduce a top-down approach for efficient debugging of micro-architectural specification and RTL implementation. Our solution is based on the formalism introduced by statistical transactional analysis that we call MAGENTA – Modeling AGENT for Transactional Analysis. To the best of our knowledge, MAGENTA based root-cause analysis pioneers in the efficient characterization of the micro-architectural design misbehavior via abstraction of validation output by transactions and micro-architectural events.

1. INTRODUCTION

Increasing design complexity and market forces driving a shorter time to market poses severe challenge for timely verification of the design implementation.

A major factor of validation crisis is the unavailability of traditional executable Design-Under-Test (DUT) during early stages of design. In addition to late readiness of DUT (i.e., Register-Transfer-Level (RTL) implementation), validation productivity is challenged by debugging complexity which consumes somewhere around half of the validation schedule.

In the recent years, Electronic System-Level Languages (ESL) based design methodologies [1,2] have been developed to model architectural and system-level features at a high-level of abstraction, independent of (and before) the RTL implementation (the traditional DUT for validation) and used for exploring the impact of various design trade-offs and verification of the functionality. ESL based design clearly attempts to enhance verification productivity by moving back the bug peak in the design validation schedule. It facilitates full-chip validation to start early at the micro-architectural specification stage rather than waiting for the arrival of the entire RTL. The major potential benefit of ESL based design is accelerated closure and built-in correctness of the architectural and micro-architectural definition and executable reference to RTL design and collaterals for RTL validation [3] (e.g., checkers, stimuli, coverage monitors). By promoting early validation and thus early maturity of micro-architectural definition, ESL based design will reduce the occurrence of bugs in definition phase; but, it cannot guarantee their total elimination and does not surely address the prevention of implementation bugs. Thus, it is critical that we not only engage with bugs up front, but also plan for them, and deal with them efficiently. This suggests that special focus needs to be given to the development of advanced debugging techniques to manage effective bug resolution.

Permission to make digital or hard copies of part or all of this work for personal or classroom use is granted without fee provided that copies are not made or distributed for profit or commercial advantage and that copies bear this notice and the full citation on the first page. To copy otherwise, to republish, to post on servers or to redistribute to lists, requires prior specific permission and/or a fee.
DAC'09, July 26-31, 2009, San Francisco, California, USA

In this paper, we introduce a top-down approach for efficient debugging of micro-architectural specification and RTL implementation adopting an ESL based design and validation methodology. Our solution is based on the formalism introduced by the statistical transactional analysis that we call MAGENTA – Modeling AGENT for Transactional Analysis. Our approach abstracts out details and compactly represents the huge amount of trace data output by formal/dynamic verification (i.e., simulation traces or counter-examples, respectively). The simplification via abstraction is achieved by (1) transactional micro-architectural event view versus bit-level RTL signal view (2) formalism to model micro-architectural event causality via data mining and machine learning techniques. For a given set of transactions and micro-architectural events, MAGENTA models the trace data canonically using formalism based on *Markov Chain* [4]. The origin of malfunction is singled out by comparing simple MAGENTA models extracted from specification (micro-architecture) and implementation (RTL) traces. In case the comparison is between two different micro-architectural or RTL configurations, the debug will involve traces from the same abstraction level (e.g., only RTL or only micro-architecture). To the best of our knowledge, MAGENTA based root-cause analysis pioneers in the analysis of both functional and performance/power bugs via abstraction of validation output and efficient characterization of the misbehavior by differentiating the misbehavior from the desired behavior.

MAGENTA *Root Cause Analysis* applies a multifaceted discipline that leads the user through the analysis of a micro-architectural misbehavior or alternatively a behavior of interest using a standard, repeatable inquiry process. This process guides the user to logically reconstruct the misbehavior from the causal facts. The user relies on the very nature of initial first line of causal facts characterized by the root-cause analysis flow and determines dynamically the second line of behaviors of interest to focus. Thus, the proposed root-cause analysis is a dynamic, systematic, iterative process that guides the user to discover the relevant facts, their causal relationship and potential solutions. The characterization flow facilitates the refinement of the causal facts and generation of detailed and informative chain of facts that can potentially describe the misbehavior. Clearly, an important objective of the root-cause analysis is a high level of confidence that the outcome of the analysis accurately describes the actual misbehavior that is being debugged. Therefore, our approach also aims to get to accurate results by facilitating the application of a *"confidence score"* to the characterization results at every stage.

The paper is organized as follows. Section 2 introduces the formalism of MAGENTA statistical transactional model. Section 3 provides an overview of use cases that MAGENTA based transactional debugging addresses. Section 4 introduces the details of the characterization flow. Section 5 presents a short overview of related work. In Section 5, we summarize the results supporting the validity of our solution based on tests from a wide variety of benchmark applications. In Section 7, we present future directions and conclude.

2. MAGENTA : <u>M</u>odeling <u>AGENT</u> for <u>T</u>ransactional <u>A</u>nalysis

Transaction-Level Modeling (TLM) [6] is a high-level approach for modeling digital systems where details of communication among modules are separated from the details of the implementation of functional units or of the communication architecture. Communication mechanisms such as busses or FIFOs are modeled as channels, and are presented to modules using interfaces. Transaction requests take place by calling interface functions of these channel models, which encapsulate low-level details of the information exchange. Moreover, in the life-time of a transaction, micro-architectural events occur.

The basis of our transactional analysis via MAGENTA is the assumption that micro-architectural events occurring in TLM can be monitored by both micro-architectural and RTL simulation through events that are instrumented to both micro-architectural transaction-level model (specification) and RTL (implementation).

When given traces in terms of transactions and events from either micro-architectural or RTL simulation runs, MAGENTA can either extract a canonical causality graph in terms of micro-architectural events of interest or a flow graph that represents all the micro-architectural paths exhibited by the trace. In case of causality graphs, the vertices correspond to the events and edges represent probabilistic causality relations. In this paper, we will mainly focus on the statistical causality modeling of MAGENTA. In this representation, the relations between the events in a transaction are learned based on the sequential occurrence of the events. The vertices (events) and edges (causality relation between the events) of MAGENTA graphs are annotated with statistical data learned from all the transactions in the trace/traces at hand. In Figure 1, we illustrate a MAGENTA transactional event causality model that represents the statistical event dependency witnessed in the trace or a set of traces at hand following the occurrence of event "*allocate*". The edges of MAGENTA graph (e.g., *allocate* → *cache_miss*) are annotated with the information on the activity count (AC), latency (min, average, max) and probability (PRB) of occurrence of *from-event* (i.e., *allocate*) followed by *to-event* (i.e., *cache_miss*).

MAGENTA event causality model is based on a stochastic process with the Markov Chain [4] property. Having the Markov property means that, *given the present state*, future states are independent of the past states. The description of the present state fully captures all the information that could influence the future evolution of the process. Future states will be reached through a probabilistic process instead of a deterministic one. In the case of MAGENTA a state is represented by the occurrence of an event. The causality model satisfies a set of rules. For a given vertex representing event$_j$,

$$\sum_{1}^{N} PRB(edge_i) = 1$$

when $1 \leq i \leq N$ and edge$_i$ is member of *fan_out*(event$_j$)

Similarly,

$$AC(event_j) = \sum_{1}^{N} AC(edge_i)$$

Since events are shared between transactions, every event is represented by only one vertex in MAGENTA causality graphs making them canonical; i.e., for a given trace and set of events of interest there is one and only one MAGENTA causality graph representation of the trace. The difference of two graphs G and H, corresponds to the graph with adjacency matrix [5] given by the difference of adjacency matrices of G and H, when the orders of G and H are the same. In the same manner, the merge of two graphs G and H, corresponds to the graph with adjacency matrix given by the merge of adjacency matrices of G and H. Therefore, using canonicity of MAGENTA causality graphs, "diff" and "merge" of the graphs can be performed in a robust manner. In case of graph merge the statistical annotation on vertices and edges are averaged.

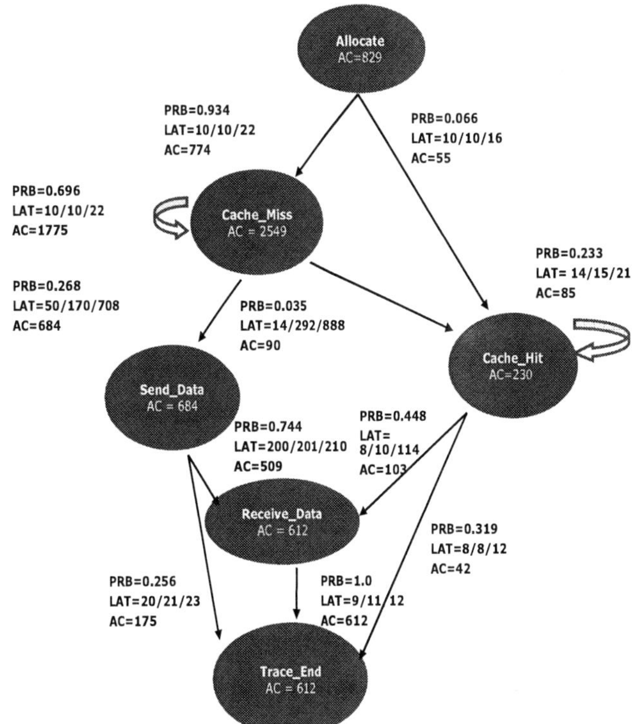

Figure 1: MAGENTA statistical transactional modeling of trace data focusing on five micro-architectural events (i.e., *allocate, cache_miss, cache_hit, send_data, receive_data*). The graph includes statistical data on activity count (AC), latency (LAT) and probability (PRB).

3. USE CASES: From Performance Verification to Functional Verification

To be effective, verification must ensure that designs are shipped without critical functional bugs and satisfy the product's aggressive timing and power targets. Thus, functional and performance/power verification need to be viewed as interdependent activities and best if addressed by a unified verification solution. For example, typically near tape out, full chip integration reveals challenging timing and power closure

978-1-60558-497-3/09 $25.00 © 2009 ACM

problems. At this point, designers may need to make micro-architectural optimizations without breaking the functional correctness. In the same manner, functional bugs (e.g., starvation) can be the cause of performance degradation or alternatively excessive power consumption.

The essence of performance verification is the comparison of one configuration of micro-architecture (e.g., including a feature set) versus another. As the design evolves from technology readiness to execution for the same micro-architectural configuration the comparison is between TLM and RTL simulations. The comparison is feasible since each event monitored by performance model is monitored through corresponding monitors in RTL simulation. During micro-architectural performance (specification) or alternatively RTL (implementation) simulation, the values of the events are traced across time (in units of cycles) along transactions.

MAGENTA's main role in performance verification is an intelligent diff between the micro-architectural performance model and the RTL implementation via compact graphs built in terms of events of interest with respect to performance. The paradigm that we applied for performance verification can be inherited by functional verification. The knowledge on micro-architectural events and their corresponding RTL implementation can be a valuable source for abstraction of RTL implementation. This paradigm anchors performance model to RTL and thus is the essential step towards RTL validation collateral generation by leveraging from the performance model. Natural collaterals in this direction are the specification of RTL validation properties and checkers in terms of micro-architectural events that are monitored by TLM simulation. Moreover, for particular properties, through extended MAGENTA graphs, RTL implementation can be verified against micro-architectural specification.

3.1 BUCKETING

One of the challenges of debugging process is the bucketing of issues to a predefined set of categories. Efficient bucketing can facilitate the efficient allocation of architects or validation engineers that have the relevant knowledge to the debugging of the respective outliers. Bucketing of issues can be achieved by differentiating the MAGENTA graph representing the specification behavior (i.e., trace output by micro-architectural simulation) against the implementation (i.e., trace output by RTL simulation). The structural difference is reported in context of vertices (i.e., micro-architectural events) and edges (i.e., causality relations between micro-architectural events). Additionally, statistical diff which is relevant in case of performance/power verification is reported by differencing the annotations on the vertices and edges (i.e., AC – activity count, PRB – Probability, and LAT – min/average/max latency). As can be seen in Figure 2, a simple category for bucketing can be the clusters of functional units of the chip. Each micro-architectural event (vertex of MAGENTA graph) belongs to a specific cluster (i.e., it represents activities in the cluster). In case of an outlier analysis, the vertices and edges of the "diff graph" represent the differentiating factors between the specification and the implementation. The outliers are attributed to the cluster or clusters that the differentiating vertices/edges belong.

3.2 MISBEHAVIOR CHARACTERIZATION

Misbehavior characterization flow can be applied to a substantial set of tests (ideally the whole study list). In case the characterization is applied to a large set of tests, the aim is to either characterize similar functional failures and thus strengthen the root-cause analysis process by more data, or fine tune a micro-architectural power/performance feature by learning how an undesired behavior can be prevented. The more common usage of the characterization flow is on a single trace characterizing the failure by tracing it to the origin of the malfunction.

4. CHARACTERIZATION FLOW

The characterization flow requires the user to describe the misbehavior in term of a checker. The flow consists of three stages:

- ***Identification of good and bad examples for characterization***: The cycles or alternatively transactions that satisfy the misbehavior (e.g., *write* operation occurs less than 3 times at a given cycle) represent the "bad" examples when the cycles/transactions that do not satisfy the misbehavior represent the "good" examples. Figure 3 represents the categorization of bad and good cycles for a transactional trace.

- ***Statistical characterization of misbehavior:*** The goal of the characterization flow is to learn the characteristics of the bad examples distinguishing them from the good ones. This means the learned characteristics should be unique to the bad examples. Since the learning data can be misbehavior specific, the user defines the instruction/cycle interval that is relevant in context of the misbehavior at hand. For example, characterization flow may be requested to learn the consecutive intervals representing the "bad" examples distinguishing them from the consecutive intervals of "good" ones. Alternatively, the characterization flow may be requested to focus on a time step some number of cycles preceding the occurrence of "good" or "bad" examples.

The characterization of "bad" examples in the trace is achieved by collecting over N uniformly distributed "learning intervals" for good and for bad. Separate MAGENTA statistical causality models are built for intervals of "good" and "bad" in term of the focus events of interest provided by the user.

The causality model of "bad" is compared with causality model of "good". The structural differences between the two models (vertices/edges that are in bad and not in good) are saved as distinguishing characteristics of "bad". Similarly, "good" is distinguished from "bad". The characterization is refined by applying a confidence score. Following several experimentations, we have converged on the following confidence metric:

$$1 - \frac{\#characterizes\_good}{\#characterizes\_bad}$$

The metric scales the ability of the characteristics to distinguish "bad" from "good". The learned characteristic is further refined statistically by prioritizing with respect to the number of times it characterizes the "bad" behavior versus the number of times it characterizes the "good" behavior.

- ***Refinement of characterization (i.e., causality of misbehavior)***: Based on the nature of first line of learned characteristics, the characterization can be further refined by the second line of behaviors of interest to focus.

5. RELATED WORK

Generally speaking, *Learning from Examples* [7] addresses the problem of extracting, for a given group of objects, a set of predefined *features* (in our case focus events of interest) differentiating the objects, and a classification of the objects as positive and negative examples, a rule that generates the classification in terms of feature values. Since this is equivalent to classification of the given objects in accordance to the set of *features*, this process (*class label* extraction) identifies machine learning based classification techniques. There has been several attempts to apply trace based machine learning to CAD domains in context dynamic invariant extraction [10,11].

Similarly, the extraction of event causality from event logs has been attempted in several applications as business management. Carmona et al. [8, 9] propose the extraction of event graphs based on traces of events and construct a transition system (i.e. a state graph with arcs labeled by events). Using a tool called "Petrify" they synthesize a Petri Net out of the transition system. Petri net is a directed bipartite graph, in which the vertices represent transitions (i.e. discrete events that may occur), places (i.e. conditions), and directed arcs (that describe which places are pre- and/or post-conditions for which transitions). For our micro-architectural design and validation use cases, we have chosen the formalism of Markov Chains instead of Petri Nets due to their simplicity and compact representation of statistical information.

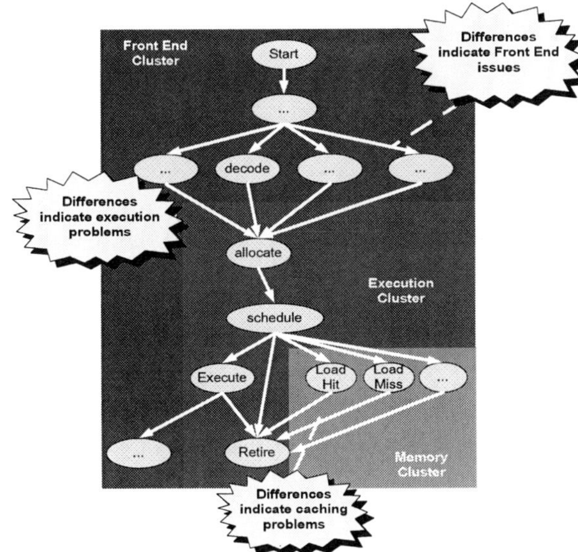

Figure 2: Bucketing outliers per cluster based on the differences on the edges of MAGENTA graphs

6. RESULTS

The characterization flow has been used in the analysis of micro-architectural outliers on a single trace or alternatively in the study and characterization of a behavior of interest over a set of traces. We report in Table 1 the results of from the analysis of traces extracted from 89 sample applications in an attempt to characterize a micro-architectural misbehavior from maximum 2000 samples of "good" and 2000 samples of "bad" from each trace. The traces have been run for 6 M transactions over 3.8 M cycles. The results, presented in Table 1, are encouraging as witnessed by the confidence score and manual review by the relevant architects. The characterization flow has learned 22 distinguishing characteristics of "bad" in the form (event$_i$ → event$_j$) when the confidence score of each trait is high since the traits of "bad" are clearly not strong distinguishers for also "good".

The characterization of the misbehavior based on 89 applications run for 6 M transactions over 20000 cycles has taken a few hours. Clearly, manual learning of this amount of data is humanly not possible.

	Tests in which characterize bad	Tests in which characterize good	Confidence Score
Chact1	38	0	1
Chact2	32	0	1
Chact3	24	1	0.96
Chact4	24	1	0.96
Chact5	20	0	1
Chact6	19	0	1
Chact7	14	1	0.92
Chact8	13	1	0.92
Chact9	12	2	0.84
Chact10	10	0	1
Chact11	8	0	1
Chact12	8	0	1
Chact13	7	1	0.86
Chact14	7	1	0.86
Chact15	7	1	0.86
Chact16	5	0	1
Chact17	5	0	1
Chact18	5	0	1
Chact19	4	0	1
Chact20	3	0	1
Chact21	3	0	1
Chact22	1	0	1

Table 1: Characterization flow has learned 22 characteristics of misbehavior in form of event $_i$ → event $_j$. The first and second column represents the number of tests in which characteristics *Chart*$_n$ characterizes "bad" samples and "good" samples, respectively.

7. CONCLUSIONS AND FUTURE WORK

In this paper, we present a new approach based on a stochastic process for abstract transactional representation of the huge amount of data in traces for the efficient root-cause analysis of functional and performance/power bugs. Through MAGENTA's new statistical analysis views, user can classify causes and major paths in which there is high discrepancy. Interactive mode makes it easier to pinpoint problematic paths to zoom in for local analysis; while the batch mode allows issues to be distributed to the appropriate owners, enabling more tests to be analyzed and solved.

The characterization flow creates, MAGENTA - statistical micro-architectural event causality graphs that helps the user explore all potential causes that result in misbehavior. As next steps we would like to facilitate the analysis of the discrete causes of misbehavior and depict how they interact to produce the overall behavior of interest. We will also research how the causality graphs can be utilized to prevent the unwanted behavior.

978-1-60558-497-3/09 $25.00 © 2009 ACM

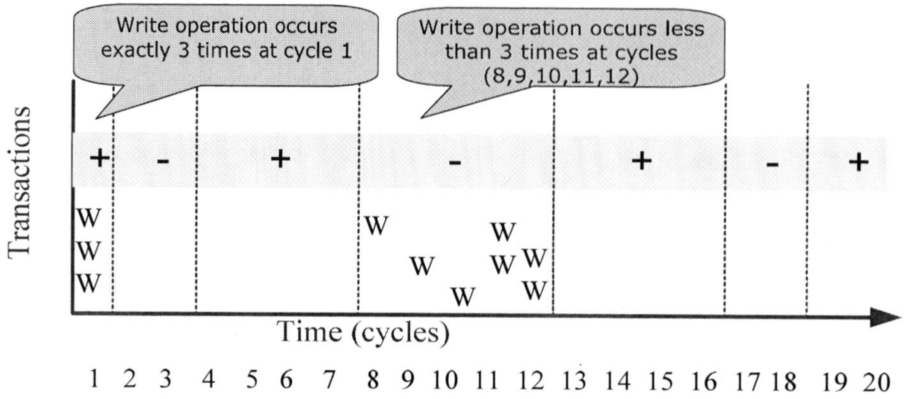

Figure 3. A trace snippet illustrating the extraction of bad and good cycles in context of the misbehavior specification at hand; i.e., "Write operation occurs less than 3 times at a given cycle". As emphasized in the call-outs, *Cycle 1* demonstrates a "good" sample and Cycles 8-12 illustrate "bad" samples. The good and bad intervals are marked with "+", and "-", respectively. The X axis specifies time in terms of cycles and Y axis specifies the transactions that have been exercised. (X, Y) entries in the figure specify the events (e.g., Write (W)) that occur. In this example, for readability the events occurring only at the two cycle intervals (1), (8-12) are highlighted; i.e., W operation occurrence for cycles 2-7 and 13-20 are not highlighted.

8. REFERENCES

[1] B. Black and J. Donovan, *SystemC: From the Ground up,* Kluwer-Academic Publishers.

[2] G. Frank, Transaction-Level Modeling with SystemC: TLM Concepts and Applications for Embedded Systems, Kluwer-Academic Publishers.

[3] A. Kasuya and T. Tesfaye, "Verification Methodologies in a TLM-to_RTL Design Flow", Proc. of Design Automation Conference, June 2007, pp. 199-204.

[4] S. P. Meyn and R.L. Tweedie, Markov Chains and Stochastic Stability,Cambridge University Press, 2008

[5] Weisstein, Eric W., "Adjacency Matrix." From MathWorld—A Wolfram Resource.http://mathworld.wolfram.com/AdjacencyMatrix.html

[6] Grotker, Liao, Martin, Swan, System Design with System C, Kluwer Academics Publishers, 2002 775-778.

[7] J. G. Carbonell, (Ed.). "Machine Learning: Paradigms and Methods", MIT Press., 1990.

[8] J. Carmona, J. Cortadella and M. Kishinevsky, "A region-based algorithm for discovering Petri nets from event logs", Business Process Management, Kluwer Academic Publishers, 2008

[9] J. Carmona, J. Cortadella, M. Kishinevsky, A. Kondratyev, L. Lavagno, A. Yakovlev, "A symbolic algorithm for the synthesis of bounded Petri nets" , in Proceedings of the 29th international conference on Applications and Theory of Petri Nets, 2008

[10] Hangal, N. Chandra, S. Narayan, S. Chakravorty, "IODINE: a tool to automatically infer dynamic invariants for hardware designs". Proc. of Design Automation Conference, 2005, pp.

[11] P.Babighian, G. Kamhi, M.Y. Vardi "PowerQuest : Trace driven data mining for power optimization". Proc.of DATE, 2007.

978-1-60558-497-3/09 $25.00 © 2009 ACM

Untwist Your Brain – Efficient Debugging and Diagnosis of Complex Assertions

Michael Siegel
OneSpin Solutions
Theresienhoehe 12
80339 Munich, Germany
Tel.: +4989990130

Michael.Siegel@onespin-solutions.com

Adriana Maggiore
OneSpin Solutions
1183 Bordeaux Drive, Suite 16,
Sunnyvale, CA 94089, USA
Tel.: +1 408 734 1900

Adriana.Maggiore@onespin-solutions.com

Christian Pichler
OneSpin Solutions
Theresienhoehe 12
80339 Munich, Germany
Tel.: +4989990130

Christian.Pichler@onespin-solutions.com

Abstract

Assertions are recognized in the industry to be a major improvement in functional RTL verification flows. Today's standard assertion languages, such as SVA [1] and PSL [2], are very expressive, capable of describing sophisticated temporal design behavior at different abstraction levels. Nevertheless, most assertion users stick to writing simple assertions because of the intricacy and effort required to debug complex assertions – one of the major bottlenecks in assertion based verification.

We present debugging and diagnosis techniques that automatically identify those parts of an assertion that cause the assertion to fail for a given design and that provide additional automation to efficiently identify the root cause of the failure. These techniques enable major effort savings when working with complex assertions, allowing engineers to use the full capabilities of assertion languages. This enables further productivity and quality improvements in functional verification by lifting mainstream assertion usage to higher abstraction levels such as efficient capture and verification of high-level design features, operations, and transactions. Advanced debugging automation is key for this progress - solving a problem that many designers and verification engineers face in their daily work.

Categories and Subject Descriptors

B.5.2 [**Register-Transfer-Level Implementation**]: Design Aids – *Verification, Simulation;* B.6.3 [**Logic Design**]: Design Aids – *Verification, Simulation;* B.7.2 [**Integrated Circuits**]: Design Aids – *Verification, Simulation*

General Terms

Algorithms, Design, Languages, Verification.

Keywords

Functional Verification, Assertions, Debugging, Root Cause Analysis, Fault Localization, SystemVerilog Assertions

Permission to make digital or hard copies of part or all of this work for personal or classroom use is granted without fee provided that copies are not made or distributed for profit or commercial advantage and that copies bear this notice and the full citation on the first page. To copy otherwise, to republish, to post on servers or to redistribute to lists, requires prior specific permission and/or a fee.
DAC'09, July 26-31, 2009, San Francisco, California, USA

1. Introduction

Since the definition of standard assertion languages such as SVA [1] and PSL [2], assertions see a rapid adoption in the industry. Adoption is fueled by broad tool support in simulators, testbench automation solutions and formal verification tools, by new design and verification methodologies emerging around assertions and a growing market of assertion-based Verification IP.

Assertions are concise, executable statements describing expected design behavior. They are used to monitor signals on interfaces, to ensure that designs behave as expected and to detect forbidden behaviors. Since assertions can be used in simulation, emulation and formal verification, they allow leveraging the complementary strengths of these technologies for block, subsystem, integration and chip-level verification.

Assertions can be used in many different ways to increase design and verification productivity and quality: documentation and sharing of design intent and integration conditions, definition of concise monitors and checkers for simulation, definition and measuring of coverage information, exhaustive verification of design features and operations/transactions using formal verification, just to name a few. While SVA and PSL offer rich sets of constructs to capture almost arbitrarily complex temporal relationships between signals, typically only a small fraction of the power of these languages is leveraged by most users. One of the main reasons is, that assertions quickly become hard to debug once they become more complex.

In this paper we present advanced debugging and diagnosis techniques for efficient root cause analysis of failing assertions. These new techniques enable efficient working with complex assertions, making advanced SVA constructs better accessible for main-stream assertion-based verification.

To illustrate the concepts we use SVA assertions of an arbiter design and demonstrate the techniques using OneSpin's 360 MV formal assertion-based verification tool [3].

2. The Intricacy of Debugging Complex SVAs

When using assertions in the verification of designs, failure of an assertion is indicated by e.g. a simulation trace that flags the assertion as violated or by a counterexample generated by a formal tool that shows a design behavior that violates the assertion. There are three possible causes why an assertion fails:

- The design has a bug, causing unexpected behaviors

- The assertion is wrong, e.g., a wrong signal is used, a signal is referenced at a wrong clock cycles, just to name a few

- The design has been triggered with an input stimulus that violates the assumption the designer made about how his block is triggered, leading to an unexpected behavior of the design that violates the assertion; in formal verification this situation corresponds to a missing input constraint

We look at some assertions for an arbiter design to illustrate how to find out which of the three cases is the root cause of the failure. The arbiter receives requests from three masters and grants the one with the highest priority – it then waits for the shared resource to become free again before a further request is granted. It is implemented by a finite state machine with states "idle", "start" and "busy". First we consider an assertion for the idle behavior: "when the arbiter is in state idle and there is no request then it will remain in state idle and will not issue a grant".

```
idle_check: assert property
   (@(posedge clk) disable iff (reset)
            state_s == idle      &&
            request_i == 3'b000
        |=> state_s == idle      &&
            grant_o == 3'b000
```

The antecedent (left hand side of the implication operator |=>) defines the trigger condition of the assertion, the consequent (right hand side of the implication) defines what the arbiter shall do when the trigger condition holds. If this simple assertion fails, it is easy to determine from the waveform displaying the failure whether the signal state_s is set differently by the design or whether unexpectedly a grant is issued. An RTL source code debugger is then sufficient to analyze the root cause of the failure.

Simple assertions are a good way of getting started with assertion-based verification. However, more complex use cases - such as the verification of complete design features or transactions - require more complex assertions to capture the expected behavior and are significantly harder to debug.

Let's look at a slightly more involved assertion expressing a full grant cycle of the arbiter: "If the arbiter is in state idle and receives at least one request, than the highest priority request is granted in the next cycle, followed by a period where no further grant is issued until the free signal is received from the shared resource, upon which the arbiter goes back to the idle state with the grant signal deasserted." To capture this requirement, we first define the function grant_to_highest_prio that determines the correct grant signal based on the pending requests:

```
function [2:0]
grant_to_highest_prio(input[2:0]request);
begin
     grant_to_highest_prio =
            request[0] ? 3'b001:
            request[1] ? 3'b010:
            request[2] ? 3'b100: 3'b000;
end
endfunction
```

Additionally we define the sequence no_grant_until_freed - it matches if the free_i signal occurs within at most 3 clock cycles and no grant signal is issued during this wait period:

```
sequence no_grant_until_freed;
     grant_o == 3'b000 throughout
     (!free_i[*0:2] ##1 free_i);
endsequence
```

The grant cycle can now be specified, using a local variable "req" to store the initial value of request_i. Note that the grant cycle is supposed to start and end in state idle according to this assertion.

```
property grant_master;
reg [2:0] req;
     (state_s == idle &&
      request_i != 3'b000,
      req = request_i)
|=>
     (grant_o ==
      grant_to_highest_prio(req)) and
     (##1 no_grant_until_freed
      ##1 grant_o == 3'b000 &&
            state_s == idle);
endproperty.
```

While assertions can become much more complex than in this example, the above example is sufficient to explain why debugging becomes intricate. As can be seen in the grant_master property, SVA assertions basically consist of conjunctions, disjunctions and intersections of nested regular expressions, functional calls, and instantiations of pre-defined sequences and properties. When such an assertion fails, its debugging requires from the user to determine why the counterexample does not match the complex regular expression defined by the property. This requires determining, amongst other things, from which clock cycle onwards the counterexample no longer matches the assertion, to find out which sub-expressions of the assertion are violated, which signals or function calls are involved in the failure, what their values are and why these signals deviate in the design and the assertion. These otherwise often manual steps are highly automated by the techniques explained in this paper.

3. Related Work

The debugging help that commercial verification tools offer spans (1) annotating counterexamples with the time points when failing assertions are triggered and when they fail, (2) RTL source code value annotations based on counterexamples, and (3) "what-if" analysis where counterexamples are modified by the user to better understand which signals contribute to the failure of an assertion.

All these techniques nevertheless require high effort for manual divide-and-conquer debugging: the user needs to manually isolate failing parts of a complex assertion by omitting parts of the assertion, rerunning the verification and checking if the assertion still fails. The main benefit of the presented techniques is, that this significant, repeated manual effort is replaced by an automatic analysis that displays the failing clause(s).

Research on fault detection and localization is presented and discussed, e.g., in [5]. This paper as well as referenced research assume that the checked assertion/property is correct and that sufficient constraints have been used. Based on this assumption, they address techniques to locate and correct design faults.

In contrast to the related work, we present techniques that (1) analyze the assertion and the counterexample and automatically identify those parts of the assertion that make it fail and (2) help to determine if the failure is caused by an error in the assertion, by missing constraints or a bug in the design. The work in [5] and the cited material on design fault localization comes into play, once

978-1-60558-497-3/09 $25.00 © 2009 ACM

the *design* has been identified to be incorrect. In this sense most academic work on fault localization is orthogonal to the approaches presented in this paper.

4. Structural Assertion Debugging

In this section we present a new diagnosis technique for SVA called *structural debugging*. Structural debugging reads the values of signals from the counterexample and performs a semantic analysis that annotates all objects at all clock cycles in the assertion (signals, local variables, functional calls, parameters, etc.) with values according to the counterexample. A second semantic analysis explores the logical and temporal structure of the assertion and evaluates all Boolean expressions at all clock cycles to find the time point(s) and the sub-expression(s) that make the assertion fail. Violated sub-expressions are marked red at the respective clock cycles as well as the path through the SVA source code leading to the sub-expression. Additionally the structural debugger analyses the hierarchical structure of the SVA (nested function calls, instantiations of named sequences and properties) and generates a hierarchical view of the assertion, that can be explored by the user – using unfolding and/or collapsing parts of the SVA.

When checking the property grant_master:

```
master_check: assert property
(@(posedge clk) disable iff (reset)
grant_master);
```

structural debugging delivers the hierarchical diagnosis information shown in the screenshot below.

Besides the waveform showing the counterexample the screenshot shows the structural debugger at the left hand side. Initially it shows the upper red line (respectively bold line in black and white printouts) indicating that the property grant_master failed. Expanding it by clicking on the symbol to the left of it shows (marked red) that there is something wrong with the last line in the consequent of the grant_master property. Further expansion shows that the counterexample does not match the no_grant_until_freed sequence. One more expansion indicates that the last free_i clause is responsible for the failure. The final view at the very bottom shows that the free_i signal is low in the

counterexample at clock cycle 4 (cf. the waveform to the right) whereas the property required it to be "1".

So the cause of the failure is found: the designer assumed that the input free_i is set by the environment of the block to "1" at most 3 cycles after the arbiter grants a request. In the counterexample this is not the case. If the assumption about the environment is indeed valid it has to be added as a new assumption to prove the grant_master property, or the assumption that the free_i is set after at most three cycles has to be relaxed in the assertion.

As explained, the structural debugger highlights the condition that fails, while conditions satisfied by the DUV are not highlighted. The SVA can then be inspected at multiple levels of hierarchy by expanding or collapsing the assertion's sub-expressions, including function calls and instantiated sequences and properties, tracing the highlighted red path to the failing sub-expression.

The algorithm – that underlies this kind of structural debugging – computes for a given counterexample instantiations of the repetition operators of an SVA that witness the failure of the SVA. This computation is similar to the techniques employed for bounded model checking of SVAs, cf. [4]. In the simplest case there exists only one such instantiation but typically there exist multiple such instantiations. Then the algorithm builds an internal tree structure that represents all the computed instantiations and determines by means of heuristics one specific instantiation for subsequent root cause analysis. The internal tree structure is mapped to the SVA source tree to annotate all hierarchical elements of the SVA with values based on the selected instantiation. Then the user can either explore the displayed instantiation or select other instantiations for debugging.

5. Temporal Fan-in Analysis

While structural debugging determines in a first step an instantiation of the failing SVA as well as all relevant sub-expression(s) and clock cycle(s) that make the instance – and thus the assertion - fail, a subsequent temporal fan-in analysis supports root analysis of the failure.

Once the problematic instantiation and clause(s) have been determined, a temporal fan-in analysis traces step-by-step the fan-

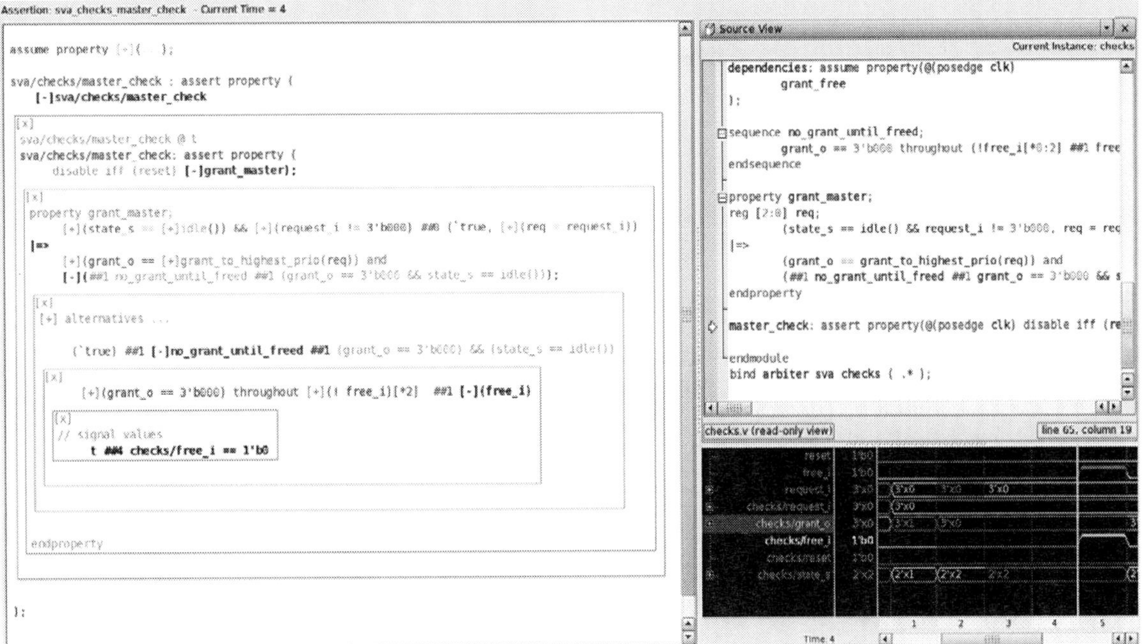

in of signals that are referenced in failing sub expressions of the SVA. The exploration is again based on value annotations from the counterexample.

We illustrate this analysis using a simple assertion that states: "When the arbiter is in state idle and receives a request from the master with index 1, it will grant this request in the next cycle":

```
property grant_master1;
reg [2:0] req;
      state_s == idle() && request_i [1]
|=>
      grant_o[1] ;
endproperty
```

The assertion fails and the structural debugger shows that the grant is not given to master 1 but to master 0 (see screenshot below). The "fan-in view" of the grant_o signal shows that its value depends on the request bits request_i[0] and request_i[1], and reminds the user that master 0 has highest priority – in this case the property was written incorrectly, there is no design bug.

So, the temporal fan-in analysis displays a signal's dependencies on other signals through the design hierarchy. The view can be expanded and collapsed to display the values of signals at different clock cycles that influence the signals in the identified failing sub-expression.

In case that the assertion is considered to be correct and the design is suspected to contain a bug, a source code debugger can be used to analyze the HDL code. The temporal fan-in analysis helps to identify the clock cycles where the drivers must be examined, guiding the HDL code analysis.

6. Applicability

Both structural debugging and temporal fan-in analysis can be efficiently applied to full SVA, including advanced constructs such as sequence intersection, local variables, and property instantiation, which are known to cause additional complexity in the verification and debug of SystemVerilog Assertions, cf. [6].

The presented techniques have been applied in numerous industrial verification projects, e.g., in the high-level formal verification of Infineon's TriCore processor [7] where they have

been key to efficiently debug complex assertions, some of them consisting of several hundred lines of nested assertion code – avoiding significant manual divide-and-conquer and re-run effort .

7. Summary and Conclusion

We have presented structural debugging and temporal fan-in analysis techniques that complement each other to support and speed up debugging of complex SystemVerilog assertions. Similar techniques can also be applied to significantly reduce debugging effort in other assertions languages such as PSL. These techniques enable efficient working with complex assertions, making the expressive power of advanced assertion languages better accessible for main-stream assertion-based verification. The presented work further improves productivity and quality of functional verification by lifting mainstream assertion usage to higher abstraction levels.

References

[1] IEEE Standard 1800-2005 SystemVerilog: Unified Hardware Design, Specification and Verification Language, USA, 2005.

[2] IEEE Standard 1850-2005 Property Specification Language (PSL), IEEE, Inc., New York, NY, USA, 2005.

[3] http://www.onespin-solutions.com/360mv.php

[4] R. Wille, G. Fey, M. Messing, R. Drechsler et. al.; Identifying a subset of SystemVerilog Assertions for Efficient Bounded Model Checking, *Conf. on Digital System Design (DSD)*, 2008.

[5] G. Fey, S. Staber, R. Bloem, R. Drechsler. Automatic fault localization for property checking. IEEE Trans. on CAD of Integrated Circuits and Systems, 27:1138-1149, 2008.

[6] D. Bustan and J. Havlicek. Some complexity results for SystemVerilog Assertions. CAV 2006, volume 4144 of *LNCS*.

[7] J. Bormann, S. Beyer, T. Blackmore, et al. Complete Formal Verification of TriCore2 and Other Processors, DVCon 2007

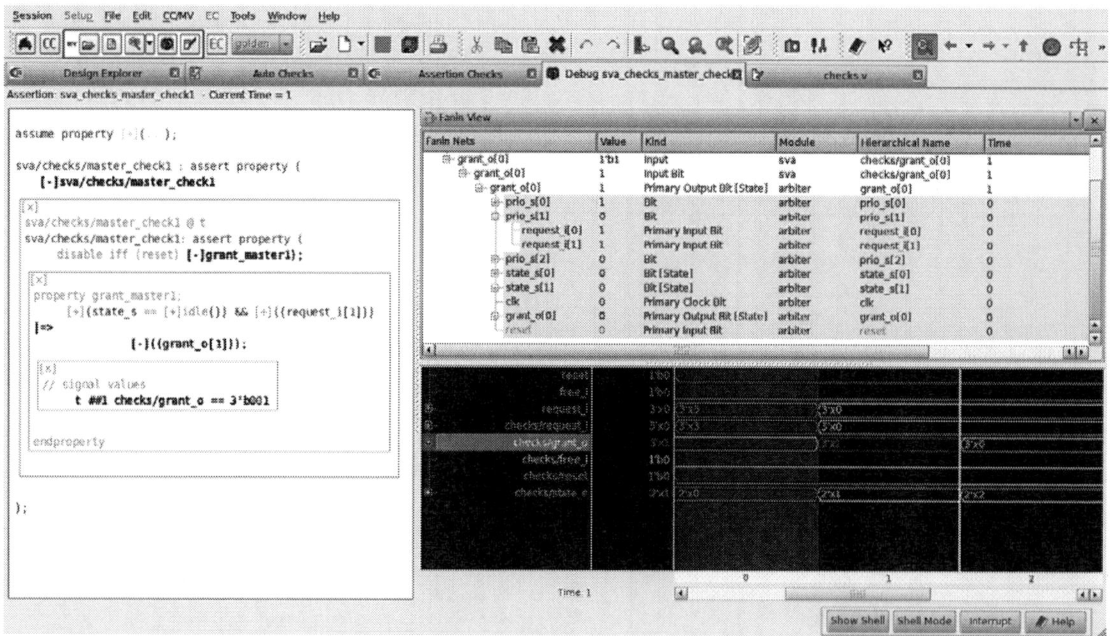

Beyond Verification: Leveraging Formal for Debugging

Rajeev K. Ranjan
Jasper Design Automation
100 View St., Suite 101
Mountain View, CA 94041

rajeev@jasper-da.com

Claudionor Coelho
Jasper Design Automation Brazil
Rua Sebastião Fabiano Dias, 210/406
Belo Horizonte - MG-30320-690

coelho@jasper-da.com

Sebastian Skalberg
Jasper Design Automation
100 View St., Suite 101
Mountain View, CA 94041

sebastian@jasper-da.com

ABSTRACT

The latest advancements in the commercial formal model checkers have enabled the integration of formal property verification with the conventional testbench based methods in the overall verification plan. This has led to significant verification productivity across the entire design flow (from architectural verification to post-silicon debugging). As verification productivity is improved, debugging efficiency has become more important than before. In this paper, we discuss how formal technology can be leveraged to bring efficiency in the debugging process. In particular, we discuss how "behavioral indexing" enables a top-down view of the counter-example and facilitates debugging by overlaying a higher abstraction view on the bit-level counter-example. We also discuss how formal technology can be leveraged to do "what-if" analysis to localize the root cause of the bug. We also discuss how formal technology supports the even more challenging task of traceless debugging (the process of debugging the "absence of witness/counter-example").

Categories and Subject Descriptors

D.5.2 [**Register-transfer-level Implementation**]: Design aids – *verification*

General Terms

Algorithms, Design, Verification.

Keywords

Formal verification, debugging, behavioral indexing, property verification, post-silicon debugging, traceless debugging

1. INTRODUCTION

Verification has traditionally been considered the longest pole in the design flow with as much as 70% of the overall design effort attributed to verification related tasks. The increasing functional complexity of the chips and SoCs being designed and the associated verification challenges have led to great deal of innovation in technology and methodology related to non-testbench based methods. In particular, of late, formal technology

Permission to make digital or hard copies of part or all of this work for personal or classroom use is granted without fee provided that copies are not made or distributed for profit or commercial advantage and that copies bear this notice and the full citation on the first page. To copy otherwise, to republish, to post on servers or to redistribute to lists, requires prior specific permission and/or a fee.
DAC'09, July 26-31, 2009, San Francisco, California, USA

has come of age and has been leveraged successfully for both implementation verification (aka equivalence checking) as well as functional verification (aka model checking). The ability of formal property verification to identify the corner case design bugs at the earliest stage as well as to establish correctness of critical features of the design is well known. In recent times, formal property verification has reached a maturity level where it is appropriately accounted for in the beginning of the overall verification planning. As verification productivity has improved with the introduction of new technologies, debugging efficiency has become more important than before.

The debugging challenges in the functional verification world span the entire design flow: from the architectural stage to RTL development to block/chip/system-level verification including the post-silicon debugging stage. The stakeholders are different at different stages of the design flow and the nature of the debugging challenge differs from one stage to another.

For an RTL developer, the key to efficiency lies in the ability to "reason" about the counter-examples without needing to change the underlying RTL. This way, the developer can be reasonably confident about the correctness of the bug fix prior to making the actual fix. For sub-system and system level verification, an additional debugging challenge comes from the fact that, more often than not, the verification task falls to verification engineers who have little visibility in the design functionality. This poses a significant barrier for the verifiers to root cause the problem. Post-silicon debugging poses significant challenges due to lack of visibility and poor controllability. Formal verification has been successfully leveraged in reducing the time in isolating the failure, root causing it, and later verifying the fix [2,3].

In this paper we will talk about novel applications of formal technology to facilitate the debugging process. The applications described in the paper target design verification at RT level, however, can be extended to support higher (high-level models) or lower (gate) levels of abstractions as well.

The paper is organized as follows: In Section 2, we give a brief overview of RTL formal function particularly in the context of debugging. In Section 3, we discuss how formal technology can be leveraged to facilitate trace-based debugging. In particular, we discuss "waveform-based" what-if analysis in Section 3.1, and "behavioral indexing" [1] technology based "transaction-annotation" in Section 3.2. In Section 4, we discuss challenges in traceless debugging and how formal technology can facilitate that process.

2. OVERVIEW OF FORMAL FUNCTIONAL VERIFICATION

In a traditional testbench setting, functional verification is carried out by specifying a design, a set of observers, and a set of input vectors, and then simply reporting if any of the observers trigger when the design is stimulated with the given input vectors. Formal functional verification produces the same kind of results, but with vastly different techniques.

The crucial difference is the specification and treatment of input vectors: In the testbench setting, the input vectors to test are specified, explicitly or implicitly, by some procedure. Formal functional verification, on the other hand, takes a list of constraints and proceeds to test all input vectors that have not been explicitly ruled out by the constraints.

The greatest advantage of describing the input vectors by what they are not is that it often allows a much more compact and readable specification. Also, due to the exhaustive nature of the formal analysis, it relieves the verification engineer of the task of figuring out all the possible corner cases in the design or relying on random processes to hit them. All in all, this means that we get higher quality verification while spending both less time and less intellectual effort in specifying the input vectors.

There are, of course, trade-offs. Apart from getting lower capacity in return for the exhaustive analysis, we also get some new debugging problems. The most immediate one is that of coverage: Did we over-constrain the verification, ruling out legal input scenarios? There are several methodologies to avoid this, the primary one being making sure that each constraint is clearly stated and supported by the specification. Also, it is good practice to only add a constraint after seeing an unrealistic trace that the constraint would rule out.

Even the best specification and most clearly written set of constraints is useless if the constraints actually express something else than intended. Since both the constraints and assertions are typically expressed using a special language or library like SVA, PSL, or OVL, which are still relatively new to many designers and verification engineers, there is the problem of debugging the constraints and assertions themselves. Interestingly, one can apply formal analysis on the constraints and assertions themselves, and use the inherent waveform generating capabilities of the tools to debug them.

3. DEBUGGING WITH A TRACE

This is the classical case of debugging, where a counter-example for the specification violation is presented by the tool (either a simulator or a model checker). Over the years, there have been a number of advances that have made debugging counter-example waveforms easier, such as co-relating the waveform with the actual source code, showing the reason why a signal has a given value in the trace, showing temporal causality information on assertions, and performing transaction-level analysis of a trace. In this section, we will discuss two new ways in which the formal technology can be leveraged to improve the debugging efficiency.

3.1 Debugging efficiency with "what-if" analysis

"What-if" analysis stems for an interactive debugging flow where the user adds on-the-fly constrains to the proof by either inserting textual or visual temporal constraints, reanalyzes the property by the formal tool, and obtains a new trace to debug.

In "what-if" analysis, the user attempts to address one or more of the following issues:

- whether the driving logic to the module under test correctly drives the design;

- which part of the design can better explain the violation trace;

- if a fix for the problem will fix the issue with respect to the trace obtained.

In both simulation and formal verification, designers usually spend a lot of time determining if they have the right set of stimuli for the design under test.

Modern commercial formal verification tools with interactive proof flows enable one to perform "what-if" analysis to check whether assumptions are correctly in place.

This can be used in a flow to interactively constrain the design. One starts with a minimal set of constraints, such that the design is under-constrained. Upon obtaining a violation trace the user determines which parts of the design need to be further constrained. This is achieved by specifying on-the-fly constraints to the proof that eventually becomes the full proof scenario.

This flow has long been successfully used when developing simulation testbenches, but only with recent advances in proof engines and interactive flows, has it been made possible in formal verification tools.

Another use for "what-if" analysis is as an isolation analysis for the root cause of the problem. Having determined that the assumptions specifying the proof scenario are not the cause for the problem, designers can focus on the design itself.

In this flow, the user determines which portions of the logic is the root cause for the violation. In many scenarios, a failure occurs when two conditions are met at the same time. For example, a cache hit is the conjunction of a tag address match and the valid entry being set. If both conditions are false, the user can check what would happen to the violation trace if only one of them were false.

This is accomplished by allowing the user to specify cycle-accurate temporal constraints that should be used by the proof tool in order to accept or refute the trace.

By performing "what-if" analysis, the user can safely and quickly determine if the violation should be expected or not.

Once a failure of the design has been identified, the user can use the same analysis technique to determine if a fix to the problem can resolves the problem.

Again, most of the problems identified can be circumvented by adding simple cycle accurate conditions phrased as constraints to the circuit behavior. For example, one such constraint could specify that a given signal should not toggle as frequently as determined by the proof engine.

In this case, by adding this additional constraint to the tool, the designer may determine that the violation trace indeed was caused by this problem, and with the fix specified by the high-level description, the problem is gone.

It is important to note that the high-level fix provided by the user will later on be coded, and that too can be used in a formal verification environment to check whether the fix was proper or not.

The overall "what-if" analysis flow is depicted in Figure 1. It is important to note that although sequential, these three steps overlap somehow, as after performing assumptions refinement, and moving to bug isolation, the user may determine that a wrong behavior is in fact caused by yet another missing assumption, in which case, he/she will have to think whether the newly added assumption will resolve the problem or whether it requires additional refinements.

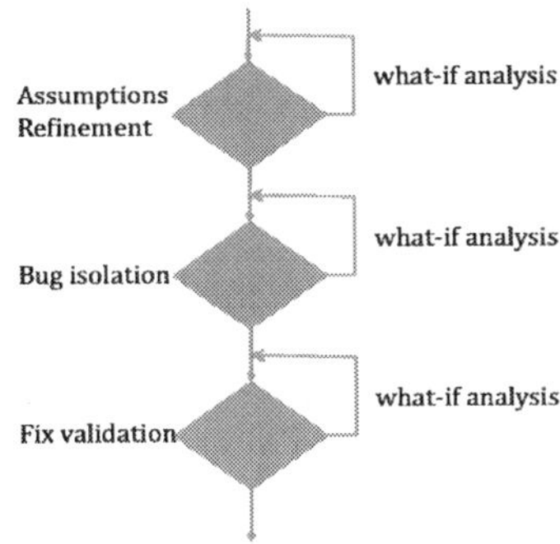

Figure 1. "What-if" based debugging flow

3.2 Debugging efficiency with "Behavioral Indexing" of designs

Typically the verification tasks verification at the system and sub-system levels (and sometimes even at the block-level) fall on a verification engineer who does not have sufficient visibility within the design functionality. As a result, when a trace capturing some form of specification violation is identified, due to lack of sufficient design knowledge, the verification engineer is handicapped, and quite often the actual debugging task is handed over to the designer. This may not be practical (or even feasible) depending upon the bandwidth availability on the designer's part.

Behavioral Indexing solves this problem by enabling automatic transaction annotation on the traces. It is the process of capturing meaningful design behaviors (temporal relationship amongst design states) in executable form and tagging them with appropriate descriptions and storing them in a database. The analysis system associated with the database is capable of analyzing a given trace and performing a pattern matching of the stored behaviors in the database against the trace. Any matched patterns can be automatically highlighted and the associated descriptions for the corresponding behaviors can be accessed from the database on-the-fly and made visually available close to the trace. For details on the behavioral indexing technology and the associated analysis system, please refer to [1].

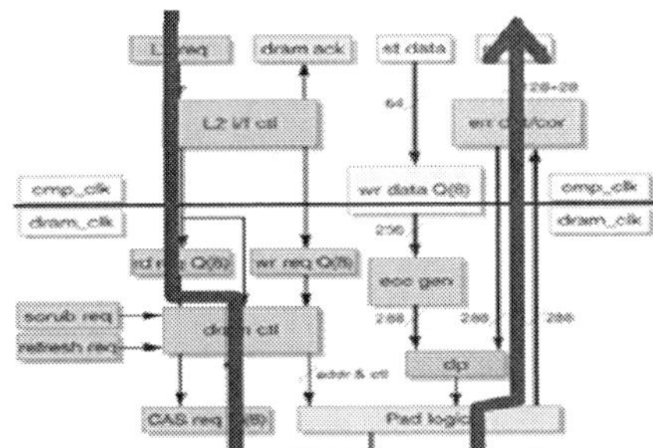

Figure 2. Counter-example at the bit-level

Consider the DRAM controller example in Figure 2. The controller accepts read/write requests from the L2 cache (the client) to access the DRAM. The functionality of a complete read cycle for this design consists of the following behaviors: 1) read request from the L2 cache 2) read request acknowledged by the controller 3) execution of specific read request from the queue 4) read data delivered to the client. A trace for this entire read sequence consisting of bit-level design signals is shown in Figure 3.

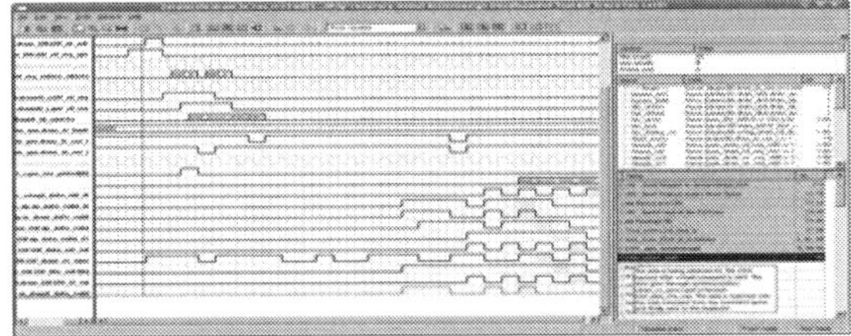

Figure 3. Read-cycle trace with bit-level activity

With appropriate behavioral indexing of the dram controller design, the same bit-level trace can be viewed as a sequence of overlapping transactions (automatically shown by the analysis system in Figure 4). For someone who is not familiar with this design, understanding the trace at the transaction level will be far easier and will lead to an efficient debugging process compared to dealing with the bit-level activity in the design. Essentially, behavioral indexing has enabled an "abstracted view" and top-down understanding of the design functionality.

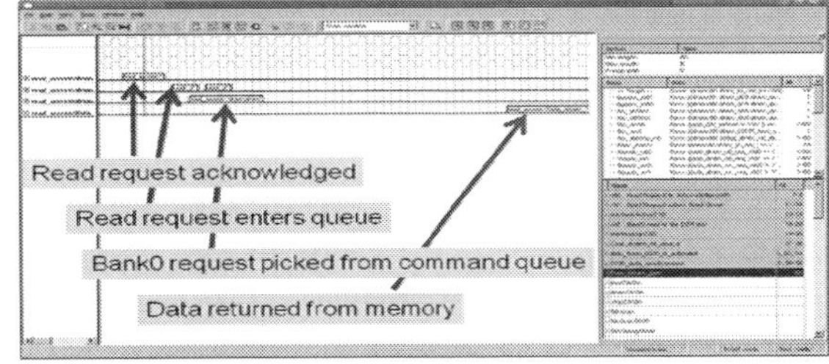

Figure 4. Read-cycle trace automatically annotated with transactions

4. DEBUGGING WITHOUT A TRACE

Most bugs found are demonstrated by a witness trace; this is true for both testbench based verification as well as formal

978-1-60558-497-3/09 $25.00 © 2009 ACM

verification. There is, however, a class of bugs that can be particularly difficult to debug: the bugs that are demonstrated without any specific trace for the verification engineer to consider. Whether this is due to the verification technology (post-silicon bugs typically come just with the observed failure, not a full trace) or due to the verification result itself (a provably unreachable cover point, for example), the immediate information provided is minimal. In this section, we will talk about the latter (debugging of provably unreachable point as determined by formal analysis).

There are many root causes for the unexpected absence of traces: a design bug might lead to dead code, the formal environment might be over-constrained, or the specification – attempting, maybe, to be overly complete – speaks of impossible scenarios. Or some other problem with the usual culprits: design, testbench, and specification. To debug such a problem, the verification engineer will naturally try to answer the following questions

- Why is a trace expected?

- What is such a trace expected to look like?

- Why does such a trace not exist?

Of course, the answers are interdependent and efficient debugging is all about making progress on the understanding of all three at any given time – until either it is decided that, in fact, no trace is expected or the bug prohibiting a trace is found.

While debugging, a number of different techniques may be used to produce a relevant trace, something concrete that can be scrutinized.

First, depending on the phrasing of the unreachable cover, it may be possible to identify a "sub-cover" that indeed can be reached, yielding a trace to debug. For example, if the tool has proven that A & B is unreachable, the verification engineer may try to cover A. If this succeeds, "what-if" analysis (as discussed in Section 3.1) can be done to attempt to understand why B can't happen concurrently with A.

Other times, such a simple decomposition of the cover statement is not possible, but the verification engineer has some intermediate events in mind that should lead to the cover. In a simulation environment, the natural thing would be to write a testbench to stimulate to design to take those steps. However, formal technology can be used very efficiently in such a scenario, to automatically produce such fulfilling input sequences.

For example, if the unreachable cover statement says that a 128 element FIFO cannot fill up, a reasonable intermediate step would be to see the FIFO as half full, with 64 elements. Depending on whether that is reachable, one can try covering less or more elements in the FIFO and establish exactly what the maximum number of elements in the FIFO is. Again, once such a trace is up, one can use "what-if" analysis to examine why one more push is not possible – or maybe why elements must start popping at this point.

The inspiration for which steps may lead to a cover can come from many sources. It may be taken from experience or common sense, as in the FIFO example, or by design knowledge learned through RTL exploration or 'what-if' analysis of related cover traces. Sometimes, however, one starts from scratch: No idea what a trace would look like, no related covers can be thought of, but

also no reason to believe that a trace should not exist. In this case, one can use Design Tunneling to produce abstract cover traces.

Design Tunneling is a JasperGold specific feature that is typically used to achieve convergence on hard-to-prove properties. At the beginning of a Design Tunneling session, the tool abstracts all registers in the design, and presents the user with the ensuing abstract trace. Using normal waveform debugging, the user identifies one or more registers whose abstraction is causing the problem and tells the tool to de-abstract them. If a trace is still possible, it is presented to the user who tries to identify more registers to de-abstract, and so forth until a proof is found. The strength of this methodology is that it helps the tool identify a minimal subset of the states that is needed for a full proof.

Each trace presented to the user in the Design Tunneling flow is a bit less abstract than the last. This very fact can be used to great effect to do traceless debugging without any prior design knowledge. As an example, take a failing cover stating that a timeout occurs. Using Design Tunneling, we immediately get a trace as the time out detection logic is abstracted. We tell the tool to de-abstract the timeout logic and are presented with a new trace where the timeout counter is started. When we de-abstract the trigger, we're told there is no trace: We now realize that the problem is that we've constrained configuration updates away and the timeout logic is disabled at reset.

Applied effectively, Design Tunneling will allow traceless debugging in a very natural, interactive flow. Depending on the amount of logic on the critical path of the cover, it can however also be quite laborious if done completely manually. While the tool does provide automatic Design Tunneling features, using these defeat the purpose of understanding why the critical path includes the logic that it does. One will therefore normally use all three techniques, swapping between them based on the knowledge gained and the current hypothesis or angle of attack; each iteration resulting in some insight that brings the user closer to resolving the issue.

5. CONCLUSION

Formal technology has emerged as a strong complementary approach to traditional testbench-based simulation for verification across the entire design flow. Formal technology can be additionally leveraged for bringing efficiency in the debugging process – for trace-based as well as traceless debugging. We present waveform-based "what-if" analysis and "behavioral indexing" based automatic transaction annotation as two applications for trace-based debugging. We also discuss how "what-if" capability enables "traceless debugging".

6. REFERENCES

[1] Rajeev Ranjan et al. 2009. Toward harnessing the true potential of IP reuse. Design Con 2009.

[2] Jamil Mazzawi et al. 2008. Formal technology in the post-silicon lab. Haifa Verification Conference 2008 (www.haifa.**ibm.**com/**conferences**/hvc2008/present/**PostSilic on**JamilRMazzawi_v1.pdf)

[3] Richard Ho et al. 2009. Post-silicon debug using formal verification waypoints. DV Con 2009.

Power Modeling of Graphical User Interfaces on OLED Displays

Mian Dong   Yung-Seok Kevin Choi   Lin Zhong
Department of Electrical & Computer Engineering, Rice University, Houston, TX 77025
{dongmian, ykc1,lzhong}@rice.edu

ABSTRACT

Emerging organic light-emitting diode (OLED)-based displays obviate external lighting; and consume drastically different power when displaying different colors, due to their emissive nature. This creates a pressing need for OLED display power models for system energy management, optimization as well as energy-efficient GUI design, given the display content or even the graphical user interface (GUI) code. In this work, we present a comprehensive treatment of power modeling of OLED displays, providing models that estimate power consumption based on pixel, image, and code, respectively. These models feature various tradeoffs between computation efficiency and accuracy so that they can be employed in different layers of a mobile system. We validate the proposed models using a commercial QVGA OLED module. For example, our statistical learning-based image-level model reduces computation by 1600 times while keeping the error below 10%, compared to the more accurate pixel-level model.

Categories and Subject Descriptors

I.6.5 [**Simulation and Modeling**]: Model Development

General Terms

Algorithms, Measurement, Human Factors

Keywords

OLED Display, Graphic User Interface, Low Power

1. INTRODUCTION

Energy consumption is an important design concern for mobile embedded systems that are battery-powered and thermally constrained. Displays have been known as one of the major power consumers in mobile systems [1-4]. Conventional liquid crystal display (LCD) systems provide very little flexibility for power saving because the LCD panel consumes almost constant power regardless of the display content while the external lighting dominates the system power consumption. In contrast, the power consumption by emerging organic light-emitting diode (OLED)-based displays [5] is highly dependent on the display content because their pixels are emissive. For example, our measurement shows that a commercial QVGA OLED display consumes 3 and 0.7 Watts showing black text on a white background and white text on a black background, respectively. Such dependence on display content leads to new challenges to the modeling and optimization of display power consumption. First, it makes it much more difficult to account display energy consumption in the operating sys-

tem for optimized decisions. Second, GUI designers will have a huge influence on the energy cost of applications, which is not their conventional concern. As a result, there is a great need of OLED display power models for use at different layers of a computing system and different stages of system design.

In this work, we provide a comprehensive treatment of OLED display power modeling. In particular, we make three contributions by addressing the following research questions.

First, *given the complete bitmap of the display content, how to estimate its power consumption on an OLED display with the best accuracy?* An accurate *pixel-level model* is the foundation for modeling OLED display power. We base our pixel-level model on thorough measurements of a commercial QVGA OLED module. In contrast to the linear model assumed by previous work [6], we show that the power consumption is nonlinear to the intensity levels of the color components. Our nonlinear power model achieves 99% average accuracy against measurement of the commercial OLED display module. This is presented in Section 4.

Second, *given the complete bitmap of the display content, how to estimate the power consumption with as few pixels as possible?* This *image-level model* is important because accessing information of a large number of pixels can be costly due to memory and processing activities. We formulate the tradeoff as a sampling problem and provide a statistical optimal solution that outperforms both random and periodical sampling methods. Our solution achieves 90% accuracy with 1600 times reduction in sampling numbers. This is presented in Section 5.

Third, *given the code specification of a GUI, how to estimate its power consumption on an OLED display?* This *code-level model* is important to GUI designers as well as application and system based energy management. We use the code specification to count pixels of various colors and calculate the power consumption of the OLED display. Our model guarantees over 95% accuracy for 10 benchmark GUIs. This is presented in Section 6.

To the best of our knowledge, this is the first public study that addresses the three research questions above. The modeling methods presented here provide powerful mechanisms for operating systems and applications to construct energy-conserving policies for OLED displays. They will also enable GUI designers of mobile systems to build adaptable and energy-efficient GUIs; and empower end users to make informed tradeoffs between battery lifetime and usability.

The rest of the paper is organized as follows. We provide background and address related work in Section 2. We describe the experimental setup used in this study in Section 3. From Sections 4 to 6, we present the power models and their experimental validations. We conclude in Section 7.

2. BACKGROUND AND RELATED WORK

How Display Works. Figure 1 illustrates the relationships between the display and the rest of the system. The *main processor*, or *application processor*, runs the operating system (OS) and

Permission to make digital or hard copies of part or all of this work for personal or classroom use is granted without fee provided that copies are not made or distributed for profit or commercial advantage and that copies bear this notice and the full citation on the first page. To copy otherwise, to republish, to post on servers or to redistribute to lists, requires prior specific permission and/or a fee.
DAC'09, July 26-31, 2009, San Francisco, California, USA

978-1-60558-497-3/09 $25.00 © 2009 ACM

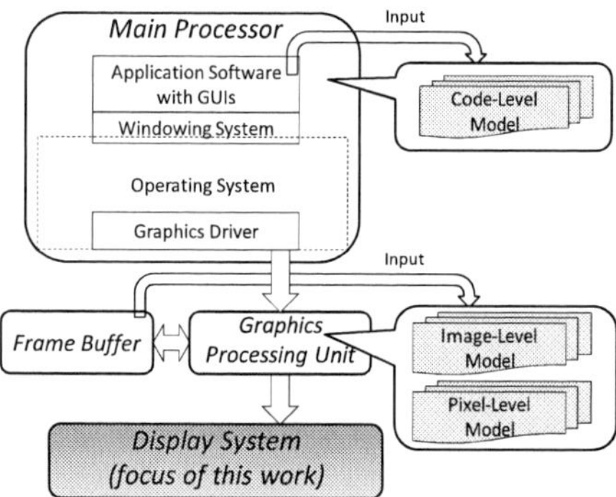

Figure 1. Display system in a typical mobile system. The three power models proposed require input of different abstractions and provide different accuracy-efficiency tradeoffs

application software with GUIs. Note the windowing system can be either part of the OS, e.g. Windows, or a standalone process, e.g. X Window under Linux. the *graphics processing unit*, often including a graphics accelerator and a LCD controller in a system-on-a-chip for mobile devices, generates the bitmap of the display content and stores it in a memory called *frame buffer*; the bitmap is sent to the display for displaying. Each unit of this bitmap is described using sRGB, or standard RGB, color space, in which a color is specified by (R, G, B), the intensity level of red, green and blue component. Because sRGB uses gamma correction to cope with the nonlinearity introduced by cathode ray tube (CRT) displays, the intensity level of each component and its corresponding luminance follow a nonlinear relation [7].

OLED Display. Organic light-emitting diode or OLED [5, 8] is an emerging display technology that provides much wider view angle and higher image quality than conventional LCDs . The key difference in power characteristics between an OLED display and a LCD is that an OLED display does not require external lighting because its pixels are emissive. Each pixel of an OLED display consists of three types of devices, corresponding to red, green and blue components, respectively. Moreover, the red, green, and blue components of a pixel have different luminance efficacies. As a result, the color of a pixel directly impacts its power consumption and GUI has a significant impact on the display power. In contrast, color only has negligible power impact on LCDs and illumination of external lighting dominates. OLED displays and LCDs have a very similar organization, including a panel of addressable pixels, LCD or OLED, control circuitry that generates the control and data signals for the panel based on display content, and interface to the graphics processing unit. In this work, we address the power consumption of the display and focus on the variance introduced by the OLED panel. Our power models take input from different places of the system and can be implemented either as a software tool, an operating system module, or an extra circuit.

We focus on the power consumption by a constant screen because of the following two reasons. First, a display spends most time displaying a constant screen, even for high-definition video that requires 30 updates per second. Second, our measurement showed that the power consumption by an OLED display during updating is close to the average of those by the constant screens before and after the updating. Therefore, the energy contribution by OLED display updating is very small and can be readily estimated from the power models of a constant screen. However, we note that

Figure 2. Measurement setup of the QVGA OLED module used in our experimental validation

screen updating may incur considerable energy overhead in graphics processing unit, frame buffer, and data buses, which are out of the scope of this work.

Related Work. HP Labs pioneered energy reduction for OLED displays [6, 9, 10]. Yet no real OLED displays were reported in the work. The power model employed was pixel-level, thus expensive to use, and incorrectly assumed a linear relationship between intensity levels of color components and power consumption. As we will show, the relationship is indeed nonlinear. The IBM Linux Wrist Watch was one of the earliest users of OLED displays [11]. The work, however, did not employ or provide a power model for the OLED display. There is also a large body of work on energy optimization of conventional LCD systems [2, 4, 12-23]. While many of the proposed techniques may be applied to OLED displays, they are orthogonal to the power modeling techniques presented in this work.

3. EXPERIMENTAL SETUP

Benchmark GUI Images. To evaluate the proposed power models, we collect 300 GUI images from three Windows Mobile-based cell phones, HTC Touch, HTC Mogul, and HTC Wizard, all with a resolution of QVGA (240×320). On each phone, we exhaust all the varying GUI screens, representative of everyday use of a smart phone (e-mail, web browser, games, etc). We also capture screens with different color themes available.

Measurement Setup. For our experimental validation, we use a 2.8" OLED QVGA display module with an integrated driver circuit, μOLED-32028-PMD3T, from 4D Systems [24]. We connect it to a PC using a micro USB interface, which also supplies power to it. Through the USB, we can send commands to the OLED module to display images. The display module employs a standard 16-bit (5,6,5) RGB setting. That is, there are five, six, and five bits to represent the intensity of red, green, and blue component, respectively. In this work, we call their numbers the *intensity level* or *value* of the three color components.

We obtain the power consumption of the OLED module by measuring the current it draws from the USB interface and its input voltage. Figure 2 shows the measurement setup with a DAQ board from Measurement Computing and the OLED module. To overcome the variance among different measurements of the same image, we take an average of 1000 measurements for each image.

4. PIXEL-LEVEL POWER MODEL

We first present a pixel-level power model that estimates the power consumption of OLED modules based on the RGB specification of each pixel. It is intended to be the most accurate and constitutes the baseline for models based on more abstract descriptions of the display content.

978-1-60558-497-3/09 $25.00 © 2009 ACM

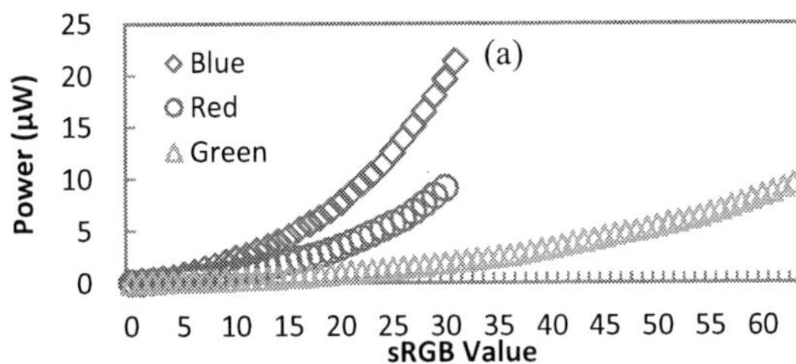

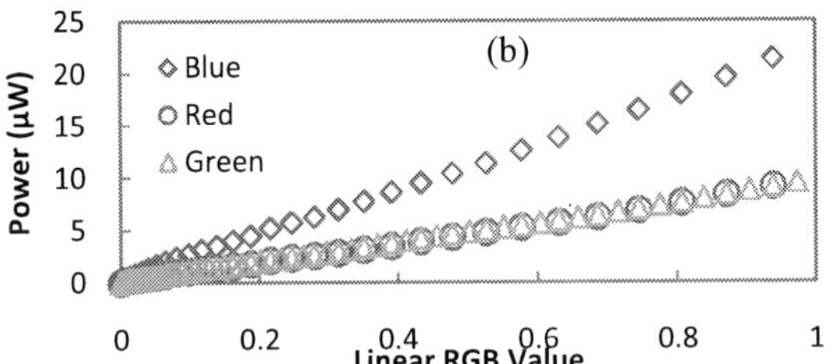

Figure 3. Intensity Level vs. Power Consumption for the R, G, and B components of an OLED pixel

Power Model of OLED Module. We model the power contributed by a single pixel, specified in (R, G, B), as

$$P_{pixel}(R, G, B) = f(R) + h(G) + k(B),$$

where $f(R)$, $h(G)$ and $k(B)$ are power consumption of red, green and blue devices of the pixel, respectively. And the power consumption of an OLED display with n pixels is

$$P = C + \sum_{i=1}^{n}\{f(R_i) + h(G_i) + k(B_i)\}.$$

Note that the model includes a constant, C, to account for static power contribution made by non-pixel part of the display, which is independent with the pixel values. This pixel model has a generic form that applies to all the colorful OLED display modules.

Power Models for RGB Components. We obtain C by measuring the power consumption of a completely black screen. To obtain $f(R)$, we fill the screen with colors in which the green and blue components are kept zero and the red component, R, varies from 0 to 31, enumerating every possible intensity level. For each measurement, we subtract out C to get just the power contribution by the red pixel component, or $f(R)$. We obtain $h(G)$ and $k(B)$ similarly. Figure 3 (a) presents the measured data for all three components. Apparently, the power contribution by pixel components is a nonlinear function, instead of a linear function as assumed in [6], of the intensity level. The nonlinearity is due to the gamma correction in sRGB standard. After transforming the intensity level into linear RGB format, which is the indication of luminance, we obtain a linear relation between pixel power consumption and intensity, as shown in Figure 3(b).

The measured data can be directly used to estimate the power consumption of displaying an image on the OLED display through simple table lookup. One can also apply curve fitting to obtain close-form functions for $f(R)$, $h(G)$, and $k(B)$.

Estimation vs. Measurement. Figure 4 shows the histogram of the error of the estimation against the measurement for the 300 benchmark images. It shows that 63% of the samples have no more than 1% error and 93% have no more than 3% errors. The average absolute error is only 1%.

It is important to note that although we derived the pixel-level power model from a specific OLED display, the methodology can

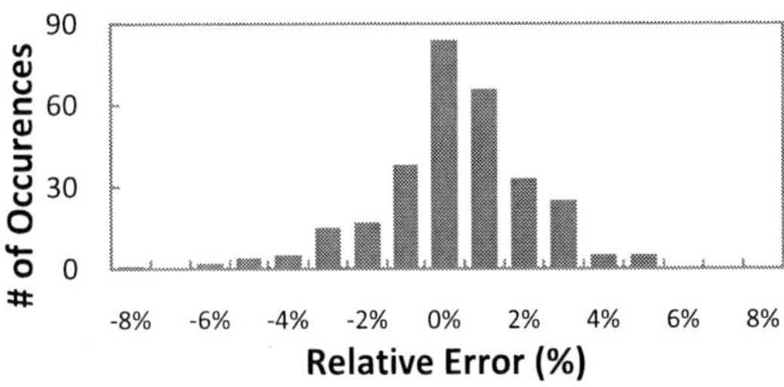

Figure 4. Histogram of percent errors observed for the 300 benchmark GUIs using our pixel-level power model

be largely extended to other OLED displays with a similar RGB organization.

5. IMAGE-LEVEL POWER MODEL

We next present models that estimate power consumption given the image to display. They are important for a system to assess the display power cost when the pixel information is known, e.g. from the frame buffer. A straightforward image-level model can be a simple application of the pixel-level model to all pixels. Such a method, unfortunately, is exceedingly expensive. Modern displays can have hundreds of thousands or millions of pixels. The cost of a large number of pixel power calculations and the overhead of accessing the frame buffer can be prohibitively high for the system. Our solution to this problem is to estimate the power based on a small subset of pixels, or *sampling*.

5.1 Problem Formulation

The power consumption by an image on an OLED display can be described by a vector of N elements, i.e., $y = (y_1, y_2, ..., y_N)^T$, in which y_n denote the power consumption of the i-th pixel, $i = 1, 2, ..., N$. The total display power can be calculated as $Power = \mathbf{1}^T y = \sum_{i=1}^{N} y_i$.

Sampling is to select K pixels out of N and use the average of K samples to approximate the average of all N pixels. For each pixel, we use a random variable to indicate whether this pixel is sampled or not, i.e., $X_i = 1$ when the i-th pixel is sampled; otherwise $X_i = 0$. Thus, the sampling of an image can be represented by a vector $X = (X_1, X_2, ..., X_N)^T$. And denote the joint probabilistic distribution of them as $f_X(x) = f_{X_1, X_2, ..., X_N}(x_1, x_2, ..., x_N)$. Given an image, y, the estimation error can be calculated as

$$\varepsilon = E_X\left[\left(\frac{X^T y}{K} - \frac{\mathbf{1}^T y}{N}\right)^2\right] = E_{X_1, X_2, ..., X_N}\left[\left(\frac{\sum_{i=1}^{i=N} X_i y_i}{K} - \frac{\sum_{i=1}^{i=N} y_i}{N}\right)^2\right]$$

$$= \int \left(\frac{\sum_{i=1}^{i=N} x_i y_i}{K} - \frac{\sum_{i=1}^{i=N} y_i}{N}\right)^2 f_{X_1, X_2, ..., X_N}(x_1, x_2, ..., x_N) dx.$$

Therefore, an optimal sampling given the image can be obtained by finding $f_X(x)$ that leads to the minimal ε, or $f_X^*(x) = \arg \min_{\forall f_X(x)} \varepsilon$.

Now, we consider all possible images by treating the power consumption of an image as a random vector, $Y = (Y_1, Y_2, ..., Y_N)^T$, and denote the joint probabilistic distribution as $g_Y(y) = g_{Y_1, Y_2, ..., Y_N}(y_1, y_2, ..., y_N)$. Then the optimal sampling for all images can be found as $f_X^*(x) = \arg \min_{\forall f_X(x)} E_Y[\varepsilon]$, i.e., finding a probability simplex $p = (p_1, p_2, ..., p_{\binom{N}{K}})^T$ such that $E_Y[\varepsilon]$ is minimal, where $p_j = prob(X = x^{(j)})$ and $x^{(j)}$ is the j-th possible combination of X. This is a linear programming problem, i.e.,

978-1-60558-497-3/09 $25.00 © 2009 ACM

654

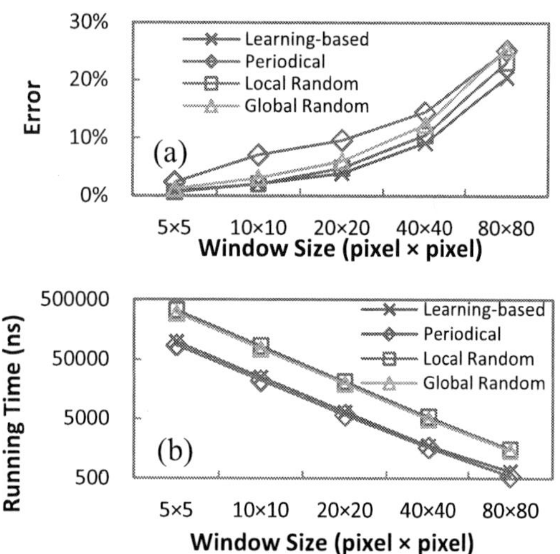

Figure 5. Comparison of four different sampling methods

$$\min \int \sum_{j=1}^{\binom{N}{K}} p_j \left(\frac{\sum_{i=1}^{i=N} x_i^{(j)} y_i}{K} - \frac{\sum_{i=1}^{i=N} y_i}{N} \right)^2 g_{Y_1, Y_2, \ldots, Y_N}(y_1, y_2, \ldots, y_N) dy$$

s.t. $\quad \sum_j p_j = 1$ and $p_j \geq 0$

It is, however, impractical to obtain the joint distribution $g_{Y_1, Y_2, \ldots, Y_N}(y_1, y_2, \ldots, y_N)$, because the dimension is too high. Therefore, we transform the objective function to eliminate the join distribution as below.

$$\int \sum_{j=1}^{\binom{N}{K}} p_j \left(\frac{\sum_{i=1}^{i=N} x_i^{(j)} y_i}{K} - \frac{\sum_{i=1}^{i=N} y_i}{N} \right)^2 g_{Y_1, Y_2, \ldots, Y_N}(y_1, y_2, \ldots, y_N) dy$$

$$= E_X \left[\left(\frac{X}{K} - \frac{\mathbf{1}}{N} \right)^T E_Y[YY^T] \left(\frac{X}{K} - \frac{\mathbf{1}}{N} \right) \right] = \sum_{j=1}^{\binom{N}{K}} p_j b^{(j)T} \bar{A} b^{(j)}$$

in which $\bar{A} = E_Y[YY^T]$ and $b_i^{(j)} = \frac{x_i^{(j)}}{K} - \frac{1}{N}$, for $i = 1, 2, \ldots, N$.

Therefore, the key to an accurate estimation is to obtain a good knowledge of the expectation of outer products of all possible images, which can be *statistically learned* from a set of training images by using the mean of the outer products of the training images to approximate the expectation $\bar{A}$.

5.2 Practical Learning-based Sampling

There are two critical barriers for the practical implementation of the statistically optimal sampling. First, the linear programming problem described above is very difficult, if possible, to solve because of the extremely high dimension, i.e. $\binom{N}{K}$ or N-choose-K. To address this issue, we divide the image into much smaller windows, and then solve the linear programming problem for each window. By preparing a training set for the window at every position within an image, we are able to obtain an optimal sampling solution for every window. Second, the solution requires generating a large number of random numbers, which is expensive for mobile embedded platforms. Therefore, instead of random sampling, we decide to use deterministic sampling method in which we seek to solve a combinatorial problem to find the samples for each window such that $b^{(j)T} \bar{A} b^{(j)}$ achieves the minimal. That is,

$$\min \left(\frac{X}{K} - \frac{\mathbf{1}}{N} \right)^T \bar{A} \left(\frac{X}{K} - \frac{\mathbf{1}}{N} \right)$$

s.t. $x_i = 1 \ or \ 0$, and $\sum_i x_i = K$

Another practical issue is the number of samples to take from each window, or K in the formulation above. Given a fixed sampling rate, or the total number of samples, there are two strategies. One can increase the window size, therefore reduce the number of windows, but allow more samples per window. Or one can reduce the window size, therefore increase the number of windows, but allow fewer samples per window. The second strategy obviously is more efficient because the computational load increases much faster with the window size than with the number of samples per window. We also experimentally compared the accuracy of these two strategies. We found that there is no difference in accuracy between them. Therefore, we take the second strategy in our image-level mode, i.e. a single sample is taken from each window.

5.3 Experimental Results

To evaluate our proposed learning-based sampling method (Learning-based), we compare its performance against three other sampling methods, described below.

- Periodical sampling: an image is divided into non-overlapping windows of the same size, exactly the same as Learning-based sampling but the bottom right pixel of each window is selected.
- Local Random sampling: one pixel is randomly selected from each window.
- Global Random sampling: pixels are randomly selected from the whole image. The number of pixels is kept the same as the number of windows in the first three methods.

Collectively these benchmark sampling methods will highlight the effectiveness of the design decisions made in Section 5.2. It is important to note that for learning-based sampling, we employ a bootstrapping method to improve the reliability of accuracy evaluation because training is involved. That is, we divide the 300 benchmark images evenly into 10 groups, i.e. 30 in each. Then we use nine groups as training set and test the 10th group. We repeat the process 10 times using different groups as the test group and the report the average accuracy.

We investigate the tradeoff between sampling rate and accuracy by varying the window size from 5×5, 10×10, 20×20, 40×40, to 80×80. This leads to a sampling rate between 1/25 and 1/6400.

Figure 5 (a) presents the tradeoffs between estimation error and window size over the 300 benchmark GUIs described in Section 3. It clearly shows that as window sizes increases, or sample rates decreases, the estimation error increases, for all four sampling methods. Figure 5 (a) clearly shows that our learning-based method achieves the best tradeoffs between accuracy and sampling rate. When window size is 40×40 and a 1600 times reduction in pixels needed for power estimation, learning based sampling method achieves accuracy of 90%.

Figure 5 (b) presents the run-time of all four sampling methods on a Lenovo T61 laptop with a Core 2 Duo T7300 2GHz processor and a 2GB memory. It clearly demonstrates the advantage in efficiency of deterministic methods, i.e. Learning-based and Periodical. With the comparable accuracy, Learning-based sampling is at least three times faster than the random methods, both Global and Local. This will lead to significant efficiency improvement when the power model is employed in mobile embedded systems for energy management and optimization.

In summary, our learning-based sampling method achieves the best tradeoff between accuracy and efficiency. We note that the learning-based method achieves this through training, which can

978-1-60558-497-3/09 $25.00 © 2009 ACM

```
for i = 2 to n
    for j = 1 to i − 1
        if O_i ∩ O_j ≠ Ø
            update(pl_j);
        end
    end
end
for k = 1 to n
    pl_GUI = merge(pl_GUI, pl_k);
end
```

Figure 6. Power estimation for a composition of GUI objects

be compute-intensive. However, training can be carried off-line and therefore does not consume any resource at run-time.

6. Code-Level Power Model

Both pixel and image-level models require that the RGB information is available for all pixels of the display content. In this section, we present a power model that is based on the code specification of the display content. This is possible because graphical user interfaces (GUIs) are usually described in high-level programming languages, e.g. C# and Java, and are highly structured and modular. The code-level model can be even more efficient than the sampling-based image-level power model because they do not need to access the frame buffer. Therefore, it can be readily employed directly by the application or the operating system. In addition, they also provide a tool for GUI designers to evaluate their designs at an early stage.

We take an object-oriented approach because modern GUIs are composed of multiple objects, each with specified properties, e.g. size, location, and color. Moreover, GUI programming is also object-oriented. Developers rarely specify the graphics details; instead they extensively reuse a "library" of customizable common objects, e.g. buttons, menus, and lists. They customize these objects by specifying their properties and their relationship with each other within a GUI. Our methodology for code-based power modeling is first estimating power contribution by individual GUI objects based on their properties and then estimating the total power based on the composition of these objects.

6.1 Power by GUI Object

We can view a GUI object as a group of pixels with designated colors. By accounting the number of pixels with each color that appear in an object, we can obtain its power consumption using the pixel-level power model. To obtain the estimated power for the object, we enhance the object class with a pixel list property that records the (R, G, B) values and pixel number N of each color. Therefore, by utilizing our pixel-level power model, we can calculate the power consumption of an object with K colors as

$$P_{object} = \sum_{i=1}^{K} N_i \times P_{pixel}(color_i).$$

In most cases, a GUI object only has three colors, i.e., border color, background color and foreground (text) color. All three can be obtained from the GUI object's properties. Knowing the geometric specification of the object, usually rectangular, we can calculate the number of pixels for both background and border colors. Then we create a pixel list of two entries for this object, which includes the (R, G, B) values of colors and their corresponding pixel numbers. By extending the pixel list, we are able to handle more complicated objects with arbitrary shapes and more colors.

Text is a special object property and needs a special treatment. Unlike the color information available from the object properties, the pixel number of text is not easy to obtain. Thus, we build a library for all the ASCII characters; the library contains the number of pixels of each character based on its font type and size. Thus, we are able to account the total number of pixels of a whole text string, which we should subtract from the background pixel number. For our experiments, we have constructed the library for Times New Roman and Tahoma of font sizes 12 and 9, which are the most common on Windows Mobile devices.

6.2 Power by Composition of Objects

A GUI usually consists of multiple objects. Simply aggregating the power of all of the objects is not accurate enough because they may overlap with each other. To consider the effect of overlapping, we maintain a pixel list (pl_{GUI}) for the whole GUI based on the pixel list of each objects. We first sort the objects from back to front in the GUI. Denote the sorted object list includes n objects $O_1, O_2, ..., O_n$, with the sequence from the back to the front, and the pixel list of O_i is pl_i. As shown in Figure 6, we go through all the objects one by one while checking their overlapping situation. If two objects overlap with each other, we update the pixel list of the object in the back by subtracting the number of pixels being covered. Finally, we merge all the pixel lists by combining the pixels with the same color together. We describe how we deal with two special effects of GUIs below.

Dynamic Objects. Some GUI objects can have multiple states. For example, a radio button has two states, checked and non-checked; a menu has even more states when different items are activated. For these objects, we maintain a pixel list for each state and treat each state as an individual object in the procedure described above.

Themes. Many operating systems support color themes that contain pre-defined graphical details, such as colors and fonts, so that GUIs of different applications will follow a coherent style. For instance, in Windows Mobile and C#, BackColor and ForeColor can be either specified using (R, G, B) or selected from a set of pre-defined colors, such as Window and ControlText. These system colors are determined by the theme of Windows Mobile. Fortunately, the color and font information can be obtained in the GUI code in the runtime for power estimation.

6.3 Experimental Results

We have implemented the object-based power estimation on .NET Compact Framework and C#, for Windows Mobile-based mobile embedded systems. In C#, most GUI objects belong to namespace System.Windows.Controls. All objects provide height, width, and background color, which are enough for power estimation. We use eight sample programs with 10 GUIs from the Windows Mobile 5.0 SDK R2 to evaluate the implementation. Using code-level model, we estimate the power consumption of the 10 GUIs, and compare the results with the estimation from the pixel-level model and measurement. Figure 7 presents the results and shows that our code-level model with text achieves higher than 95% accuracy for all the benchmark GUIs.

7. Conclusions

In this work, we provided models for efficient and accurate power estimation for OLED displays, at pixel, image, and code levels, respectively. The pixel-level model built from measurements achieves 99% accuracy in power estimation for 300 benchmark

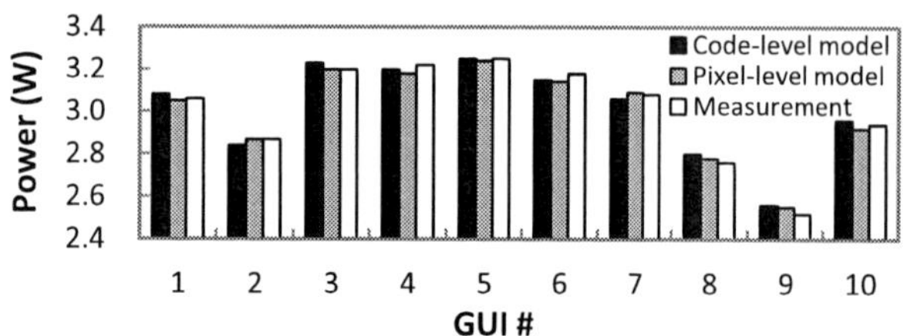

Figure 7. Power estimation comparison

images. By aggregating the power of a small group of pixels instead of all the pixels in an image, our image-level power model reduces the computation cost by 1600 times, while achieving 90% accuracy in power estimation. Our code-level model utilizes specification of the GUI objects to calculate the power consumption, which guarantees 95% accuracy.

The three power models we presented require different inputs and provide different tradeoffs between accuracy and efficiency. Therefore, we intend them for implementation and use at different system components. The pixel and image-level models are best for implementation in hardware or the operating system because they need to access the frame buffer. In contrast, the code-level model is best for implementation in application software or GUI development kits due to its dependence on knowing the composition of GUI object. All three models can provide power estimation for the system to better manage and optimize energy consumption, for the end user to make better tradeoffs between usability and battery lifetime, and for GUI designers to design energy-efficient GUIs.

8. ACKNOWLEDGMENTS

The work is supported in part by NSF Award IIS/HCC 0713249 and by the Texas Instruments Leadership University program. The authors would like to thank the anonymous reviewers whose comments helped improve the final version of this paper.

9. REFERENCES

[1] F. Gatti, A. Acquaviva, L. Benini, and B. Ricco, "Low Power Control Techniques For TFT LCD Displays," in *Proc. Int. Conf. Compilers, Architecture, and Synthesis for Embedded Systems (CASES)*, Grenoble, France, 2002.

[2] W.-C. Cheng, Y. Hou, and M. Pedram, "Power Minimization in a Backlit TFT-LCD Display by Concurrent Brightness and Contrast Scaling," in *Proc. Conf. Design, Automation and Test in Europe (DATE)*, Paris, France, 2004.

[3] L. Zhong and N. K. Jha, "Energy efficiency of handheld computer interfaces: limits, characterization and practice," in *Proc. ACM/USENIX Int. Conf. Mobile Systems, Applications, and Services (MobiSys)*, Seattle, Washington, USA, 2005.

[4] L. Cheng, S. Mohapatra, M. E. Zarki, N. Dutt, and N. Venkatasubramanian, "Quality-based backlight optimization for video playback on handheld devices," *Advanced Multimedia*, vol. 2007, pp. 4-4, 2007.

[5] S. R. Forrest, "The road to high efficiency organic light emitting devices," *Organic Electronics*, vol. 4, pp. 45-48, 2003.

[6] S. Iyer, L. Luo, R. Mayo, and P. Ranganathan, "Energy-Adaptive Display System Designs for Future Mobile Environments," in *Proc. ACM/USENIX Int. Conf. Mobile Systems, Applications, and Services (MobiSys)* San Francisco, California, USA, 2003.

[7] C. A. Poynton, *Digital Video and HDTV: Algorithms and Interfaces*. San Francisco: Morgan Kaufmann, 2003.

[8] J. Shinar, *Organic Light-Emitting Devices: A Survey*: Springer, 2004.

[9] T. Harter, S. Vroegindeweij, E. Geelhoed, M. Manahan, and P. Ranganathan, "Energy-aware user interfaces: an evaluation of user acceptance," in *Proc. ACM Conf Human Factors in Computing Systems (CHI)*, Vienna, Austria, 2004.

[10] P. Ranganathan, E. Geelhoed, M. Manahan, and K. Nicholas, "Energy-Aware User Interfaces and Energy-Adaptive Displays," *IEEE COMPUTER*, pp. 31-38, 2006.

[11] N. Kamijoh, T. Inoue, C. M. Olsen, M. T. Raghunath, and C. Narayanaswami, "Energy trade-offs in the IBM wristwatch computer," in *Proc. IEEE Int. Sym. Wearable Computers*, Zurich, Switzerland, 2001.

[12] I. Choi, H. Shim, and N. Chang, "Low-power color TFT LCD display for hand-held embedded systems," in *Proc. Int. Sym. Low Power Electronics and Design (ILSPED)*, Monterey, California, USA, 2002.

[13] N. Chang, I. Choi, and H. Shim, "DLS: dynamic backlight luminance scaling of liquid crystal display," *IEEE Trans. Very Large Scale Integration (VLSI) Systems,* vol. 12, pp. 837-846, 2004.

[14] H. Shim, N. Chang, and M. Pedram, "A Backlight Power Management Framework for Battery-Operated Multimedia Systems," *IEEE Design & Test,* vol. 21, pp. 388-396, 2004.

[15] A. K. Bhowmik and R. J. Brennan, "System-Level Display Power Reduction Technologies for Portable Computing and Communications Devices," in *Proc. IEEE Int. Conf. Portable Information Devices*, Orlando, Florida , USA, 2007.

[16] A. Iranli, H. Fatemi, and M. Pedram, "HEBS: histogram equalization for backlight scaling," in *Proc. Conf. Design, Automation and Test in Europe (DATE)*, 2005, pp. 346-351 Vol. 1.

[17] W.-C. Cheng and C.-F. Chao, "Minimization for LED-backlit TFT-LCDs," in *Proc. ACM/IEEE Design Automation Conf. (DAC)*, San Francisco, California, USA, 2006.

[18] A. Iranli and M. Pedram, "DTM: dynamic tone mapping for backlight scaling," in *Proc. ACM/IEEE Design Automation Conf.(DAC)*, Anaheim, California, USA, 2005.

[19] C.-N. Wu and W.-C. Cheng, "Viewing direction-aware backlight scaling," in *Proc. ACM Great Lakes Sym. VLSI (GLVLSI)*, Stresa-Lago Maggiore, Italy, 2007.

[20] W.-C. Cheng and C.-F. Chao, "Perception-guided power minimization for color sequential displays," in *Proc. ACM Great Lakes Sym. VLSI (GLVLSI)*, Philadelphia, PA, USA, 2006.

[21] W.-C. Cheng, C.-F. Hsu, and C.-F. Chao, "Temporal vision-guided energy minimization for portable displays," in *Proc. Int. Sym. Low Power Electronics and Design (ISLPED)*, Tegernsee, Bavaria, Germany, 2006.

[22] S. Salerno, A. Bocca, E. Macii, and M. Poncino, "Limited intra-word transition codes: an energy-efficient bus encoding for LCD display interfaces," in *Proc. Int. Sym. Low Power Electronics and Design (ISLPED)*, Newport Beach, California, USA, 2004.

[23] H. Shim, N. Chang, and M. Pedram, "A compressed frame buffer to reduce display power consumption in mobile systems," in *Proc. IEEE Conf. Asia South Pacific Design Automation (ASPDAC)*, Yokohama, Japan, 2004.

[24] 4D Systems，http://www.4dsystems.com.au/.

978-1-60558-497-3/09 $25.00 © 2009 ACM

Energy-aware Error Control Coding for Flash Memories

Veera Papirla
Arizona State University, Tempe
AZ 85287, USA
vpapirla@asu.edu

Chaitali Chakrabarti
Arizona State University, Tempe
AZ 85287, USA
chaitali@asu.edu

ABSTRACT

The use of Flash memories in portable embedded systems is ever increasing. This is because of the multi-level storage capability that makes them excellent candidates for high density memory devices. However, cost of writing or programming Flash memories is an order of magnitude higher than traditional memories. In this paper, we design an algorithm to reduce both average write energy and latency in Flash memories. We achieve this by reducing the number of expensive '01' and '10' bit-patterns during error control coding. We show that the algorithm does not change the error correction capability and moreover improves endurance. Simulations results on representative bit-stream traces show that the use of the proposed algorithm saves, on average, 33% of write energy and 31% of latency of Intel MLC NOR Flash memory, and improves the endurance by 24%.

Categories and Subject Descriptors

B.3.2 [**Memory Structures**]: Mass storage; C.4 [**Performance of Systems**]: Fault tolerance

General Terms

Algorithms, Design, Performance

Keywords

Flash memories, Error Control Coding, Low-power design, Endurance

1. INTRODUCTION

Flash memory devices are the most popular form of nonvolatile storage in embedded systems today. They are widely used for multimedia data storage such as memory sticks in digital cameras, in MP3 players, etc. As the multi-level storage per Flash memory cell continues to improve, it is even being considered as a general purpose memory solution such as disk caches [1].

Permission to make digital or hard copies of part or all of this work for personal or classroom use is granted without fee provided that copies are not made or distributed for profit or commercial advantage and that copies bear this notice and the full citation on the first page. To copy otherwise, to republish, to post on servers or to redistribute to lists, requires prior specific permission and/or a fee.
DAC'09, July 26-31, 2009, San Francisco, California, USA

Table 1: Delay and Energy measurements for Intel MLC NOR 28F256L18 Flash memory [2].

Write bit pattern	Time	Energy
00	110.00 μs	4.738 μJ
01	644.23 μs	29.531 μJ
10	684.57 μs	31.194 μJ
11	24.93 μs	0.752 μJ
word read	0.02 μs	0.001 μJ

In many modern portable embedded systems, memory devices consume significant portion of the total energy. Reducing the energy consumption of memory is therefore crucial to extending the operation lifetime of portable embedded systems. While the read energy and latency of Flash memories are same as traditional memories [2], the write energy and latency are an order of magnitude higher. In fact, the write energy and latency depend on the data values that have to be written. Table 1 gives the write latency and energy involved for all four bit patterns for one of the most popular 4-level NOR Flash memories from Intel [2]. Similar trends are present in Samsung Flash memories as well. Table 1 clearly shows the large difference in writing '01' & '10' bit-patterns compared to '00' & '11' bit-patterns. This feature was exploited in [2] along with prefix-coding to reduce write energy and latency for storing multimedia files.

Almost all Flash memories have Error Control Coding (ECC) to retrieve the correct information bits in case of error. ECC is used to generate redundant bits which along with the information bits form the codeword to be written into memory [3]. In multi-level Flash memories, the probability of error increases because of closer voltage levels assigned to consecutive logic states. The most widely used ECC in Flash memories are Hamming codes and Reed-Solomon codes [4], [5].

In this paper, we propose a method that builds on top of the ECC in Flash memories to reduce the number of '01' & '10' bit-patterns that are stored in memory. We do this by modifying the codeword if the number of '01' & '10' bit-patterns is large. We formally show that the proposed modification does not change the error correction capability of the ECC. Further more, the proposed algorithm can be used on top of any ECC. We show that this method results in significant reduction in the energy and latency of 'write' operations. Simulation results on representative bit-stream traces for a 4-level Intel MLC NOR Flash memory, show that this method, on average, reduces the energy by 33% and latency by 31%. Finally, the proposed method increases the

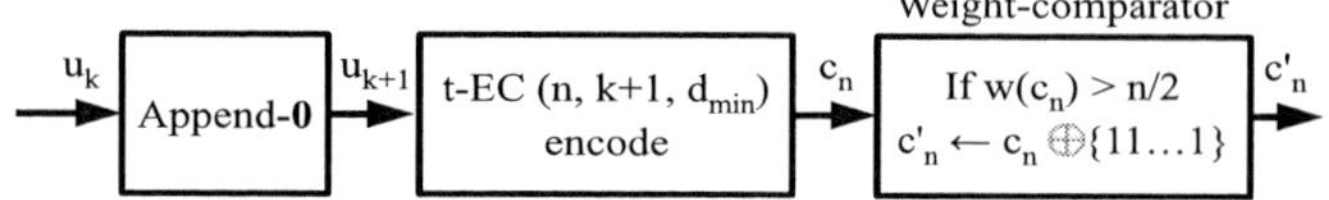

Figure 1: Block diagram of encoder for *Weight-reduction algorithm.*

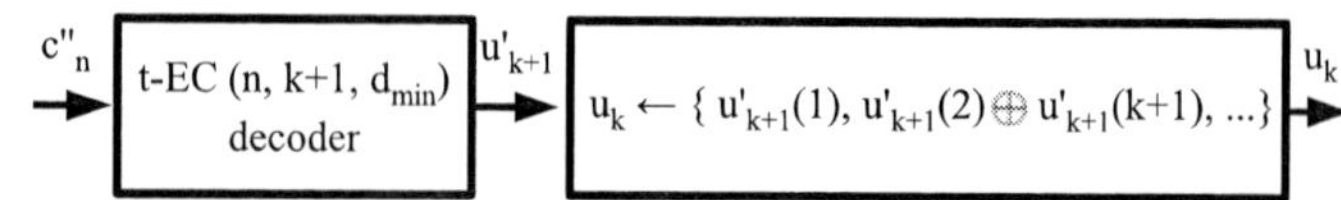

Figure 2: Block diagram of decoder for *Weight-reduction algorithm.*

endurance or lifetime of the memory by 24%. The overhead of this method is fairly small. For NOR Flash memory with block size 512 bits, the overhead is mostly contributed by a Digital Majority Voter (DMV) which cost 340ps in latency and 2.7pJ in energy. In contrast, the method in [2] requires an additional 52% memory for reducing 41% of the write energy.

The rest of the paper is organized as follows. We describe the original weight reduction that was designed for asymmetric codes in Section 2 and show how the algorithm can be modified to reduce the number of '01' & '10' bit-patterns during ECC in Section 3. We prove that the error correction performance of the proposed method is the same as the original ECC in the Appendix. We describe the circuit-level overhead of implementing the proposed method in Section 4. Simulation results showing the write energy and latency savings as well as potential increase in endurance are presented in Sections 5 and 6. The paper is concluded in Section 7.

Following are the definitions and notations which will be frequently used in the next few sections:

Definition 1. If $\mathbf{c_n}$ is an n-bit codeword of a code C, then *weight of a codeword*, $w(\mathbf{c_n})$, is the number of 1s in $\mathbf{c_n}$.

Definition 2. t-EC (n, m, d_{min}) code stands for a symmetric ECC which can correct t errors in an n-bit codeword with minimum Hamming distance of d_{min}, where $d_{min} \geq 2t + 1$. The number of information bits is m.

2. BACKGROUND: WEIGHT-REDUCTION ALGORITHM

In this section, we describe the *Weight-reduction algorithm* for reducing the *average weight* of a code. This algorithm was originally proposed in [6] for reducing weight in the design of asymmetric codes. We modify and apply this algorithm to reduce the write energy and latency for multi-level Flash memories.

2.1 Encoding

The block diagram of the encoder is given in Figure 1. The input to the encoder is a k-bit information vector $\mathbf{u_k}$ which is appended with bit '0' called the *Inverting bit* to generate the $(k + 1)$-bit vector $\mathbf{u'_{k+1}}$. The vector $\mathbf{u'_{k+1}}$ is input to a t-EC $(n, k+1, d_{min})$ code which produces $\mathbf{c_n}$, an n-bit codeword. $\mathbf{c_n}$ is input to the *Weight-comparator block* which inverts the bit-vector $\mathbf{c_n}$ if the following condition is satisfied: If the weight of the codeword, $w(\mathbf{c_n})$ is greater than half the number of bits in the codeword $\mathbf{c_n}$, i.e., if $w(\mathbf{c_n}) > \frac{n}{2}$, the bit-vector $\mathbf{c_n}$ is inverted, i.e., $\mathbf{c'_n} = \mathbf{c_n} \oplus \mathbf{f_n}$, where $\mathbf{f_n}$ is the n-bit vector $\{11...1\}$. Otherwise, if $w(\mathbf{c_n}) \leq \frac{n}{2}$, $\mathbf{c'_n}$ is the same as $\mathbf{c_n}$.

2.2 Decoding

The block diagram for the corresponding decoder is given in Figure 2. The input is an n-bit vector $\mathbf{c''_n}$ which is fed to the decoder of t-EC $(n, k + 1, d_{min})$ ECC. $\mathbf{c''_n}$ could be different from $\mathbf{c'_n}$ because of errors in the system. The decoder outputs a $(k + 1)$-bit vector $\mathbf{u'_{k+1}}$, whose $(k+1)^{th}$ bit contains the information about the conditional inversion of codeword $\mathbf{c_n}$. The original k-bit information vector $\mathbf{u_k}$ can be obtained by doing binary bit-wise XOR of the first k bits with the $(k + 1)^{th}$ bit of $\mathbf{u'_{k+1}}$. In other words,

$$\mathbf{u_k} = \{u'_{k+1}(1) \oplus u'_{k+1}(k+1), \ldots, u'_{k+1}(k) \oplus u'_{k+1}(k+1)\}$$

3. APPLICATION TO FLASH MEMORIES

In this section, we show how the *Weight-reduction algorithm* in Section 2 can be used to reduce the number of '01' & '10' bit-patterns in the data that is written in Flash memories. This algorithm is built on top of existing ECC in Flash memories. We describe the encoding and decoding process followed by simulation results comparing its error performance with that of the original ECC.

3.1 Encoding

The block diagram of the encoder for the modified *Weight-reduction algorithm* is given in Figure 3. The first two blocks are the same as in Figure 1. In addition, here a $\frac{n}{2}$-bit vector $\mathbf{x_{n/2}}$ is created, where $x_{n/2}(i) = c_n(2i - 1) \oplus c_n(2i)$, $1 \leq i \leq \frac{n}{2}$. Thus a '1' in the vector $\mathbf{x_{n/2}}$ corresponds to a '01' or '10' bit-pattern in $\mathbf{c_n}$. The modified *Weight-reduction algorithm* conditionally modifies $\mathbf{c_n}$ if the number of 1s in $w(\mathbf{x_{n/2}})$ is $> \frac{n}{4}$. The modification involves bit-wise XOR of $\mathbf{c_n}$ with a constant vector $\mathbf{g_n}$, where $\mathbf{g_n}$ is a codeword of t-EC $(n, k + 1, d_{min})$ code. $\mathbf{g_n}$ corresponds to the k-bit input $\{0101 \ldots 01\}$, when k is odd and $\{1010 \ldots 10\}$, when k is even. This choice of $\mathbf{g_n}$ guarantees that $\mathbf{c'_n}$ will always contain less number of '01' & '10' bit-patterns than $\mathbf{c_n}$. It also sets the $(k+1)^{th}$ bit of $\mathbf{c'_n}$ to '1' if there is a modification. The $(k+1)^{th}$ bit can be used to decode the modified vector back at the decoder.

3.1.1 Procedure

Step-1: Let $\mathbf{u_k}$ be the input vector. Appending bit-0 gives the $(k + 1)$-bit vector $\mathbf{u_{k+1}}$.

Step-2: The t-EC $(n, k+1, d_{min})$ ECC gives an n-bit codeword $\mathbf{c_n}$ corresponding to the input vector $\mathbf{u_{k+1}}$.

Step-3: Compute $\mathbf{x_{n/2}}$: $x_{n/2}(i) = c_n(2i - 1) \oplus c_n(2i)$, $1 \leq i \leq \frac{n}{2}$.

Step-4: If $w(\mathbf{x_{n/2}}) > \frac{n}{4}$ then $\mathbf{c'_n} = \mathbf{c_n} \oplus \mathbf{g_n}$, else $\mathbf{c'_n} = \mathbf{c_n}$. $\mathbf{g_n}$ is a codeword which corresponds to the input $\{0101 \ldots 01\}$, when k is odd, and $\{1010 \ldots 10\}$, when k is even.

3.1.2 Example

978-1-60558-497-3/09 $25.00 © 2009 ACM

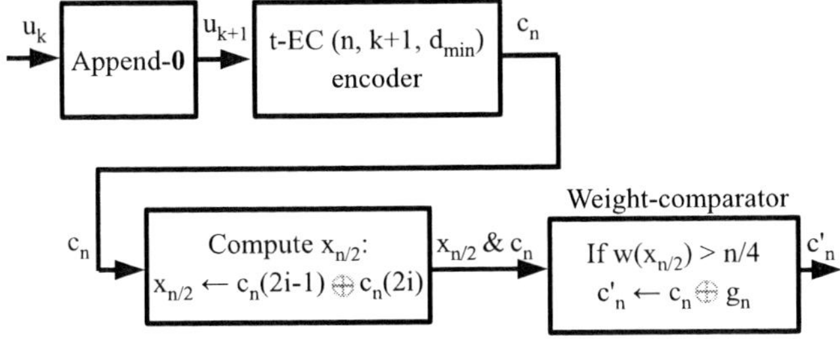

Figure 3: Block diagram of the encoder for MLC Flash memories.

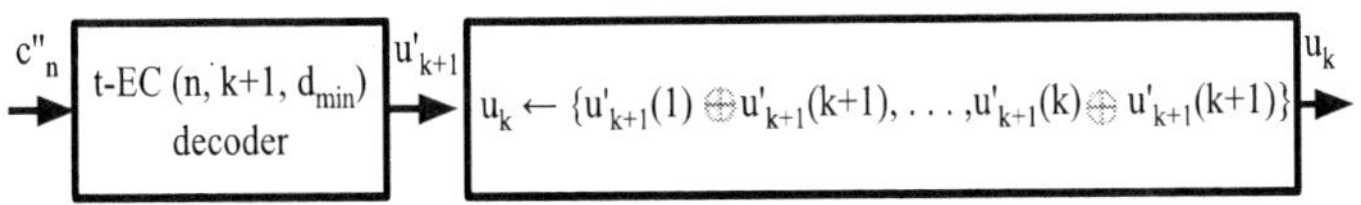

Figure 4: Block diagram of the decoder for MLC Flash memories.

Step-1: Let $\mathbf{u_3} = \{101\}$ be the input vector. Appending zero gives $\mathbf{u_4} = \{1010\}$.

Step-2: The codeword $\mathbf{c_8}$ is generated using 2-EC $(8, 4, 4)$ extended Hamming code, $\mathbf{c_8} = \{10101010\}$.

Step-3: $\mathbf{x_4}$ is $\{1111\}$.

Step-4: Since $w(\mathbf{x_4}) = 4 > \frac{8}{4}$, $\mathbf{c'_8} = \mathbf{c_8} \oplus \mathbf{g_8}$. Since $k = 3$, $\mathbf{g_8}$ is the codeword $\{01010101\}$ of 2-EC $(8, 4, 4)$ extended Hamming code corresponding to the input $\{0101\}$. Note that $\mathbf{c'_8} = \{10101010\} \oplus \{01010101\} = \{11111111\}$ has only '11' bit-patterns, which is significantly less costly to write in Flash memory.

3.2 Decoding

The block diagram of the encoder for the modified *Weight-reduction algorithm* is given in Figure 4. The n-bit vector $\mathbf{c''_n}$ is decoded to $\mathbf{u'_{k+1}}$ using the decoder for t-EC $(n, k+1, d_{min})$ code. The $(k+1)^{th}$ bit of the bit-vector $\mathbf{u'_{k+1}}$ has the information about the conditional modification at the encoder. When k is odd, the original k-bit information vector $\mathbf{u_k}$ is obtained by computing the XOR of every 2nd bit with the $(k+1)^{th}$ bit, starting with the 2nd bit. This is because at the encoder, the codeword $\mathbf{g_n}$ corresponds to the information vector $\{0101\ldots01\}$. When k is even, $\mathbf{g_n}$ corresponds to the information vector $\{1010\ldots10\}$, and the original k-bit information vector $\mathbf{u_k}$ is obtained by computing the XOR of every 2nd bit with the $(k+1)^{th}$ bit, starting with the 1st bit.

3.2.1 Procedure

Step-1: Let $\mathbf{c''_n}$ be the vector read from the memory. Decoding using the t-EC $(n, k+1, d_{min})$ code gives the error corrected vector $\mathbf{u'_{k+1}}$.

Step-2: Compute $\mathbf{u_k}$: When k is odd,
$$\mathbf{u_k} = \{u'_{k+1}(1), \ u'_{k+1}(2) \oplus u'_{k+1}(k+1), u'_{k+1}(3), \ldots, u'_{k+1}(k)\}.$$
When k is even,
$$\mathbf{u_k} = \{u'_{k+1}(1) \oplus u'_{k+1}(k+1), u'_{k+1}(2), u'_{k+1}(3) \oplus u'_{k+1}(k+1), \ldots, u'_{k+1}(k) \oplus u'_{k+1}(k+1)\}.$$

3.2.2 Example

Step-1: Let the vector read from the memory be $\mathbf{c''_8} =$

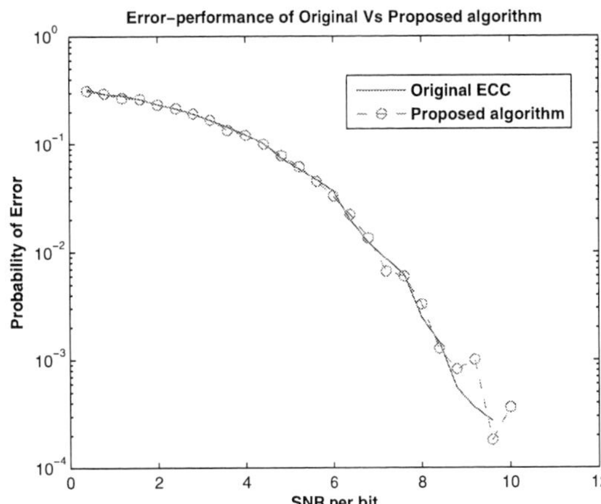

Figure 5: Error performance of the original ECC and the proposed modified *Weight-reduction algorithm*.

$\{11111011\}$ with an error in the 6th bit. The 2-EC $(8, 4, 4)$ extended Hamming code decoder corrects the error in the 6th bit and outputs $\mathbf{u'_4} = \{1111\}$.

Step-2: $\mathbf{u_3} = \{u'_4(1), u'_4(2) \oplus u'_4(4), u'_4(3)\} = \{101\}$.

3.3 Error Performance

We show through simulations that the error performance of the proposed modified *Weight-reduction algorithm* is the same as that of original ECC. Specifically, we used a (15, 11, 3) Hamming code as the original ECC. Monte Carlo simulations for 10^5 iterations were done in MATLAB. Gaussian random noise was used for generating errors. The algorithm's error performance with SNR per bit is compared with that of the original ECC. The error performance of the original ECC and that of the proposed algorithm are given in Figure 5. SNR per bit is $20 log \frac{Energy\ per\ bit}{var}$, where var is the variance of the Gaussian random noise. We see that the error performance of the proposed algorithm is exactly the same as that of the original ECC. This is formally proved in the Appendix.

4. IMPLEMENTATION ISSUES

In this section, we discuss the overhead involved in implementing the modified *Weight-reduction algorithm* in the ECC block.

4.1 Implementation of Encoder

From Figure 3, we see that the encoder for the modified *Weight-reduction algorithm* has four blocks. Out of these, the t-EC $(n, k+1, d_{min})$ encoder block for ECC already exists in the write interface of the Flash memory, so only the first, third and the fourth blocks account for the overhead.

The first block, *Append-0* adds '0' bit to the input. The implementation costs in terms of delay, area and power for this block is negligible. The third block calculates $x_{n/2}$ from $\mathbf{c_n}$ and has an overhead of $\frac{n}{2}$-XOR gates. The fourth block which checks *if* $w(\mathbf{x_{n/2}}) > \frac{n}{4}$ contributes the most to the overhead. It requires a counting circuit which counts the number of 1s in the vector $\mathbf{x_{n/2}}$. The circuit outputs logic 1 if the number of 1s are greater than the number of 0s and is referred to as majority voter. There are digital and analog

978-1-60558-497-3/09 $25.00 © 2009 ACM

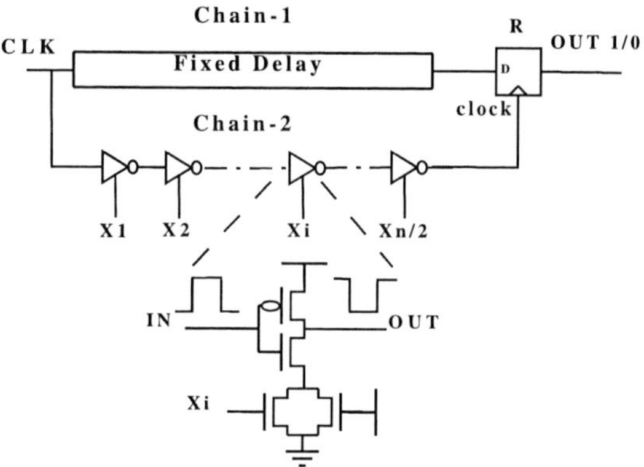

Figure 6: DMV circuit in the Weight-comparator.

majority voters in the literature. We focus our study on the digital majority voter (DMV).

4.1.1 Digital Majority Voter (DMV)

We propose a new DMV, the circuit of which is given in Figure 6. The circuit is based on a design presented in [7]. The complicated dynamic latching of the original design has been replaced by a simple register. The basic operation of the proposed DMV in Figure 6 is as follows. The input CLK signal is delayed through two paths. One path is through a chain of variable-delay inverters whose delay is controlled by the ith bit of the vector $\mathbf{x_{n/2}}$. The other path has fixed delay; here it is chosen as the delay if $\frac{n}{4}$ '1's were fed to the chain of variable-delay inverters.

To understand the working of the proposed circuit, let us assume Chain-1 has more delay than Chain-2. Then at register R, the rising edge of CLK (through Chain-1) at input D arrives later than the rising edge of CLK (through Chain-2) at the clock input of register R. That means the rising edge of input D of R is shifted right with respect to the rising edge of CLK. So, in this case R will sample a '0'. If Chain-2 has more delay than Chain-1, R will sample a '1'.

Two-stage DMV implementation: For a MLC NOR Flash memory with block size 512 bits [2], the number of parity bits required is $2x \log 512 = 18 bits$ according to method in [8], and the codeword $n = 512 + 18 = 530$ bits. Thus, we need a DMV of size 265. A 265 long chain is not very realistic, so we propose breaking the DMV into two stages. All two stage implementations result in some loss in accuracy. We considered the following decompositions: 1) Sixty six 4-bit DMVs in the first stage followed by a 67-bit DMV in the second stage, 2) thirty three 8-bit DMVs in the first stage followed by a 34-bit DMV in the second stage, and 3) Sixteen 16-bit DMVs in the first stage followed by a 25-bit DMV in the second stage. Of these, the second combination has the performance closest to a single stage 265-bit DMV. Thus for the NOR flash with $k + 1 = 512$ bits, this combination gave incorrect outputs $\approx 5\%$ of the time. For NAND flash with $k + 1 = 4096$ bits, $n = 4096 + 24 = 4120$ bits, and a two-stage DMV with thirty two 64-bit DMVs in the first stage followed by a 44-bit DMV in the second stage gave the best result with $\approx 7\%$ loss in accuracy.

The proposed DMV circuit includes a flip-flop and so metastability issues can arise. We propose using the 'metastability detector' in Razor flip-flop design [9] to detect cases when the input signal transition occurs close to the clock

Table 2: Latency and energy overhead of DMV implementations for various sized delay chains.

Size	Delay	Energy
8	0.8 ns	97 fJ
16	0.12 ns	163 fJ
32	0.26 ps	228 fJ
48	0.41 ps	294 fJ
64	0.60 ps	341 fJ

Table 3: Latency and energy overhead of the blocks in the $n = 530$, $k + 1 = 512$ encoder.

	Delay	Energy
First block	0 ps	0 pJ
Third block	11 ps	0.834 pJ
Fourth block	340 ps	2.7 pJ

edge. For such cases, the DMV or the flip-flop's output can be taken to be 1 or 0. This does not affect the performance of the weight reduction algorithm much since metastability occurs (if at all) when the weight of the input vector is close to $\frac{n}{4}$.

4.1.2 Simulations

The simulations for the encoder blocks are done using HSPICE. Bulk High-performance (HP) 45 nm Predictive Technology Model (PTM) was used. Table 2 gives delay and energy overhead for the DMV implementations.

The total energy and delay overheads for the three blocks for $n = 530$, $k + 1 = 512$ are tabulated in Table 3. The delay of the two-stage DMV (thirty three 8-bit DMVs followed by a 34-bit DMV) is $340ps$. Thus, the total delay and energy overheads are $351ps$ and $3.534pJ$ respectively. These low overheads make the implementation of the modified *Weight-reduction algorithm* a very attractive method.

The reliability of the proposed DMV circuits was also tested. The circuits were able to tolerate a $\pm 15\%$ variation from nominal 1.0 V and $\pm 10\%$ variation in transistor length for the 45 nm technology node.

4.2 Implementation of Decoder

In the decoder, the t-EC $(n, k+1, d_{min})$ Hamming decoder block already exists in the read interface of the Flash memory. The overhead is only due to the second block in Figure 4 which computes XOR of $(k + 1)^{th}$ bit with every other bit. So the decoder overhead is due to $\frac{k}{2}$ parallel XOR gates. Simulations show that the total delay and energy overheads are $11ps$ and $256fJ$ respectively, for $k + 1 = 512$.

5. SAVINGS IN ENERGY AND LATENCY

In this section, we first walk through an example for calculating write energy and latency savings. We then give detailed simulation results of the write energy and latency savings of the algorithm for various bit-streams derived from real computing sources.

5.1 Illustrative example

Let $\{101110100101\}$ be the 12 bits that have to be written into Flash memory. Let the ECC be 2-EC $(8, 4, 4)$ extended Hamming code.

978-1-60558-497-3/09 $25.00 © 2009 ACM

Table 4: Energy and latency savings for various bit-streams for Intel MLC NOR 28F256L18 Flash memory with $n = 530$, $k + 1 = 512$. In the notation A(B), A corresponds to the proposed two-stage DMV implementation and B corresponds to the ideal single-stage DMV implementation.

Bit-stream	Size	Energy savings	Latency savings
Mp3	10 MB	33% (39%)	32% (38%)
Linux	699 MB	35% (39%)	33% (37%)
Adobe PDF	2 MB	31% (36%)	30% (34%)
Images (.jpg)	5.6 MB	32% (37%)	31% (35%)
Speech (.wav)	8 MB	36% (42%)	35% (41%)
Gaussian	100 MB	30% (38%)	28% (36%)

Table 5: Reduction in number of 01 & 10 bit-patterns for NAND Flash memory with $k + 1 = 4096$. In the notation A(B), A corresponds to the proposed two-stage DMV implementation and B corresponds to single-stage DMV implementation.

Bit-stream	Size	Reduction
Mp3	10 MB	49% (54%)
Linux	699 MB	48% (54%)
Adobe PDF	2 MB	45% (50%)
Images (.jpg)	5.6 MB	47% (51%)
Speech (.wav)	8 MB	51% (55%)
Gaussian	100 MB	47% (51%)

In the traditional case, the bit-stream is divided into 3 blocks with 4 bits per block. The 3 blocks: {1011}, {1010} and {0101} are encoded by the (8,4,4) extended Hamming code to give codewords {10110100}, {10101010} and {01010101} respectively. Note that there is one '00', five '01', five '10' and one '11' bit-patterns. Using Table 1 for NOR Flash memory, the write energy required is $1 \times 4.738 + 5 \times 29.531 + 5 \times 31.194 + 1 \times 0.752 = 309.115\ \mu J$ and the write latency is $1 \times 110.00 + 5 \times 644.23 + 5 \times 684.57 + 1 \times 24.93 = 6778.93\ \mu s$.

If the proposed modified *weight reduction algorithm* is to be applied, the bit-stream is divided into 4 blocks with 3 bits per block to accommodate for the inversion bit in each block. The 4 blocks: {101}, {110}, {100} and {101} are appended with 0s to give {1010}, {1100}, {1000} and {1010} respectively, and encoded by the extended Hamming code to give codewords {10101010}, {11001100}, {10000111} and {10101010} respectively. The *Weight-comparator block* conditionally XORs the codewords with $g_8 = \{01010101\}$. In this case, only the first and the fourth vectors were XORed and the final vectors are {11111111}, {11001100}, {10000111} and {11111111}. Since there are three '00', one '01', one '10' and eleven '11' bit-patterns to be written into memory, using Table 1, the write energy required is $83.211\ \mu J$ and the write latency is $1933\ \mu s$. In this particular example, the proposed modified *weight reduction algorithm* reduces write energy and latency by 73% and 71% respectively.

5.2 Examples from real bit-streams

Industry standards for ECCs used in NAND Flash memories are described in [8], [10] and [11]. Typically NOR Flash memories do not have on-chip ECC but MLC NOR Flash now demand use of strong ECC protection due to higher probability of error [12], [11]. We consider the Hamming code as the ECC to demonstrate the write energy and latency savings in case of NOR Flash memory. The block size of the Hamming code is chosen as $k + 1 = 512$ bits.

Bit-streams from real computing sources have been chosen as input bit-streams. These include compressed files such a regular mp3 file of size 10 MB, Ubuntu-8.10 Linux zImage file of size 699 MB, an Adobe PDF of size 5.6 MB and four 256x256 JPEG images. We have also used an uncompressed .wav file and a bitstream of size 100 MB generated from concatenating 32-bit numbers that are randomly drawn from a Gaussian distribution.

We will discuss the case with mp3 file as bit-stream. Our simulations show that without the modified *Weight-reduction algorithm*, the mp3 file requires writing of 63 '00', 71 '01', 74 '10' and 57 '11' bit-patterns per 530 bits on an average over 10 MB. The write energy is $4.71\ mJ$ and latency is $104\ ms$. Using the modified *Weight-reduction algorithm*, the mp3 file requires writing of roughly 90 '00', 43 '01', 42 '10' and 95 '11' bit patterns per 530 bits on an average. The write energy is $3.16\ mJ$ and latency is $70.72\ ms$ per 530 bits resulting in significant savings.

The write energy and latency savings for Intel MLC NOR 28F256L18 Flash memory are tabulated in Table 4. The difference between two-stage and single-stage DMV implementation is due to loss of accuracy as described in Section 4.1.1. We note a loss of just 5% savings due to using the two-stage DMV. Simulations show similar results for other Flash memories like Samsung MLC NOR KAK38200AM.

Next we present results for NAND Flash memory with block size of 4096 bits [8], [10]. Table 5 gives the percentage reduction in 01s & 10s for the two-stage and single-stage DMV implementations. We did not have access to NAND Flash energy and latency measurement values and could not provide the savings results. However, we expect the savings to be significant for NAND Flash memories as well.

6. IMPROVEMENT IN ENDURANCE

In a 4-level NOR Flash memory, of the four states, '11' is called erased state since it does not require programming the memory cell, while '00', '01' & '10' are called programmed states. The endurance of a Flash memory cell strongly depends upon the number of programming operations on the cell [13], [14]. The use of modified *Weight-reduction algorithm* reduces the number of '01' & '10' bit-patterns which translates to reduction in the number of program cycles and improvement in the endurance of the Flash memory cells. Table 6 lists the average improvement in endurance for the different bit-streams. Improvement in endurance is calculated as the reduction in the average number of program cycles. We conclude from Table 6 that the endurance or life time of Flash memory cells increases by nearly 24%.

7. CONCLUSION

The energy and latency for writing data in Flash memories are an order of magnitude higher than traditional memories. Moreover, multi-level Flash memories suffer from endurance and reliability issues. In this paper, we design an algorithm to reduce the average write energy and latency in Flash memories as well as improve the endurance. The algorithm is built on top of existing ECC in Flash memories and effectively reduces the number of '01' & '10' bit-patterns in

978-1-60558-497-3/09 $25.00 © 2009 ACM

[12] B. Godard and et. al., "Hierarchical Code Correction and Reliability Management in Embedded NOR Flash memories," *European Test Symposium*, pp. 84–90, May 2008.

[13] "IEEE Standard Definitions and Characterization of Floating Gate Semiconductor Arrays," Feb 1999.

[14] N. Mielke and et. al., "Bit error rate in NAND Flash memories," *IEEE International Reliability Physics Symposium*, pp. 9 – 19, April-May 2008.

Table 6: Improvement in endurance for the various bit-streams.

Bit-stream	Size	Improvement in endurance
Mp3	10 MB	26%
Linux	699 MB	25%
Adobe PDF	2 MB	22%
Images (.jpg)	5.6 MB	22%
Speech (.wav)	8 MB	28%
Gaussian	100 MB	24%

the codeword that is written into the memory. The overhead of this algorithm is contributed primarily by a digital majority voter (DMV). The use of the proposed modified *Weight-reduction algorithm* saves 33% and 31% of write energy and latency, respectively, for Intel NOR Flash memories. It also improves the endurance of the Flash memory by 24%.

ACKNOWLEDGEMENTS

This work was supported in part by a grant from the National Science Foundation CSR-EHS 0615135.

REFERENCES

[1] T. Kgil, D. Roberts, and T. Mudge, "Improving NAND flash based disk caches," *35th International Symposium on Computer Architecture*, pp. 327–38, 2008.

[2] Y. Joo and et. al., "An energy characterization platform for memory devices and energy-aware data compression for multilevel-cell Flash memory," *ACM Transactions on Design Automation of Electronic Systems*, vol. 13, July 2008.

[3] S. Lin and D. J. Costello, *Error Control Coding*. Prentice Hall, 2 ed., 2004.

[4] S. Gregori and et. al., "On-chip error correcting techniques for new-generation Flash memories," *Proceedings of the IEEE*, vol. 91, April 2003.

[5] B. Chen, X. Zhang, and Z. Wang, "Error correction for multi-level NAND flash memory using Reed-Solomon codes," *IEEE Workshop on Signal Processing Systems*, pp. 94–99, November 2008.

[6] J. Bruck and M. Blaum, "Some new EC/AUED Codes," *The Nineteenth International Symposium on Fault-Tolerant Computing*, pp. 208–15, Nov 1989.

[7] M. Fujino and V. Moshnyaga, "An efficient Hamming distance comparator for low-power applications," *9th International Conference on Electronics, Circuits and Systems*, vol. 2, pp. 641–644, September 2002.

[8] "Hamming codes for NAND Flash memory devices," *http://www.micron.com/products/nand/technotes/*.

[9] D. Ernst and N. S. Kim, "Razor: A Low-Power Pipeline Based on Circuit-Level Timing Speculation," *36th Annual International Symposium on Microarchitecture*, pp. 7–18, Dec 2003.

[10] S. Chen, "What types of ECC Should be used on Flash memory?," *Application Note, http://www.eetasia.com*.

[11] F. Sun and et. al., "Design of on-chip error correction systems for multilevel NOR and NAND Flash memories," *IET, Circuits, Devices and Systems*, vol. 1, pp. 241–249, June 2007.

APPENDIX:PROOF OF ERROR-CORRECTION CAPABILITY

In this section, we prove that the proposed modified *Weight-reduction algorithm* does not affect the error correction capability of the ECC. Specifically, we show that the minimum Hamming distance d_{min} of the proposed system is the same as the original ECC.

PROOF. Let t-EC $(n, k+1, d_{min})$ be a t-error correcting code $\mathbf{C}$ used in step-2 of the modified *Weight-reduction algorithm*. The minimum Hamming distance $d_{min} \geq 2t+1$ by definition. The input vector in step-2 of the algorithm is the vector $\mathbf{u_{k+1}} = \{\mathbf{u_k}, 0\}$, where $\mathbf{u_k}$ corresponds to the k information bits. Since $\mathbf{C}$ is a code with minimum Hamming distance d_{min}, all the codewords in the code $\mathbf{C}$, except the all-0 vector $\mathbf{0_n}$, have a minimum weight of at least d_{min},

$$w(\mathbf{c_n}) > d_{min} \text{ for any codeword } \mathbf{c_n} \in \mathbf{C}$$

In the encoding algorithm, $\mathbf{g_n}$ is a codeword of $\mathbf{C}$ corresponding to k-bit input $\{0101\ldots01\}$ when k is odd and corresponding to the input $\{1010\ldots10\}$ when k is even. Step-4 of the encoding algorithm is given by

$$if\ w(\mathbf{x_{n/2}}) > \frac{n}{4}\ then\ \mathbf{c'_n} = \mathbf{c_n} \oplus \mathbf{g_n}\ else\ \mathbf{c'_n} = \mathbf{c_n}$$

Alternately, *if* $w(\mathbf{x_{n/2}}) > \frac{n}{4}$ *then* $\mathbf{c'_n} = \mathbf{c_n} \oplus b.\mathbf{g_n}$, where b is 1 or 0 depending on condition $w(\mathbf{x_{n/2}}) > \frac{n}{4}$, and $b.\mathbf{g_n}$ is obtained by 'AND'ing bit b with each of the bits in the vector $\mathbf{g_n}$.

Let $\mathbf{y'_n}$ & $\mathbf{z'_n}$ be any two output vectors corresponding to the any two input vectors $\mathbf{y_n}$ & $\mathbf{z_n} \in \mathbf{C}$. The Hamming distance between the two output vectors $\mathbf{y'_n}$ and $\mathbf{z'_n}$ can be written as

$$d(\mathbf{y'_n}, \mathbf{z'_n}) = d(\mathbf{y_n} \oplus e.\mathbf{g_n}, \mathbf{z_n} \oplus f.\mathbf{g_n})$$

where e and f are 1 if the number of '01' & '10' bit-patterns is larger than $\frac{n}{4}$ in $\mathbf{y_n}$ and $\mathbf{z_n}$ respectively. Since the Hamming distance between any two codewords $\mathbf{p_n}, \mathbf{q_n}$ of a code is the weight of the codeword $\mathbf{p_n} \oplus \mathbf{q_n}$, $d(\mathbf{y'_n}, \mathbf{z'_n})$ can be expressed as

$$d(\mathbf{y'_n}, \mathbf{z'_n}) = w(\mathbf{y_n} \oplus e.\mathbf{g_n} \oplus \mathbf{z_n} \oplus f.\mathbf{g_n})$$

Since $\mathbf{r_n} = \mathbf{y_n} \oplus e.\mathbf{g_n} \oplus \mathbf{z_n} \oplus f.\mathbf{g_n}$ is a linear combination of the codewords of the code $\mathbf{C}$ in the Galois Field $\mathbf{GF(2)}$, the resultant vector $\mathbf{r_n}$ is also a codeword of the code $\mathbf{C}$. By definition, $w(\mathbf{r_n}) > d_{min}$. Thus,

$$d(\mathbf{y'_n}, \mathbf{z'_n}) > d_{min}$$

Thus, the minimum Hamming distance between any two output vectors of the encoder is d_{min}, and the error correcting capability remains the same as the original ECC. □

978-1-60558-497-3/09 $25.00 © 2009 ACM

PDRAM: A Hybrid PRAM and DRAM Main Memory System

Gaurav Dhiman
gdhiman@cs.ucsd.edu

Raid Ayoub
rayoub@cs.ucsd.edu

Tajana Rosing
tajana@ucsd.edu

Department of Computer Science and Engineering
University of California, San Diego
La Jolla, CA 92093-0404

ABSTRACT

In this paper, we propose PDRAM, a novel energy efficient main memory architecture based on phase change random access memory (PRAM) and DRAM. The paper explores the challenges involved in incorporating PRAM into the main memory hierarchy of computing systems, and proposes a low overhead hybrid hardware-software solution for managing it. Our experimental results indicate that our solution is able to achieve average energy savings of 30% at negligible overhead over conventional memory architectures.

Categories and Subject Descriptors

B.7.1 [**Integrated Circuits**]: Memory Technologies

General Terms

Design, Experimentation, Performance

Keywords

Phase Change Memory, Energy Efficiency, Memory management

1. INTRODUCTION

Power consumption is a major concern in the design of modern systems. Most of the recent research in power management has focused more on dynamic management of CPU power consumption, assuming it to be the most dominant contributor to system power consumption. However, recent studies [1] have shown that in modern systems CPU is no longer the primary energy consumer. Main memory has become a significant energy consumer, contributing to as much as 30-40% of total consumption on modern server systems.

In this paper, we propose a new approach for tackling the high levels of energy dissipation in main memory while minimizing the impact on performance. We introduce a new heterogeneous organization for main memory that is composed of DRAM and PRAM memories. The properties of PRAM that we leverage are its lower read access and standby power compared to DRAM while having a comparable throughput. However, the primary challenges in using PRAM include its lower write endurance (typical mean time to failure in the range of $10^9 - 10^{12}$ cycles), and the higher power cost of write accesses compared to DRAM. These properties motivate the use of a heterogeneous memory architecture consisting of both DRAM and PRAM, which we refer to as PDRAM, enabling exploitation of positive aspects of the respective memories.

To manage the PDRAM memory organization, we propose a hybrid hardware/software solution. In order to maintain reliability for PRAM (because of write endurance problem), we introduce a cost efficient book keeping hardware technique that stores the frequency of writes to PRAM at a page level granularity. On the software side, we propose an efficient operating system (OS) level page manager that utilizes the write frequency information provided by the hardware to perform uniform wear leveling across all the PRAM pages. Wear leveling refers to the process of prolonging the lifetime of erasable storage devices with endurance problems (like Flash, PRAM etc) by ensuring uniform usage/utilization of all the storage blocks/pages of the device. Furthermore, the page manager intelligently allocates/migrates pages across DRAM/PRAM in order to minimize the impact of wear leveling on performance. The benefits of using such a hybrid approach is that the hardware can maintain and track page level accesses at a very low cost, while the software (OS) can leverage the high level observability and policies for management of free pages across DRAM/PRAM. We experiment across benchmarks with varying memory access characteristics, and show that using the PDRAM organization, we can achieve as high as 37% energy savings at negligible performance overhead over comparable DRAM organizations. Furthermore, we show that it achieves better energy and performance efficiency compared to homogeneous PRAM based memory systems as well.

Overall, the *primary contributions* of our work are: 1) We outline and evaluate the challenges in incorporating PRAM as an alternative main memory technology. 2) We propose an architecture and system policies for managing a PRAM/DRAM based main memory system. 3) We present a thorough evaluation and discussion of the system's performance, benefits and overhead.

The rest of the paper is organized as follows. In Section 2, we provide an overview of the related work. In Section 3, we explain our memory organization and the solution to manage it for higher energy efficiency, reliability and performance. In Section 4, we describe the experimental evaluation and results before concluding in Section 5.

Permission to make digital or hard copies of part or all of this work for personal or classroom use is granted without fee provided that copies are not made or distributed for profit or commercial advantage and that copies bear this notice and the full citation on the first page. To copy otherwise, to republish, to post on servers or to redistribute to lists, requires prior specific permission and/or a fee.
DAC'09, July 26-31, 2009, San Francisco, California, USA

2. RELATED WORK

Existing memory power management research has primarily focused on DRAM based systems. In [12], the authors propose power aware page allocation algorithms for DRAM power management. They assume support in memory controller for fine grained bank level power control, and show that their allocation algorithms give

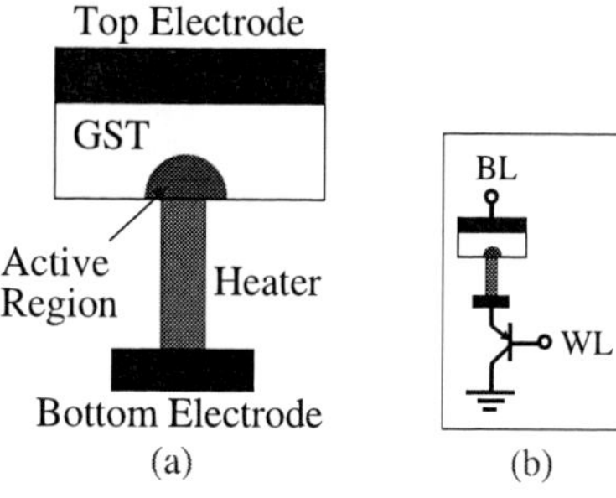

Figure 1: PRAM cell (a) and transistor (b)

greater opportunities for placing memory in low power modes. In [5, 9], the authors propose an OS level approach, where the OS maintains tables that map processes onto the memory banks they have their memory allocated in. This allows the OS to dynamically move unutilized DRAM banks into low power modes.

The possible use of alternative memory technology for improving energy efficiency has also been explored before. The authors in [10] propose NAND flash based page cache, which reduces the amount of DRAM required for system memory. This results in energy efficiency due to lower power consumption and higher density of NAND flash compared to DRAM. However, this approach is beneficial for more disk intensive applications, since flash is used only for page cache. In addition, flash also has endurance problems in terms of number of write cycles, which is tackled using wear leveling in the flash translation layer [14].

PRAM is an attractive alternative to flash, since: 1) Its endurance is higher by several orders of magnitude, 2) It is a RAM, and hence does not require an overhead of an erase before a write. In [13], the authors evaluate the challenges involved in architecting PRAM as a DRAM alternative. The authors in [15] propose a hybrid cache, that is composed of PRAM and SRAM for power savings. To solve the endurance problem of PRAM, they set a threshold on number of writes to the PRAM cache lines, beyond which they do not use those lines. However, the concern is that the typical level of conflicts and activity in cache could create reliability problems very soon on such a configuration. Extensive research has been done in modeling and understanding the basic characteristics of this technology [8, 11, 17–19].

3. DESIGN

3.1 PRAM/DRAM Background

The DRAM memory is organized as a grid of rows and columns, where each bit is stored in the form of charge in a small capacitor. As the charge can get exhausted due to leakage and frequent accesses, DRAM requires a consistent refresh operation to sustain its data. This results in a constant power consumption referred to as the *refresh* power. DRAM further consumes power for setting up its row and column for a physical address accessed, and also for closing a row if some other row needs to be accessed. This is referred to as the *activation/precharge* power. Additionally, it consumes power for the actual read/write accesses, and consistent standby power due to leakage and clock supply.

Unlike DRAM, PRAM is designed to retain its data even when the power is turned off. A PRAM cell stores information permanently in the form of the cell material state, which can be amorphous (low electrical conductivity) or crystalline (high electrical conductivity). A PRAM cell typically comprises of a chalcogenide alloy material (eg. $Ge_2Sb_2Te_5$ (GST)) and a small heater as shown

in Figure 1a. The cell can be addressed using a selection transistor (MOS or BJT) that is connected to the word-lines (WL) and bit-lines (BL) as illustrated in Figure 1b. To write to a PRAM cell, the GST state needs to be altered by injecting a large but fast current pulse (few 100ns) to heat up the GST active region. Consequently, PRAM write power is high compared to DRAM. For reading a PRAM cell, the power consumption is much lower, since no heating is involved. It is also lower than DRAM cell read power based on the measurements shown in [17]. Besides, PRAM consumes no refresh power, as it retains its information permanently, and [17] shows that it consumes much lower standby power due to its negligible leakage. However, PRAM has limited write endurance ($10^9 - 10^{12}$ cycles), which poses a reliability problem. Regarding access times, the access latency of random reads/writes on PRAM is slower compared to DRAM, although their read throughput is comparable. Thus considering all of these factors, PRAM is a promising candidate for energy savings because of its low read and standby power compared to DRAM.

3.2 Architecture

Based on the PRAM/DRAM characteristics described above, we observe that both the technologies have their respective pros and cons. This motivates us to propose a hybrid memory architecture, which consists of both DRAM and PRAM (PDRAM), for achieving higher energy efficiency. While PRAM provides low read and standby power, DRAM provides higher write endurance and lower write power.

The primary design challenge in managing a PDRAM system is to manage efficient wear leveling of PRAM pages to ensure its longer lifetime. For this purpose, we provide a hybrid hardware-software based solution. The hardware portion is based in the memory controller and manages the access information to different PRAM pages. The software portion is part of the operating system (OS) memory manager (referred to as the page manager), which performs the wear leveling by page swapping/migration. The components of the solution are described in detail below.

3.2.1 Memory controller

Figure 2 illustrates the various components and interactions of the PDRAM memory controller. The memory controller is aware of the partitioning of system memory between DRAM and PRAM. Based on the address being accessed, it is able to route requests to the required memory. To help wear leveling the PRAM, it maintains a map (access map in Figure 2) of the number of write accesses to it. This information is kept at a page level granularity, which is a function of the processor being used. For instance, the page size is 4KB for x86, 8KB for Alpha etc. We use page level granularity, since it is the unit of memory management, i.e. allocation and deallocation in the OS. If the number of writes to any PRAM page exceed a given threshold, then the controller generates a 'page swap' interrupt to the processor, and provides the page address. The OS then assumes the responsibility of handling this interrupt and performing page swapping as described in the next sub-section. The controller stores the map in the PRAM, for which it reserves space during bootup. The access map is maintained for the lifetime of the system, and after the first page swap interrupt, future interrupts are generated whenever the write access count becomes a multiple of the threshold. To maintain the map across reboots, it is stored on disk before the shutdown, and copied back into PRAM during the startup. To protect this data against crashes, it is synced with the disk periodically. If the write count for a page reaches the endurance limit (10^9 in the worst case), the controller generates a 'bad-page' interrupt for that page. This interrupt is also

978-1-60558-497-3/09 $25.00 © 2009 ACM

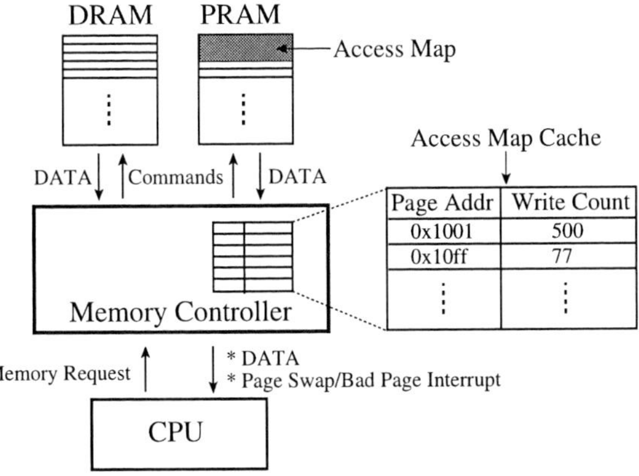

Figure 2: PDRAM Memory controller

handled by the page manager as described in the next section.

The enhancement of the controller incurs energy and memory overhead: 1) time and energy for updating PRAM access map; this involves extra accesses to PRAM, which causes extra power consumption. To avoid this overhead we introduce a small SRAM based cache in the controller (see Figure 2), which caches the updates to the map, hence reducing the consequent number of PRAM accesses. 2) memory overhead for storing the access map; the amount of memory required is proportional to the size of the PRAM used, and the size of the entry stored for each physical page in the access map. For instance, if the PRAM used is 4GB in size and the page size is 8K, the memory required for the access map would be around 4MB (each entry = 8 bytes). This is fairly small for modern systems, which have multiple GBs of memory.

3.2.2 Page Manager

The page manager is the OS level entity responsible for managing memory pages across PRAM/DRAM. It consists of two key subsystems, described below in detail, which help it perform its key tasks: Memory Allocator (page allocation/deallocation) and Page Swapper (uniform wear leveling of PRAM).

Memory Allocator. The goal of the OS memory allocator is to serve the memory allocation requests from the OS and the user processes. A list of free pages is maintained by typical allocators, from which the page request is fulfilled. Traditional memory allocators instantly mark the pages released by applications as free, and may immediately allocate them on subsequent requests. Such an approach can create hot spots of pages in memory, where the activity (reads/writes) is significantly higher compared to rest of the pages. This is fine for memories (like DRAM) that do not face endurance problems, but for memories like PRAM, it is a problem, as it may render some pages unusable very soon. To get around it, we make the PRAM memory allocator aware of these issues. The PRAM allocator maintains three lists of free pages: *free*, *used-free* and *threshold-free* list. At the startup time, all PRAM pages are in the *free* list, and the allocation requests from the applications are served from it. When the pages are freed, they go to the *used-free* list rather than the free list. If an allocated page is written to a lot by an application, and if the number of writes crosses a *'threshold'*, then as described before, the memory controller generates a page swap interrupt for that page. At this point, the page swapper (described below) handles the interrupt, and releases this page, which goes to the *threshold-free* list.

When the free-list becomes exhausted, or it lacks sufficient pages

to service a request, it is merged with the used-free list, which is then used as the source of page allocation. The user-free list will get exhausted only when all the free PRAM pages have been written to at least a *'threshold'* number of times. When this happens, the *free* and *threshold-free* list pointers are swapped to move all the free pages from the *threshold-free* list to the original *free* list. Such an approach tries to achieve wear leveling across all the PRAM pages by ensuring that all the free pages have been written to at least *'threshold'* times before getting reused. We assume that there is a separate allocator for DRAM memory, which does not need any such changes.

Page Swapper. The page swapper is responsible for managing the page swap and bad-page interrupts generated by the memory controller. It handles the page-swap interrupt by doing the following: 1) Allocates a new page from the memory allocator. The new page could be either from the PRAM or DRAM. 2) Finds the page table entry/entries (PTE/PTEs) of the physical page for which the interrupt is generated. This can be accomplished in modern systems such as Linux using the reverse mapping (RMAP), which maintains a linked list containing pointers to the page table entries (PTEs) of every process currently mapping a given physical page. 3) Copies the contents of the old page to the new one using their virtual addresses. The advantage of using virtual address for the copy as opposed to something like DMA is that it results in coherent copying of data, which ensures that the new page gets the latest data. 4) Updates the PTE/PTEs derived from the RMAP with the new physical page address. 5) Replaces the TLB entries corresponding to the old PTE/PTEs. 6) Releases the old physical page. If it is a PRAM page, it goes to the threshold-free list as described above.

The key decision for the swapper is in selection of the new page for replacing the old PRAM page, for which interrupt is generated. We implement two policies for this decision:

Uniform memory policy: This policy allocates the new page from the PRAM allocator. We introduce this as our baseline policy, which can be used in a PRAM only memory configuration as well, since it allocates only PRAM pages. This policy exploits the wear leveling mechanism of page swapping to extend PRAM endurance, but does not benefit from the memory heterogeneity of a PDRAM system.

Hybrid memory policy: This policy allocates new page from the DRAM allocator. The motivation to do so is based on the fact that there is a high probability of the page, for which page swap interrupt got generated, being very write intensive. As we show later, this has a two fold advantage: a) It reduces number of page swap interrupts, which is good for performance; b) It reduces number of writes in PRAM, which is good from perspective of both reliability and power, since PRAM writes consume higher power than DRAM writes. This policy exploits the heterogeneity of the PDRAM system, hence the name hybrid memory policy.

For the bad-page interrupt, the page swapper does exactly the same things as it does for the page-swap interrupt, except that it moves the page off its free lists and moves it into a bad-page list. The pages in the bad-page list are discarded and not used for future allocations. The list is stored reliably in a known location on PRAM.

3.3 Endurance Analysis

In this section, we analyze PRAM reliability with and without our wear leveling policies. Lets assume a system with pages of size 4K (eg. x86 systems) and 4GB of PRAM, implying there are N_p=1M pages available for allocation. We assume an application, which consistently writes to two different addresses in PRAM, that map to different PRAM rows in the same memory bank. This ensures that each row is consistently written back to the PRAM cells with alternative writes, because at a time only one row in a bank can be open. Typical latencies for such writes in PRAM is around 150ns (T_w). We assume that this application writes to these two addresses every 150ns or T_w in an alternative fashion to generate the worst case from endurance perspective. Note that this is unrealistic, since there are caches and write buffers between the CPU and memory, which will throttle the rate of writes. However, the analysis will allow us to estimate PRAM reliability under extreme cases. Lets refer to the write endurance of PRAM (10^9 write cycles in worst case) as N_w. If we do not take any wear leveling into account, then such an application can cause row failure in the page containing these addresses in $2N_w$ writes or $T_w \times 2N_w = 300s$. This is a very low time scale, and hence not acceptable for a real system deployment.

Now we analyze the case with our uniform memory wear leveling policy. Let the threshold of writes, at which the memory controller generates a page swap interrupt, to be N_t (where $N_t \ll N_w$), and the % of free PRAM pages in the system be $\alpha\%$. Lets assume, that the PRAM rows containing the two addresses being written to by the application map to the same physical page. Now as soon as the application writes N_t times, the page swapper will swap the physical page mapping these two addresses to a new PRAM page, which will correspond to different rows in the PRAM. The old physical page will then be moved into the *threshold-free* list. From our discussion in the previous section, we know that the application will not be able to write to the freed physical page (and its corresponding PRAM rows) again until both the *free* and *used-free* lists become empty and the *threshold-free* and *free* list pointers are swapped. For this to happen, the application will have to write at least N_t times to every free PRAM page ($\alpha N_p N_t$ writes) before it can access the old physical page again. In other words, to write $2N_t$ times to the old physical page, the application will do $\alpha N_p N_t + 2N_t$ writes, i.e.:

$$2N_t \rightarrow \alpha N_p N_t + 2N_t$$
$$3N_t \rightarrow 2\alpha N_p N_t + 3N_t$$
$$\beta N_t \rightarrow (\beta - 1)\alpha N_p N_t + \beta N_t \qquad (1)$$

This means that for doing N_w writes ($\beta N_t = N_w$ in equation 1):

$$N_w \rightarrow \left(\frac{N_w}{N_t} - 1\right)\alpha N_p N_t + \left(\frac{N_w}{N_t}\right)N_t$$
$$N_w \rightarrow \alpha N_w N_p + N_w \approx \alpha N_w N_p \quad (N_w \gg N_t \,;\, \alpha N_w N_p \gg N_w)$$
$$(2)$$

Based on this analysis, our application will have to do approximately $\alpha N_w N_p = \alpha 10^{15}$ writes in order to perform N_w writes to a given physical page. This means, to write N_w times to a PRAM row, it will have to perform $2N_w$ writes. Using T_w as 150ns (see the paragraph above) and $\alpha = 50\%$, this translates to around 2.4 years. Thus, with our wear leveling scheme the bounds go from order of seconds to years. It must be noted that this analysis has been done assuming an application which bypasses cache and write buffers to perform just writes. If we assume a conservative assumption of 50% cache hit for writes, the PRAM lifetime would increase to about 4.8 years. As the quality of PRAM is expected to increase to 10^{12} cycles and beyond, the bounds will be much higher. For the hybrid memory policy, the expected bounds would be even higher since the write intensive pages are moved to DRAM, where endurance is not an issue. In our experiments, we use $N_t=1000$ as a good trade-off between endurance and swapping cost.

Parameter	DRAM	PRAM
Power Characteristics		
Row read power	210 mW	78 mW
Row write power	195 mW	773 mW
Act Power	75 mW	25 mW
Standby Power	90 mW	45 mW
Refresh Power	4 mW	0 mW
Timing Characteristics		
Initial row read latency	15 ns	28 ns
Row write latency	22 ns	150 ns
Same row read/write latency	15 ns	15 ns

Table 1: DRAM and PRAM Characteristics (1Gb memory chip)

4. EVALUATION

4.1 Methodology

For our experimental evaluation we use the M5 architecture simulator [4]. M5 has a detailed DRAM based memory model, which we significantly enhance to model timing and power of a state of the art modern DDR3 SDRAM based on the data sheet of a Micron x8 1Gb DDR3 SDRAM running at 667MHz [7]. We implement a similar model for PRAM based on timing and power characteristics described in [13, 17–19]. We assume PRAM cells to be arranged in a grid of rows and columns just like DRAM. Such configuration of PRAM has been practically implemented and demonstrated by Intel [17].

The power parameters used for DRAM and PRAM are listed in Table 1. The DRAM parameters are based on 78nm technology [7]. The read-write power values for PRAM are obtained from the results in [2, 3, 13, 17]. For a fair comparison, the values are down scaled for 78nm technology based on the rules described in [16]. We can observe that read power of PRAM is around three times lower compared to DRAM, while write power is around four times more. This suggests that PRAM is not very attractive for write intensive applications both from the perspective of reliability as well as energy efficiency. The third parameter in Table 1 (Act) refers to the activation/precharge power, which is consumed in opening and closing a row in the memory array. It is higher for DRAM, since it has to refresh the row data before closing a row, which can be avoided in PRAM due to its non-volatility. The standby power, which the memory consumes when it is idle, is also lower for PRAM due to its negligible leakage power consumption. Finally, DRAM consumes refresh power for supplying sustained refresh cycles for it to retain its data. This is not required in PRAM, since it is non-volatile.

For timing, we use the Micron data sheet to get the detailed parameters for DRAM. For PRAM, we use the results in [2] to obtain the read-write latency values for 180nm technology. We scale down the read latency for 78nm, but maintain the same value for write as a conservative assumption, since the write latency is a function of the material property. Table 1 shows these values. We can observe, that for an initial read, PRAM requires almost twice the amount of time as compared to DRAM. This happens due to the higher row activation time of PRAM. Similarly, writing back or closing an open row in PRAM is around seven times more expensive than for DRAM. However, reads/writes on an open row have latency values similar to DRAM.

Besides this, we further extend M5 for incorporating the memory controller and page manager as described in section 3. We implement the access map cache in the memory controller (see section 3) as a 32 entry (each entry = 8 bytes) fully associative cache. For the page manager, we implement both the uniform and hybrid memory

978-1-60558-497-3/09 $25.00 © 2009 ACM

Benchmark	*rpi* (%)	*wpi* (%)	*# of pages*
applu	1.94	0.93	24435
bzip2	0.12	0.08	24600
facerec	0.6	0.5	2240
gcc	0.15	0.06	2781
sixtrack	0.01	0.008	7601

Table 2: Benchmark Characteristics

policies as described in section 3.

For our experiments, we assume a baseline system with 4GB of DDR3 SDRAM, which we refer to as DRAM. The 4GB memory consists of four 1GB ranks, where each rank is made up of eight x8 1Gb chips with characteristics described in Table 1. We evaluate it against two experimental systems: 1) **Hybrid system**: It comprises of 1GB DDR3 SDRAM and 3GB of PRAM, and employs *hybrid memory policy* for managing page swap requests. 2) **Uniform system**: It comprises of 4GB of PRAM, and employs *uniform memory policy* for managing page swap requests. The motivation of the comparison is to show how heterogeneity in memory organization can result in better overall performance and energy efficiency.

For workloads, we use benchmarks from the SPECCPU2000 suite, which we execute on M5 using a detailed out-of-order execution ALPHA processor running at 2.66GHz. The simulated processor has two levels of caches: 64KB of data and instruction L1 caches, and 4MB of L2 cache. We use benchmarks described in Table 2, and simulate the first five billion instructions. Table 2 illustrates the memory access characteristics of these benchmarks in terms of *rpi* (reads per instruction), *wpi* (writes per instruction), and *number of pages* (total number of pages allocated). We can see that they have varying memory access characteristics. For instance, sixtrack has very low *rpi* (0.01%) and *wpi* (0.008%), while for applu it is an order of magnitude higher (1.94% and 0.93%); facerec has high *rpi* (0.6%) and *wpi* (0.5%), while gcc and bzip2 have medium *rpi* and low *wpi*. Similarly, the working set of these benchmarks in terms of the number of pages allocated also varies from just 2240 (facerec) to as high as 24600 (bzip2).

4.2   Results

Energy Savings and Performance Overhead

Figure 3 shows the results of the hybrid and uniform memory systems baselined against the DRAM system for all the benchmarks. Figure 3b shows the overhead incurred in terms of execution time by these systems, while Figure 3a shows the reduction in memory energy consumption. Please note that the %overhead and %energy numbers in these figures include the energy and time overhead due to page migrations and accesses to the access map in the memory controller and the PRAM. We describe the details of the overhead in the following sections.

We can see in Figure 3, that on average, the hybrid system achieves around 30% energy savings for just 6% performance overhead across all the benchmarks. In contrast, the uniform system gets 30% energy savings at the cost of 31% overhead. For sixtrack, which has low *rpi* and *wpi*, the impact on performance is negligible since there are not enough accesses to expose the slower access times of PRAM. Both systems achieve high energy savings due to the lower standby power consumption of PRAM compared to DRAM (see Table 1). The energy savings for the uniform system is higher (around 49%) than hybrid system (37%) since the uniform system comprises exclusively of PRAM, while the hybrid system contains 1GB of DRAM as described in section 4.1. For gcc and bzip2, the performance overhead is higher for the uniform system (around 6%). This happens due to the relatively higher *rpi* and *wpi* of these

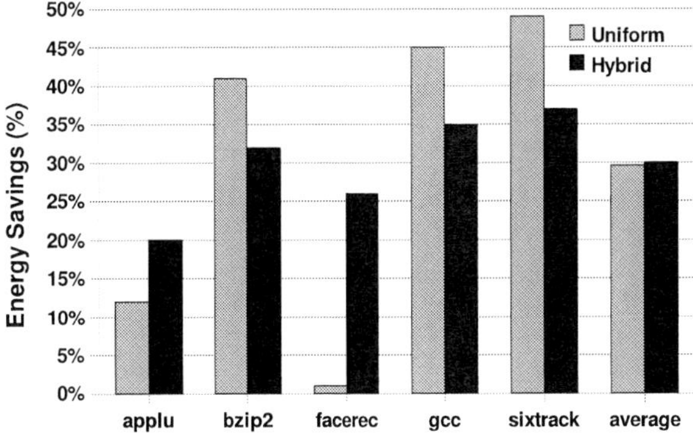

(a) Energy Savings

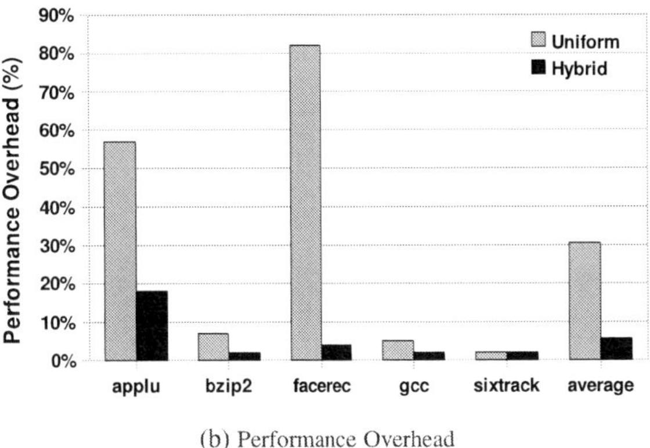

(b) Performance Overhead

Figure 3: Energy Savings and Performance Overhead Results

benchmarks compared to sixtrack (see Table 2), which exposes the slower access times of the PRAM. The overhead is lower for the hybrid system (2%), since it migrates the write intensive pages to DRAM. The energy savings is higher again for the uniform system due to the lower standby power of PRAM.

In contrast, for applu and facerec, the performance overhead of the uniform system is significantly high (57% and 82% respectively). This happens due to the higher *wpi* and *rpi* of both these benchmarks (see Table 2). The overhead for facerec is higher than applu (despite its lower *rpi* and *wpi*) due to its higher IPC (60% more than applu). This implies, facerec is more sensitive to higher memory access latencies, which results in the poor performance of the uniform system due to slower access times of PRAM. The high overhead nullifies the lower power consumption of PRAM and results in negligible energy savings. In contrast, the performance overhead of the hybrid system is very low (4%). This happens because facerec has a very high locality of read and writes to its pages, and the hybrid system migrates them to DRAM. This results in much higher energy savings as well (26%). For applu, the energy savings for the uniform system is low (around 10%) due to its high performance overhead. For the hybrid system as well, the overhead is high (around 18%) because of low locality of reads and writes. However, the energy savings is still higher (20%) compared to uniform system.

Thus, the results indicate, that in terms of comparison between the systems, the hybrid system is clearly more beneficial. It is able to exploit the read-friendliness of PRAM as well as the write friendliness of DRAM, and hence achieves better overall performance and energy efficiency.

978-1-60558-497-3/09 $25.00 © 2009 ACM

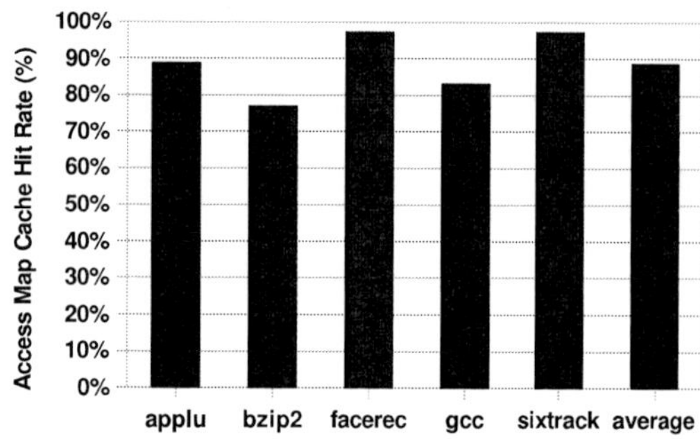

Figure 4: Access Map Cache Hit Rate (%)

Benchmark	Uniform	Hybrid	% Reduction
applu	29000	18000	38
bzip2	2400	635	74
facerec	24600	1500	94
gcc	1350	320	77
sixtrack	0	0	0

Table 3: Page Swap Interrupts

System Overhead

In this discussion, we analyze the sources of system overhead and their impact in terms of performance and energy in detail. We focus on the overhead due to accesses to the access map and its cache, and page swapping.

Access Map

As described in section 3, the access map is used to store write access information to PRAM pages. The updates to the access map and its memory controller cache thus add energy overhead in the system in terms of extra accesses to PRAM and the power consumption of the cache itself.

For understanding the extra accesses to PRAM, Figure 4 shows the hit rate of the access map cache. The hit rate was almost the same for both the hybrid and uniform systems. We can see that the hit rate is fairly high across all the benchmarks (average around 90%). This indicates that the overhead is very low, since extra PRAM accesses are done only for 10% of writes. From Table 2, we know that *wpi* of most of the benchmarks is fairly low, so in the context of overall time frame, the impact is negligible.

For the access map cache, we estimate the power consumption to be around 78mW per access using CACTI 4.1 [6]. Since the cache is accessed only for writes, in the overall time-frame, the extra energy consumption due to it is also very low. It must be noted that the extra energy consumption due to accesses to the access map and its cache is included in the results in Figure 3.

Page Swapping

The second source of overhead is page swapping interrupts and the consequent page swaps. Table 3 shows the statistics related to page swap interrupts for the uniform and hybrid system. We can observe that for most of the benchmarks, the number of interrupts drop significantly for the hybrid system. For instance, for facerec, it drops down by as much as 94%. This happens due to its high *wpi* and significant locality in its reads and writes to a small set of addresses. In the uniform system, when the pages mapping these addresses reach the *threshold*, they get mapped to a new PRAM page. However, sustained writes to these addresses generate further page swap interrupts. In contrast, with the hybrid system, once the pages mapping these addresses reach the threshold, they are mapped to DRAM pages, where they no longer generate any further page swap interrupts. For applu, the reduction is smaller (37%) due to its bigger working set, and relatively lower lack of locality of reads-writes.

In terms of time overhead of a page swap, it is fairly low since it is a quick software operation of allocating and copying the page, and modifying the page table entries. We assume it to be $5\mu s$, which is based on the estimate of timer interrupt overhead in modern systems that get generated as frequently as 1ms. Hence, overall the page swapping overhead is also minimal in terms of performance and energy. Note, the page swap overhead is also included in the results in Figure 3.

5. CONCLUSION

In this paper we propose PDRAM, a novel, energy efficient hybrid main memory system based on PRAM and DRAM. We highlight the challenges involved in managing such a system, and provide a hardware/software based solution for it. We evaluate the system using benchmarks with varying memory access characteristics and demonstrate that the system can achieve up to 37% energy savings at negligible overhead. Furthermore, we show that it provides better overall energy and performance efficiency compared to homogeneous PRAM based memory systems as well.

6. REFERENCES

[1] L. A. Barroso et al. The case for energy-proportional computing. *Computer*, 40(12):33–37, 2007.

[2] F. Bedeschi et al. An 8Mb demonstrator for high-density 1.8V phase-change memories. *In Symposium on VLSI Circuits*, pages 442–445, 2004.

[3] F. Bedeschi et al. A multi-level-cell bipolar-selected phase-change memory. *Proc ISSCC'08*, pages 427–429, 2008.

[4] N. L. Binkert et al. The m5 simulator: Modeling networked systems. *IEEE Micro*, 26(4):52–60, 2006.

[5] V. Delaluz et al. Scheduler-based dram energy management. In *Proc DAC'02*, pages 697–702, 2002.

[6] http://www.hpl.hp.com/research/cacti/.

[7] http://www.micron.com/products/dram/ddr3/.

[8] http://www.numonyx.com/Documents/WhitePapers.

[9] H. Huang et al. Design and implementation of power-aware virtual memory. In *Proc ATEC'03*, 2003.

[10] T. Kgil et al. Improving nand flash based disk caches. In *Proc ISCA '08*, pages 327–338, 2008.

[11] A. Lacaita et al. Status and challenges of pcm modeling. *Solid State Device Research Conference, 2007. ESSDERC 2007. 37th European*, pages 214–221, Sept. 2007.

[12] A. R. Lebeck et al. Power aware page allocation. *SIGOPS Oper. Syst. Rev.*, 34(5):105–116, 2000.

[13] B. Lee et al. Architecting phase change memory as a scalable dram alternative. *Proc ISCA'09*, 2009.

[14] H.-L. Li et al. Energy-aware flash memory management in virtual memory system. *IEEE Trans. Very Large Scale Integr. Syst.*, 16(8):952–964, 2008.

[15] P. Mangalagiri et al. A low-power phase change memory based hybrid cache architecture. In *Proc. GLSVLSI '08*, pages 395–398, 2008.

[16] A. Pirovano et al. Scaling analysis of phase-change memory technology. *Proc IEDM'08*, pages 29.6.1– 29.6.4, 2003.

[17] A. Pirovano et al. Phase-change memory technology with self-aligned μtrench cell architecture for 90nm node and beyond. *Solid-State Electronics*, 52(9):1467 – 1472, 2008.

[18] N. Takaura et al. A GeSbTe phase-change memory cell featuring a tungsten heater electrode for low-power, highly stable, and short-read-cycle operations. *Electron Devices Meeting, 2003. IEDM '03 Technical Digest. IEEE International*, pages 37.2.1–37.2.4, Dec. 2003.

[19] B. Yu et al. Chalcogenide-nanowire-based phase chnage memory. *IEEE Transactions on Nanotechnology*, 7(4), 2008.

978-1-60558-497-3/09 $25.00 © 2009 ACM

A Voltage-Scalable & Process Variation Resilient Hybrid SRAM Architecture for MPEG-4 Video Processors

Ik Joon Chang, Debabrata Mohapatra, Kaushik Roy

Purdue University, West Lafayette, IN 47907, USA

{ichang, dmohapat, kaushik}@purdue.edu

ABSTRACT

We present a voltage-scalable and process-variation resilient memory architecture, suitable for MPEG-4 video processors such that power dissipation can be traded for graceful degradation in "quality". The key innovation in our proposed work is a hybrid memory array, which is mixture of conventional 6T and 8T SRAM bit-cells. The fundamental premise of our approach lies in the fact that human visual system (HVS) is mostly sensitive to higher order bits of luminance pixels in video data. We implemented a preferential storage policy in which the higher order luma bits are stored in robust 8T bit-cells while the lower order bits are stored in conventional 6T bit-cells. This facilitates aggressive scaling of supply voltage in memory as the important luma bits, stored in 8T bit-cells, remain relatively unaffected by voltage scaling. The not-so-important lower order luma bits, stored in 6T bit-cells, if affected, contribute insignificantly to the overall degradation in output video quality. Simulation results show average power savings of up to 56%, in the hybrid memory array compared to the conventional 6T SRAM array implemented in 65nm CMOS. The area overhead and maximum output quality degradation (PSNR) incurred were 11.5% and 0.56 dB, respectively.

Categories and Subject Descriptors
B.3.1 [**Semiconductor Memories**]: Static memory (SRAM)

General Terms
Design

Keywords
Low power SRAM, Graceful degradation, Supply voltage over-scaling

1. INTRODUCTION
The widespread availability of 3G communication networks, coupled with the growing popularity of powerful smart-phones, has fueled the exponential growth of demand for multimedia services in wireless communications. As more and more functionality is integrated into hand held mobile devices (such as cell phones and PDAs), the corresponding power dissipation associated with these devices is increasing as well. Hence, it is of paramount importance for mobile devices supporting multimedia applications to achieve low power dissipation in order to prolong the battery life of these devices. The MPEG-4 (Moving Pictures Experts Group) [1] audio visual standard has been instrumental in bringing several multimedia applications (primarily video) to cell phones, PDAs and other hand held devices, and is the current *de-facto* standard for wireless multimedia communications. Due to its widespread prevalence, it forms the basis of our motivation to investigate low power techniques for MPEG-4 hardware.

The primary cause of power dissipation in MPEG-4 video processors is memory access power [3]. Extensive research has been conducted to reduce the memory access power from an algorithmic standpoint [3]. These algorithmic techniques aim at reducing the complexity of video algorithms by decreasing the number of basic arithmetic operations per second and/or the number of memory accesses by data reuse. From a power dissipation perspective, ($P=\alpha CV_{DD}^2 f$ where α is the activity, C is the effective switched capacitance, V_{DD} is the supply voltage and f is the operating frequency), low complexity algorithms can effectively reduce f and α in the above relation for power dissipation.

Due to the quadratic dependence of switching power dissipation on V_{DD}, we can also lower power by scaling down V_{DD} [4]. Unfortunately, failure probability of a conventional 6T SRAM bit-cell increases considerably as V_{DD} is scaled down [12], imposing a lower bound on the supply voltage. It should be noted that read failure, which occurs due to lack of read *static noise margin* (SNM), is one of the major obstacles impeding supply scaling of a 6T SRAM array [12]. Recently, L. Chang et al. proposed an 8T SRAM bit-cell [5] to overcome the above issue. In the 8T bit-cell, data memory nodes are decoupled from bitline accessing during a read, improving the read SNM significantly. The 8T bit-cell is promising for storage of CIF/QCIF [1] format data in mobile devices displaying video, where processors are operated at low frequency (less than 10 MHz) [8, 9]. In such a situation, performance degradation resulting from supply scaling can be considered as a minor issue. Hence, we can lower V_{DD} of video memories more aggressively using the 8T bit-cell. However, the 8T bit-cell incurs large area penalty (larger than 30% compared to the conventional 6T bit-cell), limiting application of the 8T bit-cell in video memory.

The key innovation in our proposed work is the concept of hybrid SRAM architecture for video memory, which is mixture of conventional 6T and 8T SRAM bit-cells. The hybrid SRAM array allows us to scale down the operating voltage of memory to as low as 600mV (~200mV more than the allowed supply voltage of conventional 6T SRAM in 65nm) without degrading the output video quality significantly. Concurrently, the hybrid nature allows us to circumvent the area overhead issue associated with 8T SRAM array. The fundamental premise of our approach lies in the fact that, human visual system (HVS) is sensitive mainly to higher order bits of luminance pixels in video data. We exploit this property of video data to implement a preferential storage policy in which the higher order luma bits are stored in robust 8T bit-

Permission to make digital or hard copies of part or all of this work for personal or classroom use is granted without fee provided that copies are not made or distributed for profit or commercial advantage and that copies bear this notice and the full citation on the first page. To copy otherwise, to republish, to post on servers or to redistribute to lists, requires prior specific permission and/or a fee.
DAC'09, July 26-31, 2009, San Francisco, California, USA

978-1-60558-497-3/09 $25.00 © 2009 ACM

cells while the lower order bits are stored in conventional 6T bit-cells. This facilitates aggressive supply voltage scaling in memory as the important luma bits, stored in 8T bit-cells, remain unaffected by voltage scaling. The not-so-important lower order luma bits, stored in 6T bit-cells, if affected by voltage scaling or mismatch due to process variations, contribute insignificantly to the overall degradation in output video quality.

The rest of the paper is organized as follows. In section 2, the impact of voltage scaling on SRAM bit-cell stability is presented. In section 3, we present a brief overview of MPEG-4 decoder architecture. Section 4 shows the worst case output quality in presence of memory failures. A detailed description of our hybrid SRAM array design, along with optimal choice of ratio between 8T and 6T bit-cells is given in section 5. Simulation results and comparison between 6T-only array and hybrid SRAM array are drawn in section 6 while section 7 concludes the paper.

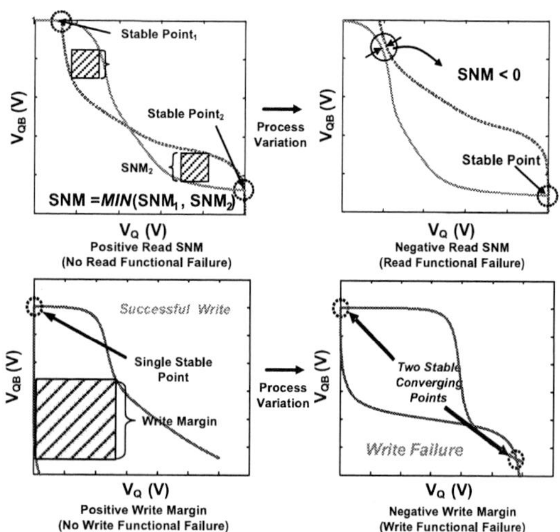

Figure 1. Read and Write Failure in a 6T SRAM Bit-cell

2. Failure Analysis in SRAM bit-cells

SRAM bit-cells experience possible failures under process variations. The failures can be broadly categorized as delay and functional failures. Since CIF/QCIF format requires low operating frequency (below 10 MHz), the delay constraint can be easily satisfied in 65nm technology (even with 600mV of V_{DD}). However, the bit-cells still suffer from large number of functional failures due to negative read SNM [6] and negative write margin [7], as shown in Fig. 1. Since read static noise margin (SNM) and write margin have conflicting design requirements [12], we can ignore the probability that both have negative values simultaneously for a bit-cell. Consequently, the failure probability (P_F) of an SRAM bit-cell can be expressed as follows:

$$P_F \approx P(R_F) + P(W_F) : P(R_F) = P(read\,SNM < 0),$$

$$P(W_F) = P(write\,margin < 0)$$

where, $P(R_F)$ represents the read failure probability and $P(W_F)$ represents the write failure probability. Using critical point sampling [6], one can accurately derive $P(R_F)$ of the 6T and the 8T bit-cells. In order to determine $P(W_F)$, we employed extensive Monte Carlo (MC) simulations. Since $P(R_F)$ and $P(W_F)$ are strongly dependent on inter-die variations, we estimated them at every process corner, which are 'Typical NMOS and Typical

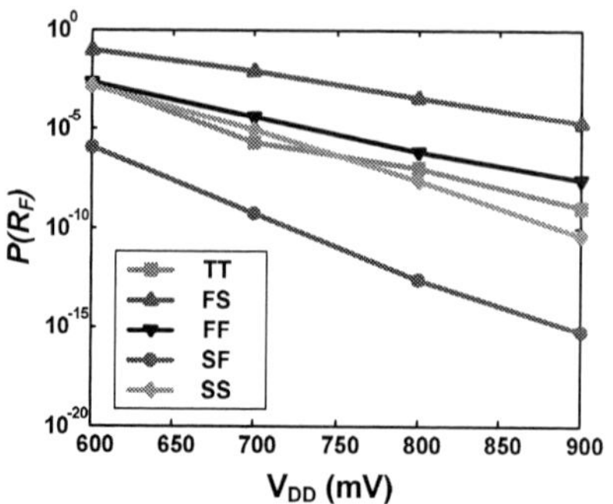

Figure 2. 6T Read Failure Probability @ T=25°C, 65nm CMOS (cell size: 0.64μm²)

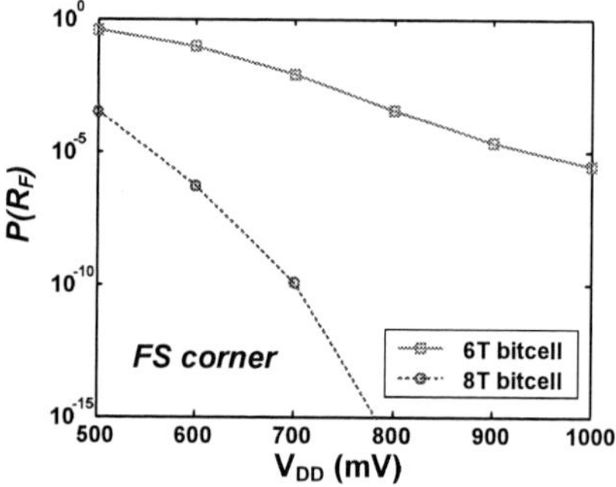

Figure 3. Read Failure Probability @ FS corner, T=25°C, 65nm CMOS (6T size: 0.64μm², 8T size: 0.832μm²)

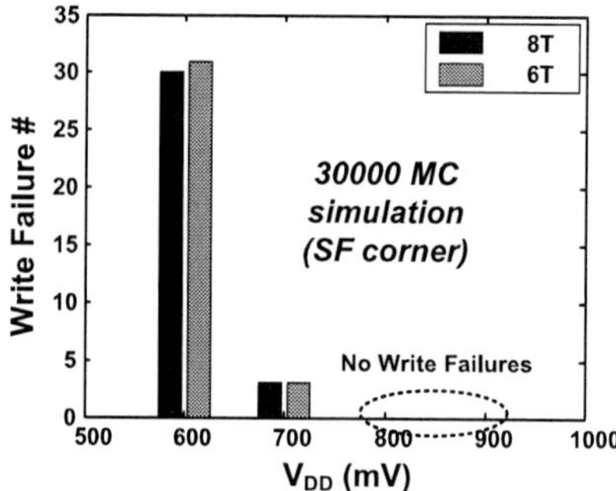

Figure 4. # of Write Failures @ SF corner, T=25°C, 65nm CMOS (6T size: 0.64μm², 8T size: 0.832μm²)

PMOS' (TT), 'Fast NMOS and Slow PMOS' (FS), 'Slow NMOS and Fast PMOS' (SF), 'Slow NMOS and Slow PMOS' (SS) and 'Fast NMOS and Fast PMOS' (FF). Fig. 2 shows $P(R_F)$ of the 6T bit-cell, where the worst corner is found to be the FS corner. In case of the 8T bit-cell, we note that memory nodes are fully decoupled from bitline access. Hence, $P(R_F)$ of the 8T bit-cell is almost negligible except in the worst corner situation, which incidentally, is also observed at the FS corner. The comparison between $P(R_F)$ of the 6T and 8T bit-cells at the FS corner is shown in Fig. 3. It demonstrates that the $P(R_F)$ of 8T bit-cell improves almost by three orders of magnitude compared to that of the 6T bit-cell at 600mV V_{DD}.

In state-of-the-art video encoder and decoder [8, 9], SRAM array size is less than 30Kbits. Hence, we consider 30000 MC simulations for the estimation of $P(W_F)$ at each process corner. The result obtained from these simulations shows that write failures occur only at SF corner. Fig. 4 shows the number of write failures at this process corner. It should be noted that $P(W_F)$ of the

978-1-60558-497-3/09 $25.00 © 2009 ACM

8T bit-cell is 0.001 at 600mV, which is significantly larger than $P(R_F)$ (=3.41×10^{-4}) of this bit-cell at the FS corner. It implies that due to improved read stability, write stability is the dominant failure mechanism in 8T bit-cell.

From the above discussion, we can calculate overall failure probability P_F for FS and SF corner. At the FS corner, $P(R_F)$ is much larger than $P(W_F)$. Hence, we can regard $P(R_F)$ as P_F. On the other hand, P_F can be approximated to $P(W_F)$ at the SF corner. It is because $P(R_F)$ is almost negligible compared to $P(W_F)$ at this corner. In the following sections, we discuss the impact of memory failures on quality of reconstructed frames in an MPEG-4 video decoder. We show how errors in individual bits of luminance pixel affects the overall quality and exploit the special characteristics of video data to implement a hybrid SRAM consisting of 6T and 8T bit-cells. As discussed above, the worst P_F of the 6T and 8T bit-cells appears at FS and SF corner, respectively. Hence, we simulate the overall quality at these process corners.

3. MPEG-4 Decoder Architecture

Before we proceed to show the impact of SRAM failures (at scaled supply voltages) on output video quality, we present a brief overview of MPEG-4 decoder architecture which is shown in Fig. 5. First, we discuss the salient features of compression on the encoder side. The MPEG encoder achieves compression by exploiting spatial and temporal redundancy existing within and between video frames. *Discrete cosine transform* (DCT), followed by quantization, is used to exploit the spatial redundancy within a frame, while *motion estimation* (ME) [2, 10] is employed for taking advantage of temporal redundancy across multiple frames. In MPEG encoding process, there are three basic types of frames, namely I, P and B frames. The I-frames are fundamental in nature, which undergo spatial compression only. The P-frames are predicted from I-frames, and B-frames are predicted from both I and P-frames by the ME process. Hence, the P and B-frames experience temporal compression. ME is done by dividing the video frame into blocks of non-overlapping 16x16 pixels known as a *macro-block* (MB). For each of the MB in the current frame, there is a search area allocated in the reference frame. The size of this search area is determined by the algorithm used and search area of 31x31 pixels is most popular [10]. The MB in the current frame is then moved within the search area in the reference frame with the objective of minimizing one of the distance criterions. The *sum of absolute difference* (SAD) is the most widely used distance criteria due to its accuracy and ease of implementation in hardware. In the SAD distance criteria, a pixel-by-pixel difference is accumulated for the 16x16 pixel MB at each position of the search area. The displacement of the MB from its position in the current frame within the search area that minimizes the SAD, yields the motion vector information (stored as displacement in x

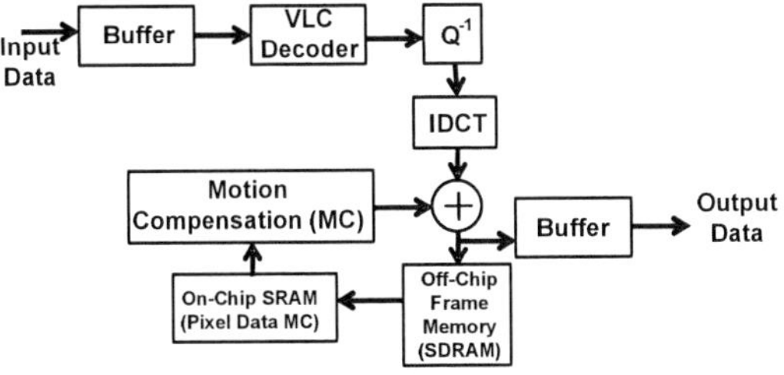

Figure 5. MPEG-4 Decoder Architecture

and y coordinates). The information stored as motion vectors is transmitted to the decoder side along with the DCT coded frame data.

The MPEG decoder has a stream buffer to store the incoming data which is then fed to the *variable length coding* (VLC) decoder. Inverse quantization and inverse DCT is performed on the output of the VLC to reconstruct the prediction error in the spatial domain. The *motion compensation* (MC) [10] block uses the past reconstructed frames stored in the frame memory and the transmitted motion vectors to compute subsequent frames. The final reconstructed frame of luma pixels is obtained by adding the motion compensated frame to the spatial prediction error (the output of IDCT block). Since MC involves reconstruction of present frames from past reconstructed frames, we need to transfer images from SDRAM to the on-chip SRAM. The size of on-chip SRAM required to accommodate a single CIF/QCIF frame is 792/198 Kbits (352x288 pixels for CIF and 176x144 pixels for QCIF), respectively. However, the SRAM array size in state-of-the-art, low power video decoder [9] is below 25Kbits. Hence, frame data has to be reloaded continuously to the on-chip SRAM array from the off-chip SDRAM in order to perform MC. The use of SRAM, prone to bit-cell failures, degrades output quality in two ways. First, it degrades the quality of the I-frame being reconstructed. Second, as P and B-frames are derived from I-frames, the quality degradation is more severe in case of reconstructed P and B-frames. This is due to the propagation of error in the reconstructed I-frames. In the following section, we show the degradation in output quality in presence of SRAM bit-cell failures at scaled voltage.

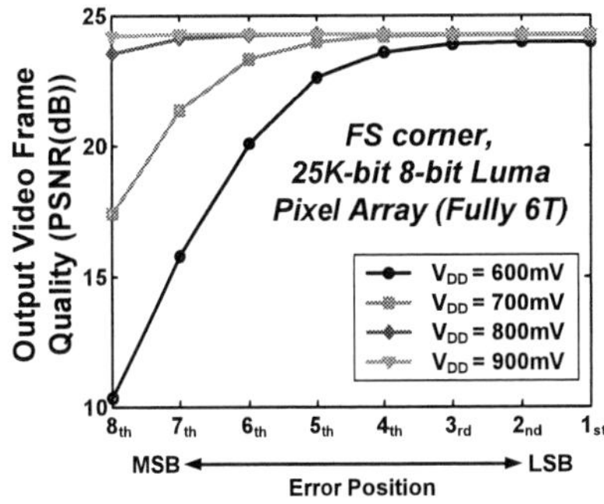

Figure 6. PSNR vs. Error Position in 6T SRAM (AKIYO)

4. Output Quality of MPEG-4 Decoder in the Presence of Memory Errors

We use P_F values obtained for 6T and 8T bit-cells in section 2 to study the effect of memory failures on output PSNR (peak-signal-to-noise ratio). For this purpose, we use 50 frames of the standard CIF video sequence "AKIYO" at 30 frames per second (fps). As discussed in section 2, FS is the worst corner for conventional 6T SRAM where read failures are dominant compared to write failures ($P(R_F) >> P(W_F)$). Under such failure probability, we simulate a hypothetical scenario in which SRAM failures occur only at one particular bit position of the 8-bit luminance pixel. This gives us an idea about the sensitivity of the output quality to errors in specific bit position of luma pixels. The simulation results are shown in Fig. 6. The key observation from Fig. 6 is that first few higher order bits contribute significantly to the output quality (PSNR) compared to the lower order bits. This

Table 1. Output PSNR as a function of # of 8T Bit-cells and V_{DD} @ FS Corner

V_{DD} (mV)	0_bit 8T	1_bit 8T	2_bit 8T	3_bit 8T	4_bit 8T	5_bit 8T	6_bit 8T	7_bit 8T	8_bit 8T
600	15.27	19.80	22.46	23.60	23.96	24.08	24.15	24.22	24.29
700	21.99	23.56	24.07	24.21	24.25	24.26	24.27	24.28	24.29
800	24.15	24.25	24.28	24.28	24.29	24.29	24.29	24.29	24.29
900	24.27	24.29	24.29	24.29	24.29	24.29	24.29	24.29	24.29
1000	24.29	24.29	24.29	24.29	24.29	24.29	24.29	24.29	24.29

Table 2. Output PSNR as a function of V_{DD} @ SF Corner

V_{DD} (mV)	PSNR (dB)
600	23.90
700	24.24

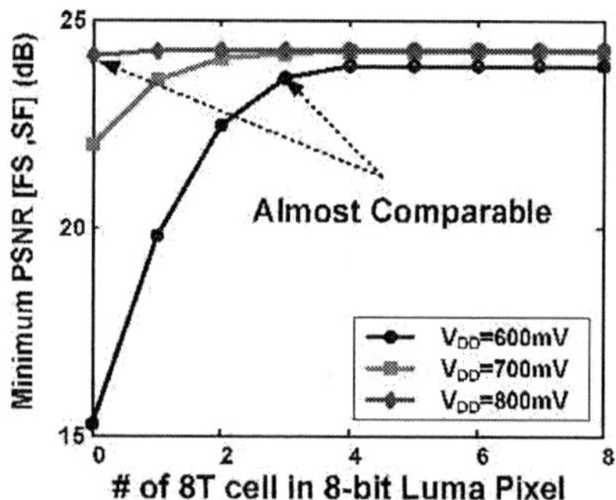

Figure 7. Worst Case PSNR in Hybrid SRAM (AKIYO)

analysis forms our basis for implementing the more significant bits (MSB) in 8T bit-cells (as $P_{8T}(R_F) << P_{6T}(R_F)$) and the less significant bits (LSB) in 6T bit-cells. However, as noted earlier in section 2, even though 8T bit-cells have excellent read stability compared to 6T bit-cells at FS corner, the write stability is degraded for 8T bit-cells at the SF corner. Hence, there is a need to jointly consider the FS and SF corner failure probabilities in hybrid SRAM.

In order to obtain the worst case PSNR, we perform simulations at both FS and SF corner. The worst case PSNR is obtained to be the minimum of values at the two corners. We assume the memory errors to be uniformly distributed throughout the array in our simulation. The PSNR results are summarized in tables 1 and 2 for FS and SF corner, respectively. Interestingly, as the number of 8T bit-cells is increased in hybrid SRAM, the worst PSNR is bounded by the write failure probability of 8T bit-cells at the SF corner. This stresses the importance of hybrid memory concept -- a fully 8T SRAM implementation may not improve the output PSNR beyond a certain point since the output PSNR is limited by write failure probability in 8T SRAM at scaled V_{DD}. Fig. 7 plots the worst case output PSNR (considering FS and SF corner), as a function of the number of 8T bit-cells (8-bit representation of luminance pixel values). The results agree fairly well with the sensitivity analysis of output quality shown in Fig. 6. It should be noted that when few MSB's of the luminance pixel are implemented as 8T bit cells, we can achieve significant improvement in output PSNR. For example, when we implement 3 MSB's in 8T bit-cells, the output PSNR at V_{DD}=600mV is comparable to that of V_{DD}=800mV of a 6T only SRAM. The key observation here is that we can over-scale the supply voltage in memory by 200mV for the same output quality achievable by 6T

SRAM while incurring roughly 11% area overhead. It is due to the fact that the proposed hybrid video memory allows memory failures to occur, but at the same time ensures that the failures minimally affect the output quality.

5. Hybrid SRAM Array Architecture
5.1 Optimal Ratio of 6T and 8T bit-cells

The most important parameter in the design of a hybrid SRAM array is the ratio of number of 6T bit-cells to 8T bit-cells in the representation of a luminance pixel. We can achieve greater robustness by implementing more MSB's as 8T bit-cells, however the area penalty associated with it increases proportionally. Thus, we need to jointly optimize the quality of output frames and the area of the hybrid SRAM as a function of number of 8T bit-cells. In order to formulate the above problem in an optimization framework, we define a new metric as follows:

$$Hybrid\ Memory\ Metic\ (HMM) = \frac{Quality(PSNR)}{Array\ Area}$$

which is also our objective function for determining the optimal number of 8T bit-cells in hybrid SRAM. The numerator is the quality of reconstructed frame in the presence of memory errors while the denominator is the area of hybrid SRAM. We denote the number of 8T bit-cells per luminance pixel as η. The optimization problem can now be formulated as:

$$\eta_{opt} = \underset{0 \le \eta \le 8}{argmax}[\min(HMM_{FS}(\eta), HMM_{SF}(\eta))]$$

where FS and SF denote the process corners, and η_{opt} is the optimal number of 8T bit-cells that maximizes the minimum of HMM values at different process corners. Since the error probabilities are different at FS (read dominated) and SF (write dominated) corners, we included both of them in our problem formulation.

In order to solve the above optimization problem, we use the following analysis. We define $Quality(PSNR) = 10 log_{10} \dfrac{255^2}{MSE}$

where MSE denotes the mean square error between the luma pixels of the original frame and the motion compensated frame with SRAM failures. We calculate the MSE by the following relation where p_{6T} and p_{8T} denote 6T and 8T bit-cell failure probability, respectively. The first term in the equation denotes the MSE due to failures in 6T bit-cells while the second term gives the MSE due to 8T bit-cell failures.

$$MSE_{pixel}(\eta) = \sum_{m=0}^{7-\eta} p_{6T} 2^{2m} + \sum_{n=8-\eta}^{7} p_{8T} 2^{2n} \ (1 \le \eta \le 7)$$

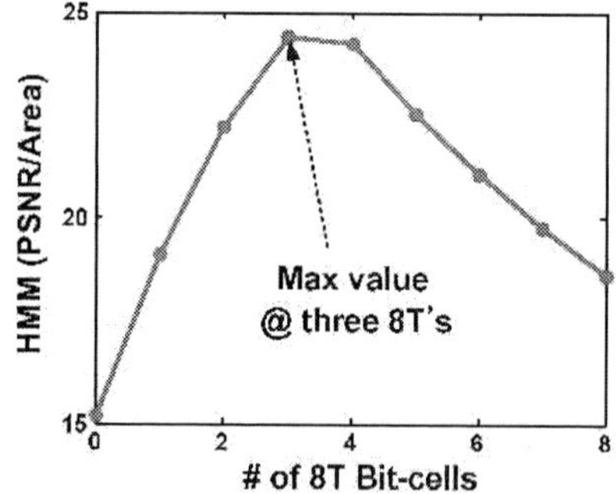

Figure 8. Optimal # of 8T Bit-cells for the hybrid SRAM array

The hybrid SRAM array area is given by the expression

$$Array\ Area = \left[\frac{8-\eta}{8} + \frac{\eta}{8}(8/6) \right]$$

which is normalized to the conventional 6T SRAM area. We use the above relations for MSE, Quality and SRAM area in the expression for *HMM*, and plot *HMM* as a function of η, which is shown in Fig. 8. We used p_{6T}=0.0899, p_{8T}=3.41×10^{-4} at FS corner, and p_{6T}=0.001, p_{8T}=0.001 at SF corner, which were obtained based on the discussion in section 2. The optimal number of 8T bit-cells is found to be three from our analysis and we used it in the hybrid SRAM design. In the next subsection, we present the integration of 6T and 8T bit-cells into a single SRAM array.

5.2 Integration of 6T and 8T Bit-cells and Array Architecture

Figure 9 shows the schematic of 6T and 8T bit-cells for the hybrid SRAM. Note that the 8T bit-cell has two word lines, which are for read (RWL) and write (WWL), respectively. For efficient integrated layout, we also split the word line of the 6T bit-cell to WWL and RWL, as illustrated in Fig. 10. Since the hybrid SRAM uses single-ended sensing, we can achieve successful read operation with one access transistor in the 6T bit-cell. The timing diagram given in Fig. 10 shows the operating principles of the bit-cells. In the read mode, only RWL is enabled while WWL remains disabled. Depending on data value of each cell, pre-charged RBL (for 8T) and BL (for 6T) are evaluated. During write process, both word lines (RWL and WWL) are enabled to turn on both access transistors of the 6T bit-cell.

The layout of one luma pixel array (3 MSB's: 8T and 5 LSB's: 6T) is shown in Fig. 11. Since both bit-cells require two poly pitches in the thin-cell layout [5], they can be integrated into one row successfully. RWL and WWL contacts are shared by neighboring cells and hence, the integration of 6T and 8T bit-cells

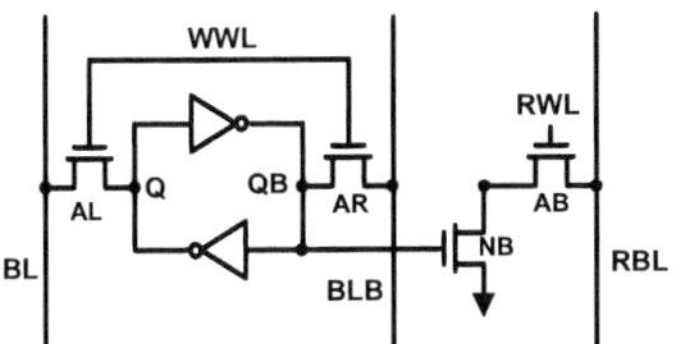

8T Bit-cell (MSB's of Hybrid SRAM)

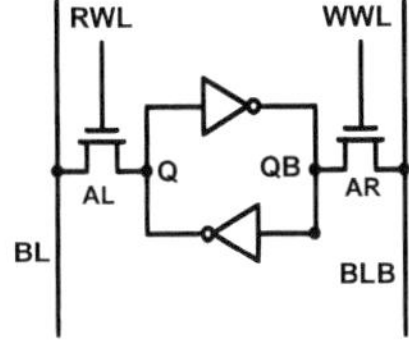

6T Bit-cell (LSB's of Hybrid SRAM)

Figure 9. 6T & 8T Bit-cell Schematic in Hybrid SRAM

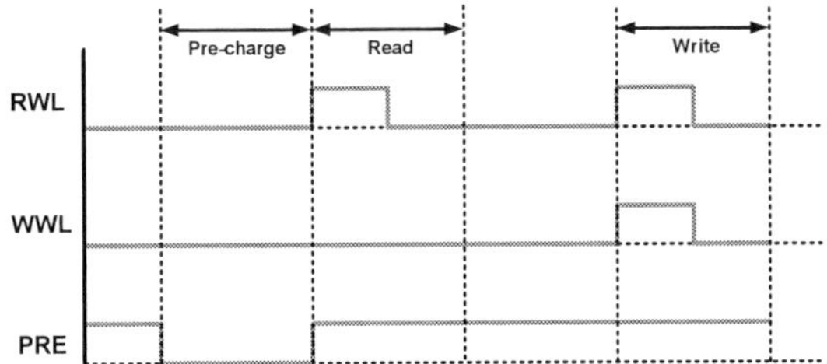

Figure 10. Timing Diagram for Read & Write in Hybrid SRAM

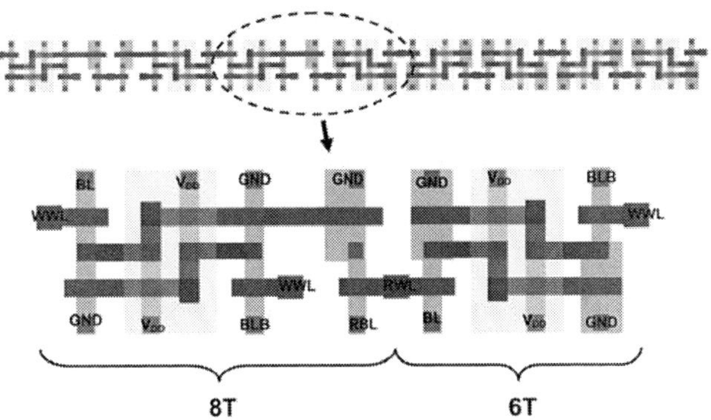

Figure 11. Hybrid SRAM Layout of 8-bit Luminance Pixel

does not require any extra area overhead except the area penalty incurred due to the extra transistors in 8T bit-cells. Unlike cache memory of high performance microprocessors, soft-errors may not be a big issue in video memories. Hence, we do not consider bit-interleaving in our implementation (which is required for soft-error immunity [11]).

To reduce area penalty of memory decoders in the short bitline architectures, we implemented four pixels (=32 bits) as one decoding block. It should be noted that in such a scheme, half selection problem [5] occurs during writing one pixel. In this work, we avoided this issue by writing the four pixels at the same time. Sequential write access of video memory allows us to implement such a writing approach easily.

6. Simulation Results

Figure 12 shows snapshots of simulated video screen under two simulation environments. For the first simulation, we assume that a 6T-only array is used for the video memory and consider two cases for simulation: V_{DD} of 600mV and 800mV, respectively. We generated errors in the memory array with P_F of FS corner, which is the worst case for the 6T bit-cell. For the other simulation, we replaced the 6T array with the proposed hybrid SRAM array (The first three MSB bits implemented as 8T bit-cells) and lowered V_{DD} to 600mV. Since the worst P_F of 6T and 8T appears at different process corners (FS and SF respectively),

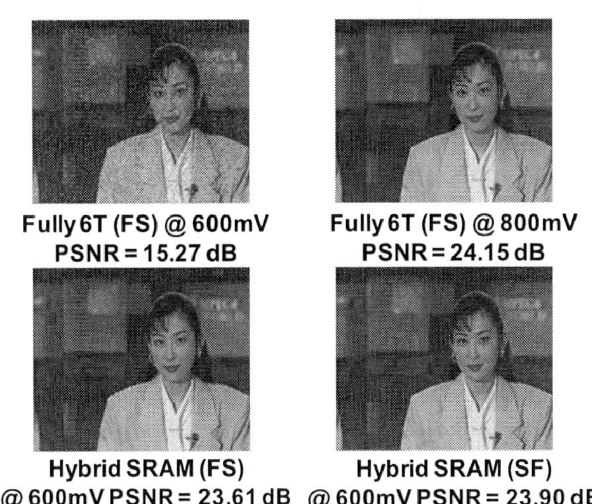

Figure 12. Quality of "AKIYO" Video Sequence in Hybrid SRAM

978-1-60558-497-3/09 $25.00 © 2009 ACM

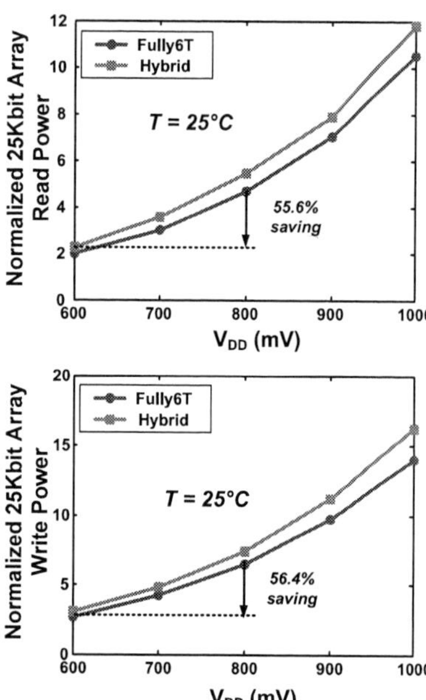

Figure 13. Read & Write Power Simulation Results

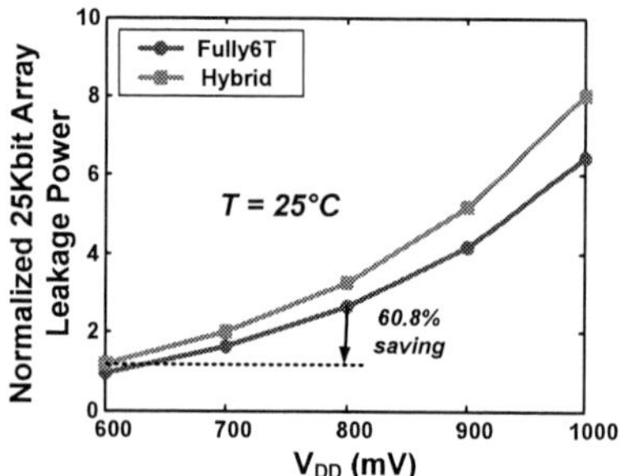

Figure 14. Leakage Power Simulation Results

we reconstructed video images at these two corners. In spite of aggressive voltage over-scaling (by 200mV), we can observe that the video image quality in the second set of simulation is as good as that of the first set at 800mV.

Figure 13 and 14 show power and leakage simulation results for the two SRAM arrays. At 800mV, the hybrid SRAM array dissipates larger power compared to the 6T array (by 14~16%). However, our hybrid SRAM array allows 200mV over-scaling of V_{DD} for similar image quality. Simulation results show that it achieves significant power saving (55~56%) for both read and write (at 10MHz frequency). Note that switching power dissipation is expressed as '$\alpha CV^2 f$'. When we assume that two SRAM arrays have same effective capacitance, the 200mV over-scaling is expected to provide 43% power reduction. However, due to low operating frequency of CIF/QCIF format (below 10MHz), the active leakage power ($I_{Leak}V_{DD}$) also forms substantial portion of the total power dissipation. As shown in Fig. 14, the 200mV over-scaling improves the leakage power by 60.8% due to exponential dependence of leakage power on V_{DD}. As a result, we obtain significant power savings.

7. Conclusion

In this paper, we presented a novel hybrid SRAM for video memory. The proposed hybrid SRAM allows aggressive scaling of V_{DD} to 600mV, while providing identical output quality to that of conventional 6T SRAM operating at 800mV. While algorithm level techniques save power by reducing the frequency of operation (reduced complexity requiring fewer computations), the proposed hybrid SRAM provides a system/circuit level technique for lowering power dissipation in video memory by allowing a graceful degradation in output quality.

8. Acknowledgement

This research was funded in part by Focused Center Research Program (GSRC) and by Semiconductor Research Corporation.

9. REFERENCES

[1] http://www.mpeg4.org

[2] P. Kuhn, "Algorithms, Complexity Analysis and VLSI architectures for MPEG-4 Motion Estimation", Kluwer Academic Publishers, 1999.

[3] C. Lin et. al. "A 5mW MPEG4 SP Encoder with 2D Bandwidth-Sharing Motion Estimation for Mobile Application", *ISSCC Dig. Tech. Papers,* Feb., 2006

[4] G. V. Varatkar, N. Shanbhag. "Variation-Tolerant Motion Estimation Architecture", IWSPS, 2007

[5] L. Chang et al., "An 8T-SRAM for Variability Tolerance and Low-Voltage Operation in High-Performance Caches", *IEEE Journal of Solid State Circuits*, Vol. 44, No. 4, pp. 956-963, 2008

[6] I. J. Chang et al., "Fast and Accurate Estimation of Nano-Scaled SRAM Read Failure Probability using Critical Point Sampling", *Custom Integrated Circuit Conference (CICC)*, pp 439-442, Sep. 2005

[7] K. Takeda et al., "Redefinition of write margin for next-generation SRAM and write-margin monitoring circuit", *ISSCC Dig. Tech. Papers*, pp.630-631, Feb. 2006

[8] Y. Lin et al., "A 242mW 10mm^2 1080p H.264/AVC High-Profile Encoder Chip", *ISSCC Dig. Tech. Papers*, Feb. 2008

[9] T. Liu et al., "A 125 µW , Fully Scalable MPEG-2 and H.264/AVC Video Decoder for Mobile applications", *IEEE Journal of Solid State Circuits*, Vol. 42, No. 1, pp. 161-169, 2007

[10] V. Bhaskaran, K.Konstantinides, *"Image and Video Compression Standards"*, Kluwer Academic Publishers, Second Edition, 1997

[11] I. J. Chang et. al, "A 32kb 10T Subthreshold SRAM Array with Bit-Interleaving and Differential Read Scheme in 90nm CMOS", *ISSCC Dig. Tech. Papers*, pp.388-389, Feb. 2008

[12] S. Mukhopadhyay, H. Mahmoodi, K. Roy, "Statistical design and optimization of SRAM cell for yield enhancement", *IEEE/ACM International Conference on Computer Aided Design*, pp. 10-13, 2004

978-1-60558-497-3/09 $25.00 © 2009 ACM

A Physical Unclonable Function Defined Using Power Distribution System Equivalent Resistance Variations

Ryan Helinski
University of New Mexico
Albuquerque, NM
helinski@unm.edu

Dhruva Acharyya
Verigy Inc.
Cupertino, CA
dhruva.acharyya@verigy.com

Jim Plusquellic
University of New Mexico
Albuquerque, NM
jimp@ece.unm.edu

ABSTRACT

For hardware security applications, the availability of secret keys is a critical component for secure activation, IC authentication and for other important applications including encryption of communication channels and IP protection in FPGAs. The vulnerabilities of conventional keys derived from digital data can be mitigated if the keys are instead derived from the inherent statistical manufacturing variations of the IC. Robust silicon-derived keys are implemented using physically unclonable functions (PUFs). A PUF consists of a specialized hardware circuit and a mechanism to retrieve a set of responses under a variety of different challenges. In this paper, we propose a PUF that is based on the measured equivalent resistance variations in the power distribution system (PDS) of an IC. The effectiveness of the PUF is evaluated on twenty-four ICs fabricated in a 65 nm technology.

Categories and Subject Descriptors

K.6.5 [**Management of Computing and Information Systems**]: Security and Protection -- *Authentication.*

General Terms

Security

Keywords

Hardware security, unique identifier, process variations

1. INTRODUCTION

Many hardware security and trust mechanisms depend on the availability of a secret key or signature, i.e., a unique, unclonable identifier that can be derived from each IC. The signature of the IC defines the basis of hardware security mechanisms implemented at high levels, e.g., those that perform encryption of data communication channels, or provide IP protection in FPGAs. Conventional IC signatures are defined using digital data stored, for example, in a flash or ROM on the chip. It is critical that access to the key remains restricted to hardware circuits on the chip. Unfortunately, since the keys always remain in digital form, they are subject to an invasive attack by adversaries who may be able to extract the key,

thereby defeating the security mechanisms. Also, once a digital key is stolen, it becomes possible to produce *clone* chips that have the same identifier. This is a problem for applications that use the key in authentication protocols.

The vulnerability of embedded digital keys to attacks can be mitigated if the keys are instead derived from the inherent statistical manufacturing variations of the IC. Physically unclonable functions (PUFs) are used to realize these silicon-variation-based keys [1]. A PUF consists of a specialized hardware circuit that is sensitive to process variations. A PUF also incorporates a mechanism to retrieve a unique set of responses from a variety of different challenges. Keys derived from PUFs possess important properties including *volatility* and *non-replicability*; properties which make it extremely difficult for the attacker to steal and/or duplicate the keys. Therefore, PUFs can revolutionize next generation security and trust infrastructures in ICs.

There are two general approaches to implementing PUFs, one that is based on the variability in passive and active devices [2-9] or leakage current [10] and one that is based on variability in only passive structures, e.g., metal wires [11]. Although process variations in active devices can be leveraged to create a diverse set of responses across ICs, performance variations in active devices are also subject to environmental variations such as temperature and noise. Therefore, such approaches must also incorporate a technique to calibrate for environmental variations otherwise the response of the PUF may depend on the conditions. Calibration complicates the design and use of the PUF and makes them less attractive for security applications.

On the other hand, a PUF that is based on the variations in passive components of the IC is less susceptible (and therefore more robust) to environmental variations. The challenge in this case is implementing the PUF such that the infrastructure which defines the key does not consume a large area overhead. We propose a PUF that leverages the inherent variations in the metal resistances that define the power grid [12]. Since the power grid is an existing, distributed resource in every design, the overhead of a power grid-derived PUF is limited to the added challenge/response circuitry. Moreover, the distributed nature of the power grid makes it more prone to larger random and systematic process variation effects. Distributed process variation effects introduce resistance variations whose magnitudes vary across different regions of the power grid. This characteristic improves the robustness of the PUF because it makes it less probable that the PUFs from two ICs will produce the same response.

In this paper, we investigate a PUF derived from the resistance variations in the power grids of chips fabricated in a 65 nm technology. The PUF's response is defined in two ways; 1) as set

Permission to make digital or hard copies of part or all of this work for personal or classroom use is granted without fee provided that copies are not made or distributed for profit or commercial advantage and that copies bear this notice and the full citation on the first page. To copy otherwise, to republish, to post on servers or to redistribute to lists, requires prior specific permission and/or a fee.

DAC'09, July 26-31, 2009, San Francisco, California, USA

978-1-60558-497-3/09 $25.00 © 2009 ACM

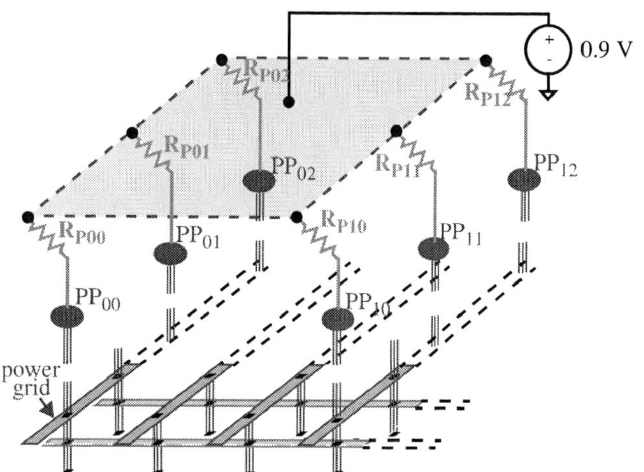

Figure 1. Power Grid Architecture

of voltage drops measured at a set of distinct locations on the power grid of the IC, and 2) as a corresponding set of equivalent resistances computed at these same locations. A distributed PUF circuit is proposed as a means of introducing a variety of stimuli (challenges) and for measuring the voltage drops (responses). A statistical analysis is carried out on the data collected from a set fabricated chips to determine the effectiveness of the PUF and to measure its susceptibility to environmental variations.

The organization of this paper is as follows. A brief background is presented in Section 2. The experimental design and test setup is described in Section 3. The PUF is described in Section 4. Section 5 describes the experiments carried out on the 65 nm chips and the experimental results. Section 6 concludes.

2.BACKGROUND

PUFs have been proposed for many applications including IC identification [13][14], addressing security in wireless sensor nodes and IC process quality control [2], hardware metering [10][15], challenge-based IC authentication [3][4][10], IP protection in FPGAs [16-18] and remote service and feature activation [19][20].

For IC authentication, a secret key is embedded that enables the IC to generate a unique response to a challenge, which is valid only for that challenge (called challenge-based IC authentication). In this manner, the key remains secret and the authentication mechanism is not vulnerable to spoofing. The authors of [7][17] propose that the same secret keys can also be used for cryptography.

The authors of [15][19] describe remote activation schemes that enable IC designers to lock each IC at start-up and then to enable it remotely, providing IP protection and hardware metering. In [19], their objectives are realized by adding states to the finite state machine (FSM) of a design and by adding control signals that are a function of the unique IDs. In effect, the hardware "locks up" waiting for a specific activation code. This offers protection against unauthorized use of Intellectual Property (IP) and hardware piracy (the illegal manufacturing of ICs).

Various PUF techniques have also been proposed that are based on mismatched delay-lines [3][6][21][22], SRAM power-on patterns [17][18], MOS device mismatch [2][13][14] and input-dependent leakage patterns [10].

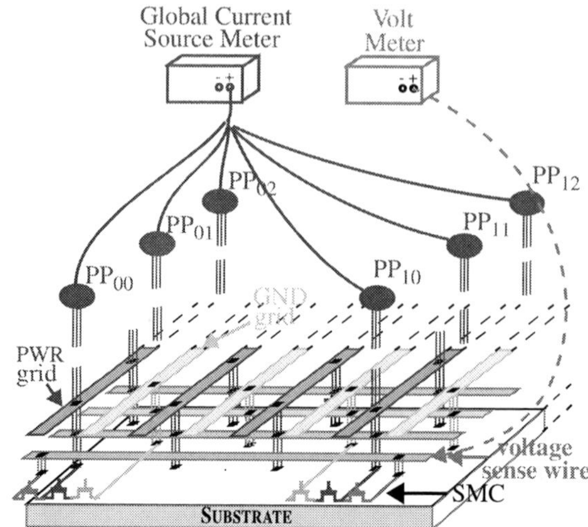

Figure 2. Instrumentation Setup

3.EXPERIMENTAL DESIGN AND SETUP

A high-level representation of the power grid architecture used in the hardware experiments is shown in Figure 1. The bottom portion shows that adjacent metal layers are routed at right angles to each other in a mesh configuration with vias placed at the intersections. The GND grid (not shown) is interleaved with the power grid and routed in a similar fashion. Both grids are routed across the 10 metal layers available in the 65 nm process. The width of the wires and the granularity of the mesh vary across the metal layers. In particular, the widths of the lower metal wires are smaller and the granularity is finer than the widths and granularity of the metal wires in the upper layers. This feature of the power grid is typical of commercial designs.

The power grid is connected to a set of six C4s or power ports (**PP**s) in the top metal layer. The PPs are shown as ovals in the figure and are labeled PP_{00} through PP_{12}. The C4s enable the power grid to be connected to the power supply, either through a membrane style probe card (during wafer probe) or through the package wiring. The finite resistance of power port connections are represented as series resistances, Rp_{xy}, in the figure.

The test jig used in the experiments is shown in Figure 2. The package pins that are connected to the PPs wire onto a printed circuit board to the *global current source meter* (GCSM). The GCSM provides 0.9 V to the power grid and can measure current at a resolution of approximately 300 nA. In addition to the global currents, our technique also requires on-chip voltage measurements. The voltage is measured in our experiments using a pin that is connected internally to a globally routed *voltage sense wire*. A voltmeter is connected to this pin off-chip, as shown in Figure 2.

The last element of the proposed infrastructure is shown along the bottom of Figure 2 and in more detail in Figure 3(b). A **Stimulus/Measure Circuit** (SMC) is inserted under each of the six C4s as shown in Figure 3(a). The SMC consists of a *shorting inverter*, a *voltage sense transistor* and a set of three *scan flip-flops* (FFs). The outputs of the FFs connect to the gates of the three transistors as shown in Figure 3(b). The shorting inverter provides a controlled stimulus, i.e., a short between the power and ground grid, when the states of FF_1 and FF_2 are set to 0. The voltage on the power grid is measured using the voltage sense transistor,

978-1-60558-497-3/09 $25.00 © 2009 ACM   677

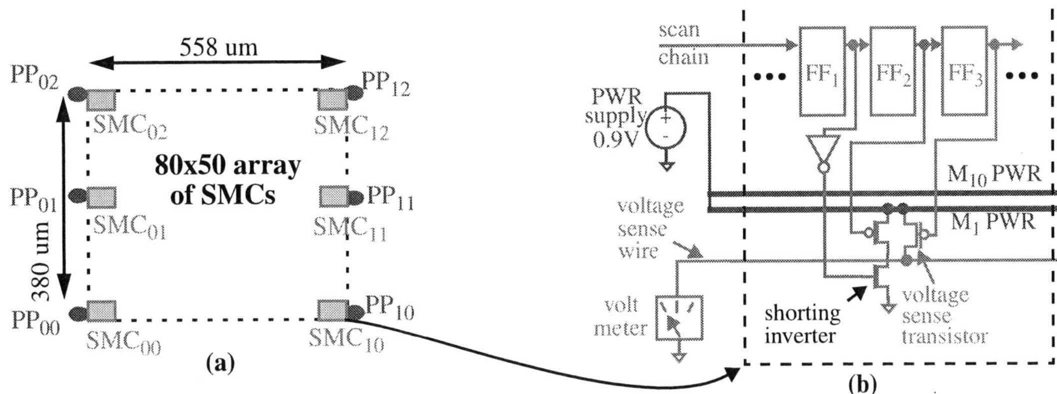

Figure 3. (a) Block diagram of the test structure and (b) details of the Stimulus/Measure Circuit (SMC).

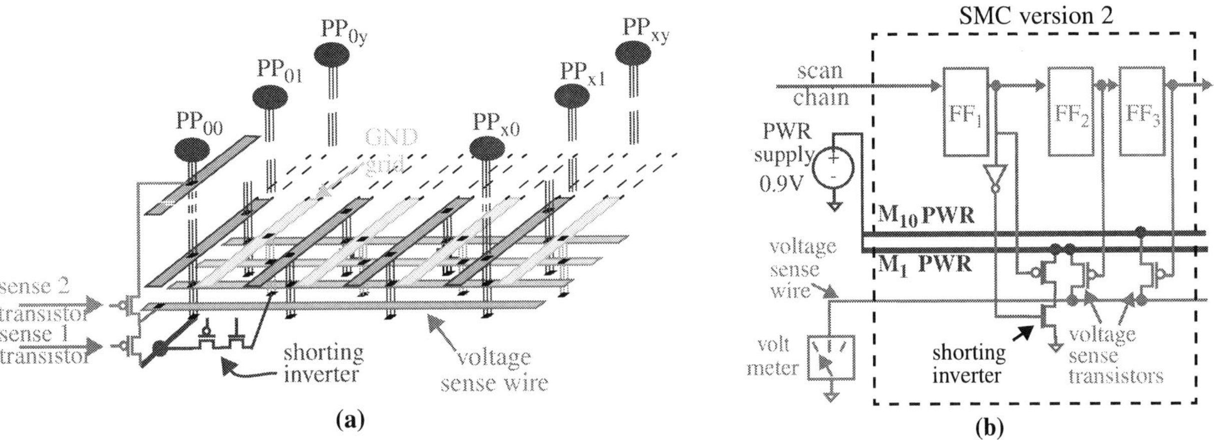

Figure 4. (a) Connections of a modified versions of SMC (b) details of the modified SMC.

enabled with a 0 in FF_3.

4. PUF SIGNATURES AND ARCHITECTURES

The PUF signature is derived using two strategies, one that is based on voltage drops and one on equivalent resistance. In either case, the signature associated with the chip is composed of six quantities, each corresponding to one of the six SMCs. The signature for a given IC under the voltage drop strategy is constructed by enabling the shorting inverters in the SMCs, one at a time, and then measuring the voltage at its source using the voltage sense transistor. A voltage drop is computed by subtracting the measured voltage from the supply voltage, 0.9 V. This process is repeated for each of the other SMCs. The resulting set of six voltages defines the signature.

The values in the voltage drop signature are affected by the magnitude of the current through the shorting inverter. The variations in the current magnitude among the shorting inverters actually adds to the 'randomness' of the PUF. However, the PUF is also more sensitive to environmental conditions, which detracts from its ability to generate the same signature (reproducibility). The equivalent resistance (ER) strategy eliminates this dependency by dividing the voltage drops by the global currents. The elimination of the current dependency makes the ER-based PUF less sensitive to environmental variations.

Bear in mind that hundreds of SMCs can be inserted into commercial power grids, which would greatly expand the complexity of the signature over that shown in these proof-of-concept experiments. Doing so is practical because the overhead of the SMC is small. For example, assuming a total of 100 SMCs, each with an area of 50 um^2 yields 5000 um^2. This is only 0.02% of the 25,000,000 um^2 area available in a 5 mm X 5 mm chip.

The PUF as described has several drawbacks. First, it is only able to produce a single signature. Second, signature generation requires the use of external instrumentation to measure the voltages and currents. Although this serves some applications, it poses problems for others that need to apply a challenge and obtain a response while operating in mission mode.

Simple modifications of the PUF architecture can address these issues. For example, the SMC shown in Figure 3(b) can be modified to incorporate more than one 'voltage sense' transistor. The left side of Figure 4 shows a modification in which a second sense transistor, 'sense 2 transistor' is added to enable the voltage to be measured in metal 10 underneath the power port. With the second sense transistor, the voltage drops between M_1 and M_{10} at different places on the power grid can be measured. This increases the number of stimulus/response pairs of the PUF from linear to quadratic because voltage drops can now be computed between any pairing of 'sense 1' and 'sense 2' transistors across the array of SMCs. The right side of the figure shows a schematic in which an additional flip-flop, FF_3, is used to control the second sense

978-1-60558-497-3/09 $25.00 © 2009 ACM

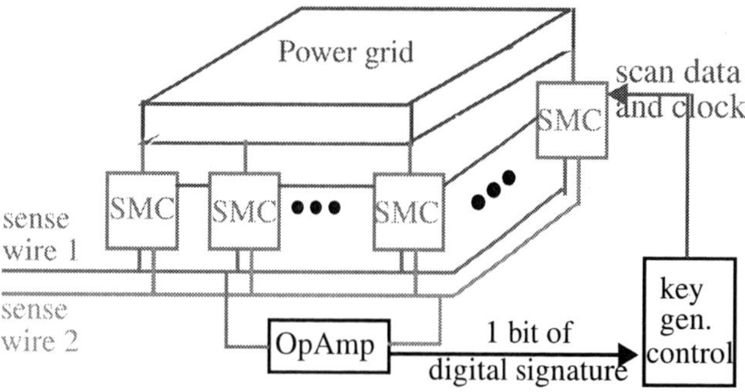

Figure 5. On-chip instrumentation for signature generation.

transistor[1].

Another strategy to increase the number of stimulus/response pairs is to allow the stimulus to be applied from more than one SMC. In these scenarios, multiple shorting inverters are enabled simultaneously at different locations and the voltage drops are measured using different combinations of sense 1 and sense 2 transistor pairs. We refer to these scenarios as multiple-on and the former as single-on. Since the power grid is a linear system, superposition applies. Therefore, to make this more resilient to attack, whereby the attacker systematically deduces the voltage drops that would occur under a multiple-on scenario by measuring the voltage drops under all single-on scenarios, this scheme can be combined with an obfuscation of the scan chain control bits. Under obfuscation, the number and position of the enabled shorting inverters are deterministically (or randomly) scrambled for a given scan chain control sequence, making it difficult or impossible to systematically apply single-on tests at known locations on the chip. We have investigated scan-chain obfuscation techniques in previous work where the objective was to prevent an adversary from using the scan chain to reverse engineer a design [23][24]. These techniques are applicable here as well. For chip-specific random scrambling, a subset of the SMCs can be used during initialization to define the state of a *selector* that controls the scan chain scrambling configuration.

The PUF as proposed requires the use of external instrumentation to measure the voltages and global currents needed to compute the IC's signature. Although this approach serves the chip authentication application well, e.g., where the objective is to periodically check the authenticity of a chip or set of chips to circumvent attempts to replace the chips with counterfeits, it is not amenable to cryptology applications that use the signature as the secret key in hardware implemented encryption/decryption algorithms. In order to serve this latter need, the signature generation process needs to occur using on-chip instrumentation.

The simplest approach to accomplishing this is shown in Figure 5. The *key generator control* unit drives the scan-in, scan-out and scan-clock signals of the SMCs with a specific pattern to enable one or more of the shorting inverters in the array of SMCs[2].

The scan pattern also enables two voltage sense transistors, one for each of the two voltage sense wires, labeled *sense wire 1* and *sense wire 2*. The two voltage sense wires are routed to the inputs of a simple differential OpAmp. The OpAmp outputs a '0' or a '1' depending on whether the voltage on 'sense wire 1' is larger or smaller than the voltage on 'sense wire 2', respectively. The 1-bit output is sent to the *key generation control* unit and the process is repeated until a sufficient number of bits are generated to realize the key. Note that this implementation is more sensitive to environmental variations because it makes use of voltages instead of equivalent resistances, as described earlier. Therefore, the response for a given chip under a given sequence of scan patterns may differ over time unless temperature and power supply noise are monitored and tightly controlled. Other more noise tolerant architectures are possible but they will increase the area overhead associated with the key generation infrastructure.

5. EXPERIMENTAL RESULTS

We carried out a set of experiments to evaluate the diversity in the voltage drops and equivalent resistances in a set of thirty-six chips. We also carried out an additional set of experiments to evaluate the stability of the PUF. The PUF stability experiments were performed on one of the chips in the set. To evaluate stability, we repeated the signature generation/measurement process seventy-two times[3]. The variation across the set of signatures from these experiments is due entirely to environmental noise and temperature variations. The stability experiments are important for determining the probability of signature aliasing, i.e., the probability that two chips from the population generate the same signature. We refer to data from the stability experiments as control data.

The experimental results for twelve of the chips from the set of thirty-six are shown in Figures 6 and 7, using the voltage drops and equivalent resistances, respectively. The left half of the figure lists the chip number along the x-axis. The right half gives twelve of the PUF stability results for one chip. The six data points defining the chip signature are displayed vertically above the chip identifier. The y axis gives the voltage drop and equivalent resistance, respectively, in each of the figures.

The diversity among the signatures in the twelve chips shown on the left side of the figures is evident in both plots. In addition to the different patterns of dispersion in the signatures, the ordering of the data points from top to bottom is also distinct across all chips. The ordering is in reference to the SMCs that each data point corresponds to. For example, SMC_{00} in Figure 3(a) is assigned 0, SMC_{01} is assigned 1, ..., SMC_{12} is assigned 5. In Figure 6, the ordering for chip 1 is 5, 1, 2, 0, 4, 3. while the ordering for chip twelve is 3, 0, 5, 1, 2, 4. Therefore, the apparent diversity among the signatures due to dispersion is actually larger because of the differences in the orderings. It is also clear from the PUF stability experiments that environmental variations have an impact on the signature and therefore, they must be taken into account.

In many cases, there are differences in the dispersion and

---

1. It is possible to replace the 'shorting inverter' with a single PFET. However, the stacked devices of the shorting inverter are more robust to defects and is proposed as a fault tolerant strategy to prevent yield loss that might result if a defect caused the stimulus transistor to remain in the on-state.

2. This scheme refers to the original SMC (Fig. 3) modified to include a second sense transistor connected between M1 and a new voltage 'sense wire 2' (Fig. 5).

3. No temperature control or specialized low noise test apparatus was used.

978-1-60558-497-3/09 $25.00 © 2009 ACM

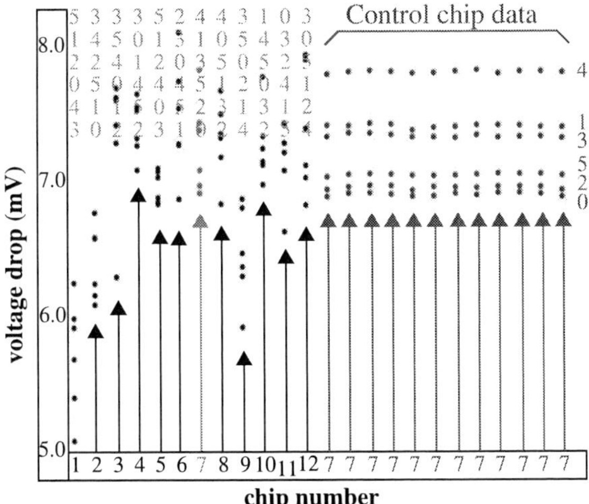

Figure 6. Voltage drop signatures for 12 chips and 12 control samples.

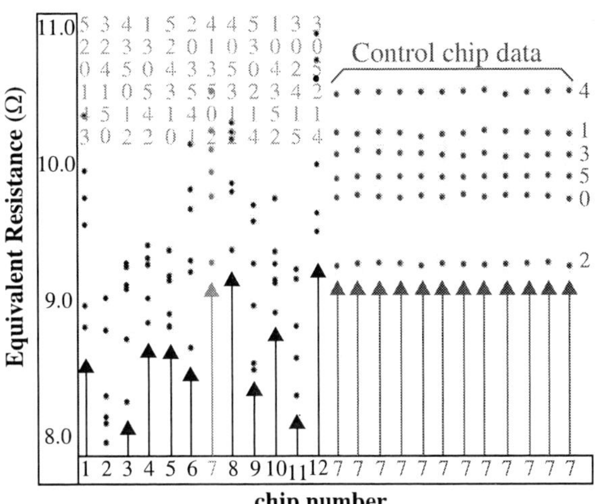

Figure 7. Equivalent resistance signatures for 12 chips and 12 control samples.

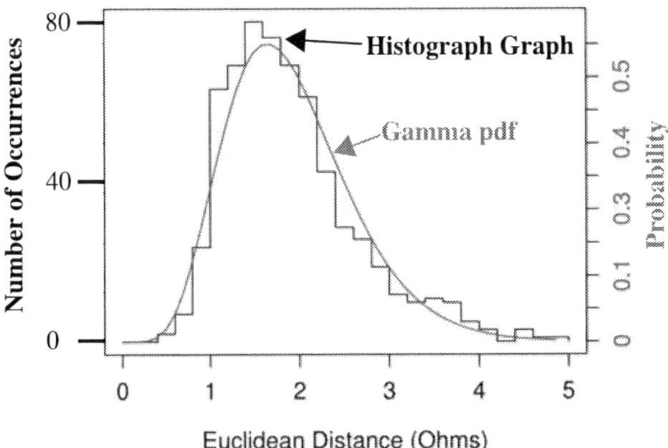

Figure 8. Gamma function fit of chip equivalent resistance histogram.

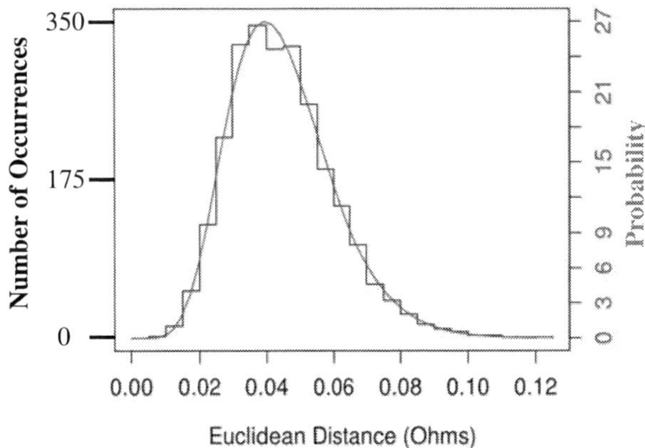

Figure 9. Gamma function fit of noise equivalent resistance histogram.

ordering of the data points for the same chip across the voltage drop and equivalent resistance analyses. This is expected because the equivalent resistance eliminates an element of the diversity introduced by variations in the magnitude of the shorting currents.

In order to quantitate the dispersion among the chip signatures, we compute the Euclidean distance between the data points and analyze their variance. The six data points in each signature can be interpreted as a single point in a six-dimensional space. The Euclidean distance between two signatures for chips x and y is given by Equation 1. The Euclidean distance is computed between

$$\text{dist} = \sqrt{(x_1 - y_1) + (x_2 - y_2) + \ldots + (x_6 - y_6)} \qquad \textbf{Eq. 1.}$$

all possible pairing of chips, i.e., (36*35)/2 = 630 combinations. The same procedure is carried out using the control data in which (72*71)/2 = 2556 combinations are analyzed.

In order to compute the probability of two chips producing the same signature given the uncertainty associated with the measurements, we first compute a histogram that tabulates the number of Euclidean distances partitioned into a set of bins for the chip and noise data sets separately. The bins in each histogram are

equal in width, with each equal to 1/25th of the total span that defines the range of Euclidean distances among the 630 and 2556 combinations of chip and noise data pairings, respectively. We then fit these histograms to gamma probability density functions (pdf). The histograms and the gamma pdfs are shown superimposed in Figure 8 (chip) and Figure 9 (noise) for the equivalent resistance analysis. In both cases, the gamma functions are a good fit to the histograms. The range of values found among the 630 chip pairings is between 0.45 and 5.0, as indicated by the x-axis, while the range for the noise analysis is between 0.01 and 0.12. Therefore, the largest value in the noise data is approximately four times smaller than the smallest value in the chip data.

We compute the probability of aliasing by first determining the Euclidean distance in the noise data that bounds 99.7% (3 sigma) of the area under the pdf. This particular Euclidean distance upper bounds the worst case noise and is equal to 0.099 for the data shown in Figure 9. We then compute the cumulative distribution function (cdf) of the chip data and use this worst case noise value to determine the probability of aliasing by looking up the y value on the chip cdf associated with this x value. This gives us the

978-1-60558-497-3/09 $25.00 © 2009 ACM

680

probability that the Euclidean distance between any pairing of two chips is less than or equal to the worst case Euclidean distance among the control data.

The results for the equivalent resistance and voltage analyses are given in Table 1. Using equivalent resistances, the probability of aliasing is 6.9e-8 or approximately 1 chance in 15 million. For the voltage analysis, the probability increases to approximately 1 chance in 28 billion. Given that the number of SMCs used to define the signature in these experiments is only six, we can expect, based on these results, that the probability would improve in a commercial design that included a larger number of SMCs.

Table 1: Probability of aliasing.

	Analysis Type	
	Volt.	Eq. Res.
Prob. Eucl. dist. of chips < 99.7% of all noise Eucl. dist.	3.5e-11	6.9e-8

6. CONCLUSIONS

Methods to provide IC security based on manufacturing variability have recently emerged for applications such as IC authentication, secure activation, encrypted communication and IP protection on FPGAs. These applications rely on intrinsic variability of the hardware to provide a signature that is probabilistically unique to each IC. Hardware circuits that leverage process variations to implement IC signatures are called physically unclonable functions (PUFs). In this paper, we propose a PUF that leverages the inherent resistance variations in the metal layers defining the power grid. Data from a set of thirty-six chips fabricated in a 65 nm technology is used to confirm the feasibility of this strategy.

7. ACKNOWLEDGEMENTS

We acknowledge Sani Nassif and Kanak Agarwal of IBM Austin Research Laboratory for their support of this research. This research was supported in part by NSF grant CNS-0716559.

8. REFERENCES

[1] R. S. Pappu, B. Recht, J. Taylor and N. Gershenfeld, "Physical One-Way Functions," *Science*, 297(6), 2002, pp. 2026-2030.

[2] Y. Su and J. Holleman and B. Otis, "A 1.6pJ/bit 96% Stable Chip ID Generating Circuit using Process Variations," *Proc. of International Solid State Circuits Conference*, 2007, pp. 406-407.

[3] B. Gassend and D. E. Clarke and M. van Dijk and S. Devadas, "Silicon Physical Unknown Functions," *Proc. of Conference on Computer and Communications Security*, 2002, 148-160.

[4] G. E. Suh and S. Devadas, "Physical Unclonable Functions for Device Authentication and Secret Key Generation", *Proc. Design Automation Conference*, 2007, pp. 9-14.

[5] D. Lim, J. W. Lee, B. Gassend, G.E. Suh, M. van Dijk and S. Devadas, "Extracting Secret Keys from Integrated Circuits, " *Trans. on Very Large Scale Integration Systems*, 13(10), Oct. 2005, pp. 1200-1205.

[6] B. Gassend, D. Clarke, M. van Dijk and S. Devadas, "Controlled Physical Random Functions," *18th Annual Computer Security Applications Conference*, 2002.

[7] B. Gassend and M. Van Dijk and D. Clarke and E. Torlak and S. Devadas and P. Tuyls, "Controlled Physical Random Functions and Applications," *ACM Transactions on Information and System Security*, Volume 10, Number 4, 2008.

[8] E. Ozturk, G. Hammouri and B. Sunar, "Physical Unclonable Function with Tristate Buffers," *Proc. International Sympo-sium on Circuits and Systems*, 2008, pp. 3194-3197.

[9] E. Ozturk, G. Hammouri and B. Sunar, "Towards Robust Low Cost Authentication for Pervasive Devices", *Proc. International Conference on Pervasive Computing and Communications*, March 2008 pp. 170-178.

[10] Y. Alkabani and F. Koushanfar and N. Kiyavash and M. Potkonjak, "Trusted Integrated Circuits: A Nondestructive Hidden Characteristics Extraction Approach," *Information Hiding*, 2008.

[11] R. Helinski, "Measuring Power Distribution System Resistance Variations for Application to Design for Manufacturability and Physical Unclonable Functions," M.S. thesis, University of Maryland, Baltimore Co., July, 2008.

[12] R. Helinski, J. Plusquellic, "Measuring Power Distribution System Resistance Variations," *Transactions on Semiconductor Manufacturing*, Volume 21, Issue 3, pp. 444-453, Aug. 2008.

[13] S. Maeda and H. Kuriyama and T. Ipposhi and S. Maegawa and Y. Inoue and M. Inuishi and N. Kotani and T. Nishimura, "An Artificial Fingerprint Device (AFD): a Study of Identification Number Applications Utilizing Characteristics Variation of Polycrystalline Silicon TFTs," *Trans. on Electron Devices*, number 50, issue 6, June, 2003, pp.1451- 1458.

[14] K. Lofstrom, W. R. Daasch and D. Taylor, "IC Identification Circuits using Device Mismatch," *Proc. of International Solid State Circuits Conference*, 2000, pp. 372-373.

[15] J. Huang and J. Lach, "IC Activation and User Authentication for Security-Sensitive Systems," *Proc. of IEEE International Workshop on Hardware-Oriented Security and Trust*, 2008, 79-83.

[16] E. Simpson and P. Schaumont, "Offline Hardware/Software Authentication for Reconfigurable Platforms," *Cryptographic Hardware and Embedded Systems*, Volume 4249, Oct., 2006, pp. 10-13.

[17] J. Guajardo, S. S. Kumar, G.-J. Schrijen and P. Tuyls, "Physical Unclonable Functions and Public Key Crypto for FPGA IP Protection," *Conference on Field Programmable Logic and Applications*, 2007, 189-195.

[18] S. S. Kumar and J. Guajardo and R. Maes and Geert-Jan Schrijen and P. Tuyls, "Extended Abstract: The Butterfly PUF Protecting IP on Every FPGA," *Proc. of IEEE International Workshop on Hardware-Oriented Security and Trust*, 2008, pp. 70-73.

[19] Y. Alkabani and F. Koushanfar and M. Potkonjak, "Remote Activation of ICs for Piracy Prevention and Digital Right Management," *Proc. of International Conference on Computer-Aided Design*, 2007, pp. 674--677.

[20] J. Guajardo, S. S. Kumar and G. Schrijen and P. Tuyls, "Brand and IP Protection with Physical Unclonable Functions," *IEEE Symposium on Circuits and Systems*, 2008, pp. 3186-3189.

[21] B. Gassend and D. Lim and D. Clarke and M. van Dijk and S. Devadas, "Identification and Authentication of Integrated Circuits, Concurrency and Computation: Practice and Experience, 2003.

[22] J. Li and J. Lach, "At-Speed Delay Characterization for IC Authentication and Trojan Horse Detection," *Workshop on Hardware-Oriented Security and Trust*, 2008, 8-14.

[23] J. Lee, Mohammad Tehranipoor, Chintan Patel and Jim Plusquellic, "Securing Designs Against Scan-Based Side-Channel Attacks," Transactions on Dependable and Secure Computing, Volume 4, Number 4, October-December 2007, pp. 325-336.

[24] J. Lee, M. Tehranipoor, J. Plusquellic, "A Low-Cost Solution for Protecting IPs against Scan-Based Side-Channel Attacks," Proc. VLSI Test Symposium, May 2006, pp. 42-47.

Hardware Authentication Leveraging Performance Limits in Detailed Simulations and Emulations *

Daniel Y. Deng
Cornell University
Ithaca, NY
deng@csl.cornell.edu

Andrew H. Chan [†]
University of California
Berkeley, CA
andrewhc@eecs.berkeley.edu

G. Edward Suh
Cornell University
Ithaca, NY
suh@csl.cornell.edu

ABSTRACT

This paper proposes a novel approach to check the authenticity of hardware based on the inevitable performance gap between real hardware and simulations or emulations that impersonate it. More specifically, we demonstrate that each processor design can be authenticated by requiring a checksum incorporating internals of complex micro-architectural mechanisms to be computed within a time limit; this checksum is different for each processor model and only authentic secure hardware can obtain the checksum fast enough. This new authentication approach provides attractive solutions to privacy, scaling, and security issues of traditional approaches that otherwise rely only on certificates. Architectural simulations and an RTL implementation show that the proposed approach is viable with very low hardware overheads.

Categories and Subject Descriptors. C.3 [**Special-Purpose and Application-Based Systems**]

General Terms. Design, Security

Keywords. Hardware authentication, Secure processors

1. INTRODUCTION

Hardware serves as a foundation for trust in software. Security mechanisms built in software can be compromised if hardware is insecure. Moreover, new hardware features are playing an increasingly important role in securing highly interconnected computer systems. Recent Intel microprocessors are enhanced with Trusted eXecution Technology (TXT) [7] and many computing systems are already equipped with a Trusted Platform Module (TPM) [15], which enable trusted applications over the network. For example, in an Internet banking application, TXT and TPM can attest software on a client system so that the bank server can ensure that there is no malicious software.

---

*This work was partially supported by the National Science Foundation under grants CNS-0746913 and SA4897-10808PG, and an equipment donation from Intel.

†The work was done while the author was at Cornell.

Permission to make digital or hard copies of part or all of this work for personal or classroom use is granted without fee provided that copies are not made or distributed for profit or commercial advantage and that copies bear this notice and the full citation on the first page. To copy otherwise, to republish, to post on servers or to redistribute to lists, requires prior specific permission and/or a fee.
DAC'09, July 26-31, 2009, San Francisco, California, USA

To be trusted, secure hardware must convince other systems that they are indeed interacting with *authentic hardware of a trustworthy design*, not a software emulator or untrustworthy hardware. As an example, without authentication of TPM hardware, a virtual machine can simply pretend to be a TPM by emulating its external behavior. This paper aims to enable bootstrapping of trust between remote systems by checking the authenticity of trusted hardware.

Today's approach to authenticate an unknown system rely on a Certificate Authority (CA) to certify a public key of the remote system. For example, TPMs must come with a manufacturer's certificate saying that it is an authentic TPM. Unfortunately, this certificate-based approach faces a number of limitations [6] when applied to hardware (processor). First, there is a serious concern for privacy; activities done with a particular processor can be linked together because one public key pair is used for authentication. Second, if a certificate is issued for a wrong public key or a private key becomes exposed, the authentication scheme is broken and an adversary can easily impersonate the corresponding secure processor. Finally, the centralized CA introduces cost and scalability issues. Today, for websites, companies such as Verisign charge website owners for their service as a trusted party. Requiring such costly services can severely limit the applications of secure processors.

This paper proposes to directly check the authenticity of hardware based on its low-level implementation details. More specifically, we focus on micro-architectural features, which are very complex and different for each processor model, of a high-end secure processor, In our approach, a secure processor is challenged to provide a checksum that depends on cycle-by-cycle activities of its internal micro-architectural mechanisms for a given code within a time limit. Even after years of research in developing fast simulation technologies for design space exploration and design validation, accurate simulation or emulation of processor micro-architecture is still extremely slow compared to real hardware. Moreover, producing a high-end microprocessor is becoming prohibitively expensive as processor complexity and fabrication costs increase. Therefore, only an authentic secure processor hardware can compute a correct checksum fast enough.

This proposed technique complements the traditional certificate approach. There is no privacy concern because the authentication process only reveals that a processor is a particular model with security features, but not which particular instance. The processor can easily have many independent key pairs. Also, impersonating a secure processor is difficult even if a legitimate private key or all of the micro-architecture details are known; an adversary still needs to compute the checksum as fast as real hardware. Finally, our approach enables secure processors to be introduced in a distributed fashion without relying on a few centralized certificate authorities.

Architectural simulations and an RTL implementation show that the proposed approach is indeed viable. The hardware extension to compute the micro-architecture checksum only requires negligible amounts of additional hardware resources. Moreover, even a small difference in the processor's micro-architecture results in significant deviation in the checksum, therefore an adversary will not be able to use other processors or a stripped down FPGA implementation to impersonate a secure processor. Also, even with recent advances in simulation technologies, simulations are still multiple orders of magnitude slower than real hardware; an adversary will not be able to use simulations to obtain valid checksums fast enough. We believe that the only viable attack on the proposed approach is to build a clone of the secure processor that has an identical micro-architecture and comparable performance but without the security features of an authentic secure processor, which is prohibitively difficult and expensive for most adversaries.

The rest of the paper is organized as follows. Section 2 describes our proposed authentication approach. Section 3 and Section 4 discuss the architectural mechanisms and challenge programs that are required to realize the approach. Section 5 evaluates the overheads and the security of our scheme through software and RTL modeling. Section 6 compares related works and Section 7 concludes.

2. AUTHENTICATION APPROACH

The goal of a verifier (V) is to check that a public key PK_T from the target system (T) belongs to an authentic hardware, not a simulator or emulator, of a particular model. Here, we assume that an authentic secure processor has a private key that is only known to itself.

Our approach leverages the complex cycle-by-cycle low-level micro-architecture state of modern high-end processors. Effectively, a microprocessor pipeline is used as a complex function that maps a set of challenges to responses based on the implementation details. In our context, a challenge is a program or a test vector, and a response is a checksum computed based on cycle-by-cycle processor operations. We call this function, Micro-architecture Signature Function (MSF).

For security, the MSF must be different for each processor model and each challenge, be efficiently computable only by an authentic processor, and be difficult to duplicate or simulate quickly. For usability, the MSF must be deterministic and consistent across all instances of the same model. To prevent an adversary from using an authentic secure processor to impersonate the processor, the interface to the MSF is restricted to the following three.

- MSF_SIGN(challenge): Produce $H(PK_T||response)$, the hash of its public key and the response for a given challenge. This operation is used by the target system.

- MSF_VERIFY(challenge, signature, PK): Check if a given signature matches $H(PK||response)$ for a given challenge. Returns TRUE or FALSE.

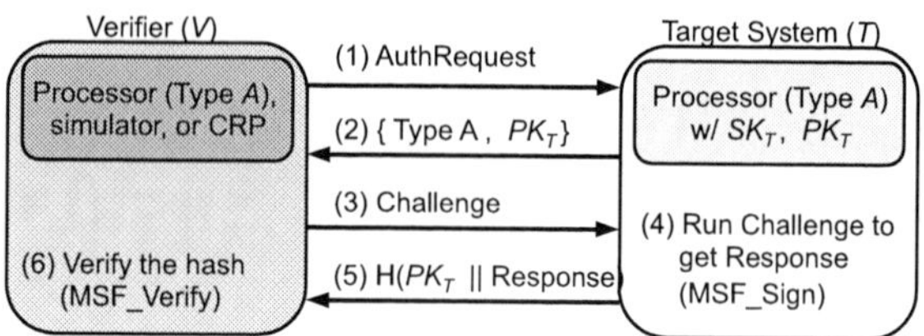

Figure 1: The authentication protocol based on micro-architectural features.

- MSF_CRP(): Return a Challenge-Response Pair (CRP) for a randomly chosen challenge.

Figure 1 shows our authentication process, with an example protocol. First, a verifier (V) initiates the process by sending an *AuthRequest* message to a target system (T), which replies with its public key PK_T and its hardware model. To validate the claim, the verifier sends a challenge to the target. The challenge could be a full program or simply a seed to a program (see Section 4). The target replies with a signature from MSF_SIGN: runs the challenge and hashes (cryptographic hash such as SHA256) the response with its public key to bind the public key with the response. The verifier checks the hash using either MSF_VERIFY or a previously obtained Challenge Response Pair (CRP).

To prevent an adversary from using a simulator to obtain the response, the verifier sets a time limit. If the response is not received within the predetermined limit, the authentication fails. The time limit is set to be the time required for an authentic hardware to run the challenge plus the worst-case network latency. Note that this limit even with the network latency can be made much smaller than the time required for the fastest simulator to compute the correct response by increasing the execution time for a challenge. The verifier changes the challenge for each authentication to ensure that an adversary cannot replay a recorded response.

To prevent denial of service attacks, the target system needs to limit the execution time of a challenge and the frequency of the authentication requests. The challenge programs must also be properly isolated so that they cannot perform sensitive system calls. Operating systems normally perform such isolation using software layers and virtual machines techniques.

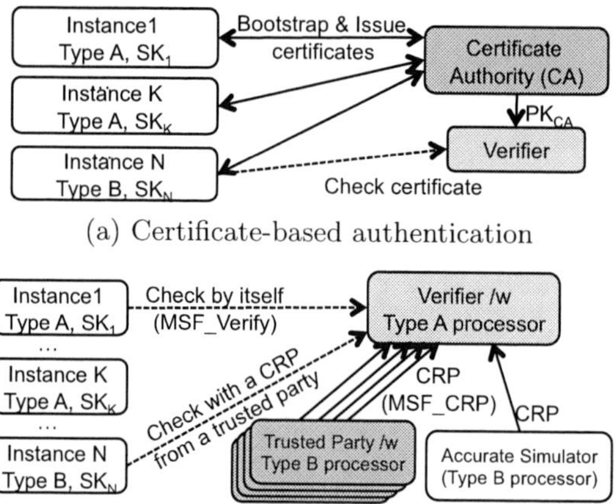

(a) Certificate-based authentication

(b) MSF-based authentication

Figure 2: The authentication infrastructures.

Figure 2 compares our approach with a traditional certificate-based approach. In the certificate-based framework, a trusted

978-1-60558-497-3/09 $25.00 © 2009 ACM

Certificate Authority (CA) issues a certificate for each instance of a secure processor. The CA must somehow establish trust with each and every processor instance before it can issue a certificate. As a result, a few centralized CAs must provide service for millions of devices.

The proposed scheme enables processor authentication in a more distributed fashion as shown in Figure 2 (b). First, a verifier can independently authenticate a target without a trusted third-party if it has access to any authentic secure processors of the target's model (use `MSF_VERIFY`). Similarly, *if* the vendor releases an accurate simulator for a particular processor model, anyone can authenticate any instance of the model in a completely distributed fashion. Even if a verifier does not have an authentic processor of the target's model, the processor vendor or *any secure processor of the same model* can provide a randomly selected challenge-response pair (CRP) using the `MSF_CRP` primitive. Unlike a traditional CA, the trusted party here does not need to bootstrap each target instance. The verifier can save and use the CRP for an authentication process at any time.

In addition to its distributed nature, the proposed scheme preserves privacy because a single processor can easily use many different public keys. Also, the security of the traditional approach is enhanced because knowing one private key does not enable an adversary to fake the response.

3. HARDWARE DESIGN

This section describes hardware extensions to construct the MSF on modern high-end processors using internal pipeline checksums. For security, the checksums must reflect details of architectural features and be sensitive to even small differences in them. For reliability, the checksums must be consistent so that all processors of the same model can generate the same response.

3.1 Checksum Computation

To incorporate effects of various micro-architectural mechanisms, our checksum includes the fetched and dispatched instructions in each cycle. While the checksums can easily include more details in each pipeline stage, we found the fetched and dispatched instructions (augmented with cycle counts) capture almost all of the plausible micro-architecture discrepancies between any two processor designs.

For example, modern processors can be largely characterized by their mechanisms for speculative and out-of-order execution, and the memory hierarchy. A cycle-by-cycle record of fetched instructions reflect the speculation mechanisms such as branch predictors and instruction cache hits/misses. Similarly, the record of dispatched instructions show the behavior of dynamic scheduling and data caches.

For the MSF checksums, a processor is augmented with a checksum unit for each fetch and dispatch slot. For example, a 32-bit four-way superscalar processor will produce four 32-bit checksums for fetched instructions and four 32-bit checksums for dispatched instructions. The checksum units are zeroed and initiated at the start of a challenge program. At the end of the challenge, all checksums are combined by a cryptographic hash to generate the final response.

Figure 3 shows how each checksum is computed, which applies to both fetched and dispatched instructions. In each cycle, the checksum circuit either ADDs or XORs the incoming 32bit value. The toggling between an ADD and an XOR allows the results of the accumulation to be strongly-ordered;

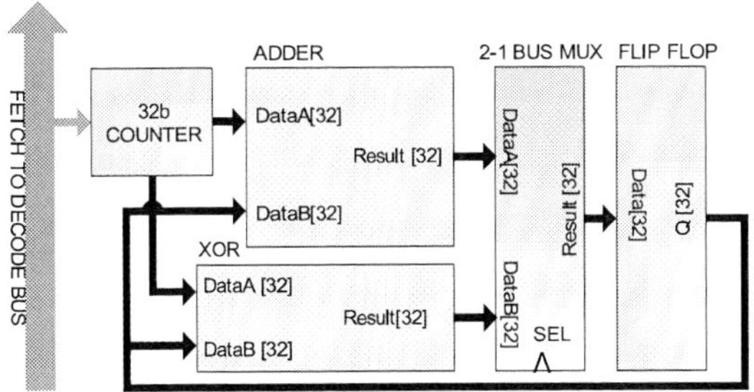

Figure 3: A checksum unit for fetched instructions.

deviation in the order of the incoming values would alter the checksum results. In the checksum, idle cycles in addition to real instructions are important because they may indicate interesting events such as a cache miss or branch misprediction. To account for these idle cycles, our checksum units include the number of idle cycles encountered since the start of a challenge instead of the incoming 32bit value on each idle cycle. Note that an adversary must know exact cycle-by-cycle pipeline operations in order to compute a correct checksum.

3.2 New Instructions

We introduce five new instructions to support the MSF operations. First, the `MSF_BEGIN` and `MSF_END` instructions have a processor internally generate a response for a challenge program. The `MSF_BEGIN` instruction prepares the processor for a deterministic execution and initiates the checksum computation. The instruction first disables interrupts, loads challenge code and data into the L2 cache, and sets the TLB to preclude off-chip accesses. Then, the processor is bootstrapped into a deterministic state by clearing any non-deterministic state including the pipeline, L1 caches, cache controller buffers, and branch predictor tables, etc., and starts the checksum units. The `MSF_END` stops the checksum units, compute a response by hashing checksums, stores the response in an internal register, and enables interrupts. Second, the `MSF_SIGN` and `MSF_VERIFY` instructions use the stored response to provide MSF operations. The `MSF_SIGN` instruction outputs the hash of the concatenation of its public key and the response. The `MSF_VERIFY` instruction allows the hash from another processor to be checked. The instruction gets a public key (PK_T) and a hash ($H(PK_T \| Response)$) as arguments, compute the hash with its own response, and checks if the two hashes match. Finally, the `MSF_CRP` instruction runs a hard-coded challenge code with a random seed and returns its response (see the next section for a challenge code design).

3.3 Non-Determinism

For reliable authentication, a processor must produce a consistent response to a challenge every time the same challenge runs. A microprocessor is essentially a finite state machine, therefore its cycle-by-cycle behavior will be deterministic if non-deterministic state and signals are eliminated. In fact, microprocessors often support short but deterministic execution for post-silicon validation. A previous study also demonstrated a longer deterministic execution for modern microprocessors [5].

In this work, to ensure consistent initial state, the processor sets up its state on `MSF_BEGIN`. During execution, there are two sources of non-determinism: off-chip memory accesses and interrupts. We eliminate the off-chip memory accesses by carefully designing the challenge program to only access on-chip memory and avoid accessing memory through off-chip buses that cross clock domains. Alternatively, we can also make off-chip accesses slow but deterministically long. The processor disables maskable interrupts and ignore I/O for the duration of challenge execution. Extremely rare exceptions such as a power failure can still occur during challenge program execution. In this case, the processor simply aborts the checksum computation and the authentication needs to be retried later. Avoiding off-chip accesses also allow the processors with one micro-architecture design to produce the same response regardless of the operating frequency.

4. CHALLENGE PROGRAM

The challenge program must satisfy several properties to be effective in distinguishing between processor models. First, to elicit unique responses, the challenge program must exercise the functionality of a processor with high coverage so that small differences in micro-architecture can be captured. Second, new challenges must be relatively easy to create so that the verifier can use a unique challenge on each authentication attempt. Thus, processor hardware is designed to be able to generate a random challenge for the MSF_CRP operation. Lastly, the challenge program must be small and self-contained to fit into on-chip caches, especially if hardware cannot support deterministic off-chip accesses.

To differentiate micro-architectures, the challenge program must present many different patterns to a micro-architecture mechanism. Also, to make checksums difficult to predict, the patterns should result in high variance in micro-architectural behavior. In our design, we make judicious use of random patterns to achieve both objectives. For example, a random traversal of data can produce access patterns that result in unpredictable hits and misses depending on the cache configuration. Similarly, branches with random targets of many different locations result in significantly varying predictions from different branch prediction algorithms.

In our design, the challenge program takes a seed for a pseudo-random number generator (PRNG) and performs repeated sequences of pseudo-random operations: memory accesses to pseudo-random locations to differentiate data memory organizations, pseudo-random branches to differentiate speculative execution mechanisms, jumps to pseudo-random locations to differentiate instruction memory organizations and BTBs, and long sequences of instructions with pseudo-random dependencies and operations to differentiate out-of-order execution schemes. The use of a pseudo-random function has an added benefit that it is very easy to generate many different challenges by simply changing the seed.

Figure 4 shows the pseudo-code of our challenge program. The challenge program takes a seed as an argument, which is used to seed the T-function, our pseudo-random number generator (PRNG). This PRNG is then used to initialize a small array, whose size can be set to control cache miss-rates. In the main body, the challenge program generates a new random number x and reads the array element determined by the random number. Following the read, the program jumps to a random location in the program based on x (the switch statement). Then, a conditional branch based on a random condition executes. Finally, a number of unique instructions are placed in both branch paths. The

```
void challenge(int seed) {
    MSF_BEGIN;              \\ prepare and start checksums
    x = seed;
    \\ create an a[] of size N
    do {
        x = x + ((x*x) | 5); \\ generate a new random number
        v = a[x & (N - 1)];  \\ load a random address
        switch(x & (M - 1)) {
        case 0:
            if((x & 32) == 0) { \\ branch on a random condition
            \\ Z random instructions with random dependencies
            } else {
            \\ Z random instructions with random dependencies
            } break;
        case 1:
            if((x & 32) == 0) { \\ branch on a random condition
            \\ Z random instructions with random dependencies
            } else {
            \\ Z random instructions with random dependencies
            } break;
            ...
        case M-1:
            if((x & 32) == 0) { \\ branch on a random condition
            \\ Z random instructions with random dependencies
            } else {
            \\ Z random instructions with random dependencies
            } break;
        }
    } while (j < N);   \\ do N times
    MSF_END; \\ signals checksum units and cycle counter to stop
}
```

Figure 4: The implemented challenge program. For our simulations, Z=20, M=1024, N=65,536.

instructions have random dependencies and operations to differentiate steering mechanisms in out-of-order processors. This read-jump-branch-compute sequence is then repeated a number of times to build a long sequence of operations and to sufficiently prolong the execution time to mask the worst-case network latency.

Note that the results of the challenge program, which are easy to compute on any processor, are unimportant in our context because the checksums collect the internal pipeline operations. We also note that the challenge code design can leverage knowledge from processor validation and testing as they face similar goals of achieving high design coverage. Randomly generated test vectors are commonly used to elicit unexpected behaviors. Test programs can also be customized for a particular processor design [2].

5. EVALUATION

5.1 Overheads

The proposed hardware extension has negligible area overheads, only adding checksum units, a cryptographic hash engine, and a cycle counter. Each checksum circuit requires a 32-bit adder, 32 XOR gates, a 32-bit 2-to-1 multiplexer, a latch, and a register. The adder is the largest and most complex element in the circuit. Prior work [16] showed that a 5GHz, 32-bit adder in the 130nm technology consumed about $0.03mm^2$ area overhead. Therefore, 8 adders $(0.24mm^2)$ represent less than 0.6% of the area of a Intel Pentium M processor $(39mm^2)$ [4]. Recent ASIC implementations [10] show that a SHA-256 engine only requires on the order of 40-50K transistors. Further, the hash engine does not need to be high performance as it is rarely used.

The proposed extension will not adversely affect performance or power consumption. The checksum units do not lie on the critical path of computation, and their added load

978-1-60558-497-3/09 $25.00 © 2009 ACM

Varied Parameters ('+': change from the baseline)		Fetch		Dispatch	
		Avg	Min	Avg	Min
L1 I-cache					
- latency (clks)	1, 2, 3, 4	22.9	22.4	23.3	22.8
- size (KB)	4,8,16,32,64	22.4	21.2	22.7	21.6
- associativity	1, 2, 4, 8, 16	21.9	21.1	22.3	21.4
L1 D-cache					
- latency (clks)	1, 2, 3, 4	22.9	22.6	23.3	22.9
- size (KB)	4,8,16,32,64	21.0	20.7	21.3	21.1
- associativity	1, 2, 4, 8, 16	21.2	21.1	21.6	21.5
L2 latency(clks)	6,8,10,12,14	22.7	21.4	23.0	21.7
BTB					
- size (entries)	512,1K,2K,4K	21.6	21.1	21.2	20.8
- associativity	1, 2, 4	21.0	20.8	21.1	20.9
Branch Predictor	YAGS,gskew, gshare,hybrid	23.7	23.3	23.4	23.0
Fetch widths	2, 3, 4, 5, 6	19.3	16.5	22.4	20.1
Retire widths	2, 3, 4, 5, 6	22.8	21.1	23.1	21.4
Issue widths	2, 3, 4, 5, 6	23.0	21.5	29.8	22.1
Int FU latency	+0,+1,+2,+3	20.9	20.0	21.2	20.3
FP FU latency	+0,+1,+2,+3	22.1	21.5	22.4	21.8
ROB size	80,96,112,128, 144,160	23.0	22.3	23.4	22.9

Table 1: Differences (%) in cycle-by-cycle fetched and dispatched instructions when varying a single micro-architecture parameter.

capacitance can be minimized with low input-capacitance buffers. Also, the added circuits can be turned off during normal operation to minimize energy overheads.

5.2 Effectiveness

To study its effectiveness in differentiating micro-architectures, the proposed scheme was implemented in superscalar processors built on a simulator (SESC) [9] and an RTL model (IVM) [17]. For each processor, we compared differences in cycle-by-cycle behavior when a single architectural parameter is changed.

Table 1 shows the percentage of cycles that a different instruction was observed by each checksum unit when a single micro-architecture parameter was varied for a 4-way superscalar processor modeled in SESC. *Fetch Avg* and *Fetch Min* show the average and the minimum differences for the fetched instructions, and *Dispatch Avg* and *Dispatch Min* show the average and the minimum for the dispatched instructions. Percentages appear low because checksum units often don't have valid instructions to process. The difference exceeds 90% if we only count cycles with valid instructions. RTL model studies show similar differences and weren't included due to space limits. The results demonstrate that each processor behaves very differently at the micro-architectural level even when only one configuration is changed.

Our authentication scheme also requires that each challenge for the same processor elicit unique responses. We ran 50,000 different challenge seeds on the same configuration and a different response was observed for each challenge.

5.3 Deterministic Execution

To confirm that the micro-architecture checksums can be reliably reproduced, we studied deterministic execution of a challenge on the Verilog model (IVM) and Pentium 4. Our RTL model implemented mechanisms described in Section 3.3; before computing checksums, the processor clears its state such as branch history tables, caches, and the processor pipeline, and pre-loads on-chip caches and TLBs to avoid off-chip accesses. In IVM, running a challenge multiple times, with arbitrary programs in between, produced

an identical response. To further verify the feasibility of getting deterministic responses, we executed our challenge code on a Pentium 4 machine running Linux, and measured the cycle count for each execution using a built-in counter. For deterministic execution, we disabled interrupts (CLI instruction), cleared the pipeline (CPUID), and initialized caches and TLB entries with multiple pre-runs before running the challenge. The experiment confirmed that a short execution [1] can be deterministic. Repeatedly running the challenge with a fixed seed returned consistent cycle counts, and different seeds returned different cycle counts.

5.4 Security Discussion

The goal of an adversary is to fool a verifier to accept a public key PK_{Fake} that does not belong to an authentic secure processor. Therefore, to be successful, the adversary must be able to present a valid hash $H(PK_{Fake}\|Response)$ for a given challenge within the time limit. Here, we discuss potential ways that an adversary can try to obtain the valid response for a given challenge.

An adversary can try to use an authentic processor to extract a response. However, none of the three MSF instructions allows an adversary to impersonate the authentic processor. The MSF_SIGN instruction gives out a hashed response, $H(PK_T\|Response)$. However, an adversary cannot replace PK_T with PK_{Fake} or obtain $Response$ from the hash given a non-malleable one-way hash function. The MSF_VERIFY instruction only reveals whether a hashed response is correct or not. Finally, the MSF_CRP does not let an adversary choose a particular challenge that is asked by the verifier. Note that different processor models cannot be used either because they generate different checksums.

Without exploiting existing authentic processors, an adversary must compute the correct response directly. However, this approach requires an adversary to reverse engineer the exact micro-architecture of a target processor, which is quite difficult and expensive. Even if the detailed micro-architecture is known, the adversary still needs to compute the exact cycle-by-cycle pipeline operations *as fast as* the authentic processor. Years of research in performance modeling and processor verification suggest that this proposition is practically infeasible.

Due to the complexity of modern processors, accurate simulations of internal microprocessor operations are extremely slow. Even low-fidelity academic simulators run 3 to 4 orders of magnitude slower than the actual processor being modeled. As an example, the SESC simulator used in our evaluations resulted in a 2200X slowdown in comparison to executing the challenge on actual hardware. More accurate performance simulators used in industry often are about a million times slower than the target [8]. Moreover, even the industry performance simulators are not sufficient to obtain accurate cycle-by-cycle checksums because they often ignore some details as their goal is to *explore* the design space. Faithful RTL simulations that are used in functional verification will be accurate enough, but are even slower (only run at tens of Hertz [14]). Also, our RTL studies showed that inputs to checksum units depend on the entire RTL model; therefore it will be very difficult to compute the checksums using a simpler sub-circuit of the processor.

Recent advances in simulation methodology propose to use Field Programmable Gate Arrays (FPGAs) for timing

---

[1]The challenge code ran from 20 milliseconds to 1 second.

models [3]. This approach has the promise of increasing simulation speeds by upwards of 100 times over that of pure software simulators. An adversary can also achieve additional speed ups by running simulations on future microprocessors and FPGAs. However, for high-performance microprocessors built on the state-of-the-art fabrication technologies, accurate simulations will still be orders of magnitude slower than authentic hardware during the processor's life cycle. For example, if a processor is used for 10 years, industry-level performance simulators will be still at least 1000 times slower.

As a result, to match the performance of actual hardware, the adversary must build a custom chip that is essentially a clone of the target microprocessor. However, such an attack will be simply too costly for most applications and adversaries as high-end microprocessors cost enormous amounts of time and money to design and produce.

6. RELATED WORK

Public Key Infrastructure (PKI) based on a Certificate Authority (CA) is the most widely used framework for remote authentication of unknown parties. As discussed before, the proposed scheme complements the weaknesses of the PKI such as privacy, security, and cost.

A simple approach to authenticate a hardware model is to embed the same private key in all instances of a design. This approach greatly simplifies binding of a key to trusted hardware because there is a single public key that's common to all instances. However, this approach suffers from serious security issues; if any device is compromised and the private key is leaked, all devices are broken.

Seshadri et al. proposed time-bound verification functions to authenticate software executing on a remote system [12, 11], which exploits the fact that malicious software will slow down an execution. However, such time bounds can only be defined for known hardware configurations. This paper aims to verify that the hardware is a trusted model so as to enable additional software-based techniques such as theirs.

A Physical Unclonable Function (PUF) is a function that maps challenges to responses using an intractably complex physical system. For example, silicon PUFs use the unique delay characteristics of each IC, which are determined by random manufacturing variations, and enables the secure authentication of each IC instance [13]. While both use a challenge-response protocol, PUFs address a different problem of the authentication of a particular IC instance *once it is registered and trusted*, whereas our scheme addresses the initial bootstrap of trust in unknown hardware.

Research by Agrawal et al. proposed to use side-channel information to detect malicious circuits in an IC [1]. This is related to our approach as this side-channel approach also tries to determine if an unknown IC can be trusted by checking implementation details. The differences are in capabilities and applications. The side-channel approach required physical access to an IC to measure power consumption, and addressed detecting malicious circuits inserted by manufacturers. On the other hand, our proposed approach is applicable *remotely* and assumes that authentic processors are manufactured by trusted sources.

7. CONCLUSION

In this paper, we presented a new challenge-response scheme to check the authenticity of a processor based on the performance gap between authentic hardware and simulations or emulations. We augmented a modern processor pipeline with modest hardware enhancements to capture internal processor state during challenge execution to produce a unique response for each processor model. Experimental results demonstrate that the results from the checksum mechanisms are different for micro-architectures varying in a single configuration parameter, and for each challenge. These experimental results suggest that the scheme is effective in authenticating processor models.

8. REFERENCES

[1] D. Agrawal, S. Baktir, D. Karakoyunlu, P. Rohatgi, and B. Sunar. Trojan detection using ic fingerprinting. In *Proceedings of IEEE 2007 Symposium on Security and Privacy*, pages 296–310, May 2007.

[2] A. Aharon, D. Goodman, M. Levinger, Y. Lichtenstein, Y. Malka, C. Metzger, M. Molcho, and G. Shurek. Test program generation for functional verification of powerpc processors in IBM. In *Design Automation Conference*, pages 279–285, 1995.

[3] D. Chiou, H. Sunjeliwala, D. Sunwoo, J. Xu, and N. Patil. FPGA-based Fast, Cycle-Accurate, Full-System Simulators. In *Proceedings of the 2nd Annual Workshop on Architecture Research using FPGA Platforms*, February 2006.

[4] H. de Vries. http://www.chip-architect.com/.

[5] B. Greskamp, S. R. Sarangi, and J. Torrelas. Designing Hardware that Supports Cycle-Accurate Deterministic Replay. In *Proceedings of 2006 Workshop on Complexity-Effective Design*, pages 22–25, June 2006.

[6] P. Gutmann. PKI: It's not dead, just resting. *Computer*, 35:41–40, August 2002.

[7] Intel. Intel Trusted Execution Technology. http://www.intel.com/technology/security/, 2007.

[8] M. Pellauer, J. Emer, and Arvind. HAsim: Implementing a Partitioned Performance Model on an FPGA. http://publications.csail.mit.edu/abstracts/abstracts07/pellauer-abstract/hasim.html.

[9] J. Renau, B. Fraguela, J. Tuck, W. Liu, M. Prvulovic, L. Ceze, S. Sarangi, P. Sack, K. Strauss, and P. Montesinos. SESC simulator, January 2005. http://sesc.sourceforge.net.

[10] A. Satoh and T. Inoue. ASIC-Hardware-Focused Comparison for Hash Functions MD5, RIPEMD-160, and SHS. In *Proceedings of the International Conference on Information Technology: Coding and Computing (ITCC'05)*, 2005.

[11] A. Seshadri, M. Luk, E. Shi, A. Perrig, L. van Doorn, and P. Khosla. Pioneer: Verifying Code Integrity and Enforcing Untampered Code Execution on Legacy Systems. In *Proceedings of the 20th Annual Symposium on Operating Systems Principles*, pages 148–162, October 2005.

[12] A. Seshadri, A. Perrig, L. van Doorn, and P. Khosla. SWATT: SoftWare-based ATTestation for Embedded Devices. In *2004 IEEE Symposium on Security and Privacy*, May 2004.

[13] G. E. Suh and S. Devadas. Physical Unclonable Functions for Device Authentication and Secret Key Generation. In *Proceedings of the 44th Design Automation Conference*, 2007.

[14] H. Tago, K. Hashimotof, N. Ikumi, M. Nagamatsu, M. Suzuoki, and Y. Yamamoto. Importance of CAD Tools and Methodologies in High Speed CPU Design. In *Proceedings of the Asia and South Pacific Design Automation Conferences (ASP-DAC 2000)*, pages 631–633, 2000.

[15] Trusted Computing Group. TCG Specification Architecture Overview. http://www.trustedcomputinggroup.com/.

[16] S. Vangal, M. A. Anders, N. Borkar, E. Seligman, V. Govindarajulu, V. Erraguntla, H. Wilson, A. Pangal, V. Veeramachaneni, J. W. Tschanz, Y. Ye, D. Somasekhar, B. A. Bloechel, G. E. Dermer, R. K. Krishnamurthy, K. Soumyanath, S. Mathew, S. G. Narendra, M. R. Stan, S. Thompson, V. De, and S. Borkar. 5-GHz 32-bit integer execution core in 130-nm dual-V/sub T/ CMOS. *IEEE Journal of Solid-State Circuits*, 37:1421–1432, November 2002.

[17] N. J. Wang, J. Quek, T. M. Rafacz, and S. J. Patel. Characterizing the effects of transient faults on a high-performance processor pipeline. In *Proceedings of the 2004 International Conference on Dependable Systems and Networks*, June 2004.

978-1-60558-497-3/09 $25.00 © 2009 ACM

Hardware Trojan Horse Detection
Using Gate-Level Characterization

Miodrag Potkonjak   Ani Nahapetian   Michael Nelson   Tammara Massey

Computer Science Department
University of California, Los Angeles (UCLA)
Los Angeles, CA 90095

{miodrag, ani, tmassey}@cs.ucla.edu

ABSTRACT

Hardware Trojan horses (HTHs) are the malicious altering of hardware specification or implementation in such a way that its functionality is altered under a set of conditions defined by the attacker. There are numerous HTHs sources including untrusted foundries, synthesis tools and libraries, testing and verification tools, and configuration scripts. HTH attacks can greatly comprise security and privacy of hardware users either directly or through interaction with pertinent systems and application software or with data. However, while there has been a huge research and development effort for detecting software Trojan horses, surprisingly, HTHs are rarely addressed. HTH detection is a particularly difficult task in modern and pending deep submicron technologies due to intrinsic manufacturing variability.

Our goal is to provide an impetus for HTH research by creating a generic and easily applicable set of techniques and tools for HTH detection. We start by introducing a technique for recovery of characteristics of gates in terms of leakage current, switching power, and delay, which utilizes linear programming to solve a system of equations created using non-destructive measurements of power or delays. This technique is combined with constraint manipulation techniques to detect embedded HTHs. The effectiveness of the approach is demonstrated on a number of standard benchmarks.

Categories and Subject Descriptors

B.7.m [**Hardware**]: Integrated Circuits – *Miscellaneous.*

General Terms

Experimentation, Security

Keywords

Hardware Trojan horses, gate-level characterization, linear programming, manufacturing variability.

Permission to make digital or hard copies of part or all of this work for personal or classroom use is granted without fee provided that copies are not made or distributed for profit or commercial advantage and that copies bear this notice and the full citation on the first page. To copy otherwise, to republish, to post on servers or to redistribute to lists, requires prior specific permission and/or a fee.
DAC'09, July 26-31, 2009, San Francisco, California, USA

1. INTRODUCTION

Since semiconductor manufacturing demands a large capital investment, the role of contract foundries has dramatically grown, increasing exposure to theft of masks, attacks by insertion of malicious circuitry, and unauthorized excess fabrication [1]. The development of hardware security techniques is exceptionally difficult due to reasons that include limited controllability and observability (50,000+ gates for each I/O pin in modern designs) [7], large size and complexity (the newest Intel processor has 2.06B transistors), variety of components (e.g., clock, clock distribution interconnect, and finite state machine), unavoidable design bugs, possibility of attacks by non-physically connected circuitry (e.g., using crosstalk and substrate noise), many potential attack sources (e.g. hardware IP providers, CAD tools, and foundries), potentially sophisticated and well-funded attackers (foundries and foreign governments), and manufacturing variability that makes each IC coming from the same design unique [5][11].

There are several broad types of malicious hardware attacks that we consider. The first is gate resizing, where the attacker intentionally changes the sizing factors of one or more gates in such a way that the circuit passes all standard timing test, but its timing for a certain inputs is incorrect or its switching or leakage power are globally or locally increased drastically. Note that many other gate sizing attacks can be envisioned, including one where the sizes of the gates are altered in such a way that the calculation of internal signals is facilitated through altered timing or switching power. In the second type of attack, the adversary adds one or more gates so that the functionality of the design is altered. It is important to observe that the gates can be added so that no timing path between primary inputs and flip-flops (FFs) and primary outputs and FFs is altered. However, leakage power is always altered because even if the attacker gates the added circuitry, the gating requires an additional gate. Our HTH detection approach is generic in a sense that it can easily be retargeted to other circuit components, such as interconnect by considering more comprehensive timing and/or power models.

The main technical obstacle to HTH detection is manufacturing variability, which may have a very significant impact on timing and power characteristics. Its difficulty is often compounded by low controllability and observability. Our approach is non-destructive and aims to minimize measurement test time. Currently, we target an off-line scenario, but the approach can be applied in the case of timing on-line. Even power-based techniques can be applied on-line by measuring the change of timing characteristics due to rise in temperature. However, for the sake of focus, we restrict our attention to the off-line case.

The basis for our approach is gate-level characterization using a set of timing and/or power measurements. The measurements are treated as a set of linear equations with imposed measurements errors. They are processed using linear programming (LP). When we address detection of additional ghost circuitry using LP, we impose additional constraints on our LP formulation in such a way that the results indicate what circuitry is added and where. Essentially, we use LP to ask if the characterization of gates is significantly more consistent under the assumption of added circuitry.

Finally, it is important to observe that gate-level timing characterization has other numerous applications beyond security. For example, it addresses in an elegant way static timing analysis and greatly facilitates dynamic timing analysis. It can be also used for tailoring post-silicon optimization. For example leakage power vectors can be calculated much more precisely when the gate sizing information is available. Also, it can be used for inexpensive silicon process characterization.

2. RELATED WORK

Most available hardware security, as well as hardware-based secure system methods, are based upon implementations of digital cryptography protocols [11]. In traditional cryptographic protocols, security is provided by trapdoor mathematical functions and digital keys, which make the protocols resilient to algorithmic attacks [17]. However, digital hardware security keys can be attacked in a number of ways including side-channel, electromigration, imaging and fault injection [5][6].

To address the above shortcomings and vulnerabilities, a new generation of security techniques based on manufacturing variability (MV) has been developed [14] [15][21]. With scaling of feature sizes, the physical limits of the devices are reached and uncertainty in the device size increases [7]. Variations in transistor feature sizes and thus, in gate characteristics are inevitable (e.g., delay or power). In present and pending technologies, the variation is large compared to the device dimensions. As a result, VLSI circuits exhibit a high variability in both delay and power consumption.

In contrast, the proposed new scheme does not require storing of secret information, does not require any exchange of secret information, and is much faster and less power expensive. Most available hardware security, as well as hardware-based secure system methods are based upon implementations of digital cryptography protocols [6]. In traditional cryptographic protocols, security is provided by trapdoor mathematical functions and digital keys, which make the protocols resilient to algorithmic attacks [10]. However, digital hardware security keys can be attacked in a number of ways including side-channel, electromigration, imaging and fault injection [3][4].

To address the above shortcomings and vulnerabilities, a new generation of security techniques based on manufacturing variability (MV) has been developed [8][9][12][15][16]. Note that the manufacturing variations are because of the intense industrial CMOS feature scaling. With scaling of feature sizes, the physical

limits of the devices are reached and uncertainty in the device size increases [5]. Variations in transistor feature sizes and thus, in gate characteristics are inevitable (e.g., delay or power). In present and pending technologies, the variation is large compared to the device dimensions. As a result, VLSI circuits exhibit a high variability in both delay and power consumption.

3. TROJAN HORSE (HTH) ATTACK
3.1 Motivation
A Hardware Trojan Horse (HTH) is an intentional hardware alteration of the design specification or of the corresponding implementation. These alterations only affect the circuit's functionality in a few specific circumstances and are hidden otherwise. HTHs are more difficult to detect, diagnose, and mask than design bugs or manufacturing faults since they are intentionally implanted to be unperceivable by the current debugging and testing methodologies and tools. The vast number of possibilities for implementing HTHs further complicates their detection.

Diagnosis of HTHs can be especially intricate since a large number of Trojans may be present simultaneously in an IC. In addition, HTHs are not necessarily present in all ICs coming from a design. Since HTHs are embedded within the circuit and are active only under certain, very rare conditions, detection methods must be complex. In addition, the introduction of new technologies, ultra large scale integration, and intrinsic MV of deep submicron technologies also increase the difficulty of HTH uncovering. Implementation of HTHs by the attackers and its detection/prevention by the designers is a battle that is only bounded by the creativity, knowledge, and skills of each side. Although the IC layout data is shared with the foundry, the designers and the tool developers have the advantage of making the first and last move. Thus, they can construct their design to be the least suitable to HTH placement, and they can also choose the type of post-silicon HTH detection procedure.

3.2 Example of HTH Attacks
Here we present specific HTH, in an attempt to describe the nature of HTH attacks in general. A simple, yet powerful HTH attack is presented in Figure 1, which shows how ghost circuitry can be activated in a cell phone when specific inputs or data are detected at specific memory locations. The unshaded portion of the circuit represents the HTH circuitry when it is activated by a HTH caller ID number. Upon activation, the attacker bitstream (ABS) is activated and the initial cell phone design is corrupted. In this example, HTHs will either cause the cell phones to malfunction or cause confidential information to be leaked. Important information can be disclosed after activation of the HTH. The exploited phone can automatically dial a hidden spy third party when certain numbers are dialed. Ghost circuitry (HTH) may be difficult to identify by traditional timing/power analysis techniques. To avoid timing analysis-based detection, an attacker only needs to ensure that no path delays between the inputs to flip-flops (FFs) or between the outputs to FFs are increased. Also, the switching power can remain stable until the trigger of the attacker's caller ID activates the HTH.

978-1-60558-497-3/09 $25.00 © 2009 ACM

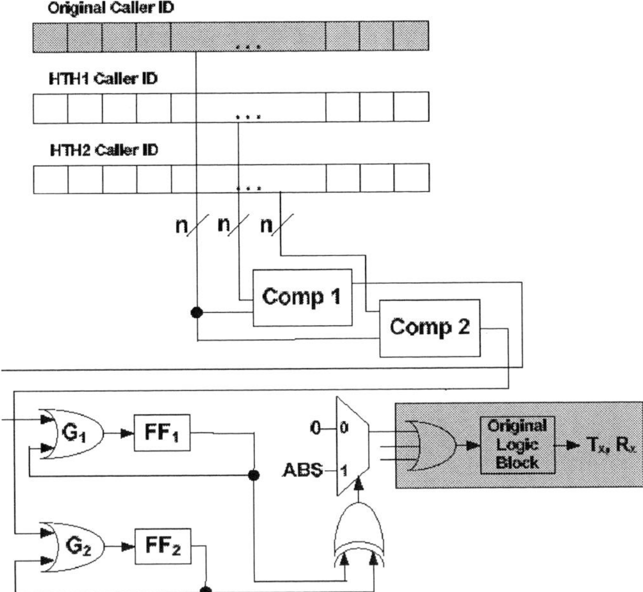

Figure 1. Example of a cell phone HTH

The crosstalk caused by HTH in Figure 2 is particularly difficult to detect since no physical connections between the ghost and the affected circuitry is necessary. Our SPICE simulations of wire crosstalk demonstrate that for two wires closer than 0.1 micrometers, the affected wire's delay can increase by more than four times, as shown in Table 1. A prevention method, during the place and route phase, is to carefully place the nonfunctional interconnects that fill the empty spaces to facilitate crosstalk.

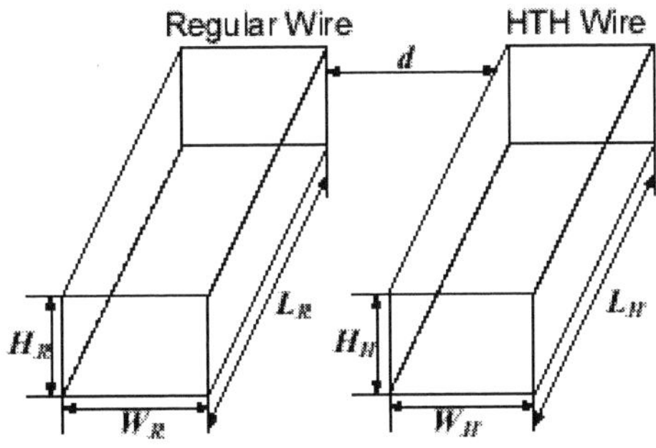

Figure 2. Example of a wire crosstalk HTH

Table 1. Interconnect capacitance vs. distance of the HTH wire

Distance	0.1	0.2	0.5	1.0
Increase (%)	420	105	34	3.2

Finally, consider an example from photonic crystals-based on-chip optical interconnects, one the future, potentially pervasive IC technologies. Due to its bandwidth, low power consumption, high speed, feature size compatibility, and easy transduction to and from electric signals, photonic crystals-based on-chip optical interconnects are one of the leading candidates for implementation of global on-chip interconnects [13]. However, if an attacker places a single transducer on the optical fiber, data

being communicated on the high bandwidth channel can be interfered/obtained with no timing or electromagnetic impact.

3.3 HTH Classification

In the modern design flows of ASICs, ASIPs, and microprocessors, most of the phases may be potentially untrusted [1]. CAD tools, hardware intellectual properties (IPs), design libraries, testing and verification tools, untrusted foundries, untrusted testing facilities, and even untrusted designers in large design teams are all possible malicious threats. With field programmable gate arrays (FPGAs), the situation is even worse because the software and configuration files can also be attacked.

Even though the number and types of potential hardware attacks are essentially unlimited, we currently classify HTH attacks into the seven following categories: (i) damage objectives; (ii) components and mechanisms of the attack; (iii) components of the IC under attack; (iv) duration and initiation mechanisms; (v) design phase implantation and usage phase; (vi) optimization level; and (vii) customization level.

The HTH attackers desiderata may include alteration of the computed results, slowing the IC, increasing the power consumption, releasing confidential data, and facilitating side-channel attacks by making the gates' power consumptions observable at output. Attackers can employ various techniques including excessive switching, interconnect resizing, and substrate noise addition. Components, such as gates, clocks, and memory are vulnerable to physical attacks. Attacks may be randomly or actively initiated, e.g., event triggered. The impact may be temporary/permanent or recoverable/unrecoverable. The effectiveness and the detection difficulty of attacks can be quantitatively optimized. The attack may be on a specific weak IC component or on a predetermined set of components. For example, incorrectly sized gates are very slow on the design's critical path. Sizing variations caused by MV in ICs may be exploited for this attack. Unfortunately, the broad spectrum of possible attacks prevents the development of universally effective defense and protection measures. To mitigate this problem, our research work develops generic hardware security techniques that can easily be retargeted to new tasks.

We focus on detecting the differences between the taped-out chip's characteristics and post-silicon tests. We also address the more difficult HTH discovery problems where the malware is implanted within the hardware intellectual property (IP), cell and module libraries, CAD tools, and FPGA configuration software.

3.4 Global Flow

To carry out HTH detection, we first carry out gate-level characterization, by using non-destructive leakage power and timing delay measurements to create linear programs. Once, the scaling factor of all or most of the gates in the IC have been approximated, then we are able to carry out a suite of statistical techniques for determining the presence of HTHs.

4. GATE-LEVEL CHARACTERIZATION

Gate-level characterization (GLC) aims to recover the post-silicon and unique properties of each IC in the presence of manufacturing variability. GLC calculates the relevant characteristics of each gate of the design using a limited number of nondestructive

978-1-60558-497-3/09 $25.00 © 2009 ACM

measurements. We currently target delay, switching power, and leakage measurements and use GLC on combinatorial gates. However, other options such temperature and response to radiation may be also used for this approach. Even though we currently consider only combinatorial gates, one can generalize our techniques and easily apply our approach to sequential elements and interconnects. Due to space limitations and for the sake of clarity, we discuss the simple structure of GLC applied to leakage power, in this section. The key observations are that each gate consumes an energy that is proportional to its manufacturing variability scaling factor, *si*. Each gate *gi* depends on its input. Table 2 demonstrates this for a two input NAND gate for 90 nanometer technologies.

Table 2. Leakage current for different inputs

Input Vector	00	01	10	11
NAND-2 Leakage	0.776	10.39	4.137	15.15

GLC leakage characterization can be formulated as a set of linear equations. We assume that the design structure is known. Given an IC with n gates and m input pins, K different input vectors can be applied ($K << 2m$). The total leakage current is measured for each input vector. The goal is to find the scaling (sizing) factor for each gate's leakage. The stated problem is an over-constrained system of linear equations. Each equation has the form:

$$\sum_{j=1}^{n} s_j I_{k,j}^{nom} = \hat{I}_k^{total}$$

where $I_{k,j}^{nom}$ is the nominal leakage power for gate j for the input vector k, $\hat{I}_k^{total}$ is the total measured leakage current for the k-th input that is the sum of the correct value and a measurement error.

Depending on the choice of optimization objective function, the above problem can be stated in a linear, convex, or nonlinear program format. First, each equation is transformed into a constraint where the difference of the left and the right side of equation is less than $|\varepsilon_k|$ for the k-th equation. The objective is to minimize an appropriate error norm defined over the K error terms, ε_k's, e.g., the l_1 error norm is $\sum_{k=1}^{k} |\varepsilon_k|$, that can be linearly stated by a set of auxiliary variables ρ_k as

$$\sum_{k=1}^{k} \partial_k \text{ , s.t., } \partial_k \geq \varepsilon_k \text{ and } \partial_k \geq -\varepsilon_k .$$

The stated system of equations can be optimally solved using linear programming. The relative advantages of linear, convex, and nonlinear programs are that they provide a fine trade-off between accuracy, run time, and modeling flexibility. While linear programming (LP) is fast, it assumes piece-wise linear error models. On the other hand, while non-linear programming is slow and has potential convergence problems, it can fit nonlinear error norms and, more importantly, handle non-linear scaling and leakage power models. There are two major potential difficulties with solving the system of equations with linear, convex, or non-linear programs. The first is that two or more scaling variables

can be correlated in all equations and thus one cannot find their actual values. The second is that the large circuits cannot be solved in a reasonable amount of time.

To resolve the above concerns and also to address several others, we created a more elaborate GLC flow. Our current GLC flow has five phases that are embedded within a loop that terminates once the user defined accuracy criteria is satisfied or the runtime limit is reached. The five phases are:
(i) measurement organization;
(ii) equation analysis and selection;
(iii) solving system of equations;
(iv) results post-processing;
and (v) results validation.

First, we select the input vectors. In the case of delay measurements, we also decide on relative input arrival times. Second, we analyze the equations (constraints) that maximize the corresponding matrix rank to solve for the maximal number of scaling variables in a numerically stable manner. Third, we derived several heuristics for the task, which we use. Fourth, after the equations are solved for different input vectors, we combine their solutions to minimize the probability of large errors. Finally, learn-and-test and resubstitution statistical validation techniques are used to estimate bounds for the calculated scaling factors.

4.1 Gate-Level Characterization using Linear Programming (LP)

In evaluating the LP formulation for gate-level characterization, we studied three properties of input variables: (i) distribution type, (ii) the mean of the absolute measurement errors, and (iii) the number of constraints.

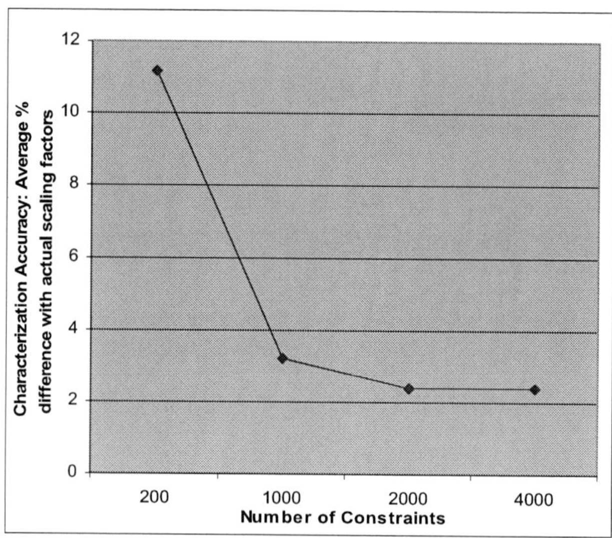

Figure 3. Accuracy of the gate-level characterization vs. the relative measurement error (%) for different number of constraints (equations) on c432 benchmarks

In Figure 3, we evaluate the accuracy of the gate characterization procedure versus the uniform measurement error and the number of constraints (NC) the c432 benchmark. The number of constraints refers to the number of equations that we used in the

978-1-60558-497-3/09 $25.00 © 2009 ACM

LP, corresponding to the number of different inputs used to measure the leakage. We see that even though increasing the NC on average improves the accuracy, there exists some cases where with NC=2000 there is more accuracy than with NC=4000, because of the random variations in the subset of used equations.

Table 3 presents the results for LP-based GLC using leakage power, delay, and both leakage power and delay for various ISCAS85 benchmarks. The average error in GLC is presented for various percentages of average measurement errors. Overall in this paper, our results demonstrate that we are able to characterize gates with an error smaller than the measurements error.

Table 3. LP-based GLC for leakage power, delay, or both, for five ISCAS85 different benchmarks

Accuracy of Solvable Scaling Factor Values				
Measure Err	1%	2%	5%	10%
c432 Power	0.48	0.92	2.02	4.57
c432 Delay	0.33	0.71	1.59	4.04
c432 Both	0.14	0.71	0.98	2.63
c499 Power	0.72	1.57	3.93	7.82
c499 Delay	0.32	0.77	1.59	1.90
c499 Both	0.06	0.24	1.00	1.22
c880 Power	0.92	1.80	3.98	8.02
c880 Delay	0.30	0.80	1.09	1.23
c880 Both	0.08	0.69	0.92	1.20
c1355 Power	0.88	1.71	3.47	6.62
c1355 Delay	0.12	0.71	0.80	1.09
c1355 Both	0.14	0.66	0.78	0.92
c1908 Power	0.94	1.74	4.02	8.83
c1908 Delay	0.19	0.32	0.79	0.85
c1908 Both	0.19	0.22	0.52	0.58

5. HTH DETECTION

The goal of HTH detection is to identify the presence of HTHs and to estimate the probability of added ghost circuitry. HTH detection problem can be stated in many different frameworks that result in sharply different difficulties and false positive/false negative likelihoods. A key observation is that certain HTHs are impossible to detect without test points or analysis methods beyond GLC. For example, if an added gate has the same inputs as a gate of the same type in the original design, solely GLC techniques will not be able to identify it. Another important observation is that the difficulty of detection depends on several factors, such as the number and characteristics of the added gates, the interconnects, the measurement errors, the original design size, the design structure, and the number of conducted measurements.

We conduct HTH detection using three novel techniques: (i) statistical analysis; (ii) constraint manipulations; and (iii) comparison with technological and physical laws. In the first technique, we analyze the variable residuals and errors in individual equations. For example, a systematic positive measurement error is a strong indicator that a ghost circuitry is added. In the second technique, we manipulate constraints and the objective function in a nonlinear program. The solver in the nonlinear program indicates if a ghost circuitry is present. For instance, we may add the same extra variable to the right side of each constraint. Now, if the gates can be characterized in a more

accurate and consistent manner, it is an indicator that a malware is present. This example assumes that only one gate is added. Finally, we also compare the GLC results to the relative characteristics of the gates with respect to the well-established physical design and technological laws. For example, if two gates are next to each other in the original layout, but the GLC determines that the gates have highly different scaling factors, we can deduce that they are placed further than the layout specification and it is likely that a ghost circuitry is placed in between them.

Once an HTH is detected, the natural next phase is to identify its location. We conduct HTH diagnosis using a two phase process. In the first phase, we identify the type of gates added and their position. After we determine which interconnects and gates receive signals and where these gates are physically located, our integer linear and nonlinear programming techniques can make the equations maximally consistent. Consistency will be achieved by adding gates that are not in the initial specification and by allowing the program to assign inputs to these gates. In the second phase, we correlate the assigned inputs with the existing input and intermediate signal locations to identify the most likely location of the added gate's signal. The high local correlation of sizing factors identifies the most likely HTH location using the maximum likelihood principle. We conducted a study of our HTH diagnosis technique by adding a single inverter to various ISCAS85 benchmarks. The inverter receives its input from a randomly selected gate in the benchmark and has no output to any of the original gates. Even if we set the measurement error to 5%, we were able to correctly detect/ diagnose the presence/ location of the inverter with higher than 98% accuracy.

Table 4. Detection accuracy using a single extra variable

Bench-marks	% of Measurement Error			
	1	2	5	10
c17	100	100	100	100
c432	100	100	100	98
c499	100	100	100	96
c880	100	100	100	95
c1355	100	100	98	95
c1908	100	100	98	94

Table 4 presents the results where an extra variable is added to the LP constraints per measurement.

Another approach we use is to examine the correlations between an input value(s) and the overall switching or leakage energy consumption and/or delay. This technique aims to leverage the observation that if a ghost gate is set to almost always use the same inputs, that input will have a significantly higher consumption than other sets of inputs. For all types of gates, this difference is a least two times higher. By fixing one or more inputs to a specific value and searching all combinations, it is very likely that there will be a subset of the inputs that impact the HTH with a higher probability.

The main technical difficulty is to account for the unknown scaling factors of the gates. We address this in the following two ways. First, we calculate the scaling factors of the gates using out linear programming maximum likelihood procedure for gate characterization. Then, we can compare the expected energy for various input vectors and find that there is a consistent bias due to the HTH. The second technique iterates the first technique by perturbing the scaling factors in order to address uncertainty about the actual gate scaling factors.

6. CONCLUSION

We have developed a system of techniques for two major classes of hardware Trojan horse detection and diagnosis: gate resizing and embedding of ghost circuitry. The techniques are non-destructive and apply a system of delay or power measurements followed by singular value decomposition and linear programming (LP) processing. The final step is either statistical data analysis or the more effective addition of LP constraints. The effectiveness of the approaches is demonstrated on a number of benchmarks.

7. REFERENCES

[1] Defense Science Board (DSB) study on high performance microchip supply, http://www.acq.osd.mil/dsb/reports/2005-02-hpms report final.pdf, 2006.

[2] D. Agrawal, S. Baktir, D. Karakoyunlu, P. Rohatgi, and B. Sunar. Trojan detection using ic fingerprinting. In *IEEE Symposium on Security and Privacy (SP)*, pages 296–310, 2007.

[3] R. Anderson, M. Bond, J. Clulow, and S. Skorobogato. Cryptographic processors-a survey. *Proceedingsof the IEEE*, 94(2):357–369, 2006.

[4] R.J. Anderson. *Security Engineering: A guide to building dependable distributed systems*. John Wiley and Sons, 2001.

[5] K. Bernstein, D.J. Frank, A.E. Gattiker, W. Haensch, B.L. Ji, S.R. Nassif, E.J. Nowak, D.J. Pearson, and N.J. Rohrer. High-performance CMOS variability in the 65-nm regime and beyond. *IBM Journal of Research and Development*, 50(4/5):433–450, 2006.

[6] D. Hwang, P. Schaumont, K. Tiri, and I. Verbauwhede. Securing embedded systems. *IEEE Security & Privacy*, 4(2):40–49, 2006.

[7] N.K. Jha and S. Gupta. *Testing of Digital Systems*. Cambridge University Press, 2003.

[8] F. Koushanfar and M. Potkonjak. CAD-based security, cryptography, and digital rights management.In *Design Automation Conference (DAC)*, 2007.

[9] K. Lofstrom, W.R. Daasch, and D. Taylor. IC identification circuits using device mismatch. In *International Solid State Circuits Conference (ISSCC)*, pages 372–373, 2000.

[10] A. Menezes, P. van Oorschot, and S. Vanstone. *Handbook Of Applied Cryptography*. CRC Press, 1997.

[11] A. Srivastava, D. Sylvester, and D. Blaauw. *Statistical Analysis and Optimization for* VLSI: *Timing and Power*. Series on Integrated Circuits and Systems. Springer, 2005.

[12] Y. Su, J. Holleman, and B. Otis. A 1.6J/bit stable chip ID generating circuit using process variations. In *International Solid State Circuits Conference (ISSCC)*, page to appear, 2007.

[13] G. Suh and S. Devadas. Physical unclonable functions for device authentication and secret key generation. In *Design Automation Conference (DAC)*, pages 9–14, 2007.

[14] E. Yablonovitch. Can nano-photonic silicon circuits become an intra-chip interconnect technology? In *IEEE/ACM International Conference on Computer-Aided Design (ICCAD)*, pages 309–309, 2007.

[15] M. Nelson, A. Nahapetian, F. Koushanfar, M. Potkonjak. SVD-based Ghost Circuitry Detection. In *Information Hiding (IH)*, 2009.

[16] R. Rad, X. Wang, J. Plusquellic, and M. Tehranipoor. Taxonomy of Trojans and Methods of Detection for IC Trust. In *International Conference on Computer-Aided Design (ICCAD)*, 2008.

978-1-60558-497-3/09 $25.00 © 2009 ACM

Process Variation Characterization of Chip-Level Multiprocessors

Lide Zhang[†]   Lan S. Bai[‡]   Robert P. Dick[‡]   Li Shang[⋆]   Russ Joseph[†]

[†]EECS Department
Northwestern University
Evanston, IL, USA
{lzh228@u,rjoseph@ece}.northwestern.edu

[‡]EECS Department
University of Michigan
Ann Arbor, MI, USA
{dickrp@eecs.,lanbai@}umich.edu

[⋆]ECE Department
University of Colorado
Boulder, CO, USA
li.shang@colorado.edu

ABSTRACT

Within-die variation in leakage power consumption is substantial and increasing for chip-level multiprocessors (CMPs) and multiprocessor systems-on-chip. Dealing with this problem via conservative assumptions is sub-optimal. Instead, operating systems may adapt task assignment and power management decisions to the variable characteristics of cores, improving system-wide power consumption and performance. Researchers have proposed such adaptation techniques. However, they rely on knowledge of CMP process variation (PV) maps. These maps are not provided by processor vendors, providing them would impose additional cost during the testing process, and static maps would not permit adaptation to aging effects. Further progress on developing and validating PV aware control techniques for CMPs requires access to PV maps for real processors. We present an online technique to extract the PV maps of CMPs. Potentially automatic temperature measurements with built-in on-die sensors during the execution of characterization workloads are used to determine variation in leakage power consumption. The proposed technique is applied to real CMPs, and the resulting PV maps are used within a PV aware task assignment and scheduling algorithm.

Categories and Subject Descriptors: B.8 [**Performance and Reliability**]: Performance Analysis and Design Aids

General Terms: Design, Verification, Performance

Keywords: Process variation, characterization, software

1. Introduction and Motivation

Process variation (PV), the deviation of process parameters from their nominal values, can be divided into three categories: die-to-die variation, within-die variation, and wafer-to-wafer variation. Ongoing technology scaling has a tendency to increase PV [1]. In this paper, we will focus on spatially-correlated within-die variation and die-to-die variation. More specifically, we focus on the within-die variation among CMP and multiprocessor system-on-chip cores.

PV has received substantial attention from researchers [2, 3], who have developed techniques that adapt to the characteristics of individual cores to achieve better power consumption and performance [2, 3]. Such techniques require a PV

This work was supported in part by the SRC under awards 2007-HJ-1593 and 2007-TJ-1589 and in part by the NSF under awards CCF-0702761, CNS-0347941, CNS-0720820, and CNS-0720691.

Permission to make digital or hard copies of part or all of this work for personal or classroom use is granted without fee provided that copies are not made or distributed for profit or commercial advantage and that copies bear this notice and the full citation on the first page. To copy otherwise, to republish, to post on servers or to redistribute to lists, requires prior specific permission and/or a fee.
DAC'09, July 26-31, 2009, San Francisco, California, USA

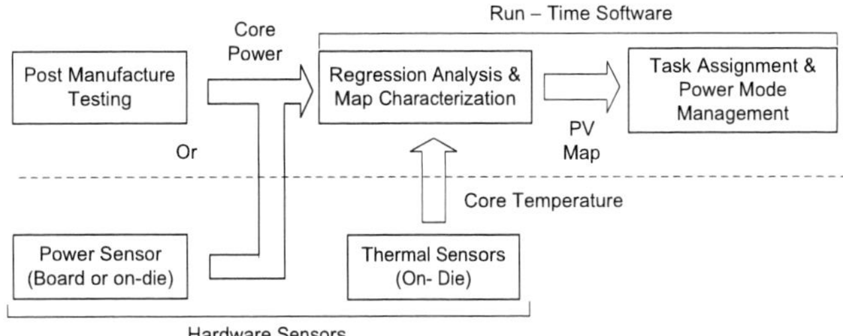

Figure 1: Generating a PV map and applying it in on-line power consumption and performance optimizations.

map as input. However, these maps are not provided by processor vendors, providing them would impose additional cost during the testing process, and in-factory characterization results might be invalidated by aging effects. Furthermore, there are not published post-testing methods for researchers or end users to derive PV maps. Borkar et al. discuss power, voltage, and temperature variations and their impact on circuits and microarchitecture [1]. They also present measurements of the leakage power consumptions and maximum frequencies of numerous microprocessors. However, these data are not provided with individual processors and their work suggests no way for end users or OSs to derive them.

We present an automatic on-line software-based technique to characterize the threshold voltages and leakage power consumptions of CMP cores, i.e., derive PV maps. These parameters are used to optimize CMP power consumption and performance. Our work makes the following main contributions: (1) it is the first on-line software-based technique for characterizing the variation in leakage and threshold voltage among CMP cores using built-in sensors; (2) we present a technique that predicts the power and thermal profiles of a given workload when run on a CMP with a particular PV map; and (3) we formulate and solve the time-constrained CMP task assignment and power management mode selection problem.

The proposed PV characterization technique derives a PV map based on temperature differences among cores. Due to PV, core leakage power coefficients (and also leakage power) may differ, resulting in temperature variation given identical workloads. We extract the PV map by performing multivariable regression analysis based on measured results, a thermal model, and a leakage model. The temperature readings come from on-die sensors present in commercially available processors. Figure 1 outlines our approach for producing a PV map, and explains how this map can be used to perform run-time optimization for task assignment and power mode selection. The characterization technique can be implemented within the OS, firmware, or hardware. It imposes low computational overhead, eliminates the need to measure the PV map during testing, and can be used to adapt to aging effects.

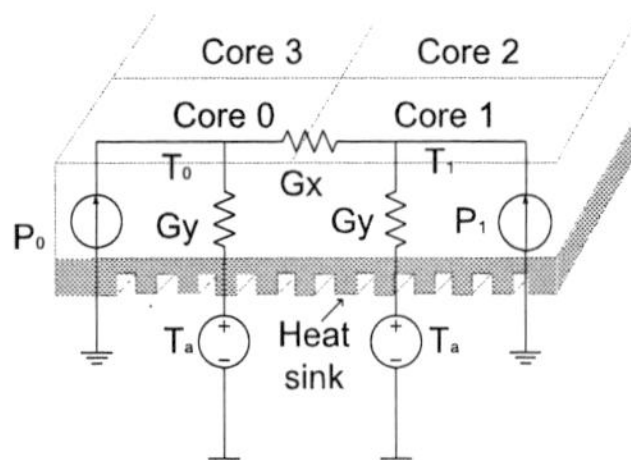

Figure 2: CMP chip-package thermal model.

2. Thermal and Leakage Models

Our approach to generating PV maps is based on two physical models: a compact CMP thermal model and a leakage power model, which are described in this section. These models will be used to derive and validate the PV map as explained in Section 3.

2.1 Thermal Model for Multi-Core Processors

To analyze heat flow, we use a discretized Fourier thermal model [4], in which heat flow is analogous to electrical current and temperature is analogous to voltage. As shown in Figure 2, multicore processors can be divided into several blocks, each representing a core. T_a is the ambient temperature of the processor. T_0 and T_1 are the temperatures for Core 0 and Core 1. G_y is the thermal conductance from the core to the ambient and G_x is the thermal conductance between cores. We assume a symmetric cooling solution for which all core-to-ambient thermal conductances are identical. P_0 and P_1 represent the power dissipations of the cores and hence correspond to the heat generated in the core active layers.

Heat flow can be modeled as follows:

$$\mathbf{C}dT(t)/dt + \mathbf{G}T(t) = Pu(t), \tag{1}$$

where N is the number of thermal elements (with at least one element per core), $\mathbf{C}$ is a diagonal $N \times N$ matrix containing the core heat capacities as the diagonal elements, $T(t)$ is an N-element vector, each element of which represents the temperature of a core as a function of time, and $\mathbf{G}$ is an $N \times N$ thermal conductance matrix where G_{ij} represents the thermal conductance between core i and core j. $u(t)$ is the $t = 0$ unit step function. Section 3.2 explains one use of heat capacity and dynamic thermal effects to assist characterization.

2.2 Leakage Power Consumption Model

Leakage power consumption is a first-order concern in CMP management, and is sensitive to the effects of inter-core PV. Subthreshold leakage power consumption is presently the major component of leakage power, and the introduction of high-κ gate dielectrics means this is likely to remain true in the near future. According to the BSIM model [5], transistor-level subthreshold leakage power can be approximated as follows:

$$P^{leak} = \eta \cdot T^2 e^{\frac{-q(V_{th}(T)+V_{off}-V_{GS})}{nkT}}, \tag{2}$$

where n ($=1.5$) is the subthreshold swing coefficient for the transistor [6], $V_{off} = 0.08\,\mathrm{V}$ is offset voltage, T is the temperature, and η is a technology-dependent parameter.

3. Process Variation Map Characterization

In this section, we formulate the PV map characterization problem based on the thermal and leakage power models described in Section 2. Our objective is to determine the threshold voltages and leakage power consumptions of individual cores from the temperature readings of all cores. For this purpose, we produce a number of different thermal profiles by using workloads with carefully controlled levels of processor utilization. For each thermal profile, the dynamic power profile is held uniform among cores. Leakage power is then estimated by doing regression analysis with the dynamic power profiles as independent variables and the thermal profile as dependent variables. The thermal model (Equation 4) and leakage power model (Equation 2) are used as regression equations.

This regression process requires several parameters including the nominal leakage power coefficient, the dynamic power of each workload, and thermal conductance characteristics of the processor. We explain our characterization technique by first developing regression model with the assumption that these parameters are known, then describe how the parameters may be obtained.

3.1 Regression Analysis

Assuming knowledge of dynamic power, the nominal leakage power coefficient η, and the thermal conductance matrix, the primary input of our regression model is a series of thermal profiles. These thermal profiles are readings from temperature sensors when all the cores are stressed with different workloads with known CPU usage. T_{ij} is the temperature of the ith processor with workload j. Given knowledge of the thermal conductance matrix $\mathbf{G}$, power and steady-state temperature are related as follows:

$$\mathbf{G} \times T_j = P_j + T_a \times G_y, \tag{3}$$

where T_j is the jth column of the T matrix, P_j is the power profile of all the cores with the jth workload, and T_a is the ambient temperature as shown in Figure 2. The power consumption is composed of dynamic power and leakage power, the equation for which follows:

$$P_j^{leak} = \mathbf{G} \times T_j - P_j^{dyn} - T_a \times G_y. \tag{4}$$

Similarly, P_j^{leak} and P_j^{dyn} are the leakage power profile and dynamic power profile of the CMP with the jth workload. ξ is an $N \times W$ matrix, in which P_j^{leak} is the jth column. W is the set of workloads available for use. Therefore, ξ_{ij} is the leakage power of the ith core at temperature T_{ij}, which is approximated based on Fourier thermal model.

Given the nominal leakage power coefficient η, we can also estimate the leakage power profile using the model shown in Equation 2. This leakage profile is subject to the threshold voltage $(V_0^{th}, V_1^{th}, ..., V_N^{th})$. Γ_{ij} is the leakage power for the ith core with T_{ij} using the leakage model. For a set of data (T_j, P_j^{dyn}), the threshold voltages can be estimated by minimizing the sum of squared errors: $\sum_{j \in W} |\xi_j - \Gamma_j|^2$.

3.2 Parameter Derivation

The proposed regression approach constructs a leakage variation map for a CMP based on knowledge of the temperature, CMP total power consumption, and the thermal conductances of the system. In practice, these quantities are not readily available. In this subsection, we describe methods for determining the vertical thermal conductance G_Y, the lateral thermal conductance G_X, nominal leakage, and core dynamic power consumption P_j^{dyn}. The nominal leakage power is the average leakage power of all chips that the manufacturer measures during testing process [1]. The dynamic power of each workload is determined by subtracting the nominal leakage power from the total power. As these values may be associated with a particular processor model, they can potentially be stored in non-volatile memory or fuses by the chip manufacturer. However, this is not now common, so we have developed a characterization technique.

Existing and upcoming processor technologies for power management support on-die power measurement. Some, such as Intel's Foxton technology, support independent power sensors on each processor core [7]. The required power measurements could instead be provided by motherboards. Even if the total power consumption of each workload on an individual core is known, separating leakage and dynamic power is challenging. To do this, we use workloads designed to control the utilization, and therefore power consumptions, of cores. We also take advantage of the dynamic thermal effect resulting from heat capacities ($\mathbf{C}$) in Equation 1 to set dynamic power and temperature independently.

978-1-60558-497-3/09 $25.00 © 2009 ACM

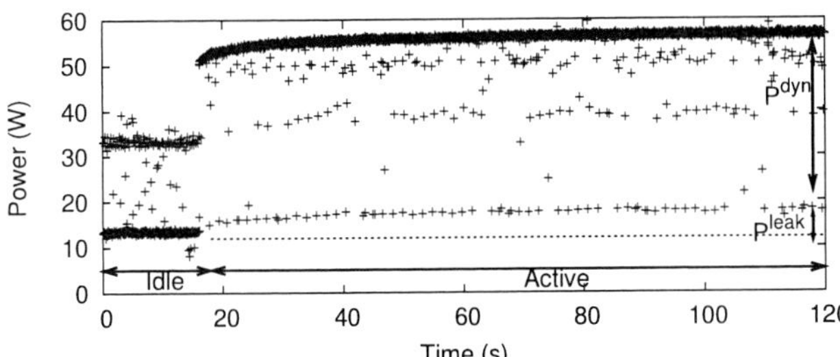

Figure 3: Power measurement of processor.

Figure 3 shows power measurements when all cores transition between idle and active workloads. Note that due to the heat capacity, the temperatures and therefore leakage power consumptions of all cores increase gradually. By stressing the processor to alternate CPU utilization between 0% and 100%, we isolate the impact of the workload on dynamic ($\mathbf{P}^{dyn}$) and change in leakage power consumption ($\mathbf{P}^{leak}$). With the knowledge of the temperature-dependent component (change in leakage power), we use regression on Equation 2 to determine the values of η and the temperature-independent component of the power consumption associated with the measured temperature-dependent component of the power consumption, temperature, and the process-dependent parameters.

In addition to the nominal leakage power and dynamic power, the vertical and lateral thermal conductances are also required. Starting from the processor power consumptions, we get the relationship between total power consumption, thermal profile, and the vertical thermal conductance from Equation 3. To give an example, we apply Equation 3 to the four-core processor shown in Figure 2 to derive the following equations. The same technique can be applied to general multi-core processors. By adding the four equations in the extension of Equation 3, we have

$$G_y \times \sum_{i=1}^{4}(T_{ij}) - 4 \times T_a \times G_y = \sum_{i=1}^{4}(P_{ij}) = P_j^{total}, \quad (5)$$

where P_j^{total} is the total power, measured using power sensors. Hence, by taking advantage of the linear relationship between $\sum_{i=1}^{4}(T_{ij})$ and P_j^{tota}, we set parameters to minimize the sum of squared errors. The vertical thermal conductance and ambient temperature can thus be derived. Similarly, we get the lateral thermal conductance by doing linear regression on thermal and power profiles. To get the relationship between the lateral thermal conductance, thermal profile, and power profile, we subtract the $i + 1$th equation from the ith in the extension of Equation 3. For example, subtracting the 2nd equation from the 1st yields the following equation:

$$(3G_x + G_y)T_{0j} - (3G_x + G_y)T_{1j} + G_x T_{2j} - G_x T_{3j} = P_{0j} - P_{1j}.$$

Hence, G_x is set to minimize the sum of squared errors in this relationship between thermal profile and its corresponding power profile. This regression is easy if per-core power sensors are available. However, if only a single sensor or single power supply network is available for the entire CMP, it still remains a challenge to isolate the power consumptions of individual cores. To achieve this, we use the nominal power coefficient and thermal profile to estimate the leakage power of each core. By adding the dynamic power profile and the leakage power profile, we approximate the total power profile yielding an initial G_x. This dynamic power profile is obtained by measuring the difference between the power consumption immediately after the power transition in Figure 3. As a result the heat capacity $\mathbf{C}$ in Equation 1, the temperature, and therefore leakage power consumption, change only gradually after the power consumption change, allowing the isolation of dynamic and leakage power consumption. This is not yet accurate because we assume the leakage power coefficients are the same for each core. To solve this, we use the initial G_x as input in the multi-variable regression and get the initial leakage power map. The initial leakage power map is then used in the estimation of G_x, iterating until G_x convergence.

4. Process Variation Aware Task Assignment and Power Management Mode Selection

In this section, the time-constrained CMP task assignment and power management mode selection problem is used to demonstrate the importance of knowing a PV map when attempting to optimize system characteristics such as power consumption and performance.

Problem statement: Given a deadline for a set of tasks, determine the assignment of tasks to cores and power management configurations to tasks. The objective is to complete all tasks within a time constraint and with minimal energy consumption. Tasks are independent. The core configuration parameters, such as supply voltage, frequency, and active cache size, are controlled independently for each core. Many of these features are supported on existing multi-core processors, e.g., the AMD's Quad-Core Opteron.

Let T be the set of tasks, C be the set of cores, P be the set of power management configurations, and B be the time constraint. Let E_{ijk} be the energy consumption of running task i on core j with configuration k. Let δ_{ijk} be the execution time of task i on core j with configuration k. Let binary variable A_{ijk} indicate whether task i is assigned to core j with configuration k. The task assignment and core configuration problem is formulated as an integer linear program (ILP).

$$\text{minimize} \sum_{i \in T, j \in C, k \in P} E_{ijk} \times A_{ijk}$$

$$\text{subject to} \max_{j \in C} \sum_{i \in T, k \in P} \delta_{ijk} \times A_{ijk} \leq B$$

$$\forall i \in T \sum_{j \in C, k \in P} A_{ijk} = 1$$

To evaluate the impact of considering PV when solving this problem, we compare optimal solutions for formulations that consider, and neglect, PV. From the perspective of problem input, the PV-unaware approach bases its decisions on the mean energy–performance relationships, while the PV-aware approach considers the different energy-performance tradeoffs of different cores. Note that we are not proposing to use the optimal ILP solver within the OS.

5. Experimental Results and Validation

This section explains our experimental setup and results, the approach used for preliminary validation of the proposed technique, and indicates the impact of considering PV during task assignment and power state control.

5.1 Experimental Setup

To validate our idea, we use Intel Core 2 Duo E6420 processors and a Shuttle SD32G2B motherboard as our experimental platform. This processor is equipped with one thermal diode per core, which were calibrated by the manufacturer to within $1\,°C$ error [8]. To reduce the impact of asymmetric cooling, we replace the original fan and heat sink with a symmetric cooling solution. Intel E6420 processors lack on-die power sensors. We therefore use a current clamp to measure CPU power consumption. Note that this is unnecessary for processors or motherboards with built-in power sensors (see Section 3.2). The voltage regulator power offset does not influence the result because it affects both ambient temperature and dynamic power estimates. The workloads used to stress the processor are controlled CPU utilization programs. By adjusting the duty cycle of the workload, we indirectly control processor temperature.

978-1-60558-497-3/09 $25.00 © 2009 ACM

Table 1: Fitted Parameters of Processor

	Processor 0		Processor 1	
V_{th}	Core 0	Core 1	Core 0	Core 1
	0.2176 V	0.2233 V	0.2672 V	0.2404 V
Thermal cond.	Vert. G_y	Lat. G_x	Vert. G_y	Lat. G_x
	0.54 W/K	0.28 W/K	0.37 W/K	0.18 W/K
Leakage const. η	0.0478 V $\cdot$ $(eV/K)^2$		0.0272 V $\cdot$ $(eV/K)^2$	

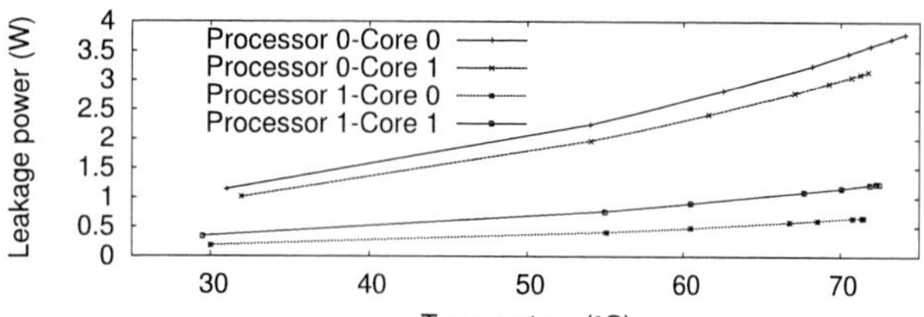

Figure 4: Leakage power of four cores.

5.2 Results and Analysis

Figure 4 illustrates the leakage power map characterized through regression. Consider processor 0. The largest difference between leakage power for two cores is 0.6 W, 20.28% of the leakage power of core 1. The total leakage power is 6.93 W at 72.9 °C, 12.57% of the total power. When the processor is idle, the leakage power is responsible for up to 16.71% of the total power. All the parameters including the threshold voltage and thermal conductances are shown in Table 1. The estimated variation in threshold voltage is 2.62%.

5.3 Validation of Characterization Technique

In the interest of validation, we use another set of workloads with different dynamic power consumptions and predict temperatures based on the extracted PV map (see Section 3). We start from dynamic power consumption and iterate until Equation 4 converges. The predicted temperatures are compared with the measured values for four cores. The average difference between the predicted and measured temperatures is 1.1 °C and the maximum is 2.14 °C. These results are a step toward validating the proposed technique. This prediction technique is also a potential use of the extracted PV maps.

5.4 Task Assignment and Power Management

We used the M5 instruction set architecture simulator running a number of SPEC CPU2000 benchmarks to generate input energy–delay relationships for different tasks. We consider the following configuration parameters: instruction cache (one-way or two-way), L1 data cache (one-way or two-way), L2 data cache (one-way to eight-way), frequency (1.0, 0.95, 0.9, 0.85× maximum frequency). The M5 simulation results were used as the input for different tasks on cores with mean PV parameters, i.e., without any PV.

After obtaining the energy–delay relationships with mean variation parameters, we applied the variation to each curve to emulate cores with PV. Based on the switching and leakage power scaling trends reported by Keshavarzi [9], the leakage power is approximately half of the total power for a 45 nm process. We use this leakage power proportion. We obtained the PV map (Figure 4) by using the characterization technique described in Section 3 on two Intel Core 2 Duo processors. The frequency variation is derived from the leakage power variation according to a function fitted to the frequency–leakage variation plot (see Borkar et al.'s Figure 1 [1]). In addition to solving the instance based on our characterized PV maps, we also solved the problem for a number of synthetic PV maps using frequency and leakage power variation distributions based on large-scale measurements [1]. The problem instances were solved with the CPLEX ILP solver [10]. Each problem instance is an assignment of 12 tasks to four cores with 128 power management modes.

The solutions yielded by the PV-unaware formulation have two disadvantages compared with those from the PV-aware

Table 2: Penalty of PV-Unaware Formulation

Time constraint (ms)	4	5	6	7	8
Energy overhead (%)	2.2	5.3	7.2	8.8	10.7
Deadline violation (%)	34.0	21.5	20.5	8.5	15.5

formulation: (1) they have higher energy consumption and (2) they sometimes violate their deadlines. The second row of Table 2 shows the energy consumption overhead of the PV-unaware formulation relative to the PV-aware formulation with various time constraints for the synthetic PV maps based on the distributions of Intel measurements [1]. They are computed by averaging the energy overheads of the 20 simulated chips, neglecting infeasible solutions. As the time constraint is relaxed, the PV-aware technique is increasingly able to adapt to CMP characteristics. On average, the PV-unaware formulation imposes a 6.8% energy penalty. For the measured PV map, the PV-unaware formulation imposes a 26.0% energy overhead. The third row of Table 2 shows the probability of PV-unaware solutions violating time constraints. Averaging over all task sets and constraints, the PV-unaware technique violates the deadline for 20% of the problems while the PV-aware technique meets all deadlines; the use of the PV-unaware formulation would require tightening timing constraints, thereby increasing energy consumption or making the problem unsolvable.

In summary, considering PV during optimization of task assignment and power mode selection can substantially improve system quality. However, for such an approach to be used, the PV maps of individual processors are required, motivating the characterization technique described in Section 3.

6. Conclusions

This paper presented a technique that uses built-in sensors to derive the leakage power PV maps for processor cores in a CMP or multiprocessor system-on-chip. This technique makes use of the interdependence between temperature and leakage power to isolate dynamic and leakage power consumption for different regions of the CMP, and uses dynamic thermal effects to independently control the temperature and dynamic power consumption of the CMP being characterized. Our characterization results are validated by comparing measured and predicted temperatures for CMP workloads. We also present a novel PV aware technique for controlling CMP task assignment and power management state in order to optimize energy consumption under hard deadlines.

7. References

[1] S. Borkar, et al., "Parameter variation and impact on circuits and microarchitecture," in *Proc. Design Automation Conf.*, June 2003, pp. 338–342.

[2] R. Teodorescu and J. Torrellas, "Variation-aware application scheduling and power management for chip multiprocessors," in *Proc. Int. Symp. Computer Architecture*, June 2008.

[3] P. Ndai, et al., "Within-die variation-aware scheduling in superscalar processors for improved throughput," in *IEEE Trans. Computers*, vol. 57, no. 7, July 2008.

[4] J. Fourier, *The Analytical Theory of Heat*, 1822.

[5] "BSIM4," http://www-device.eecs.berkeley.edu/~bsim4/bsim4.html.

[6] Z. Chen, et al., "Estimation of standby leakage power in CMOS circuits considering accurate modeling of transistor stacks," in *Proc. Int. Symp. Low Power Electronics & Design*, Aug. 1998, pp. 239–244.

[7] C. Poirier, et al., "Power and temperature control on a 90 nm Itanium–family processor," in *Proc. Int. Solid-State Circuits Conf.*, Feb. 2005, pp. 304–305.

[8] "Core 2 Quad and Duo temperature guide," http://www.tomshardware.com/forum/ 221745-29-core-quad-temperature-guide.

[9] A. Keshavarzi, "Technology scaling and low-power circuit design," in *The VLSI Handbook*, W.-K. Chen, Ed. CRC Press, 2007, ch. 21, p. 21.12.

[10] "CPLEX," ILOG, Inc., http://www.ilog.com/products/cplex.

978-1-60558-497-3/09 $25.00 © 2009 ACM

Information Hiding for Trusted System Design

Junjun Gu
Electrical and Computer Engineering
Department and the Institute for
Advanced Computer Studies
University of Maryland
College Park, MD, USA

alicegu@umd.edu

Gang Qu
Electrical and Computer Engineering
Department and the Institute for
Advanced Computer Studies
University of Maryland
College Park, MD, USA

gangqu@umd.edu

Qiang Zhou
Tsinghua National Laboratory for
Information Science and Technology
Department of Computer Science and
Technology, Tsinghua University
Beijing, P. R. China

zhouqiang@tsinghua.edu.cn

ABSTRACT

For a computing system to be trusted, it is equally important to verify that the system performs no more and no less functionalities than desired. Traditional testing and verification methods are developed to validate whether the system meets all the requirements. They cannot detect the existence or show the non-existence of the unknown undesired functionalities. In this paper, we propose a novel approach that converts this problem to a less challenging design quality measuring problem. Our approach is based on information hiding and constraint manipulation of the original system design specification. We lay out the basic requirements for our approach and demonstrate it through the popular graph coloring problem. Results show that information can be embedded into the original graph without significant impact to the solution quality. However, when the same information is added to the graph modified based on our approach, there will be noticeable drop in the solution quality.

Categories and Subject Descriptors

B.7.3 [**Integrated Circuits**]: Reliability and testing – *testability*
B.m [**Hardware**]:MISCELLANEOUS – *design management.*

General Terms

Algorithms, Measurement, Security, Legal Aspects, Verification.

Keywords

Trusted IC, constraint manipulation, information hiding, VLSI design intellectual property, graph coloring.

1. INTRODUCTION

In the past few years, there has been a fast increasing interest in building integrated circuit (IC) based trusted systems [1-5], mainly driven by the prolific applications, both military and civilian, that require security and access control. To ensure the user that the system does each of the required functionalities, the design house only needs to show that expected output will be produced under the given testing input data. However, *verifying that an IC or a system is not intentionally designed to perform any undesired functionality* is much harder. Particularly, in most cases the user may not know the undesired functionalities he wants to verify.

This has also been identified in the recent Defense Science Board study on High Performance Microchip Supply, *"Trust cannot be added to integrated circuits after fabrication; electrical testing and reverse engineering cannot be relied on to detect undesired alternations in military integrated circuits"* [1]. The Trusted Foundry Program and the complementary Trusted IC Supplier Accreditation were specifically designed for domestic fabrication, where trust and accreditation are built on reputation and partnership [2]. The notions of trusted and trustworthiness are presented in [3], where the authors discuss the challenges, opportunities, call for new initiatives and programs to establish principles, tools, and standards for building trusted hardware. Trimberger in [4] argues that trusted IC can be built on FPGA platform because it separates the manufacturing process from the design process. However, the base array still needs to be verified through manufacture and trust still needs to be built during the design process. Suh and Devadas [5] propose physical unclonable functions (PUFs) based on transistor and wire delay to uniquely identify chip. Their method authenticates ICs, but cannot be used to verify the non-existence of unwanted functionalities.

To verify the non-existence of known unwanted functionality, one can use mathematical techniques such as proof by contradiction or conduct a careful testing. But there is no systematic method to verify the non-existence of unknown functionalities. The main contribution of this paper is a general approach that detects the unwanted functionality with high probability, regardless such functionality is known or not. The proposed method can be applied at any design levels. However, we believe that it is more effective to be applied at high level. This is because that the user has the control of high level design specification and the adversary also prefers to hide extra functionality at high level. Our approach is based on formal models (such as graph and finite state machine models) and can be used for the detection of unwanted features during both hardware and software development.

The key idea behind our approach is the *conversion of the intractable verification problem to the problem of measuring a (normally simple) metric for design quality*. If the original design spec is altered, we will be able to observe changes in the design quality. More specifically, we apply information hiding and constraint manipulation techniques to modify the original design spec such that the following objectives for trusted system design can be achieved:

Permission to make digital or hard copies of part or all of this work for personal or classroom use is granted without fee provided that copies are not made or distributed for profit or commercial advantage and that copies bear this notice and the full citation on the first page. To copy otherwise, to republish, to post on servers or to redistribute to lists, requires prior specific permission and/or a fee.
DAC'09, July 26-31, 2009, San Francisco, California, USA

(1) The system does the right thing. Any design and implementation based on the modified design specification will maintain the correct functionality from the original design.

(2) The system has high quality and meet performance requirement. There exist implementations based on the modified specification which have quality as good as the optimal (or near optimal) implementations based on the original design specification.

(3) The system does only the right thing. Although we cannot guarantee this, it will become difficult, if not impossible, for others (e.g. adversary) to introduce new functionalities into the modified design specification without noticeable design quality degradation.

(4) System testing and verification cost does not increase dramatically. We use metrics that can be conveniently measured to compare the design quality with the expected design quality.

2. OVERVIEW OF THE APPROACH

We propose an information hiding and constraint manipulation based method to establish trust in systems that are designed and built through supply chain that might not be trusted. We choose to control *the design specification*, which is among the few trusted steps in today's IC design and fabrication supply chain model [1]. The rationale for making this decision is that user has the full control of the design at such early stage and will gradually lose it as design tools, third-party intellectual properties, and untrusted design teams and foundries are involved in the design and fabrication process.

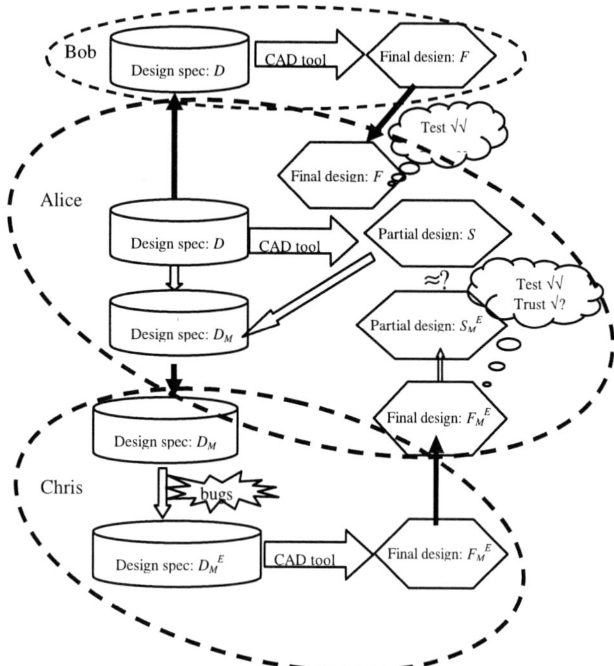

Figure 1. Overview of the information hiding and constraint manipulation based trusted IC design flow. From the original design spec, Bob's final design can be tested, but may not be trusted. Using the modified design space Chris' final design can be tested for correctness and trustworthiness.

Figure 1 illustrates how we achieve trusted IC system design. In the top half of the figure, Alice gives her design specification D to Bob's design team. Bob returns Alice the final design F. Alice can test relatively easily whether F meets all her design constraints. However, she will not be able to judge the trustworthiness of design F, for example, does F also performance any unwanted functionalities?

The bottom half of the figure shows how Alice follows our approach to ask Chris' design team. Before she gives Chris her design specification, Alice first performs some very simple and routine design using commercial CAD tools to get a partial design S. S does not need to be the final design or even close to the final design. For example, for a combinational design, S can be the two-level optimized layout; for a sequential design, S can be the finite state machine with the fewest number of states. Alice then modifies the design specification to D_M based on D and S such that S remains as a solution for D_M.

When Chris gets D_M and decides to embed some extra constraints/functions ("bugs" shown in the figure), he creates a newer design specification D_M^E. He then continues to build the design F_M^E and gives the design back to Alice. To verify whether the design is trustworthy, Alice creates the partial design S_M^E from F_M^E at the same design level that she generates S. She then compares the quality of S and S_M^E. If she observes large quality difference, she concludes that Chris' design may not be trustworthy. In summary, our proposed approach consists of

Step 1: determine a high quality (partial) solution S for the problem D.

Step 2: create a modified problem D_M, based on S and D.

Step 3: ask others to solve the modified problem D_M without revealing S and D.

Step 4: compare the quality of the (partial) solution S_M^E with the known solution S. If the qualities are similar, trust is built; otherwise, the party is not trustworthy.

There are several challenges in this approach, especially in Step 2. First, Alice should be able to *define a quality metric* at early design stage and *get a high quality partial solution S* according to that metric; second, when creating D_M, Alice needs to be assured that D_M will still *have high quality solutions* and at the same design stage, as good as S she has; third, Alice wants to make sure that if D_M is modified, it will be *hard to find design with similar high quality*; fourth, the quality metric that Alice defines at the early design stage *can be measured from a final design*.

3. INFORMATION EMBEDDING FOR THE GRAPH COLORING PROBLEM

In this section, we elaborate how to develop techniques that meet the above requirements, particularly how to meet the challenges in Step 2. We focus on the graph coloring problem for illustrative purpose. Similar examples can be derived for other abstract graph models, finite state machine, and Boolean satisfiability (SAT) that are commonly used to describe design specification.

The graph vertex coloring problem (GC) is to label the vertices of a graph with minimal number of colors such that vertices connected by an edge do not receive the same color. The quality of the solution can be conveniently measured by the number of

978-1-60558-497-3/09 $25.00 © 2009 ACM

colors used. It is considered a significant improvement of solution quality if the number of colors can be reduced by even one [6].

Consider that now Alice's goal is to modify a graph $G=(V,E)$ based on a known high quality solution S_0 to obtain a new graph $G'=(V',E')$ such that (1) a coloring solution S for $G=(V,E)$ can be obtained from any solution S' to the modified graph $G'=(V',E')$; (2) if S' is obtained by coloring $G'=(V',E')$ directly, S and S_0 will have the same or similar quality. That is, they use the same number of colors to color the original graph $G=(V,E)$; and (3) if changes were made on $G'=(V',E')$, solution S will either fails to be a solution for $G=(V,E)$ or have lower quality than S_0. The former can be verified by checking each edge in E. The latter implies that S requires more colors than S_0.

In the remainder of this section, we describe a technique that builds traps (i.e. low quality solutions) for the adversaries to fall in when they attempt to alter the modified graph $G'=(V',E')$. We assume that the adversaries intend to add new unwanted functionalities to the system, which correspond to adding more edges to the modified graph $G'=(V',E')$.

Consider Fig. 2(a), where we extract several nodes from a colored graph. Nodes B and C are connected by an edge (the solid line) and they receive different colors, green and red, respectively. Node A is not connected to either node B or node C. Now if Alice adds an edge (the dashed line) between node A and node B, the current coloring scheme remains valid because A and B have different colors. However, if an adversary, who sees the uncolored graph with the edge between A and B, adds an edge (the dashed line with arrow heads) between node A and node C, he falls into the trap Alice made by forming a triangle of nodes A, B, and C. This makes the graph harder to color.

The following is the pseudo-code for this clique trap on a pair of nodes for graph $G=(V,E)$ and a coloring solution S:

```
while (more edges need to be added)
{ pick a node v random from the graph;
    let C(v) be the color that node v receives in S;
    find a node u that receives the same color as v;
    find a node w such that (u,w)∈ E, (v,w)∉E;
    add edge (v,w) to E;
}
```

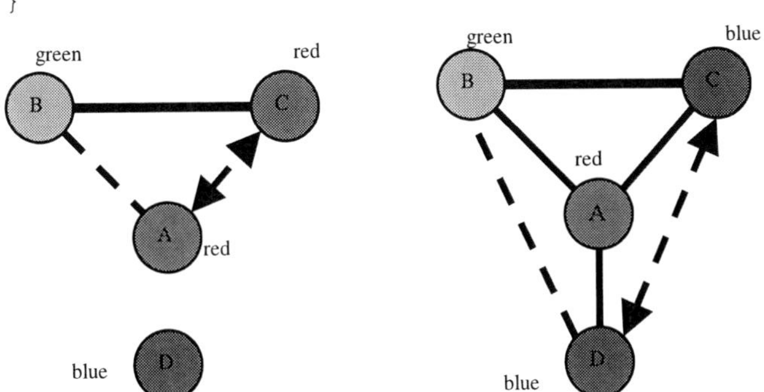

Figure 2. Clique traps: (a) on a pair of nodes B and C; (b) on a triangle of nodes A, B, and C. Solid edges are from the original graph; dashed edges are newly added edges; dashed edges with arrowheads are those added by adversaries.

Fig. 2(b) illustrates the same idea when we find a triangle ABC (a clique trap of size 3) in the original graph and another node D that receives the same color (blue) as node C. We connect nodes B and D to construct a trap for the adversary. If the adversary needs to connect nodes C and D to embed the unwanted functionality, this will form a clique of size four (nodes A, B, C, and D).

The strength of the clique trap can be enhanced by connecting a node in the clique to a node outside the clique. For example, in Fig. 2(a), if we further add edges between node A and node D (or node C and node D), node A (or C) cannot have the same color as node D and it makes the clique harder to color.

Forming one small clique may not make the graph very hard to color. However, when we introduce many such clique traps in the graph, collectively they will significantly increase the difficulty to color the graph. We will show later in the simulation results section that (1) the adversary can add many edges to the original graph while still color the graph with the minimum number of colors; however, (2) when the adversary adds the same edges to the modified graph, where there are many clique trap constructed as shown in Fig. 2, in most cases, he has to use extra colors, which is a very significant degradation of the solution quality [6].

4. EXPERIMENTAL RESULTS

To validate the effectiveness of our approach within the context of graph coloring problem, we need to show (i) *high quality solution can be found from the modified graph*; (ii) *adversary can successfully add information to the original graph without sacrificing solution quality*; (iii) *solution quality goes down if the adversary adds the same information to the modified graph.*

To achieve these goals, we perform the following sequence of experiments for a graph G = (V,E):

Step 1. color it 10 times to obtain the best and average solution;

Step 2. create a modified graph $G_M = (V, E \cup E_M)$ based on G and the best solution;

Step 3. create graph G' = (V, E \cup E') based on G with the adversary's edges E';

Step 4. create graph $G'_M = (V, E \cup E_M \cup E')$ based on G_M with the same E';

Step 5. color graphs G_M, G', and G'_M each for 10 times and record the best and average solution;

Step 6. for comparison, we also list the watermarking solution quality obtained by adding edges as reported in [7].

Table 1 reports the results of this series of experiments on the random graph G(n,p) from the DIMACS benchmark, where n is the number of nodes in the graph and p =0.1, 0.5, or 0.9 is the edge probability. The solver we use is the one reported in [6]. Rows G_b and G_a show the best and average number of colors needed to color each graph from step 1. In the next three paragraphs, we show how the three experimental goals are supported by these results.

For step 2, we add n edges, the same as the number of nodes in the graph, to the original graph to create clique traps. We observe a slight increase of the average number of colors in G_{M_a}. However, for all the graphs, we manage to find the best solution

978-1-60558-497-3/09 $25.00 © 2009 ACM

(n,p)	(250,0.1)	(500,0.1)	(125,0.5)	(250,0.5)	(500,0.5)	(125,0.9)	(250,0.9)	(500,0.9)
$|E|$	3218	12458	3891	15668	62624	6961	27897	112437
G_b	9	14	18	29	50	45	77	133
G_a	9	14	18.8	29.9	50.4	45	77	133.9
$|E_M|$	250	500	125	250	500	125	250	500
G_M_b	9	14	18	29	50	45	77	133
G_M_a	9	14	18.5	29.5	50.8	45	77.3	134.5
$|E'|$	250	500	125	250	500	125	250	500
G'_b	9	14	19	30	50	49	78	133
G'_a	9.2	14.9	19	30.5	50.9	49	78.4	134.2
G'_M_b	**10**	**15**	**20**	**31**	**53**	**50**	**78**	**136**
G'_M_a	**10.5**	**15**	**20.1**	**31.8**	**53.9**	**50.2**	**79.3**	**137.8**
G_1_b	n/a	n/a	19	30	50	n/a	n/a	n/a
G_1_a	9.1	14	19	30.2	50.4	49.1	79	n/a
G_2_b	n/a	n/a	19	30	50	n/a	n/a	n/a
G_2_a	9.8	14.2	19	30.2	50.6	52	82	n/a

Table 1. Solution quality on random graphs G(n,p). $|E|$: # of edges; G_: original graph; _b: # of colors in the best solution; _a: # of colors in the average solution; $|E_M|$=n: # of edges added to form traps; G_M_: modified graph; $|E'|$=n: # of edges added based on extra functionality; G'_: adversary's graph based on G; G'_M_: adversary's graph based on G_M; G_1_: graph with n random edges added; G_2_: graph with 2n random edges added; the last four rows are copied from [7] (n/a: data not available from [7]).

G_M_b of the same quality as G_b. This verifies goal (i): high quality solution can be found from the modified graph.

Next, we verify goal (ii). When the adversary adds n random edges to the original graph, we have similar observations. The adversary is able to obtain at least one solution of the best quality, G'_b, after adding hundreds of edges. For graphs (125,0.5) and (250,0.5), one extra color is used. This is because that we were fortunate to get the best solution in step 1. As shown below in the row G_1_b, the authors in [7] were not able to get the best solution after adding the same amount of edges for their watermark.

The most interesting and convincing results occur when we add the same adversary's edges to our modified graph G_M. As shown in the rows of G'_M_b and G'_M_a, we see significant increase in both the best and average solution. For example, five more colors are used to color (125,0.9) and three are needed for (500,0.5). This verifies goal (iii): solution quality goes down if the adversary adds the same information to the modified graph.

One may argue that this happens because we have added n edges to form G_M from G and n more to form G'_M from G_M. A total of 2n edges have been added and the corresponding graph should be harder to color. This is a valid argument. However, the key is not only the number of edges. Where the edges are added is also important. To see this, we take the results G_2_b and G_2_a from [7]. These are the best and average number of colors when 2n random edges are added to the graph. We see that G_2_b and G_2_a are very close to the results for the original graph G_b and G_a. This shows that our modified graph G_M does have the nice property that can be used to effectively detect whether extra constraints have been added or not.

5. CONCLUSION

In the paper, we propose a method to detect the potential existence of unwanted functionality in a system. The novelty of this approach is that we embed information into the system such that if extra functionality is added, the design quality will have noticeable drop. The basic idea is to modify the system spec to build some traps which an adversary will fall into and find low quality design when he tries to add extra functionality. We demonstrate the effectiveness of this approach by the example of graph coloring. This work paves a new way to how to build trusted hardware and software systems.

6. REFERENCES

[1] Report of the Defense Science Board Task Force on High Performance Microchip Supply, February 2005.

[2] B.S. Cohen. "On Integrated Circuits Supply Chain Issues in a Global Commercial Market –Defense Security and Access Concerns", March 2007.

[3] C.E. Irvine and K. Levitt. "Trusted Hardware: Can It Be Trustworthy?", ACM/IEEE Design Automation Conference, pp. 1-4, June 2007.

[4] S. Trimberger. "Trusted Design in FPGAs", ACM/IEEE Design Automation Conference, pp. 5-8, June 2007.

[5] G.E. Suh and S. Devadas. "Physical Unclonable Functions for Device Authentication and Secret Key Generation", ACM/IEEE Design Automation Conference, pp. 9-12, June 2007.

[6] D. Kirovski and M. Potkonjak. "Efficient Coloring of a Large Spectrum of Graphs", ACM/IEEE Design Automation Conference Proceedings, pp. 427-432, June 1998.

[7] G. Qu and M.Potkonjak, Hiding Signatures in Graph Coloring Solutions", Information Hiding Workshop, pp. 391-408, 1999.

978-1-60558-497-3/09 $25.00 © 2009 ACM

On Systematic Illegal State Identification for Pseudo-Functional Testing

Feng Yuan and Qiang Xu
CUhk REliable computing laboratory (CURE)
Department of Computer Science & Engineering
The Chinese University of Hong Kong, Shatin, N.T., Hong Kong
Email: {fyuan,qxu}@cse.cuhk.edu.hk

ABSTRACT

The discrepancy between integrated circuits' activities in normal functional mode and that in structural test mode has an increasing adverse impact on the effectiveness of manufacturing test. Pseudo-functional testing tries to resolve this problem by identifying illegal states in functional mode and avoiding them during the test pattern generation process. Existing methods, however, can only extract a small set of illegal states in the system due to various limitations. In this paper, we first show that illegal states in the system are mainly caused by multi-fanout nets in the circuit, and we develop efficient and effective heuristics to identify them. Experimental results on benchmark circuits demonstrate the effectiveness of our proposed systematic solution.

Categories and Subject Descriptors

B.7.3 [**Integrated Circuits**]: Reliability and Testing

General Terms

Reliability, Design.

Keywords

Pseudo-Functional Testing, Illegal States

1. INTRODUCTION

Functional testing was historically used to find manufacturing defects, but its nature of exhaustive testing makes it impractical for reasonable-sized integrated circuits (ICs) [3]. The semiconductor industry hence mainly resorts to structural testing to identify defective chips, wherein test patterns are generated based on circuit structural information and a set of fault models. With the ever increasing transistor-to-pin ratio in IC products, structural test pattern generation has become extremely complex and it is only possible with the help of dedicated design-for-testability (DfT) circuits.

Scan is the mainstream DfT technique in use today, which makes automatic test pattern generation (ATPG) viable for large ICs. Scan-based DfT technique, however, changes the circuit states in test mode, making them possibly different from that in functional mode. Consider a finite state machine encoded with one-hot code, the legal combinations of values in the circuit's storage elements are only those with a single logic '1' and all the others logic '0'. Without taking this functional constraint into consideration, traditional

Permission to make digital or hard copies of part or all of this work for personal or classroom use is granted without fee provided that copies are not made or distributed for profit or commercial advantage and that copies bear this notice and the full citation on the first page. To copy otherwise, to republish, to post on servers or to redistribute to lists, requires prior specific permission and/or a fee.
DAC'09, July 26-31, 2009, San Francisco, California, USA

ATPG tools may generate scan patterns that contain multiple logic '1's and hence are illegal in functional mode.

The discrepancy between functional mode and test mode is not a big issue for testing defects that cause slow speed failures, as it mainly affects the timing behavior of the circuit. For ICs fabricated with latest technology, however, at-speed testing is essential to ensure the quality of the shipped IC products, rendering over-testing due to such discrepancy a serious concern for the industry [16, 18, 4]. Recent design evaluations have revealed that at-speed scan patterns were up to 20% slower than any functional pattern [19]. Consequently, some good ICs that would work in application might fail at-speed delay tests, leading to unnecessary *test yield loss* (also known as *test overkill*) [15]. With today's tight profit margins, particularly for chips that go into consumer products, achieving a high manufacturing yield can mark the difference between success and failure, and just a small variation in yield percentage can translate to millions of dollars of revenue change. Therefore, how to reduce test yield loss has become a main concern for the industry.

One way to reduce test overkills is to identify structurally-testable but functionally-untestable delay faults (FU-faults) in the circuit and do not target them during test generation. A significant amount of research work has been conducted in this direction (e.g., [2, 5, 6, 8, 21]). However, FU-fault identification generally has the same complexity as that of sequential ATPG [14], which is exponential in terms of circuit's size. In addition, while test patterns are only generated for those functionally-testable faults in the above methods, it is still possible that they incidentally detect some FU-faults and hence lead to test overkills [14]. Moreover, even for patterns that detect only functionally-testable faults, as we typically let them target as many faults as possible to reduce testing time, they might lead to excessive noises that could not occur in functional mode, again, rendering possible false rejects.

From another perspective, several power-aware test generation methodologies were proposed to reduce test overkills, by reducing switching activities in scan capture mode to ensure the power integrity in manufacturing test [7, 11, 17, 22]. This, however, leads to the concern for under-testing, i.e., if we over-restrict test power, some defective chips containing speed-related defects may pass manufacturing test, leading to *test escapes* [1]. Therefore, the real question is: *How do we make sure that circuits' activities in test mode correlate well with that in functional mode so that we can exercise the worst-case timing of the circuits under test (CUTs) in their functional mode during manufacturing test?*

Lin *et al.* [13] proposed the concept of pseudo-functional testing to tackle the above problem. Instead of identifying FU-faults in the CUT, functionally-unreachable states in the circuit are extracted in [13]. These illegal states are then fed to the ATPG engine to generate *functional-like* patterns. When illegal states are available, the above constrained ATPG process can be conducted quite efficiently as in [9, 14], in which the ATPG tool backtracks immediately when illegal states are reached during test generation.

978-1-60558-497-3/09 $25.00 © 2009 ACM

Pseudo-functional testing seems to have great potential for resolving the discrepancy problem between structural test mode and functional mode. However, whether we could realize this potential highly relies on whether we could effectively identify as many illegal states as possible. That is, if the extracted illegal states are far from complete, there is still a high possibility that the generated pseudo-functional test patterns are not within the CUT's functionally-reachable space, which invalidates the objective of pseudo-functional testing.

Several approaches were proposed for illegal state identification in the literature, including SAT-based methods [13], implication-based strategies [20, 24], and mining-based techniques [23]. None of the above techniques, however, answers the fundamental question why some states are functionally-unreachable from a structural point of view. They also have their own specific limitations and can only extract a small set of illegal states in the circuit (detailed in Section 2). Consequently, for the success of pseudo-functional testing, more effective techniques are required for illegal state identification.

In this paper, we propose novel solutions to tackle the above problem. The contributions of our work include:

- we show that the illegal states in CUTs are mainly caused by the multi-fanout nets in the circuit.

- we propose novel algorithms that are able to effectively identify much more functionally-unreachable states when compared to state-of-the-art techniques.

The remainder of this paper is organized as follows. Section 2 reviews related prior work and motivates this paper. In Section 3, we study the structural root cause for illegal states. The main flow and the key concept for the proposed illegal state identification scheme are then detailed in Section 4 and Section 5, respectively. Next, Section 6 presents our experimental results on several ISCAS'89 benchmark circuits. Finally, Section 7 concludes this paper.

2. PRELIMINARIES AND MOTIVATION

Illegal state identification has been studied in the context of sequential ATPG in several earlier works [10, 12], wherein they were used to prune the search space for sequential tests. In [10], known illegal states are used to generate larger candidate illegal spaces by eliminating one assignment in the illegal states at a time and trying to justify them. [12], on the other hand, identified invalid states by exploring all valid states through simulation from an unknown initial state. These techniques used in sequential ATPG are not practically viable for today's large ICs due to their extremely high computational complexity.

In [13], Lin *et al.* used a sequential boolean satisfiability (SAT) solver to extract the functional constraints in the system. While theoretically SAT solver is able to find almost all the unreachable states in a circuit, its computational complexity is extremely high and hence cannot be applied to a large circuit. Therefore, [13] proposed to divide the flip-flops (FFs) in the circuit into a number of much smaller groups based on topological analysis and the targeted fault information, each containing few FFs only (e.g., less than 10), and run SAT solver within each group to identify illegal states. Apparently, it is possible that functional constraints exist among different groups and this method cannot identify such kind of illegal states. Moreover, the SAT solver might still abort computation even within the small group of FFs.

Zhang *et al.* [24] proposed an implication-based technique for illegal state identification. The method starts from any gate, say gate A, and finds the implications when its output value is logic '1' and logic '0', respectively. Suppose B and C are internal flip-flops in the circuit, and there exist two implications: $[A(0) \rightarrow B(1)]$ (i.e., $A = 0$ implies $B = 1$) and $[A(1) \rightarrow C(0)]$, we then have $[B(0) \rightarrow A(1)]$

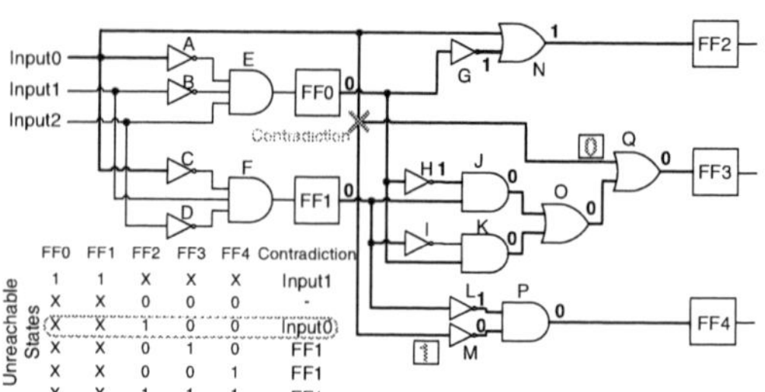

	FF0	FF1	FF2	FF3	FF4	Contradiction
	1	1	X	X	X	Input1
	X	X	0	0	0	-
Unreachable States	X	X	1	0	0	Input0
	X	X	0	1	0	FF1
	X	X	0	0	1	FF1
	X	X	1	1	1	FF1

Figure 1: Unreachable State Analysis

and $[C(1) \rightarrow A(0)]$ according to contrapositive law. Consequently, we can conclude $\{B(0), C(1)\}$ is an illegal state cube. [24] also considered the implications from impossible input-output combinations (e.g., for a 2-input AND gate, when a input is logic '0' while the output is logic '1') in their approach. To keep the computational complexity manageable, [24] implies values based on a single node in one time-frame only. This, however, significantly restricts the number of identified illegal states.

Syal *et al.* [20] considered multi-node functional constraints obtained through sequential implications, which cannot be identified within a single time-frame as in [24]. They also advocated to represent illegal states in the form of boolean expressions on arbitrary nets in the circuit instead of values on the state elements. Unlike what the authors claimed, however, using this representation has actually increased the storage overhead for the constraints significantly, because the number of arbitrary nets is much larger than the number of state elements in the circuit and lots of their obtained functional constrains are redundant in nature.

Recently, Wu and Hsiao [23] proposed a mining-based illegal state identification strategy. In this work, the circuit is expanded to multiple time-frames first and then simulated with a number of random patterns. By analyzing the obtained data, some suspicious functional constraints are extracted and they use SAT solver to verify whether they are actually functionally-unreachable. While this dynamic learning method accelerates the search procedure, due to the large amount of data, it can only check pair-wise and three-node relations within small groups of state elements and hence cannot find many illegal states in the system.

To sum up, identifying illegal states using SAT-based [13] or mining-based techniques [23] can be seen as brute-force approaches, which can only be applied to a small circuit at a time. Implication-based methods [20, 24] that consider circuit structural information to tackle this problem look to be more promising. Unfortunately, none of them answers the fundamental question why there are illegal states in the system from the structural point of view. *Finding the root cause of illegal states is extremely important for this problem, as with this information the solution exploration space can be significantly reduced without sacrificing its completeness.* This facilitate pseudo-functional testing to be applicable to real industrial designs. It is the above observation that motivates this work.

3. WHAT IS THE ROOT CAUSE OF ILLE- GAL STATES?

Let us examine an example circuit as shown in Fig. 1 to demonstrate our structural root cause analysis for illegal states. This is the gate netlist of a finite state machine that contains six illegal state cubes (see Fig. 1). A closer observation of the circuit shows that except $\{FF2(0), FF3(0), FF4(0)\}$, all the other five illegal state cubes imply logic violation at a multi-fanout net. For example, let us try to justify illegal state $\{FF2(1), FF3(0), FF4(0)\}$. First, we can learn $Input0(0)$ and $O(0)$ through OR gate Q, and $G(1)$ and $FF0(0)$ through OR gate N. We can then derive $J(0)$ and $K(0)$ through OR gate O and $H(1)$ through the inverter H. Next, $FF1(0)$ is justified through AND gate J. Finally, we can infer $Input0(1)$

978-1-60558-497-3/09 $25.00 © 2009 ACM

through AND gate P with $FF1(0)$ and $FF4(0)$, which, however, contradicts to the aforementioned $\{Input0(0)\}$ justified through another fanout going through gate Q. On the other hand, although the illegal state cube $\{FF2(0), FF3(0), FF4(0)\}$ do not imply any logic violation at a multi-fanout net directly, it actually infers another illegal state cube $\{FF0(1), FF1(1)\}$. In other words, this particular illegal state also causes logic violation on a multi-fanout net $Input1$ implicitly, which occurs at another time frame. This example motivates us to consider whether multi-fanout nets are the main root cause of illegal states.

DEFINITION 1. *Consider a circuit that does not contain any multi-fanout nets, denoted as circuit C. FF_i is the ith flip-flop in circuit C. The flip-flop set $\mathcal{F}$ contains all flip-flops that do not belong to any sequential loop. The flip-flop set $\mathcal{L}_j$ contains all flip-flops belonging to the jth sequential loop, and $\mathcal{L}$ $(\mathcal{L} = \bigcup_j \mathcal{L}_j)$ is the set of all flip-flops belonging to any sequential loops.*

DEFINITION 2. *$S(FF_i)$ is defined as the **father-cone set** of FF_i, including all state elements (including both primary inputs and flip-flops) that directly determine the next state of FF_i. **Ancestor-cone set** $\mathcal{A}(FF_i)$ includes all state elements that directly or indirectly determine the state of FF_i in one or more clock cycle(s).*

LEMMA 3. *In circuit C, $S(FF_k) \bigcap S(FF_n) = \emptyset$ when $k \neq n$.*

PROOF. If there exist a common state element within $S(FF_k)$ and $S(FF_n)$, there must be a multi-fanout net in the circuit, which is conflicted with the definition of circuit C. ∎

LEMMA 4. *The states of any elements in $\mathcal{L}_j$ cannot affect the states of elements outside of this set.*

PROOF. Suppose there is a flip-flop FF_k belonging to the flip-flop set $\mathcal{L}$, whose father-cone set $S(FF_k)$ includes at least one flip-flop FF_n belonging to $\mathcal{L}_j$. Since all flip-flops in $\mathcal{L}_j$ are connected one by one to form a sequential loop, FF_k also belongs to the father-cone set of another flip-flop FF_q in $\mathcal{L}_j$. Therefore, $S(FF_k)$ and $S(FF_q)$ have a common state element, which contradicts Lemma 3. ∎

DEFINITION 5. *The sequential level of a flip-flop FF_i in set $\mathcal{L}$ is set to be the maximal sequential level of all state elements in $S(FF_i)$ plus one, assuming the sequential level of all primary inputs is 0.*

THEOREM 6. *In a circuit C, suppose the maximum sequential level of the elements in $\mathcal{L}$ is n, any state of $\mathcal{L}$ can be reachable within n clock cycles from the primary inputs.*

PROOF. We prove this theorem by mathematical induction principle. Let us start with the case $n = 1$. Apparently, any states on Level 1 flip-flops are reachable within one clock cycle. This is because Level 1 flip-flops are directly determined by primary inputs, and the father-cone sets of any two Level 1 flip-flops are independent. Next, suppose any state for flip-flops with level k is reachable within k clock cycles, where $k \leq n$. In this case, all flip-flops with level $n+1$ are reachable within $n+1$ clock cycles. This is because, Level 1 flip-flops have essentially the same functionality as primary inputs after one clock cycle. In addition, the flip-flops with level 2 to n that directly connected with the flip-flops with level $n+1$ can also be viewed as primary inputs, without multi-fanout nets in the circuit. Therefore, any state for flip-flops with level $n+1$ is reachable. ∎

The above theorem proves that no functionally-unreachable states exists in the NLF of CS. However, it cannot guarantee the same conclusion holds true for sequential loop structure. Take the circuit shown in Fig. 2 as an example, we observe that if $FF0$ and $FF1$ are initialized as $\{FF0(0), FF1(0)\}$, it would never escape from this state even if we change the states of $FF2 - FF5$, the other

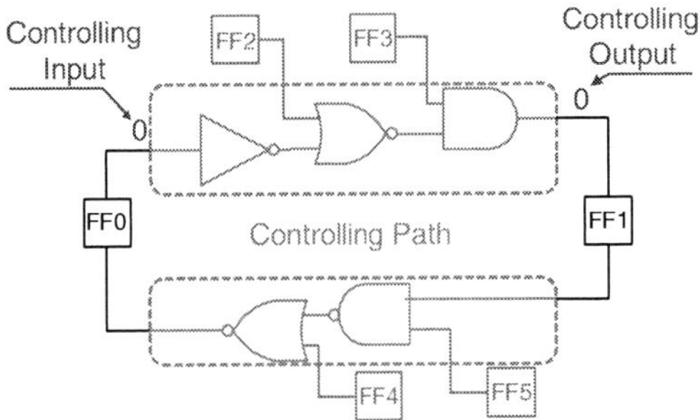

Figure 2: Sequential Loop-Induced Unreachable States

three states for $FF0$ and $FF1$, therefore, are not reachable. Such sequential loop-induced illegal states are very difficult, if not impossible, to be found by any automatic illegal state identification method. This is because: (i). sequential loop expands their effects to different sequential levels in infinite number of time frames; (ii). which states are reachable depends on the initial state of the involved flip-flops.

Fortunately, the illegal states caused by sequential loop are rare cases, because the conditions to form sequential loop-induced unreachable states are quite stringent: (i). between any connected flip-flops on the loop, there must be a controlling path as shown in Fig. 2, which is a logical path where a controlling value can be directly propagated from the beginning to the end; (ii). all controlling paths should have transitivity, i.e., the output controlling value of a path should be the input controlling value of the next path. If any one of the above two conditions is not true, any states in the sequential loop is functionally-reachable. For example, in Fig. 2, suppose we just change the NOR gate which connects with $FF0$ to an OR gate, the controlling path from $FF1$ to $FF0$ is broken. The value of $FF0$ is not solely dependent on its previous on-loop flip-flop $FF1$. Starting from this flip-flop, we first set its value by assigning $FF4 = 1$ and $FF5 = 1$ such that $FF0$ can be flipped to logic 1 and then controlling path between $FF0$ and $FF1$ will be also broken. Therefore, the state of the sequential loop in next clock cycle can be determined by $FF2 - FF5$, thus can reach any state. Moreover, even for these rare sequential loop-induced unreachable states, their impact on testing is rather limited because the involved state elements typically span a number of clock cycles while we target defects in combinational logic between adjacent cycles in testing.

Because of the above reasons, we can simply ignore the unreachable states caused by sequential loops without damaging the effectiveness of pseudo-functional testing much.

4. ILLEGAL STATE IDENTIFICATION FLOW

As discussed earlier, illegal states would imply logic violations at different branches of the same multi-fanout net, explicitly or implicitly. A first thought to generate illegal states is then to propagate contradictive logic values at different branches of multi-fanouts concurrently, and find out which state cubes can justify this combination. These state cubes can then be deemed as illegal. We initially tried out for this intuitive method and found out it is not a good solution. This is because: (i). we need to avoid the case that two fanout branches with contradictive values propagate through the same logic elements (as we cannot determine its value under such circumstances), which results in incomplete identified illegal states; (ii). as we need to propagate contradictive logic values at the branches of each multi-fanout net pairwisely, for those nets that contain a high number of branches, we need to propagate along each branch multiple times and it is a waste of computational efforts. To resolve the above problems, instead of propagating contradictive values at multi-fanout nets, we determine which state cubes

978-1-60558-497-3/09 $25.00 © 2009 ACM   704

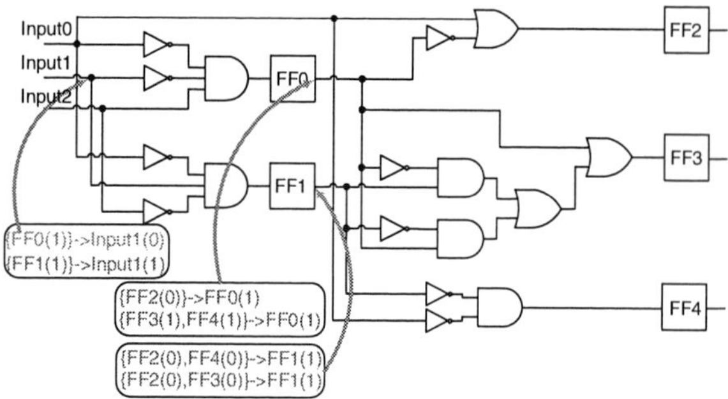

Figure 3: Example Circuit for Illegal State Identification

can justify logic '1'('0') at the multi-fanout nets independently and use this information to obtain illegal states. We use the example circuit shown in Fig. 3 to explain the main flow of the proposed illegal state identification scheme, as depicted in Fig. 4.

Let us define the so-called *justification scheme* at every circuit node in the format of $Cube0 \rightarrow 0$ and $Cube1 \rightarrow 1$, denoting that a state cube $Cube0/Cube1$ justifies logic '0/1' on this node. For example, a justification scheme $FF0(1) \rightarrow Input1(0)$ in Fig. 3 means that $FF0 = 1$ can justify logic '0' at circuit node $Input1$. All such justification schemes are systematically built for each net before unreachable state extraction (detailed in Section 5). According to previous discussion, we always start from a multi-fanout net and try to derive illegal states using contradictory justification schemes. In this example, for the multi-fanout at $Input1$, we have $\{FF0(1)\} \rightarrow Input1(0)$ and $\{FF1(1)\} \rightarrow Input1(1)$. We can therefore conclude that the state cube $\{FF0(1), FF1(1)\}$ is illegal as they justify contradictory values in this fanout. The above procedure needs to be conducted for every multi-fanout nets.

As the illegal state cubes obtained from different paths may contain redundant information (e.g., the set of illegal state cubes shown in Fig. 1 is not the most compact one), we need to remove such redundancy so that all the cubes are minimized and disjoint to each other. This step is conducted by gradually building up a hypergraph for identified illegal state cubes. That is, each flip-flop (FFx) is split into two vertices ($FFx(0)$ and $FFx(1)$) to denote its two possible logic values, and an illegal state cube can be represented as a hyperedge that connects the corresponding vertices. We define the following relationships between hyperedges A and B:

- if A connects to only a subset of vertices of B, we denote this relationship as A *dominates* B. Apparently, the corresponding illegal state cube for hyperedge A contains that of B;

- if both A and B connect to n vertices, and among them $n-1$ vertices are the same and the remaining one corresponds to the same flip-flop with different logic values, we denote this relationship as A *and* B *complement each other*. These two hyperedges can be replaced by a single hyperedge that connects to the $n-1$ common vertices according to our definition.

Every time we add a new hyperedge into the graph, we check the above relationships for its related edges and finally we get a compact set of illegal state cubes without redundancy. For example, for the illegal state cubes shown in Fig. 1, they can be finally compacted into five illegal state cubes: $\{FF0(1), FF1(1)\}$, $\{FF3(0), FF4(0)\}$, $\{FF2(0), FF3(0)\}$, $\{FF2(0), FF3(1), FF4(0)\}$, and $\{FF2(1), FF3(1), FF4(1)\}$, using the above method.

Next, we expand the current illegal state cubes to the next sequential level, again, using the justification scheme information. For the example circuit in Fig. 3, according to structural analysis, two justification schemes at $FF0$ and $FF1$ are detected. We explore

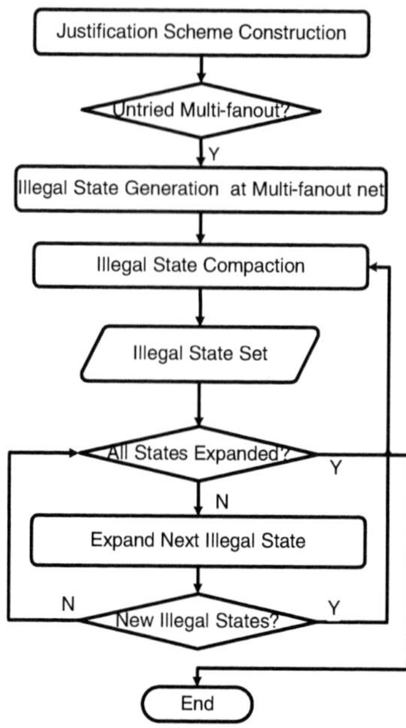

Figure 4: Flowchart for the Proposed Illegal State Identification Scheme

all possible combinations of expanded unreachable cubes, and obtain new cubes, e.g., $\{FF2(0), FF4(0)\}$ and $\{FF2(0), FF3(0)\}$. As shown in our flowchart in Fig. 4, this step is conducted in a stack-like manner. That is, whenever we generate new illegal state cubes through expansion, they will be pushed into the illegal state set for later expansion as well. Compared to prior work that explicitly expand the circuit into a few time frames for illegal state justification, the above procedure is able to implicitly walk through unlimited number of time frames of the circuit efficiently. Finally, the whole procedure for our illegal state justification terminates when there is no unexpanded illegal state.

So far we have discussed how to conduct illegal state extraction, expansion, and compaction, with available justification schemes at every circuit node. In the following section, the key issue in our algorithm, how to construct these justification schemes in an efficient and effective manner, is discussed in detail.

5. JUSTIFICATION SCHEME CONSTRUCTION

We start our justification scheme construction at the input of each flip-flop, which can be obtained directly with the state of the flip-flop. We then propagate these initial justification schemes to every circuit node based on the following propagation rules:

Rule 1: Backward Propagation

A non-controlled value at the output of a logic gate (e.g., logic '1' for a NOR gate) will imply non-controlling value for its inputs of this logic gate. Therefore, a justification scheme that justifies a non-controlled value at the output of a logic gate can be propagated to all inputs of this logic gate to justify the non-controlling value there, referred as *backward propagation*. The backward propagation of justification schemes over a two-input NOR gate is shown in Fig. 5(a) as an example.

Rule 2: Forward Propagation

A justification scheme that justifies the controlling value at any input of a logic gate also justifies the controlled value at its output (e.g., logic '1' for an OR gate). If this scenario occurs, we can propagate the justification scheme from the input to the output directly, referred as *forward propagation*. Similarly, an example for two-input NOR gate is shown in Fig. 5(b).

978-1-60558-497-3/09 $25.00 © 2009 ACM

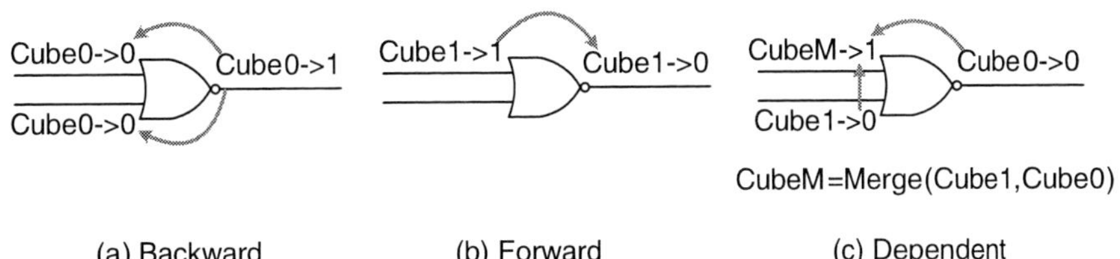

Figure 5: Propagation Rules for Justification Schemes

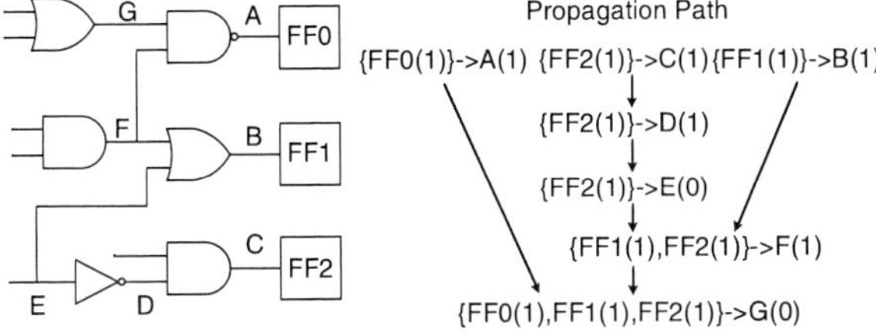

Figure 6: Generation of Sophisticated Justification Schemes

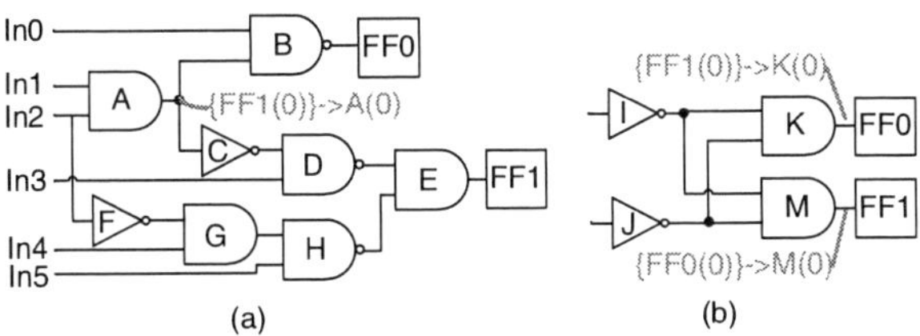

Figure 7: The Impact of Reconvergent Nodes

Rule 3: Dependent Propagation

Generally speaking, to justify a controlling value at a certain input of a logic gate, we need to justify its output to be the controlled value as well as all other inputs to be the non-controlling value. Therefore, dependency exist when propagating such justification schemes and we need to merge state cubes to obtain such justification schemes, referred as *dependent propagation*. Again, an example for NOR gate is shown in Fig. 5(c). Please note that, no justification schemes would be generated if there is any conflict when merging state cubes.

Let us use an example circuit (see Fig. 6) to show how we can build sophisticated justification schemes based on the above propagation rules. After initialization, three schemes $FF0(1) \rightarrow A(1)$, $FF2(1) \rightarrow C(1)$ and $FF1(1) \rightarrow B(1)$ are built. Then, justification scheme $FF2(1) \rightarrow C(1)$ is propagated along path $C - D - E$ and we can obtain justification scheme $FF2(1) \rightarrow E(0)$ at E. Together with $FF1(1) \rightarrow B(1)$, we have $\{FF1(1), FF2(1)\} \rightarrow F(1)$. According to the above and $FF0(1) \rightarrow A(1)$, we can finally obtain $\{FF0(1), FF1(1), FF2(1)\} \rightarrow G(0)$. The last two schemes which is generated based on dependent rule cannot be extracted by implication based technique since such method can only generate justification schemes between single flip-flop and single node.

The above propagation rules, however, are not enough for us to build complete justification schemes. For example, for the example circuit shown in Fig. 7(a), if we would like to propagate justification scheme $FF1(0) \rightarrow E(0)$ backwardly, as logic '0' is a controlled value for AND gate E, based on dependent propagation rule, we need to merge $FF1(0)$ with the cube that justifies $H(1)$ (i.e., $FF1(1)$) to obtain the cube that justifies $D(0)$. Apparently, this merging is not possible and hence we have to stop propagation through gate D. Consequently, we cannot obtain justi-

fication scheme $FF1(0) \rightarrow A(0)$ based on our earlier propagation rules. This justification scheme, however, does exist, because $E(0)$ implies $D(0)$ and/or $H(0)$ while both $D(0)$ and $H(0)$ imply $A(0)$. Similarly, the justification schemes shown in Fig. 7(b) cannot be obtained with the above propagation rules. A closer examination of the circuit shown in Fig. 7(a) reveals that the existence of the justification scheme $FF1(0) \rightarrow A(0)$ is due to the fact that gate E is a reconvergent node of multi-fanout net $In2$. That is, both inputs of gate E are affected by $In2$ and this essentially leads to the implication of $A(1) \rightarrow In2(1) \rightarrow E(1)$ and hence $FF1(0) \rightarrow E(0) \rightarrow A(0)$. For the circuit in Fig. 7(b), even though there is no reconvergent node in the traditional sense, considering logic cubes are propagated both forwardly and backwardly, gates K and M can be also deemed as *virtual* reconvergent nodes.

As the occurrence of the above missing justification schemes can be attributed to reconvergent nodes, to resolve this problem, we conduct implication at every circuit node and we record those implications that lead to the same non-controlling value of a logic gate (if not, there is no reconverging effects). For example, by doing so, we can learn $A(1)$ implies the non-controlling logic '1' at both inputs of logic gate E. Hence we have $A(1) \rightarrow E(1)$ and we would record $E(0) \rightarrow A(0)$, which can be used during propagation through gate E to obtain $FF1(0) \rightarrow A(0)$. Based on the same principle, we would record $K(0) \rightarrow M(0)$ and $M(0) \rightarrow K(0)$ for the circuit shown in 7(b), which can then be used to build the justification schemes shown in the figure. It is important to note that, those implications that do not lead to the same non-controlling value of a logic gate would be discarded, as they can be obtained based on our propagation rules.

By conducting the above implication procedure before the justification scheme propagation process, we are able to build complete justification schemes for the circuit.

6. EXPERIMENTAL RESULTS

To evaluate the effectiveness of the proposed solution, we perform experiments on several ISCAS'89 benchmark circuits, comparing against the implication-based algorithm in [24], which is shown to be able to obtain much more illegal states than the SAT-based method in [13]. We do not compare against [20] and [23] because they do not store functional constraints in the format of illegal states (in arbitrary nets instead). Our experiments are conducted on a 2GHz PC with 1GB memory.

In [24], the authors reported the total number of illegal state cubes only. Let us first compare this value obtained using the two algorithms, as shown in Table 1 (see Columns 2 and 7). Apparently, we can find much more illegal states when compared to [24]. It should be also highlighted that our algorithm are generally more effective for those large circuits.

As one 2-bit illegal state cube can be represented as two 3-bit illegal state cubes, this makes comparing the total number of illegal state cubes not quite dependable. As a result, for fair comparison, we re-implement the algorithm in [24] and all illegal state cubes detected by the two algorithms are compacted in the same manner to remove redundancy. In Table 1, we classify all the illegal state cubes according to the number of specified bits. We can observe

978-1-60558-497-3/09 $25.00 © 2009 ACM

Benchmark	[24]					Proposed							
	Total #	# of 2-bit	# of 3-bit	# of 4-bit	Running Time	Total #	# of 2-bit	# of 3-bit	# of 4-bit	# of 5-bit	# of Large	# of Expanded Cubes	Running Time
s208	16	16	0	0	0.00	23	23	0	0	0	0	0	0.01
s444	16	12	4	0	0.00	49	14	20	15	0	0	0	0.00
s953	116	101	15	0	0.05	365	153	133	78	1	0	0	12.32
s1196	35	10	25	0	0.02	138	10	43	42	20	23	0	0.68
s5378	1006	293	711	2	5.23	8516	399	2584	266	2383	2884	681	153.86
s9324	195	47	148	0	2.30	1109	47	168	108	204	582	17	19.65
s13207	2808	1222	1330	256	19.48	82651	1445	4156	14002	27760	34928	60140	151.28
s15850	507	112	392	3	20.32	5568	239	399	777	261	3892	65	211.82
s38417	988	483	496	9	56.30	90983	864	3094	16544	34641	35840	42559	397.58

Table 1: Experimental Results for Illegal State Identification.

that most illegal state cubes detected by [24] are 2-bit or 3-bit cubes (there are no cubes with size more than 4). This is because their rather simple implication scheme can only obtain relatively small-sized functional constraints. The proposed method, on the other hand, is able to generate a large amount of large-sized illegal state cubes (i.e., cube size larger than 3). More importantly, our method also obtains much more 2-bit and 3-bit cubes than [24], which is able to restrict the functional space more effectively. It can be also observed from the table that our illegal state expansion technique is quite effective. For certain benchmark circuits (e.g., s13207), the illegal cubes generated by expansion can reach nearly 80% of the total illegal states.

Finally, we carefully examine the illegal state cubes generated from the two algorithms, and we find out that all the illegal states obtained in [24] are covered in the illegal state cubes obtained using our proposed method. This further proves that multi-fanout nets are the main structural root cause for functionally-unreachable states, and gives us more confidence to realize the potential of pseudo-functional testing using the proposed methodology.

7. CONCLUSION

The discrepancy between ICs' activities in normal functional mode and that in structural test mode has an increasingly adverse impact on the effectiveness of manufacturing test with technology advancement, as reported in several industrial studies (e.g., [4, 18, 19]). Pseudo-functional testing seems to have great potential to resolve this problem, but whether we could realize this potential highly relies on whether we could effectively identify as many illegal states as possible. In this paper, we show that the main structural root cause for illegal states is the multi-fanout nets in the circuit. Based on this observation, we develop efficient and effective algorithms for illegal state identification. Experimental results on ISCAS'89 benchmark circuits demonstrate that the proposed technique is much more effective when compared to state-of-the-art solutions.

8. ACKNOWLEDGEMENTS

This work was supported in part by the General Research Fund CUHK417406, CUHK417807, and CUHK418708 from Hong Kong SAR Research Grants Council, and in part by a grant N_CUHK417/08 from the NSFC/RGC Joint Research Scheme.

9. REFERENCES

[1] Moderator: K. Butler, Organizer: N. Mukherjee. Power-Aware DFT - Do We Really Need it? Panel, International Test Conference, 2008.

[2] D. Brand and V. S. Iyengar. Identification of Redundant Delay Faults. *IEEE Transactions on Computer-Aided Design*, 13(5):553–565, May 1994.

[3] M. Bushnell and V. Agrawal. *Essentials of Electronic Testing*. Kluwer Academic Publishers, 2000.

[4] C. Shi and R. Kapur. How Power Aware Test Improves Reliability and Yield. EE Times, Sept. 15, 2004.

[5] G. Chen, S. M. Reddy, and I. Pomeranz. Procedures for Identifying Untestable and Redundant Transition Faults in Synchronous Sequential Circuits. In *Proceedings International Conference on Computer Design (ICCD)*, pages 36–41, 2003.

[6] K.-T. Cheng and H.-C. Chen. Classification and Identification of Nonrobust Untestable Path Delay Faults. *IEEE Transactions on Computer-Aided Design*, 15(8):845–853, August 1996.

[7] D. Czysz, M. Kassab, X. Lin, G. Mrugalski, J. Rajski, and J. Tyszer. Low Power Scan Shift and Capture in the EDT Environment. In *Proceedings IEEE International Test Conference (ITC)*, paper 13.2, 2008.

[8] K. Heragu, J. H. Patel, and V. D. Agrawal. Fast Identification of Untestable Delay Faults Using Implications. In *Proceedings International Conference on Computer-Aided Design (ICCAD)*, pages 642–647, 1997.

[9] M. Konijnenburg, J. van der Linden, and A. van de Goor. Test Pattern Generation with Restrictors. In *Proceedings IEEE International Test Conference (ITC)*, pages 598–605, 1993.

[10] M. Konijnenburg, J. van der Linden, and A. van de Goor. Illegal State Space Identification for Sequential Circuit Test Generation. In *Proceedings IEEE/ACM Design, Automation, and Test in Europe (DATE)*, pages 741–746, 1999.

[11] J. Li, Q. Xu, Y. Hu and X. Li. iFill: An Impact-Oriented X-Filling Method for Shift- and Capture-Power Reduction in At-Speed Scan-Based Testing. In *Proceedings IEEE/ACM Design, Automation, and Test in Europe (DATE)*, pages 1184-1189, 2008.

[12] H.-C. Liang, C. L. Lee, and J. E. Chen. Identifying Invalid States for Sequential Circuit Test Generation. *IEEE Transactions on Computer-Aided Design*, 16(9):1025–1033, Sept. 1997.

[13] Y.-C. Lin, F. Lu, and K. Cheng. Pseudofunctional Testing. *IEEE Transactions on Computer-Aided Design*, 25(8):1535–1546, August 2006.

[14] X. Liu and M. S. Hsiao. A Novel Transition Fault ATPG that Reduces Yield Loss. *IEEE Design & Test of Computers*, 22(6):576–584, Nov.-Dec. 2005.

[15] P. Maxwell, I. Hartanto, and L. Bentz. Comparing Functional and Structural Tests. In *Proceedings IEEE International Test Conference (ITC)*, pages 400–407, 2000.

[16] J. Rearick. Too Much Delay Fault Coverage Is A Bad Thing. In *Proceedings IEEE International Test Conference (ITC)*, pages 624–633, Nov. 2001.

[17] S. Remersaro, X. Lin, S. M. Reddy, I. Pomeranz, and J. Rajski. Scan-Based Tests with Low Switching Activity. *IEEE Design & Test of Computers*, 24(3):268–275, May-June 2007.

[18] J. Saxena, K. Butler, V. Jayaram, and S. Kundu. A Case Study of IR-Drop in Structured At-Speed Testing. In *Proceedings IEEE International Test Conference (ITC)*, 2003.

[19] S. Sde-Paz and E. Salomon. Frequency and Power Correlation between At-Speed Scan and Functional Tests. In *Proceedings IEEE International Test Conference (ITC)*, paper 13.3, 2005.

[20] M. Syal, K. Chandrasekar, V. Vimjam, M. S. Hsiao, Y.-S. Chang, and S. Chakravarty. A Study of Implication Based Pseudo Functional Testing. In *Proceedings IEEE International Test Conference (ITC)*, page paper 24.3, 2006.

[21] M. Syal and M. S. Hsiao. New Techniques for Untestable Fault Identification in Sequential Circuits. *IEEE Transactions on Computer-Aided Design*, 25(6):1117–1131, 2006.

[22] X. Wen, K. Miyase, T. Suzuki, S. Kajihara, Y. Ohsumi, and K. K. Saluja. Critical-Path-Aware X-Filling for Effective IR-Drop Reduction in At-Speed Scan Testing. In *Proceedings ACM/IEEE Design Automation Conference (DAC)*, pages 527–532, June 2007.

[23] W. Wu and M. S. Hsiao. Mining Sequential Constraints for Pseudo-Functional Testing. In *Proceedings IEEE Asian Test Symposium (ATS)*, pages 19–24, 2007.

[24] Z. Zhang, S. Reddy, and I. Pomeranz. On Generate Pseudo-Functional Delay Fault Tests for Scan Designs. In *Proceedings IEEE International Symposium on Defect and Fault Tolerance in VLSI Systems (DFT)*, pages 215–226, 2005.

978-1-60558-497-3/09 $25.00 © 2009 ACM

Automated Failure Population Creation for Validating Integrated Circuit Diagnosis Methods[*]

Wing Chiu Tam, Osei Poku and R.D. (Shawn) Blanton

Department of Electrical and Computer Engineering

Carnegie Mellon University, Pittsburgh, PA 15213

Email: {wtam, opoku, blanton}@ece.cmu.edu

ABSTRACT

Integrated circuit (IC) diagnosis typically analyzes failed chips by reasoning about their responses to test patterns to deduce what has gone wrong. Current trends use diagnosis as the first step in extracting valuable information from a large population of failing ICs that include, for example, design-feature failure rates and defect-occurrence statistics. However, it is difficult to examine the accuracy of these techniques because of the unavailability of sufficient fail data where such information is known. This paper describes an approach for benchmarking and verifying diagnosis techniques through failure population creation that builds on prior work in this area. Specifically, we describe how a population of realistic IC failures is created through circuit-level simulation of extracted layouts. The most novel feature of the work is that the virtual test responses produced are both a precise function of defect type and the three-dimensional location within the layout. The extended approach is demonstrated using twelve placed-and-routed circuits. An example application of the developed framework is given to illustrate the utility of having a failure population where the location and type of defect are known a priori.

Categories and Subject Descriptors

B.8.1 [**Performance and Reliability**]: Testing

General Terms

Algorithms, Reliability, Verification

Keywords

Fault Diagnosis, Failure Analysis, Verification

1. INTRODUCTION

Diagnosis of integrated circuits (ICs) is playing an increasingly important role in the IC manufacturing process. Traditionally, diagnosis is used as a fault localization step for physical failure analysis (PFA). Emerging yield learning techniques (*e.g.*, [1,2]) make use of on-going diagnoses to derive important information that includes, for example, design-feature failure rates. Recently, diagnosis is also used as a part of a yield-monitoring vehicle so that

[*] This work is supported by the National Science Foundation under award No. CCF-0541297.

Permission to make digital or hard copies of part or all of this work for personal or classroom use is granted without fee provided that copies are not made or distributed for profit or commercial advantage and that copies bear this notice and the full citation on the first page. To copy otherwise, to republish, to post on servers or to redistribute to lists, requires prior specific permission and/or a fee.
DAC'09, July 26-31, 2009, San Francisco, California, USA

defect density and size distribution can be estimated [3], and in [4], diagnosis is being used to monitor and control IC quality. However, no matter how diagnosis is applied, one important objective remains the same: a diagnosis technique should produce an outcome that correctly pinpoints the defect location in a failed IC. Unfortunately, due to the complex and unpredictable nature of defects, many diagnosis techniques are often based on justifiable assumptions and simplications [5-10]. With assumptions and simplifications come inaccuracies however. To evaluate the merit of various diagnosis techniques, it is important to understand and quantify the amount of inaccuracies introduced.

1.1 Motivation

Verifying the accuracy of a diagnosis technique and the associated follow-on methodologies (*e.g.*, [4,11,12]) is quite challenging for several reasons. Firstly, it is costly to collect and make available sufficient fail data. To have sufficient data, there must be a sufficient number of failed ICs to begin with. Moreover, because diagnosis techniques are not perfect, only a subset of the failed ICs will be successfully diagnosed. Here, diagnosis success means two criteria are satisfied: a) the set of suspect sites must be non-empty and b) the set of suspect sites must not include too many sites (*i.e.*, diagnostic resolution must be reasonably high). Successful candidates are then typically subjected to PFA which attempts to uncover the actual root cause of the located defect. Since PFA is a very time-consuming process [13], only a handful of the successful candidates are actually selected for PFA. (The selection criterion is to optimize PFA success rate.) Thus, at the end of PFA, there might not be sufficient data available to draw any statistically-sound conclusions concerning the accuracy of the diagnosis technique employed to locate the defect. To make matters worse, even if there is sufficient data, the data actually selected for the diagnosis-PFA cycle is a biased sample since only the candidates with high probability of success are chosen.

Another challenge for verifying diagnosis accuracy stems from the sheer amount of data that needs to be diagnosed. Many emerging test-data learning techniques requires on-going diagnosis of a large population of failures (*e.g.*, [1,2,4]). Certainly, PFA cannot be performed on the entire population, thus making accuracy verification in such cases virtually impossible.

Finally, verifying diagnosis accuracy can be challenging for non-technical reasons as well. Most IC failure-analysis data are foundry confidential and thus may not be available even to foundry customers. Unless the diagnosis methodology developers have a close affiliation with the foundry, it is impossible for them to obtain the necessary data to verify their techniques.

1.2 Our Contribution

To address the aforementioned problems, work in [14] has

978-1-60558-497-3/09 $25.00 © 2009 ACM

proposed a framework for creating and evaluating a diverse set of defects. Specifically, the framework is a simulation environment that allows realistic defects to be injected into a circuit through appropriate modifications to its corresponding circuit-level netlist. The "defective" circuit is simulated using a circuit simulator [15] and the "tester response" is extracted from the simulation result. This framework utilized a simple 4-bit ALU circuit with approximately one thousand realistic defects crafted specifically for it. The defects are abstracted as a set of instructions in XML syntax [16] in order to make appropriate changes to the ALU circuit-level netlist. The resulting population of ALU failures was used to examine the effectiveness of various diagnosis techniques by the authors themselves, as well as by commercial tools. The work described here is a direct but significant extension of the work in [14].

The existing framework in [14] is useful for examining diagnosis techniques at the logic level. Since this is a controlled framework, the affected (logic-level) lines where a defect is injected are precisely known. The reported diagnostic candidates, as a result, can be easily checked against the affected lines for investigating diagnosis accuracy. This process can be carried out completely in simulation and thus avoids the issues concerning data insufficiency, bias, *etc.* Since circuit-level simulation is used, the response generated is the most accurate achievable without using real silicon. However, the framework described in [14] is still lacking in several respects. Firstly, the existing defect model is accurate only for the logic level. This is true since the defect is abstracted as a set of XML instructions that modify the netlist, that is, it does not reflect the physical changes in the layout caused by the defect. For example, if an open occurs in a long interconnect connecting node *A* to node *B*, the physical location of the open along the interconnect cannot be captured by simply adding a large resistance between the two nodes. This makes the existing framework unsuitable for verifying the class of diagnosis techniques that aim for improving localization of a defect to a particular three-dimensional region [*e.g.*, 17,18]. Secondly, the framework lacks flexibility in that there is no defect generation capability. Moreover, each available defect model is tailored to the specific 4-bit ALU circuit. This means the available defect models cannot be ported to another design without significant effort. Because the 4-bit ALU circuit does not provide sufficient function variability, its usefulness as a benchmark for evaluation purposes is limited. A variety of circuit types with realistic defect types that produce behaviors as a function of their three-dimensional location are needed for such a purpose.

This paper addresses the shortcomings of the existing framework in [14] by automating (i) defect generation and injection at the layout level and (ii) re-extraction of the circuit-level netlist from the modified layout. This ensures that the defect is accurately modeled. In addition, since the defect generation and injection process are completely automated, the defect spectrum can be easily applied to a variety of designs using a variety of technologies, thus enabling the efficient creation of realistic failure populations.

Automatic defect generation at the layout level has another important application. It can aid the "debugging" of a diagnosis technique and can be used to identify the boundary cases for which a diagnosis technique begins to fail. This exposes incorrect assumptions in the technique by providing a counter-example. By understanding when and how the diagnosis technique fails, it

makes diagnosis improvement possible.

One clear disadvantage of the extended framework is that it is based on circuit-level simulation and thus does not scale well with circuit size. However, it must be emphasized that the utility of the framework does not diminish despite the scalability limitation. This is because a set of small circuits can be used to validate the accuracy of diagnosis techniques as long as the set of circuits is sufficiently rich in its layout features mimicking that of a much larger design. In other words, this framework need not and should not be applied to a multi-million gate commercial design simply to check diagnosis accuracy.

It should also be noted that despite the apparent similarity to inductive fault analysis (IFA) [19,20], our work has two major differences. Firstly, our work has a different focus from IFA. IFA is a systematic method for determining what faults are likely to occur in a given circuit [19]. This information is then used to direct test effort to faults with high probability of occurrence. In contrast, our work focuses on creating a large population of failed ICs to validate diagnosis techniques. While the scalability of IFA is certainly limited by the circuit size, our work is not because, as discussed above, a set of small circuits with rich layout features would suffice for our purpose. Secondly, the end result of IFA is typically a list of faults, ranked according to their likelihood of occurrence. The focus of IFA is not on predicting the tester response of a defect, nor the potential differences in tester response due to defect location. On the other hand, our work attempts to accurately capture tester response as a precise function of defect type and location.

The rest of this paper is organized as follows: Section II briefly reviews the key components, and the flow of the existing framework [14]. Section III describes the details of the extended framework. This is followed by an experiment carried out using the framework on twelve placed-and-routed benchmarks [21,22], and an application of the framework to validate a recently-introduced diagnosis methodology [23] in Section IV. Finally, a summary and future work are presented in Section V.

2. BENCHMARKING FRAMEWORK

In this section, we describe the structure and flow of the benchmarking framework described in [14].

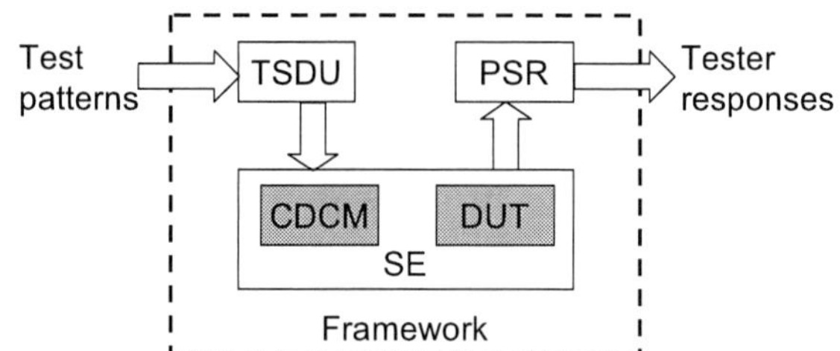

Figure 1: Block diagram that illustrates the structure of the benchmarking framework from [14].

2.1 Framework Structure

Figure 1 shows the key components of the benchmarking framework described in [14]. It is composed of a Test Set Definition Unit (TSDU), Collection of Defective Circuit Models (CDCM), a Simulation Engine (SE), a Design under Test (DUT), and a Parser of Simulation Results (PSR).

978-1-60558-497-3/09 $25.00 © 2009 ACM

2.2 Framework Flow

The TSDU converts test patterns to waveforms for use in a circuit-level simulator. SE takes three inputs, a transistor-level netlist of the DUT, the defect models (CDCM) in XML format, and the stimulus from the TSDU. The SE creates a defective circuit by carrying out the instructions specified in the defective models in CDCM. The XML instruction sets specify the addition and/or removal of components to/from the DUT. The SE then applies the stimulus to the defective circuit for simulation. The output waveforms are subsequently parsed by the PSR to generate the "tester response", that is, the logic ones and zeros that would be presumably measured by the tester in response to application of the test input. It was envisioned that users of this framework would not need to change the internals of framework and that the only "control knob" would be the test patterns.

3. FRAMEWORK EXTENSION

This section describes in detail all the extensions made to the benchmarking framework described in the prior section.

3.1 Extended Framework Structure

Figure 2 shows the key components of the extended framework described in this work. It too is composed of a Test Set Definition Unit (TSDU), a Simulation Engine (SE), and a Parser of Simulation Results (PSR), but also has a new component called the Defect Generation Unit (DGU). Comparing the extended framework with the original framework, it is clear that the DUT and the CDCM are no longer part of the framework. In other words, the framework is no longer dependent upon a particular design. This is compensated however by the addition of the DGU, which accepts a DUT, creates the defect to be injected into the DUT, and the corresponding defective circuit itself.

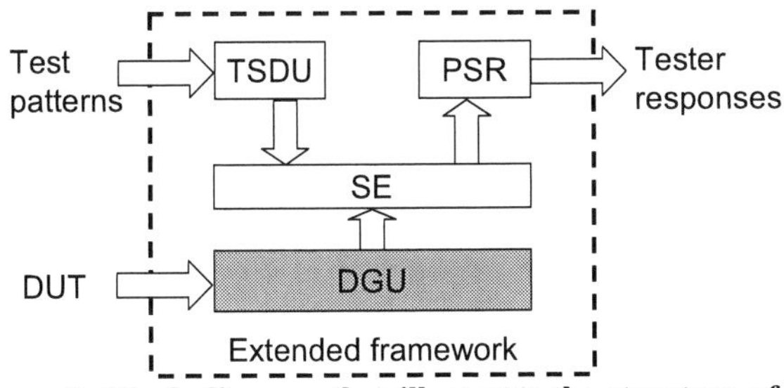

Figure 2: Block diagram that illustrates the structure of the extended benchmarking framework.

3.2 Defect Generation Flow

The TSDU, PSR and SE function in the same manner as previously described; we thus focus on the DGU only. The DGU is implemented as a Cadence SKILL [24] program. Figure 3 shows the flow of the DGU.

The DGU accepts the DUT in GDS format [25], which is converted and stored in the Cadence database. Based on the user-provided configuration, different types of defects are precisely injected into a temporary copy of the design database at the layout level. A circuit-level netlist is then extracted from the "defective" (modified) layout using the Cadence extraction engine [26]. Since the layout geometry is changed before the extraction, the extraction engine would reflect the change in the form of modified connectivity and parasitic components within the netlist. This ensures that the location of defect is accurately captured. After the extraction, the temporary Cadence database is no longer needed and is therefore discarded. The extracted netlist is then modified to account for specific defect characteristics. For example, if a bridge is injected, then its resistance value can be modulated to model the resistance of the extra conductive material at the bridged location. The information of the injected defect (*e.g.*, type, location, size, layer, *etc.*) is recorded as comments in the modified netlist. The SE then uses the netlist for simulation.

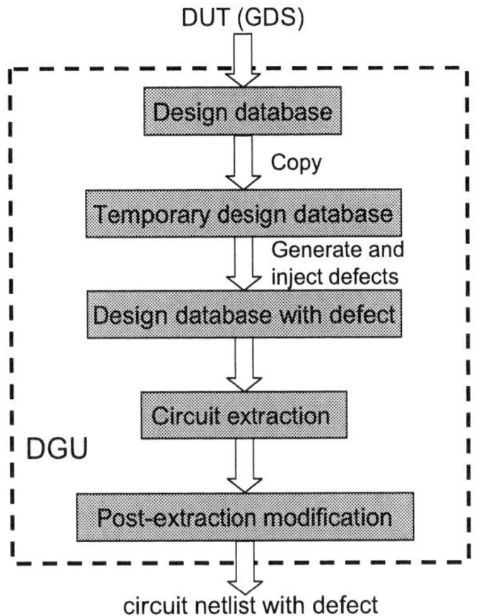

Figure 3: Flow diagram that illustrates the defect-generation process at the layout level.

The sequence of steps involving "copy-generate-inject-extract" operations can be repeated as many times as needed to create a large population of failures exhibiting a variety of defects which are random and/or systematic [27] in nature.

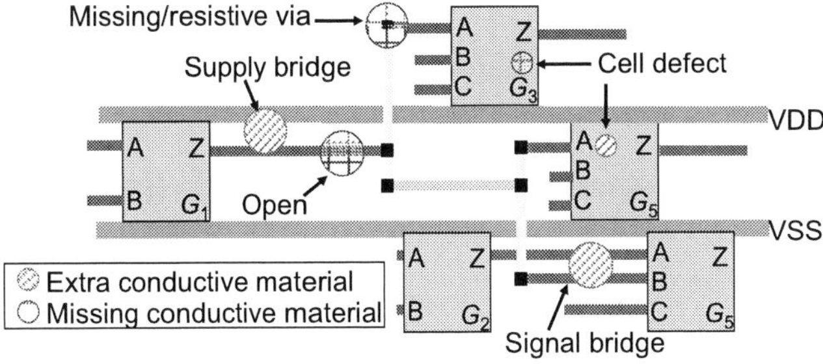

Figure 4: Illustration of defect types supported by DGU.

3.3 Defect Generation Details

Figure 4 shows the defect types currently supported by the DGU, which include opens, signal bridges, supply bridges, missing/resistive vias, and cell defects. These defect types commonly occur in a manufacturing process. If additional conductive material occurred between two neighboring wires and was large enough to electrically connect them, a short could result. Depending on the short location, a signal bridge, a supply bridge (a short between a signal and power or ground), or a cell defect could result. On the other hand, missing conductive or additional insulating material that occurred along the wire could cause an unexpectedly high impedance. Depending on the location of the high impedance, an open, a resistive via or a cell defect could result.

978-1-60558-497-3/09 $25.00 © 2009 ACM

Table 1 shows design-database modifications employed at the layout level as well as the post-extraction modifications performed at the circuit level. For example, consider an open at location (x_0, y_0) in layer z that affects net N_1. The polygon found at the specified location would be partitioned into two separate segments at the layout level. This results in two separate nodes N_{1a} and N_{1b} in the extracted netlist. To model the impedance of the missing conductive material, a resistor of resistance R (specified by the user) is added to the extracted netlist between the nodes N_{1a} and N_{1b}. Similarly, a resistive/missing via can be generated in a net N_1 by removing the via from the layout at a given location (x_0, y_0). To model the impedance of the resistive via, a resistor of resistance R (again specified by the user) is added to the extracted netlist between the disconnected nodes N_{1a} and N_{1b}. For a signal bridge defect involving two nets N_1 and N_2, each of the polygons involved at the bridge location would be partitioned into two segments. This ensures that the parasitics of the involved wires are accurately captured before and after the bridge location. Resistors of resistance R (specified by the user) are added to the extracted netlist for each of the following node pairs (N_{1b}, N_{1b}), (N_{2a}, N_{2b}), (N_{1a}, N_{2a}) to model the impedances at the bridge location. A supply bridge is simply an unwanted connection between a signal line and one of the supply rails, and is created similarly to a signal bridge.

Defect type	Original layout	Modified layout	Modification to the extracted netlist
Open	N_1	N_{1a} N_{1b}	Add a resistor of resistance R to the nodes (N_{1a}, N_{1b})
Resistive via	N_1	N_{1a} N_{1b}	Add a resistor of resistance R to the nodes (N_{1a}, N_{1b})
Signal bridge	N_1 / N_2	N_{1a} N_{1b} / N_{2a} N_{2b}	Add three resistors of resistance R to the nodes (N_{1a}, N_{1b}), (N_{2a}, N_{2b}), (N_{1a}, N_{2a})
Supply bridge	N_1	N_{1a} N_{1b}	Add two resistors of resistance R to the nodes (N_{1a}, N_{1b}), $(Vdd/Vss, N_{1a})$

Table 1: Illustration of layout-level modifications supported by DGU and the associated circuit-level modifications.

The method for generating cell defects is not listed in Table 1 since they can be generated by applying the same methods described in Table 1 at the standard-cell level instead of at the interconnect level. Cell defect types also include however the "nasty polysilicon defect" [28], transistor stuck-open, and transistor stuck-closed.

4. FRAMEWORK APPLICATION

This section describes two experiments to demonstrate the applicability and the usefulness of the extended framework.

4.1 Defect Generation Experiment

To demonstrate the extended framework, twelve ISCAS benchmarks [21,22], varying in size and complexity, are placed and routed using Cadence First Encounter [29]. The technology used is the TSMC 180 nm CMOS process from the publicly-accessible MOSIS webpage [30]. In this experiment, we are interested in creating failing tester responses for slow-speed scan test. Thus, no simulation is performed to create delay-test responses. Test sets are created for each benchmark using Atalanta [31], each having 100% stuck-at fault coverage. The critical-path delay of all the benchmark circuits does not exceed 3.6 ns. To balance simulation

time and mimic a stuck-at, scan testing environment, the applied test clock period is chosen to be 10 ns. The test patterns are applied to each ISCAS benchmark one-by-one in a single circuit-level simulation, that is, all tests for a single defective circuit are simulated in a single run.

To demonstrate the utility of the benchmarking framework, defects of the same type are injected at various locations along the *same* segment of the *same* net in the placed-and-routed benchmarks. Figure 5 illustrates this process. As shown in Figure 5, net f has five different segments (labeled B_1 to B_5). In this example, two defects, denoted by circular dots, are injected one at a time at two different locations along B_3 in two different copies of the same design database. Defects are injected in the other segments B_1, B_2, B_4 and B_5 in a similar manner. This process is repeated on all the nets for each benchmark. If the extended framework accurately captures the physical location of the defect in the extracted netlist, the simulated tester responses of the circuit should be different for some nets (*e.g.*, a long net). This experiment is repeated for three defect types that include an interconnect open, a resistive via, and a signal bridge and is summarized in Table 2. (Supply bridges and standard-cell defects are not used in this experiment because they do not exhibit location dependency at the interconnect level.) The number of net segments exhibiting a location-dependent response (N_{diff}) is recorded along with the number of net segments modified ($N_{modified}$). Similarly, the number of defective circuits that fails at least one test (D_{failed}) is recorded along with the number of defects injected ($D_{injected}$) to demonstrate that the injected defects do lead to failing circuits.

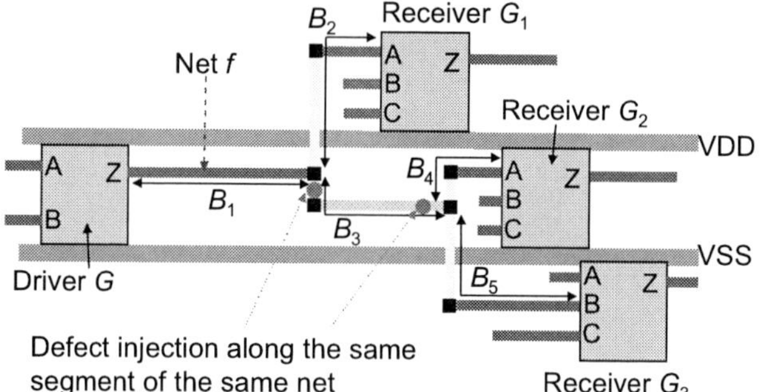

Figure 5: Illustration of defect injection along the same segment.

From these results, it is clear that even for a simple circuit like S208, which consists of only 96 gates, the framework is able to produce location-dependent tester responses for all three defect categories. The ability to capture location-dependent behavior is very important especially for opens since it is well-known that the exact location of an open can have a measurable impact on the tester response [8,32]. For each defect category in Table 2, the ratio of the number of nets with different tester response (N_{diff}) to the total number of nets modified (N_{net}) is less than the ideal value of one because the tester response is binary in nature and does not necessarily reflect the differences in the analog output signals caused by an injected defect. In addition, Table 2 shows that the responses generated for bridges at different locations are very likely to be different (82% on average), when compared to the other two defect types. This result is expected since a bridge injected at different locations along a given segment is likely to bridge to different neighboring nets, which generally results in different responses.

978-1-60558-497-3/09 $25.00 © 2009 ACM

Benchmark name	Opens				Resistive vias				Bridges			
	N_{diff}	$N_{modified}$	D_{failed}	$D_{injected}$	N_{diff}	$N_{modified}$	D_{failed}	$D_{injected}$	N_{diff}	$N_{modified}$	D_{failed}	$D_{injected}$
C432	45	336	1274	1277	13	83	646	648	109	116	435	450
C499	83	441	1586	1602	49	273	906	913	79	84	407	428
C880	92	646	2262	2262	58	301	1096	1097	166	388	1084	1163
C1355	181	960	3361	3372	81	455	1660	1664	190	238	1053	1161
S208	11	140	509	514	5	58	213	215	21	28	150	158
S382	26	263	931	932	14	132	443	444	25	36	226	251
S386	70	311	1112	1115	29	165	564	566	67	72	329	340
S444	34	319	1085	1093	15	133	481	484	47	63	328	350
S510	73	399	1436	1439	35	179	669	672	114	121	494	512
S641	65	464	1781	1783	33	212	766	767	68	76	358	387
S713	49	500	1756	1829	29	234	774	808	87	101	456	512
S953	173	727	2636	2643	101	365	1280	1284	237	251	958	1006

Table 2: Statistics of simulated tester responses that exhibit location dependency for opens, resistive vias and signal bridges.

4.2 Diagnosis Experiment

The framework is also used here to investigate the accuracy of several defect localization heuristics that are based on layout analysis. The diagnosis methodology described in [23] is applied to the population of failures described in the previous section. Layout analysis heuristics developed in [18] are used to precisely localize the x-y-z region of the failure for reducing the time needed for successful PFA. In [23], both passing and failing tester patterns are analyzed to identify potential defect site(s) and to extract *neighboring conditions* for defect/failure activation. The neighboring conditions are simply the logic values of neighboring nets (within some appropriately-deemed distance) that must be satisfied for the defective site to become faulty. The neighboring conditions are mapped into a set of one or more fault-tuple products [33], which are then disjunctively combined into a *macrofault* which is an expression that represents the logical conditions that must be satisfied for a site to become faulty. For example, the macrofault $(n_1,0,i)^c (n_2,0,i)^c (s_x,1,i)^e + (n_1,1,i)^c (n_2,1,i)^c (s_x,1,i)^e$ indicates that site s_x becomes stuck-at-1 when $n_1=n_2=0$ or $n_1=n_2=1$. (More details of fault-tuple macrofaults can be found in [33].) The resulting macrofault is then fault simulated using *all* the test patterns that were applied to the failing chip. Only those macrofaults that produce a simulation response that exactly matches the tester response are reported as an outcome of diagnosis.

Table 3 reports the accuracy of diagnosis. A diagnosis outcome is considered correct if it has at least one macrofault that contains the injected defect location. From Table 3, we can see that the diagnosis methodology described in [23] is of high accuracy and achieved 85% accuracy on average.

Benchmark name	Diagnosis accuracy	Benchmark name	Diagnosis accuracy
C432	78.4%	S386	85.4%
C499	89.8%	S444	94.9%
C880	86.3%	S510	90.0%
C1355	79.3%	S641	76.0%
S208	85.2%	S713	80.5%
S382	77.6%	S953	96.7%

Table 3: Quantification of diagnosis accuracy.

In [18], the outcome of the diagnosis is further analyzed with layout information to improve defect localization. Work in [18] describes defect-based and defect-independent approaches for improving defect localization. The defect-based approach exploits characteristics of the defect type that is inferred from the diagnosis-extracted macrofault. The defect-independent approach improves localization less but has the advantage of being more conservative and thus is less likely to incorrectly localize the position of the defect. The accuracy of both approaches are examined in this experiment.

The approach in [18] attempts to categorize failing chips by the properties of their diagnosis outcome so that defect-specific heuristic can be applied to improve defect localization. Five specific categories, namely, stuck-at (SA), branch-open (BO), two-line bridge (TB), cell defect (CD), and defect-type independent (DTI), are described. The SA category captures failing chips that exhibit stuck-at behavior for all patterns and is characterized by a diagnosis outcome that contains stuck-at fault(s) that matches all the tester patterns[1]. The BO category contains failing chips with an open occurring in a branch and is characterized by a diagnosis outcome that exhibits both stuck-at-0 and stuck-at-1 behavior at a logical branch. The TB category has failing chips with a bridge involving two signal lines and is characterized by a diagnosis outcome with two failure sites s_1 and s_2 that have this property: when s_1 exhibits stuck-at behavior, s_2 is in the neighboring condition of s_1 and vice versa. The CD category has failing chips with defective standard cells and is characterized by a diagnosis outcome that exhibits stuck-at behavior at a failure site only when the driving cell of the site has specific input values. The DTI category is a catch-all category that contains any failing chips that are not captured by the other categories. The first four categories use defect-based approaches for defect localization while the last category uses a defect-independent approach.

The correctness of a category-specific heuristic is investigated by comparing the implicated portion of the layout against the actual x-y-z location of the injected defect. In addition, to factor out the effect of diagnostic resolution on the accuracy of the heuristic, only diagnosis outcomes consisting of a single macrofault are considered. Similarly, in order to evaluate the accuracy of the heuristic alone, the diagnosis candidate is checked to ensure that it contains the defect location before performing layout analysis.

---

[1] A stuck-at fault matches all the patterns when the simulated response of the fault agrees completely with the observed response for all the tester patterns.

978-1-60558-497-3/09 $25.00 © 2009 ACM

Table 4 shows accuracy of the category-specific heuristics. The "-" in a specific category for a specific benchmark indicates that there is no diagnosis candidate. This occurs because only diagnosis outcomes consisting of a single macrofault are considered. From table 4, it is clear that both defect-based and defect-independent approaches have good accuracy. The average accuracy for the categories CD, TB, BO, SA, and DTI are 94.9%, 100.0%, 95.1%, 100.0% and 85.7%, respectively.

Benchmark name	CD (%)	TB (%)	BO (%)	SA (%)	DTI (%)
C432	-	100.0	-	100.0	91.8
C499	100.0	100.0	96.2	100.0	89.4
C880	100.0	100.0	98.9	100.0	95.8
C1355	-	100.0	99.7	100.0	80.2
S208	-	100.0	-	100.0	73.9
S382	-	-	89.7	100.0	92.3
S386	97.6	100.0	95.1	100.0	82.9
S444	94.4	-	100.0	100.0	100.0
S510	96.0	-	98.0	100.0	72.0
S641	-	100.0	98.9	100.0	67.0
S713	97.4	100.0	96.1	100.0	82.9
S953	78.9	-	78.9	100.0	100.0

Table 4: Accuracy quantification of category-specific heuristics of diagnosis outcomes with a single candidate.

5. SUMMARY AND FUTURE WORK

A benchmarking framework for producing a population of chip failures is useful for analyzing the efficiency and efficacy of various diagnosis techniques. In this work, we describe the shortcomings of an existing framework [14] and describe extensions and modifications that allow the three-dimensional location of a defect to be more precisely controlled. The generated population of chip failures reflects the exact three-dimensional position of the injected defect location in their tester responses. The utility of the framework was demonstrated by investigating the ability of two new diagnosis approaches to precisely localize defects in failing chips. The framework is able to generate scenarios when diagnosis and defect localization fail, which provides concrete examples for these diagnosis approaches to be improved. Other test and diagnosis methodologies can also be investigated in a similar manner using the benchmark framework and is the focus of future work. Also, at some point in the future, we plan to add these and other benchmarks to our website: www.ece.cmu.edu/~actl.

6. REFERENCES

[1] M. Keim, et al., "A Rapid Yield Learning Flow Based on Production Integrated Layout-Aware Diagnosis," *International Test Conference*, p. 7.1, Oct. 2006.

[2] H. Tang, et al., "Analyzing Volume Diagnosis Results with Statistical Learning for Yield Improvement," *European Test Symposium*, pp. 145-150, May 2007.

[3] J. E. Nelson, W. Maly and R. D. Blanton, "Diagnosis-Enhanced Extraction of Defect Density and Size Distributions from Digital Logic ICs," *Techcon* 2007.

[4] X. Yu, et al., "Controlling DPPM through Volume Diagnosis," *VLSI Test Symposium*, pp. 134-139, May 2009.

[5] S. D. Millman, E. J. McCluskey and J. M. Acken, "Diagnosing CMOS Bridging Faults with Stuck-At Fault Dictionaries," *Inlemafional Test Conference*, pp. 860-870, Oct. 1990.

[6] S. Venkataraman and S. B. Drummonds, "POIROT A Logic Fault Diagnosis Tool and Its Applications," *International Test Conference*, pp. 253-262, Oct. 2000.

[7] S. B. Drummonds et al., "Bridging the Gap Between Logical Diagnosis and Physical Analysis," *IEEE International Workshop on Defect Based Testing*, April 2002.

[8] Y. Sato et al., "A Persistent Diagnostic Technique for Unstable Defects," *International Test Conference*, pp. 242-249, Oct. 2002.

[9] T. Bartenstein et al., "Diagnosing combinational Logic Designs Using the Single Location At-a-Time (SLAT) Paradigm," *International Test Conference*, pp. 287-296, Oct. 2001.

[10] D. B. Lavo, I. Hartanto and T. Larrabee, "Multiplets, Models, and the Search for Meaning: Improving Per-Test Fault Diagnosis," *International Test Conference*, pp. 250-259, Oct. 2002.

[11] M. Sharma et al., "Faster Defect Localization in Nanometer Technology Based on Defective Cell Diagnosis," *International Test Conference*, p. 15.3, Oct. 2007.

[12] M. Sharma et al., "Efficiently Performing Yield Enhancements by Identifying Dominant Physical Root Cause from Test Fail Data," *International Test Conference*, p. 14.3, Oct. 2008.

[13] B. Engel et al., "The Art of Cross Sectioning," *Microelectronics Failure Analysis Desktop Reference*, p. 473, 2004.

[14] T. Vogels et al., "Benchmarking Diagnosis Algorithms With a Diverse Set of IC Deformations," *International Test Conference*, pp. 250-259, pp. 508 – 517, Oct. 2004.

[15] *The Star-HSPICE Reference Manual*, Synopsys, Mountain View, CA, www.synopsys.com.

[16] Extensible Markup Language (XML) 1.0, http://www.w3.org/XML/.

[17] C. Hora and S. Eichenberger, "Towards High Accuracy Fault Diagnosis of Digital Circuits," *International Symposium for Testing and Failure Analysis*, pp. 47 – 51, 2004.

[18] W. C. Tam, O. Poku, and R. D. Blanton, "Precise Failure Localization Using Automated Layout Analysis of Diagnosis Candidates," *Design Automation Conference*, pp. 367-372, June 2008.

[19] J. P. Shen, W. Maly, and F. J. Ferguson, "Inductive Fault Analysis of MOS Integrated Circuits," *IEEE Design and Test of Computers*, Dec. 1985.

[20] A. L. Jee and F. J. Ferguson, "CARAFE: An Inductive Fault Analysis Tool for CMOS VLSI Circuits," *VLSI Test Symposium*, pp. 92-98, April 1993.

[21] F. Brglez and H. Fujiwara, "A Neutral Netlist of 10 Combinational Benchmark Circuits and a Target Translator in Fortran," *Int. Symp. on Circuits and Systems*, 1985.

[22] F. Brglez, D. Bryan and K. Kozminski, "Combinational profiles of sequential benchmark circuits," *Int. Symp. on Circuits and Systems*, pp. 1929 - 1934, 1989.

[23] R. Desineni, O. Poku, and R. D. Blanton, "A Logic Diagnosis Methodology for Improved Localization and Extraction of Accurate Defect Behavior," *International Test Conference*, p. 12.3, Oct. 2006.

[24] *The SKILL Language Reference Manual*, Cadence Design Systems, Inc., San Jose, CA, www.cadence.com.

[25] *The Graphic Data System (GDS) II Format*, Cadence Design Systems, Inc., San Jose, CA, www.cadence.com.

[26] *The Diva Reference Manual*, Cadence Design Systems, Inc., San Jose, CA, www.cadence.com.

[27] B. Kruseman et al., "Systematic Defects in Deep-Submicron Technologies," *International Test Conference*, pp. 290 - 299, Oct. 2004.

[28] R. D. Blanton et al., "Fault Tuples in Diagnosis of Deep-Submicron Circuits," *International Test Conference*, pp. 233 – 241, Oct 2002.

[29] *The Cadence Encounter Reference Manual*, Cadence Design Systems, Inc., San Jose, CA, www.cadence.com.

[30] The MOSIS Service, Marina del Rey, CA, www.mosis.com.

[31] H.K. Lee and D.S. Ha, "Atalanta: An Efficient ATPG for Combinational Circuits," Technical Report, 93-12, Dept of Electrical Engineering, Virginia Polytechnic Institute and State University, Blacksburg, Virginia, 1993.

[32] R. Rodriguez-Montanes et al., "Diagnosis of Full Open Defects in Interconnecting Lines," *VLSI Test Symposium*, pp. 158-166, May 2007.

[33] R. D. Blanton, K. N. Dwarakanath, and R. Desineni, "Defect Modeling Using Fault Tuples," *IEEE Transactions on Computer-Aided Design of Integrated Circuits and Systems*, pp. 2450 - 2464, Nov. 2006.

978-1-60558-497-3/09 $25.00 © 2009 ACM

Fault Models for Embedded-DRAM Macros

Mango C.-T. Chao*, Hao-Yu Yang*, Rei-Fu Huang†, Shih-Chin Lin‡, Ching-Yu Chin*

*Dept. of Electronics Engineering, National Chiao-Tung University, Hsinchu, Taiwan
†MediaTek Inc., Hsinchu, Taiwan
‡ United Microelectronics Corporation, Hsinchu, Taiwan
{mango@faculty.nctu.edu.tw, max0327.eecs94@nctu.edu.tw, rf.huang@mediatek.com,
Shih_Chin_Lin@umc.com, alwaysrain.gr@gmail.com}

Abstract

In this paper, we compare embedded-DRAM (eDRAM) testing to both SRAM testing and commodity-DRAM testing, since an eDRAM macro uses DRAM cells with an SRAM interface. We first start from an standard SRAM test algorithm and discuss the faults which are not covered in the SRAM testing but should be considered in the DRAM testing. Then we study the behavior of those faults and the tests which can detect them. Also, we discuss how likely each modeled fault may occur on eDRAMs and commodity DRAMs, respectively.

Categories and Subject Descriptors

B.8.1 [**Hardware**]: Reliability, Testing, and Fault-Tolerance

General Terms

Design

Keywords

Memory testing, embedded DRAM

1. INTRODUCTION

With the continually growing need to an effective and economic embedded-memory core in the SoC era, researchers attempt to carry DRAM's advantages, such as high density, structure simplicity, low-power consumption, and low cost [1], from a commodity memory into a SoC. In the past decade, a lot research effort has been put into the area of the embedded-DRAM (*eDRAM*) technologies, such as deep-trench capacitor with bottle etch [2], planar capacitor [3] [4], shallow trench capacitor [4], and metal-insulator-metal (MIM) capacitor [3] [5], to reduce eDRAM's process adders to the CMOS process. However, few previous research works have discussed the testing strategies used for eDRAMs, which cannot be directly carried from the testing of commodity DRAMs.

The conventional DRAM testing contains two main tasks: the retention testing and the functional testing. In the retention testing, we test whether the data retention time of each

Permission to make digital or hard copies of part or all of this work for personal or classroom use is granted without fee provided that copies are not made or distributed for profit or commercial advantage and that copies bear this notice and the full citation on the first page. To copy otherwise, to republish, to post on servers or to redistribute to lists, requires prior specific permission and/or a fee.
DAC'09, July 26-31, 2009, San Francisco, California, USA

DRAM cell can meet its specification. In the functional testing, we test whether the DRAM-cell array and its peripheral circuits can function correctly at different operating modes, which combine different cycle latencies with different clock frequencies for special applications, such as the burst mode and page read/write. To cover various fault models for the DRAM array, several test algorithms, such as checkerboard, address complement, March, row/column disturb, self-refresh, XMOVI, and butterfly, need to be applied. Applying all the above algorithms at different operating modes is time-consuming, and hence, in reality, most DRAM companies price their DRAM chips differently according to the length of the applied test. With this price model, DRAM companies need to analyze their process as well as their memory design to rank the fault models by their possibility of occurrence. Then, the test engineers can choose a proper combination of test algorithms to cover the high-ranked faults as much as possible when the length of the applied test is limited.

Testing eDRAMs is quite different from testing commodity DRAMs due to the following reasons. First of all, most eDRAM macros use the SRAM interface (the so-called 1T-SRAM architecture), which consists of no address multiplexer (no CAS, RAS) and can auto-refresh. Second, unlike commodity DRAM, whose application might be unknown before the fabrication, eDRAM macros are more application-specific and hence have only one operating mode, meaning that one cycle latency at only one operating frequency needs to be tested. Due to the simplicity of eDRAM's interface, testing eDRAMs is more like testing SRAMs and requires a shorter test algorithm than testing commodity DRAMs. However, testing eDRAM is not as simple as testing SRAMs since some fault models which may not be considered in SRAM testing, such as retention faults and coupling faults, should be considered in eDRAM. Third, the process of eDRAMs is different from that of commodity DRAMs, meaning that their storage capacitors, bit lines, word lines, transistors models, number of metal layers, and wire models are all different. As a result, the possibility of a fault's occurrence for eDRAMs is different from that for commodity DRAMs as well.

In addition, when a bare DRAM die is integrated to a system-in-package, the responsibility of testing the DRAMs is on the DRAM provider. However, when an eDRAM macro is integrated to a system-on-chip (SoC), the responsibility of testing the eDRAMs is transferred to the system integrator. Testing eDRAM macros in an SoC chip relies on the BIST circuitry to enhance the test accessibility of the SoC testing [6][7][8]. Its test application time depends on the efficiency of the test scheduling among various types of embedded cores. On the other hand, testing commodity DRAMs (especially the MCM

978-1-60558-497-3/09 $25.00 © 2009 ACM

DRAM KGDs) seldom uses BIST in practice and relies more on the the parallel testing capability provided by the memory testers to shorten the average test time. Several BIST schemes are proposed for the embedded DRAM testing [9][10][11][12]. However, these previous works mainly focus on the architecture and the automatic generation of the BIST circuitry, not on the testing algorithms and methodologies.

In this paper, we would like to share our experience in testing an UMC 65nm eDRAM macro. We first introduce the overall architecture of the eDRAM macro under test and the differences between commodity-DRAMs testing and eDRAMs testing. Since an eDRAM macro uses an SRAM interface, we start from a standard SRAM algorithm and then discuss the fault models which are not covered by the SRAM algorithm but should be particularly considered when testing DRAMs. Next, we discuss the impact of those faults on both eDRAMs and commodity DRAMs, and suggest a test sequence to cover each of those faults. Also, we discuss how likely each modeled fault may occur on eDRAMs and commodity DRAMs, respectively.

2. BACKGROUND

2.1 Overview of an Embedded-DRAM Macro

Figure 1 shows the block diagram of a 16Mb eDRAM macro used in our SoC design. This eDRAM macro utilizes deep trench capacitors and is implemented in a UMC 65nm low-leakage logic process. Figure 2 shows a cross-section view of an eDRAM cell in our design. The word size on the interface of this eDRAM macro is 32 bits. Due to the use of ECC, we need to add 6-bit more memory cells to the physical array for each word, and hence the physical data stored in the memory array is 38 bits per word. The size of the eDRAM macro is around 4 mm^2, which contains two symmetric eDRAM arrays. Each array contains 128 banks, and each bank contains 64 word-lines and its own local sense amplifier. Each word-line on each array is connected to 64 half-words, and the data-width of each half-word is 19 bits. Note that the layout topology of the eDRAM array utilizes the distributed folding scheme, where the ith bit of the jth word is adjacent to the ith bit of the $(j+1)$th word, not the $(i+1)$th bit of the original jth word. A physical 38-bit word is read out from or written into the memory array through the ECC circuitry, which encodes a 32-bit word into a 38-bit word or decodes a 38-bit word to a 32-bit word. When operating at 100 MHz, the bandwidth of this eDRAM macro is 3.125 Gb/s (32 bits x 100 MHz).

In modern memory designs, scrambling techniques are commonly used to optimize memory's layout geometry, address decoder, cell area, performance, yield, and I/O pin compatibility [13]. The types of scrambling include folding, address decoder scrambling, contact and well sharing, and bit-line twisting. Figure 3 shows an exemplary scrambling used in our eDRAM design, where the ordering of word-lines in this example is arranged according to the least significant bits of the address. With an SRAM interface, eDRAM utilizes both bit-lines and bit-line-bars to distinguish the data value stored in an eDRAM cell, but a cell's data is only connected to either one of the corresponding bit-line and bit-line-bar. In this example, each word-line connects to two 4-bit words. The first word on a word line uses the 0th, 2nd, 4th, and 6th pairs of the bit-line and bit-line-bar, and the second word uses the 1st, 3rd, 5th, and 7th pairs. By proper arrangement, half of eDRAM cells are connected to bit-line, and the other half to the bit-line-bar. As a result, the physical value of those cells connected to a bit-line-bar is inverse to their logical value. In addition, the

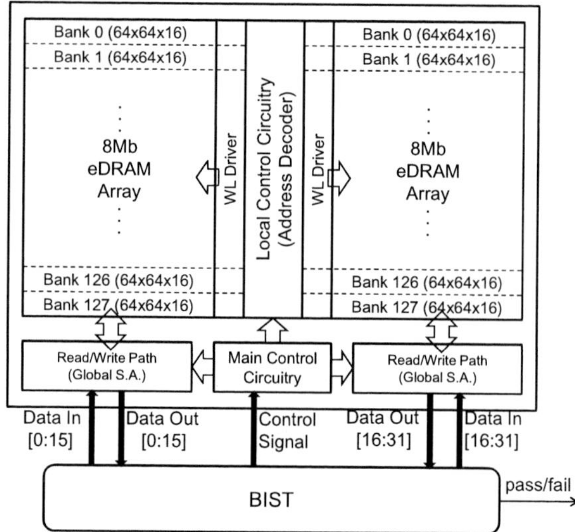

Figure 1: Architecture of a UMC 16Mb eDRAM macro.

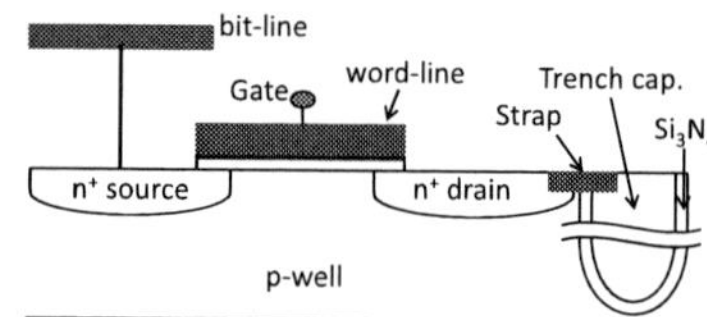

Figure 2: Illustration of an eDRAM cell.

bit-line twist for a column reverses the physical-value/logical-value relation of the cells below that twist.

During the eDRAM testing, the data background written into or read from the memory macro should represent cell's physical value instead of its logical value. Therefore, when designing the BIST circuitry, we need to build a scramble table to map the physical value described in the test algorithm to its corresponding logical value for a given address [13][14]. Those logical values then form the functional test patterns or expected responses during testing. This scramble table can be implemented by a simple two-level logic, whose inputs contain few least significant bits and most significant bits of an address. In addition, when performing March algorithm, the sequence of the activated word-lines also needs to follow the physical sequence, not logical address sequence. Thus, the BIST requires another physical-address-mapping circuitry to handle this address scrambling.

2.2 Difference between eDRAM and Commodity DRAM

Table 1 compares the eDRAMs with the commodity DRAMs. First of all, the array structure and the peripheral circuits of commodity DRAMs are both simple and hence a commodity-DRAMs process usually requires two or three metal layers. On the other hand, eDRAMs are integrated into a logic process, which usually uses five or more metal layers. Second, the storage capacitor (C_s) in commodity DRAMs is larger than that of eDRAMs since a commodity-DRAM process is developed specifically for DRAM cells. Third, the read mechanism of DRAMs is based on the charge sharing between the storage capacitor and the bit-line's capacitor (C_{bl}). Thus, the length

978-1-60558-497-3/09 $25.00 © 2009 ACM

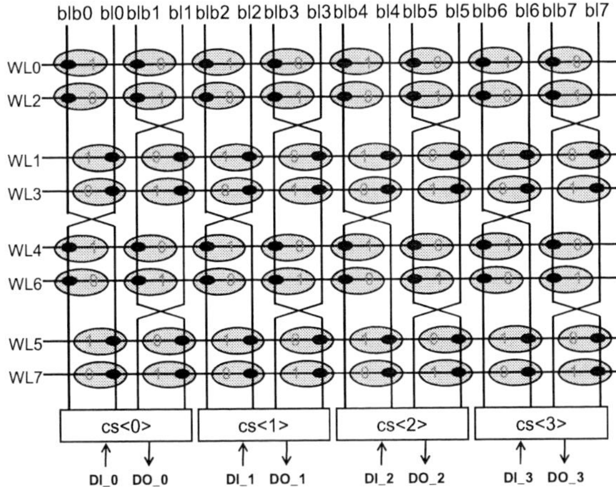

Figure 3: An exemplary array scrambling.

of the bit-line is limited by the ratio of C_{bl} over C_s such that the C_{bl} is small enough to guarantee a successful read operation through the sense amplifier. In our eDRAM design, this ratio is set to 15x. Since the C_s of commodity DRAMs is larger than that of eDRAMs, the bit-line in commodity DRAMs is also longer, meaning more bit cells per bit-line for commodity DRAMs. Also, the refresh period for commodity DRAMs is longer since a larger C_s can tolerate more leakage. In addition, the word-line, which is implemented by polysilicon, is longer in eDRAMs because the extra metal layers in the logic process can be used to reduce word-line's resistance by connecting the word-line in parallel.

	stand-alone DRAM	embedded DRAM
metal layers	$2 \sim 3$	$5 \sim 6$
C_s	$30f \sim 45f$	$7f \sim 10f$
C_{bl}	fixed ratio to C_s	fixed ratio to C_s
refresh period	$> 64ms$	$4ms \sim 16ms$
data size	$512Mb \sim 2Gb$	$2Mb \sim 64Mb$
operating modes	multiple	single
ESD	Yes	No
interface test	timing check+IO	setup/hold time
ECC	Mostly no	Yes

Table 1: Comparison between commodity DRAMs and embedded DRAMs.

From testing's point of view, the eDRAM testing is simpler than the commodity-DRAM testing. First, the data size of eDRAM is smaller, and hence requires shorter test application time. Second, commodity DRAMs have multiple operating modes, i.e., several difference latency cycles mix with several clock frequencies. All operating modes need to be tested for commodity DRAMs. On the other hand, eDRAMs uses SRAM interface which contains no address multiplexors and only one operating mode. Also, commodity DRAMs need to test their IO pads and package pins, such as the ESD testing and the DC testing (open, short, static current, etc.), whereas eDRAMs need not.

Even though testing a single eDRAM macro is much faster than testing a commodity DRAM, the test application time for an eDRAM macro is still a big concern when testing the whole SoC chip. A lot SoC designs using eDRAM macros are actually simple applications and do not contain too many IP cores. Based on our experience, if all the conventional DRAM

test algorithms are applied to an eDRAM macro, its testing application time may take much longer than testing SoC's logic circuits and dominates the total testing time of the SoC testing. Therefore, a minimal eDRAM testing algorithm is still highly demanded in industry, especially for the eDRAM providers.

3. FAULT MODELS FOR EDRAM

To design a minimal test algorithm for eDRAMs, we first start from a standard SRAM test algorithm, March C- [1], since the eDRAM uses the SRAM interface. This standard SRAM algorithm is much shorter than the conventional DRAM algorithms and can already cover stuck-at faults, transition faults, address decoder faults, inversion coupling faults, idempotent coupling faults, and state coupling faults [1]. In the remaining of this section, we will discuss the faults that are not covered by the above SRAM test algorithm but should be considered in the eDRAM testing. Then we discuss how likely these faults may occur in eDRAMs and commodity DRAMs. Based on that, when the length of the test algorithm is limited, we know which fault can be first removed from our target fault list. Also, by comparing eDRAMs to commodity DRAMs, we could discover which commodity-DRAM test patterns can be re-used in eDRAM testing under which conditions.

3.1 Retention Faults

DRAM is a volatile memory, in which a bit-cell is composed of a storage capacitor and a switch transistor (nMOS). A storage capacitor may lose its stored charges through the leakage mechanisms of the switch transistor, such as reverse-bias pn junction leakage, subthreshold leakage, direct tunneling current, etc. [15]. This charge loss may in turn change the logic value of the memory cell. To prevent this value change of a memory cell, a refresh mechanism is utilized to rewrite cell's original value back to its storage capacitor after every specified period of time. However, due to the defects or process variations, the size of a manufactured capacitor may become smaller or the leakage of a switch transistor may become larger than expected, and hence a DRAM cell can lose its stored value before the next refresh is applied. This failure behavior is defined as a retention fault.

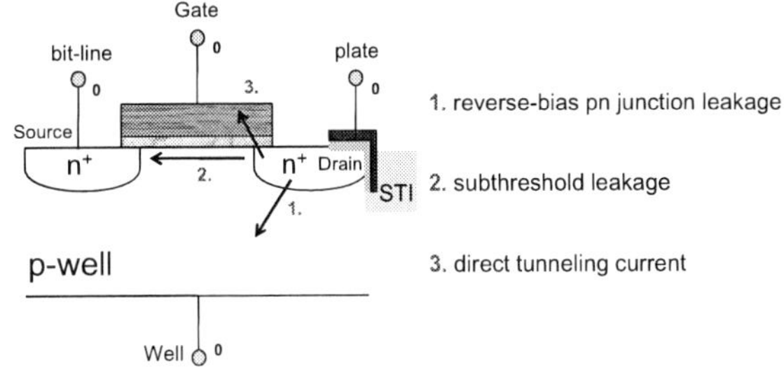

Figure 4: Main leakage sources of an eDRAM cell.

To test retention faults, we need to perform a self-refresh followed by a delay element, which delays the next operation for a specified period of time at a specified temperature. At the same time that the self-refresh is performed, the counter of the auto-refresh is reset. So after the delay element ends, the data will be auto-refreshed again. Then a read operation is performed to check if any retention fault occurs during the delay element. During this retention test, the checkerboard background should be applied because this background can exacerbate the leakage and help to catch a retention fault. Note that

978-1-60558-497-3/09 $25.00 © 2009 ACM

the checkerboard background here refers to data's physical values, not logical values. Also, we need to perform this retention test to both the data-background and its complement.

For SRAMs, retention faults may also occur due to stuck-open defects or resistive-open defects in the two cross-coupled inverters. Several techniques have been proposed to detect the SRAM retention faults, such as IDDQ monitoring [18], low-voltage write with high-voltage read [19], consecutive read with bit line pre-charge disabled [20] [21], or specialized write operations in test mode [22]. The most significant difference between a DRAM retention fault and an SRAM retention fault is that the data of a DRAM cell is supposed to be lost by its nature after a certain period of time, but the data of a SRAM cell is not. Thus, when testing SRAM retention faults, we often try to create the scenario which can exacerbate the faulty behavior as much as possible, so that the test time can be reduced. However, when testing DRAM retention faults, we cannot exacerbate the faulty behavior as much as in SRAM testing since DRAM retention faults correspond to a specified time. Any voltage lost which does not change a DRAM cell's logic value within that specified time can be saved by DRAM's refresh mechanism. Exacerbating the faulty behavior in DRAM retention testing may lead to significant over-testing.

Compared to commodity DRAMS, the possibility of the occurrence of a retention fault is higher for eDRAMs since the capacitor in eDRAMs is smaller and more susceptible to unexpected leakage. Also, eDRAMs have the function of auto-refresh but DRAMs do not. Thus, the eDRAM testing need to turn on the auto-refresh from the beginning of the test algorithm to test its functionality.

3.2 Word-line-Coupling Faults

The word-line-coupling faults can be classified into two types: (1) switching word-line-coupling fault and (2) hammering word-line-coupling fault. In the following, we detail the behavior and corresponding test strategies for each of them.

3.2.1 Switching word-line-coupling fault

For most DRAM designs, word-lines are implemented by polysilicon and hence have a large parasitical resistor. Also, DRAM arrays are usually designed dense and the spacing between two adjacent word-lines is usually small, which implies that the coupling capacitance between word-lines is large compared to SRAMs. SRAM designs utilize a lot shielding technique to reduce the potential coupling capacitance. The word-lines in SRAMs are implemented by metal, whose parasitical resistor is smaller compared to that of DRAMs. With a large parasitical RC, the word-line switching (turn on/off) in DRAM designs is slow.

A switching word-line-coupling fault occurs when an extra coupling capacitor is imposed to a word-line, such that the word-line's parasitical RC is significantly increased and hence the word-line may not fully turn off before the next word-line turns on. This faulty behavior results in that two adjacent word-lines are both on at the same time, and thus the charge sharing occurs among two storage capacitors and a bit-line capacitor, not between a storage capacitor and a bit-line capacitor as it should. As a result, a wrong data might be read from or write into the cells if the cell's data is different from the data of the cell utilizing the same bit-line on its adjacent word-line. This faulty scenario can be created by the checkerboard background.

In order to test this switching word-line-coupling fault, a Y-direction March-based algorithm with the checkerboard background is required. Note that the Y-direction sequence refers

to the physical word-line sequence, not the logical address sequence, such that two adjacent word-lines can be turned on in order. For example in Figure 3, the physical word-line sequence is "WL 0, 2, 1, 3, 4, 6, 5, 7", not "WL 0, 1, 2, 3, 4, 5, 6, 7".

3.2.2 Hammering word-line-coupling fault

A hammering word-line-coupling fault also occurs when an extra coupling capacitor is imposed but the way to discover this extra coupling capacitor is different. This extra coupling capacitor can induce a noise signal to its adjacent word-line (victim) every time when the current word-line (aggressor) turns on. This coupling noise signal then turns on the victim word-line slightly and induces the extra leakage to the storage capacitors on the victim word-line. If we repeatedly toggle the aggressor word-line, the data 1 stored in the cells of the victim word-line may be changed. Figure 5 shows the simulation result of a hammering word-line-coupling faults.

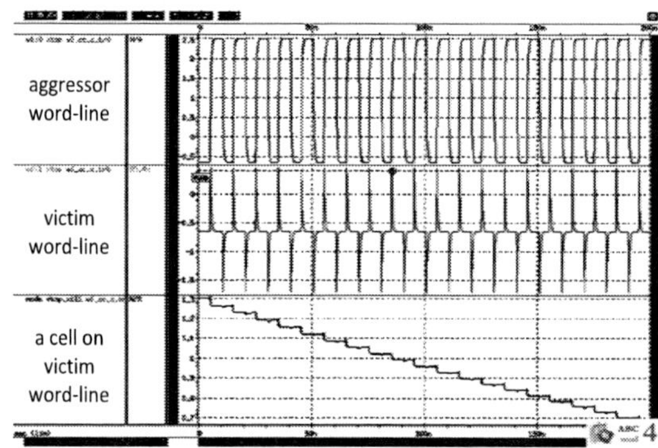

Figure 5: Simulation of hammering word-line-coupling faults.

To test hammering word-line-coupling faults, we need to repeatedly read (toggle) a word-line before reading the data from its victim word-lines (usually the adjacent word-lines) since a read operation is always followed by an automatic write-back. Also, we need to assign the data 1 (physical value) to the cells of the victim word-lines. In fact, the hammering word-line-coupling faults occur more often than the switching word-line-coupling faults since the extra imposed coupling capacitor needs not to be very large for the hammering word-line-coupling faults. The repeated read operations to the aggressor word-line can accumulate the small leakage and result in a failure.

Then, the next question is how many repeated read operations to a word-line is sufficient when testing hammering word-line-coupling faults? The most conservative way is to repeat until the next auto-refresh operation. However, the period between two auto-refresh operations is in the order of milliseconds (more specifically 32ms in our eDRAM design). Then applying repeated read operations to each word-line in our eDRAM design takes 128(# of bank)x64(# of word-line per bank)x32ms= 262 seconds, which is too long for practical eDRAM testing. Also, it could be over-testing since it is possible that no real application may read a word-line for that many consecutive times. Therefore, the actual number of repeated operations applied in our eDRAM testing is an empirical number.

The word-line coupling faults occur more often in eDRAMs than that in commodity DRAMs because the word-line in eDRAMs is usually longer than that in commodity DRAMs. Also, eDRAM's

978-1-60558-497-3/09 $25.00 © 2009 ACM   717

storage capacitor is smaller, so that it can tolerate less extra leakage resulting from the hammering word-line-coupling faults.

3.3 Bit-line-Toggling Faults

Because the DRAM array is dense, the coupling capacitance between the bit-lines cannot be neglected and may affect the sensing mechanism on the bit-lines. If the coupling capacitance is too large, the speed of the charge-sharing mechanism during a read operation will become slow. This slow charge-sharing mechanism may reduce the voltage difference between a bit-line pair captured while the sense amplifier fires. Hence its read-out data is wrong, and then the automatic write-back mechanism will write the wrong read-out data back to the bit cell. This failure behavior is called the bit-line-toggling fault.

A bit-line-toggling fault occurs when the bit-line or bit-line-bar of a cell is close to the bit-line or bit-line-bar of its adjacent cell, and these two adjacent lines have opposite data values. In order to create this scenario for each pair of adjacent cells, we need to perform the solid data-background in our March-based algorithm due to our array scrambling shown in Figure 3. The direction of the applied algorithm does not matter because all the cells on the selected word-line will automatically read out and then write back at the same time through local sense amplifiers once a read operation is applied to any word at the selected word-line.

The bit-line-toggling faults occur more often in commodity DRAMs than that in eDRAMs because a bit-line in commodity DRAMs is longer than that in eDRAMs. In fact, the possibility of the occurrence of a bit-line-toggling fault also depends on the capability of the sense amplifier in use. Again, we need not to consider a bit-line-toggling fault in SRAMs because SRAMs already apply sufficient shielding techniques to prevent this fault.

3.4 Stuck-open Faults

Stuck-open faults in DRAMs can be classified into two categories, the transistor-open faults and the resistive-open faults. Figure 6 illustrates the possible stuck-open faults in a DRAM cell. A transistor-open fault occurs when the resistance R_{wg} between the word-line and the gate of the switch transistor is too large such that the switch transistor is difficult to open and no data can be written into or read from the cell. A resistive-open fault occurs when the resistance between the bit-line and the switch transistor R_{bt}, the resistance between the switch transistor and the storage capacitor R_{tc}, or the resistance between the capacitor and the ground R_{cg} is too large.

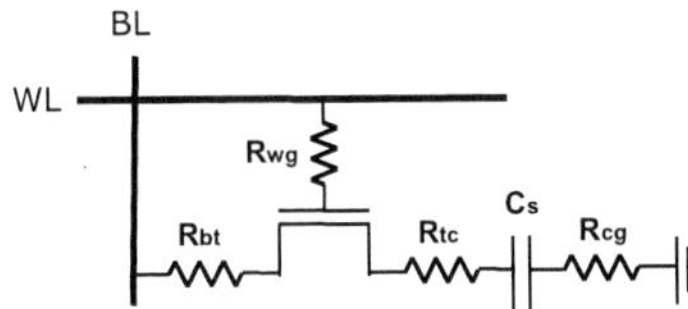

Figure 6: Illustration of stuck-open faults.

To detect transistor-open faults, a March element (read, write, read) is required if the sense amplifier is latch-based and not transparent to the stuck-open fault, which is usually the case in eDRAM designs. Otherwise, testing stuck-open faults is the same as testing stuck-at faults. The data background used in the March element (read, write, read) can be 0 or 1 for detecting transistor-open faults. To detect resistive-open faults, the data background for the March element has to be

(read0, write1, read1) [16] [17].

The transistor-open faults occur more often in commodity DRAMs than that in eDRAMs. It is because eDRAMs are implemented by a logic process, which usually has five or more metal layers. Thus, a word-line in eDRAMs can be connected to a metal wire in parallel such that a large increase of R_{wg} is difficult to occur. However, the commodity-DRAM process only uses two or three metal layers. No extra metal layers can be used for such a parallel connection.

On the other hand, the resistive-open faults occurs more often in eDRAMs than that in commodity DRAMs because of R_{cg}. The amount of R_{cg} can be affected by the doping density of the substrate. For eDRAMs, the doping density of the memory array is different from the doping density of the logic circuits, whereas only one doping density is used in commodity DRAMs. To well control both doping densities in eDRAM process (especially the boundary between eDRAM and logic) is not trivial. Thus a large R_{cg} may occurs more in eDRAM than that in DRAMs.

For SRAM testing, we rarely consider stuck-open faults because an SRAM cell supplies data on both ends of its sense amplifier (bit-line and bit-line-bar) instead of only one end as eDRAMs or commodity DRAMs. As long as one end of an SRAM cell can provide correct data, the sense amplifier can differentiate them.

3.5 Bank Faults

DRAM designs use a two-level sensing scheme to read a cell's data. The cell's data is first passed to the local sense amplifier associated with the bank and then passed to the global sense amplifier as introduced in Section 2.1. Those control signals to the farthest banks, such as bank select and word-line select, may need to be routed in a long wire path and synchronize with the clock. To design a high-speed DRAM, the master-stage of these signals is placed in the central controller (on the bottom of the memory design) while its slave-stage is placed next to the corresponding bank, such as Figure 7. Those signal paths from a master to its slave can be long and has larger parasitical RC. The timing on these signals is always critical.

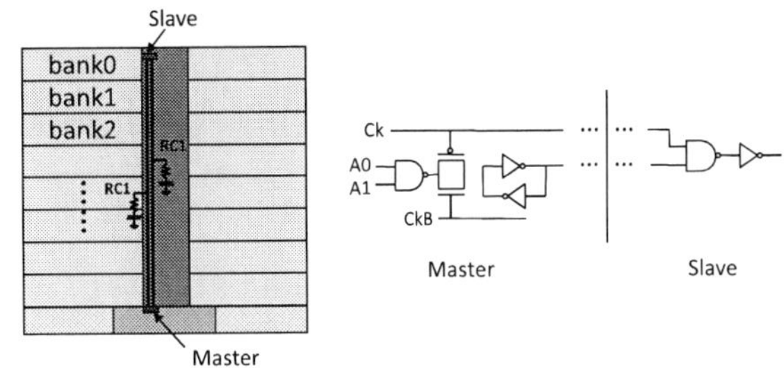

Figure 7: Illustration of a bank fault.

When a timing defect falls in the path of these signals, their signal delay may greatly increase, resulting in that two banks may turn on concurrently. Then the values from the two local sense amplifiers of these two banks will fight with each other on the global bit-line. This value-fighting from two local sense amplifiers is called the *bank fault*. To test this fault, we can apply consecutive read operations from the farthest bank to the nearest one. Each read operation reads the first word-line of a different bank. This is called the *bank-switching* pattern.

The occurrence of a bank fault is determined by the design margin and should be especially considered for high-performance designs. Both eDRAMs and commodity DRAMs could have

978-1-60558-497-3/09 $25.00 © 2009 ACM

bank faults. For SRAMs, the bank faults do not exist since SRAMs use single-level sensing and no local sense amplifier is used as shown above.

4. CONCLUSION

The eDRAM testing mixes up the techniques of SRAM testing and commodity-DRAM testing since eDRAMs use the SRAM interface and DRAM cells. In this paper, we first introduced an exemplary eDRAM design and discussed the key issues which should be emphasized in eDRAM testing by comparing to commodity-DRAM testing. Then we started from a short SRAM algorithm and discussed the fault models that are not covered by the SRAM testing but should be considered in DRAM testing. The impact of those faults on both eDRAMs and DRAMs are also discussed, based on which a test sequence to detect each modeled fault is provided. A test algorithm for eDRAMs can be constructed by combining those suggested test sequences.

5. FUTURE WORK

One important issue about eDRAM testing is its reliability testing, such as THB (temperature, humidity, bias) test [23], HAST (highly-accelerated temperature and humidity stress) test [24], and HTOL (high temperature operating life) test [25]. The objective of reliability testing is to measure the reliability or lifetime of manufactured chips by operating them under an extreme condition of certain environmental or electrical parameters. Those extreme conditions can accelerate the wear-out failure mechanisms. Hence we could use the result of the above reliability testings to foresee the product's failure rate occurring after certain months or years of usage. This failure caused by the wear-out defects is one source of product's defect level.

Unlike commodity DRAM utilizing spare rows and columns for repair, most eDRAM macros utilize ECC (error correction code) circuitry to improve their yield as well as its reliability since eDRAM's storage capacitor is small and thus more susceptible to the noise, soft errors, or wear-out defects. Because some wear-out defects might be masked by the ECC circuitry, we need to turn off the ECC circuitry and collect the raw defect statistics of the eDRAM macros when applying the reliability testing. Then, how to estimate the defect level with ECC based on the collected raw defect statistics without ECC becomes an interesting but practical problem for eDRAMs.

6. REFERENCES

[1] A. J. van de Goor, "Testing Semiconductor Memories, Theory and Practice," Gouda, The Netherlands: ComTex, 1998.

[2] G. Wang, et al., "A 0.127 μm^2 High Performance 65nm SOI Based embedded DRAM for on-Processor Applications," *International Electron Devices Meeting*, 11-13 Dec. 2006, pp. 1-4.

[3] E. Gerritsen, et al., "Evolution of Materials Technology for Stacked-Capacitors in 65 nm Embedded-DRAM," *Solid-State Electronics*, vol. 14, 2005, pp. 1767-1775.

[4] M.-E. Jones, "1T-SRAM-Q^{TM}: Quad-Density Technology Reins in Spiraling Memory Requirements," *Mosys, Inc.*, Retrieved on 2007-10-06.

[5] A. Berthelot, C. Caillat, V. Huard, S. Barnola, B. Boeck, H. Del-Puppo, N. Emonet, F. Lalanne, "Highly Reliable TiN/ZrO$_2$/TiN 3D Stacked Capacitors for 45 nm Embedded DRAM Technologies," *Proceeding of Solid-State Device Research Conference*, Sept. 2006, pp. 343-346.

[6] C. Cheng, C.-T. Huang, J.-R. Huang, C.-W. Wu, C.-J. Wey, and M.-C. Tsai, "BRAINS: A BIST compiler for embedded memories," *Proceedings of IEEE International Symposium on Defect and Fault Tolerance in VLSI Systems*, Yamanashi, Oct. 2000, pp. 299-307.

[7] J.-F. Li, R.-S. Tzeng and C.-W. Wu, "Diagnostic Data Compression Techniques for Embedded Memories with Built-In Self-Test," *J. Electronic Testing: Theory and Application*, vol.18, no.4, Aug. 2002, pp. 515-527.

[8] B. Nadeau-Dostie, A. Silburt, V.K. Agarwal, "Serial Interfacing for Embedded Memory Testing," *IEEE Design & Test of Computers*, vol. 7, no. 2, Apr 1990, pp. 52-63.

[9] C.-T. Huang, J.-R. Huang, C.-F. Wu, C.-W. Wu, and T.-Y. Chang, "A Programmable BIST Core for Embedded DRAM," *IEEE Design & Test of Computers*, vol. 16, no. 1, Jan.-Mar. 1999, pp. 59-70.

[10] J. E. Barth, et al., "Embedded DRAM Design and Architecture for the IBM 0.11-μm ASIC Offering," *IBM Journal of Research and Development*, vol. 46, no. 6, Nov. 2002, pp. 675-689.

[11] S. Miyano, K. Sato, K. Numata, "Universal Test Interface for Embedded-DRAM Testing," *IEEE Design & Test of Computers*, vol. 16, no. 1, Jan.-Mar. 1999, pp. 59-70.

[12] N. Watanabe, F. Morishita, Y. Taito, A. Yamazaki, T. Tanizaki, K. Dosaka, Y. Morooka, F. Igaue, K. Furue, Y. Nagura, T. Komoike, T. Morihara, A. Hachisuka, K. Arimoto, and H. Ozaki, "An Embedded DRAM Hybrid Macro with Auto Signal Management and Enhanced-on-Chip Tester," *Proceedings of the IEEE Int'l Solid-State Circuits Conference (ISSCC), Digest of Technical Papers*, 2001, pp. 388-389.

[13] A.J. van de Goor and I. Schanstra, "Address and Data Scrambling: Causes and Impact on Memory Tests," *Proc. 1st IEEE Int'l Workshop on Electronic Design, Test and Application* (DELTA 02), IEEE Press, 2002, pp. 128-136.

[14] K.-L. Cheng, M.-F. Tsai, and C.-W. Wu, "Neighborhood pattern-sensitive fault testing and diagnostics for random-access memories," *IEEE Trans. on Computer-Aided Design of Integrated Circuits and Systems*, vol. 21, no. 11, Nov. 2002, pp. 1328-1336.

[15] K. Roy, S. Mukhopadhyay, and H. Mahmoodi-Meimand, "Leakage Current Mechanisms and Leakage Reduction Techniques in Deep-Submicrometer CMOS Circuits," *Proceeding of the IEEE*, vol. 91, No. 2, Feb. 2003, pp. 305-327.

[16] Z. Al-Ars, A. J. van de Goor, "Approximating Infinite Dynamic Behavior for DRAM Cell Defects", *IEEE VLSI Test Symposium*, pp. 401-406

[17] Z. Al-Ars, S. Hamdioui, A.J. van de Goor, G. Mueller, "Defect Oriented Testing of the Strap Problem Under Process Variations in DRAMs", *IEEE International Test Conference*, pp.1-10, 2008.

[18] J. Castillejos, V.H. Champac, "A forced-voltage technique to test data retention faults in CMOS SRAM by I_{DDQ} testing", *Circuits and Systems*, pp.433-436, 1997.

[19] A. Meixner, J. Banik, "Weak Write Test Mode: an SRAM cell stability design for test technique", *Test Conference, Proceedings., International*, pp.1043-1052, 1997.

[20] V.H. Champac, V. Avendano, M. Linares, "Bit line sensing strategy for testing for data retention faults in CMOS SRAMs", *Electronics Letters*, pp.1182-1183, 2000.

[21] V.H. Champac, V. Avendano, "Test of data retention faults in CMOS SRAMs using special DFT circuitries", *Circuits, Devices and Systems, IEE Proceedings*, pp.78-82, 2004.

[22] B. Wang, Y. Wu, J. Yang, A. Ivanov, Y. Zorian, "SRAM retention testing: zero incremental time integration with March algorithms", *VLSI Test Symposium, Proceedings. 23rd IEEE*, pp.66-71, 2005.

[23] Electronic Industries Association and JEDEC Solid State Technology Association, "Steady State Temperature Humidity Bias Life Test," EIA/JESD22-A101-B, April 1997.

[24] Electronic Industries Association and JEDEC Solid State Technology Association, "Highly-Accelerated Temperature and Humidity Stress Test," EIA/JESD22-A110-B, June 2008.

[25] JEDEC Solid State Technology Association, "Temperature, Bias, and Operating Life," JESD22-A108C, June 2005.

Adaptive Test Elimination for Analog/RF Circuits

Ender Yilmaz
Department of Electriacal Engineering
Arizona State University
Tempe, AZ 85281
eyilmaz2@asu.edu

Sule Ozev
Department of Electriacal Engineering
Arizona State University
Tempe, AZ 85281
sule.ozev@asu.edu

ABSTRACT

In this paper, we propose an adaptive test strategy that tailors the test sequence with respect to the properties of each individual instance of a circuit. Reducing the test set by analyzing the dropout patterns during characterization and eliminating the unnecessary tests has always been the approach for high volume production in the analog domain. However, once determined, the test set remains typically fixed for all devices. We propose to exploit the statistical diversity of the manufactured devices and adaptively eliminate tests that are determined to be unnecessary based on information obtained on the circuit under test. We compare our results with other similar specification-based test reduction techniques for an LNA circuit and observe 90% test quality improvement for the same test time or 24% test time reduction for the same test quality.

Categories and Subject Descriptors

B.7 [**Integrated Circuits**]: Miscellaneous

General Terms

Algorithms

Keywords

Adaptive Testing

1. INTRODUCTION

Analog circuits are tested for a large number of specifications resulting in unreasonable test time. It is crucial to reduce test time not only to lower the cost per device but also to meet throughput requirements.

The most common practice in the industry is to analyze the dropout patterns of each test during the characterization step and eliminate tests that are unnecessary. Since specifications stem from the same underlying physical circuits, they can be highly correlated. The dropout patterns are a

Permission to make digital or hard copies of part or all of this work for personal or classroom use is granted without fee provided that copies are not made or distributed for profit or commercial advantage and that copies bear this notice and the full citation on the first page. To copy otherwise, to republish, to post on servers or to redistribute to lists, requires prior specific permission and/or a fee.
DAC'09, July 26-31, 2009, San Francisco, California, USA

way of capturing these correlations. In line with this philosophy, various methods have been proposed in the literature to provide an efficient algorithm for this selection process.

The test selection problem, as described above, can be easily modeled as a set-cover problem [12, 10]. Unfortunately, set-cover is an NP-hard problem [4], thus an optimal solution is typically elusive. Using integer linear programming (ILP) to generate a near-optimal cover has been proposed in [12, 5]. For very large test sets, greedy algorithms can be used to generate the desired test set [9]. Statistical pass/fail based decision algorithms has been proposed in [2, 3] utilizing binary decision trees and support vector machine learning respectively. In [1], an efficient method is proposed to approximately obtain a minimum covering set based on pass/fail test data. The use of a genetic algorithm for test selection purposes has also been proposed [12].

Instead of eliminating tests, ordering them such that defective devices are pruned away quickly has also been proposed [9, 6, 7]. Such an approach is particularly useful for early production cycles where the yield is low. Tests that are at the tail end can be eliminated during production ramp up.

While such test compaction methods are extremely helpful in reducing the test time, the basic philosophy has been to treat all devices in the same manner. With increasing process variations however, devices are statistically diverse. Some devices, which fall near the nominal of the distribution space do not require extensive testing while more resources are needed for marginal devices. Devoting more time to marginal devices will also help increase test quality in terms of reducing test escapes. Recently, test quality concerns have been raised in the analog domain and customers have started to demand quality metrics similar to the digital domain, such as DPPM (Defective Parts Per Million) [13, 14]. Tailoring the test set with respect to the properties of the CUT (Circuit Under Test) would not only shrink the test set, but also increase the test quality as marginal devices are more extensively tested. Along these lines, in [15] a dynamic test scheduling technique is presented.In our earlier work [15], we have aimed to tailor the test set with respect to the statistical properties of the CUT.

In this paper, we propose an efficient and adaptive test solution for analog circuits that eliminates unnecessary tests based on the collective statistical information obtained *a priori* and information obtained from the CUT on-line. As in previous approaches, we make use of collective statistical information from a sample set of devices. However, we analyze this information in more detail. Once a circuit parameter

978-1-60558-497-3/09 $25.00 © 2009 ACM

is measured, it contains much more information than a simple pass/fail answer. We exploit this invaluable information to make decisions on the so-far unmeasured specifications and decide whether any of them should be skipped during the test program flow. During the measurement process, we process the data in an interleaved fashion in the background to avoid any timing overhead.

We use an LNA circuit as an experimental venue and compare the test time and DPPM results of our technique with previously proposed test elimination techniques. For the same test time, we consistently achieve 90% reduction in DPPM. We also observe that for the same test quality, we are able to reduce the test time by 24% compared to previously proposed techniques. We also provide a parameter to trade-off test time with test quality in much finer steps.

2. TEST FLOW

The proposed method utilizes two pieces of information to eliminate unnecessary tests: (a) the collective statistical information obtained from a set of sample circuits, and (b) individual measurement information obtained from the CUT during testing. The fundamental difference of our technique is that instead of prescribing a predetermined set of tests for the circuits, we may skip any test during test application depending on the measurement results and how confident we are in terms of meeting that specification. In this regard, the test flow development is not altered.

We capture the collective statistical information in terms of a correlation matrix. However, instead of relying only on pass/fail correlations, we analyze information on how marginal *passes* may be for the circuit by dividing the acceptance range of each specification into sub-ranges, or *bins*. The flow of our test approach is shown in Figure 1. We first order the complete test set based on the coverage they provide using a greedy heuristic. This initial order is coded into the test program. We also obtain a correlation matrix from a sample set of devices which can be periodically updated during production ramp-up as new data become available.

Initially, we do not have much information on the circuit. Thus, the likelihood of passing unmeasured specifications is around 0.5. When the test program enters into the routine of measuring one specification, we first check if the likelihood is beyond a parameterizable *confidence level threshold*. If this is the case, we skip that test. Else, the test is executed. Once a measurement is taken, we have additional information on the circuit under test. Using this information, as we will explain in more detail in the following section, and the correlation information, we update all the likelihood profiles of unmeasured specifications until either the circuit fails or we reach the end of the test sequence. Since circuits are failed only upon determining, through measurements, that they violate a specification, the yield loss of our technique is zero. However, there may be some circuits which fail a skipped specification, resulting in test escapes. The confidence level parameter provides a gauge to trade-off between test time and test quality as defined in terms of DPPM.

Intuitively, since we can spend more resources on marginal devices without blowing up the test time, test quality of our approach is improved when compared with techniques which use the same test set for all devices. In our results section, we evaluate the test time and DPPM metrics and compare them with previous approaches.

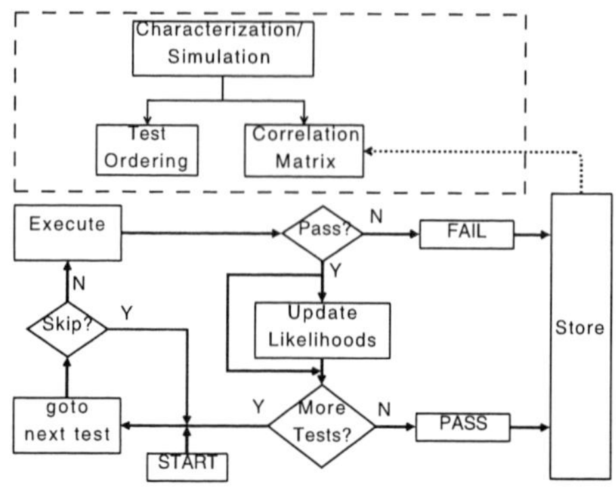

Figure 1: Adaptive Test Elimination Flow

3. CHARACTERIZATION/ SIMULATION STEP

We use a training sample set to get information about the distribution characteristics and exploit this information to improve test metrics. Traditionally, the training sample set is used to generate a minimal covering set of tests that can cover all the dropouts in the training set. Hence, test time is reduced proportionally to the size of the reduced subset. The first step in our method is to generate an order for all the tests.

3.1 Test Ordering

Since we may end up going into the tail end of the test set, we order the complete test list instead of generating a reduced test list. We use a greedy heuristic for test ordering purposes. The initial order is determined with the goal of obtaining as much information as possible from the CUT in as little time as possible. Tests are ordered descending based on their dropout coverage performances; tests that cover more dropouts are placed at the top. Failing circuits that are already covered are excluded from the list in each iteration. This proposed heuristic algorithm quickly sorts the list by scheduling the test that has the highest dropout coverage in each iteration. Although this method does not provide the optimum order, it is a good starting point for the adaptive step. Optimality of this initial test list has negligible impact on the final result, since we rely on the adaptive elimination step. During the testing step, we perform two operations. The measurement of specifications is the same process as in the traditional test flow. Once a measurement is taken, we update the likelihood profiles of the unmeasured specifications in the background.

The training set can be obtained using either early characterization data, as in [8], or using simulation data, as in [11], or a combination of the two. On the one hand, simulations can provide a way of analyzing the data *a priori*, determining optimal binning strategies, and a confidence level-threshold that is desired. Simulations can also be helpful in determining the performance of the method under various defect scenarios, while such information may not be available for the characterization step. Recently, techniques for accurate reduced-set Monte-Carlo sampling have been introduced for efficient evaluation [11]. Characterization data on the other hand, are invaluable to accurately model the inter-specification correlations, which are not captured in simulations.

3.2 Correlation Matrix

Capturing detailed inter-specification correlations are es-

978-1-60558-497-3/09 $25.00 © 2009 ACM

	Spec1	Spec2	Spec3	Spec4	Spec5
Spec1 B1					
Spec1 B2					
Spec1 B3					
Spec2 B1					
Spec2 B2					
Spec2 B3					

Figure 2: Correlation Matrix

sential for any test reduction method. We capture the correlations between the specifications by slicing each specification space into bins and calculating patterns between these bins and pass/fail correlations of other specifications. Thus, once we know in which bin a measurement result falls, we can estimate the likelihood-of-pass of the unmeasured specifications.

Correlation matrix is a $(nspec \times nbin) \times nspec$ matrix, where $nspec$ is the number of specifications and $nbin$ is the number of bins in each specification. Figure 2 illustrates sample entries of a correlation matrix. Bins can be addressed by a specification index and a bin index. For instance, $spec_i bin_j$ addresses the jth bin of the ith specification. Correlation matrix is filled by calculating the correlations of circuits that fall in each bin and good circuits of each specification. For instance, in order to calculate the correlation between $spec_i bin_j$ and $spec_k$, where $k \neq i$, two vectors are generated of size $ntrial \times 1$ each, where $ntrial$ is the size of the training set. Entries in the first vector are filled with "1" for the circuits in training set that fall into $spec_i bin_j$, and "0" elsewhere. Second vector is filled with "1" for circuits passing the specification $spec_k$, and "0" elsewhere. Correlation is simply calculated by evaluating the correlation coefficient of these vectors. The calculated correlation coefficient is used to estimate the passing likelihood of $spec_k$ given that the measurement result of specifications falls into bin_j. Similarly, we build the rest of the matrix using the same method.

4. TESTING STEP

During the testing step, we perform two operations. The measurement of specifications is the same process as in the traditional test flow. Once a measurement is taken, we update the likelihood profiles of the unmeasured specifications in the background.

4.1 Updating Likelihoods

Initially, all the specifications are assigned unbiased likelihoods of passing since we have no information about the CUT. After measuring the first specification, likelihoods of unmeasured specifications are updated using the measurement result and the correlation matrix. As an example, $spec_i$ is measured and the CUT falls into bin_j. Now we use the correlation matrix to update the likelihood of passing $spec_k$ for all k, where $k \geq i + 2$. Correlation entry ($corr$) used to relate measurement result and $spec_k$ is at the intersection of row $spec_i bin_j$ and column $spec_k$ of the correlation matrix. Likelihood (L_k) of $spec_k$ is updated as follows:

1 If $corr > 0$: likelihood (L_k) of $spec_k$ is increased since there is a positive correlation.

2 If $corr < 0$: likelihood (L_k) of $spec_k$ is reduce due to

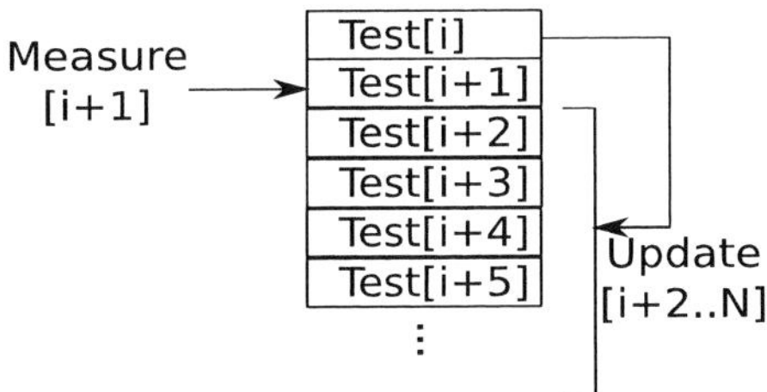

Figure 3: Test and Update

the negative correlation.

3 Else: likelihood of $spec_k$ does not change since there is no correlation.

$$
L_k = \begin{cases} 1 - (1 - L_k)(1 - corr) & , corr \geq 0 \\ (1 + corr)(L_k) & , corr < 0 \end{cases}
$$

As illustrated in Figure 3, the update procedure is designed as a pipelined process; update is performed concurrently with the measurement process to avoid any timing overhead. Therefore, there is no reason to update the likelihood of $spec_{i+1}$ since it is currently being measured.

We also introduce an update mechanism for possible variations in the process, such as lot-to-lot variations. Thus, we can adapt our test elimination mechanism to altering specification distributions. We simply keep the record of measured device specifications and update the correlation matrix incorporating the new data. It is important to note that we are able to update these correlations since we end up measuring a diverse set of specifications instead of a fixed set. Since we use spec-to-spec correlations, we can use information from any DUT for which both of the specifications are measured. Hence, missing data is not a problem for correlation matrix update as long as a diverse set of specifications is measured. In our experimental results section, we show that this is indeed the case. Inclusion of the new data reveals systematic variations in the process enabling the algorithm to fit into the changing conditions. Correlation matrix is updated on a regular basis during the testing process in the background; hence testing process is not affected.

4.2 Test Elimination

After a few iterations, likelihoods of unmeasured specifications are either raised or lowered by the update procedure. Increasing likelihoods indicate specifications that are likely to pass. Specification likelihoods that reach a certain level of confidence threshold are dropped from the test schedule. Even though the initial test list is the same for all devices, it is adapted for each DUT according to their performance. Using this test elimination scheme, we achieve test sequences that are unique for each device.

Since not all of the tests from the initial test list are applied, test time reduction is achieved. Test time is a function of the confidence level, since the confidence level changes the rate of test elimination. Hence, it is possible to reduce the test time by decreasing the confidence level. However, DPPM is inversely proportional to the confidence level. A high confidence threshold level will let fewer tests to be skipped and therefore result in better test quality. Effect of confidence level threshold on test quality (DPPM) and test time is explained in the results section.

978-1-60558-497-3/09 $25.00 © 2009 ACM

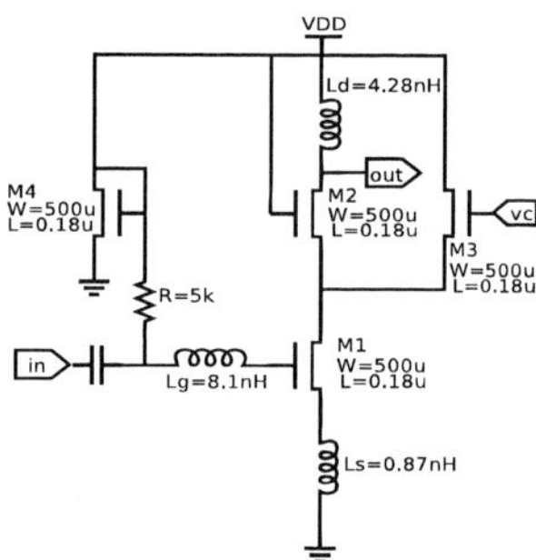

Figure 4: Test Circuit

5. RESULTS AND ANALYSIS

We demonstrate our method on a controllable gain LNA, shown in Figure 4. Randomly generated 60k instances of this circuit are used for analysis. To mimic process variations, we inject 10% variation into passive components (R, L) and 3% variation into active components (Width, Length). 12 specifications for 4 different gain levels are specified for the circuit: gain, bandwidth, center frequency, -3dB frequencies, output voltage offset, input impedance, output impedance, IIP3, 1db compression point, power consumption, and noise figure.

We have employed equal test time for each test in our method for the sake of simplicity. Nonetheless, incorporation of test time information can easily be done.

We present the efficiency and characteristics of our algorithm in the following sections. First, we show the performance of our method for various configurations and describe the effect of these parameters. Then, the performance of our method is compared to various test set reduction techniques.

5.1 Skip Histogram and Delay of Identification

As we have described in the introduction section, the premise of our method is to treat each individual circuit differently by using not only collective information but also the device specific information. Hence, it is possible to estimate the passing likelihood of the tests and eliminate some of them about which we are confident enough. Figure 5 shows the percentage of elimination of each test. The elimination rate starts to increase after the 5th test and increases as the number of applied tests increase, indicating that the greedy heuristic that we used in initial test ordering is useful in obtaining early information. The test elimination plot shows that tests are skipped with a rate of more than 50% on the average, which shows the diversity of the test for each CUT.

Figure 6 illustrates the delay of identification (DoI) histogram for bad and good circuits, where DoI refers to the number of executed tests before the circuit is labeled as pass or fail. Figure 6(b) is particularly interesting because it clearly shows that some circuits are identified very early on despite the fact that they are good circuits. We also observe that some circuits require extensive testing. Under a non-adaptive test approach, all good circuits would have the same delay of identification.

5.2 Effect of Training Set Size and Number of Bins

Two of the parameters of our adaptive method is described

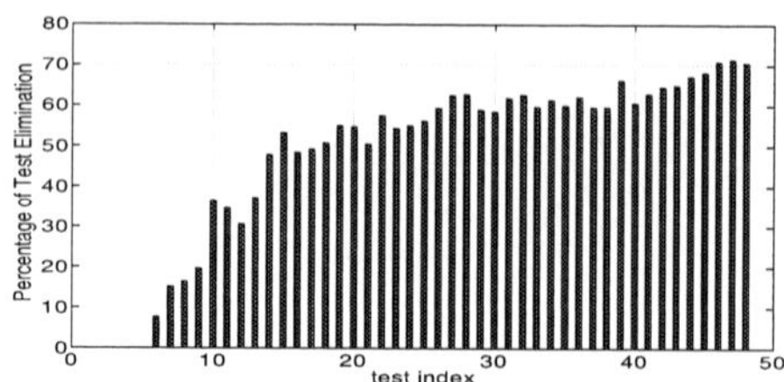

Figure 5: Test Elimination Histogram

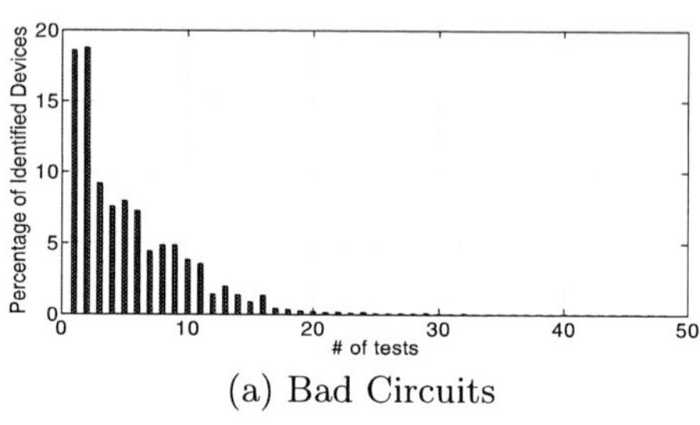

(a) Bad Circuits

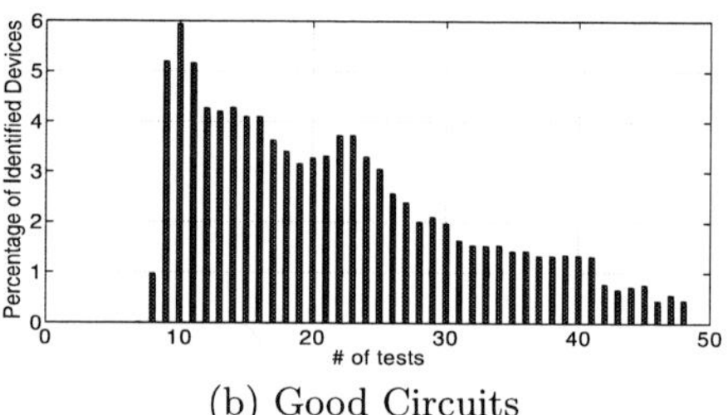

(b) Good Circuits

Figure 6: Delay of Identification Histogram

in this section: off-line training set size and number of bins used to slice distribution spaces. Figure 7(b) shows DPPM comparison for our method and set-cover method for training sizes from 1k to 10k. In this experiment, for a fair comparison, new information obtained from the measurement of additional chips is not included. In other words, the correlation matrix was fixed after the training set is analyzed.

The DPPM level reduces as expected as the training set size increases for both methods, yet adaptive method maintains an order of magnitude better performance for reasonable training set sizes. Time performance of two methods for the same configuration is given in Figure 7(a). Increasing curve of the set-cover method indicates that better test quality is achieved via increasing the number of applied tests. Time curve of our method, in comparison is almost flat, which implies that the performance of our method relies on device specific information instead of relying on collective information. Employing device specific information virtually pins the time performance of our method and does not change as the training set size increases. This also implies that as more collective information becomes available, the proposed method can improve DPPM without blowing up the test time.

The second parameter we discuss in this section is the number of bins used to slice the distribution spaces. Using a large number of bins improves the resolution and therefore gives more information about the devices. However, DPPM level does not improve after a point. In the rest of the analysis, we use 10 bins, since this setup yields a reasonable test quality and test time.

978-1-60558-497-3/09 $25.00 © 2009 ACM

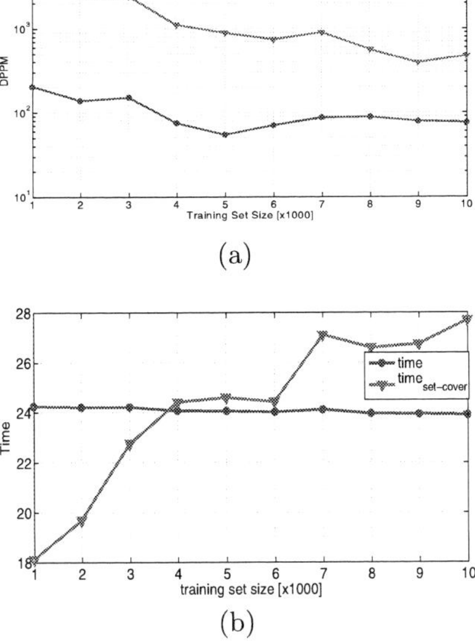

(a)

(b)

Figure 7: Effect of Training Size

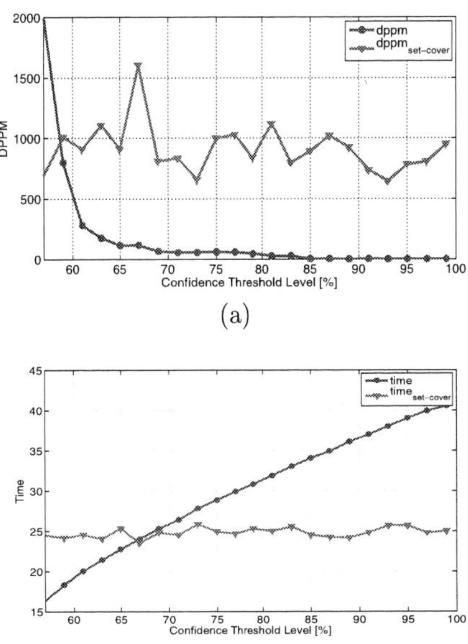

(a)

(b)

Figure 8: Effect of Confidence Threshold

5.3 Trade-off via Confidence Level Threshold

In this section, we introduce the parameter that we use to adjust the quality-time trade-off. Confidence level threshold parameter sets the rate of skipping tests in our method. Therefore, it directly affects the test quality and test time. Figure 8(a), shows the comparison of DPPM for our method and set-cover method. Note that performance levels of set-cover method is not a function of confidence level, hence they should remain fixed. Nonetheless, some variation is observed due to the randomly selected training set. Time performance is given in Figure 8(b), which shows a somewhat linear dependency between the confidence level threshold and test time. Using both relations, it is possible to adjust the test quality and test time trade-off.

5.4 Comparison

Similar work in the area of test selection has focused on developing an efficient algorithm to find a minimal cover for all the faulty circuits [12, 5, 9, 2, 3, 1, 6, 7]. Most approaches can provide near-optimal cover with reasonable run-times. In order to abstract away the implementation details from our comparison, we first find the optimal cover by exhaustively searching between lower and upper bounds for a viable cover. Test set that includes only uniquely identifying tests gives a lower bound to the minimal covering test set size. Uniquely identifying tests are defined as tests that can identify CUTs that cannot be identified using other tests. Since a complete covering set is required, uniquely identifying tests must be included in the test list. Size of this test list is smaller or equal to the size of the optimal test set. $OPT \geq n$, where OPT is defined as the size of the optimal test set and n is the size of uniquely identifying test set. Upper bound is obatined using ILP-based set covering method. Because ILP produces a sub-optimal test set, the size of test set obtained using ILP is greater or equal to the size of the opimum test set. Thus, $n \leq OPT \leq m$, where m is defined as the size of the test set obtained using ILP. Once we obtain upper and lower bounds of the optimal test set size, we

search for the optimal test set by enumerating all possible combiantions between test sizes between n to m. We have observerd that due to the high efficiency of ILP, lower and upper bounds are usually very tight and $m - n$ never exceeded 3 for the data set we used in this paper. It is very important to note that proposed optimal set covering algorithm is not a part of our adaptive test elimination method. Computational complexity of set-cover method is not a concern for our algorithm; it is included for comparison purpose only.

In this section, we compare the performance of our method with the set cover method. Since previously published methods rely on offline test compaction algorithms [9, 12, 2, 3, 1], which can achieve the performance of the set-cover method at best, we have used only set-cover method for comparison. We also compare the proposed technique with the dynamic scheduling(DS) technique in [15]. It should be noted that the technique in [15] would suffer from on-line scheduling overhead, while it would be hard to quantify. We provide average results over 10 non-overlapping runs to reduce the effect of randomness. Table 1 shows the comparison of DPPM where we used a confidence level threshold to equate the test time to the previously proposed techniques. For each run, we select a different training set of samples out of the 60k available circuit instances.

Our method improves test quality almost 10 times for the same test time and 3 times compared with our previous dynamic scheduling method. Results indicate that there is a lower limit of the performance that can be achieved using only the collective information of the devices. Our method results in a substantial improvement in test quality by utilizing device specific information. Table 1(b), shows the time reduction performance of our method for the same quality level as of the other methods. 24% reduction in test time is achieved compared to set-cover method.

5.5 Fault Analysis

Any method that reduces the test set based on learned information out of a training set runs the risk of test escapes,

978-1-60558-497-3/09 $25.00 © 2009 ACM

Table 1: Comparison With The Previous Techniques

	DPPM	Time
set-cover	927	24.88
DS [15]	310	25
Proposed	97	24.5

(a) DPPM for the same test time

	DPPM	Time
set-cover	927	24.88
DS [15]	980	20.5
Proposed	897	18.5

(b) Test time for the same DPPM level

Table 2: DPPM and Test Time Results of Injected Faults

method	DPPM		Time	
	ours	set-cover	ours	set-cover
Paremetric	0	0	3.194	4.922
Catastrophic	0	0	10.71	12.45

especially under defect scenarios. To ensure that defective devices are identified correctly, the results of the test scheme should be analyzed under fault injection. In order to facilitate this evaluation, we analyze our technique under both parametric and hard faults. Marginal parametric faults are injected into the passive component parameters (R, L) and active component parameters (Width, Length) at a distance of %30 and %10 from the nominal values for passive and active components respectively. 500 random faulty circuit instances are used for the analysis. A fault free sample set of 5k circuits are used for training the algorithm. No misclassification occurred either for our method or for set-cover based methods. Test time for defective devices is not an issue since it is much less compared with good devices.

For hard faults, we determine 8 locations for potential opens and shorts. These faults are injected on 500 circuits with parametric variation, where the variations are the same as we have used in the previous sections. Both of the methods yield zero misclassification. Test quality and test time comparison for catastrophic fault is given in Table 2.

6. CONCLUSION

We present an adaptive test elimination method for analog circuits that tailors the test set with respect to the properties of each device. Collective information from a training set is utilized to determine an initial order for the test set as well as extracting specification-based correlation information. Device-specific information together with the correlation information are used to determine which tests may be redundant. While the initial test set is the same for all the devices due to the statistical diversity of devices, their final test sets are also diverse. With this adaptive technique, less resources are devoted to well-performing circuits by passing them quickly whereas more test time is devoted to marginal devices. Our technique also enables a mechanism to trade-off test quality with test time systematically. Experimental results on an LNA circuit confirm the premise of our technique that some devices need less while others need more time for identification. These experiments also confirm that with our technique, a diverse set of specifications are measured and the test quality can be increased 10-fold without changing the test time compared with previous techniques that use an identical test set for all devices.

Acknowledgment

This work is supported by Semiconductor Research Corporation under contract number 1836.012 and by National Science Foundation under contract 0917766

7. REFERENCES

[1] S. Biswas and R. D. S. Blanton. Statistical test compaction using binary decision trees. *IEEE Design and Test of Computers*, 23(6):452–462, 2006.

[2] S. Biswas and R. S. Blanton. Test compaction for mixed-signal circuits using pass-fail test data. *VLSI Test Symposium, IEEE*, pages 299–308, 2008.

[3] S. Biswas, P. Li, R. Blanton, and L. Pileggi. Specification test compaction for analog circuits and mems [accelerometer and opamp examples]. *Design, Automation and Test in Europe, 2005. Proceedings*, 1:164–169, March 2005.

[4] T. H. Cormen, C. E. Leiserson, and R. L. Rivest. *Introduction to Algorithms*. McGraw-Hill, 1997.

[5] P. Drineas and Y. Makris. Independent test sequence compaction through integer programming. *IEEE International Conference on Computer Design*, page 380, 2003.

[6] S. Huss and R. Gyurcsik. Optimal ordering of analog integrated circuit tests to minimize test time. In *IEEE/ACM Design Automation Conference*, pages 494–499, 1991.

[7] W. Jiang and B. Vinnakota. Defect-oriented test scheduling. *IEEE VLSI Test Symposium*, pages 433–438, 1999.

[8] N. Kupp, P. Drineas, M. Slamani, and Y. Makris. Confidence estimation in non-rf to rf correlation-based specification test compaction. *European Test, 2008 13th*, pages 35–40, 2008.

[9] L. Milor. A tutorial introduction to research on analog and mixed-signal circuit testing. *IEEE Transactions on Circuits and Systems II: Analog and Digital Signal Processing*, 45(10):1389–1407, Oct 1998.

[10] C.-J. R. Shi and M. Tian. Automatic test generation of linear analog circuits under parameter variations. *IEEE/ACM Design Automation Conference*, 1998.

[11] H.-G. Stratigopoulos, J. Tongbong, and S. Mir. A general method to evaluate rf bist techniques based on non-parametric density estimation. *Design, Automation and Test in Europe*, pages 68–73, March 2008.

[12] H.-G. D. Stratigopoulos, P. Drineas, M. Slamani, and Y. Makris. Non-RF to RF test correlation using learning machines: A case study. In *IEEE VLSI Test Symposium*, pages 9–14, 2007.

[13] S. Sunter and N. Nagi. Test metrics for analog parametric faults. *IEEE VLSI Test Symposium*, pages 226–234, 1999.

[14] T. Williams and N. Brown. Defect level as a function of fault coverage. In *IEEE Transactions on Computers*, volume C-30, pages 987–988, 1981.

[15] E. Yilmaz and S. Ozev. Dynamic test scheduling for analog circuits for improved test quality. In *IEEE International Conference on Computer Design*, pages 227–233, 2008.

978-1-60558-497-3/09 $25.00 © 2009 ACM

WCET-aware Register Allocation based on Graph Coloring[*]

Heiko Falk
Computer Science 12
Technische Universität Dortmund
D - 44221 Dortmund, Germany
Heiko.Falk@tu-dortmund.de

ABSTRACT

Current compilers lack precise timing models guiding their built-in optimizations. Hence, compilers apply ad-hoc heuristics during optimization to improve code quality. One of the most important optimizations is register allocation. Many compilers heuristically decide when and where to spill a register to memory, without having a clear understanding of the impact of such spill code on a program's run time.

This paper extends a graph coloring register allocator such that it uses precise worst-case execution time *(WCET)* models. Using this WCET timing data, the compiler tries to avoid spill code generation along the critical path defining a program's WCET. To the best of our knowledge, this paper is the first one to present a WCET-aware register allocator. Our results underline the effectiveness of the proposed techniques. For a total of 46 realistic benchmarks, we reduced WCETs by 31.2% on average. Additionally, the runtimes of our WCET-aware register allocator still remain acceptable.

Categories and Subject Descriptors

D.3.4 [**Programming Languages**]: Compilers; Optimization; C.3 [**Real-time and embedded systems**]; B.3.3 [**Memory Structures**]: Worst-case analysis

General Terms

Algorithms, Performance

Keywords

WCET, Register Allocation

1. INTRODUCTION

Embedded systems are often real-time systems whose correctness depends on both the logical results and on the time at which the results are produced. A program's *worst-case*

[*]Funded by the European Community's 7th Framework Programme FP7/2007-2013 under grant agreement n° 216008.

Permission to make digital or hard copies of part or all of this work for personal or classroom use is granted without fee provided that copies are not made or distributed for profit or commercial advantage and that copies bear this notice and the full citation on the first page. To copy otherwise, to republish, to post on servers or to redistribute to lists, requires prior specific permission and/or a fee.
DAC'09, July 26-31, 2009, San Francisco, California, USA

execution time (WCET) is used to guarantee that real-time constraints are safely met. But besides safety, the market demands high performance, energy efficiency and low cost. Hence, designing such products implies solving a complex optimization problem with multiple optimization criteria. Compilers play an important role during real-time system design since they are able to apply automated optimizations improving the quality of the generated executable code.

Unfortunately, even modern optimizing compilers are often unable to quantify the effect of an optimization since they lack precise timing models [8]. Hence, simple ad-hoc heuristics are applied during optimization in the hope that these finally improve code quality. But it is well-known that this is not always true: due to the absence of precise models, optimizations may have a negative impact on code quality.

Among all optimizations studied in the past, *register allocation* is considered the most important one. It intends to use a processor's registers most efficiently in order to minimize slow main memory accesses. Due to the increasing speed gap between processors and memories, accesses to physical processor registers *(PHREGs)* are orders of magnitudes faster than memory accesses. However, memory accesses can not be totally avoided during register allocation, since the amount of temporary variables (aka. *virtual registers, VREGs*) at a certain place in a program can exceed the number of available PHREGs. In such a situation, *spill code* is inserted swapping a register out to memory and back.

Currently, register allocators usually decide heuristically where to insert spill code. Due to a lack of precise models, the compiler is unaware of the impact of generated spill code on a program's execution time. Especially in the area of real-time system design and optimization, badly placed spill code can have a dramatic impact on a program's WCET.

This paper is the first one to present a technique for a WCET-aware graph coloring register allocator. The main contributions of the proposed register allocator are that it

- explicitly uses WCET data during optimization,
- automatically updates its WCET data in the course of the optimization in order to cope with the inherent instability of the critical path defining the WCET,
- reduces average WCETs by 31.2% for 46 benchmarks while requiring only moderate optimization runtimes.

Section 2 gives a survey of related work on register allocation, WCET optimization and compilers using formal models during optimization. Section 3 presents graph coloring which is the standard technique for register allocation, followed by the WCET-aware register allocator in Section 4. Section 5 describes the benchmarking results, and Section 6 summarizes this paper and gives an outlook on future work.

978-1-60558-497-3/09 $25.00 © 2009 ACM

2. RELATED WORK

Graph coloring is the standard technique for register allocation nowadays. Due to its importance for compilers and for this paper, it is discussed in more detail in Section 3.

An optimal register allocator using *integer-linear programming (ILP)* is presented by [7]. As opposed to the motivation given in Section 1, [7] does not apply any heuristic during register allocation since ILP produces optimal results. The ILP minimizes code size, i.e. the total amount of generated spill code. Since code size is a criterion not requiring sophisticated models in a compiler, optimal results can be produced. Nevertheless, the absence of timing models implies that the impact of [7] on WCETs is fully unknown. How to make this ILP WCET-aware is unclear up to now.

A third standard register allocation technique is *linear scan* [11]. Using a linear order of all operations of a program, a life time interval is computed for each VREG. Register allocation is done by mapping each life time interval to a PHREG and by applying a simple spill heuristic if all PHREGs are in use. Compared to graph coloring, linear scan is fast but produces results of inferior quality. In addition, the code quality resulting from linear scan heavily depends on the chosen operation order and on the spill heuristic. Hence, linear scan allocation is a typical example of an optimization not guided by actual models.

Profile feedback optimization is discussed in [10] as a workaround for lacking formal models. Here, information about a program's run-time behavior is supplied to the compiler, which is collected by applying code instrumentation and run-time profiling. The profile-based register allocator of the Sun Studio compiler finally uses basic block counts from this profile data. In contrast to such profile-based approaches, our approach relies on timing models tightly integrated in a compiler making profiling runs obsolete.

One of the very few compilers featuring an actual model used during optimization is the energy-aware C compiler *encc* [13]. Its energy model reflects the energy dissipation of each operation of the ARM thumb instruction set, plus the energy consumed during various memory accesses. Using this energy model, *encc* exploits memory hierarchies by moving program code or data onto scratchpad memories and minimizes a program's energy consumption.

The compiler WCC [5] is the first fully functional compiler explicitly designed for WCET minimization. WCET timing models are integrated into WCC by coupling its backend with the static WCET analyzer aiT [1]. This way, WCC can apply static WCET analysis while optimizing and can use all the WCET-related data computed by aiT for optimization. WCC serves as technical infrastructure for the WCET-aware register allocator presented in this paper.

Compiler optimizations minimizing WCET are an emerging area of research where only few related works currently exist. In [9], a combination of procedure cloning and procedure positioning to improve worst-case I-cache performance is proposed. The authors of [4, 6, 12] propose to move parts of a program's code and data onto a scratchpad memory or onto a software-controlled cache. These papers focus on exploiting memory hierarchies outside a processor core to minimize WCETs. Exploiting the register file – which is that part of a memory hierarchy being closest to a processor – in a WCET-aware way has not yet been considered in any publication to the best of our knowledge.

3. TRADITIONAL GRAPH COLORING

Due to its relevance, traditional graph coloring is presented in-depth in this section. The overall structure of graph coloring is discussed first in Section 3.1, followed by a selection of typical spill heuristics in Section 3.2.

3.1 Overall Algorithm

Traditional graph coloring based register allocation was originally published in [3] and later improved in [2]. Its basic data structure is the *interference graph* $G = (V, E)$. For each VREG of a function and for each of the C available PHREGs, a node is added to V. An undirected edge $e = \{v, w\}$ is added to E whenever nodes v and w interfere. v and w interfere either if they represent VREGs which are simultaneously alive and thus should not share the same PHREG, or if a VREG v must not be allocated to PHREG w for architectural reasons.

Graph coloring assigns one of C colors representing the physical registers to each node $v \in V$ such that no two adjacent nodes have the same color. According to [2], this is done as follows:

Build: Construct the interference graph $G = (V, E)$.

Simplify: Successively remove each $v \in V$ from G having a degree $< C$, push v onto stack S. This step removes all nodes from G which are always colorable due to the small amount of adjacent nodes.

Spill: After *simplify*, each node v has degree $\geq C$. Select one node $v \in V$, mark v as *potential spill*, remove v from G, push v onto S.
If $V \neq \emptyset$, continue with *simplify*.

Select: Successively pop nodes v from S and re-insert them into G. If v is not a potential spill, assign a free color c to v. If v is a potential spill, there may be a free color c available for v. If this is the case, assign c to v. Otherwise, don't assign a color to v and mark v as *actual spill*.

Start over: If there are actual spills $v \in V$, insert a load operation before each use of v and a store operation after each definition of v and continue with *build*.

By inserting spill operations before uses and after definitions of an actual spill v during *start over*, the lifetime of v is split into smaller intervals in the hope that these smaller intervals will only interfere with lifetimes of fewer VREGs in the next iteration of the above algorithm. This register allocator has proven to produce results of high quality and has an overall complexity of $\mathcal{O}(n \log n)$, where n is the number of instructions in the program to be allocated.

3.2 Spill Heuristics

A crucial issue of the allocator sketched above is the question which node v to select and to mark as potential spill during the *spill* phase. In the related literature, several heuristics to select a potential spill are proposed:

- Select nodes according to the order in which registers occur in the compiler's intermediate code.

- Select the node v with highest degree, since spilling this node reduces the degree of many other nodes in G so that it is more likely to maximize the number of nodes with degree $< C$ after spilling v.

978-1-60558-497-3/09 $25.00 © 2009 ACM

- Select a node v depending on the degree of v, on the number of operations o using or defining v, on the register pressure around o and on the loop nesting level of each such operation o.

From the description of these spill heuristics, it becomes obvious that no formal timing model is used to take a spilling decision. Due to a lack of such models, these heuristics try to estimate the impact of a spilling decision on code quality. It is not surprising that in some cases, one heuristic is better, whereas another heuristic might be more appropriate in other cases. Nowadays, spill decisions are not steered by actual timing data so that their estimates may guide a register allocator into a wrong direction.

4. WCET-AWARE GRAPH COLORING

The discussion of the spill heuristics in Section 3.2 reveals that current register allocators have no direct control over where spill code is generated, since only simplified measures are used. This can have severe effects on a program's WCET. The WCET of a program P is equal to the length of the longest possible execution path from the start node to an end node in P's control flow graph *(CFG)*. For such a path, its length is the sum of the products of WCET and worst-case execution frequency for all basic blocks of the path. This longest path is also known as *Worst-Case Execution Path (WCEP)*. Traditional spill heuristics may now lead to spill code generation along this WCEP, thus increasing the WCET considerably.

Since it is intractable to compute a program's WCET in general, upper bounds of the actual WCET of P need to be estimated. The state-of-the-art technique to obtain safe and tight WCET estimates is to apply static analyses to the executable machine code of P, taking into account the influence of P on memories, caches, processor pipelines etc. Static WCET analyzers like e.g. aiT [1] basically estimate *a)* the WCET per basic block of P and *b)* how many times each basic block of the CFG is executed in the worst-case (i.e. the *worst-case execution frequency* per node). Using this data, the WCEP can be determined, since any block with non-zero worst-case execution frequency belongs to the WCEP. In the remainder of this paper, the shortcuts WCET and WCET_{est} are used synonymously – we always refer to the WCET estimates produced by a static WCET analyzer in the following.

This section presents our WCET-aware graph coloring approach. Section 4.1 discusses properties of WCETs and WCEPs that heavily influence compiler optimizations aiming at WCET reduction. Section 4.2 shows the overall workflow of our WCET-aware register allocator, followed by the discussion of our WCET-aware spill heuristic in Section 4.3.

4.1 Instable Worst-Case Execution Paths

The WCET of a program P is the maximal time P's execution can ever take. The CFG of P, whose nodes represent basic blocks and whose edges tell that one basic block can be reached from the other, reflects all possible ways of executing P. Among all paths from P's start node in the CFG to some end node, there is one longest path. This path is the WCEP and its length is equal to P's WCET.

A compiler aiming at WCET minimization must thus reduce the length of the WCEP. Assume p_1 is P's current WCEP and some disjoint path p_2 is the second longest path in the CFG. If a compiler optimization is successful in shortening p_1 by more than $|p_1| - |p_2|$ time units (where $|p|$ stands for the length of p), p_2 becomes the new WCEP after this optimization.

However, if the optimization is unaware of the WCEP change from p_1 to p_2, the compiler keeps on reducing the length of p_1. Unfortunately, this effort may be in vain since it not necessarily leads to any further WCET reduction, because the new WCEP p_2 might not be affected.

As a consequence, the following requirements have to be met by compilers aiming at WCET minimization. They must

- have detailed knowledge about the WCEP,

- apply optimizations exclusively to those parts of P lying on the WCEP, since optimizing parts of P not lying on the WCEP don't reduce the WCET at all, and

- be aware of switches of the WCEP in the course of applied optimizations.

These requirements are very challenging, because they show that it is not sufficient to equip a compiler with WCET models to obtain a WCEP, but that the influence of an optimization on the model needs to be tracked steadily. A WCET-aware register allocator is even more challenging as will be shown in the next section.

4.2 A Chicken-Egg Problem

In order to design a WCET-aware register allocator, the worst-case execution frequencies per CFG node need to be known. Unfortunately, static WCET analysis can not be applied to the program P serving as input for register allocation to obtain this data. This is because P is not an executable program since it uses VREGs instead of PHREGs. Static WCET analysis relies on executable and thus register-allocated code in order to correctly take the mutual influences of P and the processor hardware into account. Hence, there are cyclic dependencies between register allocation and WCET analysis – in addition to the requirements discussed in Section 4.1 – which need to be broken in order to obtain a WCET-aware register allocator.

Conventional register allocators follow the strategy to keep as many VREGs in PHREGs as possible, and to move a VREG to memory only if this is really necessary. The traditional way of thinking thus assumes optimistically that all VREGs fit into the physical register file and that only exceptionally, a VREG is allocated to memory. This way of thinking is also reflected by the graph coloring algorithm presented in Section 3, since it first tries to remove all colorable nodes from the interference graph, and only after that, a decision on one single potential spill is taken.

However, this traditional approach is not applicable for a WCET-aware register allocator. The intermediate code produced in the course of all the steps and iterations of traditional graph coloring is not executable and thus, no WCEP can be determined. The first stage where traditional graph coloring produces executable and thus WCET analyzable code is when register allocation is just finished. And at the end of the entire procedure, it does not make sense to reason about WCEPs, since all spill decisions are already taken.

For WCET-aware graph coloring, we propose the opposite way of thinking: we assume pessimistically that all VREGs are kept in memory. During register allocation, we thus

978-1-60558-497-3/09 $25.00 © 2009 ACM

```
1   IR WCET-GC-RA( IR P ) {
2     while ( true ) {
3       IR P' = P.copy();
4       P'.spillAllVREGs();

5       set<basicBlock> WCEP = computeWCEP( P' );
6       if ( getVREGs( WCEP ) == ∅ )
7         break;

8       basicBlock b' = getMaxSpillCodeBlock( WCEP );
9       basicBlock b = getBlockOfOriginalP( b' );

10      list<virtualRegister> vregs = getVREGs( b );
11      vregs.sort( occurrences of VREG in b );
12      traditionalGraphColoring( P, vregs );
13    }
14    traditionalGraphColoring( P, getVREGs( P ) );

15    return P;
16  }
```

Figure 1: Algorithm for WCET-Aware Graph Coloring

move VREGs from memory to PHREGs. This approach has the advantage that the intermediate code generated in the course of register allocation is always executable so that WCEPs can be determined.

4.3 A WCET-Aware Spill Heuristic

The WCET-aware spill heuristic we present in this paper bases on the following two key characteristics:

- Due to the instability of the WCEP, it has to be recomputed regularly. Since it is practically infeasible to recompute the WCEP after each individual spill decision, we propose to recompute WCEPs after deciding on the allocation and spilling of one single basic block.

- For a given WCEP, that basic block b leading to the highest execution of spill code in the worst case is chosen for allocation. All VREGs v of b are sorted by to the number of occurrences of v in b. This precedence list of VREGs is passed to a standard graph coloring register allocator selecting that register with least occurrences in the list as potential spill, if necessary.

The overall algorithm for WCET-aware graph coloring is depicted in Figure 1. It executes an optimization loop processing one basic block per iteration (lines 2-13). For an *intermediate representation (IR)* of a program P being input to register allocation, the algorithm maintains a copy P' which is fully spilled, i. e. where all VREGs of P are marked as actual spills and load / store operations are inserted for spilling (lines 3 and 4). This fully spilled IR is passed to a static WCET analyzer to obtain the current WCEP, which is feasible since P' does not contain any VREGs (line 5).

Among all basic blocks on the current WCEP, that block b' with the highest execution of spill code in the worst case is selected. Worst-case execution of spill code is computed by multiplying the number of inserted spill operations per basic block with the worst-case execution frequency per basic block as determined by the WCET analyzer (line 8). For b' within the fully spilled IR P', its counterpart b within the IR P still containing VREGs is determined (line 9).

This basic block b of P is the most critical one within the current iteration of the WCET-aware register allocator. Hence, the allocator should try to keep all VREGs of b in PHREGs. However, if register pressure is too high and some VREGs of b must be spilled, this should be done such that only a minimal amount of spill code will be executed in b in the worst case. For this purpose, all VREGs v of b are sorted by their number of occurrences within b (line 11). Since spill code generation always inserts a load operation before each use of v and a store after each definition of v, the number of occurrences of v in b correlates with the amount of spill code required in b to spill v. This sorting order models a precedence which VREGs are better spill candidates and which ones not. This precedence list of VREGs is passed to a standard graph coloring allocator (cf. Section 3) which performs the actual work of mapping these virtual to physical registers (line 12). After that, b' is put in a black-list to prevent it from being selected again by line 8. For the sake of simplicity, this black-listing is omitted in Figure 1.

If the current WCEP does not contain VREGs any more, the optimization loop terminates (lines 6 and 7). However, the IR P may still contains VREGs after leaving the optimization loop. This happens for basic blocks on other paths within the CFG which have never been the WCEP – such blocks were never considered by the optimization loop. However, they still need to be allocated. For this purpose, all remaining VREGs within P are passed to a final run of the traditional graph coloring register allocator in order to obtain a fully allocated IR (line 14).

For this final run of the standard register allocator, the applied spill heuristic does not matter. This is because even in the worst case where all remaining VREGs would be spilled, the basic blocks still containing VREGs in line 14 will never ever lie on the WCEP and thus will never influence the global WCET of P. If they lay on the WCEP, they would have been captured by the optimization loop which is in contradiction to the loop's termination condition in lines 6 and 7. Hence, it is fully sufficient to provide the standard graph coloring allocator with some arbitrary precedence list of VREGs. For the sake of simplicity, we make the allocator select nodes for spilling just in the order in which VREGs occur in P.

To sum up, the algorithm sketched in Figure 1 implements a WCET-aware spill heuristic for a traditional graph coloring register allocator. It sorts VREGs using a criterion depending on worst-case execution frequencies of spill code, stemming from static WCET analysis. This way, our spill heuristic tries to avoid spill code generation along the WCEP and especially within those blocks lying on the WCEP which are most frequently executed in the worst case.

5. EVALUATION

This section presents results obtained by applying the proposed WCET-aware graph coloring allocator to real-life benchmarks. Section 5.1 describes the experimental environment used to perform benchmarking. Sections 5.2 and 5.3 discuss benchmarking results in terms of worst-case and average-case execution times, respectively.

5.1 Experimental Environment

Our WCET-aware register allocator is integrated into a compiler for the Infineon TriCore TC1796 processor. This processor features a relatively large register file having 16 data and 16 address registers. However, not all of these 32

978-1-60558-497-3/09 $25.00 © 2009 ACM

Optimal Static WCET-aware Scratchpad Allocation of Program Code[*]

Heiko Falk
Computer Science 12
Technische Universität Dortmund
D - 44221 Dortmund, Germany
Heiko.Falk@tu-dortmund.de

Jan C. Kleinsorge
Computer Science 12
Technische Universität Dortmund
D - 44221 Dortmund, Germany
Jan.Kleinsorge@tu-dortmund.de

ABSTRACT

Caches are notorious for their unpredictability. It is difficult or even impossible to predict if a memory access will result in a definite cache hit or miss. This unpredictability is highly undesired especially when designing real-time systems where the *worst-case execution time (WCET)* is one of the key metrics. *Scratchpad memories (SPMs)* have proven to be a fully predictable alternative to caches. In contrast to caches, however, SPMs require dedicated compiler support.

This paper presents an optimal static SPM allocation algorithm for program code. It minimizes WCETs by placing the most beneficial parts of a program's code in an SPM. Our results underline the effectiveness of the proposed techniques. For a total of 73 realistic benchmarks, we reduced WCETs on average by 7.4% up to 40%. Additionally, the run times of our ILP-based SPM allocator are negligible.

Categories and Subject Descriptors

D.3.4 [**Programming Languages**]: Compilers; Optimization; C.3 [**Real-time and embedded systems**]; B.3.3 [**Memory Structures**]: Worst-case analysis

General Terms

Algorithms, Performance

Keywords

WCET, Scratchpad Allocation

1. INTRODUCTION

Embedded systems are often real-time systems whose correctness depends on both the logical results and on the time at which the results are produced. A program's *worst-case execution time (WCET)* is used to guarantee that real-time

---

[*]Funded by the European Community's 7th Framework Programme FP7/2007-2013 under grant agreement n° 216008.

Permission to make digital or hard copies of part or all of this work for personal or classroom use is granted without fee provided that copies are not made or distributed for profit or commercial advantage and that copies bear this notice and the full citation on the first page. To copy otherwise, to republish, to post on servers or to redistribute to lists, requires prior specific permission and/or a fee.
DAC'09, July 26-31, 2009, San Francisco, California, USA

constraints are safely met. But besides safety, the market demands high performance, energy efficiency and low cost. Hence, designing such products implies solving a complex optimization problem with multiple optimization criteria. Compilers play an important role during real-time system design since they are able to apply automated optimizations improving the quality of the generated executable code.

For hard real-time systems, caches are problematic. Since they are hardware controlled, it is virtually impossible to determine the latency of a memory access for many popular cache architectures. Additionally, sporadically executed code like e. g. scheduler or interrupt handlers may modify cache contents. Hence, statically determined WCET estimates may be heavily overestimated in the presence of caches. Therefore, real-time system designers tend to disable caches. Such systems suffer a low average-case performance since each memory access uses the slow main memory.

Scratchpad memories (SPMs) have both a good average and worst-case performance. This paper presents a WCET-aware SPM allocation of program code. Our algorithm determines a static SPM allocation: the SPM's contents is pre-computed at compile time and remains fixed during run time. Due to *integer-linear programming (ILP)*, our approach is optimal in that it results in a minimal WCET.

A program P's WCET is the maximal time P's execution can ever take. The *control flow graph (CFG)* of P, whose nodes represent basic blocks and whose edges indicate that one basic block can be reached from the other, reflects all possible ways of executing P. Among all paths from P's start node in the CFG to some end node, there is one longest path, called *worst-case execution path (WCEP)*, and its length is equal to P's WCET. Here, path length is the sum of the products of WCET and worst-case execution frequency for all basic blocks of the path.

A WCET minimizing compiler thus has to reduce the WCEP's length. Assume e. g. that p_1 is P's current WCEP and some disjoint path p_2 is the second longest path in the CFG. If an optimization successfully shortens p_1 by more than $|p_1| - |p_2|$ time units (where $|p|$ stands for the length of p), p_2 becomes the new WCEP after this optimization.

However, if the optimization is unaware of the WCEP change from p_1 to p_2, the compiler keeps on reducing the length of p_1. Unfortunately, this effort may be in vain since it not necessarily leads to any further WCET reduction, because the new WCEP p_2 might not be affected.

Hence, the following requirements have to be met by compilers aiming at WCET minimization. They must

- have detailed knowledge about the WCEP,
- apply optimizations exclusively to those parts of P ly-

ing on the WCEP, since optimizing parts of P not lying on the WCEP don't reduce the WCET at all, and

- be aware of changes of the WCEP in the course of applied optimizations.

These requirements are very challenging, because such instable WCEPs are difficult to consider within compiler optimizations. This paper is the first one to present an optimal WCET-aware SPM allocation technique for program code. The main contributions of the proposed approach are that it

- inherently captures a program's current WCEP and its possible switches,

- improves the current state of the art of SPM allocation of program code. Our technique is the first one to consider jump penalties and variable basic block sizes based on jump scenarios.

- achieves average WCET reductions from 7.4% up to 40% for 73 real-life benchmarks while requiring only negligible optimization run times.

Section 2 gives a survey of related work. Section 3 presents our ILP model to perform optimal WCET-aware SPM allocation of program code. The link between our ILP model and a compiler infrastructure used to extract all constants required by the ILP is described in Section 4. Section 5 describes the benchmarking results, and Section 6 gives a summary and an outlook on our future work.

2. RELATED WORK

Compiler-guided SPM allocation has been studied intensely in the past. SPMs are frequently used to reduce average-case performance or energy dissipation. Generally, global data, basic blocks, sequences of basic blocks or functions are placed in an SPM to realize a certain profit in terms of run time or energy dissipation. As pointed out in [14, 18], simply greedily selecting the SPM contents can lead to suboptimal or even degraded results. Among all proposed allocation strategies for energy dissipation or run time minimization, ILP-based approaches are most popular due to the optimality of the results and the elegance of the models.

An ILP for static SPM allocation of functions, basic blocks and data minimizing energy dissipation is proposed in [11]. It introduces multi basic blocks to model the fact that sizes and energy savings of basic blocks vary depending on a current SPM allocation due to the insertion of additional jump instructions. This concept basically fully enumerates the power set of all basic blocks in the ILP, thus leading to an exponential explosion of the ILP's size. This exponential ILP explosion is avoided in the present paper by modeling variable block sizes and gains using so-called jump scenarios.

In [16, 17], the impact of SPMs on WCET prediction is studied. Even though WCET is subject of these papers, the ILP-based SPM allocation is not WCET-aware. Instead, an energy minimizing selection algorithm is employed, and the effect of this energy reduction strategy on WCET is evaluated afterwards. Hence, that work is not a true WCET-aware optimization and does not consider WCEPs at all.

Software controlled caches allowing to load contents into a cache and to lock it afterwards, i.e. to prevent it from being replaced, behave like SPMs. All following publications explicitly focus on WCET minimization. In [2], a genetic algorithm for cache contents selection of statically locked I-caches is presented. However, this approach does not necessarily yield optimal results. An explicit search for the

WCEP within the CFG is performed in [6] and I-cache contents selection is done along the found WCEP. [6] relies on repeatedly investigating the CFG and is therefore expensive to perform. A similar iterative approach was presented in [10]: multiple optimization steps along the current WCEP are performed without recomputing the WCEP. After a certain number of optimization steps, the partially optimized program is analyzed and the resulting WCEP is computed for subsequent optimization steps.

The authors of [3] present a hybrid approach for WCET-centric dynamic SPM allocation of data. It is hybrid in that sense that it combines an ILP with an iterative heuristic. Using a static WCET analyzer, the current WCEP is computed. After that, an ILP tailored for this particular WCEP is solved that determines optimally which data is allocated to the SPM. In the next iteration, the WCEP is recomputed and some more SPM contents is determined using ILP.

In [12], a fully ILP-based solution to the problem of static allocation of data to SPMs for WCET reduction is presented. This work serves as basis for the techniques presented in the following. However, [12] is unable to allocate code onto SPMs and suffers from several limitations preventing it from being applied to real-life programs. This survey of related work shows that no WCET minimizing unified ILP-based SPM allocation scheme for program code currently exists, whereas basic techniques for program data already exist. For this reason, we focus on SPM allocation of code.

The compiler WCC [5] is the first fully functional compiler explicitly designed for WCET minimization. WCET timing models are integrated into WCC by coupling its backend with the static WCET analyzer aiT [1]. This way, WCC can apply static WCET analysis while optimizing and can use all the WCET-related data computed by aiT for optimization. WCC serves as technical infrastructure for the WCET-aware SPM allocation presented in this paper.

3. ILP FOR PROGRAM CODE SCRATCH-PAD ALLOCATION

This section presents our optimal WCET-aware SPM allocation for program code. Section 3.1 discusses those parts of the ILP modeling a function's control flow. Section 3.2 deals with SPM allocation of consecutive sequences of basic blocks. Modeling a program's global control flow, capacity constraints and objective function are subject of Sections 3.3, 3.4 and 3.5, respectively.

3.1 Modeling of a Function's Control Flow

In the following, ILP variables are represented using lowercase letters whereas constants use uppercase letters. The ILP allocating program code to the SPM uses one binary decision variable x_i per basic block b_i of a program. x_i specifies whether block b_i is allocated to the main memory (mem_{main}) or to the SPM (mem_{spm}):

$$x_i = \begin{cases} 1 & \text{if basic block } b_i \text{ is assigned to } mem_{spm} \\ 0 & \text{if basic block } b_i \text{ is assigned to } mem_{main} \end{cases} \quad (1)$$

Each basic block b_i of a function F causes some costs c_i. These costs reflect the WCET of b_i depending on whether b_i is executed from main memory or from the SPM:

$$c_i = C_{main}^i * (1 - x_i) + C_{spm}^i * x_i \quad (2)$$

For reducible CFGs, an innermost loop L of F has exactly

978-1-60558-497-3/09 $25.00 © 2009 ACM

one back-edge turning it into a cyclic graph. Not considering this back-edge turns L's CFG into an acyclic graph. This acyclic graph without L's back-edge is denoted as $G_L = (V, E)$ in the following. Without loss of generality, it can be assumed that there is exactly one basic block b_{exit}^L in G_L being the loop's unique exit node and one unique entry node b_{entry}^L. The WCET w_{exit}^L of b_{exit}^L is equal to the costs of b_{exit}^L:

$$w_{exit}^L = c_{exit}^L \qquad (3)$$

The WCET of a path leading from a node b_i of G_L different from b_{exit}^L to b_{exit}^L must be greater than or equal to the WCET of any successor of b_i in G_L, plus the costs b_i causes:

$$\forall b_i \in V \setminus \{b_{exit}^L\} : \forall (b_i, b_{succ}) \in E : w_i \geq w_{succ} + c_i \qquad (4)$$

Variable w_{entry}^L models the WCET of all paths of loop L if it is executed exactly once. To model several executions of L, all CFG nodes $v \in V$ of G_L are merged to a new super-node v_L. The costs of v_L are the product of L's WCET if executed once and L's maximal loop iteration count:

$$c_L = w_{entry}^L * C_{max}^L \qquad (5)$$

Replacing a loop L by a super-node v_L in the CFG may turn another loop L' of F directly surrounding L into an innermost loop with acyclic CFG G_L'. Hence, the constraints of Equations (3) and (4) can be formulated for L'. This way, the innermost loops of F are successively collapsed in the CFG so that ILP constraints modeling F's control flow are created from the innermost to the outermost loops.

A program's WCEP can change during optimization only at such points in the CFG where a basic block b_i has more than one successor because only there, forks in the control flow are possible. Since constraint (4) is formulated for each successor of block b_i, variable w_i always reflects the WCET of any path starting from b_i – irrespective of the fact which of the successors actually lies on the current WCEP. This way, constraint (4) realizes the implicit consideration of WCEPs and their changes in the ILP.

The structure of the ILP constraints of Equations (2) – (5) was originally proposed by [12]. However, these basic constraints of Suhendra et al. need to be refined substantially in order to obtain a functional scratchpad allocation technique for program code. Our extensions to the original ILP formulation are described in the following sections.

3.2 Allocation of Consecutive Basic Blocks

The binary decision variables x_i allow to place a basic block b_i in the SPM independent of allocation decisions concerning any other basic block within a function F. However, this independence of the allocation decisions for single basic blocks is particularly problematic for embedded processors.

If a basic block b_i is allocated to the main memory and an immediate successor b_j of b_i within the CFG is allocated to the SPM, jump instructions must ensure that b_j is reached from b_i. Due to the limited displacement which can be encoded as branch target of typical jump instructions, and due to the usually too large distance between the address spaces of SPM and main memory, a single jump instruction is often insufficient to transfer control from b_i to b_j. Frequently, the address where to jump needs to be computed and stored in an address register so that a register-indirect jump instruction can finally be issued. In such a scenario, transferring control from b_i to b_j requires several machine instructions constituting a severe jumping overhead.

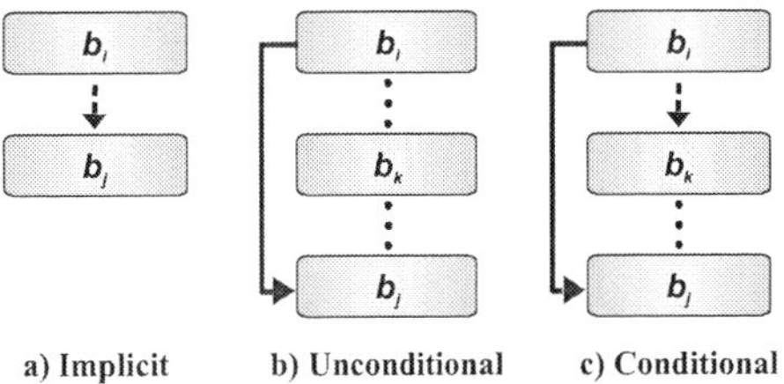

Figure 1: Typical Jump Scenarios

This jumping overhead might be avoided if both b_i and b_j are allocated to the same memory. Thus, the ILP should consider this kind of jumping overhead and should try to allocate groups of consecutive basic blocks to the same memory if this helps in reducing jumping overhead.

On typical embedded processors, three different *jump scenarios (JS)* can be found (cf. Figure 1). If control flows from basic block b_i to b_j without any jump instruction at the end of b_i, this is called *implicit jump*. If a jump instruction at the end of b_i always transfers control from b_i to b_j, this is called *unconditional jump*. Finally, a *conditional jump* transfers control conditionally from b_i to either b_j using an unconditional jump or to b_k via an implicit jump scenario.

The variables x_i, x_j and x_k for the basic blocks b_i, b_j and b_k resp. now provide the information whether jumping overhead needs to be considered within the ILP or not.

If two blocks b_i and b_j belong to the implicit jump scenario and if b_i and b_j are placed in different memories, a penalty should be added since this situation leads to a large jumping overhead to transfer control from b_i to b_j across the different memories. In contrast, no penalty needs to be considered at all if both b_i and b_j are placed in the same memory, because it is ensured that b_i and b_j are allocated adjacently so that no jump is required. We thus define a jump penalty for implicit jumps between basic blocks b_i and b_j as follows:

$$jp_{impl}^i = (x_i \otimes x_j) * P_{high} \qquad (6)$$

In Equation (6), the operator $\otimes$ represents the Boolean XOR of two binary decision variables – the Boolean XOR can be modeled within an ILP, but we omitted the listing of these constraints for the sake of simplicity. P_{high} is a constant realizing a high penalty for jumps across the different memories due to their large jumping overhead.

An unconditional jump from b_i to b_j usually bypasses a number of other basic blocks b_k (cf. Figure 1b). These other blocks b_k also need to be considered, because they determine if a jump from b_i to b_j is required at all. If b_i and b_j are allocated to different memories, the high penalty P_{high} already used in Equation (6) needs to be used. If b_i and b_j are allocated to the same memory *mem*, and if no other basic block b_k originally lying between b_i and b_j is allocated to *mem*, b_i and b_j are adjacent within *mem*. Hence, no jump from b_i to b_j is required at all and thus no penalty within the ILP is necessary. If any other basic block b_k is placed between b_i and b_j in *mem*, an unconditional jump from b_i to b_j bypassing b_k is necessary which is penalized by a constant P_{low} which is much lower than P_{high}. The jump penalty for unconditional jumps between basic blocks b_i and b_j is thus defined as follows:

$$jp_{uncond}^i = (x_i \otimes x_j) * P_{high} + \qquad (7)$$
$$\overline{(x_i \otimes x_j)} * (1 - \prod_{b_k \in \text{Figure 1b)}} (x_i \otimes x_k)) * P_{low}$$

978-1-60558-497-3/09 $25.00 © 2009 ACM

In analogy to Equation (6), we omit the description of the ILP equations to model the product of XOR terms.

Since a conditional jump can be seen as the combination of an implicit and an unconditional jump (cf. Figure 1c), the jump penalty for conditional jumps is the combination of Equations (6) and (7):

$$jp^i_{cond} = (x_i \otimes x_k) * P_{high} + (x_i \otimes x_j) * P_{high} + \qquad (8)$$
$$\overline{(x_i \otimes x_j)} * (1 - \prod_{b_k \in \text{Figure 1c}} (x_i \otimes x_k)) * P_{low}$$

Depending on the jump scenario of a basic block b_i, the overall jump penalty jp_i is defined as follows:

$$jp_i = \begin{cases} jp^i_{impl} & \text{if JS of } b_i \text{ is } implicit \\ jp^i_{uncond} & \text{if JS of } b_i \text{ is } unconditional \\ jp^i_{cond} & \text{if JS of } b_i \text{ is } conditional \\ 0 & \text{else} \end{cases} \qquad (9)$$

This jump penalty is used to extend the basic control flow constraints defined in Equations (3) and (4):

$$w^L_{exit} = c^L_{exit} + jp^L_{exit} \qquad (10)$$

$$\forall b_i \in V \setminus \{b^L_{exit}\} : \forall(b_i, b_{succ}) \in E : w_i \geq w_{succ} + c_i + jp_i \qquad (11)$$

3.3 Modeling of the Global Control Flow

Up to this point, the ILP defined in Equations (1) – (11) only models the intra-procedural control flow of a single function F. Without loss of generality, each function F has one dedicated entry block b^F_{entry}. For b^F_{entry}, the ILP variable w^F_{entry} denotes the WCET of any path starting at b^F_{entry} under the assumption that F is called exactly once.

However, some basic block b of a function F' may contain a call of a function F. In this situation, F's WCET represented by variable w^F_{entry} needs to be added to the WCET of block b. In addition, a function call penalty needs to be added to b's WCET since branching overhead similar to that one described in Section 3.2 occurs if b and b^F_{entry} are allocated to different memories. As a result, the overall function call penalty cp_i for a basic block b_i is defined as follows:

$$cp_i = \begin{cases} w^F_{entry} + (x_i \otimes x^F_{entry}) * P_{high} & \text{if } b_i \text{ calls } F \\ \quad +(1 - (x_i \otimes x^F_{entry})) * P_{low} \\ 0 & \text{else} \end{cases} \qquad (12)$$

In analogy to Section 3.2, cp_i is used to extend the control flow constraint defined in Equation (11):

$$\forall b_i \in V \setminus \{b^L_{exit}\} : \forall(b_i, b_{succ}) \in E : \qquad (13)$$
$$w_i \geq w_{succ} + c_i + jp_i + cp_i$$

Equation (13) now reflects the constraint which is finally generated for our ILP per basic block b_i and per successor b_{succ} of b_i. In Equation (13), we assume non-recursive functions. Due to the practical irrelevance of recursion for embedded real-time software [4], this assumption is valid even if support of recursion could be added to the ILP.

3.4 Scratchpad Capacity Constraints

To obtain a valid SPM allocation, it must be ensured that the size of all basic blocks allocated to the SPM does not exceed the SPM's capacity. Previous work on ILP-based SPM allocation either assumed constant basic block sizes or performed an exponential enumeration of the power set of all basic blocks to model basic block sizes.

As already discussed in Section 3.2, different kinds of jump instructions need to be issued, depending on the contents of the decision variables x_i and x_j for a basic block b_i transferring control to b_j. No jump instruction is needed in situations where b_i and b_j are adjacently placed in the same memory. A conventional jump instruction needs to be generated if b_i and b_j are allocated to the same memory but not adjacently. Finally, complex address computations and register-indirect branches need to be generated if a jump across memories needs to be performed.

Obviously, these different situations have an impact on the size of a basic block b_i. Hence, a block's size depends on the ILP decision variables in practice. In order to cope with such variable block sizes, we fall back to the jump scenarios introduced in Section 3.2 (cf. Figure 1). In the ILP, we only consider b_i's size if b_i is placed in the SPM since we assume a main memory which is large enough to hold the entire program. For a block b_i placed in the SPM, a new variable s_i denotes the growth in size of b_i in bytes if the successors b_j of b_i are kept in main memory. Depending on the jump scenario of b_i (implicit, unconditional or conditional), or if b_i contains a function call, s_i is computed as follows:

$$s_i = \begin{cases} (x_i \wedge \overline{x_j}) * S_{impl} & \text{if JS of } b_i \text{ is } implicit \\ (x_i \wedge \overline{x_j}) * S_{uncond} & \text{if JS of } b_i \text{ is } uncond. \\ (x_i \wedge \overline{x_k}) * S_{impl} + & \text{if JS of } b_i \text{ is } cond. \\ \quad (x_i \wedge \overline{x_j}) * S_{uncond} \\ (x_i \wedge \overline{x^F_{entry}}) * S_{call} & \text{if } b_i \text{ calls } F \\ 0 & \text{else} \end{cases} \qquad (14)$$

For each jump scenario, dedicated constants S_{impl}, S_{uncond} and S_{call} are used. They represent the growth of b_i in bytes for the different jump scenarios. Using s_i, the scratchpad capacity constraint ensuring the validity of an SPM allocation is defined as follows:

$$\sum_{b_i} (S_i * x_i + s_i) \leq S_{spm} \qquad (15)$$

In Equation (15), the constant S_i denotes the byte size of b_i in its original form without any cross-memory jumps. S_{spm} represents the available SPM size in bytes.

3.5 Objective Function

The overall goal of our ILP is to minimize a program's WCET by assigning basic blocks to the SPM. Due to the nature of Equations (12) and (13), variable w^F_{entry} corresponds to the WCET of function F including the WCETs of all functions called by F, plus some abstract jump penalties. Since function main is the unique entry point of an entire program, variable w^{main}_{entry} denotes the WCET of a program including all penalties. As a consequence, the value of this decision variable needs to be minimized by the ILP:

$$w^{main}_{entry} \rightsquigarrow min. \qquad (16)$$

4. COMPILER INFRASTRUCTURE

To turn the ILP model presented in Section 3 into a fully functional optimization, support by an underlying compiler infrastructure is required. In particular, we employ the infrastructure of our WCET-aware C compiler *WCC* [5] for the Infineon TriCore TC1796 processor (cf. Figure 2) to extract all the constants required by the ILP from the code currently under optimization.

978-1-60558-497-3/09 $25.00 © 2009 ACM

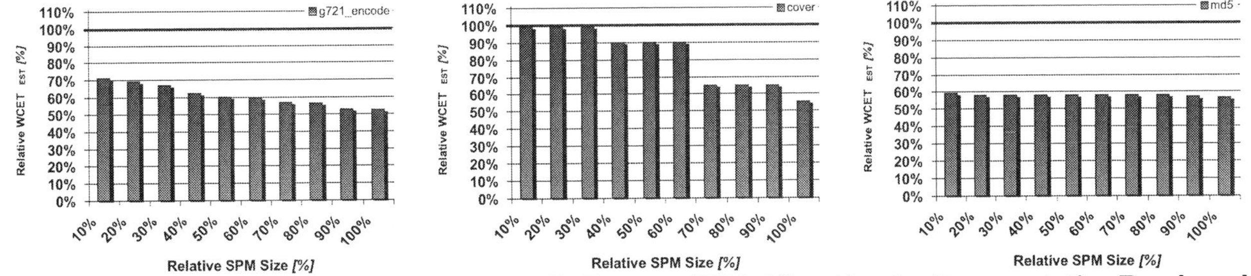

Figure 3: Relative WCET Estimates after WCET-aware SPM Allocation for Representative Benchmarks

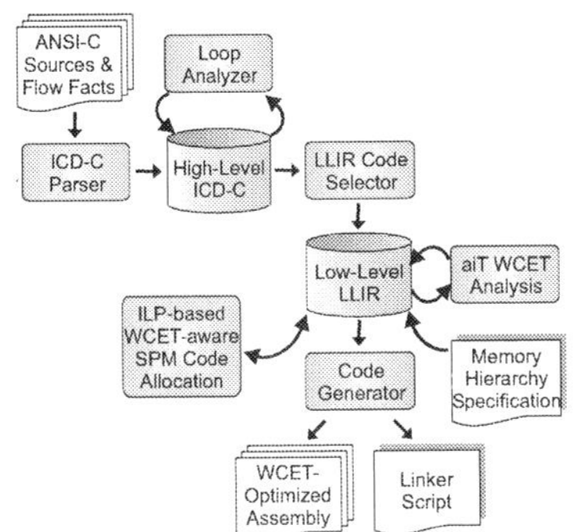

Figure 2: WCET-aware C Compiler WCC

Equation (2) depends on the constants C_{main}^i and C_{spm}^i representing the WCET of basic block b_i if it is located either in main memory or in the SPM, respectively. C_{main}^i is obtained by placing the whole program in main memory and performing a WCET analysis using the coupling of our compiler backend to the static WCET analyzer aiT [1]. In a second step, the whole program is virtually placed in the SPM using an infrastructure to model arbitrary memory hierarchies within our compiler. Another WCET analysis of this program yields the values C_{spm}^i.

Equation (5) depends on a loop's maximal iteration count C_{max}^L. In our compiler, this value can stem from user-specified flow fact annotations or it can be generated by our automatic loop analyzer [8]. Irrespective of the origin of C_{max}^L, flow fact mechanisms take care to keep the values C_{max}^L up to date during all loop optimizations of our compiler such that correct values are used by our ILP for SPM allocation.

The jump penalties P_{high} and P_{low} introduced in Section 3.2 do not rely on any part of our compiler infrastructure. WCET analyses of jumping code for the different jump scenarios revealed that the values 16 and 8 are appropriate for the considered TriCore architecture.

Equation (14) depends on constants S_{impl}, S_{uncond} and S_{call} representing the byte size of the additional code required to jump from a block b_i to b_j if b_i is allocated to the scratchpad memory but b_j is not. Due to the shape of the jumping code required for the TriCore architecture and the different jump scenarios of Equation (14), S_{impl} and S_{uncond} equal to 10 bytes and S_{call} is equal to 12 bytes.

A basic block's size S_i (cf. Equation (15)) without consideration of cross-memory jumps at the end of b_i is simply computed by accumulating the size of all instructions of b_i. The totally available SPM size S_{spm}, however, is extracted again from WCC's memory hierarchy infrastructure.

After solving the ILP, the values for the decision variables x_i determine where to place each basic block b_i. Using WCC's memory hierarchy API, the code of the program cur-

rently under optimization is finally transformed such that it reflects exactly the allocation decisions taken by the ILP. In addition to the SPM-allocated assembly code, our compiler finally emits a linker script required to generate a binary executable reflecting the ILP's SPM allocation.

5. EVALUATION

This section presents real-life benchmarking results for the proposed optimal WCET-aware SPM allocator. At optimization level *-O2*, our compiler (cf. Figure 2) applies a total of 34 different optimizations, including, among others, code reordering transformations. As very last optimization, the WCET-aware SPM allocation of program code discussed in this paper is performed. Hence, our SPM allocation is always applied to already highly optimized code.

As target architecture, the Infineon TriCore TC1796 is considered. Its memory address space is divided into 16 segments of maximum size of 256 MB each. The instruction set allows to jump within a segment with only one single machine instruction. Cross-segment jumps require more instructions (cf. Section 3.2). The TC1796 features a 48 kB large program code scratchpad mapped to segment 13. From these 48 kB, 1 kB is reserved for system code so that 47 kB remain for free use. An access to the program SPM of the TC1796 takes place within one cycle whereas accessing the program Flash serving as main memory takes 6 cycles.

Our ILP-based SPM allocation was applied to a total of 73 different real-life benchmarks from the MRTC [9], MediaBench [7], UTDSP [13] and DSPstone [15] benchmark suites. The number of basic blocks of all considered benchmarks ranges from 4 for the simplest DSPstone codes up to 585 for the most complex MediaBench applications. The code sizes of the benchmarks range from 52 bytes up to 18 kB with an average code size of 2.8 kB per benchmark.

Since the benchmarks' code sizes are considerably smaller than the totally available SPM size, we artificially limit the available SPM size for benchmarking. For every single benchmark, SPM sizes of 10%, 20%, ..., 100% of the benchmark's code size were used. The following results show the WCET estimates of all benchmarks produced by aiT resulting from our WCET-aware SPM allocator as a percentage of the WCET when not using the SPM at all.

Figure 3 shows the impact of our WCET-aware SPM allocation on the WCET estimates *(WCET_est)* of three representative benchmarks. For `g721_encode` with a total code size of 3,204 bytes, a steady decrease in terms of WCET can be observed the more scratchpad is available. Already for extremely small SPMs of only 10% of the program's code size, the WCET after our optimization amounts to 71% of the original WCET, i.e. a WCET reduction of 29% was achieved. If the benchmark fits into the SPM in its entirety,

978-1-60558-497-3/09 $25.00 © 2009 ACM

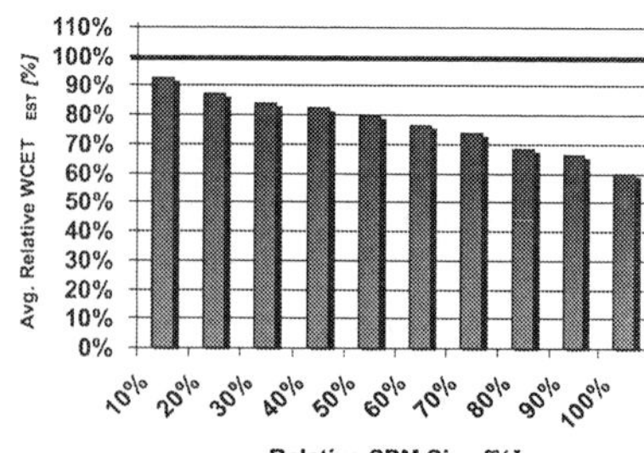

Figure 4: Average WCET Estimates after WCET-aware SPM Allocation for 73 Benchmarks

the resulting WCET is only 52.2% of the original WCET leading to savings of 47.8%.

For `cover` (code size: 2,670 bytes), a stepwise WCET reduction was observed. At 40%, 70% and 100% of SPM size, our ILP is able to move the most important loops leading to the highest WCET savings entirely onto the scratchpad memory. Thus, WCETs of 89.8%, 65.1% and 55.7% were achieved for these SPM sizes, resulting in savings of 10.2%, 34.9% and 44.3% respectively.

A quite extreme evolution of WCETs was finally observed for `md5` (6,354 bytes size). Here, an SPM size of only 10% of the total size of `md5` is enough to reduce the benchmark's WCET down to 59.1%. In absolute values, a tiny SPM of 636 bytes leads to a WCET reduction of 40.9%. This indicates that `md5` consists of one single hot spot which is executed extremely frequently and which has a very small code size. A further increase of the SPM for `md5` does not translate into any further WCET reductions. If `md5` fits entirely into the SPM (100%), an overall WCET of 56.3% was obtained which is only 2.8% off the WCET obtained for the 10% SPM. This benchmark clearly demonstrates that our ILP unerringly selects those basic blocks leading to the highest WCET reductions when being moved onto the SPM.

On average over all 73 benchmarks, we finally obtained steadily decreasing WCETs with increasing SPM sizes (cf. Figure 4). Already for small SPMs, WCETs decrease to 92.6% of the WCET without any SPM, corresponding to a WCET reduction of 7.4%. For large SPMs storing the entire benchmark, average WCETs of only 60% of the original WCET were obtained, leading to overall savings of 40%.

The complexity of our proposed ILP-based SPM allocator is of no practical relevance. For a CFG with n nodes, the ILP defined in Section 3 has a size of $\mathcal{O}(n^2)$ constraints and variables. However, the ILP solver *cplex* only takes one or two CPU seconds on an Intel Xeon running at 2.4 GHz for each of the 73 benchmarks. Compared to this, the two WCET analyses required to generate the constants C^i_{spm} and C^i_{main} (cf. Section 4) are more expensive, but they also terminate within a few CPU minutes for our largest benchmarks.

6. CONCLUSIONS

This paper is the first one to present a technique for optimal WCET-aware SPM allocation of program code. It improves the current state of the art of SPM allocation in that it performs ILP-based SPM allocation of code for the very first time under consideration of jump penalties and variable basic blocks sizes based on jump scenarios. The effectiveness of our approach is demonstrated by WCET reductions from 7.4% up to 40% for 73 different real-life benchmarks.

Our future work will concentrate on developing ILP-based scratchpad memory allocators for program data and on dynamic SPM allocation of both code and data where changing the scratchpad's content at run time will be possible.

Acknowledgments

The authors would like to thank AbsInt Angewandte Informatik GmbH for their support concerning WCET analysis using aiT (`www.absint.com/ait`).

7. REFERENCES

[1] AbsInt Angewandte Informatik GmbH. aiT: Worst-Case Execution Time Analyzers. *www.absint.com/ait*, 2009.

[2] A. M. Campoy, I. Puaut, A. P. Ivars, et al. Cache contents selection for statically-locked instruction caches: An Algorithm Comparison. In *Proceedings of ECRTS*, Palma de Mallorca, July 2005.

[3] J.-F. Deverge and I. Puaut. WCET-Directed Dynamic Scratchpad Memory Allocation of Data. In *Proceedings of ECRTS*, Pisa, July 2007.

[4] J. Engblom. Static Properties of Commercial Embedded Real-Time Programs, and Their Implication for Worst-Case Execution Time Analysis. In *Proceedings of RTAS*, Vancouver, 1999.

[5] H. Falk, P. Lokuciejewski, and H. Theiling. Design of a WCET-Aware C Compiler. In *Proceedings of ESTIMedia*, Seoul, Oct. 2006. *ls12-www.cs.tu-dortmund.de/research/activities/wcc*.

[6] H. Falk, S. Plazar, and H. Theiling. Compile Time Decided Instruction Cache Locking Using Worst-Case Execution Paths. In *Proceedings of CODES+ISSS*, Salzburg, Oct. 2007.

[7] C. Lee, M. Potkonjak, and W. H. Mangione-Smith. MediaBench: A Tool for Evaluating and Synthesizing Multimedia and Communications Systems. In *Proceedings of MICRO 30*, Washington DC, 1997.

[8] P. Lokuciejewski, D. Cordes, H. Falk, and P. Marwedel. A Fast and Precise Static Loop Analysis based on Abstract Interpretation, Program Slicing and Polytope Models. In *Proceedings of CGO*, Mar. 2009.

[9] Mälardalen WCET Research Group. WCET Benchmarks. *www.mrtc.mdh.se/projects/wcet*, Sept. 2008.

[10] I. Puaut. WCET-centric Software-controlled Instruction Caches for Hard Real-Time Systems. In *Proceedings of ECRTS*, July 2006.

[11] S. Steinke, L. Wehmeyer, B.-S. Lee, et al. Assigning Program and Data Objects to Scratchpad for Energy Reduction. In *Proceedings of DATE*, Paris, Mar. 2002.

[12] V. Suhendra, T. Mitra, A. Roychoudhury, et al. WCET Centric Data Allocation to Scratchpad Memory. In *Proceedings of RTSS*, Miami, Dec. 2005.

[13] UTDSP Benchmark Suite. *www.eecg.toronto.edu/~corinna/DSP/infrastructure/UTDSP.html*, Sept. 2008.

[14] M. Verma and P. Marwedel. *Advanced Memory Optimization Techniques for Low-Power Embedded Processors*. Springer, 2007.

[15] V. Živojnović, J. M. Velarde, C. Schläger, and H. Meyr. DSPstone: A DSP-Oriented Benchmarking Methodology. In *Proceedings of ICSPAT '94*, Dallas, 1994.

[16] L. Wehmeyer and P. Marwedel. Influence of Onchip Scratchpad Memories on WCET Prediction. In *Proceedings of WCET*, Catania, June 2004.

[17] L. Wehmeyer and P. Marwedel. Influence of Memory Hierarchies on Predictability for Time Constrained Embedded Software. In *Proceedings of DATE*, Munich, Mar. 2005.

[18] L. Wehmeyer and P. Marwedel. *Fast, Efficient and Predictable Memory Accesses – Optimization Algorithms for Memory Architecture Aware Compilation*. Springer, 2006.

978-1-60558-497-3/09 $25.00 © 2009 ACM

A Real-Time Program Trace Compressor
Utilizing Double Move-to-Front Method

Vladimir Uzelac, Aleksandar Milenkovic
The University of Alabama in Huntsville
{uzelacv, milenka}@ece.uah.edu

ABSTRACT

This paper introduces a new unobtrusive and cost-effective method for the capture and compression of program execution traces in real-time, which is based on a double move-to-front transformation. We explore its effectiveness and describe a cost-effective hardware implementation. The proposed trace compressor requires only 0.12 bits per instruction of trace port bandwidth, at the cost of 25K gates.

Categories and Subject Descriptors

B 7.2 [Integrated Circuits]: Design Aids-Verification. D.2.5: [**Testing and Debugging**]: Debugging aids, Tracing.

General Terms

Design, Verification.

Keywords

Debugging, Program Trace, Compression.

1. INTRODUCTION

Continual growth in the complexity of embedded systems-on-a-chip (SoCs) makes traditional approaches to system-level testing and debugging infeasible or impractical. For example, the development of a dedicated In-Circuit-Emulator (ICE) with additional support for debugging is cost-prohibitive; in addition, the ICE's physical characteristics such as chip floorplan, pin layout, and timing characteristics, differ from the targeted SoC. Traditional software approaches to debugging that rely on hardware and software breakpoints are often insufficient to capture the real sources of a bug. Moreover, they interfere with normal program execution, often causing the original error to disappear. This is especially important for real-time and safety-critical embedded systems that often need to be tested in real operating conditions. Last but not least, software step-by-step debugging is time consuming and places an additional strain on system developers, resulting in either poorly tested designs or product delays or both.

Embedded processor manufacturers responded to this debugging challenge by incorporating on-chip hardware resources exclusively dedicated to program tracing and debugging. For

Permission to make digital or hard copies of part or all of this work for personal or classroom use is granted without fee provided that copies are not made or distributed for profit or commercial advantage and that copies bear this notice and the full citation on the first page. To copy otherwise, to republish, to post on servers or to redistribute to lists, requires prior specific permission and/or a fee.
DAC'09, July 26-31, 2009, San Francisco, California, USA

example, ARM based embedded systems may include Embedded Trace Macrocell [1] modules to support program tracing; Altera Nios II [2] and Xilinx Microblaze [3] based systems may also include trace modules to enable real-time tracing of programs and data. Lauterbach [4] offers a number of program tracing hardware and software tools for a variety of processors. Typically, trace modules capture instruction and data traces (and possibly other bus signals), perform branch filtering, and store traces in on-chip trace buffers. The trace buffers can be read by external trace units through a JTAG interface or through the system bus. Alternatively, a trace module can send traced data directly through a trace port. The traces are then used in conjunction with program binaries to faithfully replay program execution and locate a bug source. In addition to debugging, program execution traces are also vital for workload characterization and performance tuning and optimization.

The existing commercially available trace modules rely either on large on-chip buffers to store execution traces of sufficiently large program segments or on wide trace ports that can sustain a large amount of trace data in real-time. However, large trace buffers and/or wide trace ports significantly increase the system complexity and cost. Moreover, they do not scale well, which is a significant problem in the era of multicore systems.

Compressing program execution traces at runtime in hardware can be used to reduce requirements for on-chip trace buffers and trace port communication bandwidth. Whereas commercially available trace modules typically implement only rudimentary forms of hardware compression with a relatively small compression ratio (5:1), several recent research efforts in academia propose effective trace compression techniques that can achieve compression ratios one order of magnitude higher [5-7]. For example, Kao *et al.* [5] propose an LZ-based program trace compressor that achieves a good compression ratio for a selected set of programs. However, the proposed module has a relatively high complexity (50K gates). In addition, the selected program segments are relatively small with less than 10 million instructions, so it is unclear how effective it would be in tracing more diverse programs.

In this paper we introduce a new cost-effective technique for compression of program traces in real-time. The proposed technique exploits common program characteristics and utilizes a two-level move-to-front transformation. We thoroughly explore program characteristics with regard to trace compression (Section 2), introduce a new Double Move-To-Front method (DMTF) for compression of program traces (Section 3), explore its design space (Section 4), and describe a cost-effective hardware implementation (Section 5). We also introduce two enhancements to the original method and explore their effectiveness using 17 diverse benchmarks from the MiBench benchmark suite [8]. A trace module configuration of 25,000 logic gates achieves compression ratios between 83 and 29,389, depending on the benchmark. The average weighted compression ratio is 268:1, which translates into 0.12 bits/instruction.

2. PROGRAM CHARACTERISTICS AND MOVE-TO-FRONT TRANSFORMATION

To replay a program flow offline, we only need to trace the information about program dynamic basic blocks (or streams). An instruction stream is a sequential run of instructions, from the target of a taken branch to the first taken branch in the sequence. Each instruction stream can be uniquely represented by its starting address (SA) and its length (SL). Thus, the complete trace of instruction addresses from an instruction stream can be replaced by the corresponding stream descriptor, i.e., the (SA, SL) pair. Relatively simple logic can be used to capture (SA, SL) pairs. In processors with fixed instruction word length, the current program counter (PC) is compared to the previous PC. If they differ for a value other than the instruction length, the current instruction is the beginning of a new stream. The current values of the SA and SL registers are output and the current PC is moved to the SA register to mark the beginning of a new stream. The SL register is set to 1. If the difference corresponds to the instruction length, the current value in the SL is incremented. In processors with variable instruction length, the stream detector requires an additional control line from the CPU to indicate a taken branch instruction. We introduce a slight modification to the original definition of an instruction stream. When we encounter an unconditional direct branch we do not terminate the current stream because the address of the next instruction in sequence can be inferred directly from the binary. Thus, when such a branch is identified, the SL register is just incremented as it was a non-branching instruction.

Most programs have only a small number of unique program streams, with just a fraction of them responsible for majority of program execution. Figure 1 shows some important characteristics of MiBench [8] benchmarks collected using SimpleScalar [9] while running ARM binaries. The first 4 columns *(a-d)* show the number of executed instructions (in millions), the number of unique streams, the maximum and average stream length, respectively. The number of unique streams ranges from 341 to 6871, and the average dynamic stream length is between 5.9 (*bf_e*) and 54.7 (*adpcm_c*) instructions. The fifth column *(e)* shows the number of unique program streams that constitute 90% of dynamically executed streams. This number ranges between 1 and 235, and it is 78 on average. Note that all calculations assume weighted average, where weights are determined based on the number of executed instructions, since the raw instruction address trace is directly proportional to the number of executed instructions. The maximum stream length never exceeds 256, thus we may choose to use 8 bits to represent SL. In addition to this, it can be shown that these frequently executed program streams create repeating patterns with strong local correlation. Our approach is to exploit these program characteristics in designing a cost-effective trace compressor that will achieve an excellent compression ratio with minimal storage and trace port bandwidth requirements.

Move-to-Front (MTF) [10] is an encoding of data designed to reduce the entropy of symbols in a data message by exploiting the local correlation between symbols. It is used in conjunction with the Burrows-Wheeler transform in the *bzip2* utility program [11].

The MTF algorithm encodes an input data message as follows. If an incoming input symbol is found in a history table *ht*, it is replaced with its index *i* in the *ht*, and the symbol is moved at the top of the table (the entry with index 0). The *ht* is updated by shifting down first *i-1* entries by one position, such that $ht[i]=ht[i-1]$, ..., $ht[1]=ht[0]$. To illustrate the MTF operation, let us consider an input message *AABC*, and a history table *ht*=[*C*, *B*, *A*] (symbol *C* is at the position 0). The MTF transforms the 3-symbol input message into a new 2-symbol message *2022*.

	IC (mil.)	SC	Max SL	Avg SL	CDF 90%	ht CDF 90%	ht[x] HR	ht2[0] HR
adpcm_c	733	341	71	54.7	1	1	0.99	1.00
bf_e	544	403	70	5.9	22	5	0.41	0.69
cjpeg	105	1590	239	12.3	47	11	0.46	0.90
djpeg	23	1261	206	25.1	31	11	0.69	0.87
fft	631	846	94	10.5	209	32	0.17	0.66
ghostscript	708	6871	251	10.0	67	22	0.20	0.76
gsm_d	1299	711	165	19.5	33	6	0.48	0.90
lame	1285	3229	237	32.4	235	21	0.23	0.74
mad	287	1528	206	20.7	42	25	0.30	0.75
rijndael_e	320	513	77	21.0	45	2	0.57	0.79
rsynth	825	1238	180	17.6	49	10	0.26	0.77
stringsearch	4	436	65	6.0	48	38	0.62	0.81
sha	141	519	65	15.4	10	2	0.86	0.92
tiff2bw	143	1038	43	12.8	2	1	0.97	0.99
tiff2rgba	152	1131	75	27.7	2	1	0.92	0.99
tiffmedian	541	1335	92	22.3	5	1	0.91	0.97
tiffdither	833	1777	67	14.3	63	38	0.45	0.78
Average	816	1791	145	21.6	77.8	14.5	0.46	0.82
	(a)	(b)	(c)	(d)	(e)	(f)	(g)	(h)

Figure 1. MiBench program characteristics.

The original MTF transformation can be easily extended to allow operation starting from an empty history table. The history table is searched for an incoming input symbol. If the symbol is not found in the table (we call this event an *ht* miss), the original symbol is output and the table is updated by shifting its content by one position and by placing the incoming symbol in the *ht*[0]. If the symbol is found in the history table (an *ht* hit event), its index is output and the table is updated as described above. The MTF allows for an effective encoding of frequently executed program sections. Let us consider several typical examples of program loops where symbols A, B, and C represent unique instruction streams characterized by their respective (SA, SL) pairs. For example, a program loop consisting of a sequence of two streams A and B repeating many times, illustrated as {AB}, is transformed into a hit pattern {11}; similarly, a loop with a repeating pattern {ABC} is transformed into {222}.

A relatively small history table will suffice to achieve a good hit rate due to a strong temporal locality of instruction streams in common programs. When a stream descriptor (SA, SL) is found in the history table, it is replaced with its index in the history table. Otherwise, the full stream descriptor of 40 bits is output in case that the SA is a target of an indirect branch. If the SA is a target of a direct branch, it can be inferred from the program binary, and we output only 8 bits for SL. The effectiveness of the MTF transformation on program execution traces consisting of a sequence of stream descriptors is shown in Figure 1*(f)*. We measure the frequency of the output symbols after the MTF transformation is applied. The average number of unique MTF output symbols that constitute 90% of all dynamically executed program streams is only 14.5 (down from 78 before the MTF transformation), ranging from 1 (*adpcm_c*) to 38 (*stringsearch*).

978-1-60558-497-3/09 $25.00 © 2009 ACM

Note: the experiments are conducted assuming a history table with 128 entries; the hit rate is over 97%, so a very small number of streams are not transformed with the MTF.

A perfect trace compression without stream pattern recognition would replace each stream with just a single bit. As described above, the MTF transformation significantly reduces the number of trace symbols. To come close to a one bit per stream goal, we need to identify the most frequent entry and to encode it with a single bit. Figure 1*(g)* shows the percentage of the hit events in the most frequent *ht* entry. Although this percentage is fairly high for many benchmarks (e.g., *adpcm_c*, *tiff2bw*), it is relatively modest for others (e.g, 17% for *fft*, and 46% on average across all benchmarks). An additional problem is how to identify the most frequent entry in the *ht* because it varies across benchmarks.

In order to resolve these two problems, we introduce an additional, second level move-to-front transformation. Let us consider the following repeating stream pattern {ABAC}. The hit pattern at the output from the first-level MTF is {1212}. If we supply this pattern to the second-level MTF history table, the hit pattern at the output is {1111}, with even lower entropy of information. Because the MTF transformation lowers the number of frequent symbols, the level 2 history table can be significantly smaller.

This approach can be further extended by introducing another level of MTF transformation; in general we could introduce a hierarchy of MTF history tables. The size of the MTF history tables will exponentially decrease as we move toward the upper levels. However, an increase in the number of MTF levels will reach the point of diminishing returns, and will not yield expected gains. In general, the optimal configuration is application specific. Our analysis shows that a 2-level MTF configuration appears to be optimal. Figure 1*(h)* shows a high percentage of program streams that end up in the entry 0 of the level 2 history table (*ht2*), from 66% to 100%. By encoding this entry with a single bit we approach the goal of having one bit per instruction stream. Note: the experiments are conducted using *ht2* with 16 entries achieving 94% hit rate.

3. DMTF METHOD

The analysis from the previous section suggests the use of a 2-level move-to-front transformation as optimal in compressing program instruction traces. Consequently, we design an instruction trace compressor with two history tables in sequence. We name this scheme Double Move-to-Front (DMTF). The first- and the second-level history tables are named *mtf1* and *mtf2*, respectively.

Figure 2 illustrates operation of the proposed trace compressor. When a new stream is detected, its descriptor (SA, SL) is forwarded to *mtf1*. As described before, the *mtf1* table is searched for the stream descriptor. If we find a match, we have an *mtf1* hit; the index of the matching entry is output to the next stage, and *mtf1* is updated accordingly. Otherwise, we have an *mtf1* miss; the *mtf1* content is shifted down by one position, and *mtf1[0]* is loaded with the stream descriptor. In case of an *mtf1* hit, the index *i1* is sent to the *mtf2* history table; *mtf2* is searched for the index *i1*. If we find a match in the entry 0, *mtf2[0]*, we have an *mtf2* zero entry hit. If we find a match in the remaining *mtf2* entries, we have an *mtf2* non-zero entry hit. Otherwise we have an *mtf2* miss event.

We can distinguish four different events in the DMTF scheme and they are encoded as follows. An *mtf2[0]* hit is encoded with a single bit '0'. An *mtf2* non-zero entry hit is encoded with a one-bit header '1' and the *mtf2* index *i2* ('1'+*i2*). An *mtf1* hit with a miss in *mtf2* is encoded with ('1'+*i2miss*+*i1*); note that the last index in the *mtf2* table, *i2miss*, is reserved to indicate a miss event in the *mtf2*. Finally, a miss in *mtf1* is encoded with a header ('1'+*i2miss*+*i1miss*) followed by a full or a partial stream descriptor ([SA], SL) – 40 or 8 bits. Note: the last index in the *mtf1* table, *i1miss*, is reserved to indicate a miss in the *mtf1*.

The compression ratio that can be achieved using DMTF scheme can be expressed analytically as follows. Equation 1 shows the number of bits needed to encode a single stream after DMTF compression, as a function of five parameters: *mtf2* zero-entry hit rate, *mtf2.zhr*; *mtf2* non-zero entry hit rate, *mtf2.ohr*; *mtf1* hit rate, *mtf1.hr*; *mtf1* size, *mtf1.size*; and *mtf2* size, *mtf2.size*. Note: *mtf2.hr=mtf2.zhr + mtf2.ohr*. Equation 2 shows the compression ratio as a function of the number of instructions in a program, the number of executed instruction streams, and the number of bits per one instruction stream.

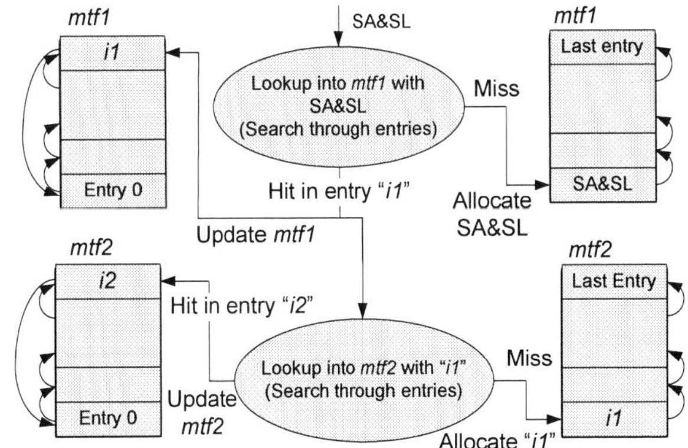

Figure 2. DMTF Operation.

$$Eq.\ 1 \quad \begin{aligned} BitsPerStream = mtf2.zhr + \\ (mtf2.ohr*(1+\log_2(mtf2.size) + \\ (mtf1.hr - mtf2.hr)*(1+\log_2 mtf2.size + \log_2 mtf1.size) + \\ (1 - mtf1.hr)*(1+\log_2 mtf2.size + \log_2 mtf1.size + 8 + [32]) \end{aligned}$$

$$Eq.\ 2 \quad CR = \frac{32*InstructionCount/StreamCount}{BitsPerStream}$$

An Example Compression/Decompression. Let us illustrate the compression flow using an example from Figure 3*(a)*. We consider the following sequence of instruction streams *ABCAABABAC*, where A, B, and C denote 3 instruction streams with distinct stream descriptors. Let us assume a 64-entry *mtf1* and an 8-entry *mtf2*. Note that the actual number of entries is 63 and 7, respectively, since the last indices are reserved to indicate miss events. First three instruction streams are not found in the *mtf1* and are output with the header '1', followed by a 3-bit index in the *mtf2* reserved for miss events ('111'), a 6-bit index in the *mtf1* reserved for miss events ('111111'), and individual stream descriptors ([SA], SL). Next, the stream A is found in *mtf1*[2], but index 2 is not found in the *mtf2* resulting in an *mtf2* miss with *mtf1* hit event; we emit a header '1'+'111' followed by the *mtf1* index

978-1-60558-497-3/09 $25.00 © 2009 ACM

'000010'. The next stream in sequence is A, resulting again in an *mtf2* miss with *mtf1* hit event; this event is encoded with '1'+'111'+'000000'. The rest of the compression flow continues as illustrated in Figure 3*(a)*.

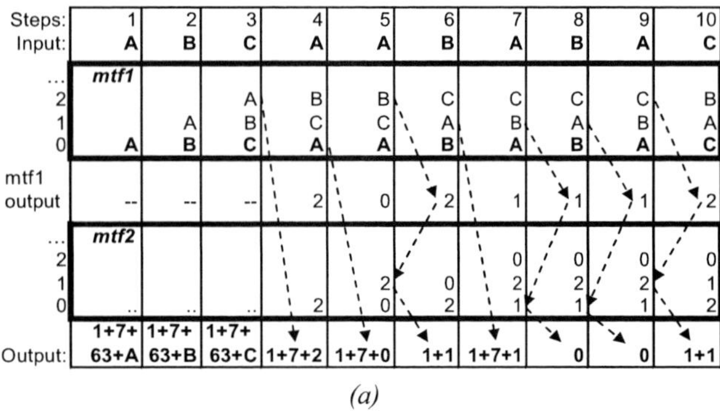

(a)

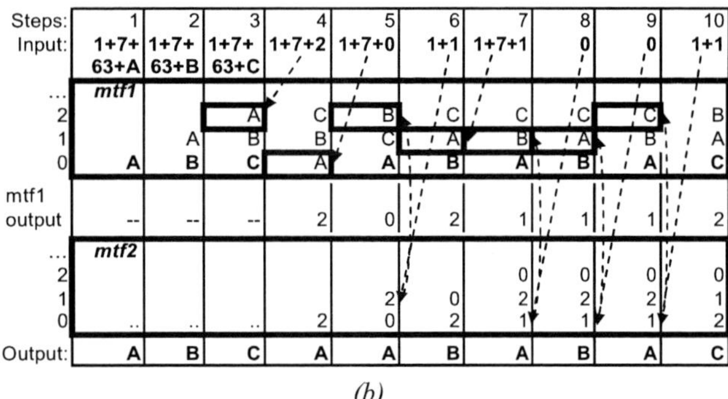

(b)

Figure 3. DMTF compression (a) and decompression (b) flow examples.

The decompression flow is a reversed compression flow and it requires the same configuration of the history tables. The compressed trace is read, headers are analyzed and the history tables updated according to the DMTF method described above. The decompression flow is illustrated in Figure 3*(b)*. The first item starts with the header '1'+'111'+'111111', which indicates that the next 40 bits represent the first stream descriptor, (SA, SL). The stream descriptor is loaded into *mtf1*[0]. The de-compressor now can recreate a complete instruction trace for this stream. The next two items in the trace are decompressed in the same way. The next trace record '1'+'111'+'000010' directs the de-compressor to find the original stream descriptor in *mtf1*[2] (instruction stream A). The following trace record '1'+'111'+'000000' directs the de-compressor to find the stream in *mtf1*[0], which is stream A. The rest of the decompression process is illustrated in Figure 3*(b)*.

Performance Analysis. Figure 4*(a)* shows the compression ratio for MiBench benchmarks; the size of the *mtf1* is fixed to 128 entries and the *mtf2* size is varied from 4 to 16 entries. The last row (*Average*) shows the total compression ratio calculated as the weighted harmonic mean of individual benchmark compression ratios. The results show that we are able to achieve an excellent compression ratio ranging from 45 to 1738. The results also indicate that a DMTF configuration with only 4-entry *mtf2* will outperform configurations with larger *mtf2*.

	CR (mtf1 size = 128)			Distribution per component			
mtf2 size	4	8	16	zht	mtf2ht	mtf1ht	mtf1mt
adpcm_c	1738	1734	1729	99%	1%	0%	0%
bf_e	110	94	84	40%	52%	8%	0%
cjpeg	250	253	252	57%	10%	31%	2%
djpeg	434	434	433	47%	12%	30%	11%
fft	45	44	43	9%	3%	10%	78%
ghostscript	107	106	107	25%	7%	55%	13%
gsm_d	358	343	327	52%	13%	4%	32%
lame	326	329	323	23%	12%	28%	37%
mad	210	210	212	24%	8%	48%	20%
rijndael_e	323	409	362	38%	17%	45%	0%
rsynth	209	199	186	29%	18%	13%	41%
strings.	81	78	75	34%	7%	58%	2%
sha	425	399	375	79%	20%	0%	0%
tiff2bw	391	390	388	95%	2%	3%	0%
tiff2rgba	852	844	834	95%	3%	0%	1%
tiffmedian	523	515	505	71%	5%	2%	22%
tiffdither	170	161	155	29%	9%	44%	18%
Average	181	175	169	42.8%	12.3%	20.0%	25.0%

(a)   *(b)*

Figure 4. Compression ratio for DMTF(128,X), X= 4-16 *(a)*. Distribution of the individual trace components *(b)*.

4. ENHANCED DMTF METHOD

The output of the DMTF trace compressor contains a lot of redundant information. We introduce two low-cost enhancements that exploit this redundancy and/or reduce complexity of the compressor implementation. The four components of the output trace, *mtf2* zero hit trace (*zht*), *mtf2* non-zero hit trace (*mtf2ht*), *mtf2* miss with *mtf1* hit trace (*mtf1ht*), and *mtf1* miss trace (*mtf1mt*) are analyzed separately. Figure 4*(b)* shows the contribution of each component to the total trace size for DMTF(128,4) (128-entry *mtf1* and 4-entry *mtf2*). The results show the *mtf1mt* component is responsible for 25% of the total size, in spite of high hit rates in the *mtf1*. Fortunately, the redundant information in this trace can be easily exploited using a simple last-value predictor on upper address bits that stay constant during program execution. This enhancement is described in Section 4.1 and also helps reduce hardware complexity of the compressor implementation. Next, the zero trace occupies 43% of the total trace. We expect it to contain long runs of '0's, and its size can be reduced by replacing them by a counter value (Section 4.2). Finally, in Section 4.3 we put both enhancements together and evaluate the effectiveness of the DMTF instruction trace compressor.

4.1 Last-Value Predictor for Upper Address Bits

The upper address bits of the starting address (SA) field in the stream descriptor rarely change during program execution. We analyzed the locality of stream starting addresses; the SA field of the incoming stream is compared bit by bit to the SA of the previous instruction stream or to SAs of the several last instruction streams. The results indicate that the upper 12 address bits, SA[31:20], stay constant during program execution in 99% of cases. Therefore, we divide the SA field into two parts: the lower 20 address bits SA[19:0] that are compressed through the DMTF, and the upper 12 bits that are handled using a simple last value predictor (*HLV*). Note: SA[1:0] is '00' for the ARM ISA and could be omitted; SA[0]= '0' for the ARM Thumb ISA, so the SA[0] could be omitted. Here we keep the whole address.

A 12-bit last value (LV) register keeps the upper 12 bits of the last stream's SA. The upper 12 address bits of an incoming stream are compared to the LV. If they match we have an *HLV* hit. The lower 20 address bits SA[19:0] and SL are used in *mtf1* lookup. We adopt a scheme where *mtf1* hits are conditional upon the corresponding *HLV* hits. An *HLV* miss will cause that a miss *trace* record is emitted regardless of mtf1 hits. When we have an *HLV* hit with *mtf1* miss event, the upper address bits are not emitted (in case that a full stream descriptor is required). Finally, in case when both *mtf1* and *HLV* have a hit, a regular DMTF record is emitted. The miss trace format is consequently extended to support these modifications.

The effectiveness of this enhancement is analyzed below. In general, it is beneficial in DMTF configurations with a relatively small *mtf1* and less so with a larger *mtf1*. The performance benefits are somewhat limited because direct branches dominate in the MiBench suite (92% of all branches on average) and all stream descriptors that start with targets of direct branches do not require the SA field. However, it significantly reduces the complexity of the DMTF implementation, as we do not need to keep upper 12 address bits in the *mtf1* history table.

4.2 Zero Hit Trace Counters

We show that the DMTF method ensures that the *mtf2* zero hit event is the most frequent one, and thus it is encoded with a single bit '0'. In many benchmarks the output trace will consist of long runs of zeros. The redundancy in this trace can be exploited by utilizing a zero-length counter (ZLC for short); it counts the consecutive zeros and replaces them with a counter value preceded by a new header. The number of bits used to encode this trace component is determined by the counter size. A longer counter can capture longer runs of zeros, but too long counter results in wasted bits. Our analysis of the *zht* trace component shows a fairly large variation in the average number of consecutive zeros, ranging from 5 in *ghostscript* and *fft* to hundreds of in *adpcm_c* and *tiff2bw*. In addition, zero runs in a program may vary across different program phases. This implies that an adaptive ZLC length method would be optimal.

The adaptive zero-length counter (AZLC) tries to dynamically adjust the ZLC size to the program flow characteristics. An additional 4-bit saturating counter monitors the *zht* component and it is updated as follows. It is incremented by 3 when the number of consecutive zeros in the trace (*mtf2*[0] hits) exceeds the current size of the ZLC. The monitoring counter is decremented by 1 when a detected zero sequence is smaller than the ZLC counter maximum value. When the monitoring counter reaches the maximum (15) or minimum (0) values, a change in the ZLC size occurs.

The AZLC requires a slight modification of the trace output format. A header bit '0' is followed by $\log_2$(AZLC Size) bits. The counter size is automatically adjusted as described above. The decompressor needs to implement the same adaptive algorithm.

4.3 Putting It All Together

Figure 5 shows a modified trace format that supports two enhancements, HLV and AZLC. Figure 6 shows the average compression ratio (CR) of several DMTF configurations as a function of the *mtf1* size (64-320). The basic DMTF (bDMTF) with *mtf2*=4 performs better than with *mtf2*=8 for any *mtf1* size as previously indicated in Figure 4. The DMTF with HLV predictor and *mtf2*=4 (hDMTF) performs better than bDMTF only for small *mtf1* sizes. When *mtf1*=192 bDMTF slightly outperforms hDMTH primarily due to significant performance degradation for lame

benchmark. Finally, the enhanced DMTF with *mtf2*=4 (eDMTF) with both improvements performs the best. For all configurations the compression ratio saturates for the *mtf1* with 256 entries, and the *mtf1* with 192 entries strikes an optimal balance between the complexity and compression ratio. Figure 6 gives a design guideline and shows how one can trade compression ratio for complexity (the most complex resource in the enhanced DMTF module is the *mtf1* table).

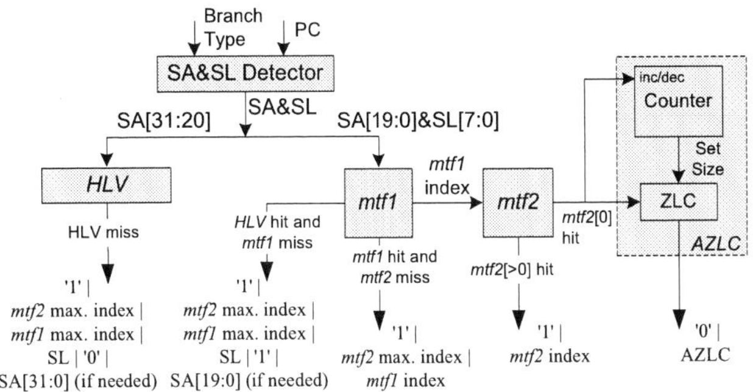

Figure 5. An Enhanced DMTF Trace Format.

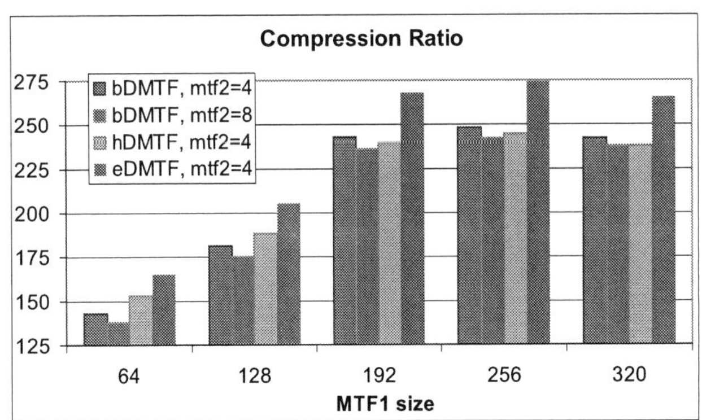

Figure 6. Compression ratio as a function of the *mtf1* size.

| | mtf1=192, mtf2=4 | | | | mtf1=64, mtf2=4 | | | |
| | CR | CR | CR | bits/inst. | CR | CR | CR | bits/inst. |
	bDMTF	hDMTF	eDMTF	eDMTF	bDMTF	hDMTF	eDMTF	eDMTF
adpcm_c	1738	1738	29389	0.001	1738	1738	29441	0.001
bf_e	108.8	108.8	112.7	0.284	110.5	110.5	114.5	0.279
cjpeg	245.3	245.7	353.1	0.091	245.6	248.7	360.5	0.089
djpeg	448.9	451.6	612.9	0.052	421.9	433.6	582.4	0.055
fft	151.7	151.9	159.1	0.201	26.9	31.1	31.6	1.012
ghostscript	104.7	106.1	104.6	0.306	104.8	108.4	106.9	0.299
gsm_d	495.4	495.4	808.4	0.040	363.2	381.2	547.8	0.058
lame	317.1	274.2	283.2	0.113	318.6	279.7	289.8	0.110
mad	202.7	208.8	217.0	0.147	219.0	227.6	237.4	0.135
rijndael_e	308.9	308.9	333.5	0.096	333.7	334.5	363.4	0.088
rsynth	307.2	307.4	296.3	0.108	159.1	176.6	173.8	0.184
strings.	77.0	77.2	82.7	0.387	32.9	37.2	38.8	0.825
sha	424.9	425.1	654.3	0.049	425.1	425.2	654.7	0.049
tiff2bw	390.2	390.4	2807.6	0.011	221.3	241.0	521.8	0.061
tiff2rgba	852.5	853.8	5334.6	0.006	348.8	393.5	637.7	0.050
tiffmedian	653.1	653.4	2712.9	0.012	386.1	418.7	819.1	0.039
tiffdither	167.8	169.6	193.2	0.166	140.9	144.8	162.7	0.197
Average	**242.5**	**239.5**	**268.1**	**0.119**	**143.1**	**153.1**	**165.1**	**0.193**

Figure 7. Compression ratios for xDMTF (x=b,h,e).

Figure 7 shows a detailed evaluation for bDMTF, hDMTF, and eDMTF with two configurations (192,4) and (64,4), x={b,h,e}. The hDMTF configuration achieves 7% higher CR than bDMTF for *mtf1*=64. This improvement is due to reducing the size of the miss trace. For *mtf1*=192, we see a decrease in hDMTF performance over bDMTF as explained above. The eDMTF configuration achieves 15% higher CR over bDMTF for *mtf1*=64 and 11% for *mtf1*=192. This improvement is unevenly distributed over benchmarks and is useful for tests such as *adpcm_c* (17x) or *tiff2rgba* (6x). The best performing configuration (eDMTF with *mtf1*=192) achieves the total weighted average bandwidth on the trace port of only 0.12 bits per instruction.

5. DMTF HARDWARE IMPLEMENTATION

The *mtf1* and *mtf2* history tables can be implemented as custom fully associative structures with a single-clock cycle lookup and additional hardware needed to support the move-to-front update operation. Instead, we propose a cost-effective implementation that combines a standard content addressable memory (CAM) and a most-recently used (MRU) stack (Figure 8). The MRU stack has the same number of entries as the history table, but its content are indices in the CAM memory. Each MRU stack entry points to a particular CAM location, and thus has [$\log_2$(MTF_Size)] bits.

The *mtf* lookup operation encompasses a lookup into the CAM with (SA, SL) pair and a lookup into the MRU stack. In case of a CAM hit, the corresponding CAM index is forwarded to the MRU stack and the MRU lookup is performed. The selected entry is moved at the top of the MRU stack, and the top *(i-1)* locations are shifted down. In case of a CAM miss, the MRU stack provides the address of the CAM location where the new stream is going to be stored (the index at the bottom of the MRU stack), and the MRU stack is updated accordingly. Figure 8 shows a block diagram of a single level MTF history buffer. The lookup and update together require only two processor clock cycles and are performed only when a new instruction stream is detected. Hence, the compression can be done at the full processor speed without ever slowing the processor.

To estimate the complexity of the proposed implementation we consider enhanced DMTF(192,4) configuration. The *mtf1* CAM memory has 191 entries, each with 28 bits (20 for SA, and 8 for SL). With 3 gates per CAM bit [12], the CAM complexity is estimated at 3x28x191 ~ 16000 logic gates. The *mtf1* MRU stack has 191 8-bit entries, plus comparators attached to each of them. Registers use latches that occupy approximately 2.5 logic gates per bit, comparators use 2.5 logic gates per bit while tri-state buffer use 0.5 logic gates per bit. The *mtf1* MRU stack size is estimated at approximately 8400 gates. Similarly the *mtf2* size is estimated to be approximately 150 logic gates. Together with the LV predictor (12-bit register + comparator) and the AZLC counter (4 bits) the total complexity of the DMTF(192,4) is less than 24,600 gates.

6. CONCLUSIONS

This paper presents the double move-to-front method for program trace compression that successfully exploits temporal and spatial locality of program streams to achieve compression ratios of two orders of magnitude. Detailed evaluation of its effectiveness on a diverse set of benchmarks shows that the compression ratio for our best performing configuration ranges between 82.7:1 and 29,389:1 (268 on average), that translates into trace port bandwidth of 0.001 to 0.39 bits/instruction (0.12 bits/instruction on average). We have introduced a cost-effective implementation

of the proposed program trace compressor. The best performing configuration has an estimated complexity equivalent to 25,000 logic gates, which is a half of the complexity reported for the LZ-based trace compressor [5].

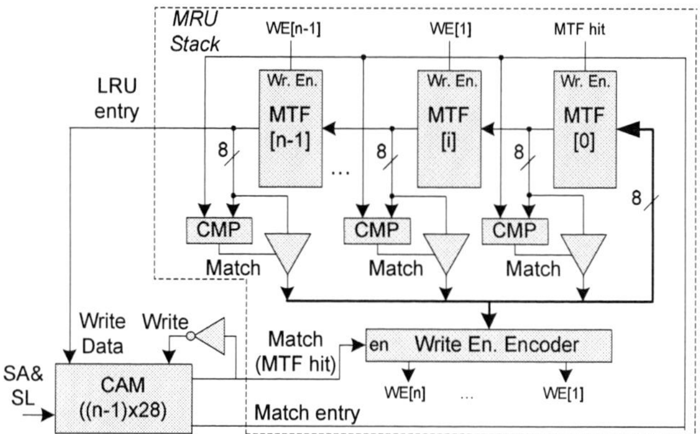

Figure 8. MTF Hardware Implementation.

The proposed trace compressor allows designers to effectively trade complexity and compression ratio, depending on application characteristics and available on-chip area for the trace module. For example, with DMTF(64, 4) we achieve 0.2 bits/instruction on average (ranging from 0.001 to 1) at the cost of 8,200 logic gates. With DMTF(128, 4) we achieve 0.16 bits/instruction on average (ranging from 0.001 to 0.6) at the cost of 16,500 logic gates.

7. REFERENCES

[1] ARM, "Embedded Trace Macrocell Architecture Specification," http://infocenter.arm.com.
[2] Altera, "Nios II Processor Reference Handbook," http://www.altera.com.
[3] Xilinx, "MicroBlaze Processor Reference Guide Embedded Development Kit EDK 10.1i," http://www.xilinx.com.
[4] "Lauterbach GmbH," http://www.lauterbach.com.
[5] C.-F. Kao, et al., "A Hardware Approach to Real-Time Program Trace Compression for Embedded Processors," *IEEE Transactions on Circuits and Systems,* vol. 54, pp. 530 - 543, 2007.
[6] M.-C. Hsieh and C.-T. Huang, "An embedded infrastructure of debug and trace interface for the DSP platform," in *45th ACM Design Automation Conference,* 2008.
[7] M. Milenkovic, et al., "Algorithms and Hardware Structures for Unobtrusive Real-Time Compression of Instruction and Data Address Traces " *Data Compression Conference,* pp. 283-292, 2007
[8] M. R. Guthaus, et al., "MiBench: A free, commercially representative embedded benchmark suite," in *Proceedings of the IEEE 4th Workshop on Workload Characterization,* 2001.
[9] T. Austin, et al., "SimpleScalar: An Infrastructure for Computer System Modeling," *Computer,* vol. 35, pp. 59-67, 2002.
[10] B. Jon Louis, et al., "A locally adaptive data compression scheme," *Commun. ACM,* vol. 29, pp. 320-330, 1986.
[11] M. Burrows and D. J. Wheeler, "A block-sorting lossless data compression algorithm," Digital SRC Research Report 1994.
[12] K. Pagiamtzis and A. Sheikholeslami, "Content-Addressable Memory (CAM) Circuits and Architectures: A Tutorial and Survey," *IEEE Journal of Solid-State Circuits,* vol. 41, 2006.

Heterogeneous Code Cache: Using Scratchpad and Main Memory in Dynamic Binary Translators

José A. Baiocchi
Department of Computer Science
University of Pittsburgh
Pittsburgh, PA 15260
baiocchi@cs.pitt.edu

Bruce R. Childers
Department of Computer Science
University of Pittsburgh
Pittsburgh, PA 15260
childers@cs.pitt.edu

ABSTRACT

Dynamic binary translation (DBT) can be used to address important issues in embedded systems. DBT systems store translated code in a software-managed code cache. Unlike general-purpose systems, embedded systems often have specialized memory resources, such as a fast scratchpad memory, that can be used to mitigate DBT performance overhead. This paper presents the *Heterogeneous Code Cache* (HCC), a code cache split among scratchpad and main memory. We explore several HCC management policies and show that, on average, an HCC outperforms a code cache allocated only to scratchpad or only to main memory.

Categories and Subject Descriptors

C.3 [**Computer Systems Organization**]: Special-purpose and application-based systems—*Real-time and embedded systems*; D.3.4 [**Programming Languages**]: Processors—*Code generation, Compilers, Incremental compilers, Interpreters, Optimization, Run-time environments*

General Terms

Measurement, Performance, Design, Experimentation

Keywords

Dynamic Binary Translation, Scratchpad, Software Caching

1. INTRODUCTION

Embedded systems continue to increase their capabilities to support more demanding applications: multimedia, image recognition, advanced online signal processing, etc. Their evolution creates new challenges (e.g., security, reliability, etc.), while traditional constraints (e.g., performance, memory, real-time, energy) are still important. Many of these challenges have been successfully addressed in general-purpose systems by Dynamic Binary Translation (DBT).

Permission to make digital or hard copies of part or all of this work for personal or classroom use is granted without fee provided that copies are not made or distributed for profit or commercial advantage and that copies bear this notice and the full citation on the first page. To copy otherwise, to republish, to post on servers or to redistribute to lists, requires prior specific permission and/or a fee.

DAC'09, July 26-31, 2009, San Francisco, California, USA

DBT is a technology that allows control over the execution of a program by processing every instruction before it is executed. DBT uses include virtualization [1], optimization [5], software caching [13] and code compression [8].

Despite the potential, DBT use in embedded systems has been limited due to performance and memory constraints. To ensure good performance, a dynamic binary translator stores translated code in a software-controlled memory buffer, called the *code cache* (CC). The CC is usually placed in main memory and large enough to capture an application's working set. An unconstrained CC minimizes DBT's *base performance overhead* – i.e., overhead in the absence of complex code transformations – to 4% on average [12]. An unconstrained CC may grow to several megabytes in size for desktop applications [10] and to hundreds of kilobytes for embedded applications [3].

This paper addresses the problem of allocating and managing a CC in an embedded system with scratchpad memory (SPM). SPM is a small on-chip SRAM with low energy consumption [6]. It is fast but must be explicitly controlled by software. We propose a new approach, the *Heterogeneous Code Cache* (HCC), that splits the CC among multiple memory levels. Unlike a traditional multi-level memory hierarchy, objects in the lower levels of the HCC are not replicated in the higher levels. The goal is to exploit the fast but small SPM, while keeping the low miss rate of a large CC. We study a two-level HCC with one level in SPM and the other in main memory. Our contributions include:

- A new code cache organization (HCC) that uses both scratchpad and main memory;

- Several new HCC management policies, including novel "SPM-aware" policies, that place the most recently translated code in the scratchpad memory;

- A new heuristic to adaptively resize the HCC to minimize translation overhead while constraining HCC size;

- A thorough evaluation of our techniques with a dynamic binary translator in a simulated embedded system.

The rest of the paper is organized as follows: Section 2 describes the framework for our work. Section 3 describes an HCC organization and several solutions to design issues that impact performance. Section 4 evaluates the proposed techniques, finds those that work best and compares them to traditional CC approaches. Section 5 describes related work and Section 6 concludes.

978-1-60558-497-3/09 $25.00 © 2009 ACM

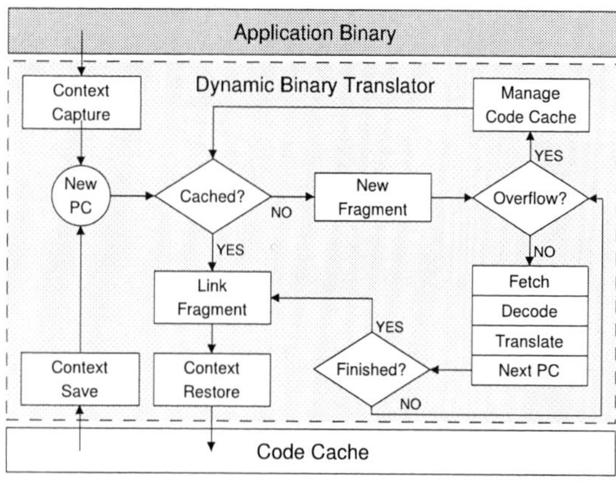

Figure 1: DBT overview

2. FRAMEWORK

We initially describe the class of embedded system targeted by our research and the operation of a typical DBT.

2.1 Target System

The techniques in this paper target embedded systems-on-chip (SoC) with a heterogeneous memory system. We assume a SoC that has a processor with L1 instruction and data caches, application-specific integrated circuits, SPM (on-chip SRAM), ROM (on-chip NOR Flash), controllers for external memories and off-chip I/O channels. External (off-chip) memories include SDRAM and NAND Flash. SDRAM is used as main memory and holds application code and data during program execution. NAND Flash is accessed through a file system and holds user files, including application binaries. System code, including the boot program and OS, is kept in ROM. SPM is *directly mapped* in the address space and has a fast access latency (1 cycle).

A dynamic binary translator can be part of the SoC's system code. It accesses application binaries from external NAND Flash. When an application is initiated, the translator loads the static data into main memory. The code is loaded on demand as execution progresses using DBT, which provides a form of *software instruction caching* [13, 4].

2.2 Dynamic Binary Translation

A high-level view of the DBT process is illustrated in Figure 1. The translator must ensure that all untranslated application instructions are loaded, examined and possibly modified prior to their execution. To do so, it must be invoked whenever new application code is requested. New code is requested when execution begins and every time a control transfer instruction (CTI) invokes an untranslated application address. Each CTI is replaced with an exit stub, called a *trampoline*, that "re-enters" the translator.

To safely re-enter the translator, the application's context must be saved to free registers for use by the translator, i.e., a *context switch* is done. The translator checks if there is translated code in the code cache (CC) for the requested application address. If so, the application context is restored and the corresponding translated code is executed. Otherwise, a new set of translated instructions, called a *fragment*, is built and saved in the CC. When translation stops, the context is restored and the new fragment is executed.

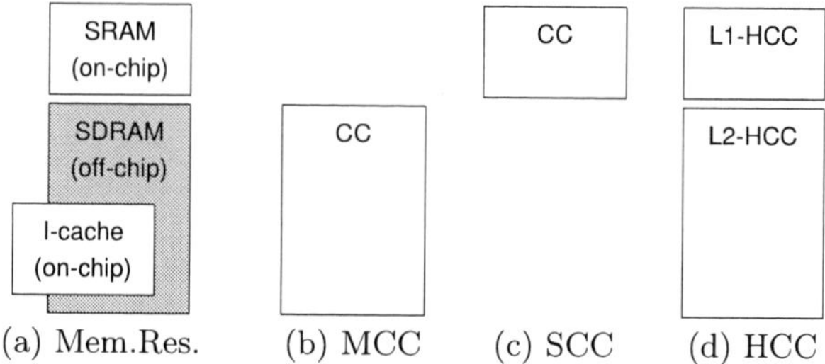

(a) Mem.Res.   (b) MCC   (c) SCC   (d) HCC

Figure 2: Code Cache Allocation Alternatives

To avoid unnecessary context switches, *fragment linking* overwrites trampolines to jump to their target fragments [5]. Indirect CTIs may not always have the same target, so they are replaced by an *Indirect Branch Translation Cache* (IBTC) lookup. The IBTC maps application addresses targeted by indirect CTIs to their translated addresses [12].

The CC is progressively filled with new fragments as execution progresses. If it is too small to hold the application's working set, a CC overflow occurs. This event is handled by a *code cache management* routine, which makes room in the CC for new fragments [10, 11, 7, 4].

3. HETEROGENEOUS CODE CACHE

In embedded systems where memory resources have different access latencies and sizes, we propose to split the CC among those resources. This arrangement introduces several CC design issues that affect performance. In this section, we describe the issues and propose solutions to address them.

3.1 Resource Allocation

We consider an embedded system with a scratchpad (on-chip SRAM) and main memory (off-chip SDRAM) serviced by an instruction cache (I-cache), as shown in Figure 2(a). The first design decision is where to allocate the CC.

One choice, shown in Figure 2(b), places the CC in main memory (MCC), so it can be relatively large to fully capture the application's translated working code set. A large CC leads to a minimal miss rate and avoids costly Flash memory reads to fetch binary code for retranslation. Another choice, shown in Figure 2(c), places the CC in SPM (SCC). The advantage to this choice is that translated instructions can be fetched in a single cycle, with the SPM serving as a *software instruction cache* [13]. The disadvantage is that the CC will be small (constrained to the SPM size), potentially leading to a high miss rate and additional Flash memory reads due to non-compulsory misses. Neither choice fully utilizes the on-chip memory resources – the first one does not use the SPM and the second one does not use the I-cache.

Our approach, the *Heterogeneous Code Cache* (HCC) exploits both SPM and I-cache. It aims to get the capacity benefit of a large CC in main memory and the latency benefit of a CC in SPM. A two-level HCC, shown in Figure 2(d), has its first level (L1-HCC) in SPM and its second level (L2-HCC) in main memory. The two levels share the lookup tables for CC metadata (translator data about fragments and trampolines). SPM is directly mapped in the address space, so the levels may not be contiguous. SPM use may help to improve the I-cache miss rate due to less potential conflicts.

978-1-60558-497-3/09 $25.00 © 2009 ACM   745

3.2 Eviction Policies

When the capacity of the HCC is exhausted and space is needed for newly translated code, an *eviction policy* is used to determine what code should be removed from the HCC to make room for new code. In general, evicting one or more fragments requires updating the CC metadata, reverting links from non-evicted fragments to evicted fragments back into trampolines (*unlinking*), and invalidating IBTC entries. An effective CC eviction policy must balance the miss rate, the frequency of evictions and the cost of unlinking [11]. We propose three eviction policies, derived from general-purpose ones, that address specific HCC challenges:

FLUSH is the simplest eviction policy. It discards the whole contents of the CC on an overflow [5]. The CC is refilled when translation is restarted after an overflow. This scheme may increase the HCC miss rate, but eviction management is greatly simplified. The CC metadata can be simply discarded and unlinking is not necessary. For the HCC, FLUSH initially emits code into L1-HCC. When it becomes full, L2-HCC is filled. When both levels are full, all fragments are evicted and translation resumes in L1-HCC.

FIFO is a fine-grained eviction policy. It treats the CC as a circular buffer, evicting only the *least recently created* (LRC) fragment when space is needed. FIFO has a similar or better CC miss rate than more complex policies (e.g., LRU, LFU), but does not suffer from fragmentation and has low management overhead [10]. FIFO requires unlinking only *backward links* that target the evicted fragment. FIFO for an HCC, like FLUSH, starts filling L1-HCC and continues with L2-HCC. However, when both levels are full, only the LRC fragment in L1-HCC is evicted (rather than evicting all the code). Translation resumes in L1-HCC after an eviction. When all fragments in L1-HCC have been replaced, the fragments in L2-HCC are evicted next. The effect of this policy is that the combined levels of the HCC form a circular buffer.

Segmented FIFO divides the HCC into several *segments* of the same size. To handle CC overflows, whole segments are evicted in FIFO order [11]. This policy tries to get both FIFO's low miss rate and FLUSH's low management overhead. Evicting a group of fragments together makes more space available at once, which reduces the frequency of evictions. With Segmented FIFO only *inter-segment links* must be unlinked on eviction; *intra-segment links* do not need to be processed. We use the same segment size in both HCC levels. First, segments in L1-HCC are filled. When the L1-HCC is full, segments in L2-HCC are in turn filled. After both levels are full, the first segment in L1-HCC is flushed when space is needed. When the code in all L1-HCC segments has been replaced, eviction continues at L2-HCC.

The choice of eviction policy affects the layout of the translated code. One reason is that space must be reserved for the trampolines created when unlinking. Another reason is that when the next translated fragment does not fit in the remaining space of a CC segment, it is stored instead into the next segment. This leaves unused space at the end of the CC segments. A third reason is retranslation. A finer eviction granularity may cause a fragment to be found in the CC, but with a coarser granularity that same fragment would be retranslated at a different location after being previously evicted. These differences affect both SPM usage (depending on fragment execution frequency) and I-cache effectiveness (due to mapping conflicts).

3.3 SPM-aware Fragment Placement

The HCC policies described to this point allow new fragments to be stored into the SPM only after the L2-HCC has been completely filled or replaced. Increasing the size of the L2-HCC reduces the chances for (first-time or retranslated) fragments to be assigned to the L1-HCC, i.e., the benefit of the SPM is reduced. To address this problem, we develop a set of new "SPM-aware" management policies. These policies force the DBT to place new fragments only into the SPM.

When all new fragments are put into the L1-HCC (SPM), they have to be moved to L2-HCC (main memory) when the L1-HCC overflows. Code relocation requires the capability to fix any links that are associated with a relocated fragment. We explore both single fragment and group relocations. Relocating fragments requires redirecting links and updating IBTC entries.

Our SPM-aware policies ensure that the SPM holds the most recently translated code. The first three policies use the eviction granularity for relocation. The fourth policy relocates one fragment at a time but uses Segmented FIFO for eviction. The policies are:

FLUSH@L1: Code is initially translated into L1-HCC. When it becomes full, *all fragments* in L1-HCC are relocated to L2-HCC and the DBT starts to fill L1-HCC again. When both levels are full, i.e., there is no space in L2-HCC to hold the contents of L1-HCC, all code in both levels is evicted.

FIFO@L1: Code is initially translated into L1-HCC and when it becomes full, fragments in L1-HCC are moved one at a time to L2-HCC in FIFO order. When L2-HCC is full, fragments are evicted from it in FIFO order. From the eviction point of view, the effect is the same as basic FIFO.

Segmented FIFO@L1: Code is initially translated into L1-HCC segments. When L1-HCC becomes full, its least recently filled segment is relocated to L2-HCC. When L2-HCC is full, its least recently added segment is evicted. From the eviction point of view, the overall effect is the same of the Segmented FIFO policy.

FIFO / Segmented FIFO: This hybrid policy combines single-fragment relocation with segmented eviction. Fragments are translated into L1-HCC and moved to L2-HCC one at a time in FIFO order. However, L2-HCC is divided in segments, which are evicted in FIFO order.

3.4 Retranslation-aware Resizing

When the HCC is too small to fully capture the translated code working set, some fragments are repeatedly evicted and retranslated, leading to excessive DBT overhead. To avoid this problem, we use a *retranslation-aware resizing heuristic*. On an overflow, the heuristic is used to decide whether to increase L2-HCC capacity (i.e., "adding" more main memory to it) or to evict code.

To make this decision, the translator monitors code that has been previously seen using the *Translation History Table* (THT). The THT records the application address (PC) of every fragment. If a PC is not found in the THT, it is recorded and the fragment is classified as "first-time"; otherwise, the fragment is classified as "retranslated". L2-HCC expansion is chosen when L2-HCC (or the segment to evict) contains more retranslated fragments than first-time fragments. The fragments that caused the expansion are marked to force eviction the next time. This strategy reduces the likelihood of keeping unneeded fragments in the HCC.

Parameter	Configuration
Fetch queue	4 entries
Branch predictor	not taken, 2 cycle mispred.penalty
Fetch/decode width	1 instr./cycle
Issue	1 instr./cycle, in order
Functional units	1 IALU, 1 IMULT, 1 FPALU, 1 FPMULT
RUU capacity	4 entries
Issue/commit width	1 instr./cycle
Load/store queue	4 entries
Memory page size	4 Kbytes
TLBs	64 entries, 2-way, 10 cycle miss
L1 D-cache	8 Kbytes, 4-way, LRU, 1 cycle
L1 I-cache	4 Kbytes, 4-way, LRU, 1 cycle
Scratchpad memory	4 Kbytes, 1 cycle
Bus width	4 bytes
Main memory latency	10 cycles first chunk, 4 cycles rest
Flash page size	512 Kbytes
Flash latency	3000 cycles first chunk, 10 cycles rest

Table 1: SimpleScalar Configuration

4. EXPERIMENTAL EVALUATION

In this section we evaluate our techniques, choose those that work best, and compare the HCC to a CC that uses only SPM or only main memory.

4.1 Methodology

We implemented our techniques in the Strata DBT [14], retargeted to run on SimpleScalar/PISA [2]. We extended the out-of-order simulator with ROM, SPM and Flash, which can have different sizes and latencies. The simulator reports the number of cycles spent executing code fetched from each kind of memory.

The simulator was configured as shown in Table 1. The resources are modeled after the ARM926EJ-S processor, used in smart phones, digital cameras, etc. PISA is an instruction set similar to MIPS, but with a 64-bit encoding to facilitate experimentation (e.g., instructions have a 16-bit annotation field and 8-bit register fields) [2]. To account for the long instructions, we doubled the size of the SPM and I-cache for the simulations but refer to the *effective size* (e.g., a 4KB SPM/I-cache is simulated with 8KB). We run programs from MiBench [9], a benchmark suite representative of embedded applications, with their large input data sets.[1]

Strata was configured to create *Dynamic Basic Blocks* (DBB), i.e., to stop fragment formation whenever a CTI is found. This configuration minimizes code duplication and speculative translation of code that may never be executed. To ensure a small translated code footprint, we enabled a set of techniques described in [3]: shadow link register trampolines, out-of-line shared IBTC lookup with shared target register copies, and prologue elimination.

4.2 Eviction Policies

To understand which eviction policy is most appropriate for the HCC, we measured their impact on program performance. Our evaluation is done with an HCC that has two levels: a 4K L1-HCC (SPM) and a L2-HCC (main memory) with an initial size of 16K. Capacity is added in 2K increments to L2-HCC using our resizing heuristic. We evaluate DBT performance with an HCC in the absence of complex code transformations. We compare FLUSH, Segmented FIFO with 2K segments (2K-Segs) and FIFO. The slowdown for the benchmarks relative to native execution with these

policies is shown in Figure 3. Our result charts show slowdown bars split into three sections: the bottom section is time spent executing code from main memory (main memory time), the middle section is time spent executing code from SPM (SPM time), and the top section is the time spent executing Strata (translation time). We report these times as a slowdown relative to native execution to facilitate comparisons. *Native execution* means running a binary fully loaded into main memory (with a 4K I-cache), without the use of DBT.

HCC resizing prevents the programs from thrashing, so translation time is generally small and does not vary much across policies, with a few exceptions (*lame*, *pgp*, *typeset*). For instance, *typeset* has slowdowns of 1.17x (FLUSH), 1.19x (2K-Segs) and 1.14x (FIFO). Its translation times are: 0.22x (FLUSH), 0.15x (2K-Segs) and 0.24x (FIFO).

For some programs, SPM usage is insignificant (e.g., *adpcm*, *crc*, *rijndael*, *sha*). In programs with significant SPM usage, an important trend arises: more frequent use of SPM helps performance. For instance, *basicmath* has slowdowns of 1.35x (FLUSH), 1.04x (2K-Segs) and 1.29x (FIFO). Its translation times are 0.17x (FLUSH, 2K-Segs) and 0.16x (FIFO) and its SPM times are 0.07x (FLUSH), 0.11x (2K-Segs) and 0.06x (FIFO). *tiffdither* has speedups of 1.25x with FLUSH and 1.10x with 2K-Segs, but a 1.04x slowdown with FIFO. Its SPM times are 0.19x (FLUSH), 0.06x (2K-Segs) and 0.03x (FIFO), and its translation time is 0.03x.

Using the SPM reduces pressure on the I-cache, which helps to improve performance. For instance, *ghostscript* has speedups: 1.16x (FLUSH), 1.15x (2K-Segs), 1.14x (FIFO). Its translation and SPM times are respectively 0.06x and 0.02x for all policies. However, the native execution has 15% I-cache miss rate, which is reduced to 9-10%.

4.3 SPM-aware Policies

Our SPM-aware policies attempt to improve SPM usage by keeping the *most recently translated code* in SPM. Although they do not guarantee that the *most frequently executed* code is assigned to the SPM, they improve performance for several benchmarks. Figure 4 shows the slowdown relative to native execution for the benchmarks with the SPM-aware policies.

For some benchmarks where no evictions occur (*adpcm*, *bitcount*, *blowfish*, *crc*, *susan.smoothing*), the SPM is heavily used, but performance is practically unaffected since the I-cache miss rate is low. For other benchmarks without evictions (*dijkstra*, *qsort*, *rijndael*, *stringsearch*) there are significant improvements. For instance, the speedup of *qsort* is increased from 1.25x to 1.54x with all policies; *stringsearch*'s slowdowns of 1.11x (FLUSH), 1.12x (2K-Segs, FIFO) are improved to speedups of 1.32x (FLUSH@L1), 1.37x (2K-Segs@L1) and 1.43x (FIFO@L1, FIFO/2K-Segs).

The SPM-aware policies increase *ghostscript*'s speedups to 2.70x (FLUSH@L1), 1.22x (2K-Segs@L1), 2.78x (FIFO@L1) and 2.56x (FIFO/2K-Segs). For *basicmath*, its 1.35x slowdown with FLUSH is reduced to 1.03x with FLUSH@L1, and its 1.29x slowdown with FIFO turns into a 1.04x speedup with FIFO@L1. However, *basicmath*'s 1.04x slowdown with 2K-Segs is increased to 1.20x with 2K-Segs@L1.

ispell is not helped. Its 1.28x speedup with FLUSH turns into a 1.25x slowdown with FLUSH@L1. Its 1.23x slowdown with 2K-Segs is increased to 1.28x with 2K-Segs@L1. Its 1.15x speedup with FIFO is reduced to 1.10x with FIFO@L1.

---

[1] We could not compile `mad`, `rsynth`, and `sphinx`.

978-1-60558-497-3/09 $25.00 © 2009 ACM

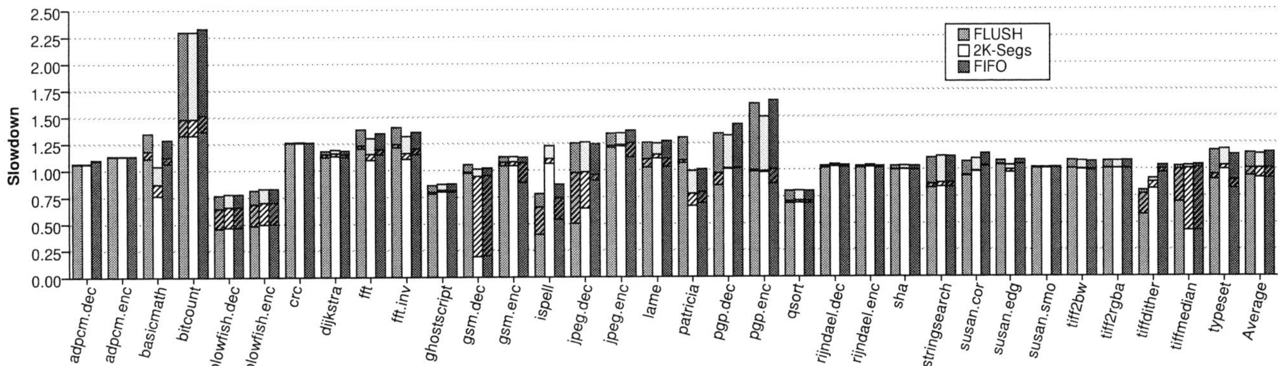

Figure 3: Slowdown relative to native execution with FLUSH, Segmented FIFO and FIFO eviction policies

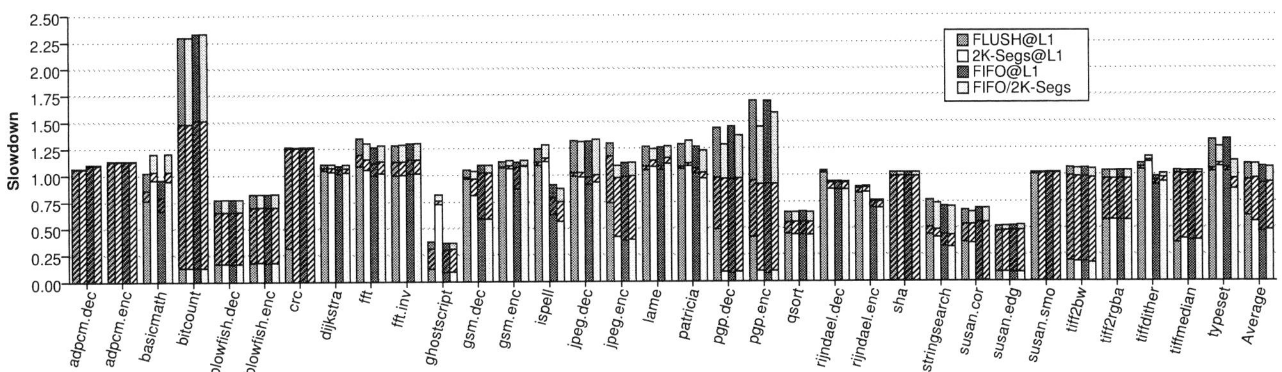

Figure 4: HCC slowdown relative to native execution with SPM-aware policies

However, with FIFO/2K-Segs, it has the same 1.15x speedup than with FIFO. In this case, the most recently translated code is not the most frequently used, and the SPM-unaware policies casually capture code with higher execution frequency in the SPM.

Interestingly, the average performance of all benchmarks with FIFO@L1 and FIFO/2K-Segs is the same: 1.07x slowdown. The average slowdown with FIFO is 1.16x. The average slowdowns with FLUSH and 2K-Segs are also improved by putting all new fragments in the SPM: from 1.15x (FLUSH) to 1.10x (FLUSH@L1) and from 1.14x (2K-Segs) to 1.10x (2K-Segs@L1).

In conclusion, the SPM-aware HCC management policies should be used instead of their SPM-unaware counterparts.

4.4 Comparison to Traditional Code Caches

We compared an HCC managed with an SPM-aware policy to a CC that uses only SPM and a CC that uses only main memory. The performance results for these three alternatives are shown in Figure 5.

SCC:FLUSH allocates the CC to the 4K SPM and handles overflows with FLUSH, without resizing. In most cases, the translated code does not fit in the small SPM, leading to unacceptable slowdowns[2]. For instance, *stringsearch* has a 15.09x slowdown, spent mostly in translation – SPM time is just 0.45x. With HCC:FLUSH@L1, it has a 1.32x speedup. Another example is *typeset*, which has a 43.93x slowdown with SCC:FLUSH, but its SPM time is just 0.61x. With

HCC:FLUSH@L1, its slowdown is only 1.32x. When the translated code working set fits in SPM, SCC outperforms the techniques that execute code in main memory. For instance, *crc* has slowdowns of 1.09x with SCC:FLUSH and 1.26x with HCC:FLUSH@L1 and MCC:FLUSH.

MCC:FLUSH initially allocates a 16K CC to main memory and uses our resizing heuristic to adaptively increase its size (2K each time). FLUSH is used for evictions. In most cases, the SPM-aware techniques outperform MCC:FLUSH thanks to their taking better advantage of the SPM. For instance, *fft* with MCC:FLUSH has a 1.48x slowdown, but only a 1.34x with HCC:FLUSH@L1. For *ghostscript*, MCC:FLUSH does reasonably well: 1.15x speedup. With HCC:FLUSH@L1, it does much better: 2.70x speedup.

On average, the HCC outperforms both SCC and MCC. HCC:FLUSH@L1 has an average slowdown of 1.10x, while MCC:FLUSH slowdown is 1.23x. SCC:FLUSH has the worst average slowdown, 11.41x, because SCC is too small.

Only some benchmarks need the CC in main memory to grow. Table 2 shows the final CC size for benchmarks where L2-HCC grows at least by 4K. HCC size includes the 4K SPM. Although no growth limit is set when enabling the resizing heuristic, the amount of main memory needed for DBT is much less than for native execution. For instance, the final CC size for *ghostscript* is less than 8% the size of the binary's code segment.

From these results, we conclude that the HCC is a good choice for a SoC with heterogeneous memory resources since it can fully use those resources to minimize DBT overhead.

---

[2] *ispell* did not run to completion.

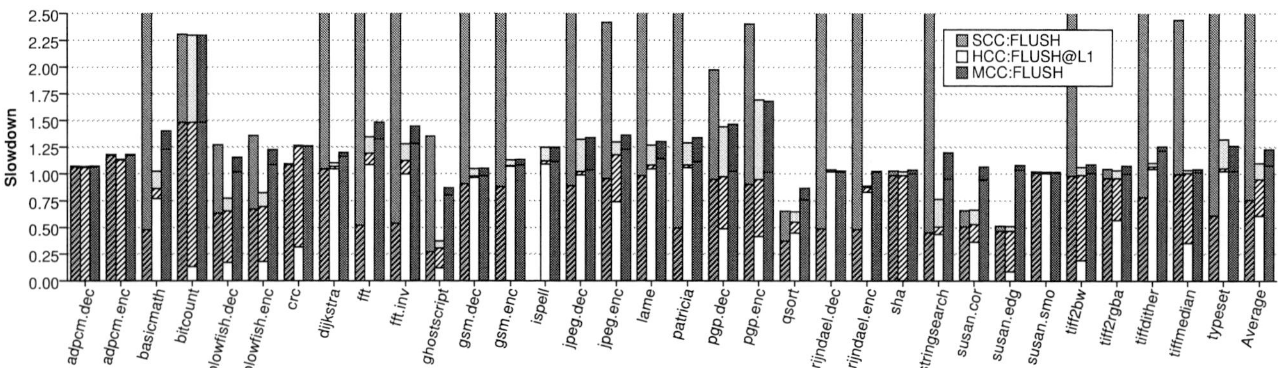

Figure 5: SCC, HCC and MCC slowdown relative to native execution with FLUSH

Benchmark	Binary	HCC:FLUSH@L1	MCC:FLUSH
ghostscript	888K	60K (6.76%)	64K (7.21%)
gsm.enc	78K	28K (35.90%)	28K (35.90%)
ispell	106K	28K (26.42%)	28K (26.42%)
lame	158K	70K (44.30%)	68K (43.04%)
patricia	58K	28K (48.28%)	26K (44.83%)
pgp.dec	228K	30K (13.16%)	30K (13.16%)
pgp.enc	228K	28K (12.28%)	26K (11.40%)
typeset	544K	80K (14.71%)	86K (15.81%)

Table 2: Final code cache size

5. RELATED WORK

CC management has been extensively studied in general-purpose DBT systems. [5] proposed using FLUSH on overflows or preemptively when a phase change is detected. [10] found that FIFO's miss rate is similar to LRU's with less management overhead and 50% better than the miss rate of FLUSH. [11] found that mid-grained evictions (Segmented FIFO) scale better than FLUSH and FIFO, and proposed a generational approach that stores short-lived and long-lived fragments in disctinct CCs to prevent premature evictions.

[7] explored expanding a bounded CC. [7] doubles the CC size when the ratio of retranslated to replaced fragments is above a threshold. To avoid exponential growth, our resizing approach expands the HCC only by a constant amount.

[13] developed software instruction caching for processors with SPM and no I-cache. [13] uses a static binary rewriter to form cache blocks. [4] showed that DBT can also treat the SPM as a software I-cache. [4] used victim compression and pinning to reduce retranslation cost when the binary is in external Flash. The CC in [13, 4] is only in SPM.

6. CONCLUSION

This paper presents the Heterogeneous Code Cache (HCC), which is a CC split among SPM and main memory. Several HCC design issues and management policies are evaluated. Our results show that an HCC outperforms a CC that uses only SPM or only main memory. The best performance is achieved when treating SPM and main memory distinctly.

Our techniques make DBT more feasible in embedded systems. HCC can be combined with compelling DBT uses, such as compressing infrequently used code and decompressing it on-demand [8]. Future HCC enhancements include promoting frequently used code from main memory to SPM.

7. ACKNOWLEDGEMENTS

This work was supported by NSF under grants CCF-0811295, CCF-0811352, CNS-0702236, CNS-0720483 and CNS-0551492.

8. REFERENCES

[1] K. Adams and O. Agesen. A comparison of software and hardware techniques for x86 virtualization. In *International Conference on Architectural Support for Programming Languages and Operating Systems*, 2006.

[2] T. Austin, E. Larson, and D. Ernst. Simplescalar: An infrastructure for computer system modeling. *Computer*, 35(2):59–67, 2002.

[3] J. A. Baiocchi, B. R. Childers, J. W. Davidson, and J. D. Hiser. Reducing pressure in bounded DBT code caches. In *International Conference on Compilers, Architecture, and Synthesis for Embedded Systems*, 2008.

[4] J. A. Baiocchi, B. R. Childers, J. W. Davidson, J. D. Hiser, and J. Misurda. Fragment cache management for dynamic binary translators in embedded systems with scratchpad. In *International Conference on Compilers, Architecture, and Synthesis for Embedded Systems*, 2007.

[5] V. Bala, E. Duesterwald, and S. Banerjia. Dynamo: a transparent dynamic optimization system. In *Conference on Programming Language Design and Implementation*, 2000.

[6] R. Banakar, S. Steinke, B.-S. Lee, M. Balakrishnan, and P. Marwedel. Scratchpad memory: design alternative for cache on-chip memory in embedded systems. In *International Conference on Hardware/Software Codesign*, 2002.

[7] D. Bruening and S. Amarasinghe. Maintaining consistency and bounding capacity of software code caches. In *International Symposium on Code Generation and Optimization*, 2005.

[8] S. Debray and W. Evans. Profile-guided code compression. In *Conference on Programming Language Design and Implementation*, 2002.

[9] M. R. Guthaus, J. S. Ringenberg, D. Ernst, T. M. Austin, T. Mudge, and R. B. Brown. Mibench: A free, commercially representative embedded benchmark suite. In *IEEE Workshop on Workload Characterization*, 2001.

[10] K. Hazelwood and M. D. Smith. Code cache management schemes for dynamic optimizers. In *Workshop on Interaction between Compilers and Computer Architectures*, 2002.

[11] K. Hazelwood and M. D. Smith. Managing bounded code caches in dynamic binary optimization systems. *ACM Trans. on Architecture and Code Optimization*, 3(3):263–294, 2006.

[12] J. D. Hiser, D. Williams, W. Hu, J. W. Davidson, J. Mars, and B. R. Childers. Evaluating indirect branch handling mechanisms in software dynamic translation systems. In *International Symposium on Code Generation and Optimization*, 2007.

[13] J. E. Miller and A. Agarwal. Software-based instruction caching for embedded processors. In *International Conference on Architectural Support for Programming Languages and Operating Systems*, 2006.

[14] K. Scott, N. Kumar, S. Velusamy, B. Childers, J. W. Davidson, and M. L. Soffa. Retargetable and reconfigurable software dynamic translation. In *International Symposium on Code Generation and Optimization*, 2003.

From Milliwatts to Megawatts :
System Level Power Challenge

Ruchir Puri (Organizer)
IBM T J Watson Research
Center, Yorktown Hts, NY USA

Eshel Haritan (Coorganizer)
Coware Inc.
San Jose, CA, USA

Stan Krolikoski (Chair)
Cadence Design Systems
San Jose, CA, USA

Jason Cong
University of California
Los Angeles, CA, USA

Tim Kogel
Coware Inc
Aachen, Germany

Bradley McCredie
IBM Corporation
Austin, TX, USA

John Shen
Nokia Research Center
Palo Alto, CA, USA

Andres Takach
Mentor Graphics Corporation
Wilsonville, OR, USA

PANEL SUMMARY

This panel discusses power optimization at the system level. What are the needs and opportunities? What are examples of successful practices? Can system-level power optimizations ever be automated? Can power optimizations be the enabling event for a methodology change in system-level design?

Categories and Subject Descriptors

B.7.1 [Integrated Circuits]: Advanced technologies, VLSI; B.7.2 [Design Aids];

General Terms

Performance, Design.

Keywords

System Level Power, System Design

PANELIST VIEWPOINTS

Jason Cong: In order to meet ever-increasing computing needs and overcome power density limitations, the computing community has halted simple processor frequency scaling and entered the *era of parallelization*, with tens to hundreds of computing cores integrated in a single processor, and hundreds to thousands of computing servers in a warehouse-scale data center. Such highly parallel general-purpose computing systems, however, still face serious challenges in terms of performance, power, heat dissipation, space, and cost. We believe the next significant opportunity is to look beyond parallelization and focus on *customization* which may bring orders-of-magnitude power-performance efficiency improvement for applications in a specific domain. We make the observation that each user or

enterprise typically has a high computing demand in one or a few application domains, while its other computing needs can be easily satisfied by COTS technologies. Therefore, it is possible to develop a customizable computing platform where computing engines and interconnects can be specialized to a particular application domain to gain significant improvement in power-performance efficiency as compared to a general-purpose architecture. A simple analogy can be made to the history of civilization, where more advanced societies have a higher degree of specialization, resulting in a much improved efficiency.

Tim Kogel: Obviously the most significant power savings can be achieved at the platform architecture level when major architecture decisions are done. Further down in the implementation flow at the transistor, gate and register-transfer level only few percent can be gained, which will not save the day. Automatic platform architecture level power optimization is still in early research, and even quantitative power analysis at that level has not been solved. Initial successes on the latter have been reported, but it still requires significant expertise to come to a reliable power characterization of the SoC components.

We see that some practical and proven Electronic Systems Virtualization methodologies are successfully deployed to drastically reduce the overall power budget:

- Reduce block-level power dissipation by optimizing the computational efficiency of all processing elements. We believe that programmable accelerators with customized instruction set are the silver bullet to optimize best flexibility/performance tradeoff

- Fine-tune the SoC interconnect and memory architecture to the specific requirements by means of an platform architecture performance model. This allows measuring the activity of the different SoC components as a metric for power.

- Use aggressive software-driven power management to shut- or scale-down clock regions wherever possible.

Permission to make digital or hard copies of part or all of this work for personal or classroom use is granted without fee provided that copies are not made or distributed for profit or commercial advantage and that copies bear this notice and the full citation on the first page. To copy otherwise, to republish, to post on servers or to redistribute to lists, requires prior specific permission and/or a fee.
DAC'09, July 26-31, 2009, San Francisco, California, USA

Again a virtual hardware platform is required to validate that that the performance requirements of the relevant application use-cases are still satisfied.

Bradley McCredie: System Power Management has become a key driver of system design in the past years. This is due to a clear mandate from systems customers and the need to be good stewards of our environment. The drive to "manage" power has touched every aspect of design from technology, to chip micro-architecture, to firmware and middle ware. It has forced changes into systems design and the leaders that will emerge in this race will be those that tie all of these disciplines together in a way that is easy for customers to use and exploit.

John Shen: "System Level Power Challenge" must be viewed at the global level. From the mobility perspective, this involves over 3B cell phones (0.2 GW), 3M cellular basestations (4.5 GW), and 10K cellular core network elements (0.1 GW), contributing over 20 MT of CO2/year. Mobility continues to grow at a rapid pace and is quickly converging with the Internet. I will present the system level power challenges from the mobility perspective and argue that the mobility domain will be the technology and innovation driver for achieving unprecedented power optimizations that will be necessary for the next decade.

Andres Takach : Optimizing a design for power requires a comprehensive approach that provides designers with tools to 1) explore and evaluate the effects on power of architectural

choices and power management strategies and 2) synthesize power-efficient implementations from abstract specifications.

At the system level, a requirement for evaluating both performance and power is the availability of fast abstract models for blocks or subsystems in the design. The TLM2.0 standard provides a common framework to build those models on, but the manual creation of models is costly and tools are needed to automate their generation. Hardware blocks are either manually created or generated using HLS tools or IP generators. Manually created RTL blocks need to have their power characterized by analyzing their protocols and major modes of operations to construct the abstract power models. For blocks generated by HLS design tools or IP generators, the expectation is that the tools will create the required TLM models with both timing and power annotations.

At the block and subsystem level, HLS tools now provide designers with the ability to rapidly explore and evaluate architectural decisions. This capability is very important as it is difficult to predict the best low-power hardware architecture. HLS design flows also present many opportunities to further improve both static and dynamic power by providing designers the ability to consider various Vth, Vdd, and power gating techniques in a more automated way. As the scope of HLS expands to larger subsystems, designers will expect more support in evaluating, synthesizing and verifying power management approaches.

978-1-60558-497-3/09 $25.00 © 2009 ACM

A Direct Integral-Equation Solver of Linear Complexity for Large-Scale 3D Capacitance and Impedance Extraction

Wenwen Chai, Dan Jiao, and Cheng-Kok Koh

School of Electrical and Computer Engineering, Purdue University, 465 Northwestern Avenue, West Lafayette, IN 47907, USA (phone: 765-494-5240; fax: 765-494-3371; e-mails: {wchai, djiao, chengkok}@purdue.edu)

ABSTRACT

State-of-the-art integral-equation-based solvers rely on techniques that can perform a matrix-vector multiplication in $O(N)$ complexity. In this work, a fast inverse of linear complexity was developed to solve a dense system of linear equations directly for the capacitance extraction of any arbitrary shaped 3D structure. The proposed direct solver has demonstrated clear advantages over state-of-the-art solvers such as FastCap and HiCap; with fast CPU time and modest memory consumption, and without sacrificing accuracy. It successfully inverts a dense matrix that involves more than one million unknowns associated with a large-scale on-chip 3D interconnect embedded in inhomogeneous materials. Moreover, we have successfully applied the proposed solver to full-wave extraction.

Categories and Subject Descriptors

B.7.2 **[Integrating Circuits]:** Design Aids - simulation, verification

General Terms

Algorithms

Keywords

Integral-equation-based methods, direct solver, capacitance extraction, full wave.

1. INTRODUCTION

Integral-equation-based (IE-based) methods have been methods of choice in extracting the capacitive parameters of 3D interconnects since they reduce the solution domain by one dimension, and they model an infinite domain without the need of introducing an absorbing boundary condition. Compared to their partial-differential-equation-based counterparts, however, IE-based methods generally lead to dense systems of linear equations. Using a naïve direct method to solve a dense system takes $O(N^3)$ operations and requires $O(N^2)$ space, with N being the matrix size. When an iterative solver is used, the memory requirement remains the same, and the time complexity is $O(N_{it}N^2)$, where N_{it} denotes the total number of iterations required to reach convergence. In state-of-the-art IE-based capacitance solvers, Fast Multipole Method (FMM) and hierarchical algorithms [1-3] were used to perform a matrix-vector multiplication in $O(N)$ complexity,

—This work was supported by NSF under award No. 0747578 and No. 0702567.

Permission to make digital or hard copies of part or all of this work for personal or classroom use is granted without fee provided that copies are not made or distributed for profit or commercial advantage and that copies bear this notice and the full citation on the first page. To copy otherwise, to republish, to post on servers or to redistribute to lists, requires prior specific permission and/or a fee.
DAC'09, July 26-31, 2009, San Francisco, California, USA

thereby significantly reducing the complexity of iterative solvers. In the limited work reported on the direct IE solutions for capacitance extraction [4], no linear complexity has been achieved. Compared to iterative solvers, direct solvers have advantages when the number of iterations is large or the number of right hand sides is large. For example, if there exist N right hand sides, each solve of which costs $O(N)$ operations, the total cost is still $O(N^2)$, which is expensive.

The contribution of this paper is the development of a linear-complexity direct IE solver that is kernel independent, and hence suitable for solving both quasi-static and full-wave problems. To be specific, the inverse of a dense system matrix arising from a quasi-static or full-wave problem is obtained in linear CPU time and memory consumption without sacrificing accuracy. Our solution hinges on the observation that the matrices resulting from an IE-based method, although dense, can be thought of as *data-sparse*, i.e., they can be specified by few parameters. There exists a general mathematical framework, called the "Hierarchical ($\mathcal{H}$) Matrix" framework [5], which enables a highly compact representation and efficient numerical computation of dense matrices. Both Storage requirements and matrix-vector multiplications using $\mathcal{H}$-matrices are of complexity $O(N\log^{\alpha}N)$. $\mathcal{H}^2$-matrices, which are a specialized subclass of hierarchical matrices, were later introduced in [6]. It was shown that the storage requirements and matrix-vector products are of complexity $O(N)$ for $\mathcal{H}^2$-based representation of both quasi-static and electrodynamic problems [7-8]. The nested structure is the key difference between $\mathcal{H}$-matrices and $\mathcal{H}^2$-matrices, since it permits an efficient reuse of information across the entire hierarchy. Solvers based on $\mathcal{H}$- and $\mathcal{H}^2$-matrices are kernel independent, and are therefore suitable for any IE-based formulation.

Although the matrix-vector product involving an $\mathcal{H}^2$-matrix can be performed in $O(N)$ complexity, the complexity of $\mathcal{H}^2$-matrix-based inverse has not been clearly established in the literature. In this work, we developed a direct IE solver of linear complexity for solving large-scale quasi-static and electrodynamic problems. The remainder of this paper is organized as follows. In Section II, IE formulations for capacitance extraction in both uniform and non-uniform materials are presented. An $\mathcal{H}^2$-matrix-based representation of the dense system matrix is constructed, and its error bound derived. We show that exponential convergence with respect to the number of interpolation points can be achieved irrespective of the problem size. In Section III, we provide the details of the linear-complexity direct inverse, which includes the orthogonalization of cluster bases, a recursive inverse formula, and fast matrix-matrix multiplication in linear complexity. In

Section IV, numerical results are given to demonstrate the accuracy and efficiency of the proposed IE solver for both capacitance and full-wave extraction. We conclude in Section V.

2. IE FORMULATION WITH $\mathcal{H}^2$ MATRIX

2.1 IE Formulation

An integral-equation-based analysis of a multi-conductor structure embedded in inhomogeneous materials results in the following linear system [3]

$$\underbrace{\begin{bmatrix} \mathbf{P}_{cc} & \mathbf{P}_{cd} \\ \mathbf{E}_{dc} & \mathbf{E}_{dd} \end{bmatrix}}_{G} \underbrace{\begin{bmatrix} q_c \\ q_d \end{bmatrix}}_{q} = \underbrace{\begin{bmatrix} v_c \\ 0 \end{bmatrix}}_{v} \qquad (1)$$

where q_c and q_d are the charge vectors of the conductor panels and dielectric-dielectric interface panels, respectively, and v_c is the potential vector associated with the conductor panels. The entries of $\mathbf{P}$ and $\mathbf{E}$ are

$$\mathbf{P}_{ij} = \frac{1}{a_i}\frac{1}{a_j} \int_{S_i} \int_{S_j} g(\vec{r},\vec{r}')dr_i dr_j$$

$$\mathbf{E}_{ij} = (\varepsilon_a - \varepsilon_b)\frac{\partial}{\partial n_a}\frac{1}{a_i}\frac{1}{a_j} \int_{S_i} \int_{S_j} g(\vec{r},\vec{r}')dr_i dr_j \qquad (2)$$

where a_i and a_j are the areas of panel S_i and S_j, respectively, $\hat{n}$ is a unit vector normal to the dielectric interface, and ε_a and ε_b are the permittivity in the two dielectric regions separated by the interface. The diagonal entries of $\mathbf{E}_{dd}$ are $e_{ij} = (\varepsilon_a + \varepsilon_b)/(2a_i\varepsilon_0)$. In a uniform dielectric, (1) is reduced to

$$\mathbf{P}_{cc}q_c = v_c . \qquad (3)$$

2.2 Cluster Tree and Block Cluster Tree

In order to capture the nested hierarchical dependence present in $\mathbf{G}$ shown in (1), we explore the use of a cluster tree and a block cluster tree. Denoting the full index set of all the panels by $\mathcal{I} := \{1, 2, ..., N\}$. A representative cluster tree $T_\mathcal{I}$ is shown in Fig. 1(a). Clusters with indices no more than *leafsize* are leaves. The set of leaves of $T_\mathcal{I}$ is denoted by $\mathcal{L}_\mathcal{I}$.

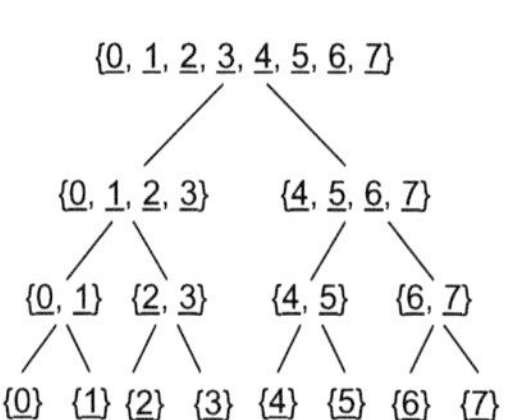
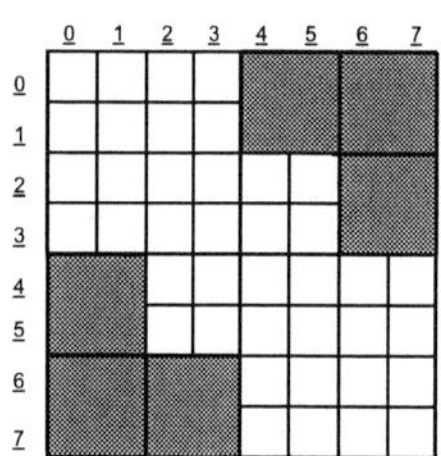

Fig. 1 (a) A cluster tree. (b) An $\mathcal{H}^2$-matrix structure.

Consider two subsets t and s of $\mathcal{I}$. We define a strong admissibility condition as follows [5]:

$$(t,s)\text{ are admissible} := \begin{cases} \text{True} & \text{if } \max\{diam(\Omega_t),diam(\Omega_s)\} \leq \\ & \eta\, dist(\Omega_t,\Omega_s) \\ \text{False} & \text{otherwise} \end{cases} \qquad (4)$$

in which Ω_t and Ω_s are the supports of the union of all the panels in t and s respectively, and η is a parameter that controls the solution accuracy. Constructing an admissible block cluster tree from the cluster trees $T_\mathcal{I}$ and $T_\mathcal{I}$ itself (the testing and basis functions are the same in Galerkin-based IE solvers) and a given admissibility condition can be done recursively [5], in which the constructing procedure results in an admissible block cluster tree which can be mapped to a matrix structure shown in Fig. 1(b). Each leaf block cluster corresponds to a matrix block. The shaded matrix blocks are admissible blocks in which the $\mathcal{H}^2$-matrix representation is used; the un-shaded ones are inadmissible blocks in which a full matrix representation is employed.

2.3 $\mathcal{H}^2$-Matrix Representation and Its Error Bound

If two subsets t and s of $\mathcal{I}$ satisfy the strong admissibility condition (4), the original kernel function $g(\vec{r}_i,\vec{r}_j)$ in (2) can be replaced by a degenerate approximation

$$\tilde{g}^{t,s}(\vec{r},\vec{r}') = \sum_{v\in K^t}\sum_{\mu\in K^s} g(\xi_v^t,\xi_\mu^s)L_v^t(\vec{r})L_\mu^s(\vec{r}') \qquad (5)$$

where $K := \{v\in\mathbb{N}^d : v_i \leq p \ \forall i\in\{1,...,d\}\} = \{1,...,p\}^d$, $d = 1, 2, 3$, for 1-, 2-, and 3-D problems, respectively; p is the number of interpolation points in one dimension; $(\xi_v^t)_{v\in K^t}$ and $(\xi_\mu^s)_{\mu\in K^s}$ are two families of interpolation points, respectively, in t and s; and $(L_v^t)_{v\in K^t}$ and $(L_\mu^s)_{v\in K^s}$ are the corresponding Lagrange polynomials. With (5), (2) are separated into two single integrals:

$$\tilde{\mathbf{P}}_{ij}^{t,s} := \sum_{v\in K^t}\sum_{\mu\in K^s} 1/(a_i a_j) \cdot g(\xi_v^t,\xi_\mu^s)\int_{S_i} L_v^t(\vec{r})dr \cdot \int_{S_j} L_\mu^s(\vec{r}')dr'$$

$$\tilde{\mathbf{E}}_{ij}^{t,s} := \sum_{v\in K^t}\sum_{\mu\in K^s} 1/(a_i a_j)\cdot(\varepsilon_a - \varepsilon_b)\frac{\partial}{\partial n_a} g(\xi_v^t,\xi_\mu^s)\int_{S_i} L_v^t(\vec{r})dr \cdot \int_{S_j} L_\mu^s(\vec{r}')dr' \qquad (6)$$

Hence, the submatrix $\tilde{\mathbf{G}}^{t,s}$ can be written in a factorized form as:

$$\tilde{\mathbf{G}}^{t,s} := \mathbf{V}^t\mathbf{S}^{t,s}\mathbf{V}^{s^T}, \ \mathbf{V}^t\in\mathbb{R}^{t\times K^t}, \mathbf{S}^{t,s}\in\mathbb{R}^{K^t\times K^s}, \ \mathbf{V}^s\in\mathbb{R}^{s\times K^s} \qquad (7)$$

$$where, \quad \mathbf{V}_{iv}^t = \int_{S_i} L_v^t(\vec{r})dr, \qquad \mathbf{V}_{j\mu}^s = \int_{S_j} L_\mu^s(\vec{r}')dr'$$

$$\mathbf{S}_{v\mu}^{t,s} = \begin{cases} g(\xi_v^t,\xi_\mu^s)/(a_i a_j) & (t\text{ contains conductor panels}) \\ (\varepsilon_a - \varepsilon_b)/(a_i a_j)\dfrac{\partial g(\xi_v^t,\xi_\mu^s)}{\partial n_a} & (t\text{ contains dielectric panels}) \end{cases}$$

for $i\in t$, $j\in s$, $v\in K^t$, and $\mu\in K^s$. The matrix $\mathbf{G}$ in (7) forms an $\mathcal{H}^2$-matrix representation if the same space of polynomials are used across t and s. After a detailed error analysis, we found that if the admissibility condition given in (4) is satisfied, the error of (5) is bounded by

$$\| g(r,r') - \tilde{g}^{(t,s)}(r,r') \|_\infty \leq \frac{4ed}{\pi}(\Lambda_p)^{2d} p \frac{1}{dist(Q_t,Q_s)}[1+\sqrt{2}\eta][1+\frac{\sqrt{2}}{\eta}]^{-p} \qquad (8)$$

where Λ_p is a constant related to p and the interpolation scheme, and $dist(Q_t,Q_s)$ is the Euclidean distance between cluster t and cluster s. Clearly, exponential convergence with respect to p can be obtained irrespective of the choice of η. The larger p is, the smaller the error is. In addition, the block entries represented by (7) can be kept to the same order of accuracy across tree levels.

978-1-60558-497-3/09 $25.00 © 2009 ACM

3. Direct Inverse of Linear Complexity

The algorithms in the proposed direct solver are outlined below:

> Direct IE solver of linear complexity
>   1. Orthogonalize the cluster basis $\mathbf{V}^t$
>   2. Compute the inverse of $\mathcal{H}^2$–based $\mathbf{G}$
>
>   • Recursive inversion
>   • Linear-time matrix multiplication
>   3. Compute the capacitance matrix by $q=\mathbf{G}^{-1}v$

Before we provide the details, we introduce the following concepts and notation: 1) For each cluster $t \in T_\mathcal{I}$, the cardinality of the sets $col(t) := \{s \in T_\mathcal{J} : (t,s) \in T_{\mathcal{I} \times \mathcal{J}}\}$ and $row(s) := \{t \in T_\mathcal{I} : (t,s) \in T_{\mathcal{I} \times \mathcal{J}}\}$ is bounded by a constant C_{sp} [5]. 2) Each non-leaf cluster t has two child nodes. 3) The rank of $\mathbf{V} = (\mathbf{V}^t)_{t \in T_\mathcal{I}}$ is denoted by k. 4) The parameter *leafsize* is denoted by $n_{\min}$, and $\#t \le n_{\min}$ if $t \in \mathcal{L}_\mathcal{I}$.

3.1 Orthogonalization of the Nested Cluster Basis

Recall that when constructing the $\mathcal{H}^2$-matrix representation of system matrix $\mathbf{G}$, we use the same space of polynomials for all clusters. Consider a cluster t', which is a child of t, $L^t_v(\vec{r})$ in (5) can be written as $L^t_v(\vec{r}) = \sum_{v' \in K^{t'}} \mathbf{T}^{t'}_{v'v} L^{t'}_{v'}(\vec{r})$, with $\mathbf{T}^{t'}_{v'v} = L^t_v(\xi^{t'}_{v'})$. As a result,

$$\mathbf{V}^t_{iv} = \int_{S_i} L^t_v(\vec{r})dr = \sum_{v' \in K^{t'}} \mathbf{T}^{t'}_{v'v} \int_{S_i} L^{t'}_{v'}(\vec{r})dr = \sum_{v' \in K^{t'}} \mathbf{T}^{t'}_{v'v} \mathbf{V}^{t'}_{iv'} = (\mathbf{V}^{t'}\mathbf{T}^{t'})_{iv},$$

where $\mathbf{T}^{t'} \in \mathbb{R}^{K^{t'} \times K^t}$ is called transfer matrix for cluster t'. Hence, assuming that $children(t) = \{t_1, t_2\}$ with $t_1 \ne t_2$, we have

$$\mathbf{V}^t = \begin{pmatrix} \mathbf{V}^{t_1}\mathbf{T}^{t_1} \\ \mathbf{V}^{t_2}\mathbf{T}^{t_2} \end{pmatrix} = \begin{pmatrix} \mathbf{V}^{t_1} & \\ & \mathbf{V}^{t_2} \end{pmatrix}\begin{pmatrix} \mathbf{T}^{t_1} \\ \mathbf{T}^{t_2} \end{pmatrix} \tag{9}$$

This means that we only need to store the matrices $\mathbf{V}^t$ for leaf clusters t and use the transfer matrices $\mathbf{T}$ to represent all other clusters. This nested property of $\mathbf{V}^t$ as shown in (9) enables $O(N)$ storage of $\mathbf{G}$ and $O(N)$ matrix vector multiplication [6-8].

To make the inverse calculation efficient, we first orthogonalize $\mathbf{V}^t$ while still preserving the nested property of $\mathbf{V}^t$.

For leaf cluster bases, we can construct an orthogonal matrix $\mathbf{Z}^t$ such that

$$\tilde{\mathbf{V}}^t = \mathbf{V}^t\mathbf{Z}^t \quad \text{and} \quad \tilde{\mathbf{V}}^{t^\mathsf{T}}\tilde{\mathbf{V}}^t = (\mathbf{V}^t\mathbf{Z}^t)^\mathsf{T}\mathbf{V}^t\mathbf{Z}^t = (\mathbf{Z}^t)^\mathsf{T}\mathbf{G}^t\mathbf{Z}^t = \mathbf{I}$$

with $\mathbf{G}^t = \mathbf{V}^{t^\mathsf{T}}\mathbf{V}^t$. To find $\mathbf{Z}^t$, we first perform Schur decomposition $\mathbf{G}^t = \mathbf{P}\mathbf{D}\mathbf{P}^\mathsf{T}$, where $\mathbf{P}$ contains the eigenvectors of $\mathbf{G}^t$ and the diagonal matrix $\mathbf{D} = diag(\lambda_1, ..., \lambda_k)$ contains the corresponding eigenvalues $\lambda_1 \ge \lambda_2 \ge ... \ge \lambda_k \ge 0$. We fix a rank $k' \in \{0, ..., k\}$ such that $\lambda_i > 0$ holds for all $i \in \{1, ..., k'\}$. Define matrix $\tilde{\mathbf{D}} \in \mathbb{R}^{k \times k'}$ by $\tilde{\mathbf{D}}_{ij} = \delta_{ij}/\sqrt{\lambda_i}$. If $\mathbf{Z}^t = \mathbf{P}\tilde{\mathbf{D}}$, we obtain

$$\tilde{\mathbf{V}}^{t^\mathsf{T}}\tilde{\mathbf{V}}^t = (\mathbf{Z}^t)^\mathsf{T}\mathbf{G}^t\mathbf{Z}^t = \tilde{\mathbf{D}}^\mathsf{T}\mathbf{P}^\mathsf{T}\mathbf{P}\mathbf{D}\mathbf{P}^\mathsf{T}\mathbf{P}\tilde{\mathbf{D}} = \tilde{\mathbf{D}}^\mathsf{T}\mathbf{D}\tilde{\mathbf{D}} = \mathbf{I}$$

Hence, $\mathbf{Z}^t = \mathbf{P}\tilde{\mathbf{D}}$ is the matrix that can orthogonalize leaf cluster bases $\mathbf{V}^t$. Based on the nested property of $\mathbf{V}^t$, the total complexity of obtaining $\mathbf{Z}^t$ using the above procedure is $O(N)$. The non-leaf clusters can be orthogonalized in a similar fashion with the nested property preserved.

3.2 Fast Inversion of Linear Complexity

(1) Recursive Inversion Equation: Casting the $\mathcal{H}^2$-matrix representation of $\mathbf{G}$ into the following form $\mathbf{G} = \begin{bmatrix} \mathbf{G}_{11} & \mathbf{G}_{12} \\ \mathbf{G}_{21} & \mathbf{G}_{22} \end{bmatrix}$. Its inverse can be recursively computed by using the equation

$$\mathbf{G}^{-1} = \begin{bmatrix} \mathbf{G}_{11}^{-1} \oplus \mathbf{G}_{11}^{-1} \otimes \mathbf{G}_{12} \otimes \mathbf{S}^{-1} \otimes \mathbf{G}_{21} \otimes \mathbf{G}_{11}^{-1} & -\mathbf{G}_{11}^{-1} \otimes \mathbf{G}_{12} \otimes \mathbf{S}^{-1} \\ -\mathbf{S}^{-1} \otimes \mathbf{G}_{21} \otimes \mathbf{G}_{11}^{-1} & \mathbf{S}^{-1} \end{bmatrix} \tag{10}$$

where $\mathbf{S} = \mathbf{G}_{22} \oplus (-\mathbf{G}_{21} \otimes \mathbf{G}_{11}^{-1} \otimes \mathbf{G}_{12})$, and $\oplus$, $\otimes$ are respectively addition and multiplication defined for the $\mathcal{H}^2$ matrix to be elaborated soon. The recursive inverse equation (10) can be realized by the pseudo-code shown below

> Recursive inverse algorithm ($\mathbf{X}$ is used for temporary storage)
> Procedure $\mathcal{H}^2$-inverse($\mathbf{G}$, $\mathbf{X}$)   ($\mathbf{G}$ is input matrix, $\mathbf{X}$ is inverse)
>   *If* matrix $\mathbf{G}$ is a non-leaf matrix block
> $\mathcal{H}^2$-inverse ($\mathbf{G}_{11}$, $\mathbf{X}_{11}$)
> $\mathbf{G}_{21} \otimes \mathbf{X}_{11} \to \mathbf{X}_{21}$, $\mathbf{X}_{11} \otimes \mathbf{G}_{12} \to \mathbf{X}_{12}$, $-\mathbf{X}_{22} \oplus (\mathbf{X}_{21} \otimes \mathbf{G}_{12}) \to \mathbf{X}_{22}$ (11)
> $\mathcal{H}^2$-inverse ($\mathbf{X}_{22}$, $(\mathbf{G}^{-1})_{22}$)
> $-(\mathbf{G}^{-1})_{22} \otimes \mathbf{X}_{21} \to (\mathbf{G}^{-1})_{21}$, $-\mathbf{X}_{12} \otimes (\mathbf{G}^{-1})_{22} \to (\mathbf{G}^{-1})_{12}$, $\mathbf{X}_{11} \oplus (-\mathbf{G}_{12} \otimes \mathbf{X}_{21}) \to (\mathbf{G}^{-1})_{11}$
>   *else*
>   Inverse ($\mathbf{G}$)   (normal full matrix inverse)

From (11), it can be seen that the computation of inverse involves a full-matrix inverse at the leaf level and a number of matrix-matrix multiplications at other levels. Hence, efficient matrix-matrix multiplication is essential to an efficient inverse in linear time, which is elaborated in next section.

(2) Fast Matrix-Matrix Multiplication in Linear Time

The fast multiplication in (11) can be done recursively. Assuming $b_1 = (t,s) \in T^G_{\mathcal{I} \times \mathcal{I}}$, $b_2 = (s,r) \in T^G_{\mathcal{I} \times \mathcal{I}}$, and the multiplication target block is $b = (t,r) \in T^G_{\mathcal{I} \times \mathcal{I}}$. Matrix blocks $\mathbf{G}^{b1}$, $\mathbf{G}^{b2}$, and $\mathbf{G}^b$ can be admissible blocks, non-admissible blocks, or non-leaf blocks. The $\mathbf{G}^{b1} \otimes \mathbf{G}^{b2} \to \mathbf{G}^b$ encountered in (11) can be divided into the following cases, each of which has a constant complexity.

	$\mathbf{G}^{b1}$	$\mathbf{G}^{b2}$	$\mathbf{G}^b$	complexity
1	admissible	admissible	admissible	$O(k_1^3)$
2	admissible	admissible	non-leaf	$O(k_1^3)$
3	admissible	non-leaf	admissible	$O(k_1^3)$
4	admissible	non-leaf	non-leaf	$O(k_1^3)$
5	admissible	non-admissible	admissible	$O(k_1^3)$
6	non-admissible (full matrix)	non-admissible (full matrix)	non-admissible (full matrix)	$O(k_1^3)$
7	non-leaf	non-leaf	admissible	$O(k_1^3)$
8	non-leaf	non-leaf	non-leaf	$O(k_1^3)$

978-1-60558-497-3/09 \$25.00 © 2009 ACM

In the Table above, $k_1 = \max(k, n_{\min})$ is a constant that is independent of N for quasi-static applications.

Next, we will use only cases 1, 2 and 3 to explain how the fast multiplication is performed and omit other cases due to space constraint.

Case 1: $\mathbf{G}^{b1} = \mathbf{V}^t \mathbf{S}_{b1} \mathbf{V}^{s^\mathrm{T}}$ and $\mathbf{G}^{b2} = \mathbf{V}^s \mathbf{S}_{b2} \mathbf{V}^{r^\mathrm{T}}$. Then,

$$\mathbf{G}^{b1} \otimes \mathbf{G}^{b2} = \mathbf{V}^t \mathbf{S}_{b1} \mathbf{V}^{s^\mathrm{T}} \otimes \mathbf{V}^s \mathbf{S}_{b2} \mathbf{V}^{r^\mathrm{T}} = \mathbf{V}^t \mathbf{S}_{b1} \mathbf{I} \mathbf{S}_{b2} \mathbf{V}^r = \mathbf{V}^t (\mathbf{S}_{b1} \mathbf{S}_{b2}) \mathbf{V}^r$$

$$\mathbf{G}_b^{new} = \mathbf{G}^b + \mathbf{G}^{b1} \otimes \mathbf{G}^{b2} = \mathbf{V}^t \mathbf{S}_b \mathbf{V}^r + \mathbf{V}^t (\mathbf{S}_{b1} \mathbf{S}_{b2}) \mathbf{V}^{r^\mathrm{T}} = \mathbf{V}^t (\mathbf{S}_b + \mathbf{S}_{b1} \mathbf{S}_{b2}) \mathbf{V}^{r^\mathrm{T}}$$

with $\mathbf{S}_b^{new} = \mathbf{S}_b + \mathbf{S}_{b1} \mathbf{S}_{b2}$. This process does not involve any approximation. Since the dimension of each of $\mathbf{S}_b, \mathbf{S}_{b1}, \mathbf{S}_{b2}$ is $k \times k$, the complexity of computing $\mathbf{S}_b^{new}$ is at most $O(k^3)$.

Case 2: $\mathbf{G}^{b1} = \mathbf{V}^t \mathbf{S}_{b1} \mathbf{V}^{s^\mathrm{T}}$ and $\mathbf{G}^{b2} = \mathbf{V}^s \mathbf{S}_{b2} \mathbf{V}^{r^\mathrm{T}}$, while $\mathbf{G}^b$ is a non-leaf block. We first compute the multiplication as in Case (1) to get an admissible block as shown in step (a) in Fig. 2. We then split the resultant admissible block into four small admissible blocks as shown in step (b). However, the sub-blocks in $\mathbf{G}^b$ are not necessarily all admissible blocks. Based on the block structure in the target matrix, we may convert an admissible block to a full matrix block as shown in step (c). We add the resultant matrix upon $\mathbf{G}^b$.

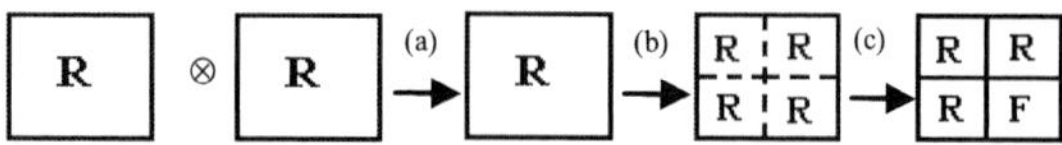

Fig.2. A scheme to compute the product of two admissible blocks and format the product to be a non-leaf. (R—an admissible block, F—an inadmissible block.

Steps (b) and (c) are called split operation and conversion operation respectively, which are performed as follows:

Split operation: A split operation is in fact a transformation from parents to children, which does not involve any approximation. For one block $b = (t,s) \in T_{\mathcal{I} \times \mathcal{I}}^G$, we perform

$$\mathbf{V}^t \mathbf{S}_b \mathbf{V}^{s^\mathrm{T}} = \begin{bmatrix} \mathbf{V}^{t1} \mathbf{T}^{t1} \\ \mathbf{V}^{t2} \mathbf{T}^{t2} \end{bmatrix} \mathbf{S}_b \begin{bmatrix} \mathbf{V}^{s1} \mathbf{T}^{s1} \\ \mathbf{V}^{s2} \mathbf{T}^{s2} \end{bmatrix}^\mathrm{T} = \begin{bmatrix} \mathbf{V}^{t1} (\mathbf{T}^{t1} \mathbf{S}_b \mathbf{T}^{s1\mathrm{T}}) \mathbf{V}^{s1\mathrm{T}} & \mathbf{V}^{t1} (\mathbf{T}^{t1} \mathbf{S}_b \mathbf{T}^{s2\mathrm{T}}) \mathbf{V}^{s2\mathrm{T}} \\ \mathbf{V}^{t2} (\mathbf{T}^{t2} \mathbf{S}_b \mathbf{T}^{s1\mathrm{T}}) \mathbf{V}^{s1\mathrm{T}} & \mathbf{V}^{t2} (\mathbf{T}^{t2} \mathbf{S}_b \mathbf{T}^{s2\mathrm{T}}) \mathbf{V}^{s2\mathrm{T}} \end{bmatrix} \quad (12)$$

where $children(t) = \{t_1, t_2\}$, and $children(s) = \{s_1, s_2\}$. It can be seen that one admissible block is divided into four admissible blocks with $\mathbf{S}_{b_{ij}} = \mathbf{T}^{ti} \mathbf{S}_b \mathbf{T}^{s_j^\mathrm{T}}$ in (12). Hence, one split operation costs $O(k^3)$.

Conversion operation: If we convert the admissible block to a full matrix block as shown in step (c), we just need to compute $\mathbf{V}^{t2} \mathbf{S}_{b22} \mathbf{V}^{r2^\mathrm{T}}$. Since the largest dimension of a full matrix block is $n_{\min} \times n_{\min}$, the cost is at most $O(n_{\min}^2 k)$. If we convert a full matrix block to an admissible block, the best approximation of $\mathbf{G}_{full}^{(t2, r2)}$ for the admissible block is $\mathbf{V}^{t2} \mathbf{V}^{t2^\mathrm{T}} (\mathbf{G}_{full}^{(t2,r2)}) \mathbf{V}^{r2} \mathbf{V}^{r2^\mathrm{T}} = \mathbf{V}^{t2} \mathbf{S}_{b22} \mathbf{V}^{r2^\mathrm{T}}$ with $\mathbf{S}_{b22} = \mathbf{V}^{t2^\mathrm{T}} (\mathbf{G}_{full}^{(t2,r2)}) \mathbf{V}^{r2}$. Hence, each conversion operation at most costs $O(n_{\min}^2 k)$. In summary, each of steps (a), (b), and (c) shown in Fig. 2 has complexity $O(k^3)$, and hence the total complexity is $O(k^3)$.

Case 3: If $\mathbf{G}^{b1}$ is an admissible block, $\mathbf{G}^{b2}$ is a non-leaf block, and $\mathbf{G}^b$ is an admissible block, we first do split operations on $\mathbf{G}^{b1}$ and get a new block $\tilde{\mathbf{G}}^{b1}$ as shown in step (a) of Fig. 3. We then use simple recursive multiplication to compute $\tilde{\mathbf{G}}^{b1} \otimes \mathbf{G}^{b2}$ and obtain a non-leaf block with four admissible sub-blocks in step (b). In step (c), a collect operation is performed to get a single admissible block, which is depicted below. After obtaining the single admissible block, we directly add it to original matrix block $\mathbf{G}^b$ and get $\mathbf{G}_b^{new}$. The only approximation in this case is from the collect operation done in step (c), the accuracy of which is controllable.

Fig. 3. A scheme to compute the product of an admissible block with a non-leaf block with the target block being an admissible block.

Collect operation: This process is a transformation from children to parents, which involves an approximation since we are not able to express the cluster bases of children in terms of the parent cluster bases. However, we can get the best approximation of the children blocks in the cluster bases corresponding to the parent using the orthogonal cluster basis.

We approximate the child matrix block b_{ij} by the parent block b. The best approximation in the cluster bases $\mathbf{V}^t$ and $\mathbf{V}^s$ is

$$\begin{aligned} & \mathbf{V}^{t} \mathbf{V}^{t^\mathrm{T}} \begin{bmatrix} \mathbf{V}^{t1} \mathbf{S}_{b11} \mathbf{V}^{s1\mathrm{T}} & \mathbf{V}^{t1} \mathbf{S}_{b12} \mathbf{V}^{s2\mathrm{T}} \\ \mathbf{V}^{t2} \mathbf{S}_{b_{21}} \mathbf{V}^{s1\mathrm{T}} & \mathbf{V}^{t2} \mathbf{S}_{b22} \mathbf{V}^{s2\mathrm{T}} \end{bmatrix} \mathbf{V}^s \mathbf{V}^{s^\mathrm{T}} \\ & = \mathbf{V}^t \left(\begin{bmatrix} \mathbf{V}^{t1} \mathbf{T}^{t1} \\ \mathbf{V}^{t2} \mathbf{T}^{t2} \end{bmatrix}^\mathrm{T} \begin{bmatrix} \mathbf{V}^{t1} \mathbf{S}_{b11} \mathbf{V}^{s1\mathrm{T}} & \mathbf{V}^{t1} \mathbf{S}_{b12} \mathbf{V}^{s2\mathrm{T}} \\ \mathbf{V}^{t2} \mathbf{S}_{b_{21}} \mathbf{V}^{s1\mathrm{T}} & \mathbf{V}^{t2} \mathbf{S}_{b22} \mathbf{V}^{s2\mathrm{T}} \end{bmatrix} \begin{bmatrix} \mathbf{V}^{s1} \mathbf{T}^{s1} \\ \mathbf{V}^{s2} \mathbf{T}^{s2} \end{bmatrix} \right) \mathbf{V}^{s^\mathrm{T}} \quad (13) \\ & = \mathbf{V}^t \left(\sum_{i \in children(t)} \sum_{j \in children(s)} \mathbf{T}^{ti^\mathrm{T}} \mathbf{S}_{b_{ij}} \mathbf{T}^{sj} \right) \mathbf{V}^{s^\mathrm{T}} \end{aligned}$$

It can be seen that four admissible blocks are collected to be one admissible block with $\mathbf{S}_b = \sum_{i \in children(t)} \sum_{j \in children(s)} \mathbf{T}^{ti^\mathrm{T}} \mathbf{S}_{b_{ij}} \mathbf{T}^{sj}$. Hence, one collect operation costs $O(k^3)$. The total complexity of steps (a)-(c) shown in Fig. 3 is $O(k_1^3)$.

3.3 Compute Capacitance Matrix

Since the inverse obtained from (11) is also an $\mathcal{H}^2$ matrix, and $\mathcal{H}^2$-matrix-vector multiplication has linear complexity [6-8], we can compute $q = \tilde{\mathbf{G}}^{-1} v$ in $O(N)$ time. By adding all the entries of q in each conductor, the capacitance matrix element can be obtained.

3.4 Complexity Analysis and Error Analysis

Complexity Analysis: The cost of orthogonalization of the cluster basis described in Section 3.1 is $O(N)$. The cost of direct inverse shown in (11) can be analyzed below

$$\begin{aligned} Comp(G^{-1}) &= \sum_{l=0}^{L} (\#blocks\ at\ level\ l) O(k_1^3) \leq \sum_{l=0}^{L} 2^l C_{sp} O(k_1^3) \\ &= C_{sp} O(k_1^3) \# T_\mathcal{I} = C_{sp} k_1^3 O(N) \end{aligned} \quad (14)$$

in which L is the number of tree levels. The inverse procedure shown in (11) essentially traverses a block cluster tree from bottom to top. At each tree level, the matrix block at that level is formed by a matrix-matrix multiplication. Since each matrix-

978-1-60558-497-3/09 $25.00 © 2009 ACM

matrix multiplication has an $O(k_1^3)$ complexity as shown in Section 3.3, and there are at most $2^l C_{sp}$ matrix blocks in level l, we obtain a linear cost for matrix inverse as shown in (14). Computing the capacitance matrix described in Section 3.3 also costs $O(N)$. Therefore, the total CPU cost of the proposed direct inverse is $O(N)$. It is worth mentioning that if the linear system is symmetric, we can compute only half of the entries in the inverse, further reducing the CPU cost.

Accuracy Analysis: In Section 3.1, orthogonal bases $\tilde{\mathbf{V}}^t$ are constructed. The best approximation of a general $\mathbf{V}^t$ in the space $\tilde{\mathbf{V}}^t$ is given by $\tilde{\mathbf{V}}^t(\tilde{\mathbf{V}}^t)^T\mathbf{V}^t$. The error of this approximation is:

$$\| \mathbf{V}^t - \tilde{\mathbf{V}}^t(\tilde{\mathbf{V}}^t)^T\mathbf{V}^t \|_2^2 = \lambda_{k^t+1} , \tag{15}$$

where λ_{k^t+1} is the $(k^t+1)th$ eigenvalue of $\mathbf{V}^{t^T}\mathbf{V}^t$, in which k^t is the rank of cluster basis $\tilde{\mathbf{V}}^t$. Clearly, if k^t is chosen to be the same as the rank of $\mathbf{V}^t$, the error of (15) is zero. Therefore $\tilde{\mathbf{V}}^t\tilde{\mathbf{V}}^{t^T}\mathbf{G}\tilde{\mathbf{V}}^s\tilde{\mathbf{V}}^{s^T}$ is the best approximation of a matrix block $\mathbf{G}^{(t,s)}$ in the bases $\tilde{\mathbf{V}}^t$ and $\tilde{\mathbf{V}}^s$. In Section 3.2, the inverse is performed by using formatted multiplication. For example, when computing $\mathbf{G}_{21} \otimes \mathbf{X}_{11} \to \mathbf{X}_{21}$ in (11), the block structure of $\mathbf{X}_{21}$ is assumed to be the same as that of $\mathbf{G}_{21}$. The goal of a formatted multiplication is to represent the $\mathcal{H}^2$ tree of $\mathbf{G}^{-1}$ by the same $\mathcal{H}^2$ tree used to represent $\mathbf{G}$. Certainly, one can assume a different tree to represent the tree of $\mathbf{G}^{-1}$, or perform unformatted multiplication, i.e. without specifying the target matrix. However, using the $\mathcal{H}^2$ tree of $\mathbf{G}$ to represent that of $\mathbf{G}^{-1}$ is an ideal choice based on physical understanding. Here, the capacitance matrix $\mathbf{G}^{-1}$ is a sparse matrix. In the $\mathcal{H}^2$ tree constructed for $\mathbf{G}$ and hence $\mathbf{G}^{-1}$, the blocks formed by clusters that satisfy admissibility condition (4) (i.e., they are far away) are represented by low rank matrices. In real $\mathbf{G}^{-1}$, these blocks can even be considered as zero. Hence, using the $\mathcal{H}^2$ tree of $\mathbf{G}$ to represent that of $\mathbf{G}^{-1}$ is indeed an ideal choice. The same argument holds true for full-wave cases. This has been verified by our numerical experiments. Therefore, the inverse performed here can be considered as an exact inverse if one neglects the round off error incurred in numerical computation.

4. Numerical Results

The first example is a $m \times m$ crossing bus structure embedded in free space or dielectric materials as shown in Fig. 4 [3]. Two

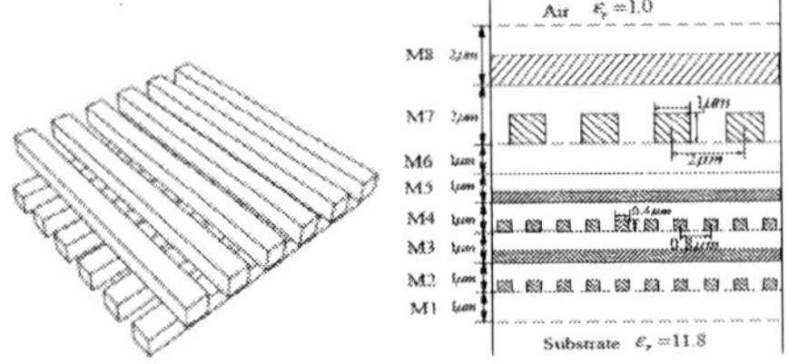

Fig. 4. (a) A bus structure. (b) An on-chip interconnect.

methods are compared: FastCap 2.0 and the proposed direct IE solver. The m in this bus structure varies from 4 to 16. The dimension of each bus is scaled to $1 \times 1 \times (2m+1)$ m^3. The distance

between buses in the same layer is 1 m, and the distance between the two bus layers is 1 m. For this bus structure, we simulated both free-space case and non-uniform dielectric case. For the case involving non-uniform dielectrics, the dielectric surrounding the upper layer conductors has relative permittivity of 3.9, and the lower layer conductors are in the dielectric having relative permittivity 7.5. Each bus is also scaled to $1 \times 1 \times (2m+1)$ m^3. The distance between buses in the same layer is 1 m, and the distance between the two bus layers is 2 m. (Note that capacitances are scalable with respect to the length unit). In the proposed solver, the parameters used to construct the cluster tree and block cluster tree are *leafsize*=10 and η =1.6. The number of interpolation points p is determined by a function $p = a + b(L-l)$, with a=2, b=1, and L being the maximum number of tree level, and l tree level. In Fig. 5(a), we plot the error of $\mathcal{H}^2$-matrix representation of system matrix $\mathbf{G}$ (so called as original matrix error), and the error of extracted capacitances with respect to the number of unknowns. The former is measured by $\| \mathbf{G} - \tilde{\mathbf{G}} \|_F / \| \mathbf{G} \|_F$, where $\tilde{\mathbf{G}}$ is shown in (7), and $\| \cdot \|_F$ is the Frobenius norm; the latter is measured by $\| C - C' \|_F / \| C \|_F$, where C is the capacitance matrix obtained from FastCap 2.0, and C' is that generated by the proposed solver. As can be seen clearly from Fig. 5(b), very good accuracy of the proposed direct solver can be observed in both $\tilde{\mathbf{G}}$ and capacitance matrix C'. In addition, the error of $\tilde{\mathbf{G}}$ reduces with the number of unknowns because of increased p and hence increased accuracy as can be seen from (8). In addition, we are able to keep the accuracy of the capacitance matrix to the same order in the entire range.

In Fig. 6, we plot the total CPU time and memory consumption of the proposed direct inverse for the $m \times m$ bus structure in free space. In Fig. 7, we plot the same for the $m \times m$ bus structure embedded in multiple dielectrics. The performance of FastCap 2.0 is also plotted for comparison, the convergence tolerance of which is set to 1%. Compared with FastCap 2.0, the proposed direct solver is 9–25 times faster and reduces memory usage by 85–95%. Dell 1950 Server was used for all simulations in this paper.

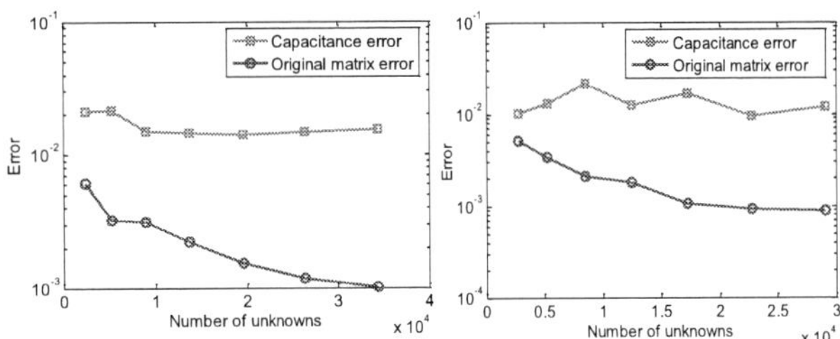

Fig. 5 Original matrix error and capacitance error with respect to N. (a) Uniform dielectric. (b) Non-uniform dielectric.

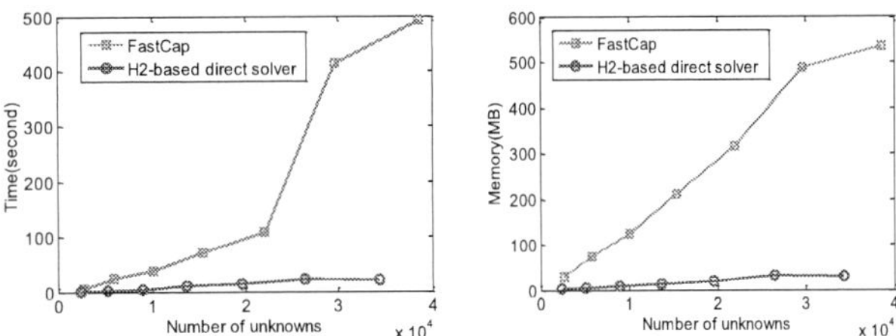

Fig. 6. Comparison of time and memory complexity in simulating the bus structure in free space.

978-1-60558-497-3/09 $25.00 © 2009 ACM

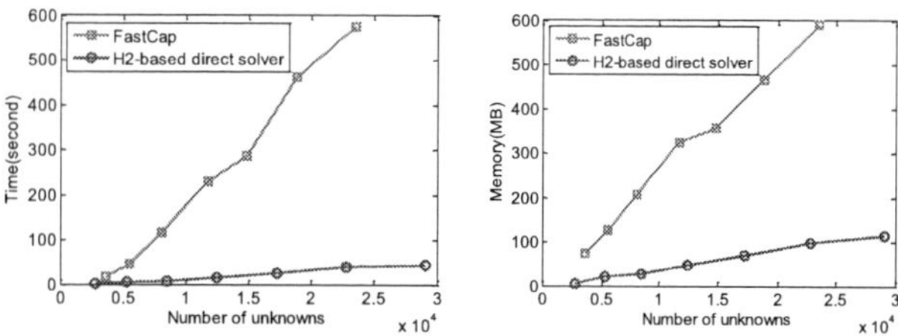

Fig. 7 Comparison of time and memory complexity in simulating the bus structure embedded in multiple dielectrics.

To test the performance of the proposed solver in simulating very large examples, we simulated a structure shown in Fig. 4(b) [3]. The relative permittivity is 3.9 in M1, 2.5 from M2 to M6, and 7.0 from M7 to M8. The discretization of this 48-conductor structure results in 25,556 unknowns. To test the large-scale modeling capability of the proposed solver, the 48-conductor structure shown in Fig. 4(b) is duplicated horizontally, resulting in 72, 96, 120, 144, 192, 240, 288, and 336 conductors, which lead to more than 1 million unknowns. The simulation parameters are chosen as *leafsize*=10, η=1, and *p*=1. The error of the $\mathcal{H}^2$-matrix representation of system matrix **G** in the maximal admissible block is shown in Fig. 8(a), with the number of

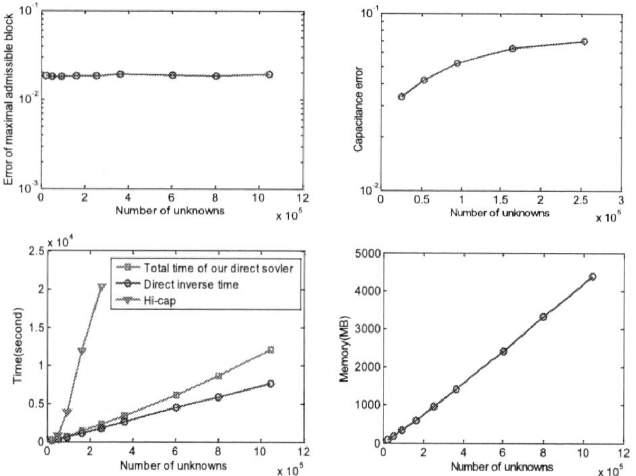

Fig. 8. (a) Error of maximal admissible block. (b) Error of Capacitance. (c) Time Complexity. (d) Memory Complexity.

unknowns varying from 25,556 to 1,047,236. Good accuracy is observed in the entire range. In Fig. 8(b), we show the capacitance error with respect to N. Again, good accuracy is observed. Since we need to use the capacitance C generated from an existing solver such as FastCap to assess the accuracy of the capacitance C' extracted by the proposed solver based on $\| C - C' \|_F / \| C \|_F$,

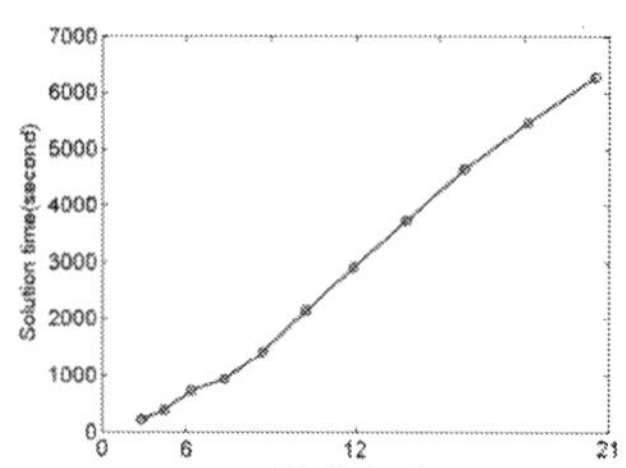

Fig. 9. Simulation of a 4λ–20λ plate.

and C is not available within feasible computational resources when the number of unknowns is too large, the error in Fig. 8(b) was only plotted up to 253792 unknowns. In Fig. 8(c), we plot the inverse time and the total CPU time of the proposed direct solver with respect to N. Clearly a linear complexity can be observed. For comparison, the solution time of the HiCap [2] is also plotted. The advantage of the proposed direct solver is clearly demonstrated even though HiCap only calculated the results for m right hand sides with m being the number of conductors, whereas the proposed solver obtained the entire inverse, i.e., the results for N right hand sides. In Fig. 8(d), we plot the memory complexity of the proposed solver, which again demonstrates a linear complexity.

The proposed $\mathcal{H}^2$-matrix-based method is kernel independent, and hence is equally applicable to electrodynamic problems. Using the proposed solver, we simulated a square plate having electric size from 4 wavelengths to 20 wavelengths. The simulation parameters were chosen as η=1, *leafsize*=20, and *p*=10. In Fig. 9, the CPU time was plotted with respect to the number of unknowns. Again, linear complexity is observed. In addition, the error $\| \mathbf{G} - \tilde{\mathbf{G}} \|_F / \| \mathbf{G} \|_F$ across all the electric sizes is smaller than 3.5e-3; and the error of the inverse matrix, which is $\| \mathbf{I} - \tilde{\mathbf{G}}^{-1}\mathbf{G} \|$, is smaller than 3.3%. For full-wave cases, the rank k used in the $\mathcal{H}^2$-based representation needs to be determined adaptively [8] in order to keep a constant order of accuracy across electric sizes without compromising the computational complexity.

5. CONCLUSIONS

A linear-complexity direct inverse was developed for fast integral-equation-based analysis of quasi-static and electro-dynamic systems. Numerical results demonstrated its superior performance.

6. REFERENCES

[1] K. Nabors and J. White, "FastCap: A multipole accelerated 3-d capacitance extraction program," *IEEE Trans. on CAD*, pp.1447–1459, 1991.

[2] W. Shi, J. Liu, N. Kakani, and T. Yu, "A fast hierarchical algorithm for 3-D capacitance extraction," *IEEE Trans. on CAD*, pp. 330–336, 2002.

[3] S. Yan, V. Saren, and W. Shi, "Sparse Transformations and Preconditioners for Hierarchical 3-D Capacitance Extraction with Multiple Dielectrics," DAC 2004, pp. 788-793.

[4] D. Gope, I. Chowdhury, and V. Jandhyala, "DiMES: Multilevel fast direct solver based on multipole expansions for parasitic extraction of massively coupled 3D microelectronic structures," DAC 2005, pp. 159-162.

[5] S. Borm, L. Grasedyck, and W. Hackbusch, "Hierarchical matrices," Lecture note 21 of the Max Planck Institute for Mathematics, 2003.

[6] S. Borm. "$\mathcal{H}^2$-matrix approximation of integral operators by interpolation," Applied Numerical Mathematics, 43: 129-143, 2002.

[7] W. Chai and D. Jiao, "An $\mathcal{H}^2$-Matrix-Based Integral-Equation Solver of Linear-Complexity for Large-Scale Full-Wave Modeling of 3D Circuits," *IEEE 17th Conference on Electrical Performance of Electronic Packaging* (EPEP), pp. 283-286, Oct. 2008.

[8] W. Chai and D. Jiao, "An $\mathcal{H}^2$-Matrix-Based Integral-Equation Solver of Reduced Complexity and Controlled Accuracy for Solving Electrodynamic Problems," accepted for publication, *IEEE Trans. Antennas Propagat.*, 2009.

Variational Capacitance Extraction of On-Chip Interconnects Based on Continuous Surface Model

Wenjian Yu[1,3], Chao Hu[2,3], Wangyang Zhang[1,3]

[1]Department of Computer Science and Technology, [2]Institute of Microelectronics, [3]Tsinghua National Laboratory for Information Science and Technology, Tsinghua University, Beijing 100084, China
E-mail: yu-wj@tsinghua.edu.cn, huc07@mails.tsinghua.edu.cn, zwy677@gmail.com

ABSTRACT

In this paper we present a continuous surface model to describe the interconnect geometric variation, which improves the currently used model for better accuracy while not increasing the number of variables. Based on it, efficient techniques are presented for chip-level capacitance extraction considering the window technique. The sparse-grid-based Hermite polynomial chaos combined with a novel weighted principle factor analysis is employed for intra-window extraction. Then, the inter-window capacitance covariance is calculated through matrix pseudo inverse. Numerical results validate the accuracy and efficiency of the proposed method, which is more than 50 times faster than the Monte-Carlo simulation with 10000 samples.

Categories and Subject Descriptors

J.6 [COMPUTER-AIDED ENGINEERING]: Computer-aided design (CAD); I.6 [Computing Methodologies]: Simulation and Modeling

General Terms

Algorithms, Theory, Design

Keywords

Geometric variation modeling, Hermite polynomial chaos method, quadratic variation model, spatial correlation, variational capacitance extraction

1. INTRODUCTION

As technology scales down to the nanometer era, the parasitic capacitance is delivering more impact on performance of integrated circuits. Delay caused by interconnect parasitics has already overtaken gate delay to become the major limiting factor of performance. At the same time, the increased mismatch between the actual fabricated interconnects and those in design brings great challenges to the computational accuracy and efficiency of parasitic capacitance extraction.

Process variations can be classified into systematic variations and random variations [1]. The systematic variations are often pattern-dependent and can be modeled with some deterministic methods [2]. In contrast, the nature of random variations is more complicated and a stochastic modeling methodology is needed for capacitance extraction. The random variation is mainly due to process uncertainty and results in non-smooth geometry surfaces

Permission to make digital or hard copies of part or all of this work for personal or classroom use is granted without fee provided that copies are not made or distributed for profit or commercial advantage and that copies bear this notice and the full citation on the first page. To copy otherwise, to republish, to post on servers or to redistribute to lists, requires prior specific permission and/or a fee.

DAC'09, July 26-31, 2009, San Francisco, California, USA

of both conductor and dielectric. Another important character of on-chip variation is the spatial correlation. Thus, a spatially correlated multivariate Gaussian distribution is often assumed for the geometry parameter variations [3-9].

A straight-forward approach for variation-aware capacitance extraction is the Monte Carlo method, which suffers from huge computing time with thousands of stochastic samplings. Although it can be accelerated with the quasi-Monte-Carlo techniques [10], the latter only generates the mean and standard deviation of capacitance, which is often not sufficient as an accurate model required by the subsequent statistical circuit analysis. A non-Monte-Carlo method was firstly proposed for variation-aware capacitance extraction [3-4], which emphasizes on the off-chip rough surface effect. The technique can only compute the mean and standard deviation of capacitance, but an accurate high-order variation model. Based on a 3-D boundary element method (BEM) capacitance solver, a perturbation method was proposed in [5] to generate a quadratic model for capacitance variation. Two efficient methods were then proposed to obtain the quadratic variational capacitance with the Hermite polynomial chaos [6, 7]. The spectral stochastic collocation method (SSCM) in [6] employs the techniques of homogenous chaos expansion and sparse grid quadrature to obtain the coefficients in the quadratic expression. The method in [7] employs the spectral stochastic method, and then solves augmented potential coefficient matrices. Considering only the first-order variation of potential coefficients, this method is very efficient, but not promising on the accuracy. Recently, a method combining the Neumann expansion and Hermite expansion was proposed for impedance extraction [8]. The computational efficiency is greatly improved for rough-surface structures with a large number of random variables.

Although the methods in [5-7] generate quadratic variation model of capacitance, which has merit for subsequent statistical circuit analysis, they all consider a simplified model of geometric variations. The variational geometry is constructed by disturbing the panels on nominal conductor surface, which produces an incomplete surface. This surface discontinuity obviously deviates from the actual situation, and is likely to induce large error for the interconnect capacitance under nanometer technology process. Another shortage of [5-7] is that they do not consider the actual chip-level capacitance extraction. Recently, techniques considering the window-based chip-level capacitance extraction were proposed in [9]. A so-called Hermite polynomial collocation (HPC) method is employed to extract the intra-window interconnects. And then, the capacitance covariances between windows and the statistical full-path capacitance are calculated. However, [9] uses an even simpler grid-based variation model than [5-7], to avoid the difficulty of random variable reduction.

In this paper, the variation-aware capacitance extraction for on-

978-1-60558-497-3/09 $25.00 © 2009 ACM

chip interconnects is considered, where the correlation length of variables are not as short as that in the rough-surface structures. We firstly present a new geometric variation model, which generates continuous surface while not increasing the number of variables. Experimental results are also presented to reveal the inaccuracy of capacitance caused by the simple geometric model in [5-7]. Based on the variation model with continuous surfaces, the techniques in [9] are extended for chip-level window-based capacitance extraction. A weighted principle factor analysis (wPFA) technique is proposed to reduce the random variables while considering their different contributions to the resulting capacitance. The pseudo inverse of matrix is utilized to derive the capacitance covariance between windows. With these techniques, both variational intra-window capacitances and full-path capacitance cross windows can be computed efficiently. Numerical results show that the proposed method is about 50 times faster than the Monte-Carlo method, while preserving high accuracy.

2. BACKGROUND

2.1 Statistical Capacitance Extraction

Random variations are usually modeled with a set of random variables ξ. For interconnect capacitance extraction, we restrict the random variables to the geometric parameters of conductor. Their variation behavior is often assumed to follow the spatially correlated multivariate Gaussian distribution:

$$\rho(\xi_i) = \frac{1}{\sqrt{2\pi}\sigma} \exp(-\frac{\xi_i^2}{2\sigma^2}) \quad , \tag{1}$$

$$\text{cov}(\xi_i, \xi_j) = \sigma^2 \exp(\frac{-|\vec{r}_i - \vec{r}_j|^2}{\eta^2}) \quad . \tag{2}$$

where $\rho(\xi_i)$ is the possibility density function (PDF) of the i'th variable ξ_i, and σ is the standard derivation (Std). The spatial correlation between two variables are reflected by the correlation function in (2), where η is called correlation length. $\vec{r}_i$ and $\vec{r}_j$ are the spatial positions associated with ξ_i and ξ_j, respectively. There are other forms of correlation function, which may be extracted through sophisticated techniques using statistical data from actual chips [1]. The larger the correlation length, the stronger the correlation between variables spatially close to each other is.

The boundary element method (BEM) is one of the main 3-D capacitance extraction approaches for deterministic interconnect structures. The surfaces of conductor are firstly dissected into 2-D panels, for each of which a uniform charge distribution is assumed. With the potential at a panel center expressed as the sum of contributions from all panel charges, a linear equation system is formed [5-7]:

$$Pq = v \quad . \tag{3}$$

Here P is called potential coefficient matrix; v consists of the voltages imposed on the conductors; q is the unknown vector of panel charges. The capacitances are computed by summing up the panel charges on each conductor.

The main object of accurate statistical capacitance extraction is to obtain a high-order capacitance representation with respect to a set of random variables. A second order (quadratic) representation is sufficient in most applications [8]. For this purpose, the methods in [5, 7] firstly expand the matrix entry of P with a second-order form of the variables ξ, like

$$\tilde{P}_{ij} = P_{ij} + \tilde{a}_{ij}^T \xi + \xi^T \tilde{A}_{ij} \xi \quad . \tag{4}$$

Then, after matrix conversion, the statistical charge vector $\tilde{q}$ and further capacitance, can be obtained. The perturbation method includes invoking d^2 times of deterministic capacitance extraction for the quadratic model, where d is the number of independent variables. This method may have large error [6][1], partially because high-order items are truncated at the both stages of forming $\tilde{P}$ and solving for $\tilde{q}$. The Hermite polynomial chaos method used in [6, 9] has the same order of complexity but higher accuracy, and is introduced in the following subsection.

2.2 Hermite Polynomial Chaos Techniques

Essentially, the HPC method in [9] is the same as the SSCM of [6]. Ref. [9] gives a concise presentation of it and points out a distinct advantage that the method can suit to any deterministic capacitance solver, even formula-based capacitance solver. Below the main techniques in [6] and [9] are briefly introduced.

The statistical capacitance $\tilde{C}$ can be expressed with the Hermite polynomial expansion:

$$\tilde{C}(\zeta) = a_0 \Psi^0 + \sum_{i_1=1}^{d} a_{i_1} \Psi^1(\zeta_{i_1}) + \sum_{i_1=1}^{d} \sum_{i_2=1}^{i_1} a_{i_1 i_2} \Psi^2(\zeta_{i_1}, \zeta_{i_2}) + \cdots \tag{5}$$

where $\zeta = (\zeta_1, \ldots, \zeta_d)$ is a set of independent Gaussian random variables, and Ψ^i are the Hermite polynomials of the i'th order. Specifically, the Hermite polynomials below order 3 are:

$$\Psi^0 = 1, \quad \Psi^1(\zeta_i) = \zeta_i, \quad \Psi^2(\zeta_i, \zeta_j) = \zeta_i \zeta_j - \delta_{ij} \quad , \tag{6}$$

where δ_{ij} is the Kronecker delta function. The Hermite polynomial expansion is guaranteed to converge for any Gaussian random process with finite second-order moments [11]. Moreover, the Askey principle [12] shows that the expansion has the optimal convergence rate.

The Hermite polynomials can further be labeled consistently to simplify (5):

$$\tilde{C}(\zeta) = \sum_{j=1}^{\infty} \tilde{c}_j \Psi_j(\zeta) \quad , \tag{7}$$

where Ψ_j is the j'th Hermite polynomials in ascending order. The Hermite polynomials satisfy the following orthogonal property:

$$< \Psi_j(\zeta), \Psi_k(\zeta) >_\rho = \alpha_j \delta_{jk}, \quad \alpha_j > 0 \tag{8}$$

where α_j are constant values, and the subscript ρ means that the variables obey the Gaussian distribution. The inner product in (8) is defined as the mathematical expectation of the product of the two stochastic functions:

$$< X, Y >_\rho = E(XY) \quad . \tag{9}$$

For quadratic capacitance model, the terms with order larger than 2 in (7) are truncated. This approximation of capacitance is denoted by $C(\zeta)$:

$$C(\zeta) = \sum_{j=1}^{M} c_j \Psi_j(\zeta) \quad , \tag{10}$$

where M is the number of Hermite polynomials, $M = d^2/2 + 3d/2 + 1$. The coefficients can be evaluated with:

$$c_j = \frac{1}{\alpha_j} < C(\zeta), \Psi_j(\zeta) >_\rho \quad , j = 1, \ldots, M \quad . \tag{11}$$

---

[1] Experiment results in [5] even show a probability density function of capacitance with both positive and negative values.

According to (9), we need to compute a d-dimensional integral:

$$< C(\zeta), \Psi_j(\zeta) >_\rho = \int C(\zeta)\Psi_j(\zeta)\rho(\zeta)d\zeta \quad . \tag{12}$$

The Gaussian-Hermite quadrature technique can be used, which has a form of weighted summation:

$$< C(\zeta), \Psi_j(\zeta) >_\rho = \sum_{i=1}^{N} w_i C(\zeta^i)\Psi_j(\zeta^i) \quad , \tag{13}$$

where ζ^i is the ith Hermite-Gaussian integral point. Therefore, the variational capacitance model can be obtained after solving N deterministic structures defined by the geometric parameters ζ^i.

To improve the computational efficiency, the number of integral points N in (13) can be further reduced with the sparse grid quadrature [13]. It is proved in [9], that level-2 sparse grid is required for the quadratic capacitance model. This makes N about $2d^2$, much less than 3^d for the case using the Gaussian-Hermite quadrature. The sparse grid based method can be accelerated also by combining coincident integral points with different weights, or adopting the technique of minimum spanning tree [6] when extracting deterministic capacitances with iterative linear equation solver.

3. CONTINUOUS SURFACE MODEL FOR GEOMETIRC VARIATIONS

3.1 Model for One-Direction Variation

The geometric variation model in [5-7] assumes each quadrilateral panel on nominal conductor surface fluctuates along the normal direction of surface separately, but keeps its shape unchanged. Thus, discontinuous conductor surfaces are generated, as that shown in Fig. 1(a). Under this model, the total area of conductor surfaces is the same as that of the nominal conductor, and the fragmented surface obviously deviates from the actual situation.

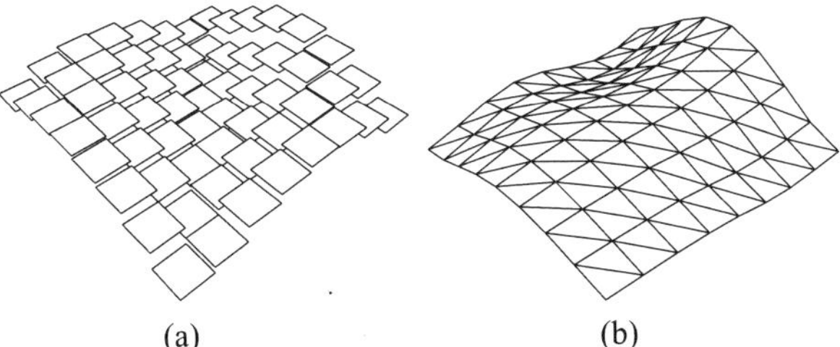

(a)   (b)

Figure 1. A variational plane generated with: (a) the model used in [5-7], (b) the continuous surface model.

A straightforward remedy to the simplified model can be made by characterizing the fluctuations of vertices of nominal panels separately. Since four vertices of a panel may not lie on a same plane, the triangular discretization is needed for the variational surface. Thus, the modified geometric variation model generates a continuous surface (as shown in Fig. 1(b)), where the random variables are imposed on the panel vertices. This variational surface model converges to a smooth shape as the triangular mesh becomes denser. The grid size of triangular mesh in practical use depends on the accuracy requirement of BEM capacitance solver and the extent of variation.

For a plane structure, the nominal vertices fluctuate along only one direction (normal to the nominal surface). Thus, one variable is associated with one nominal vertex, as one variable corresponds to one nominal panel in the simplified model. Since the number of vertices is about half of the number of panels for the triangular discretization, the number of variables will also be half of that in the simplified model if both models have same number of panels. For BEM capacitance solver, the assumption of same number of panels usually implies the comparable result accuracy.

3.2 Model for Two-Direction Variation

The above model is suitable for the cases where only one-direction variation is considered. For actual interconnect wire, modeling two-direction variations of height and width is often needed. In this case, the model for one-direction variation will encounter difficulty. Since surface fluctuation occurs along two directions, the projection of top surface on the XOY plane is different from the nominal shape. Fig. 2 shows an example of the effect of two-direction fluctuation, where the projection of arris is not a straight line. So, we do not have a certain nominal shape for the top surface and thus cannot construct the variational surface through disturbing the nominal vertices.

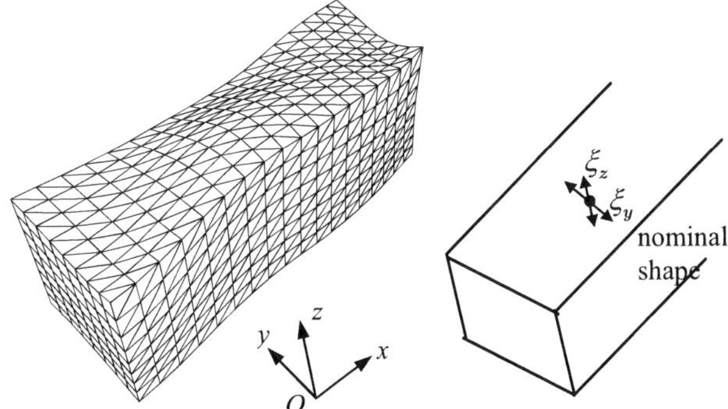

Figure 2. A metal wire with two-direction fluctuation, and the illustration of nominal shape with two-direction variables.

To model the two-direction variation, we extend the one-direction model by defining two variables for each discretization vertex on the nominal conductor. For each vertex on a wire routed along X-axis, random variables ξ_z and ξ_y are defined to represent the fluctuations along Z-axis and Y-axis, respectively (see Fig. 2). The variables for two directions form two variable sets, each of which obeys a Gaussian distribution with spatial correlation. With this model, the continuous variational surface can be easily generated. Now the number of variables approximates the number of triangular panels. So, if the continuous model employs the same number of panels as the simplified model, the number of variables will not increase.

3.3 Comparison Results

In this subsection, we compare the continuous surface model and simplified model in [5-7], through Monte Carlo (MC) simulations. For various discretization density, the MC simulation with 10000 samples is performed; the capacitances of each sample structure are extracted with FastCap 2.0 [14].

The first structure is a $1\mu m \times 1\mu m$ plane. The densest discretization mesh includes 800 panels, whose result is regarded as golden value. The simulation results for two variation models are listed in Table 1. Experimental results on the corresponding non-variation plane show that the capacitance error is less than 1% if the panel number is not less than 242. Table 1 shows that the variational capacitance with the continuous surface model has the same convergence property. But with the simplified model, a comparable panel number corresponds to a nearly 8% error of capacitance Std. With denser discretization, simulation results show the simplified model still results in large error of Std.

Table 1. Simulation results for the plane (σ=0.2μm, η=1μm)

	Continuous surface model			Simplified model	
Num. of panel	800	242	392	256	400
Mean (10^{-18}F)	41.656	41.235	41.427	41.224	41.411
Std (10^{-18}F)	0.854	0.840	0.854	0.788	0.793
Err. of Mean	--	-1.0%	-0.5%	-1.0%	-0.6%
Err. of Std	--	-1.6%	0.0%	**-7.7%**	**-7.1%**

The second structure includes two parallel wires (2-bit bus). The width, height and length of each wire are 1μm, 1μm and 4μm, respectively. The wire spacing is 2μm. The width and height variations are assumed to have same Std and correlation length. The simulation errors of both models for different discretization are listed in Table 2, where the results for 4096-panel discretization are regarded as golden value. From Table 2, we can see that the Std error of total capacitance C_{11} from the simplified model is prominent, just as the simulation results for plane structure.

Table 2. Simulation results for the 2-bit bus (σ=0.2μm, η=8μm)

		Continuous surface model			Simplified model	
Num. of panel		4096	640	1536	640	1568
Err. of C_{11}	**Mean**	--	-1.5%	-0.6%	-1.1%	-0.4%
	Std	--	0.3%	1.4%	**-6.7%**	**-5.0%**
Err. of C_{12}	**Mean**	--	2.2%	0.8%	1.7%	0.6%
	Std	--	-1.2%	1.0%	-1.1%	-0.5%

Due to variables of both side surfaces are correlated, and the wire width is the difference of two $N(0, \sigma^2)$ distribution, the Std of width in this case is actually $\sqrt{2\sigma^2(1-\exp(-1/64))}$ =0.035μm. The Std of height is the same small quantity. For larger variance of width and height, the error of simplified model would be larger.

4. VARIATION-AWARE EXTRACTION CONSIDERING WINDOW TECHNIQUE

In this section, the chip-level capacitance extraction in [9] is extended to consider the continuous surface model for geometric variations. For chip-level extraction, the window technique is necessary to limit the problem size. Only the small structures within windows are simulated with a BEM solver. Then, the capacitance elements from different windows form a distributed circuit for further analysis, or are used to assemble the total and coupling capacitances of whole critical net. Sophisticated techniques for window partition and assembling can be employed to guarantee high accuracy [15].

With consideration of window technique, the main tasks for variational capacitance extraction include the intra-window extraction, and the calculation of capacitance covariance between windows [9]. Because the continuous surface model involves much larger number of random variables than [9], efficient variable reduction is needed for the intra-window extraction. Since different windows are associated with each other through the physical variables, not the variables in capacitance model, calculating the capacitance covariance becomes an issue.

4.1 Intra-Window Extraction and Variable Reduction with Weighted PFA

For intra-window structure, we employ the techniques in Section 2.2 to obtain the quadratic capacitance expression (10). The computational time is proportional to d^2, where d is the number of independent variables. Random variable reduction must be performed for the proposed geometric variation model.

The principle factor analysis (PFA) is employed in [5-7] to reduce the random variables, which is done by eigen-decomposition on the covariance matrix $\Delta n(\xi)$ of geometric random variables ξ. If the first K largest eigenvalues are λ_1, λ_2, ..., λ_K, and e_1, e_2, ..., e_K are the corresponding eigenvectors, the variables ξ can be approximated by the linear combination of K dominant variables ζ [5-7]:

$$\xi \approx \sum_{i=1}^{K} \sqrt{\lambda_i} e_i \zeta_i \quad . \tag{14}$$

Here $\{\zeta_i\}$ is a set of independent variables with $N(0,1)$ distribution, that can be used to form the Hermite polynomial chaos (10). Thus, the intra-window extraction includes the following main steps:

1). Perform the eigen-decomposition for the covariance matrix of physical variations generated by (2) to obtain the first K largest eigenvalues and eigenvectors.

2). With the technique of sparse grid quadrature, obtain the integral points ζ^i and corresponding weights.

3). For each ζ^i, convert it to physical parameters ξ^i with (14), and then generate and extract the deterministic conductor structure with continuous surfaces. This step obtains the deterministic capacitances $C(\zeta^i)$.

4). Calculate the coefficients of the quadratic capacitance expression with (11) and (13).

It should be addressed that the number of dominant variables required in PFA mainly depends on the correlation function (2) and is independent of the BEM discretization. Another comment is that, the eigen-decomposition is not very time-consuming, since it is performed only once and the dimension of covariance matrix for intra-window structure is just moderate. If there are several groups of variables and no correlation between any two groups, the PFA shall be performed for each group separately.

The PFA method compares variable parameters one with another without considering their different impact on the output capacitance. A new variable reduction technique was proposed in [16] to take the impact on output performance into account. Inspired by [16], we propose a weighted PFA (wPFA) technique for better variable reduction.

If a weight is defined for each physical variable ξ_i to reflect its impact on output, then a set of new variables ξ^* are formed:

$$\xi^* = W \xi \quad , \tag{15}$$

where W is a diagonal matrix of weights. The covariance matrix $\Delta n(\xi^*)$ of ξ^* includes the weight information, and performing PFA based on it makes weighted variable reduction. We have

$$\Delta n(\xi^*) = E(W \xi (W \xi)^T) = W \Delta n(\xi) W^T \quad , \tag{16}$$

and denote its eigenvalues and eigenvectors by $\{\lambda_i^*\}$ and $\{e_i^*\}$.

Then, the variables ξ can be approximated by the linear combination of a set of independent dominant variables ζ^*:

$$\xi = W^{-1} \xi^* \approx W^{-1} \sum_{i=1}^{K} \sqrt{\lambda_i^*} e_i^* \zeta_i^* \quad . \tag{17}$$

For capacitance extraction, the capacitance value is the sum of panel charges, and the charge distribution is not uniform. At

978-1-60558-497-3/09 $25.00 © 2009 ACM

different positions, there is charge contribution with different significance. Based on this observation, we define the weight for variable ξ_i according to the charge density around its position. Firstly we perform capacitance extraction on the nominal structure to get a charge distribution. Then, the charge density $q_{v,i}$ for vertex i is:

$$q_{v,i} = \sum_{\vec{r}_i \in panel_j} \frac{q_j}{3S_j} \,, \tag{18}$$

where q_j and S_j are the charge and area of panel j, respectively. $\vec{r}_i$ stands for the vertex corresponding to variable ξ_i. Finally, the weight is calculated as the square ratio of vertex charge density to its minimum value:

$$w_i = \frac{q_{v,i}^2}{\min_i(q_{v,i}^2)} \,. \tag{19}$$

The matrix W is formed with w_i as diagonal entries. Except one additional extraction of nominal structure, the weighted PFA hardly increases computational expense of PFA as W is diagonal.

4.2 Inter-Window Covariance of Capacitance

Because the variable parameters belonging to different windows may be correlated, the corresponding capacitances from these windows have nonzero covariance [9]. After intra-window extraction, we suppose the quadratic capacitance expressions for the kth capacitance in windows i and j are

$$C_{i,k} = \sum_{p=1}^{M_i} c_{i,k,p} \Psi_p(\boldsymbol{\zeta}^*) \quad \text{and} \quad C_{j,k} = \sum_{p=1}^{M_j} c_{j,k,p} \Psi_p(\boldsymbol{\zeta}'^*) \,,$$

respectively. The covariance of capacitance can be evaluated with:

$$\text{cov}(C_{i,k},\, C_{j,k}) = \sum_{p=1}^{M_i} \sum_{q=1}^{M_j} c_{i,k,p} c_{j,k,q} \, \text{cov}(\Psi_p(\boldsymbol{\zeta}^*), \Psi_q(\boldsymbol{\zeta}'^*)) \,. \tag{20}$$

Here $\boldsymbol{\zeta}^*$ and $\boldsymbol{\zeta}'^*$ are dominant variable sets in windows i and j, respectively. Thus, the problem becomes calculating the covariance between Hermite polynomials.

We first determine the covariance between a variable ζ_a^* in windows i and a variable $\zeta_b'^*$ in windows j. The relationship between $\boldsymbol{\zeta}^*$ and physical variables $\boldsymbol{\xi}$ in the weighted PFA (17) can be rewritten as

$$\boldsymbol{\xi} = L_i \boldsymbol{\zeta}^* \,, \tag{21}$$

where L_i is a $n \times K$ matrix. n is the number of correlated physical variables, and K is the number of dominant factors. With the concept of pseudo inverse, we derive from (21):

$$\boldsymbol{\zeta}^* \approx G_i \boldsymbol{\xi} \,,$$

where G_i is the pseudo inverse of L_i, with dimensions of $K \times n$:

$$G_i = (L_i^T L_i)^{-1} L_i^T \,. \tag{22}$$

Now,

$$\text{cov}(\zeta_a^*, \zeta_b'^*) = \text{cov}(\sum_s g_{i,a,s} \xi_s,\ \sum_r g_{j,b,t} \xi_t') \,,$$
$$= \sum_s \sum_t g_{i,a,s} g_{j,b,t} \, \text{cov}(\xi_s, \xi_t') \tag{23}$$

where $g_{i,a,s}$ and $g_{j,b,t}$ represent matrix entries of G_i and G_j, respectively. $\text{cov}(\xi_s, \xi_t')$ can be checked out from the geometric variation model.

With the covariance between two dominant variables from different windows, we can calculate the covariance between Hermite polynomials as in [9]. For problem with large n, directly calculating (23) is time-consuming. Considering the attenuation of correlation function (2), we can ignore the covariance item of two variables physically far away from each other. This will reduce the expense of calculating (23).

The inter-window covariance of capacitance can be used to calculate the full-path capacitance. The mean values of relevant intra-window capacitances are assembled to obtain the mean of full-path capacitance, with the same manner as non-statistical extraction [15]. If considering a simple window technique with windows non-overlapping, the capacitance variance (the square of Std) of a whole net k can be calculated with [9]:

$$D(C_k) = D(\sum_i C_{ki}) = \sum_i D(C_{ki}) + 2\sum_{i \neq j} \text{cov}(C_{ki}, C_{kj}) \,, \tag{24}$$

where C_{ki} stands for the related window capacitance, whose summation approximates the capacitance C_k for net k. The technique to obtain the PDF of capacitance is also available in [9].

5. NUMERICAL RESULTS

The proposed techniques are implemented with MATLAB, which generates the deterministic sample structures and calculates the statistical capacitances. The sample structures are extracted with FastCap 2.0 [14]. For comparison, the MC simulation with 10000 samples is also performed. For each test case, the continuous geometric model with two-direction variation is assumed. Since the rough-surface effect is not prominent for on-chip interconnects, we set the correlation lengths to be 8 to 9 times of conductor width, like [6]. All experiments are carried out on a Linux server with 8 Xeon CPUs with 2.33GHz.

5.1 Small Cases

Two small structures are firstly tested. They are 2-bit bus and 1×1 crossover. The width, height and length of each wire are $1\mu m$, $1\mu m$ and $4\mu m$, respectively. The spacing of wires is $1\mu m$, the same for two cases. Each case includes 768 triangular panels. The techniques of HPC and that in Section 4.1 are used to calculate the statistical capacitances. For both PFA and wPFA, the truncation criterion of eigenvalues is 0.1%.

For the 2-bit bus structure, the correlation lengths for height variation and width variation are $8\mu m$ and $9\mu m$ respectively. The variation Std's for the two directions are $0.15\mu m$ and $0.2\mu m$. Their computational results are listed in Table 3, along with the error to the results of MC simulation. The table shows that both HPC techniques have good accuracy. For this case, wPFA reduces the number of variables to 9, instead of 11 by PFA. The integral points for sparse-grid quadrature are reduce from 276 to 191 (**30% reduction**). The speedup of HPC with wPFA to 10000-times MC simulation is about $10000/191 \approx \mathbf{52}$.

Table 3. Results for the 2-bit bus (in unit of 10^{-18}F)

		MC	HPC(PFA)	Error	HPC(wPFA)	Error
C_{11}	Mean	172.0	172.0	0.0%	172.0	0.0%
	Std	1.640	1.609	-1.9%	1.614	-1.6%
C_{12}	Mean	-82.4	-82.3	0.1%	-82.4	0.0%
	Std	1.720	1.799	4.6%	1.770	3.0%

For the 1×1 crossover structure, the computational results are listed in Table 4. The settings of correlation length and variation Std are the same as those for the 2-bit bus. Table 4 also shows the accuracy of proposed techniques. For this case, wPFA reduces the number of variables to 9, instead of 12 by PFA. The integral

978-1-60558-497-3/09 $25.00 © 2009 ACM

points are reduced by **41%** with wPFA. The speedup of HPC with wPFA to 10000-times MC simulation is about **52**.

Table 4. Results for the 1×1 crossover (in unit of 10^{-18}F)

		MC	HPC(PFA)	Error	HPC(wPFA)	Error
C_{11}	Mean	169.0	169.0	0.0%	169.0	0.0%
	Std	1.808	1.738	-3.9%	1.783	-1.4%
C_{12}	Mean	-81.1	-81.3	-0.2%	-81.2	-0.1%
	Std	1.727	1.774	2.7%	1.746	1.1%

5.2 A Large Case for Full-Path Extraction

The experiment for a large structure is carried out to validate the accuracy and efficiency of the techniques presented in Section 4.2. This case contains two parallel lines dissected into 10 windows, and parameter settings are just the same as the first case in [9]. The sample geometry of this case is constructed with the continuous surface model, rather than the grid-based simple model assumed by [9]. According to the correlated multivariate Gaussian distribution, 10000 sample structures are generated. For each one, the structure is divided into 10 windows for extraction and the capacitance results of all windows are summed up to obtain the full-path capacitances. The results for 10000 samples are then collected to get the MC simulation results. While using the proposed techniques, we extract the statistical capacitance expressions for each window separately. Then, the statistical full-path capacitances are calculated with the techniques proposed in Section 4.2. The comparison of full-path capacitances are shown in Table 5.

Table 5. Results of the full-path capacitances (in unit of 10^{-18}F)

		MC	HPC(wPFA)	Error
C_{11}	Mean	1483	1484	0.1%
	Std	8.76	8.87	1.3%
C_{12}	Mean	-511	-513	-0.4%
	Std	6.68	7.34	9.8%

From Table 5, we can see the results of proposed techniques match those from 10000-times MC simulation. This validates the accuracy of the techniques in Section 4.2. It is also found out that the mean of C_{11} has little discrepancy with that in [9], but the Std of C_{11} is much less than the latter (see Fig. 4(a) of [9]). Because [9] is based on an over-simplified model, its results obviously largely overestimate the capacitance variation.

For this case, each window includes 1024 triangular panels, and the wPFA reduces the variables to 8 dominant factors. The HPC solves 154 deterministic structures for each window. The speedup to 10000-times MC is expected to be $10000/154 \approx$ **64**. The actual breakdown of computational times is listed in Table 6. From the table we see that additional 40.5 seconds are spent on calculating the statistical full-path capacitance, mainly devoted to calculating the capacitance variance with (23) and (24). And, the actual speedup to MC simulation is still as large as **58**.

Table 6. CPU time (in second) of the full-path extraction

MC	Proposed method			Speedup
	HPC(wPFA)	Full-path variances	Total	ratio
20918	322	40.5	362.5	58

6. CONCLUSIONS

The main contributions of this work are as follows:

1) A continuous surface model is proposed for the geometric variations of on-chip interconnects. Compared with the currently used model, the new model does not increase number of variables and guarantees accurate modeling of capacitance variance.

2) Based on the continuous surface model, efficient techniques for chip-level extraction of variational capacitance are presented, including the HPC technique for intra-window extraction and the calculation of inter-window capacitance covariance.

3) A weighted PFA technique for capacitance extraction is proposed to reduce the random variables. Its advantages over the original PFA are preliminarily demonstrated by experiments.

7. ACKNOWLEDGEMENT

This work was supported by NSFC under Grant 60720106003, and the Basic Research Foundation of Tsinghua National Laboratory for Information Science and Technology (TNList).

8. REFERENCES

[1] J. Xiong, V. Zolotov and L. He, "Robust extraction of spatial correlation," IEEE Trans. CAD, vol. 26, pp. 619-631, 2007.

[2] X. Qi, A. Gyure, Y. Luo, S. C. Lo, M. Shahram, and K. Singhal, "Measurement and characterization of pattern dependent process variations of interconnect resistance, capacitance and inductance in nanometer technologies," in Proc. GLS-VLSI, pp. 14-18, 2006

[3] Z. Zhu, A. Demir, and J. White, "A stochastic integral equation method for modeling the rough surface effect on interconnect capacitance," in Proc. ICCAD, pp. 887–891, 2004

[4] Z. Zhu and J. White, "Fastsies: A fast stochastic integral equation solver for modeling the rough surface effect," in Proc. ICCAD, pp. 675–682, 2005.

[5] R. Jiang, W. Fu, J. M. Wang, V. Lin, and C. C.-P. Chen, "Efficient statistical capacitance variability modeling with orthogonal principle factor analysis," in Proc. ICCAD, pp. 683-690, 2005.

[6] H. Zhu, X. Zeng, W. Cai, J. Xue, and D. Zhou, "A sparse grid based spectral stochastic collocation method for variations-aware capacitance extraction of interconnects under nanometer process technology," in Proc. DATE, pp.1514-1519, 2007.

[7] J. Cui, G. Chen, R. Shen, S. X.-D. Tan, W. Yu, and J. Tong, "Variational capacitance modeling using orthogonal polynomial method," in Proc. GLS-VLSI, pp. 23-28, 2008

[8] T. El-Moselhy and L. Daniel, "Stochastic integral equation solver for efficient variation-aware interconnect extraction," in Proc. DAC, pp. 415-420, June 2008.

[9] W. Zhang, W. Yu, Z. Wang, Z. Yu, R. Jiang, and J. Xiong, "An efficient method for chip-level statistical capacitance extraction considering process variations with spatial correlation," in Proc. DATE, pp. 580-585, Mar. 2008.

[10] A. Singhee, S. Singhal, and R. A. Rutenbar, "Practical, fast Monte Carlo statistical static timing analysis: Why and how," in Proc. ICCAD, pp. 190-195, Nov. 2008.

[11] R. G. Ghanem and P. D. Spanos, Stochastic Finite Elements: A Spectral Approach, New York: Springer-Verlag, 1991.

[12] D. Xiu and G. E. Karniadakis, "The Wiener-Askey polynomial chaos for stochastic differential equations," in SIAM J. Sci. Comput., vol. 24, no. 2, pp. 619-644, Oct. 2002.

[13] E. Novak and K. Ritter, "Simple cubature formulas with high polynomial exactness," Constructive Approximation, vol. 15, no. 4, pp. 449-522, Dec. 1999.

[14] K. Nabors and J. White, "FastCap: A multipole accelerated 3-D capacitance extraction program,", IEEE Trans. CAD, vol. 10, pp. 1447–1459, Nov. 1991.

[15] W. Shi and F. Yu, "A divide-and-conquer algorithm for 3-D capacitance extraction," IEEE Trans. CAD, vol. 23, no. 8, pp. 1157-1163, Aug. 2004.

[16] A. Mitev, M. Marefat, D. Ma, and J. M. Wang, "Principle Hessian direction based parameter reduction with process variation," in Proc. ICCAD, pp. 632-637, Nov. 2007.

PiCAP: A Parallel and Incremental Capacitance Extraction Considering Stochastic Process Variation

Fang Gong
UCLA EE Department
Los Angeles, CA 90095
gongfang@ucla.edu

Hao Yu
Berkeley Design Automation
Santa Clara, CA 95054
hao.yu@berkeley-da.com

Lei He
UCLA EE Department
Los Angeles, CA 90095
lhe@ee.ucla.edu

ABSTRACT

It is unknown how to include stochastic process variation into fast-multipole-method (FMM) for a full chip capacitance extraction. This paper presents a parallel FMM extraction using stochastic polynomial expanded geometrical moments. It utilizes multi-processors to evaluate in parallel for the stochastic potential interaction and its matrix-vector product (MVP) with charge. Moreover, a generalized minimal residual (GMRES) method with deflation is modified to incrementally consider the nominal value and the variance. The overall extraction flow is called piCAP. Experiments show that the parallel MVP in piCAP is up to 3X faster than the serial MVP, and the incremental GMRES in piCAP is up to 15X faster than non-incremental GMRES methods.

Categories and Subject Descriptors: B.7.2[Hardware]: Integrated circuits – Design aids

General Terms: Algorithms, design

Keywords: Capacitance extraction, Stochastic geometrical moments, Parallel fast-multipole method, Incremental precondition

1. INTRODUCTION

Capacitance extraction, which considers process variations, has recently regained attention. As IC designs are approaching processes below 45nm, the fabricated interconnect and dielectric can show significant difference from the nominal shape due to large uncertainties from chemical mechanical polishing (CMP), etching and lithography [1, 2, 3, 4, 5, 6]. As a result, the value of extracted capacitance can differ from the nominal value by a large margin, which may further lead to a significant variability for the timing analysis. This leads to a need to accurately extract the capacitance with stochastic process variation.

An accurate extraction usually leads to higher computational complexity. To avoid discretizing the entire space, the boundary element method (BEM) is one approach to evaluate capacitance by discretizing the surface into panels on the boundary of the conductor and the dielectric. Though it results in a discretized system with a small dimension, the discretized system under BEM is dense. FastCap [7] employs an iterative method to solve the dense system by a generalized minimal residual (GMRES) method. Instead of performing the expensive LU decomposition, the GMRES iteratively reaches the solution with use of the matrix-vector mul-

tiplication. The computational cost of the matrix-vector-product (MVP) can be reduced by a fast-multipole-method (FMM) [7], a low-rank approximation [8], or a hierarchical-tree decomposition [9].

There are a few recent works [3, 4, 5, 6] on the interconnect extraction considering the process variation. [3] calculates the variational capacitance through a stochastic integral, where the computational cost is alleviated by the low-rank approximation. The works in [4, 5, 6] represent the variation by the stochastic orthogonal polynomial [10] when calculating a variational capacitance. [4] employs a spectral collocation, [5] applies a perturbation analysis, and [6] is in the framework of the stochastic integral similar to [3]. Since the interconnect length and cross-area are at different scales, the variational capacitance extraction is quite different between the on-chip [4, 5] and the off-chip [3, 6]. The on-chip interconnect variation from the geometrical parameters, such as width length of one panel and distance between two panels, is more dominant [4, 5] than the rough surface effect seen from the off-chip package trace. Including the stochastic integral or collocation in the full chip capacitance extraction would be computationally expensive. Instead, similar to deal with the stochastic analog mismatch for transistors [11], a full chip extraction needs to explore an explicit relation between the stochastic variation and the geometrical parameter. As such, the electrical property has an explicit dependence on geometrical parameters, and it can lead to a scalable extraction algorithm similar to [7, 8, 9].

The complexity of fast full chip extractions in [7, 8, 9] generally comes from two parts: the evaluation of MVP and the preconditioned GMRES iteration. Note that the FMM algorithm in [7] can be parallelized on the modern multi-processor system. However, it is unknown how to incorporate the stochastic process variation in the FMM. Moreover, the GMRES needs to be designed in an incremental fashion to consider the update from the process vacation.

Parallel and incremental analysis are the two fundamental concepts in reducing computational cost. In this paper, we have developed a parallel and incremental full chip capacitance extraction considering stochastic variation, namely *piCAP*. Our contributions are two-fold. First, we reveal that the potential interaction can be represented by a number of geometrical moments. As such, the process variation can be further included by expanding the geometrical moments with use of stochastic orthogonal polynomials. Therefore, the variation can be incorporated into a modified FMM algorithm to evaluate the MVP in parallel. Next, we develop a preconditioned GMRES method that can incrementally update the preconditioner with different variations. Experiments show that our method with the stochastic polynomial expansion is hundreds of times faster than the Monte-Carlo based method while maintaining a similar accuracy. Moreover, the parallel MVP in our method is up to 3X faster than the serial method, and the incremental GMRES in our method is up to 15X faster than non-incremental GMRES methods.

The rest of the paper is organized in the following manner. We first review the background of the capacitance extraction in Section 2. In Section 3, we present the parallel FMM based on the

Permission to make digital or hard copies of part or all of this work for personal or classroom use is granted without fee provided that copies are not made or distributed for profit or commercial advantage and that copies bear this notice and the full citation on the first page. To copy otherwise, to republish, to post on servers or to redistribute to lists, requires prior specific permission and/or a fee.

DAC'09, July 26-31, 2009, San Francisco, California, USA

978-1-60558-497-3/09 $25.00 © 2009 ACM

stochastic geometrical moments. In Section 4, we introduce a new incremental GMRES method and its application for solving large-augmented stochastic systems. We present experiment results in Section 5 and conclude the paper in Section 6.

2. BACKGROUND

The boundary element method (BEM) starts with an integral equation

$$\phi(r) = \int_{r' \in a'} \frac{\rho(r')}{4\pi\epsilon_0 |r - r'|} da', \qquad (1)$$

where $\phi(r)$ is the potential at the observer metal, $\rho(r')$ is the surface-charge density at the source metal, da' is an incremental area at the surface of the source metal, and r' is on da'.

By discretizing the metal surface into N panels sufficiently such that the charge-density is uniform at each panel, a linear system equation can be obtained by the *point-collocation* [7]:

$$Pq = b, \qquad (2)$$

where P is an $N \times N$ matrix of potential coefficients (or potential interactions), q is an N vector of panel charges, and b is an N vector of panel potential. By probing v iteratively with one volt at each panel in the form of $[0,...,1,...,0]$, the solved vector q is one column of the capacitance matrix.

Note that each entry P_{ij} in the potential matrix P represents the potential observed at the *observer panel* a_j due to the charge at the *source panel* a_i:

$$P_{ij} = \frac{1}{a_i} \int_{r_i \in a_i} \frac{1}{4\pi\epsilon_0 |r_i - r_j|} da_i. \qquad (3)$$

Since panels are assumed as well-separated, P_{ij} can be well approximated by $\frac{1}{4\pi\epsilon_0 |r_i - r_j|}$ [7, 5].

The resulting potential coefficient matrix P is usually dense in the BEM method. As such, directly solving (2) would be computationally expensive. FastCap [7] applies the iterative GMRES method to solve (2). Instead of performing an expensive LU decomposition of the dense P, GMRES first forms a preconditioned $W \cdot P$ to approximate P and then finds an initial solution q_0 accordingly by

$$(W \cdot P)q_0 = b.$$

Then, using either the multipole-expansion [7], low-rank approximation [8] or hierarchical-tree method [9] to efficiently evaluate the matrix-vector-product (MVP) for Pq_i, the GMRES method minimizes the residue error

$$min: \; ||b - Pq_i||$$

iteratively till converged.

Therefore, the use of GMRES requires a well-designed preconditioner and a fast matrix-vector-product. To this end, our stochastic extraction has a parallel evaluation Pq with variations presented in Section 3, and has an incremental preconditioner presented in Section 4.

3. PARALLEL FMM BY STOCHASTIC GEOMETRICAL MOMENT

A 3D FMM discretizes the conductor surface into panels and forms a cube with a finite height containing a number of panels. Then, it builds a hierarchical oct-tree of cubes and evaluates the potential interaction P at different levels. As such, the MVP of Pq can be provided efficiently during the GMRES iteration. Since such a spatial decomposition is geometrically dependent, it is helpful to express P by geometrical moments with an explicit geometry-dependence. As a result, this can lead to an efficient recursive update (M2M, M2L, L2L) of P on the oct-tree. The geometry-dependence is also one key property to preserve in presence of the stochastic variation. In this section, we first introduce a stochastic geometrical moment to calculate the potential interaction with variations. Using stochastic geometrical moments, we then illustrate how to develop a parallel FMM considering variations.

3.1 Stochastic Geometrical Moment

The process variation includes global systematic variation and local random variation. This paper focuses on the local random variation, or stochastic variations. Due to the local random variation, the width of discretized panel may show deviations from nominal values, and so does the distance between panels. Therefore, our paper considers two geometrical parameters under the stochastic vacation: panel-distance (d) and panel-width (h). With expansions in Cartesian coordinates, we can connect the potential interaction with the geometry parameter through *geometrical moments* that can be extended to consider stochastic variations.

Let the center of an observer-cube to be r_0 and the center of a source-cube to be r_c. We assume that the distance between the i-th source-panel and r_c as a vector $\mathbf{r}$ is

$$\mathbf{r} = r_x \overrightarrow{x} + r_y \overrightarrow{y} + r_z \overrightarrow{z}$$

with $|\mathbf{r}| = r$, and the distance between r_0 and r_c as a vector $\mathbf{d}$ is

$$\mathbf{d} = d_x \overrightarrow{x} + d_y \overrightarrow{y} + d_z \overrightarrow{z}$$

with $|\mathbf{d}| = d$.

In Cartesian coordinates $(x-y-z)$, when the observer is outside the source region ($d > r$), a *multipole expansion* (ME) [12, 13] can be defined as

$$\frac{1}{|\mathbf{r} - \mathbf{d}|} = \sum_{p=0} \frac{(-1)^p}{p!} \underbrace{(\mathbf{r} \cdots \mathbf{r})}_{p} \times \cdots \times \underbrace{(\nabla \cdots \nabla}_{p} \frac{1}{d})$$

$$= \sum_{p=0} M_p = \sum_{p=0} l_p(d) m_p(r) \qquad (4)$$

by expanding r around r_c, where

$$l_0(d) = \frac{1}{d}, \; m_0(r) = 1$$

$$l_1(d) = \frac{d_k}{d^3}, \; m_1(r) = -r_k$$

$$l_2(d) = \frac{3d_k d_l}{d^5}, \; m_2(r) = \frac{1}{6}(3r_k r_l - \delta_{kl} r^2)$$

$$\cdots$$

$$l_p(d) = \underbrace{\nabla \cdots \nabla}_{p} \frac{1}{d}, \; m_p(r) = \frac{(-1)^p}{p!} \underbrace{(\mathbf{r} \cdots \mathbf{r})}_{p}. \qquad (5)$$

Note that ∇ is the Laplace operator to take the spatial difference, δ_{kl} is the Kronecker delta function, and $(\mathbf{r} \cdots \mathbf{r})$ and $(\nabla \cdots \nabla \frac{1}{d})$ are rank-p tensors with $x^\alpha, y^\beta, z^\gamma$ ($\alpha + \beta + \gamma = p$) components.

Assume that there is a spatial shift at the source-cubic center r_c, for example, changing one child's center to its parent's center by $\mathbf{h}$ ($|h| = c \cdot h$), where c is a constant and h is the panel width. This leads to the following transformation for m_p in (5)

$$m'_p = \underbrace{((\mathbf{r} + \mathbf{h}) \cdots (\mathbf{r} + \mathbf{h}))}_{p}$$

$$= m_p + \sum_{q=0}^{p} \frac{p!}{q!(p-q)!} \underbrace{(\mathbf{h} \cdots \mathbf{h})}_{j} m_{p-j}. \qquad (6)$$

Moreover, when the observer is inside the source region ($d < r$), a *local expansion* (LE) under Cartesian coordinates is simply achieved by exchanging d and h in (4)

$$\frac{1}{|\mathbf{r} - \mathbf{d}|} = \sum_{p=0} L_p = \sum_{p=0} m_p(d) l_p(r). \qquad (7)$$

Also, when there is a spatial shift of the observer-cubic center r_0, the shift of moments $l_p(r)$ can be derived similarly to (6).

Clearly, both M_i, L_i and their spatial shifts show an explicit dependence on the panel-width h and panel-distance d. As such, we call M_i and L_i *geometrical moments*, and we also express P

978-1-60558-497-3/09 $25.00 © 2009 ACM

as a geometrical-dependence function $P(h,d)$ via geometrical moments. Usually, a second-order expansion can be accurate enough to calculate P_{ij}.

Moreover, assuming that the local random variations are described by two random variables: ξ_h for the panel-width h, and ξ_d for the panel-distance d, the stochastic forms of M_k and L_k become

$$\hat{M}_p(\xi_h,\xi_d) = M_p(h_0 + h_1\xi_h, d_0 + d_1\xi_d)$$
$$\hat{L}_p(\xi_h,\xi_d) = L_p(h_0 + h_1\xi_h, d_0 + d_1\xi_d) \qquad (8)$$

where h_0 and d_0 are the nominal values, and h_1 and d_1 define the perturbation range (% of nominal). Similarly, the stochastic potential interaction becomes $\hat{P}_{ij}(\xi_h,\xi_d)$.

3.2 Parallel FMM

Based on the multipole expansion (ME) (4), the local expansion (LE) (7), and the stochastic geometrical moments (8) of both, we can apply the $O(n)$ M2M, M2L and L2L operations and form a fast-multipole method (FMM) as shown in Fig. 1. We assume that there are N panels at the finest (or bottom) level. Providing depth H, we build an oct-tree with $H = \lceil log_8 \frac{N}{n} \rceil$ by assigning n panels in one cube. I.e., there are 8^h cubes at the bottom level. A parallel FMM further distributes a number of cubes into different processors to evaluate P. The decomposition of the tasks needs to minimize the communication cost and balance the workload.

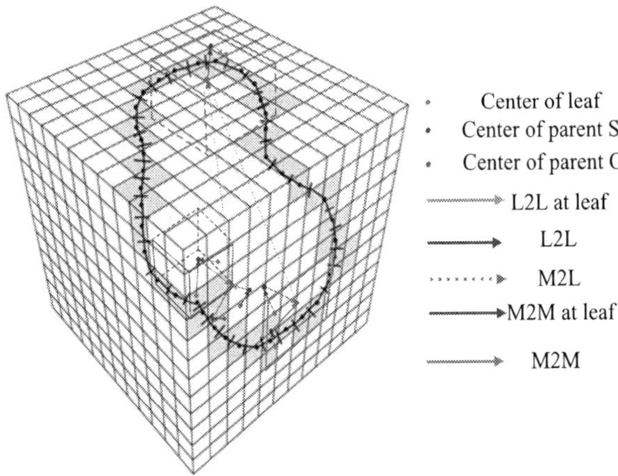

Figure 1: The 3D flow of FMM including M2M, M2L and L2L operations.

3.2.1 Upward Pass

The upward-pass manages the computation during the source expansion. It accumulates the multipole-expanded near-field interaction starting from the bottom level ($l = 0$). For each child cube (leaf) at the bottom level, it first evaluates the stochastic geometrical moment (8) for all panels in that cube. Then, it transverses to the upper level to consider the contribution from parents. The moment of a parent cube can be efficiently updated by summing the moments of its 8 children via an M2M operation. Based on (6), the M2M translates the children's $\hat{M}_p$ into their parents. The M2M operations at different parents are performed in parallel since there is no data-dependence.

3.2.2 Downward Pass

The potential evaluation at the observer is managed during a downward pass. At lth level ($l > 0$), two cubes are said to be *adjacent* if they have at least one common vertex. Two cubes are said to be *well separated* if they are not adjacent at level l but their parent cubes are adjacent at level $l - 1$. Otherwise, they

are said to be *far* to each other. The list of all the well-separated cubes from one cube at level l is called the *interaction list* of that cube.

From the top level $l = H - 1$, the interactions from cubes in the interaction list to one cube are calculated by an M2L operation at one level. Assuming that a source-parent center r_c is changed to a observer-parent's center r_0, this leads to a LE (7) using the ME (4) when exchanging the r and d. As such, the M2L operation translates the source's $\hat{M}_p$ into the observer's $\hat{L}_p$ for a number of source-parents on the interaction list of one observer-parent at the same level. Due to the use of the interaction list, the M2L operations have the data-dependence to prevent a parallel evaluation.

After the M2L operation, interactions are further distributed down to the children from their parents by a L2L operation. Assume that the parent's center r_0 is changed to the child's center r_0' by a constant $\mathbf{h}$. Identical to the M2M update by (6), an L2L operation updates $\mathbf{r}$ by $\mathbf{r}' = \mathbf{r} + \mathbf{h}$ for all children's $\hat{L}_k$s in a parallel manner.

Finally, the FMM sums the L2L results for all leaves at the bottom level ($l = 0$) and returns $\hat{P}$ and its product with q_i for the next GMRES iteration.

3.2.3 Data Sharing and Communication

The total runtime complexity for the parallel FMM with stochastic geometrical moments can be estimated by $O(N/B) + O(log_8 B) + C(N, B)$, where N is the total number of panels and B is the number of used processors. The $C(N, B)$ implies communication or synchronization overhead.

Therefore, this part discusses how to minimize the overhead of data sharing and communication during a parallel evaluation. In our parallel FMM implementations,. the message-passing-interface (MPI) is used for the data communication and synchronization between multiple processors. We notice that data dependency mainly comes from the interaction list during M2L operations. In this operation, a local cube needs to know the ME moments from cubes in its interaction list. To design a task distribution with small latency between the computation and communication, our implementation uses a complement interaction list. For a given cube, its complement interaction list records cubes that require ME moments. As such, the implemented communication protocol first anticipates which ME moments will be needed by other processes, and then distributes the ME moments prior to the computation where they will be required. Therefore, the communication overhead can be significantly reduced.

4. INCREMENTAL GMRES

The parallel FMM presented in Section 3 provides a fast matrix-vector-product for the fast GMRES iteration. As discussed in Section 2, another critical factor for a fast GMRES is the construction of a good preconditioner. In this section, to improve the convergence of GMRES iteration, we first present a deflated GMRES method with the use of a spectral precondition. Then, we discuss how to incrementally update the spectral preconditioner for the deflated GMRES, and apply such an incremental GMRES to efficiently consider the perturbations for an augmented system equation.

4.1 Deflated GMRES

The convergence of GMRES can be slow in the presence of degenerated small eigen-values of the potential matrix P. This is the case for most extraction problems with fine meshes. Constructing a preconditioner W to shift the eigen-value distribution (spectrum) of a preconditioned matrix $W \cdot P$ can significantly improve the convergence [14]. This is one of the so called *deflated GMRES* methods [15].

To avoid fully decomposing P, an implicitly restarted Arnoldi method by ARPACK [1] can be applied to find its first K eigen values $[\lambda_1, ..., \lambda_K]$ and its Kth-order Krylov subspace composed

---

[1] http://www.caam.rice.edu/software/ARPACK/

by the first K eigen-vector $V_K = [v_1, ..., v_K]$, where

$$PV_K = V_K D_K, \quad V_K^T V_K = I. \tag{9}$$

Note that D_K is a diagonal matrix composed of the first K eigen-values

$$D_K = V_K^T A V_K = diag[\lambda_1, ..., \lambda_K]. \tag{10}$$

Then, an according spectrum preconditioner is

$$W = I + \sigma(V_K D_K^{-1} V_K^T), \tag{11}$$

which leads to a shifted eigen-spectrum by

$$W P v_i = (\sigma + \lambda_i) v_i \; i = 1, ..., K. \tag{12}$$

Note that σ is the shifting value that leads to a better convergence. Moreover, as discussed below, the spectral preconditioner W can be easily updated in an incremental fashion.

4.2 Incremental Precondition

The essence of the deflated GMRES is to form a preconditioner that shifts degenerated small eigen-values. For a new P with updated $P^{(1)}$, the distribution of the degenerated small eigen-values changes accordingly. Therefore, given a preconditioner W for the nominal system with the potential matrix $P^{(0)}$, it would be expensive by another native Arnoldi iteration to form a new preconditioner W' for a new P with updated $P^{(1)}$. Instead, we show that W can be incrementally updated as follows.

If there is a perturbation δP ($P^{(1)}$) in P, the perturbation δv_i of ith eigen-vector v_i ($k = 1, ..., K$) can be given by [16]

$$\delta v_i = V_i B_i^{-1} V_i^T \delta P v_i. \tag{13}$$

Note that V_i is the subspace composed of

$$[v_1, ..., v_j, ..., v_K]$$

and B_i is the perturbed spectrum

$$diag[\lambda_i - \lambda_1, ..., \lambda_i - \lambda_j, ..., \lambda_i - \lambda_K]$$

($j \neq i$, $i, j = 1, ..., K$). As a result, δV_K can be obtained similarly for K eigen-vectors.

Assume that the perturbed preconditioner is W'

$$\begin{aligned} W' &= (I + \sigma V_K'(D_K')^{-1}(V_K')^T) \\ &= W + \delta W \end{aligned} \tag{14}$$

where

$$V_K' = V_K + \delta V_K, \quad D_K' = (V_K')^T P V_K'. \tag{15}$$

After expanding V_K' by V_K and δV_K, the incremental change in preconditioner W can be obtained by

$$\delta W = \sigma(E_K - V_K D_K^{-1} F_K D_K^{-1} V_K) \tag{16}$$

where

$$E_K = \delta V_K D_K^{-1} V_K^T + (\delta V_K D_K^{-1} V_K^T)^T, \tag{17}$$

and

$$F_K = \delta V_K^T V_K D_K + (\delta V_K^T V_K D_K)^T. \tag{18}$$

Note that all the above inverse operations only deal with the diagonal matrix D_K and hence the computational cost is low.

As there is only one Arnoldi iteration to construct a nominal spectral preconditioner W, it can be efficiently updated when $P^{(1)}$ changes. For example, $P^{(1)}$ is different when one alters the perturbation range h_1 of panel-width, or changes the variation type from panel-width h to panel-distance d. We call this deflated GMRES method with the incremental precondition an *iGMRES method*.

4.3 iGMRES for Stochastic Extraction

We further discuss how to apply iGMRES for our stochastic capacitance extraction in this part. For a full-chip extraction, simultaneously considering variations from all kinds of geometrical parameters would significantly increase model complexity, if not impossible. In this paper, we study the stochastic variation contributed by each parameter individually in an incremental fashion. Together with the incremental GMRES discussed above, the computational cost can be dramatically reduced for a large-scale extraction.

For example, to study the variation caused by the panel-distance d, one can assume that the potential coefficient matrix by

$$\hat{P}(\xi_d) = P^{(0)}(d_0) + P^{(1)}(d_1)\xi_d. \tag{19}$$

Note that as each panel center can experience a variation, $P^{(1)}$ is also a dense matrix with the same dimension as $P^{(0)}$.

Then, instead of performing an expensive Monte-Carlo simulation on ξ_d, one can span the distribution of ξ_d by an orthogonal polynomial [10, 4, 5, 6]. Based on the *Askey scheme* [10], a Gaussian distribution of ξ_d can be spanned by Hermite polynomials

$$\Phi(\xi_d) = [1, \; \xi_d, \; \xi_d^2 - 1, ...,]^T. \tag{20}$$

Accordingly, the charge-density vector q becomes

$$q(\xi_d) = \sum_{i=0} \alpha_j \Phi_j(\xi_d). \tag{21}$$

Usually, the expansion order is 2 to calculate the mean and the variance. As such, by further applying an inter-product with Φ_j ($i, j = 0, 1, 2$) to minimize the residue,

$$\langle \hat{P}(\xi_d) q(\xi_h) - b, \; \Phi_j \rangle = 0 \tag{22}$$

we have

$$\begin{bmatrix} P^{(0)} & P^{(1)} & 0 \\ P^{(1)} & P^{(0)} & 2P^{(1)} \\ 0 & 2P^{(1)} & P^{(0)} \end{bmatrix} \begin{bmatrix} \alpha_0 \\ \alpha_1 \\ \alpha_2 \end{bmatrix} = \begin{bmatrix} b \\ 0 \\ 0 \end{bmatrix} \tag{23}$$

due to the orthogonality of the Hermite polynomials.

By solving α_0, α_1 and α_2, the mean and the variance can be obtained from

$$E(q(\xi_d)) = \alpha_0$$
$$Var(q(\xi_d)) = \alpha_1^2 Var(\xi_d) + \alpha_2^2 Var(\xi_d^2 - 1) = \alpha_1^2 + 2\alpha_2^2,$$

where the parallel FMM discussed in Section 3 is applied to calculate the MVPs among $P^{(0)}$, $P^{(1)}$ and α_0, α_1, α_2 with the use of the spectrum preconditioner.

The above procedure can be similarly applied to calculate the variance and the mean for the geometrical parameter h. Clearly, the stochastic orthogonal expansion leads to an augmented system with perturbed blocks in the off-diagonal. It increases the computational cost for any GMRES method, and remains an unresolved issue in the previous applications of the stochastic orthogonal polynomial [10, 4, 5, 6]. In addition, when variation changes, a different $P^{(1)}$ results. Forming a new preconditioner to consider the augmented (23) would therefore be expensive.

To efficiently solve the augmented (23) under different $\delta P(h)$ or $\delta P(d)$, we first analyze an augmented nominal system with

$$\mathcal{W} = diag[W, \; W, \; W]$$
$$\mathcal{P} = diag[P^{(0)}, \; P^{(0)}, \; P^{(0)}]$$
$$\mathcal{D}_K = diag[D_K, \; D_K, \; D_K]$$
$$\mathcal{V}_K = diag[V_K, \; V_K, \; V_K],$$

which are all block diagonal. Hence there is only one preconditioning cost from the nominal block $P^{(0)}$.

The variation contributes to the perturbation matrix by

$$\delta \mathcal{P} = \begin{bmatrix} 0 & P^{(1)} & 0 \\ P^{(1)} & 0 & 2P^{(1)} \\ 0 & 2P^{(1)} & 0 \end{bmatrix}. \tag{24}$$

Table 1: Accuracy and Runtime(s) Comparison between MC(3000), *piCap*.

Table 1: Accuracy and Runtime(s) Comparison between MC(3000), *piCap*.

2 panels, $d_0 = 7.07\mu m$, $h_0 = 1\mu m$, $d_1 = 20\%d_0$		
	MC	piCAP
$C_{ij}(fF)$	-0.3113	-0.3056
Runtime (s)	10.786037	0.008486

2 panels, $d_0 = 11.31\mu m$, $h_0 = 1\mu m$, $d_1 = 10\%d_0$		
	MC	piCAP
$C_{ij}(fF)$	-0.3861	-0.3824
Runtime (s)	10.7763	0.007764

2 panels, $d = 4.24\mu m$, $h_0 = 1\mu m$, $d_1 = 20\%d_0$, $h_1 = 20\%$		
	MC	piCAP
$C_{ij}(fF)$	-0.2498	-0.2514
Runtime (s)	11.17167	0.008684

Based on (16) we can do an incremental update of the preconditioner W to consider a new variation $P^{(1)}$ when changing the perturbation range of h_1 or d_1. Moreover, we can also make an incremental update of W when changing the variation type from $P^{(1)}(h)$ to $P^{(1)}(d)$. This can dramatically reduce the cost when applying the deflated GMRES during the variational capacitance extraction. The same procedure can be easily extended for the high-order expansion with stochastic orthogonal polynomials.

5. EXPERIMENT RESULTS

We have implemented our *piCap* algorithm in c++. The experiments were carried out on Linux network servers with Xeon processors (2.4GHz CPU and 2GB memory). We first show the accuracy of using stochastic geometrical moments when compared to the Monte-Carlo integral. We then study the parallel runtime scalability when building the potential interaction and its MVP with charge. Finally, we show the impact of using the incremental GMRES preconditioner.

5.1 Accuracy Validation

The first example is two square panels with introduced random variation for their distance d and width h. In the experiment, we use a set of different perturbation ranges for d and h. First, Monte-Carlo method is used to calculate their C_{ij} 3000 times, and each time the variation is added to d randomly with a normal distribution. As such, we can evaluate the mean value μ and standard deviation σ. Then we introduce the same random variation to geometric moments in (8) with stochastic polynomial expansion. Because of an explicit dependence on geometrical parameters according to (19), we can efficiently calculate $\hat{C}_{ij}$s from (23). Table I shows the C_{ij} value and runtime using the aforementioned two approaches. The comparison in Table I shows that stochastic geometric moments can not only keep high accuracy, which yields an average error of 1.8%, but also be up to $\sim 1000X$ faster than the Monte-Carlo method.

5.2 Speed Validation

In this part, we study the runtime scalability using a few large examples to show both the advantage of the parallel FMM for MVP, and the advantage of the deflated GMRES with incremental precondition.

5.2.1 Parallel Fast Multipole Method

The four large examples are comprised of 20, 40, 80 and 160 conductors, respectively. For the two-layer example with 20 conductors, each conductor is of size $1\mu m \times 1\mu m \times 25\mu m$ (width×thick×len), and *piCap* employs a uniform $3\times3\times50$ discretization. Fig.2 shows its structure and surface discretization.

For each example, we use a different number of processors to calculate the MVP of $P \times q$ by the parallel FMM. Here we assume that only d has 10% perturbation range. As shown in Table 2, the runtime of the parallel MVP decreases evidently when more processors are involved. Due to the use of the complement interaction list, the latency of communication is largely reduced and

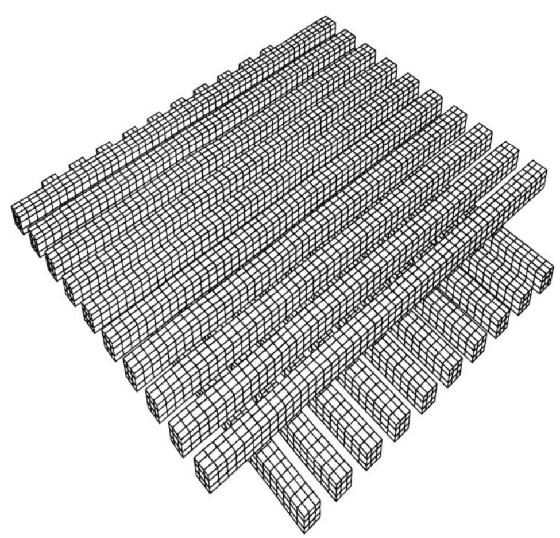

Figure 2: The structure and discretization of two-layer example with 20 conductors.

Table 2: MVP Runtime (seconds)/Speedup Comparison for four different examples

#wire	20	40	80	160
#panels	12360	10320	11040	12480
1 proc	0.737515/1.0	0.541515/1.0	0.605635/1.0	0.96831/1.0
2 procs	0.440821/1.7X	0.426389/1.4X	0.352113/1.7X	0.572964/1.7X
3 procs	0.36704/2.0X	0.274881/2.0X	0.301311/2.0X	0.489045/2.0X
4 procs	0.273408/2.7X	0.19012/2.9X	0.204606/3.0X	0.340954/2.8X

the runtime shows a good scalability vs. the number of processors. For all examples, the runtime with four processors is about $3X$ faster on average when compared to the runtime of the single processor.

It is worth mentioning that MVP needs to be performed many times in the iterative solver such as GMRES. Hence, even a small reduction of MVP runtime can lead to essential impact on the total runtime of the solution, especially when the problem size increases rapidly.

5.2.2 Deflated GMRES

piCap has been used to perform analysis for three different structures as shown in Fig.(3). The first is a plate with size $32\mu m \times 32\mu m$ and be discretized as 16×16 panels. The other two examples are Cubic capacitor and Bus2x2 cross-over structures. For each example, we can obtain two stochastic equation systems in (23) by considering variations separately from width h of each panel and from the centric distance d between two panels, both with 20% perturbation range from their nominal values.

To demonstrate the effectiveness of the deflated GMRES with a spectral preconditioner, two different algorithms are compared in Table 3. In the baseline algorithm (column "diagonal prec."), it constructs a simple preconditioner using diagonal entries. As the fine mesh structure in the extraction usually introduces degenerated or small eigen values, such a preconditioning strategy within the traditional GMRES usually needs much more iterations to converge. In contrast, since the deflated GMRES employs the spectral preconditioner to shift the distribution of non-dominant eigen values, it accelerates the convergence of GMRES leads to a reduced number of iterations. As shown by Table 3, the deflated GMRES consistently reduces the number of iterations by 3X on average.

5.2.3 Incremental Preconditioner

With the spectral preconditioner, an incremental GMRES can be designed easily to update the preconditioner when considering different stochastic variations. It quite often happens that a change occurs in the perturbation range of one geometry param-

Table 3: Runtime and Iteration Comparison for different Examples.

	#panel	#variable	diagonal prec.		spectral prec.	
			# iter	time	# iter	time
single plate	256	768	29	24.594	11	8.625
cubic	864	2592	32	49.59	11	19.394
cross-over	1272	3816	41	72.58	15	29.21

eter, or in the variation type from one geometry parameter to the other. As the system equation in (23) is augmented to 3X larger than the nominal system, it becomes computationally expensive to apply any non-incremental GMRES methods whenever there is a change from the variation. As shown by the experiments, the incremental preconditioning in the deflated GMRES can reduce the computation cost dramatically.

As described in Section 4, iGMRES only needs to perform the precondition one time for the nominal system, and to update the preconditioner with perturbations from matrix block $P^{(1)}$. In order to verify the efficiency of such an incremental preconditioner strategy, we apply two different perturbation ranges for h_1 for panels of the two-layer 20 conductors shown in Fig. 2, and then compare the total runtime of the iGMRES and GMRES, both with the deflation. The results are shown in Table 4.

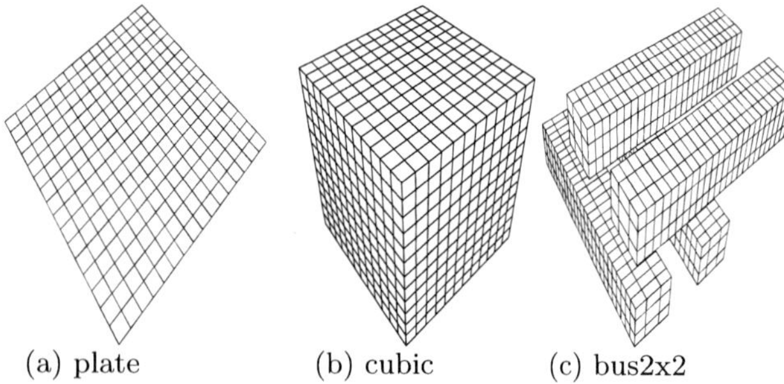

Figure 3: Test structures:(a)plate;(b)cubic;(c)crossover2x2

Table 4: Total Runtime (seconds) Comparison for 2-layer 20-conductor by different methods

discretization w×t×l	#panel	#variable	Total Runtime(s)	
			non-incremental	incremental
$3 \times 3 \times 7$	2040	6120	419.438	81.375
$3 \times 3 \times 15$	3960	11880	3375.205	208.266
$3 \times 3 \times 24$	6120	18360	-	504.202
$3 \times 3 \times 50$	12360	37080	-	3637.391

From Table 4, we can see that a non-incremental approach needs to constructs its preconditioner whenever there is an update of variations, which is very time consuming. For our proposed *iGMRES*, it can reduce much CPU time during the construction of the preconditioner by only updating the nominal spectral preconditioner incrementally with (16). The result of iGMRES shows a speedup up to $15X$ over non-incremental algorithms and only iGMRES can finish all large-scale examples up to 12360 panels.

6. CONCLUSIONS

To consider local random variations for a full chip capacitance extraction, this paper presents a fast parallel FMM method with an incremental precondition, namely piCAP. With the use of stochastic-polynomial expanded geometrical moments, the parallel FMM performs a fast evaluation of both potential interaction and its variation, and hence dramatically reduces the computational cost of the matrix-vector product (MVP). Moreover, the incrementally preconditioned GMRES is developed to consider the different update of variations with an improved convergence by deflation. Experiments on a few different large examples show that piCAP is hundreds of times faster than the Monte-Carlo based evaluation of variation with a similar accuracy, is up to 3X faster than the serial method in MVP, and is also up to 15X faster than non-incremental GMRES methods.

Future studies are planned to show the impact to the distribution of the eigenvalues when adding variations, and to identify the important parameters that lead to significant variations.

7. REFERENCES

[1] Y. Liu, S. Nassif, L. Pileggi, and A. Strojwas, "Impact of interconnect variations on the clock skew of a gigahertz microprocessor," in *Proc. ACM/IEEE Design Automation Conf. (DAC)*, 2000.

[2] R. Chang, Y. Cao, and C. Spanos, "Modeling the electrical effects of metal dishing due to CMP for on-chip interconnect optimization," *IEEE Trans. on Semi. Manufacturing*, pp. 1577–1583, 2004.

[3] Z. Zhu and J. White, "Fastsies: A fast stochastic integral equation solver for modeling the rough surface effect," in *Proc. IEEE/ACM Int. Conf. Computer-aided-design (ICCAD)*, 2005.

[4] H. Zhu, X. Zeng, W. Cai, J. Xue, and D. Zhou, "A sparse grid based spectral stochastic collocation method for variations-aware capacitance extraction of interconnects under nanometer process technology," in *Proc. IEEE/ACM Design, Automation, and Test in Europe (DATE)*, 2007.

[5] J. Cui, G. Chen, R. Shen, S. X.-D. Tan, W. Yu, and J. Tong, "Variational capacitance modeling using orthogonal polynomial method," in *Proc. ACM/IEEE Great Lake VLSI*, 2008.

[6] T. El-Moselhy and L. Daniel, "Stochastic integral equation solver for efficient variation-aware interconnect extraction," in *Proc. ACM/IEEE Design Automation Conf. (DAC)*, 2008.

[7] K. Narbos and J. White, "FastCap: A multipole accelerated 3D capacitance extraction program," *IEEE Tran. on Computer-aided-design (TCAD)*, pp. 1447–1459, 1991.

[8] S.Kapur and D.Long, "Ies3: a fast integral equation solver for efficient3-dimensional extraction," in *Proc. IEEE/ACM Int. Conf. Computer-aided-design (ICCAD)*, 1997.

[9] W. Shi, J. Liu, N. Kakani, and T. Yu, "A fast hierarchical algorithm for 3-d capacitance extraction," in *Proc. ACM/IEEE Design Automation Conf. (DAC)*, 1998.

[10] S. Vrudhula, J. M. Wang, and P. Ghanta, "Hermite polynomial based interconnect analysis in the presence of process variations," *IEEE Tran. on Computer-aided-design (TCAD)*, pp. 2001–2011, 2006.

[11] M. Pelgrom, A. Duinmaijer, and A. Welbers, "Matching properties of MOS transistors," *IEEE J. of Solid State Circuits*, pp. 1433–1439, 1989.

[12] J. D. Jackson, *Classical Electrodynamics*. John Wiley and Sons, 1975.

[13] C. Brau, *Modern Problems In Classical Electrodynamics*. Oxford Univ. Press, 2004.

[14] L. Giraud, S. Gratton, and E. Martin, "Incremental spectral preconditioners for sequences of linear systems," *Appl. Num. Math.*, pp. 1164–1180, 2007.

[15] V. Simoncini and D. Szyld, "Recent computational developments in Krylov subspace methods for linear systems," *Num. Lin. Alg. with Appl.*, pp. 1–59, 2007.

[16] G. W. Stewart, *Matrix algorithms (Volume II): Eigensystems*. SIAM, 2001.

978-1-60558-497-3/09 $25.00 © 2009 ACM

An Efficient Resistance Sensitivity Extraction Algorithm for Conductors of Arbitrary Shapes

Tarek El-Moselhy
MIT
Cambridge, MA 02139
tmoselhy@mit.edu

I. M. Elfadel
IBM
Yorktown Heights, NY 10598
elfadel@us.ibm.com

Bill Dewey
IBM
East Fishkill, NY 12533
bdewey@us.ibm.com

ABSTRACT

Due to technology scaling, integrated circuit manufacturing techniques are producing structures with large variabilities in their dimensions. To guarantee high yield, the manufactured structures must have the proper electrical characteristics despite such geometrical variations. For a designer, this means extracting the electrical characteristics of a whole family of structure realizations in order to guarantee that they all satisfy the required electrical characteristics. Sensitivity extraction provides an efficient algorithm to extract all realizations concurrently. This paper presents a complete framework for efficient resistance sensitivity extraction. The framework is based on both the Finite Element Method (FEM) for resistance extraction and the adjoint method for sensitivity analysis. FEM enables the calculation of resistances of interconnects of arbitrary shapes, while the adjoint method enables sensitivity calculation in a computational complexity that is independent of the number of varying parameters. The accuracy and efficiency of the algorithm are demonstrated on a variety of complex examples.

Categories and Subject Descriptors. J.6 [**Computer Aided Engineering**]: Computer-aided design (CAD)

General Terms. Algorithms

Keywords. resistance extraction, finite-element method, sensitivity, adjoint method, shape variations

1. INTRODUCTION

Resistance calculation is a fundamental component of any standard VLSI layout extraction flow. Several algorithms have been proposed for resistance calculation, ranging from simple analytical formulae to complex field solvers [1, 2, 3, 4, 5, 6].

Among the latter, more emphasis has been placed recently on the finite element method (FEM) [4, 5, 6]. This emphasis is justified by the fact that rigorous and accurate resistance calculation is the result of solving a Laplace equation in a *closed* domain that may enclose a conducting medium with a non-homogeneous resistivity distribution. FEM enables solving such equation even if the boundaries of the closed domain are irregular or fall outside a pre-determined Manhattan grid. Common instances of such boundaries are the

ones outlining the contours due to lithographic processing. Common instances of non-homogeneous resistivity distributions include low-resistivity on-chip copper interconnect and vias surrounded by high-resistivity liners. FEM is also very well adapted to the specific boundary conditions of the resistance calculation problem, namely, the mixed Neumann-Dirichlet boundary conditions. These features of resistance calculation have to be contrasted with those of capacitance calculation which is the result of solving a Laplace equation in an *open* domain with Dirichlet boundary conditions. Such calculation is better handled with a boundary element method. The reader is referred to [7] for more details about the fundamentals of FEM and to [8] for its application to resistance calculation.

Despite a fair amount of research directed towards resistance calculation and its incorporation in the VLSI layout extraction flow, the impact of manufacturing variability on the extracted resistance has been barely addressed. The several sources of variability, whether systematic and random, will combine to impact the boundaries of shapes printed on the wafer, thus leading to variations in their electrical properties. Lithography, etching, and chemical-mechanical polishing are such sources. With the decrease in feature sizes, the radius of mutual interaction between shapes is scaling up, thus complicating even further the correct prediction of the ultimate boundaries of a given shape. This uncertainty on the shape boundary is at the root of a pressing need to come up with efficient resistance calculation methods that can extract the resistance of a given shape for a whole family of boundary realizations.

Accounting for variability in extraction has focused on the capacitance calculation problem where the proposed solutions have ranged from very simple analytical correction formulas to very complex stochastic integral methods. Very little work has been done for resistance calculation [9], perhaps because of the latter's deceptive simplicity when it deals with wire-like patterns. Yet, non wire-like geometric patterns such as bends, jogs, corners, steps, pads, ports, fingers, diffusions, contacts, vias, and all their lithographic variations require very special attention when it comes to resistance extraction. Furthermore, in a variation-aware CAD context, even the resistance of wire-like patterns now require special attention for the simple reason that these patterns are no longer rectangular. The work described in this paper aims at addressing the resistance "gap" in the variability-aware extraction flow by proposing a rigorous resistance variation analysis based on sensitivity calculation. Among the range of variational methods, sensitivity-based ones strike a needed balance between accuracy and computational efficiency. Such methods have been extensively applied to capacitance calculation in its various formulations [10, 11, 12, 13] but are yet to be derived, implemented, and validated for resistance extraction. In a variation-aware VLSI extraction flow, one can make use of both the nominal resistance and the sensitivity of the resistance

Permission to make digital or hard copies of part or all of this work for personal or classroom use is granted without fee provided that copies are not made or distributed for profit or commercial advantage and that copies bear this notice and the full citation on the first page. To copy otherwise, to republish, to post on servers or to redistribute to lists, requires prior specific permission and/or a fee.
DAC'09, July 26-31, 2009, San Francisco, California, USA

978-1-60558-497-3/09 $25.00 © 2009 ACM

to the geometrical variations to predict the resistance of a slightly perturbed shape. More precisely, one may approximate the resistance function using a multivariate, first-order Taylor expansion

$$R = R_0 + \sum_m \frac{\partial R}{\partial p_m} \Delta p_m \tag{1}$$

where Δp_m is the perturbation around a nominal value of the p_m parameter and $\frac{\partial R}{\partial p_m}$ is the sensitivity (expressed as a partial derivative) of the resistance with respect to such perturbation. This paper is devoted to presenting an algorithm for computing such sensitivities for conductors of arbitrary shapes. In particular, we show how a pre-existing tool for FEM resistance calculation can be augmented with a sensitivity calculation capability using adjoint variational analysis.

It is important to note that in a VLSI layout extraction flow, FEM may be used in two different ways. The first way is in computing accurate look-up tables of specific wiring patterns for which the wire-like resistance formula will fail. Such computations are done off-line and do not impact the CPU cost of the resistance extraction step. The second way is on-line where FEM is applied very selectively using rules similar to those given in [1]. In this latter approach, caching and pattern recognition are extensively employed to reduce the number of times the FEM solver is actually called. Another important aspect is that in the full layout extraction flow, the overall performance is gated by the capacitance extraction phase rather than the resistance extraction phase. Our work on FEM-based resistance sensitivities should be placed within this overall VLSI layout extraction context.

This paper is organized as follows. Section 2 is a review of both FEM for resistance calculation and adjoint sensitivity analysis. Section 3 is the core of our theory and algorithmic contributions in this paper, namely how FEM and the adjoint analysis can be combined to produce an efficient algorithm for resistance sensitivity extraction. Section 4 illustrates such an algorithm on a set of numerical examples of industrial relevance. These results have been obtained by coding our sensitivity algorithm within an industrial FEM resistance calculator.

2. BACKGROUND

2.1 Resistance Calculation

The problem of resistance calculation is governed by the following partial differential equation

$$\begin{aligned}
\nabla \cdot (\sigma(r)(-\nabla\phi(r))) &= 0 & r &\in D \\
\sigma(r)(-\nabla\phi(r)) \cdot \hat{n} &= 0 & r &\in \partial D_{nc} \\
\phi(r) &= \phi_0 & r &\in \partial D_c
\end{aligned} \tag{2}$$

where $\phi(r)$ is the electric potential, $\sigma(r)$ is the electric conductivity of the material, D is the *closed* domain of the problem, ∂D_{nc} is the union of the boundary segments which are not assigned a particular potential (referred to as non-contact), ∂D_c is the union of the boundary segments which are assigned a specific potential (referred to as contact), and $\hat{n}$ is the normal to the boundary surface. The second equation is the Neumann boundary condition at the non-contact boundary of the problem and indicates that the current does not flow outside of the metal (the perpendicular current component vanishes). The last equation is the Dirichlet boundary condition at the contacts with prescribed potential ϕ_0. For a set of N_c contacts the resistance between contacts p and q is computed by assigning unit potential to contact p, i.e., $\phi_p = V_p (\equiv 1)$, and assigning zero potential to all other contacts. One then solves (2) to find $\phi(r)$ everywhere in D and subsequently computes the total current

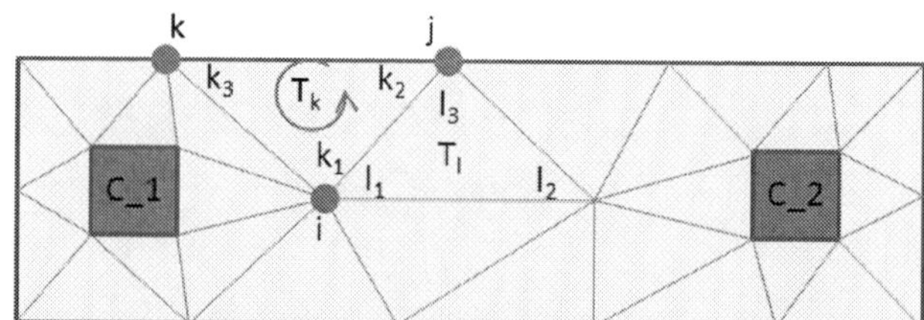

Figure 1: Triangular mesh of a 2-D structure. Triangular element local indexing is *anti* clockwise. Each node has a unique global index and multiple local indices.

entering contact q as

$$I_q = \int_{\partial D_{cq}} \sigma(r)(-\nabla\phi(r)) \cdot \hat{n} dS$$

where ∂D_{cq} is the boundary of port q.

The required resistance is then $R(p,q) = V_p/I_q$. The previous discussion makes it very clear that only one simulation is needed to compute an entire row of the resistance matrix $R(p,x) : 1 \le x \le N_c$.

2.2 Finite Element Method (FEM)

For the sake of simplicity, we will present the basics of FEM in 2D. The extension to 3D is straightforward. Since Problem (2) has mixed Dirichlet-Neumann boundary conditions, the continuous FEM formulation is derived from minimizing the following energy functional

$$E(\phi(x,y)) = \int_D \sigma(x,y)\nabla\phi(x,y) \cdot \nabla\phi(x,y)dxdy \tag{3}$$

To solve for $\phi(r)$, the geometry is first subdivided into smaller elements (Fig. 1). For 2D structures, the Delaunay triangulation is used for the discretization, since it tends to guarantee triangles of reasonable aspect ratios [7]. The resulting elements are described by the coordinates of their nodes. Each node has a unique global index to identify it in the mesh as well as a local index within each triangle it belongs to. Clearly, a given node may have more than one local index, since it may belong to more than one triangle. The local nodes of triangle T_k are labeled k_1, k_2 and k_3 (Fig. 1). The potential within each triangular element is then approximated using a basis of polynomial functions. For simplicity, this basis is taken to be that of first-order polynomials, so that

$$\phi(x,y) = \beta_x x + \beta_y y + \beta_0 \tag{4}$$

Consequently, the potential of every element T_k is described by the three unknown potentials of its nodes $\phi(k_1)$, $\phi(k_2)$ and $\phi(k_3)$ and three coefficients β_x, β_y, and β_0 of (4). The gradient of the potential in (4) can be rewritten in terms of the nodal potential as

$$\begin{aligned}
\nabla\phi &= \sum_{i=1}^{3} \frac{1}{\alpha_{T_k}} A(k_i)\phi(k_i) \\
A(k_i) &= \left(y_{k_{i+1}} - y_{k_{i-1}}\right)\hat{x} + \left(x_{k_{i-1}} - x_{k_{i+1}}\right)\hat{y}
\end{aligned}$$

where the coordinates of the node k_i are (x_{k_i}, y_{k_i}), α_{T_k} is the area of the triangle T_k, and $\hat{x}$ and $\hat{y}$ are the unit vectors of the x and y axis. Substituting in (3) we obtain the discretized quadratic form

$$E(\phi) = \phi_n^T A \phi_n \tag{5}$$

where ϕ_n is the vector of all node potentials in the mesh and A is the system matrix. The elements of A are written in a compact form

$$A(i,j) = \sum_{k:(i,j)\in T_k} \frac{\sigma(k)}{\alpha_{T_k}} A(k_i) \cdot A(k_j) \tag{6}$$

978-1-60558-497-3/09 $25.00 © 2009 ACM

where i and j are the global indices of the nodes, k is the index of the triangle, $(i,j) \in T_k$ means that i and j belong to a triangle T_k, and $\sigma(k)$ is the conductivity of the region bounded by triangle k. It is further assumed that the local indices of i and j in T_k are k_i and k_j, respectively. Equation (5) is then rewritten as

$$\phi_n^T \begin{pmatrix} A_{11} & A_{12} \\ A_{21} & A_{22} \end{pmatrix} \phi_n = \phi_1^T A_{11} \phi_1 + \phi_2^T A_{21} \phi_1 + \phi_1^T A_{12} \phi_2 + \phi_2^T A_{22} \phi_2 \tag{7}$$

where ϕ_1 is the vector of the unknown potential (potential of all N non-contact nodes), ϕ_2 is the vector of the known fixed potential (potential of N_f nodes on the contacts), A_{11} represents the self interaction of the non-contact nodes, $A_{12} = A_{21}^T$ mutual interaction of contact and non-contact nodes, and A_{22} self interactions of the contact nodes. Equation (7) is then minimized with respect to the unknown potential vector ϕ_1 to obtain

$$A_{11}\phi_1 = -A_{12}\phi_2 \tag{8}$$

Equation (8) is cast in the standard compact form $M\phi = b$, where $M = A_{11}$, $\phi = \phi_1$ and $b = -A_{12}\phi_2$. This linear system is then solved for ϕ to obtain the potential everywhere inside the domain D. Because of our interest in sensitivity calculation, the dependence of M, ϕ, and b on the problem parameters has to be made explicit, so the above linear system is written as

$$M(\mathbf{P})\phi(\mathbf{P}) = b(\mathbf{P}) \tag{9}$$

where $\mathbf{P}$ is a vector of geometrical parameters, such as the dimensions of the structure and the relative position of the contacts within it. The goal is to efficiently evaluate the impact of the variation of such parameters on the computed resistances. This we undertake in the next subsection.

2.3 Adjoint Sensitivity Analysis

The adjoint sensitivity computation is a very efficient algorithm for finding the sensitivity of a given vector $f(\mathbf{P}, \phi(\mathbf{P}))$ of length n_o with respect to a parameter vector $\mathbf{P}$ of length n_p. In this subsection, we summarize the simple derivation of the adjoint method as given in [14].

Taking the total derivative of $f(\mathbf{P}, \phi(\mathbf{P}))$ with respect to $\mathbf{P}$, we get

$$\frac{df(\mathbf{P}, \phi(\mathbf{P}))}{d\mathbf{P}} = \frac{\partial f(\mathbf{P}, \phi(\mathbf{P}))}{\partial \mathbf{P}} + \frac{\partial f(\mathbf{P}, \phi(\mathbf{P}))}{\partial \phi(\mathbf{P})} \frac{d\phi(\mathbf{P})}{d\mathbf{P}} \tag{10}$$

where $\frac{df(\mathbf{P}, \phi(\mathbf{P}))}{d\mathbf{P}}$, $\frac{\partial f(\mathbf{P}, \phi(\mathbf{P}))}{\partial \mathbf{P}}$ are matrices of size $n_o \times n_p$, $\frac{\partial f(\mathbf{P}, \phi(\mathbf{P}))}{\partial \phi(\mathbf{P})}$ is a matrix of size $n_o \times N$ and $\frac{d\phi(\mathbf{P})}{d\mathbf{P}}$ is a matrix of size $N \times n_p$.

Direct sensitivity methods are based on computing $\frac{d\phi(\mathbf{P})}{d\mathbf{P}}$ using a finite difference (FD) perturbation for each component of $\mathbf{P}$, which is computationally very expensive since it requires n_p independent system solves. However, taking the derivative of linear system (9) with respect to $\mathbf{P}$, we get

$$\frac{\partial M(\mathbf{P})\phi(\mathbf{P})}{\partial \mathbf{P}} + M(\mathbf{P})\frac{d\phi(\mathbf{P})}{d\mathbf{P}} = \frac{db(\mathbf{P})}{d\mathbf{P}} \tag{11}$$

Solving for $\frac{d\phi(\mathbf{P})}{d\mathbf{P}}$ and substituting into (10), we get

$$\frac{df(\mathbf{P}, \phi(\mathbf{P}))}{d\mathbf{P}} = \frac{\partial f(\mathbf{P}, \phi(\mathbf{P}))}{\partial \mathbf{P}} + \frac{\partial f(\mathbf{P}, \phi(\mathbf{P}))}{\partial \phi} M(\mathbf{P})^{-1} \left(\frac{db(\mathbf{P})}{d\mathbf{P}} - \frac{\partial M(\mathbf{P})\phi(\mathbf{P})}{\partial \mathbf{P}} \right) \tag{12}$$

Defining now the *adjoint* vector Λ as the solution of the *adjoint*

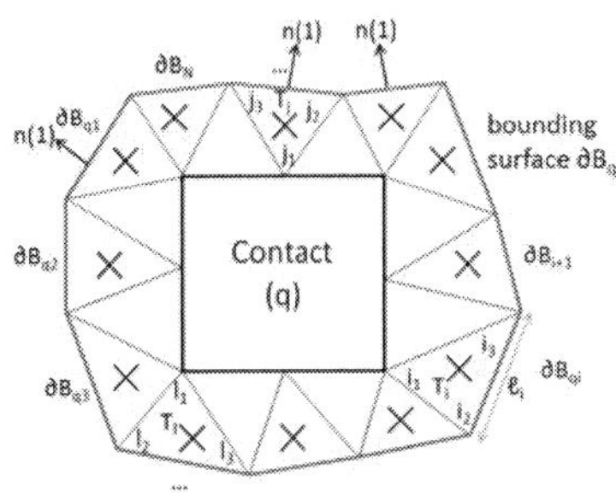

Figure 2: Hypothetical boundary surrounding the contact. Such boundary is useful for current computation.

linear system

$$M(\mathbf{P})^T \Lambda = \left(\frac{\partial f(\mathbf{P}, \phi(\mathbf{P}))}{\partial \phi(\mathbf{P})} \right)^T \tag{13}$$

we get the governing equation of the adjoint sensitivity method

$$\frac{df(\mathbf{P}, \phi(\mathbf{P}))}{d\mathbf{P}} = \frac{\partial f(\mathbf{P}, \phi(\mathbf{P}))}{\partial \mathbf{P}} + \Lambda^T \left(\frac{db(\mathbf{P})}{d\mathbf{P}} - \frac{\partial M(\mathbf{P})\phi(\mathbf{P})}{\partial \mathbf{P}} \right) \tag{14}$$

The main advantage of the adjoint method is that with only two system solves, one for the nominal system (9) and one for the adjoint system (13), we can obtain the sensitivity of $f(\mathbf{P}, \phi(\mathbf{P}))$ with respect to an arbitrary number of parameters n_p. In other words, when compared with the standard direct sensitivity method the time complexity of the adjoint method is independent of the number of parameters. Furthermore, its accuracy is independent of numerical differentiation.

3. RESISTANCE SENSITIVITY EXTRACTION

3.1 Defining the Output Function

Recall that the resistance $R_{pq} = \frac{V_p}{I_q}$ is a function of the total current I_q at a particular port q, which in turn is a linear function of the potential $\phi(r)$. The total derivative of the resistance with respect to the parameter vector $\mathbf{P}$ is given by

$$\frac{dR_{pq}}{d\mathbf{P}} = -\frac{V_p}{I_{pq}^2} \frac{dI_q}{d\mathbf{P}} \tag{15}$$

Consequently, the quantity of interest to extract R_{pq} is the total current at port q due to a unit potential $V_p = 1$ excitation at port p, i.e., the q-th component of the output vector is given by $f(\mathbf{P}, \phi(\mathbf{P}))(q) = I_q$. Let ∂B_q be a hypothetical closed boundary surrounding port q (Fig. 2). By current continuity we know that the net current flowing through port q is the same as the net current flowing through ∂B_q. The latter is computed from the relation between the current density and the potential

$$I_q = - \int_{\partial B_q} \sigma(r) \nabla \phi(r) \cdot \hat{n} d\ell$$

where ∂B_q is constructed as the union of the sides of the triangles touching the contact q at a single point (marked with an "X" in Fig. 2). Without loss of generality, the local numbering of these triangles is made such that the point touching the contact is always numbered l_1, where l is the index of the triangle. Let the side of triangle T_l belonging to ∂B_q be referred to as ∂B_{ql}. Note that the local normal to the boundary is the unit vector in direction of $A(l_1)$,

978-1-60558-497-3/09 $25.00 © 2009 ACM

772

i.e., $\hat{n} = \hat{n}(l_1) = \hat{A}(l_1)$. The I_q integral is discretized as

$$I_q = \sum_{T_l} \int_{\partial B_{ql}} \sum_{k=1}^{3} \frac{\sigma(l)}{\alpha_{T_l}} A(l_k)\phi(l_k) \cdot \hat{A}(l_1) d\ell$$

$$= \sum_{T_l} \sum_{k=1}^{3} \frac{\sigma(l)}{\alpha_{T_l}} A(l_k)\phi(l_k) \cdot \hat{A}(l_1) \int_{\partial B_{ql}} d\ell$$

$$= \sum_{T_l} \sum_{k=1}^{3} \frac{\sigma(l)}{\alpha_{T_l}} A(l_k) \cdot A(l_1)\phi(l_k)$$

where we have used the relation $A(l_1) = \hat{A}(l_1) \int_{\partial B_{ql}} d\ell$.

Note that due to the assumed local indexing of triangle T_l, node l_1 is on the boundary of the contact and $\phi(l_1) = 0$. This leads to

$$I_q = \sum_{T_l} \sum_{k=2}^{3} \frac{\sigma(l)}{\alpha_{T_l}} A(l_k) \cdot A(l_1)\phi(l_k)$$

which is simply the addition of the contributions of all the points on the boundary ∂B_q that are connected to points on the contact boundary. A careful investigation of this formula reveals that with the aid of (6) it can be cast in the following compact form

$$I_q = S_q \left(A_{21}\phi_1 + A_{22}\phi_2 \right) \tag{16}$$

where S_q is a row vector of zeros and ones and the rest of the notation is as in (7). S_q has ones at columns corresponding to the global indices of the nodes representing port q. Equation (16) indicates that the total current depends linearly on the potential of *any* point connected to a boundary point through a common triangle. More importantly, (16) indicates that the entries of the output matrices $S_q A_{21}$ and $S_q A_{22}$, along with those of both the system matrix M and the RHS vector b all share the same formulas, i.e., they all rely on elements of the form (6). This will become very useful later on when we compute derivatives of such elements with respect to geometrical variations. Finally, (16) can be cast in a more compact linear relation between the current and potential

$$I(\phi(\mathbf{P}), \mathbf{P}) = C_1^T(\mathbf{P})\phi(\mathbf{P}) + C_2^T(\mathbf{P})\phi_2 \tag{17}$$

where $\phi(\mathbf{P})$ is the unknown potential of the non-contact nodes, and ϕ_2 is the vector of fixed potentials of the contact nodes and is of length N_f, while $C_1(\mathbf{P})$ and $C_2(\mathbf{P})$ are known parameter-dependent matrices of size $N \times n_o$ and $N_f \times n_o$, respectively. Note that the above derivation is valid *only* for ports that are assigned zero potential, i.e., $q \neq p$, where p is the index of the excited port. However this is not a limitation since only one port is assigned a nonzero potential and the self-resistance of such port is given by the sum of all the mutual resistances of the port $R_{pp} = \sum_{q=1, q \neq p}^{N_c} R_{pq}$.

Finally, the derivatives of the current function required for the adjoint equations (13) and (14) are given by

$$\frac{\partial I(\phi(\mathbf{P}), \mathbf{P})}{\partial p_i} = \frac{dC_1^T(\mathbf{P})}{dp_i}\phi(\mathbf{P}) + \frac{dC_2^T(\mathbf{P})}{dp_i}\phi_2$$

$$\frac{\partial I(\phi(\mathbf{P}), \mathbf{P})}{\partial \phi} = C_1^T(\mathbf{P})$$

where p_m is the m-th component of $\mathbf{P}$.

3.2 Computing Derivatives w.r.t. P

In this subsection we are interested in computing the derivatives of the different matrix and vector entries with respect to the parameter vector $\mathbf{P}$. To do so we first recall that the matrix elements of the system matrix M, the right hand side vector b and the output matrices C_1 and C_2 are all computed from (6) and therefore computing

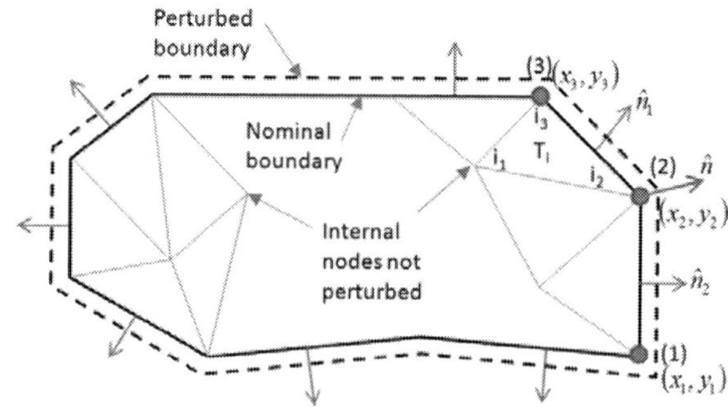

Figure 3: Typical variations defined by boundary node perturbations.

the derivative of (6) with respect to the geometrical parameters covers all the required derivatives. Second we make the observation that since all the entries of the matrices and vectors depend solely on the coordinates of the nodes, one can define all geometrical perturbations by their effect on such nodal coordinates. Using (6) the derivative of any matrix element with respect to the geometrical parameter p_m is given by

$$\frac{dA(i, j)}{dp_m} = \sum_{k:(i,j)\in T_k} \sum_{z_\ell \in T_k} \frac{\partial a(k_i, k_j)}{\partial z_\ell} \frac{dz_\ell}{dp_m} \tag{18}$$

$$a(k_i, k_j) = \frac{\sigma(k)}{\alpha_{T_k}} A(k_i) \cdot A(k_j)$$

where z_ℓ is one of the six coordinates (x and y coordinates of the three nodes of the triangle) on which the term $A(i, j)$ depends. The global indices of the nodes of triangle T_k are (i, j, l) and the corresponding local indices are (k_i, k_j, k_l). The computation of $\frac{\partial a(k_i, k_j)}{\partial z_\ell}$ is illustrated by the following example. The goal is to compute $\frac{\partial a(k_i, k_j)}{\partial z_\ell}$ where $z_\ell = x_l$, the x coordinate of the l-th global node in triangle T_k. This is achieved through the following algebra steps

$$a(k_i, k_j) = \frac{\sigma(k)}{\alpha_{T_k}} \left(\left(y_{k_j} - y_{k_l} \right)\left(y_{k_l} - y_{k_i} \right) + \left(x_{k_l} - x_{k_j} \right)\left(x_{k_i} - x_{k_l} \right) \right)$$

$$\frac{\partial a(k_i, k_j)}{\partial x_l} = \frac{\sigma(k)}{\alpha_{T_k}} \left(-2x_{k_l} + x_{k_j} + x_{k_i} \right) + \frac{-\sigma(k)}{\alpha_{T_k}} \frac{\partial \alpha_{T_k}}{\partial x_{k_l}} a(k_i, k_j)$$

$$\alpha_{T_k} = 0.5 \left(\left(x_{k_j} - x_{k_i} \right)\left(y_{k_l} - y_{k_i} \right) - \left(x_{k_l} - x_{k_i} \right)\left(y_{k_j} - y_{k_i} \right) \right)$$

$$\frac{\partial \alpha_{T_k}}{\partial x_{k_l}} = -0.5 \left(y_{k_j} - y_{k_i} \right) \tag{19}$$

Next we illustrate how to compute the chain rule factor $\frac{dz_\ell}{dp_m}$ in (18). We do so using an instance of a generic perturbation that implements uniform shape changes such as expansion or shrinking, as shown in (Fig. 3). Since this type of perturbations affects only the boundary of the structure, all the internal nodes will remain unchanged, i.e., $\frac{dz_\ell}{dp_m} = 0$ for any z_ℓ coordinate of an internal node. Only nodes defining the outer boundary will change. The direction of the boundary node perturbation is in the average direction of the normals to both boundary segments connected through the node. This is direction $\hat{n}$ in Fig. 3. As an example, we will compute $\hat{n}$, the direction of perturbation of (x_2, y_2)

$$\hat{n} = \frac{1}{||\hat{n}_1 + \hat{n}_2||} \left(\hat{n}_1 + \hat{n}_2 \right)$$

where $||v||$ is the length of vector v and

$$\hat{n}_1 = \frac{1}{\sqrt{(y_3 - y_2)^2 + (x_2 - x_3)^2}} \left((y_3 - y_2)\hat{x} + (x_2 - x_3)\hat{y} \right)$$

978-1-60558-497-3/09 $25.00 © 2009 ACM

$$\hat{n}_2 = \frac{1}{\sqrt{(y_2 - y_1)^2 + (x_1 - x_2)^2}} \left((y_2 - y_1)\hat{x} + (x_1 - x_2)\hat{y} \right)$$

Consequently, the sensitivities of coordinates (x_2, y_2) to a small node perturbation p_i along the normal $\hat{n}$ are given by

$$\frac{dx_2}{dp_i} = \hat{n} \cdot \hat{x} \qquad \frac{dy_2}{dp_i} = \hat{n} \cdot \hat{y} \qquad (20)$$

The mechanism suggested above for defining a perturbation is in fact general and can be used to model any geometric perturbation of either the domain boundaries or the contact locations. All that is required is to determine the set of nodes defining the perturbation, determine the changes in the coordinates of these nodes in response to a unit variation, and finally determine the partial derivatives. This process is summarized in Algorithm 1.

Algorithm 1 Efficient Assembly of Derivative Terms

1: Determine the set of nodes S_1 defining the perturbation
2: **for all** nodes in set S_1 **do**
3:     create a list of the parameters on which the node depends
4:     determine the partial derivatives of the node coordinates with respect to every parameter on which it depends $\frac{dz_k}{dp_i}$ (similar to (20))
5: **end for**
6: **for all** parameter p_i in parameter set $\mathbf{P}$ **do**
7:     create a set of nodes S_2 (by global index) that depend on the parameters
8: **end for**
9: When filling matrices M, b, C_1 and C_2, assemble $\frac{dM}{dp_i}$, $\frac{db}{dp_i}$, $\frac{dC_1}{dp_i}$ and $\frac{dC_2}{dp_i}$

$$\frac{d\left(M, b, C_1, C_2\right)}{dp_i} = \sum_{z_k \in S_2} \frac{d\left(M, b, C_1, C_2\right)}{dz_k} \frac{dz_k}{dp_i} \qquad (21)$$

3.3 Complexity Analysis of Sensitivity Extraction

It is well known that the FEM system matrix M is symmetric and very sparse. Moreover, by proper numbering of the nodes in the FEM mesh one can generate a banded system matrix M [7]. The maximum bandwidth B of the matrix is the maximum difference between the global indices of any interacting non-contact nodes (i.e., nodes that share a common triangle). In the remainder of this subsection, B is assumed a constant much smaller than N but in the order of both n_p and n_o. The most important observation is that the adjoint system matrix in (13) is the transpose of the symmetric matrix M. Consequently, both the linear system and the adjoint system share the same system matrix. Following all the previous observations the complete set of equations can be summarized as

$$M(\mathbf{P}) \begin{bmatrix} \phi(\mathbf{P}) & \Lambda \end{bmatrix} = \begin{bmatrix} b(\mathbf{P}) & C_1(\mathbf{P}) \end{bmatrix}$$
$$\frac{df(\mathbf{P}, \phi(\mathbf{P}))}{d\mathbf{P}} = \frac{dC_1(\mathbf{P})^T \phi(\mathbf{P})}{d\mathbf{P}} + \frac{dC_2(\mathbf{P})^T \phi_2}{d\mathbf{P}}$$
$$+ \Lambda^T \left(\frac{db(\mathbf{P})}{d\mathbf{P}} - \frac{\partial M(\mathbf{P})\phi(\mathbf{P})}{\partial \mathbf{P}} \right)$$

The complexity of solving the first equation is that of solving the same nominal sparse linear system with multiple right hand sides. The number of right hand sides is equal to $1 + n_o$. Therefore, the complexity of solving all systems concurrently using Gaussian elimination is $O(B^2 N)$. In other words, the complexity of our method is independent of the number of outputs, and indeed,

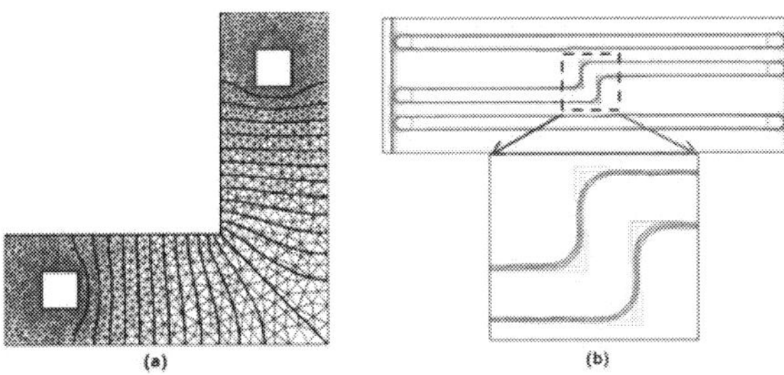

Figure 4: a) A simple 2-port rectangular corner. Equipotential contours clearly visible. b) 2-port jog. Different litho generated contours clearly visible.

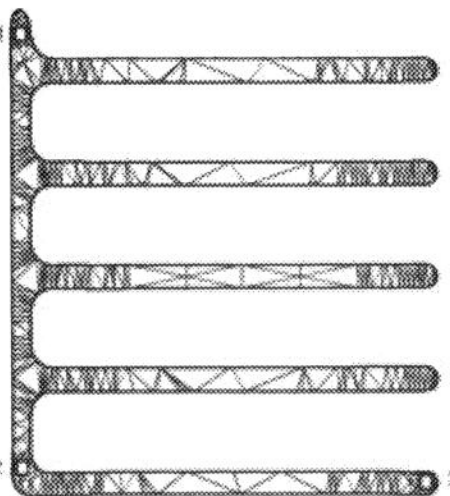

Figure 5: Three-port comb structure.

we have avoided one of the main pitfalls of the adjoint sensitivity method, namely, the linear growth of the complexity as a function of the number of outputs. The Gaussian elimination complexity is inherited from the solution of the nominal system and therefore the incremental complexity of solving both systems as compared to solving only the original system is insignificant. The added complexity of forming $\frac{dC_{1,2}(\mathbf{P})^T \phi(\mathbf{P})}{d\mathbf{P}}$, which is n_o matrix-vector products, is $O(n_o n_p)$ due to the sparsity of matrices C_1 and C_2. Finally, the complexity of forming the term $\frac{\partial M(\mathbf{P})\phi(\mathbf{P})}{\partial \mathbf{P}}$, which is n_p sparse matrix-vector products, is $O(N n_p)$. The total complexity is $O(B^2 N + n_o n_p + N n_p)$, which is $O(B^2 N)$, i.e., it is the exact same complexity as solving only the nominal system.

The memory complexity can also be shown to be $O(B^2 N)$, i.e., exactly the same memory complexity required

4. RESULTS

All results are obtained from a C implementation of the algorithm on a PowerPC workstation running at slightly more than 1 GHz with 16GB of RAM.

4.1 Accuracy Validation: Rectangular Corner

The first example is that of a rectangular corner Fig. 4.a. Each side of the corner is made of 9×3 squares, where the side of the square is equal to the minimum width per the design rules. The port size is 1 square and is centered at a point 1.5 square away from the corners. In this calculation the objective is to illustrate the accuracy of our FEM solver. The structure is discretized with 5000 triangles. The calculated resistance between the two ports of the corner is 0.807Ω. Such values are correlated with hardware using delay measurements on ring oscillator structures. Fig. 4.a shows the structure, the discretization and some of the equipotential surfaces.

4.2 Jog and Multiport Comb Structures

The second and third examples are that of a 2-port jog (Fig. 4.b) and a 3-port 5-finger comb structure (Fig. 5), respectively. Note that the structure boundaries do not follow a Manhattan pattern. They are rather described by piecewise linear approximations. The jog is

978-1-60558-497-3/09 $25.00 © 2009 ACM

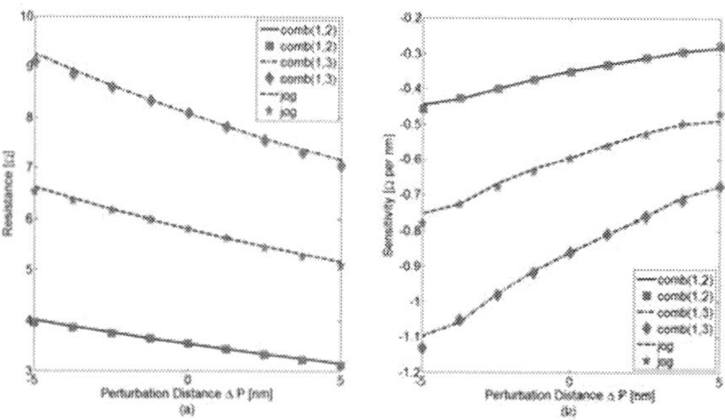

Figure 6: a) Resistance of the comb and jog structures as a function of the perturbation distance. The lines represent reference FEM results, the dots represent resistance calculated with extracted FEM sensitivities. b) Sensitivity of the resistances of the comb and jog structures w.r.t. structural perturbations. The lines represent sensitivities calculated using direct FD method, the dots represent extracted FEM sensitivities.

discretized using 5000 triangles, while the comb is discretized using 15,000 triangles. We extract the resistance between the ports of the jog and the resistance from port 1 to the other two ports (2 and 3) of the comb structure. We also extract the sensitivity of such resistances with respect to orthogonal uniform shape variations, i.e., variations which result in the boundary contours being uniformly offset from their nominal location along the normal to the boundary by distances ranging from -5nm to +5nm. This corresponds to 13% relative variation in the minimum dimension. This type of variation is illustrated in Fig. 3.b and was described mathematically by (20). The total extraction time (including resistances and sensitivities) for the jog and the comb drive is 0.8sec and 2.9sec, respectively. We further extract the resistances and sensitivities of both structures for 9 different boundary contour realizations corresponding to -5, -3.75, -2.5, -1.25, 0, 1.25, 2.5, 3.75, 5 nm perturbation from the nominal geometry. [1] The resistances of such different realizations are used to approximate the derivatives via finite difference approximations.

Fig. 6.a shows the excellent agreement between the resistances computed with the first order gradient model (1) and the FEM extracted sensitivity on the one hand (*red dots*) and on the other hand the resistances extracted directly from the finite element program (*blue lines*) for the structures under test. The maximum relative discrepancy was about 2 % and attained as expected at the end points of the perturbation interval. As apparent from Fig. 6.a, the resistance function can be accurately approximated using a first-order gradient model even for relative variations as high as 13%. This result is a nice illustration of the value of sensitivity extraction.

Fig. 6.b also shows the excellent agreement between the FEM-based sensitivities (*red dots*) and the sensitivities computed with an FD formula using the 9 sample points mentioned above (*blue lines*). Finally, it is worth noting that the cost of the first order gradient model is 1.5× the nominal extraction, while that of the full FD alone is 9× the nominal extraction.

5. CONCLUSIONS

In this paper we have presented a complete framework for extracting the resistance sensitivities to geometrical variations of conductors of arbitrary shapes. The sensitivity algorithm is based on the adjoint method and has been fully integrated into an industrial

---

[1] This way of perturbing the structure is physically meaningful as it mimics the way lithographic process variations impact the shape of nominal boundaries of the conductors.

FEM resistance solver. Due to the sparsity and symmetry of the FEM formulation, the complexity of the sensitivity extraction is overshadowed by that of the resistance calculation. The overall CPU complexity remains therefore unchanged. Furthermore, because of the sparsity of the FEM method and the ease of defining perturbations through the exclusive usage of boundary elements, there is negligible memory utilization associated with our algorithm. We have demonstrated the validity and efficiency of our algorithm on a variety of examples of industrial relevance. Future work will address how such resistance sensitivities can be used in a litho-aware VLSI extraction flow.

Acknowledgments

The authors would like to acknowledge very helpful discussions with Aditya Bansal, Koushik Das, Fook-Luen Heng, Mark Lavin, Rama Singh, and Amith Singhee from IBM Research. T. El-Moselhy would like to acknowledge IBM financial support in the form of a PhD Fellowship from IBM Research.

6. REFERENCES

[1] M. Horowitz, and R. W. Dutton, "Resistance Extraction from Mask Layout Data", *IEEE Trans. on Computer-Aided Design of Integrated Circuits and Systems*, Vol. 2, Issue 3, Jul. 1983, pp. 145–150.

[2] A. Vithayathil, X. Hu, and J. White, "Substrate Resistance Extraction using a Multi-Domain Surface Integral Formulation" *International Conference on Simulation of Semiconductor Processes and Devices, SISPAD 2003*, pp. 323-326.

[3] X. Wang, W. Yu, and Z. Wang, " Efficient Direct Boundary Element Method for Resistance Extraction of Substrate With Arbitrary Doping Profile" *IEEE Trans. on Computer-Aided Design of Integrated Circuits and Systems*, Vol. 25, Issue 12, Dec. 2006, pp. 3035-3042.

[4] B. Yang, and H. Murata, "A Finite Element-Domain Decomposition Coupled Resistance Extraction Method with Virtual Terminal Insertion" *Midwest Symposium on Circuits and Systems, MWSCAS 2007*, pp. 1425-1428.

[5] S. Rajagopalan, and S. Batterywala, "A 3-Dimensional FEM Based Resistance Extraction" *20th International Conference on VLSI Design, 2007*, pp. 565-570.

[6] M. Oppeneer, P. Sumant, and A. Cangellaris, "Robust Iterative Finite Element Solver for Multi-Terminal Power Distribution Network Resistance Extraction" *Electrical Performance of Electronic Packaging IEEE-EPEP, 2008*, pp. 181-184.

[7] M. Sadiku, *Numerical Techniques in Electromagnetics*, Boca Raton : CRC Press, 2000.

[8] C. Sakkas, "Potential Distribution and Multi-Terminal DC Resistance Computations for LSI Technology" *IBM Journal of Research and Development*, Vol. 23, No. 6, Nov. 1979, pp. 640-651.

[9] Q. Chen, and N. Wong, "A Stochastic Integral Equation Method for Resistance Extraction of Conductors with Random Rough Surfaces" *ISPACS '06*, pp. 411-414.

[10] I. Park, J. Coulomb, and S. Hahn, "Implementation of Continuum Sensitivity Analysis with Existing Finite Element Code" *IEEE Trans. on Magnetics*, Vol. 29, No.2, pp. 1787-1790, 1993.

[11] H. Qu, L. Kong, Y. Xu, X. Xu and Z. Ren, "Finite-Element Computation of Sensitivities of Interconnect Parasitic Capacitance to the Process Variation in VLSI" *IEEE Trans. on Magnetics*, Vol. 44, No.6, pp. 1386-1389, 2008.

[12] T. El-Moselhy, I. Elfadel and D. Widiger, "Efficient Algorithm for the Computation of On-Chip Capacitance Sensitivities with respect to a Large Set of Parameters" *ACM/IEEE Design Automation Conference, DAC 2008*, pp. 906-911.

[13] Y. Bi, N. van der Meijs and D. Ioan, "Capacitance Sensitivity Calculation for Interconnects by Adjoint Field Technique" *12th IEEE Workshop on Signal Propagation on Interconnects, 2008*, pp. 1-4.

[14] F. H. Branin, Jr., "Network Sensitivity and Noise Analysis Simplified" *IEEE Trans. Circuit Theory*, vol. CT-20, pp. 285-288, 1973.

Throughput Optimal Task Allocation under Thermal Constraints for Multi-core Processors *

Vinay Hanumaiah[‡], Ravishankar Rao[§], Sarma Vrudhula[‡], and Karam S. Chatha[‡]
[‡]Computer Science and Engineering Department, Arizona State University, Tempe, AZ 85287, USA
[§]Synopsys, Inc., Mountain View, CA 94043, USA
[‡]{vinayh, vrudhula, kchatha}@asu.edu
[§]ravirao@asu.edu

ABSTRACT

It is known that temperature gradients and thermal hotspots affect the reliability of microprocessors. Temperature is also an important constraint when maximizing the performance of processors. Although DVFS and DFS can be used to extract higher performance from temperature and power constrained single core processors, the full potential of multi-core performance cannot be exploited without the use of thread migration or task-to-core allocation schemes. In this paper, we formulate the problem of throughput-optimal task allocation on thermally constrained multi-core processors, and present a novel solution that includes optimal speed throttling. We show that the algorithms are implementable in real time and can be implemented in operating system's dynamic scheduling policy. The method presented here can result in a significant improvement in throughput over existing methods (5X over a naive scheme).

Categories and Subject Descriptors

C.4 [**Computer Systems Organization**]: Performance of Systems—*modeling techniques, performance attributes*; B.8.2 [**Performance and Reliability**]: Performance Analysis and Design Aids

General Terms

Algorithms, Performance, Design, Theory

Keywords

Thermal management, multi-core processors, task allocation, optimal throughput, thread migration

1. INTRODUCTION

*This work was supported in part by NSF grant CSR-EHS 0509540, Consortium for Embedded Systems grant DWS 0086, and by a grant from Science Foundation Arizona (SFAz) and Stardust Foundation.

Permission to make digital or hard copies of part or all of this work for personal or classroom use is granted without fee provided that copies are not made or distributed for profit or commercial advantage and that copies bear this notice and the full citation on the first page. To copy otherwise, to republish, to post on servers or to redistribute to lists, requires prior specific permission and/or a fee.
DAC'09, July 26-31, 2009, San Francisco, California, USA

While the processor industry is aggressively scaling the number of cores on processors [3], power and thermal constraints pose a significant challenge to future multi-core processors. Currently, there are up to 64 cores available commercially on a single die [6]. The trend towards multi-core strategy is motivated by the fact that multi-cores can achieve higher performance at reduced core speeds by exploiting thread-level parallelism within a given power budget. However there are many challenges that need to be addressed in order to have hundreds or even thousands of cores in the near future [3]. The die temperature has to be maintained below the specified maximum temperature for safe and reliable operation of the processor. Cost and volume constrains the packaging and cooling solutions, and this necessitates the need to incorporate the Dynamic Thermal Management (DTM) techniques.

The traditional focus of performance optimization has been to maximize performance subject to constraints on power or energy consumption. This is a different and easier problem than optimizing performance under thermal constraints because the power budget is a single upper bound on the total power consumed by all units in all cores, whereas the thermal constraint is an upper bound that must be satisfied by each and every unit. Moreover, the relation between power and speed is a much simpler one (simple algebraic) than the temperature-speed relation (requires solution to hundreds of equations). Thus the traditional power management techniques such as Dynamic Frequency Scaling (DFS) and Dynamic Voltage and Frequency Scaling (DVFS) must also account for temperature.

1.1 Related Work

While there are DTM methods to guarantee optimal performance in case of single core processors, the same is not true in the case of multi-core processors, because of the larger dimension of the problem. Local optimization for a single core is not necessarily optimal for a multi-core processor [20]. Apart from speed and voltage control, task migration or task-to-core allocation provides an additional dimension for performance optimization. Although a significant body of research has recently started on the problems of performance optimization of multi-cores under thermal constraints [5, 8, 10, 13], there are no optimal methods that incorporate task-to-core allocation.

A detailed comparison of various techniques aimed at general purpose, single core processors appears in [4]. The problem of optimal speed control for a sequence of tasks on a single core was addressed in [19]. DVFS with thermal con-

978-1-60558-497-3/09 $25.00 © 2009 ACM

straints for a periodic sequence of tasks on a single core was addressed in [22]. In [12], the problem of *off-line* multi-core speed control under thermal and power constraints is solved through convex optimization. In [12, 22] the leakage dependence on temperature, which substantially complicates the temperature-power relation, is not considered. In [20], the authors provided an online computational model to calculate the speeds of cores, for both transient and steady-state cases. Leakage dependence on temperature (LDT) is considered in [20] and an accurate thermal model (HotSpot [14]) is included. A comprehensive summary of thread migration techniques is provided in [11, 13]. The authors of [8] proposed a scheme called *heat-and-run* which moves threads from over heated SMT cores to cooler cores for maximizing performance. While this technique works when the number of tasks is less than number of cores and for processors that have a temperature slack, it may not be optimal for high performance processors where most of the cores operate close to thermal maximum. In [5] various thermal management techniques are studied, including OS based migration controllers. They make use of multi-loop control, wherein thread migration controls the outer loop, and DVFS makes up the inner loop.

Previous studies on task migration methods [8, 11, 13] used detailed cycle accurate simulators to determine the migration policies. The simulation times can be as large as tens of hours. These are not suitable for thread assignment in real time. Hence, there is a need to make use of accurate architectural thermal models with suitable approximations so that the run time of the algorithm is close to the migration interval.

1.2 Main Contributions and Outline

In this paper we present a complete formulation of the throughput-optimal task-to-core allocation problem under thermal constraints. This turns out to be a computationally expensive non-linear optimization problem, that cannot be solved in real time. We then present a novel simplification of the problem that allows us to compute the optimal task assignment and throughput-optimal speed at the start of each migration interval. This is then combined with a throughput-optimal continuous speed function over the migration interval. The problem formulation includes a detailed thermal model that accounts for thermal characteristics of the functional units within each core, the thermal interface material, the package and the intra-core and inter-core heat flow. We demonstrate, through simulation, that throughput improvements over a naive scheme can be as high as 5X. The proposed algorithm is feasible for on-line computation as it can be executed within a reasonable OS scheduling interval of $\approx 5ms$, for a large number of cores and tasks (as much as 128 cores and 128 tasks).

The paper is organized as follows: Section 2 introduces the system models and notations. It explains the HotSpot model [14] and the piece-wise linear (PWL) model for decoupling LDT. The problem of transient throughput optimization is introduced in Section 3. We also explain the structure of the problem and assumptions made in order to reduce the problem complexity. Section 4 contains a polynomial time algorithm for maximizing the instantaneous throughput. Finally Sections 5 and 6 present results and conclusion respectively.

2. SYSTEM MODELS AND NOTATIONS

2.1 Performance Model

We consider a heterogeneous multi-core processor with a queue containing n_t non-identical tasks to be run on n cores. We assume that each core i can be assigned an independent speed s_i, which is normalized and can be continuously varied over $[0, 1]$. We do not use dynamic voltage scaling in this work for the following reasons: (i) speed control through clock gating or fetch throttling can be activated with less overhead, and (ii) supply voltages are already small and are not expected to scale any further in future generations, and this leaves a very small margins for scaling [21].

In our work we do not assume SMT (simultaneous multi-threading) for the reasons of simplicity and also due to the fact that simple single-context cores are expected to be more power efficient [3]. We define the throughput of a multi-core processor as the weighted sum of the speeds of all cores, i.e. $S = \int_0^t \Sigma_{i=1}^n w_i s_i(t) dt$. The weights could be the IPC (instructions per clock cycle), the task priorities or any other performance criterion that varies with tasks. For inactive cores (those cores without tasks assigned), the weights $w_i = 0$. When IPCs are used as weights, the throughput reduces to instructions per second.

We assume that the threads run for a duration of at least the die thermal constant (around few milliseconds). Class A and Class B of NAS benchmarks [1] fall into this category. This allows the thermal die capacitances to saturate and hence they can be ignored without any significant loss of accuracy.

2.2 Power and Thermal Models

We use the latest HotSpot thermal RC circuit model [14] to model the thermal behavior of a processor. HotSpot uses the duality between heat flow and electrical circuit phenomena. Heat transfer and retaining capacity are represented by resistances and capacitances respectively, while the heat generating sources are modeled through current sources. Figure 1 taken from [18] shows HotSpot thermal model for a four core processor. Each core on the die is divided into several blocks (4 shown), with the same division being applied to the Thermal Interface Material (TIM). The heat spreader and the heat sink are divided into a number of blocks as shown. For a processor with n cores and m functional units for each core, there will be nm blocks in the die, another nm blocks in the TIM and 14 blocks in the package according to HotSpot-4 model [14], for a total of $N = 2nm + 14$ blocks.

Execution of various tasks on cores creates a spatial and temporal distribution of temperature on the die. Spatial variations arise due to different circuit styles, their size, different core speeds, activity factor of tasks, etc., whereas, temporal variations arise due to the time-varying nature of the code and core speeds, context switching among the threads, etc. The system power and temperature vectors[1] are represented by $\mathbf{P}$ and $\mathbf{T}$ respectively. The power $\mathbf{P}$ is the sum of the dynamic power component $\mathbf{P}_d$ and the leakage power component $\mathbf{P}_s$. The dynamic power depends only on speed $\mathbf{s}(t)$, whereas the leakage power depends only on temperature $\mathbf{T}(t)$. Hence $\mathbf{P} = f(\mathbf{s}(t), \mathbf{T}(t))$.

The relationship between the temperature vector $\mathbf{T}$ and

---

[1]All vectors are assumed to be column vectors and variables in bold represent matrices.

978-1-60558-497-3/09 $25.00 © 2009 ACM

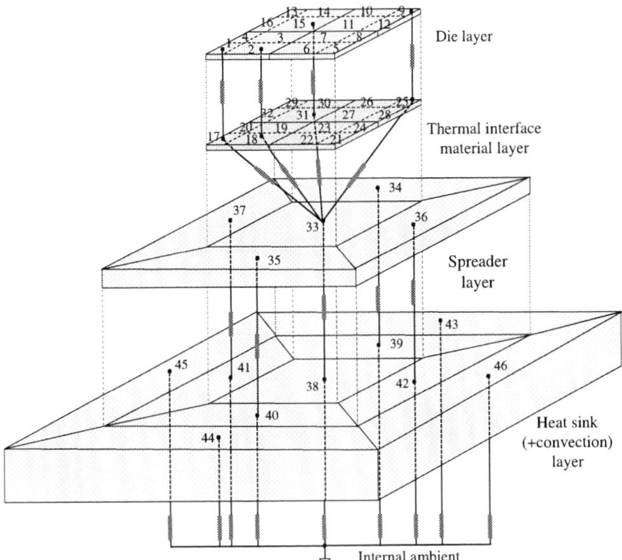

Figure 1: HotSpot thermal model for a four core processor [18].

the power vector $\mathbf{P}$ is given by the following dynamic state-space differential equation [7, 16]:

$$\frac{d\mathbf{T}}{dt} = \mathbf{AT} + \mathbf{BP}(\mathbf{s}, \mathbf{T}, t) \qquad (1)$$

Given a floorplan of a multi-core processor, the $N \times N$ ($N = 2nm + 14$) matrices, $\mathbf{A}$ and $\mathbf{B}$ are determined from HotSpot [7, 14] model. The vectors $\mathbf{P}$ and $\mathbf{T}$ are of length N. However $\mathbf{P}$ only has nm non-zero components because the remaining blocks of the TIM and the package do not generate heat.

In order to decouple the cyclic dependency between power and temperature, we need to linearize the non-linear relation between leakage and temperature. From [20], we note that the leakage power can be represented as $\mathbf{P}_s = \mathbf{P}_{s0} + \mathbf{G}_{\mathrm{LDT}}\mathbf{T}$, where $\mathbf{G}_{\mathrm{LDT}}$ is a matrix whose elements represent the slopes of the power-temperature curve. Let us define $\mathbf{P}_d^{max}$ as the vector of maximum power dissipation when $\mathbf{s} = \mathbf{1}_{n \times 1}$ (constant vector of 1's). The dynamic power is given by $\mathbf{P}_d(\mathbf{s}) = diag(\mathbf{P}_d^{max})\mathbf{Xs}$, where $\mathbf{X}$ is given by,

$$X(i,j) = \begin{cases} 1, & \text{if block } i \text{ belongs to core } j \text{ and } i \leq nm, \\ 0, & \text{otherwise,} \end{cases}$$

The state space equation (1) can now be re-written in standard state space form as follows:

$$\begin{aligned} \frac{d\mathbf{T}}{dt} &= \mathbf{AT} + \mathbf{B}(\mathbf{P}_d(\mathbf{s}) + \mathbf{P}_{s0} + \mathbf{G}_{\mathrm{LDT}}\mathbf{T}) \\ &= \hat{\mathbf{A}}\mathbf{T} + \mathbf{B}(\mathbf{P}_d(\mathbf{s}) + \mathbf{P}_{s0}) \end{aligned} \qquad (2)$$

where $\hat{\mathbf{A}} \triangleq \mathbf{A} + \mathbf{BG}_{\mathrm{LDT}}$.

3. PROBLEM DEFINITION

Given a heterogeneous multi-core processor with n cores, a weight vector $\mathbf{w}$, power vectors $\mathbf{P}_d^{max}$ (of all tasks) and $\mathbf{P}_{s0}$, conductance matrix $\mathbf{G}$ and spreader temperature T_{spr}, find an optimal allocation of n_t (non-identical) tasks on n cores with optimal speeds of operation such that the total

throughput defined by $\frac{1}{t_s}\int_0^{t_s} \mathbf{w}^T\mathbf{s}(t)dt$ is maximized over the migration interval $[t_0, t_s]$, subject to the constraints that the temperature of no functional block is greater than the allowed maximum temperature T_{max} and frequency ranges over $[0, 1]$.

$$\max_{\mathbf{s}(t),\mathbf{M}} \quad \frac{1}{t_s}\int_0^{t_s} \mathbf{w}^T\mathbf{s}(t)\, dt, \qquad (3)$$

$$s.t. \quad \frac{d\mathbf{T}}{dt} = \hat{\mathbf{A}}\mathbf{T} + \mathbf{B}(\mathbf{P}_{dt}^{max}\mathbf{M}^T\mathbf{s}(t) + \mathbf{P}_{s0}), \qquad (4)$$

$$\sum_{i=1}^{n} \mathbf{M}(i,j) = 1, j \in \{1, \ldots, n_t\}, \qquad (5)$$

$$\sum_{j=1}^{n_t} \mathbf{M}(i,j) = 1, i \in \{1, \ldots, n\}, \qquad (6)$$

$$\mathbf{M}(i,j) \in \{0,1\}, i \in \{1, \ldots, n\}, j \in \{1, \ldots, n_t\}, \quad (7)$$

$$\mathbf{T}(0) = \mathbf{T_0}, \qquad (8)$$

$$\mathbf{T}(t) \leq \mathbf{T}_{max}, \ \forall \ 0 \leq t \leq t_s, \qquad (9)$$

$$\mathbf{0}_{n \times 1} \leq \mathbf{s}(t) \leq \mathbf{1}_{n \times 1}, \ \forall \ 0 \leq t \leq t_s \qquad (10)$$

where, $\mathbf{P}_{dt}^{max}$ is a matrix whose columns contain $\mathbf{P}_d^{max}$ of n_t tasks and is of dimension $N \times n_t$. $\mathbf{T}_0$ is the vector of initial temperatures. $\mathbf{M}$ is the mapping/allocation matrix of size $n \times n_t$. The elements of $\mathbf{M}$ are defined as follows:

$$M(i,j) = \begin{cases} 1, & \text{if task } j \text{ is assigned to core } i, \\ 0, & \text{otherwise,} \end{cases}$$

Equation (4) represents a time-dependent non-linear constraint involving variables $\mathbf{s}(t)$ and $\mathbf{M}$. Performing non-linear optimization in real time is computationally not practical. This motivates us to simplify the problem as follows.

Problem Simplification: Owing to penalty of high migration overhead, the task allocation is done once in tens of milliseconds. A significant simplification is obtained if we restrict the problem to finding an allocation that maximizes the *instantaneous* throughput at the start of the migration interval. Then with those corresponding initial speeds, we can find the throughput-optimal speed curve over the migration interval. Letting $\mathbf{s}_0$ denote the vector of initial speeds at the start of a migration interval, the objective will be to maximize the weighted sum of initial speeds, $\mathbf{w}^T\mathbf{s}_0$. In order to solve for the initial speeds $\mathbf{s}_0$, we need to decouple the dependence of $\mathbf{M}$ on $\mathbf{s}_0$ and vice-versa. In the following sections we show how to decouple the above dependency by making a few simple, but realistic assumptions.

4. OPTIMAL PLACEMENT ALGORITHM

4.1 Structure of HotSpot Conductance Matrix

We build multi-core floorplan from single core floorplans with L2 caches surrounding these cores. Intel Core 2 Duo processor has a similar floorplan. Given the geometric floorplan, one can calculate the conductance matrix $\mathbf{G}$ using HotSpot thermal circuit model. Figure 2 taken from [18] shows the sparsity plot of the conductance matrix for the dual core Alpha processor constructed as explained above. A dot in the plot corresponds to a non zero entry in the matrix. We can see in the figure that most of the entries are concentrated along the diagonal blocks of the core and the TIM. The size of the blocks correspond to the number of functional blocks m. These square diagonal blocks represent the lateral resistances of the core and TIM. The sets of

978-1-60558-497-3/09 $25.00 © 2009 ACM

dots running in parallel to the main diagonal represent the vertical resistances connecting the die and the TIM layers. The vertical resistances between the TIM and the spreader are shown by the vertical and the horizontal strip of dots in the figure.

Caches separate the cores and they tend to be the coolest parts of the die as they have larger die area and the resulting low power density and temperature [9]. This results in smaller lateral resistances between cores as seen in Figure 2.

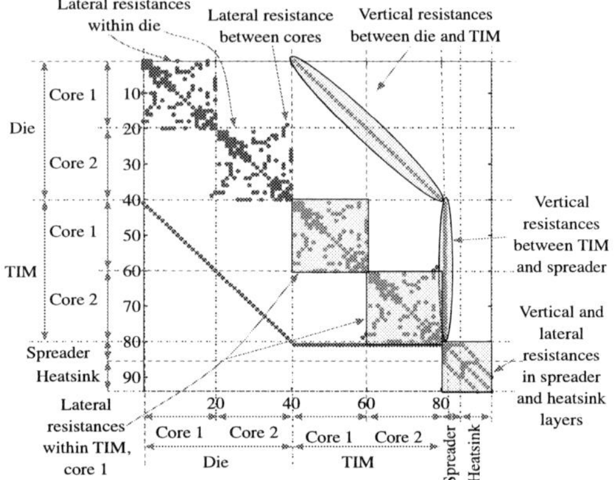

Figure 2: Plot of conductance matrix sparsity for the constructed dual core Alpha floor plan [18].

Figure 3 taken from [18] shows the various components of the conductance matrix shown in Figure 2. The diagonal components of the matrix are named $\mathbf{G}_{\text{die}}$ and $\mathbf{G}_{\text{tim}}$ corresponding to the die and the TIM sections of the chip. The vertical conductances connecting the die to the TIM are denoted by $\mathbf{G}_{\text{die-tim}}$ and $\mathbf{G}_{\text{tim-die}}$, which are identical as it is the same conductance that connects both the die and the TIM layer. $\mathbf{G}_{\text{tim-spr}}$ and $\mathbf{G}_{\text{spr-tim}}$ denote the conductances connecting the spreader and the TIM layer. Finally the conductances in the package are denoted by $\mathbf{G}_{\text{pkg}}$.

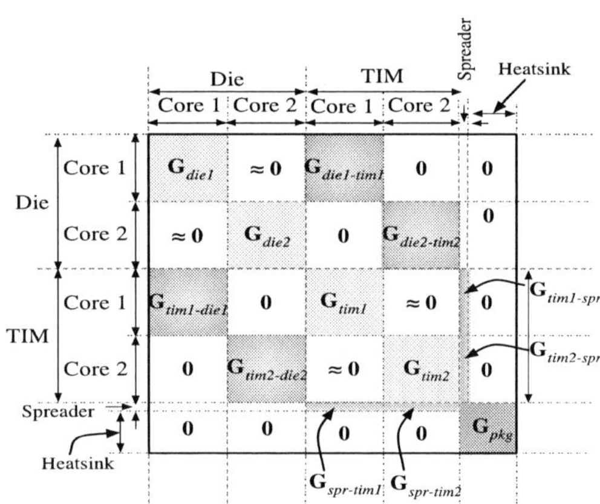

Figure 3: Components of the thermal conductance matrix G for the constructed Alpha dual core [18].

We make following observations based on the conductance matrix shown in Figure 3:

- There are fewer lateral resistances between cores and since the cores are surrounded by caches, the heat flow

between the cores can be neglected (as caches are the cooler parts of the chip).

- The lateral thermal resistances are nearly four times higher than the vertical resistances in the die and the TIM layers [19]. Thus majority of the heat flow occurs through the vertical resistances. Also since the vertical thermal gradient is larger between the spreader and the TIM than across the die, we can ignore the lateral resistances.

- The TIM has lower thermal conductance than the silicon, which further reduces the heat flow in that layer.

Based on the above observations, we neglect the sparse blocks of the matrix corresponding to inter-core die and TIM resistances.

4.2 Computation of Die Temperatures

The total power dissipation in the processor is given by:

$$\mathbf{GT} = \mathbf{P}_d(\mathbf{s}) + \mathbf{G}_{LDT}\mathbf{T} + \mathbf{P}_{s0} \tag{11}$$

Using Kirchhoff's current law for the die and TIM layers, the following equation can be derived for core i running task j:

$$\mathbf{G}_{\text{die,i}}\mathbf{T}_{\text{die,i}} + \mathbf{G}_{\text{die-tim,i}}\mathbf{T}_{\text{tim,i}} = s_{ij}\mathbf{P}_{d,j}^{max} + \mathbf{G}_{\text{LDT,i}}\mathbf{T}_{\text{die,i}} + \mathbf{P}_{s0,i} \tag{12}$$

$$\mathbf{G}_{\text{tim-die,i}}\mathbf{T}_{\text{die,i}} + \mathbf{G}_{\text{tim,i}}\mathbf{T}_{\text{tim,i}} + \mathbf{G}_{\text{tim-spr,i}}T_{\text{spr}} = 0. \tag{13}$$

T_{spr} is the scalar temperature of the spreader center. Given $\mathbf{P}_{d_i}^{max}$ for task j and T_{spr}, we can find the maximum operational frequency s_{ij} of any core i. Let,

$$\mathbf{K}_{g,i} = \mathbf{G}_{\text{die,i}} - \mathbf{G}_{\text{die-tim,i}}\mathbf{G}_{\text{tim,i}}^{-1}\mathbf{G}_{\text{die-tim,i}} - \mathbf{G}_{\text{LDT,i}} \tag{14}$$

$$\mathbf{K}_{p,i} = \mathbf{G}_{\text{die-tim,i}}\mathbf{G}_{\text{tim,i}}^{-1}\mathbf{G}_{\text{tim-spr,i}}T_{\text{spr}} + \mathbf{P}_{s0,i}. \tag{15}$$

Let $\mathbf{T}_{\text{die,ij}}$ be the die temperature of core i when executing task j at the maximum speed and is given by:

$$\mathbf{T}_{\text{die,ij}} = \mathbf{K}_{g,i}^{-1}\mathbf{K}_{p,i} + \mathbf{K}_{g,i}s_{ij}\mathbf{P}_{d,j}^{max}. \tag{16}$$

Note that we need not be concerned with T_{tim}, since the hottest blocks will be part of the die. Now the problem stated in Section 3 can be re-stated as follows:

$$\max_{\mathbf{s},\mathbf{M}} \qquad \mathbf{w}^T\mathbf{s}, \tag{17}$$

$$s.t. \qquad \mathbf{T}_{\text{die,i}} = \mathbf{K}_{g,i}^{-1}\mathbf{K}_{p,i} + \mathbf{K}_{g,i}s_i\mathbf{P}_{dt}^{max}\mathbf{M}(i)^T,$$
$$\forall i \in \{1,\ldots,n\} \tag{18}$$

$$\sum_{i=1}^{n}\mathbf{M}(i,j) = 1, j \in \{1,\ldots,n_t\}, \tag{19}$$

$$\sum_{j=1}^{n_t}\mathbf{M}(i,j) = 1, i \in \{1,\ldots,n\}, \tag{20}$$

$$\mathbf{M}(i,j) \in \{0,1\}, i \in \{1,\ldots,n\}, j \in \{1,\ldots,n_t\}, \tag{21}$$

$$T_{spr} = T_{spr,0}, \tag{22}$$

$$\mathbf{T}_{\text{die,i}} \leq \mathbf{T}_{max}, \tag{23}$$

$$\mathbf{0}_{n\times 1} \leq \mathbf{s} \leq \mathbf{1}_{n\times 1} \tag{24}$$

where, $\mathbf{M}(i)$ refers to the ith row of the matrix $\mathbf{M}$.

Equation (18) allows us to compute the speed s_i for a core i running task j independent of other tasks running on the remaining cores. The maximum of $\mathbf{T}_{\text{die,ij}}$ should be less than the maximum allowed temperature T_{max}. Thus

978-1-60558-497-3/09 $25.00 © 2009 ACM

the hottest functional unit determines the maximum speed of operation for a given task j on core i. The hottest unit h is given by the row corresponding to the $row_h(\mathbf{T}_{\text{die},ij}) = max(\mathbf{T}_{\text{die},ij})$. The speed s_{ij} is given by:

$$
s_{ij} = \begin{cases} \frac{T_{max} - row_h(\mathbf{K}_{g,i}^{-1}\mathbf{K}_{p,i})}{\mathbf{K}_{g,i}\mathbf{P}_{d,j}^{max}}, & max(\mathbf{T}_{\text{die,ij}}^{max}) \geq T_{max}, \\ 1, & \text{otherwise}, \end{cases}
\tag{25}
$$

This removal of dependency allows us to compute the speeds of a core for all available tasks and similarly for all other cores. Thus we can construct a matrix of speeds $\mathbf{S}$ whose elements are s_{ij}, where i, j refer to the core and task number respectively. We refere to this matrix as the *speed matrix*.

Given the above *speed matrix* we need to find an optimal task-to-core allocation. This problem can be formally stated as shown below:

$$
\max_{\mathbf{M}} \qquad \sum \mathbf{M}^T \mathbf{S},
\tag{26}
$$

$$
s.t. \qquad \sum_{i=1}^{n} M(i,j) = 1, j \in \{1, \ldots, n_t\},
\tag{27}
$$

$$
\sum_{j=1}^{n_t} M(i,j) = 1, i \in \{1, \ldots, n\},
\tag{28}
$$

$$
M(i,j) \in \{0,1\}, i \in \{1, \ldots, n\}, j \in \{1, \ldots, n_t\}.
\tag{29}
$$

This is a linear assignment problem and can be efficiently solved in polynomial time using *Munkres* algorithm [17]. The time complexity of this algorithm is $O((nn_t)^3)$. Once the allocation is known along with the optimal initial speeds $\mathbf{s}_0$, the speed profile for the core i for the migration interval $[t_o, t_f]$ is given by [20]:

$$
s_i(t) = s_{i0}e^{-t/\tau_i} + s_{i,ss}[1 - e^{-t/\tau_i}].
\tag{30}
$$

where the τ_i is the time constant for the core i. The steady state speeds $s_{i,ss}$ and time constants τ_i are determined by solving the linear and non-linear program described in [20].

5. RESULTS

We used HotSpot thermal model to model the thermal behavior of the dual core Alpha 21264 processor constructed as described in Section 4.1. To construct the thermal model for the multi-core Alpha processor, we replicate the thermal model of one single core processor to desired number of cores by placing them side-by-side. The power numbers for the tasks were generated by uniformly distributing the power dissipation values obtained from SPEC benchmarks [2] using PTscalar tool [15]. We allowed a maximum temperature of $110\,^{\circ}$C and 130 W of dynamic power and 60 W of leakage power. The maximum frequency of operation was set to 4 GHz. Spatial variation among cores is created by choosing a mean leakage power and varying it over a uniform distribution.

Chaparro et al. [10] have shown that performing thread assignment often can degrade performance. In fact they show that when the migration penalty is more than 2% of the total performance, the throughput starts degrading. The migration cost is assumed to be 40,000 cycles. For a 4GHz processor this approximates to 10 μs. In order to keep the migration penalty low and also since our tasks are assumed to run for at least the die thermal time constant, we set the migration interval to 10 ms or 40 million clock cycles . To

the best of our knowledge there are no available literature on task-to-core allocation for thermally over constrained multi-core processors and hence chose to compare with round robin scheme.

5.1 Performance Comparison against Round Robin Scheme

Figure 4 shows the plot of the throughput of optimal allocation scheme versus round robin allocation for a four core processor with eight tasks. The simulation is run for a duration of 200 seconds with initial spreader temperature of $55\,^{\circ}$C. The throughput between migrations is obtained by exponentially throttling the initial speed as described in [20]. We see a large improvement in throughput achieved by our optimal allocation algorithm. The throughput drops in both procedures after a few seconds as the package temperature increases necessitating the throttling of the cores and continues to drop until the cores attain the steady state speeds after package time constant (≈ 100 s).

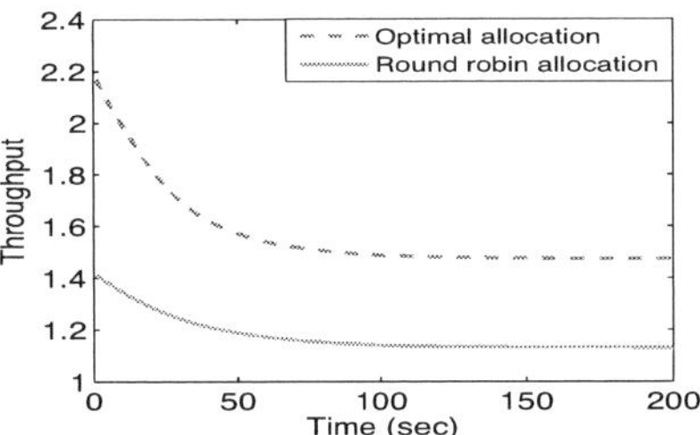

Figure 4: Temporal performance of optimal allocation and round robin allocation schemes.

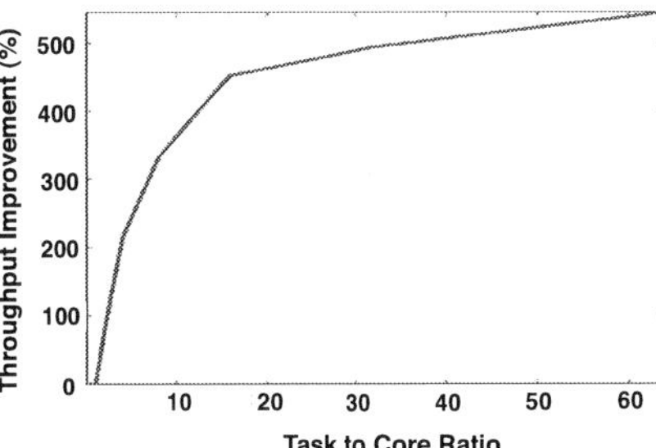

Figure 5: Plot of performance of the algorithm versus processor utility factor

5.2 Effect of Number of Cores and Tasks

Figure 5 shows the performance of the algorithm in optimally determining the allocation of tasks to cores against a random allocation. The improvement is described in terms of percentage throughput improvement and is plotted for several task to core ratios. We see that our algorithm shows a very high improvement when the processor utilization factor is very high. We see that the improvement can be as high as 5X.

5.3 Effect on Temperature Gradient

Figure 6 shows the spatial spatial distribution of temperature for a four core processor running optimal alloca-

978-1-60558-497-3/09 $25.00 © 2009 ACM

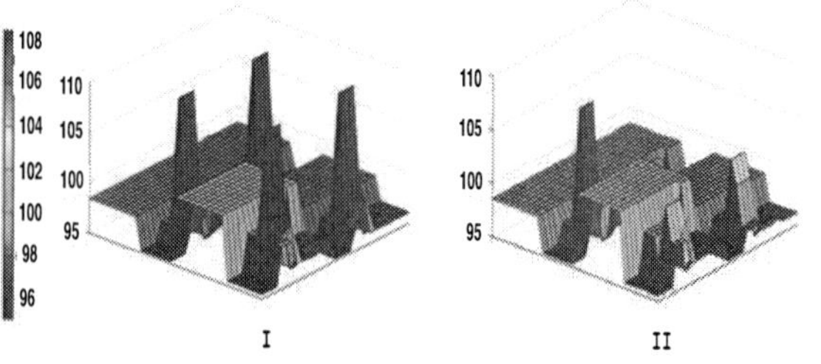

Figure 6: Spatial thermal map of 4-core processor utilizing I. optimal, II. random allocation algorithm

tion algorithm versus random allocation. To obtain maximum performance the total power dissipation has to be minimized and this fact is exploited by our procedure in seeking higher performance, yet minimize power dissipation and hence hotspots. The peaks in the plot represent the hotspots. This example explains the effectiveness of the optimal against random allocation algorithm in reducing the temperature deviation across the die.

5.4 Computation Time

Figure 7 shows the plot of the computation time for various size of the *speed matrix*. The algorithm is implemented in Matlab and is run on Intel core 2 duo at 2.16 GHz with 2GB RAM. The number of cells in the matrix is given by the product of number of cores and number of tasks. We see that the computation time is proportional to the number of cells. (**Note:** The x-axis is logarithmic, while the y-axis is linear). For the largest size of 16384 cells (128 cores and 128 tasks), the time needed to determine the optimal allocation is around 200 ms. Assuming that if there is a core dedicated to determine the allocation and the algorithm executed in binary format, we can fairly assume an 100X speedup over the Matlab implementation. This shows that even for a large number of cores and tasks it is possible to determine the allocation within the OS scheduling interval (≈ 5 ms).

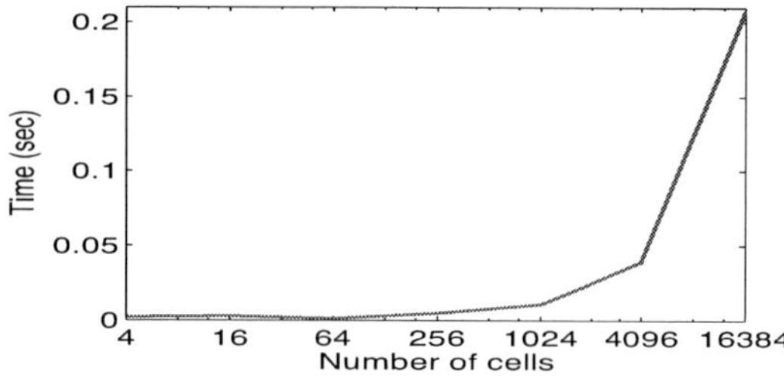

Figure 7: Plot of computation times as a function of number of cells in the *speed matrix*

6. CONCLUSION

Although multi-core processors can deliver higher performance, as they grow from tens to hundreds of cores, their power dissipation will also increase proportionately. With such large numbers of cores, it is unlikely that packaging alone will be sufficient to remove the heat, and dynamic thermal management will have to be employed far more frequently than with single core processors. In this

work, we have used the frequency scaling and task-to-core allocation schemes of thermal management to optimally increase the performance of multi-core processors under thermal constraints. We formulated the combined problem as a linear assignment problem and showed that significant improvement in performance can be achieved over existing techniques (as high as 5X). The algorithm determines the optimal task-to-core allocation and subsequent speed control, accounting for the dependence of leakage on temperature and using an accurate thermal model (HotSpot). Finally, the proposed algorithm is sufficiently fast to be implementable in real time, within the OS.

7. REFERENCES

[1] NAS Parallel Benchmarks. http://www.nas.nasa.gov/Resources/Software/npb.html.
[2] SPEC CPU2000 Benchmarks. http://www.spec.org/benchmarks.html.
[3] S. Borkar. Thousand core chips: A technology perspective. In *DAC*, pages 746–749, 2007.
[4] D. Brooks and M. Martonosi. Dynamic thermal management for high-performance microprocessors. In *Proc. HPCA*, pages 171–182, 2001.
[5] J. Donald and M. Martonosi. Techniques for multicore thermal management: Classification and new exploration. *SIGARCH Comput. Archit. News*, 34(2):78–88, 2006.
[6] D. Wentzlaff et al. On-chip interconnection architecture of the Tile Processor. *IEEE Micro*, 27(5):15–31, 2007.
[7] K. Skadron et al. Control-theoretic techniques and thermal-RC modeling for accurate and localized dynamic thermal management. In *Proc. HPCA'02*, pages 17–28, 2002.
[8] M. D. Powell et al. Heat-and-run: Leveraging SMT and CMP to manage power density through the operating system. *SIGOPS Oper. Syst. Rev.*, 38(5):260–270, 2004.
[9] M. Monchiero et al. Power/performance/thermal design-space exploration for multicore architectures. *IEEE Trans. Parallel Distrib. Syst.*, 19(5):666–681, 2008.
[10] P. Chaparro et al. Understanding the thermal implications of multicore architectures. *TPDS*, 18(8):1055–1065, 2007.
[11] P. Michaud et al. A study of thread migration in temperature-constrained multicores. *ACM Trans. Archit. Code Optim.*, 4(2):9–1–9–28, 2007.
[12] S. Murali et al. Temperature-aware processor frequency assignment for MPSoCs using convex optimization. In *Proc. CODES+ISSS*, pages 111–116, 2007.
[13] T. Constantinou et al. Performance implications of single thread migration on a chip multi-core. *ACM SIGARCH*, 33(4):80–91, 2005.
[14] W. Huang et al. An improved block-based thermal model in hotspot 4.0 with granularity considerations. In *WDDD*, 2007.
[15] W. Liao et al. Temperature and supply voltage aware performance and power modeling at microarchitecture level. *TCAD*, 24(7):1042–1053, 2005.
[16] Y. Han et al. Temptor: A lightweight runtime temperature monitoring tool using performance counters. In *TACS*, pages 17–28, 2006.
[17] J. Munkres. Algorithms for the assignment and transportation problems. *Journal of the Society for Industrial and Applied Mathematics*, 5(1):32–38, 1957.
[18] R. Rao. *Fast and accurate techniques for early design space exploration and dynamic thermal management of multi-core processors*. PhD thesis, Arizona State University, 2008.
[19] R. Rao and S. Vrudhula. Performance optimal processor throttling under thermal constraints. In *Proc. CASES*, pages 257–266, 2007.
[20] R. Rao and S. Vrudhula. Efficient online computation of core speeds to maximize the throughput of thermally constrained multi-core processors. In *Proc. ICCAD*, pages 537–542, 2008.
[21] Y. Taur. CMOS design near the limit of scaling. *IBM J. Res. and Dev.*, 46(23):213–222, 2002.
[22] S. Zhang and K. S. Chatha. Approximation algorithm for the temperature-aware scheduling problem. In *Proc. ICCAD*, pages 281–288, 2007.

978-1-60558-497-3/09 $25.00 © 2009 ACM

An Adaptive Scheduling and Voltage/Frequency Selection Algorithm for Real-time Energy Harvesting Systems

Shaobo Liu, Qing Wu, and Qinru Qiu
Department of Electrical and Computer Engineering
Binghamton University, State University of New York
Binghamton, New York 13902, USA
{sliu5, qwu, qqiu}@binghamton.edu

Abstract – In this paper we propose an adaptive scheduling and voltage/frequency selection algorithm which targets at energy harvesting systems. The proposed algorithm adjusts the processor operating frequency under the timing and energy constraints based on workload information so that the system-wide energy efficiency is achieved. In this approach, we decouple the timing and energy constraints and simplify the original scheduling problem by separating constraints in timing and energy domains. The proposed algorithm utilizes maximum task slack for energy saving. Experimental results show that the proposed method improves the system performance in remaining energy, deadline miss rate and the minimum storage capacity requirement for zero deadline miss rate. Comparing to the existing algorithms, the new algorithm decreases the deadline miss rate by at least 23%, and the minimum storage capacity by at least 20% under various processor utilizations.

Categories and Subject Descriptors

B.8.2 [**Performance and Reliability**]: Performance Analysis and Design Aides

General Terms

Algorithms, Design, Performance

Keywords

Energy harvesting, Dynamic voltage and frequency selection

I. Introduction

The energy constraint remains a major issue for battery powered devices, despite that a lot of researchers have been working actively to solve this problem. Generally the research activities can be grouped into two categories: one is to focus on the reduction of the power consumption of the battery powered device, such as dynamic power management (DPM) [1-3] and dynamic voltage and frequency selection (DVFS) [4-6]. The other is to focus on seeking new energy sources for the device, such as energy harvesting [7-9]. Although both DPM and DVFS techniques are able to effectively reduce the power consumption of a device, the limited energy in the battery will be exhausted eventually; and then the battery has to be either recharged or replaced before the device can continue to function.

However, in some applications, neither recharging nor replacing batteries is practical. One example is the sensor nodes that are deployed in the radioactive surroundings and they are networked together for environment surveillance. In order to increase the lifespan of such application, the energy harvesting technologies [7-9] have been actively explored recently. Energy harvesting is considered as a promising method for overcoming the energy limitation for battery-powered systems and it could let systems achieve energy autonomy. Simply speaking, the energy harvesting system is a system that draws parts or all of its operating energy from its physical surroundings. Several prototypes have been proposed to demonstrate the effectiveness of energy harvesting system such as *Heliomote* [8] and *Prometheus* [9].

Several research works have been carried out in power minimization techniques for energy harvesting systems. An offline algorithm using dynamic voltage and frequency selection (DVFS) is proposed in [10] that targets at real-time tasks. The optimization is done by assuming that harvested energy from the ambient energy source is constant, which is not the case in real applications. The work in [11] chooses the solar power as the harvesting energy source and models it as time-variant. The energy source is assumed to work in two modes: daytime and nighttime. A lazy scheduling algorithm (LSA) is proposed in [12] that executes task as late as possible at full speed, in which the task slack is not exploited for energy savings.

In order to utilize the task slack for energy saving, the authors of [13] proposed an energy-harvesting-aware dynamic voltage and frequency selection (EA-DVFS) algorithm. The proposed algorithm slows down the task execution if the system does not have sufficient available energy; otherwise, the tasks are executed at the full speed. The main shortcomings of this work are:

1) The "sufficient available energy" is defined based on a single current task. As long as the remaining operation time of system at the full speed is more than the relative deadline of the task, then the system considers it has sufficient energy. However, there may be just as little as 1% energy left in the energy storage while the system can operate at full speed for more than the relative deadline of a task. Then EA-DVFS algorithm schedules the task at full speed. That is not the desired behavior.

2) When tasks are scheduled and operating voltages are selected, the EA-DVFS algorithm only considers one task instead of considering all tasks in the ready task queue. This results in that the task slacks are not fully exploited for energy savings.

In this paper we propose an adaptive task scheduling and DVFS algorithm for real-time energy harvesting systems. The goal of the proposed algorithm is to schedule all tasks in the ready queue at the lowest possible speed and allocate the workload to the processor as evenly as possible. The evenly distributed workload not only reduces the overhead from processor voltage and operating frequency switches, but also achieves system-wide energy efficiency [14]. The proposed algorithm also adaptively updates the scheduling and voltage/frequency selection when a new task arrives at the task ready queue. The main features of our approach can be summarized as follows,

1) It decouples the energy constraints and timing constraints for the real-time energy harvesting system so that the scheduling problem subjected to constraints both in timing domain and energy domain can be easily handled.

2) It fully explores the possibility of trading the task slack for

Permission to make digital or hard copies of part or all of this work for personal or classroom use is granted without fee provided that copies are not made or distributed for profit or commercial advantage and that copies bear this notice and the full citation on the first page. To copy otherwise, to republish, to post on servers or to redistribute to lists, requires prior specific permission and/or a fee.
DAC'09, July 26-31, 2009, San Francisco, California, USA

energy saving by adaptively solving the problem when considering multiple tasks in the queue at the same time.

Comparing to the EA-DVFS algorithm, the proposed algorithm fully exploits the task slack for energy savings under timing and energy constraints. As long as the task can be slowed down for energy saving under given timing and energy constraints, the task is executed at a lower speed. The proposed algorithm results in more available/stored energy at any time, comparing to the EA-DVFS and LSA algorithms. Experimental results also show that, the proposed adaptive algorithm can significantly reduce the deadline miss rate under various processor utilizations; it also requires considerably less storage capacity for zero deadline miss rate, comparing to EA-DVFS and LSA algorithms.

The rest of this paper is organized as follows. The energy harvesting system model and some assumptions are presented in Section II. The proposed adaptive scheduling algorithm is described in Section III. Simulation results and discussions are presented in Section IV. Finally Section V gives the conclusions.

II. System Model and Assumptions

As shown in Figure 1, the energy harvesting system we consider in this paper consists of four major modules: energy source module, energy storage module, the uniprocessor module and the real-time task queue module. The energy source module harvests energy and feeds into the energy storage at power $Ps(t)$ at time instance t. The energy storage is the place to store energy and its capacity is denoted as C; the stored energy at time t is denoted by $E_C(t)$. When the stored energy reaches the capacity C, the incoming energy harvesting overflows the energy storage, so we have

$$0 \leq E_C(t) \leq C \quad \forall t \tag{1}$$

When the processor executes the real-time task, it draws energy from energy storage. If the energy storage is empty, the processor stops functioning.

2.1. Energy source

We denote $P_S(t)$ as the net power that the energy source feeds into the storage. The harvested energy $E_S(t_1, t_2)$ at time interval $[t_1, t_2]$ can be calculate by integration:

$$E_S(t_1, t_2) = \int_{t_1}^{t_2} P_S(t)dt \tag{2}$$

The power output of the energy source is a function of time, thus $P_S(t)$ can not be determined ahead. But we can predict it by tracing the energy source profile [15].

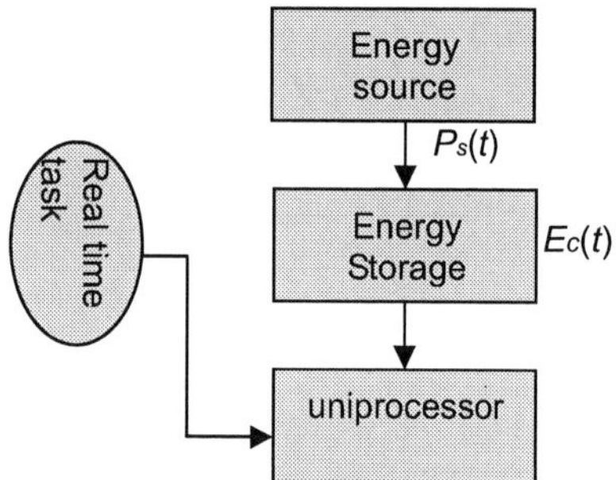

Figure 1 A real-time system with energy harvesting module.

2.2. Energy storage

The energy storage is assumed to be ideal. It can be fully charged and also fully discharged no matter how many charge/discharge cycles it has gone through. The energy that the processor demands only comes from the storage and harvested energy. Let $E_D(t_1, t_2)$ denote the processor energy dissipation from time t_1 to t_2, then we have:

$$E_D(t_1, t_2) \leq E_C(t_1) + E_S(t_1, t_2) \quad \forall t_1 < t_2 \tag{3}$$

Meanwhile, the stored energy should be the surplus that available energy deducts the energy dissipation by the processor if no overflow occurs, so we have:

$$E_C(t_2) \leq E_C(t_1) + E_S(t_1, t_2) - E_D(t_1, t_2) \quad \forall t_1 < t_2 \tag{4}$$

2.3. DVFS-enabled processor and real-time tasks

Assume the DVFS-enabled processor has N discrete operating frequencies f_n: $\{ f_n \mid 1 \leq n \leq N, \quad f_{\min} = f_1 < f_2 < \ldots < f_N = f_{\max} \}$; and the power consumption with regards to f_n is denoted as P_n.

We define a slowdown factor S_n as the normalized frequency of f_n with respect to the maximum frequency $f_{\max}$, that is:

$$S_n = f_n / f_{\max} \tag{5}$$

For the sake of convenience, we use notations f_n, $f(n)$ interchangeably in this paper. Similarly for notations P_n & $P(n)$, and S_n & $S(n)$.

The triplet (a_m, d_m, w_m) is used for characterizing a real-time task τ_m, where a_m, d_m, w_m indicate the arrival time, the relative deadline and the worst case execution time of task τ_m, respectively. Before the real-time task τ_m is released, the triplet (a_m, d_m, w_m) is unknown. Once the task τ_m is released, the triplet is finalized, and τ_m is pushed into the ready task queue Q.

If task τ_m is stretched by a slowdown factor S_n, then its actual execution time at frequency f_n is w_m/S_n. All tasks are scheduled based on earliest deadline first (EDF) policy. The system is considered to be preemptive. The task with the earliest deadline has the highest priority and should be executed first; and it preempts any other task if needed.

III. Adaptive Scheduling Algorithm

In this section we will introduce the proposed adaptive scheduling and voltage/frequency selection algorithm for real-time systems with energy harvesting module. This algorithm dynamically adjusts the processor speed to achieve system-wide energy efficiency based on the workload and available energy information. The key point is that the proposed algorithm decouples the energy constraints and timing constraints originated from a real-time system so that the problem can be easily tackled. The framework of the proposed algorithm consists of three steps:

1) Create an initial schedule for all tasks in the ready task queue; that schedule is based on the lazy scheduling policy with tasks having earlier deadline having higher priority. This step guarantees that timing constraints of the real-time system are met.

2) Distribute the workload as evenly as possible on the processor; dynamic voltage and frequency selection (DVFS) policy is used for slowing down the processor so that the system power consumption goes down under given timing and performance constraints.

3) Tune up the scheduling from step (2) by taking into account the energy constraints. The schedule from step (2) is the energy efficient schedule under the timing constraints [14], but it does not consider the available energy for energy-harvesting system. If the schedule from step (2) is invalidated due to energy shortage, we do not simply remove the tasks. Instead an adaptive policy is adopted: if the system is able to harvest enough energy to finish the task under its given timing constraints, then the task is delayed until the system has sufficient energy; otherwise, the task is removed. Removing the task gives the system a chance to purely accumulate energy by harvesting, which improves the available energy for future tasks.

In the following part, we will explain each step in more details.

3.1 Generate an initial schedule

All tasks in the ready task queue Q are sorted ascendingly in terms

978-1-60558-497-3/09 $25.00 © 2009 ACM

of the task deadline. The task with earliest deadline is put in the head of the queue, and the one with latest deadline in the tail of the queue. In the initial schedule, all tasks are executed at full speed. Then the lazy policy is used to schedule tasks in Q and tasks are always executed as late as possible. In other words, the task in the tail gets executed right at its deadline, and it starts being executed at the time instance when its deadline minuses its worst case execution time.

Assuming that there are M tasks in the task queue, and the first task is located in the head, the last one (M-th) in the tail. In order to get the initial schedule easily, the initial starting time (ist_m) and initial finishing time (ift_m) of each task τ_m ($m=1,2,\ldots, M$) are calculated in a reversed order . Hence, ist_M and ift_M are calculated first, while ist_1 and ift_1 last.

Based on lazy scheduling policy, for the last task τ_M, we have:

$$ift_M = a_M + d_M \tag{6}$$

$$ist_M = a_M + d_M - wcet_M \tag{7}$$

For all other tasks left, the initial schedule is easily obtained by the following equations,

$$ift_m = \min(a_m + d_m, ist_{m+1}) \tag{8}$$

$$ist_m = \min(\max(a_m + d_m - wcet_m, \ a_m), \ ist_{m+1} - wcet_m) \tag{9}$$

where index variable m ranges from M-1 to 1. In order to make the schedule practical, the ist_m can not be smaller than a_m. Note that $a_m + d_m - wcet_m$ is no less than a_m; otherwise task τ_m is not schedulable under the given timing constraint; so we have

$$ist_m = \min(a_m + d_m - wcet_m, \ ist_{m+1} - wcet_m) \tag{10}$$

The indication of the above equation is clear that task τ_m starts being executed either at time instance $a_m + d_m - wcet_m$, when its deadline minuses its worst case execution time, or at the time instance, $ist_{m+1} - wcet_m$, when the starting executing time of its next task ist_{m+1} minus its worst case execution time $wcet_m$, no matter which one is earlier. That scheduling is justified by the following facts: 1) task τ_m is delayed as much as possible so that system may have more energy to execute task by energy harvesting; 2) the timing constraints of task τ_m is guaranteed.

3.2 Balance workload

As long as each task (τ_m) is finished at its initial finishing time (ift_m), the timing constraint is met. However, based on the initial schedule, all tasks are executed at the full speed of the processor, which is not an energy-efficient scheme. We need to make use of the task slacks for energy saving. The dynamic voltage and frequency selection (DVFS) [13] is applied to stretch the execution time of each task and slow down the processor.

The DVFS-enabled processor has multiple operating voltage and frequency levels. In order to achieve the maximum power savings, all tasks should be stretched uniformly [14]. In other words, the processor should avoid operating frequency switches as much as possible.

In terms of the initial schedule, all tasks are executed at the full speed, with the same slowdown factor index SI_m equal to N. Then all tasks in the ready queues are stretched by N rounds of DVFS policy, where N is equal to the number of available operating frequencies to the processor. The effort of using N round of DVFS policy is to make all tasks executed in the same frequency level so that the switch activity of the processor is minimized and the system-wide energy efficiency is maximized.

For a given round, the starting time (st_m) of task τ_m for execution is determined by:

$$st_m = \begin{cases} \max(a_1, current\_time), & m=1 \\ \max(a_m, \ ft_{m-1}), & m=2,\cdots,M \end{cases} \tag{11}$$

where m is from 1 to M.

However, its finishing time (ft_m) is more complicated to obtain. Before calculating ft_m, two questions need to be answered. First,

check if the slowdown factor index SI_m for task τ_m can be reduced further. If the following inequality holds,

$$st_m + wcet_m/S(SI_m\text{-}1) < ift_m \tag{12}$$

then the timing constraint is still met after further stretching task τ_m. Second, check if the slowdown factors for tasks indexed from $m+1$ to M is still valid. If the answers to these two questions are yes, then SI_m for task τ_m is decremented by 1; in other words, the operating frequency of task τ_m is reduced to $f(SI_m\text{-}1)$ from $f(SI_m)$. Otherwise, SI_m is kept as it is.

The slowdown factor S_n is called valid for a given task τ_m if task τ_m can be executed by the processor at frequency f_n subjected to the timing constraints.

Now ft_m can be easily calculated as:

$$ft_m = st_m + wcet_m/S(SI_m). \tag{13}$$

The workload balance algorithm is shown in Figure 2. Line 10 in Figure 2 tells us that the slowdown index SI_m for each task τ_m is decreased at most by 1 at a given round of DVFS. The meaning is two-fold: 1) each task has the same opportunity to be stretched, which avoids some tasks getting overstretched by squeezing out the slack of other tasks; 2) the slack time of tasks is sufficiently exploited for energy savings.

1.	**Require**: get the initial schedule for M tasks in queue Q
2.	**for** $n = 1{:}N$ **do**
3.	**for** $m =1{:}M$ **do**
4.	**if** $m == 1$, **then**
5.	$st_m = \max(a_m, current\_time)$
6.	**else**
7.	$st_m = \max(a_m, ft_{m-1})$
8.	**endif**
9.	**if** $st_m + wcet_m/S(SI_m\text{-}1) < ift_m$ && *the slowdown factors for tasks with lower priority is valid*, **then**
10.	$SI_m = SI_m - 1$
11.	**end if**
12.	$ft_m = st_m + wcet_m/S(SI_m)$
13.	**end for**
14.	**end for**

Figure 2: Workload balance algorithm.

3.3 Check energy availability and tune up schedule

One of the features of the energy harvesting systems is that the available energy is limited by the energy storage capacity. The available energy dynamically fluctuates with time in two opposite directions: increasing or decreasing. Therefore we need to tailor the workload-balanced schedule to that feature.

When tasks are scheduled based on the workload-balanced algorithm in Section 3.2, the energy constraint is not considered. If the energy availability invalidates the schedule, then the processor has to stop the task execution before the task can be finished. In order to overcome that problem, we have to check the energy availability after getting the workload balanced schedule; then tune up the schedule.

If the workload-balanced schedule is invalidated by energy shortage, tasks are not directly removed from the ready task queue. Instead the task execution is first delayed.

For example, if we define that mth task τ_m in Q is the first task whose schedule is invalidated by the energy shortage, so we have:

$$E_C(st_m) + E_S(st_m, ft_m) < E_D(st_m, ft_m) \tag{14}$$

where $E_S(st_m, ft_m)$ is the harvested energy between st_m and ft_m, and it can be estimated based on the profile of energy-harvesting source. Then task τ_m is rescheduled by delaying dl_m until the following equality holds,

$$E_C(st_m) + E_S(st_m, ft_m+dl_m) = E_D(st_m+dl_m, \ ft_m+dl_m) \tag{15}$$

If the deadline of task τ_m is not violated, that is:

$$ft_m + dl_m \leq a_m + d_m \tag{16}$$

978-1-60558-497-3/09 $25.00 © 2009 ACM

and the slowdown factors for tasks indexed by $m+1, \ldots, M$ are still valid, then task τ_m is executed at time interval $[st_m+dl_m, ft_m+dl_m]$ at the frequency $f(SI_m)$, obtained from workload-balanced schedule; and the schedule for tasks with lower priority is updated, as shown in lines 7~10 in Figure 3; otherwise, task τ_m is simply removed from task ready queue, as shown in line 12 in Figure 3.

Note that the tune-up algorithm presented in Figure 3 is executed on the fly and the scheduler has to check the energy availability before the task's execution. We would like to give an example to explain how the tune-up algorithm works. Assuming that the DVFS-enabled processor has 4 operating frequency levels with slowdown factor 1, 0.6, 0.4 and 0.15; and the corresponding power levels are 32, 10, 4, and 0.8. Also assume that there are 2 tasks τ_1 and τ_2 in Q, and they are scheduled by the workload-balanced schedule with $(st_1, ft_1, deadline_1)=(50, 56, 59)$, and $(st_2, ft_2, deadline_2)=(56, 62, 68)$. Both tasks are scheduled to execute at the lowest speed, and the power consumption of the processor is 0.8 at lowest operating frequency.

1.	**Require**: the workload balanced schedule for M tasks in Q
2.	**if** $E_C(st_m)+E_S(st_m,ft_m) < E_D(st_m,ft_m)$, **then**
3.	calculate dl_m from equation (16)
4.	**if** $ft_m+dl_m \leq d_m$ && *the slowdown factors for tasks with lower priority is valid*, **then**
5.	$st_m = st_m+dl_m$
6.	$ft_m = ft_m+dl_m$
	//update schedule for tasks with lower priority in **for** loop
7.	**for** $i = m+1:M$ **do**
8.	$st_i = max(st_i, ft_{i-1})$
9.	$ft_i = st_i + exe_i;$
10.	**endfor**
11.	**else**
12.	remove task τ_m from queue Q
13.	**endif**
14.	**endif**

Figure 3: The tune-up schedule algorithm to guarantee the energy availability.

The available energy in the storage at time instance 50 is set to 1. The harvesting power from time instance 50 to 68 is set to 0.5. If the system executes those two tasks based on the workload balance schedule, then the execution of both tasks will be suspended due to the energy shortage. The following calculation verifies our conclusion: The total energy the system provides at time instance 56 is $1+6*0.5=4$; and the total energy demand for executing task τ_1 is $6*0.8=4.8$. So the energy shortage forces the processor to stop running at time instance 53.3 and the schedule for task τ_1 can not be carried out, shown by the "lime" color long dash line in Figure 4.

On the other hand, before running task τ_1, the energy availability is checked by equation (15), and then task will be delayed by 2 time units; accordingly task τ_1 is executed between time interval [52, 58], and the schedule for task τ_2 is updated as $(st_2, ft_2, deadline_2)=(58, 64, 68)$. After finishing task execution, the remaining energy is 0.2 shown by the "lime" color solid line in Figure 4; energy is not a concern any more for the schedule of task τ_1. The similar argument holds for task τ_2.

3.4 Put all together

As we stated earlier, the proposed algorithm comprises three steps:
1) generate the initial schedule;
2) balance workload for the processor;
3) tune up the schedule under the constraints of energy availability.

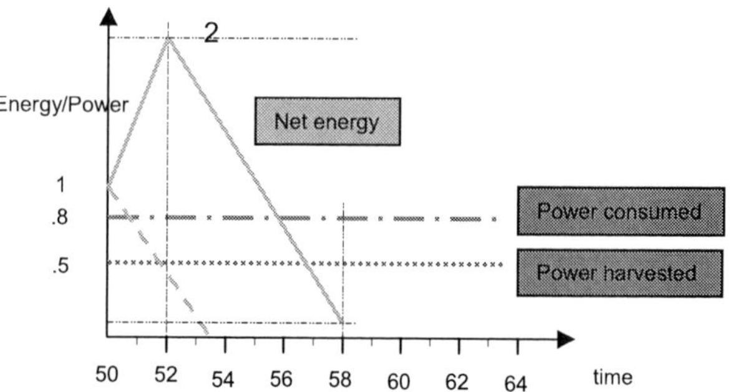

Figure 4: Tuning up scheduling algorithm illustration.

1.	**Require**: maintain a ready task queue Q
2.	set task queue Q empty
3.	**while** (true) **do**
4.	**if** new task coming, **then**
5.	push new task in Q,
6.	sort all tasks ascendingly in Q based on the deadline
7.	get initial schedule for tasks in Q,
8.	balance the workload based on algorithm shown in Figure 2
9.	tune up scheduling, shown in Figure 3
10.	**endif**
11.	execute task
12.	**if** the task is finished, **then**
13.	remove task from the ready task queue Q
14.	**end if**
15.	**end while**

Figure 5: The proposed adaptive and voltage/frequency selection algorithm for real-time energy harvesting system.

In this section, we put the prior discussions together, and construct a complete adaptive scheduling and voltage/frequency algorithm for real-time energy harvesting system, presented in Figure 5.

In the beginning, we assume the ready task queue Q is empty, shown in line 2. Every time the new task comes, it is pushed into Q, as shown in line 5, and then all tasks in Q are sorted ascendingly based on their deadlines.

The event that the new task comes triggers rescheduling all tasks in Q, so that the task with higher priority (the earlier deadline) in the new task queue is scheduled to run earlier, as shown from lines 6~9 in Figure 5.

If there is no new task coming, the processor executes tasks based on the schedule obtained before, as shown in line 11. Once the task is finished, it is removed from Q, as shown in line 13. The key of the proposed algorithm is in lines 6~9. The initial schedule guarantees that tasks meet the timing requirement. The workload balance scheduling algorithm achieves the system-level energy efficiency by two ways: 1) trading the task slack for energy savings by slowing down processor execution speed; and 2) the balanced workload reduces the frequency switch activities for the processor and then the correspondent overhead goes down. The tune-up algorithm makes sure the system has needed energy to execute each task based on the schedule.

IV.   Simulations and Discussions

In this section, we evaluate the performance of the proposed scheduling algorithm based on simulations. We have developed a discrete event-driven simulator in C++ and implement the proposed scheduling algorithm. For comparison purposes, the lazy scheduling algorithm (LSA) in [12, 16] and energy aware DVFS algorithm (EA-DVFS) in [13] are also implemented as benchmarks.

We design three sets of experiments. The first set is designed to show how the system remaining energy is improved by the proposed scheduling algorithm, comparing to LSA and EA-DVFS. The second set of experiments reports how much the deadline miss rate is reduced based on the proposed algorithm, comparing to the other two; the third set of experiments shows the minimum energy storage capacity requirement in order to maintain zero deadline miss rate for three different scheduling algorithms.

4.1 Simulation setup

We choose the solar energy as the energy harvesting source in our simulations. In order to exhibit the stochastic and periodic characteristics of the solar energy source, we multiply a random number generation function with two deterministic cosine functions together [13, 17], as shown in:

$$P_S(t) = \left| 10 \cdot N(t) \cdot \cos(\frac{t}{70\pi}) \cdot \cos(\frac{t}{120\pi}) \right| \qquad (17)$$

where $N(t)$ is a random number generator and it is subject to the normal distribution with mean 0 and variance 1.

A DVFS-enabled processor similar to Intel's Xscale processor [18] is used in the simulations. The processor has five operating frequencies: 150MHz, 400MHz, 600MHz, 800MHz and 1000MHz. Accordingly, the processor also has 5 power levels: 80mW, 400mW, 1000mW, 2000mW and 3200mW, each corresponding to a specific operating frequency. The overhead from the processor operating frequency switching is ignored in the simulations.

The task set consists of limited specific periodic tasks and the cardinality of the task set is arbitrary. The period of the specific task is uniformly drawn from a set {10, 20, 30, ..., 120}; the relative deadline of the task is set to its period; and the worst case execution time is calculated based on its period and harvesting power. Assume that the average harvesting power is $\overline{P_S}$, and the task period is p, the worst case energy consumption e of the task is uniformly drawn from interval $[0,\ \overline{P_S} \cdot p]$; so e is a sample of a uniform-distributed random variable with distribution $[0,\ \overline{P_S} \cdot p]$. Then the worst case execution time of the task can be computed as $e/P_{\max}$.

We define the processor utilization U as

$$U = \sum_m \frac{w_m}{p_m} \qquad (18)$$

where w_m is the worst case execution time of task τ_m, and p_m is the period. The processor utilization stands for the ratio of its busy time over the summation of its busy time plus its idle time when the processor operates at full speed. So the utilization U cannot be larger than 1. To obtain a specific U, we need scale the worst case execution time of each task in a task set in the same ratio in some cases.

The energy storage is assumed to be full in the beginning of the simulation. The simulation terminates after 10,000 time units. For a specific utilization, we repeat experiments for 5,000 task sets.

4.2 Remaining energy comparison

In this set of experiments, we focus on comparing the remaining energy for systems using proposed scheduling algorithm, against systems using two baseline scheduling algorithms, LSA and EA-DVFS. In order to have fair comparison, all simulations are performed under the same condition, except for scheduling algorithms.

From the discussion of the proposed scheduling algorithm, we know that the processor utilization has significant impact on the remaining energy. We do experiments sweeping utilization U from 0.2 to 0.8 with a step of 0.2.

When the utilization U is given, all experiments show the similar trend in the remaining energy, we present the results with 6 periodic tasks in a task set. Note that the reported remaining energy is normalized to the storage capacity.

The simulation results with 4 different processor utilization ratios are plotted in Figure 6. As shown in Figure 6, the proposed algorithm has the most available remaining energy in various processor utilization, comparing to the other algorithms.

When utilization is set to 0.2, the proposed algorithm is able to keep the energy storage almost full for most of the time shown in Figure 6(a). The reason is that most of tasks are scheduled at the lowest speed and the energy consumption goes down considerably; furthermore, the harvesting energy replenishes to the energy storage. The EA-DVFS algorithm only slows down task execution when the system available energy is not sufficient. Therefore, some tasks are executed at full speed, and the others at reduced speed, which leads to the EA-DVFS-based system has less available energy. LSA-based system always executed tasks at full speed, and it accordingly has least available energy at any time.

With the processor utilization increasing from 0.4 to 0.8, the proposed algorithm has less task slack to take advantage for energy savings, therefore the available energy difference among LSA, EA-DVFS and the proposed algorithm decreases, as shown in Figure 6(b-d).

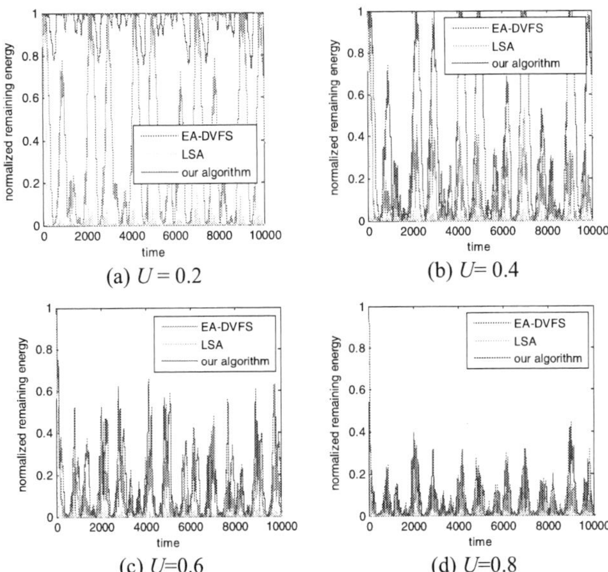

(a) $U = 0.2$ (b) $U = 0.4$

(c) $U=0.6$ (d) $U=0.8$

Figure 6: Remaining energy with different utilizations.

4.3 Deadline miss rate comparison

From the last sub-section, we know that the system based on the proposed scheduling algorithm has significantly more available energy than LSA-based/EA-DVFS-based systems no matter the processor utilization is low or high. The more available energy helps reduce the deadline miss rate due to energy shortage for future tasks. Hence, the proposed algorithm significantly decreases the deadline miss rate, comparing to LSA-based/EA-DVFS-based system, as shown in Figure 7, when U is set to 0.4 and 0.8, respectively.

Note that when utilization is set to 0.8, EA-DVFS performs the same as LSA, as shown in Figure 7(b); nevertheless our algorithm is able to reduce the deadline miss rate by at least 23% on average, comparing to LSA and EA-DVFS. When utilization is high, EA-DVFS schedules almost all of tasks at the full speed, and it can not achieve much energy efficiency. However, the proposed algorithm is able to squeeze out the task slack for energy savings and still improves the system energy efficiency. Hence, the proposed algorithm decreases the deadline miss rate even when the processor utilization is high.

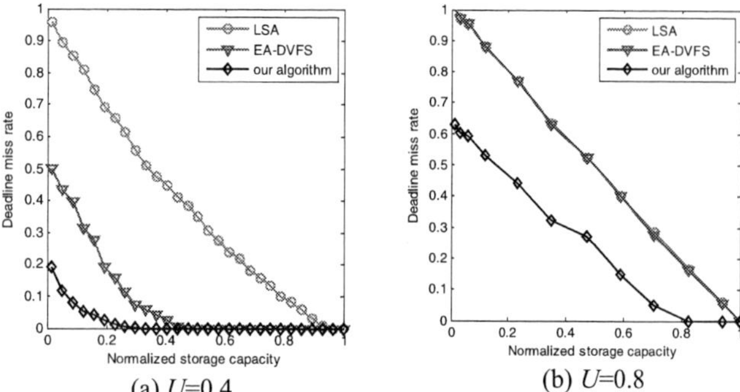

(a) U=0.4   (b) U=0.8

Figure 7: Deadline miss rate with two workloads.

4.4 Storage capacity comparison

Finally, we compare the minimum storage capacity requirement for three scheduling algorithms in order to avoid any deadline miss for any task. Notations $C_{min\text{-}LSA}$, $C_{min,EA\text{-}DVFS}$ and $C_{min,our}$ represent the minimum storage requirement for LSA, EA-DVFS and our algorithm, respectively. They are all normalized to $C_{min\text{-}LSA}$. We sweep the processor utilization U from 0.2 to 0.8 to run experiments with a step 0.2. The results are reported in table 1.

Table 1 Normalized storage capacities.

U	0.2	0.4	0.6	0.8
$C_{min\text{-}LSA}$	1	1	1	1
$C_{min,EA\text{-}DVFS}$	0.43	0.73	0.92	0.98
$C_{min,our}$	0.04	0.34	0.62	0.79

To visualize the storage capacity difference, the normalized storage capacities for LSA , EA-DVFS and the proposed algorithm, are presented in a bar chart in Figure 8.

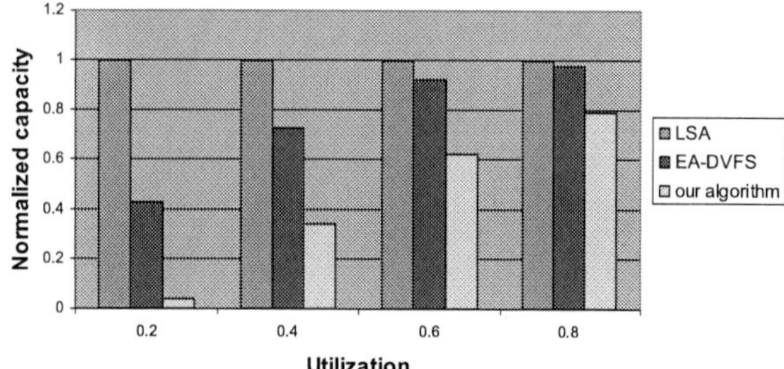

Figure 8: Normalized storage capacity comparison

Observing the results in Table 1 and in Figure 8, we can see that the proposed algorithm requires less storage capacity to achieve zero deadline miss rate in all cases. When utilization is set to 0.2, our algorithm needs as little as 1/10 of the storage capacity that EA-DVFS needs, and as little as 1/23 of the storage capacity that LSA needs. With the utilization going up, the difference between $C_{min\text{-}LSA}$, $C_{min,EA\text{-}DVFS}$ and $C_{min,our}$ reduces. Note that when utilization is set to 0.8, $C_{min,our}$ is about 21% less than $C_{min\text{-}LSA}$, while $C_{min\text{-}LSA}$ and $C_{min,EA\text{-}DVFS}$ are almost the same.

Whether the processor utilization is high or low, the proposed algorithm aggressively trades task slack for energy savings under timing and energy constraints. When utilization is low, the processor are scheduled to run task at lower speed; while utilization high, the task are executed at higher speed but not full speed. Under any circumstances the proposed algorithm saves more energy than LSA and EA-DVFS algorithms. That's why the proposed algorithm requires less storage capacity in all cases.

V. Conclusions

In this paper we have proposed an adaptive scheduling and voltage/frequency selection algorithm targeting at real-time systems with energy harvesting capability. The proposed algorithm consists of three steps:1) generate initial schedule; 2) balance workload; and 3) check energy availability for each scheduled task and tune up the schedule. The first step is to guarantee the timing constraints are met; the second step is to trade task slack for energy savings and the third step is to make sure the energy constraints are met. By dividing the original scheduling problem into three steps, we separate the constraints in timing and energy domains. So the problem can be easily handled.

Experimental results show that, the proposed algorithm aggressively trade the task slacks for energy saving. Hence, no matter the processor utilization is high or low, the proposed algorithm, comparing to LSA and EA-DVFS, increases the system available energy, decreases the deadline miss rate and reduces the energy storage capacity requirement for zero deadline miss rate.

References

1] Lu, Y.-H., Benini, L., and De Micheli, G, "Low-power task scheduling for multiple device", Proc. Int. Workshop HW/SW Co-design, Mar.2000, pp. 39–43

2] Mishra, R., Rastogi, N., Zhu, D., Mosse, D., and Melhem, R, "Energy aware scheduling for distributed real-time systems", Proc. Int. Parallel & Distributed Processing Symp., Apr. 2003

3] S. Liu, Q. Qiu, Q. Wu, "Task merging for dynamic power management of cyclic applications in real-time multi-processor systems", in Proc of ICCD, 2006.

[4] F. Yao, A. Demers, et al,"A scheduling model for reduced CPU energy," in IEEE symposium on Foundations of Comp. Science, 1995.

5] I. Hong, D. Kirovski, G. Qu, M. Potkonjak, and M. B. srivastava. "Power Optimization of Variable-Voltage Core-Based systems," IEEE Trans. On Computer-Aided Design, 1999.

[6] J. Luo and N. K. Jha, "Static and dynamic variable voltage scheduling algorithms for real-time heterogeneous distributed embedded systems," In Proc. Of Int. Conf. on VLSI Design, pp.719-726, 2002

[7] S. Roundy, et.al, "Power sources for wireless sensor networks,". In Proc. Of Wireless Sensor Networks, First Europeean Workshop, 2004.

8] V. Raghunathan, A. Kansal, et al, "Design considerations for solar energy harvesting wireless embedded systems", In Proc. of the International Symposium on Information Processing in Sensor Networks, 2005

9] X. Jiang, J. Polastre, and D. E. Culler, "Perpetual environmentally powered sensor networks", In Proc. of the International symposium on Information Processing in Sensor Networks , 2005

[10] A. Allavena and D. Mosse, "Scheduling of frame-based embedded systems with rechargeable batteries," In Workshop on Power Management for Real-time and Embedded Systems, 2001

11] C. Rusu, R. G. Melhen, and D. Mosse, "Multi-version scheduling in rechargeable energy-aware real-time systems", In 15th Euromicro Conference on Real-time systems, ECRTS 2003, , 2004.

12] C. Moser, D. Brunelli, L. Thiele, and L. Benini, "Lazy scheduling for energy-harvesting sensor ndoes," in Fifth Working Conference on Distributed and Parallel Embedded Systems , 2006

[13] S. Liu, Q. Qiu, Q. Wu, "Energy Aware Dynamic Voltage and Frequency Selection for Real-Time Systems with Energy Harvesting", In Proc. of DATE 2008, 236-241.

14] C. Xian, Y. Lu, Z. Li, "Energy-Aware Scheduling for Real-Time Multiprocessor Systems with Uncertain Task Execution Time", In Proc. of DAC 2007: 664-669

15] A. Kansal, J. Hsu, S. Zahedi and M. Srivastava, "Power Management in Energy Harvesting Sensor Networks," In ACM Transactions on Embedded Computing Systems (in revision) , 35 pages , May 2006.

16] C. Moser, D. Brunelli, L. Thiele, and L. Benini, "Real-time scheduling with regenerative energy," in Proc. of the 18th Euromicro Conference on Real-time Systems (ECRTS06),, 2006.

17] C. Moser, D. Brunelli, L. Thiele, and L. Benini, "Real-Time Scheduling for Energy Harvesting Sensor Nodes,"MICS Scientific Conference and SNF Panel Review, 2006.

18] Intel-Xscale Micro-architecture, available at http://www.intel.com

978-1-60558-497-3/09 $25.00 © 2009 ACM   787

Software-Assisted Hardware Reliability:
Abstracting Circuit-level Challenges to the Software Stack

Vijay Janapa Reddi, Meeta S. Gupta,
Michael D. Smith, Gu-yeon Wei, David Brooks
Harvard University
Cambridge, USA
{vj, meeta, smith, guyeon, dbrooks}@eecs.harvard.edu

Simone Campanoni
Politecnico di Milano
Milano, Italy
simone.campanoni@mail.polimi.it

ABSTRACT

Power constrained designs are becoming increasingly sensitive to supply voltage noise. We propose a hardware-software collaborative approach to enable aggressive operating margins: a checkpoint-recovery mechanism corrects margin violations, while a run-time software layer reschedules the program's instruction stream to prevent recurring margin crossings at the same program location. The run-time layer removes 60% of these events with minimal overhead, thereby significantly improving overall performance.

Categories and Subject Descriptors

C.0 [**Computer Systems Organization**]: General—
Hardware/Software interfaces and System architectures.

General Terms

Performance, Reliability.

Keywords

Runtime Optimization, Hardware Software Co-Design.

1. INTRODUCTION

Power supply noise impacts the robustness and performance of microprocessors. In order to meet power requirements, processor designs are relying on ever lower supply voltages and aggressive power management techniques such as clock gating that can cause large current swings. These current swings when coupled with the parasitic inductances in the power-delivery subsystem can cause voltage fluctuations that violate the processor's operating margins—a significant drop in the voltage can lead to timing-margin violations, due to slow logic paths. On the other hand, significant overshoots in the voltage can also cause long-term degradation in transistor characteristics. For reliable and correct operation of the processor, large voltage swings, also called *voltage emergencies*, should be avoided.

The traditional way of dealing with voltage emergencies has been to over-design the system to accommodate the worst-case voltage swing. A recent paper analyzing supply noise in a Power6 processor [15] shows the need for operating margins greater than 20% of the nominal voltage (200mV for a nominal voltage of 1.1V). Conservative processor designs with large timing margins ensure robustness. However, conservative designs either lower the operating frequency or sacrifice power efficiency.

Alternatively, researchers have proposed designing for the average case operating conditions while providing a "fail-safe" hardware mechanism that guarantees correctness in the presence of voltage emergencies. Such a fail-safe mechanism enables more aggressive timing margins in order to maximize performance, but

Permission to make digital or hard copies of part or all of this work for personal or classroom use is granted without fee provided that copies are not made or distributed for profit or commercial advantage and that copies bear this notice and the full citation on the first page. To copy otherwise, to republish, to post on servers or to redistribute to lists, requires prior specific permission and/or a fee.
DAC'09, July 26-31, 2009, San Francisco, California, USA

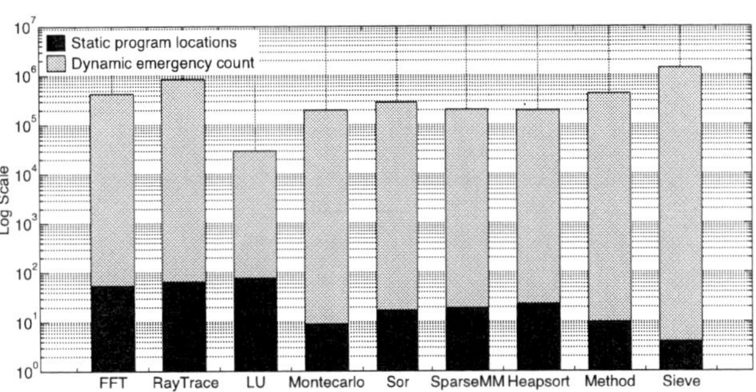

Figure 1: Fewer than 100 static program addresses are responsible for the large number of voltage emergencies. We assume a 4% operating margin, but this trend is the same across margins.

at the expense of some performance penalty. Architecture- and circuit-level techniques either proactively take measures to prevent a potentially impending voltage emergency [11, 16, 22, 23], or operate reactively by recovering a correct processor state after an emergency corrupts machine execution [13].

Traditional hardware techniques do not exploit the effect of program structure on emergencies. Figure 1 shows the number of unique static program addresses responsible for emergencies,[1] and the total number of emergencies they contribute over the lifetime of a program. The stacked log-scale distribution plot indicates that on average fewer than 100 static program addresses are responsible for several hundreds of thousands of emergencies. Even an ideal oracle-based hardware technique will need to activate its fail-safe mechanism once per emergency. Additionally, hardware-based schemes must ensure that performance gains from operating at a reduced margin outweigh the fail-safe penalties. They therefore rely on tuning the fail-safe mechanism to the underlying processor and power delivery system specifics [13]. When combined with implementation costs, potential changes to traditional architectural structures, and challenges like response-time delays [13], design, testing, validation and wide-scale retargetability all become increasingly difficult.

In this paper, we present a hardware-software collaborative approach for dealing with voltage emergencies. Hazelwood and Brooks *et al.* [14] suggest the potential for a collaborative scheme, but we demonstrate and evaluate a full-system implementation design. The collaborative approach relies on a general-purpose fail-safe mechanism as infrequently as possible to handle emergencies, by having a software layer dynamically smooth bursty machine activity via code transformation to prevent frequently occurring emergencies. Ideally, the fail-safe mechanism activates only once per static emergency location, and therefore only a few times in all, as shown in Figure 1.

Dynamic optimization systems [5] are well suited for scenarios where "90% of the execution time is spent in 10% of the code". Figure 1 shows similar behavior with respect to emergencies. A compiler-assisted scheme, in contrast to hardware techniques, can exploit the fact that programs have so few static emergency-

---

[1] We use the event categorization algorithm described by Gupta *et al.* [12] to identify the instruction that gives rise to an emergency.

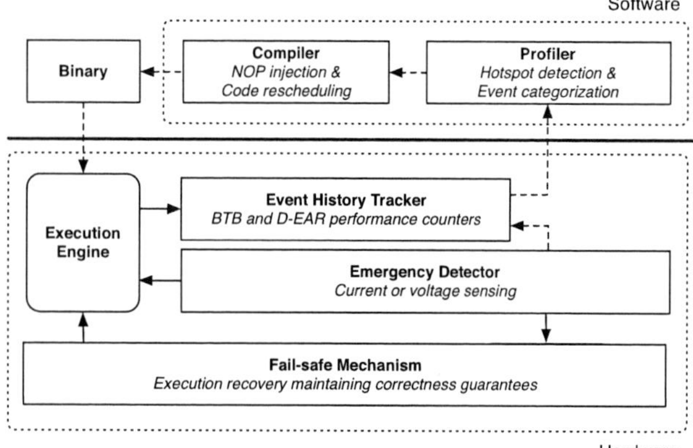

Figure 2: Workflow diagram of the proposed software-assisted hardware-guaranteed approach to deal with emergencies.

prone hot spots. In our scheme, a dynamic compiler eliminates a large fraction of the Dynamic emergency count. We demonstrate a compiler-based issue rate staggering technique that reduces emergencies by applying transformations such as code rescheduling or injecting new code into the dynamic instruction stream of a program. Our primary contributions are as follows:

1. Design and implementation of a dynamic compiler-based system for suppressing recurring voltage emergencies.

2. An instruction rescheduling algorithm that prevents voltage emergencies by staggering the issue rate.

3. Demonstration and evaluation of general purpose checkpoint-recovery hardware to handle voltage emergencies even at aggressive operating margins with our software co-design.

The rest of the paper is organized as follows: Section 2 presents an overview of the proposed hardware-software collaborative approach along with design details for the hardware and software components. Section 3 discusses performance results, and Section 4 concludes the paper.

2. A COLLABORATIVE SYSTEM

The benefits of a collaborative hardware-software approach are twofold: First, recurring emergencies are avoidable via software code transformation. Second, a collaborative scheme allows hardware designers to relax worst-case timing margin requirements because of the reduced number of emergencies. The net effect is better energy efficiency or improved performance. In this section, we first present an overview of how our collaborative architecture works and highlight the critical components. Following that, we present details about each of the hardware and software components.

2.1 Overview

Figure 2 illustrates the operational flow of our system. An Emergency Detector continuously monitors execution. When it detects an emergency, it activates the hardware's Fail-safe Mechanism. We assume that a general-purpose checkpoint-recovery mechanism restores execution to a previously known valid processor state whenever an emergency is detected. After recovery, the detector notifies the software layer of the voltage emergency.

The software operates in *lazy* mode; it waits for emergency notifications from the hardware. Whenever a notification arrives, the software's Profiler extracts information about recent processor activity from the Event History Tracker, which maintains information about cache misses, pipeline flushes, and so on. The Profiler uses this information to identify the code region corresponding to an emergency. Subsequently, the Profiler calls a run-time Compiler to alter the code responsible for causing the emergency in an attempt to prevent future emergencies at the same location.

2.2 Hardware Design

Emergency detector. To detect operating margin violations, we rely on a voltage sensor. The detector invokes the fail-safe mechanism when it detects an emergency. After recovery, the detector invokes the software layer for profiling and code transformation to eliminate subsequent emergencies.

Fail-safe mechanism. Our scheme allows voltage emergencies to occur in order to identify emergency-prone code regions for software transformation. We therefore require a mechanism for recovering from a corrupt processor state. We use a recovery mechanism similar to that found in reactive techniques for processor error detection and correction that have been proposed for handling soft errors [1, 26]. These are primarily based on checkpoint and rollback. Checkpoints can be made either explicitly [2, 17, 18] or they can be saved implicitly [13]. We only evaluate the explicit scheme since it is already shipping in production systems [3, 10] and does not require modifications to traditional microarchitectural structures. The interval between checkpoints is just tens of cycles.

Several researchers have proposed a variety of diverse applications using checkpoint-recovery hardware [17, 18, 20, 24, 26]. Our use of checkpoint-recovery for handling inductive noise in collaboration with software is another novel application of this general-purpose hardware. However, explicit-checkpointing by itself cannot be used to handle voltage emergencies because the performance penalties are too large (as discussed in Section 3.2).

Event history tracker. The software layer requires pertinent information to locate the instruction sequence responsible for an emergency in order to do code transformation. For this purpose, we require the processor to maintain two circular structures similar to those already found in existing architectures. The first is a *branch trace buffer (BTB)*, which maintains information about the most recent branch instructions, their predictions, and their resolved targets. The second is a *data event address register (D-EAR)*, which tracks recent memory instruction addresses and their corresponding effective addresses for all cache and TLB misses. The software extracts this information whenever it receives a notification about an emergency.

2.3 Software Design

Profiler. The profiler is notified whenever an emergency occurs. The profiler identifies emergency-prone program locations for the compiler to optimize. It records the time and frequency of emergency occurrences in addition to recent microarchitectural event activity extracted from the performance counters. Using this information the profiler locates the instruction responsible for an emergency using an *event categorization* algorithm [12]. We refer to this problematic instruction as the *root-cause* instruction. Event categorization identifies root-cause instructions based on the understanding that microarchitectural events along with long-latency operations can give rise to pipeline stalls. A burst of activity following the stall can cause the voltage to drop below the minimum operating margin due to a sudden increase in current draw. Such a violation of the minimum voltage margin is by definition a voltage emergency.

Figure 3(a) illustrates a scenario where a data dependence on a long-latency operation stalls all processor activity. When the operation completes, issue rate increases rapidly as several dependent instructions are successively allocated to different execution units. This gives rise to a voltage emergency because of the sudden increase in current draw. The categorization algorithm associates the long-latency operation as the root cause since it caused the burst of activity that gave rise to an emergency. The algorithm also takes into account other processor activity such as cache and TLB misses and branch mispredictions, as they can also cause emergencies.

978-1-60558-497-3/09 $25.00 © 2009 ACM

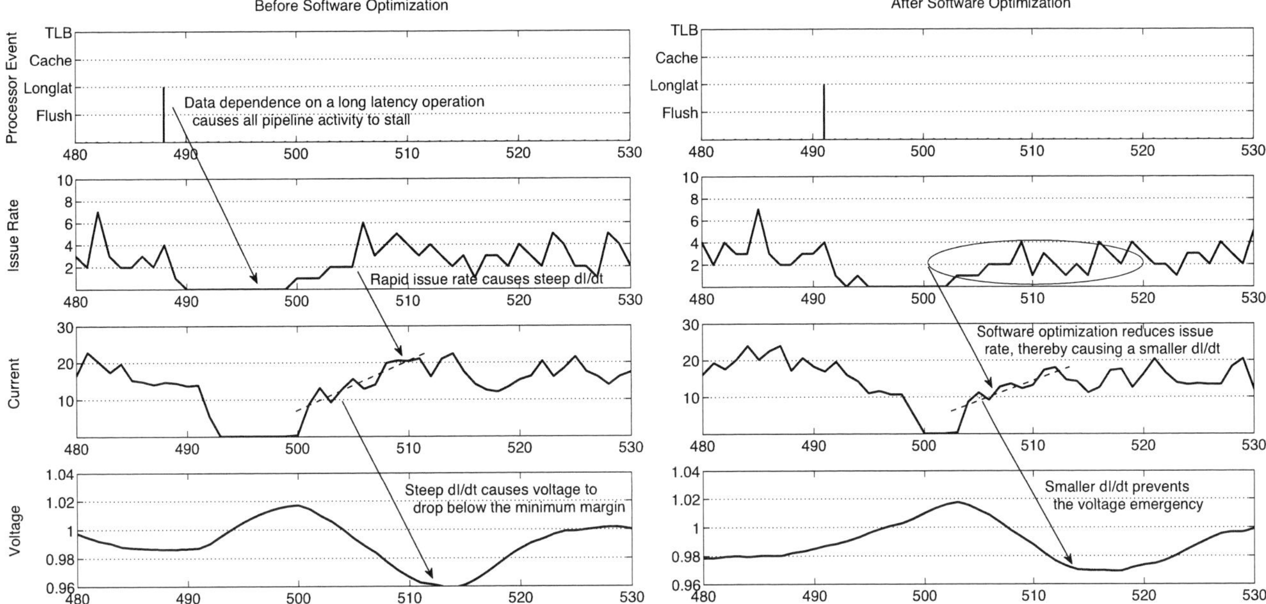

Figure 3: Snapshot of benchmark *Sieve* showing the impact of a pipeline stall due to data dependency. An operating margin of 4% is assumed (i.e., max. of 1.04V and min. of 0.96V). (a) **Before Software Optimization** shows how a stall triggers an emergency as the issue rate ramps up quickly once the long-latency operation completes. (b) **After Software Optimization** demonstrates how compiler-assisted code rescheduling slows the issue rate to prevent the emergency illustrated in (a).

Compiler. Figure 3(a) illustrates that voltage emergencies are dependent on the issue rate of the machine. Therefore, slowing the issue rate at the appropriate point can prevent voltage emergencies. Hardware-based solutions have been proposed that prevent emergencies by altering machine behavior via execution throttling [11,16,22,23] or staggering the issue rate [21,22]. But as high issue rate alone is insufficient to cause emergencies, throttling in all cases can penalize performance unnecessarily.

Alternatively, Toburen [25] and Yun and Kim [27] demonstrate static compiler techniques that target voltage emergencies. However, emergencies are the result of complex interactions between the application, the execution engine, and the power delivery subsystem. Therefore, these static optimizations are not easily retargetable across different combinations of platform and application. In other words, the emergency-prone static program locations discussed in Figure 1 differ depending on platform specifics.

In contrast, our software approach prevents emergencies by altering the program code that gives rise to emergencies at execution time and does so without slowing down the machine. The compiler tries to exploit pipeline delays by rescheduling instructions to decrease the issue rate close to the root-cause instruction. Pipeline delays exist because of NOP instructions or read-after-write (RAW), write-after-read (WAR), or write-after-write (WAW) dependencies between instructions. Hardware optimization techniques like register renaming in superscalar machines can optimize away WAR and WAW dependencies, so a RAW dependence is the only kind that forces the hardware to execute in sequential order. The compiler tries to exploit RAW dependencies that already exist in the program to slow the issue rate by placing the dependent instructions close to one another.

NOP injection. The compiler can slow down pipeline activity by inserting NOP instructions specified in the instruction set architecture into the dynamic instruction stream of a program. However, modern processors discard NOP instructions at the decode stage. Therefore, the instruction does not affect the issue rate of the machine. Instead of real NOPs, the compiler can generate a sequence of instructions containing RAW dependencies that have no effect. But since these *pseudo-NOP* instructions perform no real work, this approach degrades performance.

Code rescheduling. A better way reduce processor activity is to exploit RAW dependencies already existing in the original control flow graph (CFG) of the program. The compiler attempts to relocate RAW dependencies to a point following the root cause of an emergency, thereby constraining the burst of activity after the stall and consequently preventing the emergency.

Whether the compiler can successfully move instructions to create a sequence of RAW dependencies depends on if moving the code violates either control dependencies or data dependencies. The compiler's instruction scheduler does not break data dependencies, but it works around control dependencies by cloning the required instructions and moving them around the control flow graph carefully such that the original program semantics are still maintained. Aggressive cloning can potentially impact the performance of a program by increasing the total number of instructions executed at run time. For this reason, our scheduler does not migrate instructions if the estimated instruction count of the program is likely to increase from cloning.

For illustrative purposes, we present in Figure 4(a) a simplified sketch of the code corresponding to the activity shown in Figure 3(a). The long-latency operation illustrated in Figure 3 corresponds to the *divide* instruction shown in basic block 4 of Figure 4. An emergency repeatedly occurs in basic block 3 along the dotted loop backedge path $4 \rightarrow 1 \rightarrow 2 \rightarrow 3$. The categorization algorithm identifies the *divide* instruction corresponding to $C \leftarrow A \ / \ B$ in basic block 4 as the root-cause instruction. The compiler identifies the control flow path using the branch history information extracted by the profiler from the BTB counters, and recognizes that moving instruction $A \leftarrow B$ from basic block 1 to 2 will constrain the issue rate of the machine because of a tighter sequence of RAW dependencies. But the compiler also recognizes that the result of $A \leftarrow B$ is live along edge $1 \rightarrow 3$, so it clones the instruction into a new basic block (basic block 5) along that edge to ensure correctness.

The resulting effect (on the more complex actual code sequence) after rescheduling is illustrated in Figure 3(b). The slight change in current activity between cycles 490 and 500 is a result of code rescheduling. After dependent instructions are packed close to one another in basic block 2, the issue rate in Figure 3(b) does not spike as high as it does in Figure 3(a) once

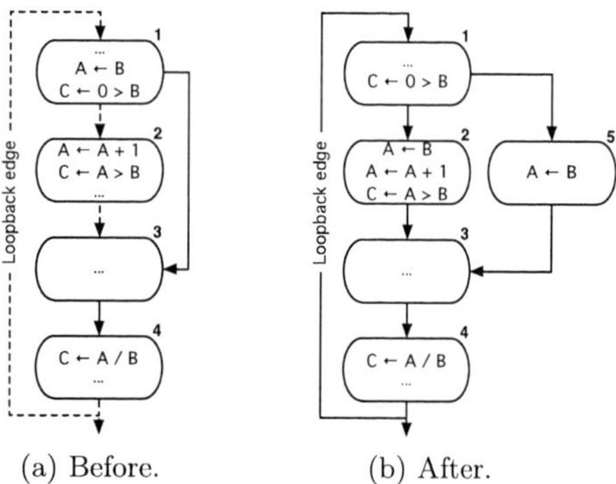

(a) Before.   (b) After.

Figure 4: Effect of code rescheduling on an emergency-prone loop from benchmark *Sieve*. (a) Emergencies consistently occur in basic block 3 along the dotted loop backedge path $4 \to 1 \to 2 \to 3$. (b) Moving instruction $A \leftarrow B$ from block 1 to block 2 puts dependent instructions closer together, thereby constraining the issue rate. This prevents all subsequent emergencies in basic block 3.

pipeline activity resumes after the stall.

Code rescheduling alters the current and voltage profile. Therefore, the scheduler must be careful not to simply displace emergencies from one location to another by arbitrarily moving code from far away regions. To retain the original activity, the code rescheduling algorithm searches for RAW dependencies starting with the basic block containing the root-cause instruction. As the program is running, the profiler tracks the instructions that are giving rise to voltage emergencies. Using this information, the compiler computes a set of program instructions, P, for rescheduling. For each instruction in P, the scheduler searches for RAW dependencies starting from the root-cause instruction. The scheduler enlarges its search window iteratively over the CFG until it finds a RAW dependence to exploit or it reaches the scope of a function body, at which point it gives up.

Out-of-order execution complicates instruction rescheduling, as the machine can bypass the RAW dependence chain generated by the compiler if there is enough other code available for execution in the hardware's scheduling window. The scheduler handles this by choosing a RAW candidate from a set C_1 of candidates by computing the subset $C_2 \subseteq C_1$ such that each element of C_2 has the longest RAW dependence chain after moving the instructions to the required location. By targeting long RAW dependence chains, the compiler causes the machine's scheduling window to fill up with dependent code that reduces the issue rate. Otherwise, the compiler must generate multiple sets of smaller RAW dependence chains. But the more code the compiler moves around, the higher the chances that it will render optimization ineffective because of statically unpredictable interactions among the dependence chains.

3. RESULTS

Evaluation of our system demonstrates the effectiveness of the compiler at reducing voltage emergencies and shows the impact of its code changes on performance. After showing in Section 3.1 that the compiler can reduce over 60% of emergencies, we present a performance study in Section 3.2 showing that our software-assisted scheme overcomes the challenges of existing hardware techniques effectively.

Hardware simulator. We use SimpleScalar/x86 to simulate a Pentium 4 with the characteristics shown in Table 1. The modified 8-way superscalar $x86$ SimpleScalar gathers detailed cycle-accurate current profiles using Wattch [7]. Voltage variations are calculated by convolving the simulated current profiles with an impulse response of the power delivery subsystem [16,23]. In this work, we focus on a power delivery subsystem model based on the characteristics of the Pentium 4 package [4], which exhibits a mid-frequency resonance at 100MHz with a peak impedance of $5m\Omega$. Finally, we assume peak current swings of 16–50A.

Clock Rate	3.0 GHz	RAS	64 Entries
Inst. Window	128-ROB, 64-LSQ	Branch Penalty	10 cycles
Functional Units	8 Int ALU, 4 FP ALU, 2 Int Mul/Div, 2 FP Mul/Div	Branch Predictor BTB	64-KB bimodal gshare/chooser 1K Entries
Fetch Width	8 Instructions	Decode Width	8 Instructions
L1 D-Cache	64 KB 2-way	L1 I-Cache	64 KB 2-way
L2 I/D-Cache	2MB 4-way, 16 cycle latency	Main Memory	300 cycle latency

Table 1: Architecture parameters for SimpleScalar.

Software infrastructure. We use the ILDJIT [9] CIL compiler as our framework for optimizing emergencies at run time. The compiler dynamically generates native $x86$ code from CIL byte code, which it then executes directly on the simulator. We extended the ILDJIT compiler to include the code injection and scheduling algorithms described in Section 2.3. The compiler has access to metadata such as the complete control flow graph and data flow graph, all of which is utilized at run time for optimization. The C# benchmarks evaluated in this paper come from the Java Grande benchmark suite [8,19].

Due to space constraints, we omit discussion about the negligible overheads of run-time code transformation. Figure 1 shows that the number of static emergency-prone program locations (root-cause instructions) is fewer than a hundred. Therefore, our compiler is rarely invoked during execution to transform the code. When combined with the fact that the ILDJIT compiler does its profiling and code transformation in a separate thread on a separate core, performance overhead on the original program is negligible. Also, in our experiments we observe that the fraction of emergencies encountered during ILDJIT's own execution is around 1% across all benchmarks. Since ILDJIT cannot recompile itself, we incur rollback penalties during the compiler's execution. But because the fraction of emergencies is so small, the rollback overhead is insignificant.

3.1 Compiler Efficiency

The goal of our software-based voltage emergency elimination is to: (1) reduce the number of voltage emergencies, and (2) ensure that performance does not suffer as a result of our code transformations. In the next section, we factor in all costs to evaluate full-system performance. In this section, we evaluate the effectiveness of NOP injection and code rescheduling. The main observations are that (1) the choice of transformation affects performance, and that (2) the transformation itself can introduce new emergencies if the scheduler is not careful.

NOP injection. In this transformation, the compiler modifies the original program to contain new instructions that simulate a NOP instruction immediately following the root-cause instruction. The effectiveness of the transformation is shown by the left bar in Figure 5(a). The bar shows the fraction of emergencies remaining after the compiler has attempted to prevent emergencies by injecting pseudo-NOP code. The number of emergencies is reduced by ~50% in benchmarks *FFT*, *RayTrace*, *Method*, *Sieve*, and *Heapsort*, which shows that the transformation can be effective. However, the transformation is ineffective across the remaining benchmarks *LU*, *Montecarlo*, *Sor* and *SparseMM*.

Analysis reveals that pseudo-NOP injection does reduce the original program's emergencies, but the transformation itself gives rise to new emergencies. The compiler might occasionally have to spill and fill registers to generate pseudo-NOP code. This has the adverse effect of not only increasing the number of instructions needed to simulate the NOP, but also potentially causing architectural events like cache misses (from the spill and fill code) that dramatically alter the current and voltage profile. These side effects depend on the number of registers available for use, the properties of the original instruction schedule, and other conditions. It is hard to predict what current and voltage activity will result from injecting new code, so it may give rise to new emergencies. That is what we observe with the poorly performing

978-1-60558-497-3/09 $25.00 © 2009 ACM

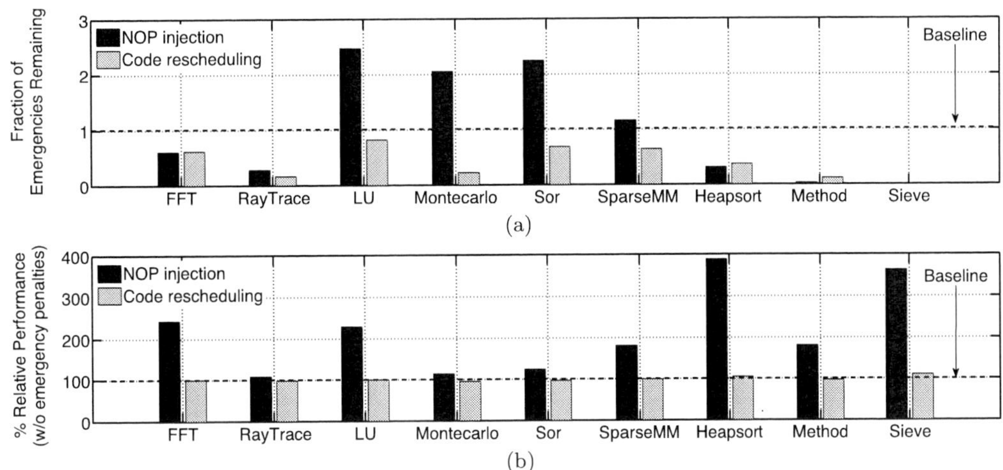

Figure 5: (a) Fraction of emergencies remaining after code transformation. (b) Code performance after transformation. The cost for handling emergencies is not shown in this plot to isolate the effect of code transformation on the run-time performance. Section 3.2 evaluates overall performance after factoring in code performance costs, along with penalties for handling emergencies.

benchmarks *LU*, *Montecarlo*, *Sor*, and *SparseMM*.

Additionally, the run-time performance of the original program suffers with the injection of pseudo-NOP code, as the new code does not serve the original program's purpose. The left bar in Figure 5(b) shows execution performance of the program with the injected code. The data indicates that the effect of simply adding new code to prevent emergencies can be severely detrimental to performance. In the case of benchmarks *Heapsort* and *Sieve* performance degrades by as much as 300%. Large execution overheads indicate that while a transformation can be very effective at reducing voltage emergencies (e.g., benchmark *Sieve* has fewer than 10 emergencies remaining), the compiler must be sensitive to its run-time performance implications.

Code rescheduling. A compiler approach that relocates RAW dependencies following the root-cause instruction does not suffer from the severely unpredictable behavior of injecting code to prevent emergencies. Code rescheduling is superior to simple NOP injection for the following reasons. First, it successfully reduces more emergencies across all the benchmarks (illustrated by the bars on the right in Figure 5(a)). Second, it does so without dramatically increasing the execution time of a program (as shown in Figure 5(b)). Our analysis also shows that it does not introduce new emergencies, as the compiler does not inject new code that significantly alters the current and voltage profile.

For instance, consider benchmark *FFT*. The NOP injection transformation and the code rescheduling transformation eliminate approximately the same number of emergencies. However, the effect on performance between the two transformations is substantially different. The NOP injection transformation causes the original program to take twice as long to execute, whereas code rescheduling has a negligible effect on the original program's performance. That is because the NOP code wastes processor cycles, while the rescheduled instructions are real program code that is simply restructured to prevent emergencies.

Overall, changes in the run-time performance of the rescheduled code are negligible across all the benchmarks, and the reduction in emergencies averages ~61%. Reductions are smaller over benchmarks *LU*, *Sor* and *SparseMM* (around 30%) because the compiler could not find enough RAW dependencies that it could relocate to slow the issue rate at the frequently occurring root-cause locations.

3.2 Full-System Performance Evaluation

Reducing operating voltage margins allows for frequency improvements or improved energy efficiency. However, there are fail-safe mechanism penalties associated with handling voltage emergencies at tighter margins. In this section, we demonstrate that by using our dynamic compilation strategy, it is possible to leverage general-purpose checkpoint-recovery for voltage emergencies at very aggressive margins. Performance gains for our collaborative approach are within 4 percentage points of an oracle-based throttling scheme. Results are presented in Table 2.

Bowman et al. show that removing a 10% operating voltage margin leads to a 15% improvement in clock frequency [6]. This indicates a scaling factor of 1.5 from operating voltage margin to clock frequency. We assume an aggressive operating margin of 4% in our experiments as compared to a 18% worst-case margin[2]. Based on the 1.5x scaling factor, the 4% operating voltage margin assumed in this paper corresponds to a 6% loss in frequency. Similarly, a conservative voltage margin of 18%, sufficient to cover the worst-case drops observed, leads to 27% lower frequency. If we take this conservative margin as the baseline for comparisons and the 18% margin can reduce to 4% while avoiding voltage emergencies, the resulting clock frequency increase could be ~29%. This sets the upper bound on frequency gains achievable. We make the simplifying assumption that frequency improvements translate to higher overall system performance.

Fail-safe mechanism. An explicit-checkpointing scheme recovers from an emergency by rolling back execution. The explicit-checkpoint scheme suffers from the penalty of rolling back useful work done whenever a voltage emergency occurs. The restart penalty is a direct function of the sensor delay in the system, i.e., the time required to detect a margin violation. An explicit-checkpoint scheme incurs additional overhead associated with restoring the registers (assumed to be 8 cycles, for 32 registers with 4 write ports) and memory state (when volatile lines are flushed, additional misses can occur at the time of rollback).

Assuming a 50-cycle rollback penalty per recovery, an explicit-checkpoint scheme incurs an average increase of 25% in CPI over the set of benchmarks evaluated in Figure 5. Performance gains from scaling the operating margin down to 4% are negligible at only 3%. This minimal improvement in performance implies that explicit-checkpointing by itself cannot handle voltage emergencies successfully at aggressive margins.

Fail-safe mechanism with code rescheduling. While the performance gains using only explicit-checkpointing are minimal, the gains are larger when the fail-safe mechanism is combined with a software counterpart (as proposed in Section 2 and illustrated in Figure 2). Of the two compiler transformations discussed in Section 2.3 we evaluate the code rescheduling transfor-

---

[2]The worst voltage drop we observe for our power delivery package is 18%.

978-1-60558-497-3/09 $25.00 © 2009 ACM

Scheme	CPI Overhead	Performance Gain
Fail-safe mechanism	25.0%	3.0%
Fail-safe mechanism with code rescheduling	7.6%	19.8%
Oracle-based throttling	4.0%	23.8%

Table 2: Increase in CPI to handle voltage emergencies, and net performance improvement after scaling the operating margin and factoring in the overheads. The upper bound on performance improvement is 29% assuming the margin is scaled from 18% to 4%. These results are the average measured across all benchmarks.

mation only, since its changes effectively reduce the number of emergencies (as discussed in Section 3.1), but are not detrimental to program performance.

The profiler identifies root-cause instructions as the fail-safe checkpoint scheme initiates rollbacks. So there is some amount of rollback penalty associated with initially discovering root-cause instructions for transformation. Thereafter, however, the compiler optimizes the root-cause instructions to prevent subsequent occurrences of emergencies at the same program location. If the rescheduling algorithm is ineffective at fixing certain emergency points, rollback penalties may still arise at those points (as shown in Figure 5(a) and discussed in Section 3.1). Combining explicit checkpointing with compiler assistance reduces checkpointing overhead substantially, from 25% to 7.6%. This translates to a net performance gain of $\sim$20%.

Comparison to other schemes. Several researchers have proposed mechanisms that spread out a sudden increase in current via execution throttling. Several kinds of throttling have been proposed [11,16,22,23]. For evaluation purposes, we compare the performance of our scheme against a frequency throttling mechanism that quickly reduces current load. The frequency of the system is halved whenever throttling is turned on, which results in performance loss.

We compare against an oracle-based throttling scheme, which throttles once per emergency and always successfully prevents the emergency. As a result, an oracle scheme does not suffer from rollback costs, nor does it suffer from performance loss due to throttles that cannot prevent emergencies. Oracle-based throttling enables $\sim$24% improvement in performance for tightened margins, which is just 4 percentage points better than our scheme. Of course, our scheme represents a practical design.

While an oracle-based scheme always successfully prevents emergencies, it is important to remember that realistic sensor-based implementations suffer from a tight feedback loop that involves detecting an imminent emergency and then activating the throttling mechanism in a timely manner to avoid the emergency. The detectors are either current sensors or voltage sensors that trigger when a certain threshold is crossed, indicating that a violation is likely to occur. Unfortunately, the delay required to achieve acceptable sensor accuracy inherently limits the effectiveness of these feedback-loop schemes, and operating margins must remain large enough to allow time for the loop to respond [13].

In contrast, our collaborative approach does not suffer from the limitations of sensor-based schemes. It leverages general-purpose checkpointing hardware that is already shipping in production systems [3,10] to reduce voltage emergencies at very aggressive margins that enable significant performance gains.

4. CONCLUSION

The primary contribution of this work is a full system implementation design for a hardware-software collaborative approach to handle voltage emergencies. The collaborative approach reduces hardware penalties associated with handling voltage emergencies by having the software (a dynamic compiler) permanently fix the code region responsible for emergencies. The hardware provides fail-safe guarantees via a checkpoint-recovery mechanism, while the software layer identifies the emergency-prone code regions and reschedules that code to prevent further emergencies. The compiler eliminates over 60% of the emergencies on

average, and therefore dramatically reduces the recurring overhead of the fail-safe mechanism. We show that by scaling the operating margin down from a conservative 18% to an aggressive 4% setting, we can achieve $\sim$20% higher performance, which is within 4 percentage points of an oracle-based throttling scheme.

Acknowledgments

We are grateful to Glenn Holloway and the anonymous reviewers for their comments and suggestions. This work is supported by gifts from Intel Corporation, National Science Foundation grants CCF-0429782 and CSR-0720566 and in part by ST Microelectronics and the European Commission under Framework Programme 7 (the OpenMedia Platform project).

5. REFERENCES

[1] S. Agarwal et al. Adaptive incremental checkpointing for massively parallel systems. In *ICS*, 2004.

[2] H. Akkary, R. Rajwar, and S. Srinivasan. Checkpoint processing and recovery: Towards scalable large instruction window processors. In *MICRO-36*, 2003.

[3] H. Ando and et al. A 1.3 GHz Fifth-Generation SPARC64 Microprocessor. In *Design Automation Conference*, 2003.

[4] K. Aygun et al. Power delivery for high-performance microprocessors. *Intel Technology Journal*, 9, 2005.

[5] V. Bala, E. Duesterwald, and S. Banerjia. Dynamo: a transparent dynamic optimization system. In *PLDI*, 2000.

[6] K. A. Bowman et al. Energy-efficient and metastability-immune timing-error detection and instruction replay-based recovery circuits for dynamic variation tolerance. In *ISSCC*, 2008.

[7] D. Brooks, V. Tiwari, and M. Martonosi. Wattch: A framework for architectural-level power analysis and optimizations. In *ISCA-27*, 2000.

[8] M. Bull, L. Smith, M. Westhead, D. Henty, and R. Davey. Benchmarking java grande applications. In *The Practical Applications of Java*, 2000.

[9] S. Campanoni, G. Agosta, and S. C. Reghizzi. A parallel dynamic compiler for cil bytecode. *SIGPLAN Not.*, 2008.

[10] S. et al. Ibm's s/390 g5 microprocessor design. *Micro, IEEE*, 1999.

[11] E. Grochowski et al. Microarchitectural simulation and control of di/dt-induced power supply voltage variation. In *HPCA-8*, 2002.

[12] M. S. Gupta et al. Towards a software approach to mitigate voltage emergencies. In *ISLPED*, 2007.

[13] M. S. Gupta et al. DeCoR: A delayed commit and rollback mechanism for handling inductive noise in processors. In *HPCA-14*, 2008.

[14] K. Hazelwood and D. Brooks. Eliminating Voltage Emergencies via Microarchitectural Voltage Control Feedback and Dynamic Optimization. In *ISPLED*, 2004.

[15] N. James et al. Comparison of split-versus connected-core supplies in the POWER6 microprocessor. In *ISSCC*, 2007.

[16] R. Joseph, D. Brooks, and M. Martonosi. Control techniques to eliminate voltage emergencies in high performance processors. In *HPCA-9*, 2003.

[17] N. Kirman, M. Kirman, M. Chaudhuri, and J. Martinez. Checkpointed early load retirement. In *HPCA-11*, 2005.

[18] J. F. Martínez, J. Renau, M. C. Huang, M. Prvulovic, and J. Torrellas. Cherry: Checkpointed early resource recycling in out-of-order microprocessors. In *MICRO-35*, 2002.

[19] J. A. Mathew, P. D. Coddington, and K. A. Hawick. Analysis and development of java grande benchmarks. In *Java Grande Conference*, 1999.

[20] S. Narayanasamy, G. Pokam, and B. Calder. BugNet: Continuously Recording Program Execution for Deterministic Replay Debugging. In *ISCA*, 2005.

[21] M. D. Pant et al. An architectural solution for the inductive noise problem due to clock-gating. In *ISLPED*, 1999.

[22] M. D. Powell and T. N. Vijaykumar. Pipeline muffling and a priori current ramping: architectural techniques to reduce high-frequency inductive noise. In *ISLPED*, 2003.

[23] M. D. Powell and T. N. Vijaykumar. Exploiting resonant behavior to reduce inductive noise. In *ISCA-28*, 2004.

[24] S. Shyam, K. Constantinides, S. Phadke, V. Bertacco, and T. Austin. Ultra Low-Cost Defect Protection for Microprocessor Pipelines. In *ASPLOS*, 2006.

[25] M. Toburen. Power Analysis and Instruction Scheduling for Reduced di/dt in the Execution Core of High-Performance Microprocessors. Master's thesis, NC State University, USA, 1999.

[26] N. J. Wang and S. J. Patel. ReStore: Symptom-based soft error detection in microprocessors. *TDSC.*, 2006.

[27] H.-S. Yun and J. Kim. Power-aware Modulo Scheduling for High-Performance VLIW Processors. In *ISLPED*, 2001.

978-1-60558-497-3/09 $25.00 © 2009 ACM

Simultaneous Clock Buffer Sizing and Polarity Assignment for Power/Ground Noise Minimization[*]

Hochang Jang
School of Electrical Engineering and Computer Science
Seoul National University
Seoul, Korea
potato99@snu.ac.kr

Taewhan Kim
School of Electrical Engineering and Computer Science
Seoul National University
Seoul, Korea
tkim@ssl.snu.ac.kr

ABSTRACT

This work addresses the problem of minimizing power/ground noise in the clock tree synthesis. Contrary to the previous approaches which only make use of assigning polarities to clock buffers to reduce power/ground noise, our approach solves a new problem of simultaneous consideration of assigning polarities to clock buffers and determining buffer sizes to fully exploit the effects of buffer sizing together with polarity assignment on the minimization of power/ground noise while satisfying the clock skew constraint. Through experiments with MCNC benchmark circuits, it is shown that the proposed solution produces designs with 19.1% less power and 16.2% less ground noise as well as 15.6% less total peak current over that by the conventional method.

Categories and Subject Descriptors

B.7.2 [**Integrated Circuits**]: Design Aids—*Layout, Placement and routing*

General Terms

Design, Reliability

Keywords

Clock synthesis, buffer insertion, power/ground noise

1. INTRODUCTION

[*]The work was supported by Nano IP/SoC Promotion Group of Seoul R&BD Program, IT-SoC Program, System IC2010 project of Korea Ministry of Knowledge Economy, the Korea Science and Engineering Foundation (KOSEF) grant funded by the Korea government (No. R01-2007-000-20891-0), and the Korea Ministry of Knowledge Economy (MKE) under the Information Technology Research Center (ITRC) support program supervised by the Institute for Information Technology Advancement (IITA) (IITA-2008-C1090-0804-0009).

Permission to make digital or hard copies of part or all of this work for personal or classroom use is granted without fee provided that copies are not made or distributed for profit or commercial advantage and that copies bear this notice and the full citation on the first page. To copy otherwise, to republish, to post on servers or to redistribute to lists, requires prior specific permission and/or a fee.
DAC'09, July 26-31, 2009, San Francisco, California, USA

As the supply voltage decreases in the modern VLSI design, the power and ground noise has a crucial effect on the circuit performance, such as the delay of switching signal [1, 2]. The maximum voltage drop of power/ground line is determined by the current peak [3], and the high current peak is driven by wrong or inappropriately dimensioned power/ground routings or peak current sources. The amplitude of peak current increases when numerous signals driven by neighboring sources switch simultaneously. In a synchronous circuit, the buffered clock tree incessantly consumes a considerable amount of current. In addition, the amount of resulting clock power on the clock distribution network and the clocked loads typically accounts for one third to one half of the total chip power dissipation [4]. Since the clock buffers consume the current at the clock edges, a large amount of current is generated around the clock edges, which makes the clock buffers be one of the major sources of power/ground noise.

There have been proposed several works which have tried to reduce the peak current around the clock edges and the resulting power/ground noise. Benini, Vuillod, Bogliolo, and De Micheli [3] proposed to schedule the clock arrival times of FFs (flip-flops) in order to disperse the peak current. Vittal, Ha, Brewer, and Sadowska [5] then formulated the clock arrival time scheduling problem as a 0-1 integer linear program. Later, Huang, Chang, and Nieh [6] proposed a refined technique to reduce the computational expense of the 0-1 integer linear program. On the other hand, rather than scheduling the clock arrival times, Nieh, Huang, and Hsu [7] first proposed to assign positive polarity onto a half of clock buffers and negative polarity onto the remaining half of the clock buffers.[1] They equally partitioned the clock tree into two subtrees and replaced the buffer at the root of one subtree with an inverter so that when the clock signal switches from 0 to 1 (or 1 to 0), all the buffers on the one subtree charge (or discharge) current from VDD (or to GND) while all the buffers on the other subtree discharge (or charge) current to GND (or from VDD). Note that the FFs connected to the sink buffers in the one subtree should be replaced with negative-edge triggered FFs. As implied by the clock tree structure, even this simple modification can reduce the total peak current over the chip up to the limit, it is not able to effectively reduce the power/ground noise in local

---

[1]It is said that a buffer is assigned with a *positive polarity* or a *negative polarity* if its output switches in the same direction as or in the opposite direction to that of the clock source, respectively.

regions. To overcome this limitation, Samanta, Venkataraman, and Hu [8] used the physical placement information of the buffering elements in determining buffers and inverters so that for local regions, roughly half of the buffering elements are assigned with positive polarity and the other half with negative polarity. Note that even this work is able to reduce the power/ground noise greatly, sometimes it is likely to cause a long clock skew because the effect of the different delays of inverters and buffers on the clock skew have not been taken into account.

Chen, Ho, and Hwang [9] observed that the peak current occurs at the time when the clock signal arrives at the buffering elements (called sinks) that are directly incident to FFs, as validated by SPICE simulation. Thus, they proposed a method to assign polarities to the sinks, using the physical placement information of the sinks, with the objective of minimizing the power/ground noise while satisfying a minimum clock skew constraint. Even though the optimization problem described in the work of [9] is well defined and the proposed solution is effective to overcome the limitations of the prior works ([7, 8]), the proposed approach is practically less acceptable in that it assumes that only single (fixed) types of buffers and inverters are to be used, and the currents and behaviors by the allocated buffers and inverters are all identical. Recently, Ryu and Kim [11] attempted a new design flow that placed more importance on minimizing the power/ground noise than minimizing the total wirelength of clock tree, proposing a two-step approach, i.e., clock buffer polarity assignment followed by clock tree generation. Even though the approach is effective in reducing power/ground noise, the wirelength of clock tree may unnecessarily increase to meet the clock skew constraint. Moreover, it have not taken into account the exploitation of buffer sizing on reducing power/ground noise. This work aims to completely eliminate the limitations of the previous works. Precisely, this work *addresses a new problem of simultaneous consideration of assigning polarities to the sink buffering elements and determining the sizes of buffers and inverters to fully exploit the effects of the polarity assignment and buffer/inverter sizing on the minimization of power/ground noise* while satisfying the clock skew constraint.

2. PROBLEM FORMULATION

Since the power/ground noise is a local effect, we assume to have as input a set of sub-areas in a chip in which the peak current caused by the current flows through the sink buffering elements in each sub-area should be minimized. (The generation of sub-areas will be discussed in section 3.) Then, solving the problem of minimizing power/ground noise on a chip corresponds to solving the problem of minimizing the peak current on each sub-area.

Problem 1 (<u>PEAK-min</u>): (*Polarity assignment and buffer / inverter sizing for noise minimization*) *For a sub-area that contains a set L of sink buffering elements, a buffer type set B, an inverter type set I, and clock skew bound κ, find a mapping function $\phi : L \mapsto \{B \cup I\}$ that minimizes the quantity of*

$$\max\left\{ \sum_{\phi(e_i)\in B} peak(\phi(e_i))), \quad \sum_{\phi(e_i)\in I} peak(\phi(e_i)) \right\} \quad (1)$$

$$s.t. \max_{i=1,\cdots,|L|}\left(arr\_max(\phi(e_i))\right) - \min_{i=1,\cdots,|L|}\left(arr\_min(\phi(e_i))\right) \leq \kappa$$

where $arr\_max(\phi(e_i))$, $arr\_min(\phi(e_i))$, and $peak(\phi(e_i))$ represent the latest arrival time and the earliest arrival time from the clock source to FFs that are connected directly to $\phi(e_i)$, and the amount of peak current on $\phi(e_i)$, respectively.

In the following, we confirm that PEAK-min is NP-complete by reducing an NP-complete partitioning problem (PARTITION) to PEAK-min, and propose an optimal algorithm that solves the PEAK-min problem in pseudo-polynomial time.

Problem 2 (<u>PARTITION</u>): *For a finite set A and a 'size' $s(a) \in \mathbf{Z}^+$ for each $a \in A$, is there a subset $A' \subseteq A$ such that*

$$\sum_{a \in A'} s(a) = \sum_{a \in A - A'} s(a)?$$

Theorem 1: PARTITION *is NP-complete.* [13]

Problem 3 (<u>decision-PEAK-min</u>): *For a PEAK-min instance with (L, B, I, κ) and constant c, is there a mapping ϕ such that the value of expression (1) is less than or equal to c?*

Theorem 2: decision-PEAK-min *is NP-complete.* (Proof: It is showed that PARTITION can be reduced to decision-PEAK-min.)

Our approach to solving an instance of PEAK-min first divides the problem into many subproblems (discussed in section 3.1). Then, each subproblem is tackled optimally (discussed in section 2), from which a globally optimal solution is derived (discussed section 3.2). The generation of subproblems is based on constraining the arrival times of clock signals, thus restricting the types of buffers and inverters to be allocated to the sinks.

More specifically, let us consider a signal arrival time interval $H(t) = [t - \kappa, \ t]$. Let C_i denote the set of buffers and inverters such that the values of $arr\_max(\cdot)$ and $arr\_min(\cdot)$ for their assignments to sink $e_i \in L$ are in $H(t)$. If C_i contains more than one buffer type, remove all the buffer types from C_i except the one with the lowest peak current. Similarly, if C_i has more than one inverter type, remove all the inverter types except the one with the lowest peak current.[2] (Recall that we want to minimize the total peak current by allocating buffers and inverters to sinks.) Then, we classify the sinks into four groups according to C_i's:

- $G1 = \{e_i \in L | \ |C_i| = 2, \ C_i$ has a buffer and an inverter,$\}$
- $G2 = \{e_i \in L | \ |C_i| = 1, \ C_i$ has a buffer,$\}$
- $G3 = \{e_i \in L | \ |C_i| = 1, \ C_i$ has an inverter,$\}$
- $G4 = \{e_i \in L | \ |C_i| = 0.\}$

Definition 1 (feasible time interval): *A signal arrival time interval is called* feasible *if each sink has at least one buffer or inverter type to be assigned such that the resulting earliest and latest arrival times are in the arrival time interval, i.e., $G4 = \phi$.*

Note that since the numbers of buffers and inverters available in B and I are bounded by constants, the number of feasible time intervals is bounded by $|L| \cdot (|B| + |I|)$, where the worst case happens when all the arrival times of buffer/inverter mapping to sinks are different.

---

[2] We often simply say 'buffer' for buffer type and 'inverter' for inverter type if it incurs no confusion.

Once the feasible time intervals are produced, as a next step, we focus on solving the PEAK-min problem for each of the feasible arrival time intervals, which we call PEAK-min-interval problem. Precisely, PEAK-min-interval can be formulated as:

Since the sink buffering element $e_i \in G2$ should be assigned to a buffer type, and $e_i \in G3$ should be assigned to an inverter type, we define constants P_f^+ and P_i^- as follows:

$$P_f^+ = \sum_{e_i \in G2} peak(\phi(e_i)), \quad P_f^- = \sum_{e_i \in G3} peak(\phi(e_i))$$

We introduce constant notations p_i^+ and p_i^-, and 0-1 integer variables x_i and y_i, for each sink $e_i \in G1$:

$$p_i^+ = peak(\phi(e_i)), \quad \text{for } \phi(e_i) \in C_i \cap B,$$
$$p_i^- = peak(\phi(e_i)), \quad \text{for } \phi(e_i) \in C_i \cap I,$$
$$x_i = \begin{cases} 1 \text{ if } \phi(e_i) \in B \\ 0 \text{ otherwise} \end{cases}, \quad y_i = \begin{cases} 1 \text{ if } \phi(e_i) \in I \\ 0 \text{ otherwise} \end{cases}$$

where $x_i + y_i = 1$. Then, the (minimization) cost function in expression (1) can be reformulated as minimizing

$$\max \left\{ \sum_{e_i \in G1} p_i^+ x_i + P_f^+, \ \sum_{e_i \in G1} p_i^- y_i + P_f^- \right\}. \quad (2)$$

We formulate the problem of determining the values of variables x_i and y_i that minimizes the quantity of expression (2) into the KNAPSACK problem [14]. The KNAPSACK problem is stated as:

Problem 4 (<u>KNAPSACK</u>): *Given a set of n items with a 'gain' $g_i \in \mathbf{Z}^+$ and a 'value' $v_i \in \mathbf{Z}^+$ for item i, and a capacity constraint $W \in \mathbf{Z}^+$, select a subset of the items so as to*

$$maximize \quad \sum_{i=1}^{n} g_i z_i$$
$$subject\ to \quad \sum_{i=1}^{n} v_i z_i \leq W$$
$$where \quad z_i = \begin{cases} 1 & if\ item\ i\ is\ selected \\ 0 & otherwise. \end{cases}$$

We divide expression (2) into two cases :

$$case1: \quad \sum_{leaf_i \in G1} p_i^+ x_i + P_f^+ \geq \sum_{e_i \in G1} p_i^- y_i + P_f^-$$

$$case2: \quad \sum_{leaf_i \in G1} p_i^+ x_i + P_f^+ < \sum_{e_i \in G1} p_i^- y_i + P_f^-$$

For *case1*, by replacing x_i with $1 - y_i$, the PEAK-min-interval problem becomes:

$$maximize \quad \sum_{e_i \in G1} p_i^- y_i$$
$$subject\ to \quad \sum_{e_i \in G1} (p_i^+ + p_i^-) y_i \leq \sum_{e_i \in G1} p_i^+ + P_f^+ - P_f^-,$$
$$y_i = 0 \text{ or } 1,$$

Then, the *case1* problem can be reduced to the KNAPSACK in problem 4 by setting y_i to z_i, p_i^- to g_i, $p_i^+ + p_i^-$ to v_i,

$\sum_{e_i \in G1} p_i^+ + P_f^+ - P_f^-$,[3] to W, and $|G1|$ to n. The *case2* of the PEAK-min-interval problem is also similarly transformed to the KNAPSACK problem. Consequently, an instance of the PEAK-min-interval problem can be solved by solving two instances of the KNAPSACK problem and selecting the solution with the smaller cost value.

We solve each instance of the KNAPSACK problem by formulating it into a dynamic programming (DP) [14]. The subproblems to be solved are, for $j = 1, \cdots, n$, the forms of $K(w, j)$, which designates the maximum value achievable using a knapsack of capacity w and items $1, \cdots, j$. Then, the answer we look for is $K(W, n)$. We can express subproblem $K(w, j)$ with smaller subproblems:

$$K(w, j) = max\{K(w - w_j, j - 1) + v_j, \ K(w, j - 1)\}. \quad (3)$$

As a result, our dynamic programming algorithm, called PEAK-min-DP, consists of filling a two-dimensional table of $W + 1$ rows and $n + 1$ columns. Each entry takes constant time, and overall time takes $O(nW)$. The initializations are $K(0, j) = 0$, $j = 0, 1, 2, \cdots, n$ and $K(w, 0) = 0$, for $w = 0, 1, 2, \cdots, W$.

Theorem 2: PEAK-min-interval *problem can be solved in pseudo-polynomial time.*

3. ALGORITHM FOR NOISE MINIMIZATION

The proposed algorithm called CLK-NOISE accepts, as input, a synthesized clock routing tree that contains buffering elements. The objective of CLK-NOISE is to minimize the power/ground noise on the clock tree by considering buffer sizing and polarity assignment to the sink buffering elements while satisfying the clock skew constraint. CLK-NOISE is performed in two phases. In the first phase, a minimal set of feasible signal arrival time intervals is extracted. In the second phase, CLK-NOISE searches a solution of buffer sizing and polarity assignment with the lowest power/ground noise by exploiting the PEAK-min-DP algorithm repeatedly on the feasible time intervals obtained in the first phase.

3.1 Phase 1: Generation of a Minimal Set of Feasible Time Intervals

Since selecting a buffer or an inverter from the libraries and assigning it to a sink determine the minimal and maximal arrival times from the clock source to the FFs connected to the sink, we are interested in finding all the *feasible* arrival time intervals of length κ . In other words, we remove from our consideration the time intervals for which there exists at least one sink that is able to be assigned to none of buffers and inverters in the libraries. For example, consider the sink buffering element $e_1, \cdots, e_{36}$ that have been disposed in Figure 1(a). When we assume that the buffer library B and inverter library I have four types of buffers and inverters, respectively, Figure 1(b) illustrates the arrival times for all possible assignments of buffers and inverters to sinks where the left and right end-points of each line segment in a sink indicate the minimum and maximum arrival times of clock signal to FFs through the corresponding buffer or inverter assigned to the sink. The assigned buffers and inverters on the left side in the arrival times of Figure 1(b)

---

[3]All p_i^+, p_i^-, P_f^+, and P_f^- values are scaled to be positive integer numbers.

978-1-60558-497-3/09 $25.00 © 2009 ACM

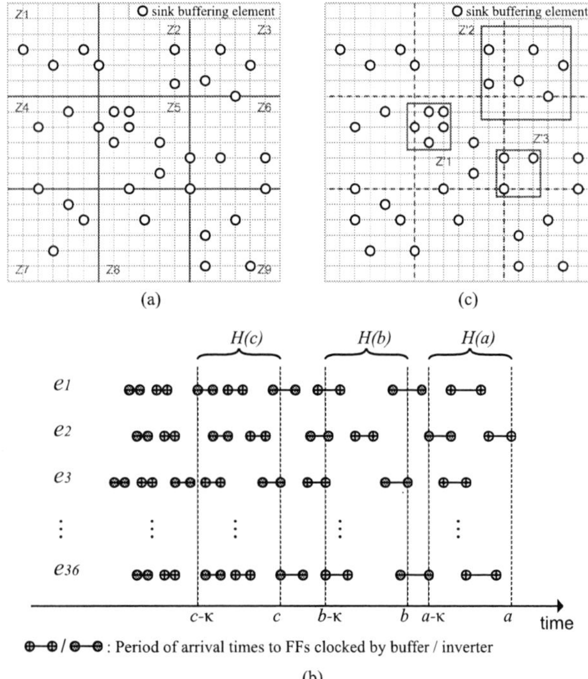

(b)

$\oplus\!\!-\!\!\oplus\,/\,\ominus\!\!-\!\!\ominus$: Period of arrival times to FFs clocked by buffer / inverter

Figure 1: Design example for explaining the proposed algorithms. (a) A disposition of sink buffering elements and uniformly distributed zones (i.e., sub-areas). (b) The arrival times intervals for sinks with various allocation of buffers and inverters. (c) User specified zones.

are relatively faster, but larger in size and consume more current than those on the right. We search through, from right to left, feasible time intervals (e.g., starting from the interval $H(a)$ in Figure 1(b)). The number of distinct time intervals is bounded by the sum of the numbers of line segments because setting the time corresponding to the right end-point, say t_i, of each segment uniquely determines one arrival time interval $H(t_i) = [t_i - \kappa,\ t_i]$, and any of other time intervals can be reduced to one of $H(t_i)$'s. Thus, the number of time intervals to check is $|L| \cdot (|B| + |L|)$. For each time interval $H(t_i)$, we check if every sink has at least one line segment that is contained in $H(t_i)$. That is, we check if $G4$ is empty or not. If $G4$ is empty, we include $H(t_i)$ in the set of feasible arrival time intervals. For example, $H(a)$ and $H(c)$ are feasible arrival time intervals, but $H(b)$ is not because there is no possible assignment for sink e_1. The following theorem claims that the search process of feasible time intervals can be greatly simplified by cutting-off the feasible time intervals that will cause higher power/ground noises than some of the others.

Definition 2 (completely feasible): *An arrival time interval is called* completely feasible *if it is feasible and its $G1$ contains all sinks, i.e., $|G1| = |L|$.*

For example, the arrival time interval $H(c)$ in Figure 1(b) is completely feasible, but $H(a)$ is not completely feasible.

Theorem 3: *If $H(t_i)$ is completely feasible, then for any t_j such that $t_j < t_i$ the peak current on $H(t_j)$ is always greater than or equal to the minimum peak current on $H(t_i)$.* (Proof

is omitted due to space limitation.)

According to theorem 3, we can terminate the search process when a completely feasible time interval is encountered.

Table 1: An illustration of selecting a feasible interval with minimum noise in phase 2 for the example in Figure 1(c).

zones		Feasible time intervals in $\mathcal{H}$ ($|\mathcal{H}| = n$)					current
		$H(t_1)$	$H(t_2)$	$H(t_3)$	$\cdots$	$H(t_n)$	constr.
Z'	z'_1	26	<u>19</u>	24	$\cdots$	<u>16</u>	≤ 20
	z'_2	39	<u>29</u>	34	$\cdots$	<u>27</u>	≤ 31
	z'_3	17	<u>13</u>	15	$\cdots$	<u>12</u>	≤ 15
Z	z_1	-	12	-	$\cdots$	10	
	z_2	-	10	-	$\cdots$	13	
	$\vdots$	$\vdots$	$\vdots$	$\vdots$	$\vdots$	$\vdots$	
	z_5	-	**42**	-	$\cdots$	**37**	
	$\vdots$	$\vdots$	$\vdots$	$\vdots$	$\vdots$	$\vdots$	
$p_{H(\cdot)}^{max}$		-	42	-	$\cdots$	37	

3.2 Phase 2: Finding a Feasible Interval of Minimum Noise

Since the power/ground noise is a local effect, it is assumed that designer is given a set of circuit sub-areas, which we call *zones*, for which their peak currents should be minimized. We consider two strategies of zone generation: (1) *generating zones by uniform partitioning*, which partitions the circuit into small pieces of equal size with no overlapping, and (2) *specifying zones by designer*, in which according to the designer's experience and knowledge on designing the circuit, she or he locates zones where the peak currents should be minimized. For example, Figure 1(a) shows a uniform partition of circuit with nine zones. On the other hand, Figure 1(c) shows designer-specified three zones (Note that some zones may be overlapped. In that case, priorities among the zones for minimizing peak currents should be given. In addition, a peak current constraint (i.e., current bound) on each specified zone is given.) First, we consider the simple strategy of uniform partition. Handling the strategy of user-specified zones will be described subsequently.

• *Handling zones generated by uniform partitioning*: The objective of CLK-NOISE is to find a feasible interval that leads to a minimum among the maximum values of peak currents of zones of feasible intervals. For a feasible interval $H(t)$, the peak current values of zones can be obtained by applying PEAK-min-DP to each zone, from which the maximum peak current value, $p_{H(t)}^{max}$, can be determined. Then, CLK-NOISE chooses the feasible interval with the smallest value of $p_{H(\cdot)}^{max}$. Note that the selection of feasible interval and the application of PEAK-min-DP solve the combined problem of buffer sizing and polarity assignment accordingly.

• *Integrating designer-specified zones*: For each feasible interval, PEAK-min-DP is applied to the designer-specified zones first. Then, CLK-NOISE selects the feasible intervals for which the peak currents of the designer-specified zones all satisfy their peak current constraints. For example, the three rows labeled z'_1, z'_2, and z'_3 in Table 1 illustrate the peak current values for the zones z'_1, z'_2, and z'_3 in Figure 1(c) for the feasible intervals $H(t_1), \cdots, H(t_n)$ obtained in phase

978-1-60558-497-3/09 $25.00 © 2009 ACM

1. The values in the last column show the peak current constraints of the zones. The values with underscore indicate that the corresponding zones satisfy the peak current constraints. For example, feasible intervals $H(t_1)$ and $H(t_3)$ violate the peak current constraints, but $H(t_2)$ and $H(t_n)$ can be considered as candidates for further noise minimization. If there are additional zones to reduce noise, the process is repeated for the set of selected feasible intervals while preserving the results of buffer sizing and polarity assignment done in the previous iteration(s). For example, the rows labeled $z_1, \cdots, z_9$ show the peak current values for the uniform partitioned zones for the selected feasible intervals. If the zones are from uniform partition of circuit, CLK-NOISE computes $p_{H(\cdot)}^{max}$ values for the feasible intervals, and chooses the feasible interval with minimum value of $p_{H(\cdot)}^{max}$. For the example in Table 1, $H(t_n)$ is selected.

4. EXPERIMENTAL RESULTS

The proposed algorithm CLK-NOISE for solving the combined problem of polarity assignment and buffer sizing with the objective of minimizing power/ground noise has been implemented in C on a Linux machine and tested on ISCA89 benchmark circuits. We obtained the locations of the FFs of the circuits in [15], by performing synthesis using Berkeley SIS and placement using UCLA Dragon. The clock trees were then generated by using the algorithm in [16]. We combined the cluster based algorithm in [17] with the clock tree generation to produce sink buffering elements. We used four pairs of buffer and inverter types taken from the UMC $(0.13\mu m)$ standard cell library. The pairs have different driving strength levels G, H, I, and J. (Level-J indicates the fastest delay but consumes the largest current whereas level-G indicates the slowest delay but consumes the least current.) The model parameters of the buffers and inverters were taken from [18]. For the SPICE simulation to measure the peak current, we used the power grid model in [12]. We compared the results produced by CLK-NOISE with that by the approach, we named Polarity-only, proposed by Chen, Ho, and Hwang [9], which heuristically solved the polarity assignment problem only, under the assumption that the peak currents of a buffer and an inverter are the same.

Table 2 summarizes the simulation results of the designs produced by Polarity-only which uses level-J BUF and INV and the designs by CLK-NOISE which uses all four levels of BUF and INV. Since Polarity-only produces minimum skews of 19.2ps - 19.9ps when the level-I BUF and INV are used, we set the skew budget to 20ps in CLK-NOISE for a fair comparison. The column $|L|$ represents the number of sink buffering elements on each circuit. The last three columns in each column section labeled Polarity-only and CLK-NOISE represent the values of total peak current, maximum power noise, and maximum ground noise of the designs produced by Polarity-only and CLK-NOISE, respectively. Column $|Z|$ indicates the number of zones that are used in CLK-NOISE. The last three columns labeled *Improvement* show the improvements by CLK-NOISE over Polarity-only. We can see that the average improvements of the maximum power and ground noises by CLK-NOISE are 19.1% and 16.2%, respectively, which clearly reveals that considering polarity assignment together with buffer sizing is effective in reducing the power/ground noise. Figures 2(a) and (b) show the power noise maps of the designs produced by Polarity-only and CLK-NOISE for s35932. (s35932 has the largest number of sinks in ISCA89 benchmarks.)

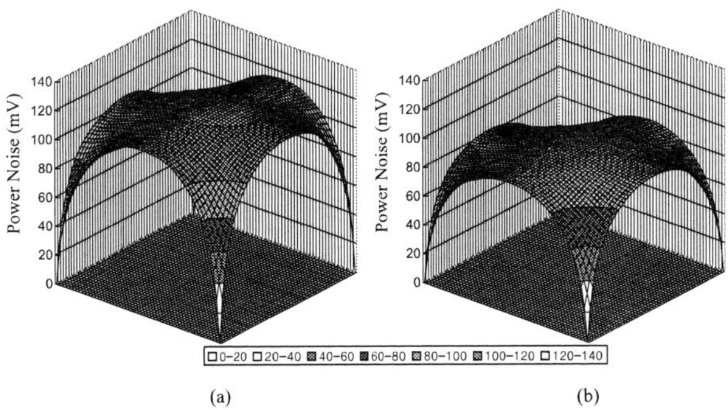

Figure 2: Comparison of the power noise map for s35932 **in Table 2. (a) Power noise map by** Polarity-only. **(b) Power noise map by** CLK-NOISE.

We constrained the buffer and inverter library to have level-I BUF and INV only for both Polarity-only and CLK-NOISE. Since level-I BUF and INV are slower than that of level-J BUF and INV that was used in Table 2, the skew bound was also relaxed to 30ps. Table 3 summarizes the simulation results. We can see the improvements of power and ground noises are 13.1% and 11.1% on average, respectively. This means that by comparing the improvement results in Table 2 with the improvements in Table 3, the contribution to the power/ground noise reduction by exploiting buffer/inverter sizing by CLK-NOISE can lead to 6% (= 19.1-13.1) power noise and 5.1% (= 16.2-11.1) ground noise reductions.

5. CONCLUSIONS

This work presented a comprehensive solution to the integrated problem of buffer sizing and polarity assignment for minimizing power/ground noise. The contributions were a precise estimation of peak current by clock buffers/inverters, a practically efficient optimal algorithm based on dynamic programming for the problem, and a systematic design flow for reducing power/ground noise using two types of 'zone' concept. Through experiments with MCNC benchmark circuits, we confirmed that the proposed algorithm was able to produce designs with 19.1% less power noise, 16.2% ground noise, and 15.6% less total peak current over that by the conventional method which solved the problem of polarity assignment only.

6. REFERENCES

[1] S. Chowdury and J. Barkatullah, "Estimation of maximum currents in MOS IC logic circuits," *IEEE Transaction on CAD*, vol. 9, no. 6, pp. 642-654, 1990.

[2] L. H. Chen, M. M.-Sadowska, and F. Brewer, "Buffer delay change in the presence of power and ground noise," *IEEE Transaction on VLSI Systems*, vol. 11, no. 3, pp. 461-473, June 2003.

[3] L. Benini, P. Vuillod, A. Bogliolo and G. D. Micheli, "Clock skew optimization for peak current reduction," *Journal of VLSI Signal Processing*, vol. 16, 1997.

[4] N. H. E. Weste and D. Harris, *CMOS VLSI DESIGN, A Circuits and Systems Perspective*, ed. 3, Addison-Wesley, 2005.

Table 2: Comparison of results (Polarity-only[9]: using level-J BUF/INV, CLK-NOISE: using all 4 levels of BUF/INV, skew_ budget < 20ps).

ISCA89		Polarity-only[9] (using level-J)				CLK-NOISE (using level-G, H, I, J)						Improvement		
Circuit	$\|L\|$	Skew (ps)	Peak curr. (mA)	Power noise (mV)	Ground noise (mV)	Skew (ps) ($\leq$20)	$\|Z\|$	Run time (sec)	Peak curr. (mA)	Power noise (mV)	Ground noise (mV)	Peak curr. (%)	Power noise (%)	Ground noise (%)
s5378	40	19.5	19.4	14.2	14.1	19.8	12	0.02	14.8	9.8	10.0	23.7	31.0	29.1
s9234	50	19.6	23.6	18.3	18.2	18.9	16	0.02	20.8	15.2	15.1	11.9	16.9	17.0
s13207	134	19.7	54.8	47.7	47.8	19.7	36	0.19	50.1	40.3	40.3	8.8	15.5	15.7
s15850	133	19.2	56.7	51.5	52.9	19.9	49	0.12	46.3	39.1	41.8	18.3	24.1	21.0
s35932	407	19.9	129.7	122.4	111.6	19.9	121	1.26	105.1	96.0	98.9	19.0	21.6	11.4
s38417	343	19.9	113.7	100.1	112.8	19.8	81	1.20	100.4	95.0	97.8	11.7	5.0	13.3
s35584	330	19.7	119.8	113.1	99.8	19.9	100	0.80	100.9	90.8	93.9	15.8	19.7	5.9
Average												**15.6**	**19.1**	**16.2**

Table 3: Comparison of results (Polarity-only[9]: using level-I BUF/INV, CLK-NOISE: using level-I BUF/INV, skew_ budget < 30ps).

ISCA89		Polarity-only[9] (using level-I)				CLK-NOISE (using level-I)						Improvement		
Circuit	$\|L\|$	Skew (ps)	Peak curr. (mA)	Power noise (mV)	Ground noise (mV)	Skew (ps) ($\leq$20)	$\|Z\|$	Run time (sec)	Peak curr. (mA)	Power noise (mV)	Ground noise (mV)	Peak curr. (%)	Power noise (%)	Ground noise (%)
s5378	40	28.9	15.2	10.7	11.3	28.9	12	<0.01	13.2	9.6	9.9	13.2	10.3	12.4
s9234	50	28.9	19.9	16.5	16.8	29.3	16	<0.01	18.5	13.6	14.0	7.0	17.6	16.7
s13207	134	29.3	50.7	41.6	42.1	29.5	36	0.01	46.4	36.6	37.9	8.5	12.0	10.0
s15850	133	26.9	49.4	41.5	43.2	28.4	49	0.02	42.3	36.9	36.7	14.4	11.1	15.0
s35932	407	29.7	114.1	114.0	101.0	29.8	121	0.04	97.0	90.0	92.8	15.0	21.1	8.1
s38417	343	29.8	101.7	94.1	103.0	29.9	81	0.05	92.9	87.3	92.0	8.7	7.2	10.7
s35584	330	29.6	105.5	95.2	91.0	29.7	100	0.04	96.0	83.5	86.5	9.0	12.3	4.9
Average												**10.8**	**13.1**	**11.1**

[5] A. Vittal, H. Ha, F. Brewer, and M. M.-Sadowska, "Clock skew optimization for ground bounce control," *Proc. of IEEE/ACM International Conference on Computer-Aided Design,* pp. 395-399, 1996.

[6] S.-H. Huang, C.-H. Chang, and Y.-T. Nieh, "Fast multi-domain clock skew scheduling for peak current reduction," *Proceedings of IEEE/ACM Asia and South Pacific Design Automation Conference,* pp. 254-259, 2006.

[7] Y.-T. Nieh, S.-H. Huang, and S.-Y. Hsu, "Minimizing peak current via opposite-phase clock tree," *Proc. of IEEE/ACM Design Automation Conference,* pp. 182-185, 2005.

[8] R. Samanta, G. Venkataraman, and J. Hu, "Clock buffer polarity assignment for power noise reduction," *Proc. of IEEE/ACM International Conference on Computer-Aided Design,* pp. 558-562, 2006.

[9] P.-Y. Chen, K.-H. Ho, and T. Hwang, "Skew aware polarity assignment in clock tree," *Proc. of IEEE/ACM International Conference on Computer-Aided Design,* pp. 376-379, 2007.

[10] K. D. Bosse, A. B. Kahng, "Zero-skew clock routing with minimum wirelength," *Proc. of IEEE International ASIC Conference,* pp. 111-115, 1992.

[11] Y. Ryu and T. Kim, "Clock buffer polarity assignment combined with clock tree generation for power/ground noise minimization," *Proc. of IEEE/ACM International Conference on Computer-Aided Design,* pp. 416-419, 2008.

[12] Q. Zhu, *Power Distribution Network Design for VLSI,* John Wiley & Sons, 2004.

[13] M. R. Garey and D. S. Johnson, *Computers and Intractability, A Guide to the Theory of NP-Completeness,* W. Freeman and Company, 1979.

[14] S. Martello and P. Toth, *Knapsack Problems,* John Wiley & Sons, 1990.

[15] Placement of ISCAS89 Benchmark Circuits *http://www.ece.wisc.edu / vlsi/tools/iscas-placement/index.html*

[16] R. Chaturvedi and J. Hu, "Buffered clock tree for high quality IC design," *Proc. of IEEE/ACM International Symposium on Quality Electronic Design,* pp. 381-386, 2004.

[17] M. Edahiro, "A clustering-based optimization algorithm in zero-skew routing," *Proc. of IEEE/ACM Design Automation Conference,* pp. 612-616, 1993.

[18] *FSC0H_D 0.13μm Standard Cell Databook,* Faraday Technology, 2004.

An SDRAM-Aware Router for Networks-on-Chip

Wooyoung Jang and David Z. Pan

Department of Electrical and Computer Engineering

University of Texas at Austin

wyjang@cerc.utexas.edu, dpan@ece.utexas.edu

ABSTRACT

In this paper, we present an NoC (Networks-on-Chip) router with an SDRAM-aware flow control. Based on a priority-based arbitration, it schedules packets to improve memory utilization and reduce memory latency. Moreover, our multi-scheduling scheme performed by the multiple SDRAM-aware routers helps to achieve better SDRAM performance and save the hardware cost of NoC platform. Experimental results show that our SDRAM-aware router improves memory latency by 18% and memory utilization by 4.9% on average with over 42% saving of gate count of the NoC platform with dual memory subsystem.

Categories and Subject Descriptors

C.2.1 [**COMPUTER-COMMUNICATION NETWORKS**]: Network Architecture and Design - Packet-switching networks.

General Terms

Algorithms, Performance, Design.

Keywords

Networks-on-Chip, router, flow control, memory

1. INTRODUCTION

An NoC (Networks-on-Chip) has been proposed as a scalable solution to complex on-chip interconnection problems [1, 2]. Recently, one inter-layer interconnection comes to extend the NoC into three dimensions by TSV (Through-Si-Vias) technology [3]. Especially, a 3D NoC embedded with multiple SDRAMs (Synchronous Dynamic Random Access Memory) for L2 or L3 caches on top of processing elements in different layers achieves higher memory bandwidth, shorter latency and more reliable electrical features than a conventional 2D NoC only interfacing with one or two off-chip SDRAMs [4].

A memory subsystem managing an SDRAM is one of the most important components in most 2D/3D NoC designs since the performance of the whole system is considerably sensitive to the performance of the memory subsystem. However, a memory subsystem frequently underperforms due to characteristic operation flows of an SDRAM [5] and dynamic accesses by various processing components. For example, DDR (Double Data Rate) II SDRAM utilization (defined as the number of clock cycles transferred valid data divided by the total number of clock cycles required to access data) becomes much worse by 55% in DTV (Digital Television) application [6]. Moreover, the corresponding number of memory subsystem is also equipped to manage the SDRAMs. Since the gate count of a single memory subsystem occupies over 35% of the entire 3×3 NoC platform consisting of a single memory subsystem, nine routers and twelve links from our experimental results, the NoC design cost highly increases depending on the number of memory subsystem. Therefore, considerable attention has shifted toward memory-aware NoC exploration to improve memory utilization and latency and save the design cost of the NoC platform.

In this paper, we propose an SDRAM-aware NoC router improving total SDRAM utilization and latency and decoupling the NoC design cost from the number of SDRAM. Our key ideas are two-fold. First, if an NoC router schedules packets to access SDRAM efficiently, the packets arrive at a memory subsystem into the order that is more friendly to SDRAM operations. At a result, the complexity of a memory subsystem considerably decreases while the memory performance is more improved. Since our SDRAM-aware router uses existing resources to schedule packets, e.g. input buffers for storing blocked packets or other flow-control mechanisms, the additional design cost is tiny. Instead, heavy reordering buffers and a scheduler are removed in a memory subsystem. Second, a multi-scheduling scheme performed by multiple SDRAM-aware routers outperforms a single-scheduling scheme performed by a single memory subsystem. The reason is that the performance of a single scheduling scheme is mainly limited by the depth of reordering buffers in a memory subsystem. However, our multi-scheduling scheme uses all input buffers in multiple routers to reorder packets. Based on these ideas, the major novelty and contribution of this paper include the following.

- We propose an NoC router with an SDRAM-aware flow control. It schedules a packet to access an SDRAM instead of a memory subsystem.

- We employ both a priority-based arbitration based on SDRAM-awareness and a multi-scheduling scheme that are performed by multiple NoC routers.

- We show that the NoC design with our SDRAM-aware router can achieve higher memory utilization, shorter memory latency and cheaper hardware cost than the conventional NoC design with an SDRAM-unaware router.

The rest of this paper is organized as follows. In the next section, we survey related works. In section 3, SDRAM operation principle and SDRAM scheduling are reviewed. In section 4, a problem in NoC design with a memory subsystem and our basic solution are present. Section 5 presents a detailed description of the SDRAM-aware router. Experimental results are shown in Section 6. Finally, section 7 concludes the paper.

Permission to make digital or hard copies of part or all of this work for personal or classroom use is granted without fee provided that copies are not made or distributed for profit or commercial advantage and that copies bear this notice and the full citation on the first page. To copy otherwise, to republish, to post on servers or to redistribute to lists, requires prior specific permission and/or a fee.
DAC'09, July 26-31, 2009, San Francisco, California, USA

2. RELATED WORKS

A 3D NoC embedded with multiple SDRAMs provides more bandwidth, shorter latency and less number of timing-critical paths. However, the corresponding number of an expensive memory subsystem is also equipped to manage the SDRAMs. A memory subsystem usually consists of three parts, i.e., a buffer, a SDRAM scheduler and a SDRAM interface signal generator, where the depth of buffer and the scheduler for reordering memory accesses are a key factor of high memory utilization and short memory latency. Memory access scheduler proposed in [7] supports preemption and reordering to optimize bandwidth and average latency. The schedulers discussed in [8, 9] support preemption for high-priority requestors to decouple latency and data rate. PREDATOR proposes grouping memory accesses and a predictable arbiter for this group in [10]. Memory scheduler proposed in [11] adopts the adaptive history-based algorithm.

Flow control in an NoC router focuses on how network resources, e.g. channel bandwidth, buffer capacity and control state, are allocated to packets traversing a network. A decentralized control system and a predictive explicit-rate control are developed in [12, 13]. A predictive flow controller that controls the packet injection rate is proposed in [14]. To improve the overall execution time and link utilization, optimal link scheduler and shared buffer router architecture are proposed in [15]. An open-loop flow control scheme is proposed in [16] to reduce the conflicts of data transfers from multiple memory modules to the same masters. Our SDRAM-aware flow-controller schedules packets to improve the utilization and latency of SDRAM based on priority-based arbitration and a multi-scheduling scheme, which can work together with other flow controls and routing schemes.

3. PRELIMINARY

3.1 Basic SDRAM Operation

An SDRAM has a three dimensional structure, i.e., a bank, a row and a column as shown in Fig. 1. Basic commands to access the SDRAM are ACT (Activate), R/W (Read/Write) and PRE (Precharge), where the ACT is executed with a BA (Bank Address) and an RA (Row Address), the R/W is executed with a BA and a CA (Column Address) and the PRE is executed with a BA. In Fig. 1, when any idle bank is activated by ACT, one row data of the bank move to a row buffer of the bank. It takes t_{RCD} to complete the ACT. Timing parameters of a DDR II SDRAM used in this paper is shown in Table 1. Then, R/W is executed on the filled row buffer. After read latency called CL or write latency called t_{DQSS}, successive data go from or to the SDRAM. Finally, PRE is executed to deactivate the active row buffer in the bank. It also takes t_{RP} to become the idle state again.

3.2 Memory Scheduling

An SDRAM consists of independent multiple banks while address and data resources serialize access to different banks, as shown in Fig. 1. Its benefit is that pin/wire resources are saved and commands to a different bank are pipelined, i.e. while data is transferred to or from one bank, the rest of banks become idle and active for a later request. Nevertheless, the improvement of SDRAM utilization is still limited due to wasted cycles caused by the above characteristic operation flows of an SDRAM and dynamic accesses by various processing components. Main factors

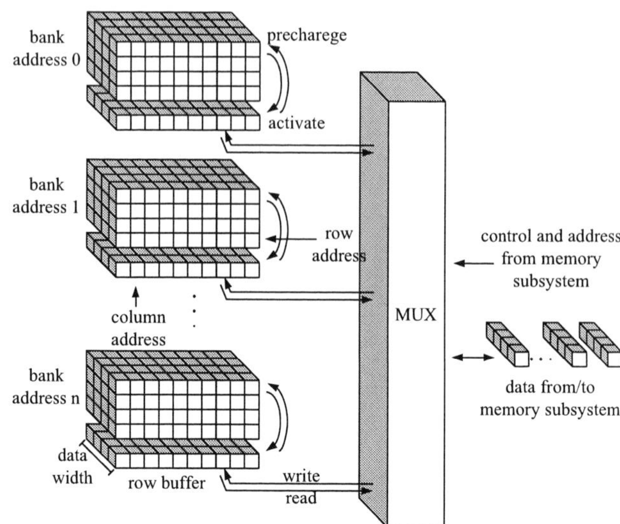

Figure 1: SDRAM architecture

Table 1: Timing parameter of DDR II SDRAM @333MHz [5]

CL	CAS (Column Access Strobe) latency (4 clocks)
t_{RCD}	RAS (Row Access Strobe) to CAS delay (4 clocks)
t_{RP}	Row precharge time (4 clocks)
t_{WR}	Write recovery time (5 clocks)
t_{WTR}	Internal write to read command delay (3 clocks)
t_{DQSS}	DQS (Data Strobe) latching rising transitions to associated clock edges (1 clock)

that deteriorate the SDRAM utilization are the following or their combinations:

- Continuous accesses of the same bank with different RAs.

- Write (read) access followed by any read (write) access.

The first factor called *bank conflict* is the most critical to SDRAM utilization since a bank activated by the former access should get idle and then active for the later access. In Fig. 2, there are two SDRAM schedulers that reorder four read accesses, i.e. read 1 (RA 0, BA 0, CA 0), read 2 (RA 1, BA 0, CA 0), read 3 (RA 0, BA 1, CA 0) and read 4 (RA 1, BA 1, CA 0). As shown in Fig. 2(a), let them scheduled in the order, read 1, read 2, read 3 and read 4 by scheduler 1. After completing read 1 as mentioned in section 3.1, read 2 cannot be immediately executed since the row buffer of bank 0 is already occupied by the data of RA 0. Therefore, PRE should be executed to release the filled row buffer of bank 0 and then ACT should be executed to fill the row buffer of bank 0 with the data of RA 1. On the other hand, read 3 can be pipelined, that is called *bank interleaving* since it has the different BA from the BA of read 2. At a result, the data 3 accessed by read 3 are generated with no loss of clock cycles as shown in Fig. 2(a). The last read 4 conflicts with read 3 since they have the same BAs, but the different RAs. On the other hand, scheduler 2 changes the order of four reads: read 1, read 3, read 2, and read 4 as shown in Fig. 2(b). This order does not cause any bank conflict such that all of the read accesses are pipelined. Consequently, the second SDRAM scheduler lets all SDRAM accesses completed faster than the first SDRAM scheduler. In this example, memory utilization of the first scheduler is 9.5% (= 4 data / 42 clock cycles) and that of the second scheduler is 13.3% (= 4 data / 30 clock cycles).

978-1-60558-497-3/09 $25.00 © 2009 ACM

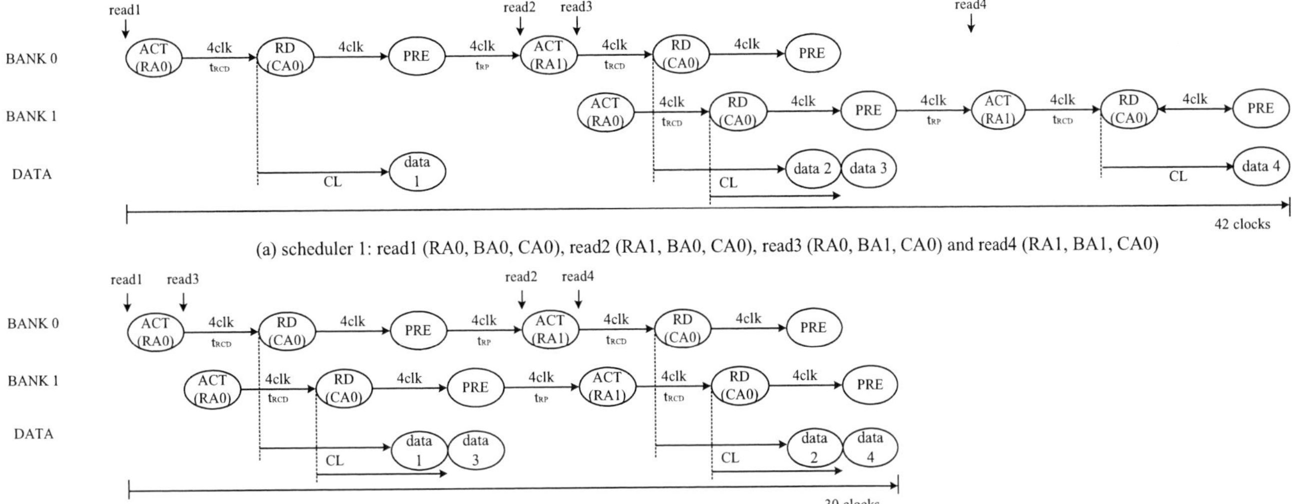

(a) scheduler 1: read1 (RA0, BA0, CA0), read2 (RA1, BA0, CA0), read3 (RA0, BA1, CA0) and read4 (RA1, BA1, CA0)

(b) scheduler 2: read1 (RA0, BA0, CA0), read3 (RA0, BA1, CA0), read2 (RA1, BA0, CA0) and read4 (RA1, BA1, CA0)

Figure 2: Examples of SDRAM operation depending on schedulers

The second factor is called *data contention*. Data pins or wires in an SDRAM are bidirectional while control and address pins or wires are unidirectional. Thus, input data can be collided with output data. To prevent it, at least one clock cycle should pass when a write access after a read access is accessed. A write access followed by a read access also waits for write recovery time (t_{WTR}) after the write access, which degrades the memory performance. Therefore, continuous read or write accesses are encouraged.

4. NOC DESIGN WITH SDRAM
4.1 Problem Description

The bank conflict and data contention frequently happen in the conventional NoC design due to the limited resources in the memory subsystem. Fig. 3 shows a simple example of bank conflict in the 2x3 NoC design. It includes one memory subsystem that schedules packets to avoid the bank conflict and data contention. In Fig. 3, *RxBy* means that an RA and a BA of packet are *x* and *y*, respectively and the arrow means a packet will move at the next cycle. We assume that the length of packet is 1, the memory subsystem includes two buffers to store two packets and its scheduler makes one of two stored packets executed every cycle (although actual execution time is longer than one cycle). Round-robin arbitration [17] as a flow control of NoC routers is adopted to assign a channel and a buffer of the next node to one packet among several competing packets. At cycle 0, three packets, R2B0, R2B1 and R3B0 get competition for advance to the router interconnecting with the memory subsystem and then R2B0 wins. R0B1 is executed in the memory subsystem. At cycle 1, R2B0 advances to the router interconnecting with the memory subsystem and then R3B1 also advances to the empty router by

the advance of R2B0. Then three packets, R2B1, R3B0 and R3B1 also get competition such that R3B0 wins. R0B0 but not R1B1 is executed for avoiding bank conflict in the memory subsystem. At cycle 2, by round-robin arbitration R3B0 advances. Then, the rest of two packets get competition such that R2B1 wins. R1B1 is executed in the memory subsystem. At cycle 3, bank conflict happens in the memory subsystem since current execution is a bank 0 access and two buffers are also filled with bank 0 accesses, where all RAs are different. It is difficult to avoid bank conflict completely in the conventional NoC with the small depth of buffer.

4.2 Basic Idea of SDRAM-Aware NoC Router

In our NoC design, the packet scheduling for an SDRAM access is performed by multiple SDRAM-aware routers. Consequently, the possibility of bank conflict becomes lower since packets arrive at the memory subsystem into the order that is friendly to the SDRAM operation. Fig. 4 shows how an NoC with our SDRAM-aware router works. At the first competition (cycle 0), the winner that advances to the router interconnecting with the memory subsystem is R2B1 that accesses bank 1 since the former access, R1B0 accessed bank 0. The rest of two packets may cause bank conflict since they have the same BAs but different RAs from the former access. At cycle 1, R2B1 advances and both competing packets, R2B0 and R2B1 can be a winner for advance but, R2B0 is chosen in this example. At cycle 2, R2B0 advances and R3B1 wins. Finally, R3B1 advances and R3B0 wins at cycle 3. Therefore, the proposed router prevents bank conflict better.

In addition, while it is desirable to use multiple SDRAMs for high performance, it is not desirable to use a corresponding number of memory subsystems. The reason is that the memory subsystem in

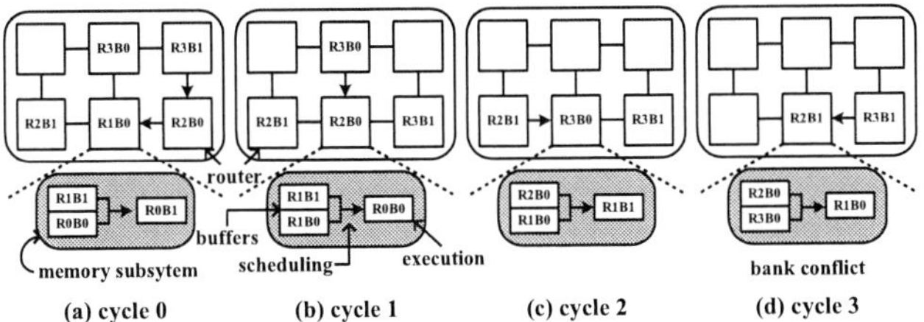

Figure 3: Conventional NoC with round-robin flow control

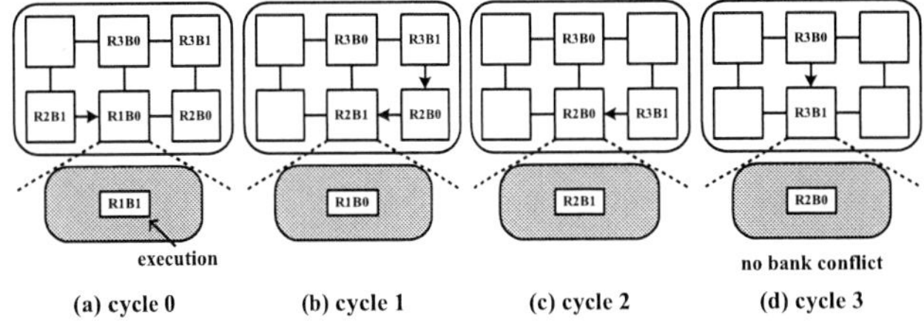

Figure 4: Novel NoC with SDRAM-aware flow control

Fig. 3 is too high in design cost due to a lot of buffers and a scheduler. Due to no buffer and no scheduler in the memory subsystem as shown in Fig. 4, our NoC design not only improves memory utilization and latency but also saves NoC design cost.

5. SDRAM-AWARE NOC ROUTER

Rather than a detail implementation, the proposed NoC router is about a novel paradigm for SDRAM-aware NoC exploration, which has a flow-control mechanism improve memory utilization and latency. Indeed, based on our idea, any deterministic and adaptive routing scheme can be combined to form our SDRAM-aware router. Another flow-control mechanism mentioned in section 2 can also be combined to avoid deadlock and livelock [17], to make traffic load balanced on a network [12-14] and to manage buffers and channel bandwidth [15].

5.1 Router Description

Our NoC router consists of input buffers, routing logics, SDRAM-aware flow controllers and an output schedulers as shown in Fig. 5. A packet is split into so-called flits (flow control digits) which are then routed and stored in a pipelined fashion. The input buffers are managed by wormhole flow control or virtual-channel flow control and a backpressure is used to inform the upstream nodes when they must stop transmitting flits because all of the downstream flit buffers are full. For our experiment, the wormhole flow control is implemented due to its simplicity and wide popularity [17] and an on/off flow control is adopted to avoid the loss of flits as the backpressure. Our SDRAM-aware router can be implemented to both deterministic and adaptive routers according to a routing logic that guarantees deadlock and livelock freeness. The approaches to solve deadlock are to use virtual channels [17] and deterministic dimension-ordered routings (e.g. XY routing, odd-even routing) [17]. We implement the XY routing that is a deterministic and minimal path routing algorithm, i.e. free deadlock and livelock. Simply to avoid another deadlock condition mentioned in [16], a master can send requests to another slave only after finishing requests for one slave.

In this router, over two flits arriving on different inputs buffers at the same time may both desire the same channel, where the final destination of flits is an SDRAM subsystem. In this case, the SDRAM-aware flow-control mechanism resolves this contention, allocating the channel to one packet and dealing with the others, blocked packets. Our SDRAM-aware flow control adopts winner-take-all bandwidth allocation that allocates all of the bandwidth to one packet until it is finished or blocked before serving the other packets [17]. Output scheduling detects if the buffers of the next node are available or expects when the buffers are available such that each SDRAM-aware flow control can choose the best path for low latency when multiple paths are given by the routing logic. In next section, a detail SDRAM-aware flow-control mechanism that uses a priority-based arbitration is present.

5.2 SDRAM-aware Flow Control

Our flow control allocates a channel to one of the competing flits whose destination is a memory subsystem controlling an SDRAM. Therefore, our flow-control mechanism performs an arbitration to determine which flit gets the channel it has requested. After the arbitration, the winning flit advances over this channel. Our arbitration algorithm also decides how to dispose of any flits that do not get their requested destination.

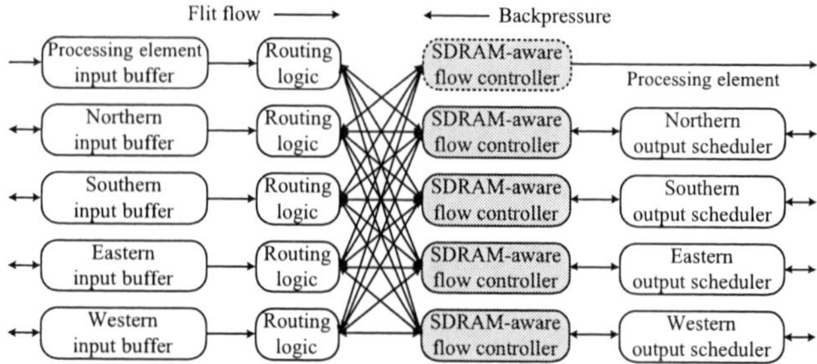

Figure 5: The proposed NoC router in mesh architecture

As shown in Algorithm 1, our arbitration is a priority-based algorithm, where the priority is composed from SDRAM awareness. The priority is assigned to all competing head flits which destination is a memory subsystem. Let $h(n)$ be the head flit of packet, which is already allocated a channel by the SDRAM-aware flow control at nth arbitration. Body and tail flits are assigned the same channel as their head flit. Let $h_i(n+1)$ be one of all competing head flits (I) that may be allocated the same channel as $h(n)$ at $(n+1)$th arbitration, where $i \in I$. The head flits, $h(n)$ and $h_i(n+1)$ contain addresses and command to access an SDRAM, denoted by (RA_n, BA_n, R/W_n) and ($RA_{n+1,i}$, $BA_{n+1,i}$, $R/W_{n+1,i}$), respectively, where the notations are (row address, bank address, read/write). At $(n+1)$th arbitration, all competing $h_i(n+1)$ are compared to $h(n)$ and then given a delay point from Table 2 (line 7) that is composed from DDR II SDRAM operating at 333MHz clock cycle [5] and that is changed according to the kind of SDRAM and operating speed.

Table 2 shows how many clock cycles are wasted when $h_i(n+1)$ is accessed in an SDRAM after $h(n)$. In Table 2, both case 1 and 10 have no loss of clock cycle since $h_i(n+1)$ gets the same R/W, BA and RA as $h(n)$, which means that data already stored in the row buffer of the same bank are accessed again by the same command. In case 3 and 12, data can be also accessed continuously since bank interleaving is completely performed as mentioned in section 3.2. Case 4 and 6 have one clock loss between the former read and the latter write access to avoid data contention. Case 7 and 9 waste 7 cycles to read data after writing data. Although bank interleaving is completely performed in case 7 and the continuous access of the same row buffer are performed in case 9, the later read command can be accepted t_{WRT} (internal write to read command delay) after finishing the former write. Then read data go out from an SDRAM after CL. In case 5, after reading data, its row buffer should be idle and then active since a write access with the same BAs but different RAs are executed. Then, data are written after write latency. Case 2 is already explained in section 3.2. Case 11 is a write-to-write access with the same BAs but the

---

Algorithm 1 SDRAM-Aware Flow Control Algorithm
Input: $h(n)$, $h_i(n+1)$ and table 2
1:    **for** each $h_i(n+1)$, $i \in I$ **do**
2:       **if** $h_i(n+1)$ is a new packet entering to the router **then**
3:          $w_i = 0$;
4:       **else**
5:          $w_i = w_i$ + waiting cycles from the previous arbitration(n);
6:       **end if**
7:       d_i = delay cycle between $h(n)$ and $h_i(n+1)$ from table 2;
8:       $p_i = w_i - d_{i_i}$;
9:    **end for**
10:   $h_i(n+1)$ with maximum(p_i) is allocated to a channel;
Output: $h(n+1)$

978-1-60558-497-3/09 $25.00 © 2009 ACM

Table 2: Execution delay between $h(n)$ and $h_i(n+1)$

Relation of $h(n)$ with RW_n, BA_n, and RA_n and $h_i(n+1)$ with $RW_{n+1,i}$, $BA_{n+1,i}$, and $RA_{n+1,i}$				delay cycle (d_i) @ 333MHz DDR II SDRAM	case
$RW_n=R$	$RW_{n+1,i}=R$	$BA_n=BA_{n+1,i}$	$RA_n=RA_{n+1,i}$	0 cycle	1
			$RA_n \neq RA_{n+1,i}$	t_{RP} (4 cycles) + t_{RCD} (4 cycles) + CL (4 cycles) = 12 cycles	2
		$BA_n \neq BA_{n+1,i}$		0 cycle	3
	$RW_{n+1,i}=W$	$BA_n=BA_{n+1,i}$	$RA_n=RA_{n+1,i}$	1 cycle (timing gap between input and output not to collide)	4
			$RA_n \neq RA_{n+1,i}$	t_{RP} (4 cycles) + t_{RCD} (4 cycles) + t_{DQSS} (1 cycle) = 9 cycles	5
		$BA_n \neq BA_{n+1,i}$		1 cycle (timing gap between input and output not to collide)	6
$RW_n=W$	$RW_{n+1,i}=R$	$BA_n=BA_{n+1,i}$	$RA_n=RA_{n+1,i}$	t_{WTR} (3 cycles) + CL(4 cycles) = 7 cycles	7
			$RA_n \neq RA_{n+1,i}$	t_{WR} (5cycles) + t_{RP} (4 cycles) + t_{RCD} (4 cycles) + CL (4 cycles) = 17 cycles	8
		$BA_n \neq BA_{n+1,i}$		t_{WTR} (3 cycles) + CL (4 cycles) = 7 cycles	9
	$RW_{n+1,i}=W$	$BA_n=BA_{n+1,i}$	$RA_n=RA_{n+1,i}$	0 cycle	10
			$RA_n \neq RA_{n+1,i}$	t_{WR} (5cycles) + t_{RP} (4 cycles) + t_{RCD} (4 cycles) + t_{DQSS} (1 cycle) = 14 cycles	11
		$BA_n \neq BA_{n+1,i}$		0 cycle	12

different RAs. After the first writing, write recovery time (t_{WR}) is needed. Then, its row buffer is released (t_{RP}), filled with new data (t_{RCD}), and finally wrote after write latency (t_{DQSS}). Case 8 is the worst case since a read access with the same BAs but different RAs happens after writing data. The write access needs t_{WR} and the row buffer should be released, filled with new data and finally read after read latency (CL).

Our priority-based arbitration guarantees the upper bound latency since the low priority of packet may last for a long time. For example, let a packet with case 11 lose a competition against a packet with case 10. Nevertheless, since the priority condition is not changed at the next arbitration, the defeated packet keeps losing the next competition if it meets another packet with case 10. Therefore, the packet should escape from this competition after several defeats. To solve this starvation problem, our flow control counts the number of clock cycles passed from the first competition to the current competition (line 5) for each defeated packet. Then, this waiting time is subtracted by the delay cycle obtained in Table 2 (line 8). By this operation, any packet delayed for 17 clocks does not get a lower priority than a new packet entered in the router.

Finally, the packet with the maximum p_i is allocated the channel (line 10) and the blocked packets wait for the next competition or get another competition at a different SDRAM-aware flow controller if multiple routing paths are allocated by a routing logic. Thus, if an adaptive router instead of a deterministic router is employed in a routing logic, the performance would be better.

5.3 Hardware Implementation

Simple logics are added in the SDRAM-aware router to compute the delay (d_i) and the waiting time (w_i) while the buffers and schedulers of memory subsystem are removed as shown in Fig. 4. The buffers in the memory subsystem are used to store several packets and then to reorder the packets such that an SDRAM is accessed as fast as possible. However, since the massive size of packet is generated in a GPU (Graphics Processing Unit) and a high-definition video system, the size of buffers also becomes larger and larger. Our NoC embedded with SDRAMs does not require the buffers in the memory subsystem since the scheduling is performed in multiple NoC routers. Furthermore, the size of input buffer in the router does not increase since the maximum four input buffers per router in a mesh network substitute for the buffers of memory subsystem by wormhole flow control and our multi-scheduling scheme gets the similar effect to the increase of input buffers. Consequently, the distinguished decrease of buffers

in the memory subsystem exceeds the increase by the complexity of arbiter in multiple routers such that total chip area decreases. In addition, each router can get a different scheduler for a faster SDRAM access or for another purpose, which is one of the most important benefits of a multi-scheduling scheme.

6. EXPERIMENTAL RESULTS

An NoC with our SDRAM-aware router is implemented in Verilog HDL (Hardware Description Language). A memory subsystem operates with a DDR II SDRAM at 333MHz [5] and is implemented by the design concept from Sonics' MemMax [18] and Denali's Databahn [19], where MemMax is an SDRAM scheduler with four 32-flit buffers and Databahn is an SDRAM signal generator. Both are included in a conventional NoC design with a round-robin flow control based router. This is compared to our NoC design that includes multiple SDRAM-aware routers and a SDRAM signal generator instead of a full memory subsystem.

6.1 DTV Application

Our SDRAM-aware NoC router is applied to a Samsung DTV system that consists of 9 subsystems, i.e., an ARM9, an MPEG (Moving Picture Experts Group) decoder, a DNIE (Digital Natural Image Engine), a GPU, an audio decoder, a TS (Transport Stream) decoder, an AV (Audio and Video) format converter, a channel decoder and a memory subsystem, which are mapped to 3×3 mesh network. A conventional NoC router with round-robin flow control is gradually replaced by our SDRAM-aware router where the router that is the closest to a memory subsystem is replaced first and the router that is the farthest away from memory subsystem is replaced last. Fig. 6 shows the results according to the number of SDRAM-aware router placed to the order. In Fig. 6(a), the memory utilization of our NoC design starts 57% in case of no buffer, no memory scheduler and no SDRAM-aware router. When three SDRAM-aware routers are placed, its memory utilization increases by 72% (4.8% higher than the conventional NoC design). The reason why it is no longer improved in the case that over three SDRAM-aware routers are employed is that the bank conflicts and data contentions are almost removed. In Fig. 6(b), service latency is also shortened by 79 cycles (33% shorter than the conventional NoC design) since the high memory utilization lets a packet accessed faster with short waiting time and our flow control mechanism manages the upper bound latency. Our SDRAM-aware NoC design and the conventional NoC design are also synthesized by DesignVision from Synopsys with TSMC130LV library. The gate count of SDRAM-aware NoC design is 24.8% smaller as shown in Fig. 6(c) even if all

978-1-60558-497-3/09 $25.00 © 2009 ACM

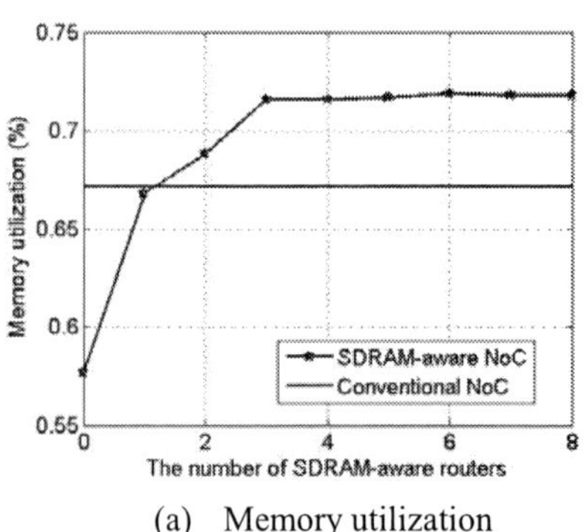

(a) Memory utilization

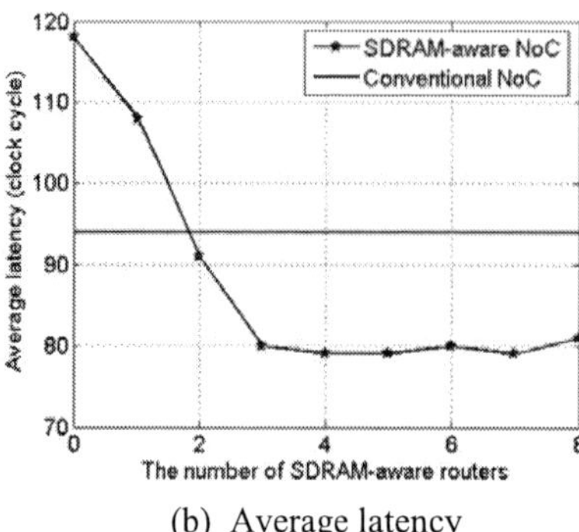

(b) Average latency

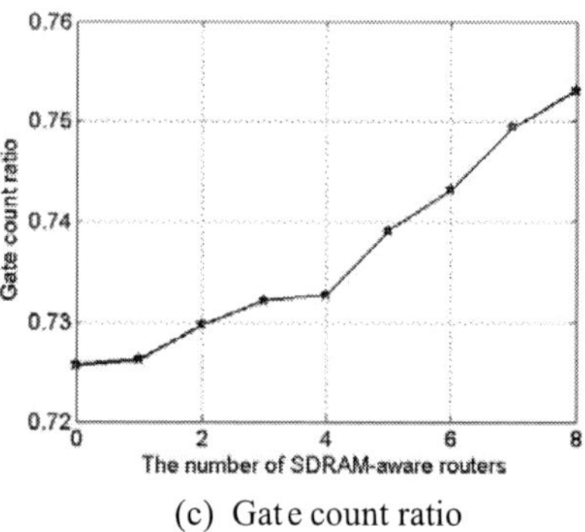

(c) Gate count ratio

Figure 6: Comparison in DTV application according to the number of SDRAM-aware routers

conventional NoC routers are replaced by our SDRAM-aware router. The reason is that large buffers and a memory scheduler in a memory subsystem are removed while the gate count increase caused by our flow-control mechanism is minimal. Thus, considering both the performance and the hardware cost, it is the best choice to replace three conventional routers to the SDRAM-aware routers in our DTV application. Our SDRAM-aware NoC router is also applied into dual DTV model [20] containing two memory subsystems. As the result, over 42% gate count is saved.

6.2 Experiments with Synthetic Benchmark

We evaluate the improvement of memory utilization and latency using 10 randomly generated applications based on industrial IPs (Intellectual Properties). The IPs mapped into 3×3 to 6×6 mesh topology generate 4 to 32 flits per packet at dynamic intervals. Each simulation runs for one million cycles. Table 3 shows our SDRAM-aware NoC improves memory utilization by 4.9% and latency by 18% on average compared to the conventional NoC. Especially, the improvement is higher in 6x6 NoC than in 3x3 NoC since packets passing through more SDRAM-aware routers are scheduled better for SDRAM operations. Thus, the improvement would be greater in a more complex NoC.

Table 3: Comparison in synthetic benchmarks

Content	router	3x3 NoC	6x6 NoC	average
memory utilization	SDRAM-aware NoC	63.7%	61.2%	64.6
	conventional NoC	59.4%	53.2%	59.7
	improvement	3.3%	8.0%	4.9%
memory latency	SDRAM-aware NoC	59 cycles	71 cycles	68 cycles
	conventional NoC	65cycles	99 cycles	83 cycles
	improvement	9.2%	28.3%	18%

7. CONCLUSION

We propose a NoC router including an SDRAM-aware flow-control mechanism. The proposed scheme allocates one of the competing packets toward an SDRAM, to a channel in our multiple routers based on SDRAM-awareness. Thus, the memory utilization and latency are more improved and the expensive buffers and scheduler of memory subsystem are removed. Furthermore, our SDRAM-aware router achieves better performance when it is employed in a more complex NoC design or when its routing scheme is adaptive. Our basic idea can be also extended to all SoCs with a flow controller and an SDRAM.

8. REFERENCES

[1] Luca Benini and Giovanni De Micheli, "Network on chips: a new SoC paradigm," Computer, 2002, 35, pp. 70-78.

[2] W. J. Dally and B. Towles, "Route Packets, Not wires: On-Chip Interconnection Networks," In Proc. Design Automation Conf., 2001.

[3] A. W. Topol, et al., "Three-dimensional integrated circuits," IBM J. Research and Development, vol. 4, 2006.

[4] F. Li, C. Nicopoulos, T. Richardson, Y. Xie, N. Vijaykrishnan, and M. Kandemir, "Design and Management of 3D Chip Multiprocessors Using Network-in-Memory," In Proc. International Symposium on Computer Architecture (ISCA), 2006.

[5] "Samsung DDR II SDRAM. Device operations & timing diagram," http://www.samsung.com/global/business/semiconductor.

[6] L. A. Vervoort, Philip Yeung, and Anil Reddy, "Addressing memory bandwidth needs for next-generation digital TVs with cost-effective, high-performance Consumer DRAM Solutions," http://www.rambus.com.

[7] S. Rixner, W. J. Dally, U. J. Kapasi, P. Mattson, and J. D. Owens, "Memory access scheduling," In Proc. ISCA, 2000.

[8] S. Heithecker and R. Ernst, "Traffic shaping for an FPGA based SDRAM controller with complex QoS requirements," In Proc. Design Automation Conference, 2005.

[9] W. D. Weber, "Efficient shared DRAM subsystem for SoCs," Sonics Inc., 2001, White paper.

[10] B. Akesson, K. Goossens and M. Ringhofer, "Predator: a predictable SDRAM memory controller," In Proc. CODES+ISSS, 2007.

[11] I. Hur and C. Lin, "Memory scheduling for modern microprocessors," ACM Trans. on Computer Systems, vol. 25, no. 4, Dec. 2007.

[12] F. Paganini, J. Doyle, and S. Low, "Scalable laws for stable network congestion control," In Proc. IEEE Conference on Decision and Control, 2001.

[13] Dongyu Qiu and Ness B. Shroff, "A Predictive flow control scheme for efficient network utilization and QoS," IEEE Trans. on Networking, vol. 12, no. 1, Feb. 2004.

[14] U. Y. Ogras and R. Marculescu, "Prediction-based flow control for network-on-chip traffic," In Proc. Design Automation Conf., 2006.

[15] W. C. Kwon, S. M. Hong, S. Yoo, B. Min, K. M. Choi, and S. K. Eo, "An open-loop flow control scheme based on the accurate global information of on-chip communication," In Proc. DATE, 2008.

[16] W. C. Kwon, S. Yoo, S. M. Hong, B. Min, K. M. Choi, and S. K. Eo, "A practical approach of memory access parallelization to exploit multiple off-chip DDR memories," In Proc. DAC, 2008.

[17] W. J. Dally and B. Towles, "Principles and practices of interconnection networks," Morgan Kaufmann, 2004.

[18] "Sonics MemMax," http://www.sonicsinc.com.

[19] "Denali Databahn: DRAM Memory Controller IP," http://www.denali.com.

Multiprocessor System-on-Chip Designs with Active Memory Processors for Higher Memory Efficiency

Junhee Yoo
School of Electrical Engineering
and Computer Science
Seoul National University

ihavnoid@poppy.snu.ac.kr

Sungjoo Yoo
Department of Electronic and
Electrical Engineering
POSTECH

sungjoo.yoo@postech.ac.kr

Kiyoung Choi
School of Electrical Engineering
and Computer Science
Seoul National University

kchoi@snu.ac.kr

ABSTRACT

Memory access latency and memory-related operations are often the performance bottleneck in parallel applications. In this paper, we present a concept of *active memory operations* which is an on-chip network transaction that operates based on the microcode provided by the software designer. Utilizing the active memory operation, we can replace multiple transactions of memory accesses over the on-chip network and related local processing element computation with a smaller number of high-level transactions and near-memory computation. We implemented a processor called active memory processor which is located near the memory and executes the active memory operations. In our case studies, we applied the concept to three real-world applications (parallelized JPEG, FFT, and text indexing for data mining) running on a 36-tile architecture with 32 cores and 4 memories and found that the programmable transaction approach can improve performance by 34.3% to 618% at the cost of additional design effort.

Categories and Subject Descriptors

C.1.4 [**Computer System Organization**]: Parallel Architectures – *Distributed architectures.*

General Terms

Performance, Design

Keywords

Network-on-chip, Computer Architecture, System-on-Chip

1. INTRODUCTION

As the number of transistors available on a chip increases, we are experiencing a large boost in the number of processors on a single design. Due to the larger number of processors, modern SoC designs require more sophisticated communication methods. Traditional shared-wire buses evolved into crossbar-based designs, and are further evolving into on-chip networks. Although on-chip networks have alleviated bandwidth bottlenecks and long wire delays, on-chip networks suffer from large communication latencies. Thus, they require better methods for reducing the

Permission to make digital or hard copies of part or all of this work for personal or classroom use is granted without fee provided that copies are not made or distributed for profit or commercial advantage and that copies bear this notice and the full citation on the first page. To copy otherwise, to republish, to post on servers or to redistribute to lists, requires prior specific permission and/or a fee.
DAC'09, July 26-31, 2009, San Francisco, California, USA

latency between processors and memories. Although there have been many methods proposed to reduce on-chip communication latency, there is an inherent limitation in reducing it. This is because latency is dependent on the communication distance, which is an unavoidable physical constraint.

In this paper, we propose a different approach, which tries to reduce the number of transactions over the on-chip network, rather than reducing the latency itself. In short, the proposed method combines many simple memory read/write operations into one high-level operation, called active memory operation.

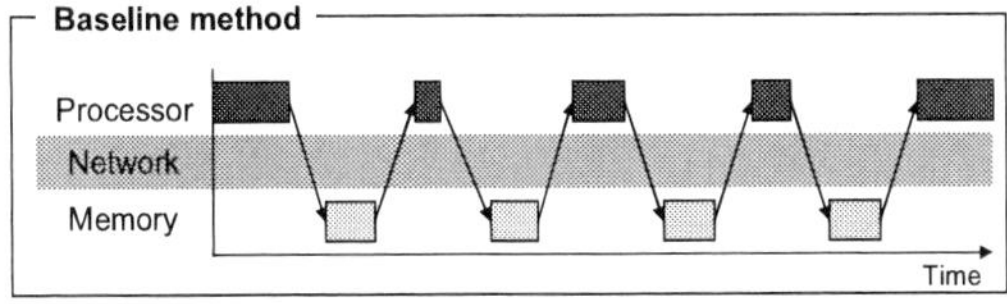

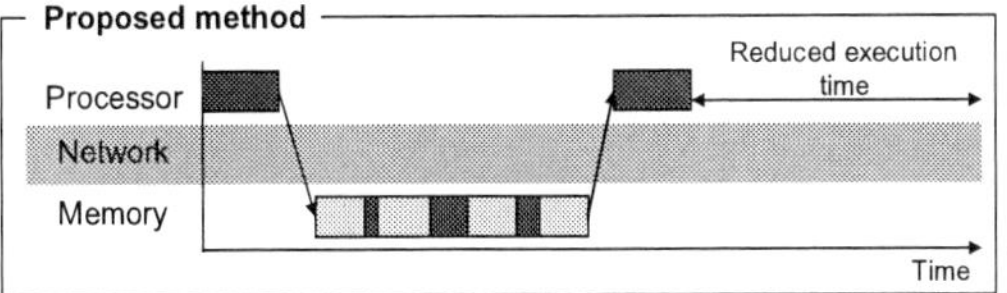

Figure 1. Benefit of active memory operations.

Figure 1 illustrates the benefit of active memory operations. In the baseline method, when a given code is executed only on the processor side, four memory accesses are required to complete the function, where each memory access incurs network latency thereby increasing the execution cycle. Our idea is to reduce the number of network transactions. To do that, we propose moving near the memory side the computation (the given code) which generates such frequent (and possibly irregular) memory accesses. By doing that, we can reduce the number of transactions on the on-chip network from four to one as Figure 1 shows. Note that the remaining single transaction orders the original computation (consequently, the related memory accesses) to be performed on the memory side. This scheme can reduce both the memory stall time of the target function and the overall traffics on the on-chip network. The active memory operation should be reprogrammable, so that different active memory operations can be used for different applications.

Figure 2 illustrates how the active memory operation works. A transaction is invoked by the master (typically, a software program running on the processor) as a function, such as 'search for the largest integer on the linked list' (step 1 in the figure). Then, the master-side network interface (NI) generates a request packet, which contains the starting address of the slave-side NI's microcode and additional parameters (step 2). Using the packet, the slave-side NI processes the high-level request by executing the microcode and accessing the nearby memory (step 3). Then, it generates a response packet which is returned to the master-side NI (step 4). The master-side NI forwards the result from the packet to the master.

In this approach, the slave-side NI can be specialized to execute instructions which are critical to memory accesses. For example, the NI can have instructions for complex address and data manipulation which do not exist on general-purpose processors. Thus, the slave-side NI becomes a specialized processor for dealing with complex memory operations.

We expect our method to be useful at least for functions which (1) load/store data in irregular access sequences, 2) perform intensive memory operations, and 3) manipulate high-level data structures (which may also involve many memory accesses and heavy use of synchronization primitives). In this paper, as an implementation of the proposed approach, we present an *active memory processor* (AMP), which is integrated into the slave-side NI for implementing user-provided, custom operations.

Original code:

```
void fx(int x, int y) {
  p = mem_a[x].v1;
  q = mem_b[p+y];
  r = mem_c[q*3].right;
  s = mem_d[r];
  return s;
}
```

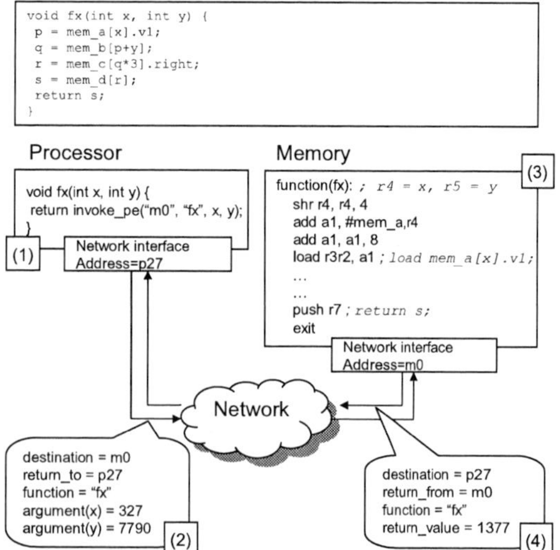

Figure 2. Outline of our proposed communication method.

2. RELATED WORK

There have been many studies on efficient data structures which are aware of the underlying memory hierarchy. The concept of cache-oblivious algorithms was proposed on [6][7][8]. Our method does not eliminate the usefulness of cache-oblivious (and cache-aware) algorithms, but complements them by allowing the designer to implement memory-intensive algorithms on the lower level of the memory hierarchy (e.g., the memory controller, or the L2 cache controller) for further optimization.

Additionally, there have been many approaches relying on cache prefetches [3][4][5]. Although prefetching is a useful method, bandwidth is wasted when the prefetched data 'misses' the expectation. Our approach may reduce the necessity of prefetches since the memory latency critical operations, which might require prefetch previously, can be performed directly on the memory side. On the memory side, many studies have focused on intelligent cache and memory controllers. [1] presents an intelligent memory controller, which optimizes the memory architecture both for conventional cache line fills, and vector-style scatter/gather operations. This is achieved by implementing configurable physical address remapping for higher locality, and simultaneously operating bank controllers for vectorized accesses. In [2], the authors present a method to reduce DRAM latency by optimizing the SDRAM controller's prefetching behavior specifically for the L2 cache controller. However, our work involves microcode-controlled processors for custom types of transactions which operate on the memory side.

In addition to efficient architectures, there also have been various researches on system design flows for low latency, low power, and low area in the memory hierarchy. These studies include scratchpad memory management [11], memory hierarchy customization [12], and cache management [13]. Our method complements these methods. Thus, they can be used together for co-optimizations.

The concept of active memory operations is similar to active messages [9] [14], in the sense that the transactions contain user-defined control information for manipulating the state of packet destination. Our method can be considered to be a special case of active messages, which is designed for efficient communication between processors and memories over the on-chip network, where the memory blocks have dedicated processors for handling the custom transactions. Unlike previous active messages, which are implemented for high-performance supercomputing, active memory operations over the on-chip network have tighter resource constraints. For example, on-chip designs have limited memory for code and data in the AMP near the memory. In addition, there can be a limitation in the number of AMPs considering their costs. Thus, we consider our method as an application-specific active message for many-core architectures. The approach from [14] uses the active message concept for manipulating memory contents, but the support is limited to two kinds – scalar and stream operations.

3. ACTIVE MEMORY PROCESSOR ARCHITECTURE

In the viewpoint of software, a memory operation is similar to a function call. Thus, in the request packet of transaction, the following information is encoded:

- Destination address of the memory block where the target AMP is located,
- Microcode address of the code to be executed,
- Source address, and
- Additional parameters, which are equivalent to function arguments

In the viewpoint of AMP, the contents of request packet act as the initial state of the AMP's execution flow, which includes the PC and registers. The microcode is already stored in the AMP, where it can be reconfigured by the application. Typically, the AMP's microcode is configured during system initialization.

The AMP operates as a processor which is tightly integrated into the NI of a memory tile. Not only the AMP handles active memory operations, but also handles ordinary reads and writes to the underlying memory. The AMP needs to meet the following requirements:

- The AMP must be capable of maximizing memory access throughput. To do that, at least one memory access operation needs to be able to be executed every cycle. Thus, it requires that at least four operations to be done in parallel - one memory access, one address computation, one conditional branch, and one address comparison for boundary checking.
- The AMP should have an abundant set of instructions related to data accessing, while having less instructions on data computation. For example, floating-point computation, multiplication, or division can be omitted, while complex address computation methods or bitwise operations need to be implemented.

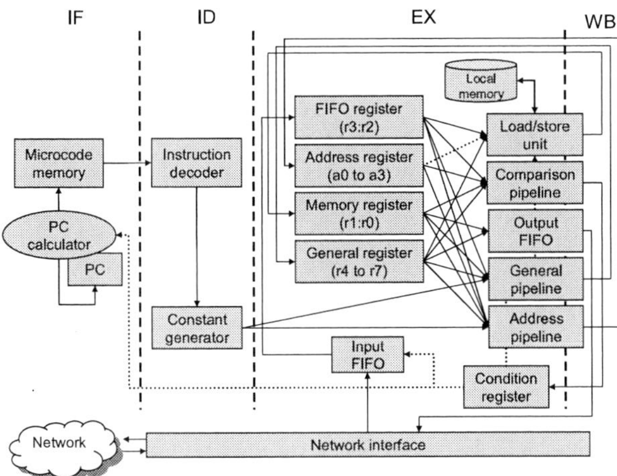

Figure 3. The active memory processor architecture.

To address these requirements, we designed the AMP with the following features:

- The AMP is designed using a custom 64-bit wide VLIW instruction set, which can issue, on each cycle, one memory access, one data computation, one address computation, one branch, and one network read/write operation. Each operation can be conditionally executed.
- The AMP can perform address computation and comparison with one instruction. By combining these with conditional branches, zero-overhead loops are supported.
- The AMP supports instructions for bitwise operations and byte-addressed operations. These are for dealing with data structures with bitwise accesses efficiently.

Figure 3 outlines our four-stage pipelined AMP architecture. The AMP is a typical in-order VLIW processor. The AMP contains four 32-bit address registers (A0 ~ A3) and four 32-bit data registers (R4~R7). The address registers are intended to be used for pointers, counters, and other control-related variables. Additionally, there are three condition flags that can be modified by comparison operations, two 32-bit registers for memory data loading (R3 and R2), and two 32-bit registers for network buffering (R1 and R0).

The AMP is connected to its local memory by a 64-bit wide memory interface. The local memory is accessed by conventional load and store instructions. The AMP is connected externally via a NI which initializes the AMP's state according to the contents of transaction request packet, and generates the response packet after completing the service of the request. The AMP's NI is controlled using FIFO instructions. The processor gets the function pointer (i.e., microcode address) and arguments by executing a *pop* instruction on the input FIFO where the incoming packets are stored, and puts return values into the output FIFO by executing a *push* instruction.

The response packet contains the function's 'return value' and the information of *(src, dest)* pair and function pointer that the request packet provides. As the AMP receives a request packet, the AMP generates the response's header, which contains the identification information (src, dest, function pointer) of the response packet. Subsequent flits are generated as the return value by the *push* instructions, which move a 64-bit data from a pair of data registers to the output FIFO as the payload of the response packet.

4. SYSTEM ARCHITECTURE

Figure 4 shows our 36-tile architecture. The 32 processing element (PE) tiles and 4 memory tiles (3D stacked SRAM) are connected via two 6x6 mesh networks, where one is used for request packets, and the other for response packets. Both of them have 32-bit wide flits, but can operate on different clock speeds. The networks use a simple XY routing algorithm.

Each PE has a RISC processor with the performance comparable to an ARM7 processor. Figure 5 shows the block diagram of each PE. Each PE has a 4KB I-cache and 4KB D-cache, both of which are four-way write-back caches with 32-byte line size. Each PE also has its own 32KB local memory which can be used as a local scratchpad memory. A master NI is attached to the local scratch pad memory for generating transactions for the AMP.[1] The PEs are connected to the outer world via a 64-bit wide interface, which has asynchronous clock bridges and data width converters, so that the PE can operate on a clock frequency different from that of the network.

Each memory tile contains one AMP connected to a 64KB SRAM memory block which is shared among the 32 PEs.[2] Each AMP has its own code memory which can be programmed by the PEs. The full system, including AMPs, PEs, and the network is implemented in VHDL.

5. DESIGN METHODOLOGY

Identifying active memory operations, i.e., which part of the application code to implement on the AMPs may be quite difficult, since each AMP is a resource shared among all the PEs. In this section, we explain our design flow for assisting the software designer in implementing software with the AMP.

5.1. Overall Flow

Figure 6 shows the design flow. First, we assume an initial parallelized software implementation of the application, which runs on the system without using the AMPs. The parallelized software is executed and profiled using conventional software profiling methods. The recorded information consists of not only the execution time, but also the number and latency of memory accesses and the average waiting time on each synchronization primitive, such as lock and barrier. This is because the AMP's purpose is to minimize communication latency, especially, on code regions and data which are heavily locked and shared.

Based on the profiling results, the designer selects candidate code blocks (arbitrary code sections, functions, and/or loops) for implementing on the AMP, based on the ratio of total memory stall time and/or synchronization waiting time to total execution cycle. The selected candidates are then profiled with a higher precision, and the branch probability of all branches in the selected candidates is also measured for more accurate estimation. Then, the full application's execution time is estimated using our execution time re-estimation flow, described in subsection 5.2. Based on the estimation results, the designer selects the code blocks that give the maximum reduction in the execution time of candidate code blocks.

5.2. Execution Time Estimation

The execution time of the full application is estimated by calculating the new execution time of the candidate code block to be obtained when executed on the AMP. This is done by

---

[1] Cache misses are (de)packetized in the cache controllers.

[2] In our examples, four 64KB SRAMs were enough to cover the entire data set. There is no limitation in supporting larger memory in our architecture.

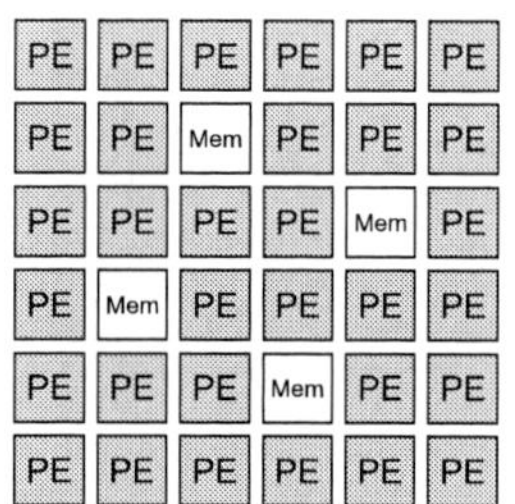
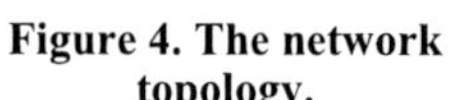

Figure 4. The network topology.

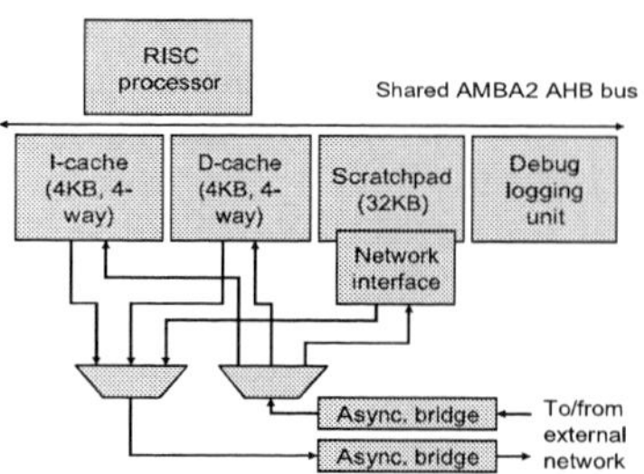

Figure 5. The processing element (PE) architecture.

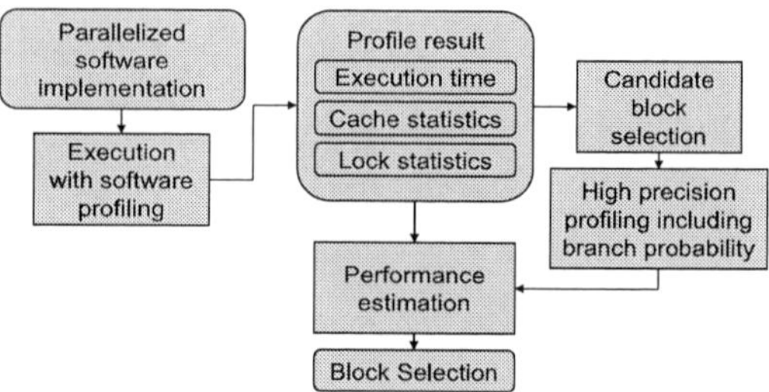

Figure 6. AMP operation implementation flow.

estimating 1) the execution time of the candidate code block on the AMP without considering congestions, and 2) the congestions on the AMP and locks.

5.2.1. Execution Time of Candidate Code Block

The execution time T_p of a code block on the PE can be formulated as:

$$T_p = E_p + M_p$$

where E_p is the computation time of the code itself on the PE, and M_p is the memory stall time. Likewise, the execution time T_t of the code block, when executed on the AMP can be formulated as:

$$T_t = E_t + M_t + X_t$$

where E_p is the computation time of the code on the AMP, M_t is the memory stall time, and X_p is the delay of processing the operation and its response (including the execution time at the PE, the round-trip delay on the on-chip network, and the execution time at the AMP). E_t and M_t can be estimated by re-compiling the code block for the AMP, and then running monte-carlo simulations using the branch probability profiles. M_t can be calculated by identifying the amount of data required to be transmitted from the PE and to the AMP.

Based on the measured result, if $T_p \leq T_t$ - that is, if there is no reduction in the execution time when implemented on the AMP - it is unlikely that the AMP implementation will benefit the application performance. Thus, the estimation stops here. Otherwise, we continue the estimation with the congestion modeling on the AMP and locks.

5.2.2. Congestion and Lock Modeling

Since implementing an active memory operation on the AMP will add workload to the AMP, all memory accesses through the AMP will experience additional memory latency. The additional latency is calculated using a simple M/M/1 queuing model.

When the execution time of a critical section reduces, the waiting period of locks protecting that section also reduces. To estimate waiting periods when the lock is accessed by the AMP, not by the PE, each synchronization primitive, e.g., a lock, is modeled as a queue according to its behavior. For example, a lock is modeled as an M/M/1 queuing model, where the service time and the arrival time obtained from the profiled information are used to estimate the lock waiting time.

6. CASE STUDIES

We have implemented three examples on our 36-tile architecture. The Fast Fourier Transform example enhances performance by implementing irregular memory access sequences on the AMP. The JPEG encoder example takes advantage of the memory-specific instructions implemented on the AMP. The text indexing example gains performance by implementing high-level data

structures directly on the AMP. For all experiments, the on-chip networks run at 200MHz, and PEs run at 125MHz. All experiments were done using a commercial VHDL simulator on an eight-core Xeon workstation, at the speed of approximately 400 PE cycles per second.

In these case studies, we have implemented the code for the AMPs by manually writing assembly code, but we believe that it is quite straightforward to design a compiler for it, using conventional compiler techniques for VLIW processors.

6.1. 4096-point Fast Fourier Transform

We used a 4096-point Fast Fourier Transform (FFT). Each point is a complex number, which consists of two 16-bit fixed point integers. Each of the 32 processors loads 128 points (64-points x2) of data from the shared memory on the memory tile, performs two 64-point FFTs which are implemented using 48 radix-4 FFTs, multiplies the appropriate twiddling factors, and stores the 128 points back to the shared memory. The whole loop is executed three times, of which the last iteration rearranges the output sequence.

In our baseline case, each PE performs non-sequential memory accesses for data reading without any help from the AMPs, only relying on the PE's data cache. 64 doubleword data have poor spatial locality, one data per cache line on average. Thus, most (3/4) of communication bandwidth is wasted to yield poor performance. The improved case, however, executes the non-sequential accesses directly from the slave NI at the memory tile, thus loading only 512B. This gives double benefits: reduction in both on-chip network traffics and memory accesses.

In this example, the code size of the active memory operation on the AMP is 48B and the sizes of request and response packets are 8B and 136B, respectively.

Figure 7 shows the execution time of the FFT for different number of processors. To prevent transient effects, we have run the 4096-point FFT four times, while measuring only the last three runs. For the 32 processor case, the execution time of the *improved* case decreases by 25.6% (1,537us to 1,144us) compared to the *baseline* case, which translates to 34.3% higher performance. This is due to the decrease in the number of data access transactions, which has dropped to 30.6%. Although the number of transactions for loading the PE's instruction cache has increased approximately by 37% due to the additional code for invoking the AMP, the performance improvement was large enough to compensate the additional instruction accesses.

As the number of used processors increases, the benefit of the AMP becomes more significant. It is because the baseline case suffers from more congestion on the on-chip network due to the large number of network traffics. On the contrary, the *improved* case reduces network traffics significantly (down to 30.6%), which contributes to avoiding network congestion.

978-1-60558-497-3/09 $25.00 © 2009 ACM

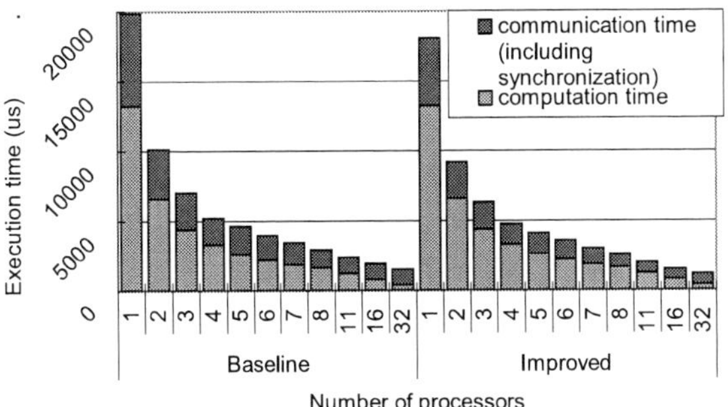

Figure 7. Experiment results of the FFT example.

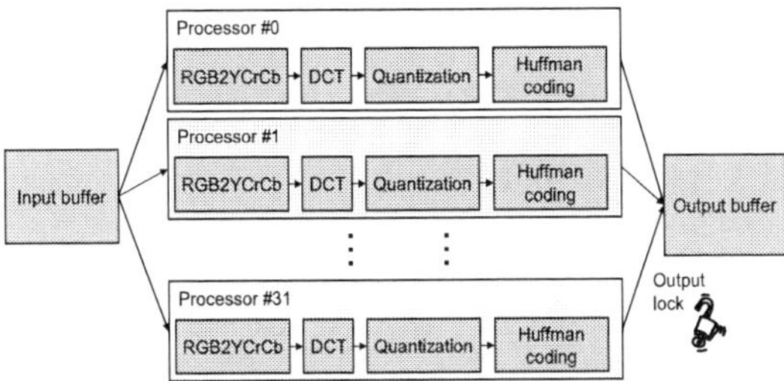

Figure 8. The JPEG encoder design.

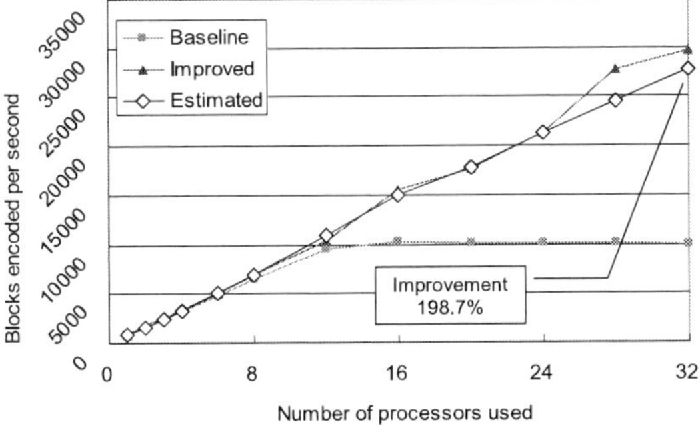

Figure 9. Encoding performance of the JPEG example.

6.2. Parallelized JPEG Encoder

For our second example, we implemented a JPEG encoder which can run on all 32 cores in parallel. Each processor takes, as the input, three 16x16 blocks (each for R, G, and B), converts the color space to YUV, runs a discrete cosine transform (DCT), quantizes it, and then encodes the data using Huffman coding. The encoded data is stored to the output buffer. Figure 8 shows the block diagram of the JPEG encoder example.

The output buffer can be accessed by only one processor at a time. Many bit packing operations are required to write data on the buffer. These involve many shifts and bitwise operations. In this example, the *baseline* case packs the output bitstream using the RISC processor in the PE, while the *improved* case uses the AMP for the packing.

Figure 9 shows the performance of the baseline and improved cases. It shows that for the *baseline* case, as the number of processors increases, performance improvement starts to saturate near 12 processors. This is because the bit packing function is becoming the performance bottleneck, making all the other processors wait for the shared lock. For example, when all 32 processors were used, each processor had to wait for the lock for an average of 1.87ms per access. However, the *improved* case does not show that saturation, because the bit packing function

running on the AMP takes much less execution time compared to the *baseline* case. As a result, there has been a 198.7% performance improvement in the *improved* case for the 32 processor case.

Figure 9 also shows the estimation results of our estimation flow. The estimation results follow the actual measured results fairly well, with an average error rate of 5.65%. Based on our observations, the errors are caused by non-ideal queuing models, and due to not considering the reduced number of instruction cache misses.

Table 1 shows the breakdown of the bit packing routine's execution time. In the baseline case, during the bit packing routine, there are 61 transactions (=45 instruction cache misses and 16 data cache misses). The transaction of an I-cache miss takes on average 72 cycles, and that of a D-cache miss 50 cycles. This is mainly due to the network delay and asynchronous bridge delay for handshaking. However, in the *improved* case, there is only one transaction for sending the data to the AMP. Thus, we could reduce the overhead of cache misses significantly.

Table 1. Breakdown of bit packing routine's execution time

Baseline		Improved	
Operation	Execution time (ns)	Operation	Execution time (ns)
Handling I-cache misses	20688	Preparing for transaction	4632
Handling D-cache misses	4304	Bit-packing computation	11864
Bit-packing computation	76328	Finalizing computation	2392
Total	101320	Total	18888

Table 1 also shows that a significant part of execution time reduction comes from the reduction in the execution time of bit-packing computation. It is because bit-packing computation benefits from the data structure manipulation instructions of the AMP that the RISC processor in the PE does not support. In addition, the AMP can issue multiple instructions each cycle. Our measurements show that the IPC (instruction per cycle) of bit-packing routine is 1.96 when it is executed on the AMP. Additionally, the 64-bit wide bit manipulation instructions on the AMP contribute to the performance gain.

In this example, the code size of the active memory operation on the AMP is 736B and the sizes of request and response packets are 274B and 8B, respectively.

6.3. Text Indexing

For our third example, we have implemented text indexing, which is a key algorithm for various data mining applications. This involves tokenizing the given text, categorizing the parsed words, and then inserting the *(word, location_of_word)* pair into one of the three global hashtables, depending on the hash value.

In our implementations, the *baseline* case implements the hashtable purely on the RISC processor side, and the *improved* case implements the hashtable on the AMP. The *baseline* case protects the hashtable using 32 locks per memory block, where each lock protects 1/32 of the whole table. The improved case handles each hashtable access as one memory transaction, and all locking is done directly by the AMP.

The *baseline* case requires at least 10 memory accesses (over the on-chip network) per hashtable insertion, which involves acquiring and releasing the locks, modifying the memory allocator data structure, the chaining data structure, and the hashtable itself. However, the *improved* case performs all of those operations in one transaction, thus significantly reducing the number of total transactions over the on-chip network.

978-1-60558-497-3/09 $25.00 © 2009 ACM

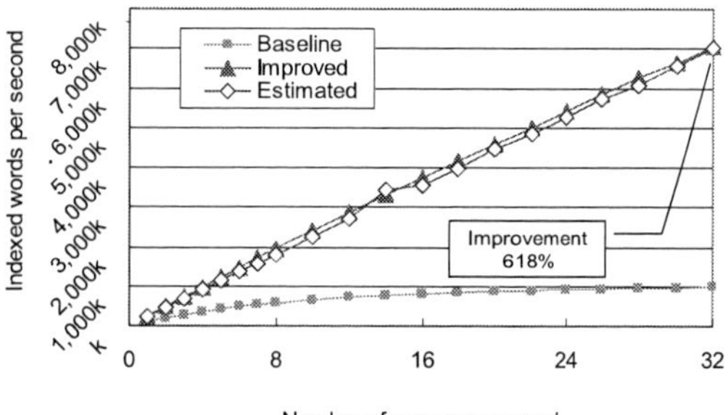

Figure 10. Indexing performance of the text indexing example.

Figure 10 shows the performance of the *baseline* and the *improved* cases in terms of the number of words processed per second. The *baseline* case has its performance saturate around 1,000k words per second, while the *improved* case is capable of processing up to 7,000k words per second, when all 32 processors are used. This results in 618% performance improvement. The main reason is because the *baseline* case requires frequent cache update and flushing over the on-chip network when performing heavy lock accesses and the modification of global data structure. However, the *improved* case makes the AMP perform all those operations by issuing one transaction over the on-chip network. Thus, memory accesses no longer become a bottleneck of system performance. Figure 10 also shows the estimation results of our performance estimation flow. It shows that our estimation flow predicts the performance fairly accurately, where the average error rate is 5.32%.

In this example, the code size of the active memory operation on the AMP is 186B and the sizes of request and response packets are 32B and 8B, respectively.

6.4. FPGA Implementation

We have implemented PE and memory tiles and the on-chip network on an Altera Stratix 3 FPGA [10], where the PEs are confirmed to operate on 50MHz, while the network is confirmed to operate on 100MHz.

Table 2. Resource consumption of PE and memory tiles

	ALUTs used	Memory used		ALUTs used	Memory used
Processing element			Memory block		
RISC processor	4314	0	Buffer and async. bridge	1424	0
Scratchpad & network interface	413	32KB	Active memory processor	5747	4KB
I-cache	1297	4KB	Local SRAM		64KB
D-cache	1333	4KB			
Local bus	445	0			
Network logic (including async. Bridge)	911	0			
Debugging logic	494	0			
Total PE tile size	9207		**Total memory tile size**	7171	

Table 2 shows the resource consumption of one PE and one memory tile. The table shows that the resource consumption of the AMP is comparable to that of the RISC processor in the PE.

7. CONCLUSION

In this paper, we proposed the concept of active memory operations, and the active memory processor as an implementation.

Experimental results with a 36-tile architecture show that our approach can improve by 34.3% ~ 618% the performance of real-life applications with a moderate area overhead on the network interface of memory tile.

As our future work, we will work on the integration of the AMP with DDR memory controller and the investigation of increasing the computing power of the AMP via multi-threading, e.g., SMT or multiple AMPs per memory tile. Additionally, based on our performance estimation flow, we will also work on the design method to explore the design space to find code sections, functions, and loops suited to active memory operations.

8. ACKNOWLEDGMENTS

This work was supported in part by KOSEF under NRL Program Grant (ROA-2008-000-20126-0) funded by MEST, Korea, and in part by IITA under ITRC support program (IITA-2009-C1090-0902-0024) funded by MKE, Korea.

9. REFERENCES

[1] B. K. Mathew, S. A. McKee, J. B. Carter, A. Davis, "Design of a parallel vector access unit for SDRAM memory systems," in *Proc. 6th International Symposium on High-Performance Computer Architecture*, pp.39-48, Jan 2000

[2] Wei-fen Lin, Steven K. Reinhart, D. Burger. "Reducing DRAM latencies with an integrated memory hierarchy design," in *Proc. 7th International Symposium on High-Performance Computer Architecture*, pp.301, Jan 2001

[3] A. Roth and G. S. Sohi. "Effective jump-pointer prefetching for linked data structures," in *Proc. 26th International Symposium on Computer Architecture*, May 1999.

[4] M. Karlsson, F. Dahlgren, P, Stenstrom. "A prefetching technique for irregular accesses to linked data structures," in *Proc. 6th International Symposium on High-Performance Computer Architecture*, 2000.

[5] S. P. Vanderwiel , D. J. Lilja, "Data prefetch mechanisms," *ACM Computing Surveys*, v.32 n.2, p.174-199, June 2000.

[6] M. Frigo, C. E. Leiserson, H. Prokop, S. Ramachandran, "Cache-oblivious algorithms," in *Proc. 40th Annual Symposium on Foundations of Computer Science*, 1999.

[7] M. Bender, E. Demaine, M. Farach-Coltom. "Cache-oblivious B-trees," in *Proc. 41st Annual Symposium of Foundations of Computer Science*, 2000.

[8] L. Arge , M. Bender , E. Demaine, B. Holland-Minkley , J. Ian Munro, "Cache-oblivious priority queue and graph algorithm applications," in *Proc. 34th annual ACM Symposium on Theory of Computing*, May 2002.

[9] T. v. Eicken, D. E. Culler, S.C. Goldstein, K.E. Schauser, "Active messages: a mechanism for integrated communication and computation," in *Proc. 19th Annual Internation Symposium on Computer Architecture*, 1992.

[10] Stratix III FPGA Device Family Overview, http://www.altera.com/products/devices/stratix-fpgas/stratix-iii/overview/st3-overview.html

[11] L. Li, L. Gao, J. Xue, "Memory coloring : a compiler approach for scratchpad memory management," in *Proc. of 14th International Conference on Parallel Architectures and Compilation Techniques*, pp. 329-338, 2005.

[12] I. Issenin, E. Brockmeyer, B. Durinck, N. Dutt, "Multiprocessor system-on-chip data reuse analysis for exploring customized memory hierarchies," in *Proc. of 43rd Design Automation Conference*, pp. 49-52, July 2006.

[13] L. Rudolph, P. Jain, S. Devadas, D. Chiou, "Application-specific memory management for embedded systems using software-controlled caches," in *Proc. of 37th Design Automation Conference*, pp. 416-419, June 2000.

[14] Z. Fang, L. Zhang, J. B. Carter, A. Ibrahim, M. A. Parker, "Active Memory Operations", in *Proc. of 21st Annual International Conference on Supercomputing*, pp. 232-241, July 2007.

978-1-60558-497-3/09 $25.00 © 2009 ACM

Vicis: A Reliable Network for Unreliable Silicon

David Fick, Andrew DeOrio, Jin Hu, Valeria Bertacco, David Blaauw, and Dennis Sylvester
Department of Electrical Engineering and Computer Science
University of Michigan, Ann Arbor, MI 48109
{dfick, awdeorio, jinhu, valeria, blaauw, dmcs}@umich.edu

ABSTRACT

Process scaling has given designers billions of transistors to work with. As feature sizes near the atomic scale, extensive variation and wearout inevitably make margining uneconomical or impossible. The ElastIC project seeks to address this by creating a large-scale chip-multiprocessor that can self-diagnose, adapt, and heal. Creating large, flexible designs in this environment naturally lends itself to the repetitive nature of network-on-chip (NoC), but the loss of a single link or router will result in complete network failure. In this work we present Vicis, an ElastIC-style NoC that can tolerate the loss of many network components due to wearout induced hard faults. Vicis uses the inherent redundancy in the network and its routers in order to maintain correct operation while incurring a much lower area overhead than previously proposed N-modular redundancy (NMR) based solutions. Each router has a built-in-self-test (BIST) that diagnoses the locations of hard fault and runs a number of algorithms to best use ECC, port swapping, and a crossbar bypass bus to mitigate them. The routers work together to run distributed algorithms to solve network-wide problems as well, protecting the networking against critical failures in individual routers. In this work we show that with stuck-at fault rates as high as 1 in 2000 gates, Vicis will continue to operate with approximately half of its routers still functional and communicating.

Categories and Subject Descriptors

B.4.5 [**Hardware**]: Input/Output and Data Communications—*Reliability, Testing, and Fault-Tolerance*

General Terms

Algorithms, Design, Reliability

Keywords

Network-on-Chip, Fault Tolerance, Hard Faults, N-Modular Redundancy, Reconfiguration, Torus, Built-in-Self-Test

1. INTRODUCTION

Silicon processes have continued to improve in transistor density and speed due to aggressive technology scaling. However, each generation increases variability and susceptibility to wearout as silicon features approach the atomic scale. It has been predicted that future designs will consist of hundreds of billions of transistors, where upwards of 10% of them will be defective due to wear-out and variation [5]. At that point we must learn to design reliable

Permission to make digital or hard copies of part or all of this work for personal or classroom use is granted without fee provided that copies are not made or distributed for profit or commercial advantage and that copies bear this notice and the full citation on the first page. To copy otherwise, to republish, to post on servers or to redistribute to lists, requires prior specific permission and/or a fee.
DAC'09, July 26-31, 2009, San Francisco, California, USA

systems from unreliable components, managing both design complexity and process uncertainty [6].

The ElastIC project approaches this challenge by having a large network of processing cores that are individually monitored for various wearout mechanisms and operating conditions [20]. Each processing core can be tested, repaired, powered down, or quarantined individually. With proper lifetime management, the performance of some cores can even be boosted for short periods of time when needed [14]. The interconnect architecture becomes a single point of failure as it connects all other components of the system together. A faulty processing element may be shut down entirely, but the interconnect architecture must be able to tolerate partial failure and operate with partial functionality. Network-on-chip provides opportunities to address this issue, as redundant paths exist from point to point, potentially allowing for reconfiguration around failed components.

Network-on-chip is a distributed, router-based interconnect architecture that manages traffic between IPs (intellectual properties, *e.g.*, processing cores, caches, *etc.*) [4]. Traffic sent through the network is organized into packets, of arbitrary length, which are broken down into smaller, link-sized chunks called flits (FLow unITS). This style of interconnect scales bandwidth with network size, making it suitable for systems with many more communicators than what could be supported by bus-based architectures. Recent commercial designs include the Tilera TILE64 [2], Intel Polaris [21], and Ambric AM2045 [1, 12]. NoCs do not inherently support fault tolerance - the first link or router failure in the network will cause the entire network to deadlock.

The reliability of NoC designs is threatened by transistor wearout in aggressively scaled technology nodes. Wear-out mechanisms such as oxide breakdown and electromigration become more prominent in these nodes as oxides and wires become thinned to the physical limits. These breakdown mechanisms occur over time, so traditional post burn-in testing will not capture them. Additionally, so many faults are expected to occur that *in situ* fault management will be economically beneficial. In order to address these faults, we propose a framework with the following steps: 1) error detection, 2) error diagnosis, 3) system reconfiguration, and 4) system recovery. Error detection mechanisms notices new faults by invariance checks such as cyclic redundancy code checks on packets. Error diagnosis determines the location of the new fault via a built-in-self-test. System reconfiguration disables faulty components and configures functional components to work around the faulty ones. Lastly, system recovery restores the system to a previously known good state or checkpoint.

Network-on-chip provides inherent structural redundancy and interesting opportunities for fault diagnosis and reconfiguration. Each router is comprised of input ports, output ports, decoders and FIFOs, which are duplicated for each channel. By leveraging the redundancy that is naturally present in routers and the network, Vicis is able to provide robustness at a low area cost of 42% while exhibiting greater fault tolerance than the previously proposed N-modular redundancy based solutions. We experimentally show that Vicis is able to sustain as many as one stuck-at fault per every 2000 gates, and still maintain half of its routers.

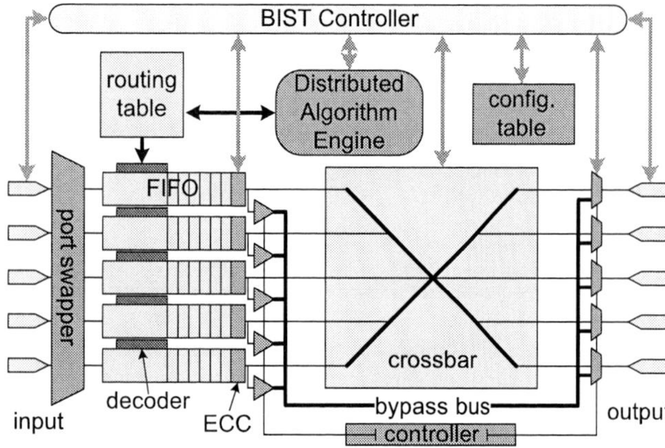

Figure 1: **Architecture of the Vicis router.** Enhancements include a port swapper, ECC, crossbar bypass bus, BIST, and distributed algorithm engine.

2. RELATED WORK

Reliable network design was first introduced by Dally *et al.* with a reliable packet-switching network, focusing on protecting against faulty board traces due to physical and thermal stress [8]. This network used link-level monitoring and retransmission to allow for the loss of a single link or router anywhere in the network, without interruption of service.

Constantinides *et al.* demonstrated the BulletProof router, which efficiently used NMR techniques for router level reliability [7]. However, NMR approaches are expensive, as they require at least N times the silicon area to implement. Additionally, network level reliability needs to be considered since some logic is impossible or expensive to duplicate (*e.g.*, clock tree) or spares may run out, resulting in the loss of a router, and subsequently the network.

Pan *et al.* explored another strategy that exploits redundant components [16]. Instead of using NMR, inputs are sent through two copies of the same component and the outputs are then compared. If they do not match, then an error is detected and can be repaired. In addition, they included on-line testing and a built-in-self-test (BIST) for correction.

A recent architecture was proposed by Park *et al.* that focused on protecting the intra-router connections against soft errors [17]. They explored a hop-by-hop, flit retransmission scheme in case of a soft error with a three-cycle latency overhead. Though each component has localized protection, there is no guarantee that the network will be operational if certain hard faults occur. Researchers have also explored efficient bus encoding schemes to protect against transient errors [3, 24].

Routing algorithms for fault-tolerant networks have been extensively explored for network level reconfiguration [9–11, 13, 18, 19, 22, 23]. These algorithms direct traffic around failed network components in a way that avoids network deadlock. Although Vicis must also accomplish this, it additionally diagnoses the location of hard faults and reconfigures around them using other methods as well. Vicis adopts an implementation of the routing algorithm described in [9] for part of its network level reconfiguration.

3. THE VICIS NETWORK

Vicis is able to maintain and prolong its useful lifetime by accurately locating faults, reconfiguring around them at the router level, and then reconfiguring at the network level. The following section is split into three parts, giving a top down view of Vicis. The first section discusses how Vicis reconfigures around failed routers and

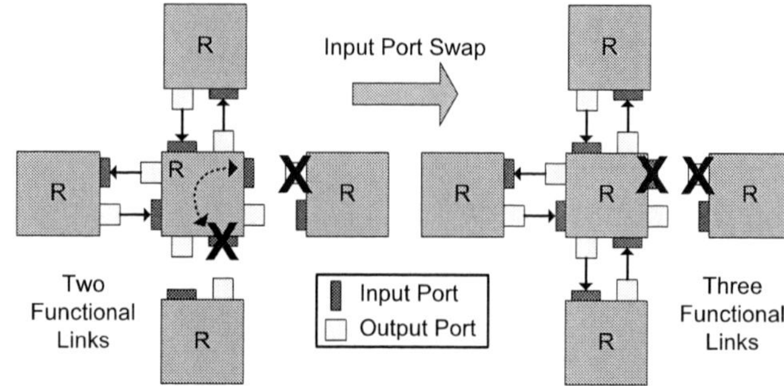

Figure 2: **An example of port swapping.** On the left side a failed input port and a failed output port are part of two different links. By swapping the failed input port to the link connected to the failed output port, Vicis increases the number of functional links from two to three.

links at the network level. The second section looks at how Vicis uses ECC and a crossbar bypass bus to tolerate faults internally. The last section discusses the built-in-self-test and how it is able to accurately diagnose faults within the router with 100% coverage.

3.1 Network Level Reconfiguration

Network level reconfiguration addresses network visible problems by running distributed algorithms to optimize connectivity and redirect traffic. The input port swapping algorithm runs first in order to increase the number of available links in the network. The network routing algorithm then rewrites the static routing table (shown in Figure 1) to detour traffic around failed network links and routers. These algorithms use fault information provided by the router level reconfiguration discussed in Section 3.2.

3.1.1 Input Port Swapping

The routing algorithm needs fully functional bidirectional links in order to safely route through the network. Each link is comprised of two input ports and two output ports, all four of which must be functional to establish a link. If one of these four ports fail, then there are still three functional ports that might be used in other places (three spares).

As shown in Figure 1, the input ports are comprised of a FIFO and a decode unit which are identical for each lane of traffic. Vicis includes a port swapper to the input of the FIFOs in order to change which physical links are connected to each input port. A port swapper is not included for the output ports due to their relatively small area. The area of the input ports, however, is majority of the total area of the router and will therefore attract the most faults, so adding an input port swapper gives Vicis great ability to move faults around the router. Since the outputs of the input ports are connected to a crossbar, they do not need a second swapper.

Figure 2, shows an example of the input port swapper in use. On the left side there is a failed input port and a failed output port. Since these two failed ports are on different links, both links fail and become unusable by the network routing algorithm. One of the failed ports is a input port of the router in the center, so the physical channel that is connected to can be changed. The port swapping algorithm reconfigures the port swapper so that the failed input port is instead connected to the neighboring failed output port, as shown on the right side. By doing this, Vicis takes advantage of the inherent redundancy in the router, and increase the number of functional links from two to three.

The port swapper does not need to be fully connected, that is, not every input port needs to be connected to every physical link. In this implementation of Vicis, the link to the local network adapter

978-1-60558-497-3/09 $25.00 © 2009 ACM

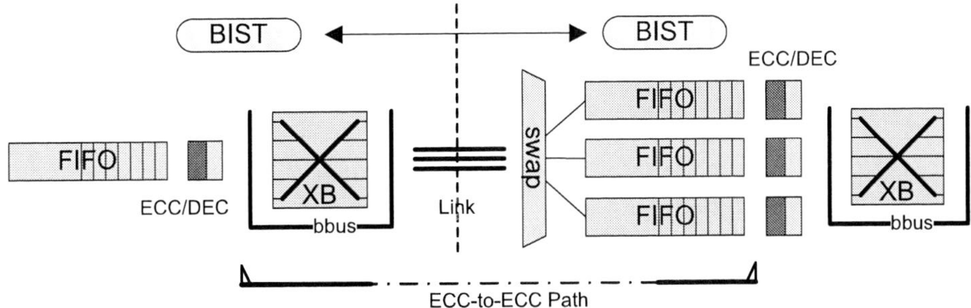

Figure 3: Convergence path of ECC-mitigated faults. Faults in the datapath can be corrected using ECC as long as no more than one is encountered between corrections. The bypass bus and port swapper provide alternate paths between routers to reduce the number of faults in each ECC-to-ECC path. There are six available paths in this example.

is able to connect to three different input ports, and the other links are able to connect to two input ports. The port swapping algorithm is a greedy algorithm that takes into account the ECC related information described in Section 3.2.2. It also gives priority to the local network adapter link, making sure that it is always connected when possible.

After the port swapping algorithm finishes running, each router knows which of its links are functional, and provides that information to the routing algorithm, which completes the network reconfiguration.

3.1.2 Network Rerouting

For a non-fault tolerant network, or a network that supports only router level reliability, the first link or router failure will cause the entire network to fail. Traffic that gets directed to that link or router gets lost, and eventually the network will deadlock when traffic flow tokens fail to be back propagated. In order to detour traffic around faulty links and routers, Vicis uses a routing algorithm to rewrite the routing tables shown in Figure 1. We selected the algorithm described by Fick *et al.* in [9], since it does not need virtual channels, it supports link-level faults, and it has a low overhead implementation. Since it is a distributed algorithm, the loss of any number of links or routers will not prevent it from running, and the remaining routers will have correctly written routing tables.

3.2 Router Level Reconfiguration

Router level reconfiguration is used to contain faults within the router so that they are not visible at network level in the form of failed links and routers. The Vicis architecture, shown in Figure 1, includes a crossbar bypass bus to protect against crossbar failures, and single error correct, single error detect error correction codes (SEC-SED ECC) to protect datapath elements. Information about the location of faults is provided by the hard fault diagnosis stage, described in Section 3.3.

3.2.1 Crossbar Bypass Bus

The bypass bus, parallel to the crossbar, allows flits to bypass a faulty crossbar (see Figure 1). The crossbar controller is configured by the BIST to direct traffic to either the crossbar or to the bypass bus on a packet by packet basis. If multiple flits needs to use the bypass bus (simultaneously), then one flit is chosen to proceed first while the others wait until a later cycle. This spare path allows Vicis to maintain operation when multiple faults appear in the crossbar, trading off performance to maintain correct operation.

3.2.2 Error Correcting Codes (ECC)

ECC-protected units allow each datapath component to tolerate a small number of faults while maintaining correct functionality at the cost of minimal overhead. Any fault that manifests along an

ECC-guarded datapath can be corrected when the flit passes though an ECC unit. In order to take full advantage of ECC, the BIST tracks the location of every datapath fault and ensures that every ECC-to-ECC path has at most one hard fault.

In Figure 3 we show the ECC-to-ECC path in Vicis. After encoding, a flit travels through the crossbar and bypass bus, the link between the two routers, the input port swapper, and finally the FIFO. At each mentioned unit along the path, the faults (if any) are diagnosed and cataloged by the two BISTs. If two faults accumulate in the same datapath, the bypass bus and port swapper provide alternative configurations to either avoid one of the faults or move one of the faults to a different datapath, respectively.

For example, one fault could be in the crossbar, another in a link, and a third in the default FIFO for the datapath. The ECC implementation in Vicis can only correct one of these faults, so the crossbar bypass bus and input port swapper need to mitigate two of them. The bypass bus will be used to avoid the crossbar fault, potentially resulting in a loss of performance. The input port swapper will be used to swap in a fault-free input port to the datapath, moving the single-fault input port to another physical link that does not have any faults. By doing this, full functionality is maintained, even with three faults originally in the same datapath.

The FIFOs were designed to take advantage of ECC as well. The FIFO is implemented as a circular queue, which results in a given flit only ever entering a single line of the FIFO. In this way, a FIFO could have upwards of one fault per entry while remaining operational, as it effectively adds only a single fault to the ECC-to-ECC path.

3.3 Hard Fault Diagnosis

The techniques in Vicis rely on knowing precisely where faults exist. A port cannot be swapped if it is not known what ports are faulty and which are functional. The use of ECC requires that Vicis knows precisely how many faults are in each part of the datapath. In this section we discuss the strategies that Vicis uses to locate faults with low cost, yet obtain 100% coverage.

3.3.1 Built-in-Self-Test (BIST)

The built-in-self-test (BIST) unit controls all of the router diagnosis and reconfiguration. It is power gated during normal operation in order to protect it from wearout. Because of this, the BIST controls other units through a configuration table, which stays powered on at all times. The configuration table is tested as well, as part of testing the units that it controls. Non-wear-out faults in the BIST need to be found after fabrication, and the router disabled by the manufacturer.

Figure 4 shows the built-in-self-test procedure that occurs when the BIST is activated. One of network adapters will detect a network failure by using a cyclic redundancy check or another invari-

978-1-60558-497-3/09 $25.00 © 2009 ACM

```
 1: (*)   Propagate Error Status (Error Unit)
 2: ( )   Power up BIST
 3: (*)   Synchronize Network
 4: (D)   Test Router Configuration Table
 5: (P)   Test Error Propagation Unit
 6: (P)   Test Crossbar Controller
 7: (P)   Test Routing Table
 8: (P)   Test Decode/ECC Units
 9: (P)   Test Output Ports
10: (P)   Test FIFO Control Logic
11: (D)   Test FIFO Datapath
12: (D*)  Test Links
13: (D*)  Test Swapper
14: (D)   Test Crossbar
15: (D)   Test Bypass Bus
16: (*)   Communicate ECC-to-ECC Path Information
17: (*)   Run Port Swapping Algorithm
18: (*)   Run Network Rerouting Algorithm
19: ( )   Power Down BIST
20: ( )   Resume Operation
```

Figure 4: Built In Self Test procedure. Distributed algorithms are marked (*), datapath testing is marked (D), and pattern based testing is marked (P).

ance check. Once a network adapter detects an error, it will send an error status bit to its connected router, at which point the routine in Figure 4 begins.

The network first broadcasts an error status bit via a low-overhead error unit. The error unit also powers up the BIST for that router. Once the BIST has been powered up, it performs a distributed synchronization algorithm so that every BIST in the network runs the rest of the routines in lock-step. Once synchronization completes, each unit is diagnosed for faults. The diagnosis step does not use information from previous runs of the routine, or from the detection mechanism that started the routine, so all faults will be diagnosed regardless of whether they caused the error or not. Once all units have been tested, distributed configuration algorithms run, and finally the BIST powers down and normal operation resumes.

3.3.2 *Functional Unit Wrappers*

Each functional unit has a test wrapper so that the built-in-self-test can control it during fault diagnosis. The wrapper consists of multiplexers for each of the functional unit inputs, which switch between the normal unit inputs, and a testing bus coming from the BIST. A signature bus takes a copy of each of the outputs back to the BIST for analysis. The wrappers are interlocked to guarantee fault-coverage is not lost between each of the units. Although a hard fault on a wrapper gate may be caught trivially, a hard fault may occur on the gate driving the multiplexer select bit, forcing it to always be in test mode. By reading the inputs after the multiplexer to the next unit, this case will also be caught. Additionally, synthesis might insert buffers between where the signature bus connects and the other units connecting to the output - those buffers would not be covered without interlocking wrappers. Cooperative testing between routers is needed to test links, the input port swapper, and their associated wrappers.

3.3.3 *Datapath Testing*

Lines marked with (D) in Figure 4 are part of a datapath test. Units in the datapath need an exact count of ECC errors for each unit so that the maximum number of errors is not exceeded for any ECC-to-ECC path. The tests each send all 1's or all 0's through the datapath, looking for bit-flips faults. A specially designed bit-flip count unit determines if the datapath has zero, one, or more bit flips so that multiplexers are not needed to inspect each bit individually.

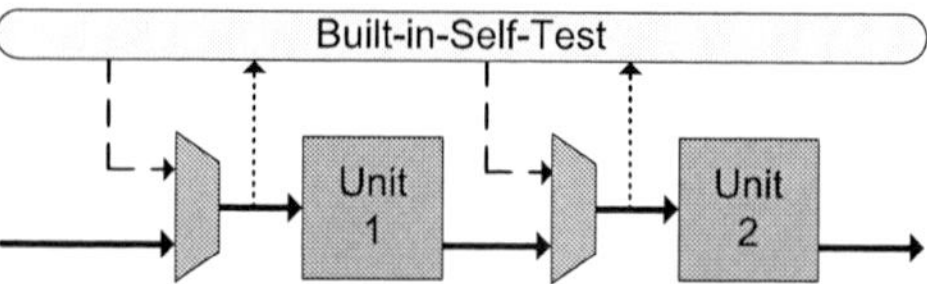

Figure 5: Functional Unit Wrappers. Vicis uses interlocking functional unit wrappers to provide the BIST with access to each unit for hard-fault diagnosis. Incoming patterns from the BIST are shown with long dash arrows, outgoing signatures are shown with short dash arrows, and normal operation paths are shown with solid arrows.

Units like the FIFO reuse the same test for each instance, which helps maintain a low overhead.

3.3.4 *Pattern Based Testing*

Lines marked with (P) in Figure 4 are part of a pattern based test. These units are implemented with synthesized logic, so a hand-written test would not have full coverage. To test these units, Vicis uses a linear feedback shift register (LFSR) to generate a number of unique patterns, and a multiple input signature register (MISR) to generate a signature. Each unit tested with the pattern based test receives the same sequence of patterns from the LFSR, but each has its own signature. Identical units, such as all of the decoders, have the same signature. A signature mismatch will flag the corresponding unit as unusable, and the unit is then disabled by the BIST. Implementation of the pattern based test is lightweight due to the simplicity of the LFSR and MISR.

Pattern based testing dominates the total runtime of the BIST, taking approximately 200,000 cycles when 10,000 patterns are used per test. The time between faults should be measured in hours, however, so this is a relatively short amount of time.

4. EXPERIMENTAL RESULTS

In this section, we discuss Vicis' effectiveness in improving network reliability. First we discuss the experimental setup used to evaluate Vicis. Next, we look at network reliability and compare Vicis to a TMR based implementation. Then we look at router reliability and its *Silicon Protection Factor*. Finally, we show results for network performance degradation.

4.1 Experimental Setup

We implemented Vicis in a 3x3 torus topology, as well as a baseline router design. Each design routes flits with a 32-bit data portion, and have 32-flit input port buffers, which is consistent with the Intel Polaris router [21]. The baseline router has the same functionality as Vicis, except for the reliability features. Both Vicis and the baseline NoC were written in Verilog and were synthesized using Synopsis Design Compiler in a 42nm technology. Compared against the baseline NoC, Vicis has an area overhead of 42%, which is much lower than the 100+% needed for NMR.

To test the reliability features of Vicis, we randomly injected stuck-at faults onto gate outputs in the synthesized netlist. Selection of the gates was weighted based the area of the gate, which is consistent with the breakdown patterns found experimentally in Keane *et al.* [15].

Test packets were generated at each network-adapter. It was verified that each packet made it to the correct destination with the correct data. For each error combination, we ran 10,000 packets of a random uniform traffic distribution, each packet having between 2 and 11 flits.

Network throughput was measured using a custom made, cycle accurate C++ simulator. Profiles of the network were gen-

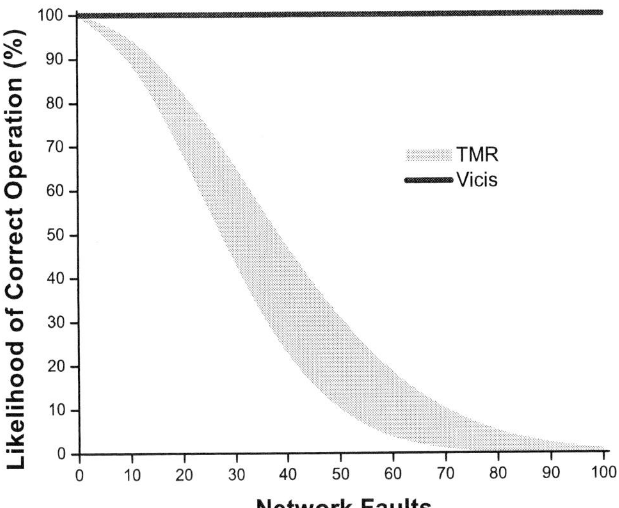

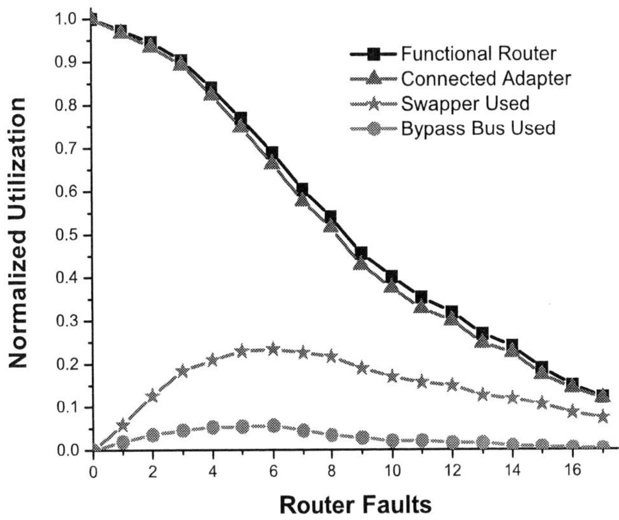

Figure 6: Reliability Comparison with TMR. TMR has probabilistic reliability while maintaining performance, where Vicis has probabilistic performance while maintaining reliability (see Figure 8.)

Figure 7: Router reliability with increasing faults. Reliability of an individual router as it receives stuck-at faults inside of a faulty network. Utilization of individual features is also shown. *Connected Adapter* shows how many functional routers also have a functional link to its adapter.

erated with the Verilog simulations described above, which were then given to the C++ simulator to model longer simulation traces. Throughput was measured using random uniform traffic for this test as well.

4.2 Network Reliability

To test the reliability of the simple network, we randomly injected faults into the system and monitored the packet traffic. Although the full system is prone to wear-out, we primarily focused on injecting faults into the router - a total of 20,413 gates. Since we only consider wear-out induced faults, we power-gated the BIST, thus making it immune to fault injection.

To rigorously test the system, we considered eleven different cases with varying simultaneous faults: 1, 10, 20, 30, 40, 50, 60, 70, 80, 90, and 100, 100 being one fault for every 2000 gates. For each number of simultaneous faults, we considered 1200 different (random) error combinations.

In this series of tests there was a single failure that manifested at 30-40 faults, but disappeared at 50 faults. The disappearance of the failure is likely due to new faults being injected into the same unit as the fault(s) that were previously overlooked by the BIST. We expect the failure to be an implementation error, as opposed to a problem with the described techniques. Either way, the reliability of the network is dominated by the routing algorithm for larger networks, as described in [9].

We compare the reliability of our design to a TMR-based implementation. Note that TMR only provides probabilistic reliability - since the voter takes the most common signal of the three replicated units, two well-placed faults could cause the system to fail. In the worst case, a single fault could cause system failure if it occurred in a clock tree or another non-replicable cell. Unlike BulletProof [7] and other prior work that relies on maintaining total functionality, Vicis is able to tolerate multiple, simultaneous faults, including ones that render entire routers useless. Thus, Vicis is able to maintain 100% reliability regardless of the number or occurrence of faults, although performance becomes degraded.

In Figure 6 we show the reliability of TMR, which is based on analysis from [7]. Where TMR provides 100% performance and degrading reliability for 200+% area overhead, Vicis provides

100% reliability and degrading performance for 42% overhead. Comparing TMR reliability degradation to Vicis performance degradation, TMR degrades much more quickly - at 100 network faults, TMR's reliability approaches zero whereas Vicis continues to operate with half of its routers still functional and communicating.

4.3 Router Reliability

In Figure 7, we present the different internal router statistics based on the tests described above. To fairly evaluate and collect results for each router component, we injected faults on standard-cell outputs in a network level netlist in order to consider the states of the surrounding routers. A router is considered operational if it has at least two functional bidirectional ports connected to either other routers or network adapters.

4.3.1 Input Port Swapper

We found that the input port swapper was very successful at keeping the cores (IPs) connected to the network. As shown in Figure 7, only eight percent of the available routers not have a functional network adapter. The swapper had a high utilization, being used nearly 24% of the time for routers with seven faults.

4.3.2 Crossbar Bypass Bus

At seven faults, the crossbar bypass bus was used less than six percent of the time. This is due to two reasons: 1) the crossbar is relatively small - less than five percent of the total area of the router, and 2) the crossbar is protected by both the input port swapper and ECC as well. We expect that an improved implementation of the bypass bus will have an greater effect, and are pursuing that in future work.

4.3.3 Silicon Protection Factor (SPF)

Constantinides *et al.* introduce the concept of *Silicon Protection Factor* for router level reliability, defined as the number of faults a router can tolerate before becoming inoperable, normalized by the area overhead of the technique [7]. The normalization step is to take into account the increase of gates since having more gates means that the design will also experience more faults.

978-1-60558-497-3/09 $25.00 © 2009 ACM

From Figure 7 we can interpolate to get the median value, which is 9.3 faults before the system fails. Normalized to our area overhead of 42%, this gives Vicis an SPF of 6.55. In comparison, the best SPF provided by the Bulletproof router is 11.11, which incurs a 242% area overhead. The lowest area overhead Bulletproof configuration is 52% overhead, but provides an SPF of only 2.07.

Actual network reliability cannot be compared since a router failure in Bulletproof causes full network failure, whereas a router failure in Vicis only renders that single router inoperable. Additionally, Bulletproof does not give a breakdown of what faults were critical (first fault causes a failure) versus cumulative (multiple faults interact to cause a failure). The number of critical failure points in the Bulletproof router would place a limit on overall network reliability, where that is not the case in Vicis.

4.4 Network Performance

In Figure 8, we demonstrate how network level performance gracefully degrades as the number of faults in the network increases. The black line with squares shows the number of available IPs connected through the network to at least one other IP. At 90 faults, which is more than 10 faults per router, or one fault per 2000 gates, over 50% of the original IPs are still available. Since faults are injected at the gate level, there would not likely be any functional IPs to connect, as they would be experiencing the same fault levels.

In Figure 8, we also show the normalized network throughput across different number of simultaneous hard faults (data points as triangles). For the first 30-40 faults, the network experiences a loss in normalized throughput. This is due to link failures (2+ faults within links), forcing packets to take longer paths and avoid network obstructions. After 40 faults, however, the network loses enough routers to effectively create a smaller network that intrinsically supports higher normalized throughput. The light shading behind the data line shows 5th-95th percentiles of normalized throughput - the line itself is the median.

5. CONCLUSIONS

In this work we presented Vicis, a network-on-chip that is highly resilient to hard-faults. By using the inherent redundancy in the network and its routers, Vicis can maintain higher reliability than NMR based solutions while incurring only a 42% overhead. Each router uses a built-in-self-test (BIST) to diagnose the locations of hard faults and runs a number of algorithms to best use ECC, port swapping, and a crossbar bypass bus to mitigate them. The routers work together to run distributed algorithms to solve network-wide problems as well, protecting the networking against critical failures in individual routers. Ultimately, we show that with stuck-at fault rates as high as 1 in 2000 gates, Vicis will continue to operate with approximately half of its routers still functional and communicating. Additionally, we provide results detailing the utilization of some of the architectural features, as well as a reliability comparison with triple modular redundancy and prior work.

6. ACKNOWLEDGMENT

This research was funded in part by the Gigascale Systems Research Center and the United States National Science Foundation. We would like to thank Synopsys for their generous support of this project, and the reviewers for their detailed feedback.

7. REFERENCES

[1] Massively Parallel Processing Arrays Techhnology Overview. *Ambric Technology Overview*, 2008.

[2] S. Bell et al. TILE64 processor: A 64-core SoC with mesh interconnect. *Proc. ISSCC*, 2008.

[3] D. Bertozzi, L. Benini, and G. De Micheli. Low power error resilient encoding for on-chip data buses. *Proc. DATE*, 2002.

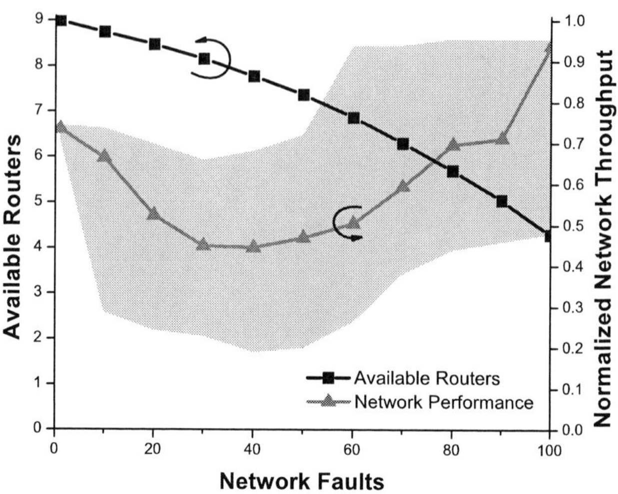

Figure 8: Network performance with increasing faults. Normalized network throughput is shown as the number of faults in the network increases. Throughput is normalized to the bandwidth of the remaining network adapter links. The shaded region show the 5th-95th percentiles, while the line is the median.

[4] T. Bjerregaard and S. Mahadevan. A survey of research and practices of network-on-chip. *ACM Computer Survey*, 2006.

[5] S. Borkar. Microarchitecture and design challenges for gigascale integration. *Proc. Micro, keynote address*, 2004.

[6] S. Borkar. Designing reliable systems from unreliable components: the challenges of transistor variability and degradation. *Proc. Micro*, 2005.

[7] K. Constantinides, S. Plaza, J. Blome, B. Zhang, V. Bertacco, S. Mahlke, T. Austin, and M. Orshansky. BulletProof: a defect-tolerant CMP switch architecture. *Proc. HPCA*, 2006.

[8] W. J. Dally, L. R. Dennison, D. Harris, K. Kan, and T. Xanthopoulos. The reliable router: A reliable and high-performance communication substrate for parallel computers. *Proc. PCRCW*, 1994.

[9] D. Fick, A. DeOrio, G. Chen, V. Bertacco, D. Sylvester, and D. Blaauw. A highly resilient routing algorithm for fault-tolerant NoCs. *Proc. DATE*, 2009.

[10] C. J. Glass and L. M. Ni. Fault-tolerant wormhole routing in meshes without virtual channels. *IEEE Trans. on Parallel and Distributed Systems*, 1996.

[11] M. E. Gomez et al. An efficient fault-tolerant routing methodology for meshes and tori. *IEEE Computer Architecture Letters*, 2004.

[12] T. R. Halfhill. Ambric's New Parallel Processor: Globally Asynchronous Architecture Eases Parallel Programming. *Microprocessor Report*, 2006.

[13] C.-T. Ho and L. Stockmeyer. A new approach to fault-tolerant wormhole routing for mesh-connected parallel computers. *IEEE Trans. on Computers*, 2004.

[14] E. Karl, D. Blaauw, D. Sylvester, and T. Mudge. Reliability modeling and management in dynamic microprocessor-based systems. *Proc. DAC*, 2006.

[15] J. Keane, S. Venkatraman, P. Butzen, and C. H. Kim. An array-based test circuit for fully automated gate dielectric breakdown characterization. *Proc. CICC*, 2008.

[16] S.-J. Pan and K.-T. Cheng. A framework for system reliability analysis considering both system error tolerance and component test quality. *Proc. DATE*, 2007.

[17] D. Park, C. Nicopoulos, and J. K. N. V. C. Das. Exploring fault-tolerant network-on-chip architectures. *Proc. DSN*, 2006.

[18] V. Puente, J. A. Gregorio, F. Vallejo, and R. Beivide. Immunet: A cheap and robust fault-tolerant packet routing mechanism. *ACM SIGARCH Computer Architecture News*, 2004.

[19] S. Rodrigo, J. Flich, J. Duato, and M. Hummel. Efficient unicast and multicast support for CMPs. *Proc. Micro*, 2008.

[20] D. Sylvester, D. Blaauw, and E. Karl. ElastIC: An Adaptive Self-Healing Architecture for Unpredictable Silicon. *IEEE Design & Test*, 2006.

[21] S. R. Vangal et al. An 80-tile sub-100w teraflops processor in 65-nm cmos. *IEEE Journal of Solid-State Circuits*, 2008.

[22] J. Wu. A fault-tolerant and deadlock-free routing protocol in 2D meshes based on odd-even turn model. *IEEE Trans. on Computers*, 2003.

[23] J. Zhou and F. C. M. Lau. Multi-phase minimal fault-tolerant wormhole routing in meshes. *Parallel Computing*, 2004.

[24] H. Zimmer and A. Jantsch. A fault model notation and error-control scheme for switch-to-switch buses in a network-on-chip. *Proc. CODES+ISSS*, 2003.

978-1-60558-497-3/09 $25.00 © 2009 ACM

Technology-Driven Limits on DVFS Controllability of Multiple Voltage-Frequency Island Designs: A System-Level Perspective[*]

Siddharth Garg   Diana Marculescu   Radu Marculescu
Dept. of ECE
Carnegie-Mellon University
{sgarg1,dianam,radum}@ece.cmu.edu

Umit Ogras
Strategic CAD Labs
Intel Corp.
umit.y.ogras@intel.com

ABSTRACT

In this paper, we consider the case of network-on-chip (NoC) based multiple-processor systems-on-chip (MPSoCs) implemented using multiple voltage and frequency islands (VFIs) that rely on fine-grained dynamic voltage and frequency scaling (DVFS) for run-time control of the system power dissipation. Specifically, we present a framework to compute theoretical bounds on the performance of DVFS controllers for such systems under the impact of three important technology driven constraints: (i) reliability and temperature driven upper limits on the maximum supply voltage; (ii) inductive noise driven constraints on the maximum rate of change of voltage/frequency; and (iii) increasing manufacturing process variations. Our experimental results show that, for the benchmarks considered, any DVFS control algorithm will lose up to 87% performance, measured in terms of the number of steps required to reach a reference steady state, in the presence of maximum frequency and maximum frequency increment constraints. In addition, increasing process variations can lead to up to 60% of fabricated chips being unable to meet the specified DVFS control specifications, irrespective of the DVFS algorithm used.

Categories and Subject Descriptors

B.7.2 [**Hardware**]: Integrated Circuits—*Design Aids*

General Terms

Performance

Keywords

Networks-on-chip, power management, performance bounds

1. INTRODUCTION

With increased levels of integration in scaled technologies, novel on-chip communication architectures that use a Network-on-Chip approach have emerged as a scalable alternative to traditional bus-based or point-to-point communication solutions. Furthermore, due to increased power density and energy consumption, NoCs implemented using a multiple Voltage Frequency Island design style have become an attractive alternative to single-clock, single-voltage designs [2, 7]. Each island in a VFI system is locally clocked and has an independent voltage supply, while inter-island communication is orchestrated via mixed-clock, mixed-voltage FIFOs. The power savings result from the fact that the voltage of each island can be independently tuned to minimize the system power dissipation under performance constraints.

To cope with *run-time* variations in the workload or power characteristics of VFI systems, the voltage and frequency of each island can be dynamically scaled to exploit the slack in the application

to reduce the power dissipation under a performance budget [6]. Not surprisingly, designing appropriate dynamic voltage and frequency scaling (DVFS) control algorithms for run-time control of VFI systems is a matter of great importance. While this problem has been addressed before by a number of authors [7, 9, 6], no attention has been given to analyzing the *fundamental limits* of the capabilities of DVFS controllers for multiple VFI systems. Starting from these overarching ideas, in this paper, we specifically focus on three technology driven constraints that we believe have the most impact on DVFS controller characteristics: (1) reliability and temperature constrained upper-limits on the maximum voltage and frequency at which any VFI can operate; (2) inductive noise driven limits on the maximum rate of change of voltage and frequency; and (3) the impact of manufacturing process variations. We note that each of these factors is becoming increasingly important with technology scaling.

Given the broad range of proposed DVFS control algorithms proposed in literature, we believe that it is insufficient to merely analyze the performance limits of a specific control strategy. The only assumption we make, which is common to many of the DVFS controllers proposed in literature, is that the goal of the control algorithm is to ensure that the occupancies of a pre-defined set of mixed-clock, mixed-voltage queues in the NoC are controlled to remain at pre-specified reference values[1]. We then define the performance of a controller to be its ability to bring the queues, starting from an arbitrary initial state, back to their reference utilizations in a *desired but fixed number of control intervals*. Given the technology constraints, our framework is then able to provide a *theoretical guarantee* on the existence of a controller that can meet this specification.

2. RELATED WORK AND PAPER CONTRIBUTIONS

Power management of multiple VFI-based MPSoCs has been the subject of extensive prior research. [6] presents a Lagrange optimization based approach to perform DVFS in multiple VFI systems, while in [9], the authors propose a PID DVFS controller to set the occupancies of the queues in a multiple clock-domain processor to reference values. Finally, [7] presents a state-space model of the queue occupancies in an NoC with multiple VFIs and proposes a formal feedback control algorithm to control the queues based on the state-space model. We note that, compared to prior work, we focus on the fundamental limits of controllability of DVFS enabled multiple VFI systems and not on a specific control algorithm. Our results are, therefore, equally applicable to any of the control techniques proposed before.

As compared to previous work we make the following novel contributions: (i) we propose a computationally efficient framework to analyze the impact of three major technology-driven constraints on the performance of DVFS control algorithms for multiple VFI networks-on-chip; and (ii) starting from a formal state-space representation of the queues in an NoC, we provide theoretical bounds on the performance capabilities of *any* DVFS control algorithm (i.e., the proposed bounds are *DVFS control algorithm agnostic*).

[*]This research was funded in part by SRC Grant 2008-HJ-1800

Permission to make digital or hard copies of part or all of this work for personal or classroom use is granted without fee provided that copies are not made or distributed for profit or commercial advantage and that copies bear this notice and the full citation on the first page. To copy otherwise, to republish, to post on servers or to redistribute to lists, requires prior specific permission and/or a fee.
DAC'09, July 26-31, 2009, San Francisco, California, USA

---

[1]While we explicitly control the NoC queues in this work, it can easily be applied to other systems with queue-based communication, such as point-to-point queues or even software queues.

3. PRELIMINARIES AND ASSUMPTIONS

We begin by briefly reviewing the state-space modeled developed in [7] to model the controlled queues in a multiple VFI system. We start with a design with N interface queues and M VFIs. An example of such a system is shown in Figure 1, where $M = 3$ and $N = 2$. Furthermore, without any loss of generality, we assume that the system is controlled at discrete intervals of time, i.e., the kth control interval is the time period $[kT, (k + 1)T]$, where T is the length of a control interval.

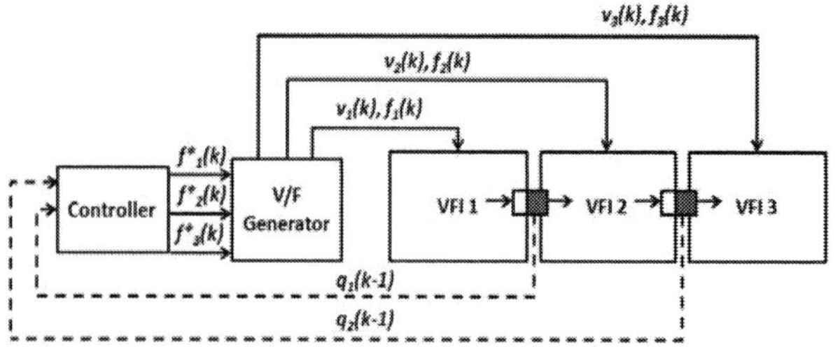

Figure 1: VFI system with three islands and two queues.

The following notation can now be defined:

- The vector $Q(k) \in \mathbb{R}^N = [q_1(k), q_2(k), \ldots, q_N(k)]$ represents the queue occupancies in the kth control interval.

- The vector $F(k) \in \mathbb{R}^M = [f_1(k), f_2(k), \ldots, f_M(k)]$ represents the frequency of each VFI in the kth control interval.

- λ_i and μ_i ($i \in [1, N]$) represent the average arrival and service rate of queue i, respectively. These parameters are workload dependent and can be obtained by probabilistic workload characterization.

- The system matrix $B \in \mathbb{R}^{M \times N}$ is defined such that the $(i, j)^{th}$ entry of B is the rate of write (read) operations at the input (output) of the ith queue due to the activity in the jth VFI. We refer the reader to [7] for a detailed example on how to construct the system matrix.

The state-space equation that represents the queue dynamics can now simply be written as [7]:

$$Q(k + 1) = Q(k) + TBF(k) \tag{1}$$

The key observation is that given the applied frequency vector $F(k)$ as a function of time, this equation describes completely the evolution of queue occupancies in the system. In this equation, we have implicitly assumed that the occupancy feedback from the queues is available to the centralized controller within a single control interval - this is a reasonable assumption, since, in practice, a control interval is of the order of micro- to milli-seconds [7].

As shown in Figure 1, we also introduce an additional vector $F^*(k) = [f_1^*(k), f_2^*(k), \ldots, f_M^*(k)]$, which represents the *desired* control frequency values at control interval k. For a perfect system, $F^*(k) = F(k)$, i.e., the desired and applied control frequencies are the same. However, due to the technology driven constraints, the applied frequencies may deviate from the frequencies desired by the control, for example if there is a limit on the maximum frequency at which a VFI can be operated. The technology driven deviations between the desired and actual frequency will be explained in greater detail in the next section.

4. LIMITS ON DVFS CONTROL

We now present the proposed framework to analyze the limits of performance of DVFS control strategies in the presence of technology driven constraints. To describe more specifically what we mean by performance, we define $Q_{ref} \in \mathbb{R}^N$ to be the desired *reference queue occupancies* that have been set by the designer. Furthermore, we assume that as a performance specification, the designer also sets a *limit*, J, that specifies the *maximum* number of control intervals that the control algorithm should take to bring

the queues back from an arbitrary starting vector of queue occupancies, $Q(0)$, back to their reference occupancy values. Note that the time index 0 here refers to a time instant at which the queue occupancies deviate from their reference values due to workload variations, and not to the start-up state of the system.

Given this terminology, using Equation 1, we can write the queue occupancies at the Jth control interval as [4]:

$$Q(J) = Q(0) + TB \sum_{k=0}^{J-1} F(k) \tag{2}$$

Since we want $Q(J) = Q_{ref}$, we can write:

$$(TB) \sum_{k=0}^{J-1} F(k) = (Q_{ref} - Q(0)) \tag{3}$$

4.1 Limits on Maximum Frequency

In a practical scenario, reliability concerns and peak thermal constraints impose an **upper limit** on the frequencies at which the VFIs can be clocked. For now, let us assume that each VFI in the system has a maximum frequency constraint $f_{MAX}^i (i \in [1, M])$. Therefore, we can write:

$$f_i(k) = min(f_{MAX}^i, f_i^*(k)) \quad \forall i \in [1, M] \tag{4}$$

Consequently, the system can be returned to its required state Q_{ref} in at most J steps *if and only if* the following system of linear equations has a feasible solution:

$$(TB) \sum_{k=0}^{J-1} F(k) = (Q_{ref} - Q(0)) \tag{5}$$

$$0 \leq f_i(k) \leq f_{MAX}^i \quad \forall k \in [0, J - 1], \forall i \in [1, M] \tag{6}$$

Note that this technique only works for a *specific* initial vector of queue occupancies $Q(0)$; for example, $Q(0)$ may represent an initial condition in which all the queues in the system are full. However, we would like the system to be controllable in J time steps for a *set* of initial conditions, denoted by R_Q.

Let us assume that the set of initial conditions for which we want to ensure controllability is described as follows: $R_Q = \{Q(0) : A_Q Q(0) \leq B_Q\}$, where $A_Q \in \mathbb{R}^{P \times N}$ and $B_Q \in \mathbb{R}^P$. Clearly, the set R_Q represents a bounded *closed convex polyhedron* in $\mathbb{R}^N$. We will now show that to ensure controllability for all points in R_Q, it is sufficient to show controllability for each vertex of R_Q. In particular, without any loss of generality, we assume that R_Q has V vertices given by $\{Q^1(0), Q^2(0), \ldots, Q^V(0)\}$. First, due to the Krein-Milman theorem [8], we get the following result [2]:

LEMMA 1. *Any $Q(0) \in R_Q$ can be written as a convex combination of the vertices of R_Q, i.e., $\exists \{\alpha_1, \alpha_2 \ldots \alpha_V\} \in \mathbb{R}^N$ s.t. $\sum_{i=1}^V \alpha_i = 1$ and $Q(0) = \sum_{i=1}^V \alpha_i Q^i(0)$.*

LEMMA 2. *The set of all $Q(0)$ for which Equation 5 and Equation 6 admit a feasible solution is convex.*

Finally, based on Lemma 1 and Lemma 2, we can show that:

THEOREM 1. *Equation 5 and Equation 6 have feasible solutions $\forall Q(0) \in R_Q$ if and only if they have feasible solutions $\forall Q(0) \in \{Q^1(0), Q^2(0), \ldots, Q^V(0)\}$.*

Significance Theorem 1 establishes necessary and sufficient conditions to *efficiently* verify the ability of a DVFS controller to bring the system back to its reference state, Q_{ref}, in J control intervals starting from a large set of initial states, R_Q, without having to independently verify that *each* initial state in R_Q can be brought back to the reference state; instead, it is sufficient to verify the controllability for only the set of initial states that form the vertices of R_Q. This significantly reduces the computational cost of the proposed framework.

---

[2]For brevity, all results in the paper are stated without proof. Detailed proofs can be found in [1].

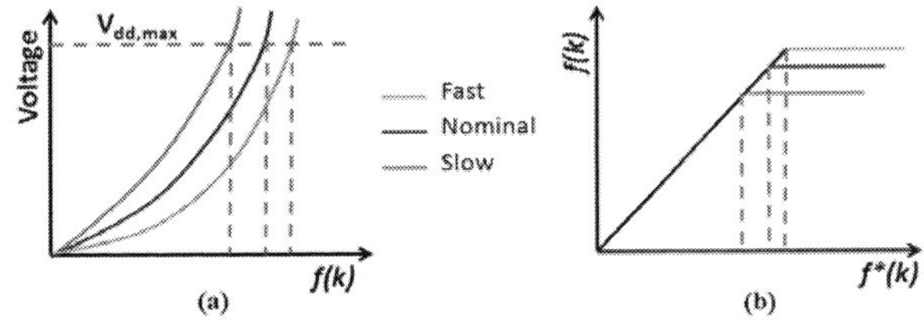

Figure 2: Process variation impact on DVFS controller.

In practice, the region of initial states R_Q will depend on the behavior of the workload, for example, a bursty read or a bursty write. While it is possible to obtain R_Q from extensive simulations of real workloads, R_Q can be defined conservatively as follows: $R_Q = \{Q(0) : 0 \leq q_i(0) \leq q^i_{MAX}\}, \forall i \in [1, N]$, where q^i_{MAX} is the physical queue length of the ith queue in the system.

Finally, we note that minimum frequency constraints are not considered in this work because we assume that the frequency of each VFI can safely be reduced to zero, for example, using clock gating. However, minimum frequency constraints can easily be included in the framework if required.

4.2 Frequency Increment Constraints

A major consideration for the design of systems that support dynamic voltage and frequency scaling is the resulting inductive noise in the power delivery network due to sudden changes in the power dissipation and current requirement of the system. While there exist various circuit-level solutions to the inductive noise problem, it may be necessary to additionally constrain the maximum frequency increment within a control interval to minimize inductive noise. Furthermore, the limited slew rate of off-chip voltage controllers also constrains the maximum frequency change possible in a control interval [3]. We note that while there exist techniques to scale frequency on the order of a few clock cycles, both voltage *and* frequency scaling are required to obtain maximum energy efficiency from the system.

Frequency increment constraints can be modeled in the proposed framework as follows:

$$|f_i(k + 1) - f_i(k)| \leq f^i_{step} \quad \forall i \in [1, M], \forall k \in [0, J - 1] \quad (7)$$

where f^i_{step} is the maximum frequency increment allowed in the frequency of VFI i. Equation 7 can further be expanded as linear constraints as follows:

$$f_i(k + 1) - f_i(k) \leq f^i_{step} \quad \forall i \in [1, M], \forall k \in [0, J - 1] \quad (8)$$

$$-f_i(k + 1) + f_i(k) \leq f^i_{step} \quad \forall i \in [1, M], \forall k \in [0, J - 1] \quad (9)$$

Together with Equations 5 and 6, Equations 8 and 9 define a linear program that can be used to determine the existence of a time-optimal control strategy.

Finally, we note that for Theorem 1 to hold, we need to ensure that Lemma 2 is valid with the additional constraints introduced by Equation 7. We show that this is indeed the case.

LEMMA 3. *The set of all $Q(0)$ for which Equation 5, Equation 6 and Equation 7 admit a feasible solution is convex.*

Significance Lemma 3 ensures that Theorem 1 still remains valid after the inductive noise constraints given by Equation 7 are added to the original set of linear constraints.

4.3 Process Variation Impact

To demonstrate the effect of process variations on DVFS control, we plot the voltage-frequency curves for the *slow, nominal* and *fast* process corners of a VFI in Figure 2(a). As mentioned before, we assume that due to reliability concerns, the process allows a maximum voltage of $V_{dd,max}$. As it can be seen, for a given $V_{dd,max}$, we obtain three different values of f_{MAX} for the *slow, nominal* and *fast* VFIs. Figure 2(b) shows the relationship between desired and applied frequency values in the presence of process variations - from the figure it is clear that under the impact

of process variations, we must think of f^i_{MAX} as *random variables*, not fixed upper limits on the operating frequency of each VFI.

As a result, the linear programming framework described in the previous sections will now have a certain probability of being feasible, i.e., there might exist values of f^i_{MAX} for which it is not possible to bring the system back to steady state within J control intervals. We will henceforth refer to the probability that a given instance of a multiple VFI system can be brought back to the reference queue occupancies in J time steps as the *probability of controllability* (PoC).

In this paper, we use Monte Carlo simulations to estimate the PoC, i.e., in each Monte Carlo run, we obtain a sample of the maximum frequency for each VFI, f^i_{MAX}, and check for the feasibility of the linear program defined by Equations 5, 6, 8 and 9. Furthermore, we are able to exploit the specific structure of our problem to *speed up the Monte Carlo simulations*. In particular, we note that if a given vector of upper bounds, $f^{i,1}_{MAX}(i \in [1, M])$, has a feasible solution, then another vector, $f^{i,2}_{MAX}(i \in [1, M])$, where $f^{i,2}_{MAX} \geq f^{i,1}_{MAX} \forall i \in [1, M]$ must also have a feasible solution. Therefore, we do not need to explicitly check for the feasibility of the upper bound $f^{i,2}_{MAX}$, thereby saving significant computational effort. This provides significant speed-up over a naive Monte Carlo implementation.

5. EXPERIMENTAL RESULTS

To validate the theory presented in this paper, we experiment on two benchmarks: (1) *MPEG*, is a distributed implementation of an MPEG-2 encoder with six ARM7-TDMI processors that are partitioned to form a 3 VFI system, as shown in Figure 3(a); and (2) *Star*, a 5 VFI system organized in a star topology as shown in Figure 3(b). The MPEG encoder benchmark was profiled on a cycle-accurate MPSoC simulator to obtain the average rates at which the VFIs read and write from the queues, as tabulated in Figure 3(a). The arrival and service rates of the *Star* benchmark are randomly generated.

To begin, we first compute the nominal frequency values f^i_{NOM} of each VFI in the system to ensure stable queues for the nominal workload values. The maximum frequency constraint, f^i_{MAX} is then set using a parameter $\gamma = \frac{f^i_{MAX}}{f^i_{NOM}}$ - in our experiments we use three values of $\gamma = \{1.1, 1.25, 1.5\}$. Finally, we allow the maximum frequency increment per control interval to vary from 5% to 20% of the nominal frequency.

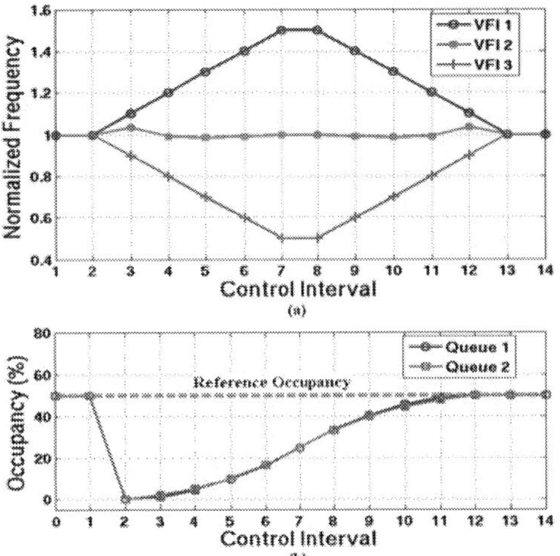

Figure 4: (a) Response of proposed controller to a deviation from the reference queue occupancies. (b) Evolution of queue occupancies in the system.

Figure 3(c) shows the obtained results as γ and the maximum frequency step are varied for the *MPEG* benchmark. The results for *Star* benchmark are quantitatively similar, so we only show the graph for *MPEG* benchmark in Figure 3(c). As it can be seen, the

978-1-60558-497-3/09 \$25.00 © 2009 ACM

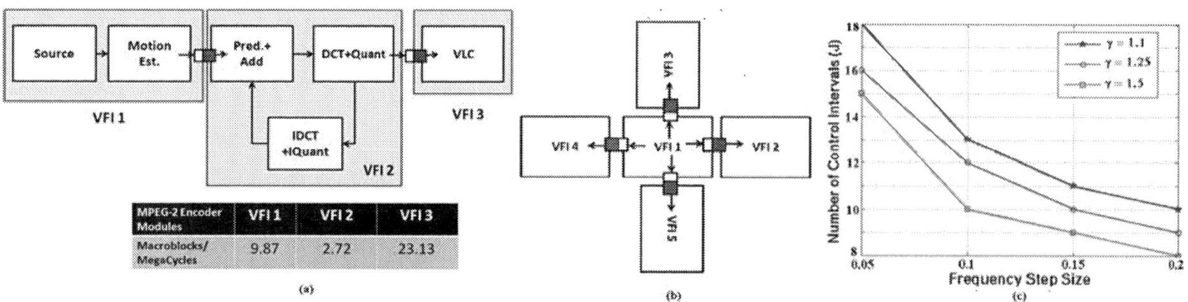

MPEG-2 Encoder Modules	VFI 1	VFI 2	VFI 3
Macroblocks/ MegaCycles	9.87	2.72	23.13

(a)

(b)

(c)

Figure 3: (a) Topology and workload characteristics of the *MPEG* benchmark. (b) Topology of the *Star* benchmark. (c) Impact of γ and maximum frequency increment on the minimum number of control intervals, J.

frequency step size has a significant impact on the controllability of the system, in particular, for $\gamma = 1.5$ we see an 87% increase in the number of control intervals required to bring the system back to reference queue occupancies, J, while for $\gamma = 1.1$, J increases by up to 80%. The impact of γ itself is slightly more modest - we see a 20% to 25% increase in J as γ increases from 1.1 to 1.5.

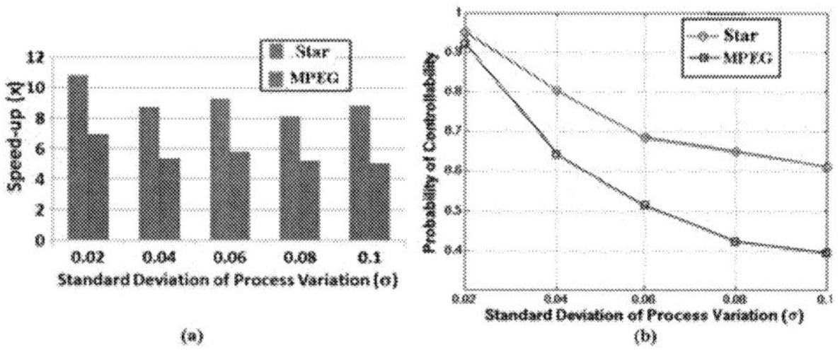

Figure 5: (a) Speed-up ($\times$) of the efficient Monte Carlo technique compared to naive Monte Carlo. (b) PoC as a function of increasing process parameter variations.

In Figure 4, we plot the response of the time-optimal control strategy for the *MPEG* benchmark when the queue occupancies of the two queues in the system drop to zero at the control interval 2. As a result, the applied frequency values are modulated to bring the queues back to their reference occupancies within $J = 10$ control intervals. From Figure 4(a), we can clearly observe the impact of both the limit on the maximum frequency, and the limit on the maximum frequency increment, on the time-optimal control response. Figure 4(b) shows how the queue occupancies change in response to the applied control frequencies, starting from 0% occupancy until they reach their reference occupancies.

Next, we investigate the impact of process variations on the PoC of DVFS enabled multiple VFI systems. For this experiment, we model the maximum frequency of each VFI as an independent normal distribution, and increase the standard deviation (σ) of the distribution from 2% to 10% of the maximum frequency. Finally, we use 5000 runs of both naive Monte Carlo simulations and the proposed efficient Monte Carlo simulations (see Section 4.3) to obtain the PoC for various values of σ and for both benchmarks. From Figure 5(a), we can see that the proposed efficient version of Monte Carlo provides **significant speed-up** over the naive Monte Carlo implementation - on average, a $9\times$ speed-up for the *MPEG* benchmark and a $5.6\times$ speed-up for the *Star* benchmark.

From the estimated PoC values in Figure 5(b), we can see that the PoC of both *MPEG* and *Star* benchmarks are significantly impacted by process variations, though *MPEG* sees a greater degradation in the PoC, decreasing from 92% for $\sigma = 2\%$ to only 40% for $\sigma = 10\%$. On the other hand, the PoC of *Star* drops from 95% to 62% for the same values of σ. To explain the significance of these results, we point out that a PoC of 40% implies that, on average, 60% of the fabricated circuits will *not* be able to meet the DVFS control performance specification, irrespective of the control algorithm that is used. Of note, while the specific parameters used in the Monte Carlo simulations (for example, the value of γ at various technology nodes) are implementation dependent and

may cause small changes in the PoC estimates in Figure 5, the fundamental predictive nature of this plot will remain the same.

We now briefly discuss the integration of the proposed framework in a power-aware NoC design methodology. First, given a specific implementation of a DVFS controller, for example, the ones proposed in [7] or [9], the framework can be used to determine how far away a particular implementation of the controller in Figure 1 is from an optimal solution. Second, based on the application, designers may be willing to trade-off reliability, peak temperature or maximum inductive noise for a better DVFS controller. In this scenario, the proposed approach can be used to efficiently explore the trade-offs between these technology driven factors and the upper bound on performance of DVFS control, similar to the analysis presented in Figure 3(c).

6. CONCLUSION

In this paper, we presented a theoretical framework to efficiently obtain the limits on the controllability and performance of DVFS controllers for multiple VFI based networks-on-chip. Using a computationally efficient implementation of the framework, we present results, using both real and synthetic benchmarks, that explore the impact of three major technology driven factors - temperature and reliability constraints, maximum inductive noise constraints and process variations - on the performance bounds of DVFS control strategies. Our experiments demonstrate the importance of considering the impact of these three factors on DVFS controller performance, particularly since all three factors are becoming increasingly important with technology scaling. As future work, we plan to consider explicit control of power consumption instead of purely time optimal control.

7. REFERENCES

[1] S. Garg et al. Technology-Driven Limits on DVFS Controllability of Multiple Voltage-Frequency Island Designs. Technical report, CSSI, CMU, 2009.

[2] W. Jang, D. Ding, and D.Z. Pan. A Voltage-Frequency Island aware energy optimization framework for networks-on-chip. In *Proceedings of ICCAD*, 2008.

[3] W. Kim et al. System Level Analysis of Fast, Per-Core DVFS using On-Chip Switching Regulators. In *Proceedings of HPCA*, 2008.

[4] B.C. Kuo. *Digital Control Systems*. Oxford University Press, Inc. New York, NY, USA, 1992.

[5] D. Marculescu and S. Garg. System-level process-driven variability analysis for single and multiple voltage-frequency island systems. 2006.

[6] K. Niyogi and D. Marculescu. Speed and voltage selection for GALS systems based on voltage/frequency islands. In *Proceedings of ASP-DAC*, 2005.

[7] U.Y. Ogras et al. Variation-Adaptive Feedback Control for Networks-on-Chip with Multiple Clock Domains. In *Proceedings of DAC*, 2008.

[8] H.L. Royden. *Real analysis*. Macmillan New York, 1968.

[9] Q. Wu et al. Formal online methods for voltage/frequency control in multiple clock domain microprocessors. *Proceedings of ASPLOS*, 2004.

NoC Topology Synthesis for Supporting Shutdown of Voltage Islands in SoCs

Ciprian Seiculescu*, Srinivasan Murali§ *, Luca Benini‡, Giovanni De Micheli*
* LSI, EPFL, Lausanne, Switzerland,{ciprian.seiculescu, giovanni.demicheli}@epfl.ch
§ iNoCs, Lausanne, Switzerland, murali@inocs.com
‡ DEIS, Univerity of Bologna, Bologna, Italy, luca.benini@unibo.it

ABSTRACT

In many *Systems on Chips (SoCs)*, the cores are clustered in to *voltage islands*. When cores in an island are unused, the entire island can be shutdown to reduce the leakage power consumption. However, today, the interconnect architecture is a bottleneck in allowing the shutdown of the islands. In this paper, we present a synthesis approach to obtain customized application-specific *Networks on Chips (NoCs)* that can support the shutdown of voltage islands. Our results on realistic SoC benchmarks show that the resulting NoC designs only have a negligible overhead in SoC active power consumption (average of 3%) and area (average of 0.5%) to support the shutdown of islands. The shutdown support provided can lead to a significant leakage and hence total power savings.

Categories and Subject Descriptors

B.4.3 [**INPUT/OUTPUT AND DATA COMMUNICATIONS**]: Interconnections (Subsystems)—*topology*

General Terms

Design

Keywords

NoC, voltage islands, shutdown, leakage power, topology

1. INTRODUCTION

Power management is a challenge for modern Systems on Chips (SoCs), as many of them are destined for the embedded market and have to operate with low power consumption. With technology scaling, the leakage power consumption is increasing rapidly as a fraction of the total power consumption. In fact, leakage power can be responsible for 40% or more of the total system power [6].

In order to reduce the leakage power consumption, cores that are not used by an application can be shutdown or placed in sleep mode, while the other cores can be operational. For example, power gating using sleep transistors is a popular way to shutdown cores [6]. To achieve power gating, the sleep transistors are added between the actual ground lines and the circuit ground (also called the virtual ground) [6], which are turned off in the sleep mode to cut-off the leakage path. Due to routing restrictions, separate VDD and ground lines cannot be used for each core. Instead, cores are grouped in to *Voltage Islands (VIs)*, with cores in an island using the

same VDD and ground lines [5]-[8]. When all the cores in an island are unused for an application, the entire island can be shutdown. For example, the IBM fabrication processes CU-08, CU-65HP and CU-45HP all support the partitioning of chips into multiple VIs and power gating of the VIs [4].

In today's SoCs, the interconnect architecture is a bottleneck in allowing the shutdown of the islands. There are several approaches presented to synthesize application-specific Networks on Chips (NoCs) [12]-[15]. However, none of them consider the issue of shutdown of VIs. These approaches cannot be directly extended to design NoCs for SoCs with voltage islands. If the NoC is designed using such approaches, either the whole NoC should be placed in a separate VI or the islands cannot be shutdown.

Placing the entire NoC in a separate VI is not a feasible solution. The NoC switches are usually spread across the chip, connecting the different cores. If the entire NoC is in the same island, it is difficult to route the VDD and ground lines for the NoC across the chip. On the other hand, if all the NoC switches are physically clustered and placed in the center of the chip, then long wires are needed to connect all the cores to the NoC island. Thus, the routing congestion would be enormous and the solution is not scalable. Moreover, additional resources for routing the additional voltage and ground lines may not even be available in the design. If the switches are spread across the different VIs and if a VI needs to be shutdown, then packets between cores on the other VIs that use the switches in this VI cannot be transmitted. This will prevent the shutdown of the entire island.

The concept of voltage island should be considered during the NoC topology synthesis phase itself. In this work, we present a synthesis approach to determine the best NoC topology points that are tailored to meet the application performance constraints, minimizing power consumption and supporting the ability to shutdown voltage islands. To the best of our knowledge, this is the first work that addresses custom NoC topology synthesis for supporting the shutdown of voltage islands.

2. RELATED WORK

Power gating of designs has been widely applied in many SoCs [5]. In [5], the authors present the importance of partitioning cores in to voltage islands for power reduction. Several methods have been presented to achieve shutdown of islands [5]-[8]. Any of these methods can be used in conjunction with our topology synthesis process to achieve the actual shutdown of cores.

A description of the NoC paradigm with the related benefits and issues is presented in [1]-[3]. Many works have been presented on synthesizing bus based systems [16]-[18]. In [9]-[11], algorithms for mapping application to regular NoC topologies are presented. In [12]-[15], methods for designing application specific NoCs are described. However, none of these works address the issue of sup-

Permission to make digital or hard copies of part or all of this work for personal or classroom use is granted without fee provided that copies are not made or distributed for profit or commercial advantage and that copies bear this notice and the full citation on the first page. To copy otherwise, to republish, to post on servers or to redistribute to lists, requires prior specific permission and/or a fee.
DAC'09, July 26-31, 2009, San Francisco, California, USA

978-1-60558-497-3/09 $25.00 © 2009 ACM

porting shutdown of voltage islands on the chip.

In [24] an architecture for *Globally Asynchronous Locally Synchronous (GALS)* NoC is presented. In [23] the authors present a physical implementation of multi-synchronous NoC. In [22], the authors present a methodology to partition a NoC into multiple voltage islands. This work is complementary to ours, as we present a methodology to design a custom NoC topology with VIs. In [19], the authors present an approach to design NoCs with voltage islands. However, the designs produced by the method do not support the shutdown of the islands.

In [20], the authors present approaches to route packets even when parts of the NoC have failed. A similar approach can be used for handling NoC components that have been shutdown. However, such methods do not guarantee the availability of paths when elements are shutdown. Moreover, mechanisms for re-routing and re-transmission can have a large area-power overhead on the NoC [21] and are difficult to design and verify.

3. PROBLEM DESCRIPTION

In this section, we describe the architectural features of the NoC and the synthesis problem.

3.1 Architecture Description

An example of the architecture for which our custom NoC synthesis algorithm is designed is presented in Figure 1. The cores of the design are assigned to different VIs, which is given as an input to our method. The cores in a VI have the same operating voltage (same power and ground lines), but could have different operating frequencies. In order to have a scalable solution, we build NoC systems, where cores in a VI are connected to switches in the same VI. We follow an approach similar to the *GALS* approach, where the NoC components in a VI are synchronous and operate at the same frequency. Having a locally synchronous design eases the integration of the NoC with standard back-end placement&routing tools and industrial flows. Moreover, if the different switches in a VI operate at different frequencies, power and latency hungry synchronizers are needed to connect them.

The cores are connected to the NoC switches by means of Network Interfaces (NIs) that convert the protocol of the cores to that of the network. The NIs also perform clock frequency conversion, if the cores are running at different frequencies than the switches in the VI. When a switch in one VI is to be connected to a switch in another VI, we use a bi-synchronous FIFO to connect them together. The FIFO takes care of the voltage and frequency conversion across the islands. The frequency conversion is needed because the switches in the different VIs could be operating at different frequencies. Even if they are operating at the same frequency, the clock tree is usually built separately for each VI. Thus, there may be a clock skew between the two synchronous islands. We use over the cell routing with unpipelined links to connect switches across different VIs, as the wires could be routed on top of other VIs.

3.2 Synthesis Problem

In our synthesis procedure, we generate switches in each VI to connect the cores in the VI. Optionally, our method can explore solutions where a separate NoC VI can be created. We take the availability of power and ground lines for the intermediate VI as an input, and our method will use the intermediate island, only if the resources are available. Our method produces topologies such that a traffic flow across two different VIs can be routed in two ways: (i) the flow can go either directly from a switch in the VI containing the source core to another switch in the VI containing the destination core, or (ii) it can go through a switch which is placed in the intermediate NoC VI, if the VI is available. The switches in the intermediate VI are never shutdown. The method will automatically explore both alternatives and choose the best one for meeting the application constraints.

The objective of our synthesis method is to determine the number of switches needed in each VI, the size of the switches, their operating frequency and routing paths across the switches, such that application constraints are satisfied and VIs can be shutdown, if needed. Our method determines if an intermediate NoC VI needs to be used to connect the switches in the different VIs and if so, the number of switches in the intermediate island, their sizes, frequency of operation, connectivity and paths. The synthesis method can be plugged in our design flow presented in [15] in order to generate fully implementable NoCs

Our method produces several design points that meet the application constraints with different switch counts, with each point having different power and performance values. The designer can then choose the best design point from the trade-off curves obtained.

4. TOPOLOGY SYNTHESIS APPROACH

The synthesis algorithm is explained in detail in this section. From the input specifications, we construct the VI communication graph defined as follows:

DEFINITION 1. *A VI Communication Graph (VCG(V, E, isl)) is a directed graph, each vertex $v_i \in V$ represents a core in the VI denoted by isl and the directed edge (v_i, v_j) representing the communication between the cores v_i and v_j. The bandwidth of traffic flow from cores v_i to v_j is represented by $bw_{i,j}$ and the latency constraint for the flow is represented by $lat_{i,j}$. The weight of the edge (e_i, e_j), defined by $e_{i,j}$, is set to a combination of the bandwidth and the latency constraints of the traffic flow from core v_i to v_j: $h_{i,j} = \alpha \times bw_{i,j}/max\_bw + (1 - \alpha) \times min\_lat/lat_{i,j}$, where max_bw is the maximum bandwidth value over all flows, min_lat is the tightest latency constraint over all flows and α is a weight parameter.*

The value of the weight parameter α can be set experimentally or obtained as an input from the user, depending on the importance of performance and power consumption objectives.

Algorithm 1 Core-to-switch connectivity

1: Determine the frequency at which the NoC will operate in each VI and $max\_sw\_size_j, \forall j \in [1 \cdots N_{VI}]$
2: $min\_sw_j = |VCG(V, E, j)|/max\_sw\_size_j, \forall j$
3: {Vary number of switches in each VI}
4: **for** $i = 1$ to $max_{\forall j \in 1 \cdots N_{VI}} |V_j|$ **do**
5:     **for** $j = 1$ to N_{VF} **do**
6:         **if** $i + min\_sw_j < |V_j|$ **then**
7:             $k = i + min\_sw_j$
8:         **else**
9:             $k = |V_j|$
10:         **end if**
11:         Perform k min-cut partitions of $VCG(V, E, j)$.
12:     **end for**
13:     {Vary number of switches in intermediate NoC VI}
14:     **for** $k = 0$ to $max_{\forall j \in 1 \cdots N_{VI}}$ **do**
15:         Compute least cost paths for inter-switch flows. Choose flows in bandwidth order and find the paths.
16:         If paths found for all flows save design point
17:     **end for**
18: **end for**

The algorithm for topology synthesis is presented in Algorithm 1. In the first step of the algorithm, the frequencies of operation

978-1-60558-497-3/09 $25.00 © 2009 ACM

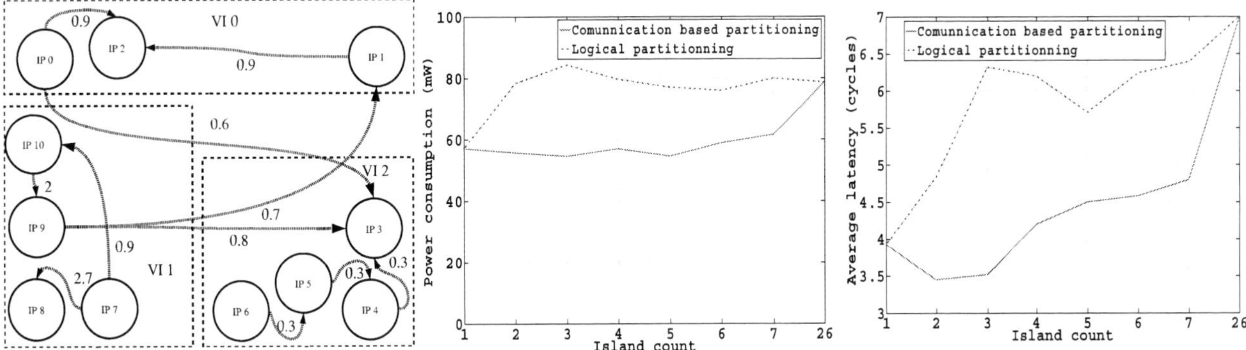

<table>
<tr><td>Figure 1: Example Input</td><td>Figure 2: VI count vs. power</td><td>Figure 3: VI count vs. latency</td></tr>
</table>

of the switches in each of the islands are determined. In our NoC design, a core is connected to only one switch, through a NI. The links connecting the NI and the switch determine the frequency at which the NoC elements have to run in an island. The bandwidth available on a link is a product of the link data width and the frequency. In our synthesis procedure, without loss of generality, we fix the data width of the NoC links to a user-defined value. Please note that it could be varied in a range and more design points could be explored, which does not affect the algorithm steps. For a fixed data width, the frequency of the switches in an island is determined by the link that has to carry the highest bandwidth from or to a core in the island.

A larger switch will have a longer critical path in the crossbar and therefore will have to operate at a smaller frequency. The frequency at which a switch has to operate determines the maximum size (number of inputs and outputs) of the switch that can be allowed, denoted as $max\_sw\_size_j$. As the switches in the different VIs can operate at different frequencies, the maximum switch size is different for the different VIs. Once the maximum switch sizes are determined, based on the number of cores in each VI, the minimum number of switches required for each island is determined in step 2 of the algorithm. Let N_{VI} denote the total number of VIs in the design. In steps 4 to 10 of the algorithm, the number of switches in each island is varied from the minimum value (computed in step 2) to the maximum number of cores in the island. In step 11, for the current switch count of the VI, that many min-cut partitions of the VCG corresponding to the VI are obtained. Cores in a partition share the same switch. As min-cut partitioning is used, cores that communicate heavily or that have tighter latency constraints would be connected to the same switch, thereby reducing the power consumption and latency.

At this point, the connectivity of the cores with the different NoC switches is obtained. We still need to connect the switches together and find paths for the inter-switch traffic flows. If the switches from a VI are directly connected to the switches on the other VIs, then several switch-to-switch links would be needed. This may lead to large switch sizes, which may lead to violation of the $max\_sw\_size_j$ constraint. By using switches in an intermediate NoC island, the number of switch-to-switch links can be reduced. These switches act as indirect switches, as they are not directly connected to the cores, but only connect other switches. If the design constraints permit the usage of another VI, then we explore the solution space (step 14) with varying number of switches on the intermediate NoC VI.

For each combination of direct and indirect switches, the cost of opening links is calculated and the minimum cost paths are chosen for all the flows (step 15). The traffic flows are ordered based on the bandwidth values and the paths for each flow in the order is

computed. The cost of using a link is a linear combination of the power consumption increase in opening a new link or reusing an existing link and the latency constraint of the flow. When opening links, we ensure that the links are either established directly across the switches in the source and destination VIs or to the switches in the intermediate NoC island.

If for all the flows, paths that do not violate the latency constraints are found, then the design point is saved. Finally, for each valid design point, the NoC components are inserted on the floorplan and the wire lengths, wire power and delay are calculated. The time complexity of our algorithm is $O(V^2 E^2 ln(V))$, however in practice the algorithm runs quite fast as the input graphs typically are not fully connected.

5. EXPERIMENTAL RESULTS

Experiments are performed using the power, area and latency models for the NoC components based on the architecture from [25]. The models are built for 65nm technology node. We extended the library with models for the bi-synchronous voltage and frequency converters.

When the NoC has to be designed to support power gating of islands, there is an additional overhead on the dynamic power consumption of the NoC. To study the impact of the overhead and to see the impact of different core assignment to islands and different number of islands, we consider a case-study on a realistic SoC benchmark. The SoC design is used for mobile communication and multimedia applications. The benchmark has 26 cores, consisting of several processors, DSPs, caches, DMA controller, integrated memory, video decoder engines and a multitude of peripheral I/O ports.

We consider two ways of assigning the cores to different VIs. One way, designated as *logical partitioning*, is based on the functionality of the cores. For example, shared memories are placed in the same VI, as they have the same functionality and therefore are expected to operate at the same frequency and voltage. The island with the shared memories is also expected not to be shutdown, since memories are shared and should be accessible at any time and this is another reason to cluster them in the same VI. Similar reasoning was used to partition all the cores for the case of *logical partitioning*. Another way we considered for partitioning is based on communication and is called *communication based partitioning*. In this case, cores that have high bandwidth communication with one another will be placed in the same VI. Please note that the assignment of cores to the VIs is an input to our synthesis algorithm.

In Figure 2, we show how the dynamic power consumption of the NoC varies when the cores are partitioned in to different number of VIs. The power consumption values comprise the consumption on

978-1-60558-497-3/09 $25.00 © 2009 ACM

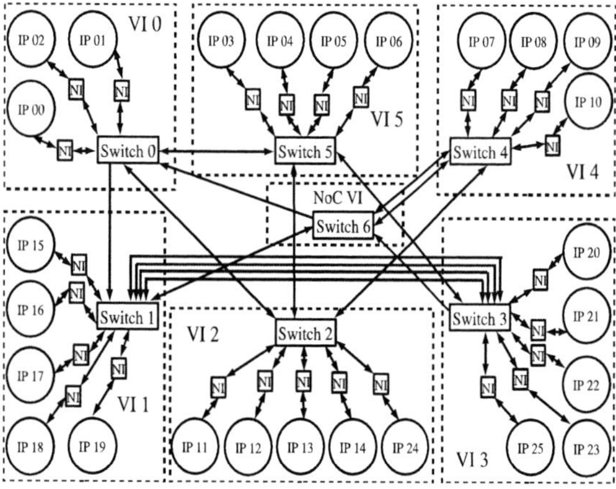

Figure 4: Topology example

switches, links and the synchronizers. In the x-axis, the first point (1 island) is actually a design point with all the cores in the same island, which is the reference point. The last point on the graph corresponds to 26 VIs, which is the point when each core is in its own island. It can be seen that in the case of *logical partitioning*, we have to pay a some overhead in NoC dynamic power, as there are more high bandwidth flows that have to go across islands. In the case of the *communication based partitioning*, the NoC consumes less power than the reference point with 1 island, as the NoC can run at a slower frequency in some of the islands. In this case most of the high bandwidth flows are inside an island, so the power overhead is less. In Figure 3, we show the average packet latencies for the different design points. The latency quoted is the number of cycles needed to transfer a single chunk of the packet from the output of the source NI until the input of the destination NI under zero-load conditions. When packets cross the islands, a 4 cycle delay is incurred on the voltage-frequency converters. Thus, with increasing number of islands, the latencies increase. A topology for the 6 VI *logic partitioning* case is shown in Figure 4 and a floorplan example is presented in Figure 5.

We studied the support of voltage islands on a variety of SoC benchmarks. For the different SoC benchmarks, we found that the topologies synthesized to support multiple VIs incur a 3% overhead on the total system's dynamic power. We found that the area overhead is also negligible, with less than 0.5% increase in the total SoC area. In many SoCs, the shutdown of cores can lead to large reduction in leakage power, leading to even 25% or more reduction in overall system power [6]. Thus, compared to the power savings achieved, the penalty incurred in the NoC design is negligible.

The exploration of the design points for all the benchmark took only a few hours on a 2 GHz Linux machine. To be noted that the synthesis process is only run once at design time and therefor the computational time required by the algorithm is negligible.

6. CONCLUSIONS

Stand by and leakage power consumption of the SoC is becoming a large fraction of the total power consumption. Clustering of cores in to voltage islands and shutdown of unused islands is an effective way to reduce the leakage power consumption. The system interconnect has to be designed to ensure proper operation when shutting down voltage islands. In this paper, we presented an approach to synthesize application specific *Networks on Chip (NoC)* interconnects that can effectively support shutdown of islands. The topologies synthesized by our methods have negligible power and area overhead (3% power and 0.5% area, on average), in order to support shutdown. The presented approach also allows the design

Figure 5: Floorplan example

space exploration of NoCs with different power-performance values that meet the application constraints.

7. ACKNOWLEDGMENT

We would like to acknowledge the financial contribution of CTI under project 10046.2 PFNM-NM and the ARTIST-DESIGN Network of Excellence.

8. REFERENCES

[1] L.Benini and G.De Micheli, "Networks on Chips: A New SoC Paradigm", IEEE Computers, pp. 70-78, Jan. 2002.

[2] P.Guerrier, A.Greiner,"A generic architecture for on-chip packet switched interconnections", Proc. DATE, pp. 250-256, March 2000.

[3] G. De Micheli, L. Benini, "Networks on Chips: Technology and Tools", Morgan Kaufmann, First Edition, July, 2006.

[4] IBM ASIC Solutions, www.ibm.com

[5] D. Lackey et al, "Managing power and performance for System-on-Chip designs using Voltage Islands", ICCAD 2002.

[6] F. Fallah and M. Pedram, "Standby and active leakage current control and minimization in CMOS VLSI circuits.",IEICE Trans. on Electronics, 2005.

[7] A. Sathanur et al., "Multiple power-gating domain (multi-VGND) architecture for improved leakage power reduction", ISLPED 2008.

[8] Q. Ma, E. F. Y. Young, "Voltage Island Driven Floorplanning", ICCAD 2007.

[9] J. Hu et al., 'Exploiting the Routing Flexibility for Energy/Performance Aware Mapping of Regular NoC Architectures', Proc. DATE, March 2003.

[10] S. Murali, G. De Micheli, "SUNMAP: A Tool for Automatic Topology Selection and Generation for NoCs", Proc. DAC 2004.

[11] S. Murali, G. De Micheli, "Bandwidth Constrained Mapping of Cores on to NoC Architectures", Proc. DATE 2004.

[12] A.Pinto et al., "Efficient Synthesis of Networks on Chip", ICCD 2003, pp. 146-150, Oct 2003.

[13] W.H.Ho, T.M.Pinkston, "A Methodology for Designing Efficient On-Chip Interconnects on Well-Behaved Communication Patterns", HPCA, 2003.

[14] J. Xu et al., "A design methodology for application-specific networks-on-chip", ACM TECS, 2006.

[15] S. Murali et al., "Designing Application-Specific Networks on Chips with Floorplan Information", pp. 355-362, ICCAD 2006.

[16] K. Ryu, V. Mooney, "Automated Bus Generation for Multiprocessor SoC Design", Proc. DATE, pp. 282-287, March 2003.

[17] K.Lahiri et al., "Design Space Exploration for Optimizing On-Chip Communication Architectures", IEEE TCAD, pp. 952- 961, June 2004.

[18] S. Pasricha et al., "Floorplan-aware automated synthesis of bus-based communication architectures", Proc. DAC '05.

[19] L. Leung, C. Tsui, "Energy-Aware Synthesis of Networks-on-Chip Implemented with Voltage Islands", DAC 2007.

[20] T. Dumitras et al., "Towards on-chip fault-tolerant communication", ASPDAC 2003.

[21] S. Murali et al., "Analysis of error recovery schemes for Networks on Chips", IEEE D&T, 2005.

[22] U. Y. Ogras et al., ' Voltage-Frequency Island Partitioning for GALS-based Networks-on-Chip ', in Proc. DAC, June 2007.

[23] Miro-Panades, I. et al., "Physical Implementation of the DSPIN Network-on-Chip in the FAUST Architecture", NoC Symposium , 2008.

[24] Bjerregaard, T. et al. "An OCP Compliant Network Adapter for GALS-based SoC Design Using the MANGO Network-on-Chip", Proc. SoC 2005

[25] S. Stergiou et al., "×pipesLite: a Synthesis Oriented Design Library for Networks on Chips", pp. 1188-1193, Proc. DATE 2005.

978-1-60558-497-3/09 $25.00 © 2009 ACM

Hierarchical Reconfigurable Computing Arrays
For Efficient CGRA-based Embedded Systems

Yoonjin Kim
Dept. of Computer Science and Engineering
Texas A&M University, College Station
TX 77843
ykim@cse.tamu.edu

Rabi N. Mahapatra
Dept. of Computer Science and Engineering
Texas A&M University, College Station
TX 77843
rabi@cse.tamu.edu

ABSTRACT

Coarse-grained reconfigurable architecture (CGRA) based embedded system aims at achieving high system performance with sufficient flexibility to map variety of applications. However, significant area and power consumption in the arrays prohibits its competitive advantage to be used as a processing core. In this work, we propose *hierarchical reconfigurable computing array architecture* to reduce power/area and enhance performance in configurable embedded system. The CGRA-based embedded systems that consist of hierarchical configurable computing arrays with varying size and communication speed were examined for multimedia and other applications. Experimental results show that the proposed approach reduces on-chip area by 22%, execution time by up to 72% and reduces power consumption by up to 55% when compared with the conventional CGRA-based architectures.

Categories and Subject Descriptors

C.3[Special-Purpose and Application-Based Systems]: *Microprocessor and microcomputer applications, real-time and embedded systems, signal processing systems.*

General Terms

Algorithms, Design, Performance, Experimentation, Verification.

Keywords

Embedded Systems, Coarse-Grained Reconfigurable Architecture (CGRA), Computing Hierarchy.

1. INTRODUCTION

With the growing demand for high quality multimedia, especially over portable media, there has been continuous development on more sophisticated algorithms for audio, video, and graphics processing. These algorithms have the characteristics of complex data-intensive computation. In addition, as the market pressure of embedded systems compels the designer to meet tighter constraints on cost, performance, and power, the application specific optimization of a system becomes inevitable. On the other hand, the flexibility of a system is also important to accommodate rapid-

Permission to make digital or hard copies of part or all of this work for personal or classroom use is granted without fee provided that copies are not made or distributed for profit or commercial advantage and that copies bear this notice and the full citation on the first page. To copy otherwise, to republish, to post on servers or to redistribute to lists, requires prior specific permission and/or a fee.
DAC'09, July 26-31, 2009, San Francisco, California, USA

ly changing consumer needs. To accommodate these incompatible demands while efficiently supporting the complex embedded-applications, domain-specific design has emerged as a suitable solution for embedded systems. CGRA-based embedded system is the very domain-specific design in that it can boost the performance by adopting specific hardware engines while it can be reconfigured to adapt to ever-changing characteristics of the applications.

In spite of the above advantages, the use of CGRA-based design has been prohibitive due to its significant area/power consumption. The area and power overheads are caused by large memory components and the computing block of many processing elements including ALU, multiplier and divider, etc. There is also a performance bottleneck due to the conventional communication structure between processor and reconfigurable computing block that cannot adaptively support various applications. Therefore, reducing area/power and improving performance of CGRA-based system has been a serious concern. For efficient domain specific designs, this paper provides a new hierarchical CGRA computing array structure and its hardware implementation.

The paper has following contributions:

- A new reconfigurable computing hierarchy has been proposed to design cost-effective CGRA-based embedded systems.

- Efficient communication structure between processor and reconfigurable computing blocks is introduced to reduce performance bottleneck in the CGRA-based architecture.

- Gate level design and simulation were carried out with real applications and benchmarks to demonstrate the cost-effectiveness of the proposed approach in reducing both the area and power consumptions while enhancing its performance.

This paper is organized as follows. After the related work in Section 2, we briefly describe coarse-grained reconfigurable architecture as preliminary in Section 3. In Section 4, we present the motivation of our approach. Then we propose the new reconfigurable computing hierarchy for CGRA in section 5 and the experimental results are given in Section 6. Finally we conclude the paper in the Section 7.

2. RELATED WORKS

In [1], Hartenstein summarized many CGRAs that had been suggested until 2001. Since then, many more new CGRAs have been continuously proposed and evolved. Most of them comprise of a fixed set of specialized processing elements (PEs) and interconnection fabrics between them. The run-time control of the operation of each PE and the interconnection provides the reconfigurability. However, such fixed architecture has limitations in optimizing the cost and performance for various applications. For example, Morphosys [2] consists of 8x8 array of Reconfigurable Cell coupled with Tiny_RISC processor through system bus. It shows good performance for regular code segments in computation intensive domains but requires large amount of area and power consumption. XPP configurable system-on-chip architecture [3] is another example. XPP has 4 x 4 or 8 x 8 reconfigurable array and LEON processor with AMBA bus architecture. A processing element of XPP is composed of an ALU and some registers. Since the processing elements do not include heavy resources, the total area cost is not high but the range of applicable domains is restricted. In addition, XPP shows significant communication overhead between the processor and RAA through the system bus. REMARC [5] is reconfigurable Multimedia Array Coprocessor that consists of a global control unit and an 8x8 array of nano processors. The nano processors do not also include heavy resources like XPP but it also restricts the range of applicable domains. However, the communication with main processor is faster than [2] or [3] because the processor can access the register-set by coprocessor data transfer instructions. However, limited size of the register-set causes heavy registers-array traffic restricting performance enhancement. ADRES [4] tightly couples a VLIW processor and a reconfigurable matrix through shared register file. The reconfigurable matrix is used to accelerate the dataflow-like kernels in a highly parallel way, whereas the VLIW processor executes the non-kernel code by exploiting instruction-level parallelism. Even though it also provides the fast communication speed between VLIW and the matrix but the entire structure is very dependent on VLIW processor architecture and it require huge register file for the communication. Therefore, the performance is limited by size of the register file. In [12], authors have also used reconfigurable computing cache that is different from our proposed RCC. Their RCC block is LUT-based and used for both memory and computing units.

3. PRELIMINARY

Typically, a CGRA-based system consists of a main processor, a Reconfigurable Array Architecture (RAA), and their interface. The RAA has identical processing elements (PEs) containing functional units and a few storage units. The PEs in the array are connected to the nearest neighbor PEs In addition, they have limited interconnections between non-neighboring PEs in order to perform column-wise or row-wise data-transfer efficiently. These interconnections are reconfigurable because input multiplexers in each PE have inputs from other PEs. The data buffer provides operand data to PE array through a high-bandwidth data bus. The configuration cache (or context memory) is composed of cache elements (CEs) and each CE provides context word to configure each PE.

4. MOTIVATION

4.1 Limitation of Existing Processor-RAA Communication Structures

A typical coarse-grained reconfigurable architecture consists of a microprocessor, a Reconfigurable Array Architecture (RAA), and their interface. We can consider three types of organizations in connecting RAA to the processor. First, the array can be connected to the processor through a system bus as an 'Attached IP'[2][3][8] shown in Figure 1 (a). In this case, the main benefit of

this organization is the ease of constructing such a system using a standard processor without modifying the processor and its compiler. In addition, large data buffer of RAA can be used to support applications having large inputs/outputs. However, the speed improvement using the RAA may have to compensate for significant communication overhead between the processor and

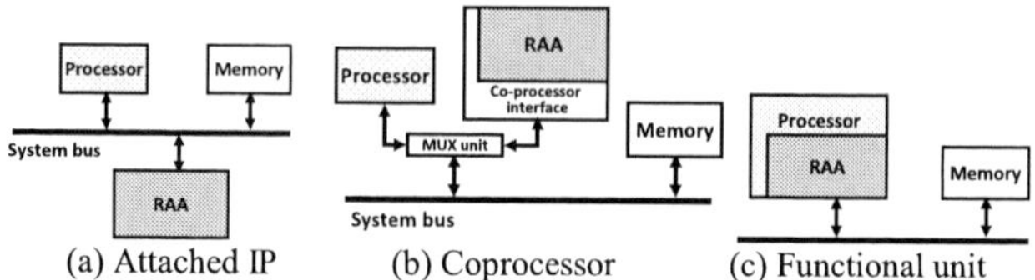

Figure 1. Basic types of RAA coupling.

RAA through system bus as well as SRAM-based large data buffer in RAA consumes much power. Second type of organization involves the array connected with the processor as a 'Coprocessor'[5][6][7] shown in Figure 1 (b). In this case, the standard processor does not change and the communication is faster than 'Attached IP' type interconnects because the coprocessor register-set is used as data buffer of the RAA and the processor can access the register-set by coprocessor data transfer instructions. In addition, the register-set consumes less power than the data buffer of 'Attached IP'. Since the size of the register-set is fixed by the processor ISA, it creates performance bottleneck for registers-PE array traffic due to applications having large inputs/outputs run on the RAA. In the third type of organization, the array is placed inside the processor like a 'FU (Functional Unit)'[4][10][11] as shown in Figure 1 (c). In this case, the instruction decoder issues special instructions to perform specific functions on the RAA as if it were one of the standard functional units of the processor. In this case, the communication speed is faster than 'Coprocessor' and power consumption of the data storage is less than 'Attached IP' because the processor register-set is used as data buffer of the RAA and the processor can directly access the register-set by the processor instructions. However, standard processor needs to be modified for due to integration with RAA and its compiler should be also changed. The performance bottleneck is caused by limited size of the processor registers as in the case of 'Coprocessor' type organization. Table 1 shows a summary about advantage and disadvantage of three coupling types.

Table 1. Comparison of the basic coupling types

Coupling type	'Comm' power	'Comm' speed	Performance Bottleneck	Application feasibility
Attached IP	high	slow	communication through system bus	large size of input/output
Coprocessor	low	fast	limited size of coprocessor register-set	small size of input/output
Functional unit	low	very fast	limited size of processor registers	small size of input/output

*'Comm' power: power consumption by data-storage (data buffer or registers)
**'Comm' speed: Communication speed between processor and RAA

4.2 RAA-based Computing Hierarchy

As mentioned in the previous section, basic three types of RAA organizations show advantage and disadvantage according the input/output size of the applications. It shows the existing coupling structure with a conventional RAA cannot be flexible to support various applications with sacrificing performance. In addition, such an RAA structure cannot efficiently utilize PE arrays and data buffers leading to high power consumption.

We hypothesize that if CGRA can maintain a computing hierarchy of its RAAs having difference size and communication speed, the CGRA-based embedded system can be optimized for its performance and power. It is because such a hierarchical ar-

978-1-60558-497-3/09 $25.00 © 2009 ACM

rangement of the RAA can optimize the communication latency and efficiently utilize functional resources of PE array in various applications. In this paper we propose a new CGRA-based architecture that supports such a RAA-based computing hierarchy.

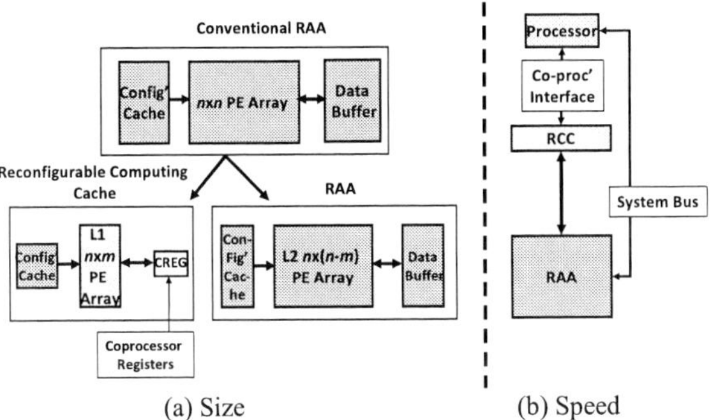

(a) Size   (b) Speed

Figure 2. Computing hierarchy of CGRA.

5. COMPUTING HIERARCY in CGRA

In order to implement efficient CGRA-based embedded systems[1], we propose a new computing hierarchy consisting of two computing blocks using two types of coupling structures together – 'Attached IP' and 'Coprocessor'. In this organization, a general RAA having large size PE array is connected to a system bus and another is a small RAA composed of small PE array coupled with a processor through coprocessor interface. We call the small RAA *reconfigurable computing cache* (RCC) because it plays important role in enhancing performance and power of the entire CGRA like data cache. The RCC and the RAA share critical resources and such a sharing structure provides efficient communication interface between two computing blocks. The propose approach ensures that the RCC and the RAA are efficiently utilized to support variable size of inputs and outputs for variety of applications. In subsection 5.1 and 5.2, we describe computing hierarchy and resource sharing in RCC and RAA in detail. Then we show how to optimize computing flow based on reconfigurable computing cache according to the applications in subsection 5.3.

5.1 Computing Hierarchy – Size and Speed

A CGRA-based computing hierarchy is formed by splitting a conventional computing RAA block into two computing blocks – RCC with small PE array and RAA having large PE array as shown in Figure 2 (a). The RCC is coupled with coprocessor interface and the RAA is attached to a system bus as shown in Figure 2 (b). The RCC provides fast communication with the processor and offers low power consumption by using coprocessor register-set and small array size. Therefore the RCC can enhance performance and reduce power consumption when small applications run on CGRA. If RCC is not sufficient to support computing requirements of in applications, intermediate data from the RCC can be moved to the RAA through the interconnections as shown in Figure 3. Such interconnections between the two blocks offer flexibility in migrating computing demands from one to another. Such computing flow may help to optimize performance and

power for the applications having various sizes of inputs /outputs whereas the existing models show performance bottlenecks caused by the communication overheads or the limited size of the data-storages as shown in Table 1. We have described the computing flow optimization in detail in subsection 5.3.

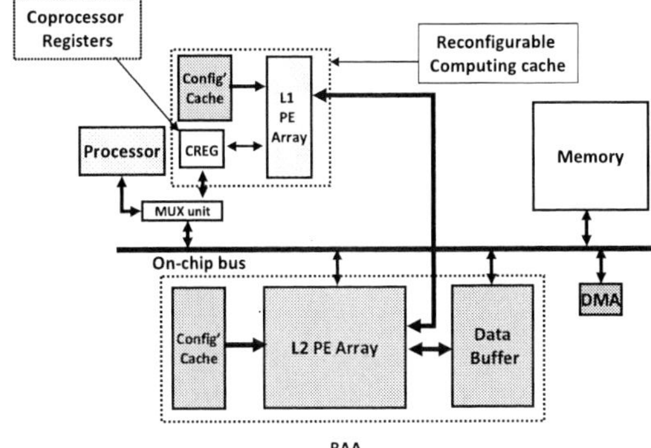

Figure 3. CGRA configuration with RCC and RAA.

5.2 Resource Sharing in RCC and RAA

We have so far presented two factors (speed and size) in building computing hierarchy for CGRAs similar to memory hierarchy. It seems a small portion of RAA has been detached from large CGRA block and placed as the fast RCC block adjacent to the processor coupled with coprocessor interface. However, only considering two factors is not sufficient to design compact RCC for power and area benefits. This is because computing blocks can have diverse functionality which affects the system capabilities. The functionality of computing blocks is specified by functional resources of its PE such as adder, multiplier, shifter, logic operations etc. Therefore, it is necessary to examine how to select the functionalities of RCC and RAA. This leads to further studies on resource assignment/sharing between RCC and RAA.

First of all, we can classify the functional resources into two groups: primitive resources and critical resources. Primitive resources are basic functional units such as adder/subtractor and logical operators. Critical resources are area/delay-critical ones such as multiplier and divider. Based on the classification, let us consider two cases of the functional resource configurations as shown in Figure 4. Figure 4 (a) shows hierarchical functionality that indicates L1 PE array has primitive resources and L2 PE array includes critical resources as well as primitive resources. The Figure 4 (b) shows identical functionalities both in the L1 and L2 PE arrays. In the case of Figure 4 (a), the RCC with L1 PE array is relatively lightweight computing block compared to the RAA with L2 PE array. Therefore, the RCC can perform small applications having only primitive operations with low power consumption. However, it causes 'lack of resource' problem when applications demand critical operations. In Figure 4 (b) L1 and L2 PE arrays have identical functionality with area and power overheads.

To prevent such extreme cases, we propose resource sharing for the RCC and the RAA based on [9]. L1 and L2 PE array have the same primitive resources and shared the pipelined critical resources as shown in Figure 5. Here the RCC and the RAA basically perform the primitive operations and their functionality will include the critical operations using the shared resources. Figure 6 shows interconnection structure with shared critical resources along with RCC and RAA. PEs in the same row of the L1 and L2 array share the pipelined critical resources in the same manner as [9]. Such a structure avoids the 'lack of resource' problem in Figure 4 (a) and this structure is more area and power-efficient than

---

[1] We exclude the type of 'Functional Unit' from the proposed computing hierarchy because it requires the modification of processor and its compiler and entire CGRA is heavily dependent on the specific processor architecture. Our proposed approach aims to implement entire systems allowing any combination of standard processors and RAAs.

978-1-60558-497-3/09 $25.00 © 2009 ACM   828

Figure 4 (b) because the number of critical resources is reduced and the critical resources taken out of L1 and L2 PE array are not affected by unnecessary switching activity caused by other resources. In addition, interconnections for resource sharing can be also utilized for communication interface between the RCC and the RAA by adding multiplexer and de-multiplexer between front and end of the critical resources as shown in Figure 6 (b).

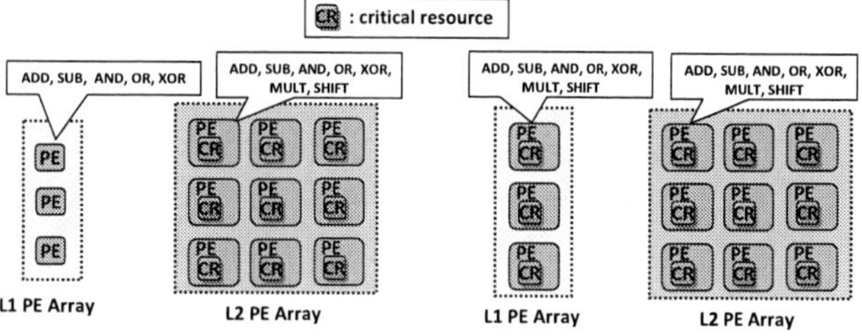

(a) Hierarchical functionality (b) Identical functionality

Figure 4. Two cases of functional resource assignment.

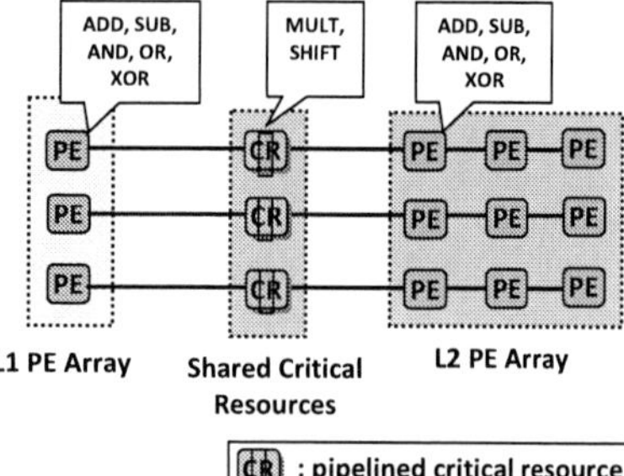

Figure 5. Critical resource sharing and pipelining in L1 and L2 PE Array.

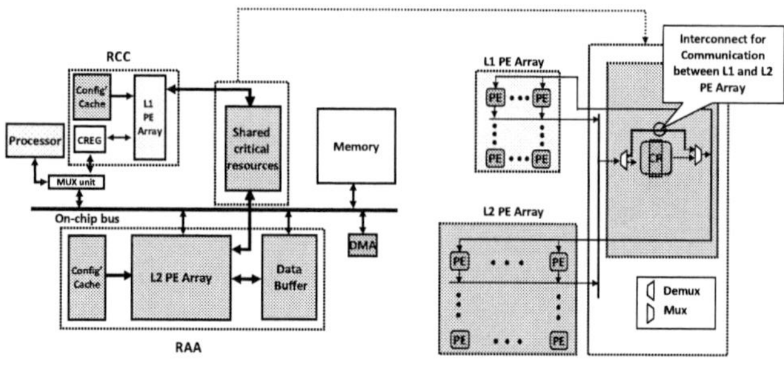

(a) Entire structure (b) Interconnection structure

Figure 6. Interconnection structure among RCC, shared critical resources and L2 PE Array.

5.3 Computing Flow Optimization

Based on the proposed CGRA structure, we can classify four cases of optimized computing flow to achieve low power and high performance. Figure 7 shows such four computing flows on the proposed CGRA according to variance of input and output size of applications – In subsection 6.1.2, Table 3 shows that we can select the optimal case among the proposed computing flows for several applications with variance in their input/output size. All of the cases show that shared critical resources are used as needed because they are only utilized when applications have the operations requiring the critical resources. Figure 7 (a) shows computing flow when application has the smallest inputs and outputs. In this case, only RCC functional units are used to execute the application while the RAA is disabled to reduce power consumption. However, if the application has larger inputs and outputs than Figure 7 (a), the computing flow can be extended to L2 PE array

as shown in Figure 7 (b). Even though L2 PE array is used for this case, data buffer of the RAA is not used because the coprocessor register-set (CREG) is sufficient to save the all of the inputs or outputs. The next case is that when RAA is used with RCC because of large inputs and small outputs as shown in Figure 7 (c). In this case, data buffer of the RAA receives inputs using DMA which is more efficient for overall performance than CREG. This is because insufficient CREG resource for large inputs causes performance bottleneck with heavy registers-PE array traffic. Therefore, the L2 PE array may be used first for running such application and the L1 PE array can be utilized for enhancing parallelized execution as needed. However, the outputs are stored on CREG because their size is small. Finally, Figure 7 (d) shows a case of RAA used with L1 PE array with large inputs and outputs. To avoid heavy registers-PE array traffic by the large input/output size, the data buffer with DMA is used and L1 PE array can be optionally utilized for enhancing parallelized execution.

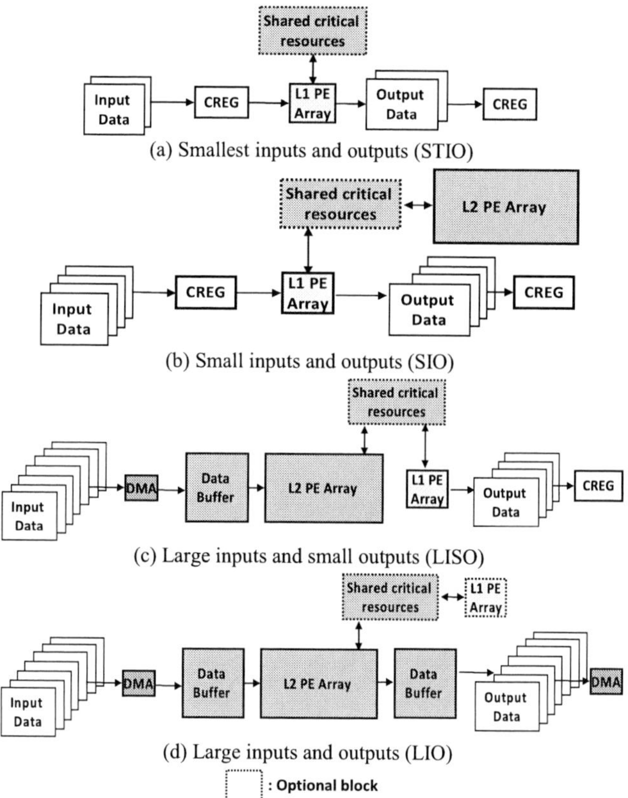

Figure 7. Four cases of computing flow according to the input/output size of application.

In summary, the computing flow on the proposed CGRA can be adapted according to the input/output size of applications. It is more power-efficient than using a conventional CGRA by separated computing blocks with sharing critical resources. This way is only necessary computing blocks are utilized. In addition, computing flow with supporting two communication interfaces reduces power and enhances performance.

978-1-60558-497-3/09 $25.00 © 2009 ACM

6. EXPERIMENTS AND RESULTS

6.1 Experimental Setup

6.1.1 Architecture implementation

To demonstrate the effectiveness of the proposed RCC-based CGRA, we have designed three different organizations of CGRA with RT-level implementation using VHDL as shown in Table 2.

Table 2. Comparison of the basic coupling types

CGRA	PE array	Data storage
Attached IP	8x8 PE array	6KB data buffer
Coprocessor	8x8 PE array	256-byte coprocessor register-set
Proposed RCC-based	8x2 L1 PE array and 8x6 L2 PE array	4KB data butter and 256-byte coprocessor register set

(Leon2 processor [15] is used as main processor)

In addition, for resource sharing of RCC-based CGRA, two pipelined multipliers and two shifters are shared by PEs in the same row of L1 and L2 PE array whereas conventional two types of CGRA do not support such a resource sharing and pipelining.

The architectures have been synthesized using Synopsys Design Compiler with TSMC 0.18 μm technology. Synopsys PrimePower tools have been used for gate-level simulation and power estimation. To obtain the power consumption data, we have used the applications in Table 2 for simulation with operation frequency of 70MHz and typical case of 1.8V Vdd and 27℃.

6.1.2 Evaluated applications

Evaluated applications are composed of real multimedia applications and benchmarks. We have analyzed the input/output size and operation-types in the applications to identify specific computing flow in Figure 7. Table 3 shows the selected applications and the optimal computing flows for them.

Table 3. Applications characteristics

Real Applications	SHR	Computing Flow	Benchmarks	SHR	Computing Flow
(H.263) 8x8 DCT	✓	SIO	*256-point FFT	✓	LISO
(H.263) 8x8 IDCT	✓	SIO	*256-tap FIR	✓	LISO
(H.263)8x8 QUANT	✓	SIO	*Complex Mult	✓	LISO
(H.263) 8x8 DEQUANT	✓	SIO	**State	✓	STIO
(H.263) SAD	-	LISO	**Hydro	✓	STIO
(H.264) 4x4 ITRANS	✓	STIO	**Tri-Diagonal	✓	LIO
(H.264) MSE	✓	LISO	**First-Diff	-	STIO
(H.264) MAE	-	LISO	**ICCG	✓	STIO
(H.264) 16x16 DCT	✓	LISO	**Inner Product	✓	LIO
8x8*8x1 Matrix-Vector Multiplication	✓	SIO	*: DSPstone benchmarks [13]		
16x16*16x1Matrix-Vector Multiplication	✓	LISO	**: Livermore loop benchmarks [14] SHR:'✓'means critical resources are used for the application.		
8x8 Matrix Multiplication	✓	SIO	STIO: smallest inputs and outputs SIO: small inputs and outputs		
16x16 Matrix Multiplication	✓	LISO	LISO: large inputs and small outputs LIO: large inputs and outputs		

Table 4. Area cost comparison

PE Array	No' of PEs	No' of MULTs	No' of SHTs	Gate Equivalent			Reduction (%)
				Interconnect	Logic	Total	
Base 8x8	64	64	64	151234	478908	630143	-
Proposed	64	16	16	161311	325908	487219	22.68

6.2 Results

6.2.1 Area cost evaluation

Table 4 shows area cost evaluation for the two cases. 'Base 8x8' means 8x8 PE array included in 'Attached IP' and 'Coprocessor' type CGRA. 'Proposed' means L1 and L2 PE array included in the proposed RCC-based CGRA. Even though interconnection area of the proposed model increases because of resource sharing structure, entire area of the proposed one is reduced by 22.68% because it has less critical resources than base 8x8 PE array.

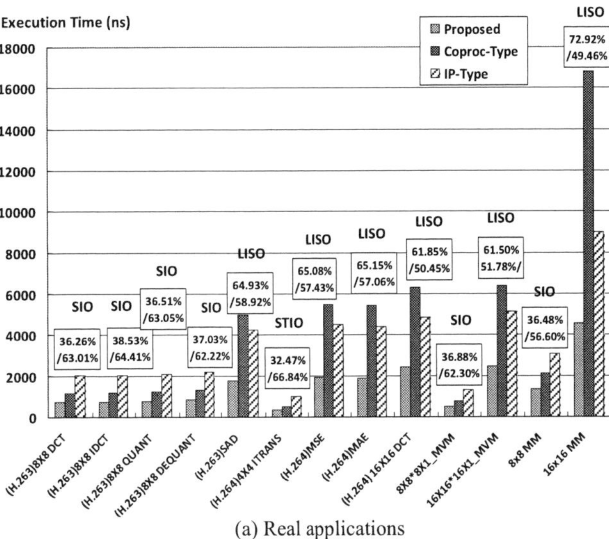

(a) Real applications

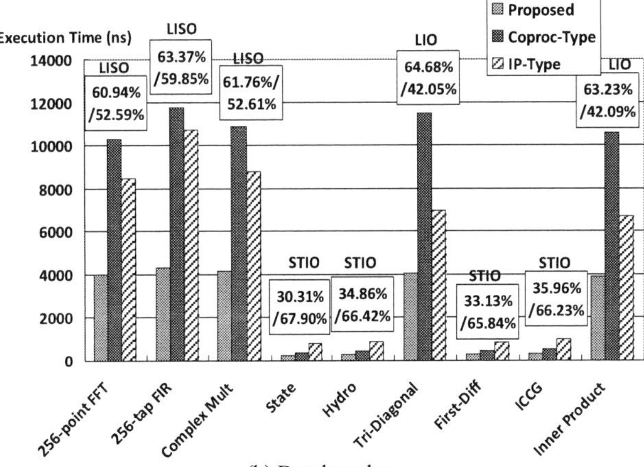

(b) Benchmarks

A%/B%: A% means reduced execution time ratio compared with Coproc-Type and B% means reduced execution time ratio compared with IP-Type.

Figure 8. Performance comparison.

6.2.2 Performance evaluation

The synthesis results show that the proposed PE array has reduced critical path delay (5.12 ns) compared to the base PE array (8.96 ns). This is because pipelined multipliers are excluded from the original set of critical paths. Based on the synthesis results, we evaluate execution times of the selected applications on three cases of CGRA as shown in Figure 8. The execution times include communication time between memory/processor and the RAA or RCC. Each application is executed on the RCC-based CGRA in the manner of selected computing flow as shown in Table 3 – all of the applications are classified under 4 cases of computing flow (STIO, SIO, LISO and LIO). In the case of STIO and SIO, performance improvement compared with 'Coprocesssor' type is relatively less (30.31%~37.03%) than LIO and LISO (60.94%~72.92%). This is because the improvements of STIO and SIO are achieved by only reduced critical path delay whereas the improvements of LIO or LISO are achieved by avoiding heavy coprocessor registers-PE array traffic as well as reduced critical path delay. However, compared with 'Attached-IP' type, STIO and SIO achieve much more performance improvement (56.60%~67.90%) whereas LISO and LIO show the improvement of (42.05%~59.85%). This is because STIO and SIO do not use data buffer of the RAA causing communication overhead on system bus.

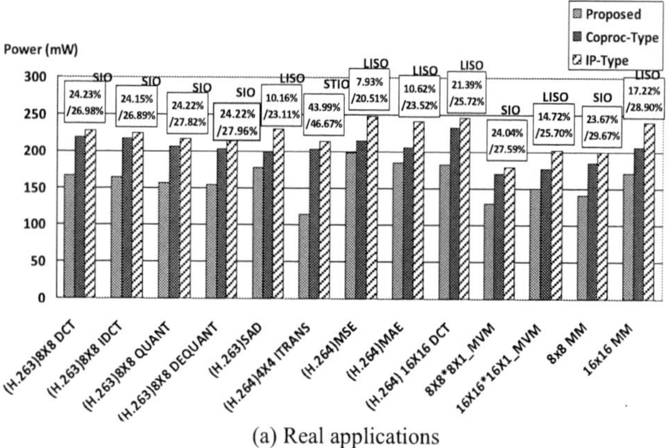

(a) Real applications

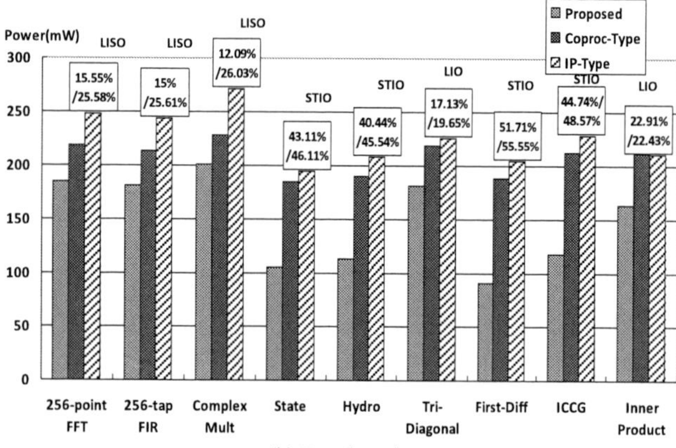

(b) Benchmarks

A%/B%: A% means power saving ratio compared with Coproc-Type and B% means power saving ratio compared with IP-Type.

Figure 9. Power comparison.

6.2.3 Power evaluation

Figure 9 shows the comparison of power consumptions in three different organizations of CGRA. First of all, the proposed L1 and L2 PE array is more power-efficient than the base PE array because of the reduced critical resources. With such a power-efficient PE array, the amount of power saving depends on the selected computing flow for the application. The most power-efficient computing flow is STIO that shows relatively much power saving (40.44%~55.55%) compared to other cases (7.93%~29.67%) because the STIO does not use the RAA - specially, 'First_Diff' shows the highest power saving ratio of 51.71%/55.55% because of not using the shared critical resources. The next power-efficient model is SIO showing power saving (23.67%~29.67%). This is because the SIO computing flow does not use data buffer of the RAA whereas LISO (7.93%~26.03%) and LIO (17.13%~22.91%) utilizes the data buffer for input data or output data. Finally, power saving of LISO and LIO is mostly achieved by reduced critical resources and by not activating L1 PE array.

7. CONCLUSION

Coarse-grained reconfigurable architectures have emerged as a suitable solution for embedded systems because it aims to achieve high performance and flexibility. However, the CGRA has been considered as prohibitive one due to its significant area/power overhead and performance bottleneck. This is because existing CGRAs cannot efficiently utilize large reconfigurable computing block of many processing elements and data buffer wasting area and power. In addition, fixed communication structure between the computing block and processor can not guarantee good performance for various applications. To overcome the limitations, in this paper, we proposed a new computing hierarchy consisting of two reconfigurable computing blocks with two types of communication structure together. In addition, the two computing blocks have shared critical resources. Such a sharing structure provides efficient communication interface between them with reducing overall area. Based on the proposed architecture, optimized computing flows have been implemented according to the varying applications for low power and high performance. Experiments with implementation of several applications on the new hierarchical CGRA demonstrate the effectiveness of our proposed approach. The proposed approach reduces area by up to 22.68%, execution time by up to 72.92% and power by up to 55.55.% when compared with the existing undivided arrays in CGRA architectures.

REFERENCES

[1] Reiner Hartenstein, "A decade of reconfigurable computing: a visionary retrospective," in *Proc. of Design Automation and Test in Europe Conf.*, pp. 642-649, Mar. 2001.

[2] Hartej Singh, et al, "MorphoSys: an integrated reconfigurable system for data-parallel and computation-intensive applications," *IEEE Trans. on Computers*, vol. 49, no. 5, pp. 465-481, May 2000.

[3] A. Deledda, et al, "Design of a HW/SW communication infrastructure for a heterogeneous reconfigurable processor," in *Proc. of Design, Automation, and Test in Europe Conf.*, pp.1352-1357, Mar. 2008.

[4] C. Arbelo, et al, "Mapping Control-Intensive Video Kernels onto a Coarse-Grain Reconfigurable Architecture: the H.264/AVC Deblocking Filter," in *Design Automation and Test in Europe Conf.*, pp. 642-649, Mar. 2007.

[5] T. Miyamori, et al, "A quantitative analysis of reconfigurable coprocessors for multimedia applications," in *Proc. of IEEE Symposium on FPGAs for Custom Computing Machines*, pp 15-17, April 1998.

[6] Michalis D, Galanis, et al, "Speedups in embedded systems with a high-performance coprocessor datapath," *ACM Transactions on Design Automation of Electronic Systems*, vol.12, no 35, Aug. 2008.

[7] T.J. Callahan, et all, "The Garp Architecture and C Compiler", *IEEE Computer*, Vol. 33, No. 4, pp. 62.69, Apr. 2000.

[8] Ada S. Y. Poon, "An Energy-Efficient Reconfigurable Baseband Processor for Wireless Communications," *IEEE Trans. on Very Large Scale Integration Systems*, vol. 15, no. 3, pp. 319-327, Mar. 2007.

[9] Yoonjin Kim, et al, "Resource Sharing and Pipelining in Coarse-Grained Reconfigurable Architecture for Domain-Specific Optimization," in *Design Automation and Test in Europe Conf.*, pp. 642-649, Mar. 2005.

[10] Francisco Barat, et al, "Low power coarse-grained reconfigurable instruction set processor," in *Proc. of Int. Conf. on Field Programmable Logic and Applications*, pp. 230-239, Sept. 2003.

[11] Marco Lanuzza, et al, "Cost-effective low-power processor-in-memory-based reconfigurable datapath for multimedia applications," in *Proc. of Int. Symp. on Low Power Electronics and Design*, pp. 161-166, Aug. 2005.

[12] Rama Sangireddy, et al, "Low-Power High-Performance Reconfigurable Computing Cache Architectures," *IEEE Trans. on Computers*, vol. 53, no. 10, pp. 1274-1290, Oct. 2004.

[13] http://www.ert.rwth-aachen.de/Projekte/Tools/DSPSTONE

[14] http://www.netlib.org/benchmark/livermorec

[15] Gaisler Research; http://www.gaisler.com/cms

Multicore Parallel Min-Cost Flow Algorithm for CAD Applications

Yinghai Lu[1], Hai Zhou[2,1], Li Shang[3], Xuan Zeng[1*]
[1]State Key Lab of ASIC & System, Microelectronics Dept., Fudan University, China
[2]EECS, Northwestern University, U.S.A., [3]ECEE, University of Colorado, Boulder, U.S.A.

Abstract—Computational complexity has been the primary challenge of many VLSI CAD applications. The emerging multicore and many-core microprocessors have the potential to offer scalable performance improvement. How to explore the multicore resources to speed up CAD applications is thus a natural question but also a huge challenge for CAD researchers. Indeed, decades of work on general-purpose compilation approaches that automatically extracts parallelism from a sequential program has shown limited success. Past work has shown that programming model and algorithm design methods have a great influence on usable parallelism. In this paper, we propose a methodology to explore concurrency via nondeterministic transactional algorithm design, and to program them on multicore processors for CAD applications. We apply the proposed methodology to the min-cost flow problem which has been identified as the key problem in many design optimizations, from wire-length optimization in detailed placement to timing-constrained voltage assignment. A concurrent algorithm and its implementation on multicore processors for min-cost flow have been developed based on the methodology. Experiments on voltage island generation in floorplanning demonstrated its efficiency and scalable speedup over different number of cores.

Categories and Subject Descriptors:
J.6 [Computer-Aided Engineering]: Computer-Aided Design
D.1.3 [Programming Techniques]: Concurrent Programming – Parallel programming
General Terms: Algorithms, Performance
Keywords: Min-cost flow, Multicore, Parallel programming

I. INTRODUCTION

VLSI computer-aided design (CAD) software for multi-billion transistor IC design has become increasingly complex and requires more and more computation resources. This challenge can potentially be mitigated by emerging multicore and manycore systems. Since 2004, multicore microprocessor has become the main engine of mainstream servers and personal computers [6], [7]. Nowadays, it is rare to see uni-core processors even in laptop computers, and servers often come with eight cores on one or two CPUs. Therefore, it is natural to hope that manycores available in modern computers may be effectively utilized to speed up CAD programs. However, numerous unsuccessful past attempts have shown that, programming model has great influence on usable parallelism; without exploring concurrency in algorithm design, it is impossible to achieve reasonable speedup in multicore or manycore systems.

Recently, multicore parallel CAD has drawn significant attention in the design automation field [2], [5], [21], [30]. Various existing techniques to explore program concurrency have been borrowed for parallel CAD programming.

Automated parallelization is a compilation approach that extracts parallelism from a sequential program. It has been extensively investi-

gated for many years, but has shown limited success [26]. The general consensus in the community is that a program's automatically-exploitable concurrency is generally fixed by the programming or the programmer's way of thinking. Conventional sequential programming heavily limits a program's usable concurrency.

Message passing approaches such as MPI [24] explicitly implement a computation by multiple processes that work in separate memory spaces and synchronize via passing messages. It is easy to understand. However, the programming model is at a low abstraction level, closer to the physical platform. Similar to assembly language programming, it requires the programmer to think in physical details and the (even more difficult) concurrent execution of processes. Furthermore, such a program needs to be redesigned for different generations of many-core processors.

Threading (or multithreading) implements a computation using multiple threads that share a common memory space and can be executed concurrently. Thread synchronizations are most commonly achieved by locking. However, coarse-grain locking does not perform well, while fine-grain locking is error-prone. Common problems in fine-grain locking include deadlock and the inability to compose program fragments that are correct in isolation [9]. In addition, it is not known how a programmer can come up with a multithreaded program with correctness guarantee.

Transactional memory is a shared memory model proposed by Herlihy et al. [10]. Instead of locking a set of memory elements before accessing them, a program using transactional memory marks some blocks of instructions *transactions*. A transaction is a state change that happens *atomically*. A runtime implementation of transactional memory, either via software or hardware, may speculatively execute many transactions in different threads. If there is no conflict among these transactions, concurrency is effectively explored. Otherwise, only one of the conflicting transactions is permitted to take effect and the others are aborted.

In this paper, we identify the nondeterministic transactional programming model (as in UNITY [3] and TLA+ [16]) as the most effective algorithm design approach for exploring concurrency. There is a systematic algorithm design methodology for such a model that is natural for problem solving and constructs an algorithm together with its correctness proof. More importantly, the correctness of the algorithm allows nondeterminacy in execution order of the commands. Two such commands can be executed concurrently if there is no conflict. We believe that the nondeterministic atomic command model is perhaps the closest to concurrency while still manageable in our brain. We also advocate a design principle to create many small local commands for concurrency exploration. Philosophically, program parallelization is a trade-off between communication and load balancing. Optimizing both is often possible, if we create many short actions with nondeterministic order. Starting with such an algorithm, we also develop an approach to programming it in such a way that we only need to program and compile it once to run on machines with different cores. We believe such a feature is important to achieve scalable performance speedup on multicore architectures which will double the core number in each generation.

To test and demonstrate such a multicore CAD programming method, we select the min-cost network flow problem and develop a multicore parallel program for it. There are tens of different CAD

---

*Corresponding author. E-mail: xzeng@fudan.edu.cn

Permission to make digital or hard copies of part or all of this work for personal or classroom use is granted without fee provided that copies are not made or distributed for profit or commercial advantage and that copies bear this notice and the full citation on the first page. To copy otherwise, to republish, to post on servers or to redistribute to lists, requires prior specific permission and/or a fee.
DAC'09, July 26-31, 2009, San Francisco, California, USA

978-1-60558-497-3/09 $25.00 © 2009 ACM

problems that can be formulated as min-cost flow problem or solved via min-cost flow as the key subroutine. These include voltage assignment [20], gate sizing [28], clock skew optimization [18], retiming [29], floorplan area minimization [19], placement wire minimization [27], etc. In general, many CAD problems need to minimize a weighted summation of element costs under the timing constraint in terms of the longest path delay or the maximal delay-to-register ratio on the cycles. Such a problem is very close in structure to the dual problem of the min-cost flow, as will be explained in Section II. There was theoretical study of parallel algorithms for min-cost flow problem [22]. However, practical multicore parallel program is still needed.

One of the most recent applications of the min-cost flow technique to CAD is the voltage assignment in voltage island floorplanning [20], [17], [12]. Different supply voltages need to be assigned to different blocks to minimize the total power consumption under the constraint that the longest delay is upper bounded. Such a problem can be formulated as the dual problem of the convex-cost network flow, which is then translated into a min-cost flow problem [20]. We apply our multicore parallel program to the voltage island floorplanning problem and demonstrate the effectiveness and scalability of our approach.

The contributions of the paper include a nondeterministic transactional algorithm design method for exploring concurrency, a systematic approach to program such an algorithm for multicore platforms, their application in developing an efficient multicore program for the min-cost flow problem, and the application of the program to speed up voltage island floorplanning. Furthermore, such an approach is applicable to other algorithmic problems in CAD.

The rest of the paper is organized as follows. In Section II, the voltage island assignment problem is formulated as the dual of min-cost flow problem. A nondeterministic transactional algorithm for min-cost flow problem is developed in Section III, and the general methodology for mapping such a transactional algorithm to parallel program on a multicore platform is described in IV. Speedup improvement techniques for solving the voltage assignment problem on multicore is introduced in V. The effectiveness of the proposed methodology is demonstrated through experiments in Section VI. Finally, the paper is concluded in Section VII.

II. Problem Formulation

Many CAD problems, especially timing-constrained design optimization problems, center around minimizing implementation cost under various timing constraints. When signal arrival time in the design is considered, and the implementation cost of an element is a function of its delay, such a problem can usually be formulated as the following mathematical program.

$$Min \quad \sum_{(i,j)\in E} cost_{ij}(d(i,j))$$
$$s.t. \quad \forall(i,j) \in E : p(i) + d(i,j) \leq p(j) \quad (1)$$

where the decision variables are p (arrival time) and d (element delay). The function $cost_{ij}(d)$ computes the element cost to achieve a delay of d. It is a non-increasing function, and is usually convex, meaning that speeding up an element is more difficult at higher speed zone.

Recently, voltage island generation has become an important CAD problem because of the thermal issues in modern VLSI designs [20]. Multiple different supply voltages will be applied to different blocks to minimize the power consumption while satisfying the timing constraints. It is also desirable that blocks with the same supply voltages be placed together in order to minimize the power/ground network. The central problem in voltage island generation is the **timing-constrained voltage assignment problem** which can be

formulated as follows.

$$Min \quad \sum_{(i,j)\in E} power_{ij}(v(i,j))$$
$$s.t. \quad \forall(i,j) \in E : p(i) + d_{ij}(v(i,j)) \leq p(j) \quad (2)$$
$$\forall i \in V : 0 \leq p(i) \leq \phi \quad (3)$$
$$\forall(i,j) \in E : v(i,j) \in Voltage \quad (4)$$

where v is the supply voltage, and both the delay d and the power consumption $power$ are its functions. When v is treated as (inverse) function of d, $power$ can be represented as a function of d; the above formulation becomes very close to our general formulation. The only exceptions are the bound constraints on the arrival time p and the discrete voltage requirements in the last two formulas. The bound constraints can be subsumed into the difference inequalities with an introduced ground node. The idea is to treat the lower bound as $p(O) + 0 \leq p(i)$ and the upper bound as $p(i) - \phi \leq p(O)$, where $p(O)$ for the ground node O is always 0. The discrete requirement will be first relaxed to get a continuous solution, which will then be rounded by a heuristic.

The continuous timing-constrained voltage assignment problem and the general formulation as discussed above can be translated into the following formulation, which is the dual of the min-cost network flow problem.

$$Max \quad \sum_{(i,j)\in E} c(i,j)d(i,j)$$
$$s.t. \quad \forall(i,j) \in E : p(i) + d(i,j) - w(i,j) \leq p(j) \quad (5)$$

Its dual, the min-cost flow problem, is given as follows.

$$Min \quad \sum_{(i,j)\in E} w(i,j)f(i,j)$$
$$s.t. \quad \forall(i,j) \in E : 0 \leq f(i,j) \leq c(i,j) \quad (6)$$
$$\forall j \in V : \sum_{(i,j)\in E} f(i,j) = \sum_{(j,k)\in E} f(j,k) \quad (7)$$

The Karush-Kuhn-Tucker condition [25] for both the primal and dual problems is the same and given as follows.

$$P0 \quad \stackrel{\triangle}{=} \quad \forall(i,j) \in E : 0 \leq f(i,j) \leq c(i,j) \quad (8)$$
$$P1 \quad \stackrel{\triangle}{=} \quad \forall j \in V : \sum_{(i,j)\in E} f(i,j) = \sum_{(j,k)\in E} f(j,k) \quad (9)$$
$$P2 \quad \stackrel{\triangle}{=} \quad \forall(i,j) \in E : (f(i,j) < c(i,j) \Rightarrow p(i) - w(i,j) \leq p(j))$$
$$\wedge (f(i,j) > 0 \Rightarrow p(i) - w(i,j) \geq p(j)) \quad (10)$$

They are necessary and sufficient conditions for both the primal and the dual problems. Any correct algorithm need to satisfy them as its post-conditions, which must be true at the end of the algorithm.

III. Nondeterministic Transactional Algorithm Design for Concurrency

We propose to use a **nondeterministic transactional programming** method to explore concurrency in algorithm design. Such a method can be traced back to Dijkstra's guarded commands [4], which had been later developed into UNITY [3]. In UNITY, every algorithm is composed of an initialization followed by a loop of guarded commands. No order is imposed on the commands. When the condition (i.e., guard) is valid for a command, it can be selected for execution. The correctness of the algorithm does not depend on the execution order of the commands, but depends on the atomic execution of each command. Lamport further demonstrated in his TLA [16] that such a model can be used to naturally specify any system including reactive systems.

We believe nondeterministic transactional programming is suitable for multicore algorithm design, based on the following reasons.

978-1-60558-497-3/09 $25.00 © 2009 ACM

First, thinking and reasoning about arbitrary asynchronous concurrent actions are difficult for human brains, since the possibilities are exponential. Fortunately, reasoning on isolated actions without restricting their ordering is within our capabilities, thanks to the concept of invariant and mathematical induction. Second, there is a systematic algorithm design method for this model that employs assertional proof techniques [23], [15] to guarantee correctness, as the fruit of many years' research on programming theory. Finally, high performance can be achieved by concurrently executing many actions, as long as conflicts are rare, which is common in practice and has been confirmed by transactional memory research [10]. Therefore, *one principle we use to explore concurrency in algorithm design is to produce as many small actions as possible.*

We will now develop an algorithm for the min-cost flow problem using the nondeterministic transactional programming method to explore concurrency. The post-condition of the algorithm is $P0 \wedge P1 \wedge P2$ as defined in the previous section. An important design decision in the method is to select a predicate within the post-condition as the invariant and use the remaining as the loop goal. Here we simply select $P0$ as the invariant which can be easily satisfied by an initialization $f := 0$.

When $P1$ is not true, there must be at least two nodes whose in-flows are not equal to their out-flows. The node excess is defined as

$$X(j) \triangleq \sum_{(i,j) \in E} f(i,j) - \sum_{(j,k) \in E} f(j,k).$$

A node with positive excess is defined to be active. Under a given f, the residual edges $E(f)$ are defined as the edges where an extra flow can be added. Decreasing a flow on an edge is equal to adding the flow on the reverse direction. Formally,

$$E(f) \triangleq \{(i,j)|(i,j) \in E \wedge f(i,j) < c(i,j)\}$$
$$\cup \{(j,i)|(i,j) \in E \wedge f(i,j) > 0\}.$$

The residual capacity $c_f(i,j)$ of an residual edge (i,j) is defined as $c(i,j) - f(i,j)$ if it is in E (i.e., forward edge) or $f(j,i)$ if it is not (i.e., backward edge). For an active node i, we want to push flows from i over residual edges to its neighbors. But such an operation may introduce a residual edge in the reverse direction. We define the reduced cost of an edge (i,j) as

$$w^p(i,j) \triangleq w(i,j) - p(i) + p(j).$$

With the definitions, the post-condition $P2$ can be simplified as

$$P2 = \forall(i,j) \in E(f) : w^p(i,j) \geq 0.$$

To satisfy $P1$ while not to violate $P2$, we only push flow over (i,j) with $w^p(i,j) < 0$, which is called an admissible edge, giving the first guarded command in the algorithm (Figure 1). When there is no admissible edge from an active i, we will relabel it by increasing $p(i)$ by $\epsilon/2$, giving the second command. When $P2$ is not true, there will be a residual edge (i,j) with $w^p(i,j) < -\epsilon$. We can simply remove such an edge by filling its capacity, giving the third command.

Now a tricky problem is how to decide the potential changing step ϵ in the second command. A larger step will render more valid residual edges for pushing flows out of i, but may give more residual edges violating $P2$. As a trade-off, we can gradually reduce ϵ until $\epsilon < 1/|V|$, giving the fourth command. In summary, the whole algorithm is given in Figure 1, where

$$P2(\epsilon) \triangleq \forall(i,j) \in E(f) : p(i) - w(i,j) \leq p(j) + \epsilon.$$

It is actually Goldberg's min-cost flow algorithm, and its correctness and complexity are given in the following theorem. Such an algorithm exposes much concurrency since it has $2|E| + |V| + 1$ actions.

Theorem 1: Goldberg's algorithm in Figure 1 is correct under nondeterministic atomic execution of the guarded commands. The number of iterations is upper bounded by $O(|V|^2|E|log(|V|\max_{(i,j) \in E}|w(i,j)|))$.

$$f,p,\epsilon := 0,0,\max_{(i,j)\in E}|w(i,j)|$$

do $\{P0\}$
    $\exists(i,j) \in E(f) : X(i) > 0 \wedge -\epsilon \leq w^p(i,j) < 0$
        $\rightarrow push(i,j)$
    $\exists i \in V : X(i) > 0 \wedge \forall(i,j) \in E(f) : w^p(i,j) \geq 0$
        $\rightarrow p(i) := p(i) + \epsilon/2$
    $\exists(i,j) \in E(f) : w^p(i,j) < -\epsilon$
        $\rightarrow f(i,j) := f(i,j) + c_f(i,j)$
    $P1 \wedge P2(\epsilon) \wedge \epsilon \geq 1/|V| \rightarrow \epsilon := \epsilon/2$
od $\{P0 \wedge P1 \wedge P2(\epsilon) \wedge \epsilon < 1/|V|\}$

Fig. 1.   Nondeterministic transactional algorithm for min-cost flow.

IV. Multicore Programming of Nondeterministic Transactional Algorithm

Even though a nondeterministic transactional algorithm can be naturally designed and formally proved correct, as demonstrated in the previous section, no current existing programming platform supports it. In this section, we will leverage existing mechanism and develop a systematic approach for mapping such an algorithm to a multithreading program for multicore platforms. An outstanding feature of the approach is that the program only needs to be developed once and can then take advantage of different number of cores in a platform.

A. General Methodology

It is tempting to create a thread for each guarded command in the transactional algorithm; the many number of threads will be automatically scheduled by the OS to available cores, and the atomicity can be enforced with simple locking or transactional memory. However, since the OS has no knowledge of the operations in the threads, it might preempt a thread doing useful thing by a busy-waiting one. Even worse, a thread could be preempted in the middle of a (atomic) transaction, increasing the chance of conflict or abortion, not to mention the overhead of scheduling and managing the threads. Another plausible approach is to have one thread take care of a group of guarded commands. For example, four threads may be created for the four groups of commands in Goldberg's algorithm in Figure 1. However, due to instance-dependent dynamics of execution, the number of iterations for each command is unknown and could change dramatically, making load balance hard to achieve, if not impossible. The same is true for static task partitioning based on data (for example, on V or E in Figure 1).

After careful examination of many plausible ideas, we propose the following general method to implement a nondeterministic transactional algorithm on a multicore platform: Each core will have all the code and be able to execute every guarded command; which command is executed will depend on which data are available and controlled by each core. The data (or their tokens) will be moved dynamically among the cores. Because of the universality of the cores, no data need to be moved if not for the purpose of load balancing. Data affinity is thus preserved.

We leverage the mechanism of multithreading to create a thread for each core. Such a thread is created at the beginning of the program and will last till the end of the program. It is ideal for each thread to keep running on a core during the program execution. From now on, core and thread will be used interchangeably. In order to manage the control of data (or tasks) among the threads, a queue will be maintained as a globally shared data structure. Each thread will fetch control tokens from the queue when it finishes its current work, and will release tokens to the queue based on its schedule. Dynamic load balancing is thus attainable through self-discipline. Since the execution of a command will change the state thus may render more valid guards, all threads need to be synchronized in order to detect that all guards are false thus the program can be terminated.

Applying the general method, we can map Goldberg's algorithm in Figure 1 into a multithreaded program with the identical thread program in Figure 2. We maintain a global queue Q to hold the

978-1-60558-497-3/09 $25.00 © 2009 ACM

nodes whose excess flow or reduced cost condition will enable the corresponding push/relabel commands. A thread repeatedly tries to fetch a bunch of nodes into its local input buffer q_{in}. Then, it exams the nodes in q_{in} and their associated edges one by one to see whether they enable the first three groups of commands in Figure 1. If a guarded command is enabled, the thread carries out the corresponding action such as push, relabel or fill on the node or edge. The actions, as are parenthesized in Figure 2, should be executed as an atomic transaction, in order to guarantee the correctness of the program. The atomicity can be achieved by using conventional mutual exclusion or modern transactional memory [10]. An action such as push will probably introduce more valid guarded commands because pushing a flow toward a node may make it active. The newly active nodes are stored in the local output buffer q_{out} and are later on flushed to Q.

If a thread fails to fetch any active nodes from Q, which means that Q is currently empty, it becomes idle. Then it tries to synchronize with other threads in order to know their status. The detail of global synchronization (*Sync on idle* in Figure 2) will be discussed in the next subsection. When all the threads are idle, the fourth guarded command is enabled, and ϵ is reduced. With reduced ϵ, the first three group of guarded commands may becomes valid again. To enable their examination, all nodes will be activated and added to Q. Each thread will repeat the process until $\epsilon < 1/|V|$. Then all threads terminate and the program completes.

---

while $\epsilon > 1/|V|$
    if get some active nodes V_a
        for $i \in V_a$
            for $(i, j) \in E(f)$
                {**if** $(w^p(i,j) < -\epsilon)$ $f(i,j) := f(i,j) + c_f(i,j)$
                elseif $(w^p(i,j) < 0)$ $push(i,j)$}
            end for
            if $(X(i) > 0)$ {$relabel(i)$}
        end for
    elseif *Sync on idle*
        $\epsilon := \epsilon/2$
        activate V
end while

Fig. 2. The program for each core/thread.

Our implementation of transactional algorithm on multicore machine overcomes many disadvantages of other possible approaches discussed above. Firstly, with long-living threads, the overhead of thread creation and termination is largely reduced. Second, since each thread is bound with a core, it is much less likely to be preempted during execution, reducing overhead and half-complete transactions. Furthermore, instead of being idle, each thread (core) is constantly making progress or checking for new valid guarded commands. All threads terminate almost simultaneously when there are no valid guarded commands. The flexibility of task-based programming facilitates load balance tuning of our program using the technique introduced later.

B. Global Synchronization

According to previous discussions, the fourth command in Figure 1 become valid only when all the other three guarded commands are not enabled. However, a thread cannot check the validation of the fourth command by simply checking the status of global queue to see whether it is empty, because some other threads may still be processing and their operations may re-enable some guarded commands. In the context of Goldberg's algorithm, a push operation will add flow to the target node and may make it active. So besides the status of global queue, we need to introduce a global synchronization mechanism that tells each thread the status of all other threads. In out implementation, a modified version of termination detection barrier (TDBarrier) in [11] is used to do the job.

A TDBarrier contains a counter, implemented by an atomic integer register, and listens to the status of each thread. Before all the threads are launched, a TDBarrier is created and the counter is initialized to be zero. Each time a thread sets itself as idle, it decreases the counter by one; and each time it sets itself as active, it increases the counter by one. When a thread asks TDBarrier about the global thread status, the TDBarrier returns *true* if the counter equals to zero, indicating that all the threads are idle. Otherwise, the TDBarrier returns *false*, informing the querying thread that there are still other active threads and it should keep on checking the global queue for potential newly-added active nodes.

Note that the TDBarrier only does half of the job since each thread has to register its status to the TDBarrier at a proper phase of its own computation. To achieve correct synchronization, each thread registers itself as active before fetching the active nodes from Q, and as idle if the fetching fails.

Theorem 2 ([11]): The global synchronization mechanism with termination detection barrier described above guarantees correct synchronization of the threads in Figure 2.

C. Load Balancing

We have introduced a local input buffer q_{in} and a local output buffer q_{out} to hold active nodes (valid commands) in order to reduce the access to global Q. Moreover, the local buffers can be dynamically adjusted to balance the workload among different threads [1]. Due to the heterogeneity of the flow network, it is not unusual that other threads have exhausted all the active nodes in Q and are waiting for one busy thread to flush its output buffer to the global queue so that they can fetch again. In this case, it makes sense for the busy thread to shrink its q_{out} and flushes new active nodes to global queue. Conversely, if other threads are all busy and there are plenty of active nodes in Q, the size of q_{out} should grow back in order to reduce the frequency of accessing Q. The local input buffers are adjusted accordingly. Let b_k be the original buffer size for thread k, L be the size of global queue, n_{active} and n_{total} be the number of active and total threads, the dynamic load balance adjustment works as follows:

$$\text{if } n_{active} \leq n_{total} \times 0.75$$
$$b_k = b_k/2$$
$$\text{else if } n_{active} + L/b_k \geq n_{total}$$
$$b_k = b_k \times 2$$

The above adjustment is carried out by each thread after processing every 100 valid guarded commands.

V. Performance Improvements

We have implemented the multicore min-cost flow solver using the techniques described in the last section, and tested it on the general graph benches [14]. The speedup in terms of the number of cores is satisfactory. However, when our algorithm is applied to solving the voltage island assignment problem, the speedup is not so good, especially in the 4-core (4C) case. Detailed examination shows that this speedup problem is due to the existence of the ground node, which is introduced to enforce bound constraints (Inequality 3) on the nodes. Figure 3(a) shows the extreme imbalance between the ground node and other nodes in terms of their connecting edges and the number of operations executed on them. Because of the high connectivity and large operation numbers of the ground node, it is highly possible that multiple cores compete for the the ground node at the same time to execute their own commands, which causes heavy contention and slows down our multicore program. Increasing the number of working cores makes the contention problem even worse.

Knowing the cause, we propose a two-fold ground network solution to address the heavy contention problem of our multicore min-cost flow solver on the voltage island assignment problem. First, as described in Section II, the function of the ground node is to

---

978-1-60558-497-3/09 $25.00 © 2009 ACM

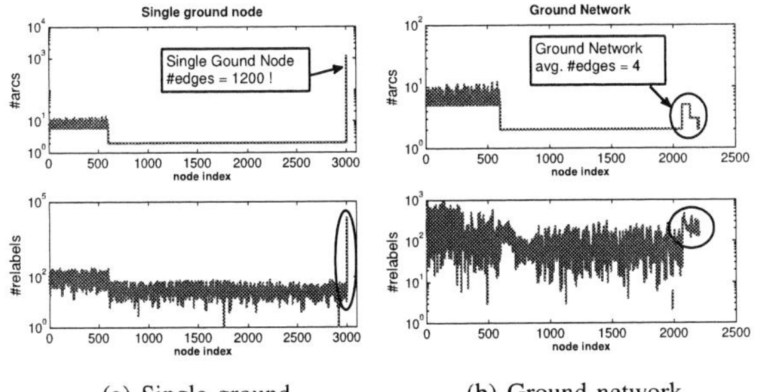

(a) Single ground (b) Ground network

Fig. 3. Performance profiling of case n300.

assert a bound on the arrival time of each block. However, it is not necessary to add such constraint to every node since a bound on the primary inputs/outputs plus the constraint in Inequality 2 guarantees that every internal block satisfies the timing constraint. So we can prune redundant edges between the ground node and the internal nodes, which reduces the number of edges and operations on the ground node.

To further balance the constraint graph, we apply a node splitting technique. The idea is to apportion the connections of IO nodes to the original single ground node to a group of ground nodes. The single ground node is first split into several ground nodes, with the connections of primary IO nodes evenly assigned to them. Then a new ground node is used to bridge all the split nodes to form a ground network with infinite capacity and zero cost. This process is demonstrated in Figure 4. Every ground node in the constructed ground network has only a few edges. When the number of primary nodes is large, we extend this idea and build a tree-structured ground network.

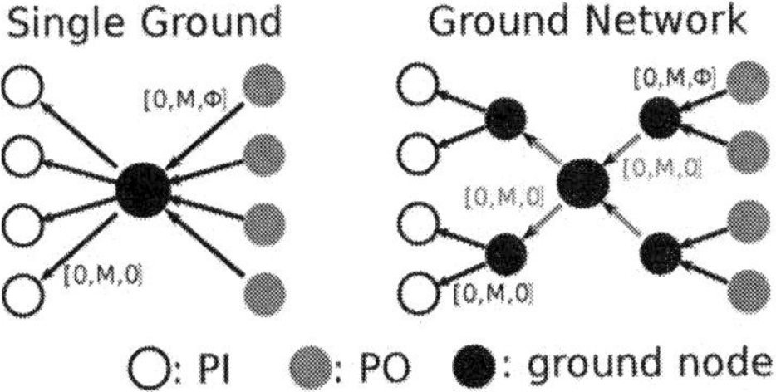

Fig. 4. Node splitting. Edge parameter: [lower cap., upper cap., cost].

After applying redundant edge pruning and node splitting, the structure and operation statistics on the constraint graph is shown in Figure 3(b). The reduction in the number of nodes is due to redundant edge pruning. When converting the convex cost flow problem into the min-cost problem, we introduce one edge for each piecewise linear part of the cost function and attach an auxiliary node with it. When the number of edges is reduced, so is the number of nodes. The circled part in Figure 3(b) corresponds to the ground network. As we can see, the number of edges and relabel operations on these ground nodes are well balanced with the other nodes in the graph. Using the proposed techniques, the heavy contention problem is solved with the balanced constraint flow graph and the multicore min-cost flow solver regains its performance on the voltage assignment problems as shown in Section VI.

VI. Experiment Results

We have implemented the multicore min-cost flow solver in C++ programming language with Intel Threading Building Blocks [13]. All the experiments are carried out on a Linux server with two dual-core 3.0GHz CPUs and 2GB RAM, which supports up to 4-core parallelism. The multicore program is compiled once and runs with a user-specified number of cores.

First, we demonstrate the effectiveness of performance improvement techniques described in Section V. Using the 4-core (4C) min-cost flow solver, we compare the average speedup and contention of the program in solving voltage assignment problems with single ground node and the proposed ground network technique. The test cases are GSRC benchmarks with additional delay-power information provide by the authors of [20] and there are four legal working voltages for each block. Because of nondeterminacy in runtime, the program is run for 10 times on every test case and the results are reported in Table I. The speedup rate is computed against the single-core program. The number of contentions is defined as the average number of data conflicts seen in a core when it commits guarded commands. Table I shows, especially for the large benchmarks, significant reduction in contention and increased speedup of the proposed multicore solver when the single ground node is replaced by the ground network.

TABLE I
EFFECTIVENESS OF THE GROUND NETWORK

Cases	Single Ground		Ground Network	
	#Contentions	4C Speedup	#Contentions	4C Speedup
n10	0.00	1.25	0.00	0.93
n30	58.50	1.03	4.50	1.25
n50	196.75	1.25	5.00	1.42
n100	908.75	1.31	51.75	1.46
n200	6111.00	1.07	94.75	2.26
n300	8809.00	1.02	116.50	1.90

In the second experiment, we modified the MSV-driven floorplanner developed in [20] by replacing the voltage assignment module with the proposed multicore solver. Using the same set of GSRC benchmarks, we compare the modified floorplanner with the original floorplanner. The original voltage assignment module is implemented using CS2 [8], a well-developed sequential min-cost flow solver. Performance comparison between the modified floorplanner and the original one (denoted as [20]) is given in Table II. All the results denoted as *Ours* are averaged over the 1-core (1C), 2-core (2C) and 4-core (4C) versions of our floorplanner. It can be seen that our results are very close to those in [20] on all design parameters. Especially, the power cost (P), which is the objective of the voltage assignment problem, gives almost the same results as in [20]. The tiny differences of the power cost in cases n50 to n300 are caused by the difference of total number of level shifters inserted by the two floorplanners, which consume a small amount of power themselves. With the same number of level shifters inserted, as in the cases of n10 and n30, the power cost is the same, witnessing the correctness of our multicore solver.

The running time comparisons are shown in Table III. The speedup of our program is computed against [20]. For smaller benchmark cases such as n10 to n50, the speedup is small, which is not unexpected due to the implementation overhead such as thread scheduling and maintenance of the global queue. The multi-core programs begin to gain significant speedup against original CS2 on the larger cases such as n100 to n300. Up to 2X speedup of our 4C program against [20] is achieved. We also notice the speed leap from 2C to 4C program, which shows the power of multicore programming when the number of cores increases and the parallel overhead evens out.

TABLE III
COMPARISON OF RUNTIME AND SPEEDUP WITH PREVIOUS WORK

Cases	Run Time				Speedup Rate		
	[20]	1C	2C	4C	1C	2C	4C
n10	2.28	2.82	2.44	2.57	0.81	0.93	0.89
n30	18.90	19.06	15.42	15.27	0.99	1.23	1.24
n50	55.42	70.55	55.76	49.93	0.78	0.99	1.11
n100	223.14	230.12	172.48	150.40	0.96	1.29	1.48
n200	960.31	976.86	657.21	470.01	0.98	1.46	2.04
n300	2032.86	2281.80	1625.99	1087.52	0.89	1.25	1.87

In the last experiment, we further inspect the scalability of our multicore min-cost flow solver on even larger cases, which are constructed by combining and duplicating the smaller ones. We fix the position of the blocks and run voltage assignment once instead of running the whole floorplanning because it is too time consuming to be finished by the original single-threaded program. More specifically,

978-1-60558-497-3/09 $25.00 © 2009 ACM

TABLE II

COMPARISON OF PERFORMANCE WITH PREVIOUS WORK

Cases	Max Power (MaxP)	Power Cost with LS (P)		Power Saving (%)		Power Network Resource		LS Number		Dead Space (%)		Wire Length	
		[20]	Ours	[20]	Ours	[20]	Ours	[20]	Ours	[20]	Ours	[20]	Ours
n10	216841	167012	167012	22.98	22.98	1643	1734	9	9	6.01	6.9	6818	6866
n30	205650	142717	142717	30.6	30.6	2323	2582	37	37	13.4	13.48	32102	29636
n50	195146	145911	143562	25.23	26.43	2297	2784	48	43	16.59	16.42	67611	68003
n100	180028	126209	126442	29.89	29.77	2257	2669	106	104	14.62	15.11	126918	127447
n200	177647	133081	134091	25.09	24.52	2543	2680	165	169	16.59	16.86	228890	231119
n300	273527	171134	170232	37.43	37.76	2971	3096	146	145	24.67	23.85	254085	264317
Average	-	147677	147342	28	28	2339	2590	85	84	15.31	15.43	119404	121231
Difference	-	1	0.99	1.00	1.01	1	1.11	1	0.99	1	1.01	1	1.02

TABLE IV

SPEEDUP RATES OF MULTICORE PROGRAMS ON LARGER CASES

Cases	Constraint Graph		1C Time (s)	Speedup Rate of 2C			Speedup Rate of 4C		
	#Nodes	#Arcs		AVG.	MIN.	MAX.	AVG.	MIN.	MAX.
n200	1344	2329	0.25	1.61	1.40	1.81	2.26	1.99	2.96
n300	2209	3834	0.48	1.44	1.17	1.84	1.90	1.31	2.44
n600	4414	7662	1.15	1.46	1.26	1.60	2.24	1.87	2.64
n800	5376	9322	2.08	1.73	1.52	1.99	2.78	2.32	3.31
n900	6619	11490	1.99	1.44	1.15	1.97	2.15	1.65	2.51
n1000	6720	11653	2.45	1.76	1.51	2.02	2.92	2.36	3.30
n1200	8824	15318	3.00	1.53	1.27	1.95	2.54	2.17	3.41
n1400	9410	16319	4.20	1.83	1.67	2.03	3.16	2.86	3.44
n1600	10752	18646	4.17	1.57	1.47	1.69	2.72	2.30	3.05
Average	-	-	-	1.59	1.38	1.88	2.52	2.09	3.01

the original voltage assignment program from [20] cannot deal with cases larger than n600 due to the overflow problem. Thus, we use our single-core program (1C) as the baseline of comparison. Experiments are repeated for 10 times for each case and the average, minimum and maximum speedup rates of the 2C and 4C programs against the 1C are reported in Table IV. An average speedup rate of 2.52 and a maximum of 3.01 are achieved, showing smooth scalability of our multicore min-cost flow solver when handling large problems.

VII. CONCLUSION AND FUTURE WORK

It is desperately needed for computationally intensive CAD algorithms to speed up with the increasing number of cores in each generation of microprocessors, since their operating frequencies are largely flattened. We proposed in this paper to use nondeterministic transactional programming method to explore concurrency in algorithm design, and developed an approach to program such an algorithm on multicore platforms. By applying the method to the min-cost flow problem, one of the most important problems in CAD with tens of applications, we developed a multicore parallel program for the problem. The efficiency and the smooth scalability of the program with increasing number of cores are demonstrated in the important voltage assignment and island floorplanning problem. The design process also convinced us that such a method can be deployed to other problems. We are currently developing other multicore algorithms for CAD applications.

ACKNOWLEDGMENT

This research is supported partially by NSFC research project 60676018 and 60806013, China National Basic Research Program under the grant 2005CB321701, China National Major Science and Technology special project 2008ZX01035-001-06 during the 11th five-year plan period, the doctoral program foundation of Ministry of Education of China under 200802460068, the International Science and Technology Cooperation program foundation of Shanghai under 08510700100, the program for Outstanding Academic Leader of Shanghai, NSF under CNS-0613967, and SRC under 2007-HJ-1593.

REFERENCES

[1] R. J. Anderson and J. C. Setubal. On the parallel implementation of goldberg's maximum flow algorithm. In *SPAA*, 1992.

[2] B. Catanzaro, K. Keutzer, and B. Y. Su. Parallelizing CAD: A timely research agenda for EDA. In *DAC*, 2008.

[3] K. M. Chandy and J. Misra. *Parallel Program Design: A Foundation*. Addison-Wesley Publishing Company, 1988.

[4] E. W. Dijkstra. Guarded commands, nondeterminacy, and the formal derivation of programs. *Commun. ACM*, 8:453–457, 1975.

[5] W. Dong, P. Li, and X. Ye. Wavepipe: Parallel transient simulation of analog and digital circuits on multi-core shared-memory machines. In *DAC*, 2008.

[6] J. F. et al. Design of the Power6™ microprocessor. In *ISSCC*, 2007.

[7] U. G. et al. An 8-core 64-thread 64b power-efficient SPARC SoC. In *ISSCC*, 2007.

[8] A. V. Goldberg. An efficient implementation of a scaling minimum-cost flow algorithm. *Journal of Algorithms*, 22:1–29, 1997.

[9] M. Herlihy. The multicore revolution. In *FSTTCS 2007: Foundations of Software Technology and Theoretical Computer Science, 27th International Conference*, pages 1–8, 2007.

[10] M. Herlihy and J. E. B. Moss. Transactional memory: Architectural support for lock-free data structures. In *ISCA*, pages 289–300, 1993.

[11] M. Herlihy and N. Shavit. *The Art of Multiprocessor Programming*. Morgan Kaufmann, 2008.

[12] H.Wu, D. Wong, and I.-M. Liu. Timing-constrained and voltage-island-aware voltage assignment. In *DAC*, 2006.

[13] Intel. Threading building blocks. http://www.threadingbuildingblocks.org/.

[14] D. Klingman, A. Napier, and J. Stutz. Netgen: A program for generating large scale capacitated assignment, transportation, and minimum cost flow network problems. *Management Science*, 20(5):814–821, 1974.

[15] L. Lamport. Proving the correctness of multiprocess programs. *IEEE Transactions on Software Engineering*, SE-3(2):125–143, Mar. 1977.

[16] L. Lamport. *Specifying Systems: The TLA+ Language and Tools for Hardware and Software Engineers*. Addison-Wesley Publishing Company, 2002.

[17] W.-P. Lee, H.-Y. Liu, and Y.-W. Chang. An ILP algorithm for post-floorplanning voltage-island generation considering power network planning. In *ICCAD*, 2007.

[18] C. Lin and H. Zhou. Clock skew scheduling with delay padding for prescribed skew domains. In *ASPDAC*, 2007.

[19] C. Lin, H. Zhou, and C. Chu. A revisit to floorplan optimization by lagrangian relaxation. In *ICCAD*, 2006.

[20] Q. Ma and E. F. Y. Young. Network flow-based power optimization under timing constraints in MSV-driven floorplanning. In *ICCAD*, 2008.

[21] T. Mattson and M. Wrinn. Parallel programming: Can we please get it right this time? In *DAC*, 2008.

[22] J. B. Orlin and C. Stein. Parallel algorithms for the assignment and minimum-cost flow problems.

[23] S. Owicki and D. Gries. Verifying properties of parallel programs: An axiomatic approach. *Commun. ACM*, 19:279–285, May 1976.

[24] P. S. Pacheco. *Parallel Programming with MPI*. Morgan Kaufmann, 1997.

[25] R. L. Rardin. *Optimization in Operations Research*. Prentice Hall, 1998.

[26] J. P. Shen and M. H. Lipasti. *Modern Processor Design: Fundamentals of Superscalar Processors*. McGraw-Hill Professional, 2005.

[27] X.-P. Tang, R.-Q. Tian, and D. F. Wong. Minimizing wire length in floorplanning. *IEEE Trans. on CAD*, 25(9):1744–1753, 2006.

[28] J. Wang, D. Das, and H. Zhou. Gate sizing by lagrangian relaxation revisited. In *ICCAD*, 2007.

[29] J. Wang and H. Zhou. An efficient incremental algorithm for min-area retiming. In *DAC*, 2008.

[30] X.-J. Ye, W. Dong, P. Li, and S. Nassif. MAPS: multi-algorithm parallel circuit simulation. In *ICCAD*, 2008.

978-1-60558-497-3/09 $25.00 © 2009 ACM

FPGA-Targeted High-Level Binding Algorithm for Power and Area Reduction with Glitch-Estimation[*]

Scott Cromar, Jaeho Lee, Deming Chen
Department of Electrical and Computer Engineering
University of Illinois, Urbana-Champaign

ABSTRACT

Glitches (i.e. spurious signal transitions) are major sources of dynamic power consumption in modern FPGAs. In this paper, we present an FPGA-targeted, glitch-aware, high-level binding algorithm for power and area reduction, accomplished via dynamic power estimation and multiplexer balancing. Our binding algorithm employs a glitch-aware dynamic power estimation technique derived from the FPGA technology mapper in [6]. High-level binding results are converted to VHDL, and synthesized with Altera's Quartus II software, targeting the Cyclone II FPGA architecture. Power characteristics are evaluated with the Altera Power-Play Power Analyzer. The binding results of our algorithm are compared to LOPASS, a state-of-the-art low-power high-level synthesis algorithm for FPGAs. Experimental results show that our algorithm, on average, reduces toggle rate by 22% and area by 9%, resulting in a decrease in dynamic power consumption of 19%. To the best of our knowledge this is the first high-level binding algorithm targeting FPGAs that considers glitch power.

Categories and Subject Descriptors

B.6.3 [**Hardware**]: Design Aids—*optimization*

General Terms

Algorithms, Design, Measurement, Performance

Keywords

FPGA, high-level synthesis, glitch power, power reduction

1. INTRODUCTION

FPGAs hold significant promise as a fast-to-market replacement for ASICs in many applications. As the price of single-purpose chip development skyrockets in each successive technology iteration, the relative price of the FPGA architecture becomes more and more attractive. This, coupled with the many other advantages of FPGAs, such as rapid prototyping and field reprogrammability, makes them a more and more viable alternative in many current ASIC applications. Unfortunately, the advantages of FPGAs are

[*]This work is partially supported by NSF CCF 07-02501 and a research grant from Altera. We used machines donated by Intel. Jaeho Lee participated in the project when he was an undergraduate student in ECE at University of Illinois. Email contact: scromar2@illinois.edu or dchen@illinois.edu

Permission to make digital or hard copies of part or all of this work for personal or classroom use is granted without fee provided that copies are not made or distributed for profit or commercial advantage and that copies bear this notice and the full citation on the first page. To copy otherwise, to republish, to post on servers or to redistribute to lists, requires prior specific permission and/or a fee.
DAC'09, July 26-31, 2009, San Francisco, California, USA

offset in many cases by high power consumption and large area. In fact, it has been shown that FPGAs can be up to 40 times larger and consume up to 12 times more dynamic power than the equivalent ASIC implementation [13].

In FPGAs there are two sources of power consumption: *static power* and *dynamic power*. Static power is power consumed when the circuit is either active or idle. Unless power gating or other transistor-level techniques are built in the chip, static power cannot be easily reduced. Dynamic power, on the other hand, is power consumed when a signal transition occurs at gate outputs, and, being a characteristic of the design implemented on the FPGA, is more easily mitigated. Signal transitions make up the switching activity (SA) of a circuit, and can be classified into two types: *functional transitions*, and *glitches*. Functional transitions are the signal transitions necessary to perform the required logic function, while glitches are spurious transitions: unnecessary signal transitions that occur due to unbalanced path delays at the inputs of a gate.

Dynamic power consumption can be estimated as $P_d = 0.5 \times SA \times C \times V_{dd}^2 \times f$, where SA is the switching activity of the circuit, C is the effective capacitance, V_{dd} is the supply voltage, and f is the operating frequency. Reducing any of these factors will reduce the dynamic power of a circuit. In Altera's Stratix II FPGAs in the 90 nm process technology, dynamic power is the dominant type of power consumed. Further, glitches can account for up to 19% of the total power consumed in FPGAs, and if considering only dynamic power the percentage is even higher [16]. In this work, we focus on reducing SA (through a glitch-aware SA estimator and multiplexer balancing) and C (by indirectly reducing multiplexer area), for power minimization.

High-level synthesis—the mapping of a behavioral description of a circuit to RTL—is a well studied topic consisting of three steps: scheduling, allocation, and binding. Scheduling determines when an operation will take place; allocation determines how many of each resource is needed; and binding assigns operations, or variables, to the resources.

Our FPGA-targeted, glitch-aware, high-level binding algorithm for power and area reduction is unique in that it can connect to the gate-level implementation and employ a glitch-aware dynamic power estimation to guide the synthesis process. The dynamic power estimation is accomplished using a low-power FPGA technology mapper [6], which makes use of a switching activity estimation model considering glitches and has been shown to be effective at capturing glitch power. Reduction of switching activity, and thus dynamic power, is futher targeted via the balancing of multiplexers on the two input ports of a functional unit. The Altera Cyclone II FPGA is used as a testbed FPGA architecture in our experiments (due to its support of a large number of I/O pins that can accommodate large benchmarks), and our binding results are verified using Altera's gate-level

978-1-60558-497-3/09 $25.00 © 2009 ACM

power estimator, Quartus II PowerPlay Analyzer. To the best of our knowledge this is the first FPGA-targeted high-level binding algorithm that considers glitch power.

The rest of this paper is organized as follows: In Section 2 we discuss related work. In Section 3 we present the problem formulation. In Section 4 we describe the technique used for switching activity estimation considering glitches. In Section 5 we describe the binding algorithm, herein referred to as HLPower, in detail. In Section 6 we present experimental results. In Section 7 the paper is concluded.

2. RELATED WORK

Much effort has gone into evaluating and reducing dynamic power in FPGAs. Recent work has included improved SA estimation tools and architectural changes to reduce glitches [15]. Additionally, binding for ASICs is a well-studied topic. The work in [18] proved that the problem of resource binding for multiplexer reduction is NP-complete. Work in the area of low-power binding has included a bipartite graph formulation for multiplexer reduction [11], low-power register binding through a network-flow formulation [1], simultaneous register and resource binding and scheduling algorithms [9], generalized low-power binding formulated as an ILP problem with heuristic speed-ups [10], and early evaluation of DFGs for low-power binding [14], among many others. Reference [10] provides a good overview of the previous work in low-power binding.

Despite this previous work, low-power high-level synthesis and binding for FPGAs is a relatively new area of research. In [20], the switching activity characteristics of the functional units were pre-characterized and used during low-power synthesis targeting FPGAs. In [5], a low-power, simultaneous resource allocation and binding algorithm for FPGAs was presented, and included a high-level power estimator. In [3] and [4], the authors presented a simulated annealing-based algorithm which carried out high-level synthesis subtasks simultaneously, targeting FPGAs for low-power, called LOPASS. Their binding algorithm initially used minimum weight bipartite matching, and then was enhanced using a network flow approach presented in [2] that binds all the resources simultaneously. We compare our own algorithm to LOPASS and show that an iterative approach enables greater power savings.

HLPower, the FPGA targeted, low-power binding algorithm we present here, sets itself apart from previous work in that it considers low-level glitches in its power estimation. This allows HLPower to target a major contribution to dynamic power during high-level synthesis that has not been considered before. We think the main reason this has been missing is because of the difficulty of estimating glitches during high-level synthesis. We present a unique way to address this problem in this paper.

3. PROBLEM FORMATION

The input to our binding algorithm is a scheduled CDFG, a resource constraint, and a resource library. The problem to be solved involves the allocation and assignment of registers to variables, and functional units to operations. Efficient sharing of functional units by operations, and registers by variables, in order to reduce power and area, are the challenges of binding. The binding problem can be formulated as follows:

Given: A scheduled CDFG, a resource constraint, and a resource library.

Tasks: Allocate and bind registers to variables, and allocate and bind functional units to operations.

Objectives: Produce a valid binding solution while meeting the resource constraint and optimizing the solution for power and area on the targeted FPGA.

4. SWITCHING ACTIVITY ESTIMATION

As dynamic power estimation is a central driver of our binding algorithm, the way this is accomplished is described in this section. Dynamic power is estimated in the form of a switching activity model based on probabilistic techniques developed originally in [17], extended in [7], and further developed to target FPGA mapping and include glitches in [6].

In [17] the ideas of *transition density* (also referred to as *toggle rate* or *switching activity*) and *signal probability* are initially developed. The *transition density* of a logic signal is defined as the average number of transitions per unit time, while the *signal probability* is defined as the fraction of the time that the logic signal is in the 1 state (i.e. the average value of the logic signal over all time). An efficient technique is presented that allows for the calculation of the total circuit switching activity by means of propagation of transition densities and signal probabilities from input nodes to output nodes. For a node y with independent fanin nodes $x_1, x_2, \ldots, x_n$, and given the transition density (switching activity) $s(x_i)$ of each fanin node x_i, the transition density of node y, $s(y)$ can be computed using the Boolean difference $(\partial y / \partial x)$ of y with respect to x_i:

$$s(y) = \sum_{i=1}^{n} P(\frac{\partial y}{\partial x_i})s(x_i) \tag{1}$$

where $P(\partial y/\partial x_i)$ is the signal probability of the Boolean difference.

This technique was extended in [7] to take into account simultaneous switching, something that the technique in [17] lacked. Let y be a Boolean expression, $y(t)$ be its value at time t, $P(y)$ be the signal probability of y, and $s(y)$ now be the normalized switching activity of y. $s(y)$ is the probability of y having different values at time t and $t + T$, where T is a unit time period, and is thus given by $s(y) = P(y(t)\overline{y(t+T)}) + P(\overline{y(t)}y(t + T))$. Additionally, note that $P(y(t)\overline{y(t+T)}) = P(\overline{y(t)}y(t+T))$. Thus, $P(y(t)\overline{y(t+T)}) = P(\overline{y(t)}y(t + T)) = 1/2 \cdot s(y)$. Since $P(y(t)) = P(y(t)y(t+T)) + P(y(t)\overline{y(t+T)})$, we find

$$s(y) = 2(P(y(t)) - P(y(t)y(t+T))) \tag{2}$$

And, as noted in [6], the term $P(y(t)y(t+T))$ can be calculated from the probabilities and switching activities of fanin nodes of y using the procedure in [7].

Finally, the technique for switching activity estimation was applied to FPGA technology mapping in [6]. The algorithm in [6] reads in a netlist, and uses a cut-enumeration technique [8] to select K-input cuts that will be mapped to the FPGA (K-input look-up tables). Primary inputs are assumed to have signal probabilities and switching activities of 0.5. For each node, the signal probability of all of the K-input feasible cuts of that node are computed using the weighted averaging algorithm from [12]. When calculating the switching activities for each cut, the widely accepted unit delay model is assumed for the FPGA look-up tables. This means that signal transitions are assumed to happen

978-1-60558-497-3/09 $25.00 © 2009 ACM

Algorithm 1 HLPower Binding Algorithm

1: **Input:** Scheduled CDFG, library, resource constraint
2: **Output:** Scheduled and bound CDFG
3: precalc SA values for all FU & MUX combinations
4: bind registers according to [11]
5: traverse CDFG, select nodes for set U
6: put remaining nodes in set V
7: **while** resource constraint is not met **do**
8:     initialize bipartite graph $G = (U, V, E)$
9:     **for all** edges in E **do**
10:        calculate input MUX sizes (if nodes were combined)
11:        look up SA value for particular FU & MUXs
12:        calculate edge weight
13:     **end for**
14:     solve G for maximum weight
15:     combine matched nodes & allocate functional units
16: **end while**

only at discrete time units: $1, 2, \ldots, D(C)$, where $D(C)$ is the depth of the cut. The transition that takes place at time $D(C)$ is considered the functional transition, while the transitions that occur at the other time steps are considered glitches. Switching activities are then calculated and propagated through the cut according to Equation (2). For a given cut, the effective switching activity is a summation of the switching activities at each time step. An example of how this is accomplished can be found in [6].

The best cuts, those with the lowest switching activities, are then chosen for implementation of the node in the FPGA. Summing up the switching activities, sa_i, for all of the selected cuts, $1, 2, \ldots n$, provides the *total estimated switching activity*, SA, for the netlist:

$$SA = \sum_{i=1}^{n} sa_i \qquad (3)$$

The *total estimated switching activity*, SA, is used in the binding algorithm. This technique for switching activity estimation has the advantages over previous techniques of being mapping-aware and considering glitches.

5. BINDING ALGORITHM

The HLPower binding algorithm proceeds in two major parts. First, registers are allocated and bound, and second, functional units are allocated and bound. In this paper we focus on the functional unit binding. The functional unit binding proceeds in an iterative fashion until the resource constraint is met, driven by the estimated dynamic power usage and multiplexer sizes of various operation-to-functional unit bindings, as will be explained below. Algorithm 1 provides a summary of the HLPower binding algorithm.

5.1 Register Binding

Register binding is accomplished in a manner similar to that described in [11], where variables are bound by solving a weighted bipartite graph. An allocated set of registers is determined by counting the number of variables present in the control step with the largest number of variables with overlapping lifetimes. This set of registers is allocated, and a cluster of mutually unsharable variables (meaning the lifetimes of these variables are overlapping) is bound at a time, by way of a weighted bipartite graph, sorted in ascending order according to their birth times. Operator ports are randomly bound during this step.

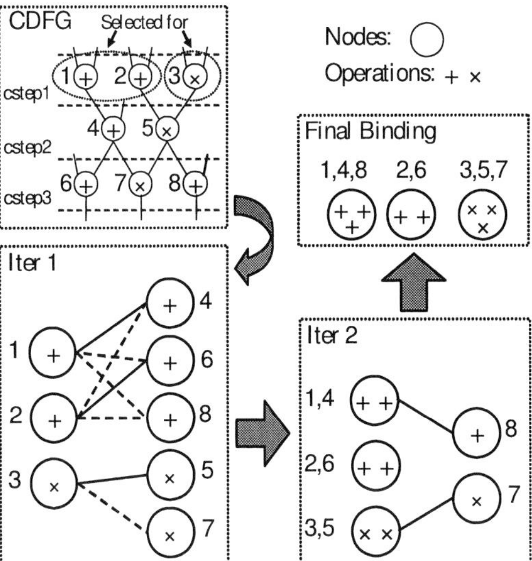

Figure 1: An example of the functional unit binding.

5.2 Functional Unit Binding

5.2.1 Algorithm Overview

Functional unit binding iteratively constructs weighted bipartite graphs, finds a maximum matching, and combines nodes that are matched. Before the first iteration of the functional unit binding, the scheduled CDFG is traversed, and for each operation type, the control step with the largest number of operations of that type is found. This gives a lower bound on the possible resource constraint. These operations are selected to make up one set of vertices (or nodes), U, in the bipartite graph. The second set of vertices, V, includes all of the other nodes. See Figure 1.

During functional unit binding, the nodes of the graph are each considered an allocated functional unit. Initially, as none of the operations have been bound to functional units, every operation is considered to be bound to its own functional unit, and each is represented by an individual node of the graph. On subsequent iterations, each node (functional unit) of the graph may contain more than one operation. Edges (making up the set E) are created between compatible nodes in the graph. Two nodes are compatible if they meet the following two criteria:

1. They perform the same type of operation, e.g. are both multiplications.
2. They do not contain any operations that have overlapping lifetimes in the schedule.

Each edge of the graph represents a possible binding of two sets of operations to the same functional unit. Edge weights are then assigned as described in the next section. This formulation is similar to binding that works with a compatibility graph, but not all nodes will be bound in a single iteration.

Figure 1 illustrates the bipartite graph formulation. In iteration one, add operations 1 and 2, and mult operation 3 are selected for set U, because they come from the control steps of maximum density for their respective types in the scheduled CDFG. (Note that, alternatively, any of the mult operations could have been chosen, or add operations 6 and 8 could have been chosen.) Solid edges represent those selected in the maximum weighted matching. Nodes are combined, and in iteration two nodes are further combined.

978-1-60558-497-3/09 $25.00 © 2009 ACM

In iteration three there is no longer any compatibility between the nodes, and the algorithm is completed. The final allocation is 2 adders and 1 multiplier.

THEOREM 1. *A weighted bipartite graph $G = (U, V, E)$, containing the single-cycle operations of a scheduled CDFG, if iteratively generated and solved, combining matching nodes in each iteration (as previously described), guarantees that the minimum possible resource constraints can be met.*

PROOF. Suppose on the contrary that the minimum resource constraint cannot be met. This would mean that there exists a node in the set V that is incompatible with all nodes in the set U. Since this incompatibility could not be due to compatibility criterion 1 given above—the non-existence of a compatible operation type (if, for example, there was only one operation of a particular type in a CDFG, then it would already lie in set U)—it must be due to criterion 2—operations that have overlapping lifetimes. That would imply that there were more operations in the incompatible operation's control step than were in the control step chosen initially for set U. This could not be the case due to the selection criteria for set U. □

Theorem 1 guarantees that, for a library of single-cycle resources, the minimum resource constraint for the given scheduled CDFG can be met. Although no similar guarantee can be made for multi-cycle resources, our experiments show that the algorithm is nonetheless effective in achieving a minimum resource allocation in most cases. The runtime complexity of the algorithm is $O(|N|^2 * (|E| + |N|log|N|))$, where $|N|$ is the total number of nodes in the CDFG. This is because the number of bipartite graphs solved is, in the worst case, linear with the number of nodes in the CDFG.

5.2.2 Edge Weight Calculation

Edge weights for each graph edge are calculated as follows:

1. The sizes of the input multiplexers to the functional unit to which the operations connected by the edge would be bound (if the matching included the given edge) are found. This is possible because the registers have already been assigned, enabling the calculation of the exact multiplexer sizes. This combination of multiplexers and a functional unit make up a partial datapath.

2. A gate-level netlist of the partial datapath is generated in .blif format [19]. This is accomplished by creating a new .blif file with proper input and output ports, importing existing instantiations of the multiplexers and functional units, and making the necessary connections. See Figure 2.

3. The switching activity is estimated for the gate-level netlist (.blif), based on the technique described in Section 4. This produces an estimate of the dynamic power, including glitch power, which will be used to estimate part of the cost of this particular binding of operations to functional unit.

4. The weight on the edge is computed as follows:

$$w(e_{i,j}) = \alpha \times \frac{1}{SA} + (1-\alpha) \times \frac{1}{(muxDiff + 1) \times \beta} \quad (4)$$

where SA is the *total estimated switching activity* as defined in Equation (3), α is a weighting coefficient, β is a value used to adjust the size of the $muxDiff$ factor relative to SA (based on emperical study $\beta \approx 30$

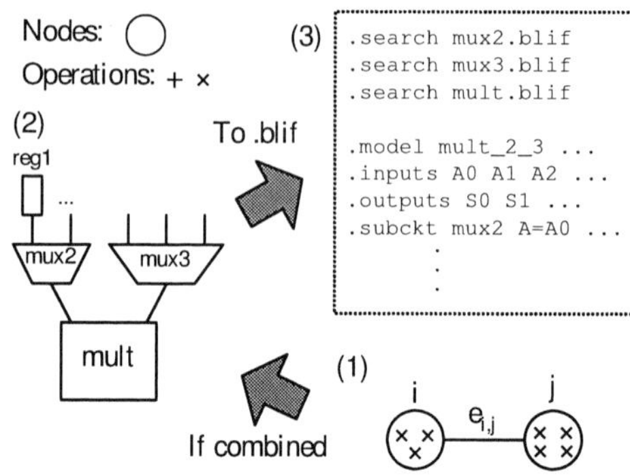

Figure 2: Gate-level partial data-path netlist generation. Based on the register binding, and operations assigned to the edge nodes (1), it is determined that a 2-input and a 3-input MUX are needed (2). The .blif netlist is then generated (3).

for add operations, and 1000 for mult), and $muxDiff$ is defined as the absolute difference in the sizes of the two multiplexers that input to the functional unit. $((muxDiff + 1)$ is used so the equation will be valid if $muxDiff = 0$.)

Equation (4) uses the weighting coefficient α to balance the contribution to the weight of two important factors for the reduction of glitches and switching activity: the *total estimated switching activity*, or SA, and the difference in size between the two multiplexers, or $muxDiff$.

SA provides a *low-level* consideration of the circuit. It explicitly estimates dynamic power usage (including glitches) at the gate-level, taking into account the multiplexers in the partial datapath. It also implicitly considers area through the number of look-up tables required to implement the partial datapath, because a larger area correlates with a higher SA. $muxDiff$, on the other hand, provides a *high-level* consideration of the circuit. It explicitly considers multiplexer balancing, which would have a direct impact on glitch reduction, even if the SA estimation is not 100% accurate. This combination of *high-level* multiplexer balancing, and *low-level* SA estimation, work together to select the best matches for power and area reduction in each iteration of the binding algorithm.

As dynamic calculation of the switching activities for each edge during the binding iterations can be time consuming, in our experiments we precalculate the switching activities for all combinations of multiplexers and functional units. This is done by generating the gate-level netlists for the partial data-path of each combination of functional unit and multiplexers, and running the SA estimation on each. The calculated SA values are then stored in a text file. A hash table is then generated when HLPower is initially run by reading in the precalculated values from the text file. This allows fast look-up of the estimated SA value for a particular combination of input multiplexer sizes, and functional unit. Experimental results show that this method provided us with the same results as running the algorithm with dynamic SA estimation, but with a much shorter run time.

The iterative approach to functional unit binding allows the multiplexer size to be better controlled than is possible with single iteration approaches, such as with a network flow algorithm. By iteratively building up the numbers of operations assigned to allocated functional units, the multiplexer sizes, balance among multiplexers, and the contributions of

Table 1: Benchmark Profiles.

Bench-marks	No. of PIs	No. of POs	No. of Adds	No. of Mults	Total No. of Edges
chem	20	10	171	176	731
dir	8	8	84	64	314
honda	9	2	45	52	214
mcm	8	8	64	30	252
pr	8	8	26	16	134
steam	5	5	105	115	472
wang	8	8	26	22	134

Table 2: Resource Constraints, Scheduling Length, and Number of Registers used for both LOPASS and HLPower binding. Identical schedules and register bindings were used by both LOPASS and HLPower.

Benchmarks	Add	Mult	Cycle	Reg	HLPower Runtime (s)
chem	9	7	39	70	812
dir	3	2	41	25	56
honda	4	4	18	13	14
mcm	4	2	27	54	16
pr	2	2	16	32	2
steam	7	6	28	39	189
wang	2	2	18	39	2

the multiplexers to dynamic power (including glitch power) consumption can be carefully controlled and evaluated.

6. EXPERIMENTAL RESULTS

6.1 Experimental Setup

Our experiments are carried out on a 2.8 GHz Intel Pentium 4 Linux machine, with 2 GB of memory. A number of data-intensive benchmarks are used. The benchmark CDFGs include several different DCT algorithms including *pr*, *wang*, and *dir*, and several DSP programs including *chem*, *steam*, *mcm* and *honda*. The benchmarks are profiled in Table 1. Each node in the benchmarks is either an addition/subtraction or a multiplication.

We compare our binding algorithm, HLPower, to a state-of-the-art low-power high-level synthesis algorithm for FPGAs, LOPASS [3] [4], which can perform scheduling, allocation, and binding. We use a resource library containing single-cycle resources, including a multiplier, an adder, a register, and multiplexers. The benchmarks are first run through LOPASS, and a binding solution is obtained. Then they are run through HLPower with the same schedule, register allocation, and resource constraints, to obtain the HLPower binding solution. Table 2 summarizes the scheduled benchmark characteristics used for both the LOPASS and the HLPower solutions.

The binding solutions, in CDFG format, are then converted to RTL design in VHDL with a CDFG to VHDL tool. To verify our results using a commercial tool, the designs are put into Quartus II for RTL synthesis, placement and routing, timing analysis, simulation, and power analysis. This is done by first building a project on each benchmark's VHDL, setting the device family to Cyclone II, and selecting the same device for each benchmark. We use the Quartus II vector waveform file (.vwf) editor to generate 1000 random input vectors for each benchmark. We also set the simulator settings *glitch filtering* to *never*, *max balancing dsp blocks* to *0*, *wysiwyg remap* to *on*, *optimization technique* to *speed*, and *synthesis effort* to *fast*. These settings help to ensure that

Table 4: Mean and variance of $muxDiff$ across all allocated resources for the final binding solutions.

Bench-marks	LOPASS	HLPower $\alpha = 1$	HLPower $\alpha = 0.5$	
	mean/var	mean/var	mean/var	# muxes
chem	7.4/16.1	4.6/9.8	2.4/5.3	16
dir	5.4/12.2	4/11.2	4.2/3.8	5
honda	3.1/11.1	3.9/6.4	3/6.3	8
mcm	1/0.3	1.8/0.5	0.5/0.3	6
pr	0.8/0.2	0.3/0.2	0.8/0.2	4
steam	8.1/56.1	6.8/29.9	5.8/26.7	8
wang	1.3/0.7	0.8/0.2	1.8/0.7	4
average	3.9/13.8	3.2/8.3	2.6/6.2	

the benchmarks for both LOPASS and HLPower are synthesized in the same way without Quartus II optimizations that would invalidate the power results produced by both algorithms. The same .vwf file is used for both LOPASS and HLPower. Then we run the command *quartus_sh –flow compile* (which runs the synthesis, placement and routing, and timing analysis), *quartus_sim* (which makes use of the .vwf file, and generates a switching activity file, .saf), and *quartus_pow* (which makes use of the .saf file). The command *quartus_pow* runs PowerPlay Power Analyzer, and reports the dynamic power consumption.

6.2 FPGA Area and Power Reduction Results

Table 3 summarizes the synthesis and power analysis results for both LOPASS and HLPower (with $\alpha = 0.5$ from Equation (4)), for each benchmark. There was an average reduction in dynamic power of 19.3%, and area (in the form of LUTs) of 9.1%. These reductions came at the expense of 0.6% of the clock period, on average. For comparison, an $\alpha = 1$ yielded an average power reduction of 6.5%, clock period increase of 3.5%, and an area reduction of 5.1%. (See below for a further discussion of the choice of α.)

Table 3 columns 5, 6, 10, and 11 show the multiplexer reduction results. In the table, *Largest MUX* is the largest multiplexer needed to implement the binding solution, while *MUX length* is a measure of the total number of multiplexers implemented, and is calculated by adding up the total number of multiplexer inputs (sizes).

HLPower reduced the largest multiplexer size by an average of 2.6, and the length an average of 7.2%, over LOPASS. Additionally, Table 4 shows the change in the mean and variance of $muxDiff$ across all allocated resources for the final binding solutions of LOPASS, HLPower with $\alpha = 1$, and HLPower with $\alpha = 0.5$. The variation from $\alpha = 1$ (no $muxDiff$ influence in the weighting equation) to $\alpha = 0.5$ (equal weighting of SA and $muxDiff$) shows a clear reduction in both the mean and variance of $muxDiff$. (Note that the same number of multiplexers were allocated in each of the solutions.) This better balancing of multiplexers on the resource inputs contributes to a power reduction by balancing paths and eliminating extra glitch transitions.

Evidence for the decrease in switching activity, and indirectly glitches, is given in Figure 3. Average toggle rate is defined as number of transitions per second, and is a number reported by Quartus II. HLPower with $\alpha = 1$ reduces the toggle rate for most benchmarks (averaging 8.4%), while $\alpha = 0.5$ produces reductions for all benchmarks, averaging 21.9%. Again we see that SA estimation alone (in Equation (4)) effectively helps reduce toggling, but the combina-

978-1-60558-497-3/09 $25.00 © 2009 ACM

Table 3: Power, Clock Period, Number of LUTs, and Multiplexer Results for LOPASS and HLPower Bindings.

	LOPASS/HLPower					Change				
Bench-marks	Dynamic Power (mW)	Clk Per. (ns)	LUTs	Largest MUX	MUX Length	Dynamic Pow.(%)	Clk Per. (%)	LUTs (%)	Lrgst MUX	MUX Len.(%)
chem	1602.3/1468.6	26.0/27.5	9,806/9,613	26/23	672/637	-8.35	5.67	-1.97	-6	-5.2
dir	709.1/405.8	23.8/24.2	4,527/3,453	18/15	167/157	-42.78	2.04	-23.72	-3	-6.0
honda	658.7/534.1	23.5/23.2	3,352/3,057	15/13	165/162	-18.92	-1.40	-8.80	-2	-1.8
mcm	351.3/208.7	24.1/24.2	3,274/2,548	17/14	159/153	-40.60	0.38	-22.17	-3	-3.8
pr	232.7/192.9	20.9/21.7	1,714/1,732	11/8	70/57	-17.09	3.60	1.05	-3	-18.6
steam	729.6/690.6	24.4/23.6	5,121/4,469	19/22	429/321	-5.35	-3.32	-12.73	3	-25.2
wang	161.5/158.5	20.5/19.9	1,697/1,775	12/8	69/76	-1.85	-2.88	4.60	-4	10.1
Average						-19.28	0.58	-9.11	-2.6	-7.2

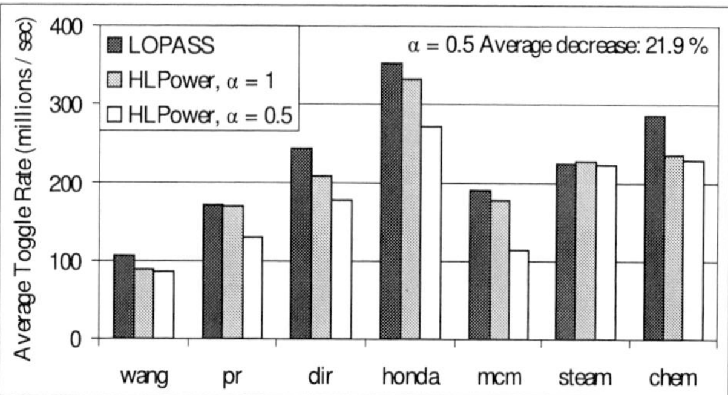

Figure 3: Average Toggle Rate.

tion of SA and $muxDiff$ is more effective. The combination of area savings through multiplexer reduction, and reduced glitching (as evidenced by the significant decrease in toggle rate), produces an aggregate reduction in dynamic power, for a negligible change in clock period.

7. CONCLUSIONS AND FUTURE WORK

In this paper we have presented a new high-level binding algorithm for power and area reduction, targeting FPGAs. The binding algorithm, HLPower, is based on weighted bipartite matching, and makes use of glitch power-aware, dynamic power estimation. HLPower successfully reduces the switching activity and area of a design, producing savings in dynamic power consumption. Future work will include integrating HLPower into a complete high-level synthesis algorithm that includes scheduling and module selection, while providing better support for multi-cycle resources.

8. REFERENCES

[1] J.-M. Chang and M. Pedram. Register allocation and binding for low power. *DAC*, 1995.

[2] D. Chen and J. Cong. Register binding and port assignment for multiplexer optimization. *ASP-DAC*, 2004.

[3] D. Chen, J. Cong, and Y. Fan. Low-power high-level synthesis for FPGA architectures. In *ISLPED*, 2003.

[4] D. Chen, J. Cong, Y. Fan, and L. Wan. LOPASS: A low-power architectural synthesis system for FPGAs with interconnect estimation and optimization. *IEEE Transactions on Very Large Scale Integration (VLSI) Systems*, to be published.

[5] D. Chen, J. Cong, Y. Fan, and Z. Zhang. High-level power estimation and low-power design space exploration for FPGAs. In *ASP-DAC*, 2007.

[6] L. Cheng, D. Chen, and M. D. F. Wong. GlitchMap: an FPGA technology mapper for low power considering glitches. In *DAC*, 2007.

[7] T.-L. Chou and K. Roy. Estimation of activity for static and domino CMOS circuits considering signal correlations and simultaneous switching. *IEEE Transactions on Computer-Aided Design of Integrated Circuits and Systems*, 15(10):1257–1265, Oct. 1996.

[8] J. Cong, C. Wu, and Y. Ding. Cut ranking and pruning: enabling a general and efficient FPGA mapping solution. In *FPGA*, 1999.

[9] A. Dasgupta and R. Karri. Simultaneous scheduling and binding for power minimization during microarchitecture synthesis. In *ISLPED*, 1995.

[10] A. Davoodi and A. Srivastava. Effective techniques for the generalized low-power binding problem. *ACM Trans. Des. Autom. Electron. Syst.*, 11(1):52–69, 2006.

[11] C.-Y. Huang, Y.-S. Chen, Y.-L. Lin, and Y.-C. Hsu. Data path allocation based on bipartite weighted matching. *DAC*, 1990.

[12] B. Krishnamurthy and I. Tollis. Improved techniques for estimating signal probabilities. *IEEE Transactions on Computers*, 38(7):1041–1045, Jul. 1989.

[13] I. Kuon and J. Rose. Measuring the gap between FPGAs and ASICs. In *FPGA*, 2006.

[14] E. Kursun, A. Srivastava, S. O. Memik, and M. Sarrafzadeh. Early evaluation techniques for low power binding. In *ISLPED*, 2002.

[15] J. Lamoureux, G. G. Lemieux, and S. J. E. Wilton. GlitchLess: an active glitch minimization technique for FPGAs. In *FPGA*, 2007.

[16] F. Li, Y. Lin, L. He, D. Chen, and J. Cong. Power modeling and characteristics of field programmable gate arrays. *IEEE Transactions on Computer-Aided Design of Integrated Circuits and Systems,*, 24(11):1712–1724, Nov. 2005.

[17] F. Najm. Transition density: a new measure of activity in digital circuits. *IEEE Transactions on Computer-Aided Design of Integrated Circuits and Systems*, 12(2):310–323, Feb. 1993.

[18] B. Pangrle. On the complexity of connectivity binding. *IEEE Transactions on Computer-Aided Design of Integrated Circuits and Systems*, 10(11):1460–1465, Nov. 1991.

[19] E. Sentovich, et al. SIS: A system for sequential circuit synthesis. Technical report, UCB/ERL Memorandum M89/49, Department of EECS, University of California, Berkeley, Nov. 1992.

[20] F. Wolff, M. Knieser, D. Weyer, and C. Papachristou. High-level low power FPGA design methodology. *NAECON*, 2000.

978-1-60558-497-3/09 $25.00 © 2009 ACM

FPGA-based Accelerator for the Verification of Leading-Edge Wireless Systems

Amirhossein Alimohammad
Saeed F. Fard
Ukalta Engineering
Edmonton, Alberta, Canada
amir@ukalta.com, saeed@ukalta.com

Bruce F. Cockburn
Department of ECE
The University of Alberta
Edmonton, Alberta, Canada
cockburn@ualberta.ca

ABSTRACT

The design of communication systems becomes increasingly challenging as product complexity and cost pressures increase and as the time-to-market is shortened more than ever before. This paper presents a bit error rate tester (BERT) for the hardware-based verification of the physical layer (PHY) layer of emerging wireless systems. We integrate fundamental modules of a typical PHY layer along with the channel simulator onto a single field-programmable gate array (FPGA). For a proof-of-concept, we present the results of a FPGA-based performance verification exercise for a multiple antenna system. The proposed BERT system significantly decreases the test time compared to conventional software-based verification, hence increasing designer productivity.

Categories and Subject Descriptors

B.8.2 [**Performance and Reliability**]: Performance Analysis and Design Aids

General Terms

Verification, Performance, Measurement

Keywords

Bit error rate, Wireless communications

1. INTRODUCTION

In the rapidly evolving area of wireless digital communications, new wireless systems must be developed in very short design cycles. Realizing an efficient and reliable design despite frequently changing standards in a short amount of time poses major challenges for both communication system designers and test equipment vendors. One efficient approach to reducing the design verification time is to decouple the verification of the baseband layer from that of the RF layer. The physical (PHY) or baseband layer resides between the transmitter/receiver frontend (RF layer)

Permission to make digital or hard copies of part or all of this work for personal or classroom use is granted without fee provided that copies are not made or distributed for profit or commercial advantage and that copies bear this notice and the full citation on the first page. To copy otherwise, to republish, to post on servers or to redistribute to lists, requires prior specific permission and/or a fee.
DAC'09, July 26-31, 2009, San Francisco, California, USA

and the medium access control (MAC) layer. Following this strategy of concurrent development, the design effort for the baseband layer and RF chipsets can be overlapped.

At the PHY layer, the bit error rate (BER) performance metric is widely used to measure the reliability of communication systems. Since BER properties are not in general amenable to analysis, Monte Carlo (MC) simulation techniques have been widely used to generate BER versus signal-to-noise ratio (SNR) characteristics. However, the execution times of software-based MC simulations of the baseband layer on workstations can be extremely long, especially for increasingly complex communication systems with computationally-intensive signal processing. For example, decoding of received signals that are transmitted over a multiple-input multiple-output (MIMO) wireless fading channel is a computationally-daunting process. In addition, the joint number of test cases and operating modes can be very large. For example, a wireless system should operate reliably in various radio propagation environments (e.g., indoor and outdoor) and adapt to unpredictable channel conditions using different operating modes (e.g., different modulation constellations and code rates). As the number of possible configurations in emerging standards is increasing (e.g., more than 300 modulation and coding schemes are present in the $802.11n$ standard), software simulation has become the bottleneck to timely design and verification.

The most accurate measurements of receiver performance are performed in-the-field. However, the expensive and time-consuming manual test setup and execution of typical over-the-air (OTA) tests and field measurements combined with the large number of test cases make this conventional approach increasingly time-consuming and expensive. A more practical approach for measuring the error rate performance of complex communication systems is to use a hardware-accelerated verification system at the baseband. This system integrate computationally-intensive baseband signal processing algorithms on a hardware platform. Fortunately, as we will show, the design and verification of the baseband layer can be accelerated by several orders of magnitude using hardware-based prototyping [14, 9].

To provide a realistic wireless channel model and repeatable experimental results in a laboratory settings, a baseband fading channel simulator can accurately reproduce the statistical properties of propagation environments, such as those captured in standard radio channel models [12]. Available off-the-shelf channel simulators offer standard interfaces and support different numbers of channels, numbers of fading paths per channel, maximum Doppler frequencies, max-

Fading channel simulator	Number of channels	Number of paths per channel	Max. Doppler (Hz)	Max. delay spread (msec)	Baseband bandwidth (MHz)
ARC SamrtSim [4]	64	36	2777	–	20
ACE 400NB [5]	32	18	6	0.0012	20
PROPsim C8 [8]	16	48	22000	6.4	50
N5115A [18]	2	48	2400	10	15
R&S ABFS [7]	2	12	1600	1.6	5
NJZ-1600B [13]	2	12	800	0.2	10
SR5500 [17]	2	24	2000	2	13
Sofi05 [16]	2	24	4000	25	8

Table 1: Commercially-available fading channel simulators

imum delays and bandwidths. Table 1 shows the characteristics of some typical commercially-available fading channel simulators. In addition to being costly (as much as half a million dollars), commercially-available simulators often cannot provide all of the required features of the latest wireless communication standards.

We developed a parameterizable hardware-based baseband fading channel simulator that can be implemented on a small fraction of a field-programmable gate array (FPGA). To further accelerate performance verification at the PHY layer, other computationally-intensive signal processing modules are also implemented on the same FPGA. This parameterizable BERT system allows evaluation of the baseband layer of emerging communication systems under various channel conditions at hardware speeds. The rest of this paper is organized as follows. Section 2 presents an efficient model for accurate radio propagation modeling. Section 3 details our bit error rate tester for the performance verification of state-of-the-art communication systems such as MIMO systems. And finally, Section 4 make some concluding remarks.

2. BASEBAND CHANNEL SIMULATORS

Typically, a signal travels from the transmitter to the receiver along different propagation paths with a range of possible delays, called the delay spread [12]. Thus the receiver experiences a superposition of multiple copies of the transmitted signal with different attenuations, phases and delays. This scenario is called multipath fading where the delay spread of the paths varies with time and space. In many emerging standards there are multiple antennas at both transmitter and receiver. Figure 1 shows a MIMO channel with two transmit antennas and two receiver antennas. Each sub-channel between a transmitter antenna and a receiver antenna is modeled as a multipath fading channel that is usually constructed using a tapped-delay line model, where τ_ℓ, $\ell = 1, \ldots, L$, is the delay of the ℓ-th path, and $c_\ell(t)$ is the complex-valued fading coefficient of path ℓ. The noise at the receiver, which is commonly modeled as additive white Gaussian noise (AWGN) [2], is superimposed on the channel symbols to produce the received baseband signal.

The Rayleigh fading model is a reasonably accurate mathematical representation of a wireless environment [12]. This model is widely used to represent an environment where neither a line-of-sight nor a strong specular path is present from the transmitter to the receiver. One efficient approach to generate a Rayleigh fading profile for each path is based on the sum-of-sinusoid (SOS) channel model [6]. In the SOS approach the complex-valued fading process $c(t) = c_i(t) + jc_q(t)$ is formed by superimposing sinusoidal waveforms with amplitudes, frequencies and phases that are se-

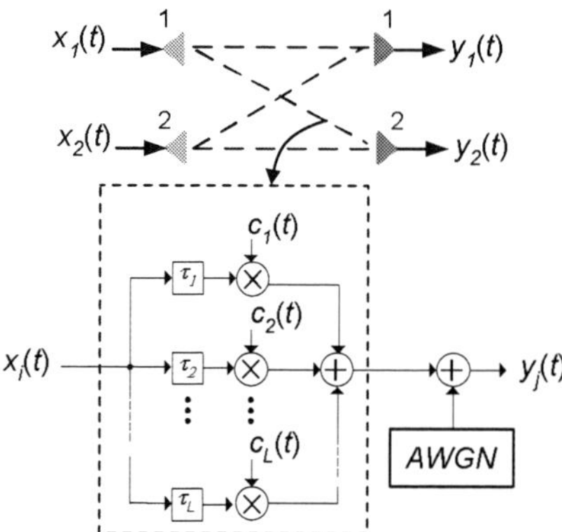

Figure 1: MIMO channel model.

lected appropriately to generate the desired statistical properties. In this model, the real and imaginary parts of the complex Gaussian process $c(t)$, $c_i(t)$ and $c_q(t)$, are independent zero-mean Gaussian [11] with equal variance. The resulting envelope $|c(t)| = \sqrt{c_i(t)^2 + c_q(t)^2}$ follows the desired Rayleigh distribution.

In this work, we used a recently improved Rayleigh fading channel model from [3]. In this model, each complex discrete-time Rayleigh fading process is described as $c[m] = c_i[m] + jc_q[m]$, where

$$c_i[m] = \sqrt{\frac{2}{N}} \sum_{n=1}^{N} \cos\left(2\pi f_D T_s \cos(\alpha_n[m])m + \varphi_n[m]\right),$$

$$c_q[m] = \sqrt{\frac{2}{N}} \sum_{n=1}^{N} \cos\left(2\pi f_D T_s \sin(\alpha_n[m])m + \psi_n[m]\right),$$

and $\alpha_n[m] = (2\pi n - \pi + \theta[m])/(4N)$. Here, $m = 0, 1, 2, \cdots$ is the discrete time index, $n = 1, \cdots, N$, where N is the number of sinusoids, f_D is the maximum Doppler frequency, T_s is the symbol period, $\alpha_n[m]$ is the angle of arrival of the n-th sinusoid, and $\varphi_n[m]$ and $\psi_n[m]$ are the phases of the in-phase and quadrature components, respectively of the n-th sinusoid. The $\psi_n[m]$ $\varphi_n[m]$ and $\theta[m]$ are mutually-independent random processes for all n.

Conventional software simulation of a wireless channel is a computationally-daunting process and inconveniently slow when implemented on a workstation. This is mainly because (i) the components of the fading samples, $c_i(t)$ and $c_q(t)$, must be Gaussian. Therefore, according to the Central Limit Theorem, the number N of sinusoids should be relatively large (e.g., 16 or 32), which increases the com-

978-1-60558-497-3/09 $25.00 © 2009 ACM

putational requirements correspondingly; (ii) the number of supported channels and paths in the emerging wireless networks (such as MIMO systems and ad-hoc networks) is significantly larger than the single antenna systems; (iii) in a typical MIMO scenario, the fading sequences usually exhibit spatial correlations [10]. Implementing spatial correlation among the sub-channels of a MIMO system requires a linear transformation (i.e., complex-valued matrix multiplications), which is a computationally-intensive process; (iv) wireless systems should be tested over as many channel conditions as possible. For example, in indoor and outdoor environments a wide variety of paths and delay spreads must be considered. Therefore, the number of test cases is very large.

We have developed a parameterizable hardware-based baseband fading channel simulator. The implemented fading channel simulator is compact enough to support a complete 4×4 MIMO channel using a small fraction of a single FPGA. As an example, the 4×4 MIMO fading channel simulator uses only 14% of the configurable slices and 2% of the on-chip block memories (BRAMs) on a Virtex-4 4VSX55FF1148-12 FPGA, while operating at 185 MHz. Figure 2 shows excellent agreement between the probability density function (PDF) of 10^7 generated fixed-point fading samples using $N = 16$ sinusoids and the theoretical PDF curve computed using double-precision floating-point arithmetic.

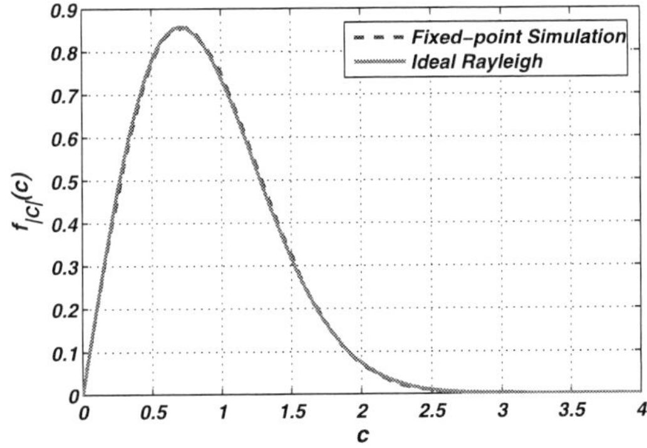

Figure 2: PDF of the fading envelope $|c(t)|$.

3. HARDWARE-ACCELERATED BASEBAND ERROR-RATE MEASUREMENT

Figure 3 shows the cascaded components of the baseband layer of a typical MIMO communication system. As shown in Fig. 3, the binary information sequence generated by the digital source SRC passes through various blocks such as a channel encoder, interleaver and modulator. The serial-to-parallel (S/P) block distributes the bit stream over the available transmitting antennas. The parallel-to-serial (P/S) block rejoins the substreams after detection. The MIMO fading channel attenuates and distorts the transmitted message symbols. The receiver performs the reverse operation of the transmitter. Finally, the decoded bit stream will be compared with the reference sequence to calculate the BER. Note that the baseband layer of different communication systems could have different numbers and types of signal processing modules, perhaps including additional blocks such as an encryptor/decryptor pair or an FFT/IFFT pair [15]. A

pre-existing flexible interface scheme from [1] was utilized to configure the components in the layered architecture shown in Fig. 3.

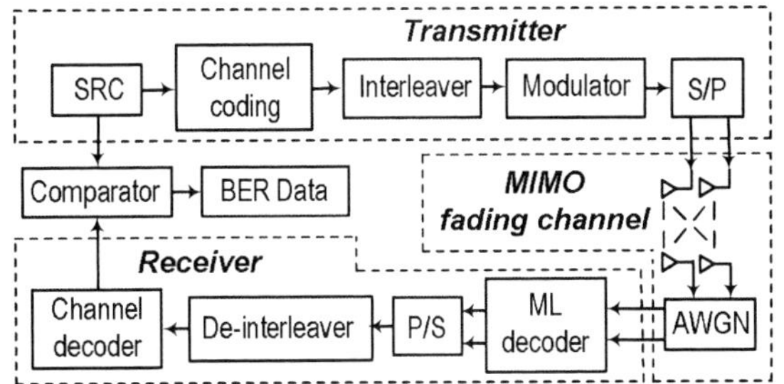

Figure 3: Layered architecture of a digital baseband BER system.

We developed a parameterizable BERT for a MIMO system on a single Virtex-4 4VSX55FF1148-12 FPGA. Figure 4 shows the test setup for evaluating the baseband performance of a 2×2 MIMO system. The system can be scaled to support various numbers of transmitter and receiver antennas. Using a custom graphical user interface (GUI), the BERT can be configured for different test scenarios. The BERT allows us, for example, to rapidly evaluate the impact of Doppler frequency, number of paths, different values of delay spreads, and spatial correlation between antennas of a MIMO system on the error rate of the system. We can also rapidly simulate different signal processing algorithms to quantify the BER performance versus system resource trade-offs. Table 2 gives the hardware costs and performance

Figure 4: Test setup for evaluating the baseband performance of MIMO systems on an FPGA board.

of the implemented modules.

Figure 5 shows the simulated BER performance of a 2×2 MIMO system for both the uncoded and a coded cases assuming either (a) independent channels, where there is no correlation between sub-channels of the MIMO system, or (b) a spatially-correlated MIMO channel. The channel state information is assumed to be available at the receiver, hence no channel estimation was performed. The channel coding scheme was Golay code $(24, 12)$, the length of the convolutional interleaver was 16384, the 4-QAM modulation scheme was selected, the sample rate was set to 50 MHz and the Doppler frequency was set to $f_D = 50$ Hz. Received samples are detected using a maximum likelihood (ML) detector.

978-1-60558-497-3/09 $25.00 © 2009 ACM

Table 2: Characteristics of different modules on a Xilinx Virtex-4 4VSX55FF1148-12 FPGA

Module	Slices	BRAMs	Freq. (MHz)
SRC	37 ($< 1\%$)	–	777
Channel encoder	52 ($< 1\%$)	–	421
Interleaver	49 ($< 1\%$)	4 (1%)	336
Fading channel	3522 (14%)	7 (2%)	185
AWGN	632 (2%)	5 (1%)	303
ML Detector	939 (3%)	–	188
De-interleaver	51 ($< 1\%$)	4 (1%)	332
Channel decoder	407 (1%)	–	191

At least 5000 errors were found when calculating the BER value for each SNR value in the design range. BER averaging was performed over a minimum of 60 seconds, to ensure statistically meaningful measurements over the time-varying channel. Using the single-chip BERT, we have accelerated

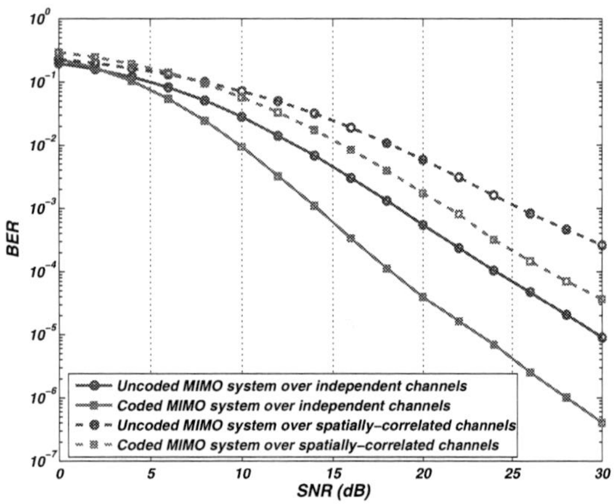

Figure 5: Measured BER performance of a 2×2 MIMO system over independent channels and spatially-correlated correlated channels.

significantly the performance verification of both single and multiple antenna systems. For example, 10 seconds of BER measurements using our hardware platform takes more than three days of bit-true software simulation running on a 3.6-GHz dual-core Pentium processor, producing a speed-up of over $25,000$.

4. CONCLUSION

Hardware prototyping of the physical layer is becoming an indispensable technique in the design and verification of rapidly-evolving modern wireless systems. We implemented an entire baseband layer of a 2×2 MIMO communication system on a single Virtex-4 4VSX55FF1148-12 FPGA. The implemented bit error rate tester system uses 13247 (53%) of slices and 44 (13%) on-chip block memories of the FPGA, and operates at 174 MHz. Using a developed graphical user interface, the error rate performance of the system over a wide range of parameters can be quickly evaluated. This BERT system is flexible enough to be easily reconfigured

to adapt to the new channel specifications of emerging standards and is scalable to support various numbers of channels and multipath configurations.

5. REFERENCES

[1] A. Alimohammad, S. F. Fard, and B. F. Cockburn. Hardware-based error rate testing of digital baseband communication cystems. In *IEEE Int. Test Conference*, pages 1–10, 2008.

[2] A. Alimohammad, S. F. Fard, B. F. Cockburn, and C. Schlegel. A compact and accurate Gaussian variate generator. *IEEE Trans. Very Large Scale Integr. (VLSI) Syst.*, 16(5):517–527, 2008.

[3] A. Alimohammad, S. F. Fard, B. F. Cockburn, and C. Schlegel. A novel technique for efficient hardware simulation of spatiotemporally correlated MIMO fading channels. In *IEEE ICC*, pages 718–724, 2008.

[4] ARC Seibersdorf Research GmbH, Vienna, Austria. *MIMO Channel Simulator*, 2007.

[5] Azimuth Systems Inc., Acton, MA, USA. *ACE 400NB MIMO Channel Emulator*, 2006.

[6] R. H. Clarke. A statistical theory of mobile-radio reception. *Bell System Technical Journal*, 47:957–1000, 1968.

[7] Data sheet, Rohde & Schwarz. *Baseband Fading Simulator ABFS, Saving costs through baseband simulation*, 2004.

[8] Elektrobit Inc., Oulunsalo, Finland. *Multi-Channel Emulator, EB Propsim C8*, 2008.

[9] Y. Guo and D. McCain. Compact hardware accelerator for functional verification and rapid prototyping of 4G wireless communication systems. In *Asilomar Conference on Signals, Systems and Computers*, pages 767–771, 2004.

[10] M. Kiessling and J. Speidel. Statistical transmit processing for enhanced MIMO channel estimation in presence of correlation. In *IEEE Global Telecommunications Conference*, pages 2411–2415, 2003.

[11] R. N. Kolte, S. C. Kwatra, and G. H. Stevens. Computer controlled hardware simulation of fading channel models. In *Proc. IEEE Intl. Conference on Communications*, pages 1646–1650, 1998.

[12] M. Pätzold. *"Mobile Fading Channels"*. West Sussex, U.K.:Wiley, 2002.

[13] Product Manual, NJZ-1600B, Japan Radio Co. *Multipath Fading Simulator*, 2005.

[14] V. Singh, A. Root, E. Hemphill, N. Shirazi, and J. Hwang. Accelerating bit error rate testing using a system level design tool. In *IEEE Symposium on FCCM*, pages 62– 68, 2003.

[15] B. Sklar. *"Digital Communications: Fundamentals and Applications"*. Prentice Hall, 2001.

[16] Sofimation, Helsinki, Finland. *SOFI05 Radio Channel Simulator*, 2006.

[17] Technical Document, SR5500, Spirent Communications. *Wireless Channel Emulator*, 2006.

[18] Technical Overview, N5115A, Agilent Technologies Inc. *Baseband Studio for Fading*, 2005.

Transmuting Coprocessors: Dynamic Loading of FPGA Coprocessors

Chen Huang and Frank Vahid*

Dept. of Computer Science and Engineering, Univ. of California, Riverside

{chuang/vahid}@cs.ucr.edu, http://www.cs.ucr.edu/{~chuang/~vahid}

*Also with the Center for Embedded Computer Systems, UC Irvine

ABSTRACT

Field-programmable gates arrays (FPGAs) are increasingly used in general-purpose computing platforms to augment microprocessors, enabling runtime loading of coprocessors customized to speed up some applications. Such transmuting coprocessors create new dynamic management problems involving decisions as to when to load a coprocessor, where to place the coprocessor in the FPGA, or which resident coprocessor to replace. We define a transmuting coprocessor problem based on Intel's FSB-FPGA architecture, with attention on communication and memory contention. We develop an online algorithm to manage coprocessor loading, the *AG* algorithm, which uses *aggregated gains* to guide coprocessor load, placement, replacement, and wait decisions. Experiments using embedded system applications, for random, biased, and periodic input application sequences, a range of reconfiguration times, and different FPGA types with different numbers of partial reconfigurable regions, demonstrate that the AG algorithm is robust across a variety of situations. The AG algorithm results are within 15% of an unlimited-size FPGA on average, exhibit a small standard deviation, and show a 1.4x speedup versus a static coprocessor loading approach and a 3x speedup over execution on a microprocessor-only solution.

Categories and Subject Descriptors

C.1.3 [**Processor Architectures**]: Adaptable architectures, heterogeneous systems.

General Terms

Algorithm, Performance, Design.

Keywords

FPGAs, dynamic optimization, runtime configuration, coprocessing, acceleration, online algorithms.

1. INTRODUCTION

FPGAs have extended beyond their original role as application-specific integrated circuit (ASIC) prototypes, to ASIC replacements in end-products, and more recently to reprogrammable processing chips in compute platforms such as Intel's QuickAssist, SGI's Altix, and Cray's XD1. For example, Intel's QuickAssist technology [8] enables tight incorporation of FPGAs onto the Front Side Bus (FSB) of Intel processing systems, resulting in fast communication. FPGAs in supercomputing, desktop, or embedded compute platforms typically implement custom coprocessors that speed up particular applications, akin to existing math, graphics, encryption, or signal processing coprocessors but specialized to one application. For example, an image processing application's custom FPGA coprocessor may perform dozens of matrix operations in parallel on different regions of data and a bioinformatics application's coprocessor may perform DNA substring matching.

The increased use of FPGAs to implement different coprocessors at different times introduces new problems, including managing coprocessor loading, illustrated in Figure 1. Some modern FPGA

Permission to make digital or hard copies of part or all of this work for personal or classroom use is granted without fee provided that copies are not made or distributed for profit or commercial advantage and that copies bear this notice and the full citation on the first page. To copy otherwise, to republish, to post on servers or to redistribute to lists, requires prior specific permission and/or a fee.
DAC'09, July 26-31, 2009, San Francisco, California, USA

devices have independently reconfigurable regions, which supports coprocessing in programmable platforms. At any time, an FPGA may have different loaded coprocessors based on the system's present application workload, leading to *transmuting coprocessors*. Loading a coprocessor into an FPGA involves time overhead, so coprocessor load decisions must minimize total system execution time (or energy), respecting FPGA size limits. Much previous FPGA acceleration work statically pre-determined the coprocessors. The distinction is analogous to cache memory and scratchpad memory; the latter excels when applications are known a priori, but the former is common when application executions are dynamically determined.

We define a particular transmuting coprocessor problem. While dynamic task scheduling and placement algorithms for FPGAs have been previously developed, our contribution is in creating an online problem definition, and developing a robust "aggregated gain" algorithm that excels across different application sequence scenarios, reconfiguration overheads, numbers of processor cores, and types of FPGAs with different numbers and sizes of reconfigurable regions.

2. RELATED WORK

Runtime reconfiguration has been studied extensively. Early work included a time-multiplexed FPGA storing multiple configurations in on-chip memory which could be swapped in at runtime [14]. Partial reconfiguration is another important FPGA feature to support fast runtime reconfiguration. Horta [6] presented a partially reconfigurable architecture in which reconfiguration is partially done within the FPGA, to reduce reconfiguration time and energy. McMillan [11] introduced a partial reconfiguration tool JRTR for the Xilinx Virtex family [16]. Runtime reconfigurable systems have been developed such as PipeRench[3] and Chimaera [5].

Online algorithms for dynamic reconfiguration have also been emphasized. Bauer [1] presented a runtime instruction set selection

Figure 1: Illustrated an example for transmuting coprocessor. Five applications a1-a5 with FPGA coprocessors are installed. The past sequence D of application executions is shown. Presently, a5 is running, with its coprocessor c5 in FPGA Region3, and a3 is running with c3 in *Region1*. *c1* is loaded but presently idle. *a4* has arrived for execution—based on the execution history and reconfiguration overhead, would the time of loading *a4*'s coprocessor *c4* be worthwhile? If so, should *c4* replace the idle *c1*, or should it wait for a coprocessor in one of the smaller regions to become idle and replace that one? If *c4* is placed in *Region2* and then a smaller region becomes idle, should it be moved to the smaller region?

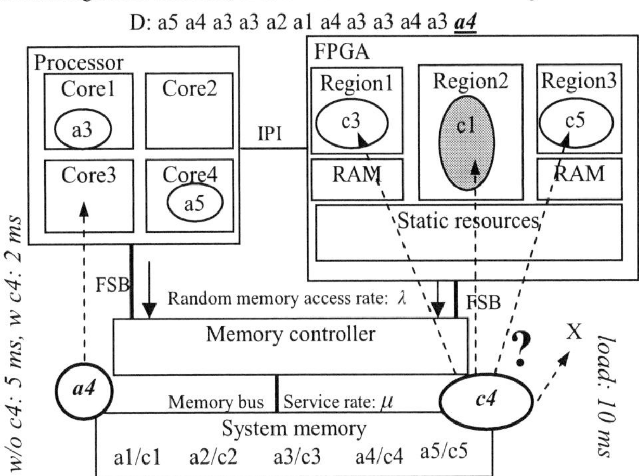

algorithm to explore the design space, with a reconfigurable fabric tightly coupled with a processor. Walder [15] introduced an online task placement algorithm using a hash matrix data structure. Noguera [12] introduced multi-tasking scheduling techniques for FPGAs. Our work differs by using an online algorithm that makes decisions based on past applications, and considers various reconfiguration overheads, as we did in [7]. The current work incorporates a memory contention model and pre-defined reconfigurable regions that involve key interface and routing issues.

3. PROBLEM DEFINITION

3.1 The System and FPGA architecture

We use an Intel FSB-FPGA architecture [8], illustrated in Figure 1. The architecture has a multi-core processor, an FPGA attached on the Front Side Bus (FSB), and a shared system memory. The processor and FPGA share data through the system memory, and can also communicate through inter-processor interrupt (IPI). The FPGA includes a static region that serves as the interface and that performs other basic functions (e.g. routing), an on-chip RAM, and a dynamic region that can be divided into several partial reconfiguration regions (PRRs) [9]. Coprocessors can be loaded into PRRs at runtime to improve performance. The FPGA supports partial reconfiguration, meaning that reconfiguring one PRR does not interfere with running coprocessors in other PRRs.

Each PRR supports only one coprocessor, due to interface and routing issues. Because coprocessor sizes vary, different PRR types also exist to reduce wasted space that would occur if a small coprocessor were placed in a large PRR. Each coprocessor has a corresponding PRR type (small, medium, large) that is the smallest PRR type into which the coprocessor can be placed. A PRR is *vacant* if no coprocessor is loaded. A PRR/coprocessor is *idle* if a coprocessor is loaded but is not presently utilized.

3.2 Communication overhead

This work assumes little communication between applications, as with user applications running on an iPhone. Heavily communicating tasks would be represented as one application. This work *does* consider the important communication that occurs *within* an application, including communication between processor and FPGA.

We capture communication overhead via system memory contention, as shown in Figure 1. Because the memory bus is the bottleneck, FSB contention is omitted. We use queuing theory to estimate average memory access delay, and model memory contention by the M/M/1 queue [4]. The applications in the microprocessor and FPGA generate memory access requests. We define the following:

- Random memory access rate: $\lambda = Sum(\lambda i)$. λi is the memory access rate of running application i.
- Bus service rate: μ. E.g. Assuming a 400Mhz memory bus, one access takes 20 cycles, so μ=20M/s.
- Average delay: According to queuing theory, average delay for one access is $D=\lambda/(\mu(\mu-\lambda))$. System delay: *delay = Dλ*.

3.3 Problem definition

Applications are assumed to be written for a particular compute platform with custom coprocessors for the platform's FPGA. Each application has two possible running modes—*M-only:* The application runs entirely on the platform's processor, or *M+CP:* The application runs on the processor using its FPGA coprocessor to speed up execution.

Applications can run concurrently on different cores in different modes. Multiple coprocessors may exist for a single application (which we collectively refer to as one coprocessor for the application). We define the *Transmuting Coprocessors* (TC) problem as follows. In addition to the definitions in the previous section, we are given:

- An application set $A=\{a1,a2,...,an\}$ containing the n application types that will run on the platform.
- A set of execution times $Tp=\{tp1,tp2,...,tpn\}$ describing the execution time of each application type running in *M-only* mode.
- A set of execution times $Tc=\{tc1,tc2,...,tcn\}$ for each application type i when the application is running in *M+CP* mode without communication and reconfiguration overhead.
- A set of sizes $S=\{s1,s2...,sn\}$ giving the size of each application type's coprocessor in terms of equivalent FPGA LUTs.
- The total size capacity SF of the FPGA, in equivalent LUTs.
- The time for reconfiguration TR per LUT of the FPGA, from which the total coprocessor loading time, written as *loading time(i)=TR*si*, can be computed for a coprocessor of a given size. Unloading a coprocessor takes negligible time, as it consists of invalidating an FPGA region.
- The dynamic input to the problem is an application sequence, such as *<a2, a1, a4, a2, a1, a1....>*, that lists the application instances that run on the platform in the order of arrival time.

The *transmuting coprocessors* problem is an online problem: For each application in the application sequence, using only knowledge of *prior* and *current* applications in the sequence, when a processor becomes available to execute the current application, determine whether to load that application's coprocessor, in which PRR to load that coprocessor, and whether to replace other coprocessors in the FPGA, such that the execution time for the *entire* application sequence (including future applications) is minimized. A coprocessor in the FPGA is said to be *FPGA resident*. The *current application* is the application that at a given time is to be executed next and for which the coprocessor load determination must be made.

We assume $tci < tpi$; else application i is removed from consideration. Each application instance is either run in *M-only* or *M+CP* mode. Running FPGA coprocessors cannot be unloaded.

Limitations: We consider inter-application communication to be negligible; while suitable for many scenarios, others may involve extensive communication. Cache coherency is not considered. We assume each application has one coprocessor; a more general formulation would consider more coprocessors that tradeoff size and performance. FPGA capacity is in LUTs, while modern FPGAs have hard-core resources like multipliers, block RAM, and input/output pins. The current execution model does not allow an application instance's mode to change once the instance has begun executing. Finally, the formulation does not support running the application during reconfiguration. Future work may address above points.

4. TC ALGORITHMS

We now describe several algorithms for the TC problem. We define the FPGA as having N PRRs.

4.1 Upper and lower bounds

An upper bound on total execution time can be determined by running every application in *M-only* mode, referred to as the *microprocessor-only* algorithm. That total time includes the communication overhead with memory. A lower bound can be determined by running every application in *M+CP* mode, assuming unlimited FPGA, and ignoring all communication and reconfiguration overhead, referred to as the *big FPGA/no comm* algorithm. Another interesting comparison is running every application in *M+CP* mode, assuming a big FPGA, but in this case considering communication overhead, referred to as the *big FPGA* algorithm.

4.2 Statically preloaded

To see the advantage of dynamically reconfiguring the FPGA during runtime, we compare to a *statically preloaded* approach, which assumes the FPGA is not reconfigured during runtime. We sort the coprocessors by how much time each could save for one application execution (*tpi-tci*), and load the coprocessors according to the sorted list until no space remains in the FPGA. Communication overhead is included.

4.3 Greedy algorithm

A greedy algorithm can be defined that always tries to load the application's coprocessor into the FPGA before executing the application. In particular, assuming the current application is *ai*, if an *idle* coprocessor *i* exists in the FPGA, run *ai* in *M+CP* mode; else if there is one *vacant* or *idle* PRR that can hold coprocessor *i*, load coprocessor *i* and run *ai* in *M+CP* mode; else run *ai* in *M-only* mode. The time complexity of the Greedy algorithm is *O(N)*.

4.4 Aggregate gain algorithm (AG)

The Aggregate Gain algorithm (*AG*) uses the history of application executions to attempt to predict future executions and hence to predict when reconfiguration overhead is worthwhile. The AG algorithm considers the reconfiguration and communication overhead.

4.4.1 Aggregate gain table and fading process

We define gain as *tpi-tci*, representing the time saved by running an application in *M+CP* mode instead of *M-only* mode. The algorithm maintains an *aggregate gain table*, containing one entry per application *i*. Initially, all entries are 0. When presented with a current application *i* in the sequence, the algorithm updates the table for entry *i* using the equation: *ag(i) = ag(i) + (tpi – tci)*.

Real application sequences often exhibit temporal locality—recently-executed applications are more likely to execute in the near future than are applications from long ago. One way to account for temporal locality is to apply a "fading process" to the aggregate gain table. When a new application arrives, the process multiplies all table entries by the fading factor *f*, *0 < f < 1*. Thus, the aggregate gain values for applications not executed for a long time will approach 0.

The choice of a good *value* for *f* depends in part on the overhead of reconfiguration. If overhead is small, fading should be more aggressive (meaning a small *f*) because reconfigurations can be done freely with little overhead and so current applications should be preferred. Conversely, if reconfiguration overhead is large, fading should be more conservative. We thus define *f* to be proportional to the average reconfiguration time versus average time saved running in *M+CP* mode, namely *f = min{TR*(Σsi/n)/(Σ(tpi-tci)/n), 1}*.

4.4.2 Overhead of reconfiguration

Reconfiguring the FPGA at runtime causes two types of overhead. **FPGA reconfiguration overhead (RO)**: Loading coprocessor *i* into FPGA involves writing bits into SRAM, during which time the current application waits before executing. We define *RO = TR*si*. **Communication overhead (CO)**: Loading coprocessor *i* into FPGA causes extra memory contention delay. We define *current memory access rate:* $\lambda c=Sum(\lambda j)$, where *j* corresponds to currently running applications, and *new memory access rate:* $\lambda n =\lambda c+\lambda i$. Then:

$$CO = (\lambda n)^2/(\mu(\mu-\lambda n)) - (\lambda c)^2/(\mu(\mu-\lambda c)) * tci.$$

4.4.3 AG objective and decision policy

The AG algorithm's objective is to maximize the current system gain. We define *current system gain as: SG=Sum(ag(i))-RO-CO*, where *i* corresponds to all applications whose coprocessor is currently resident in the FPGA. The intuition of the objective is that the future application sequence will be similar to the past. We thus developed the following loading and replacement policy. Assume coprocessor *i* is being considered for loading. If the FPGA has one vacant PRR that can hold coprocessor *i*, the algorithm only loads coprocessor *i* when *ag(i) – RO – CO >= 0.*

The replacement policy has a similar intuition. The policy searches for an idle coprocessor *j* resident in a PRR that can hold

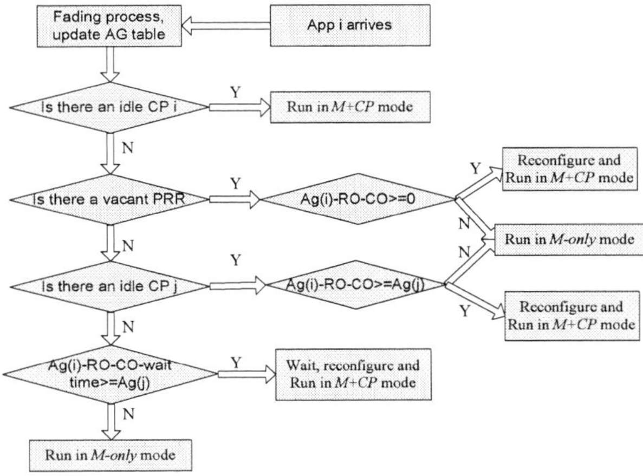

Figure 2: Aggregate gain (AG)

coprocessor *i* and having minimal aggregate gain. The policy is to only replace coprocessor *j* when *ag(i) - RO - CO >=ag(j)*.

If no vacant or idle PRR exists for a current application's coprocessor, we can decide whether to run the current application in *M-only* mode or let the current application wait until a PRR becomes idle. Such a wait may be worthwhile if *M+CP* mode has very high gain. The algorithm can be improved by letting current application wait when *ag(i) – RO- CO – wait time(j) > ag(j)*, where *j* is the next coprocessor that will end execution and whose PRR can hold the new coprocessor. The load, replacement and wait policies guarantee that after making the decision the *system gain* will not decrease.

Figure 2 shows the program flow for the AG algorithm. The time complexity for AG is *O(N)*, where *N* is the number of PRRs.

5. EXPERIMENTS

5.1 Framework

We developed a simulator in C++ to test our algorithms, and applied the simulator to benchmarks from EEMBC [2] and Powerstone [10]. To evaluate the algorithms across a spectrum of application sequence scenarios, we created a generator capable of creating three categories of application sequences. **Random:** Applications are randomly inserted into the sequence with the same possibility. **Biased:** A small number of applications appear most of the time. We defined two percentages *A* and *B*, and then generated the sequence such that *A* percent of the applications executed *B* percent of the time. For our experiments, we used *A*=20% and *B*=80%. **Periodic:** We defined a length *T*, and generated a random subsequence of length *T* that then repeats. For our experiments, we used *T=15*.

Each sequence's length was 1,000. For all experiments, because sequences involve some random ordering, we generated 20 sequences, and report the arithmetic average and standard deviation. Execution time data includes the time to run the algorithm itself. We tested the algorithms for three different platforms; the bus services rate is 20M/sec for all platforms. **Small platform**: A two-core processor with a small FPGA (Xilinx Virtex5 LX30 with 1 large PRR, 1 medium PRR, 1 small PRR). **Medium platform**: A four-core processor with a medium size FPGA (Virtex5 LX50 with 1 large PRR, 2 medium PRRs, 3 small PRRs). **Large platform**: An eight-core processor with a large FPGA (Virtex5 LX85 with 2 large PRRs, 4 medium PRRs, 8 small PRRs.)

5.2 Benchmarks

We obtained data from Stitt [13] and other sources for nine embedded system benchmarks from Powerstone and EEMBC. The data is from earlier experiments seeking to show energy advantages of partitioning applications among microprocessor and FPGA. The data included execution time and coprocessor size for each benchmark, running on

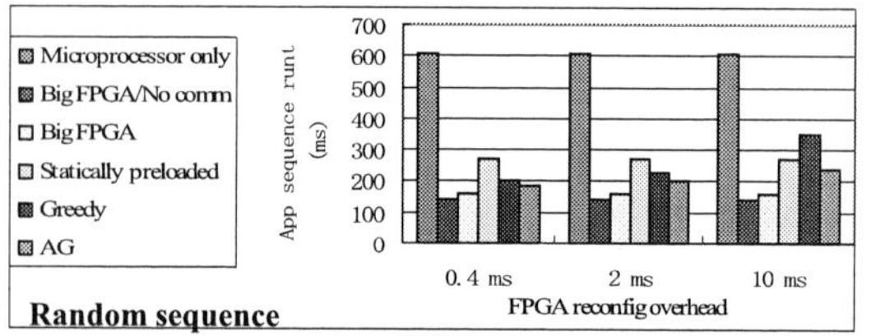
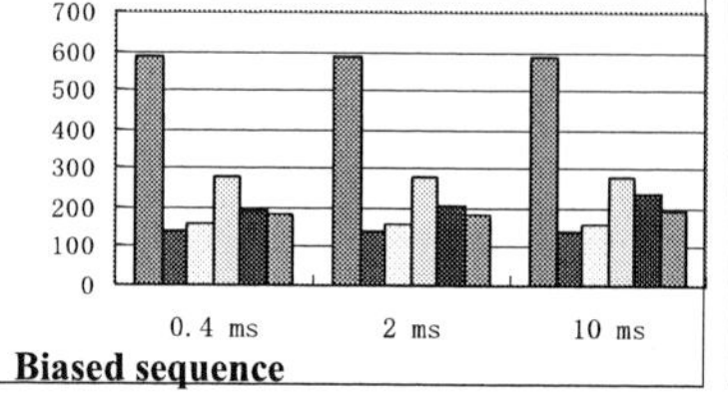
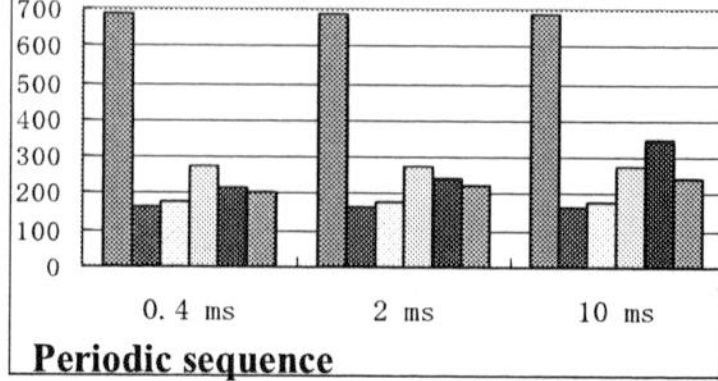

Figure 3: Total execution times (ms) of the entire application sequence for the different algorithms for the Medium platform, for random, biased, and periodic application sequences, for whole FPGA reconfiguration times ranging from 0.4 ms to 10 ms.

a microprocessor alone or partitioned to use a coprocessor on a particular Xilinx FPGA device. We measured the communication overhead and determine the corresponding PRR type for each coprocessor based on the coprocessor size. Benchmark execution time on a microprocessor (a 1.6GHz processor) alone averaged 2.4 milliseconds, reduced to 0.56 milliseconds when using a coprocessor, for an average speedup of 6.6x. The number of equivalent LUTs used on the FPGA per benchmark was 3,105. The magnitudes of the runtimes are inflated by each benchmark internally being iterated a constant number of times.

5.3 Evaluation

Figure 3 shows total execution time of the application sequence for different algorithms for the Medium platform (other platforms had similar results and are omitted). The static FPGA approach yields >2x speedup versus a microprocessor only solution. The dynamic approach (greedy or AG algorithm) yields further speedup; the additional reconfiguration overhead is outweighed by more applications benefiting from coprocessing. Note that greedy algorithm results worsen for larger reconfiguration times. In contrast, note that the AG algorithm does well in all examined situations, including different application sequences and reconfiguration times, because AG adapts to both such items. The execution time obtained by AG is 1.4x and 1.2x faster than statically preloaded and greedy solutions, respectively. AG yields 3x speedup versus a microprocessor only solution.

Comparing with the Big FPGA lower bound (i.e., all coprocessors fit in the FPGA) shows that the AG algorithm obtains execution times on average within 16% of this lower bound, indicating that an improved algorithm could not improve beyond 16% (and in reality less as the lower bound is not generally achievable).

Figure 4 shows the decomposition of the runtimes for different algorithms for the large platform with the random input sequence. We logged the communication, reconfiguration, and algorithm overheads during simulation. The Big FPGA solution has the highest communication overhead, as all the applications run in *M+CP* mode and thus generate more memory accesses due to microprocessor/FPGA communication than when running in *M-only* mode (whose communication is not seen in the plot due to being off the top of the chart). Greedy has much reconfiguration overhead

Figure 4: Runtime overhead breakdown of algorithms for the Large platform and the random input sequence scenario (SRO, MRO, LRO: small, medium and large FPGA reconfiguration overhead).

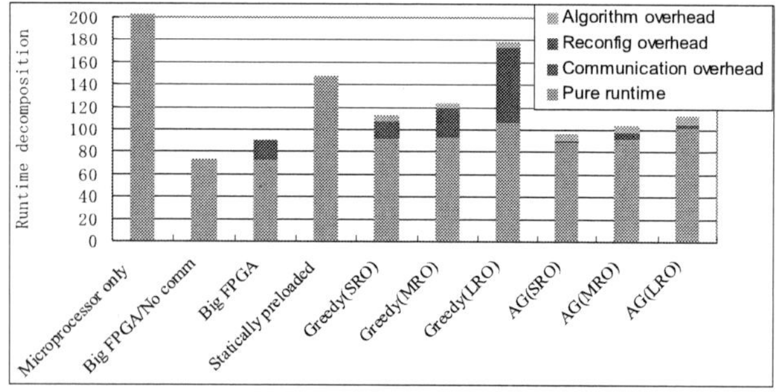

when the FPGA reconfiguration time is large, due to always trying to load a coprocessor. AG adapts to the reconfiguration time, and thus AG reconfigures less when reconfiguration overhead is large. AG also resists loading coprocessors with large communication overhead, resulting in low communication overhead. AG's runtime overhead was 5% of the total runtime for EEMBC/Powerstone benchmarks which usually lasted for several milliseconds. For longer running applications, AG's runtime overhead becomes negligible. Our results were averaged over 20 runs; AG's standard deviation was consistent with the Big FPGA solution and is stable across different FPGA reconfiguration times.

6. CONCLUSIONS

We defined the transmuting coprocessors (TC) problem and developed an intuitive online algorithm, Aggregate Gain (AG), to dynamically manage the loading of coprocessors into an FPGA. The AG algorithm is robust across various platforms, reconfiguration times, and input application sequences.

Acknowledgements: This research was supported in part by the National Science Foundation (CNS-0614957).

7. REFERENCES

[1] L. Bauer, M. Shafique, J. Henkel. Run-time instruction set selection in a transmutable embedded processor. DAC 2008.

[2] Embedded Microprocessor Benchmark Consortium. EEMBC. http://www.eembc.org/home.php

[3] S.C. Goldstein, H. Schmit, M. Budiu, S. Cadambi, M. Moe, R.R. Taylor, R. Laufer. PipeRench: a coprocessor for streaming multimedia acceleration. Computer Architecture, 1999.

[4] D. Gross, C.M. Harris. Fundamentals of queueing theory. John Wiley & Sons, Inc. New York, NY, USA. 1985.

[5] S. Hauck, T.W. Fry, M.M. Hosler, J.P. Kao The Chimaera reconfigurable functional unit. Very Large Scale Integration (VLSI) Systems, IEEE. 2004.

[6] E.L. Horta, J.W. Lockwood, D.E. Taylor and D. Parlour. Dynamic Hardware Plugins in an FPGA with Partial Run-time Reconfiguration. Design Automation Conference (DAC), 2002.

[7] C. Huang and F. Vahid. Dynamic Coprocessor Management for FPGA-Enhanced Compute Platforms. IEEE/ACM Int. Conf. on Compilers, Architectures, and Synthesis for Embedded Systems (CASES), Oct 2008.

[8] Intel QuickAssist Technology, http://www.intel.com/technology/platforms/quickassist/index.htm

[9] P. Lysaght, B. Blodget, J. Mason, J. Young, B. Bridgford. Invited paper: enhanced architectures, design methodologies and CAD tools for dynamic reconfiguration of Xilinx FPGAs. Field Programmable Logic and Applications, 2006.

[10] A. Malik, B. Moyer, D. Cermak. A Lower Power Unified Cache Architecture Providing Power and Performance Flexibility. Int. Symp. on Low Power Electronics and Design, 2000.

[11] S. McMillan, S.A. Guccione. Partial Run-Time Reconfiguration Using JRTR. Lecture Notes in Computer Science, 2000.

[12] J. Noguera, R.M. Badia. HW/SW codesign techniques for dynamically reconfigurable architectures. IEEE Trans. on Very Large Scale Integration (VLSI) Systems, 2002.

[13] G. Stitt, F. Vahid. Energy advantages of microprocessor platforms with on-chip configurable logic. Design & Test of Computers, IEEE, 2002.

[14] S. Trimberger. Scheduling designs into a time-multiplexed FPGA. Proceedings of the 1998 ACM/SIGDA sixth international symposium on Field programmable gate arrays, 1998.

[15] H. Walder, C. Steiger, M. Platzner. Fast Online Task Placement on FPGAs: Free Space Partitioning and 2D-Hashing. Parallel and Distributed Processing Symposium, 2003.

[16] Xilinx Virtex-5 FPGAs, http://www.xilinx.com

Dynamic Thread and Data Mapping for NoC Based CMPs

Mahmut Kandemir
Pennsylvania State University
University Park, PA
kandemir@cse.psu.edu

Ozcan Ozturk
Bilkent University
Turkey
ozturk@cs.bilkent.edu.tr

Sai P. Muralidhara
Pennsylvania State University
University Park, PA
smuralid@cse.psu.edu

ABSTRACT

Thread mapping and data mapping are two important problems in the context of NoC (network-on-chip) based CMPs (chip multiprocessors). While a compiler can determine suitable mappings for data and threads, such static mappings may not work well for multithreaded applications that go through different execution phases during their execution, each phase with potentially different data access patterns than others. Instead, a dynamic mapping strategy, if its overheads can be kept low, may be a more promising option. In this work, we present dynamic (runtime) thread and data mappings for NoC based CMPs. The goal of these mappings is to reduce the distance between the location of the core that requests data and the core whose local memory contains that requested data. In our experiments, we evaluate our proposed thread mapping and data mapping in isolation as well as in an integrated manner.

Categories and Subject Descriptors

D.3.4 [**Programming Languages**]: Processors-compilers

General Terms

Algorithms, Performance

Keywords

Thread, data, dynamic, mapping, NoC, CMP

1. INTRODUCTION

It is clear today that large, complex uniprocessors are no longer scaling performance-wise. This is because there is only a limited amount of parallelism that can be extracted from a single thread of execution, and even extracting this limited parallelism requires sophisticated superscalar instruction fetch, dependence check and instruction execution mechanisms. Unfortunately, it is not possible to continuously increase clock frequency of these architectures as it is well known that, beyond a point, power dissipation becomes a major bottleneck. In addition to these constraints, with extremely large numbers of transistors available on today's microprocessor chips, it is too costly to design and debug ever-larger and ever-complex processors in every two or three years. All these trends clearly

Permission to make digital or hard copies of part or all of this work for personal or classroom use is granted without fee provided that copies are not made or distributed for profit or commercial advantage and that copies bear this notice and the full citation on the first page. To copy otherwise, to republish, to post on servers or to redistribute to lists, requires prior specific permission and/or a fee.
DAC'09, July 26-31, 2009, San Francisco, California, USA

point that chip multiprocessors (CMPs) are in very good position to become the next generation mainstream computer architecture. The fact that there are many applications from diverse set of application domains that can take advantage of thread-level parallelism (e.g., embedded multi-media codes, data-intensive simulation programs) further motivates for CMPs.

Since future technologies offer the promise of being able to integrate billions of transistors on a chip, the prospects of having hundreds to thousands of cores on a single chip along with an underlying memory hierarchy and an interconnection system is entirely feasible. Once the number of cores on one chip passes some technology and architecture dependent threshold, point-to-point buses will no longer be a sufficient interconnect structure. Thus, it is reasonable to assume that future CMPs will employ an NoC (network-on-chip) in order to be able to handle the required communications between the cores in a scalable, flexible, programmable, and reliable fashion.

Clearly, assigning (mapping) threads and data manipulated by threads (of a multithreaded application) to CMP nodes is a challenging task for extracting maximum performance. One option is to assign threads and data to CMP nodes statically at compile-time and not to change this assignment throughout the execution of the application. While an optimizing compiler can certainly perform thread-data assignments, it is not possible to capture dynamic modulations in data access patterns when such static mappings are employed. Instead, a dynamic (runtime) mapping strategy, if its overheads can be kept low, may be more promising for data-intensive applications which go through multiple phases during execution.

Data and computation mapping for generic multiprocessors have been widely studied. From the data mapping angle, previously proposed approaches range from generic data partitioning to techniques specifically target NoC based systems. Balasundaram et al [3] propose a static performance estimator to guide data partitioning decisions for generic data partitioning. In [12], authors present a data layout selection tool that generates data layout specifications of distributed-memory machines automatically. Kim et al [13] introduce a dynamic and static algorithms to partition the cache among the threads. Panda et al [16] implement a data partitioning algorithm in embedded processor-based systems. In the multi-core domain, Guz et al [8] propose Nahalal, an architecture which partitions data according to its usage as shared versus private data, and locates the private data close to each processor. Kandemir [10] present an interprocessor data reuse vector based data locality optimization scheme for CMPs. Jin et al [9] implement a page-level data to L2 cache slice mapping for CMPs. Chishti et al [6] propose a data mapping scheme which utilizes replication and capacity allocation to neighboring processors within a CMP.

Pop and Kumar [17] use multi-threaded processors as resources in NoC based CMPs. They map the concurrent applications to multi-threaded processors of the CMP off-line. Chen et al [5] compare two thread scheduling techniques on CMPs, namely, Parallel Depth First and Work Stealing. In

978-1-60558-497-3/09 $25.00 © 2009 ACM

[2], authors propose a scheduling method for real-time systems targeted on CMPs. Chou et al [7] propose a run-time strategy for allocating the application tasks to platform resources in an NoC. Nuzzo et al [20] present task allocation strategies based on bin-packing algorithms. Kempf et al [11] propose and evaluate a task mapping framework in the context of heterogeneous chip multiprocessors.

In this work, our goal is to reduce data access latency by reducing the distance between the core that requests the data and the memory that holds the requested data. For this purpose, we first present an application-specific, dynamic thread assignment strategy for NoC based CMP systems. In this strategy, the job of thread-to-core assignment is given to a helper thread which carries out this task on behalf of the application and in parallel with the execution of application threads. Consequently, the resulting thread mappings are customized for an application and can change at runtime to adapt dynamic data sharing patterns exhibited by the threads of the application. We implemented this thread mapping strategy using a simulation platform and compared it to a static mapping scheme. The results from our experimental evaluation show that application-specific dynamic thread mapping can bring significant improvements in performance for all CMP sizes and applications we ran. More specifically, we achieve an average execution latency saving of 16.3% over static mapping.

Secondly, we present a dynamic data-to-core mapping scheme, also implemented using a helper thread. The results collected show that dynamic data mapping alone brings an improvement of 8.2% on average. More importantly, our integrated thread-data mapping scheme, which alternates between thread mapping and data mapping in successive mapping periods (epochs), takes these savings (over the static mapping) to 29.1% when all the applications are considered. Our results clearly demonstrate that to achieve the best performance both thread mapping and data mapping should be considered together. The results also indicate that the integrated scheme achieves consistent savings under different thread counts and CMP sizes.

2. EXECUTION MODEL

We assume an NoC based CMP architecture in which each node has a processor core, a private L1 cache (instruction and data) and a portion of the shared on-chip main memory space. Note that, while the on-chip memory space is distributed across CMP nodes, the entire memory space is logically shared, i.e., a core can access the memory attached to any node. It should be emphasized however the distance between the node of the requesting core and the node that holds the requested data element in its memory can make significant difference in performance (in data access latency), and in fact, our main goal in this work is to reduce this distance through dynamic thread mapping and data mapping across CMP nodes.

We assume that, when requested by an application, operating system (OS) gives a set of nodes (called *allocation* in this paper) to the requesting application and the application has full control in assigning its threads and data anywhere within its allocation. We further assume that the allocation for an application does not change during its execution, until it finishes. We also assume the existence of two types of migration support in this architecture: *thread migration* and *data migration*. The former moves a thread from one location in the allocation to another, whereas the latter moves a data block from one location to another within the allocation. While it is conceptually possible to modify an application code (i.e., thread bodies) to insert explicit "thread move" and "data move" instructions, this can have significant degradation in performance of the application. Therefore, in this work, we explore a different option which employs *helper threads* to migrate threads and data blocks across CMP nodes. Specifically, we use two helper threads, which are created at the time the

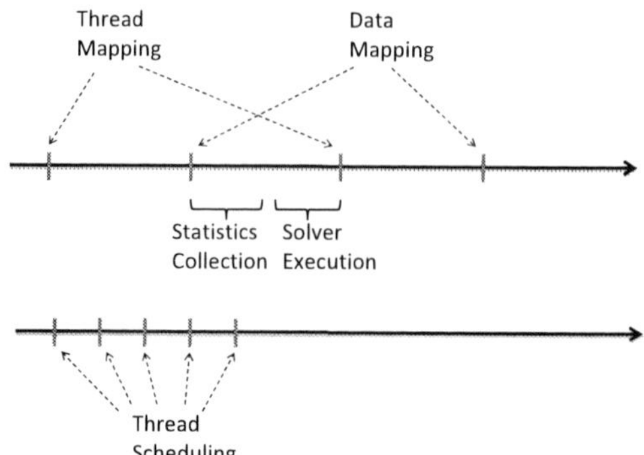

Figure 1: Thread migrations, data migrations and thread schedulings. Time flows from left to right.

application has been initiated. One of these threads moves application threads across the CMP nodes at runtime and the other moves data blocks. Note that a helper thread can execute in parallel with application threads. The overall goal of these helper threads is to reduce the distance between the core that requests a data block and the memory that holds the requested data block. While we focus on a two-dimensional (2D) mesh-based NoC (i.e., CMP nodes form a 2D mesh), our approach is applicable to other types of on-chip communication network topologies as well, as long as the target topology is exposed to our scheme. In the rest of this paper, we use the terms "mapping" and "migration" interchangeably. Basically, what we mean by "migrating a thread/data" is to "re-map" it from one location to another.

Figure 1 depicts how thread migrations, data migrations and thread schedulings take place on the timeline. Typically, thread scheduling (performed by the process scheduler in that node) is invoked much more frequently; in comparison, data and thread migrations (carried out by helper threads as explained above) are less frequent and they alternate (in our integrated scheme)—i.e., a set of thread migrations followed by a set of data migrations, which in turn are followed by a set of thread migrations, and so on. In each thread (or data) migration interval, the first half of the interval is used to collect runtime statistics, which indicate the frequency of accesses to each data block from each core. In the second half, an ILP (integer linear program) solver is invoked and the new thread (or data) mapping is determined and implemented. Note that the new mappings take place only after the ILP solver returns a solution, and at each step, the number of threads (or data blocks) to be migrated is dictated by the changes in the data access patterns of the threads between the previous interval and current interval.

In the following two sections, we describe the thread migration and data migration components of our approach. As mentioned above, in our integrated scheme, thread migrations and data migrations interleave at an (mapping) interval granularity. That is, at the end of the first interval (see Figure 1), threads are migrated; at the end of the second interval, data blocks are migrated; at the end of the third, threads are migrated again, and so on. For deciding both thread and data migrations (at each interval), we use the ILP based approaches explained below.

3. DYNAMIC THREAD MAPPING

We now present our ILP based formulation of dynamic thread mapping targeting 2D mesh NoC based CMP architecture. Our goal is to assign these threads to the CMP nodes such that the overall distance-to-data is reduced across all threads. To achieve this, our approach tries to place the threads that share a lot of data between them in close-by locations (cores) in the CMP.

Constant	Definition
$X \times Y$	Number of CMP nodes
T	Number of threads
$G(T)$	Graph representing thread affinities
$E(T)$	Edges in graph $G(T)$
$V(T)$	Vertices in graph $G(T)$
w_e	Weight of edge e or degree of affinity between the corresponding threads
C_{comm}	Unit cost of data access in the CMP

Table 1: The constant terms used in our dynamic thread mapping formulation. These are either architecture specific or program specific. $G(T)$ is built by collecting statistics during execution at each interval.

An integer linear program, or ILP for short, tries to solve a linear objective function via linear functional constraints along with integer solution variables. In 0-1 ILP, however, each solution variable is restricted to be either 0 or 1 [15]. Table 1 gives the constant terms used in our ILP formulation of the thread mapping problem. Although choice of the ILP tool is orthogonal to the focus of this paper, we formulated our ILP problem using a commercial tool, *Xpress-MP* [1]. Recall that in our execution model, the ILP solver is invoked by the helper thread. For the sake of our formulation, we view the CMP as a 2D grid, and assign the threads to the nodes of this 2D grid. Assume that the target CMP is composed of $X \times Y$ nodes, where X denotes the number of nodes on the x-axis and Y denotes the number of nodes on the y-axis. We denote a specific node in this grid using (x, y).

We use 0-1 integer variables R to indicate where the threads are mapped (once our solver returns a solution). More specifically,

- $R_{i,j,k}$: indicates whether thread i is running on (j, k).

The distance between two threads is captured using D_{t_1,t_2}. Specifically, we have:

- D_{t_1,t_2} : indicates the distance between threads t_1 and t_2.

We next give our constraints based on these definitions. A thread denoted by t can be assigned to a unique node in the CMP:

$$\sum_{j=1}^{X} \sum_{k=1}^{Y} R_{t,j,k} = 1, \quad \forall t. \tag{1}$$

We also need to make sure that only one thread is mapped to a node (x, y) (this assumption can be dropped if there are more threads than CMP nodes, or can be limited by an upper bound; in our implementation, we have a parameter that limits the number of threads per node, which is set to 4 in our default configuration):

$$\sum_{i=1}^{T} R_{i,x,y} \leq 1, \quad \forall x, y. \tag{2}$$

We use the Manhattan Distance to capture the cost of the placement of threads that share data. Manhattan Distance is given as the distance between two points measured along axes at right angles. In a plane with p_1 at (x_1, y_1) and p_2 at (x_2, y_2), it is $|x_1 - x_2| + |y_1 - y_2|$. We assume this distance to be the cost of the relative placements (on the CMP) of thread t_1 and thread t_2. To capture the Manhattan Distance, we use the variable, D_{t_1,t_2} and employ the following constraints:

$$D_{t_1,t_2} \geq (R_{t_1,j_1,k_1} + R_{t_2,j_2,k_2} - 1) \times (|j_1 - j_2| + |k_1 - k_2|) \tag{3}$$
$$\forall t_1, t_2, j_1, j_2, k_1, k_2.$$

Using the above constraint, we capture the distance between thread t_1 and thread t_2. Note that, R_{t_1,j_1,k_1} is 1 if t_1 is running on CMP node (j_1, k_1), whereas R_{t_2,j_2,k_2} is 1 if t_2 is running on CMP node (j_2, k_2). The role of the first term in the above constraint is to perform a logical 'and' on R_{t_1,j_1,k_1} and R_{t_2,j_2,k_2}. More specifically, if both of these 0-1 variables are equal to 1, then the first term in the constraint is equal to 1. We multiply

this term with the absolute distance between the two nodes given with (j_1, k_1) and (j_2, k_2). As can be observed from this expression, the first term given with $(R_{t_1,j_1,k_1} + R_{t_2,j_2,k_2} - 1)$ is going to be either 0 or negative if threads t_1 and t_2 are not running on the indicated CMP nodes. However, our ILP solver will pick the maximum value among all the possible D_{t_1,t_2} in order to satisfy all the constraints. Hence, we will have the distance in variable D_{t_1,t_2}.

Our helper thread employs a *weighted directed graph* to indicate the data sharing volume between threads. In this graph, each vertex represents a thread in the program and an edge between two vertices indicates the data sharing between the corresponding threads. An edge is annotated with a *weight* which is multiplication of the amount of data shared between the threads and the total number of references made to these shared data elements by the two threads. Note that this weight in a sense represents the *affinity* between the two threads. Clearly, two threads with high affinity should be as close to each other as possible in the 2D grid. In other words, placing two threads with high affinity in locations (on the CMP) far away from each other can incur high access latencies to data which should be avoided.

In mathematical terms, we map the given threads, T, to its graph representation $G(T)$, where $G(T) = V(T) \cup E(T)$ and $E(T) \subseteq V(T) \times V(T)$.

As stated above, each edge carries a weight value between the vertices of the graph. Since we have at this point covered mapping constraints and distance constraints, we can now give our objective function. Our objective function actually targets minimizing the distance between threads of high affinity and does not consider specific optimization problems such as power and performance. However, this formulation, if desired, can easily be converted to serve such specific goals. We express our objective function using the aforementioned graph structure as an input. More specifically, the cost can be captured, for each edge e in $G(T)$, using the distance between the two threads designated by the vertices connected by this edge. This distance is multiplied with the weight of the corresponding edge along with the unit data access cost, which is a constant. We can express this as follows:

$$Comm = \sum_{\forall e \in E(T)} w_e \times D_{ev1,ev2} \times C_{comm}. \tag{4}$$

In the above expression, we assume that C_{comm} is the unit cost of data access within the NoC. We multiply this with the volume of the data access, which is specified as w_e, i.e., the weight of the edge between two threads. Distance, $D_{ev1,ev2}$, captured in our previous constraints, is used to obtain the overall cost of placement of two threads. It is to be noted that e_{v1} and e_{v2} are the two vertices connected by edge e which actually represent the two threads. Based on these constraints, we can express our objective function as min $\quad Comm$.

4. DYNAMIC DATA MAPPING

Our goal in this section is to present an ILP formulation of the problem of minimizing data access latencies by determining the optimal placement of data blocks in the CMP. A data block in this section corresponds to a set of consecutive cache lines. Each dataset is assumed to be divided into *blocks* of equal size, which is the granularity of migration in our scheme across the memories attached to CMP nodes. Unless stated otherwise, we allow at most one copy of each data block to exist within the cumulative on-chip memory space (data cache can have its own copy). As previously stated, we assume a 2D grid CMP composed of $X \times Y$ nodes, and we represent a specific node in this 2D grid using (x, y). Table 2 gives the constant terms used in our ILP formulation of the data mapping problem.

We use A to indicate where a data block is assigned within the CMP (once the ILP solver return a solution). More specifically,

978-1-60558-497-3/09 $25.00 © 2009 ACM

Constant	Definition
$X \times Y$	Number of nodes
D	Number of data blocks
l_m	Size of a local memory of a core
$BSize$	Size of a data block
$CT_{i,j,k}$	Number of accesses to data block i by processor (j,k)
C_{tr}	Cost of moving a data block to an adjacent core
C_{acc}	Cost of accessing a data block from the local memory
$C_{off-chip}$	Cost of bringing a data block from off-chip memory

Table 2: The constant terms used in our dynamic data mapping formulation. These are either architecture specific or program specific. The values of $CT_{i,j,k}$ are obtained by collecting statistics at runtime.

- $A_{i,j,k}$: indicates whether data block i is in core (j,k).

Similarly, M is used in our formulation to identify whether a data block is in off-chip memory. Observe that the cumulative on-chip memory space may not be sufficient, in general, to store all the data blocks manipulated by the multithreaded program.

- M_i : indicates whether data block i is in off-chip memory.

The distances between a core and a data block is captured using $D_{i,j,k}$. Specifically:

- $D_{i,j,k}$: indicates the distance between data block i and core (j,k).

Based on these definitions, we can start describing our constraints. A data block given by d needs to be assigned to a unique core or off-chip memory:

$$M_d + \sum_{j=1}^{P} \sum_{k=1}^{Q} A_{d,j,k} = 1, \quad \forall j, k. \tag{5}$$

As before, the Manhattan Distance is a factor in determining the cost when data block d is accessed by processor core (x, y). This is also referred to as the *data access cost* in this section, and is the metric whose value we want to minimize. We want to remind the reader that a data block can be shared across multiple cores. To capture the Manhattan Distance, we use variables $D_{d,x,y}$ and accommodate the following constraints:

$$D_{d,x,y} \geq \sum_{j=1}^{X} \sum_{k=1}^{Y} A_{d,j,k} \times (|x - j| + |y - k|) \forall j, k. \tag{6}$$

In the above constraint, we capture the distance of data block d to the corresponding core denoted using (x, y). Note that, $A_{d,j,k}$ is used to indicate the location of the data block and we take the vertical and horizontal distances with the core (x, y). As can be seen from this expression, in the case of data not being stored in the on-chip memory, the distance value given with $D_{d,x,y}$ will be returned as 0. However, as we will explain in more detail, the off-chip memory access costs will be captured separately.

We also need to make sure that the local memory size (capacity), which is given by l_m, is not exceeded. This can be captured using the following constraint:

$$\sum_{i=1}^{D} A_{i,x,y} \times BSize \leq l_m \forall x, y. \tag{7}$$

This expression sums up the total amount of space required to keep the assigned data blocks within the local memory of core (x, y). $BSize$ indicates the data block size being used and may vary depending on the implementation and the data block granularity. Although it is possible to add a constraint to check the overall on-chip memory size, this is not necessary as we do it for individual local memories.

We next discuss our objective function. For clarity reasons we give our objective function in two main components. First, on-chip data access cost can be captured, for each data block access, using the distance between the location of that data

block and the location of the core accessing it multiplied with the cost of moving this data block within the chip, C_{tr}. On top of this cost, we have the local access cost which is given by C_{acc}. This will then be multiplied by the frequency of these accesses. As a result, we have:

$$TC_{on-chip} = \sum_{i=1}^{D} \sum_{j=1}^{X} \sum_{k=1}^{Y} ((D_{i,j,k} \times C_{tr} + C_{acc}) \times CT_{i,j,k}). \tag{8}$$

In a similar fashion, we can compute the off-chip memory access cost using the following expression:

$$TC_{off-chip} = \sum_{i=1}^{D} \sum_{j=1}^{X} \sum_{k=1}^{Y} ((M_i \times C_{off-chip} + C_{acc}) \times CT_{i,j,k}). \tag{9}$$

In the above expression, we assume that $C_{off-chip}$ captures the cost of bringing the data block to the local memory of the requesting core. That is, we assume that each core can access the off-chip memory with uniform latency. Alternatively, if the accesses are not uniform for some reason, one can introduce individual off-chip access costs for each core such as $C_{x,y}$ for off-chip access cost of core (x, y). Thus, our overall objective function can be written as:

$$\min \quad TC = TC_{on-chip} + TC_{off-chip}. \tag{10}$$

Within each interval (at the end of which we want to migrate data blocks), the helper thread collects access statistics for each data block, build the ILP formulation summarized above, and invokes the ILP solver to solve it. Based on this solution, the required data block migrations are performed by the helper thread.

Data Replication. Recall that so far in our ILP formulation, we assumed that there is only one copy of each data block, either on-chip memory or off-chip memory. We can expand the scope of our formulation by allowing data block replication within the local memories if this helps to reduce the total data access costs. We replicate a data block only if the block is read-only (to avoid coherence-related issues). In order to reflect these changes on our baseline formulation, we modify the constraint given in Expression 5 as follows:

$$M_d + \sum_{j=1}^{X} \sum_{k=1}^{Y} A_{d,j,k} \geq 1, \quad \forall j, k. \tag{11}$$

This modified constraint enables us have multiple copies of the data block in different local memories, i.e., for a specific data block d, we can have multiple $A_{d,j,k} = 1$. However, this requires additional modifications such as selecting the closest data block in case of multiple data blocks residing within the local memories. In order identify the closest data block, we define an additional 0-1 variable, $N_{d,j,k,l,m}$. More specifically:

- $N_{i,j,k,l,m}$: indicates whether core (j,k) can access the closest data block i from core (l, m).

A core (j, k) can only access a data block d if it resides in the said core (x, y), therefore:

$$N_{d,j,k,x,y} \leq A_{d,x,y}, \quad \forall j, k. \tag{12}$$

Also, we have to make sure that there is one and only one closest data block d for a core (j, k). Note that, if the data block is not located within local memories, then the closest location is the off-chip memory, which can be expressed as M_d. In mathematical terms:

$$M_d + \sum_{l=1}^{X} \sum_{m=1}^{Y} N_{d,j,k,l,m} = 1, \quad \forall j, k. \tag{13}$$

In addition to these constraints, we have to modify Expres-

NoC topology	5×5 2D mesh
Core (fetch,issue,retire width)	(4,2,2)
L1 cache	16KB per node, 2-way, 2 cycle
On-chip memory	256KB per node, banked, 10 cycle
Scheduling frequency	50 msec
Mapping frequency	500 msec
Data block size	512 bytes

Table 3: Our simulation parameters and their default values.

Program	Application Domain	Dataset Size (MB)	Execution Latency (sec)
radiosity	Graphics	6.44	7.21
radix	General	5.68	6.17
raytrace	Graphics	4.96	5.89
volrend	Graphics	7.26	8.73
water	Simulation	5.17	5.37

Table 4: Benchmark codes used in our evaluation. The last column gives the execution latency values for the baseline static (thread and data) mapping scheme.

sion 6 accordingly:

$$D_{d,x,y} \geq \sum_{l=1}^{X} \sum_{m=1}^{Y} N_{d,x,y,l,m} \times (|x - l| + |y - m|) \forall l, m. \quad (14)$$

As explained earlier, we only allow a certain number of copies for a given data block, which is τ. This requirement can be enforced using:

$$\sum_{j=1}^{X} \sum_{k=1}^{Y} A_{d,j,k} \leq \tau, \quad \forall d. \quad (15)$$

In the above expression, the total number of copies of data block d stored within the on-chip memory space is captured with the *sum*. Finally, our objective function give earlier remains the same since both $TC_{on-chip}$ and $TC_{off-chip}$ are the same as in our original ILP formulation.

5. EXPERIMENTAL EVALUATION

The baseline thread and data mapping used in our experiments is a static one, where a compiler determines appropriate data-thread mapping statically and these mappings do not change during the course of execution. Note that, this static scheme is in fact derived from the ILP based formulations explained in Sections 3 and 4. Specifically, we analyze the entire application and record the number of accesses made by threads to data blocks and the affinities between thread pairs (for the entire execution). Once these counts have been obtained, we first execute our thread mapping scheme and then our data mapping scheme. Note that, this baseline static mapping scheme can also be seen as a specific instance of the dynamic mapping scheme where the mapping interval is set to the entire execution period. In the results presented in the next subsection, all values are given as *normalized* with respect to this static scheme. In addition to this static scheme, we conducted experiments with the following dynamic schemes:

- TM. In this scheme, data mapping is decided statically by the compiler (as in the static mapping case) and only dynamic thread mapping is applied at runtime.

- DM. This is in a sense the opposite of scheme TM. In this scheme, thread mapping is fixed at compile time (as in the static mapping case) and only dynamic data mapping is exercised at runtime.

- TM+DM (C). This is our integrated scheme which exercises both data and thread mapping at runtime. The initial data and thread mappings are decided by the compiler. And at runtime, we use data and thread mappings in alternate fashion as explained earlier.

- TM+DM (R). This is similar to the previous scheme except that the initial thread and data mappings are random, that is, threads and data blocks are assigned to the CMP nodes randomly at compile time but we employ the integrated scheme at runtime.

All these mapping schemes are implemented on top of SIM-ICS [14] and have been tested using the platform summarized in Table 3. Recall that the target architecture is a shared memory based one, i.e., although each core has a portion of the on-chip memory space (as its local memory), all on-chip memory space are accessible (albeit with different costs) by all cores. Table 4 on the other hand lists the benchmark codes used in this study. We selected five parallel benchmark codes from

the Splash-2 benchmark suite [19] and executed them under all the mapping schemes discussed above. For each benchmark code in our experimental suite, 100 threads are generated and mapped onto 25 CMP nodes (we later present results with different number of threads and different number of CMP nodes). Finally, in all the experiments we made, the default node-level thread scheduling scheme is round-robin (clearly, if there is only one thread per node, no scheduling is needed).

Figure 2 gives the normalized execution latencies under different mapping policies. Our first observation is that while both TM and DM generate better results than the static thread-data mapping, the integrated scheme generates the most savings for all the benchmarks. The second observation is that the difference between TM+DM (C) and TM+DM (R) is not significant, meaning that initial mapping may not matter too much if the dynamic re-mapping is activated at runtime. We also see that TM performs better than DM. This is mainly because, in general, a data block can be shared by more than one thread and positioning threads with respect to that data turns out to be easier than finding a good location for that data. The average execution latency improvements brought by TM, DM, TM+DM (C) and TM+DM (R) over the static scheme are 16.3%, 8.2%, 29.1% and 26.7%, respectively.

The sensitivity of these mapping schemes to the number of threads is presented in Figure 3. Each bar in this plot represents the average (normalized) execution latency across all five benchmarks. Recall that the default number of threads was 100 for our benchmarks. We see from these results that the integrated scheme consistently generates good results for all the thread counts considered. The results are a bit higher with larger thread counts as it gives more flexibility to our mapping scheme. Figure 4 on the other hand gives the sensitivity of the mapping schemes with respect to the number of nodes. In addition to our default 5×5 mesh, we experimented with 4×4 and 6×6 mesh sizes (and the number of threads is fixed at 100 in this experiment). The average execution latency results are presented in Figure 4 indicate that all dynamic mapping schemes achieve slightly better results with larger meshes as larger meshes give more flexibility to a dynamic approach in moving data and threads across CMP nodes.

We next study the sensitivity of our dynamic mapping schemes to the mapping frequency. Recall from Table 3 that the mapping frequency used in experiments so far was 500 msec. The results with mapping frequencies of 100 msec, 200 msec, 800 msec, and 1000 msec are shown in Figure 5. Maybe the most important observation from these results is that working with very low or very high frequencies may not be the best option. Though the results change from one benchmark to another, we see that, for each benchmark, there is an ideal value (among the frequency values tested) that generates the maximum improvement in execution latency.

So far in our experiments no data block is replicated across CMP nodes, i.e., each data block has exactly one copy in the on-chip memory space. Figure 6 plots the results when each data block is replicated n times (x-axis), where $1 \leq n \leq 5$. Note that, when n is 1, we have no replication. We note that, as may be expected, the DM scheme takes advantage of block replication. This also reflects on, to varying extent, other schemes as well. However, we also witness a reduction in improvements (in some cases) as we move from $n = 4$ to $n = 5$. This is because too much replication reduces effective

978-1-60558-497-3/09 $25.00 © 2009 ACM

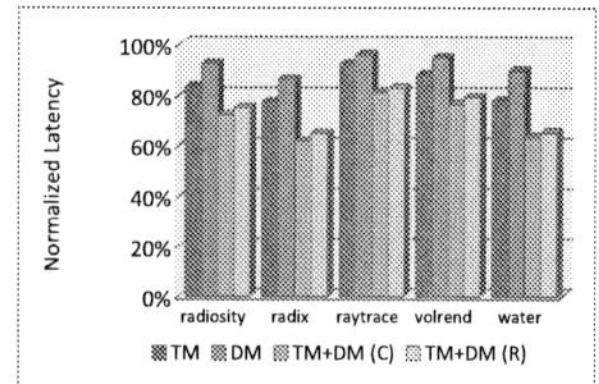

Figure 2: Normalized execution latencies under different mapping policies.

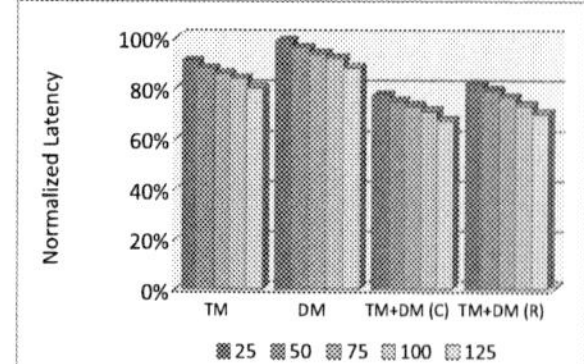

Figure 3: Sensitivity of the mapping schemes to the number of threads.

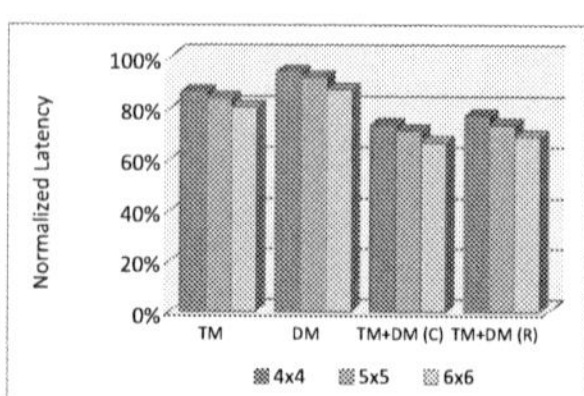

Figure 4: Sensitivity of the mapping schemes to the number of nodes.

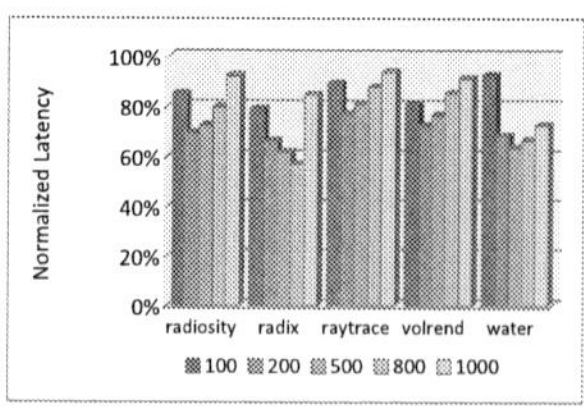

Figure 5: Sensitivity of the dynamic mapping schemes to the mapping frequency.

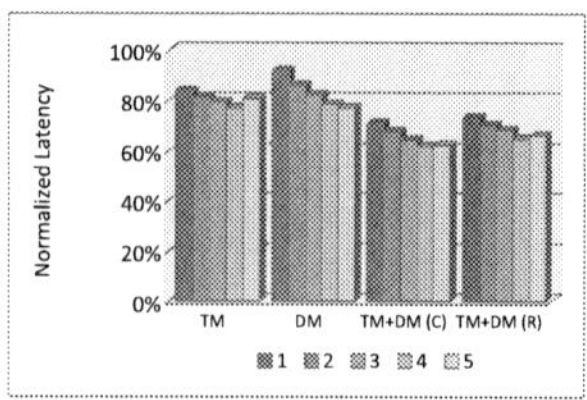

Figure 6: Results when each data block is replicated n times (x-axis), where $1 \leq n \leq 5$.

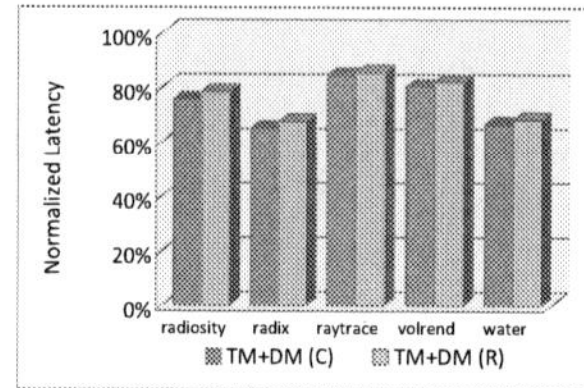

Figure 7: Results with an alternate mapping invocation scheme.

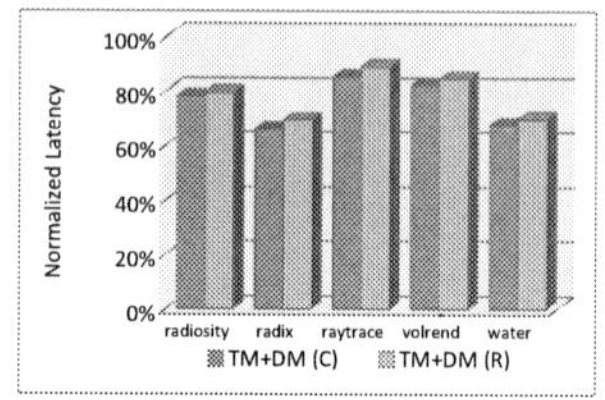

Figure 8: Impact of the shape of the allocation.

capacity of on-chip memory space, and this increases average access latency. Ideally, one may want a dynamic scheme that selects the best n value at runtime; exploring such a scheme is in our future research agenda.

We now change the order in which thread mapping and data mapping are invoked at runtime. Recall that our default approach is to invoke first thread migration and then data migration, and continue with this pattern until the end of execution. Figure 7 illustrates the results with an alternate scheme which invokes thread mapping twice followed by two invocations of data mapping, which in turn is followed by two invocations of thread mapping, and so on. Compared to the results in Figure 2, the results in Figure 7 are slightly worse. Clearly, there are many other potential invocation patterns and we plan to explore this issue further in our future work. We want to mention here however invoking both thread mapping and data mapping in the same interval did not help too much as it caused conflicting migrations for threads and some data blocks.

In our final set of experiments, we study the impact of the shape of the allocation (as defined in Section 2). All the results presented so far have been collected assuming a 2D mesh allocation (5×5 by default). We also performed experiments when the set of nodes allocated to an application is selected randomly. More specifically, we assumed a set of 25 nodes are assigned to an application randomly from a 6×6 mesh based CMP. The values of all the other experimental parameters are as given in Table 3. The results presented in Figure 8 show that the integrated scheme generates significant improvements in this scenario as well. In fact, compared to the savings shown in Figure 2, those in Figure 8 are only slightly worse.

6. CONCLUSIONS

As processor design has become severely power limited, it is now commonly accepted that staying on the current performance trajectory will come about through the integration of multiple processors on a chip rather than through increases in the clock rate of single processors. As a result, several manufacturers already have CMP architectures on the market, and we can expect future CMPs to employ NoC for scalable and reliable communication. In this paper, we have presented dynamic thread and data mapping schemes for NoC based CMPs and evaluated their performance. The results from our experimental evaluation show that application-specific thread mapping can bring significant improvements in performance for all CMP sizes and applications we tested.

7. ACKNOWLEDGMENTS

This work is supported in part by NSF grants 0811687, 0720645, 0720749, 0702519, a grant from Microsoft Research and a grant from GSRC.

8. REFERENCES

[1] Xpress-mp. *http://www.dashoptimization.com/pdf/Mosel1.pdf*, 2002.

[2] J. H. Anderson and J. M. Calandrino. Parallel real-time task scheduling on multicore platforms. In *Proc. of RTSS '06*.

[3] V. Balasundaram et al. A static performance estimator to guide data partitioning decisions. In *Proc. of PPOPP '91*.

[4] G. Chen et al. Application mapping for chip multiprocessors. In *Proc. of DAC '08*.

[5] S. Chen et al. Scheduling threads for constructive cache sharing on cmps. In *Proc. of SPAA '07*.

[6] Z. Chishti et al. Optimizing replication, communication, and capacity allocation in cmps. In *Proce. of ISCA '05*.

[7] C.-L. Chou and R. Marculescu. User-aware dynamic task allocation in networks-on-chip. In *DATE*, pages 1232–1237, 2008.

[8] Z. Guz et al. Utilizing shared data in chip multiprocessors with the nahalal architecture. In *Proc. of SPAA '08*.

[9] L. Jin et al. A flexible data to l2 cache mapping approach for future multicore processors. In *Proc. of MSPC '06*.

[10] M. Kandemir. Data locality enhancement for cmps. In *Proc. of ICCAD '07*.

[11] T. Kempf et al. A modular simulation framework for spatial and temporal task mapping onto multi-processor soc platforms. In *DATE*, pages 876–881, 2005.

[12] K. Kennedy and U. Kremer. Automatic data layout for distributed-memory machines. *ACM TOPLAS*, 20(4):869–916, 1998.

[13] S. Kim et al. Fair cache sharing and partitioning in a chip multiprocessor architecture. In *Proc. of PACT '04*.

[14] P. S. Magnusson et al. Simics: A full system simulation platform. *IEEE Computer*, 35(2):50–58, 2002.

[15] G. L. Nemhauser and L. A. Wolsey. *Integer and combinatorial optimization*. Wiley-Interscience, New York, NY, USA, 1988.

[16] P. R. Panda et al. On-chip vs. off-chip memory: the data partitioning problem in embedded processor-based systems. *ACM Trans. Des. Autom. Electron. Syst.*, 5(3):682–704, 2000.

[17] R. Pop and S. Kumar. Mapping applications to noc platforms with multithreaded processor resources. *NORCHIP Conference, 2005. 23rd*, pages 36–39, Nov. 2005.

[18] V. Suhendra et al. Integrated scratchpad memory optimization and task scheduling for mpsoc architectures. In *Proc. of CASES '06*.

[19] S. C. Woo et al. The splash-2 programs: characterization and methodological considerations. In *Proc. of ISCA '95*.

[20] J. Zhuo and C. Chakrabarti. System-level energy-efficient dynamic task scheduling. In *Proc. of DAC '05*..

978-1-60558-497-3/09 $25.00 © 2009 ACM

A Commitment-based Management Strategy for the Performance and Reliability Enhancement of Flash-memory Storage Systems

Yuan-Hao Chang
Graduate Institute of Networking and Multimedia
Department of Computer Science and
Information Engineering
National Taiwan University
Taipei 106, Taiwan (R.O.C.)
d93944006@ntu.edu.tw

Tei-Wei Kuo*
Department of Computer Science and
Information Engineering
Graduate Institute of Networking and Multimedia
National Taiwan University
Taipei 106, Taiwan (R.O.C.)
ktw@csie.ntu.edu.tw

ABSTRACT

Cost has been a major driving force in the development of the flash memory technology, but has also introduced serious challenges on reliability and performance for future products. In this work, we propose a commitment-based management strategy to resolve the reliability problem of many flash-memory products. A three-level address translation architecture with an adaptive block mapping mechanism is proposed to accelerate the address translation process with a limited amount of the RAM usage. Parallelism of operations over multiple chips is also explored with the considerations of the write constraints of multi-level-cell flash memory chips.

Categories and Subject Descriptors

D.4.2 [**Operating Systems**]: Storage Management—*Secondary storage*; D.4.7 [**Operating Systems**]: Organization and Design—*Real-Time Systems and Embedded Systems*

General Terms

Design, Experimentation, Management, Measurement, Performance, Reliability

Keywords

Secondary storage, embedded systems, flash memory, performance, reliability

1. INTRODUCTION

*This work was supported in part by the NSC under grant No. 97-EC-17-A-01-S1-034 and 97-2221-E-002-206-MY3, by the Excellent Research Projects of National Taiwan University under grant No. 97R0062-01, and by the Ministry of Economic Affairs under grant No. 96-EC-17-A-01-S1-034 in Taiwan.

Permission to make digital or hard copies of part or all of this work for personal or classroom use is granted without fee provided that copies are not made or distributed for profit or commercial advantage and that copies bear this notice and the full citation on the first page. To copy otherwise, to republish, to post on servers or to redistribute to lists, requires prior specific permission and/or a fee.
DAC'09, July 26-31, 2009, San Francisco, California, USA

Flash memory is widely adopted in various storage systems, and its applications have grown much beyond its original designs[1]. Well-known examples are flash-memory cache of hard drives (known as TurboMemory), fast booting devices for Microsoft Windows Vista, and solid-state disks (SSD) (for the replacement of hard drives). In recent years, multiple-level-cell (MLC) flash memory has become a major player in the market of (lower cost) large-scale storage systems, despite its weakness on reliability and performance (compared to single-level-cell (SLC) flash memory)[2]. While the capacity of MLC flash memory grows dramatically, the performance and reliability problems of MLC flash memory are also quickly exaggerated. In addition, the appearing of new write constraints of MLC flash memory, e.g., the prohibition of partial programming, results in efficiency (and even feasibility) problems of many flash-translation-layer (FTL) designs. Such observations motivate this work in the proposing of new management strategies for performance and reliability enhancement.

A NAND flash memory chip might consist of multiple planes, and every plane is composed of blocks with one read/write buffer. Each block is of a fixed number of pages, and a block is the smallest unit for erase operations. A page, the basic unit for read and write operations, contains a user area and a spare area, where the user area is for data storage, and the spare area stores *out of band (OOB) data*, e.g., logical block addresses (LBAs), status flags, and error detection code (EDC) and/or error correction code (ECC). When a page is written, the space is no longer available unless it is erased. As a result, "out-place update" is adopted so that data are always written onto free pages. Pages that contain the effective(/latest) copy of data are considered as "live" (or valid), and those with old data versions are "dead" (or invalid).

In the past decade, researchers have explored different system architectures and layer designs [14, 18], such as NAND flash memory with a SRAM cache. Some implemented native file systems such as JFFS2 and YAFFS over raw flash-memory media [9, 17]. Some others proposed to adopt a write buffer in flash-memory devices to improve the write performance [12, 19]. Note that write buffers might damage the integrity of file systems, due to power failures. Different from the adoption of write buffers, Chang et al. [7] proposed an adaptive striping architecture (referred to as ASA) to access different flash planes or chips in parallel so as to improve the read and write performance over flash memory. However, the

---

[1]In this paper, we focus our discussions on NAND flash.
[2]Each cell of SLC(/MLC$_{\times n}$) flash memory contains 1(/n)-bit information.

proposed translation mechanism is based on page-level mapping and might need a large amount of RAM for the maintenance of the translation table, and it does not consider the reliability problems and the write constraints of MLC flash memory.

This work is motivated by the needs in the designs of flash-memory management strategies that meet the reliability and performance requirements of flash-memory storage systems. In particular, we propose a commitment-based management strategy to resolve the reliability problem of many flash-memory products. A three-level address translation architecture with an adaptive block mapping mechanism is proposed to accelerate the address translation process with a limited amount of the RAM usage. The proposed mechanism could also support crash recovery over MLC flash memory without the support of version number, i.e., a unique serial number for each write operation or each version of an LBA. Parallelism of operations over multiple chips is explored with the considerations of the write constraints of advanced multi-level-cell flash memory. The proposed approach is then evaluated with the distributions of bad blocks and bit errors. A series of experiments is conducted based on realistic workloads. We show that the reliability and performance of a multi-chipped flash memory storage system can be significantly improved with limited RAM usage, compared to existing popular flash-translation-layer designs.

The rest of this paper is organized as follows: Section 2 presents the system architecture and the motivation of this work. In Section 3, a commitment-based management strategy is proposed. Section 4 summarizes the experiment results on the performance and reliability enhancement with the RAM-usage analysis. Section 5 is the conclusion.

2. SYSTEM ARCHITECTURE AND MOTIVATION

Flash memory has different block size configurations: Each page of small-block(/large-block) SLC flash memory can store 512B(/2KB) of data and 16B(/64B) of spare information, and there are 32(/64) pages per block. For small-block(/large-block) MLC$_{\times 2}$ flash memory, each page can store 2KB(/4KB) of data and 64B(/128B) of spare information, and each block consists of 128 pages. A low-cost flash-memory device usually adopts MLC$_{\times n}$ flash memory chips as its storage media because of its low unit price (Please see Table 1). However, *pages in MLC$_{\times n}$ flash memory usually can only be written sequentially in a block, while pages in SLC flash memory can be written randomly within a block. Furthermore, MLC$_{\times n}$ flash memory usually makes partial page write/programming impossible even though it can be done over SLC flash memory [10], so that it is impossible to mark a page of MLC flash memory as invalid by modifying the flag in its spare area.* MLC$_{\times 2}$ flash memory has lower endurance and has a higher possibility to develop additional bad blocks than SLC flash memory. The write and random read performance of MLC$_{\times n}$ flash memory is also much worse than that of SLC flash memory, as shown in Table 1.

Type & Part Number	SLC	MLC$_{\times 2}$
Endurance (Erase Cycles)	100,000	10,000
Price (US dollars/GB, 2008 Q1)[11]	7.10	2.48
Serial Access	25ns	25ns
Random Read	20μs	60μs
Write/Program	200μs	800μs
Erase	1.5ms	1.5ms
Bit error probability	1×10^{-9}	1×10^{-6}
Error rate	1 bit every 119MB	1 bit every 100KB

Table 1: Comparisons Between 8Gbit SLC and 8Gbit MLC Flash.

A typical system architecture of flash-memory storage systems has two major layers, as shown in Figure 1: The Flash Translation Layer (FTL) driver and the Memory Technology Device (MTD) driver. The FTL driver usually consists of *Allocator, Cleaner,* and *Wear Leveler* for flash memory management: The Allocator handles address translations between Logical Block Addresses (LBA) and Physical Block Addresses (PBA). LBAs are addresses of sectors as requested by the operating system, where the size of each sector is usually 512B. In a multi-chipped flash-memory storage system, each PBA has three parts ($chip, block, page$), where $chip$ denotes its chip number. The Cleaner is to do garbage collection to reclaim invalid pages. The Wear-leveler is an optional component that is to do wear leveling so as to extend the lifetime of flash memory. Note that native flash file systems, e.g., JFFS2 or YAFFS, combine the file system layer and the FTL driver together. The MTD driver provides primitive functions, such as read, write, and erase, over flash memory. A multi-chipped MTD driver does not block and wait for the completion of any read, write, or erase operation, since several chips are available and can be accessed concurrently. Instead of adopting several independent chips in a multi-chipped flash-memory device, low-cost flash-memory devices(/cards) usually adopt one large-scale chip, which is composed of several flash-memory sub-chips stacked together with chip selects, and an access to one sub-chip does not interfere with any operations to the other sub-chips in the same chip.

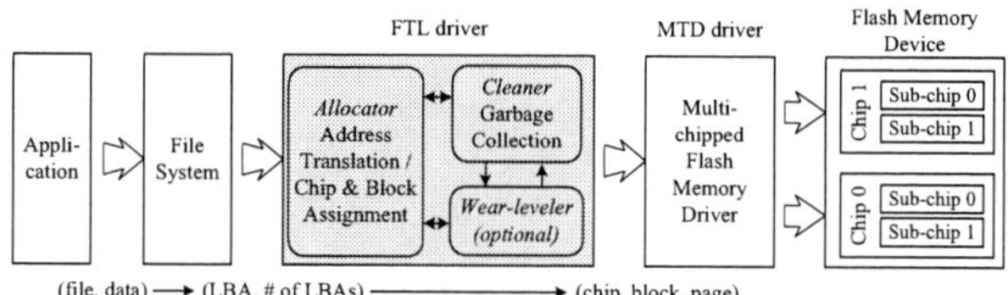

Figure 1: System Architecture

There are three major implementations of the flash-translation-layer (or its management strategies): *FTL, BL,* and *NFTL* [1, 4, 2, 3, 13]. FTL adopts a page-level address translation mechanism, where an address translation table is used to map any given LBA to its corresponding PBA [2, 4]. BL adopts a block-level translation mechanism. Each LBA is divided into a virtual block address (VBA) and a block offset (page number) under BL, and is mapped to its PBA by a translation table. Data are written to pages of their corresponding PBAs and block offsets. If any of the corresponding pages of a block is already used, or the MLC write constraints are violated, then BL must allocate a new block to write the corresponding data and copy all of the valid data of the previous block to the new block. Similar to BL, each LBA under NFTL is also divided into a VBA and a block offset [3]. A VBA can be translated to a (primary) physical block address by the block-level address translation. When a write request is issued, the data of the write request is written to the page with the corresponding block offset in the primary block. Since the following write requests can not overwrite the same pages in the primary block, a replacement block is needed to handle subsequent write requests, and the data of the (overwritten) write requests are sequentially written to the replacement block.

Although MLC flash memory is especially suitable for large-scale storage systems due to its cost-effectiveness, such as those up to 16GB or above, it suffers from worse read and write performance, compared to SLC flash memory. The situation is exaggerated with the appearing of MLC$_{\times 4}$ flash-memory. Most of the existing designs of FTL/MTD drivers, such as well-known NFTL

978-1-60558-497-3/09 $25.00 © 2009 ACM

and FTL [1, 2, 3], often invalidate a page by updating the status bit in the space area of a page. It is usually assumed that data can be written to pages randomly within a block. Such assumptions and designs are no longer possible because of new constraints on the write operations of MLC flash memory, due to its unique characteristics. Furthermore, the error rate of MLC flash memory is around 1000 times of that of SLC flash memory, as shown in the last two rows of Table 1. Such a phenomenon requires better design on error correction and data recovery for MLC-based flash-memory storage systems.

This research is motivated by the needs of performance and reliability enhancement for low-cost flash-memory storage systems with the considerations of physical constraints of MLC flash memory. Our goal is to consider the designs of low-end products in which there could be even no hardware multi-channels or hardware designs for advanced error correction coding (ECC). At the same time, we should also consider the scalability issue due to the fast-growing capacity of flash memory, where serious challenges in restricted main-memory usage and limited computing power of embedded systems are faced because of cost considerations. We are interested in low-cost multi-chipped MLC flash-memory storage systems that are considered being disposable by users. The technical problem is how to explore the parallelism of multiple chips with the considerations of the write constraints and the low reliability problem of MLC flash memory.

3. COMMITMENT-BASED MANAGEMENT STRATEGY

In this section, a commitment-based management strategy is proposed to improve the performance and reliability of flash-memory storage systems. In particular, we propose a three-level address translation architecture with a log-based strategy for each read or write request to locate its corresponding physical page sets (Section 3.1). An adaptive block mapping mechanism between (virtual) blocks seen by the users and physical blocks of the flash memory is presented to adapt to random and sequential accesses (Section 3.2).

3.1 A Three-level Address Translation

3.1.1 A Log-based Strategy and Its Flash Layout

We propose to partition a multi-chipped flash-memory storage system into a fixed number of physical regions and a spare-block area, as shown in Figure 2. *Physical regions* are used to store data, and spare blocks are reserved to replace bad blocks of physical regions with good blocks whenever it is needed. Each physical region contains sets of physical blocks, referred to as *physical block sets*, where a physical block set is the set of physical blocks with the same offset of their corresponding chips. Each physical block set is partitioned exclusively into sets of physical pages, referred to as *physical page sets*, where physical pages of the same physical page set have the same offset in their corresponding blocks. The size of a physical page set is the unit of data written to or read from the flash memory. The physical pages of the last physical page set of each physical block set are referred to as *summary pages*, and other physical pages are referred to as *data pages*.

Each physical page set is stored with data of consecutive LBAs, and the physical page sets of a physical block set are sequentially written from the first physical page set to the last one. When all of the data pages of a physical block set are written, the summary pages of the physical block set are written to save the LBA translation information and the parity-check information. The purpose of summary pages is to enhance the performance on address trans-

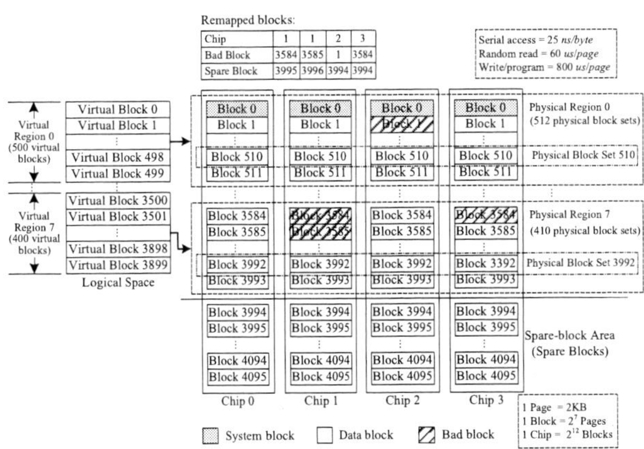

Figure 2: Mapping Between Logical Addresses and Physical Addresses

lation and the ability to recover from flash-cell failures (with log-based writes). However, if a physical block set is not filled up before the system turns to write other LBAs, data of the later LBAs are written to another empty physical block set. The LBA translation information and the parity-check information of the former physical block set are cached in the main memory until the former physical block set is filled up, or the cache is full. If the former physical block set is filled up, then the corresponding summary pages are written accordingly. If the cache is full before the former physical block set is filled up, then the rest of the former physical block set is left empty, and the corresponding summary pages are written accordingly. When the summary pages of a physical block set are written, the physical block set is said being "*committed*". Note that the spare area of each data page and each summary page is to store the house-keeping information, including *LBA*, *old block set*, *ECC*, and *EDC*. The LBA field stores the first LBA of an LBA range of the stored data. The old block set field is to save the information of the previous physical block set allocated for the same LBA range of this physical block set, so that the proposed strategy can support crash recovery without the help of version numbers (Please see Section 3.2.1). ECC is to correct errors in a page. EDC is to detect errors in the house-keeping information to facilitate the reconstruction of translation tables.

The logical LBA space of a multi-chipped flash-memory storage system is also divided into *virtual regions*. Each virtual region is partitioned into *virtual blocks*, and there is a one-to-one mapping from virtual regions to physical regions based on their physical locations on the chips (and their LBAs). In order to improve the write performance, each virtual block could be mapped to at most two physical block sets, where they correspond to the current and previous versions (Please see Section 3.2.1 for details). A virtual block consists of a fixed number of *virtual pages*, and the number of virtual pages of a virtual block is equal to the number of data page sets of a physical block set. The number of LBAs in a virtual page is equal to the number of sectors that can be stored in the data pages of a physical page set. Data of each virtual page of a virtual block can be stored in any data page set of the corresponding physical block sets. Note that the number of virtual blocks of a virtual region might not be the same as the number of physical block sets of the corresponding physical region because some bad physical blocks must be masked by spare physical blocks, and extra physical blocks are needed to facilitate data updates over blocks for better performance. As shown in Figure 2, the flash-memory storage system consists of four 1GB MLC flash-memory chips. Each phys-

978-1-60558-497-3/09 $25.00 © 2009 ACM

ical region, except the last one, is composed of 512 physical block sets because some spare physical blocks are reserved (to replace bad physical blocks in physical regions) right after the last physical region. The block set 0 is reserved to store the system information because the first block of a flash-memory chip is 100% guaranteed being a valid block with at least one-thousand erase cycles in many products. In the logical LBA space, each virtual region except the last region consists of 500 virtual blocks and is mapped to a physical region, and each virtual block can be mapped to any physical block set of the corresponding physical region.

3.1.2 Address Translation Architecture

Given an LBA, the physical location is derived by a three-level address translation architecture, as shown in Figure 3: The virtual block address (VBA) of a given LBA is obtained by dividing the LBA with the number of LBAs per virtual block, and the virtual page address (VPA) is the quotient in dividing the remainder of the former division with the number of LBAs per virtual page. The virtual region address (VRA) of the LBA is derived by dividing its VBA with the number of virtual blocks per virtual region. The remainder of the last division is called the offset of the corresponding virtual block, referred to as its virtual block offset (VBO), relative to the first virtual block of its virtual region. As shown in Figure 3, three types of translation tables are used to serve as an index table to look up the address translation information of an given LBA in a multi-level paging fashion, referred to as the *virtual region table*, *virtual block table*, and *virtual page table*, respectively. In addition to the translation tables, each physical region is associated with a *block remapping table* (to replace bad physical blocks with spare physical blocks) and a *block remapping list* (to accelerate the finding of remapped physical blocks). Here a *bad physical block set* is a physical block set that contains one or more bad physical blocks.

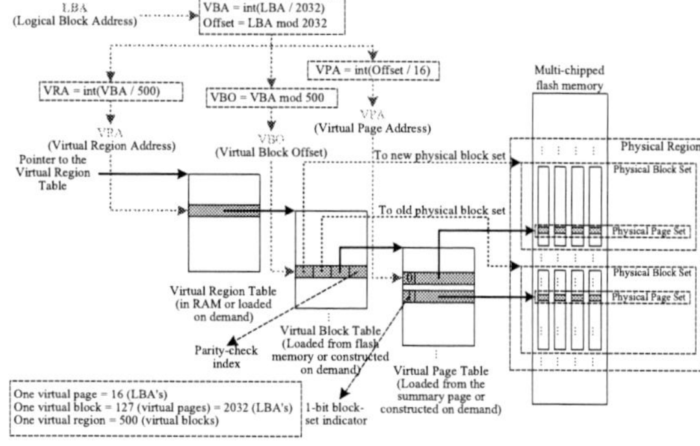

Figure 3: An Example of the Address Translation

The virtual region table is constructed and stored in RAM during the system's start-up. Each entry of the virtual region table is an index to the virtual block table of the corresponding virtual region. A virtual block table can be reconstructed by scanning the spare area of anyone physical page of each physical block set of the corresponding physical region when needed. Each entry of a virtual block table consists of four fields, as shown in Figure 3: The first two fields are the offsets of its two corresponding physical block sets, *i.e.*, its new and old physical block sets (Please see Section 3.2.1). The third field is an index to the virtual page table of the corresponding virtual block. The fourth field points to the parity-check information of its new physical block set. A virtual page table consists of 127 one-byte entries. Each entry is composed of a 1-bit *block-set indicator* and a 7-bit *page-set index*. A one(/zero) value for the block-set indicator denotes that the virtual

page is stored in the old(/new) physical block set of the corresponding virtual block, and the page-set index points to the physical page set that stores the virtual page. A virtual page table is written to the summary page of the new physical block set of the corresponding virtual block when it is removed from the SRAM cache. If a virtual page table is crashed, it can be reconstructed in an on-demand fashion. Therefore, *the system can be fast initialized (even after system crashes), in terms of address translation information.*

In order to improve the space utilization and lifetime of flash memory, each physical region is associated with one block remapping table to replace bad physical blocks with spare physical blocks, where spare physical blocks are only used as replacements for bad physical blocks of the same chip. In the block remapping table, each entry records the offset of a bad physical block set (relative to the first physical block set of its physical region) and its remapping information. Entries of the block remapping table are sorted in an increasing order of the offsets of bad physical block sets so that some binary search is possible. When several consecutive read/write requests to the same physical block set are received, we only need to look up this table for the first read/write request. When the performance is a critical issue, each physical region should have its own block remapping list to accelerate the identification of a bad physical block set because of the reducing in the frequency of binary searches over the block remapping table. The list can be implemented as a bit-array, and each bit indicates the status of one physical block set of the physical region. Whenever the corresponding bit of a physical block set indicates the block set as bad, the corresponding block remapping table is binary-searched for the remapping information of the physical block set. In this way, the block remapping table is seldom searched because the percentage of bad physical blocks of an MLC flash-memory chip is usually less than 2%.

3.2 An Adaptive Block Mapping Mechanism

3.2.1 Block Management

In order to improve the write performance of the proposed management strategy, a virtual block is mapped to at most two physical block sets, referred to as the new physical block set and the old physical block set, where the new and old physical block sets correspond to the currently and previously allocated physical block sets, respectively. Initially, each virtual block is not mapped to any physical block set. When a virtual block is written for the first time, then a free block set is allocated and becomes its *new physical block set*. Note that data are always written to a physical block set sequentially from the first page set. If the new physical block set of the virtual block is filled up or committed, and there is any further write to the virtual block, then another free physical block set is allocated as the new physical block set. The "original" new physical block set becomes the old physical block set of the virtual block. Notice that pages of new/old physical block sets might be valid or invalid, depending on the access patterns to their corresponding virtual block. Consider a virtual block that is already associated with a new physical block set and an old one. Suppose that its new physical block set is filled up or committed, and there is another write to the virtual block. The old physical block set is set as one of the "original" new and old physical block sets that has the largest number of valid data pages. Valid data pages of the other "original" physical block set (referred to as the discarded physical block set) are copied to the newly allocated physical block set (referred to as the new physical block set) together with the newly arrived data. *The discarded physical block set is then erased immediately.*

As shown in Figure 4, the old physical block set of a virtual block has more valid data pages than its corresponding new physical block set, and both of them are filled up or committed. Once there is a new write to the virtual block, the new physical block set is discarded, and the "original" old physical block set is still the old physical block set of the virtual block. The valid data pages (i.e., data pages of virtual pages 1 and 2) in the discarded physical block set are copied to the newly allocated physical block set that is set as the new physical block set. Conversely, if the new physical block set of a virtual block has more valid data pages than its corresponding old physical block set, then the old physical block set is discarded, and the "original" new physical block set becomes the old physical block set of the virtual block. Note that the number of physical block sets in a physical region should be larger than the number of virtual blocks in a virtual region because a virtual block could be mapped to two physical block sets.

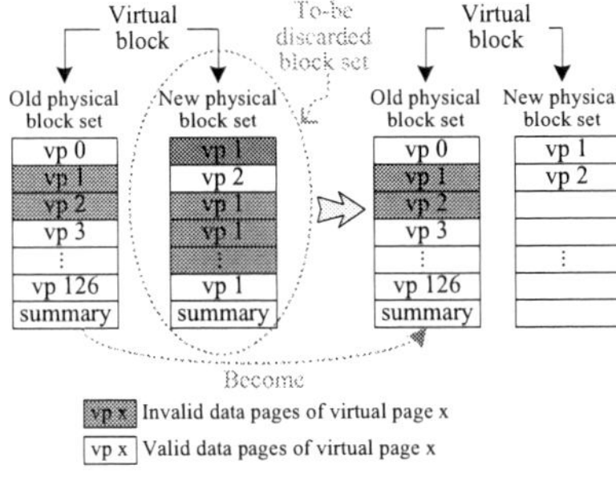

Figure 4: An Example of Block-set Replacement

The above block management strategy aims at providing adaptive mapping for both random and sequential accesses. When sequential writes are observed (such as those for many multimedia files), data are always written to the newly allocated physical block sets, and the previous copy in the old physical block sets are obsoleted naturally. Random updates to file attributes or small-size writes always incur the minimum overheads on any potential data page copying because the discarded physical block set is the one with the smallest number of valid pages. In the spare area of each physical page (Please see Section 3.1.1), the "LBA" field can help in identifying the corresponding virtual block, and the "old block set" field can help in locating the other physical block set of the corresponding virtual block. A physical block set is the old physical block set of its corresponding virtual block if it is pointed by the "old block set" field of some other physical block set; otherwise, it is the new physical block set of its corresponding virtual block. However, when the system crashes before some discarded physical block set is erased, there could exist three physical block sets that are mapped to the same virtual block. This problem could be easily resolved because the physical block set that is not committed should be the new physical block set, and the physical block set that is not discarded must be pointed by the new physical block set.

3.2.2 Free Block Allocation and Garbage Collection

In order to accelerate the performance of free block allocation, each physical region is associated with its own list of free physical block sets. During the reconstruction of a virtual block table, its corresponding list of free physical block sets can also be reconstructed at the same time (so that the time overhead on the reconstruction of the list is limited). This list can be stored to the spare physical blocks together with its corresponding virtual block table and be loaded in an on-demand fashion. This list is a bit-array, and each bit indicates the status of one physical block set. Whenever a

free physical block set in the physical region is needed, the list is scanned in a circular chain fashion until a free physical block set is found. *Note that the circular scanning of physical block sets in the selection of free physical block sets can be very effective in the implementation. The design is close to a random selection policy in reality because each virtual block could be stored in any physical block set of its residing physical region.*

Once there is only one free physical block set in the physical region, the garbage collection is activated to reclaim physical block sets. Similar to the way to locate a free physical block set, the garbage collector goes over its corresponding virtual block table in a circular chain fashion until it finds a virtual block that has its own old and new physical block sets. The two physical block sets of the selected virtual block are then merged and replaced with a free physical block set. Since the garbage collector can generate one additional free block set after the merging of the two physical block sets of the selected virtual block, the worst-case time on free block allocation can be bounded and estimated. *This property is critical to real-time applications.* To achieve the wear-leveling efficiently and effectively, the garbage collector can reclaim a physical block set that has not been erased over a predetermined time, or it can randomly reclaim a physical block set whenever a fixed number of block erases has been done [5, 6, 8, 16].

4. PERFORMANCE EVALUATION

4.1 Performance Metrics and Experiment Setup

The purpose of this section is to evaluate the capability of the proposed commitment-based management strategy, in terms of reliability, performance, and main-memory requirements (Section 4.2). The proposed commitment-based management strategy (referred to as *CBMS*) was evaluated, compared with *BL*, *NFTL*, *FTL*, and *ASA* strategies, where ASA was to scatter write requests over chips simultaneously [7]. ASA could be classified into ASA_s (to store data of each LBA to a dedicated chip) and ASA_d (to distribute hot(/cold) data to chips that have smaller(/larger) block erase counts). BL didn't need a garbage collector since any replaced block in BL was erased immediately. All of the other evaluated strategies adopted the same greedy policy in their garbage collector.

# of chips	4	Write	$800\mu s$
Chip Size	16384 blocks (4GB)	Erase	$1.5ms$
Block size	128 pages (256KB)	Random read	$60\mu s$
Page size	2KB + 64B	Serial access	$25ns/byte$
Bad blocks	200 per chip	ECC	4-bit correction
Erase cycles	10,000	Bir error rate	10^{-6}

Table 2: The Evaluated MLC Flash-memory Storage System

As shown in Table 2, a 16GB four-chipped MLC flash-memory storage system was under investigation, and 80% of the storage system was initially stored with data [15, 10]. The capability of the management strategies was evaluated through a realistic trace. The trace was collected over a desktop PC with a 160GB hard disk (by NTFS) for a month. The workload was mainly on daily activities, such as document editing, email accessing, file uploading/downloading, and web surfing. Accesses within the first 16GB of the trace were used for the performance evaluation. In the experiment, the proposed management strategy partitioned the 16GB four-chipped MLC flash memory storage system into 31 regions. Each physical(/virtual) region consisted of 512(/500) physical(/virtual) blocks. There were 512 spare physical blocks in the spare area of each chip and about 200 initial bad blocks in each chip. The RAM

978-1-60558-497-3/09 $25.00 © 2009 ACM

cache could cache at most 4 virtual block tables(/block remapping lists/lists of free physical block sets) and 8 virtual page tables(/parity-check information of physical block sets).

4.2 Experimental Results

Figure 5(a) shows the reliability of the storage system with different management strategies, in terms of the lifetime. The lifetime of the storage system under CBMS was 4.21 and 7.92 times of those under BL and NFTL, respectively. However, the lifetime of the storage system under CBMS was shorter than that of FTL, because FTL could store data of an LBA to any physical place, and it introduced very limited live-page copyings and block erases during garbage collection (even though FTL seemed infeasible for large-scaled flash memory). Furthermore, CBMS outperformed ASA_s and ASA_d for 12% and 32%, respectively. That was because ASA_s assigned each logical address to a specific chip, so that the chip that stored more frequently updated data was worn-out faster and had a higher possibility to encounter bit errors that could not be corrected by the ECC. ASA_d separated hot data from cold data, and stored them in different blocks so that the blocks that stored hot data tended to fail faster than other blocks. Although the lifetime of the storage system under FTL and ASA seemed long, they were not practical in large-scale or high bit-error-rate flash memory. The reason was that they needed large main-memory space to maintain the address translation table, and the error probability increased much faster than the bit error rate.

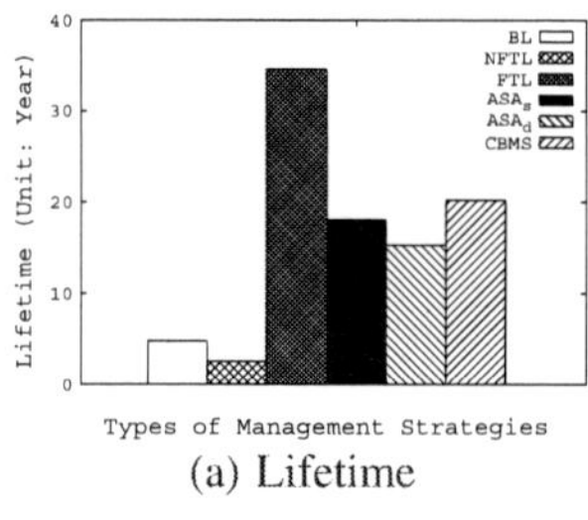

(a) Lifetime

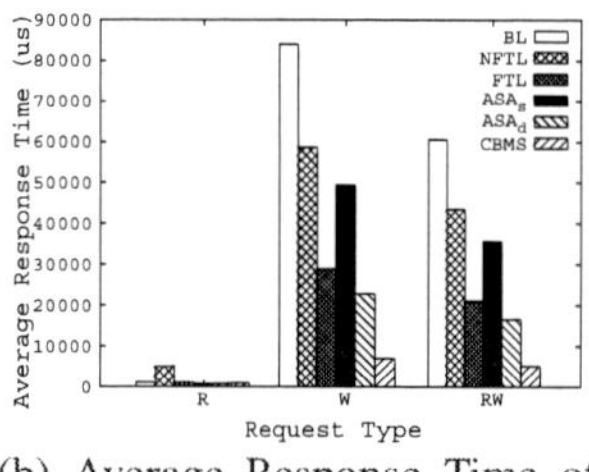

(b) Average Response Time of Read/Write Requests

Figure 5: Lifetime and Access Performance

Figure 5(b) shows that CBMS could greatly improve the access performance of the flash-memory storage system, compared to BL, NFTL, FTL, and ASA, where the x-axis denotes the request type (i.e., read or write), and the y-axis denotes the average response time of each request. As shown Figure 5(b), the average read response time of CBMS could outperform that of NFTL for more than 400% since NFTL needed to scan the spare area of pages for the latest version. CBMS could still outperform both BL and FTL for 16.8% on the average read response time. The less improvement was because the time on transmitting the data of pages dominated the time on the serving of read requests. In addition, CBMS read data in the unit of one page set and needed to reconstruct or load translation tables in an on-demand fashion. Due to the above overheads of CBMS, the average read response time of CBMS was slightly longer than that of ASA. For the average response time on write requests, the performance of CBMS was 12.08 and 8.45 times of those of BL and NFTL, respectively. The huge improvement was because BL needed to copy all of the live data to another free block on each write request; NFTL needed to scan the spare area of the replacement blocks for the latest version and had low space utilization in the primary block of virtual blocks due to the sequential write constraint of MLC flash memory. The average write response time of CBMS only outperformed that of FTL for 3.15 times, since FTL adopted a page-level address translation mechanism. It was also interesting to see that the average write response time of CBMS outperformed that of ASA_s for 5.86 times. That was because the garbage collector of ASA_s was lack of efficiency. Comparatively, the average write response time of CBMS only outperformed that of ASA_d for 2.18 times. The reason was that ASA_d stored hot data and cold data in different blocks, so that the efficiency of the garbage collector could be improved.

The main-memory requirement is a main factor to the scalability of a flash-memory management strategy, and the translation tables of a management strategy are its major RAM space overheads. As shown in Figure 3, the main-memory requirements of FTL, ASA, BL, and NFTL were proportional to the capacity of the storage system, where FTL and ASA adopted page-level translation mechanisms, and BL and NFTL adopted block-level translation mechanisms. Comparatively, CBMS needed very small RAM space, and its main-memory requirement grew slowly. For example, the main-memory requirements of CBMS were 12.867KB and 15.617KB when the system's capacities were 1GB and 1TB, respectively.

Capacity	BL	NFTL	FTL	ASA_s	ASA_d	RBMS
1GB	6	12	1216	1216	1234	12.867
32GB	272	544	49152	49152	49170	12.914
1TB	11264	22528	1900544	1900544	1900562	15.617

Table 3: Sizes of Translation Tables. (Unit: KB)

5. CONCLUSION

This work is motivated by a strong demand on the performance and reliability degradation problem, due to the development of flash-memory technology and new write constraints of MLC flash memory. In this paper, a commitment-based management strategy is proposed to improve the reliability and performance of MLC flash memory with the considerations of the limited RAM consumption and the new write constraints of MLC flash memory. For future research, we shall further explore the tradeoff between the RAM usage and the performance improvement. The proposed strategy can be further extended for the joint designs of management software and multi-channel hardware architectures.

6. REFERENCES

[1] Flash File System. US Patent 540,448. In *Intel Corporation*.

[2] FTL Logger Exchanging Data with FTL Systems. Technical report, Intel Corporation.

[3] Flash-memory Translation Layer for NAND flash (NFTL). *M-Systems*, 1998.

[4] Understanding the Flash Translation Layer (FTL) Specification. Technical report, Intel Corporation, Dec 1998.

[5] A. Ban. Wear Leveling of Static Areas in Flash Memory. US Patent 6,732,221. *M-systems*, 2004.

[6] A. Ben-Aroya and S. Toledo. Competitive Analysis of Flash-memory Algorithms. In *the 14th Conference on Annual European Symposium (ESA)*, pages 100–111, 2006.

[7] L.-P. Chang and T.-W. Kuo. An Adaptive Striping Architecture for Flash Memory Storage Systems of Embedded Systems. In *the IEEE Real-Time and Embedded Technology and Applications Symposium (RTAS)*, pages 187–196, 2002.

[8] Y.-H. Chang, J.-W. Hsieh, and T.-W. Kuo. Endurance Enhancement of Flash-Memory Storage Systems: An Efficient Static Wear Leveling Design. In *the 44th ACM/IEEE Design Automation Conference (DAC)*, June 2007.

[9] A. O. Company. Yet Another Flash Filing System.

[10] R. Dan and R. Singer. *Implementing MLC NAND Flash Memory for Cost-Effective, High-capacity Memory*. M-Systems, September 2003.

[11] DRAMeXchange. *Flash Contract Price,http://www.dramexchange.com/*, 03 2008.

[12] H. Kim and S. Ahn. BPLRU : A Buffer Management Scheme for Improving Random Writes in Flash Storage. In *the 6th USENIX Conference on File and Storage Technologies (FAST)*, pages 239–252, 2008.

[13] T.-W. Kuo, Y.-H. Chang, P.-C. Huang, and C.-W. Chang. Special Issues in Flash. In *the IEEE/ACM International Conference on Computer-Aided Design (ICCAD)*, 2008.

[14] J.-H. Lin, Y.-H. Chang, J.-W. Hsieh, T.-W. Kuo, and C.-C. Yang. A NOR Emulation Strategy over NAND Flash Memory. In *the 13th IEEE International Conference on Embedded and Real-Time Computing Systems and Applications (RTCSA)*, pages 95–102, 2007.

[15] Samsung Electronics. *K9GAG08U0M 2G x 8bit NAND Flash Memory Data Sheet*, Sep. 2006.

[16] M. Spivak and S. Toledo. Storing a Persistent Transactional Object Heap on Flash Memory. In *the 2006 ACM Conference on Language, Compilers, and Tool Support for Embedded Systems (LCTES)*, pages 22–33, 2006.

[17] D. Woodhouse. JFFS: The Journalling Flash File System. In *Ottawa Linux Symposium*, 2001.

[18] C.-H. Wu and T.-W. Kuo. An Adaptive Two-level Mnagement for the Flash Translation Layer in Embedded Systems. In *the IEEE/ACM Iinternational Conference on Computer-Aided Design (ICCAD)*, pages 601–606, 2006.

[19] S. yeong Park, D. Jung, J. uk Kang, J. soo Kim, and J. Lee. CFLRU: a Replacement Algorithm for Flash Memory. In *the International Conference on Compilers, Architecture and Synthesis for Embedded Systems (CASES)*, 2006.

Quality-Driven Synthesis of Embedded Multi-Mode Control Systems

Soheil Samii
Dept. of Computer and
Information Science
Linköping University
Linköping, Sweden
sohsa@ida.liu.se

Petru Eles
Dept. of Computer and
Information Science
Linköping University
Linköping, Sweden
petel@ida.liu.se

Zebo Peng
Dept. of Computer and
Information Science
Linköping University
Linköping, Sweden
zpe@ida.liu.se

Anton Cervin
Dept. of Automatic
Control
Lund University
Lund, Sweden
anton@control.lth.se

ABSTRACT

At runtime, an embedded control system can switch between alternative functional modes. In each mode, the system operates by using a schedule and controllers that exploit the available computation and communication resources to optimize the control performance in the running mode. The number of modes is usually exponential in the number of control loops, which means that all controllers and schedules cannot be produced in affordable design-time and stored in memory. This paper addresses synthesis of multi-mode embedded control systems. Our contribution is a method that trades control quality with optimization time, and that efficiently selects the schedules and controllers to be synthesized and stored in memory.

Categories and Subject Descriptors

C.3 [**Special-Purpose and Application-Based Systems**]: *process control systems, real-time and embedded systems*; D.4.1 [**Operating Systems**]: Process Management—*scheduling*; J.6 [**Computer-Aided Engineering**]: *computer-aided design*; J.7 [**Computers in Other Systems**]: *industrial control*

General Terms

Algorithms, Design, Performance, Theory

Keywords

control performance, embedded control, multi-mode systems

1. INTRODUCTION AND RELATED WORK

In systems that control several physical plants, mode changes occur at runtime either as responses to external events or at predetermined moments in time. In an operation mode, the system controls a subset of the plants by executing several control applications concurrently—one for each controlled plant in the mode. The application tasks execute periodically (reading sensors, computing control signals, and writing to actuators) on a platform comprising several computation nodes connected to a bus. Such systems have complex timing behavior that leads to bad control performance if not taken into account during controller design [10]. To achieve good control performance in a certain mode, system scheduling must be integrated with controller design (periods and control laws). A mode change means that some of the running control loops are deactivated, some are activated, or both. This leads to a change in the execution and communication demand and, consequently, a new schedule and new controllers must be used to achieve best possible control performance by an efficient use of the resources.

Permission to make digital or hard copies of part or all of this work for personal or classroom use is granted without fee provided that copies are not made or distributed for profit or commercial advantage and that copies bear this notice and the full citation on the first page. To copy otherwise, to republish, to post on servers or to redistribute to lists, requires prior specific permission and/or a fee.

DAC'09, July 26-31, 2009, San Francisco, California, USA

Control–scheduling co-design methods in the literature focus on single-mode control systems. Sampling-period optimization for mono-processor systems with multiple control loops and priority-based scheduling was first studied by Seto et al. [9]. Bini and Cervin improved their work by considering controller delays in the optimization process [2]. Rehbinder and Sanfridson proposed optimal control strategies for static scheduling of controllers on a mono-processor system [7]. In our previous work, we proposed a framework [8] for the scheduling and synthesis of controllers on distributed execution platforms.

The contribution of this paper is a scheduling and synthesis method for embedded multi-mode control systems. The number of modes to be considered in the synthesis is usually exponential in the number of control loops and leads to two problems: (1) all modes cannot be synthesized in affordable time, and (2) all synthesized controllers and schedules cannot be stored in memory. With the objective of optimizing the control performance in a multi-mode control system, we address these problems by a limited exploration of the set of modes, followed by a selection of the produced schedules and controllers to store in memory.

2. MULTI-MODE SYSTEMS

Given is a set of plants $\mathbf{P}$, indexed by $\mathcal{I}_{\mathbf{P}}$, where each plant P_i ($i \in \mathcal{I}_{\mathbf{P}}$) is described by a continuous-time linear model [11]

$$\dot{\boldsymbol{x}}_i = A_i \boldsymbol{x}_i + B_i \boldsymbol{u}_i + \boldsymbol{v}_i, \ \boldsymbol{y}_i = C_i \boldsymbol{x}_i + \boldsymbol{e}_i. \tag{1}$$

The vectors $\boldsymbol{x}_i$ and $\boldsymbol{u}_i$ are the plant state and controlled input, respectively, and the vector $\boldsymbol{v}_i$ models plant disturbance as a continuous-time white-noise process with intensity R_{1i}. The continuous-time output $\boldsymbol{y}_i$ is measured and sampled periodically and is used to produce the control signal $\boldsymbol{u}_i$. The measurement noise $\boldsymbol{e}_i$ in the output is modeled as a discrete-time white-noise process with variance R_{2i}. The control signal is updated at discrete time instants and is held constant between two updates (zero-order hold [11]). As an example, let us consider three inverted pendulums $\mathbf{P} = \{P_1, P_2, P_3\}$. Each pendulum P_i ($i \in \mathcal{I}_{\mathbf{P}} = \{1, 2, 3\}$) is modeled according to Equation 1, with $A_i = \begin{bmatrix} 0 & 1 \\ g/l_i & 0 \end{bmatrix}$, $B_i = \begin{bmatrix} 0 & g/l_i \end{bmatrix}^\mathsf{T}$, and $C_i = \begin{bmatrix} 1 & 0 \end{bmatrix}$, where $g \approx 9.81$ m/s^2 and l_i are the gravitational constant and length of pendulum P_i, respectively (all pendulums have equal length, $l_i = 0.2$ m). For the plant disturbance and measurement noise, we have $R_{1i} = B_i B_i^\mathsf{T}$ and $R_{2i} = 0.1$.

The execution platform comprises a set of nodes $\mathbf{N}$, indexed by $\mathcal{I}_{\mathbf{N}}$, which are connected by a communication controller to a bus; common communication protocols in automotive systems are TTP [5], FlexRay [4], and CAN [3], which all are supported by our framework. On the execution platform runs a set of applications $\mathbf{\Lambda}$, indexed by the set $\mathcal{I}_{\mathbf{\Lambda}} = \mathcal{I}_{\mathbf{P}}$. Application $\Lambda_i \in \mathbf{\Lambda}$ ($i \in \mathcal{I}_{\mathbf{\Lambda}}$) controls plant P_i and is modeled as a directed acyclic graph $\Lambda_i = (\mathbf{T}_i, \mathbf{\Gamma}_i)$, where the nodes $\mathbf{T}_i$, indexed by $\mathcal{I}_i$, represent computation tasks and the edges $\mathbf{\Gamma}_i \subset \mathbf{T}_i \times \mathbf{T}_i$ represent data dependencies between tasks. Figure 1 shows two nodes $\mathbf{N} = \{N_1, N_2\}$ connected to a bus. Three applications $\mathbf{\Lambda} = \{\Lambda_1, \Lambda_2, \Lambda_3\}$ are mapped to the nodes and control

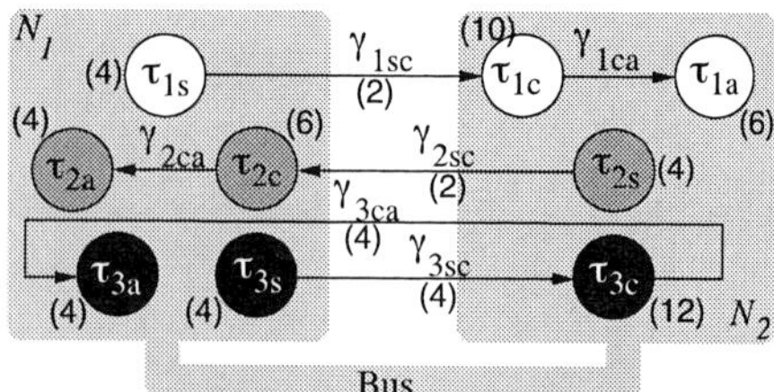

Figure 1: Motivational example

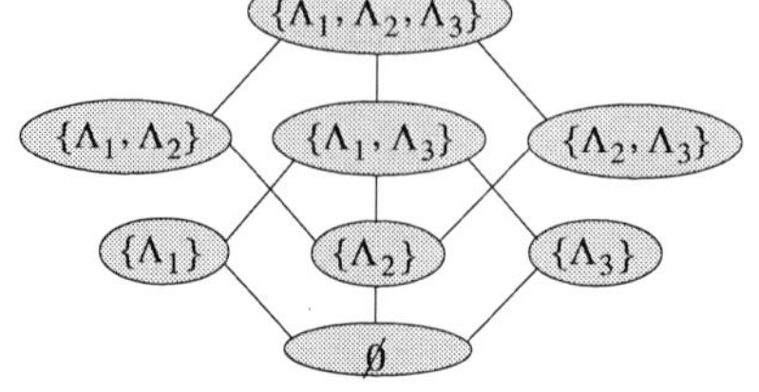

Figure 2: Hasse diagram of modes

Table 1: Individual control costs

Mode $\mathbf{M}$	$J_1^{\mathbf{M}}$	$J_2^{\mathbf{M}}$	$J_3^{\mathbf{M}}$
$\{\Lambda_1, \Lambda_2, \Lambda_3\}$	2.98	1.19	2.24
$\{\Lambda_1, \Lambda_2\}$	1.60	1.42	-
$\{\Lambda_1, \Lambda_3\}$	1.60	-	1.95
$\{\Lambda_2, \Lambda_3\}$	-	1.36	1.95
$\{\Lambda_1\}$	1.60	-	-
$\{\Lambda_2\}$	-	1.14	-
$\{\Lambda_3\}$	-	-	1.87

the three pendulums P_1, P_2, and P_3 in the example before. For this example, the task set of application Λ_i is $\mathbf{T}_i = \{\tau_{is}, \tau_{ic}, \tau_{ia}\}$ ($\mathcal{I}_i = \{s, c, a\}$). The data dependencies are given by the edges $\mathbf{\Gamma}_i = \{\gamma_{isc} = (\tau_{is}, \tau_{ic}), \gamma_{ica} = (\tau_{ic}, \tau_{ia})\}$. The sensor task is τ_{is}, the controller task is τ_{ic}, and the actuator task is τ_{ia}.

The system can run in different operation modes, where each mode is characterized by a set of running applications. A mode is a subset $\mathbf{M} \subseteq \mathbf{\Lambda}$, indexed by $\mathcal{I}_{\mathbf{M}} \subseteq \mathcal{I}_{\mathbf{\Lambda}} = \mathcal{I}_{\mathbf{P}}$, that contains the applications that are running in that mode. The complete set of modes is $\mathcal{M} = 2^{\mathbf{\Lambda}}$ and its cardinality is $|\mathcal{M}| = 2^{|\mathbf{\Lambda}|}$. The set of modes $\mathcal{M}$ is a partially ordered set under the subset relation $\subset$. For our example with the three applications in Figure 1, we show in Figure 2 the Hasse diagram of the partially ordered set of modes $\mathcal{M} = 2^{\mathbf{\Lambda}} = 2^{\{\Lambda_1, \Lambda_2, \Lambda_3\}}$. A mode $\mathbf{M}' \in \mathcal{M}$ is called a *submode* of $\mathbf{M}$ if $\mathbf{M}' \subset \mathbf{M}$. Mode $\mathbf{M}$ is called a *supermode* of mode $\mathbf{M}'$. We define the set of submodes of $\mathbf{M} \in \mathcal{M}$ as $\underline{\mathcal{M}}(\mathbf{M}) = \{\mathbf{M}' \in \mathcal{M} : \mathbf{M}' \subset \mathbf{M}\}$. Similarly, the set of supermodes of $\mathbf{M}$ is denoted $\overline{\mathcal{M}}(\mathbf{M})$. The idle mode is the empty set $\emptyset$, which indicates that the system is inactive, whereas the mode $\mathbf{\Lambda}$ indicates that all applications are running. It can be the case that certain modes do not occur at runtime—for example, because certain plants are never controlled concurrently. Let us therefore introduce the set of *functional* modes $\mathcal{M}^{\text{func}} \subseteq \mathcal{M}$ that can occur during execution. Modes $\mathcal{M}^{\text{virt}} = \mathcal{M} \setminus \mathcal{M}^{\text{func}}$ do not occur at runtime and are therefore called *virtual* modes. In a given mode $\mathbf{M} \in \mathcal{M}$, an application $\Lambda_i \in \mathbf{M}$ ($i \in \mathcal{I}_{\mathbf{M}}$) releases jobs for execution periodically with the period $h_i^{\mathbf{M}}$. Job q of task τ_{ij} is denoted $\tau_{ij}^{(q)}$. For a message $\gamma_{ijk} = (\tau_{ij}, \tau_{ik}) \in \mathbf{\Gamma}_i$, the message instance produced by job $\tau_{ij}^{(q)}$ is denoted $\gamma_{ijk}^{(q)}$. An edge $\gamma_{ijk} = (\tau_{ij}, \tau_{ik}) \in \mathbf{\Gamma}_i$ means that the earliest start time of a job $\tau_{ik}^{(q)}$ is when $\tau_{ij}^{(q)}$ has completed its execution and the produced data (i.e., $\gamma_{ijk}^{(q)}$) has been communicated to $\tau_{ik}^{(q)}$. We define the hyperperiod $h_{\mathbf{M}}$ of $\mathbf{M}$ as the least common multiple of the periods $\{h_i^{\mathbf{M}}\}_{i \in \mathcal{I}_{\mathbf{M}}}$.

Each task is mapped to a node; in Figure 1, task τ_{1s} is mapped to N_1 and τ_{1c} is mapped to N_2. A message between tasks that are mapped to different nodes is sent on the bus. For a message γ_{ijk} between two nodes, we denote with c_{ijk} the communication time when there are no conflicts on the bus. We model the execution time of task τ_{ij} as a stochastic variable c_{ij} with probability function $\xi_{c_{ij}}$. We assume that the execution time of τ_{ij} is bounded by given best-case and worst-case execution times. In Figure 1, the execution times (constant in this example) and communication times for the tasks and messages are given in milliseconds in parentheses. We support static-cyclic and priority-based scheduling of tasks and messages [5]. In static-cyclic scheduling, the period of the schedule is the hyperperiod $h_{\mathbf{M}}$ of the applications in the running mode $\mathbf{M}$. The schedule determines the start times of the jobs and message instances that are released within a hyperperiod of the mode. Further, the schedule must satisfy precedence constraints, which are given by the data dependencies, and account for the worst-case execution times of the tasks and the communication times of the messages. For preemptive priority-based scheduling, the tasks and messages are scheduled at runtime based on fixed priorities that are decided at design time.

As an example of a static-cyclic schedule, let us consider the system in Figure 1 and the schedule in Figure 3. All times are given in milliseconds in the discussion that follows. The schedule is constructed for mode $\mathbf{M} = \mathbf{\Lambda}$ (i.e., for the mode in which all three applications are running concurrently) and for the periods $h_1^{\mathbf{M}} = 40$, $h_2^{\mathbf{M}} = 20$, and $h_3^{\mathbf{M}} = 40$. The period of the schedule is $h_{\mathbf{M}} = 40$ (the hyperperiod of $\mathbf{M}$). Considering the periods of the applications, the jobs to be scheduled on node N_1 are $\tau_{1s}^{(1)}$, $\tau_{2c}^{(1)}$, $\tau_{2c}^{(2)}$, $\tau_{2a}^{(1)}$, $\tau_{2a}^{(2)}$, $\tau_{3s}^{(1)}$, and $\tau_{3a}^{(1)}$. The jobs on node N_2 are $\tau_{1c}^{(1)}$, $\tau_{1a}^{(1)}$, $\tau_{2s}^{(1)}$, $\tau_{2s}^{(2)}$, and $\tau_{3c}^{(1)}$. The message transmissions on the bus are $\gamma_{1sc}^{(1)}$, $\gamma_{2sc}^{(1)}$, $\gamma_{2sc}^{(2)}$, $\gamma_{3sc}^{(1)}$, and $\gamma_{3ca}^{(1)}$. The schedule in Figure 3 is shown with three rows for node N_1, the bus, and node N_2, respectively. The small boxes depict task executions and message transmissions. The white, grey, and black boxes show the execution of applications Λ_1, Λ_2, and Λ_3, respectively. Each box is labeled with an index that indicates a task or message, and with a number that specifies the job or message instance. The black box labeled $c(1)$ shows that the execution of job $\tau_{3c}^{(1)}$ on node N_2 starts at time 8 and completes at time 20. The grey box labeled $sc(2)$ shows the second message between the sensor and controller task of Λ_2.

An *implementation* of a mode $\mathbf{M} \in \mathcal{M}$ comprises the period $h_i^{\mathbf{M}}$ and control law $\boldsymbol{u}_i^{\mathbf{M}}$ of each control application $\Lambda_i \in \mathbf{M}$ ($i \in \mathcal{I}_{\mathbf{M}}$), and the schedule (or priorities) for the tasks and messages in the mode. We denote with $\mathsf{mem}_d^{\mathbf{M}}$ the memory consumption on node $N_d \in \mathbf{N}$ ($d \in \mathcal{I}_{\mathbf{N}}$) of the implementation of $\mathbf{M}$. An implementation of a mode can serve as an implementation of all its submodes. This means that the system can run in a submode $\mathbf{M}' \subset \mathbf{M}$ ($\mathbf{M}' \neq \emptyset$) with the same controllers (periods and control laws) and schedule as for mode $\mathbf{M}$—for example, by not running the applications in $\mathbf{M} \setminus \mathbf{M}'$ or by not writing to their outputs. However, to achieve better performance in mode $\mathbf{M}'$, a customized set of controllers and schedule for submode $\mathbf{M}'$ can exploit the available computation and communication resources that are now not used by the other applications $\mathbf{M} \setminus \mathbf{M}'$. To have a correct implementation of the whole system, for each functional mode $\mathbf{M} \in \mathcal{M}^{\text{func}}$, there must exist an implementation in memory, or there must exist an implementation of at least one of its supermodes. The set of implemented modes is $\mathcal{M}^{\text{impl}} \subseteq \mathcal{M} \setminus \{\emptyset\}$.

3. CONTROL QUALITY AND SYNTHESIS

The quality of a controller Λ_i for plant P_i (Equation 1) is given by the quadratic cost [11]

$$J_i = \lim_{T \to \infty} \frac{1}{T} \mathrm{E}\left\{\int_0^T \begin{bmatrix} \boldsymbol{x}_i \\ \boldsymbol{u}_i \end{bmatrix}^{\top} Q_i \begin{bmatrix} \boldsymbol{x}_i \\ \boldsymbol{u}_i \end{bmatrix} dt\right\}. \quad (2)$$

The matrix Q_i, which is given by the designer, is a positive semi-definite matrix with weights for the magnitude of the plant states and the control signals ($\mathrm{E}\{\cdot\}$ denotes the expected value of a stochastic variable). For a given sampling period h_i and a given constant sensor–actuator delay (i.e., the time elapsed between sampling the output $\boldsymbol{y}_i$ and updating the controlled input $\boldsymbol{u}_i$), it is possible to find the control law $\boldsymbol{u}_i$ that minimizes J_i [11]. The cost J_i of a controller is increased (its quality is

978-1-60558-497-3/09 $25.00 © 2009 ACM

865

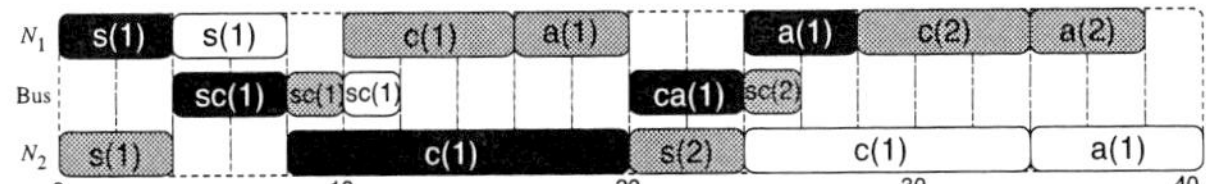

Figure 3: Schedule for mode $\{\Lambda_1, \Lambda_2, \Lambda_3\}$ with periods $h_1 = 40$, $h_2 = 20$, and $h_3 = 40$

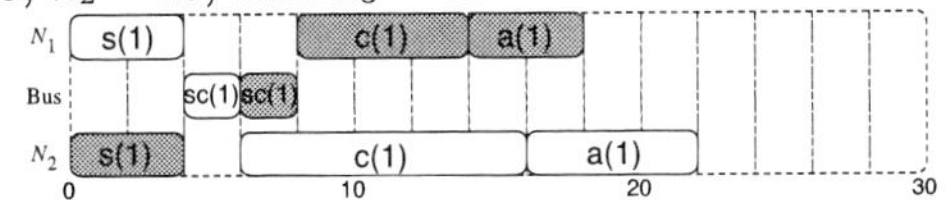

Figure 4: Schedule for mode $\{\Lambda_1, \Lambda_2\}$ with periods $h_1 = 30$ and $h_2 = 30$

degraded) if the sensor–actuator delay at runtime is different from what was assumed during control-law synthesis, or if this delay is not constant (i.e., there is jitter). Given the delay characteristics, we compute this cost J_i with Jitterbug [6].

We have developed a framework [8] to synthesize controllers and schedule their execution and communication for single-mode systems. In this paper, we use the framework to obtain an implementation of a certain mode $\mathbf{M} \in \mathcal{M} \setminus \{\emptyset\}$ such that the total control cost of the mode $J^{\mathbf{M}} = \sum_{i \in \mathcal{I}_{\mathbf{M}}} J_i^{\mathbf{M}}$ is minimized, where $J_i^{\mathbf{M}}$ is the cost of controller Λ_i. The produced mode implementation comprises the periods $h_i^{\mathbf{M}}$, control laws $u_i^{\mathbf{M}}$, and system schedule (schedule table or priorities). We also obtain the memory consumption $\mathsf{mem}_d^{\mathbf{M}} \in \mathbb{N}$ of the implementation on each computation node $N_d \in \mathbf{N}$ ($d \in \mathcal{I}_{\mathbf{N}}$).

4. MOTIVATIONAL EXAMPLE

In this section, we shall highlight the two problems that are addressed in this paper: (1) the time complexity of the synthesis of embedded multi-mode control systems, and (2) the memory complexity of the storage of produced mode implementations. Let us consider our example in Section 2 with three applications $\mathbf{\Lambda} = \{\Lambda_1, \Lambda_2, \Lambda_3\}$ that control the three inverted pendulums $\mathbf{P} = \{P_1, P_2, P_3\}$. For the computation of the control cost in Equation 2, we have used $Q_i = \mathrm{diag}(C_i^{\mathsf{T}} C_i, 0.002)$ as the weight matrix for each inverted pendulum P_i.

By using our framework for controller scheduling and synthesis [8], we synthesized an implementation of mode $\mathbf{M}_{123} = \mathbf{\Lambda}$ with the periods $h_1^{\mathbf{M}_{123}} = 40$, $h_2^{\mathbf{M}_{123}} = 20$, and $h_3^{\mathbf{M}_{123}} = 40$ and the schedule in Figure 3. The outputs of the plants are sampled periodically without jitter (e.g., by some mechanism that stores the sampled data in buffers). The actuations of Λ_1 and Λ_3 finish at times 40 and 28, respectively. Because the schedule is periodic, the delay from sampling to actuation is 40 in each instance of Λ_1 (i.e., constant). Similarly, application Λ_3 has the constant sampling–actuation delay 28. Two instances are scheduled for application Λ_2. The first actuation finishes at time 20, whereas the second actuation finishes at time 38. By considering the sampling period 20, we note that the sampling–actuation delay of Λ_2 during periodic execution of the schedule is either 20 or 18. With this implementation, we obtained the individual control costs $J_1^{\mathbf{M}_{123}} = 2.98$, $J_2^{\mathbf{M}_{123}} = 1.19$, and $J_3^{\mathbf{M}_{123}} = 2.24$. Table 1 contains the individual controller costs of the modes considered in this section.

Let us now consider mode $\mathbf{M}_{12} = \{\Lambda_1, \Lambda_2\}$ in which Λ_3 is not executing. The system can operate in the new mode by using the schedule and control laws from mode $\mathbf{M}_{123}$. This can be done by not writing to the actuators of Λ_3 or by omitting the execution and communication of Λ_3. By using the implementation of mode $\mathbf{M}_{123}$, the overall control cost of mode $\mathbf{M}_{12}$ is $J_1^{\mathbf{M}_{123}} + J_2^{\mathbf{M}_{123}} = 2.98 + 1.19 = 4.17$. This cost can be reduced because, compared to mode $\mathbf{M}_{123}$, there is now more computation and communication power available for applications Λ_1 and Λ_2. Thus, it is worth investigating whether a better implementation (e.g., with reduced periods and delays) can be pro-

duced for mode $\mathbf{M}_{12}$. By running the synthesis of $\mathbf{M}_{12}$, we obtained an implementation with the periods $h_1^{\mathbf{M}_{12}} = h_2^{\mathbf{M}_{12}} = 30$ and the schedule in Figure 4. Note from the new schedule that both the period and delay of Λ_1 have been reduced. The costs of Λ_1 and Λ_2 with the new implementation are $J_1^{\mathbf{M}_{12}} = 1.60$ and $J_2^{\mathbf{M}_{12}} = 1.42$ (Table 1). The cost of Λ_1 is reduced significantly as a result of the reduction in sampling period and delay. The sampling period of Λ_2 is increased, which results in a small increase in the cost of Λ_2. This cost increase is accepted because it makes possible a significant quality improvement of Λ_1, which leads to a significant cost reduction of mode $\mathbf{M}_{12}$ to 3.02.

To achieve the best control performance, implementations of all functional modes have to be produced at design time. However, the number of modes is exponential in the number of control loops that run on the platform. Thus, even if some modes are functionally excluded, implementations of all possible functional modes cannot be produced in affordable time (except cases with small number of control loops). Let us consider that we can only run the synthesis of three modes at design time. Modes $\mathbf{M}_{123}$ and $\mathbf{M}_{12}$ have already been discussed. Considering the third mode to be $\mathbf{M}_{13}$, the synthesis resulted in the costs $J_1^{\mathbf{M}_{13}} = 1.60$ and $J_3^{\mathbf{M}_{13}} = 1.95$ (Table 1). When using the implementation of $\mathbf{M}_{123}$, the costs of Λ_1 and Λ_3 are 2.98 and 2.24, respectively. By running $\mathbf{M}_{13}$ with the new implementation thus leads to a significant improvement in control performance, compared to when using the implementation of $\mathbf{M}_{123}$. At runtime, the system has implementations of the modes $\mathcal{M}^{\mathrm{impl}} = \{\mathbf{M}_{123}, \mathbf{M}_{12}, \mathbf{M}_{13}\}$ in memory. For modes that are not implemented, the system chooses an implementation at runtime based on the three available implementations. For example, mode $\mathbf{M}_2 = \{\Lambda_2\}$ does not have an implementation but can run with the implementation of either $\mathbf{M}_{12}$ or $\mathbf{M}_{123}$. The cost of Λ_2 when running with the implementation of $\mathbf{M}_{12}$ is $J_2^{\mathbf{M}_{12}} = 1.42$, whereas it is $J_2^{\mathbf{M}_{123}} = 1.19$ for the implementation of $\mathbf{M}_{123}$. Thus, at runtime, the implementation of $\mathbf{M}_{123}$ is chosen to operate the system in mode $\mathbf{M}_2$.

We now consider that memory limitations in the platform imply that we can only store implementations of two modes out of the three modes in $\mathcal{M}^{\mathrm{impl}}$. We cannot use the implementation of $\mathbf{M}_{12}$ or $\mathbf{M}_{13}$ to run the system in mode $\mathbf{M}_{123}$ and thus we cannot remove its implementation. As we discussed in the beginning of this section, the total control cost when running mode $\mathbf{M}_{12}$ with the implementation of $\mathbf{M}_{123}$ is 4.17, compared to the total cost 3.02 when running with the implementation of $\mathbf{M}_{12}$. The implementation of $\mathbf{M}_{12}$ thus gives a total cost reduction of 1.15. If, on the other hand, $\mathbf{M}_{13}$ runs with the implementation of $\mathbf{M}_{123}$, the total cost is $J_1^{\mathbf{M}_{123}} + J_3^{\mathbf{M}_{123}} = 2.98 + 2.24 = 5.22$. If $\mathbf{M}_{13}$ runs with its produced implementation, the total cost is $J_1^{\mathbf{M}_{13}} + J_3^{\mathbf{M}_{13}} = 1.60 + 1.95 = 3.55$. This gives a total cost reduction of 1.67, which is better than the reduction obtained by the implementation of $\mathbf{M}_{12}$. Thus, in the presence of memory limitations, the implementations of $\mathbf{M}_{123}$ and $\mathbf{M}_{13}$ should be stored in memory to achieve the best control performance. In this discussion, we have assumed that $\mathbf{M}_{12}$ and $\mathbf{M}_{13}$ are equally important. However, if $\mathbf{M}_{12}$ occurs more frequently than $\mathbf{M}_{13}$, the cost improvement of the implementation of $\mathbf{M}_{12}$ becomes more significant. In this case, the best selection could be to exclude the implementation of $\mathbf{M}_{13}$ and store implementations of $\mathbf{M}_{123}$ and $\mathbf{M}_{12}$ in memory.

As the last example, let us consider that $\mathbf{M}_{123}$ is a virtual mode (i.e., it does not occur at runtime). Let $\mathcal{M}^{\mathrm{impl}} = \{\mathbf{M}_{12}, \mathbf{M}_{13}, \mathbf{M}_{23}\}$ be the set of modes with produced implementations. We have run the synthesis of mode $\mathbf{M}_{23} = \{\Lambda_2, \Lambda_3\}$ and obtained a total cost of 3.31. Let us assume that the three produced implementations of the modes in $\mathcal{M}^{\mathrm{impl}}$ cannot be all stored in memory. If the implementation of $\mathbf{M}_{23}$ is removed,

978-1-60558-497-3/09 $25.00 © 2009 ACM

for example, there is no valid implementation of the functional mode $\mathbf{M}_{23}$ in memory. To solve this problem, we implement the virtual mode $\mathbf{M}_{123}$. Its implementation can be used to run the system in all functional modes—however, with degraded control performance. The available memory allows us to further store the implementation of one of the modes $\mathbf{M}_{12}$, $\mathbf{M}_{13}$, or $\mathbf{M}_{23}$. We choose the mode implementation that gives the largest cost reduction, compared to the implementation with $\mathbf{M}_{123}$. By looking in Table 1 and with the cost reductions in our discussions earlier in this example, we conclude that $\mathbf{M}_{13}$ gives the largest cost reduction among the modes in $\mathcal{M}^{\mathrm{impl}}$. The best control performance under these tight memory constraints is achieved if the virtual mode $\mathbf{M}_{123}$ and functional mode $\mathbf{M}_{13}$ are implemented and stored in memory. This shows that memory limitations can lead to situations in which virtual modes must be implemented to cover a set of functional modes.

5. PROBLEM FORMULATION

In addition to the plants, control applications, and their mapping to an execution platform (Section 2), the inputs to the problem that we address are

- a set of functional modes[1] $\mathcal{M}^{\mathrm{func}} \subseteq \mathcal{M}$ that can occur at runtime (the set of virtual modes is $\mathcal{M}^{\mathrm{virt}} = \mathcal{M} \setminus \mathcal{M}^{\mathrm{func}}$),
- a weight[2] $w_{\mathbf{M}} > 0$ for each mode $\mathbf{M} \in \mathcal{M}^{\mathrm{func}}$ ($w_{\mathbf{M}} = 0$ for $\mathbf{M} \in \mathcal{M}^{\mathrm{virt}}$), and
- the available memory in the platform, modeled as a memory limit $\mathrm{mem}_d^{\max} \in \mathbb{N}$ for each node N_d ($d \in \mathcal{I}_{\mathbf{N}}$).

The outputs are implementations of a set of modes $\mathcal{M}^{\mathrm{impl}} \subseteq \mathcal{M} \setminus \{\emptyset\}$. For each functional mode $\mathbf{M} \in \mathcal{M}^{\mathrm{func}} \setminus \{\emptyset\}$, there must exist an implementation (i.e., $\mathbf{M} \in \mathcal{M}^{\mathrm{impl}}$) or a supermode implementation (i.e., $\mathcal{M}^{\mathrm{impl}} \cap \overline{\mathcal{M}}(\mathbf{M}) \neq \emptyset$).

If $\mathbf{M} \in \mathcal{M}^{\mathrm{impl}}$, the cost $J^{\mathbf{M}}$ is given from the scheduling and synthesis step that produces the implementation (Section 3). If $\mathbf{M} \notin \mathcal{M}^{\mathrm{impl}}$ and $\mathbf{M} \neq \emptyset$, the cost $J^{\mathbf{M}}$ is given by the available supermode implementations: Given an implemented supermode $\mathbf{M}' \in \mathcal{M}^{\mathrm{impl}} \cap \overline{\mathcal{M}}(\mathbf{M})$ of $\mathbf{M}$, the cost of $\mathbf{M}$ when using the implementation of $\mathbf{M}'$ is

$$J^{\mathbf{M}}(\mathbf{M}') = \sum_{i \in \mathcal{I}_{\mathbf{M}}} J_i^{\mathbf{M}'} = J^{\mathbf{M}'} - \sum_{i \in \mathcal{I}_{\mathbf{M}'} \setminus \mathcal{I}_{\mathbf{M}}} J_i^{\mathbf{M}'}. \quad (3)$$

At runtime, we use the supermode implementation that gives the smallest cost, and thus

$$J^{\mathbf{M}} = \min_{\mathbf{M}' \in \mathcal{M}^{\mathrm{impl}} \cap \overline{\mathcal{M}}(\mathbf{M})} J^{\mathbf{M}}(\mathbf{M}').$$

The cost of the idle mode $\emptyset$ is defined as $J^{\emptyset} = 0$. The objective is to find a set of modes $\mathcal{M}^{\mathrm{impl}}$ and synthesize them such that their implementations can be stored in the available memory of the platform. The cost to be minimized is

$$J_{\mathrm{tot}} = \sum_{\mathbf{M} \in \mathcal{M}^{\mathrm{func}}} w_{\mathbf{M}} J^{\mathbf{M}}. \quad (4)$$

6. SYNTHESIS APPROACH

Our synthesis approach consists of two sequential parts. First, we synthesize implementations for a limited set of functional modes (Section 6.1). Second, we select the implementations to store under given memory constraints, and if needed, we produce virtual-mode implementations (Section 6.2).

---

[1]Due to the large number of modes, it can be nonpractical (or even impossible) to explicitly mark the set of functional modes. In such cases, however, the designer can indicate control loops that cannot be active in parallel due to functionality restrictions. Then, the set of functional modes $\mathcal{M}^{\mathrm{func}}$ includes all modes except those containing two or more mutually exclusive controllers. If no specification of functional modes is made, it is assumed that $\mathcal{M}^{\mathrm{func}} = \mathcal{M}$.

[2]It can be impossible for the designer to assign weights to all functional modes explicitly. An alternative is to assign weights to some particularly important and frequent modes (all other functional modes get a default weight). Another alternative is to correlate the weights to the occurrence frequency of modes (e.g., obtained by simulation).

Initialization:
1. $\mathcal{M}^{\mathrm{impl}} := \emptyset$
2. **list** edges := **empty**

For each mode $\mathbf{M} \in \mathcal{M}_{\uparrow}^{\mathrm{func}}$, perform Steps 3–5 below.

3. Synthesize $\mathbf{M}$ and set $\mathcal{M}^{\mathrm{impl}} := \mathcal{M}^{\mathrm{impl}} \cup \{\mathbf{M}\}$

4. For each $\mathbf{M}' \in \mathcal{M}^-(\mathbf{M})$, add the edge $(\mathbf{M}, \mathbf{M}')$ to the beginning of the list edges

5. **while** edges $\neq$ **empty**
    (a) Remove edge $(\mathbf{M}, \mathbf{M}')$ from the beginning of edges
    (b) **if** $\mathbf{M}' \in \mathcal{M}^{\mathrm{virt}}$ and $\mathbf{M}' \neq \emptyset$, for each $\mathbf{M}'' \in \mathcal{M}^-(\mathbf{M}')$, add $(\mathbf{M}, \mathbf{M}'')$ to the beginning of edges
    (c) **else if** $\mathbf{M}' \in \mathcal{M}^{\mathrm{func}} \setminus \mathcal{M}^{\mathrm{impl}}$ and $\mathbf{M}' \neq \emptyset$
        i. Synthesize $\mathbf{M}'$ and set $\mathcal{M}^{\mathrm{impl}} := \mathcal{M}^{\mathrm{impl}} \cup \{\mathbf{M}'\}$
        ii. $\Delta J^{\mathbf{M}'}(\mathbf{M}) := \left(J^{\mathbf{M}'}(\mathbf{M}) - J^{\mathbf{M}'} \right) \big/ J^{\mathbf{M}'}(\mathbf{M})$
        iii. Compute the average weight w^{avg} of the immediate functional submodes of $\mathbf{M}'$
        iv. **if** $\Delta J^{\mathbf{M}'}(\mathbf{M}) \geqslant \frac{\lambda}{w^{\mathrm{avg}}}$, for each $\mathbf{M}'' \in \mathcal{M}^-(\mathbf{M}')$, add $(\mathbf{M}', \mathbf{M}'')$ to the beginning of edges

Figure 5: Synthesis with improvement factor λ

6.1 Control-Quality versus Synthesis Time

Our heuristic is based on a limited depth-first exploration of the Hasse diagram of modes. We use an improvement factor $\lambda \geqslant 0$ to limit the exploration and to trade quality with optimization time. Given a set of modes $\mathcal{M}' \subseteq \mathcal{M}$, let us introduce the set $\mathcal{M}'_{\uparrow} = \{\mathbf{M} \in \mathcal{M}' : \overline{\mathcal{M}}(\mathbf{M}) \cap \mathcal{M}' = \emptyset\}$, which contains the modes in $\mathcal{M}'$ that do not have supermodes in the same set $\mathcal{M}'$. Such modes are called *top* modes of $\mathcal{M}'$. For example, mode Λ is the only top mode of $\mathcal{M}$ (i.e., $\mathcal{M}_{\uparrow} = \{\Lambda\}$). We introduce the set of immediate submodes of $\mathbf{M} \in \mathcal{M}$ as $\mathcal{M}^-(\mathbf{M}) = \{\mathbf{M}' \in \underline{\mathcal{M}}(\mathbf{M}) : |\mathbf{M}| - 1 = |\mathbf{M}'|\}$, and the set of immediate supermodes as $\mathcal{M}^+(\mathbf{M}) = \{\mathbf{M}' \in \overline{\mathcal{M}}(\mathbf{M}) : |\mathbf{M}| + 1 = |\mathbf{M}'|\}$—for example, $\mathcal{M}^+(\{\Lambda_2\}) = \{\{\Lambda_1, \Lambda_2\}, \{\Lambda_2, \Lambda_3\}\}$.

Figure 5 outlines our approach. In the first and second step, the set of modes with produced implementations is initialized to the empty set and an empty list is initialized. In this first part of the synthesis, we consider only implementations of functional modes. Virtual modes are implemented only as a solution to memory constraints (Section 6.2). Note that we must implement the top functional modes (i.e., functional modes that do not have any functional supermodes). For each such mode $\mathbf{M}$, we perform Steps 3–5. In Step 3, we run the synthesis for mode $\mathbf{M}$ to produce an implementation (periods, control laws, and schedule) [8]. After the synthesis of a mode, the set $\mathcal{M}^{\mathrm{impl}}$ is updated. We proceed, in Step 4, by finding the edges from mode $\mathbf{M}$ to its immediate submodes. These edges are added to the beginning of the list edges, which contains the edges leading to modes that are chosen for synthesis.

As long as the list edges is not empty, we perform the substeps in Step 5. First, in Step (a), an edge $(\mathbf{M}, \mathbf{M}')$ is removed from the beginning of edges. In Step (b), if $\mathbf{M}' \neq \emptyset$ is a virtual mode, we do not consider it in the synthesis and resume the exploration with its immediate submodes. In Step (c), if an implementation of the functional mode $\mathbf{M}' \neq \emptyset$ has not yet been synthesized, we perform four steps (Steps i–iv). We first run the synthesis for mode $\mathbf{M}'$. Based on the obtained cost, we decide whether or not to continue synthesizing modes along the current path (i.e., synthesize immediate submodes of $\mathbf{M}'$). To take this decision, we consider the cost improvement of the implementation of mode $\mathbf{M}'$ relative to the cost when using the implementation of mode $\mathbf{M}$ to operate the system in mode $\mathbf{M}'$. We also consider the weights of the immediate submodes of $\mathbf{M}'$. In Step ii, we compute the relative cost improvement $\Delta J^{\mathbf{M}'}(\mathbf{M})$. The cost $J^{\mathbf{M}'}$ of the synthesized mode $\mathbf{M}'$ was de-

---

1. **if** $\sum_{\mathbf{M} \in \mathcal{M}_{\uparrow}^{\mathrm{impl}}} \mathrm{mem}_d^{\mathbf{M}} > \mathrm{mem}_d^{\max}$ for some $d \in \mathcal{I}_{\mathbf{N}}$

   (a) Find $\mathcal{M}' \subseteq \bigcup_{\mathbf{M} \in \mathcal{M}_{\uparrow}^{\mathrm{impl}}} \mathcal{M}^+(\mathbf{M})$ with smallest size such that, for each $\mathbf{M} \in \mathcal{M}_{\uparrow}^{\mathrm{impl}}$, $\mathcal{M}' \cap \overline{\mathcal{M}}(\mathbf{M}) \neq \emptyset$

   (b) For each $\mathbf{M}' \in \mathcal{M}'$, synthesize $\mathbf{M}'$ and set $\mathcal{M}^{\mathrm{impl}} := \mathcal{M}^{\mathrm{impl}} \cup \{\mathbf{M}'\}$

   (c) Repeat Step 1

2. Find $b_{\mathbf{M}} \in \{0,1\}$ for each $\mathbf{M} \in \mathcal{M}^{\mathrm{impl}}$ such that the cost in Equation 5 is minimized and the constraint in Equation 6 is satisfied for each $d \in \mathcal{I}_{\mathbf{N}}$ and $b_{\mathbf{M}} = 1$ for $\mathbf{M} \in \mathcal{M}_{\uparrow}^{\mathrm{impl}}$

Figure 6: Mode-selection approach

fined in Section 3, whereas the cost $J^{\mathbf{M}'}(\mathbf{M})$ of mode $\mathbf{M}'$ when using the implementation of mode $\mathbf{M}$ is given in Equation 3. In Step iii, we compute the average weight of the immediate functional submodes of $\mathbf{M}'$ as

$$w^{\mathrm{avg}} = \sum_{\mathbf{M}'' \in \mathcal{M}^-(\mathbf{M}') \cap \mathcal{M}^{\mathrm{func}}} w_{\mathbf{M}''} \Big/ \left| \mathcal{M}^-(\mathbf{M}') \cap \mathcal{M}^{\mathrm{func}} \right|.$$

Based on the given improvement factor $\lambda \geqslant 0$, the relative improvement $\Delta J^{\mathbf{M}'}(\mathbf{M})$, and the average mode weight w^{avg}, we decide in Step iv whether to consider the immediate submodes $\mathcal{M}^-(\mathbf{M}')$ in the continuation of the synthesis. Note that, in this way, the submodes with larger weights are given higher priority in the synthesis. If $\mathcal{M}^-(\mathbf{M}') \cap \mathcal{M}^{\mathrm{func}} = \emptyset$ (i.e., all immediate submodes are virtual), the average weight is set to $w^{\mathrm{avg}} = \infty$, which means that all immediate submodes are added in Step iv.

The parameter $\lambda \geqslant 0$ is used to tune the exploration of the set of functional modes $\mathcal{M}^{\mathrm{func}}$; for example, a complete synthesis of the functional modes corresponds to $\lambda = 0$. The results of this first part of the synthesis are implementations of a set of functional modes $\mathcal{M}^{\mathrm{impl}} \subseteq \mathcal{M}^{\mathrm{func}}$. Note that all top functional modes are synthesized, which means that the system can operate in any functional mode.

6.2 Control-Quality versus Memory Consumption

Given are implementations of functional modes $\mathcal{M}^{\mathrm{impl}}$ as a result of the first part. We shall now discuss how to select the implementations to store such that given memory constraints are satisfied and the system can operate in any functional mode with the stored implementations. Note that the set of top functional modes $\mathcal{M}_{\uparrow}^{\mathrm{func}}$ must have implementations in memory. These implementations can be used to operate the system in any of the other functional modes. If the implementations of the top functional modes can be stored in the available memory, we do not need to implement virtual modes. The remaining problem is to additionally select the remaining implementations to store in memory. If, however, the implementations of the top functional modes cannot be stored, we must produce implementations of virtual modes. These implementations should cover a large amount of functional modes. An implementation of a virtual supermode can replace several implementations and, therefore, save memory space but with degraded control performance in the modes with replaced implementations.

To select the mode implementations to store, we propose the two-step approach outlined in Figure 6. We first check whether the implementations of the functional top modes $\mathcal{M}_{\uparrow}^{\mathrm{impl}} = \mathcal{M}_{\uparrow}^{\mathrm{func}}$ can be stored in the available memory on each computation node $N_d \in \mathbf{N}$. If not, we proceed by selecting virtual modes to implement. In Step (a), we find among the immediate supermodes of the top modes $\mathcal{M}_{\uparrow}^{\mathrm{impl}}$ (the supermodes are virtual modes) a subset $\mathcal{M}'$ with minimal size that contains a supermode for each top mode. In Step (b), we produce implementations of the chosen virtual modes $\mathcal{M}'$ and add them to the set of implemented modes $\mathcal{M}^{\mathrm{impl}}$. The top modes of $\mathcal{M}^{\mathrm{impl}}$

is now $\mathcal{M}'$. We go back to Step 1 and check whether the implementations of the new top modes can be stored in memory. If not, we repeat Steps (a) and (b) and consider larger virtual modes until we find a set of virtual modes that cover all functional modes and that can be implemented in the available memory. After this step, we know that the implementations of $\mathcal{M}_{\uparrow}^{\mathrm{impl}}$ can be stored in the available memory and that they are sufficient to operate the system in any functional mode.

The second step selects mode implementations to store in the available memory such the control quality is optimized. This selection is done by solving a linear program with binary variables: Given the set of implemented modes, possibly including virtual modes from Step 1 in Figure 6, let us introduce a binary variable $b_{\mathbf{M}} \in \{0,1\}$ for each mode $\mathbf{M} \in \mathcal{M}^{\mathrm{impl}}$. If $b_{\mathbf{M}} = 1$, it means that the implementation of $\mathbf{M}$ is selected to be stored in memory, whereas it is not stored if $b_{\mathbf{M}} = 0$. For a correct implementation of the multi-mode system, we must store the implementation of each top mode $\mathbf{M} \in \mathcal{M}_{\uparrow}^{\mathrm{impl}}$, and thus $b_{\mathbf{M}} = 1$ for those modes. To select the other mode implementations to store, we consider the obtained cost reduction by implementing a mode. Given $\mathbf{M} \in \mathcal{M}^{\mathrm{impl}} \setminus \mathcal{M}_{\uparrow}^{\mathrm{impl}}$, this cost reduction is

$$\Delta J^{\mathbf{M}} = \min_{\mathbf{M}' \in \mathcal{M}^{\mathrm{impl}} \cap \overline{\mathcal{M}}(\mathbf{M})} \Delta J^{\mathbf{M}}(\mathbf{M}'),$$

where $\Delta J^{\mathbf{M}}(\mathbf{M}')$ is given in Step ii in Figure 5. By removing a mode $\mathbf{M} \in \mathcal{M}^{\mathrm{impl}} \setminus \mathcal{M}_{\uparrow}^{\mathrm{impl}}$ (i.e., setting $b_{\mathbf{M}} = 0$), we lose this cost improvement. The cost to be minimized is the overall loss of cost reductions for mode implementations that are removed from $\mathcal{M}^{\mathrm{impl}} \setminus \mathcal{M}_{\uparrow}^{\mathrm{impl}}$. This cost is formulated as

$$\sum_{\mathbf{M} \in \mathcal{M}^{\mathrm{impl}} \setminus \mathcal{M}_{\uparrow}^{\mathrm{impl}}} (1 - b_{\mathbf{M}}) w_{\mathbf{M}} \Delta J^{\mathbf{M}}. \tag{5}$$

The memory consumption of an implementation of mode $\mathbf{M}$ is $\mathrm{mem}_d^{\mathbf{M}}$ on node $N_d \in \mathbf{N}$. This information is given as an output of the controller scheduling and synthesis (Section 3). The memory constraint on node $N_d \in \mathbf{N}$ is

$$\sum_{\mathbf{M} \in \mathcal{M}^{\mathrm{impl}}} b_{\mathbf{M}} \mathrm{mem}_d^{\mathbf{M}} \leqslant \mathrm{mem}_d^{\max}, \tag{6}$$

considering a memory limit $\mathrm{mem}_d^{\max}$ on node N_d. After solving the linear program in Step 2, we store the implementations of the selected modes $\{\mathbf{M} \in \mathcal{M}^{\mathrm{impl}} : b_{\mathbf{M}} = 1\}$ and the multi-mode system synthesis is completed.

7. EXPERIMENTAL RESULTS

We have conducted experiments to evaluate our proposed approach. We created 100 benchmarks using a database of inverted pendulums, ball and beam processes, DC servos, and harmonic oscillators [11]. Such plants are acknowledged as representatives of realistic control problems and are used extensively for experimental evaluation. We considered 10 percent of the modes to be virtual for benchmarks with 6 or more control loops. We generated 3 to 5 tasks for each control application; thus, the number of tasks in our benchmarks varies from 12 to 60. The tasks were mapped randomly to platforms comprising 2 to 10 nodes. Further, we considered uniform distributions of the execution times of tasks. The best-case and worst-case execution times of the tasks were chosen randomly from 2 to 10 milliseconds and the communication times of messages were chosen from 2 to 4 milliseconds.

The first set of experiments evaluate the synthesis heuristic in Section 6.1. The synthesis was run for each benchmark and for different values of the improvement factor λ. All experiments were run on a PC with a quad-core CPU at 2.2 GHz, 8 Gb of RAM, and running Linux. For each value of λ and for each benchmark, we computed the obtained cost after the synthesis. This cost is $J_{\mathrm{tot}}^{(\lambda)} = \sum_{\mathbf{M} \in \mathcal{M}^{\mathrm{func}}} w_{\mathbf{M}} J^{\mathbf{M}}$ (Equation 4) and, because of the large number of modes, we could compute

978-1-60558-497-3/09 $25.00 © 2009 ACM

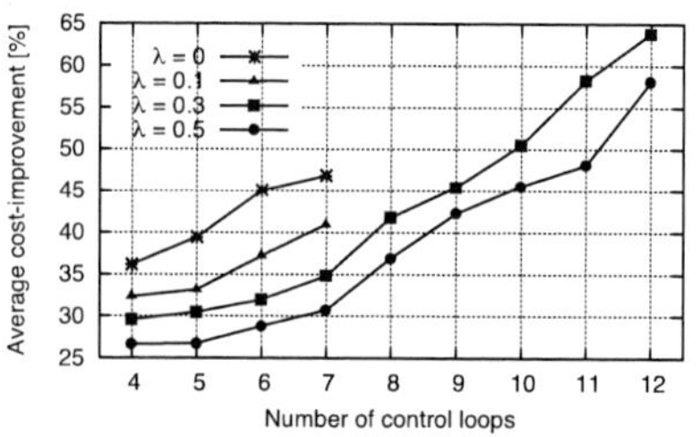

Figure 7: Control-performance improvements

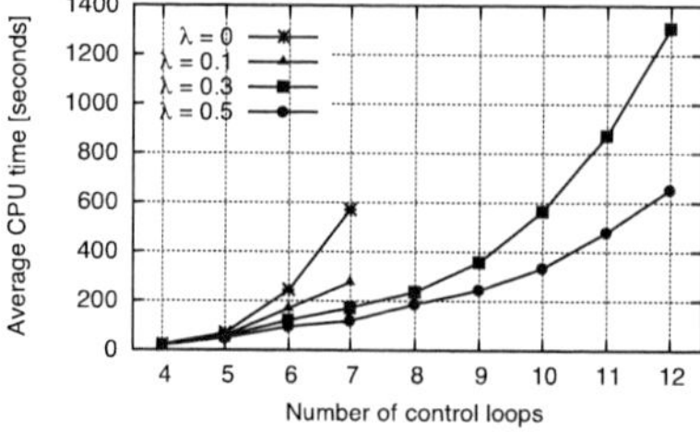

Figure 8: CPU times for mode synthesis

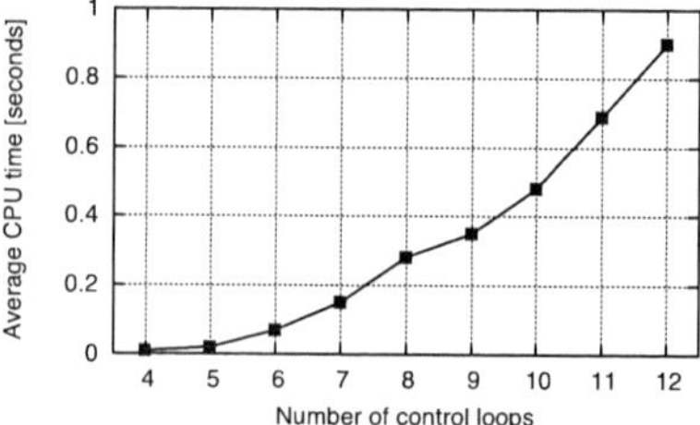

Figure 9: CPU times for mode selection ($\lambda = 0.3$)

it in affordable time for benchmarks with at most 12 control loops. It is very important to note that the time-consuming calculation of the cost is only needed for the experimental evaluation and is not part of the heuristic. We used the values 0, 0.1, 0.3, and 0.5 for the improvement factor λ. Because of long runtimes, we could run the synthesis with $\lambda = 0$ (exhaustive synthesis of all functional modes) and $\lambda = 0.1$ for benchmarks with at most 7 control loops. As a baseline for the comparison, we used the cost $J_{\text{tot}}^{\uparrow}$ obtained when synthesizing only the top functional modes $\mathcal{M}_{\uparrow}^{\text{func}}$. For each benchmark and for each value of λ, we computed the achieved cost improvement as the relative difference $\Delta J_{\text{tot}}^{(\lambda)} = \left(J_{\text{tot}}^{\uparrow} - J_{\text{tot}}^{(\lambda)} \right) \big/ J_{\text{tot}}^{\uparrow}$. Figure 7 shows the obtained cost improvements. The vertical axis indicates the average value of the relative cost-improvement $\Delta J_{\text{tot}}^{(\lambda)}$ for benchmarks with number of control loops given on the horizontal axis. The results show that the achieved control performance becomes better with smaller values on λ (i.e., a more thorough exploration of the set of modes). We also observe that the improvement is better the larger number of control loops in the multi-mode system. This demonstrates that for large number of control loops, it is important to additionally synthesize other modes than the top functional modes. The experiments also show that good quality results can be obtained also with larger values of λ, which provide affordable runtimes for even larger examples. The runtimes for the synthesis heuristic are shown in Figure 8. We show the average runtimes in seconds for the different values of the improvement factor λ and for each dimension of the benchmarks (number of control loops). Figures 7 and 8 illustrate how the designer can trade off quality of the synthesis with optimization time, depending on the size of the control system. To further illustrate the scaling of synthesis time, we mention that a system consisting of 20 control loops on 12 nodes has been synthesized with $\lambda = 0.5$ in 38 minutes.

In the second set of experiments, we evaluate the mode-selection approach. We have run the mode-synthesis heuristic (Section 6.1) with $\lambda = 0.3$. Let us denote with $\text{mem}_d^{\text{req}}$ the amount of memory needed to store the generated mode implementations on each node N_d. We have run the mode selection (Section 6.2) considering a memory limitation of $0.7\text{mem}_d^{\text{req}}$ for each node. To solve the linear program given by Equations 5 and 6, we used the `eplex` library for mixed integer programming in ECL^{i}PS^e [1]. The average runtimes are shown in seconds in Figure 9. Note that the mode selection is optimal if all top functional modes can be stored in memory. The selection of modes to store in memory is performed only once as a last step of the synthesis, and as the figure shows, the optimal selection is found in neglectable runtime compared to the overall runtime of the multi-mode system synthesis. To study the quality degradation as a result of memory limitations, we used a benchmark with 12 control loops running on 10 nodes. We have run the mode synthesis for three scenarios: First, we considered no memory limitation in the platform, which resulted in a cost improvement of 48.5 percent, relative to the cost obtained with the baseline approach (synthesis of only top functional modes). Second, we considered that only 70 percent of the memory required by the produced implementations can be used. As a result of the mode selection, all top functional modes could be stored, leading to a cost improvement of 41.4 percent (a degradation of 7.1 percent compared to the first scenario without memory constraints). For the third and last scenario, we considered memory limitations such that the implementations of the top functional modes cannot be all stored in memory. After the mode-selection approach, including the synthesis of virtual modes, we obtained a solution with a cost improvement of only 30.1 percent (a degradation of 18.4 percent compared to the case without memory constraints).

8. CONCLUSIONS

We addressed control-performance optimization for embedded multi-mode control systems. The main design difficulty is raised by the potentially very large number of possible modes. In this context, an appropriate selection of the actual modes to be implemented is of critical importance. We presented our synthesis approach that produces schedules and controllers for an efficient deployment of embedded multi-mode control systems.

9. REFERENCES

[1] K. R. Apt and M. G. Wallace. *Constraint Logic Programming using ECL^{i}PS^e*. Cambridge University Press, 2007.

[2] E. Bini and A. Cervin. Delay-aware period assignment in control systems. In *Proc. of the 29th Real-Time Systems Symposium*, pp. 291–300, 2008.

[3] R. Bosch GmbH. *CAN Specification Version 2.0*. 1991.

[4] FlexRay Consortium. *FlexRay Communications System. Protocol Specification Version 2.1*. 2005.

[5] H. Kopetz. *Real-Time Systems—Design Principles for Distributed Embedded Applications*. Kluwer Academic, 1997.

[6] B. Lincoln and A. Cervin. Jitterbug: A tool for analysis of real-time control performance. In *Proc. of the 41st Conference on Decision and Control*, pp. 1319–1324, 2002.

[7] H. Rehbinder and M. Sanfridson. Integration of off-line scheduling and optimal control. In *Proc. of the 12th Euromicro Conference on Real-Time Systems*, pp. 137–143, 2000.

[8] S. Samii, A. Cervin, P. Eles, and Z. Peng. Integrated scheduling and synthesis of control applications on distributed embedded systems. In *Proc. of the Design, Automation and Test in Europe Conference*, pp. 57–62, 2009.

[9] D. Seto, J. P. Lehoczky, L. Sha, and K. G. Shin. On task schedulability in real-time control systems. In *Proc. of the 17th Real-Time Systems Symposium*, pp. 13–21, 1996.

[10] K. E. Årzén, A. Cervin, J. Eker, and L. Sha. An introduction to control and scheduling co-design. In *Proc. of the 39th Conference on Decision and Control*, pp. 4865–4870, 2000.

[11] K. J. Åström and B. Wittenmark. *Computer-Controlled Systems*. Prentice Hall, 1997.

978-1-60558-497-3/09 $25.00 © 2009 ACM

Context-Sensitive Timing Analysis of Esterel Programs

Lei Ju[1]   Bach Khoa Huynh[1]   Samarjit Chakraborty[2]   Abhik Roychoudhury[1]

[1]Department of Computer Science, National University of Singapore
[2]Institute for Real-Time Computer Systems, TU Munich, Germany

{julei,huynhbac}@comp.nus.edu.sg, samarjit@tum.de, abhik@comp.nus.edu.sg

ABSTRACT

Traditionally, synchronous languages, such as Esterel, have been compiled into hardware, where timing analysis is relatively easy. When compiled into software – e.g., into sequential C code – very conservative estimation techniques have been used, where the focus has only been on obtaining safe timing estimates and not on the cost of the implementation. While this was acceptable in avionics, efficient implementations and hence tight timing estimates are needed in more cost-sensitive application domains. Lately, a number of advances in Worst-Case Execution Time (WCET) analysis techniques, coupled with the growing use of software in domains such as automotives, have led to a considerable interest in timing analysis of code generated from Esterel specifications. In this paper we propose techniques to obtain tight estimates on the processing time of input events by sequential C code generated from Esterel programs. Execution of an Esterel program – as in all other synchronous languages – is logically made up of a sequence of *clock ticks*. In reality, they take non-zero time which depends on the generated C code as well as the underlying hardware platform on which this code is executed. Apart from exploiting the specific structure of this C code to obtain tight WCET estimates, we capture program-level *contexts* across ticks in order to obtain tight estimates on response times of events whose processing spans across multiple clock ticks. Such tighter estimates immediately translate into more cost-effective implementations. Our experimental results with realistic case studies show 30% reduction in timing estimates when program level context information is taken into account.

Categories and Subject Descriptors

C.3 [**SPECIAL-PURPOSE AND APPLICATION-BASED SYSTEMS**]: Real-time and embedded systems

General Terms

Design, Languages, Performance

Keywords

Esterel, Synchronous programming, Worst-case Execution Time (WCET) analysis

Permission to make digital or hard copies of part or all of this work for personal or classroom use is granted without fee provided that copies are not made or distributed for profit or commercial advantage and that copies bear this notice and the full citation on the first page. To copy otherwise, to republish, to post on servers or to redistribute to lists, requires prior specific permission and/or a fee.
DAC'09, July 26-31, 2009, San Francisco, California, USA

1. INTRODUCTION

Generating implementations from synchronous language specifications is widely practiced in safety-critical domains such as avionics where certification of the generated implementation is essential. Programs in synchronous languages are considered to execute as a sequence of *clock ticks*. The processing of all events arriving within a single tick are assumed to happen instantaneously or in zero time, which is referred to as the *perfect synchrony hypothesis*. This logical view of program execution offers a clean semantics and greatly simplifies the programming and verification of concurrent reactive systems. An implementation generated from a synchronous language specification is said to follow the synchrony hypothesis if all events that are logically assumed to be processed instantaneously are processed before the next set of events arrive. Hence, for the synchrony hypothesis to hold, the estimated Worst-case Execution Time (WCET) associated with the processing of events should be less than the minimum separation time between the arrival of sets of events (that are assumed to be processed instantaneously).

Verifying the synchrony hypothesis when a synchronous language program is compiled into hardware is relatively straightforward. When such languages are compiled into software – e.g., into a sequential C code – the associated WCET analysis is more complicated and depends both on the generated code, as well as on the micro-architecture of the platform executing this code. For an arbitrary C program running on a modern processor, determining its execution time accurately is an intractable problem and also requires a substantial amount of user intervention. As a result, when performing timing analysis of software generated from synchronous languages, it is difficult to produce very tight timing estimates. This is acceptable in domains such as avionics where the only focus is on safety and formal certification.

However, with the growing use of software in domains such as automotives – which are significantly more cost-sensitive – there is an increasing need to develop timing analysis techniques that return *safe* as well as *tight* timing estimates. Tighter timing estimates clearly lead to more cost-effective, resource-saving designs. There has also been a renewed interest in timing analysis of software generated from synchronous languages, because of the recent advances in WCET analysis techniques and the availability of industry-strength tools (e.g., [1, 9]). Very recently, plans to integrate the synchronous language SCADE from Esterel Technologies with the aiT WCET analyzer from AbsInt GmbH [1], targeting general-purpose processors, was reported in [7].

Our contributions: In this paper we propose techniques for tight response time analysis of sequential C code generated from the popular synchronous language Esterel [3]. Our techniques are fairly general and would extend to other synchronous languages as well (such as Lustre/SCADE) with minor modifications. Note that the

978-1-60558-497-3/09 $25.00 © 2009 ACM

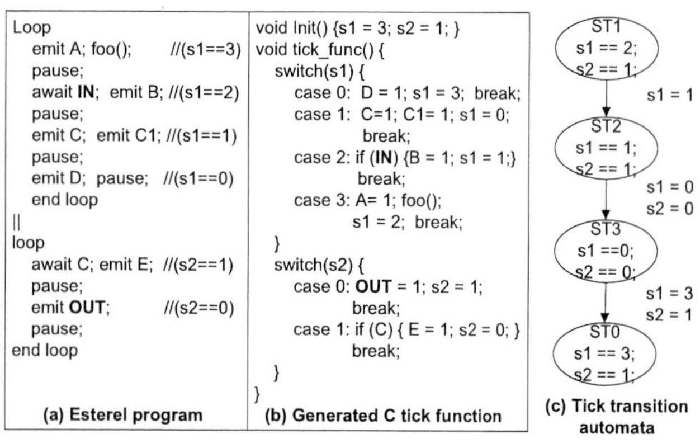

Figure 1: An Esterel program, compiled C tick function, and Tick Transition Automata

central problem here is to verify whether the synchrony hypothesis holds for a particular C code and underlying processor architecture. Towards this, we systematically remove infeasible program paths in the generated C code by exploiting the syntax and semantics of the source Esterel program. Further, we capture program contexts at the start of any clock tick in order to further refine our execution time estimates. Program contexts might comprise values of certain variables at the start of a clock tick, which might help in ruling out certain execution paths in the code to be executed within this tick. Capturing such program contexts lead to significantly improved response time estimates of events whose processing spans multiple clock ticks.

Related work: Analysis and debugging of timing properties of synchronous language specifications (targeting general-purpose processors) has been somewhat ignored until recently. High-level timing analysis of Esterel programs have been studied in [8, 15], where the problem was to compute the number of transitions in the underlying automata (encoding an Esterel specification) in response to different input events. In other words, the problem is that of computing the number of Esterel clock ticks, rather than the execution time of code within a tick. The timing analysis problem where the states of the automata have been annotated with WCET estimates has been discussed in [11].

Low level WCET analysis for a single Esterel tick is solved in [2] for a special Esterel processor, where the instruction set and microarchitecture are different from a general-purpose processor. In [14], the problem of infeasible paths in the generated code is mentioned and timing analysis of the whole Esterel program is studied. However, the methodology is restricted, since it requires two separate codes to be generated from the synchronous program — one on which the WCET analysis is performed, and one which guides the analysis. Again, the issue of capturing context information has not been addressed.

Finally, very recently, [10] and [7] report preliminary results on integrating stand-alone WCET analyzers with synchronous language compilers to provide real-time guarantees on the generated C code. The framework we propose here follows a similar direction, and we believe that our techniques can be integrated with the works in [10] and [7] to obtain even tighter timing estimates.

2. OVERVIEW OF ESTEREL

Figure 1(a) shows a toy Esterel program which will be used as an illustrative example in this paper. It consists of two concurrently running processes. The input event IN is consumed by the first process in the second logical tick, and the corresponding output event OUT is emitted by the second process in the fourth tick. Thus, it takes three logical ticks to produce the output. The first tick

in the first process performs some other operations (e.g., system initialization) that are irrelevant to the computation of IN. Details of the syntax and semantics of Esterel may be found in [3, 6].

Compiling Esterel: In our problem setting, we assume an Esterel program to be compiled into C code and executed on a single-core general purpose processor. Various techniques exist for compiling Esterel into C programs [13]. In this paper, we focus our discussion on the control flow graph-based Esterel compilation, which normally produce fast and small C code. Figure 1(b) shows (a simplified version of) the generated C code from the example Esterel program using the open source Columbia Esterel Compiler (CEC) [5]. The compiled C code is in the form of a tick function, such that the set of (feasible) execution paths of the tick function represent all possible computations in an Esterel tick.

3. TICK TRANSITION AUTOMATA

Esterel language is finite state in nature, that is, a finite-state automata can capture the behavior of an Esterel program. The full automata corresponding to an Esterel program has many uses, such as in compilation and/or program property verification. However, the combinatorial explosion in the number of states of the full automata is well-known. Instead, for our response-time calculation, we construct a smaller automata called the Tick Transition Automata (TTA for convenience, but not to be confused with Time triggered architectures). The states of this automata capture only the global control flow of an Esterel program — data variable values do not appear in the states.

As already mentioned above, compiling an Esterel program generates a tick function which captures all possible Esterel executions in a single tick. In other words, the Esterel program execution corresponds to repeated executions of this tick function. Naturally, the tick function needs auxiliary variables to capture the progress in control flow in each process (or thread) of the Esterel program. For each process i, a state variable s_i is introduced. Different values of a state variable s_i correspond to the different ticks that process i could execute.

Control states: The states of the TTA correspond to valuations of the s_i variables. A transition in TTA corresponds to assignment of one or more s_i variables. In the example shown in Figure 1, two state variables $s1$ and $s2$ are introduced to encode the tick transition information of the Esterel program. The states of the TTA correspond to the valuations of $s1, s2$. We call a valuation of the s_i variables as a *global control state* since it captures the progress in control flow of all the threads of an Esterel program. The individual s_i variables will be called as *control state variables*.

Formal definitions: Formally, a TTA identifies all paths in the Esterel program that can be executed between an input event IN and its output OUT. It can be defined as a 5-tuple $\langle Q, \Sigma, \delta, Q_0, F \rangle$, where

- Q is the set of all TTA states. A TTA state is a global control state capturing the progress in control flow in all the program threads.
- Σ is a finite set of symbols, where each symbol represents a value assignment on one or more state variables.
- δ is the transition function, such that $\delta : Q \times \Sigma \rightarrow Q$. Each transition in the automata represents an execution of a tick in the Esterel program.
- Q_0, F are the set of initial/final states of the TTA. An initial state is a global control state of the Esterel program where the input signal IN is ready to be consumed. A final state is one where the output signal OUT is produced.

978-1-60558-497-3/09 $25.00 © 2009 ACM

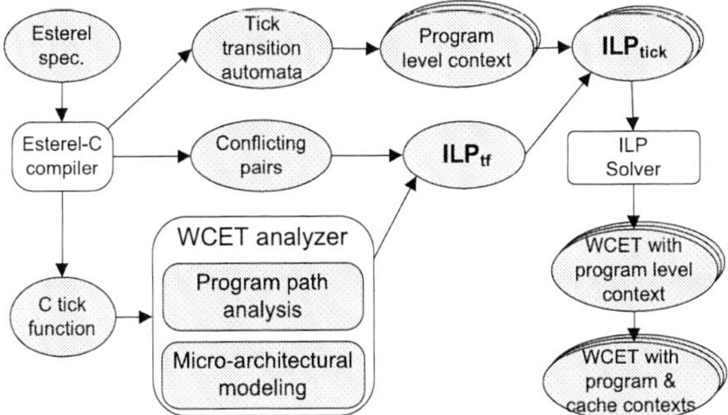

Figure 2: Framework to compute WCET for a single tick

Figure 1(c) shows the TTA between the input event *IN* and its output *OUT* for the example Esterel program and generated C tick function. In between the initial state $ST1$ and final state $ST0$, there is only one possible execution path which consists of three ticks.

4. WCET ESTIMATION

We first describe how to compute WCET of a single Esterel tick in the compiled C program. Then we show how to incorporate program contexts captured by TTA into the analysis to obtain tighter WCET estimate. An overview of the analysis framework is shown in Figure 2.

4.1 WCET of a Single Tick

Since the time required to produce a designated output *OUT* in response to a given input *IN* may involve several ticks, the first step is to tightly bound the Worst-case Execution Time (WCET) of a single tick.

The WCET of any single Esterel tick can be calculated using a static analysis based WCET estimation on the generated C tick function. The generated C code normally contains huge number of infeasible paths that do not correspond to any concrete program execution traces. In order to obtain a tighter estimate, specifically designed infeasible path analysis can be used to effectively exclude infeasible paths from being considered as the WCET path. Readers are referred to [10] for a detailed description of the single tick WCET analysis. Note that such an analysis is context-insensitive, due to the lack of consideration of program states before execution of a particular Esterel tick. In this following discussion, let us refer to this estimate as $WCET_{tf}$.

4.2 Handling Program Contexts

So far, we have elaborated on the WCET estimation of *any* given tick. Our goal now is to study whether our estimation method is *context-aware*. In other words, we want to estimate the WCET of a tick $ST_i \rightarrow ST_j$ (where ST_i and ST_j are states of the TTA) by considering the program states in which the tick is executed. The program-level context with which $ST_i \rightarrow ST_j$ is executed is captured by ST_i. We describe how such context information is integrated into our WCET estimation for a single tick.

As mentioned in the previous section, we assume that each tick between input *IN* and output *OUT* takes time $WCET_{tf}$ (obtained by solving ILP_{tf}, see Figure 2). However, this clearly leads to a gross over-estimation. To accurately estimate the worst-case response time between *IN* and *OUT*, our first step is to accurately estimate the WCET of each individual tick between *IN* and *OUT*. Thus, we accurately estimate the WCET of each transition in the TTA. This is achieved by generating additional constraints for each specific transition, and integrating them with the tick function's WCET

constraints ILP_{tf} to build a new ILP formulation. This yields an ILP formulation ILP_{tick} for each tick in the TTA (see Figure 2). Solving ILP_{tick} will produce the accurate WCET estimate of the specific tick in question.

We now explain how the additional ILP constraints for a specific tick transition $ST_i \rightarrow ST_j$ are generated. The key difficulty here is that the ILP constraints refer to occurrences of code fragments in the generated C code – they *do not* refer to occurrences of specific ticks at the Esterel program level. Hence, constraints resulting from the occurrence of a specific tick $ST_i \rightarrow ST_j$ need to be expressed in terms of *occurrences of nodes in the Sequential Control flow graph (SCFG) of the code generated from Esterel*.

Recall that ST_i and ST_j correspond to valuations of control state variables $s_1, \ldots, s_n$ where n is the number of threads in the Esterel program. Now, let the value of a state variable s_k in ST_i be v, and suppose it is assigned to a new value v' in the tick transition $ST_i \rightarrow ST_j$. For each edge $B \rightarrow B'$ from node B to B' in the sequential Control Flow Graph (SCFG) of the generated C code, if B contains a test on s_k and the edge can be taken only when "$s_k ! = v$", the following ILP constraint is generated: $\{E_{B \rightarrow B'} = 0\}$. Here, $E_{B \rightarrow B'}$ is the number of times control flows through the SCFG edge $B \rightarrow B'$. These path constraints ensure that the tick execution that corresponds to $ST_i \rightarrow ST_j$ takes only the SCFG path where "$s_k == v$" whenever state variable s_k is tested, for each state variable s_k in the TTA state ST_i.

Moreover, there can be multiple outgoing tick transitions from a TTA state ST_i. Suppose the control state variable s_k is assigned to a new value v' ("$v! = v'$") in the tick transition $ST_i \rightarrow ST_j$. To calculate the WCET for a particular tick from $ST_i \rightarrow ST_j$, we simply add ILP constraints to ensure that state variable s_k is assigned to v' during the tick execution. Let $\mathcal{B}$ be the set of SCFG nodes that contain the assignment "$s_k = v'$". We set $\{\sum_{B \in \mathcal{B}} N_B > 0\}$ where N_B is the execution count of a node $B \in \mathcal{B}$.

5. WCRT ESTIMATION

Our TTA captures all execution paths between the consumption of a given input *IN* and the production of an output *OUT*. Given a TTA and tight WCET values for tick transitions in it, we now need to compute the worst case response time (WCRT) between an input signal *IN* and output signal *OUT*.

Since the execution count of each tick transition is an integer, we employ an Integer Linear Programming (ILP) approach to compute the WCRT. We solve the following ILP optimization problem. This problem uses the WCET values of the individual ticks as constants.

$$\text{maximize} \sum_{ST_i \rightarrow ST_j \in \mathcal{T}} Cnt_{i,j} \times wcet'_{i,j}$$

where $Cnt_{i,j}$ and $wcet'_{i,j}$ are the execution count and WCET of tick transition $ST_i \rightarrow ST_j$ in TTA $\mathcal{T}$.

The linear constraints on $Cnt_{i,j}$ are developed from the TTA's control flow. Since we are calculating the WCRT between an input signal and its output, only one of the initial transitions is allowed to take place. This is captured by the constraints

$$\sum_{ST_i \rightarrow ST_j \in \mathcal{T} \wedge ST_i \in Q_0} Cnt_{i,j} = 1$$

where Q_0 is the set of initial states of TTA $\mathcal{T}$. Furthermore, for each state ST_i in the TTA, the aggregate execution counts of all incoming tick transitions should be equal to that of the outgoing transitions. Thus,

$$\sum_{ST_j \rightarrow ST_i \in \mathcal{T}} Cnt_{j,i} - \sum_{ST_i \rightarrow ST_k \in \mathcal{T}} Cnt_{i,k} = 0$$

978-1-60558-497-3/09 $25.00 © 2009 ACM

Over and above the constraints given in the preceding, we need to bound the number of iterations of every cycle in the TTA. This is a difficult task. The Esterel program may contain loops in one or more processes in the computation between the input signal *IN* and output *OUT*. Even if we know the loop bounds of these Esterel level loops, we cannot directly use them to bound the cycles of the TTA. While constructing the TTA from the Esterel program, an Esterel level loop is often partially unrolled, or fused with other loops.

We bound the number of executions of each TTA cycle as follows. Recall that each state in the TTA is a valuation of control state variables $s_1, \ldots, s_n$ – each variable s_i corresponds to a thread or process in the Esterel program. Now, for each loop L in the Esterel program, we first find the control state variable s_k that captures the control flow of the process p containing L. Since the loop defines repeated execution of certain local control states of the process p, and the variable s_k simply encodes the progress in control flow of p – we can always find a value v of s_k that appears exactly once in each iteration of the loop L. Such a value v corresponds to a control state of p lying inside the loop L. We now generate the following ILP constraints to incorporate the loop bound B_L for each loop L in the Esterel program

$$\sum_{ST_i \rightarrow ST_j \in T \wedge (s_k==v) \in ST_i} Cnt_{i,j} <= B_L$$

where s_k is the control state variable for the process containing loop L, v is a value of s_k that holds once in each iteration of L.

6. CASE STUDY

In this section, we first discuss some implementation details of our response time estimation framework (as shown in Figure 2).

In our experiments, an Esterel program is compiled into C by the Columbia Esterel Compiler (CEC). Instrumentation code is added into CEC to build a mapping between Esterel statements, intermediate representation's (e.g., abstract syntax tree, SCFG) node ID, and the generated C code. Chronos [9] – a static analysis based popular WCET estimation tool – is used to generate the WCET objective function, micro-architectural models and control flow constraints of the compiled C tick function (ILP_{tf} in Figure 2). In this paper, we use the default architecture configuration of Chronos, which assumes a direct mapped L1 instruction cache, dynamic 2-level branch predictor, 5-staged pipeline, and an instruction dispatch queue size of 4. We consider an L1 instruction cache with 256 cache blocks, where each block's size is 8 bytes. In order to compute a tight WCET of each single tick in the TTA, constraints due to program level contexts are added into ILP_{tf} to restrict the path can be taken in the tick function, which forms the ILP problem ILP_{tick}). ILP_{tick} is solved by ILOG CPLEX to produce WCET value of each individual tick.

To illustrate our response time estimation technique, we used the TURBOchannel Interface (TcInt) benchmark from the Estbench Esterel Benchmark Suite [4]. It contains 684 lines of Esterel code and 2192 lines of compiled C code. TURBOchannel is an I/O interconnect that allows several I/O options to connect to one system [12]. In our case study, we consider the "ReadROM" operation which takes 9 ticks between the input event and the final output.

The WCET of the generated tick function is $10,949$ cycles without considering any tick path information from TTA. Since the operation takes a maximum of 9 ticks to finish, a pessimistic estimation of the response time is $9 \times 10,949 = 98,541$ clock cycles. One the other hand, by utilizing program path contexts, the calculated response time is $68,103$ clock cycles, which gives a 30.9% reduction over the previous result.

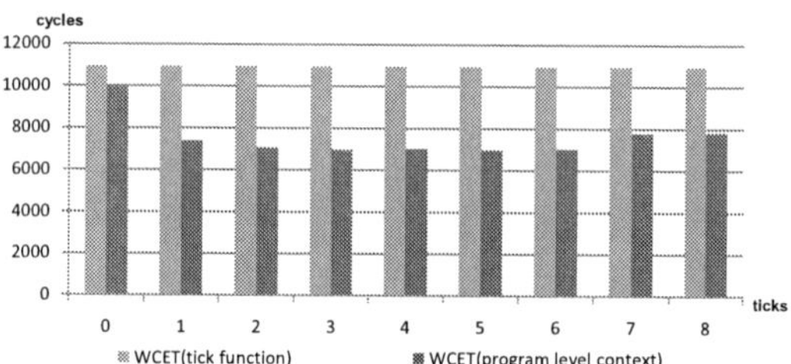

Figure 3: Tick WCET results from different calculation approaches

Finally, Figure 3 compares the WCET values calculated for the individual ticks which appear in the longest path of the TTA – ticks responsible for the worst-case response time of a ReadROM operation. For each tick, (i) the left bar shows the WCET value without any context information, (ii) the right bar shows the WCET value considering program level contexts. In our experiment, the ILP solving time for WCET of each individual tick takes 2-3 seconds, while ILP solving time to compute WCRT of multiple ticks takes less than 0.1 second.

7. CONCLUDING REMARKS

In this paper, we have proposed a context-sensitive timing analysis for Esterel programs, which is useful for obtaining tight estimates on the response times of input events being processed by the program. Towards this goal, we captured control flow contexts at the beginning of each Esterel clock tick.

In future, we will also consider micro-architectural context (e.g., cache states) to obtain even tighter WCET estimates. Note that we have assumed a single-core processor architecture in this paper. It would be interesting to investigate possible extensions of this work to multi-cores.

Acknowledgments

This work was partially supported by NUS URC grant R252-000-321-112.

8. REFERENCES

[1] AbsInt GmbH, http://www.absint.com/.

[2] M. Boldt, C. Traulsen, and R. von Hanxleden. Worst Case Reaction Time Analysis of Concurrent Reactive Programs. *Electronic Notes in Theoretical Computer Science (ENTCS)*, 203(4):65–79, 2008.

[3] F. Boussinot and R. de Simone. The Esterel language. *Proceedings of the IEEE*, 9(79):1270–1282, 1991.

[4] S.A. Edwards. The Estbench Esterel Benchmark Suite. *http://www1.cs.columbia.edu/ sedwards/software.html*, 2003.

[5] S.A. Edwards and J. Zeng. Code Generation in the Columbia Esterel Compiler. *EURASIP Journal on Embedded Systems*, 2007.

[6] A. Benveniste *et al.* The synchronous languages 12 years later. *Proceedings of the IEEE*, 91(1):64–83, 2003.

[7] R. Heckmann *et al.* Combining a high-level design tool for safety-critical systems with a tool for WCET analysis on executables. In *4th European Congress on Embedded and Real Time Software (ERTS)*, 2008.

[8] V. Bertin *et al.* TAXYS = Esterel + Kronos. A tool for verifying real-time properties of embedded systems. 2001.

[9] X. Li *et al.* Chronos: A timing analyzer for embedded software. *Science of Computer Programming*, 69(1-3), 2007. http://www.comp.nus.edu.sg/~rpembed/chronos.

[10] L. Ju, B.K. Huynh, A. Roychoudhury, and S. Chakraborty. Performance debugging of Esterel specifications. In *International Conference on Hardware Software Codesign and System Synthesis (CODES-ISSS)*, 2008.

[11] G. Logothetis, K. Schneider, and C. Metzler. Generating formal models for real-time verification by exact low-level runtime analysis of synchronous programs. In *RTSS*, 2003.

[12] M.J.K. Nielsen. TURBOchannel. In *36th IEEE Computer Society International Conference, COMPCON*, 1991.

[13] D. Potop-Butucaru, S.A. Edwards, and G. Berry. *Compiling ESTEREL*. Springer, 2007.

[14] T. Ringler. Static worst-case execution time analysis of synchronous programs. In *5th Ada-Europe International Conference*, LNCS 1845, 2000.

[15] R. K. Shyamasundar and J. V. Aghav. Realizing real-time systems from synchronous language specifications. In *RTSS Work-in-Progress Session*, 2000.

978-1-60558-497-3/09 $25.00 © 2009 ACM

Scheduling the FlexRay Bus Using Optimization Techniques

Haibo Zeng
GM R&D, Palo Alto, CA
haibo.zeng@gm.com

Wei Zheng
Univ. of California, Berkeley
zhengwei@eecs.berkeley.edu

Marco Di Natale
Scuola Superiore S. Anna, Pisa
marco@sssup.it

Arkadeb Ghosal
GM R&D, Warren, MI
arkadeb.ghosal@gm.com

Paolo Giusto
GM R&D, Palo Alto, CA
paolo.giusto@gm.com

Alberto Sangiovanni-Vincentelli
Univ. of California, Berkeley
alberto@eecs.berkeley.edu

ABSTRACT

FlexRay is a new communication protocol for automotive systems, providing support for transmission of periodic messages in static segments and priority-based scheduling of event-triggered messages in dynamic segments. The design of a FlexRay schedule is not an easy task because of protocol constraints and demands for extensibility and flexibility. We study the problem of FlexRay bus scheduling from the perspective of the application designer, interested in optimizing the performance of application related timing metrics or extensibility. We provide solutions for different task scheduling policies on existing industry standards based on a mixed integer linear programming (MILP) framework.

Categories and Subject Descriptors

C.3 [**Special-purpose and application-based systems**]: Real-time and embedded systems; G.1.6 [**Optimization**]: Integer programming

General Terms

Algorithms, Design

Keywords

FlexRay, MILP, scheduling, automotive, real-time systems

1. INTRODUCTION

The development of new by-wire functions with stringent requirements for determinism and short latencies, and the upcoming active safety functions, characterized by large volumes of data traffic, generated by $360°$ sensors around the vehicle, are among the motivations for the definition of the FlexRay standard. FlexRay is being developed by a consortium (www.flexray.org) that includes a few core members, namely, BMW, Daimler-Benz, General Motors, Freescale, NXP, Bosch and Volkswagen/Audi, as a new communication standard for highly deterministic and high speed communication. The stated objective is *to support cost-effective deployment of distributed by-wire controls*. In FlexRay, the communication speed is defined at 10 Mb/s.

Permission to make digital or hard copies of part or all of this work for personal or classroom use is granted without fee provided that copies are not made or distributed for profit or commercial advantage and that copies bear this notice and the full citation on the first page. To copy otherwise, to republish, to post on servers or to redistribute to lists, requires prior specific permission and/or a fee.
DAC'09, July 26-31, 2009, San Francisco, California, USA

The available bus bandwidth is assigned in a time-triggered pattern, which is divided in communication cycles and each communication cycle contains up to four segments *Static*, *Dynamic*, *Symbol* and *Network idle time* (Nit). Clock synchronization for the purpose of communication is embedded in the standard, using part of the Nit segment. The *static* segment of the communication cycle enables the transmission of time critical messages according to a periodic cycle, in which a time slot, of fixed length and in a given position in the cycle, is always reserved to the same node. The dynamic segment allows for flexible communications. Transmission of messages in the dynamic part is arbitrated by identifier priority (the lowest identifier messages are transmitted first). The time-triggered model of FlexRay not only allows for time determinism, but is considered as a paradigm for composability and extensibility. Each node only needs to know the time slots for its outgoing and incoming communications. The specification of these time slots is kept in local scheduling tables. No global description exists and each nodes executes with respect to its own (synchronized) clock. As long as the local tables are kept consistent, no timing conflicts nor interferences arise. Slots that are free in the (virtual) global table from the composition of the local tables can be used for future extensions. Clock synchronization and time determinism on the communication channel allow the implementation of end-to-end computations in which the data generation, data consumption and communication processes are temporally aligned, avoiding the sampling delays that are characteristic of composable periodic activation semantics, in exchange of determinism. Also, system-level time-triggered schedules allow the semantics-preserving implementation of distributed control models (including synchronous reactive models in tools such as Simulink).

In this work, we explore two options for CPU (task) scheduling, based on the two existing standards for RTOS. In one, the time-triggered OSEKTime standard is used. The other option is based on the fully-preemptive, priority-based scheduling of the OSEK standard. The main focus of this work is the development of a methodology, based on an MILP formulation of the FlexRay scheduling problem that can

- accommodate legacy components with pre-existing task and message schedules;
- deal with end-to-end deadline constraints as well as generic types of timing constraints, including jitter constraints and output coherency constraints;
- accommodate different task scheduling options.

Our method allows optimizing the scheduling configuration with respect to a number of metric functions.

State of the art In automotive systems, computation and communication functions can be time- or event-triggered. In the first case, task activations and message transmissions happen at predefined points in time. Support for this type of communication is provided by protocols that schedule messages based on tables that define transmission times. TTCAN [1] and the Time-Triggered Protocol (TTP) [7] are examples. TTP uses a generalized time-division multiple-access (TDMA) scheme with variable sized slots, in which each node has only one opportunity to transmit for each cycle. In FlexRay, slots have the same size, but a node can have more than one transmission opportunity for each cycle. Scheduling techniques for the static segment [8] [6] have been developed by extending the work for scheduling messages in a TDMA bus [3]. [4] considers a hard real-time application implemented on an FlexRay-based architecture. Messages are scheduled in the static segment, and the method reuses scheduling techniques developed for the TDMA bus [3].

In order to accommodate dynamically activated traffic, flexibility can be added with a dedicated transmission window. This is the case of Byteflight [2], introduced by BMW and later superseded by FlexRay. In [8] timing analysis of applications communicating over FlexRay is presented. The authors first present a static cyclic scheduling technique for messages in static segments. Then, they develop a worst-case response time analysis for event-based transmissions in dynamic segments. These techniques are integrated in a holistic schedulability analysis method that computes the worst-case response times of tasks and messages. [5] presents a compositional analysis framework to compute worst-case end-to-end delays for FlexRay-based architectures. *All the proposed schedulers are based on heuristics and consider only the feasibility problem with respect to deadlines.*

2. SYSTEM MODEL AND NOTATIONS

In this paper, we consider a *dataflow* model of the system. The vertices represent the *tasks* and the edges represent the *signals* communicated among them. A task τ_i is characterized by $(e_i, T_i, \Phi_i, C_i, d_i)$, where e_i is the ECU resource it needs to execute, T_i its period, Φ_i its initial phase, C_i its worst case execution time, and d_i its deadline. Each task runs an infinite sequence of instances or *jobs*. The *application cycle* or *hyperperiod* H is defined as the least common multiple of the periods of all tasks. Inside the hyperperiod, each job is considered as an individual scheduling entity and denoted as t_i. Jobs can also be denoted with reference to their task. In this case, $t_{k,j}$ denotes the j-th job of task τ_k.

Each periodic task reads its inputs at the beginning and writes its results at the end of its execution. A task pair (τ_i, τ_j) may communicate by exchanging a signal $\sigma_{i,j}$ characterized by a bit width $b_{i,j}$, represented by a directed edge from τ_i to τ_j in the dataflow graph (Figure 1). For simplicity, signals will also be denoted by a single index as σ_i. Each signal also carries a precedence constraint in the execution of the sender and receiver job. Also, signals may optionally be delivered with a *unit delay*, e.g. the signals from τ_8 to τ_2 and τ_6 in Figure 1. If the signal is delivered without a unit delay, the successor must be executed after the sender job instance activated immediately before it, but before the following one; otherwise, the successor will use the signal value produced by the previous job of the sender (see Figure 2).

The *arrival time* of a job instance t_i is denoted as a_i or, using the instance index notation as $a_{k,j}$, with $a_{k,j} = \Phi_k + (j-1) \times T_k$. The *release time* of a job is A_i, the *start of*

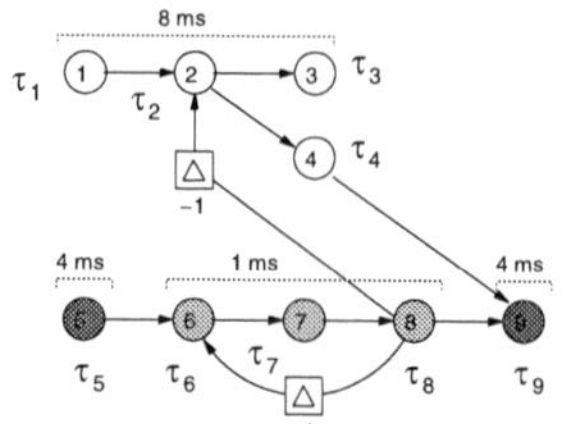
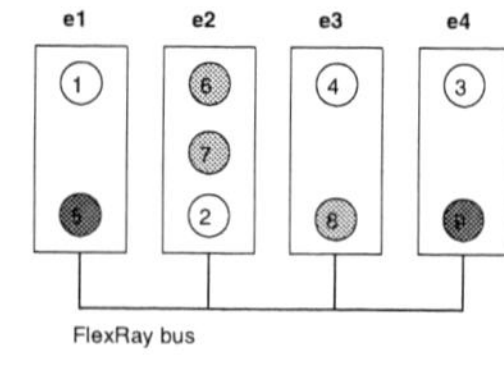

Figure 1: System model with tasks, and signals

execution time is s_i and the *finish time* f_i. The *response time* r_i of a job t_i is defined as the time interval from its arrival to its termination, i.e., $r_i = f_i - a_i$. The worst case task response time R_i is defined as the maximum of the response times $r_{i,j}$ of all its jobs $t_{i,j}$.

The set of all the task instances transmitted in the application cycle defines the *application instance graph*, as in Figure 2. Signals transmitted by tasks on the same node may be mapped to the data content of the same message. The scheduling problem of FlexRay consists of the mapping of the tasks and signals defined in the application cycle into a set of communication cycle instances (up to 64 FlexRay cycles for an application cycle). We assume a FlexRay bus configuration $(H, l_{\text{comm}}, n_{\text{slot}}, l_{\text{slot}}, b_{\text{slot}})$, where l_{comm} is the length of the *FlexRay cycle* or *communication cycle*, n_{slot} the number of slots in the static segment, l_{slot} the length of the slot, and b_{slot} the slot size in bits. Then, based on this configuration, we apply a mathematical programming framework to encode the problem and synthesize the slot ownership, the signal to slot mapping, and the message and task scheduling. This two step approach is consistent with the typical design flow in the auto industry. The application cycle is based on the application, but the communication cycle and the slot size are defined based on the need to reuse legacy components and standardize configurations. We denote the start time of the j-th slot inside FlexRay cycle k as $s_{k,j}^s$, and its finishing time $f_{k,j}^s$.

A *path* from τ_i to τ_j, or $P_{i,j}$, is a sequence $P = [\tau_i, \ldots, \tau_j]$ of tasks such that there is a link between any two consecutive tasks. For example, in Figure 1 a path exists between the tasks τ_1 and τ_9. The *latency of path* $P_{i,j}$ is defined as the time interval between the arrival of one instance of τ_i and the completion of the instance of τ_j that produces a result dependent on the output of τ_i.

3. PROBLEM FORMULATION

We use an MILP formulation to find a solution to the FlexRay scheduling problem.

Activation, release and deadline constraints T_i and d_i are given parameters as inputs, while Φ_i, and consequently a_i, A_i, s_i (for OSEKTime scheduling only) and f_i are variables for the optimization framework. Given that each task is periodic with an initial phase, its arrival time must be

$$\forall t_{i,k}, a_{i,k} = \Phi_i + kT_i \tag{1}$$

$$0 \leq \Phi_i < T_i \tag{2}$$

All tasks must finish before their deadlines.

$$\forall t_i, f_i \leq a_i + d_i \tag{3}$$

In *OSEKTime*, jobs are dispatched and executed according to a time table. We assume dispatching delays is negligible, hence $a_i = A_i$. The job start times must be larger than the corresponding activation times.

$$\forall t_i, A_i \leq s_i \tag{4}$$

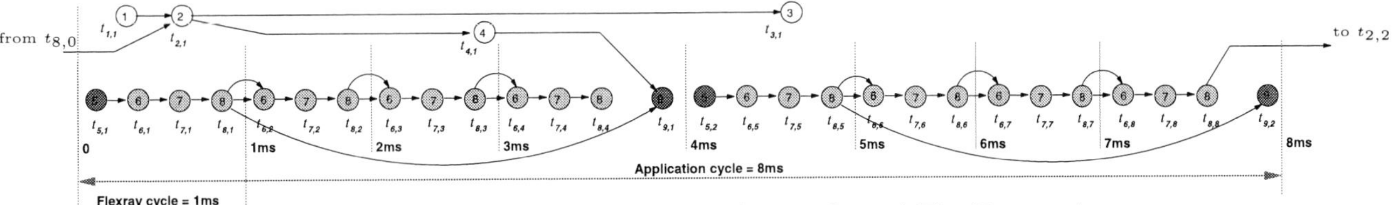

Figure 2: Instance graph, application cycle and FlexRay cycle.

In *OSEK*, tasks are activated periodically by an alarm and scheduled according to their priorities. The response time of the scheduler introduces jitter J_i to the activation of τ_i.

$$\forall t_{i,k}, A_{i,k} - a_{i,k} \geq J_i \tag{5}$$

OSEKTime start times and preemption A binary variable is used to define the order of job start times.

$$y_{i,j} = \begin{cases} 0, & \text{if } s_i < s_j \\ 1, & \text{otherwise} \end{cases} \tag{6}$$

y variables should be consistent with the definition (M is a number larger than any input parameter in the problem).

$$s_i < s_j + y_{i,j} \times M \tag{7}$$

$$s_j < s_i + (1 - y_{i,j}) \times M \tag{8}$$

A binary variable $p_{i,j}$ is used to encode preemption between t_i and t_j on the same node with no data dependency

$$p_{i,j} = \begin{cases} 0, & \text{if job } t_i \text{ is not preempted by job } t_j \\ 1, & \text{otherwise} \end{cases} \tag{9}$$

A set of constraints apply to the preemption variables. To start, mutual preemption is not allowed.

$$p_{i,j} + p_{j,i} \leq 1 \tag{10}$$

If t_i starts later than t_j, t_j doesn't need to preempt t_i. This leads to an additional constraint between y and p

$$p_{i,j} \leq 1 - y_{i,j} \tag{11}$$

If job t_i starts before t_j and is not preempted by it, then t_i finishes execution before the start time of t_j. However, if t_i is preempted by t_j, then t_i must finish after t_j.

$$f_i \leq s_j + y_{i,j} \times M + p_{i,j} \times M \tag{12}$$

$$f_j < f_i + y_{i,j} \times M + (1 - p_{i,j}) \times M \tag{13}$$

Schedulability constraints The schedulability constraints are to model the rules for computing the job start and finish times. In *OSEKTime*, we define θ as the set of job pairs (t_i, t_j) where t_i and t_j are mapped onto the same node with no data dependency and $[a_i, d_i] \cap [a_j, d_j] \neq \emptyset$. Equation 14 calculates the delay from the start to the finish of job t_i

$$f_i = s_i + C_i + \sum_{(i,j)\in\theta} p_{i,j} \times C_j \tag{14}$$

In *OSEK*, the worst case response time r_i is computed for task τ_i and applies to all its jobs (where $hp(i)$ is the set of tasks with priority higher than τ_i).

$$R_i = J_i + C_i + \sum_{j\in hp(i)} \left\lceil \frac{R_i - J_i + J_j}{T_j} \right\rceil C_j \tag{15}$$

Since all the parameters in Equation 15 except R_i are known, R_i is calculated beforehand and used as an input parameter. The finish time f_i is calculated as

$$f_i = a_i + R_i \tag{16}$$

FlexRay protocol rules The mapping of signal to slot is encoded in a binary variable

$$A_{i,j,k} = \begin{cases} 1, & \text{if } \sigma_i \text{ is mapped to cycle } j, \text{ slot } k \\ 0, & \text{otherwise} \end{cases} \tag{17}$$

If σ_i is mapped to the kth slot of the jth cycle, s_i is constrained to $s^s_{j,k}$, where $s^s_{j,k} = l_{comm} \times (j-1) + l_{slot} \times (k-1)$.

$$s^s_{j,k} < s_i + (1 - A_{i,j,k}) \times M \tag{18}$$

$$s_i \leq s^s_{j,k} + (1 - A_{i,j,k}) \times M \tag{19}$$

$$f_i = s_i + l_{slot} \tag{20}$$

Each signal can only be mapped to one slot and the sum of the payloads over all mapped signals with one specific slot will be bounded by the slot size

$$\sum_{s^s_{j,k} \leq t_{\max}} A_{i,j,k} = 1 \tag{21}$$

$$\sum_{i\in\text{signals}} b_i \times A_{i,j,k} \leq b_{\text{slot}} \tag{22}$$

where $t_{\max}$ is the maximum time span over which the schedule must be computed (see Section 3.1).

A set of binary variable encodes the ownership of each slot

$$A_{e_i,j} = \begin{cases} 1, & \text{if slot } j \text{ is owned by ECU } e_i \\ 0, & \text{otherwise} \end{cases} \tag{23}$$

The slot-to-ECU assignment applies to all the communication cycles. If signal σ_i is mapped to slot k of communication cycle j, then Equation 24 requires that slot k is owned by ECU e_i where the source task of σ_i is allocated. Equation 25 ensures that each slot is owned by at most one ECU. If no signal is mapped to slot k in any communication cycle, Equation 26 sets the slot ownership to null (n_a is the number of communication cycles in $[0, t_{\max}]$, see Section 3.1).

$$A_{i,j,k} \leq A_{e_i,k} \tag{24}$$

$$\sum_{e_p\in\text{ECUs}} A_{e_p,k} \leq 1 \tag{25}$$

$$A_{e_p,k} \leq \sum_{i\in\text{signals},j\leq n_a} A_{i,j,k} \tag{26}$$

Data dependencies If there is a data dependency between two jobs or between a job and a signal, the successors must start later than the predecessors. The following inequalities encode these constraints, where $\gamma_{i,j}$ is the worst case delay for σ_j to be written into the transmit register at the FlexRay adapter, and $\gamma_{j,l}$ is the copy time from the adapter to the input variable for the receiving job.

$$f_i \leq s_j - \gamma_{i,j} \wedge f_j \leq a_l - \gamma_{j,l} \tag{27}$$

If the communication is between two jobs t_i and t_l on the same node, a similar condition is enforced between the f_i and a_l for the case of OSEKTime.

$$f_i \leq a_l \tag{28}$$

Equation 28 is overly constrained for OSEK tasks. If the sender t_i has a higher priority than t_l, then it is sufficient that t_i arrives before t_l.

$$\begin{cases} f_i \leq a_l, & \text{if } P_i < P_l \\ a_i \leq a_l, & \text{otherwise} \end{cases} \tag{29}$$

To avoid overwriting, σ_j needs to start transmission before the arrival of the next job of the sender task.

$$s_j \leq a_{i+1} \tag{30}$$

Legacy components For legacy components, the schedule of tasks and signals are pre-defined, thus the values of design variables such as the task phase and signal to slot mapping (Equation 17) are given.

3.1 Scheduling domain

The scheduling of the tasks and the FlexRay bus must be performed beyond the first application cycle, that is, until $H + \max_i(\Phi_i)$. Since Φ_i is an optimization variable, the upper bound on the previous formula is $t_{\max} = H + \max_i(T_i)$. In the interval $[0, t_{\max}]$ there are $n_i = \lfloor t_{\max}/T_i \rfloor$ jobs for each task τ_i. However, not all these jobs can be scheduled

978-1-60558-497-3/09 $25.00 © 2009 ACM

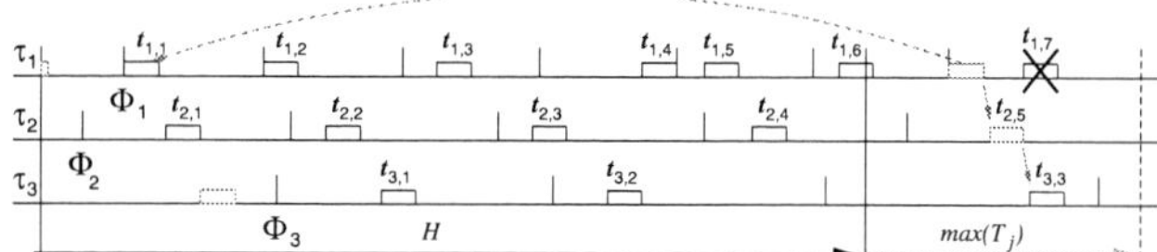

Figure 3: Schedule beyond the first hyperperiod

freely. In the example of Figure 3, this is true for $t_{3,3}$, but $t_{1,7}$ must be scheduled in a position defined by $t_{1,1}$ given that they are actually the same instance in the hyperperiod cycle. This translates into a constraint on their finish times.

$$f_{i,q} = f_{i,k} + H \text{ where } q = nc_i + k \qquad (31)$$

For time-triggered (OSEKTime) schedulers, it must also be

$$s_{i,q} = s_{i,k} + H \text{ where } q = nc_i + k \qquad (32)$$

Here $nc_i = H/T_i$ is the number of jobs of τ_i in one hyperperiod. Similar constraints exist on the scheduling of the FlexRay slots. We need to schedule beyond one application cycle, up to n_a number of cycles where $n_a = \lceil t_{\max}/l_{\text{comm}} \rceil$. If the sender and receiver jobs of signals σ_i and σ_j are exactly one application cycle away, i.e.,

$$\begin{aligned} \text{src}(\sigma_i) &= t_{k,l} \wedge \text{dst}(\sigma_i) = t_{m,n} \\ \wedge \quad \text{src}(\sigma_j) &= t_{k,p} \wedge \text{dst}(\sigma_j) = t_{m,q} \\ &\text{where } p = l + nc_k \text{ and } q = n + nc_m \end{aligned} \qquad (33)$$

then the two signals are actually the same and must be allocated to the corresponding slots (with a distance of H). We denote this relationship as $\sigma_i = \sigma_j$. For each cycle k and slot index l and, for each pair $\sigma_i = \sigma_j$, it must be

$$A_{j,m,l} = 1 \text{ if and only if } A_{i,k,l} = 1 \wedge m = k + H/l_{\text{comm}} \qquad (34)$$

3.2 Objective functions

In addition to satisfying the constraints, we seek optimality with respect to a cost function. One objective, related to extensibility, is to minimize the number of used slots

$$\text{minimize} \sum_{e_i \in \text{ECUs}, j < n_{\text{slot}}} A_{e_i, j} \qquad (35)$$

4. CASE STUDY

Table 1 describes a prototypical X-by-Wire application from General Motors. The application has 10 ECUs interconnected by a FlexRay bus. There are 47 tasks and 132 signals with periods of $1ms$, $4ms$ and $8ms$.

τ_i	e_i	$T_i(\mu s)$	$C_i(\mu s)$	τ_i	e_i	$T_i(\mu s)$	$C_i(\mu s)$
τ_8	e_9	8000	810	τ_{20}	e_{10}	8000	230
τ_9	e_9	8000	550	τ_{21}	e_5	1000	25
τ_{11}	e_9	8000	100	$\tau_{22}/\tau_{26}/\tau_{30}/\tau_{34}$	$e_5/e_6/e_7/e_8$	1000	60
τ_{12}	e_9	8000	770	$\tau_{25}/\tau_{29}/\tau_{33}$	$e_6/e_7/e_8$	1000	30
τ_{13}	e_9	8000	200	$\tau_{24}/\tau_{28}/\tau_{32}/\tau_{36}$	$e_5/e_6/e_7/e_8$	1000	20
τ_{14}	e_9	8000	110	$\tau_{23}/\tau_{27}/\tau_{31}/\tau_{35}$	$e_5/e_6/e_7/e_8$	1000	40
τ_{15}	e_9	8000	550	$\tau_{37}/\tau_{42}/\tau_{47}/\tau_{52}$	$e_1/e_2/e_3/e_4$	8000	1000
τ_{16}	e_{10}	8000	780	$\tau_{38}/\tau_{43}/\tau_{48}/\tau_{53}$	$e_1/e_2/e_3/e_4$	8000	500
τ_{10}	e_{10}	8000	510	$\tau_{41}/\tau_{46}/\tau_{51}/\tau_{56}$	$e_1/e_2/e_3/e_4$	8000	350
τ_{17}	e_{10}	8000	190	$\tau_{39}/\tau_{44}/\tau_{49}/\tau_{54}$	$e_1/e_2/e_3/e_4$	8000	1500
τ_{18}	e_{10}	8000	260	$\tau_{40}/\tau_{45}/\tau_{50}/\tau_{55}$	$e_1/e_2/e_3/e_4$	4000	1300
τ_{19}	e_{10}	8000	100				

Table 1: Tasks of the X-by-wire example

No end-to-end delay constraints are defined for this case, and the objective is to find the schedule with the minimum number of assigned slots. We tried two FlexRay configurations. In both cases, $H = 8000\mu s$. For configuration 1 (confg1), $l_{\text{comm}} = 1000\mu s$ with $n_{\text{slot}} = 22$, and the slot size $l_{\text{slot}} = 200$ bits, or $35\mu s$. In confg2 it is $l_{\text{comm}} = 8000\mu s$. The slot size is unchanged and there are $n_{\text{slot}} = 222$ slots in the cycle. Also, we tried three scheduling policies: OSEK, and OSEKTime with and without preemption (all $p_{i,j}{=}0$).

The problem is modeled in AMPL and solved using CPLEX with a time limit of one hour. If the problem formulation in Section 3 is applied "as is" to the case studies, the solver can find a feasible solution, but not the optimal one within a

hour of run time (Method (1) in Table 2). For example, for the OSEK scheduling with bus configuration 1 (top left of Table 2), there are 32672 binary variables after the AMPL pre-solve phase. CPLEX finds a feasible solution with 13 assigned slots, but cannot guarantee optimality.

We leverage additional knowledge on the problem to further restrict the search space (Method (2) in Table 2). For each case study, the optimal solution can be found within an hour, with the minimum number of used slots the same as of Method (1). In another word, these reduction techniques preserve the optimality of the solutions of the case studies. Please contact the authors for details of these techniques.

	OSEK+confg1				OSEK+confg2			
Method	# Binary	Time(s)	result	Opt	# Binary	Time(s)	result	Opt
(1)	32672	3600	13	No	42916	3600	46	No
(2)	25886	311.2	13	Yes	34412	2099.7	44	Yes
	OSEKTime+confg1				OSEKTime+confg2			
Method	# Binary	Time(s)	result	Opt	# Binary	Time(s)	result	Opt
(1)	47878	3600	13	No	62472	3600	45	No
(2)	32048	564.2	13	Yes	42314	3018.4	44	Yes
	OSEKTime+Preem+confg1				OSEKTime+Preem+confg2			
Method	# Binary	Time(s)	result	Opt	# Binary	Time(s)	result	Opt
(1)	50313	3600	N/A	No	64907	3600	50	No
(2)	34909	771.1	13	Yes	45175	1952.7	44	Yes

Table 2: Results on the X-by-wire Example

5. CONCLUSIONS

We present an optimization framework based on MILP for the scheduling of real-time applications on FlexRay-based systems. We provide solutions for task scheduling policies based on existing industry standards and optimality with respect to an extensibility metric. Optimal solutions can be found in reasonable time for a case study taken from an X-by-wire system, which proves the efficiency of our formulation and interaction with the MILP solver.

6. REFERENCES

[1] Road vehicles-controller area network (can) part 4: Time-triggered communication. *ISO Standard-11898-4, International Standards Organisation (ISO)*, 2004.

[2] J. Berwanger, M. Peller, and R. Griessbach. Byteflight - a new high-performance data bus system for safety related applications. http://www.byteflight.com, 2004.

[3] P. Caspi, A. Curic, A. Maignan, C. Sofronis, S. Tripakis, and P. Niebert. From simulink to scade/ lustre to tta: a layered approach for distributed embedded applications. *SigPlan Not.*, 38(7):153–162, 2003.

[4] S. Ding, N. Murakami, H. Tomiyama, and H. Takada. A ga-based scheduling method for flexray systems. In *EMSOFT '05: Proceedings of the 5th ACM international conference on Embedded software*, pages 110–113, 2005.

[5] A. Hagiescu, U. D. Bordoloi, S. Chakraborty, P. Sampath, P. V. V. Ganesan, and S. Ramesh. Performance analysis of flexray-based ecu networks. In *DAC '07: Proceedings of the 44th annual conference on Design automation*, pages 284–289, 2007.

[6] A. Hamann and R. Ernst. Tdma time slot and turn optimization with evolutionary search techniques. In *DATE '05: Proceedings of the conference on Design, Automation and Test in Europe*, pages 312–317, 2005.

[7] H. Kopetz and I. G. Bauer. The time-triggered architecture. In *Proceedings of the IEEE*, pages 112–126, 2003.

[8] P. Pop, P. Eles, and Z. Peng. Schedulability-driven communication synthesis for time-triggered embedded systems. *Real-Time Systems*, 24:297–325, 2004.

978-1-60558-497-3/09 $25.00 © 2009 ACM

The Wild West: Conquest of complex Hardware-dependent Software design

Hiroyuki Yagi (Organizer)
Sony Corp.
Atsugi, Kanagawa, Japan

Wolfgang Rosenstiel (Chair)
University of Tubingen
Tubingen, Germany

Jakob Engblom
Virtutech AB
Stockholm, Sweden

Jason Andrews
Cadence Design Systems, Inc.
Ham Lake, MN, USA

Kees Vissers
Xilinx, Inc.
San Jose, CA, USA

Marc Serughetti
CoWare, Inc.
San Jose, CA, USA

PANEL SUMMARY

Embedded SW design can be compared to the lawless wild west. With no clear methodology and no standard multi core platform modeling environment every company has to improvise its own solution. The problems facing embedded software users are becoming more complex since:

- Hardware platforms become very complicated with many heterogeneous processors, complicated memory structure and interconnect
- Multiple platform configurations and platform migrations drive an explosion of the number of version that need to be developed/maintained/ported
- Processors are becoming very complicated and hard to program
- Semiconductor vendors can not justify the investment required to provide a complete Hardware Abstraction Layer (HAL), RTOS, etc.
- It is hard to put together an embedded software development environment (platform model + HAL+ development tools)
- The speed required for efficient software development is very high

This panel composed of experts in the problem and solution domains will discuss the current problems and potential solutions.

Categories and Subject Descriptors

D.2.11 **[SOFTWARE ENGINEERING]**: Software Architectures –*Domain-specific architectures*.

C.3 **[SPECIAL-PURPOSE AND APPLICATION-BASED**

Permission to make digital or hard copies of part or all of this work for personal or classroom use is granted without fee provided that copies are not made or distributed for profit or commercial advantage and that copies bear this notice and the full citation on the first page. To copy otherwise, to republish, to post on servers or to redistribute to lists, requires prior specific permission and/or a fee.

DAC'09, July 26-31, 2009, San Francisco, California, USA

SYSTEMS]: Microprocessor/microcomputer applications

General Terms

Algorithms, Performance, Design, Reliability, Standardization, Languages, Theory, Verification.

Keywords

Hardware-dependent Software, Multiprocessors, Multi-core, Many-core, MPSoC, programming model, ESL, Virtual Prototyping, Virtual platform, Virtualization, heterogeneous/homogenous multi-core, symmetric/asymmetric multi-core.

PANELIST VIEWPOINTS

Jakob Engbolm: List first panelist view point here, in non-bold. The problem of software development for complex software-intensive multicore hardware is not just one of tools but also one of process. Virtual platforms is the core enabler for process change, serving as the natural hub of the system development project. With a virtual platform, software teams can decouple from hardware development schedules, and also gain superior control, insight, debuggability (reverse execution and perfect repeatability are particularly valuable for multicore), improved communication between teams, fault-injection and error provocation.

To be useful, virtual platforms have to model a full end-user system, not just an SoC. The platform has to be neutral in terms of vendors for the hardware modeled as well as the software tools attached to the virtual platform, combining SoCs and cores from many different vendors under a single roof. The virtual platform should not be tied to an SoC development process, but rather be considered as a key enabler of the system project.

Virtual platforms need to execute fast and be developed quickly, which means using loosely-timed or software-timed models as the default. Virtual platforms should be used as executable specifications and experiment carriers, expressing and testing system design assumptions for timing, features, programming models, and core counts as easily-changed variables. Detailed timing is introduced later in the system design process, when and where needed.

Jason Andrews: There is a positive trend showing software developers moving to virtual platforms for hardware-dependent software. The current generation of tools use techniques to run target code on the host machine at or above the speed of the target system. This is enabling much more software to be developed on virtual platforms much earlier in the design process. Another positive trend is the improvement in debugging. Virtual platforms have much better control and visibility compared to physical hardware and are a big improvement in overcoming issues that were difficult to reproduce. In summary, the execution and debugging of hardware dependent software using virtual platforms are positive trends.

There are two main areas for improvement. Help is coming, but adoption will take time. The first area is to improve the connection between the hardware dependent software and the hardware design process. The hardware design flow must be connected to software development. Software engineers cannot use models developed on the side from paper specifications while hardware engineers work independently on implementation. A common golden source for both hardware implementation and software development on virtual platforms is needed. The second area for improvement is in mixed hardware/software verification. Just running software does not equal verification. Better methodology to verify both hardware and software together is needed. The methodology should cover verification planning, automated environment creation, sequence and test creation, functional coverage, code coverage, and test results mapped back to the verification plan. This automation will improve quality and better leverage the benefits of virtual platforms.

Kees Vissers: *Statement not available.*

Marc Serughetti: *Statement not available.*

Nobu Matsumoto: It has been a cost-effective solution for quite some time to customize architecture for each application task. Configurable processors are one of the best approaches for the customization and provide extreme examples. At present, heterogeneous multi-core architectures are in widespread use. However, as software development dominates the total cost for SoC designs, regularly structured architectures suitable for software reuse are needed although they are still categorized as heterogeneous ones.

Since regularly structured and symmetric multi-cores are subject to particular difficulties in programming, solid programming models are required for minimizing the difficulties. On the other hand, from the viewpoint of market competitiveness, power and/or performance efficiency are strongly demanded and the large overheads associated with existing multi-core programming schemes are unacceptable. In order to satisfy these conditions, each developer pioneers its own platform consisting of symmetric yet heterogeneous multi-cores amid intensifying competition.

Thus, the use of "Wild West" in the title is apposite. Although pioneers often face difficulties, it should be emphasized that solutions definitely exist. Furthermore, differentiation is enabled by flexibility in design at the present day. This flexibility should be maintained in upcoming standardization of parallel programming from the user's standpoint. Once, the EDA industry greatly contributed to realization of design flexibility by providing ASIC solutions. Can one anticipate a similar contribution in the field of multi-core programming?

As a consequence of the widespread adoption of multi-cores, software development has become increasingly dependent on simulation techniques.

Racing detection and performance analysis based on simulation are useful measures in the field of embedded software development. Leading-edge simulation techniques presented by the EDA industry is expected to enhance the usefulness and productivity for developers.

Internet-in-a-Box: Emulating Datacenter Network Architectures using FPGAs

Jonathan D. Ellithorpe
Computer Science Division
EECS Department
UC Berkeley
Berkeley, CA 94720-1776

jde@cs.berkeley.edu

Zhangxi Tan
Computer Science Division
EECS Department
UC Berkeley
Berkeley, CA 94720-1776

xtan@cs.berkeley.edu

Randy H. Katz
Computer Science Division
EECS Department
UC Berkeley
Berkeley, CA 94720-1776

randy@eecs.berkeley.edu

ABSTRACT

In this paper we describe the Internet-in-a-Box datacenter network emulator, an FPGA-based tool for researchers to rapidly experiment with O(10,000) node datacenter network architectures. Our basic approach to emulation involves constructing a model of the target architecture by composing simplified hardware models of key datacenter building blocks, including switches, routers, links, and servers. Since models in our system are implemented in programmable hardware, designers have full control over emulated buffer sizes, line rates, topologies, and many other network properties. Full system control also gives researchers a significant degree of system visibility. Additionally, because our node model emulates servers using a full SPARC v8 ISA compatible processor, each node in the network is capable of running real applications. This allows researchers to study a network under complex real-world workloads at a scale that matches that of a large datacenter today. Moreover, because the system is a private testbed, experiments can be deterministic and therefore reproduced by other researchers. Lastly, the system is cost effective for designers, and we show that using FPGA technology on the market today we can actually emulate a network of 256-nodes for about $2,000.

Categories and Subject Descriptors

C.3 Special-purpose and application-based systems

General Terms

Design, Experimentation, Performance

Keywords

Datacenter Networking, Hardware Modeling

1. INTRODUCTION

In recent years, datacenters have been growing rapidly to scales of up to 100,000 servers [1]. Many key technologies make this possible, including modular datacenter design and server virtualization. The changes in scale enabled by these technologies

Permission to make digital or hard copies of part or all of this work for personal or classroom use is granted without fee provided that copies are not made or distributed for profit or commercial advantage and that copies bear this notice and the full citation on the first page. To copy otherwise, to republish, to post on servers or to redistribute to lists, requires prior specific permission and/or a fee.
DAC'09, July 26-31, 2009, San Francisco, California, USA

draws great attention to datacenter networking architecture. To explore this large design space, researchers typically use software simulation on virtualized servers, which can only reach a scale of O(10)~O(100) nodes [2][3]. In addition to having such limited scale, these approaches often fail to represent the true timing characteristics of the target network architectures, thus leading to less creditable results. Recently, however, cloud computing vendors have offered pay-per-use services to enable users to share their datacenter infrastructure at an O(1,000) node scale, e.g., Amazon EC2, Microsoft Azure and Google App Engine [4][5][6]. Such services, however, provide almost no visibility into the network infrastructure and provide no mechanisms for experimenting with new networking elements or devices. Also, it is expensive to run continuous experiments on thousands of nodes in a cloud.

To address the above limitations, we propose an inexpensive and reconfigurable emulation testbed for datacenter network research, called "Internet-in-a-Box", that is scalable to O(10,000) nodes using an array of FPGAs. Instead of fully implementing the target system, we build several smaller models targeting key datacenter components and compose them together in hardware. The way it works is analogous to a software based computer architecture simulator. With the latest Xilinx Virtex 5 FPGA, we can build a laptop adapter powered emulation system capable of emulating 256 nodes on a single chip at over 100MHz for about $2,000. A 10,000 node system can then be constructed from a rack of multi-FPGA boards, e.g., the BEE3 [7] system. Additionally, each node in the network is able to run real applications and OSes. The performance may be slower than a real implementation, but is still several orders of magnitude faster compared to event-driven software simulators. Furthermore, logging can be done at the hardware level, giving researchers clear visibility into the network for little or no performance penalty.

2. EMULATION METHODOLOGY

FPGAs and similar programmable ASICs have been widely used in the field of scalable system design verification, such as Cadence Palladium [8] and MentorGraphics Veloce [9]. These multi-million dollar boxes allow designers to directly implement a fully functional model of the target chip and run it at a frequency of around 1~2 MHz. We define the *Target* here to mean the system we want to emulate, while the *Host* refers to its physical representation on FPGAs. Unlike those systems, however, Internet-in-a-Box is designed to emulate a much larger target system, and therefore we do not implement a fully functional target. Neither do we implement a complex and monolithic model of the target. Rather, we have developed simplified hardware models of key datacenter building blocks, including switches, routers, links, and servers, and allow a designer or researcher to compose them together to then construct a model of the desired network architecture. In this paper we describe our node and switch model, and show how they can be used to emulate a target system consisting of a 64 node rack of machines connected to a top of rack switch. Before going through this example and our current system architecture, however, we must first describe two important concepts that define our basic approach to building hardware models and emulating the target system: a) the functional and timing split model-building methodology, and b) model virtualization.

2.1 Functional and Timing Split

Every model is composed of two parts: a *functional model* and a *timing model*. Both models are implemented in hardware. A functional model is composed of the hardware necessary to implement the basic functionality of the target, regardless of timing or performance. The timing model then orchestrates the functional model to execute operations in time. For example, say that our target is a single cycle access SRAM buffer. We can use a slower DRAM buffer as the functional model for the target in the host, and design the timing model to tick time as if we were using an SRAM (one tick for each access). The advantages of the functional and timing split approach to hardware modeling, when compared to implementing a fully functional target or a monolithic model of the target with inseparable functional and timing parts, are the following:

1) Reducing the hardware cost on FPGAs compared to a full system implementation. Because the functional

model alone lacks any notion of timing, we can neglect any performance optimizations in the target (as in the previous example).

2) Timing models can be interchanged, allowing a designer to change system characteristics like link speeds and delays without modifying the functional model.

3) The timing model only encodes the timing characteristics of the target, and can be as simple as a set of counters, which can be seen from the example in Section 3.

2.2 Virtualization

One design goal of Internet-in-a-Box is to maximize emulation throughput per FPGA. One naïve way to do this is to replicate our models to fill up the chip. Our previous work has shown that this is suboptimal [10], and that by virtualizing a host processor the overall emulation throughput in MIPS/FPGA is twice that of using the naïve model replication method. By "virtualization" we mean the use of multiple hardware contexts sharing a model's core data-path in time (analogous to hardware multithreading in modern processors). The increase in emulation throughput as a result of virtualization is due to its ability to hide emulation latencies and maximize memory bandwidth utilization. Therefore, we introduce the idea to our hardware models to improve the emulation throughput. Leveraging the timing model, then, we can patch the serialization introduced by virtualization and preserve the parallel nature of the target. Our result is that we are able to increase emulation capacity to 256 nodes on a single Xilinx Virtex 5 LX110T FPGA, as shown in Section 4.1.

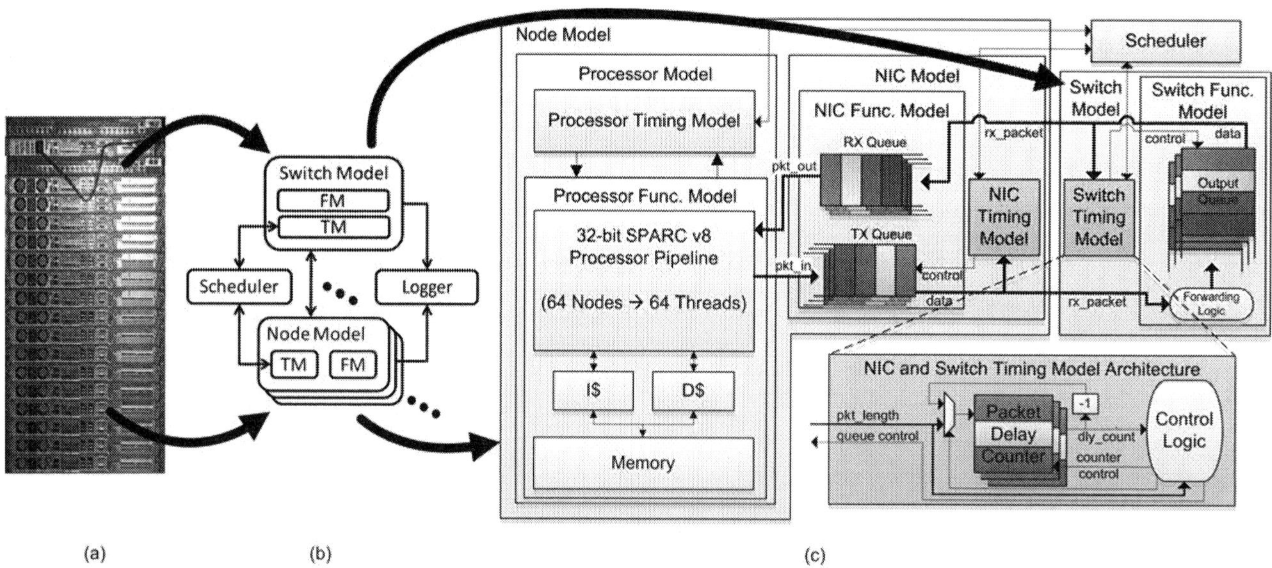

Figure 1. Mapping a target system into our emulation architecture. (a) The target is a 64 node datacenter rack with a 64-port output buffered layer-2 switch. (b) Diagram describing the logical setup and componentry of our emulation system. (c) Mapping the logical emulation setup (b) into our emulation architecture.

978-1-60558-497-3/09 $25.00 © 2009 ACM   881

3. EMULATION ARCHITECTURE

In this section we walk through and describe our emulation architecture by mapping a target system into it, as illustrated in Figure 1. Our target in this example is a 64 node datacenter rack equipped with a 64-port, output buffered layer-2 switch. We first translate the target system into abstract emulation components as in (b), and then into a detailed emulation architecture in (c). In this example, our abstract emulation setup consists of 64 node models, a switch model, a scheduler, and a logger. The node and switch models are split into their timing and functional components, as described in Section 2. Model synchronization is done by the scheduler, which maintains a bound on the spread in emulation progress between models through regular communication with and control of the timing models. Lastly, the logger is responsible for monitoring interesting places in the system and streaming event records to off-chip storage. The following sections describe the node and switch model architectures, and how they are used to emulate the target system.

3.1 Node Model

Our node model targets a single datacenter server, and as illustrated in Figure 1(c) is actually composed of two more specific models: one for the node's processor and one for the node's Network Interface Card (NIC). The processor's functional model implements the 32-bit SPARC v8 ISA, and is a single issue in-order pipeline that runs at over 100MHz. The model has been verified against the certification test suite donated by SPARC International, and can therefore run unmodified SPARC binaries. A single functional model pipeline is virtualized using *host multithreading*, which refers to the usage of N hardware threads to implement N functional models. We have chosen N to be 64 for FPGA hardware mapping efficiency. Therefore, we can use a single pipeline to emulate the 64 nodes in the target system and run real applications on each of them. If more nodes are to be emulated, we would replicate the multithreaded processor pipeline appropriately. The processor timing model tracks the number of completed target cycles for each processor functional model, and statically schedules the threads in round-robin order for design simplicity. In the event that the functional models get out of sync, the timing model can inject nop instructions into the faster threads' instruction stream while letting the slower threads catch up.

The second component of the node model is a NIC which has been virtualized, similar to the processor, to support one NIC for every thread. Our NIC model supports sending and receiving of 64-bit layer-2 packet descriptors, illustrated in Figure 2. The packet payload is stored in memory, and can be looked up using the TAG field by any component in the system (e.g., by the receiving node model to transfer the full packet into his address space).

SRC MAC	DST MAC	LENGTH	TAG

Figure 2. 64-bit IIAB Packet Descriptor

Our use of packet descriptors is motivated by the limits of FPGA memory resources (less than 1MB of on-chip memory for an LX110T FPGA). Upon receipt of a packet from the processor, the NIC's functional model observes the source node and places that packet into the node's transmit queue. Similarly, packets destined for node X from the network are stored in node X's receive queue. Because we have separated the functional and timing models, it is not necessary for us to implement 64 independent queues. Instead we have implemented each set of 64 queues using a single-port

memory and access them serially to save FPGA resources. We let the timing model control the completion of target cycles to reflect the parallel and independent nature of the queues in the target system.

The timing model for the NIC is illustrated in Figure 1(c) and determines at what target cycle the packet at the head of each TX queue should be transmitted. It does this by using its Packet Delay Counters which count down the number of target cycles for transmission. The timing model therefore has the ability to control the line rate of each port. For example, if the NIC's timing model specifies a 1Gbps link and 1ns target cycles, a 64B packet translates into a 512-cycle delay in the target. When the counter reaches zero the packet is sent and a new value is calculated and loaded for the next packet in the queue, if any. Similar to the queues described above, the Packet Delay Counters are implemented using a single-port memory to save FPGA resources. Thus, in 64 host cycles the timing model will have checked, decremented, or loaded 64 counter values one after the other.

3.2 Switch Model

Our switch model targets a 64-port, 1Gbps, output buffered layer-2 switch, and has a very similar architecture to that of the NIC model. The functional model is composed of 64 queues implemented in a single-port memory, and forwarding logic to direct incoming packets to the correct output queue. The queues may be sized appropriately to model, in approximation, the size of the target buffers. We say approximate because the switch model stores a fixed number of packets, not bytes. The timing model controls the transmission of these packets across a single on-chip link that is multiplexed in time to emulate 64 physical links in the target. In our current implementation the switch stores 64 packets per output buffer using on-chip SRAM. If more buffer storage is needed than is afforded by on-chip resources then external DRAM can also be used. In this case the timing model can slow the passage of target cycles for the switch in the event of a high-latency access to DRAM.

4. DISCUSSION

4.1 Resource Consumption

Table 1 outlines the resource consumption for our system on a Virtex5 LX110T FPGA. Given these resource requirements, we can instantiate 4 processor pipelines, 4 NICs, and 5 switches each with a 7,680 packet buffering capacity on an LX110T FPGA. Since each packet descriptor represents a real packet of size between 64B and 1500B, 7,680 packets translates to between 0.5MB and 11MB of emulated capacity.

Table 1. FPGA resource consumption on an LX110T.

	Reg. Bits	LUTs	BRAM (18kb)
Processor	4886 (7.1%)	2839 (4.1%)	30 (10.1%)
Switch	53(0.1%)	137 (0.2%)	30 (10.1%)
NIC	77 (0.1%)	250 (0.4%)	4 (1.4%)
Mem. Ctrl.	2472 (3.6%)	1975 (2.9%)	10 (3.4%)

As mentioned before, if more memory is needed then DRAM can also be used. A single $2,000 FPGA platform [11] is therefore capable of emulating a 256-node system with a network of 5 switches.

978-1-60558-497-3/09 $25.00 © 2009 ACM

Looking at Table 1, it is clear that our limiting resource is on-chip memory. Thus, projecting forward to the largest 40nm Xilinx Virtex6 FPGA (SX475T), we will have 6x more resources to either spend on more models or increase the complexity of our models. In terms of on-chip packet-buffering capacity in the network, that translates to about 500,000 in-flight packets.

4.2 Slowdown

Our emulation system runs slower than the target system in terms of wall clock time. There are two contributing factors to this slowdown: a) FPGA clock speeds, and b) virtualization. Since the FPGA runs at ~100MHz, emulating a 1GHz target system contributes 10x to our slowdown. Also, because we virtualize our hardware models into 64 time slices, there is an additional slowdown factor of 64. However, host multithreading on the processor can hide emulation latencies and offset this slowdown. In fact, recent results have shown that our processor model is able to achieve more than 60 MIPS in the host running at 100 MHz and all while the timing model synchronizes threads on every target cycle. This translates to approximately 1 MIPS per thread. If the target, then, is a 1 GHz processor with a CPI of 2, our slowdown factor is approximately 530, which is less than our upper bound on slowndown (640) and is attributable to the benefits of virtualization. Our actual slowdown factor has yet to be determined, since our processor currently runs at more than 130 MHz and a more permissive timing model may not synchronize threads every target cycle, reducing the synchronization overhead.

4.3 Scaling with Multiple FPGAs

Although our discussion thus far has focused on a single FPGA, partitioning the emulation across multiple FPGAs is fairly straightforward. This is because the functional and timing model split methodology makes each model inherently asynchronous and loosely coupled to the rest of the system. Furthermore the host multithreading technique will help to hide the inter-chip communication latencies.

4.4 Use for Energy Modeling

In our system it is easy to monitor any component and record any event. Therefore, energy consumption can be estimated by gathering the appropriate data and plugging it into the right power models. Coupled with various other system measurements, it is possible to generate a complete system picture and easily identify places for potential optimizations.

5. FUTURE WORK

Our future work primarily involves continuing to develop the system's functionality and scaling across many FPGAs. Validating our switch model using real workloads is also an important next step, as well as understanding the performance bottlenecks in the system when scaling. Finally, we would like to use our system to validate the datacenter networking architecture research of others and explore ideas of our own. At present we are working towards using the system to study the TCP Incast problem [12]. In particular, we are using a ported TCP implementation to allow nodes to establish connections in a many-to-one fashion over a single switch, and seeing how the resulting throughput collapse unfolds at both the hardware and software levels. Since we have full control of the hardware, we may then investigate the effect of various switch architectures and features on the system's behavior.

6. CONCLUSION

In this paper, we have described the Internet-in-a-Box datacenter network emulator and demonstrated how to map a rack of 64 machines with a 64-port rack switch into our emulation architecture on a single FPGA. Additionally, we have shown that our system is capable of emulating 256 nodes connected in a network of 5 switches, also on a single FPGA system costing $2,000. The relative cost of ownership for such a system compared to a 256-node cluster makes our platform attractive for researchers. Moreover, every researcher can own one on his or her desktop for themselves as a private research testbed, as opposed to time sharing hundreds of machine in a large machine room. We believe that Internet-in-a-Box will be a powerful CAD tool for exploring datacenter networking architectures in the future.

7. REFERENCES

[1] R. H. Katz, "Tech Titans Building Boom: The Architecture of Internet Datacenters," *IEEE Spectrum,* February 2009

[2] Mohammad Al-Fares, Alex Loukissas, and Amin Vahdat, A Scalable, Commodity, Data Center Network Architecture, Proceedings of the ACM SIGCOMM Conference, Seattle, WA, August 2008.

[3] Dilip Joseph, Arsalan Tavakoli, Ion Stoica, *A Policy-aware Switching Layer for Data Centers*, SIGCOMM, August 2008.

[4] http://aws.amazon.com/ec2/

[5] http://www.microsoft.com/azure/default.mspx

[6] http://code.google.com/appengine/

[7] BEE3 System, http://research.microsoft.com/en-us/projects/bee3/default.aspx

[8] http://www.cadence.com/products/sd/palladium_series/pages/default.aspx

[9] http://www.mentor.com/products/fv/emulation-systems/

[10] Zhangxi Tan, Krste Asanović, and David A. Patterson, "An FPGA Host-Multithreaded Functional Model for SPARC v8", 3rd Workshop on Architectural Research Prototyping, at 35th International Symposium on Computer Architecture, June 2008.

[11] Digilent Virtex-5 OpenSPARC Evaluation Platform, http://www.digilentinc.com/Products/Detail.cfm?NavTop=2&NavSub=599&Prod=XUPV5

[12] E. Krevat, V. Vasudevan, A. Phanishayee, D. Andersen, G. Ganger, G. Gibson, S. Seshan. On Application-level Approaches to Avoiding TCP Throughput Collapse in Cluster-Based Storage Systems. Proceedings of the 2nd international Petascale Data Storage Workshop held in conjunction with Supercomputing '07. November 11, 200

978-1-60558-497-3/09 $25.00 © 2009 ACM

Sustainable Data Centers: Enabled by Supply and Demand Side Management

Prith Banerjee, Chandrakant D. Patel, Cullen Bash, Parthasarathy Ranganathan

Hewlett-Packard Laboratories, 1501 Page Mill Road, Palo Alto, CA-94304

{firstname.lastname}@hp.com

ABSTRACT

The environmental impact of data centers is significant and is growing rapidly. Servers alone in the US consumed 1.2% of the nation's energy in 2005, according to the EPA. In the following year, the EPA found that the cost of energy rose by 10%. However, there are many opportunities for greater efficiency through integrated design and management of data center components. To that end, we propose a sustainable data center that replaces conventional resource delivery models with a framework centered around the supply and demand side management of *all* data center resources including IT, power and cooling. We have identified five elements for achieving this vision: data center scale lifecycle design, flexible and configurable building blocks, pervasive cross-layer sensing, knowledge discovery and visualization, and autonomous control. We describe these principles and provide selected results that quantify the potential for savings.

Categories and Subject Descriptors

C.0 [**Computer System Designs**]: General – *system architectures;* K.4 [**Computing milieux**]: General – *computers and society.*

General Terms

Design, Management, Economics

Keywords

Sustainability, Exascale, Data Centers

1. INTRODUCTION

A recent study found that IT is responsible for about 2% of global greenhouse gas emissions [1], about as much as the aviation industry. Furthermore, it projected that that this share would double by the year 2020. Increasing environmental concern and regulatory action will soon force a paradigm shift in how IT solutions are designed and managed across their lifecycles. Data centers are a prominent component of this impact, as well as the fastest growing.

To turn this crisis into an opportunity, we propose the development of a suite of technologies for a *sustainable data center* (SDC). Specifically, we propose developing technologies

to reduce the environmental footprint of the data center to such an extent that the services offered from such a facility would be more environmentally friendly than conventional services offered within the physical infrastructures. The SDC would be enabled through supply and demand side resource management:

Supply Side

- Design of physical infrastructure with focus on lifecycle engineering and management, and the available energy required to extract, manufacture, operate and reclaim components;

- Utilization of local resources to minimize destruction of available energy in transmission, and construction of transmission infrastructure.

Demand Side

- Provisioning data center resources based on the needs and service level agreement of the user through use of flexible building blocks, pervasive sensing, knowledge discovery and policy based control

Developing and demonstrating SDC requires the multi-disciplinary collaboration of mechanical engineers, electrical engineers, computer scientists, and others. The *hardware* infrastructure of the data center consists of thousands of servers hosting revenue-generating services, interconnected with each other and the outside world via networking equipment, and relying on storage devices for persistent data. These hardware elements are managed by a datacenter-wide *software* stack that spans the platforms and virtualization layers. The data center also

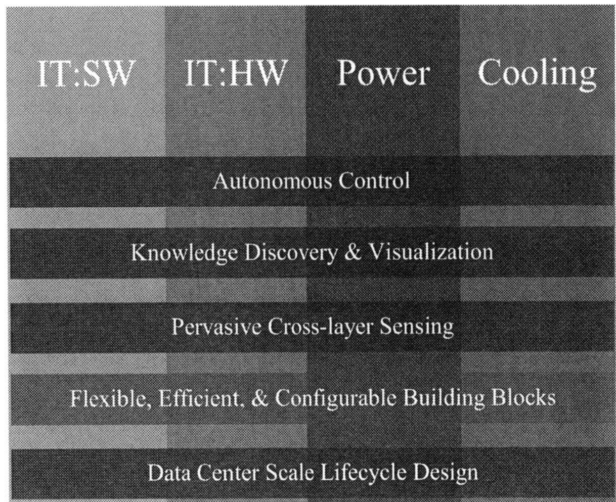

Figure 1: Sustainable Data Center Key Elements

Permission to make digital or hard copies of part or all of this work for personal or classroom use is granted without fee provided that copies are not made or distributed for profit or commercial advantage and that copies bear this notice and the full citation on the first page. To copy otherwise, to republish, to post on servers or to redistribute to lists, requires prior specific permission and/or a fee.
DAC'09, July 26-31, 2009, San Francisco, California, USA
Copyright 2009 ACM 978-1-60558-497-3/09/07...$10.00

has a *power* infrastructure that feeds electricity to all of the equipment, and a *cooling* infrastructure that removes heat from the equipment. The economic and environmental burden of the latter two infrastructures often equals or exceeds that of the compute infrastructure [2].

Figure 1 shows the five key elements of SDC along with the four infrastructure verticals with which they interact. The foundation of a sustainable data center is *lifecycle-based design* based on sustainability, service-level agreement (SLA) and total cost of ownership (TCO) goals through the entire data center lifecycle (i.e. build, operation and end-of-life). The data center is designed with *flexible efficient elements* that enable the manipulation of resources within each infrastructure. Examples can include a multiplicity of micro-grids for power generation, variable air-conditioning systems for cooling resource generation and server architectures that are scalable and power proportional. *Pervasive sensing* that cross-cuts the platforms and management layers is necessary in order to gather the information necessary during operation. *Knowledge discovery and data analytics* can then be employed to evaluate resource needs within each infrastructure and ensure the operational requirements are being met. The final element is an *autonomous controller* that holistically optimizes the energy for a given end-user SLA from both the supply and demand side. Finally, aside from cutting across all infrastructures, each element also cuts across multiple academic disciplines and will require close collaboration with historically separate research groups.

2. DATA CENTER LIFECYCLE DESIGN

Sustainable operation must begin with sustainable design. As the first and foundational element of SDC, we propose a Data Center Synthesizer to automate and optimize data center design to facilitate meeting sustainability, SLA and TCO goals. This process begins with a description of the services to be provided and any constraints on the design. At HP Labs, we have developed an SLA decomposition approach that can transform Service Level Objectives (SLOs) for multiple multi-tier applications into the computing, power and cooling resources required to deliver those services, a complex and challenging task since the joint requirements are typically less than the sum of the individual requirements [3-5]. Power distribution and cooling equipment are selected to meet the operating characteristics of the computing equipment, taking into consideration real-time performance management that will be incorporated. The next step is projection of a candidate data center solution based on the selected computing, power and cooling equipment. This step results in a complete model and physical layout of the data center, incorporating all computing, networking and storage devices and the power and cooling infrastructure required to support them. Future enhancements will include provisions to specify anticipated growth rates to enable the equipment selection to be optimized to accommodate that growth, plus additional specifications and operational policies that might affect equipment selection.

The candidate data center is analyzed for sustainability criteria, performance, availability, thermal characteristics and TCO. The sustainability criteria include power consumption for computing and cooling, carbon emissions, heat load and embedded exergy, which accounts for the loss in exergy (usable energy) for all components and manufacturing processes associated with each

piece of equipment, from raw materials extraction to finished products. If the design does not meet key customer requirements or if the design can be improved, a re-synthesis process uses the results to modify the specifications and generate another candidate data center. Iteration continues in a fully-automated loop until an optimal solution is reached.

3. FLEXIBLE BUILDING BLOCKS

As an example of innovations in the architectural building blocks driving energy efficiency and sustainability, we next discuss our work on microblades and megaservers [6]. Specifically, our work seeks to understand and design energy-efficient next-generation servers for emerging cloud workloads. We put together a detailed

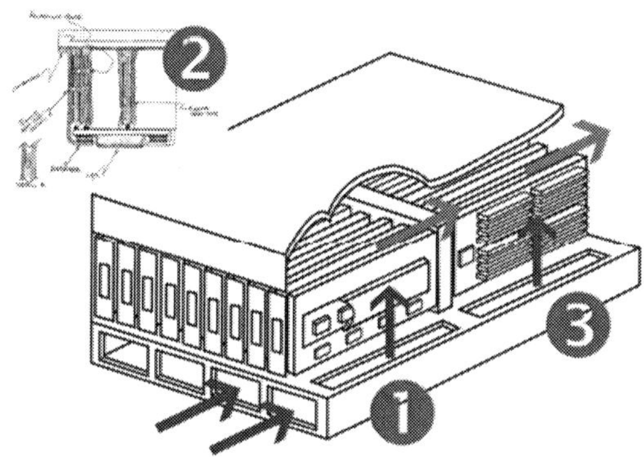

Figure 2: Microblade server.

evaluation infrastructure including a new benchmark suite for internet sector workloads, and detailed performance, cost, and power models to quantitatively characterize bottlenecks. Based on the insights from a systematic cost analysis, we propose a new design involving modular cost-effective server blocks ("microblades") to build large powerful computing environments ("megaservers").

Our work incorporates volume non-server-class components (CPU, flash) in disaggregated building blocks with ensemble-level provisioning (memory, storage) and novel form-factor, interface, infrastructure optimizations, and novel packaging of power and cooling at the micro-blade level. Our evaluation shows that this approach has the potential for dramatic improvements compared to the state-of-the-art, improving energy efficiency by factors of 4 to 6 in some cases.

As ongoing work, we are currently looking at extending the ideas proposed in this work further to look at co-designed packaging and system architecture, specifically exploring disaggregated blade designs for future dematerialized datacenters.

4. PERVASIVE SENSING

The next element discussed in Figure 1 is pervasive cross-layer sensing and monitoring. In current data centers, the degree of sensing is not well distributed across the infrastructure layers. While the IT layers (particularly the software layer) contains an ubiquitous sensing network, sensing in the power and cooling layers is sparse. We have shown that pervasive sensing in the environmental layers, particularly the cooling layer, is necessary for efficient resource distribution [15][16].

978-1-60558-497-3/09 $25.00 © 2009 ACM

Furthermore, aggregation and coordination of data across the infrastructure layers is required for integrated operation. A key element of this is an underlying architecture for coordinated monitoring and management. To enable this, we have proposed vManage [7], a solution to loosely couple platform and virtualization management and facilitate monitoring and management coordination in data centers. Our solution is comprised of registry and proxy mechanisms that provide unified monitoring and actuation across platform and virtualization domains, and coordinators that provide policy execution for better VM placement and runtime management, including a formal approach to ensure system stability from inefficient management actions. Figure 3 illustrates our approach.

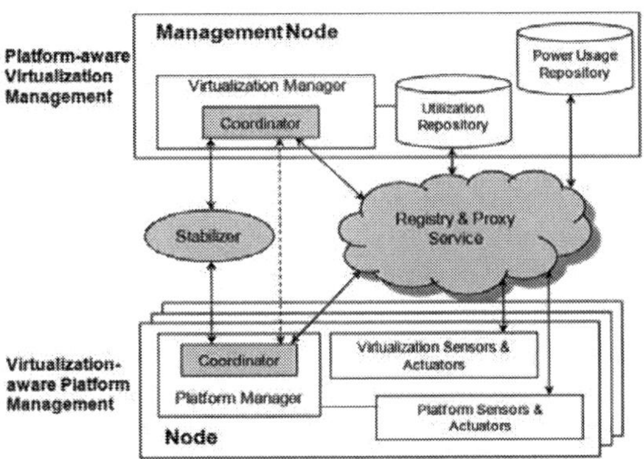

Figure 3: Cross-layer monitoring and management

5. KNOWLEDGE DISCOVERY

Knowledge discovery and data analysis tools are utilized to evaluate the data generated during operation. Knowledge discovery techniques generate insights and knowledge from disparate data sources across the facility. Data center facilities and systems produce huge amounts of data, on the order of megabytes per server per day, including environmental sensor data (e.g., temperature), operational states (e.g., system utilization), and workload information (e.g., user requests). Since the sheer volume of data precludes manual inspection, automated data mining and knowledge discovery techniques are applied to identify trends, patterns, and models for more efficient operation of a data center. In particular, we focus on three key areas: (1) detecting uncorrelated data center events using Principal Component Analysis [8]; (2) visual analytics for thermal state management [9], and (3) temporal data mining of common motifs or patterns for enhancing operational efficiency [10]. Among other benefits, the tools will help optimize resource utilization, predict events, manage growth and improve reliability.

6. AUTONOMOUS CONTROL

The final element of Fig. 1 considers the autonomous control of resources within the infrastructure verticals. Since available energy supply has direct impact on the exchequer, it is sensible to manage the available energy supply based on demand and improve the effectiveness of utilization of available energy for the total IT delivery process at each step. Managing onsite power generation, power delivery, power storage and cooling

infrastructure operation to provide critical services to the IT demand layer are critical to sustainable operation.

Figure 4 shows the sustainable data center ecosystem. During runtime, the data center is managed to reduce resource consumption from the IT, power and cooling infrastructures. Service Level Agreements (SLAs) are used to define the operational requirements of the infrastructure based on performance, TCO and sustainability criteria. These criteria are applied to all aspects of data center operation from, for example, workload placement and server consolidation to the distribution of cooling resources based on the provisioned workload. Apart from the distribution of resources like power and cooling, the generation of these resources is also considered in order to take advantage of efficient sources that may have time-varying attributes (like photovoltaics for power generation, or bringing in outside air for cooling during certain times of the day or year).

A key characteristic of autonomous control in SDC is that control is integrated across the infrastructure verticals. Since sensing is pervasive and information is shared, resource control decisions are made that consider the entire data center ecosystem rather than on a single portion of that ecosystem. Workload schedulers provide an example. Currently, workload may be scheduled

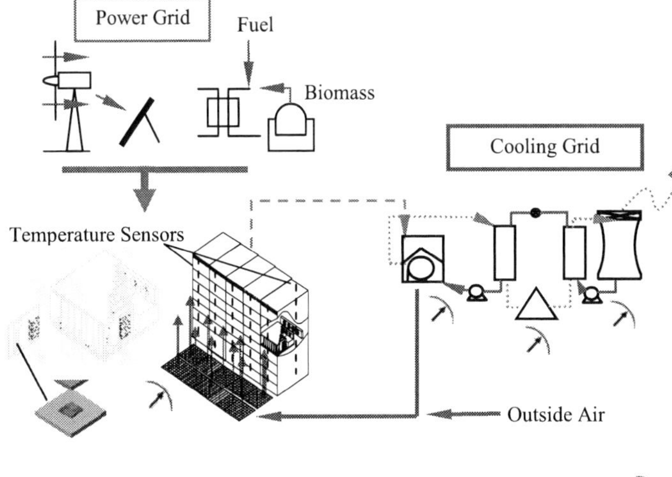

Figure 4: Supply and demand side management

according to performance and availability requirements with supply side attributes (i.e. power and cooling generation and distribution) ignored. However, workload placement location can impact overall data center operational cost. We have developed a method for ranking workload placement according to cooling efficiency in the data center. We've incorporated this ranking into both batch and real-time workload schedulers [11]. We have experimental evidence that quantifies the impact of this integration and shows how the automated migration of a virtualized workload from one set of nodes located in an inefficient thermal zone to a more efficient zone reduced data center energy usage by 27% and improved available thermal capacity by 22% [12].

Furthermore, the flexible building blocks described in Section 2 provide sufficient granularity to provide resources where and when they are needed rather than relying on over-provisioning to meet operational requirements. We recently consolidated 14 laboratory data centers into one large site in Bangalore, India, and

978-1-60558-497-3/09 $25.00 © 2009 ACM

applied a dynamic control system that adjusts the utilization of the air conditioning system based on 7,500 temperature sensors deployed throughout the data center. The 40 per cent savings in cooling power consumption achieved in this facility translates to annual savings of approximately $1.2 million, when compared to the conventional approach [13].

Finally, while several past solutions have individually evaluated different techniques to address separate aspects of this problem, in hardware and software, and at local and global levels, there has been not much corresponding work on architectural approaches to coordinating all these solutions. In the absence of such coordination, these solutions are likely to interfere with one another, in unpredictable (and potentially dangerous) ways. We discuss our solution that addresses this problem [14]. We make two key contributions. First, we propose and validate a power management solution (illustrated in Figure 5) that coordinates different individual approaches. Using simulations based on 180 server traces from nine different real-world enterprises, we demonstrate the correctness, stability, and efficiency advantages of our solution. Second, using our unified architecture as the base, we perform a detailed quantitative sensitivity analysis and draw conclusions about the impact of different architectures, implementations, workloads, and system design choices.

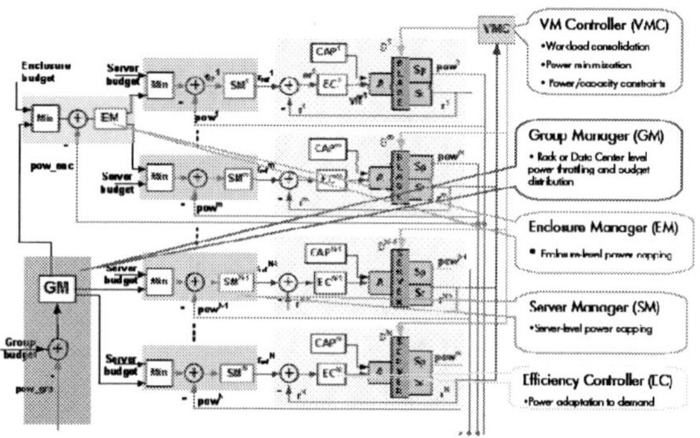

Figure 5: Coordinated policy management across different layers

7. CONCLUSIONS

The environmental impact of IT is a growing concern worldwide. Data centers contribute a large fraction of this impact. This concern, coupled with increasing government interest in regulating data center resource consumption, is resulting in a new approach to management of data centers. This paper describes an integrated life-cycle based approach to the design and management of sustainable data centers enabled by supply and demand side management of resource consumption.

8. ACKNOWLEDGEMENTS

The authors would like to collectively acknowledge the members of the Sustainable IT Ecosystem and Exascale Computing Laboratories within Hewlett-Packard Labs, many of whom contributed directly to the research described herein.

9. REFERENCES

[1] Mingay, S., 2008, "IT's Role in a Low Carbon Economy," Keynote Address, *Greening the Enterprise 2.0 Conf.*

[2] Belady, C. L., 2007, "In the Data Center, Power and Cooling Costs More Than the IT Equipment It Supports," Electronics Cooling Mag., **13**(1)

[3] G. (John) Janakiraman, Jose Renato Santos, Yoshio Turner, "Automated Multi-Tier System Design for Service Availability," HP Labs Technical Report, HPL-2003-109.

[4] G. (John) Janakiraman, Jose Renato Santos, Yoshio Turner, "Automated System Design for Availability," HP Labs Technical Report, HPL-2004-54.

[5] Yuan Chen, et al., "A Systematic and Practical Approach to Generating Policies from Service Level Objectives," Proceedings of the 11th IFIP/IEEE International Symposium on Integrated Network Management (IM 2009), accepted.

[6] Understanding and Designing New Server Architectures for Emerging Warehouse-Computing Environments, Kevin Lim, Parthasarathy Ranganathan, Jichuan Chang, Chandrakant Patel, Trevor Mudge, Steve Reinhardt, Proceedings of the International Conference on Computer Architecture, 2008

[7] vManage: Loosely Coupled Platform and Virtualization Management in Data Centers, S. Kumar et al, Proceedings of the International Conference on Autonomic Computing, June 2009

[8] Sharma, Marwah, Lugo, Bash, "Application of data analytics to heat transfer phenomena for optimal design and operation of complex systems" 2009 ASME summer Heat Transfer Conference, HT2009-88550

[9] Hao, Sharma, Dayal, Patel, Kleim, "Application of Visual Analytics For Thermal State Management in Large Data Centers", submitted to Eurographics/ IEEE-VGTC Symposium on Visualization (2009)

[10] M. Marwah, R. Sharma, L. Bautista, and W. Lugo. Stream mining of sensor data for anomalous behavior detection in datacenters. Technical Report HPL-2008-40, HP Labs, 2008.

[11] Bash, C.E., Forman, G., "Cool Job Allocation: Measuring the Power Savings of Placing Jobs in Cooling-Efficient Locations in the Data Center", Usenix, San Jose, 2007.

[12] Bash, C.E., Hyser, C., Hoover, C., "Thermal Policies and Active Workload Migration Within Data Centers", InterPack, San Francisco, 2009.

[13] R.K. Sharma, R. Shih, C. Bash, C. Patel, P. Varghese, M. Mekanapurath, S. Velayudhan, V. Manu Kumar (2008), "On Building Next Generation Data Centers: Energy Flow in the Information Technology Stack," Proceedings of 2008 ACM Compute Conference, Bangalore, India.

[14] R. Raghavendra, P. Ranganathan, V. Talwar, Z. Wang, and X. Zhu."No "power" struggles: Coordinated multi-level power management for the data center" In Proc. of the 13th International ASPLOS, pages 48–59, Seattle, WA, 2008.

[15] C. D. Patel, C. E. Bash, R. Sharma, M. Beitelmam, and R. J. Friedrich. "Smart cooling of data centers" InterPack, pages 129–137, Kauai, Hawaii, July 2003.

[16] Cader, T., Sharma, R., Bash, C.E., Fox, L., Bhatia, V., Mekanapurath, M. "Operational Benefits of real time monitoring and visualization in data centers", InterPack, San Francisco, CA, 2009.

978-1-60558-497-3/09 $25.00 © 2009 ACM

Green Data Centers and Hot Chips

Dilip D. Kandlur
IBM Austin Research Laboratory
11501 Burnet Road
Austin, TX 78758
+1 512 286 9996

kandlur@us.ibm.com

Tom W. Keller
IBM Austin Research Laboratory
11501 Burnet Road
Austin, TX 78758
+1 512 286 6943

tkeller@us.ibm.com

ABSTRACT

Today's computing environment is changing, with growing emphasis on reducing both operational energy costs and the high capital costs to reliably deliver power and cooling to systems in factory-sized data centers. The "greening" of data centers is challenging microprocessor and system designs, including server, storage, and network designs, to deliver ever higher performance within strict power and cooling constraints. This paper outlines some of the challenges, research approaches and holistic solutions to more energy efficient data centers, starting with microprocessor and system design and extending to the broader environment.

Categories and Subject Descriptors

C.0 [Computer Systems Organization] – System architectures.

General Terms

Algorithms, Management, Measurement, Performance, Design, Economics, Reliability, Experimentation, Standardization.

Keywords

Data Centers, Architecture, Energy Management.

1. EXTENDED ABSTRACT

Today's computing environment is changing, with growing emphasis on reducing both operational energy costs and the high capital costs to reliably deliver power and cooling to systems in factory-sized data centers. Data centers consume US electricity at a rate that is doubling every 5 years and presently at 2% [1]. The expense of generating and delivering this energy has become a national concern. Ever increasing wattage requirements of servers and storage systems translate into the commonly voiced complaint by operators that their data centers are out of power and/or out of cooling capacity. The drive for increased efficiency has generated a new metric for data center energy efficiency, namely Power Usage Effectiveness (PUE), with better numbers being published for recognition among companies. New designs

Permission to make digital or hard copies of part or all of this work for personal or classroom use is granted without fee provided that copies are not made or distributed for profit or commercial advantage and that copies bear this notice and the full citation on the first page. To copy otherwise, to republish, to post on servers or to redistribute to lists, requires prior specific permission and/or a fee.
DAC'09, July 26-31, 2009, San Francisco, California, USA

by Google and Microsoft capture attention for mega-scale and highly efficient data centers filled with fleets of brand new, homogeneous, specially configured systems to support cloud computing. In contrast, existing data centers are very heterogeneous in equipment population and design, with multiple types of cooling systems and power distribution systems that provide varying levels of reliability and redundancy for all types of computing equipment over a range of ages.

New data center design trends such as modularization provide for just-in-time, just-in-need power and cooling resources to support varying power densities across the machine room floor. Data centers are also experimenting with "just open the windows" approaches to cooling to reduce energy usage [3]. Computer system vendors are starting to relax input temperature requirements by up to 10 degrees F [2], allowing warmer machine rooms and consequently greater cooling tower, chiller and machine room air handler efficiencies. This "greening" of data centers is challenging microprocessor and system designs, including server, storage, and network designs, to deliver ever-higher performance within strict power and cooling constraints. This means system-level power is now a first-class design constraint along with performance and cost.

In this paper we present IBM's perspective on some of the challenges, research approaches and solutions to more energy efficient data centers, spanning microprocessor and system design, novel packaging and cooling, to data center level measurements and optimization.

The demand for increased power and cooling efficiencies has obvious impacts on microprocessor architectures and design methodologies. The multi-core POWER6 and POWER7 microprocessor designs represent an approach that balances performance with power in multi-core systems. Simple power control techniques, such as fetch throttling have been combined with the more complex technique of Dynamic Voltage/Frequency Scaling [6]. Global power management solutions have been found to outperform local per-core approaches, with improvements in power savings found at the finest control period granularities. Coordinating power/performance tradeoffs is best achieved with an on-chip power management unit, responding to a power budget [7].

Power and performance management at the system level, not just the chip microprocessor level, is necessary. The POWER6 was the first IBM microprocessor to support a system-wide power/performance scheme based on on-chip, high-resolution, real-time measurements of temperature, performance and power.

EnergyScale™ supported new capabilities such as power capping and dynamic power savings, as well as enforcing thermal limits [8]. The implementation of the POWER6 and POWER7-based systems employs not only on-chip features but also board-level hardware, firmware and systems software [9]. Today's systems deploy power capping, power supply oversubscription protection, and "turbo" frequency increases to deliver significant power savings. The implementation is "out-of-band," thus insulated from operating systems and hypervisor coordination, and uses one or more service processors executing management logic and coordinated over dedicated communication paths.

At the system and data center level, a new approach to cooling pioneered by the IBM Zurich Research Lab, coupled with the commercial use of waste heat holds the promise of low, if not free, utility cooling costs. The enabler of this "Zero Emission Data Center" is a high-performance, hot-liquid cooling solution that can successfully cool chips with 80° C water which is only heated by 5°C (to the maximum sustainable junction temperature of microprocessors) before being either employed in neighborhood heating systems or being inexpensively cooled [5]. This has been made possible by dramatically reducing the convective thermal resistance in the heat sink by the introduction of liquid microscale cooling, permitting liquid cooling to remove heat from enclosures with power densities as high as 60 W/cm². A modified rack of blades incorporating this technique is scheduled for demonstration this year.

The techniques described above will be applicable to new systems and data centers. However, we also need to address the needs of current data centers which contain a diversity of equipment from different vendors. In this context, we have developed cost-effective instrumentation and measurement methodology for both IT and non-IT equipment in the data center, and control/optimization software for data center level energy optimization.

Hamann et. al. have used high resolution 3D thermal scanning coupled with permanent sensors to produce 3D models of a data center's "weather" [4]. Their "Mobile measurement technology" (MMT) employs three stages to improve the energy efficiency of data centers. First, a cart of sensors enables an advanced metrology technique for rapid data collection at high spatial resolution to drive thermal models which spotlight cooling imbalances. The tool "digitizes" the machine room by scanning and quickly logging parameters such as temperature, flow, humidity and rack spatial dimensions. This data is then supplemented with a network of permanent sensors left in place for dynamic updating of the static model. The combination of both static and dynamic data, combined with new modeling techniques, results in orders-of-magnitude speed improvements over computational fluid dynamics models. IBM has applied this approach to rebalance cooling flow and turn off unneeded cooling units in our data centers, with savings of 20% of the cooling costs being common.

Another facet of our approach is that closer coupling of facilities and systems instrumentation is a requirement to improve cooling efficiency and reduce wasted provisioning of power. "Stranded" power is a costly phenomenon of today's data centers in which all power distribution capacity is statically provisioned based on worst-case or peak power requirements, resulting in the data center being "out of power" even though the actual power consumed by the deployed equipment is much lower. A recent study by one of the authors in a very large, modern data center showed that the amount of actual power consumed was but a small fraction of the designed-for power, and that it could easily be doubled by improving the rack equipment density.

Lastly, since systems power does not yet track systems utilization, with today's best systems requiring around one-half the energy at idle as at peak load, real work per watt efficiencies are being achieved by machine consolidation via virtualization and consolidation, with the goal of deploying fewer running systems. A related approach to minimizing the number of running systems is to power on and off systems dynamically in response to load. Recent experiments have shown 25% power savings using this approach under very conservative assumptions without incurring performance penalties for typical daily and weekly periodic demands [10]. Practical implementations may require multi-agent management, one for power and one for performance, if power and performance are to be woven into the fabric of today's software systems [11]. We are now extending this control approach and combining it with the instrumentation from MMT to create a highly dynamic, energy-optimized data center.

2. REFERENCES

[1] U.S. Environmental Protection Agency, "Report to Congress on Server and Data Center Efficiency (Public Law 109-431)," ENERGY STAR Program, Aug. 2007.

[2] http://www.rackable.com/cloudrackC2

[3] J. Atwood, and Don G. Miner, "Reducing Data Center Cost with an Air Economizer," Intel Online Brief, Aug. 2008.

[4] H. F. Hamann, T. G. van Kessel, M. Iyengar, J.-Y. Chung, W. Hirt, M. A. Schappert, A. Claassen, J. M. Cook, W. Min, Y. Amemiya, V. Lopez, J. A. Lacey, and M. O'Boyle, "Uncovering energy-efficiency opportunities in data centers," IBM Journal of Research and Development, vol. 53, no. 3, paper 10, 2009.

[5] T. Brunschwiler, B. Smith, E. Ruetsche, and B. Michel, "Toward zero-emission data centers through direct reuse of thermal energy," IBM Journal of Research and Development, vol. 53, no. 3, paper 11, 2009.

[6] J. Sharkey, A. Buyuktosunoglu, and P. Bose, "Evaluating design tradeoffs in on-chip power management for CMPs," In *Proceedings of the 2007 International Symposium on Low Power Electronics and Design (ISLPED '07),* pp. 44-49, Portland, OR, USA, August 27 - 29, 2007.

[7] R. Bergamaschi, G. Han, A. Buyuktosunoglu, H. Patel, I. Nair, G. Dittmann, G. Janssen, N. Dhanwada, Zhigang Hu, P. Bose, and J. Darringer, "Exploring power management in multi-core systems," In *Proceedings of the Asia and South Pacific Design Automation Conference (ASPDAC 2008),* pp. 708 – 713, Seoul, Korea, March, 2008.

[8] M. S. Floyd, S. Ghiasi, T. W. Keller,, K. Rajamani, F. L. Rawson, J. C. Rubio, and M. S. Ware, "System power management support in the IBM POWER6 microprocessor," IBM Journal of Research and Development, vol. 51, no. 6, November 2007.

[9] H.-Y. McCreary, M. A. Broyles, M. S. Floyd, A. J. Geissler, S. P. Hartman, F. L. Rawson, T. J. Rosedahl, J. C. Rubio, and M. S. Ware, "EnergyScale for IBM POWER6 processor-

based systems, IBM Journal of Research and Development, vol. 51, no. 6, November 2007.

[10] R. Das, J. O. Kephart, C. Lefurgy, G. Tesauro, D. W. Levine, and H. Chan, "Autonomic multi-agent management of power and performance in data centers," In *Proceedings of the 7th International Joint Conference on Autonomous Agents and Multiagent Systems (AAMAS '08)*, pp. 107-114, Estoril, Portugal, May 12-16, 2008.

[11] J. O. Kephart, Hoi Chan, R. Das, D. W. Levine, G. Tesauro, F. Rawson, and C. Lefurgy, "Coordinating multiple autonomic managers to achieve specified power-performance tradeoffs," In *Proceedings of the Fourth International Conference on Autonomic Computing*, 2007 (ICAC '07), pp. 24-24, Jacksonville, Florida, USA, June 11-15, 2007.

978-1-60558-497-3/09 $25.00 © 2009 ACM

Optimum LDPC Decoder: A Memory Architecture Problem

Erick Amador
EURECOM
06904 Sophia Antipolis, France
erick.amador@eurecom.fr

Renaud Pacalet
TELECOM ParisTech
06904 Sophia Antipolis
renaud.pacalet@telecom-paristech.fr

Vincent Rezard
Infineon Technologies France
06560 Sophia Antipolis
vincent.rezard@infineon.com

ABSTRACT

This paper addresses a frequently overlooked problem: designing a memory architecture for an LDPC decoder. We analyze the requirements to support the codes defined in the IEEE 802.11n and 802.16e standards. We show a design methodology for a flexible memory subsystem that reconciles design cost, energy consumption and required latency on a multistandard platform. We show results after exploring the design space on a CMOS technology of $65nm$ and analyze various use cases from the standardized codes. Comparisons among representative work reveal the benefits of our exploration.

Categories and Subject Descriptors

B.3.2 [**Memory Structures**]: Design Styles

General Terms

Algorithm, Design

Keywords

LDPC codes, low power architectures, memory optimization

1. INTRODUCTION

Low-density parity-check (LDPC) codes [4] are among the best known error-correcting codes because of their capacity-approaching performance and the explicit parallelism exhibited by their iterative decoding algorithm. These codes have already been adopted by several wireless communication standards like IEEE 802.11n [5] and 802.16e[6] among others. Such standards define several coding rates and codeword lengths in order to provide different levels of link adaptability, consequently an implementation of a decoder should be flexible to support various modes of operation. Furthermore, if the target application entails mobile terminals then ultra-low power operation is mandatory. Standardized LDPC codes embed a particular structure into the parity-check matrix that reduces the interconnection complexity between the processing nodes and enables the realization of semi-parallel architectures. These type of architectures are amenable for flexibility, i.e. supporting different parity-check matrices that arise from different codeword lengths and coding rates. Decoding of LDPC codes comprises a memory intensive kernel, rendering the memory subsystem as the main source of energy consumption. Although this subsystem is an important factor for the overall performance of the decoder, its design is very frequently overlooked. Even though there are numerous published works on LDPC decoders design, to the best of our knowledge the memory subsytem design exploration has been omitted in most cases.

Several works have presented the tradeoff between throughput and area, finding the right compromise between the level of processing parallelism and implementation area. Representative state-of-the-art works like [1] and [9] present an exploration of the design space that spans from processing parallelism, the processing units approximation up to the decoding schedules.

In [3] the memory allocation problem is investigated to avoid memory conflicts, but like the other works lacks an exploration in terms of energy efficiency and area cost. While these works explore the fundamental tradeoff between processing parallelism, operating frequency and implementation area, they usually result in overdimensioned designs. The resulting memory organizations are costly in terms of area per bit, since fewer bigger memories are preferable to several small memories for on-chip SRAM.

In this work we concentrate on the exploration of the memory architecture to optimally implement an LDPC decoder that supports the codes defined in the IEEE 802.11n and 802.16e standards. The paper is organized as follows: structured LDPC codes and the implemented decoding algorithm are presented in Section 2. Section 3 presents the required storage components of the decoder and section 4 presents the real time requirements. Section 5 presents the methodology we propose and results from the design space exploration, as well as comparisons among similar works. Section 6 concludes the paper.

2. STRUCTURED LDPC CODES

LDPC codes are linear block codes described by a sparse parity-check matrix $H_{M \times N}$ that defines M parity-check constraints over N codeword symbols. The number of non-zero elements in a column and in a row determine the degree of the column and the row respectively. The code can be described by a bipartite graph, where columns of H are mapped to *variable* nodes and rows are mapped to *check* nodes. A non-zero element in H represents an edge be-

Permission to make digital or hard copies of part or all of this work for personal or classroom use is granted without fee provided that copies are not made or distributed for profit or commercial advantage and that copies bear this notice and the full citation on the first page. To copy otherwise, to republish, to post on servers or to redistribute to lists, requires prior specific permission and/or a fee.
DAC'09, July 26-31, 2009, San Francisco, California, USA

978-1-60558-497-3/09 $25.00 © 2009 ACM

tween the corresponding variable and check node. *Structured* LDPC codes are composed of several layers of non-overlapping rows, where these layers are formed by $Z \times Z$ sub-matrices. Z is an *expansion factor* that shows the degree of available parallelism as Z non-overlapping rows can be processed concurrently. Each of these sub-matrices can be either a zero matrix or a circularly right-shifted version of the identity matrix. Furthermore, these codes are constructed such that their encoding process exhibits linear complexity, identifying two sections within H: a random part for the systematic information and a prearranged part for the parity information. Figure 1 shows the H matrix and the bipartite graph of an example structured code, where $H_{M \times N} = [H^i_{n \times k} | H^p_{n \times n}]$, with $M = n$ and $N = n + k$. Indeed, H consists of an array $m_b \times n_b$ of $Z \times Z$ blocks.

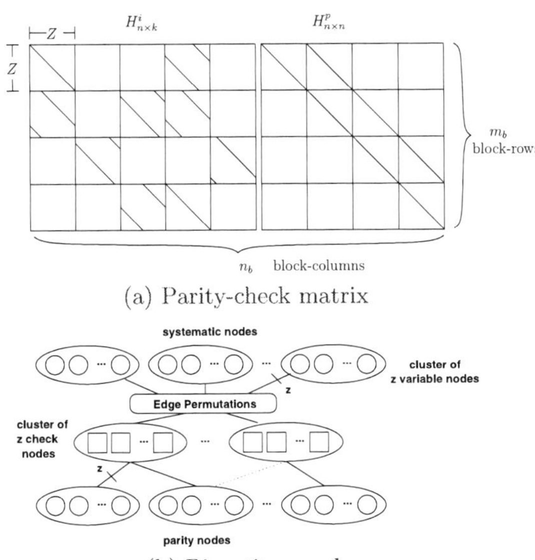

(a) Parity-check matrix

(b) Bipartite graph

Figure 1: Structured LDPC code

The graph representation in figure 1(b) contains edges of width Z, grouping Z nodes into clusters. This grouping enables the possibility to instantiate a subset of the processing nodes, representing a clear advantage in terms of flexibility and silicon area as well as reducing the complexity of the interconnection network (edge permutations).

LDPC codes are decoded iteratively using a two-phase message-passing algorithm [4], where check and variable nodes exchange extrinsic reliability values associated with each codeword symbol. Node operations are independent and may be executed in parallel, allowing the possibility to use different scheduling techniques for various purposes. The turbo-decoding message-passing (TDMP) schedule [8] shows important improvements over the two-phase schedule: a faster convergence and reduced memory requirements.

In TDMP the check nodes of the graph are evaluated sequentially, updating and propagating more reliable messages along the graph. In the following we summarize this algorithm from [8]. Initial channel observations per codeword symbol are given as log-likelihood ratios in an initial vector $\delta = [\delta_1, \ldots, \delta_N]$. Each row i in H has a corresponding vector of *extrinsic* messages $\lambda^i = [\lambda^i_1, \ldots, \lambda^i_{c_i}]$, where c_i is the degree of row i. Let I_i denote the set of c_i elements in row i. A vector $\gamma = [\gamma_1, \ldots, \gamma_N]$ contains the sum of all messages generated in the rows for each codeword symbol, thereby encompassing *posterior* messages. Hard decisions are taken by slicing the vector γ to obtain the decoded message. The decoding of the ith row involves these steps:

1. Read λ^i and the c_i elements of γ that participate in the row, denoted as $\gamma(I_i)$.

2. Generate a vector ρ of prior messages given by:
$\rho = \gamma(I_i) - \lambda^i$.

3. Process ρ with a soft-input soft-output (SISO) decoding algorithm. Denote the result as $\Lambda = SISO(\rho)$. Any SISO algorithm may be implemented with different levels of hardware complexity and error-correcting performance. A practical choice commonly found is the *Min-Sum* algorithm [2] and its variants.

4. Update the previously read vectors according to these rules: $\lambda^i \leftarrow \Lambda$; $\gamma(I_i) \leftarrow \rho + \Lambda$

This method merges check and variable updates in one step reducing the memory requirements when compared to the traditional two-phase method. Another important advantage is the reduction in the number of iterations by up to 50%, indeed saving energy by the same proportion.

3. MEMORY ARCHITECTURE

TDMP decoding requires the following storage elements: posterior messages memory (γ) and extrinsic messages memory (λ). It is important to note that as described in [8], with TDMP decoding the vector γ is initialized with the channel observations δ, such that δ does not play a role in the iterative process other than during initialization. Besides these memories, storage is required for the structure of the parity-check matrix, involving the location of the non-zero sub-matrices and their shift value.

3.1 Parity-check matrix storage

The IEEE 802.11n and 802.16e standards define different parity-check matrices for different use cases, where a use case is defined by a codeword length and a coding rate. In 802.11n there are a total of 12 matrices for 12 use cases, and in 802.16e there are 6 matrices for 19 use cases. The structure of each of these matrices follows the structure shown in figure 1(a). As described in the previous section, for decoding the ith row it is necessary to retrieve the set I_i of c_i elements from γ. $H_{M \times N}$ consists of an array $m_b \times n_b$ of $Z \times Z$ blocks, where for both standards $n_b = 24$ and $m_b \in \{4, 6, 8, 12\}$ which depends upon the use case. Storing H involves saving the shift value of the non-zero blocks and the block-column position. Since for Z rows c_i remains constant, the decoding of Z rows involves retrieving c_i shift values from H. These values can be encoded in a read-only memory by concatenating in one word the shift value and column position. The shift values for 802.11n are taken directly from the specified matrices, whereas the shift values for 802.16e follow a precalculation dependent upon the use case. For all cases the shift value $s \leq Z$, where $Z_{11n} = \{27, 54, 81\}$ and $Z_{16e} = \{24, 28, 32, \ldots, 96\}$. In this way one 12-bit word encodes both the shift value and the column position (7-bits for shift value and 5-bits for the column number). It is evident that only the random part of H, $H^i_{n \times k}$ needs to be stored. For the 18 matrices this translates into a total of 1295 values.

978-1-60558-497-3/09 $25.00 © 2009 ACM

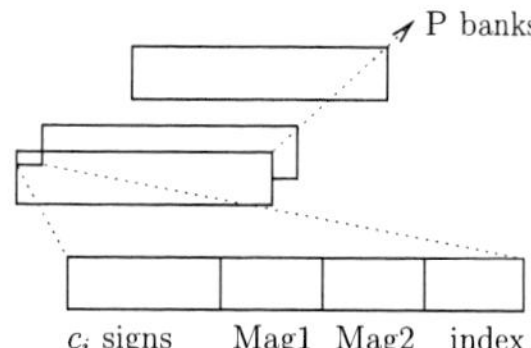

Figure 2: Extrinsic memory

3.2 Posterior messages memory

This memory is initialized with the log-likelihood ratio values δ taken from the channel observations and updated after each row processing. This memory contains as many values as the codeword length. Hard decisions on the stored vector produce a valid codeword after successful decoding. One property provided by the structure of the codes is that when processing each block-row of H only a subset of the block-columns is used. This property can be exploited when mapping the data to the memory architecture, which we explain in section 5.

3.3 Extrinsic messages memory

This memory stores the messages that are obtained after each decoding round per row, and are used as input prior messages to decode subsequent rows. The dimension of this memory corresponds to the number of non-zero elements in H, or equivalently to the number of edges in the code graph. While this is the largest of the two memories it is easier to design, since the data allocation is straightforward by following the flow in TDMP decoding: the processing of each row uses always the same values from the same locations. The data does not have to be neither shifted nor distributed to different processing units if each row decoding is bound to a specific processing unit. The choice of *Min-Sum* as the SISO algorithm has further consequences on this memory. This algorithm takes c_i inputs and produces c_i outputs, with each having a sign and one of two possible magnitudes: the first or second minimum value out of the c_i inputs. So instead of storing c_i messages it is only necessary to store c_i signs, two magnitudes and the index where the first minimum is to be found. For our case this represents a reduction in memory size of up to 65% when using messages of 8-bits. Figure 2 shows the storage for these values.

Figure 3(a) shows a top level view of the architecture for a TDMP LDPC decoder. Shuffling units π and π^{-1} are used to distribute the data to and from P computation units. These units perform the processing of the rows in H, figure 3(b) shows the dataflow of the processing pipeline.

4. BANDWIDTH REQUIREMENTS

The codes defined in the 802.11n and 802.16e standards exhibit very tight and demanding timing requirements. Table 1 shows the main characteristics defined in the standards for both the most and less demanding cases (4 out of 31 cases shown). The decoding latency corresponds to the maximum permissible time in order to perform the decoding task, such that the baseband processing respects the time allotted for generating negative or positive acknowledgement messages.

From this table we can estimate the sizes of the memories for posterior and extrinsic messages and their respective bandwidth requirements. In terms of memory size the 802.16e case is used as it requires the most number of sam-

Table 1: Specifications for standardized LDPC codes

Use case	Most demanding		Less demanding	
Standard 802.xx	11n	16e	11n	16e
Codeword length	1944	2304	648	576
H dimensions	12x24	12x24	4x24	4x24
($m_b \times n_b$)				
Z value	81	96	27	24
Latency	$8\mu s$	0.25ms	$8\mu s$	0.25ms
Row degree c_i	7,8	6,7	22	20
Rows in H	972	1152	108	96
Graph edges	6966	7296	2376	1920

Table 2: Required frequency and memory width

Bus width (samples)	12	16	32
Average freq. [MHz]	580.50	435.38	217.69
Peak freq. [MHz]	648.00	486.00	243.00

ples, but for the timing constraints it is clear that 802.11n has tighter deadlines. From this point on we use a maximum number of 8 decoding iterations and 8-bits long messages. Initially we can fix a hypothetical operating frequency to observe the minimum required bandwidth for the memory architecture. This bandwidth is obtained by:

$$BW = \frac{total\ \ samples \times iterations}{decoding\ \ latency \times frequency} \qquad (1)$$

In order to manage realistic and practical widths on the memory organization, it is possible to fix the memory bus width after observing the initial estimations provided by table 1 and obtain the required operating frequency. It is worth noting from table 1 that when decoding different rows the number of required samples may vary, this given by the degree of the rows having different values, e.g. $c_i \in \{7, 8\}$ for the 802.11n most demanding case. From this we define an average and a peak operating frequency, where the average frequency considers the actual cases where different row degrees are used, and where the peak frequency considers the maximum row degree only. By introducing a peak operating frequency, we introduce several dead cycles or equivalently a decrease in the utilization of the memory bandwidth. Table 2 shows the obtained operating frequencies for different bus widths.

The decoding throughput can be estimated by the number of decoded bits, the operating frequency and the number of computation cycles. For our case we use P serial processing units that decode a row of degree c_i in $2 \times c_i$ cycles (including read and writeback stages). The throughput T is given by:

$$T = \frac{(n_b - m_b) \times Z}{v \times m_b \times 2c_i \times t} \times f_{clock} \qquad (2)$$

where t is the number of iterations and $v = Z/P$ shows the level of parallelism used.

5. DESIGN STRATEGY

The memory architecture has to be dimensioned to handle several use cases with differences in workload. We concentrate on a worst case analysis to observe the greatest impacts on energy, so we use the case of a memory bandwidth of 12 samples/cycle. For the 802.11n most demanding case 8

978-1-60558-497-3/09 $25.00 © 2009 ACM

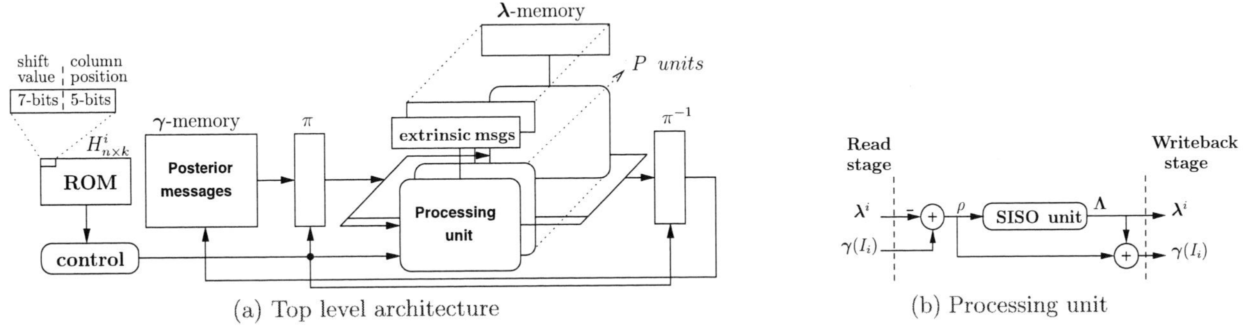

(a) Top level architecture (b) Processing unit

Figure 3: TDMP LDPC Decoder

samples are required per decoding of a row operating at the peak frequency. If 12 samples are available per read cycle this suggests either the parallel processing of 1.5 rows per cycle, or a combination of decoding a number of rows in a number of cycles: e.g. 3 rows in 2 cycles. In order to avoid cooperative processing between SISO units and take advantage of the structure in H, we assign the decoding of a row to a processing unit. Notice that we are dealing with the most demanding case (biggest H) such that for the less demanding cases there is more time available to fetch the samples. As mentioned in section 3.3 the extrinsic messages memory data allocation is straightforward and it is conveniently distributed among different processing units. In section 5.2 we explore how to further partition this memory to capitalize on energy scalability. Next we elaborate on the particular challenges for designing the posterior messages memory.

5.1 Data mapping and allocation

With a bandwidth requirement and a cycle budget, we design the number of banks and the dimensions of the posterior messages memory by exploiting the structure of the codes. From this we identify clearly two types of data organization and placement:

1. Micro-organization: provided by the shifted identity matrices that make up the block structure of H. This structure allows the processing of non-overlapping rows, so that consecutive samples are distributed to several processing units by means of a shuffling unit. From this it follows that for distributing samples to P units each memory location could store exactly P samples, such that all are accessible per read operation. Figure 4 shows this organization for $P = 3$ and $Z = 81$ for one block-column. The codeword length is in this case $24 \times Z = 1944$, a use case from 802.11n. The micro-organization is shown for the block-column that corresponds to the codeword symbols $\{324, 325, \ldots, 404\}$. Due to the structure of the $Z \times Z$ blocks all Z samples will be used when processing a block-row. In figure 4 we further show how 3 samples that are required are read from two memory locations, and how the remaining samples of the read accesses can be stored as they will be used in subsequent rows' processing.

2. Macro-organization: c_i block-columns are used per block-row processing. For the most demanding case (802.11n) when operating on a given block-row it is required only to have access to either 7 or 8 block-columns. From this we define the macro-organization as the mapping of block-columns to several banks such that the required blocks are all accessible per read operation.

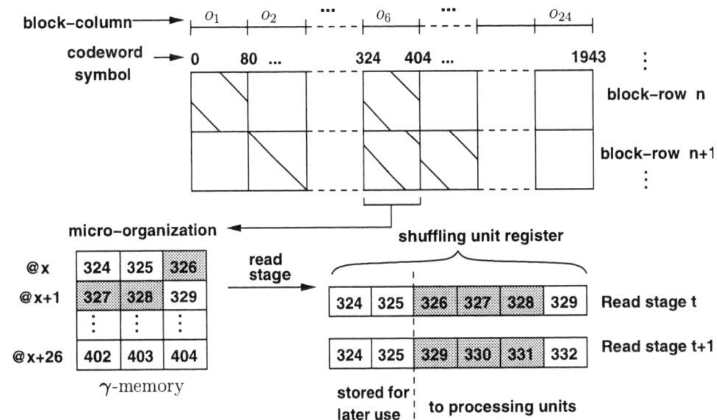

Figure 4: Micro-organization example

This constitutes a problem of allocating static objects (block-columns) to memory banks. As shown in [7] this problem is *NP-complete* and it maps naturally onto the well-known graph colouring problem. This introduces another dimension to the design criteria to organize the memory architecture, since not only does the bandwidth have to be achieved, but the data allocation required has to be possible. This situation is formalized as follows: a set of static objects $O = \{o_1, o_2, \ldots, o_{24}\}$ is to be mapped to a set of memory banks B. Objects in O *conflict* when they need to be accessed at the same time. The memory allocation problem is solved by finding an allocation function $alloc: \; O \to B$ that avoids as much conflicts as possible. A conflict represents a loss in execution speed, introducing stall cycles and bubbles into a pipelined implementation.

From the most demanding case in table 1, we analyze the mapping of data in the posterior messages memory for the case in 802.11n at rate $1/2$ and codeword length of 1944. From [5] we take the LDPC matrix definition and build *conflict graphs* $G = (V, E)$, where V is the set of vertices representing the block-columns used on a given block-row and E is the set of edges between the nodes of V that incur in a conflict. The chromatic number $\chi(G)$ of a conflict graph determines the minimum number of colors needed such that all vertices of the graph are coloured and adjacent vertices do not share the same color. In other words, the chromatic number determines the minimum number of memories needed to provide the required bandwidth. There should be as many conflict graphs as block-rows in H, but we consider now the bandwidth requirement and the degree of the rows in H to observe the cycle budget for reading the required samples. The case of 12 samples/cycle and the concurrent processing of 3 rows every 2 cycles leads to ac-

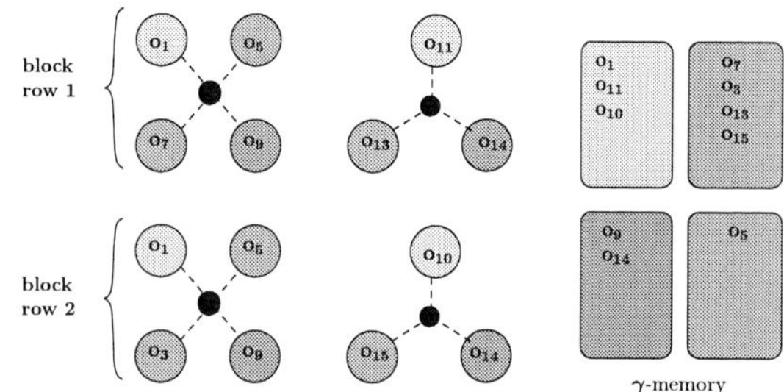

Figure 5: Macro-organization by conflict graphs

Table 3: Energy from H matrix storage

Use case	Energy/iteration [nJ]
$R = 1/2, N = 2304$ (16e)	0.47
$R = 5/6, N = 576$ (16e)	0.64
$R = 5/6, N = 648$ (11n)	0.70

cessing as much as 4 block-columns per read access. This produces 2 conflict graphs per block-row, where the chromatic number is bounded, $\chi(G) \leq 4$. Figure 5 shows the situation for the first 2 block-rows of H for this use case, using $B = 4$ memory banks that provide a bandwidth of 12 samples/cycle to $P = 3$ processing units.

This mapping is performed offline and is solvable by the well known graph colouring techniques. By analyzing the parity-check matrix the conflict graphs can be constructed and coloured such that the block-columns are allocated to the memory banks and the number of conflicts is reduced or even avoided.

5.2 Exploration

One important parameter for the cost of a memory architecture is the area per bit. By reducing the number of banks also interconnection links and control circuitry are reduced. This impacts not only cost but also power consumption, so it is fundamental to perform a study on the memory architecture candidates to achieve an efficient and economical design. From the area/bit point of view there are two extreme cases: using one single large memory or a maximum number of memories that guarantees the required number of parallel accesses. Compared to small memories, large memories consume more energy per access because of the longer word and bit lines. But distributing the data along several small memories leads also to relatively high energy consumption due to the added interconnection lines and decoding circuitry overhead.

Using an in-house memory exploring tool and *Synopsys Primetime*, we investigate candidate memory architectures where the data mapping has been shown to be possible. The target technology is a $65nm$ typical CMOS process with $V_{dd} = 1.32V$.

The exploration for the storage of H is limited since it requires non-volatile storage in principle. We synthesized this memory as a [324x48] low power ROM. In table 3 we show the energy consumed per decoding iteration when reading this control data for the corner use cases, defined by the number of non-zero blocks in each H matrix.

The exploration of the extrinsic messages memory involves inquiring about the possibility of partitioning each of the memories that are bound to each processing unit. This memory stores the compressed row vector shown in figure

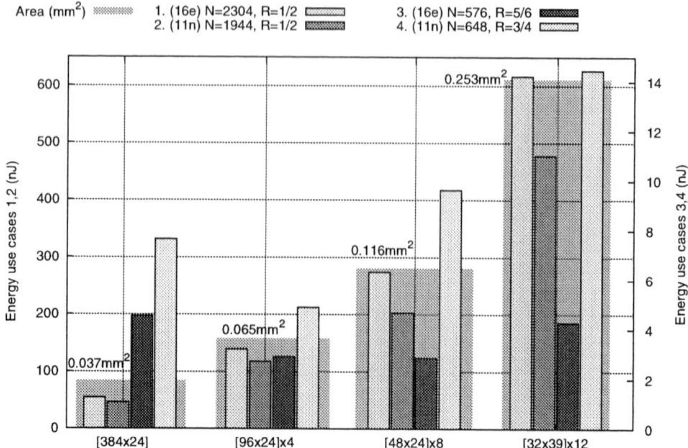

Figure 6: λ-memory exploration

2. Each of the 31 use cases has different requirements. The corner cases are defined by the number of edges in the code graph and the degree of the rows. In figure 6 we explored the area and average energy per iteration consumed in the extrinsic messages memory for the two corner cases (both from 16e) and two intermediate cases with several memory configurations, using low power dual-port SRAMs clocked at 648MHz. For the less demanding cases the unused partitions are powered off. This simple power management strategy allows to observe the energy cost behavior according to the memory partitions, where an optimal configuration exists for a particular use case. Once an optimal choice is taken more elaborated power management techniques (voltage/frequency scaling) can further improve the energy efficiency.

For the case of the posterior messages memory it is compelling to follow suit and partition further this memory while maintaining the required bandwidth and achieving gains on energy by exploiting the flexibility of the architecture. We explored several configurations for bandwidths of 12 and 32 samples/cycle using low power dual-port SRAMs clocked at 648MHz and 243MHz respectively. The corner use cases explored are characterized by the number of memory accesses, which depends upon the degree of the block-rows. For example the case $R = 1/2$ and $N = 2304$ in 802.16e requires 76 accesses per decoding iteration to the 24 blocks of 96 samples each stored in this memory. Figure 7 shows the average energy consumed per iteration in this memory for several partitions.

For both use cases shown we can observe that the configurations for 32 samples/cycle are more energy efficient than the 12 samples/cycle ones, but more costly in terms of area/bit.

With this exploration an appropriate selection of the optimum memory configurations is possible. Figure 8 shows the breakdown of the estimated implementation area [1] and average energy consumed per iteration on different use cases for the following realization, λ-memory: [96x24]x4, γ-memory: [192x48]x2 and $P = 3$. While the use case dictates which of the memories (λ- or γ-memory) dominates the energy consumption, it is clear that at least 85% of the energy is spent on the memory architecture.

Although there are numerous published works on LDPC

---

[1] Min-Sum used as SISO kernel and Benes networks as shuffling units

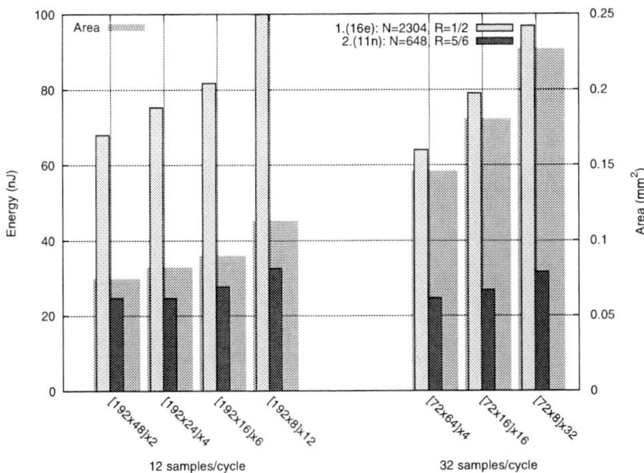

Figure 7: γ-memory exploration

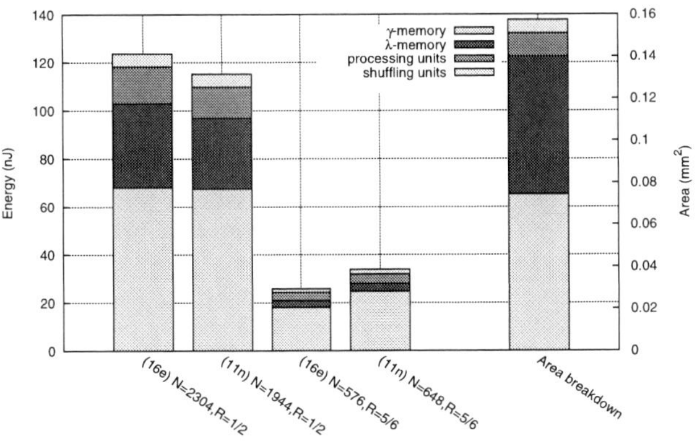

Figure 8: Area and use case energy breakdown

decoder design, it is hard to find a thorough report on the breakdown of implementation complexity and energy on a use case basis when addressing standardized codes. In table 4 we show a comparison with representative state-of-the-art works. All works have been implemented on $65nm$ CMOS technologies. In this table our work corresponds to the realization λ-memory: [96x24]x4 and γ-memory: [72x64]x4. Even though it is difficult to draw comparisons among multi-objective explorations we can observe that our approach avoids the usual overdimensions (in terms of throughput) and focuses instead on meeting the requirements (latency) while choosing an optimal memory architecture. In our case optimality being energy and area costs. Only [9] reports the breakdown of energy for a use case, so we show in table 4 the total energy consumed on λ- and γ-memories. While [9] averages over three different SNRs, ours is a worst case estimation at the maximum number of iterations.

6. CONCLUSIONS

In this work we explored the frequently overlooked problem of finding an optimal memory configuration on a memory-driven application. We have presented how the memory architecture alone represents one of the most challenging parts in the design of an LDPC decoder. Our design space exploration emphasizes on implementation area cost and energy consumption on a use case basis. The design steps followed comprise the analysis of the real-time requirements, an initial estimation on dimensions and bandwidth of the memories, and a further pruning of the search space by considering

Table 4: Memory area and energy comparison

Work	this work	[1]		[9]
Standard 802.xx	11n/16e	11n	16e	11n
Operating point	1.32V 243MHz	N/A V 400MHz		1.2V 240MHz
Messages length	8-bit	6-bit		5 and 8-bit
Memory area [mm^2]	0.211	0.467	0.551	0.265
Throughput [$Mbps$]	92-162	108-337	96-399	175-292
Latency [μs]	3.0-7.5	5.8-6.0	5.7-6.0	5.65-5.98
Memory energy Use case 802.11n: R=1/2,N=648	275.71 nJ	N/A	N/A	590.81 nJ

the cost merits of different realizations in terms of area and energy per use case. We detailed the challenges of designing and partitioning the posterior messages memory by identifying two levels of data organization. Albeit it remains challenging to implement LDPC decoders, our work reveals that a thorough exploration yields more attractive results in terms of cost of the storage elements and energy scalability according to the supported use cases.

7. REFERENCES

[1] T. Brack, M. Alles, T. Lehnigk-Emden, F. Kienle, N. Wehn, N. E. L'Insalata, F. Rossi, M. Rovini, and L. Fanucci. Low Complexity LDPC Code Decoders for Next Generation Standards. In *Proc. of the Conference on Design, Automation and Test in Europe*, pages 331–336, San Jose, USA, 2007.

[2] J. Chen, A. Dholakia, E. Eleftheriou, M. Fossorier, and X. Hu. Reduced-Complexity Decoding of LDPC Codes. *IEEE Trans. Comms.*, 53:1288–1299, August 2005.

[3] J. Dielissen and A. Hekstra. Non-fractional Parallelism in LDPC Decoder Implementations. In *Proc. 2007 Design, Automation and Test in Europe*, Nice, France, 2007.

[4] R. Gallager. Low-Density Parity-Check Codes. *IRE Trans. Inf. Theory*, 7:21–28, January 1962.

[5] IEEE-802.11n. Wireless LAN Medium Access Control and Physical Layer Specifications: Enhancements for Higher Throughput. *P802.11n/D3.07*, March 2008.

[6] IEEE-802.16e. Air Interface for Fixed and Mobile Broadband Wireless Access Systems. *P802.16e-2005*, October 2005.

[7] P. Keyngnaert, B. Demoen, B. De Sutter, B. De Sus, and K. De Bosschere. Conflict Graph Based Allocation of Static Objects to Memory Banks. *Informal proceedings of the First workshop on Semantic, Program Analysis, and Computing Environments*, pages 131–142, September 2001.

[8] M. Mansour. A Turbo-Decoding Message-Passing Algorithm for Sparse Parity-Check Matrix Codes. *IEEE Trans. on Signal Processing*, 54(11):4376–4392, November 2006.

[9] M. Rovini, G. Gentile, F. Rossi, and L. Fanucci. A Scalable Decoder Architecture for IEEE 802.11n LDPC Codes. In *Proc. of IEEE GLOBECOMM*, pages 3270–3274, 2007.

978-1-60558-497-3/09 $25.00 © 2009 ACM

A DVS-based Pipelined Reconfigurable Instruction Memory

Zhiguo Ge, Tulika Mitra, and Weng-Fai Wong
School of Computing
National University of Singapore
{gezhiguo, tulika, wongwf}@comp.nus.edu.sg

ABSTRACT

Energy consumption is of significant concern in battery operated embedded systems. In the processors of such systems, the instruction cache consumes a significant fraction of the total energy. One of the most popular methods to reduce the energy consumption is to shut down idle cache banks. However, we observe that operating idle cache banks at a reduced voltage/frequency level along with the active banks in a pipelined manner can potentially achieve even better energy savings. In this paper, we propose a novel DVS-based pipelined reconfigurable instruction memory hierarchy called PRIM. A canonical example of our proposed PRIM consists of four cache banks. Two of these cache banks can be configured at runtime to operate at lower voltage and frequency levels than that of the normal cache. Instruction fetch throughput is maintained by pipelining the accesses to the low voltage banks. We developed a profile-driven compilation framework that analyzes applications and inserts the appropriate cache reconfiguration points. Our experimental results show that PRIM can significant reduce the energy consumption for popular embedded benchmarks with minimal performance overhead. We obtained 56.6% and 45.1% energy savings for aggressive and conservative V_{DD} settings, respectively, at the expense of a 1.66% performance overhead.

Categories and Subject Descriptors

B.3.2 [**Design Styles**]: Cache memories—*Memory structures*

General Terms

Algorithm, Design, Performance

Keywords

Instruction cache, low power, reconfigurable memory.

1. INTRODUCTION

Power consumption has become a major design concern with the proliferation of portable consumer electronics devices. As instructions are fetched in nearly every processor cycle, the instruction delivery system accounts for a significant fraction of the total energy consumption. Pose et al. [3] reported that the front-end instruction delivery path (consisting of the fetch, decode, rename, dispatch and issue stages) consumes about 20% of the overall system power and about 35% power of each processor core. Brooks et al. [4] found that instruction fetch consumes 22.2% of power in the Intel Pentium Pro processor. According to [15], the instruction cache of the StrongARM 110 processor accounts for 27% of total power and is higher than the energy consumption of the data cache.

Several approaches have been proposed to reduce the energy consumption of caches in embedded processors. For example, application specific customizations allow the cache parameters, such as associativity, line size, cache size, to be adapted to specific applications and/or their execution phases [1, 18, 7]. Dynamic voltage scaling (DVS) [6] is another effective technique for power reduction where the supply voltage and clock frequency are dynamically adjusted in accordance to the need of the application.

DVS-based techniques are especially useful in applications that have multiple phases of varying resource requirements. Resources that are under-utilized at a certain phases can be exploited to further reduce energy consumption. There are two different classes of techniques for doing this. The first class of approaches are *reactive* in nature in that they use DVS to slow down or completely shut down a resource when it becomes under-utilized. The second category of approaches *predict* the resource utilization and employ DVS to exploit the idle capacity of the under-utilized resources to save energy while maintaining performance. Proposals for each of these two methods have been made for computational units [13, 2, 9]. For example, an aggressive technique proposed in [6] replicates a logic block number of times with each instance operating at a lower supply voltage and frequency. In the context of memory hierarchies, most of the previously proposed techniques are reactive in nature. Examples include selective cache ways [1], drowsy cache [11] and cache decay [10] mechanism. They generally disable the under-utilized memory resources to save energy.

This paper presents a novel predictive DVS-based energy savings technique for the memory hierarchy that uses idle banks rather than shutting them down. We propose the *pipelined reconfigurable instruction memory hierarchy* (PRIM) for power constrained embedded processors. Our canonical example of PRIM consists of an instruction cache with four banks for data storage, where two banks can be dynamically reconfigured to DVS mode. In DVS mode, the supply voltage and frequency of the two banks are decreased. Moreover, the low-voltage banks are operated as scratchpad memory that are explicitly controlled through software. In particular, frequently executed instructions are loaded and locked into the two DVS banks. Finally, instruction throughput is maintained by pipelining and alternating the instruction fetches between these two banks. To the best of our knowledge, this is the first work

Permission to make digital or hard copies of part or all of this work for personal or classroom use is granted without fee provided that copies are not made or distributed for profit or commercial advantage and that copies bear this notice and the full citation on the first page. To copy otherwise, to republish, to post on servers or to redistribute to lists, requires prior specific permission and/or a fee.

DAC'09, July 26-31, 2009, San Francisco, California, USA

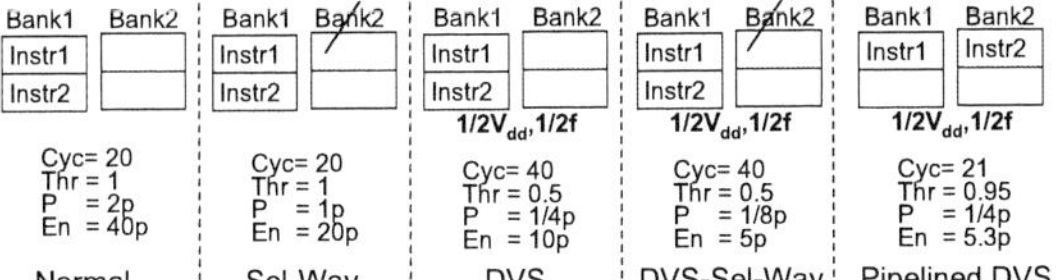

Figure 1: Comparison of cache energy savings techniques.

that proposes taking advantage of under-utilized storage resources to obtain more energy savings instead of disabling them.

Motivating Example. Let us illustrate the intuition behind PRIM and its advantage over existing cache energy savings techniques with a motivating example. Consider the execution of a small loop containing only two instructions from a 2-way set associative cache for 10 iterations. For simplicity of exposition, let us assume that each cache block contains only one instruction. Finally, both the instructions are assumed to be present in the same cache bank. Figure 1 shows a comparison among the different schemes based on (1) total cycles for instruction fetch (*Cyc*), (2) instruction fetch throughput (*Thr*), (3) power (*P*) and (4) total energy consumption (*En*) for one complete execution of the loop. We assume 1-cycle cache latency in normal mode and 2-cycle cache latency in DVS mode (as DVS mode runs at half the frequency). The dynamic power of each normal cache bank is $p = CV^2 f$.

In normal mode, it will take 20 cycles to fetch 20 instructions with a throughput of 1 instruction/cycle. As the normal cache has two banks, $P = 2p$ and $En = 2p \times 20$. In the *selective-way cache*, Bank2 is switched off as all the accesses go to Bank1. So the selective-way scheme can achieve 50% power and energy reduction compared to normal cache without sacrificing performance.

A more aggressive technique applies *DVS* to both the cache banks. The power requirement of DVS scheme is $P = 2 \times C \times (\frac{V}{2})^2 \times \frac{f}{2} = \frac{p}{4}$. However, as frequency is halved, the instruction fetch latency increases to 2 cycles and throughput drops to 50% of the normal cache. The overall energy consumption of DVS scheme is $40 \times (\frac{p}{4})$, which is 50% savings compared to selective-way cache. To save additional power, the unused bank can be switched off in DVS mode as well leading to *DVS selective way* cache. Clearly, DVS selective way has the same performance as DVS scheme but reduces energy consumption further by 50%. Our key observation is that in DVS mode, one cache bank is not utilized and it either sits idle or gets switched off. Instead of powering down one cache bank for this loop, PRIM (a) distributes the instructions to the two cache banks and (b) powers up both the cache banks in DVS mode. We then propose the use of pipelined instruction fetches to hide the latency introduced by DVS thereby achieving the one instruction fetch per cycle throughput of the normal cache. An example of this pipelined access is shown next.

	Cycle :	$C1$	$C2$	$C3$	$C4$	$C5$	$C6$	...
Bank1 :		$i1$		$i1$		$i1$		...
Bank2 :			$i2$		$i2$		$i2$	...

Clearly, the throughput of 1 instruction per cycle is sustained. However, an accurate next instruction address prediction scheme needs to be in place in order for this to succeed.

In summary, PRIM has the same power consumption as the DVS scheme. However, the careful overlapping of accesses to the two banks reduces the total cycles requirement to 21 (1 additional cycle is required for pipeline initialization). Thus the total energy consumption for this scheme is $En = \frac{p}{4} \times 21 = 5.3p$, which is similar to DVS selective way scheme (the best scheme in terms of energy) while the performance is similar to that of the normal cache. Moreover, with larger loop sizes and greater number of iterations, the pipeline initialization overhead will become negligible.

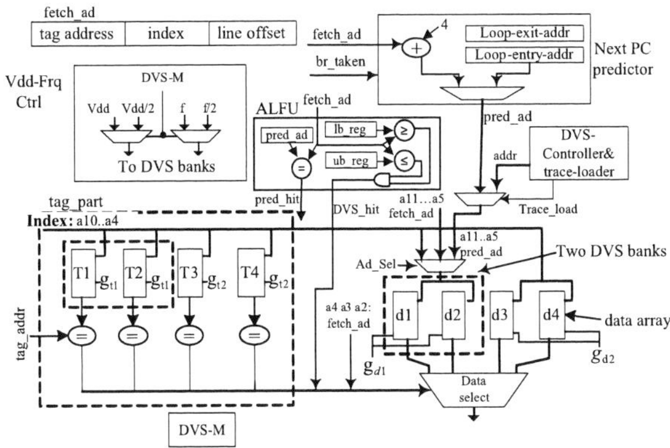

Figure 2: PRIM architecture.

The motivating example illustrates that PRIM needs co-operation from the compiler in statically identifying program phases with under-utilized cache banks. Once these program fragments have been identified, the compiler inserts appropriate instructions to load and lock the content to the cache banks and switch the banks to DVS mode. At runtime, the execution follows these cache configuration hints to switch the cache between normal and DVS modes.

2. THE PRIM ARCHITECTURE

Architectural considerations: We present the PRIM architecture based on a 4-way set associative cache as shown in Figure 2. PRIM architecture can achieve the following functionalities: (1) run in DVS mode and normal cache mode according to execution needs, (2) dynamically switch running modes, (3) reload instruction blocks to DVS banks at runtime, and (4) fetch instructions from DVS banks in pipelined fashioned to maintain throughput. Here we only consider uninterrupted execution of a single application that controls PRIM throughout its execution.

Control of DVS mode and normal cache mode: The cache consists of four data banks, the DVS control logic, the voltage-frequency controller (Vdd-Frq Ctrl), the address lookup function unit (ALFU), and the next-PC predictor. Unlike a conventional cache, the supply voltage of banks d1 and d2 can be dynamically changed by the DVS controller. When a DVS reconfiguration instruction is encountered, the DVS controller loads the appropriate instructions to the DVS banks. DVS controller is essentially a component of Direct Memory Access (DMA) which is controlled by processor. It has two inputs: initial address of an instruction block and number of instructions to be loaded. After loading the instructions, DVS controller sets the DVS-M register, which in turn changes the supply voltage and frequency of d1 and d2 to $\frac{V_{dd}}{2}$ and $\frac{f}{2}$, effectively reconfiguring the banks to DVS mode.

Instruction searching: The ALFU (*address lookup functional unit*) determines if an instruction is residing in the DVS banks or not. When two banks are configured to DVS mode, the tags corresponding to these two banks are completely disabled. In other words, the DVS banks act as software controlled memory. The address matching is accomplished inside the ALFU instead of tag matching in conventional caches. The ALFU contains two address registers and two parallel comparators. The two registers, ub_reg and lb_reg, hold respectively the upper and lower bound addresses for the instruction block that currently resides in the DVS banks. The values in ub_reg and lb_reg are updated when the DVS controller loads a block of instructions into the DVS banks. In the current PRIM design, there is only one pair of ub_reg and lb_reg. As a result, only one block of instructions may reside in the DVS banks at any point of time. If the address of the instruction

978-1-60558-497-3/09 $25.00 © 2009 ACM

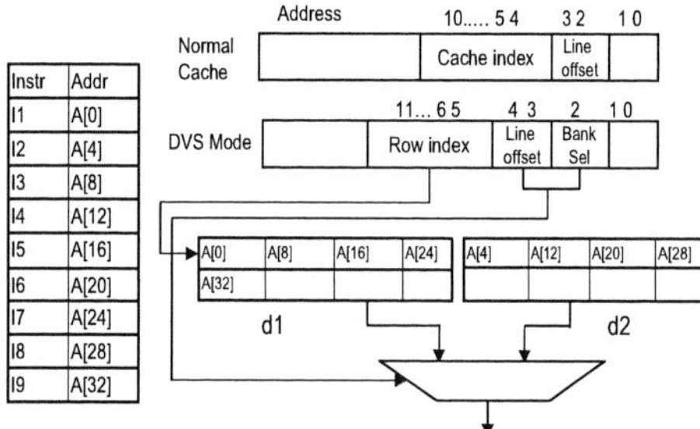

Figure 3: The layout of instructions

to be fetched falls within the range of these two registers, then it resides in the DVS banks and the ALFU generates `DVS_hit` signal. This signal controls the selection and gating of the tag and the data banks. To save dynamic energy, the two normal cache banks and their tags are not searched if the `DVS_hit` signal is asserted. On the other hand, the DVS banks are not searched if `DVS_hit` signal is not asserted. This is accomplished by clock gating of the tag and data banks. The gating signals g_{t1}, g_{d1} and g_{t2}, g_{d2} are used to gate the clock of DVS banks and non-DVS banks, respectively:

$$g_{t1} = \text{DVS\_mode} \qquad g_{t2} = \text{DVS\_mode} \wedge \text{DVS\_hit}$$
$$g_{d1} = \text{DVS\_mode} \wedge \sim \text{DVS\_hit} \qquad g_{d2} = \text{DVS\_mode} \wedge \text{DVS\_hit}$$

As a result, in DVS mode, only two banks – either DVS banks or normal cache banks – are accessed at any point of time.

Pipelined parallel instruction fetches: In order to maintain the throughput of instruction fetches in DVS mode, when an instruction is fetched from one DVS bank, a speculative fetch is also started in the other DVS bank. The ALFU will later check to see if the speculation was successful. If the speculated PC equals that requested by the processor, the address prediction hit (pred_hit) is asserted to indicate a successful speculation. Otherwise, it is necessary to redo the instruction fetch, thereby incurring an overhead of twice the normal cache's latency.

>From the above description, one can easily see why it is crucial to carefully layout the instructions in DVS banks so that sequentially executed instructions can be fetched in parallel from the two DVS banks. Figure 3 shows an example of the instruction layout. The left side of the figure is the instruction layout of a loop containing six instructions in main memory. Suppose the sequence of instruction executed is $1 - 2 - 3 - 4 - 5 - 6 - 1 \cdots$. To be able to fetch consecutive instruction in parallel, the odd and even numbered instructions need to be in different banks. We propose to layout the instructions in the way shown in right hand side of Figure 3. Bits 11...5 in Figure 3 are used to address the row of the DVS banks, while the column selection is determined by the least significant bit (bit 2 in Figure 3) of cache block index. The instruction selection from a cache line is decided by two bits 3 and 4.

As the two DVS banks can be configured as two different types of instruction buffer, namely conventional cache and software controlled memory, the address of two DVS banks need to be selectable from different address sources according to their mode. This is achieved by selection signal Ad_Sel. When applications try to load instruction block into DVS mode, Ad_Sel will select the address generated by 'DVS-Controller&trace-loader' component. If they are configured as DVS mode and operate on instruction fetching state, address a11...a5 will be chosen. Otherwise, the address (i.e cache index) will be selected for conventional cache mode.

The 'Next PC predictor' component of PRIM predicts the address of the next instruction to be fetched. If a branch is taken, we assume that we have encountered a loop and the address boundary will be dynamically written to registers 'Loop-entry-addr' and 'Loop-exit-addr'. If the current fetch address is not equal to the value in 'Loop-exit-addr', the next fetch address will be speculated to be the increment of the current fetch address. Otherwise, the next fetch address will be the value in 'Loop-entry-addr'.

3. COMPILER SUPPORT

PRIM requires compiler support to achieve dynamic reconfiguration. The compiler decides which program regions are suitable for running in DVS mode, inserts appropriate instructions into the application to switch the two designated banks to DVS mode, and dynamically loads the selected instructions into the DVS banks.

Our algorithm works on the Loop-Procedure Hierarchy Graph (LPHG) [14] of the program. The nodes of this graph represent procedures and loops while the edges between the nodes denote the call relationship between them.

Algorithm 1: Algorithm for determining dynamic reconfiguration and DVS instruction load points

---

Input: $Proc\_list$: Procedure list whose procedures have been intra-procedural optimized

Output: $Basic\_block\_list$: list of instruction blocks of basic blocks assigned to DVS banks

BOOL $conflict$: if severe cache conflict incurred;
1 Build Loop-Procedure Hierarchy Graph(LPHG)
2 Get_sizes_of_all_loops();
3 $list\_loops \longleftarrow$ all leaf loops;
4 **foreach** $loop\ l\ in\ list\_loops$ **do**
    if l $is\ leaf\ loop$ $\bigwedge$ $\#iterations\ of\ l \geq Thresh\_hold$ **then**
        | Annotate_reconfig_point_and_instrs_partitioned(l);
5   **else if** l $is\ non\text{-}leaf\ loop$ **then**
6         $list\_child\_loops \longleftarrow$ all child loops of l;
7         $Update\_iteration\_num(list\_child\_loops)$;
8         $conflict$=evalConflict($DVSmode, child\_loops\_of\_l \cup l$);
9   **if** $!conflict$ **then**
10 | Instr_alloc($list\_child\_loops \cup l$, DVS-banks);
        end
11   **if** $!list\_loops.contain(l)$ **then**
12 | $list\_loops.push\_back(l)$;
    end
end
13 Hoist_reconfig_position();
14 Insert_reconfig_and_code_loading_instructions();
15 return $Proc\_list$;

Our proposed algorithm is shown in Algorithm 1. We assume that most of the energy consumption due to instruction fetch occur inside the loops. The intuition is that if the time spent in a loop is long enough to hide the overhead of reconfiguration and instructions loading, then that loop is a candidate for allocation into the DVS banks. Thus we look for loops where the number of iterations exceeds a threshold. In this paper, we empirically set the threshold value to be 20 iterations. If the loop size is too big to fit into the DVS banks, then the cache is used to buffer the rest of for the loop.

In a LPHG, the deeper a loop is, the higher is its execution frequency. The algorithm therefore starts from the leaf loops and works toward the root node. After a loop has been examined, its parent loop is added to $list\_loops$ (line 12). We may at some later point examine the parent node for further opportunities. If a loop is an internal node (line 5), then the algorithm evaluates whether it is beneficial to configure the cache to DVS mode for its child loops (line 8). The evaluation function we use is conservative and simple. If the allocation of a child loop L to DVS banks does not adversely affect the miss rate of the other children, then the allocation is considered beneficial. In that case, the cache is configured to DVS mode at the entry point of the parent loop.

Figure 4 is an example of how the algorithm resolves conflicts, determines reconfiguration points, and selects the instructions for

978-1-60558-497-3/09 $25.00 © 2009 ACM

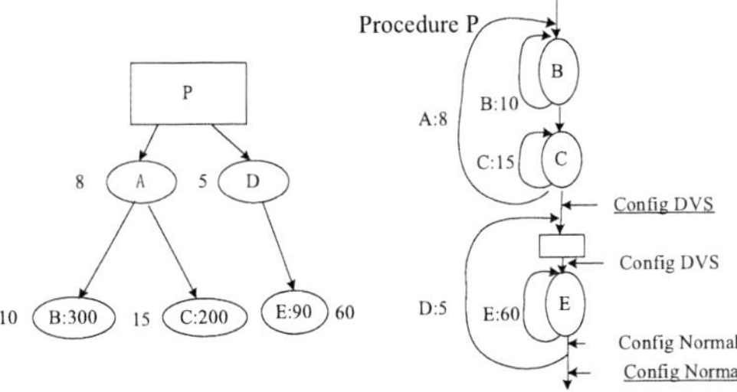

Figure 4: Code transformation for reconfiguration.

DVS banks. The left part of Figure 4 is a sample program represented in LPHG, while the right part is its corresponding CFG. The LPHG consists of the procedure P and five loops A, B, C, D, E. The numbers beside the loops are the number of loop iterations while the numbers inside the loops are their sizes in terms of the number of instructions. Each of the four cache banks can hold up to 64 instructions. The algorithm first selects the three leaf loops. Only the number of iteration of E is larger than the threshold value (20). So the cache is configured to DVS mode at the entry point of E. We then traverse upward to the parent loops, A and D. As the number of iterations of D is smaller than the threshold, we will not use DVS mode for it. Loop A has two children, B and C, and the total number of iterations of B and C from the point of the entry of A is now larger than 20. The algorithm then checks whether there will be severe cache conflicts if B or C is placed in the DVS banks. As the sizes of B and C are both larger than the remaining two normal cache banks, allocating either loop to the DVS banks will cause severe cache conflict to the other loop. It is therefore prudent to operate A in DVS mode.

The instruction allocation function assigns the frequently executed instructions inside the selected loop to the DVS banks (line 10). If the loop size exceeds the capacity of the DVS banks, then only the most frequently executed instructions are allocated to DVS banks.

After determining the reconfiguration points and the instructions selection for the DVS banks, the number of reconfigurations can be reduced by hoisting the reconfiguration point from the inner loop to the outer loop (line 13) if L is the only loop selected among its siblings. Figure 4 shows an example where the DVS configuration instructions are hoisted to where they are underlined.

4. EXPERIMENTAL EVALUATION

4.1 Experimental Methodology

We used the Simplescalar-PISA 3.0d simulator [5] for our experiments. The full-featured simulator `sim-outorder` was modified to support PRIM. The cache line used in the simulation corresponds to 32 bytes. The instruction memory hierarchy consists of PRIM and the main memory. Our PRIM implementation has four banks of data storage, each capable of holding 64 instructions. The latency of accessing L1 cache is 1 cycle. The main memory is assumed to be pipelined. The latency of the first access to the main memory is 10 cycles, while that of the subsequent accesses is 2 cycles. In our simulation, the Trace_load instructions used to load instruction blocks to DVS banks are compiled to binary code. We include the overhead of execution time and cache miss rate increase caused by these instructions.

We use six benchmarks from the MediaBench and MiBench suites. We compared the energy consumption and performance across three different configurations: (1) a baseline system comprising of a tra-

ditional 4-way associative instruction cache, (2) a simple low V_{DD} scheme (Low-V_{DD}) (3) a selective way cache [1] (Sel-Way), and (4) PRIM. We use two voltage settings, $0.5 \times V_{DD}$ and $0.8 \times V_{DD}$, for different process technologies. We chose selective way cache for comparison because it disables and puts under-utilized memory banks into low power mode to save energy.

In our experiments, Sel-Way gates the clock of a certain number of cache banks if it is under-utilized. To do this, we analyze the LPHG and put frequently executed loops into certain cache banks and gate the rest of the cache banks. The instructions of the frequently executed loops are locked for a certain time interval to avoid unintentional replacement. Low-V_{DD} scheme simply halves supply voltage and clock frequency of L1 instruction cache to save energy while the access latency to the instruction cache is doubled.

4.2 Energy Calculations

We model the energy consumption of the memory hierarchy using the CACTI [17] model for $0.13\mu m$ technology. In this work, our focus is on reducing dynamic energy consumption. To calculate the energy consumption of PRIM, we include the extra logic elements not found in a normal cache, namely the ALFU, the Next PC predictor and the two extra multiplexers. We assume that the energy consumed by the PRIM's comparators and multiplexers are the same as the multiplexers and comparators components of the normal cache assumed by CACTI. We approximate the energy consumption of an adder or subtractor in ALFU as that of a multiplexer. These hardware components are powered at regular voltage.

According to CACTI, the total per-access energy of a n way cache is $e_C = e_t + e_d + e_{com} + e_{mux} + e_{out}$ where e_t, e_d, and e_{com} stand for the energy per access to all tags, data storages, and comparators, respectively. e_{mux} is the energy consumption of the multiplexer, while e_{out} is the energy of the output driver. We rewrite this formula as $e_C = n \times e_b + e_{mux} + e_{out}$ where e_b is the energy per access for one cache bank. This includes the per-access energy for the data and tag of one bank, and a comparator.

According to general energy equation, $E = C \times V^2 \times f \times t$, when supply voltage and frequency is decreased to V', the energy consumption becomes $(\frac{V'}{V})^2 \times 2f$ (we define this factor as C_f) of the original value. We assume that the energy of the data output drivers of PRIM will be the same as that of a normal cache since this component has to be powered at the regular voltage while that for the cache banks, comparators and multiplexer is decreased. There are three ways in which PRIM is accessed. When PRIM is in DVS mode and the instruction is in one of the DVS banks, PRIM can deliver it directly from the DVS bank without searching the other DVS bank, the energy of an instruction fetch is: $e_D = \frac{e_b + e_{mux}}{C_f} + e_{out} + e_O$, where e_O is the energy overhead per access due to the extra hardware components described above. It is about 3.5% of the total per-access energy of the baseline cache. When PRIM is in DVS mode and the instruction is not in a DVS bank, the other two cache banks powered at the regular V_{DD}, will be searched. The per-access energy for this is: $e_{\neg D} = 2 \times e_b + e_{mux} + e_{out} + e_O$. If PRIM is in normal cache mode, energy consumption is simply the energy of the baseline cache plus the overhead: $e_N = e_C + e_O$.

The last two energy components of PRIM are the energy consumed during the loading of traces, E_L, into the DVS banks, and the energy caused by cache misses. E_L is approximated by:

$$E_L = \sum_{l=0}^{L} \frac{s_l}{B} \times e_M$$

where L is the total number of traces loaded, s_l is the size of trace l and B is the size of a cache block. e_M is the energy penalty of a

978-1-60558-497-3/09 $25.00 © 2009 ACM

base cache(e_C)	Low-V_{DD}	PRIM				Sel-Way			
		$e_{\neg D}$	e_N	e_D ($0.5 \times V_{DD}$)	e_D ($0.8 \times V_{DD}$)	1way	2way	3way	4way
0.538	0.144	0.296	0.557	0.065	0.165	0.146	0.277	0.407	0.538

Table 1: Per-access energy consumption (in nJ).

cache miss. Its value is assumed 50 times of energy consumption of one access to four way baseline cache [18]. Since the DVS controller is only active during instruction trace loading to PRIM, and because the energy of a cache miss is orders of magnitude greater, we did not include it in the energy calculations. The total overall energy consumption due to memory accesses to PRIM is given by:

$$E_{\text{PRIM}} = A_D \times e_D + A_{\neg D} \times e_{\neg D} + A_N \times e_N + A_M \times e_m + E_L$$

where A_D and $A_{\neg D}$ are the numbers of accesses to the DVS banks and two cache banks respectively when PRIM is in DVS mode, while A_N is the number of accesses to PRIM when it is in normal cache mode. A_M is the number of cache misses.

As with PRIM, for Low-V_{DD} cache, the voltage of data output driver is assumed normal and the energy consumption of this component remains same as that of baseline cache. For Sel-Way cache, the energy is depend on the number of active cache banks, n, which is shown as $e_S = n \times e_b + e_{mux} + e_{out}$

Table 1 shows the energy consumption per access to different architectures. The result of our energy calculation is consistent with the energy consumption of 32K L1 cache in [12] in which SPICE was used to simulate the energy consumption. According to [12], the total energy consumption of 32K L1 cache powere at 1V is 1.055nJ while the energy at 0.5V is 0.1250nJ.

4.3 Experiment results

Performance: The performance results are shown in Table 2. 'fetch miss' represents the miss rate of instruction fetch while 'perform overhd' stands for the performance overhead compared to the baseline. 'Pred-miss' is the miss rate of address prediction of the next speculatively fetched instruction in DVS mode. Compared to the baseline cache configuration, we can see some miss rate reduction achieved by PRIM and Sel-Way schemes for certain benchmarks. We attribute this to the locking of frequently executed instructions, thereby reducing conflict misses. The miss rate of address prediction ranges from 0.63% to 8.45% with an average value of 2.85%. The 'pred-miss' rate for mpeg2-dec and pegwit are quite high. The loops in mpeg2-dec have many procedure calls resulting in poor locality. For pegwit, there are many control flow transfers inside the loops which is detrimental to the prediction process.

The performance overhead varies across different applications. There are mainly four factors affecting the overhead: the latency of cache accesses, the cache miss rate, the next address prediction miss rate, and the overhead of cache reconfigurations. Low-V_{DD} suffers from a high performance overhead that averages 125.6% because of the doubling of the latency in the instruction cache. PRIM, on the other hand, has a low performance overhead ranging from -6.65% to 2.82%. For the benchmark susan-edge, performance is in fact improved because of the reduction in the cache miss rate. On the other hand, for the benchmark pegwit, the poorer performance is due to the penalty of pred-miss and reconfiguration exceeding the gains from cache miss rate reduction.

Energy consumption: The total dynamic energy consumption of the four instruction memory hierarchies are shown in Table 3. We evaluated two PRIM configurations – one that halves V_{DD} and one that lowers V_{DD} by 20%. In both these configurations, the operating frequency is halved, doubling access times to these banks. All the schemes achieve significant energy savings compared to the baseline. The reduction in energy consumption achieved by the Sel-Way cache is 39.0% for the benchmarks, while that for PRIM is 56.6% at $0.5 \times V_{DD}$ and 45.1% at $0.8 \times V_{DD}$. In other words, PRIM at $0.5 \times V_{DD}$ and $0.8 \times V_{DD}$ achieved 17.6% and 6.1% more energy saving than Sel-Way respectively. Energy savings of PRIM at $0.8 \times V_{DD}$ is smaller than PRIM at $0.5 \times V_{DD}$ because the supply voltage scaling down is more conservative.

Low-V_{DD} achieved an average 56.5% energy savings which is almost the same as PRIM at $0.5 \times V_{DD}$. However, Low-V_{DD} comes with significant performance loss as shown in Table 2 because of the doubling of the cache hit latency. As the result, the additional energy consumed by other parts of a processor can surpass the energy savings obtained by instruction memory hierarchy.

Energy savings come mainly from DVS mode access, clock gating of under-utilized ways, and cache miss rate reduction. Operating in the DVS mode, PRIM can save more energy than the clock gating used in Sel-Way. This is evident in the benchmark FFT where the improvement of PRIM over Sel-Way scheme is very significant. In the case of FFT, the cache miss rate is very low and so cache misses contribute only a small amount of total energy consumption. The main contributor to the total energy is the on-chip cache access energy which is significantly reduced in PRIM. For the benchmark susan-edge, the energy improvement of PRIM over Sel-Way is small. This benchmark has a big loop with a large number of iterations that is larger than the cache. As a result, it is beneficial to allocate as many cache entries as possible and lock the instructions of this loop in these cache entries to avoid cache conflict. Sel-Way allocates and locks three cache banks for the instructions in the loop, while our PRIM configuration can only allocate two cache banks in DVS mode. Sel-Way is therefore able to avoid more cache conflicts in this benchmark.

Energy delay product: 'En*delay' in Table 3 represents normalized energy delay product relative to the baseline cache. PRIM at $0.5 \times V_{DD}$, PRIM at $0.8 \times V_{DD}$, Low-V_{DD} and Sel-Way achieved average normalized energy delay product value 0.43, 0.54, 0.93, and 0.61 respectively. Although Low-V_{DD} achieves significant energy savings, the energy delay product is about the same as the baseline cache due to the large performance overhead. PRIM at $0.5 \times V_{DD}$ and $0.8 \times V_{DD}$ achieved the best energy-delay product since it can effectively decreased energy consumption with only a slight performance loss.

5. RELATED WORK

A lot of research have been done on the memory hierarchy with the aim of saving energy. Schemes were proposed to shut down certain number of cache ways when the cache is under-utilized [1, 16]. Zhang [18] proposed a highly reconfigurable cache whose associativity can be dynamically changed.

Many schemes and algorithms utilizing dynamic voltage and frequency scaling have also been developed for low energy processors. In [13, 2], the worst-case slack time and workload-variation slack time were exploited using DVS. When the utilization computed based on the WCETs of tasks is lower than 1, it may be possible to run some of tasks at a slower clock and a lower supply voltage without delaying their deadlines. Some researchers [2, 8] developed compilation flows that detect code regions for DVS

978-1-60558-497-3/09 $25.00 © 2009 ACM

Benchmark	Baseline	Low-V_{DD}	PRIM			Sel-Way	
	fetch miss(%)	perform overhd(%)	fetch miss(%)	pred-miss (%)	perform overhd(%)	fetch miss(%)	perform overhd(%)
mpeg2-dec	1.35	107.0	1.30	4.8	-0.03	1.49	3.07
gsm-dec	0.42	121.8	0.40	0.71	0.20	0.41	0.07
susan-edge	2.76	80.9	2.36	1.11	-6.65	2.15	-13.5
FFT	0.06	146.5	0.09	1.42	2.82	0.09	0.85
pegwit	0.38	45.8	0.35	8.45	0.78	0.38	3.54
sha	0.054	252.1	0.35	0.63	1.66	0.046	3.85
Average	-	125.6	-	2.85	-	-	-

Table 2: Miss rate and performance overhead

Benchmark	Baseline	Low-V_{DD} ($0.5 \times V_{DD}$)			PRIM($0.5 \times V_{DD}$)			PRIM($0.8 \times V_{DD}$)			Sel-Way		
	En(mJ)	En(mJ)	impr(%)	En*delay	En(mJ)	impr(%)	En*delay	En(mJ)	impr(%)	En*delay	En(mJ)	impr(%)	En*delay
mpeg2-dec	34.5	19.43	43.7	1.17	25.6	25.8	0.74	27.55	20.1	0.80	30.26	12.3	0.90
gsm-dec	8.10	3.20	60.5	0.88	3.66	54.7	0.45	4.61	43.0	0.57	4.35	46.3	0.54
susan-edge	2.92	2.02	30.8	1.25	1.89	35.2	0.60	2.00	31.4	0.64	1.98	32.3	0.59
FFT	1.06	0.31	71.0	0.71	0.21	80.0	0.21	0.40	62.0	0.39	0.51	51.5	0.49
pegwit	19.78	7.63	61.5	0.56	7.78	60.7	0.40	10.19	48.5	0.52	11.21	43.4	0.59
sha	7.53	2.16	71.3	1.01	1.25	83.4	0.17	2.58	65.8	0.35	3.89	48.4	0.54
Average	-	-	56.5	0.93	-	56.6	0.43	-	45.1	0.54	-	39.0	0.61

Table 3: Energy consumption.

and determine the optimal frequency scaling. There are also many proposals to apply DVS to the memory hierarchy. Kim et. al. [11] proposed the drowsy cache architecture where cache lines will be put into drowsy mode if they are not accessed for a certain time interval. This technique is quite similar to the selective way cache [1] in which under-utilized cache lines are set to a low power mode. The main difference between them is that the drowsy cache can hold data when switching between low and normal power modes. All these previous works tried to disable under-utilized memory resources to prevent these parts from consuming energy. Instead PRIM tries to make use of these resource in DVS mode so as to retain the energy savings while recovering the performance loss.

6. CONCLUSION

In this paper, we propose PRIM, a pipelined, DVS-based dynamically reconfigurable instruction memory hierarchy. A number of ways in a PRIM cache can be reconfigured as low-V_{DD}, low frequency loop buffer. This achieves better energy savings over methods that disable unused resources for energy reduction. Pipelined speculative reads are used to maintain the throughput of instruction fetch in the low-V_{DD} buffer. To support PRIM, we also developed an algorithm to partition instructions and determine reconfiguration points. Our experimental results showed significant energy savings while performance is affected only marginally.

7. REFERENCES

[1] David H. Albonesi. Selective cache ways: On-demand cache resource allocation. In *Proceedings of MICRO*, 1999.

[2] Ana Azevedo et al. Profile-based dynamic voltage scheduling using program checkpoints. In *Proceedings of DATE*, 2002.

[3] P. Bose et al. Early-stage definition of LPX: A low power issue-execute processor prototype. In *Proceedings of HPCA Workshop on Power-Aware Computer Systems*, 2002.

[4] David Brooks, Vivek Tiwari, and Margaret Martonosi. Wattch: A framework for architectural-level power analysis and optimizations. In *Proceedings of ISCA*, 2000.

[5] Doug Burger and Todd M. Austin. The Simplescalar tool set, version 2.0. *Technical Report #1342, University of Wisconsin-Madison Computer Sciences Department*, May 1997.

[6] A. Chandrakasan and R. W. Brodersen. Lower power digital CMOS design. In *Kluwer Academic Publishers*, 1995.

[7] Zhiguo Ge, Weng-Fai Wong, and Hock Beng Lim. DRIM: A low power dymaically reconfigurable instruction memory hierarchy for embedded systems. In *Proceedings of DATE*, 2007.

[8] Chung-Hsing Hsu, Ulrich Kremer, and Michael S. Hsiao. Compiler-directed dynamic voltage/frequency scheduling for energyreduction in microprocessors. In *Proceedings of ISLPED*, 2001.

[9] Zhigang Hu et al. Microarchitectural techniques for power gating of execution units. In *Proceedings of ISLPED*, 2004.

[10] Stefanos Kaxiras, Zhigang Hu, and Margaret Martonosi. Cache decay: Exploiting generational behavior to reduce cache leakage power. In *Proceedings of ISCA*, 2001.

[11] Nam Sung Kim et al. Drowsy instruction caches: Leakage power reduction using dynamic voltage scaling and cache sub-bank prediction. In *Proceedings of MICRO*, 2002.

[12] Cheng-kok Koh, Weng-Fai Wong, Yiran Chen, and Hai Li. VOSCH: Voltage Scaled Cache Hierarchies. In *Proceedings of ICCD*, 2007.

[13] Soongsoo Lee and Takayasu Sakurai. Run-time voltage hopping for low-power real-time systems. In *Proceedings of DAC*, 2000.

[14] Yanbing Li et al. Hardware-software co-design of embedded reconfigurable architectures. In *Proceedings of DAC*, 2000.

[15] James Montanaro et al. A 160-MHz, 32-b, 0.5-W CMOS RISC Microprocessor. *Digital Technical Journal*, 9(1), 1997.

[16] Michael Powell et al. Gated-Vdd: A circuit technique to reduce leakage in deep-submicron cache memories. In *Proceedings of ISLPED*, 2000.

[17] Steven J. E. Wilton and Norman P. Jouppi. CACTI: An enhanced cache access and cycle time model. *IEEE Journal of Solid-State Circuits*, 31(5):677–688, 1996.

[18] Chuanjun Zhang, Frank Vahid, and Walid Najjar. A highly configurable cache architecture for embedded systems. In *Proceedings of ISCA*, 2003.

978-1-60558-497-3/09 $25.00 © 2009 ACM

LICT: Left-uncompressed Instructions Compression Technique to Improve the Decoding Performance of VLIW Processors

Talal Bonny and Jörg Henkel
University of Karlsruhe
Chair for Embedded Systems
Karlsruhe, Germany
{bonny,henkel}@informatik.uni-karlsruhe.de

ABSTRACT

Compressing program code compiled for VLIW processors to reduce the amount of memory is a necessary means to decrease costs. The main disadvantage of any code compression technique is the system performance penalty because of the extra time required to decode the compressed instructions during run time. In this paper we improve the performance of decoding compressed instructions by using our novel compression technique ($LICT$: $Left$-$uncompressed$ $Instruction$ $Technique$) which can be used in conjunction with any compression algorithm. Furthermore, we adapt a new code compression approach called Burrows-Wheeler (BW) [9] which has been used before in data compression. It significantly reduces the code size compared to state-of-the-art approaches for VLIW processors. Using our $LICT$ in conjunction with the BW algorithm improves the performance explicitly (2.5x) with little impact on the compression ratio (only 3% compression ratio loss).

Categories and Subject Descriptors

B.3 [**Hardware**]: Memory Structures

General Terms

Performance, Design

Keywords

Code compression, embedded systems, Huffman Coding

1. INTRODUCTION AND RELATED WORK

Embedded applications are growing rapidly due to the attractive price/performance ratio of embedded systems. Since memory of the embedded system must be small according to the constraints on the embedded market, a key challenge in designing high volume and cost effective embedded systems is to host this vast amount of software in an efficient

Permission to make digital or hard copies of part or all of this work for personal or classroom use is granted without fee provided that copies are not made or distributed for profit or commercial advantage and that copies bear this notice and the full citation on the first page. To copy otherwise, to republish, to post on servers or to redistribute to lists, requires prior specific permission and/or a fee.
DAC'09, July 26-31, 2009, San Francisco, California, USA

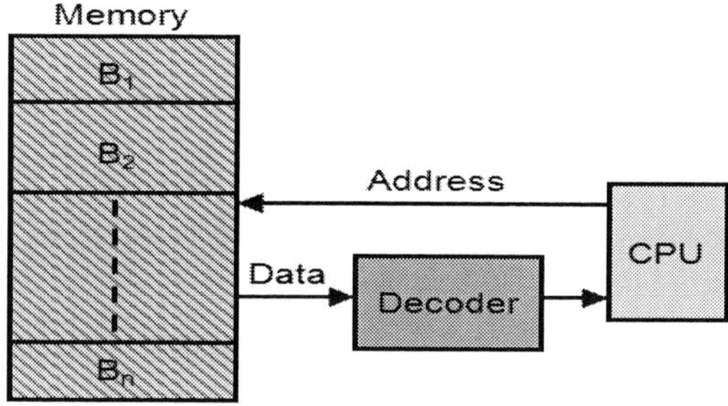

Figure 1: Traditional decoding techniques.

way. This can be accomplished by deploying code compression which, in addition to memory size reduction, may also reduce the power consumption [4, 2].

In code compression techniques, the time required to execute the compressed instructions includes the time of decoding the compressed instructions plus the time of executing the decompressed instructions. As the time of decoding and executing the instructions is always longer than the time of executing the original ones, this results in a performance penalty in the decoding. We therefore improve the performance of decoding compressed instructions by using our novel compression technique called $LICT$: $Left$-$uncompressed$ $Instruction$ $Technique$ which can be used in conjunction with any compression algorithm to improve the performance. In this paper, we use $LICT$ along with the Burrows-Wheeler (BW) compression algorithm [9] and we show that it results in a high compression ratio (58%) without loss in performance. To show the orthogonality of our technique, we use it also in conjunction with another compression algorithm (called FBT: $Filled$ $Buffer$ $Technique$) from [3] and we show that our technique explicitly improves the performance ratio with only a slight impact on the compression ratio.

In **Related Work**, various compression techniques have been studied for VLIW processors. Seong and Mishra [10, 11] used a bitmask technique to compress the program code. This technique is based on the Hamming distance method and can create more matching patterns that improves the compression ratio to 60%. The fetch packet can be decoded in parallel but there was no information about the performance ratio. In [6], the authors proposed a variable-sized block code compression method based on LZW for VLIW processors. They achieved an average compression ratio of 75.2% targeting the TMS320C6x VLIW processor. Their method provides 8 bytes peak decompression bandwidth and 1.82 bytes in average. Ros and Sutton ([7] and [8]) developed a dictionary compression technique based on vector Hamming distances for the complete 32-bit instructions. The authors achieved compression ratios between 72.1% to 80.3%. There was no information about the performance ratio. Bonny et al. [3] proposed a new technique called Filled Buffer Technique to improve the compression ratio when it is applied to any Lempel-Ziv family algorithm. When they applied their technique to the "Deflate" compression algorithm, they achieved an average compression ratio of 61% targeting the TMS320C62x VLIW processor and an average performance of 4 Bytes per cycle.

The rest of the paper is organized as follows. In Section 2 we preset the traditional code compression techniques. Our compression technique *LICT* and the compression algorithm *BW* are presented in Section 3 and 4, respectively. The experimental results are presented in Section 5. We conclude this paper with Section 6.

2. TRADITIONAL CODE COMPRESSION TECHNIQUES

In most of the previous work, a data compression algorithm is selected and applied to each compression block separately (to ensure random access). The compressed blocks are then stored contiguously in memory with a constraint that the beginning of each compressed block should be aligned to the memory boundary so that the decoder can quickly access the compressed blocks.
The decompression (see Fig. 1), hardware decoder is implemented and placed between the memory and the CPU. Each compressed block is **decoded** first using the hardware decoder and then **executed** by the CPU. The time required to complete this process is computed as follow:

$$T_{block\_total} = T_{block\_dec} + T_{block\_exe}$$

One of the disadvantages of using code compression is the performance loss due to the extra time required to decode the compressed code before execution. Interestingly, the entire research in the area has always focused on achieving a better code compression ratio without explicitly targeting the performance loss problem of the hardware decoder. We solved this problem by using our *LICT*

3. LEFT-UNCOMPRESSED INSTRUCTION COMPRESSION TECHNIQUE (LICT)

In this paper, we extend the size of a compression block to include more instructions than the *Basic Blocks* (as we did in [3]). The new compression blocks are called *Extended Blocks*. In our *LICT*, we conduct the following steps:
I. The compression blocks (*Extended Blocks*) of the program code are specified and extracted.
II. Each block is split into two parts. "P1" contains the first *n1* instructions of the compression block and "P2" contains the remaining instructions *n2* in the compression block.
III. The second part "P2" of each compression block is compressed using a data compression algorithm. The instructions *n1* in the first part "P1" of the compression block are

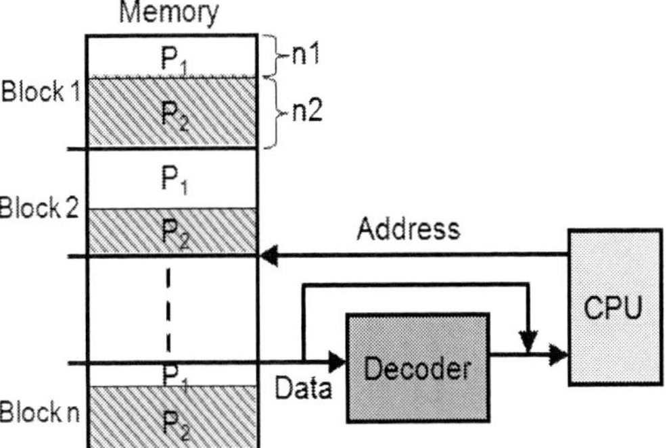

Figure 2: Our compression technique (LICT)

left uncompressed. *n1* is selected such that the time required to execute the instructions in "P1" is equal to the time required to decode the compressed instructions in "P2". For that, the time required to decode the compressed instructions will be hidden by the time required to execute the uncompressed instructions and no performance loss will occur during the decoding.
Fig. 2 shows the same example used in Fig. 1 but using our *LICT*. The time required to complete this process is computed as following:

$$T_{block\_total} = T_{P1\_exe} + T_{P2\_exe}$$

Such that $T_{P1\_exe} = T_{P2\_dec}$
T_{P1_exe}: Time to execute uncompressed instructions "n1".
T_{P2_dec}: Time to decode the compressed instructions "n2".
T_{P2_exe}: Time to execute the decoded instructions "n2".

Algorithm 1 LICT: Left-uncompressed Instructions Compression Technique

---

/* "B": Compression Block */
/* "P1": First part of the compression Block "B" */
/* "P2": Second part of the compression Block "B" */
/* n1: Number of instructions in "p1" */

1: **for all** Compression Blocks "B" in the application **do**
2:    "B"→"P2" {Move all instructions in "B" to "P2"}
3:    0→ "P1" {Reset "P2"}
4:    Move one instruction from "P2" to "P1"
5:    Compress "P2" {Using the compression algorithm}
6:    Compute the time of decoding "P2" {i.e. T_{P2_dec}}
7:    Compute the time of executing "P1" {i.e. T_{P1_exe}}
8:    **if** $T_{P1\_exe} < T_{P2\_dec}$ **then**
9:      GOTO 4
10:    **else** {Times are equal}
11:      Return(n1) {This is the optimal number of uncompressed instructions}
12:    **end if**
13: **end for**

Algorithm 1 shows how to determine the instructions *n1* in the first part "P1" which have to be left uncompressed. This number will keep the time of executing the uncompressed instructions equal to the time of decoding the compressed instructions. The algorithm starts by moving all instructions of the compression block to its second part "P2" and resetting "P1" (lines 2-3).

978-1-60558-497-3/09 $25.00 © 2009 ACM

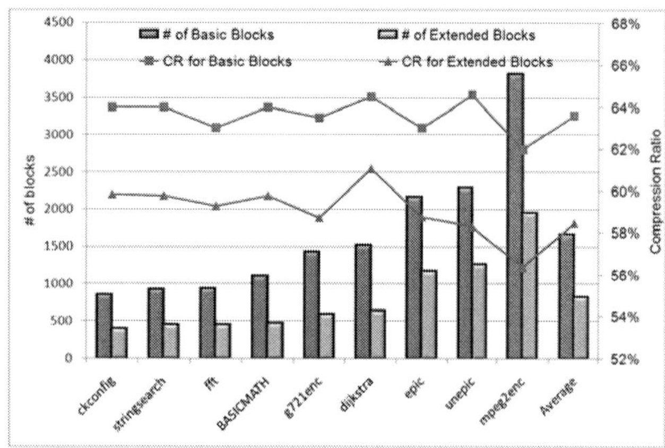

Figure 3: Compression results of the *BW* algorithm

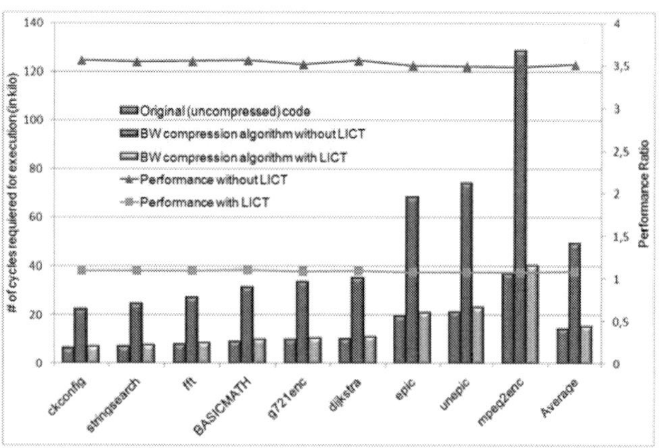

Figure 4: Performance ratio for different size of benchmarks

4. BW CODE COMPRESSION ALGORITHM

To compress the compression blocks (i.e. *Extended Blocks*) we use the Burrows-Wheeler (*BW*) compression algorithm [9] to compress the second part "P2" of each compression block. The *BW* algorithm works in a *block mode* where each block is encoded separately as one string. The main idea of the *BW* algorithm is to start with a string "S" of "n" symbols and to scramble them into another string "L" such that Any region of "L" tends to have a concentration of just a few symbols. This can be done by constructing an n x n matrix. In this matrix, the encoder stores the string "S" in the top row, followed by "n-1" copies of S, each rotated one symbol to the left. The matrix is then sorted lexicographically by rows. The permutation "L" selected by the encoder is the last column of the sorted matrix. After the string "L" is created, it is compressed using the Move-to-front (MTF) method [9] followed by Huffman Coding [1]. Using MTF method increases the concentrations of identical symbols in the String "L" and improves the compression results when the Huffman Coding is used afterwards.

5. EXPERIMENTS AND RESULTS

All experiments are conducted for the TMS320C62x VLIW processor (from Texas Instruments). The applications are compiled and linked using the Code-Composer-Studio (CCS)

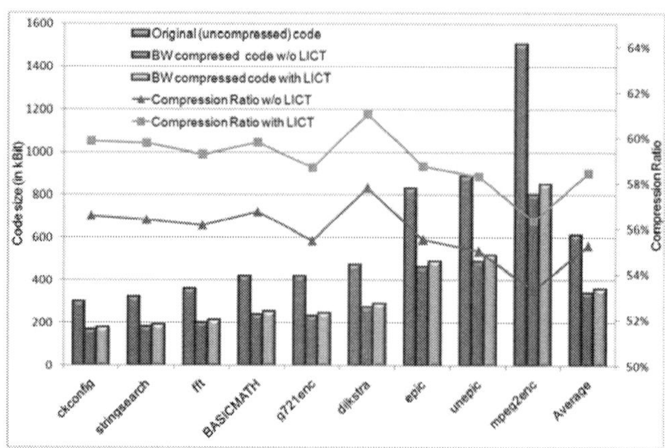

Figure 5: Compression ratio for different size of benchmarks

and simulated using the simulator "c6xsim" [5]. Fig. 3 shows the number of the *Basic Blocks* and *Extended Blocks* for different benchmarks. As shown in this figure, the number of *Extended Blocks* is much smaller than the number of *Basic Blocks* for the same benchmark (in average it may be 48% smaller). This will result in a better compression ratio when the compression technique is applied to the *Extended Blocks* other than the *Basic Blocks*. Fig. 3 shows also the compression ratios of different benchmarks when the *BW* compression algorithm is applied to the *Extended Blocks* (58%) and the *Basic Blocks* (63%). For that, we use the *Extended Blocks* as the compression blocks in all of our experimental results. Fig. 4 shows number of static cycles required to execute different benchmarks of the original applications (first bar) and the compressed applications using the *BW* compression technique (second bar). The average performance ratio obtained in this case is 3.5 The third bar shows the number of decoding and executing cycles when *LICT* is used. Fig. 4 shows that using our compression technique improves the performance in average by 2.5x. For that, there will be no performance loss in the hardware decoder. Our compression technique (*LICT*) does not entirely come for free. Small increase of compressed code size is the price to pay (see Fig. 5). The average compression ratio achieved using the *BW* compression algorithm is 55%. When *LICT* is used, the average compression ratio is degraded to 58%.

Figure 6 shows the trade-off between the performance and compression ratios for the "mpeg2enc" benchmark. The range of the uncompressed instructions is presented in this figure between 1 and 20 instructions. From Fig. 6 we can observe and conclude that by increasing the number of uncompressed instructions in each compression block, the number of instructions that are required to be compressed will be decreased and consequently the compression ratio will be decreased, i.e. unimproved (the compression ratio is decreased from 55% by 1 uncompressed instruction to 80% by 20 uncompressed instructions). On the other hand, increasing the number of uncompressed instructions reserves more time to the decoder to decode the compressed instructions during the execution of the uncompressed instructions. This will improve the performance ratio from 3.47 (by 1 uncompressed instruction) to 1.0 (by 20 uncompressed instructions), i.e. 2.47x.

978-1-60558-497-3/09 $25.00 © 2009 ACM

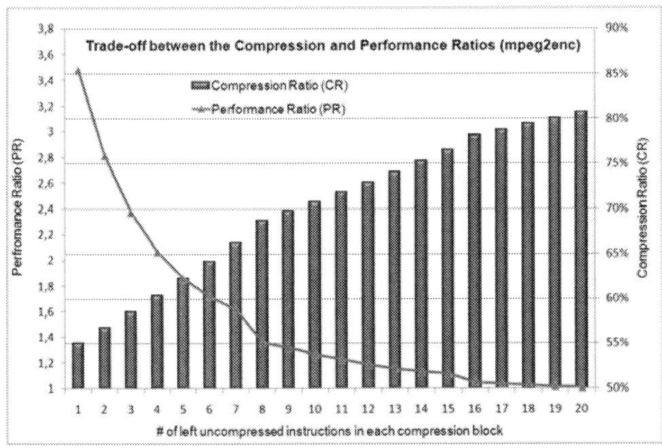

Figure 6: Trade-off between the compression ratio and performance ratio

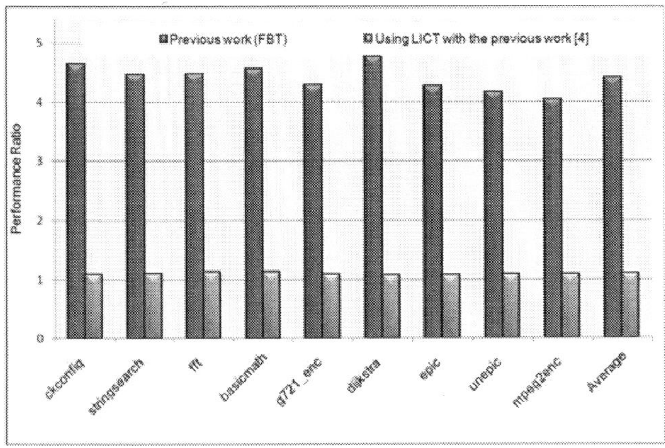

Figure 7: Performance ratio when *LICT* is applied to our previous work [3]

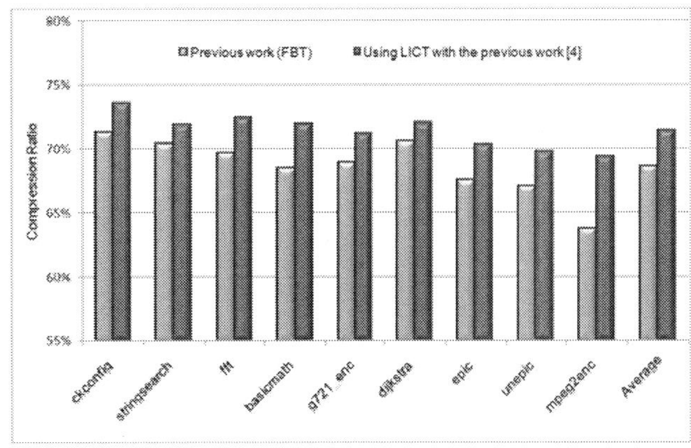

Figure 8: Compression ratio when *LICT* is applied to our previous work [3]

Figures 7 and 8 show the results of applying *LICT* to our previous work (Filed Buffer Technique) [3]. In Figures 7 and 8, the first bar in each benchmark shows the results of the sole Filed Buffer Technique, i.e. without applying *LICT* to it. The second bar shows the results of the Filed Buffer Technique in conjunction with *LICT*. Fig. 7 shows that using *LICT* improves the performance ratio of the decoder in average by 3.3x. That means the decoder decodes the compressed instruction without (or with a very slight) performance penalty. On the other hand, a slight degrading in the compression ratio is observed (from 68% to 71%) because of the left uncompressed instructions (see Fig. 8).

6. CONCLUSION

We have presented a new compression technique for embedded system. We have showed that the performance ratio of any compression algorithms may be improved by using our *Left-uncompressed Instructions Technique* with a slight degrading on the compression ratio. Our Technique can be used with any previous work to improve their performance. It is independent from the processor architecture. We achieved an average compression ratio of 58% without (or very slight) performance penalty for the TMS320C62x VLIW processor.

7. REFERENCES

[1] T. Bell and et al. Text Compression. *Prentice-Hall, Englewood Cliffs*, 1990.

[2] L. Benini and et al. Hardware-assisted data compression for energy minimization in systems with embedded processors. *Proc. of the Design, Automation and Test in Europe conf.(DATE)*, pp. 449-453, 2002.

[3] T. Bonny and J. Henkel. FBT: Filled Buffer Technique to reduce code size for VLIW processors. *Proceedings of international conference on CAD (ICCAD)*, pp. 549-554, 2008.

[4] M. Collin and M. Brorsson. Low Power Instruction Fetch using Profiled Variable Length Instructions. *Proceedings of the IEEE International SoC Conference*, pp. 183-188, 2003.

[5] V. Cuppu. Cycle Accurate Simulator for TMS320C62x, http://www.cs.cmu.edu/afs/cs/academic/class/15745-s07/www/c6xref.

[6] C. Lin, Y. Xie, and W. Wolf. LZW-Based Code Compression for VLIW Embedded Systems. *Proc. of the Design, Automation and Test in Europe conf.(DATE04)*, 2004.

[7] M. Ros and P. Sutton. A hamming distance based VLIW/EPIC code compression technique. *Proc. of the 2004 inter. conf. on Compilers, architecture and synthesis for embedded systems*, pp. 132-139, 2004.

[8] M. Ros and P. Sutton. A post-compilation register reassignment technique for improving hamming distance code compression. *Proc. of the 2005 inter. conf. on Compilers, architecture and synthesis for embedded systems*, pp. 97-104, 2005.

[9] D. Salomon. Data compression: The complete reference. 2007.

[10] S. Seong and P. Mishra. A bitmask-based code compression technique for embedded systems. *Proceedings of international conference on CAD (ICCAD)*, pp. 251-254, 2006.

[11] S. Seong and P. Mishra. An efficient code compression technique using application-aware bitmask and dictionary selection methods. *Proc. of the conf. on Design, automation and test in Europe (DATE)*, pp. 582-587, 2007.

978-1-60558-497-3/09 $25.00 © 2009 ACM

Hierarchical Architecture of Flash-based Storage Systems for High Performance and Durability

Sanghyuk Jung[†]   Jin Hyuk Kim[‡]   Yong Ho Song[†]

[†]School of Electronics and Computer Engineering, Hanyang University, Korea

[‡]Samsung Electronics. Co., Ltd. Korea

{shjung, yhsong}@enc.hanyang.ac.kr[†], jh7711.kim@samsung.com[‡]

ABSTRACT

The use of NAND flash memory for building permanent storage has been increasing in many embedded systems due to properties such as non-volatility and low energy consumption. The persistent requirements for high storage capacity have given rise to the increase of bit density per cell as in multi-level cells but this has come at the expense of performance and has resulted in degradation of durability. In this paper, we introduce a complementary approach to boost the performance and durability of MLC-based storage systems by employing a non-volatile buffer that temporarily holds the data heading to MLCs. We also propose algorithms to efficiently eliminate unnecessary write and erase operations in MLCs by performing a pre-merge in the buffer. Our experiments show that the proposed approach can increase performance by up to 4 times and durability by 4 times by adding only a small hardware cost.

Categories and Subject Descriptors

C.3 [**Computer Systems Organization**]: Special-purpose and Application based Systems, Real-time and embedded systems; D.4.2 [**Operating Systems**]: Storage Management: Secondary storage

General Terms

Algorithms, Design, Management, Performance

Keywords

Storage Systems, Flash Memory, Flash Translation Layer

1. INTRODUCTION

The increasing demand for high capacity and low cost has led to the changes in the architecture of NAND flash memory and storage systems based on this. First, to increase the storage density and reduce the cost per bit, multi-level cell (MLC) technology has been applied to NAND flash devices. A typical MLC flash device stores two bits per cell by using four different voltage ranges. However, this capacity enhancement comes at the expense of performance and durability degradation [1].

Second, to effectively increase the peak bandwidth of read and/or write operations, NAND-based storage systems such as SSDs often implement interleaving techniques in such a way that a long-latency operation is divided into multiple sub-operations which are processed in parallel by many devices. This approach is now common in most high-performance SSDs because the capacity and performance of an individual flash device remain still insufficient. The interleaving technique is also useful to increase the durability of flash devices because write operations are distributed over multiple devices which can reduce write counts at each device. Although the capacity problem can be mitigated by the use of MLCs, the performance degradation is an obstacle to the wide use of MLCs in SSDs.

Permission to make digital or hard copies of part or all of this work for personal or classroom use is granted without fee provided that copies are not made or distributed for profit or commercial advantage and that copies bear this notice and the full citation on the first page. To copy otherwise, to republish, to post on servers or to redistribute to lists, requires prior specific permission and/or a fee.

DAC'09, July 26-31, 2009, San Francisco, California, USA

There have been many research activities in the attempt to address performance and durability problems for MLC-based storage systems. One recent effort is to build a hierarchical structure by using both SLCs and MLCs in such a way that SLCs are used to hold small random write data and MLCs for large sequential write data. Considering that SLC has better endurance and that small random data tends to be updated more frequently, the approach of storing frequently updated data in SLC could be effective in increasing the durability of the storage system. However, the performance of such storage can still be limited by those of MLCs, especially when long sequential data writes occur to the storage frequently. Moreover, write operations can be poorly balanced between SLCs and MLCs due to the uneven amount of sequential and random accesses to storage systems, possibly resulting in the reduction of the storage lifetime.

In this paper, we propose a new complementary approach to boost the performance and durability of MLC-based SSDs by temporarily depositing write data in a non-volatile buffer consisting of SLC flash memories. We also propose algorithms to efficiently eliminate unnecessary write and erase operations in MLCs. In this architecture, the performance of storage system on write operations can be improved because all the write data are first stored into SLC buffers regardless of whether they are random or sequential. In addition, the data in SLCs can be *pre-merged* to a data block of an SLC before being merged to MLCs, which could contribute to the reduction of the effective write counts to MLCs and, thus, to the lengthening of their lifetime. Our experiments show that the proposed approach can increase performance by up to 4 times and durability by up to 4 times by adding a small hardware cost, as compared with MLC-only storage systems.

The remainder of this paper is organized as follows: Section 2 summarizes the previous approaches which explain the architecture and a mapping algorithm used in our system. Section 3 presents our proposed architecture and software algorithm. Section 4 provides experimental results, and the paper concludes in Section 5.

2. BACKGROUND AND RELATED WORK
2.1 Mapping Algorithm

The lack of in-place update capability in flash memories makes data update an expensive operation. Many flash translation layers (FTLs) handles this problem by providing a mapping mechanism from a logical sector address to a physical flash address. There are three types of mapping techniques: block-level mapping, page-level mapping, and hybrid mapping. The block-level and page-level mapping techniques maintains a mapping entry for each block and page, respectively. Unlike these, hybrid mapping mechanisms partition blocks in flash memory into log and data blocks, and use page-level mapping for log blocks and a block-level mapping for data blocks [2][3][4][5][6][7].

FAST is a well-known example of hybrid mapping techniques [3]. It partitions log blocks into a sequential log block and multiple random log blocks and uses them to store write data separately depending on access behavior. Each random log block is fully associated with all the data blocks in flash memories. That is, it can store write (or update) data to *any* data blocks in flash. Later, the data in a log block are merged into their associated data block by allocating new data blocks and copying up-to-date data from log and data blocks into the new blocks. Afterwards, the log block and previously used data blocks are to be erased for further use. This type of merge is called *Full Merge*. By associating one log block with multiple data blocks, FAST enhances the utilization of log blocks, and thus reduces erase operations. However, the time to merge a log block greatly varies depending on the associativity.

FAST uses sequential log block to store write data to consecutive memory location. Once the log block is completely filled with sequential data at consecutive address locations, it becomes a new data block and a free block is designated to a sequential block. This is called *Switch Merge*. However, when a sequential access starts at a different memory location before the sequential log block is completely filled, the log block should be rendered to the incoming sequence. If this happens, the sequential log block becomes a new data block and the remaining empty pages of the new data block are copied from the current data block. This type of merge is called *Partial Merge*, which, if happens frequently, degrades the log block utilization and durability [3].

2.2 Hybrid NAND Storage Architecture

Recently, many FTL mapping schemes classify incoming data into *hot* and *cold* based on the access frequency and size. If a data is updated frequently, it is to be *hot*, and otherwise *cold*. This classification is to improve the performance and durability of flash memories by applying different management policies suitable for access patterns. One of the ways of using hot/cold property is to store hot data in a relatively faster and durable memory device to enhance overall performance and durability.

The mechanism proposed in [8] is a good example of such technique which uses SLC as a faster and durable memory. Figure 1(a) illustrates the organization of such hybrid architecture. In this architecture, a hot/cold detection module separates hot data from cold ones dynamically and directs them either to SLC or MLC based on the decision. It uses one SLC along with multiple MLCs in such a way that all the blocks in SLC are used for hot data at the page level. However, before the SLC runs out of free blocks, it performs garbage collection on log blocks to merge valid data in the SLC to the corresponding data blocks of the MLCs. This operation puts the SLC in busy state until finished. Therefore, if a new request arrives at the SLC from a host during the state, the handling of the request could be delayed until the SLC becomes available after returning to idle state because the SLC has no support for multi-port parallel accesses. In some applications, such conflicts on the SLC could be a performance-limiting factor [9][10].

It is possible to use a volatile memory as a faster and more durable memory device, but with the risk of losing data on sudden power disruption. An example of such an idea is presented in [11]. It uses a DRAM to hold hot data. From the viewpoint of performance and durability, DRAM has much better characteristics than SLC or any other non-volatile memories. However, data safety is not guaranteed by storage.

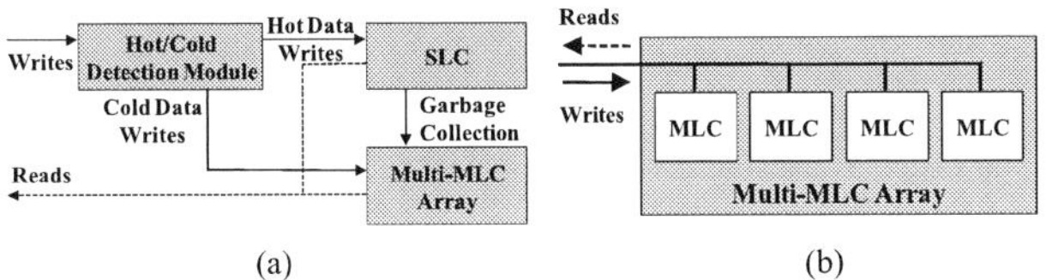

Figure 1. (a) Existing SLC/MLC hybrid architecture (b) Multi-MLC flash only architecture

3. PROPOSED ARCHITECTURE
3.1 Idea in a Nutshell

In the proposed architecture, more than one SLC is used as the non-volatile write buffer in the round-robin fashion. The idea behind this is to efficiently overlap the buffering operations for the data coming from a host with the merge operations for the buffered data to MLCs. Among SLCs, one is selected to hold data from the host until it gets full, called a *foreground* operation, while the others perform time-consuming merge and garbage collection operations with MLCs, called a *background* operation.

All the data written to the storages are firstly directed to the foreground SLC irrespective of whether the data access is sequential or random. However, in order to improve buffering efficiency, the SLC partitions its blocks into three exclusive regions: a sequential log block, random log blocks, and data blocks, as in the FAST algorithm, which is explained later. If the log blocks of the foreground SLC are full, the storage system selects the next SLC in the round-robin order for foreground operations, and the SLC used so far starts to work in background mode.

For the support of foreground and background operations to be processed in parallel, the proposed architecture provides a *Data Migration Controller (DMC)*, which is responsible for establishing independent data channels for both foreground and background operations, as illustrated in Figure 2. This is to avoid unnecessary conflicts on data channels which could lead to the performance degradation of storage systems. For example, if SLCs share the same data channel, a data transfer over the channel for a background merge operation could block a foreground operation, possibly increasing response latency.

The number of SLCs used in this architecture is a design parameter which should be explored to determine a cost-effective value. When the number increases, it is more likely that a frequent update to hot data invalidates the previous copy in SLCs before being merged to MLCs, which contributes to eliminate unfruitful writes or merges to MLCs. On the contrary, if the number decreases, write operations can quickly overflow the log blocks in SLCs and then make merge operations to MLCs occur frequently.

3.2 Hierarchical Architecture

The storage system proposed in this architecture consists of three components: SLCs, MLCs, and DMC, as shown in Figure 2. SLCs are used as a data buffer and all the data from a host will be eventually stored into MLCs unless they are removed from the storage during the stay in SLCs. The DMC is responsible for making data paths for foreground and background operations. It has a direct connection to each SLC memory to support parallel accesses, and a bus-based connection to MLCs.

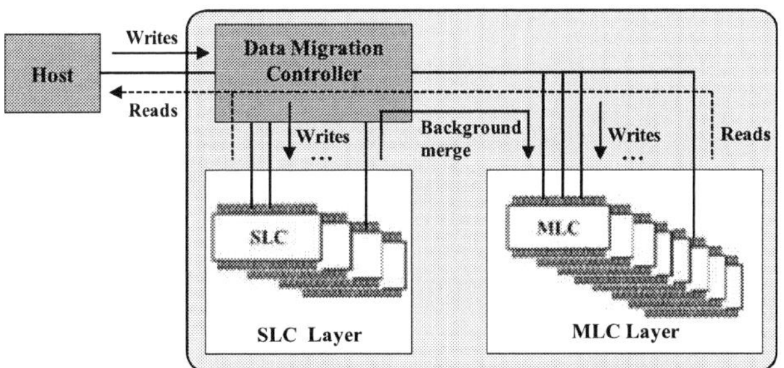

Figure 2. Hierarchical architecture with parallel multi-SLC

Upon the reception of a write data from the host, the DMC determines which SLC is in the foreground and directs the data to the device. Before storing the data, it checks whether the access is sequential or random. For this, it compares the starting logical page number and count of the current access against those of the previous access stored in the sequential log block of the SLC. If it is subsequent to the previous access or if its starting logical page number falls on the starting page number of the data blocks, it is considered as a sequential access and stored in the sequential log block. Otherwise, it is considered as a random access and stored in one of the random log blocks. When log blocks become full, they are merged to data blocks in the same SLC. If the data blocks become full, no more merge operations can occur. The DMC then puts the next SLC into foreground mode. The DMC is also capable of performing DMA-like operations to move pages from SLC in background to MLCs for merge operations.

The blocks in SLCs are partitioned into three groups, but there is no partition in MLCs. The reasons to provide data blocks in the SLC are twofold. First, when log blocks become full, they are merged into data blocks in the same SLC as long as there are free data blocks available.

978-1-60558-497-3/09 $25.00 © 2009 ACM

908

Later, when an SLC enters into the background, the data blocks are copied to MLCs without expensive merge operations. Considering that the time to merge in SLCs is smaller than that in MLCs, the data blocks in SLCs are used for such *pre-merge* operations. It is possible to bring pages from MLCs to SLCs during the pre-merge operations. However, since the performance difference in read operations at SLCs and MLCs are narrowed, the pre-merge overheads become less significant.

Figure 3 shows the algorithms of foreground write operations (Algorithm). In Algorithm, when the log block of the foreground SLC has enough free pages to hold the write data, the algorithm returns successfully after the store operation completes. However, if the log block is full, the algorithm merges the log block into data blocks in the same SLC. This can repeat as long as there are free data blocks available. Later, if no more free data blocks are available in the foreground SLC, the next SLC needs to be selected to buffer the data in the foreground. If there are a sufficient number of SLCs in the storage system, it is probable that the next SLC is ready and available for further foreground operations. But if there is only small number of SLCs provided in the storage, the next SLC could remain busy still being involved in background merge operations with MLCs. In this case, the current SLC remains in the foreground and the log block is merged directly to the corresponding data block in MLCs until the next SLC becomes ready.

Algorithm. Foreground page write algorithm

Input: page address, page number
Output: none
Procedure FTL foreground write (page_addree, page_num, data)
begin
  while (page_num > 0) **do**
    write a page to the current SLC log block;
    if (random log blocks full || sequential log block full || sequential fail) **then**
      if (the current SLC data blocks are full) **then**
        if (the next SLC background operation is complete) **then**
          chip_selection++;
          current_SLC = next_SLC;
          background operation trigger;
        else
          if (sequential log block full) **then**
            switch merge with MLC; erase current sequential log block;
          else if (sequential fail) **then**
            partial merge with MLC; erase current sequential log block;
          else then
            random merge with MLC; erase current random log blocks;
          end if
        end if
      else
        if (sequential log block full) **then**
          switch merge with SLC; allocate new sequential log block in SLC;
        else if (sequential fail) **then**
          partial merge with SLC; allocate new sequential log block in SLC;
        else then
          random merge with SLC; allocate new random log blocks in SLC;
        end if
      end if
    end if
    page_num--;
  end while
end

Figure 3. Algorithms of Foreground and Background Operations

When a sequential write arrives from a host, there could be a case where the access itself is sequential but not subsequent to the previous access in the sequential log block. Now the new write conflicts with the sequential log block with the previous access. This is denoted as a *sequential fail* in Algorithm. This conflict is resolved by merging the incomplete sequential access in the log block to a data block and rendering the log block to the new access, as in FAST.

Once an SLC enters into the background mode, the log blocks begin to be merged to MLCs and the data blocks to be copied to MLCs. A partial merge is performed on the sequential log block such that the incomplete log block is copied to a data block in an MLC and the remaining pages are copied from the previous data block. Note that all the random log blocks in an SLC are scanned to collect up-to-date pages belonging to the same data block and merged together at the same time. By doing this, we can increase the efficiency of merge operations. Otherwise, multiple merge operations can take place on the same data block. We call this *collective random merge,* which is different from a merge on a single random log block, called a *single random merge.* Figure 4 illustrates the two different merge operations. After completing all the background operations, the entire blocks of the SLC are erased to be free.

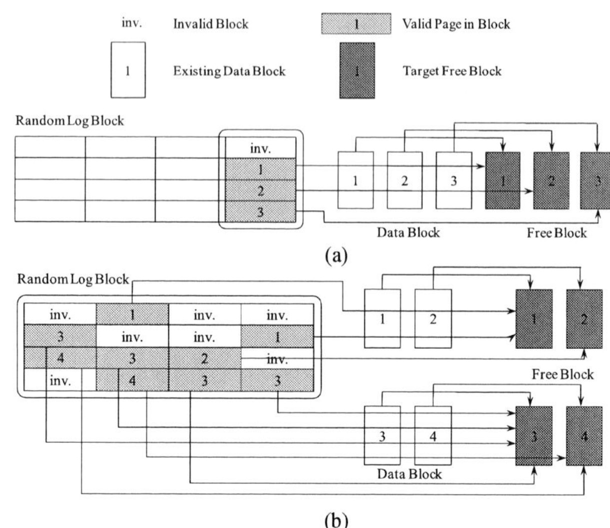

Figure 4. (a) Single random merge and (b) Collective random merge

3.3 Mapping Algorithm

The mapping algorithm used in our architecture, named *FAST-MS (FAST with Multi-SLC)*, is an extension of FAST. Because the SLCs are responsible for handling both random and sequential write accesses, the page-level mapped log block scheme of FAST fits well to our architecture. The extension includes the following. First, collective random merge is introduced in FAST-MS to efficiently merge pages in random log blocks to the same data block. Second, when performing a single random merge in a foreground SLC, FAST-MS may bring pages from a data block in an MLC. Third, log blocks are provided over multiple SLCs even though only one of the SLCs is in the foreground mode at a given time.

4. EXPERIMENTAL RESULTS
4.1 Simulation Environment

In order to perform quantitative analysis of the proposed architecture, we developed a trace-driven simulator with the architectural and timing parameters summarized in [1]. We assume that both SLC and MLC are of the same size, 256 MBytes per device. The write response time of SLC is assumed to be 0.2 ms as described in the specification while that of MLC is 0.8 ms.

The benchmark application traces used in our simulation study are summarized in Table 1. The first two application traces, ATTO and Sandra, are gathered by running benchmark tools. ATTO [12] generates the same amount of I/O requests irrespective of storage capacity. It makes sequential requests more than random ones, and most of the random requests are the accesses to small number of sectors. The Sandra file system [13] generates sequential and random requests, most of which are 2048 sectors long. The next two application traces, MP3 Small Data Copy and Large File Data Copy, are done by copying small MP3 files and big files of size of 3 GB to disks, respectively. The last one is a trace gathered from a PC running Windows XP on the NTFS file system. These traces are gathered by Disk Monitor [14], a disk I/O trace tool.

Table 1. Benchmark applications for simulation

Benchmark	Description	Random/Sequential ratio	Note
ATTO	0.5KB ~ 1MB Measure Test in Size Unit	40% / 60%	< 16 Sector Request 80%
Sandra Filesystem	Sequential, Random, Update Measure Test	55% / 45%	2048 Sector Request 95%
MP3 Small Data Copy	3 ~ 5MB MP3 File Copy	25% / 75%	128 Sector Request 90%
Large File Data Copy	3GB Large Data File Copy in same Disk	0.1% / 99.9%	128 Sector Request 99.9%
System Disk I/O	Windows XP System Only I/O	80% / 20%	OS Filesystem I/O

4.2 Simulation Results

4.2.1 Average Write Response Time

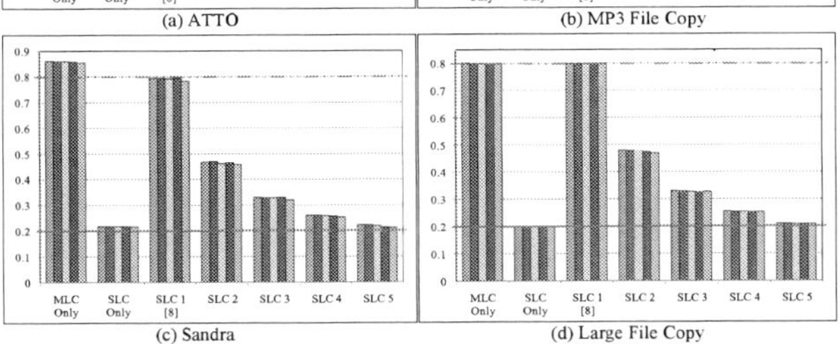

Figure 5. Response time per a page write request

In order to compare the performance of our architecture against one of the previously proposed schemes, a set of simulations was performed using the four benchmark inputs. Figure 5 illustrates the time with other architectural combinations including the average response time to write requests when one SLC is used to store only random requests as in [8] and the one proposed in this paper. The dotted lines and solid lines in the figures represent the program latency in MLC and SLC, respectively. As we can see in the figures, the average response time of this architecture is much longer than the SLC program latency. When the storage consists of SLCs only, its performance is almost equal to or better than the other cases. Note that both MLC only and SLC only architectures do not have background operations performed in parallel. This explains why the SLC only performs slightly poorer than SLC*x* for ATTO and MP3 File Copy. For the simulated benchmark traces, the storages with 4 or 5 SLCs perform almost comparable to the SLC only configuration, and speed up the write operations by up to about 4 times compared with the MLC only.

In an ATTO benchmark, because almost all of the write requests are performed on a limited set of logical pages, they tend to update existing pages. Therefore, there are many merges taking place to the data blocks in the foreground SLC, which provides enough time for the background SLCs to finish merges to MLCs. In this case, we have enough speed up even when only SLCs are used. In the trace of Large File Copy, 99.9% of write operations are sequential. Thus, the SLC data blocks, once in the background mode, are simply copied to MLCs without merge operations and the foreground SLC has only page write operations again without any random merge operations.

4.2.2 Durability

We use the erase count as an index of durability for the configurations on which simulations have been performed. Figure 6 shows the erase count for the configurations that have one (the leftmost set) and two SLCs (the right five sets) based on variation in the number of log blocks. Note that erase count includes the number of erased blocks and blocks yet to be erased.

In our proposed architecture, the erase count has a loose correlation to the number of SLCs, and the total erase count is the same at any configuration. In the case of MP3 File Copy, there is about 4 times improvement in the MLC erase count. The things to note from the figure are that the erase count of SLCs is much higher than that of MLCs, and that the erase count of MLCs decreases when two SLCs are used. Considering that SLCs are 10 times more durable compared with MLCs, it is desirable that SLCs would be able to offload the erase count from MLCs, as long as they do not become worn out too fast.

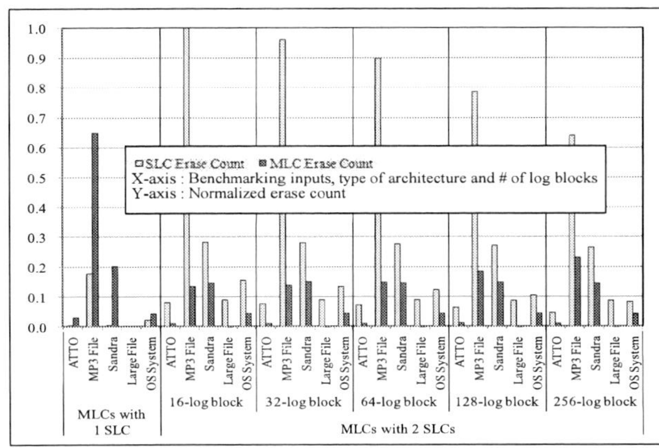

Figure 6. Erase count for durability

5. CONCLUSION

In this paper, we proposed a new storage organization with SLCs as a temporal buffer and analyzed its impact on the performance and durability. The new architecture, along with a proposed mapping algorithm, contributes to significantly enhance the performance and durability of MLC-based storage systems. The simulation study shows that the proposed technique can increase performance by up to 4 times and durability by 4 times with only a small hardware cost of adding a data migration controller.

6. ACKNOWLEDGMENTS

This work was developed within the scope of Human Resource Development Project for IT SoC Architecture. & This research paper has been supported by 「Nano IP/SoC Promotion Group」 of 「Seoul R&BD Program」 in 2009.

7. REFERENCES

[1] Samsung flash memory chip SLC/MLC Spec, Samsung K9F2G08UXA, Samsung K9GAG08B0M, Datasheet.

[2] J. Kim, J. Kim, S. Noh, S. Min and Y. Cho, A Space-Efficient Flash Translation Layer for CompactFlash Systems, *IEEE Transactions on Consumer Electronics*, Vol.48, No.2, May 2002.

[3] S. Lee, D. Park, T. Chung, D. Lee, S. Park, and H. Song, A Log Buffer-Based Flash Translation Layer Using Fully-Associative Sector Translation, *ACM Transactions on Embedded Computing Systems*, Vol. 6, No.3, Article 18, July 2007.

[4] C. Park, W. Cheon, J. Kang, K. Roh, W. Cho and J. Kim, A Reconfigurable FTL (Flash Translation Layer) Architecture for NAND Flash-Based Applications, *ACM Transactions on Embedded Computing Systems*, Vol. 7, No.4, Article 38, July 2008.

[5] S. Lee, B. Moon, Design of Flash-Based DBMS: An In-Page Logging Approach, 27th *ACM SIGMOD International Conference on Management of Data* (SIGMOD'07), June 11-14, 2007.

[6] A. Gupta, Y. Kim, B. Urgaonkar, DFTL: A Flash Translation Layer Employing Demand-based Selective Caching of Page-level Address Mappings, 14th *Architectural Support for Programming Language and OS* (ASPLOS'09), March 7-11, 2009.

[7] J. Kim, S. Jung, Y. Song, Cost and Performance Analysis of NAND Mapping Algorithms in a Shared-bus Multi-chip Configuration, 3rd *International Workshop on Software Support for Portable Storage* (IWSSPS'08), October 23, 2008.

[8] L. Chang, Hybrid Solid-State Disks: Combining Heterogeneous NAND Flash in Large SSDs, *Asia and South Pacific Design Automation Conference* (ASPDAC'08), January 21-24, 2008.

[9] N. Duann, SLC & MLC Hybrid, *Flash Memory Summit 2008*, August 12-14, 2008.

[10] R. Fisher, Optimizing NAND Flash Performance, *Flash Memory Summit 2008*, August 12-14, 2008.

[11] K. Yim, A Novel Memory hierarchy for Flash Memory Based Storage System, *Journal of Semiconductor Technology and Science*, Vol. 5, No. 4, December 2005.

[12] ATTO Windows Disk Benchmark 2.02, http://www.attotech.com

[13] Sandra File System Benchmark, http://www.sisoftware.co.uk

[14] Disk Monitor for Windows v2.01, http://technet.microsoft.com/en-us/sysinternals/bb896646.aspx

Reduction Techniques for Synchronous Dataflow Graphs

Marc Geilen
Eindhoven University of Technology
Department of Electrical Engineering
Den Dolech 2, 5600 MB, Eindhoven, The Netherlands
m.c.w.geilen@tue.nl

ABSTRACT

The Synchronous Dataflow (SDF) model of computation is popular for modelling the timing behaviour of real-time embedded hardware and software systems and applications. It is an essential ingredient of several automated design-flows and design-space exploration tools. The model can be analysed for throughput and latency properties. Although the SDF model is fairly simple, the analysis algorithms are often of high complexity and the models that need to be analysed may be fairly large. This paper introduces two graph transformations for reducing large SDF graphs into simpler, smaller ones that can be analysed more efficiently and give a conservative and often tight estimation of the timing of the original model and hence of the hard real-time system. We can make SDF based methods more efficient and prove that analyses that were done manually in an ad-hoc fashion in the past, can be done automatically and with guaranteed correctness. Additionally we introduce a novel conversion from SDF to Homogeneous SDF, a step applied in many analysis methods for SDF, which yields an up to 250X improvement on the number of actors, thus mitigating the problems with the size explosion observed in the traditional conversion.

Category 1.1 System specification, modeling, simulation, verification, and performance analysis

General Terms Algorithms, Performance, Languages, Theory

Keywords Synchronous Dataflow Graphs, model-based design, reduction techniques

1. INTRODUCTION

For many modern embedded applications, real-time constraints are important, while at the same time, the amount of resources that are spent on an application need to be minimised. Worst-case execution time models help to analyse the timing behaviour of an application, mapped on a specific platform with given resources and can serve as the basis for an automated design-flow. A popular modelling formalism for such timing analysis are Synchronous Dataflow (SDF) graphs [11]. SDF graphs consist of actors that periodically execute a function (sometimes extended with information capturing their execution time), while communicating tokens of information between each other. These actors can be used to model

Permission to make digital or hard copies of part or all of this work for personal or classroom use is granted without fee provided that copies are not made or distributed for profit or commercial advantage and that copies bear this notice and the full citation on the first page. To copy otherwise, to republish, to post on servers or to redistribute to lists, requires prior specific permission and/or a fee.
DAC'09, July 26-31, 2009, San Francisco, California, USA

the algorithms, networks, arbiters, memories and other components of a system under study in a conservative fashion [3, 16, 13, 4, 15, 2]. SDF graphs with timing annotations moreover allow for off-line analysis or conservative estimation of their timing behaviour. In this way it can be verified at design time whether a particular design will be able to meet its deadlines at run-time. [1, 13, 15].

Although the SDF model is fairly simple, some of the actual analysis algorithms are of high complexity. Some require the transformation of the SDF graph into an equivalent homogeneous SDF graph (explained later). The size of the graph may however expand exponentially with this conversion [11, 15]. In combination with that, some of the modelling techniques that allow embedded systems to be analysed with SDF graphs, lead to graphs with large numbers of actors. However, these graphs often have a regular or almost regular structure. These observations have lead us to investigate firstly the possibilities of transforming such graphs into smaller and hence easier to analyse, equivalent or almost equivalent models that will still allow us to conclude that a system will meet its real-time deadlines and secondly to improve on the traditional algorithm to convert SDF graphs to homogeneous SDF graphs.

The former problem is illustrated by the (homogeneous) SDF graph depicted in Figure 1(a), which has a typical structure of e.g., the prefetching of data from a remote memory for some block based image processing application. It is clear that all actors Ai serve a similar role, as well as all actors Bi and their dependencies are fairly regular, except for the beginning and the end of for instance a video frame. Such behaviour is typical for processing of a finite amount of data such as an audio or video frame consisting of smaller blocks, where only start or end or borders have slightly different dependencies. The size of this type of graph is determined by the number of blocks in a frame and can be large. We argue that the behaviour of the graph can be modelled by a simpler graph like the one in Figure 1(b). How we arrive at this graph is explained in detail in Section 4.1. We show in this paper how this graph can be automatically derived and proved to be a conservative estimate of the worst-case execution time of the original graph.

The rest of the paper is organised as follows. Section 2 discusses related work. Section 3 provides preliminary definitions w.r.t. SDF graphs. The abstraction method is introduced in Section 4. In Section 5 we argue that the method is sound and gives a conservative estimate of the throughput of the original model. Section 6 presents a novel transformation from SDF to homogeneous SDF. Experimental results are described in Section 7 and Section 8 concludes.

2. RELATED WORK

SDF is a model of computation that is popular to study scheduling and performance of embedded systems and software. The model is introduced by Lee [11] and further studied in several other pub-

978-1-60558-497-3/09 $25.00 © 2009 ACM

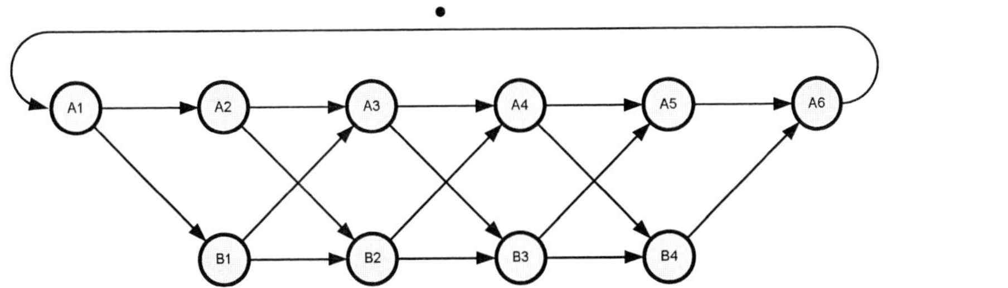

Figure 1: Example regular (a) and abstracted (b) SDF graph

lications, among which the following [4, 15, 1]. [4, 15] study the use of SDF for software synthesis and scheduling. [1] describes a more general mathematical framework called max-plus algebra in which SDF graphs can be described.

A recent trend is to not only describe the embedded software and software synthesis in terms of SDF graphs, but also the underlying hardware platform and multiprocessor systems-on-a-chip, in order to be able to derive guaranteed bounds on execution times and throughput for real-time applications [15, 3, 2, 13, 16].

SDF analysis has been used for throughput analysis [11, 15, 8], latency analysis [15, 9] or buffer sizing by heuristics [19] or exact analysis [18]. Complexity and efficiency of various approaches of throughput analysis based on translation from SDF to HSDF have been studied in [5]. Many of the analysis problems are coming within reach of exact analysis algorithms, although they may still fail on tough cases where heuristics are required.

To the best of our knowledge, the only conversion algorithm to convert SDF to HSDF is the traditional one given in [11, 15]. [10] considers the equivalence between the SDF graph and its corresponding HSDF in more detail. The algorithm in this paper is the first to provide an improved conversion algorithm, although the equivalence with the original graph is less tight.

3. PRELIMINARIES

We start with some definitions concerning SDF graphs, that will allow us to develop the abstraction and reduction mechanisms of this paper. Figure 2 shows an example of an SDF graph. Circles denote *actors*, which can repeatedly *fire*. This models for instance an amount of data processing. A firing consumes and produces at constant rates, *tokens* which are communicated over the edges (denoted by arrows) of the graph, which act as unbounded FIFO channels. Tokens represent data- or other causal dependencies between actor firings. Tokens are indicated by dots. The amount of tokens consumed and produced per firing is constant. Formally, an SDF graph is defined as follows.

DEFINITION 1. (SDF GRAPH) *An SDF graph (SDFG) is a pair (A, D) consisting of a set A of actors and a set $D \subseteq A \times A \times \mathbb{N}^3$ of dependency edges. A dependency (a, b, p, c, d) represents a dependency of actor b on actor a with production rate p of a and a consumption rate c of b and an initial delay of d tokens.*

Because of the constant consumption and production rates, the execution of an SDFG has a periodic behaviour. A graph is *consistent* [11], if there exists fixed numbers of firings for each of the actors such that execution the corresponding numbers of firings brings the graph back into its original state, i.e. with the same distribution of tokens. A vector denoting this number for each actor is called the *repetition vector*. The total number of firings in such an iteration depends on the rates of the edges and can therefore be exponential

in terms of the number of actors of the graph. Because many algorithms need to deal with iterations of the graph, complexity of the analysis can sometimes be high.

A special kind of SDF graphs are *homogeneous SDFGs* (HSD-FGs) where all production and consumption rates equal 1. In case of an HSDFG, the length of an iteration is equal to the number of actors of the graph. In literature, there is a classical conversion algorithm to go from an SDFG to an equivalent HSDFG [11, 15]. However, this conversion will always increase the number of actors to exactly the length of an iteration of the graph which, as mentioned before, can be exponential in terms of the size of the graph. In Section 6 we introduce a novel transformation, which yields a graph that is at most quadratic in the number of initial tokens of the graph. Next, we add timing information to SDF graphs. The execution time of an SDF graph (A, D) is a mapping $T : A \to \mathbb{N}$, that assigns to every actor $a \in A$ the time it takes to execute the actor once, the time that elapses between consumption of the input tokens and production of the output tokens.

DEFINITION 2. (TIMED SDF GRAPH) *A timed SDF graph is a triple (A, D, T) consisting of an SDF graph (A, D) and a corresponding execution time T.*

For lack of space we cannot formalise the semantics of timed SDF graphs in this paper. We assume standard *self-timed execution* [1, 4]. The performance of an SDF graph is characterised by the number of times actors can fire per unit of time, its *throughput* [11, 8].

To study graph reductions later on we make the following observation to compare two graphs. If a timed SDFG contains at least all the actors and all the dependency edges of another graph (and possibly more), its execution times are at least as long and it has not more initial tokens, then the throughput of the former graph is a lower bound for the throughput of the latter.

PROPOSITION 1. *If (A, D, T) and (B, E, U) are two timed SDF graphs with $A \subseteq B$, the execution times T and U are such that $T(a) \leq U(a)$ for all $a \in A$ and for every $(a, b, p, c, d) \in D$, there is some $(a, b, p, c, d') \in E$ such that $d' \leq d$, then the throughput of (A, D, T) is at least as high as the throughput of (B, E, U).*

The intuitive reason for this is that the actors are slower and there are more and stricter dependencies, which will slow the graph down. Proofs in the paper are omitted for space reasons, see [6].

4. ABSTRACTION

Important aspects of SDF graphs, such as their throughput or buffer sizes can be analysed in principle. However, some analyses have a high complexity [12, 15, 18]. In this section, we present an abstraction method to transform an SDF graph into a smaller SDF graph from which a conservative and often reasonably tight estimate of the behaviour of the original graph can be obtained.

978-1-60558-497-3/09 $25.00 © 2009 ACM

4.1 Example

We return to the example of Figure 1(a). It shows an almost regular SDF graph. This type of graph may be obtained from modelling a memory pre-fetching mechanism [16]. We can make an abstract version of it by taking all the Ai actors together into one actor A and similarly put the Bi actors together in one actor B. This results in the abstract graph of Figure 1(b). Firings of the abstract actor model subsequently firings of $A1$, $A2$, ..., $A6$ and then back to $A1$. Note that because actor firing of the same actor can be concurrent in SDF, this does not automatically mean in general that the new graph is much slower than the original. With this connection between the original graph and the abstract graph in mind, we translate the dependency edges of the original graph into dependency edges of the abstract graph. The cycle of edges Ai to $A(i+1)$ and back to $A1$, can be captured by the self edge on A, forcing them to be sequential. Similar for the edges between Bi, although here the original graph does not have the edge back from $B4$ to $B1$. All edges Ai to Bi are captured by the singe edge A to B. The edges back from Bi to $A(i+2)$ result in the edge from B to A. It gets two initial tokens, because the edges connect B's with A's from a different step in the cycle of 6.

Assume that the execution times of the actors are as follows: $A1$, $A2$: 2 time units, $A3$ and $A4$: 5 time units and $A5$ and $A6$: 3 time units. $B1$ up to $B4$ all consume 4 time units. To be conservative, the abstract actor A then takes 5 time units and B takes 4, being the maximum of the firings they represent. It is straightforward to check that a single execution of the graph of Figure 1(a) takes 23 time units. The throughput is $\tau(a) = \frac{1}{23}$ for every actor $a \in A$. In general, for a graph with n copies of the Ai actor, the throughput is: $\frac{1}{5n-7}$. The throughput of the abstract graph is $\frac{1}{5}$ and hence it estimates the throughput of the original graph with n copies as $\frac{1}{5n}$. The estimate is conservative ($\frac{1}{5n-7} \geq \frac{1}{5n}$) and as n gets larger, the relative error made by the abstract graph decreases and it provides a better approximation of the throughput of the large graph.

4.2 The abstraction method

We give the precise definition of the transformation, illustrated by the example in Figure 2(a). The abstraction groups the Ai actors together and the Bi actors together. The result of the abstraction is then the graph depicted in Figure 2(b). The executions of the A actor model subsequent executions of $A1$, $A2$, $A3$, $A1$, etcetera. The executions of abstract actor B models executions of $B1$, $B2$ and some 'dummy' actor to account for the fact that the numbers of actors that have been abstracted are not equal.

The principle behind the construction is as follows. A group of actors *with identical firing rates* will be ordered and represented by a single new actor. If the groups of actors that are combined are not of equal size, then the abstract actor may have a number of 'dummy' firings that do not correspond to firings of the original actors. The firing time of the new, abstract actor is adapted to the slowest of the actors that it represents. We first define an abstraction. It tells us which actors are going to be grouped together and in what order.

DEFINITION 3. (ABSTRACTION) *An abstraction (α, I) of a consistent SDF graph (A, D) consists of a function $\alpha : A \to B$ that maps actors $a \in A$ to abstract actors $\alpha(a) \in B$. The function $I : A \to \mathbb{N}$ assigns to every actor $a \in A$ an index number. α and I are further such that:*

- *if $\alpha(a_1) = \alpha(a_2)$ then $I(a_1) \neq I(a_2)$ and $\gamma(a_1) = \gamma(a_2)$ where γ is the repetition vector of the graph.*

- *for every edge $(a, b, p, c, d) \in D$, $I(a) \leq I(b)$ or $d > 0$.*

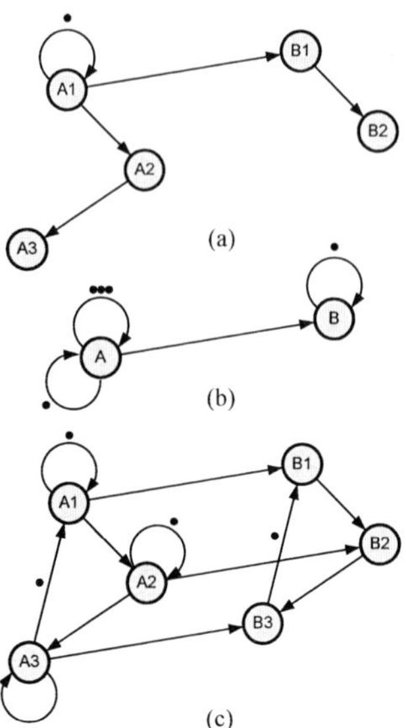

Figure 2: Abstracting and unfolding a graph

Given an SDF graph and an abstraction, we can determine the resulting smaller SDF graph. We consider here for clarity the case that the original graph is homogeneous, but the method can be extended to non-homogeneous graphs as well.

DEFINITION 4. (ABSTRACT GRAPH) *The abstract timed graph of (A, D, T) according to the abstraction (α, I) is the graph $(A, D, T)^{\alpha, I} = (A', D', T')$, where (with $N = \max_{a \in A} I(a)$)*

- $A' = \{\alpha(a) \mid a \in A\}$

- $D' = \{(\alpha(a_1), \alpha(a_2), p, c, I(a_2) - I(a_1) + Nd) \mid (a_1, a_2, p, c, d) \in D\}$

- $T'(b) = \max_{a \in A, \alpha(a) = b} T(a)$.

The new set of actors consists of all abstractions $\alpha(a)$ for the original actors $a \in A$. An edge is introduced for every edge in the original graph and the number of initial tokens is adapted to the number of initial tokens in the original graph (which must be multiplied by N, because firings of the new actor representing firings of the original actor occur only once every N firings) and the relative index ($I(a_2) - I(a_1)$) of the actors. Finally, the execution time is set to the maximum of the execution times of the actors mapped to it. Note that according to the definition, every dependency edge in the original graph results in an edge in the abstract graph. For a regular graph, suitable for the abstraction, many edges of the original graph are mapped onto the same edge of the new, reduced graph. Still, this might lead to having several edges between two actors in the abstract graph. Such a set of edges can always be pruned to only the one with the smallest number of initial tokens. E.g., in Figure 2, the abstract graph, the self-edge on actor A with three initial tokens is redundant because there is another one with only one token and it could have been removed. For the unfolding this would mean that the self-edges on the A actors disappear, without changing its throughput.

5. CONSERVATIVITY

In this section, we give a brief sketch of the proof that every behaviour of the original graph is mimicked by the new graph in a conservative fashion and hence that the method is sound to verify that real-time constraints are met. This is shown by the following steps. First we define an unfolding of the abstract graph, splitting every abstract actor back into N individual actors. Figure 2(c) shows this for the example. It gives a graph that has more actors and more dependencies and larger execution times than the original graph (compare Figure 2(a) and Figure 2(c)) and according to Proposition 1, its throughput is a lower bound on the throughput of the original graph and all individual firing times are conservative approximations of the original firing times. Note that how tight the conservative bounds are depends strongly on how well-suited the original graph is for abstraction and how well-chosen the abstraction functions are. We define an unfolding of an SDF graph, which will be used to unfold the abstract graph to be able to directly compare it to the original graph using Proposition 1. Formally, the unfolding procedure is defined as follows.

DEFINITION 5. *Let (A, D, T) be a timed SDF graph. The N-fold unfolding unf$(A, D, T, N) = (B, E, U)$ where*

- $B = \{a_i \mid a \in A, 0 \leq i < N\}$.

- *for every edge $(a, b, p, c, d) \in D$, there are precisely N edges in E. For every $0 \leq i < N$, let $j = (i + d) \mod N$ and let $d' = d \operatorname{div} N + t$, where $t = 1$ if $j < i$ and $t = 0$ otherwise, then $(a_i, b_j, p, c, d') \in E$.*

- $U(a_i) = T(a)$ *for every $a \in A$ and $0 \leq i < N$.*

The timed SDF graph and its unfolding mimic each other exactly. If actor a can execute for the i'th time in the original graph, then actor $a_{i \mod N}$ can execute for the i div n 'th time in the unfolded graph. They have the same throughput up to the factor N.

PROPOSITION 2. *Let (A, D, T) be a timed SDF graph and $(B, E, U) = $ unf(A, D, T, N) its N-fold unfolding. Let τ be the throughput of (A, D, T) then τ', with $\tau'(a_i) = \tau(a)/N$ is the throughput of (B, E, U).*

For the remainder of this section, assume that we have a timed SDF graph (A, D, T), an abstraction (α, I) for this graph and the abstract graph (A', D', T'). We further compute the N-fold unfolding of the abstract graph as (B, E, U), where $N = \max_{a \in A} I(a)$. We investigate the relationship between the graphs (A, D, T) and (B, E, U). With every actor $a \in A$ we associate an image actor in B as follows. $\sigma(a) = a_i'$ if $\alpha(a) = a'$ and $I(a) = i$. Note that such $\sigma(a)$ exists for every $a \in A$. One can show that the actor $\sigma(a)$ conservatively mimics a. Firstly, it actor has at least the same amount of execution time.

PROPOSITION 3. *If $a_i' = \sigma(a)$ then $T(a) \leq U(a_i')$.*

For every edge, there is a corresponding one with at most the same amount of tokens.

PROPOSITION 4. *Let $(a, b, p, c, d) \in D$, then there is some $(a_i', b_j', p, c, d') \in E$ such that $a_i' = \sigma(a)$, $b_j' = \sigma(b)$ and $d' \leq d$.*

As such, σ represents a one-to-one mapping from the original SDF graph to the unfolding of the abstract graph. This shows that the throughput of the original graph is at least as high as the throughput of the unfolding.

THEOREM 1. *Let (A, D, T) be a timed SDF graph with throughput τ and (α, I) an abstraction. Then the throughput of (A, D, T) can be conservatively estimated from $(A, D, T)^{\alpha, I}$.*
$\tau(a) \geq \tau^u(\sigma(a)) = \tau(\alpha(a))/N$.

The throughput of the abstracted graph is equivalent to its unfolding by Proposition 2, and the unfolding conservatively models the original SDF, because it follows from Propositions 3 and 4 that the conditions of Proposition 1 are met.

Algorithm 1 Convert SDFG to HSDFG

1: CONVERTTOHSDF$(G = (A, D, T))$
2: $V \leftarrow \{(t_k, \bar{i}_k) \mid t_k \in InitialTokens(D)\}$
3: $\overline{\gamma} \leftarrow$ REPETITIONVECTOR$((A, D))$
4: $\sigma \leftarrow$ SEQUENTIALSCHEDULE$((A, D), \overline{\gamma})$
5: **for** $j = 1$ to LENGTH(σ) **do**
6:     Actor $a \leftarrow \sigma[j]$ /* a the j'th actor in the schedule */
7:     fire a consuming tokens $W \subseteq V$
8:     $V \leftarrow V \backslash W$
9:     produce output tokens with time-stamp
       $\overline{g}_p \leftarrow \max \{\overline{g}(t) \mid t \in W\} + T(a)$
10:    add new tokens to V with time stamp $\overline{g}_{[}$
11: **end for**
12: create graph according to coefficients of vectors $\overline{g}_k = [g_{j,k}]$ and structure of Figure 4

---

6. A NOVEL HSDF CONVERSION

In this section we introduce a novel transformation from an SDFG to an equivalent HSDFG, which is in most cases considerably smaller than the HSDFG obtained from the classical transformation of [11, 15]. However, a note is in order to explain what we mean by 'equivalent'. In the traditional transformation, every actor is duplicated a number of times, according the number of firings of the actor in a single iteration. The firing times of the actors correspond one-to-one to the firing times of the corresponding firing in an iteration. Such a direct correspondence is not maintained by our new transformation. We seek to obtain a graph which has the same throughput and latency as the original graph. When specific actor firings within an iteration are of particular interest, for instance, a dedicated 'output actor' of the system, then it is straightforward to include this information in the constructed graph.

In order to derive an equivalent HSDFG, we consider a symbolic execution of the original graph. Recent developments in throughput calculation of SDFGs [8] have shown that the most efficient method for throughput calculation is to do an execution of the graph until a recurrent state is found and the behaviour becomes periodic, then the throughput is obtained from examining the throughput achieved in the periodic part of the behaviour. Although the number of states that need to be explored in the behaviour is in the same order as the number of actors in the equivalent HSDFG according to the traditional transformation, the main strength of the algorithm comes from the fact that hardly any of these states need to be stored in memory and used later on.

In the same spirit, we use state-space exploration to determine a reduced HSDFG. Instead however of doing an explicit exploration, we apply the method *symbolically*. If a firing actor consumes n tokens from various channels and we know that these tokens have become available at times $t_1, t_2, \ldots, t_n$ respectively, then we know that the actor starts its firing at: $t = \max_i t_i$ If the duration of the firing of that actor is E, then the actor ends it firing at: $t' = E + \max_i t_i = \max_i t_i + E$. This is also the time when all tokens produced by that firing become available to actors that depend on such tokens. We will show that if we continue execution like this, the time at which the token becomes available can always be expressed using an equation of the form:

$$t = \max_i (t_i + g_i)$$

using suitable constants g_i. Assuming the original t_i are fixed, we can characterise such a *symbolic time stamp* as a vector $\overline{g} = [g_i]$.

We illustrate this with the simple graph of Figure 3. An iteration consist of three firings, two of the left and one of the right

978-1-60558-497-3/09 $25.00 © 2009 ACM

actor. In the initial state (a), the symbolic time stamps of the initial tokens in the graph are simply t_1, t_2, t_3 and t_4. The left actor fires consuming tokens labelled t_1 and t_2. Hence the firing takes place at time $\max(t_1, t_2)$ and ends at $\max(t_1 + 3, t_2 + 3)$, which is the symbolic time stamp of the newly produced tokens. Towards state (c), the left actor fires another time, consuming tokens t_3 and $\max(t_1 + 3, t_2 + 3)$, so it starts at $\max(t_1 + 3, t_2 + 3, t_3)$ and ends at $\max(t_1 + 6, t_2 + 6, t_3 + 3)$. Finally, the right actor fires, bringing the graph back in the original state, but with symbolic tokens representing the impact of a single iteration.

In general, if an actor fires consuming tokens with a start time equation characterised by $\overline{g}_n = [g_{n,i}]$ then its starting time $t = \max_j(\max_i(t_i + g_{j,i})) = \max_i(t_i + (\max_j g_{j,i}))$ In other words, the starting time can be characterised by the vector $\overline{g} = \max_j \overline{g}_j := [\max_j g_{i,j}]$. The actor ending time, which is then equal to the time stamps of the new tokens is equal to: $\max_j \overline{g}_j + E$. This is again of the same form, it shows that we can characterise the time of *any* token in the graph symbolically by an expression of this form.

We mimic the state-space exploration algorithm using symbolic time stamps instead of concrete time stamps. After a single iteration of the graph (executing every actor a number of times in accordance with the repetition vector, all tokens are back at their original places and labelled with a time stamp which expresses an equation relating their production time to the production times of the original, initial tokens.

Now it is straightforward to construct an HSDFG which mimics exactly this behaviour. If $t'_k = \max_j g_{j,k} + t_j$, then t'_k needs to respect certain minimum distances ($g_{j,k}$) to the previous token times t_j. This can be enforced by placing a single actor between them. Because these distances are enforced pair-wise, additional multiplexing and de-multiplexing of tokens is required. This ultimately leads to a homogeneous graph of the form depicted in Figure 4. In the middle there is a matrix of actors for the coefficients $g_{j,k}$. Not all tokens depend on each other. If there is no dependence, the actor need not be present, as illustrated by the gray actors in Figure 4. In practice, this matrix is often quite sparse. To the left and at the bottom we use actors with execution time 0, which act as multiplexors and de-multiplexors to the tokens on the channels between them. These actors only need to be present if there is actually more than one actor that needs the token or multiple actors from which the tokens need to synchronise.

The algorithm for calculating the graph is straightforward and illustrated in Algorithm 1. (The algorithm is derived from an algorithm to convert an SDFG into a MaxPlus matrix [8, 7].) Firstly, the repetition vector of the graph is computed and an arbitrary sequential schedule for one iteration of the graph, using well-known methods [11, 15]. The initial tokens in the graph are initialised with the corresponding vector representations. t_1 is represented by $\overline{i}_1 := [0; \ -\infty; \ -\infty; \ldots]$, t_2 by $\overline{i}_2 := [-\infty; \ 0; \ -\infty; \ \ldots]$ and so on. $-\infty$ is used here to denote when there is no dependency on the corresponding token. ($-\infty$ is the neutral element of the max operator and zero element of addition in Max-Plus algebra which formalises these types of equations [1].) Next, the schedule is executed and with every actor firing, the symbolic input tokens are read, the symbolic starting time and ending time of the actor are computed and new tokens are produced carrying this symbolic ending time of the actor. In this way, the iteration is completed and returns the graph to its original token distribution. By reading the vector $[g_{j,k}]$ corresponding to token k, we get the coefficients for the execution times in the k-th column of the matrix of Figure 4. If the entry in the vector is $-\infty$, no actor is created. Finally, the (de-)multiplexing actors are created when necessary and all edges are created including the edges with initial tokens between the (de-

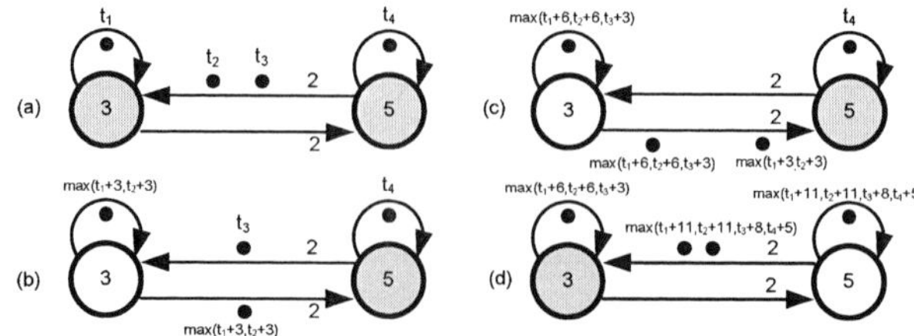

Figure 3: Example symbolic execution

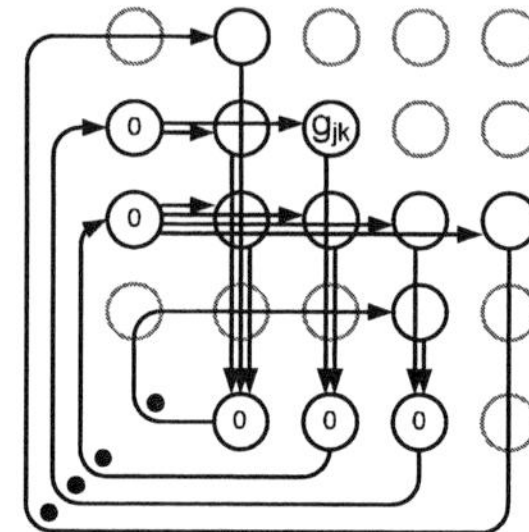

Figure 4: Structure of HSDF conversion

)multiplexing actors.

The Max-Plus matrix is a square matrix and the number of rows and columns corresponds to the number (N) of initial tokens in the graph. From Figure 4 we see that the resulting graph has at most $N(N + 2)$ actors, $N(2N + 1)$ edges and N initial tokens.

7. EXPERIMENTS

The method described in this paper was inspired by practical work in which SDF models were used for worst case timing analysis of multiprocessor system-on-chip realisations of multimedia applications [16, 13]. Here, informal arguments were used to defend that the very large SDF models that resulted from the analysis methods could be replaced by simpler models, which could be analysed by hand. The results of this paper allow one to ensure oneself that this method is sound. [16] investigates the embedding of large data structures in the memory of a multiprocessor system with network-on-chip. The algorithm mapped onto the platform is a full-search block matching algorithm for detecting motion vectors in video sequences, typically used in video encoding algorithms such as H.263 and MPEG-2. For efficient operation of the algorithm, data is to be pre-fetched from a frame-memory residing in a different tile on the SoC. The data has to be transported over the network-on-chip. The result of the analysis is essentially the model depicted in Figure 5, where the CA actors represent the communication assist components on either side of the network. The A actors take care of the pre-fetch requests and the actual computations. In total, 1584 of such computations (all taking the same amount of time) have to be performed within the duration of a single video frame. By the obvious abstraction, the graph is reduced to the model on the right of Figure 5, which in this case, has exactly the same throughput as the original graph. Note that in the abstract graph of 5 redundant edges have been pruned for readability. We do not show any experimental results for the abstraction technique, because they are not intended to be used on arbitrary graphs, but rather specific graphs which show a specific regular structure. Results for arbitrary graphs would not be good and regular graphs can

978-1-60558-497-3/09 $25.00 © 2009 ACM

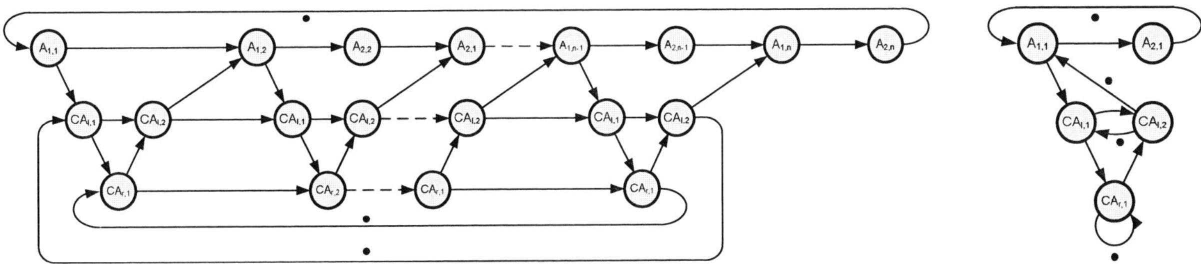

Figure 5: Remote memory access model and abstraction

Table 1: HSDF Transformations Compared

test case	Traditional conversion number of actors	new conversion number of actors	ratio
1. h.263 decoder	1190	10	119
2. h.263 encoder	201	11	18.3
3. modem	48	210	0.23
4. mp3 dec. block par.	911	8	114
5. mp3 dec. granule par.	27	8	3.38
6. mp3 playback	10601	38	279
7. sample rate conv.	612	31	19.7
8. satellite	4515	217	20.8

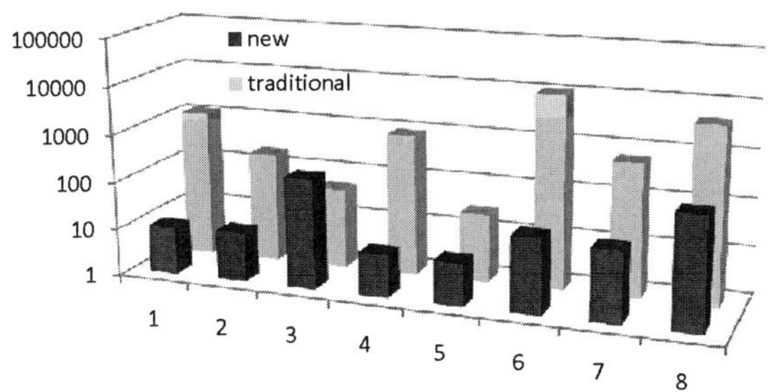

Figure 6: Experimental results

be constructed for which the abstraction returns small graphs with a perfectly accurate prediction of performance.

The novel HSDF conversion algorithm has been implemented as an extension to the SDF³ tool set available from [17]. For a benchmark of available SDFGs of various applications [14] we compare the sizes of the HSDFGs in terms of their numbers of actors. The run-time of the algorithms is a few milliseconds. The results in Table 1 and visually presented in Figure 6, show that in all but one case, the new conversion algorithm yields much smaller graphs, The differences in size vary from a small factor up to a factor of 279 times fewer actors. Only for the case of the modem graph [11], the result is actually larger. This is a graph which is itself 'almost HSDF' with only few rates different from 1 and with a large number of initial tokens. Because the size of the traditional HSDF is exactly predictable and a bound on the size of the new method can be estimated from the number of initial tokens, it is possible to assess beforehand when this might occur.

8. CONCLUSIONS

We have presented two SDF graph reduction techniques. A transformation that allows simplification of SDF models to allow for a more efficient analysis with guaranteed conservative results. Experiments show that the results can be used in practice assure that transformations that are often used manually in practice, are in-

deed correct. Secondly a novel conversion algorithm has been introduced to transform SDF graphs into HSDF graphs.

9. REFERENCES

[1] F. Baccelli, G. Cohen, G. Olsder, and J.P.Quadrat. *Synchronization and Linearity*. John Wiley & Sons, 1992.

[2] N. Bambha, V. Kianzad, M. Khandelia, and S. Bhattacharya. Intermediate representations for design automation of multiprocessor DSP systems. *Design Automation for Embedded Systems*, 7:307–323, 2002.

[3] M. Bekooij, O. Moreira, P. Poplavko, B. Mesman, M. Pastrnak, and J. V. Meerbergen. Predictable multiprocessor system design. In *SCOPES 2004, 8th Int. Workshop on Software and Compilers for Embedded Systems*, 2004.

[4] S. Bhattacharyya, P. Murthy, and E. Lee. *Software Synthesis from Dataflow Graphs*. Kluwer Academic Publishers, 1996.

[5] A. Dasdan, S. S. Irani, and R. K. Gupta. Efficient algorithms for optimum cycle mean and optimum cost to time ratio problems. In *DAC '99: Proceedings of the 36th ACM/IEEE conference on Design automation*, pages 37–42, New York, NY, USA, 1999. ACM.

[6] M. Geilen. Reduction techniques for synchronous dataflow graphs. Technical Report ESR-2009-01, Electronic Systems Group, Dept. of Electrical Engineering, Eindhoven University of Technology, 2009.

[7] M. Geilen. Synchronous data flow scenarios. *Transactions on Embedded Computing Systems, Special issue on Model-driven Embedded-system Design, to be published*, 2009.

[8] A. Ghamarian, M. Geilen, S. Stuijk, T. Basten, A. Moonen, M. Bekooij, B. Theelen, and M. Mousavi. Throughput analysis of synchronous data flow graphs. In *Proceedings of the 6th International Conference on Application of Concurrency to System Design (ACSD'06)*, pages 27–30. IEEE Computer Society Press, Los Alamitos, CA, USA, 2006, 2006.

[9] A. Ghamarian, S. Stuijk, T. Basten, M. Geilen, and B. Theelen. Latency minimization for synchronous data flow graphs. *Digital Systems Design, Euromicro Symposium on*, pages 189–196, 2007.

[10] R. Govindarajan and G. R. Gao. Rate-optimal schedule for multi-rate dsp computations. *J. VLSI Signal Process. Syst.*, 9(3):211–232, 1995.

[11] E. Lee and D. Messerschmitt. Synchronous data flow. *IEEE Proceedings*, 75(9):1235–1245, Sept. 1987.

[12] P. K. Murthy. *Scheduling Techniques for Synchronous and Multidimensional Synchronous Dataflow*. PhD thesis, University of California, Berkeley, December 1996.

[13] P. Poplavko, T. Basten, and J. van Meerbergen. Execution-time prediction for dynamic streaming applications with task-level parallelism. In *DSD '07: Proceedings of the 10th Euromicro Conference on Digital System Design Architectures, Methods and Tools*, pages 228–235, Washington, DC, USA, 2007. IEEE Computer Society.

[14] SDF³. http://www.es.ele.tue.nl/sdf3.

[15] S. Sriram and S. S. Bhattacharyya. *Embedded Multiprocessors: Scheduling and Synchronization*. Marcel Dekker, Inc., New York, NY, USA, 2000.

[16] S. Stuijk, T. Basten, B. Mesman, and M. Geilen. Predictable embedding of large data structures in multiprocessor networks-on-chip. In *Proc. DSD, 8th Euromicro Conference, DSD'05*, pages 388–395. IEEE Computer Society Press, 2005.

[17] S. Stuijk, M. Geilen, and T. Basten. SDF³: SDF For Free. In *Application of Concurrency to System Design, 6th International Conference, ACSD 2006, Proceedings*, pages 276–278. IEEE Computer Society Press, Los Alamitos, CA, USA, June 2006.

[18] S. Stuijk, M. Geilen, and T. Basten. Throughput-buffering trade-off exploration for cyclo-static and synchronous dataflow graphs. *IEEE Trans. Comput.*, 57(10):1331–1345, 2008.

[19] M. Wiggers, M. Bekooij, and G. J. M. Smit. Efficient computation of buffer capacities for cyclo-static dataflow graphs. In *DAC*, pages 658–663, 2007.

978-1-60558-497-3/09 $25.00 © 2009 ACM

A Parameterized Compositional Multi-dimensional Multiple-choice Knapsack Heuristic for CMP Run-time Management [*]

Hamid Shojaei[1,2], AmirHossein Ghamarian[2], Twan Basten[2,3], Marc Geilen[2],
Sander Stuijk[2], Rob Hoes[2]

[1] University of Wisconsin-Madison, Madison, WI
[2] Eindhoven University of Technology, Eindhoven, the Netherlands
[3] Embedded Systems Institute, Eindhoven, the Netherlands
shojaei@wisc.edu

ABSTRACT

Modern embedded systems typically contain chip-multiprocessors (CMPs) and support a variety of applications. Applications may run concurrently and can be started and stopped over time. Each application may typically have multiple feasible configurations, trading off quality aspects (energy consumption, audio-visual quality) with resource usage for various types of resources. Overall system quality needs to be guaranteed and optimized at all times. This leads to the need for a run-time management solution that selects an appropriate system configuration from all the application configurations of active applications. This run-time management problem can be phrased as a multi-dimensional multiple-choice knapsack (MMKP) problem. We present a compositional heuristic to solve MMKP, that due to the compositionality is better suited to CMP run-time management than existing heuristics that are all not compositional. Our heuristic outperforms the best-known heuristic to date. The heuristic is parameterized, leading to the additional advantage that it allows to trade off execution time vs. solution quality, and to bound the time needed to compute a solution. The latter makes it particularly well-suited for resource-constrained embedded platforms.

Categories and Subject Descriptors

C.3 [**Special-purpose and Application-based Systems**]: Realtime and embedded systemsC.5.4Computer System ImplementationVLSI Systems
General Terms: Algorithms, Design, Performance
Keywords: MMKP, CMP run-time management, Pareto algebra

1. INTRODUCTION

Complex dynamic applications are becoming increasingly dominant in different types of embedded systems, and ever more embedded systems have a chip-multiprocessor (CMP) as the compute platform. When (parts of) applications may start and stop over time, one of the most challenging issues is to select at run time a cost-effective mapping of the applications on these platforms.

Selecting an optimal configuration for each application requires optimization of power consumption, perceived quality, timeliness, and so on. This needs to be done within available platform resources and within performance constraints, and requires optimization of processing, memory, and communication resources of various types. This leads to a multi-dimensional design-space exploration problem. For dynamic systems with applications starting and stopping over time, this exploration needs to be done at run-time, within the resource-constrained embedded platform.

To alleviate run-time decision making, the available configurations of the various applications themselves may be obtained from an off-line Pareto analysis in the multi-dimensional quality/costs/resources trade-off space. The selection of a combination of the application configurations at run-time then requires composition of feasible Pareto-optimal configurations of applications into feasible Pareto-optimal configurations of a system, ultimately selecting one configuration per application, optimizing the overall cost and quality (in terms of e.g, power, perceived quality, etc.).

Figure 1 gives an illustrative example. Consider a system with three applications A, B, and C, that need to be mapped onto a systemwith 6 processing elements (PEs), allowing two clock speeds (ck1 and ck2 with ck1<ck2). Potential configurations are determined off-line. Some of the configurations are infeasible because of a latency constraint. The objective is to map applications onto the system so that latency and resource constraints are met and the energy consumption is minimal. Applications may start and stop at arbitrary times. When application A starts, a 4-PE configuration is selected with slow clock speed ck1. At time t1, application B starts and among all possible combinations, a configuration with fast clock ck2, 2 PEs assigned to each application, and energy consumption as low as possible is selected. This configuration is still valid when application C starts, which is assigned two additional PEs. Conceptually, in each step, the configurations of the newly arriving task are joined with the configurations of the already active tasks, making sure that clock speeds match (Join in the figure). Infeasible configurations, denoted in grey, and configurations that are guaranteed to be worse than others (dominated configurations) are removed (Const resp. Min in the figure).

In general, application configurations can be assumed to have a *value* (that takes into account all optimization objectives) and a *multi-dimensional resource usage*. We need

[*]This work was supported in part by the EC through FP7 IST project 216224, MNEMEE.

Permission to make digital or hard copies of part or all of this work for personal or classroom use is granted without fee provided that copies are not made or distributed for profit or commercial advantage and that copies bear this notice and the full citation on the first page. To copy otherwise, to republish, to post on servers or to redistribute to lists, requires prior specific permission and/or a fee.
DAC'09, July 26-31, 2009, San Francisco, California, USA
Copyright 2009 ACM 978-1-60558-497-3/09/07 ...$10.00.

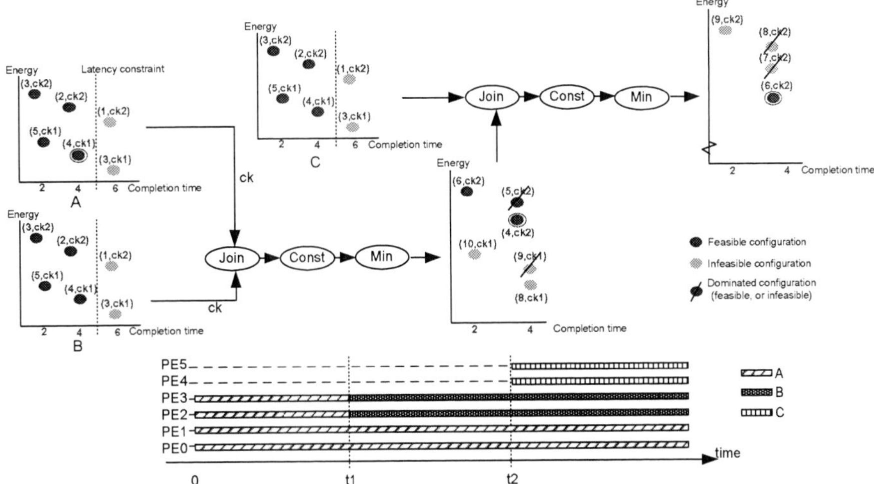

Figure 1: Configuration selection (adapted from [12]).

to choose exactly one feasible configuration per application, maximizing the total value of the selected configurations, without exceeding the resource constraints in any dimension. We do not explicitly consider other types of constraints (like the latency constraint in the example), because these can be treated in the same way as resource constraints. This leads to an optimization problem that is known as the Multi-dimensional Multiple-choice Knapsack Problem (MMKP, [8]). This optimization problem needs to be solved each time something changes in the set of active applications, i.e., each time an application starts or stops executing. We do not consider dynamic behavior within one application. This type of dynamic behavior can be handled independently, by other means like scenario-based design [3].

Let N be the number of applications, and N_i the number of configurations of application i $(0 \leq i < N)$. Let R be the number of resources. The amount of available resources of type k $(0 \leq k < R)$ is given by R_k. Suppose that configuration j of application i has value v_{ij} and requires resources r_{ijk} $(0 \leq k < R)$. MMKP is then defined as follows, where the x_{ij} are binary-valued variables:

- *Maximize*

$$\sum_{0 \leq i < N} \sum_{0 \leq j < N_i} x_{ij} v_{ij}$$

- *Subject to*

$$- \ x_{ij} \in \{0,1\}, 0 \leq i < N, 0 \leq j < N_i$$

$$- \sum_{0 \leq j < N_i} x_{ij} = 1, 0 \leq i < N$$

$$- \sum_{0 \leq i < N} \sum_{0 \leq j < N_i} x_{ij} r_{ijk} \leq R_k, 0 \leq k < R.$$

MMKP is an NP-hard problem. For CMP run-time management as sketched above, the MMKP problem needs to be solved at run-time, at a resource-constrained platform, each time the active application set changes. This puts severe constraints on the efficiency of the solution. There already exist various exact and heuristic algorithms for solving MMKP. We propose a compositional heuristic solution that follows the conceptual approach sketched in Figure 1. To the best of our knowledge, our solution is the first compositional solution to MMKP.

Our heuristic uses Pareto algebra [2], which is an algebraic framework for compositional calculation of Pareto-optimal solutions in multi-dimensional optimization problems. Pareto optimality is used as a central criterion to compare configurations and discard the ones that cannot be optimal. The practical complexity of the heuristic is determined by the number of configurations of the (sets of) applications considered in each optimization step. This provides a parameter to control the execution time of the heuristic. This parameter allows to trade off execution time and quality of the end result. It is furthermore possible to bound in practice the execution time of the heuristic for a given problem in terms of this parameter. This allows in turn to bound the amount of time allowed for the computation. This is particularly convenient in a context with hard or firm real-time constraints. The possibilities to trade off quality for execution time and to budget the execution time are unique to our heuristic.

The experimental results indicate that we achieve the same results in terms of speed and obtained value compared to the fastest non-compositional heuristic ([12]) known to date when assuming that all applications are started at a single point in time. If applications start and stop at different points in time, our heuristic shows execution times that are independent of the number of active tasks, in contrast to the execution times of non-compositional approaches that depend on the number of active tasks. The compositional approach is then generally much faster than the fastest heuristic so far.

The rest of the paper is organized as follows. The next section provides an overview of related work. Section 3 gives an introduction to Pareto algebra. Our Pareto-algebra-based heuristic is introduced in Section 4. Experimental results are presented in Section 5. Section 6 concludes.

2. RELATED WORK

Finding an optimal solution for an MMKP instance is not applicable for CMP run-time management, because the execution times increase dramatically with increasing problem sizes (see e.g. [4]). Also general heuristic search strategies such as genetic algorithms, applied to a knapsack variant in

978-1-60558-497-3/09 $25.00 © 2009 ACM

[6], are not sufficiently fast.

The literature also describes some fast dedicated heuristic methods [1, 5, 8, 10]. Important ideas to improve efficiency originate from those papers, such as to consider the aggregate resource consumption of an application configuration, the idea to reduce the multi-dimensional search space into a two-dimensional search space, and the idea to reduce complexity by considering the configurations from each application that are on the convex hull. However, none of the heuristics is sufficiently fast for run-time management in a CMP context.

Researchers from IMEC propose in [12] a greedy approach to find near-optimal solutions for MMKP instances for CMP run-time management. This heuristic is by far the fastest among all heuristics to date. It is sufficiently fast for run-time management, finding a solution typically in the order of 1 ms, and providing results of good quality close to the best heuristics mentioned above. All (Pareto-optimal) application configurations are sorted according to value vs aggregate resource-usage ratio in a two-dimensional search space. Starting from the configurations with the lowest resource usage, a solution is constructed in a greedy way in a single traversal of this list by swapping configurations each time such a swap increases the value.

None of the mentioned approaches is compositional. In this paper, we propose a compositional approach building upon some of the ideas of [12] and earlier papers on MMKP. Besides the advantage of fast incremental decision making, our approach allows to trade off quality vs. execution time, and to bound the time spent on computing a solution.

3. PARETO ALGEBRA

This section provides an overview of Pareto algebra [2]. We briefly define the concepts and operations for multi-dimensional optimization in this algebra, to the extent that they are relevant for solving MMKP. We use the case study of Fig. 1 as a running example.

A *quantity* is a parameter, quality metric, or any other quantified aspect of a system. It is represented as a set Q with a partial order $\preceq_Q$ that denotes preferred values (lower values being preferred). The subscript is dropped when clear from the context. If $\preceq_Q$ is total, the quantity is *basic*; if $\preceq_Q$ is the identity, the quantity is *unordered*. In Fig. 1, *Energy* and *Completion time* are two basic quantities with total order $\leq$. Also the number of PEs in use is a basic quantity with order $\leq$. The quantity of clock speeds {ck1,ck2} is an unordered quantity, because there is no inherent preference for a low or high clock speed; the clock speed is a parameter of the system. Ultimately, the combination of clock speed and used number of PEs translates to energy usage and completion time, where lower values are preferred.

A configuration space $\mathcal{S}$ is the Cartesian product $Q_1 \times Q_2 \times \ldots \times Q_n$ of a finite number n of quantities and a configuration $\bar{c} = (c_1, c_2, \ldots, c_n)$ is an element of such a space.

A dominance relation $\preceq \subseteq \mathcal{S}^2$ defines configurations that are preferred over others. If $\bar{c}_1, \bar{c}_2 \in \mathcal{S}$, then $\bar{c}_1 \preceq \bar{c}_2$ iff for every quantity Q_k of $\mathcal{S}$, $\bar{c}_1(Q_k) \preceq_{Q_k} \bar{c}_2(Q_k)$. If $\bar{c}_1 \preceq \bar{c}_2$, then $\bar{c}_1$ *dominates* $\bar{c}_2$, expressing that $\bar{c}_1$ is in all aspects at least as good as $\bar{c}_2$. Dominance is reflexive, i.e., a configuration dominates itself. The irreflexive strict dominance relation is denoted $\prec$. A configuration is a *Pareto point* of a configuration set iff it is not strictly dominated by any other configuration. Configuration set $\mathcal{C}$ is *Pareto minimal* iff it contains only Pareto points, i.e., for any $\bar{c}_1, \bar{c}_2 \in \mathcal{C}$, $\bar{c}_1 \not\prec \bar{c}_2$. Fig. 1 shows the Pareto minimal configuration sets of the applications A, B, and C in our example.

A crucial observation is that, by allowing partially ordered quantities, quantities, configuration sets, and configuration spaces become essentially the same concepts. Pareto algebra exploits this to define operations on configuration sets that can be used to compute or reason about Pareto points for composite systems. We define those operations that are relevant for this paper. Let $\mathcal{C}$ and $\mathcal{C}_1$ be configuration sets of space $\mathcal{S}_1 = Q_1 \times Q_2 \times \ldots \times Q_m$ and $\mathcal{C}_2$ a configuration set of space $\mathcal{S}_2 = Q_{m+1} \times \ldots \times Q_{m+n}$. Let $\bar{c} \cdot \bar{d} = (c_1, \ldots, c_m, d_1, \ldots, d_n)$ for $\bar{c} = (c_1, \ldots, c_m) \in \mathcal{C}_1$ and $\bar{d} = (d_1, \ldots, d_n) \in \mathcal{C}_2$, and $k \in \{1, \ldots, m\}$.

- $\min(\mathcal{C}) \subseteq \mathcal{S}_1$ is the set of Pareto points of $\mathcal{C}$: $\min(\mathcal{C}) = \{\bar{c} \in \mathcal{C} \mid \neg(\exists \bar{c}' \in \mathcal{C} : \bar{c}' \prec \bar{c})\}$.
- $\mathcal{C}_1 \times \mathcal{C}_2 \subseteq \mathcal{S}_1 \times \mathcal{S}_2$ is the (free) product of $\mathcal{C}_1$ and $\mathcal{C}_2$.
- $\mathcal{C} \cap \mathcal{C}_1 \subseteq \mathcal{S}_1$ is the $\mathcal{C}_1$-constraint of $\mathcal{C}$.
- $\mathrm{Join}(\mathcal{C}_1, \mathcal{C}_2, Q) = (\mathcal{C}_1 \times \mathcal{C}_2) \cap \mathcal{D}$, where $\mathcal{S}_1$ and $\mathcal{S}_2$ both include unordered quantity Q and constraint $\mathcal{D}$ is defined by $\mathcal{D} = \{\bar{c}_1 \cdot \bar{c}_2 \mid \bar{c}_1 \in \mathcal{S}_1, \bar{c}_2 \in \mathcal{S}_2, \bar{c}_1(Q) = \bar{c}_2(Q)\}$.

Fig. 1 illustrates the compositional computation of Pareto-optimal system configurations from Pareto-optimal configurations of the applications. The Join operation in each step enforces that the applications should run in configurations with the same clock speed; the constraint operation (Const in the figure; set intersection in Pareto Algebra) applies resource bounds for each dimension and removes infeasible configurations. The minimization operation removes dominated configurations. If we assume that the value of a configuration is inversely proportional to energy consumption, then the sketched (compositional) computation finds an optimal solution to the run-time configuration problem, maximizing value at all times. Note that the MMKP formulation assumes that all configurations of two different applications are always compatible, in contrast to the example where clock speeds need to match. An exact solution to MMKP can be formulated in terms of Pareto algebra in a straightforward way, following the example. Our MMKP heuristic given in the next section can easily be adapted to incorporate constraints like the need to match parameter settings such as the clock speed, making it only more efficient because less configurations need to be considered.

4. THE PARETO-ALGEBRA SOLUTION

This section introduces our Pareto-algebra-based solution for CMP run-time management, focusing on efficiently solving the underlying MMKP instances. It follows the approach of Figure 1. Each time the active application set changes, an MMKP instance needs to be solved. Consider the case that we have K active applications and M newly starting applications. The CMP run-time management approach sketched in [12] then takes an MMKP instance with $N = K + M$ applications. The MMKP instance that we need to solve takes $N = 1 + M$ applications, the M newly starting ones, and one 'application' that corresponds to the total set of active applications.

The $M+1$ configuration sets in our MMKP instance are all $R + 1$-dimensional, 1 dimension corresponding to the value and R dimensions corresponding to the resources. Conceptually, the solution takes the Cartesian product of these sets,

978-1-60558-497-3/09 $25.00 © 2009 ACM

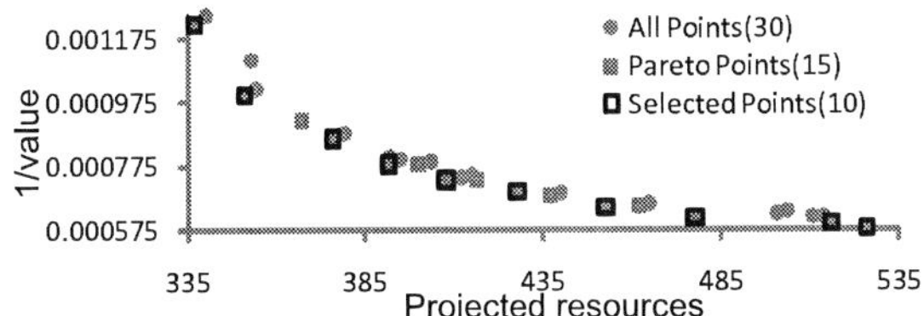

Figure 2: Minimization and reduction in a two-dimensional configuration space.

removes infeasible configurations, and removes dominated configurations. In this way, theoretically, all feasible configurations are computed, and a maximum-value configuration can be selected. In practice, this is intractable. To speed up the computation, we follow other MMKP heuristics by transforming all the configuration sets into 2-dimensional spaces. Dominance is checked in this 2-dimensional space. Furthermore, we avoid the explicit enumeration of the Cartesian product by interleaving the computation with the feasibility and dominance checks. Finally, we introduce a parameter L, which corresponds to the maximum number of Pareto-optimal configurations per application that we consider in each step. The projected 2-dimensional spaces are filtered to make sure that at most L configurations remain.

Given an MMKP instance I with N applications, it turns out the execution time $T(I)$ of our heuristic is bounded in practice as follows:

$$T(I) \leq C \cdot N \cdot L^2, \qquad (1)$$

where C is some constant depending on the processor used. From this bound, it follows that we are able to dynamically at run-time bound the time budget we want to spend on solving an MMKP instance. Furthermore, lower values for L, meaning that less Pareto-optimal configurations are kept in each step of the heuristic, typically result in lower quality of the final solution found. This allows (dynamic) trade offs between quality and execution time.

Before presenting the algorithm in detail, we first discuss the projection of a configuration set into a 2-dimensional space, and the selection of L points from such a space. Fig. 2 gives an example from one of the benchmarks, I6, introduced and studied in detail in the next section.

One dimension in the 2-dimensional space is the value dimension, where we take the inverse value for illustration purposes, so that less is better in the figure. The second dimension is a projected resource usage. Ref. [1] suggests the transformation of the R resource dimensions to a single dimension by taking the distance to the origin in the resource subspace (that omits value). To avoid expensive square and square-root computations, we simply take the sum of all resources: if $\bar{r}$ is a configuration $(r_0, \ldots, r_{R-1})$ in the resource subspace, then the projected resource value $r^*(\bar{r})$ is defined as $\sum_{0 \leq k < R} r_k$. In Fig. 2, the blue points represent the resulting configurations in the derived 2-dimensional space.

In Fig. 2, the blue points represent the resulting configurations in the derived two-dimensional space.

As mentioned, Pareto points are determined in this 2-dimensional space. The red configurations in Fig. 2 are the Pareto points in the example.

It remains to filter out L Pareto points from this space, where L is assumed to be at least 2. Suppose points in the 2-dimensional space are represented in terms of variables v (value) and r (projected resource). For any given constant

c, $v \cdot r = c$ then provides a line through the origin of the space. We select our L points, by first taking the 2 Pareto points $(r_{\max}, v_{\max})$ and $(r_{\min}, v_{\min})$ that result in two lines $v_{\max} \cdot r_{\max} = c_{\max}$ and $v_{\min} \cdot r_{\min} = c_{\min}$ with maximal and minimal values for the constant c, respectively. We then divide the range $c_{\max}..c_{\min}$ in $L - 1$ equally large parts, resulting in $L - 2$ new lines through the origin of the graph. For each of these lines, we take a Pareto point closest (Euclidean distance) to the line. In this way, we select the extreme Pareto points in the space, plus a good spread of Pareto points over the remainder of the space. This provides the highest possible flexibility in selecting fitting configurations for an application.

Algorithm 1 provides pseudo-code for our MMKP heuristic. The algorithm gets as input an N-sized vector of configuration sets (S), an R-sized vector of resource bounds (Rb), and the value for our parameter L. The first step of the algorithm uses the min operation to find Pareto-optimal configurations of each application (Lines 2, 3). The algorithm then transforms the MMKP problem to linear programming constraints and uses lp_solve [7] to compute an upper bound for the objective function (value, Lines 5, 6). This upper bound is added to the resource bounds to consider as a bound for the value dimension in the $Constrain$ function (Line 7).

The compositional part of the algorithm is implemented from lines 8 to 18. The for-loop in Line 10 iterates over applications and in each step configurations of the new application are combined with already found configurations. In each iteration, the algorithm first limits the size of configuration sets C_i and Cm to L, following the procedure given above. We then use the $SumBoundMin$ function for combining configurations. This function gets two configuration sets and a multi-dimensional resource bound. It computes the feasible, Pareto-optimal configurations in the Cartesian product of the two configuration sets. In each step of this

---

Algorithm 1 Pareto-Algebra heuristic for MMKP

```
 1: MMKP(S, Rb, L)
 2: for all  C_i ∈ S do
 3:     min(C_i);
 4: end for
 5: lp = Convert_MMKP_to_lp
 6: vb = lp_solve(lp);
 7: append vb to Rb;
 8: ConfSet Cm = S[0];
 9: S = S − S[0];
10: for all  C_i ∈ S do
11:     if size(Cm) > L then
12:         reduce(Cm, L);
13:     end if
14:     if size(C_i) > L then
15:         reduce(C_i, L);
16:     end if
17:     Cm = SumBoundMin(Cm, C_i, Rb);
18: end for
19: return  maxVal(Cm);
    SumBoundMin(Cm, C, Rb)
20: ConfSet res = empty;
21: for all  c_i ∈ Cm do
22:     for all  c_j ∈ C do
23:         if Constrain(c_i, c_j, Rb) then
24:             AddAndMin(res, product(c_i, c_j));
25:         end if
26:     end for
27: end for
28: return  res;
```

978-1-60558-497-3/09 $25.00 © 2009 ACM

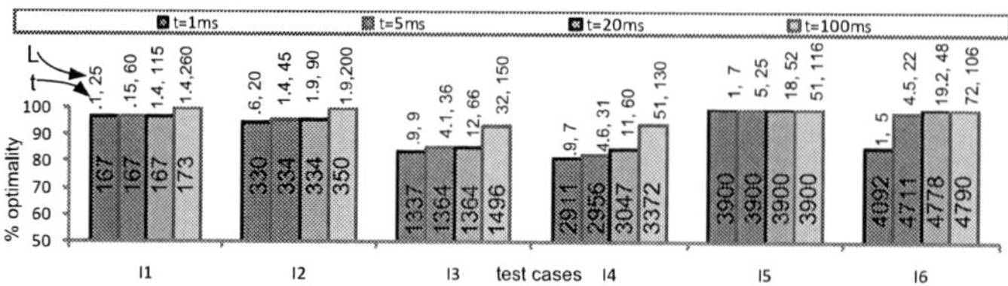

Figure 3: Controlling time budget.

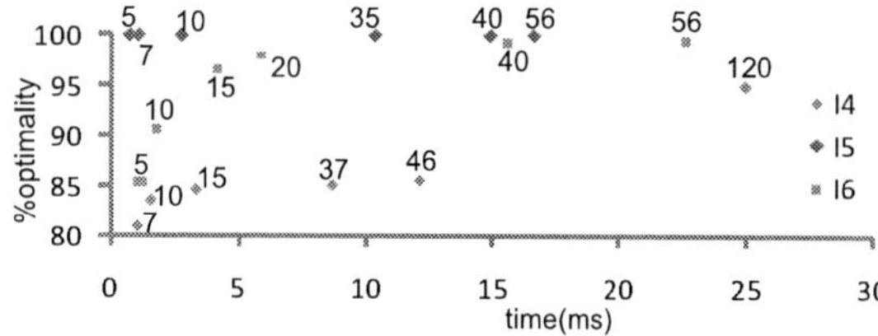

Figure 4: Trading off value vs. execution time.

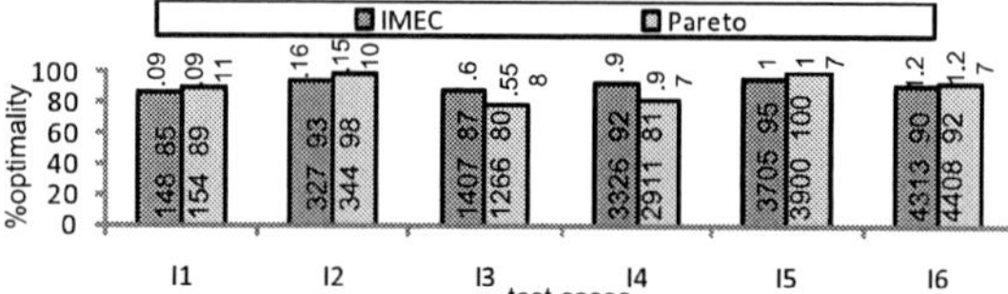

Figure 5: Comparison with the IMEC heuristic.

computation, the function sums all corresponding dimensions from two considered configurations elementwise, and it keeps only the resulting configurations that meet the *multi-dimensional* resource bound and are at that moment in time Pareto-optimal (Lines 23, 24).

After all compositional steps over the applications, Cm contains all feasible configurations of the system that are Pareto-optimal in the projected 2-dimensional space. The algorithm then returns the configuration with the maximum value among those configurations (Line 19).

It is interesting to consider the theoretical worst-case complexity of the algorithm. The complexity of the first part of the algorithm up to Line 8 is determined by the linear programming call of Line 6. The complexity in each step of the compositional part depends on the number of configurations in the considered configuration sets (Cm and C_i), and on parameter L. In each step of the optimization, the algorithm first limits the size of configuration sets C_i and Cm to L; the second part invokes SUMBOUNDMIN on the reduced Cm and C. The two size-limiting phases each require a sorting of the configuration list and a walk through the input configuration list to pick L configurations. Therefore, observing that the size of any of the intermediate Cm results is bounded by L^2 and assuming N_{max} is the maximum number of configurations in any of the input sets in S, then the complexity of these two reductions is $O(L^2 \log L)$ and $O(N_{max} \log N_{max})$. In the second part, the complexity of SUMBOUNDMIN is $O(L^4)$, as it has two nested loops of size at most L and the statement of Line 24 has complexity $O(L^2)$. Since $N - 1$ compositional steps are made by Algorithm 1, the overall complexity of the compositional part of the algorithm, which is the part that is executed online, becomes $O(N(\max(N_{max} \log N_{max}, L^4)))$. In practice, the running time of the algorithm typically never shows its worst-case behavior, because the Pareto-dominance checks

and the checks on the resource and value bounds prune the total search space, and because N_{max} is often (much) smaller than L. In the experiments in the next section, the execution-time bound of Eq. (1) that is quadratic in L is always conservative. If necessary, the bound can be tuned to specific cases.

In any case, by increasing the value of L we can change the behavior of the algorithm from a very short running time and solutions with potentially limited quality towards longer running times and solutions close to the optimum solution of the MMKP instances. By storing Pareto-optimal feasible configurations of the active applications in a system, the complexity of the sketched CMP run-time management approach building upon the heuristic becomes independent of the number of active applications, and it depends only on the number of new applications.

5. EXPERIMENTAL EVALUATION

In this section, we experimentally investigate the various aspects of our approach, and compare our heuristic to the fastest state-of-the-art heuristic of [12], which is the only one sufficiently fast for CMP run-time management. We implemented this heuristic and our compositional heuristic, using the C programming language. We tested the heuristics on the cycle-accurate SimIt-ARM simulator [11] for the StrongARM architecture running at 206 MHz. The results we achieve for the IMEC heuristic [12] are in line with the results reported in [12] for the same platform.

Experiments are reported for the standard MMKP benchmarks available via [9]. The thirteen benchmarks, I1-I13, have values for N (number of applications), N_i (number of configurations per application i), and R (number of resource dimensions) up to 400, 10, and 10, respectively. In line with [12], we divide the benchmark problems in two groups. Instances I1-6 are representative for CMP problem sizes with $N \leq 30$, $N_i \leq 10$, and $R \leq 10$. The second group I7-I13 provides larger test cases. We focus our experiments on the 1-I6 benchmarks, and only briefly investigate the scalability of our approach to larger problem instances.

An important strength of our heuristic is the possibility to bound the time needed to compute a solution, via Equation (1). In the first set of experiments, we bounded the computation time to 1, 5, 20, and 100 ms. To compute L from a time budget, we use Eq. (1) with C equal to 0.0003 for the considered StrongArm configuration. Fig. 3 illustrates for the CMP problem sizes the value of parameter L, the real execution time t, and the quality of the solution in each case (given both as a bar on a normalized scale with the optimum value as a reference, and as an absolute number in that bar). For some problems, the size of intermediate configuration sets are never close to the computed L. Consequently, the real execution time is far below the considered bound. All the results confirm that the bound of Eq. (1) is

978-1-60558-497-3/09 $25.00 © 2009 ACM

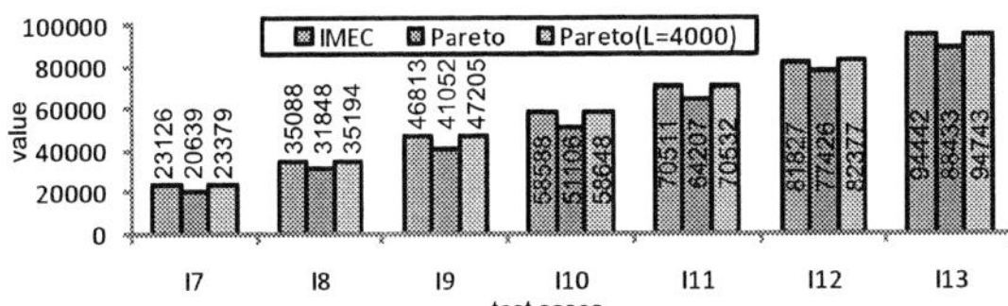

Figure 6: Large MMKP instances

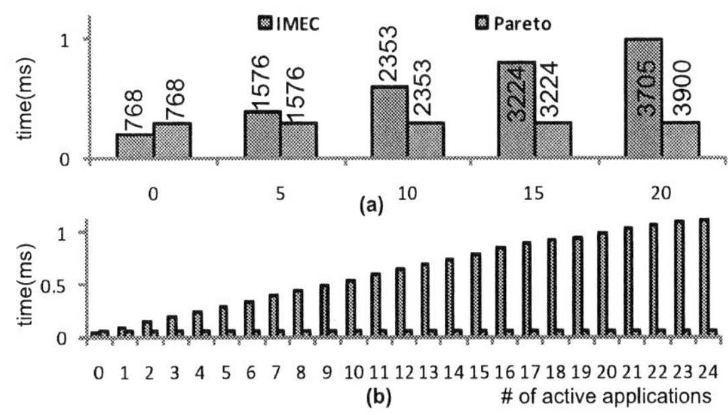

Figure 7: Compositionality

conservative.

Fig. 4 illustrates the trade-off between quality of the solution and the computation time for our heuristic. The results are reported for the three largest CMP benchmarks, I4-I6, with different values for L, given as annotations in the figure. The results indicate that the trade-off can be controlled by L. For I5, the resource bounds are very tight and the number of feasible configurations after each step of the compositional computation remains very low. So in this case, we can have an exact solution with a small L. The resource bounds in I4 are more relaxed and we should consider larger values for L to find a good-quality solution.

An important issue is how our heuristic compares to the IMEC heuristic for the same problem. Fig. 5 shows execution times for the CMP benchmark in the worst-case situation for our approach in which all applications start simultaneously without any active applications. For this experiment, we considered the execution time needed by the IMEC heuristic as a bound for the computation time for our heuristic, deriving L from this bound. In Fig. 5, the numbers on the bars indicate the measured computation time and, for our heuristic also L. The execution times for our heuristic are always within the bound determined by the IMEC heuristic. The numbers inside the bars give obtained absolute and relative (%) values. On average, we may conclude that our obtained solution quality is as good as that of the IMEC heuristic.

It is also interesting to consider the large test cases, despite the fact that these are not the target for our heuristic. Fig. 6 shows experiments comparing the solution quality of our heuristic to the solution quality of the IMEC heuristic. When taking the time needed by the IMEC heuristic on a 2 GHz laptop with 2 GB main memory and running Windows XP, as the budget for our heuristic, we obtain $L = 10$ in all cases. For those cases, we obtain a lower quality than IMEC, illustrating that our heuristic scales worse in terms of quality given a fixed time budget. When increasing the value of L, however, we can obtain better quality in all test cases, at the expense of extra execution time. For these large cases, the projection of resources into one dimension leads to poor results, so Pareto points are computed in the multi-dimensional space.

Finally, we evaluate the impact of compositionality, a second important strength of our heuristic. Fig. (7) compares the performance of the compositional heuristic with the IMEC heuristic, for benchmark I5 with $L = 10$ and assuming that applications start in groups of 5 or 1. The results confirm that the performance of our heuristic does not depend on the number of active applications. The IMEC heuristic needs to solve the complete MMKP instance for all applications each time, which results in increasing execution times when more applications are active. It is clear that our heuristic outperforms the IMEC heuristic when applications start at different points in time.

6. CONCLUSION

We have presented a novel parameterized compositional heuristic to solve MMKP. This heuristic may form the basis of a run-time quality and resource manager for CMPs. It allows to select application configurations at run-time, in such a way that overall system quality is optimized. The heuristic outperforms the best known heuristic to date (that is not compositional). The parametrization furthermore allows to trade off the time needed to find a solution for solution quality, and to bound the time used for finding a solution. These properties imply that it fits well with the constraints imposed by embedded applications and resource-constrained embedded platforms. In future work, we plan to consider the actual reconfiguration of a CMP when new applications enter or leave the system, ultimately aiming at the development of a prototype run-time quality and resource manager for CMPs.

7. REFERENCES

[1] M. Akbar et al. Solving the multidimensional multiple-choice knapsack problem by constructing convex hulls. *Computers and Operations Research*, 33(5):1259-1273, 2006.

[2] M. Geilen et al. An algebra of Pareto points. *Fundamenta Informaticae*, 78(1):35–74, 2007.

[3] S.V. Gheorghita et al. Application Scenarios in Streaming-Oriented Embedded-System Design. *IEEE Design and Test of Computers*, 25(6):581-589, 2008.

[4] S. Khan et al. The utility model for adaptive multimedia systems. *Int. Workshop on Multimedia Modeling*, p. 111–126, 1997.

[5] S. Khan et al. Solving the knapsack problem for adaptive multimedia systems. *Studia Informatica Universalis*, p. 161–182, 2002.

[6] S. Khuri et al. The zero/one multiple knapsack problem and genetic algorithms. *ACM Symp. of applied computation*, 1994.

[7] Lp_solve. http://lpsolve.sourceforge.net.

[8] M. Moser et al. An algorithm for the multidimensional multiple-choice knapsack problem. *IEICE Trans. on Fundamentals of Electronics*, E80-A(3):582-589, 1997.

[9] MMKP Problems. ftp://cermsem.univparisl.fr/pub/CERMSEM/hifi /MMKP/MMKP.html.

[10] R. Parra-Hemandez and N. Dimopoulos. A new heuristic for solving the multichoice multidimensional knapsack problem. *IEEE Trans. on Systems, Man, Cybernetics*, 35(5):708-717, 2005.

[11] SimIt-ARM. http://simit-arm.sourceforge.net.

[12] Ch. Ykman-Couvreur et al. Fast Multi-Dimension Multi-Choice Knapsack Heuristic for MP-SoC Run-Time Management. In *SoC 2006*, p. 1–4. IEEE, 2006.

978-1-60558-497-3/09 $25.00 © 2009 ACM

Mode Grouping for More Effective Generalized Scheduling of Dynamic Dataflow Applications

William Plishker, Nimish Sane, and Shuvra S. Bhattacharyya
Electrical and Computer Engineering Department and Institute for Advanced Computer Studies
University of Maryland
College Park, Maryland, USA
{plishker, nsane, ssb}@umd.edu

ABSTRACT

For a number of years, dataflow concepts have provided designers of digital signal processing systems with environments capable of expressing high-level software architectures as well as low-level, performance-oriented kernels. To apply these proven techniques to new complex, dynamic applications, we identify repetitive sequences of atomic, repeatable actions ("modes") inside dynamic actors to expose more of the static nature of the application. In this work, we propose a mode grouping strategy that aids in the decomposition of a dynamic dataflow graph into a set of static dataflow graphs that interact dynamically. Mode grouping enables the discovery of larger static subgraphs improving scheduling results. We show that grouping modes results in improved schedules with lower memory requirements for implementations by up to 37% including a common imaging benchmark with dynamic behavior: 3D B-spline interpolation.

Categories and Subject Descriptors

D.4.1 [**Operating Systems**]: Process Management

General Terms

Algorithms, Management

Keywords

dataflow, scheduling, mode grouping

1. INTRODUCTION

Dataflow models have become both an important formal underpinning for tools and a convenient way of reasoning about applications in a precise manner for application areas such as digital signal processing (DSP). Their graph-based formalisms allow natural and yet semantically rigorous application descriptions. Such a semantic foundation enables a variety of analysis tools, including determining buffer bounds and efficient scheduling [6]. As a result, dataflow

Permission to make digital or hard copies of part or all of this work for personal or classroom use is granted without fee provided that copies are not made or distributed for profit or commercial advantage and that copies bear this notice and the full citation on the first page. To copy otherwise, to republish, to post on servers or to redistribute to lists, requires prior specific permission and/or a fee.
DAC'09, July 26-31, 2009, San Francisco, California, USA

languages are increasingly popular. Their diversity, portability, and intuitive appeal have extended them to many application areas with a variety of targets. As system complexity increases in digital signal processing platforms, designers are expressing more types of behavior in dataflow languages. While the semantic range of DSP-oriented dataflow models has expanded to cover dynamic interactions, many of the powerful static optimizations algorithms have become less applicable. This is especially problematic in scheduling, which impacts key implementation metrics for embedded software systems, like memory size, performance, and power consumption (e.g., see [9]).

Even though new applications are dynamic, they still exhibit static behavior during execution. This may be in the form of static subsystems or repetitious sequences of actions that occur during otherwise dynamic execution. While current techniques already address the former, the latter is unexploited. Traditional static scheduling techniques are potentially applicable to these application-wide static sequences, but the behavior must be identified and properly modeled based on the original application description.

In this paper, we extend a scheduling algorithm [7] that takes a dynamic application described in the functional dataflow interchange format formalism [8] and decomposes it into a set of static dataflow graphs. In particular we introduce a notion of grouping actor modes, which allows certain static sequences of fixed actions to be exposed to a scheduler. This exploits the statically known behavior of an otherwise dynamic actor resulting in a more efficient schedule for the overall application.

2. BACKGROUND

2.1 Dataflow Modeling

Modeling DSP applications through coarse-grain dataflow graphs is widespread in the DSP design community, and a variety of dataflow models have been developed for dataflow-based design. A growing set of DSP design tools support such dataflow semantics. Designers are expected to be able to find a match between their application and one of the well-studied models, including cyclo-static dataflow (CSDF), synchronous dataflow (SDF), single-rate dataflow, homogeneous synchronous dataflow (HSDF), or a more complicated model such as boolean dataflow (BDF).

Common to each of these modeling paradigms is the representation of computational behavior as a dataflow graph. A dataflow graph G is an ordered pair (V, E), where V is a set of vertices (or nodes), and E is a set of directed edges.

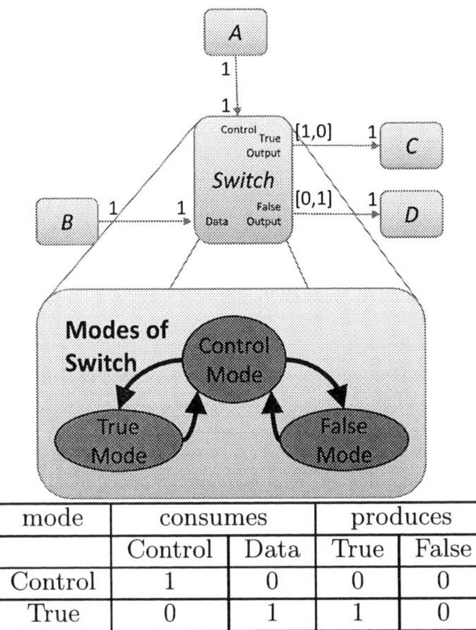

mode	consumes		produces	
	Control	Data	True	False
Control	1	0	0	0
True	0	1	1	0
False	0	1	0	1

Figure 1: Boolean dataflow *Switch* in CFDF.

A directed edge $e = (v_1, v_2) \in E$ is an ordered pair of a source vertex $v_1 \in V$ and a sink vertex $v_2 \in V$. Nodes or *actors* represent computation while edges represents a FIFO communication links between them.

One of the semantic foundations for a structured dynamic dataflow formalism is *core functional dataflow* (CFDF) [8], which is capable of expressing deterministic, dynamic dataflow applications. In this formalism, each actor $a \in V$ has a set of *modes*, M_a, in which it can execute. Each mode, when executed, consumes and produces a fixed number of tokens. For example, consider the *Switch* actor and four SDF actors in Figure 1 with the mode behavior of the *Switch* actor shown below it.

2.2 Dataflow Interchange Format

To describe dataflow applications for the wide range of dataflow modeling techniques that are relevant to DSP system design, application developers can use the dataflow interchange format (DIF) [3], a standard approach founded in dataflow semantics and tailored for DSP system design. It provides an integrated set of syntactic and semantic features that can fully capture essential modeling information of DSP applications without over-specification. DIF is designed to describe mixed-grain graph topologies and hierarchies as well as to specify dataflow-related and actor-specific information. The DIF package transforms DIF descriptions into an internal representation and contains graph utilities, optimization engines, and other utilities. *Functional DIF* also supports CFDF making it an effective environment for modeling dataflow applications, providing interoperability with other design environments, and developing new tools.

2.3 Generalized Scheduling

While there exist techniques for finding static islands in a dynamic application, more static behavior can be found in dataflow graphs described in CFDF. A generalized scheduling approach can utilize the fact that every mode has fixed production and consumption behavior and decompose a dy-

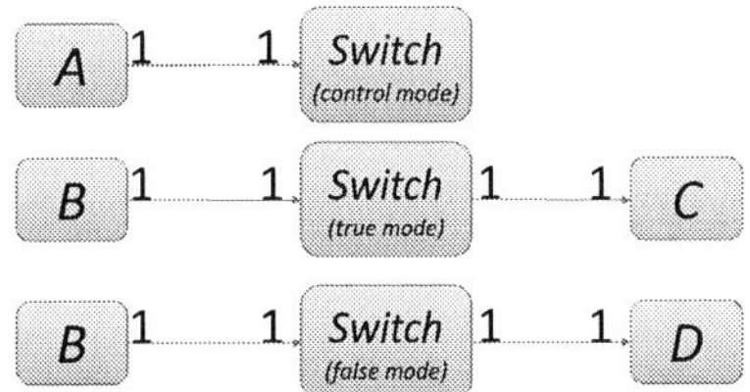

Figure 2: Mode decomposition of switch application

namic dataflow graph into a set of static subgraphs [7]. To construct a single static subgraph, it finds the combination of modes in which one mode from each actor in the subgraph is producing or consuming on an edge that has a consuming or producing mode respectively at the other end of the edge. For example, consider the decomposition that results from Figure 1 as shown in Figure 2. From these static subgraphs, generalized schedule trees (GSTs) are produced [5]. A GST is a schedule representation based on ordered trees where leaf nodes represent the actors of a dataflow graph and internal nodes represent loop counts.

2.4 Related Dynamic Dataflow Work

Besides CFDF, there are a variety of other dataflow models and dataflow based design environments that capture and simulate dynamic applications. In the Stream Based Function (SBF) model of computation [4], actors are represented by a set of functions, a controller, state, and transition function. Ptolemy II encompasses a diversity of dataflow-oriented and other kinds of models of computation [2]. Developers employ a "director" that controls the communication and execution schedule of an application graph.

Beyond modeling, dataflow has been used in the optimization of dynamic systems. For example, systems with data dependent communication can be modeled with variable rate dataflow [10]. When dynamic behavior is modeled as such, buffer capacities can be computed based on throughput constraints in the application. By contrast, our work focuses on identifying more static behavior in a dynamic application so that traditional scheduling techniques can be used. While Functional DIF is used as our demonstration vehicle for extending generalized scheduling, the techniques and results should be applicable across many such tools.

3. MODE GROUPING

While static subgraphs can be successfully found by a generalized scheduling approach to dynamic applications [7], some static behaviors are not considered. For example, in the decomposition of our switch application from Figure 2 the *True* and *False* modes act predictably, always returning to *Control*, which is the mode that transitioned to them in the first place. The repeatable nature of these branches are the kind of static behavior that is exploitable but is not found by existing generalized scheduling approaches.

To this end, we augment our actor description with the concept of *mode grouping*, in which application writers can refine their original application by grouping modes together. For an actor a with modes M_a, we define a mode grouping, $D_a \subseteq M_a$ as a set of modes with a static relationship. The static mode behavior we expose in this work is cyclic mode transitions in which all modes in the grouping return exactly one mode as the next mode, except for one mode, called the

978-1-60558-497-3/09 $25.00 © 2009 ACM

Table 1: The actors in the cubic B-spline application ("control" corresponds to c* edges and "point" corresponds to d* edges)

Actors	Modes	consumed			prod
		control	point	data	out
Grid Lookup	Normal	NA	3	NA	192
Interp Plane	Control	1	0	0	0
	New Cube	0	1	192	48
	Recalc	0	1	0	48
Interp Row	Control	1	0	0	0
	New Plane	0	1	48	12
	Recalc	0	1	0	12
Interp Point	Control	1	0	0	0
	New Row	0	1	12	3
	Recalc	0	1	0	3
Print	Normal	NA	NA	3	NA

entrance mode. The entrance mode may have multiple transitions out, as it marks the single point of dynamic behavior in the grouping, but after it is fired, the modes that follow it do so in a static sequence. The mode grouping can be considered by the scheduler as a single mode that has production and consumption behavior equal to the sum of the individual modes in it. The resulting schedule then includes a repeated firing the size of the mode grouping.

In our switch example, two mode groupings are $D_a = \{\{control, true\}, \{control, false\}\}$, each with *Control* as the entrance mode. This exposes that *Control* always precedes a *True* or a *False*, allowing a larger schedule tree to be formed. For this small example performance benefits are slight, but for more complex applications, the assertion that a set of modes execute in a static sequence can lead to notably smaller buffer requirements.

4. CUBIC B-SPLINE INTERPOLATION

Splines are commonly used in image processing for applying smooth, non-linear transformation to images. Such deformation fields may be modeled by piecewise polynomial interpolation from *basis functions* based on a set of *control points*. They can be computationally efficient by restricting the impact of a control point to a local region, but still be globally smooth. In this work we focus on *cubic B-splines* which use third order polynomials as their basis functions and use contributions from four neighboring control points. A popular way of transforming a 3-dimensional set of control points into deformation field is through B-spline interpolation in three different axes, each time adding the contributions from two control points before and two control points after the voxel in question.

A dataflow implementation of this application (shown in Figure 3) avoids re-computation and additional memory access to the intermediate results of interpolation. Mode behaviors are described in Table 1 and Table 2. Each interpolating actor may accept new inputs for its interpolation step or it may compute another voxel on the shared support of the control points at that step. We exploit grouping in this application by pairing each *Recalc* and *New Cube/Plane/Row* mode with the *Control* mode, creating a total of 6 new modes to the datapath actors.

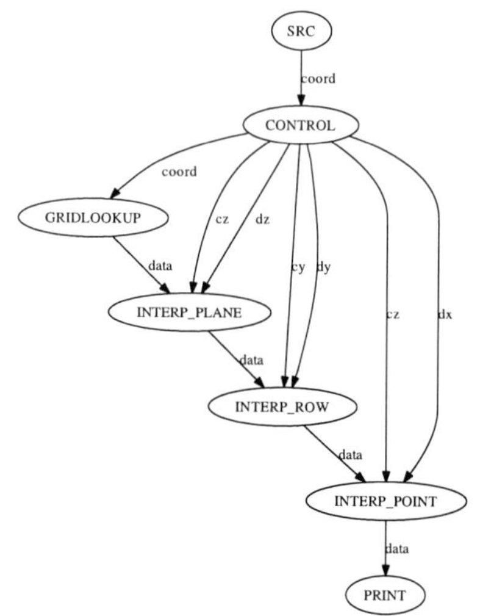

Figure 3: Dataflow graph of 3D cubic B-spline

Table 2: The CONTROL actor modes in B-spline

Modes	cons	produces						
	coord	coord	cz	dz	cy	dy	cx	dx
Control	3	0	0	0	0	0	0	0
New Cube	0	3	1	1	1	1	1	1
New Plane	0	0	1	1	1	1	1	1
New Row	0	0	0	0	1	1	1	1
New Point	0	0	0	0	0	0	1	1

5. RESULTS

To evaluate the benefits of mode grouping, we started with a set of both static and dynamic applications with actors that had mode groupings to exploit: the B-spline application presented in the previous section, a CSDF data distribution of audio streams to be sample-rate-converted, a polyphase decimated DFT filter bank, and an application with multiple polynomial evaluation accelerators (PEAs). For each, we ran the generalized scheduler (using APGAN as the static scheduler [1]) with and without mode groupings that were manually added. The resulting schedule trees were balanced and ordered based on the known input conditions, be it static patterns or probability distributions. In the case of the B-spline application, the steps through the different dimensions were known at run-time, and schedule trees were balanced accordingly. Figure 4 shows these five unique schedule trees (each of which has a root at the second level of the tree) that were replicated and ordered with a modified iteration count to account for the number of voxels to be processed at each step. Also for the B-spline application, a valid schedule could not be found by our generalized scheduler without groups. Instead, we generated the flat schedule balanced with equivalent execution rates as shown in Figure 5.

We pushed hundreds of input data tokens through each of these implementations to monitor the amount of buffering consumed by the application with that schedule as shown in Table 3. The two purely static applications showed no benefit of using mode grouping. While mode groups were identified in CSDF actors, the original generalized scheduler performed equally well with and without groups. Once any

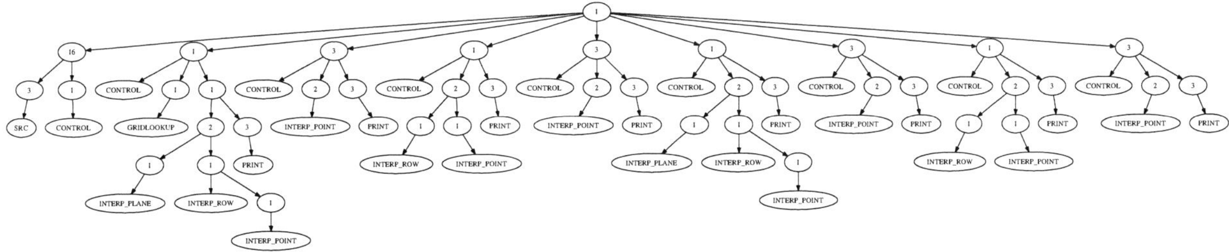

Figure 4: Schedule derived from mode grouping for B-spline application

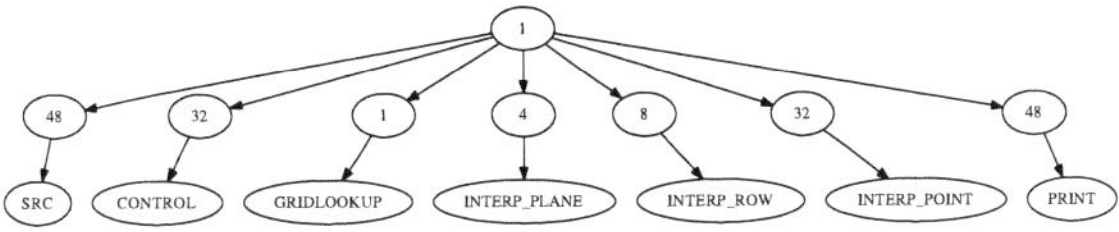

Figure 5: Rate balanced schedule for B-spline

Table 3: Total buffer size requirements with and without mode grouping

Application	Without Groups	With Groups	Percentage Improvement
BSpline Interp	479	304	37%
Sample Rate Conv	2,278	2,278	0%
PolyPhase	24	24	0%
Multi-PEA	3,802	2,976	22%

dynamic behavior was inserted (i.e. the B-spline controller and the PEA dynamic switch and select pair), mode grouping showed a significant improvement finding more (and larger) static schedule trees, which provided a direct savings in buffering by more optimal actor firings. Generalized scheduling with and without groups for each of these examples took less than 5 seconds on a modern CPU.

6. CONCLUSIONS

The rise of complex, dynamic applications has hindered the applicability of traditional static dataflow analysis. In this paper, we have reclaimed some of this ground by presenting a new extension to a generalized scheduling approach that exposes more static behavior of a dynamic application graph. By identifying static groups of "modes" inside actors, we expose more of the static nature of the application, allowing traditional scheduling techniques to improve on memory requirements by up to 37%. We plan to extend this work by developing dynamic schedule tree selector so that a simulator or a final implementation may strategically switch between the known static behaviors at run-time.

Acknowledgments

This research was sponsored in part by the U.S. National Science Foundation (Grant number 0325119), and the US Army Research Office (Contract number TCN07108, administered through Battelle-Scientific Services Program).

7. REFERENCES

[1] S. S. Bhattacharyya, P. K. Murthy, and E. A. Lee. *Software Synthesis from Dataflow Graphs.* Kluwer Academic Publishers, 1996.

[2] J. Eker, J. Janneck, E. A. Lee, J. Liu, X. Liu, J. Ludvig, S. Neuendorffer, S. R. Sachs, and Y. Xiong. Taming heterogeneity - the Ptolemy approach. *Proceedings of the IEEE, Special Issue on Modeling and Design of Embedded Software*, 91(1):127–144, January 2003.

[3] C. Hsu, I. Corretjer, M. Ko., W. Plishker, and S. S. Bhattacharyya. Dataflow interchange format: Language reference for DIF language version 1.0, userŠs guide for DIF package version 1.0. Technical Report UMIACS-TR-2007-32, Institute for Advanced Computer Studies, University of Maryland at College Park, June 2007. Also Computer Science Technical Report CS-TR-4871.

[4] B. Kienhuis and E. F. Deprettere. Modeling stream-based applications using the SBF model of computation. In *Proceedings of the IEEE Workshop on Signal Processing Systems*, pages 385–394, September 2001.

[5] M. Ko, C. Zissulescu, S. Puthenpurayil, S. S. Bhattacharyya, B. Kienhuis, and E. Deprettere. Parameterized looped schedules for compact representation of execution sequences. In *Proceedings of the International Conference on Application Specific Systems, Architectures, and Processors*, pages 223–230, Steamboat Springs, Colorado, September 2006.

[6] E. A. Lee and D. G. Messerschmitt. Static scheduling of synchronous dataflow programs for digital signal processing. *IEEE Transactions on Computers*, February 1987.

[7] W. Plishker, N. Sane, and S. S. Bhattacharyya. A generalized scheduling approach for dynamic dataflow applications. In *Proceedings of the Design, Automation and Test in Europe Conference and Exhibition*, Nice, France, April 2009. 6 pages in electronic proceedings.

[8] W. Plishker, N. Sane, M. Kiemb, K. Anand, and S. S. Bhattacharyya. Functional DIF for rapid prototyping. In *Proceedings of the International Symposium on Rapid System Prototyping*, pages 17–23, Monterey, California, June 2008.

[9] S. Sriram and S. S. Bhattacharyya. *Embedded Multiprocessors: Scheduling and Synchronization.* Marcel Dekker, Inc., 2000.

[10] M. H. Wiggers, M. J. G. Bekooij, and G. J. M. Smit. Computation of buffer capacities for throughput constrained and data dependent inter-task communication. In *DATE '08: Proceedings of the conference on Design, automation and test in Europe*, pages 640–645, New York, NY, USA, 2008. ACM.

Efficient Program Scheduling for Heterogeneous Multi-core Processors

Jian Chen and Lizy K. John

ECE Department, University of Texas at Austin

Austin, TX 78712, USA

chenjian@mail.utexas.edu, ljohn@ece.utexas.edu

ABSTRACT

Heterogeneous multicore processors promise high execution efficiency under diverse workloads, and program scheduling is critical in exploiting this efficiency. This paper presents a novel method to leverage the inherent characteristics of a program for scheduling decisions in heterogeneous multicore processors. The proposed method projects the core's configuration and the program's resource demand to a unified multi-dimensional space, and uses weighted Euclidean distance between these two to guide the program scheduling. The experimental results show that on average, this distance based scheduling heuristic achieves 24.5% reduction in energy delay product, 6.1% reduction in energy, and 9.1% improvement in throughput when compared with traditional hardware oblivious scheduling algorithm.

Categories and Subject Descriptors: C.1.3 [**Processor Architectures**]: Other Architecture Styles – *Hybrid Systems*

General Terms: Design, Performance

Key Words: Heterogeneous Multi-core, Energy-Delay Product, Program Scheduling

1. INTRODUCTION

Heterogeneous multicore processors (HMP) have been demonstrated to be an attractive design alternative to its homogeneous counterpart, as it has a unique advantage in improving both system throughput and execution efficiency. Although most of the existing multicore processors are homogeneous, this design paradigm leads to an inevitable dilemma. That is, replicating smaller cores compromises the throughput of the high-complexity single-threaded applications; whereas replicating larger cores sacrifices the execution efficiency of the low-complexity low-priority threads. The HMP, however, integrates cores of different types or complexities in a single chip, and hence is able to address both throughput and efficiency for various workloads by matching execution resources to each application's needs. As a result, HMP is gaining preference in both industry (e.g. IBM's CELL [12]) and academia (e.g. Core Fusion [9], TFlex [7]).

While HMP is capable of addressing diverse workload demands, it relies on an appropriate program scheduling scheme to unleash its architectural potential for energy efficient computing. The successful program scheduler must be able to find the match between programs and cores with minimum cost in performance and power, which is nontrivial and remains challenging. A straightforward scheduling policy proposed by Kumar et.al uses trial-and-error approach to find the match between programs and cores [3]. The problem with this method is that it incurs significant energy overhead in context switching [15]. Becchi et al extends Kumar's work by measuring IPC ratios between two different cores to migrate applications [17]. Essentially, this method uses pair-wise program swapping to reduce

Permission to make digital or hard copies of part or all of this work for personal or classroom use is granted without fee provided that copies are not made or distributed for profit or commercial advantage and that copies bear this notice and the full citation on the first page. To copy otherwise, to republish, to post on servers or to redistribute to lists, requires prior specific permission and/or a fee.

DAC'09, July 26-31, 2009, San Francisco, California, USA

the number of tentative runs.

These existing methods rely on monitoring the exhibited performance metric during tentative runs to identify the matching program-core pair. This paper, however, attempts to leverage the fundamental relationship between the inherent program characteristics and the corresponding resource demands for program scheduling. This paper, for the first time, presents a framework that addresses three aspects of a program scheduler in the HMP context: understanding the physical configurations of the core supply, estimating the resource demands of the running applications, and identifying the program-core matching for a given criteria [1]. The proposed method projects the core configurations and the program's resource demands into a unified multi-dimensional space, where the program-core matching can be easily identified with Weighted Euclidean Distances (WED). We demonstrate that the WED is strongly correlated with EDP, hence can be used to guide program scheduling in HMPs. Compared with traditional hardware oblivious scheduling, the distance based scheduling heuristic achieves an average of 24.5% reduction in EDP, and 6.1% reduction in energy, and improves the throughput by 9.1%.

The rest of the paper is organized as follows: Section 2 summarizes the related work. Section 3 describes the framework for multidimensional program-core matching. Section 4 shows the projection functions used in the proposed framework. Section 5 presents the scheduling heuristics. Section 6 gives the setup of the experimental environment. Section 7 discusses the experimental results and Section 8 concludes the paper.

2. RELATED WORK

Prior research on program scheduling in heterogeneous systems mainly focuses on optimizing the scheduling of subtasks. The widely adopted approach is to decompose the application into several dependent subtasks, formulate the constraint graph based on the inter-subtask dependencies, and partition the graph to minimize the overall execution time [13] or the data transfer on buses [18][19]. This scheduling approach requires the performance/power of a workload be known a priori or easy to predict, which fundamentally limits its applicability. Our method, on the contrary, does not need the microarchitecture dependent performance/power data as a priori information for scheduling.

Recently, Kumar et al. [3] propose a dynamic program scheduling approach based on the sampled EDP during tentative runs. Becchi et al. [17] extend this method by using the measured IPC ratios between two programs for program migrations. Gulati et al. [8] uses efficiency threshold to dynamically allocate processor for the given task. All of these methods exploit intra-program diversity, and could adapt to program phase changes. Our scheduling scheme exploits inter-program diversity and statically allocates programs to cores by analyzing inherent program characteristics.

Chen and John [6] employ fuzzy logic to calculate the program-core suitability, and use that to guide the program scheduling. However, their method is not scalable since the complexity of fuzzy logic increases exponentially as the number of characteristics increases. Our method, however, is scalable and can be easily extended to the case where the number of characteristics is four or beyond.

3. FRAMEWORK

The idea of multi-dimensional program-core matching stems from the observation that both programs and cores can be described with a set of orthogonal characteristics. By projecting these characteristics from both program side and core side to a unified multi-dimensional space, we are able to visualize the correlation between the program and the core, and simplify the program-core matching.

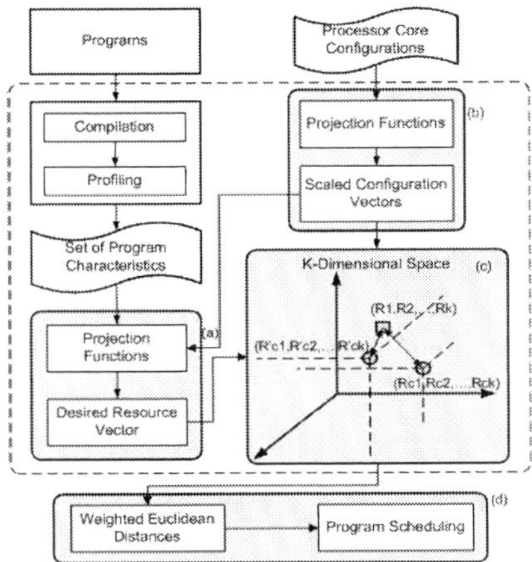

Fig 1. Framework for multidimensional program-core matching

Figure 1 shows the proposed framework for multidimensional program-core matching. The programs are first compiled and profiled to obtain K sets of inherent program characteristics: $\overrightarrow{X_1}, \overrightarrow{X_2}, ..., \overrightarrow{X_K}$, where $\overrightarrow{X_i} = (x_{i0}, x_{i1}, x_{i2}, ..., x_{in})$ is the vector that describes program characteristic i (i=1..K). Define g_i as the projection function that transforms characteristic $\overrightarrow{X_i}$ to the program's desired resource demand R_{pi}, i.e. $R_{pi} = g_i(\overrightarrow{X_i}) = g_i(x_{i0}, x_{i1}, x_{i2}, ..., x_{in})$. We have the program's desired resource vector: $\overrightarrow{R_D} = (R_{p1}, R_{p2}, ..., R_{pK}) = (g_1(\overrightarrow{X_1}), g_2(\overrightarrow{X_2}), ..., g_K(\overrightarrow{X_K}))$. This vector points to the program's desired configuration node in the K-dimensional space, as shown in Figure 1 (c). On the other hand, each processor core has a configuration vector $(C_1, C_2, ..., C_K)$ with element C_i correspondent to the program characteristic $\overrightarrow{X_i}$ (i=1..K). Similarly, this vector can also be transformed by a set of projection functions to a scaled configuration vector $\overrightarrow{R_C} = (R_{c1}, R_{c2}, ..., R_{ck})$ in the K-dimensional space. Once the desired resource vector $\overrightarrow{R_D}$ and the configuration vector $\overrightarrow{R_C}$ have been projected to the same space, the distance between these two becomes the natural measurement for the degree of match between the program and the core. Specifically, larger distance leads to less compatibility between the program and the core, and hence less execution efficiency. Note that not every dimension of the vector contributes equally to the degree of match. Therefore, we use WED in the K-dimensional space as the metric for the program-core matching, as shown in equation (1):

$$D^2 = \sum_{i=1}^{k} w_i (R_{pi} - R_{ci})^2 \tag{1}$$

where w_i is the weight coefficient for the i-th dimension (i=1..K), and D is the weighted Euclidean distance. This distance could guide the program scheduler to identify the matching program-core pair.

4. PROJECTION FUNCTIONS

The projection functions are used to map program characteristics and core configurations to the unified K-dimensional space. On one hand, the projection functions need to interpret and quantify the implications of program's inherent characteristics on its hardware resource demand. On the other hand, the projection functions need to scale the raw hardware configurations in accordance to the diminishing return effect. This section gives the detailed description of the projection functions in this study.

4.1 Projection Functions for Core Configurations

Generally speaking, hardware resources suffer from the diminishing return effect, that is, the additional hardware resource yields less than proportional increase in the marginal benefit. This diminishing return effect has its implication on the core configuration projection: the spacing between adjacent configurations decreases as the value of the configuration increases. To capture this effect, we use the reciprocal function to scale the inter-configuration distance, which is formulated as follows:

$$R_X(i) = c \left(\frac{1}{X_{min}} - \frac{1}{X_i} \right) \tag{2}$$

where X_{min} is the minimum value of the configuration X among the cores, X_i is the value of configuration X of core i (i=1..n), and c is a normalizing factor. While the proposed framework supports K (K$\geq$3) different hardware configurations, this paper only examines three configurations, i.e., issue width, branch predictor size and L1 data cache size.

4.2 Projection Functions for Programs Characteristics

In accordance with the hardware configurations, this paper investigates three important program characteristics: instruction level parallelism (ILP), branch predictability, and data locality. The projection functions for these characteristics are used to identify the desired issue width, branch predictor size, and data cache size.

4.2.1. Desired Issue Width

The desired issue width represents the pipeline width that a processor core needs to efficiently exploit the amount of ILP in the program. We use the instruction dependency distance [5] to capture the program's ILP. Typically, for a given dependency distance distribution, the more instructions with long dependency distance, the more ILP the program has, and hence the desired issue width is larger. Therefore, to obtain the desired issue width, we cluster the instructions into four groups with respect to their dependency distances, i.e., group 1 with distance of 1, group 2 with distance of 2-3, group 3 with distance of 4-7, and group 4 with distance of 8 and beyond. Let $X_{dep,i}$ represent the percentage of instructions whose dependency distance falls in group i (i=1..4). We have the program's dependency distance vector $(X_{dep,1}, X_{dep,2}, X_{dep,3}, X_{dep,4})$. Each element in this vector has its most suitable issue width, that is, 1-way issue for $X_{dep,1}$, 2-way for $X_{dep,2}$, and so on. Therefore, the mass center (or the weighted average) of the distribution indicates on average the issue width demand of the program, hence:

$$R_{issue} = \frac{\sum_{i=1}^{4} X_{dep,i} * W_i}{\sum_{i=1}^{4} X_{dep,i}} \tag{3}$$

where W_i, i=1..4, are the projected coordinates in the issue width dimension of the space, representing the issue width from 1 to 8.

4.2.2. Desired Branch Predictor Size

The desired branch predictor size represents the branch predictor size that a processor core needs to efficiently exploit branch predictability in the program. We use branch transition rate [14] to capture the branch predictability of the program. Generally speaking, the branch instructions with extremely low or extremely high transition rate are easy to predict, and as the transition rate approaches 50%, branches become harder to predict. Based on this observation, we evenly divide the transition rates into 10 buckets: [0, 0.1], [0.1, 0.2],...,[0.9, 1.0]. Let $X_{br,i}$ be the amount of branch instructions whose transition rate falls in the bucket i (i=1..10). We have the branch transition rate vector $(X_{br,1}, X_{br,2}, ..., X_{br,10})$. Since the branch instructions in the buckets [0.4, 0.5] and [0.5, 0.6] are the hardest to predict, they are associated with the largest branch predictor. The branch instructions in the buckets [0.3, 0.4] and [0.6, 0.7] are relatively easier to predict, and hence are associated with a smaller branch predictor. Same applies in buckets [0.2, 0.3] and [0.7, 0.8], and buckets [0.1, 0.2] and [0.8, 0.9]. Therefore, the mass center

978-1-60558-497-3/09 $25.00 © 2009 ACM

of the transition rate distribution indicates on average the demand of branch predictor size, hence:

$$R_{branch} =$$
$$\frac{(B_1*(X_{br,2}+X_{br,9})+B_2*(X_{br,3}+X_{br,8})+B_3*(X_{br,4}+X_{br,7})+B_4*w*(X_{br,5}+X_{br,6}))}{\sum_{i=2}^{4} X_{br,i}+\sum_{i=7}^{9} X_{br,i}+w*\sum_{i=5}^{6} X_{br,i}} \quad (4)$$

where B_i, $i=1..4$, are the coordinates in the branch predictor dimension of the space ($B_1 < B_2 < B_3 < B_4$). Buckets [0, 0.1] and [0.9, 1] are not considered because branch instructions in this range are very easy to predict, and even the smallest branch predictor in this study would be more than enough for them. The parameter w is employed to tune the weight of the largest branch predictor, and is set to $\alpha \times P_{cond}$. The parameter α is an empirically determined value, and is in proportional to the instruction issue width. It is used to keep track of the fact that as the issue width gets wider the branch misprediction penalty also increases, and hence a larger branch predictor with higher prediction accuracy is more desirable. P_{cond} is the percentage of the conditional branches in the instruction mix. A large P_{cond} may lead to a large number of hard-to-predict branches; hence the weight of large branch predictor should be high.

4.2.3. Desired L1 Data Cache

Similarly, the desire L1 data cache is the data cache size that a processor core needs to efficiently exploit the data locality of the program. This paper uses Mattson's stack distance distribution [16] to measure the program's data locality. Let H_i ($i = 1..4$) be the possible L1 data cache sizes with $H_1 < H_2 < H_3 < H_4$, and let $X_{stk,i}$ ($i = 1..4$) be the number of accesses whose stack distance is within the range of $[0, H_1]$, $(H_1, H_2]$, $(H_2, H_3]$, $(H_3, H_4]$ respectively. Note that $X_{stk,1}$ can be most efficiently hold by the cache with size H_1, and $X_{stk,2}$ can be most efficiently hold by the cache with size H_2, and so on. Similarly, the desired L1data cache size of the program can be calculated as:

$$R_{cache} = \frac{\sum_{i=1}^{4} X_{stk,i}*H_i'}{\sum_{i=1}^{4} X_{stk,i}} \quad (5)$$

where H_i' represents the scaled coordinates for the data cache size $H_i, i = 1..4$.

Since the issue width has the highest impact on the program-core matching, followed by cache size and branch predictor size, the highest weight is assigned to the issue width dimension, with the second highest to the cache size dimension, and the lowest to the branch predictor dimension. In this paper, we empirically assign the weights to be 0.8, 0.2, and 0.1 to demonstrate the concept.

5. SCHEDULING HEURISTICS

The WED represents the match between the program's desired resource demand and the core's hardware resource supply. The smaller the distance is, the better the match is, and hence the higher the execution efficiency would be. Therefore, the distance can be treated as a proxy of the program's execution efficiency on a certain core. Given that, the optimum scheduler shall minimize the total distance of the scheduled program-core pairs. However, such scheme is NP-complete (O(n!) with a naïve implementation), and becomes impractical for large number of programs.

Let P_j be the j-th program in the program queue (j=1..M).
Let C_i be the processor core i. (i=1..N)
for j (1 .. M)
　for i (1.. N)
　　if (C_i is available && min_dist > distance(P_j,C_i))
　　　min_dist = distance(P_j,C_i);
　　　k=i;
　　end if
　end for
　schedule P_j to C_k;
　mark C_k as unavailable;
end for

Fig 2.Pseudo code of the distance based scheduling heuristic.

This section presents a scalable scheduling heuristic based on the distance. As shown in Figure 2, this heuristic schedules the program on a first-come, first-served (FCFS) basis, and for each program, it allocates the available core with the minimum distance to that program. The complexity of this scheme is the same as the complexity of finding the minimum distance, which is O(n). In addition, this paper also examines the following scheduling algorithms for comparison:

Hardware Oblivious Scheduling (baseline): This scheduling scheme is unaware of the hardware substrate, and schedules the program on a FCFS basis. Specifically, the first in program queue is schedule to core 1, the second to core 2, and so on.

Min EDP Scheduling: This scheduling method attempts to schedule the programs such that the overall EDP is minimized. It assumes that the EDP of each program-core pair is known a priori, and hence sets the *best case* scenario the overall EDP.

Max EDP Scheduling: This scheduling method attempts to maximize the overall EDP. It also assumes the EDP of each program-core pair is known a priori, and provides the *worst case* scenario in the overall EDP for comparison.

6. EXPERIMENTAL SETUP

To validate the WED model, we vary the instruction issue width, L1 data cache size and branch predictor size of an out-of-order superscalar processor. The configuration options of these parameters are shown in Table I. To evaluate the proposed scheduling heuristics, we create a hypothetical heterogeneous single-ISA quad-core processor. The detailed configurations of these cores are listed in Table II. Like Cortex-A9 multicore processor [2], each core is an out-of-order processor, and there is no on-chip L2 cache. Other parameters not shown in the table are chosen in a way that the design of the core is balanced.

Table I.　Configuration Options

Items	Configuration Options
Issue Width	single-issue, 2-issue, 4-issue, 8-issue
L1 D-Cache	8KB, 4-way, block size 64byte, 16KB, 4-way, block size 64byte, 32KB, 4-way, block size 64byte, 64KB, 4-way, block size 64byte
Branch Predictor	1K Gshare, 2K Gshare, 4K Gshare, 8K Gshare

Table II.　Configurations of Each Core

Items	Configurations
Core 1	Out-of-order, single-issue, Gshare(1k), 8KB 4-way L1d-cache 64byte, 16k 2-way i-cache 64byte
Core 2	Out-of-order, 2-issue, Gshare(2k), 8 KB 4-way L1 d-cache 64byte, 16k 2-way i-cache 64byte
Core 3	Out-of-order, 2-issue, Ghsare(2k), 64 KB 4-way L1 d-cache 64byte, 16k 2-way i-cache 64byte
Core 4	Out-of-order, 4-issue, Gshare(4k), 16 KB 4-way L1 d-cache 64byte, 16k 2-way i-cache 64byte

The workload of the experiment is composed of benchmark programs from Mibench, MediaBench and SPEC CPU2000. Each program is compiled to Alpha-ISA with peak configurations. In order to count in the effect of different input datasets, we use the test/training input datasets to profile the programs, and use the profiled characteristics to schedule the programs with other input datasets. In this study, we assume the workloads are independent with each other, and we do not consider the hardware mechanisms for core-level communication because these mechanisms are symmetric and have similar impact on independent workloads.

We use extensively modified SimProfile from Simplescalar tools [10] to profile programs and collect the aforementioned program characteristics. We employ Wattch [11] to obtain the performance and power data for each benchmark program. To demonstrate the effectiveness of the framework, we evaluate three aspects of the

978-1-60558-497-3/09 $25.00 © 2009 ACM

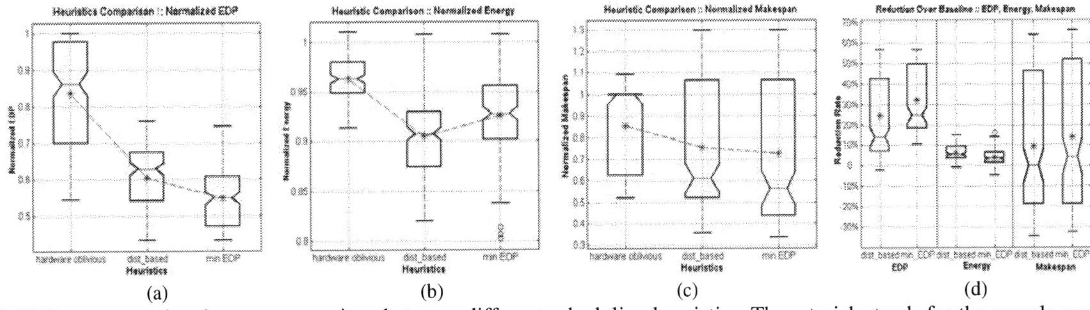

(a)   (b)   (c)   (d)

Fig3. EDP, energy and makespan comparison between different scheduling heuristics. The asterisk stands for the sample average.

proposed scheduling scheme: EDP, energy, and makespan. EDP takes into account both energy and speed, and is widely adopted as a metric for the execution efficiency. Makespan is the time between the start and finish of a group of programs, and is used as the metric for throughput [8].

7. RESULTS & ANALYSIS

To demonstrate that the WED is the appropriate proxy for the program's execution efficiency on a certain core, we need to show the correlation between EDP and WED. To do that, we measure the EDPs of each program on each of the 64 processor cores, and calculate the distance of every program-core pair according to the proposed framework. Table III shows the Pearson's correlation coefficient for each benchmark program included in this study. Note that for each program, we use one set of input for EDP measurement and a different set of input for program profiling. Therefore, the coefficients in Table III not only demonstrate that the WED is strongly correlated with EDP, but also show that the input set has negligible influence on the program's inherent characteristics.

Table III. Correlation Coefficient between EDP and Distance

Benchmarks	Coeff.	Benchmarks	Coeff.
susan	0.91544	ghostscript	0.95708
qsort	0.94244	epic	0.94086
fft	0.81584	g271encode	0.89022
rawdaudio	0.83496	mcf	0.87527
blowfish	0.93837	gcc-166	0.80145
bitcnts	0.81135	gcc-expr	0.88296
mpeg2decode	0.95841	gcc-scilab	0.80598
cjpeg	0.94807	twolf	0.81261

To evaluate the proposed program scheduling schemes, we select programs with different characteristics from the benchmarks in table III to compose 150 diverse program combinations, with each one containing 4 programs. Figure 3 shows the boxplots [4] of the scheduling results of these workloads. According to these boxplots, the proposed distance based scheduling has statistically significant improvement over the baseline scheduling with an average of 24.5% reduction in EDP, 6.1% reduction in energy, and 9.1% reduction in makespan. Compared with min EDP scheduling, the distance based scheduling achieves higher reduction in energy (6.1% vs 3.9%), but less reduction in makespan (9.1% vs 13.8%). However, the difference in makespan reduction is not statistically significant because the notches of the two boxplots overlap, as shown in figure 3 (d). Overall, the distance based scheduling achieves a good balance between throughput and energy.

8. CONCLUSION

This paper presents a framework of multidimensional program-core matching for program scheduling in heterogeneous multiprocessors. The proposed method projects the core's configuration and the program's desired resource demand to a unified multi-dimensional space, and uses weighted Euclidean distance between these two to guide the program scheduling. The experimental results show that the distance based scheduling heuristic achieves an average of 24.5% reduction in EDP, 6.1% reduction in energy, and 9.1% improvement in throughput when compared with traditional hardware oblivious scheduling algorithm.

References

[1] F. A. Bower, et al. "The Impact of Dynamically Heterogeneous Multicore Processors on Thread Scheduling", *IEEE Micro, pp 17-25, May 2008.*

[2] The ARM Cortex-A9 Processor, the ARM white paper. http://www.arm.com/pdfs/ARMCortexA-9Processors.pdf

[3] R. Kumar, et al, "Single-ISA heterogeneous multi-core architectures: the potential for processor power reduction", *Micro-36, pp 81-92, Dec. 2003.*

[4] Yoav Benjamini. "Opening the Box of a Boxplot". *The American Statistician. Vol 42 (4), pp257–262, Nov. 1988.*

[5] A. Phansalkar, et al, "Measuring Program Similarity: Experiments with SPEC CPU Benchmark Suites," *ISPASS'05, pp 10-20, Mar.2005*

[6] J. Chen and L. K. John,"Energy aware program scheduling in a heterogeneous multicore system", *IISWC'08, pp.1-9, Sept. 2008.*

[7] Changkyu Kim, et.al., "Composable Lightweight Processors," *Micro-40, pp.381-394, Dec. 2007*

[8] D.P.Gulati et al., "Multitasking Workload Scheduling on Flexible Core Chip Multiprocessors", *PACT'08, pp187-196, Oct. 2008.*

[9] E. İpek, et al., "Core Fusion: Accommodating software diversity in chip multiprocessors". *ISCA-34, pp 186-197, June 2007.*

[10] D. Burger and T. M. Austin, "The simplescalar tool set version 3.02" , http://www.simplescalar.com/

[11] David Brooks, et al. "Wattch: A Framework for Architectural-Level Power Analysis and Optimizations", *ISCA-27, pp 83-94, June, 2000*

[12] H.P Hofstee, "Power efficient processor architecture and the CELL processor", *HPCA-11. pp. 258-262, Feb. 2005*

[13] M.Maheswaran and H.J.Siegel, "A Dynamic Matching and Scheduling Algorithm for Heterogeneous Computing Systems", *Proc. Heterogeneous Computing Workshop, pp. 57-69, 1998*

[14] M.Haungs, et.al, "Branch transition rate: a new metric for improved branch classification analysis", *HPCA-6, pp.241-250, Feb.2000*

[15] C. S. Ballapuram, A. Sharif and H. S. Lee, "Exploiting Access Semantics and Program Behavior to Reduce Snoop Power in Chip Multiprocessors", *ASPLOS XIII, pp 60-69, Mar 2008*

[16] R.L.M. et al. "Evaluation Techniques for Storage Hierarchies". *IBM Systems Journal, pp 78-117. 1970*

[17] M. Becchi, Patrick Crowley, "Dynamic thread Assignment on Heterogeneous Multi-processor Architectures", *Computing Frontiers 2006, pp 29-40, May 2006*

[18] J. Liu, et al., "Power-aware scheduling under timing constraints for mission-critical embedded systems", *DAC-38, pp840-845, July, 2001.*

[19] M. Ruggiero, et al. "Communication aware system-on-chip allocation and scheduling framework for stream-oriented multi-processor", *DATE, pp3-8, Apr.2006.*

978-1-60558-497-3/09 $25.00 © 2009 ACM

Polynomial Datapath Optimization Using Partitioning and Compensation Heuristics

O. Sarbishei[†], B. Alizadeh[‡] and M. Fujita[‡]

[†]Department of Electrical Engineering, Sharif University of Technology, Tehran, Iran
[‡]VLSI Design and Education Center (VDEC), University of Tokyo and CREST, Tokyo, Japan
sarbishei@ee.sharif.edu, alizadeh@cad.t.u-tokyo.ac.jp, and fujita@ee.t.u-tokyo.ac.jp

ABSTRACT

Datapath designs that perform polynomial computations over Z_2^n are used in many applications such as computer graphics and digital signal processing domains. As the market of such applications continues to grow, improvements in high-level synthesis and optimization techniques for multivariate polynomials have become really challenging. This paper presents an efficient algorithm for optimizing the implementation of a multivariate polynomial over Z_2^n in terms of the number of multipliers and adders. This approach makes use of promising heuristics to extract more complex common sub-expressions from the polynomial compared to the conventional methods. The proposed algorithm also utilizes a canonical decision diagram, Horner-Expansion Diagram (HED) [1] to reduce the polynomial's degree over Z_2^n. Experimental results have shown an average saving of 27% and 10% in terms of the number of logic gates and critical path delay respectively compared to existing high-level synthesis tools as well as state of the art algebraic approaches.

Categories and Subject Descriptors

B.5.1 [**Register Transfer Level Implementation**]: Design – *Datapath design*

General Terms

Algorithms

Keywords

Polynomial Datapath, High-Level Synthesis, Modular HED.

1. INTRODUCTION

As the modern embedded applications continue to grow in size and complexity, a growing demand has led to design hardware at higher levels of abstraction targeting faster design adjustments and higher simulation speed. Regarding this trend, providing efficient implementation of high-level descriptions for polynomial datapaths has become a challenging process. Designers usually optimize such

Permission to make digital or hard copies of part or all of this work for personal or classroom use is granted without fee provided that copies are not made or distributed for profit or commercial advantage and that copies bear this notice and the full citation on the first page. To copy otherwise, to republish, to post on servers or to redistribute to lists, requires prior specific permission and/or a fee.
DAC'09, July 26-31, 2009, San Francisco, California, USA

polynomial functions manually to achieve efficient Register-Transfer-Level (RTL) implementations. However this process can be both time consuming and error prone. As a result, the development of high-level optimization or synthesis tools has been necessitated to automate the design of custom datapaths from a behavioral description.

On the other hand, it is evident that RTL models are implemented with fixed-bit-width datapath architectures, so that polynomial computations are carried out over *n-bit* integers where the sizes of the entire datapaths are kept constant by the way of signal truncation. Hence polynomial optimizations over Z_2^n should be supported. Several optimization techniques have been proposed for multivariate polynomials. The mathematical concepts of finite-ring algebra and modulo optimizations can be used to reduce the degree of polynomials over Z_2^n. A canonical decision diagram called *modular Horner-Expansion Diagram* (HED) [1] and the approaches in [3,4] can perform such modulo optimizations. These algorithms can reduce the complexity of the polynomial over Z_2^n in terms of multipliers and adders. As an example $f_1 = 4x^2 y^2 + x^2 y + 4xy^2 + 21xy + 4$ can be reduced to $f_1 = -3x^2 y + xy + 4$ over Z_2^4.

The Horner form is a popular representation of polynomial expressions and is the most straightforward way of evaluating polynomials approximations of trigonometric functions in many libraries. This method transforms the expression into a sequence of nested additions and multiplications, which are suitable for sequential machine evaluation using Multiply-Accumulator (MAC) instructions. In spite of the advantage of the Horner form in sequential implementations, it does not provide an efficient optimization for combinational multivariate polynomials. As an example if we use the Horner form, the polynomial $f_2 = 2x^2 z + 6xyz$ will be represented by $f_2 = x(2xz + 6yz)$.

Factorization with CSE [5], and symbolic computer algebra based manipulation [2-4] are much better approaches in optimizing polynomials compared to the Horner. For instance, the function $f_2 = 2x^2 z + 6xyz$ is reduced to $f_2 = xz(2 + 6y)$, if we use the CSE method. We can enhance this approach with a coefficient factorization to obtain $f_2 = 2xz(1 + 3y)$. Moreover, CSE can be combined with *modular HED* to provide more efficient polynomials over Z_2^n. In spite of the advantages of the CSE method, it is unable to efficiently factorize some polynomials. As an example, consider the arithmetic

978-1-60558-497-3/09 $25.00 © 2009 ACM

function $f_3 = x^2 + 6xy + 9y^2$. If we employ the kernel/co-kernel extraction with CSE [5-6], this function is factorized as $f_3 = x(x + 6y) + 9y^2$. It is obvious that the better factorization for this polynomial is $f_3 = (x + 3y)^2$.

The approximate factorization algorithm in [7] is an efficient approach in representing an arithmetic function f as a product of sub-functions: $f = f_1 f_2 ... f_N$, where f_i is a multivariate polynomial. This approach can not deal with a sub-function f_i with a degree higher than one. An example is the polynomial $f_3 = (x + 3y)^2$, which includes a sub-function with the degree of two. Regarding such polynomials, the method in [7] has to be enhanced with techniques to initially reduce the degree of all the sub-functions to one. Another important drawback of the method presented in [7] is that it cannot deal with polynomials that are not reducible to $f = f_1 f_2 ... f_N$. As an example the function $f_4 = x^2 + 6xy + 9y^2 + 2z$ cannot be reduced to the function $f_4 = (x + 3y)^2 + 2z$ using [7], because we need to *reject* the *monomial z*.

Definition 1 (poly/monomial): The multivariate polynomial, *poly*, is in the form of $poly = \sum_{i=1}^{M} k_i m_i$, where k_i is the coefficient of the monomial m_i and it is a constant integer value. The above *poly* consists of M monomials. Each monomial is also in the form of $m_i = \prod_{j=1}^{N} x_j^{p_{ij}}$, where x_j is an input variable and p_{ij} is the degree of x_j in m_i and it is a non-negative integer value.

As an example consider the irreducible polynomial $f_5 = x^2 + 2xy + y^2 + 3x + 2y + z + 1$, which consists of 3 multipliers and 7 adders. If we put $3x=2x+x$ and reject z, we obtain $f_5 = (x + y + 1)^2 + x + z$, which includes 1 multiplier and 4 adders. In other words, the two monomials z and x in f_5 need to be *rejected* and *compensated*, respectively to obtain the above reduced f_5. Note that the conventional factorization approaches do not cover such coefficient compensations and monomial rejections.

In this paper we present an efficient algorithm for optimizing a multivariate polynomial, *poly* over Z_2^n. In the initial phase of the algorithm we use modulo optimizations with modular HED in [1] to reduce the degree and coefficients of the polynomial over Z_2^n. Then we perform two efficient heuristics called the *partitioning* and *coefficient compensation* algorithms. These steps are briefly described below:

1) Partitioning poly in the form of poly=$p_1 p_2$+p_3, where p_3 is a low complexity polynomial and it consists of some rejected monomials obtained from poly.

The partitioning heuristic aims to realize all the monomials in *poly* within the product form of $p_1 p_2$. In this way, we expect to have $p_3=0$. However, in some polynomials we may need to reject a number of monomials from *poly* and send them to p_3. As an example for the function $f_6 = x^2 + 2xy + y^2 + 3x + 2y + 4z + 1$, the *partitioning* heuristic tries to obtain $p_1 = (a_1 x + a_2 y + a_3)$ and

$p_2 = (b_1 x + b_2 y + b_3)$, where a_i and b_j are unknown integer coefficients in the partitioning step. Under such conditions we will have $p_3=4z$.

2) Choose the most suitable coefficients (a_i,b_j) for the monomials in p_1 and p_2 so that it minimizes the complexity of compensated monomials transferred to p_3.

For instance, in the previous example, the unknown coefficients in p_1 and p_2 are a_i and b_i ($i=1,2,3$), respectively. The *coefficient compensation* algorithm in this step starts evaluating a bounded range of values for a_i and b_i. It will be explained that using the coefficient compensation approach, we obtain the most suitable compensation of $3x=2x+x$, in which $2x$ can be realized within $p_1 p_2$, but x will be sent to p_3.

Using the above compensation approach, we obtain $f_5 = (x + y + 1)^2 + x + 4z$, where $p_1 = p_2 = (x + y + 1)$ and $p_3 = x + 4z$. After this reduction, the above two steps are performed for p_1, p_2 and p_3, again to reduce them as well. In our example none of these polynomials are reducible and the algorithm converges to $f_5 = (x + y + 1)^2 + x + 4z$.

The proposed algorithm can improve the implementation of several polynomials, which cannot efficiently be optimized by the conventional methods. For example the algorithm can reduce the arithmetic function $f_5 = x^2 + 258xy + y^2 + 3x + 2y + z + 1$ to $f_5 = (x + y + 1)^2 + x + z$ over Z_2^8. Our main contributions in this paper are as follows:

- Proposing an efficient optimization algorithm for multivariate polynomials using *partitioning* and *coefficient compensation* approaches.

- Providing experimental results on a set of benchmark applications to show the superiority of our approach in extracting more complex common sub-expressions in comparison with the traditional techniques.

The rest of the paper is organized as follows. Section 2 presents the proposed algorithm. The experimental results are addressed in Section 3. Finally, Section 4 concludes the paper.

2. THE PROPOSED POLYNOMIAL OPTIMIZATION ALGORITHM

This section addresses the proposed algorithm in optimizing a multivariate polynomial, *poly*. At first the algorithm performs a degree reduction procedure over Z_2^n using *modular HED* [1]. Then we perform the *partitioning* and *coefficient compensation* approaches. The rest of this section is organized as follows. In Section 2.1 basic definitions are presented. Section 2.2 presents the *partitioning* algorithm to obtain the monomials in p_1, p_2 and p_3 in the first main step of the algorithm, where *poly*=$p_1 p_2$+p_3. Section 2.3 addresses the *compensation* heuristic to obtain the most suitable coefficients of the monomials in p_1 and p_2. Finally, Section 2.4 presents the whole polynomial optimization algorithm.

2.1 Definitions

In this subsection basic definitions, which are used throughout the rest of the paper are presented.

978-1-60558-497-3/09 $25.00 © 2009 ACM

Definition 2 (a|b): If a divides b we represent it by $a|b$. Note that a,b can be integer coefficients or monomials.

Definition 3 (monomial complexity, C(m)): We represent the complexity of the monomial $m_i = \prod_{j=1}^{N} x_j^{p_{i,j}}$ by $C(m_i) = \sum_{j=1}^{N} p_{i,j}$. This is due to the fact that variable degrees of m_i represent the number of multipliers in its realization.

Definition 4(sub-polynomial): If we choose L monomials and their coefficients from $poly = \sum_{i=1}^{M} k_i m_i$, where $2 \leq L \leq M-1$, a sub-polynomial is obtained.

As an example $S_1 = x^2 + y^2 + 3x + 1$ is a sub-polynomial of $f_6 = x^2 + 2xy + y^2 + 3x + 2y + 4z + 1$.

Definition 5(gcm(poly),gcc(poly)): The greatest common monomial and the greatest common coefficient in *poly* are represented by *gcm(poly)* and *gcc(poly)* respectively.

For instance, we have $gcm(2x^2 + 4xy^2 + 6x) = x$ and $gcc(2x^2 + 4xy^2 + 6x) = 2$.

2.2 Partitioning

In this subsection we present the partitioning algorithm to represent *poly* as $poly = p_1 p_2 + p_3$, where p_3 should be a low-cost polynomial. The pseudo-code of the partitioning algorithm for the input polynomial, *poly* is depicted in "Fig. 1". At first all the monomials in *poly* are sorted with respect to their complexity, *i.e.* $C(m_i)$, in a descending order. After sorting a *poly* with M monomials, we obtain a set of M monomials in which we have $C(m_i) \geq C(m_{i+1})$ for $i=1,...,M-1$. In the next phase we try to rewrite all the monomials in *poly* to the product form of $p_1 p_2$. Note that the monomials with higher complexity are more prior to be written as a product form of $p_1 p_2$, if possible. In this way based on the sorted monomial complexities, we evaluate each monomial to see if it is possible to show it as a form of $p_1 p_2$ or not. If this is not achievable the monomial stays in the rejection list of p_3.

In order to evaluate the possibility of each monomial to be represented in the form of $p_1 p_2$ we need to have non-zero initial values for p_1 and p_2. In the proposed algorithm we evaluate all the possible initial values for p_1 and p_2 so that $p_1 p_2$ realizes a sub-polynomial of *poly* and then for each initialization we check the possibility of all the other monomials in *poly* to be represented in the form of $p_1 p_2$. Finally, the best initialization, which constitutes the lowest complexity p_3, is chosen as the most suitable partitioning.

In order to provide all the initializations for p_1, p_2 we consider all the sub-polynomials of *poly*, *i.e.* $S_k = \sum_{j=1}^{L} k'_j m'_j$, so that $gcm(S_k)$

$\neq 1$. Then we put:

/* Input: $poly = \sum m_i k_i$ */
Partition(poly)
1 $poly = Sort(poly)$;
2 For those sub-polynomials of *poly*, S_k, so that $gcm(S_k) \neq 1$
{
   3 Build p_1, p_2 in (1) and (2);
   4 For all the monomials m_i in S_k
   {
     5 If (conditions in Lemma1 are satisfied)
       update p_1 as in (3);
     6 Else $p_3 = p_3 + k_i m_i$;
   }
   7 If a better p_3 is obtained choose this partition of p_1, p_2, p_3;
}

Figure 1. Pseudo-code of the partitioning algorithm.

$$p_1 = a_1 gcm(S_k) \qquad (1)$$
$$p_2 = (b_1 m'_1 / gcm(S_k) + b_2 m'_2 / gcm(S_k) + ... + b_L m'_L / gcm(S_k)) \qquad (2)$$

where a_1 and b_j $(j=1,...,L)$ are unknown integer coefficients. Note that in the partitioning algorithm the coefficients are ignored and we only evaluate the monomials in p_1 and p_2. For each initialization of p_1, p_2 in (1) and (2) we start evaluating the rest of monomials in $(poly-S_k)$ to see if they can be inserted into p_1 or not. This process is performed from the highest complexity monomial in $(poly-S_k)$ to the lowest complexity one based on the following lemma.

Lemma 1: A monomial m_i in $(poly-S_k)$ can be added to the partitioning of p_1 in (1) if both of the following conditions are satisfied:

   1)   *There exists a monomial m_P in p_2 that divides m_i.*

   2)   *All the monomials produced by $(m_i/m_P) \times p_2$ are found in poly.*

Under such conditions m_i and all the monomials in $(poly-S_i)$ that also exist in $(m_i/m_P) \times p_2$ can be inserted into the product form of $p_1 p_2$ if we update p_1 as follows:

$$p_1 = p_1 + a_2(m_i/m_P) \qquad (3)$$

Note that if more than one monomial m_P in p_2 satisfies both of the above conditions we choose the highest complexity monomial for the updating process in (3).

On the other hand, if the monomial m_i under evaluation does not satisfy at least one of the conditions in Lemma 1, it is added to the rejection list of p_3. This process is performed until all the monomials in $(poly-S_k)$ are evaluated.

After considering all the initializations in (1)/(2) and building a rejection list of p_3 for each of them by evaluating all the monomials based on Lemma 1, we choose the partitioning of p_1 and p_2, which constitutes the lowest-complexity p_3.

Example 1: We evaluate the partitioning algorithm in the following polynomial, which is sorted w.r.t monomial complexities:

$$poly = w^4 - x^3 - x^2 y - 5x^2 + 2xy + 2y^2 + 3zx + 3zy + 14y + 5z + 3$$

978-1-60558-497-3/09 $25.00 © 2009 ACM   933

The most suitable initialization is when we choose the sub-polynomial $S=-x^3-x^2y-5x^2$. Under such initialization we have: $p_1=a_1x^2$, $p_2=(b_1x+b_2y+b_3)$.

We then start evaluating each monomial in (poly-S) based on Lemma 1. The first monomial w^4 does not satisfy the first condition in Lemma 1 and as a result it is added to p_3. The next monomial is xy, which divides $m_P=x$ in p_2. This value of m_P also satisfies the second condition in Lemma 1. Therefore, based on (3) we update p_1 as follows:

$$p_1= p_1+a_2(xy/x)= a_1x^2+a_2y$$

Then the monomials y^2 and y are eliminated from the evaluation process, because p_1p_2 produces them.

The next monomial is zx, which divides $m_P=x$. It also satisfies the second condition in Lemma 1 and as a result we update p_1 as follows:

$$p_1= p_1+a_3(zx/x)= a_1x^2+a_2y+a_3z$$

We also eliminate the monomials zy and z from the evaluation process. The final monomial is "1", which does not satisfy the first condition in Lemma 1. Therefore, it is added to p_3. In this way the following partitioning is obtained:

$$p_1= a_1x^2+a_2y+a_3z, \ p_2=(b_1x+b_2y+b_3), \ p_3=w^4+3$$

2.3 Coefficient Compensation

When the partitioning algorithm is done, we need an algorithm to find the unknown values of a_i and b_j in p_1 and p_2. One way would be to solve the nonlinear equations of $poly - p_3 = p_1p_2$. However, solving those equations are not easy. Moreover, the equations may not have an answer. In this way we propose a coefficient compensation algorithm, which finds the most suitable values of a_i and b_j in p_1 and p_2 so that we have to transfer the lowest complexity polynomial p back to p_3.

The pseudo code of the coefficient compensation algorithm is shown in "Fig. 2". Consider that $(poly - p_3)$ consists of L monomials. In order to have $poly - p_3 = p_1p_2$, we can write L non-linear equations. We consider a subset of these equations so that at least one set of answers exists for those equations. These equations under evaluation are called the *accepted list* of equations. The other equations are considered as a *compensation list*. The *accepted list* of equations refers to the monomials in *poly* that their coefficients cannot be changed. However, the *compensation list* of equations indicates the monomials in *poly* that may need a coefficient-compensation. We build the *accepted list* (step#1 in "Fig. 2") by taking the following step:

If the equation is in the form of a_ib_j = constant, and all the previous equations in the list does not include a_i or b_j, then add the equation to the list (perform this step for all the equations)

For example consider the following set of equations:

$a_1b_1=2, a_1b_2=3, a_1b_3=5, a_2b_2+a_3b_3=-3, a_2b_3=6, a_3b_1=2, a_3b_2+a_2b_1=4$

After performing the above procedure we obtain the following accepted list:

Accepted list=$\{a_1b_1=2, a_2b_3=6\}$

Theorem 1: The accepted list has at least two equations (four variables).

Compensation(p_1,p_2,p_3)
1 Build the *accepted list* of equations;
2 Build the *compensation list* of equations;
3 For all the answers of the *accepted list* {
    4 Choose the answer with minimum inequalities in the
       compensation list;
}
5 Update p_1,p_2,p_3;

Figure 2. Coefficient compensation algorithm.

Proof: consider that the product form of p_1p_2 is in the following form:

$$p_1p_2 =(a_{n1}x^{n1}f_{n1} +....+a_1xf_1 +a_0f_0)(b_{n2}x^{n2}g_{n2} +....+b_1xg_1 +b_0g_0)$$

where f_i and g_i are polynomials that do not include the variable x. In the above equation the monomials produced by $a_{n1}b_{n2}x^{n1+n2}f_{n1}g_{n1}$ and $a_0b_0f_0g_0$ cannot be realized by the multiplication of other terms in p_1 and p_2. This is due to the fact that they include the highest/lowest degree terms w.r.t the variable x. Therefore the relevant equations for these monomials obtained from *poly-$p_3=p_1p_2$* are in the form of $a_ib_j=cte$. In the minimum-case both $a_{n1}b_{n2}x^{n1+n2}f_{n1}g_{n1}$ and $a_0b_0f_0g_0$ include one monomial and as a result at least two equations exist in the accepted list.

Lemma 2: For an accepted list with l equations in the form of $A_iB_i=C_i$ $(i=1,...,l)$, where C_i is a constant integer, there exists n sets of answers for A_i,B_i, where $n = \prod_{i=1}^{l} Div(C_i)$.

Note that $Div(C_i)$ is the number of integers like "a" so that $a|C_i$. For example $Div(4)=6$, because $\{\pm1, \ \pm2, \ \pm4\}$ divide 4.

After building the accepted list we evaluate all of its possible answers (step#3 in "Fig. 2"). These answers can be obtained by calculating all the divisors of the constant values in each equation. For example the answers of the above accepted list are:

For all a_1 and a_2 so that $a_1|2$, $a_2|6$

we have: $b_1=2/a_1$, $b_3=6/a_2$

At this point we put all the above sets of answers into the compensation list of equations. Then we choose the set of answers with the fewest inequalities in the compensation list of equations (step#4 in "Fig. 2"). In other words, the most suitable answer for the accepted list is the one that results in the lowest complexity polynomial of p to be added to the rejections list of p_3 obtained from the partitioning algorithm.

Example 2: We evaluate the coefficient compensation algorithm for the polynomial in Example 1.

The unknown variables are a_1,a_2,a_3,b_1,b_2,b_3. The accepted list after performing step#1 in "Fig. 2" is:

Accepted list=$\{a_1b_1=-1, a_2b_2=2, a_3b_3=5\}$

The answers for the accepted list are:

For all a_1, a_2 and a_3 so that $a_1|-1$, $a_2|2$, $a_3|5$

we have: $b_1=-1/a_1$, $b_2=2/a_2$, $b_3=5/a_3$

978-1-60558-497-3/09 $25.00 © 2009 ACM

We evaluate all of the above answers and put them into the compensation list of equations. The best answer is when we have $a_1=-1, a_2=2, a_3=1$. Under such conditions, only one of the compensation equations will not be satisfied. This equation takes place for the monomials *14y*, where we should have: $a_2b_3=14$. However, by using the values of a_2 and b_3 from the accepted list we obtain $a_2b_3=10$. In this way this monomial needs compensation and we put $p=(14-10)y=4y$. This function is added to the rejection list of p_3 obtained from the partitioning algorithm. In this way the following final answer is obtained:

$$poly=(-x^2+2y+z)(x+y+5)+w^4+4y+3$$

2.4 Polynomial Optimization Algorithm

This subsection explains the whole multivariate polynomial optimization algorithm. The pseudo-code of the proposed algorithm is shown in "Fig. 3". The algorithm starts from a pre-processing degree reduction procedure for the input polynomial, *poly*, over Z_2^n using *modular HED*. Note that *n* is the bit-length of *poly* and it is an input value for the algorithm. Then the polynomial optimization algorithm is performed. If *poly* consists of one or two monomials, it can easily be reduced by the procedure in step#1 and step#2. We then extract the *gcm* and *gcc* of *poly* (step#3). In the next step the algorithm performs the partitioning and coefficient compensation approaches for the polynomial to obtain p_1, p_2 and p_3 (step#4 and step#5). At this point the new polynomials p_1, p_2, p_3 are optimized independently using the partitioning and coefficient compensation algorithms (step#6-8). This process is performed until all the resulting polynomials cannot be partitioned anymore.

Example 3: Consider the following polynomial:

$$poly = w^4z^3 - 3w^4z^2 - 2w^4z - 6w^3z^3 - 18w^3z^2 - 12w^3z$$
$$- 11w^2z^3 - 33w^2z^2 - 19w^2z - 6wz^3 - 15wz^2 - 12wz$$
$$+ z^3 + w^3 + x^2 + y^2 + 52xy + 69$$

The goal is to optimize this function over Z_2^4. First, the preprocessing with *modular HED* results in $poly = z^3 + 3z^2w + 3zw^2 + w^3 + x^2 + y^2 + 4xy + 5$. Then the partitioning and compensation algorithms are performed to obtain $poly = (z+w)(z^2 + 2zw + w^2) + x^2 + 4xy + y^2 + 5$. The two polynomials $p_2 = (z^2 + 2zw + w^2)$, $p_3 = x^2 + 4xy + y^2 + 5$ can also be optimized and finally we obtain:

$$poly = (z+w)^3 + (x+y)^2 + 2xy + 5$$

3. EXPERIMENTAL SETUP AND RESULTS

In this section, we provide data from the experiments we conducted with our implementation of the proposed algorithms. In our experiments we have employed several polynomials extracted from real embedded systems such as DSP, computer graphics and automotive applications. The algorithm has been implemented in C++ and carried out on an Intel 2.1GHz Core 2 and 1 GByte of main memory running Windows XP. Table 1 tabulates the results where Column 1 lists the benchmarks which include a multivariate

```
/* Inputs: n,poly=∑kᵢmᵢ */
poly=HED(poly,n);
PolyOptimize(poly)
1 If (M==1) return poly; /* M=#of monomials in poly */
2 If (M==2) return
   GCD(k₁m₁,k₂m₂)× ((k₁m₁+k₂m₂)/GCD(k₁m₁,k₂m₂));
3 If gcm(poly)!=1 | gcc(poly)!=1 return
   gcm*gcc*PolyOptimize(poly/[gcm*gcc]);
4 Partition(poly);
5 Compensate(p₁,p₂,p₃);
6 PolyOptimize(p₁);
7 PolyOptimize(p₂);
8 PolyOptimize(p₃);
```

Figure 3. Our proposed polynomial optimization algorithm.

Cosine Wavelet and Quartic-Spline from [5], Savitzky-Golay filters from [8], Quadratic filters from [9], two benchmarks from [10] to be prevalent in signal processing, computer graphics and automotive applications. We have synthesized the polynomials using Leonardo synthesis tool in $0.25\mu m$ CMOS technology. The important parameters of the circuits including #of logic gates (instances) as well as critical path delay (*ns*) are addressed in Table 1. Column 3 (*M/V/D/n*) in this table lists the number of monomials/number of variables/the highest degree/bit-vector size of each variable. We have compared the savings in #of gates and delay obtained by our technique over Horner form, CSE in [5] and the enhanced CSE method in [11]. We obtained an average savings of 49.2%, 41.4% and 26.4% of logic gates over the Horner form, CSE in [5] and its extension in [11], respectively. Most of the improvement is due to efficient factorization of the polynomial expressions and from finding multiple term common sub-expressions, which effectively reduces the number of multiplications and additions. Note that extracting common sub-expressions may impact the parallelism of a multi-polynomial system. In this way the maximum delay may negligibly be impacted when we use our approach. This issue can be seen in some test-cases in Table 1.

The CPU run-time of the algorithms is also shown in Table 1. The run-time values addressed in Table 1 are in seconds. The run-time bottlenecks of the proposed algorithm are the exhaustive check of all the sub-polynomials produced by *poly* in the partitioning heuristic with the time complexity of $O(M2^M)$, where M is the number of monomials in *poly* and evaluating all the answers for the *accepted list* in the compensation algorithm with the time complexity of $O(D^L)$, where D is the number of divisors for the constant integer in an equation of the accepted list and L is the number of equations in the list ($L<M$). Note that M is typically low, while D is typically high in DSP applications. In this way although the run-time complexity of the proposed heuristics are exponential w.r.t the variables M and L, they do not become much slow in optimizing polynomial datapaths. This issue is explored in Table 1 for several test-cases. As can be seen, the first *Image Processing* test-case with 10 monomials constitutes the worst case in terms of run-time. Note that regarding the CSE approaches in [5] and [11], if we prefer to choose the most suitable common sub-expression extraction we also need to evaluate different sub-polynomials in *poly*, which constitutes an exponential time-complexity. This issue implies that the run-time complexity of the proposed algorithm is not much worse than the CSE methods in [5] and [11].

Table 1. Comparison of the proposed algorithm with other decompositions

Application	Function	M/V/D/n	Horner			CSE in [5]			Enhanced CSE in [11]			Our Approach		
			#Gates	Delay	CPU Time	#Gates	Delay	CPU Time	#Gates	Delay	CPU Time	#Gates	Delay	CPU Time
Graphics	Cosine-Wavelet	9/2/3/16	7850	23.73	0.04	6809	21.26	3.26	5109	18.26	3.47	3678	18.46	6.52
Image Processing	Savitzky-Golay 1	10/2/3/16	7218	20.7	0.04	6057	20.81	3.78	3757	22.74	4.75	2879	17.8	14.9
Image Processing	Savitzky-Golay 2	5/2/2/16	1697	20.76	0.03	1565	18.14	0.46	1433	18.08	0.48	1057	16.19	1.63
Filter	Quad 1	5/2/2/16	2737	17.65	0.03	2375	16.43	0.63	2269	16.43	0.63	1763	17.12	1.78
Filter	Quad 2	5/2/2/16	2569	16.43	0.03	2090	16.44	0.63	2032	16.44	0.63	1571	15.31	1.56
Automotive	Mibench	9/3/2/16	2058	15.65	0.04	2046	15.63	0.46	2046	15.63	0.46	1303	14.12	8.75
Average Saving w.r.t Horner			0%	0%	-	13.2%	5.4%	-	31%	6.3%	-	49.2%	13.8%	-

4. CONCLUSIONS AND FUTURE WORKS

In this paper we have proposed an optimization algorithm to reduce the complexity of polynomial datapaths in terms of multipliers and adders over Z_2^n. The modularity optimizations have been fulfilled by making use of finite ring algebra like the approaches in [1,3,4], while two efficient heuristics called *partitioning* and *coefficient compensation* have been applied to reduce and factorize the polynomials, which may even be impossible to be represented in the product form of $poly = f_1 f_2 ... f_N$, where f_i is also a polynomial. Up to our knowledge none of the conventional factorization and optimization algorithms can handle such polynomials. Another advantage of the proposed algorithm is that unlike the conventional factorization algorithms it does not perform the time-consuming processes of GCD extraction or multivariate polynomial division. We can also benefit from the proposed algorithm in optimizing a system of different polynomials. In experimental results we have run several benchmarks and evaluated the efficiency of the proposed algorithm compared to the conventional approaches.

In spite of the superiority of the proposed algorithm over conventional methods in reducing polynomials that cannot be factorized, it faces a limitation in the partitioning procedure when dealing with the *hidden* monomials, which may be used in factorization. As an example $f_1 = x^2 - y^2$ needs a hidden monomial of xy as follows: $f_1 = x^2 - y^2 - xy + xy$. Unless we revive this monomial, the proposed partitioning algorithm would not be able to factorize f_1 as $f_1 = (x-y)(x+y)$. Note that the conventional CSE algorithms cannot handle such polynomials either. As a future work we are going to enhance the proposed partitioning algorithm by considering *hidden* monomials so that more efficient partitions of p_1, p_2, p_3 can be obtained. Moreover, a synergism between the proposed techniques and the conventional logic synthesis methods can be explored to optimize arithmetic circuits and their control units.

5. REFERENCES

[1] B. Alizadeh and M. Fujita, "Modular-HED: A Canonical Decision Diagram for Modular Equivalence Verification of Polynomial Functions", *in the fifth Workshop on Constraints in Formal Verification (CFV)*, pp. 22-40, 2008.

[2] A. Peymandoust and G. DeMicheli, "Application of Symbolic Computer Algebra in High-Level Data-Flow Synthesis", *IEEE Trans. CAD*, vol. 22, pp. 1154–11656, 2003.

[3] Sivaram Gopalakrishnan and Priyank Kalla, "Optimization of polynomial datapaths using finite ring algebra", *ACM Trans. Design Automation of Electronics Systems*, vol. 12, pp. 49, 2007.

[4] S. Gopalakrishnan, P. Kalla, Brandon Meredith, and F. Enescu, "Finding linear building-blocks for RTL synthesis of polynomial datapaths with fixed-size bit-vectors", *in ICCAD*, pp. 143-148, 2007.

[5] A. Hosangadi, F. Fallah, R. Kastner, "Factoring and Eliminating Common Subexpressions in Polynomial Expressions", in *ICCAD*, pp. 169-174, 2004.

[6] JuanCSE, "Extensible, programmable and reconfigurable embedded systems group", *Available at: http://express.ece.ucsb.edu/suif/cse.html.*

[7] E. Kaltofen, J. P. May, Z. Yang, L. Zhi, "Approximate Factorization of Multivariate Polynomials Using Singular Value Decomposition", *Journal of Symbolic Computation,* pp. 359-376, 2008.

[8] J. Krumm, "Savitzky-Golay Filters for 2D Images", *Available at: http://homepages.inf.ed.ac.uk/rf/CVonline/LOCAL_COPIES/KRU MMI/SavGol.htm.*

[9] V. J. Mathews and G. L. Sicuranza, *Polynomial Signal Processing*, Wiley-Interscience, 2000.

[10] M. R. Guthaus and et al., "Mibench: A Free, Commercially Representative Embedded Benchmark Suite", *in IEEE 4th Annual Workshop on Workload Characterization*, pp. 3-14, 2001.

[11] A. Hosangadi, F. Fallah, and R. Kastner, "Optimizing polynomial expressions by algebraic factorization and common subexpression elimination", *IEEE Trans. CAD*, vol. 25, pp. 2012-2022, Oct 2006.

Register Allocation for High-Level Synthesis Using Dual Supply Voltages

Insup Shin
Dept. of Electrical
Engineering, KAIST
Daejeon 305-701, Korea

Seungwhun Paik
Dept. of Electrical
Engineering, KAIST
Daejeon 305-701, Korea

Youngsoo Shin
Dept. of Electrical
Engineering, KAIST
Daejeon 305-701, Korea

ABSTRACT

Reducing the power consumption of memory elements is known to be the most influential in minimizing total power consumption, since designs tend to use more memories these days. In this paper, we address a problem of high-level synthesis with the objective of minimizing power consumption of storage using dual-V_{dd}. Specifically, we propose a complete design framework that starts from dual-V_{dd} scheduling, dual-V_{dd} allocation, and controller synthesis down to the final layout. Its main feature is dual-V_{dd} register allocation, which exploits timing slacks left in the data-path after operation scheduling. In experiments on benchmark designs implemented in 1.08 V (with V_{ddl} of 0.8 V), 65-nm CMOS technology, both switching and leakage power were reduced by 20% on average, respectively, compared to data-path with dual-V_{dd} applied to functional units alone. Detailed analysis of slack histogram, area, wirelength, and congestion were performed to assess feasibility of the design framework.

Categories and Subject Descriptors: B.5.2 [**Register-transfer-level implementation**]: Design Styles—*Automatic synthesis*

General Terms: Algorithms, Design, Performance

Keywords: High-level synthesis, register allocation, dual supply voltage, low power

1. INTRODUCTION

Reducing supply voltage (V_{dd}) is the most effective way to reduce switching power due to its quadratic dependence on V_{dd}; it is also effective in reducing subthreshold leakage, which is roughly proportional to V_{dd}^3, and in reducing gate leakage, which is proportional to V_{dd}^4 [1]. Since reducing single overall V_{dd} increases circuit delay, multiple-V_{dd} technique, which applies higher V_{dd} to timing-critical elements and lower V_{dd} to elements that are not critical to timing, is employed instead to maintain clock frequency. Due to complexity of building power network [2], two V_{dd}s (dual-V_{dd}) are typically used at gate-level of granularity, i.e. V_{dd} is assigned gate by gate; when more than two V_{dd}s are used, they are used at block-level, where each block is put in its own voltage island [3].

Several algorithms have been proposed to determine V_{dd}-type (high- or low-V_{dd}) at gate-level netlist [4–6]; these algorithms are applied after netlist is determined, while netlist itself is derived without dual-V_{dd} in mind (e.g. only high-V_{dd} is assumed during logic synthesis), which limits the scope low-V_{dd} can be applied.

Permission to make digital or hard copies of part or all of this work for personal or classroom use is granted without fee provided that copies are not made or distributed for profit or commercial advantage and that copies bear this notice and the full citation on the first page. To copy otherwise, to republish, to post on servers or to redistribute to lists, requires prior specific permission and/or a fee.
DAC'09, July 26-31, 2009, San Francisco, California, USA

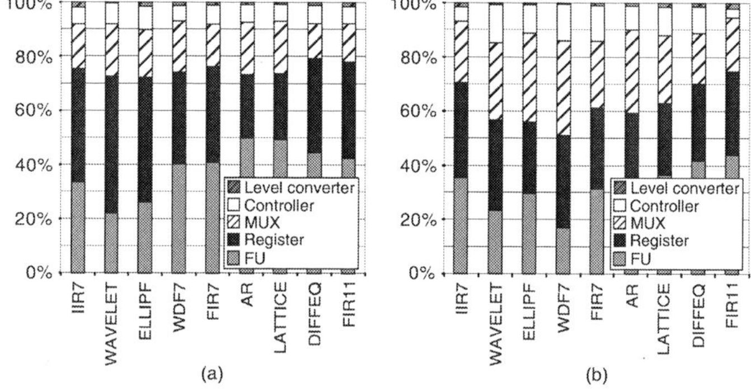

Figure 1: Distribution of (a) switching- and (b) leakage-power of several benchmark circuits when dual-V_{dd} is applied to functional units alone.

Therefore, it is natural to consider dual-V_{dd} at earlier design stage such as during architectural design. Many algorithms have been proposed to determine V_{dd}-type of functional units (such as adders and multipliers) during high-level synthesis (HLS) [7–12]; the problem can be cast to conventional operation scheduling except that the same functional unit with different V_{dd}-type is considered as different units. Surprisingly, however, no research effort has been made toward dual-V_{dd} register allocation, even though the proportion of memory elements (in recent complex designs such as SoCs and microprocessors) is substantial, about 80% of total transistors [13].

To assess the importance of storage (together with multiplexers that are located between storage and functional units) in power consumption, we took a set of of behavioral benchmark designs. Each design was transformed into a DFG (data flow graph); dual-V_{dd} scheduling [11] was applied to determine V_{dd}-type of each operation, which was followed by allocating functional units; remaining steps of HLS were performed when we assume that only high-V_{dd} can be used for registers, multiplexers, and controller; RTL design was then synthesized to obtain a gate-level netlist, which was taken for fast SPICE simulation [14]; simulation was repeated 100 times using different input vectors in commercial 65-nm technology to obtain both switching- and leakage-power consumption. Figure 1(a) shows that the registers contribute, on average, 36% to the total switching power of these circuits (multiplexers contribute 17%), which is substantial; they contribute 30% on average to the total leakage power (26% for multiplexers) as shown in Figure 1(b).

Selecting V_{dd}-type of register and functional unit affects each other when they are on the same timing path from register to register. Since register allocation is usually performed after scheduling, in this paper, we take a scheduled DFG, where V_{dd}-type of each operation is already determined, and try to determine V_{dd}-type of variables, which are then mapped to registers. Central to this approach is how we exploit timing slacks that are left in the data-path

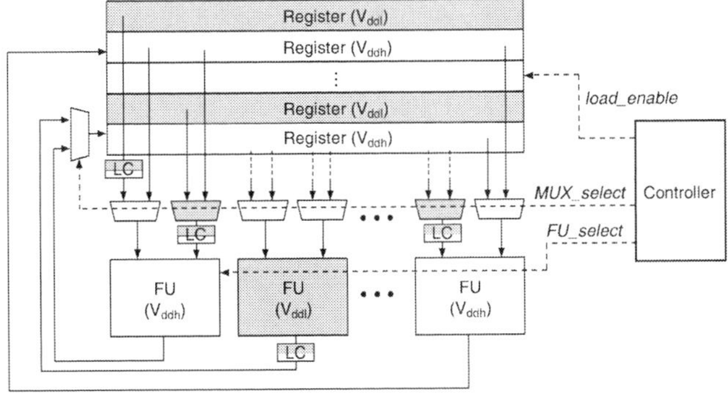

Figure 2: Target architecture for dual-V_{dd} HLS.

after dual-V_{dd} scheduling is performed, so that we can respect the latency and the number of functional units, which are either given by designers as constraints or determined as a result of scheduling. Our main contributions are:

- A complete framework for dual-V_{dd} HLS that covers design processes from scheduling, allocation, and controller synthesis to the final circuit layout (Section 2.2).

- A heuristic solution to dual-V_{dd} register allocation, which is a main feature of the framework, minimizing the number of high-V_{dd} registers as well as total number of registers (Section 3); a heuristic for assigning V_{dd}-type to multiplexers (Section 4).

- Extensive experimental results from commercial 65-nm technology applied to behavioral benchmark designs to assess effectiveness of dual-V_{dd} register allocation on power, and to assess the framework on timing and physical design aspects (Section 5).

2. PRELIMINARIES

2.1 Target Architecture for HLS

We consider a register file-based architecture as a target of dual-V_{dd} HLS. As shown in Figure 2, it consists of a data-path (functional units, registers, and their connections) and a controller. Each functional unit is assigned either high-V_{dd} (V_{ddh}) or low-V_{dd} (V_{ddl}) as a result of scheduling and allocating functional units [11]. For simplicity of presentation, the controller is assumed to be assigned V_{ddh}, which eliminates the need for level converters between data-path and controller.

The delay of timing path across data-path has several components:

$$D_{dp} = T_{cq} + \alpha \cdot T_{mux} + \beta \cdot T_{lc} + T_{FU} + T_{su}, \qquad (1)$$

where T_{cq} and T_{su} are clock-to-Q delay and setup time of a register, respectively, and T_{FU} is a (maximum) delay through a functional unit; T_{mux} and T_{lc} denote the delay of a multiplexer and a level converter, respectively. The number of multiplexers, α, and the number of level converters, β, on the timing path are determined after resource allocation and dual-V_{dd} assignment on all data-path components. The controller delay can also be factored in (1) through estimating its delay [15]; or, it can be assumed to be negligible if all control signals (see Figure 2) are ready before data actually arrive (e.g. if MUX_select arrives before multiplexer inputs do), which can be done via adjusting clock arrival times to controller.

Note that (1) is needed during operation scheduling, i.e. the delay (as a number of control steps) required for executing i-th operation (when it is executed on a resource having T_{FU} delay) is given

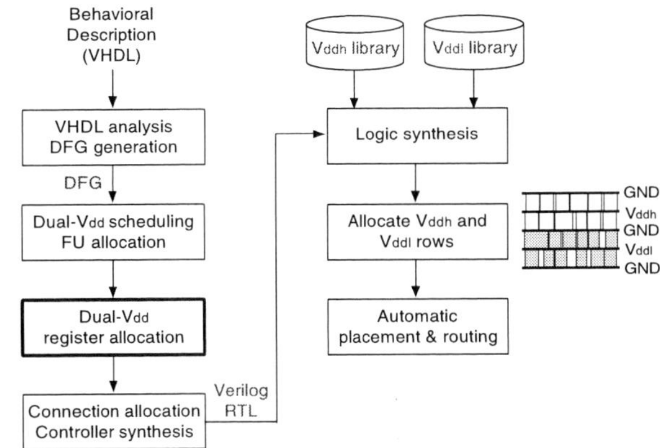

Figure 3: Overall design flow based on dual-V_{dd} HLS.

by

$$d_i = \left\lceil \frac{D_{dp}}{P_c} \right\rceil, \qquad (2)$$

where P_c is a designer-specified clock period. The V_{dd}-type of a functional unit can be determined during scheduling by assuming two T_{FU}s (one for V_{ddh} and the other for V_{ddl}). However, the values of parameters in (1) except for T_{FU} are unknown when they are needed in (2), since resource allocation is typically performed after scheduling. Therefore, during scheduling, we assume that V_{ddh} is initially assigned to all registers, which gives us the values for T_{cq} and T_{su}, and $\alpha=2$ (one before each FU input and the other after FU output); we also assume V_{ddh} for multiplexers[†]; $\beta=1$ is assumed when functional unit is assigned V_{ddl}, i.e. we assume a level converter after the output of functional unit.

Due to the way operation delay is defined as in (2), some timing paths across data-path may have timing slacks that can be utilized to assign V_{ddl} to some registers (thus increasing T_{cq} and T_{su}) during dual-V_{dd} register allocation in Section 3.

2.2 Dual-V_{dd} Design Framework

The overall design flow based on dual-V_{dd} HLS is shown in Figure 3. A behavioral description written in VHDL is first analyzed [16] and is then transformed into a DFG [17]. The DFG is an input to dual-V_{dd} operation scheduling [11], which, together with subsequent step of allocating functional units, determines V_{dd}-type of each functional unit, while minimizing latency under resource constraints (or its dual). The scheduled DFG will then be an input to register allocation, which will be described in Section 3. The connection allocation, which uses multiplexers to connect registers to functional units and functional units to registers, is formulated as vertex coloring of a connection conflict graph [18], which is performed by left edge algorithm [19]. The FSM controller is synthesized as a hard-wired sequential circuits; thus, it is described as a state transition graph [18]. Allocation and control synthesis generates the data-path and controller as a Verilog HDL, which goes through a standard logic synthesis [20] to create a gate-level netlist. Two libraries of gates were created and used: one for V_{ddh} (1.08 V in our experiment) and the other for V_{ddl} (0.80 V).

For the physical design of dual-V_{dd} netlist, we first determine the number of V_{ddh} and V_{ddl} placement rows that will be required,

---

[†]We assume 10-to-1 multiplexer during scheduling. The exact number of inputs of each multiplexer can only be determined after register allocation and connection allocation. The difference of delay arising from this assumption is utilized when we determine V_{dd}-type of multiplexers in Section 4.

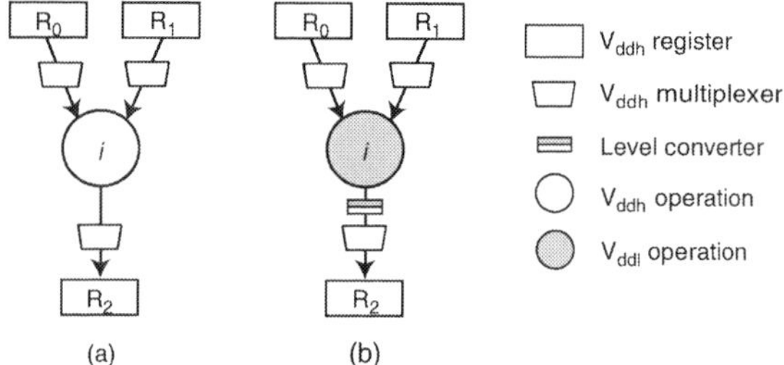

(a)   (b)

Figure 4: Data-path after dual-V_{dd} scheduling: operation is assigned (a) V_{ddh} and (b) V_{ddl}.

based on the area of the cells to be placed in each type of row. This is followed by decision of how to interleave them, i.e. how many V_{ddh}-rows are followed by how many V_{ddl}-rows. In order to apply double-back layout pattern, which helps reduce area, we try to group even number of consecutive V_{ddh}- or V_{ddl}-rows, e.g. every four V_{ddh}-rows followed by every two V_{ddl}-rows, which is then followed by automatic placement and routing [21] of the netlist.

3. DUAL-V_{dd} REGISTER ALLOCATION

We accept a scheduled DFG, where each operation is assigned either V_{ddh} or V_{ddl}, as an input to dual-V_{dd} register allocation, which consists of the following three steps. We use Figure 5 as an example to illustrate each step.

3.1 Find Feasible V_{dd} Combinations of Operands

After dual-V_{dd} scheduling, data-path can be either in the form of Figure 4(a) when operation i is assigned V_{ddh} or Figure 4(b) when it is assigned V_{ddl}. In this data-path, we want to determine whether input or output operands can be assigned V_{ddl}[†]. We compute slack along each timing path from input operand to output operand. In Figure 4, we first compute the timing slack from R_0 to R_2,

$$S_l = d_i \cdot P_c - D_{dp,l}, \qquad (3)$$

where $D_{dp,l}$ is the delay of timing path across data-path from R_0 to R_2 ($D_{dp,r}$ is the delay from R_1 to R_2; thus D_{dp} in (1) represents the maximum of $D_{dp,l}$ and $D_{dp,r}$). We also compute the amount of time required to assign V_{ddl} to R_0,

$$\Delta T_{cq} = T_{cq}(V_{ddl}) - T_{cq}(V_{ddh}) + T_{lc}[‡], \qquad (4)$$

and the amount of time required to assign V_{ddl} to R_2,

$$\Delta T_{su} = T_{su}(V_{ddl}) - T_{su}(V_{ddh}). \qquad (5)$$

We then compare S_l to ΔT_{cq}, ΔT_{su}, and $\Delta T_{cq} + \Delta T_{su}$ to see if we can assign V_{ddl} to R_0, R_2, or both of them. We repeat the process for the path from R_1 to R_2.

In Figure 5(b), combinations of operand V_{dd}-type are illustrated within each operation, where the first two letters correspond to input operands (for binary operations) and the last letter to output operand. Since V_{dd}-type of each operand is determined independently in each operation and operand can be shared in different operations, there can be inconsistencies of V_{dd}-type. In Figure 5(b),

---

[†]V_{dd}-type is assigned to functional unit (not to operation) and register (not to operand). Assigning V_{dd} to operation and operand implies assignment before allocation.

[‡]Note that we need a level converter to use V_{ddl} for R_0, since multiplexer is initially assigned V_{ddh}. We may instead try to locate level converter at the output of multiplexer while we use V_{ddl} for multiplexer; we postpone the decision of V_{dd}-type of multiplexer (see Section 4), since exact number of multiplexer inputs can only be determined after connection allocation.

z_1 always takes V_{ddh} in op_1, which is why the third combination in op_4 (where z_1 is V_{ddl}) is crossed out; we cross out the second combination in op_6 because z_2 has to be assigned V_{ddh} in op_2.

3.2 Categorize Variables

3.2.1 Initial Categorization

Once we derive candidate V_{dd}-types of operands, we categorize operands (or variables) into three groups: V_{ddh}-variables, independent variables, and dependent variables. V_{ddh}-variables (HV) are those whose V_{dd}-types are fixed to V_{ddh} (note that there are no V_{ddl}-variables, because if any variable can be assigned V_{ddl}, it can be assigned V_{ddh} as well): z_1, z_2, and y_1 belong to this category as shown in Figure 5(c). Independent variables (IV) can take either V_{ddh} or V_{ddl}, and the decision does not affect V_{dd}-type of other variables; if the decision affects other variables (even though they can take both types of V_{dd}), they are called dependent variables (DV). In Figure 5(c), x_1 and x_2 belong to IV, because op_1 has all four combinations of V_{dd}-type (HH, HL, LH, and LL) for them, while x_2 is fixed to V_{ddh}; z_3 is DV because if it is assigned V_{ddl}, x_3 has to be assigned V_{ddh} (see op_4).

The decision of IV or DV can be made as follows: for each variable (which is not HV), we check each operation that takes the variable as one of its operands one by one; in each operation, we remove the letters corresponding to HVs; if there are all possible combinations of V_{dd}-type for the remaining variables, the variable is IV in that particular operation; if the variable is IV in all the operations, it is IV, otherwise it is DV. In op_4 of Figure 5(c), we remove the letters (H) for z_1 that leaves us three combinations (HH, LH, and HL), thus both x_3 and z_3 are DV; z_6 is IV in op_7 but it is DV in op_5, thus it is DV.

3.2.2 Refinement by Extending Operation Delay

Our objective of dual-V_{dd} register allocation is to minimize the number of registers assigned to V_{ddh} as well as the total number of registers. We have better chance to achieve this goal (during the next step of register allocation) when we have less number of HVs and less number of DVs (note that allocating V_{ddl} to some DVs may cause other DVs to be allocated to V_{ddh}). This can be accomplished by extending the delay of operations having some HV or DV operands, which do not have any successor operations scheduled directly in the next control step (thus have a room for extending its delay): in Figure 5(d), delays of op_2 and op_6 are extended; as a result, z_2 becomes IV (while it was HV in Figure 5(c)) and z_5 also becomes IV (it was DV).

Note that allocation of functional units remains intact: in Figure 5(a), we need 1 V_{ddh} adder (for op_1, op_2, and op_7), 1 V_{ddh} multiplier (for op_4 and op_6), and 1 V_{ddl} multiplier (for op_3 and op_5); the same is true in Figure 5(d) because we simply delay loading the results of op_2 and op_6 to output registers by one cycle, which can be done by generating `load_enable` signals (during controller synthesis) one cycle later. This can be checked by counting the number of operations of the same type in the control step where operation will be extended and comparing that to resource constraint; in Figure 5(a), there are no additions in the third control step, which allows op_2 to be extended.

Extending operation delay increases the lifetime interval of input operands (and decreases that of output operands), which may increase the total number of registers. This is simply checked by counting the maximum number of lifetimes that are overlapped in any control steps, which serves as a lower bound of total number of registers, before and after extending operation delay; if the number increases, extending operation delay is not performed.

978-1-60558-497-3/09 $25.00 © 2009 ACM

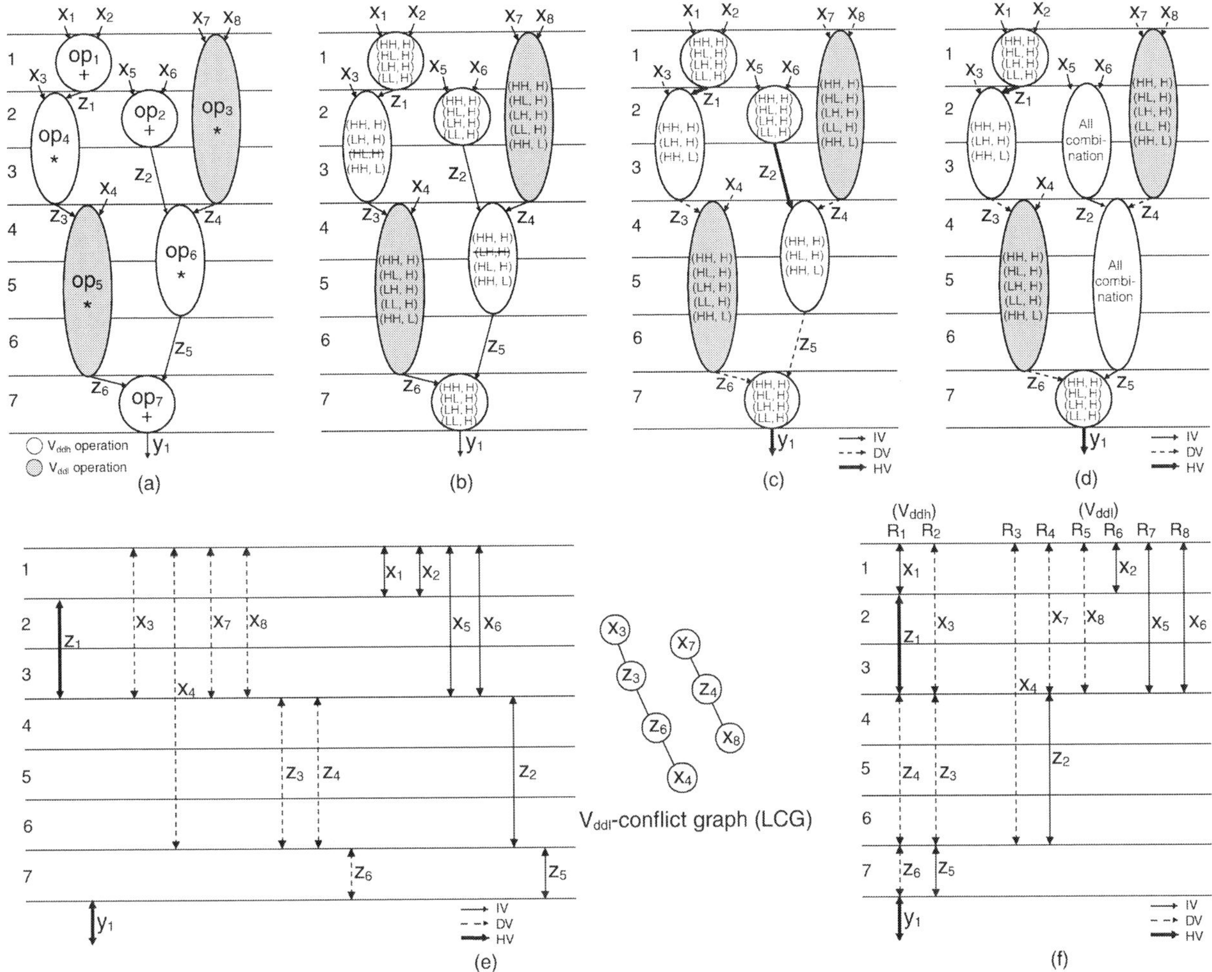

Figure 5: (a) An initial DFG, which is scheduled with dual-V_{dd}, (b) finding feasible V_{dd} combinations of operands, (c) categorizing variables as HV, DV, and IV, (d) re-categorizing variables after extending delay of some operations, (e) variable lifetimes and V_{ddl}-conflict graph of DVs, and (f) result of register allocation.

3.3 Register Allocation

Once we categorize variables as HV, DV, and IV, we derive their lifetimes as shown in Figure 5(e). We also derive a V_{ddl}-conflict graph (LCG), where each vertex corresponds to DV and an edge signifies conflict of allocating V_{ddl}, e.g. if x_3 is allocated to V_{ddl}-register, z_3 has to be allocated to V_{ddh}-register (see Figure 5(e) and (d)).

Since we want to minimize the number of V_{ddh}-registers, we first perform register allocation using HVs alone (by using left edge algorithm [19, 22] if lifetime conflict graph is interval graph [18] or heuristic vertex coloring algorithm [23] otherwise). In Figure 5(f), z_1 and y_1 are assigned to R_1. Since R_1 is already V_{ddh}-register, we want to allocate as many (DV and IV) variables as possible to it (in general, to all V_{ddh}-registers). We first consider DVs and their LCG; in Figure 5(e), z_3, z_4, and z_6 are candidates because their lifetimes do not overlap with those of z_1 and y_1, which are already in R_1. Since only one of z_3 and z_4 can be allocated to R_1, we select the one that removes the most number of edges in LCG when it is taken away, which is why z_4 (together with z_6) is allocated to R_1. Note that once z_4 and z_6, with their edges, are removed from LCG, x_4, x_7, and x_8 become IVs, thus are removed from LCG as well. We continue to pick candidates from IVs that can be allocated to R_1; we select x_1 in Figure 5(f).

If we still have some DVs left in LCG, we need more V_{ddh}-registers. We remove the node having the most number of edges in LCG, and allocate it to a new V_{ddh}-register; we then take DVs and IVs in turn until the register is filled in; we repeat the process until LCG becomes empty; in Figure 5(f), x_3, z_3, and z_5 are allocated to V_{ddh}-register R_2. Remaining variables are all IVs, which are submitted to standard register allocation minimizing the number of V_{ddl}-registers.

4. DUAL-V_{DD} ALLOCATION FOR MULTI-PLEXERS

Even after dual-V_{dd} register allocation, there may be some slacks left in the data-path, which can be used to assign V_{ddl} to multiplexers. This is formulated as a problem of finding maximum-weight independent set (MWIS) from a conflict graph $G_{mux} = (V, E)$; each vertex in V correspond to a multiplexer that can be assigned V_{ddl} (after checking timing slack similar to what we did in Section 3.1) and there is an edge in E between two multiplexers if assigning V_{ddl} to one multiplexer has a conflict with assigning V_{ddl} to the other. Each vertex has a weight, which reflects the amount of power saving from using V_{ddl}-multiplexer; weight varies from multiplexer to multiplexer depending on the number of inputs and possibility of

978-1-60558-497-3/09 $25.00 © 2009 ACM

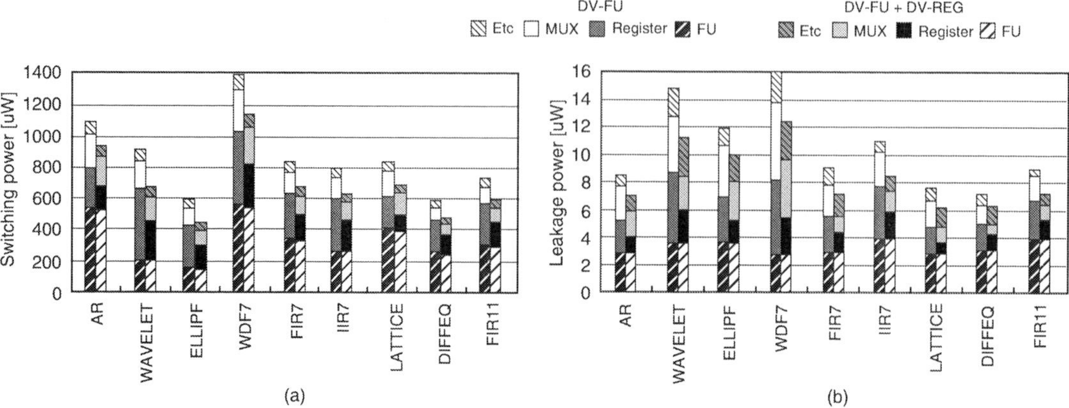

Figure 6: Comparison of (a) switching- and (b) leakage-power between DV-FU (left-hand bars) and DV-FU+DV-REG (right-hand bars).

Table 1: Area and delay of (32-bit) circuit elements used in the experiment. Register delay is denoted by T_{cq}/T_{su}.

Circuits		Area (um^2)	Delay (ps)
Adder/subtractor	V_{ddh}	387	6880
	V_{ddl}	387	27320
Multiplier	V_{ddh}	3593	6950
	V_{ddl}	3593	27500
Register	V_{ddh}	378	230/150
	V_{ddl}	378	700/630
10-to-1 MUX	V_{ddh}	933	500
	V_{ddl}	933	2130
Level converter		82	350

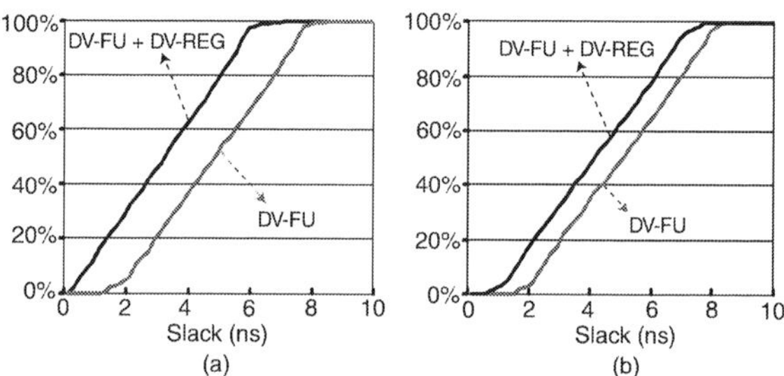

Figure 7: Cumulative slack histogram of (a) LATTICE and (b) IIR7 in DV-FU+DV-REG and DV-FU.

reducing the number of level converters when level converters are moved from inputs to output (details are omitted). MWIS problem is solved by employing a heuristic algorithm [24].

5. EXPERIMENTAL RESULTS

We carried out experiments on a set of behavioral benchmark designs to assess the effectiveness of dual-V_{dd} register allocation (DV-REG) in reducing switching as well as leakage power consumption. For dual-V_{dd} scheduling and allocation of functional units, the algorithm presented in [11] (referred to as DV-FU) was used; it was also used as a reference of comparison, i.e. we compared DV-FU and DV-FU+DV-REG. DV-REG was implemented in C under Centos 5.0 Linux machine. Each design was synthesized in commercial 1.08V, 65-nm bulk CMOS technology; V_{ddl} was set to 0.8V (74% of V_{ddh} of 1.08) based on the observation that V_{ddl} should be set to about 70% of V_{ddh} for maximum saving of power [25]. For physical design presented in Section 5.3, we forced at least 70% of the placement region to be occupied by cells during automatic placement; metal layers up to M5 were allowed for routing.

Table 1 summarizes circuit elements used in the experiment, each one with its area and delay; single-supply level converter [5] was designed and used. For all benchmark designs, we used 10 ns of clock period (see P_c in (2)); this is slightly larger than maximum delay along the data-path (multiplier, register, and two multiplexers; see Figure 2), which is $6950 + 210 + 700 + 2 \cdot 500 = 8860$ ps when all circuit elements are implemented in V_{ddh}.

5.1 Power Consumption

We compared results from DV-FU+DV-REG against those from DV-FU alone (thus, registers and multiplexers remain in V_{ddh}) un-

der the same latency and the same number of functional units used. The comparisons are illustrated in Figure 6; Figure 6(a) for switching power and Figure 6(b) for leakage power. Each bar consists of four components: power consumption from FUs, registers, multiplexers, and the remaining circuit elements such as level converters, and controller. Both switching and leakage power were obtained by simulating each circuit with fast transistor-level circuit simulator [14], which was repeated 100 times using different random input vectors and the average was taken as a result.

To summarize Figure 6, switching power is reduced by 20.4% on average; leakage power is reduced by 19.9%. If we consider registers and multiplexers alone, switching power is reduced by 43.6%, and leakage power is reduced by 37.3%.

5.2 Slack Histogram

In two implementations we compared in Figure 6, clock period (10 ns), as well as latency, is the same. This implies that DV-REG utilizes timing slacks that are left in the implementation from DV-FU. In Figure 7, we compare DV-FU+DV-REG and DV-FU of the benchmark designs LATTICE and IIR7 in terms of cumulative slack (e.g. about 50% of register-to-register timing paths in the data-path have slacks less than 5 ns in DV-FU implementation of LATTICE). From DV-FU to DV-FU+DV-REG, we see the clear shift of histogram to the left, which shows that we indeed use the slacks left from DV-FU implementation.

Note that even after DV-REG consume some slacks left by DV-FU, there are still plenty of slacks left in the data-path. This can be understood as follows. During scheduling, operation delay (see (2)) is computed from maximum delay through combinational path, D_{dp}, which is based on the maximum delay values in Table 1; maximum delay of adder/subtractor (in V_{ddh}) is 6880 ps while minimum

Table 2: Comparison of area, total wirelength, and average congestion of DV-FU and DV-FU+DV-REG

Circuits	% of V_{ddl}-rows		Area			Wirelength			Average congestion		
	DV-FU	DV-FU+DV-REG	DV-FU (um^2)	DV-FU+DV-REG (um^2)	Inc. (%)	DV-FU (mm)	DV-FU+DV-REG (mm)	Inc. (%)	DV-FU (%)	DV-FU+DV-REG (%)	Inc.
AR	18	57	20834	21325	2.4	224	270	20.6	44	51	7
WAVELET	11	34	33526	34837	3.9	340	454	33.5	42	53	11
ELLIPF	15	51	25191	26338	4.6	284	318	12.2	46	50	4
WDF7	10	51	35400	37448	5.8	477	555	16.4	55	61	6
FIR7	18	59	20588	21244	3.2	235	280	19.2	46	53	7
IIR7	27	68	27560	28543	3.6	246	348	41.1	38	53	15
LATTICE	20	59	17977	18930	5.3	195	213	12.0	44	46	2
DIFFEQ	22	49	16754	17410	3.9	168	223	33.1	41	51	10
Average	18	54			4.1			23.5			8

is 50 ps (adder sum bit of LSB). This also applies to register allocation (see (3)) when we use $D_{dp,l}$. In other words, bit-level of granularity is exposed in the slack histograms shown in Figure 7, while scheduling and register allocation do not perform in that granularity (e.g. scheduling considers a multiplier as a single entity). Therefore, the netlist we obtain after performing DV-FU and DV-REG may be submitted to gate-level dual-V_{dd} allocation for further reducing power, which is left for future work.

5.3 Physical Design

Physical design aspects of DV-FU+DV-REG are compared to DV-FU in Table 2. The area, which is the sum of the areas of all the cells in the design, increases by 4.1% on average; more use of level converters due to DV-REG is a main reason for the increase. The total wirelength, reported after running detailed routing [21], increases by 23.5% on average. This relatively large increase of wires can be understood from the percentage of V_{ddl}-rows; it occupies 18% on average in DV-FU (second column in Table 2) and 54% on average in DV-FU+DV-REG (third column). As V_{ddh}- and V_{ddl}-rows are more evenly used, we can expect more connections between cells placed in each type of rows; placement is more constrained, which makes shorter connections more difficult to be achieved. Due to increased wirelength, average congestion also increases by 8% on average, as reported in the last three columns of Table 2.

6. CONCLUSION

We have presented dual-V_{dd} register allocation, focusing on the problem of minimizing power consumption of storage. Dual-V_{dd} HLS framework includes the complete design flow from dual-V_{dd} scheduling to circuit layout, which was tested using commercial 65-nm technology. Main feature of the framework is a heuristic solution to the problem of finding a register allocation, which minimizes the number of high-V_{dd} registers as well as total number of registers. Dual-V_{dd} allocation for multiplexers was formulated as MWIS problem, which was then solved by using a heuristic algorithm. Experiments on benchmark designs showed that DV-REG can reduce switching- and leakage-power by 20% on average, respectively.

References

[1] R. K. Krishnamurthy et al., "High-performance and low-power challenges for sub-70nm microprocesssor circuits," in *Proc. CICC*, May 2002, pp. 125–128.

[2] C. Yeh et al., "Layout techniques supporting the use of dual supply voltages for cell-based designs," in *Proc. DAC*, June 1999, pp. 62–67.

[3] J. Hu et al., "Architecting voltage islands in core-based system-on-a-chip designs," in *Proc. ILSPED*, Aug. 2004, pp. 180–185.

[4] K. Usami et al., "Automated low-power technique exploiting multiple supply voltages applied to a media processor," *JSSC*, vol. 33, no. 3, pp. 463–472, Mar. 1998.

[5] R. Puri et al., "Pushing ASIC performance in a power envelope," in *Proc. DAC*, June 2003, pp. 788–793.

[6] S. Kulkarni, A. Srivastava, and D. Sylvester, "A new algorithm for improved VDD assignment in low power dual VDD systems," in *Proc. ILSPED*, Aug. 2004, pp. 200–205.

[7] S. Raje and M. Sarrafzadeh, "Variable voltage scheduling," in *Proc. Int. Symp. on Low Power Design*, Apr. 1995, pp. 9–14.

[8] Y. R. Lin et al., "Scheduling techniques for variable voltage low power designs," *TODAES*, vol. 2, no. 2, pp. 81–97, Apr. 1997.

[9] M. C. Johnson and K. Roy, "Datapath scheduling with multiple supply voltages and level converters," *TODAES*, vol. 2, no. 3, pp. 227–248, July 1997.

[10] J. Chang and M. Pedram, "Energy minimization using multiple supply voltages," *TVLSI*, vol. 5, no. 4, pp. 436–443, Dec. 1997.

[11] W. Shiue and C. Chakrabarti, "Low-power scheduling with resources operating at multiple voltages," *TCAS-II*, vol. 47, no. 6, pp. 536–543, June 2000.

[12] A. Manzak and C. Chakrabarti, "A low power scheduling scheme with resources operating at multiple voltages," *TVLSI*, vol. 10, no. 1, pp. 6–14, Feb. 2002.

[13] K. Nose and T. Sakurai, "Optimization of VDD and VTH for low-power and high-speed applications," in *Proc. ASP-DAC*, Jan. 2000, pp. 469–474.

[14] Synopsys, "NanoSim User Guide," Dec. 2007.

[15] G. Gupta et al., "Rapid estimation of control delay from high-level specifications," in *Proc. DAC*, July 2006, pp. 455–458.

[16] T. Ahn et al., "Incremental analysis and elaboration of VHDL description," in *Proc. APCHDL*, Jan. 1996, pp. 128–131.

[17] J. Jeon et al., "High-level synthesis under multi-cycle interconnect delay," in *Proc. ASP-DAC*, Jan. 2001, pp. 662–667.

[18] G. De Micheli, *Synthesis and Optimization of Digital Circuits*, McGraw-Hill, Inc., 1994.

[19] A. Hashimoto and J. Stevens, "Wire routing by optimizing channel assignment within large apertures," in *Proc. Design Automation Workshop*, June 1971, pp. 155–169.

[20] Synopsys, "Design Compiler User Guide," Mar. 2007.

[21] Synopsys, "Astro User Guide," June 2006.

[22] F. J. Kurdahi and A. C. Parker, "REAL: a program for register allocation," in *Proc. DAC*, June 1987, pp. 210–215.

[23] D. Brelaz, "New methods to color the vertices of a graph," *Comm. of the ACM*, vol. 22, no. 4, pp. 251–256, Apr. 1979.

[24] D. Kagaris and S. Tragoudas, "Maximum independent sets on transitive graphs and their applications in testing and CAD," in *Proc. ICCAD*, Nov. 1997, pp. 736–740.

[25] T. Kuroda and M. Hamada, "Low-power CMOS digital design with dual embedded adaptive power supplies," *JSSC*, vol. 35, no. 4, pp. 652–655, Apr. 2000.

978-1-60558-497-3/09 $25.00 © 2009 ACM

GPU-Based Parallelization for Fast Circuit Optimization *

Yifang Liu
Department of ECE
Texas A&M University
College Station, TX
yliu@ece.tamu.edu

Jiang Hu
Department of ECE
Texas A&M University
College Station, TX
jianghu@ece.tamu.edu

ABSTRACT

The progress of GPU (Graphics Processing Unit) technology opens a new avenue for boosting computing power. This work is an attempt to exploit GPU for accelerating VLSI circuit optimization. We propose GPU-based parallel computing techniques and apply them on simultaneous gate sizing and threshold voltage assignment, which is often employed in practice for performance and power optimization. These techniques are aimed to fully utilize the benefits of GPU through efficient task scheduling and memory organization. Compared to conventional sequential computation, our techniques can provide up to $56\times$ speedup without any sacrifice on solution quality.

Categories and Subject Descriptors

B.7 [**Integrated Circuits**]: Design Aids

General Terms

Algorithms, Design

Keywords

parallel computing, GPU, VLSI circuit optimization

1. INTRODUCTION

Fast circuit optimization technique is an increasingly compelling need for chip designs. While the pressure of time-to-market is almost never relieved, design complexity keeps growing along with transistor count. In addition, more and more issues need to be considered – from conventional objectives like performance and power to new concerns like process variability and transistor aging. On the other hand, the advancement of chip technology opens new avenues for boosting computing power. One example is the amazing progress of GPU (Graphics Processing Unit) technology. In the past 5 years, the computing performance of GPU has grown from about the same as CPU to about ten times of CPU in term of GFLOPS [1]. GPU is particularly good at data-intensive parallel computations. Recently, GPU-based parallel computation has been successfully applied for the speedup of fault simulation [2] and power grid simulation [3].

---

*This work is partially supported by Intel.

Permission to make digital or hard copies of part or all of this work for personal or classroom use is granted without fee provided that copies are not made or distributed for profit or commercial advantage and that copies bear this notice and the full citation on the first page. To copy otherwise, to republish, to post on servers or to redistribute to lists, requires prior specific permission and/or a fee.
DAC'09, July 26-31, 2009, San Francisco, California, USA

In this work, we propose GPU-based parallel techniques for simultaneous gate sizing and threshold voltage assignment. We will focus on discrete algorithm, which faces two inter-dependent difficulties. First, the underlying topology of a combinational circuit is typically a DAG (Directed Acyclic Graph). The path reconvergence of DAG makes it difficult to carry out systematic solution search like dynamic programming (DP). Second, the size of a combinational circuit can be very large, sometimes with dozens of thousands of gates. As a result, most of existing methods are simple heuristics [4–6]. Recently, a Joint Relaxation and Restriction (**JRR**) algorithm [8] is proposed to handle the path reconvergence problem and enable a DP-like systematic solution search. Indeed, the systematic search [8] remarkably outperforms its previous work. To address the large problem size, a grid-based parallel gate sizing method is introduced in [7]. Although it can obtain high solution quality with very fast speed, it concurrently uses 20 computers and entails significant network bandwidth. In contrast, GPU-based parallelism is much more cost-effective. The expense of a GPU card is only a few hundreds of dollars and the local parallelism obviously causes no overhead on network traffic.

It is not straightforward to map a conventional sequential algorithm onto GPU computation and achieve desired speedup. In general, parallel computation implies that a large computation task needs to be partitioned to multiple threads. For the partitioning, one needs to decide its granularity levels, balance the computation load and minimize the interactions among different threads. Data sharing and memory management are also very important. One needs to properly allocate the storage to facilitate date sharing. Apart from general parallel computing issues, the characteristics of GPU should be taken into account. For example, the parallelization should be SIMD (Single Instruction Multiple Data) in order to better exploit the advantages of GPU. In this work, we propose task scheduling and memory management techniques for performing gate sizing/V_t assignment on GPU. To the best of our knowledge, this is the first work on GPU-based combinational circuit optimization. In the experiment, we compared our parallel version of the joint relaxation and restriction algorithm [8] and its original sequential implementation. The results show that our parallelization achieves speedup of up to $56\times$ and $39\times$ on average. At the same time, our techniques can retain the exactly same solution quality as [8]. Such speedup will allow many systematic optimization approaches, which were slow and previously regarded as impractical, to be widely adopted in realistic applications.

2. ALGORITHM OF SIMULTANEOUS GATE SIZING AND V_T ASSIGNMENT

We briefly review the simultaneous gate sizing and V_t assignment algorithm proposed in [8], since the parallel techniques proposed here are built upon this algorithm. This algorithm has two phases: relaxation phase and restriction phase. It is called Joint Relaxation and Restriction (**JRR**). Each phase consists of two or multiple circuit traversals. Each traversal is a solution search in the same spirit as dynamic programming. For the ease of description, we call a combination of certain size and V_t level as an implementation of a gate (or a node in the circuit graph).

The relaxation phase includes two circuit traversals: history consistency relaxation and history consistency restoration. The history consistency relaxation is a topological order traversal of the given circuit. In the traversal, a set of partial solutions are propagated, which is similar to dynamic programming based buffer insertion algorithm [9]. However, the topology here is a DAG as opposed to a tree in buffer insertion. Due to reconvergent paths in DAGs, history consistency constraint applies. When DP-like algorithm is performed directly on a DAG, this constraint requires to keep or trace all history information and consequently causes substantial computation or memory overhead. To overcome this difficulty, the work of [8] suggests to relax this constraint in the initial traversal, i.e., solutions are allowed to be merged even if their ancestor solution are inconsistent. The resulting solutions, which may be not legitimate, provides a useful global view (and a lower bound to the delay at each node) for subsequent solution search.

In the second traversal of the relaxation phase, any history inconsistency resulted from the first iteration is solved in a reverse topological order traversal of the circuit. When a node is visited in the traversal, only one implementation is selected for the node. Hence, no history inconsistency should exist after the traversal is completed. At each node, the gate implementation is selected to minimize the cost function according to the problem formulation.

The solution at the end of the relaxation phase can be further improved. After relaxation phase, the constraints are enforced in restriction phase thereafter. Solution propagation in topological order and solution selection in reverse topological order are conducted one after the other iteratively, while history consistency is maintained. The iterative solution search in restriction phase is shown to monotonically improve the solution.

3. GPU-BASED PARALLELIZATION

3.1 Background on GPU

A GPU is usually composed of an array of multiprocessors and each multiprocessor consists of multiple processing units (or ALUs). Each ALU is associated with a set of local registers and all the ALUs of a multiprocessor share a control unit and some shared memory. A typical GPU may have over one hundred ALUs. GPU is designed for SIMD (Single Instruction Multiple Data) parallelism. As such, an ideal usage of GPU is to execute identical instructions on a large volume of data.

The software program applied on GPU is called kernel function, which is executed on multiple ALUs in the basic unit of thread. The threads are organized in a two-level hierarchy. Multiple threads form a warp and a set of warps constitute a block. For more details on the organization and relation between warp, thread block, and multiprocessors, readers are referred to [1]. The global memory for a GPU is usually a DRAM off the GPU chip but on the same board as GPU. The latency of data access on global memory and kernel function loading can be very high. In order to improve the efficiency of GPU usage, one needs to load data infrequently and make the ALUs dominate the overall program runtime.

3.2 Two-Level Task Scheduling

Exploring GPU-based parallelism for gate sizing and V_t assignment is motivated by the observation that the algorithm repeats a few identical computations on many different gates. When evaluating the effect of an implementation (a specific gate and V_t level) for a gate, we compute its corresponding delay, AT/RAT, and power. These a few computations are repeated for many gates, sometimes hundreds of thousands of gates, and for multiple iterations [8]. Evidently, such repetitive computations on a large number of objects fit very well with SIMD parallelism.

We propose a two-level task scheduling method which allocates the computations to multiple threads. At the first level, the computations for multiple gates are allocated to a set of thread blocks. The algorithm, software and hardware unit of this level are gate, thread block and multiprocessor, respectively. Since different implementations of a gate share the same context (fanin and fanout characteristics), such allocation matches the shared context information with the shared memory for each thread block. At the second level, the evaluations of multiple gate implementations are assigned to a set of threads. The algorithm, software and hardware unit of this level are gate implementation, thread and ALU, respectively. For each gate, the evaluations for all of its implementation options are independent of each other. Hence, it is convenient to parallelize them on multiple threads. The two-level task scheduling will be described in details in the subsequent sections.

3.3 Gate-Level Task Scheduling

We describe the gate-level task scheduling in the context of topological order traversal of the circuit. The techniques for reverse topological order traversal are almost the same. In a topological order traversal, a gate is a *processed gate* if the computation of delay/power for all of its implementations is completed. A gate is called a *current gate* if all of its fanin gates are processed gates. We term a gate as a *prospective gate* when all of its fanin gates are either current gates or processed gates. In GPU-based parallel computing, multiple current gates can be processed at the same time because there is no inter-dependency among their computations. A set of current gates are called *independent current gates* (ICG), since there is no computational interdependency among them. Due to the restriction of GPU computing bandwidth, the number of current gates which can be processed at the same time is limited. A critical problem here is how to select a subset of independent current gates for parallel processing. This subset is designated as *concurrent gate group*, which has a maximum allowed size.

The way of forming a concurrent gate group may greatly affect the efficiency of utilizing the GPU-based parallelism. This can be illustrated by the example in Figure 1. The processed gates, independent current gates and prospective gates are represented by dashed, grey and white rectangles, respectively. If the maximum concurrent gate group size is 4, there could be at least two different ways of forming a concurrent gate group for the scenario shown in Figure 1(a). As in Figure 1(b), $\{G1, G2, G3, G4\}$ can be selected to be the concurrent gate group. After they are processed, any four gates among $\{G5, G6, G7, G8, G9\}$ may become the next concurrent gate group. Alternatively, we can choose $\{G2, G3, G4, G5\}$ as in Figure 1(c). However, after $\{G2, G3, G4, G5\}$ are processed, we can include at most three gates $\{G1, G9, G10\}$ in the next concurrent gate group. The selection of concurrent gates in Figure 1(b) is preferred over that in Figure 1(c), since it better utilizes the bandwidth of concurrent group.

The problem of finding concurrent gate group among a set of independent current gates can be formulated as a max-throughput problem, which maximizes the minimum size of all concurrent gate groups. The max-throughput problem is very difficult to solve. Therefore, we will focus on a reduced problem: max-succeeding-group. Given a set of independent current gates, the max-succeeding-group problem asks to choose a subset of them as

978-1-60558-497-3/09 $25.00 © 2009 ACM

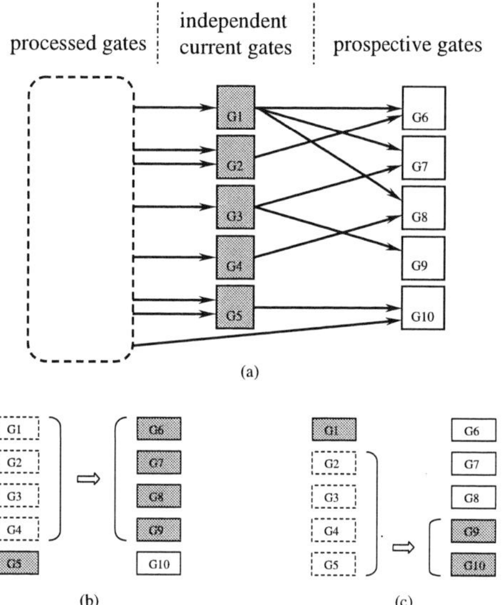

Figure 1: Processed gates: dashed rectangles; independent current gates: grey rectangles; prospective gates: solid white rectangles. Given the scenario in (a), one can choose at most 4 independent current gates for parallel processing. In (b): G1, G2, G3 and G4 are selected, we may choose another 4 gates in G5, G6, G7, G8 and G9 for the next parallel processing. In (c): G2, G3, G4 and G5 are selected, only 3 independent current gates G1, G9 and G10 are available for the next parallel processing.

the concurrent gate group such that the size of the succeeding independent gate group is maximized. Unfortunately, even the reduced problem - max-succeeding-group - is NP-complete.

Input : $current\_grp, prospective\_grp$
Output: $concurrent\_grp$

1   $concurrent\_grp \leftarrow \emptyset$;
2 **repeat**
3     $prop\_gate \leftarrow gate(min\_ICGfanin)$;
4     $prospective\_grp \leftarrow prospective\_grp - prop\_gate$;
5     $concurrent\_grp \leftarrow$
     $concurrent\_grp \bigcup (inputs(prop\_gate) \bigcap current\_grp)$;
6     $current\_grp \leftarrow$
     $current\_grp - inputs(prop\_gate) + prop\_gate$;
7   update_ICGfanin $(outputs(inputs(prop\_gate)))$;
8 **until** $|concurrent\_grp| \geq concurrent\_budget$;

Algorithm 1: Concurrent Gate Group Selection

Since the max-succeeding-group problem is NP-complete, we propose a linear-time heuristic to solve it. This heuristic iteratively examines the prospective gates and puts a few independent current gates into the concurrent gate group. For each prospective gate, we check its fanin gates that are independent current gates. The number of such fanin gates is called ICG (independent current gate) fanin size. Each time, the prospective gate with the minimum ICG fanin size is selected. Then all of its ICG fanin gates are put into the concurrent gate group. After this, the selected prospective gate will no longer be considered in subsequent iterations. At the same time, the selected ICG fanin gates are not counted in the ICG fanin size of the remaining prospective gates.

In the example of Figure 1, the prospective gate with the minimum ICG fanin size is $G9$. When it is selected, gate $G3$ is put into the concurrent gate group. Then, the ICG fanin size of $G7$ becomes 1, which is the minimum. This requires that gate $G1$ is put into the

concurrent gate group. Next, any two of $G2$, $G4$ and $G5$ can be selected to form the concurrent gate group of size 4.

The rationale behind this heuristic is as follows. The maximum allowed size of concurrent gate group can be treated as a budget. The goal is to maximize the number of succeeding ICGs. If a prospective gate has a small ICG fanin size, selecting its ICG fanin gates can increase the number of succeeding ICGs with the minimum usage of concurrent gate group budget.

This heuristic is performed on CPU once. The result, which is the gate-level scheduling, is saved since the same schedule is employed repeatedly in the traversals of the JRR algorithm (see Section 2). The pseudo code for the concurrent gate selection heuristic is given in Algorithm 1. The minimum ICG fanin size is updated each time an ICG fanin size is updated, so the computation time is dominated by fanin size updating. If the maximum fanin size among all gates is F_i, each gate can be updated on its ICG fanin size for at most F_i times. Thus, the time complexity of this heuristic is $O(|V|F_i)$, where V denotes the set of nodes in the circuit.

3.4 Parallelization for Gate Implementation Evaluation

It is not difficult to see that the evaluations for different implementations of the same gate are independent of each other. Therefore, we can allocate these evaluations into multiple threads without worrying interdependency. These evaluations include a few common computations: the timing and power estimation for each implementation. Different implementations of the same gate also share some common data such as the parasitics of the fanin and fanout. According to this observation, the evaluations of implementations for the same gate are assigned to the same thread block and the same multiprocessor. Since all the ALUs of a multiprocessor have access to a fast on-chip shared memory, the shared data can be stored in the shared memory to reduce communication cost.

In order to facilitate simultaneous memory access, the shared memory is usually divided into equal-sized banks. In order to reduce the chance of bank conflict and avoid the cost of serialized access, we store all information of a gate in the same bank and separate it from that of other gates in other banks.

GPU global memory often has memory coalescing mechanism for improved access efficiency. we save gate information of sibling nodes in the circuit adjacent to each other in the global memory whenever possible, so that global memory requests from different threads in a warp have a greater chance to be coalesced and the access latency is thereby reduced.

Loading the kernel function to GPU can be very time consuming. Since the computation operations for all gates are the same, we load the computation instructions only once and apply them to all of the gates in the circuit. The gate-level task schedule is computed before the circuit optimization and saved in the GPU global memory. Once the kernel function is called, the optimization follows the schedule saved on GPU and no kernel reloading is needed.

To reduce the idle time of the multiprocessors during memory access, we assign multiple thread blocks to a multiprocessor for concurrent execution. This arrangement has positive impact on the performance, because memory access takes a large portion of the total execution time of the kernel function.

4. EXPERIMENTAL RESULTS

In our experiment, the testcases are the ISCAS85 combinational circuits. The cell library is based on $70nm$ technology. Each gate has 6 different sizes and 4 V_t levels so that it has 24 options of implementation.

In order to test the runtime speedup of our parallel techniques,

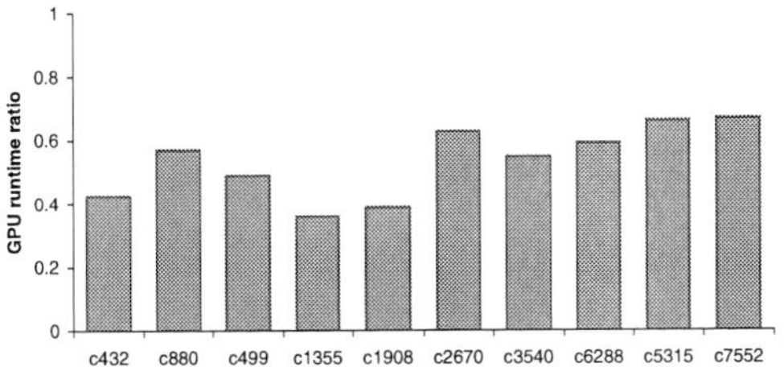

Figure 2: The ratio of GPU runtime over overall runtime.

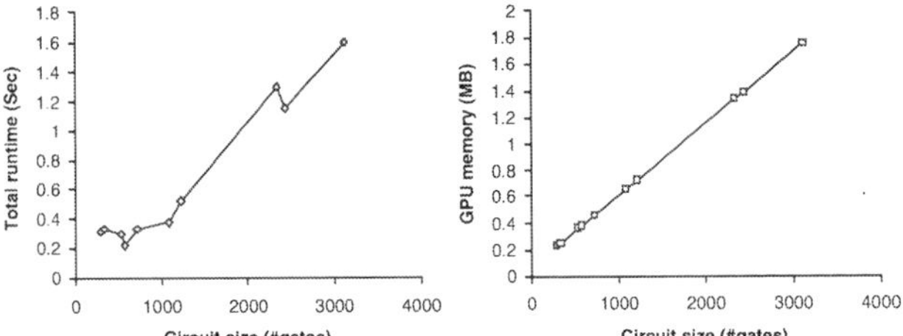

Figure 3: Runtime and GPU memory scalability.

Circuit	SA [10]		Sequential JRR [8]		Parallel JRR	
	power	runtime	power	runtime	runtime	speedup
c432	703	1.7	701	3.25	0.317	10×
c499	1669	4.9	1590	6.27	0.295	21×
c880	1817	5.1	1050	3.61	0.328	11×
c1355	1385	3.3	1076	7.36	0.218	34×
c1908	2502	10.7	2296	9.20	0.327	28×
c2670	3412	18.6	2509	15.70	0.376	42×
c3540	4645	22.3	3830	21.30	0.515	41×
c5315	8406	26.8	5023	64.88	1.156	56×
c6288	13685	19.2	12356	53.47	1.295	41×
c7552	9510	46.1	5949	67.44	1.595	42×
Average	4773	15.87	3638	25.25	0.64	39×
Norm.	1.0		0.76			

Table 1: Comparison on power (μW) and runtime (seconds). All solutions satisfy timing constraints.

we compare our parallel version of the JRR algorithm [8] with its original sequential implementation. Regarding solution quality, we compare the results with another previous work [10] while ensuring that the parallel JRR solutions are identical with those of sequential JRR. The method of [10] is based on slack allocation (SA). The problem formulation for all these methods is to minimize total power dissipation subject to timing constraints. The power dissipation includes both dynamic and leakage power. The timing accounts for both gate and wire delay. Since we do not have lookup table based timing and power information for the cell library, we use analytical model for power [8] and the Elmore model for delay computation. However, the JRR algorithm and our parallel techniques can be easily applied with lookup table based models.

The SA method and sequential JRR algorithm are implemented in C++. The parallel JRR implementation includes two parts: one part is in C++ and runs on the host CPU; the other part runs on the GPU through CUDA. CUDA (Compute Unified Device Architecture) is a parallel programming model and interface developed by NVIDIA [11]. The major components of the parallel programming model and software environment include thread groups, shared memory and thread synchronization. The experiment was performed on a Windows XP based machine with an Intel core 2 duo CPU of 2.66GHz and 2GB memory. The GPU is NVIDIA GeForce 9800GT, which has 14 multiprocessors, which of each has 8 ALUs. The GPU card has 512MB off-chip memory. We set the maximum number of gates being parallel processed to 4. There-

fore, at most 96 gate implementations are evaluated at one time.

The main results are listed in Table 1. Since all of these methods can satisfy the timing constraints, timing results are not included in the table. The solution quality can be evaluated by the results of power dissipation. One can see that JRR can reduce power by about 24% on average when compared to SA [10]. The parallel JRR achieves exactly the same power as the sequential JRR. Our parallel techniques provide runtime speedup from 10× to 56×. One can see that the speedup tends to be more significant when the circuit size grows. One of the reasons is that both small and large circuits have similar overhead for setup.

In Figure 2, we depict the ratio between the GPU runtime and the total runtime. The ratio is mostly between 0.4 and 0.6 among all circuits. Usually, larger circuits have higher GPU runtime percentage. This may be due to the higher parallel efficiency of larger circuits.

In Figure 3, the total runtime and GPU memory usage versus circuit size are plotted to show the runtime and memory scalability of our techniques. The main trend of the runtime curve indicates a linear dependence on circuit size. There are a few non-monotone parts in the curve which can be explained by the fact that the runtime depends on not only the circuit size but also circuit topology. The memory curve exhibits a strong linear relationship with circuit size. At least for ISCAS85 benchmark circuits, we can conclude that our techniques scale well on both runtime and memory.

5. CONCLUSIONS AND FUTURE WORK

It has long been a challenge to optimize combinational circuit in a systematic yet fast manner due to its topological reconvergence and large size. A recent progress [8] suggests an effective solution to the reconvergence problem. This work addresses the large problem size by exploiting GPU-based parallelism. The proposed parallel techniques are integrated with the state-of-the-art gate sizing and V_t assignment algorithm [8]. These techniques and the integration effectively solves the challenge of combinational circuit optimization. A circuit with thousands of gates can be optimized with high quality in less than 2 seconds. The parallel techniques provide up to 56× runtime speedup. They also show an appealing trend that the speedup is more significant on large circuits. In future, we will test these techniques on larger circuits, for example, circuit with hundreds of thousands of gates.

6. ACKNOWLEDGEMENTS

The authors appreciate Zhuo Feng's help on GPU programming.

7. REFERENCES

[1] J. Owens, D. Luebke, N. Govindaraju, M. Harris, J. Krüger, A. Lefohn, T. Purcell. A Survey of General-Purpose Computation on Graphics Hardware. In *Proceedings of Eurographics*, 2005.

[2] K. Gulati and S. Khatri. Towards Acceleration of Fault Simulation using Graphics Processing Units. In *Proceedings of the ACM/IEEE DAC*, 2008.

[3] Z. Feng and P. Li. Multigrid on GPU: Tackling Power Grid Analysis on Parallel SIMT Platforms. In *Proceedings of the ACM/IEEE ICCAD*, 2008.

[4] O. Coudert. Gate sizing for constrained delay/power/area optimization. In *IEEE Trans. VLSI*, 1997.

[5] L. Wei, Z. Chen, K. Roy and V. De. Design and Optimization of Dual Threshold Circuits for Low Voltage Low Power Application. In *IEEE Trans. VLSI*, 1999.

[6] S. Sirichotiyakul, T. Edwards, C. Oh, J. Zuo, A. Dharchoudhury, R. Panda and D. Blaauw. Stand-by power minimization through simultaneous threshold voltage selection and circuit sizing. In *Proceedings of the ACM/IEEE DAC*, 1999.

[7] T. Wu and A. Davoodi. PaRS: Fast and Near-Optimal Grid-Based Cell Sizing for Library-Based Design. In *Proceedings of the ACM/IEEE ICCAD*, 2008.

[8] Y. Liu and J. Hu. A New Algorithm for Simultaneous Gate Sizing and Threshold Voltage Assignment. In *Proceedings of the ACM ISPD*, 2009.

[9] L.P.P.P. van Ginneken. Buffer Placement in Distributed RC-Tree Networks for minimal Elmore Delay. In *IEEE ISCS*, 1990.

[10] D. Nguyen, A. Davare, M. Orshansky, D. Chinnery, B. Thompson and K. Keutzer. Minimizion of Dynamic and Static Power Through Joint Assignement of Threshold Voltages and Sizing Optimization. In *ISLPED*, 2003.

[11] NVIDIA CUDA homepage. *http://www.nvidia.com/object/cude_home.html*.

Architectural Assessment of Design Techniques to Improve Speed and Robustness in Embedded Microprocessors

Thomas Baumann[1,2], Doris Schmitt-Landsiedel[2], Christian Pacha[1]
[1] Infineon Technologies, Munich, Germany [2] Technical University of Munich, Munich, Germany
Thomas.Baumann1@infineon.com

ABSTRACT

This work investigates the interrelation of performance and robustness against variability in industrial microprocessor designs. A novel analysis technique for variation-sensitive hardware and two figures of merit to quantify the robustness of a design against variations are proposed. Together with a multi-stage STA this enables an efficient application of low-V_T cell insertion and pulsed latch design to compensate for within-die delay variations. For the same speed margin of 5% on design level, a pulsed latch design of an ARM926 microprocessor shows a 2.5x higher robustness compared to a MS-FF design with selective low-V_T cell insertion.

Categories and Subject Descriptors

B.7.1 [**Integrated Circuits**]: Types and Design Styles; B.8.2 [**Performance and Reliability**]: Performance Analysis and Design Aids

General Terms

Performance, Design, Reliability.

Keywords

Variability-aware design, robustness, micro-architecture.

1. INTRODUCTION

The increasing computational requirements in mobile communication drive the introduction of deeper pipelining and superscalar architectures in embedded microprocessors. However, their implementation efficiency in scaled CMOS technologies is strongly depending on the required timing margins caused by process, environmental voltage and temperature variations (PVT). Even though recent indications show that process variations are manageable down to 45nm [1], the delay sensitivity of CMOS logic significantly increases mainly due to V_{DD} scaling down to 0.8-1V under constants V_T constraints. While global Die-to-Die (D2D) process variations can be compensated by speed binning, V_{DD} trimming and adaptive voltage scaling [2], within-die (WID) variations are typically handled by timing margins [3] derived by classical OCV or statistical design methodologies [4]. In this work we introduce a complementary approach which explicitly considers micro-architectural aspects such as path topologies and critical hardware within the variation sensitive timing regions. Due to the interdependency of micro-architecture, circuit and technology level, this approach is a prerequisite for the implementation of compensation techniques for WID delay variations.

First step is to analyze the impact of micro-architectural trends on performance and variability for the ARM processor family in section 2 using an in-house microprocessor model. In section 3 we introduce

Permission to make digital or hard copies of part or all of this work for personal or classroom use is granted without fee provided that copies are not made or distributed for profit or commercial advantage and that copies bear this notice and the full citation on the first page. To copy otherwise, to republish, to post on servers or to redistribute to lists, requires prior specific permission and/or a fee.
DAC'09, July 26-31, 2009, San Francisco, California, USA

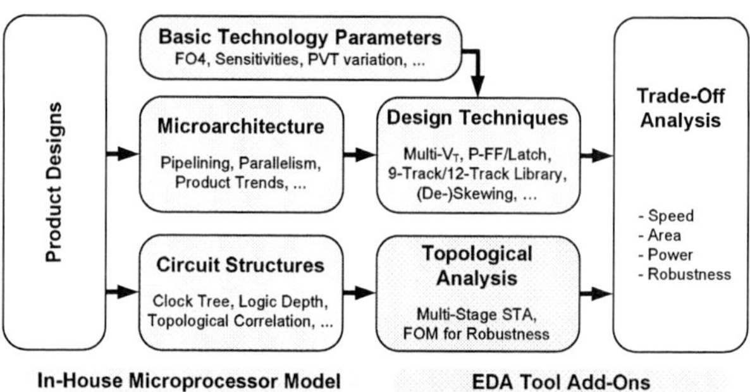

Figure 1. Unified approach for design space exploration.

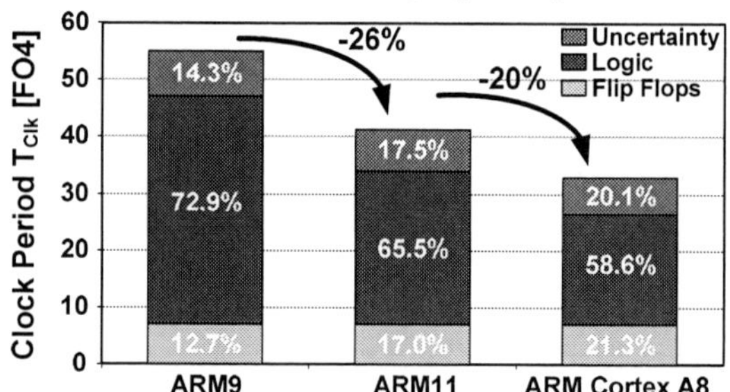

Figure 2. Delay contributions in different µ-processor cores.

two figures of merit (FOM) to assess topological correlation of speed critical paths and the robustness of circuit designs. A multi-stage STA analysis is proposed in section 4 to facilitate the application of time borrowing techniques in semicustom-design. Section 5 contains an assessment of selective low-V_T cell insertion and pulsed FF/latches using the derived FOMs. Trade-offs between speed, area, and robustness are determined to quantify costs and benefits from an industrial point of view.

2. MICROPROCESSOR SCALING TREND

For design space exploration in new CMOS technologies we use an in-house microprocessor model (Fig. 1). The model includes key timing contributions, such as flip flop timing overhead, combinational logic delay, and WID variation induced timing uncertainties. This way, architectural parameters are combined with basic technology parameters and speed critical circuit topologies, which are derived from productive designs using STA/OCV-based timing sign-off. Main timing uncertainties are clock skew and jitter, WID, and local variations of supply voltage and temperature. The derived timing contributions are quantified in terms of the technology-independent fan-out-4 inverter timing metric [5].

The application of the model to the ARM microprocessor family (Fig. 2) shows the well-known increase of flip flop overhead as well as an increasing timing uncertainty in regard to the clock period. Obviously, this limits the efficiency of deeper pipelining for embedded microprocessors from a theoretical value of 38% to 26% for the migration from the 5-stage ARM9 to 8-stage ARM11 core. The migration from ARM11 to 13-stage ARM Cortex A8 core is even less efficient since

Table 1. Topological correlation factors of an ARM926 core.

Delay Region	Upper 3%	Upper 5%	Upper 7%
Avg. Correl. κ_{Top}	88%	96%	98%
#Independent paths	8	31	80
#Total paths	182	4612	47906

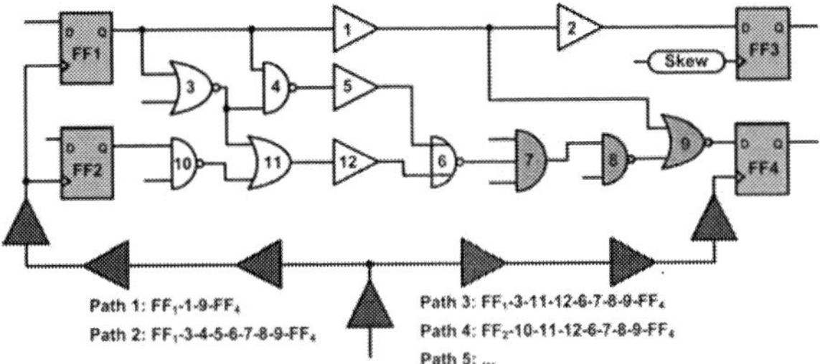

Figure 3. Generic pipeline stage with different paths.

the gain is reduced from 38% to 20%. In high-speed microprocessors clock skew reduction techniques and high-speed flip flops are used to preserve the efficiency of pipelining. Since physical synthesis tools are emerging to enable e.g. the use of pulsed latches in semi-custom design, we quantify their cost and benefit trade-offs for the ARM microprocessor family.

3. CRITICAL HARDWARE ANALYSIS

To assess the impact of variations on microprocessor core level the precise knowledge of circuit structures, such as number and type of cells in critical path regions is mandatory. Starting points of this analysis are conventional STA timing reports of an ARM926 microprocessor product design (140k cells) in 90nm single reg-V_T CMOS technology [6] generated by the STA tool PrimeTime SI including in-house OCV methodology. Here all cells are assigned to the most critical path comprising this cell. Since the sum of systematic and random WID variations is on the order of 5-8% [1] we focus our investigations on all critical paths arriving at the receiving flip flop within the upper 10% of the clock period. Unlike a simple delay check where the timing analysis of the most critical paths ending at a certain flip flop is sufficient, the variation-aware topology analysis demands the total investigation of all critical paths ending at each FF cell. For the ARM design the average number of paths per receiving flip flops is about 500 resulting in timing reports of about 16GB.

To extract the essential information about divergent and re-convergent circuit topologies out of this huge database we apply an in-house STA add-on software. Typical circuit topologies are illustrated in Fig. 3, where a topological correlation among several paths is shown due to the common use of gates 7, 8, 9. Obviously, any WID variation affecting these gates will have a strong impact since the performance of a large number of paths is potentially affected. Previous investigations of the impact of variations on maximum clock frequency oversimplify the underlying path topologies by assuming independent paths. However, topological correlations are quite typical for industrial designs and determine the efficiency of statistical static timing analysis [4] as well as circuit level techniques to optimize the speed and robustness of a design. Therefore, to quantify the degree of correlation within a certain path delay window $\Delta T_{Var}=Z*\Delta t$ (e.g. 5% of the T_{Clk}), we define an average topological correlation factor as follows:

$$\kappa_{Top}\left(\Delta T_{Var}\right)=1-\sum_{i=1}^{Z}\frac{n_{cells}^{i}}{n_{cells,avg}^{i}}\bigg/\sum_{i=1}^{Z}n_{paths}^{i} \qquad (1)$$

Here, $n_{cells,i}$ is the number of cells, $n_{paths,i}$ the number of paths and $n_{cells,avg,i}$ the average number of cells per path in the time interval Δt of

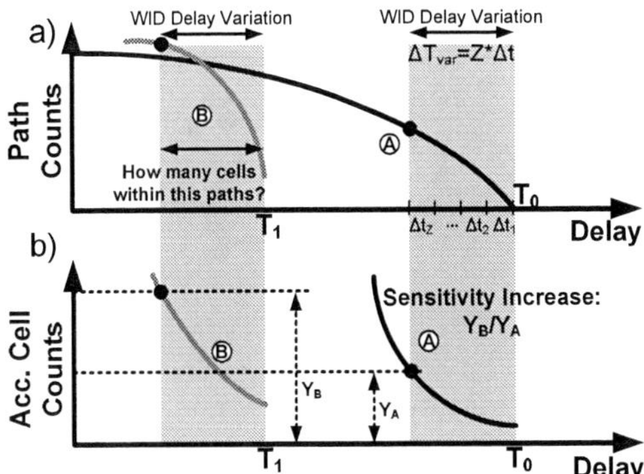

Figure 4. Illustration of path and cell distributions.

interest. Z is the number of time intervals within the delay window ΔT. Idea of this approach is to combine the information of path and cell distributions in a single FOM, to have an evaluation factor for hardware sharing within the critical path delay region.

Table 1 shows the correlation factors, the number of independent paths and total number of paths in certain delay regions of the ARM926 design. Apparently, the majority of paths within the sensitive 7% delay region share a significant number of cells and less than 100 paths are completely independent.

This insight has two consequences: first, any optimization technique should focus on these highly shared cells to gain speed and robustness without major area and leakage impact. Second, a definition of robustness solely based on path distributions is insufficient. Therefore we propose an additional FOM to quantify the amount of cells within the sensitive delay region. To assess the robustness of a circuit to PVT variations we propose a metric for circuit sensitivity which can be gained by the output data of state-of the art STA tools.

Figure 4a schematically illustrates two path delay and cell distributions of an original design A and a speed optimized design B. Depending on the topological correlation the build-up of the timing wall is often coupled to a strong increase of critical hardware, i.e. the number of cells Y_B within the variation critical delay region increases compared to Y_A (Fig. 4b). Even though performance of design B is improved by $\Delta T_{OPT}=T_0-T_1$ the robustness of the design may be reduced. To distinguish between raw performance gain and robustness we define the following circuit sensitivity factor:

$$s=\frac{Y}{N_{cells,tot}}=\frac{1}{N_{cells,tot}}\cdot\sum_{i=1}^{Z}n_{cells}^{i} \qquad (2)$$

The number of cells within a certain time interval is denoted by $n_{cells,i}$. $N_{cells,tot}$ is the total number of cells in the design. Here, Z is again the number of time intervals Δt within the delay variation sensitive path region ΔT_{var}. This definition enables a straightforward evaluation of circuit robustness and is later applied to compare different design options using state of the art design tools.

4. MULTI-STAGE BASED STA

As illustrated in section 2, the efficiency of deeper pipelining of high performance processors relies on high-speed flip flops and skew reduction techniques which are mainly based on time-borrowing (TB) techniques. A key challenge for transferring these techniques to a semi-custom design is to detect successive critical path configuration comprising several pipeline stages. Especially, loops of pipelined critical paths are fundamental to all processor designs and its number increases with deeper pipelining [7]. Therefore, we introduce a multi-stage extended STA for a post-processing of the conventional timing

978-1-60558-497-3/09 $25.00 © 2009 ACM

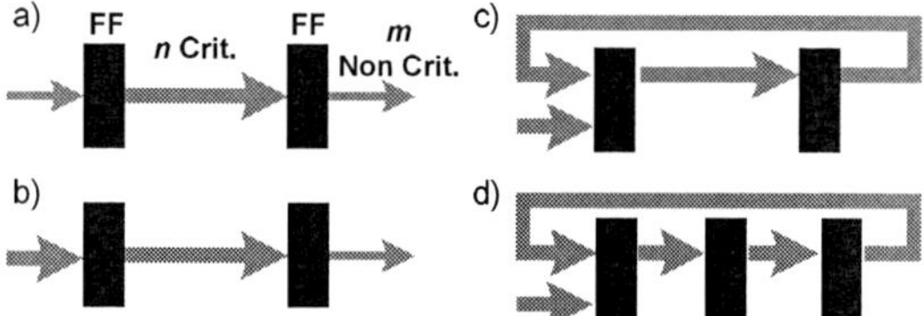

Figure 5. Performance critical multi-pipeline stage paths.

report. Our algorithm maps the critical hardware to a set of frequently occurring multi-pipeline stage configurations as illustrated in figure 5.

The most simple path configuration is an isolated critical path. Here, time-borrowing from the successive non-critical path is possible at the receiving FF without creating a setup-time violation (Fig. 5a). For successive critical path configurations (Fig. 5b) straight forward application of TB is limited due to the balancing of pipeline stages. Most challenging configurations are pipelined loops (fig. 5c&5d). Any design technique based on TB e.g. useful skew, pulsed flip flops or post-layout skew adaptation, is finally prevented by this path configuration. Here TB requires the speed-up of the combinational logic between the FF cells. Figure 5 also shows the contributions of path configurations within the upper 5% of path delay region for the investigated ARM926 design. Among these, 20% are loop internal paths and 55% are successive critical paths. Overall this corresponds to 1791 combinational logic cells. These cells are candidates for speed-up using local design techniques.

5. P-LATCH AND LOW-V_T INSERTION

In this section the previously introduced FOMs and the multi-stage based STA are combined with circuit-level design techniques to improve the robustness of a design against WID variations. D2D variations affect each device on the die in the same way and are treated as global offset by an appropriate PVT-corner strategy. Our intention is to compensate for 5-8% WID delay variations. The straightforward approach is to apply selective, speed increasing techniques in variation-sensitive parts of the design. Two promising techniques in this context are low-V_T cell insertion in the identified sensitive circuit topologies and pulsed latch (P-L) design. The performance improvement and the topological changes of the ARM926 design due to the applied design techniques is assessed by STA runs. Furthermore the respective trade-off between raw performance gain and circuit sensitivity is determined.

5.1 Selective Low-V_T Cells Insertion

Selective insertion of low-V_T cells in the original reg-V_T design is in general limited by an 8-10x leakage increase per cell. Therefore, we derived a replacement algorithm exploiting the high topological correlation within the sensitive delay region. Basic idea is to replace cells in similar path topologies as cells 7, 8, and 9 in Fig. 2 using in-house developed STA add-on software. After the initial place and route based on industry standard design flow all cells with major timing impact are identified. Therefore we apply a weighting factor based algorithm, which considers cell delay, path delay, the previously derived topological correlation factor, as well as the position of the cell within the path configuration. This restricts the number of low-V_T cells to a minimum and prevents an increase of leakage, in contrast to existing mixed low/regular-V_T design with a typically much higher leakage increase of up to 2-3x. Moreover, after cell replacement, the circuit sensitivity factor for each optimization run is determined.

Figure 6 shows the path delay distributions and the corresponding sensitivity factors (2) after STA re-runs for designs with cell replacement resulting in 5% and 7% total speed gain. As expected, the shift

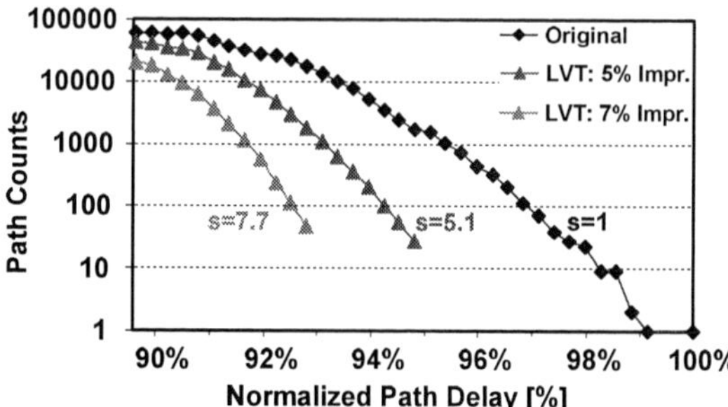

Figure 6. Path distribution and normalized circuit sensitivity.

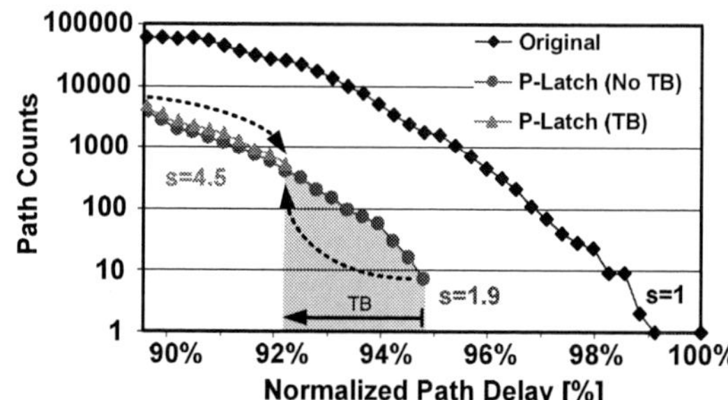

Figure 7. Path delay distribution for P-Latch designs.

of paths to the lower delay region causes an accumulation of critical paths at the new path delay limit. This build-up of the timing wall is reflected in an increase of the circuit sensitivity factor by x7.7 for the design with 7% speed gain.

Since the replacement algorithm excludes cells in hold-time critical paths, the cell placement within the design is not changed by low-V_T cell insertion. The leakage increase of all low-V_T insertions for the ARM926 is less than 3% and hence negligible. However, we see that the single selective insertion of low-V_T cells is insufficient if a FOM such as the circuit sensitivity factor is considered as design optimization criterion.

5.2 Pulsed Latch Design

Pulsed Latches (P-L) and pulsed FFs (P-FF) enable the reduction of FF timing overhead at the expense of increased hold-times [8, 9]. The main speed improvement is achieved by the significant reduction of setup-time to t_{SU}=0ps. Target pulse-widths are 5-10% of cycle time corresponding to 3-6 FO4 delays. This design point extends the classical edge-triggered FF design approach by a soft-edge property. Similar to high-performance full-custom designs, this soft-edge property allows for time-borrowing (TB). Time-borrowing provides a certain elasticity to compensate for design and timing uncertainties due to variations on individual path level. To quantify the amount of TB, we apply the multi-stage STA in section 4.

In this section we investigate the global replacement of MS-FFs by P-Ls using an external pulse generator cell. In contrast to a P-FF concept with cell-internal pulse generator (PG), here the PG is used as local clock buffer (LCB) and is shared among 10-20 P-Ls.

P-Ls in 90nm and 65nm CMOS, are optimized to achieve nearly the same clock-to-Q delay t_{CLK-Q} as MS-FFs, as required in successive and loop internal critical paths. Robust pulse distribution i.e. full swing signals for all design corners, is guaranteed for a minimum pulse-width of 3 FO4 delays and steep slopes. Therefore, we use a factor of 1.3x larger output driving capability for the PG output stage compared to a conventional LCB. The upper bound of the pulse-width is limited by the area overhead due to additional hold-time fixing. For

978-1-60558-497-3/09 $25.00 © 2009 ACM

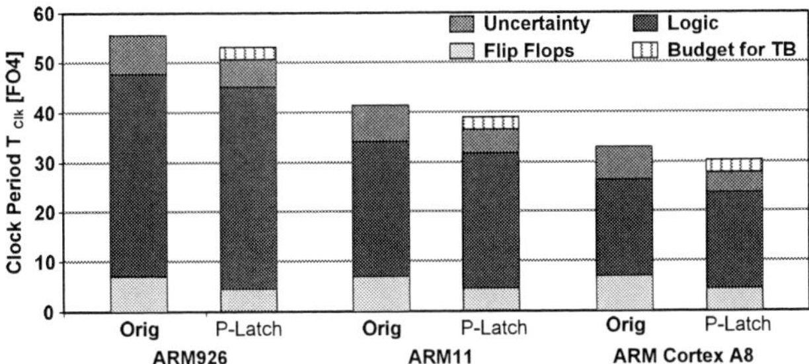

Figure 8. Timing contributions for P-Latch designs in ARM µPs.

the investigated ARM926 core, this leads to a pulse-width of 5 FO4 delays.

Figure 7 shows the reshaping of the path delay distribution and the speed improvement for P-L design after multi-stage STA timing re-calculation. The P-L design results in a speed gain of about 5% due to setup-time reduction without time-borrowing. For the ARM926 de-sign static time-borrowing further reduces maximum path delays by 2.5% of clock period without causing any new setup-time violation. The budget for time-borrowing is indicated by the shaded region. Depending on the activation of successive critical paths, even the total pulse-width can dynamically be used for TB. If a path borrows time from a sub-critical successive path, the delay of both paths is bal-anced as indicated by the arrows in figure 7.

As expected the benefit of P-L designs increases for deeper pipelined microprocessors (Fig. 8 / Table 2), since the contribution of FF over-head and timing uncertainty significantly decreases compared to an original design based on standard MS-FFs. According to our micro-processor model, about 40% of the timing uncertainty can be compen-sated by TB for an ARM Cortex A8 processor.

Compared to the original design, the use of P-Ls causes only a slight increase of sensitivity factor. The use of time-borrowing results in a very fast build-up of a timing wall since the paths of the 1st stage are shifted from higher to lower delay regions. In addition, the effective clock period of the 2nd stage is reduced and sub-critical paths become more critical. Especially for a large TB, i.e. for designs with large pulse widths >6 FO4, the number of interacting paths significantly increases. This interaction between two subsequent pipeline stages is a basic, micro-architectural limitation of TB and therefore demands a thorough multi-stage STA timing closure.

5.3 Impact on Circuit Robustness

Figure 9 illustrates the trade-off between speed improvement and circuit sensitivity for the investigated ARM926 design. With increas-ing speed gain, the circuit sensitivity factor significantly increases for all investigated design techniques. Compared to the low-V_T design the P-L designs show about 2.5x higher robustness due to the lower cir-cuit sensitivity factors for the same performance improvement of 5% and 7.5%, respectively.

So far dynamic time-borrowing in P-Ls, i.e. the capability to sample a late arriving signal within a critical path, has not been considered here. This property is indicated by the horizontal bars. For both P-L designs the soft-edge property enables to compensate for variation induced timing uncertainty up to 2% and 4.5% of clock period. In contrast to the pure static techniques investigated before, this intrinsic elasticity is hardly to predict without precise information about dy-namic effects such as clock jitter and activation of successive critical paths.

Table 2. Area and performance of pulsed latch designs.

		ARM926	ARM1176	Cortex A8
Speed Gain (*:TB= 2.5%)	No TB	+4.7%	+6.5%	+8.1%
	S-TB	+7.5%	+9.0%*	+10.6%*
Area		-5.6%	-3.6%	-1.3%

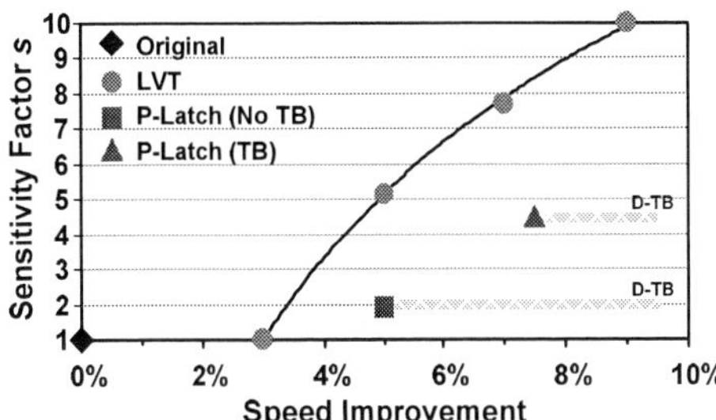

Figure 9. ARM926 results: sensitivity vs. speed gain (=margin).

6. CONCLUSION

The performance impact of 5% to 8% due to WID variations can be compensated in embedded microprocessor by means of circuit-specific speed enhancements with no area overhead. We introduce a topological correlation factor to identify critical hardware regions. This enables a cost-efficient selective low-V_T cell insertion. To reduce the flip flop overhead and WID induced timing uncertainties, a multi-stage STA is applied to facilitate pulsed-latch designs. To quantify the response of cell and path distributions on the investigated compensa-tion techniques, i.e. the build-up of a timing wall, a circuit sensitivity factor is introduced.

For a 90nm CMOS ARM926 design the use of pulsed latches shows a performance increase of 5%-7.5%. This speed gain is a sufficient margin to compensate for WID variations. For the same timing mar-gin, selective low-V_T cell insertion results in a less robust design due to a 2.5x higher circuit sensitivity factor. Time borrowing further increases the robustness of deeply pipelined processors, such as the ARM Cortex A8. Here, up to 40% of the total timing uncertainty can be compensated at constant circuit sensitivity and a net timing margin of 8%.

7. ACKNOWLEDGMENTS

This work was supported by the German BMBF research project SIGMA65, FKZ 01M3080.

8. REFERENCES

[1] K. Kuhn et al., "Managing Process Variation in Intel's 45nm technology", Intel Tech. J., Vol. 12, 2008, pp. 93-109.

[2] G. Gammie et al., "A 45nm 3.5G Baseband-and-Multimedia Application Proc-essor using Adaptive Body-Bias and Ultra-Low-Power Techniques", ISSCC 2008, pp. 258-259.

[3] M. Horowitz et al., "Scaling, Power, and the Future of CMOS", IEDM 2005, pp. 9-15.

[4] D. Blaauw et al., "Stat. Timing Analysis: From Basic Principles to State of the Art", Trans. on CAD, 2008, pp. 589-607.

[5] D. Brooks et al., "New methodology for early-stage, micro-architectural-level power-performance analysis of microprocessors", IBM J. of R&D, Sep./Nov. 2003, pp. 585-598.

[6] T. Lueftner et al., "A 90-nm CMOS low-power GSM/EDGE multi-media-enh. baseband processor with 380-MHz ARM926 core and mixed-signal exten-sions", J. Solid-State Circuits, Vol. 42, No.1, Jan. 2007, pp.1-12.

[7] E. Borch et al., "Loose Loops Sink Chips", Int. Symp. on High-Performance Comp. Arch., Feb. 2002, pp. 299-310.

[8] T. Baumann et al., "Perf. Improv. of Embedded Low-Power Microprocessor Cores by Selective Flip Flop Replacement", ESSCIRC 2007, pp. 308-311.

[9] J. Tschanz et al., "Comparative delay and energy of single-edge triggered & dual-edge triggered pulsed flip-flops for high-performance microprocessors", ISLPED 2001, pp. 147-152.

978-1-60558-497-3/09 $25.00 © 2009 ACM

ARMS - Automatic Residue-Minimization based Sampling for Multi-Point Modeling Techniques

Jorge Fernández Villena [*]
INESC ID / IST - TU Lisbon
Rua Alves Redol 9, 1000-029 Lisbon, Portugal
jorge.fernandez@inesc-id.pt

L. Miguel Silveira
INESC ID / IST - TU Lisbon / Cadence Labs
Rua Alves Redol 9, 1000-029 Lisbon, Portugal
lms@inesc-id.pt

ABSTRACT

This paper describes an automatic methodology for optimizing sample point selection for using in the framework of model order reduction (MOR). The procedure, based on the maximization of the dimension of the subspace spanned by the samples, iteratively selects new samples in an efficient and automatic fashion, without computing the new vectors and with no prior assumptions on the system behavior. The scheme is general, and valid for single and multiple dimensions, with applicability on rational nominal MOR approaches, and on multi-dimensional sampling based parametric MOR methodologies. The paper also presents an integrated algorithm for multi-point MOR, with automatic sample and order selection based on the transfer function error estimation. Results on a variety of industrial examples demonstrate the accuracy and robustness of the technique.

Categories and Subject Descriptors

B.7.2 [**Integrated Circuits**]: Design Aids—*simulation,verification*

General Terms

Algorithms, Verification

Keywords

Model Order Reduction, Multi-Dimensional Parametric Sampling

1. INTRODUCTION

It is well known in the EDA industry that the detailed models representing physical devices, obtained after the modeling and extraction steps, are in general too large for efficient simulation and verification. Reducing their order or dimension, while guaranteeing that the input-output response is accurately captured, is crucial to enabling the simulation and verification of those large systems [20].

This is the realm of *Model Order Reduction* (MOR). The methods for linear model reduction have greatly evolved and can be broadly characterized into two types: those based on subspace generation and projection [7, 14], and those based on balancing techniques [13, 16]. Multi-point based approaches [6, 8, 17] have recently gathered renewed attention due to their robustness and reliability. In particular, some of them [15] have been extended to tackle the issue of *Parametric Model Order Reduction* (pMOR), in which the system behavior depends on a set of parameters, modeling process and environmental variations, and the frequency. In

[*]Acknowledges NXP Semiconductors support via the EMonIC project.

Permission to make digital or hard copies of part or all of this work for personal or classroom use is granted without fee provided that copies are not made or distributed for profit or commercial advantage and that copies bear this notice and the full citation on the first page. To copy otherwise, to republish, to post on servers or to redistribute to lists, requires prior specific permission and/or a fee.
DAC'09, July 26-31, 2009, San Francisco, California, USA

this context, the multi-dimensional sample-based techniques [15], seem less sensitive to the number of parameters, and become a good alternative to the multi-dimensional moment matching approaches, [4, 11, 12, 21], which seem unable to deal with what has been denoted as the *curse of dimensionality*. Such techniques generate oversized models when the number of parameters and the accuracy require matching a large set of moments. However, the multi-dimensional sampling can be expensive if there is no good indication of where to place the sample points. Random based sampling can defeat the reliability of these algorithms, whereas trying to cover the complete subspace with a linear scheme is overwhelming beyond the single dimension.

In this single dimensional framework, some effort has been devoted to the sampling selection issue. Some schemes, such as Adaptive Frequency Sampling [9], are broadly applied on interpolation and rational based procedures. However these schemes start from tabulated data to build a compact model, and thus they do not take into account the data generation (i.e. to solve the system for different points), a costly issue in MOR of large systems. The work in [10] aims at automatically finding the optimal $\|H\|_2$ model for a fixed q-th order, based on a refinement of the interpolation points. However the algorithm can only be applied on single dimension linear systems, and needs to solve the original system at q points at each refinement iteration, which makes the procedure expensive. Resampling plans have been proposed in [18] to provide guidance for future sample point placement in linear MOR, based on variance measurements of the reduced transfer function. However, it requires the evaluation of several reduced models, which may be expensive, and, in addition, seem awkward to extend to multi-dimensional cases.

In this paper we will try to fill this gap, by proposing a novel methodology for sampling selection in single and multi-dimensional spaces. The procedure, which is directly related to the algorithm in [17], is based on maximizing the subspace spawned by the sampling set, in order to obtain a good approximation of the dominant subspace in the orthonormalization step. To avoid falling into inefficiency, **the algorithm will iteratively provide a good new sample point without solving the system**, and assuming **no prior knowledge of the system**. The method will also attemp to **avoid oversampling by estimating the error on the fly**, using for this the norm associated with the vectors generated. It will decide at run time whether to generate a new sample, and where to place it. Therefore, the proposed methodology will present a **fully automated procedure**, which will try to overcome the lack of automation in existing (sample-based) MOR procedures, and in particular in the pMOR scenario.

Although beyond the scope of this paper, the procedure is general enough to be applied to different fields, such as non-linear MOR in which the matrices exhibit non-linear dependences with the parameters or time [20] (for example, to determine suitable new trajectory

978-1-60558-497-3/09 $25.00 © 2009 ACM

points in Trajectory Piecewise-Linear approaches [3, 19]).

The paper is structured as follows: in Section 2 an overview of the MOR paradigm is presented, along with a discussion of existing sampling schemes. In Section 3 the new scheme is introduced, along with a study of its complexity and computational issues. In this section, an efficient integration into a multi-point model order reduction algorithm is also discussed and presented. In Section 4 several examples are shown that illustrate the efficiency of the proposed technique, and in Section 5 conclusions are drawn.

2. BACKGROUND

2.1 Model Order Reduction

The main techniques in MOR are geared toward the reduction of a state space linear time-invariant system, obtained by some modeling methodology, and representing a physical system. In such representation, the output y is related to the input u via some inner states x. When parametric variations are taken into account, the system is represented as a parametric state-space descriptor,

$$C(\lambda)\dot{x}(\lambda) + G(\lambda)x(\lambda) = Bu$$
$$y(\lambda) = Lx(\lambda) \tag{1}$$

where $C, G \in \mathbb{R}^{n \times n}$ are respectively the dynamic and static matrices, $B \in \mathbb{R}^{n \times m}$ is the matrix that relates the input vector $u \in \mathbb{R}^m$ to the inner states $x \in \mathbb{R}^n$ and $L \in \mathbb{R}^{p \times n}$ is the matrix that links those inner states to the outputs $y \in \mathbb{R}^p$. We assume here, as is common, that the elements of C and G, as well as the states x, depend on a set of P parameters $\lambda = [\lambda_1, \lambda_2, \ldots, \lambda_P] \in \mathbb{R}^P$ which model the effects of the uncertainty. Usually the system is formulated so that the input (B) and output (L) matrices do not depend on the parameters. This time-domain description yields a frequency response modeled via the transfer function

$$H(s, \lambda) = L(sC(\lambda) + G(\lambda))^{-1}B, \tag{2}$$

The objective of pMOR techniques is to generate a reduced order approximation, able to accurately capture the input-output behavior of the system for any point in the joint frequency-parameter space,

$$\hat{H}(s, \lambda) = \hat{L}(s\hat{C}(\lambda) + \hat{G}(\lambda))^{-1}\hat{B}, \tag{3}$$

where $\hat{C}, \hat{G} \in \mathbb{R}^{q \times q}$, $\hat{B} \in \mathbb{R}^{q \times m}$, and $\hat{L} \in \mathbb{R}^{p \times q}$, are the reduced set of matrices, with $q \ll n$ the reduced order.

In general, one attempts to generate a *Reduced Order Model* (ROM) whose structure is as similar to the original as possible, i.e. exhibiting a similar parametric dependence allowing more control within analysis and optimization frameworks, in order to facilitate further simulations. The most common procedure to achieve this goal is to use a sensitivity-based Taylor Series Representation [4], combined with some form of orthogonal projection scheme [14].

2.2 Multi-point Projection based Approaches

Standard pMOR methodologies rely on the generation of a suitable low order subspace (spanned by the basis $V \in \mathbb{R}^{n \times q}$), in which the original system matrices $C(\lambda)$, $G(\lambda)$, B and L are projected. This step generates a set of new matrices $\hat{C}(\lambda) = V^T C(\lambda)V$, $\hat{G}(\lambda) = V^T G(\lambda)V$, $\hat{B} = V^T B$ and $\hat{L} = LV$, that define the parametric ROM.

The goal of Multi-point approaches is to generate the basis either by generating the transfer function moments from multiple expansion points (s_k, λ_k), or from solving the system at different sample points on the relevant frequency plus parameter space,

$$z_k = z(s_k, \lambda_k) = (s_k C(\lambda_k) + G(\lambda_k))^{-1}B, \tag{4}$$

where $z(s_k, \lambda_k)$ is the sample vector generated at the sample point (s_k, λ_k), i.e. the zero order moment at that point. In this frame-

work, the approach in [15] presents a statistical interpretation of the algorithm in [17], and enhances its applicability. In that work, the Gramian is presented in an integral form, which is numerically approximated by a quadrature scheme. The most relevant vectors among those generated by such quadrature are selected via *Singular Value Decomposition* (SVD) in order to build the projection matrix V. An *a posteriori* expected error bound, based on the study of the singular values of the SVD, can be used, with rather good results.

This approach, less sensitive to the number of parameters, is more reliable but, on the other hand, depends on a good sampling selection scheme. However, it was argued in [17] that the efficiency of the method does not depend on the "quality" of the quadrature of the integral representation of the Gramian, but on the approximation of the subspace. Regrettably, and although good results are reported with very simplistic sampling schemes, no methodology is provided for sample placement. Furthermore very little automation has been reported on pMOR methodologies, since the number of samples must be decided beforehand. Bad or poor sampling may lead to inaccurate results, whereas an excessive number of samples can lead to inefficiency.

2.3 Sample Selection Schemes

Little work has been devoted to the issue of sample selection within the framework of MOR. The authors in [10] present an iterative procedure for selecting the q best interpolation points that generate an optimal $\|H\|_2$ model of order q. The algorithm generates q initial random samples, used to obtain a q-th order model. The eigenvalues of this reduced model are obtained (this operation is relatively cheap on the reduced model), and these eigenvalues are used as a new set of q sampling points. The procedure is repeated until the eigenvalues (i.e. next iteration sampling points) converge, generating the $\|H\|_2$ optimal model. However, there are some obvious drawbacks to this approach: it is only applicable on single dimensional linear systems, and it needs to solve the **original system (an expensive operation) at q points on each iteration**. Also, increasing q (the reduced order) implies complete model recalculation, and q has a dramatic impact on accuracy (and, of course, the required value of q for a good approximation is not known beforehand). Furthermore, the initial guess of the points is relevant in terms of the stability of the model and the efficiency, due to convergence of the eigenvalues. This means that some a priori knowledge is needed for a good initial guess.

The work presented in [18] is explicitly aimed at sample point selection for optimizing multi-point MOR schemes. There, the idea followed is to generate R ROMs of identical reduced size q, resampled from the same large set, $Z_1 \ldots Z_R$, with an associated transfer function $H_k(s)$ (with $k = 1 \ldots R$). If the various ROMs do not give similar results, that means that the places where the transfer functions disagree the most are likely to be good places to put a new sample point. As a metric of model uncertainty, a variance-like quantity is proposed

$$var(s) = \frac{1}{R} \sum_1^R \|H_k(s) - H_{ref}(s)\|^2, \tag{5}$$

where H_{ref} is the transfer function of a reference reduced model, obtained by using all the available samples (i.e. from $Z = \cup_1^R Z_k$). Points in regions with larger variance are good candidates for new sample points, and thus the procedure gives not only indication of which are the best points (and thus vectors) in the generated set, but also can guide the process of finding new sample points for generating new vectors. Although the work presents an efficient implementation to build the various ROMs and to obtain the variance, the method still has some drawbacks: one is that it requires

978-1-60558-497-3/09 $25.00 © 2009 ACM

the evaluation of the reduced models' frequency transfer function, and another is that it requires one to compute an initial population, i.e. an initial set of samples (and vectors) to start working with (and with no guarantee that these first samples are a good set). These issues are aggravated in multi-dimensional spaces: how to compute efficiently the multi-dimensional variance? and how to set a first sample population in this case?

3. AUTOMATIC SAMPLING PROCEDURE

This section introduces the proposed sampling scheme. As already pointed out, it is based on trying to find the minimum set of samples that will spawn a subspace that generates a good ROM.

Before presenting it, we have to settle onto two basic premises. If we define $A_j = s_j C(\lambda_j) + G(\lambda_j)$, $A_j \in \mathbb{R}^{n \times n}$ the system matrix for the j-th sampling point (s_j, λ_j), we note that:

- **To compute the system matrix for a sampling point is cheap**[1]. We have to obtain A_j, and the matrices C_j and G_j are sparse, so the sums and products of these matrices come at a cost of $O(n^\alpha)$, with $1 \leq \alpha \leq 1.2$.

- **To solve the system for a sampling point is expensive**. We have to obtain the sample vector $z(s_j, \lambda_j) = A_j^{-1}B$. And thus, we must compute the LU factorization of A, at a cost of $O(n^\beta)$, with $1.2 \leq \beta \leq 1.5$, and a set of sums and products at a cost of $O(n^\alpha)$.

This means that we should avoid computing the sample vectors $z(s_j, \lambda_j)$ unless it is strictly necessary.

It is clear that any sample will give a good vector for approximating the system, at least in the neighborhood of the point, but on the other hand, a new vector may add information that is already enclosed in the subspace generated so far (it can be well approximated by the already obtained vectors), and thus it is not a good new sampling choice. Therefore, a good new vector is one that is as much different from the ones we already have as possible, and thus, cannot be well approximated by the currently available subspace.

However, this unleashes a question: how can we determine if a candidate sample point (s_j, λ_j) will generate a vector that adds rank to our set of vectors $Z = \{z_1 \cdots z_k\}$ *without computing it*? Furthermore, how can we know if this new vector will help to *minimize the number of samples needed to obtain a good ROM*?

The answer is simple, we cannot. But on the other hand, we can do some simple and efficient computations that will provide a good estimate of the best candidate sample point to solve for.

3.1 Iterative Sample Selection

We start from a set $\Psi = \{\psi_1 \ldots \psi_k\}$, in which each ψ_j is a point in the space of interest. Ψ is a set that covers our space of interest, and at each step, we will select from it the most appropriate point for our goals. This will be done by trying to find which point has an associated vector that is less similar to the vectors we have already computed. Although it is not a perfect indicator, this will maximize our chances of adding "relevant" rank to the subspace, and we can obtain such information in a cheap fashion.

We take the j-th candidate sample point, ψ_j, with an associate matrix $A_j = A(\psi_j)$. If we have a vector $z_i \in \mathbb{R}^n$, we can determine if it is a good solution for our system at this j point by computing the norm of the residue,

$$\|r_{i,j}\| = \|B - A_j z_i\|, \tag{6}$$

---

[1] we envision the application of AMRS in conjunction with an extraction methodology that provides variational information [1, 5], which makes A_j evaluation for different parameter settings inexpensive.

Algorithm 1 Automatic Residue Minimization Sampling - ARMS

Start from the system matrices C, G, B, L, and a set of candidate sample points $\Psi = \{\psi_1 \ldots \psi_k\}$.

1: Select randomly the first sample ψ_j: $A_j = A(\psi_j)$, $z_j = A_j^{-1}B$
2: Orthonormalize the vector $v_j = QR(z_j)$
3: $V = [V\ v_j]$, and $Z = [Z\ z_j]$
4: Withdraw sample: $i = [1, \ldots (j-1),\ (j+1), \ldots k]$
5: $\forall \psi_i$, generate residues $r_i = B \perp A_i Z$
6: Find index j for which $\|r_j\| = max(\|r_i\|)$
7: Compute $z_j = A_j^{-1}B$
8: Compute $v_j = RRQR([V\ v_j])$
9: if $rank([V\ v_j]) \neq rank(V)$ goto *step 3*
10: Apply V in a congruence transformation on the system,
$\quad \hat{C} = V^T C V \quad \hat{G} = V^T G V \quad \hat{B} = V^T B \quad \hat{L} = L V$

---

where $r_{i,j}$ is the residue of the vector z_i for the system $\{A_j, B\}$. The cost of this operation is relatively cheap since A_j is sparse. If the norm is small, it means that the system $\{A_j, B\}$ is well approximated by vector z_i, so in this case ψ_j is not a good new point.

To see if the system $\{A_j, B\}$ is well approximated by a set of vectors Z, we simply orthogonalize vector B against the set of vectors $A_j Z$, with a resulting vector r_j with an associated norm.

$$r_j = B \perp A_j Z, \tag{7}$$

where r_j is the residue after the orthogonalization ($\perp$) of B against the set of vectors $A_j Z$. If the norm of r_j is small, the system $\{A_j, B\}$ is well approximated by the set of vectors Z, and thus, the vector associated with the candidate point ψ_j, will probably add "small" rank to the subspace spanned by the already existing set of vectors Z.

We can then repeat the operation for all the candidate sample points, to know, for each of them, whose solution is the worst approximated by the set of vectors. This information can be obtained by finding the maximum among the norm of the residues given by (7) for each candidate point j. Since such sample point is not well approximated by the currently available subspace, it is likely that a new vector generated by this point will add rank to the subspace, and furthermore, be a relevant point to include a sample.

Only once we have selected the best suited candidate ψ_j, we solve the system to generate the vector z_j, and withdraw the sample point for the candidate set, $\Psi = \{\psi_1 \ldots \psi_{j-1}\ \psi_{j+1} \ldots \psi_k\}$. Then we repeat the procedure for finding the next point.

3.2 Efficient Implementation

Some non-trivial issues arise from the computations outlined in the previous section. One is that since the operation in (7) must be repeated for each sample point in the candidate set, we are not interested in having an extremely fine set of samples covering the space under study. So how to generate the candidate set of points, and in particular how to generate it in a large dimensional space?

There is no good answer to this question, but in [18] it was shown that linear sampling is a good scheme. In addition, the PMTBR based procedure is quite robust, and able to generate good results from small number of samples. Therefore we propose a relatively coarse linear mesh, covering the space of interest. A finer mesh may provide better candidates, and thus a final more compressed model, but on the other hand would increase the complexity of the model generation. Also, there is nothing to prevent us from adding new sample points to Ψ at a later stage (or even on the fly) if we determine that a certain region is under-represented in the set.

Another issue is when to stop looking for new samples. In this case, the answer is simple: when the latest generated vector does not add rank. An efficient method for obtaining such information

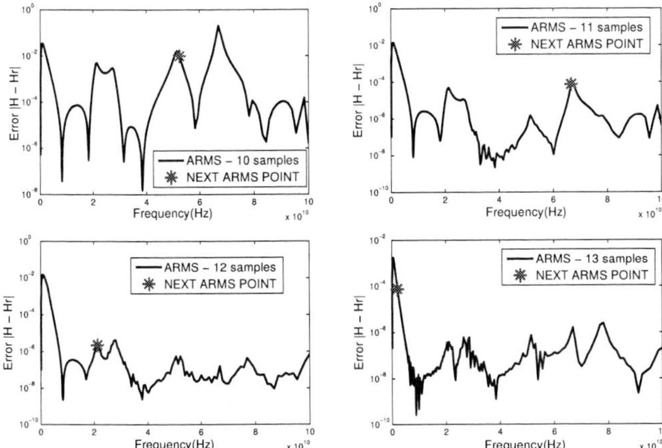

Figure 1: Sample Point Selection in Ex-B example: Transfer function error with 10, 11, 12 and 13 samples, and placement of the 11, 12, 13 and 14-th points respectively.

is to use an incremental QR decomposition, for example a Rank Revealing (RR) QR [2]. Every time a new sample vector is generated, we orthonormalize it against the already orthonormalized set of vectors, and check its norm against the tolerance desired. It must be noticed that although not a perfect indicator, this tolerance is correlated with the error, and can be used to automate the procedure. This step can be integrated with the sampling scheme into a complete multi-point MOR projection based algorithm, as the orthonormalized set of vectors provides the basis of the subspace into which the matrices are going to be projected. Algorithm 1 gives a more clear depiction of the complete methodology.

The last issue is related to the applicability on multiple-input multiple-output (MIMO) systems. Let us take the case in which $B = L^T$, i.e. all the inputs are also outputs of the system, for simplicity, and without loss of generality. It is important to notice that the procedure tries to find new points for which the available subspace does not provide good approximations. This can be extrapolated to the MIMO case in a straightforward manner. The only difference is that in (7), B and r_j are matrices with m (m the number of inputs) columns. The norm must be obtained for a matrix instead for a vector, which is slightly more expensive, but on the other hand each step generates m new vectors, that maximize the subspace growth, and thus fewer samples will be required.

3.3 Weighted Approaches

Some applications may require different levels of accuracy in different regions of the sampling space. For example, some regions may be characterized in detail, whereas other regions may not need a very fine approximation. This can be easily achieved with the proposed methodology by controlling the sample population used. For example, let us assume we have an I region Ψ_I in which we desire a fine approximation, and a II region Ψ_{II} which needs to be coarsely approximated. We can then use the procedure on a candidate set that covers the I region, with a small tolerance, to generate a good model in such region. Once this model is generated, we add the samples of the candidate set that covers the II region. Then the methodology can be applied with this new set, with a less strict tolerance, in order to upgrade the model generated in the first step to obtain a coarse accuracy in the new region II, while ensuring that the accuracy in region I is not degraded (to add new vectors to a basis will not degrade the ROM's existing accuracy).

This weighting scheme can be used in parametric approaches, where the more probable region of the parameters are covered using an initial Ψ set. Once this region is well approximated, points

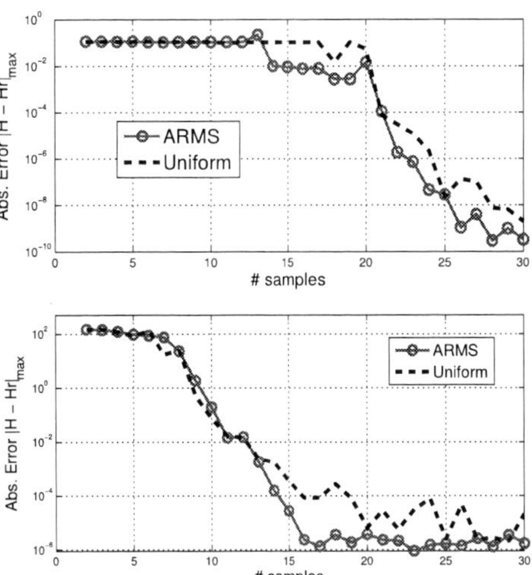

Figure 2: Maximum error versus number of samples for uniform sampling and ARMS, for (top) Ex-A and (bottom) Ex-B examples.

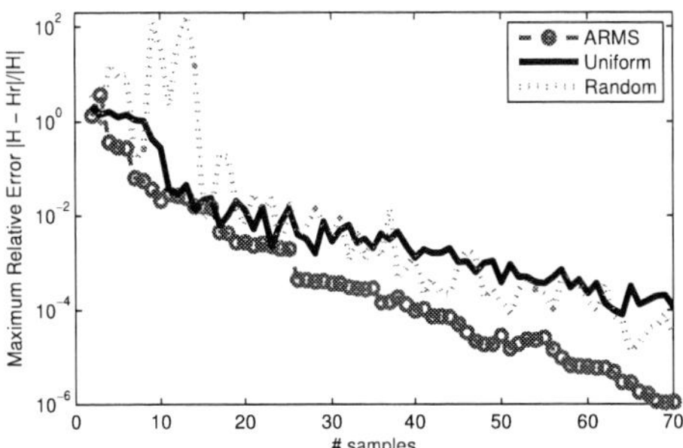

Figure 3: pEx-C: Maximum error from an MC analysis with 159720 evaluation points versus the number of samples.

belonging to less probable regions can be included in the remaining set, and this can be used to refine the first obtained model. This procedure also allows us to reduce the number of samples in the candidate set every time the algorithm is applied: less probable regions can be coarsely populated with a smaller number of points in the candidate set. This will lead to better efficiency, although it may degrade the optimality of the sample selection, since we are not taking into account the complete domain, and thus are selecting new samples based on a local criterion.

4. SIMULATION RESULTS

In this section we present the capabilities of the proposed algorithm (denoted as ARMS) for multiple different benchmarks. For comparison, we will use a PMTBR approach based on other sampling schemes. In the single dimensional case (frequency), uniform sampling will be used. For the multi-dimensional (parametric) cases, a pure random and a pseudo-uniform sampling will be used. The pseudo-uniform sampling consist in covering the frequency range with uniform samples (i.e. with zero variation), and around each of those samples take a set of parametric random samples. As benchmarks we will use two examples depending solely on the frequency, and two parametric systems. The first (denoted as Ex-A) is a well known 304-states PEEC example, which has already appeared in the literature [7], and which has very sharp resonances. The second (Ex-B) is a frequency dependent example obtained with an RCK-based EM extraction method. It corresponds to a strip-line whose model has 9172 states. The third (pEx-C) is a 334-states parametric RLC system obtained from an industrial

978-1-60558-497-3/09 $25.00 © 2009 ACM

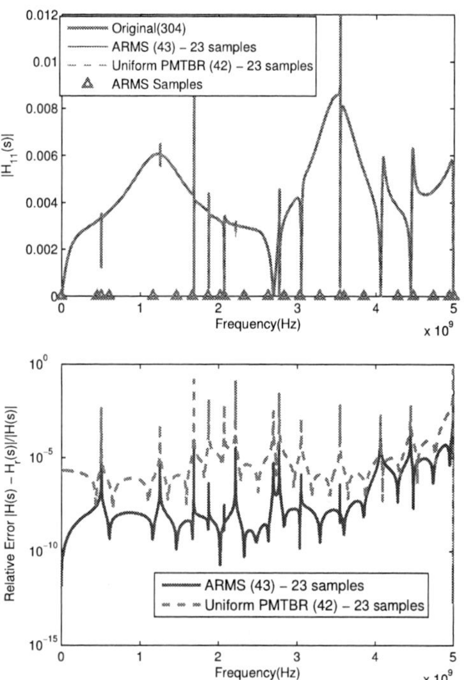

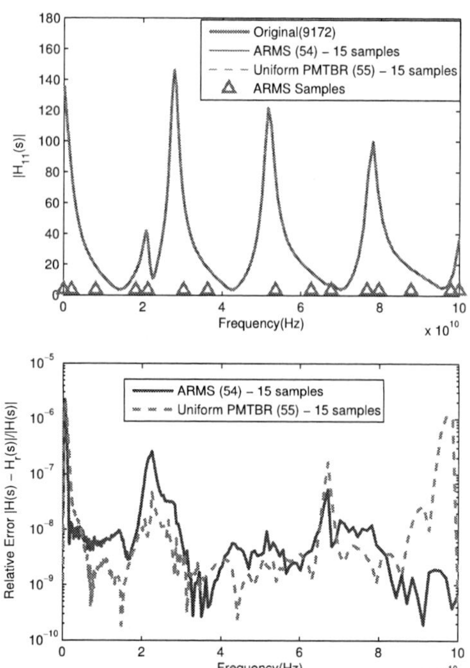

Figure 4: Ex-A: Transfer function and relative error w.r.t. the original system, for the proposed ARMS procedure, and the uniform sampling, with the same number of samples (note the scale on the Y-axis). ARMS sample points are also shown.

Figure 5: Ex-B: Transfer function and relative error w.r.t. to the original system for the proposed ARMS procedure, and uniform sampling, with the same number of samples. ARMS sample points are also shown.

three-metal-layer interconnect network. Its response will depend on the frequency and on a set of 3 parameters, which model the width of each metal layer. The fourth (pEx-D) is a Full Wave EM model of an industrial integrated planar spiral connected to a Metal-Insulator-Metal capacitor. It has 11005 states and depends on the frequency and two very different parameters, one process parameter modeling the width of the capacitor insulator, and the other a design parameter modeling the length of the side of the spiral square (with quite a large range of variation).

4.1   Point Selection and Error Convergence

In this section we demonstrate the capabilities of our point selection method, and the fast convergence to a good solution of the proposed methodology. In Figure 1 we show a set of consecutive sample points automatically selected by the ARMS algorithm, without any knowledge of the system, and an initial population of 100 candidate samples between the minimum and maximum frequencies. The figure shows, for the Ex-B example, the transfer function absolute error with respect to the original system for a given number of consecutive samples and, marked with a star, the frequency where ARMS is going to place the next sampling point. We can see that the selection points are placed in regions of large error, or in regions in which the sample will significantly reduce the overall error. Next, in Figure 2 we show the convergence of the error in the transfer function versus the number of samples for Ex-A and Ex-B examples. It can be seen that the models generated with ARMS exhibit a faster convergence of the error. In the case of multi-dimensional environments, we show the same plot in Figure 3 for the pEx-C example, with the results of pure random, pseudo-uniform sampling (denoted as uniform) and ARMS methodologies. The results are obtained from a Monte-Carlo (MC) simulation with 159720 evaluation points finely covering the complete space of interest. The convergence of the error in the ARMS approach is much faster even in this relatively simple test case.

4.2   Algorithm Accuracy and Automation

In this section we discuss the algorithm's features in terms of accuracy and automation. For the non-parametric systems, the ARMS algorithm will be compared against a uniformly sampled PMTBR. It is important to notice that the comparison is not fair (to our disadvantage), since the uniform PMTBR approach needs to know the number of samples beforehand, whereas the ARMS approach does the selection of the number and placement of the samples at run time. For ARMS, we only fix the minimum and maximum frequencies, and the tolerance. The reduction with PMTBR is done with **the same number of samples** reported by the ARMS method, and with **the same tolerance**. First we show the transfer functions and error for ARMS and PMTBR, versus the original response, for the non-parametric examples (Ex-A and Ex-B), in Figures 4 and 5 respectively. In all cases the accuracy is very good, and the responses of the original and ROM models appear indistinguishable, but the sample points selected by the ARMS procedure (also shown) are placed in regions of large variation of the transfer function, or near sharp resonances, yielding a more accurate model than the PMTBR approach. Figure 6 shows the absolute error versus the tolerance given by the ARMS algorithm for the Ex-A and Ex-B examples. It can be seen that although there is not a perfect match, the tolerance gives a good estimator of the error. Figure 7 shows the relative error distribution for the more complex parametric pEx-D benchmark. Results are shown for the Variational PMTBR based on Pseudo Uniform and Random sampling, and for the proposed ARMS algorithm. Notice that the X-axis scale is the same. For the same number of samples, the ARMS methodology outperforms the other methodologies. In this more complex case, in which the parameters have very different range of variations, ARMS provides a better guidance for sample selection. Notice that the deviation around the mean error is much smaller in the ARMS case. This is a good indicator that the procedure selects new samples in the areas of larger error, in order to minimize the maximum.

978-1-60558-497-3/09 $25.00 © 2009 ACM

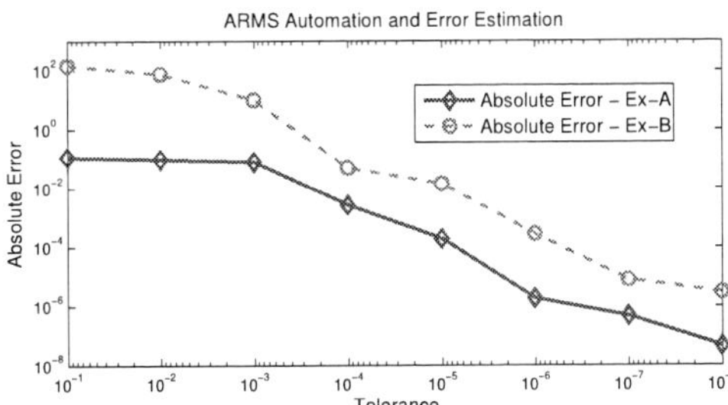

Figure 6: Maximum absolute error $|H_r - H|$ versus the tolerance given to the ARMS algorithm, for Ex-A and Ex-B.

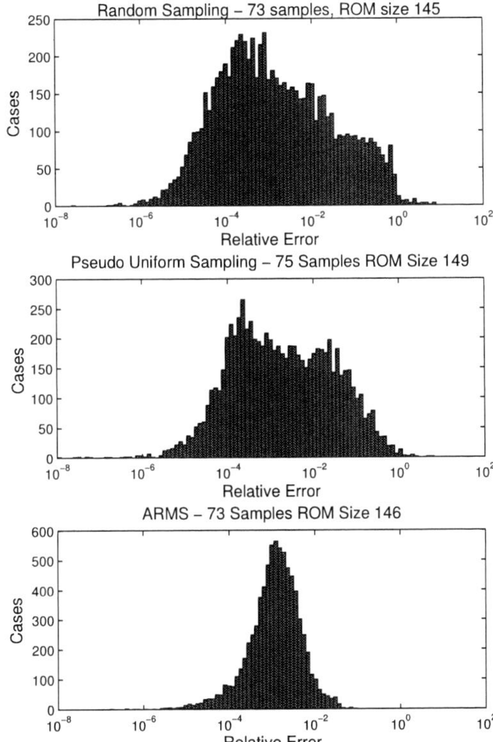

Figure 7: pEx-D: Distribution of the Relative error from an MC simulation with 7700 evaluation points in the space of interest, for (Top) Random sampling, (Center) Uniform sampling, and (Bottom) ARMS.

5. CONCLUSIONS

This paper presents an efficient algorithm for automatic selection and minimization of the number of samples to use in a MIMO multi-point MOR framework. The methodology, based on maximizing the spanned subspace, iteratively provides new sample points. It is general and hence easily applicable to multi-dimensional problems. The procedure was also integrated into an efficient MOR algorithm that controls the number of samples generated, and provides a good convergence of the error. It was shown that it offers a reliable, robust and accurate framework able to deal with different cases. Results demonstrate that it outperforms other sampling schemes both in single and multiple dimensions.

6. REFERENCES

[1] Y. Bi, K.-J. van der Kolk, D. Ioan, and N. van der Meijs. Sensitivity computation of interconnect capacitances with respect to geometric parameters. In *IEEE International Conference on Electrical Performance of Electronic Packaging*, San Jose, CA, October 2008.

[2] C. H. Bischof and G. Quintana-Ortí. Computing rank-revealing QR factorizations of dense matrices. *ACM Trans. Math. Softw.*, 24(2):226–253, 1998.

[3] B. N. Bond and L. Daniel. A piecewise-linear moment-matching approach to parameterized model-order reduction for highly nonlinear systems. *IEEE Trans. Computer-Aided Design*, 26(12):2116–2129, December 2007.

[4] L. Daniel, O. C. Siong, S. C. Low, K. H. Lee, and J. K. White. A multiparameter moment-matching model-reduction approach for generating geometrically parametrized interconnect performance models. *IEEE Trans. Computer-Aided Design*, 23:678–693, May 2004.

[5] T. El-Moselhy, I. Elfadel, and L. Daniel. A capacitance solver for incremental variation-aware extraction. In *IEEE/ACM International Conference on CAD*, San Jose, CA, November 2008.

[6] I. M. Elfadel and D. L. Ling. A block rational arnoldi algorithm for multipoint passive model-order reduction of multiport rlc networks. In *International Conference on Computer Aided-Design*, pages 66–71, San Jose, California, November 1997.

[7] P. Feldmann and R. W. Freund. Efficient linear circuit analysis by Padé approximation via the Lanczos process. *IEEE Transactions on Computer-Aided Design of Integrated Circuits and Systems*, 14(5):639–649, May 1995.

[8] K. Gallivan, E. Grimme, and P. V. Dooren. Multi-point padé approximants of large-scale systems via a two-sided rational krylov algorithm. In 33rd *IEEE Conference on Decision and Control*, Lake Buena Vista, FL, December 1994.

[9] J. D. Geest, T. Dhaene, N. Faché, and D. D. Zutter. Adaptive CAD-model building algorithm for general planar microwave structures. *IEEE Trans. On Microwave Theory and Techniques*, 47(9):1801–1809, September 1999.

[10] S. Gugercin, A. Antoulas, and C. Beattie. A rational krylov iteration for optimal H_2 model reduction. In *Proceedings of the 17th International Symposium on Mathematical Theory of Networks and Systems*, pages 1665–1667, Orlando, Florida, USA, July 2006.

[11] P. Li, F. Liu, X. Li, L. Pileggi, and S. Nassif. Modeling interconnect variability using efficient parametric model order reduction. In *Proc. Design, Automation and Test in Europe Conference and Exhibition*, February 2005.

[12] Y. Li, Z. Bai, Y. Su, and X. Zeng. Model order reduction of parameterized interconnect networks via a two-directional arnoldi process. *IEEE Trans. Computer-Aided Design*, 27(9):1571–1582, September 2008.

[13] B. Moore. Principal Component Analysis in Linear Systems: Controllability, Observability, and Model Reduction. *IEEE Transactions on Automatic Control*, AC-26(1):17–32, February 1981.

[14] A. Odabasioglu, M. Celik, and L. T. Pileggi. PRIMA: passive reduced-order interconnect macromodeling algorithm. *IEEE Trans. Computer-Aided Design*, 17(8):645–654, August 1998.

[15] J. Phillips. Variational interconnect analysis via PMTBR. In *International Conference on Computer Aided-Design*, pages 872–879, San Jose, CA, USA, November 2004.

[16] J. Phillips, L. Daniel, and L. M. Silveira. Guaranteed passive balancing transformations for model order reduction. *IEEE Trans. Computer-Aided Design*, 22(8):1027–1041, August 2003.

[17] J. R. Phillips and L. M. Silveira. Poor Man's TBR: A simple model reduction scheme. *IEEE Trans. Computer-Aided Design*, 24(1):43–55, Jan. 2005.

[18] J. R. Phillips and L. M. Silveira. Resampling plans for sample point selection in multipoint model order reduction. *IEEE Trans. Computer-Aided Design*, 25(12):2775–2783, December 2006.

[19] M. Rewienski and J. White. A trajectory piecewise-linear approach to model order reduction and fast simulation of nonlinear circuits and micromachined devices. *IEEE Trans. Computer-Aided Design*, 22(2):155–170, Feb. 2003.

[20] W. H. A. Schilders, H. A. van der Vorst, and J. R. (Eds.). *Model Order Reduction: Theory, Research Aspects and Applications*. Mathematics in Industry, The European Consortium for Mathematics in Industry , Vol. 13, Springer, 2008.

[21] Z. Zhu and J. Phillips. Random sampling of moment graph: a stochastic krylov-reduction algorithm. In *Proc. Design, Automation and Test in Europe Conference and Exhibition*, pages 1502–1507, Nice, France, April 2007.

An Efficient Passivity Test for Descriptor Systems Via Canonical Projector Techniques

N. Wong
Department of Electrical and Electronic Engineering
The University of Hong Kong
Pokfulam Road, Hong Kong
nwong@eee.hku.hk

ABSTRACT

An efficient passivity test based on matrix chain and canonical projector techniques is proposed for descriptor systems (DSs) commonly encountered in VLSI modeling. The test entails a natural flow that first evaluates the index of a DS, followed by possible decoupling into its proper and improper (if any) subsystems. Compared to the recent DS passivity tests using linear matrix inequality (LMI)-based extended positive real lemma or skew-Hamiltonian/ Hamiltonian (SHH) transformations, numerical examples show that the proposed projector approach is significantly more efficient. Explicit state space formulations for the decoupled subsystems that facilitate further processing such as passivity enforcement and model order reduction are also derived.

Categories and Subject Descriptors

I.6.5 [**Simulation and Modeling**]: Model Development—*modeling methodologies*

General Terms

Algorithms

Keywords

Passivity test, descriptor system, canonical projector, spectral projector

1. INTRODUCTION

Passivity of a linear system is a crucial property for ensuring the stability of global simulation [1], namely, when the system is connected to other system models for a full-system simulation. Specifically, a passive system is one that does not generate energy internally, and a strictly passive system means a dissipative system. In VLSI circuit and multibody dynamics macromodeling, models are often rendered as descriptor systems (DSs) (also known as singular systems) and represented in the linear descriptor state space format [2–5]. For example, the widely used modified nodal analysis (MNA) representation of electrical circuits, as well as the recent

attempts to couple the linear RLC network to transistors or RF objects for direct system-level co-simulation [6–8], naturally give rise to DSs.

The passivity test for a DS, however, is far less developed than its regular (or nonsingular) state space counterparts whose passivity can be readily derived from the positive real lemma [9] or through frequency sweeping [10]. An extension of the positive real lemma to DSs has been given in [11], but its numerical implementation results in prohibitive computation. Although a recent skew-Hamiltonian/Hamiltonian (SHH) transformation has been proposed for testing DS passivity with higher efficiency [4, 5], it still requires complicated subspace identification and construction (mainly singular value decompositions (SVDs) and Gram-Schmidt orthogonalizations), and faces a bottleneck that effectively amounts to the expensive solution of an algebraic Riccati equation plus a Lyapunov equation.

To overcome the computational hurdle, the emerging canonical projector technique is visited [12–14] and utilized to formulate a highly efficient passivity test for DSs. It is shown that the canonical projectors give rise to a spectral projector that provides a natural and conceptually simple way to decouple a DS into its proper (impulse-free) subsystem and improper (impulsive) subsystem, if any. Such decoupling is considered the major difficulty in the passivity tests for DSs [4, 5].

A major difference that marks the efficiency of the proposed method over existing ones is the early knowledge of the matrix pencil index, which quickly manifests as a direct consequence of constructing the matrix projector chain. Given the fact that a minimal passive system can at most be of index 2, this quickly rules out nonpassive systems after two initial matrix chain iterations. Then, if the proper and improper parts are to be decoupled, the algorithm can proceed to form the spectral projector for doing so.

We further show that only one spectral projector, either the left or right (spectral) projector, is all that is needed for decoupling a DS. Explicit construction of state spaces for the proper and improper subsystems of a DS, together with analytical expressions for the derivatives of a DS, are detailed. This leads to a highly efficient DS decomposition and characterization procedure, which in turn facilitates further processing such as model order reduction (MOR) and/or passivity enforcement of DSs using standard regular state space procedures, e.g., [9, 10].

2. BASICS OF PASSIVITY

We study a minimal linear time-invariant (LTI) DS [12–14] in the state space format

$$E\dot{x} = Ax + Bu, \tag{1a}$$

$$y = Cx + Du, \tag{1b}$$

Permission to make digital or hard copies of part or all of this work for personal or classroom use is granted without fee provided that copies are not made or distributed for profit or commercial advantage and that copies bear this notice and the full citation on the first page. To copy otherwise, to republish, to post on servers or to redistribute to lists, requires prior specific permission and/or a fee.
DAC'09, July 26-31, 2009, San Francisco, California, USA

978-1-60558-497-3/09 $25.00 © 2009 ACM

where $E, A \in \mathbb{R}^{n \times n}$ and $B, C^T \in \mathbb{R}^{n \times m}$. Also, $u, y \in \mathbb{R}^m$ and $x \in \mathbb{R}^n$ are the input, output and state vector, respectively. The matrix E is generally singular with $\text{rank}(E) \leq n$. We assume a regular matrix pencil $\{E, A\}$, namely, there exists an $s_0 \in \mathbb{C}$ such that $\det(s_0 E - A) \neq 0$. Then, there always exist nonsingular $W, T \in \mathbb{R}^{n \times n}$ that transform E and A into the so-called *Weierstrass form* [12]:

$$E = W \begin{bmatrix} I_q & 0 \\ 0 & N \end{bmatrix} T \text{ and } A = W \begin{bmatrix} J & 0 \\ 0 & I_{n-q} \end{bmatrix} T, \quad (2)$$

where I_k denotes an identity matrix of order k (though the order is sometimes omitted when the dimension is obvious). J corresponds to the finite eigenvalues of $(sE - A)$ whereas N is nilpotent and corresponds to the infinite eigenvalues. The matrix pencil is stable if all eigenvalues of J are stable (i.e., having negative real parts). The nilpotency index μ of N, viz. $N^{\mu-1} \neq 0$ and $N^\mu = 0$, is called the *index* of the matrix pencil $\{E, A\}$. Expressing the pencil $\{E, A\}$ in Weierstrass form, the DS transfer function of (1) is obtained as

$$
\begin{aligned}
G(s) &= D + C(sE - A)^{-1}B \\
&= D + C(sW \begin{bmatrix} I_q & \\ & N \end{bmatrix} T - W \begin{bmatrix} J & \\ & I_{n-q} \end{bmatrix} T)^{-1} B \\
&= D + CT^{-1} \begin{bmatrix} (sI_q - J)^{-1} & \\ & (sN - I_{n-q})^{-1} \end{bmatrix} W^{-1} B \\
&= D + \begin{bmatrix} C_p & C_\infty \end{bmatrix} \begin{bmatrix} (sI_q - J)^{-1} & \\ & -(I_{n-q} - sN)^{-1} \end{bmatrix} \begin{bmatrix} B_p \\ B_\infty \end{bmatrix} \\
&= \underbrace{D - C_\infty B_\infty + C_p(sI_q - J)^{-1}B_p}_{G_p(s)} \\
&\quad \underbrace{-sC_\infty N B_\infty - s^2 C_\infty N^2 B_\infty - s^3 C_\infty N^3 B_\infty - \cdots}_{G_\infty(s)} \quad (3)
\end{aligned}
$$

which consists of the proper part $G_p(s)$ (viz. bounded as $s \to \infty$) and improper part $G_\infty(s)$ (viz. unbounded as $s \to \infty$). The definitions $\begin{bmatrix} C_p & C_\infty \end{bmatrix} = CT^{-1}$ and $\begin{bmatrix} B_p \\ B_\infty \end{bmatrix} = W^{-1}B$ are conformal to the partitions in the Weierstrass form.

For an LTI system, (strict) passivity is equivalent to the transfer function being (strictly) *positive real*.

THEOREM 1. *[11] A rational matrix-valued transfer function $G(s) \in \mathbb{C}^{m \times m}$ is positive real (strictly positive real) if*

1. *$G(s)$ is analytic in $\mathbb{C}_+ = \{s \in \mathbb{C} | \text{Re}(s) > 0\}$;*

2. *$\Phi(j\omega) = G(j\omega) + (G(j\omega))^* \geq 0 \; (> 0 \text{ for strict positive realness}), \forall \omega \in \mathbb{R}, \text{ with } j\omega \text{ not a pole of } G(s)$;*

3. *If $j\omega_0$ or ∞ is a pole of $G(s)$, then it is a simple pole and the $m \times m$ residue matrix is positive semidefinite.*

Alternatively, if $G(s)$ is decomposed as

$$G(s) = \overbrace{\underbrace{G_{sp}(s)}_{\text{strictly proper}} + M_0}^{G_p(s)} + sM_1 + \underbrace{\sum_{k=2}^{\infty} s^k M_k}_{G_\infty(s)}, \quad (4)$$

then $G(s)$ is positive real if and only if $G_p(s)$ is positive real and $M_1 \geq 0$, $M_k = 0$, $k \geq 2$. Comparing (4) and (3), it is obvious that (1) is passive if and only if $G_p(s)$ in (3) is passive and $-C_\infty N B_\infty \geq 0$, while $C_\infty N^i B_\infty = 0$, $i = 2, 3, \cdots$.

In the rest of the paper, we assume that $\{E, A\}$ is regular and stable, i.e., the finite eigenvalues of J in (2) are in the open left half plane. Such stability assumption is a practical one for dissipative linear circuits, say, the RLC networks obtained from VLSI parasitic extraction. Therefore, the first condition in Theorem 1 is always true. However, we make no assumption about the improper part of $G(s)$, namely, it may have nonzero M_i, $i = 1, 2, \cdots$.

Referring to the Weierstrass form in (2), we define the left and right (spectral) projectors, P_l and P_r, respectively, as

$$P_l = W \begin{bmatrix} I_q & 0 \\ 0 & 0 \end{bmatrix} W^{-1} \text{ and } P_r = T^{-1} \begin{bmatrix} I_q & 0 \\ 0 & 0 \end{bmatrix} T. \quad (5)$$

Obviously, P_l and P_r ($Q_l = I - P_l$ and $Q_r = I - P_r$) project onto the left and right deflating subspaces, respectively, corresponding to the finite (infinite) eigenvalues.

3. MATRIX PROJECTOR CHAIN

For a regular index-μ matrix pencil $\{E, A\}$, setting $E_0 := E$ and $A_0 := A$ (':=' denotes assignment), a projector Q_0 onto $\ker E_0$ is constructed (i.e., $Q_0^2 = Q_0$ and $\text{im } Q_0 = \ker E_0$). Subsequently, a chain of matrix projectors are formed:

$$E_{j+1} := E_j + A_j Q_j \quad \text{and} \quad A_{j+1} := A_j P_j \quad (6)$$

for $j = 0, 1, \cdots$, where $Q_j^2 = Q_j$ is a projector *onto* $\ker E_j$ and $P_j = I - Q_j = P_j^2$ is a projector *along* $\ker E_j$. Obviously, $P_j Q_j = Q_j P_j = 0$. Then, we have the important theorem:

THEOREM 2. *[12] For a regular index-μ matrix pencil $\{E, A\}$, $E_0, \cdots, E_{\mu-1}$ are singular while E_μ is nonsingular.*

Consequently, for the matrix chain formed from the matrix pencil of a DS:

- If E_1 is nonsingular [w.r.t. (2), $\mu = 1$ or $N = 0$], the system (1) corresponds to a non-impulsive system [4, 5] whose proper part is extracted and tested for passivity as in Section 4.1.

- If E_1 is singular but E_2 is nonsingular, we then proceed to form the spectral projector P_r (or P_l) through constructing canonical projectors, which then allows the decoupling of the proper and improper subsystems (viz. $G_p(s)$ and $G_\infty(s)$).

- If E_2 is singular ($\Rightarrow N^2 \neq 0$), minimality of the DS dictates $M_2 \neq 0$ [cf. (4)] so the system is nonpassive and the passivity test is complete. Therefore, μ is at most 2 for passive systems.

Some matrix chain properties that will be utilized later are highlighted:

1. Post-multiplying Q_j and P_j, respectively, to the first equation of (6) yields

$$E_{j+1} Q_j = A_j Q_j, \quad (7a)$$
$$E_{j+1} P_j = E_j, \quad (7b)$$

for $j = 0, 1, \cdots, \mu - 1$.

2. When the projectors $Q_0, \cdots, Q_{\mu-1}$ satisfy $Q_j Q_i = 0$, $j > i$, they are called *admissible projectors*. Many new properties arise from using admissible projectors. For example, for $j = 0, \cdots, \mu - 1$,

$$A_{j+1} = A - E_\mu(Q_0 + \cdots + Q_j), \quad (8a)$$
$$E_\mu Q_j = A_j Q_j. \quad (8b)$$

3. *Canonical projectors* are admissible projectors that further satisfy, for $j = 0, \cdots, \mu - 2$,

$$
\begin{aligned}
Q_j &= Q_j P_{j+1} \cdots P_{\mu-1} E_\mu^{-1} A_j \\
&= Q_j P_{j+1} \cdots P_{\mu-1} E_\mu^{-1} A, \quad (9\text{a})
\end{aligned}
$$

whereas $Q_{\mu-1} = Q_{\mu-1} E_\mu^{-1} A_{\mu-1}$

$$
= Q_{\mu-1} E_\mu^{-1} A. \quad (9\text{b})
$$

The second equality signs in (9a) and (9b) can be established from (8a).

4. In the matrix chain (6), Q_j is generally not unique as only its range is constrained. Suppose $\bar{Q}_j = \bar{Q}_j^2$ is another projector onto ker E_j, we must have $\bar{Q}_j Q_j = Q_j$ and also $Q_j \bar{Q}_j = \bar{Q}_j$ (from which we can show $-\bar{Q}_j P_j = Q_j \bar{P}_j$). This permits a relationship between the E_{j+1}'s generated by different projectors:

$$
\begin{aligned}
\bar{E}_{j+1} &= E_j + A_j Q_j \bar{Q}_j = E_j + A_j Q_j (I - \bar{P}_j) \\
&= (E_j + A_j Q_j)(I + \bar{Q}_j P_j) \\
&= E_{j+1}(I + \bar{Q}_j P_j). \quad (10)
\end{aligned}
$$

The inverse of $\bar{E}_{j+1}$, if exists, is readily shown to be $\bar{E}_{j+1}^{-1} = (I - \bar{Q}_j P_j) E_{j+1}^{-1}$.

In the following, we elaborate the formation of canonical and spectral projectors for the non-trivial case $\mu = 2$. We also remark that the illustration given here is much more compact and easier to follow than that in [13].

3.1 Constructing canonical projectors

Setting $E_0 := E$ and $A_0 := A$, the matrix chain in the case of $\mu = 2$ is formed, namely,

$$
\begin{aligned}
E_1 &:= E_0 + A_0 Q_0 \text{ and } A_1 := A_0 P_0, \quad (11\text{a}) \\
E_2 &:= E_1 + A_1 Q_1, \quad (11\text{b})
\end{aligned}
$$

where E_2 is nonsingular by Theorem 2. Q_0 and Q_1 in (11) are generally not canonical or admissible, but a new matrix chain with canonical (and therefore admissible) projectors can be constructed from (11) with smart reformulation of projectors and careful algebras.

We begin by making the projector in (11b) canonical. This is done by replacing Q_1 with

$$
Q_1' := Q_1 E_2^{-1} A_1. \quad (12)
$$

To see Q_1' is a valid projector onto ker E_1, we note that $Q_1' Q_1 = Q_1 E_2^{-1} A_1 Q_1 = Q_1 E_2^{-1} E_2 Q_1 = Q_1$ and $Q_1 Q_1' = Q_1'$, which implies $Q_1'(= Q_1'^2)$ is a projector onto the same range as Q_1. To show Q_1' and Q_0 are admissible, we have $Q_1' Q_0 = Q_1 E_2^{-1} A_1 Q_0 = Q_1 E_2^{-1} A_0 P_0 Q_0 = 0$. Using Q_1' in place of Q_1, (11) is updated to

$$
\begin{aligned}
E_1 &:= E_0 + A_0 Q_0 \text{ and } A_1 := A_0 P_0, \quad (13\text{a}) \\
E_2' &:= E_1 + A_1 Q_1'. \quad (13\text{b})
\end{aligned}
$$

Note that Theorem 2 again guarantees E_2' to be nonsingular. Now to verify the canonicity of Q_1' in (13b), we make use of the property in (10) to get $E_2' = E_2(I + Q_1' P_1)$ and have

$$
\begin{aligned}
Q_1' E_2'^{-1} A_1 &= Q_1'(I - Q_1' P_1) E_2^{-1} A_1 \\
&= Q_1' Q_1 E_2^{-1} A_1 = Q_1 E_2^{-1} A_1 = Q_1', \quad (14)
\end{aligned}
$$

thus completing the proof. Next, we set

$$
Q_0' := Q_0 P_1' E_2'^{-1} A_0. \quad (15)
$$

Noting $A_0 Q_0 = E_2' Q_0$, it follows that $Q_0'^2 = Q_0'$ is a projector onto ker E_0. Recompute (13a) with this new projector to give

$$
E_1' := E_0 + A_0 Q_0' \text{ and } A_1' := A_0 P_0'. \quad (16)
$$

Again, from (10), $E_1' = E_1(I + Q_0' P_0)$ and it turns out Q_1' is also a projector onto ker E_1', namely,

$$
\begin{aligned}
E_1' Q_1' &= E_1(I + Q_0' P_0) Q_1' = E_1(I + Q_0 P_1' E_2'^{-1} A_0 P_0) Q_1' \\
&= E_1 Q_1' + E_1 Q_0 P_1' E_2'^{-1} A_1 Q_1' = 0,
\end{aligned}
$$

since $A_1 Q_1' = E_2' Q_1'$. With Q_0', (13) is further updated as

$$
\begin{aligned}
E_1' &:= E_0 + A_0 Q_0' \text{ and } A_1' := A_0 P_0', \quad (17\text{a}) \\
E_2'' &:= E_1' + A_1' Q_1'. \quad (17\text{b})
\end{aligned}
$$

Because $Q_1' Q_0 = 0$, so is $Q_1' Q_0' = 0$ (i.e., admissible) and

$$
\begin{aligned}
Q_0' Q_1' &= Q_0 P_1' E_2'^{-1} A_0 Q_1' = Q_0 P_1' E_2'^{-1} A_0 (Q_0 + P_0) Q_1' \\
&= Q_0 P_1' E_2'^{-1} E_2' Q_0 Q_1' + Q_0 P_1' E_2'^{-1} A_1 Q_1' \\
&= Q_0 Q_1' + Q_0 P_1' E_2'^{-1} E_2' Q_1' = Q_0 Q_1'. \quad (18)
\end{aligned}
$$

This also implies $P_0' Q_1' = P_0 Q_1'$ and $A_1' Q_1' = A_1 Q_1'$, from which a link can be found between E_2'' and E_2' as

$$
\begin{aligned}
E_2'' &= E_1(I + Q_0' P_0) + A_1 Q_1' \\
&= E_2' P_1'(I + Q_0' P_0) + E_2' Q_1' = E_2'(I + Q_0' P_0). \quad (19)
\end{aligned}
$$

Still, we need to show both Q_0' and Q_1' in (17) are indeed canonical. Utilizing properties (18), (19) and noting that $A_1 = A_0 P_0 = A_0 P_0 P_0' = A_1' - E_2' Q_0 P_0'$, we get

$$
\begin{aligned}
Q_0' P_1' E_2''^{-1} A_0 &= Q_0'(I - Q_1')(I - Q_0' P_0) E_2'^{-1} A_0 \\
&= (Q_0' - Q_0' Q_1' - Q_0' P_0) E_2'^{-1} A_0 \\
&= (Q_0' Q_0 - Q_0' Q_1') E_2'^{-1} A_0 \quad [\text{cf. (18)}] \\
&= Q_0 P_1' E_2'^{-1} A_0 = Q_0' \quad [\text{cf. (15)}] \\
Q_1' E_2''^{-1} A_1' &= Q_1'(I - Q_0' P_0) E_2'^{-1} A_1' \\
&= Q_1' E_2'^{-1}(A_1 + E_2' Q_0 P_0') \\
&= Q_1' E_2'^{-1} A_1 = Q_1'. \quad [\text{cf. (14)}]
\end{aligned}
$$

In practice, only the canonical projectors are required for computing the spectral projectors. To summarize the implementation, an initial chain (11) is first formed, followed by computing the canonical Q_1' and Q_0' via

$$
\begin{aligned}
Q_1' &:= Q_1 E_2^{-1} A_1, \quad (20\text{a}) \\
Q_0' &:= Q_0 P_1' E_2'^{-1} A_0 = Q_0 P_1'(I - Q_1' P_1) E_2^{-1} A_0 \\
&= Q_0 P_1' E_2^{-1} A_0 = Q_0(I - Q_1 E_2^{-1} A_1) E_2^{-1} A_0, \quad (20\text{b})
\end{aligned}
$$

so that the matrix chain updates from (11) to (13) to (17) never need to be computed explicitly.

3.2 Constructing spectral projectors

Following from above, suppose the canonical projectors Q_0' and Q_1' are constructed. An important result for canonical projectors is that the (right) spectral projector P_r is readily given by

$$
P_r = P_0' P_1' = (I - Q_0')(I - Q_1'). \quad (21)
$$

The proof is shown in Appendix A.

978-1-60558-497-3/09 $25.00 © 2009 ACM

4. PASSIVITY TEST

The canonical projector technique provides a natural way to decouple the proper (non-impulsive) and improper (impulsive) parts of the DS in (1), which translates into a highly effective way for checking passivity. Moreover, all that is needed is one, either the left or right, spectral projector. Without loss of generality, we assume the availability of P_r in (5). The case when P_l is available follows analogously. In the following, we see how P_r permits the simple and explicit construction of regular state spaces for the proper and improper subsystems.

4.1 Proper subsystem

We first project E onto its finite deflating subspace via $E_f := EP_r$, which effectively reduces (1) to an index-1 DS with $N = 0$. The resulting transfer function is then [cf. (3)]

$$
\begin{aligned}
G_1(s) &= D + C(sE_f - A)^{-1}B \\
&= D + C(sW\begin{bmatrix} I_q & \\ & 0 \end{bmatrix}T - W\begin{bmatrix} J & \\ & I_{n-q} \end{bmatrix}T)^{-1}B \\
&= D + \begin{bmatrix} C_p & C_\infty \end{bmatrix}\begin{bmatrix} (sI_q - J)^{-1} & \\ & -I_{n-q} \end{bmatrix}\begin{bmatrix} B_p \\ B_\infty \end{bmatrix} \\
&= G_p(s).
\end{aligned}
\tag{22}
$$

In terms of state space, since now we know the system with E_f represents an impulse-free system, a reduced-order system of order q thus follows. Specifically, performing the singular value decomposition $E_f = U\begin{bmatrix} \Sigma \\ 0 \end{bmatrix}V^T$ (here $\Sigma > 0$ must contain q positive singular values due to the deterministic rank of E_f) and letting $\tilde{x} = V^Tx$, we readily derive the state space corresponding to $G_p(s)$:

$$
\begin{bmatrix} \Sigma & \\ & 0 \end{bmatrix}\begin{bmatrix} \dot{\tilde{x}}_1 \\ \dot{\tilde{x}}_2 \end{bmatrix} = \begin{bmatrix} A_{11} & A_{12} \\ A_{21} & A_{22} \end{bmatrix}\begin{bmatrix} \tilde{x}_1 \\ \tilde{x}_2 \end{bmatrix} + \begin{bmatrix} B_1 \\ B_2 \end{bmatrix}u,
$$
$$
y = \begin{bmatrix} C_1 & C_2 \end{bmatrix}\begin{bmatrix} \tilde{x}_1 \\ \tilde{x}_2 \end{bmatrix} + Du,
\tag{23}
$$

where A_{22} is invertible due to its impulse-free nature [4, 5], which then implies $\tilde{x}_2 = -A_{22}^{-1}A_{21}\tilde{x}_1 - A_{22}^{-1}B_2u$. This reduces (23) to a qth-order state space,

$$
\tilde{E}\dot{\tilde{x}}_1 = \tilde{A}\tilde{x}_1 + \tilde{B}u
$$
$$
y = \tilde{C}\tilde{x}_1 + \tilde{D}u
\tag{24}
$$

with $\tilde{E} = \Sigma$, $\tilde{A} = A_{11} - A_{12}A_{22}^{-1}A_{21}$, $\tilde{B} = B_1 - A_{12}A_{22}^{-1}B_2$, $\tilde{C} = C_1 - C_2A_{22}^{-1}A_{21}$ and $\tilde{D} = D - C_2A_{22}^{-1}B_2$. Also, $G_p(s) = \tilde{D} + \tilde{C}(s\tilde{E} - \tilde{A})^{-1}\tilde{B}$. The passivity test of $G_p(s)$ can then make use of the Hamiltonian matrix eigenvalue test for regular systems [9] (since $\tilde{E}$ is nonsingular) or the fast frequency sweeping approach in [10, 15].

4.2 Improper subsystem

To extract $G_\infty(s)$ in (3), we first project E onto its infinite deflating subspace via $E_{inf} := E(I - P_r)$ such that the transfer function becomes

$$
\begin{aligned}
G_2(s) &= D + C(sE_{inf} - A)^{-1}B \\
&= D + C(sW\begin{bmatrix} 0 & \\ & N \end{bmatrix}T - W\begin{bmatrix} J & \\ & I_{n-q} \end{bmatrix}T)^{-1}B \\
&= D + \begin{bmatrix} C_p & C_\infty \end{bmatrix}\begin{bmatrix} -J^{-1} & \\ & (sN - I_{n-q})^{-1} \end{bmatrix}\begin{bmatrix} B_p \\ B_\infty \end{bmatrix} \\
&= \underbrace{D - C_\infty B_\infty - C_p J^{-1} B_p}_{G_p(0)} \\
&\quad - sC_\infty NB_\infty - s^2C_\infty N^2B_\infty - s^3C_\infty N^3B_\infty - \cdots.
\end{aligned}
\tag{25}
$$

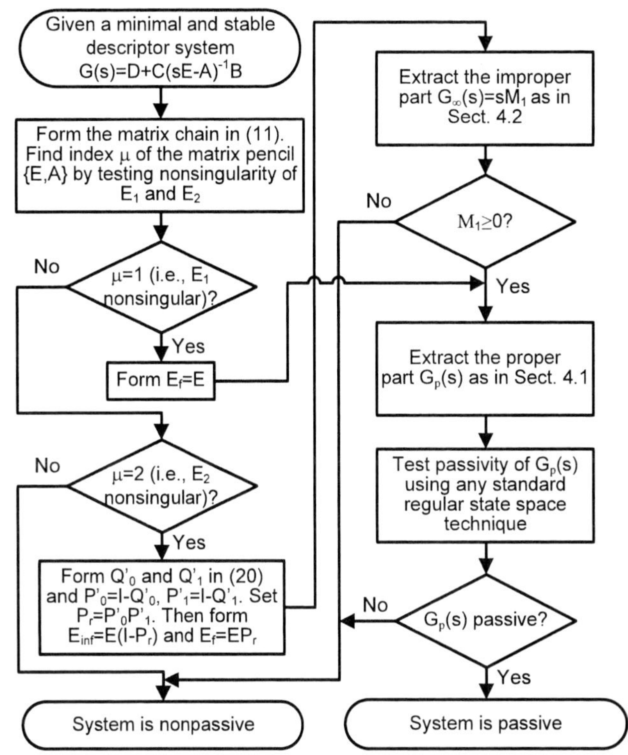

Figure 1: Flowchart for the proposed DS passivity test (can alternatively be viewed as pseudocodes).

Contrasting (3) and (4) it follows that

$$
M_1 = \frac{dG_2(s)}{ds}\bigg|_{s=0} = -C_\infty NB_\infty.
\tag{26}
$$

Therefore, the remaining task is to find the derivative of $G_2(s)$ with respect to s where A is invertible by assumption. With some care, it can be shown that $d(sE_{inf} - A)^{-1}/ds = -(sE_{inf} - A)^{-2}AEA^{-1}$ which results in

$$
M_1 = -CA^{-1}E_{inf}A^{-1}B,
\tag{27}
$$

whose positive semidefiniteness necessary for a passive DS can be easily tested. The flow of the proposed projector-based DS passivity test is summarized in Fig. 1.

4.3 Reconstruction into a DS

Finally, if $G_p(s)$ is passive and $M_1(\in \mathbb{R}^{m \times m}) \geq 0$ (possibly after passivity enforcement), a passive DS corresponding to the transfer function $G_p(s) + sM_1$ can be reconstructed for export to a simulator. To do so, the positive semidefiniteness of M_1 allows a Cholesky factorization $M_1 = LL^T$ where $L \in \mathbb{R}^{m \times m}$. Consequently, the DS state space can be formulated as

$$
\begin{bmatrix} \tilde{E} & & \\ & 0 & I_m \\ & 0 & 0 \end{bmatrix}\dot{x} = \begin{bmatrix} \tilde{A} & \\ & I_{2m} \end{bmatrix}x + \begin{bmatrix} \tilde{B} \\ 0 \\ L^T \end{bmatrix}u,
$$
$$
y = \begin{bmatrix} \tilde{C} & -L & 0 \end{bmatrix}x + \tilde{D}u,
\tag{28}
$$

whose transfer function is easily checked to be $G_p(s) + sM_1$.

4.4 Complexity Analysis

DS passivity test by the extended positive real lemma in [11] via solution of LMIs is impractical due to its $O(n^6)$ (n: order of

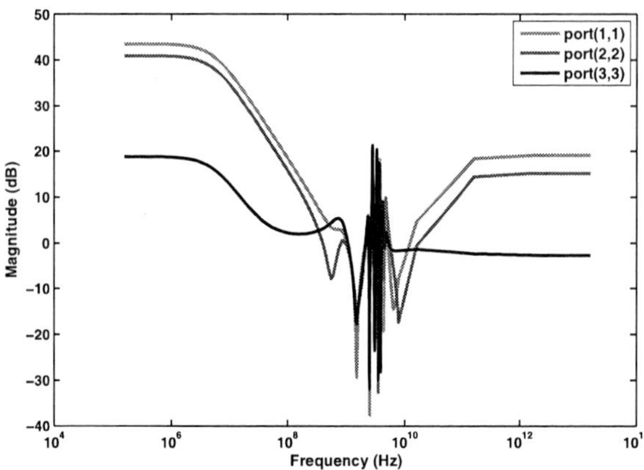

Figure 2: Some reflection responses of the RF circulator.

Examples	Proposed	SHH	GUPTRI
RLC100	0.22(0.09)	1.47	0.43
RLC200	1.64(0.58)	20.17	26.45
RLC300	5.61(2.20)	77.25	73.55
RLC400	13.09(5.17)	207.7	417.75
Touchstone500	13.05(2.47)	352.81	589.10

Table 1: CPU times in seconds of various passivity tests (bracketed numbers in the second column denote times up to M_1 extraction)

state vector) complexity. Also, direct transformation to the Weierstrass form (2) can be numerically unstable and expensive (e.g., the GUPTRI algorithm for doing so has a worst-case $O(n^4)$ complexity [16]). Consequently, the much more efficient SHH transformation technique is proposed in [4, 5] to decouple the original DS into its proper and improper subsystems, followed by standard regular system passivity tests such as the Hamiltonian matrix eigenvalue test (of $O((2n)^3)$ work) or fast frequency sweeping (of $O(m^3)$ work, where usually the number of ports $m \ll n$) [9, 10]. However, a closer inspection of [4, 5] reveals that, up to subsystem decoupling, the SHH approach requires $4\times$SVDs, one QR decomposition, one Arnoldi algorithm, and the solution of one algebraic Riccati equation plus one Lyapunov equation, each having $O((2n)^3)$ work, whereas the proposed projector approach, in its standard implementation, requires mainly $3\times$SVDs of $O(n^3)$ work. This leads to drastic computational savings, as will be seen in the numerical examples.

5. NUMERICAL EXAMPLES

The proposed DS passivity test is contrasted against the SHH transformation [4, 5] and the Weierstrass tests, where the public-domain routine GUPTRI [16] is used in the latter. The extended LMI test in [11] has been shown by [4, 5] to be computationally impractical and therefore not included. All algorithms are coded in Matlab and executed on a 3GHz PC with 3Gb memory. For fairness, in testing the passivity of the extracted proper subsystem, the conventional $O((2n)^3)$ Hamiltonian matrix eigenvalue test is used in all algorithms. First, the tests are run on several RLC interconnect models with orders from 100 to 400 generated from port-to-port response fitting with their $M_1 \geq 0$ but proper parts nonpassive, so that the proposed and the SHH tests are run till the end of their test flow for maximum CPU times. The dimension of N in these

examples ranges from 2 to 10. The results are listed in Table 1 which shows the projector decoupling framework is significantly faster than existing DS tests.

We then experiment with another touchstone example of a real world 3-port RF counterclockwise circulator given in 500th-order singular admittance state space. Fig. 2 shows some frequency responses of the circulator. Again, the last row of Table 1 reveals the remarkable efficiency of the projector-based passivity test for DSs.

Several remarks are in order:

1. In the proposed projector-based test, if non-passivity is detected due to M_1 not being positive semidefinite, the CPU times would be further reduced (see Table 1). In fact, the computation bottleneck lies in the passivity test of the extracted proper subsystem. For even faster testing, the frequency sweeping test for regular systems can be employed, e.g., [10].

2. Compared to the SHH transformation DS passivity test, the proposed algorithm is conceptually much simpler and straightforward. As described in Section 4.4, the SHH technique deals with matrices of order $2n$ from the onset, while the projector approach deals only with matrices of order n in forming the projectors. To the knowledge of the authors, this projector-based decoupling passivity test flow for DSs is by far the fastest flow in identifying the proper and improper subsystems.

3. In our examples, standard techniques like SVD has been used for computing the null spaces and projectors. Since physical matrix pencils almost always come with structures like sparsity or bands, structure-aware algorithms for the fast computation of matrix null space or inverse may provide significant acceleration (say, from $O(n^3)$ to $O(n^2)$ or even less) in spectral projector construction. The closed forms in (20) also allow matrix algebra speedup in certain scenarios such as MNA pencils. Findings along this line, however, are not in the central scope of this paper and will be reported elsewhere.

4. By virtue of the explicit formulation of regular state spaces for the proper and improper subsystems in Section 4, further processing such as MOR and passivity enforcement can readily be exercised, e.g., [1, 9, 10, 15]. The reconstructed passive DS in Section 4.3 can also be exported as a macromodel to circuit simulator for coupled simulation, e.g., [6–8].

6. CONCLUSION

This paper has presented a canonical projector-based passivity test for DS models commonly found in VLSI modeling. Compared to the existing LMI and SHH transformation approaches for testing DS passivity, the proposed test is theoretically straightforward and enjoys simple codings and significantly faster computation. The index of a DS comes at an early stage which immediately reveals the possible passivity of a DS. Closed-form construction of the spectral projector then allows efficient decomposition of the DS into its proper and improper subsystems, whose individual passivity can be efficiently evaluated with conventional regular system techniques. State spaces for these subsystems have also been explicitly formulated for utilization in subsequent passivity enforcement or MOR exercise.

7. REFERENCES

978-1-60558-497-3/09 $25.00 © 2009 ACM

[1] J. R. Phillips, L. Daniel, and L. M. Silveira, "Guaranteed passive balancing transformations for model order reduction," *IEEE Trans. Comput.-Aided Design Integr. Circuits Syst.*, vol. 22, no. 8, pp. 1027-1041, Aug. 2003.

[2] A. Varga, "A descriptor systems toolbox for MATLAB," in *Proc. IEEE Intl. Symp. Computer-Aided Control System Design*, 2000, pp. 150-155.

[3] P. Benner, V. Mehrmann, and D. C. Sorensen, *Dimension Reduction of Large-Scale Systems*, ser. Lecture Notes in Computational Science and Engineering. Springer Verlag, 2005.

[4] N. Wong and C. K. Chu, "A fast passivity test for descriptor systems via structure-preserving transformations of skew-Hamiltonian/Hamiltonian matrix pencils," in *Proc. IEEE Design, Automation Conf.*, Jul. 2006, pp. 261-266.

[5] ——, "A fast passivity test for descriptor systems via skew-Hamiltonian/Hamiltonian matrix pencil transformations," *IEEE Trans. Circuits Syst. I*, vol. 55, no. 2, pp. 635-643, Mar. 2008.

[6] M. S. Soto and C. Tischendorf, "Numerical analysis of DAEs from coupled circuit and semiconductor simulation," *Applied Numerical Mathematics*, vol. 53, no. 2, pp. 471-488, May 2005.

[7] M. S. Soto, *An index analysis from coupled circuit and device simulation*, ser. Scientific Computing in Electrical Engineering. Berlin/Heidelberg: Springer-Verlag, May 2005, vol. 9.

[8] M. Bodestedt and C. Tischendorf, *Numerical Simulation of High-Frequency Circuits in Telecommunication*, ser. Mathematics Ú Key Technology for the Future. Berlin/Heidelberg: Springer-Verlag, 2008.

[9] S. Grivet-Talocia, "Passivity enforcement via perturbation of Hamiltonian matrices," *IEEE Trans. Circuits Syst. I*, vol. 51, no. 9, pp. 1755-1769, Sep. 2004.

[10] D. Saraswat, R. Achar, and M. S. Nakhla, "Fast passivity verification and enforcement via reciprocal systems for interconnects with large order macromodels," *IEEE Trans. VLSI Syst.*, vol. 15, no. 1, pp. 48-59, Sep. 2007.

[11] R. W. Freund and F. Jarre, "An extension of the positive real lemma to descriptor systems," *Optimization Methods and Software*, vol. 19, no. 1, pp. 69-87, Feb. 2004.

[12] E. Griepentrog and R. März, "Basic properties of some differential-algebraic equations," *Zeitschrift für Analysis und ihre Anwendungen*, vol. 8, no. 1, pp. 25-40, 1989.

[13] R. März, "Canonical projectors for linear differential algebraic equations," *Computers Math. Applications*, vol. 31, no. 4, pp. 121-135, Feb. 1995.

[14] ——, "Projectors for matrix pencils," Humboldt-Universität zu Berlin, Tech. Rep. 04-24, 2004. [Online]. Available: http://www.mathematik.hu-berlin.de/publ/pre/2004/P-04-24.ps

[15] Y. Liu and N. Wong, "Fast sweeping methods for checking passivity of descriptor systems," in *Proc. IEEE Asia Pacific Conf. on Circuits and Systems*, Dec. 2008, pp. 1794-1797.

[16] Z. Bai, J. Demmel, J. Dongarra, A. Ruhe, and H. van der Vorst, *Templates for the solution of Algebraic Eigenvalue Problems: A Practical Guide*. Philadelphia: SIAM, 2000.

APPENDIX

A. SPECTRAL PROJECTORS FROM CANONICAL PROJECTORS

We give a constructive proof for $P_r = P_0' P_1'$ for the case $\mu = 2$. This complements, and is also in contrast to, the indirect argument towards spectral projector presented in [13]. Taking the Weierstrass viewpoint for E_0 and A_0 (wherein $N^2 = 0$) and assuming canonical Q_0' and Q_1', we can easily show

$$Q_0' = T^{-1} \begin{bmatrix} 0 & 0 \\ 0 & \widehat{Q}_0' \end{bmatrix} T \text{ and } Q_1' = T^{-1} \begin{bmatrix} 0 & 0 \\ 0 & \widehat{Q}_1' \end{bmatrix} T, \quad (29)$$

where $\widehat{Q}_0', \widehat{Q}_1' \in \mathbb{R}^{(n-q) \times (n-q)}$ project onto $\ker N$ and $\ker(N + \widehat{Q}_0')$, respectively. Moreover,

$$N_2 := N + \widehat{Q}_0' + \widehat{P}_0' \widehat{Q}_1' = N + I - \widehat{P}_0' \widehat{P}_1' \quad (30)$$

is nonsingular due to Theorem 2. By canonicity,

$$\widehat{Q}_0' = \widehat{Q}_0' \widehat{P}_1' N_2^{-1} , \ \widehat{Q}_1' = \widehat{Q}_1' N_2^{-1} \text{ and } \widehat{Q}_1' \widehat{Q}_0' = 0. \quad (31)$$

Next, we prove $\widehat{P}_0' \widehat{P}_1' = 0$. First,

$$N N_2 = N^2 + N \widehat{Q}_0' + N \widehat{P}_0' \widehat{Q}_1' = N \widehat{Q}_1'. \quad (32)$$

Post-multiplying N_2^{-1} to (32) and using (31) we have $N = N \widehat{Q}_1'$ or equivalently $N \widehat{P}_1' = 0$. Next, recognizing $\widehat{P}_0' \widehat{P}_1'$ is a projector by itself, i.e., $(\widehat{P}_0' \widehat{P}_1')^2 = \widehat{P}_0' \widehat{P}_1'$, and post-multiplying it to (30), we obtain

$$N_2 \widehat{P}_0' \widehat{P}_1' = N \widehat{P}_0' \widehat{P}_1' = N \widehat{P}_1' = 0. \quad (33)$$

Nonsingularity of N_2 then implies $\widehat{P}_0' \widehat{P}_1' = 0$. Consequently, the right (spectral) projector P_r in (5) is given by

$$P_r = P_0' P_1' = T^{-1} \begin{bmatrix} I_q & 0 \\ 0 & \widehat{P}_0' \widehat{P}_1' \end{bmatrix} T = T^{-1} \begin{bmatrix} I_q & 0 \\ 0 & 0 \end{bmatrix} T. \quad (34)$$

To obtain the left projector, canonical projectors are constructed from the matrix chain starting with $E_0 := E^T$ and $A_0 := A^T$ instead. It is then easily verified that

$$P_l = (P_0' P_1')^T = W \begin{bmatrix} I_q & 0 \\ 0 & 0 \end{bmatrix} W^{-1}. \quad (35)$$

978-1-60558-497-3/09 $25.00 © 2009 ACM

A Parameterized Mask Model for Lithography Simulation

Zhenhai Zhu
Cadence Research Labs, Berkeley, CA 94704
zhzhu@cadence.com

1. ABSTRACT

We formulate the mask modeling as a parametric model order reduction problem based on the finite element discretization of the Helmholtz equation. By using a new parametric mesh and a machine learning technique called Kernel Method, we convert the nonlinearly parameterized FEM matrices into affine forms. This allows the application of a well-understood parametric reduction technique to generate compact mask model. Since this model is based on the first principle, it naturally includes diffraction and couplings, important effects that are poorly handled by the existing heuristic mask models. Further more, the new mask model offers the capability to make a smooth trade-off between accuracy and speed.

Categories and Subject Descriptors

B.7.2 [**Integrated Circuits**]: Design Aides—*Simulation*

General Terms

Algorithms, Performance, Design

Keywords

Lithography, Mask Model, Parameterized Model Order Reduction

2. INTRODUCTION

From lithography simulation point of view, two classes of designs pose particularly challenging problems. The first class is the memory design. With just six or seven transistors, each memory cell has small layout. The carefully designed cell is duplicated hundred millions to billions of times. Since the manufacturing defects in one cell affect the entire row of cells, the memory design is highly sensitive to such defects. Therefore, high accuracy lithography simulation is mandatory for the memory design. The second class is the custom logic design. With a finite number of unique standard cells as the building blocks, the layout is typically large, in the order of $1 \times 1mm^2$. Therefore, highly efficient lithography simulation is mandatory for the custom logic design. Since timing and power are the main concerns, the accuracy of lithography simulation has to be good enough to model the impact of manufacturing imperfection to the timing and power. Clearly, a good lithography simulation tool should inherently have the capability to make smooth trade-off between accuracy and speed for various designs. And it should be based on rigorous mathematical foundation to ensure its robustness.

Permission to make digital or hard copies of part or all of this work for personal or classroom use is granted without fee provided that copies are not made or distributed for profit or commercial advantage and that copies bear this notice and the full citation on the first page. To copy otherwise, to republish, to post on servers or to redistribute to lists, requires prior specific permission and/or a fee.
DAC'09, July 26-31, 2009, San Francisco, California, USA

There are three main steps in lithography simulation [1]: photo mask modeling, aerial image simulation and photo resist simulation. This paper focuses on the photo mask modeling. The goal of this step is to compute the near field, the electric field at the bottom of the computational domain that contains the mask pattern. The existing approaches can be categorized into two groups: field solvers and heuristic models.

The field solver approach is to solve the Maxwell's equations using well-known numerical techniques such as finite difference time domain (FDTD) [2] or finite element method (FEM) [3]. This is the most accurate and robust approach. But experience indicates that even the state-of-the-art field solvers are too slow or memory-bounded to handle a medium-size mask pattern.

In practice, the heuristic models are used instead to obtain the approximate solution. A commonly used model is based on the so-called Kirchhoff approximation [4]: if there is a mask opening, the light shines through it without any change in magnitude and phase; otherwise, light is completely blocked. This approximation neglects the effects such as diffraction, polarization and coupling. Attempts have been made to obtain a modified mask model to improve the accuracy [5, 6]. The piecewise constant curve fitting approach in [5, 6] is simple and efficient but not accurate and robust. For example, it has been shown in [7] that such model can result in wrong wafer imaging prediction.

The field solvers in [2, 3] and the heuristic models in [5, 6] sit at the opposite corners in the accuracy-speed trade-off matrix and it is difficult to smoothly trade accuracy with speed. In our early work [8], the mask modeling was formulated as a parametric model order reduction problem based on the finite element discretization of the Helmholtz equation. The discretization in [8] uses a uniform and rectangular mesh. This natually leads to the affine parametric form for the stiff and the mass matrices in FEM and hence the well-understood parametric reduction technique [9, 10] can be directly applied to generate the compact mask model. However, the non-uniform and unstructure mesh is indispensable to handle complicated mask patterns. Unfortunately, as will be shown in section 6, a direct application of the technique in [9] could result in large mask models.

In this paper, we present a new approach to approximate the nonlinearly parameterized FEM matrices with affine forms. This allows the direct application of the parametric reduction in [9, 10] again. Though we demonstrate this new technique for the Helmholtz equation that governs the photo mask modeling, it should be straightforward to apply the same approach to other partial differential equations in the context of the parametric model order reduction.

3. PROBLEM FORMULATION

For the sake of simplicity, we present the problem formulation based on the 2D example shown in Fig 1. The extension to 3D cases is straightforward.

Assuming an S-polarization (TE) case, the governing 2D Helmholtz

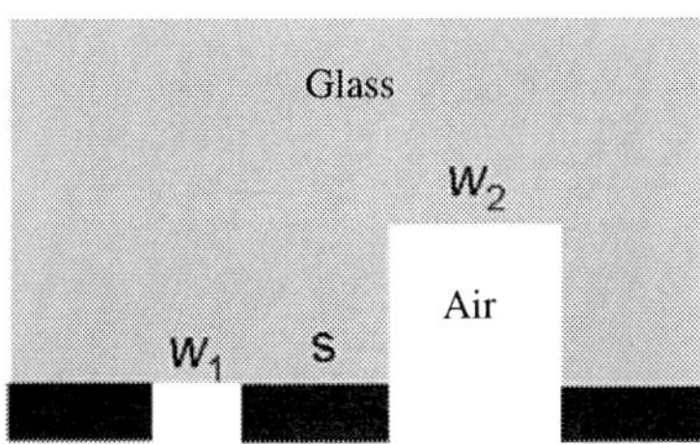

Figure 1: The structure of the 2D phase-shift mask. Variables w_1, w_2 and s are the parameters in the reduced mask model.

equation is

$$\nabla^2 u - \omega^2 \varepsilon \mu u = \nabla^2 u + k^2 u = 0 \tag{1}$$

where $u(x,y)$ is the z-component of the total electric field, ω is the frequency, ε and μ are respectively the dieletric constant and permeability. Following the standard FEM procedure [11], we obtain the weak form

$$\int_\Omega ds \nabla v \cdot \nabla u - \omega^2 \int_\Omega ds \varepsilon \mu v u - \int_{\partial\Omega} dl v \mathrm{DtN}(u) = -2 \int_{\partial\Omega} dl v \mathrm{DtN}(u_{\mathrm{in}}), \tag{2}$$

where Ω and $\partial\Omega$ are respectively the computational domain and its boundary, v is the testing function, u_{in} is the known incidence plane wave and the $\mathrm{DtN}()$ operator defines the transparent boundary condition [12]. In this paper, we assume the periodic boundary condition at the east and the west side of the computational domain and transparent or non-reflecting boundary condition on the north and the south side of the computational domain.

Using the standard FEM piecewise polynomial basis functions [11] to discretize (2), we arrive at the parameterized system equation

$$[S(\bar{w},\bar{s}) - M(\bar{w},\bar{s}) - B]\bar{u} = \bar{r}, \tag{3}$$

where vector $\bar{w}$ and $\bar{s}$ respectively contain the widths and the spacings in the layout, the stiff matrix $S(\bar{w},\bar{s})$ corresponds to the first term in (2), the mass matrix $M(\bar{w},\bar{s})$ corresponds to the second term in (2), the matrix B corresponds to the third term in (2), and vector $\bar{r}$ corresponds to the right-hand-side term in (2).

For mask patterns with fixed topology but different width and spacing values, computing the near field involves solving equation (3) for different $\bar{w}$ and $\bar{s}$. This kind of multiple-inquery scenario is precisely what the parametric model order reduction approach is designed for.

4. PARAMETRIC MODEL ORDER REDUCTION

A parametric model order reduction technique called Reduced Basis method has been developed by the finite element research community [9]. A similar idea has also been independently proposed in the area of parameterized model order reduction for circuit simulation [10]. As will be shown later, the Reduced Basis is a very powerful idea on top of which we can build our new mask model to achieve the desirable accuracy and speed trade-off. Here we summarize its key steps. Please refer to [9] for more details.

Suppose the parameterized governing equation for the problem at hand is

$$A(\bar{\sigma})\bar{u} = \left[A_0 + \sum_i f_i(\bar{\sigma})A_i\right]\bar{u} = \bar{r}, \tag{4}$$

where the scalar function $f_i(\bar{\sigma})$ can be arbitrary and the size of vector $\bar{u}$, $\bar{r}$ and constant matrix A_i is $N \times 1$, $N \times 1$ and $N \times N$, respectively. In practical applications, N can be as large as a few millions

for a medium-sized 3D structure. The Reduced Basis method in [9] has two stages: the off-line pre-characterization and the on-line evaluation.

Off-line pre-characterization stage. We randomly generate a set $\bar{\sigma}^k = \{\sigma_1^k, \sigma_2^k, ...\}$ using the given ranges of σ_i and solve (4) for $\bar{u}_k$. After a few sampling solves, we collect all solutions into the projection matrix

$$P = [\bar{u}_1, \bar{u}_2, ..., \bar{u}_M] \tag{5}$$

and perform projection

$$\hat{A}_i = P^T A_i P; \quad \hat{r} = P^T \bar{r}. \tag{6}$$

Now we arrive at the reduced governing equation

$$\left[\hat{A}_0 + \sum_i f_i(\bar{\sigma})\hat{A}_i\right]\hat{u} = \hat{r}, \tag{7}$$

where the size of matrix $\hat{A}_i$ is $M \times M$ and M is the number of sampling solves.

Similar to the standard procedure in the Model Order Reduction [10], the columns in the projection matrix P are orthogonalized using techniques such as incremental QR decomposition. This makes the matrix P well conditioned. In addition, both theoretical and practical ways to estimate the error of the reduced model are readily available [9, 10]. Hence the off-line model generation can be made incremental.

On-line evaluation stage. We substitute the given set $\bar{\sigma}^*$ into (7) and solve for $\hat{u}$. The approximate solution to equation (4) is obtained from

$$u = P\hat{u}. \tag{8}$$

The key observation here is that the CPU time of the on-line stage is only related to M, not to N in (4). And M is typically many orders of magnitude smaller than N, as shown by the extensive experiments in [9, 10]. Hence equation (7) is a much more efficient but still accurate reduced model than the original model in (4). However, this dramatic efficiency gain critically depends on the fact that matrix A_i in (4) is not a function of $\bar{\sigma}$. Otherwise, the projection step in (6) involves calculating $A_i(\bar{\sigma})$ at a particular value $\bar{\sigma}^k$. This essentially means that the CPU time used by the reduced model in (7) is related to the original problem size N and hence we have gained no efficiency at all [9, 13]. This issue of representing the potentially arbitrary nonlinearity in $A(\bar{\sigma})$ in the form amenable to the projection framework in (6) is one of the main challenges in the nonlinear Model Order Reduction [14].

The main contribution in this paper is to show how to convert $S(\bar{w},\bar{s})$ and $M(\bar{w},\bar{s})$ into the appropriate form similar to that in equation (4) so that we can apply the congruence projection in (6). This is done in two steps: parameterization of mesh and parameterization of the FEM matrices.

5. PARAMETERIZE MESH

In this section, we show an effective technique to parameterize the unstructured triangular mesh in an affine form of the geometry parameters $\bar{w}$ and $\bar{s}$. This is an important first step toward parameterizing the stiff matrix $S(\bar{w},\bar{s})$ and the mass matrix $M(\bar{w},\bar{s})$ in (3).

When the size of a geometric feature changes, say w_2 in Fig 1, the mesh points surrounding the feature will move as well. However, to capture the changes in the resulting electric field, not all mesh points in the computational domain have to be moved. Only mesh points within a certain distance from the changing geometric feature need to be moved. To measure such a distance, we borrow a well-established concept called distance function from the celebrated Level Set Method [15].

978-1-60558-497-3/09 $25.00 © 2009 ACM

5.1 The Distance Function

We use a simple example to explain the basic ideas in the distance function. Let (x_1, y_1) and (x_2, y_2) be the lower-left and upper-right corner of a rectangle, respectively. The distance from a point (x, y) to such a rectangle is defined as

$$d(x,y) = -min(min(y-y_1, y_2-y), min(x-x_1, x_2-x)), \quad (9)$$

where function $min(a,b)$ returns the smaller value of the two variables. Fig. 2 shows the contour plot of the distance function for a square where $x_1 = y_1 = 5$ and $x_2 = y_2 = 10$. It should be noted that the zero level set corresponds to the boundary of the square. All points outside of the square have positive distance and all points inside the square have negative distance.

5.2 The Blending Function

The movement of mesh points due to geometry changes depend on the distance of the mesh points to the changing geometry. Intuitively, the function that characterizes such movement should satisfy the following constrains

$$
\begin{aligned}
b(\eta) &= 1, \quad if \ \eta = 0 & (10) \\
\frac{db}{d\eta} &< 0, \quad if \ 0 < \eta < 1 & (11) \\
b(\eta) &= 0, \quad if \ \eta \geq 1 & (12) \\
\frac{db}{d\eta} &= 0, \quad if \ \eta = 1 & (13)
\end{aligned}
$$

where η is the normalized distance defined as

$$\eta = \frac{d(x,y)}{B}, \quad (14)$$

B is a user-defined radius of influence, and $d(x,y)$ is the distance function like the one defined in (9). Equation (10) means that the mesh points on the geometry move by the same amount as that of the moving geometry features. Equation (11) means that the movement magnitude decreases monotonically as the mesh points are away from the geometry. Equation (12) means that the movement magnitude is zero if the mesh point is not inside the influence region. Equation (13) ensures a smooth transition of the movement at the boundary of the influence region. In spirit, function $b(\eta)$ is similar to the so-called blending function in [16] used to generate the mesh for the computational domain with moving boundaries. In this paper we have chosen the following function as the blending function

$$b(\eta) = 2(1 - p(\frac{\eta}{2} + 0.5)), \quad \eta \in [0,1] \quad (15)$$

where $p(\eta)$ is a third-order polynomial

$$p_3(\eta) = 3\eta^2 - 2\eta^3. \quad (16)$$

5.3 The Parametric Form of Mesh

Armed with the blending function in (15) based on the distance function like the one defined in (9), we are ready to parameterize the movement of the mesh points due to the change of geometric features. Without loss of generality, we use the parameter w_2 in Fig. 1 as an example. Suppose w_2 is changed by δw_2. Furthermore, again without loss of generality, suppose that the center of the opening does not move, only the left and the right wall move by $-\frac{\delta w_2}{2}$ and $\frac{\delta w_2}{2}$, respectively. Within the influence region around the left and the right side wall, the location (x,y) of a mesh point can be expressed as

$$
\begin{aligned}
x &= x_0 + \frac{b_R - b_L}{2} \delta w_2 = x_0 + b_2 \delta w_2, & (17) \\
y &= y_0 & (18)
\end{aligned}
$$

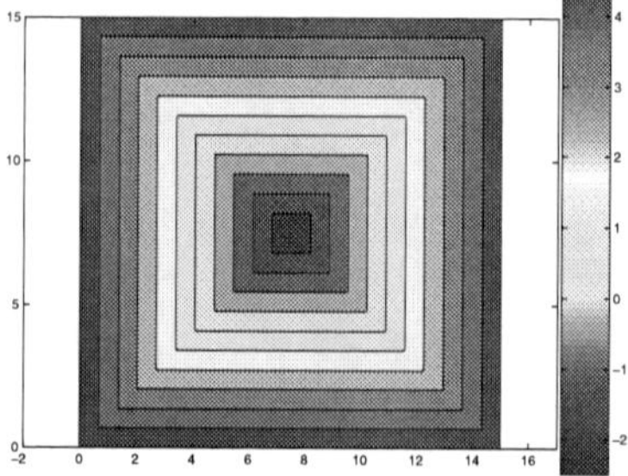

Figure 2: Contour plot of the distance to a square.

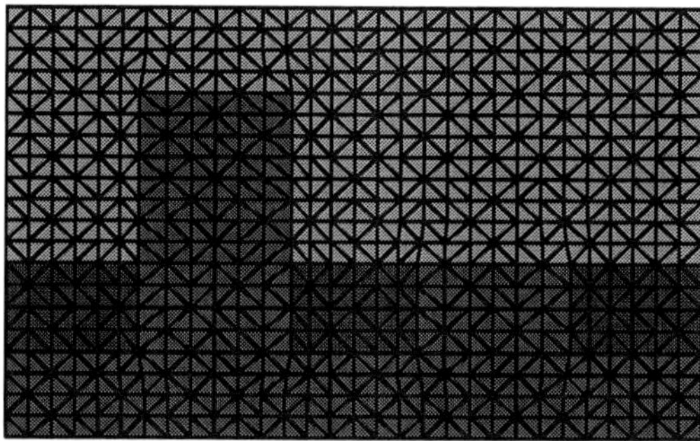

Figure 3: Triangular mesh for the 2D mask in Fig. 1. Notice the deformation of the 90-degree triangles due to the change in w_1 and w_2.

where

$$b_L = b(\frac{d_L(x_0, y_0)}{B}), \quad b_R = b(\frac{d_R(x_0, y_0)}{B}) \quad (19)$$

(x_0, y_0) is the initial location, and $d_L(x_0, y_0)$ and $d_R(x_0, y_0)$ are its distance to the left and the right side wall, respectively.

Using linear superposition, one can easily generalize (17) to the multiple parameter cases

$$x = x_0 + <\bar{b}, \bar{\sigma}> \quad (20)$$

where $<;>$ is the inner product, and b_i is the blending term for parameter σ_i which represents the change of w_1, w_2 or s in Figure 1. Clearly the parametric form in (20) is an affine function of the perturbation in each geometry parameter. It should be pointed out that the simple parametric form in (17) and (18) can be easily extended to handle parameter changes along arbitrary directions. But for the purpose of modeling masks, it is probably sufficient to consider just the cases where the parameters are changing only along the horizontal direction (x-direction for 2D). This will be the case in the remaining part of this paper. Figure 3 shows the deformation in the triangular mesh for the mask shown in Fig. 1 due to the change in w_1 and w_2.

6. PARAMETERIZE THE FEM MATRICES

For the isoparametric linear triangular elements defined on the triangle element with vertexes $(x_i, y_i), i = 1, 2, 3$, the element stiff matrix and mass matrix are respectively [11]

$$S^e = W \frac{\bar{\delta}_x \bar{\delta}_x^T + \bar{\delta}_y \bar{\delta}_y^T}{4A} W^T \tag{21}$$

and

$$M^e = 4AM_c \tag{22}$$

where A is the area of the triangle and can be written as

$$A = \frac{1}{2}|det([\bar{\delta}_x, \bar{\delta}_y])|, \tag{23}$$

$$\bar{\delta}_x = [x_1 - x_3, x_2 - x_1]^T, \tag{24}$$

$$\bar{\delta}_y = [y_3 - y_1, y_1 - y_2]^T, \tag{25}$$

$$W = \begin{bmatrix} -1 & -1 \\ 1 & 0 \\ 0 & 1 \end{bmatrix}, \tag{26}$$

$$M_c = \frac{1}{24}\begin{bmatrix} 2 & 1 & 1 \\ 1 & 2 & 1 \\ 1 & 1 & 2 \end{bmatrix}. \tag{27}$$

6.1 The Parametric Element FEM Matrices

Suppose a triangle element is within the region of the influence defined in section 5, in view of (20) and (18), (24) and (25) become

$$\bar{\delta}_x = \begin{bmatrix} x_1 - x_3 \\ x_2 - x_1 \end{bmatrix} = \begin{bmatrix} x_{1,0} - x_{3,0} + <\bar{d}_{13}, \bar{\sigma}> \\ x_{2,0} - x_{1,0} + <\bar{d}_{21}, \bar{\sigma}> \end{bmatrix} \tag{28}$$

$$\bar{\delta}_y = \begin{bmatrix} y_{3,0} - y_{1,0} \\ y_{1,0} - y_{2,0} \end{bmatrix} \tag{29}$$

where $(x_{i,0}, y_{i,0})$ is the nominal position for the node-i, $d_{ij}^{(k)} = b_i^{(k)} - b_j^{(k)}$ and $b_i^{(k)}$ is the k-th component of the vector $\bar{b}$ in (20) for node-i. For the sake of clarity, we assume that the triangle is a 90-degree triangle that satisfies

$$\begin{bmatrix} x_{1,0} - x_{3,0} \\ x_{2,0} - x_{1,0} \end{bmatrix} = \begin{bmatrix} 0 \\ h_x \end{bmatrix}, \quad \begin{bmatrix} y_{3,0} - y_{1,0} \\ y_{1,0} - y_{2,0} \end{bmatrix} = \begin{bmatrix} h_y \\ 0 \end{bmatrix}, \tag{30}$$

then we have

$$\bar{\delta}_x = h_x \begin{bmatrix} <\frac{\bar{d}_{13}}{h_x}, \bar{\sigma}> \\ 1 + <\frac{\bar{d}_{21}}{h_x}, \bar{\sigma}> \end{bmatrix} = h_x \begin{bmatrix} Y \\ 1+Z \end{bmatrix} \tag{31}$$

$$\bar{\delta}_y = h_y \begin{bmatrix} 1 \\ 0 \end{bmatrix}, \tag{32}$$

$$A = 0.5 h_x h_y (1+Z) \tag{33}$$

where $Y =< \frac{\bar{d}_{13}}{h_x}, \bar{\sigma} >$ and $Z =< \frac{\bar{d}_{21}}{h_x}, \bar{\sigma} >$. Substituting (31), (32) and (33) into (21), we obtain the parametric element stiff matrix

$$
\begin{aligned}
S^e &= \frac{h_x}{2h_y} W \begin{bmatrix} \frac{Y^2}{1+Z} & Y \\ Y & 1+Z \end{bmatrix} W^T + \frac{h_y}{2h_x} \frac{1}{1+Z} W \begin{bmatrix} 1 & 0 \\ 0 & 0 \end{bmatrix} W^T \\
&= \frac{h_x}{2h_y}\left(W \begin{bmatrix} 0 & 0 \\ 0 & 1 \end{bmatrix} W^T + W \begin{bmatrix} 0 & Y \\ Y & Z \end{bmatrix} W^T \right) \\
&\quad + \frac{Y^2}{1+Z}\frac{h_x}{2h_y} W \begin{bmatrix} 1 & 0 \\ 0 & 0 \end{bmatrix} W^T + \frac{1}{1+Z}\frac{h_y}{2h_x} W \begin{bmatrix} 1 & 0 \\ 0 & 0 \end{bmatrix} W^T \\
&= S_0^e + \sum_{k=1}^{N_p} \sigma_k S_k^e + \frac{Y^2}{1+Z} S_{N_p+1}^e + \frac{1}{1+Z} S_{N_p+2}^e
\end{aligned} \tag{34}
$$

where

$$S_0^e = \frac{h_x}{2h_y} W \begin{bmatrix} 0 & 0 \\ 0 & 1 \end{bmatrix} W^T, \quad S_k^e = \frac{1}{2h_y} W \begin{bmatrix} 0 & d_{13}^{(k)} \\ d_{13}^{(k)} & d_{21}^{(k)} \end{bmatrix} W^T \tag{35}$$

$$S_{N_p+1}^e = \frac{h_x}{2h_y} W \begin{bmatrix} 1 & 0 \\ 0 & 0 \end{bmatrix} W^T, \quad S_{N_p+2}^e = \frac{h_y}{2h_x} W \begin{bmatrix} 1 & 0 \\ 0 & 0 \end{bmatrix} W^T. \tag{36}$$

In view of (33) and (22), the parametric element mass matrix is

$$M^e(\bar{\sigma}) = 2h_x h_y (1+Z) M_c = M_0^e + \sum_{k=1}^{N_p} \sigma_k M_k^e \tag{37}$$

where $M_0^e = 2h_x h_y M_c$ and $M_k^e = 2h_y M_c d_{21}^{(k)}$.

6.2 The Assembled FEM Matrices

The standard way of generating matrices $S(\bar{w}, \bar{s})$ and $M(\bar{w}, \bar{s})$ is by a procedure called stamping. The 3×3 element stiff matrix $S^e(\bar{w}, \bar{s})$ and the 3×3 element mass matrix $M^e(\bar{w}, \bar{s})$ are first generated for each element. Using the local-to-global vertex index map, the entries of S^e and M^e are directly added (stamped) to the large matrices $S(\bar{w}, \bar{s})$ and $M(\bar{w}, \bar{s})$, respectively. The parametric stiff matrix is

$$S(\bar{\sigma}) = S_0 + \sum_{k=1}^{N_p} \sigma_k S_1^{(k)} + \sum_{i=1}^{N_1} f_i(\bar{\sigma}) S_2^{(i)} + \sum_{j=1}^{N_2} g_j(\bar{\sigma}) S_3^{(j)} \tag{38}$$

where matrices S_0, $S_1^{(k)}$, $S_2^{(i)}$ and $S_3^{(j)}$ are respectively the reults of the stamping of the element matrices S_0^e, S_k^e, $S_{N_p+1}^e$ and $S_{N_p+2}^e$ in (34),

$$f_i(\bar{\sigma}) = \frac{Y_i^2}{1+Z_i}, \quad g_j(\bar{\sigma}) = \frac{1}{1+Z_j}, \tag{39}$$

$$Y_i =< \frac{\bar{d}_{13}^{(i)}}{h_x}, \bar{\sigma} >, \quad Z_i =< \frac{\bar{d}_{21}^{(i)}}{h_x}, \bar{\sigma} >, \tag{40}$$

N_1 and N_2 are respectively the number of unique $f_i(\bar{\sigma})$ and $g_j(\bar{\sigma})$ for all the triangle elements in the computation domain, N_p is the number of parameters in the vector $\bar{\sigma}$. Similarly, the parametric mass matrix is

$$M(\bar{\sigma}) = M_0 + \sum_{k=1}^{N_p} \sigma_k M_1^{(k)}, \tag{41}$$

where matrices M_0 and $M_1^{(k)}$ are respectively the reults of the stamping of the element matrices M_0^e and M_k^e in (37).

6.3 Regression Using the Kernel Method

The parametric forms in (38) and (41) are very similar to that in (4). But since N_1 and N_2 are related to the number of elements in the computational domain, so is the size of the reduced model. Hence we have not obtained a mask model that is independent of the original problem size yet. In fact, this is the main reason why we can not directly apply the technique in [9] to generate the reduced mask model. The next critical task is to approximately represent the large number of $f_i(\bar{\sigma})$ and $g_j(\bar{\sigma})$ by the linear combination of a small number of the so-called Kernel functions that are independent of the original problem size. Essentially we seek to approximate a high-dimension space where the nonlinear function f_i and g_j reside with a much lower dimension space. This kind of approximation has been well-studied in the Machine Learning literatures and the so-called Kernel Method has been shown to be effective [17].

The gist of the Kernel Method is the following. Consider representing a scalar function $f(\bar{x}): R^n \to R$ of multi-variable argument. An interesting class of approximations are of the form

$$f(\bar{x}) = \sum_{k=1}^{N_k} \alpha_k K(\bar{x}, \bar{x}_k) \tag{42}$$

978-1-60558-497-3/09 $25.00 © 2009 ACM

where $\bar{x} \in R^n$ is the evaluation point, $K(\bar{x},\bar{y}): R^n \times R^n \to R$ is the Kernel, and the N_k vectors $\bar{x}_k \in R^n$ are denoted as the "support vectors". The basic kernel methods [17] use simple function forms for the kernels and pick the coefficient α_k based on certain loss functions.

In this paper, we choose the kernel form to be the same as the nonlinear functions to be approximated. So the kernel regression form is

$$f_i(\bar{\sigma}) \approx \sum_{k=1}^{N_k} \alpha_k^{(i)} K_1(\xi_k), \quad g_j(\bar{\sigma}) \approx \sum_{k=1}^{N_k} \beta_k^{(j)} K_2(\xi_k), \quad (43)$$

where

$$K_1(\xi_k) = \frac{\xi_k^2}{1+\xi_k}, \quad K_2(\xi_k) = \frac{1}{1+\xi_k}, \quad (44)$$

and $\xi_k = <\bar{d}_k, \bar{\sigma}>$, N_k is the number of kernels necessary to achieve the desired accuracy. As shown in section 8, 10 to 15 kernels can achieve 3 to 4 digits of accuracy. The coefficients $\alpha_k^{(i)}$ and $\beta_k^{(j)}$ are obtained by the simple Least Square Fitting, the simplest loss function form [17]. The support vectors $\bar{d}_k$ are selected from all possible values of $\bar{d}_{13}^{(i)}$ and $\bar{d}_{21}^{(i)}$ in (40) by the K-mean Clustering Methods [17].

It should be noted that due to the locality of the parametric mesh, each mesh node is influenced by a small number of parameters, typically 2 in the 2D cases and 4 in the 3D cases. This means that the length of the vector $\bar{d}_{13}^{(i)}$ and $\bar{d}_{21}^{(i)}$ is 2 for 2D cases and 4 for 3D cases, respectively. Essentially the number of clusters is independent of the number of the parameters in the mask structures. This is a key benefit from using the parametric mesh in section 5. Covering a relatively low-dimensional space with the K-mean clustering usually means small cluster number N_k in (43). Hence the mask model size is largely determined by the number of sampling solves in (5).

Substituting (43) into (38), we obtain the final parametric form of the stiff matrix

$$S(\bar{\sigma}) \approx S_0 + \sum_{k=1}^{N_p} \sigma_k S_1^{(k)} + \sum_{k=1}^{N_k} K_1(\xi_k) S_2^{(k)} + \sum_{k=1}^{N_k} K_2(\xi_k) S_3^{(k)} \quad (45)$$

where $S_2^{(k)}$ and $S_3^{(k)}$ are the results of the stamping $\alpha_k^{(i)} S_{N_p+1}^e$ and $\beta_k^{(j)} S_{N_p+2}^e$, respectively.

7. PARAMETRIC MASK MODEL

Now we are ready to put everything together to present the final mask model. Substituting (45) and (41) into (3), we obtain

$$\left[A_0 + \sum_{k=1}^{N_p} \sigma_k A_1^{(k)} + \sum_{k=1}^{N_k} K_1(\xi_k) A_2^{(k)} + \sum_{k=1}^{N_k} K_2(\xi_k) A_3^{(k)} \right] \bar{u} = \bar{r}, \quad (46)$$

where $A_0 = S_0 - M_0 - B$, $A_1^{(k)} = S_1^{(k)} - M_1^{(k)}$, $A_2^{(k)} = S_2^{(k)}$, and $A_3^{(k)} = S_3^{(k)}$. Using the congruence projection in (6), we obtain the reduced model

$$\left[\hat{A}_0 + \sum_{k=1}^{N_p} \sigma_k \hat{A}_1^{(k)} + \sum_{k=1}^{N_k} K_1(\xi_k) \hat{A}_2^{(k)} + \sum_{k=1}^{N_k} K_2(\xi_k) \hat{A}_3^{(k)} \right] \hat{u} = \hat{r}. \quad (47)$$

The usage of this mask model is similar to that described in section 4. But there is a key difference. In lithography simulation flow, we only care about the near field at the bottom of the mask. In view of (8), this can be accomplished by using

$$\bar{u}_n = \Lambda \bar{u} = \Lambda P \hat{u} = P_n \hat{u}, \quad (48)$$

where $\bar{u}_n$ is the portion of $\bar{u}$ that corresponds to the near field at the bottom of the mask, and the sparse matrix Λ selects those entries

from $\bar{u}$. An important benefit from (48) is that we only need to store $P_n = \Lambda P$ in the mask model. Matrix P_n is much smaller than matrix P. The algorithmic details for the offline and the online stage are shown in Algorithm 1 and Algorithm 2, respectively. We want to emphasize again that the cost of online stage is $O(N_s^3 + (N_p + N_k)N_s^2)$. This is clearly independent of the orginal problem size N.

Algorithm 1: Offline Stage to generate mask model

Input: Range of parameters in $\bar{\sigma}$; mesh; N_s: number of samplings; *tol*: truncation tolerance for rank revealing QR

Output: $\hat{A}_0, \hat{A}_1^{(k)}, \hat{A}_2^{(k)}, \hat{A}_3^{(k)}; \hat{r}; P_n$

(1)   Form matrices $A_0, A_1^{(k)}, A_2^{(i)}, A_3^{(j)}$ and $\bar{r}$ in (46)

(2)   **foreach** $k = 1 : N_s$

(3)   Randomly sample $\bar{\sigma}^{(k)}$ using the given ranges

(4)   Solve (46) for $\bar{u}^{(k)}$

(5) $P = [P \ \ \bar{u}^{(k)}]$

(6)   Run rank-revealing incremental QR to obtain rank r and the full-rank columns $P_1, P_2, ..., P_r$ using the given truncation tolerance *tol*

(7)   **if** $r < k$, i.e., P is rank deficient

(8) $P = [P_1, P_2, ..., P_r]$

(9)   Exit the sampling loop

(10)   Use P as projector and compute $\hat{A}_0, \hat{A}_1^{(k)}, \hat{A}_2^{(i)}, \hat{A}_3^{(j)}, \hat{r}$ as shown in (6)

(11) $P_n = \Lambda P$ as shown in (48)

Algorithm 2: Online Stage to evaluate mask model

Input: $\bar{\sigma}*; \hat{A}_0, \hat{A}_1^{(k)}, \hat{A}_2^{(k)}, \hat{A}_3^{(k)}; \hat{r}; P_n$

Output: $\bar{u}_n$: the near field at the bottom of the mask

(1)   Instantiate the compact mask model in (47) using $\bar{\sigma}*$

(2)   Solve (47) for $\hat{u}$

(3) $\bar{u}_n = P_n \hat{u}$ as shown in (48)

---

8. EXPERIMENTAL EVIDENCE

We use the mask in Figure 1 to validate our ideas. For the $32nm$ node, the nominal values of w_1, w_2 and s are assumed to be $128nm$. They have identically independent distribution with the variance being 10%. This 20% variation range is probably more than sufficient for practical consideration. It should be noted that if we add correlation among these parameters, it will only change the sampling but not the main steps in algorithm 1 and 2. With a randomly generated parameter set $\bar{\sigma}*$, we substitute (38) and (41) into (3) and solve (3) for $\bar{u}_1$. This is treated as the accurate solution. This way, we can clearly see the error introduced separately by the Kernel-based fitting in (43) and the model order reduction in Algorithm 1.

A) Kernel-based Regression Accuracy   With the same parameter set $\bar{\sigma}*$ mentioned above, we solve (46) for $\bar{u}_2$ and then compute the relative L_2 norm error $\frac{\|\bar{u}_1 - \bar{u}_2\|_2}{\|\bar{u}_1\|_2}$. The relative error vs. the number of kernels is shown in Fig. 4.

B) Model Order Reduction Accuracy   In this experiment, $N_k = 20$ kernels are used in (47). The corresponding small error in kernel fitting ensures that the MoR error in (47) dominates the overall error, as is clearly seen in Figure 5. With the increase of the number

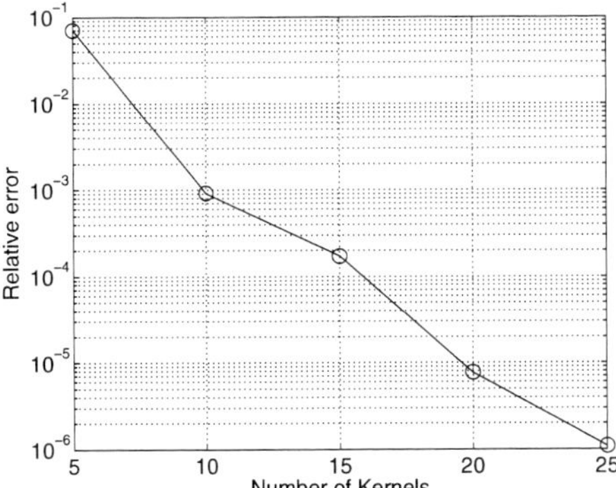

Figure 4: Relative error in regression.

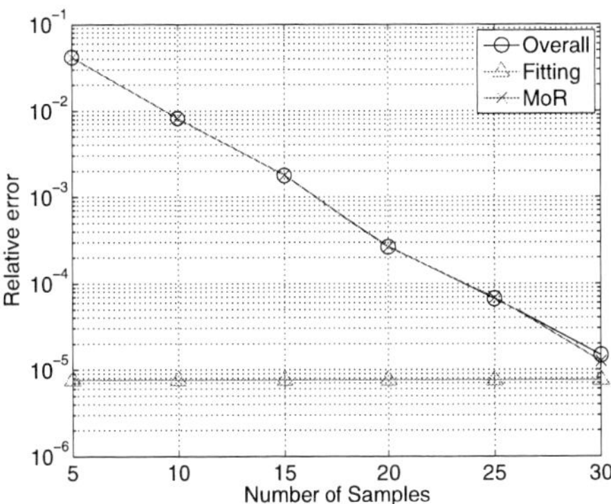

Figure 5: Convergence of the L2 norm relative error in field distribution for the mask shown in Figure 1. Each parameter varies between -10% and $+10\%$ from its nominal value of $128nm$. "MoR" refers to the error caused by the model reduction stage. "Fit" refers to the error caused by the regression fit in (43). "Overall" refers to the overall error due to both.

of samples N_s in Algorithm 1, the relative error due to model order reduction decreases rapidly in Figure 5.

C) Mask Model Accuracy Figure 5 also shows the overall error in the approximate solution provided by the new mask model. With about 8 kernels and 10 samples, the mask model can achieve a 1% overall relative error or 2-digits of accuracy.

D) Mask Model Speed The main cost of using the reduced mask model is to instantiate the small dense matrices in (47) and then invert one small dense matrix. Instantiating a few 10×10 dense matrices and then inverting a 10×10 dense matrix takes about $1e - 4$ second. For the simple two-dimensional phase-shift mask shown in Fig. 1, direct use of FEM approach would result in a sparse matrix $A(\bar{\sigma})$ in (46) with N being around 5000. The CPU time used by a direct sparse solver for such a system would be in the order of a few seconds on a desktop PC. So we see a 4-order of magnitude improvement in speed.

In addition, since the CPU time at the online stage directly relates to the number of samples in Algorithm 1, the smooth trade-off between the accuracy and the CPU time is clearly demonstrated in Figure 5. Further more, our empirical studies indicate that number of samples is a weak function of the original problem size or the number of parameters. This is certainly a very attractive feature of the new mask model proposed in this paper.

9. CONCLUSIONS

In this paper, we propose two key new ideas to improve our work on the parametric mask model in [8]: 1) Parametric unstructured mesh using the distance function and the blending function; 2) Kernel-based regression to significantly reduce the number of parametric terms in the final system. Numerical experiments demonstrate that the new mask model offers smooth accuracy-speed trade-off and hence can be tuned to different design applications. The model generation in Algorithm 1 is inherently incremental and both theoretical and practical ways to estimate the model error are readily available. In addition, the parametric mesh allows the decoupling between the mesh generation and the parametrization of the FEM matrices, a considerable simplification from implementation point of view.

10. ACKNOWLEDGMENTS

We want to thank Dr. Joel Phillips from Cadence Research Labs for bringing the Kernel Methods in [14, 17] to our attention. We want to thank Prof. Per-olof Persson from University of California at Berkeley for bringing to our attention the use of blending function in [16]. And finally, we want to thank Dr. Apo Sezginer from Cadence Design Systems for the valuable discussions.

11. REFERENCES

[1] A.K.T. Wong, *Resolution Enhancement Techniques in Optical Lithography*, SPIE, Bellingham, WA, 2001.

[2] A. K. Wong and A.R. Neureuther, "Rigorous three-dimensional time-domain finite-difference electromagnetic simulation for photolithograph applications," in *IEEE Trans. on Semiconductor Manufacturing*, Nov. 1995, vol. 8, p. 419.

[3] S. Burger, R. Kohle, L. Zschiedrich, H. Nguyen, F. Schmidt, R. Marz, and C. Nolscher, "Rigorous simulation of 3D masks," in *Proc. SPIE*, 2006, vol. 6349.

[4] J.W. Goodman, *Introduction to Fourier Optics*, Roberts and Companies, Greenwood Village, CO, 2005, third edition.

[5] Jaione Tirapu-Azpiroz; Paul Burchard; Eli Yablonovitch, "Boundary layer model to account for thick mask effects in photolithography," in *Advanced Lithography, Proc. SPIE*, Feb. 2003, p. 1611.

[6] K. Adam and A.R. Neureuther, "Domain decomposition methods for the rapid electromagnetic simulation of photomask scattering," in *J. Microlitho., Microfab., and Microsys.*, Oct. 2002, vol. 14, p. 253.

[7] V. Singh, B. Hu, K. Toh, S. Bollepalli, S. Wagner, and Y. Borodovsky, "Making a tillion pixels dance," *Advanced Lithography, Proc. SPIE*, vol. 6924, Feb. 2008.

[8] Zhenhai Zhu and Frank Schmidt, "An efficient and robust mask model for lithography simulation," in *Advanced Lithography, Proc. SPIE*, Feb. 2008, vol. 6925, p. 126.

[9] G. Rozza, D.B.P. Huynh, and A. Patera, "Reduced basis approximation and a posteriori error estimation for affinely parameterized elliptic coercive partial differential equations," *Arch. Comput. Methods Eng.*, vol. 15, pp. 229–275, 2008.

[10] J.R. Phillips, "Variational interconnect analysis via PMTBR," in *International Conference on CAD*, San Jose, CA, Nov. 2004, p. 872.

[11] J.M. Jin, *The Finite Element Method in Electromagnetics*, John Willeys and Sons Inc, New York, 2002, second edition.

[12] F. Ihlenburg, *Finite Element Analysis of Acoustic Scattering*, Springer, 1998.

[13] J. Phillips, "Projection-based approaches for model reduction of weakly nonlinear, time-varying systems," *IEEE Trans. on CAD*, vol. 22, pp. 171, Feb. 2003.

[14] J.R. Phillips, J. Afonso, A. Oliveira, and L.M. Silveira, "Analog macromodeling using kernel methods," in *International Conference on CAD*, San Jose, CA, Nov. 2003, p. 446.

[15] J.A. Sethian, *Level Set methods and Fast Marching Methods*, Cambridge University Press, 1999.

[16] P. Persson, J. Peraire, and J. Bonet, "Discontinuous galerkin solution of the navier-stokes equations on deformable domains," in *Proc. 45th AIAA Aerospace Sciences Meeting and Exhibit*, Jan. 2007.

[17] T. Hastie, R. Tibshirani, and J.H. Friedman, *Elements of Statistical Learning*, Springer, 2003.

978-1-60558-497-3/09 $25.00 © 2009 ACM

CALL FOR CONTRIBUTIONS

47th DESIGN AUTOMATION CONFERENCE®
Anaheim Convention Center, Anaheim, CA • June 14 - 18, 2010
FOR INFORMATION CALL: +1-303-530-4333

With a heritage of excellence for almost half a century, DAC continues to be the premier conference devoted to Electronic Design Automation (EDA) and the application of EDA tools in designing advanced electronic systems. DAC 2010 is seeking papers that deal with tools, algorithms, and design techniques for all aspects of electronic circuit and system design. We invite submissions in the following categories: research papers, User Track papers, "Wild and Crazy Ideas" (WACI) topic papers, suggestions for special sessions, panels, and tutorials, and proposals for workshops and colocated events. Submissions must be made electronically at the DAC website. Detailed guidelines for all categories are available on the DAC website.

RESEARCH PAPERS DUE BEFORE 5:00pm MT, November 19, 2009
Research paper submissions MUST (1) be in PDF format only, (2) contain an abstract of approximately 60 words clearly stating the significant contribution, impact, and results of the submission, (3) be no more than six pages (including the abstract, figures, tables, and references), double columned, 9-pt or 10-pt font, and (4) MUST NOT identify the author(s) by their name(s) or affiliation(s) anywhere on the manuscript or abstract, with all references to the author(s)'s own previous work or affiliations in the bibliographic citations being in the third person. Preliminary submissions will be at a disadvantage. Format templates are available on the DAC website. Submissions not adhering to these rules will be rejected. DAC will compare each submission against a vast database and any paper with significant similarity to previously published works or with papers that are simultaneously under review with other conferences and symposia, will be rejected. All research papers will be reviewed as finished papers. Authors of accepted papers must sign a copyright release form for their paper. Acceptance notices will be available by logging in to the DAC website after Feb. 25, 2010. Complete author kits will be available on the DAC website by March 19, 2010.

USER TRACK PAPERS - EXTENDED ABSTRACTS DUE BEFORE 5:00pm MT, November 9, 2009
DAC's User Track addresses the real-life issues facing IC designers, application engineers, and design flow developers. It provides valuable insights and experiences with in-house or commercial EDA tool flows. User Track papers may describe the application of EDA tools to the design of a novel electronic system, or the integration of EDA tools within a design flow or methodology to produce such systems. A User Track paper may be problem-specific in scope (e.g., analyzing substrate coupling during floorplanning) or may address a specific application domain (e.g., designing wireless handsets). Submissions are in the form of an extended abstract. For more details, please see the separate User Track Call for Papers on the DAC website: *http://www.dac.com/47th/UTinfo.html*

WILD AND CRAZY IDEAS (WACI) PAPERS DUE BEFORE 5:00pm MT, November 19, 2009
The "Wild and Crazy Ideas" sessions cover interesting activities on a wide variety of topics that do not fit in the conventional mold. The WACI track features novel (and even preliminary or unproven) technical ideas. The aim of WACI is to promote revolutionary and way-out ideas that inspire discussion among conference attendees, create a buzz, and get people talking. Submissions to the "Wild and Crazy Ideas" track should not exceed two pages, but must otherwise follow the above rules and deadlines for the research papers. Unlike a DAC research paper that explores a specific technology problem and proposes a complete solution to it, with a full table of results, a WACI paper could present less developed but highly innovative ideas related to areas relevant to DAC. All WACI accepted papers will be required to post a two-minute video describing the work as part of the acceptance process.

SPECIAL SESSION SUGGESTIONS DUE BEFORE 5:00pm MT, November 2, 2009
Special session suggestions must include descriptions of the proposed papers and speakers, and the importance of the special session to the DAC audience. As the term implies, a special session is devoted to a topic of strong contemporary or future interest. The topic must represent an emerging area that does not yet receive sufficient focus from research papers. A submission must list at least three inspiring speakers who address the topic from different angles. DAC reserves the right to restructure all special session suggestions.

PANEL and **TUTORIAL** SUGGESTIONS DUE BEFORE 5:00pm MT, November 2, 2009
Panel and tutorial suggestions should not exceed two pages, should describe the topic and intended audience, and should include a list of suggested participants. Tutorial suggestions must include a bulleted outline of covered topics. DAC reserves the right to restructure all panel and tutorial suggestions. Please see the website for further details, or call Kevin Lepine, Conference Manager, at +1-303-530-4333.

WORKSHOP SUBMISSIONS DUE BEFORE 5:00pm MT, January 29, 2010
DAC invites you to organize a workshop on topics related to design, design methodologies, and design automation. DAC provides the financial and organizational support, including attendee registration, rooms at the conference center and audio visual equipment. Please see the website for further details, or call Kevin Lepine, Conference Manager at +1-303-530-4333.

COLOCATED EVENT PROPOSALS DUE BEFORE 5:00pm MT, January 29, 2010
DAC invites you to colocate your conference, meeting or other special event with DAC. We will provide you with meeting rooms at the conference center at no cost. Your event will be financed and otherwise organized by you. Please see the website for further details, or call Kevin Lepine, Conference Manager, at +1-303-530-4333.

SUBMISSION CATEGORIES FOR RESEARCH PAPERS
Authors of research papers are required to specify a category from the following list:

1. **System-Level Design and Codesign**
   1.1 System specification, modeling, simulation, verification, and performance analysis
   1.2 Scheduling, HW/SW partitioning, HW/SW interface synthesis
   1.3 IP and platform-based design, IP protection
   1.4 System-On-Chip (SOC) and multiprocessor System-On-Chip (MPSOC)
   1.5 Application-specific processor design tools
2. **System-Level Communication and Networks-On-Chip**
   2.1 Modeling and performance analysis
   2.2 Communications-based design, communication and network synthesis
   2.3 Architectural synthesis, mapping, routing, scheduling
   2.4 Optimization for energy, fault-tolerance, reliability
   2.5 Interfacing and software issues, beyond-the-die communication
   2.6 NOC design methodologies, case studies and prototyping
3. **Embedded Hardware Design and Applications**
   3.1 Case studies of embedded system design
   3.2 Flows and methods for specific applications and design domains
4. **Embedded Software Tools and Design**
   4.1 Retargetable compilation
   4.2 Memory/cache optimization
   4.3 Software for single-/multi-processor, multicore, GPU systems
   4.4 Real-time operating systems
   4.5 Verification of embedded software
5. **Power Analysis and Low-Power Design**
   5.1 System-level power design and thermal management
   5.2 Embedded low-power approaches: partitioning, scheduling, and resource management
   5.3 High-level power estimation and optimization
   5.4 Gate-level power analysis and optimization
   5.5 Device and circuit techniques for low-power design
   5.6 Power-aware and energy-efficient wireless protocols, algorithms and associated design techniques and methodologies
6. **Verification**
   6.1 Functional, transaction-level, RTL, and gate-level modeling and verification of hardware design
   6.2 Dynamic simulation, equivalence checking, formal (and semiformal) verification model and property checking
   6.3 Emulation and hardware simulators or accelerator engines
   6.4 Modeling languages and related formalisms, verification plan development and implementation
   6.5 Assertion-based verification, coverage analysis, constrained-random testbench generation
   6.6 Verification techniques for software correctness

7. **High-Level Synthesis, Logic Synthesis and Circuit Optimization**
   7.1 Combinational, sequential, and asynchronous logic synthesis
   7.2 Library mapping, cell-based design and optimization
   7.3 Transistor and gate sizing, resynthesis
   7.4 Interactions between logic design and layout or physical synthesis
   7.5 High-level, behavioral, algorithmic, and architectural synthesis, "C" to gates tools and methods
   7.6 Resource scheduling, allocation, and synthesis

8. **Circuit, Interconnect and Manufacturing Simulation and Analysis**
   8.1 Electrical-level circuit simulation
   8.2 Model order reduction methods for linear systems
   8.3 Interconnect and substrate modeling and extraction
   8.4 High-frequency and electromagnetic simulation of circuits
   8.5 Thermal and electrothermal simulation
   8.6 Technology CAD and fab automation

9. **Timing Analysis**
   9.1 Process technology characterization, and modeling
   9.2 Deterministic static timing analysis and verification
   9.3 Statistical performance analysis and optimization

10. **Physical Design and Manufacturability**
    10.1 Floorplanning, partitioning, placement
    10.2 Buffer insertion, routing, interconnect planning
    10.3 Physical verification and design rule checking
    10.4 Automated synthesis of clock networks
    10.5 Reticle enhancement, lithography-related design optimizations
    10.6 Design for manufacturability, yield, defect tolerance, cost issues, and DFM impact
    10.7 Physical design of 3-D integrated circuits
    10.8 System-in-package design, package-board codesign
    10.9 Design for resilience under manufacturing variations

11. **Signal Integrity and Design Reliability**
    11.1 Signal integrity, capacitive and inductive crosstalk
    11.2 Reliability modeling and analysis
    11.3 Novel clocking and power delivery schemes
    11.4 Power grid robustness analysis and optimization
    11.5 Soft errors
    11.6 Thermal reliability

12. **Analog, Mixed-Signal, and RF**
    12.1 Analog, mixed-signal, and RF design methodologies
    12.2 Automated synthesis
    12.3 Analog, mixed-signal, and RF simulation
    12.4 High-frequency design and advanced antenna design for wireless design

13. **FPGA Design Tools and Applications**
    13.1 Rapid prototyping
    13.2 Logic synthesis and physical design techniques for FPGAs
    13.3 Configurable and reconfigurable computing

14. **Testing**
    14.1 Test quality/reliability, current-based test, delay test, low-power test
    14.2 Digital fault modeling, automatic test generation, fault simulation
    14.3 Digital design for test, test data compression, built-in self test
    14.4 Memory test and repair, FPGA testing
    14.5 Fault-tolerance and online testing
    14.6 Analog/mixed-signal/RF testing, system-in-package (SIP) testing
    14.7 Board- and system-level test, system-on-chip (SOC) testing
    14.8 Silicon debug and diagnosis, post-silicon design validation

15. **New and Emerging Design Technologies**
    (including but not restricted to)
    15.1 MEMS, sensors, actuators, imaging devices
    15.2 Nanotechnologies, nanowires, nanotubes
    15.3 Quantum computing
    15.4 Synthetic and systems biology, biologically-based or biologically-inspired systems
    15.5 Non-CMOS 3-D design, new transistor structures and devices, design in new or radical process technologies
    15.6 Optical devices and communication

ALL SUBMISSIONS MUST BE MADE ELECTRONICALLY AT THE DAC WEBSITE: WWW.DAC.COM.
ACM and IEEE reserve the right to exclude a paper from archival distribution after the conference if the paper is not presented at the conference, or in other exceptional cases.

AUTHOR INDEX

AUTHOR INDEX